MW00909196

Computer-Aided Design of Analog Integrated Circuits and Systems

IEEE Press
445 Hoes Lane, P.O. Box 1331
Piscataway, NJ 08855-1331

Technical Reviewers

Rick Carley, *Carnegie Mellon*
John Cohn, *IBM*
Tamal Mukherjee, *Carnegie Mellon*
Emil Ochotta, *Xilinx*
Willy Sansen, *Katholieke Universiteit Leuven*

Computer-Aided Design of Analog Integrated Circuits and Systems

Edited by

Rob A. Rutenbar

Georges G. E. Gielen

Brian A. Antao

A Selected Reprint Volume

IEEE Press

A JOHN WILEY & SONS, INC., PUBLICATION

This text is printed on acid-free paper. ♾

For ordering and customer service, call 1-800-CALL-WILEY.

ISBN 0-471-22782-X

Printed in the United States of America.

10 9 8 7 6 5 4 3 2 1

Contents

Preface

Ten years ago, analog seemed to be a dead-end technology, or at best a "niche" area to be tolerated as a necessary evil. The world was inevitably becoming digital—what do we need from analog? Today, the world is amazingly digital, and yet the inevitable consequence is an explosion of analog design interest. An ever-increasing level of integration is the hallmark of today's IC market. Complete systems that previously occupied one or more boards are now integrated on a few chips or even on one chip — a so called *System-on-Chip* (SoC). Although most functions in such integrated systems are implemented with digital or digital signal processing (DSP) circuitry, the circuits needed at the interface between the electronics and the continuous-valued outside world are analog circuits. These analog interface components are being integrated on the same die for reasons of cost and performance. Modem SoC designs are therefore increasingly mixed-signal designs, and this will even be more prevalent in the future.

Interestingly, this has put significant pressure on the CAD tools and methodologies that support the analog parts of these system designs. Although analog circuits typically occupy only a small fraction of the total area of mixed-signal ICs, their design is often the bottleneck in mixed-signal systems, in design time and effort as well as test cost. Analog components are embarrassingly often the ones responsible for chip-level design errors and expensive reruns. And the situation is getting worse. Just as deeply scaled CMOS technologies have complicated the design of digital functions with issues such as complexity management and predictable timing closure, likewise these deeply scaled technologies complicate the analog subsystems on SoC designs. Since analog circuits exploit (rather than abstract away) the low-level physics of the fabrication process, they remain difficult and costly to design, validate, and reuse.

In the digital domain, we have come to depend on the ability of CAD tools to synthesize our logic, map it onto arbitrary libraries, transform logic to layout, verify layout against logic, and optimize both to meet tight timing/power goals. Indeed, the ASIC revolution of the 1990s would have been impossible without these design automation tools.

So, where are the *analog* CAD tools? This volume is our attempt to answer the question. Although the most common strategy for analog is still hand-crafted full-custom design one transistor at a time, one polygon at a time—the situation is now changing rapidly. Today, new circuit and physical synthesis tools are designing practical analog circuits; new modeling and analysis tools are allowing rapid exploration of system-level alternatives; new simulation tools are giving accurate answers for analog circuit behaviors and interactions that were considered impossible to handle only a few years ago. Our goal for this volume has been to collect in one place the essential set of analog CAD papers that form the foundation for the wealth of new analog design automation work we see today.

Some of these papers have been selected to give the reader a better understanding of the history of the field, while others really present a snapshot of state of the art in a particular topic. We have organized the papers into eight areas:

I. **Introduction to Analog CAD:** two general survey papers, giving a broad overview of the field.
II. **Analog Synthesis:** techniques to transform performance specifications into sized circuits.
III. **Symbolic Analysis:** techniques to extract symbolic performance equations from analog circuits.
IV. **Analog Layout:** techniques to transform from circuit to mask, for cells and for mixed-signal systems.
V. **Analog Modeling Analysis:** techniques to fit reduced/simplified macromodels for analog circuits, and analysis methods for common analog concerns such as coupling and noise isolation.
VI. **Specialized Analog Simulation:** techniques to simulate over-sampled converters, switched-capacitor filters, mixed analog/digital circuits, behaviorally represented circuits, and radio frequency (RF) designs.
VII. **Analog Centering and Yield Optimization:** techniques to optimize analog designs for manufacturing yield.
VIII. **Analog Test:** techniques to test analog and mixed-signal designs.

Of course, given this extremely broad range of topics, space limitations have forced many difficult trade-offs; many worthy papers could not be included. As a result, we have chosen to emphasize in this volume the synthesis related topics (parts II — IV), and then to specialize the modeling (part V), simulation (part VI) and optimization (part VII) areas to problems unique to analog macromodeling and analog verification, and to limit the test work (part VIII) to only a few broad survey papers. With the material contained in this volume, the energetic reader should be able to gain a solid understanding of the historical progress in this area, and a good global picture of the current state of the art in analog CAD. This book is therefore intended not only for researchers in the field, but also CAD professionals responsible for implementing or maintaining analog / mixed-signal tools, and the circuit designers who depend on the resulting tools.

ROB A. RUTENBAR
GEORGES G. E. GIELEN
BRIAN A. ANTAO

May 2002

Acknowledgments

The editors would like to thank Professor Resve Saleh, under whose initial impetus this collected volume got its start. They also gratefully acknowledge the many useful comments, suggestions, and critiques from several generous colleagues: Rick Carley of Carnegie Mellon, John Cohn of IBM, Tamal Mukherjee of Carnegie Mellon, Emil Ochotta of Xilinx, Willy Sansen of Katholieke Universiteit Leuven. We would also like to acknowledge the efforts of Hong Zhou Liu and Amit Singhee of Carnegie Mellon, for their careful, proofeading of the final manuscript. Finally, we would like to thank our IEEE editor, John Griffin, whose patience and persistence were heroic and essential to the completion of this volume.

Computer-Aided Design of Analog Integrated Circuits and Systems

PART I

Introduction to Analog CAD

Computer-Aided Design of Analog and Mixed-Signal Integrated Circuits

GEORGES G. E. GIELEN, SENIOR MEMBER, IEEE, AND ROB A. RUTENBAR, FELLOW, IEEE

Invited Paper

This survey presents an overview of recent advances in the state of the art for computer-aided design (CAD) tools for analog and mixed-signal integrated circuits (ICs). Analog blocks typically constitute only a small fraction of the components on mixed-signal ICs and emerging systems-on-a-chip (SoC) designs. But due to the increasing levels of integration available in silicon technology and the growing requirement for digital systems to communicate with the continuous-valued external world, there is a growing need for CAD tools that increase the design productivity and improve the quality of analog integrated circuits. This paper describes the motivation and evolution of these tools and outlines progress on the various design problems involved: simulation and modeling, symbolic analysis, synthesis and optimization, layout generation, yield analysis and design centering, and test. This paper summarizes the problems for which viable solutions are emerging and those which are still unsolved.

Keywords**—***Analog and mixed-signal computer-aided design (CAD), analog and mixed-signal integrated circuits, analog circuit and layout synthesis, analog design automation, circuit simulation and modeling.*

I. Introduction

The microelectronics market and, in particular, the markets for application-specific ICs (ASICs), application-specific standard parts (ASSPs), and high-volume commodity ICs are characterized by an ever-increasing level of integration complexity, now featuring multimillion transistor ICs. In recent years, complete systems that previously occupied one or more boards have been integrated on a few chips or even one single chip. Examples of such **systems on a chip (SoC)** are the single-chip TV or the single-chip camera [1] or new generations of integrated telecommunication systems that include analog, digital, and eventually radio-frequency (RF) sections on one chip. The technology of choice for these systems is of course CMOS, because of the good digital scaling, but also BiCMOS is used when needed for the analog or RF circuits. Although most functions in such integrated systems are implemented with digital or digital signal processing (DSP) circuitry, the **analog circuits needed at the interface between the electronic system and the "real" world** are also being integrated on the same die for reasons of cost and performance. A typical future SoC might look like Fig. 1, containing several embedded processors, several chunks of embedded memory, some reconfigurable logic, and a few analog interface circuits to communicate with the continuous-valued external world.

Despite the trend previously to replace analog circuit functions with digital computations (e.g., digital signal processing in place of analog filtering), there are **some typical functions that will *always* remain analog**.

- The first typically analog function is on the input side of a system: signals from a sensor, microphone, antenna, wireline, and the like, must be sensed or received and then amplified and filtered up to a level that allows digitization with sufficient signal-to-noise-and-distortion ratio. Typical analog circuits used here are low-noise amplifiers, variable-gain amplifiers, filters, oscillators, and mixers (in case of downconversion). Applications are, for instance, instrumentation (e.g., data and biomedical), sensor interfaces (e.g., airbag accelerometers), process control loops, telecommunication receivers (e.g., telephone or cable modems, wireless phones, set-top boxes, etc.), recording (e.g., speech recognition, cameras), and smart cards.
- The second typically analog function is on the output side of a system: the signal is reconverted from digital to analog form and it has to be strengthened so that it can drive the outside load (e.g., actuator, antenna, loudspeaker, wireline) without too much distortion. Typical analog circuits used here are drivers and buffers, filters, oscillators and mixers (in case of upconversion). Applications are, for instance, process control

Manuscript received February 4, 2000; revised July 14, 2000.

G. G. E. Gielen is with the Electrical Engineering Department, Katholieke Universiteit Leuven, Leuven, Belgium.

R. A. Rutenbar is with the Electrical and Computer Engineering Department, Carnegie Mellon University, Pittsburgh, PA 15213 USA.

Publisher Item Identifier S 0018-9219(00)10757-1.

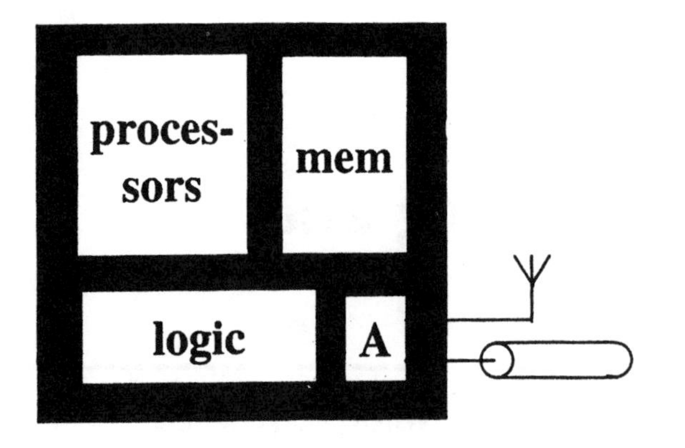

Fig. 1. Future system-on-a-chip.

loops (e.g., voltage regulators for engines), telecommunication transmitters, audio and video (e.g., CD, DVD, loudspeakers, TV, PC monitors, etc.), and biomedical actuation (e.g., hearing aids).

- The third type of blocks are the true mixed-signal circuits that interface the above analog circuits with the DSP part of the system. Typical circuits used here are the sample-and-hold circuits for signal sampling, analog-to-digital converters for amplitude discretization, digital-to-analog converters for signal reconstruction, and phase-locked loops and frequency synthesizers to generate a timing reference or perform timing synchronization.
- In addition, the above circuits need stable absolute references for their operation, which are generated by voltage and current reference circuits, crystal oscillators, etc.
- Finally, the largest analog circuits today are high-performance (high-speed, low-power) digital circuits. Typical examples are state-of-the-art microprocessors, which are largely custom sized like analog circuits, to push speed or power limits.

Clearly, analog circuits are indispensable in all electronic applications that interface with the outside world and will even be more prevalent in our lives if we move toward the intelligent homes, the mobile road/air offices, and the wireless workplaces of the future.

When both analog (possibly RF) and digital circuits are needed in a system, it becomes obvious to integrate them together to reduce cost and improve performance, provided the technology allows us to do so. The growing market share of integrated mixed-signal ICs observed today in modern electronic systems for telecommunications, consumer, computing, and automotive applications, among many others, is a direct result of the need for higher levels of integration [2]. Since the early 1990s, the average growth rate of the mixed-signal IC market has been between 15% and 20% per year, and this market is predicted to surpass $22 billion by 2001. Recent developments in CMOS technology have offered the possibility to combine good and scalable digital performance with adequate analog performance on the same die. The shrinking of CMOS device sizes down to the deep submicrometer regime (essentially in line with, or even ahead of, the predicted technology roadmap [3]) makes higher levels of system integration possible and also offers analog MOS transistor performance that approaches the performance of a bipolar transistor. This explains why CMOS is the technology of choice today, and why other technologies like BiCMOS are only used when more aggressive bipolar device characteristics (e.g., power, noise, or distortion) are really needed. The technology shift from bipolar to CMOS (or BiCMOS) has been apparent in most applications. Even fields like RF, where traditionally GaAs and bipolar were the dominant technologies, now show a trend toward BiCMOS (preferably with a SiGe option) and even plain CMOS for reasons of higher integration and cost reduction. These higher levels of mixed-signal integration, however, also introduce a whole new set of problems and design effects that need to be accounted for in the design process.

Indeed, together with the increase in circuit complexity, the design complexity of today's ICs has increased drastically: 1) due to integration, more and more transistors are combined per IC, performing both analog and digital functions, to be codesigned together with the embedded software; 2) new signal processing algorithms and corresponding system architectures are developed to accommodate new required functionalities and performance requirements (including power) of emerging applications; and 3) due to the rapid evolution of process technologies, the expectation for changing process technology parameters needs to be accounted for in the design cycle. At the same time, many ASIC and ASSP application markets are characterized by shortening product life cycles and tightening time-to-market constraints. The time-to-market factor is very critical for ASICs and ASSPs that eventually end up in consumer, telecom, or computer products: if one misses the initial market window relative to the competition, prices and, therefore, profit can be seriously eroded.

The key to managing this increased design complexity while meeting the shortening time-to-market factor is the **use of computer-aided design (CAD) and verification tools**. Today's high-speed workstations provide ample power to make large and detailed computations possible. What is needed to expedite the analog and mixed-signal design process is a structured methodology and supporting CAD tools to manage the entire design process and design complexity. CAD tools are also needed to assist or automate many of the routine and repetitive design tasks, taking away the tedium of manually designing these sections and providing the designer with more time to focus on the creative aspects of design. ICs typically are composed of many identical circuit blocks used across different designs. The design of these repetitive blocks can be automated to reduce the design time. In addition, CAD tools can increase the productivity of designers, even for nonrepetitive analog blocks. Therefore, analog CAD and circuit design automation are likely to play a key role in the design process of the next generation of mixed-signal ICs and ASICs. And although the design of mixed-signal ASICs served as the initial impetus for stepping up the efforts in research and development of analog design automation tools, the technology trend toward integrating complete **systems on a chip** in recent years has provided yet another driving force to bolster analog CAD efforts. In addition, for such systems new design paradigms are being developed that greatly affect how we will design analog blocks. One example is the macrocell design reuse methodology of assembling a system by reusing soft or hard macrocells ("virtual components") that are available on the intellectual property (IP) market and that can easily be mixed and matched in the "silicon board" system if they comply with the virtual socket inferface (VSI) standard [4]. This methodology again poses many new constraints, also on the analog blocks. Platform-based design is another emerging system-level design methodology [5].

In the digital domain, CAD tools are fairly well developed and commercially available today, certainly for the lower levels of the design flow. First, the digital IC market is much larger than the analog IC market. In addition, unlike analog circuits, a digital system can naturally be represented in terms of Boolean representation and programming language constructs, and its functionality can easily be represented in algorithmic form, thus paving the way for a logical transition into automation of many aspects of digital system design. At the present time, many lower-level aspects of the digital design process are fully automated. The hardware is described in a hardware description language (HDL) such as VHDL or Verilog, either at the behavioral level or most often at the structural level. High-level synthesis tools attempt to synthesize the behavioral HDL description into a structural representation. Logic synthesis tools then translate the structural HDL specification into a gate-level netlist, and semicustom layout tools (place and route) map this netlist into a correct-by-construction mask-level layout based on a cell library specific for the selected technology process. Research interest is now moving in the direction of system synthesis where a system-level specification is translated into a hardware–software coarchitecture with high-level specifications for the hardware, the software, and the interfaces. Reuse methodologies and platform-based design methodologies are being developed to further reduce the design effort for complex systems. Of course, the level of automation is far from the push-button stage, but the developments are keeping up reasonably well with the chip complexity offered by the technology.

Unfortunately, the story is quite different on the analog side. There are not yet any robust commercial CAD tools to support or automate analog circuit design apart from circuit simulators (in most cases, some flavor of the ubiquitous SPICE simulator [6]) and layout editing environments and their accompanying tools (e.g., some limited optimization capabilities around the simulator, or layout verification tools). Some of the main reasons for this lack of automation are that analog design in general is perceived as less systematic and more heuristic and knowledge-intensive in nature than digital design, and that it has not yet been possible for analog designers to establish a higher level of abstraction that shields all the device-level and process-level details from the higher level design. Analog IC design is a complex endeavor, requiring specialized knowledge and circuit design skills acquired through many years of experience. The variety of circuit schematics and the number of conflicting requirements and corresponding diversity of device sizes is also much larger. In addition, analog circuits are more sensitive to nonidealities and all kinds of higher order effects and parasitic disturbances (crosstalk, substrate noise, supply noise, etc.). These differences from digital design also explain why analog CAD tools cannot simply adapt the digital algorithms, but why specific analog solutions need to be developed that are targeted to the analog design paradigm and complexity. The analog CAD field, therefore, had to evolve on its own, but it turned into a niche field as the analog IC market was smaller than the digital one. As a result, due to the lack of adequate and mature commercial analog CAD tools, analog designs today are still largely being handcrafted with only a SPICE-like simulation shell and an interactive layout environment as supporting facilities. The **design cycle for analog and mixed-signal ICs remains long and error-prone**. Therefore, although analog circuits typically occupy only a small fraction of the total area of mixed-signal ICs, their design is often the bottleneck in mixed-signal systems, both in design time and effort as well as test cost, and they are often responsible for design errors and expensive reruns.

The economic pressure for high-quality yet cheap electronic products and the decreasing time-to-market constraints have clearly **revealed the need in the present microelectronics industry for analog CAD tools** to assist designers with fast and first-time-correct design of analog circuits, or even to automate certain tasks of this design process where possible. The push for more and more integrated systems containing both analog and digital circuitry heavily constrains analog designers. To keep pace with the digital world and to fully exploit the potential offered by the present deep submicrometer VLSI technologies, boosting analog design productivity is a major concern in the industry today. The design time and cost for analog circuits from specification to successful silicon has to be reduced drastically. The risk for design errors impeding first-pass functional (and possibly also parametrically correct) chips has to be eliminated. Second, analog CAD tools can also help to increase the quality of the resulting designs. Before starting detailed circuit implementation, more higher-level explorations and optimizations should be performed at the system architectural level, preferrably across the analog–digital boundary, since decisions at those levels have a much larger impact on key overall system parameters such as power consumption and chip area. Likewise, designs at lower levels should be "automated" where possible. Designers find difficulty in considering multiple conflicting tradeoffs at the same time—computers do not. Computers are adept at trying out and exploring large numbers of competing alternatives. Typical examples are fine-tuning through optimization of an initial handcrafted design and improving design robustness with respect to operating parameter variations (temperature, supply voltage) and/or with respect to manufacturing tolerances and

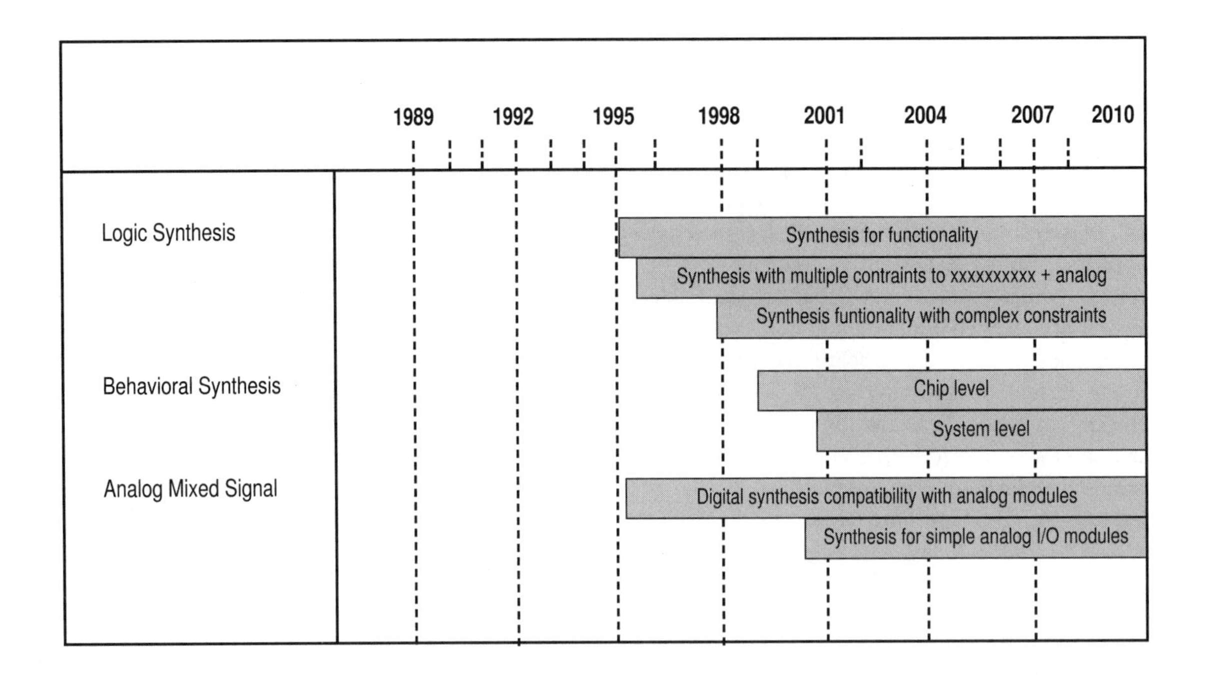

Fig. 2. SIA synthesis potential solutions roadmap [3].

mismatches. Third, the continuous pressure of technology updates and process migrations is a large burden on analog designers. CAD tools could take over a large part of the technology retargeting effort, and could make analog design easier to port or migrate to new technologies. Finally, the SoC design reuse methodology also requires executable models and other information for the analog macrocells to be used in system-level design and verification. Tools and modeling techniques have to be developed to make this possible. This need for analog CAD tools beyond simulation has also clearly been identified in the SIA roadmap, as indicated in Fig. 2, where analog synthesis is predicted to take off somewhere beyond the year 2000 [3].

Despite the lack of commercial analog CAD tools, analog CAD and design automation over the past 15 years has been a field of profound academic and industrial research activity, although with not quite as many researchers as in the digital world, resulting in a slow but steady progress [7]. Some of the aspects of the analog CAD field are fairly mature, some are ready for commercialization, while others are still in the process of exploration and development. The simulation area has been particularly well developed since the advent of the SPICE simulator, which has led to the development of many simulators, including timing simulators in the digital field and the newer generation of mixed-signal and multilevel commercial simulators. Analog circuit and layout synthesis has recently shown promising results at the research level, but commercial solutions are only starting to appear in the marketplace. The development of analog and mixed-signal hardware description languages like VHDL-AMS [8] and Verilog-A/MS [9] is intended to provide a unifying trend that will link the various analog design automation tasks in a coherent framework that supports a more structured analog design methodology from the design conceptualization stage to the manufacturing stage. They also provide a link between the analog and the digital domains, as needed in designing mixed analog–digital ICs and the SoC of the future.

In this survey, the relevant developments to date in analog and mixed-signal CAD will be covered in a general overview. The paper is organized as follows. Section II describes the analog and mixed-signal integrated system design process, as well as a hierarchical design strategy for the analog blocks. Section III then describes general progress and the current status in the various fields of analog CAD: simulation and modeling, symbolic analysis, circuit synthesis and optimization, layout generation, yield analysis and design centering, and test and design for testability. This is illustrated with several examples. Most of the emphasis will be on circuit and layout synthesis as it is key to analog design automation, while other topics such as test will only be covered briefly in this paper. For the sake of completeness, we did not want to omit those topics, but they require overview papers of their own for detailed coverage. Conclusions are then provided in Section IV, and an extensive list of references completes the paper.

II. Analog and Mixed-Signal Design Process

We will now first describe the design flow for mixed-signal integrated systems from concept to chip, followed by the description of a hierarchical design methodology for the analog blocks that can be adopted by analog CAD systems.

A. Mixed-Signal IC Design Flow

Fig. 3 illustrates a possible scenario for the design flow of a complex analog or mixed-signal IC. The various stages that are traversed in the design process are as follows.

1) Conceptual Design: This is typically the product conceptualization stage, where the specifications for a design

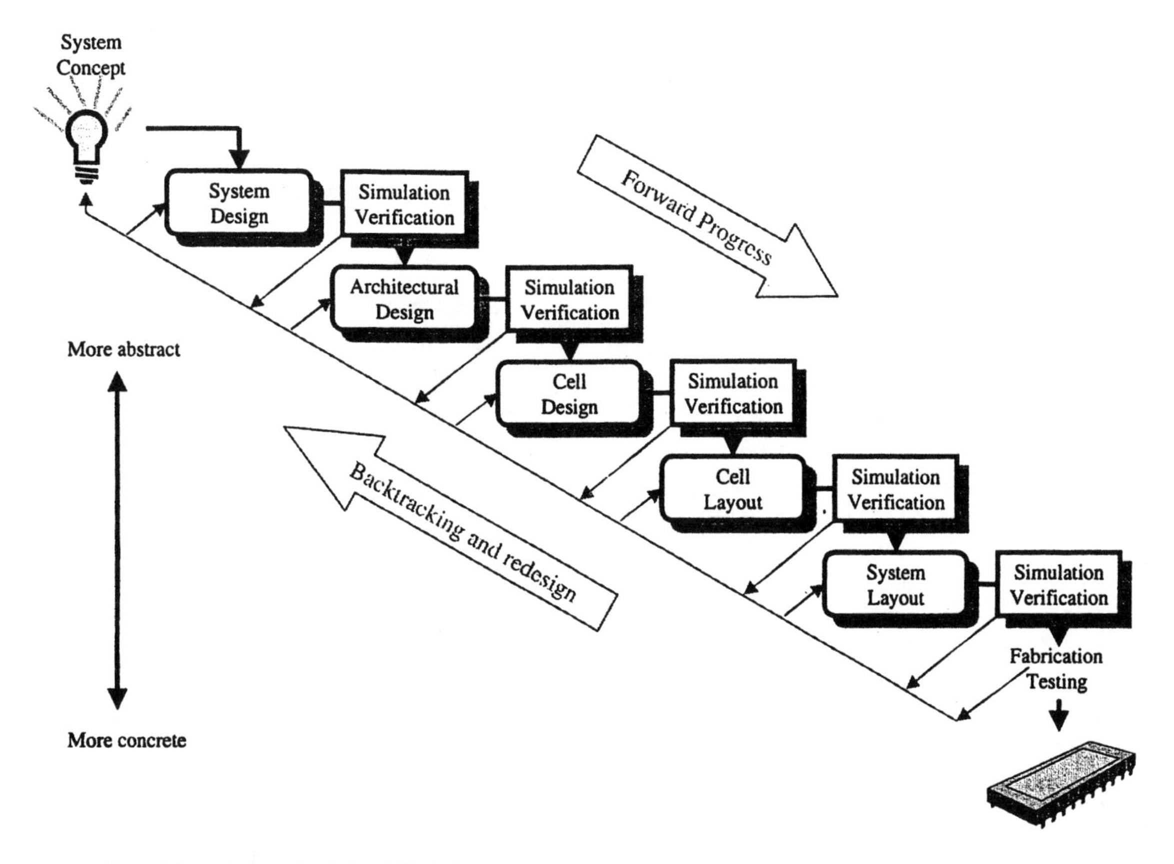

Fig. 3. High-level view of the analog or mixed-signal IC design process.

are gathered and the overall product concept is developed. Careful checking of the specifications is crucial for the later success of the product in its application context. Mathematical tools such as Matlab/Simulink are often used at this stage. This stage also includes setting project management goals such as final product cost and time-to-market, project planning, and tracking.

2) System Design: This is the first stage of the actual design, where the overall architecture of the system is designed and partitioned. Hardware and software parts are defined and both are specified in appropriate languages. The hardware components are described at the behavioral level, and, in addition, the interfaces have to be specified. This stage includes decisions about implementation issues, such as package selection, choice of the target technology, and general test strategy. The system-level partitioning and specifications are then verified using detailed cosimulation techniques.

3) Architectural Design: This stage is the high-level decomposition of the hardware part into an architecture consisting of functional blocks required to realize the specified behavioral description. This includes the partitioning between analog and digital blocks. The specifications of the various blocks that compose the design are defined, and all blocks are described in an appropriate hardware description language (e.g., VHDL and VHDL-AMS). The high-level architecture is then verified against the specifications using behavioral mixed-mode simulations.

4) Cell Design: For the analog blocks, this is the detailed implementation of the different blocks for the given specifications and in the selected technology process, resulting in a fully sized device-level circuit schematic. The stage encompasses both a selection of the proper circuit topology and a dedicated sizing of the circuit parameters. Throughout this process, more complex analog blocks will be further decomposed into a set of subblocks. This whole process will be described in more detail in Section II-B. Manufacturability considerations (tolerances and mismatches) are taken into account in order to guarantee a high yield and/or robustness. The resulting circuit design is then verified against the specifications using SPICE-type circuit simulations.

5) Cell Layout: This stage is the translation of the electrical schematic of the different analog blocks into a geometrical representation in the form of a multilayer layout. This stage involves area optimization to generate layouts that occupy a minimum amount of chip real-estate. The layout is followed by extraction of layout parasitics and detailed circuit-level simulations of the extracted circuit in order to ensure that the performance characteristics do not deviate on account of layout parasitics.

6) System Layout: The generation of the system-level layout of an IC not only includes system-level block place and route, but also power-grid routing. Crosstalk and substrate coupling analysis are important in mixed-signal ICs, and proper measures such as shielding or guarding must also be included. Also, the proper test structures are inserted to make the IC testable. Interconnect parasitics are extracted and detailed verification (e.g., timing analysis) is performed. Finally, the system is verified by cosimulating the hardware part with the embedded software.

7) Fabrication and Testing: This is the processing stage where the masks are generated and the ICs fabricated. Testing is performed during and after fabrication in order to reject defective devices.

Note that any of the many simulation and verification stages throughout this design cycle may detect potential problems with the design failing to meet the target requirements. In that case, **backtracking or redesign** will be needed, as indicated by the upward arrow on the left-hand side of Fig. 3.

B. Hierarchical Analog Design Methodology

This section focuses on the design methodology adopted for the design of analog integrated circuits. These analog circuits could be part of a larger mixed-signal IC. Although at the present time there is no clear-cut general design methodology for analog circuits yet, we outline here the hierarchical design methodology prevalent in many of the emerging experimental analog CAD systems [10]–[15]. For the design of a complex analog macroblock such as a phase-locked loop or an analog-to-digital converter, the analog block is typically decomposed into smaller subblocks (e.g., a comparator or a filter). The specifications of these subblocks are then derived from the initial specifications of the original block, after which each of the subblocks can be designed on its own, possibly by further decomposing it into even smaller subblocks. In this way, constraints are passed down the hierarchy in order to make sure that the top-level block in the end meets the specifications. This whole process is repeated down the decomposition hierarchy until a level is reached that allows a physical implementation (either the transistor level or a higher level in case analog standard cells or IP macrocells are used). The top–down synthesis process is then followed by a bottom–up layout implementation and design verification process. The need for detailed design verification is essential since manufacturing an IC is expensive, and a design needs to be ensured to be fully functional and meet all the design requirements within a window of manufacturing tolerances, before starting the actual fabrication. When the design fails to meet the specifications at some point in the design flow, redesign iterations are needed.

Most experimental analog CAD systems today use a **performance-driven design strategy within such analog design hierarchy**. This strategy consists of the alternation of the following steps in between any two levels i and $i+1$ of the design hierarchy (see Fig. 4):

1) **Top–down path**:
   a) topology selection;
   b) specification translation (or circuit sizing);
   c) design verification.

2) **Bottom–up path**:
   a) layout generation;
   b) detailed design verification (after extraction).

Topology selection is the step of selecting the most appropriate circuit topology that can best meet the given specifications out of a set of already known alternative topologies.

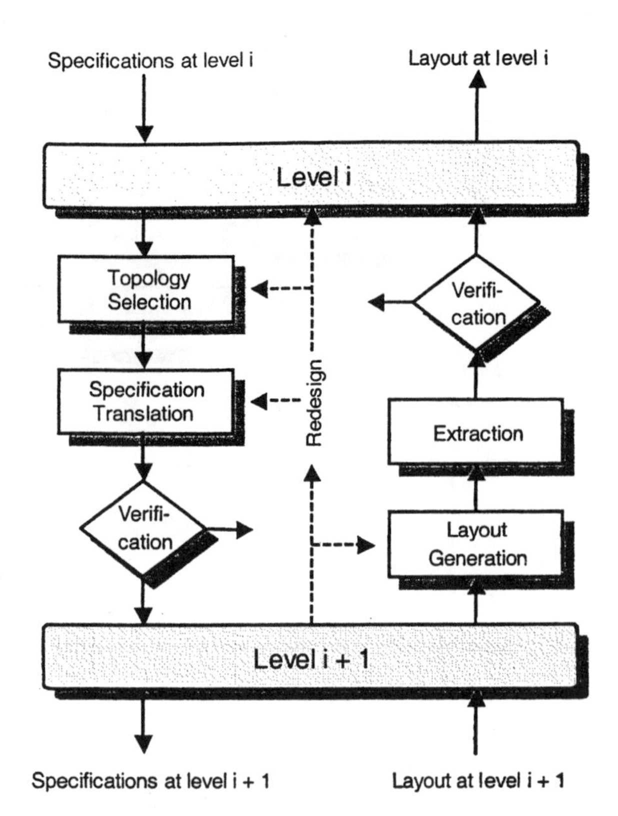

Fig. 4. Hierarchical design strategy for analog circuits.

(An alternative is that the designer develops his/her own new topology.) A topology can be defined hierarchically in terms of lower-level subblocks. For an analog-to-digital converter, for instance, topology selection could be selecting between a flash, a successive approximation, a $\Delta - \Sigma$ or any other topology that can best realize the specifications.

Specification translation is then the step of mapping the specifications for the block under design at a given level (e.g., a converter) into individual specifications for each of the subblocks (e.g., a comparator) within the selected block topology, so that the complete block meets its specifications, while possibly optimizing the design toward some application-specific design objectives (e.g., minimal power consumption). The translated specifications are then verified by means of (behavioral or circuit) simulations before proceeding down in the hierarchy. Behavioral simulations are needed at higher levels in the design hierarchy (when no device-level implementation is available yet); circuit simulations are used at the lowest level in the design hierarchy. At this lowest level, the subblocks are single devices and specification translation reduces to **circuit sizing** (also called circuit dimensioning), which is the determination of all device sizes, element values, and bias parameters in the circuit tuned to the given specifications.

Layout generation is the step of generating the geometrical layout of the block under design, by assembling (place and route) the already generated layouts of the composing subblocks. At the lowest level, these subblocks are individual devices or selected device groupings, which themselves are generated by parameterized procedural device layout generators. Also, power, ground, and substrate connection routing is part of the layout generation step. This step is followed by extraction and, again, detailed verification and simulation to check the impact of the layout parasitics on the overall circuit performance.

The above methodology is called **performance-driven or constraint-driven**, as the performance specifications are the driving input to each of the steps: each step tries to perform its action (e.g., circuit sizing or layout generation) such that the input constraints are satisfied. This also implies that throughout the design flow, constraints need to be propagated down the hierarchy in order to maintain consistency in the design as it evolves through the various design stages and to make sure that the top-level block in the end meets its target specifications. These propagated constraints may include performance constraints, but also geometrical constraints (for the layout), or manufacturability constraints (for yield), or even test constraints. Design constraint propagation is essential to ensure that specifications are met at each stage of the design, which would also reduce the number of redesign iterations. This is the ultimate advantage of top–down design: **catch problems early in the design flow and, therefore, have a higher chance of first-time success, while obtaining a better overall system design.**

Ideally, one would like to have one clean top–down design path. However, this rarely occurs in practice, as a realistic design needs to account for a number of sometimes hard-to-quantify second-order effects as the design evolves. For instance, a choice of a particular topology for a function block may fail to achieve the required specifications or performance specifications may be too tight to achieve, in which case a redesign step is necessary to alter the block topology or loosen the design specifications. In the above top–down/bottom–up design flow, **redesign or backtracking iterations** may therefore be needed at any point where a design step fails to meet its input specifications. In that case, one or more of the previously executed steps will have to be redone, for example, another circuit topology can be selected instead of the failing one, or another partitioning of subblock specifications can be performed. One of the big differences between analog or mixed-signal designs and the more straight top–down digital designs is exactly the larger number of design iterations needed to come to a good design solution. The adoption of a top–down design methodology is precisely intended to reduce this disadvantage.

A question that can be posed is why the analog circuits need to be redesigned or customized for every new application. The use of a library of carefully selected analog standard cells can be advantageous for certain applications, but is in general inefficient and insufficient. Due to the large variety and range of circuit specifications for different applications, any library will only have a partial coverage for each application, or it will result in an excess power and/or area consumption that may not be acceptable for given applications. Many high-performance applications require an optimal design solution for the analog circuits in terms of power, area, and overall performance. A library-based approach would require an uneconomically large collection of infrequently used cells. Instead, analog circuits are better custom tailored toward each specific application and tools should be available to support this. In addition, the porting of the library cells whenever the process changes is a serious effort, that would also require a set of tools to automate.

The following section in this survey paper will describe the progress and the current state of the art in CAD tool development for the main tasks needed in the above analog design methodology: simulation and modeling, symbolic analysis, circuit synthesis, layout generation, yield estimation and design centering, test, and design for testability.

III. Current Status for the Main Tasks in Analog Design Automation

A. Numerical Simulation of Analog and Mixed-Signal Circuits

A key to ensuring design correctness is the use of simulation tools. Simulation tools have long been in use in the IC design process and provide a quick and cost-effective means of design verification without actual fabrication of the device. The most widely used analog CAD tool today, therefore, is a circuit simulator that numerically calculates the response of the circuit to an input stimulus in the time or frequency domain. In the design methodology of Fig. 4, simulation plays a key role. First of all—and this has been its traditional role—simulation is critical for detailed verification after a design has been completed (before layout as well as after extraction from the layout). Analog integrated circuits are typically impacted by many higher order effects that can severely degrade the circuit performance once fabricated, if the effects are not properly accounted for during the design process. Circuit simulation is a good design aid here by providing the capability of simulating many of the higher order effects and verifying circuit performance prior to fabrication, provided the effects are modeled properly. Second, a result of adopting the top–down design paradigm, simulation is needed to explore tradeoffs and verify designs at a high level, before proceeding with the detailed implementation of the lower-level subblocks. The latter also implies a higher level of modeling for the analog blocks. Finally, executable simulation models are also part of the interface needed to enable the integration of complex systems on a chip by combining IP macrocells.

1) Circuit Simulation: Circuit simulation began with the early development of the **SPICE simulator** [6], [16], and [17], which spawned many of the CAD and IC design efforts and has been the cornerstone of many of today's IC designs. The SPICE simulator is to an analog designer what a calculator is to an engineering school sophomore. Advances in mathematics and the development of many new and efficient numerical algorithms as well as advances in interfaces (e.g., user interfaces, waveform displays, script languages, etc.) have over the years contributed to a vast number of commercial CAD tools. Many variants of the SPICE simulator are now marketed by a number of CAD vendors; many of the IC manufacturers have in-house versions of the SPICE simulator that have been adapted to their own proprietary processes and designs. These simulators have been fine-tuned to

Table 1
Different Analog Hardware Description Levels

	continuous time	discrete time
functional level	signal flow diagram with blocks described by mathematical equations	signal flow diagram with blocks described by mathematical equations
behavioral level	block diagram with building blocks described by differential-algebraic equations and/or s-domain transfer functions	block diagram with sampled-data blocks described by difference-algebraic equations and/or z-domain transfer functions
macro level	circuit with basic elements and equivalent (non)linear controlled sources	circuit with switches, capacitors and equivalent controlled sources
circuit level	circuit with basic elements (transistors, diodes, capacitors, resistors, inductors, etc.)	

meet the convergence criteria of the many difficult-to-simulate ICs. SPICE or its many derivatives have evolved into an established designer utility that is being used both during the design phase (often in a designer-guided trial-and-error fashion) and for extensive postlayout design verification.

A problem that has frustrated analog designers for many years is the limited accuracy of the semiconductor device models used in these simulators, especially for small-signal parameters and on the boundary between different operating regions of the devices (where the earlier models had discontinuities). Fortunately, recent models such as BSIM3 v3, Philips model 9 or EKV look more promising for analog design by providing smooth and continuous transitions between different operating regions [18]. For RF applications, however, even these models are not accurate enough, and the latest research work concentrates on analyzing and modeling the extra effects that become important at higher operating frequencies (e.g., the distributed gate, the resistive bulk, and nonquasi-static effects) [19].

With the explosion of mixed-signal designs, the need has also arisen for simulation tools that allow not only simulation of analog or digital circuits separately, but also simulation of truly mixed analog–digital designs [20]. Simulating the large digital parts with full SPICE accuracy results in very long overall simulation times, whereas efficient event-driven techniques exist to simulate digital circuits at higher abstraction levels than the transistor level. Therefore, mixed-mode simulators were developed that glue together an accurate SPICE-like analog simulator to an efficient digital simulator. These so-called **glued mixed-mode simulators** address the conversions of the signals between analog and digital signal representations and of the appropriate loading effects by inserting interface elements at the boundaries between analog and digital circuitry. Also, the synchronization between the analog kernel with its tiny integration steps and the digital kernel with its events determines the efficiency of the overall simulation. Such synchronization is needed at each time point when an event crosses the boundary between analog or digital. Today, the trend clearly is toward a more unified level of algorithm integration with single-kernel multiple-solver solutions, and commercial solutions following that line have recently appeared in the marketplace.

2) Circuit Modeling: In recent years, the need has also arisen for **higher levels of abstraction to describe and simulate analog circuits**. There are three reasons for this. In a top–down design methodology at higher levels of the design hierarchy, where the detailed lower-level circuit implementations are yet unknown, there is a need for higher-level models describing the pin-to-pin behavior of the circuits rather than the (yet unknown) internal structural implementation. Second, the verification of integrated mixed-signal systems also requires higher description levels for the analog sections, since such integrated systems are computationally too complex to allow a full simulation of the entire mixed-signal design in practical terms. Third, when providing or using analog IP macrocells in a SoC context, the virtual component has to be accompanied by an executable model that efficiently models the pin-to-pin behavior of the virtual component. This model can then be used in system-level design and verification, even without knowing the detailed circuit implementation of the macrocell.

To solve those three problems, modeling paradigms and languages from the digital world have migrated to the analog domain. For this reason, macro, behavioral, and functional simulation levels have been developed for analog circuits besides the well-known circuit level [21]. For a commercial simulator to be useful in current industrial mixed-signal design practice, it therefore has to be capable of simulating a system containing a mix of analog blocks described at different levels and in different domains, in combination with digital blocks. This requires a true mixed-signal, multilevel, mixed-domain simulator.

Table 1 gives an overview of the different analog description levels, both for continuous-time and discrete-time analog circuits [21]. In a *macromodel*, an equivalent but computationally cheaper circuit representation is used that has approximately the same behavior as the original circuit. Equivalent sources combine the effect of several other elements that are eliminated from the netlist. The simulation speed-up is roughly proportional to the number of nonlinear devices that can be eliminated. In a *behavioral or functional model*, a purely mathematical description of the input–output behavior of the block is used. This typically will be in the form of a set of differential-algebraic equations (DAE) and/or transfer functions. At the behavioral level, conservation laws still have to be satisfied on the pins connecting different blocks. At the functional level, this is no longer the case and the simulated system turns into a kind of signal-flow diagram. Fig. 5 shows an example of the output response of a CMOS current-steering digital-to-analog converter, modeled at the full device level Fig. 5(b) and at the behavioral level Fig. 5(a). The responses are quite similar (the error between the two time-domain responses for the same input signal is less than 1%), while the behavioral model simulates about 1000 times faster.

To allow an easy exchange of these models across different simulators and different users, the need arose for **standardized analog hardware description languages** in which to describe these higher-level models. These language standards have to provide a consistent way of representing and sharing design information across the different design tasks and across the design hierarchy, and, therefore, provide a unifying trend to link the various tools in a global analog CAD system. For mixed-signal designs, the analog HDLs had to be compatible with the existing digital HDLs (such as VHDL and Verilog). Several parallel analog or mixed-signal HDL language standardization efforts, therefore, have been initiated, recently resulting in the standardized languages VHDL-AMS [8] and Verilog-A/MS [9]. The VHDL-AMS language targets the mixed-signal domain and is a superset of the digital VHDL language. Verilog-A for the analog part and Verilog-MS for the mixed-signal part target compatibility with the Verilog language. Recently, also, the standardization of an extension of VDHL-AMS toward RF has been started.

One of the remaining difficulties with higher-level analog modeling is the automatic characterization of analog circuits and more particularly the **automatic generation of analog macromodels or behavioral models from a given design**. This is a difficult problem area that needs to be addressed in the near future, as it might turn out to be the biggest hurdle for the adoption of these high-level modeling methodologies and AHDLs in industrial design practice. Current approaches can roughly be divided into fitting approaches and constructive approaches. In the fitting approaches, a parameterized model (for example, a rational transfer function, a more general set of equations, or even a neural network model) is first proposed by the model developer and the values of the parameters are then fitted by some least-square error optimization so that the model response matches as closely as possible the response of the real circuit [22]–[24]. The problem with this approach is that first a good model template must be proposed. The second class of methods, therefore, tries to generate or build a model from the underlying circuit description. One approach, for instance, uses symbolic simplification techniques to simplify the physical equations that describe the circuit up to a maximum error bound [25]. Up until now, however, the gains in CPU time were not high enough for practical circuits. More research in this area is definitely needed.

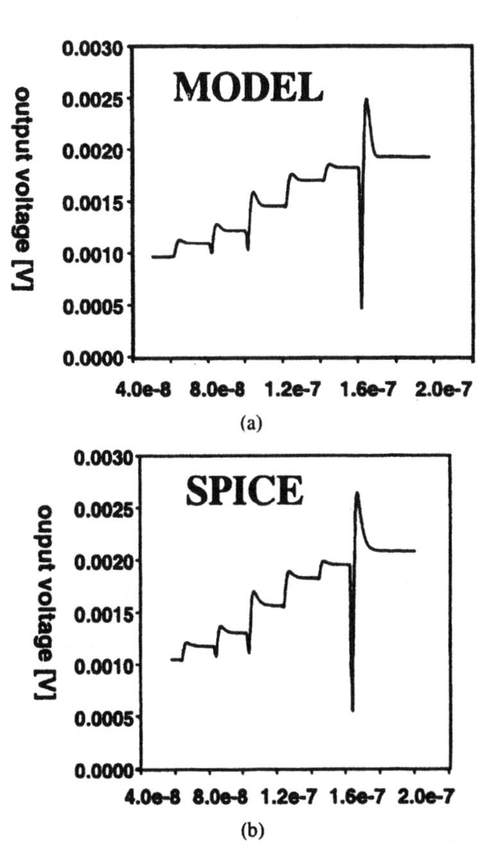

Fig. 5. Comparison of the output response to the same input waveform of a digital-to-analog converter modeled at the behavioral level (a) and at the circuit level (b). The horizontal axis is time in seconds.

3) Dedicated Simulation Techniques: In addition to the above general-purpose simulation tools for analog and mixed-signal circuits, other techniques or tools have been developed for dedicated purposes. An important class of circuits that are used in many signal processing and communication systems are the switched circuits, like switched-capacitor and, more recently, switched-current circuits. Their switched nature, with the resulting switching transients, requires many small numerical integration steps to be simulated within each clock phase if a standard SPICE simulator is used. On the other hand, advantage can be taken of the periodically switched nature of the circuits and the fact that in a time-discrete circuit the signals are only important and, thus, only have to be calculated at specific time points (e.g., the end of each clock phase). This is exploited in several switched-capacitor simulation tools like SWITCAP [26], [27] and SWAP [28] but also in dedicated tools like TOSCA that analyzes switched-capacitor-based $\Delta - \Sigma$ converters [29].

Another important domain is **RF simulation**, needed for instance when developing circuits for wireless applications, where modulated signals have to be simulated and effects like noise, distortion, and intermodulation become important. Here, techniques have been developed to directly simulate the steady-state behavior of these circuits without having to wait for the decay of the initial transients [30], [31]. In the time domain, shooting methods are used for this, which tend to be more suited for strongly nonlinear circuits. In the frequency domain, harmonic balance methods are used, which allow a simulation of the steady-state behavior of nonlinear circuits driven by one- or two-tone signals but which historically required large CPU times and memory sizes for large circuits or for strong nonlinearities. Recently, the implicit matrix technique in combination with both shooting or harmonic balance methods extended the range of these methods to much larger circuits [32]. In parallel, other techniques have been developed such as the envelope simulation technique [33], which combines time and frequency domain simulation to efficiently calculate the circuit's response to truly modulated signals by separating the carrier from the modulation signal. Other dedicated simulation algorithms have been developed for specific applications such as the high-level analysis of entire analog RF receiver front ends in the ORCA tool [34], or for the analysis of nonlinear noise and phase noise in both autonomous and driven circuits such as oscillators, mixers, and frequency synthesizers [35], [36].

An important problem in deep submicrometer technologies where interconnect delays are exceeding gate delays is the **analysis of interconnect networks** during postlayout timing verification. Accurate models for each wire segment and the driving gates are needed, which makes the overall interconnect network too complex to simulate. Therefore, recent developments try to improve the efficiency of timing verification while keeping the accuracy by using piecewise-linear models for gates and model-order reduction techniques for the interconnect network [37]. The complexity of the interconnect network can be reduced by techniques such as asymptotic waveform evaluation (AWE) [38] or related variants such as Padé via Lanczos (PVL), that use moment matching and Padé approximation to generate a lower order model for the response of a large linear circuit like an interconnect network. The early AWE efforts used explicit moment matching techniques, which could generate unstable reduced-order models. Subsequent developments using iterative methods resulted in methods like PVL that overcome many of the deficiencies of the earlier AWE efforts, and stability is now guaranteed using techniques like Arnoldi transformations [39]. The interconnect delay problem has become so important that it is now driving the layout generation to get in-time timing closure, and that it even is becoming essential for synthesis (where, of course, estimation techniques must be used) [40].

An important problem in mixed-signal ICs is **signal integrity analysis**: *the analysis of crosstalk and couplings such as capacitive or inductive interconnect couplings or couplings through the supply lines or the substrate.* Crosstalk can be a limiting factor in today's high-speed circuits with many layers of interconnect. Substrate or supply coupling noise is particularly important for analog circuits, especially where they have to sense small input signals, such as in receiver front ends. Research has been going on to find efficient yet accurate techniques to analyze these problems, which depend on the geometrical configuration and, therefore, are in essence three-dimensional field-solving problems. Typically, finite difference methods or boundary element methods are used to solve for the substrate potential distribution due to injected noise sources [41]–[44]. Recently, these methods have been speeded up with similar acceleration techniques as in RF or interconnect simulation, e.g., using an eigendecomposition technique [45]. Their efficiency even allows one to perform some substrate design optimizations [46]. A problem is that the noise-generating sources (i.e., the switching noise injected by the digital circuitry) are not accurately known, but vary with time depending on the input signals or the embedded programs, and, therefore, have to be estimated statistically. Some attempts to solve this problem characterize every cell in a digital standard cell library by the current they inject in the substrate due to an input transition, and then calculate the total injection of a complex system by summing the contributions of all switching cells over time [47].

B. Symbolic Analysis of Analog Circuits

Analog design is a very complex and knowledge-intensive process, which heavily relies on circuit understanding and related design heuristics. Symbolic circuit analysis techniques have been developed to help designers gain a better understanding of a circuit's behavior. A symbolic simulator is a computer tool that takes as input an ordinary (SPICE-type) netlist and returns as output (simplified) analytic expressions for the requested circuit network functions in terms of the symbolic representations of the frequency variable and (some or all of) the circuit elements [48], [49]. They perform the same function that designers traditionally do by hand analysis (even the simplification). The difference is that the analysis is now done by the computer, which is much faster, can handle more complex circuits, and does not make as many errors. An example of a complicated BiCMOS opamp is shown in Fig. 6. The (simplified) analytic expression for the differential small-signal gain of this opamp has been analyzed with the SYMBA tool [50] and is shown below

$$A_{V0} = \frac{g_{m,M2}}{g_{m,M1}} \frac{g_{m,M4}}{\left(\dfrac{g_{o,M4} g_{o,M5}}{g_{m,M5}+g_{mb,M5}} + \dfrac{G_a + g_{o,M9} + g_{o,Q2}}{\beta_{Q2}}\right)}.$$

The symbolic expression gives a better insight into which small-signal circuit parameters predominantly determine the gain in this opamp and how the user has to design the circuit to meet a certain gain constraint. In this way, symbolic circuit analysis is complementary to numerical (SPICE) circuit simulation, which was described in the previous section. Symbolic analysis provides a different perspective that is more suited for obtaining insight in a circuit's behavior and for circuit explorations, whereas numerical simulation is more appropriate for detailed design validation once a design point has been decided upon. In addition, the generated symbolic

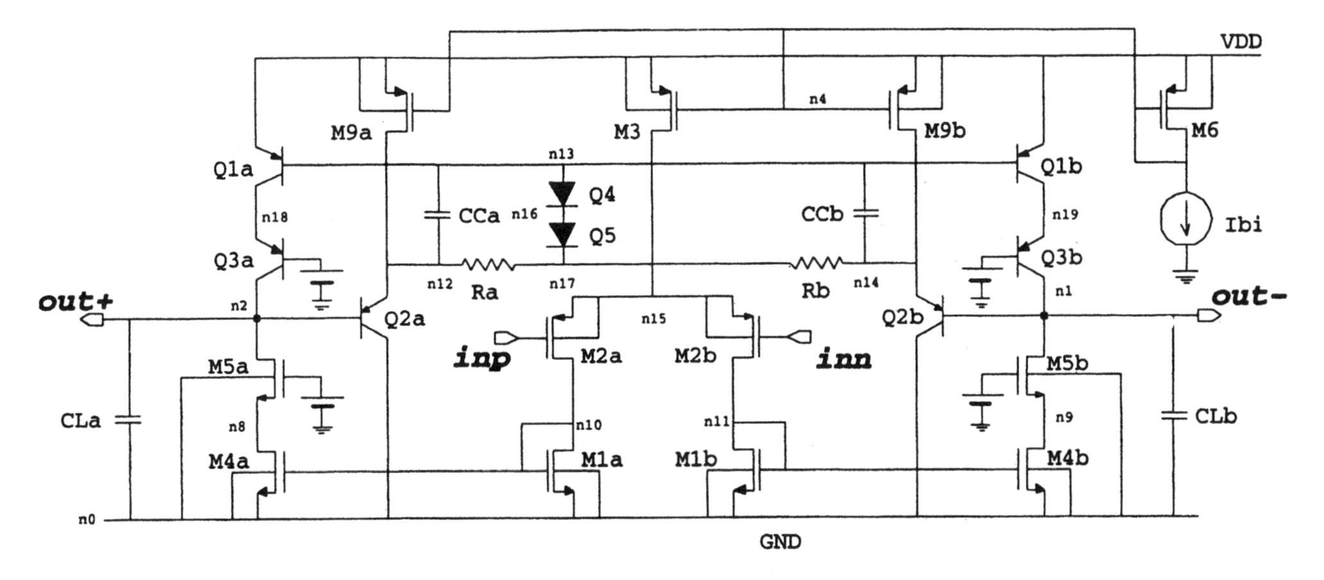

Fig. 6. BiCMOS operational amplifier to illustrate symbolic analysis.

design equations also constitute a model of the circuit's behavior that can be used in CAD tasks such as analog synthesis, statistical analysis, behavioral model generation, or formal verification [48].

At this moment, only symbolic analysis of linear or small-signal linearized circuits in the frequency domain is possible, both for continuous-time and discrete-time (switched) analog circuits [48], [49], [51]. In this way, symbolic expressions can be generated for transfer functions, impedances, noise functions, etc. In addition to understanding the first-order functional behavior of an analog circuit, a good understanding of the second-order effects in a circuit is equally important for the correct functioning of the design in its system application later on. Typical examples are the PSRR and the CMRR of a circuit, which are limited by the mismatches between circuit elements. These mismatches are represented symbolically in the formulas. Another example is the distortion or intermodulation behavior, which is critical in telecom applications. The technique of symbolic simulation has been extended to the symbolic analysis of distortion and intermodulation in weakly nonlinear analog circuits where the nonlinearity coefficients of the device small-signal elements appear in the expressions [52].

Exact symbolic solutions for network functions, however, are too complex for linear(ized) circuits of practical size, and even impossible to calculate for many nonlinear effects. Even rather small circuits lead to an astronomically high number of terms in the expressions, that can neither be handled by the computer nor interpreted by the circuit designer. Therefore, since the late 1980s, and in principle similar to what designers do during hand calculations, dedicated symbolic analysis tools have been developed that use heuristic simplification and pruning algorithms based on the relative importance of the different circuit elements to reduce the complexity of the resulting expressions and retain only the dominant contributions within user-controlled error tolerances. Examples of such tools are ISAAC [51], SYNAP [53], and ASAP [54] among many others. Although successful for relatively small circuits, the fast increase of the CPU time with the circuit size restricted their applicability to circuits between 10 and 15 transistors only, which was too small for many practical applications.

In recent years, however, an algorithmic breakthrough in the field of symbolic circuit analysis has been realized. The techniques of simplification before and during the symbolic expression generation, as implemented in tools like SYMBA [50] and RAINIER [55], highly reduce the computation time and, therefore, enable the symbolic analysis of large analog circuits of practical size (like the entire 741 opamp or the example of Fig. 6). In simplification before generation (SBG), the circuit schematic, or some associated matrix or graph(s), are simplified before the symbolic analysis starts [56], [57]. In simplification during generation (SDG), instead of generating the exact symbolic expression followed by pruning the unimportant contributions, the desired simplified expression is built up directly by generating the contributing dominant terms one by one in decreasing order of magnitude, until the expression has been generated with the desired accuracy [50], [55].

All these techniques, however, still result in large, expanded expressions, which restricts their usefulness for larger circuits. Therefore, for really large circuits, the technique of hierarchical decomposition has been developed [58], [59]. The circuit is recursively decomposed into loosely connected subcircuits. The lowest-level subcircuits are analyzed separately and the resulting symbolic expressions are combined according to the decomposition hierarchy. This results in the global nested expression for the complete circuit, which is much more compact than the expanded expression. The CPU time increases about linearly with the circuit size, provided that the coupling between the different subcircuits is not too strong. Another compact representation of symbolic expressions was presented recently. Following the use of binary decision diagrams in logic synthesis, determinant decision diagrams (DDD) have been proposed as a technique to canonically represent determinants in a compact nested format [60]. The advantage is that all operations on these DDDs are linear with the size of the DDD, but the DDD itself is not always linear with the size of the circuit. This technique has been combined with hierarchical analysis in [61]. Further investigation will have to prove the usefulness of this technique in practice.

Based on the many research results in this area over the last decade, it can be expected that symbolic analysis techniques will soon emerge in the commercial EDA marketplace and that they will soon be part of the standard tool suite of every analog designer. In the meantime, new (possibly heuristic) algorithms for the symbolic analysis of transient and large-signal circuit characteristics are currently being developed in academia.

C. Analog Circuit Synthesis and Optimization

The first step in the analog design flow of Fig. 4 is analog circuit synthesis, which consists of two tasks: topology selection and specification translation. Synthesis is a critical step since most analog designs require a custom optimized design and the number of (often conflicting) performance requirements to be taken into account is large. Analog circuit synthesis is the inverse operation of circuit analysis. During analysis, the circuit topology and the subblock parameters (such as device sizes and bias values) are given and the resulting performance of the overall block is calculated, as is done in the SPICE simulator. During synthesis, on the other hand, the block performance is specified and an appropriate topology to implement this block has to be decided first. This step is called topology selection. Subsequently, values for the subblock parameters have to be determined, so that the final block meets the specified performance constraints. This step is called specification translation at higher levels in the design hierarchy, in which case performance specifications of subblocks have to be determined, or circuit sizing at the device level, in which case the sizes and biasing of all devices have to be determined. See Fig. 7 for an illustration of this flow for low-level cells. The inversion process inherent to synthesis, however, is not a one-to-one mapping, but typically is an underconstrained problem with many degrees of freedom. *The different analog circuit synthesis systems that have been explored up till now can be classified based on how they perform topology selection and how they eliminate the degrees of freedom during specification translation or circuit sizing.* In many cases, the initial sizing produces a near optimal design that is further fine-tuned with a circuit optimization tool. The performance of the resulting design is then verified using detailed circuit simulations with a simulator such as SPICE, and when needed the synthesis process is iterated to arrive at a close-fit design. We will now discuss the two basic steps in more detail.

1) Topology Selection: Given a set of performance specifications and a technology process, a designer or a synthesis tool must first select a circuit schematic that is most suitable to meet the specifications at minimal implementation cost (power, chip area). This problem can be solved by selecting a schematic from among a known set of alternative topologies such as stored in a library (topology selection), or by generating a new schematic, for example by modifying an existing schematic. Although the earliest synthesis approaches

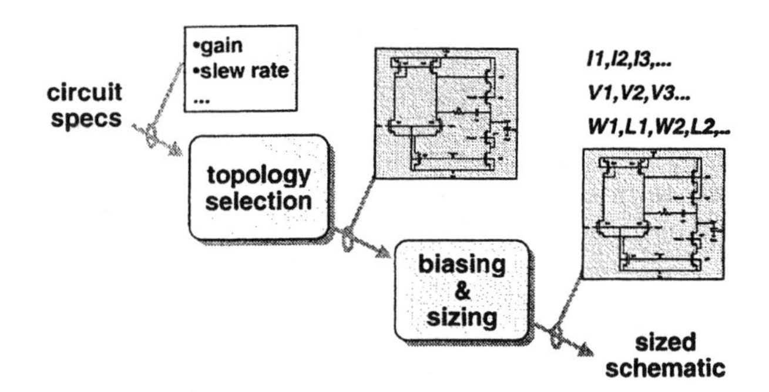

Fig. 7. Basic flow of analog circuit synthesis for a basic cell: topology selection and circuit sizing.

considered topology selection and sizing together, the task of topology selection has received less attention in recent years, where the focus was primarily on the circuit sizing. Finding the optimal circuit topology for a given set of performance specifications is rather heuristic in nature and brings to bear the real expert knowledge of a designer. Thus, it was only natural that the first topology selection approaches like in OASYS [10], BLADES [62], or OPASYN [63] were rather heuristic in nature in that they used rules in one format or another to select a proper topology (possibly hierarchically) out of a predefined set of alternatives stored in the tool's library.

Later approaches worked in a more quantitative way in that they calculate the feasible performance space of each topology that fits the structural requirements, and then compare that feasible space to the actual input specifications during synthesis to decide on the appropriateness and the ordering of each topology. This can for instance be done using interval analysis techniques [64] or using interpolation techniques in combination with adaptive sampling [65]. In all these programs, however, topology selection is a separate step. There are also a number of optimization-based approaches that integrate topology selection with circuit sizing as part of one overall optimization loop. This was done using a mixed integer-nonlinear programming formulation with Boolean variables representing topological choices [66], or by using a nested simulated evolution/annealing loop where the evolution algorithm looks for the best topology and the annealing algorithm for the corresponding optimum device sizes [67]. Another approach that uses a genetic algorithm to find the best topology choice was presented in DARWIN [68]. Of these methods, the quantitative and optimization-based approaches are the more promising developments that address the topology selection task in a deterministic fashion as compared to the rather ad-hoc heuristic methods.

2) Analog Circuit Sizing: Once an appropriate topology has been selected, the next step is specification translation, where the performance parameters of the subblocks in the selected topology are determined based on the specifications of the overall block. At the lowest level in the design hierarchy, this reduces to circuit sizing where the sizes and biasing of all devices have to be determined such that the final circuit meets the specified performance constraints. This mapping from performance specifications into proper, preferrably optimal, device sizes and biasing for a selected analog circuit topology

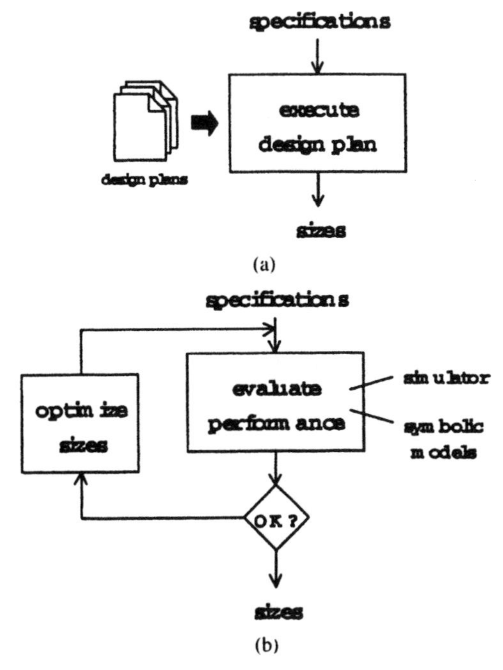

Fig. 8. The two basic approaches toward analog circuit synthesis: (a) the knowledge-based approach using procedural design plans, and (b) the optimization-based approach.

in general involves solving the set of physical equations that relate the device sizes to the electrical performance parameters. However, solving these equations explicitly is in general not possible, and analog circuit sizing typically results in an underconstrained problem with many degrees of freedom. The two basic ways to solve for these degrees of freedom in the analog sizing process are either by exploiting analog design knowledge and heuristics, or by using optimization techniques. These two basic methods, which are schematically depicted in Fig. 8, correspond to the two broad classes of approaches adopted toward analog circuit synthesis, i.e., the knowledge-based approaches and the optimization-based approaches [7], [69].

a) Knowledge-Based Analog Sizing Approaches: The first generation of analog circuit synthesis systems presented in the mid to late 1980s were **knowledge-based**. Specific heuristic design knowledge about the circuit topology under design (including the design equations but also the design strategy) was acquired and encoded explicitly in some computer-executable form, which was then executed during the synthesis run for a given set of input specifications to directly obtain the design solution. This approach is schematically illustrated in Fig. 8(a). The knowledge was encoded in different ways in different systems.

The IDAC tool [70] used manually derived and prearranged design plans or design scripts to carry out the circuit sizing. The design equations specific for a particular circuit topology had to be derived and the degrees of freedom in the design had to be solved explicitly during the development of the design plan using simplifications and design heuristics. Once the topology was chosen by the designer, the design plan was loaded from the library and executed to produce a first-cut design that could further be fine-tuned through local optimization. The big advantage of using design plans is their fast execution speed, which allows for fast-performance space explorations. The approach also attempts to take advantage of the knowledge of analog designers. IDAC's schematic library was also quite extensive, and it included various analog circuits such as voltage references, comparators, etc., besides operational amplifiers. The big disadvantages of the approach are the lack of flexibility in the hardcoded design plans and the large time needed to acquire the design equations and to develop a design plan for each topology and design target, as analog design heuristics are very difficult to formalize in a general and context-independent way. It has been reported [71] that the creation of a design script or plan typically took four times more effort than is needed to actually design the circuit once. A given topology must therefore at least be used in four different designs before it is profitable to develop the corresponding design plan. Considering the large number of circuit schematics in use in industrial practice, this large setup time essentially restricted the commercial usability of the IDAC tool and limited its capabilities to the initial set of schematics delivered by the tool developer. Also, the integration of the tool in a spreadsheet environment under the name PlanFrame [72] did not fundamentally change this. Note that due to its short execution times, IDAC was intended as an interactive tool: the user had to choose the topology him/herself and also had to specify values for the remaining degrees of freedom left open in the design plan.

OASYS [10] adopted a similar design-plan-based sizing approach where every (sub)block in the library had its own handcrafted design plan, but the tool explicitly introduced hierarchy by representing topologies as an interconnection of subblocks. For example, a circuit like an opamp was decomposed into subcircuits like a differential pair, current mirrors, etc., and not represented as one big device-level schematic as in IDAC. OASYS also added a heuristic approach toward topology selection, as well as a backtracking mechanism to recover from design failures. As shown in Fig. 9, the complete flow of the tool was then an alteration of topology selection and specification translation (the latter by executing the design plan associated with the topology) down the hierarchy until the device level is reached. If the design does not match the desired performance characteristics at any stage in this process, OASYS backtracks up the hierarchy, trying alternate configurations for the subblocks. The explicit use of hierarchy allowed to reuse design plans of lower-level cells while building up higher-level-cell design plans and, therefore, also leveraged the number of device-level schematics covered by one top-level topology template. Although the tool was used successfully for some classes of opamps, comparators and even a data converter, collecting and ordering all the design knowledge in the design plans still remained a huge manual and time-consuming job, restricting the practical usefulness of the tool. The approach was later adopted in the commercial MIDAS system [71], which was used successfully in-house for certain types of data converters. Also, AZTECA [73] and CATALYST [74] use the design-plan approach for the high-level design of successive-approximation and high-speed CMOS data converters, respectively. Inspired

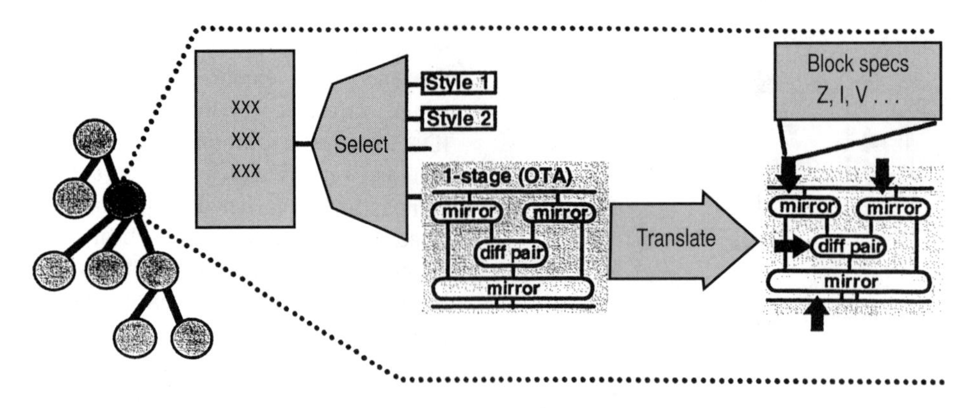

Fig. 9. Hierarchical alternation of topology selection and specification translation down the design hierarchy.

by artificial intelligence research, also other ways to encode the knowledge have been explored, such as in BLADES [62], which is a rule-based system to size analog circuits, in ISAID [75], [76] or in [77].

In all these methods, the heuristic design knowledge of an analog designer turned out to be difficult to acquire and to formalize explicitly, and the manual acquisition process was very time consuming. In addition to analytic equation-based design knowledge, procedural design knowledge is also required to generate design plans, as well as specialized knowledge to support tasks such as failure handling and backtracking. The overhead to generate all this was too large compared to a direct design of the circuit, restricting the tools basically to those circuits that were delivered by the tool developers. Their coverage range was found to be too small for the real-life industrial practice and, therefore, these first approaches failed in the commercial marketplace.

b) Optimization-Based Analog Sizing Approaches: In order to make analog synthesis systems more flexible and extendible for new circuit schematics, an alternative approach was followed starting in the late 1980s. This research resulted in a second generation of methods, the optimization-based approaches. These use numerical optimization techniques to implicitly solve for the degrees of freedom in analog design while optimizing the performance of the circuit under the given specification constraints. These strategies also strive to automate the generation of the required design knowledge as much as possible, e.g., by using symbolic analysis techniques to automatically derive many of the design equations and the sizing plans, or to minimize the explicitly required design knowledge by adopting a more equation-free simulation-oriented approach. This optimization-based approach is schematically illustrated in Fig. 8(b). At each iteration of the optimization routine, the performance of the circuit has to be evaluated. Depending on which method is used for this performance evaluation, two different subcategories of methods can be distinguished.

In the *subcategory of equation-based optimization approaches*, (simplified) analytic design equations are used to describe the circuit performance. In approaches like OPASYN [63] and STAIC [78], the design equations still had to be derived and ordered by hand, but the degrees of freedom were resolved implicitly by optimization. The OPTIMAN tool [79] added the use of a global simulated annealing algorithm, but also tried to solve two remaining problems. Symbolic simulation techniques were developed to automate the derivation of the (simplified) analytic design equations needed to evaluate the circuit performance at every iteration of the optimization [69]. Today, the ac behavior (both linear and weakly nonlinear) of relatively large circuits can already be generated automatically. The second problem is then the subsequent ordering of the design equations into an application-specific design or evaluation plan. Also, this step was automated using constraint programming techniques in the DONALD tool [80]. Together with a separate topology-selection tool based on boundary checking and interval analysis [64] and a performance-driven layout generation tool [81], all these tools are now integrated into the AMGIE analog circuit synthesis system [82] that covers the complete design flow from specifications over topology selection and circuit sizing down to layout generation and automatic verification. An example of a circuit that has been synthesized with this AMGIE system is the particle/radiation detector front end of Fig. 10, which consists of a charge-sensitive amplifier (CSA) followed by an n-stage pulse-shaping amplifier (PSA). All opamps are complete circuit-level schematics in the actual design as indicated in the figure. A comparison between the specifications and the performances obtained by an earlier manual design of an expert designer and by the fully computer-synthesized circuit is given in Table 2. In the experiment, a reduction of the power consumption with a factor of 6 (from 40 to 7 mW) was obtained by the synthesis system compared to the manual solution. Also, the final area is slightly smaller. The layout generated for this example is shown in Fig. 11.

The technique of equation-based optimization has also been applied to the high-level synthesis of $\Delta-\Sigma$ modulators in the SD-OPT tool [83]. The converter architecture is described by means of symbolic equations, which are then used in a simulated-annealing-like optimization loop to derive the optimal subblock specifications from the specifications of the converter. Recently, a first attempt was presented toward the full behavioral synthesis of analog systems from an (annotated) VHDL-AMS behavioral description. The VASE tool follows a hierarchical two-layered optimization-based design-space exploration approach to

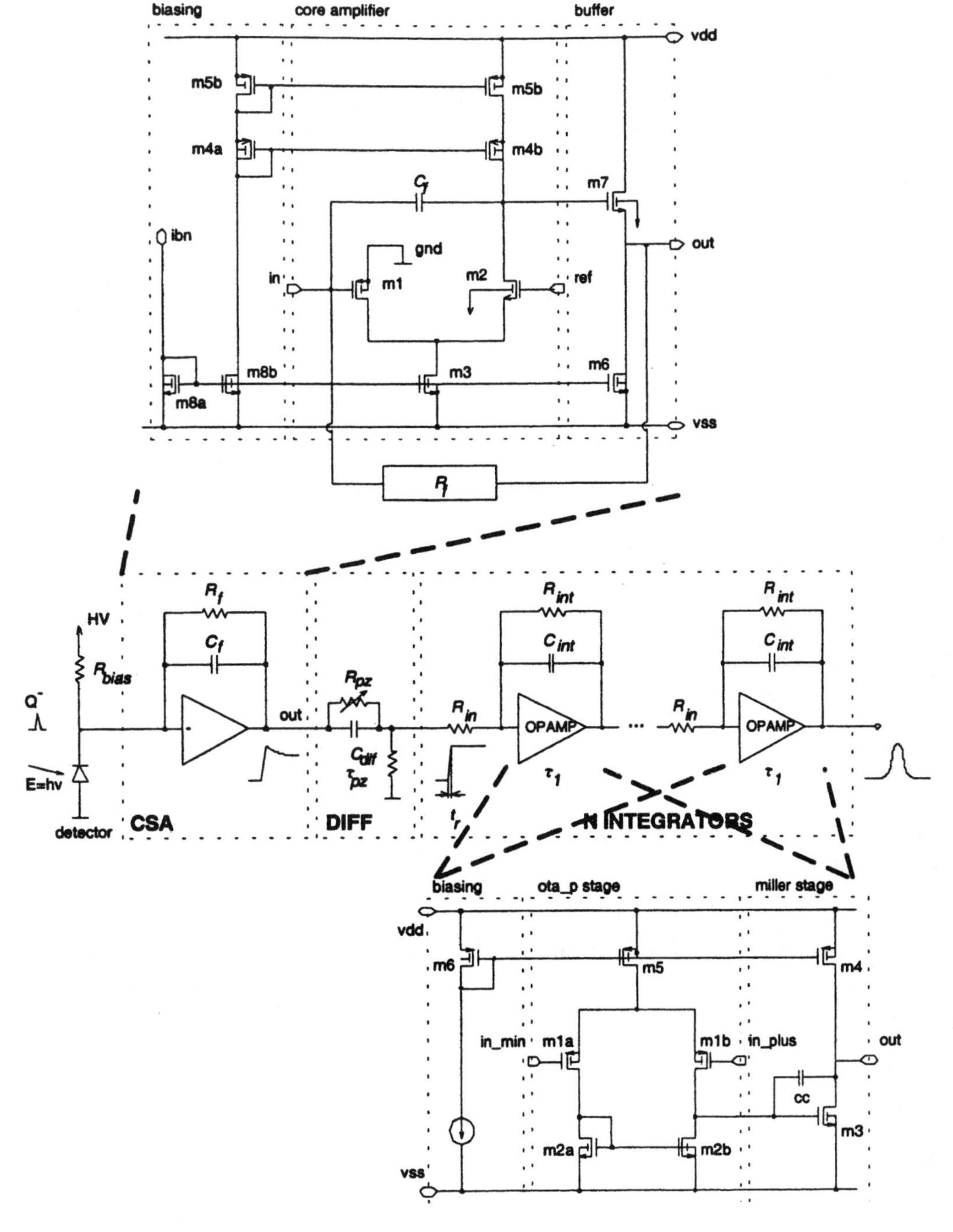

Fig. 10. Particle/radiation detector front end as example for analog circuit synthesis. (The opamp and filter stage symbols represent full circuit schematics as indicated.)

produce sized subblocks from behavioral specifications [84]. A branch-and-bound algorithm with efficient solution-space pruning first generates alternative system topologies by mapping the specifications via a signal-flow-graph representation onto library elements. For each resulting topology a genetic-algorithm-based heuristic method is then executed for constraint transformation and subblock synthesis, which concurrently transforms system-level constraints into subblock design parameters (e.g., a bias current) and fixes the topologies and transistor sizes for all subblocks. For reasons of efficiency the performances at all levels in the considered hierarchy are estimated using analytic equations relating design parameters to performance characteristics. The genetic algorithms operating at the different levels are speeded up by switching from traditional genetic operators to directed-interval-based operators that rely on characterization tables with qualitative sensitivity information to help focusing the search process in promising local regions.

In general, the big advantages of these analytic approaches are their fast evaluation time and their flexibility in manipulation possibilities. The latter is reflected in the freedom to choose the independent input variables, which has a large impact on the overall evaluation efficiency [85], as well as the possibility to perform more symbolic manipulations. Recently, it has been shown that the design of CMOS opamps can be formulated (more precisely, it can be fairly well ap-

Table 2
Example of Analog Circuit Synthesis Experiment with the AMGIE System

performance	specification	manual design	automated synthesis
peaking time	< 1.5 ms	1.1 ms	1.1 ms
counting rate	> 200 kHz	200 kHz	294 kHz
noise	< 1000 RMS e-	750 RMS e-	905 RMS e-
gain	20 V/fC	20 V/fC	21 V/fC
output range	> -1..1 V	-1..1 V	-1.5..1.5 V
power	minimal	40 mW	7 mW
area	minimal	0.7 mm^2	0.6 mm^2

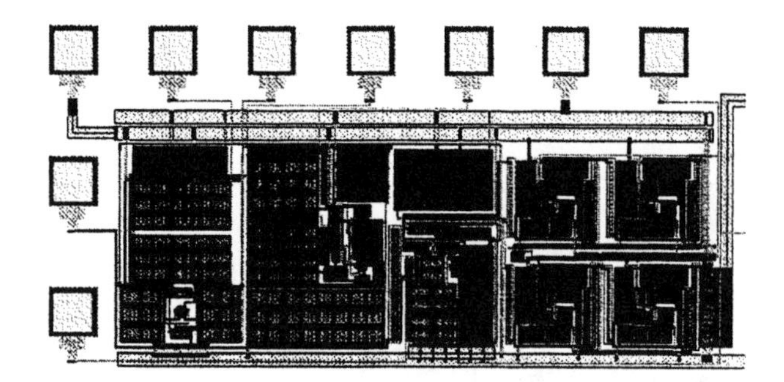

Fig. 11. Layout of the particle/radiation detector front end generated with the AMGIE analog synthesis system.

proximated) as a posynomial convex optimization problem that can then be solved using geometric programming techniques, producing a close-by first-cut design in an extremely efficient way [86]. The initial optimization time literally reduces to seconds. The same approach has been applied to some other circuits as well [87]. The big drawback of the analytic methods, however, is that the design equations still have to be derived and, despite the progress in symbolic circuit analysis, not all design characteristics (such as transient or large-signal responses) are easy to capture in analytic equations with sufficient accuracy. For such characteristics either rough approximations have to be used, which undermines the sense of the whole approach, or one has to fall back on numerical simulations. This problem has sparked research efforts to try to develop equation-free approaches.

Therefore, in recent years and with improving computer power, a second *subcategory of simulation-based optimization approaches* toward analog circuit synthesis has been developed. These methods perform some form of full numerical simulation to evaluate the circuit's performance in the inner loop of the optimization [see Fig. 8(b)]. Although the idea of optimization-based design for analog circuits dates back at least 30 years [88], it is only recently that the computer power and numerical algorithms have advanced far enough to make this really practical. For a limited set of parameters circuit optimization was already possible in DELIGHT.SPICE [89]. This method is most favorable in fine-tuning an already designed circuit to better meet the specifications, but the challenge in automated synthesis is to solve for all degrees of freedom when no good initial starting point can be provided. To this end, the FRIDGE tool [90] calls a plain-vanilla SPICE simulation at every iteration of a simulated-annealing-like global optimization algorithm. In this way, it is able to synthesize low-level analog circuits (e.g., opamps) with full SPICE accuracy. Performance specifications are divided in design objectives and strong and weak constraints. The number of required simulations is reduced as much as possible by adopting a fast cooling schedule with reheating to recover from local minima. Nevertheless, many simulations are performed, and the number of optimization parameters and their range has to be restricted in advance by the designer. The introduction of a new circuit schematic in such an approach is relatively easy, but the drawback remains the long run times, especially if the initial search space is large.

An in-between solution was therefore explored in the ASTRX/OBLX tool [91] where the simulation itself is speeded up by analyzing the linear (small-signal) characteristics more efficiently than in SPICE by using Asymptotic Waveform Evaluation [38]. For all other characteristics equations still have to be provided by the designer. So this is essentially a mixed equation-simulation approach. The ASTRX subtool compiles the initial synthesis specification into an executable cost function. The OBLX subtool then numerically searches for the minimum of this cost function via simulated annealing, hence determining the optimal circuit sizing. To achieve accuracy in the solution, encapsulated industry-standard models are used for the MOS transistors. For efficiency the tool also uses a dc-free biasing formulation of the analog design problem, where the dc constraints [i.e., Kirchhoff current law (KCL) at every node] are not imposed by construction at each optimization iteration, but are solved by relaxation throughout the optimization run by adding the KCL violations as penalty terms to the cost function. At the final optimal solution, all the penalty terms are driven to zero, thus resulting in a KCL-correct and thus electrically consistent circuit in the end. ASTRX/OBLX has been applied successfully to a wide variety of cell-level designs, such as a 2× gain stage [92], but the CPU times remain large. The tool is also most suited only when the circuit behavior is relatively linear, because the other characteristics still require equations to be derived.

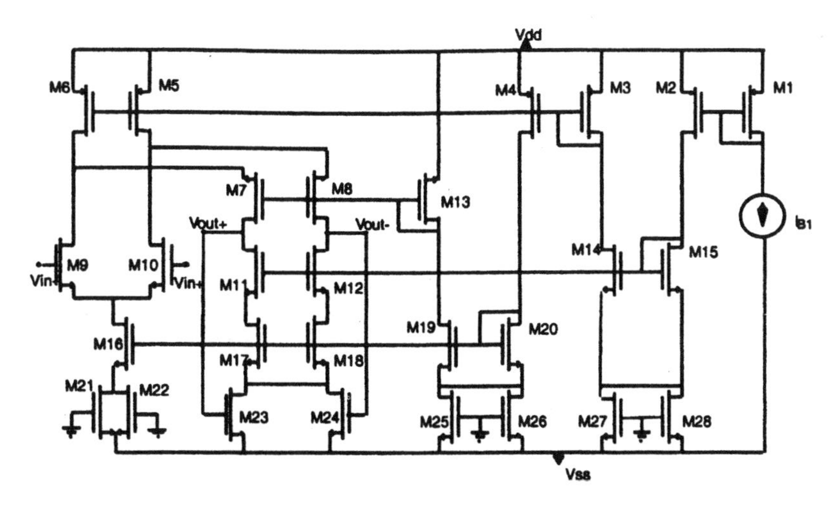

Fig. 12. Example circuit for analog circuit synthesis.

Table 3
Example of Analog Circuit Synthesis Results with FRIDGE and MAELSTROM

performance	specification	FRIDGE	MAELSTROM
nr of variables	-	10	31
CPU time	-	45 min	184 min
DC gain	≥ 70 dB	79 dB	76 dB
UGF	≥ 30 MHz	35 MHz	31 MHz
phase margin	≥ 60°	66°	61°
output swing	≥ 3.0 V	3.2 V	2.9 V
settling time	minimal	90 ns	62 ns

In the quest for industry-grade quality, most recent approaches, therefore, use complete SPICE simulations for all characteristics. To cut down on the large synthesis time, more efficient optimization algorithms are used and/or the simulations are executed as much as possible in parallel on a pool of workstations. In [93], the generalized boundary curve is used to determine the step length within an iterative trust-region optimization algorithm. Using the full nonlinear cost function based on the linearized objectives significantly reduces the total number of iterations in the optimization. The ANACONDA tool, on the other hand, uses a global optimization algorithm based on stochastic pattern search that inherently contains parallelism and, therefore, can easily be distributed over a pool of workstations, to try out and simulate 50 000 to 100 000 circuit candidates in a few hours [94]. MAELSTROM is the framework that provides the simulator encapsulation and the environment to distribute both the search tasks of the optimization algorithm as well as the circuit evaluations at every iteration of the optimization over parallel workstations in a network [95]. It uses another parallel global algorithm, a combined annealing-genetic algorithm, to produce fairly good designs in a few hours. These brute-force approaches require very little advance modeling to prepare for any new circuit topology and have the same accuracy as SPICE. Fig. 12 shows an example of an opamp circuit that has been synthesized with FRIDGE and MAELSTROM. The results are summarized in Table 3. In [96], ANACONDA/MAELSTROM, in combination with macromodeling techniques to bridge the hierarchical levels, was applied to an industrial-scale analog system (the equalizer/filter front end for an ADSL CODEC). The experiments demonstrated that the synthesis results are comparable to, or sometimes better than, manual design. Although appealing, these methods still have to be used with care by designers because the run times (and, therefore, also the debug time) remain long, and because the optimizer may easily produce improper designs if the right design constraints are not added to the optimization problem. Reducing the CPU time remains a challenging area for further research.

Other simulation-based approaches can be found in tools such as OAC [97], which is a specific nonlinear optimization tool for operational amplifiers and which is based on redesign starting from a previous design solution stored in the system's database. It also performs physical floorplanning during the optimization, which accounts for physical layout effects during circuit synthesis. A recent application of the simulation-based optimization approach to the high-level optimization of analog RF receiver front ends was presented in [98]. A dedicated RF front-end simulator was developed and used to calculate the ratio of the wanted signal to the total power of all unwanted signals (noise, distortion, aliasing, phase noise, etc.) in the frequency band of interest. An optimization loop then determines the optimal specifications for the receiver subblocks such that the desired signal quality for the given application is obtained at the lowest possible power consumption for the overall front-end topology. Behavioral models and power estimators are used to evaluate the different front-end subblocks at this high architectural level.

In summary, the initial design systems like IDAC were too closed and restricted to their initial capabilities and, therefore, failed in the marketplace. The current trend is toward open analog design systems that allow the designer to easily extend and/or modify the design capabilities of the system without too much software overhead. Compared to the initial knowledge-based approaches, the big advantages of the more recent optimization-based approaches are their high flexibility and extendibility, both in terms of design objectives (by altering the cost function) and in terms of the ease to add new circuit schematics. Although some additional research is still needed, especially to reduce the CPU times, it can be concluded that a lot of progress has been achieved in the field of analog circuit synthesis during the past ten years. This has resulted in the development of several experimental analog synthesis systems, with which several designs have successfully been synthesized, fabricated, and measured. This has been accomplished not only for opamps, but also for filters [99] and data converters [100]. Based on these recent methods, several commercial tools are currently being developed that will be introduced in the marketplace in the near future.

Finally, it has to be added that industrial design practice not only calls for fully optimized nominal design solutions, but also expects high robustness and yield in the light of varying operating conditions (supply voltage or temperature variations) and statistical process tolerances and mismatches [101]. Techniques to analyze the impact of this on the yield or the capability index Cpk of the circuit [102] after the nominal design has been completed will be discussed in detail in Section III-E. Here, we briefly describe the efforts to integrate these considerations in the synthesis process itself. Yield and robustness precautions were already hardcoded in the design plans of IDAC [70], but are more difficult to incorporate in optimization-based approaches. Nevertheless, first attempts in this direction have already been presented. The ASTRX/OBLX tool has been extended with manufacturability considerations and uses a nonlinear infinite programming formulation to search for the worst case "corners" at which the evolving circuit should be evaluated for correct performance [103]. The approach has been successful in several test cases but does increase the required CPU time even further (roughly by $4\times$–$10\times$). Also, the OPTIMAN program has been extended by fully exploiting the availability of the analytic design equations to generate closed-form expressions for the sensitivities of the performances to the process parameters [104]. The impact of tolerances and mismatches on yield or Cpk can then easily be calculated at each optimization iteration, which then allows to synthesize the circuits simultaneously for performance and for manufacturability (yield or Cpk). The accuracy of the statistical predictions still has to be improved. The approach in [93] uses parameter distances as robustness objectives to obtain a nominal design that satisfies all specifications with as much safety margin as possible for process variations. The resulting formulation is the same as for design centering and can be solved efficiently using the generalized boundary curve. Design centering, however, still remains a second step after the nominal design. More research in this direction, therefore, is still needed.

D. Analog and Mixed-Signal Layout Synthesis

The next important step in the analog design flow of Fig. 4 after the circuit synthesis is the generation of the layout. The field of analog layout synthesis is more mature than circuit synthesis, in large part because it has been able to leverage ideas from the mature field of digital layout. Yet, real commercial solutions are only beginning to appear in recent years. Below we distinguish analog circuit-level layout synthesis, which has to transform a sized transistor-level schematic into a mask layout, and system-level layout assembly, in which the basic functional blocks are already laid out and the goal is to floorplan, place, and route them, as well as to distribute the power and ground connections.

1) Analog Circuit-Level Layout Synthesis: The earliest approaches to analog cell layout synthesis relied on **procedural module generation**, like in [105], in which the layout of the entire circuit was precoded in a software tool that generates the complete layout at run time for the actual parameter values entered. This approach is today frequently used during interactive manual layout for the single-keystroke generation of the entire layout of a single device or a special group of (e.g., matched) devices by means of parameterized procedural device generators. For circuits, however, the approach is not flexible enough, and large changes in the circuit parameters (e.g., device sizes) may result in inefficient area usage. In addition, a module generator has to be written and maintained for each individual circuit.

A related set of methods are called **template driven**. For each circuit, a geometric template (e.g., a sample layout [71] or a slicing tree [63]) is stored that fixes the relative position and interconnection of the devices. The layout is then completed by correctly generating the devices and the wires for the actual values of the design according to this fixed geometric template, thereby trying to use the area as efficiently as possible. These approaches work best when the changes in circuit parameters result in little need for global alterations in the general circuit layout structure, which is the case for instance during technology migration or porting of existing layouts, but which is not the case in general.

In practice, changes in the circuit's device sizes often require large changes in the layout structure in order to get the best performance and the best area occupation. The performance of an analog circuit is indeed impacted by the layout. Parasitics introduced by the layout, such as the parasitic wire capacitance and resistance or the crosstalk capacitance between two neighboring or crossing wires, can have a negative impact on the performance of analog circuits. It is, therefore, of utmost importance to generate analog circuit layouts such that: 1) the resulting circuit still satisfies all performance specifications; and 2) the resulting layout is as compact as possible. This requires full-custom layout generation, which can be handled with a **macrocell-style layout strategy**. The terminology is borrowed from digital floorplanning algorithms, which manipulate flexible layout blocks (called "macros"), arrange them topologically and then route them. For analog circuits, the entities to be handled are structural groups of one single or a special grouping of devices (e.g., a matching pair of transistors). These "device-level macros" are to be folded, oriented, placed, and interconnected to make up a good overall layout. Note that many analog devices and special device groupings, even for the same set of parameters, can be generated in different geometrical variants, e.g., two matching devices can be laid out in interdigitated form, or stacked, or in a quad-symmetric fashion, etc. For each variant of each macrocell structure used, procedural module generators have to be developed to generate the actual layouts of the cells for a given set of parameter values. A drawback is that these generators have to be maintained and updated whenever the technology process changes, which creates some pressure to limit the number of different generators. Whatever the macrocells considered in a custom analog circuit layout synthesis tool, a placement routine optimally arranges the cells, while also selecting the most appropriate geometrical variant for each; a router interconnects them, and sometimes a compactor compacts the resulting layout, all while taking care of the many constraints like symmetry and matching typical for analog circuits, and also attending to the numerous parasitics and couplings to which analog circuits (unfortunately) are sensitive. This general analog circuit layout synthesis flow is shown in Fig. 13.

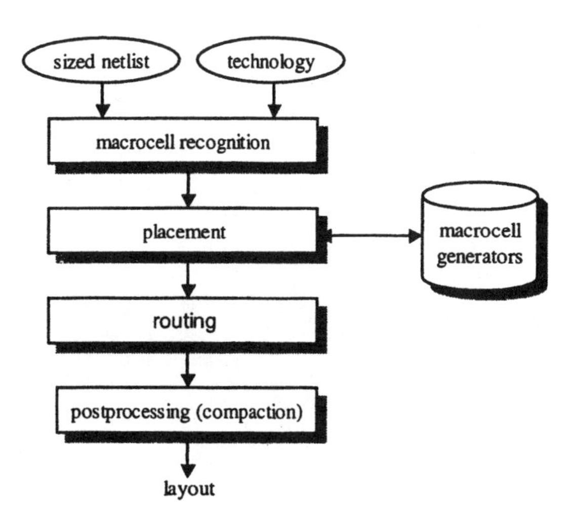

Fig. 13. General flow of an analog circuit layout synthesis tool.

The need to custom optimize analog layouts led to the **optimization-based macrocell-place-and-route layout generation approaches** where the layout solution is not predefined by some template, but determined by an optimization program according to some cost function. This cost function typically contains minimum area and net length and adherence to a given aspect ratio, but also other terms could be added, and the user normally can control the weighting coefficients of the different contributions. The advantage of the optimization-based approaches is that they are generally applicable and not specific to a certain circuit, and that they are flexible in terms of performance and area as they find the most optimum solution at run time. The penalty to pay is their larger CPU times, and the dependence of the layout quality on the set-up of the cost function. Examples of such tools are ILAC [106] and the different versions of KOAN/ANAGRAM [107]. ILAC borrowed heavily from the best ideas from digital layout: efficient slicing-tree floorplanning with flexible blocks, global routing via maze routing, detailed routing via channel routing, area optimization via compaction [106]. The problem with the approach was that it was difficult to extend these primarily-digital algorithms to handle all the low-level geometric optimizations that characterize expert manual design. Instead, ILAC relied on a large, very sophisticated library of device generators.

ANAGRAM and its successor KOAN/ANAGRAM II kept the macrocell style, but reinvented the necessary algorithms from the bottom up, incorporating many manual design optimizations [107]–[109]. The device placer KOAN relied on a very small library of device generators and migrated important layout optimizations into the placer itself. KOAN, which was based on an efficient simulated annealing algorithm, could dynamically fold, merge, and abut MOS devices and, thus, discover desirable optimizations to minimize parasitic capacitance on the fly during optimization. Its companion, ANAGRAM II, was a maze-style detailed area router capable of supporting several forms of symmetric differential routing, mechanisms for tagging compatible and incompatible classes of wires (e.g., noisy and sensitive wires), parasitic crosstalk avoidance and over-the-device routing. Also, other device placers and routers operating in the macrocell-style have appeared (e.g., LADIES [110] and ALSYN [111]). Results from these tools can be quite impressive. For example, Fig. 14 shows two versions of the layout of an industrial 0.25-μm CMOS comparator. On the left is a manually created layout, on the right is a layout generated automically with a commercial tool operating in the macrocell style. The automatic layout compares well to the manual one.

An important improvement in the next generation of optimization-based layout tools was the shift from a rather qualitative consideration of analog constraints to an explicit quantitative optimization of the performance goals, resulting in the **performance-driven or constraint-driven approaches**. For example, KOAN maximized MOS drain-source merging during layout and ANAGRAM II minimized crosstalk, but without any specific, quantitative performance targets. The performance-driven approaches, on the other hand, explicitly quantify the degradation of the performance due to layout parasitics and the layout tools are driven such that this extra layout-induced performance degradation is within the margins allowed by the designer's performance specifications [112]. In this way, more optimum solutions can be found as the importance of each layout parasitic is weighed according to its impact on the circuit performance, and the tools can much better guarantee by construction that the circuit

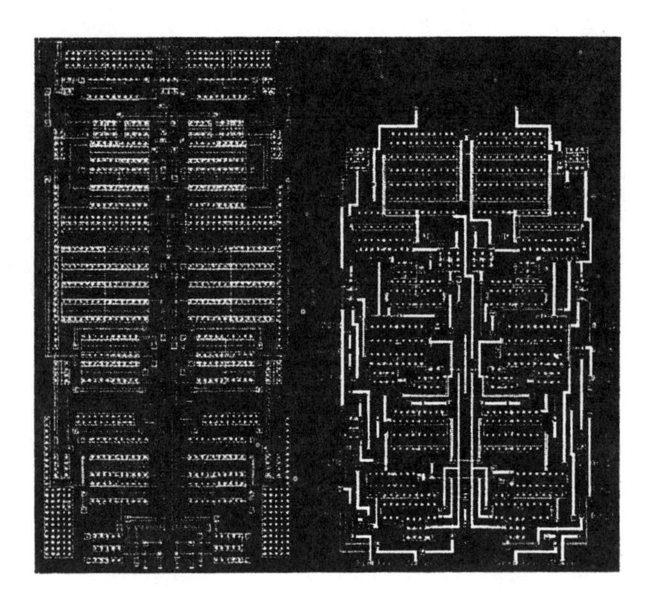

Fig. 14. Manual (left) versus automatic (right) layout for an industrial 0.25 μm CMOS analog comparator.

will meet the performance specifications also after the layout phase (if possible).

Tools that adopt this approach include the area router ROAD [113], the placement tool PUPPY-A [114] and the compaction tool SPARCS-A [115]. The routers ROAD [113] and ANAGRAM III [116] have a cost function that drives them such that they minimize the deviation from acceptable bounds on wire parasitics. These bounds are provided by designers or derived from the margins on the performance specifications via sensitivities. The LAYLA system [81], [117], [118] consists of performance-driven analog placement and routing tools that penalize excess layout-induced performance degradation by adding the excess degradation directly as an extra term to the cost function. Effects considered include for instance the impact of device merging, device mismatches, parasitic capacitance and resistance of each wire, parasitic coupling due to specific proximities, thermal gradients, etc. The router can manage not just parasitic wire sensititivies, but also yield and testability concerns. A layout generated by means of LAYLA was shown in Fig. 11.

In all the above tools, sensitivity analysis is used to quantify the impact on final circuit performance of low-level layout decisions and has emerged as the critical glue that links the various approaches being taken for circuit-level layout and for system assembly. An influential early formulation of the sensitivity analysis problem was [119], which not only quantified layout impacts on circuit performance, but also showed how to use nonlinear programming techniques to map these sensitivities into maximum bounds on parasitics, which serve as constraints for various portions of the layout task. Later approaches [117], however, showed that this intermediate mapping step may not be needed. Other work [120] showed how to extract critical constraints on symmetry and matching directly from a device schematic.

A recent innovation in CMOS analog circuit layout generation tools is the idea of separating the device placement into two distinct tasks: device stacking followed by stack placement. By rendering the circuit as an appropriate graph of connected drains and sources, it is possible to identify natural clusters of MOS devices that ought to be merged, called stacks, to minimize parasitic capacitance, instead of discovering these randomly over the different iterations of the placement optimization. The work in [121] presented an exact algorithm to extract all the optimal stacks, and dynamically choose the right stacking and the right placement of each stack throughout the placement optimization. Since the underlying algorithm has exponential time complexity, enumerating all stacks can be very time consuming. The work in [122] offered a variant that extracts one optimal set of stacks very fast. The idea is to use this in the inner loop of a placer to evaluate fast trial merges on sets of nearby devices.

One final problem in the macrocell place-then-route style is the separation of the placement and routing steps. The problem is to estimate how much wiring space to leave around each device for the subsequent routing. Too large estimates result in open space, too small estimates may block the router and require changes to the placement. One solution is to get better estimates by carrying out simultaneous placement and global routing, which has been implemented for slicing-style structures in [123]. An alternative is to use dynamic wire space estimation where space is created during routing when needed. Another strategy is analog compaction, where extra space is left during placement, which after routing is then removed by compaction. Analog compactors that maintain the analog constraints introduced by the previous layout steps were for instance presented in [115] and [124]. A more radical alternative is to perform simultaneous device place and route. An experimental version of KOAN [125] supported this by iteratively perturbing both the wires and the devices, but the method still has to be improved for large practical circuits.

Performance-driven macrocell-style custom analog circuit-level layout schemes are maturing nowadays, and the first commercial versions already started to be offered. Of course, there are still problems to solve. The wire space problem is one. Another open problem is "closing the loop" from circuit synthesis to circuit layout, so layouts that do not meet the specifications can, if necessary, cause actual circuit design changes (via circuit resynthesis). Even if a performance-driven approach is used, which should generate layouts correct by construction, circuit synthesis needs accurate estimates of circuit wiring loads to obtain good sizing results, and circuit synthesis needs to leave sufficient performance margins for the layout-induced performance degradation later on. How to control this loop, and how to reflect layout concerns in synthesis and synthesis concerns in layout remain difficult.

2) Mixed-Signal System Layout Assembly: A mixed-signal system is a set of analog and digital functional blocks. System-level layout assembly means floorplanning, placement, and global and detailed routing (including the power grid) of the entire system, where the layouts of the individual blocks are generated by lower-level tools such as discussed in the previous section for the analog circuits. In

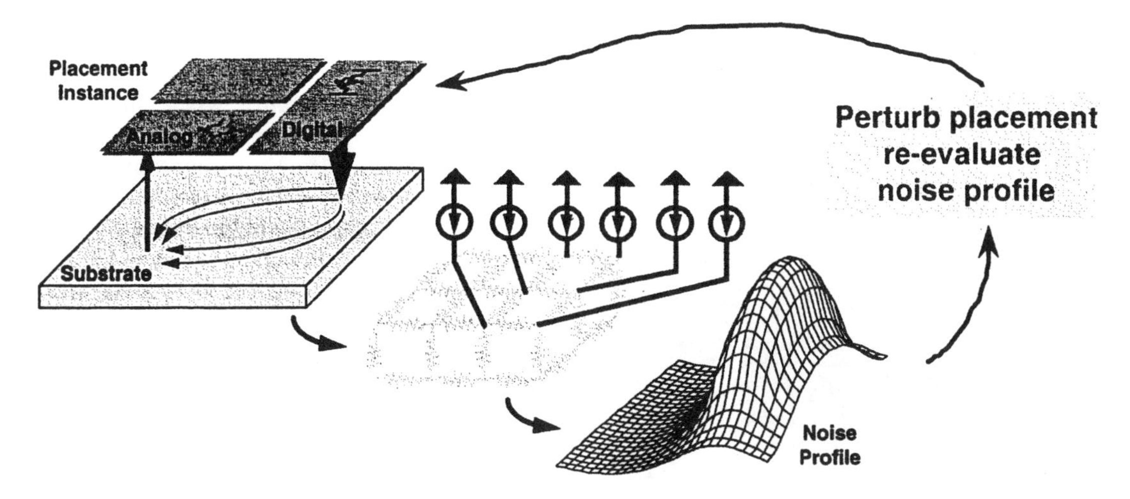

Fig. 15. Principal flow of the WRIGHT mixed-signal floorplanner that incorporates a fast substrate noise coupling evaluator.

addition to sensitivities to wire parasitics, an important new problem in mixed-signal systems is the coupling between digital switching noise and sensitive analog circuits (for instance, capacitive crosstalk or substrate noise couplings).

As at the circuit level, procedural layout generation remains a viable alternative for well-understood designs with substantial regularity (e.g., switched-capacitor filters [126]). More generally though, work has focused on custom placement and routing at the block level. For row-based layout, an early elegant solution to the coupling problem was the segregated-channels idea of [127] to alternate noisy digital and sensitive analog wiring channels in a row-based cell layout. The strategy constrains digital and analog signals never to be in the same channel, and remains a practical solution when the size of the layout is not too large. For large designs, analog channel routers were developed. In [128], it was observed that a well-known digital channel routing algorithm, based on a gridless constraint-graph formulation, could easily be extended to handle critical analog problems that involve varying wire widths and wire separations needed to isolate noisy and sensitive signals. The channel router ART extended this strategy to handle complex analog symmetries in the channel, and the insertion of shields between incompatible signals [129].

The WREN [130] and WRIGHT [131] tools generalized these ideas to the case of arbitrary layouts of mixed functional blocks. WREN comprises both a mixed-signal global router and channel router [130]. The tool introduced the notion of signal-to-noise ratio (SNR)-style constraints for incompatible signals, and both the global and detailed routers strive to comply with designer-specified noise rejection limits on critical signals. WREN incorporates a constraint mapper that transforms input noise rejection constraints from the across-the-whole-chip form used by the global router into the per-channel per-segment form necessary for the channel router. WRIGHT on the other hand uses simulated annealing to floorplan the blocks, but with an integrated fast substrate-noise-coupling evaluator so that a simplified view of substrate noise influences the floorplan [131]. Fig. 15 shows the flow of this tool. Accurate methods to analyze substrate couplings have been presented in Section III-A. In the frame of layout synthesis tools, however, the CPU times of these techniques are prohibitive and there is a need for fast yet accurate substrate-noise-coupling evaluators to explore alternative layout solutions.

Another important task in mixed-signal system layout is power grid design. Digital power grid layout schemes usually focus on connectivity, pad-to-pin ohmic drop, and electromigration effects. But these are only a small subset of the problems in high-performance mixed-signal chips, which feature fast-switching digital systems next to sensitive analog parts. The need to mitigate unwanted substrate interactions, the need to handle arbitrary (nontree) grid topologies, and the need to design for transient effects such as current spikes are serious problems in mixed-signal power grids. The RAIL system [132], [133] addresses these concerns by casting mixed-signal power grid synthesis as a routing problem that uses fast asymptotic-waveform-evaluation-based [38] linear system evaluation to electrically model the entire power grid, package and substrate in the inner loop of grid optimization. Fig. 16 shows an example RAIL redesign of a data channel chip in which a demanding set of dc, ac, and transient performance constraints were met automatically.

Most of these mixed-signal system-level layout tools are of recent vintage, but because they often rely on mature core algorithms from similar digital layout problems, many have been prototyped both successfully and quickly. Although there is still much work to be done to enhance existing constraint mapping strategies and constraint-based layout tools to handle the full range of industrial concerns, the progress obtained opens possibilities for commercialization activities in the near future to make these tools available to all practicing mixed-signal designers.

E. Yield Estimation and Optimization

The manufacturing of integrated circuits suffers from statistical fluctuations inherent to the fabrication process itself [101]. These fluctuations can be local (e.g., lithographic spots) or global (e.g., gate oxide gradients). These

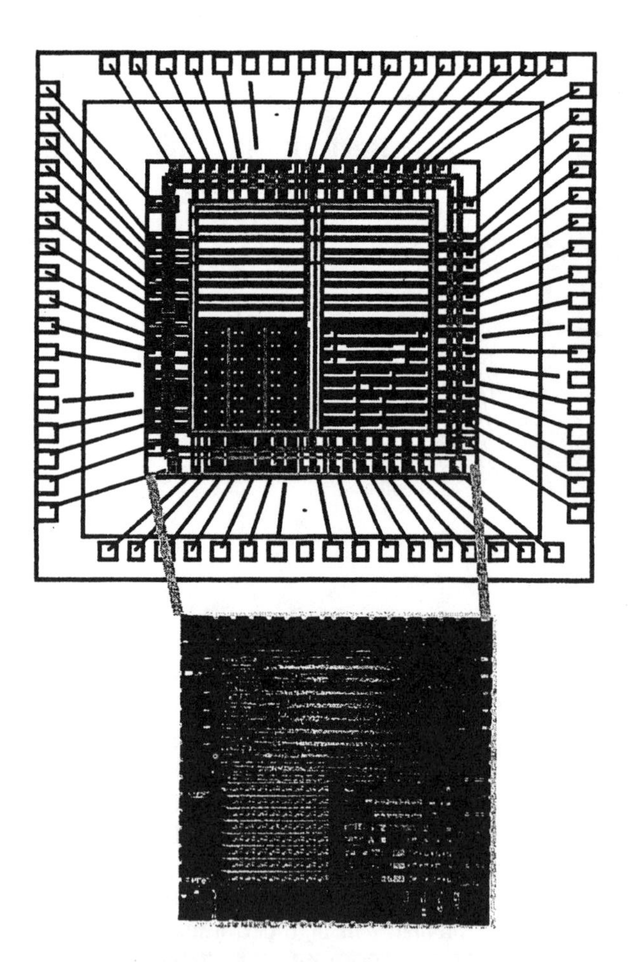

Fig. 16. RAIL power grid design for IBM data channel chip.

process-induced deformations result in deviations of the actual performances of the fabricated ICs from the expected performances. Some of these deformations result in the fabricated IC not having the expected functionality. This is then called a *structural failure*. Others cause the fabricated IC to have, despite the correct functionality, observed performances far from the targeted values. This is then called a *hard performance failure*. Both failures are also termed *catastrophic failures*, caused by a catastrophic fault. Other process deformations result in the observed performances deviating only slightly around the targeted values due to the statistical tolerances (interdie variations) and mismatches (intradie variations) of the device parameters. This is then called a *parametric* (*or soft*) *performance failure*, caused by a parametric fault.

Both catastrophic and parametric faults cause a fraction of the fabricated ICs to have performances that do not meet the required specifications. The ratio of accepted to the total number of fabricated ICs is called the **yield**. If the yield is significantly less than 100%, this implies a financial loss to the IC manufacturer. Therefore it is important to already calculate and maximize the manufacturing yield during the design stage. This is called **design for manufacturability**, which implies techniques for yield estimation and yield optimization. Even more expensive are the field failures that show up when the IC is in use in a product in the field, for instance when the IC is used under extreme operating conditions like high temperatures. To try to avoid this, the design has to be made as robust as possible. This is called **design for robustness or design for quality**, which implies techniques for variability minimization and design centering. Both aspects are of course interrelated and can be captured in a characteristic like the capability index **Cpk** [102].

Due to the fluctuations of the device parameters, the performance characteristics of the circuit will show fluctuations. The corresponding parametric yield is the integral over the acceptability region (i.e., the region where all specifications are satisfied) of the joint probability density function of the fluctuating parameters. This yield can be calculated in both the device parameter space or the circuit performance space. Its calculation, however, is complicated by the fact that in either space one of the two elements is not known explicitly: the statistical fluctuations are known in the device parameter space but not in the circuit performance space, whereas the acceptability region is known in the performance space but not in the parameter space [134]. The relation between the device parameter space and the circuit performances depends on the nominal design point chosen for the circuit, but is in general a nonlinear transformation that is not known explicitly. This relation can for instance be derived on a point by point basis using SPICE simulations. All this makes yield estimation a difficult task where the different techniques trade off accuracy versus CPU time in a different way.

A simple approach often used in practice is **worst case analysis** [102]. Instead of doing a true statistical analysis, some worst case combinations (e.g., slow p-type devices, fast n-type devices, etc.) of the device parameters are used to calculate the "worst case" corners of the performances. Problems are that this approach can lead to large overestimations and that for analog circuits it is usually not known *a priori* which combinations of the device parameters result in the worst case corners. A deterministic technique based on sequential quadratic programming (SQP) to calculate the worst case device model parameters and worst case operating conditions for each performance separately has been described in [135]. The corresponding worst case distances can be used to measure the yield and the robustness of the design.

The most straightforward statistical approach is to use **Monte-Carlo simulations**, in which a large number of samples of the device parameters are generated according to the given statistics. The samples must include the correlations, but principal component analysis (PCA) techniques can be used to generate uncorrelated parameter sets. For each generated sample set a SPICE simulation is performed and all resulting performance data are combined to derive the statistics of the circuit performances. Since this process requires many circuit simulations, especially when the confidence factor on the yield or Cpk estimate has to be small, this Monte-Carlo approach is very time-consuming.

The CPU time can be reduced with the response surface method [136], which works in two steps. First, the parameter space is sampled with controlled simulations according to some design-of-experiments (DOE) scheme. For each performance characteristic a response surface is then constructed by fitting a simple function (e.g., a first- or second-order polynomial) of the device parameters to the simulated performance data. By initial screening, unimportant device parameters can be eliminated from each model. Then, in the second step, the evaluation of these simple response surface models replaces full circuit simulations during the yield or Cpk calculation, for instance using the Monte-Carlo method. The only limitation of this approach is the accuracy of the response surface models. Nevertheless, various forms of Monte-Carlo simulation—with or without response surface acceleration technique—remain staples for almost all analog design methodologies. A rigorous formulation to estimate the parametric yield of a circuit affected by mismatches has been presented in [137]. The approach, however, does not include yield due to defects.

Finally, if the calculated yield or Cpk is not sufficient, then the design will have to be altered in order to increase the yield or the design robustness [102], [134]. By changing the nominal design point, either the probability density function in the performance space, or the acceptability region in the device parameter space will be modified, and hence the yield/Cpk. The goal is then to modify the nominal design point such that the yield is maximized, or the variability of the circuit performance is minimized, or the design is centered. The most direct approach is to put one of the above yield/Cpk estimation routines in an optimization loop and directly try to maximize the yield/Cpk. Needless to say, this is extremely time consuming. Here again, the response surface technique can be used to reduce the number of required simulations, by using the technique in two subsequent phases and generating an explicit response surface model for the yield or Cpk as a function of the nominal design point parameters. An alternative is the technique of design centering where the design is centered as much as possible within the acceptability region. Geometric approximation techniques such as simplicial approximation, etc., have to be used to approximate the acceptability region in the device parameter space, but these techniques suffer from a bad computational complexity. The approach in [93] uses the worst case distances as robustness objectives during design centering, and applies the generalized boundary curve technique to reduce the number of simulations required. The challenge is also to incorporate yield and robustness optimization as an integrated part of circuit synthesis, instead of considering it as an separate step to be performed after a first nominal-only circuit sizing. First attempts in this direction have already been presented in Section III-C [103], [104], but the execution times still have to be reduced further.

F. Analog and Mixed-Signal Testing and Design for Testability

The final step in the IC production cycle is the testing of the IC [138], [139]. The objective of testing is to verify whether the fabricated IC meets the specifications it was designed for or not. The cost of testing is a considerable part of the final cost of an IC, and the later in the production cycle a certain defect is detected, the more expensive the financial loss. The effective cost of a product is therefore related to the quality of the applied tests and the time necessary to generate and apply the tests. In mixed-signal designs, an additional complication is that, although analog circuits constitute only a small fraction of the die area, they are responsible for the largest fraction of the overall test cost. This is not only because they require more expensive mixed- signal test equipment, but also because they require longer test times than the digital circuitry and because there are no structured test waveform generation tools nor structured design for test methodologies.

The testing of an IC is complicated by the limited pin count, which means that only a limited number of nodes can be accessed from the outside. In order to increase the controllability and observability of internal nodes, design for testability (DfT) measures have to be included in the design and layout of the IC. A second problem critical in analog designs is the presence of statistical process parameter fluctuations, which make multiple fabricated samples of the same circuit showing a statistical spread of responses around the nominal response (called a *tolerance box*). Due to this spreading, the tolerance boxes of the fault-free and certain faulty circuits may overlap for the given test set, creating so called ambiguity regions where the given test cannot uniquely distinguish between fault-free and faulty devices. The test may therefore result in some percentage of undecisive or wrong test decisions, which of course has to be reduced as much as possible. Fortunately, the situation is changing and the field of analog and mixed-signal testing is characterized by many interesting developments, which will only be touched upon here briefly. The reader is referred to [138]–[140] for more details.

For **fault detection** in go/no-go production testing, test cost and test time are critical. Two basic approaches can be distinguished. The traditional test approach is **specification based**, also called functional testing. Tests are applied to check the functionality of the circuit under test against the specifications. For instance, for an analog-to-digital converter, these are the traditional tests such as the histogram test to derive the integral and differential nonlinearity (INL and DNL) and the effective number of bits (ENOB). The problems with this approach are that there is no quantification of how necessary and how sufficient these tests are for detecting whether a circuit is fault-free or not, and that this test generation process is difficult to automate. A more recent approach, therefore, is **defect-based** analog testing, mimicking the approach in use for a long time in digital testing. In this case, tests are applied to detect whether specific faults are present in the fabricated circuit structure or not. This requires that the faults that are being tested for are collected in advance in a so-called fault dictionary, which implies the use of a certain fault model and of fault simulation. Unlike in the digital world, however, there is no common agreement in the analog design and test community on the fault models to be used. For instance, a spot defect shorting two neighboring wires can be represented as an ideal short of the two corresponding nodes, or as a resistor between the two nodes with a value of $0.1\ \Omega$, or a resistor with a value of $1\ \Omega$, etc. All this makes it hard to calculate a "fault coverage" characteristic to qualify test sets. In addition, not all possible faults can occur in practice. For instance, assuming all possible shorts between any two nodes in the circuit and all possible opens for all interconnections in the circuit may lead to unrealistically large fault lists. A more efficient technique is inductive fault analysis [140] that generates a list of physically realistic faults to be tested for by sprinkling defects across the actual layout and extracting possible faults from this. In any case, whatever the fault list, each faulty circuit considered has to be simulated during the design stage to build up the fault dictionary, which is extremely time consuming. Therefore, techniques to speed up fault simulations have received large attention in recent years [141].

During **fault diagnosis** in the initial IC prototype characterization and validation phase the test problem is even more complicated: not only must it be detected whether there is a fault or not in the circuit, but also the location of the fault has to be identified to diagnose the error. This is more difficult as there may also be ambiguity regions between the responses of circuits with different faults, making it impossible to distinguish between the ambiguous cases for the given test set. Two basic approaches are possible [138]. In **simulation before test** (SBT), the fault-free and faulty circuits are simulated beforehand and the responses are stored in a fault dictionary. During testing the measured response is then compared to the signatures in the fault dictionary to identify the most likely case. The second approach is **simulation after test** (SAT), where the circuit is measured first and then the circuit parameters are reconstructed from the measurement results and compared to the fault-free values. It is not always possible, however, to uniquely solve the value of each element out of a given measurement set, making it necessary to use optimization techniques to estimate the most likely element values from the measurements.

Concerning the tests themselves, both dc, ac (frequency), or transient tests can be carried out, and the measured signals can be node voltages, pin currents (e.g., the power-supply current—the equivalent of digital IDDQ testing) or derived characteristics. The goal of **test waveform generation** is to determine which tests have to be carried out (which stimuli to be applied, which signals to be measured) in order to have maximum fault coverage at a minimum test cost. In general, this has to be formulated as an optimization problem and some interesting approaches have been presented in recent years [142], [143]. In order to overcome the problem of the limited accessibility in ICs, **design for testability** measures have to be taken to improve the controllability and observability by propagating test signals from the external IC pins to and from internal, otherwise inaccessible nodes [140]. These measures can for instance take the form of extra test pins or extra test structures [e.g., (de)multiplexers] to isolate and individually test the different analog blocks in the system [144], analog scan path techniques to read out selected test node signals into a serial scan path [145], or extra circuitry to reconfigure the circuit in the test mode, e.g., use a feedback structure to reconfigure the circuit into an oscillator mode where the oscillation frequency then indicates whether the circuit is faulty or fault-free [146].

With the move toward systems on a chip where the analog circuitry is deeply embedded into the system, the limited accessibility problem will become more and more stringent. An appealing alternative for such systems is the use of **built-in self test** (BIST). In this case, the generation of the test waveforms, the capturing of the test responses, and their comparison to the expected good responses are all performed on-chip, resulting in the circuit autonomously returning a pass or fail signal, at the expense of some chip area overhead. The response comparison has to consider tolerances both on the stored good signature and on the measured signals in order to avoid false rejections and escapes. Although most mixed-signal BIST schemes today have a restricted applicability, some noteworthy approaches have already been presented, most of them exploiting the presence of on-chip data converters and surrounding digital circuitry. In the MADBIST scheme, after self-testing the digital circuitry, the D/A converter is configured in oscillator mode to test the A/D converter, after which the D/A converter and then the other analog circuitry is tested [147]. In the HBIST scheme for so-called "discretized analog" applications (i.e., applications consisting of the sequence: analog in–ADC–DSP–DAC–analog out), the analog output in test mode is fed back to the analog input, to create a loop that starts and ends at the central DSP part [148]. After first self-testing the DSP part, the analog loop can then be tested while test signals are provided and processed by the DSP part. With the move toward systems on a chip and the paradigm of building systems by mixing and matching IP blocks, the role of analog BIST schemes will become more and more important in the future, and more research in this direction is to be expected. Also, the IEEE P1149.4 standardized mixed-signal test bus may create opportunities here [149].

IV. CONCLUSION

The increasing levels of integration in the microelectronics industry, now approaching the era of systems on a chip, has also brought about the need for systematic design methodologies and supporting CAD tools that increase the productivity and improve the quality in the design of analog and mixed-signal integrated circuits and systems. This survey paper has presented an overview of the current context and state of the art in the field of analog and mixed-signal CAD tools. After introducing the industrial context and the design flow, an overview has been given of the progress in analog and mixed-signal simulation and modeling, symbolic analysis, analog circuit synthesis and optimization, analog and mixed-signal layout generation, yield analysis and design centering, as well as analog and mixed-signal test.

Most progress has been obtained in the field of analog, mixed analog–digital and multilevel analog simulation, where many commercial solutions are available. Also, the standardization of analog and mixed-signal hardware description languages is approaching completion. In the field of custom analog circuit and layout synthesis, substantial progress at the research level has been achieved over the past decade, despite the dearth of commercial offerings yet. Cast mostly in the form of numerical and combinatorial optimization tasks, linked by various forms of sensitivity analysis and constraint mapping, leveraged by ever faster workstations, some of these tools show glimmers of practical application, and commercial startups have embarked to bring these to industrial practice. Also, in the field of analog and mixed-signal test new ideas have been developed, but they are striving for industrial acceptance now.

One conclusion is clear: In order to meet the economic constraints (time to market, cost, quality) of future semiconductor products, analog design will have to be carried out in a much more systematic and structured way, supported by methodologies and tools that fit in the overall system design flow. Although research in academia has not fully solved all the relevant problems yet, this paper has shown that real progress has been made over the last decade. Therefore, given the current market pressures and given the existence of core design technology in academia, we can hope that commercial offerings will soon follow and that analog designers will finally get the boost in productivity needed to take them from the dark ages of analog black magic to the bright future of integrated systems on a chip. In the emerging era of combined information and communication technologies and ubiquitous computing, we need good tools for analog circuits more than ever.

References

[1] "Systems-on-a-chip," in *IEEE Int. Solid-State Circuits Conf. (ISSCC)*, 1996.

[2] J. Liang, "Mixed-signal IC market to surpass $10 billion in 1997 and $22 billion by 2001," Dataquest Rep., Jan. 1998.

[3] *The National Technology Roadmap for Semiconductors*: Semiconductor Industry Association (SIA), 1994.

[4] *Virtual Socket Interface Architecture Document*: Version 1.0, VSI Alliance, 1997.

[5] H. Chang *et al.*, *Surviving the SOC Revolution—A Guide to Platform-Based Design*. Norwell, MA: Kluwer, 1999.

[6] L. Nagel, "SPICE2: A computer program to simulate semiconductor circuits," Electronics Research Lab., Univ. Calif., Berkeley, Memo UCB/ERL M520, May 1975.

[7] L. R. Carley, G. Gielen, R. Rutenbar, and W. Sansen, "Synthesis tools for mixed-signal ICs: Progress on frontend and backend strategies," in *Proc. ACM/IEEE Design Automation Conf. (DAC)*, 1996, pp. 298–303.

[8] *IEEE Standard VHDL 1076.1 Language Reference Manual—Analog and Mixed-Signal Extensions to VHDL 1076*: IEEE 1076.1 Working Group, July 1997.

[9] *Verilog-A: Language Reference Manual: Analog Extensions to Verilog HDL*: Version 0.1, Open Verilog International, Jan. 1996.

[10] R. Harjani, R. Rutenbar, and L. R. Carley, "OASYS: A framework for analog circuit synthesis," *IEEE Trans. Computer-Aided Design*, vol. 8, pp. 1247–1265, Dec. 1989.

[11] G. Gielen, K. Swings, and W. Sansen, "Open analog synthesis system based on declarative models," in *Analog Circuit Design*, J. Huijsing, R. van der Plassche, and W. Sansen, Eds. Norwell, MA: Kluwer, 1993, pp. 421–445.

[12] S. Donnay *et al.*, "Using top-down CAD tools for mixed analog/digital ASICs: A practical design case," *Kluwer Int. J. Analog Integrated Circuits Signal Processing (Special Issue on Modeling and Simulation of Mixed Analog-Digital Systems)*, vol. 10, pp. 101–117, June–July 1996.

[13] H. Chang *et al.*, "A top-down, constraint-driven design methodology for analog integrated circuits," in *Proc. IEEE Custom Integrated Circuits Conf. (CICC)*, 1992, pp. 8.4.1–8.4.6.

[14] E. Malavasi *et al.*, "A top-down, constraint-driven design methodology for analog integrated circuits," in *Analog Circuit Design*, J. Huijsing, R. van der Plassche, and W. Sansen, Eds. Norwell, MA: Kluwer, 1993, ch. 13, pp. 285–324.

[15] H. Chang *et al.*, *A Top-Down, Constraint-Driven Design Methodology for Analog Integrated Circuits*. Norwell, MA: Kluwer, 1997.

[16] L. Nagle and R. Rohrer, "Computer analysis of nonlinear circuits, excluding radiation (CANCER)," *IEEE J. Solid-State Circuits*, vol. SSC-6, pp. 166–182, Aug. 1971.

[17] A. Vladimirescu, *The SPICE Book*. New York: Wiley, 1994.

[18] D. Foty, *MOSFET Modeling with SPICE*. Englewood Cliffs, NJ: Prentice-Hall, 1997.

[19] C. Enz, "MOS transistor modeling for RF integrated circuit design," in *Proc. IEEE Custom Integrated Circuit Conf. (CICC)*, 2000, pp. 189–196.

[20] R. Saleh, B. Antao, and J. Singh, "Multi-level and mixed-domain simulation of analog circuits and systems," *IEEE Trans. Computer-Aided Design*, vol. 15, pp. 68–82, Jan. 1996.

[21] A. Vachoux, J.-M. Bergé, O. Levia, and J. Rouillard, Eds., *Analog and Mixed-Signal Hardware Description Languages*. Norwell, MA: Kluwer, 1997.

[22] G. Casinovi and A. Sangiovanni-Vincentelli, "A macromodeling algorithm for analog circuits," *IEEE Trans. Computer-Aided Design*, vol. 10, pp. 150–160, Feb. 1991.

[23] Y.-C. Ju, V. Rao, and R. Saleh, "Consistency checking and optimization of macromodels," *IEEE Trans. Computer-Aided Design*, pp. 957–967, Aug. 1991.

[24] B. Antao and F. El-Turky, "Automatic analog model generation for behavioral simulation," in *Proc. IEEE Custom Integrated Circuits Conf. (CICC)*, May 1992, pp. 12.2.1–12.2.4.

[25] C. Borchers, L. Hedrich, and E. Barke, "Equation-based behavioral model generation for nonlinear analog circuits," in *Proc. IEEE/ACM Design Automation Conf. (DAC)*, 1996, pp. 236–239.

[26] S. Fang, Y. Tsividis, and O. Wing, "SWITCAP: A switched-capacitor network analysis program—Part I: Basic features," *IEEE Circuits Syst. Mag.*, vol. 5, pp. 4–10, Sept. 1983.

[27] S. Fang, Y. Tsividis, and O. Wing, "SWITCAP: A switched-capacitor network analysis program—Part I: Advanced applications," *IEEE Circuits Syst. Mag.*, vol. 5, pp. 41–46, Sept. 1983.

[28] J. Vandewalle, H. De Man, and J. Rabaey, "Time, frequency, and Z-domain modified nodal analysis of switched-capacitor networks," *IEEE Trans. Circuits Syst.*, vol. CAS-28, pp. 186–195, Mar. 1981.

[29] V. Dias, V. Liberali, and F. Maloberti, "TOSCA: A user-friendly behavioral simulator for oversampling A/D converters," in *Proc. IEEE Int. Symp. Circuits Syst. (ISCAS)*, 1991, pp. 2677–2680.

[30] K. Kundert, "Simulation methods for RF integrated circuits," in *Proc. IEEE/ACM Int. Conf. Computer-Aided Design (ICCAD)*, 1997, pp. 752–765.

[31] R. Telichevesky, K. Kundert, I. Elfadel, and J. White, "Fast simulation algorithms for RF circuits," in *Proc. IEEE Custom Integrated Circuits Conf. (CICC)*, 1996, pp. 437–444.

[32] K. Kundert, "Introduction to RF simulation and its applications," *IEEE J. Solid-State Circuits*, vol. 34, pp. 1298–1319, Sept. 1999.

[33] P. Feldmann and J. Roychowdhury, "Computation of circuit waveform envelopes using an efficient, matrix-decomposed harmonic balance algorithm," in *Proc. IEEE/ACM Int. Conf. Computer-Aided Design (ICCAD)*, 1996, pp. 295–300.

[34] J. Crols, S. Donnay, M. Steyaert, and G. Gielen, "A high-level design and optimization tool for analog RF receiver front-ends," in *Proc. IEEE/ACM Int. Conf. Computer-Aided Design (ICCAD)*, 1995, pp. 550–553.

[35] A. Demir and A. Sangiovanni-Vincentelli, *Analysis and Simulation of Noise in Nonlinear Integrated Circuits and Systems*. Norwell, MA: Kluwer, 1998.

[36] J. Phillips and K. Kundert, "Noise in mixers, oscillators, samplers, and logic: An introduction to cyclostationary noise," in *Proc. IEEE Custom Integrated Circuits Conf. (CICC)*, 2000, pp. 431–439.

[37] L. Pileggi, "Coping with RC(L) interconnect design headaches," in *Proc. IEEE/ACM Int. Conf. Computer-Aided Design (ICCAD)*, 1995, pp. 246–253.

[38] L. Pillage and R. Rohrer, "Asymptotic waveform evaluation for timing analysis," *IEEE Trans. Computer-Aided Design*, vol. 9, pp. 352–366, Apr. 1990.

[39] L. Silveira *et al.*, "A coordinate-transformed Arnoldi algorithm for generating guaranteed stable reduced-order models of RLC circuits," in *Proc. IEEE/ACM Int. Conf. Computer-Aided Design (ICCAD)*, 1996, pp. 288–294.

[40] M. Kamon, S. McCormick, and K. Shepard, "Interconnect parasitic extraction in the digital IC design methodology," in *Proc. IEEE/ACM Int. Conf. Computer-Aided Design (ICCAD)*, 1999, pp. 223–230.

[41] N. Verghese and D. Allstot, Rapid simulation of substrate coupling effects in mixed-mode IC's," in *Proc. IEEE Custom Integrated Circuits Conf. (CICC)*, 1993, pp. 18.3.1–18.3.4.

[42] R. Gharpurey and R. Meyer, "Modeling and analysis of substrate coupling in integrated circuits," *IEEE J. Solid-State Circuits*, vol. 31, pp. 344–353, Mar. 1996.

[43] N. Verghese, T. Schmerbeck, and D. Allstot, *Simulation Techniques and Solutions for Mixed-Signal Coupling in Integrated Circuits*. Norwell, MA: Kluwer, 1995.

[44] T. Blalack, "Design techniques to reduce substrate noise," in *Advances in Analog Circuit Design*, Huijsing, van de Plassche, and Sansen, Eds. Norwell, MA: Kluwer, 1999, pp. 193–217.

[45] J. Costa, M. Chou, and L. Silveira, "Efficient techniques for accurate modeling and simulation of substrate coupling in mixed-signal ICs," *IEEE Trans. Computer-Aided Design*, vol. 18, pp. 597–607, May 1999.

[46] E. Charbon, R. Gharpurey, R. Meyer, and A. Sangiovanni-Vincentelli, "Substrate optimization based on semi-analytical techniques," *IEEE Trans. Computer-Aided Design*, vol. 18, pp. 172–190, Feb. 1999.

[47] M. van Heijningen, M. Badaroglu, S. Donnay, M. Engels, and I. Bolsens, "High-level simulation of substrate noise generation including power supply noise coupling," in *Proc. IEEE/ACM Design Automation Conf. (DAC)*, 2000, pp. 446–451.

[48] G. Gielen, P. Wambacq, and W. Sansen, "Symbolic analysis methods and applications for analog circuits: A tutorial overview," *Proc. IEEE*, vol. 82, pp. 287–304, Feb. 1994.

[49] F. Fernández, A. Rodríguez-Vázquez, J. Huertas, and G. Gielen, *Symbolic Analysis Techniques—Applications to Analog Design Automation*. Piscataway, NJ: IEEE Press, 1998.

[50] P. Wambacq, F. Fernández, G. Gielen, W. Sansen, and A. Rodríguez-Vázquez, "Efficient symbolic computation of approximated small-signal characteristics," *IEEE J. Solid-State Circuits*, vol. 30, pp. 327–330, Mar. 1995.

[51] G. Gielen, H. Walscharts, and W. Sansen, "ISAAC: A symbolic simulator for analog integrated circuits," *IEEE J. Solid-State Circuits*, vol. 24, pp. 1587–1596, Dec. 1989.

[52] P. Wambacq, G. Gielen, P. Kinget, and W. Sansen, "High-frequency distortion analysis of analog integrated circuits," *IEEE Trans. Circuits Syst. II*, vol. 46, pp. 335–345, Mar. 1999.

[53] S. Seda, M. Degrauwe, and W. Fichtner, "A symbolic analysis tool for analog circuit design automation," in *Proc. IEEE/ACM Int. Conf. Computer-Aided Design (ICCAD)*, 1988, pp. 488–491.

[54] F. Fernández, A. Rodríguez-Vázquez, and J. Huertas, "Interactive ac modeling and characterization of analog circuits via symbolic analysis," *Kluwer Int. J. Analog Integrated Circuits Signal Processing*, vol. 1, pp. 183–208, Nov. 1991.

[55] Q. Yu and C. Sechen, "A unified approach to the approximate symbolic analysis of large analog integrated circuits," *IEEE Trans. Circuits Systems I*, vol. 43, pp. 656–669, Aug. 1996.

[56] W. Daems, G. Gielen, and W. Sansen, "Circuit complexity reduction for symbolic analysis of analog integrated circuits," in *Proc. IEEE/ACM Design Automation Conf.*, 1999, pp. 958–963.

[57] J. Hsu and C. Sechen, "DC small signal symbolic analysis of large analog integrated circuits," *IEEE Trans. Circuits Systems I*, vol. 41, pp. 817–828, Dec. 1994.

[58] J. Starzyk and A. Konczykowska, "Flowgraph analysis of large electronic networks," *IEEE Trans. Circuits Syst.*, vol. CAS-33, pp. 302–315, Mar. 1986.

[59] O. Guerra, E. Roca, F. Fernández, and A. Rodríguez-Vázquez, "A hierarchical approach for the symbolic analysis of large analog integrated circuits," in *Proc. IEEE Design Automation and Test in Europe Conf. (DATE)*, 2000, pp. 48–52.

[60] C.-J. Shi and X.-D. Tan, "Canonical symbolic analysis of large analog circuits with determinant decision diagrams," *IEEE Trans. Computer-Aided Design*, vol. 19, pp. 1–18, Jan. 2000.

[61] X.-D. Tan and C.-J. Shi, "Hierarchical symbolic analysis of analog integrated circuits via determinant decision diagrams," *IEEE Trans. Computer-Aided Design*, vol. 19, pp. 401–412, April 2000.

[62] F. El-Turky and E. Perry, "BLADES: An artificial intelligence approach to analog circuit design," *IEEE Trans. Computer-Aided Design*, vol. 8, pp. 680–691, June 1989.

[63] H. Koh, C. Séquin, and P. Gray, "OPASYN: A compiler for CMOS operational amplifiers," *IEEE Trans. Computer-Aided Design*, vol. 9, pp. 113–125, Feb. 1990.

[64] P. Veselinovic *et al.*, "A flexible topology selection program as part of an analog synthesis system," in *Proc. IEEE Eur. Design Test Conf. (ED&TC)*, 1995, pp. 119–123.

[65] R. Harjani and J. Shao, "Feasibility and performance region modeling of analog and digital circuits," *Kluwer Int. J. Analog Integrated Circuits Signal Processing*, vol. 10, pp. 23–43, Jan. 1996.

[66] P. Maulik, L. R. Carley, and R. Rutenbar, "Simultaneous topology selection and sizing of cell-level analog circuits," *IEEE Trans. Computer-Aided Design*, vol. 14, pp. 401–412, Apr. 1995.

[67] Z. Ning *et al.*, "SEAS: A simulated evolution approach for analog circuit synthesis," in *Proc. IEEE Custom Integrated Circuits Conf. (CICC)*, 1991, pp. 5.2.1–5.2.4.

[68] W. Kruiskamp and D. Leenaerts, "DARWIN: CMOS opamp synthesis by means of a genetic algorithm," in *Proc. ACM/IEEE Design Automation Conf. (DAC)*, 1995, pp. 550–553.

[69] G. Gielen and W. Sansen, *Symbolic Analysis for Automated Design of Analog Integrated Circuits*. Norwell, MA: Kluwer, 1991.

[70] M. Degrauwe *et al.*, "IDAC: An interactive design tool for analog CMOS circuits," *IEEE J. Solid-State Circuits*, vol. 22, pp. 1106–1115, Dec. 1987.

[71] G. Beenker, J. Conway, G. Schrooten, and A. Slenter, "Analog CAD for consumer ICs," in *Analog Circuit Design*, J. Huijsing, R. van der Plassche, and W. Sansen, Eds. Norwell, MA: Kluwer, 1993, pp. 347–367.

[72] R. Henderson *et al.*, "A spreadsheet interface for analog design knowledge capture and reuse," in *Proc. IEEE Custom Integrated Circuits Conf. (CICC)*, 1993, pp. 13.3.1–13.3.4.

[73] N. Horta, J. Franca, and C. Leme, "Automated high level synthesis of data conversion systems," in *Analogue–Digital ASICs—Circuit Techniques, Design Tools and Applications*, Soin, Maloberti, and Franca, Eds. Stevenage, U. K.: Peregrinus, 1991.

[74] J. Vital and J. Franca, "Synthesis of high-speed A/D converter architectures with flexible functional simulation capabilities," in *Proc. IEEE Int. Symp. Circuits Syst. (ISCAS)*, 1992, pp. 2156–2159.

[75] C. Toumazou and C. Makris, "Analog IC design automation—I: Automated circuit generation: New concepts and methods," *IEEE Trans. Computer-Aided Design*, vol. 14, pp. 218–238, Feb. 1995.

[76] C. Makris and C. Toumazou, "Analog IC design automation—II: Automated circuit correction by qualitative reasoning," *IEEE Trans. Computer-Aided Design*, vol. 14, pp. 239–254, Feb. 1995.

[77] B. Sheu, A. Fung, and Y.-N. Lai, "A knowledge-based approach to analog IC design," *IEEE Trans. Circuits Syst.*, vol. 35, pp. 256–258, Feb. 1988.

[78] J. Harvey, M. Elmasry, and B. Leung, "STAIC: An interactive framework for synthesizing CMOS and BiCMOS analog circuits," *IEEE Trans. Computer-Aided Design*, vol. 11, pp. 1402–1416, Nov. 1992.

[79] G. Gielen, H. Walscharts, and W. Sansen, "Analog circuit design optimization based on symbolic simulation and simulated annealing," *IEEE J. Solid-State Circuits*, vol. 25, pp. 707–713, June 1990.

[80] K. Swings and W. Sansen, "DONALD: A workbench or interactive design space exploration and sizing of analog circuits," in *Proc. IEEE Eur. Design Automation Conf. (EDAC)*, 1991, pp. 475–479.

[81] K. Lampaert, G. Gielen, and W. Sansen, *Analog Layout Generation for Performance and Manufacturability*. Norwell, MA: Kluwer, 1999.

[82] G. Gielen *et al.*, "An analog module generator for mixed analog/digital ASIC design," *Wiley Int. J. Circuit Theory Applicat.*, vol. 23, pp. 269–283, July–Aug. 1995.

[83] F. Medeiro, B. Pérez-Verdú, A. Rodríguez-Vázquez, and J. Huertas, "A vertically-integrated tool for automated design of $\Sigma\Delta$ modulators," *IEEE J. Solid-State Circuits*, vol. 30, pp. 762–772, July 1995.

[84] A. Doboli, A. Nunez-Aldana, N. Dhanwada, S. Ganesan, and R. Vemuri, "Behavioral synthesis of analog systems using two-layered design space exploration," in *Proc. ACM/IEEE Design Automation Conf. (DAC)*, 1999, pp. 951–957.

[85] F. Leyn, G. Gielen, and W. Sansen, "An efficient dc root solving algorithm with guaranteed convergence for analog integrated CMOS circuits," in *Proc. IEEE/ACM Int. Conf. Computer-Aided Design (ICCAD)*, 1998, pp. 304–307.

[86] M. Hershenson, S. Boyd, and T. Lee, "GPCAD: A tool for CMOS op-amp synthesis," in *Proc. IEEE/ACM Int. Conf. Computer-Aided Design (ICCAD)*, 1998, pp. 296–303.

[87] M. Hershenson, S. Mohan, S. Boyd, and T. Lee, "Optimization of inductor circuits via geometric programming," in *Proc. IEEE/ACM Design Automation Conf. (DAC)*, 1999, pp. 994–998.

[88] S. Director and R. Rohrer, "Automated network design—The frequency domain case," *IEEE Trans. Circuit Theory*, vol. 16, pp. 330–337, Aug. 1969.

[89] W. Nye, D. Riley, A. Sangiovanni-Vincentelli, and A. Tits, "DELIGHT.SPICE: An optimization-based system for the design of integrated circuits," *IEEE Trans. Computer-Aided Design*, vol. 7, pp. 501–518, Apr. 1988.

[90] F. Medeiro *et al.*, "A statistical optimization-based approach for automated sizing of analog cells," in *Proc. ACM/IEEE Int. Conf. Computer-Aided Design (ICCAD)*, 1994, pp. 594–597.

[91] E. Ochotta, R. Rutenbar, and L. R. Carley, "Synthesis of high-performance analog circuits in ASTRX/OBLX," *IEEE Trans. Computer-Aided Design*, vol. 15, pp. 273–294, Mar. 1996.

[92] E. Ochotta, T. Mukherjee, R. Rutenbar, and L. R. Carley, *Practical Synthesis of High-Performance Analog Circuits*. Norwell, MA: Kluwer, 1998.

[93] R. Schwencker, F. Schenkel, H. Graeb, and K. Antreich, "The generalized boundary curve—A common method for automatic nominal design and design centering of analog circuits," in *Proc. IEEE Design Automation and Test in Europe Conf. (DATE)*, 2000, pp. 42–47.

[94] R. Phelps, M. Krasnicki, R. Rutenbar, L. R. Carley, and J. Hellums, "ANACONDA: Robust synthesis of analog circuits via stochastic pattern search," in *Proc. Custom Integrated Circuits Conf. (CICC)*, 1999, pp. 567–570.

[95] M. Krasnicki, R. Phelps, R. Rutenbar, and L. R. Carley, "MAELSTROM: Efficient simulation-based synthesis for custom analog cells," in *Proc. ACM/IEEE Design Automation Conf. (DAC)*, 1999, pp. 945–950.

[96] R. Phelps, M. Krasnicki, R. Rutenbar, L. R. Carley, and J. Hellums, "A case study of synthesis for industrial-scale analog IP: Redesign of the equalizer/filter frontend for an ADSL CODEC," in *Proc. ACM/IEEE Design Automation Conf. (DAC)*, 2000, pp. 1–6.

[97] H. Onodera, H. Kanbara, and K. Tamaru, "Operational-amplifier compilation with performance optimization," *IEEE J. Solid-State Circuits*, vol. 25, pp. 466–473, Apr. 1990.

[98] J. Crols, S. Donnay, M. Steyaert, and G. Gielen, "A high-level design and optimization tool for analog RF receiver front-ends," in *Proc. Int. Conf. Computer-Aided Design (ICCAD)*, Nov. 1995, pp. 550–553.

[99] J. Assael, P. Senn, and M. Tawfik, "A switched-capacitor filter silicon compiler," *IEEE J. Solid-State Circuits*, vol. 23, pp. 166–174, Feb. 1988.

[100] G. Gielen and J. Franca, "CAD tools for data converter design: An overview," *IEEE Trans. Circuits Systems II:*, vol. 43, pp. 77–89, Feb. 1996.

[101] S. Director, W. Maly, and A. Strojwas, *VLSI Design for Manufacturing: Yield Enhancement*. Norwell, MA: Kluwer, 1990.

[102] J. Zhang and M. Styblinski, *Yield and Variability Optimization of Integrated Circuits*. Norwell, MA: Kluwer, 1995.

[103] T. Mukherjee, L. R. Carley, and R. Rutenbar, "Synthesis of manufacturable analog circuits," in *Proc. ACM/IEEE Int. Conf. Computer-Aided Design (ICCAD)*, Nov. 1995, pp. 586–593.

[104] G. Debyser and G. Gielen, "Efficient analog circuit synthesis with simultaneous yield and robustness optimization," in *Proc. IEEE/ACM Int. Conf. Computer-Aided Design (ICCAD)*, Nov. 1998, pp. 308–311.

[105] J. Kuhn, "Analog module generators for silicon compilation," in *VLSI System Design*, 1987.

[106] J. Rijmenants *et al.*, "ILAC: An automated layout tool for analog CMOS circuits," *IEEE J. Solid-State Circuits*, vol. 24, pp. 417–425, Apr. 1989.

[107] J. Cohn, D. Garrod, R. Rutenbar, and L. R. Carley, *Analog Device-Level Layout Generation*. Norwell, MA: Kluwer, 1994.

[108] D. Garrod, R. Rutenbar, and L. R. Carley, "Automatic layout of custom analog cells in ANAGRAM," in *Proc. ACM/IEEE Int. Conf. Computer-Aided Design (ICCAD)*, Nov. 1988, pp. 544–547.

[109] J. Cohn, D. Garrod, R. Rutenbar, and L. R. Carley, "KOAN/ANAGRAM II: New tools for device-level analog placement and routing," *IEEE J. Solid-State Circuits*, vol. 26, pp. 330–342, Mar. 1991.

[110] M. Mogaki *et al.*, "LADIES: An automated layout system for analog LSI's," in *Proc. ACM/IEEE Int. Conf. Computer-Aided Design (ICCAD)*, Nov. 1989, pp. 450–453.

[111] V. Meyer zu Bexten, C. Moraga, R. Klinke, W. Brockherde, and K. Hess, "ALSYN: Flexible rule-based layout synthesis for analog IC's," *IEEE J. Solid-State Circuits*, vol. 28, pp. 261–268, Mar. 1993.

[112] E. Malavasi, E. Charbon, E. Felt, and A. Sangiovanni-Vincentelli, "Automation of IC layout with analog constraints," *IEEE Trans. Computer-Aided Design*, vol. 15, pp. 923–942, Aug. 1996.

[113] E. Malavasi and A. Sangiovanni-Vincentelli, "Area routing for analog layout," *IEEE Trans. Computer-Aided Design*, vol. 12, pp. 1186–1197, Aug. 1993.

[114] E. Charbon, E. Malavasi, U. Choudhury, A. Casotto, and A. Sangiovanni-Vincentelli, "A constraint-driven placement methodology for analog integrated circuits," in *Proc. IEEE Custom Integrated Circuits Conf. (CICC)*, May 1992, pp. 28.2.1–28.2.4.

[115] E. Malavasi, E. Felt, E. Charbon, and A. Sangiovanni-Vincentelli, "Symbolic compaction with analog constraints," *Wiley Int. Journal Circuit Theory Applicat.*, vol. 23, pp. 433–452, July–Aug. 1995.

[116] B. Basaran, R. Rutenbar, and L. R. Carley, "Latchup-aware placement and parasitic-bounded routing of custom analog cells," in *Proc. ACM/IEEE Int. Conf. Computer-Aided Design (ICCAD)*, Nov. 1993.

[117] K. Lampaert, G. Gielen, and W. Sansen, "A performance-driven placement tool for analog integrated circuits," *IEEE J. Solid-State Circuits*, vol. 30, pp. 773–780, July 1995.

[118] ——, "Analog routing for performance and manufacturability," in *Proc. IEEE Custom Integrated Circuits Conf. (CICC)*, May 1996, pp. 175–178.

[119] U. Choudhury and A. Sangiovanni-Vincentelli, "Automatic generation of parasitic constraints for performance-constrained physical design of analog circuits," *IEEE Trans. Computer-Aided Design*, vol. 12, pp. 208–224, Feb. 1993.

[120] E. Charbon, E. Malavasi, and A. Sangiovanni-Vincentelli, "Generalized constraint generation for analog circuit design," in *Proc. ACM/EEE Int. Conf. Computer-Aided Design (ICCAD)*, Nov. 1993, pp. 408–414.

[121] E. Malavasi and D. Pandini, "Optimum CMOS stack generation with analog constraints," *IEEE Trans. Computer-Aided Design*, vol. 14, pp. 107–122, Jan. 1995.

[122] B. Basaran and R. Rutenbar, "An O(n) algorithm for transistor stacking with performance constraints," in *Proc. ACM/IEEE Design Automation Conf. (DAC)*, June 1996.

[123] J. Prieto, A. Rueda, J. Quintana, and J. Huertas, "A performance-driven placement algorithm with simultaneous place&route optimization for analog IC's," in *Proc. IEEE Eur. Design Test Conf. (ED&TC)*, 1997, pp. 389–394.

[124] R. Okuda, T. Sato, H. Onodera, and K. Tamuru, "An efficient algorithm for layout compaction problem with symmetry constraints," in *Proc. ACM/IEEE Int. Conf. Computer-Aided Design (ICCAD)*, Nov. 1989, pp. 148–151.

[125] J. Cohn, D. Garrod, R. Rutenbar, and L. R. Carley, "Techniques for simultaneous placement and routing of custom analog cells in KOAN/ANAGRAM II," in *Proc. ACM/IEEE Int. Conf. Computer-Aided Design (ICCAD)*, Nov. 1991, pp. 394–397.

[126] H. Yaghutiel, A. Sangiovanni-Vincentelli, and P. Gray, "A methodology for automated layout of switched-capacitor filters," in *Proc. ACM/IEEE Int. Conf. Computer-Aided Design (ICCAD)*, Nov. 1986, pp. 444–447.

[127] C. Kimble *et al.*, "Analog autorouted VLSI," in *Proc. IEEE Custom Integrated Circuits Conf. (CICC)*, June 1985.

[128] R. Gyurcsik and J. Jeen, "A generalized approach to routing mixed analog and digital signal nets in a channel," *IEEE J. Solid-State Circuits*, vol. 24, pp. 436–442, Apr. 1989.

[129] U. Choudhury and A. Sangiovanni-Vincentelli, "Constraint-based channel routing for analog and mixed analog/digital circuits," *IEEE Trans. Computer-Aided Design*, vol. 12, pp. 497–510, Apr. 1993.

[130] S. Mitra, S. Nag, R. Rutenbar, and L. R. Carley, "System-level routing of mixed-signal ASICs in WREN," in *ACM/IEEE Int. Conf. Computer-Aided Design (ICCAD)*, Nov. 1992.

[131] S. Mitra, R. Rutenbar, L. R. Carley, and D. Allstot, "Substrate-aware mixed-signal macrocell placement in WRIGHT," *IEEE J. Solid-State Circuits*, vol. 30, pp. 269–278, Mar. 1995.

[132] B. Stanisic, N. Verghese, R. Rutenbar, L. R. Carley, and D. Allstot, "Addressing substrate coupling in mixed-mode ICs: Simulation and power distribution synthesis," *IEEE J. Solid-State Circuits*, vol. 29, Mar. 1994.

[133] B. Stanisic, R. Rutenbar, and L. R. Carley, *Synthesis of Power Distribution to Manage Signal Integrity in Mixed-Signal ICs*. Norwell, MA: Kluwer, 1996.

[134] S. Director, P. Feldmann, and K. Krishna, "Optimization of parametric yield: A tutorial," in *Proc. IEEE Custom Integrated Circuits Conf. (CICC)*, 1992, pp. 3.1.1–3.1.8.
[135] K. Antreich, H. Graeb, and C. Wieser, "Circuit analysis and optimization driven by worst-case distances," *IEEE Trans. Computer-Aided Design*, vol. 13, pp. 57–71, Jan. 1994.
[136] C. Guardiani, P. Scandolara, J. Benkoski, and G. Nicollini, "Yield optimization of analog IC's using two-step analytic modeling methods," *IEEE J. Solid-State Circuits*, vol. 28, pp. 778–783, July 1993.
[137] M. Conti, P. Crippa, S. Orcioni, and C. Turchetti, "Parametric yield formulation of MOS IC's affected by mismatch effect," *IEEE Trans. Computer-Aided Design*, vol. 18, pp. 582–596, May 1999.
[138] J. Huertas, "Test and design for testability of analog and mixed-signal integrated circuits," in *Selected Topics in Circuits and Systems*, H. Dedieu, Ed. Amsterdam, The Netherlands: Elsevier, 1993, pp. 77–156.
[139] B. Vinnakota, Ed., *Analog and Mixed-Signal Test*. Englewood Cliffs, NJ: Prentice-Hall, 1998.
[140] M. Sachdev, *Defect Oriented Testing for CMOS Analog and Digital Circuits*. Norwell, MA: Kluwer, 1998.
[141] C. Sebeke, J. Teixeira, and M. Ohletz, "Automatic fault extraction and simulation of layout realistic faults for integrated analogue circuits," in *Proc. IEEE Eur. Design Test Conf. (ED&TC)*, 1995, pp. 464–468.
[142] G. Devarayanadurg and M. Soma, "Dynamic test signal design for analog ICs," in *Proc. IEEE/ACM Int. Conf. Computer-Aided Design (ICCAD)*, 1995, pp. 627–630.
[143] W. Verhaegen, G. Van der Plas, and G. Gielen, "Automated test pattern generation for analog integrated circuits," in *Proc. IEEE VLSI Test Symp. (VTS)*, 1997, pp. 296–301.
[144] K. Wagner and T. Williams, "Design for testability of mixed signal integrated circuits," in *Proc. IEEE Int. Test Conf. (ITC)*, 1988, pp. 823–829.
[145] C. Wey, "Built-in self-test (BIST) structure for analog circuit fault diagnosis," *IEEE Trans. Instrum. Meas.*, vol. 39, pp. 517–521, June 1990.
[146] K. Arabi and B. Kaminska, "Oscillation-test strategy for analog and mixed-signal integrated circuits," in *Proc. IEEE VLSI Test Symp. (VTS)*, 1996, pp. 476–482.
[147] G. Roberts and A. Lu, *Analog Signal Generation for Built-In Self-Test of Mixed-Signal Integrated Circuits*. Norwell, MA: Kluwer, 1995.
[148] M. Ohletz, "Hybrid built-in self test (HBIST) for mixed analogue/digital integrated circuits," in *Proc. IEEE Eur. Test Conf. (ETC)*, 1991, pp. 307–316.
[149] *IEEE 1149.4 Standard for Mixed-Signal Test Bus*, 1997.

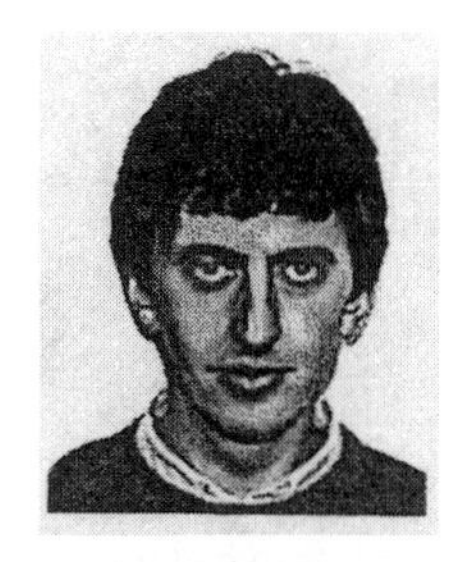

Georges G. E. Gielen (Senior Member, IEEE) received the M.Sc. and Ph.D. degrees in electrical engineering from the Katholieke Universiteit Leuven, Leuven, Belgium, in 1986 and 1990, respectively.

In 1990, he was appointed as a Postdoctoral Research Assistant and Visiting Lecturer at the Department of Electrical Engineering and Computer Science, University of California, Berkeley. From 1991 to 1993, he was a Postdoctoral Research Assistant of the Belgian National Fund of Scientific Research at the ESAT-MICAS Laboratory of the Katholieke Universiteit Leuven. In 1993, he was appointed as a Tenure Research Associate of the Belgian National Fund of Scientific Research and, at the same time, as an Assistant Professor at the Katholieke Universiteit Leuven. In 1995, he was promoted to Associate Professor and in 2000 to full-time Professor at the same university. His research interests are in the design of analog and mixed-signal integrated circuits and especially in analog and mixed-signal CAD tools and design automation (modeling, simulation and symbolic analysis, analog synthesis, analog layout generation, analog and mixed-signal testing). He is the Coordinator or Partner of several (industrial) research projects in this area. He has authored or coauthored two books and more than 100 papers in edited books, international journals, and conference proceedings.

Dr. Gielen was the 1997 Laureate of the Belgian Royal Academy of Sciences, Literature and Arts, in the category of engineering sciences. He also received the 1995 Best Paper Award from the John Wiley international journal on *Circuit Theory and Applications*. He is a regular member of the Program Committees of international conferences (ICCAD, DATE, CICC, etc.), he is an Associate Editor of the IEEE TRANSACTIONS ON CIRCUITS AND SYSTEMS, PART II, and is a Member of the Editorial Board of the Kluwer international journal on *Analog Integrated Circuits and Signal Processing*. He is a Member of the Board of Governors of the IEEE Circuits and Systems (CAS) Society and is the Chairman of the IEEE Benelux CA CAS Chapter.

Rob. A. Rutenbar (Fellow, IEEE) received the Ph.D. degree from the University of Michigan, Ann Arbor, in 1984.

He subsequently joined the faculty of Carnegie Mellon University (CMU), Pittsburgh, PA. He is currently Professor of Electrical and Computer Engineering, and (by courtesy) of Computer Science. From 1993 to 1998, he was Director of the CMU Center for Electronic Design Automation. He is Cofounder of NeoLinear, Inc., and served as its Chief Technologist on a 1998 leave from CMU. His research interests focus on circuit and layout synthesis algorithms for mixed-signal ASICs, for high-speed digital systems, and for FPGAs.

Dr. Rutenbar received a Presidential Young Investigator Award from the National Science Foundation in 1987. He has won Best/Distinguished paper awards from the Design Automation Conference (1987) and the International Conference on CAD (1991). He has been on the program committees for the IEEE International Symposium on FPGAs, and the ACM International Symposium in Physical Design. He also served on the Editorial Board of IEEE Spectrum. He was General Chair of the 1996 ICCAD. He Chaired the Analog Technical Advisory Board for Cadence Design Systems from 1992 through 1996. He is a Member of the ACM and Eta Kappa Nu.

Design of Mixed-Signal Systems-on-a-Chip

Ken Kundert, *Member, IEEE*, Henry Chang, *Member, IEEE*, Dan Jefferies, Gilles Lamant, *Member, IEEE*, Enrico Malavasi, *Member, IEEE*, and Fred Sendig

***Abstract*—The electronics industry is increasingly focused on the consumer marketplace, which requires low-cost high-volume products to be developed very rapidly. This, combined with advances in deep submicrometer technology have resulted in the ability and the need to put entire systems on a single chip. As more of the system is included on a single chip, it is increasingly likely that the chip will contain both analog and digital sections. Developing these mixed-signal (MS) systems-on-chip presents enormous challenges both to the designers of the chips and to the developers of the computer-adided design (CAD) systems that are used during the design process. This paper presents many of the issues that act to complicate the development of large single-chip MS systems and how CAD systems are expected to develop to overcome these issues.**

***Index Terms*—Design automation, design methodology, hardware design languages, integrated circuit layout, integrated circuit modeling, mixed analog-digital integrated circuits, simulation, testing.**

I. Introduction

INCREASING time-to-market (TTM) pressures due to the continued consumerization of the electronics market place and the availability of shrinking process technologies are the two fundamental forces driving designers, design methodologies, and electronic design automation (EDA) tools and flows today. This is illustrated in Fig. 1.

On one hand, TTM pressures, along with the added integration afforded by newer process technologies, have forced a move to higher levels of abstraction to cope with the added complexity in design. This can already be seen in the digital design domain space, where cell based design is rapidly moving to intellectual property (IP), re-use based or block-based design methodologies [4]. On the other hand, shrinking process technologies have also caused a move in the opposite direction: because of the increasing significance of physical effects, there has been a need to observe lower levels of detail. Signal integrity, electromigration, and power analysis are now adding severe complications to design methodologies already stressed by the increasing device count. This is true for both analog and digital design. The total range of design abstractions encountered in a single design flow is continually growing, and pulling in opposite directions (abstraction versus detail). Managing this increasing range, and insuring that the system definitions (constraints) are preserved and verified (or verifiable) through all levels of abstractions and between different levels is where one becomes acutely aware of

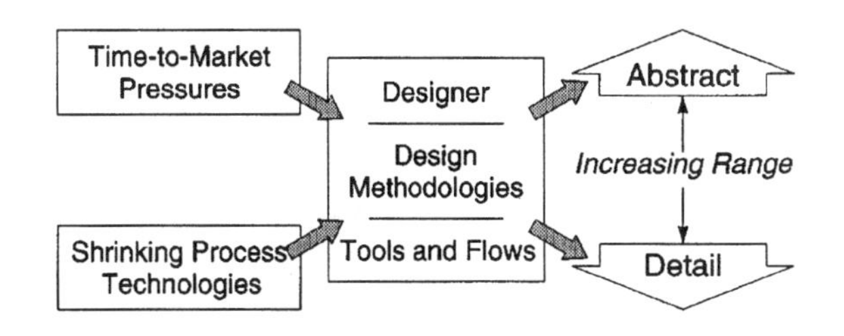

Fig. 1. Design drivers and design methodology gaps.

the widening gaps in today's design methodologies. However, to meet TTM needs, it is imperative that these be kept under control.

The stresses caused by this wide abstraction range and the increasing complexity of design at each level of abstraction uncover significant methodology gaps. These occur both between abstraction layers as well as within them. Design methodologies, tools, and flows, evolve to try to hold the design "system" together. However, what we see today is just the beginning of what is to come, with the new, even smaller, process technologies.

Stressed by cost and performance objectives resulting from the consumerization of electronics, designers are driven to take advantage of the smaller process technologies, putting entire systems on chips. Two basic types of systems-on-a-chip (SOC) exist—one that has grown from the application-specific integrated circuit (ASIC) world, and the other from the custom integrated circuit (IC) world. An example of the former is shown in Fig. 2. This is a design that is mostly digital. It is a programmable system that integrates most of the functions of the end product. It contains processors. It has embedded software, peripherals both analog and digital, and has a bus-based architecture. Analog and mixed-signal (MS) design blocks are only integrated if they can be in a reasonable time and at a reasonable cost. For example, high-frequency radio frequency (RF) remains as a separate chip for this type of design. For this type of design, the integrator is a digital designer and increasingly, the cost is in the development of the embedded software rather than in the hardware design of the IC.

The other type of design, which we will, henceforth, refer to as *AMS-SOC*, is shown in Fig. 3. This is a design that began in the realm of custom MS designs. These are designs that are both high in performance and have complex signal paths through both analog and digital components. Examples of these designs include PRML disk drive controllers, xDSL front-ends, 10/100 base-T physical layers and RF front-ends. This era of process technology has also allowed analog and MS designers to begin to integrate significant amounts of the functionality of the entire systems onto a single chip. However, unlike the case of the ASIC design moving to SOC, the analog/MS design is not an

Manuscript received January 3, 2000; revised May 24, 2000. This paper was recommended by Associate Editor M. Pedram.

K. Kundert, H. Chang, E. Malavasi, and F. Sendig are with Cadence Design Systems, San José, CA 95134-1937 USA.

G. Lamant is with Cadence Design Systems, Tokyo, Japan.

D. Jefferies is with Cadence Design Systems, Columbia, MD 21046 USA.

Publisher Item Identifier S 0278-0070(00)10456-7.

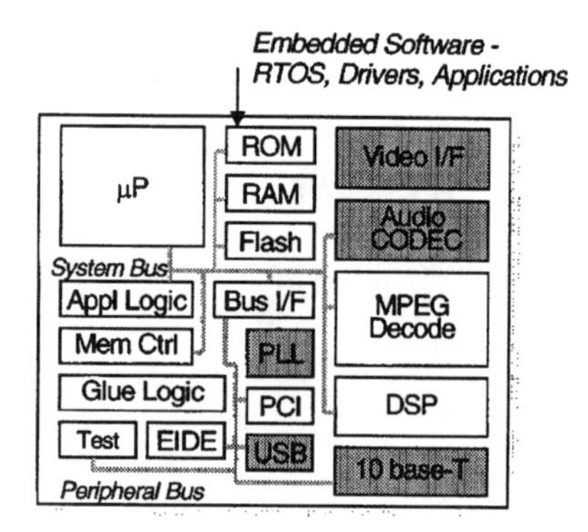

Fig. 2. An ASIC-SOC example.

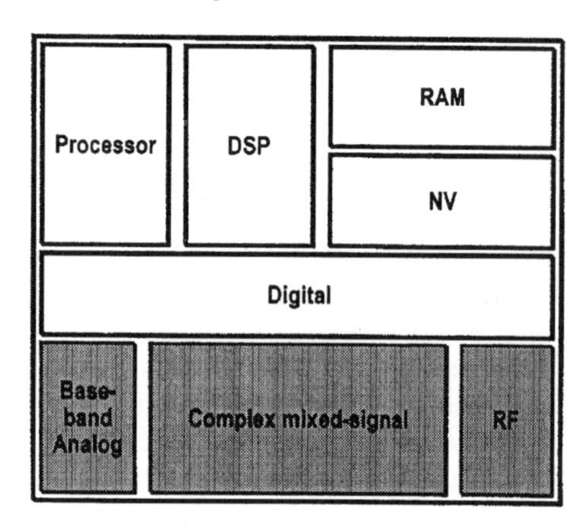

Fig. 3. An AMS-SOC example.

"option." It is the critical and probably the differentiating part of the ICs with the digital part optional as to whether or not it is integrated. In this case, at today's process technology, embedded software is not yet a significant issue. The most significant issues lie around the design and integration of digital and analog/MS blocks.

These are the designs that are the main focus of this paper. The highest level of abstraction is the system level. Thus, in this case, the range of abstraction levels spans from the device level (including parasitic devices) through to the system level.

Due to the complex feedback loops that involve signal paths crossing the interface between digital and analog blocks multiple times, as well as less obvious physical effects between the analog and digital blocks, we believe that we are reaching a point where ad-hoc "patching" of the design process will not hold it together anymore and allow meeting TTM objectives for this type of design.

The design methodology needed for the design of AMS-SOCs dictates a design flow that can be broken down into a set of design stages as shown in Fig. 4. Section II explores this design methodology and each of the design stages. But not all aspects of design can be neatly separated into these stages. There are certain design capabilities and tool requirements that span the design process. In many ways, these are the more difficult for EDA tool vendors to address as they are not contained wholly within the expertise of a specific design stage, and of necessity require interaction across the designers and design tools at each of these stages. Section III explores these complex intraflow issues and how they might be addressed.

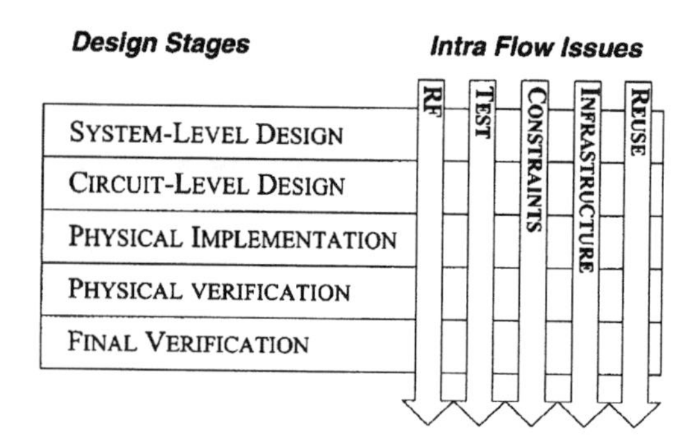

Fig. 4. The design flow.

II. THE DESIGN FLOW

In this section, we analyze the main areas that must be addressed to provide a workable solution to the problem of developing successful AMS-SOC designs.

The solution must consist of a set of design methodologies, tool flows, as well as an appropriate and cohesive set of tools. All of these are necessary to create a complete solution. While a specific design group may not have the desire or need for all three, all three must be considered in concert to develop the complete solution.

To provide a framework for what the solution entails, we have selected specific aspects of the design process and provided an overview of some of what is needed in each area.

The design flow of a complex AMS-SOC starts with an idea and ends with a layout. In between is a series of refinement and verification steps. First, the idea is refined to a series of specifications, which are verified by talking to potential customers. Then the specifications are refined to a functional description or an algorithm, which is verified with system simulators. The functional description is refined to an architecture, which is verified by simulators that interpret MS hardware description languages (MS-HDLs), see Section III-E). The blocks are then refined to the transistor level and are verified with timing simulators or with SPICE. This represents the electrical design process.

A similar process occurs for the physical design. The architecture is converted to a floorplan, which is then refined until the blocks are laid-out and routed. Verification of the layout involves checking the layout to assure it matches the schematic and that it satisfies all manufacturability rules. Final verification involves extracting the circuit from the layout, including layout parasitics, and simulating it with transistor-level simulators.

In high-performance analog and MS designs, the physical implementation often has such an impact on the circuit performance that circuit and layout issues must be considered together [44]. As a consequence the physical design is intertwined with circuit design and optimization, and the physical implementation is subject to frequent extraction, analyzes and engineering change orders, or incremental modifications. These frequent disruptions of the design flow are characteristic of AMS circuit development, and often account for a good portion of the overall time to market.

Once layout is complete, the whole design is represented in fine detail and the simulations are quite expensive. This prevents all but basic functionality from being verified at this point. Thus, the design process itself must assure, with a high degree of confidence, that the design functions properly in all situations and meets its performance requirements. This requires the following.

- A formal verification plan be developed and followed throughout the design process [18]. The plan must assure that the design be verified continually along the design process.
- The ability to co-simulate blocks at different levels of abstraction so that the design can be continuously verified as it progresses from an abstract to detailed levels of representation.
- Constraint definition, translation, and verification from the architecture level through functional, circuit and physical levels.
- Reliable and easy communication of connectivity, constraints, parasitics and models between systems, circuit and layout designers.
- Extraction of models for each block that faithfully represent its behavior and performance as implemented. These models are used with system or HDL simulators to verify the design from the bottom up.

A. Top-Down Design

Most analog chips at one time were designed to be general-purpose building blocks optimized for performance, cost, or low power dissipation. This involved precision work at the transistor level by a specialist. Design exploration and circuit function and performance verification occurred more or less together. For small performance-critical analog and MS ICs, this remains the dominant design style. For large designs, however, this approach has several problems, including at least the following two fundamental ones.

1) Simulations take so long that comprehensive analysis of the design in manageable time frames becomes problematic. Because of this, projects may be delayed because of the need for extra silicon prototypes caused by inadequate verification.
2) For large designs, improvements made at the architectural level generally provide the greatest impact on the performance, cost and functionality of the chip. By the time the development reaches the circuit level, meaningful improvements are often very expensive.

In order to address these challenges, many design teams are either looking to, or else have already implemented, top-down methodologies [3]. In a basic top-down approach, the architecture of the chip is defined as a block diagram and simulated and optimized using either a MS-HDL simulator or a system simulator. From the high-level simulation, requirements for the individual circuit blocks are derived. Circuits are then designed individually to meet these specifications. Finally, the entire chip is laid out and verified against the original requirements.

A few of the key characteristics of these design styles are as follows.

- Design exploration and verification are somewhat separate. The combination of greater simulation speed from the use of high-level behavioral models and the ability to perform parametric design make MS-HDL simulation appropriate for design exploration. The use of transistor-level simulation becomes more focused on verifying that the blocks match the intent of the high-level design.
- Parametric design at the system level. MS-HDLs provide users great flexibility in modeling. However, since a fundamental objective of the block-level analysis is to develop specifications for the block implementation, good top-down practice is to write the MS-HDL models so that their key performance characteristics are specified using parameters and so can be easily adjusted.
- Mixed-level simulation. In general, it is much faster to verify the functionality and performance of a specific block against its specifications within an MS-HDL representation of the system than it is to verify the entire design "flat" at the circuit level.

In practice, a final verification of the entire design at the circuit level, in SPICE, may still be desirable for verification of connectivity, proper startup, and the performance of critical paths. However, a major objective of most top-down approaches is to eliminate the need to do this more than once per project.

These practices require substantial attention early in the design process. This is the essential tradeoff of top-down methodologies: more analysis early in design to avoid problems later on.

The main objectives of top-down approaches are to optimize globally the performance of the design, and to increase the general predictability of the design schedule. They also make it easier to coordinate the efforts of multiple designers working in parallel on different parts of the design at once.

The principal drawbacks are the need for rigor in the design process, and the need for designers to learn an MS-HDL, which presently few have significant familiarity with. Some of the early proprietary languages acquired a perhaps-justified reputation as difficult to learn and use. However, modern MS-HDLs like Verilog-AMS are better. Furthermore, our experience is that the effort required to make MS-HDL models is not only worthwhile, but also drops dramatically over the first few projects, as engineers learn the methodology and begin to reuse their existing models.

Top-down design represents a substantial shift from the way most people design today and there is considerable inertia that acts to slow its adoption. Those that have moved to a top-down design style have seen dramatic improvements in time-to-market and the ability to handle complexity. The best way to overcome the inertia that prevents top-down design from being adopted is to teach the art of top-down design and behavioral modeling in the universities.

B. System-Level Design

System-level design is generally performed by system architects. Their goal is to find an algorithm and architecture that implement the required functionality while providing adequate performance at minimum cost. They use system-level simulators, such as Matlab [26] or SPW [48], that allow them to explore various algorithms and evaluate tradeoffs quickly. These tools are preferred because they represent the design as a block diagram and have large libraries of predefined blocks for common application areas.

Once the algorithm is chosen, it must be mapped to a particular architecture. Thus, it must be refined to the point where the blocks used at the system level accurately reflect the way the circuit is partitioned for implementation. The blocks must represent sections of the circuit that are to be designed and verified as a unit. Furthermore, the interfaces must be chosen carefully to avoid interaction between the blocks that are hard to predict and model, such as loading or coupling. The primary goal at this phase is the accurate modeling of the blocks and their interfaces. This contrasts with the goal during algorithm design, which is to quickly predict the output behavior of the entire circuit with little concern about matching the architectural structure of the chip as implemented. As such, MS-HDLs such as Verilog-AMS [51] or VHDL-AMS [6], [22], [52] become preferred during this phase of the design because they allow accurate modeling of the interfaces and support mixed-level simulation (discussed in Section II-D).

The transition between algorithm and architecture design currently represents a discontinuity in the design flow. The tools used during algorithm design are different from the ones used during architecture design, and they generally operate off of different design representations. Thus, the design must be re-entered, which is a source of inefficiencies and errors. It also prevents the test benches and constraints used during the algorithm design phase from being used during the rest of the design.

On the digital side, tools such as SPW do provide paths to implementation via Verilog and VHDL generation. However, as of today, they have yet to be tightly integrated into the remainder of the design flow. Similar capabilities do not yet exist for the analog or MS portions of the design. An alternative is to use Verilog-AMS or VHDL-AMS for both algorithm and architecture design. This has not been done to date because simulators that support these languages are just now becoming available. As such, there is a dearth of application specific libraries.

C. Analog Synthesis

The ability to automatically convert a high-level specification of a block to a circuit-level implementation is referred to as synthesis. While synthesis is well established in the digital world, for analog or MS circuits it is only available in special cases, such as for filters. Research into analog synthesis has developed over the last two decades in many directions, from early work on knowledge-based module compilation [2], [8] to more recent optimization intensive approaches [7], [29], [36]. Optimization is based either on numerical simulation [35] or on analytic models [28]. To help in the development of analytic models, a significant research effort went into exploration of symbolic analysis [12], [45]. Beyond model building, symbolic analysis was also applied to more ambitions goals, such as topology exploration with interesting results [27], whose applicability unfortunately is limited to selected categories of analog circuits.

Many attempts at building analog design automation systems have been made. The most important ones are probably ADAM [8], [9], a commercial product developed at CSEM, and ACACIA [5], [36], developed at CMU, which recently expanded to include, among other things, RF design [1]. Several good survey papers provide insight into the extensive research production in this field [13], [43], [44].

Some commercial offerings in this space have appeared recently. Noticeable among others are NeoCell [37], a system for analog cell design automation, which leverages in part from the technology developed for ACACIA; and Picasso Op-Amp [38] and Dali RF Tool Suite [39], web-based tools providing on-demand circuit topology selection and sizing based on geometric programming [17]. However, we believe that the large variety and complexity of analog cells makes it unlikely that a general solution for the problem of analog synthesis will be available in the near future. Instead, it is likely that a variety of design aids and very specific module generators will become available for an increasing variety of analog cells and blocks to ease the transition from high-level specification to circuit-level implementation.

D. Mixed-Level Simulation

Using a top-down design methodology is expected to become the norm for designing complex MS circuits. As such, the system architecture will be fully explored and verified using either a MS-HDL or a system simulator before individual blocks are designed. However, once the blocks are designed, they must be verified in the context of the system to assure that they will operate properly within the system. At this point, it must be possible to co-simulate behavioral models and transistor-level circuits together. The block-diagram used in the simulation of the architecture must be refined to the point where each block represents a relatively independent circuit that would be designed as a single unit. Pin-accurate MS-HDL models are developed for each block and the system is verified using these models.

The block designers then take the HDL models and a series of specifications as input and produce the transistor-level schematic and layout, which are passed back to the system engineers for integration and verification with the rest of the system. Using the ability to co-simulate transistor-level and behavioral-level descriptions of the blocks, the system is repeatedly verified by replacing one-by-one the HDL model of one block with the transistor-level implementation to verify the functionality of the block and its interfaces. This approach greatly reduces the cost of each simulation and increases the chance that miscommunications concerning block interfaces are caught early in the design process.

E. Physical Implementation

Physical implementation corresponds to a variety of tasks that can be grouped into two major areas:

- block authoring;
- block/chip assembly.

These two areas are deeply intertwined as most design flows require a mix of top-down and bottom-up approaches with a combination of soft and hard blocks, and behavioral, logical and physical representations of different parts [4] (p. 189ff.).

Therefore, any solution needs to incorporate a seamless flow including access to both authoring and assembly of complex blocks.

For block authoring successful commercial tools, methodologies and flows have been developed over the last few years. However, the coordination of different design approaches into consistent flows and adequate solutions for block assembly is still under investigation, especially for MS applications in a SOC environment. Assembly and authoring need to be addressed simultaneously, since a design environment for large MS applications requires cooperation between an interactive editing environment and reentrant automation. A full custom implementation is also required for most analog portions of the design. Finally, critical issues such as IP reuse [4], [24], power management and signal integrity [49] are key to the success of large SOC designs.

A key component required to guarantee a good integration between assembly and authoring is the floorplanner. Commercial floorplanning technology today is focused on digital designs, and it is often poorly equipped to handle AMS issues such as noise and signal integrity. While encapsulation of AMS blocks is available in most commercial tools, it does not allow a correct representation of the interactions between blocks such as intermodulation, cross talk and substrate noise, and top-level MS interconnections require special modeling and planning [16]. The organization of power distribution is significantly different when analog and digital supply lines are used, with severe impact on substrate noise [46]. Finally, the design flow often follows a combination of top-down and bottom up steps very tightly interleaved.

For example, a design may be partitioned into various digital, analog, and MS blocks. Not only may these blocks be designed concurrently, they will be physically realized at different times and using different implementation flows. While large digital blocks can be created using a semi-automated design flow, other custom portions often need to be carefully hand crafted. Manual design does not scale well with circuit size, and custom blocks become the bottleneck of the entire flow, even though advanced assisted custom design methodologies have become available recently [33]. Finally, during chip assembly the communication between blocks will be subject to timing, power or signal integrity constraints. Coupling between blocks sometimes determines the inner structure of the blocks themselves. In order to meet these constraints, top-down modifications are forced upon the blocks. These operations must be kept consistent with the specific design flow, often bottom up, used for the authoring of each block. The need for reentrant and interoperable environments for the authoring of ASIC-style digital blocks, custom analog blocks, and the assembly of all these parts is a major paradigm shift that characterizes complex AMS-SOCs.

For the complex ASIC MS SOC designs, a physically aware automated synthesis to silicon flow such as being delivered as part of Envisia PKS [15] may also be utilized. This flow, while currently targeting complex deep submicrometer designs, needs to be extended in order to be able to read AMS designs. While not necessarily implementing the analog portions, this tool suite needs to become aware of them and to take them into account during physically aware synthesis.

F. Physical Verification

Very powerful technology for physical verification has been developed in recent years by the EDA industry. Commercial tools, such as Cadence's Assura, Avant!'s Hercules and Mentor Graphics's Calibre, often include hierarchical capability for verification and extraction, and present various levels of integration with the block authoring tool suites. The resulting design cycle improvements have proven not only the importance of an efficient verification tool, but also the criticality of a solution flow where such a tool is well integrated with all the other applications:

- using a common database;
- using a common set of interactive commands for browsing and fixing errors;
- using a common user interface look-and-feel;
- supporting the same set of constraints.

Some of the proprietary netlist-based integration methodologies are shared by most commercial tools. Digital description languages used for simulation, such as Verilog and VHDL have been extended for AMS designs [22], [51], and commercial verification tools will soon be required to support these AMS extensions.

From a strategic point of view, the verification phase must be tied more closely with the physical design cycle. New constraint-driven layout applications, able to enforce physical and electrical constraints, have recently become available [14]. These must be matched by corresponding new capabilities in the verification phase, which will have to become cognizant of the same set of AMS constraints. Substrate coupling noise verification, currently heavily limited for capacity reasons to circuit level within small blocks, must be used to optimize the distribution of guard rings and to drive block placement in mixed signal systems.

Yield has a parametric dependency on AMS performance functions and measurements, which can be captured through behavioral and stochastic models. The support of electrical and design constraints derived from these models will enable the physical verification phase to help in the design centering analysis.

With respect to manufacturing, the verification tools must also support the increasingly common post-layout processing techniques such as optical proximity correction (OPC) and new subwavelength lithography processes such as phase shift mask (PSM) [23].

G. Final Verification

Final verification is performed by using a physical verification tool to extract a netlist of the circuit, including parasitics, from the final layout. Of course, such circuits are very large and only limited verification is possible. With an AMS-SOC, it is often possible to do some transistor-level full chip simulation. Typically only areas of special concern that cannot be sufficiently verified using mixed-level simulation are considered. Examples include power-up behavior and timing of the critical paths.

In digital blocks, final verification is often performed using timing simulators, such as Synopsys's TimeMill or Avant!'s Star-Sim. Relative to circuit simulators such as Spice, timing simulators trade accuracy and generality for speed. They generally provide at least 10× in speed and capacity over SPICE but are suitable only for estimating the timing of MOS digital circuits, and can generally be counted on to produce timing numbers that are accurate to within 5% on these circuits. Analog circuits or circuits constructed with bipolar transistors often confuse timing simulators, causing them to run slowly and give incorrect results. Cadence's ATS overcomes this problem by combining a circuit simulator with a timing simulator and so can handle large digital MOS circuits that contain some analog or bipolar circuitry.

ASIC-SOCs are usually too large to be verified using any type of transistor-level simulation. Instead bottom-up verification is required. With bottom-up verification, individual blocks are extracted and characterized, macromodels are created that exhibit the behavior and performance of the block as implemented, and the macromodels are combined and simulated using a fast high-level simulator, such as a SPW or an AMS simulator. In practice, this is done by refining the models for the blocks used during the top-down design. To reduce the chance of errors, it is best done during the mixed-level simulation procedure. Thus, the verification of a block by mixed-level simulation becomes a three step process. First, the proposed block functionality is verified by including an idealized model of the block in system-level simulations. Then, the functionality of the block as implemented is verified by replacing the idealized model with the netlist of the block. This also allows the effect of the block's imperfections on the system performance to be observed. Finally, the netlist of the block is replaced by an extracted model. By comparing the results achieved from simulations that involved the netlist and extracted models, the functionality and accuracy of the extracted model can be verified. From then on, mixed-level simulations of other blocks are made more representative by using the extracted model of the block just verified rather than the idealized model. The extracted model may also be used to support reuse of the block.

III. Intra Flow Design Issues

The previous section described how the design process is partitioned into tasks that support the refinement of complex systems from a top level architectural concept to a working physical implementation. In this section, we will analyze some design issues that cannot be addressed by enhancements to any particular point tool in a flow. Instead they require a holistic approach, where every task in the flow must participate in a comprehensive solution to the design problem. A new design methodology, better suited for more complex design objectives or for more aggressive time to market, can be made possible by the coordinated operation of all design phases.

The first consideration is the frequency range of the components of the SOC. If one or more RF components are present, simulation, verification and physical implementation must all include very different sets of models, parasitics and performance measurements.

Significant advantage in terms of cost and risk reduction for the entire design can be achieved by adopting a constraint-driven approach. However, this requires transformation techniques, and every tool must understand, enforce and verify constraints.

In a similar fashion, testability considerations must be carried along the design flow and every single application must be able to understand and improve, or at least not reduce, testability of the entire chip.

Finally, the communication between tools in a complex design flow, data integrity, constraint transformations and the simulta-neous use of multiple models and levels of abstraction requires a strong software infrastructure with standards and interfaces to which all applications must adhere.

A. RF

The addition of RF to a MS chip adds considerable risk and so is done sparingly today. It is common to find the RF transceiver paths combined with a frequency synthesizer, but it is unusual to see the baseband processing or the micro-controller combined with the RF sections. This is expected to change with the development of relatively low performance RF systems such as Bluetooth and HomeRF. Here, the large volumes and low costs make a single chip implementation compelling, while the low performance makes it feasible. Once success is achieved here, higher performance systems such as PCS and 3G phones are expected to be implemented on a single chip. The wireless market will be an important technology driver for MS-SOCs, and of course, including the RF sections is crucial.

There are several aspects of RF that make this a challenge. First and foremost, RF circuits operate at very high frequencies, typically between 1–5 GHz. Wires that carry RF signals must be short and carefully placed to avoid interference. Floorplanning, layout, and packaging must take this into account. Accurate models are needed for the active devices, the interconnect, the package, and passive components, both on and off chip. For example, spiral inductors are used on chip and ceramic or SAW resonators are used off chip. Often exotic process are used, such as SiGe or SOI, which affects both the active and passive models. Links to field solvers and the ability to read in files of S-parameters is necessary to assure adequate verification.

Another important challenge is that RF circuits can be sensitive to interference from signals generated in the digital portion of the circuit. Signals at the input of a receiver can be as small as 1 μV. Any signals that couple into the front-end of a receiver through the substrate, supplies, interconnect, or package degrades its sensitivity. The ability to accurately model these portions of the circuit and predict coupling is important.

A third challenge is that in the RF section of a transceiver, the information signal is present as a relatively low frequency modulation on a high-frequency carrier. Simulating these circuits is expensive because the high-frequency carrier necessitates a small time step while the low frequency modulation requires a long simulation interval. RF simulators provide special analyzes that are designed to efficiently simulate these circuits, but they are incapable of including the non-RF sections [30]. One possible solution to these problems is to use the RF simulator to extract macro-models of the RF blocks that can be efficiently evaluated in an AMS simulator [40]–[42].

B. Constraint Management

Especially in the design of an SOC, several levels of abstraction are used in different phases and using different models. The formulation of constraints, their management [34] along every phase in the design, their validation, verification and enforcement are extremely critical to the consistency of the design flow. Furthermore, the design of analog and M/S systems is a process of progressive constraint refinement, where data tolerances and their level of confidence change at every step.

Physical constraints apply to the physical entities used to implement the layout. Examples are distances, area and aspect ratio, alignment between instances etc. Some commercial tools used for physical implementation such as IC-Craftsman [14] have achieved good results in enforcing physical constraints within the context of their specific application. Academic research has also devoted considerable attention to physical constraints, especially for analog design applications [5], [32]. Some physical constraints such as distances, are routinely used to enforce timing and cross-talk specifications during placement and routing.

Electrical constraints apply to specific signals in the circuit. Hence, these constraints require a register transfer level or schematic level representation of the circuit where nets and devices are identifiable. Examples are timing, parasitics, voltage drop, crosstalk noise, substrate coupling noise, and electromigration. Because of their extraordinary importance in the design of digital circuits, timing constraints need to be handled by all synthesis and physical tools. As mentioned above, special transformations into physical constraints such as net length or spacing between devices have been commonly adopted by physical tools. In the case of more complex constraints, analysis and design tools might need to be entirely redesigned to properly take them into account. An example is the case of power and ground routing with mixed analog and digital supply lines [46], [47].

Finally, *design constraints* are used to characterize the behavior of individual components in terms of their I/O signals and performance. Examples are throughput, slew rate, bandwidth, gain, phase margin, power dissipation, jitter, etc. These can be specified on a circuit characterized by a model at any level of abstraction, from behavioral to physical. With complex AMS chips, design constraints might include specifications on sophisticated measurements such as distortion, noise and frequency response.

So far design constraints have not been handled adequately by commercial applications. The main reason is that their enforcement is usually impractical, since they require a transformation into a set of electrical or physical constraints in order to be handled by automatic applications. Another reason is the lack of a standard to represent these measurements and their constraints, consistent with the high-level behavioral modeling language. Such a standard should provide a description of the dependencies between electrical and design constraints when such transformation is actually performed.

A constraint management system, therefore, must have the following characteristics.

- It must be able to handle constraints of different types (design, electrical and physical) in a consistent way with a language applicable to all relevant description models.
- It must provide a way to facilitate transformations [31], which can be fairly complex especially for MS applications where the behavioral description of blocks might be quite abstract from the actual implementation. This includes mapping of digital-to-analog and vice versa, as well as generation of noise constraints from coupling between interconnections or through the substrate. It also includes use of behavioral and stochastic models to generate electric constraints for design centering.
- It must be able to provide a consistency check to validate constraints and detect infeasible specifications and over-constraints as early as possible in order to reduce the number of design iterations.
- Constraints must be verifiable. That means that analysis and verification tools must be able to access the definition of measurements and evaluate the corresponding performance functions using the appropriate models.

C. AMS Test

Generally, the last thing done for a design before it is sent to manufacturing is test program development. Verifying the test program involves running it on a working model of the chip, which is only available late in the design process. This is costly in two important ways. First, MS testers are very expensive, and test development can tie up these machines for long periods of time. Second, starting test development at the end of the design process greatly prolongs the time-to-market. If instead of running the test program on an actual chip, it can be run on a simulated version of the chip, then it is possible to address both of these issues [25].

If a top-down design methodology is used, then a system-level model of the chip exists early in the design process. This system-level model can be used during the development of the test program. Thus, the test engineers can become involved with the project much earlier, and like the block designers, are given a working virtual prototype of the chip in the form of a system-level model [10], [11]. This improves communication between the test and design engineers, acts to greatly reduce the cost of test development, allows the test programs to be more thoroughly verified, and permits the test programs to be developed concurrently with the chip. All of which helps to insure that the test program is available as soon as the chip is ready to be manufactured. In addition, involving the test engineers while the design is ongoing allows fault simulation and design for test to be attempted [21].

Commercial tools are available that allow test development on virtual prototypes of the chip, but they do not as yet support Verilog-AMS or VHDL-AMS [50], [55].

D. Infrastructure

The AMS-SOC design stages we have described are frequently addressed by specific tools, or mini flows, in isolation by existing EDA vendors. This is not surprising as most vendors have a rather small subset of the tools required in a complex AMS-SOC solution flow. This correspondingly restricts how much of the problem they are able to address. Without access to the internals of the tools within the flow, and without co-operation between vendors, problems cannot be addressed where they are best addressed. This leads to a patched together rats nest of tools, which can be, with a lot of wasted time, manual user intervention, and design iterations, used to create chips that eventually work. Such patching together of tools can never succeed in creating an efficient design environment capable of the fast time to market that is needed in today's AMS-SOC market.

Further, many of the tools in use today were not designed for the complexities and sizes implicit in AMS-SOCs. This manifests itself both in the analog and the digital design tools from front end through physical realization. On the analog side, most tools still target traditional transistor-based bottom-up design methodologies. The capacity of such capture and analysis tools is inherently limited. Further, the physical realization of such designs is a largely manual process. On the digital side, the tools currently do not take sufficient account of the physical affects during the logical and planning design phases. This results in designs that cannot meet the constraints when physically realized, and thereby require costly design rework (silicon iterations). But worse than these specific limitations that are being addressed at a localized level is the interaction of the digital and analog design processes. Not only do the digital and analog design tools tend to be targeted only to their specific design methodologies, they frequently do not take into account the effects of their counterparts. Even the communication of these tools between the digital and analog domain tends to be in different forms than the other expects, thereby making it difficult, or impossible, to create an efficient design flow that ensures data integrity.

To create a truly efficient design environment for AMS-SOC design, we need to start from scratch. First, the AMS-SOC design methodology needs to be defined. Given that methodology, a design flow can be specified. This flow will then clearly dictate the necessary tools, design representations and data formats that are needed to convey all needed data both within a design stage, and across design stages. Such a definition is the contract, or infrastructure, to which all tools must conform.

Given this infrastructure it then becomes possible to design tools that will not only have the necessary functionality, but will by definition be plug-and-play in an efficient solution. Thus, by restricting the data locations that all tools must both read and write data to, as well as the allowed formats, it becomes possible to insert tools as needed into the flow without requiring a redesign of other components. It also becomes possible to create utilities that perform design integrity checks. With a restricted set of formats dictated by the needed abstraction levels, as opposed to the eccentricities and whims of a tool designer, the types of checks needed is greatly reduced and confined to what is required by the design methodology.

Perhaps the most import of these formats will be the MS-HDLs. They are expected to be used as a common language for representing the design and will be understood by most tools, even those from competing vendors. As such, tools other than simulators are expected to be extended to support one or both of the MS-HDLs. The MS-HDLs are also open standard languages, which means there will be greater willingness by the design and EDA communities to invest in developing model libraries and support tools for these languages. MS-HDLs are also likely to develop into a medium of exchange between block authors and block integrators.

E. Mixed-Signal Hardware Description Languages

Both Verilog-AMS and VHDL-AMS have been defined and simulators that support these languages are becoming available. These languages are expected to have a big impact on the design of MS systems because they provide a single language and a single simulator that are shared between analog and digital designers. It will be much easier to provide a single design flow that naturally supports analog, digital and MS blocks, making it simpler for these designers to work together. It also becomes substantially more straight-forward to write behavioral models for MS blocks. Finally, the AMS languages bring strong event-driven capabilities to analog simulation, allowing analog event-driven models to be written that perform with the speed and capacity inherited from the digital engines.

It is important to recognize that the AMS languages are primarily used for verification. Unlike the digital languages, the AMS languages will not be used for synthesis in the foreseeable future because the only synthesis that is available for analog circuits is very narrowly focused.

1) Verilog-AMS: Verilog-A is an analog HDL patterned after Verilog-HDL [19]. Verilog-AMS combines Verilog-HDL and Verilog-A into a MS-HDL that is a super-set of both seed languages [51]. Verilog-HDL provides event-driven modeling constructs, and Verilog-A provides continuous-time modeling constructs. By combining Verilog-HDL and Verilog-A it becomes possible to easily write efficient MS behavioral models. Verilog-AMS also provides automatic interface element insertion so that analog and digital models can be directly interconnected even if their terminal/port types do not match. In addition, it provides support for real-valued event-driven nets and back annotating interconnect parasitics.

A commercial version of Verilog-AMS that also supports VHDL is available from Cadence Design Systems.

2) VHDL-AMS: VHDL-AMS [6], [22], [52] adds continuous time modeling constructs to the VHDL event-driven modeling language [20]. Like Verilog-AMS, MS behavioral models can be directly written in VHDL-AMS. Unlike with Verilog, there is no analog-only subset.

VHDL-AMS inherits support for configurations and abstract data types from VHDL, which are very useful for top-down design. However, it also inherits the strongly typed nature of VHDL, which is a serious issue for MS designs. Within VHDL-AMS you are not allowed to directly interconnect digital and analog ports, and there is no support for automatic interface element insertion built-in to the language. In fact, you are not even allowed to directly connect ports from an abstract analog model (a signal flow port) to a port from a low-level analog model (a conservative port). This makes it difficult to support mixed-level simulation. These deficiencies have to be overcome by a simulation environment, making VHDL-AMS much more dependent on its environment. This should slow deployment of effective VHDL-AMS-based flows.

A commercial version of VHDL-AMS that also supports Verilog is available from Mentor Graphics [53]. A VHDL-AMS simulator is also expected soon from Avanti [53].

F. Design Reuse

The push to reduce costs for the consumer market place by increasing integration will result in larger and more complete systems on chip. Once MS circuits exceed a certain size, a full-custom design style becomes impractical. With circuits of this size, the AMS-SOCs described above become blocks that are combined with very large digital blocks such as microcontrollers to form ASIC-SOCs. In this case, the intended complexity of the interaction between MS blocks is relatively low and a top-down design style that includes the MS blocks is usually not necessary. The MS blocks can generally be designed with little interaction from the system engineer.

It is hoped that the MS blocks could be designed in advance as relatively generic components and incorporated into many designs. To support this, the MS blocks must be designed for reuse. At a minimum this implies that certain documentation be available that describes the block. Standards that specify what type of documentation is required have been set by the Virtual Socket Interface Alliance (VSIA) [54]. In addition, if the block is large it may be required to be embedded in special interface collars to make it easier to import them into an ASIC-SOC. These collars provide a standard interface and guard-banding to provide some degree of isolation from the rest of the circuit.

With the rapid changes in technology, and with the difficulty of migrating MS blocks to a new technology, it is generally not possible to reuse a single block design more than a few times. Thus, preparing a design for reuse must take significantly less effort than redesigning the block for a new application. An important task when preparing a block for reuse is generating a high-level model of the block that captures its essential behavior. This model is used to evaluate the suitability of the block for use in follow-on projects. It must capture the significant imperfections of the block, and must be generated as a bi-product of the block design with little extra effort by a person with limited modeling skills.

Even though the design community has become familiar with these issues and understands well the advantages, reuse today is still used infrequently. Organizations such as VSIA have taken on significant roles to define standards and methodologies. However, more work needs to be done, especially to reduce the overhead on designers, and to define widely accepted practices for design for reuse and IP interchange.

The main areas for improvement are:

- design methodologies that improve the chances a block can be reused such as interface-based design for digital blocks;
- interface verification and ip qualification and certification;
- tools that help create matching behavioral blocks for actual analog implementations;
- formal and robust techniques to associate constraint sets to the behavioral description of MS blocks

Once blocks are designed and made available for reuse, it is also necessary for them to be easily accessible to other designers. As such, the ability to automatically generate datasheets for the blocks and publish them on the web so that they are easily searched and browsed by other designers. These datasheets should include an accurate high-level model that can be used to audition the block in the intended system.

IV. SUMMARY

In this paper, we have analyzed the problems that must be addressed in the immediate future to handle the complexity of system on a chip designs for MS applications. Our analysis shows that in many areas improvements are required not only in the tools, but in the entire design methodology. A solution for AMS-SOC requires advanced tools, well defined flows, an infrastructure supporting design reuse and excellent communication between the interacting resources participating in the design flow. It also requires designers that are willing and able to change the way that they design. To change, they must have a broader set of skills, such as a understanding of modeling and a familiarity with MS-HDLs. Graduate and continuing education should be expanded to provide these skills.

A significant market is opening up for large MS consumer applications using SOC devices in the next few years. Some major EDA vendors are already positioning themselves to provide technology and comprehensive services in this arena. This effort will have to include not only large scale tools for specific tasks such as a mixed signal floor planning, but also a consistent representation for the characterization of MS behavioral data (and measurements) at all levels of abstraction and for constraints. Finally, it will require utilities for design integrity checking, constraint validation, and manufacturing sign-off.

REFERENCES

[1] M. Aktuna, R. A. Rutenbar, and L. R. Carley, "Device-level early floorplanning algorithms for RF circuits," *IEEE Trans. Computer-Aided Design*, vol. 18, pp. 375–388, Apr. 1999.

[2] L. R. Carley and R. A. Rutenbar, "How to automate analog IC designs," *IEEE Spectrum, Mag.*, vol. 25, no. 8, pp. 26–30, Aug. 1988.

[3] H. Chang, E. Charbon, U. Choudhury, A. Demir, E. Felt, E. Liu, E. Malavasi, A. Sangiovanni-Vincentelli, and I. Vassiliou, *A Top-Down Constraint-Driven Design Methodology for Analog Integrated Circuits*. Norwell, MA: Kluwer Academic, 1997.

[4] H. Chang, L. Cooke, M. Hunt, G. Martin, A. McNelly, and L. Todd, *Surviving the SOC Revolution: A Guide to Platform Based Design*: Kluwer Academic Publishers, 1999.

[5] J. M. Cohn, D. J. Garrod, R. A. Rutenbar, and L. R. Carley, *Analog Device Level Layout Automation*. Norwell, MA: Kluwer Academic, 1994.

[6] E. Christen and K. Bakalar, "VHDL-AMS—A hardware description language for analog and mixed-signal applications," *IEEE Trans. Circuits Syst. II*, vol. 46, pp. 1263–1272, Oct. 1999.

[7] G. Debyser and G. Gielen, "Efficient analog circuit synthesis with simultaneous yield and robustness optimization," in *Proc. IEEE/ACM Int. Conf. Computer-Aided Design, Digest of Technical Papers (ICCAD'98)*, Nov. 1998, pp. 308–311.

[8] M. Degrauwe *et al.*, "IDAC: An interactive design tool for analog CMOS circuits," *IEEE J. Solid-State Circuits*, vol. 22, pp. 1106–1116, Dec. 1987.

[9] M. G. R. Degrauwe, B. L. A. G. Goffart, C. Meixenberger, M. L. A. Pierre, J. B. Litsios, J. Rijmenants, O. J. A. P. Nys, E. Dijkstra, B. Joss, M. K. C. M. Meyvaert, T. R. Schwartz, and M. D. Pardoen, "Toward an analog system design environment," *IEEE J. Solid-State Circuits*, vol. 24, pp. 659–671, June 1989.

[10] C. Force and T. Austin, "Testing the design: The evolution of test simulation," presented at the Int. Test Conf., Washington, DC, 1998.

[11] C. Force, "Integrating design and test using new tools and techniques," *Integrated Syst. Design*, Feb. 1999.

[12] G. G. E. Gielen, H. C. C. Walscharts, and W. M. C. Sansen, "ISAAC: A symbolic simulator for analog integrated circuits," *IEEE J. Solid-State Circuits*, vol. 24, pp. 1587–1597, Dec. 1989.

[13] G. Gielen, P. Wambaq, and W. Sansen, "Symbolic analysis methods and applications for analog circuits: A tutorial overview," *Proc. IEEE*, vol. 82, pp. 287–304, Feb. 1994.
[14] R. Goering, "Cadence ties routing to RC extraction," *Electron. Eng. Times*, Sept. 28, 1998.
[15] ——, "Cadence claims synthesis coup," *Electron. Eng. Times*, July 12, 1999.
[16] R. S. Gyurcsik and J. C. Jean, "A generalized approach to routing mixed analog and digital signal nets in a channel," *IEEE J. Solid-State Circuits*, vol. 24, pp. 436–442, Apr. 1989.
[17] M. D. M. Hershenson, A. Hajimiri, S. S. Mohan, S. P. Boyd, and T. H. Lee, "Design and optimization of LC oscillators," in *Dig. Tech. Papers IEEE/ACM Int. Conf. Computer-Aided Design (ICCAD'99)*, Nov. 1999, pp. 65–69.
[18] J. Holmes, F. James, and I. Getreu, "Mixed-signal modeling for ICs," *Integrated Syst. Design Mag.*, June 1997.
[19] *Standard Description Language Based on the Verilog™ Hardware Description Language*, IEEE Standard 1364, 1995.
[20] *VHDL Language Reference Manual*, IEEE Standard 1076, 1993.
[21] *IEEE Standard for a Mixed-Signal Test Bus*, IEEE Standard 1149 4, 1999.
[22] *Definitions of Analog and Mixed-Signal Extensions to IEEE Standard VHDL*, IEEE Standard 1076.1, 1999.
[23] A. B. Kahng and Y. C. Pati, "Subwavelength lithography and its potential impact on design and EDA," in *Proc. 36th Design Automation Conf. (DAC'99)*, June 21–25, 1999, pp. 799–810.
[24] M. Keating and P. Bricaud, *Reuse Methodology Manual for System-on-a-Chip Designs*. Norwell, MA: Kluwer Academic, 1998.
[25] N. Khouzam, "Simulating mixed-signal tests to reduce time-to-market," *Integrated Syst. Design*, Apr. 1997.
[26] A. Knight, *Basics of Matlab and Beyond*. Boca Raton, FL: CRC, 1999.
[27] A. Konczykowska and M. Bon, "Automated design software for switched-capacitor ICs with symbolic simulator SCYMBAL," in *Proc. 25th Design Automation Conf. (DAC'88)*, 1988, pp. 363–368.
[28] H. Y. Koh, C. H. Sequin, and P. R. Gray, "OPASYN: A compiler for CMOS operational amplifiers," *IEEE Trans. Computer-Aided Design*, vol. 9, pp. 113–125, Feb. 1990.
[29] M. Krasnicki, R. Phelps, R. A. Rutenbar, and L. R. Carley, "MAELSTROM: Efficient simulation-based synthesis for custom analog cells," in *Proc. 36th Design Automation Conf. (DAC'99)*, June 1999, pp. 945–950.
[30] K. Kundert, "Introduction to RF simulation and its application," *IEEE J. Solid-State Circuits*, vol. 34, pp. 1298–1319, Sept. 1999.
[31] E. Malavasi and E. Charbon, "Constraint transformation for IC physical design," *IEEE Trans. Semiconduct. Manufact.*, vol. 12, pp. 386–395, Nov. 1999.
[32] E. Malavasi, E. Charbon, E. Felt, and A. Sangiovanni-Vincentelli, "Automation of IC layout with analog constraints," *IEEE Trans. Computer-Aided Design*, vol. 15, pp. 923–942, Aug. 1996.
[33] E. Malavasi, D. Guilin, K. Jones, and W. Kao, "Layout acceleration for IC physical design," in *Proc. DATE-99 (User Forum)*, Mar. 1999, pp. 7–11.
[34] E. Malavasi and W. H. Kao, "Constraint-driven physical design issues for a mixed-signal flow," in *Proc. ED&TC-97 (User Forum)*, Mar. 1997, pp. 63–67.
[35] W. Nye, D. C. Riley, A. Sangiovanni-Vincentelli, and A. L. Tits, "DELIGHT-SPICE: An optimization-based system for the design of integrated circuits," *IEEE Trans. Computer-Aided Design*, vol. 7, pp. 501–519, Apr. 1988.
[36] E. S. Ochotta, R. A. Rutenbar, and L. R. Carley, "ASTRX/OBLX: Tools for rapid synthesis of high-performance analog circuits," *IEEE Trans. Computer-Aided Design*, vol. 15, Mar. 1996.
[37] S. Ohr, "Cell builder tool anticipates analog synthesis," *Electron. Eng. Times*, Nov. 9, 1998.
[38] ——, "Pay per use op-amp synthesizer hits the net," *Electron. Eng. Times*, Apr. 10, 2000.
[39] ——, "Barcelona adds RF passives to online tool suite," *Electron. Eng. Times*, May 1, 2000.
[40] J. Phillips, "Model reduction of time-varying linear systems using approximate multipoint Krylov-subspace projectors," in *Dig. Tech. Papers IEEE/ACM Int. Conf. Computer-Aided Design (ICCAD'98)*, Nov. 1998, pp. 96–102.
[41] J. R. Phillips, "Automated extraction of nonlinear circuit macromodels," presented at the 2000 IEEE Custom Integrated Circuits Conf. (CICC'00), May 2000.
[42] J. Roychowdhury, "Reduced-order modeling of linear time-varying systems," in *Dig. Tech. Papers IEEE/ACM Int. Conf. Computer-Aided Design (ICCAD'98)*, Nov. 1998, pp. 92–95.
[43] R. A. Rutenbar, "Analog design automation: Where are we? Where are we going?," in *Proc. 1993 IEEE Custom Integrated Circuits Conf. (CICC'93)*, May 1993, pp. 13.1.1–13.1.7.
[44] R. A. Rutenbar and J. M. Cohn, "Layout tools for analog ICs and mixed-signal SoCs: A survey," in *Proc. Int. Symp. Physical Design*, Apr. 2000, pp. 76–82.
[45] S. J. Seda, M. G. R. Degrauwe, and W. Fichtner, "A symbolic analysis tool for analog circuit design automation," in *Dig. Tech. Papers IEEE/ACM Int. Conf. Computer-Aided Design (ICCAD'88)*, Nov. 1988, pp. 488–491.
[46] B. R. Stanisic, N. K. Verghese, R. A. Rutenbar, L. R. Carley, and D. J. Allstot, "Addressing noise decoupling in mixed signal ICs: Simulation and power distribution synthesis," *IEEE J. Solid-State Circuits*, vol. 29, pp. 226–238, Mar. 1994.
[47] B. R. Stanisic, R. A. Rutenbar, and L. R. Carley, "Addressing noise decoupling in mixed-signal ICs: Power distribution design and cell customization," *IEEE J. Solid-State Circuits*, vol. 30, pp. 321–326, Mar. 1995.
[48] *Signal-Processing Worksystem User's Guide*. San Jose, CA: Cadence Design Systems.
[49] B. R. Stanisic, R. A. Rutenbar, and L. R. Carley, *Synthesis of Power Distribution to Manage Signal Integrity in M/S ICs*. Norwell, MA: Kluwer Academic, 1996.
[50] SpectreVX and SaberVX virtual test environments. [Online]. Available: http://www.teradyne.com
[51] Verilog-AMS Language Reference Manual: Analog & Mixed-Signal Extensions to Verilog HDL version 2.0. Open Verilog International [Online]. Available: www.ovi.org
[52] VHDL-AMS [Online]. Available: http://www.vhdl.org/analog
[53] VHDL-AMS simulators [Online]. Available: http://www.vhdl-ams.com
[54] Virtual Socket Interface Alliance Official Web Page [Online]. Available: http://www.vsia.com
[55] Dantes virtual test environment [Online]. Available: http://www.virtual-test.com

Ken Kundert (S'88–M'89) received the B.S., M.Eng., and Ph.D. degrees in electrical engineering and computer sciences from the University of California, Berkeley in 1979, 1983, and 1988, respectively.

He specialized in circuit simulation and analog circuit design. He is a Fellow with Cadence Design Systems, San Jose, CA and is the Principal Architect of the Spectre circuit simulation family. As such, he has led the development of Spectre, SpectreHDL, and SpectreRF. He also played a key role in developing Hewlett-Packard's MNS harmonic balance simulator as well as the Verilog-AMS and VHDL-AMS analog hardware description languages. Finally, he is the author of two books on circuit simulation, *Steady-State Methods for Simulating Analog and Microwave Circuits* (Norwell, MA: Kluwer Academic, 1990) and *The Designer's Guide to Spice and Spectre* (Norwell, MA: Kluwer Academic, 1995).

Henry Chang (M'94) received the Sc.B. degree in electrical engineering from Brown University, Providence, RI, in 1989, and the M.S. and Ph.D. degrees in electrical engineering from the University of California at Berkeley in 1992 and 1994, respectively.

Since 1995, he has been at Cadence Design Systems, San José, CA, working on system-on-a-chip activities. He is currently the chair of the VSI Alliance Mixed-Signal Development Working Group. At Cadence, he works on SOC design environments, methodologies, capabilities, tools, and standards both looking at the general SOC problem as well as focusing on mixed-signal design issues. He is also the author of two books, *A Top-Down, Constraint-Driven Design Methodology for Analog Integrated Circuits* (Norwell, MA: Kluwer Academic, 1997) and *Surviving the SOC Revolution: A Guide to Platform-Based Design* (Norwell, MA: Kluwer Academic, 1999).

Dan Jefferies received the B.S. degree in electrical engineering from the University of Illinois, Urbana, in 1982.

He worked for 14 years designing a wide range of mixed-signal devices, mostly for military application. Since 1996, he has been with Cadence's Mixed-Signal Design Factories, Columbia, MD, focusing on design methodology implementation and improvements. He is currently responsible for all aspects of the design flow used in design and implementation of state of the art, mixed-signal chips. In this position, he has been responsible for numerous advances in methodology that help drive both the design practices and tool innovations.

Gilles Lamant (M'99) received the engineering degree from Ecole Supérieure d'Ingénoieurs en Electronique et Electronique (E.S.I.E.E.), Paris, France, in 1991.

Since 1993, he has been working with Cadence Design Systems, focusing on working directly with customers and improving their designs flows. In that context, and since 1994, he has focused on the mixed-signal flows, covering all aspects from the front-end to the back end. He is currently a Senior Project Manager with the Cadence Service organization, and involved with the R&D organization for the definition of the future infrastructure. He is currently based in Tokyo, Japan.

Enrico Malavasi (M'97) graduated in electrical engineering at the University of Bologna, Italy, in 1984. He received the M.S. in electrical engineering from the University of California at Berkeley in 1993.

Between 1986 and 1989, he worked at the Department of Electrical Engineering and Computer Science (DEIS) at the University of Bologna on computer—adided design for analog circuits. In 1989, he joined the *Dipartimento di Elettronica ed Informatica* of the University of Padova, Padova, Italy, as Assistant Professor. Between 1990 and 1995, he has collaborated with the CAD group of the Department of EECS of the University of California at Berkeley where he has carried out research on performance-driven CAD methodologies for analog design. In July 1995, he joined Cadence Design Systems, San Jose, CA, where he has been working as architect in the physical design and mixed signal groups. He has been driving the development of advanced tools for the automation and acceleration of constraint-driven mixed-signal integrated circuit layout design and mixed signal floorplanning. He has authored or coauthored more than 40 papers in international journals and conferences, two books, and one U.S. patent.

Fred Sendig received the B.S. degree in electrical engineering and computer science from the University of California at Berkeley in 1984.

Since 1985, he has been at Cadence Design Systems, San José, CA, working on front-end design, mixed-signal design environments, infrastructure and corporate architecture. He is currently a Sr. Architect in the mixed signal design group.

PART II

Analog Synthesis

IDAC: An Interactive Design Tool for Analog CMOS Circuits

MARC G. R. DEGRAUWE, MEMBER, IEEE, OLIVIER NYS, MEMBER, IEEE, EVERT DIJKSTRA, JEF RIJMENANTS, MEMBER, IEEE, SERGE BITZ, BERNARD L. A. G. GOFFART, ERIC A. VITTOZ, SENIOR MEMBER, IEEE, STEFAN CSERVENY, MEMBER, IEEE, CHRISTIAN MEIXENBERGER, G. VAN DER STAPPEN, AND HENRI J. OGUEY, MEMBER, IEEE

***Abstract* — A design system has been developed which is able to design transconductance amplifiers, operational amplifiers, low-noise BIMOS amplifiers, voltage and current references, quartz oscillators, comparators, and oversampled A/D converters including their digital decimation filter starting from building-block and technology specifications.**

I. Introduction

ALTHOUGH analog circuits take up only a minor part of most ASIC's, their design time and cost is very important. In order to reduce this design effort, analog standard cell libraries can be used. However, since the circuits are then not tailored to their application, an optimum solution, with respect to power dissipation and area, is not obtained. Furthermore, such libraries, which typically have required more than 20 man years of design effort, very rapidly become obsolete due to technology evolution.

Rather than investing all of this man power in the design of only one analog cell library in geometric form, we have accumulated the design knowledge in a software program. This design system, called (Interactive Design for Analog Circuits (IDAC), is able to size a library of analog schematics (actually more than 40) as a function of technology (p-well and n-well CMOS) and desired building-block specifications. IDAC also generates a complete data sheet, an input file for SPICE2, and an input file for the analog layout program ILAC [1].

In the next section, the design system as seen by the user, as well as how it works, will be discussed. Then the design capabilities are illustrated with the design of oversampled A/D converters. Further, experimental results are given. Finally, program limitations and further developments are discussed.

II. Overview of the Design System IDAC

IDAC is able to design transconductance amplifiers [2]–[5], operational amplifiers [6], low-noise BIMOS amplifiers [7], voltage and current references [8], quartz oscillators [9], [10], comparators, and oversampled A/D converters including their digital decimation filter [11], [12].

After the technology parameters have been entered, the user has to select the kind of circuits he wants to design. For each of these families of circuits, the program is structured as shown in Fig. 1. The different inputs and outputs of the program and the design procedure will be discussed hereunder in detail.

A. Input Specifications

The user has to specify the technology, the desired building-block specifications, and design options. Furthermore one or more schematics of the library need to be selected.

In order to do a worst-case design, the program asks information concerning the following:

a) minimum and maximum value of the electrical parameters of MOS transistors, lateral bipolar devices, poly and well resistors, and capacitors;
b) the matching of components;
c) layout rules needed to calculate drain and source capacitances; and
d) extreme temperature and supply voltage values.

Besides the desired electrical building-block specifications, the user also has to specify some degrees of freedom which exist in almost all designs. For example, for the operational transconductance amplifier (OTA) of Figure 2, these degrees of freedom are the gain in the first stage, the current mirror ratio B, and the degree of inversion of the input devices (weak, moderate, or strong). For the degrees of freedom, no technology- and application-independent design strategy could be found. Therefore we give the opportunity to the designer to use the default values, which are a convenient choice for a very large domain of applications and for various technologies or to interactively change these parameters to evaluate other possible solutions.

With the design options, the user can further tailor his circuit. He can, for instance, ask to put input transistors in

Manuscript received May 2, 1987; revised July 15, 1987.
The authors are with the Centre Suisse d'Electronique et de Microtechnique S.A. (CSEM), Maladière 71, CH-2000 Neuchâtel 7, Switzerland.
IEEE Log Number 8716796.

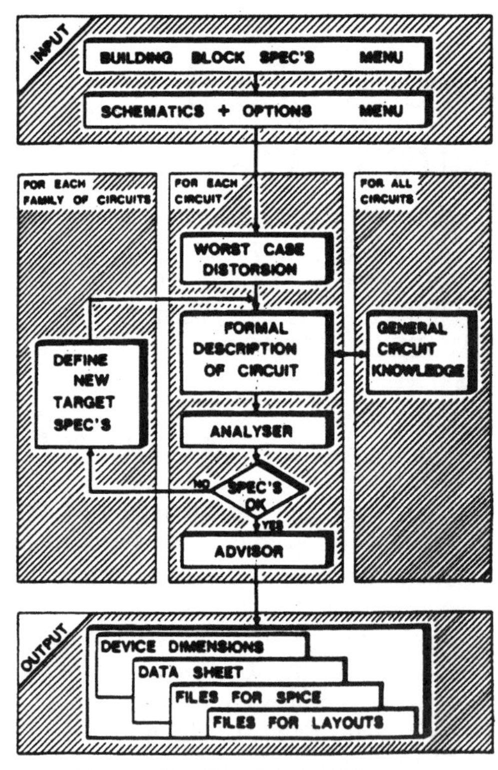

Fig. 1. Overview of the IDAC program.

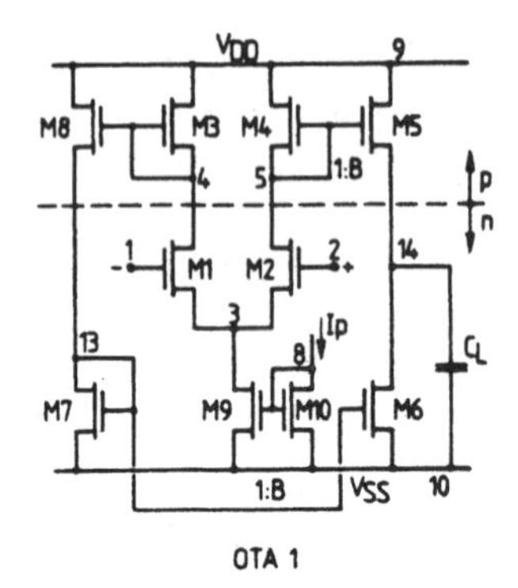

Fig. 2. Operational transconductance amplifier.

a separate well, to use n- or p-channel transistors at the input, to consider circular or comb-like layout of the devices, etc.

Finally, the user has to select one or more schematic of the library. Since the optimum solution depends on the required performances, technology specifications and parameters such as chip area, power consumption, and minimum required supply voltage, "the best schematic" does not exist. Therefore, we have included for each type of building block various schematics.

B. Output Results

As output the designer can ask to obtain the following:

a) a detailed input listing for documentation purposes;

b) a comparative table of the most important characteristics of various schematics;

c) detailed results concerning each schematic including the circuit performances at reference temperature for slow–slow electrical parameters, the dc voltages on each node of the circuit, the device dimensions and the operating point information, as well as a complete data sheet;

d) a frequency response (for amplifiers and filters), a transient response (for the comparators), the temperature behavior (for voltage references) or the signal-to-noise ratio as a function of the input signal (for A/D converters), as well as an input file for PPR-G (trademark of Sylvar-Lisco) containing all this information;

e) a SPICE2 input file;

f) an input file for the layout program ILAC [1]; this file contains the net list, information concerning the shape of the transistors, the different basic structures (differential pair, current mirror, etc), the devices which need to be matched, and the sensitive and noise nodes of the circuit; with the ILAC program, it is then possible to generate the layout of the circuit; and

g) an input file for PPR-G representing the data sheet.

C. Design Procedure

To size a schematic, IDAC makes use of three types of knowledge, i.e., knowledge specific to the schematic, general circuit knowledge (for instance, how to size cascode devices,) and knowledge common for a family of circuits (for example, how to stabilize an amplifier, how to improve the slew rate, etc.).

We will illustrate the working of the program with the design procedure of the OTA of Fig. 2 and the input specifications given in Fig. 3.

In order to take into account temperature and bias current variations, a worst-case distortion of the building-block specifications is performed. Therefore we assume that the bias current I_p is given by

$$\frac{I_p}{I_{p_0}} = \left(\frac{T}{T_{\text{ref}}}\right)^a \cdot (1 \pm X \text{ percent}) \tag{1}$$

where I_{p0} is the nominal bias current at reference temperature T_{ref} and X expresses the accuracy of I_{p0}. Furthermore we assume that the mobility is given by

$$\mu = \mu_0 \cdot \left(\frac{T}{T_{\text{ref}}}\right)^{-p} \tag{2}$$

By using (1) and (2), the variation of the building-block specifications can be easily calculated. For instance, if the input transistors of the OTA operate in weak inversion, the

```
**********************************************************************
*      Example                                         IDAC CSEM    V *
*      Input summary                                    8-JUL-87    2 *
*      OTA specifications                               15:34:31    D *
**********************************************************************

Gain ......................... 1.000E+03  -
GBW .......................... 2.000E+05 Hz
Phase margin ................. 6.000E+01 Degree
Slew rate .................... 1.000E-01 V/usec
Max. white noise ............. 1.000E-07 V/sqrt(Hz)
Max. 1/f noise at 1 KHz....... 5.000E-07 V/sqrt(Hz)
CMRR ......................... 6.000E+01 dB
Max. capacitive load ......... 5.000E-12 Farad
Min. capacitive load ......... 5.000E-12 Farad
Temp. exp. of bias current .. 1.000E+00  -
Max. variation of nominal Ip. 2.000E+01  %
Gain in the first stage ..... 2.000E+00  -
Input inversion coefficient . 2.500E-01  -
Current mirror ratio B ...... 1.000E+00  -
```

Fig. 3. Example of input specifications of OTA_1 which need to be satisfied in the commercial temperature range (0–70°C). The inversion coefficient of the input devices is defined as the ratio $I/(2\beta n(kT/q)^2)$ where n is the slope factor in weak inversion.

```
**********************************************************************
*      Example                                         IDAC CSEM    V *
*      OTA1                                             8-JUL-87    2 *
*      Required spec. at Tref                           15:34:34    D *
**********************************************************************

Gain....  1.133E+03  -              Gbw.....  2.508E+05  Hz
PhMar...  6.430E+01  Deg            S.R. ...  1.373E-01  V/usec
Vnwhit..  7.336E-08  V/sqrt(Hz)     Cmrr....  6.192E+01  dB
```

Fig. 4. Required specifications at reference temperature.

gain–bandwidth product GBW is given by

$$GBW = \frac{g_{m1} \cdot B}{2\pi C_L} \tag{3}$$

$$= \frac{I_p \cdot B}{2n \cdot \frac{kT}{q} \cdot 2\pi \cdot C_L} \tag{4}$$

$$= GBW_0 \cdot \left(\frac{T}{T_{\text{ref}}} \right)^{a-1} \cdot (1 \pm X \text{ percent}) \tag{5}$$

where GBW_0 is the GBW at reference temperature for nominal bias conditions. From (5), the nominal GBW_0 can then be evaluated which is needed to satisfy the GBW input specifications over the whole temperature and bias range. The same kind of derivations can also be made for the gain, the slew rate, and the noise and phase margins. At reference temperature, the specifications given in Fig. 4 are required.

Once the new "worst-case condition" specifications have been calculated, the program starts to execute the formal description of the circuit which will result in the size and bias conditions for each device. This description is based on an in-depth analytic study of the characteristics of the circuit which reveals the interrelationship of these characteristics as well as their dependence on device dimensions, bias current, and technology parameters. The formal description of the OTA (Fig. 2) is given in the Appendix.

After this sizing phase, the program will use a built-in analyzer to verify the correctness of the design. For amplifiers, this analyzer includes the evaluation of the transfer function to verify, for instance, the phase margin. For the comparators, a high-speed transient analyzer is used which is about 50 times faster than SPICE2 when exactly the same transistor model is used. This transient simulator is a voltage-controlled simulator [19] which has been adapted to be suitable for analog circuits [20].

If the analyzer detects that some specifications are not satisfied, then a new set of target specifications is defined and the formal description is re-executed. If all specifications are satisfied, then the program checks if the used device models are valid for the particular design. The program warns if one has potential problems with leakage currents, high-frequency effects in the transistors [13], short-channel effects [14], high or low injection in bipolar devices, too high base or emitter resistances, and a too low transient time of the bipolar devices. Finally all outputs are generated. For the design example of Fig. 3, the obtained results are summarized in Figs. 5 and 6. The data sheet is obtained starting from the results of the built-in simulator by introducing "distortions" as described. If the program does not find a solution, it tells the user why. For instance, it might indicate that the supply voltage is too small, or that the phase margin cannot be respected. When possible, IDAC gives a hint of how a solution can be obtained.

Typical design times are a few CPU seconds on a VAX780, which makes it possible to evaluate different designs in a very short time.

III. Design of Oversampled A/D Converters

With the IDAC program it is possible to design first- and second-order oversampled A/D converters including

```
************************************************************************
*     Example                                          IDAC CSEM    V *
*     OTA1                                             8-JUL-87     2 *
*     characteristics (Slow-Slow , Tref)               15:34:43     D *
************************************************************************

Gain....   1.453E+03   -             GBW.....   2.508E+05   Hz
PhMar...   6.700E+01   Deg           Vdyn....   4.097E+00   V
Ctot....   5.000E-12   F             Cadd....   0.000E+00   F
Gm......   8.910E-06   A/V           Gout....   6.133E-09   A/V
Ip......   9.737E-07   A             Itot....   1.947E-06   A
Vcas1...   0.000E+00   V             Vcas2...   0.000E+00   V
Vcmin...  -1.203E+00   V             Vcmax...   1.716E+00   V
Vomin...  -2.049E+00   V             Vomax...   2.048E+00   V
Vn.....w   7.296E-08   V/sqrt(Hz)    Vn...1/f   3.983E-07   V/sqrt(Hz)
Gamma..W   4.286E+00   -             Gam..1/f   2.867E+00   -
Ran Voff   4.120E-04   V             sys Voff  -7.798E-04   V
Cin.....   2.835E-13   F             Area....   5.853E+03   um**2
S.R. ...   1.937E-01   V/usec
```

(a)

	TYPE	W μm	L μm
M1,M2	N	90	5,5
M3,M4,M5,M8	P	26	22,5
M6,M7	N	8,5	19
M9,M10	N	5	5

(b)

```
************************************************************************
*     Example                                          IDAC CSEM    V *
*     OTA1                                             8-JUL-87     2 *
*     Data sheet                                       15:35:08     D *
************************************************************************

OTA characteristics given for Load  Cap. = 5.000E-12 Farad

                                    Min. Temp. = 0.000E+00 Deg C
                                    Max  Temp. = 7.500E+01 Deg C

                                    Vdd        = 2.250E+00 V
                                    Vss        =-2.250E+00 V

    Specification            Minimum value  Maximum value    Units

D.C. volt. gain                1.282E+03      2.815E+03       -
Gain Bandwidth                 2.000E+05      3.486E+05       Hz
Phase margin                   6.246E+01      7.177E+01       Deg
Slew rate                      1.411E-01      2.696E-01       V/usec
Pos. output swing              2.027E+00      2.066E+00       V
Neg. output swing             -2.033E+00     -2.063E+00       V
Random offset volt.            3.184E-04      5.081E-04       V
White noise                    5.641E-08      9.521E-08       V/sqrt(Hz)
1/f.Noise at 1KHz              3.744E-07      4.422E-07       V/sqrt(Hz)
Bias current                   7.090E-07      1.355E-06       A
current consump.               1.418E-06      2.710E-06       A
Power consumption              6.381E-06      1.219E-05       W
```

(c)

Fig. 5. Output results generated by IDAC (dc solution and operating point information are not shown): (a) characteristics at T_{ref}; (b) device sizes; and (c) data sheet.

the digital decimation filter [21]. This filter can be realized as a single equiripple FIR filter [12], [15] or as a cascade of (square or cube) Comb filters [22] followed by a truncation filter [16] and a final decimation filter [11]. The designer thus has a choice between various A/D architectures. Without the IDAC program, a global optimum solution can only be obtained by an experienced analog and digital designer who has a lot of time to spend. There is indeed a trade-off to be made between the design complexity of the analog and digital parts. The optimal solution will depend on the available technology, power consumption, and chip area.

With the IDAC program the job of the designer is significantly simplified. For input specifications the designer has to specify: a) the converter specifications (number of bits, gain error, offset error, maximum frequency of input signal, output rate, etc.); b) the boundaries of the obtainable performances of the used amplifiers, comparator, and voltage reference; and c) the loss of the signal-to-noise ratio, with respect to the ideal case, due to imperfections of the analog and digital circuits. Those last input specifications are in fact degrees of freedom which influence the block frequency and the design complexity of the analog and digital parts of the converter. By manipulating those the designer can scan various possible solutions and pick out the best. Starting from those input specifications, the program calculates the required system specifications and those of all the circuits used in the converter which can then be designed with the other molecules of the program. The program thus allows an hierarchical design style. The converter will, for instance, specify the characteristics of the switched-capacitor integrator(s) in the analog loop. Starting from these specifications, the program can then dimension the required transconductance amplifier [17]. To obtain a fast response time, the design phase is also based on an in-depth analytical study of the A/D converter. This analysis takes into account nonidealities such as finite gain, slew rate and gain bandwidth of the amplifiers, offset and hysteresis of the comparator, nonlinearity of capacitors, output impedance and bandwidth of the voltage reference, noise of the analog circuits, truncation noise in digital filters, etc.

In Fig. 7 the predicted and measured signal-to-noise ratios of a 10.5 significant-bit first-order A/D converter

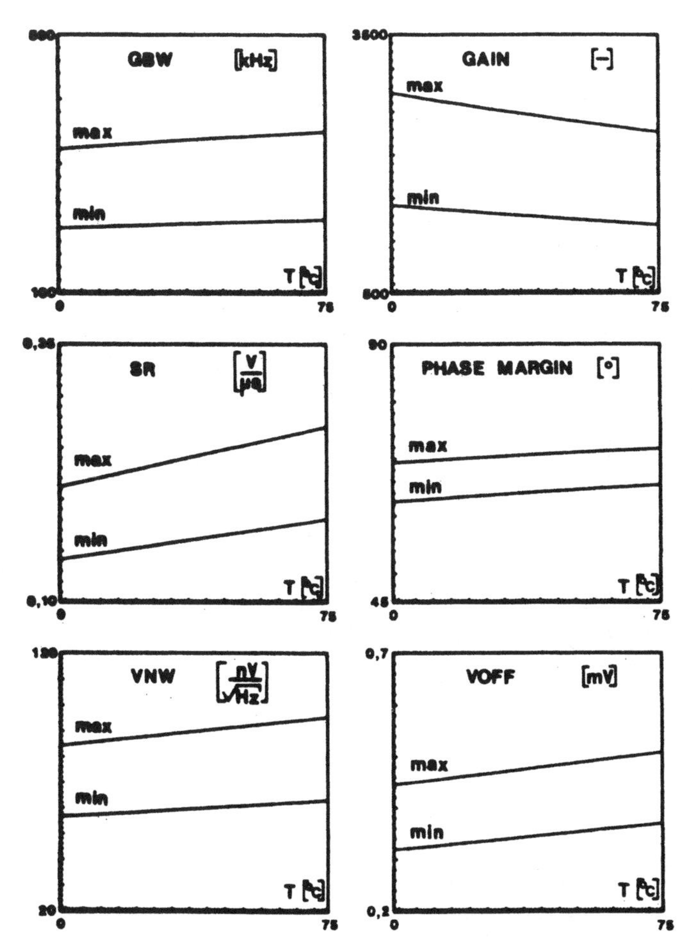

Fig. 6. Plot of variation of several OTA characteristics in the commercial temperature range. For each parameter, minimum and maximum values (due to technology and current variations) are given.

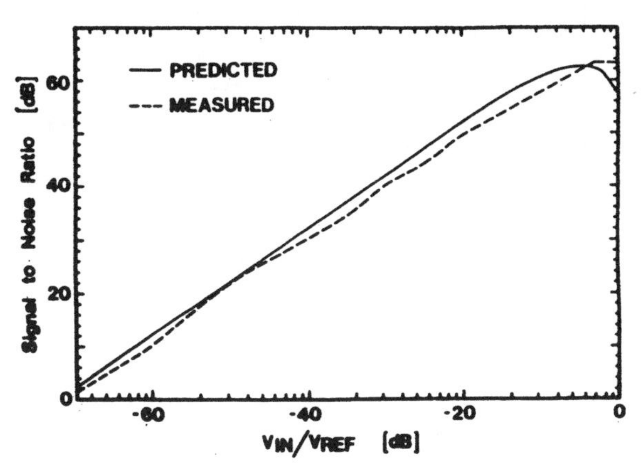

Fig. 7. Measured [15] and predicted signal-to-noise ratio of a 10.5 significant-bit A/D converter. Since calculations were done for dc signals while measurements were done with ac signals, the calculated curve should be shifted 3 dB to the right, which shows even better the close correspondence between measurements and theory.

[15] are compared. One can notice that even the overload region can be accurately predicted.

The required signal-to-noise ratio of a converter is equal to

$$\mathrm{SNR} = \mathrm{SNR}/_{\text{ideal}} - \Delta\mathrm{SNR}/_{\substack{\text{analog}\\ \text{imperfections}}} - \Delta\mathrm{SNR}/_{\substack{\text{digital}\\ \text{imperfections}}} \tag{6}$$

The less imperfections tolerated, the harder the design of the circuits. Too many imperfections will, however, require a much larger ideal SNR and thus clock frequency which can also make the design of the circuits very tedious. Some possible designs for a 12-bit converter are shown in Fig. 8. It is seen that if a single decimation filter is used, then a second-order $\Sigma-\delta$ converter will yield a less complex digital filter. This is mainly due to the much smaller ratio between the clock frequency and the transition band of the digital decimation filter. A more in-depth description of this module will be published in the near future.

IV. Experimental Results

In order to validate the program, besides a lot of SPICE simulations of each schematic two particular circuits in two different technologies are realized. The close correspondence between the measurement results and predicted values by IDAC for OTA's have already been published [2]. In this section, first the results of a bandgap reference are given. Thereafter, those of a low noise preamplifier are discussed.

A. Bandgap Voltage Reference

The bandgap references which can be realized with the program are insensitive to the low alfa and beta current gains of the lateral bipolar transistors [18] and to any amplifier offset [8]. One of them is shown in Fig. 9. A detailed description can be found in [8].

In Fig. 10(a) and (b), the measurement results of a circuit integrated in a p-well 3-μm CMOS process (Fig. 11) are compared with the characteristics predicted by IDAC. From Fig. 10(b) it is seen that even output voltage variations due to technology imperfections (ΔV_{Be}, ΔV_T, mismatch of resistors, etc.) are accurately predicted.

B. Low-Noise BIMOS Preamplifier

The schematic of one of the low-noise amplifiers, which can be realized with IDAC, is shown in Fig. 12. The amplifier consists of a current source (M_8-M_{10}, Q_9-Q_{10}, R_7), a first gain stage realized with lateral bipolar transistors (Q_1, Q_2) and resistors (R_1, R_2) where transistor M_1 is only used as a level shifter, and a second stage transconductance amplifier which is resistively loaded. The total gain of the amplifier is controlled by the ratio of the resistors and is independent of the alpha current gains of the bipolar transistors (role of Q_7 and Q_8). The resistors R_3 and R_4 are used to limit the overall distortion.

This circuit has also been integrated in a 3-μm p-well CMOS technology (Fig. 13). From Fig. 14, in which measurement results are compared with the performances predicted by IDAC, it is seen that the program is very accurate.

		1 order			2 order	
Loss [dB]	1	7	13	1	7	13
δp --	1.02 10^{-2}	4.1 10^{-2}	4.8 10^{-2}	9.9 10^{-3}	4.1 10^{-2}	4.9 10^{-2}
Δf [Hz]	579	4288	9592	365	2649	4988
δs --	1.198 10^{-5}	3.85 10^{-5}	4.8 10^{-5}	6.9 10^{-6}	1.9 10^{-5}	3.1 10^{-5}
N --	22180	3731	2532	10460	1524	1002
f_c [kHz]	2688	4264	6752	768	1016	1336

Fig. 8. Trade-offs for first- and second-order converters with a single FIR decimation filter (δ_s = stopband ripple, Δf = frequency transition, N = order of the filter, f_c = clock frequency).

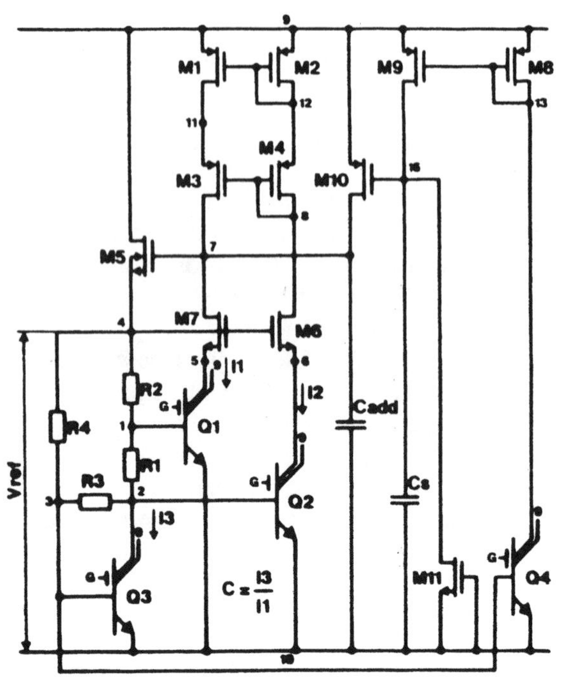

Fig. 9. Bandgap voltage reference.

	PROGRAM	MEASURED	Units
Reference voltage	1.273	1.28	Volt
standard deviation	±9,4	±12,3	mV
Minimum needed VDD	2.755	2.7	Volt
Supply regulation	1.49	0.95	mV/V
Load regulation	66	68.7	mV/mA
Total current cons.	2.82	3.05	μA
Start-up time	12	8	msec
1/f noise at 10Hz	6,61	7,24	μV/√Hz
White noise	1.67	1.8	μV/√Hz

(a)

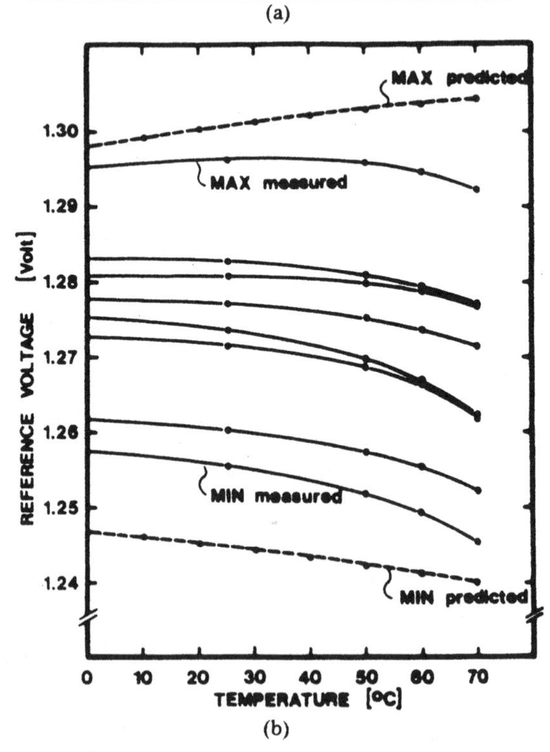

(b)

Fig. 10. (a) Characteristics of the voltage reference. (b) Predicted and measured temperature behavior of various samples.

V. Program Limitations and Further Developments

Often people wonder whether the IDAC system can make as good a circuit as an experienced analog designer. The answer to this question—for so far the analog designer has restricted himself to using the schematics available in the IDAC library—is probably yes. First of all, the IDAC system includes both knowledge we developed with several people inside the company who have extensive analog experience [24], [25] and knowledge we gathered from discussions with several well-known analog designers

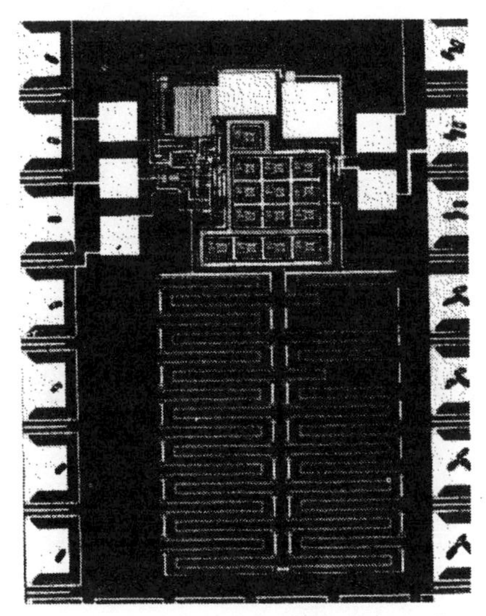

Fig. 11. Chip photograph of the voltage reference.

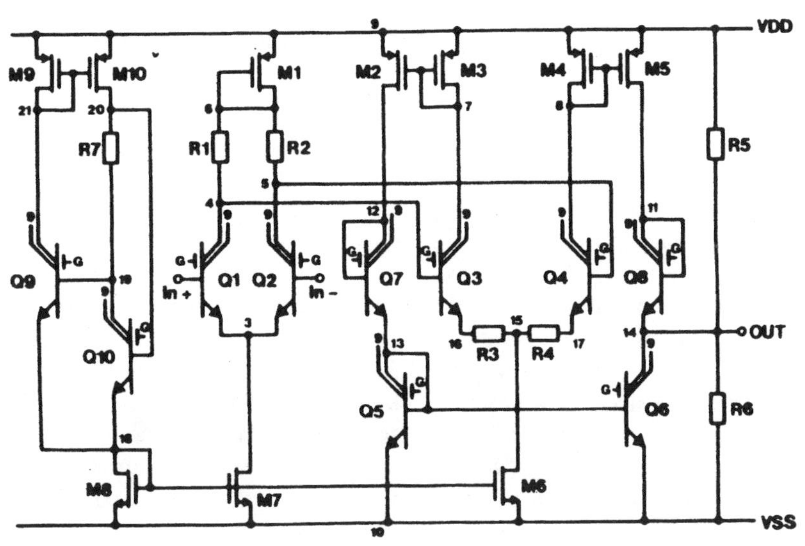

Fig. 12. Low-noise BIMOS amplifier.

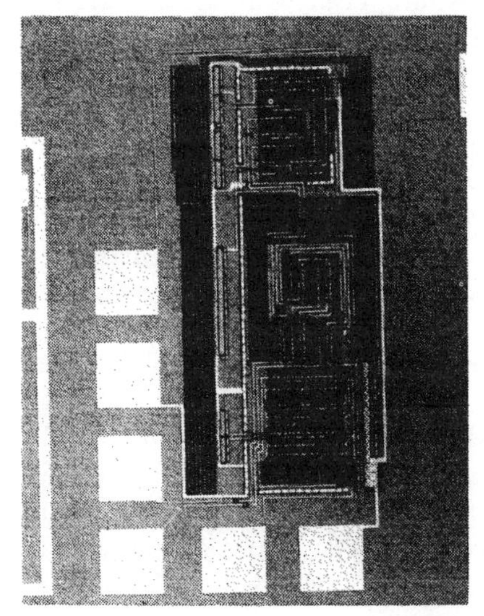

Fig. 13. Chip photograph of the preamplifier.

	PROGRAM	MEASURED	Units
Gain	38	39	dB
Total current cons.	0.7	0.65	mA
1/f noise at 10 Hz	<58	50	nV/√Hz
White noise	<10	9	nV/√Hz

Fig. 14. Predicted and measured results of the preamplifier.

such as P. Gray. Y. Tsividis, and E. Swanson. Second, due to the fast response time of IDAC, it is possible for experienced designers to converge towards a better solution than with a manual design since they can explore and take advantage of the degrees of freedom. In order to keep the IDAC library attractive, we also scan the literature to detect interesting schematics. Actually we include some new amplifiers [26], [27], as well as interleaved A/D converters [28], D/A converters, and Q oscillators.

```
{**********************************************************
Voltage reference - REF2
**********************************************************}
VERSION( 0, 5 );
CELL( 'REF2' ) ;
TECHNOLOGY( 'mosaic$demo:cmn20a.drd' );

PMOS( M1, PIN(4, 4, 9, 9), 1.500E-05, 4.000E-06,, 9 );
PMOS( M2, PIN(5, 4, 9, 9), 1.500E-05, 4.000E-06,, 9 );
PMOS( M3, PIN(6, 5, 9, 9), 9.900E-04, 4.000E-06 );
NMOS( M4, PIN(6, 6,10,10), 3.600E-05, 2.000E-06 );
NMOS( M5, PIN(3, 6,10,10), 3.600E-05, 2.000E-06 );
NMOS( M6, PIN(2, 6,10,10), 3.600E-05, 2.000E-06 );
LATPNP( QL1, PIN(4, 9, 1, 3,99), 1 );
LATPNP( QL2, PIN(5, 9, 2, 3,99), 8 );
LATPNP( QL3, PIN(9, 9, 9,13,99), 1 );
RES( R1, PIN(1, 2), SQUARES(350.00), POLY );
RES( R2, PIN(13, 1), SQUARES(3500.0), POLY );
CAP( CADD, PIN(5, 9), AREA(27128UU), METAL, POLY );

MIRROR( M4, M5, M6 );
CENTER( QL1, QL2 );

CONNECTOR( VDD, 10 );
CONNECTOR( VSS, 9 );
CONNECTOR( VREF, 1 );
CONNECTOR( VG, 99 );
END.
```

Fig. 15. File created by IDAC describing the layout of a voltage reference.

For system designers, however, the degrees of freedom which subsist for each design seem to be a handicap. Therefore we foresee the possibility of letting IDAC manage the degrees of freedom in order to minimize cost function taking into account the area and power consumption of the building blocks.

Recently we designed a 13-bit A/D converter with the program optimized with respect to power consumption. The total design time with IDAC (system design, voltage reference, comparator, amplifiers, and digital decimation filter) took about 4-h elapse time. The manual layout took about four to five man weeks and for final simulations another week was needed. This example clearly shows that the layout is responsible for the bottleneck in our design methodology. However, with the ILAC program [1], which takes a file generated by IDAC as input (see Fig. 15), we can now generate a layout (see Fig. 16) in a few minutes there where by hand we needed several days.

During the design phase the IDAC program takes into account the drain and source capacitances of the devices. Those capacitances are exactly known since IDAC tells the layout program ILAC how to make the device layout. The capacitances associated with the interconnections are not taken into account by IDAC. These capacitances are, however, typically only a few percent of the drain capacitances of the devices and therefore have only a small effect on the performances. We foresee extracting all information from the layout and feeding it back to IDAC in order to generate a final data sheet. For the time being, we do not think it will be necessary to bring the layout phase "into the design loop" of IDAC.

One of the drawbacks of the design methodology is that much more knowledge is needed to realize the program than to design an analog standard cell library. This means that a much larger initial (financial) effort is needed and it is questionable whether it can pay off. However, the derivation of a formal description of the actual available circuits has shown us the way to obtain, in an automated way, a formal description of circuits. Actually we are realizing, together with W. Fichtner, an "Analog Assistant" who will generate in an interactive way code for IDAC starting from a schematic entry. With this tool, the development time of new modules will be significantly reduced.

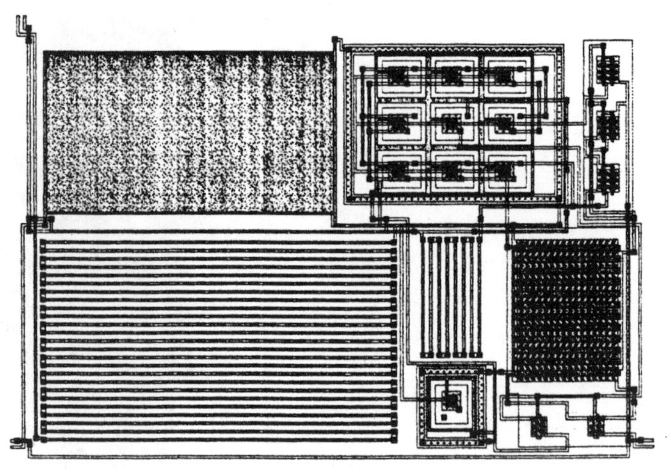

Fig. 16. Layout of voltage reference generated by ILAC starting from the description of Fig. 15.

VI. Conclusions

A standard cell approach is not optimum for the analog part of ASIC's since the circuit performances need to be tailored to their application. Therefore we propose a standard schematic approach. This consists of having a library of schematics of proven analog concepts and a corresponding formal description of the circuits. Starting from desired behavioral and technology specifications, this formal description is used to generate the electrical description, files for layout, and simulation. During synthesis, a built-in verification system is used and the validity range of the used device models is verified in order to ensure the correctness of the obtained circuits.

The program has been used to design circuits in several technologies. Since measurement results and predicted performances are always in close correspondence, the program has proven to be a reliable design tool.

With the implementation in the program of the converter design, it has been shown that an in-depth analytic study of the behavior of a building block is not restricted to circuits with only a few transistors, but is also possible for larger systems.

The use of the IDAC program makes analog design accessible to digital and junior designers. Through interactive use it is possible to deduce the knowledge embedded in the program. This makes it possible for a large group of engineers to become reasonably good analog designers in an easy-to-learn way. With the use of the design system, analog designers can enhance their productivity by more than an order of magnitude without sacrificing circuit performance. Disposing of this design tool allows them to evaluate even more possibilities, in an economic way, resulting in better customized solutions. The analog designers will certainly not lose their jobs, but in the future they will be free to do mainly innovative work.

Appendix
Formal Description of OTA

In this appendix, a summary of the formal description for the OTA of Fig. 2 is given. The symbol I_p is the bias current of the input stage, B stands for a current mirror ratio, G_1 stands for the gain in the first stage, and C_L is for the load capacitance.

Step 1: Determine the required bias current.

a) The GBW specifications, together with the parameter B and the load capacitance, allow the calculation of the minimum required g_m of the input transistors:

$$g_{m_1} \geqslant GBW \cdot 2 \cdot C_L / B. \tag{7}$$

b) The slew-rate specification SR, together with the parameter B and the load capacitance, allow the calculation of the minimum required bias current I_p:

$$I_p \geqslant SR \cdot C_L / B. \tag{8}$$

c) The white-noise specification ($v_{n,w}^2$), the gain in the first stage G_1, and the current mirror ratio B also impose a minimum value on the g_m of the input transistors:

$$g_{m_1} \geqslant \frac{16\,kT}{3 \cdot v_{n,w} 2} \cdot \left(1 + \frac{2 + 1/B}{G_1}\right). \tag{9}$$

d) By taking into account the operation region of the input devices (weak, moderate, or strong inversion), the required bias current I_p, g_m, and the W over L ratio of the input transistors can then be determined.

Step 2: Calculate W/L ratio of the other devices.

a) From the g_m of the input devices and the gain in the first stage, the required g_m of transistors M_3 and M_4 can be calculated. Since also the current through these devices is known, the W/L ratio of $M_3 - M_5$ and M_8 can be easily calculated:

$$g_{m_3} = g_{m_1} / G_1. \tag{10}$$

b) The g_m of the devices M_5 and M_6 are chosen equal and therefore the W/L of M_6 and M_7 are also known.

c) The transistors M_9 and M_{10} can be sized starting from CMRR specifications and input common-mode swing considerations.

d) Finally the gain specification, the g_m of the input devices, and the current mirror ratio B allow the calculation of the required output conductance of devices M_5 and M_6 and thus the minimum required effective length of these devices.

Step 3: $1/f$ noise consideration.

a) By using the weight of each noise source in the circuit and the $1/f$ noise specification, the minimum required area of each transistor can be calculated in such a way that the total gate area is minimal.

Step 4: Determining W and L of each device.

a) Starting from the W/L ratios and the $W \cdot L$ products and taking into account the matching considerations, the W and L of each device can be calculated.

During the design phase, the appropriate technology parameters are used in order to guarantee that the input specifications are met regardless of the technology variations. Offset is predicted according to the formulas proposed in [23].

Acknowledgment

The authors wish to express their gratitude to M. Schwoerer, V. von Kaenel, P. Deck, M. Meyvaert, and J. Sanchez for their help with the layout and testing of the circuits and for their help with the coding of the program.

References

[1] R. Zinszner, J. Rijmenants, T. Schwarz, and J. Litsios, "A layout program for analog circuits," submitted for publication.
[2] M. Degrauwe and W. Sansen, "The current efficiency of MOS transconductance amplifiers," *IEEE J. Solid-State Circuits*, vol. SC-19, pp. 349–359, June 1984.
[3] M. Degrauwe, J. Rijmenants, E. Vittoz, and H. De Man, "Adaptive biasing CMOS amplifiers," *IEEE J. Solid-State Circuits*, vol. SC-17, pp. 522–528, June 1982.
[4] F. Krummenacher, E. Vittoz, and M. Degrauwe, "Class AB CMOS amplifier for micropower SC-filters," *Electron. Lett.*, vol. 17, no. 13, pp. 433–435, June 25, 1981.
[5] M. Degrauwe and W. Sansen, "A synthesis program for operational amplifiers," in *Proc. ISSCC'84* (San Francisco, CA), Feb. 1984, pp. 18–19.
[6] P. R. Gray and R. G. Meyer, "MOS operational amplifier design—A tutorial overview," *IEEE J. Solid-State Circuits*, vol., SC-17, no. 6, pp. 969–982, Dec. 1982.
[7] M. Degrauwe *et al.*, "A analog design expert system," in *Proc. ISSCC'87* (New York, NY), Feb. 1987, pp. 212–213.
[8] M. Degrauwe, O. Leuthold, E. Vittoz, H. Oguey, and A. Descombes, "CMOS voltage references using lateral bipolar transistors," *IEEE J. Solid-State Circuits*, vol. SC-20, no. 6, pp. 1151–1157, Dec. 1985.
[9] E. Vittoz, "Quartz oscillators for watches," presented at the Int. Congress of Chronometry, Geneva, Switzerland, 1979.
[10] S. Bitz, M. Degrauwe, and E. Vittoz, "A very precise 2 MHz micropower quartz oscillator," to be published in *ESSCIRC'87*, Sept. 1987.
[11] O. Nys, M. Degrauwe, and E. Dijkstra, "A CAD tool for oversampling A/D converters," in *Proc. Symp. VLSI Circuits* (Karuizawa Japan), May 1987, pp. 113–114.
[12] E. Dijkstra, M. Degrauwe, J. Rijmenants, and O. Nys, "A design methodology for decimation filters in sigma-delta A/D converters," in *Proc. ISCAS* (Philadelphia, PA), 1987, pp. 479–482.
[13] M. Bagheri and Y. Tsividis, "A small signal dc-to-high frequency nonquasistatic model for the four-terminal MOSFET valid in all regions of operation," *IEEE Trans. Electron Devices*, vol. ED-32, no. 11, pp. 2383–2391, Nov. 1985.
[14] J. Brews, W. Fichtner, E. Nicollian, and S. Sze, "Generalized guide for MOSFET miniaturization," *IEEE Electron Device Lett.*, vol. EDL-1, no. 1, pp. 2–4, Jan. 1980.
[15] M. Hauser, P. Hurst, and R. Brodersen, "MOS-ADC filter combination that does not require precision analog components," in *Proc. ISSCC'85*, pp. 80–81.
[16] J. Candy and An-Ni-Huynh: "Double interpolation for digital-to-analog conversion," *IEEE Trans. Commun.*, vol. COM-34, no. 1, pp. 77–81, Jan. 1986.

[17] M. Degrauwe and F. Salchli, "A multi-purpose micropower SC-filter," *IEEE J. Solid-State Circuits*, vol. SC-19, pp. 343–348, June 1984.

[18] E. Vittoz, "MOS transistors operated in the lateral bipolar mode and their application in CMOS technology," *IEEE J. Solid-State Circuits*, vol. SC-18, pp. 273–279, June 1983.

[19] D. Tsao and C. Chen, "A fast-timing simulator for digital MOS circuits," *IEEE Trans. Computer Aided Des.*, vol. CAD-5, no. 4, pp. 536–540, Oct. 1986.

[20] M. Degrauwe, C. Meixenberger, and G. van der Stappen, "A silicon compiler for voltage comparators," to be published.

[21] O. Nys, E. Dijkstra, and M. Degrauwe, "The front-end of a silicon compiler for $\Sigma\delta$ A/D converters," submitted to *IEEE J. Solid-State Circuits*.

[22] S. Chu and C. S. Burrus, "Multi-rate filter designs using comb filters," *IEEE Trans. Circuit Syst.*, vol. CAS-31, no. 1, Nov. 1984.

[23] K. Lakshmikumar, R. Hadaway, and M. Copeland, "Characterization and modeling of mismatch in MOS transistors for precision analog design," *IEEE J. Solid-State Circuits*, vol. SC-21, no. 5, pp. 1057–1066, Dec. 1986.

[24] E. Vittoz and J. Fellrath, "CMOS analog integrated circuits based on weak inversion operation," *IEEE J. Solid-State Circuits*, vol. SC-12, no. 3, pp. 224–231, June 1977.

[25] H. Oguey and B. Gerber, "MOS voltage reference based on polysilicon gate work function difference," *IEEE J. Solid-State Circuits*, vol. SC-15, pp. 264–269, June 1980.

[26] D. Monticelli, "A quad CMOS single-supply op amp with rail-to-rail output swing," *IEEE J. Solid-State Circuits*, vol. SC-21, pp. 1026–1034, Dec. 1986.

[27] M. Armstrong, H. Ohara, H. Ngo, C. Rahim, and A. Grossman, "A CMOS programmable self-calibrating 13 b light-channel analog interface processor," in *Proc. ISSCC 1987* (New York, NY), Feb. 1987, pp. 46–47.

[28] Y. Matsuya *et al.*, "A 16 b oversampling A/D conversion technology using triple integration noise shaping," in *Proc. ISSCC 1987* (New York, NY), Feb. 1987, pp. 48–49.

Marc G. R. Degrauwe (S'78–M'84) was born in Brussels, Belgium, on August 16, 1957. He received the engineering degree in electronics and the Ph.D. degree in applied sciences from the Katholieke Universiteit Leuven (KUL), Leuven, Belgium, in 1980 and 1983, respectively.

During the summer of 1980, he was on leave at Centre Electronique Horloger S.A. (CEH), Neuchâtel, Switzerland. From autumn 1980 to 1983, he was associated with KUL where he worked on the design of micropower amplifiers and sampled data filters. In July 1983 he returned to CEH. In 1984, when CEH was reorganized into the Swiss Centre of Electronics and Microtechnics (CSEM), he became Head of the Circuits Department. His actual field of interest is design automation of analog circuits. He is also lecturing on analog circuit design at the University of Neuchâtel.

Olivier Nys (S'73–M'85) was born in Tournai, Belgium, in 1960. In 1984 he graduated as an electrical engineer from the Catholic University of Louvain-La-Neuve.

He then joined the Center Suisse d'Electronique et de Microtechnique (CSEM), Neuchâtel, Switzerland. He was first involved with optimization of ROM's and RAM's with respect to access time, power consumption, and testability. His research interest is oversampled A/D converters.

Evert Dijkstra was born in Heerenveen, The Netherlands, on January 1, 1960. He received the M.S.E.E. degree (cum laude) from Twente University of Technology, The Netherlands.

He completed his technical training period for Twente University of Technology at the Centre Electronique Horloger S.A., Neuchâtel, Switzerland, from June 1983 to December 1984. He is presently at the Centre Suisse d'Electronique et de Microtechnique (CSEM), Neuchâtel, Switzerland, involved in the design of digital signal processing systems, CMOS design methodologies, and computer-aided design techniques.

Jef Rijmenants (S'80–M'81) was born in Nijlen, Belgium, on May 30, 1956. He received the degree of Burgerlijk Ingenieur Electronica from the Katholieke Universiteit Leuven (KUL), Leuven, Belgium, in 1979.

After graduating he spent two years at ESAT Laboratories, Leuven, Belgium as a Research Assistant working in the field of micropower sampled data circuits. Later he joined Silvar-Lisco, Leuven, Belgium where he worked as a Software Engineer on a symbolic layout editor integrated with a compactor, a graphics postprocessor, and database interfaces. He joined the Centre Suisse d'Electronique et de Microtechnique (CSEM), Neuchâtel, Switzerland, in 1985, where he is currently working on automated layout techniques for analog integrated circuits.

Serge Bitz was born in Sion, Switzerland, on November 30, 1959. He received the M.S. degree in electrical engineering from the Swiss Federal Institute of Technology (ETHZ), Zurich, in 1985.

He joined the Centre Suisse d'Electronique et de Microtechnique (CSEM), Neuchâtel, Switzerland, in 1985. Since then he has been engaged in the research and development of CMOS analog integrated circuits.

Bernard L. A. G. Goffart was born in Brussels, Belgium, on May 17, 1962. He received the engineering degree in electronics in applied science from the Université Catholique de Louvain (UCL), Louvain-La-Neuve, Belgium, in 1985.

From the fall of 1985 to March 1986 he was employed as an assistant at the Université Catholique de Louvain. Now he is working at the Centre Suisse d'Electronique et de Microtechnique S.A. (CSEM), Neuchâtel, Switzerland, on the design automation of analog building blocks.

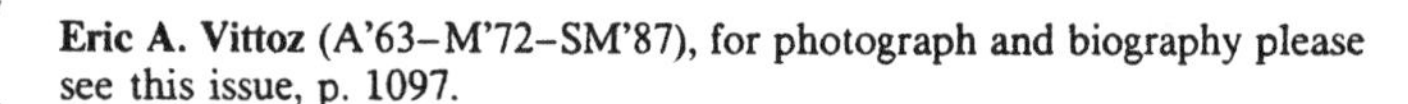

Eric A. Vittoz (A'63–M'72–SM'87), for photograph and biography please see this issue, p. 1097.

Stefan Cserveny (M'86) was born in Resita, Romania, on January 14, 1941. He received in 1962 the Dipl. Ing. degree in electronics from the Polytechnical Institute of Bucharest, Romania, and in 1973 the M.S. degree in electrical engineering from the University of California, Berkeley.

Between 1962 and 1979 he lectured on electronic devices and circuits at the Polytechnical Institute of Bucharest and at the Telecommunications Institute of Oran, Algeria, writing several textbooks and papers. In 1979 he joined the Centre Electronique Horloger S.A. in Neuchâtel, Switzerland, which in 1984 became the Centre Suisse d'Electronique et de Microtechnique (CSEM), where he is involved in device modeling, analog circuit design for intelligent sensors, and EEPROM.

Christian Meixenberger was born in Bern, Switzerland, on October 11, 1960. He received the engineering degree in electronic physics from the University of Neuchâtel in 1986.

Now he works at the Centre Suisse d'Electronique et de Microtechnique (CSEM) in Neuchâtel, Switzerland, where he is involved in analog circuit design.

G. van der Stappen, photograph and biography not available at time of publication.

Henri J. Oguey (M'60) was born in Geneva, Switzerland, on December 10, 1928. He received the M.E. degree in 1952 and the Ph.D. degree in 1956, both from the Swiss Federal Institute of Technology.

From 1956 to 1963 he worked for the IBM Research Laboratories in Zurich, Switzerland and Yorktown Heights, NY. Then he joined the Centre Electronique Horloger, Neuchâtel, Switzerland. His past experience covers, among others, thin magnetic films applications, micromotors, the development of the first quartz wristwatch, and the design of digital and analog CMOS circuits. He has been granted more than 20 patents in these various fields. He also lectured on electromechanics and IC design at the University of Neuchâtel. He is now Manager of the Integrated Devices Department at the Centre Suisse d'Electronique et de Microtechnique (CSEM), Neuchâtel, Switzerland. His present activities include device modeling and characterization, CMOS compatible smart sensors, and nonvolatile memories.

OPASYN: A Compliler for CMOS Operational Amplifiers

HAN YOUNG KOH, MEMBER, IEEE, CARLO H. SÉQUIN, FELLOW, IEEE, AND PAUL R. GRAY, FELLOW, IEEE

***Abstract*—A silicon compilation system for CMOS operational amplifiers (OPASYN) has been developed. The synthesis system takes as inputs system level specifications, fabrication-dependent technology parameters, and geometric layout rules. It produces a design-rule-correct compact layout of an optimized op amp. The synthesis proceeds in three stages: 1) heuristic selection of a suitable circuit topology, 2) parametric circuit optimization based on analytic models, and 3) mask geometry construction using a macro cell layout style. The synthesis process is fast enough for the program to be used interactively at the system design level by system designers who are inexperienced in op amp design.**

I. INTRODUCTION

THE pervasive trend in recent years towards the integration of whole systems into single-chip VLSIC's requires that both digital functional units and dedicated analog interface subsystems (such as A/D converters and filters) be implemented on the same chip. Many digital parts of such chips can nowadays be synthesized rapidly and reliably using CAD tools developed for semicustom design methods such as gate arrays, standard cells, and macro cells. On the other hand, analog interface subsystems still need to be entirely handcrafted by a specialist. Therefore, the design time and cost associated with dedicated analog interface components often constitute a bottleneck in semicustom design of mixed analog/digital systems such as voice-band data modems and high speed data transceivers [1].

For semicustom design of analog circuits, building blocks may be stored in the form of parameterized generators or as entries in macro cell or standard cell libraries [2]-[5]. As in the digital domain, the usage of libraries of predefined analog building blocks can shorten the design period significantly. However, in the analog domain it is difficult to configure a rich enough set of library cells for the wide spectrum of possible applications. Performance specifications for analog building blocks are much more diverse and complicated than those for digital blocks. For a digital inverter, delay and area are the major performance specifications. On the other hand, specifications for an operational amplifier (op amp) may include power dissipation, small signal dc gain, bandwidth, phase margin, slew rate, settling time, output voltage swing, $1/f$ noise, area, and so forth. Some of these specifications, such as the small signal dc gain of a CMOS op amp, may vary over a wide range, depending on the application and the chosen system architecture. Furthermore, analog library entries become obsolete even more quickly when technology or design rules change. A performance characteristic such as $1/f$ noise of a CMOS op amp degrades as technology scales down while that for small signal dc gain improves. As a result, *cell generators* that operate at the circuit or netlist level are more flexible and can be useful over a much larger domain of applications than fixed library cells. With such generators, design parameters (such as device sizes or bias currents) of a building block can be individually optimized for a particular application and for the particular technology to be used. The physical layout geometry of the block with its optimal design parameter values is then also produced by the generator.

The most often used analog building block is the operational amplifier. It is at the heart of many interface circuits, in particular, A/D and D/A converters, and filters. An efficient design of optimal op amps is thus a cornerstone of a design environment for many applications. As discussed above, op amp specifications for different applications vary so widely that it is impractical to store op amps as library cells for all applications. Using one of a few "standard" library cells that are poorly matched to a particular application is unacceptable; if the performance of the op amp falls below a certain "threshold," the quality of the overall system will suffer.

Designing a good op amp is a rather complicated multifacet task [6], [7]. An op amp topology appropriate for the given specifications must be chosen. Then its design parameters such as device sizes and bias currents must be adjusted under multiple design objectives and constraints. The many degrees of freedom in parameter space as well as the need for repeated circuit performance evaluation make this a lengthy and tedious process. This circuit optimization can be carried out by a computer but the process is still expensive and requires a skilled operator. The

Manuscript received March 16, 1989. This work was supported by DARPA under Contract DARPA VLSI-N00039-87-C-0812 and under Contract SRC 82-11-008. This paper was recommended by Editor M. R. Lightner.

H. Y. Koh was with the University of California, Berkeley, CA 94720. He is now with Cadence Design Systems, Inc., 2475 Augustine Drive, Santa Clara, CA 95054-3082.

C. H. Séquin and P. R. Gray are with the Department of Electrical Engineering and Computer Sciences, University of California, Berkeley, CA 94720.

IEEE Log Number 8931875.

optimized circuit needs then to be transformed into mask level geometries. The layout process is critical for good performance since op amps like other analog circuits are sensitive to parasitic elements, to process/thermal gradients, and to noise. In CMOS technology, device sizes in op amps vary over a large range of values. Some devices may have many alternative shapes, and terminal configurations. Thus, most of the classical layout generation methods developed for digital IC's are insufficient for CMOS op amps [8]–[12]. As a result, work on automatic layout of monolithic switched capacitor filters typically uses op amps as fixed cells from a library [13]–[15]. For many applications, this approach does not give enough flexibility and performance.

Most of the work published on the automated synthesis of op amps concern the schematic generation and parametric optimization of op amp *circuits*. The most common approaches are either based on optimization procedures or on an expert systems approach, or a combination of both.

The optimization-based approach uses various optimization algorithms combined with circuit simulation techniques to produce general purpose parametric optimization tools [16]–[18]. Since no a priori circuit design knowledge is necessary, this approach is applicable to a broad range of analog circuits. High degrees of optimality can be obtained through the use of robust and elaborate optimization algorithms. However, optimization tools based on this approach are not tailored to be used by novice designers. For instance, a system designer who wants to design an op amp for his/her A/D converter, has to provide the system with detailed design knowledge on op amps such as how to calculate slew rate from a particular node voltage. Besides, the designer must choose a good initial guess for each design parameter. Without such a good starting point, an optimization run may converge very slowly or converge to a local minimum whose performance is significantly worse than the circuit's best capabilities [17]. This approach is normally very costly in CPU time because repeated circuit simulation is performed in the inner loop of the optimization step.

In the knowledge-based approach, domain-specific knowledge has been used to produce synthesis tools that emulate expert designer's design processes [19]–[23], [48]. The domain specific design knowledge integrated into these systems permits novice designers to specify simply their design requirements. Fine tuning of the system for a specific application area can also be achieved easily. But the large and complicated search space corresponding to the complete design of an op amp circuit and a lack of an efficient performance estimation method make it difficult and inefficient to perform design optimization using this approach. In addition, *a priori* design knowledge has to be integrated into the system whenever a new type of building block is added. A general synthesis system that starts from first principles of circuit design such as Kirchhoff's current and voltage laws does not yet exist!

Alternatively, advantages of both approaches can be combined to devise synthesis tools whose optimization is done in an algorithmic way but substitutes expensive circuit simulation with algebraic evaluation of analytic design equations acquired from domain specific design knowledge [24]. This is the approach taken by OPASYN described in this paper.

Automatic layout generation for CMOS op amps has been less widely explored [23], [25], [26]. The approaches rely either on dedicated slicing trees [23], [25] or use a general-purpose layout program based on simulated annealing and constraint-based routing [26].

This paper describes a design synthesis tool for monolithic CMOS op amps (OPASYN) which starts from a set of performance specifications (open-loop gain, bandwidth, slew rate, etc.) and produces design-rule-correct mask geometries in a macro cell layout style. The following sections explain the framework of OPASYN, the circuit selection scheme, the parametric circuit optimization, and the layout generation method. Implementation of the system and experimental results are discussed, and a synthesis example is provided to illustrate the design data flow in the system.

II. OPASYN Framework

The OPASYN system consists of an internal database and three functional modules; a circuit selection module, a parametric circuit optimization module, and a layout generation module. The database contains the necessary design knowledge for each op amp topology available to the users of the system; it includes [24], [25] the following:

- a decision tree for topology selection,
- analytic circuit models for parametric circuit optimization,
- slicing tree descriptions for floorplanning,
- netlist descriptions for routing.

Device parameters and design rules for different process technologies are retrieved from a technology library. Synthesis flow and its interface to other physical design tools in the U.C., Berkeley CAD environment are shown in Fig. 1.

Synthesis starts from a set of op amp performance specifications. Based on the general domain of the specifications, the program selects an appropriate option out of a small database of generic, widely applicable op amp circuit topologies, unless the user explicitly specifies a particular circuit topology to be used. For the chosen circuit, optimal values for its design parameters will be determined to meet the objectives implied by the given specifications [25]. The SPICE interface program is called by this module to verify the optimization result through extensive circuit simulation [27], [28]. Finally, the layout generation module takes the netlist of the sized circuit schematic from the previous synthesis phase together with a specification of a desired aspect ratio or of a vertical/horizontal size constraint, and a reference to a file of

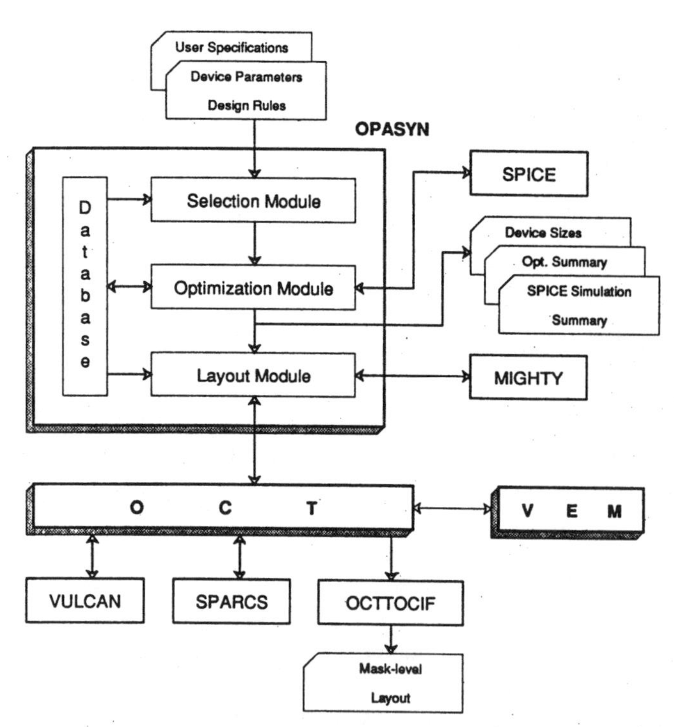

Fig. 1. OPASYN framework and its interface to the Berkeley CAD environment.

geometrical design rules, and produces a symbolic layout of the circuit using a macro cell style. The symbolic layout is then properly spaced by a spacing program to produce design-rule-correct mask geometries. Design data produced at each phase of the layout generation is hierarchically managed by the data manager OCT [29], [30].

The modular architecture and flexible functional modules used in OPASYN make it possible to extend the system to synthesize other analog building blocks such as output buffers, comparators, and bandgap references with minimal alterations to the program. Various technologies such as CMOS, bipolar and BiMOS can be also accommodated. Up to now, CMOS technology has been employed exclusively because of its prevailing usage in dedicated analog interface component designs.

III. Circuit-Selection based on Heuristic Pruning

The first step in op amp design is to obtain a promising circuit topology fitting the given design requirements. Expert designers perform this task based on their design experience. They may select an existing topology that, based on prior experience, fits the new application. Or they may modify one that almost fits by enhancing specific performance characteristics by trial and error, which often requires many design cycles. Creating an entirely "new" topology is quite difficult. The circuit should be designed to cancel out first-order variations in design parameters to yield stability and immunity against a rather wide range of spreads in active and passive integrated circuit component values. The tight and often intricate coupling between functional modules in typical op amp circuits makes hierarchical decomposition of functionality difficult. For instance, the phase margin of a two-stage op amp must be determined by considering the frequency response of the entire circuit. Propagating the constraints imposed by the phase margin of the entire circuit to its consitituent modules, namely input stage, output stage, and compensation stage is rather infeasible. As a result, most synthesis techniques for analog circuits are based on analysis; an appropriate known circuit topology is selected and its component values are then adjusted to meet the overall design requirements by analyzing the circuit. There have been some activities in knowledge-based topology creation [19], [22] but so far success has been quite limited. Practical op amp topologies are still being invented by expert designers.

The circuit selection strategy used in OPASYN has been devised based on the aforementioned observations. A decision tree (shown in Fig. 2) has been defined based on some key design specifications such as general application area, open-loop gain, power supply rejection ratio (PSRR), or fully differential topology requirement. The leaf nodes in this tree correspond to proven op amp topologies commonly used in many applications, and corresponding design knowledge is stored in the database of OPASYN. Currently five different circuit topologies have been fully incorporated and they are denoted in Fig. 2 as SFC, FFC, BTS, CTS, and OPB. Searching for a suitable topology starts at the root of the decision tree; nodes of this tree are checked in turn whether some subtrees can be pruned away (eliminated from further consideration) based on the range of the given specifications. The unpruned leaf nodes are forwarded to the optimization module. The following example illustrates the basic idea.

Example:

Let us assume that the application demands an "ordinary" op amp with an open-loop gain of 10 000 and a PSRR of 70 dB at 1 kHz using the MOSIS 3-μm process. The decision process starts at the top and immediately eliminates all the "special-purpose" op amps since none of these special characteristics are being called for. Among the "standard" op amps, the subtree of two-stage op amps is selected since the specified open-loop gain of 10 000 is higher than the gain limit of single-stage op amps for the MOSIS 3-μm process (which is set to 5000 in the technology library). The particular decision path for this example is shown in Fig. 2 with dotted arrows; "default" represents the default choice and "> limit" means that the branch is to be taken if the corresponding design specification at the node is greater than the limit value retrieved from the technology library.

The discussed circuit selection scheme has been implemented in Common Lisp using a rule-based approach. This has the advantage that the decision tree can be readily updated whenever new circuits are added to the database or when the pruning heuristics are modified. Furthermore, the decision process has been made visible to the user. The program is capable of explaining its decisions and behavior to the user in response to such queries as "why" and "how" [31]–[34].

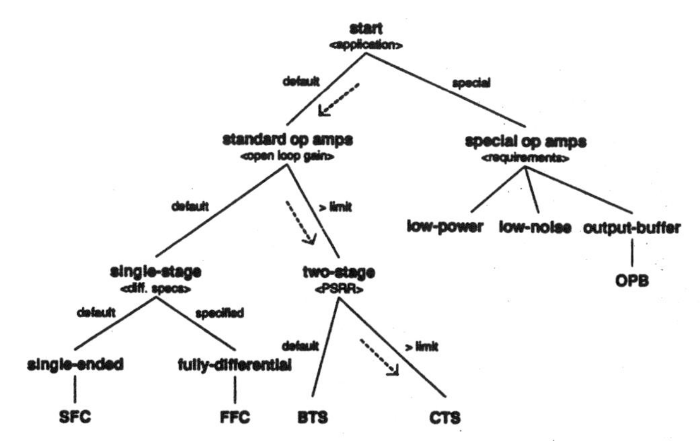

Fig. 2. Decision tree and heuristic pruning for circuit selection.

If the specification-based decision criteria are loose enough, more than one leaf node may be returned, and the optimization process is carried out for all of these nodes. In this case, the user makes the final selection among the alternatives. Often, the given specifications can be met with a "standard" op amp as well as with a "special-purpose" op amp. The default selection is to use the "standard" op amp since it is simpler and smaller. The user should only ask for a "special-purpose" op amp if none of the "standard" op amps can meet the specifications.

IV. Parametric Optimization

Once a promising op amp topology has been selected, the circuit synthesis problem is reduced to a task of parametric component optimization. A set of parameter values needs to be found that optimizes some performance attributes while meeting all explicitly stated specifications. For op amps, key component values are widths and lengths of MOSFET's, bias currents, and compensation capacitor values (if any). However, all these values need not be independently adjusted. Sound design strategies link many of these values into groups that depend on just a few independent design parameters. Performance of an op amp is then a nonlinear function of these design parameters through some circuit properties. For instance, the small signal dc gain of a two-stage op amp is determined by such circuit properties as transconductances (g_m) and output conductances (g_o) of certain transistors in the circuit, where g_m and g_o are in turn functions of device sizes and bias currents of the transistors. Therefore, evaluation of the cost function and of its gradient requires the solution of a very large set of simultaneous nonlinear differential equations and dictates the usage of a circuit simulation program such as SPICE in the inner loop of the optimization process. As a result, conventional optimization programs are very expensive in terms of CPU time.

4.1. Analytic Circuit Models

The parametric optimization process in OPASYN relies on analytic models specially developed for each of the op amp circuit topologies in the database [24]. An analytic model of a circuit is a set of analytic expressions representing sound *a priori* design decisions, and macro behavior of the circuit. The analytic circuit model for a given circuit typically includes:

- netlist description of the circuit,
- declaration of the independent design parameters,
- reasonable upper and lower bounds for the design parameter values,
- analytic design equations to express dependency of circuit performance on design parameters.

In order to reduce the dimensionality and size of the search domain, one should define a minimal set of independent design parameters and set reasonable upper and lower bounds on their range. Our investigations have shown that using expert design knowledge and first-order circuit analyses, the number of independent design parameters of the circuit can be greatly reduced. In the basic two-stage (BTS) op amp shown in Fig. 3 which consists of 13 transistors, a capacitor and a current source, we have found that the number of design parameters can be reduced to only five: $(W/L)_1$, $(W/L)_5$, $(W/L)_6$, C_f, and I_{bias}. Similarly, the fully differential folded cascode (FFC) op amp shown in Fig. 4 with over 60 devices has only 7 independent design parameters. The upper and lower bounds for the design parameter values can be set to further restrict the search space. For instance, the compensation capacitor value in the BTS op amp has been limited to the range from 2 to 20 pF since we know that values outside this range are not useful in practical applications.

The analytic design equations were derived by using first-order circuit analysis techniques and topology-specific approximations [6], [7], [35]–[39]. For most dc characteristics, these computed approximations are excellent. For highly non-linear specifications such as small signal dc gain, phase margin, and settling time of the circuit, fitting parameters have been introduced to obtain more accurate prediction of specific performance characteristics. Here is an example of such analytic design equations:

$$a_v = cf_{\text{gain}} \times \frac{g_{m2} g_{m6}}{(g_{o2} + g_{o4})(g_{o6} + g_{o7})}$$

$$g_m = \sqrt{2 k_p \frac{W}{L} I_d}$$

$$g_o = \lambda I_d$$

where a_v is the small signal dc gain of the BTS op amp shown in Fig. 3, g_m is a transconductance, g_o is an output conductance of a transistor, and cf_{gain} is a fitting parameter on the order of 1.0. All the conductances are dependent on transistor sizes and bias currents. The fitting parameters are being updated as the system acquires more information from repeated synthesis and subsequent analysis with SPICE.

Analytic models can be developed not only for op amps but also for any type of analog functional blocks such as

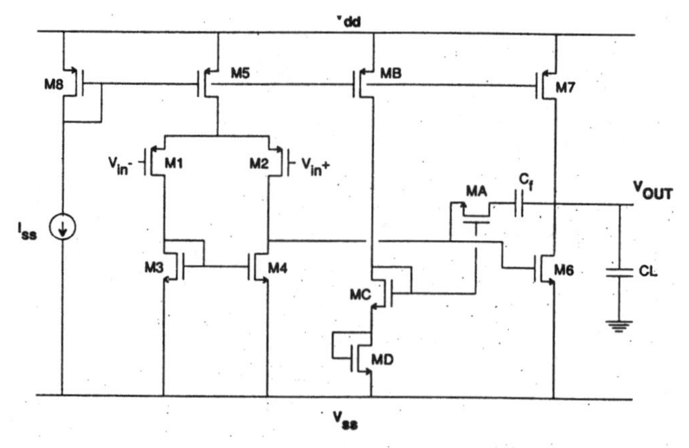

Fig. 3. Basic two-stage (BTS) op amp.

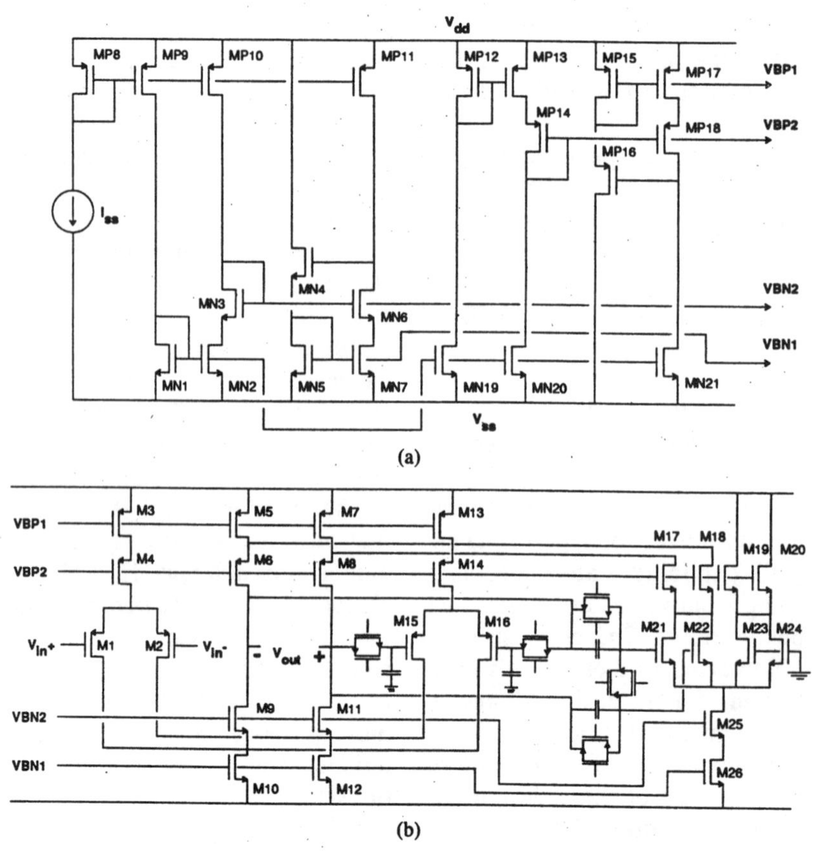

(a)

(b)

Fig. 4. Fully differential folded cascode (FFC) op amp. (a) Bias circuit. (b) Amplifier.

comparators, bandgap references, and output buffers. However, to derive an analytic model for these circuits demands the attention of experienced analog circuit designers. The one time design efforts to add our most complicated op amp topology, the FFC op amp (shown in Fig. 4) was two weeks and only a few days for the simpler circuits. Model acquisition process can be summarized as follows. A model developer, who is typically an expert circuit designer, determines a set of independent design parameters of the circuit topology and sets lower and upper bounds on each design parameters based on first-order manual circuit analyses as well as his/her design experience. Then the next step is to derive an analytic expression for each performance characteristic of the circuit again using first-order circuit analyses and designer's expertise. Even though performance characteristics predicted by first-order analytic expressions derived in this manner show less than 20-percent discrepancy from SPICE simulation results in most cases, our experiments revealed that such discrepancy can be as much as 200 percent for such performance characteristics. To enhance the accuracy of analytic expressions, various fitting parameters have been used for such highly non-linear circuit characteristics as settling time, phase margin, and small signal dc-gain (see Section 4.5). Recently, a symbolic analysis tool has been developed to generate analytic de-

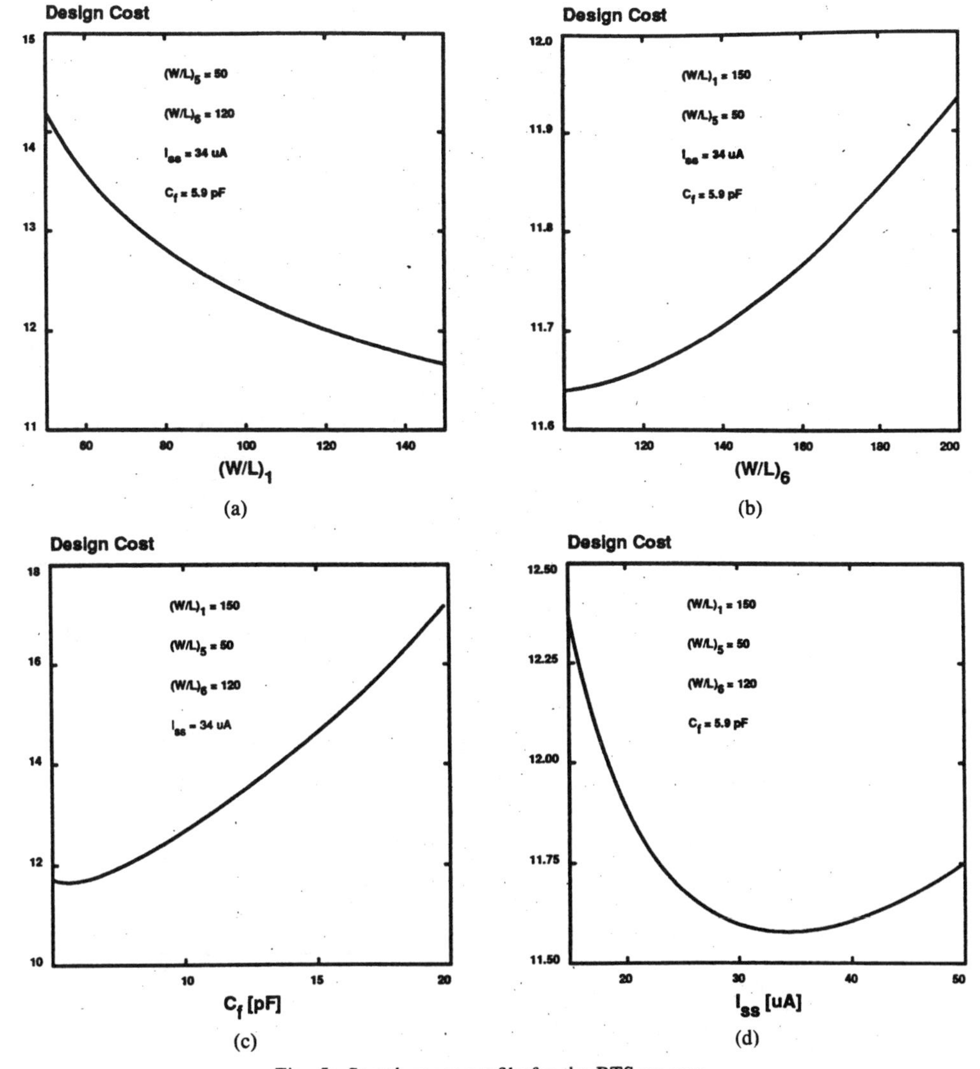

Fig. 5. Search space profile for the BTS op amp.

sign equations for analog circuits [40]. As such symbolic circuit analysis techniques mature, the effort to introduce new circuits into OPASYN may become significantly smaller.

4.2. *Objective Function*

Based on the analytic design equations and the user-defined design targets, total design cost (C) which represents a relative figure-of-merit for any particular combination of design parameter values is computed as follows:

$$C = \sum_{1}^{n} f_p \left(\frac{w_i(\text{spec}_i - p_i)}{\text{spec}_i} \right)$$

where n is the number of circuit performance parameters considered in the program, p_i is the ith performance parameter, w_i is a relative design priority of p_i, and spec_i is the corresponding design specification. We use two different types of objective functions $f_p(\,)$ depending on the type of specifications given: "center value" or "limit value." The limit values are treated by an exponential function ($f_p = \exp(\cdot)$). Note the exponential function not only produces smooth search spaces but also prevents the penalty for violating any limit specification from being compensated by overly satisfying other specifications. The center values are taken care of with a quadratic function ($f_p = (\cdot)^2$). In addition to the two types of constraints, the user can also express other design goals through the use of relative design priority, w_i, associated with each performance parameter. The resulting search spaces are smooth enough (see below) so that simple numerical optimization algorithms such as a steepest descent method can be used effectively [41], [42].

4.3. *Search Space*

Figs. 5 and 6 show search space profiles for the BTS op amp and the FFC op amp, respectively. The total design cost is plotted as one of the design parameters changes over a wide range of values and all the other design parameters remain fixed at nominal design values. The profiles show that the search space is indeed very smooth. The profiles have the same appearance for simple circuits such as the BTS op amp with 13 devices as well as for the FFC op amp with over 60 devices!! The

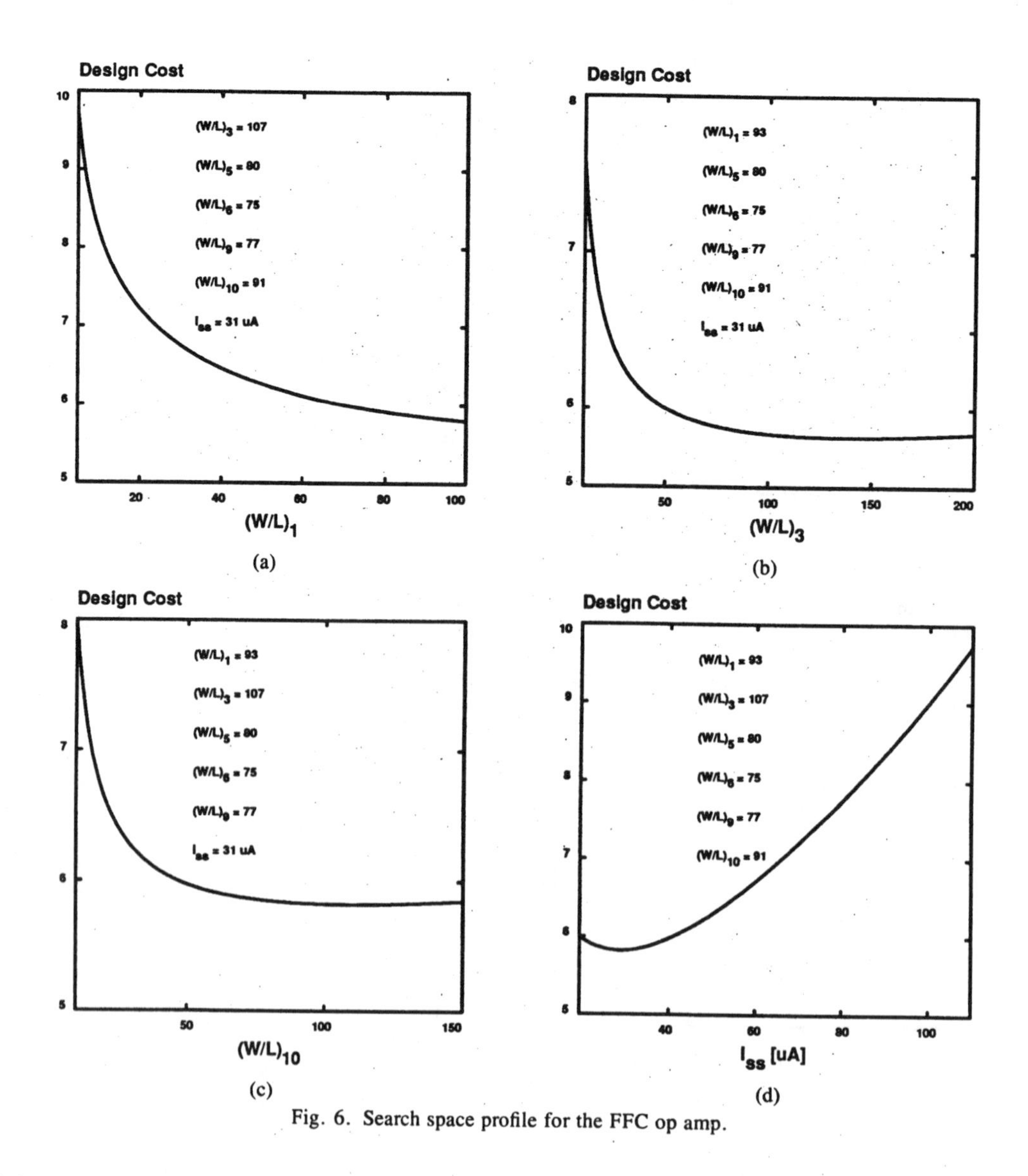

Fig. 6. Search space profile for the FFC op amp.

smoothness of the solution space results from the fact that both the cost function and the analytic models are composed of continuously differentiable functions.

4.4. Optimization Strategy

The search for an optimal solution starts with a coarse grid sampling through the bounded domain of all the independent design parameters. To avoid poor locally optimal solutions, a user specifiable number of grid points (typically about ten points) are randomly selected as starting points. A steepest descent algorithm then finds the optimal solution in the neighborhood of each of these points. OPASYN reports all the different solutions (if any) that meet specifications to the user who selects one for further processing.

The steepest descent algorithm used is simple and efficient but it requires a differentiable search space with a continuous first-derivative [41]. Furthermore, it provides only limited constraint handling capability, i.e., a set of specifications that has to be met in a limited subrange of some design parameters cannot be accommodated. To make the search procedure more robust and to handle more complicated design requirements, a more sophisticated method such as "The Phase I–II–III Method of Feasible Directions" could be used [43]. However, the simplicity and the approximate nature of the analytic models limit the effectiveness of such a powerful optimization method—the predictions of the first-order analytic models may deviate from the actual circuit response by as much as 20 percent. It appears that most design requirements for basic analog functional blocks (op amps, comparators, voltage references, output buffers, etc.) can be accommodated by the simple steepest descent method.

The described optimization strategy is independent of any specific device technology or circuit topology. Different types of circuits or process technologies can be accommodated by providing the corresponding analytic circuit models and technology files.

4.5. Fine Tuning the Analytic Models

As mentioned in Section 4.1, preparing an analytic model for a new circuit requires not only to create the proper expressions that describe the functional dependencies in the circuit but also to determine right values for the various fitting parameters used in the expressions. Ideally, they are determined by a substantial number of simulation runs and eventual comparison with fabricated circuits. However, it is normally too expensive and too time-consuming to fine-tune the fitting parameter values using discrepancy between the predictions and the results of fabrication runs. The best we can do is to compare the predictions of OPASYN with the results of SPICE simulations, and to adjust the fitting parameter values in an analytic model using zeroth- or first-order curve fitting techniques. Thus the predictions from the program can eventually reach SPICE accuracy—even without using SPICE in the inner loop of the optimization step.

Fitting parameters are also dependent on fabrication technology. Upon updating fabrication technology, fitting parameters in each analytic model should be updated based on discrepancy between the predictions and the SPICE simulation results. However, this process can be easily automated. A set of data points required for the update are first generated by running SPICE while varying the design parameters of the corresponding circuit. Then the produced data points are used to find the right values of the fitting parameters.

V. Layout Generation

Producing a good layout for an analog circuit is typically more difficult than to lay out a digital circuit with the same number of components owing to the following reasons.

- To guarantee the required performance, parasitic side effects such as parasitic capacitance and resistance should be minimized; use of finger structures for large devices and terminal sharing are two common techniques. Some nets are so critical that they must be routed in metal to minimize routing capacitance and resistance.
- The laid-out circuit should be least sensitive to process variations, thermal gradients, and noise. As a result, certain groups of devices are clustered not only because they are strongly connected but also in order to match their dependencies on and sensitivities to variations and gradients. This applies, for instance, to two transistors forming a differential input pair or a current mirror.
- Undesirable couplings between critical signal paths should be avoided. For instance, an input signal net of an op amp must be routed away from an output signal net. Otherwise, the large changes in the output signal may be coupled into the input, resulting in undesired feedback.

In addition, for CMOS op amps another difficulty arises from the fact that device sizes in the same circuit can vary over wide ranges. Depending on the performance specifications, it is not uncommon to see two orders of magnitude variations in device sizes. These large devices can be designed in many different shapes and/or be divided into sub-devices if necessary, and their terminal configurations may vary, too. These many degrees of freedom make the optimization process even more difficult. As a result, the layout process for analog circuits is time consuming and demands a fair amount of experience.

We have developed a layout strategy using a macro cell style to automatically produce good op amp layouts. The layout process divides into the three well-known phases, floorplanning, routing, and layout spacing, and makes use of many layout heuristics used by experienced designers. Each of these phases will be explained in the following subsections.

5.1. Floorplanning

Floorplanning is the first phase of the layout process. The layout floor is divided into as many partitions as the number of building blocks in the chip to be laid out, with each partition assigned to one specific building block. There exist many floorplanning algorithms developed for digital circuits such as min-cut, simulated annealing, clustering, constructive placement and so forth [8], [9], [11]. Most of these algorithms will work efficiently only if all the blocks have fixed shapes and fixed pin locations. As mentioned earlier, some of the devices in op amps can take many different shapes. In addition, certain pairs of devices should be placed as close as possible and have the same orientation and the same dimensions to prevent process and thermal gradients from affecting the devices in different ways. Min-cut and clustering algorithms are not adequate to handle such constraints. Simulated annealing can in principle handle all possible constraints but it would require very long run times due to the many degrees of freedom resulting from variable shapes and different terminal configurations. Top-down constructive placement seems to be the right way to arrive at a good layout topology, but placement methods based on simple net-length minimization cannot deal with variable shapes and geometrical matching requirements for the blocks.

Based on the above observations, an efficient floorplanning strategy has been developed using slicing trees and a set of parameterized leaf cell generators. For each of the generic op amp topologies in the database, a slicing tree has been designed that specifies a sound topological arrangement for the building blocks of the circuit based on traditionally accumulated design experience. As an example, the slicing tree associated with the basic two-stage op amp is shown in Fig. 7. The entire layout floor is first partitioned into three horizontal slices, $H1$, $H2$, and $H3$, and each of the horizontal slices is in turn divided into several vertical slices corresponding to one or two devices.

Small circuit elements such as transistors, transistor pairs, and capacitors are produced by parameterized leaf cell generators. The parameterized generators can run in two modes. In *analysis mode*, they take the electrically relevant device parameters of a cell as inputs and quickly report the range of geometrical dimensions possible for the cell without going through the mask level synthesis process. In *synthesis mode*, a generator takes the electrical parameters and the desired aspect ratio of a cell to produce the closest possible detailed mask level layout.

Layout generation, conceptually illustrated in Fig. 8, starts by examining the possible shapes of each leaf cell in the circuit by running the corresponding cell generator

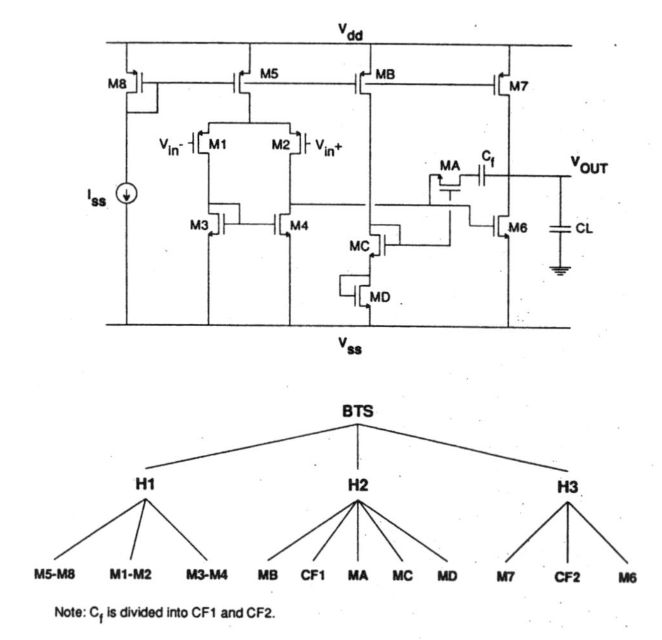

Fig. 7. BTS op amp and corresponding slicing tree.

in analysis mode. Given a set of appropriate specifications, i.e., width and length for a transistor, capacitance value for a capacitor, etc., a cell generator reports a list of possible size options for the cell (Fig. 8(a)). Based on this information, the shape constraint relation [44] is computed for each leaf node. Then all of the ancestors of the leaf nodes are annotated in bottom-up fashion with the cumulative shape constraint relation using Stockmeyer's algorithm (Fig. 8(b)) [45]. In a second, top-down traversal, the user-specified aspect ratio or size constraints (1 : 1 in Fig. 8(c)) are converted into actual x- and y-dimensions for each sub-tree, and finally for each leaf cell. It must be noted that Stockmeyer's algorithm, used in the above top-down constraint propagation process, selects the optimal shapes of the leaf cells which together best fit the given floor shape with the minimal area. Between two alternatives (5-by-4 or 4-by-5), the 5-by-4 floor shape is arbitrarily selected. Then this constraint is propagated to its descendant nodes until x- and y-dimensions of all the leaf cells are determined. Next, the cell generators are called once again, this time in synthesis mode, to produce the best matching mask layout for each cell. These layouts are assembled in a bottom-up manner according to the structure of the slicing tree (Fig. 8(d)).

From the above discussion, the algorithm for floorplanning can be summarized as described in the floorplanning algorithm *OPASYN_FLOORPLAN* (see Fig. 9).

5.2. Routing

For analog circuits, routing has additional requirements that go beyond those of digital circuits.

- Routing capacitance and resistance should be minimized in signal paths, since routing parasitics will affect the ac and transient characteristics of the performance.
- The resistance of some nets are more critical to the circuit performance than others, so they should be routed

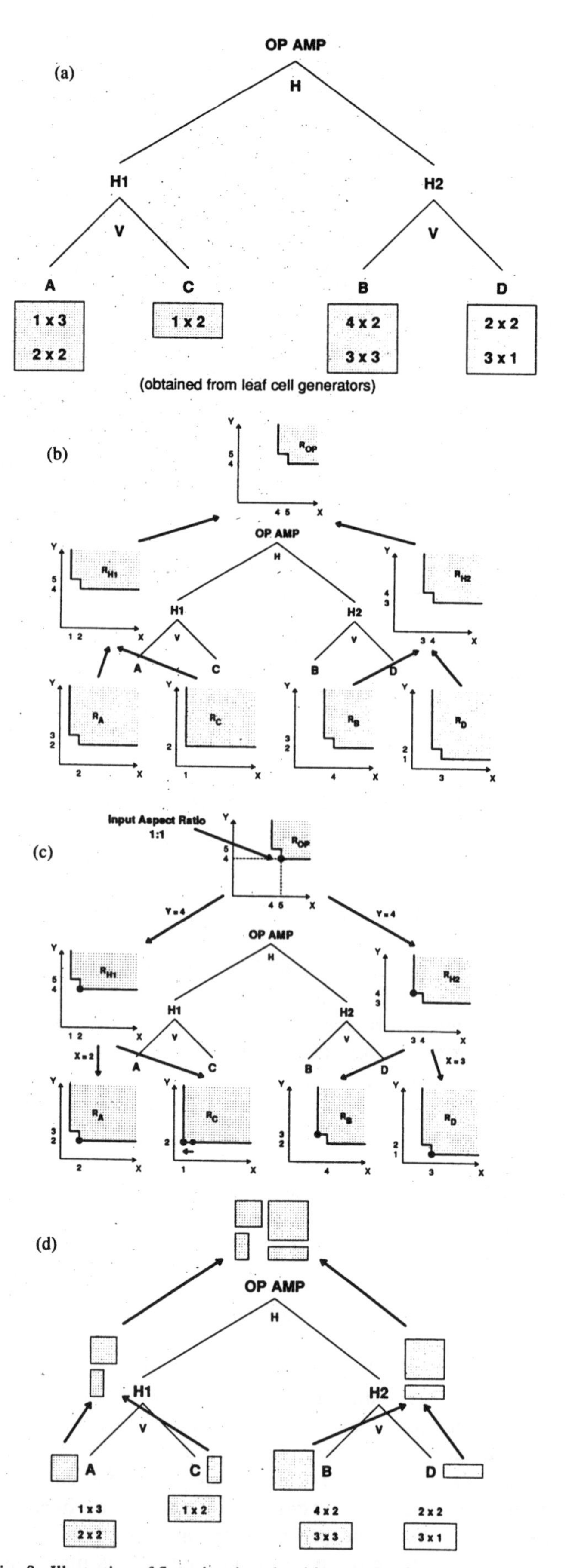

Fig. 8. Illustration of floorplanning algorithm. (a) Leaf cell shape inquiry. (b) Bottom-up shape annotation. (c) Cell selection by top-down constraint propagation. (d) Bottom-up cell assembly.

Floorplanning Algorithm: *OPASYN_FLOORPLAN*

Inputs:
- Sized circuit schematic
- Slicing tree (T_s) description
- Size constraints
- Layout design rules

Outputs:
- X-y dimensions of the layout floor
- Optimal shape selection for each leaf cell
- An unrouted symbolic layout

Algorithm:

Step 1. Leaf cell shape inquiry

- For all leaf cells in T_s do {
    1. Call the corresponding generator in analysis mode with appropriate specifications (W, L, C).
    2. Store the list of possible size options returned from the generator.

  }

Step 2. Bottom-up shape annotation

- For all nodes in T_s do {

  if (node == leaf node)

  Generate a shape constraint relation (R) using a possible shape list.

  else if (node == vertical-cut node with n-children)

$$R_{node} = \left\{(y,x) \mid (y_1,x) \in R_1 \ \& \ (y_2,x) \in R_2, \ \cdots \ (y_n,x) \in R_n \ \& \ y = y_1 + y_2 + \cdots + y_n \right\}.$$

  else if (node == horizontal-cut node with n-children)

$$R_{node} = \left\{(y,x) \mid (x_1,y) \in R_1 \ \& \ (x_2,y) \in R_2, \ \cdots \ (x_n,y) \in R_n \ \& \ x = x_1 + x_2 + \cdots + x_n \right\}.$$

  }

Step 3. Top-down constraint propagation

- Using R_{root} and the the input aspect ratio (otherwise the width or height constraint), determine x- and y-dimensions of the entire floor.
- Do until all x- and y-dimensions of the leaf cells are known {

  if (node has vertical-cut parent node) {
    1. Inherit x-dimension from the parent.
    2. Identify the y-dimension on R_{node} that corresponds to the x-dimension inherited from the parent node.

  } else if (node has horizontal-cut parent node) {
    1. Inherit y-dimension from the parent.
    2. Identify the x-dimension on R_{node} that corresponds to the y-dimension inherited from the parent node.

  }

Step 4. Cell synthesis

- Using the shape of a leaf cell determined at step 3, implement the cell by running the corresponding cell generator in synthesis mode.

Step 5. Bottom-up cell assembly

- Assemble the generated leaf cells according to T_s.

Fig. 9. Floorplanning algorithm OPASYN_FLOORPLAN.

all in metal rather than in poly silicon. For instance, if source terminals of a differential pair are connected in poly silicon, the small-signal dc gain of the pair degrades due to the source regeneration effect.

- A certain set of nets should be matched in terms of routing parasitics and/or their geometries.
- The crosstalk between signal nets should be avoided. Therefore, noise-sensitive nets should be buffered from noisy nets such as clock lines and output nets.

In OPASYN, routing is done in several steps. All the nets of each circuit are pre-assigned to a few classes with different routing priorities. Then each class of nets is routed based on its priority; e.g., the input connections are placed in the first priority class and are wired early, to make it easy to obtain a symmetrical routing with reasonably matched parasitics. Within each class, the nets are routed one at a time with the switch-box router MIGHTY [46]. When a particular net runs into a bottleneck produced by another net, the latter may undergo some local rip-up and rerouting. Once all the nets belonging to a certain class have been routed, they constitute obstacles for the nets in the lower priority classes. As a result, the whole circuit area is considered as a routing area littered with many obstacles, of which some are building blocks of the circuit and others are pre-placed nets. This approach works well since op amps have normally less than 100 nets and cover only a limited area.

5.3. Layout Spacing

The final step in the layout process is to re-space the symbolic layout to produce a compact, design-rule-correct mask level layout of an op amp. The routed layout in a symbolic format, together with additional geometrical constraints such as symmetry and location of I/O pins is sent to a spacing program (compactor), such as the one-dimensional graph-based compactor SPARCS [47]. This program will not change the layout of the leaf cells themselves, but will simply adjust their separations and compact the wiring. The leaf cells produced from the cell generators are already in a compact, design-rule-correct physical format. In this way, the complexity of the spacing problem is substantially reduced. The output from the compactor undergoes a few postprocessing steps to straighten out unnecessary jogs and to remove redundant vias.

5.4. Design Data Management

The object-oriented data manager OCT developed at Berkeley [29], [30] has been used to manage the design data. During various phases of the layout process, different views of the circuit such as a "physical view," an "unspaced view," and a "spaced view" are stored in a hierarchical manner in this database. Leaf cells are represented in a physical view with fully specified geometry for all devices and connections. They observe all specified design rules and need no further spacing. The entire layout of an op amp, on the other hand, is represented in an unspaced view where only relative placement, block sizes and shapes, and the general routing of the interconnections are recorded. This level of the layout hierarchy needs to be subject to a spacing program where all the design rules are checked and placements of the leaf cells are suitably adjusted. The result is a design-rule-correct physical view of the entire op amp.

The different views communicate with one another through suitably abstracted "interface facets" that store overall cell shapes, terminal locations, and protection frames. They contain the minimal amount of information about a block necessary for manipulation of the entire block or for interfacing to it from the outside. The detailed information about the block is stored in a "contents facet" which is not directly visible from the outside. By observing a set of pre-defined policies, our layout program can communicate with other existing tools with minimal overhead and with no need for intermediate conversions.

VI. Implementation and Results

The OPASYN modules shown in Fig. 1 have been implemented in Common Lisp and in C. Among those modules, the circuit selection module and the layout module are written in Common Lisp. The parametric optimization module is written in C because it is iteratively used in an inner loop of the optimization step, and efficiency of execution is most important. The leaf cell generation pro-

User: Performance Specifications			
Standard OP Amp; V_{dd} = 2.5 V, V_{ss} = -2.5 V; Load Cap. = 5 pF; MOSIS 3 μm Process			
power diss. (mW)	< 1	settling time (nsec)	< 500
output swing (V)	> 1.5 / < -1.5	total gate area (mil²)	MINIMIZE
open loop gain	> 5,000	1/f noise (V/√Hz)	< 1E-7
unity gain bw. (MHz)	> 4	PSRR-dc (dB)	> 70
slew rate (V/μsec)	> 2.5	PSRR-1kHz (dB)	> 40

(a)

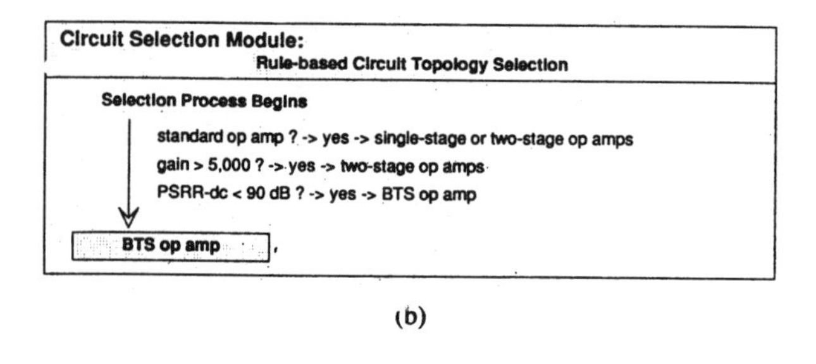

(b)

Parametric Optimization Module: Optimal Device Sizes	
Current Sources	
Device Name	Size
I_{ss}	16 μA
Capacitors	
Device Name	Size
C_f	8.9 pF
MOSFETs	
Device Name	Width (μm) / Length (μm)
M1 = M2	470 / 3
M3 = M4	84 / 6
M5	250 / 4.5
M6	890 / 6
M7	1300 / 4.5
MA	90 / 3
MB	170 / 4.5
MC	12 / 3
MD	56 / 3

(c)

BTS OP Amp - Performance Summary			
V_{dd} = 2.5 V, V_{ss} = -2.5 V; Load Cap. = 5 pF; MOSIS 3 μm Process			
Performace	Specifications	OPASYN	SPICE
power diss. (mW)	< 1	0.6	0.6
output swing (V)	> 1.5 / < -1.5	2.3 / -2.2	2.3 / -2.2
open loop gain	> 5,000	5,200	4,900
unity gain bw. (MHz)	> 2	3.1	3.0
slew rate (V/μsec)	> 1	1.4	1.4
settling time (nsec)	< 1000	980	1230
total gate area (mil²)	MINIMIZE	42	N/A
1/f noise (V/√Hz)	< 1E-7	1.1E-7	NOT AVAILABLE
PSRR-dc (dB)	> 70	78	78
PSRR-1kHz (dB)	> 40	63	70

(d)

Layout Generation Module: Mask Level Layouts of Size Circuit Schematics		
BTS OP Amp; MOSIS 3 μm Process		
Vital Data	Layout (Fig. 10f)	Layout (Fig. 10g)
input aspect ratio	1.0	2.0
achieved aspect ratio	0.9	1.8
total area (mil²)	255	278
routing area (mil²)	53 (20 %)	68 (24 %)

(e)

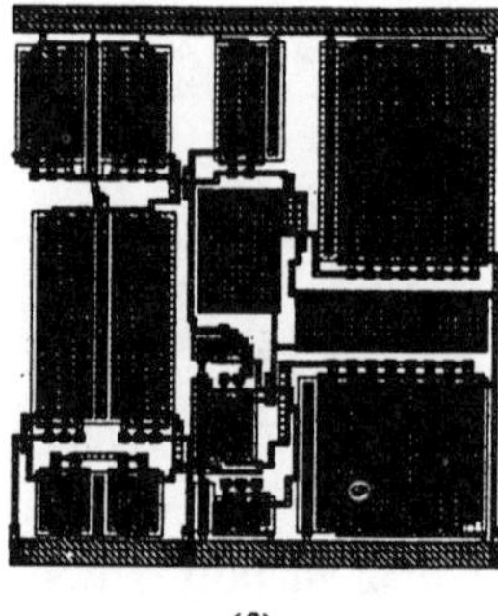

(f)

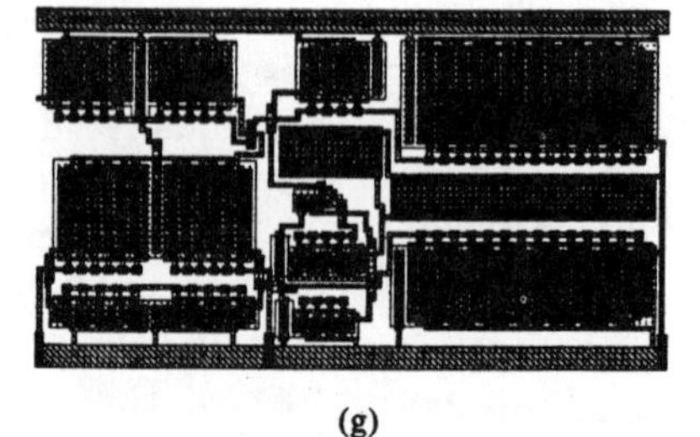

(g)

Fig. 10. A synthesis example—from specifications to silicon. (a) A summary of performance specifications. (b) Circuit topology selection process. (c) Parametric optimization results—optimal device sizes. (d) A summary of expected circuit performance. (e) Vital data for layouts shown in (f) and (g). (f) A layout with an input aspect ratio of 1. (g) A layout with an input aspect ratio of 2.

grams are also coded in C. The synthesis system is running under an Ultrix operating system on a VAX 8800. None of these modules have any process technology dependent codes; all technology information is stored in the database. The user simply specifies which fabrication process is to be used for the synthesis. Two different process technologies (MOSIS 3 μm and GE 1.5 μm processes) have been used in our experiments so far. Four commonly used op amp topologies and an output buffer have been incorporated into the database; they are the BTS op amp (shown in Fig. 3), the FFC op amp (shown in Fig. 4), a single-stage folded cascode op amp (SFC), a two-stage op amp with cascoded input stage (CTS), and a complementary Class B output buffer (OPB).

6.1. Run Times

The entire synthesis, from specifications to layout, takes under 5 min on a VAX 8800. The exact CPU time varies with the difficulty of the user specifications, the complexity of the circuit, and the degree of optimality to be achieved. More than 80 percent of the CPU time is currently spent in the layout generation module which is written in Lisp. Reimplementation of this module in C together with further code optimization could lead to much shorter CPU times, which would make it quite practical to include OPASYN as a module generator in an interactive system planning tool for mixed A/D circuits.

6.2. *Quality of Results*

The quality of the achieved designs is quite high. The selection decisions made are consistent with those made by expert human designers. The circuit performance predictions made by OPASYN are in good agreement with SPICE simulation results [24]. For dc characteristics, excellent agreement is generally achieved. For ac and transient characteristics such as phase margin, gain, and settling time, the deviations are less than 20 percent. The layout results also appear satisfactory. The generated layouts observe most of the heuristic layout rules used by expert designers and show high area usage; they are about 20–30 percent bigger than the sum of the areas of the individual devices [25]. Further improvements seem possible by improving the interface to the spacing program. The other key feature of the layout program is that it can rapidly generate a layout to some desired aspect ratio. The deviations observed vary over a 15 percent range compared to the specified target. The simulation results based on the extracted layouts demonstrate that the produced layouts also have acceptable parasitic overhead.

6.3. *An Example*

Fig. 10 presents a synthesis example from its performance specifications to the complete design-rule-correct mask level layout of an op amp. In addition to the performance specifications, Fig. 10(a) also shows the optimization priorities; in this case, the top priority is given to minimizing the total gate area. Based on these specifications, the BTS op amp (shown in Fig. 3) has been selected by the circuit selection module. The program's reasoning process is displayed in Fig. 10(b). After the parametric optimization phase, optimal values of the design parameters are determined. These values are summarized in Fig. 10(c). The expected circuit performance from OPASYN and that from the SPICE simulation runs are compared in Fig. 10(d). Mask level layouts with specified aspect ratios of 1 and 2, respectively, are shown in Fig. 10(f) and Fig. 10(g), and some quantitative data are given in Fig. 10(e).

VII. Conclusions

An analog silicon compilation system for CMOS op amps (OPASYN) has been developed. The synthesis starts from a set of system level performance specifications. The program selects the most promising op amp circuit in its internal database based on design requirements implied by the specifications. This circuit is parametrically optimized to determine optimal values for its design parameters. Design-rule-correct mask geometries are then constructed using a macro cell layout style. The integration of the whole layout subsystem could be accomplished in a relatively short period of time, since it has been built on top of the Berkeley CAD environment, reusing existing tools as much as possible.

Experimental results show that OPASYN produces practical circuit configurations and layouts. The entire synthesis process is fast enough to be interactively used at the system design level. The program can be easily used by system engineers who are inexperienced in op amp design. They simply specify their design requirements and optimization priorities at the system level; a few minutes later, after minimal interaction, the system offers one or more design-rule-correct mask layouts that meet functional specifications except for the most stringent designs.

Neither the proposed optimization algorithm nor the layout method make any assumption on what kind of circuits are represented by the analytic models or by the slicing trees. Thus the demonstrated approach can be applied for the synthesis of other analog building blocks such as output buffers, comparators, or bandgap references even though OPASYN has so far only been used as an op amp compiler. In fact, an output buffer is already fully incorporated into the OPASYN database. Work is in progress to integrate OPASYN as a module compiler into higher level subsystem synthesis programs such as compilers for A/D converter and active filters.

Acknowledgment

The authors would like to thank Jeff L. Burns for helping them with his spacing program, SPARCS, Gani Jusuf for incorporating a fully differential folded cascode op amp, and Douglas Dramgoole for incorporating an output buffer into the OPASYN database. They also gratefully acknowledge the contributions of many graduate students at U.C. Berkeley.

References

[1] P. G. Gray, "Analog IC's in the submicron era: Trends and perspectives," in *IEEE IEDM Dig. Techn. Papers*, pp. 5-9, 1987.

[2] P. E. Allen, "A tutorial—computer aided design of analog integrated circuits," in *Proc. IEEE CICC*, pp. 608-616, 1986.

[3] D. C. Stone *et al.*, "Analog CMOS building blocks for custom and semicustom applications," *IEEE J. Solid-State Circuits*, vol. SC-19, pp. 55-61, Feb. 1984.

[4] T. Pletersek, J. Trontelj, and L. Trontelj, "Analog LSI design with CMOS standard cells," in *Proc. IEEE CICC*, pp. 479-483, 1985.

[5] G. Kelson, "Design automation techniques for analog VLSI," *VLSI Des.*, pp. 78-82, Jan. 1985.

[6] P. R. Gray and R. G. Meyer, *Analysis and Design of Analog Integrated Circuits*. New York: Wiley, 1984.

[7] ——, "MOS operational amplifier design—A tutorial overview," *IEEE J. Solid-State Circuits*, vol. SC-17, pp. 969-982, Dec. 1982.

[8] S. Kirkpatrick, C. D. Gelatt Jr., and M. P. Vecchi, "Optimization by simulated annealing," *Science*, vol. 220, May 1983.

[9] M. A. Breuer, "Min-cut placement," *J. Design Fault-Tolerant Comput.*, vol. 1, no. 4, pp. 343-362, Oct. 1977.

[10] C. M. Fodiccoa and R. M. Mattheyses, "A linear-time heuristic for improving network partitions," in *Proc. 19th ACM/IEEE Design Automation Conf.*, pp. 175-181, 1982.

[11] D. M. Schuler and E. G. Ulrich, "Clustering and linear placement," *Proc. 9th Design Automation Workshop*, pp. 50-56, 1972.

[12] P. R. Suaris and G. Kedem, "Standard cell placement by quadrasection," Dep. of Comp. Sci., Duke Univ., Durham, NC.
[13] H. Yaguthiel, A. Sangiovanni-Vincentelli, and P. R. Gray, "A methodology for automated layout of switched-capacitor filters," in *Proc. IEEE ICCAD*, pp. 444-447, 1986.
[14] D. Lucas, "Analog silicon compiler for switched capacitor filters," in *Proc. IEEE ICCAD*, pp. 506-509, 1987.
[15] R. P. Sigg *et al.*, "An SC filter compiler: Fully automated filter synthesizer and mask generator for a CMOS gate-array-type filter chip," in *Proc. IEEE ICCAD*, pp. 510-513, 1987.
[16] R. K. Brayton, G. D. Hachtel, and A. Sangiovanni-Vincentelli, "A survey of optimization techniques for integrated-circuit design," *Proc. IEEE*, vol. 69, pp. 1334-1362, Oct. 1981.
[17] B. Nye, D. Riley, and A. Sangiovanni-Vincentelli, "DELIGHT.SPICE User's Guide," Dep. EECS, Univ. of California, Berkeley, May 1984.
[18] J. Shyu and A. Sangiovanni, "ECSTASY: A new environment for IC design environment," in *Proc. IEEE ICCAD*, pp. 484-487, 1988.
[19] R. Harjani, R. A. Rutenbar, and L. R. Carley, "A prototype framework for knowledge-based analog circuit synthesis," in *Proc. 24th ACM/IEEE Design Automation Conf.*, pp. 42-49, 1987.
[20] M. G. DeGrauwe, "A synthesis program for operational amplifiers," in *IEEE ISSCC Dig. Tech. Papers*, pp. 18-19, 1984.
[21] P. E. Allen and H. Nevarez-Lozano, "Automated design of MOS OP amps," in *Proc. IEEE ISCAS*, pp. 1286-1289, 1983.
[22] R. J. Bowman and D. J. Lane, "A knowledge-based system for analog integrated circuit design," in *Proc. IEEE ICCAD*, pp. 210-212, 1986.
[23] E. Berkcan, M. d'Abreu, and W. Laughton, "Analog compilation based on successive decompositions," in *Proc. 25th ACM/IEEE Design Automation Conf.*, pp. 369-375, 1988.
[24] H. Y. Koh, C. H. Séquin, and P. R. Gray, "Automatic synthesis of operational amplifiers based on analytic circuit models," in *Proc. IEEE ICCAD*, pp. 502-505, 1987.
[25] H. Y. Koh, C. H. Séquin, and P. R. Gray, "Automatic layout generation for CMOS operational amplifiers," in *Proc. IEEE ICCAD*, pp. 548-551, 1988.
[26] D. J. Garrod, R. A. Rutenbar, and L. R. Carley, "Automatic layout of custom analog cells in ANAGRAM," in *Proc. IEEE ICCAD*, pp. 544-547, 1988.
[27] A. Vladimirescu, K. Zhang, A. R. Newton, D. O. Pederson, and A. Sangiovanni-Vincentelli, "SPICE Version 2G—User's Guide," Dep. EECS, Univ. of California, Berkeley, 1981.
[28] L. W. Nagel and D. O. Pederson, "Simulation program with integrated circuit emphasis," in *Proc. Sixteenth Midwest Symposium on Circuit Theory*, Waterloo, Ont., Canada, Apr. 1973.
[29] D. Harrison, P. Moore, R. L. Spickelmier, and A. R. Newton, "Data management and graphics editing in the Berkeley design environment," in *Proc. IEEE ICCAD*, 1986.
[30] P. Moore, "OCT database programmer's manual," Internal Memo., Univ. of California, Berkeley, 1986.
[31] I. Bratko, *PROLOG—Programming For Artificial Intelligence*. Reading MA: Addison-Wesley, 1986.
[32] E. Charniak, *Artificial Intelligence*. Reading, MA: Addison-Wesley, 1986.
[33] P. H. Winston, *Artificial Intelligence*. Reading, MA: Addison-Wesley, 1984.
[34] P. H. Winston and B. K. P. Horn, *LISP*. Reading, MA: Addison-Wesley, 1984.
[35] B. K. Ahuja, "An improved frequency compensation technique for CMOS operational amplifiers," *IEEE J. Solid-State Circuits*, vol. SC-18, pp. 629-633, Dec. 1983.
[36] C. T. Chuang, "Analysis of the settling behavior of an operational amplifier," *IEEE J. Solid-State Circuits*, vol. SC-17, pp. 74-80, Feb. 1982.
[37] B. Y. Kamath, R. G. Meyer, and P. R. Gray, "Relationship between frequency response and settling time of operational amplifiers," *IEEE J. Solid-State Circuits*, vol. SC-9, pp. 347-352, Dec. 1974.
[38] M. R. Haskard and I. C. May, *Analog VLSI Design—nMOS and CMOS*. Englewood Cliffs, NJ: Prentice-Hall, 1988.
[39] A. B. Grebene, *Bipolar and MOS Analog Integrated Circuit Design*. New York: Wiley, 1984.
[40] S. J. Seda, M. G. DeGrauwe, and W. Fichtner, "A symbolic analysis tool for analog circuit design automation," in *Proc. IEEE ICCAD*, pp. 488-491, 1988.
[41] D. G. Luenberger, *Linear and Nonlinear Programming*. Reading, MA: Addison-Wesley, 1984.
[42] W. H. Press *et al.*, *Numerical Recipes in C—The Art of Scientific Computing*. Cambridge, U.K.: Cambridge Univ. Press, 1988.
[43] W. T. Nye and A. L. Tits, "An application-oriented, optimization-based methodology for interactive design of engineering systems," *Int. J. Contr.*, vol. 43, no. 6, pp. 1693-1721, 1986.
[44] R. Otten, "Laout Compilation," in *Design Systems for VLSI Circuits—Logic Synthesis and Silicon Compilation*, (Eds. G. De Micheli, A. Sangiovanni-Vincentelli, and P. Antognetti). Dordrecht, The Netherlands: Martinus Nijhoff, 1987.
[45] L. Stockmeyer, "Optimal orientations of cells in slicing floor-plan designs," *Inform. Contr.*, vol. 59, pp. 91-101, 1983.
[46] H. Shin and A. Sangiovanni-Vincentelli, "MIGHTY: A "Rip-up and reroute" detailed router," in *Proc. IEEE ICCD*, pp. 10-13, 1986.
[47] J. L. Burns and A. R. Newton, "SPARCS: A new constraint-based IC symbolic layout spacer," in *Proc. IEEE CICC*, pp. 534-539, 1986.
[48] Mark, G. R. DeGrauwe *et al.*, "IDAC: An interactive design tool for analog CMOS circuits," *IEEE J. Solid-State Circuits*, vol. SC-22, pp. 1106-1116, Dec. 1987.

*

Han Young Koh received the B.S. degree from Seoul National University, the M.S. degree from Georgia Institute of Technology, both in electrical engineering, and the Ph.D. degree in electrical engineering and computer sciences from University of California, Berkeley, in 1980, 1984, and 1989, respectively.

From 1984 to May 1989 he worked at University of California, Berkeley, in the Electronic Research Laboratory on the design synthesis of monolithic CMOS operational amplifiers. Currently, he is a senior member of technical staff at Cadence Design Systems, Inc. in Santa Clara, CA. His research interests now are focused on design synthesis of mixed analog/digital systems.

*

Carlo H. Séquin (M'71-SM'80-F'82) received the Ph.D. degree in experimental physics from the University of Basel, Switzerland in 1969. His subsequent work at the Institute of Applied Physics in Basel concerned interface physics of MOS transistors and problems of applied electronics in the field of cybernetic models.

From 1970 till 1976 he worked at AT&T Bell Telephone Laboratories, Murray Hill, NJ, in the MOS Integrated Circuit Laboratory on the design and investigation of charge-coupled devices for imaging and signal processing applications. In 1977 he joined the faculty as a professor in the Department of EECS, University of California, Berkeley. He was head of the Computer Science Division from 1980 till 1983. His research interests now are concentrated on computer graphics and solids modeling and CAD tools for mechanical and electrical design.

Dr. Séquin is a member of ACM and the Swiss Academy of Engineering Sciences.

*

Paul R. Gray (S'65-M'69-SM'76-F'81) received the B.S., M.S., and Ph.D. degrees from the University of Arizona, Tucson, in 1963, 1965, and 1969, respectively.

In 1969 he joined the Research and Development Laboratory, Fairchild Semiconductor, Palo Alto, California, where he was involved in the application of new technologies for analog integrated circuits. In 1971 he joined the Department of Electrical Engineering and Computer Sciences, University of California, Berkeley, where he is now a Professor. His research interests during this period have included bipolar and MOS analog circuit design, device modeling, CAD, and analog-digital interfaces in VLSI systems. He served as editor of the IEEE Journal of Solid-State Circuits from 1977 through 1979, and as Program Chairman of the 1982 International Solid State Circuits Conference. He has acted as consultant to a number of electronics firms over the past 20 years.

Dr. Gray is a member of Eta Kappa Nu and Sigma Xi.

OASYS: A Framework for Analog Circuit Synthesis

RAMESH HARJANI, MEMBER, IEEE, ROB A. RUTENBAR, MEMBER, IEEE, AND L. RICHARD CARLEY, MEMBER, IEEE

Abstract—**A hierarchically structured framework for analog circuit synthesis is described. Analog circuit topologies are represented as a hierarchy of templates of abstract functional blocks (called *design styles*) each with associated detailed design knowledge. This hierarchical structure has two important features: it decomposes the design task into a sequence of smaller tasks with uniform structure, and it simplifies the reuse of design knowledge. Mechanisms are described to select from among alternate design styles, and to translate performance specifications from one level in the hierarchy to the next lower, more concrete level. A prototype implementation, OASYS, synthesizes sized transistor schematics for CMOS operational amplifiers from performance specifications and process parameters. Measurements from detailed circuit simulation, and from actual fabricated analog IC's based on OASYS-synthesized designs demonstrate that OASYS is capable of synthesizing functional circuits.**

I. Introduction

DESIGN automation ideas from digital IC design have only recently begun to migrate into analog circuit design. In part, this reflects the inherent complexities of the analog design process. However, it also reflects the success with which some classical analog applications have been supplanted by digital techniques; for example, digital signal processing is frequently used in place of analog signal processing. However, there are limits to this replacement, and indeed, there has been a recent resurgence of interest in analog circuit design. Analog circuitry is widely used in systems applications such as telecommunications and robotics, where analog interfaces to an external environment are coupled with digital signal processing systems. Many board-level mixed analog/digital systems have begun to migrate down toward single chip implementations, and it has become clear that design of the analog parts of these chips is often the critical time-limiting bottleneck in the overall design process. Outside of conventional analog/digital systems, there has recently been great interest in the design of parallel analog VLSI signal processing architectures [1], [2]. Hence, it is clear that CAD tools must be developed to cope with both the complexity of large-scale analog circuit designs, and with the requirement for rapid design times.

Unfortunately, the state of analog circuit synthesis tools is quite primitive in comparison to digital synthesis tools. In the digital domain, structured abstractions and hierarchy are commonplace, and are relied upon to make seemingly large synthesis tasks tractable by breaking them into smaller steps. Such abstractions and hierarchy do not currently play a central role in analog design. Some ideas from digital design methodologies, such as standard cell libraries and module generators, have recently been applied to analog design tasks. However, such techniques usually have several drawbacks, e.g., libraries allow the designer to make only crude tradeoffs among performance specifications, and they become obsolete rapidly in the face of technological evolution. Custom analog circuits are still designed, largely by hand, by experts intimately familiar with nuances of the target application and the IC fabrication process. Analog design is commonly perceived to be one of the most knowledge-intensive of design tasks: the techniques needed to build good analog circuits seem to exist solely as expertise invested in individual designers.

This paper describes a hierarchically-structured framework for an analog circuit synthesis tool. The framework is knowledge-intensive in that it relies heavily on the codification of mature analog design expertise. What our framework provides is a consistent organization for applying such knowledge to hierarchically designed circuits. Perhaps more importantly, what the framework really demonstrates is the feasibility of attacking tightly-coupled analog design problems in a highly stylized, hierarchical fashion. We attack the behavior-to-structure portion of the design task; our goal is to produce circuit schematics including device sizes from performance specifications for common analog functional blocks. This approach is motivated by the lack of tools to support the design of custom analog circuits.

The paper is organized as follows. Section II contrasts the analog and digital domains, and summarizes related synthesis research. Section III first describes the basic structure of our synthesis framework, how it is motivated by traditional analog design approaches, and places it in context with other synthesis approaches. It then describes in detail the architecture and implementation of OASYS, a prototype designed to test the feasibility of this organization. We introduce here the critical role of hierarchy in

Manuscript received March 21, 1988; revised February 10, 1989 and June 21, 1989. This work was supported in part by the Semiconductor Research Corporation, by a grant from the Gould Foundation, and by the National Science Foundation under Grants ENG-8451496 and MIP-8657369. This paper was recommended by Associate Editor M. R. Lightner.

R. Harjani was with the Department of Electrical and Computer Engineering, Carnegie Mellon University, Pittsburgh, PA. He is now with Mentor Graphics, San Jose, CA 95112.

R. A. Rutenbar and L. R. Carley are with the Department of Electrical and Computer Engineering, Carnegie Mellon University, Pittsburgh, PA 15213.

IEEE Log Number 8930646.

analog circuits, and describe methods to structure and exploit analog design knowledge. Examples are given of how analog design knowledge is codified within OASYS. Section IV describes the domain in which OASYS operates, then evaluates the performance of OASYS, and its overall capabilities. We examine some automatically synthesized circuits OASYS has produced and present measurements from fabricated analog IC's based on OASYS designs. Finally, Section V presents some concluding remarks.

II. Background

Before describing the components of our framework for analog circuit synthesis, we review the salient differences between analog and digital design problems, and some previous approaches to analog circuit synthesis.

A. Analog Design Versus Digital Design

Consider the task of designing a functional block to be implemented as a single, perhaps large, cell in a VLSI circuit. The overall high-level synthesis task, for either analog or digital circuits, is to interconnect a set of appropriately designed primitive components (e.g., transistors) to produce the correct behavior for the cell. We informally partition the differences between analog and digital design tasks into four categories: size, hierarchy, process, and performance constraints.

- The *size* difference is easily stated: analog circuits tend to have fewer transistors than digital circuits, but these individual devices are much more difficult to design. Analog circuits often exploit the full spectrum of capabilities inherent in the physics of individual devices, so a few transistors often suffice to perform complex functions.
- Analog and digital circuits each employ *hierarchy*, but in substantially different ways. In digital circuits, there is tacit agreement on the abstraction levels through which a design must pass, and these abstractions play a central role in the organization of synthesis tools, which usually help translate downward, level to level. There is not such a well-developed hierarchy for analog circuits, but hierarchy is important nevertheless: complex, system-level analog designs are not attacked in a transistor-by-transistor fashion.
- Conceptually below the levels of hierarchy just described is the level of the *fabrication process*. At the higher levels of digital synthesis, process constraints appear in highly simplified forms, e.g., known constraints on attainable clock frequencies or drive capabilities. In analog circuit synthesis, such process constraints appear in far greater detail, even during high-level design. A common example is the level of precision in device-level component matching that a particular fabrication process permits.
- *Performance constraints* on the behavior of analog circuits also differ radically from those of digital circuits. Digital circuits are often specified using a behavioral language such as ISPS [3], which can capture the dataflow for digital quantities moving through functional blocks and storage elements. For common analog circuits, the qualitative behavior is often known implicitly: an A/D converter digitizes continuous signals; a phase-locked loop synchronizes the phase of different signals. The specification may take the form of a set of performance parameters that must be achieved, such as gain, bandwidth, input noise, or phase margin.

Practical analog synthesis tools must deal with all these concerns, i.e., unlike digital tools, they must handle performance parameters that constrain continuous quantities (e.g., voltages or currents), depend intimately on the fabrication process, and depend on the careful design of several mutually interacting devices at potentially different levels of the analog hierarchy.

B. Previous Approaches to Analog Design

The overall goal of analog design tools is to reduce high-level specifications to IC masks. However, tools for this task often break the process into two parts: reduce specifications to circuit schematics, then reduce schematics to masks. An important tradeoff that many tools make is *flexibility*: some approaches limit the freedom to reach arbitrary designs in favor of automating a more constrained view of the design process. We distinguish here two broad approaches to tools for analog synthesis: bottom-up layout-based approaches, and top-down knowledge-based approaches.

The bottom-up layout-based approaches show the most direct influence of digital design ideas. Semi-custom analog styles, such as transistor arrays (analogous to gate arrays) and analog standard cells provide a rapid path to silicon for analog functions already designed to the level of the primitive devices available in the layout style [4]–[7]. Transistor arrays provide the most design flexibility, since they provide a small set of atomic primitives from which many circuits can be built. However, layout density is very poor, and flexibility is still compromised with respect to full custom design because of the limited range of device sizes. Standard cells provide more density, and placement and routing tools that accommodate the sensitive electrical characteristics of these devices can help lay these out [8]. Unfortunately, these seem to offer even less flexibility, since designers are limited to a library of fixed circuit blocks. Both approaches constrain the circuit design itself: device parameters are not continuously variable because only a limited set of device/circuit types is available. Both approaches also track technology changes rather poorly, since they require redesign of cell layouts. A higher level approach here is the use of analog module generators. These fix some portion of a circuit's topology and parameterize the remainder into a limited number of choices. The result is a mask generator for a particular circuit topology. Regular structures, such as switched capacitor filters, are particularly amenable to this approach [9], [10]. Some systems, such as AIDE2 [11], use standard cells as the basis for building module generators; experiences with a successive approximation A/D converter generator designed in this style appear in [12]. Other compiler systems, such Seattle Silicon Technologies' CONCORDE, include a set of analog module generators based on parameterized custom layouts for common analog functions [13]. However, this approach also has several drawbacks: there may be many sets of reasonable design specifications which cannot be satisfied by the available parameter choices; it may permit the designer to make only crude tradeoffs among performance specifications; and some generators may still rapidly become obsolete in the face of technological evolution.

In contrast, top-down approaches typically strive to transform specifications to circuit schematics first, and then attack layout second. In their most general form, knowledge-based approaches attempt to reason about analog circuits, either from first principles, or from detailed domain knowledge. Analog circuits have actually provided a useful domain for testing many ideas about constraint propagation [14], causal models, and qualitative reasoning [15], [16]. Several attempts have been made to synthesize analog circuits using AI ideas, e.g., using symbolic algebra [17] and search ideas [18]. However, these early approaches focused more on AI than on practical circuit synthesis. More recent attempts along these lines have focused more on expert systems, notably rule-based systems [19]. For example, Bowman [20] describes a rule-based system that assembles op amps from specialized component pieces, but it is unclear if this methodology can be used in other circuit synthesis tasks. The BLADES system [21] proposes a framework in which individual subcircuit experts are coupled by a design manager to build higher level circuits. BLADES does use hierarchy, in the form of subcircuit experts, but does not suggest any mechanism by which such experts can be coordinated to actually perform high-level synthesis.

None of the above approaches deals adequately with hierarchy as a critical characteristic of analog design; therefore, none seem to admit straightforward generalizations to different types of analog circuits. Moreover, they do not incorporate adequate process dependence for detailed analog design, nor do they provide mechanisms to attack complex performance concerns such as noise or stability. There are few reports of synthesized circuits that have been functionally verified, either by detailed simulation with respect to a reasonable fabrication process, or by fabrication itself.

Recently, new approaches to analog synthesis problems have begun to appear. Notably, these include the following tools: IDAC [22], [23], OASYS [24], [25], which is the subject of this paper, OPASYN [26], [27], An_Com [28], and CAMP [29]. Before comparing the differing aims and approaches of these systems, we first describe in the following section the underlying synthesis framework of OASYS, its architecture and implementation mechanisms. We then return to this comparison in Section III-F.

III. OASYS Analog Synthesis Framework: Architecture and Implementation

This section presents a framework for organizing the component pieces of a hierarchically-structured analog circuit synthesis tool. We begin by outlining the synthesis task, and placing the work in the larger context of frameworks for automating design. We then describe the central components of the proposed organization: hierarchy, design style selection, and translation. Because a critical feature of analog synthesis tasks—one that distinguishes it from better understood digital synthesis tasks—is the nature of analog design knowledge, we also describe the implementation mechanisms for codifying and applying such knowledge in this framework. To make the discussion of these ideas more concrete, we introduce a simplified hierarchical circuit design example, and use it throughout this section. Finally, we examine the relationship of OASYS to other recent tools for analog circuit synthesis.

A. Overview and Justification

Our intent is to support high-level circuit synthesis for specific classes of analog functions. From an input consisting of detailed performance specifications and process specifications, we want to produce a sized, transistor-level circuit schematic. Our approach is guided by the design process employed by human analog circuit designers. Like human designers, we use approximate models for device behavior in order to simplify the analytical formulation of behavior. Our goal is to achieve a good device-level topology with correct device sizes to meet all performance specifications. Numerical optimization tools which employ a detailed circuit simulator in an optimization loop may then be applied, if desired, to further fine-tune device sizes based on specific goals, e.g., performance enhancement [30].

Previous approaches to synthesis have noted the difference between *topological design*, which interconnects devices, and *sizing*, which specifies the performance of individual devices. We make a similar distinction. The proposed framework is based on three critical ideas.

- Circuit topologies are *selected* from among fixed alternatives; they are not constructed transistor-by-transistor for each new design. The process of choosing from among these fixed alternatives for the design of a circuit topology is called *design style selection* (after the analogous digital synthesis task [31]).
- The fixed alternatives for circuit topologies are specified *hierarchically*. A topology for a high-level module (e.g., an A/D converter) is specified as an interconnection of sub-blocks, *not* as an interconnection of transistors. That the topology is fixed implies only that this arrangement of sub-blocks is fixed; the detailed design of the individual sub-blocks is not specified here. Because of this explicit hierarchy, one

high-level topology of blocks can specify *many* device-level topologies.

- After selecting a topology to accommodate a set of performance specifications given at one level of the hierarchy, we *translate* these higher level specifications into specifications for the performance of each sub-block of the topology. Informally, we are given the behavior of the interconnected sub-blocks taken as a whole, and we must deduce the specifications for each sub-block required to achieve this overall behavior. Device sizing, in the conventional sense, occurs when this process of translation reaches the bottom of the hierarchy and specifies the behavior of individual transistors.

This basic framework has many similarities to some knowledge-based design ideas. However, an important clarification here is that our framework is targeted *solely* at analog circuit synthesis. There is considerable work in progress aimed at developing general "paradigms" for engineering design (see [32] for a survey), i.e., general-purpose theories of how design activities ought to be organized, which are intended to apply across diverse engineering applications. In such work, it is typical to propose a general design paradigm, and then illustrate it with an implementation targeted at a specific—usually modest—design problem. This is the *not* the style of the work described herein. Rather, the framework described in this paper evolved from a sequence of synthesis-tool implementations each aimed specifically at analog circuit design. Specifically, our notions of a framework for circuit synthesis were abstracted from our experiences with several earlier attacks on the problem of high-level analog synthesis, and then used as the basis for an experimental implementation. Thus the framework arose specifically to accommodate problems unsolved by earlier attempts at analog design.

In the area of VLSI CAD, our framework is similar in spirit to that proposed by Brewer and Gajski [33] for digital design; it also uses the well-known idea from digital design of exploiting known design styles [31]. However, our framework was influenced by the peculiar characteristics of the analog domain, e.g., the fact that reuse of partial circuit topologies is common, contributes directly to our notions of hierarchy and design style selection. In the area of knowledge-based design paradigms, our framework bears a close resemblance to the more abstract paradigm for routine design suggested by Brown, and tested to date primarily in the area of mechanical design [34]. However, unlike some design applications in which it is immediately obvious that a particular, established design automation approach is workable, e.g., some mechanical CAD tasks [34], the general analog synthesis problem has, to date, resisted most attempts at formulating a practical, general-purpose synthesis approach. Hence, we believe that a critical contribution of the present work is simply the demonstration that the analog behavior-to-structure synthesis problem can be recast in a highly-structured form, with hierarchy as the key organizing principle here. Indeed, our aggressive use of hierarchy is clearly at odds with the conventional wisdom that tightly-coupled circuit subproblems cannot be attacked with other than flat, numerical optimization techniques tied to a circuit simulator.

Reliance on explicit hierarchy in analog design has two advantages. First, it permits the design process to be recast as a sequence of smaller design tasks, alternating between design style selection and translation. Second, it provides a measure of generality, in that sub-blocks can be reused in different contexts. Note, the *choice* of the hierarchy is regarded as expert design knowledge and is selected to minimize the information required by each sub-block. In the analog domain, there is no single, accepted hierarchical structure agreed upon by the community at large; rather, the best hierarchical decomposition for a particular analog circuit is part of the domain knowledge for the design of that circuit, i.e., it is derived from the expertise of a designer.

A hierarchical representation makes the synthesis task more tractable, but has one disadvantage. By recasting circuit design as a sequence of alternating topology selection and translation steps, we lose the easy ability to implement design tricks that jump across many levels of the hierarchy. Thus we lose a "flattened" view of the design in which all details of individual devices are simultaneously exposed, and independently changeable. Expert circuit designers occasionally employ such tricks to push circuits close to the limits of achievable performance (simultaneously exploiting tricks involving topologies, sizing, layout, and process, as in [35]). Since the hierarchy explicitly prevents the design plan for one module from depending on the details of how other modules are implemented, our approach may not be able to reach such extremal points in the design space of a given block.

The two-step topology selection and translation process is illustrated for an abstract block in Fig. 1. A particular topology is chosen because it is the best candidate to match the specified block-level behavior; the translation step then decides how its sub-blocks must behave to meet the block-level specification. An important advantage of recasting analog synthesis as a sequence of selection and translation steps is that the design task is *uniform* at all levels of the hierarchy: topology selection and translation occur when designing a complex high-level function, and when assembling a few transistors. Our original motivation for using separate selection and translation steps was to avoid the need *simultaneously* to design the interconnection and electrical characteristics of sub-blocks. Restricting the design of sub-block interconnections to a choice among fixed alternatives allows us to concentrate on specifying the electrical characteristics of the connected sub-blocks, a more tractable synthesis task. A hierarchical representation of topologies vastly simplifies the translation task because it tends to reduce the number of sub-blocks and simplify their connections: it is generally easier to synthesize the specifications for five con-

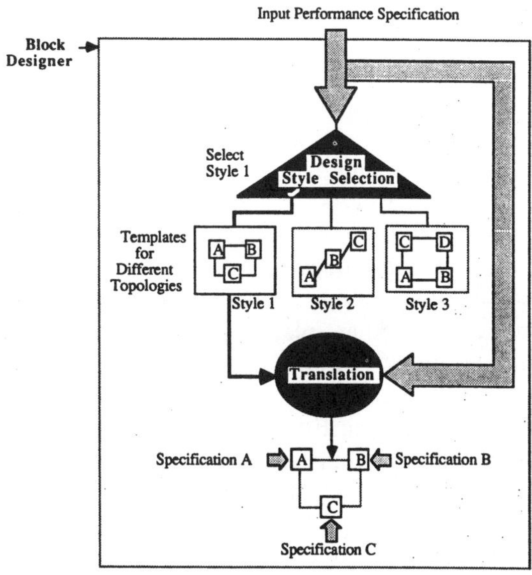

Fig. 1. Topology selection and translation processes.

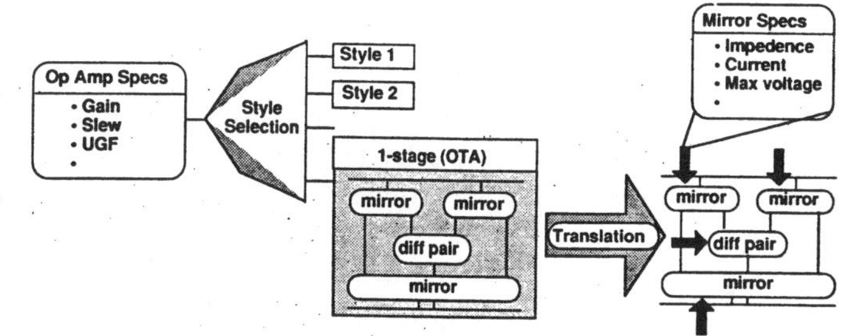

Fig. 2. Translation example: from op amp to sub-blocks.

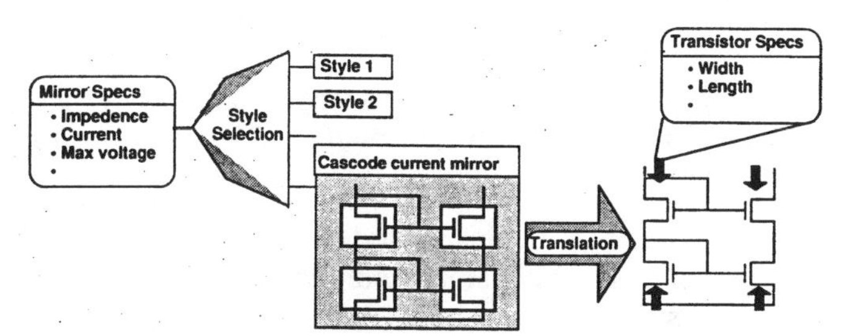

Fig. 3. Translation example: from current mirror to transistors.

nected blocks than to synthesize individual parameters for each of 50 connected transistors. Hierarchy also simplifies the selection task, because we do not require a vast number of nearly identical topologies, differing only in low-level details; only the lowest level design styles in the hierarchy actually specify transistors.

The idea of selecting from among mature design styles follows directly from standard manual design practices. Analog designers usually attempt first to find a known, mature topology to fit a given specification, and only embark on the much more difficult task of designing a new topology if this search fails to provide any reasonable candidates. One advantage of our proposed framework is that it nicely compartmentalizes such *topology-specific* design expertise. Much of the expertise in analog circuit design can be characterized as knowing what design strategies are applicable when faced with a known circuit topology and a set of performance requirements. Such clean compartmentalization implies that it should be (at least conceptually) straightforward to add new design styles to an existing synthesis tool: we add a new template for a topology, modify the selection process to incorporate a new alternative, and provide a new translation function to generate specifications for the sub-blocks of the new topology.

B. Mechanisms to Support Translation in OASYS

Translation involves knowledge of how performance specifications for a high-level block (a design style) should be transformed into specifications for each sub-block. As an example of this process, consider the simple OTA style op amp shown in Fig. 2. At this level of the design, we have no details of the transistor-level design of this op amp. The op amp appears as a fixed connection of building blocks such as current mirrors, differential pairs, bias networks, and loads. To get to transistors, we will use two levels of translation: first, op amp performance specifications will be used to design the sub-blocks; second, the (new) specifications for each sub-block will be used to design the transistors that comprise each of the sub-blocks. It is important to recognize that "to design" in this translation process means "to produce performance specifications for." For example, in Fig. 2, typical performance specifications that form the *input* to this op amp-level translation task will include dc gain, slew rate, phase margin, etc. The *output* of this translation task is a set of designed sub-blocks, specifically, a set of *input* specifications for each of these sub-blocks. The process then repeats inside each sub-block.

To continue this process down the hierarchy, consider again one of the current mirror sub-blocks appearing in Fig. 2, and shown in expanded detail in Fig. 3. Assume now a style has been chosen for the mirror, a cascoded current mirror topology. Just like the op amp, this mirror has a set of input performance specifications, including: output resistance, reference current, current gain, and output voltage range. These are the target values that the op amp translation process will produce. The mirror also has its own translation task to perform; it must translate its input specifications into width (W) and length (L) specifications for its internal MOS devices. However, when we reach this point in the sequence of translation steps, we are done, since W and L specifications for each device constitute a finished, sized design.

We observe that good analog designers exhibit two characteristics relevant to this sort of design task. First, they are adept at choosing highly simplified models of devices and device interactions to guide their choices for tradeoffs. For example, analog designers work routinely with simple algebraic descriptions (linear and nonlinear) of the relationships between component values and performance specifications; expert designers choose the right descriptions, and prune away the relationships irrelevant to the task at hand. These device models, though simple, are informed by detailed knowledge of how subtleties of the fabrication process and the desired performance parameters will interact in a specific circuit topology. Second, analog designers have a basic plan of attack for designing common functional blocks. That is, for a specific topology of abstract blocks, and well-understood performance goals, good designers do not go about the translation task of specifying sub-block constraints in a purely random order. Good designers understand how, over the course of the entire design task, design difficulties can be minimized or eliminated by attacking the right sub-problems, with the right simplifications, in the right order. These two observations suggest a way to implement translation tasks in the overall framework, which we can summarize as follows.

- Topologies, i.e., design styles, are implicitly represented as statically stored templates of connected sub-blocks.
- The process by which a high-level block specification is translated into sub-block specifications is implemented—albeit rather loosely—in the style of a planning system [36], [37]. A design plan is associated with each fixed topology and executed when the topology is instantiated.

Because we view block-level topologies as static templates, we can easily represent all relevant, useful circuit relationships within the design plan for the template, for use during the translation step. (Note, systems which dynamically design topologies must also rediscover these equations dynamically [17].) Unfortunately, just storing basic analytical circuit relationships does not completely solve the problem. Real design problems are usually so under-constrained that many heuristic choices must also be made to achieve a unique design that matches the desired performance specifications. Knowledge of these "good" heuristics is an important component of analog expertise; the critical question then is how to codify such knowledge.

In OASYS, design knowledge is codified in a style influenced by the organization of simple planning systems used in some knowledge-based programs. In this context, planning is usually defined as the process of deciding what overall course of action to take to meet some specified set of goals. Most generally, this may involve decomposing the problem into sub-goals, ordering these sub-goals, proposing individual actions to meet these sub-goals, and changing the plan if some goals cannot be met. However, OASYS uses static, essentially fixed plans that are stored with each design style template. In its most general form, planning can involve the dynamic construction of plans to meet goals, e.g., as in the ADAM digital design system [38]; we currently regard the analog synthesis problem as being too intractable for this approach. Hence, we simply attach a fixed, *default* design plan to each style template. Another fact we must deal with is that there may be many ways of meeting the required performance goals, each stressing different optimizations, e.g., design for minimum power, or area, or noise, etc. In OASYS, we handle this by treating different design plans for the same circuit topology as *different* design styles. For example, if we have both a general-purpose plan as well as a special-purpose noise-optimizing plan for some op amp topology, we store the topology twice, one copy with each plan; we regard one as a general-purpose op amp, and the other as a special noise-optimized op amp.

OASYS' translation mechanisms are most similar to the hierarchical refinement planning mechanism suggested in Brown's notion of routine design [34]. We also find some similarity to the earlier notion of skeletal planning introduced by Iwasaki and Friedman [37]. The important idea behind all these systems is that some design domains are characterized by the existence of mature plans of attack for broad classes of problems. By organizing the overall design task as levels of coarse-to-fine refinement, i.e., by organizing the problem hierarchically, we can bring these mature plans to bear on problems too complex to solve in a single refinement step. Hierarchical refinement is not a new idea, but in the analog domain in general and circuit synthesis in particular, it has long been dogmatic to assume that all representations for design tasks are necessarily flat. A major motivation behind the OASYS framework was our desire to develop an alternative to flat representations which appear to grow increasingly intractable as one attacks larger circuits.

However, there are some important differences in how OASYS adapts simple planning ideas to handle hierarchical refinement for analog design. These differences center on the granularity of the individual design steps in each plan. In [34], [37], the grain-size of each plan step appears to be fairly large. There, many plan steps commit to "large" actions, such as choosing a mechanical component or dimension [34], or a style for a genetics experiment [37]. In contrast, analog design has an explicitly fine-grained *numerical* optimization character. Neither of these earlier systems has to deal with multivariable numerical optimization issues. Accordingly, the execution of design plans in OASYS is qualitatively different. First, individual plan steps involve much more fine-grain numerical computation. Second, and perhaps more important, fixed-point iteration of plan steps is introduced as a mechanism for resolving, i.e., optimizing, numerical constraints. To understand how this optimization works, we first need to describe what actually constitutes a plan in OASYS, and how design failures during plan execution are handled.

In OASYS, each topology currently has one design plan associated with it, and this plan is a linear sequence of executable steps called plan steps. Each plan step has a set of sub-goals it expects to achieve, and methods to achieve them. Plan steps themselves perform three sorts of activities.

1) **Heuristics:** Plan steps often need to make decisions based on expertise and on incomplete knowledge, in order to advance the design state. Usually at some

later stage of the design, i.e., in some later plan step, we verify whether each heuristic decision was a viable one. A common example is a computation that estimates a parasitic capacitance *before* enough of the design is complete to *know* this value. Another example is the partitioning of gain between the stages of an amplifier; many partitionings will work, but there are good heuristics to help choose a reasonable initial value.

2) **Computation:** Plan steps frequently need to compute quantities like currents or voltages when there is sufficient information to do so, and when these quantities must be known for the design to proceed. These computations may be purely algebraic, if there is a known closed-form analytical equation for the desired result, or they may require solution of simultaneous (linear or nonlinear) equations.

3) **Refinement:** Some plan steps invoke the selection and translation mechanisms for a lower level sub-block, once sufficient information has been deduced by the current design plan. These steps link one level in the analog hierarchy with the level below it. A refinement step from our op amp example is the step that passes newly designed current mirror specifications to the mirror selection/translation code. This step invokes the design of the requested mirror, then waits for parameters to be returned from the mirror designer that inform it about the actual performance of the mirror that was synthesized.

Worst-case design is also a problem in analog design that must be handled during the translation process. Process specifications always include variances, and performance specifications are themselves often specified as ranges of acceptable performance. OASYS currently deals with worst-case design in an ad hoc way, essentially in the manner of human designers. For example, if a plan step needs to compute the current I_{DS} through a MOS transistor, and the relative constraint is to ensure some maximum current I_{DS}^{max}, then OASYS will use the relevant extremal values and compute $I_{DS}^{max} < 1/2 K^{max} (V_{GS} - V_T^{min})^2$. Of course, this can result in extremely conservative designs. However, on the other hand, this typifies the sort of approach used in manual design: good designers know when to include obvious or critically important worst-case effects, and ignore the rest. Such knowledge is simply compiled into all OASYS plans.

C. Mechanisms to Support Selection in OASYS

In the OASYS hierarchy, each circuit block is allowed to have different styles, where a style is just a different interconnection of lower level building blocks. For example, there is only one block called "op amp" in OASYS, but it is available in different circuit styles: operational transconductance amplifier (OTA), Miller-compensated two-stage, etc. Each of these styles *is not* a transistor schematic; rather, it is an interconnection of lower level blocks like current mirrors, differential pairs, transconductance amplifiers, bias chains, and so forth. Before we can invoke the translation process to refine op amp specifications into the mirror, differential pair, etc., specifications, we must select one of these op amp styles. Like translation, selection also happens at all levels of the OASYS hierarchy. For example, mirrors, differential pairs, and transconductance amplifiers all come in various styles as well, although now the sub-blocks in each of these styles are actually transistors.

Although a planning-style mechanism works well for translation tasks, it is inappropriate for this selection task. The basic reason is that selection involves a more qualitative style of decision making. Together, we can regard selection and translation as a search process through a tree of design style choices. Each node in the tree is a set of styles for a given functional block. During design, when we visit this node we must make a decision as to which specific design style to pursue, and then move down the tree as further sub-blocks are refined. If we make an incorrect style choice, some plan step will fail, and we may return to this node in order to make an alternative style choice.

Observations of expert designers lead us to believe that there are three rather broad strategies for selecting design styles.

1) **Structural Discrimination:** Some design style choices simply do not meet implied structural constraints. For example, if we know that an amplifier with differential inputs is required, we cannot pursue any design style that lacks differential inputs. This is the simplest kind of discrimination to apply.

2) **Heuristic Discrimination:** Simply put, experienced designers develop a feel for a mature fabrication process, and can often predict performance limits for certain topology choices implemented in this process. This is the hardest type of discrimination knowledge to capture: it is entirely qualitative, has a rather limited analytical foundation, and varies among designers.

3) **Generate-and-Test:** When all else fails, we can try to complete the design using all the candidate design styles: some might work; some might not. This reflects the fact that individual design styles can be coerced to work successfully in radically different design situations. Thus even if a style choice is suboptimal, it may still be viable. To determine which style is best, design in several styles and select the best one.

OASYS relies mainly on generate-and-test and structural discrimination, but also uses some heuristics. The generate-and-test style seems rather naive, but it is actually quite natural at the lower levels of the hierarchy. For example, when discriminating among transistor-level styles for a current mirror, accurate analytical expressions are available describing the performance of each style after it is designed. It is much more natural simply to compare fully designed objects than to craft heuristics to guess *a priori* which will work best. Even at slightly higher levels of detail, this strategy, combined with simple structural discrimination to prune obviously inappropriate candidates, works well.

The heuristics used for selection are based on the idea of threshold hunting described in detail in Section IV. To summarize, for a given fabrication process, we use OASYS to run a large set of controlled synthesis experiments. These experiments are meant to provide a rough estimate of the shape of OASYS' feasible region in multidimensional parameter space, i.e., the surface that bounds the set of the designs that OASYS can actually produce. These estimates are crude since, for example, they focus on only a few critical performance parameters. Nevertheless, the resulting data still suffice for rough discrimination among different styles. Simple estimates (curve fits) to the shape of this surface are stored, and used for comparison against input specifications. For example, if we are examining the gain specification for an amplifier, and are concerned about the likely area consumed by a particular amplifier style, we look at the stored shape of the area versus gain curve to get some rough insight (rough because many other parameters were held constant to generate this curve) into the appropriateness of a particular style.

This current scheme also leads naturally to a promising extension based on a more rigorous statistical approach. Again, we model each design style as a black box that transforms performance specifications into performance predictions (by involving translation down the hierarchy). We then build regression models for relevant predictions like area or power consumption. Such models can accommodate more variable parameters than the simple scheme we currently use. Still, these models will necessarily be coarse, but we only really need to rank the candidates from best to worst, and then attempt designs in rank order. By building up from the bottom of the hierarchy, higher level style selection tasks can use the regression models for the lower level styles. An important characteristic of our current simple scheme, and this proposed extension, is that the construction of the surfaces used for selection can be automated, since these surfaces are built from data generated over a statistically significant number of runs of the synthesis tool. Thus although they are approximate, they are not ad hoc. Additional discussion of these ideas appears in [39].

D. Failure Handling and Optimization During Synthesis

Unfortunately, there is no *a priori* guarantee that this descent through a hierarchy of style selection and translation by plan execution tasks will succeed in meeting all design goals. The design plan associated with each topology (one design style) is a default plan, representing essentially a best guess about the order in which to attack subproblems. These plans are tuned to work as well as possible overall, but it is almost always true that some feature of a new design problem makes it impossible to meet some subgoal within the default plan. Hence, we require a mechanism for plan modification. In OASYS, whenever a plan step finds that it has been unable to meet its goals, control is passed outside the plan to a set of failure handlers, called *plan-fixers*, attached to this plan. Each plan-fixer tries to diagnose the problem and correct it. Some plan-fixers are simply composed of many IF-THEN rules that try to match the diagnosed failure and take some corrective action. Others are more algorithmic, and resemble basic plan steps themselves. Plan-fixers can either modify circuit specifications in ways that are beyond the limited scope of individual plan steps, or can modify the flow of execution through the plan, to avoid the problems previously encountered. The feasibility of fixing plans in this way derives from our conjecture that good plans have predictable failure modes; i.e., very often, when an analog expert fails at a particular step in his overall plan of attack, the failure can be easily located because there are (relatively) few things that can go wrong, especially when the design has been constrained to a fixed block-level topology.

Fig. 4 illustrates how failure handling can modify the design process during translation and style selection down the hierarchy. It is important to note that failures during the design of lower level topologies can cause the design of higher level topologies to change. For example, in Fig. 4, a failure in the design of a mirror inside an op amp can force changes in both translation and style selection activities, locally or globally in the hierarchy. Specifically, a failure during mirror design here will invoke plan-fixers that try first to remedy the problem locally, at the level of the mirror itself, and then globally, at the level of the op amp. Typically, the following remedies would be tried, in order: change the plan of attack for refining this mirror itself; change to a different style (topology) of mirror; change the plan of attack for the op amp that invoked the mirror, to produce a different set of specifications for a more designable mirror; change the high-level op amp design style itself, if the current style is just incapable of specifying a practical mirror.

Plan steps and plan-fixers are separated in this fashion (the executable steps invoked by plan-fixers are not simply placed inside plan steps themselves) because this division reduces the complexity of developing and maintaining the implementation. Many failure-correcting actions actually do appear inside steps; it would make the code extremely cumbersome were it necessary to invoke an elaborate plan-fixer mechanism for each tiny error. We place such actions inside individual plan steps when the corrective action does not require global knowledge of design decisions, design variables, or design state that is not stored locally within this plan step. Similarly, we place corrective actions *outside* the plan itself when such actions need access to global state information about the past history of the evolving design. Were individual plan steps allowed freely to reach inside other plan steps to modify old design decisions or variables, not only would this vi-

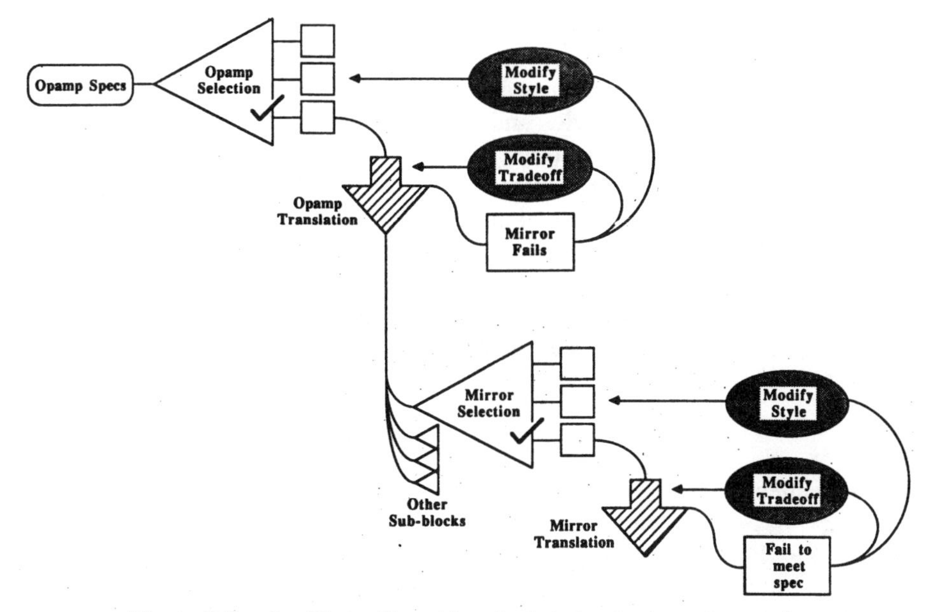

Fig. 4. Failure handling up/down hierarchy in style selection and translation.

olate our notions of a strong hierarchy, but also it would be extremely difficult to keep track of the state of the plan. This is especially critical when a corrective action needs to stop the plan, change some earlier heuristic decision, e.g., revise a parasitic estimate in light of more detailed design knowledge, and restart the plan at some earlier plan step. In essence, the separation of plan steps and plan fixers is an implementation discipline intended to prevent subtle corruption of the state of the evolving design.

This form of failure handling also permits a straightforward form of optimization. We can now describe how numerical optimization is implemented in this planning style. During circuit design, some numerical quantities can be computed directly, from first principles. However, some values that must be known to advance the design are only determined accurately after the circuit has been completely designed. OASYS uses a fixed point iteration style to "negotiate" values for such quantities. An example is most succinct here. Assume that in plan step K, we require a parasitic capacitance C to proceed with a design, i.e., to execute the rest of the plan, steps $K + 1$, $K + 2$, etc. Using whatever heuristics are available (and these may require substantial computation themselves), we produce a first guess, denoted $C_1(\textit{guess})$ and proceed with the design. Some time later, in plan step $K + N$, we know enough to revisit this guess to see if it was accurate, measured against the first computed value of this capacitance, denoted $C_1(\textit{computed})$. If these two values are considered "close enough," again determined heuristically, the design can proceed with the next plan step. If not, we jump to a plan-fixer for this situation. The likely correction will be to compute a new, second guess for the parasitic $C_2(\textit{guess})$ based on $C_1(\textit{guess})$, $C_1(\textit{computed})$, and any other useful circuit information, such as circuit sensitivity to this parasitic. Then, control will be passed back to the plan at step K: we will restore the state of the plan back at step K, update the appropriate variables involving this capacitance, and execute again the intervening design steps between K and $K + N$. This iteration has a fixed-point style: from an initial guess, we use the plan to compute a better value, generate a new improved guess, and iterate again. This will continue as long convergence appears likely, i.e., guessed and computed values are converging with adequate speed. Divergence here is simply treated as another ordinary failure: we invoke an appropriate plan-fixer to diagnose the problem and provide a remedy.

It is also important to note that we have not in any way constrained the plan steps between K and $K + N$. These steps may invoke refinements down several layers of the design hierarchy. For example, if we are looking at a parasitic capacitance in a high-level topology, we may have to translate downward a few levels to know enough device-level details to compute the actual effect of this parasitic. Moreover, fixed point iteration may happen anywhere, even across levels of the hierarchy. Indeed, between plan steps K and $K + N$, we have no guarantee that this sort optimization by iteration will not happen several times to resolve other circuit values. The key to making such a scheme work is to acquire the right design expertise, and to codify it as executable heuristics for design and for failure handling.

E. Examples of Selection and Translation

To make the ideas of hierarchy selection, and translation concrete, this section presents an extremely simplified example of OASYS-style synthesis for an opamp. For the sake of brevity, pieces of the overall synthesis process have been selected to illustrate the process. This example is divided into two parts. In the first, we assume that a specific (highly simplified) block-level topology for an op amp has been selected and we present a simplified plan for translation of op amp specifications into sub-block specifications. Then in the second part, for one specific sub-block, the current mirror, we illustrate the process of selection between two different current mirror styles and the process of translation to device sizes for each style.

We consider first the higher level translation task for the op amp. Note that we are assuming that the selection process has already occurred, and a highly simplified OTA style has been selected (shown in Fig. 5).

The following is a simplified plan for translation of this op amp. Assume that the simple op amp design plan accepts as input six performance specifications:

1) slew rate, SR;
2) load capacitance, C_L;
3) unity gain crossover frequency, ω_c;
4) dc open loop gain, A_0;
5) maximum output voltage, $V^{\max}_{o-\mathrm{OA}}$ (the "OA" subscript denotes "op amp" and is used to avoid confusion when we need output voltages for sub-blocks later.);
6) minimum output voltage, $V^{\min}_{o-\mathrm{OA}}$.

Note, throughout this plan we will attempt to minimize power and active area to remove extra degrees of freedom in the design space. The design plan is as follows.

1) Choose bias current I to achieve desired slew rate SR, using the fact that $SR \leq (1/C_L)I/2$ and minimize power by choosing the smallest possible bias current.
   Therefore, $I = 2C_L SR$
2) Choose the transconductance of the differential-pair sub-block, G_m, to achieve the desired unity gain crossover frequency, ω_c, using the fact that $\omega_c = G_m/C_L$. (Notice, G_m is an *output* of this translation process, but is itself an *input* performance specification for the differential-pair sub-block.)
   Therefore, $G_m = C_L \omega_c$
3) Choose α, where $r_{o-M} = \alpha r_{o-B}$, and r_{o-M}, r_{o-B} are the output resistance of the current mirror and output bias current source sub-blocks, respectively. In this case there is insufficient information to make an optimal choice for α, therefore, we make a *heuristic* assumption based on experience.
   For simplicity, we *assume*: $\alpha = 1$.
4) Choose r_{o-M} and r_{o-B} using the fact that $A_o = G_m(r_{o-M} \| r_{o-B})$.
   Therefore: $r_{o-M} = \alpha r_{o-B} = (A_o/G_m)(1 + \alpha)$.
5) Design the current mirror sub-block using the following input performance specifications:

- mirror current, $I_{o-M} = I$;
- minimum output resistance, $R^{\min}_o = r_{o-M}$;
- minimum output voltage, $V^{\min}_{o-M} = V_{\mathrm{DD}} - V^{\max}_{o-\mathrm{OA}}$. (As before, we use a subscript, M to denote parameters for the mirror.)

Failure Recovery: If current mirror design fails (as will be described later in details for the mirror design plan) then increase α and restart plan at step 4.

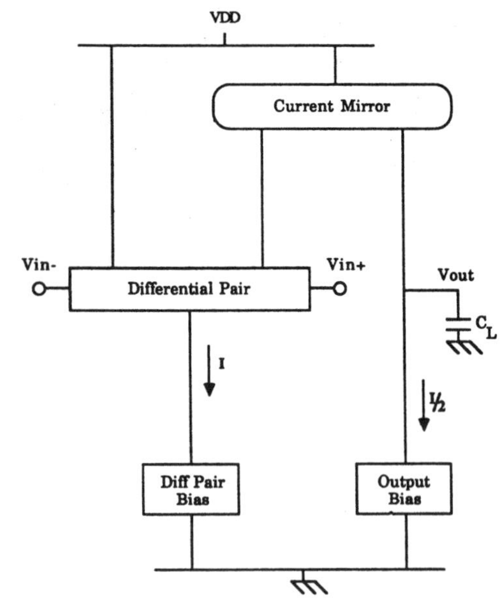

Fig. 5. Highly simplified OTA design style.

6) Design the output bias current source, the differential-pair, and the differential-pair bias current source (these are similar in style to the previous steps; we omit details.).
7) Adjust α in order to decrease total active area of all modules—note, only the output bias current source and the current mirror are affected by the choice of α. Compute sensitivities of mirror area, A_m, and bias area A_b with respect to α. If $\partial A_m/\partial \alpha < -\partial A_b/\partial \alpha$ then increase α, else decrease α. Then iterate the plan back to step 4.

Note that steps 5 and 7 in the plan use a simple form of fixed-point iteration; step 5 iterates to find a workable value for α, while step 7 iterates to optimize area.

Although the method of adjusting α is not described in detail here, a variety of numerical methods, or heuristic techniques based on designer expertise, could be employed. For example, a valid heuristic might simply be to adjust α by some constant amount. The critical point to note is the role of hierarchy in making traditional "flat-circuit" optimization impossible here. Because we refine to blocks like mirrors, *not* transistors, it is not possible to develop a set of flat equations that relate overall performance to device sizes and bias currents, and to apply optimization techniques that manipulate device parameters. Indeed, in this hierarchy, we do not even know yet the device-level implementations of the sub-blocks.

Within the OASYS implementation of real op amps, design plans are considerably more complicated than indicated by this simple illustration, and handle many additional parameters, e.g., noise, frequency dependence, and phase margin. Also, there are limits on the range of validity of the models used to generate these plans, hence, the planning process must verify that the results produced are consistent with modeling assumptions, and correct the design if they are not.

Translation for an op amp refines op amp specifications into sub-block specifications. We next consider translation, and now selection as well, at a lower level. Again, this is best illustrated by presenting a simplified example. We use the current mirror block appearing in the op amp as an example of a "lowest level" building block, i.e., a block where translation produces actual transistor sizes.

Consider a simple MOS current mirror design plan which accepts three performance specifications (as described in the previous op amp design plan) and four process specifications:

1) maximum output mirror current, I_{o-M};
2) minimum output resistance, $R_o^{\min}$;
3) minimum output voltage, $V_{o-M}^{\min}$;
4) K' for the process;
5) V_T for the process;
6) $L_{\min}$, the minimum allowable device length in the process;
7) λ for the process, where the device output resistance is given by $(1/\lambda I_D)(L/L_{\min})$.

Assume only two possible design styles for a current mirror are available: Widlar and cascode styles, as shown in Fig. 6. For this simplified example, generate-and-test is most natural for selection. We discriminate based only on the active area required by the two topologies, and choose the topology requiring the smaller area. The active area is determined by performing the translation process, which generates the device sizes, for both mirror design styles.

The translation process for the Widlar current mirror is completely algorithmic under the assumption that the smallest possible device width W and device length L will be chosen in order to minimize area.

1) Choose L to achieve desired mirror output resistance R_o, using the fact that $R_o = (1/\lambda I_D)(L/L_{\min})$.
   Therefore, $L = L_{\min} R_o^{\min} \lambda I_{o-M}$.
   If $L < L_{\min}$ then $L = L_{\min}$.
2) Choose W to achieve desired output voltage $V_o^{\min}$, using the fact that $I_D = K' W(V_{GS} - V_T)^2/2L$.
   Therefore, $W = 2LI_{o-M}/K'(V_{o-M})^2$.
3) **Failure:** we could define a maximum dimension for L and W, which if exceeded, would make the plan a "failure" by our definition, i.e., the input specifications could only be refined to an "unreasonable" design.

Because many possible combinations of device sizes can achieve the same performance specifications, the translation process for the cascode current mirror employs a heuristic in order to narrow the design space. In the example, W and L of all four devices are assumed equal. Note, many other heuristics are possible, e.g., the length of the two cascode devices could be set to their minimum size and the width of all four devices could be equal. The translation process proceeds as follows.

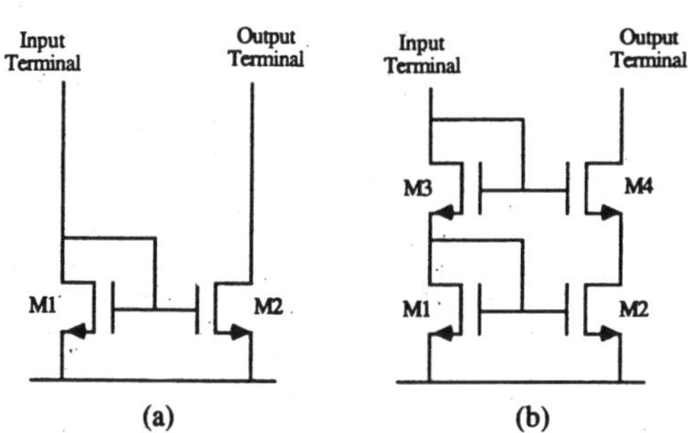

Fig. 6. Current mirror design styles. (a) Widlar mirror. (b) Cascoded mirror.

1) Choose W/L ratio to achieve desired output voltage $V_o^{\min}$, using the following facts (and neglecting the body effect):

$$\Delta V_{GS} = (V_o^{\min} - V_T)/2$$

$$I_D = K' W \Delta V_{GS}^2 / 2L.$$

   Therefore: $W/L = 2I_{o-M}/K'(\Delta V_{GS})^2$.
2) Choose L to achieve desired output resistance R_o, using the following facts (again neglecting the body effect):

$$R_o = g_{m4} r_{o4} r_{o2}$$

$$R_o = (L/\lambda I_D L_{\min})^2 K'(W/L)\Delta V_{GS4}.$$

   Therefore: $L = 2(L_{\min}\lambda I_{o-M})^2 R_o / K'W(V_o^{\min} - V_T)$.

   Failure: could be handled as with the Widlar style.
3) If $L > L_{\min}$ then $L = L_{\min}$.
4) Finally, choose W by substituting this computed L into the equation for W/L given in step 1 above.

As for the op amp plan, in the actual OASYS implementation, the current mirror design plans are more complicated than indicated here, since they must handle device-level issues such as the body effect and matching, circuit-level issues such as noise and bandwidth, and the usual problems of verifying the validity of device models over different ranges of operation.

F. Related Work

With OASYS now described, we can return to the comparison with some related synthesis tools: IDAC [22], [23] OPASYN [26], An_Com [28], and CAMP [29]. All these systems perform behavior to structure synthesis, translating specifications into sized circuit schematics. However, the tools differ substantially in their detailed aims, approaches, and implementations.

IDAC is a commercial synthesis system with an architecture rather different from OASYS. IDAC provides a set of design styles for several kinds of blocks, e.g., op amps, comparators, voltage and current references, etc., and can size these circuits to meet user specifications. Unlike OASYS, low-level blocks are fixed circuit schematics, with no topological changes possible during design. IDAC uses some limited internal hierarchy, but hierarchy is not the same strong, central organizing principle it is in OASYS. Synthesis in IDAC is a two-phase process. First, an algorithmic design strategy is executed to size the devices in the fixed schematic; this is a coarse design. If this design is deemed close enough to specifications, a second, numerical optimization phase (a downhill-style descent on the circuit's multidimensional performance-surface) is performed to tune the final circuit. A custom optimization phase is included with each circuit block in the IDAC repertoire. OASYS also does coarse design, but because it can do some limited numerical optimization during refinement, it does not include a separate, internal optimization phase. Methods for handling design failures also differ. Because IDAC does not use the same strong, information-hiding hierarchy OASYS uses, more flat circuit details are exposed for tuning to correct failures. IDAC also simply tightens specifications when the coarse design phase fails, e.g., if the achieved gain is too small, IDAC reruns the entire design with a larger requested gain. The coarse design strategies in IDAC appear to do less of the aggressive iteration done in OASYS, probably because there is a separate numerical optimization phase that follows. In contrast, OASYS backtracks and strives to correct failures locally, at each level of the design, in each sub-block of the design; this allows failure detection to be incremental, tightly localized to particular design decisions and goals in OASYS plans. Another major difference is philosophical: OASYS is intended to be an open framework for capturing the process of analog circuit design; IDAC is a set of individual programs for designing specific circuit blocks.

The OPASYN tool also performs specification-to-structure synthesis, but in a style different from either OASYS or IDAC. OPASYN currently designs only op amps. OPASYN models the overall synthesis task as a large set of tightly coupled nonlinear equations, and relies primarily on numerical techniques for solution. Like IDAC, it is uses flat, fixed topology blocks, and requires a separate program for each circuit schematic it can design. In OPASYN there is only a very simple rule-based style selection mechanism [27], no hierarchy, and apparently no direct reusability of previously coded design knowledge in new design tasks. OPASYN is primarily doing translation, but because of its models, it is also simultaneously doing detailed numerical performance optimization on approximate analytical models. OPASYN appears to produce optimized designs for a small domain of styles, but is limited to designing only those topologies in its flat schematic library.

An_Com and CAMP, which appeared after OASYS, somewhat resemble OASYS in their synthesis philosophies. Each uses hierarchy to some extent, and incorporates some notion of translation or refinement. However, OASYS appears to be a much more general synthesis framework than either of these tools. CAMP has only a limited notion of hierarchy involving transistor-level subgroups, and does no real optimization in a numerical sense. Interestingly, it uses a detailed circuit simulator (SPICE [40]) to incrementally evaluate its progress, but it appears to rely on only a rather small, simple rule-base for design knowledge. An_Com is more like OASYS, in that it has a general hierarchy, search mechanisms to refine down this hierarchy, and failure-handling mechanisms. However, it is unclear if An_Com is capable of negotiating tradeoffs iteratively, up and down the hierarchy, to optimize performance. We are unaware of any large population of aggressive designs completed by either of these systems and verified by simulation or real fabrication, that could be used to judge their relative merits.

IV. OASYS: Peformance Evaluation

This section describes the domain in which OASYS currently operates, CMOS op amps, and then evaluates the performance of the OASYS tool at designing these circuits. The basic goal here is to verify that OASYS can indeed design functional circuits. However, a synthesis tool such as OASYS can be used for two related goals. The first is conventional synthesis, in which we regard performance and process specifications as fixed, and require the tool simply to produce the best possible design to meet these inflexible specifications. A second, and closely related application is exploration, in which we use the tool to explore tradeoffs in the design space. In practice, we expect actual designers to use OASYS in both styles, exploring broad options and then focusing on particulars as the system-level design evolves. The following sections describe the capabilities of OASYS for conventional synthesis and exploration.

A. Implementation and Applications

Op amps were chosen as an initial test domain for OASYS because they are ubiquitous components in many system-level designs, and because they appear to be the favored first target of other earlier synthesis approaches [20], [18], [41]. We restricted the designs to a single technology, in this case CMOS, only because we have access to CMOS fabrication. The OASYS hierarchy was chosen to explore two levels of design style selection and translation:

1) select op amp design style;
2) translate op amp specifications into specifications for sub-blocks such as current mirrors, differential pairs, etc.;
3) select design styles for each sub-block;
4) translate each sub-block specification into device interconnections and sizes.

This reflects our basic goal to incorporate enough hierarchy in the prototype to build a working system.

OASYS currently incorporates two topologies for op amps: a standard Miller-compensated two stage style, and a one-stage OTA style (see Fig. 7). Several lower level blocks are also available, each currently in a few different styles: current mirrors, differential pairs, transconductance amplifiers, level shifters, compensation schemes, and bias blocks. Style selection and translation activities occur at each level of this hierarchy; the number of style choices is small at each level, but translation is still complex. Notice that even with these limited style choices, the hierarchical composition of styles yields several

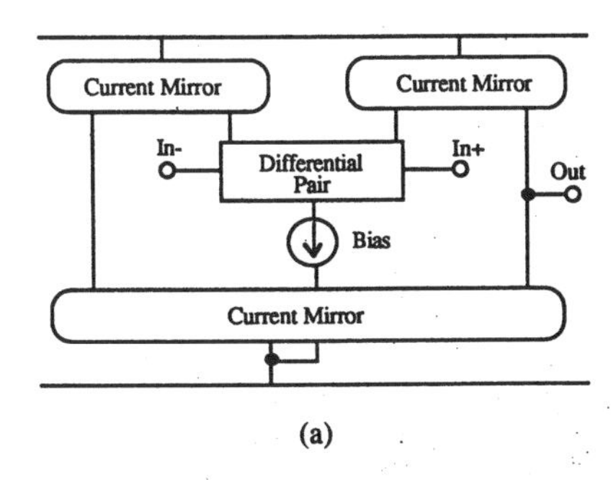

(a)

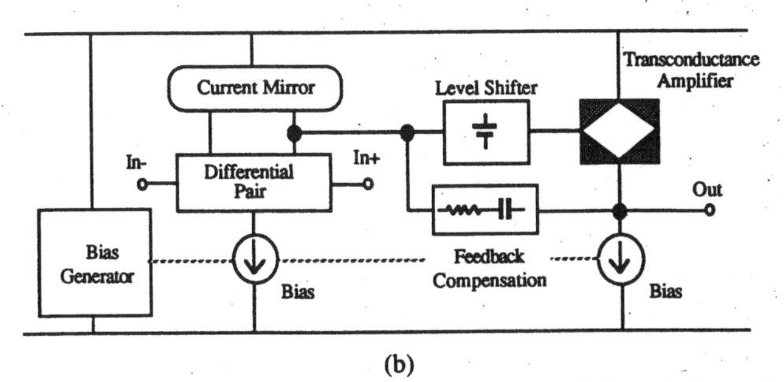

(b)

Fig. 7. OASYS op amp design styles. (a) One-stage (OTA) design style. (b) Two-stage design style.

hundred different transistor-level op amp topologies, of which about 70 are practical circuits commonly seen in real applications.

Most of these design styles are not specific to a particular topology: they are based on their own independent templates and plans, and are fully reusable as parts of other higher level designs. The only exceptions are the feedback compensation blocks, which are tightly bound to a particular parent style, and are rarely useful in other higher level design styles. However, all the other lower level blocks can be used directly in any new design plan, e.g., in a folded-cascode op amp, a voltage reference, a comparator, etc. To support this assertion, consider the time required to develop the two OASYS op amp styles. The two-stage op amp style required 18 months to implement, primarily because it was the first op amp style. To build two-stage op amps, we had first to implement design styles for all the required sub-blocks, such as mirrors, differential pairs, level shifters, etc. In addition, the framework itself was evolving as this design knowledge was being captured. In comparison, after the time spent deducing the appropriate hierarchy and heuristics for the OTA op amp design style, a design plan was coded in roughly one month. This is because all the sub-blocks were in place, and the framework itself was more stable. Any subsequent improvements in the design plan for a particular block expand the range of performance specifications that can be attained by any higher level topologies of which this block was a part. For example, an improvement in the current mirror designer would extend the range of specifications achievable by all our op amp design plans.

OASYS requires as input a description of the fabrication process and a set of op amp performance specifications. To keep pace with the rapid evolution of process technology, OASYS simply reads process parameters from a technology file; the parameters currently specifiable in OASYS appear in Table I. Because OASYS uses design equations in which process parameters are variables, it adapts easily to different processes. Only when the process is modified to such an extent that new device models are needed must the OASYS design plans be modified, and even then only in the lowest level design styles. Table II gives the performance parameters that OASYS can currently design to meet. From these inputs, OASYS produces a sized transistor-level circuit schematic in the form of an annotated SPICE file [40].

TABLE I
SPECIFIABLE PROCESS PARAMETERS FOR OASYS

Process Parameters
1. Threshold Voltage, p and n (V)
2. K′, p and n ($\mu A/V^2$)
3. Process Minimum Width and Length (μm)
4. Built-in Voltage, p and n (V)
5. Minimum Drain Diffusion Width (μm)
6. Supply Voltage, V_{DD} and V_{SS} (V)
7. Oxide Thickness (nm)
8. Mobility, p and n (cm^2/V–s)
9. C_{gso}, p and n ($fF/\mu m^2$)
10. C_{gdo}, p and n ($fF/\mu m$)
11. C_{gb}, p and n ($fF/\mu m$)
12. C_{jsw0}, p and n ($fF/\mu m$)
13. C_{j0}, p and n ($fF/\mu m^2$)
14. Temperature (K)
15. Flicker Noise Coefficient ($V^2 m^2$rad/s)
16. Channel Length Modulation, p and n, f_0 (V^{-1}), f_1 ($\mu m/V$), λ (V^{-1}), p and n for $\lambda = f_0 + f_1/Length$

TABLE II
SPECIFIABLE PERFORMANCE PARAMETERS FOR OASYS

Performance Parameters
1. Voltage Gain (dB)
2. Slew Rate ($V/\mu s$)
3. C_{load} (pF)
4. Power Dissipation (mW)
5. Unity Gain Frequency (MHz)
6. Phase Margin (degrees)
7. Common Mode Rejection Ratio (dB)
8. Power Supply Rejection Ratio (dB)
10. Offset Voltage (mV)
11. Active Area (μm^2)
12. Settling Time (μs)
13. Input Common Mode Range (V)
14. Output Range (V)
15. Input Referred Flicker Noise Voltage ($nV/\sqrt{Hz}$) at Specified Flicker Noise Frequency (Hz)
16. Broadband Noise Voltage ($nV/\sqrt{Hz}$)

The version of OASYS of interest in this paper, OASYS 0.3, comprises about 11 000 lines of Franz LISP running under UNIX.[1] Details about the evolution of this implementation appear in [24].

B. Conventional Synthesis

To verify that OASYS can synthesize functional circuits, we consider several sets of performance specifications and examine the circuits synthesized by OASYS to meet these specifications. We employ process parameters from the MOSIS 3-μm bulk CMOS process. All designs were synthesized on a VAXstation[2] II/GPX workstation running UNIX 4.3bsd.

[1]UNIX is a trademark of AT&T Bell Laboratories.
[2]VAXstation II/GPX is a trademark of Digital Equipment Corp.

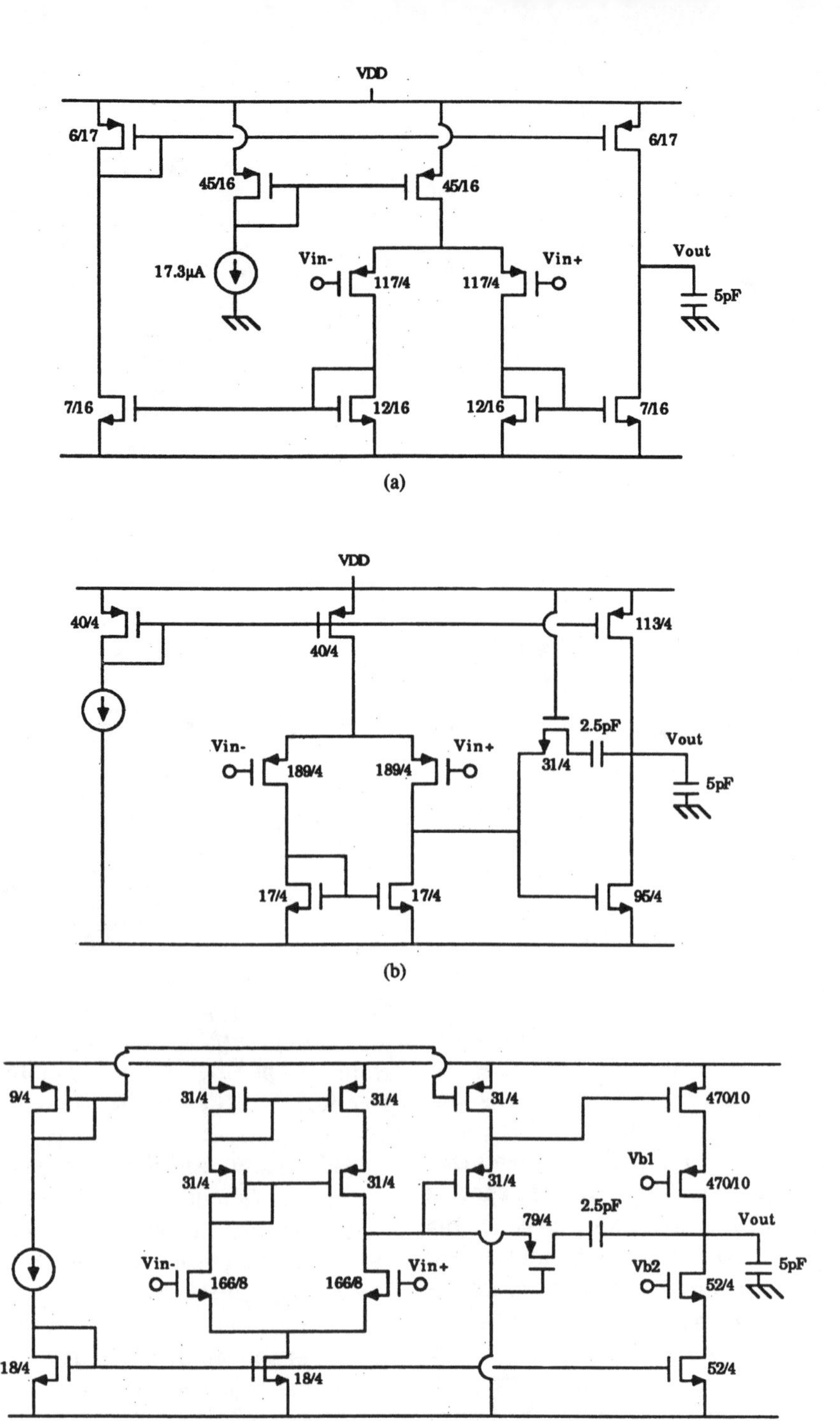

Fig. 8. OASYS-synthesized circuit schematics for test cases A, B, C.

TABLE III
INPUT SPECIFICATIONS AND SIMULATION RESULTS FOR TEST CASES, A, B, C

Parameter	Units	Circuit A Style: OTA		Circuit B Style: Simple 2-Stage		Circuit C Style: Cascoded 2-Stage	
		Spec	SPICE Result	Spec	SPICE Result	Spec	SPICE Result
Gain	dB	≥ 60	62	≥ 60	76	≥ 90	96
Unity Gain Freq.	MHz	≥ 2	2.03	≥ 10	11.2	≥ 10	12.6
Phase Margin	Degrees	≥ 45	83	≥ 45	46	≥ 50	47
Slew Rate	V/μs	≥ 2.0	+2.08/ -2.10	≥ 2.0	+13.4/ -14.6	≥ 2.0	+22.31/ -19.6
Load Capacitance	pF	5	-	5	-	5	-
Supply Voltage	V	±2.5	-	±2.5	-	±2.5	-
CMRR	dB	-	93	-	83	-	119
Power	mW	-	.193	-	1.12	-	1.53
Estimated Area	μm²	-	3020	-	6200	-	16200
Input Devices (selected by OASYS)	-	-	PMOS	-	PMOS	-	NMOS

We first consider three sets of closely related specifications intended to demonstrate that topology style selection and translation are tightly coupled in OASYS. A change in an input specification can result not only in different device sizing, but in an entirely different circuit topology as well. Table III shows three sets of performance specifications, and SPICE simulation results for the circuit synthesized by OASYS for each. Fig. 8 shows the resulting sized circuit schematics. The specification for circuit A is ordinary in that it makes no unusual performance demands. The synthesized circuit for A is a simple

TABLE IV
INPUT SPECIFICATIONS AND SIMULATION RESULTS FOR AGGRESSIVE TEST CASES

Parameter	Units	Goal: High Slew, UGF Style: 2-stage		Goal: High Gain Style: 2-stage		Goal: Low Noise Style: 2-stage	
		Spec	SPICE Result	Spec	SPICE Result	Spec	SPICE Result
Gain	dB	≥ 60	74.1	≥ 100	109	≥ 60	81.7
Unity Gain Freq.	MHz	≥ 22	23.6	≥ 5	5.6	≥ 2	1.995
Phase Margin	Degrees	≥ 45	44	≥ 45	51	≥ 45	48
Slew Rate	V/μs	≥ 40	+40.31/ -49.6	≥ 12	+13.6/ -11.8	≥8	+8.6/ -9.0
Load Capacitance	pF	5	-	5	-	5	-
Supply Voltage	V	±2.5	-	±2.5	-	±2.5	-
CMRR	dB	≥ 60	90.9	60	86	≥ 60	100
Noise (Broadband)	nV/√Hz	-	8.31	-	14.7	≤10	9.57
Noise (Broadband+Flicker)	nV/√Hz (at 100Hz)	-	60.46	-	34.93	≤ 14.14	14.19
Power	mW	-	3.84	-	1.10	-	0.83
Estimated Area	μm^2	-	14870	-	44690	-	26360
Input Devices (selected by OASYS)	-		PMOS		NMOS		PMOS

one-stage OTA. For circuit *B*, however, the unity gain frequency specification was increased from 2 to 10 MHz. OASYS determined that the OTA style was no longer the optimal style, and chose instead a rather simple two-stage Miller-compensated style to meet this new requirement. Circuit *C* then increased the gain from 60 to 90 dB, and OASYS changed topologies again, this time by cascoding many of the internal sub-blocks in this two-stage style. Notice that for each specification, the simulation results suggest that the performance is quite close to the requested performance.

As a more aggressive test, Table IV summarizes the results from OASYS designs to a set of three aggressive performance specifications. Each set of specifications stress some combination of performance parameters such that the design would be fairly difficult to complete manually. Note again that in all cases, the simulation results are quite close to the required performance values.

Table V shows how different design plans for the same circuit topology style can be used to achieve different results. One set of performance specifications was supplied, but two OASYS design plans, a default area-optimizing plan, and a newer, noise-optimizing plan, were each allowed to synthesize a circuit. Both succeeded, but notice that the noise-optimized design has slightly larger area and power, but a substantially reduced noise figure.

For all these test cases, the CPU time is very modest, usually less than 5-CPU s per circuit, and all the resulting designs are close to the specified performance parameters. The ability of OASYS to meet specifications closely occurs for three reasons. First, human designers who rely on approximate models must use fairly simple models for manual calculation. In contrast, OASYS models need only be tractable, in an analytical sense, and not necessarily simple. Thus OASYS can capture more accurately many design constraints. Second, OASYS is completely consistent in applying its models to all design situations, and updating all variables and constraints as the design evolves. Human designers often fail in such "book-keeping" tasks, taking attractive short-cuts that are sometimes incorrect. Finally, OASYS does a simple form of design optimization, fixed point numerical iteration over plan

TABLE V
INPUT SPECIFICATIONS AND SIMULATION RESULTS FOR AREA- AND NOISE-OPTIMIZING DESIGN PLANS

Parameter	Units	Spec	Plan: Area-Optimize Style: 2-stage	Plan: Noise-Optimize Style: 2-stage
			SPICE Result	SPICE Result
Gain	dB	≥ 80	81.9	82.2
Unity Gain Freq.	MHz	≥ 2	1.993	1.995
Phase Margin	Degrees	≥ 45	46	43
Slew Rate	V/μs	≥ 2	+3.33/ -3.59	+3.23/ -3.72
Load Capacitance	pF	5	-	-
Supply Voltage	V	±2.5	-	-
Noise (Broadband)	nV/√Hz	-	26.63	25.27
Noise (Broadband+Flicker)	nV/√Hz (at 100Hz)	-	102.5	48.38
Power	mW	-	22.9	24.0
Estimated Area	μm^2	-	4630	4744
Input Devices (selected by OASYS)	-		PMOS	PMOS

steps, which allows it to cope with situations for which analytical equations are intractable. Indeed, OASYS resembles human designers in this respect, except that it is more tenacious in its application of this idea. From experience, we have found this simple form of optimization to be essential in producing working designs. For example, earlier attempts at op amp design with only moderately aggressive performance goals [24] often failed to meet specifications, missing by up to 30 percent. OASYS was failing because it was being lax about verifying parasitic capacitance estimates made earlier in the design, and not iterating to correct these when early estimates differed substantially from later computed values. A more accurate parasitic model, additional verification steps and failure diagnostics, and a more aggressive iteration strategy were added to the relevant op amp design plans, which subsequently corrected this.

Fig. 9 illustrates the idea of iteration in OASYS plan execution. To show iteration over plan steps, we have uniquely numbered each plan step in the plans associated with one OASYS op amp style. As a plan executes, it visits individual steps, and we plot the plan step number as a function of the time at which that plan step was started. If no iteration occurred, these plots, called *plan traces*, would appear as simple, monotonically increasing lines, since each step is visited just once. However, in real designs, iteration occurs, and we see discontinuities on these traces: a low-numbered step early in a plan is

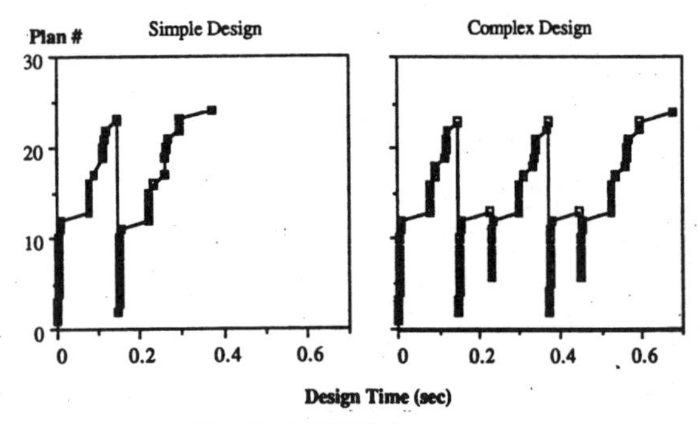

Fig. 9. OASYS plan traces.

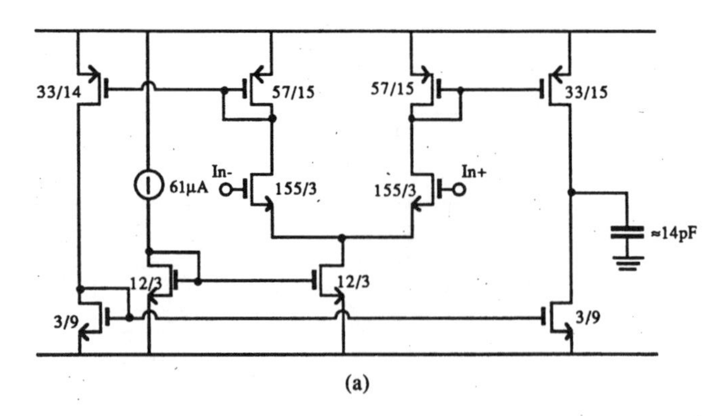

(a)

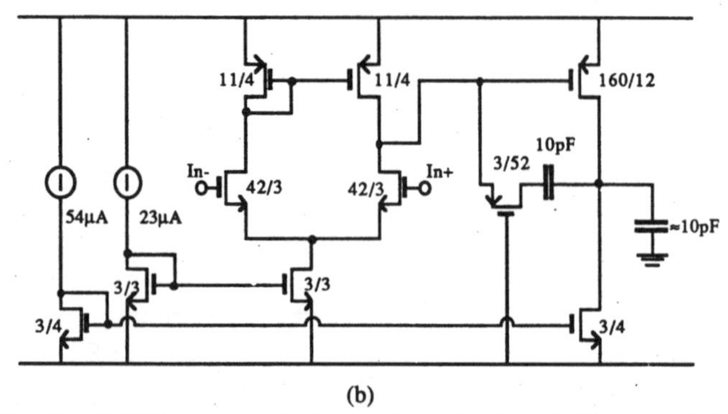

(b)

Fig. 10. OASYS-synthesized circuit schematics for (a) OTA-1 and (b) 2STAGE-1.

visited *after* a high-numbered step appearing late in a plan. Fig. 9 shows traces for two designs: a simple design, and more difficult design. Note that for the simple design shown, OASYS iterates just once (at the peak of the trace), when an early parasitic capacitance estimate failed to be sufficiently close to a final estimate. For this design, this final parasitic verification came close to the end of the design process, when even transistor-level blocks had already been designed. For the complex design shown, iteration occurs in several different places, actually across plan steps at different levels of the hierarchy. In fact, it is precisely this need for extensive iteration—to correct for unmet goals due to inaccurate heuristic guesses—that distinguishes the complex design from the simple one.

To further verify the ability of OASYS to design op amps, we have fabricated two test circuits using the MOSIS facility. The OASYS designs were laid out manually. The chip was fabricated using the 3-μm bulk CMOS process. The input performance specifications, SPICE simulation predictions, and measured performance for the two circuits, OTA-1 and 2STAGE-1, appear in Table VI; as the names suggest, these are one-stage and two-stage designs, respectively. Circuit schematics appear in Fig. 10; a photograph of the fabricated die appears in Fig. 11. There are several important observations to make about these results. First, no attempt was made after synthesis to further tune the OASYS circuit designs, either manually or with automatic optimization tools. Second, both circuits lack output buffers, and are designed for driving loads on chip only.

With respect to our measurement environment,[3] the worst deviation from specifications across the two designs is 7 percent. Given that these are first-cut designs, and the fluctuations in final circuit performance attributable to process variations, we regard this as a successful demonstration of OASYS.

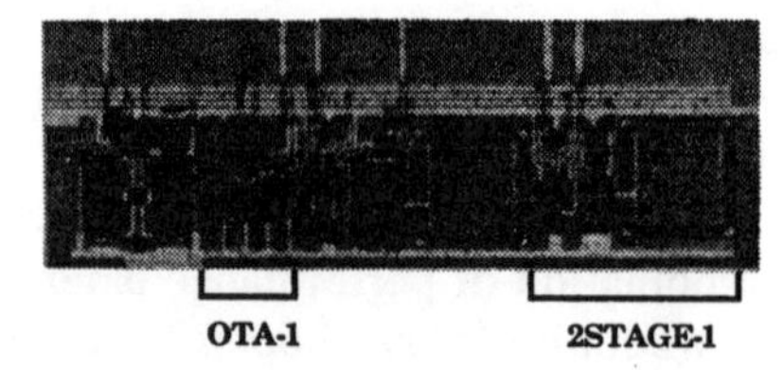

Fig. 11. Die photograph for OTA-1 and 2STAGE/1 circuits.

TABLE VI
INPUT SPECIFICATIONS AND RESULTS FOR FABRICATED OP AMPS

Parameter	Units	Circuit OTA-1 Spec	SPICE Results	Chip Results	Circuit 2Stage-1 Spec	SPICE Results	Chip Results
Gain	dB	≥ 60	60	69.5	≥ 45	45.2	48.4
Unity Gain Freq.	MHz	≥ 1.0	1.25	0.993	≥ 2.8	2.52	2.6
Phase Margin	Deg.	≥ 45	46	43	≥ 60	82	85
Load Capacitance	pF	10	10	~ 14	14	14	~ 14
Supply	V	± 2.5	± 2.5	± 2.5	± 2.5	± 2.5	± 2.5
Slew Rate	V/μS	≥ 2.0	2.1	-1.6/+2.0	≥ 2.5	2.42	-3.0/+4.5
Power	mW	≤ 3.0	0.63	0.62	-	0.56	0.56
Input Offset	mV	-	6.6	11.4	-	16	18.4

C. Exploration

The ability to synthesize complete circuits rapidly, as demonstrated in the previous section, also brings with it the ability to explore design tradeoffs rapidly. This section surveys several styles of exploration that are feasible using OASYS.

One option is simply to explore tradeoffs among performance specifications, by fixing some parameters, varying others, and examining performance variations across the range of resulting circuits. A simple example is illustrated in Fig. 12. For a two-stage design style, estimated active area and power dissipation are plotted over a range of bandwidth specifications; all other parameters remain fixed. Notice that power and estimated area both increase monotonically. This is not surprising, and simply reflects the way this two-stage design plan absorbs extra degrees

[3]Because the OTA circuit is externally compensated, it was thus designed as though it was driving a representatively small, 10-pF load. Unfortunately, available measuring equipment loaded the circuit with a minimum of 14-pF, thus degrading its measured bandwidth, phase margin, and slew rate. If we correct for this by assuming a 14-pF load in the design, the implemented circuit meets its specifications. The two-stage circuit was (fortuitously) designed for a 14-pF load.

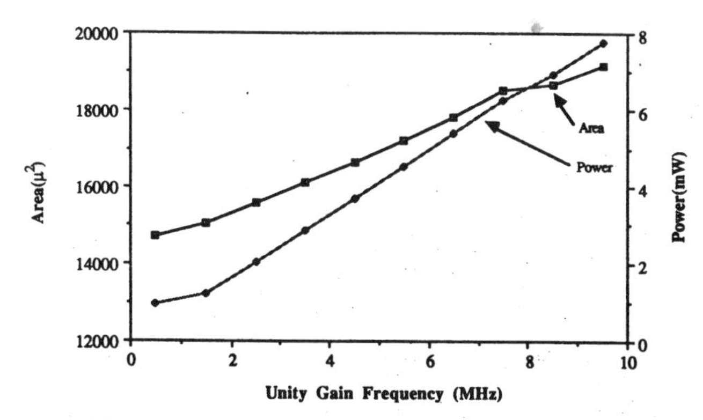

Fig. 12. Simple exploration of performance tradeoffs.

of freedom in the translation process. The plan used for this experiment optimizes primarily for power, with area as a secondary optimization criterion. However, other design strategies are also available here, e.g., OASYS has noise-optimizing plans as well. Note that the qualitative characteristics of these curves might change with a different strategy for absorbing extra degrees of freedom. For example, by optimizing for area first, it may be possible to reduce the slope of the area curve in Fig. 12, but likely at the cost of increasing the power requirements, i.e., both the slope and the actual value of the power for any given bandwidth specification would increase.

A second example of exploring performance tradeoffs is shown in Fig. 13. The two design surfaces show estimated power dissipation and estimated area over a range of bandwidth and slew rate specifications; all other parameters are fixed. Necessary changes in design styles (one-stage or two-stage), are deduced automatically by OASYS and are also shown in the figure. However, many individual designs on these surfaces have different topologies even though they share the same style: recall that OASYS refines styles hierarchically, and can reach a broad range of target topologies from any given higher level style. On the area surface, note that the one-stage styles offer considerable area savings, but cannot be coerced to work with aggressive combinations of bandwidth and slew rate specifications. On the power surface, note that at low slew rates, any increase in bandwidth requires an essentially linear increase in power; but at high slew rates, sufficient power is already available, and some increases in bandwidth can be met with negligible power increases. Although we would expect an experienced designer to have a qualitative knowledge of what these sorts of tradeoffs should yield, this experiment, which required about 3-CPU min, gives a more exact quantitative picture of these tradeoffs.

A related style of exploration is threshold hunting, in which we seek to determine the actual surface that separates feasible and infeasible designs. Fig. 14 shows, for the two-stage design style, the region of feasible gain specifications over a range of bandwidth (unity gain frequency) specifications; all other parameters are fixed. Given one known point on this curve, the procedure is to increase the bandwidth parameter to the next desired point, and then search in the vicinity of the previous maximum gain value for the new gain value above which OASYS first fails to complete the design task. In this experiment, hunting for each point on the surface required design of several individual op amps, but the total design time was only a few CPU minutes.

Finally, it is also possible to explore the impact on circuit designs of changes in the fabrication process. In OASYS, process parameters and performance specifications are treated identically. This sort of exploration is valuable, for example, in very high volume applications, where the process may be tuned to maximize circuit performance. In addition, fast synthesis gives process designers the opportunity to judge the impact, across a range of typical circuit design tasks, of introducing a modification to a fabrication process.

Although this seems to be exactly the same style of exploration as varying performance specifications, there is a subtle, but important difference here: performance parameters can be set to their respective values independently, whereas process parameters *cannot*. This is because the parameters of use during synthesis are *derived* from the interacting effects of more fundamental parameters determined by the manufacturing process. For example, the p- and n-type threshold voltages, V_T, for MOS devices are the result of several independent fabrication steps. The implication of this is that if we explore over a set of process parameters, we must be exceedingly cautious to ensure that each set of parameters we choose does not contain infeasible values. Note that this is in contrast to exploration over performance specifications: we can choose *any* set of values there, but OASYS may simply be unable to design the requested circuit.

Bearing this dependence in mind, we considered the following process exploration. For a two-stage design style with fixed performance specifications, threshold voltage (with $|V_T|$ for p and n devices assumed equal) was varied from 0.4 to 1.4 V assuming a 5-V supply; all other process parameters were fixed. This is an approximation, but a reasonable one. Circuit-level sensitivity to V_T is large for MOS devices, since device currents are directly proportional to $(V_{GS} - V_T)^2$. Thus effects dependent on the actual value of V_T dominate over other, more subtle effects that depend on nuances of the fabrication steps that lead to a particular value for V_T. Our only aim here is to examine the large-scale effects of variation in $|V_T|$. Results appear in Fig. 15, where estimated active area and power dissipation are plotted versus $|V_T|$. Note that near 0.9 V, and again to a lesser extent at 1.1 V, significant area increases occur. These result from required transistor-level topology changes (marked with crosses in Fig. 15). Specifically, at lower threshold voltages, cascoded design styles, which employ more devices but have relatively lower total area requirements, can be employed for some sub-blocks like current mirrors. However, as threshold voltage increases, the larger voltage drop of the series devices in the cascode style precludes its use. Simpler design styles with fewer devices are sub-

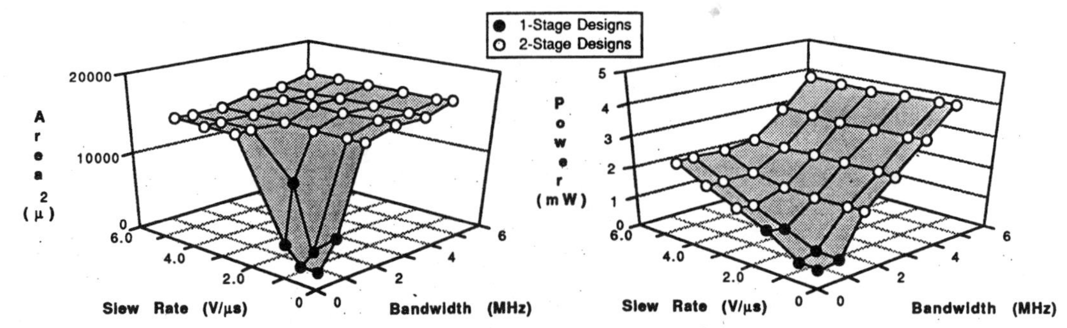

Fig. 13. Complex exploration of performance tradeoffs.

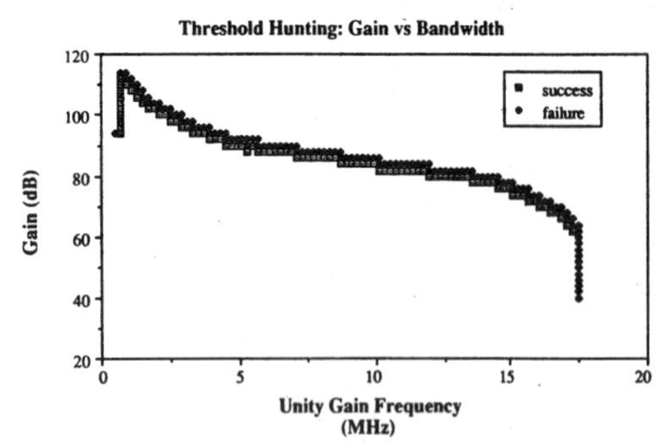

Fig. 14. Exploration by threshold hunting.

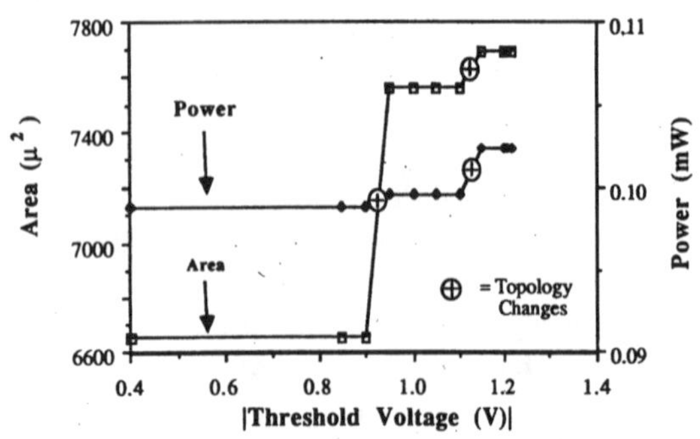

Fig. 15. Exploration of impact of process variation.

stituted, but at the cost of a large area increase to meet the performance provided by the cascode style.

V. Conclusions

We have presented a framework to support behavior-to-structure synthesis for analog circuits. The framework relies on a hierarchical representation of circuits, chooses from among a set of fixed design styles at each level of the hierarchy, and translates performance specifications down this hierarchy until individual devices are sized. Implementation mechanisms for managing the hierarchy, style selection, and translation were described. A prototype system, OASYS, can design CMOS op amps from performance specifications and process specifications; the correctness of these designs has been verified both by detailed circuit simulation, and by actual fabrication of IC's that meet their performance specifications. These experiences with the OASYS system are encouraging, and suggest that a hierarchy of selection/translation steps is a workable approach to the analog synthesis task.

Our ongoing work has concentrated on four areas: expanding the breadth and depth of the OASYS circuit repertoire, automating acquisition of design knowledge for OASYS, automating the layout of OASYS-synthesized netlists, and integrating OASYS into a complete analog design system. We have recently succeeded in adding a high-speed regenerative CMOS comparator design style to OASYS. Despite the importance of op amps in analog design, these are not sufficiently broad targets with which to fully test the generality of the OASYS synthesis framework; comparators offer a broader range of "interesting" analog problems to attack, including designs with more levels of hierarchy. Preliminary details of functional, OASYS-synthesized comparators appear in [42]; additional circuit and synthesis details appear in [39]. A longer range goal is to look at functional blocks with a few more layers of hierarchy, in particular system-level designs such as data acquisition circuits for which there is a wide range of design styles and with which we already have considerable design experience.

Knowledge acquisition is also a critical research problem for tools such as OASYS. It is essential that analog designers themselves be able to add knowledge to OASYS, without knowledge of the implementation of the OASYS framework. Toward this end, we have recently completed a preliminary design for a new version of the synthesis framework itself, called the OASYS virtual machine (OASYS-VM). OASYS-VM is structured as a runtime environment that hides all details of the machine on which it is running, and provides, in a formal and semantically precise fashion, the resources necessary to build and execute plans, plan-fixers, style selectors, and so forth. For portability, OASYS-VM is written in C. One op amp design style has already been successfully ported to OASYS-VM; details appear in [43]. OASYS-VM is actually structured as the runtime environment for a compiler. The compiler is for a language, now under development at CMU, to allow analog designers easily to write plans, plan-fixers, and so forth. Together the compiler and OASYS-VM are intended to form the basis for a knowledge acquisition environment for OASYS-style analog synthesis.

We have also recently completed a prototype layout tool for OASYS. This tool, ANAGRAM, takes OASYS-synthesized netlists and reduces them to masks. The tool operates in the style of a macrocell place and route system. Low-level primitives such as FET's and capacitors, or blocks with special matching requirements are produced using module generation techniques. An annealing algorithm then places these blocks, and an analog sensitive router connects them, while managing problems such as crosstalk, clock feedthrough, and ground loops. A variety of OASYS designs have been successfully laid out and extracted in ANAGRAM; preliminary details appear in [44]. We are currently investigating how best to couple OASYS and ANAGRAM. For example, OASYS should be able to invoke the module generators in ANAGRAM to obtain more accurate device-level parasitics during circuit synthesis.

OASYS and the other tools cited above have recently been integrated into a large analog design called ACACIA. Part of ACACIA is a graphics interface that allows analog designers to use OASYS to explore the analog design space, and also navigate graphically through tools for synthesis, schematic display, simulation, and mask layout. Discussion of a preliminary version of ACACIA appears in [45]; additional details appear in [46].

REFERENCES

[1] M. A. Sivilotti, M. A. Mahowald, and C. A. Mead, "Real-time visual computations using analog CMOS processing arrays," in *Proc. 1987 Stanford Conf. Very Large Scale Integration*, pp. 295–312, Mar. 1987.

[2] C. A. Mead, *Analog VLSI and Neural Systems*. Reading, MA: Addison-Wesley, 1987.

[3] M. R. Barbacci, "Instruction set specifications (ISPS): The notation and its applications," *IEEE Trans. Comput.*, vol. C-20, Jan. 1981.

[4] T. W. Pickerrell, "New analog capabilities on semi-custom CMOS," in *Proc. Custom Integrated Circuit Conf.*, 1983.

[5] T. Pletersek *et al.*, "Analog LSI design with CMOS standard cells," in *Proc. IEEE Custom Integrated Circuit Conf.*, 1985.

[6] G. I. Serhan, "Automated design of analog LSI," in *Proc. IEEE Custom Integrated Circuit Conf.*, 1985.

[7] R. L. Hedman *et al.*, "A device-level auto place and wire methodology for analog and digital masterslices," in *Dig. Tech. Papers, IEEE Int. Solid-State Circuits Conf.*, Feb. 1988.

[8] C. D. Kimble, A. E. Dunlop, G. F. Gross, V. L. Hein, M. Y. Luong, K. J. Stern, and E. J. Swanson, "Autorouted analog VLSI," in *Proc. Custom Integrated Circuit Conf.*, 1985.

[9] W. J. Helms and K. C. Russel, "Switched capacitor filter compiler," in *Proc. Custom Integrated Circuit Conf.*, 1986.

[10] H. Yaghutiel, A. Sangiovanni-Vincentelli, and P. R. Gray, "A methodology for automated layout of switched capacitor filters," in *Proc. Int. Conf. CAD (ICCAD86)*, Nov. 1986.

[11] P. E. Allen and E. R. Macaluso, "AIDE2: An automated analog IC design system," in *Proc. IEEE Custom Integrated Circuits Conf.*, 1985.

[12] P. E. Allen and P. R. Barton, "A silicon compiler for successive approximation A/D and D/A converters," in *Proc. IEEE Custom Integrated Circuits Conf.*, 1986.

[13] J. Kuhn, "Analog module generators for silicon compilation," *VLSI Systems Design*, May 1987.

[14] G. J. Sussman and R. M. Stallman, "Heuristic techniques in computer-aided circuit analysis," *IEEE Trans. Circuits Sys.*, vol. CAS-22, Nov. 1975.

[15] J. De Kleer and G. J. Sussman, "Propagation of constraints applied to circuit synthesis," *Circ. Theory Appl.*, vol. 8, 1980.

[16] B. C. Williams, "Qualitative analysis of MOS circuits," Master's thesis, Massachusetts Institute of Technology, 1984.

[17] G. L. Roylance, "A simple model of circuit design," Master's thesis, Massachusetts Institute of Technology, 1980.

[18] A. Ressler, "A circuit grammer for operational amplifier design," Ph.D. dissertation, Artificial Intelligence Lab., Massachusetts Institute of Technology, 1984.

[19] L. Brownston, R. Farrel, E. Kant, and N. Martin, *Programming Expert Systems in OPS5: An Introduction to Rule-Based Programming*. Reading, MA: Addison-Wesley, 1985.

[20] R. J. Bowman and D. J. Lane, "A knowledge-based system for analog integrated circuit design," in *Proc. IEEE Int. Conf. Computer-Aided Design*, 1985.

[21] F. M. El-Turky and R. A. Nordin, "BLADES: An expert system for analog circuit design," in *Proc. IEEE Int. Symp. Circuits Systems*, 1986.

[22] M. DeGrauwe, O. Nys, E. Vittoz, E. Dijkstra, J. Rijmenants, S. Cserveny, and J. Sanchez, "An analog expert design system," in *Proc. 1987 Int. Solid-State Circuit Conf.*, pp. 212–213, Feb. 1987.

[23] M. G. R. DeGrauwe *et al.*, "IDAC: An interactive design tool for analog CMOS circuits," *IEEE J. Solid-State Circuits*, vol. SC-22, Dec. 1987.

[24] R. Harjani, R. A. Rutenbar, and L. R. Carley, "A prototype framework for knowledge-based analog circuit synthesis," in *Proc. 24th ACM/IEEE Design Automation Conf.*, June 1987.

[25] ——, "Analog circuit synthesis and exploration in OASYS," in *Proc. 1988 IEEE Int. Conf. Computer Design (ICCD)*, Oct. 1988.

[26] H. Y. Koh, C. H. Sequin, and P. R. Gray, "Automatic synthesis of operational amplifiers based on analytic circuit models," in *Proc. 1987 IEEE Int. Conf. CAD*, Nov. 1987.

[27] H. Y. Koh, C. H. Sequin, and P. R. Gray, "Automatic layout generation for CMOS operational amplifiers," in *Proc. 1988 IEEE Int. Conf. CAD*, Nov. 1988.

[28] E. Berkcan, M. d'Abreu, and W. Laughton, "Analog compilation based on successive decompositions," in *Proc. 1988 ACM/IEEE Design Automation Conf.*, June 1987.

[29] A. H. Fung, D. J. Chen, Y. N. Lai, and B. J. Sheu, "Knowledge-based analog circuit synthesis with flexible architecture," in *Proc. 1988 IEEE Int. Conf. Computer Design (ICCD)*, Oct. 1988.

[30] B. Nye, A. Sangiovanni-Vincentelli, J. Spoto, and A. Tits, "DELIGHT.SPICE: An optimization-based system for the design of integrated circuits," in *Proc. Custom Integrated Circuit Conf.*, 1983.

[31] D. Thomas, "The automatic synthesis of digital systems," *Proc. IEEE*, vol. 69, Oct. 1981.

[32] C. Tong, "Toward an engineering science of knowledge-based design," *Int. J. AI Eng.*, vol. 2, no. 3, July 1987.

[33] F. D. Brewer and D. Gajski, "An expert-system paradigm for design," in *Proc. 23rd ACM/IEEE Design Automation Conf.*, 1986.

[34] D. C. Brown and B. Chandrasekaran, "Knowledge and control for a mechanical design system," *IEEE Comput.*, July 1986.

[35] R. Widlar and M. Yamatake, "A 150W Opamp," in *Dig. Tech. Papers, Int. Solid-State Circuits Conf., Feb. 1985.*

[36] M. Stefik, "Planning with constraints (MOLGEN: Part 1)," *Artificial Intelligence*, vol. 16, 1981.

[37] P. E. Friedland and Y. Iwasaki, "The concept and implementation of skeletal plans," *J. Automated Reasoning*, vol. 1, no. 2, 1985.

[38] D. W. Knapp and A. C. Parker, "A design utility manager: The ADAM planning engine," in *Proc. 23rd ACM/IEEE Design Automation Conf.*, June 1986.

[39] R. Harjani, "OASYS: A framework for analog circuit synthesis," Ph.D. dissertation, Carnegie Mellon Univ., 1989.

[40] L. W. Nagel, "SPICE2: A computer program to simulate semiconductor circuits," ERL Memo ERL-M520, Univ. Calif., May 1975.

[41] M. G. R. Degrauwe and W. M. C. Sanse, "The current efficiency of MOS transconductance amplifiers," *IEEE J. Solid-State Circuits*, vol. SC-19, June 1984.

[42] R. Harjani, R. A. Rutenbar, and L. R. Carley, "Analog circuit synthesis for performance in OASYS," in *Proc. 1988 IEEE Int. Conf. CAD*, Nov. 1988.

[43] E. Ochotta, "The OASYS virtual machine: Formalizing the OASYS analog synthesis framework," Master's thesis, Carnegie Mellon Univ., 1989.

[44] D. Garrod, R. A. Rutenbar, and L. R. Carley, "Automatic layout of custom analog cells in ANAGRAM," in *Proc. 1988 IEEE Int. Conf. CAD*, Nov. 1988.

[45] L. R. Carley and R. A. Rutenbar, "How to automate analog IC design," *IEEE Spectrum*, Aug. 1988.

[46] L. R. Carley, D. Garrod, R. Harjani, J. Kelly, T. Lim, E. Ochotta, and R. A. Rutenbar, "ACACIA: The CMU analog design system," in *Proc. 1989 IEEE Custom Integrated Circuits Conf.*, May 1989.

Ramesh Harjani (S'87–M'89) received the B.Tech degree in electrical engineering in 1982 from the Birla Institute of Technology and Science, Pilani, India, the M.Tech degree in electronics in 1984 from the Indian Institute of Technology, New Delhi, India, and the Ph.D. degree in computer engineering from Carnegie Mellon University, Pittsburgh, PA, in 1989.

He is currently working at Mentor Graphics Corporation, San Jose, CA. His research interests include design aids for analog circuit synthesis, analog circuit design, circuit simulation, and computer-aided design.

Dr. Harjani received a Best Paper Award at the 1987 IEEE/ACM Design Automation Conference. He is a member of ACM and Sigma Xi.

*

Rob A. Rutenbar (S'78–M'84) received the B.S. degree in electrical and computer engineering from Wayne State University, Detroit, in 1978, and the M.S. and Ph.D. degrees in computer engineering (CICE) from the University of Michigan, Ann Arbor, in 1979 and 1984, respectively.

In 1984, he joined the faculty of Carnegie Mellon University, Pittsburgh, PA, where he is currently an Associate Professor of Electrical and Computer Engineering, and of Computer Science. His research interests include VLSI layout algorithms, parallel CAD algorithms, and applications of automatic synthesis techniques to VLSI design, in particular, synthesis of analog integrated circuits, and synthesis of CAD software.

Dr. Rutenbar received a Presidential Young Investigator Award from the National Science Foundation in 1987. At the 1987 IEEE-ACM Design Automation Conference, he received a Best Paper Award. In 1987 he also received the George Tallman Ladd Award for Excellence in Research from the College of Engineering at Carnegie Mellon. In 1989, he received his department's Eta Kappa Nu Excellence in Teaching Award. Dr. Rutenbar is a member of ACM, Eta Kappa Nu, Sigma Xi, and AAAS.

*

L. Richard Carley (S'77–M'84) received the S.B., M.S., and Ph.D. degrees from the Massachusetts Institute of Technology, in 1976, 1978, and 1984, respectively.

He has worked for MIT's Lincoln Laboratories and has acted as a consultant in the area of analog circuit design and design automation for Analog Devices and Hughes Aircraft. In 1984, he joined Carnegie Mellon University, Pittsburgh, PA, where he is currently an Associate Professor of Electrical and Computer Engineering. His research interests are in the area of analysis, design, automatic synthesis, and simulation of mixed analog/digital systems.

Dr. Carley received the National Science Foundation Presidential Young Investigator Award in 1985, and a Best Paper Award at the 1987 Design Automation Conference.

Analog Circuit Design Optimization Based on Symbolic Simulation and Simulated Annealing

GEORGES G. E. GIELEN, STUDENT MEMBER, IEEE, HERMAN C. C. WALSCHARTS, AND WILLY M. C. SANSEN, SENIOR MEMBER, IEEE

Abstract—**A methodology for the automatic design optimization of analog integrated circuits is presented. A non-fixed-topology approach is realized as a result of combining the optimization program OPTIMAN with the symbolic simulator ISAAC. After selecting a circuit topology, the user invokes the symbolic simulator ISAAC to model the circuit. ISAAC generates both exact and simplified analytic expressions, describing the circuit's behavior. The model is then passed to the design optimization program OPTIMAN. This program is based on a generalized formulation of the analog design problem. For the selected topology, the independent design variables are automatically extracted and OPTIMAN sizes all elements to satisfy the performance constraints, thereby optimizing a user-defined design objective. The global optimization method being applied on the analytic circuit models is simulated annealing. Practical examples show that OPTIMAN quickly designs analog circuits, closely meeting the specifications, and that it is a flexible and reliable design and exploration tool.**

I. Introduction

IN PRESENT VLSI systems, the few analog circuits require an increasing share of the total design time. Therefore the automation of analog circuit design is a field of growing interest. More and more analog design strategies, frameworks, and programs are being published [1]–[7]. In each of these programs, the analog circuits are resized for every application instead of using fixed cells. This guarantees an optimal solution in terms of area, power, and overall performance, especially for ASIC's. The user first selects a particular circuit topology from the program's library. The program then sizes all circuit elements, to satisfy the performance constraints, possibly also optimizing an objective function.

The programs [2]–[4] all use some analytic description (design equations) of the circuit to perform the sizing, and all exploit expert designer knowledge to some extent. However, the main drawback of these tools is that they are fixed-topology systems, limited to a fixed set of circuit topologies (constructed hierarchically or not). None of them is able to automatically generate the appropriate analytic equations for a new topology. And the manual derivation of design equations is a tedious and error-prone job, especially for large circuits.

In IDAC [2] the circuit equations are derived after an in-depth analytic study and solved explicitly by experienced designers, to yield a straightforward step-by-step procedure. This is a time-consuming job, to be repeated for every circuit in the program's database. In addition, the circuit-specific procedures are directed towards minimum power and area consumption, not considering other performance characteristics (such as noise, distortion, etc.) to be optimized.

OASYS [3] uses design plans to successively select topologies and translate specifications downwards in a knowledge-based synthesis framework. The exploitation of hierarchy in analog circuits is a valuable feature. However, the construction of a design plan for a new topology requires an explicit representation of knowledge about the circuit behavior, heuristic design decisions, and performance trade-offs, which complicates the inclusion of new topologies. Besides, no optimization is performed on any level in the hierarchy. Only some limited form of iteration is included in the design plans.

Other expert system approaches [5]–[7] store human designer knowledge and exploit this knowledge for circuit design. However, none of them indicates how the knowledge is extracted from the human designers and how long it takes to include new topologies.

In this paper, a general approach towards automated analog circuit design is presented, which combines optimization and symbolic simulation [1]. To speed up the optimization, the circuits are characterized by an analytic model. However, whereas in OPASYN [4] the analytic model for a new topology must first be created by a good analog designer, this model is automatically generated in our approach by means of the symbolic simulator ISAAC [8]–[10]. This allows the automatic inclusion of new

Manuscript received September 4, 1989; revised January 22, 1990.
G. G. E. Gielen is a research assistant with the National Fund of Scientific Research (Belgium) at the Laboratory Elektronika, Systemen, Automatisatie, en Technologie (ESAT), Departement Elektrotechniek, Katholieke Universiteit Leuven, B-3030 Heverlee, Belgium.
H. C. C. Walscharts and W. M. C. Sansen are with the Laboratory Elektronika, Systemen, Automatisatie, en Technologie (ESAT), Departement Elektrotechniek, Katholieke Universiteit Leuven, B-3030 Heverlee, Belgium.
IEEE Log Number 9034698.

topologies. Based on the analytic model, the OPTIMAN (OPTIMization of ANalog circuits) program then sizes the circuit topology to satisfy all performance constraints. The remaining degrees of freedom are used to optimize a general cost function. The optimization is based on simulated annealing, a statistical method to search the global optimum of a function [11]. OPASYN [4], on the other hand, uses a multi-start steepest-gradient descent algorithm.

The paper is organized as follows. Section II gives a general overview of the OPTIMAN program. The implemented design methodology is discussed and the link to symbolic simulation is indicated. Section III illustrates the generation of an analytic circuit model by ISAAC. Section IV describes how analog designs are formulated as an optimization of analytic circuit models. Section V presents the simulated annealing routine and other algorithmic details of OPTIMAN. Section VI gives examples and experimental results. Concluding remarks are provided in Section VII.

II. Analog Circuit Design Methodology Based on Analytic Models

A straightforward approach to analog circuit design optimization is the combination of algorithmic optimization and numerical simulation [12]. This approach is applicable to a broad range of analog circuits and produces near optimal results. However, it is inefficient and costly in terms of CPU time, due to the full simulation at every iteration. On the other hand, a dedicated optimization program [2], [13] may be very fast, but is only applicable to one particular circuit.

The alternative, proposed in this paper, is to replace the simulation by an evaluation of analytic design equations, which model the circuit behavior. This is based on the observation that most circuit characteristics, such as the gain or the phase margin of an op amp, are influenced by a small number of parameters only. All other circuit parameters have only a negligible influence, even over a broad range of values. Moreover, the design equations are automatically derived by a circuit modeling routine, which calls the symbolic simulator ISAAC [8]–[10] to generate the ac characteristics for the circuit. ISAAC then returns expressions for transfer functions, PSRR, CMRR, impedances, noise, etc. with the frequency and the circuit elements represented by symbols. These expressions can be simplified up to a user-defined error percentage based on the order of magnitude of all elements. As shown in [10], the simplified expressions are still a good approximation for the circuit behavior. In this way, the user may trade off CPU-time consumption during optimization against accuracy, by selecting more or less accurate models.

The analog circuit design methodology introduced in this paper is schematically shown in Fig. 1. Generally, the sizing of an analog circuit from performance and technology specifications is formulated as a constrained optimization problem, based on an analytic description of the circuit behavior and simulated annealing. The user first enters the performance specifications and the technology process. Next, he hierarchically builds up a circuit topology by interacting with an advising topology selection expert system, which can also run fully automatically [14], [15]. The selected circuit topology is then looked up in the database. If it is not present, an analytic circuit model is automatically generated by calls to the symbolic simulator ISAAC, and the new model is stored in the database. As compared to fully knowledge-based systems [3], [6], the inclusion of new topologies is much easier and errors are avoided.

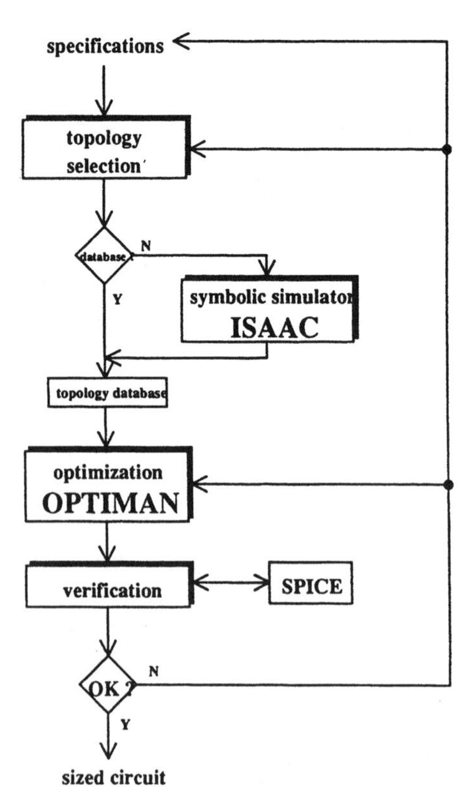

Fig. 1. Diagram of the non-fixed-topology analog design strategy.

Next, OPTIMAN sizes all circuit elements based on the analytic model and the technology data, to satisfy the performance specifications. The degrees of freedom in the design are used to minimize a user-supplied cost function. For example, biomedical applications (such as data acquisition, intelligent sensors, etc.) require minimal power consumption and minimal noise. Output buffers usually require large swing, low quiescent power, fast settling, and low distortion.

The design is then verified. The use of simplified analytic models speeds up the optimization and produces good, near-optimal results. However, a final performance check is carried out with the SPICE simulator. If deviations are noticed, the optimization routine is recalled with the deviating specifications modified. If the design is

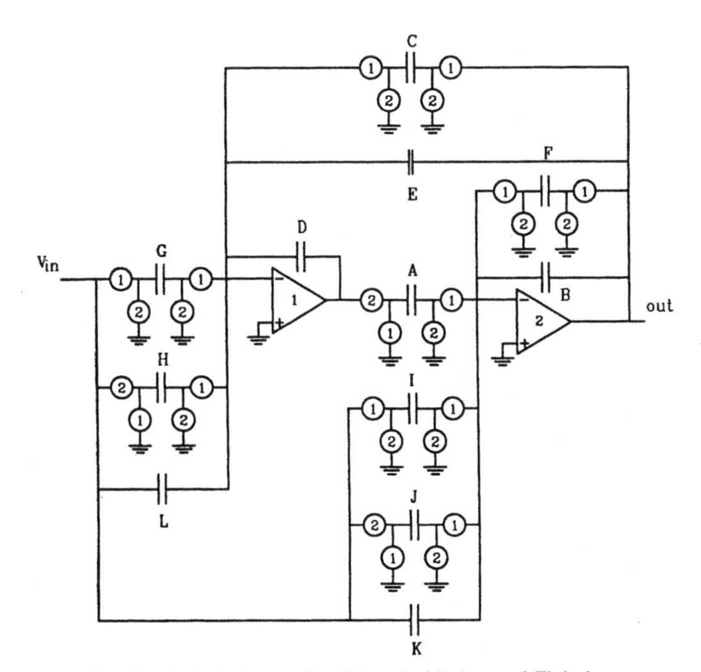

Fig. 2. Switched-capacitor biquad of Laker and Fleischer.

accepted, the circuit schematic and the device sizes are returned to the user. If no acceptable solution can be found, the program indicates which characteristics are responsible for the failure. In the future, an automatic backtrace is planned for the topology selection routine, in order to automatically select a new topology.

Summarizing, the main features of the new design methodology and the OPTIMAN program are:

1) it is independent of circuit topology and technology;
2) it allows the automatic inclusion of new topologies, due to the link with the symbolic simulator ISAAC;
3) it is general and robust: the program can optimize any circuit that is described by an analytic model. This will be illustrated by the examples of Section VI. Moreover, the use of simulated annealing guarantees finding the global optimum in moderate CPU times, as well as for circuits of practical size;
4) it is very flexible in the definition of the goal function and the constraints: OPTIMAN enables the user to optimize and explore a circuit in all dimensions of the analog design space.

III. Analog Circuit Modeling with Symbolic Simulator ISAAC

In the design methodology presented in the previous section, the symbolic simulator ISAAC [8]–[10] is used for the analytic modeling of the analog circuit. ISAAC generates the fully symbolic expressions that model the circuit behavior during the optimization. This modeling ability is now illustrated for the switched-capacitor biquad of Laker and Fleisher, shown in Fig. 2. Several analyses with ISAAC on this circuit are presented.

A. Signal Transfer Functions

With 0.8 s of CPU, the transfer function V_{out}/V_{in} is found:

$$\frac{\begin{array}{l} DK - AL - AH + DJ \\ +Z\ \ (AG - 2DK + AL - DI - DJ) \\ +Z^2\ \ D(I+K) \end{array}}{\begin{array}{l} -BD + AE \\ +Z\ \ (-AC + 2BD - AE + DF) \\ -Z^2\ \ D(B+F) \end{array}}. \quad (1)$$

However, a more readable expression is obtained if all capacitors are ratioed by the corresponding integration capacitor, B or D, respectively. This is done with a mixed symbolic–numeric analysis [10] and putting B and D equal to 1. All symbols in the expression are now to be interpreted as capacitor ratios:

$$\frac{\begin{array}{l} K - AL - AH + J \\ +Z\ \ (AG - 2K + AL - I - J) \\ +Z^2\ \ (I+K) \end{array}}{(-1+AE) + Z(-AC + 2 - AE + F) - Z^2(1+F)}. \quad (2)$$

Formulas (1) and (2) are valid if the input signal is sampled at clock phase Φ_1 and held during the full clock period, and if the output is observed during Φ_1. More generally, if input and output are considered in phases i and j, respectively, the transfer function $H_{ij}(z)$ is obtained. For example, $H_{21}(z)$ is given by

$$\frac{(AEK - AL - AH + J) + Z(-AEK + AL - J)}{(-1+AE) + Z(-AC + 2 - AE + F) - Z^2(1+F)}. \quad (3)$$

Finally, any node voltage can be taken as output variable. For instance, taking the output of op amp 1 as the circuit output yields $H'_{12}(z)$:

$$\frac{\begin{array}{l} +Z\ \ (-EFK + EI + CK - G) \\ +Z^2\ \ (FG + EFK - CK - EI - CI + G) \end{array}}{(-1+AE) + Z(-AC + 2 - AE + F) - Z^2(1+F)}. \quad (4)$$

B. Impedances

Impedances can be calculated in the s and z domains [10]. With a z-domain analysis, charges flowing in one clock phase as a result of voltages applied in other phases can be calculated. With an s-domain analysis of a phase,

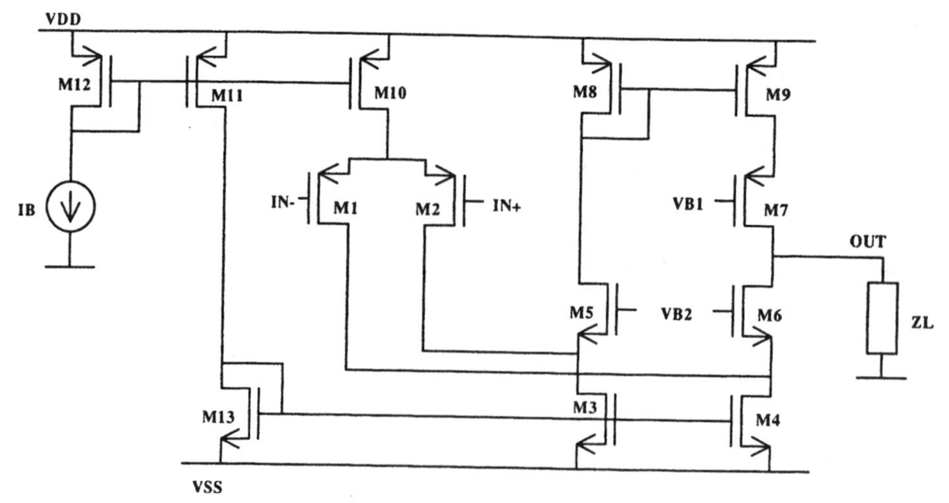

Fig. 3. CMOS folded-cascode OTA.

the influence of switch on-resistances and op-amp gain–bandwidth on impedances can be simulated. Another example of an s-domain impedance is the load capacitance of op amp 1 of Fig. 2 in phase 1:

$$\frac{(C+E+G+H+L)D}{C+E+G+H+L+D}. \tag{5}$$

C. PSRR

The rejection of capacitive power feed-in on the virtual nodes of op amps 1 and 2 in Fig. 2 is given by

$$\frac{\begin{array}{l} K-AL-AH+J \\ +Z\ \ (AG-2K+AL-I-J) \\ +Z^2\ (I+K) \end{array}}{(-AN+M)+Z(AN-2M)+Z^2M} \tag{6}$$

where M and N are the feed-in capacitors for op amps 1 and 2, respectively.

D. Noise

In phase 2, the s-domain transfer of the noise spectrum of one of the switches connected between capacitor G and ground in Fig. 2, to a voltage across capacitor G is given by

$$\frac{GON}{GON+S2G}. \tag{7}$$

This transfer causes an error charge on capacitor G, which is then transferred to the output in the z domain according to

$$\frac{AZ}{(-1+AE)+Z(-AC+2-AE+F)-Z^2(1+F)}. \tag{8}$$

Doing this for all switches and both op amps yields the total output noise of the biquad in both phases.

E. Other Analyses

More complicated analyses have been performed. For example, the influence of the finite op-amp gains can be calculated in the z domain. In this case, the lengthy exact result must be approximated. CPU times for all above analyses are below 1 s on a one-MIPS machine.

Of course, ISAAC can also be used for the modeling of operational amplifiers by generating symbolic expressions for the differential gain, CMRR, PSRR, input and output impedance, input-referred noise, etc. Detailed examples can be found in [16].

Another application for the symbolic simulator is the macromodeling of analog building blocks for system-level (eventually mixed analog–digital) simulation. For this application, the design of the building blocks has already been completed. Hence, all circuit parameters are determined and the circuit can be characterized by analytic expressions containing only the frequency variable s or z as a symbol.

IV. The Design Formulation in OPTIMAN

The optimization in OPTIMAN is based on an analytic circuit model. The ac model obtained with ISAAC is part of the OPTIMAN model, which also contains large-signal information. Actually, the model contains a description of the circuit topology, a summary of the independent design variables, relationships between dependent and independent design variables, design equations relating the circuit performance to the design variables, additional analog expert knowledge about the circuit, and other design constraints.

To illustrate these ideas, some simplified design equations for the CMOS folded-cascode OTA of Fig. 3 are

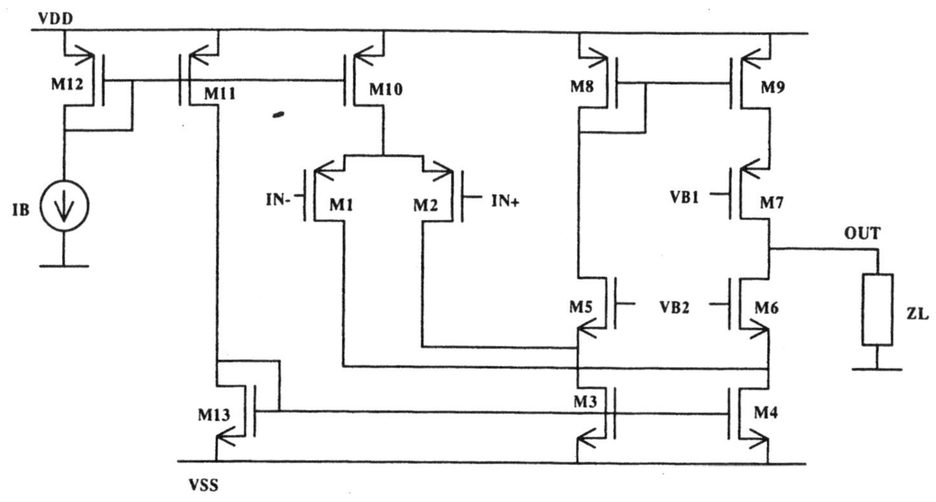

Fig. 3. CMOS folded-cascode OTA.

the influence of switch on-resistances and op-amp gain–bandwidth on impedances can be simulated. Another example of an s-domain impedance is the load capacitance of op amp 1 of Fig. 2 in phase 1:

$$\frac{(C+E+G+H+L)D}{C+E+G+H+L+D}. \tag{5}$$

C. PSRR

The rejection of capacitive power feed-in on the virtual nodes of op amps 1 and 2 in Fig. 2 is given by

$$\frac{\begin{matrix} K-AL-AH+J \\ +Z\ \ (AG-2K+AL-I-J) \\ +Z^2\ (I+K) \end{matrix}}{(-AN+M)+Z(AN-2M)+Z^2M} \tag{6}$$

where M and N are the feed-in capacitors for op amps 1 and 2, respectively.

D. Noise

In phase 2, the s-domain transfer of the noise spectrum of one of the switches connected between capacitor G and ground in Fig. 2, to a voltage across capacitor G is given by

$$\frac{GON}{GON+S2G}. \tag{7}$$

This transfer causes an error charge on capacitor G, which is then transferred to the output in the z domain according to

$$\frac{AZ}{(-1+AE)+Z(-AC+2-AE+F)-Z^2(1+F)}. \tag{8}$$

Doing this for all switches and both op amps yields the total output noise of the biquad in both phases.

E. Other Analyses

More complicated analyses have been performed. For example, the influence of the finite op-amp gains can be calculated in the z domain. In this case, the lengthy exact result must be approximated. CPU times for all above analyses are below 1 s on a one-MIPS machine.

Of course, ISAAC can also be used for the modeling of operational amplifiers by generating symbolic expressions for the differential gain, CMRR, PSRR, input and output impedance, input-referred noise, etc. Detailed examples can be found in [16].

Another application for the symbolic simulator is the macromodeling of analog building blocks for system-level (eventually mixed analog–digital) simulation. For this application, the design of the building blocks has already been completed. Hence, all circuit parameters are determined and the circuit can be characterized by analytic expressions containing only the frequency variable s or z as a symbol.

IV. The Design Formulation in OPTIMAN

The optimization in OPTIMAN is based on an analytic circuit model. The ac model obtained with ISAAC is part of the OPTIMAN model, which also contains large-signal information. Actually, the model contains a description of the circuit topology, a summary of the independent design variables, relationships between dependent and independent design variables, design equations relating the circuit performance to the design variables, additional analog expert knowledge about the circuit, and other design constraints.

To illustrate these ideas, some simplified design equations for the CMOS folded-cascode OTA of Fig. 3 are shown below:

$$A_0 = \frac{g_{m1}(g_{m6}+g_{mb6})g_{m7}}{g_{m7}g_{o6}(g_{o4}+g_{o1})+(g_{m6}+g_{mb6})g_{o7}g_{o9}}$$

$$\mathrm{GBW} = \frac{g_{m1}}{2\pi C_L}$$

$$\mathrm{PM} = 90° - bgtg\left(\frac{g_{m1}(C_{DB1}+C_{DB3}+C_{GS6}+C_{SB6})}{(g_{m6}+g_{mb6})C_L}\right)$$

$$\mathrm{POWER} = 2I_B \cdot (V_{DD}-V_{SS})$$

$$\mathrm{SR} = \frac{I_B}{C_L}$$

$$V_{\mathrm{out,max}} = V_{B1}+V_{T7}$$

$$V_{\mathrm{out,min}} = V_{B2}-V_{T6}$$

$$V_{B1}+V_{GS7} \leqslant V_{DD}-(V_{GS}-V_T)_9$$

$$V_{B2}-V_{GS6} \geqslant V_{SS}+(V_{GS}-V_T)_4$$

$$V_{os,\mathrm{syst.}} = \frac{g_{o9}}{g_{m1}}[V_{GS8}-(V_{DD}-(V_{B1}+V_{GS7}))]$$

$$dv_n^2 = \frac{g_{m1}}{2C_L}\left[dv_{n1}^2 + dv_{n3}^2\left(\frac{g_{m3}}{g_{m1}}\right)^2 + dv_{n8}^2\left(\frac{g_{m8}}{g_{m1}}\right)^2\right]. \quad (9)$$

In general, more complicated expressions are included in the analytic model.

The performance specifications are treated in two different ways in OPTIMAN.

1) Performance objectives are incorporated in the goal function. They are minimized (or maximized). For op-amp designs this function can be the power consumption, the chip area, the noise, the frequency range (bandwidth), or any weighted combination of these four basic design dimensions. The user then has to supply the weighting coefficients and the maximum (or minimum) values allowed. For example, a mixed power–area minimization may be performed with maximum values of 10 mW and 10 mm^2.

2) For other performance parameters, an inequality or an equality constraint can be specified. For example, it can be required to have a gain A_0 of minimum 70 dB or of exactly 6 dB, or a systematic offset voltage $V_{os,\mathrm{syst}}$ between -1 and $+1$ mV.

All node voltages, currents, and element values in the circuit are taken as design variables. These variables are automatically reduced to an independent set, by means of general constraints (such as Kirchoff's laws), circuit specific constraints (such as matching information), and designer constraints (such as offset reduction rules). For the design of operational amplifiers, a useful set of design variables is the variables which determine the dc operating point of the circuit: bias currents, transistor saturation voltages, bias voltages (and transistor widths or lengths). For example, a possible set of independent design variables for the CMOS folded-cascode OTA of Fig. 3 is: I_1, $(V_{GS}-V_T)_1$, $(V_{GS}-V_T)_3$, $(V_{GS}-V_T)_5$, $(V_{GS}-V_T)_7$, $(V_{GS}-V_T)_8$, $(V_{GS}-V_T)_{10}$, V_{B1}, and V_{B2}. From this dc operating point, all transistor dimensions W and L can be calculated, using a general transistor model. This transistor model may be a first-order model or a more sophisticated one, and is stored as a separate program module. For example, the gain–bandwidth is determined by g_{m1}. However, the relationship between g_{m1}, I_1, and $(V_{GS}-V_T)_1$ is part of the transistor model. The optimization routine operates independently of this transistor model.

V. Algorithmic Aspects

An analog circuit is regarded as a multidimensional space, in which the independent variables are the dimensions. The valid design space for a particular application consists of those points which satisfy the design constraints. OPTIMAN searches in this valid design space for the point which optimizes the goal function.

The optimization algorithm is simulated annealing [11]. This is a general and robust method, based on random move generation and statistical move acceptance. Moves which lower the goal function are always accepted. Moves which increase the goal function are statistically accepted. The acceptance probability of uphill moves is large in the beginning of the optimization, when the controlling temperature is high, but decreases with decreasing temperature.

Simulated annealing is a general and robust optimization method. It does not rely on any rules nor on special properties of the goal function and its derivatives. Since it also allows uphill moves, the probability of getting trapped in a local optimum is minimized and a solution close to the global optimum will be found at the expense of a large number of function evaluations. The simulated annealing implementation used is SAMURAI [17], an efficient kernel with fully adaptive temperature scaling and move range limitation and a novel inner-loop criterion. These features keep the CPU-time consumption still acceptable for large circuits.

The general optimization loop is then as follows. The program statistically selects new values for the independent variables in a "move range" around the present value. It calculates all dependent variables, checks if the corresponding design satisfies all boundary conditions and constraints, calculates the goal function, and statistically accepts or rejects the new state. This loop is executed until convergence occurs at lower temperatures, while the move range is gradually decreased with decreasing temperature.

At the initialization, the independent variables are gridded over their initial range. An important factor to reduce the CPU time is the limitation of this range. A program has been developed, that derives simple initial boundaries for the variable values from a graph representation of the relationships between specifications and variables and from the numerical specification values.

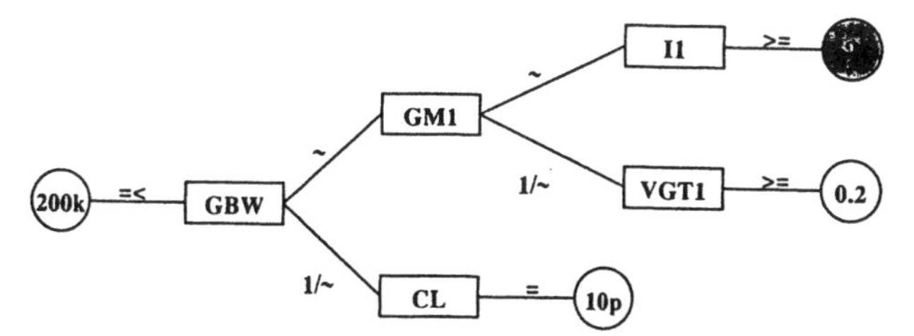

Fig. 4. Graph representation of the relationships between specifications and design variables.

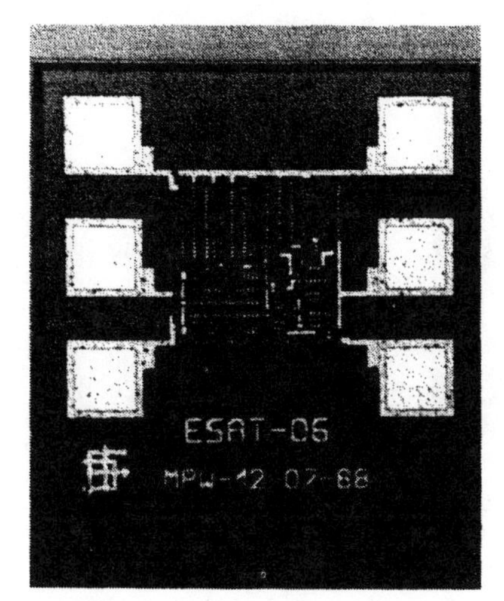

Fig. 5. Die photograph of the realized folded-cascode op amp.

This is illustrated in Fig. 4 for the gain–bandwidth of the folded-cascode OTA. The requirements $\mathrm{GBW} \geqslant 200$ kHz, $C_L = 10$ pF, and $(V_{GS} - V_T)_1 \geqslant 0.2$ V impose the following lower limit for the current I_1:

$$I_1 \geqslant \pi \ \mathrm{GBW} \ C_L(V_{GS} - V_T)_1 = 1.9 \ \mu\mathrm{A}. \tag{10}$$

The OPTIMAN program is written in VAX Pascal and runs on VAX under VMS. The topology database contains models for several frequently used CMOS op amps and their subblocks, but can easily be extended.

VI. Practical Examples

Example 1: CMOS Folded-Cascode OTA

Consider the CMOS folded-cascode OTA of Fig. 3. The technology used is 3-μm CMOS N-well. A sample has been designed towards minimum power consumption. Fig. 5 shows the die photograph. The design specifications are compared to OPTIMAN data, SPICE simulations, and measurement results in Table I. Notice the close correspondence between all values. Table II compares the final optimization variables for power minimization, noise minimization, and gain–bandwidth maximization of the folded-cascode OTA. All these designs require less than 1 min on a VAX750.

TABLE I
Comparison of Specifications, OPTIMAN Data, SPICE Simulations, and Measurement Results for a Minimal Power Design of the Folded-Cascode OTA

specification	spec	OPTIMAN	SPICE	measurement
GBW [kHz]	> 200	223	224	220
power [mW]	< 1	0.140	0.142	0.15
gain [dB]	> 60	83	97	86
phase margin [°]	> 60	89	86	83
slew rate [V/μs]	> 0.13	0.28	0.28	0.22
noise [μV_{RMS}]	< 20	16.1	17.6	17.0
output range [V]	> 1	1.0	1.07	1.12
input range [V]	> 1	1.0	1.09	1.14
offset [mV]	< 0.5	0.14	0.1	0.2

TABLE II
Optimization Variables for Power Minimization, Noise Minimization, and GBW Maximization of the Folded-Cascode OTA

variable	power	noise	GBW
I_1	13 μA	323 μA	500 μA
$(V_{GS}-V_T)_1$	0.20 V	0.20 V	0.20 V
$(V_{GS}-V_T)_3$	0.36 V	0.50 V	0.43 V
$(V_{GS}-V_T)_5$	0.20 V	0.27 V	0.27 V
$(V_{GS}-V_T)_7$	0.27 V	0.20 V	0.35 V
$(V_{GS}-V_T)_8$	0.34 V	0.50 V	0.33 V
V_{B1}	0.3 V	0.2 V	0.1 V
V_{B2}	-0.1 V	-0.2 V	-0.6 V

TABLE III
Comparison of Capacitor Values (in picofarads) for a Switch Noise Minimization of the Laker and Fleischer Biquad by OPTIMAN and BISON

capacitor	OPTIMAN	BISON
A	5.329	5.493
B	13.174	12.250
C	16.489	18.090
D	19.093	20.420
E	14.929	10.130
F	2.015	3.812
G	15.115	16.420
H	0.150	0
I	1.297	1.654
J	0.019	0
K	11.371	11.740
L	0.877	0
noise	19.49 μV	19.43 μV

Example 2: Switched-Capacitor Biquad of Laker and Fleischer

The program's generality is illustrated by the switch noise minimization of the switched-capacitor biquad of Laker and Fleischer, shown in Fig. 2. The analytic description can be found in [13]. The design variables are the capacitor values, with the optimization variables being a subset (since a fixed transfer function has to be realized). Table III compares the resulting capacitor values from OPTIMAN and BISON [13] for the same filter function. BISON obtains these results in a shorter time (a few seconds), but is dedicated to this single biquad topology and relies on expert rules. OPTIMAN, on the other hand, is general and performs a global optimization. These results demonstrate the usefulness and reliability of the program, and also verify the rules used in BISON. Note that the capacitor values in Table III are optimization results (with a grid of 1 fF) and still have to be rounded off for any physical capacitor implementation.

VII. Conclusions

An efficient non-fixed-topology analog design tool, OPTIMAN, has been developed. The program sizes all circuit elements in order to satisfy the performance constraints, thereby optimizing a user-defined design objective. OPTIMAN is based on analytic circuit models and applies simulated annealing for the global optimization. Models for new topologies are automatically generated by means of the symbolic simulator ISAAC.

OPTIMAN has successfully been applied to the design of CMOS op amps and switched-capacitor circuits. The examples demonstrate the usefulness and reliability of the approach. Inexperienced op-amp designers can easily use the program, fully in batch mode. Expert circuit designers can easily insert new topologies.

Acknowledgment

The authors are grateful to C. Verdonck and G. Van Hecke for their contributions to the OPTIMAN program, and to Philips Industries, The Netherlands, for the logistic support.

References

[1] G. Gielen, H. Walscharts, and W. Sansen, "Analog circuit design optimization based on symbolic simulation and simulated annealing," in *Proc. ESSCIRC*, 1989, pp. 252–255.
[2] M. Degrauwe *et al.*, "IDAC: An interactive design tool for analog integrated circuits," *IEEE J. Solid-State Circuits*, vol. SC-22, no. 6, pp. 1106–1116, Dec. 1987.
[3] R. Harjani, R. A. Rutenbar, and L. R. Carley, "OASYS: A framework for analog circuit synthesis," *IEEE Trans. Computer-Aided Design*, vol. 8, no. 12, pp. 1247–1266, Dec. 1989.
[4] H. Y. Koh, C. H. Séquin, and P. R. Gray, "Automatic synthesis of operational amplifiers based on analytic circuit models," in *Proc. ICCAD*, 1987, pp. 502–505.
[5] E. Berkcan, M. d'Abreu, and W. Laughton, "Analog compilation based on successive decompositions," in *Proc. Design Automation Conf.*, 1988, pp. 369–375.
[6] F. El-Turky and E. E. Perry, "BLADES: An artificial intelligence approach to analog circuit design," *IEEE Trans. Computer-Aided Design*, vol. 8, no. 6, pp. 680–692, June 1989.
[7] R. J. Bowman and D. J. Lane, "A knowledge-based system for analog integrated circuit design," in *Proc. ICCAD*, 1986, pp. 210–212.
[8] W. Sansen, G. Gielen, and H. Walscharts, "Symbolic simulator for analog circuits," in *ISSCC Dig. Tech. Papers*, 1989, pp. 204–205.
[9] H. Walscharts, G. Gielen, and W. Sansen, "Symbolic simulation of analog circuits in *s*- and *z*-domain," in *ISCAS Dig. Tech. Papers*, 1989, pp. 814–817.
[10] G. E. Gielen, H. C. C. Walscharts, and W. M. C. Sansen, "ISAAC: A symbolic simulator for analog integrated circuits," *IEEE J. Solid-State Circuits*, vol. 24, no. 6, pp. 1587–1597, Dec. 1989.
[11] S. Kirkpatrick, C. D. Gelatt Jr., and M. P. Vecchi, "Optimization by simulated annealing," *Science*, vol. 220, no. 4598, pp. 671–680, May 13, 1983.
[12] W. Nye *et al.*, "DELIGHT.SPICE: An optimization-based system for the design of integrated circuits," *IEEE Trans. Computer-Aided Design*, vol. 7, no. 4, pp. 501–519, Apr. 1988.
[13] H. Walscharts, L. Kustermans, and W. Sansen, "Noise optimization of switched-capacitor biquads," *IEEE J. Solid-State Circuits*, vol. SC-22, no. 3, pp. 445–447, June 1987.
[14] G. Gielen, K. Swings, and W. Sansen, "An intelligent design system for analogue integrated circuits," in *Proc. European Design Automation Conf.*, 1990.
[15] K. Swings, G. Gielen, and W. Sansen, "An intelligent analog IC design system based on manipulation of design equations," in *Proc. Custom Integrated Circuits Conf.*, 1990.
[16] G. Gielen, "Computer-aided design techniques for the automatic synthesis of analogue integrated circuits," Ph.D. dissertation, Katholieke Universiteit Leuven, Heverlee, Belgium, to be published in 1990.
[17] F. Catthoor, J. Vandewalle, and H. De Man, "SAMURAI: A general and efficient simulated-annealing schedule with fully adaptive annealing parameters," *Integration, VLSI J.* (North Holland), vol. 6, no. 2, pp. 147–178, July 1988.

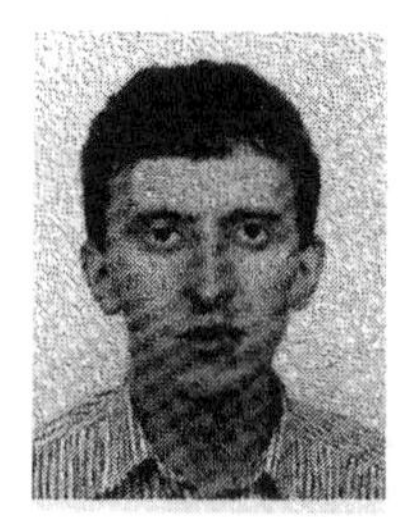

Georges G. E. Gielen (S'87) was born in Heist-op-den-Berg, Belgium, on August 25, 1963. He received the engineering degree in electronics from the Katholieke Universiteit Leuven, Heverlee, Belgium, in 1986.

In the same year he was appointed by the National Fund of Scientific Research (Belgium) as a Research Assistant at the Laboratory Elektronika, Systemen, Automatisatie, en Technologie (ESAT), Katholieke Universiteit Leuven, where he is now working towards the Ph.D. degree in electronics. His research interests are in design of analog integrated circuits and in analog design automation (synthesis, optimization, and layout).

Herman C. C. Walscharts was born in Antwerpen, Belgium, on July 22, 1961. He received the engineering degree in electronics from the Katholieke Universiteit Leuven, Heverlee, Belgium, in 1984.

In the same year he joined the Laboratory Elektronika, Systemen, Automatisatie, en Technologie (ESAT), Katholieke Universiteit Leuven, where he is currently working towards the Ph.D. degree in electronics. In the period from 1985 to 1988 he was supported by the Belgian I.W.O.N.L. Recently, he joined Bell Telephone M.C., Belgium. His interests are in design of analog integrated circuits and in analog design automation.

Willy M. C. Sansen (S'66–M'72–SM'86) was born in Poperinge, Belgium, on May 16, 1943. He received the engineering degree in electronics from the Katholieke Universiteit Leuven, Heverlee, Belgium, in 1967 and the Ph.D. degree in electronics from the University of California, Berkeley, in 1972.

In 1968 he was employed as an Assistant at the Katholieke Universiteit Leuven. In 1971 he was employed as a Teaching Fellow at the University of California. In 1972 he was appointed by the N.F.W.O. (Belgian National Fund for Scientific Research) as a Research Associate at the Laboratory Elektronika, Systemen, Automatisatie, en Technologie (ESAT), Katholieke Universiteit Leuven, where he has been full Professor since 1981. Since 1984 he has been the Head of the Department of Electrical Engineering. In 1978 he spent the winter quarter as a Visiting Assistant Professor at Stanford University, Stanford, CA; in 1981 at the Technical University of Lausanne; and in 1985 at the University of Pennsylvania, Philadelphia. His interests are in device modeling, in design of integrated circuits, and in medical electronics and sensors.

Dr. Sansen is a member of the Koninklijke Vlaamse Ingenieurs Vereniging (K.V.I.V.), the Audio Engineering Society (A.E.S.), the Biotelemetry Society, and Sigma Xi. In September 1969 he received a CRB Fellowship from the Belgian American Educational Foundation, in 1970 a G.T.E. Fellowship, and in 1978 a NATO Fellowship.

STAIC: An Interactive Framework for Synthesizing CMOS and BiCMOS Analog Circuits

J. Paul Harvey, *Student Member, IEEE*, Mohamed I. Elmasry, *Fellow, IEEE*, and Bosco Leung, *Member, IEEE*

***Abstract*—STAIC is an interactive design tool that synthesizes CMOS and BiCMOS analog integrated circuits that conform to specified performance constraints. STAIC features an input modeling language for entering hierarchical circuit descriptions and a symbolic/numeric solve unit that dynamically integrates analytical model equations across hierarchical boundaries. Output of the solver is a "flattened" homogeneous model that is customized to a user specified topology and set of performance specifications. The output is thus tailored for optimization and other numerically intense design exploration procedures. All model descriptions include physical layout so that important net parasitics may be fully accounted for during design evaluation. Synthesis proceeds via a *successive solution refinement* methodology. Multilevel models of increasing sophistication are used by scan and optimization modules to converge to what is likely a globally optimal solution. Design experiments have shown that STAIC can produce satisfactory results.**

I. Introduction

THE ANALOG design medium has proven to be a formidable domain to its digital counterpart for many high performance integrated circuit applications. However, despite its many advantages, the analog design medium remains largely unexploited due to the relative lack of analog design aids. The problem of analog synthesis may be defined as one of selecting a circuit topology, sizing the composite components, and laying out the structure in a manner that realizes the required functionality and meets the desired performance criteria. While the objective is easily stated, the design of analog integrated structures that meet such criteria is not. To appreciate the complexity of analog designs and the automation thereof, it is constructive to compare analog and digital design mediums. The following observations may be made:

1) In the digital domain, there are typically only three main performance measures of interest, namely area, power, and delay. In contrast, performance attributes typical of analog circuits are numerous and depend on the functional block of interest.

2) The digital signal may be simply characterized as having two unique logic states. A consequence of these binary discrete levels is that the digital signal is subject to formal mathematical treatment using modulo 2 arithmetic and state variables. The analog signal, however, can carry information in a variety of forms. The amplitude of the incoming signal carries vital information in the case of an analog-to-digital converter. Alternatively, phase lock loop systems process phase information and are indifferent to signal amplitudes.

3) In the digital domain, system performance may be readily expressed as a linear function of sub-system performance with little loss in modeling accuracy. In the analog domain, system performance is typically a nonlinear function of lower level attributes.

4) With the inclusion of simple back annotation schemes, digital design methods such as standard cell and gate array permit layout to be considered independent of circuit topology selection and component sizing. In the analog domain, net parasitics can play a dominant role in determining attributes of high performance analog blocks. The use of analog standard cell libraries is not practical due to the many divergent performance measures typical of analog systems. Critical layout parasitics must, therefore, be considered commensurate with topology selection and component sizing.

Several CAD tools have emerged from both industry and academia which address various aspects of the analog synthesis problem. Each of their synthesis methodologies can be broadly contrasted by placing them on a one-dimensional graph with two extreme points (Fig. 1). At the leftmost extreme lie algorithmic-based synthesis methodologies that are characterized as being numerically intensive and robust in the sense that they make few assumptions about the behavior of the circuit undergoing design. DELIGHT.SPICE [1] is one such example that epitomizes the algorithmic approach to circuit synthesis. DELIGHT.SPICE is a software package dedicated to optimizing component values of fixed topology circuits. DELIGHT.SPICE employs an enhanced version of SPICE2G6 to provide transient, ac, and dc operating point values and their corresponding sensitivities. While this approach can yield accurate results, run times are excessive, thus limiting its practicality to component tweaking during the final design stage.

The OPASYN [2] system assumes a *synthesis by analysis* approach similar to that of DELIGHT.SPICE. Optimization, however, is based strictly on analytical models rather than simulation. Run times are therefore consider-

Manuscript received April 18, 1991; revised December 9, 1991. This work was supported in part by MICRONET, NSERC, and ITRC. This paper was recommended by Associate Editor R. Brayton.

The authors are with the VLSI Group, Department of Electrical and Computer Engineering, University of Waterloo, Ont., Canada N2L 3G1.

IEEE Log Number 9200813.

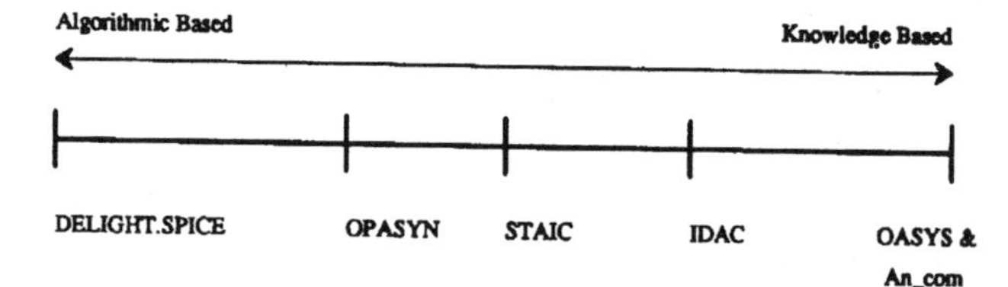

Fig. 1. Contrast of CAD synthesis methodologies.

ably faster. A course grid point search conducted over a domain space defined by the independent variable boundaries yields an initial solution for subsequent optimization. Layout in OPASYN is based on a single fixed floorplan arrangement which captures all special analog layout considerations.

A drawback of the OPASYN approach is that library circuit topologies and layout styles are nonhierarchical. As a result, the swapping and subsequent evaluation of functionally equivalent sub-blocks is not permitted. A CMOS differential pair in an opamp, for example, cannot be exchanged for a bipolar differential pair. A complete and separate opamp description must be present for each feasible combination of sub-blocks.

The IDAC [3] system uses predefined *synthesis strategies* to efficiently size components of fixed library topologies. In this context, a *synthesis strategy* is defined as a collection of analytical synthesis equations sequentially ordered to form a single-pass algorithm. While rapid circuit evaluation is possible, considerable effort is required by the expert analog designer (super-user) to develop the *strategies*. For a given circuit, multiple strategies are required to account for different combinations of input specifications.

Layout in the IDAC system is produced independently by ILAC [4], a complementary analog place and route package. The decoupling of layout from the rest of the design phase makes it difficult to model critical net parasitics and trade off electrical performance for smaller silicon real estate.

At the opposite end of the spectrum lie analog CAD tools whose synthesis methodologies are knowledge-based. The OASYS [5] compiler is one such example. The OASYS methodology employs *hierarchical decomposition with backtracking* to effect circuit synthesis. Heuristics and simple model equations referred to as "plans" are executed to recursively translate higher level specifications to lower level sub-blocks. While this method permits the swapping of functionally equivalent sub-blocks, the knowledge-based approach is limited to finding feasible rather than globally optimal circuit solutions.

The An_com [6] system adopts a similar methodology as OASYS, but also considers floorplan and routing at each decompositional step. Real estate and parasitic modeling are, therefore, accommodated. In contrast, layout in OASYS is performed by a separate analog place and route package [7] similar to ILAC.

Each of the synthesis tools presented above have made important advances to further the state of the art of analog synthesis. The IDAC package, for example, has demonstrated that analytic modeling can be an effective technique to characterize circuit performance. The OASYS package has shown the importance of hierarchy in order to accommodate circuit topology variations in an efficient manner. OPASYN and DELIGHT.SPICE have shown that numerical techniques can be applied to successfully optimize the performance of fixed topology circuits. An_com and OAC [8] have illustrated that layout-dependent phenomena can be considered in conjunction with device attributes when predicting overall performance. No tool, however, has yet attempted to define an Analog Hardware Description Language that would allow an expert analog designer to enter new circuit descriptions without having intimate knowledge of the software internals. Also, no tool has yet integrated the benefits of analytical modeling, layout dependent modeling, hierarchical circuit representation, and numerical optimization. It is these vital issues that we address.

In this paper we discuss a new synthesis tool, STAIC [9], that not only builds on the strengths of its predecessors, but also adds important features lacking in its predecessors. Similar to many analog CAD tools, STAIC is an open and interactive synthesis environment that accepts structural and performance specifications as input and generates a module generator layout description, SPICE net list, and data sheet as the final output. Unlike its predecessors, however, STAIC provides the super-user with a hardware description language for entering hierarchical CMOS and BiCMOS circuit descriptions. A unified treatment of modeling equations by STAIC allows the super-user to describe all relevant performance and layout aspects in a straightforward manner. Since layout forms part of the circuit description, important layout dependent attributes and parasitics may be incorporated into the synthesis basis when choosing an appropriate topology, layout style, and component sizing arrangement. A distinct division between the knowledge base of hierarchical circuit descriptions and the kernel of STAIC permit entry of circuit descriptions without intervention on the part of the CAD tool developer.

Pivotal to the success of STAIC is a central solve unit that dynamically assembles fragmented hierarchical design equations of selected circuit topologies to produce a homogeneous "flat" model description suitable for practical optimization and other numerical methods. Unlike its predecessors, STAIC is thus able to apply numerically intense design exploration techniques to hierarchical knowledge base representations.

STAIC is committed to finding global optimal solutions. To this end, STAIC employs a *successive solution refinement* synthesis methodology. Multilevel model descriptions of increasing complexity may be used by a variety of interactive design exploration tools to systematically converge to what is likely a global optimal solution. A summary of the salient features of STAIC and its competitors is provided in Table I.

TABLE I
SUMMARY OF SALIENT FEATURES OF STAIC AND OTHER CAD TOOLS

Attribute	STAIC [Waterloo]	IDAC & ILAC [SCEM]	OPASYN [Berkeley]	OASYS & ANAGRAM [CMU]	An_com [GE]
Synthesis Methodology	*Successive Solution Refinement*—dynamically constructed "flat" multilevel models of increasing sophistication are used to converge to what is likely a global optimal solution.	*Multiple Synthesis Strategies*—each strategy consists of "synthesis" equations ordered to form a single pass algorithm. Hierarchical decomposition at system design level.	*Synthesis by Analysis using Optimization*—problem is formulated as one in nonlinear constrained optimization.	*Hierarchical Decomposition with Backtracking*—heuristic rules propagate block performance specifications to equivalent sub-block performances.	*Successive Decomposition*—similar to OASYS, but includes structural decomposition at each level in the design hierarchy.
Analog Hardware Description Language	Expert analog designer fills "empty" templates with model descriptions using predefined language constructs and "objects."	—	—	—	—
Modeling of Layout Dependent Phenomena	Alternate layout style descriptions present at each level in the design hierarchy contain relevant model equations.	Separate treatment of layout precludes possibility. Feedback loop permits spec. update after layout.	Not reported.	Separate treatment of layout precludes possibility.	Hierarchical Decomposition includes floor planning. Layout dependent modeling is therefore accommodated.
Topology Selection & Design Space Exploration	Interactive swapping of alternate topologies and layout styles at any level in the design hierarchy. Optimization and scan modules work with multilevel model descriptions.	User may choose from library of hierarchical circuits. Optimization module works with built-in simulators.	Simple decision tree facilitates selection from among functionally equivalent "flat" circuit descriptions. Optimization module works with analytical model description.	Heuristic rules facilitate automatic topology selection at each level in the design hierarchy. No optimization.	Heuristic rules facilitate automatic topology and floorplan selection at each level in the design hierarchy. No optimization.
Layout Generation	Module generator description created from solved geometry and place and route directives embedded in selected layout style descriptions.	Created by separate place and route package tailored for analog applications.	Single slicing tree arrangement captures all special layout considerations.	Created by separate place and route package tailored for analog applications.	Floorplan is solved during synthesis. Routing is preformed by a separate s/w package.
Solution Verification	Data sheet and SPICE net list is output for post verification.	Dedicated built-in simulators.	Data sheet and SPICE net list is output for post verification.	Data sheet and SPICE net list is output for post verification.	Macro-model simulation at each hierarchical decomposition step.
Technology Support	CMOS, Bipolar, and BiCMOS	CMOS, Bipolar, and BiCMOS	CMOS	CMOS	CMOS, Bipolar

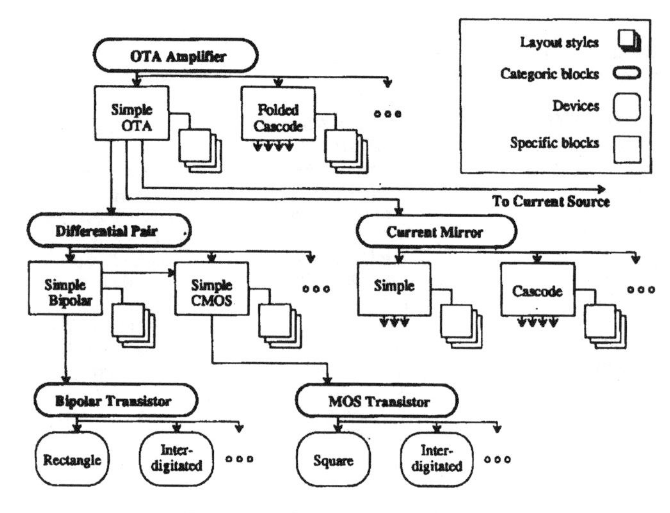

Fig. 2. Design description hierarchy.

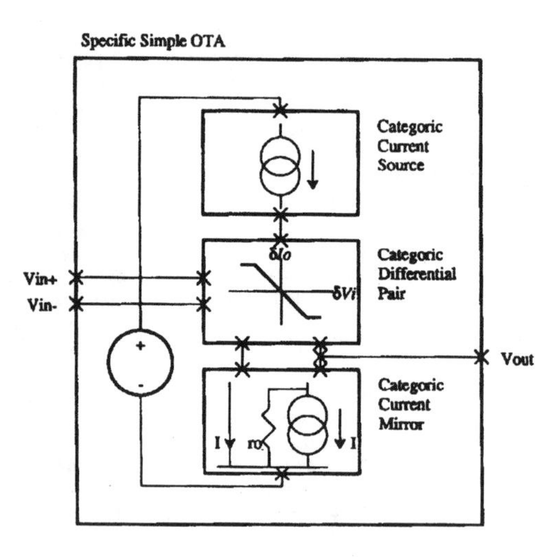

Fig. 3. Simple OTA components and interconnections.

II. Analog Hardware Descriptions

Circuit descriptions in STAIC, as illustrated in Fig. 2, can be classified as being of *categoric, specific, device*, or *layout* type. A *categoric* block declares a set of attributes (variables) that are common to all alternate implementations of a given generic function. Common attributes of a functional generic operational transconductance amplifier (OTA), for example, would include open loop dc gain, phase margin, and output voltage swing. The *categoric* block essentially maintains a well-defined interface to which all functionally equivalent blocks must conform.

A *specific* block is a particular and unique realization of a generic function. A folded cascode OTA would therefore be classified as a *specific* block since it is one of many possible OTA realizations. Each *specific* block is in turn defined by an interconnection of lower level *categoric* components along with an associated set of modeling equations that relate the *categoric* attributes of the block to the *categoric* attributes of each of its sub-blocks. Each sub-block again may have a number of alternate implementations.

A *layout* style description contains layout directives and model equations relevant to a particular floorplan and routing scheme. Each *specific* block can have a number of alternate *layout* style descriptions to accommodate a variety of application requirements. The benefits of using layout descriptions for analog circuits are twofold. First, since the floorplan and routing scheme are known in advance, mismatch and important parasitics can be accurately modeled. Second, expert domain knowledge can be exploited to yield customized compact layout. As an example, one possible floorplan arrangement for the *specific* simple OTA is as shown in Fig. 3 where each of the sub-blocks are stacked upon each other. Using a bounding box approach, the port positions, (denoted by X) and cell dimensions of the simple OTA may be expressed as a function of port positions and cell dimensions of its sub-blocks. Net parasitics can be expressed as a function of rectilinear distances between connected ports.

At the lowest level of hierarchy, *specific* blocks are supplanted by *devices*. Devices are similar to *specific* blocks in that they too have modeling equations relevant to the device. Unlike *specific* blocks, however, each *device* has an associated module generator layout description.

A. Language Constructs

The entry of a new *categoric* circuit description begins with an "empty" predefined circuit template or form that is to be filled with port and performance attribute (variable) declarations. The fundamental syntax is that of the C++ [10] object oriented programming language. All ports and variables are "objects" in the object oriented paradigm. Members of the port "object" include x and y variables for port positioning, and a capacitance variable to model interconnect capacitances. Each variable maintains an array of values along with their corresponding soft and hard upper and lower boundaries. *Specific* templates are "derived" from an appropriate *categoric* template and filled with port interconnection, analytic model descriptions, variable boundary estimates, and local equation and variable declarations. Fig. 4 depicts compilable code fragments of the *specific* simple OTA comprised of a differential pair, current source, and current mirror (Figs. 2 and 3).

Five language constructs find repeated use in the *specific* template. The *connect*( ) statement declares two or more ports to be in the same net. Upon circuit instantiation, entries are made into the hierarchical net list module of Fig. 5. The *alias_as*( ) statement declares two variables to be the same. Dominant and recessive variable designations are used to decide which variable's members are to be adopted. The *bind_lin*( ), *bind*1( ), and *bind*2( ) language constructs facilitate model equation entry into the linear, first-order, and second-order equation data-

```
// Port and net assignments
  connect( cur_mir.ref & diff.op1 );
  connect( cur_mir.out & diff.op2 );
  connect( cur_source.iout & diff.bias );
  .
  .
// Variable aliases
  cur_source.i_bias.alias_as( diff.i_bias );
  cur_mir.i_ref.alias_as( diff.i_o );
  .
  .
// Linear constraints
  bind_lin( VDD * diff.i_bias > static_power.hard_low() );
  bind_lin( v_swing + cur_source.vsat + diff.vsat + cur_mir.vsat = VDD );
  .
  .
// First order model equations
  bind1( Rl = diff.rout || cur_mir.rout );
  bind1( dc_gain = diff.gm * Rl );
  bind1(settle_eq, settle_time, pole1 & dc_gain & v_frac );
  bind1( slew = diff.i_bias / (cur_mir.out.cap + diff.cout + cur_mir.cout + c_ext) );
  .
  .
```

Fig. 4. Code fragments of the *Specific* simple OTA.

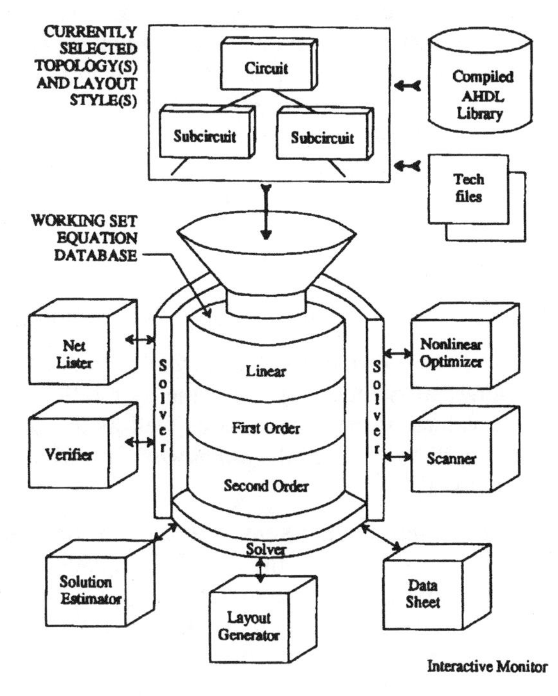

Fig. 5. Schematic of STAIC's framework.

```
// Layout directives
  diff.place.northof( cur_mir ).by( min_sep ).justify( CENTER );
  cur_source.place.northof( diff ).by( min_sep ).justify( CENTER );
  cur_mir.out.route.northto( diff.op2 );
  .
  .
// Linear constraints
  bind_lin( cell_width - diff.cell_width > 0.0 );
  bind_lin( cell_width - cur_mir.cell_width > 0.0 );
  bind_lin( diff.cell_width + cur_source.cell_width + cur_mir.cell_width - cell_width > 0.0 );
  .
  .
// First order model equations
  bind1( cur_mir.out.cap += min_sep * MIN_METAL * METAL_CAP );
  bind1( cell_height = diff.cell_height + cur_source.cell_height + cur_mir.cell_height + 2*min_sep );
  .
  .
```

Fig. 6. Code fragments of simple OTA *layout style*.

bases of Fig. 5, respectively. Two forms of binding are supported. In the normal or intrinsic mode, equations are entered directly using simple mathematical and circuit operators such as ×, +, −, /, and ‖. The *bind1 (Rl = diff.rout // cur_mir.rout)* instruction of Fig. 4, for example, implicitly instantiates an equation object of type "parallel" which operates on two input variables and yields a single output. Upon circuit instantiation, the intermediate simple OTA *specific* variable, *Rl*, the differential pair *categoric* output resistance attribute, *diff.rout*, and the current mirror *categoric* output resistance attribute, *cur_mir.rout*, are dynamically bound to the parallel equation, and an entry is made into the first-order database. Upon circuit deletion, all corresponding database entries are removed.

In the second form, equation descriptions are explicitly bound to variables. This endows the super-user with the ability to write customized equations. *Settle_eq* is an externally declared equation in the instruction *bind1 (settle_eq, settle_time, pole1 & dc_gain & v_frac)* of Fig. 4. Similar to circuit descriptions, a generic equation template is provided to ease the description process.

Layout templates are "derived" from an appropriate *specific* template and filled with layout dependent modeling equations and directives. Fig. 6 illustrates a compilable code fragment of a layout style for the simple OTA of Fig. 3. The *bind* statements convey model information while the *place* and *route* statements direct module generator code development. Blocks may be relatively placed, rotated, and justified in any Manhattan direction. Blocks may also be grouped into clusters and reflected to maintain layout symmetry.

B. Model Development

STAIC employs a multilevel modeling scheme. Three levels of model descriptions of increasing sophistication are possible for each *specific*, *layout*, and *device* block. In the simplest model level, circuits are described as a collection of linear constraints that serve to indicate feasible solution regions. Some relationships among block attributes are naturally linear, and hence may be directly loaded into the linear database. Series resistance and parallel capacitance calculations are two such examples. Many relationships, however, are highly nonlinear and alternate approaches must be pursued. The "parallel" equation $R_t = r_1 \| r_2$, for example may be approximated as

$$R_t < r_1,\ R_t < r_2$$

$$R_t > \frac{r_1 \cdot \text{hard\_low}(\,) \times r_2 \cdot \text{hard\_low}(\,)}{r_1 \cdot \text{hard\_low}(\,) + r_2 \cdot \text{hard\_low}(\,)}.$$

Alternatively, artificial variables constrained to lie between zero and one may be introduced to regionally linearize equations [11].

The first-order and second-order model relationships are described by continuous nonlinear equations and their corresponding continuous derivatives. Equations of arbitrary complexity may be employed, provided the continuity conditions are respected. The first-order model typically resembles that of a hand calculation and ignores second-order effects, while the second-order model includes all phenomena that have marginal impact on performance. Symbolic computation tools such as MAPLE [12] and ISAAC [13] can play an important role in aiding in the development of first-order and second-order models. MAPLE, in fact, was used to develop CAD delay models for sample-and-hold times of a high-speed comparator described in STAIC. The polynomial curve fitting capability of MAPLE was also exploited to develop a continuous and smooth version of the settling time model presented in [14]. This model is subsequently used in the first-order model of a folded cascode description in STAIC.

III. Synthesis Strategy

Intrinsic to each of the analog synthesis CAD tools considered is a synthesis strategy that simplifies the problem in some manner to cope with design complexity. For example, hierarchical decomposition in OASYS and An_com permit the synthesis problem to be partitioned into a number of smaller, more manageable, semi-independent design problems. STAIC has adopted a *synthesis by analysis* approach using a *successive solution refinement* technique to cope with design complexity. The *successive solution refinement* technique exploits the multilevel analytical model descriptions to systematically attain what is likely to be a global optimal solution. In early stages of the design process, simplified models that capture major design tradeoffs are used to efficiently generate a solution. Many local optimal points that would otherwise be present in a more exact treatment are masked by the simplified approach. Stepping to an advanced model permits a more realistic account of circuit behavior. In the quest for a global optimal solution, the choice of topology, layout style, and component geometries made as a result of design exploration using the simpler first-order model serves as an initial starting point for design exploration using the more advanced model.

Like OPASYN and OASYS, modeling in STAIC is based on analytical descriptions. Frequency- and time-dependent model descriptions that require simulation are not *globally* supported. The equations that comprise the model description, however, can in fact be dedicated macromodel simulators. The purpose of STAIC is to provide a framework that enables the designer to explore all important performance tradeoffs within reasonable design time. While modeling should reflect these tradeoffs, it need not yield the exact solution. The intent is that an optimal solution generated as a result of design exploration in STAIC be within close proximity to the true optimal solution. Final component tweaking is left to tools such as DELIGHT.SPICE. To this end, STAIC provides an output circuit net list description targetable to different circuit simulators.

IV. STAIC Framework

The architecture of STAIC is as depicted schematically in Fig. 5. At the heart of STAIC is a multilevel "working set" model equation database surrounded by a symbolic/numeric solve unit which, from an expert systems perspective, may be viewed as the inference engine. Upon interactive selection of a promising circuit candidate from the compiled design hierarchy, analytic design equations contained within the candidate's *device, specific*, and *layout style* templates are loaded into a central equation database. On invocation of a design exploration module, the solve unit proceeds to dynamically select and assemble relevant design equations from the database to yield a complete and customized "flat" analytical model relating user specified high level performance attributes to low-level geometric attributes. Using this derived homogeneous model, optimization and other numerically intense procedures may be readily and effectively applied to explore the design space of circuit descriptions. STAIC thus enjoys the advantages of a "flat" homogeneous model for efficient numerical design evaluation without sacrificing the benefits of hierarchical representation for minimizing the duplication of circuit description effort.

The solver consists of two sub-modules. One is designed to work with linear constraint systems while the other operates on nonlinear equalities.

A. Nonlinear Solver

The objective of the nonlinear solver is as follows:

> Given a list of *known* attributes (variables) and a list of *unknown* attributes, solve for the symbolic and numeric values of the *unknowns* as well as the corresponding derivatives of the *unknowns* with respect to the *knowns*.

We begin by introducing some useful notation. Let $\Gamma = \{f_1, f_2, \cdots, f_m\}$ denote the set of all equations contained in a nonlinear database such that $\boldsymbol{F}_{DB}(\boldsymbol{X}_{DB}) = [f_1(\boldsymbol{X}_{DB}), f_2(\boldsymbol{X}_{DB}), \cdots, f_m(\boldsymbol{X}_{DB})]^T = \boldsymbol{0}$, where $\boldsymbol{X}_{DB} = [x_1, x_2, \cdots, x_n]^T$ is an array of all variables bound to the database entries whose members form the set $\Psi = \{x_1, x_2, \cdots, x_n\}$. Furthermore, let $\Lambda = \{f_1, f_2, \cdots, f_p\}$ and $\Theta = \{x_1, x_2, \cdots, x_q\}$ be a minimum subset of equations and variables required to solve all *unknowns* of a particular design problem such that $\boldsymbol{F}_M(\boldsymbol{X}_M) = [f_1(\boldsymbol{X}_M), f_2(\boldsymbol{X}_M), \cdots, f_p(\boldsymbol{X}_M)]^T = \boldsymbol{0}$, where $\boldsymbol{X}_M = [x_1, x_2, \cdots, x_q]^T$. Put succinctly, $\Lambda \subseteq \Gamma$ and $\Theta \subseteq \Psi$. Finally, we define $\boldsymbol{X}_M = [\boldsymbol{U}, \boldsymbol{V}, \boldsymbol{C}]^T$ where $\boldsymbol{C}$ is the array of *known* variables, $\boldsymbol{U}$ is the array of *unknown* variables, and $\boldsymbol{V}$ is the array of intermediate variables which are involved in the basis to map $\boldsymbol{C}$ to $\boldsymbol{U}$ and must therefore be found commensurate with $\boldsymbol{U}$ when solving $\boldsymbol{F}_M(\boldsymbol{X}_M) = \boldsymbol{0}$.

The principle tasks of the solve unit may now be summarized as follows:

a) Identify the minimum subset of equations, Λ, that form the solution basis, $\boldsymbol{F}_M(\boldsymbol{X}_M) = \boldsymbol{0}$.

b) Recursively decompose $\boldsymbol{F}_M(\boldsymbol{X}_M) = \boldsymbol{0}$ into simple back substitution and minimal systems of simultaneous equations to be solved using multivariate root finding techniques.
c) Generate a symbolic solution consisting of an assemblage of op-code instructions which when executed yield a numerical solution for $\boldsymbol{U}$ and $\boldsymbol{V}$ at a given point $\boldsymbol{C}_o$.
d) Find $d\boldsymbol{U}/d\boldsymbol{C}$ at the given point $\boldsymbol{C}_o$.

In STAIC, the derivation of the symbolic solution begins by forming a directed dependency graph from all Γ equations in the database. In the graphical realm, a node is a variable, while an arc implies node connectivity through an equation. The arc directivity indicates whether a variable is an input or an output of an equation. Beginning with an output node, the dependency graph is systematically traversed in compliance with a set of rules [15]. As each node is successfully visited, a trail of "bread crumbs" is left behind to form an intermediate heterogeneous tree structure containing the $\boldsymbol{F}_M(\boldsymbol{X}_M)$ basis equations.[1] If integral model equations are missing from the database, or the number of specified *knowns* are too few for the requested set of *unknowns*, then the solver will be unable to formulate the intermediate tree. In this case, an underconstrained condition exists and trace information may be interactively output to help the designer resolve the problem. A successfully formed intermediate tree may be easily interpreted by STAIC to generate a set of op-code instructions analogous to the way in which a language compiler generates assembly code.

For illustrative purposes, consider equations $\Gamma = \{f_1, f_2 \cdots, f_7\}$ bound to variables $\Psi = \{x_1, x_2, \cdots, x_{13}\}$ and loaded into the first-order equation database (Fig. 7(a)). The solver problem is: Given $\boldsymbol{C} = [x_1, x_2, x_3, x_4, x_8, x_9]^T$, find a solution for $\boldsymbol{U} = [x_5, x_{12}]^T$. Figure 7(b) illustrates the resultant intermediate tree structure. The interpretation of the rightmost section of the tree is straightforward since no systems of equations are involved. The tree is interpreted by performing a depth-first, left-to-right traversal. The intermediate variable x_{11} is first solved by reading x_1, x_3, and x_4, firing equation f_3, and writing x_{11}. x_{10} is next solved by reading x_{11}, x_1, and x_2, firing equation f_2, and writing x_{10}. Finally x_{10} and x_{11} are read, f_1 is fired, and x_{12} is written. The corresponding op-code generated by STAIC is illustrated in the right column of Fig. 7(c).

Interpretation of the left branch is more involved. On the way down the tree, a local list of unknown variables is maintained at each *newexpression* node. On the way back up, each *newexpression* variable is pulled off all local lists. Any locally remaining unknowns are propagated up to the parent node. A system of equations is identified

[1]Some have posed this as a classical problem in maximal matching [16]. This solving technique, however, requires that all X_{DB} variables be matched to $F_{DB}(X_{DB})$ equations. Our method begins with the set of *unknown* variables and traces a solution path back to the *known* inputs, thereby ignoring equations that are disjoint to the particular problem at hand.

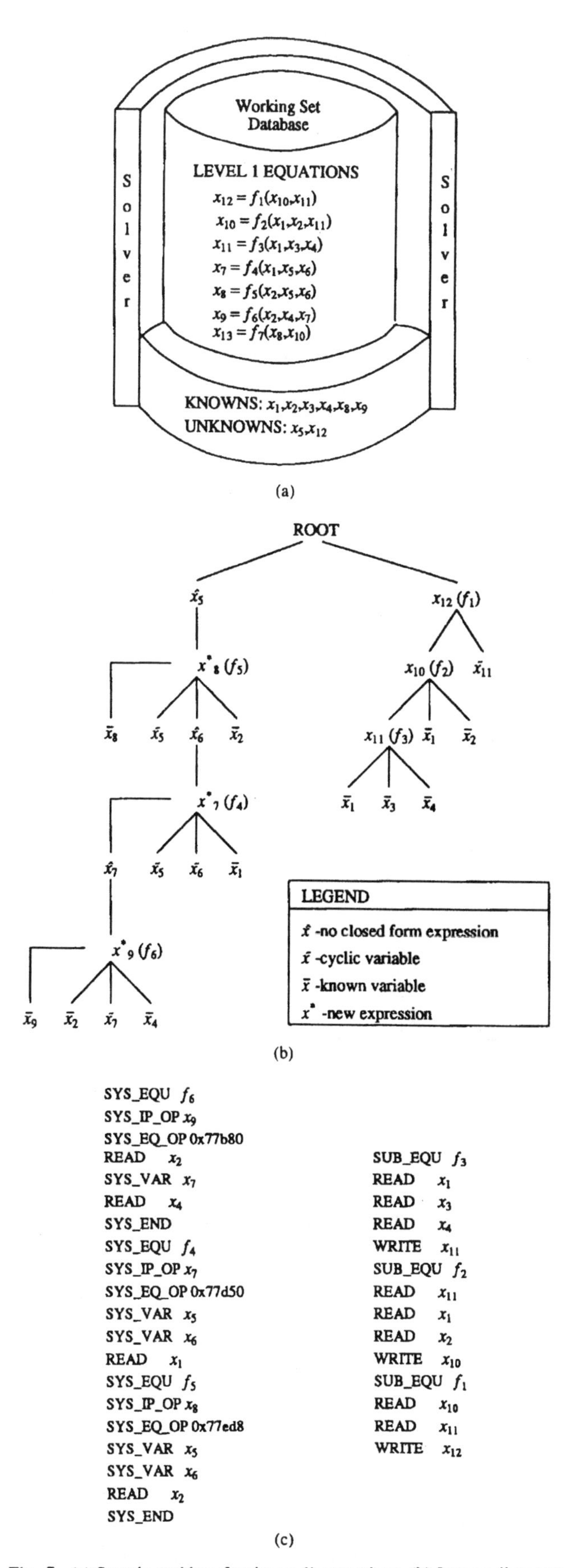

Fig. 7. (a) Sample problem for the nonlinear solver. (b) Intermediate tree output. (c) Op-code assembly instructions.

when the local unknown list is empty. For the left branch of Fig. 7(b), a system of three equations ($f_4 - f_6$) and three unknowns ($x_5 - x_7$) is readily identified. The corresponding op-code generated by STAIC is illustrated in the left column of Fig. 7(c). The actual numerical evaluation of the system of equations is performed by a commercial routine [17] that uses the Newton–Raphson technique with derivatives for finding multivariate roots. It should be noted that the solver also identifies that $\Lambda = \{f_1, f_2, \cdots, f_6\}$ with equation f_7 omitted from the solution basis.

The op-code instructions of Fig. 7(c) represent the symbolic solution. Since the symbolic solution is not tied to numeric values, it may be executed any number of times to yield new output values when the input values are changed. To avoid recomputing the symbolic solution for every similar solve request, each symbolic solution is placed in a hash table (Fig. 8). When a solve request is made, the hash table is first checked to see if a symbolic solution already exists. If it does, then the solution is pulled from the table, executed, and returned.

i. Finding the derivative: The derivative of the solver *unknowns* with respect to the solver *knowns*, dU/dC, is computed through implicit application of the chain rule. We begin with the derivative of the basis equations, $F_M(X_M)$ with respect to C *knowns*:

$$\frac{dF_M(X_M)}{dC} = \frac{\partial F_M(X_M)}{\partial X_M}\frac{dX_M}{dC} = 0.$$

Substituting $[U, V, C]^T$ for X_M in dX_M and ∂X_M above yields

$$\left[\frac{\partial F_M(X_M)}{\partial U}, \frac{\partial F_M(X_M)}{\partial V}, \frac{\partial F_M(X_M)}{\partial C}\right]\begin{bmatrix}\frac{dU}{dC}\\ \frac{dV}{dC}\\ \frac{dC}{dC}\end{bmatrix} = 0.$$

Recognizing that dC/dC is the identity matrix I, the above equation may be rewritten as

$$\left[\frac{\partial F_M(X_M)}{\partial U}, \frac{\partial F_M(X_M)}{\partial V}\right]\begin{bmatrix}\frac{dU}{dC}\\ \frac{dV}{dC}\end{bmatrix} =. -\frac{\partial F_M(X_M)}{\partial C}.$$

The left matrix is the Jacobian of $F_M(X_M)$ with C held constant, while the right matrix is the Jacobian of $F_M(X_M)$ with U and V held constant. Defining

$$J_{UV} = \left[\frac{\partial F_M(X_M)}{\partial U}, \frac{\partial F_M(X_M)}{\partial V}\right] \quad \text{and} \quad J_C = \frac{\partial F_M(X_M)}{\partial C}$$

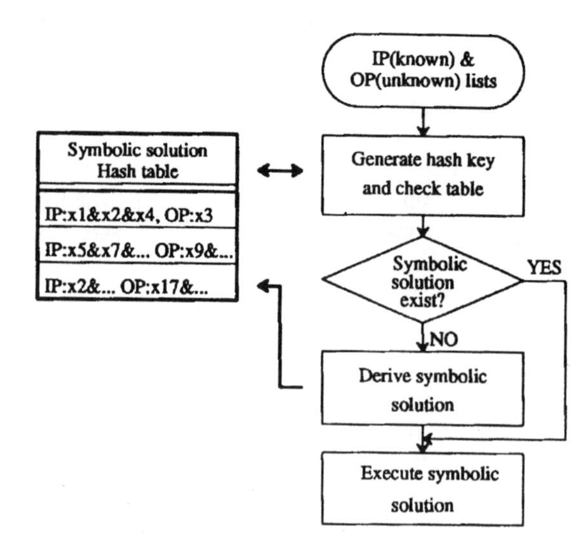

Fig. 8. Hashing of symbolic solutions.

then we may finally write

$$J_{UV}\begin{bmatrix}\frac{dU}{dC}\\ \frac{dV}{dC}\end{bmatrix} = -J_C.$$

Both J_{UV} and J_C are explicitly known as a byproduct of solving $F_M(X_M) = 0$ in the last section. The desired derivative, dU/dC, is embedded in the middle matrix of the above equation and is efficiently solved using a commercial routine employing a sparse variation on Gaussian elimination [18].

A condition code accompanies the solution to indicate the status of the linear system. A bad condition number indicates that the J_{UV} matrix is close to singular. Since systems of nonlinear dependent equations yield systems of linearly dependent partial derivatives, a bad condition number is an excellent indicator of either an improper model description or an infeasible symbolic solution. Work is underway to use this indicator in a feedback loop to find a new symbolic solution in the event that more than one solution exists.

ii. Nonlinear solve unit run time: For a typical design problem involving 740 database entries, 5 CPU seconds on a Sun Sparc 2 workstation are required to arrive at a symbolic solution involving 236 basis equations (i.e., cardinality of Λ is 236). Execution of the symbolic solution to yield a numeric value for U and dU/dC takes 1.5 CPU seconds. The system derivative information is not required for all exploration modules and, therefore, the dU/dC calculation may be optionally turned off to save further on computation time.

B. Linear Solver

Linear constraint-based systems are considerably easier to deal with than nonlinear systems. Unlike the nonlinear case, the presence of redundant and extraneous equations is inconsequential to the solution sought. The same degree of care, therefore, need not be taken in identifying the subset of database equations relevant to the particular problem. In addition, linear systems are more amenable to mathematical transformation and evaluation than their nonlinear counterparts. The function of the linear solver, therefore, is essentially reduced to that of providing an interface to established linear programming packages such as MINOS [19].

V. User Input

Three critical pieces of information are required as input to STAIC. The user must indicate the targeted process technology, the functional (*categoric*) block to be synthesized, and the performance desired of the functional block of interest. STAIC maintains a fair degree of technology independence through the use of technology files containing fundamental SPICE-like model parameters and layout design rules. The user is free to select from among any predescribed CMOS, bipolar, or BiCMOS technologies as well as to modify existing parameter values. Only the super-user can add new technology parameters often required when more advanced processes involving extra layers of interconnect are to be supported. Each circuit description in the design hierarchy maintains a list of compatible technologies.

Performance specifications take the form of equality and inequality constraints imposed on the attributes of the *categoric* block of interest. Each constraint has an associated normalized weighting factor, and multiple constraints may be applied to each and every performance attribute. A zero weighting factor indicates that the constraint is of no consequence and may be freely violated, while a one weighting factor indicates that the constraint must be satisfied at all costs. A weighting factor between zero and one indicates preferences in satisfying one constraint over another. Fig. 9 illustrates a sample input description that is used to synthesize the second example of the results section. In this example, the 0.1% settling time of the OTA block with a 3-V step and unity feedback must not exceed 300 ns. Preference, however, is given to a settling time of 200 ns with a weighting factor 0.8.

```
technology:
CMOS nt3

options:
warn on
trace optimizer on

template:
Ota_P

constants:
c_load 5.0[pF]

constraints:
phase_margin > 1.0[rad/s]; 0.5
dc_gain > 1500; 0.5
v_out_swing > 3.0[V]; 1.0
v_out_swing > 3.5[V]; 0.8
slew_rate > 10.0[V/µs]; 1.0
settle_time < 0.3[µs]; 1.0
settle_time < 0.2[µs]; 0.8
cell_area < 50000.0[µm^2]; 0.7
static_power < 6000.0[W]; 1.0
cmrr ; 0.0
v_offset ; 0.0
v_noise ; 0.0
```

Fig. 9. Sample OTA input specification.

A. Objective Function Formulation

Soft constraints, which indicate performance preferences, may be captured in an objective function to provide a measure of optimality. STAIC adopts a formulation similar to that originally presented in [20]. For each "less than" constraint, i, a term of the form $(e^{\zeta_i \text{attribute}_i - \text{spec}_i/\text{spec}_i} - 1)$ is added, and for each "greater than" constraint, j, a term $(e^{\zeta_j (\text{spec}_j - \text{attribute}_j)/\text{attribute}_j} - 1)$ is introduced, where ζ_{ij} is the weighting factor associated with meeting spec_{ij}. Both equations are combined to express equality constraints. The exponential nature of the cost function prevents the penalty for violating any specification from being compensated by overly satisfying other specifications.

VI. Design Space Exploration

In the current version of STAIC, the selection of circuit topologies and layout styles at each level of hierarchy is strictly a manual process. However, at any level in the design hierarchy, alternate functionally equivalent *specific* blocks and layout styles may be interactively substituted at will. The equation database is automatically updated to reflect the change. STAIC's architecture is open and general enough to permit the future integration of an automatic topology selector, as well as other application modules. Design exploration and evaluation are accommodated by the scanner and optimizer application modules. A brief description of each of these two blocks follows.

A. Scanner Module Applications

The scanner serves as a coarse grid point search tool with two modes of operation. As a tool to generate an initial solution, the scanner performs a coarse grid point search for a solution that is in the neighborhood of a global optimal point. Constant and independent variable declarations made within circuit descriptions are designated as *knowns* to the solve unit. Independent variable designations are typically a set of "key" component geometries and bias conditions.

The set of user imposed hard constraints and the objective value expressed above constitute the solver *unknowns*. Ideally, the *known* variables would be varied over a domain space defined by all first-order nonlinear design equations and hard variable boundaries. In practice, however, explicit determination of this nonlinear domain space is impossible for all but the simplest of problems. Instead, the independent design variables are varied over a polytope sub-space that is known to contain the nonlinear domain space. This sub-space is defined by the *linear* constraints in the linear equation database along with all hard

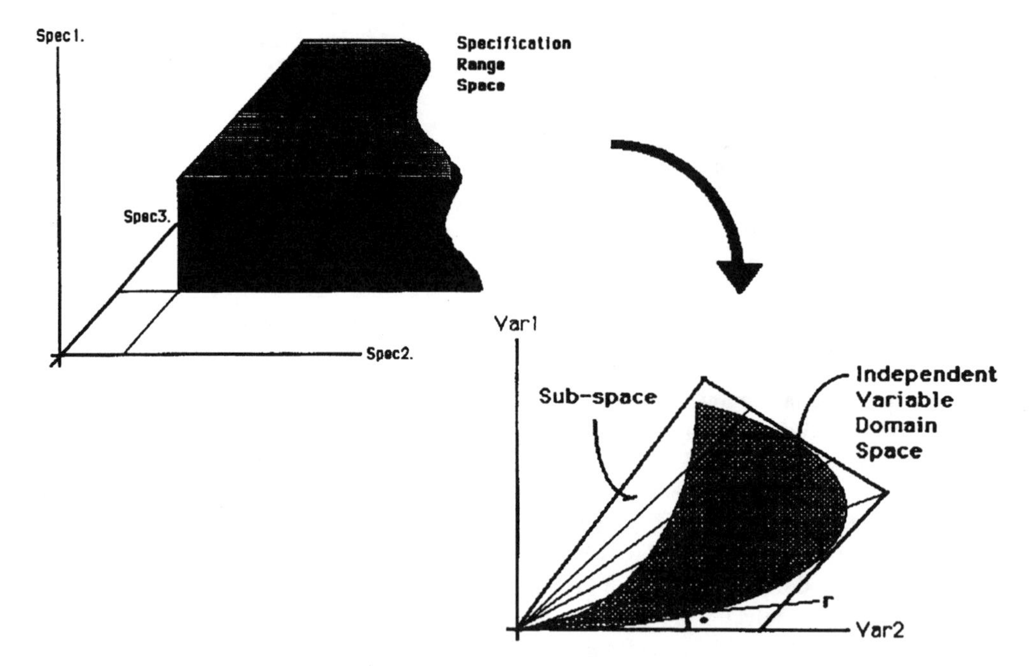

Fig. 10. Illustration of the scan operation.

upper and lower boundaries of independent and intermediate variables used by the solver in arriving at a symbolic solution.

For illustrative purposes, a situation is depicted in Fig. 10 with the sub-space of two independent variables r and θ being radially searched. In this example, θ is varied incrementally from 0 to 60 deg. At each θ_i increment the *linear solver* is used to find the point of intersection of the radial line with the sub-space boundary. This defines the radial line length r_i. The nonlinear domain space is subsequently determined by evaluating the *unknowns* (via the nonlinear solver) at gridpoints sampled along each radial line r_i, and discarding those points that fail to fall within the known range space. The domain point that evaluates to the best objective value is retained as an initial solution.

In a second mode of operation, the scanner may be used to interactively explore design tradeoffs. In this mode, the user is free to declare any set of independent and dependent variables within the design hierarchy of the currently selected circuit for gridpoint evaluation.

B. Optimization

The task of the optimizer is to adjust the set of designated independent variables so as to facilitate objective value minimization while ensuring that the final solution lies within the domain space. The optimizer interacts with the solver in a manner similar to that of mode one of the scanner. The designated independent variables for the optimizer, however, are typically more numerous than those designated for the scanner, since optimization search times are not so critically dependent on the variable count. Numeric optimization is performed by WATOPT [21], a nonlinear constrained optimizer that uses Jacobian data supplied by the solver to achieve superlinear convergence rates.

VII. Layout Generation

The function of the layout generation module depicted in Fig. 5 is to translate finalized geometrical dimensions, placement directives, and routing directives of the selected layout style into equivalent ICEWATER [22] source code. ICEWATER is a procedural layout language used for writing module generator descriptions. Fig. 11 illustrates the steps taken to produce a layout instance. Finalized component geometries are passed in as input to the parameterized layout descriptions of the composite *devices*. In many cases, considerable freedom exists for the physical shape devices may assume. *Device* descriptions typically include models that permit adjustment of the component cell dimensions to yield compact layout as part of the design exploration process. Fig. 12 shows three layout instances of a fingered MOS device produced by the same ICEWATER module generator description. The description is parameterized with respect to channel width, channel length, and cell width. All three devices are nominally equivalent in that their effective channel dimensions are the same.

From a modeling perspective, tradeoffs among device and cell dimensions of the MOS module generator layout of Fig. 12 may be succinctly described in STAIC's analog hardware description language (AHDL) by the relation:

$$Cell\_height = h\_cap + (h\_inc + L) * \cdot \mathrm{INT}\left(\frac{W}{Cell\_width - w\_cap} + 1\right)$$

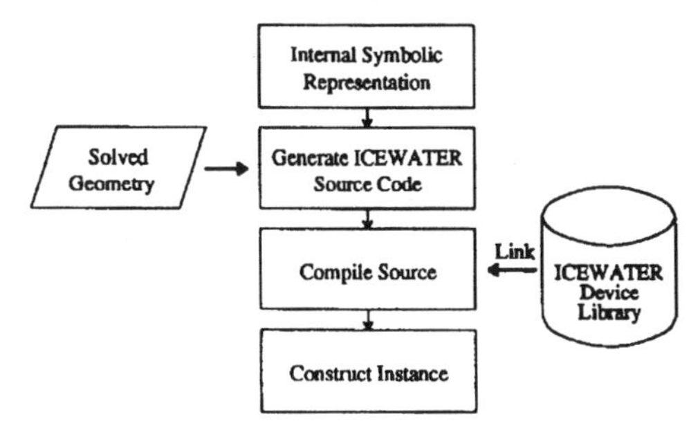

Fig. 11. Flow of control for generating layout.

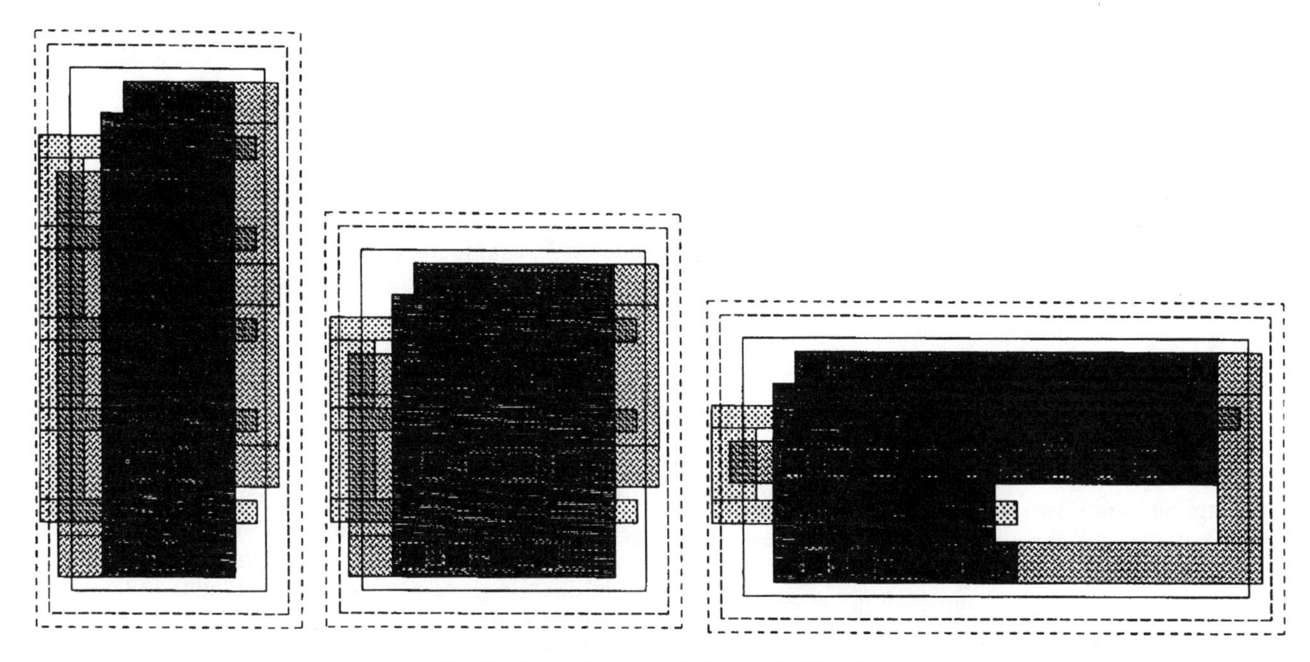

Fig. 12. Layout instances of a fingered MOS device.

where **INT**( ) yields the integer part of its argument, *w_cap* and *h_cap* are the fixed cell width and cell height contributions to the cell dimensions, *h_inc* is the incremental cell height contribution and W and L parameters are the MOS channel width and length, respectively. To satisfy nonlinear equation continuity conditions, the **INT**( ) function is approximated by a finite sum of $\tanh(x)$ functions offset by integer quantities in the x direction. Similar modeling techniques are applied to bipolar, capacitor, and resistor layout descriptions.

In general, layout models will only approximate the algorithms implicit to module generator descriptions. Consequently, to ensure that layout is correct by construction, all components and routing in STAIC are described in terms relative to bounding box attributes of previously placed ports and blocks (Fig. 6). Placement of components at absolute coordinates is prohibited. Final CIF [23] layout is created by compiling the ICEWATER source code, linking it with the device module generator library, and running the resultant executable.

VIII. Experimental Results

STAIC is programmed in the object-oriented C++ [10] language and presently comprises 40 000 lines of source code, 10 000 lines of which are AHDL circuit descriptions. To date, three hierarchical descriptions have been entered into the STAIC framework. One is a description of a fully differential folded cascode circuit depicted in Fig. 13, one is the simple OTA that has been referenced throughout this paper, and the third is a high speed regenerative comparator suitable for flash A/D converters. Both CMOS and BiCMOS subcircuit variations of each OTA and comparator are supported.

A. Simple OTA Example

The first design example entails the synthesis of a simple OTA circuit targeted for Northern Telecom's three micron double layer metal CMOS process [24]. The current design hierarchy limits topology selection to the circuit of Fig. 14. While the simple OTA is not a particularly

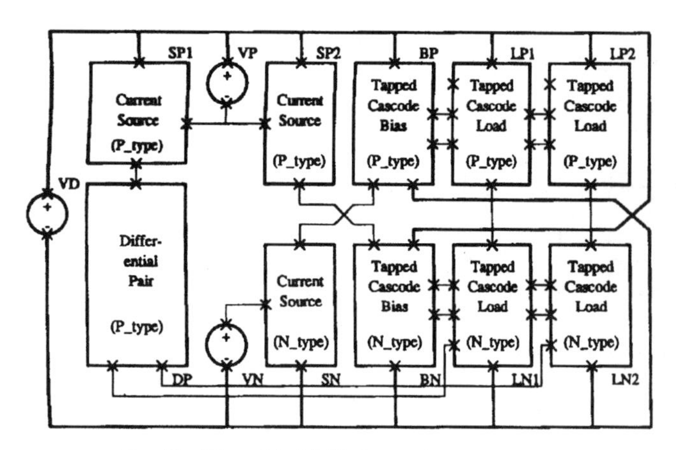

Fig. 13. Hierarchical folded cascode representation.

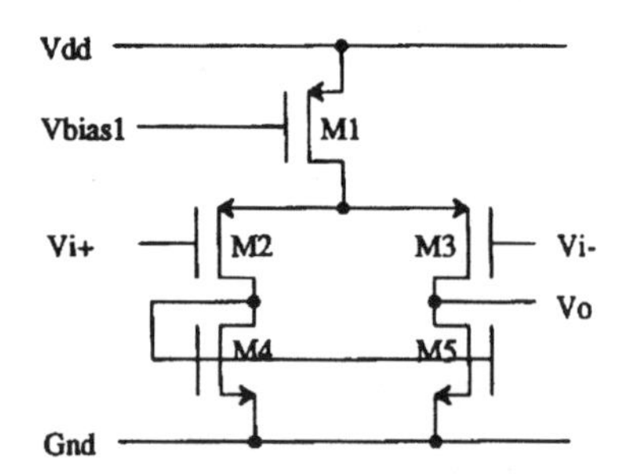

Fig. 14. CMOS simple OTA instance.

useful amplifier on its own, it does constitute an important building block in the construction of practical two-stage class *AB* opamps [25]. Low input referred noise, high ft, and small silicon real estate consumption are considered important performance traits for this example. A hard upper boundary of 0.01 $\mu v^2/Hz$ sets the maximum acceptable noise floor. An ft of 100 MHz for a 1-pf load is desirable, but not at the expense of excessive silicon area consumption. Finally, we target a reasonably square layout for compatibility with pre-existing layout.

Table II illustrates input specifications and the resultant circuit performance at various phases in the synthesis process. Design exploration begins with the scanner. Four "key" independent variables are permitted ten increments per search direction over the polytope region defined by the linear equation database. The preliminary solution of column 4 is obtained in 120 CPU seconds on a MIPS 2000 machine. The next step in the synthesis phase involves optimization of the first-order model. A larger set of independent variables are defined for optimization purposes, and WATOPT is iteratively called. The solution of column 5 in Table II is obtained in 76 CPU seconds. A SPICE net list is created by STAIC and HSPICE [26] level 2 simulations are performed to verify the predicted performance. A comparison of the results in column 5 and 6 indicates that results agree reasonably well within a worst-case error of 24%. The corresponding layout of the synthesized simple OTA is illustrated in Fig. 15. For matching purposes, the differential pair and current mirror are placed symmetrically about a north-south axis in the center of the design.

Evidently, the channel lengths of the NMOS current mirror devices at the bottom of Fig. 15 are much larger than minimum size. In light of the analysis presented in [25], these device sizes are a manifestation of the scan and optimization procedures in meeting the noise floor specification of Table II.

The effectiveness of the multilevel modeling system in addressing global optimality is exemplified in the synthesis of the simple OTA. If the scan step of this design example had been omitted, then the optimizer module would have used a default initial solution that would have led to a poorer locally optimized solution. In this case the final objective function cost would have been 47% larger.

B. CMOS Folded Cascode Example

A second example entails the synthesis of a folded cascode OTA, again targeted for Northern Telecom's CMOS process. The input specifications shown in Table III place considerable emphasis on the synthesis of a design that exhibits fast settling time over a large voltage swing without consuming excessive static power.

Column 5 of Table III illustrates the results of an optimization run based on the dynamically constructed first-order model of the instantiated folded cascode topology in Fig. 16. The initial solution supplied to the optimizer was generated by the scanner. A two-pole model of the transfer function was used as a basis in the first-order model for the ac small signal computations [27]. A comparison of HSPICE level 2 simulation and STAIC predictions indicate good agreement. The worst-case deviation of the results in Table III is 17%. Approximately 600 CPU seconds were required to generate the solution on a Sun Sparc 2 workstation.

C. BiCMOS Folded Cascode Example

A third and final example entails the synthesis of a BiCMOS instance of the folded cascode OTA. In this example, bipolar versions of the LN load elements of Fig. 13 are instantiated to improve high frequency performance (Fig. 17). A 0.8-μm BiCMOS process [28] employing a vertical n-p-n device is targeted. In this example, we are willing to trade off higher static power consumption for a smaller settling time over a minimal 3-V output swing.

Column 4 of Table IV illustrates the results of a synthesis run based on the first-order model description. For identical loading and approximately equal output voltage swings, both the settling time and cell area of the BiCMOS implementation (Table IV) is one-quarter that of the CMOS version (Table III). HSPICE level 3 simulation results presented in column 5 of Table IV indicate satisfactory agreement with a worst-case deviation of

TABLE II
SIMULATED AND PREDICTED PERFORMANCE OF CMOS SIMPLE OTA EXAMPLE

Attribute	Spec.	Wgt.	STAIC (Scanner)	STAIC (Optimizer)	HSPICE & ICEWATER
Vdd	= 5.0 V	1.0	5.0 V	5.0 V	5.0 V
$Cload_{ext}$	= 1.0 pF	1.0	1.0 pF	1.0 pF	1.0 pF
$Vout_{swing}$ (antisymmetric)	> 2.0 V	1.0	3.8 V	2.0 V	2.2 V
dc gain	> 48 dB	0.7	38.6 dB	39.2 dB	37.1 dB
slew rate	> 5.0 V/μs	0.8	39 V/μs	78.3 V/μs	76.1 V/μs
ft	> 100.0 MHz	0.8	21.0 MHz	38.4 MHz	31.0 MHz
cell area	< 1.0e4 μm^2	0.7	5.2e4 μm^2	2.92e4 μm^2	3.26e4 μm^2
cell aspect	= 1.0	0.8	0.98	1.0	0.85
V_{noise} (@ 1 kHz)	< 1e-2 μv^2/Hz	1.0	0.74e-2 μv^2/Hz	0.21e-2 μv^2/Hz	0.21e-2 μv^2/Hz
CMRR	—	—	82.9 dB	60.1 dB	63.5 dB
$Power_{static}$	—	—	711 mW	775 mW	730 mW

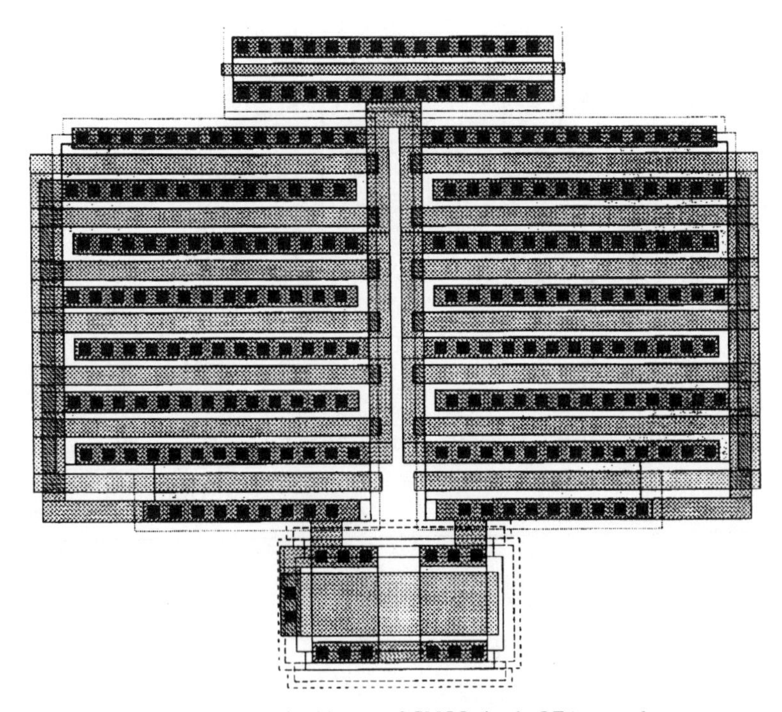

Fig. 15. Synthesized layout of CMOS simple OTA example.

TABLE III
SIMULATED AND PREDICTED PERFORMANCE OF CMOS FOLDED CASCODE EXAMPLE

Attribute	Spec.	Wgt.	STAIC (Scanner)	STAIC (Optimizer)	HSPICE & ICEWATER
Vdd	= 5.0 V	1.0	5.0 V	5.0 V	5.0 V
$Cload_{ext}$	= 5.0 pF	1.0	5.0 pF	5.0 pF	5.0 pF
$Vout_{swing}$	> 3.0 V	1.0	4.2 V	3.3 V	3.2 V
(single ended)	> 3.5 V	0.8			
dc gain	> 63.5 dB	0.5	81.0 dB	70.8 dB	69.0 dB
(single ended)					
slew rate	> 10.0 V/μs	1.0	7.2 V/μs	14.5 V/μs	17.1 V/μs
settle time	< 300 ns	1.0	428 ns	252 ns	260 ns
(0.1%, unity gain feedback, 3.2 V step)	< 200 ns	0.8			
phase margin	> 60°	0.5	73°	74°	75°
$power_{static}$	< 6.0 mW	1.0	5.8 mW	5.1 mW	5.1 mW
cell area	< 5e4 μm^2	0.7	54.6e4 μm^2	20.0e4 μm^2	19.4e4 μm^2
V_{noise} (@ 1 kHz)	—	—	3.8e-3 μv^2/Hz	7.7e-3 μv^2/Hz	6.6e-3 μv^2/Hz
CMRR	—	—	50.0 dB	44.5 dB	42.0 dB

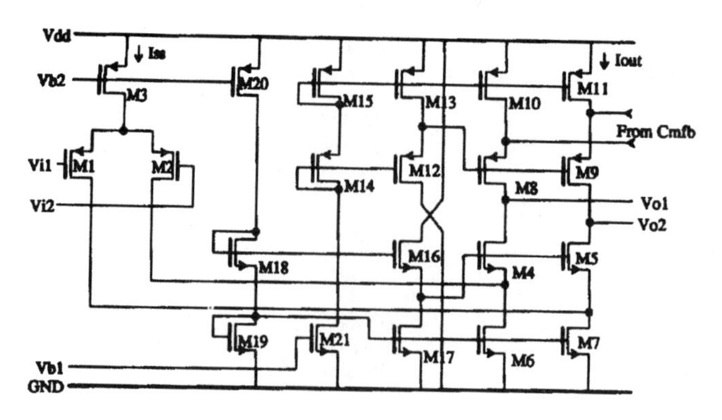

Fig. 16. CMOS folded cascode instance.

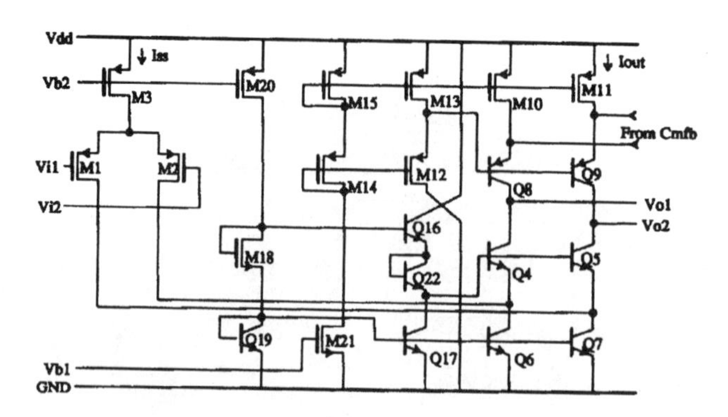

Fig. 17. BiCMOS folded cascode instance.

TABLE IV
SIMULATED AND PREDICTED PERFORMANCE OF BICMOS FOLDED CASCODE EXAMPLE

Attribute	Spec.	Wgt.	STAIC (Optimizer)	HSPICE & ICEWATER
Vdd	= 5.0 V	1.0	5.0 V	5.0 V
$Cload_{ext}$	= 5.0 pF	1.0	5.0 pF	5.0 pF
$Vout_{swing}$	> 3.0 V	1.0	3.2 V	3.0 V
(single ended)	> 3.5 V	0.8		
dc gain (single ended)	> 69.5 dB	0.5	75.6 dB	77.3 dB
slew rate	> 30.0 V/μs	1.0	82.2 V/μs	84.1 V/μs
settle time (0.1%	< 100 ns	1.0	52 ns	65 ns
unity gain feedback, 3.0 V step)	< 20 ns	0.8		
phase margin	> 60°	0.5	89°	88°
$power_{static}$	—	—	23.1 mW	24.0 mW
cell area	3e4 μm^2	0.8	5.7e4 μm^2	4.9e4 μm^2
CMRR	—	—	52.3 dB	48.7 dB

20%. The complete design was synthesized in 680 CPU seconds on a Sun Sparc 2 workstation.

The corresponding layout of the synthesized BiCMOS circuit is illustrated in Fig. 18. The main differential amplifier layout is fully symmetric and is bordered top and bottom by bias circuitry. The routing channels on either side of the main amplifier are intentionally made wide in order to reduce the adverse effects of antisymmetric temperature gradients emanating from the antisymmetric bias blocks. It may be observed that some additional area saving could be realized if the n-p-n elements shown on the lower left and lower right of the main differential amplifier were rotated −90 and +90 deg, respectively. The simple addition of an alternate *layout style* for the composite load block may be readily added to embrace such a change.

IX. CONCLUSIONS

An interactive environment for synthesizing analog circuits of medium complexity has been developed. The program, STAIC, accepts performance specifications in the

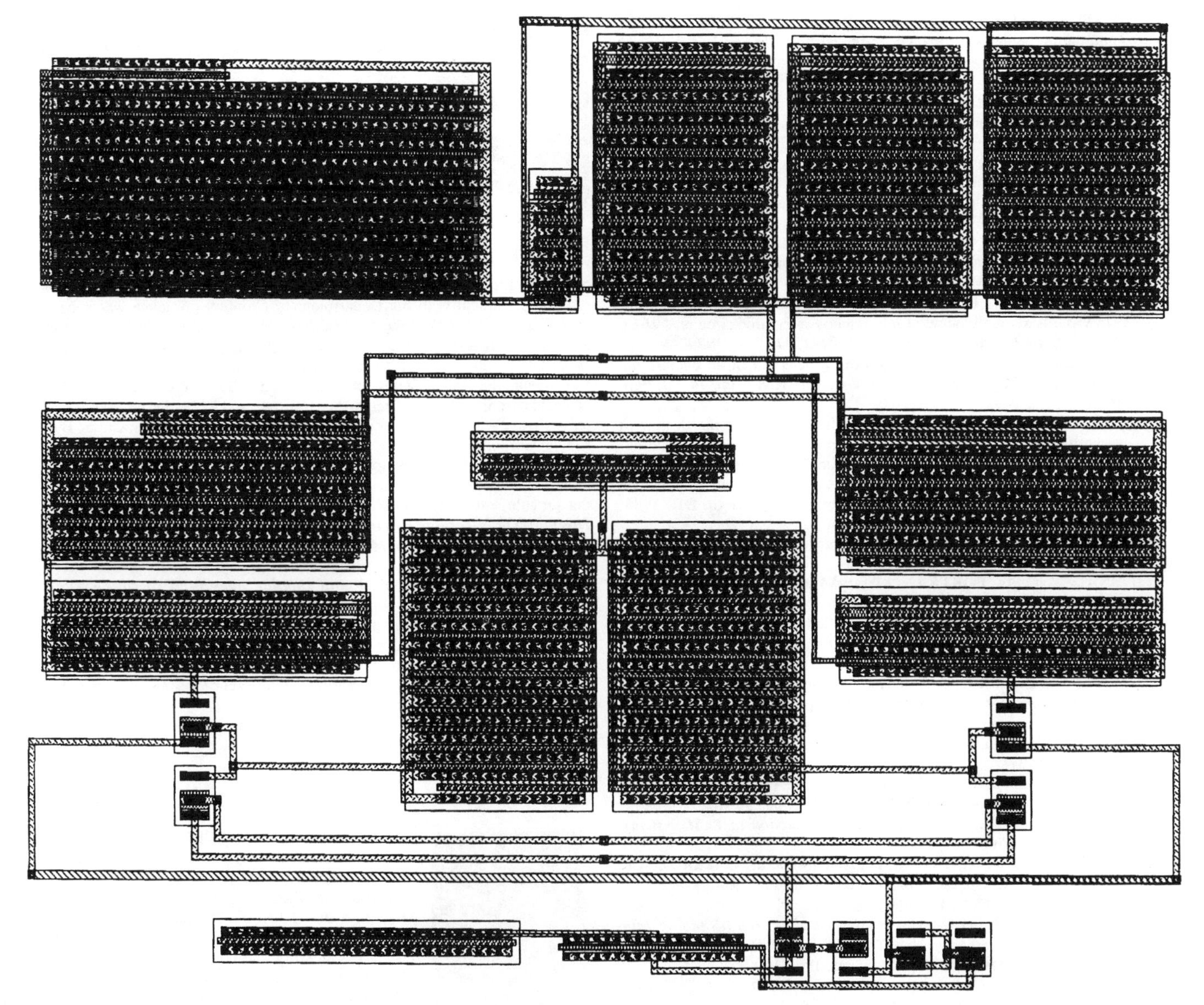

Fig. 18. Synthesized layout of BiCMOS folded cascode example.

form of constraints as input and generates a module generator layout description, SPICE file, and data sheet as output. STAIC provides the expert analog designer with an input description language for entering hierarchical circuit descriptions. A multilevel modeling database perpetuates a *successive solution refinement* methodology. Design exploration is facilitated by optimizer and scanner modules that interact extensively with a central equation database through a numeric/symbolic solve unit. A simple OTA and two realistic folded cascode design examples have been synthesized by STAIC. The results agree favorably with HSPICE simulation and generated layout.

Acknowledgment

The authors thank Dan R. Madill and Sundeep Rangan for their efforts in adding bipolar circuit descriptions and interactive commands into the STAIC framework.

References

[1] W. Nye, D. C. Riley, A. Sangiovanni-Vincentelli, and A. L. Tits, "DELIGHT.SPICE: An optimization-based system for the design of integrated circuits," *IEEE Trans. Computer-Aided Design*, vol. 7, pp. 501-518, Apr. 1988.

[2] H. Y. Koh, C. H. Sequin, and P. R. Gray, "OPASYN: A compiler for CMOS operational amplifiers," *IEEE Trans. Computer-Aided Design*, vol. 9, pp. 113-125, Feb. 1990.

[3] M. G. R. DeGrauwe *et al.*, "IDAC: An interactive design tool for analog CMOS circuits," *IEEE J. Solid-State Circuits*, pp. 1106-1116, Dec. 1987.

[4] J. Rijmenants *et al.*, "ILAC: An automated layout tool for analog CMOS circuits," *IEEE Custom Integ. Circuits*, pp. 7.6.1-7.6.4, 1988.

[5] R. Harjani, R. A. Rutenbar, and R. Carley, "Analog circuit synthesis and exploration in OASYS," in *Proc. ICCAD-88*, 1988, pp. 44-47.

[6] E. Berkcan, M. d'Abreu, and W. Laughton, "Analog compilation based on successive decompositions," in *Proc. 25th DAC*, 1988, pp. 369-375.

[7] R. Harjani, R. A. Rutenbar, and L. R. Carley, "Analog circuit synthesis for performance in OASYS," in *Proc. ICCAD-88*, 1988, pp. 492-495.

[8] H. Onodera, H. Kanabara, and K. Tamaru, "Operational-amplified compilation with performance optimization," *IEEE J. Solid-State Circuits*, vol. 25, pp. 466–473, Apr. 1990.
[9] J. P. Harvey, M. I. Elmasry, and B. Leung, "STAIC: A synthesis tool for CMOS and BiCMOS analog integrated circuits," in *Proc. IEEE ISCAS91*, June 1991, pp. 2004–2007.
[10] S. B. Lippman, *C++ Primer*. Don Mills, Canada: Addison-Wesley, 1989.
[11] D. R. Madill, "Design synthesis of analog integrated circuits employing piecewise-linear approximations," Waterloo, Ont., Canada, Univ. of Waterloo Work Rep., Jan. 1991.
[12] B. W. Char, K. O. Geddes, G. H. Gonnet, M. B. Monagan, and S. M. Watt, *MAPLE Reference Manual*, 5th edition. Waterloo, Ont. Canada: Waterloo Maple, 1990.
[13] G. G. E. Gielen, H. C. C. Walscharts, and W. M. C. Sansen, "ISAAC: A symbolic simulator for analog integrated circuits," *IEEE J. Solid-State Circuits*, vol. 24, pp. 1587-1597, Dec. 1989.
[14] C. T. Chuang, "Analysis of the settling behavior of an operational amplifier," *IEEE J. Solid-State Circuits*, vol. SC-17, pp. 74-80, Feb. 1982.
[15] J. P. Harvey, "STAIC: A synthesis framework for the design of monolithic analog circuits," M.S. dissertation, Dept. of Elect. Comp. Eng., Univ. Waterloo, Ont. Canada, May 1991.
[16] D. Serrano and D. Gossard, "Constraint management in MCAE," in *Artificial Intelligence in Engineering: Design*, J. S. Gero, Ed., 1988, pp. 217-240.
[17] "PORT Mathematical Subroutine Library," AT&T Bell Laboratories, 1984.
[18] "NAG Mark 14 FORTRAN Library Manual," Numerical Algorithms Group, 1990.
[19] B. A. Murtagh and M. A. Saunders, "MINOS 5.0 User's Guide," Stanford Univ., Stanford, CA, Tech. Rep. SOL 83-20, 1983.
[20] H. Y. Koh, C. H. Sequin, and P. R. Gray, "Automatic synthesis of operational amplifiers based on analytic circuit models," in *Proc. ICCAD-87 Computer Aided Design*, 1987, pp. 502-505.
[21] R. Chadha, *et al.*, "WATOPT An optimizer for circuit applications," *IEEE Trans. Computer-Aided Design*, vol. CAD 6, pp. 472-479, May 1987.
[22] P. A. D. Powell and M. I. Elmasry, "The ICEWATER language and interpreter," in *Proc. 21st DAC*, June, 1984, pp. 98-102.
[23] C. A. Mead and L. A. Conway, *An Introduction to VLSI Systems*. Reading, MA: Addison-Wesley, 1980.
[24] *Guide to The Integrated Circuit Implementation Services Of The Canadian Microelectronics Corporation*, Canadian Microelectronics Corp., Kingston, Ont., Canada, 1987.
[25] P. R. Gray and R. G. Meyer, *Analysis and Design of Analog Integrated Circuits*, 2nd ed. New York: Wiley, 1984.
[26] *HSPICE Users Manual*, Meta Software, Inc., 1990.
[27] J. P. Harvey *et al.*, "Design exploration of a class of CMOS op-amps," in *CCVLSI-89*, 1989, pp. 131-138.
[28] S. H. K. Embabi, A. Bellaouar, and M. I. Elmasry, "Analysis and optimization of BiCMOS digital circuit structures," *IEEE J. Solid-State Circuits*, vol. 26, pp. 676-679, Apr. 1991.

J. Paul Harvey received the B.Sc. degree in electrical engineering in 1988 and the M.Sc. degree in electrical engineering in 1991 from the University of Waterloo, Ontario, Canada.

Since May 1991, he has been working as a research associate for the VLSI Group, department of electrical engineering, University of Waterloo. His research interests include the design and design automation of analog and digital circuits and systems.

Mohammed I. Elmasry (S'69–M'73–SM'79–F'88) was born in Cairo, Egypt on Dec. 24, 1943. He received the B.S. degree from Cairo University, Cairo, Egypt, in 1965, and the M.S. and Ph.D. degrees, all in electrical engineering, from the University of Ottawa, Ottawa, Ont., Canada, in 1970 and 1974, respectively.

He has worked in the area of digital integrated circuits and system design for the last 25 years. He worked for Cairo University from 1965 to 1968 and for Bell-Northern Research, Ottawa, Canada, from 1972 to 1974. Since 1974, he has been with the Department of Electrical and Computer Engineering, University of Waterloo, where he is a professor and founding director of the VLSI Research Group. He has a cross appointment with the Department of Computer Science where he is a professor. He has held the NSERC/BNR Research Chair in VLSI design at the same university since 1986. He has served as a consultant in the area of LSI/VLSI digital circuit/subsystem design to research laboratories in Canada and the United States, including AT&T Bell Labs, GE, CDC, Ford Microelectronics, Linear Technology, Xerox, and BNR. During sabbatical leaves from Waterloo he was at the Micro Components Organization, Burroughs Corporation (Unisys), San Diego, CA, Kuwait University, Kuwait, and the Swiss Federal Institute of Technology, Lausanne, Switz. He has authored and co-authored over 150 papers on integrated circuit design and design automation. He has several patents to his credit. He is the editor of *Digital MOS Integrated Circuits* (IEEE Press, 1981), *Digital VLSI Systems* (IEEE Press, 1985), *Digital MOS Integrated Circuits II* (IEEE Press, 1991), and *Design and Analysis of BiCMOS Integrated Circuits* (IEEE Press, 1993). He is also author of *Digital Bipolar Integrated Circuits* (Wiley, 1983), and co-author of *Digital BiCMOS Integrated Circuits* (Kluwer, 1992). He has served many professional organizations in different offices, including chairman of the Technical Advisory Committee of the Canadian Microelectronics Corporation. He is a founding member of the Canadian VLSI Conference and the International Conference on Microelectronics, and he is founding president of Pico Electronics Inc.

Dr. Elmasry is a member of the Association of Professional Engineers of Ontario.

Bosco H. Leung (S'84–M'87) received the B.Sc. degree in electrical engineering from the Rensselaer Polytechnic Institute, Troy, NY, in 1979. He received the M.Sc. degree in electrical engineering from the California Institute of Technology, Pasadena, in 1980 and the Ph.D. degree in electrical engineering and computer science from the University of California, Berkeley, in 1987.

From 1980 to 1982, he was with the Northern Telecom Digital Switching Division in Calgary, Canada, working on the testing of digital central switching office. From 1982 to 1983, he was with the Northern Telecom Business Product Division in Calgary, Canada, working on the analog design of the speakerphone. In 1983 he was awarded the Bell Northern Research Scholarship for his Ph.D. studies. Since 1988 he has been an assistant professor with the Electrical and Computer Engineering Department of the University of Waterloo, Ont., Can. His main research interest is in CMOS and BICMOS analog integrated circuits, in particular for data acquisition applications. He has published over 15 technical papers in the area. He is a principal investigator in the Information Technology Research Center, a center of excellence funded by the Province of Ontario, Canada. He is also an associated researcher in Micronet, a network of centers of excellence focusing on microelectronics research, funded by the Canadian government.

Dr. Leung is a member of the Association of Professional Engineers of Ontario.

Integer Programming Based Topology Selection of Cell-Level Analog Circuits

Prabir C. Maulik, L. Richard Carley, *Senior Member, IEEE,* and Rob A. Rutenbar, *Senior Member, IEEE*

***Abstract*—A new approach to cell-level analog circuit synthesis is presented. This approach formulates analog synthesis as a Mixed-Integer Nonlinear Programming (MINLP) problem in order to allow simultaneous topology and parameter selection. Topology choices are represented as binary integer variables and design parameters (e.g., device sizes and bias voltages) as continuous variables. Examples using a *Branch and Bound* method to efficiently solve the MINLP problem for CMOS two-stage op amps are given.**

I. Introduction

As the number of transistors which can be packed onto a single integrated circuit (IC) increases, it becomes economically desirable to design systems composed of one, or a few, application specific IC's (ASIC's). Many board-level systems have already been converted into single chip implementations, and this trend will only accelerate. It is important to note that many of these system applications require analog circuits in addition to digital ones.

In the design of ASIC's, profit depends on minimizing the time between the definition of an application and completion of a working chip. Therefore, the use of sophisticated automatic synthesis techniques is desirable for ASIC design. While advanced synthesis techniques exist to facilitate rapid design of digital circuitry, only limited progress has been made toward automating the analog circuit design task in general. For example, although a typical mixed-signal ASIC design may consist of 90% digital circuitry and only 10% analog circuitry, the time required for the analog circuit design task typically exceeds that required for the digital one. Therefore, there is clear need for improved CAD tools to support analog circuit design.

The development of analog design automation tools has lagged far behind that of digital tools, in part because analog circuit design has often been regarded as a creative, highly unstructured synthesis problem. Experts rely on years of experience to guide the selection of circuit topologies and parameters for those topologies. Considerable work in the area has been devoted to issues of *parameter selection*, i.e., selecting optimum device sizes and bias points to meet specific performance targets. However, *topology selection* is also crucial in designing high-performance analog circuits; some performance specifications are vastly easier to meet if we are allowed to change not only device sizes, but also the number, type, and interconnection of devices. To date, proposed analog synthesis strategies have relied on a limited set of approaches to perform topology selection: force the user to make the topology selection, employ heuristics to guide topology selection, search exhaustively through all available topologies, or alternate iteratively between topology selection steps and parameter selection steps for various portions of the circuit.

In this paper, we present a new synthesis methodology, based on Mixed-Integer Nonlinear Programming (MINLP) methods, by which topology selection and parameter selection can be carried out *simultaneously*. By *simultaneous* we mean that topology selection and parameter selection are integrated into a single optimization framework. This results in a more efficient method than exhaustive search and a more accurate method than the heuristic and iterative approaches. Our problem formulation uses integer variables to model topology choices and continuous variables for the design parameters such as device sizes and bias voltages. Though less widely applied in VLSI or ASIC applications, it is nevertheless interesting to note that MINLP methods are actually the method of choice for some critical synthesis problems in other engineering disciplines [1]. MINLP techniques offer a natural formulation for synthesis problems that have both a *structural* component (in our case, devices and their interconnections) and a *parametric* component (in our case, the dimensions and dc operating point of each device).

II. Background

Early work [2]–[4] in the area of analog circuit synthesis addressed parameter selection for a fixed analog circuit topology using numerical optimization techniques. In general, these optimization strategies rely on circuit simulation in order to predict the performance of an analog circuit, given a set of design parameters. Current examples include DELIGHT.SPICE [5] and its successors [6] from the University of California at Berkeley and OASYS [7] from Texas Instruments. Although this approach has proven successful for some highly structured analog synthesis problems (e.g., analog filter synthesis), it

Manuscript received June 1, 1993; revised October 5, 1994. This work was supported in part by the Semiconductor Research Corporation under Contract 91-DC-068. This paper was recommended by Associate Editor J. White.

P. C. Maulik was with the Department of Electrical and Computer Engineering, Carnegie Mellon University, Pittsburgh, PA 15213 USA. He is now with Crystal Semiconductor Corporation, Austin, TX 78760 USA.

L. R. Carley and R. A. Rutenbar are with the Department of Electrical and Computer Engineering, Carnegie Mellon University, Pittsburgh, PA 15213 USA.

IEEE Log Number 9409638.

has been unable to address the general problem of selecting transistor sizes and bias points for analog circuits. In part, this is because of the large computational cost of circuit simulation. These circuit simulation-based approaches are typically used to optimize a good hand design or to redesign an existing circuit for a slightly different fabrication process.

A new generation of circuit synthesis tools have emerged that, instead of using circuit simulation, employ analytical equations as behavioral models to predict the performance of an analog circuit [8]. IDAC [9], the first commercially available synthesis tool for general analog cells like op amps and comparators, requires the human designer to choose one from a set of available topologies and then follows a procedural routine created by an expert designer to perform coarse parameter selection. Fine parameter selection uses numerical optimization based on topology-specific circuit simulation. IDAC does not address the problem of automating topology selection. A hallmark of the expert analog designer is the ability to choose from a wide variety of topologies in response to different performance requirements. Therefore, it is very important to consider design automation methods that can address the topology selection problem.

OASYS [10], developed at Carnegie Mellon University, approaches the goal of simultaneous topology selection and device sizing. It walks down a hierarchical decomposition of an analog cell, alternately selecting topologies and translating performance specifications from each level to the next lower level, and backtracking to handle design failures. This results in a depth-first search through the space of possible topologies. Although OASYS can automatically design a wide variety of topologies, creating the topology specific design knowledge—especially the knowledge required for backtracking—is very difficult and time consuming.

OPASYN [11] uses heuristic rules to choose a topology from a set of available device-level schematics and then uses numerical optimization to perform parameter selection using equations that predict circuit and subcircuit performance. HECTOR [12], [13] uses a "hierarchical, interactive and expert-system assisted" approach to construct circuit topologies from a library of basic building blocks. In both of these approaches, topology selection employs heuristics which must be provided by an expert analog designer. However, it can be very difficult to create rules that accurately predict the best topology choice without actually carrying out, at least partially, the parameter selection process.

SEAS [14], like OASYS and HECTOR, uses a hierarchical description of a variety of analog circuit topologies. It uses simulated evolution to change from one topology to another, based on a score generated after selection of parameters for the current topology. Although this approach avoids the need for expert designers to provide heuristics for topology selection, the methods for scoring subblocks in a topology must be defined by the expert designer. And, more important, choice of a new topology is made only after complete parameter selection of the current topology.

All of the above approaches separate topology selection and parameter selection. In some cases there is a fine-grained separation (e.g., OASYS) while in most of the other cases the separation is coarse grained (e.g., SEAS). The *principal contribution* of the work presented in this paper is the elimination of this separation; i.e. *simultaneous* topology selection and parameter selection.

In analog cells, the same functionality can be implemented using different connections of different numbers of components. Analog cells have a hierarchical nature to them in that they can be decomposed into subcells, which can be further decomposed into components like transistors. Each subcell can be implemented in various ways. The total number of topologies, therefore, can be combinatorially large, because a topology choice for one subcircuit can be combined with topology choices for other subcircuits. Which topology is optimal or even feasible will depend on the sizes of the devices chosen for that topology, and this in turn will depend on the specifications. It is hard therefore to make conclusions about a particular topology without having determined the sizes of the devices of that topology. In the worst case, it might be necessary to size all the topologies to determine the one that is most suitable for a particular set of specifications. It is therefore clear that any method which addresses the problems of topology selection and sizing simultaneously will be a more efficient method of solving the analog circuit synthesis problem. The following sections present how simultaneous topology selection and parameter selection can be formulated as a *Mixed-Integer Nonlinear Programming* (MINLP) problem.

III. Symbols and Notation

Before formulating our optimization problem we first present some notation. Throughout this paper, constraints and objective functions will be expressed in terms of small-signal and large-signal parameters of MOS transistors. The circuit variables used in the optimization problems described in this paper are generally the widths, lengths, gate-to-source, drain-to-source, and body-to-source voltages of MOS transistors. The following notation will be used in this paper.

- W_i — width of transistor i.
- L_i — length of transistor i.
- V_{DSi} — Drain to source voltage of device i.
- V_{GSi} — Gate to source voltage of device i.
- V_{BSi} — Body to source voltage of device i.

The items in the list above are the independent parameters of the MOS transistor i which in turn determine its small-signal and large-signal parameters *viz.*

- g_{mi} — Transconductance of device i.
- r_{oi} — Output resistance of device i.
- g_{oi} — Output conductance of device i.
- I_{Di} — Quiescent drain current of device i.

The exact functions corresponding to each of the above depend on the device model being used. For simplicity, parasitic capacitances have been ignored in the formulation examples in this paper. They are, however, included in the actual implementation.

For the supply voltages, the following notation is used.

- V_{dd} — Positive supply voltage.
- V_{ss} — Negative supply voltage.

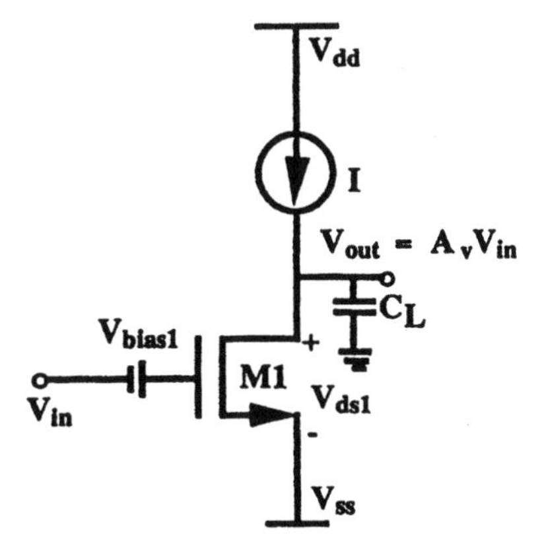

Fig. 1. Simple common-source amplifier.

Lastly, the area of a MOS transistor will be calculated as follows:

$$Area_i = W_iL_i + 2W_{diff}W_i$$

where W_{diff} is the width of the drain and source diffusions.

IV. Problem Formulation

Analog circuits can typically be represented as a sequence of subcircuit choices. Binary (0/1) variables can be used to represent the inclusion or exclusion of a subcircuit topology from the final circuit. The design variables like transistor sizes, etc., have to be represented by continuous variables. The objective function and constraints will typically be nonlinear. The formulation to be described will therefore be a MINLP problem. We illustrate the formulation with a simple example.

Fig. 1 shows a simple common source (CS) amplifier. Equations can be derived for the various performance parameters of this amplifier like gain, bandwidth, etc., in terms of design variables $\mathbf{x}$, which, in this case, are device sizes and bias voltages. Using the performance equations, we can formulate the design of this amplifier as a *constrained optimization problem* [15]–[17] as follows:

$$\min_{\mathbf{x}} Area(\mathbf{x})$$

subject to

$$Gain(\mathbf{x}) \geq \text{GAIN} \quad (1)$$
$$SlewRate(\mathbf{x}) \geq \text{SLEW RATE} \quad (2)$$
$$Unity\text{-}GainFreq.(\mathbf{x}) \geq \text{UGF} \quad (3)$$
$$Phasemargin(\mathbf{x}) \geq \text{PHASE MARGIN} \quad (4)$$
$$I_{d1}(\mathbf{x}) = I \quad (5)$$
$$X_L \leq \mathbf{x} \leq X_U. \quad (6)$$

The specifications are indicated in capital letters on the right-hand side (RHS) of the constraints[1]. Constraints (1) to (4) are examples of *performance constraints* while (5) is an example of a *functional constraint* which assures that the circuit satisfies KCL. The performance functions like $Gain(\mathbf{x})$, etc. can be written as algebraic equations in terms of small-signal parameters like $g_{m1}(\mathbf{x})$ and $g_{ds1}(\mathbf{x})$.

[1] This convention will be followed throughout the rest of the paper.

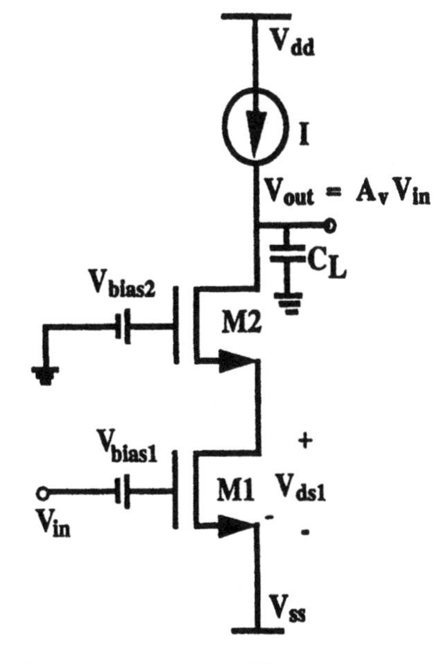

Fig. 2. Cascoded common-source amplifier.

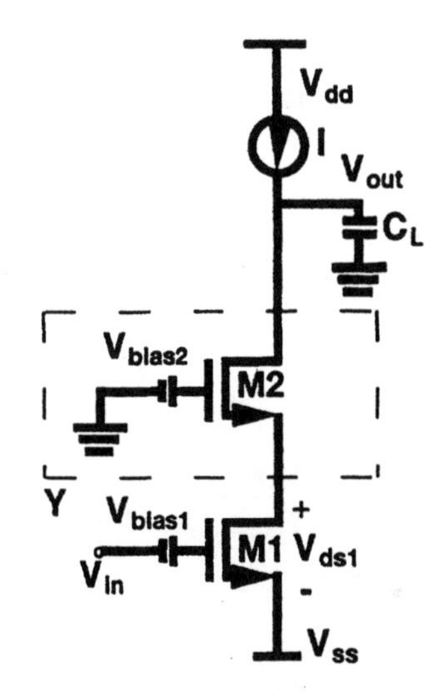

Fig. 3. Supercircuit of common-source topologies.

Fig. 2 shows an alternate topology for the common source amplifier in which a new transistor M2 has been added. This transistor, known as a cascode transistor, serves to increase the output impedance and therefore the gain of this amplifier. In order to select between these two topologies, we introduce a binary variable Y. When Y is 0 the simple topology is selected and when Y is 1, the cascoded topology is chosen. Fig. 3 shows a combined circuit from which the transistor M2 is switched in or switched out depending on the binary variable Y. We call such a combined circuit a *supercircuit*.

In order to illustrate how the binary variable is used, we show how it is incorporated into the performance and functional constraints of our supercircuit. Let us first consider the gain constraint. Let R be the output impedance of our supercircuit and let $R^s_{out}(\mathbf{x})$ and $R^c_{out}(\mathbf{x})$ denote the output impedances of the simple and cascoded cases respectively. We write an equality constraint of the form

$$R = (1 - Y)R^s_{out}(\mathbf{x}) + YR^c_{out}(\mathbf{x}). \quad (7)$$

We can see from the above equality constraint that R takes on the appropriate value depending on the value of Y. However, we cannot use the constraint in the above form because the multiplication of Y with a function such as $R^c_{out}(\mathbf{x})$ may introduce additional local minima which might adversely affect our solution method (this will be explained in a later section). In order to assure that the binary variable Y occurs only in linear terms we implement the same functionality using four inequality constraints as follows:

$$R - R^s_{out}(\mathbf{x}) + Y\Gamma_i \geq 0 \quad (8)$$
$$R - R^s_{out}(\mathbf{x}) - Y\Gamma_j \leq 0 \quad (9)$$
$$R - R^c_{out}(\mathbf{x}) + (1-Y)\Gamma_k \geq 0 \quad (10)$$
$$R - R^c_{out}(\mathbf{x}) - (1-Y)\Gamma_l \leq 0. \quad (11)$$

It can be seen from the above four constraints that if the Γ_i to Γ_l constants are chosen correctly, then, when Y is 0, the first two constraints will assure that R is equal to R^s_{out}, the third and fourth constraints being met trivially. Conversely, when $Y = 1$, the third and fourth constraints will assure that R is equal to R^c_{out}, the first two constraints being met trivially in this case. And having assured that R takes on the correct value, we can now use it in the gain constraint as follows:

$$g_{m1}R \geq GAIN \quad (12)$$

Let us next consider a functional constraint like the KCL constraint. We know that the dc bias current of transistor M2 needs to be equal to that of M1 only when M2 is present in our supercircuit, i.e., only when $Y = 1$. This can be implemented by the following equality constraint:

$$I_{D2}(\mathbf{x}) = YI_{D1}(\mathbf{x}) \quad (13)$$

For reasons mentioned before, we do not use the constraints in this form, but instead use a pair of inequality constraints of the form:

$$I_{D1}(\mathbf{x}) - I_{D2}(\mathbf{x}) + (1-Y)\Gamma_m \geq 0 \quad (14)$$
$$I_{D1}(\mathbf{x}) - I_{D2}(\mathbf{x}) - (1-Y)\Gamma_n \leq 0. \quad (15)$$

In an optimization problem, when the constraints are such that only one of a pair, or, more generally, k of m sets of constraints must hold, the constraints are said to be *disjunctive*. The constraints presented so far fall into this class. Also, this formulation is not restricted to binary variables. For example, if there were five kinds of common-source amplifiers, we would have had to use an integer variable which could assume values between 0 and 4.

The complete problem formulation is shown below.

$$\min(Area_1 + Area_2 - (1-Y)Area_{min})$$

subject to **Performance Constraints**

$$R - R^s_{out}(\mathbf{x}) + Y\Gamma_1 \geq 0 \quad (16)$$
$$R - R^s_{out}(\mathbf{x}) - Y\Gamma_2 \leq 0 \quad (17)$$
$$R - R^c_{out}(\mathbf{x}) + (1-Y)\Gamma_3 \geq 0 \quad (18)$$
$$R - R^c_{out}(\mathbf{x}) - (1-Y)\Gamma_4 \leq 0 \quad (19)$$
$$g_{m1}R \geq GAIN \quad (20)$$
$$\frac{I_{D1}(\mathbf{x})}{C_L} \geq SLEWRATE \quad (21)$$
$$\frac{g_{m1}(\mathbf{x})}{2\pi C_L} \geq UGF \quad (22)$$

KCL Constraints

$$I_{D1}(\mathbf{x}) = I \quad (23)$$
$$I_{D1}(\mathbf{x}) - I_{D2}(\mathbf{x}) + (1-Y)\Gamma_5 \geq 0 \quad (24)$$
$$I_{D1}(\mathbf{x}) - I_{D2}(\mathbf{x}) - (1-Y)\Gamma_6 \leq 0 \quad (25)$$

KVL Constraints

$$V_{C1}(\mathbf{x}) - V_{DS1}(\mathbf{x}) + (1-Y)\Gamma_7 \geq 0 \quad (26)$$
$$V_{C1}(\mathbf{x}) - V_{DS1}(\mathbf{x}) - (1-Y)\Gamma_8 \leq 0 \quad (27)$$
$$V_{C2}(\mathbf{x}) - (-V_{ss}) + (1-Y)\Gamma_9 \geq 0 \quad (28)$$
$$V_{C2}(\mathbf{x}) - (-V_{ss}) - (1-Y)\Gamma_{10} \leq 0 \quad (29)$$
$$V_{DS1}(\mathbf{x}) - (-V_{ss}) + Y\Gamma_{11} \geq 0 \quad (30)$$
$$V_{DS1}(\mathbf{x}) - (-V_{ss}) - Y\Gamma_{12} \leq 0 \quad (31)$$

Variable Range Constraints

$$W_{min} \leq W_1 \leq W_{max} \quad (32)$$
$$L_{min} \leq L_1 \leq L_{max} \quad (33)$$
$$W_2 \leq Y(W_{max} - W_{min}) + W_{min} \quad (34)$$
$$L_2 \leq Y(L_{max} - L_{min}) + L_{min} \quad (35)$$
$$W_1 \geq W_{min} \quad (36)$$
$$L_1 \geq L_{min} \quad (37)$$
$$Y \in \{0,1\} \quad (38)$$

where

$$R^s_{out}(\mathbf{x}) = r_{o1}(\mathbf{x}) \quad (39)$$
$$R^c_{out}(\mathbf{x}) = g_{m2}(\mathbf{x})r_{o1}(\mathbf{x})r_{o2}(\mathbf{x}) \quad (40)$$
$$V_{C1}(\mathbf{x}) = V_{bias2} - V_{GS2}(\mathbf{x}) \quad (41)$$
$$V_{C2}(\mathbf{x}) = V_{DS1}(\mathbf{x}) + V_{DS2}(\mathbf{x}) \quad (42)$$
$$Area_1(\mathbf{x}) = W_1L_1 + 2W_1W_{diff} \quad (43)$$
$$Area_2(\mathbf{x}) = W_2L_2 + 2W_2W_{diff} \quad (44)$$
$$Area_{Min} = W_{Min}L_{Min} + 2W_{Min}W_{diff} \quad (45)$$
$$\Gamma_1 = 0 \quad (46)$$
$$\Gamma_2 = g_{mMax}R^2_{outMax} \quad (47)$$
$$\Gamma_3 = g_{mMax}R^2_{outMax} \quad (48)$$
$$\Gamma_4 = 0 \quad (49)$$
$$\Gamma_5 = I_{DMax} - I_{DMin} \quad (50)$$
$$\Gamma_6 = I_{DMax} - I_{DMin} \quad (51)$$
$$\Gamma_7 = V_{dd} \quad (52)$$
$$\Gamma_8 = V_{dd} \quad (53)$$
$$\Gamma_9 = 0 \quad (54)$$
$$\Gamma_{10} = 2V_{dd} \quad (55)$$
$$\Gamma_{11} = -V_{ss} \quad (56)$$
$$\Gamma_{12} = 0. \quad (57)$$

The superscripts s and c denote simple and cascode, respectively. The Γ_i's are suitable positive constants. R_{outMax}, R_{outMin}, g_{mMax}, g_{mMin}, I_{DMax}, and I_{DMin} are the maximum and minimum values of R_{out}, g_m, and I_D, respectively[2]. The functions $R^s_{out}(\mathbf{x})$, $I_{D1}(\mathbf{x})$, $g_{m1}(\mathbf{x})$, etc., are not shown here because they are evaluated by calling an *encapsulated device evaluator* [17]. In the actual implementation constraints and variables are scaled for numerical stability and efficiency [15], [17]. The nonlinear solver used was NPSOL [18].

The KCL and KVL constraints can be generated algorithmically given a set of topology choices. The performance constraints, however, are generated with the help of circuit analysis. This requires the involvement of a circuit designer in the development of this kind of a synthesis tool. IDAC [9] has attempted to automatically generate the analysis equations with the help of symbolic analysis. A more recent approach, somewhere between the equation-based and simulation-based approaches is ASTRX/OBLX [19]–[21]. ASTRX/OBLX uses an extremely efficient method [22] to automatically determine all small-signal performance measures, thus eliminating the effort required to create equations for predicting performance. To date, ASTRX/OBLX has only been applied to the parameter selection problem. However, the MINLP formulation presented in this paper can easily be incorporated into the ASTRX/OBLX framework.

A. Choosing the Γ's

The value of Γ_1 is chosen to be zero because R will always be equal to or greater than $R^s_{out}(\mathbf{x})$. This is because $R^s_{out}(\mathbf{x})$ is always smaller than $R^c_{out}(\mathbf{x})$. The value of Γ_2 can be chosen to be equal to the maximum possible value of $R^c_{out}(\mathbf{x})$ because that is the maximum possible positive value that $R - R^s_{out}(\mathbf{x})$ can evaluate to. The maximum value of $R^c_{out}(\mathbf{x})$ can be estimated from the maximum length of the MOS transistors and the minimum value of currents in the circuit. The minimum value of currents in the circuit can be estimated from a slew rate specification. The values of Γ_5 and Γ_6 in the KCL constraints can be chosen by assessing the maximum possible difference in currents between the two transistors. The minimum current can be estimated, as mentioned before from the slew rate specification, and the maximum current can be estimated from a power specification. The values of Γ_i for the KVL constraints can be estimated from the minimum and maximum node voltages in the circuit which can be assumed to be equal to the positive and negative supply voltages respectively. It can be seen that the values of Γ_i are chosen very conservatively. For example, the value of the maximum output impedance of any transistor in the circuit might be much lower than R_{outMax} calculated from the minimum possible value of current and the maximum length of the transistor. The reason for choosing the Γ_i's conservatively will become clear in a later section.

B. Objective Function

The objective function is the "effective" area of our supercircuit. It is formulated in such a way that it is equal to $Area_1$ when $Y = 0$ and equal to $Area_1 + Area_2$ when $Y = 1$. The variable range (34) and (35) set W_2 to W_{min} and L_2 to L_{min} when $Y = 0$, which in turn assures that $Area_2$ is set to $Area_{min}$ when $Y = 0$.

[2] For simplicity, parasitic capacitances are ignored in this example.

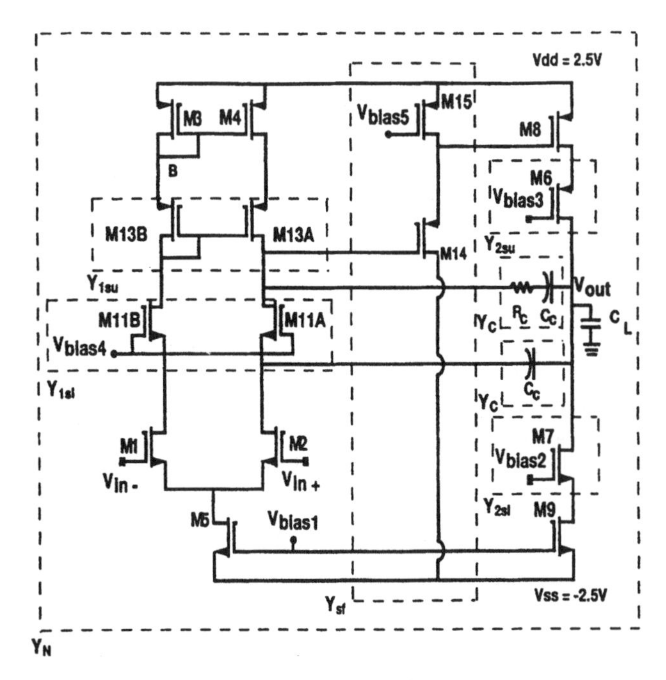

Fig. 4. Two-stage CMOS op amp Supercircuit.

C. MINLP Formulation for a CMOS Two-Stage Amplifier

We next consider the MINLP formulation for the design of a more complex circuit: a two-stage CMOS op amp. The topology choices and the binary variables used to represent them are as follows.

1) Differential pair, simple or cascoded (Y_{1sl}).
2) Current mirror, simple or cascoded (Y_{1su}).
3) Source follower between first and second stages (Y_{sf}).
4) Transconductance amplifier, simple or cascoded (Y_{2su}).
5) Load of transconductance amplifier, simple or cascoded (Y_{2sl}).
6) Resistance compensation or cascode compensation (Y_c).
7) NMOS or PMOS input with all other transistors of corresponding polarity (Y_N).

Fig. 4 shows the supercircuit of this design problem. The boxed subcircuits can be selectively included depending on the value of the binary variable indicated at the bottom left corner of each box.

Cascoding allows higher gain at the cost of common mode range and output swing depending on whether it is in the first or second stage, respectively. The source follower between the two stages can be used only if the current mirror of the first stage is cascoded, in order to achieve a lower systematic offset voltage and to allow a higher output voltage swing at the cost of extra area and power. Also, the cascode compensation can be used only if the differential pair is cascoded. Thus even though we have 7 binary variables, the total number of possible topologies is 64. In the next section we show how infeasible combinations of binary variables can be avoided

TABLE I
LOGICAL OPERATORS AND CORRESPONDING LINEAR EXPRESSIONS

Logical operation	Linear expression
$\overline{Y}$	1 - Y
$Y_i \wedge Y_j$	$Y_i \geq 1$ $Y_j \geq 1$
$Y_i \vee Y_j$	$Y_i + Y_j \geq 1$
$Y_i \oplus Y_j$	$Y_i + Y_j = 1$
$Y_i \Rightarrow Y_j$	$(1 - Y_i) + Y_j \geq 1$

using logical relations. Some details of the MINLP formulation for the CMOS two-stage op amp are given in Appendix A. The formulation consists of 7 binary variables, 39 continuous variables, 24 linear constraints, and 58 nonlinear constraints.

D. Logical Relations as Linear Inequalities

There can be logical relations between the binary variables. For example, we know that the source follower should be used only when the current mirror in the first stage is cascoded. This can be enforced by the logical relation $Y_{sf} \Rightarrow \overline{Y_{1su}}$. All logical relations can be modeled as linear constraints using the conversion relations shown in Table I [23].

V. SOLVING THE MINLP PROBLEM

MINLP problems are difficult to solve and fall into the class of NP-hard problems [24]. Some existing solution methods are Branch and Bound [25], Generalized Bender's Decomposition [26] and Outer Approximation [27]. We choose Branch and Bound in this work because it is simple, and for the problem size of our example, performs well enough to bring out the viability of the MINLP formulation. However, it must be remembered that the other two methods mentioned above are more sophisticated and may perform significantly better for large problems in terms of computation speed and robustness.

A. Branch and Bound

The Branch and Bound algorithm works by generating a tree (Fig. 5), using the binary variables to do branching [23]. When the integrality constraints are removed from an integer programming problem, the problem is said to be *relaxed*. Branch and Bound solves a sequence of relaxed problems The solution at any node provides a lower bound on the nodes below it. Any integer feasible solution is an upper bound on the final solution, and can be used to prune nodes which have lower bounds higher than the current upper bound. The method terminates when there are no more nodes to branch on. On termination, the current upper bound gives the final solution. The efficiency of the method depends on the number of nodes evaluated and this in turn depends on the branching heuristics used and the quality of the lower bounds. We adopt standard branching heuristics such as branching on the node with the lowest lower bound and using the variable furthest from integer as the next branching variable [25], [28]. The quality of the lower bounds depends on the value of the Γ_i constants. This is illustrated by considering the following

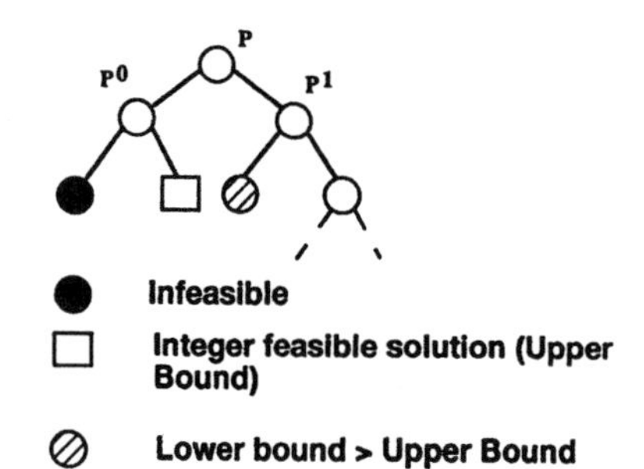

Fig. 5. Branch and Bound Tree.

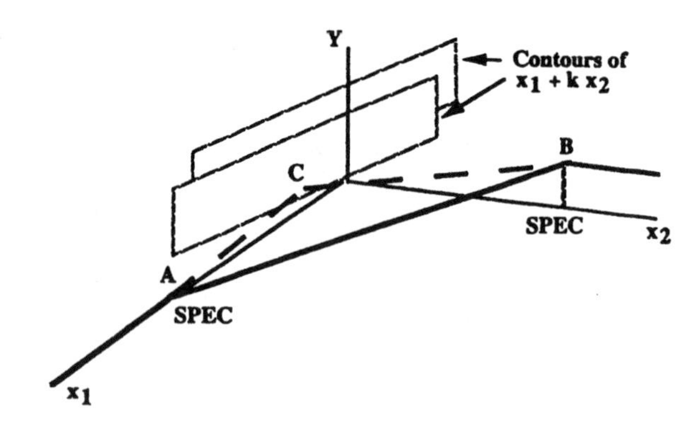

Fig. 6. Solution space of example problem.

simple problem:

$$\min \; x_1 + kx_2$$

subject to

$$x_1 + Y\Gamma_1 \geq SPEC \tag{58}$$
$$x_2 + (1 - Y)\Gamma_2 \geq SPEC \tag{59}$$
$$x_1 \leq (1 - Y)X_U \tag{60}$$
$$x_2 \leq YX_U \tag{61}$$
$$x_1, x_2 \geq 0. \tag{62}$$

The solution space of this problem is shown in Fig. 6. Since the objective function and constraints in this example are linear, this is actually a Mixed-Integer Linear Programming (MILP) problem.

The two points A and B form the boundary of the feasible region of the problem above. The line AB is the boundary of the feasible region of the relaxed problem when $\Gamma_1 = \Gamma_2 = SPEC$. One of the two points A and B will be the optimum point, depending on the value of k. This is the tightest relaxation we can possibly have, because any value of Γ_1 or Γ_2 lower than SPEC is going to render the problem infeasible at $Y = 1$ or $Y = 0$, respectively.

Now, suppose we increase the value of Γ_2 to 2 $\times SPEC$. The boundary of the feasible region of the relaxed problem now becomes ACB where C is the point (0.5, > 0, > 0.5). We can see, from Fig. 6 that C, which has a noninteger value of Y, now becomes the optimal point. The value of the objective function at C is larger than that at A, and therefore, the lower bound obtained by solving this problem has diminished in "quality."

When the functions are nonlinear, however, we might need large values for the Γ_i's. For example, let us assume that $g(\mathbf{x})$ is nonlinear in the following constraint:

$$g(\mathbf{x}) + Y\Gamma \geq SPEC. \tag{63}$$

Suppose we choose Γ based on the minimum value of $g(\mathbf{x})$, i.e., $\Gamma = SPEC - g(\mathbf{x})_{MIN}$. It is known that NPSOL linearizes nonlinear constraints in its subproblems [29]. The minimum value of the linearization of $g(\mathbf{x})$ at a certain point, may be smaller than $g(\mathbf{x})_{MIN}$. Therefore, with this choice of Γ, infeasibilities might occur in NPSOL subproblems. If an infeasibility is obtained where a feasible point exists, this is equivalent to a wrong lower bound, which may result in Branch and Bound giving a wrong answer. Hence, the Γ_i's are chosen conservatively. This might result in inefficiency, in terms of the total number of nonlinear problems solved, but correctness is more important than inefficiency. Conversely, the Γ_i's cannot be made too large, because that would make one term in the constraints very large compared to the others which might make the problem ill-conditioned [15]. Examples of rules for choosing the Γ_i's were presented in Section IV-A.

The Branch and Bound method requires that a global optimum be reached at each node [24]. Since we have a nonconvex optimization problem, this guarantee cannot be given [15]. We try to make our problem "well-behaved" by doing the following.

- Reduce nonlinearities in our formulation by transforming constraint functions [28].
- Reduce nonlinearities in our formulation by approximating constraint functions [28].
- Introduce additional constraints to ease the search for a feasible point [28].
- Choose the values of the Γ constants conservatively in order to avoid subproblem infeasiblities [28].
- Use simple device models. This can be justified by the fact that designers use simple models to decide on a topology and the intricacies of the more complex models do not greatly influence the topology choice. The sensitivity of the design to a specific device model can be reduced by allowing sufficient margin for model parameter variation during the formulation of the constraints. After the topology has been chosen, the circuit can be re-sized using more sophisticated device models [17]. If after resizing the selected topology does not turn out to be suitable, branch and bound can be run again with this topology eliminated from the list of choices. This whole process can be repeated till a suitable topology is obtained. Elimination of a topology can be easily incorporated by means of linear constraints.

Our experiments have shown that on using these precautions, Branch and Bound works correctly and gives the same values as obtained from exhaustive search for a wide range of specifications.

VI. Topology Selection Experiments

A program was written to implement the MINLP formulation of 64 topologies of a CMOS two-stage op amp. The program takes a set of specifications and generates a sized schematic, i.e., selects a topology and sizes the transistors of that topology. In this section we present some results of op amps designed using this program.

Several design experiments were done using the program. All the designs were evaluated based on efficiency and correctness. For each selected topology, the circuit rationale behind the choice of topology is presented.

The following are the performance specifications we used in our experiments.

- Gain-bandwidth product = 10 MHz.
- Phase Margin = 45°.
- Slew rate = 10 V/μs.
- Load Capacitance = 1 pF.
- Output Swing = $\pm$ 3 V.
- Flicker Noise = 200 nV/$\sqrt{\text{Hz}}$ at 1 kHz.
- Common Mode Range = $\pm$ 1 V.
- Q factor of 2nd order loop = 0.5 [28], [30].
- Input-referred Systematic Offset Voltage = 5 mV.
- Maximum width of MOS transistors = 400 μm.
- Maximum length of MOS transistors = 10 μm.
- Technology: 2 μm CMOS.

The precise value of one or more specifications was varied from experiment to experiment. The objective function used in all cases was total device area which is calculated as the sum of gate, source, and drain areas. In each experiment, four designs are attempted with gain specifications of 66 dB, 80 dB, 94 dB, and 114 dB respectively. The acronym "TA" is used for the transconductance amplifier of the second stage which consists of transistor M8 for the simple case and transistors M8 and M6 for the cascoded case. The acronym SF is used for the source follower. Power is calculated by multiplying the total quiescent current with the power supply voltage. The assumption made in this method to calculate power is that the dynamic power dissipation is dominated by the load capacitance and hence is independent of the topology. All experiments were run on a DEC 3100[3] workstation.

A. Experiment 1

In the first experiment we use the specifications as shown above. The results are shown in Table II. For the 80 dB design, the differential pair is cascoded. This is because this cascoding is cheapest in terms of area since the current in the first stage is lower than that in the second stage, thus allowing the first stage devices to be smaller. However, on increasing the gain specification further, a double cascoding becomes necessary. The current mirror in the first stage cannot be cascoded because of the common mode range specification and hence both the transconductance amplifier and load of the second stage are cascoded.

B. Experiment 2

In the second experiment we increase the output swing specification to $\pm$ 4.4 V. All other specifications remained the same. The results are shown in Table III. As in the previous

[3]Trademark of Digital Equipment Corporation.

TABLE II
EXPERIMENT 1

Gain(dB) Spec/Design	Area (μ^2)	Power (μW)	No. of nodes	CPU time (secs)	Topology
66/77	2907	164	14	957	All simple
80/87	2979	164	18	1375	Diff. pair cascoded
94/101	3161	144	20	1651	TA and Load cascoded
114/118	3539	119	22	1422	TA and Load cascoded

TABLE III
EXPERIMENT 2

Gain(dB) Spec/Design	Area (μ^2)	Power (μW)	No. of nodes	CPU time (secs)	Topology
66/77	2907	164	14	957	All simple
80/87	2980	165	14	901	Diff. pair cascoded
94/95	5779	108	24	2175	TA cascoded
114/123	8456	502	22	1738	Current mirror cascoded, SF TA and Load cascoded

TABLE IV
EXPERIMENT 3

Gain(dB) Spec/Design	Area (μ^2)	Power (μW)	No. of nodes	CPU time (secs)	Topology
66/87	3270	137	16	684	Diff pair cascoded
–	–	–	16	848	Infeasible

TABLE V
EXPERIMENT 4

Gain(dB) Spec/Design	Area (μ^2)	Power (μW)	No. of nodes	CPU time (secs)	Topology
66	–	–	2	53	Infeasible

TABLE VI
EXPERIMENT 5

Gain(dB) Spec/Design	Area (μ^2)	Power (μW)	No. of nodes	CPU time (secs)	Topology
66/69	2269	991	18	1349	Diff pair cascoded
80/81	3065	594	18	1611	Diff pair and TA cascoded
94/106	4248	534	36	2501	Diff pair, TA, Load cascoded
114/120	5950	871	36	3187	Diff pair, Curr. mirr., TA cascoded

TABLE VII
EXPERIMENT 6

Gain(dB) Spec/Design	Area (μ^2)	Power (μW)	No. of nodes	CPU time (secs)	Topology
66/74	1938	98	6	472	Diff pair cascoded Cascode compensation
80/80	2040	97	10	548	Diff pair and TA cascoded Cascode compensation
94/102	2163	96	10	591	Diff pair, TA and Load cascoded Cascode compensation
114/119	2452	124	16	1409	Diff pair, Curr Mirr and TA cascoded Source follower, Cascode compensation

experiment, the differential pair is cascoded for moderate values of the gain specification. However for the 94 dB design cascoding the transconductance amplifier and increasing the length of the load transistor becomes more efficient. The 114 dB design requires 3 cascodings and has to use the source follower as a level-shifter between the first and the second stage in order to meet the high output swing specification.

C. Experiment 3

In this experiment we increase the output swing specification even further to 4.6 V. All other specifications remained the same as in Experiment 2. The results are shown in Table IV. In this experiment only the low gain op amp could be designed. This is because the second stage cannot be cascoded at all because of the high output swing specification, and the current mirror in the first stage cannot be cascoded because of the common mode range specification.

D. Experiment 4

In this experiment the output swing specification is brought back to $\pm$ 3 V but the slew rate specification is increased to 50 V/μs. The results are shown in Table V. In this case, even the low gain op amp could not be designed. The high slew rate enforces a high V_{GS} on the input transistors M1 and M2 and this conflicts with the common mode range specification [31].

E. Experiment 5

In this experiment the specifications are the same as the previous experiment except that the common mode range specification is reduced to 0 V. This is typical of switched-capacitor circuits where the positive input of the op amp is normally tied to ac ground [32]. The results are shown in Table VI. All the op amps could be successfully designed in this case. This experiment brings out the fact that it is difficult to devise heuristics to do topology selection. A combination of the slew rate and common mode range specification caused the difference between this and the previous experiment. If we compare the designs in Table VI with the designs in Table II we find that more cascodings are needed in the former for the same gain specifications. This is because the higher currents demanded by the higher slew rate specification reduces the gain of the MOS transistors [31].

F. Experiment 6

In this experiment we specify the high frequency power supply rejection specification to be −40 dB. The common mode range specification is 0 V. The Q specification of the complex pole pair of the high PSR topology is set to be a maximum of 0.6 [28], [30]. The results are shown in Table VII. Since the resistive zero-cancellation method of compensation cannot achieve a high-frequency PSR better than 0 dB, only the topologies which have cascode compensation ($Y_c = 1$) were explored in this experiment.

G. Experiment 7

This experiment has the same specifications as Experiment 6 except that the Q specification of the complex pole pair is reduced to 0.5. The topology choices are very similar to the previous experiment but the areas of the circuits synthesized are larger. The results are shown in Table VIII.

TABLE VIII
EXPERIMENT 7

Gain(dB) Spec/Design	Area (μ^2)	Power (μW)	No. of nodes	CPU time (secs)	Topology
66/72	2225	111	4	379	Diff pair cascoded Cascode compensation
80/80	2369	111	8	790	Diff pair and Load cascoded Cascode compensation
94/113	2886	140	14	1629	Diff pair, Curr. mirr. and TA cascoded Source follower, Cascode compensation
114/119	2951	138	14	1620	Diff pair, Curr Mirr and TA cascoded Source follower, Cascode compensation

H. Evaluation of Branch and Bound Performance

In the previous sections we have presented a novel formulation for doing topology selection of cell-level analog circuits. The obvious alternative to adopting this formulation is to evaluate all the topologies separately and choose the best one, i.e., do exhaustive search. It must be remembered that a) Branch and Bound in the worst case will evaluate $2n-1$ nodes where n is the total number of topologies; and b) each individual topology solved separately constitutes a smaller optimization problem than a Branch and Bound node. Hence if the Branch and Bound strategy fails to *prune* a significant number of nodes from the tree, then exhaustive search might turn out to be computationally cheaper.

The CPU time required by the tool to determine the correct topology has to be compared to the CPU time required to exhaustively design all the topologies. The correct way of doing this would be to write computer programs to design all the topologies separately and then compare both in terms of the correctness of the topology selected and the computation cost. Because of the effort involved in carrying out this task, we adopted a simplified method to carry out this experiment. We wrote a program for the simplest topology, i.e., the topology with the smallest number of variables and equivalently a program for the most complicated topology and took the average of the computation times. We multiplied this average by the total number of topologies to get an estimate of the computation time of exhaustive search. We found that, for an average set of specifications, the time required to design the simplest topology was about 10 CPU s. and the time required to design the most complicated topology was about 120 s, giving an average of about 65 CPU s. for each topology. Thus the total time to design all topologies would be about 4160 CPU s. We also found that the average time to evaluate a node in our Branch and Bound tree was about 100 CPU s. Hence if less than about 41 nodes in the Branch and Bound tree are evaluated, then the total CPU time will be better than that of exhaustive search[4].

In Table IX, we compare the CPU times of the Branch and Bound experiments shown in the previous section with estimates of the CPU times for exhaustive search for the same set of specifications. It can be seen that the speedups obtained varied from 1.3 to 32. The method was especially efficient at identifying impossible specifications. This will be very useful if such a cell synthesis tool is used inside a system-level synthesis tool.

[4]There are exceptions to this method of evaluating the CPU time for exhaustive search. For example, when there is a PSR specification, the resistance compensation topologies should not be evaluated and this is taken into account in determining the exhaustive search time.

TABLE IX
CPU TIME COMPARISON (APPROXIMATE)

B & B (secs.)	Exhaustive Search(secs.)	B & B (secs.)	Exhaustive Search(secs.)
957	4160	1611	4160
1375	4160	2501	4160
1651	4160	3187	4160
1422	4160	472	2130
957	4160	548	2130
901	4160	591	2130
2175	4160	1409	2130
1738	4160	379	2130
684	4160	790	2130
848	4160	1629	2130
53	1696	1620	2130
1349	4160	–	–

To verify the optimality of the topology selected, all the topologies were designed. But since CPU times were not of concern in this case, a topology was designed by setting the binary variables appropriately in the MINLP program and solving it. This process was repeated till all the topologies were designed. The best topology was then compared with the one obtained by applying the Branch and Bound algorithm. In all cases Branch and Bound came up with the right topology choice and the right sizes. Lastly all circuits synthesized were simulated using HSPICE [33] to verify their proper functionality. There was very little discrepancy between the predicted performance and the simulation results.

VII. DISCUSSION

In the previous sections we have seen that an MINLP formulation together with a Branch and Bound algorithm can be used successfully to do simultaneous topology selection and sizing of analog circuits. The method was found to efficiently arrive at the correct topology choices for a large number of designs. However to ensure that MINLP formulations can be widely applied, various formulational and algorithmic steps need to be taken.

A. Global Optimization

As mentioned earlier, Branch and Bound requires that a global optimum be found at every node. In this work we have not attempted any global optimization. In the early stages of this research we found that in some cases, an infeasible point was obtained at a particular node in the Branch and Bound tree, even when there was a feasible topology in the subtree below it. This problem has been mitigated by transformation of the constraints, conservative choices of the values of the Γ constants in the MINLP formulation, and addition of constraints based on circuit knowledge to guide the search for the feasible region. However, we can try to make the formulation as convex as possible, i.e., to make sure that the objective function and constraints are convex functions [15], [34]. Reformulation and transformation of variables can be done to transfer as many of the nonlinearities to the objective function as possible. This would avoid infeasibilities due to linearization of the constraints, and thus would allow the constraints to be formulated more tightly. It might seem at first that such reformulation would require the implementer to actually manipulate the circuit and device equations, which conflicts with some other goals of analog synthesis, such as encapsulation of device models and reduction in the amount of circuit knowledge required [19]. However, there have already been attempts at automating these reformulation methods [35]. Also, global optimization algorithms should be investigated. Global optimization is currently a very active area of research [36]. With advancement in global optimization techniques, the power of such an MINLP formulation will be firmly established.

B. Alternate Formulations

It can be seen that many of the constraints in the MINLP formulation were expressed as disjunctive constraints[5]. Instead of incorporating all disjunctive constraints into one large MINLP, we can formulate two or more MINLP's each with a smaller number of disjunctive constraints. For example, as can be seen from Appendix A, the choice of NMOS and PMOS input was handled by formulating two separate MINLP's. Similarly we can split resistance compensation op amps and cascode compensation op amps into two separate MINLP's. The decision about where to draw the line of separation would depend on the following.

- The algorithm being used to solve the MINLP: A large number of disjunctive constraints might be really disadvantageous for a Branch and Bound algorithm because of the increase in the size of the nonlinear programming problems corresponding to each node. But disjunctive constraints might not be as bad for other algorithms like Generalized Benders Decomposition (GBD) [26] and Outer Approximation (OA) [27]. Both these methods alternate between solving a nonlinear programming subproblem (with all binary variables fixed) and a Mixed-Integer Linear Programming (MILP) master problem. The nonlinear programming (NLP) subproblems provide upper bounds to the final solution, and the MILP master problems provide lower bounds. The number of NLP subproblems solved can be much fewer than the number of NLP problems solved in Branch and Bound. Furthermore, since the NLP subproblems in GBD and OA have the integer variables fixed, these subproblems are of the same size as the optimization problem of the topology corresponding to the integer variables. Also MILP problems can be solved efficiently [23]. Hence these methods might turn out to be much more efficient for solving MINLP formulations of the kind presented in this paper. However, these algorithms have stricter requirements on convexity than Branch and Bound which only requires the finding of a global minimum. On the other hand decomposing the MINLP into a number of separate MINLP's might be good for Branch and Bound because branching can be done on a number of trees at the same time with upper bounds obtained from one tree being used to prune branches in another tree. But in other algorithms, each MINLP will need to be solved separately.
- Ease of automation of the formulation: Let us suppose we want to choose between various topologies of two-stage op amps and folded-cascode op amps, respectively. If this is formulated as a single MINLP, there would be a large number of disjunctive constraints. Therefore, if Branch and Bound is used, it would be better to formulate this problem as two separate MINLP's. Also, any circuit designer would find it much easier to model two-stage's and folded-cascode's as two separate MINLP's. However, as mentioned before, this might not be the most appropriate thing to do if other algorithms are used.
- The tightness of the formulation: With disjunctive constraints it is hard to make the formulations tight because the values of the Γ's will have to take into account the extreme values of all the disjunctive constraints. We observe that in most of the design experiments, very few of the NMOS-input topologies were evaluated. Formulating the NMOS and PMOS choice as two separate MINLP's resulted in the formulation being very tight with respect to the Y_N binary variable. Hence the lower bound corresponding to the NMOS choice always turned out to be much worse than that of the PMOS choice. Thus the decomposition of a problem into separate MINLP's might result in increased efficiencies because the lower bounds obtained in Branch and Bound will be of better quality.

It must be also pointed out that the formulation procedure described in this paper is not the only possible formulation for modeling topology selection of analog cells. Other supercircuits and other formulations for the same supercircuit may be tried [28]. For example, the disjunctive constraints for the gains of Fig. 3 can be formulated as

$$Gain_s(\mathbf{x}) \geq (1 - Y)GAIN \quad (64)$$

$$Gain_c(\mathbf{x}) \geq (Y)GAIN \quad (65)$$

where $Gain_s(\mathbf{x})$ and $Gain_c(\mathbf{x})$ are the gains corresponding to the simple and cascode case respectively. This results in fewer constraints than the formulation shown in Section IV. However if the same formulation method is tried for the two-stage op amp, it results in a larger number of constraints than the present formulation. Thus it is obvious there can be a number of ways of formulating the same topology selection problem.

VIII. Conclusion

In this paper we have presented a new formulation for analog circuit synthesis that casts it as an MINLP problem. And, using that formulation we have demonstrated *simultaneous* topology selection and parameter selection of analog circuits. Branch and Bound was used to solve the MINLP problem. Topology selection experiments were done for a two-stage CMOS op amp with 64 possible topologies. The Branch

[5] Disjunctive constraints were defined in Section IV.

and Bound Method performed with excellent accuracy and efficiency, even with simple branching heuristics.

APPENDIX
NOTES ON MINLP FORMULATION OF CMOS TWO-STAGE OP AMP

A. Variables

All device sizes and dc bias voltages are not used as variables in the MINLP formulation. Circuit assumptions are used to eliminate design variables [17]. For example transistors M3 and M4 in Fig. 4 are assumed to be perfectly matched and therefore W_3 and L_3 are eliminated from the list of design variables. Similarly, assumptions on pole-zero placement are used to eliminate L_{14}, V_{GS14}, W_{15} and L_{15} [28].

B. Linear Constraints

These consist of logical relations, size constraints and KVL constraints [28].

C. Nonlinear Constraints

Some of the nonlinear constraints are a little involved. We describe the gain constraint in detail as an example.

Gain constraint: Constraint sets similar to the constraint set (9) to (11) are used for each possible cascoding in our CMOS two-stage amplifier supercircuit (Fig. 4). Since we have four possible cascodings in our supercircuit we have 16 such constraints using four equivalent impedance variables, R_{1l}, R_{1u}, R_{2l}, and R_{2u}, respectively, where the variables denote the output impedances of various branches of our supercircuit. The equivalent output impedance of the first stage is the parallel combination of R_{1l} and R_{1u} and the equivalent output impedance of the second stage is the parallel combination of R_{2l} and R_{2u}. Since we do not use variables for the source follower, we assume its gain to be equal to 0.7. This is in accordance with the fact that $\frac{g_{msf}}{g_{mbsf}}$ is approximately equal to 2.5 and the gain of the source follower is approximately $\frac{g_{msf}}{g_{msf}+g_{mbsf}}$. We can now formulate our gain constraint as follows:

$$g_{m2}(\mathbf{x})(R_{1l}\|R_{1u})g_{m8}(\mathbf{x})(R_{2l}\|R_{2u}) \geq \text{GAIN}\left(1 + Y_{sf}\left(\frac{g_{msf} + g_{mbsf}}{g_{msf}} - 1\right)\right) \quad (66)$$

$$\Rightarrow \quad g_{m2}(\mathbf{x})(R_{1l}\|R_{1u})g_{m8}(\mathbf{x})(R_{2l}\|R_{2u}) \geq (1 + 0.43Y_{sf})\text{GAIN}. \quad (67)$$

Other nonlinear constraints incorporated are frequency response constraints, PSRR and noise constraints, as well as KCL, strong inversion, and saturation constraints.

D. Choice Between PMOS and NMOS

The choice between P and N type transistors for the input is incorporated in a slightly different manner. Instead of evaluating one Branch and Bound tree we evaluate two trees, one for N-transistor input op amps and the other for P-transistor input op amps. We however do them simultaneously using upper bounds from one tree to prune the other. We have the added advantage of having one less variable in the MINLP formulation of each tree.

REFERENCES

[1] I. E. Grossmann, "MINLP optimization strategies and algorithms for process synthesis," in *Proc. FOCAPD Meeting*, 1989, pp. 105–132.

[2] R. A. Rohrer, "Fully automated network design by digital computer," *Proc. IEEE*, vol. 55, pp. 1929–1939, Nov. 1967.

[3] S. W. Director and R. A. Rohrer, "Automated network design: The frequency domain case," *IEEE Trans. Circuit Theory*, vol. CT-15, pp. 330–337, June 1968.

[4] R. K. Brayton, G. D. Hachtel, and A. L. Sangiovanni-Vincentelli, "A survey of optimization techniques for integrated-circuit design," *Proc. IEEE*, vol. 69, pp. 1334–1362, Oct. 1981.

[5] W. Nye, D. C. Riley, A. Sangiovanni-Vincentelli, and A. L. Tits, "DELIGHT.SPICE: An optimization-based system for the design of integrated circuits," *IEEE Trans. Computer-Aided Design*, vol. 7, pp. 501–519, Apr. 1988.

[6] J.-M. Shyu and A. Sangiovanni-Vincentelli, "ECSTASY:A new environment for IC design optimization," in *Proc. ICCAD*, 1988, pp. 484–487.

[7] R. Arora, U. Dasgupta, D. Hocevar, and L. Goff, "OASYS: A tool for aiding in design of high performance linear circuits," in *Proc. 1990 IEEE Int. Symp. Circuits Systems*, IEEE, May 1990, pp. 1911–1914.

[8] L. R. Carley and R. A. Rutenbar, "How to automate analog IC design," *IEEE Spectrum*, pp. 26–30, Aug. 1988.

[9] M. G. R. Degrauwe, O. Nys, E. Dijkstra, J. Rijmenants, S. Bitz, B. L. A. G. Goffart, E. A. Vittoz, S. Cserveny, C. Meixenberger, G. V. der Stappen, and H. J. Oguey, "IDAC: An interactive design tool for analog CMOS circuits," *IEEE J. Solid-State Circuits*, vol. SC-22, pp. 1106–1116, Dec. 1987.

[10] R. Harjani, R. A. Rutenbar, and L. R. Carley, "OASYS: A Framework for Analog Circuit Synthesis," *IEEE Trans. Computer-Aided Design*, vol. 8, pp. 1247–1266, Dec. 1989.

[11] H. Y. Koh, C. H. Sequin, and P. R. Gray, "OPASYN: A compiler for CMOS operational amplifiers," *IEEE Trans. Computer-Aided Design*, vol. 9, pp. 113–125, Feb. 1990.

[12] K. Swings, S. Donnay, and W. Sansen, "HECTOR: A hierarchical topology-construction program for analog circuits based on a declarative approach to circuit modeling," in *Proc. CICC*, 1991, pp. 5.3/1–4.

[13] G. G. E. Gielen, H. C. C. Walscharts, and W. M. C. Sansen, "ISAAC: A symbolic simulator for analog integrated circuits," *IEEE J. Solid-State Circuits*, vol. 24, pp. 1587–1597, Dec. 1989.

[14] Z.-Q. Ning, T. Mouthaan, and H. Wallinga, "SEAS: A simulated evolution approach for analog circuit synthesis," in *Proc. CICC*, 1991, pp. 5.2/1–4.

[15] R. Fletcher, *Practical Methods of Optimization*. New York: Wiley, 1987.

[16] P. C. Maulik and L. R. Carley, "Automating analog circuit design using constrained optimization techniques," in *Proc. ICCAD*, Nov. 1991, pp. 390–393.

[17] P. C. Maulik, L. R. Carley, and D. J. Allstot, "Sizing of cell-level analog circuits using constrained optimization techniques," *IEEE J. Solid-State Circuits*, vol. 28, pp. 233–241, Mar. 1993.

[18] P. Gill, W. Murray, M. Saunders, and M. Wright, "User's Guide for NPSOL (Version 4.0)," Dept. of Operation Res., Stanford Univ., Stanford, CA, Tech. Rep. SOL 86-2, Jan. 1986.

[19] E. S. Ochotta, R. A. Rutenbar, and L. R. Carley, "Equation-free synthesis of high-performance linear analog circuits," in *Proc. 1992 Brown/MIT Conf. Advanced Res. in VLSI and Parallel Syst.*, Mar. 1992, pp. 129–143.

[20] E. S. Ochotta, L. R. Carley, and R. A. Rutenbar, "Analog circuit synthesis for large, realistic cells: Designing a pipelined A/D converter with ASTRX/OBLX," in *IEEE Custom Integrated Circuits Conf.*, May 1994, pp. 365–368.

[21] E. S. Ochotta, R. A. Rutenbar, and L. R. Carley, "ASTRX/OBLX: Tools for rapid synthesis of high-performance analog circuits}," in *ACM/IEEE Design Automation Conf.*, June 1994, pp. 24–30.

[22] L. T. Pillage and R. A. Rohrer, "Asymptotic waveform evaluation for timing analysis," *IEEE Trans. Computer-Aided Design*, vol. 9, pp. 352–366, Apr. 1990.

[23] R. Garfinkel and G. L. Nemhauser, *Integer Programming*. New York: Wiley, 1972.

[24] G. L. Nemhauser and L. A. Wolsey, *Integer and Combinatorial Optimization.* New York: Wiley, 1988.
[25] O. K. Gupta, "Branch and bound experiments in nonlinear integer programming," Ph.D. dissertation, Purdue University, Oct. 1980.
[26] A. M. Geoffrion, "Generalized benders decomposition," *J. Optimization Theory Applicat.*, vol. 10, no. 10, pp. 237–260, 1972.
[27] M. A. Duran and I. E. Grossmann, "An outer-approximation algorithm for a class of mixed-integer nonlinear programs," *Math. Programming*, vol. 36, pp. 307–339, 1986.
[28] P. C. Maulik, "Formulations for optimization-based synthesis of analog cells," Ph.D. dissertation, Dept. of Elect. and Comp. Sci., Carnegie Mellon Univ., Oct. 1992.
[29] P. Gill, W. Murray, and M. Wright, *Practical Optimization.* London, U.K.: Academic, 1989.
[30] D. B. Ribner and M. A. Copeland, "Design techniques for cascoded CMOS op amps with improved PSRR and common-mode input range," *IEEE J. Solid-State Circuits*, vol. SC-19, pp. 919–925, Dec. 1984.
[31] P. R. Gray and R. G. Meyer, *Analysis and Design Of Analog Integrated Circuits, Third Edition.* New York: Wiley, 1993.
[32] R. Gregorian and G. C. Temes, *Analog MOS Integrated Circuits for Signal Processing.* New York: Wiley, 1986.
[33] *HSPICE Manual.* Campbell, CA: Meta-Software Inc., 1990.
[34] M. S. Bazaraa and C. M. Shetty, *Nonlinear Programming Theory and Algorithms.* New York: Wiley, 1979.
[35] R. J. Amarger, L. T. Biegler, and I. E. Grossmann, "An automated modelling and reformulation system for design optimization," *Computers Chemical Eng.*, vol. 16, no. 16, pp. 623–636, 1992.
[36] A. H. G. R. Kan and G. T. Timmer, "Global optimization," in *Handbooks in OR and MS*, vol. 1, G. L. Nemhauser *et al.*, Eds. North-Holland: Elsevier Science, 1989, pp. 631–662.

Prabir C. Maulik received the B.Tech. degree in electronics and electrical communication engineering from the Indian Institute of Technology, Kharagpur, India, the M.E. degree in electrical engineering from the University of Florida, Gainesville, and the Ph.D. degree in electrical engineering from Carnegie Mellon University, Pittsburgh, PA.

He is currently employed at Crystal Semiconductor Corporation, Austin, TX. His interests are in analog circuit design, digital signal processing, computer-aided design for VLSI, and optimization algorithms.

L. Richard Carley (S'77–M-81–SM'90) received the S.B. degree from the Massachusetts Institute of Technology, Cambridge, MA, in 1976 and was awarded the Guillemin Prize for the best EE Undergraduate Thesis; he received the M.S. degree in 1978 and the Ph.D. degree in 1984, also from MIT.

He is currenly a Professor of Electrical and Computer Engineering at Carnegie Mellon University. He has worked for MIT's Lincoln Laboratories and has acted as a Consultant in the area of analog circuit design and design automation for Analog Devices and Hughes Aircraft, among others. In 1984 he joined Carnegie Mellon, and in 1992 he was promoted to Full Professor. His current research interests include the development of CAD tools to support analog circuit design, the design of high-performance analog signal processing IC's, and the design of low-power high-speed magnetic recording channels.

Dr. Carley received a National Science Foundation Presidential Young Investigator Award in 1985, a Best Technical Paper Award at the 1987 Design Automation Conference, and a Distinguished Paper Mention at the 1991 International Conference on Computer-Aided Design.

Rob A. Rutenbar (S'77–M'78–SM'90) received the Ph.D. degree in computer engineering (CICE) from the University of Michigan, Ann Arbor, in 1984.

He then joined the Faculty of Carnegie Mellon University, Pittsburgh, PA, where he is currently Professor of Electrical and Computer Engineering, and of Computer Science, and Director of the Semiconductor Research Corporation-CMU Center of Excellence in CAD and IC's. His research interests focus on circuit and layout synthesis for mixed-signal ASIC's, for high-performance digital IC's, and for FPGA's.

Dr. Rutenbar received a Presidential Young Investigator Award from the National Science Foundation in 1987. At the 1987 IEEE-ACM Design Automation Conference, he received a Best Paper Award for work on analog circuit synthesis. He is currently on the Executive Committee of the IEEE International Conference on CAD, and Program Committees for the ACM/IEEE Design Automation Conference and European Design & Test Conference. He is on the Editorial Board of IEEE SPECTRUM, and chairs the Analog Technical Advisory Board for Cadence Design Systems. He is a member of ACM, Eta Kappa Nu, Sigma Xi, and AAAS.

ARCHGEN: Automated Synthesis of Analog Systems

Brian A. A. Antao, *Member, IEEE,* and Arthur J. Brodersen, *Senior Member, IEEE*

***Abstract*—High-level design of analog systems is an open area that needs to be addressed with the emerging trend of integrating mixed analog-digital systems. Design methods compatible across the analog-digital boundaries would expedite the design process, and in this paper we address analog high-level design issues. An approach for systems-level synthesis of a class of analog systems is presented. A behavioral level for the analog domain is characterized in terms of state equations and transfer functions in the continuous and discrete domains. State-space representations are generated from transfer function specifications that exhibit system level characteristics such as controllability and observability as well as decoupled and parallel architectures. These state-space representations are synthesized into behavioral-level, technology-independent architectures composed of analog functional components. An intermediate architecture in a circuit implementation technology is synthesized from the behavioral architecture. The various algorithmic procedures for synthesis are implemented in the program ARCHGEN. Behavioral simulation is used for architecture verification and design space exploration.**

I. Introduction

BEHAVIORAL-LEVEL design and architecture synthesis has primarily been a design activity of the digital domain. The need for design and integration of analog-digital systems requires compatible design solutions in the analog domain. In this paper we introduce a methodology for behavioral level design and architecture synthesis of analog systems. While the approach is broad based in nature, the immediate focus is on the synthesis of various categories of analog filter systems. Filters compose a large portion of analog sections along with data converters in mixed analog-digital systems. The methodology can be extended to include the synthesis of specific analog functions such as data converters, frequency synthesis and locking (phase-locked loops) and other nonlinear analog systems. Some of the more complex and esoteric analog systems can be decomposed into subsystems that can be synthesized in this framework.

Earlier efforts in analog design automation have focused mainly on the circuit-level, in attempting to automate the synthesis of analog cells [1]–[5]. Another thrust has been the automation of analog layout design, where the layout process involves additional constraints such as crosstalk avoidance and parasitic minimization that can have a severe impact on the circuit performance [6]. Analog system-level design has received very little attention, though some attempts have been made at realizing analog systems using a circuit-level approach [8]. In this paper we develop a *systems-level* approach for synthesis of analog systems on the lines of the behavioral level synthesis efforts of the digital domain, thus opening an avenue for high-level synthesis of mixed analog-digital systems.

Three tasks are essential for system-level design:

1) Architecture synthesis,
2) Behavioral modeling, and
3) Architecture verification.

The focus of this paper is primarily on synthesis and some aspects of modeling, with details on verification being presented in a companion publication [9]. The objective of this research is to address the synthesis of mixed analog-digital systems through a synergy of design tasks across analog and digital boundaries. Thus methods across the analog and digital domains need to be compatible for greater efficiency in the design process. Traditional analog design has largely been heuristic in nature often relying on designers intuition and expertise. Whereas deterministic and algorithmic methods have been developed for digital systems. An additional motivation, is to research similar methods for analog systems. Furthermore, since the analog domain is composed of many different categories of systems with idiosyncratic characteristics, specialized design techniques are required for each type of system. The research results presented in this paper describe the first steps taken at developing systematic methodologies for high-level design of analog systems, with an implementation in the form of an automated synthesis tool ARCHGEN (ARCHitecture GENerator).

Traditional filter designers often start with a preference for a certain filter configuration such as leap-frog and a specific implementation technology such as switched-capacitor. Often this preference is based on the designer's experience with having designed many such filters in a particular implementation technology. Concepts such as *design space exploration* for architecture selection and *implementation technology analysis* have not been used as the process involves considerable effort and time. More so, since disparate design methods have to be used for designing each architecture and developing realizations in various technologies. Thus arises the need for unifying the process of synthesizing various architectures on a common basis and developing prototypical realizations in various implementation technologies for analysis. To address this specific need, we have utilized the state-space formulation within which most traditional and nontraditional filter archi-

Manuscript received December 16, 1993; revised September 8, 1994. This work was supported in part by the Semiconductor Research Corporation under Grants 93-DP-109 and 94-DP-109.

B. Antao is with the Coordinated Science Laboratory, University of Illinois at Urbana-Champaign, Urbana, IL 61801 USA.

A. Brodersen is with the Department of Electrical Engineering, Vanderbilt University, Nashville, TN 37235 USA.

IEEE Log Number 9410835.

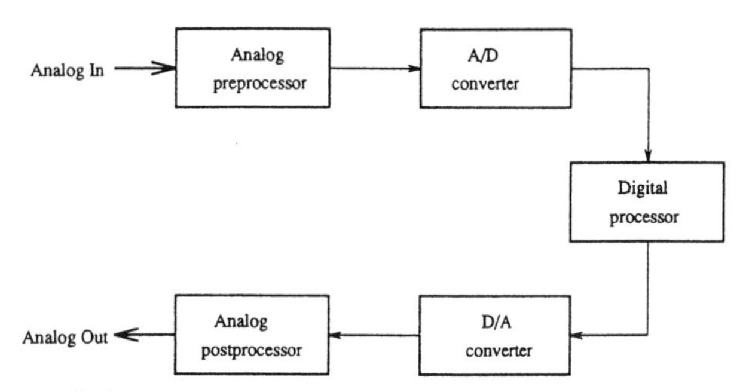

Fig. 1. Typical analog-digital processing system.

tectures are generated. The *intermediate architecture* stage is introduced for synthesizing and evaluating architectures in various implementation technologies.

The analog behavioral level is characterized in terms of mathematical expressions in the time and frequency domains, which provide input to the architecture synthesis process. The behavioral specifications are synthesized into functional architectures, independent of any circuit implementation technology. Implementation style specific intermediate architectures are then synthesized from the behavioral architecture. The intermediate architecture serves the purpose of design space exploration—evaluating various implementation styles; and serves as a bridge between system-level design and detailed circuit-level synthesis.

The paper is organized as follows: In Section II we discuss system-level design issues with a review of the digital design process, and draw out essential features to develop a compatible approach for analog high-level synthesis. In Section III we develop various synthesis formulations for analog systems both in the continuous-time and discrete-time domains along with the definition of the intermediate architecture level in Section IV. The implementation of the synthesis methodology in the form of a Computer-aided-design (CAD) tool is described in Section V. Experimental results of synthesis examples are presented in Section VI, followed by conclusions in Section VII.

II. System-Level Design Issues

A. Analog-Digital Systems

A motivation for developing the behavioral-level and a synthesis methodology for high-level design in the analog domain, is to address the design of complex analog-digital systems that can be integrated as a single chip. Though many data and signal processing tasks are carried out in the digital domain, the real world from which signals originate is analog in nature. Thus critical components of a complex system that interface with the external world are analog interface components, which include sensors, analog signal processing elements and analog-digital data converters.

Fig. 1 shows a typical system configuration that processes analog inputs and generates analog outputs. At the core of the system is a digital signal processor (DSP). Typical analog-digital systems are of the type shown in Fig. 1. The design of such a system at the system level requires the ability of being able to design and verify each of the major functional blocks and their interface. High-level synthesis in the digital domain provides for design of the DSP components. However there is a lack of a compatible methodology for architecture design in the analog domain. The behavioral level that we have characterized is aimed at providing a compatible behavioral level solution in the analog domain, and should integrate with behavioral level design phase of the digital domain, thus providing an unified behavioral level platform for the design of analog-digital systems.

B. Digital High-Level Design

Development of the synthesis efforts in the digital domain have been fairly matured with the advent of numerous formal methods as well as fairly robust tools. The behavioral level or *algorithmic level* characterizes the behavior of a system in the form of an algorithm that carries out a specific processing task. The *register-transfer (RT) level* provides the highest level structural description of a system. An RT level description is the functional structure of the system being designed, and is composed of functional blocks such as registers to hold variables specified in the algorithm, and various operators such as adders, bit shifts etc., as well as multiplexers and data busses that interconnect the elements.

High-level synthesis is the generation of a register-transfer level structural design from a given algorithmic behavioral specification [10]. This synthesis process is carried out through various phases of transformations: Behavioral compilation, scheduling, allocation and module binding. Fig. 2 depicts the digital high-level design process. The behavioral model is first compiled into a suitable internal representation, using graph theoretic forms. Compiler based optimization is applied to the internal representation at this stage to optimize the behavioral model, while maintaining its original characteristics. During this phase, the datapath and the controller are identified from the behavioral specification [11]. The data path models the flow of data through the various steps (operations) of the algorithm. The controller (or control sequencer) sequences the operations in the data path and specifies the order in which the operations are evoked when the system is in operation. Synthesis of datapaths is then completed in the scheduling and allocation phases. If a hardwired controller is desired, the realization is in the form of a finite-state machine for which logic synthesis is employed to generate the appropriate combinational logic and memory elements of the controller.

The next stage is scheduling or state synthesis, here a detailed schedule of the various control-steps or clock cycles is generated. The various data operations in the algorithm are assigned to control-steps (machine states). A finite state machine is commonly used to implement the control sequencer in hardware. A control-step then corresponds to a state in the finite state machine. The size of the clock cycle taken by a state varies depending on the number of operations scheduled within each control step. In the *scheduling process*, the operations in the data path are scheduled or assigned to control steps. The schedule is then optimized subject to specified constraints such as the number of control-steps,

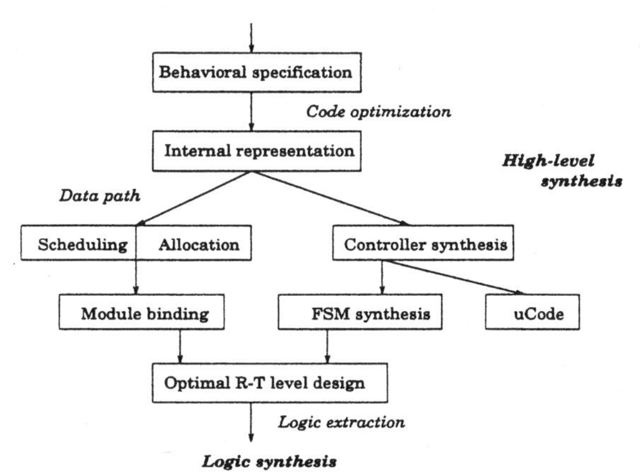

Fig. 2. Tasks in high-level digital design.

hardware resources etc. Once an optimal schedule is obtain, the *allocation* task attempts to assign appropriate functional blocks to the various operators and resources for the data path according to the schedule generated by the scheduling phase. Allocation tries to minimize the hardware resources utilized. Allocation of the functional blocks occurs in tandem with scheduling, so that effects of resource sharing and delays can be considered in developing the schedule [12]. The resources utilized are categorized as functional units (operators that implement the various operations specified in the behavioral specification), memory elements and communication paths. Optimal allocation involves making tradeoffs between the amount of hardware that is used from each of these categories, e.g., minimize functional blocks or memory elements at the expense of an increased number of interconnections and communication paths.

The allocation stage generates an abstract structure of the data path in terms of functional blocks. In this phase, for an optimal allocation, resources are shared between multiple identical operations and data variables. Depending on the schedule, multiple addition operations are assigned to the same physical functional adder, similar strategy is used in allocating registers to different data variables where a physical register is shared between variables whose lifetimes in the algorithm do not overlap, so that hardware utilities used are minimized. The resources utilized are categorized as the functional units (operators that implement the various operations specified in the behavioral specification), memory elements and communication paths. Optimal allocation involves making tradeoffs between the amount of hardware used from each of these categories, e.g., minimize functional blocks or memory elements at the expense of an increased number of interconnections and communication paths. The choice of a structural configuration for each functional block is a one-to-many mapping presenting various alternatives in the design space. Allocation of each class of hardware occurs as a subtask of the overall allocation process, entailing special considerations. During allocation a common tradeoff occurs between the number of functional units used and the amount of interconnection hardware.

The *module binding* task then makes explicit assignments of hardware components to the functional blocks. Decisions made by the module binding process involve choosing appropriate components from libraries, or invocation of lower level logic synthesis tools for full custom synthesis [13]. In module binding, selection of physical hardware structures is made and assigned to the abstract functional blocks. The physical hardware structures can be chosen from a library, or lower level synthesis tools may be evoked for full custom design.

C. The Analog Synthesis Methodology

The behavior of an analog system can be expressed in the time domain or the frequency domain. In the time domain, continuous systems are mathematically characterized by differential equations and discrete systems by difference equations.[1] In the frequency domain, the corresponding representations are algebraic expressions in terms of the Laplace variable s, and the discrete transform (z-transform) variable z. Frequency domain behavior is expressed in terms of s or z *transfer functions*. Since an analog system has to process signals of varying frequencies and magnitudes, it is essential that the frequency domain behavior of the system be consistent over the range of signal frequencies that need to be processed. Besides the frequency spectrum, the transfer functions also provide information about the system poles and zeros that indicate stability of the system.

Thus a suitable representation for analog behavior is the transfer function. Additionally, the behavior can also be specified in the time domain as differential/difference equations. The Laplace (or z) and the inverse Laplace (or inverse z) transforms can be applied to derive the equivalent representation in the time or frequency domains. Thus our goals for architectural synthesis are to suitably synthesize an architecture from a given behavioral description as a transfer function.

General behavioral level representation for continuous systems is expressed as a s-domain transfer function

$$H(s) = K\frac{b_m s^m + b_{m-1}s^{m-1} + \cdots + b_1 s + b_0}{a_n s^n + a_{n-1}s^{n-1} + \cdots + a_1 s + a_0} \qquad m \leq n. \tag{1}$$

Transfer functions specify the input-output behavior (or terminal behavior) of a system. A transfer function in the frequency domain can be transformed into a time domain equivalent, the *state-variable description* . The *state-variable description* of a system consists of a set of differential equations that describe the internal and the terminal behavior of the system [14]. A state-variable description in the a general continuous domain is:

$$\dot{\mathbf{x}}(t) = \mathbf{A}(t)\mathbf{x}(t) + \mathbf{B}(t)\mathbf{u}(t) \tag{2}$$

$$\mathbf{y}(t) = \mathbf{C}(t)\mathbf{x}(t) + \mathbf{D}(t)\mathbf{u}(t). \tag{3}$$

Here, $\mathbf{x}(t)$ is the internal state vector. Equation (2) is the internal description of the system and (3) specifies the system output in terms of the internal states and the input.

Behavioral representation for discrete systems is the z-domain transfer function

$$H(z) = K\frac{b_m z^{-m} + b_{m-1}z^{-(m-1)} + \cdots + b_1 z^{-1} + b_0}{a_n z^{-n} + a_{n-1}z^{-(n-1)} + \cdots + a_1 z^{-1} + a_0}$$

[1] Algebraic-differential or algebraic-difference equations, or explicit algorithms can be used as well.

$$m \leq n. \tag{4}$$

As in the continuous domain, a state-variable description in the discrete domain is given by

$$\mathbf{x}(k+1) = \mathbf{A}\mathbf{x}(k) + \mathbf{B}\mathbf{u}(k) \tag{5}$$

$$\mathbf{y}(k) = \mathbf{C}\mathbf{x}(k) + \mathbf{D}\mathbf{u}(k). \tag{6}$$

Here $\mathbf{x}(k)$ is the discrete state vector, $\mathbf{u}(k)$ is the discrete input vector, and $\mathbf{y}(k)$ is the discrete output vector of the system.

Equations (2), (3), (5), and (6) constitute the *dynamical equation description* or the state-variable description of a system. Once the state-variable model is known for a particular system, a structural realization can be obtained using an analog computation model. This model of a state-variable description leads to a structure consisting of weighted adders, integrators, delay elements, etc. This structure represents a functional architecture of an analog system. The internal states of a state-space model can be viewed as the analog domain counterparts of the Boolean variables that describe the behavior of digital systems. The state-variable model can be compared to the finite-state machine model used in the behavioral synthesis of sequential logic circuits. Various types of logic gates are used to synthesize digital circuits from Boolean algebraic descriptions, the process being defined by various methods developed in the field of logic synthesis. Using the state-variable basis, the synthesis of analog systems can be cast in a unified framework. Synthesis of various analog filter architectures is cast in this basis, through the use of various numerical methods. These methods are used for systematic synthesis of analog systems in a methodical fashion from state-variable models. The focus is on single-input single-output (SISO) systems. These concepts can be extended in general to multiple-input multiple-output (MIMO) systems.

In order to synthesize an architecture that realizes the transfer function, the equivalent time domain behavioral representation is required. The state-space formulation is used to derive the time domain behavioral representation from the frequency domain transfer function specification. The state-space model in the time domain also provides a basis for exploring various architectural possibilities for implementation. The method of derivation of the state-variable model is influenced by system characteristics such as *controllability* and *observability* as well as parallel, ladder and cascaded configurations.

The analog architecture synthesis methodology is illustrated in Fig. 3. The first phase is behavioral synthesis and design space exploration, where a suitable functional architecture to realize the behavioral specifications is generated. In the next phase, *Implementation technology analysis*, a feasibility analysis of implementing a functional architecture in one of the circuit-design styles, such as switched-capacitor is undertaken. *Behavioral synthesis* takes the form of architecture synthesis where a functional architecture is synthesized to implement the specified behavior. First a state-variable model is derived from the transfer function specifications. The state-variable model is then used as a basis for generating a functional architecture. Architectural verification in the form of simulation is used to verify the characteristics of the synthesized architecture, and compare the various architectures that can be synthesized. Circuit-level synthesis can be invoked to generate a complete circuit-level design by successively replacing each functional block with an equivalent circuit in a specified design style and fabrication technology. This top-down synthesis methodology starting from a high-level transfer function specification, provides a structured framework for the synthesis of a class of analog systems, such as filtering applications. More complex systems can be decomposed into units representable by their transfer functions and appropriately synthesized.

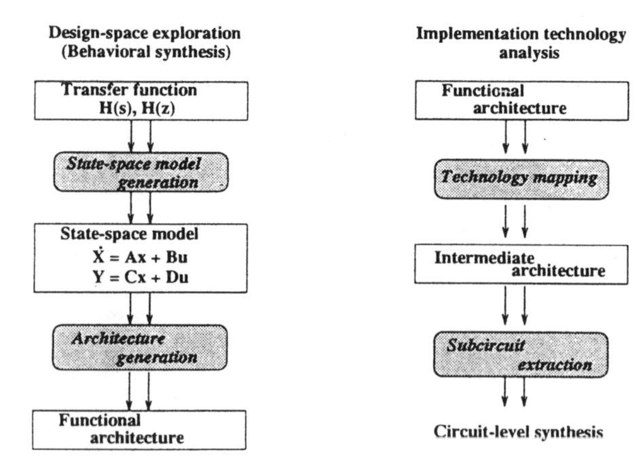

Fig. 3. Analog architecture synthesis methodology.

III. Synthesis Formulation for Analog Systems

Analog systems can be further differentiated into continuous-time and discrete-time systems. The behavioral specifications are in the form of a s-domain or z-domain transfer function, the goal of the synthesis process is to generate a suitable physical architecture that implements the transfer function characteristics. In a high level design phase the objective is to synthesize a functional architecture independent of a particular circuit implementation. This allows a designer to explore the use of specific design styles such as *MOSFET-C*, $G_m - C$, switched-capacitor etc. Hence we use the state-space approach in order to synthesize circuit design-style independent architectures.

A. Architecture Synthesis in the Continuous Domain

In the continuous analog domain behavioral synthesis starts with a Laplace (s-domain) transfer function specification. The transfer function is the frequency domain characterization of the system behavior. The state-space representation in the continuous time domain is derived from the Laplace transfer function specification. The state-space models are derived by using the *phase-variable* method for the controllable form, and the direct integration method for the observable form. These two methods are applied to the differential equations obtained from the s-domain transfer function. The decoupled architecture is derived by applying partial fraction expansion into first order factors to the transfer function. Additionally, the controllable form (also called the phase-variable canonical form) can be used as a basis for generating other architectures by applying suitable transformations. The state variable models are derived for the general case, the transfer function specification being a rational function of nth order polynomials. In typical analog filter specifications, the coefficients of the transfer function denormalized to an edge frequency are large and tend to swamp out the input signal. Hence, in these state-variable models, a scaling factor k is introduced to normalize the coefficients in the companion matrix $\mathbf{A}$ of (2), so that the internal states are at identical levels as the input signal $u(t)$ of (2). The scaling factor is defined as $a_0 k^n = 1$ for denormalized transfer functions, and $k = (1/2\pi f_c)$ for normalized transfer functions. Where f_c is the specified edge frequency. The resultant state-variable models for the continuous case are as follows; The transfer function $H(s)$ can be formulated as a state equation yielding these three forms:

1) Controllability Form:

$$\begin{bmatrix} \dot{x}_1(t) \\ \dot{x}_2(t) \\ \vdots \\ \vdots \\ \dot{x}_n(t) \end{bmatrix} = \begin{bmatrix} 0 & 1 & 0 & \cdots & 0 \\ 0 & 0 & 1 & \cdots & 0 \\ \vdots & \vdots & \vdots & \cdots & \vdots \\ 0 & 0 & 0 & \cdots & 1 \\ -a_0 k^n & -a_1 k^{n-1} & -a_2 k^{n-2} & \cdots & -a_{n-1} k \end{bmatrix} \cdot \begin{bmatrix} x_1(t)/k \\ x_2(t)/k \\ \vdots \\ \vdots \\ x_n(t)/k \end{bmatrix} + \begin{bmatrix} 0 \\ 0 \\ \vdots \\ \vdots \\ k^n K \end{bmatrix} u(t) \tag{7}$$

$$y(t) = \left[(b_0 - b_n a_0) \quad \frac{(b_1 - b_n a_1)}{k} \quad \cdots \quad \frac{(b_{n-1} - b_n a_{n-1})}{k^{n-1}} \right] \frac{\overline{X}(t)}{k} + [b_n K] u(t). \tag{8}$$

The observability form is essentially the transpose of the controllability form [14], [15]. If $\mathbf{A}$, $\mathbf{B}$, $\mathbf{C}$, and $\mathbf{D}$ are the system matrices of the phase-variable form, and $\mathbf{A}_{obs}$, $\mathbf{B}_{obs}$, $\mathbf{C}_{obs}$, and $\mathbf{D}_{obs}$ are the system matrices of a corresponding observability form, then the two forms are related as follows

$$\mathbf{A}_{obs} = \mathbf{A}^T; \qquad \mathbf{B}_{obs} = \mathbf{C}^T$$
$$\mathbf{C}_{obs} = \mathbf{B}^T; \qquad \mathbf{D}_{obs} = \mathbf{D}$$

2) Observability Form:

$$\begin{bmatrix} \dot{x}_1(t) \\ \dot{x}_2(t) \\ \vdots \\ \vdots \\ \dot{x}_n(t) \end{bmatrix} = \begin{bmatrix} 0 & 0 & 0 & \cdots & -a_0 k^n \\ 1 & 0 & 0 & \cdots & -a_1 k^{n-1} \\ 0 & 1 & 0 & \cdots & -a_2 k^{n-2} \\ \vdots & \vdots & \vdots & \cdots & \vdots \\ 0 & 0 & 0 & \cdots & -a_{n-1} k \end{bmatrix} \cdot \begin{bmatrix} x_1(t)/k \\ x_2(t)/k \\ \vdots \\ \vdots \\ x_n(t)/k \end{bmatrix} + \begin{bmatrix} (b_0 - b_n a_0) \\ \frac{(b_1 - b_n a_1)}{k} \\ \vdots \\ \vdots \\ \frac{(b_{n-1} - b_n a_{n-1})}{k^{n-1}} \end{bmatrix} u(t) \tag{9}$$

$$y(t) = [0 \quad 0 \quad \cdots \quad k^n K] \frac{\overline{X}(t)}{k} + [b_n K] u(t). \tag{10}$$

3) The Decoupled or Diagonal Form:

$$\begin{bmatrix} \dot{x}_1(t) \\ \dot{x}_2(t) \\ \vdots \\ \vdots \\ \dot{x}_n(t) \end{bmatrix} = \begin{bmatrix} p_1 & 0 & 0 & \cdots & 0 \\ 0 & p_2 & 0 & \cdots & 0 \\ 0 & 0 & p_3 & \cdots & 0 \\ \vdots & \vdots & \vdots & \cdots & \vdots \\ 0 & 0 & 0 & \cdots & p_n \end{bmatrix} \cdot \begin{bmatrix} x_1(t)/k \\ x_2(t)/k \\ \vdots \\ \vdots \\ x_n(t)/k \end{bmatrix} + \begin{bmatrix} 1 \\ 1 \\ \vdots \\ \vdots \\ 1 \end{bmatrix} u(t) \tag{11}$$

$$y(t) = [r_1 \quad r_2 \quad \cdots \quad r_n] \frac{\overline{X}(t)}{k} + [K] u(t). \tag{12}$$

here, $p_1, p_2, \cdots, p_n$ are the poles of the transfer function and $r_1, r_2, \cdots, r_n$ are the residues of the partial fraction expansion of the transfer function. In practice, the roots and residues of the transfer function are very likely to be complex in nature. To obtain a suitable physical realization the decoupled form with complex roots is converted into a real valued block diagonal form. This block diagonal form is more easily obtained by applying a similarity transform to the phase-variable canonical form [15]. The block diagonal form also has interesting characteristics in that each complex conjugate pole pair can be implemented as a biquadratic section. The overall architecture is thus composed of decoupled biquadratic sections for complex pole pairs, and single integrator sections for real poles. If more than one real pole exists, then the real poles can be grouped in pole-pairs and implemented as biquadratic sections. For generating the block diagonal form, we define a block transformation matrix $\mathbf{M}_b$ such that a new state vector $\mathbf{x}$ is introduced, where $\mathbf{x} = \mathbf{M}_b \mathbf{x}$. If $\mathbf{M}$ is the system modal matrix, which is essentially the matrix of the eigenvectors of $\mathbf{A}$. The transformation matrix $\mathbf{M}_b$ is obtained from the system modal matrix $\mathbf{M}$ such that

$$\mathbf{M}_b = [\mathrm{Re}\{\xi_1\}, \mathrm{Im}\{\xi_1\}, , \mathrm{Re}\{\xi_3\}, \mathrm{Im}\{\xi_3\}, \cdots, \mathrm{Re}\{\xi_m\}, \mathrm{Im}\{\xi_m\}, \xi_{m+2}, \cdots, \xi_n]. \tag{13}$$

Here, ξ_i are the eigenvectors that make up the system modal matrix, the eigenvectors $\xi_1, \cdots, \xi_m$ correspond to the complex conjugate eigenvalues, and the eigenvectors $\xi_{m+2}, \cdots, \xi_n$ correspond to the real eigenvalues. If $\lambda_1 = \sigma_1 + j\omega_1, \lambda_2 = \sigma_2 + j\omega_2, \cdots \lambda_{m+2}, \cdots, \lambda_n$ are the eigenvalues of $\mathbf{A}$, the block diagonal form obtained by applying the similarity transformation is

$$\Lambda_b = \mathbf{M}_b^{-1} \mathbf{A} \mathbf{M}_b = \begin{bmatrix} \sigma_1 & \omega_1 & 0 & 0 & \cdots & 0 & \cdots & 0 \\ -\omega_1 & \sigma_1 & 0 & 0 & \cdots & 0 & \cdots & 0 \\ 0 & 0 & \sigma_3 & \omega_3 & \cdots & 0 & \cdots & 0 \\ 0 & 0 & -\omega_3 & \sigma_3 & \cdots & 0 & \cdots & 0 \\ \vdots & \vdots & \vdots & \vdots & \cdots & \vdots & \cdots & \vdots \\ 0 & 0 & 0 & 0 & \cdots & \lambda_{m+2} & \cdots & 0 \\ 0 & 0 & 0 & 0 & \cdots & 0 & \cdots & \lambda_n \end{bmatrix} \tag{14}$$

Each block of the diagonal represents a complex conjugate pole-pair and can be realized using a biquadratic section.

In order to implement an architecture that has a ladder structure, also commonly referred to as the *Leapfrog architecture*, a similarity transform is applied to the phase-variable canonical form to obtain an equivalent tridiagonal form. The procedure for obtaining a tridiagonal form is based on application of the Routh–Hurwitz criterion used in the stability analysis of linear systems [16]–[18]. We define a similarity transform $\mathbf{Q}$ such that

$$\mathbf{QAQ}^{-1} = \mathbf{R}.$$

Here, $\mathbf{R}$ is the resultant system matrix, that has a particular tridiagonal structure.

In the generation of cascade architectures, the numerator and denominator of the transfer function is factored into poles and zeros. The complex roots are ordered as complex conjugate pairs to realize second-order or biquadratic sections. After factorization, a sequence of three tasks, pole-zero pairing, section ordering and gain assignment are carried out such that the dynamic range of the architecture is optimized. Each of these tasks are implemented by using some form of specialized optimization algorithms [19]. During pole-zero pairing an optimal pole-zero assignment for each biquadratic section is sought. Following this stage the order in which the biquadratic sections are cascaded is determined; finally the gain, K, of the overall transfer function is distributed across each of the sections. Once an optimal cascade decomposition is obtained a state space model is derived, and synthesis proceeds as in the previously described architectures.

The various state-space models that were derived, serve as the basis for generating functional architectures in the continuous domain. In terms of high-order filter terminology, the controllable architecture corresponds to the Follow-the-Leader (FLF) topology, and the tridiagonal or ladder architecture corresponds the Leapfrog (LF) topology. These architectures are composed of continuous analog functional blocks such as adders, integrators, amplifiers etc. The structure of a physical architecture depends on the composition of the state model. The architecture has to realize the internal states defined by the differential state equation $\dot{\mathbf{x}} = \mathbf{Ax} + \mathbf{Bu}$, and produce an output defined by the output equation $\mathbf{y} = \mathbf{Cx} + \mathbf{Du}$. The first step in the synthesis process is to implement the internal states. Each internal state requires a single continuous integrator. For the controllable form, the complete state model is given by expressions (7) and (8), the internal states are defined by differential equations as follows:

$$\begin{aligned}\dot{x}_i(t) &= x_{i+1}(t)\\ \dot{x}_n(t) &= -a_0k^{n-1}x_1(t) - a_1k^{n-2}x_2(t)\\ &\quad \cdots - a_{n-1}x_n(t) + k^nKu(t)\\ &\quad \text{for } i = 1, \cdots, n-1.\end{aligned}$$

The internal states are obtained by integrating the above set of differential equations leading to,

$$\begin{aligned}x_i(t) &= \int \dot{x}_i(t)\\ &= \int x_{i+1}(t)\end{aligned}$$

$$\begin{aligned}x_n(t) &= \int \dot{x}_n(t)\\ &= \int [-a_0k^{n-1}x_1(t) - a_1k^{n-2}x_2(t)\\ &\quad \cdots - a_{n-1}x_n(t) + k^nKu(t)]\\ &\quad \text{for } i = 1, \cdots, n-1.\end{aligned}$$

We need n integrators to realize the internal states and a $n+1$ input adder to realize the state $x_n(t)$. The next step is to implement the output definition. For the controllable form the output equation is given by expression (8). The output is defined as the weighted sum of the internal states and the input. We can implement this operation by an $n+1$ input weighted adder at the output end of the architecture. In general, the state space model would have the form

$$\begin{bmatrix}\dot{x}_1(t)\\ \dot{x}_2(t)\\ \dot{x}_3(t)\\ \vdots\\ \dot{x}_n(t)\end{bmatrix} = \begin{bmatrix}a_{11} & a_{12} & a_{13} & \cdots & a_{1n}\\ a_{21} & a_{22} & a_{23} & \cdots & a_{2n}\\ a_{31} & a_{32} & a_{33} & \cdots & a_{3n}\\ \vdots & \vdots & \vdots & \cdots & \vdots\\ a_{n1} & a_{n2} & a_{n3} & \cdots & a_{nn}\end{bmatrix} \cdot \begin{bmatrix}x_1(t)\\ x_2(t)\\ x_3(t)\\ \vdots\\ x_n(t)\end{bmatrix} + \begin{bmatrix}b_1\\ b_2\\ b_3\\ \vdots\\ b_n\end{bmatrix} u(t) \quad (15)$$

$$y(t) = [c_1 \quad c_2 \quad c_3 \quad \cdots \quad c_n]\overline{X}(t) + [d]u(t). \quad (16)$$

Equation (15) defines the internal states, and (16) defines the output in terms of the internal states and the input. Each state is realized by an adder and an integrator to implement the state definition

$$\dot{x}_i = a_{i1}x_1 + a_{i2}x_2 + a_{i3}x_3 + \cdots + a_{in}x_n. \quad (17)$$

Fig. 4(a) shows the realization of each state. An adder is required to realize the final output, depicted in Fig. 4(b). For architectures where the structure of the state-space model is known, specialized synthesis algorithms that exploit the structure of the state space models are implemented. Since they utilize information about the structure of the model, these methods are more efficient than the general algorithm. The general synthesis algorithm is used in the case of synthesizing cascaded architectures and other special-purpose architectures whose state space model may be specified directly. Fig. 5 shows the complete architecture for the controllable form.

Proceeding along similar lines, architectures for the observable, decoupled and tridiagonal forms can be generated from their state-space models. The architecture for the observable form described by the state (9) and (10) is depicted in Fig. 6. For the parallel or decoupled form, when the poles are complex the block diagonal form is used. Corresponding to each block of the diagonal of the companion matrix $\mathbf{A}$ given by (14), two single integrator sections in the ordinary decoupled form are realized by a single *biquadratic* section composed of two integrators. Each biquadratic section implements a complex-conjugate pole pair. Fig. 7 shows the functional architecture of a parallel form realization of a block-diagonal state model.

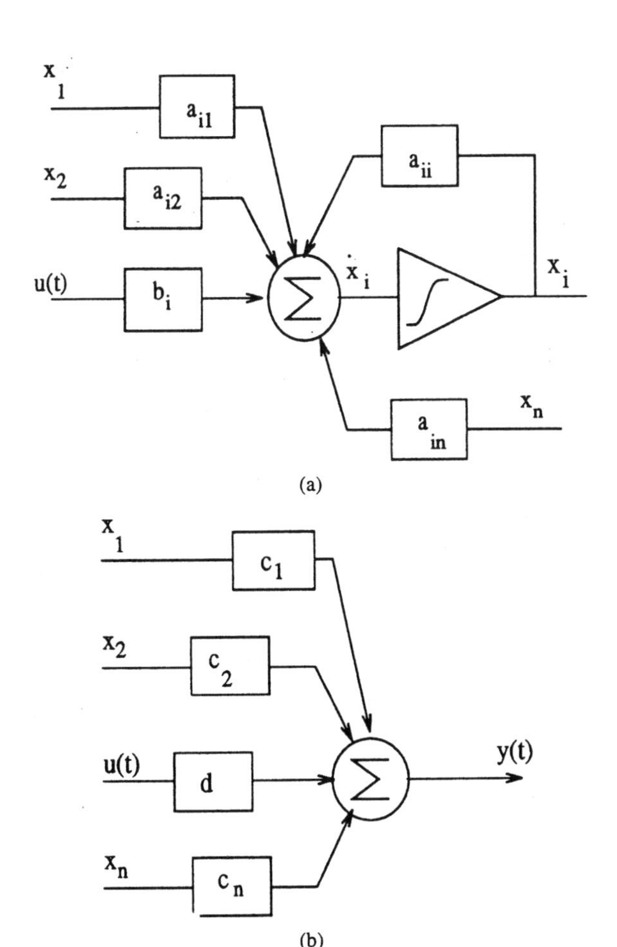

Fig. 4. Architectural realization of a general state-space model. (a) General state realization. (b) Output realization.

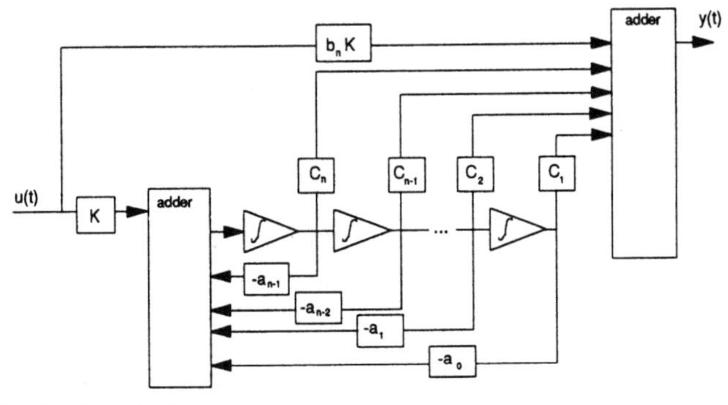

Fig. 5. A controllable form architecture.

Fig. 8 shows the functional architecture of the tridiagonal state model.

B. Architecture Synthesis in the Discrete Domain

In the discrete domain the behavioral specification is in the form of a z-domain transfer function. To obtain a physical realization of the discrete system specified by a z-domain transfer function we need an equivalent time domain representation. We use the state-variable formalism to generate a time-domain behavioral representation that can be systematically implemented as a physical architecture. The system characteristics, *controllability* and *observability*, determine the nature of the physical architecture. The state-variable representations

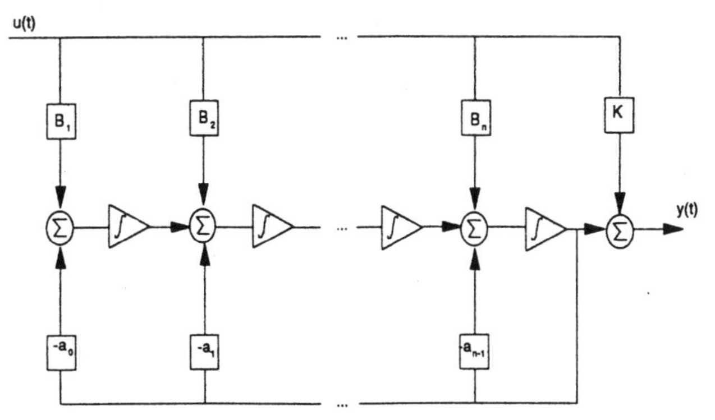

Fig. 6. An observable form architecture.

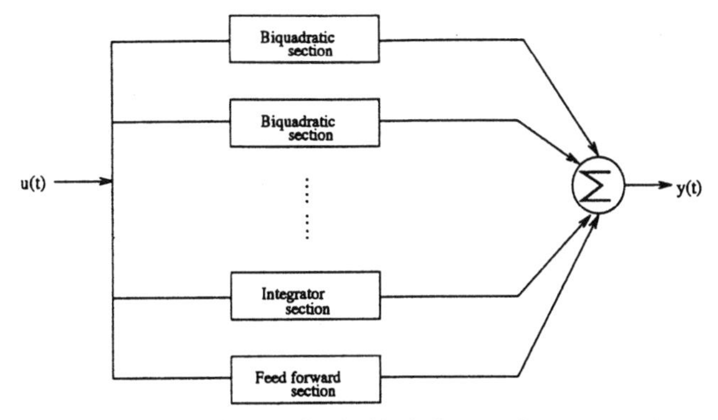

Fig. 7. A parallel architecture for the block-diagonal form.

are derived from the difference equations obtained from the z-domain transfer function specification.

For the z-domain transfer function:

$$H(z) = \frac{Y(z)}{U(z)} = K\,\frac{b_m z^{-m} + b_{m-1}z^{-(m-1)} + \cdots + b_1 z^{-1} + b_0}{a_n z^{-n} + a_{n-1}z^{-(n-1)} + \cdots + a_1 z^{-1} + a_0} \quad m \le n. \tag{18}$$

The state-space controllability form is:

$$\begin{bmatrix} x_1(k+1) \\ x_2(k+1) \\ \vdots \\ \vdots \\ x_n(k+1) \end{bmatrix} = \begin{bmatrix} -a_1 & -a_2 & \cdots & -a_{n-1} & -a_n \\ 1 & 0 & \cdots & 0 & 0 \\ 0 & 1 & \cdots & 0 & 0 \\ \vdots & \vdots & \cdots & \vdots & \vdots \\ 0 & 0 & \cdots & 1 & 0 \end{bmatrix} \cdot \begin{bmatrix} x_1(k) \\ x_2(k) \\ \vdots \\ \vdots \\ x_n(k) \end{bmatrix} + \begin{bmatrix} K \\ 0 \\ \vdots \\ \vdots \\ 0 \end{bmatrix} u(k) \quad \text{if } m = n, \tag{19}$$

$$y(k) = [(b_1 - b_0 a_1) \quad (b_2 - b_0 a_2) \quad \cdots \quad (b_n - b_0 a_n)]\,\overline{X}(k) + [b_0 K]u(k). \tag{20}$$

Equations (19) and (20) constitute the state-variable model of a discrete controllable system. Expression (19) is the internal

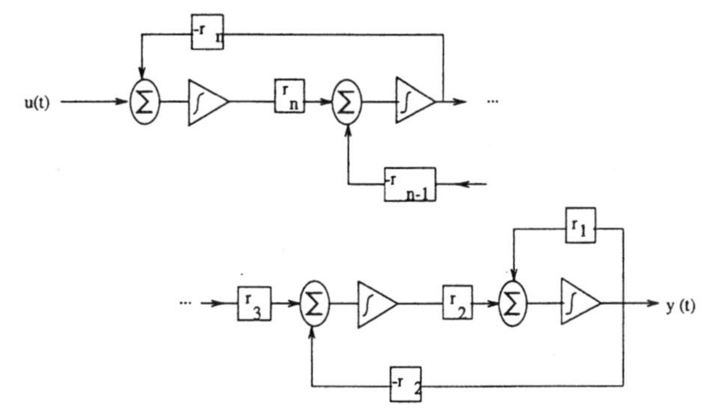

Fig. 8. A ladder architecture for the tridiagonal form.

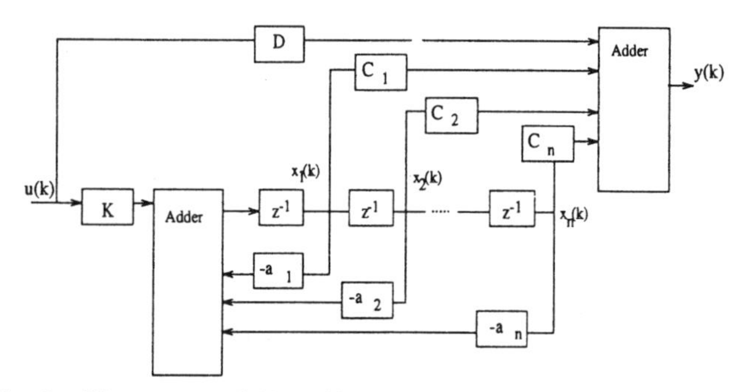

Fig. 9. Discrete controllable architecture.

description of the system and (20) defines the output of the system.

The observable form can be obtained directly from the transpose of the controllable form resulting in the following discrete observable state model:

$$\begin{bmatrix} x_1(k+1) \\ x_2(k+1) \\ \vdots \\ \vdots \\ x_n(k+1) \end{bmatrix} = \begin{bmatrix} -a_1 & 1 & 0 & \cdots & 0 \\ -a_2 & 0 & 1 & \cdots & 0 \\ \vdots & \vdots & \vdots & \cdots & \vdots \\ -a_{n-1} & 0 & 0 & \cdots & 1 \\ -a_n & 0 & 0 & \cdots & 0 \end{bmatrix} \cdot \begin{bmatrix} x_1(k) \\ x_2(k) \\ \vdots \\ \vdots \\ x_n(k) \end{bmatrix} + \begin{bmatrix} b_1 - b_0 a_1 \\ b_2 - b_0 a_2 \\ \vdots \\ \vdots \\ b_n - b_0 a_n \end{bmatrix} u(k) \quad (21)$$

$$y(k) = [K \quad 0 \quad 0 \quad \cdots \quad 0]\overline{X}(k) + [b_0 K]u(k). \quad (22)$$

The similarity transform which was defined in the continuous domain to obtain the block diagonal form from the phase-variable canonical form is also applicable in the discrete case, and yields the block diagonal discrete form as follows:

$$\mathbf{x}(k+1) = \mathbf{M}_b^{-1}\mathbf{A}\mathbf{M}_b\mathbf{x}(k) + \mathbf{M}_b^{-1}\mathbf{B}\mathbf{u}(k) \quad (23)$$

$$\mathbf{Y}(k) = \mathbf{C}\mathbf{M}_b\mathbf{x}(k) + \mathbf{D}\mathbf{u}(k). \quad (24)$$

Here $\mathbf{x}(k)$ is the equivalent state vector of the block-diagonal form.

The tridiagonal discrete form can be derived by applying the similarity transform $\mathbf{QAQ}^{-1} = \mathbf{R}$, defined in the continuous domain, to the discrete phase-variable canonical form. Application of the similarity transform $\mathbf{Q}$ to the discrete case results in the tridiagonal discrete form expressed as

$$\mathbf{x}'(k+1) = \mathbf{QAQ}^{-1}\mathbf{x}'(k) + \mathbf{QBu}(k) \quad (25)$$

$$\mathbf{Y}(k) = \mathbf{CQ}^{-1}\mathbf{x}'(k) + \mathbf{Du}(k) \quad (26)$$

$\mathbf{x}'(k)$ is the equivalent state vector of the tridiagonal form.

The above class of discrete systems are also called infinite impulse response (IIR) systems. A special class of discrete systems are the finite impulse response (FIR) systems which are described by z-domain transfer functions of the following form [20]

$$H(z) = \sum_{i=1}^{n} b_i z^{-i}. \quad (27)$$

These systems can be synthesized by considering the above transfer function to be a special case of the general z-domain transfer function with the coefficients $a_n, a_{n-1}, \cdots, a_0, b_0 = 0$. The resulting state-variable model is

$$\begin{bmatrix} x_1(k+1) \\ x_2(k+1) \\ \vdots \\ \vdots \\ x_n(k+1) \end{bmatrix} = \begin{bmatrix} 0 & 0 & \cdots & 0 & 0 \\ 1 & 0 & \cdots & 0 & 0 \\ 0 & 1 & \cdots & 0 & 0 \\ \vdots & \vdots & \cdots & \vdots & \vdots \\ 0 & 0 & \cdots & 1 & 0 \end{bmatrix} \cdot \begin{bmatrix} x_1(k) \\ x_2(k) \\ \vdots \\ \vdots \\ x_n(k) \end{bmatrix} + \begin{bmatrix} 1 \\ 0 \\ \vdots \\ \vdots \\ 0 \end{bmatrix} u(k) \quad (28)$$

$$y(k) = [b_1 \quad b_2 \quad \cdots \quad b_n]\,\overline{X}(k). \quad (29)$$

The procedure for generating the physical architectures is similar to that outlined for the continuous systems. However in the discrete domain, realization of each internal state requires an unit delay operation which implements an internal state defined as

$$x_i(k) = x_i(k+1)z^{-1}. \quad (30)$$

The state transformation operation in discrete systems is the delay operator that implements a state definition given by (30). Fig. 9 shows the discrete controllable architecture. The other architectures are similar to those derived for the continuous domain with the integrator functional block replaced by a unit delay operator.

IV. Intermediate Architecture Design

The behavioral synthesis phase generates a functional architecture that can realize a set of behavioral specifications. This architecture needs to be implemented in a specific circuit implementation technology. By introducing the intermediate architecture design stage, it is now possible to explore various implementation technologies for realizing a design. In the existing design methodology, the designer works from a constrained space where decisions on the implementation technology and even the system architecture are made a priori, and the entire design process follows. By introducing the behavioral synthesis and intermediate architecture design stages, and implementing these in the form of computer aided design (CAD) tools, architecture design space exploration and implementation technology assessment are now feasible design tasks.

An *intermediate architecture* is a functional structure composed of implementation style specific functional units. An intermediate architecture is synthesized by replacing each functional component at the architecture level with an implementation style specific realization. This process is reminiscent of the *module binding* phase in the high-level synthesis of digital systems. At the intermediate architecture level, implementations in competing technologies can be generated and evaluated during design space exploration. Thus an additional degree of freedom arises at the architecture design stage, in the form of a choice of an appropriate implementation style most appropriate for realizing the system requirements. The intermediate architecture is composed of the functional blocks that are implementation specific. For example in the $gm - C$ style, each component of the behavioral architecture will be replaced by the $gm - C$ equivalent, the intermediate architecture will be composed wholly of transconductances and capacitors. An adder in a behavioral architecture would be replaced by its $gm - C$ equivalent realization in the $gm - C$ specific intermediate architecture. The intermediate architecture is then simulated by using behavioral models for transconductances. Here the contribution of the various second order effects of the transconductance elements on the system performance can be studied.

The rationale for the *intermediate architecture* is

1) Explore the various implementation technologies.
2) Feasibility analysis of implementation technologies.
3) System-level verification in a specific implementation technology.
4) Bridge between the architecture level and the circuit level.
5) Translating architecture level specifications into specifications for a particular implementation technology.

The behavioral synthesis process generates a functional architecture from behavioral specifications. The functional architecture is then implemented in one of the various integrated circuit implementation technologies. For example, an integrator can be implemented using an operational amplifier and RC elements, an opamp with MOSFETS, or using transconductances and capacitors [21], [22]. Fig. 10 shows a $g_m - C$ adder and an integrator. For the adder:

$$I_s = I_1 + I_2 = g_{m1}V_1 + g_{m2}V_2$$

and $V_o = 1/g_{ms}I_s$, thus

$$V_o = \frac{g_{m1}}{g_{ms}}V_1 + \frac{g_{m2}}{g_{ms}}V_2.$$

For the integrator,

$$V_o = \frac{1}{C}\int I_o\,dt$$

$$V_o = \frac{1}{C}\int g_m V_i\,dt$$

or

$$V_o(s) = \frac{g_m}{C}V_i(s).$$

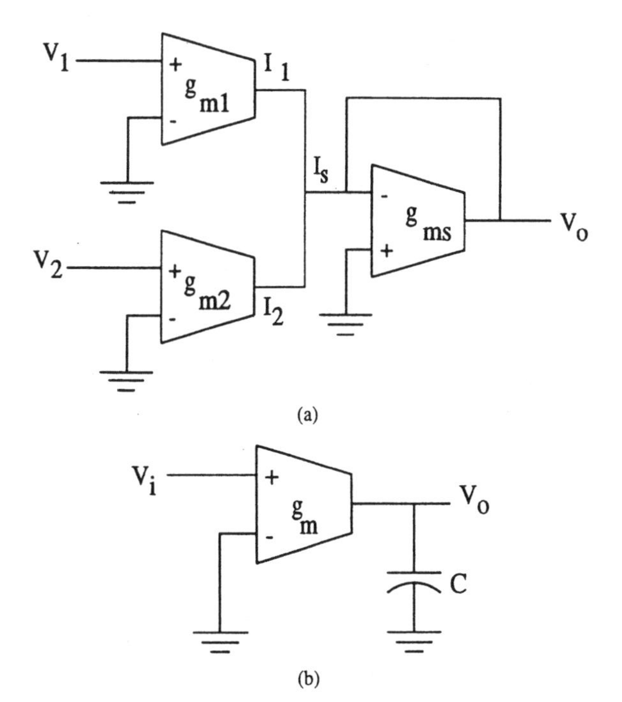

Fig. 10. $g_m - C$ (a) adder and (b) integrator.

Intermediate architecture design is essential to verify if the functional architecture can be realized in a particular implementation prior to actual circuit synthesis. If a selected implementation style is infeasible, an alternate implementation technology needs to be explored.

In the synthesis process from the behavioral architecture to the intermediate architecture stage, specifications for various components in an implementation technology are generated. The components in the intermediate architecture correspond to subcircuits that can be further synthesized using circuit-level synthesis tools, into a full custom CMOS implementation. Thus the intermediate architecture also serves as a bridge between the architecture design and circuit-level synthesis stages.

V. Implementation Description

The implementation of the synthesis methodology is composed of an architecture representation, an architecture specification language (ASL), and the synthesis module ARCHGEN.

A. Design Representation

The various architectures and functional elements that compose the architectures are represented in an object-oriented paradigm and implemented in the C++ programming language [23]. Base classes for representing the architectures and the various components are defined and during synthesis, instantiations of these base-classes are created. The base class *ArchitectureModel* represents the behavioral architecture, and the *CircuitModel* class represents the intermediate architecture. The *CircuitModelH* class is defined to implement the synthesis and technology mapping routines to generate intermediate architectures. Fig. 11 shows the behavioral architecture representation objects. Each functional block in the architecture is

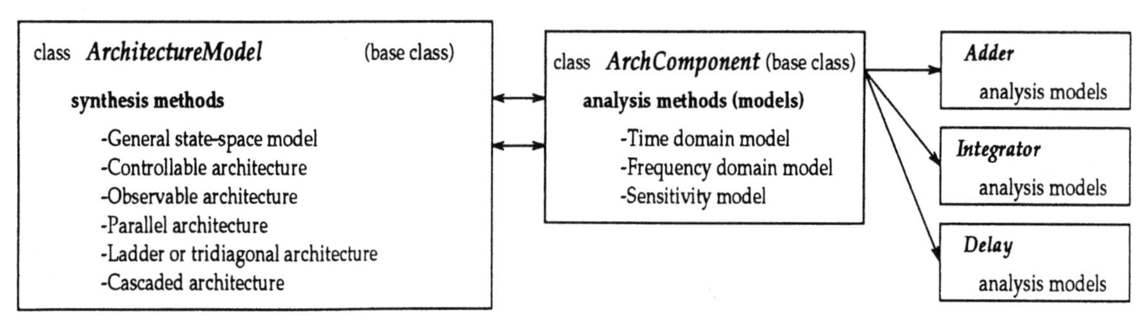

Fig. 11. Behavioral architecture representation.

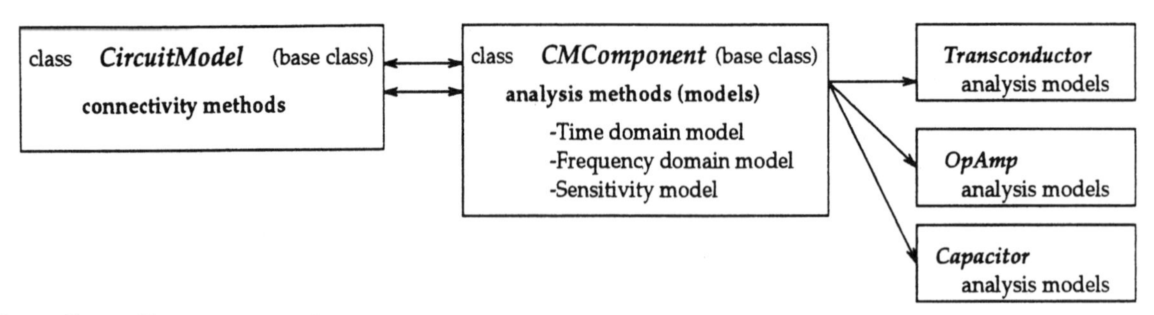

Fig. 12. Intermediate architecture representation.

```
#include <archModel.h>
class LowPassFilter5 :public ArchitectureModel {
  public:
    Node N1, N2, N3, N4, N5, N6, N7, N8, N9, N10, N11;
    Adder add1, add2, add3, add4, add5;
    Integrator integ1, integ2, integ3, integ4, integ5;
    ExternInput in1;
    ExternOutput out1;
    LowPassFilter5()
     : ArchitectureModel( "LowPassFilter5" )
      , N1( "1" ), N2( "2" ), N3( "3" ), N4( "4" ), N5( "5" )
      , N6( "6" ), N7( "7" ), N8( "8" ), N9( "9" )
      , N10( "10" ), N11( "11" )
      , in1( "in1", N1)
      , out1( "out1", N11)
      , add1( "add1", 2, N1, N5, N2, 1.23131, -0.887338)
      , add2( "add2", 2, N3, N7, N4, 0.887338, -0.817369)
      , add3( "add3", 2, N5, N9, N6, 0.817369, -0.876224)
      , add4( "add4", 2, N7, N11, N8, 0.876224, -1.27794)
      , add5( "add5", 2, N9, N11, N10, 1.27794, -1.65421)
      , integ1( "integ1", N10, N11, 5.61359e-05)
      , integ2( "integ2", N2, N3, 5.61359e-05)
      , integ3( "integ3", N4, N5, 5.61359e-05)
      , integ4( "integ4", N6, N7, 5.61359e-05)
      , integ5( "integ5", N8, N9, 5.61359e-05)
    { compile(); }
};
```

Fig. 13. Architecture specification language (ASL) description of a 5th order low pass filter behavioral architecture.

represented by the *ArchComponent* base class, with functionality specific classes defined for each component such as an Adder, Integrator, Delay etc., as shown in the figure. Similarly, Fig. 12 shows the intermediate architecture representation objects with the *CMComponent* class used to define functional blocks, with specific subclasses defined for transconductances, Opamps, etc. The object-oriented representation allows various attributes to be assigned to each object. Analysis models and routines are defined for each functional block. This allows tight integration between the synthesis and verification modules

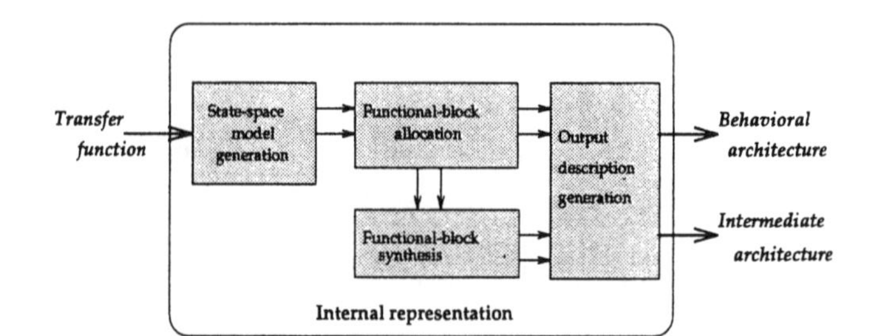

Fig. 14. Functional organization of ARCHGEN.

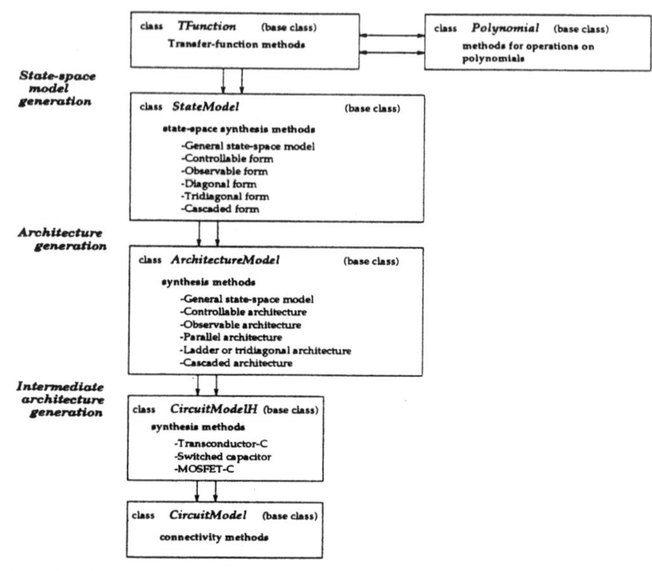

Fig. 15. Object generation through the synthesis process.

[7], [9]. The verification module ARCHSIM, provides various analysis modes such as transient analysis, frequency response

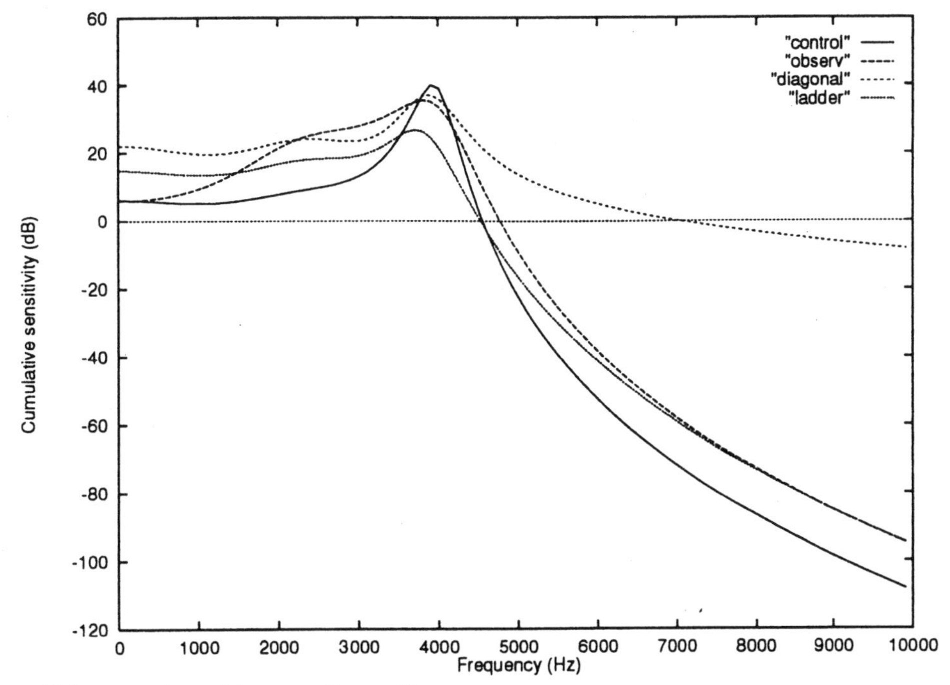

Fig. 16. Cumulative sensitivity comparison of low-pass filter architectures.

computation, sensitivity analysis, steady-state analysis and Monte-Carlo analysis through which the synthesized architectures can be validated, as well as metrics for design space exploration generated.

B. Architecture Specification Language

As a hardware description language for documentation, as well as external interface to the synthesis module and the verification module, a customized C++ based language, architecture specification language (ASL), was developed [7]. Since C++ is the implementation language, using it as a basis for the description language has the advantage of efficient compiled approach for the synthesis and verification tasks, as the language compiler is used for parsing the input description. An ASL description is in the format of a C++ class definition, as an example, the description of a 5th order Chebyshev filter is shown in Fig. 13. Description generators for emerging hardware languages, such as MHDL [24] can be integrated as a postprocessing stage.

C. ARCHGEN

The behavioral transformations applied sequentially to a transfer function specification are 1) state-variable model generation, 2) behavioral architecture synthesis and 3) Intermediate architecture generation. The program ARCHGEN is implemented to generate an architectural description in the form of an internal design representation. Fig. 14 shows the functional architecture of the program ARCHGEN. The internal design representation is composed of instances of objects that represent the functional blocks; adders, integrators, etc. Fig. 15 shows the complete set of description objects as their instances evolve through the analog systems-level synthesis process. A matrix representation of the desired state-space model is algorithmically generated and stored as an instance of the *StateModel* class. The state-space representation of a system is implemented in the class *StateModel.* The synthesis methods that derive a state-space model from a transfer function specification are implemented under this class. The next task is behavioral architecture synthesis, here a specified architecture is synthesized from the state-space model, the internal representation is an instance of the *ArchitectureModel* base class. An intermediate architecture, represented as an instance of *CircuitModel* base class, is then generated through a process of module binding and technology mapping.

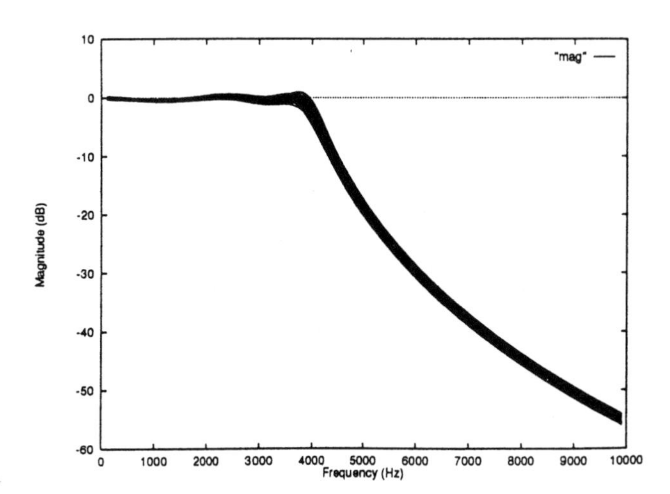

Fig. 17. Monte Carlo analysis of low pass filter.

VI. Experimental Results

In this section, we present a continuous-time and a discrete-time domain synthesis examples. For the continuous-time,

```
#include <cktmodel.h>
class LPF5GmC :public CircuitModel {
  public:
    Node N0, N1, N11, N10, N2, N3, N4, N5, N6, N7, N8, N9;
    OTASubCkt ota1, ota2, ota3, ota4, ota5, ota6, ota7, ota8,
              ota9, ota10, ota11, ota12, ota13, ota14, ota15,
              ota16, ota17, ota18, ota19, ota20;
    CapSubCkt C1, C2, C3, C4, C5;
    InputSubCkt inp1;
    OutputSubCkt out1;
    LPF5GmC()
     : CircuitModel( "LPF5GmC" )
       , N0( "0" ), N1( "1" ), N11( "11" ), N10( "10" ), N2( "2" )
       , N3( "3" ), N4( "4" ), N5( "5" ), N6( "6" )
       , N7( "7" ), N8( "8" ), N9( "9" )
    //------------ Integrator
       , ota1( "ota1", N10, N0, N11, 0.001)
       , C1( "C1", N11, N0, 5.61359e-08)
    //------------ Integrator
       , ota2( "ota2", N2, N0, N3, 0.001)
       , C2( "C2", N3, N0, 5.61359e-08)
    //------------ Integrator
       , ota3( "ota3", N4, N0, N5, 0.001)
       , C3( "C3", N5, N0, 5.61359e-08)
    //------------ Integrator
       , ota4( "ota4", N6, N0, N7, 0.001)
       , C4( "C4", N7, N0, 5.61359e-08)
    //------------ Integrator
       , ota5( "ota5", N8, N0, N9, 0.001)
       , C5( "C5", N9, N0, 5.61359e-08)
    //------------ Adder
       , ota6( "ota6", N1, N0, N2, 0.00123131)
       , ota7( "ota7", N0, N5, N2, 0.000887338)
       , ota8( "ota8", N0, N2, N2, 0.001)
    //------------ Adder
       , ota9( "ota9", N3, N0, N4, 0.000887338)
       , ota10( "ota10", N0, N7, N4, 0.000817369)
       , ota11( "ota11", N0, N4, N4, 0.001)
    //------------ Adder
       , ota12( "ota12", N5, N0, N6, 0.000817369)
       , ota13( "ota13", N0, N9, N6, 0.000876224)
       , ota14( "ota14", N0, N6, N6, 0.001)
    //------------ Adder
       , ota15( "ota15", N7, N0, N8, 0.000876224)
       , ota16( "ota16", N0, N11, N8, 0.00127794)
       , ota17( "ota17", N0, N8, N8, 0.001)
    //------------ Adder
       , ota18( "ota18", N9, N0, N10, 0.00127794)
       , ota19( "ota19", N0, N11, N10, 0.00165421)
       , ota20( "ota20", N0, N10, N10, 0.001)
    // ------------------------
       , inp1( "inp1", N1)
       , out1( "out1", N11)
    { compile();
      tc = 5.61359e-05;
      clk = 0; }
};
```

Fig. 18. Architecture specification language (ASL) description of a 5th order low pass filter at the intermediate architecture level in a $g_m - C$ realization.

a 5th order Chebyshev low-pass filter, normalized transfer function, chosen from the filter tables. The transfer function is shown next in (31).

$$H(s) = [0.1789]/[s^5 + 1.17251s^4 + 1.9374s^3 + 1.3096s^2 + 0.7525s + 0.1789. \quad (31)$$

The characteristics have a 0.5 dB ripple content in the pass band. This specification needs to be denormalized to a desired edge frequency, which is done by specifying the required frequency as an input parameter to ARCHGEN itself. In this experiment, a 4 kHz edge frequency was chosen. ARCHGEN is then used to synthesize different architectures: 1) controllable, 2) observable, 3) diagonal, and 4) ladder. Each of these architectures is analyzed during the synthesis process by ARCHSIM, the integrated behavioral simulator that provides various analysis modes [9]. One key metric used at this stage in choosing between architectures is sensitivity analysis. Fig. 16 shows the cumulative sensitivities of the four architectures that were synthesized from the above specifications. The ladder architecture has the least cumulative sensitivity and hence is chosen as the most appropriate architecture. Fig. 17 shows the results of the Monte Carlo analysis of the least sensitive architecture that can realize these specifications. The ASL description of the synthesized controllable architecture was shown in Fig. 13. The intermediate architecture generated in the $g_m - C$ technology for this specification is shown in Fig. 18.

The second example is a discrete filter specification, given by the z-domain transfer function

$$H(z) = [0.02014z^{-3} - 0.01896z^{-2} - 0.01896z^{-1} + 0.02014]/[-0.8552z^{-3} + 2.6828z^{-2} - 2.8252z^{-1} + 1].$$

The above filter specification characterizes a low pass filter with an edge frequency of 3.2 kHz and sampling clock frequency of 128 kHz. Using ARCHGEN, four different architectures were synthesized. The various simulation results, through an integrated analysis using ARCHSIM are shown in Fig. 19. The transient analysis and frequency response results shown in Fig. 19(a), and 19(b), verify the functionality of the synthesized architectures. The sensitivity analysis measurements are shown in Fig. 19(c), that compare the cumulative sensitivity of the four architectures. Here the cumulative sensitivity metric is used to choose the least sensitive architecture to parameter variations, in this case the diagonal architecture. A further analysis on the effects of parameter variation on the selected architecture is carried out through a Monte Carlo analysis in ARCHSIM. The results of this analysis are shown in Fig. 19(d), which indicate that the performance of this architecture, subject to parameter variations expected in the manufacturing stage are within a window of acceptability in the specifications.

VII. SUMMARY

The design of mixed analog-digital systems is a complex task, requiring efficient methodologies that address design issues across the two domains. Computer-aided design at the systems and high-level stages of digital systems has been an active field, with numerous techniques being developed and applied to the design process. However there is a lack of equivalent tools, techniques and methodologies in the analog domain, resulting in the analog sections posing bottlenecks to the design process of mixed analog-digital systems. In this paper we have specifically addressed the problems and issues of high-level design of analog systems, and developed a methodology which is further implemented as the CAD tool ARCHGEN. Our focus for this development has been on continuous-time and discrete-time filter architectures, which form a large component of analog functionality. Unlike digital systems, different classes of analog systems have disparate

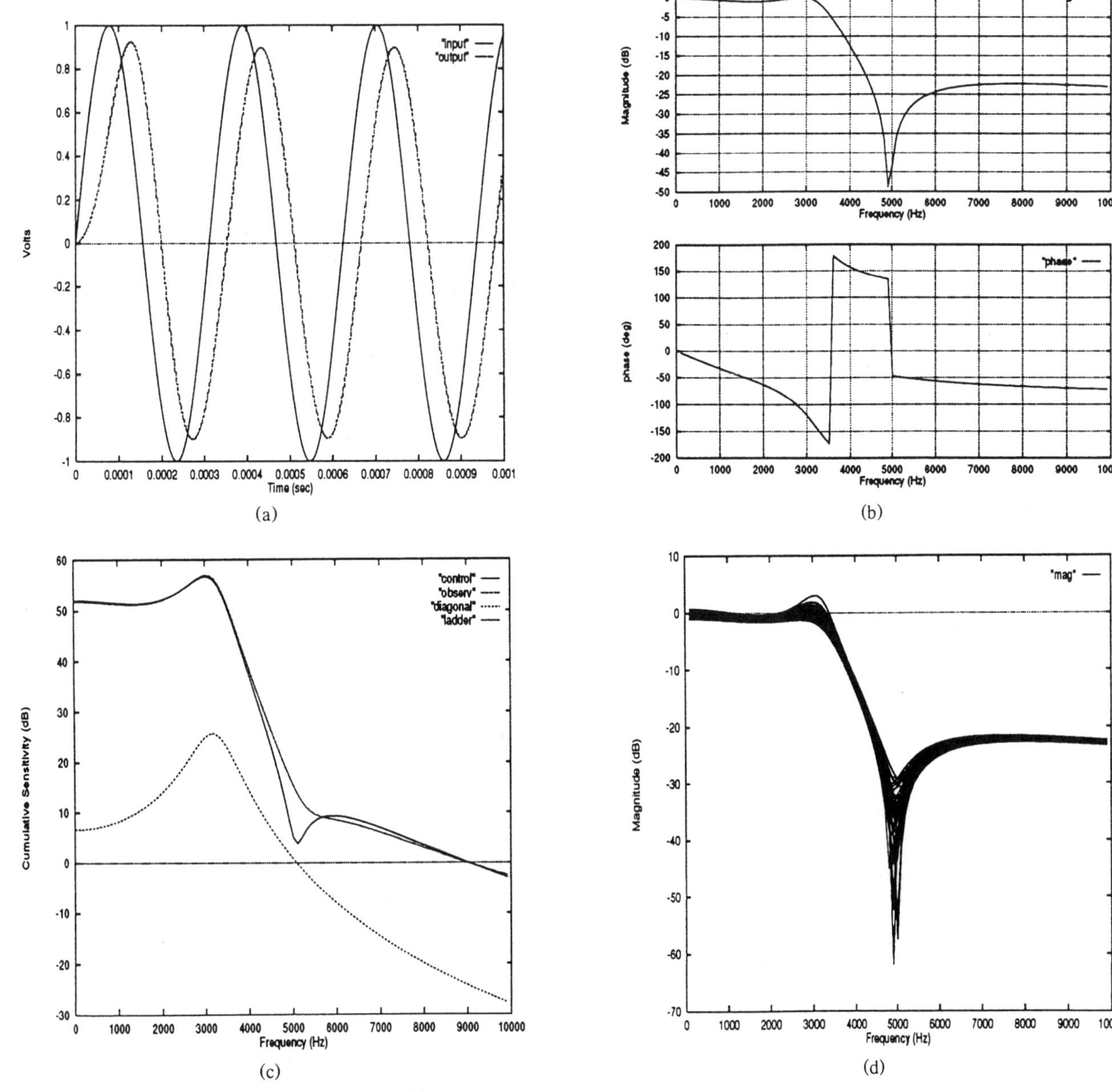

Fig. 19. Architecture verification results for discrete filter example.

characteristics, requiring specialized design methods. Or current efforts are now directed in extending the scope of the methodology presented in this paper to other analog systems, particularly data converters and phase-locked systems. Data converters are widely used to convert signals across the analog-digital boundaries and phase-locked systems are used for a wide range of tasks from tuning to clock synthesis.

ACKNOWLEDGMENT

The authors would like to thank J. Khoury for his comments and suggestions in the early stages of this research, and discussions with F. El-Turky have been helpful. The comments and constructive suggestions of the referees have helped in improving this manuscript to a large extent.

REFERENCES

[1] F. El-Turky and E. E. Perry, "BLADES: An artificial intelligence approach to analog circuit design," *IEEE Trans. Comput.-Aid. Des.*, vol. 8, pp. 680–691, June 1989.

[2] R. Harjani, R. A. Rutenbar, and L. R. Carley, "OASYS: A framework for analog circuit synthesis," *IEEE Trans. CAD,* vol. 8, pp. 1247–1265, Dec. 1989.

[3] H. Y. Koh, C. H. Sequin, and P. R. Gray, "OPASYN: A compiler for CMOS operational amplifiers," *IEEE Trans. Comput.-Aid. Des.*, vol. 9, no. 2, pp. 113–125, Feb. 1990.

[4] B. A. A. Antao and A. J. Brodersen, "Techniques for synthesis of analog integrated circuits," *IEEE Des. Test Comput.*, pp. 8–18, Mar. 1992.

[5] R. A. Rutenbar, "Analog design automation: Where are we? Where are we going?" in *IEEE Custom Integr. Circ. Conf. (CICC)*, May 1993, pp. 13.1.1–13.1.8.

[6] J. M. Cohn, D. J. Garrod, R. A. Rutenbar, and L. R. Carley, "KOAN/ANAGRAM II: New tools for device-level analog placement and routing," *IEEE J. Solid-State Circ.*, vol. 26, no. 3, pp. 330–342, Mar. 1991.

[7] B. A. A. Antao and A. J. Brodersen, "A framework for synthesis and verification of analog systems," *Analog Integr. Circ. Sign. Process.*, to be published.
[8] G. Jusuf, P. R. Gray, and A. L. Sangiovanni–Vincentelli, "CADICS-Cyclic analog-to-digital converter synthesis," *IEEE ICCAD*, pp. 286–289, 1990.
[9] B. A. A. Antao and A. J. Brodersen, "Behavioral simulation for analog system design verification," *IEEE Trans. VLSI Syst.*, 1995, to be published.
[10] M. C. McFarland, A. C. Parker, and R. Camposano, "The high-level synthesis of digital systems," *Proc. IEEE,* vol. 78, pp. 301–317, Feb. 1990.
[11] D. E. Thomas *et al.*, "Automatic data path synthesis," *IEEE Comput.*, pp. 59–70, Dec. 1983.
[12] S. Devadas and A. R. Newton, "Algorithms for hardware allocation in data path synthesis," *IEEE Trans. Comput.-Aid. Des.*, vol. 8, pp. 768–781, July 1989.
[13] R. K. Brayton, R. Rudell, A. Sangiovanni–Vincentelli, and A. R. Wang, "MIS: A multiple-level logic optimization system," *IEEE Trans. Comput.-Aid. Des.*, vol. CAD-6, no. 6, pp. 1062–1081, Nov. 1987.
[14] C. T. Chen, *Linear System Theory and Design.* New York: Holt, Rinehart, and Winston, 1984.
[15] J. G. Reid, *Linear System Fundamentals: Continuous and Discrete, Classic and Modern.* New York: McGraw-Hill, 1983.
[16] H. M. Power, "The companion matrix and Liapunov functions for linear multivariable time-invariant systems," *J. Franklin Inst.*, vol. 283, no. 3, pp. 214–234, Mar. 1967.
[17] N. N. Puri and C. N. Weygandt, "Calculation of quadratic moments of high-order linear systems via Routh canonical transformation," *IEEE Trans. Applic. Ind.*, vol. 83, no. 75, pp. 428–433, Nov. 1964.
[18] B. A. A. Antao, "Synthesis and verification of analog integrated circuits," Ph.D. dissertation, Dep. Elec. Eng., Vanderbilt Univ., Dec. 1993.
[19] R. Schaumann, M. S. Ghausi, and K. R. Laker, *Design of Analog Filters: Passive, Active RC, and Switched Capacitor.* Englewood Cliffs, NJ: Prentice-Hall Inc., 1990.
[20] A. V. Oppenheim and R. W. Schafer, *Digital Signal Processing.* Englewood Cliffs, NJ: Prentice-Hall Inc., 1975.
[21] Y. P. Tsividis, "Integrated continuous-time filter design—An overview," *IEEE J. Solid-State Circ.*, vol. 29, pp. 166–176, Mar. 1994.
[22] R. W. Brodersen, P. R. Gray, and D. A. Hodges, "MOS switched-capacitor filters," *Proc. IEEE,* vol. 67, no. 1, pp. 61–75, Jan. 1979.
[23] M. A. Ellis and B. Stroustrup, *The Annotated C++ Reference Manual.* Reading, MA: Addison Wesley, 1990.
[24] *MHDL Language Reference Manual,* Intermetrics Inc. and Army Research Laboratory, Mar. 1994.

Brian A. A. Antao (S'85–M'94) received the B.E. degree in electrical engineering from the University of Bombay (V.J.T.I.) in 1986, and the M.S. and Ph.D. degrees in electrical engineering from Vanderbilt University in 1988 and 1993.

Currently, he is a Member of the Research Faculty in the Coordinated Science Laboratory at the University of Illinois at Urbana-Champaign. He has held Summer Research Positions at AT&T Bell Laboratories, working on behavioral modeling and mixed-mode simulation. His research is in computer-aided design for analog circuits and systems, and design of application specific integrated circuits. Specific areas of focus at present include high-level analog synthesis and optimization, modeling and mixed-mode simulation.

Dr. Antao is the Guest Editor of the forthcoming special issue of the *Analog Integrated Circuits and Signal Processing Journal* on modeling and simulation of mixed analog-digital systems. He is also a member of the ACM and Tau Beta Pi.

Arthur J. Brodersen (SM'76) received the B.S., M.S. and Ph.D. degrees from the University of California, Berkeley in 1961, 1963, and 1966, respectively, all in electrical engineering.

He has been at Vanderbilt University since 1974, and has served as Chairman of Electrical and Biomedical Engineering, and Associate Dean of The School of Engineering. He is currently Professor of Electrical and Computer Engineering, Director of the Computer Engineering Program, and Associate Director of the Center for Innovation in Engineering Education. His research is in the modeling and computer simulation of integrated electronic circuits, particularly the design of linear circuits. In recent years, his research interests have been centered on the uses of computer and communications technology to enhance the undergraduate education of engineering students.

DARWIN: CMOS opamp Synthesis by means of a Genetic Algorithm

Wim Kruiskamp and Domine Leenaerts
Eindhoven University of Technology, Faculty of Electrical Engineering
P.O. Box 513, 5600 MB Eindhoven, the Netherlands
e-mail: M.W.Kruiskamp@ele.tue.nl

Abstract—DARWIN is a tool that is able to synthesize CMOS opamps, on the basis of a genetic algorithm. A randomly generated initial set of opamps evolves to a set in which the topologies as well as the transistor sizes of the opamps are adapted to the required performance specifications. Several design examples illustrate the behavior of DARWIN.

I. INTRODUCTION

The analog part of a complex mixed-signal system is often small compared to the digital part. However, the design of the analog part is often the most time consuming part in the entire design. The reason for this is that analog library cells are usually not suitable for all required analog functions [1]. The remaining analog circuits have to be designed by hand or preferable with the help of analog synthesis tools.

Analog circuit synthesis can be divided into two strongly related tasks: The selection of a suitable circuit topology and circuit sizing. During the topology selection, it is very useful if the topologies are already sized. The sized topologies, with known bias currents, capacitor values and transistor dimensions can then be evaluated and the best can be selected. An example of this approach is IDAC [2]. IDAC sizes several topologies and the user selects the sized circuit with the best performance. The main drawback of this approach is the fact that a lot of topologies have to be sized completely, of which only one will be used. To avoid this computational overhead, other tools, like OASYS [3] and OPASYN [4], select on forehand one topology, based on heuristics. If the tool can not size the selected topology correctly, other heuristics are used to redo the topology selection. Unfortunately, these heuristics are very difficult to create and there is a risk that a non-optimal topology is selected. In SEAS [5], the knowledge intensive topology selection heuristics are avoided by using an evolution algorithm to modify the topology. However, the several intermediate topologies still have to be sized completely in SEAS, using a time consuming simulated annealing algorithm. The first methodology that handles topology selection and circuit sizing simultaneously was presented by P. Maulik [6]. In this approach, topology selection is embedded in the circuit sizing problem. Therefore the risk of selecting a non-optimal topology is reduced, without the need of sizing many topologies. However, a lot of expert design knowledge is required to generate the essential design equations.

32nd Design Automation Conference®

DAC 95 - 06/95 San Francisco, CA, USA
©1995 ACM 0-89791-725-1/95/0006..$3.50

In this paper, we present a prototype synthesis tool DARWIN, which is based on a genetic algorithm [7, 8]. The genetic algorithm maintains a population of possible solutions, of which the topologies as well as the parameters gradually evolve to the final solution. The main characteristics of DARWIN are:

- Simultaneous topology selection and circuit sizing.
- Topologies are build up from basic building blocks by the program itself.
- Topology selection and circuit sizing are performed by means of a genetic algorithm.
- Only a little amount of expert design knowledge is required in the program.
- At the beginning of the synthesis, a set of constraints is solved for each building block to ensure that all intermediate generated circuits behave properly.

The organization of the rest of the paper is as follows. Section 2 describes the circuit description that is used in the genetic algorithm. Section 3 describes the genetic algorithm itself. Section 4 illustrates the behavior and performance of DARWIN by means of several design examples. Concluding remarks are provided in section 5.

II. CIRCUIT REPRESENTATION

Genetic algorithms are based on evaluation and generation processes, applied on a certain problem representation. When a sized netlist of a circuit is available, the evaluation of that circuit can be performed by means of simulations. The generation of new circuits however, will be much more complicated. For example in the domain of opamps, hundreds of suitable topologies are known, but the number of topologies that will not be able to behave as an opamp is countless. Furthermore, an unlucky chosen set of transistor dimensions and bias currents can make a circuit useless as an opamp. Therefore some restrictions must be defined to ensure that all intermediate generated circuits behave as an opamp. In section A, the topology aspects of this problem will be discussed, while the sizing aspects will be discussed in section B.

A. Topology description

The topology of an opamp can be separated into three building blocks: an input stage, an optional second gain stage and an optional output buffer. Several different topologies are possible for each building block.

On the basis of these three sets of building blocks, a large variety of topologies can be constructed. However, only a limited set of these topologies will be useful as opamp topology. By means of the connection matrix of Fig. 1, valid topologies can be recognized. A cross in entry (i, j) of the matrix indicates that block i might be succeeded by block j. As can be seen in the connection matrix, there are currently available: four input stages (element 2 ... 5), four second stages (element 6 ... 9), and three output buffers (element 10 ... 12). Element 1 and element 13 represent the input connection respectively the output connection of the opamp. Together, this results in 24 valid topologies, varying from just a simple input stage {1-2-13} as depicted in Fig. 4, to an opamp consisting of a folded cascode input stage, a second gain stage, and a class AB output buffer {1-4-8-12-13}. It is obvious that the connection matrix is easily to extend with new building blocks.

B. Building block description

In order that each building block behaves correctly, a set of simple constraints can be derived for the separate building block topologies, on the basis of the process parameters and the required specifications. An example of such a constraint is that all transistors in the input stage have to operate in saturation mode. Furthermore, some simple building block constraints can be derived out of the overall opamp specifications. For example, if the overall DC-gain has to be at least 80 dB, an input stage with a DC-gain of only 30 dB will not likely be part of an optimal design. Therefore when we require the DC-gain of the input stage to be at least 30 dB, we will limit the design freedom for the circuit parameters without loosing an optimal solution. Furthermore, since the proposed constraints will only be defined for the separate building blocks, the constraints will be much simpler than constraints for an entire opamp, as used in [6].

For the solution method we use, equations may only be linear. However, most of the equations that map the circuit parameters on the building block specifications are composed out of multiplications and square roots of the parameters. These constraints can therefore be written as linear combinations of the logarithms of the parameters. The only problems are combinations of products and summations. However, when the resulting space is convex, this problem can be overcome by approximating the (nonlinear) constraint by a piecewise linear space.

	2	3	4	5	6	7	8	9	10	11	12	13
1	×	×	×	×								
2						×			×			×
3					×					×		×
4							×			×		×
5								×	×			×
6									×	×	×	×
7									×	×	×	×
8									×	×	×	×
9									×	×	×	×
10												×
11												×
12												×

1 = input
2 = nMOS diff. pair (simple)
3 = pMOS diff. pair (simple)
4 = nMOS folded cascode
5 = pMOS folded cascode
6 = nMOS Com. Source (CS)
7 = pMOS CS
8 = nMOS CS with level shift
9 = pMOS CS with level shift
10 = nMOS-source follower
11 = pMOS-source follower
12 = class AB output buffer
13 = output

Fig. 1. Connection matrix

The expression for the DC-gain of a simple differential input stage with nMOS input transistors, as depicted in Fig. 4, is derived as follows:

$$\text{DCgain} \geq 30\text{dB} \rightarrow \begin{cases} \sqrt{\dfrac{I_{ss}}{L_1 \cdot W_1}} \leq \dfrac{\sqrt{K_n}}{31.62 \cdot \lambda_n} \\ L_1 \cdot \lambda_p = L_2 \cdot \lambda_n \end{cases} \Rightarrow \cdot \tag{1}$$

$$\begin{cases} \frac{1}{2}\log(I_{ss}) - \frac{1}{2}\log(L_1) - \frac{1}{2}\log(W_1) \leq \log\left(\dfrac{\sqrt{K_n}}{31.62 \cdot \lambda_n}\right) \\ \log(L_1) - \log(L_2) = \log(\lambda_n / \lambda_p) \end{cases}$$

As in (1), all constraints which ensure that a particular building block operates correctly, can derived. This results in a set of linear constraints for each building block in the format:

$$\begin{cases} \mathbf{A} \cdot \mathbf{x} \leq \mathbf{b} \\ \mathbf{C} \cdot \mathbf{x} = \mathbf{d} \end{cases}, \text{with } \mathbf{x} = \left(\log(W_1), \log(L_1), \cdots\cdots, \log(I_{ss})\right) \cdot \tag{2}$$

The solution of this set of constraints is a convex space of which the corner points $\mathbf{p}_i$ can be calculated with the algorithm of Tschernikow [9, 10]. The convex space given by (2) can then completely be described in the format:

$$\mathbf{x} = \left(\sum_{i=1}^{k} \alpha_i \cdot \mathbf{p}_i\right) \Big/ \sum_{i=1}^{k} \alpha_i \quad , \quad \forall_i \alpha_i \geq 0 \cdot \tag{3}$$

Expression (3) states that any k-dimensional vector α without negative entries α_i results in a vector $\mathbf{x}$ that meet all the requirements of (2). In DARWIN, buildingblocks are described by their name and a non-negative k-dimensional vector α. On behalf of computational simplicity, the values for the elements α_i are restricted to the values '0' and '1'. Due to this restriction, only 2^k points out of the solution space can be reached for each stage. In most cases however, this number will be large enough for satisfactory synthesis results. An example of this description is depicted in (4):

$$\begin{pmatrix} W_1 \\ L_1 \end{pmatrix} = \frac{\alpha_1 \cdot \begin{pmatrix} 1 \\ 1 \end{pmatrix} + \alpha_2 \cdot \begin{pmatrix} 3 \\ 1 \end{pmatrix} + \alpha_3 \cdot \begin{pmatrix} 4 \\ 4 \end{pmatrix} + \alpha_4 \cdot \begin{pmatrix} 3 \\ 5 \end{pmatrix} + \alpha_5 \cdot \begin{pmatrix} 1 \\ 3 \end{pmatrix}}{\sum_i \alpha_i} \quad , \quad \forall \alpha_i \in \{0,1\}$$

$$\alpha = (0,\ 1,\ 1,\ 1,\ 0) \Rightarrow \quad (W_1, L_1) = (3\tfrac{1}{3},\ 3\tfrac{1}{3}) \tag{4}$$

III. GENETIC ALGORITHM

Genetic algorithms or evolution programs [7, 8] can be used to find a near optimal solution for a wide variety of problems. Genetic algorithms maintain a population of individuals **P**(t) for iteration t. Each individual is an abstract representation of a potential solution of the problem at hand. During each generation, all individuals are evaluated to give some measure of their fitness. On the basis of this fitness, part of the population is selected to maintain for the next generation. In this selection, the more fit individuals have more chance to survive. The vacant places in generation $t+1$ are filled up by new individuals, generated by means of cross-over applied on the selected individuals. After a number of generations, the population converges to a population in which the best individual hopefully represents the optimal solution. This flow is depicted in Fig. 2.

In the synthesis tool DARWIN, this basic algorithm is implemented as follows. To ensure that the initial population and each following population contain at least consistent solutions to the problem (but not all of them are necessarily the best in context to the fitness function), a set of simple constraints is solved which ensure that all transistors operate in their proper region and that the transistor sizes are between maximal and minimal values. The solution is for each building block a set of vectors **p**, as discussed in section B. This prevents the algorithm to come up with circuits which are not behaving as opamps.

In the initial population $\mathbf{P}(t_0)$, the topology for each of the opamps is picked randomly out of the 24 possible opamp topologies. The entries in the vectors α for each opamp are set randomly to the value '0' or '1' (each entry in the vectors α corresponds to a vector **p**, as discussed in section B). A circuit description for each opamp can be obtained by a linear combination of the vectors **p**. All circuits in the initial population will now behave as opamps and the separate stages in each opamp will meet a minimum set of requirements. Nevertheless, the overall specifications of the opamps can still be far away from the required specifications.

```
procedure evolution program{
        t = 0;
        initialize P(t);
        evaluate P(t);
        while (not finished){
                t = t + 1;
                select P(t) from P(t-1);
                make new members with cross-over;
                insert new members in P(t);
                mutate P(t);
                evaluate P(t);
        }
}
```

Fig. 2. Genetic algorithm

It is now easy to derive the sized netlist for each opamp out of the circuit description. The performance of each opamp is then evaluated by means of a small signal equivalent circuit and analytical expressions. A fitness value is assigned to each opamp. When a specification of an opamp does not meet the requirements, this results in a negative contribution to the fitness of that opamp, proportional to the gap between specification and requirement. When all requirements are satisfied, the fitness function is defined by an optimization function based on power dissipation.

When all opamps are evaluated, their fitness values are linearly scaled in a way that the maximal fitness is equal to 1 and the minimal fitness is equal to 0. Several of the opamps are then removed out of the population. The number of removed opamps is equal to the population size, multiplied by the cross-over rate. For each opamp, the chance to survive is proportional to its fitness. Newly generated opamps fill up the resulting vacant places in the population. Each new opamp is generated out of two 'parent' opamps which are selected with a chance that is again proportional to their fitness. The new opamp is made by means of the cross over operation, as depicted in Fig. 3. The new topology is constructed by mixing the parents' building blocks. The first block of parent *A* is copied to the new topology. If, according to the connection matrix of Fig. 1, the second block of parent *B* might be placed behind the first block of *A*, this will be done. Finally, the last building block will be copied from the parent who did not deliver the second building block, if allowed by the connection matrix. Furthermore, the circuit parameters of the two parents are mixed if the parents contain similar building block(s). This is the case for the input stage of the example in Fig. 3. In that case, the α-vectors of the two parents are mixed. The α-vectors are cut at two random points and the separate parts are use in the new α-vector. In the case that the two parents have three stages in common (the same type of input stage, second stage and buffer stage),

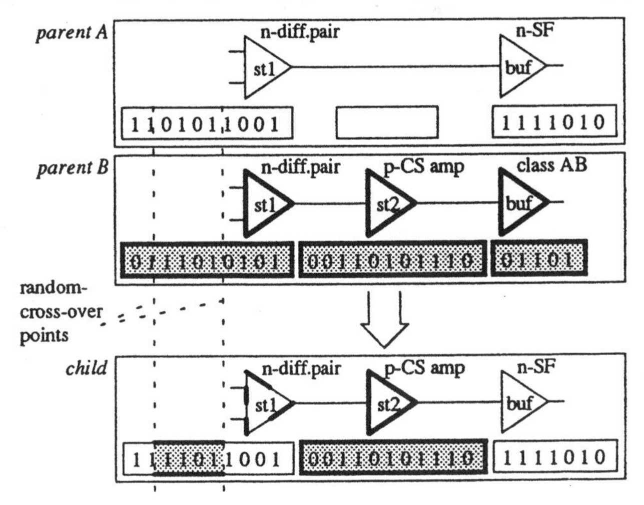

Fig. 3. Cross-over operation

this cross-over operation is performed on each of the three pairs of α-vectors.

The last modification process is mutation. This process changes the value of a small number of the entities of each α-vector. Each entity (a '0' or a '1' in Fig. 3) has a chance equal to the mutation rate to be changed in value.

It is very important here to notice that because we apply the genetic operators 'cross-over' and 'mutation' on the α-vectors, we always generate correct opamps in terms of the design constraints. Therefore, we do not have to verify whether a certain mutation is legal or not, this in contrast to genetic algorithms as in [11]. When we would perform the genetic operators directly on the circuit parameters (transistor widths and lengths), meaningless circuits can be generated, for example opamps of which the input transistors are out of saturation.

This evolution loop continues until a stop criterion has reached. In DARWIN, we use the number of generations as the stop criterion.

IV. EXPERIMENTAL RESULTS

To illustrate the synthesis process and to verify that DARWIN can synthesize functional opamps, we will discuss several circuits that were generated by DARWIN. For all opamps, we used the MIETEC 2.4 μm n-well CMOS process parameter set. The three design examples (A, B, and C) differed in the required specification for the DC gain. The optimization function was defined by the power dissipation. Table I depicts the performance specifications, the obtained results as simulated with the build in simulator and the SPICE simulation results. The component area in table I is the size of the transistors in addition to the size of the miller capacitance but without wiring and component spacing. For the SPICE simulations in table I, we used the 'typical' transistor parameters for examples A and C and the 'slow', 'typical' and 'fast' transistor parameters for example B. The sized schematics of the design example A, B, and C are depicted in Fig. 4, 5 and 6 respectively. The paracitic capacitor in Fig. 5 and 6 represent the paracitic capacitance due to the miller capacitor.

As can be seen in table I, the synthesized opamps meet all specifications according to the estimations of DARWIN and most of the specifications according to the SPICE simulations. The differences between the results as predicted by DARWIN and the SPICE simulation results are mainly due to the simplified transistor models that are used by DARWIN. The open loop transfer function of example B, as derived by DARWIN and by SPICE, is plotted in Fig. 7. In Fig. 7, 'slow', 'typical' and 'fast' refer to SPICE simulations, while 'darwin' refers to estimated transfer function by DARWIN. The estimations are good enough to be useful in DARWIN, while at the same time the evaluation of a single opamp only requires in the order of tens of milliseconds on a HP 700 workstation. In this way it is possible to evaluate an entire population of hundred opamps in only a few seconds, resulting in a complete opamp synthesis in a few minutes. When we should have used SPICE as the simulator for the intermediate evaluations, a small improvement in accuracy would have been achieved at the expense of an enormous increase in computation time.

We used the genetic algorithm with a population size of 100 opamps, a maximum of 150 generations, a cross-over rate of 0.2, and a mutation rate of 0.01. Due to the genetic algorithm, the inintial populations evolved to populations in which all opamps were of the same topology. This 'topology selection' can be seen in the area charts of Fig. 8, where the topology distribution is depicted during the first tens of generations of the synthesis process. The abbreviations that are used in the legend of Fig. 8 are explained as follows: 1st = one gain stage, 2st = two gain stages, fc = folded cascode, sf = source follower, AB = class AB buffer, n = nMOS input transistor, p = pMOS input transistor.

In design example A, the nMOS differential pair (1st_n, area marked with the number 1) dominated the population. This is not surprisingly, since the pMOS differential pair is

TABLE I.
REQUIRED SPECIFICATIONS, DARWIN ESTIMATIONS, AND SPICE SIMULATION RESULTS

Parameter	example A (60 dB)			example B (80 dB)					example C (100 dB)		
	Spec	darwin	SPICE (typ.)	Spec	darwin	SPICE (slow)	SPICE (typ.)	SPICE (fast)	Spec	darwin	SPICE (typ.)
DC-gain [dB]	≥ 60	60	54	≥ 80	93	75	87	96	≥ 100	144	133
gain bandwidth [MHz]	≥ 3	4.8	4.3	≥ 3	3.0	4.4	4.7	3.7	≥ 3	3.1	3.7
unity gain freq. [MHz]	≥ 3	3.8	5.3	≥ 3	3.2	10.1	7.4	6.8	≥ 3	3.1	6.8
phase margin [°]	≥ 60	72	74	≥ 60	72	30	63	67	≥ 60	60	47
slew rate [V/μs]	≥ 3	3.2	-4.0 +4.0	≥ 3	3.9	-7.9 +2.9	-6.7 +2.1	-5.6 +1.4	≥ 3	7.7	-7.9 +4.6
load capacitance [pF]	2	2	2	2	2	2	2	2	2	2	2
supply voltage [V]	± 2.5	± 2.5	± 2.5	± 2.5	± 2.5	± 2.5	± 2.5	± 2.5	± 2.5	± 2.5	± 2.5
input voltage range [V]	± 1	-1.3 +1.9	-1.2 +2.5	± 1	-1.8 +1.0	-2.2 +0.7	-2.2 +1.0	-2.2 +1.3	± 1	-1.0 +1.8	-1.0 +2.5
output voltage range [V]	± 1	-1.9 +2.1	-1.3 +2.2	± 1	-2.2 +2.0	-2.4 +2.4	-2.4 +2.4	-2.4 +2.4	± 1	-2.2 +2.0	-2.4 +2.4
power dissip. [μW]	low	31	52	low	89	157	124	96	low	251	308
component area [μm^2]	-	6500	-	-	3600	-	-	-	-	8100	-
technology	MIETEC 2.4 μm n-well CMOS										

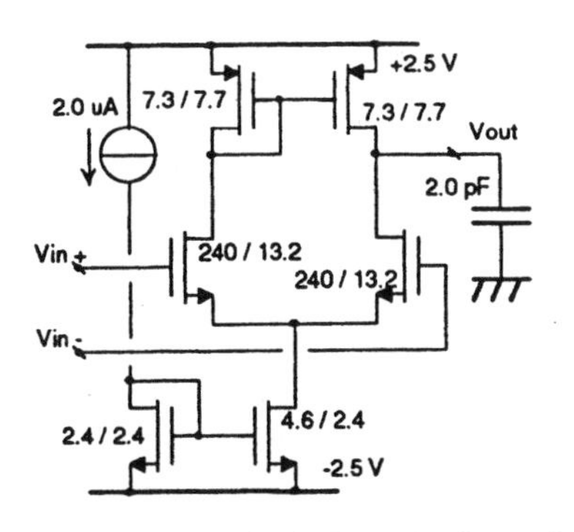

Fig. 4. Sized circuit schematic of example A

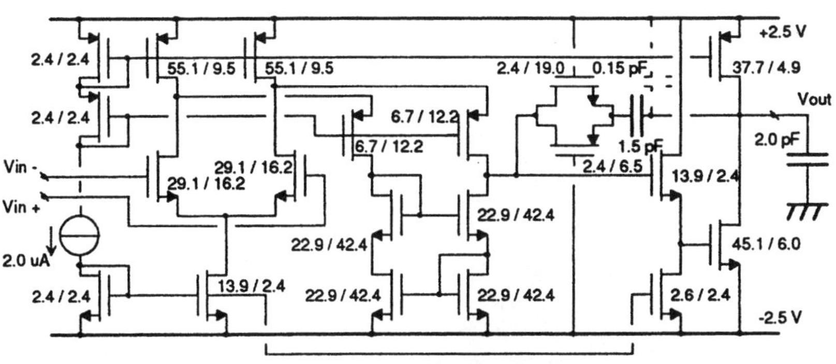

Fig. 6. Sized circuit schematic of example C

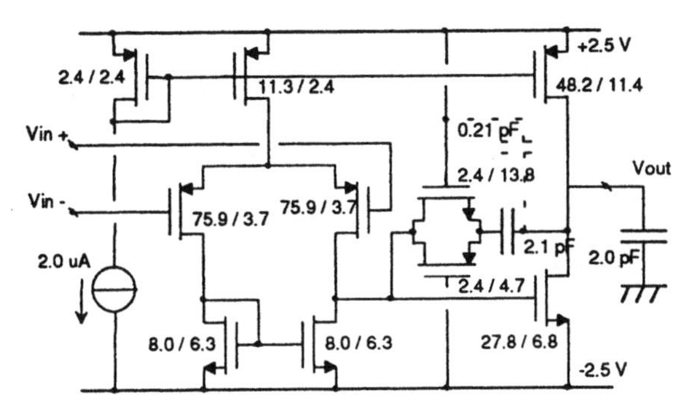

Fig. 5. Sized opamp schematic of example B

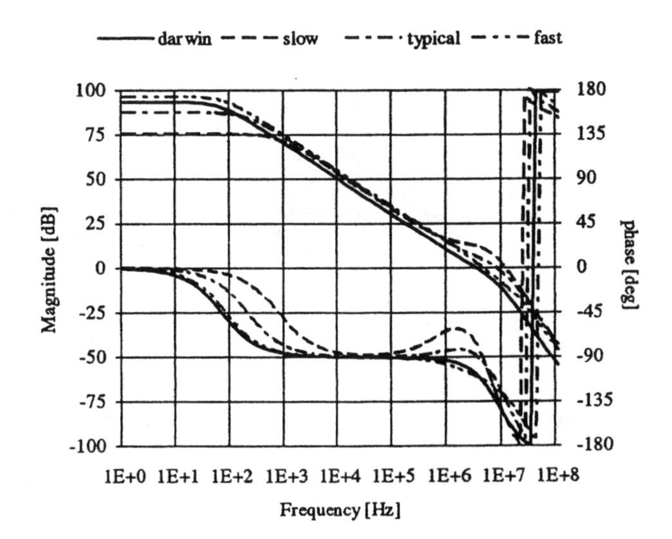

Fig 7 Bode plot of the open loop transfer function of example B

not able to have a DC-gain of 60 dB (in combination with the other specifications), and all other topologies have more stages and are therefore likely to dissipate more power.

In design example B, the required DC-gain was increased to 80 dB. All one-stage opamps died out in the first 20 generations, but several different topologies seemed to be good candidates after 20 generations. From that moment on, the two-stage opamp with pMOS input transistors (2st_n, area marked with the number 2) appeared to be slightly fitter than the other remaining topologies and the number of opamps with that specific topology increased until they dominated the population. Again the final choice of the topology is logical, since the DC-gain can not be achieved by a single differential pair, while at the other hand a folded cascode topology is not really required. Another interesting event in design example B is the fact that some topologies disappeared out of the population and reappeared some generations later. This happened for example with the topology, marked with the number 6. The fact that a new opamp can have a different topology than its two parents makes the algorithm robust, since a died-out topology can get a second chance to prove its suitability.

In design example C, the required DC-gain was further increased to 100 dB. An opamp with such a set of specification requirements can only be achieved by using a cascoded gain stage. It is therefore logical that the opamps with two gain stages, of which the first is a nMOS folded cascode stage, appeared to be more suitable than the other opamps. The areas in Fig. 11, marked with the numbers 3, 4, 5, and 6 all had a nMOS folded cascode input stage followed by a second gain stage. They only differed in their output buffer. of those four topologies, the version without a buffer became the eventual winner.

Simultaneously with this topology selection, the transistor sizes change as well during the synthesis process. The genetic algorithm will increase the maximal as well as the average fitness of the population during the synthesis process, indicating that the transistor sizes are adjusted in a way that the opamps in the population become more suitable to meet the required specifications.

The three design examples show that DARWIN is capable to select suitable topologies and to size the transistor sizes, without much design knowledge.

V. Conclusions

We described a way to synthesize analog circuits, based on genetic algorithms. An initial population of circuits evolves to a population in which the circuits are adapted to the required performance specifications. During each generation, all circuits are evaluated by means of rough simulations. On the basis of this evaluation, a fitness value is assigned to

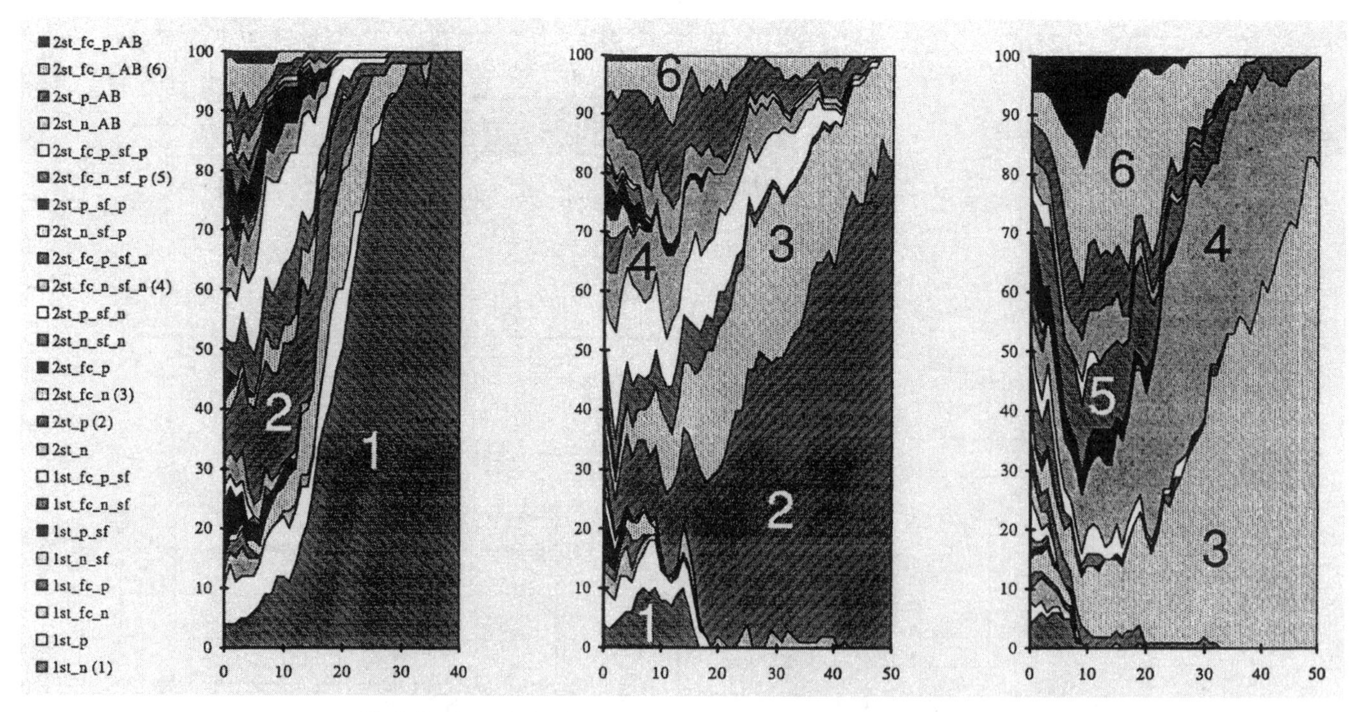

Fig. 8. Topology distribution versus generation, for design example A (left), B (centre), and C (right)

each of the circuits. Some of the ill-suited circuits are removed out of the population, while new circuits are generated out of well-suited circuits. Furthermore, the circuits are slightly changed by means of a mutation operation. The genetic algorithm does not require much design knowledge, but is nevertheless able to synthesize suitable circuits for a wide range of performance specifications.

The circuit representation that is used in the genetic algorithm allows us to put restrictions to the circuits' building blocks. In this way, it is ensured that all generated circuits posses certain qualities, required for a correct functioning of the circuit. A prototype tool, DARWIN, which is based on the proposed algorithm, can synthesize CMOS opamps from performance specifications and technology parameters. Several divergent design examples, which are verified by means of SPICE simulations, indicate that the program is capable of synthesizing the desired circuits.

Particularly in the beginning of the synthesis process, when a topology is selected and the transistor dimensions are roughly sized, the genetic algorithm showed to be very useful. A large number of different topologies can be explored, without the need of time-consuming transistor sizing algorithms. In the final stage of the opamp synthesis, gradient based local search algorithms will probably be more efficient in optimizing the transistor dimensions, although it is possible with genetic algorithms. We believe that the proposed approach is specially suited for the synthesis of circuits or systems that can have many different types of topologies and for which it is difficult to capture the topology selection knowledge in rules or equations.

REFERENCES

1. R. A. Ruthenbar, 'Analog design automation: Where are we? Where are we going?', *Proc. 15 th IEEE CICC*, San Diego, CA, 1993, IEEE, New York, pp. 13.1.1–13.1.8 (1993).
2. M. Degrauwe, et al., 'IDAC: An interactive design tool for analog integrated circuits', *IEEE J. Solid-State Circuits*, **SC-22**, 1106–1116 (1987).
3. R. Harjani, R. A. Rutenbar, and L. R. Carley, 'OASYS: A framework for analog circuit synthesis', *IEEE Trans. on Computer Aided Design*, **CAD-8**, 1247–1266 (1989).
4. H. Y. Koh, C. H. Sequin, and P. R. Gray, 'OPASYN: A compiler for MOS operational amplifiers', *IEEE Trans. on Computer Aided Design*, **CAD-9**, 113–125 (1990).
5. Z. Ning, M. Kole, T. Mouthaan, and H. Wallinga, 'Analog circuit design automation for performance', *Proc. 14 th IEEE CICC*, Boston, MA, 1992, IEEE, New York, pp. 8.2.1–8.2.4 (1992).
6. P. C. Maulik, L. R. Carley, and R. A. Rutenbar, 'A mixed-integer nonlinear programming approach to analog circuit synthesis', *Proc. 29 th ACM / IEEE DAC*, Anaheim, CA, 1992, IEEE, Los Alamitos, pp. 698–703 (1992).
7. Z. Michalewicz, *Genetic algorithms + data structures = evolution programs*, Springer Verlag, New York, 1992.
8. D. E. Goldberg, *Genetic algorithms in search, optimization and machine learning*, Addison-Wesley, New York, 1989.
9. D. M. W. Leenaerts, 'Applications of interval analysis to circuit design', *IEEE Trans. on Circuits and Systems*, **CAS-37**, 803–807 (1990).
10. D. M. W Leenaerts, and J. A. Hegt, 'Finding all solutions of piecewise linear functions and the application to circuit design', *Int. J. Cir. Theor. Appl.*, **19**, 107–123 (1991).
11. Z. Michalewicz, and C. Z. Janikow, 'Handling constraints in genetic algorithms', *Proc. 4 th int. conf. on Genetic Algorithms*, San Diego, CA, 1991, Morgan Kaufmann publishers, San Mateo, 1991, pp. 151–157 (1991).

AMGIE : a Synthesis Environment for CMOS Analog Integrated Circuits

Geert Van der Plas, Geert Debyser, Francky Leyn, Koen Lampaert, Jan Vandenbussche,
Georges Gielen, Willy Sansen,
K. U. Leuven, Department of Electrical Engineering, ESAT-MICAS
Kasteelpark Arenberg 10, B-3001 Leuven-Heverlee, Belgium

Petar Veselinovic, Domine Leenaerts
Eindhoven University of Technology, Faculty of Electrical Engineering, microlectronics group,
EEB P.O.Box 513, 5600 MB Eindhoven, The Netherlands

Abstract

A synthesis environment for analog integrated circuits is presented that is able to drastically increase design and layout productivity for analog blocks. The system covers the complete design flow from specification over topology selection and optimal circuit sizing down to automatic layout generation and performance characterization. It follows a hierarchical refinement strategy for more complex cells and is process independent. The sizing is based on an improved equation-based optimization approach, where the circuit behavior is characterized by declarative models that are then converted in a sequential design plan. Supporting tools have been developed to reduce the total effort to set up a new circuit topology in the system's database. The performance-driven layout generation tool guarantees layouts that satisfy all performance constraints. Redesign support is included in the design flow management to perform backtracking in case of design problems. The experimental results illustrate the productiveness and efficiency of the environment for the synthesis and process tuning of frequently used analog cells.

Keywords: Analog synthesis, Transistor sizing, Analog design, Performance optimization, Layout, Design reuse

1. Introduction

In recent years there is an increasing tendency in the electronics market to integrate complete systems, which before occupied one or more boards, onto a single chip or multi-chip module. This is the evolution towards Systems on a Chip (SoC) or Systems on a Package (SoP) [1]. A typical example is the highly competitive telecommunications market where cost and performance are strongly driving factors. Technologically this integration has been made possible because of the increasing miniaturization of the VLSI technology.

Most of the functions in such an integrated system are performed with digital circuitry that perform digital signal processing. Analog circuits however are always needed at the interface between the electronic system and the outer world. Nature is analog, and interaction with nature or transportation of signals is therefore inevitably through analog interface circuits.

All this explains the booming market share of mixed-signal ICs seen on the market today, with reported average growth rates well above 20% and therefore well above industry average. On the other hand, although the analog circuits occupy only a small part of the area in these mixed-signal ICs, they require an inversely large part of the design time and cost and are often responsible for design errors and expensive redesign iterations. Most steps in an analog design are basically still handcrafted, ranging from extensive and repeated SPICE simulation runs through manual place and route with the assistance of parameterized device generators. All this does not fit well with the short design cycles of time-to-market critical applications. Clearly there is an industrial need to increase analog design productivity and to lower the design risk. Therefore it is necessary to develop computer-aided design (CAD) tools that assist designers with the design of analog and mixed-signal integrated systems, and eventually automate (large parts of) it. Only in this way can the analog design community fully take advantage of the capabilities offered by the technology. This is even more true when looking at future systems on a chip, where most of the differentiation or added value will be achieved at the system level and the design at the lowest levels in the design hierarchy will mostly be automated through synthesis or through reuse, also for analog circuits.

Research in analog design automation however has been relatively slow, lagging far beyond its digital counterpart. An overview of the recent state of the art was presented in [2]. The earliest analog synthesis tools like IDAC [3], OASYS [4] or BLADES [5] relied very much on design knowledge, captured and hardcoded in expert-system-like structures such as design plans or rules. This required an expert designer to formalize and encode the design plan, with a large amount of heuristics being thrown in to come to a unique design solution, possibly using some restricted iterations and backtracking. Although this approach delivers fast results once the plan is built, the time needed to derive and craft a plan is long and the flexibility in changing the plan to a new specification target set is low. This restrained the industrial use of these early approaches. The second generation of experimental synthesis tools therefore relied on numerical optimization techniques, which allows more flexibility and user control and reduces the set-up time at the expense of larger run-times. The optimization techniques used, solve for the degrees of freedom in the design such that the performance specifications are satisfied and some user-defined design objectives (e.g. minimum power consumption) are minimized. Differences between the different approaches basically mount to the way how the circuit performance is evaluated at each iteration of the optimization loop. A first subgroup of methods use analytic models that describe the basic performance relations in the circuit by a set of (explicit or implicit) simplified symbolic equations. These methods are relatively fast but their accuracy is higher for rather linear (e.g. small-signal) characteristics. Examples of such systems include for instance OPASYN [6], OPTIMAN [7], STAIC [8], ISAID [9] and GPCAD [10] at the level of basic analog blocks, and SD-OPT [11] at the level of Δ-Σ converters. A second subgroup of methods use numerical simulation in the inner loop of the optimization. These methods have the full SPICE accuracy, but are extremely slow (unless the optimization space considered is very restricted) and still require many design constraints (e.g. stability constraints) to be incorporated into the optimization problem in order to guarantee a properly functioning circuit. Examples of such systems include for instance FRIDGE [12], MAELSTROM [13], ANACONDA [14] and ASTRX/OBLX [15]. The former three use plain-vanilla SPICE; the latter uses asymptotic waveform evaluation to boost the linear (small-signal) simulations but requires the designer to provide analytic formulas for all other characteristics. MAELSTROM and ANACONDA try to reduce the CPU time by spreading the optimization over a pool of parallel workstations.

Most of the above tools however cover only part of the entire analog design flow and therefore only partially fulfill the increasing need observed in industry today for an integrated synthesis environment that increases analog design productivity from specification to layout. For instance SD-OPT [11] and FRIDGE [12] can be used in combination to cover the higher-level respectively the lower-level design of Δ-Σ converters, but no layout solution is provided. ASTRX/OBLX [15] on the other hand has a companion analog layout synthesis tool KOAN/ANAGRAM [16] but no solution is offered for selecting a good topology from a library. In this paper a fully integrated analog synthesis environment, called AMGIE, is presented that implements a top-down hierarchical refinement design strategy for analog designs, and that covers the full design path from specifications over best topology selection and optimum circuit sizing down to automated layout generation followed by automatic verification and datasheet extraction. The system is also CMOS process independent and has an open library interface to add new circuit topologies into the library relatively easily. As every new circuit requires some amount of modeling, the approach is best suited for topologies that will be used in several designs. For instance in many companies always the same basic limited set of opamp structures is used for most applications, e.g. in Σ/Δ-converters, hence their design could be entirely automated while guaranteeing a fully customized solution for every set of specifications and for every selected CMOS process. In this way also technology porting of a basic cell library can be automated. On the other hand, some of the individual tools of the AMGIE system can also be used as point tools to leverage the design of the so-called "once-in-a-lifetime" circuits, as for instance is the case in domains like telecom and RF where the topologies tend to change with every new design.

In section 2 an overview will be given of the functionalities and design flow of the AMGIE analog synthesis environment. Also the software architecture of the entire system will be described. In section 3 the different design steps will then be discussed and the corresponding design tools and their underlying algorithms will be explained in detail. In section 4 the productiveness and efficiency of the system will be demonstrated by means of representative experimental results. Conclusions will be formulated in section 5.

2. Overview of the AMGIE analog synthesis environment

In this section first the functionality and supported design flow of the AMGIE system are described, followed by a description of the software architecture.

2.1. Design flow of the analog synthesis environment

The design strategy implemented in the AMGIE analog synthesis environment is a performance-driven top-down bottom-up strategy, which consists of the sequence of design steps shown in Fig. 1. Each of these steps is implemented in a separate software tool: either a new tool that has been developed or an existing commercial tool (esp. for simulation and back-end tasks) that has been integrated in the system. Their details will be described in section 3. The design flow is now described in more detail.

1) The user first has to select the desired type of circuit that (s)he wants to synthesize (the so-called "function block", e.g. an opamp, - the definition of a function block is an implicit behavior with a defined set of parameters, i.e. the specifications) and then has to enter the desired specifications for this circuit (including the performance specifications, the optimization targets and a choice of the technology process).

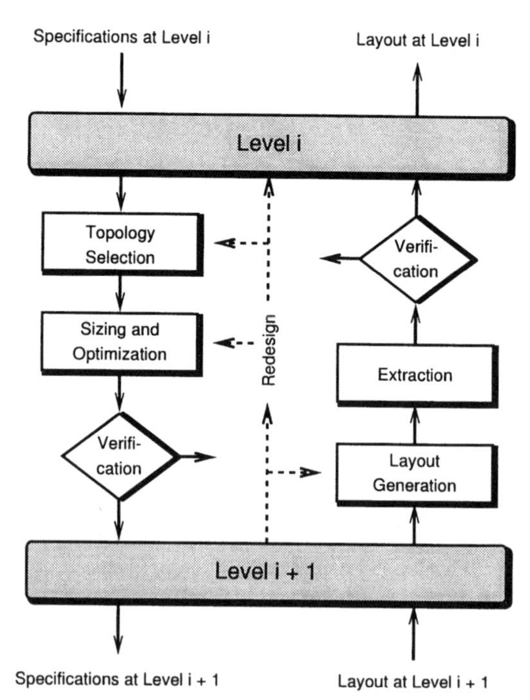

Fig. 1. Hierarchical design flow implemented in the AMGIE system.

1) $\varphi \in [\varphi_{lb}, \varphi_{ub}]$: the performance value is constrained in an interval, possibly the upper or lower bound is $\pm\infty$. The interval can be open or closed (inequality constraint), or be a single value (equality constraint). The majority of the performance specifications belong to this category. For instance the gain-bandwidth should be higher than or equal to 10 MHz.

2) $\varphi \in [\varphi_{lb}, \varphi_{ub}] @ \lambda$: the performance value is constrained in an interval at a certain parameter value. For instance the settling time should be smaller than 10 μs for settling to within 0.1%. Both the settling time value and the settling accuracy are to be specified.

2) the most promising topology for the function block capable to meet these specifications is chosen from the system's cell library by the *topology selection* tool;

3) then this topology is optimally sized such that all specifications are satisfied and the optimization targets are optimized by the *sizing and optimization* tool;

4) the resulting design is then automatically extensively simulated in the verification tool in order to verify that the design meets all performance specifications. If design failures or specification violations are noticed during this verification step, then previous design steps have to be changed and part of the design flow has to be iterated through backtracking (this is called redesign);

5) if the selected topology is defined in terms of non-primitive subblocks (e.g. an opamp within a filter), the system then first synthesizes these subblocks at this point in the design flow, by applying the complete design flow to every such subblock. When all non-primitive subblocks are synthesized, the design flow of the higher-level block continues; Note that hierarchy is used only when a decomposition in more or less non-interacting subblocks is possible.

For example, in our approach an opamp is sized as one block and is not further decomposed into subblocks like differential pairs, current mirrors, etc. because it is impossible to easily distribute specifications such as power-supply rejection ratio or settling time over the different subblocks.

6) an optimal layout is then generated or assembled by the layout generation tool;

7) the circuit is then extracted from the layout (including all layout parasitics) and again extensively simulated in the verification tool and a datasheet is generated. In case of design errors or specification violations, the same redesign procedure as outlined in step 4 is followed. If not, the design is successfully completed. The output of the system is the layout, the annotated schematic, the datasheet and design documentation.

Low-level circuits (such as opamps) are handled in a flat way as fully expanded transistor-level circuit schematics in the above design flow (step 5 is then trivial). For circuits of a higher complexity level (e.g. analog-to-digital converters) this is no longer possible, and the above design flow is executed in an hierarchical way. According to the divide and conquer strategy, hierarchy is used when the circuit is too complex to be designed as one block, and therefore is decomposed in easier to design subblocks (like a comparator in the converter). As hierarchical split-up is complicated by the interactions between the different subblocks (e.g. there is a large interaction at the level of differential pairs and current mirrors in an opamp), our rule of practice is to introduce hierarchy only when the resulting subblocks can be designed separately much easier than the overall block as a whole; if not it is preferred to design the block (e.g. an opamp) without decomposition. In the hierarchical case, sizing and optimization (step 3) then becomes specification translation which maps the specifications of the top-level cell to the individual specifications of each of the subcells (such as the comparator in the analog-to-digital converter). The subcells are then synthesized separately (step 5), one by one, using the same design flow, and the resulting subblock layouts are returned up the hierarchy to assemble the layout of the top-level cell (bottom-up layout assembly in step 6). If needed, redesign iterations are carried out across the design hierarchy.

The system supports two different design styles: 1) standard cells, and 2) custom cells. Standard cells are cells which have been completely designed before within the AMGIE system or added to the AMGIE cell library by a silicon foundry, third party or the user himself. In this way the system thus automatically supports the reuse of previous, possibly silicon-proven designs.

The above design flow has been encoded in a software module called the Design Controller. This module is the central "brain" of the AMGIE system and controls the execution of every synthesis run, according to the embedded design flow. The next step in the design flow can only be executed if the previous step has been terminated successfully. In this way the system can also guarantee full consistency in the data during the design, since no step can be executed before all required data have been correctly created before. Note however that the designer at any time can redo a previous step or - at his own responsibility - change the design after any step before proceeding. Part of the design flow is also the handling of redesign in the case of design failures. In the present AMGIE version, the Design Controller does not yet automatically impose the corrective action to be taken in case of redesign. As this largely depends on the actual design data, it would require sophisticated design expertise to be built in the design controller. Instead, the system includes a redesign wizard which is automatically invoked when a redesign action is needed and which proposes to the designer some possible context-specific alternative actions. The designer is then ultimately responsible for deciding on the actual backtracking action that has to be taken in order to correct the problem. It must be noted though that redesign iterations seldomly occur.

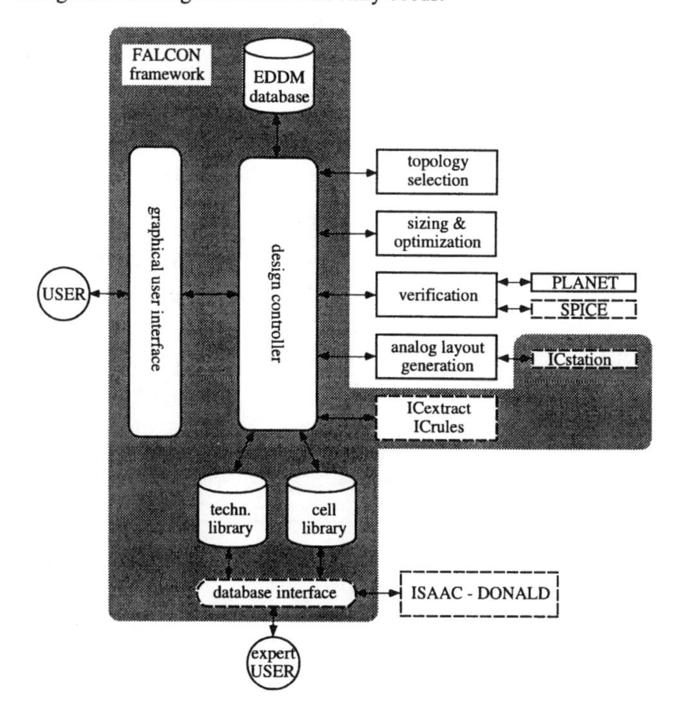

Fig. 2. Software architecture of the AMGIE system.

2.2. Software architecture of the AMGIE system

The overall software architecture of the AMGIE system is shown in Fig. 2 and has been developed in a very modular way. The system consists of a number of separate subtools, which are implemented as UNIX processes. The user interacts with the Design Controller by means of the Graphical User Interface. The individual subtools (such as the topology selection tool, the sizing and optimization tool, the verification tool with links to different simulators and the layout generation tool) are connected to the Design Controller and pass in this way design data between them. AMGIE stores data in a run-time database for the cell that is under design. Two libraries, the technology library and the cell library, store the design knowledge that is used by the design tools. The complete system has been integrated in a commercial EDA framework. The actual design database is implemented in this framework's database. The integration within a commercial framework also offered the advantage that existing commercial tools could be used for simulation, extraction, layout viewing, schematic capture and framework services. All data exchange occurs through a procedural interface, called Data Representation Interface (DRI). In this way the implementation details of the libraries and run-time database are hidden from the subtools and this modular design makes extension or modification of the libraries possible without changing the subtools. This should also make it possible to move to another EDA framework with a relatively small effort, or to extend AMGIE relatively easily with new software pieces.

The AMGIE system needs 2 libraries for its operation :

1) *cell library :* this library contains all information about the analog cells needed for topology selection, sizing and optimization, verification and layout generation. The AMGIE system can only design circuits for which the corresponding data are included in the cell library. In order to allow the designer to trade off short design time for flexibility and optimized performance depending on his

particular application, the cell library contains both standard and custom cells.

2) *technology library* : this library contains all technology process specific information that is needed by the different tools in the system.

There is also a library interface (bottom of Fig. 2) where the designer or library developer can include new cells and processes into the above libraries in a relatively easy way, so that the capabilities of the AMGIE system can grow beyond the initial capabilities delivered by the tool developers.

3. Detailed description of the analog design steps

In this section the core steps in the analog design flow and the technical and algorithmic details of the corresponding tools of the AMGIE system will be described in detail.

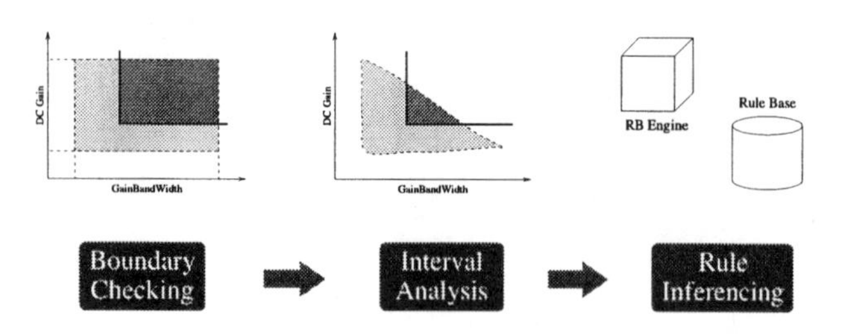

Fig. 3. Filter sequence implemented in the topology selection program

3.1. Topology selection

The topology selection tool selects from all topologies in the AMGIE cell library those that are able to satisfy the specifications of the function block as determined in the specification sheet and ranks them in order of preference. The topology selection tool actually works by eliminating inappropriate topologies from all possible candidates stored in the library and by ranking the remaining ones (see Fig. 3). Elimination of a circuit topology can be because the topology has not the correct functionality, does not fit to the selected process, or is incapable to meet the required specifications. The latter elimination is being carried out by applying a sequence of three consecutive filters on the list of candidate topologies [17]: first a boundary checking filter is applied, then an interval analysis based filter and finally a rule inferencing based ranking filter. The resulting list is presented in order of preference to the user through a graphical user interface. The user can accept the proposed topology, or force the AMGIE system to choose any other topology from the list. The three filters are now discussed in detail.

3.1.1. Boundary checking filter

The first two filters use quantitative information about the feasible performance space of a topology. This means that it is calculated what the achievable performances of each topology in the selected technology process are calculated, given the acceptable ranges in biasing values and device sizes, and that it is checked whether the specified performances are included in this performance space or not. Since the calculation of the performance space is a time-consuming process, this step is split in two consecutive filters. The first filter only calculates the multidimensional boundary box of the feasible performance space and checks whether this box overlaps with the space determined by the input specifications. This is called boundary checking (first filter in Fig. 3). If we denote the performance specification for performance j as $\varphi_j \in [\varphi_{lb_j}, \varphi_{ub_j}]$, and its feasible range for topology k as $\psi_{kj} \in [\psi_{lb_{kj}}, \psi_{ub_{kj}}]$, then topology k is an acceptable topology if :

$$\forall j : [\varphi_{lb_j}, \varphi_{ub_j}] \cap [\psi_{lb_{kj}}, \psi_{ub_{kj}}] \neq 0 \tag{1}$$

The feasible performance interval values are calculated from the declarative model that characterizes every circuit in the library (see next subsection). Since the boundary checking filter essentially considers each performance parameter independently of all the others, boundary checking is simple and fast, and can already eliminate most of the unfit topologies from the library. The disadvantage however is that no interdependencies between different performances are taken into account, and therefore that some topologies may pass the boundary checking filter but cannot meet the specifications in the end. Eliminating those topologies is exactly the task of the second filter. Some heuristics based on the relative position of the specification values within the feasible performance intervals or on the maximum size of the intervals' intersection can be applied to perform already an initial ranking of the surviving topologies. In our case the ranking is based on the ranking value rv_j which for a topology j is calculated according to :

$$rvj = \sum_{i=1}^{N} \left(\frac{2 * \left\| parameter.\max_i \right| - \left| parameter.\min_i \right\|}{\left\| parameter.\max_i \right| + \left| parameter.\min_i \right\|} \right) \tag{2}$$

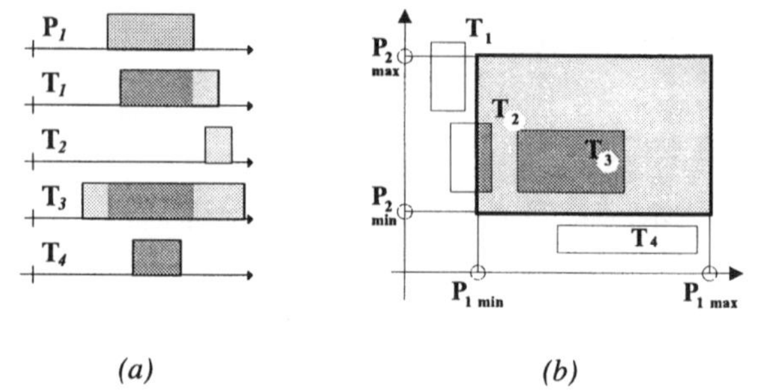

Fig. 4. Boundary checking illustrated for (a) one and (b) two performance characteristics, respectively.

Example

Fig. 4a shows a comparison of four topologies (T_1 till T_4) based on the feasible interval for one performance parameter i with a specification range P_i. The selection process rejects T_2 whose feasible interval does not overlap with P_i, and ranks the remaining topologies as T_3, T_1, T_4 depending on the size of the intersection region. Fig. 4b depicts another comparison of four topologies, this time based on feasible ranges for two performance parameters P_1 and P_2. T_1 and T_4 are rejected and the ranking of the remaining topologies is T_3, T_2.

3.1.2. Interval analysis filter

The second filter (see Fig. 3) takes the interdependencies between the different performances into account in order to calculate the complete feasible performance space more accurately, and to eliminate more inappropriate topologies that have passed the first filter. This means solving all (in)equalities simultaneously for the performance variables bounded by the required specifications. In order to do so, techniques from interval analysis are used in combination with Chernykov's algorithm that can solve all systems of nonlinear equations, including inequalities, using a piecewise-linear approximation for all nonlinear functions [17]. This system of equations is derived from the declarative model that is stored with every topology in the library (see next subsection). Topology selection now consists of checking if the calculated solution space constrained by the specifications is empty or not. If not, the topology is accepted; otherwise it is rejected. The drawback of the technique is its exponential computational complexity behavior with the size of the problem. Therefore only the most important specifications are taken into account in this filter, making the CPU time still acceptable in practice. Note also that this filter is only applied to topologies surviving the preceding boundary checking filter.

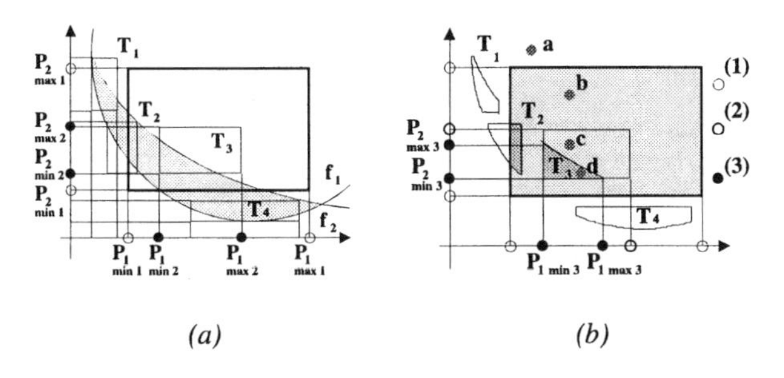

Fig. 5. Feasibility check with the relations between two parameters (a) and the result of the combination of both filters (b).

Example

The improvement that this filter brings over boundary checking is depicted in the examples of Fig. 5a and 5b. Fig. 5 differs from Fig. 4b in that the interdependency (functions f_1 and f_2) between the performance parameters P_1 and P_2 has been added. For the topology T_3 and for two performance parameters P_1 and P_2, Fig. 5b shows the user-specified specification intervals (1), the initial boundary intervals (2) as stored in the cell library, together with the remaining solution space (3) for those parameters after a calculation took place taking into account the relationship between P_1 and P_2. Point d for example belongs to the solution space of the topology T_3 for the given specifications, while point c is outside of it, in contrast with simple boundary checking that would accept both c and d.

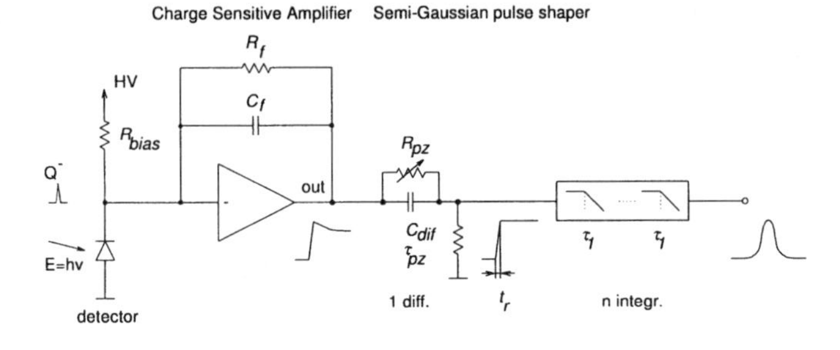

Fig. 6. Block diagram of the particle detector front-end (both the CSA opamp and the PSA integrators represent full transistor circuits).

Practical Example

Let us now consider the practical example of a particle/radiation detector front-end consisting of a charge-sensitive amplifier (CSA) and a nth-order pulse-shaping amplifier (PSA) as shown in Fig. 6. Note that the opamps are depicted symbolically here, but are in reality implemented as full transistor schematics. If we denote by A the specifications, the topology's design parameters and the technology parameters, then conceptually the model used for interval analysis looks as follows :

(speed requirement) BandWidth= $f_1(A)$ (3)

(noise requirement) EquivalentNoiseCharge= $f_2(A)$ (4)

(power requirement) DissipatedPower = $f_3(A)$ (5)

(design equations) $f_i(A)=0 \quad i = 4,\ldots,m$ (6)

The total number of model variables in the used model was larger than 25. Eight different CSA-PSA topologies were considered : either PMOS or NMOS input transistors for the CSA, each in combination with 4 different orders for the PSA. The resulting order of the topologies depends on the specifications, as shown in Fig. 7. The figure shows in the vertical axis the rank of the 8 topologies as a function of the requirement.

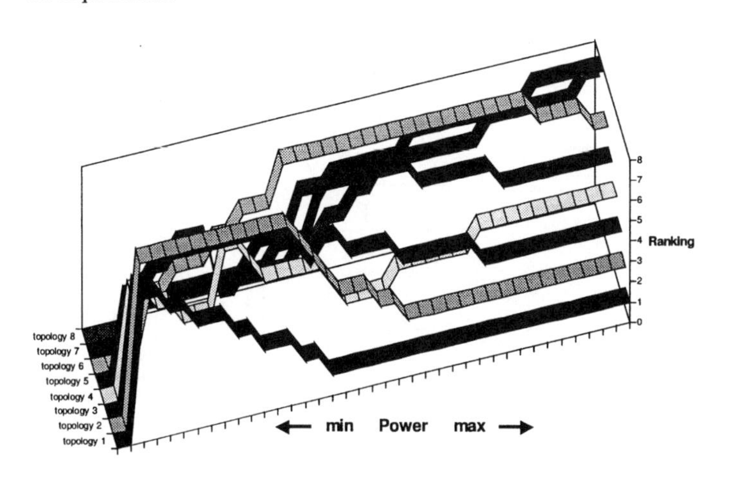

Fig. 7. Results of topology selection : order of the 8 different topologies as a function of the power requirement.

3.1.3. Rule-based ranking filter

The last filter that – if desired – can be applied to (re)rank the remaining list of candidate topologies is a heuristic rule-based filter approach, where an inference engine executes a number of rules stored in a database, to decide on the final ranking of the remaining topologies. The rules can encode both general heuristics as designer-specific preferences. They are implemented in *if-then* form.

An example of such rule based on cell attributes is :

IF {nr_integr.cell_A < nr_integr.cell_B} THEN cell_A before cell_B; (7)

3.2. Sizing and optimization

After topology selection an unsized schematic is available. The next step in the design flow is sizing and optimization where - for the lowest-level cells - the optimal device sizes and biasing will be determined for the selected topology to meet the performance specifications in the target technology process while minimizing some cost function (e.g. power consumption). At higher levels in the design hierarchy the tool searches for the optimum subblock parameters. In this case equations describe the performance of the higher-level blocks in terms of the parameters of the lower-level subblocks. Power and area estimators of these subblocks are used to assess the implementation cost.

The most difficult problem in circuit sizing is to solve for the degrees of freedom in the design, while managing the many conflicting performance trade-offs. The approach taken for the circuit sizing in the AMGIE system is improved equation-based circuit optimization. The use of optimization provides flexibility and easy set-up time; the use of equations provides speed of execution.

It alleviates many of the drawbacks of the traditional equation-based approaches, by using techniques of symbolic analysis for declarative model derivation and of constraint satisfaction for design plan generation on the one hand, and encapsulated device models to obtain high accuracy on the other hand. Also, an operating-point driven formulation of the design problem is used to speed up the evaluations. This will now be explained in more detail in the next subsections.

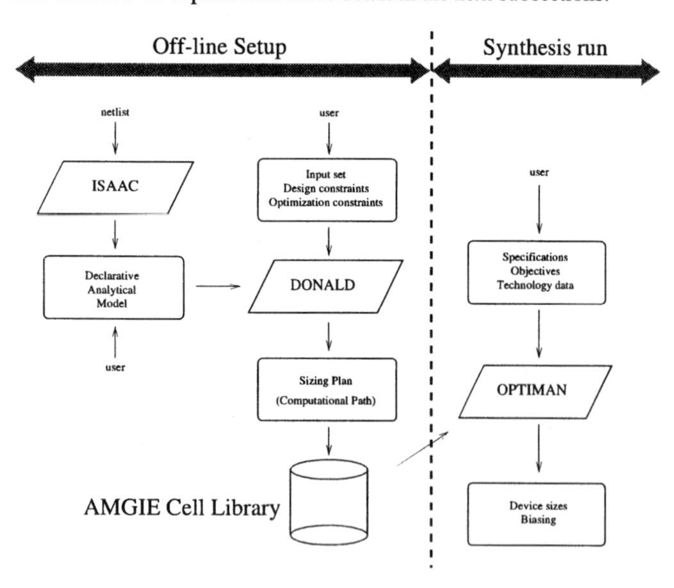

Fig. 8. Sizing model generation procedure and usage.

3.2.1. *Sizing model generation*

The general flow for generating a sizing model is shown in Fig. 8 [18]. The circuit behavior is first characterized by a declarative model that contains all the expressions (DC, AC, transient...) that fully describe the relationships between the circuit behavior and the circuit parameters. These equations are declarative, i.e. they only specify relationships that must hold simultaneously between different variables, they don't describe a direction nor sequence of solution (they are not assignments). DC equations are derived automatically from the circuit topology; AC equations are derived by means of symbolic analysis techniques like with the ISAAC [19] or SYMBA [20] tools; transient and other equations to date still have to be provided by the designer. In this way most of a declarative model can be generated automatically. The resulting model however is still declarative and therefore not yet suited for computer execution. The equation manipulation tool DONALD [21] is therefore used to automatically determine the degrees of freedom in the design, then to choose a set of independent input variables (equal to the number of degrees of freedom), and then to turn the undirected declarative model into a directed sequential computation plan, which indicates how (by means of which equations, in which direction and in which sequence) all the dependent variables have to be calculated from the values of the independent ones. DONALD uses techniques of constraint satisfaction to determine the ordering of the computation plan, and has a built-in algorithm to find a computation plan free of equation clusters if possible. The computation or design plan is then written out in C code, compiled and stored in the AMGIE cell library, ready to be used for circuit sizing and optimization by the OPTIMAN tool during an actual AMGIE synthesis run. All this model generation is done off line and thanks to the supporting tools drastically reduces the set-up time needed to include a new schematic in the AMGIE system. With this approach setup times of less than 8 hours have been achieved for moderate-complexity circuits [22]. Considering that the whole design flow can be executed in less than 20 minutes, the proposed environment also allows quick sizing plan evaluation and debugging, which is not true for any other previous approach, since either the setup time is much higher or the optimization time is much larger.

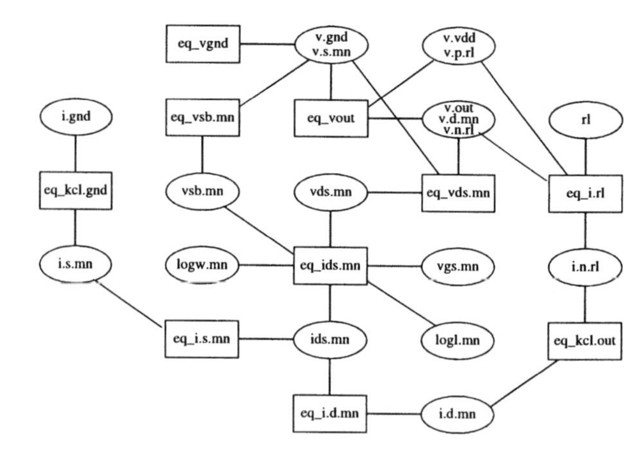

Fig. 9. Undirected bipartite graph.

Example

The approach is now illustrated for a simplified example to demonstrate the concepts. Fig. 9 shows part (the DC part) of the declarative model that corresponds to a common-source single-transistor amplifier with resistive load as shown in Fig. 10. The model is represented as a bipartite graph containing two different types of vertices: ovals for the variables, rectangles for the constraining equations. Note that the graph is undirected. The declarative equations are :

eq_vgnd : $$v.gnd = 0 \tag{8}$$

eq_vout : $$vout = \frac{v.vdd - v.gnd}{2} \tag{9}$$

eq_vsb.mn : $$vsb.mn = v.s.mn - v.b.mn \tag{10}$$

eq_vds.mn : $$vds.mn = v.d.mn - v.s.mn \tag{11}$$

eq_ids.mn : $$ids.mn = f(logl.mn, logw.mn, vgs.mn, vds.mn, vsb.mn) \tag{12}$$

eq_i.rl : $$i.rl = \frac{v.p.rl - v.n.rl}{rl} \tag{13}$$

eq_kcl.out : $$i.d.mn + i.n.rl = 0 \tag{14}$$

eq_i.d.mn : $$i.d.mn = ids.mn \tag{15}$$

eq_i.s.mn : $$i.s.mn = -ids.mn \tag{16}$$

eq_kcl.gnd : $$i.s.mn = i.gnd \tag{17}$$

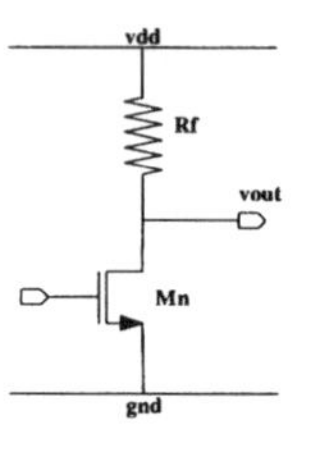

Fig. 10. Example circuit to illustrate design plan generation.

In total there are 14 variables constrained by 10 independent equations. Hence, there are 4 degrees of freedom in this simplified example. (In reality there are many more equations capturing also the AC and transient behavior of the circuit). This means that 4 independent variables can be selected as input variables. Many combinations are possible. If we choose for instance the variables {*v.vdd*, *vgs.mn*, *logw.mn*, *logl.mn*} as input set, then the originally undirected bipartite graph can be directed using constraint propagation techniques, as shown in Fig. 11, indicating the direction and order in which the equations have to be solved in order to calculate the values of the remaining 10 dependent variables (single line ovals) out of the values of the independent variables (double line ovals). The latter information can then be written out in C code as a procedural design plan and compiled for optimal evaluation speed.

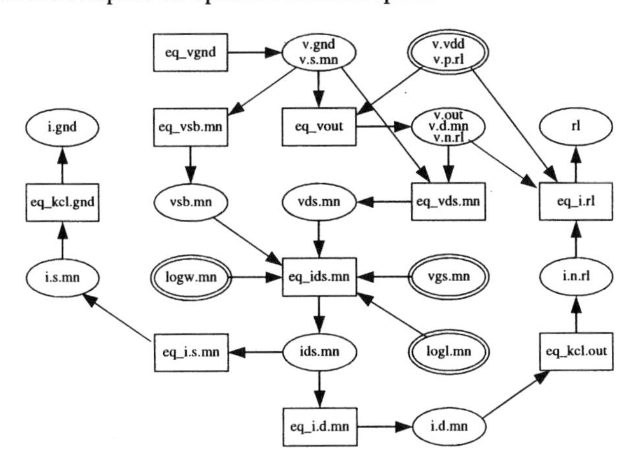

Fig. 11. Directed bipartite graph.

3.2.2. Operating-point driven formulation

The choice of the independent input variables in the computational plan, which of course are then also the optimization variables in the circuit optimization, is a very important factor in the performance of the sizing tool. As shown in [23], the choice of variables directly controlling the operating point of all MOS devices is to be preferred over all other input sets. In the presented approach the voltages at all nodes and currents in all branches are therefore specified as input variables (of course taking into account the physical dependencies resulting from the Kirchoff laws to obtain independent variables only). As a result of this, the time consuming DC operating point calculations can be avoided as all devices in the circuit can be solved independently, and operating point convergence problems often encountered in numerical simulation are avoided. Solving a device in our case requires calculating the value of W for which :

$$I_{ds} = f_I(V_{gs}, V_{ds}, V_{bs}, W, L, \vartheta, T) \tag{18}$$

in which I_{ds} is the drain current, V_{gs}, V_{ds}, V_{bs} are the operating-point voltages applied to the transistor, W and L are the width and length of the transistor, ϑ is the device model's parameter set and T is the temperature of the device.

Some simple device models, like SPICE MOS level 1, allow an explicit solution of this equation :

$$W = f_W(I_{ds}, V_{gs}, V_{ds}, V_{bs}, L, \vartheta, T) \tag{19}$$

For deep submicron technologies more advanced device models must be used, such as BSIM3v3, Philips MOS model 9 or the EKV model [24]. In this case the device equation must be iterated according to equation (18), for instance using a bisection method, which converges always since the function f_I typically is monotonic. These device models are therefore called as encapsulated functions in the sizing tool and are not hard-coded in the design plans.

3.2.3. Device model and technology parameters

The whole sizing approach is technology process independent through the use of a technology meta-model, i.e. all technology parameters are represented as variables in the equations used in the sizing plan, and their actual values during sizing are read in from the technology file in the technology library corresponding to the selected technology process specified by the user at the start-up of a synthesis session.

Four categories of parameters have been identified :

1) The *SPICE device model parameters*: these are the well known parameters used for simulation of circuits in any SPICE-like simulator. They are used to determine the operating current and voltages and small-signal parameters of each device.

2) The *mismatch model parameters*: these model the statistical intra-die differences between nominally identical devices, which heavily influence characteristics such as offset voltage or power-supply rejection ratio. These mismatches are modeled using the model of [25].

3) The *geometric model parameters*: these define the layout implementation of the device. Since in general there are multiple ways of laying out a given sized device (e.g. the normal and the fingered variants of a MOS transistor), the Geometry Calculation Method (GCM) parameter is used to indicate the correct layout variant during sizing. This GCM parameter triggers the proper set of equations to calculate or estimate the geometric dimensions of every device from its device sizes in the selected technology. For example, in the case of a fingered transistor the source and drain areas are shared, resulting in smaller overall source and drain area and junction capacitance. The GCM parameter determines which MOS variant and therefore which drain/source area function is used. For example, the total area occupied by a simple MOS transistor (GCM=1) is given by :

$$estimated\_area = \sqrt{(mos\_area + routing\_area)^2 + MINIMAL\_AREA^2} \tag{20}$$

where the core MOS area and the routing space are estimated as :

$$mos\_area = active\_area + area\_gate\_strap + area\_source\_drain \tag{21}$$

$$routing\_area = 4 * \left(\sqrt{mos\_area} + ROUTING\_SPACE\right) * ROUTING\_SPACE \tag{22}$$

Also other device geometry parameters like *ps*, *pd*, *nrs*, *nrd* are derived in this way. These are essential to estimate parasitics of the devices like the bulk capacitances, extrinsic drain/source resistances, etc.

4) The *technology info model parameters*: these define extra technology specific information used to characterize the devices. For example, the width and length of MOS transistors are constrained between a minimal and possibly maximal value, and they need to be snapped to grid. These technology constants are implemented through the process specific parameters LMIN, WMIN, LGRID, WGRID, LMAX and WMAX. By specifying the length and width of the MOS transistors as a ratio to their minimal value according to :

$$\begin{aligned} W &= WMIN * 10^{logW}, \quad logW = \log_{10}\left(\frac{W}{WMIN}\right) \\ L &= LMIN * 10^{logL}, \quad logL = \log_{10}\left(\frac{L}{LMIN}\right) \end{aligned} \tag{23}$$

and by using the *logW* and *logL* design model variables instead of the process-dependent W and L, an important technology dependence can be removed from the design plans which makes them more generic.

3.2.4. Hierarchical circuits: power and area estimators

When sizing hierarchical blocks, the performance specifications of the subblocks has to be determined. To associate a cost with a specification choice estimators are used. They link the subblock's performance with its feasibility, power and area:

$$\begin{aligned} feasibility &= f_{FB}(specifications) \\ power &= g_{FB}(specifications) \\ area &= h_{FB}(specifications) \end{aligned} \quad (24)$$

These estimators can be implemented with (1) manually derived equations, (an example for an analog sensor interface [26]), (2) by fitting to (automatically generated) design points [27,28] or (3) by empirically derived formulas [29]. Using optimization the top-level block parameters are then translated to subblock specifications. With high quality estimators an optimal trade-off in terms of power and area of the overall system is achieved.

3.2.5. Circuit optimization formulation

Once the design plan is available in the library, the OPTIMAN program [9] can then perform the actual circuit sizing and optimization as part of an AMGIE synthesis run, as shown in Fig. 8. The compiled sizing plan of the selected circuit schematic is retrieved from the cell library and linked to an optimization algorithm to tune the circuit towards the user defined specifications while optimizing some user defined design target, e.g. minimum power consumption.

The OPTIMAN tool is very modular in that the same optimization problem can be solved with several different optimization algorithms. In the AMGIE system the user can select the optimization algorithm as one of the options to be chosen in the specification sheet window. At this moment both global optimization algorithms like Very Fast Simulated Reannealing [30], as well as local optimization algorithms like Hooke-Jeeves [31], minimax [32] or Sequential Quadratic Programming [33] can be chosen. The range of all optimization variables as well as a default initial solution is provided for every schematic in the cell library, but can also be modified by the designer. After the sizing optimization in the AMGIE system the resulting optimal device sizes are automatically back-annotated onto the schematic of the circuit under design.

The used formulation of the analog circuit sizing as an optimization problem is as follows :

$$\begin{aligned} &\text{Find min } F(X) \\ &X \in D \\ \text{s.t.: } &g_j(X) \le 0 \quad \text{for j = 1..n} \\ &h_i(X) = 0 \quad \text{for i=1..m} \end{aligned} \quad (25)$$

The optimization variables X are the independent input variables of the stored design plan of the selected schematic. For optimization algorithms that cannot handle the constraints directly, like for instance Very Fast Simulated Reannealing (VFSR), penalty functions are added to the optimization target. The cost function used during optimization is then as follows :

$$\Psi(\mathbf{X}) = F(X) + u(G(X)) + u(H(X)) \quad (26)$$

where the penalty terms are added to handle the equality and inequality constraints. Different choices of penalty functions are possible, such as :

$$\begin{aligned} u_j(g_j(X)) &= w_j g_j^2(X), \; g_j(X) \ge 0 \\ &= 0, \; g_j(X) < 0 \end{aligned} \quad (27)$$

$$\begin{aligned} u_j(g_j(X)) &= w_j \ln(1 + \frac{g_j(X)}{g_{jo}}), \; g_j(X) \ge 0 \\ &= 0, \; g_j(X) < 0 \end{aligned} \quad (28)$$

For optimization algorithms that can handle constraints directly like Sequential Quadratic Programming (SQP), of course no penalty terms are added to the cost function.

Originally only the nominal performance could be optimized. Recently, however, the approach was extended to include also the impact of process parameter and operating parameter variations on the circuit performance, allowing to simultaneously optimize the circuit performance and the design yield and robustness [34].

The developed sizing approach results in a very fast sizing process. CPU times between a few minutes (for low-complexity circuits of about 10 MOS transistors) to one hour (for complex circuits of approximately 100 MOS transistors) are reached on a standard SUN Ultra 1-170 workstation, while at the same time achieving high accuracy.

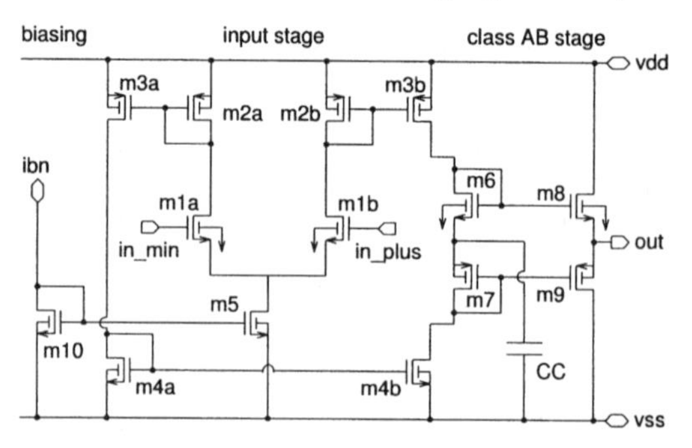

Fig. 12. Schematic of the symmetrical OTA with class AB output buffer.

Practical Example

Let us now consider the practical example of a symmetrical OTA with class AB output buffer as shown in Fig. 12. The specification set that has been limited to the following specs: Low frequency gain (Av0), Gain-bandwidth product (GBW), Slew Rate (SR), Phase margin (PM), Random Offset Voltage (Voff) and Capacitive Load (Cload). Using Donald and ISAAC and adhering to the principles proposed previously in this section, a fast compiled design plan has been created. It uses the black box device models, DC operating point formulation, mismatch model and analytical equations linking the specifications to the device's parameters. In addition to these equations design constraints have been added that generate a penalty when transistors leave preferred operating point regions (overdrive voltage limits, saturation limits, etc.). The thus generated design plan can now be used to perform optimizations.

TABLE 1
SPECIFICATIONS OF THE SYMMETRICAL OTA WITH CLASS AB OUTPUT BUFFER.

Specification		Value	Unit
Low Frequency Gain	>	1000	-
Phase Margin	>	65	*degrees*
Random Offset Voltage	<	3	*mV*
Slew Rate	>	1	V/us
Gain-bandwidth	>	1	MHz
Load Capacitance	=	100	*pF*
Power	MIN	1	*mW*
Area	MIN	0.05	mm^2

The specification values used for optimizations are summarized in Table 1. Since at first no good design point is known (to start of the optimization), the global optimization algorithm (VFSR) is employed.

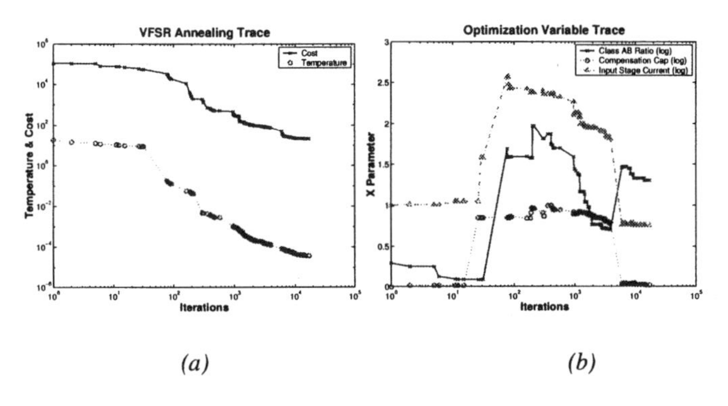

Fig. 13. Trace of VFSR optimization: (a) Cost and temperature (b) important optimization variables.

In Fig. 13a both the temperature and best cost achieved during a typical optimization run is shown. The corresponding trace of the most important optimization (independent) variables is shown in Fig. 13b. It can be clearly seen that after a short period of random exploration the annealing algorithm finds a design point that conforms to the constraints (specifications, extra design constraints), and that subsequently the power and area target are optimized. However in the latter optimization stage, large changes in the optimization variables are still possible. As can be seen on Fig. 13b at about iteration 5000 the three most important design variables change simultaneously. The input stage current drops in value, the compensation capacitance is decreased correspondingly and the class AB ratio (i.e. driver to output stage) is increased. In fact the optimization algorithm has found that decreasing the input stage current and compensation and keeping the output stage constant is advantageous to reduce the overall power consumption, while still fulfilling all specifications. Since we now have a good design point we can calculate the power-area trade-off curve using the local optimization algorithm, SQP. While maintaining the specifications set forth in Table 1, the power and area weights are modified to look for alternative designs.

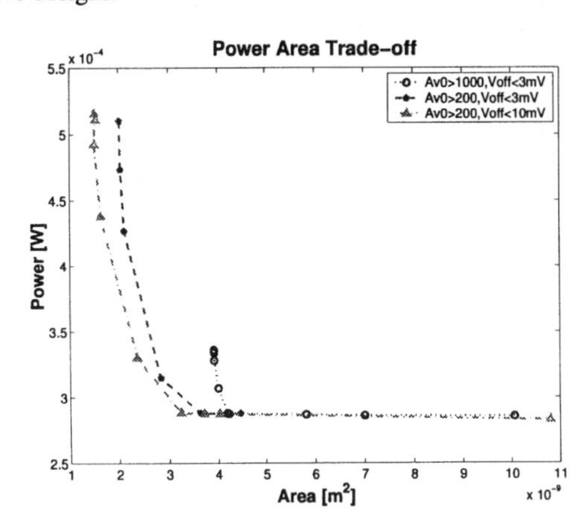

Fig. 14. Power area trade of symmetrical OTA with class AB buffer stage.

In Fig. 14 the resulting power-are points have been plotted, for the original specification set and two modified sets (low frequency gain and random offset voltage). As can be seen from the figure, the power-area trade-off curve is flat and curls up. This is explained by the fact that a power minimum does allow large transistors to be created, while an area minimum shrinks all transistors to their minimal sizes and thus allows only one power point. Furthermore, relaxing the low frequency gain and random offset voltage specifications influences the lowest possible area, but not the power (this is because power is largely GBW, Cload determined). In addition, relaxing specifications grows the trade-off curve considerably.

3.3. Layout generation

After sizing the circuit performance is verified (see next section). If this verification is passed successfully, a fully customized layout of the circuit is automatically generated by the LAYLA tool [35]. This tool implements a direct performance-driven place & route methodology [36]. The performance degradation due to layout-induced effects is quantified for every layout solution, and the placement and routing routines are driven in such a way that this performance degradation ΔP_j for every performance P_j does not exceed the user defined performance margins $\Delta P_{j,-}$ and $\Delta P_{j,+}$ in the final layout solution :

$$\Delta P_{j,-} \leq \Delta P_j \leq \Delta P_{j,+} \tag{29}$$

The allowed performance margins for layout can be derived from the specifications as follows :

$$\begin{aligned} \Delta P_{j,-} &= P_{j,spec,\min} - P_{j,nom} + 6\sigma(P_{j,tol}) \\ \Delta P_{j,+} &= P_{j,spec,\max} - P_{j,nom} - 6\sigma(P_{j,tol}) \end{aligned} \tag{30}$$

The performance degradation is calculated using a first-order linear approximation using the sensitivities of the performances to the different layout parasitics y_i :

$$\Delta P_j = \sum_i S_{y_i}^{P_j} y_i \tag{31}$$

resulting in :

$$\Delta P_{j,-} \leq \Delta P_j = \sum_i S_{y_i}^{P_j} y_i \leq \Delta P_{j,+} \tag{32}$$

The values of the layout parasitics y_i are extracted (calculated and/or estimated) for every intermediate layout solution. The sensitivities $S_{y_i}^{P_j}$ are obtained from numerical simulations that are performed only once at the beginning of the layout generation process.

In addition to performance constraints, additional geometrical requirements can be enforced by the designer, such as symmetry requirements both for devices as for nets. Also the orientation of devices can be fixed. Busses and pins can be constrained to be placed on a specific side of the circuit layout, etc.

The layout generation is split up in two parts: first placement, followed by routing. The placement algorithm [37] uses simulated annealing to place the devices. Symmetry constraints are enforced in the move set. Other constraints are enforced through the performance-driven mechanism : any violation of the performance margins for a layout solution according to equation (32) is penalized via an extra penalty term that is added to the cost function. A first effect included in this way is the performance degradation caused by (estimated) interconnect parasitics (capacitances associated with every wire and resistances associated with every wire segment in the layout) according to :

$$\begin{aligned} C_p &= C_{junction} + C_{wire} \\ R_{pi} &= \rho_{wire} \frac{L_{wire}/n}{W_{wire,i}} \\ \Delta P_j &= \sum_{k=1}^{m} \left(S_{C_{p,k}}^{P_j} C_{p,k} + \sum_{i=1}^{n_k} S_{R_{p,ki}}^{P_j} R_{p,ki} \right) \end{aligned} \tag{33}$$

The inclusion of junction capacitances in the above formula favors on-the-fly device merges at the diffusion level.

A second effect included is the matching of devices. Matching devices are handled simultaneously in the move set (same orientation and variant of the devices), but as it is not possible to put all matching devices exactly next to each other while also satisfying all other constraints, their distance is determined by the performance-driven mechanism according to the impact of their mismatch on the performance :

$$\sigma^2(V_{T0}) = \frac{A^2_{V_{T0}}}{WL} + S^2_{V_{T0}} D^2$$
$$\sigma^2(\beta) = \frac{A^2_{\beta}}{WL} + S^2_{\beta} D^2 \qquad (34)$$
$$\Delta P_j = \sum_{k=1}^{m} \left(\left| S^{P_j}_{\Delta V_{T0,k}} \right| \left(3\sigma(V_{T0})_k\right) + \left| S^{P_j}_{\Delta \beta_k} \right| \left(3\sigma(\beta)_k\right) \right)$$

Other effects have been included following the same mechanism, for example thermal effects [38].

The resulting placement is then interconnected by a performance-driven router [39]. The performance degradation caused by the actual routing parasitics is quantified and constrained in the same way as during placement by including any excess performance degradation in the cost function of the line expansion algorithm. In addition, as a post operation, the router can also trade off any remaining slacks on the performance degradation for an improved yield with respect to local catastrophic defects (pinholes and spot defects). By ripping up and rerouting nets according to the modified cost function, the layout can be made less sensitive to these defects, always without exceeding the original performance degradation margins. In this way a fully customized circuit layout is obtained that satisfies all specifications and has a high robustness.

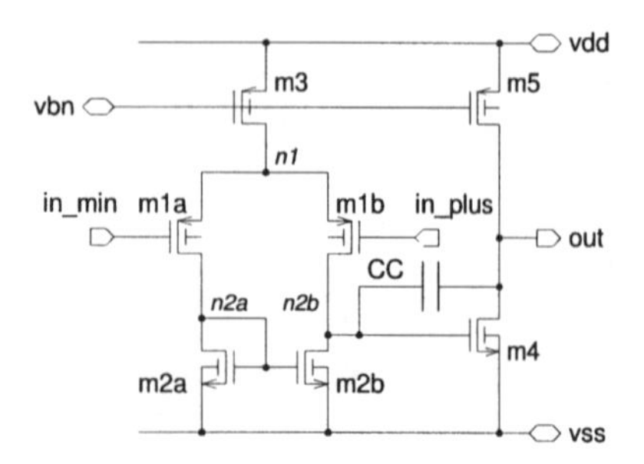

Fig. 15. Schematic of the Miller Compensated OTA with PMOS input stage.

Practical Example

In Fig. 15 the schematic of a Miller compensated OTA is shown. With the discussed performance driven place and route approach the layout of Fig. 16 has been generated. In this (limited) example, two performance specifications have been taken into account: the phase margin (PM) and the Gain-bandwidth (GBW). In Table 2 the sensitivities of these specifications with respect to the internal node capacitances are summarized, as are the estimated degradations of the placement and routing phase. The final performance degradation remains within the requested values, as shown in Table 3.

The resulting layout is then checked for layout rule violations (DRC) and compared with the schematic (LVS). If these checks don't return any errors, the actual parasitic elements of the circuit are extracted from the mask layout and back-annotated on the schematic to allow a detailed verification of the circuit performance.

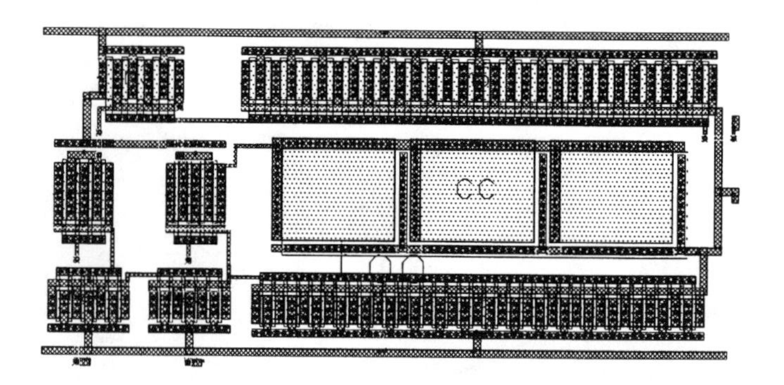

Fig. 16. Performance driven layout of the Miller Compensated OTA.

TABLE 2:
SENSITIVITY AND PERFORMANCE DEGRADATION OF THE MILLER COMPENSATED OTA.

Net	$S^{PM}_{cap_{net}}$ (*deg/F*)	ΔPM_{placed} (*deg*)	ΔPM_{routed} (*deg*)	$S^{GBW}_{cap_{net}}$ (*Hz/F*)	ΔGBW_{placed} (*Hz*)	ΔGBW_{routed} (*Hz*)
n1	-1e10	-0.00009	1e-4	-	-	-
n2a	-8.1e12	-0.12	-0.13	-6.9e17	-10300	-11000
n2b	-6.8e12	-0.24	-0.20	-1.6e18	-55400	-45800
out	-2e12	-0.063	-0.09	-4.2e17	-13200	-11020
vbn	7e10	0.0007	0.0012	-1e16	-104	-184
Total	-	-0.428	-0.421	-	-79100	-76200

TABLE 3:
PERFORMANCE OF THE MILLER COMPENSATED OTA.

Performance	Requested	Nominal	$\Delta Perf_{placed}$	$\Delta Perf_{routed}$	Extracted
PM (*deg*)	> 60	61.2	-0.428	-0.421	60.7
GBW (*Hz*)	> 30meg	30.28meg	-79100	-76200	30.2meg

3.4. Verification

Detailed verification of the circuit performance is performed twice in the design flow: a first time in the top-down path to verify the circuit sizing without the layout-induced degrading effects, and the second time in the bottom-up path after layout extraction to verify the circuit with inclusion of the actual layout-induced degrading effects. Verification implies a number of checks to be performed on the circuit as well as the execution of a simulation script to automatically simulate and extract all actual performance values obtained by the design. For this a link to existing numerical simulators is provided; the actual simulator used can be chosen by the user. The resulting performance values are then compared to the specifications and the datasheet of the design is generated, indicating whether all specifications have been met or not.

A generic verification script has been defined for every type of circuit, e.g. an opamp. As shown in Fig. 17, during verification every circuit is considered as a black box that has a specific functionality. Except for the external input and output pins, no signals are monitored inside the circuit. The verification script verifies the requested performance behavior of the function block, independent of the actual circuit implementation of that block. Details about the actual implementation and its special properties are therefore encapsulated in a verification harness that is predefined for every schematic. This includes for instance the correct biasing, which varies from schematic to schematic, the clocking in case of clocked circuits, etc.

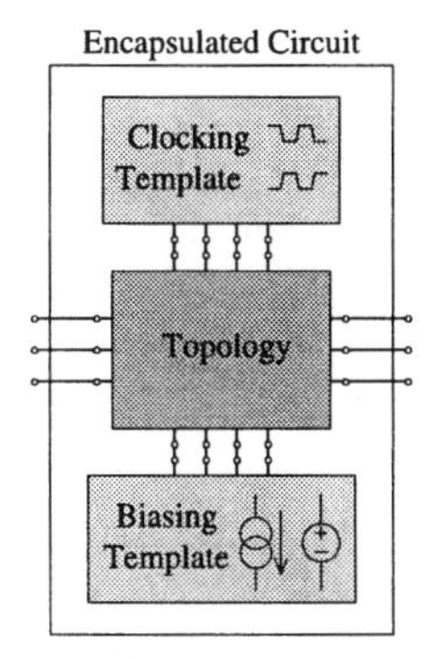

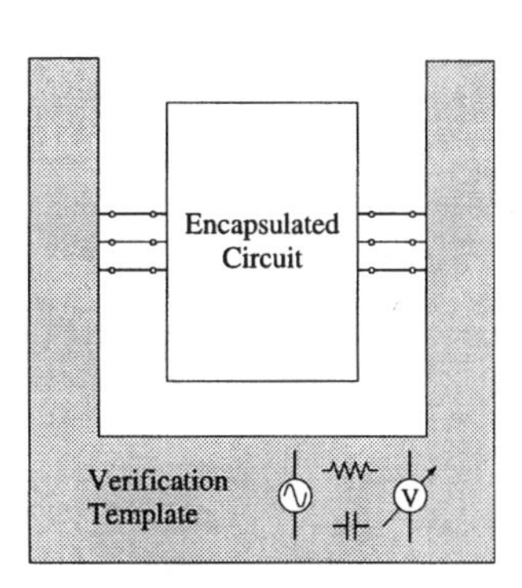

Fig. 17. The circuit under test, the encapsulated circuit with biasing and clocking templates (topology specific) and the verification template (functionblock and verification task specific) are applied to generate a test harness.

It will now be explained how the verification is performed. The verification process is fully automated, and requires no intervention from the user.

The nominal performance of the circuit can be verified with only a limited number of simulation jobs. A number of circuit performances are however influenced by device mismatches, which is a statistical phenomenon. Commercial simulators provide Monte-Carlo type of statistical simulation capabilities, but no commercial simulator provides an integrated statistical mismatch model. Therefore, a circuit preprocessor has been developed, called MMPRE [40], that replaces all MOS transistors in the netlist by an equivalent statistical mismatch model according to the Pelgrom mismatch model [25] as shown in Fig. 18. The mismatch-dependent performance specifications, as for instance the random part of the input offset voltage of an opamp, are then verified by invoking statistical simulations on this modified netlist in the corresponding simulation step in the verification script.

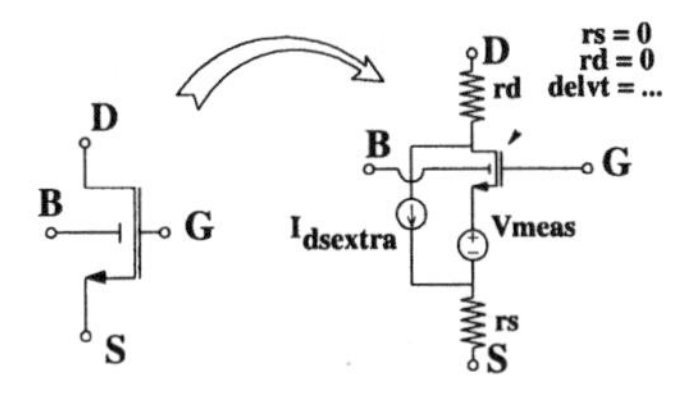

Fig. 18. For mismatch simulations every MOS transistor is replaced by an equivalent statistical model.

Example

An example is now provided to illustrate the general concept of the verification script. The example presented is the extraction of the slew rate of an opamp. The script looks as follows.

1. The topology is encapsulated by a biasing and clocking circuit (see Fig. 17). The biasing is realized through current sources and mirror transistors to generate all biasing voltages and currents. The values used have been calculated during sizing and are inserted from the database.
2. The resulting generic opamp is placed in a DC feedback loop, and loaded with the load specified in the specification sheet (see Fig. 19). For a slew rate extraction a square wave is applied. The parameters of the square wave (amplitude, rise time, period) are derived from the slew rate specification. The transient simulation mode is selected. The simulation time is set to one and a half period of the square wave.
3. The transient simulation job is run in batch mode

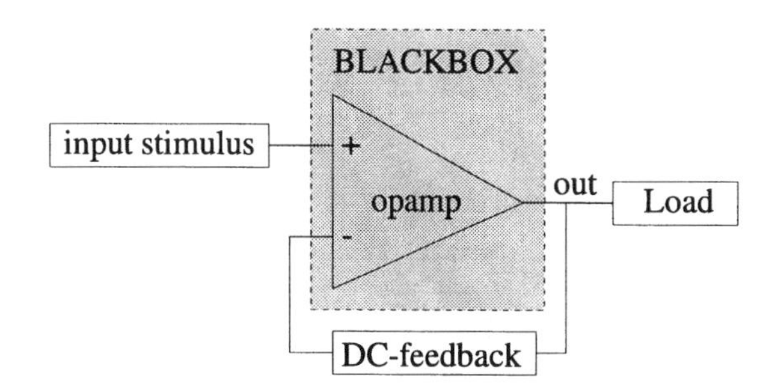

Fig. 19. The blackbox opamp in its test harness for slew rate analysis.

4. The slew rate specification is extracted from the signal on the output node. The positive slew rate is defined as the slope of the output signal between 20 and 80% of its full swing. The negative slew rate is the slope between 80 and 20% in the negative transition. The slew rate of an opamp is the minimum of the absolute value of negative and positive slew rate:

$$slewrate_- = \frac{\Delta V}{t_-}$$
$$slewrate_+ = \frac{\Delta V}{t_+} \tag{35}$$
$$slewrate = \min(|slewrate_-|, |slewrate_+|)$$

The average slopes are extracted by measuring the time elapsed between subsequent threshold passes. This is shown in Fig. 20.

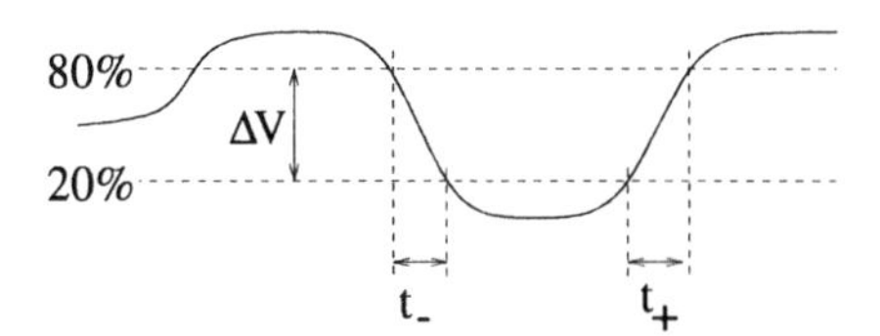

Fig. 20. Output signal and measurements.

5. The extracted value is stored in the specification sheet and compared to the specifications.

To avoid an expensive simulation run for every extracted specification, all transient and small-signal jobs are combined into two simulation jobs.

3.5. Redesign wizard

Although a performance-driven design methodology is adopted in the AMGIE system, in some cases the tool may not reach a design solution that satisfies all specifications. This may be because the specifications are too tough for the selected circuit schematic, or because simplified models used in earlier design steps incorrectly predicted the actual circuit behavior. Any such specification violations are detected by the tools themselves (for instance the sizing tool flags that it is incapable of meeting the specifications) or by one of the two detailed verification steps. In all these cases the redesign wizard is automatically started. This tool scans the status and the history of the design that failed. It then checks this against a redesign database where different predefined redesign scenarios, which contain procedures to remedy and restart the design process, have been stored. The (possibly multiple) scenarios applicable to the actual design problem are then presented to the designer. The designer selects an appropriate scenario and is guided step by step to put the design back on track. At present no automatic redesign mechanism has yet been implemented that has the built-in intelligence to automatically choose the best redesign scenario for every possible situation. The redesign scenarios in the database have been added by experienced designers. They are generic, i.e. circuit independent. It is possible for a designer to add his own scenarios to a private database.

Example

Examples of implemented scenarios are :

- If a design fails during sizing of the cell and if there are alternative topologies, select the next most promising topology. (This is the most straightforward redesign scenario).
- If a design fails during sizing of the cell, increase the maximum bounds on allowed power consumption and/or chip area. It is often found that a selected topology can reach the requested performances, but not within the (often arbitrarily) specified power or area bounds.
- If a design failure is detected during verification after sizing and the performance specification causing the failure has been calculated during sizing with an inaccurate approximate formula, restart the sizing with the specification value for this performance tightened with the same amount by which the simulated performance differs from the calculated one.
- If the verification after layout extraction fails, redo the circuit sizing and optimization after updating the estimated layout parasitics with the actually extracted parasitics of the extraction step.

4. Experimental results

The capabilities of the AMGIE system are now illustrated with practical synthesis examples, some of which have also been fabricated and measured. The AMGIE system can be used both by novice and experienced designers. The examples in this section have been generated by both types of designers. The first example set, the design of an Operational Transconductance Amplifier (OTA), has been generated in an analog circuit design class project. The second example has been generated by an experienced analog designer. Both examples illustrate that a synthesis system like AMGIE allows to generate high-performance analog designs in a very short total design time, from specification to verified layout. Hence such a system increases both analog design optimality and design productivity.

TABLE 4: SPECIFICATIONS COMMON TO ALL OTA DESIGNS.

Specification		Value	Predicted*	Simulation*	Unit
Low Freq. Gain	>	2000	6000	9400	-
Phase Margin	>	70	83	85	*degrees*
Offset Voltage	<	10	9.9	10	*mV*
Output Swing	>	70	70	73	*% of supply*
Settl. time [0.1%]	<	10	0.075	0.072	μs
Load Resistance	=	1	1	1	*GOhm*
Power	<	20	2.3	2.1	*mW*
Area	<	0.1	0.024	0.031	mm^2

4.1. Operational transconductance amplifier

The design of OTAs is a frequently occurring task. The AMGIE cell library contains a number of topologies implementing the OTA function. With this library a class of EE Master students has generated a number of designs. They were divided in 9 groups, each of which received a different requirement of gain-bandwidth (5, 15, 45 MHz) and load capacitance (5, 10, 20 pF). The specifications that were common to all projects are summarized in the second column of Table 4.

The exercise was performed on HP 712/80 UNIX workstations. All groups succeeded in the task without redesign iterations. The fastest design time was 1 hour, the slowest 2 hours and 15 minutes. The time spent getting acquainted with the tool is included in these reported times. An experienced tool user can accomplish the same task in 40 minutes from specification to layout.

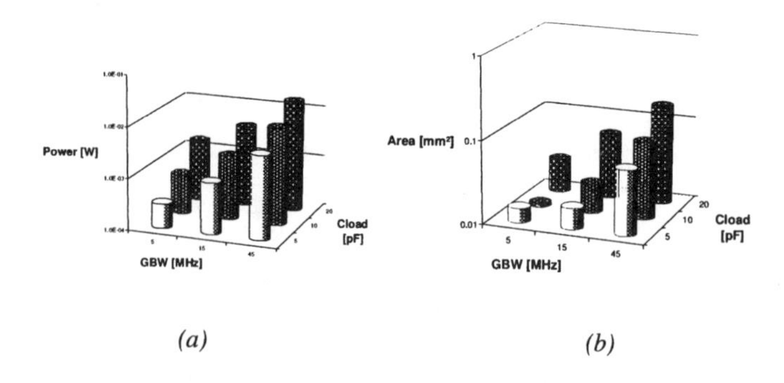

Fig. 21. Power and Area results of OTA example.

Fig. 21a and b show the achieved power consumption and area of the 9 designs. Fig. 22 shows the schematic of the typically selected topology, while Fig. 23 shows a typical layout result. The simulated performance is verified against the specifications in the third column of Table 4, and all specifications are satisfied. None of the designs encountered redesign iterations. This project showed that even less experienced designers can generate good designs in a short time with the AMGIE system. The only limitation is that sufficient cells must be available in the library to have sufficient coverage of typical performance ranges encountered in practical applications.

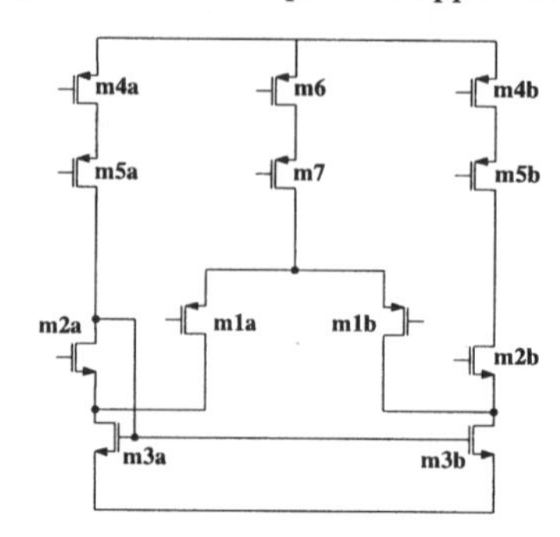

Fig. 22. Schematic of the high-speed OTA.

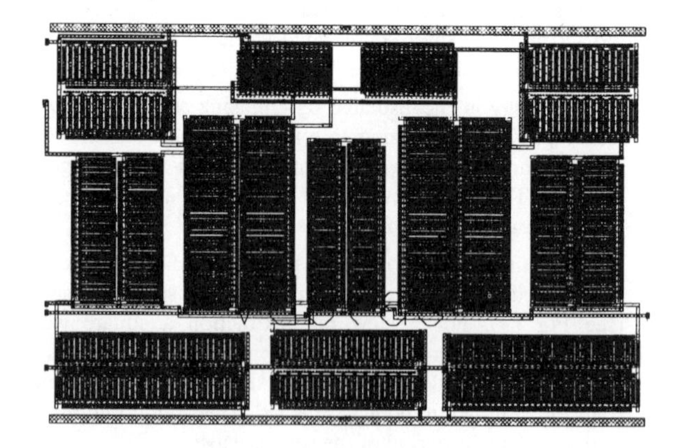

Fig. 23. Typical layout generated during the class project.

* values extracted for the design with GBW > 15MHz and C_{load} = 10pF

4.2. Particle detector front-end

A second example is the design of a particle detector front-end (PDFE) which is used in nuclear physics or space satellite experiments to measure the energy of infalling particles or radiation. The block diagram of a typical PDFE is shown in Fig. 6 [41]. An incident particle will generate a charge pulse on the detector proportional to its energy. The purpose of the PDFE is to transform this charge pulse in an electronically measurable quantity. This is achieved by an analog preprocessing chain comprising a charge sensitive amplifier (CSA) and a pulse shaping amplifier (PSA). The charge packet is integrated onto a capacitor by the CSA and then shaped by the PSA to a gaussian pulse whose peak amplitude is proportional to the original energy. This pulse is created by differentiating the integrated charge, and subsequently integrating the spike n times. The value of n is allowed to vary between 1 and 4 and will be determined by the synthesis tool depending on the required specifications (n is thus a topological variable). The most important specifications of this front-end for a certain ESA experiment are summarized in Table 5.

TABLE 5:
SPECIFICATIONS OF THE PARTICLE DETECTOR FRONT-END, PERFORMANCES OBTAINED BY A MANUAL REFERENCE DESIGN, PERFORMANCES OBTAINED BY THE AMGIE SYSTEM AND THE MEASURED PERFORMANCE.

	Specification	Reference	Synthesized	Measured	Unit
Peaking time	< 1.5	1.1	1.1	1.18	µs
Counting rate	> 200	200	294	250	kHz
Noise	< 1000	1000	905	952	e^-_{rms}
Gain	= 20	20	21	19	mV/fC
Output voltage range	= 2	2	2	2	V
Power	< 40	40	7	10	mW
Area	< 5	0.6	0.7	0.7	mm^2
Detector Capacitance	= 80	80	80	80	pF

The peaking time is defined as the time delay between the incidence of a particle and the top of the gaussian pulse. The counting rate defines the maximum number of particles that may occur before the tails of the gaussian pulses cause a measurement error. The noise of the system is expressed as an equivalent number of electron charges, which is directly related to the lowest possible particle energy that is still detectable. The gain and the output swing define the amplification and range of signals at the output of the system. The value of the detector capacitance is required to optimally (noise) match the front-end to the detector. The power and area must be minimized. The technology used is 1P2M 0.7 µm CMOS.

The custom CSAPSA cell in the AMGIE system has been implemented as an hierarchical topology. The full schematic of the topology is shown in Fig. 24. The CSAPSA topology uses 4 function blocks: the charge sensitive amplifier (CSA), the active feedback resistance (RF), the differentiator with pole-zero cancellation (PSA_DIF) and the integrator (PSA_INT). The topologies available for these four function blocks are all flat (i.e. specified at the device level).

Starting from the specifications of Table 5, the tool has selected the custom CSAPSA topology (with variable number of integrators n). The optimization of the topology uses power and area estimators to determine the optimal specifications of the composing subblocks [26]. The optimization time is typically 20 minutes on a Sun Ultra 1-170 workstation. The CSAPSA is subsequently verified using behavioral simulation. Using the derived subblock's specifications, a topology is selected for each of the subblocks and the chosen topologies are sized.

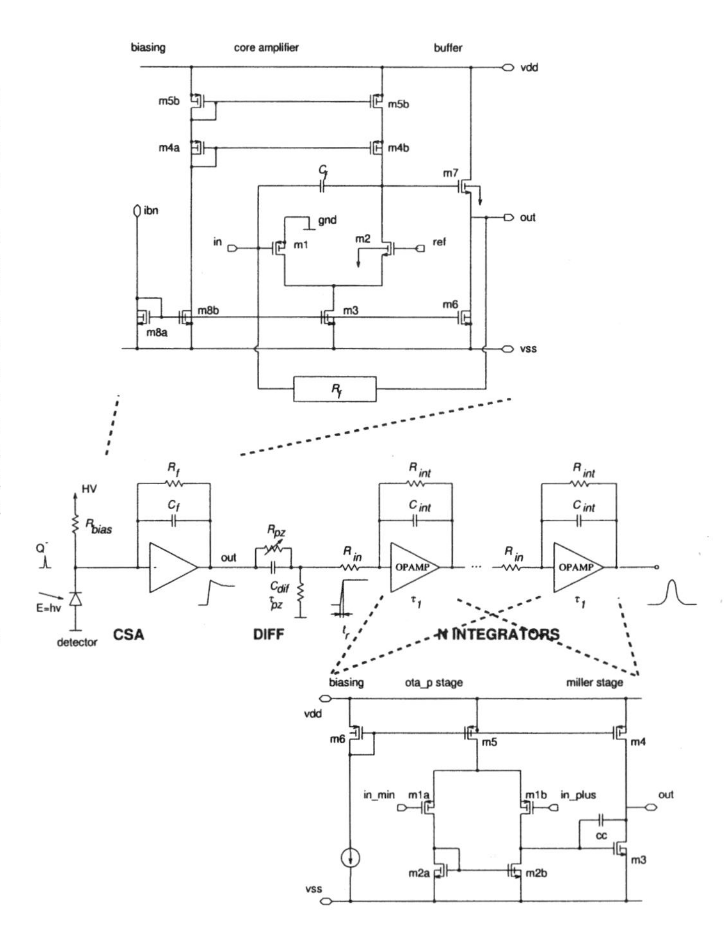

Fig 24. Hierarchical structure of the CSAPSA custom cell.

The resulting performance after sizing (down to the device level) is summarized in the fourth column of Table 5. All specifications are satisfied. The width of the input transistor is 1623 µm, its length 1.3 µm. The integration capacitor of the CSA is 215 fF. Compared to an existing reference design (third column of Table 5), the tool has been able to aggressively reduce the power consumption from 40 mW down to 7 mW, without sacrificing any extra area. The layout has been generated with LAYLA, following the hierarchy of the CSAPSA front-end.

The resulting layout is shown in Fig. 25 and is fairly dense. This design has also been fabricated [42] and the measurement results are included in the fifth column of Table 5. The results correspond very well to the predictions. A measured response of the CSAPSA to an incident particle is shown in Fig. 26. The total design time for this design was 1 week, which is far better than what can be obtained by hand.

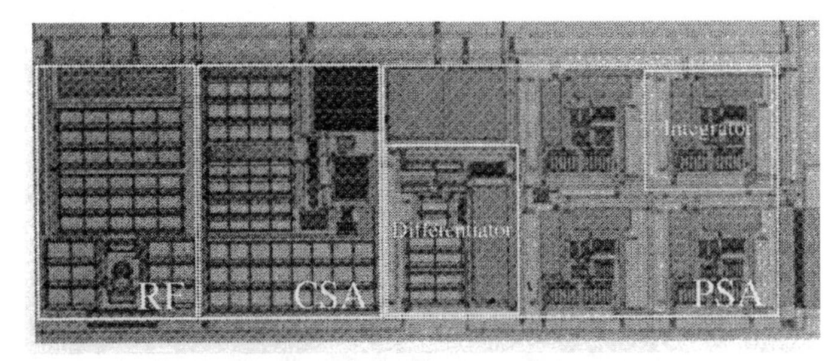

Fig. 25. Generated layout of the CSAPSA.

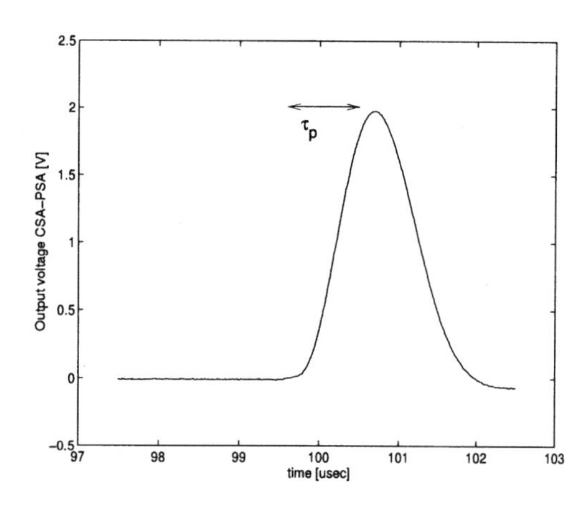

Fig. 26. Measured time response of the CSA-PSA circuit to an incident particle of 100 fC.

5. Conclusions

A synthesis environment for analog integrated circuits has been presented that is able to drastically increase design and layout productivity for analog blocks. The system covers the complete design flow from specification over topology selection and optimal circuit sizing down to automatic layout generation and performance characterization. It follows a hierarchical refinement strategy for complex cells and is process independent. The sizing is based on an improved equation-based optimization approach, where the circuit behavior is characterized by declarative models that are then converted in a sequential design plan. Supporting tools have been developed to reduce the total effort to set up a new circuit topology in the system's library. The performance-driven layout generation tool guarantees layouts that satisfy all performance constraints. The experimental results illustrate the productiveness and efficiency of the environment for the synthesis of frequently used analog cells.

Acknowledgements

This research has been supported in part by the ESPRIT project ADMIRE and by the European Space Agency (ESA) and the Belgian DWTC under the ASTP4 and GSTP2 programs.

References

[1] H. Chang, L. Cooke, M. Hunt, G. Martin, A. McNelly, Lee Todd, "Surviving the SOC Revolution A Guide to Platform-Based Design", Kluwer Academic Publishers (ISBN 0-7923-8679-5), November 1999.

[2] L. R. Carley, G. Gielen, R. Rutenbar, W. Sansen, "Synthesis tools for mixed-signal ICs: progress on front-end and back-end strategies," proceedings IEEE/ACM Design Automation Conference (DAC), pp. 298-303, June 1996.

[3] M. Degrauwe, et al., "IDAC: an interactive design tool for analog CMOS circuits," IEEE Journal of Solid-State Circuits, Vol. SC-22, No. 6, pp. 1106-1116, December 1987.

[4] R. Harjani, R. Rutenbar, L. R. Carley, "OASYS: a framework for analog circuit synthesis," IEEE Transactions on Computer-Aided Design, Vol. 8, No. 12, pp. 1247-1266, December 1989.

[5] F. El-Turky, E. Perry, "BLADES: an artificial intelligence approach to analog circuit design," IEEE Transactions on Computer-Aided Design, Vol. 8, No. 6, pp. 680-692, June 1989.

[6] H. Koh, C. Séquin, P. Gray, "OPASYN: a compiler for CMOS operational amplifiers," IEEE Transactions on Computer-Aided Design, Vol. 9, No. 2, pp. 113-125, February 1990.

[7] G. Gielen, H. Walscharts, W. Sansen, "Analog circuit design optimization based on symbolic simulation and simulated annealing," IEEE Journal of Solid-State Circuits, Vol. SC-25, No. 3, pp. 707-713, June 1990.

[8] J. P. Harvey, et al., "STAIC: An interactive framework for synthesizing CMOS and BiCMOS analog circuits," IEEE Transactions on Computer-Aided Design, Vol. 11, No. 11, pp. 1402-1415, November 1992.

[9] C. Toumazou, C. Makris, "Analog IC design automation: Part I - Automated circuit generation: new concepts and methods," IEEE Transactions on Computer-Aided Design, Vol. 14, No. 2., pp. 218-238, February 1995.

[10] M. del Mar Hershenson, S. P. Boyd, T. H Lee, "GPCAD: A Tool for CMOS Op-Amp Synthesis," proceedings ACM/IEEE International Conference on Computer-Aided Design (ICCAD), pp. 296-303, November 1998.

[11] F. Medeiro, B. Perez-Verdu, A. Rodriguez-Vazquez, J. L. Huertas, "A vertically-integrated tool for automated design of ΣΔ modulators," IEEE Journal of Solid-State Circuits, Vol. SC-30, No. 7, pp. 762-772, September 1994.

[12] F. Medeiro, et al., "A statistical optimization-based approach for automated sizing of analog cells," proceedings ACM/IEEE International Conference on Computer-Aided Design (ICCAD), pp. 594-597, November 1994.

[13] M. Krasnicki, R. Phelps, R. A. Rutenbar, L. R. Carley, "MAELSTROM: Efficient Simulation-Based Synthesis for Custom Analog Cells," proceedings IEEE/ACM Design Automation Conference (DAC), pp. 951-957, June 1999.

[14] R. Phelps, M. Krasnicki, R. Rutenbar, L.R. Carley, J. Hellums, "ANACONDA: Robust Synthesis of Analog Circuits via Stochastic Pattern Search," proceedings IEEE Custom Integrated Circuits Conference (CICC), pp. 26.2.1-26.2.4, May 1999.

[15] E. Ochotta, R. Rutenbar, L. R. Carley, "Synthesis of high-performance analog circuits in ASTRX/OBLX," IEEE Transactions on Computer-Aided Design, Vol. 15, No. 3, pp. 273-294, March 1996.

[16] J. Cohn, D. Garrod, R. Rutenbar, L. R. Carley, "Analog Device-Level Layout Automation," Kluwer Academic Publishers (ISBN 0-7923-9431-3), January 1994.

[17] P. Veselinovic, et al., "A flexible topology selection program as part of an analog synthesis system," proceedings European Design and Test Conference (ED&TC), pp. 119-123, 1995.

[18] G. Gielen, W. Sansen, "Symbolic analysis for automated design of analog integrated circuits," Kluwer Academic Publishers, 1991.

[19] G. Gielen, H. Walscharts, W. Sansen, "ISAAC: a symbolic simulator for analog integrated circuits," IEEE Journal of Solid-State Circuits, Vol. SC-24, No. 6, pp. 1587-1597, December 1989.

[20] P. Wambacq, F. Fernandez, G. Gielen, W. Sansen, A. Rodriguez-Vazquez, "Efficient symbolic computation of approximated small-signal characteristics," IEEE Journal of Solid-State Circuits, Vol. SC-30, No. 3, pp. 327-330, March 1995.

[21] K. Swings, G. Gielen, W. Sansen, "An intelligent analog IC design system based on manipulation of design equations," proceedings IEEE Custom Integrated Circuits Conference (CICC), pp. 8.6.1-8.6.4, May 1990.

[22] G. Gielen, et al., "Comparison of analog synthesis using symbolic equations and simulation," proceedings European Conference on Circuit Theory and Design (ECCTD), pp. 79-82, August 1995.

[23] F. Leyn, G. Gielen, W. Sansen, "An efficient DC root solving algorithm with guaranteed convergence for analog integrated CMOS circuits," proceedings ACM/IEEE International Conference on Computer-Aided Design (ICCAD), pp. 304-307, November 1998.

[24] D. Foty, "MOSFET modeling with SPICE: principles and practice," Prentice-Hall, 1997.

[25] M. Pelgrom, A. Duinmaijer, A. Welbers, "Matching properties of MOS transistors", IEEE Journal of Solid-State Circuits, Vol. SC-24, No. 5, pp. 1433-1439, 1989.

[26] J. Vandenbussche, S. Donnay, F. Leyn, G. Gielen and W. Sansen, "Hierarchical Top-Down Design of Analog Sensor Interfaces: From System-Level Specifications Down to Silicon", Proc. on the IEEE Design, Automation and Test in Europe Conference, pp. 716-720, February 1998.

[27] R. Harjani and J. Shao, "Feasibility and Performance Region Modeling of Analog and Digital Circuits", Analog Integrated Circuits and Signal Processing, Vol. 10, pp. 23-43, 1996.

[28] G. Van der Plas, J. Vandenbussche, G.Gielen and W. Sansen, "EsteMate: a Tool for Automated Power and Area Estimation in Analog Top-Down Design

and Synthesis", proceedings IEEE Custom Integrated Circuits Conference (CICC), pp. 139-142, May 1997.

[29] E. Lauwers, G. Gielen, "A power estimation model for high-speed CMOS A/D Converters", Proc. on the IEEE Design, Automation and Test in Europe Conference, pp. -, February 1999.

[30] L. Ingber, "Very fast simulated re-annealing, Mathl. Comput. Modelling," Vol. 12, No. 8, pp. 967-973, 1989.

[31] R. Hooke, T. A. Jeeves, "Direct Search," Solution of Numerical and Statistical Problems. JACM Vol. 8, No. 2, pp. 212-229, 1961.

[32] F. Leyn, W. Daems, G. Gielen, W. Sansen, "Analog circuit sizing with constraint programming modeling and minimax optimization", proceedings International Symposium on Circuits and Systems (ISCAS), pp. 1500-1503, June 1997.

[33] R. Fletcher, "Practical methods of optimization," second edition, Chichester and New York: John Wiley and Sons, 1987.

[34] G. Debyser, G. Gielen, "Efficient analog circuit synthesis with simultaneous yield and robustness optimization", proceedings ACM/IEEE International Conference on Computer-Aided Design (ICCAD), pp. 308-311, November 1998.

[35] K. Lampaert, G. Gielen, W. Sansen, "Analog layout generation for performance and manufacturability," Kluwer Academic Publishers (ISBN 0-7923-8479-2), April 1999.

[36] E. Malavasi, E. Charbon, E. Felt, A. Sangiovanni-Vincentelli, "Automation of IC layout with analog constraints," IEEE Transactions on Computer-Aided Design, Vol. 15, No. 8, pp. 923-942, August 1996.

[37] K. Lampaert, G. Gielen, W. Sansen, "A performance-driven placement tool for analog integrated circuits", IEEE Journal of Solid-State Circuits, Vol. SC-30, No. 7, pp. 773-780, July 1995.

[38] K. Lampaert, G. Gielen, W. Sansen, "Thermally constrained placement of analog and smart power integrated circuits," proceedings European Solid-State Circuits Conference (ESSCIRC), pp. 160-163, September 1996.

[39] K. Lampaert, G. Gielen, W. Sansen, "Analog routing for manufacturability," proceedings Custom Integrated Circuits Conference (CICC), pp. 9.4.1-9.4.4, May 1996.

[40] W. Verhaegen, G. Van der Plas, G. Gielen, "Automated test pattern generation for analog integrated circuits," proceedings IEEE VLSI Test Symposium (VTS), pp. 296-301, May 1997.

[41] Z. Chang, W. Sansen, "Low-noise wide-band amplifiers in bipolar and CMOS technologies," Kluwer Academic Publishers (ISBN 0-7923-9096-2), November 1990.

[42] J. Vandenbussche, F. Leyn, G. Van der Plas, G. Gielen, W. Sansen, "A fully integrated low-power CMOS particle detector front-end for space applications," IEEE Transactions on Nuclear Science, Vol. 45, No. 4, pp. 2272-2278, August 1998.

A High-Level Design and Optimization Tool for Analog RF Receiver Front-Ends

Jan Crols*, Stéphane Donnay, Michiel Steyaert** and Georges Gielen**
Katholieke Universiteit Leuven, ESAT-MICAS,
Kardinaal Mercierlaan 94, B-3001 Heverlee, Belgium

* supported with a fellowship of the Flemish Institute for Promotion of Scientific-Technological Research in Industry (IWT)
** research associates of the Belgian National Fund of Scientific Research (NFWO)

Abstract

This paper presents a high-level analysis and optimization tool for the design of analog RF receiver front-ends, which takes all design parameters and all aspects of performance degradation (noise, distortion, self-mixing...) into account. The simulations are performed in the spectral domain with a behavioral model library for the RF building blocks. The tool allows to explore alternative RF receiver topologies as well as to investigate design trade-offs within each topology. By having integrated the performance analysis routine within a simulated annealing optimization loop, the tool can also perform an optimal high-level synthesis of a given topology towards a specific application. It then determines the optimal specifications for the RF building blocks such that the required receiver signal quality is met while the overall power and/or area consumption is minimized.

1. Introduction

The market of wireless applications is booming nowadays. The use of digital signal processing and digital signal control has enabled wireless communications with low transmission bit-error rates. Equally important however are the analog transceiver front-ends [1]. The design of RF front-ends was until now always a very intensive process, based on the knowledge and skills of experienced RF designers. But now, the market situation is changing rapidly. The number of wireless applications is growing at a very fast pace and, as time-to-market becomes highly important, this requires the implementation on ever shorter terms of circuits operating at higher frequencies, with higher degrees of integration, lower operating voltages and lower power consumption. In addition, the optimization of each individual building block alone no longer suffices to meet the specifications. Instead, new RF transceiver topologies have to be explored and optimized at the architectural level, before designing the individual building blocks. A design tool which allows such explorations and the fast evaluation of high-level trade-offs in analog RF transceiver design is, however, not yet available today.

This paper presents a high-level design and optimization methodology for analog RF receiver front-ends. This methodology has been implemented as the ORCA (Optimizer for ReCeiver Architectures) prototype tool. Whereas transmitters deal with a well known signal and are therefore relatively straightforward to design, this is not the case with receiver design. Receivers have to handle a highly random and variable antenna signal of which the wanted signal is only a small part. An optimal receiver design is therefore highly dependent on the application, the used topology and the performance of the different types of building blocks used in this topology. The design methodology presented in this paper takes all these aspects into account and allows to evaluate the performance of an RF receiver topology as well as to automatically translate high-level system specifications into a set of specifications for each building block in the topology such that the overall power and/or area consumption of the receiver is minimized.

In section 2, the systematic modeling and analysis methodology for RF receiver topologies is introduced, including effects such as noise, distortion and aliasing. Next, section 3 describes the performance simulation method which operates on signal spectra in the frequency domain, and the optimization loop built around the simulator that allows automatic high-level synthesis of a receiver topology. Section 4 then presents experimental results. Conclusions are provided in section 5.

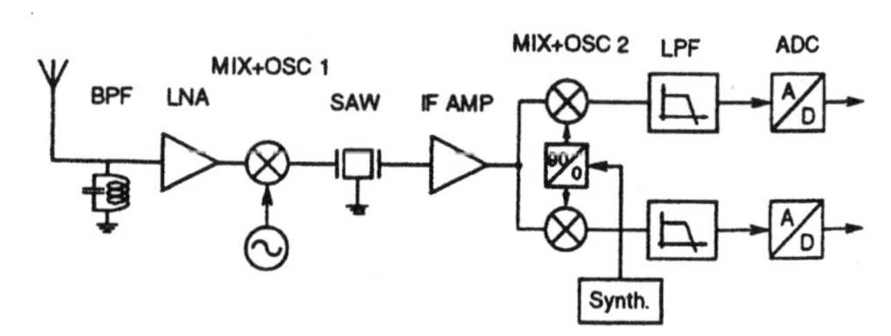

Fig. 1. A combined IF zero-IF receiver topology.

2. RF receiver front-end design

2.1. Receiver performance modeling

Fig. 1 shows an example of an RF receiver topology, a combination of an IF and a zero-IF receiver [2,3]. In successive filtering, amplification and downconversion stages the unwanted neighbor signals are further and further suppressed and the wanted signal is brought down to lower and lower frequencies until the final, low-dynamic-range, low-frequency signal can be sampled with a fairly simple A/D converter. Trade-offs have to be made in the receiver design, because it is not possible to do all the filtering and downconversion in once stage.

The performance of a receiver is defined as the output SUSR (Signal to Unwanted Signal Ratio) which is the ratio between the power of the undemodulated wanted signal at the output (after the A/D converter) and the total power of all the unwanted signals that are located at the same frequencies as the wanted signal. The unwanted signals can be subdivided in three categories:

- NOISE: all signals generated in a building block that are not correlated to any other signal (thermal noise, shot noise, but also DC offset voltages . . .).
- Distortion: all signals related to (a power of) the input signal (main sources are second- and third-order harmonic distortion and intermodulation, but also self-mixing products.
- ALIASING: all frequency-translated versions of the input signal that did not undergo the wanted frequency translation (e.g. aliasing components in A/D converters, mirror signals in downconverters, phase noise, etc.).

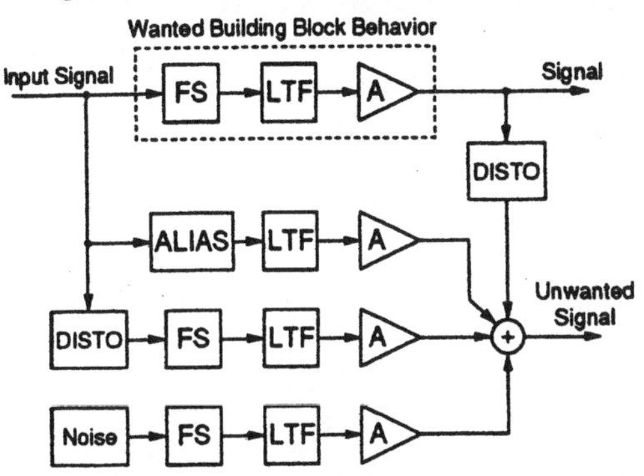

Fig. 2. The processing of wanted and unwanted signals in a building block.

As shown in Fig. 2, the signal processing operations of a receiver building block are completely defined by specifying on one hand the wanted frequency translation (FS) and linear transfer function (the filter characteristic LTF and amplification A) and on the other hand the noise sources (NOISE), the distortion levels (DISTO) and the unwanted frequency translations (ALIAS). Each block thus generates 4 different types of output signals from 1 input signal, which are also handled separately in the RF performance simulations. Distortion is modelled on both the input and the output signal.

Most properties, like self-mixing, aliasing components and the shape of the transfer function, depend on the type of the building block and the chosen IC implementation method. Only a limited number of parameters can be varied freely between certain boundaries during design. Each building block has the following set of parameters :

- BW_i : the bandwidth of the linear transfer function.
- A_i : the overall amplification in the block's passband.
- DR_i : the noise level expressed via the dynamic range
- F_i : the operating frequency (center frequency for filters, local oscillator frequency for mixers).

Only this set of building block specifications is relevant for the trade-offs in receiver design, and are therefore the variables that can be entered by a designer or that are optimized during an optimization run. The power and/or area consumption and the unwanted signal behavior of each building block will have to be modelled as a function of this parameter set.

2.2. Input conditions

The optimization results for a given topology depend highly on the possible input signal conditions, which are represented as a set of power spectral densities S_{0j}, each representing an input situation type that can occur for a certain application, and the signal to unwanted signal ratio $SUSR_j$ required for each situation. The distribution of the amplification and filtering over the different stages highly depends on these input spectra and the required $SUSR_j$. Preventing saturation for the worst condition, the relationship between the dynamic ranges and bandwidths of two consecutive stages $i-1$ and i can be determined :

$$DR_i = DR_{i-1} \cdot \frac{1}{A_i^2} \cdot \frac{BW_{i-1}}{BW_i} \tag{1}$$

A higher dynamic range is equivalent to lower noise levels and this can only be realized at the cost of a higher power and area consumption. It is thus important to find, for a given set of input conditions and a desired $SUSR_j$ for each condition, a set of bandwidths BW_i, gains A_i and center frequencies F_i which give the minimal power and/or area consumption.

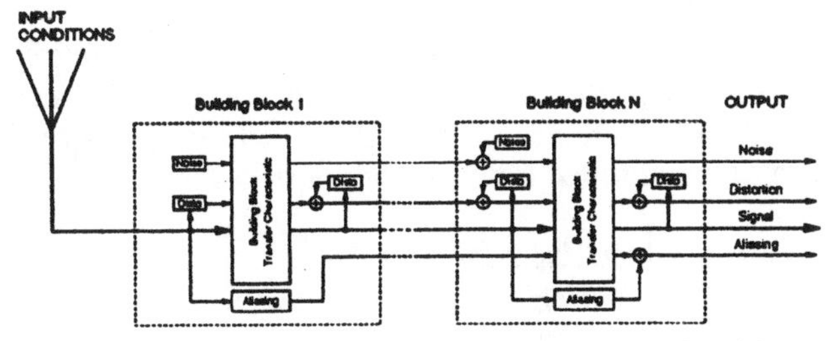

Fig. 3. Signal flow in the performance simulation algorithm.

3. The simulation and optimization method

3.1. Simulation method

Fig. 3 shows the flow diagram of the implemented performance simulator for receiver architectures. Each building block converts, as shown in Fig. 2, an input signal $S_{(i-1)j}$ into an output signal S_{ij} via a frequency translation, a filter operation and an amplification. The unwanted noise, distortion and aliasing signals, however, are processed separately in our simulator in parallel to the wanted signal. It is only after the simulation, at the output of the receiver, that they are combined into the $SUSR_j$ over the wanted output signal's passband. In this way, all contributions can be monitored separately on each intermediate node. Also, nonideal operations on the unwanted signals generated in previous building blocks can be omitted since this would only result in small higher-order corrections.

The simulator has to keep track of :

- the power spectral density distribution at each node and the signal transfer between the different receiver nodes
- the frequency shifting and aliasing effects
- the distortion and intermodulation effects

Classical SPICE AC analysis cannot be used for this purpose. Transient analysis could theoretically be used. A data sequence, similar to a sampled version of the actual antenna signal, could be applied at the input and the power spectral density distribution on each node could be determined by taking the FFT of the time domain signals. However, this would require impractically large sets of data points, resulting in massive memory and CPU time consumption. Indeed, the signal spectra range from DC to several GHz. The resolution has to be about 1 kHz, since the wanted signal is only a very small part (a few hundred kHz) of the total power spectrum. A data point representation of the signal would then require more than 1 million points. Another problem is how to discriminate between the wanted and the unwanted (noise, distortion, aliasing) signals in the resulting output spectrum.

The harmonic balance simulation technique [4] allows to get a good insight in the different types of nonlinear behavior in RF circuits. The effects of distortion and aliasing can be observed quite accurately, but it is not practical to evaluate with this technique the actual overall performance reduction due to these effects. This would again require the use of more than 1 million data points (now in the frequency domain).

Important for the simulator is thus the representation of the signal spectra. In our approach, the power spectral density distributions S_{ij} are represented symbolically as a sum of bandlimited rational polynomials RP :

$$S_{ij}(f) = \sum \left(\frac{n_0 + n_1 f + n_2 f^2 + \ldots + n_k f^k}{d_0 + d_1 f + d_2 f^2 + \ldots + d_l f^l} \Bigg|_{f_{begin}}^{f_{end}} \right) = \sum RP \qquad (2)$$

This representation technique gives a high flexibility and a very low memory consumption compared to a data point representation. Most power spectral density shapes in practice are of this form and those which are not can be represented by fitting the actual shape in small intervals to limited-order rational polynomials.

All operations on these signals are implemented as formula manipulations which is easy and fast. For example, filtering is multiplying rational polynomials with rational polynominals, resulting again in a rational polynomial :

$$LTF(S_{ij}(f)) = \left(\frac{a_0 + a_1 f + \ldots + a_k f^k}{b_0 + b_1 f + \ldots + b_l f^l} \right) \cdot \sum \left(\frac{n_0 + n_1 f + n_2 f^2 + \ldots + n_k f^k}{d_0 + d_1 f + d_2 f^2 + \ldots + d_l f^l} \Bigg|_{f_{begin}}^{f_{end}} \right)$$

$$= \sum RP \qquad (3)$$

while a frequency translation is obtained by replacing f with $f - F_i$ and calculating the new coefficients :

$$FT(S_{ij}(f)) = \sum \left(\frac{n_0 + n_1 (f - F_i) + n_2 (f - F_i)^2 + \ldots + n_k (f - F_i)^k}{d_0 + d_1 (f - F_i) + d_2 (f - F_i)^2 + \ldots + d_l (f - F_i)^l} \Bigg|_{f_{begin} - F_i}^{f_{end} - F_i} \right)$$

$$= \sum RP \qquad (4)$$

By taking into account both the even and odd terms in the power spectral density representation (2), quadrature signals and quadrature operations can also be handled correctly. The noise power spectra $N_i(f)$ and the unwanted aliasing spectra $AL_{ij}(f)$ can also be represented with formula (2). The distortion spectrum $D_{ij}(f)$ on the other hand is stored as a sum of two or more convolved rational polynomials, which only has to be evaluated at the output of the receiver in a small passband (the bandwidth of the wanted signal).

3.2. Optimization method

The above simulation technique has been integrated within an overall optimization loop, as shown in Fig. 4. This optimization tool determines the optimal set of building block specifications (A_i, BW_i, DR_i, F_i) such that the overall receiver satisfies the required $SUSR_j$ for all input conditions at the smallest overall power and/or area consumption. The performance of the given topology for the specified input condition set is evaluated at each iteration for a different set of building block specifications with the simulator discussed in the previous section.

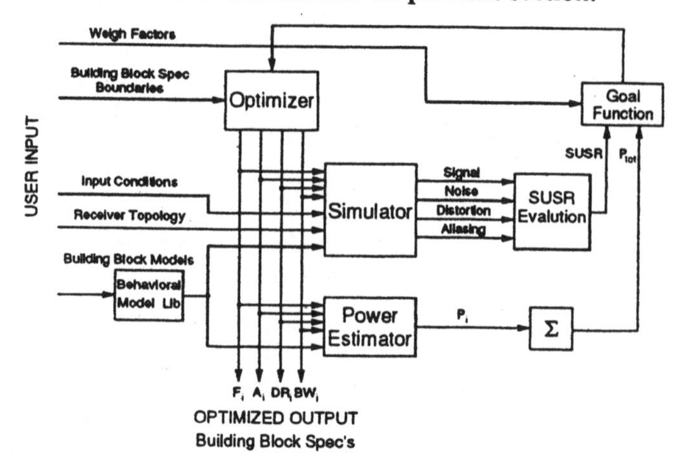

Fig. 4. The ORCA simulation and optimization tool.

A behavioral model library contains high-level models for the different types of receiver building blocks. The models contain the wanted and unwanted (noise, aliasing, distortion) signal operations of each block, and the relationships between the block's specifications and the corresponding power and area consumption. For example, a quite general model for the power consumption P_i of a block with specifications A_i, BW_i, DR_i is given by :

$$P_i = \frac{DR_i \cdot A_i^2 \cdot BW_i \cdot kT}{\eta_i} \qquad (5)$$

where the power efficiency η_i is the ratio between the theoretical and actual power needed to obtain these specifications. η_i depends on the building block type, the IC implementation technique and on some of the specifications of the building block. The power efficiency has been calculated for a large number of real-life building blocks in the library.

The optimization itself is performed by means of a simulated annealing algorithm. Optimization boundaries and weight factors are initially set by the behavioral models and the algorithm, but they can be changed interactively by the user, allowing full control over the design and optimization process.

4. Experimental results

The methodology described in the previous sections has been prototyped in the ORCA (Optimizer for ReCeiver Architectures) tool. ORCA can be used interactively, as a simulator and exploration tool, or as an automatic optimization tool. Results of both cases will be presented. For all the examples in this section the combined IF zero-IF topology of Fig. 1 was used.

When ORCA is used as a simulator, all building block specifications (BW_i, A_i, ...) have to be provided by the user. In this mode the RF designer can quickly compare different topologies or evaluate the influence of different building block specifications on the overall performance of the receiver. Simulation times range from less than 1 minute to several minutes (on a SUN Sparc10) depending on the number of different input conditions that are specified by the user. A typical output of a simulation run is shown in Fig. 5.

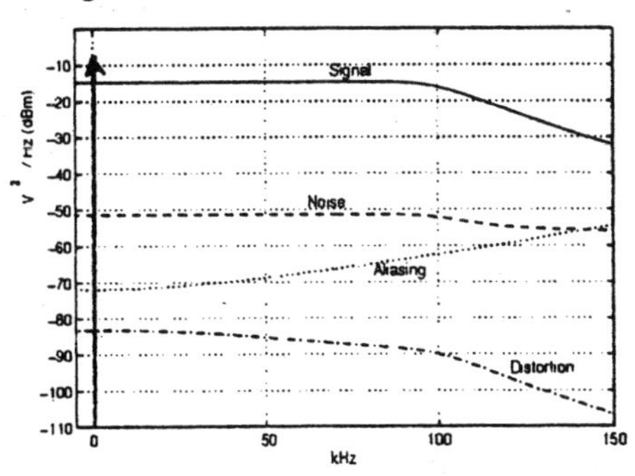

Fig. 5. A typical ORCA simulation result.

The signal band of interest is 100 kHz. The frequency distribution of the wanted and the different unwanted signals (noise, aliased signals and distortion) are displayed separately. Notice the large noise peak at DC due to LO to RF crosstalk in the second mixer, which is typical for zero-IF topologies.

ORCA can also be used in automatic mode, as an optimization tool. In that case the RF designer has to specify the optimization variables (the building block specifications that have to be varied during optimization, e.g. the BW of the bandpass filter BPF), the boundaries for each optimization variable (e.g. 20 - 800 MHz for BW_{BPF}) and the cost function (a weighted sum of the total power consumption and the deviation from the SUSR specification). Due to the nature of simulated annealing, one optimization run will typically take several hours.

Table 1 shows the results obtained from an ORCA optimization run for a specified SUSR of 40 dB. The total power required by the receiver was 50 mW. All italic numbers are optimization results. Finally, Fig. 6 shows the minimal power versus SUSR obtained from several optimization runs with different SUSR specifications.

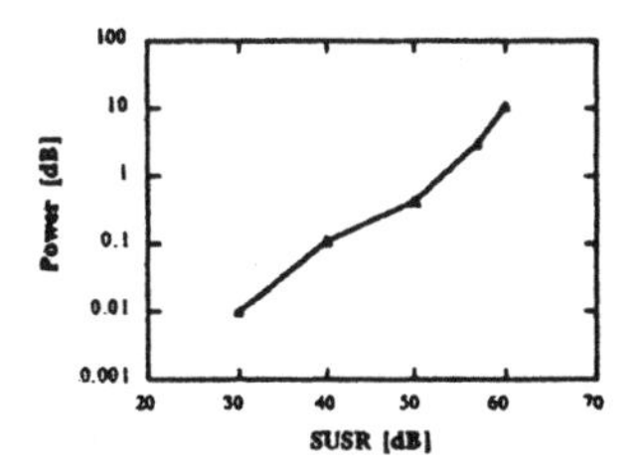

Fig. 6. Optimimal power consumption versus SUSR .

T_i	F_i [MHz]	BW_i [MHz]	A_i [dB]	DR_i [dB]	P_i [mW]
BPF	910.0	*54.0*	0.0	∞	0.0
LNA	0.0	*2680.0*	*28.0*	*50.0*	*30.5*
MIX 1	*793.0*	*1738.0*	*3.4*	*55.0*	*14.5*
SAW	*117.0*	*55.0*	0.0	∞	*0.0013*
IF AMP	0.0	*410.0*	*14.0*	*53.0*	*2.8*
MIX 2	*117.0*	*384.0*	*14.0*	*45.0*	*1.1*
LPF	0.0	*1.23*	*1.2*	*74.0*	*0.126*
ADC	0.0	*8.61*	0.0	*65.0*	*0.216*

Table 1. ORCA optimization results for a SUSR of 40 dB.

5. Conclusions

A methodology for the high-level analysis and optimization of analog RF receiver front-ends has been presented, which allows to explore alternative RF receiver topologies as well as to investigate design trade-offs within each topology. The simulation method separately determines the wanted and all unwanted signals (noise, distortion, aliasing) and all operations are performed in the spectral domain. A built-in optimization loop allows the tool to automatically perform high-level synthesis of RF receivers by translating application-specific system-level specifications, such as the antenna signal spectra and the required output signal quality, into optimum specifications for each building block which minimize the overall power and/or area consumption. Experimental results have been presented that show the practical usefulness of the tool.

References

[1] J. Sevenhans et al., "An analog radio front-end chip set for a 1.9 GHz mobile radio telephone application," *proc. ISSCC*, pp.44-45, Feb. 1994.

[2] C. Marshall et al., "A 2.7V GSM transceiver ICs with on-chip filtering," *proc. ISSCC*, pp.148-149, Feb. 1995.

[3] T. Stetzler et al., "A 2.7V to 4.5V single-chip GSM transceiver RF integrated circuit," *proc. ISSCC*, pp.150-151, Feb. 1995.

[4] K. Kundert, A. Sangiovanni-Vincentelli, "Simulation of nonlinear circuits in the frequency domain," *IEEE Trans. on Computer-Aided Design*, Vol. 5, No. 4, pp. 521-535, Oct. 1986.

A Statistical Optimization-Based Approach for Automated Sizing of Analog Cells

F. Medeiro, F. V. Fernández, R. Domínguez-Castro and A. Rodríguez-Vázquez

Dept. of Analog Circuit Design, Centro Nacional de Microelectrónica, Sevilla, SPAIN

Abstract

This paper presents a CAD tool for automated sizing of analog cells using statistical optimization in a simulation based approach. A nonlinear penalty-like approach is proposed to define a cost function from the performance specifications. Also, a group of heuristics is proposed to increase the probability of reaching the global minimum as well as to reduce CPU time during the optimization process. The proposed tool sizes complex analog cells starting from scratch, within reasonable CPU times (approximately 1hour for a fully differential opamp with 51 transistors), requiring no designer interaction, and using accurate transistor models to support the design choices. Tool operation and feasibility is demonstrated via experimental measurements from a working CMOS prototype of a folded-cascode amplifier.

1: Introduction

Most previously reported approaches for automated analog cell design are *closed* systems where the knowledge of the available topologies is provided as *analytical* design equations. The design equations associated to new topologies must be generated -- a task for only real analog design experts. Also, sizing is carried out using simplified analytical descriptions of the blocks and thus, manual fine-tuning using an electrical simulator and detailed MOS transistor models may be necessary once rough automated sizing is complete. These drawbacks are overcome in the so-called *simulation-based* systems [1], which reduce sizing to a constrained optimization problem and aim to solve it by following an iterative procedure built around an electrical simulator, with no design equations required. A representative example is DELIGHT.SPICE [2] where DELIGHT (a general algorithmic optimizer) and SPICE are combined. Also, advanced electrical simulators, like HSPICE [3], incorporate optimization routines. However, the optimization routines in both tools search for *local* solutions, and are consequently inappropriate to size analog cells from scratch.

This paper presents a simulation based approach for *global* sizing of *arbitrary* topology analog cells using *statistical* optimization. We demonstrate that by combining proper cost function formulation and innovative optimization heuristics, complex cells are designed starting from arbitrary initial points within reasonable CPU times and requiring no designer interaction -- a very appealing feature for ASIC applications.

2: Cost function formulation

Three different specification classes are considered:

- **Strong restrictions**. No relaxation of the specified value is allowed. Hence, if any of the design parameters (equivalently, any point of the design parameter space) does not satisfy one strong restriction, it must be rejected immediately.
- **Weak restrictions**. These are the typical performance specifications required of analog building blocks, i.e. $A_o > 80\text{dB}$. Unlike strong restrictions, weak restrictions allow some relaxation of the target parameters.
- **Design objectives**. Stated as the minimization/maximization of some performance features,

$$minimize \qquad y_{\Psi_i}(\mathbf{x}) \qquad 1 \le i \le P \tag{1}$$

where $\mathbf{x}$ is the vector of design parameters.

Mathematically, the fulfillment of these specifications can be formulated as a multi-objective constrained optimization problem,

$$\begin{array}{lllll} & minimize & y_{\Psi_i}(\mathbf{x}) & & ,1 \le i \le P \\ & & subjected & to & \\ \{ & y_{sj}(\mathbf{x}) \ge Y_{sj} & or & y_{sj}(\mathbf{x}) \le Y_{sj} & ,1 \le j \le Q \\ & y_{wk}(\mathbf{x}) \ge Y_{wk} & or & y_{wk}(\mathbf{x}) \le Y_{wk} & ,1 \le k \le R \end{array} \tag{2}$$

where y_{Ψ_i} denotes the value of the *i-th* design objective; y_{sj} and y_{wk} denote values of the circuit specifications (subscripts s and w denote strong and weak specifications respectively); and Y_{sj} and Y_{wk} are the corresponding targets (for instance, $A_o \ge 80\text{dB}$, settling time $\le 0.1\mu\text{s}$).

A cost function, which transforms the constrained optimization problem into an unconstrained one, is defined in the *minimax* sense as,

$$minimize\Phi(\mathbf{x}) = max\{F_{\Psi}(y_{\Psi i}), F_{sj}(y_{sj}), F_{wk}(y_{wk})\} \quad (3)$$

where the *partial* cost functions $F_{\Psi}(\bullet)$, $F_{sj}(\bullet)$, and $F_{wk}(\bullet)$ are defined as,

$$F_{\Psi}(y_{\Psi i}) = -\sum_i w_i \log(|y_{\Psi i}|) \qquad F_{sj}(y_{sj}) = K_{sj}(y_{sj}, Y_s)$$
$$F_{wk}(y_{wk}) = -K_{wk}(y_{wk}, Y_{wk}) \log\left(\frac{y_{wk}}{Y_{wk}}\right) \quad (4)$$

where w_i (called weight parameters for the design objectives) is a positive (alternatively negative) real number if $y_{\Psi i}$ is positive (alternatively negative), and for $K_{sj}(\bullet)$ and $K_{wk}(\bullet)$ we have,

$$K_{sj}(y_{sj}, Y_{sj}) = \begin{cases} -\infty & \text{if strong restriction holds} \\ \infty & \text{otherwise} \end{cases}$$
$$K_{wk}(y_{wk}, Y_{wk}) = \begin{cases} \infty\,\mathrm{sgn}(k_k) & \text{if weak rest holds} \\ k_k & \text{otherwise} \end{cases} \quad (5)$$

where k_k (weight parameters assigned to weak restrictions) is a positive (alternatively negative) real number if the weak specification is of ≥ (alternatively ≤) type. Weight parameters are used to give priority to the associated design objectives and weak specifications. As shown in the cost function formulation, only relative magnitude of the weight parameters of the same type makes sense. In (5) weak specifications are assumed positive. Sign criteria is inverted for negative specifications.

3: Process management

Fig. 1 shows a block diagram illustrating the operation flow in the proposed methodology. An updating vector $\Delta\mathbf{x}_n$, is *randomly* generated at each iteration. Strong restrictions are then checked. If any of them is not met, the corresponding movement is rejected. Otherwise, weak restrictions are examined. Only if all of them are fulfilled, the design objectives are included in the cost function. The value of the cost function is calculated at the new point and compared to the previous one. The new point is accepted if the cost function has a lower value. Otherwise, it may also be accepted according to a *probability* function,

$$P = P_o e^{-\frac{DF}{T}} \quad (6)$$

depending on a *control* parameter, *T*.

3.1: Cooling schedule

Unlike classical simulated annealing algorithms [4], where *T* in (6) decreases monotonically during the process, our tool uses a composed temperature parameter

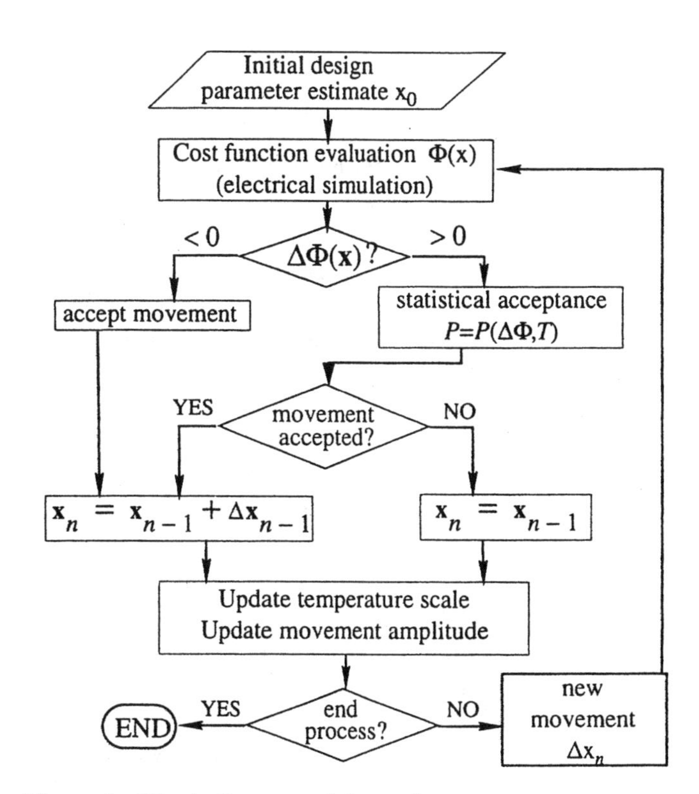

Figure 1: Block diagram of the tool

$$T = \alpha(\mathbf{x})\,T_o(n) \quad (7)$$

where *n* denotes the iteration count; $T_o(n)$ (the *normalized* temperature) is a function of *n*; and $\alpha(\mathbf{x})$ (the temperature *scale*) is a function of the position in the design parameter space. Our tool incorporates heuristics to choose T_o and α for increased convergence speed, namely:

- **Non-monotonic and adaptive normalized temperature**. Instead of a conventional slow monotonically decreasing temperature [5], a sequence of fast coolings and reheatings is used. This enables obtaining feasible designs for low iteration counts for circuits with not very demanding specifications. In the more general case, this strategy reduces iteration count by an average factor of 6. Two different evolutionary laws for the normalized temperature are incorporated in the tool: *exponential* decreasing and linear decreasing. Initial and final temperatures, number of coolings, decreasing law and rate, etc. are completely controlled by the user. An alternative cooling schedule makes T_o change as a function of the percentage of accepted movements:

$$T_o(n) = T_o(n-1) + \beta\left(1 - \frac{\rho}{\rho_s(n)}\right) \quad (8)$$

where ρ is calculated as the ratio of accepted movements to the total number of movements during the last *M* iterations, where *M* is a heuristic variable whose typical value is around 25; β in (8) controls the rate of temperature change and has a typical value around 0.1; and

$\rho_s(n)$ is a prescribed acceptance ratio, which can be fixed or vary with some given law. This schedule provides very good results for practical circuits, rendering the outcome of the optimization process somewhat independent of the specified values of the initial and final temperature.

- **Nonlinear scale.** This is done to compensate the large differences that may eventually appear in the increments of the cost function in the different regions. Thus, no temperature definition is used for those regions where strong restrictions do not hold, due to the fact that any design entering this region is automatically rejected. On the other hand, in regions where some weak specifications are violated, temperature is given as,

$$T = T_o|k_{max}| \Rightarrow \alpha(\mathbf{x}) = k_{max} \tag{9}$$

where k_{max} is the weight associated to the maximum among the $F_w(\bullet)$'s in (4), and T_o is the normalized temperature at the current iteration. Finally, if both strong and weak restrictions hold, temperature is given as,

$$T = T_o \sum |w_i| \Rightarrow \alpha(\mathbf{x}) = \sum |w_i| \tag{10}$$

where w_i is the weight associated to the *i-th* design objective.

3.2: Parameter updating

Three kinds of heuristics have been adopted:

- **Temperature-dependent amplitude.** At high *T*, large amplitude movements are allowed as they are likely to be accepted and favor wide exploration of the design parameter space. On the contrary, at low *T*, acceptance probability decreases and hence, only small movements are performed (equivalent to fine-tuning the design).
- **Logarithmic scales for independent variables.** This avoids underexploring the design space of design parameters which vary over several decades, for example, bias currents.
- **Discretization of the design parameter space.** With this partitioning, the parameter space can be viewed as a collection of *hypercubes*. Only movements over vertices of this multidimensional grid are allowed, being marked when they are visited. Thus, if during the optimization process one vertex is revisited the corresponding simulation need not be performed. Hence, an important number of simulations is avoided. When this optimization process ends, a local optimization starts within a multidimensional cube around the optimum vertex for fine tuning of the design. In this local optimization, design variables recover their continuous nature or their original grid size.

Large efficiency enhancements are also achieved by proper control of the DC electrical simulator routines. A dynamic, adaptive, DC initialization schedule is implemented which uses operating point information of previous iterations to increase convergence speed of the simulator. This significantly reduces CPU time, especially at low temperatures.

3.3: Heuristics comparison

The proposed heuristics have been tested using the function

$$f(x) = K \cdot min\{A \prod_{k=1.N} \cos(x_k - d), A \prod_{k=1.N} \cos(x_k + d) + \gamma\} \tag{11}$$

$$A = -e^{-\xi \sum_{k=1.N} (x_k - d)^2}$$

where K, ξ, d, and γ are constants. It has one absolute minimum (of value $-K$) and many local minima, whose count increases linearly with the number of variables. Thus, the complexity of the optimization process is determined exclusively by the number of variables, and not by structural changes in the cost function. Fig.2 shows this function for two independent variables.

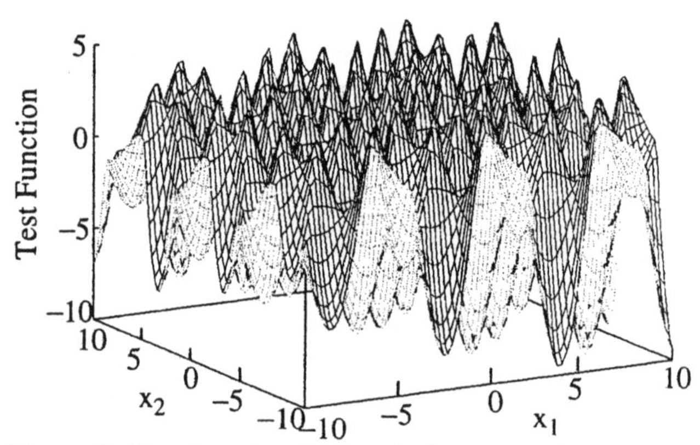

Figure 2: Test function for heuristics comparison.

The test procedure consisted in the repeated execution of the different heuristics on the test function, starting from random points of the parameter space and with a fixed iteration count. The best achieved minimum for each of these executions was stored. Experimental results from these tests are shown in the three-dimensional plots in Fig.3. In order to get better insight into the test results, the plot of the test function is allowed to assume only integer values. Hence, the minimum achieved at each test execution is represented by its closest integer value. The *X*-axis in Fig.3 represents the magnitude of the achieved minimum (its closest integer value). The *Y*-axis corresponds to the number of independent variables in the test function $f(\bullet)$, and the *Z*-axis represents the percentage of iterations that achieved that minimum. Fig.3a corresponds to a conventional cooling schedule. It had a single cooling with fixed scale in variable movements and variable Markov chain length [4]. For a function with a small number of variables, most iterations provided the global minimum of the function but this percentage decreased rapidly when the number of variables was increased. Fig.3b corresponds to our improved cooling schedule with the same number of iterations. The cooling schedule used had four successive coolings and reheatings, variable scale, and a Markov chain length equal to 1. Most iterations provided the global minimum of the function, even when the number of independent variables was increased.

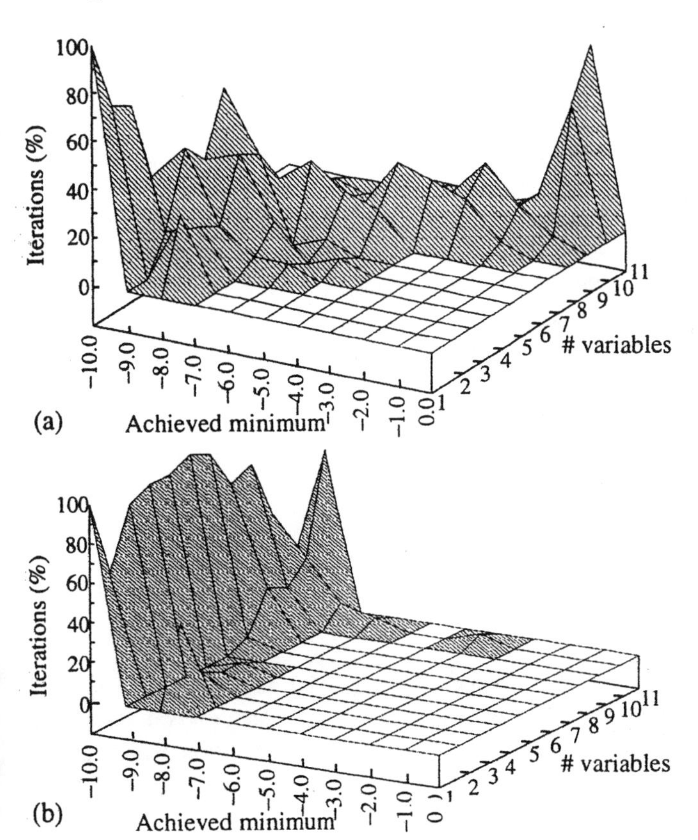

Figure 3: Cooling schedule heuristics comparison.

4: Practical results

Let us consider the folded-cascode fully-differential opamp of Fig.4, which displays the sizes provided by the tool. These sizes were obtained for the specifications needed in a 17bit@40KHz fourth order ΣΔ modulator. The specifications are given in the first column of Table 1. The power consumption was the only design objective. The optimization process started from scratch on a 10-dimension design space and required about 45min. CPU time on a 100MIPS Sparcstation. Program results for the sized circuit, corresponding to the electrical simulator output, are shown in the second column of Table 1. The opamp has been integrated in a CMOS 1.2μm double poly technology. Experimental results are given in third column of Table 1. The final ΣΔ modulator prototype displayed 16.8bit@40Khz.

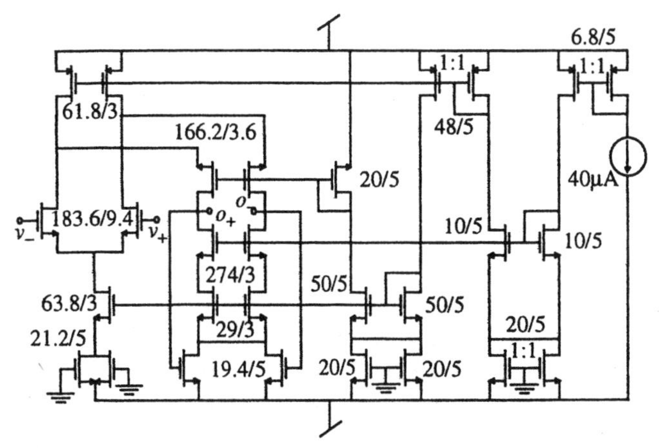

Figure 4: Fully-differential folded-cascode opamp.

These results compare advantageously to equation-based design systems. These typically spend a few seconds or minutes for the design of similar analog cells. But the effort to generate the knowledge required for new topologies varies between several weeks and 12 months. On the contrary, input file preparation in our tool requires no more than one hour of a SPICE user.

Table 1. Simulated and measured results for Fig.4.

	Specs	Simulated	Measured	Units
A_0	≥ 70	78.52	76.01	dB
GBW (1pF)	≥ 30	34.88	-	MHz
GBW(12pF,1MΩ)		4.17	4.21	MHz
PM(1pF)	≥ 60	66.28	-	o
PM(12pF, 1MΩ)		87.2	86.8	o
Input white noise	≤ 12	13.53	-	nV/√Hz
SR	≥ 70	74.81	70.5	V/μs
OS	≥ ±3	± 3.2	± 3.0	V
Offset	-	–	3.35	mV
Power	minimize	1.95	1.93	mW

5: References

[1] G. Gielen and W. Sansen: *"Symbolic Analysis for Automated Design of Analog Integrated Circuits"*. Kluwer, 1991.

[2] W. Nye et al.: "DELIGHT.SPICE: An Optimization-Based System for the Design of Integrated Circuits". *IEEE Transactions on Computer-Aided Design*, Vol. 7, pp. 501-519, April 1988.

[3] "HSPICE User Manual". Meta Software Inc. 1988.

[4] P.J.M. van Laarhoven and E.H.L. Aarts: *"Simulated Annealing: Theory and Applications"*, Kluwer Academic Pub., 1987.

[5] R. A. Rutenbar: "Simulated Annealing Algorithms: An Overview". *IEEE Circuits and Devices Magazine*, Vol. 5, pp. 19-26, January 1989.

Synthesis of High-Performance Analog Circuits in ASTRX/OBLX

Emil S. Ochotta, *Member, IEEE*, Rob A. Rutenbar, *Senior Member, IEEE*, and L. Richard Carley, *Senior Member, IEEE*

***Abstract*—We present a new synthesis strategy that can automate fully the path from an analog circuit topology and performance specifications to a sized circuit schematic. This strategy relies on asymptotic waveform evaluation to predict circuit performance and simulated annealing to solve a novel unconstrained optimization formulation of the circuit synthesis problem. We have implemented this strategy in a pair of tools called ASTRX and OBLX. To show the generality of our new approach, we have used this system to resynthesize essentially all the analog synthesis benchmarks published in the past decade; ASTRX/OBLX has resynthesized circuits in an afternoon that, for some prior approaches, had required months. To show the viability of the approach on difficult circuits, we have resynthesized a recently published (and patented), high-performance operational amplifier; ASTRX/OBLX achieved performance comparable to the expert manual design. And finally, to test the limits of the approach on industrial-sized problems, we have synthesized the component cells of a pipelined A/D converter; ASTRX/OBLX successfully generated cells 2–3× more complex than those published previously.**

I. Introduction

A SURPRISING number of technologies that most people consider hallmarks of the *digital* revolution actually rely on a core of *analog* circuitry; cellular telephones, magnetic disk drives, and compact disc players are just a few such examples. Many of tomorrow's products—e.g., neural networks, speech recognition systems, and personal digital assistants—will also require analog circuitry. Unfortunately, the present state of analog CAD tools makes it difficult to quickly and cost effectively design the new analog circuitry that these new technologies will require. To conserve space and save money, it is now commonplace to implement entire mixed analog/digital systems on a single Application Specific Integrated Circuit (ASIC). But, to maximize profit, mixed analog/digital ASIC designers must also minimize design-time and thus time-to-market. Digital CAD tools facilitate this by providing a rapid path to silicon for the large digital component of these designs. Unfortunately, the analog component of these designs, although small in size, is still designed manually by experts using time-consuming techniques that have remained largely unchanged in the past 20 years [1], [2]. With the advent of logic synthesis tools [3] and semicustom layout techniques [4] to automate much of the digital design process, the analog section may consume 90% of the overall design time, while consuming only 10% of the ASIC's die area.

Manuscript received February 15, 1995; revised July 21, 1995. This work was supported in part by the Semiconductor Research Corporation, The National Science Foundation, Harris Semiconductor, and Bell Northern Research. This paper was recommended by Associate Editor R. A. Saleh.

E. S. Ochotta was with the Department of Electrical and Computer Engineering, Carnegie Mellon University, Pittsburgh, PA 15213 USA. He is now with Xilinx, Inc., San Jose, CA 95124 USA.

R. A. Rutenbar and L. R. Carley are with the Department of Electrical and Computer Engineering, Carnegie Mellon University, Pittsburgh, PA 15213 USA.

Publisher Item Identifier S 0278-0070(96)03471-9.

This paper describes a new approach to *analog circuit synthesis*, i.e., translating performance specifications into a circuit schematic with sized devices, thereby automating part of the analog design process. The scope of this paper is synthesis for *cell-level* (less than 100 devices) circuits. Starting from a transistor schematic, we seek both to design a dc bias point and size all devices to meet performance targets such as gain and bandwidth. Our approach combines the following ideas into an analog synthesis methodology.

- A novel *unconstrained optimization formulation* to which the circuit synthesis problem is mapped;
- *Simulated annealing* to solve the resulting optimization problem;
- *Asymptotic Waveform Evaluation* (AWE) to simulate circuit performance—the key component of the function to optimize;
- A compiled database of industrial quality, nonlinear device models, called *encapsulated device evaluators*, to provide the accuracy of detailed simulation while making the synthesis tool independent of low-level device modeling concerns;
- A *relaxed-dc formulation* of the nonlinear device simulation problem to avoid a CPU intensive complete dc operating point solution for each circuit simulation; and, finally,
- A separate *compilation phase* to translate the synthesis problem from a description convenient to the designer into an executable program that designs the circuit via optimization.

Although analog synthesis via simulated annealing is not new [5], the use of AWE and the added power of a separate compilation phase are completely novel. We believe the result is a *usable* synthesis system.

Throughout this paper, we will measure the effectiveness of our new analog circuit synthesis formulation and compare it to prior systems based on the five critical metrics for analog synthesis tools.

- **Accuracy**: the discrepancy between the synthesis tool's internal performance prediction mechanisms and those of a detailed circuit simulator that uses realistic device models;

- **Generality**: the breadth of the circuits and performance specifications that can be successfully handled by the synthesis tool;
- **Complexity**: the largest circuit synthesis task that can be successfully completed by the synthesis tool;
- **Synthesis time**: the CPU time required by the synthesis tool;
- **Preparatory effort**: the designer-time/effort required to render a new circuit design in a form suitable for the tool to complete.

An ideal system maximizes accuracy, generality, and complexity, while minimizing synthesis time and preparatory effort. Note that these metrics are not always easy to quantify. For example, the complexity of a synthesis task can be affected by many factors including the number of designable parameters (element values and device sizes), the number and difficulty of the performance specifications, the number of components in the circuit, and the inherent difficulty of evaluating the performance of the circuit. For these cases, where the definition of the term is qualitative, we select specific concrete metrics that provide a good indication of the underlying factor we wish to measure. For example, as the metric for complexity, we use the number of designable parameters the designer wishes the tool to determine plus the number of components in the circuit. This is easy to quantify and relates complexity to both the problem and circuit size. In addition to the five metrics above, we shall also use one additional term, *automation*, which we define as the ratio of the time it takes to design a new circuit for the first time manually to the time it takes with the synthesis tool. When comparing synthesis tools, manual design time will be the same for a given circuit, so maximizing automation is equivalent to minimizing the sum of preparatory time and synthesis time.

To provide a concrete set of synthesis tasks to compare our approach to that of other tools, we have generated synthesis results over a large suite of analog cells. This suite includes three classes of synthesis results.

- A suite of benchmark circuits that shows the generality of our approach by blanketing essentially all previous analog cell synthesis results;
- A redesign of a recently published manually designed analog cell that shows the ability of our approach to handle difficult circuits;
- A pipelined A/D converter that includes the most complex synthesized cells of which we are aware and shows the ability of our approach to handle large, realistic designs.

In comparison to prior approaches, our approach typically predicts circuit performance more accurately, yet requires 2–3 orders of magnitude less preparatory effort by the designer. In exchange for these substantial improvements, a small price is paid in synthesis time: our approach can require several hours of CPU time on a fast workstation, instead of seconds or minutes. This is an acceptable trade-off because automation is improved when designing a new circuit for the first time, i.e., spending these hours of a computer's time can save the months of designer's time required to complete the design manually or with other analog synthesis tools.

The remainder of this paper is structured as follows. Section II reviews prior approaches to analog circuit synthesis. Section III presents the basic ideas underlying our new formulation of the analog synthesis problem, while Section IV presents a circuit synthesis example to show how these ideas are applied to a real synthesis task. In Section V, we present the formulation in detail, and in Section VI, we revisit the few related approaches to synthesis and compare them to our approach. Section VII describes synthesis results and again compares to those from other approaches. Finally, Section VIII offers concluding remarks.

II. Review of Prior Approaches

Previous approaches to analog circuit synthesis [5]–[18] have failed to make the transition from research to practice. This is due primarily to the prohibitive one-time effort required to derive the complex equations that drive these synthesis tools. Because they rely on a core of equations, we refer to these previous approaches to synthesis as equation-based, and discuss their architecture in terms of the simplified search loop shown in Fig. 1. At each step in the synthesis process (each pass through this loop), a search mechanism perturbs the element values, transistor dimensions and other variable aspects of the circuit in an attempt to arrive at a design that more closely meets performance specifications and other objectives. Performance equations are used to evaluate the new circuit design, determine how well it meets its specifications, and provide feedback to the search mechanism to guide its progress. Because of their reliance on equations, these systems are still limited in the crucial areas of accuracy and automation. Let us examine these issues in greater detail.

- **Accuracy**: Equation-based approaches rely heavily on simplifications to circuit equations and device models. Consequently, the performance of the synthesized circuit often reflects the limitations of the simplified equations used to model it, rather than the inherent limitations of the circuit topology or underlying fabrication technology. The need for designs that push the limits of circuit topologies and use the latest technologies invalidates the use of many of these simplifications. For example, in a 3μm MOS process, $I_{\rm DS} = K'W/2L(V_{\rm GS} - V_T)^2$ is a workable model of the current-voltage relationship for a device, and equation-based approaches take advantage of the fact that it can be inverted to allow either voltage or current as the independent variable. This simply inverted equation allows an equation-based tool to quickly solve for a circuit's dc operating point, but it can yield grossly inaccurate performance predictions for a device with a submicron channel length. The need to support complex device models and high-performance circuits is fundamentally at odds with equation-based strategies that rely on these simple, easily inverted equations.
- **Automation**: Equation-based tools appear to design circuits quickly. But, the run-times of these tools are not an accurate measure of automation because they do not consider the preparatory time required to derive the circuit equations. Even for a relatively simple analog circuit,

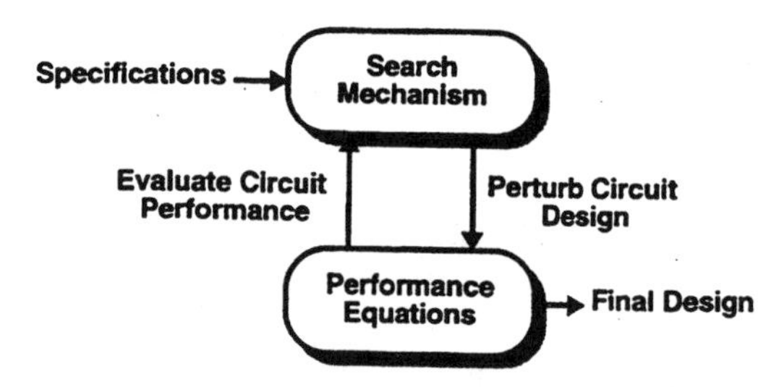

Fig. 1. Search process used in equation-based analog synthesis tools.

these equations are very complex, require considerable analog design expertise to derive, and must be entered as thousands of lines of program code. For a textbook design, this process can take weeks [6], while for an industrial design it can take several designer-years [7], and the process must be performed for each new circuit topology added to the synthesis tool's library. Moreover, adding these equations typically requires a user who is a programmer, an analog designer, and an expert intimate with the internal architecture of the tool. As a result, it is almost always easier for an industrial mixed-signal ASIC designer to design circuits manually rather than dedicate the effort required to teach these tools to do it.

However, researchers are aware of these accuracy and automation problems and several different techniques have evolved in an effort to address them. Early analog synthesis tools, such as IDAC [8] and OPASYN [6], used direct numerical means to optimize the analog circuit performance equations to meet specifications. Others, such as BLADES [9], attempted to achieve greater flexibility by using ruled-based strategies to organize the analog circuit equations. The automation problem inherent with equations was first addressed by OASYS [10], which eased the burden of equation derivation by introducing an aggressive hierarchical structure into circuit performance equations in an attempt to provide reusable circuit building blocks. This hierarchical structure became the core of later tools, such as CAMP [11] and An_Com [12]. The ability to reuse circuit equations led to the desire to be able to more easily edit and add to existing libraries of analog circuit design expertise. In early synthesis tools, analog circuit equations were hard-coded in the tool as part of the underlying solution strategy. OASYS VM [13] provided a first step toward an open system, by decoupling the expertise from the solution strategy. More recently, tools such as STAIC [14] and IDAC [15], allowed the user to specify analog circuit design equations directly using programming-like languages. Despite substantial early progress with these lines of research, the difficulty of deriving, coding, and testing performance equations still remains a daunting task, and researchers have made few strides with this style of synthesis in recent years.

The second major innovation that aimed at reducing preparatory effort was symbolic simulation [19], [20]. This technique was first introduced as a tool to aid manual design and education, and it was first integrated directly into a synthesis tool with ARIADNE [5]. Symbolic simulation generates analytical transfer functions for small linear or linearized circuits. For many important classes of circuits, such as operational amplifiers, most of the important performance specifications are linear in nature and are amenable to this kind of analysis. Moreover, fairly accurate equations for many of the remaining nonlinear specifications can be derived by inspection, whereas equations for the linear specifications are much more involved. Thus, in theory there is a great deal of leverage that can be gained from symbolic analysis. However, symbolic simulation has yet to overcome substantial technical obstacles before it can fully automate performance equation derivation for large, high-performance circuits. Applied blindly, symbolic simulation of a circuit linearized from a handful of devices can generate an expression with tens of thousands of terms, and the number of terms grows exponentially with circuit size. Because of memory and CPU time concerns, generating exact symbolic expressions for all but the smallest circuits is impractical. As a result, practical symbolic simulation algorithms generate *pruned* expressions, but this pruning leads to accuracy problems. If device models are pruned before symbolic analysis, the resulting expressions are more compact but lack accuracy for high-performance designs. If the final equations are pruned, symbolic simulation still suffers from memory and CPU time problems, and the result is faithful only to some performance concerns. For example, pruning terms whose magnitude is small typically distort phase information, on which the circuit's performance may critically depend. Very recently, strategies have been developed for effectively reducing the number of terms during simulation, and symbolic simulation is now efficient enough with computer resources to be applied to medium-sized opamps [21], [22]. However, even with these recent innovations, the pruned expressions are valid in only a very limited region of the achievable design space for the circuit, and would have to be frequently regenerated if the designable circuit parameters were varied significantly. The ability to do this in a manner efficient enough for synthesis has not been demonstrated to date, although in future these techniques may yet provide a completely automatic path to equation-generation.

Fewer technical innovations have been made to improve the *accuracy* of analog synthesis techniques. One recent area of improvement has been the incorporation of realistic device models. In [16], Maulik incorporated complete BSIM [23] device models from SPICE 3 [24] into a special purpose synthesis tool. The use of these models substantially complicates solving for dc operating points in an evolving circuit because the models cannot be inverted analytically. To address this problem, Maulik formulated the circuit synthesis problem as a constrained optimization problem and enforced Kirchhoff's current law by explicitly writing dc operating point constraints. This technique combines simultaneous circuit performance optimization with dc operating point solution. As discussed in Section III, our relaxed-dc formulation evolved from this paper.

Although many innovations have been made during the evolution of equation-based analog synthesis tools over the past decade, the combined problems of accuracy and automation (i.e., preparatory effort) have never been adequately addressed in a single cohesive approach. A new strategy is needed that addresses both these shortcomings.

One simple solution is to replace the equations with a direct simulation technique. This is the basic approach that was first proposed for analog circuit *optimization* decades ago [25], rediscovered when faster computers and improved simulators made it practical for research [26]–[28], and is now making its way into industrial CAD systems. For example, DELIGHT.SPICE [26] follows the basic structure of Fig. 1, where the search is performed by the method of feasible directions, a gradient-based optimization technique, and the performance equations have been replaced by SPICE [29], [30]. Because SPICE is a detailed circuit simulator, no designer supplied equations are required (except to extract the performance specifications from simulation results), and the performance prediction is very accurate. Unfortunately, as the core of an optimization loop, SPICE-class simulators are slow. So slow, in fact, that DELIGHT.SPICE is an *optimization* tool, not a *synthesis* tool. The key hurdle that has not been overcome to make this transition from *optimization* to *synthesis* is that optimization requires a good initial starting point to find an excellent final answer, while synthesis requires no special starting point information. This critical distinction is more carefully explained as follows:

- **Efficiency/Starting Point Sensitivity**: Because SPICE-class simulators are slow, the search mechanism must invoke the simulator as *in*frequently as possible. As a result, simulation-based methods use local optimization techniques that require few iterations to converge. These techniques must be primed with a good initial circuit design, otherwise, an optimization may not converge or may converge to a local minima significantly worse than the circuit's best capabilities [31]. In circuit synthesis, a local optimizer is not practical because the search space contains many local but non-global minima [14], [26] and because—even with a good rough design—it is only luck if optimizing from the user's initial circuit design leads to the globally optimal final solution.

The accuracy and reduced preparatory effort that comes with simulation-based optimization are the two characteristics that have been substantially lacking from equation-based systems. One approach to incorporate simulation into an equation-based system, as taken in OAC [17], is to run a simulation-based optimizer as a post-processor. This improves accuracy, but there is no guarantee that the circuit generated by the equation-based synthesis will be the starting point needed by the simulation-based optimizer to find the globally optimal circuit. Furthermore, because of the extensive simulation run-times, only the performance specifications that can be validated with ac and dc analyses are optimized using these techniques [17]. And, perhaps most importantly, the months of preparatory effort required to derive analog circuit performance equations are still required by the equation-based part of the overall design process.

A solution to the problem of preparatory effort dictates that the user not derive, code, or prune analog circuit equations. A simulation-based approach meets this criteria, but requires innovations to avoid problems with efficiency due to circuit simulation and starting point dependency due to optimization. We are aware of two analog synthesis tools that meet these criteria: ASTRX/OBLX [32]–[35], which is the subject of this paper, and a more recent tool presented in [36]. Before comparing the differing approaches of these two tools, we first describe the architecture and underlying ideas behind ASTRX/OBLX in the following sections. We return to this comparison in Section VI.

III. Basic Synthesis Formulation

In this section, we present our basic analog circuit synthesis formulation. We begin with the specific design goals that guided the evolution of this formulation and the key ideas that form its foundation. We then outline its architectural aspects.

A. Design Goals

Our design goals for a new analog circuit synthesis architecture are to directly address the automation, accuracy, and efficiency problems we identified with previous approaches. First, to streamline the path from a circuit idea to a sized circuit schematic, our new architecture should require only hours rather than weeks/months of preparatory effort to design a new circuit. Second, the system should find high-quality circuit design solutions without regard to starting point rather than getting trapped in the nearest local minima. Third, our new system should yield accurate performance predictions for high-performance circuits rather than suffer from problems due to device model or performance equation simplifications. And, finally, the system must be able to design the complex, high-performance circuits required in modern products.

Realistically, we cannot hope to achieve progress in all these areas without making some trade-offs. The first concession we are willing to make is increased run-time. This is because our primary goal here is maximal automation, which is the sum of preparatory and run-times. Equation-based synthesis tools use only minutes of CPU time but require the designer to spend months deriving, coding, and testing equations. We believe the following scenario is more appealing: after an afternoon of effort, a circuit designer goes home while the synthesis tool completes the design overnight. Realizing this scenario is our primary goal. We are also willing to make two additional concessions for our initial implementation of our new formulation. The first of these is to exclude automatic topological design. We believe that sizing/biasing is the correct starting point for a new synthesis strategy, and a suitable mechanism for choosing among topological variants can be added later. Moreover, a tool that finds optimal sizes for a single user-supplied topology is still directly usable by analog designers. The third concession is to exclude operating range and manufacturablility concerns, and—like most previous synthesis tools—the work presented here performs only nominal circuit design. However, since the conclusion of our initial work, this formulation has been augmented with the ability to handle operating range and manufacturing concerns and preliminary results appear in [37].

B. Underlying Ideas

To achieve our goals, our circuit synthesis strategy relies on five key ideas: synthesis via optimization, AWE, simulated annealing, encapsulated device evaluators, and the relaxed-dc numerical formulation. We describe these ideas below.

Synthesis via Optimization: We perform fully automatic circuit synthesis using a constrained optimization formulation, but solved in an unconstrained fashion. As in [6], [16], and [26], we map the circuit design problem to the constrained optimization problem of (1). Here $\underline{x}$ is the set of independent variables—geometries of semiconductor devices or values of passive circuit components—for which we wish to find appropriate values; $\underline{f}(\underline{x})$ is a set of objective functions that codify performance specifications that the designer wishes to optimize, e.g., power or bandwidth; and $\underline{g}(\underline{x})$ is a set of constraint functions that codify specifications must be beyond a specific goal, e.g., gain $\geq$ 60 dB. Scalar weights w_i balance competing objectives

$$\underset{\underline{x}}{\text{minimize}} \sum_{i=1}^{k} w_i \cdot f_i(\underline{x}) \quad \text{s.t.} \quad \underline{g}(\underline{x}) \leq 0. \tag{1}$$

To allow the use of simulated annealing, we perform the standard conversion of this constrained optimization problem to an unconstrained optimization problem with the use of additional scalar weights. As a result, the goal becomes minimization of a scalar cost function, $C(\underline{x})$, defined by

$$C(\underline{x}) = \sum_{i=1}^{k} w_i f_i(\underline{x}) + \sum_{j=1}^{l} w_j g_j(\underline{x}). \tag{2}$$

The key to this formulation is that the minimum of $C(\underline{x})$ corresponds to the circuit design that best matches the given specifications. Thus, the synthesis task becomes two more concrete tasks: 1) evaluating $C(\underline{x})$ and 2) searching for its minimum. However, performing these tasks is not easy. In equation-based synthesis tools, evaluating $C(\underline{x})$ is done using designer-supplied equations. To achieve our automation goals, we must avoid the large preparatory effort it takes to derive these equations. Moreover, in searching for the minimum, we must address the issues of starting point independence and global optimization, since $C(\underline{x})$ may have many local minima.

Asymptotic Waveform Evaluation: To evaluate circuit performance, i.e., $C(\underline{x})$, without designer supplied equations, we rely on an innovation in simulation called AWE [38], [39]. AWE is an efficient approach to analysis of arbitrary linear circuits that is several orders of magnitude faster than SPICE for ac analysis. By matching the initial boundary conditions and the first $2q-1$ moments of the actual circuit transient response to a reduced q-pole model, AWE can predict small-signal circuit performance using a reduced complexity model. AWE is a general simulation technique that can be applied to any linear or linearized circuit and yields accurate results without manual circuit analysis. Thus, for linear performance specifications, AWE replaces performance equations, but does so at a fraction of the run-time cost of SPICE-like simulation.

Simulated Annealing: We have selected simulated annealing [40] as the optimization engine that will drive our search for the best circuit design in the solution space defined by $C(\underline{x})$. This method provides the potential for global optimization in the face of many local minima. Simulated annealing has a theoretically proven ability to find a global optimum under certain restrictions [41]. Although these restrictions are not enforceable for most industrial applications, the proofs suggest an algorithmic robustness that has been validated in practice [42]. Because annealing incorporates controlled *hill-climbing*, it can escape local minima and is starting-point independent. Annealing has other appealing properties including its ability to optimize without derivatives. Furthermore, although annealing typically requires more function evaluations than local optimization techniques, it is now achieving competitive run-times on problems for which tuned heuristic methods exist [43]. Because annealing directly solves *un*constrained optimization problems, we require the scalar cost function of (2).

Encapsulated Device Evaluators: To model active devices, we rely on a compiled database of industrial models we call *encapsulated device evaluators*. These provide the accuracy of a general-purpose simulator while making the synthesis tool independent of low-level device modeling concerns. As with any analysis of a circuit, we use models to linearize nonlinear devices, generating a small signal circuit that can be passed to AWE. In a practical synthesis system, it is no longer a viable alternative to use one- or two-equation approximations instead of the hundreds of equations used in industrial device models. Unlike equation-based performance prediction, where assumptions about device model simplifications permeate the circuit evaluation process, with encapsulated device evaluators all aspects of the device's representation and performance are hidden and obtained only through requests to the evaluator. In this manner, the models are completely independent of the synthesis system and can be as complex as required. For our purposes, we rely entirely on device models adopted from detailed circuit simulators such as Berkeley's SPICE 3 [24].

Relaxed-DC Formulation: To avoid a CPU intensive dc operating point solution after each perturbation of the circuit design variables, we rely on a novel recasting of the unconstrained optimization formulation for circuit synthesis we call the *relaxed-dc formulation*. Supporting powerful device models is not easy within a synthesis environment because we cannot arbitrarily invert the terminal relationships of these models and choose which variables are independent and which are dependent. This critical simplification enables equation-based approaches to solve for the dc bias point of the circuit analytically and, as a result, very quickly. In contrast, when the models must be treated numerically, as in circuit simulation, an iterative algorithm such as Newton–Raphson is required. For synthesis, this approach consumes a substantial amount of CPU time that we would prefer not to waste on intermediate circuit designs that are later discarded. Instead, following Maulik, [16], we explicitly formulate Kirchhoff's laws, which are solved implicitly during dc biasing, and include them in $\underline{g}(\underline{x})$, the constraint functions in (2). Just as we must formulate optimization goals such as meeting gain or bandwidth constraints, we now formulate dc-correctness as yet another goal to meet. Of course, the idea of relaxing the dc constraints in this manner is not new, e.g., an analogous formulation for microwave circuits is discussed in [44]; however, it has been controversial [45].

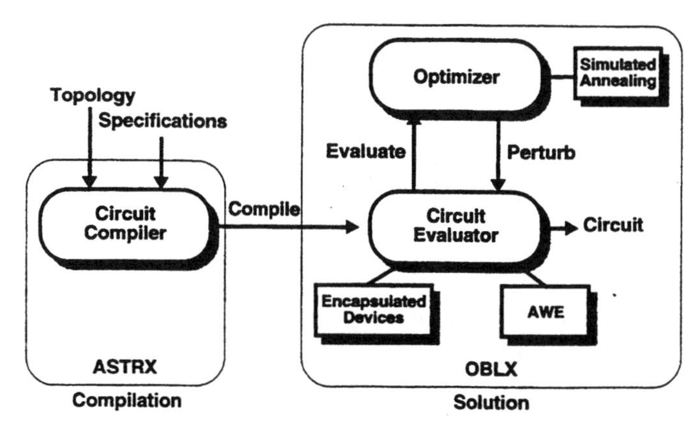

Fig. 2. New synthesis architecture.

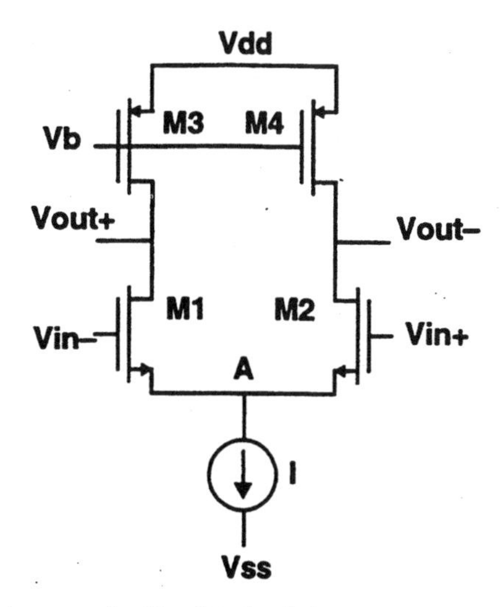

Fig. 3. Design example: Circuit under design.

C. System Architecture

We combine these five ideas to create an architecture that provides a fully automated path from an unsized circuit topology and a set of performance specifications to a completed, synthesized circuit (see Fig. 2). This path is comprised of two phases.

- **Compilation**: For each new circuit synthesis task, compilation generates code that implements the cost function, $C(\underline{x})$. To evaluate this cost function, the compiler will generate the appropriate links to the encapsulated device evaluators and AWE. Because of our relaxed-dc formulation, the compiler must also derive the dc-correctness constraints (KCL at each node) that will enforce Kirchhoff's laws and encode them in the cost function.
- **Solution**: This cost function code is then compiled and linked to our solution library, which uses simulated annealing to numerically find its minimum, thereby designing the circuit.

IV. Synthesis Example

In the previous section, we briefly introduced the concepts that underlie our new analog circuit synthesis formulation. In this section, we present a small but complete synthesis example to make concrete the entire path from problem to solution. Assume we wish to size and bias the simple differential amplifier topology shown in Fig. 3 to meet the specifications given in Table I. The topology of the circuit under design and the performance specifications the completed design must achieve—essentially the information in Fig. 3 and Table I—are the information the designer must supply to the compiler to generate the cost function and complete the synthesis process. In this section, we shall see exactly what information about the topology and specifications is required by the circuit compiler by showing how it uses this information to create $C(\underline{x})$, the cost function that is optimized (defined in (2)).

The first component of $C(\underline{x})$ is the set of independent variables, $\underline{x}$. These variables are readily apparent from the description of the circuit topology. Assume that for our example, the sizes for M3 and M4 are given as constants, the transistors M1 and M2 are matched to preserve circuit symmetry, and the rest of the component values are allowed to vary. Then, $\underline{x} = \{W, L, I, Vb\}$, where W and L represent dimensions for both M1 and M2. However, as we shall see in a few paragraphs, because of the relaxed-dc formulation, $\underline{x}$ is not yet complete.

TABLE I
DESIGN EXAMPLE: SPECIFICATIONS

Attribute	Specification
differential gain, A_{dm}	↑[a]
gain bandwidth, UGF	1 MHz
slew rate, SR	1 V/μs

a. ↑ means maximize.

Recall from (2), that $C(\underline{x})$ is composed of objective functions $\underline{f}(\underline{x})$ and constraint functions $\underline{g}(\underline{x})$. Since, for our example, the only objective is to maximize A_{dm}, $\underline{f}(\underline{x})$ contains only $f_{A\mathrm{dm}}(\underline{x})$, which calculates A_{dm}. We provide a *simulation-oriented* definition of $f_{A\mathrm{dm}}(\underline{x})$ since AWE will be used to simulate the circuit's performance. This consists of a *test jig*, a set of measurements to make on the jig, and any simple arithmetic that must be performed to calculate the values we are interested in from the simulation results. The test jig is important because it supplies the circuit environment (stimulus, load, supplies, *et cetera*) in which the circuit under design is to be tested. We use the test jig in Fig. 4 to measure A_{dm}. Following [26], we also require a *good* and *bad* value for each objective to transform $f_{A\mathrm{dm}}(\underline{x})$, from the user-supplied function to a function more amenable to optimization (see Section V for further details). The resulting function is (3), where tf is the transfer function from input to output. The nodes at which to evaluate the transfer function must be specified by the user, but the transfer function itself is obtained for each new circuit by using AWE, and the function to calculate dc gain from a transfer function is predefined within the compiler

$$f_{A\mathrm{dm}}(\underline{x}) = \frac{\mathrm{dc\_gain(tf)} - \mathrm{good}}{\mathrm{bad} - \mathrm{good}}. \tag{3}$$

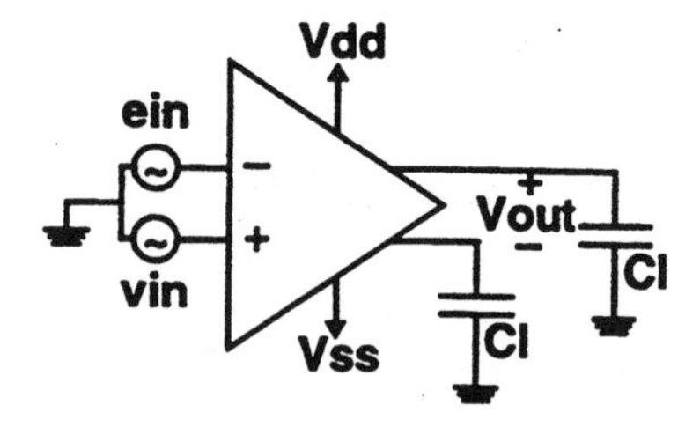

Fig. 4. Design example: Test jig for A_{dm}, UGF.

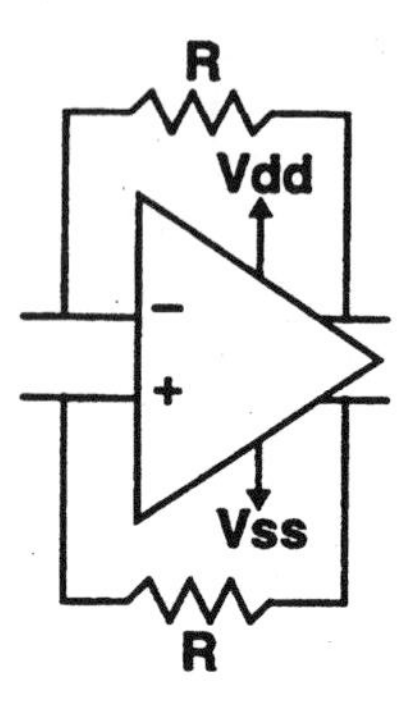

Fig. 5. Design example: Bias circuit.

The next step in creating $C(\underline{x})$ is to complete the definition of $\underline{x}$. To understand why we must add variables to $\underline{x}$, we trace the information required to calculate $f_{A\text{dm}}(\underline{x})$. Here, AWE determines the needed transfer function, but AWE requires a linearized (i.e., small-signal equivalent) circuit. Like a detailed circuit simulator, we rely on device models to linearize our circuits—in our case, these models are encapsulated device evaluators. An evaluator converts the dimensions and port voltages of each device into a set of linear elements that models the device's behavior at that operating point. After replacing each transistor with its model, we can then use AWE to evaluate the circuit's performance. What we have not discussed so far is how we obtain the port voltages of each device. In a circuit simulator, these voltages must be explicitly solved for using a time-consuming iterative procedure such as Newton–Raphson. In contrast, in our relaxed-dc formulation, we simply include these voltages as additional variables in $\underline{x}$. The compiler includes these voltage variables automatically based on a circuit analysis. To provide greater flexibility, these dc bias concerns can be separated from the small-signal test jigs—a technique familiar to analog designers. Thus dc voltage variables are obtained from a bias circuit provided by the user. For our example, an analysis of the bias circuit of Fig. 5 yields $\underline{x} = \{W, L, I, Vb, V\text{out}+, V\text{out}-, V_A\}$. This completes $\underline{x}$, but it is only half of the relaxed-dc formulation.

We complete the relaxed-dc formulation by forcing the node voltages to take values such that Kirchhoff's laws are obeyed. This is accomplished by using members of the constraint functions, $\underline{g}(\underline{x})$, the other component of $C(\underline{x})$. To begin, we replace the transistors in the circuit with large-signal models returned by the device evaluators. Simplifying the device models for the sake of clarity gives the circuit of Fig. 6. Kirchhoff's current law can then be written at each node in the circuit, e.g., at node A: $I - Id1 - Id2 = 0$. To ensure this KCL equation is met when our optimization is complete, we

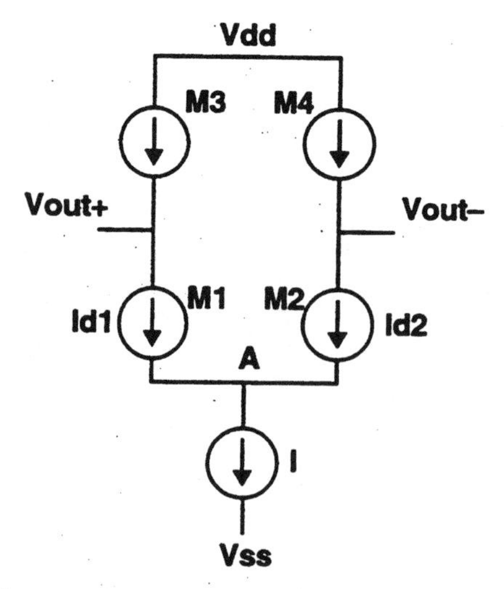

Fig. 6. Design example: Large-signal equivalent circuit.

include $g_A(\underline{x})$ from (4) as a member of $\underline{g}(\underline{x})$

$$g_A(\underline{x}) = \max(0, |I - Id1 - Id2| - \tau_{\text{abs}}). \tag{4}$$

$g_A(\underline{x})$ contributes a penalty to the cost function whenever the KCL error at node A is larger than some numeric tolerance, τ_{abs}.[1] We formulate the other KCL equations in the same fashion, creating $g_{V\text{out}+}(\underline{x})$ and $g_{V\text{out}-}(\underline{x})$. Together, these three constraints enforce KCL, thereby completing the relaxed-dc formulation.

For our example, the other members of $\underline{g}(\underline{x})$ correspond to the other performance specifications for which we wish to design. Thus, we need constraint functions for UGF and SR (see Table I). These are formulated from the user provided expressions much like $f_{A\text{dm}}(\underline{x})$ was formulated in (3); however, for specifications, the *good* value is a hard boundary and improvements beyond this value are not reflected in the cost function. The specification for UGF can be written in a simulation-based style, using the same test jig as was used for A_{dm}. This is not the case with the slew rate because measuring slew rate would require a transient simulation, which is not straightforward with AWE. However, unlike gain and unity gain frequency, slew rate is described with an easily derived expression. If we assume that we are interested only in the rate at which the output slews downwards, we can write this expression by inspection as $\text{SR} = I/(2(Cl + Cd))$, where Cd is the capacitance at the output node due to the transistors. Thus, our new formulation supports a mix of simulation- and equation-based specifications. The decision of which method to use depends on the kind of specification. As experiments with symbolic simulation have shown, equations for linear specifications such as A_{dm} and UGF can be huge, e.g., 10 000 + terms for a circuit with ten devices, and the number of terms grows exponentially with circuit size. In contrast, simulation with AWE uses a numeric technique and can evaluate A_{dm} and UGF in a few tens of milliseconds for circuits of this size. Moreover, AWE's algorithmic complexity is roughly that of an LU factorization,

[1] The large-signal model and this formulation of the KCL constraint are somewhat simplified for clarity. For more detail, see Section V.

approximately $O(n^{1.4})$ where n is the number of nodes in the circuit. The speed of AWE and the ability to describe linear performance specifications without deriving circuit equations are two of the chief advantages of our new formulation over previous synthesis approaches.

Including these final terms, we obtain (5), the final form of the cost function the compiler generates and that must be minimized to complete the circuit design

$$C(\underline{x}) = w_{\text{Adm}} f_{\text{Adm}}(\underline{x}) + w_{\text{UGF}} g_{\text{UGF}}(\underline{x}) + w_{\text{SR}} g_{\text{SR}}(\underline{x}) + w_A g_A(\underline{x}) + w_{V\text{out}+} g_{V\text{out}+}(\underline{x}) + w_{V\text{out}-} g_{V\text{out}-}(\underline{x}). \quad (5)$$

In this section, we have used an example to sketch the process the compiler must follow to map a synthesis problem into a cost function. The compiler generates this cost function as C code that is complied and linked to the optimization library, creating an executable program that performs the synthesis task specified by the user.

V. Design Details

Now that we have explained the basic ideas behind our new analog circuit synthesis formulation, and used an example to show how these ideas can be applied to analog circuit synthesis, in this section we present a more complete look at the algorithms used. We then address two issues unique to this formulation: the implications the relaxed-dc formulation has on circuit synthesis and the practicality of the formulation in terms of automation and numerical robustness.

A. Algorithmic Aspects of the Compiler

As can be seen in the example, the circuit compiler performs a number of tasks when translating the user's problem description into a cost function. These can be summarized as: a) determine the set of independent variables ($\underline{x}$), b) generate large-signal equivalent circuits for biasing, c) write KCL constraints for the large-signal circuits, d) generate small-signal equivalent circuits for AWE, e) generate cost terms for each circuit performance metric specified by the user, and f) write all the code that describes the cost function for this circuit synthesis problem.

The majority of these tasks are algorithmically straightforward, or involve algorithms that are well understood by compiler writers [46]. However, one somewhat subtle aspect of compilation that deserves mention is the task of determining the set of independent variables, $\underline{x}$. The user specifies most of these, but the compiler must find a set of independent node voltages to include in $\underline{x}$ as part of the relaxed-dc formulation. To do so, the compiler performs a symbolic tree-link analysis [47] of the large-signal equivalent circuit, which is built from the input netlist with the help of device templates provided by the encapsulated device evaluators. A path is then traced from each node to ground. Whenever a node voltage cannot be trivially determined, its value becomes another variable in $\underline{x}$.

B. The Simulated Annealing Algorithm

As described in Section III, our optimization engine employs the simulated annealing algorithm [40], [41] to solve the analog synthesis problem via unconstrained optimization. This algorithm can be described with the following pseudocode

$\underline{x} = \underline{x}_0, T = T_{\text{HOT}}$
while not *frozen*$(\underline{x}, T)$
  while not *done_at_temperature*$(\underline{x}, T)$
    $\Delta\underline{x} =$ *generate*$(\underline{x})$
    if *accept*$(\underline{x}, \underline{x} + \Delta\underline{x}, T)$ **then**
      $\underline{x} = \underline{x} + \Delta\underline{x}$
  $T =$ *update_temp*(T).

To describe the details of how our optimization engine works, we must describe each of the components of this algorithm. To begin, for all simulated annealing implementations, the *accept* function is called the Metropolis criterion [48], and is defined by (6), where *random()* returns a uniformly distributed random number on [0, 1]

$$random() < e^{[\frac{C(\underline{x}) - C(\underline{x}+\Delta\underline{x})}{T}]}. \quad (6)$$

Next, we describe annealing's four problem specific components.

- The problem *representation*, which determines how the present state of the circuit being designed is mapped to $\underline{x}$, the present state manipulated by the annealer;
- The *move-set*, which determines how the *generate* function perturbs the present circuit state $\underline{x}$ to create the new state $\underline{x} + \Delta\underline{x}$;
- The *cost function* $C(\underline{x})$ which determines how the cost of each visited circuit configuration is calculated;
- and the *cooling schedule*, which controls T, directing the overall cooling process. This defines the initial temperature, T_{HOT}, and the functions *frozen, done at temperature*, and *update temp*.

C. Problem Representation

The problem representation appears straightforward: the variables in $\underline{x}$ map to aspects of the evolving circuit design, such as device sizes and node voltages. However, there are two concerns here. The first is that the user defines a set of variables, then writes expressions to map these variables to circuit component values. As in the example of Section IV, these are typically identity relations, but may be arbitrarily complex to allow complex matchings and inter-relationships. For example, the expression '$2 * L$,' might be used in a bias circuit where one device length must always be twice that of another device.

The second concern is that we do not represent all the variables as continuous values. Node voltage values must clearly be continuous to determine an accurate bias point. Device sizes, however, can be reasonably regarded as discrete quantities, since we are limited by how accurately we can etch a device. Moreover, there is considerable advantage to be had from properly discretizing device sizes: the coarser the discretization, the smaller the space of reachable sizes that must be explored. Because small changes in device sizes make proportionally less difference on larger devices, we typically use a logarithmically spaced grid. The choice to discretize variables and the coarseness of the grid is made by the user and specified in the input description file. Thus, there may be three kinds of variables in $\underline{x}$: 1) user-specified discrete variables, e.g., a transistor width; 2) user-specified continuous variables, e.g., a current source value; and 3) automatically created continuous variables, e.g., a node voltage added to implement the relaxed-dc formulation.

D. Move-Set

Given the present state $\underline{x}$ the *move-set* $\Delta\underline{x}$ is the set of allowable perturbations on it, which are implemented by the *generate* function. The issue is the need for an efficient mechanism to generate each perturbation, and this is substantially complicated because we use a mix of discrete and continuous variables. For discrete variables, the problem is simpler because there is always a smallest allowable move, an *atomic* perturbation. The two issues here are what larger moves should be included in the move-set for efficiency and how to decide when to use these larger moves. We address these issues by defining a set of *move classes* for each variable. Each class is a tuple (x_i, r) where x_i is the variable to be perturbed by this move class and r is a range $[r^{\min}, r^{\max}]$, which is related to the range of allowable values for x_i. For example, for a transistor width variable that has been discretized such that it can take on 100 possible values, we might create three move classes with ranges $[-1,1]$, $[-10,10]$, and $[-50,50]$. The idea is that during annealing, we will randomly select a move class. This determines not just x_i, the variable to perturb, but r the range with which to bound the perturbation. Once selected, we generate Δx_i as an integral uniform random number in $[r^{\min}, r^{\max}]$. Finally, Δx_i may be adjusted to ensure the new variable value, $x_i + \Delta x_i$, lies within the allowable values for x_i.

To improve annealing efficiency, we wish to optimize the usefulness of the move-set by favoring the selection of move classes that are most effective at this particular state of the annealing process. To do this, we use a method based on the work of Hustin [49]. We track how each move has contributed to the overall cost function, and compute a *quality factor* that quantifies the performance of that move class. For move class i, the quality factor Q_i is given by (7)

$$Q_i = (1/\|G_i\|) \sum_{j \in A_i} |\Delta C_j|. \tag{7}$$

Here, $\|G_i\|$ is the number of generated move attempts that used class i over a window of previous moves and A_i is the accepted subset of those moves (i.e., $A_i \subseteq G_i$). Furthermore, $|\Delta C_j|$ is the absolute value of the change in the cost function due to accepted move j. Q_i will be large when the moves of this class are accepted frequently ($\|A_i\|$ is large), and/or if they change the value of the overall cost function appreciably (some $|\Delta C_j|$ are large). We can then compute the probability that a particular move class should be selected. If Q is the sum over all the quality factors, $Q = \sum_i Q_i$

$$p_i = \frac{Q_i}{Q} \tag{8}$$

can then be regarded as the fraction of moves that should be dedicated to move class i. Initially, when almost any move is accepted, large moves will change the cost function the most, giving them the largest quality factors and likelihood of being selected. When the optimization is almost complete, most large moves will not be accepted, so more small moves will be accepted and their quality factors and probabilities will increase. Using this scheme, we automatically bias toward the moves that are most effective during a particular phase of the annealing run.

For continuous variables, the situation is more complex. For an n-dimensional real-valued state, $\underline{x} \in R^n$, it is difficult to determine the correct *smallest* $\Delta\underline{x}$ because adjustments in real values may be infinitesimally small. We may need to explore across a voltage range of several volts and then converge to a dc bias point with an accuracy of a few microvolts. We are aware of several attempts to generalize simulated annealing to problems with real-valued variables [50]–[52], and each presents methods of controlling the move-set. These methods show promise, but require complex time-consuming matrix manipulations, large amounts of memory, or gradient information to determine the appropriate move sizes. Fortunately, for analog circuit synthesis, we can take advantage of problem-specific information to aid in selecting the correct largest and smallest moves. For user-specified variables, such as the size of a resistor or capacitor, the precision with which the value can be set—and the atomic move size—is a function of the underlying fabrication process and readily available to the designer. For node voltages, the smallest moves are dictated by the desired accuracy of performance prediction. Duplicating the method used in detailed circuit simulators, these tolerances can be specified with an absolute and relative value and the smallest allowable move derived from them. Thus, we use problem specific information to determine minimum move sizes and create move classes, allowing these continuous variables to be treated in the same fashion as the discrete variables.

To further assist in manipulating continuous node voltages, in addition to the purely random *undirected* move classes found in most annealers, we augment the annealer's move-set with *directed* move classes that follow the dc gradient. This technique is similar to the theoretically-convergent continuous annealing strategy use by Gelfand and Mitter [52]. They use gradient-based moves within an annealing optimization framework, and they prove that this technique converges to a global minimum given certain restrictions on the problem. Our addition of dc gradient moves has the same flavor. We incorporate gradient-based voltage steps as part of our set of move classes. Using KCL-violating currents calculated at each node, Δi, and the factored small-signal admittance matrix, Y^{-1}, we can calculate the voltage steps to apply at each node, Δv, as

$$\Delta\underline{v} = \alpha(Y^{-1}\Delta\underline{i}) \tag{9}$$

where α is a scaling factor that bounds the range of the move. Thus, this move effects all the node voltage variables simultaneously. Equation (9) also forms the core of a Newton–Raphson iterative dc solution algorithm—the technique used in most circuit simulators to solve for the dc operating point—so we incorporate the complete algorithm as an additional move class in the move-set. Because of the complexity of performing Newton–Raphson within a simulated annealing algorithm, we have no theoretical proof of convergence. However, in practice, this technique allows the annealer to converge to a dc operating point at least as reliably as a detailed circuit simulator.

Finally, we also combine directed and undirected moves into a single move class to further augment the move-set. These combination moves are designed to alter the circuit component values, and thereby circuit performance, while maintaining a correct dc operating point. Without combination moves, when the optimization is nearly frozen, it is difficult for the annealer to adjust circuit component values—possibly improving circuit performance—without incurring a large penalty in the cost function as a result of dc operating point errors.

In summary, the complete move-set for the annealer comprises move classes of these types

- **Undirected moves**: where we modify a single variable (discrete or continuous) and generate Δx_i as a uniform random number in $[r^{\min}, r^{\max}]$;
- **Directed moves**: where the Newton–Raphson algorithm is used to perturb all the node voltage variables simultaneously;
- **Combination moves**: that perturb a user-specified variable in an undirected fashion and then immediately perform a Newton–Raphson solve.

E. Cost-Function

The heart of the formulation and the next problem specific component of the annealing algorithm is the circuit specific cost function $C(\underline{x})$ generated by the compiler, which maps each visited circuit configuration $\underline{x}$ to a scalar cost. The cost function was described by example in Section IV. In general, it has the form

$$C(\underline{x}) = \underbrace{C^{\text{obj}}(\underline{x})}_{\text{objective}} + \underbrace{C^{\text{perf}}(\underline{x}) + C^{\text{num}}(\underline{x}) + C^{\text{dev}}(\underline{x}) + C^{\text{dc}}(\underline{x})}_{\text{penaltyterms}} \qquad (10)$$

where each term in (10) represents a group of related terms in a particular compiler-generated cost function. There are two distinct kinds of terms: the *objective* terms correspond to $\underline{f}(\underline{x})$ in (2) and must be minimized, while the *penalty* terms correspond to $\underline{g}(\underline{x})$ and must be driven to zero. Here, C^{obj} and C^{perf} are generated from the user-supplied performance objectives and specifications; C^{num} penalizes regions of the solution space that may lead to numerically ill-conditioned circuit performance calculations; C^{dev} forces active devices to be in particular user-specified regions of operation; and C^{dc} implements the relaxed-dc formulation. In the following paragraphs, we describe the most interesting of these cost-function components in greater detail.

Objective Terms: Following [26], circuit performance and figures of merit such as power and area can be specified as falling into one of two categories: an objective or a constraint. Regardless of the category, the user is also expected to provide a *good value* and a *bad value* for each specification. These are used both to set constraint boundaries and to normalize the specification's range. Performance targets that are given as objectives become part of C^{obj}, which we define as follows. Formally, let

$$\Omega^{\text{obj}} = \{f_i(x) \mid 1 \le i \le k\} \qquad (11)$$

be the set of k performance objective functions provided by the user. Then Ω^{obj} is transformed into C^{obj} as follows:

$$C^{\text{obj}}(\underline{x}) = \frac{1}{\|\Omega^{\text{obj}}\|} \cdot \sum_{i \in \Omega^{\text{obj}}} \hat{f}_i(\underline{x}) \qquad (12)$$

where

$$\hat{f}_i(\underline{x}) = \frac{f_i(\underline{x}) - \text{good}_i}{\text{bad}_i - \text{good}_i} \qquad (13)$$

and $f_i(\underline{x})$ is the ith specification function provided by the user, and bad_i and $good_i$ are the *bad* and *good* normalization values specified for function i. The normalization process of (13) has these advantages. First, it provides a natural way for the designer to set the relative importance of competing specifications. Second, it provides a straightforward way to normalize the range of values that must be balanced in the cost function. Note that the range of interesting values between *good* and *bad* maps to the range [0, 1], and regardless of whether the goal is maximizing or minimizing $f_i(\underline{x})$, optimizing toward *good* will always correspond to minimizing the normalized function $\hat{f}_i(\underline{x})$. Finally, this normalization and the inclusion of the normalizing factor $(1/\|\Omega^{\text{obj}}\|)$ helps keep the cost function formulation robust over different problems, by averaging out the effect of having a large number of objectives for one circuit and a small number for another. The user has a wide range of predefined functions at his disposal to create $f_i(\underline{x})$. These include node voltages, device model parameters, and functions to derive linear performance characteristics such as gain and bandwidth from the transfer functions continually being derived by AWE.

Performance Specifications: The performance specifications provided by the user make up C^{perf}, and are quite similar to objectives with two exceptions. First, a specification is a hard constraint so circuit performance better than the *good* value does not contribute to C^{perf}. Second, an additional scalar weight biases the contribution of each performance specification to the overall cost function. These weights are determined algorithmically during the annealing process and should not need to be adjusted by the user. We defer detailed discussion of the weight control algorithm until Section V.H.

Operating Point Terms: The most unusual component of the cost function is the C^{dc} penalty term. This term implements the relaxed-dc formulation. Recall that we explicitly formulate Kirchhoff's current law constraints for each linearly independent node in all the bias circuits. At the end of the optimization, these C^{dc} terms will be zero and Kirchhoff's laws will be met. C^{dc} includes two views of Kirchhoff's current law (KCL), a *current view* C^{KCL}, and a *voltage* view C^{DV}. To show how these terms are calculated, (14) defines the sum of the currents leaving node n via its incident branches B^n, and (15) defines the average magnitude of currents flowing through the branches at this node

$$\Delta i_n = \sum_{b \in B^n} I_b \tag{14}$$

$$\text{mag}_n = \frac{1}{\|B^n\|} \cdot \sum_{b \in B^n} |I_b|. \tag{15}$$

The KCL term for node n is C_n^{KCL} which can then be calculated as follows:

$$\begin{aligned} \text{err}_n &= |\Delta i_n| - (\tau_{\text{rel}} \cdot \text{mag}_n + \tau_{\text{abs}}) \\ C_n^{\text{KCL}} &= \max(0, \text{err}_n). \end{aligned} \tag{16}$$

As in detailed circuit simulation, the tolerances τ_{abs} and τ_{rel} ensure that numerical truncation when adding the magnitude of a small current to the magnitude of a large one does not adversely effect the measure of KCL correctness.

The C^{DV} terms present an alternative view of KCL. They measure the voltage *change* required at each node to make the circuit satisfy KCL. By this definition, calculating the DV terms in C^{dc} exactly would require actually solving for the correct circuit node voltage values, which is what we are trying to avoid with the relaxed-dc formulation. Instead, we estimate the DV error terms by multiplying the factored small-signal nodal admittance matrix, Y^{-1}, by the error currents

$$\Delta \underline{v} = Y^{-1} \Delta \underline{i}. \tag{17}$$

Since the circuit will typically contain nonlinear devices, the Y matrix will be a linearized estimate of the admittance taken at the present bias point and the DV errors first-order approximations to the actual voltage error. As the annealing state freezes, the DV terms will be quite accurate, but during the early stages of the anneal, the KCL terms will be the only reliable estimates of dc correctness because they are not approximations. The addition of the DV terms takes the impedance into account and provides more useful feedback to designers, since they tend to have better intuition regarding voltage errors than current errors.

Calculation of the Cost Function: The terms in $C(\underline{x})$ are calculated from the present problem state following the flow in Fig. 7. Each time a variable in $\underline{x}$ changes, circuit component values may change, the encapsulated device models be reevaluated, AWE used to recompute transfer functions, and cost function terms recomputed. Fortunately, this analysis is extremely fast, allowing a complete reevaluation for each circuit visited. The cell-level analog circuits we target have less than 100 devices, and produce small-signal models with less than 1000 elements. The bulk of the work is performed by AWE, which reduces such a small-signal model to an accurate low-order transfer function in 10 to 100 ms on a fast workstation. Previous synthesis strategies were unable to determine both a dc operating point and performance characteristics automatically.

F. Annealing Control Mechanisms

The final customizable component of the annealing algorithm is the cooling schedule. We have implemented the general purpose cooling schedule of Lam [43] as modified by Swartz [53]. This specifies T_{HOT}, *done at temperature* and *update temperature*. Our freezing criteria, the *frozen* function, which determines when the annealing has completed, was developed specifically for our analog synthesis application. The design is complete when both the discrete variables have stopped changing and the changes in the continuous variables are within their specified relative tolerances.

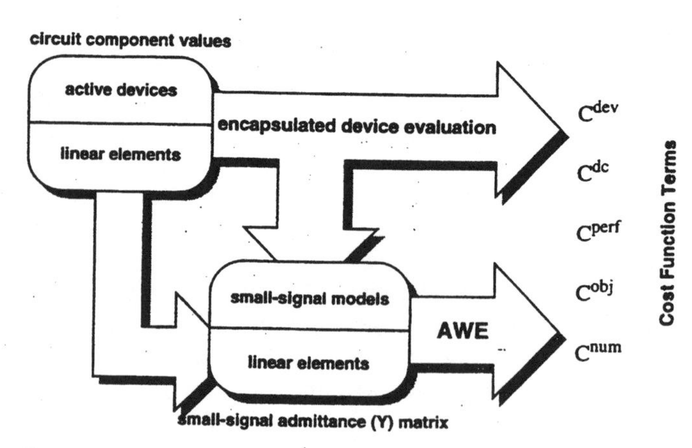

Fig. 7. Evaluation of the cost function.

G. Implications of the Relaxed-DC Formulation

Even after a complete description of our new analog circuit synthesis formulation, a few design decisions usually require further explanation. The first of these is the relaxed-dc formulation. As a result of the relaxed-dc formulation, early in the optimization process, the sum of the currents entering a given node has a significant nonzero value. An important issue to address is what it means to evaluate the performance of a circuit that is not dc-correct. One way to view this circuit is to imagine an additional current source at each node. This current source sinks the current required to ensure dc-correctness. Then, the goal of the Kirchhoff's law constraints C^{dc} is to reduce the value of these current sources to zero. When evaluating circuit performance, the fact that these current sources will not be in our final design means that our predicted performance will differ slightly from the final performance. This error factor allows us to visit many more possible circuit configurations within a given period of time, albeit evaluating each a little less accurately. As the optimization proceeds, the current sunk by these sources goes to zero and the performance prediction becomes completely accurate. This evolution is shown in Fig. 8. By the end of the annealing process, the circuit will be dc-correct within tolerances on the order of those used in circuit simulation.

A second analogy that can be used to understand how the relaxed-dc formulation behaves is to simply pretend a dc operating point solution is performed at each iteration. However, the tolerance for dc correctness is very relaxed early in the optimization process and evolves toward that of typical detailed simulations as the optimization proceeds. All simulators have numerical tolerances on dc correctness, and as a result, there is always numerical error in circuit simulation.

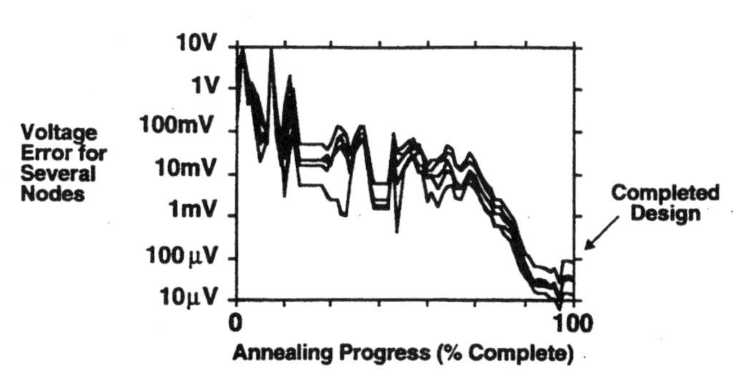

Fig. 8. Discrepancy from KCL correct voltages during optimization.

The relaxed-dc formulation simply trades increased error for increased speed of evaluation early in the optimization process.

These analogies point out a seeming inconsistency in our overall synthesis formulation. We clearly do not intend that each annealing move visits a dc-correct circuit, i.e., where the current sources ensuring dc-correctness are zero-valued. Nevertheless, we evaluate each circuit using detailed device models and highly accurate AWE techniques. Clearly this accuracy is not fully exploited early in the optimization when these Kirchhoff current law errors are substantial. Simpler models could be used in these early stages, but, practically, knowing when and how to switch from simple to accurate models would be difficult and it is unlikely significant changes in run-time would be achieved. Moreover, this accuracy is not entirely wasted because the annealer still learns much from these early circuits. For example, if we need to achieve more gain in our circuit, we probably need to increase the g_m on some critical device. We do not need a precise dc bias point to know that we must either increase that device's width, its current, or both. The optimizer can successfully make coarse decisions such as this even in the presence of the slight noise introduced by the relaxed-dc formulation.

H. Reliability Without User-Supplied Constants

The final aspect of our formulation that requires further explanation is the issue of numeric constants. Numeric algorithms in general require constants that tune the algorithm such that it reliably produces high quality solutions, and our annealer is no exception. If a numeric algorithm and the design automation tool of which it is a part are poorly designed, the constants will need to be adjusted for each new problem solved. As a result, the tool will not do a very good job of automation, since a user of the tool will spend much of his time adjusting constants, rather than designing circuits. Coupled with this automation problem is a robustness problem inherent in the choice of simulated annealing as the core of our optimization engine. Simulated annealing is a stochastic process, and each time the annealer is run it may find a slightly different trade-off between its constraints. As a result, it is not possible to guarantee that a single run will provide the best answer. However, it is important that the tool is still robust, i.e., we wish to be confident that running the annealer several (5–10) times will provide several high-quality solutions from which to choose.

Substantial effort has been spent designing our formulation and its implementation such that it is truly a robust automation tool, i.e., the user is not required to provide problem-specific constants and the tool produces a high percentage of top quality solutions. One key aspect of this algorithmic design is the use of adaptive algorithms to replace the majority of the numeric constants that would otherwise be needed within the annealer. For example the penalty terms in $C(\underline{x})$ require scalar weights to balance their contributions to the cost function. Similar to the strategies used in [54], these weights have been replaced with an adaptive algorithm. The adaptive algorithm in the annealer is based on the observation that all the penalty terms, $\underline{g}(\underline{x})$, must be zero at the end of a successful annealing run. Thus, we can use a simple feedback mechanism to force each penalty term to follow a trajectory that leads from its initial value to zero; i.e., if a penalty term presently exceeds its expected value as given by its trajectory, the weight is increased and the natural annealing action focuses on reducing that penalty term in subsequent annealing steps. With this adaptive algorithm, the problem is then one of determining what the best trajectories are and setting their values. The advantage of this technique is that, as we have shown over the circuit synthesis problems we have solved with our implementation, a single set of trajectories is much more robust than a single set of individual weights. As a result, an analog circuit designer can use our analog circuit synthesis system without understanding the internal architectural details needed to adjust components of its cost function.

To set the trajectories, we treated them as independent variables in a large optimization problem solved using Powell's algorithm [55]. We refer to the large problem as "optimizing the optimizer" because the annealer was executed as the inner loop of the Powell optimization process. Each time a trajectory was perturbed, the annealer was run and the quality of the resulting circuit solution fed back to the Powell optimizer. However, because of the stochastic nature of annealing, it is insufficient to characterize a set of trajectories on a single annealing run. Instead, using a large network of workstations, the annealer was run 200 times on the same synthesis problem and a statistical analysis performed to determine the quality of a particular set of trajectories. Because of the large amount of CPU time involved, we optimized the trajectories for a small circuit problem and validated them over our benchmark suite. Verifying that this single set of trajectories provided good results across all our circuits—and will likely do so for new circuits—was essential because optimizing the optimizer consumed approximately four years of CPU time. Comparing our initial "best guess" trajectories to those that resulted from optimizing the optimizer, the number of top quality solutions for a typical synthesis problem increased from about 30% to about 80%. As a result, the careful design and tuning of dynamic algorithms within the annealer have freed the user from providing algorithmic constants and greatly improved the overall robustness of the tool. See [33] for more details on the design and optimization of these trajectories.

VI. Comparison to Related Work

In this section, we compare and contrast our new analog circuit synthesis formulation to recent related work. Our formulation is unique among analog synthesis systems in several key ways. Architecturally, it is unique because of its two-step synthesis process. It includes a complete compiler that translates the problem from a compact user-oriented description into a cost function suitable for subsequent optimization. Compilation provides significant flexibility. It provides the opportunity for the user to interact with the tool in a language that is familiar to designers, yet, because the compiler produces an executable designed specifically to solve the user's problem, it also provides optimal run-time efficiency. Our formulation is also unique in its use of AWE for analog circuit synthesis. Although recently other researchers have used AWE for the synthesis of power distribution networks [56], to the best of our knowledge ours was the first tool to employ AWE for performance evaluation within an optimization loop.

Of course, our paper uses and builds upon the successes and failures of previous approaches to analog circuit synthesis. As discussed in Section II, previous systems are equation-based and thus suffer from problems with accuracy and automation. Because it is a simulation-based approach, our paper is a substantial departure from these previous approaches. However, the encapsulated device models and relaxed-dc formulation are based on similar ideas used by Maulik [16]. The key distinction is that Maulik still relies on hand-crafted equations to compute circuit performance from the parameters returned by the device models, and these equations must be hand-coded into the system for each new circuit topology.

Our formulation also borrows ideas from—but differs from—simulation-based optimization systems. As discussed in Section II, these tools are not true synthesis tools because they are starting point dependent. Our formulation avoids this problem by using simulated annealing, which is *not* starting point dependent, as the optimization algorithm and avoids the resulting efficiency problem by using AWE.

There is one other simulation-based synthesis tool of which we are aware. The tool described in [36] appeared shortly after the initial publication of our paper [32], [34]. It uses several of the same strategies: simulation is used to evaluate circuit performance and a form of simulated annealing is used for optimization. However, it relies on SPICE for the simulation algorithm which is typically 2–3 orders of magnitude slower than AWE for ac analysis [39]. To achieve reasonable run-times, it does not use transient analysis within SPICE (which is even slower than ac analysis) and relies on several heuristics for small-signal specifications. The first heuristic is to substantially reduce the number of iterations that would normally be required for optimization with simulated annealing. Results generated with this heuristic adaptation seem reasonable, but comparisons to difficult manual designs have not been presented so the efficacy of this heuristic is unclear. The second heuristic is to save the dc operating point from the previous SPICE run and use it as the starting point for the dc analysis of the next SPICE run. This is the first step toward a relaxed-dc formulation, where the dc operating point is maintained as part of the overall design state. They do not, however, take the final step toward relaxed-dc by loosening the dc requirements early in the design process. As a result, they cannot afford the CPU time to explore as large a region of the design space as is possible with our system.

VII. Results

In this section, we first describe our implementation of our new analog circuit synthesis formulation. We then present circuit synthesis results and compare them with those of previous approaches.

A. Implementation

To show the viability of our new formulation for analog circuit synthesis, we have implemented the ideas described in this paper as a pair of synthesis tools called ASTRX and OBLX. ASTRX is the circuit compiler that translates the user's circuit design problem into C code that implements $C(\underline{x})$. ASTRX then invokes the C compiler and the linker to link this cost function code with the OBLX solution library, which contains the annealer, AWE, and the encapsulated device library. Invoking the resulting executable completes the actual design process. Translation of the user's problem into C code requires only a few seconds, so run-times are completely dominated by optimization time. ASTRX and OBLX are themselves implemented as approximately 125 000 lines of C code.

The syntax used to describe the synthesis problem is the SPICE format familiar to analog designers, with the addition of a few cards to describe ASTRX/OBLX specific information, such as specifications. For our example of a simple differential amplifier design (Section IV), the complete input file (excluding process model parameters) is shown in Fig. 9. A complete description of the format can be found in [57].

A. Circuit Benchmark Suite

The primary goal of ASTRX/OBLX is to reduce the time it takes to size and bias a new circuit topology to meet performance specifications. Fig. 10 summarizes a representative selection of previous analog circuit synthesis results.[2] Here, each symbol represents synthesis results for a single circuit topology. The length of the symbol's "tail" represents the complexity of the circuit, which, as described in Section I, we quantify as the sum of the number of devices in the circuit and the number of variables for which the user asks the synthesis tool to determine values. The other axes represent metrics for accuracy and automation. The prediction error axis measures accuracy by plotting the worst case discrepancy between the synthesis tool's circuit performance predication and the predictions of a circuit simulator. The time axis measures automation by plotting the sum of the preparatory time spent by the designer and CPU time spent by the tool to synthesize a circuit for the first time. Fig. 10 reveals three distinct classes of synthesis results. The first class is on the right and contains the majority of previously published papers. Here, the synthesis tool predicts performance with

[2]Not all prior approches could be included in Fig. 10 because of the unavailability of the necessary data. In [14] where preparatory time was not published, we equated 1000 lines of circuit-specitic custom code to a month of effort.

```
Design Example                                                              1
                                                                            2
.param cl = 1pF                                                             3
.param vddval=2.5                                                           4
.param vssval=-2.5                                                          5
                                                                            6
.subckt oa      inpos   inneg   outpos outneg nvdd    nvss                  7
M2      outpos inneg   A       A      Ne      w='W'   l='L'                 8
+       constraint1 = 'vgs - von > 0.2' constraint2 = 'vds - vdsat > 0.05'  9
M1      outneg inpos   A       A      Ne      w='W'   l='L'                 10
+       constraint1 = 'vgs - von > 0.2' constraint2 = 'vds - vdsat > 0.05'  11
M3      outpos nvb     nvdd    nvdd   Pe      w=2u    l=1.2u                12
M4      outneg nvb     nvdd    nvdd   Pe      w=2u    l=1.2u                13
vvb     nvb     nvss    'Vb'                                                14
ib      A       nvss    'I'                                                 15
.ends                                                                       16
                                                                            17
.VAR W RANGE=(1.8u,500u) GRID=1                                             18
.VAR L RANGE=(1.2u,100u) GRID=1                                             19
.VAR I RANGE=(1uA,1mA)   RES=0.001                                          20
.VAR Vb RANGE=(0,5)      RES=0.001                                          21
                                                                            22
.SYNTH example                                                              23
                                                                            24
.OblxCkt self_bias bias                                                     25
xamp    inpos   inneg   outpos outneg nvdd    nvss    oa                    26
vdd     nvdd    0       vddval                                              27
vss     nvss    0       vssval                                              28
rf1     inneg   outpos 1e6                                                  29
rf2     inpos   outneg 1e6                                                  30
.spec SR value 'I/(2*(cl+xamp.m1.cdd+xamp.m3.cdd))' good = 1Meg bad = 10k   31
.endOblxCkt                                                                 32
                                                                            33
.OblxCkt openloop awe                                                       34
.bias self_bias                                                             35
xamp    inpos   inneg   outpos outneg nvdd    nvss    oa                    36
vdd     nvdd    0       vddval                                              37
vss     nvss    0       vssval                                              38
vin     inpos   0       0      ac      1                                    39
ein     inneg   0       0      inpos   1                                    40
cl1     outpos 0        Cl                                                  41
cl2     outneg 0        Cl                                                  42
.pz tf V(outpos) vin                                                        43
.spec   ugf     value 'unity_gain_freq(tf)'  good = 1Meg   bad = 10k        44
.obj    Adm     value 'dc_gain(tf)'          good = 1000   bad = 10         45
.endOblxCkt                                                                 46
                                                                            47
.END                                                                        48
```

Fig. 9. Design example: ASTRX input file. (Process parameters are not included.)

reasonable accuracy, but only because a designer has spent months to years of preparatory time deriving the circuit performance equations. The second group of results, those on the left, trade reduced preparatory effort for substantially reduced circuit performance prediction accuracy. Finally, the center group of results is for ASTRX/OBLX. In contrast to the other two groups, generating each new design with ASTRX/OBLX typically involved an afternoon of preparation followed by 5–10 annealing runs performed overnight, yet produced designs that matched simulation with at least as much accuracy as the best prior approaches.

For all eight circuit topologies discussed in this paper, Table II lists information about the input file used to describe the problem to ASTRX and the resulting cost function and C code generated. Note that five of these topologies (Simple OTA, OTA, Two-Stage, Folded-Cascode, and Comparator) form our benchmark suite, and cover essentially all[3] synthesis results published at the time of this writing. The limited range and performance of these prior results is perhaps the best indicator that the first-time effort to design a circuit has always been a substantial barrier to obtaining a broader range of results. The first two rows of the table give the number of lines required for each synthesis problem description. The number of lines is reported as two separate values: 1) the lines required for the netlists of the circuit under design and the test jigs (about the number of lines that would be required to simulate the circuit with SPICE), and 2) the lines for the independent variables and performance specifications (lines specific to ASTRX/OBLX). Note that this is a modest amount of input, most of which is the netlist information required for any simulation deck. The synthesis-specific information is predominantly a list of the variables the user wishes ASTRX/OBLX to determine and the performance specifications. For small circuits, creating the input to ASTRX usually takes a few hours (compared to the weeks/months required to add a new circuit to other synthesis tools). For

[3] We make the reasonable assumption that a circuit topology can represent topologies that vary in only minor detail.

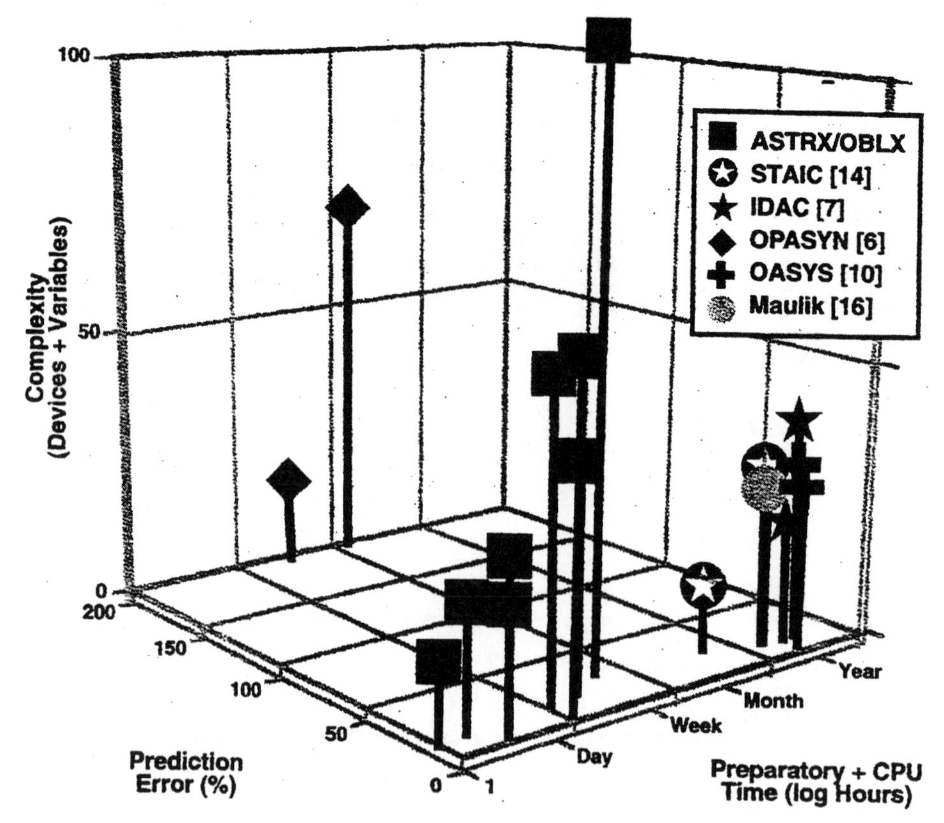

Fig. 10. Complexity, error, and first time design effort. ASTRX/OBLX compared with prior approaches.

TABLE II
RESULT OF ASTRX'S ANALYSES

		Circuit							
		Simple OTA	OTA	Two-Stage	Folded Cascode	Comparator	BICMOS Two-Stage	Novel Folded Cascode	2x Gain Stage
Input (lines)	Netlist/Models	30	34	43	65	131	39	68	189
	Synth. Specific	28	33	40	56	68	33	51	65
$\mathbf{x}$	User-Supplied	7	11	19	28	19	12	27	39
	Node Voltages	14	24	26	70	57	26	84	215
$C(\mathbf{x})$	Terms	56	85	88	212	169	86	246	483
	Lines of C	1443	1809	1894	3408	3088	1723	3960	7173
Circuits	Circuit Type[a]: nodes, elements	B: 20, 31 A: 20, 67	B: 28, 49 A: 29, 114	B: 34, 54 A: 33, 118	B: 75, 138 A: 75, 324	B: 65, 126 A: 63, 265 A: 64, 266 A: 29, 115	B: 33, 54 A: 32, 105	B: 90, 167 A: 90, 395	B: 219, 450 A: 152, 693 A: 152, 693 A: 90, 395

a. Type 'A' is a linearized, small-signal AWE circuit. Type 'B' is a bias circuit.

the 2× gain stage, creating and checking the topology, and specifications took two to three days. The third and fourth rows of Table II show the number of independent variables that must be manipulated by OBLX during optimization. Recall that since the relaxed-dc formulation requires Kirchhoff's laws to be explicitly constrained, OBLX must also manipulate most of the circuit node voltages. As shown in the fourth line of Table II, as a result of internal nodes in the device models we employ, these added variables typically outnumber the user-specified variables. This dimensional explosion is substantially alleviated by the inclusion of Newton–Raphson moves as one of the circuit perturbations OBLX can use during simulated annealing (see Section V.D.). The fifth and sixth rows of the table show the other results of ASTRX's analysis of the input description: the number of terms in the cost function OBLX will optimize and the number of lines of C code automatically generated by ASTRX for this synthesis problem. Recall that ASTRX compiles the problem description, generating the cost function as C code then compiling and linking it with OBLX. Finally, Table II shows the size of the linearized small-signal test jig circuit(s) generated by ASTRX to be evaluated by AWE for each new circuit configuration and the size of the bias circuits generated by ASTRX using the large-signal models for the nonlinear devices.

TABLE III
BASIC SYNTHESIS RESULTS, BSIM AND GUMMEL-POON MODELS, 1.2 μm PROCESS

	Specification: OBLX / Simulation				
Attribute	**Simple OTA**	**OTA**	**Two-Stage**	**Folded Cascode**	**BiCMOS Two-Stage**
Cload (pF)	1	1	1	1.25	1
Vdd	5	5	5	5	5
dc gain (dB)	↑[a]: 36.6 / 36.6	↑: 40.4 / 40.2	≥60: 66.4 / 66.4	≥70: 70.1 / 70.1	↑: 99.1 / 99.1
gain bandwidth (MHz)	≥50: 50.1 / 50.6	≥25: 25.0 / 25.4	≥10: 10.6 / 10.6	↑: 72.4 / 72.1	≥50: 73.7 / 75.1
phase margin (°)	≥60: 71.4 / 74.8	≥45: 57.9 / 57.8	≥45: 87.3 / 86.5	≥60: 80.0 / 80.0	≥45: 45.2 / 49.6
PSRR (Vss)	≥20: 21.9 / 21.9	≥40: 42.1 / 42.0	≥20: 31.0 / 30.9	≥105: 107 / 107	≥60: 78.9 / 79.0
PSRR (Vdd)	≥20: 36.8 / 36.8	≥40: 52.8 / 52.8	≥40: 45.8 / 45.8	≥105: 125 / 125	≥40: 52.2 / 52.2
output swing (V)	≥2.3: 3.7 / 3.6	≥2.5: 4.0 / 4.0	≥2: 2.7 / 2.8	≥1.0: ±1.5 / ±1.5	≥2: 3.3 / 4.0
slew rate (V/μs)	≥10: 130 / 131	≥10: 51.6 / 48.2	≥2: 3.8 / 4.0	≥50: 67 / 57	≥10: 10 / 9.5
active area ($10^3\mu^2$)	↓: 2.8	↓: 0.9	↓: 2.1	↓: 46	↓: 11.9
static power (mW)	≤1: 0.72 / 0.72	≤1: 0.33 / 0.34	≤1: 0.16 / 0.16	≤15: 10 / 10	≤20: 1.3 / 1.5
time/ckt. eval (ms)	36	37	38	116	38
CPU time (min./run)	6	9	16	120	12

a. ↑ means maximize, while ↓ means minimize.

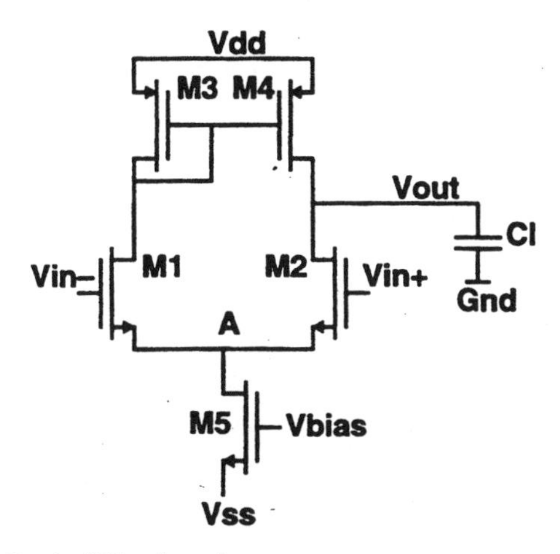

Fig. 11. Simple OTA schematic.

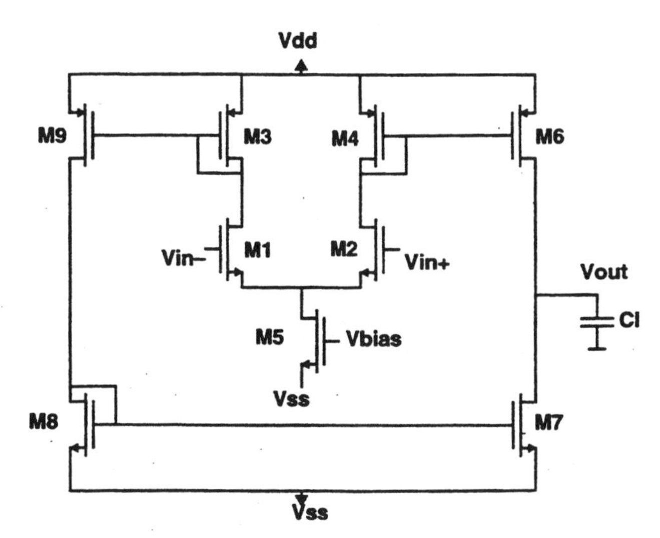

Fig. 12. OTA schematic.

The schematics for all our circuits are shown in Figs. 11–18 and the basic synthesis results for the benchmark suite are in Table III. CPU times given are on an IBM RS/6000-550 (about 60 MIPS). It is impossible to compare directly circuit performance of these ASTRX/OBLX synthesized circuits with that of circuits synthesized with other tools because of the unavailability of device model parameters to describe the processes for which they were designed. As a result, we compare the *accuracy* of ASTRX/OBLX to that of previous approaches. Because of the simplifications made during circuit analysis, results from equation-based synthesis tools differ from simulation by as much as 200% (see Fig. 10). In contrast, for the small-signal specifications where AWE predicts performance, ASTRX/OBLX results match simulation almost exactly. By simulating linearized versions of these circuits, we have determined that these minor differences are due to differences between the models used during simulation with HSPICE [58] and our models adopted from SPICE 3 [24]. For nonlinear performance specifications, such as slew rate, we used first-order equations written by inspection. This was done to validate our assertion that most large-signal specifications can be readily, yet realistically, handled despite the lack of transient simulation within the present implementation of ASTRX/OBLX. The accuracy of these is better than similar equations in completely equation-based techniques because circuit parameters used to write the equations, such as branch currents and device parasitics, are updated automatically by OBLX as the circuit evolves.

Presently ASTRX/OBLX employs three different encapsulated device models: the level 3 MOS model from SPICE [24], the BSIM MOS model [23] and the Gummel–Poon model for BJT devices [59]. Because the encapsulated model interface in OBLX was designed to be compatible with SPICE, adopting these models from SPICE source code required little more than the addition of topological information. To demonstrate the importance of supporting different models, we synthesized the same circuit (Simple OTA) with three different model/process combinations: BSIM/2μ, BSIM/1.2μ, and MOS3/1.2μ. AS-

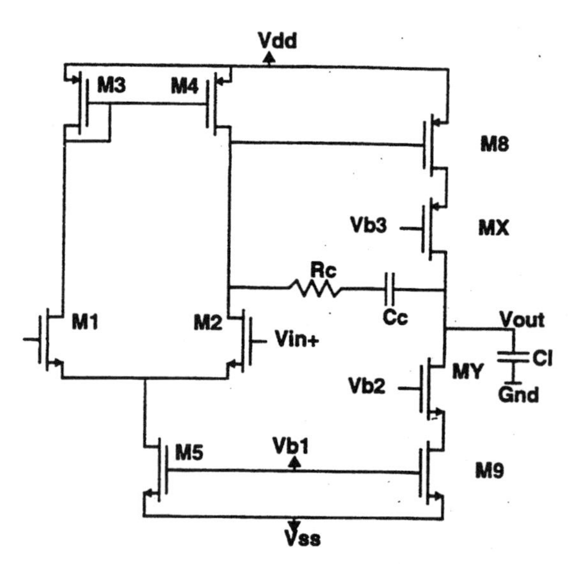

Fig. 13. Two-stage schematic.

TABLE V
COMPARATOR SYNTHESIS RESULTS

Key Attribute	Specification: OBLX/Simulation	
Vdd		5V
stage 1 settling, 0.1%, (ns)	≤5:	4.5 / 4.7
stage 2 settling, 0.1%, (ns)	≤5:	4.6 / 4.8
stage 1 slew rate (V/μs)	≥800:	800 / 750
stage 2 slew rate (V/μs)	≥800:	800 / 620
latch 1, positive pole (MHz)	≥200:	430 / 450
latch 2, positive pole (MHz)	≥200:	360 / 330
error rate, (1 in years)	≥10:	8E+24
active area ($10^3\mu^2$)	↓[a]:	1.6
static power (mW)	↓:	2.7 / 2.7
time/ckt eval (ms)		136
CPU time (min. / run)		97

a. ↓ means minimize.

TABLE IV
COMPARISON WITH MANUAL DESIGN FOR CIRCUIT NOVEL FOLDED CASCODE

Attribute	Manual Design	Automatic Re-Synthesis Spec:OBLX / Sim	
Cload (pF)	1		1
Vdd (V)	5		5
dc gain (dB)	71.2	≥71.2:	82 / 82
gain bandwidth (MHz)	47.8	↑[a]:	89 / 89
phase margin (°)	77.4	≥60:	91 / 91
PSRR (Vss)	92.6	≥93:	112 / 112
PSRR (Vdd)	72.3	≥73:	77 / 77
output swing (V)	± 1.4	≥±1.4:	1.4 / 1.3
slew rate (V/μs)	76.8	≥76:	92 / 87
active area ($10^3\mu^2$)	68.7	↓:	56
static power (mW)	9.0	≤25.0:	12 / 12
time/ckt. eval (ms)			83
CPU (min. / run)			116

a. ↑ means maximize, while ↓ means minimize.

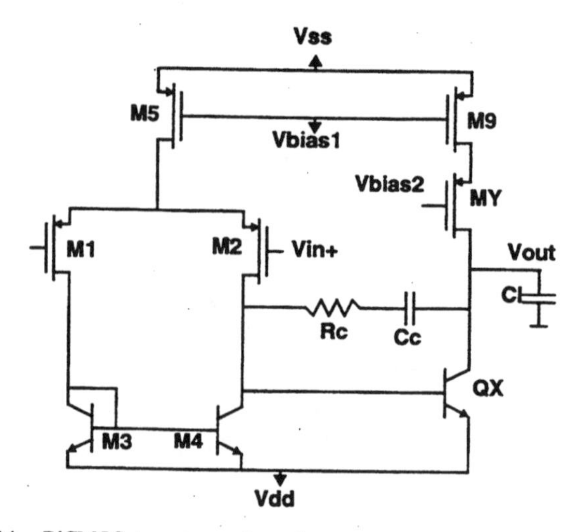

Fig. 14. BiCMOS two-stage schematic.

TRX/OBLX was given (and achieved) the same specifications for each, but was told to minimize active area. The resulting areas are shown in Fig. 20. As expected, the BSIM/2μ design required the largest area (580μ^2). But, surprisingly, the two designs for the *same* 1.2μ process also differed substantially in area: 300μ^2 for BSIM and 140μ^2 for MOS3. These models differ in their performance predictions, even though they are both intended to model the same underlying fabrication process. Clearly the choice of device model greatly effects circuit performance prediction accuracy. As a final experiment to show the utility of encapsulated device models, we designed a BiCMOS two-stage amplifier, which shows the ability of ASTRX/OBLX to handle a mix of MOS and bipolar devices. Synthesis and simulation results appear as the last column of Table III. Here again, performance predictions match detailed circuit simulations, confirming the importance of encapsulated devices for both accuracy and generality.

B. Comparison to Manual

Our next example shows the ability of ASTRX/OBLX to design difficult circuits and achieve results similar to those obtained by manual design. This circuit, a novel folded cascode fully differential opamp, shown in Fig. 18, is a new high-performance design recently published in [60] and as such is a significant test for any synthesis tool because the performance equations cannot be looked up in a textbook. Moreover, the performance of the circuit is difficult to express analytically, and as many as six poles and zeros may nontrivially effect the frequency response near the unity gain point. Table IV is a comparison of a redesign of this circuit using ASTRX/OBLX with the highly optimized manual design for the *same* 2μ process. Surprisingly, ASTRX/OBLX finds a design with higher nominal bandwidth at the cost of less area. Although we are pleased with the ability of OBLX to find this corner of the design space, this does not mean that ASTRX/OBLX outperformed the manual designer. In fact, the manual designer was willing to trade *nominal* performance for better estimated yield and performance over varying operating conditions.

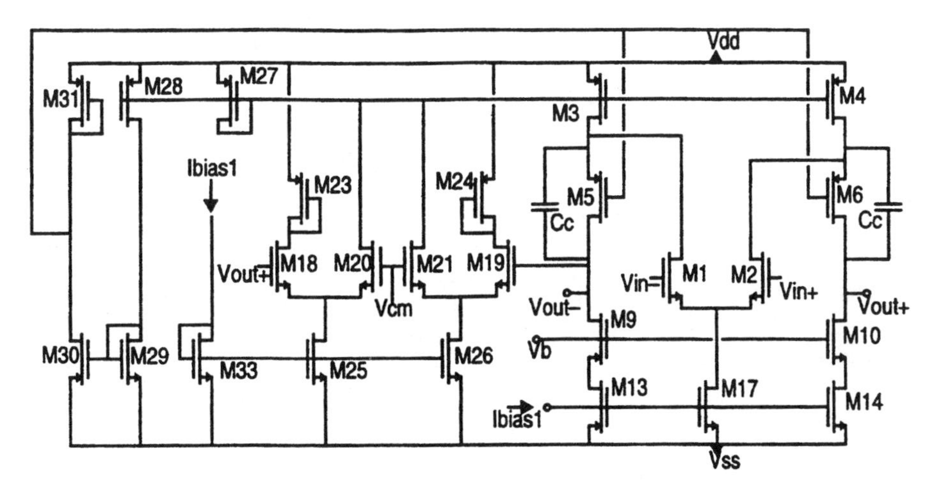

Fig. 15. Folded cascode schematic.

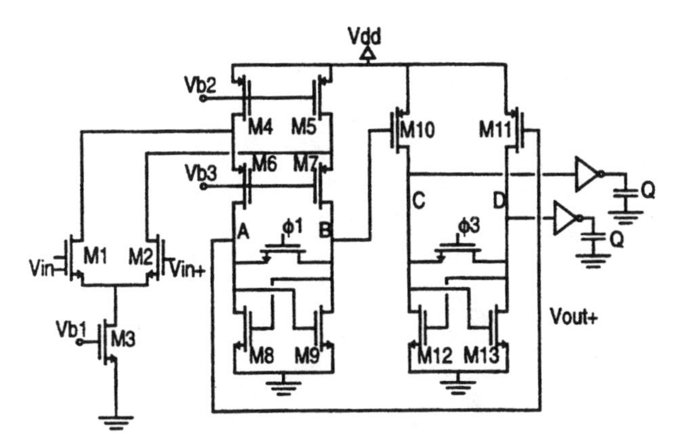

Fig. 16. Comparator schematic.

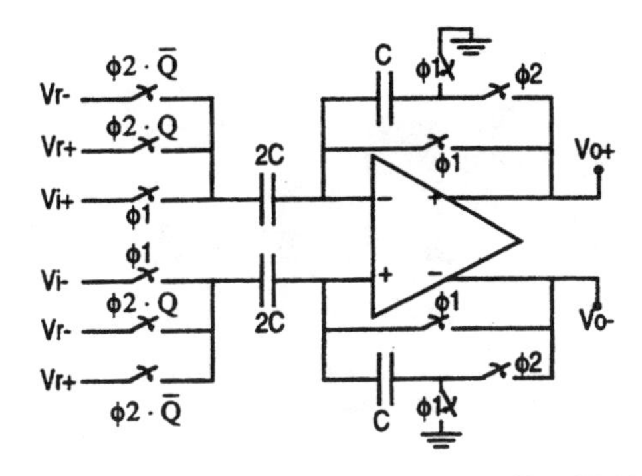

Fig. 17. $2\times$ gain stage schematic. The amplifier is the Novel Folded Cascode (Fig. 18), and the switches are MOS pairs.

Adding this ability to ASTRX/OBLX is the subject of ongoing research and preliminary results are reported in [37].

C. Pushing the Limits of Complexity and Generality

Our final result shows the utility of ASTRX/OBLX when confronted with the much more complex task of designing a pipelined A/D converter. This is an important test because it addresses the issues of generality and complexity. The pipelined A/D topology converter we selected, Fig. 19, employs two cells: a comparator (Fig. 16) and a $2\times$ gain stage (Fig. 17). This is a test of generality because these cells display important nonlinear behavioral characteristics whose modeling in ASTRX/OBLX is not as straightforward as the performance characteristics of amplifiers and other linear cells. This is a test of complexity because, using the metric of devices plus designable parameters, these cells are 2–$3\times$ more complex than those published previously. To aid noise rejection in a mixed-signal environment, the A/D converter uses a fully differential structure. The $2\times$ gain stage is implemented as a switched-capacitor circuit, and the input switches provide a convenient method to perform the multiplexing and subtraction needed to complete the design. The $2\times$ gain stage also employs an operational amplifier (Fig. 18) [60]. It is important to note that ASTRX/OBLX optimizes the *entire* gain cell as a *single* circuit. This allows the optimizer to explore crucial tradeoffs between the sizes of the switches, capacitors and amplifier devices. This also yields a very large optimization problem, which contributes to the lack of published synthesis results of this complexity. However, the ability to handle problems of this magnitude is fundamental for industrial design situations where the designer naturally works with cells of this complexity. Other characteristics of the two circuit design problems are shown in Table II.

Sample ASTRX/OBLX synthesis results for a comparator appear in Table V and results for a sample $2\times$ gain stage appear in Table VI. These results are for a 1.2 μm MOS process

TABLE VI
SWIT-CAP $2\times$ GAIN STAGE SYNTHESIS RESULTS

Key Attribute	Specification:	OBLX/Simulation
Vdd		5V
settling at output (ns)	$\leq$100:	93 / 98
settling at amp inputs (ns)	$\leq$100:	42 / 61
input range (V)	$\geq\pm$0.5:	1.6 / 1.6
output range (V)	$\geq\pm$0.5:	0.8 / 0.7
common mode gain (dB)	$\leq$-10:	-33 / -33
active area ($10^3\mu^2$)	$\downarrow$[a]:	58
static power (mW)	$\downarrow$:	21 / 21
time/ckt eval (ms)		158
CPU time (hrs. / run)		11.8

a. $\downarrow$ means minimize.

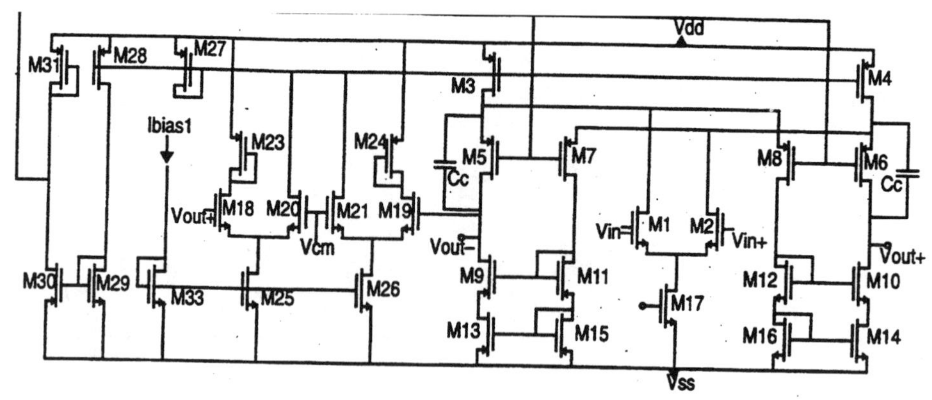

Fig. 18. Novel folded cascode schematic [60].

and use a BSIM model for the devices. Reported run-times are again on an IBM RS/6000-550 (about 60 MIPS), and simulation results were obtained with HSPICE. As before, there is a close correspondence between OBLX prediction and simulation.

One key benefit of circuit synthesis is the ability to explore the design space for a given circuit and process, quantifying the interactions between competing circuit performance constraints. Fig. 21 shows the results of several ASTRX/OBLX runs for the comparator of Fig. 16. Each point on the graph is a different complete circuit design, obtained by increasing the clock frequency specification and asking ASTRX/OBLX to minimize static power. The graph shows the expected increase in static power consumption as a function of the circuit's maximum clock frequency.

Finally, Fig. 22 shows the simulated input/output response at 3 MHz for a single stage of our completed A/D converter formed by connecting the two synthesized cells. The input–output response follows the saw-tooth pattern expected of an A/D converter stage. The 3 MHz performance is a few years behind the state of the art, but we consider this perhaps modest performance to be quite acceptable for the first fully automatic design of cells of this complexity.

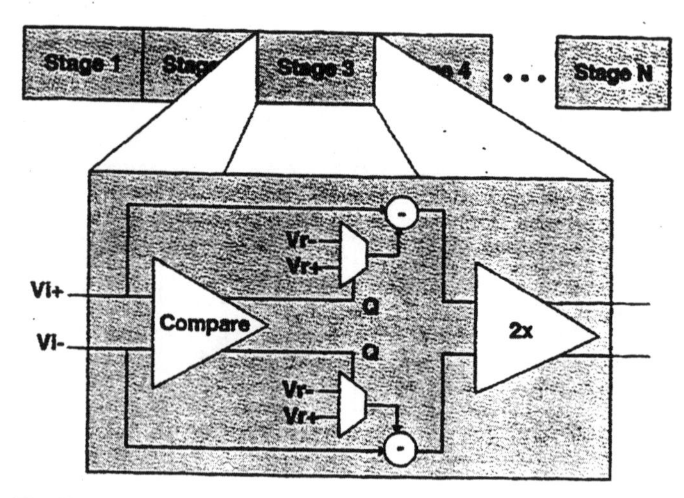

Fig. 19. Pipelined A/D converter topology.

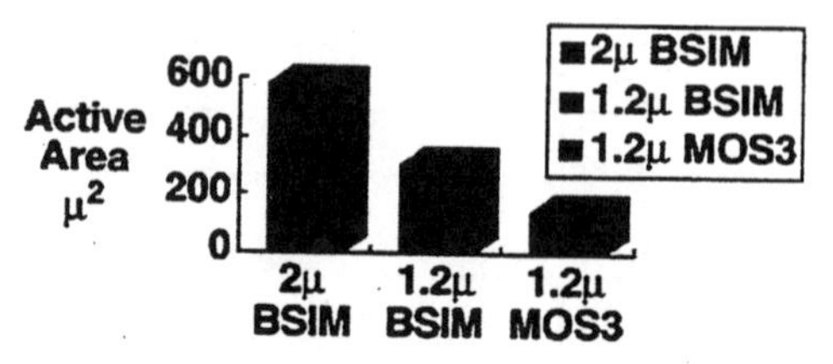

Fig. 20. Active area for the circuit simple OTA synthesized for three different process/model combinations.

VIII. Conclusion and Future Work

Our experience with ASTRX/OBLX as it has evolved over the past few years has given us considerable insight into the practical aspects of the system's use and the strengths and weaknesses of both its underlying ideas and their implementation. In practice, since the input format was designed to be very familiar to users of SPICE, analog designers need little assistance before they can begin experimenting with ASTRX/OBLX, and several colleagues at Carnegie Mellon have successfully used the tool. However, obtaining usable circuits generally requires some experience and patience. The typical failure mode is that the synthesis problem is underconstrained, and the optimizer finds a circuit that meets all the specifications given and yet is not usable. For example, if no output swing specification is given, OBLX will exploit this omission and design an amplifier that meets all the given specifications, but whose output devices will be pushed out of saturation by almost any change in the input. Correctly specifying the circuit synthesis task is usually overcome with a few iterations through the synthesize/verify cycle. This process of tuning specifications could likely be enhanced by some form of debugger that could point toward the cause of input specification problems, but as with conventional programming languages, creation of sophisticated development tools will follow only when the language is well established.

There are several ways that ASTRX/OBLX itself could be extended, and some of these are the focus of ongoing research. Three general areas of improvement are the speed and scope of circuit performance evaluation, automatic topology selection, and hierarchical design. Faster and more powerful circuit simulation is itself an open research area. Fast evaluation with AWE makes ASTRX/OBLX practical, yet it limits the information a

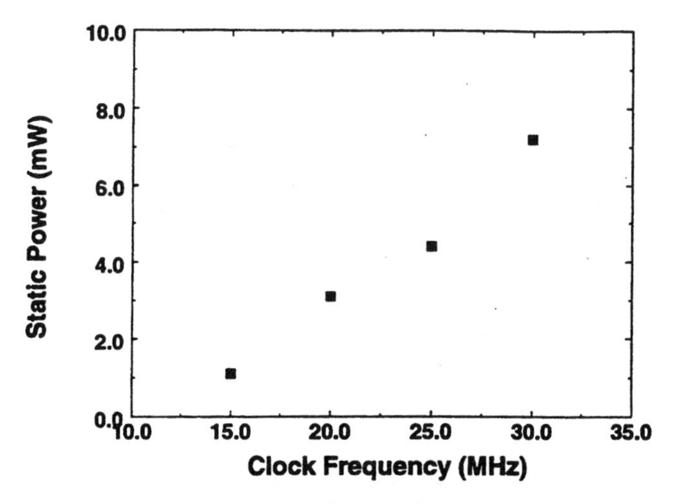

Fig. 21. Clock frequency versus static power for the comparator of Fig. 16.

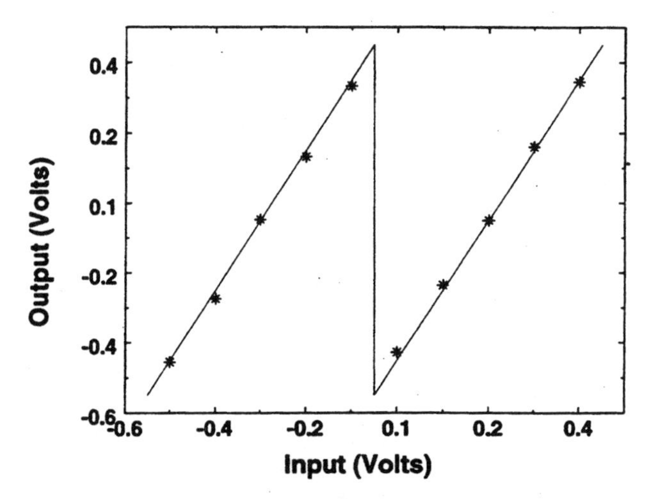

Fig. 22. Input/output response of pipeline stage.

designer can obtain about the circuit under design. The AWE techniques we employ do not work for transient specifications, and conventional transient simulation is impractical because it is many orders of magnitude too slow. This forces the designer to write some equations. Although first-order equations here are not difficult to derive, if fast transient simulation were possible, it would be a better solution. In the near term, an area in which the scope of ASTRX/OBLX circuit evaluation can be improved is manufacturablility. Since manufacturablility concerns are a critical aspect of any industrial circuit design, adding manufacturablility evaluation to ASTRX/OBLX is a critical aspect of creating a synthesis tool usable in an industrial setting. Ongoing work in this area is described in [37]. Unfortunately, adding more power and flexibility to circuit evaluation within ASTRX/OBLX exacts a penalty in run time, making synthesis of large circuits less practical. In part, the continuing trend of ever-increasing CPU speeds will alleviate this problem and make more sophisticated circuit evaluation techniques more practical in a desktop environment in the next few years.

Determining the topology—the interconnection of transistors and other circuit devices—is the other component of analog circuit design, and automating this task is another area of ongoing work with ASTRX/ OBLX. The key challenge is to extend ASTRX/OBLX while continuing to ensure that it does not contain any hard-coded circuit-specific knowledge. Our strategy is to perform simultaneous topology selection and sizing via optimization, as first introduced in [16].

Finally, a completely different way in which ASTRX/OBLX could be expanded would be to support larger circuits via hierarchy and macromodeling. Basically, this would involve using a macromodeling technique to convert a complete circuit such as an opamp into the equivalent of an encapsulated device model [61]. These macromodeled circuits would then become the atomic building blocks of a larger circuit structure such as a filter or A/D converter.

We have presented ASTRX/OBLX, tools that accurately size high-performance analog circuits to meet user-supplied specifications, but do not require prohibitive preparatory effort for each new circuit topology. For a suite of benchmark analog circuits that covers nearly all previously published synthesis results, we have validated our formulation by showing that ASTRX/OBLX requires several orders of magnitude less preparatory effort, yet can predict results more accurately. By comparing to a novel manual design, we have also shown that ASTRX/OBLX can handle difficult-to-design circuits and produce circuits comparable to those designed manually. Finally, by designing the cells of a pipelined A/D converter, we have shown that ASTRX/OBLX can successfully generate designs for problems of industrial complexity.

Acknowledgment

The authors wish to thank their colleagues whose critical analyses and insightful discussions have helped shape this paper. In particular, the authors wish to thank S. Kirkpatrick of IBM, and their coresearchers at Carnegie Mellon: R. Rohrer and his AWE group, R. Harjani, P. C. Maulik, B. Stanisic, and particularly, T. Mukherjee.

References

[1] R. Harjani, "OASYS: A framework for analog circuit synthesis," Ph.D. dissertation, Carnegie Mellon University, Pittsburgh, PA, 1989.

[2] P. G. Gray, "Analog IC's in the submicron era: Trends and perspectives," *IEEE IEDM Dig. Tech. Papers*, pp. 5–9, 1987.

[3] R. K. Brayton, A. L. Sangiovanni-Vincentelli, and G. D. Hachtel, "Multilevel logic synthesis," *Proc. IEEE*, vol. 78, pp. 264–300, Feb. 1990.

[4] E. S. Kuh and T. Ohtsuki, "Recent advances in VLSI layout," *Proc. IEEE*, vol. 78, pp. 237–263, Feb. 1990.

[5] G. Gielen, H. C. Walscharts, and W. C. Sansen, "Analog circuit design optimization based on symbolic simulation and simulated annealing," *IEEE J. Solid-State Circuits*, vol. 25, pp. 707–713, June 1990.

[6] H. Y. Koh, C. H. Sequin, and P. R. Gray, "OPASYN: A compiler for MOS operational amplifiers," *IEEE Trans. Computer-Aided Design*, vol. 9, 113–125, Feb. 1990.

[7] M. G. R. DeGrauwe, B. L. A. G. Goffart, C. Meixenberger, M. L. A. Pierre, J. B. Litsios, J. Rijmenants, O. J. A. P. Nys, E. Dijkstra, B. Joss, M. K. C. M. Meyvaert, T. R. Schwarz, and M. D. Pardoen, "Toward an analog system design environment," *IEEE J. Solid-State Circuits*, vol. 24, pp. 659–672, June 1989.

[8] M. G. R. DeGrauwe, O. Nys, E. Dijkstra, J. Rijmenants, S. Bitz, B. L. A. G. Goffart, E. A. Vittoz, S. Cserveny, C. Meixenberger, G. van der Stappen, and H. J. Oguey, "IDAC: An interactive design tool for analog CMOS circuits," *IEEE J. Solid-State Circuits*, vol. SC-22, pp. 1106–1116, Dec. 1987.

[9] F. El-Turky and E. E. Perry, "BLADES: An artificial intelligence approach to analog circuit design," *IEEE Trans. Computer-Aided Design*, vol. 8, pp. 680–691, June 1989.
[10] R. Harjani, R. A. Rutenbar, and L. R. Carley, "OASYS: A framework for analog circuit synthesis," *IEEE Trans. Computer Aided-Design*, vol. 8, pp. 1247–1266, Dec. 1989.
[11] B. J. Sheu, A. H. Fung, and Y. Lai, "A knowledge-based approach to analog IC design," *IEEE Trans. Circuits Syst.*, vol. 35, pp. 256–258, 1988.
[12] E. Berkcan, M. d'Abreu, and W. Laughton, "Analog compilation based on successive decompositions," in *Proc. 25th ACM/IEEE Design Automation Conf.*, 1988, pp. 369–375.
[13] E. S. Ochotta, "The OASYS virtual machine: Formalizing the OASYS analog synthesis framework," M.S. thesis, Rep. #89-25, Carnegie Mellon University, Pittsburgh, PA, Mar. 1989.
[14] J. P. Harvey, M. I. Elmasry, and B. Leung, "STAIC: An interactive framework for synthesizing CMOS and BiCMOS analog circuits," *IEEE Trans. Computer-Aided Design*, vol. 12, pp. 1402–1418, Nov. 1992.
[15] J. Jongsma, C. Meixenberger, B. Goffart, J. Litsios, M. Pierre, S. Seda, G. Di Domenico, P. Deck, L. Menevaut, and M. Degrauwe, "An open design tool for analog circuits," in *Proc. IEEE Int. Symp. Circuits Syst.*, June 1991, pp. 2000–2003.
[16] P. C. Maulik, L. R. Carley, and R. A. Rutenbar, "A mixed-integer nonlinear programming approach to analog circuit synthesis," in *Proc. Design Automation Conf.*, June 1992, pp. 693–703.
[17] H. Onodera, H, Kanbara, and K. Tamaru, "Operational-amplifier compilation with performance optimization," *IEEE J. Solid-State Circuits*, vol. 25, pp 466–473, Apr. 1990.
[18] C. Toumazou, C. A. Makris, and C. M. Berrah, "ISAID-A methodology for automated analog IC design," in *Proc. IEEE Int. Symp. Circuits Syst.*, 1990, pp. 531–533.
[19] S. Seda, M. DeGrauwe, and W. Fichtner, "Lazy-expansion of symbolic expression approximation in SYNAP," in *Proc. IEEE/ACM Int. Conf. CAD*, Nov. 1992, pp. 310–317.
[20] G. Gielen, H. C. Walscharts, and W. C. Sansen, "ISAAC: A symbolic simulation for analog integrated circuits," *IEEE J. Solid-State Circuits*, vol. 24, pp. 1587–1597, Dec. 1989.
[21] Q. Yu and C. Sechen, "Approximate symbolic analysis of large analog integrated circuits," in *Proc. IEEE Int. Conf. CAD*, Nov. 1994, pp. 664–672.
[22] P. Wambacq, F. V. Fernandez, G. Gielen, and W. Sansen, "Efficient symbolic computation of approximated small-signal characteristics," in *Proc. Custom Integrated Circuits Conf.*, vol. 21, no. 5, May 1994, pp. 1–4.
[23] B. Sheu, J. H. Shieh, and M. Patil, "BSIM: Berkeley short-channel IGFET model for MOS transistors," *IEEE J. Solid-State Circuits*, vol. SC-22, pp. 558–566, Aug. 1987.
[24] B. Johnson, T. Quarles, *et al.*, "SPICE version 3e user's manual," Univ. California, Berkeley, Tech. Rep., Apr. 1991.
[25] R. A. Rohrer, "Fully automatic network design by digital computer, preliminary considerations," *Proc. IEEE*, vol. 55, pp. 1929–1939, Nov. 1967.
[26] W. Nye, D. C. Riley, A. Sangiovanni-Vincentelli, and A. L. Tits, "DELIGHT.SPICE: An optimization-based system for the design of integrated circuits," *IEEE Trans. Computer-Aided Design*, vol. 7, pp. 501–518, Apr. 1988.
[27] J.-M. Shyu and A. Sangiovanni-Vincentelli, "ECSTASY: A new environment for IC design optimization," in *Proc. IEEE Int. Conf. CAD*, Nov. 1988, pp. 484–487.
[28] D. E. Hocevar, R. Arora, U. Dasgupta, S. Dasgupta, N. Subramanyam, and S. Kashyap, "A usable circuit optimizer for designers," in *Proc. IEEE Int. Conf. CAD*, Nov. 1990, pp. 290–293.
[29] L. Nagle and R. Rohrer, "Computer analysis of nonlinear circuits, excluding radiation (CANCER)," *IEEE J. Solid-State Circuits*, vol. SC-6, pp. 166–182, Aug. 1971.
[30] L. Nagle, "SPICE2: A computer program to simulate semiconductor circuits," Univ. California, Berkeley, Memo. UCB/ERL-M520, May 1975.
[31] W. Nye *et al.*, "DELIGHT.SPICE user's guide," Dept. EECS, Univ. California, Berkeley, May 1984.
[32] E. S. Ochotta, R. A. Rutenbar, and L. R. Carley, "Equation-free synthesis of high-performance linear analog circuits," in *Proc. Brown/MIT Adv. Res. VLSI and Parallel Syst.* Providence, RI: MIT, Mar. 1992, pp. 129–143.
[33] E. S. Ochotta, "Synthesis of high-performance analog cells in ASTRX/OBLX," Ph.D. dissertation, Carnegie Mellon University, Pittsburgh, PA, Feb. 1994.
[34] E. S. Ochotta, L. R. Carley, and R. A. Rutenbar, "Analog circuit synthesis for large, realistic cells: Designing a pipelined A/D converter with ASTRX/OBLX," in *Proc. Custom Integrated Circuits Conf.*, vol. 15, no. 4, May 1994, pp. 1–4.
[35] E. S. Ochotta, R. A. Rutenbar, and L. R. Carley, "ASTRX/OBLX: Tools for rapid synthesis of high-performance analog circuits," in *Proc. IEEE/ACM Design Automation Conf.*, June 1994, pp. 24–30.
[36] F. Medeiro, F. V. Fernandez, R. Dominguez-Castro, and A. Rodriguez-Vazquez., "A statistical optimization-based approach for automated sizing of analog circuits," in *Proc. IEEE Int. Conf. CAD*, Nov. 1994, pp. 594–597.
[37] T. Mukherjee, L. R. Carley, and R. A. Rutenbar, "Synthesis of manufacturable analog circuits," in *Proc. IEEE Int. Conf. CAD*, Nov. 1994, pp. 586–593.
[38] L. T. Pillage and R. A. Rohrer, "Asymptotic waveform evaluation for timing analysis," *IEEE Trans. Computer-Aided Design*, vol. 9, pp. 352–366, Apr. 1990.
[39] V. Raghavan, R. A. Rohrer, M. M. Alabeyi, J. E. Bracken, J. Y. Lee, and L. T. Pillage, "AWE inspired," in *Proc. IEEE Custom Integrated Circuits Conf.*, vol. 18, May 1993, pp. 1–8.
[40] S. Kirkpatrick, C. D. Gelatt, and M. P. Vecchi, "Optimization by simulated annealing," *Science*, vol. 220, no. 4598, pp. 671–680, May 13, 1983.
[41] F. Romeo and A. Sangiovanni-Vincintelli, "A theoretical framework for simulated annealing," *Algorithmica*, vol. 6, pp. 302–345, 1991.
[42] J. M. Cohn, D. J. Garrod, R. A. Rutenbar, and L. R. Carley, "KOAN/ANAGRAM II: New tools for device-level analog placement and routing," *IEEE J. Solid-State Circuits*, vol. 26, pp. 330–342, Mar. 1991.
[43] J. Lam and J. M. Delosme, "Performance of a new annealing schedule," in *Proc. 25th ACM/IEEE Design Automation Conf.*, 1988, pp. 306–311.
[44] G. W. Rhyne and M. B. Steer, "Comments on 'Simulation of nonlinear circuits in the frequency domain," *IEEE Trans. Computer-Aided Design*, vol. 8, pp. 927–928, Aug. 1989.
[45] K. S. Kundert and A. Sangiovanni-Vincentelli, "Reply to: Comments on 'Simulation of nonlinear circuits in the frequency domain,'" *IEEE Trans. Computer-Aided Design*, vol. 8, pp. 928–929, Aug. 1989.
[46] A. V. Aho, R. Sethi, and J. D. Ullman, *Compilers: Principles, Techniques, and Tools.* Reading, MA: Addison-Wesley, 1987.
[47] J. Vlach and K. Singal, *Computer Methods for Circuit Analysis and Design.* Princeton, NJ: van Nostrand Reinhold, 1983.
[48] N. Metropolis, A. W. Rosenbluth, M. N. Rosenbluth, and A. H. Teller, "Equation of state calculations by fast computer machines," *J. Chem. Phys.*, vol. 21, p. 1087, 1953.
[49] S. Hustin and A. Sangiovanni-Vincentelli, "TIM, a new standard cell placement program based on the simulated annealing algorithm," presented at the IEEE Physical Design Workshop on Placement and Floorplanning, Hilton Head, SC, Apr. 1987.
[50] D. Vanderbilt and G. Louie, "A Monte Carlo simulated annealing approach to optimization over continuous variables," *J. Comput. Phys.*, vol. 56, pp. 259–271, 1984.
[51] G. L. Bilbro and W. E. Snyder, "Optimization of functions with many minima," *IEEE Trans. Syst., Man, Cybern.*, vol. 21, pp. 840–849, July/Aug. 1991.
[52] S. B. Gelfand and S. K. Mitter, "Simulated annealing type algorithms for multivariate optimization," *Algorithmica*, vol. 6, pp. 419–436, 1991.
[53] W. Swartz and C. Sechen, "New algorithms for the placement and routing of macrocells," in *Proc. IEEE Int. Conf. CAD*, Nov. 1990, pp. 336–339.
[54] C. Sechen and K. Lee, "An improved simulated annealing algorithm for row-based placement," in *Proc. IEEE/ACM Int. Conf. CAD*, 1987, pp. 478–481.
[55] M. J. D. Powell, "A view of minimization algorithms that do not require derivatives," *ACM Trans. Math Software*, pp. 197–107, 1975.
[56] B. R. Stanisic, R. A. Rutenbar, and L. R. Carley, "Mixed-signal noise-decoupling via simultaneous power distribution design and cell customization in rail," in *Proc. Custom Integrated Circuits Conf.*, vol. 24, no. 2, May 1994, pp. 1–4.
[57] E. S. Ochotta, "User's Guide to ASTRX/OBLX," ECE Dept., Carnegie Mellon University, Pittsburgh, PA, Rep. CAD 94-36, July 1994.
[58] HSPICE manual, Metasoft Corp., 1990.
[59] H. K. Gummel and H. C. Poon, "An integral charge control model of bipolar transistors," *Bell Syst. Tech. J.*, pp. 827–851, May 1970.
[60] K. Nakamura and L. R. Carley, "A current-based positive-feedback technique for efficient cascode bootstrapping," in *Proc. VLSI Circuits Symp.*, June 1991, pp. 107–108.
[61] J. Shao and R. Harjani, "Macromodeling of analog circuits for hierarchical circuit design," in *Proc. 1994 IEEE Int. Conf. CAD*, Nov. 1994, pp. 656–663.

Emil S. Ochotta (S'88–M'92) received the B.Sc. degree in computer engineering from the University of Alberta, Edmonton, Canada, in 1987. He received the M.S. and Ph.D. degrees in electrical and computer engineering from Carnegie Mellon University, Pittsburgh, PA, in 1989 and 1994, respectively.

He is presently a Senior Systems Engineer at Xilinx, Inc., where he is developing new ideas for hardware and software in the field programmable gate array industry. His research interests include programming and hardware description languages, biomedical engineering, and CAD tools to support analog design and FPGA's.

Dr. Ochotta received a Best Paper award at the Semiconductor Research Corporation TECHCON conference in 1993. He was awarded the APEGGA (Association of Professional Engineers, Geologists, and Geophysicists of Alberta) Gold Medal for graduating first in his class and is a member of ACM and Sigma Xi.

Rob A. Rutenbar (S'78–M'84–SM'90) received the Ph.D degree in computer engineering (CICE) from the University of Michigan, Ann Arbor, in 1984.

He is currently a Professor of electrical and computer engineering and of computer science, and Director of the Semiconductor Research Corporation—CMU Center of Excellence in CAD and IC's with Carnegie Mellon University, Pittsburgh, PA. His research interests focus on circuit and layout synthesis for mixed-signal ASIC's, high-performance digital IC's, and FPGA's.

Dr. Rutenbar received a Presidential Young Investigator Award from the National Science Foundation in 1987. At the 1987 EIII-ACM Design Automation Conference, he received a Best Paper Award for work on analog circuit synthesis. He is currently on the Executive Committee of the IEEE International Conference on CAD, and Program Committees for the ACM/IEEE Design Automation Conference and European Design & Test Conference. He is on the Editorial Board of *IEEE Spectrum*, and chairs the Analog Technical Advisory Board for Cadence Design Systems. He is a member of ACM, Eta Kappa Nu, Sigma Xi, and AAAS.

L. Richard Carley (S'77–M'81–SM'90) received the S.B., M.S., and Ph.D. degrees from the Massachusetts Institute of Technology, Cambridge, in 1976, 1978, and 1984, respectively.

He is a Professor of electrical and computer engineering with Carnegie Mellon University, Pittsburgh, PA. He was with MIT's Lincoln Laboratories and has acted as a consultant in the area of analog circuit design and design automation for Analog Devices and Hughes Aircraft, among others. In 1984, he joined Carnegie Mellon, and in 1992 he was promoted to Full Professor. His current research interests include the development of CAD tools to support analog circuit design, design of high-performance analog signal processing IC's, and the design of low-power high-speed magnetic recording channels.

Dr. Carley received a National Science Foundation Presidential Young Investigator Award in 1985, a Best Technical Paper Award at the 1987 Design Automation Conference, and a Distinguished Paper Mention at the 1991 International Conference on Computer-Aided Design. He was also awarded the Guillemin Prize for the best EE Undergraduate Thesis.

MAELSTROM: Efficient Simulation-Based Synthesis for Custom Analog Cells

Michael Krasnicki, Rodney Phelps, Rob A. Rutenbar, L. Richard Carley
Department of Electrical and Computer Engineering
Carnegie Mellon University
Pittsburgh, Pennsylvania 15213
{kraz, rodneyp, rutenbar, carley}@ece.cmu.edu

Abstract

Analog synthesis tools have failed to migrate into mainstream use primarily because of difficulties in reconciling the simplified models required for synthesis with the industrial-strength simulation environments required for validation. MAELSTROM is a new approach that synthesizes a circuit using the same simulation environment created to validate the circuit. We introduce a novel genetic/annealing optimizer, and leverage network parallelism to achieve efficient simulator-in-the-loop analog synthesis.

I. INTRODUCTION

Mixed-signal designs are increasing in number as a large fraction of new ICs require an interface to the external, continuous-valued world. The digital portion of these designs can be attacked with modern cell-based tools for synthesis, mapping, and physical design. The analog portion, however, is still routinely designed by hand. Although it is typically a small fraction of the overall design size (*e.g.*, 10,000 to 20,000 analog transistors), the analog partition in these designs is often the bottleneck because of the lack of automation tools.

The situation appears to be worsening as we head into the era of System-on-Chip (SoC) designs. To manage complexity and time-to-market, SoC designs require a high level of reuse, and cell-based techniques lend themselves well to a variety of strategies for capturing and reusing digital intellectual property (IP). But these digital strategies are inapplicable to analog designs, which rely for basic functionality on tight control of low-level device and circuit properties that vary from technology to technology. The analog portions of these systems are still designed by hand today. They are even routinely ported by hand as a given IC migrates from one fabrication process to another.

A significant amount of research has been devoted to cell-level analog synthesis, which we define as the task of sizing and biasing a device-level circuit with 10 to 50 devices. However, as noted in [1], previous approaches have failed to make the transition from research to practice. This is due primarily to the prohibitive effort needed to reconcile the simplified circuit models needed for synthesis with the "industrial-strength" models needed for validation in a production environment. In digital design, the bit-level, gate-level and block-level abstractions used in synthesis are faithful to the corresponding models used for simulation-based validation. This is not the case for analog synthesis.

Fig. 1 illustrates the basic architecture of most analog synthesis tools. An *optimization engine* visits candidate circuit designs and adjusts their parameters in an attempt to satisfy designer-specified performance goals. An *evaluation engine* quantifies the quality of each circuit candidate for the optimizer. Most research here focuses on trade-offs between the optimizer (which wants to visit many circuit candidates) and the evaluator (which must itself trade accuracy for speed to allow sufficiently vigorous search). Much of this work is really an attempt to evade a harsh truth--that analog circuits are difficult and time-consuming to evaluate properly. Even a small cell requires a mix of ac, dc and transient analyses to correctly validate. In modern design environments, there is enormous investment in simulators, device models, process characterization, and "cell sign-off" validation methodologies. Indeed, even the sequence of circuit analyses, models, and simulation test-jigs is treated as valuable IP. Given these facts, it is perhaps no surprise that analog synthesis strategies that rely on exotic, nonstandard, or fast-but-incomplete evaluation engines have fared poorly in real design environments. To trust a synthesis result, one must first trust the methods used to quantify the circuit's performance *during* synthesis. Most prior work fails here.

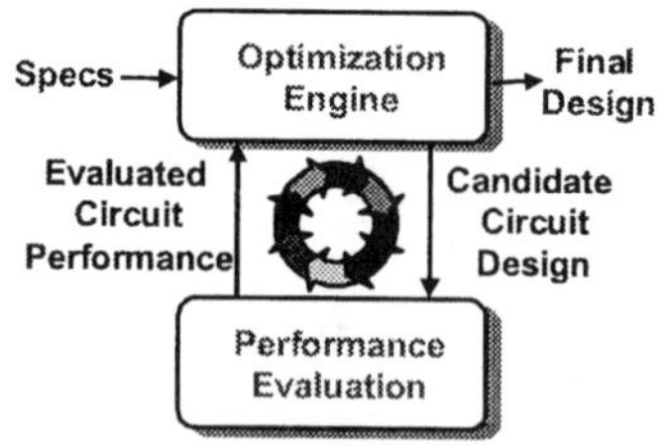

Fig. 1 Abstract Model of Analog Synthesis Tools.

Given the complexity of, investment in, and reliance on simulator-centric validation approaches for analog cells, we argue that for a synthesis strategy to have practical impact, it *must* use a simulator-based evaluation engine that is *identical* to that used to validate ordinary manual designs. This, however, poses significant challenges. For example, commercial circuit simulators are not designed to be invoked 50,000 times in the inner loop of a numerical optimizer. And, of course, the CPU time to visit and simulate this many solution candidates may be unacceptable.

In this paper we develop a new strategy to support efficient simulator-in-the-loop analog synthesis. The approach relies on three key ideas. First, we *encapsulate* commercial simulators so that their implementation idiosyncrasies are hidden from our search engine. Second, we use a novel combined *genetic/annealing optimization algorithm* that is robust in finding workable circuits, and avoids the starting-point dependency problems of gradient and other down-hill search methods. Third, we exploit *network-level workstation parallelism* to render the overall computation times tractable. Our new optimization algorithm was designed to support transparent distribution of *both* the search tasks and the circuit evaluation tasks across a network.

We have implemented these ideas in a tool called MAELSTROM. MAELSTROM has been successfully run on networks of 10 to 30 SUN or IBM UNIX workstations, and currently runs Cadence Design System's Spectre simulator [2] as its evaluation engine. In this paper we describe the basic algorithms underlying MAELSTROM, and present a set of experimental synthesis results that suggest that simulator-in-the-loop synthesis can be made both practical and efficient. The remainder of the paper is organized as follows. Section II briefly re-

DAC 99, New Orleans, Louisiana

views prior work. Section III gives a complete formulation of the synthesis problem. Section IV offers experimental results on circuits. Finally, Section V offers some concluding remarks.

II. Review Of Prior Approaches

Referring again to Fig. 1, we can broadly categorize previous work on analog synthesis by how it searches for solutions and how it evaluates each visited circuit candidate. See [3] for a more extensive survey.

Early work on synthesis used simple procedural techniques [4], rendering circuits as explicit scripts of equations whose direct evaluation completed a design. Although fast, these techniques proved to be difficult to update, and rather inaccurate. Numerical search has been used with equation-based evaluators [5], [6], [7], and even combinatorial search over different circuit topologies [8],[9], but equation-based approaches remain brittle in the face of technology changes. Hierarchical systems [10], [11], [12], [13] introduced compositional techniques to assemble equation-based subcircuits, but still faced the same update/accuracy difficulties. Some of these systems can manipulate circuit equations automatically to suit different steps of the synthesis task [6]. Qualitative and fuzzy reasoning techniques [14], [15] have been tried to capture designer expertise, but with limited success. Equation-based synthesis offers fast circuit evaluation, and is thus well suited to aggressive search over solution candidates. However, it is often prohibitively expensive to create these models--indeed, often more expensive than manually designing the circuit. Also, the simplifications required in these closed-form analytical circuit models necessarily limit their accuracy and completeness.

Symbolic analysis techniques, which have made significant strides of late[16],[17],[18],[7] offer an automated path to obtaining some of these design equations. These techniques automatically derive reduced-order symbolic models of the linear transfer function of a circuit. The resulting symbolic forms can be obtained fairly quickly, offer good accuracy, and can thus serve as evaluation engines, *e.g.*, [6]. However, they are strictly limited to linear performance specifications. Even a small analog cell may require a wide portfolio of dc, ac, and transient simulations to validate it. Symbolic analysis is a valuable but incomplete approach to circuit evaluation.

The synthesis systems most relevant to the ideas we develop in this paper are ASTRX/OBLX [1],[3] and the system from Seville [19]. In ASTRX/OBLX, we attacked the fundamental problem of tool usability with a compile-and-solve methodology. ASTRX starts from a SPICE deck describing an unsized circuit and desired performance specifications. ASTRX compiles this deck into a custom C program that implements a numerical cost function whose minimum corresponds to a good circuit solution for these constraints. OBLX uses simulated annealing [20] to solve this function for a minimum. This custom-generated cost code evaluates circuit performance via model-order reduction [21] for linear, small-signal analysis, and user-supplied equations for nonlinear specifications. ASTRX/OBLX was able to synthesize a wide variety of cells, but was still limited to essentially linear performance specifications. [19] similarly uses annealing for search, but actually runs a SPICE-class simulator in its annealer. However, this tool appears to employ a simulator customized for synthesis, only evaluates a few thousand circuit candidates in a typical synthesis run (in contrast, OBLX evaluates 10^4 to 10^5 solutions), and has only been demonstrated attacking problems with a small number of independent design variables.

Finally, we also note that there are several circuit *optimization* attacks that rely on simulator-based methods (*e.g.*, [22]). For circuit optimization we assume a good initial circuit solution, and seek to improve it. This can be accomplished with gradient and sensitivity techniques requiring a modest number of circuit evaluations. In contrast, in circuit *synthesis* we can assume nothing about our starting circuit (indeed, we usually have *no* initial solution). This scenario is much more difficult as a numerical problem, and requires a global search strategy to avoid being trapped in poor local minima that happen to lie near the starting point.

The problem with all these synthesis approaches is that they use circuit evaluation engines different from the simulators and simulation strategies that designers actually use to validate their circuits. These engines trade off accuracy and completeness of evaluation for speed. We argue that this is no longer an acceptable trade-off.

III. Synthesis Formulation

In this section, we present the full synthesis formulation of MAELSTROM. Our circuit synthesis strategy relies on three key ideas: simulator encapsulation, a novel combined genetic/annealing global optimizer, and scalable network parallelism. We describe these ideas below, beginning with a review of our basic synthesis-via-optimization formulation.

A. Basic Optimization Formulation

We use the basic synthesis formulation from OBLX [1], which we review here. We begin with a fixed circuit topology that we seek to size and bias. We approach circuit synthesis using a constrained optimization formulation, but solve it in an unconstrained fashion. We map the circuit design problem to the constrained optimization problem of (1), where x is the set of independent variables—geometries of semiconductor devices or values of passive circuit components—we wish to change to determine circuit performance; $f(\underline{x})$ is a set of objective functions that codify performance specifications the designer wishes to optimize, *e.g.* power or bandwidth; and $g(\underline{x})$ is a set of constraint functions that codify specifications that must be beyond a specific goal, *e.g.*, (gain > 60dB). Scalar weights, w_i, balance competing objectives.

$$\underset{\underline{x}}{\text{minimize}} \sum_{i=1}^{k} w_i \cdot f_i(\underline{x}) \qquad \text{s.t.} \quad g(\underline{x}) \le 0 \tag{1}$$

Formulation of the individual objective $f(\underline{x})$ and constraint $g(\underline{x})$ functions adapts ideas from [22]. The user is expected to provide a *good* value, and a *bad* value for each specification. These are used both to set constraint boundaries and to normalize the specification's range. For example, a single objective $f_i(\underline{x})$ is internally normalized as:

$$\hat{f}_i(\underline{x}) = \frac{f_i(\underline{x}) - good_i}{bad_i - good_i} \tag{2}$$

This normalization process provides a natural way for the designer to set the relative importance of competing specifications, and it provides a straightforward way to normalize the range of values that must be balanced in the cost function.

To support the genetic/annealing optimizer we shall introduce in Section IIIC, we perform the standard conversion of this constrained optimization problem to an unconstrained optimization problem with the use of additional scalar weights. As a result, the goal becomes minimization of a scalar cost function, $C(\underline{x})$, defined by (3).

$$C(\underline{x}) = \sum_{i=1}^{k} w_i \hat{f}_i(\underline{x}) + \sum_{j=1}^{l} w_j \hat{g}_j(\underline{x}) \tag{3}$$

The key to this formulation is that the minimum of $C(x)$ corresponds to the circuit design that best matches the given specifications. Thus, the synthesis task becomes two more concrete tasks: evaluating $C(x)$ and searching for its minimum. Neither of these are simple. Our major contributions in this paper are an algorithm for global search that is efficient enough to allow use of commercial circuit simulators to evaluate $C(x)$, and a methodology for encapsulating simulators to hide unnecessary details from this search process. We treat the encapsulation methodology next.

B. Simulator Encapsulation for Simulation-Based Evaluation

Our overall goal is to be able to the use the simulation methods trusted by designers--but *during* analog cell synthesis. This means invoking a sequence of detailed circuit simulations for each evaluation of $C(x)$ during numerical search. Although different SPICE-class simulation engines share core mechanisms and offer similar input/output formats, they remain highly idiosyncratic in many features. In our experience, the mechanics of embedding a simulator inside a numerical optimizer are remarkably untidy. This is a real problem since we seek a strict separation of the circuit optimization and circuit evaluation engines, and would like ultimately to be able to "plug in" different simulators. We handle this problem using a technique we refer to as *simulator encapsulation*.

Simulator encapsulation hides the details of a particular simulator behind an insulating layer of software. This software "wrapper" renders the simulator an object with a set of methods, similar to standard object-oriented programming ideas. The simulator appears to the optimization engine as an object with methods to invoke a simulation, to change circuit parameters, to retrieve simulation results as a simple vector of numbers, and so forth. Clearly one major function of this encapsulation is to hide varying data formats from the optimizer; this engine need not concern itself with the details of how to invoke or interpret an ac, dc, or transient analysis in the simulator.

A more subtle function of encapsulation is to insulate the optimization engine from "unfriendly" behavior in the simulator. Most simulators are designed either for batch-oriented operation, or for interactive schematic-update-then-simulate operation. In the latter, the time scales are optimized for humans--overheads of a few seconds per simulation invocation are negligible. But inside a numerical optimizer that seeks to run perhaps 50,000 simulations, these overheads are magnified. Our ideal is a simulator which can be invoked once, and, remaining live, can interpret quickly a stream of requests to modify circuit values and resimulate. Few simulators approach this ideal. For example, some insist on rechecking a licence manager key (possibly located remotely on a network) for every new simulation request; others flush all internal state or drop myriad temporary files in the local file system. Of course, the maximally difficult behavior exhibited by a simulator is a crash, an occurrence far from rare even in commercial offerings. This is especially problematic in synthesis, since the optimization engine may often visit circuit candidates with highly nonphysical parameter values, which occasionally cause simulator failure. Our encapsulation not only detects the crash but also restarts and reinitializes the simulator, all transparent to the optimizer. All these difficult behaviors can be hidden via appropriate encapsulation.

C. Combined Genetic/Annealing Optimization: PRSA

As in OBLX [1], we again favor global, stochastic search algorithms for the optimization engine because of their empirical robustness in the face of highly nonlinear, nonconvex cost functions. However, in OBLX we made an explicit trade-off to use a customized, highly tuned, very fast circuit evaluator to permit search over a large number of solution candidates. When we replace this custom evaluator with commercial circuit simulation, we are faced with a 10X to 100X increase in CPU time. The central question we address in this section is how to retain the virtues of global, stochastic search, but deal with the runtime implications of simulator-in-the-loop optimization.

Before we describe our new optimizer, it is worth justifying our choice of stochastic optimization. Given a good implementation of simulator encapsulation, we can replace the custom circuit evaluation used in OBLX with full, detailed simulation. We have rewritten the core annealing engine of OBLX in the form of a new, component-based optimization library called ANNEAL++ [23]. ANNEAL++ offers a range of annealing cooling schedules, move selection techniques, and dynamic updates on cost function weights, based on the ideas in [3]. As an experiment, we encapsulated the Cadence Spectre circuit simulator and used it with ANNEAL++ to resynthesize the custom folded-cascode opamp from [24]. The circuit has 32 devices and 27 designable variables; the circuit appears in Fig. 2, results appear in Table 1.

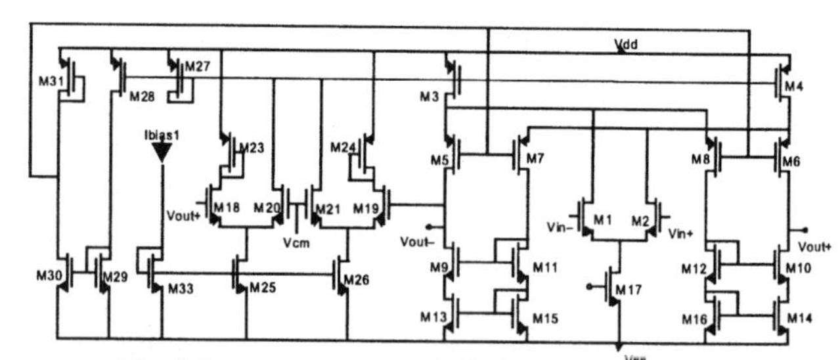

Fig. 2 Custom Folded Cascode OpAmp Circuit [24]

Table 1. Simple Synthesis Result for Circuit of Fig. 2, on a 55MHz IBM Power2

Attribute	Manual Design	Auto-Synthesis: Spec	Result
CLoad (pF)	1.25	1.25	
Vdd (V)	5	5	
DC Gain (dB)	71.2	≥ 71:	91
UGF (MHz)	47.8	≥48:	55
Phase Margin (deg)	77.4	≥ 77:	83
PSRR - Vss (dB)	92.6	≥ 93:	119
PSRR - Vdd (dB)	72.3	≥ 72:	92
Output Swing (V)	± 1.4	± 1.4:	± 1.4
Settling Time (ns)	-	↓[a]:	47
Active Area ($10^3\mu^2$)	68.7	↓:	28
Circuits Evaluated			17,100
CPU (hours)			11

a. ↑ means maximize, while ↓ means minimize.

This rather straightforward synthesis strategy yields a surprisingly reasonable result, albeit somewhat slowly. Fig. 3 shows a set of sampled cross-sections from the cost-surface for this annealing-style synthesis formulation. At an intermediate point in the synthesis, we stopped the optimizer, and then iteratively stepped each independent variable over its range, while freezing all other variables. At each step point we evaluated the synthesis cost function using Spectre. Fig. 3 shows a few of these resulting cross-sections, suitably normalized for comparison. The mix of gently sloping plateaus and jagged obstacles is typical of these landscapes. Annealing style algorithms are a good choice here because of their hill-climbing abilities.

However, annealing algorithms have a reputation for slow execution because of the large number of solution candidates that must be visited. This is greatly exacerbated when we choose to fully simulate

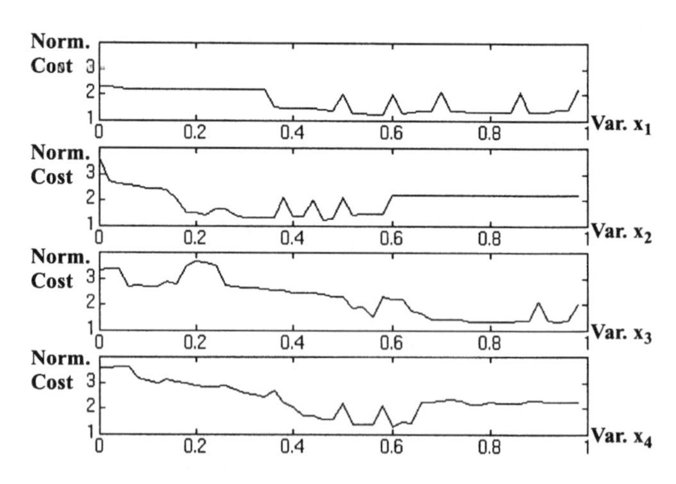

Fig. 3 Four 1-dimensional normalized cross-sections of the cost-surface for a typical simulation-based synthesis problem

each solution candidate. There are three broad avenues of solution here:

1. **Less search:** attempt to sample the cost function at fewer points. This is essentially the approach taken by [19], which uses an unusual, truncated annealing schedule with some of the character of a random multistart approach. However, in our experience, wider search always yields better solutions and a more robust tool.

2. **Parallel circuit evaluation:** each visited circuit candidate usually requires more than one circuit simulation to evaluate it. We can easily distribute these over a network to parallel workstations. Indeed, our implementation supports this simple parallelism. For example, if we resynthesize the opamp of Fig. 2, but distribute the 5 simulations required to evaluate each circuit across 3 IBM workstations, the 11 hour sequential time drops to 192 minutes. This is a useful form of parallelism to exploit, but it is strictly limited.

3. **Parallel circuit search:** what we really seek is a technique to allow multiple, concurrent points of the cost landscape to be searched in parallel, but synchronized in some manner that guarantees convergence to a final circuit or set of circuits of similar quality.

Unfortunately, annealing *per se* does not easily support parallel search. An annealing-based optimizer generates a serial stream of proposed circuit perturbations, and relies on statistics from previous circuits to adjust its control parameters. To parallelize search itself, an obvious set of methods to consider here are the genetic algorithms [25], whose population-based evolution models distribute over parallel machines more naturally. However, we do not wish to abandon the direct hill-climbing of annealing, which has empirically performed well in this task. Goldberg [26] suggests a solution here: *parallel recombinative simulated annealing* (PRSA)

PRSA, which has its roots in genetic algorithms, can be regarded as a strategy for synchronizing a population of annealers as they cooperatively search a cost surface. The idea is conceptually simple. Suppose in a serial annealer we would expect to visit 10,000 circuit candidates. To distribute this over 10 CPUs, we begin by creating 10 separate *PRSA-nodes*, each of which simply runs a standard annealing optimization (ANNEAL++ in our case) but with a schedule truncated to 10,000/10=1000 visited circuits. Obviously, the solution found by each of these 10 independent nodes will be very poor. To synchronize these nodes, we regard each annealer itself as one element of a larger population of evolving solutions, and allow annealers to exchange results among themselves. Thus, after generating a new candidate circuit solution, each annealer randomly communicates its result to a subset of the other PRSA-nodes. Each PRSA-node maintains a queue for these shared results, which represent samples of the cost surface visited by *other* annealers in the population. When generating a new circuit candidate, each annealer makes one of two choices:

1. **Perturbation:** the annealer can simply select its previously generated solution and *perturb* its element values. This is the traditional mechanism by which an annealer evolves a solution.

For all parallel PRSA nodes : P_i, ($i = 1$ to n)
(A) Set annealer temperature T = hot
(B) Generate random initial circuit solution $\underline{x}_{Pi}$.
(C) Repeat until equilibrium:
(C1) Send current circuit solution $\underline{x}_{Pi}$ to other randomly selected PRSA node
(C2) Receive migrants from other PRSA nodes
(C3) Apply perturbation or crossover to generate $\underline{x}_{Pi}^{new}$ from $\underline{x}_{Pi}$
(C4) Evaluate $\underline{x}_{Pi}^{new}$
(C5) $\Delta C = \mathrm{Cost}(\underline{x}_{Pi}^{new}) - \mathrm{Cost}(\underline{x}_{Pi})$
(C6) If $\Delta C < 0$
Replace $\underline{x}_{Pi}$ with $\underline{x}_{Pi}^{new}$ with probability 1.
(C7) Else
Replace $\underline{x}_{Pi}$ with $\underline{x}_{Pi}^{new}$ with probability $e^{-[(\Delta C)/T]}$
(D) If not frozen, lower T, goto (C)

Fig. 4 Pseudo-code for optimization in one PRSA-node.

2. **Recombination:** the annealer can *recombine* its previously generated solution with the solution on the top of its queue. This is the *crossover* (mating) operation from genetic algorithms, which randomly combines the features of two *parent* solutions into a single, new *offspring* solution.

Because circuit solution candidates are simply vectors of real numbers for us (e.g., MOSFET lengths and widths), crossover is simple to implement. We use a so-called *single-point* crossover scheme. Given two parent solutions $\underline{x} = [x_1, x_2, \ldots, x_n]$ and $\underline{y} = [y_1, y_2, \ldots, y_n]$, we combine by randomly selecting $r \in [1,n]$ and generate the offspring:

$$\underline{s} = [x_1, x_2, \ldots, x_r, y_{r+1}, y_{r+2}, \ldots, y_n] \quad (4)$$

Pseudo-code for the algorithm in each PRSA-node appears in Fig. 4.

In practice, we find that PRSA works extremely well to synchronize parallel annealers. In particular, good solutions found by one node quickly diffuse through the population, and drive annealers stuck in unpromising local minima toward better global solutions. Fig. 5. illustrates this synchronization effect by plotting the annealing cost value as a function of circuits visited in each of 10 parallel PRSA nodes during a sample circuit synthesis. Each PRSA-node visits roughly 2000 circuit candidates; the population of annealers visits 20,000, each evaluated via Spectre simulation. The curves demonstrate empirically how each annealing process is coordinated into searching for circuits of similar cost at similar times in the run.

Finally, we note that parallel circuit evaluation and parallel PRSA search are othogonal: we can do both. Each PRSA node can manage a set of independent evaluation nodes to perform the multiple simula-

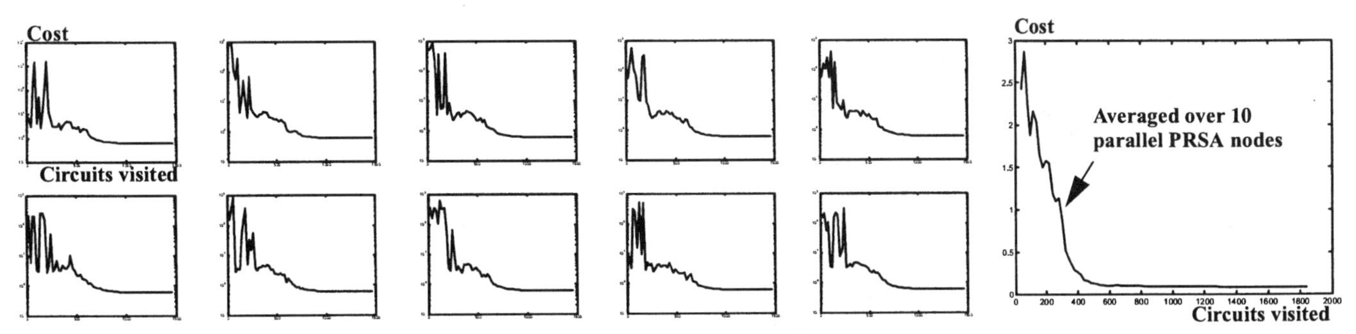

Fig. 5 Synchronized search behavior, cost versus circuits visited, for 10 parallel PRSA nodes.

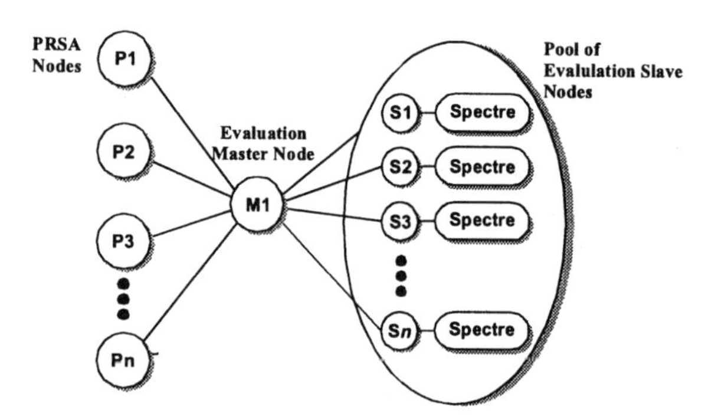

Fig. 6 Network architecture for MAELSTROM using DISTRIBUTEDPRSA

tions necessary to evaluate each solution candidate. We discuss this in the next section. We believe the capability to distribute both circuit evaluations and the optimization process itself is a significant contribution of this work.

D. Network Architecture: Distributed Search and Evaluation

Our implementation distributes all computation over a pool of workstations. At the lowest level, we manage concurrency and interprocessor communication using the publicly available PVM library [27]. We have implemented on top of this a general framework for optimization called DISTRIBUTEDPRSA. Fig. 6 shows a topological overview of DISTRIBUTEDPRSA. This library coordinates the interaction of the three concurrent tasks that comprise our synthesis tool:

1. **PRSA Node:** We use ANNEAL++ to implement a PRSA computational node, as discussed in the previous section. The DISTRIBUTEDPRSA library implements a mechanism that allows each PRSA node to send its current solution to another randomly selected PRSA node for use in crossover. In turn, each PRSA node keeps a small FIFO queue of recently received circuit solution candidates. This transfer of state information is a peer-to-peer transaction between the PRSA nodes and does not involve the evaluation master.

2. **Evaluation Master:** Each *evaluation master* schedules evaluation requests from some number of PRSA nodes across a pool of evaluation slaves. The cost calculation for each candidate circuit solution may require several Spectre simulation analyses. Each of these analyses can be performed in parallel on different machines. Thus, each evaluation master has one or more slaves for each analysis type. Currently, evaluation slaves are assigned to machines statically, based upon a configuration file. In the future, the evaluation master will dynamically reassign evaluation slaves across a pool of available workstations. The goal of this mechanism is to dynamically detect available processor time and to utilize it to expedite the synthesis process.

3. **Evaluation Slave:** An *evaluation slave* uses the simulator encapsulation library to perform one or more simulation analyses, *i.e.*, the slaves actually invoke the necessary circuit simulation tasks, with the encapsulation library serving as the interface to the simulator. If there are insufficient machines, one machine can be used to run multiple evaluation slaves.

IV. EXPERIMENTAL RESULTS

We have implemented these ideas in a tool called MAELSTROM, which currently runs on networks of SUN Solaris and IBM AIX nodes. In this section we present three results to demonstrate both the feasibility and efficiency of our synthesis strategy.

A. Custom Opamp Circuit

We have resynthesized the custom opamp [24] shown originally in Fig. 2, but now using the fully distributed version of MAELSTROM. Table 2 shows the desired specifications and the final synthesis results obtained with our tool. The optimization task had 27 independent variables that specified all device dimensions, capacitor sizes, and bias currents. Each of the variables had a broad (yet reasonable) range: all variables had a design range of at least one order of magnitude, many have ranges of two orders of magnitude. The process is 1.2μm CMOS. Note not only that we meet all specifications, but this result is significantly better than the earlier sequential synthesis shown in Table 1. The improved runtime is due to the large-scale parallelism; the improved solution is a result of allowing more search. The run in Table 1 searched only 17,000 circuits, we allowed this run to search 70,000 circuits.

Table 2. MAELSTROM Synthesis Result for Custom Opamp Circuit of Fig. 2

Attribute	Manual Design	Auto-Synthesis: Spec.	Result
CLoad (pF)	1.25	1.25	
Vdd (V)	5	5	
DC Gain (dB)	71.2	≥ 71:	110
UGF (MHz)	47.8	≥48:	70
Phase Margin (deg)	77.4	≥ 77:	84
PSRR - Vss (dB)	92.6	≥ 93:	131
PSRR - Vdd (dB)	72.3	≥ 72:	108
Output Swing (V)	± 1.4	± 1.4:	± 1.45
Settling Time (ns)	-	↓:	29
Active Area ($10^3\mu^2$)	68.7	↓:	23
Circuits Evaluated		70,000	
CPU Time (minutes)		219	

This result was obtained in 219 minutes across 15 140Mhz SUN Ultra-1 workstations. The run consisted of 10 PRSA nodes, 1 evaluation master, and 15 evaluation slaves. (Note that physical CPUs actually share search, control, and evaluation tasks concurrently.) Each PRSA node examined approximately 7000 candidate solutions across the duration of the run. Evaluating each candidate solution required 5 separate Spectre circuit simulations.

B. Basic Folded Cascode Op-amp

Fig. 7 shows a basic fully differential folded cascode circuit, again to be sized in a 1.2μm CMOS process. This is illustrative of the sort

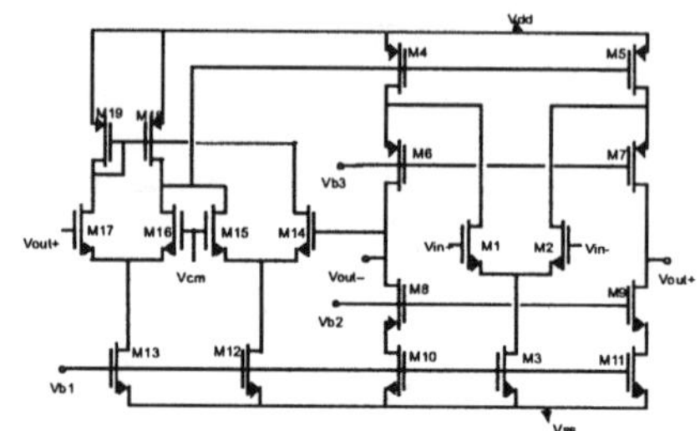

Fig. 7 Basic Folded Cascode Circuit

Table 3. MAELSTROM Result for Basic Folded Cascode Opamp Circuit in Fig. 7

Attribute	Auto-Synthesis: Spec.	Result
CLoad (pF)	1	
Vdd (V)	5	
DC Gain (dB)	≥ 70:	71.4
UGF (MHz)	≥10:	24.3
Phase Margin (deg)	≥ 60:	69
PSRR - Vss (dB)	≥ 40:	111
PSRR - Vdd (dB)	≥ 40:	132
Output Swing (V)	± 1.35:	± 1.37
Settling Time (ns)	≤ 100 :	50
Active Area ($10^3\mu^2$)	≤ 68 :	11
Circuits Evaluated	60,000	
CPU (minutes)	152	

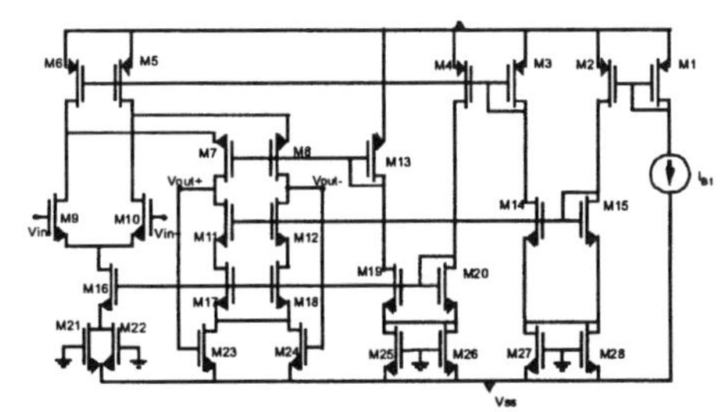

Fig. 8 Seville Benchmark Circuit

Table 4. MAELSTROM Result for Seville Benchmark Circuit of Fig. 8

Attribute	Auto-Synthesis: Spec.	Result
CLoad (pF)	1	
Vdd (V)	5	
DC Gain (dB)	≥ 70:	70
UGF (MHz)	≥30:	47
Phase Margin (deg)	≥ 60:	60
PSRR - Vss (dB)	≥ 40:	·71
PSRR - Vdd (dB)	≥ 40:	94
Output Swing (V)	± 1.5:	± 1.5
Settling Time (ns)	≤ 80 :	68
Static Power (mW)	≤ 2.1 :	1
Active Area ($10^3\mu^2$)	≤ 68 :	38
Circuits Evaluated	70,000	
CPU (minutes)	190	

of routine redesign problems faced when common analog blocks are retargeted to new applications. Table 3 shows the desired specifications and the final synthesis result. This optimization task had 21 independent design variables and was again run on 30 Ultra-1 workstations with the same PRSA configuration.

C. Seville Benchmark Circuit

Fig. 8 shows the opamp benchmark circuit used in [19]. We have synthesized this result to the specifications from [19] in a 1.2µm process. (The specification for slew rate to exceed 70 V/µs was translated to a constraint of settling time below 80ns). This optimization task had 22 independent design variables, in contrast to the formulation in [19] which had 10. This represents the trade-off between up front manual design (to determine a subset of critical designable devices) versus simply allowing the optimization tool to search a larger solution space. The circuit meets all its specifications, and is comparable to the results from [19]. This synthesis was run on 18 SUN Ultra-1 workstations.

V. CONCLUSIONS

We described a new cell-level analog synthesis strategy that evaluated each proposed solution candidate using the same simulation methods relied on by designers to validate manual circuit designs. Our approach relies on three key ideas: simulator encapsulation to hide low-level details of specific simulators; a combined genetic/annealing algorithm for robust global search of the solution space; and network parallelism to render execution times short enough to make synthesis practical. MAELSTROM, a preliminary implementation of these ideas, has been run successfully on networks of up to 30 UNIX workstations, and can explore 10^4 to 10^5 circuit candidates in a few hours. Preliminary results suggest the approach is workable for many of the routine, cell-level, nominal sizing/biasing tasks that analog designer currently perform by hand.

Our current work focuses on tuning to support usage modes where designers seek only a "quick" approximate solution to explore the feasibility of a particular circuit topology, and support for evaluation across manufacturing corners [28].

Acknowledgment: This work was funded by the Semiconductor Research Corp. and by Rockwell and Texas Instruments. We thank Dave Guilleau of CMU, Emil Ochotta of Xilinx, Chris Wolff of Rockwell, Gary Richey of TI, and Al Dunlop and Mean-Sea Tsay of Lucent for valuable discussions about this work.

REFERENCES

[1] E. Ochotta, R.A. Rutenbar, L.R. Carley, "Synthesis of High-Performance Analog Circuits and ASTRX/OBLX," *IEEE Trans. CAD*, vol. 15, no. 3, March 1996.

[2] K.S. Kundert, *The Designer's Guide to SPICE & SPECTRE*, Kluwer Academic Publishers, Kluwer Academic Publishers, 1995.

[3] E. Ochotta, T. Mukherjee, R.A. Rutenbar, L.R. Carley, *Practical Synthesis of High-Performance Analog Circuits*, Kluwer Academic Publishers, 1998.

[4] M. Degrauwe *et al.*, "Towards an analog system design environment," *IEEE JSSC*, vol. sc-24, no. 3, June 1989.

[5] H.Y. Koh, C.H. Sequin, and P.R. Gray, "OPASYN: a compiler for MOS operational amplifiers," *IEEE Trans. CAD*, vol. 9, no. 2, Feb. 1990.

[6] G. Gielen, *et al.*, "Analog circuit design optimization based on symbolic simulation and simulated annealing," *IEEE JSSC*, vol. 25, June 1990.

[7] F. Leyn, W. Daems, G. Gielen, W. Sansen, "A Behavioral Signal Path Modeling Methodology for Qualitative Insight in and Efficient Sizing of CMOS Opamps," *Proc. ACM/IEEE ICCAD*, 1997.

[8] P. C. Maulik, L. R. Carley, and R. A. Rutenbar, "Integer Programming Based Topology Selection of Cell Level Analog Circuits," *IEEE Trans. CAD*, vol. 14, no. 4, April 1995.

[9] W. Kruiskamp and D. Leenaerts, "DARWIN: CMOS Opamp Synthesis by Means of a Genetic Algorithm," *Proc. 32nd ACM/IEEE DAC*, 1995.

[10] R. Harjani, R.A. Rutenbar and L.R. Carley, "OASYS: a framework for analog circuit synthesis," *IEEE Trans. CAD*, vol. 8, no. 12, Dec. 1989.

[11] B.J. Sheu, *et al.*, "A Knowledge-Based Approach to Analog IC Design," *IEEE Trans. Circuits and Systems*, CAS-35(2):256-258, 1988.

[12] E. Berkcan, *et al.*, "Analog Compilation Based on Successive Decompositions," *Proc. of the 25th IEEE DAC*, pp. 369-375, 1988.

[13] J. P. Harvey, *et al.*, "STAIC: An Interactive Framework for Synthesizing CMOS and BiCMOS Analog Circuits," *IEEE Trans. CAD*, Nov. 1992.

[14] C. Makris and C. Toumazou, "Analog IC Design Automation Part II--Automated CIrcuit Correction by Qualitative Reasoning," *IEEE Trans. CAD*, vol. 14, no. 2, Feb. 1995.

[15] A. Torralba, J. Chavez and L. Franquelo, "FASY: A Fuzzy-Logic Based Tool for Analog Synthesis," *IEEE Trans. CAD*, vol. 15, no. 7, July 996.

[16] G. Gielen, P. Wambacq, and W. Sansen, "Symbolic ANalysis Methods and Applications for Analog Circuits: A Tutorial Overview, " Proc. IEEE, vol. 82, no. 2, Feb., 1990.

[17] C.J. Shi, X. Tan, "Symbolic Analysis of Large Analog Circuits with Determinant Decision Diagrams," *Proc. ACM/IEEE ICCAD*, 1997.

[18] Q. Yu and C. Sechen, "A Unified Approach to the Approximate Symbolic Analysis of Large Analog Integrated Circuits," *IEEE Trans. Circuits and Sys.*, vol. 43, no. 8, August 1996.

[19] F. Medeiro, F.V. Fernandez, R. Dominguez-Castro and A. Rodriguez-Vasquez, " A Statistical Optimization Based Approach for Automated Sizing of Analog Cells," *Proc. ACM/IEEE ICCAD*, 1994.

[20] S. Kirkpatrick, C.D. Gelatt, M.P. Vecchi, "Optimization by simulated annealing," *Science*, vol. 220, no. 4598, 13 May 183.

[21] L. T. Pillage and R.A. Rohrer, "Asymptotic Waveform Evaluation for Timing Analysis," IEEE Trans. CAD, vol. 9. no. 4, April 1990.

[22] W. Nye, *et al.*, "DELIGHT.SPICE: an optimization-based system for the design of integrated circuits," *IEEE Trans. CAD*, vol. 7, April 1988.

[23] M. Krasnicki, "Generalized Analog Circuit Synthesis," M.S. Thesis, Dept. of ECE, Carnegie Mellon, Dec. 1997.

[24] K. Nakamura and L.R. Carley, "A current-based positive-feedback technique for efficient cascode bootstrapping," *Proc. VLSI Circuits Symposium*, June 1991.

[25] J.H. Holland. *Adaptation in Nature and Artificial Systems*, University of Michigan Press, Ann Arbor, 1975.

[26] S. W. Mahfoud and D.E. Goldberg, "Parallel Recombinative Simulated Annealing: A Genetic Algorithm," *Parallel Computing*, vol. 21, 1995.

[27] A. Geist, A. Beguelin, J. Dongarra, W. Jiang, R. Manchek, V. Sunderam. *PVM: Parallel Virtual Machine A User's Guide and Tutorial for Network Parallel Computing.* MIT Press, 1994.

[28] T. Mukherjee, L.R. Carley, R.A. Rutenbar, "Synthesis of Manufacturable Analog Circuits," *Proc. ACM/IEEE ICCAD*, 1994.

Anaconda: Simulation-Based Synthesis of Analog Circuits Via Stochastic Pattern Search

Rodney Phelps, *Student Member, IEEE*, Michael Krasnicki, Rob A. Rutenbar, *Fellow, IEEE*, L. Richard Carley, and James R. Hellums, *Senior Member, IEEE*

***Abstract*—Analog synthesis tools have traditionally traded quality for speed, substituting simplified circuit evaluation methods for full simulation in order to accelerate the numerical search for solution candidates. As a result, these tools have failed to migrate into mainstream use primarily because of difficulties in reconciling the simplified models required for synthesis with the industrial-strength simulation environments required for validation. We argue that for synthesis to be practical, it is essential to synthesize a circuit using the same simulation environment created to validate the circuit. In this paper, we develop a new numerical search algorithm efficient enough to allow full circuit simulation of each circuit candidate, and robust enough to find good solutions for difficult circuits. The method combines the population-of-solutions ideas from evolutionary algorithms with a novel variant of pattern search, and supports transparent network parallelism. Comparison of several synthesized cell-level circuits against manual industrial designs demonstrates the utility of the approach.**

***Index Terms*—Algorithms, analog synthesis, mixed-signal design, pattern search.**

I. Introduction

MIXED-SIGNAL designs are increasing in number as a large fraction of new integrated circuits (IC's) require an interface to the external, continuous-valued world. The digital portion of these designs can be attacked with modern cell-based tools for synthesis, mapping, and physical design. The analog portion, however, is still routinely designed by hand. Although it is typically a small fraction of the overall design size (e.g., 10 000–20 000 analog transistors), the analog partition in these designs is often the bottleneck because of the lack of automation tools.

The situation appears to be worsening as we head into the era of System-on-Chip (SoC) designs. To manage complexity and time-to-market, SoC designs require a high level of reuse, and cell-based techniques lend themselves well to a variety of strategies for capturing and reusing digital intellectual property (IP). But these digital strategies are inapplicable to analog designs, which rely for basic functionality on tight control of low-level device and circuit properties that vary from technology to technology. The analog portions of these systems are still designed by hand today. They are even routinely ported by hand as a given IC migrates from one fabrication process to another.

A significant amount of research has been devoted to cell-level analog synthesis, which we define as the task of sizing and biasing a device-level circuit with 10–50 devices. However, as noted in [1], previous approaches have failed to make the transition from research to practice. This is due primarily to the prohibitive effort needed to reconcile the simplified circuit models needed for synthesis with the "industrial-strength" models needed for validation in a production environment. In digital design, the bit-level, gate-level and block-level abstractions used in synthesis are faithful to the corresponding models used for simulation-based validation. This is not the case for analog synthesis.

Fig. 1 illustrates the basic architecture of most analog synthesis tools. An *optimization engine* visits candidate circuit designs and adjusts their parameters in an attempt to satisfy designer-specified performance goals. An *evaluation engine* quantifies the quality of each circuit candidate for the optimizer. Most research here focuses on tradeoffs between the optimizer (which wants to visit many circuit candidates) and the evaluator (which must itself trade accuracy for speed to allow sufficiently vigorous search). Much of this work is really an attempt to evade a harsh truth—that analog circuits are difficult and time-consuming to evaluate properly. Even a small cell requires a mix of ac, dc, and transient analyzes to correctly validate. In modern design environments, there is enormous investment in simulators, device models, process characterization, and "cell sign-off" validation methodologies. Indeed, even the sequence of circuit analyzes, models, and simulation test-jigs is treated as valuable IP here. Given these facts, it is perhaps no surprise that analog synthesis strategies that rely on exotic, nonstandard, or fast-but-incomplete evaluation engines have fared poorly in real design environments. To trust a synthesis result, one must first trust the methods used to quantify the circuit's performance *during* synthesis. Most prior work fails here.

Given the complexity of, investment in, and reliance on simulator-centric validation approaches for analog cells, we argue that for a synthesis strategy to have practical impact, it *must* use a simulator-based evaluation engine that is *identical* to that used to validate ordinary manual designs. This, however, poses significant challenges. For example, commercial circuit simulators are not designed to be invoked 50 000 times in the inner loop of

Manuscript received October 1, 1999. This work was supported in part by the Semiconductor Research Corporation (SRC) under contract 068, by the National Science Foundation (NSF), and grants from Texas Instruments and Rockwell Semiconductor. This paper was recommended by Associate Editor E. Charbon.

R. Phelps, M. Krasnicki, and L. R. Carley are with the Department of Electrical and Computer Engineering, Carnegie Mellon University, Pittsburgh, PA, 15213 USA.

R. A. Rutenbar is with the Department of Electrical and Computer Engineering, Carnegie Mellon University, Pittsburgh, PA, 15213 USA (e-mail: rutenbar@ece.cmu.edu).

J. R. Hellums is with the Mixed Signal Products Division, Texas Instruments Inc., Dallas, TX 75243 USA.

Publisher Item Identifier S 0278-0070(00)05346-X.

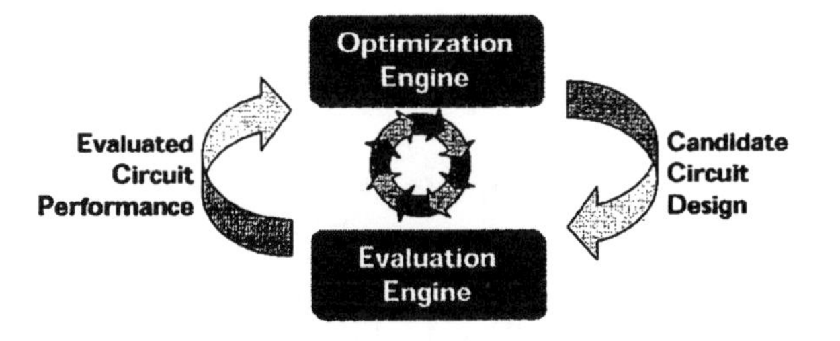

Fig. 1. Abstract model of analog synthesis tools.

a numerical optimizer. And, of course, the CPU time to visit and simulate this many solution candidates may be unacceptable.

In this paper, we develop a new strategy to support efficient simulator-in-the-loop analog synthesis. The approach relies on three key ideas. First, we *encapsulate* commercial simulators so that their implementation idiosyncrasies are hidden from our search engine. Second, we develop a novel global optimization algorithm called *stochastic pattern search*, combining ideas from evolutionary algorithms and pattern search, that is robust in finding workable circuits, and avoids the starting-point dependency problems of gradient and other down-hill search methods. Third, we exploit *network-level workstation parallelism* to render the overall computation times tractable. Our new optimization algorithm was designed to support transparent distribution of *both* the search tasks and the circuit evaluation tasks across a network.

We have implemented these ideas in a tool called ANACONDA. ANACONDA uses framework components from a companion synthesis tool, MAELSTROM [2]. ANACONDA has been successfully run on networks of 10–20 UNIX workstations, and currently runs Texas Instrument's proprietary TISpice circuit simulator as its evaluation engine [3]. In this paper, we extend the original treatment, describe in more detail the basic algorithms underlying ANACONDA, and present an expanded set of experimental synthesis results that demonstrate that simulator-in-the-loop synthesis can be made both practical and efficient for industrial designs. The remainder of the paper is organized as follows. In Section II, we briefly review prior work. In Section III, we formulate the overall synthesis problem and focuses on our global optimization algorithm, which we call *stochastic pattern search*. In Section IV, we offer experimental results on circuits. Finally, we offer some concluding remarks in Section V.

II. Review of Prior Approaches

Referring again to Fig. 1, we can broadly categorize previous work on analog synthesis by how it searches for solutions and how it evaluates each visited circuit candidate. See [4] for a more extensive survey.

Early work on synthesis used simple procedural techniques [5], rendering circuits as explicit scripts of equations whose direct evaluation completed a design. Although fast, these techniques proved to be difficult to update, and rather inaccurate. Numerical search has been used with equation-based evaluators [6]–[8], and even combinatorial search over different circuit topologies [9], [10], but equation-based approaches remain brittle in the face of technology changes. Recent work here has focused on coercing the required equations into a form more amenable to optimization; [11] shows results from rendering the equations as posynomials, thus, creating a convex optimization problem solvable via geometric programming. Hierarchical systems [12]–[15] introduced compositional techniques to assemble equation-based subcircuits, but still faced the same update/accuracy difficulties. Some of these systems can manipulate circuit equations automatically to suit different steps of the synthesis task [7]. Qualitative and fuzzy reasoning techniques [16], [17] have been tried to capture designer expertise, but with limited success. Equation-based synthesis offers fast circuit evaluation and is, thus, well suited to aggressive search over solution candidates. However, it is often prohibitively expensive to create these models—indeed, often more expensive than manually designing the circuit. Also, the simplifications required in these closed-form analytical circuit models necessarily limit their accuracy and completeness.

Symbolic analysis techniques, which have made significant strides of late [7], [8], [18]–[21] offer an automated path to obtaining some of these design equations. These techniques automatically derive reduced-order symbolic models of the linear transfer function of a circuit. The resulting symbolic forms can be obtained fairly quickly, offer good accuracy and can, thus, serve as evaluation engines. However, they are strictly limited to linear performance specifications, or at most some weakly nonlinear specifications [22]. Even a small analog cell may require a wide portfolio of dc, ac, and transient simulations to validate it. Symbolic analysis is a valuable but incomplete approach to circuit evaluation.

The synthesis systems most relevant to the ideas we develop in this paper are ASTRX/OBLX [1], [4] and the FRIDGE system from Seville [23]. In ASTRX/OBLX, we attacked the fundamental problem of tool usability with a compile-and-solve methodology. ASTRX starts from a SPICE deck describing an unsized circuit and desired performance specifications. ASTRX compiles this deck into a custom C program that implements a numerical cost function whose minimum corresponds to a good circuit solution for these constraints. OBLX uses simulated annealing [24] to solve this function for a minimum. This custom-generated cost code evaluates circuit performance via model-order reduction [25] for linear, small-signal analysis, and user-supplied equations for nonlinear specifications. ASTRX/OBLX was able to synthesize a wide variety of cells, but was still limited to essentially linear performance specifications. FRIDGE similarly uses annealing for search, but actually runs a SPICE-class simulator in its annealer. However, this tool appears to employ a simulator customized for synthesis, only evaluates a few thousand circuit candidates in a typical synthesis run (in contrast, OBLX evaluates 10^4–10^5 solutions), and has only been demonstrated solving problems with a small number of independent design variables.

Finally, we also note that there are several circuit *optimization* techniques that rely on simulator-based methods (e.g., [26]–[28]). For circuit optimization we assume a reasonable initial circuit solution, and seek to improve it. This can be accomplished with gradient and sensitivity techniques requiring a modest number of circuit evaluations. In contrast, in circuit *synthesis* we can assume nothing about our starting circuit (indeed, we usually have *no* initial solution). This scenario is much more difficult as a numerical problem, and requires a global search strategy to avoid being trapped in poor local minima that happen to lie near the starting point.

The problem with all these synthesis approaches is that they use circuit evaluation engines different from the simulators and simulation strategies that designers actually use to validate their circuits. These engines tradeoff accuracy and completeness of evaluation for speed. We argue that this is no longer an acceptable tradeoff.

III. Synthesis Formulation

In this section, we present the full synthesis formulation of ANACONDA. Our circuit synthesis strategy relies on three key ideas: simulator encapsulation, a novel combination of population methods and pattern search, and scalable network parallelism. We describe these ideas below, beginning with a review of our basic synthesis-via-optimization formulation.

A. Basic Optimization Formulation

We use the basic synthesis formulation from OBLX [1], which we review here. We begin with a fixed circuit topology that we seek to size and bias. We approach circuit synthesis using a constrained optimization formulation, but solve it in an unconstrained fashion. We map the circuit design problem to the constrained optimization problem of (1), where is the set of independent variables—geometries of semiconductor devices, device multiplicities, and values of passive circuit components—we wish to change to determine circuit performance; $f(x)$ is a set of objective functions that codify performance specifications the designer wishes to optimize, e.g., power or bandwidth; and $g(x)$ is a set of constraint functions that codify specifications that must be beyond a specific goal, e.g., (gain >60 dB). Scalar weights, w_{fi}, balance competing objectives

$$\min_{x} \sum_{i=1}^{k} w_{fi} \cdot f_i(x) \quad \text{s.t. } g(x) \leq 0. \tag{1}$$

Formulation of the individual objective $f(x)$ and constraint $g(x)$ functions adapts ideas from [26]. We require a *good* value, and a *bad* value for each specification. These are used both to set constraint boundaries and to normalize the specification's range. For example, a single objective $f_i(x)$ is internally normalized as

$$\hat{f}_i(x) = \frac{f_i(x) - \text{good}}{\text{bad}_i - \text{good}_i}. \tag{2}$$

This normalization process provides a natural way for the designer to set the relative importance of competing specifications, and it provides a straightforward way to normalize the range of values that must be balanced in the cost function.

To support the stochastic pattern search optimizer we introduce in Section IV, we perform the standard conversion of this constrained optimization problem to an unconstrained optimization problem with the use of additional scalar weights. As a result, the goal becomes minimization of a scalar cost function, $C(x)$, defined by

$$C(x) = \sum_{i=1}^{k} w_{fi}\hat{f}_i(x) + \sum_{j=1}^{l} w_{gj}\hat{g}_j(x). \tag{3}$$

The key to this formulation is that the minimum of $C(x)$ corresponds to the circuit design that best matches the given specifications. Thus, the synthesis task becomes two more concrete tasks: evaluating $C(x)$ and searching for its minimum. Neither of these are simple. Our major contributions here are a new algorithm for global search that is efficient enough to allow use of commercial circuit simulators to evaluate $C(x)$, and a methodology for encapsulating simulators to hide unnecessary details from this search process. We treat the encapsulation methodology next.

B. Simulator Encapsulation for Simulation-Based Evaluation

Our overall goal is to be able to use the simulation methods trusted by designers—but *during* analog cell synthesis. This means invoking a sequence of detailed circuit simulations for each evaluation, $C(x)$, during numerical search. Although different SPICE-class simulation engines share core mechanisms and offer similar input/output formats, they remain highly idiosyncratic in many features. In our experience, the mechanics of embedding a simulator inside a numerical optimizer are remarkably untidy. This is a real problem since we seek a strict separation of the circuit optimization and circuit evaluation engines, and would like ultimately to be able to "plug in" different simulators. We handle this problem using a technique we refer to as *simulator encapsulation*.

Simulator encapsulation hides the details of a particular simulator behind an insulating layer of software. This software "wrapper" renders the simulator an object with a set of methods, similar to standard object-oriented programming ideas. Class members are invoked to perform a simulation, to change circuit parameters, to retrieve simulation results, and so forth. Clearly one major function of this encapsulation is to hide varying data formats from the optimizer; this engine need not concern itself with the details of how to invoke or interpret an ac, dc, or transient analysis.

A more subtle function of encapsulation is to insulate the optimization engine from "unfriendly" behavior in the simulator. Most simulators are designed either for batch-oriented operation, or for interactive schematic-update-then-simulate operation. In the latter, the time scales are optimized for humans—overheads of a few seconds per simulation invocation are negligible. But inside a numerical optimizer that seeks to run perhaps 100 000 simulations, these overheads are magnified. Our ideal is a simulator which can be invoked once, and, remaining live, can interpret quickly a stream of requests to modify circuit values and resimulate. Few simulators approach this ideal. For example, some insist on rechecking a licence manager key (possibly located remotely on a network) for every new simulation request; others flush all internal state or drop myriad temporary files in the local file system. Of course, the maximally difficult behavior exhibited by a simulator is a crash, an occurrence far from rare even in commercial

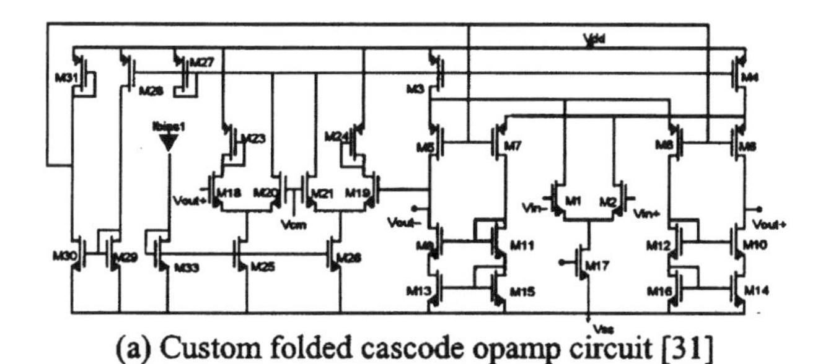

(a) Custom folded cascode opamp circuit [31]

(b) Simulation-based synthesis result for above circuit, on a 55MHz IBM Power2

Attribute	Manual Design	Auto-Synthesis: Spec	Result
CLoad (pF)	1.25	1.25	
Vdd (V)	5	5	
DC Gain (dB)	71.2	≥ 71:	91
UGF (MHz)	47.8	≥48:	55
Phase Margin (deg)	77.4	≥ 77:	83
PSRR - Vss (dB)	92.6	≥ 93:	119
PSRR - Vdd (dB)	72.3	≥ 72:	92
Output Swing (V)	± 1.4	± 1.4:	± 1.4
Settling Time (ns)	-	↓[a]:	47
Active Area ($10^3\mu^2$)	68.7	↓:	28
Circuits Evaluated			17,100
CPU (hours)			11

a. ↑ means maximize, while ↓ means minimize.

Fig. 2. Custom folded cascode opamp circuit and basic simulation-based annealing synthesis result.

offerings. This is especially problematic in synthesis, since the optimization engine may often visit circuit candidates with highly nonphysical parameter values, which occasionally cause simulator failure. Our encapsulation not only detects the crash but also restarts and reinitializes the simulator, all transparent to the optimizer. All these difficult behaviors can be hidden via appropriate encapsulation.

C. Global Optimization Issues

As in OBLX [1], we again favor global, stochastic search algorithms for the optimization engine because of their empirical robustness in the face of highly nonlinear, nonconvex cost functions. However, in OBLX we made an explicit tradeoff to use a customized, highly tuned, very fast circuit evaluator to permit search over a large number of solution candidates. When we replace this custom evaluator with commercial circuit simulation, we are faced with a 10× to 100× increase in CPU time. The central question we address in this section is how to retain the virtues of global, stochastic search, but deal with the runtime implications of simulator-in-the-loop optimization.

Before we describe our new optimizer, it is worth justifying our focus on global, rather than gradient-style local optimization. Given a good implementation of simulator encapsulation, we can replace the custom circuit evaluation used in OBLX with full, detailed simulation. We have rewritten the core annealing engine of OBLX in the form of a new, component-based optimization library called ANNEAL++ [29]. ANNEAL++ offers a range of annealing cooling schedules, move selection techniques, and dynamic updates on cost function weights. As an experiment, we encapsulated the Cadence Spectre circuit simulator [30] and used it with ANNEAL++ to resynthesize a standard OBLX benchmark circuit: the custom folded-cascode opamp from [31]. The circuit has 32 devices and 27 designable variables; the circuit schematic and associated synthesis results appear in Fig. 2.

This rather straightforward synthesis strategy yields an adequate result, albeit somewhat slowly. (In this simplified example, manufacturability concerns were ignored, which is the source of the extreme performance/area results; for annealing-style synthesis, these concerns can be addressed; we will return to these issues in Section IV.) Fig. 3 shows a set of sampled cross sections from the cost-surface for this annealing-style synthesis formulation. At an intermediate point in the synthesis, we stopped the optimizer, and then iteratively stepped each independent variable over its range, while freezing all other variables. At each step point we evaluated the synthesis cost function using Spectre. Fig. 3 shows a few of these resulting cross sections, suitably normalized for comparison. The mix of gently sloping plateaus and jagged obstacles is typical of these landscapes. We require global optimization algorithms because of their potential to avoid many of these inferior local minima.

However, annealing algorithms in particular have a reputation for slow execution because of the large number of solution candidates that must be visited. This is greatly exacerbated when we choose to fully simulate each solution candidate. There are three broad avenues of solution here.

1) **Less search:** attempt to sample the cost function at fewer points. This is essentially the approach taken by [23], which uses an unusual, truncated annealing schedule with some of the character of a random multistart approach. However, in our experience, wider search always yields better solutions and a more robust tool.
2) **Parallel circuit evaluation:** each visited circuit candidate requires more than one circuit simulation to evaluate it. We can easily distribute these over a network to parallel workstations. Indeed, our implementation supports this simple parallelism. For example, if we resynthesize the circuit of Fig. 2, but distribute the five simulations required to evaluate each circuit across three workstations, the 11-h sequential time drops to 192 min. This is a useful form of parallelism to exploit, but it is strictly limited.
3) **Parallel circuit search:** what we really seek is a technique to allow multiple, concurrent points of the cost landscape to be searched in parallel, but synchronized in some manner that guarantees convergence to a final circuit or set of circuits of similar quality.

Unfortunately, annealing *per se* does not easily support parallel search. An annealing-based optimizer generates a serial stream of proposed circuit perturbations, and relies on statistics from previous circuits to adjust its control parameters. To parallelize search itself, an obvious set of methods to consider here are the genetic [32], and evolutionary algorithms [33], [34], whose population-based evolution models distribute over parallel machines more naturally. We focus on a novel population-based strategy in Section II-D.

D. Global Optimization via Stochastic Pattern Search

We have developed two separate attacks on the global optimization problem. We initially focused on extending the annealing paradigm (which has empirically performed very well on the circuit synthesis task) to support an aggressively scalable form of parallelism. We employed an idea from Goldberg [35] called *parallel recombinative simulated annealing* (PRSA). PRSA, which has its roots in genetic algorithms, can be regarded as a strategy for synchronizing a population of annealers as they cooperatively search a cost surface. When generating a new circuit candidate, each annealer in the population makes one of two choices: perturb an element value in its previously generated solution, or recombine its previously generated solution with a solution obtained from another annealer in the population. Recombination is the *crossover* (mating) operator from genetic algorithms, which randomly combines features of two parent solutions into a single new offspring solution. The technique works extremely well to synchronize global search. In particular, good solutions found by one annealer quickly diffuse through the population, and drive annealers stuck in unpromising local minima toward better global solutions. Measured performance scaled essentially linearly out to 30 parallel UNIX CPU's, with Cadence Spectre [30] as the evaluation engine. Experimental results from our implementation of these ideas, called MAELSTROM, appear in [2].

Annealing is one member of a general class of optimization techniques for nondifferentiable cost functions. When cost surfaces are convex or nearly so, or we can assume we start optimization close to an acceptable final solution, gradient and sensitivity-based optimization techniques work well. However, these are generally unworkable in synthesis when we often have no feasible starting solution, many local minima, and gradients are either unreasonably expensive to compute (recall that we insist on full circuit-level simulation to evaluate solution candidate) or, more likely, numerically unreliable. In these cases, researchers have favored global optimization techniques based on sampling of the cost surface, with decisions for search based on the properties of previous samples. Annealing in particular has long been a favored approach here.

However, there are alternatives. An obvious class of methods are the so-called *direct-search* techniques, which sample the cost in a deterministic locus around a given solution point, and use this sample to construct a deterministic direction and distance to a conjectured better solution. As the optimization proceeds, the shape of the locus changes to reflect the success or failure of samples from previously visited regions of the cost surface. Ultimately, the locus converges to a single final locally optimal minimum. Coordinate search, Box search [36], Hook–Jeeves [37], and Nelder–Mead Simplex [38] are all variants of this idea. However, Torczon has recently suggested a unified formulation that renders many of these classical methods as particular cases of a more general method called *pattern search* [39], [40]. Surprisingly, Torczon was able to show pattern search to have provable convergence properties. The theoretical interest generated by Torczon's results motivates us to reconsider direct search ideas.

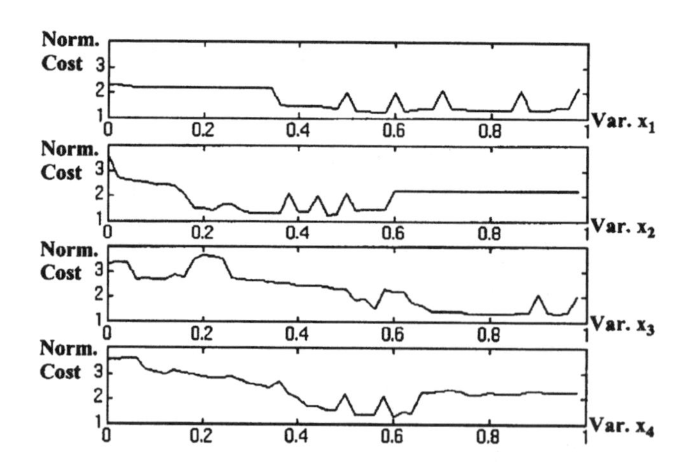
Norm. Cost
3
2
1
0
0.2
0.4
0.6
0.8
1
Var. x_1
Var. x_2
Var. x_3
Var. x_4

Fig. 3. Four one-dimensional normalized cross sections of the cost-surface for a typical simulation-based synthesis problem.

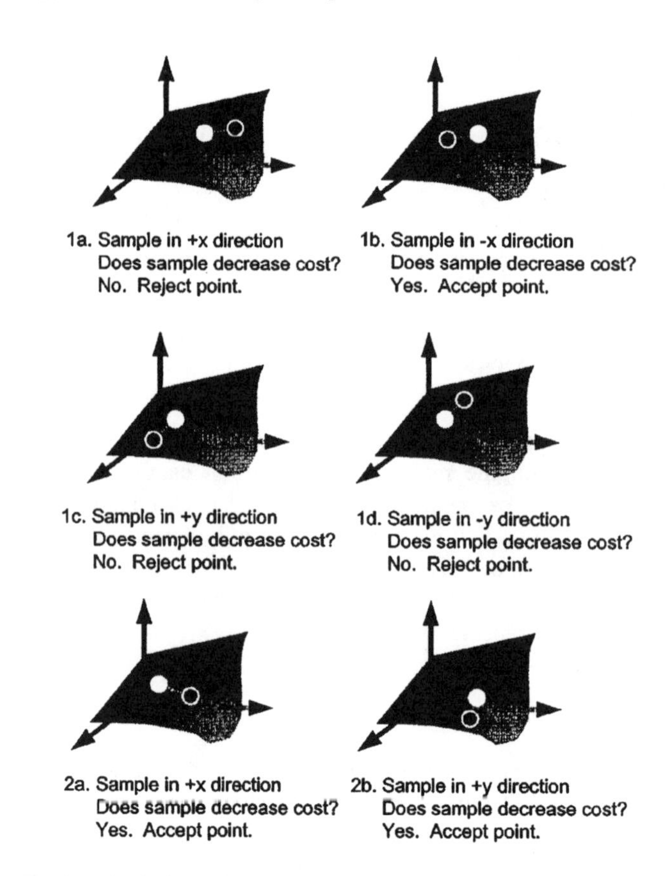
1a. Sample in +x direction
Does sample decrease cost?
No. Reject point.
1b. Sample in -x direction
Does sample decrease cost?
Yes. Accept point.
1c. Sample in +y direction
Does sample decrease cost?
No. Reject point.
1d. Sample in -y direction
Does sample decrease cost?
No. Reject point.
2a. Sample in +x direction
Does sample decrease cost?
Yes. Accept point.
2b. Sample in +y direction
Does sample decrease cost?
Yes. Accept point.

Fig. 4. Classical coordinate search algorithm.

Consider first a classical coordinate search algorithm, illustrated in Fig. 4, for a simple two-dimensional problem. Coordinate search visits in fixed order the individual coordinates x, y of the current solution, perturbs each in a deterministic pattern, and accepts only perturbations that decrease the cost. By making the starting perturbation suitably large, some local minima can be avoided: we "skip over" the hills, rather than accepting directly an uphill move, as in annealing. By shrinking the perturbation size appropriately over the course of the search, we can localize the search to converge on (we hope) a good local minimum.

Nelder–Mead Simplex has been tried in some circuit synthesis and optimization tools. Results to date have been negative: for very small problems with very simply cost surfaces

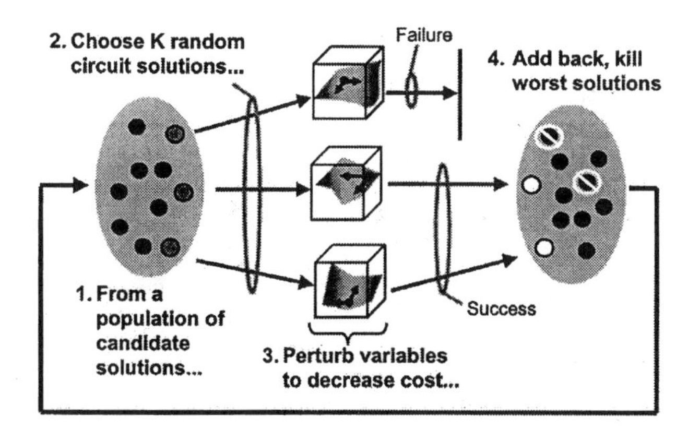
2. Choose K random circuit solutions...
Failure
4. Add back, kill worst solutions
1. From a population of candidate solutions...
3. Perturb variables to decrease cost...
Success

Fig. 5. A flow representation of the stochastic pattern search algorithm. Steps 1–4 are performed, and repeated until the population's cost can no longer be decreased. Step 3 corresponds to the subroutine decrease_cost() in the pseudocode of Fig. 7.

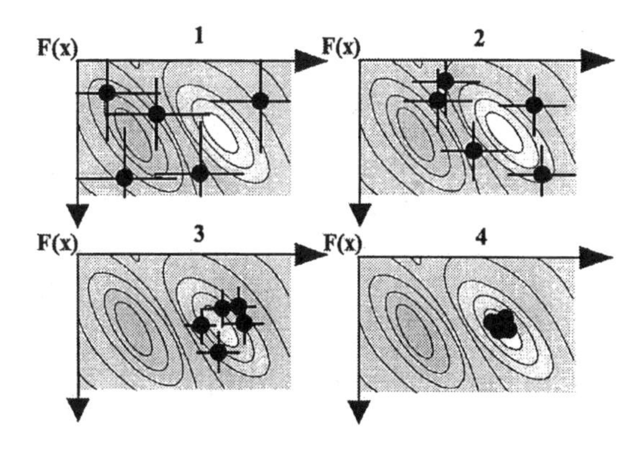
F(x)
1
F(x)
2
F(x)
3
F(x)
4

Fig. 6. Convergence to a cost-surface minimum during stochastic pattern search. The population model allows us to search multiple samples of the cost surface concurrently, all of which ultimately converge to solutions (circuits) of similar quality.

they perform adequately; large and difficult problems, they perform poorly. For example, in our preliminary comparisons to our own Nelder-Mead implementation, ANNEAL++ was always superior—although it obtained its solutions at the cost of considerably more CPU time.

To overcome these problems, we suggest coupling pattern search strategies with the population-of-circuits idea from MAELSTROM, and add a random component to the pattern search itself. The overall architecture is conceptually simple.

1) **Population of partial circuit solutions:** we maintain a large population of partial solutions. It is the population itself which combats the problem of cost surfaces with local minima. Each element is one sample of that cost surface. By maintaining a suitably diverse set of samples, and preferentially updating the population so that lower-cost samples survive and higher-cost samples are culled, we avoid the need to do explicit hill-climbing.
2) **Simple population update:** from a population of P partial solutions, we select k candidates, apply a short pattern search improvement process to each candidate, then replace these in the population. From this updated population of $P + k$ solutions, we remove the k solutions of highest (worst) cost.
3) **Evolution by random pattern search:** the update process for a candidate selected from the population is a truncated pattern search in the style of coordinate search, but with the key difference that the search pattern (direction and step size) are themselves randomized.

We evolve the population via short, randomized pattern searches, and allow only superior solutions to survive any update. The overall flow is illustrated in Fig. 5. The convergence of a population of cost samples to a set of solutions of similar numerical quality is illustrated in Fig. 6. Our intuition here is that pattern searches work well in the neighborhood of minima, but that the reliance on evolving only one solution, and doing so in a deterministic pattern, does not provide a sufficiently vigorous search. A population of solutions partially combats the problem of local minima. Evolving several elements of the population via short bursts of randomized downhill pattern search provides a suitably vigorous pressure to explore promising local minima. Favoring only improved solutions in the population coerces this search to converge to a set of solutions of similar quality, yet the combination of the population model and the randomized pattern searches helps the optimization swarm over a suitably diverse set of samples of the cost surface. This gives the technique its name: *stochastic pattern search.*

Torczon-style pattern search admits parallelism in the obvious form of concurrent exploration of different search patterns [41]. In our case, the population model allows us to perform multiple pattern searches in parallel. A detailed description of the stochastic pattern search algorithm in the form of pseudocode appears in Fig. 7. We begin with a simple coordinate search algorithm, which is shown in pseudocode beginning on line 25 of Fig. 7. The main loop of the subroutine *decrease_cost()* perturbs each independent variable (coordinate) of the current solution vector; we refer to the process of perturbing all coordinates as one atomic *pattern-search*. However, the order in which each variable is perturbed is random, and the perturbation amount is also random, although bounded. Each variable may be perturbed in both the positive and negative direction: the goal is to find a perturbation that decreases the cost. Perturbation bounds decrease as the algorithm progresses, shrinking around the evolving solution. One nonrandom component is the acceptance criterion for each coordinate perturbation: we only accept solutions that decrease the cost. Unlike annealing, we do not accept uphill moves; however, because coordinate search can take very large steps, we can "skip over" hills.

The algorithm uses a master-slave organization for managing the population. The code for each parallel node begins on line 1 of Fig. 7. We start with a population of P circuit solutions, sorted on cost. We randomly select a subset of k of these solutions, with some bias toward those of lower cost. We perform one pattern-search on each of these k candidates, add them back to the population, re-sort, and then cull out the worst k solutions. This repeats until no more improvements can be found in any elements of the population, i.e., all circuits have essentially similar cost. We can distribute the pattern-search for each of the k candidates in parallel, and each individual coordinate search itself requires t multiple circuit simulations, allowing us to easily exploit kt parallel workstations. So, as noted earlier, we can distribute both search and evaluation.

```
1   if this node is the master
2       spawn m slaves
3       initialize the population, P, with n random points
4       while the population cost is still decreasing
5           foreach slave, s
6               choose a point, p, from the population, P
7               send p to s
8           end
9           wait for a single point from each slave, m total points
10          foreach point, p, received from the slaves
11              if p decreases the cost of P
12                  insert p into P
13                  remove the highest cost point from P
14              end
15          end
16      end
17  else this node is a slave
18      wait for a point, p, from the master
19      p_new = decrease_cost(p)
20      send p_new to the master
21      adjust step size bounds
22  end
23
24  subroutine decrease_cost(p)
25      p_proposed = p
26      min_cost = cost of p
27      while there is a coordinate of p_proposed that has not been visited
28          choose a random index, i, that has not been chosen yet
29          p_proposed[i] += random_petrurbation
30          if the cost of p_proposed is less than min_cost
31              min_cost = the cost of p_proposed
32          else
33              p_proposed[i] = p[i]
34          end
35      end
36      if min_cost < the cost of p
37          return proposed_p
38      else
39          return p
40      end
41  end subroutine
```

Fig. 7. Pseudocode for parallel stochastic pattern search.

E. Network Architecture

Our implementation distributes all computation over a pool of workstations. We use software components and organizational ideas from MAELSTROM, which at the lowest level manages concurrency and interprocessor communication using the publicly available PVM library [42]. The overall network architecture is illustrated in Fig. 8. The important point to make is that the master-slave structure implied by the stochastic pattern search pseudocode of Fig. 7 is only a *logical* organization for the required parallel computation, not a *physical* organization. In other words, we do not bind the master and individual slave nodes each to a separate physical CPU. In our terminology, the process of selecting and updating k candidates in the population is handled by k *slaves*. But each slave can use as many physical CPU's as necessary. Indeed, the actual mechanics of managing the population and updating the required state information consume a negligible fraction of the overall execution time, which is wholly dominated by the circuit simulations. The two physical characteristics of the method that we must deal with are as follows.

1) Support for an arbitrary number of available workstations. We should be able to use as many machines as we have available, or from which we can harvest spare cycles. But, we must be able to make progress on synthesis even if we have fewer machines than we would prefer.
2) Support for intelligent scheduling of the simulation tasks among available workstations. Different simulation tasks (dc, ac, transient, test-jig setup, early parameter estima-

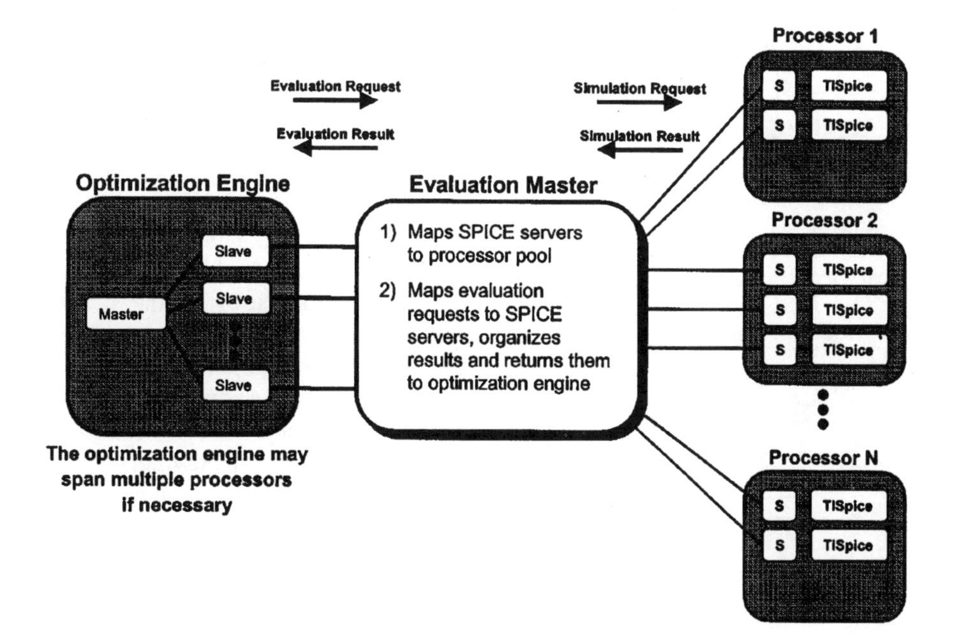

Fig. 8. ANACONDA network architecture.

tion, etc.) can each require vastly different amounts of time. Assuming that each simulation needs its own CPU is not only impractical, but inefficient. Our progress would always be limited by the slowest simulation.

The architecture in Fig. 8 addresses these concerns via a separate layer of software called the *evaluation master*. It brokers requests for simulations (from concurrent pattern searches on elements of the evolving population) among a set of available CPU's (slaves) that perform actual simulations. Any individual CPU may be performing several simulations in series. The evaluation master dynamically schedules simulation tasks to maximize the throughput on the available machines, using simple bin-packing heuristics. Thus, one CPU might be tasked to run several small simulations, each of which is expected to complete quickly, while another CPU might run only one simulation, which is expected to be long. The evaluation master tracks completion times for each simulation task on each node, and uses this information to periodically reschedule all tasks. Simulation times can vary over the course of a simulation both because of transient changes in machine loading, and because circuit candidates migrate around in the solution space, and not all solutions are equivalently easy to simulate. Dynamic scheduling makes for efficient use of the available computational resources.

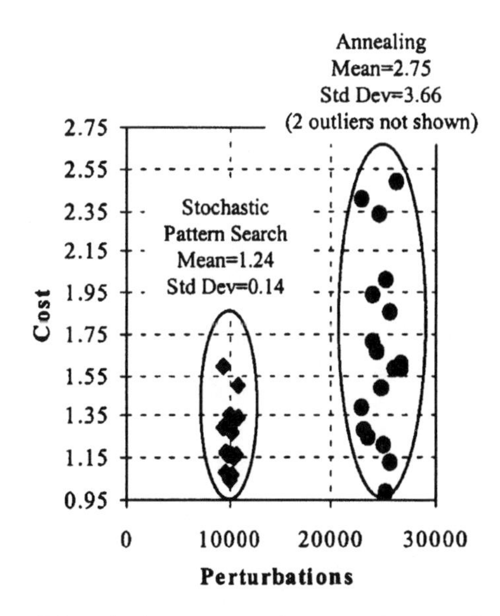

Fig. 9. Parallel stochastic pattern search compared to a well-tuned serial annealer for synthesis of the circuit in Fig. 10(c). The x axis is the number of serial perturbations. The total number of design points evaluated by the pattern search algorithm was 10× the number of serial perturbations, or approximately 100 000. The total number of design points evaluated by the annealer is 25 000 because it is a serial annealer.

At this point, we can offer a simple experiment that shows the merits of our approach. We revisited the simulator-in-serial-annealer approach used in Fig. 2, and compared it to stochastic pattern search when applied to the task of synthesizing the industrial circuit (to be described in Section IV) of Fig. 10(c). Twenty synthesis runs were performed using each algorithm, and the results are shown in Fig. 9. We show a scatter plot of cost [the OBLX-form cost-function being minimized, from (3)] versus "perturbations." For the annealer, this is the number of circuit candidates visited sequentially. For stochastic pattern search, each perturbation actually visits k new circuits, and we evaluate these in parallel across a pool of ten workstations. For this result, the parallel pattern search algorithm actually visited roughly 100 000 circuits, but with a CPU time proportional to only 10 000 circuit simulations.

Note that stochastic pattern search is producing both a better average answer (i.e., finding a better expected minimum, which corresponds to a better circuit solution), and a tighter spread of answers. We believe this is a result of our parallel population model, which not only allows us to visit more solutions in a reasonable amount of time, but also does a better job of pruning weak solution candidates before search becomes trapped in poor local minima. This is critical since any stochastic optimizer produces a spread of solutions on repeated runs; tighter spreads mean a higher likelihood of finding a good solution on each invocation of the tool.

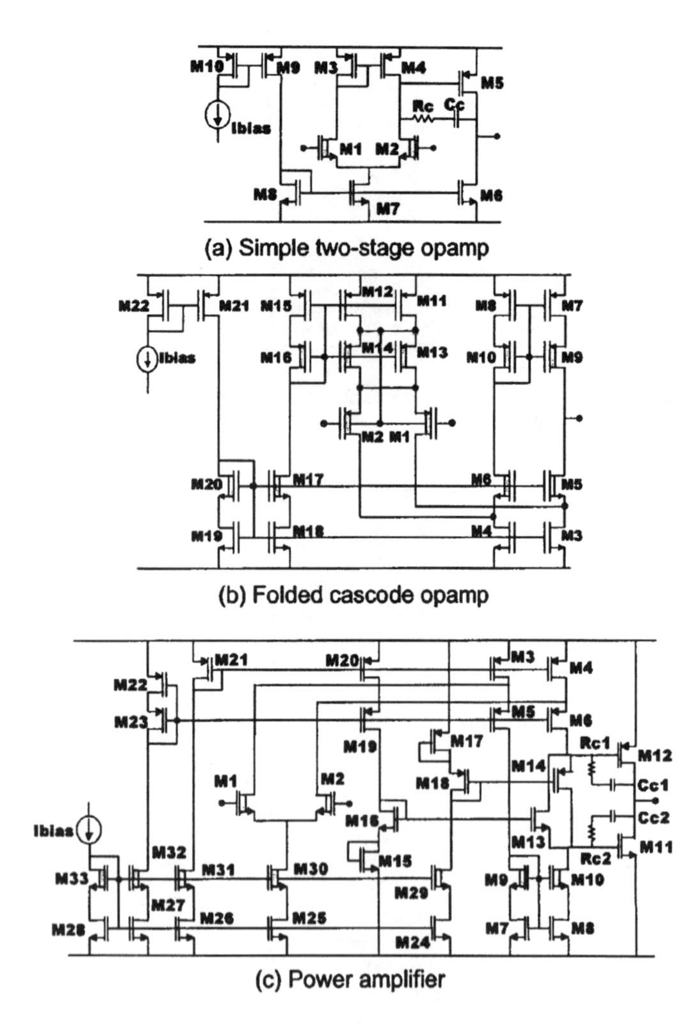

(a) Simple two-stage opamp

(b) Folded cascode opamp

(c) Power amplifier

Fig. 10. Industrial test circuits for our synthesis experiments.

IV. RESULTS

The ideas presented in this paper have been implemented in a tool called ANACONDA and tested on site at Texas Instruments in Dallas, TX. We benchmarked ANACONDA on the three opamps, shown in Fig. 10, that are all examples of production circuits for which we also have complete TI manual designs. The power amp was designed in a 0.8-μm CMOS process and the other two opamps were designed in a 0.6-μm CMOS process.

ANACONDA consists of a parallel stochastic pattern search algorithm coupled (via encapsulation) to an industrial circuit simulator that evaluates each candidate solution. The simulator used was TI's proprietary TISpice. In addition, all three circuits were synthesized using the same parameters for the pattern search algorithm: the population size was set to 200, and ten logical slaves (i.e., we select and replace ten elements of the population concurrently) were used to evaluate circuits in the population. Simulation tasks were dynamically rescheduled every 100 simulations. Also, the same weightings in the cost function from (3) were used for all three circuits. The constraints for each design had a weighting of one while the objectives had a weighting of 0.5.

Figs. 12 and 13 offer some insight into the population dynamics as these sets of benchmark circuits evolved. Fig. 12 plots the mean cost of each population of 200 circuits for each of the five synthesis runs, for each of the three benchmarks. An immediate and striking feature of the data is how closely the individual populations track each other, even across separate synthesis runs. We take this as indication that the population update/replacement strategies are performing well: they are maintaining an appropriately diverse sample of the cost surface, while encouraging convergence to solutions of similar cost by culling always the weakest solutions. Fig. 13 plots the maximum, mean, and minimum cost samples seen in each population for one sample run for each of the three circuits. As expected, there is significant disparity between the minimum and maximum values at the start of synthesis, but these rapidly converge as weaker solutions are culled and the solution candidates cluster around a single best value. Perhaps more interesting is the qualitative difference between the dynamics of the smaller two-stage and folded-cascode circuits, which show somewhat larger final spreads in contrast to the larger, more difficult power amp circuit. We interpret this as related to the size of the design—the number of degrees of freedom (DOF). The power amp is a large design; we believe there are simply more nearby solutions of equivalent quality than for the other smaller designs. For the smaller designs, there simply appear to be fewer good final solutions that are unique, hence the final population retains a larger fraction of weaker designs. A contributing factor here is also that, in industrial practice, device sizes are not continuous, but are restricted to some discrete (though closely spaced) values. This discretization affects the smaller designs somewhat more noticeably. Finally, it is worth noting that the cost scale is logarithmic: even though the spreads are visible in the plots, in reality the costs are extremely close. Table I gives the input performance constraints, simulation environment, and runtime statistics for each synthesized circuit. For our experiments, the performance constraints were set by running nominal simulations on the original hand designs. Each design was synthesized using a pool of 16 300-MHz Ultra 10's and four 300-MHz dual-processor Ultra 2's. Note that although we only update ten circuits in the population in parallel, each circuit required between five and seven simulations to evaluate. Thus, we can make use of the entire pool of 24 CPU's. Because runtime is highly dependent on the size of the circuit and the type of simulation being performed, there was significant variation among the circuits. For example, for the power amp, each total harmonic distortion (THD) analysis took a few CPU seconds and, consequently, that circuit had significantly longer synthesis times. Overall, each synthesis task required between a few hours and overnight on a pool of (typically) 20 available workstations.

Fig. 11 shows the results of five consecutive synthesis runs for each circuit. Given the large number of performance specifications for each circuit, we summarize these results as power-versus-area scatter plots. For each design, the objectives were to minimize area and static power dissipation. In all but one case, the synthesized circuits met *all* of the performance constraints specified in Table I. Several of these designs are in fact superior to their manual counterparts. And, it is worth noting again

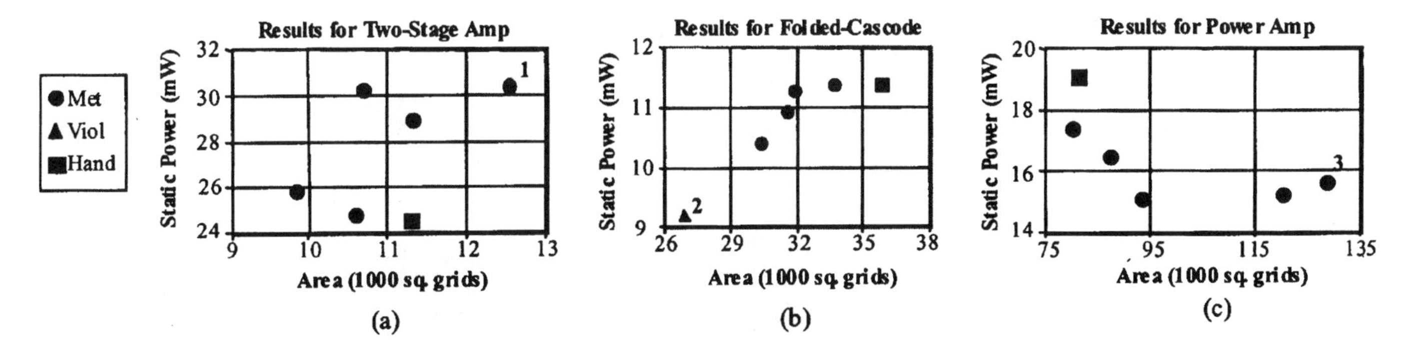

Fig. 11. Static power versus area scatter plots for five consecutive synthesis runs for each test circuit. Each scatter includes a manual circuit design represented by a square point. Designs such as "1" in result (a) that have significantly higher static power dissipation typically are over-designed for some specification; for result "1" the settling time was slightly smaller than necessary. Design "2" in (b) had a UGF of 159 MHz, and this was slightly less than the 162 MHz required. Designs such as "3" in (c) that have significantly larger area again represent over-design in a performance constraint; for result "3," the THD was 15% smaller than necessary.

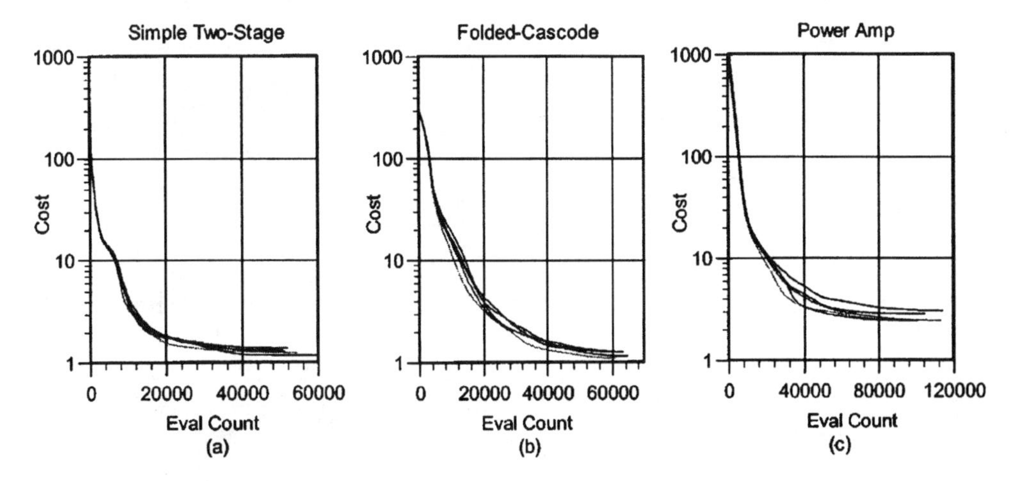

Fig. 12. Evolution of the population's mean cost over five consecutive synthesis runs, for the synthesis results shown in Fig. 11. The x axis is the total number of design points that have been evaluated.

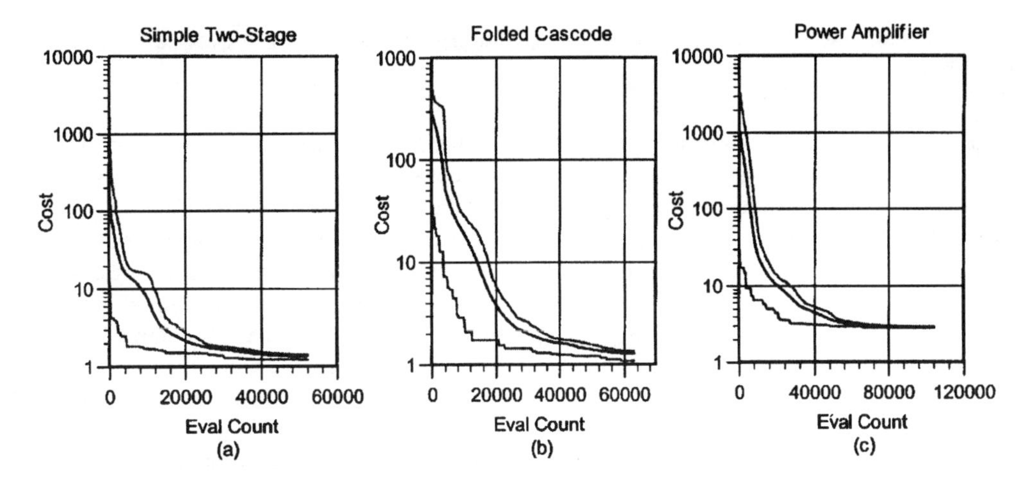

Fig. 13. A sample synthesis run for each test circuit from Fig. 10, illustrating how the minimum, mean and maximum cost of candidates in the population evolve over time. The x axis is the total number of evaluated design points.

that the quality metrics—meeting specifications under detailed circuit simulation—are *identical* to those used during manual design.

Our results so far suggest that ANACONDA is effective for nominal cell-level analog synthesis. But we must also make some efforts to address manufacturing process variations and environmental operating range constraints. We know from experience that we cannot ignore these issues, since a well-performing synthesis algorithm can push a final design very close to the edge of the feasible region where all constraints are met. Even modest changes in process or operating conditions can then render the circuit nonfunctional [43]. We have two broad classes of solutions: add first-order constraints to the synthesis task, mimicking the "conservative" design practices of skilled designers, or fold a numerical manufacturability optimization into the synthesis process itself [44]. We choose the former

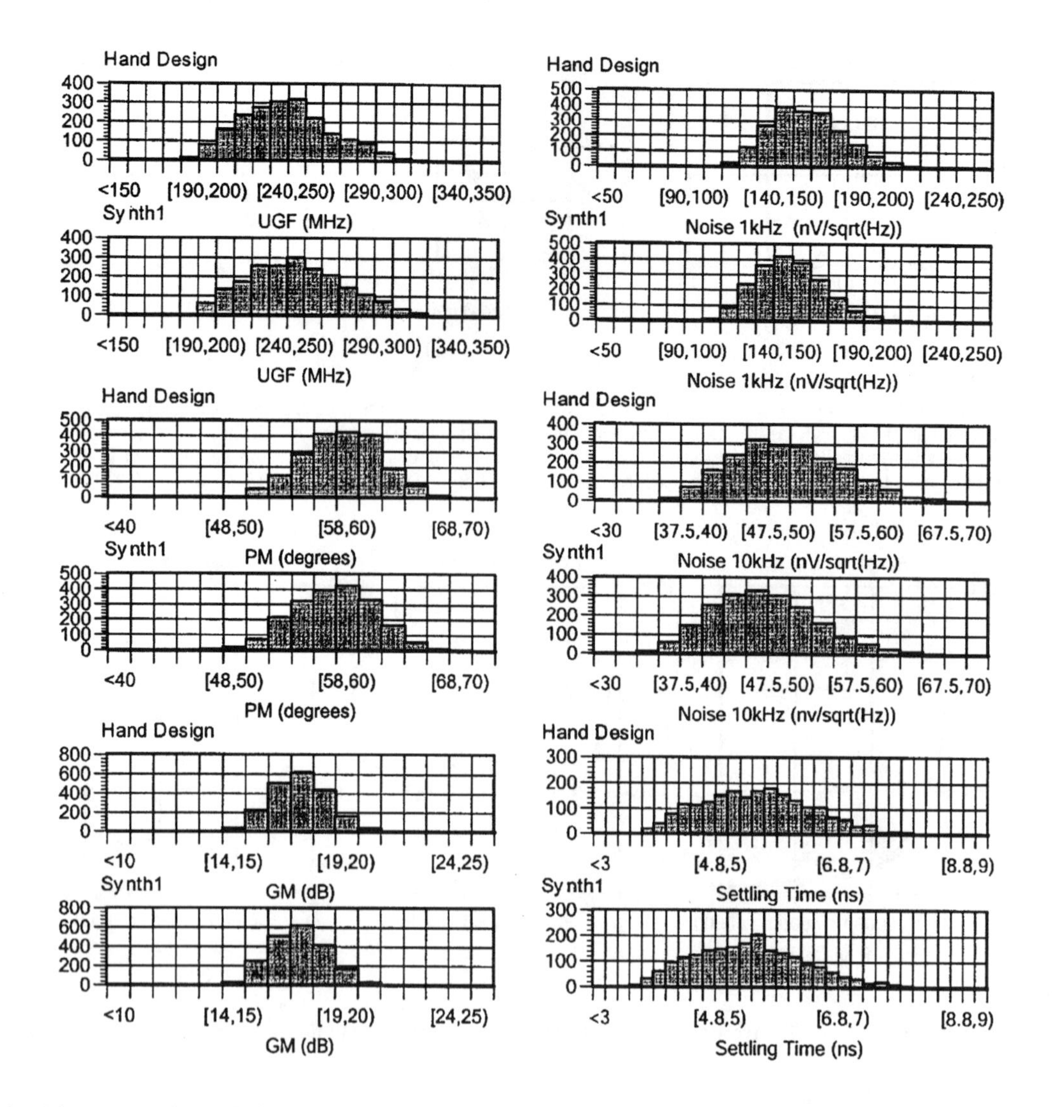

Fig. 14. A sample of the monte-carlo results for the simple two-stage opamp. Histograms for the performance specs are shown first for the hand design and second for the synthesized design.

option here, and focus on how to properly constrain a circuit so that synthesis yields a result with design centering comparable to an expert manual design. In ANACONDA, the two types of constraints supported for this purpose are *operating region constraints* and *component constraints*. These adapt ideas from ASTRX/OBLX.

Operating region constraints simply specify that individual devices should be "far enough" into the desired region that process or environmental variations cannot force the device out of this region. If operating region constraints are not used, the optimizer may choose to bias a transistor on the edge of the active region for nominal performance gains. To avoid such a situation, we can designate how large we would like the *effective voltage*, i.e., how far above V_T we require V_{GS} to be for a MOSFET, for individual devices in the design. For example, a typical number when designing CMOS opamps in micron-scale processes is simply to fix this effective voltage to 250 mV; we do not expect individual devices each to have precisely tailored, unique voltage constraints. When designing by hand the value for the effective voltage is set using first order equations. Unfortunately, the simulated value may be significantly different than the hand calculated value. One advantage of using simulator-in-the-loop synthesis is that when the constraint is met one can be assured that the effective voltage will be the value specified.

Designers can eliminate unnecessary or possibly dangerous DOF by specifying *component constraints*. As an example, consider the simple two-stage opamp in Fig. 10(a). Clearly, a reasonable design would have the input differential pair matched. This could be specified using the simple constraints $W1 = W2$ and $L1 = L2$. Such constraints are trivial to accommodate. But in real designs, many component values are determined parametrically, as functions of other designables in the circuit. For example, it is common practice to set the compensation resistor R_C in the circuit of Fig. 10(a) to the value

$$R_C = \frac{1+\delta}{gm_5} \tag{4}$$

where δ is a designer-input constant reflecting the degree of overdesign desired, and gm_5, the transconductance of another device in the circuit, requires a *separate* SPICE simulation to compute. Indeed, it is not uncommon to require a series of these "setup" simulations to first create the proper component relationships *before* real evaluation of the circuit can begin. Sim-

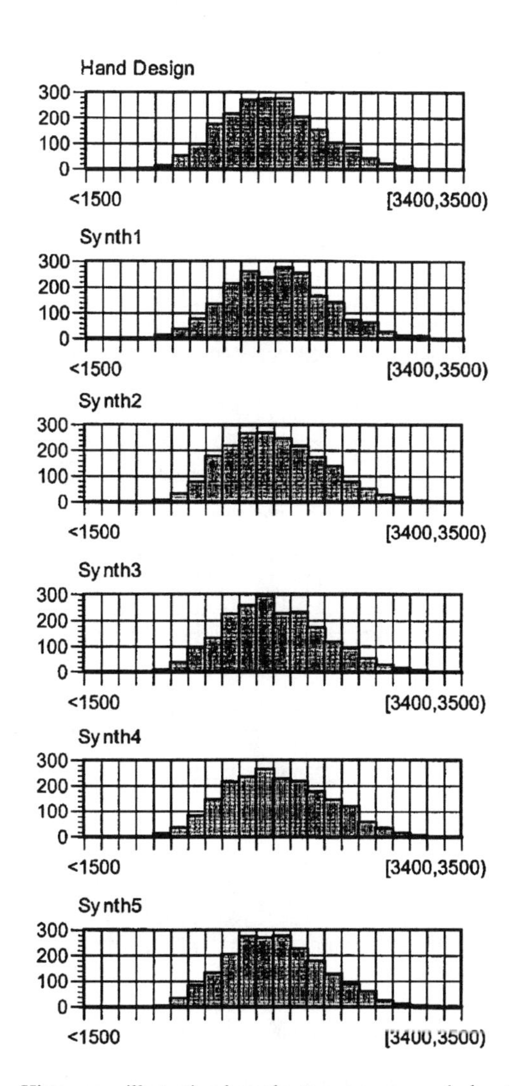

Fig. 15. Histograms illustrating how the two-stage opamp's low-frequency gain (x axis) varies across process variations. Results from the hand design are shown first, followed by results for the five consecutive synthesis runs. These histograms illustrate the consistent quality across consecutive synthesis runs.

TABLE I
DETAILED PERFORMANCE SPECIFICATIONS, SIMULATION ENVIRONMENT, AND SYNTHESIS RUNTIME RESULTS FOR THE THREE TEST CIRCUITS OF FIG. 10

Environment Spec		*Two-Stage*	*Folded-Cascode*	*Power Amp*
T=25°C				
Vdd	V	2.7	2.7	5
C_L	pF	1.5	1.75	100
R_L	Ω			25
Performance Constraint				
Gain	dB	≥ 68	≥ 71	≥ 92
UGF	MHz	≥ 260	≥ 162	≥ 0.6
PM	deg.	≥ 56	≥ 52	≥ 84
GM	dB	≥ 17		
Noise 1kHz [1]	nV $Hz^{-0.5}$	≤ 145	≤ 70	≤ 52
Noise 10kHz	nV $Hz^{-0.5}$			≤ 40
Noise 10MHz	nV $Hz^{-0.5}$	≤ 3	≤ 4	
CMRR	dB	≥ 76	≥ 108	> 138
PSRR (Vss)	dB	≥ 85	≥ 89	≥ 90
PSRR (Vdd)	dB	≥ 70	≥ 72	≥ 94
Settling Time	ns	≤ 4.6[4]	≤ 16[5]	
THD[2]	%			≤ 0.06
THD[3]	%			≤ 0.1
Runtime Info				
Independent Variable Count		15	13	20
Approx. Evaluation Count[6]		57,000	65,000	102,000
Ave. Runtime [7] (hrs)		2.8	1.8	10

1 – All noise specs are input referred and as measured in an ac analysis
2 – 4.0V p-p 1kHz input
3 – 2.6V p-p 1kHz input, R_L=5Ω
4 - 0.5V input step, 10-Bit accurate final voltage
5 - 0.1V input step, 10-Bit accurate final voltage
6 – Number of circuit evaluated in series is approx the evaluation count/10
7 – Using a pool of 16 300MHz SUN Ultra10 and 4 dual 300MHz Ultra2 workstations running Solaris 2.5.1.

ulation-based synthesis algorithms like ANACONDA (and also MAELSTROM) extend gracefully to handle these sorts of practical usage scenarios.

The circuits from Fig. 10 and results from Fig. 11 were obtained using these common-practice, first-order constraints. To assess the robustness of the approach, each of the synthesized designs was verified by performing Monte Carlo simulations with 3σ process, 10% voltage supply, and 0 °C to 100 °C temperature variations. Selected results are illustrated in Figs. 14–16. Fig. 14 shows histograms for most of the simple two-stage opamp's performance specifications. A single synthesis result, synthesis run 1, is compared to the hand design. From the histograms we see that the synthesized design performs as well as the hand design across process and environment variations for each of the performance specifications. Figs. 15–17 shows sample histograms for the low-frequency gain of the three test designs. Notice that all five consecutive synthesis runs are shown. This illustrates how the robustness of the circuits varies from one synthesis run to the next. For the simple two-stage and folded cascode opamps there is very little variation. However, for the power amplifier, Fig. 17, we see that four of the five synthesized designs actually have tighter spreads than the hand design. The main reason for these overall tight spreads is the rigorous enforcement of the operating region and component constraints. These sorts of constraints are a standard part of good manual design practice—but not always enforced with the discipline we can achieve in a numerical synthesis tool.

These results are noteworthy in several respects. First, these are production-quality industrial analog cells with difficult performance specifications. Second, our synthesis approach is using as its evaluation engine the *identical* simulation environment used by TI's designers to validate their manual designs. As a result, we can deal accurately with difficult design specification such as noise, settling time, and THD, which require detailed simulations to evaluate complex nonlinear effects. Finally, nearly all of these synthesized designs compare favorably with their manually designed counterparts, both in performance and in robustness across manufacturing and environmental corners. We believe this is a significant advance in demonstrating how an analog synthesis tool can attack realistic industrial circuit designs.

V. CONCLUSION

We have presented a novel synthesis strategy for custom analog cells. Our central contribution is stochastic pattern search, a parallelizable global optimization algorithm that

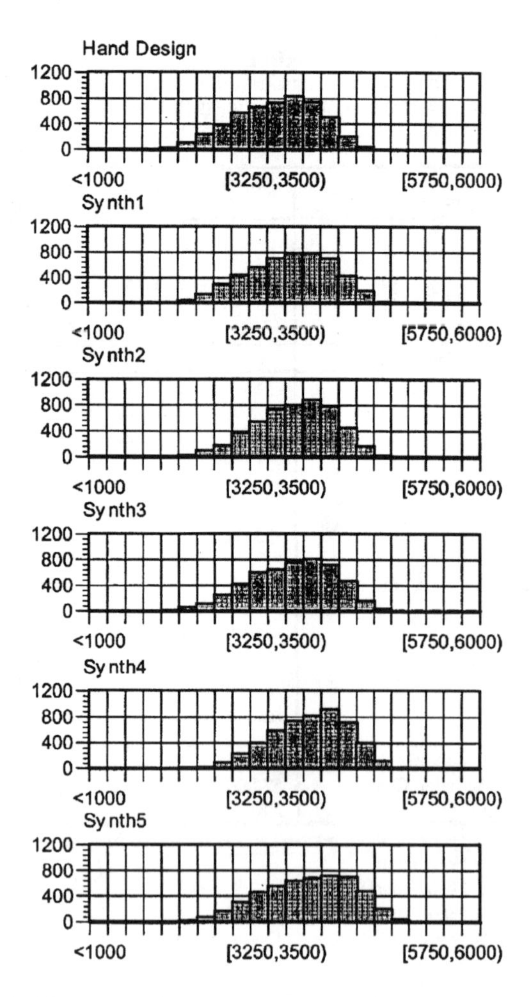

Fig. 16. Histograms illustrating how the folded cascode opamp's low-frequency gain (x axis) varies across process variations. Results from the hand design are shown first, followed by results for the five consecutive synthesis runs. These histograms illustrate the consistent quality across consecutive synthesis runs.

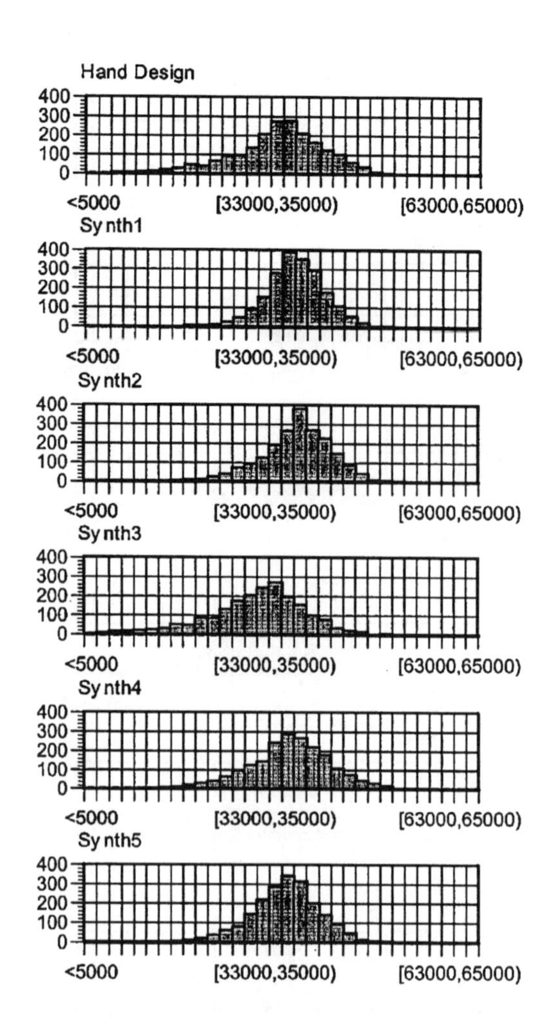

Fig. 17. Histograms illustrating how the power amp's low-frequency gain (x axis) varies across process variations. Results from the hand design are shown first, followed by results for the five consecutive synthesis runs. These histograms illustrate the consistent quality across consecutive synthesis runs.

combines ideas from evolutionary algorithms and numerical pattern search. By encapsulating commercial circuit simulators and distributing the search across a pool of workstations, we can visit 50 000–100 000 circuit candidates, and fully simulate each, in a few hours. ANACONDA, an implementation of these ideas, has successfully synthesized several difficult industrial cells. We believe analog synthesis is a necessary component of any strategy for reusing and retargeting analog circuits. We are currently working to apply ANACONDA to much larger designs, in particular, system-level designs. Other current work focusses on comparing the relative merits of the genetic-annealing approach used in MAELSTROM with the population-pattern approach used in ANACONDA. In addition, the practical ability to synthesize analog circuits raises a host of questions about analog intellectual property—how it should be archived, what information model is necessary for practical reuse, how much effort is required to "package" a cell for future use, etc. We are currently evolving the MAELSTROM framework to address these issues.

ACKNOWLEDGMENT

The authors would like to thank E. Ochotta (Xilinx) for discussions about OBLX, B. Bearden (TI) for help with TISpice and G. Richey and F. James (TI) for valuable discussions about this work.

REFERENCES

[1] E. Ochotta, R. A. Rutenbar, and L. R. Carley, "Synthesis of high-performance analog circuits and ASTRX/OBLX," *IEEE Trans. Computer-Aided Design*, vol. 15, pp. 237–294, Mar. 1996.

[2] M. Krasnicki, R. Phelps, R. A. Rutenbar, and L. R. Carley, "MAELSTROM: Efficient simulation-based synthesis for analog cells," in *Proc. ACM/IEEE Design Automation Conf.*, June 1999, pp. 945–950.

[3] R. Phelps, M. Krasnicki, R. A. Rutenbar, and L. R. Carley, "ANACONDA: Robust synthesis of analog circuits via stochastic pattern search," in *Proc. IEEE Custom Integrated Circuits Conf.*, May 1999, pp. 567–570.

[4] E. Ochotta, T. Mukherjee, R. A. Rutenbar, and L. R. Carley, *Practical Synthesis of High-Performance Analog Circuits*. Norwell, MA: Kluwer Academic, 1998.

[5] M. Degrauwe *et al.*, "Toward an analog system design environment," *IEEE J. Solid-State Circuits*, no. 3, p. 24, June 1989.

[6] H. Y. Koh, C. H. Sequin, and P. R. Gray, "OPASYN: A compiler for MOS operational amplifiers," *IEEE Trans. Computer-Aided Design*, vol. 9, pp. 113–125, Feb. 1990.

[7] G. Gielen *et al.*, "Analog circuit design optimization based on symbolic simulation and simulated annealing," *IEEE J. Solid-State Circuits*, vol. 25, pp. 707–713, June 1990.

[8] F. Leyn, W. Daems, G. Gielen, and W. Sansen, "A behavioral signal path modeling methodology for qualitative insight in and efficient sizing of CMOS opamps," in *Proc. ACM/IEEE ICCAD*, 1997, pp. 374–381.

[9] P. C. Maulik, L. R. Carley, and R. A. Rutenbar, "Integer programming based topology selection of cell level analog circuits," *IEEE Trans. Computer-Aided Design*, vol. 14, pp. 401–412, Apr. 1995.

[10] W. Kruiskamp and D. Leenaerts, "DARWIN: CMOS Opamp synthesis by means of a genetic algorithm," in *Proc. 32nd ACM/IEEE DAC*, 1995, pp. 433–438.

[11] M. Hershenson, S. Boyd, and T. Lee, "GPCAD: A tool for CMOS Op-Amp synthesis," in *Proc. ACM/IEEE ICCAD*, Nov. 1998, pp. 296–303.

[12] R. Harjani, R. A. Rutenbar, and L. R. Carley, "OASYS: A framework for analog circuit synthesis," *IEEE Trans. Computer-Aided Design*, vol. 8, pp. 1247–1266, Dec. 1989.

[13] B. J. Sheu *et al.*, "A knowledge-based approach to analog IC design," *IEEE Trans. Circuits and Systems*, vol. 35, no. 2, pp. 256–258, 1988.

[14] E. Berkcan *et al.*, "Analog compilation based on successive decompositions," in *Proc. 25th IEEE DAC*, 1988, pp. 369–375.

[15] J. P. Harvey *et al.*, "STAIC: An interactive framework for synthesizing CMOS and BiCMOS analog circuits," *IEEE Trans. Computer-Aided Design*, pp. 1402–1417, Nov. 1992.

[16] C. Makris and C. Toumazou, "Analog IC design automation part II—Automated circuit correction by qualitative reasoning," *IEEE Trans. Computer-Aided Design*, vol. 14, pp. 239–254, Feb. 1995.

[17] A. Torralba, J. Chavez, and L. Franquelo, "FASY: A fuzzy-logic based tool for analog synthesis," *IEEE Trans. Computer-Aided Design*, vol. 15, pp. 705–715, July 1996.

[18] G. Gielen, P. Wambacq, and W. Sansen, "Symbolic analysis methods and applications for analog circuits: A tutorial overview," *Proc. IEEE*, vol. 82, pp. 287–304, Feb. 1990.

[19] C. J. Shi and X. Tan, "Symbolic analysis of large analog circuits with determinant decision diagrams," in *Proc. ACM/IEEE ICCAD*, 1997, pp. 366–373.

[20] Q. Yu and C. Sechen, "A unified approach to the approximate symbolic analysis of large analog integrated circuits," *IEEE Trans. Circuits and Sys.*, vol. 43, pp. 656–669, Aug. 1996.

[21] W. Daems, G. Gielen, and W. Sansen, "Circuit complexity reduction for symbolic analysis of analog integrated circuits," in *Proc. ACM/IEEE Design Automation Conf.*, June 1999, pp. 958–963.

[22] P. Wambacq, J. Vanthienen, G. Gielen, and W. Sansen, "A design tool for weakly nonlinear analog integrated circuits with multiple inputs (mixers, multipliers)," in *Proc. IEEE CICC*, San Diego, CA, May 1991, pp. 5.1.1–5.1.4.

[23] F. Medeiro, F. V. Fernandez, R. Dominguez-Castro, and A. Rodriguez-Vasquez, "A statistical optimization based approach for automated sizing of analog cells," in *Proc. ACM/IEEE ICCAD*, 1994, pp. 594–597.

[24] S. Kirkpatrick, C. D. Gelatt, and M. P. Vecchi, "Optimization by simulated annealing," *Science*, vol. 220, no. 4598, pp. 45–54, May 1983 .

[25] L. T. Pillage and R. A. Rohrer, "Asymptotic waveform evaluation for timing analysis," *IEEE Trans. Computer-Aided Design*, vol. 9, pp. 352–356, Apr. 1990.

[26] W. Nye *et al.*, "DELIGHT.SPICE: An optimization-based system for the design of integrated circuits," *IEEE Trans. Computer-Aided Design*, vol. 7, Apr. 1988.

[27] K. Saab, D. Marche, N. N. Hamida, and B. Kaminska, "LIMSoft: Automated tool for sensitivity analysis and tesst vector generation," *Inst. Elect. Eng Proc. Circuits Devices Systems*, vol. 143, no. 6, pp. 386–392, Dec. 1996.

[28] A. R. Conn, R. A. Haud, C. Viswesvariah, and C. W. Wu, "Circuit optimization via adjoint lagrangians," in *Proc. ACM/IEEE ICCAD*, Nov. 1997, pp. 281–288.

[29] M. Krasnicki, "Generalized analog circuit synthesis," masters thesis, Dept. of ECE, Carnegie Mellon, Pittsburgh, PA, Dec. 1997.

[30] K. S. Kundert, *The Designer's Guide to SPICE & SPECTRE*. Norwell, MA: Kluwer Academic, 1995.

[31] K. Nakamura and L. R. Carley, "A current-based positive-feedback technique for efficient cascode bootstrapping," in *Proc. VLSI Circuits Symp.*, June 1991, pp. 107–108.

[32] J. H. Holland, *Adaptation in Nature and Artificial Systems*. Ann Arbor: Univ. Michigan Press, 1975.

[33] D. B. Fogel, "An introduction to simulated evolutionary optimization," *IEEE Trans. Neural Networks*, vol. 5, pp. 3–14, Jan. 1994.

[34] ——, *Evolutionary Computation: The Fossil Record*, D. B. Fogel, Ed. Piscataway, NJ: IEEE Press, 1998.

[35] S. W. Mahfoud and D. E. Goldberg, "Parallel recombinative simulated annealing: A genetic algorithm," *Parallel Computing*, vol. 21, pp. 1–28, 1995.

[36] G. E. P. Box, "Evolutionary operation: A method for increasing industrial productivity," *Appl. Statist.*, vol. 6, pp. 81–101, 1957.

[37] R. Hooke and T. A. Jeeves, "'Direct search' solution of numerical and statistical problems," *ACM JACM*, vol. 8, pp. 219–229, 1961.

[38] J. A. Nelder and R. Mead, "A simplex method for function minimization," *Comput. J.*, vol. 7, pp. 308–313, 1965.

[39] V. Torczon, "On the convergence of the multidirectional search algorithm," *SIAM J. Optimization*, vol. 1, no. 1, Feb. 1991.

[40] ——, "On the convergence of pattern search algorithms," *SIAM J. Optimization*, vol. 7, no. 1, Feb. 1997.

[41] ——, "PDS: Direct search methods for unconstrained optimization on either sequential or parallel machines," Dept. of Mathematical Sciences, Rice Univ., Houston, TX, Tech. Report 92-9, 1992.

[42] A. Geist, A. Beguelin, J. Dongarra, W. Jiang, R. Manchek, and V. Sunderam, *PVM: Parallel Virtual Machine A User's Guide and Tutorial for Network Parallel Computing*. Cambridge, MA: MIT Press, 1994.

[43] T. Mukherjee, L. R. Carley, and R. A. Rutenbar, "Synthesis of manufacturable analog circuits," in *Proc. ACM/IEEE ICCAD*, 1994, pp. 586–593.

[44] G. Debyser and G. Gielen, "Efficient analog circuit synthesis with simultaneous yield and robustness optimization," in *Proc. ACM/IEEE Int. Conf. Computer-Aided Design (ICCAD)*, Nov. 1998, pp. 308–311.

Rodney Phelps (S'96) received the B.S.E.E degree with university honors from Carnegie Mellon University, Pittsburgh, PA, in 1994. In 1996, he returned to Carnegie Mellon University where he is currently completing the Ph.D. degree.

From 1994–1996, he worked at Cirrus Logic, Fremont, CA, in the corporate EDA group. He worked at Texas Instruments, Dallas, TX, from June 1998 to August 1999, in the Mixed-Signal EDA group where he focused on synthesis algorithms for analog circuits and systems. In January 2000, he joined Neo Linear, Pittsburgh, PA.

Mr. Phelps is a member of Eta Kappa Nu and Tau Beta Pi.

Michael Krasnicki received the B. S. degree with highest distinction from the University of Virginia, Charlottesville, in 1995. He received the M. S. degree in electrical and computer engineering from Carnegie Mellon University, Pittsburgh, PA, in 1995. He is currently working toward the Ph. D. degree at Carnegie Mellon University.

He is a Semiconductor Research Corporation (SRC) Fellow. He currently works at Texas Instruments, Dallas, TX. His research focus is performance-driven synthesis of analog cells with the use of industrial quality simulation.

As an undergraduate, Mr. Krasnicki received the William L. Everett Student Award for Excellence from the Department of Electrical Engineering.

Rob A. Rutenbar (S'77–M'84–SM'90–F'98) received the Ph.D. degree from the University of Michigan, Ann Arbor, in 1984.

He joined the faculty of Carnegie Mellon University (CMU), Pittsburgh, PA, where he is currently Professor of Electrical and Computer Engineering, and (by courtesy) of Computer Science. From 1993 to 1998, he was Director of the CMU Center for Electronic Design Automation. He is a cofounder of NeoLinear, Inc., and served as its Chief Technologist on a 1998 leave from CMU. His research interests focus on circuit and layout synthesis algorithms for mixed-signal ASIC's, for high-speed digital systems, and for FPGA's.

In 1987, Dr. Rutenbar received a Presidential Young Investigator Award from the National Science Foundation. He has won Best/Distinguished paper awards from the Design Automation Conference (1987) and the International Conference on CAD (1991). He has been on the program committees for the IEEE International Conference on CAD, the ACM/IEEE Design Automation Conference, the ACM International Symposium on FPGA's, and the ACM International Symposium on Physical Design. He also served on the Editorial Board of IEEE SPECTRUM. He was General Chair of the 1996 ICCAD. He chaired the Analog Technical Advisory Board for Cadence Design Systems from 1992 through 1996. He is a member of the ACM and Eta Kappa Nu.

L. Richard Carley received the S.B. degree in 1976, the M.S. degree in 1978, and the Ph.D. degree in 1984, all from the Massachusetts Institute of Technology (MIT), Cambridge.

He joined Carnegie Mellon University, Pittsburgh, PA, in 1984. In 1992, he was promoted to Full Professor of Electrical and Computer Engineering. He was the Associate Director for Electronic Subsystems for the Data Storage Systems Center [a National Science Foundation (NSF) Engineering Research Center at CMU] from 1990–1999. He has worked for MIT's Lincoln Laboratories and has acted as a Consultant in the areas of analog and mixed analog/digital circuit design, analog circuit design automation, and signal processing for data storage for numerous companies; e.g., Hughes, Analog Devices, Texas Instruments, Northrop Grumman, Cadence, Sony, Fairchild, Teradyne, Ampex, Quantum, Seagate, and Maxtor. He was the principal Circuit Design Methodology Architect of the CMU ACA-CIA analog CAD tool suite, one of the first top-to-bottom tool flows aimed specifically at design of analog and mixed-signal IC's. He was a co-founder of NeoLinear, a Pittsburgh, PA-based analog design automation tool provider; and, he is currently their Chief Analog Designer. He holds ten patents. He has authored or co-authored over 120 technical papers, and authored or co-authored over 20 books and/or book chapters.

Dr. Carley was awarded the Guillemin Prize for best Undergraduate Thesis from the Electrical Engineering Department, MIT. He has won several awards, the most noteworthy of which is the Presidential Young Investigator Award from the NSF in 1985 He won a Best Technical Paper Award at the 1987 Design Automation Conference (DAC). This DAC paper on automating analog circuit design was also selected for inclusion in 25 years of Electronic Design Automation: A Compendium of Papers from the Design Automation Conference, a special volume, published in June of 1988, including the 77 papers (out of over 1600) deemed most influential over the first 25 years of the Design Automation Conference.

James R. Hellums (S'75–M'77–SM'96) received the B.S.E.E. (highest honors) and M.S.E.E. degrees from the University of Texas at Arlington in 1976 and 1983, respectively. He is working toward the Ph.D. degree at University of Texas at Dallas with a research topic on quantum transport theory.

In January 1978, he joined MOSTEK as an Integrated Circuit Design Engineer where he worked on five MOS analog IC's for telecommunications. He left in August 1981 to co-found Nova Monolithics where he was involved in consulting and custom IC designs which included mixed-signal chips with analog filtering and A/D conversion. He joined Texas Instruments, Dallas, TX, in May 1984 as a Senior IC Design Engineer. Since joining TI he has worked on the design of 24 analog and mixed-signal IC's of which 17 required analog-to-digital conversion. Jim has authored or coauthored 21 papers, given five conference talks, holds ten US patents, seven foreign patents with 12 patents pending.

Mr. Hellums was elected an MGTS in 1987, an SMTS in 1989, a DMTS in 1996 and a TI Fellow in 1997. He is a member of the American Physical Society, Eta Kappa Nu, Tau Beta Pi, and Alpha Chi.

A Case Study of Synthesis for Industrial-Scale Analog IP: Redesign of the Equalizer/Filter Frontend for an ADSL CODEC

Rodney Phelps, Michael J. Krasnicki, Rob A. Rutenbar, L. Richard Carley, James R. Hellums*

Department of Electrical and Computer Engineering, Carnegie Mellon University
Pittsburgh, Pennsylvania 15213 USA

*Mixed Signal Products, Texas Instruments Incorporated
Dallas, Texas 75243 USA

Abstract: *A persistent criticism of analog synthesis techniques is that they cannot cope with the complexity of realistic industrial designs, especially system-level designs. We show how recent advances in simulation-based synthesis can be augmented, via appropriate macromodeling, to attack complex analog blocks. To support this claim, we resynthesize from scratch, in several different styles, a complex equalizer/filter block from the frontend of a commercial ADSL CODEC, and verify by full simulation that it matches its original design specifications. As a result, we argue that synthesis has significant potential in both custom and analog IP reuse scenarios.*

I. Introduction

Modern fabrication technologies support the integration of many formerly discrete functions onto a single die. To manage complexity and time-to-market, these *system-on-chip* (SoC) designs require a high degree of reuse, a situation that has generated growing interest in techniques for creating silicon *intellectual property* (IP), and an evolving marketplace for IP. However, the vast majority of efforts in IP creation and packaging address only digital logic. This is an alarming situation because a large fraction of new ICs require an interface to the external, continuous-valued analog world.

IP and reuse strategies for the digital portion of these mixed-signal designs can successfully exploit cell-based tools for synthesis, mapping, and physical design. IP offered as a *soft* netlist can be retargeted via logic synthesis; IP offered as a *hard* layout can by integrated via physical synthesis. Unfortunately, with respect to current logic-centric design flows, analog blocks still fit poorly. Although a small fraction of the overall design size (*e.g.*, 10,000 to 20,000 analog transistors is typical), the analog portion of these systems remains a real design challenge. Worse, these analog blocks are still designed by hand, usually one transistor and one rectangle at a time.

The widening gap between reuse-centric digital design and manual analog design has been addressed in two very different ways. One option is *hard analog IP* in the form of layouts for common system-level analog blocks, *e.g.*, data conversion, network physical layers, phase lock loops, *etc.* Limited (but growing) selections of such analog IP blocks are now available from larger foundries, and from a new generation of third-party IP providers who target the most widely used foundries. (See [1] for a recent survey.) This is an appealing model for analog reuse--but a limited one. The problem is that the analog space cannot be fully covered by any finite library of blocks. There are hundreds of circuit-level topologies for cells (10-50 devices) and systems (10-100 cells) in use today, and each performance parameter is *continuous.* To take a crude example: a cell with 10 continuous parameters has roughly 1000 library variants even if we limit each parameter to only "low" and "high," and ignore completely that device-level circuit/layout decisions must change as we move from process to process. Routine functions can, of course, be created and rendered as part of a useful analog IP portfolio. But the analog side of mixed-signal SoCs often comprises more than just "routine" blocks connecting to "routine" interfaces. Analog functions are often *aggressively* custom, to deal with next-generation interfaces, application-specific sensors or transducers, or nontraditional trade-offs between analog and digital signal processing. Indeed, on many such designs, this custom analog is a large component of the value added by a single-chip implementation.

The alternative approach is *analog synthesis*, which seeks to transform some abstract description into a working circuit. There is an extensive literature on cell-level analog synthesis, e.g., [2]-[13], but considerably fewer attempts at analog system synthesis, e.g., [14]-[17]. None of these techniques is in widespread use today. We believe that these fail in practice because of the difficulties in reconciling the simplified models required for synthesis with the industrial-strength simulation environments required for validation. Many techniques tradeoff accuracy in the heuristic evaluation of circuits in favor of speed, to support a more vigorous numerical search. To date, strategies that rely on circuit evaluators unrelated to (often inconsistent with) current simulation-based practice have not been regarded as trustworthy by working designers, who rely on simulation to correctly assess critical second-order circuit effects. There is enormous investment in modeling, characterization, test harnesses, and simulation-based sign-off for validation in current analog design flows; synthesis strategies that cannot leverage this investment face significant obstacles.

In [18],[19] we introduced new techniques for efficient *simulation-based synthesis* for custom analog cells. By combining novel global numerical search algorithms, workstation-level parallelism, and software encapsulation methods that insulate search from the idiosyncrasies of circuit simulators, we were able to synthesize a range of custom cells competitive with industrial designs, using full "SPICE-in-the-loop" search. Our goal in this paper is to demonstrate that some *system-level* analog designs are also within the reach of these techniques. The central problem is how to retarget simulation-based synthesis to systems where the cost of a full circuit-level simulation of a design candidate is so large as to render optimization intractable. The key idea is a hierarchical decomposition in which cell-level macromodels are used to search for an optimal system-level design, while *concurrently* a full transistor-level design evolves for each cell, matching it to the abstracted cell behaviors evolving at system level.

To counter the persistent criticism that analog design--especially design beyond basic cells--is intrinsically "too hard" to make synthesis practical, we present a case study of synthesis for one significant block, an equalizer/filter (EQF), in the frontend of a state-of-the-art CODEC for a full-rate ADSL modem introduced by Texas Instruments in [20] in 1999. This design is "aggressively custom" in the sense argued above: it uses neither routine system nor cell-level circuit topologies, and meets difficult analog performance goals. Our study takes the form of a legacy IP reuse scenario: we have full access to a working design and the usual documentation archived with such designs, but *no access to the designer* (who no longer works in this group). We develop hierarchical simulation-based synthesis techniques that are successfully able to *redesign* this entire block's sizing/biasing to meet its original commercial specifications.

The remainder of the paper is organized as follows. Sec. II briefly surveys prior synthesis work. Sec. III introduces the EQF benchmark circuit. Sec. IV develops our hierarchical synthesis formulation. Sec. V presents experimental results. Sec. VI offers concluding remarks.

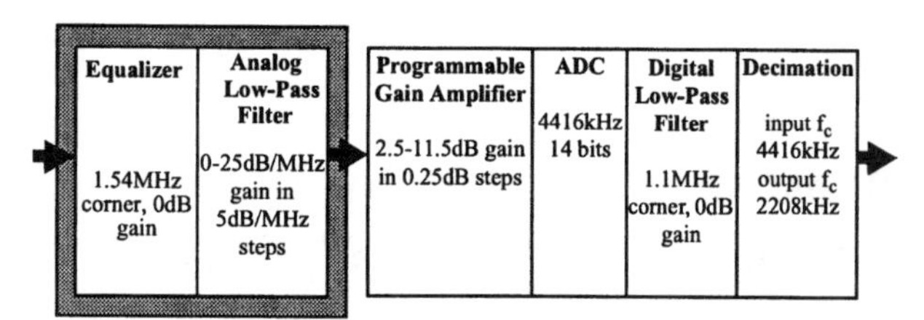

Fig. 1 Architecture for Remote Modem CODEC Receiver. The section targeted (shaded) is the analog equalizer & low-pass filter.

II. Prior Work

There is an extensive literature on cell-level synthesis, covering both generation of circuit topologies [2],[3] and sizing and biasing. The earliest approaches used procedural scripting [4], which proved fragile in the face of circuit and especially process changes. Hierarchical attacks [5-7] allowed composing of reusable subcircuits, but remained limited by their script-based underpinnings. Equation-based approaches substituted numerical search for simple scripting, and were able to attack a wider set of designs [8-10]; however, they still suffered from accuracy problems and limitations imposed by the need for closed-form models. Symbolic analysis techniques [11] can automatically extract some, but not yet all of these required equations. Attacks based on custom, lightweight simulators coupled with numerical search proved yet more capable [12],[13], but still lacked the ability to evaluate some circuit performance specifications.

The literature on system-level synthesis for analog is considerably thinner. Macromodeling, *e.g.*, [14], plays a central role, since many system-level designs are intractable to simulate flat, at the device level. There is a wider variety of attacks on topology generation, *e.g.*, via templates, pattern matching, scripting, hierarchical performance prediction [15-17], since systems have more degrees of freedom than cells. Hierarchical composition in the style of [5] plays a central role, since systems negotiate specifications with sub-blocks to achieve overall goals. The added degrees of freedom inherent in these more complex designs have limited analog systems synthesized to date to very modest size and performance.

None of these techniques is widely used. In [18],[19] we argued that the necessary remedy is efficient *simulation-based analog synthesis.* These techniques are the starting point for our work.

III. Overview of EQF Circuit Benchmark

Digital Subscriber Line (DSL) technologies combine sophisticated analog and digital signal processing to deliver high-speed digital data and conventional analog voice data over existing copper telephone wires. *Asymmetric* variants like ADSL offer full-duplex communication, but bias bandwidth usage toward data downloads (*e.g.*, video on demand provided at several Mbps), rather then uploads (*e.g.*, email, web clicks, etc., supported at a few hundred Kbps). ADSL connections require a pair of modems, one at each end of the copper line. We focus on the *remote modem* at the user's end. The *CODEC* is the interface between the modem's DSP core and the line itself; its architecture (from[20]) appears in Fig. 1. We focus on a complex subsystem at the front-end of the analog signal path, the *equalizer filter* (EQF). Copper transmission presents significant design challenges: signals attenuate strongly with increasing frequency and line length, and typical cable bundles introduce considerable crosstalk. The equalizer amplifies the attenuated line signal which is subsequently extracted by the filter. The combined EQF must do this under stringent noise and area constraints set by the overall CODEC.

The EQF itself is shown in Fig. 2 and comprises five identical low noise operational amplifiers (LNAs) connected via R's, C's and CMOS switches. The equalizer consists of opamp1 and shares opamp2 with a fourth-order elliptical low pass continuous-time filter. Typically, the equalizer would require two opamps, but a novel circuit architecture merges equalizer and filter, eliminating one opamp.

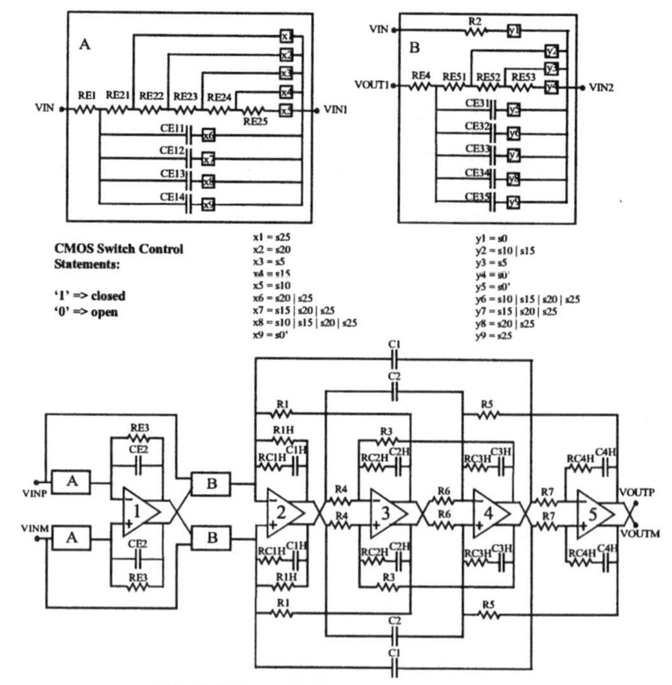

Fig. 2 Schematic for the Equalizer/Filter

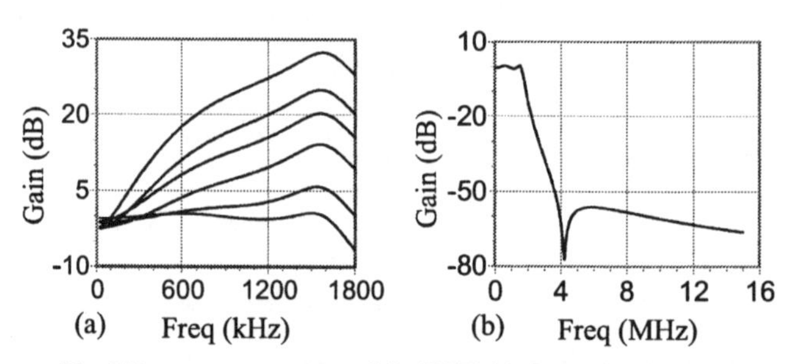

Fig. 3 Frequency response of the EQF. (a) shows the passband response for each of the six equalizer settings. (b) shows the frequency response with the equalizer set to 0dB gain.

The equalizer has six separate modes (Fig. 3a) to compensate for high frequency line attenuation. As shown in Fig. 2 the control signals on the switches program the gain to one of {0, 5, 10, 15, 20, 25}dB across the frequency range of interest, 25kHz to 1104kHz. The low pass filter itself is a standard design, with the exception of resistors RC1H-RC4H in the feedback path. Ideally, these would not be required, but because the opamp has finite bandwidth and the passband is relatively wide, these resistors compensate for peaking near the cutoff frequency. In other words, during initial design, it was decided that these small resistors were a better choice than to increase the performance of the amplifier. Fig. 3 shows the spectral mask for the EQF.

IV. Synthesis Formulation

A. Synthesis Styles for System-Level Designs

Following [13],[18],[19], we fix the topology of the EQF and formulate synthesis as the task of designing parameter values to meet performance specifications. This does simplify the problem, but also respects the fact that "libraried" analog blocks are most likely to be stored as topologies that can be re-parameterized to handle new specifications, or fabrication processes. Moreover, expert designers routinely choose good topologies to optimize gross system function, and then spend enormous effort iteratively resizing them; the problem is to determine if a proposed sizing can realize the specified performance in the face of many interacting second-order circuit effects. Hence, it is this sizing and biasing we seek to automate.

We formulate synthesis as cost-based numerical search: a minimum of an appropriately constructed cost function corresponds to a "best" circuit. We use the functional form developed in [13],[18],[22]. Creating an appropriate cost function is mostly mechanical, and we can automate much of the process; solving for a useful minimum is not.

In simulation-based synthesis, we simulate *each* design candidate during numerical search. The new problem we face is that system-level blocks are often vastly more expensive to simulate at device-level (if they can simulate *at all*) than cells. This can defeat our preferred simulator-in-the-loop formulation. We suggest three alternative strategies for coping with system-level complexity:

- **Flat synthesis** chooses to ignore the hierarchical system-level structure, flattening it down to a single, potentially large circuit, and treating it just as cell-level synthesis. Unfortunately this approach does not always work. Not only are the simulation times for large circuits problematic, but simulator convergence also becomes an issue. The reason is that numerical search often visits *exotically* parameterized designs--circuits with behaviors that deviate widely from the norms for which commercial simulators are designed.
- **Iterative-sequential synthesis** mimics top-down design practice. At the top level of our design, we replace subsystems with simplified behavioral macromodels, guess appropriate model parameters for these subsystems, and formulate synthesis as the task of choosing the remaining top-level component values to satisfy system-level goals. This is a straightforward simulation-based synthesis task since our "simulations" are just evaluations of analytical models or very simple circuits. The real problem is the need to move down the design hierarchy one level at a time, and deal with the fact that even given good macromodels, predicting feasible trade-offs among the parameters for a subsystem can be difficult. We must avoid the situation where the system design "works"--but only if its components comply with unachievable performance goals.
- **Concurrent synthesis** is a novel alternative to the above two strategies. The system-level design and its component subsystems evolve simultaneously. Unlike a fully flattened design, the system still uses macromodels for its components, and synthesis sets their input parameters. In contrast, the component cells use complete device-level models and detailed circuit simulation. We link the two synthesis processes into a single numerical problem via a transformation of the cost function. We add terms that coerce agreement between the macromodel *parameters* evolving at the top of the design hierarchy, and the actual simulated *behaviors* of the device-level components at the bottom of the hierarchy. We refer to such specifications as being *dynamically set* since they evolve naturally as a negotiation between the system and its components. The virtue of the concurrent approach is that it reduces iteration steps, and automatically avoids designs in which macromodel parameters and device-level simulated behaviors disagree.

The flat and iterated-sequential synthesis styles can be accommodated with no problems in our existing simulation-based synthesis framework. They differ only in the nature of the simulation, and the number of separate synthesis tasks to be undertaken. The concurrent style, however, requires a modification of our numerical formulation.

We map the circuit design problem to a constrained optimization problem, where $\underline{x}$ is the set of independent design variables; $\{f(\underline{x})\}$ is a set of objective functions that codify performance specifications to optimize, *e.g.*, noise; and $\{g(\underline{x})\}$ is a set of constraint functions that codify hard specifications to meet, *e.g.*, *gain* > 60dB. We perform the standard conversion to an unconstrained optimization problem, substituting suitable normalizing and penalty functions for each term [13]. Scalar weights balance competing objectives. As a result, the goal becomes minimization of a scalar cost function, $C(\underline{x})$, defined by (1).

$$(\underline{x}) = \sum_{i=1}^{k} w_{1,i}\hat{f}_i(\underline{x}) + \sum_{j=1}^{l} w_{2,j}\hat{g}_j(\underline{x}) \quad (1)$$

Suppose now that we have n subsystems in our system, and our independent variables $\underline{x}$ include both model parameters for the macromodel of each subsystem, and actual device-level parameters for the detailed design of each subsystem. We force the model parameters and simulated performance of each subsystem to *converge* by adding a set of penalty functions, $\{p(\underline{x})\}$, to the cost, as shown in (2). (3) shows the penalty function for a single parameter. The *measured* value is the value obtained via SPICE simulation of the actual subsystem; *modelparam* is the current macromodel input, set as an independent variable.

$$C(\underline{x}) = \sum_{i=1}^{k} w_{1,i}\hat{f}_i(\underline{x}) + \sum_{j=1}^{l} w_{2,j}\hat{g}_j(\underline{x}) + \sum_{m=1}^{n} w_{3,m}\hat{p}_m(\underline{x}) \quad (2)$$

$$\hat{p}_m(\underline{x}) = |modelparam - measured| \quad (3)$$

This formulation allows us to mix higher-level macromodels and lower-level detailed models, yet treat the overall synthesis task as a single, simulation-based numerical search. The added penalty functions coerce the consistency we need between different models.

B. Global Numerical Search Algorithm

This cost function creates a difficult, highly nonlinear, discontinuous optimization problem. We attack this by combining the population-based search ideas from ANACONDA [19] with some of the annealing ideas from MAELSTROM [18]. The architecture is shown in Fig. 4.

1. **Population of partial circuit solutions:** we maintain a large population of partial solutions. The population itself helps combat the problem of cost surfaces with local minima. Each element is one sample of that cost surface. We maintain a suitably diverse set of samples, and preferentially update the population so that lower-cost samples survive and higher-cost samples are culled.
2. **Limited population update:** from a population of P partial solutions, we select k candidates, apply a short annealing improvement process to each candidate, then replace these in the population. Note that as a result, individual elements may improve or degrade on cost after annealing. From this updated population of $P+k$ solutions, we remove the k solutions of highest (worst) cost.
3. **Evolution by problem-shared annealing:** the update process for a selected candidate is a *shared* annealing. We maintain k parallel annealers (k=3 in Fig. 4), and each selected candidate undergoes a small number of annealing perturbations before being returned to the overall population. Each annealer sees locally a single numerical optimization problem; in reality, it sees a sequence of snapshots of independent optimizations, each sampling different regions of the same underlying cost surface. The annealers each run a global, fixed-length cooling schedule [13]. The open question here is what is the cost C that is the current *state* of annealing?

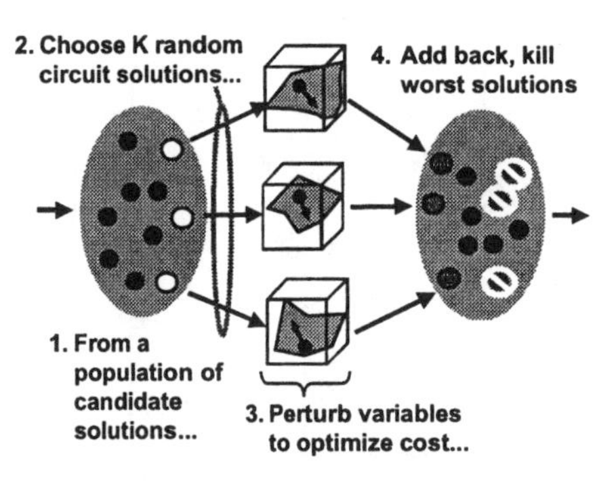
2. Choose K random circuit solutions...
4. Add back, kill worst solutions
1. From a population of candidate solutions...
3. Perturb variables to optimize cost...

Fig. 4 Overview of population-based global numerical optimization.

Suppose after move Δ_i we have annealing cost C_i, and we then swap in a new circuit candidate to perturb, which has its own current annealing cost D_i. We maintain a form of thermal equilibrium by requiring each annealer to *choose* which of these is the cost of the annealing state *before* the next perturbation Δ_{i+1}. To do this, the annealer makes a Metropolis-weighted probabilistic choice between C_i and D_i; either can be the next new state. This is essentially the notion of parallel quasi-equilibrium from [21]. See [22],[23] for additional details about this annealing process.

The population model supports significant practical parallelism: we routinely run on 20-30 workstations. In concert, the population model plus the shared-annealing update are effective as a global optimizer for simulation-based synthesis. These ideas have been implemented in an updated version of ANACONDA.

V. Experimental Synthesis Results

We describe experiments in the three synthesis styles discussed in Sec. IV. Experiments were done at TI using TISpice as the simulation engine, on a compute farm of 20 to 30 Sun UltraSparcs.

As a system, EQF has one layer of hierarchy, and five instances of a single component, the LNA circuit [20] of Fig. 5. It was designed in one of TI's 0.6μm CMOS processes. At the top level, the EQF has 46 R's, 32 C's and 36 CMOS switches. Pole and zero location constraints from the transfer function set a large number of the R's and C's. The LNA is itself a complex cell, and has 20 independent variables. As we shall see, the number of optimizable degrees of freedom depends on the synthesis style we choose.

A. Flat Synthesis for EQF

As expected, attempts here were unsuccessful. A good design for EQF can be simulated at device level, but not inside an optimization loop that seeks to visit 10,000 to 100,000 candidate solutions. In addition, we experienced simulator convergence difficulties for the "exotic" parameterizations visited early in synthesis.

B. Iterated Sequential Synthesis for EQF

We undertook several synthesis experiments in this style, including both exploratory synthesis (top-level only, no device details), and fully detailed synthesis.

In the initial stages of design it is convenient to be able to evaluate a system-level choice before each analog cell is fully designed. This allows design dead-ends, especially poor topology choices, to be quickly recognized. To evaluate the EQF's system-level topology we performed a high-level exploration that determined the relationship between noise and area. Because EQF's top-level R's are small but noisy and its C's are large but noiseless, noise versus area is a fundamental trade-off for the EQF system. To accomplish this, we replaced each LNA with a simple two-pole opamp model, and used a symbolic package (MATLAB) to derive a symbolic expression for the output noise via nodal analysis on EQF's adjoint network. At this stage in the design, we assumed that each opamp was noiseless, and ignored the

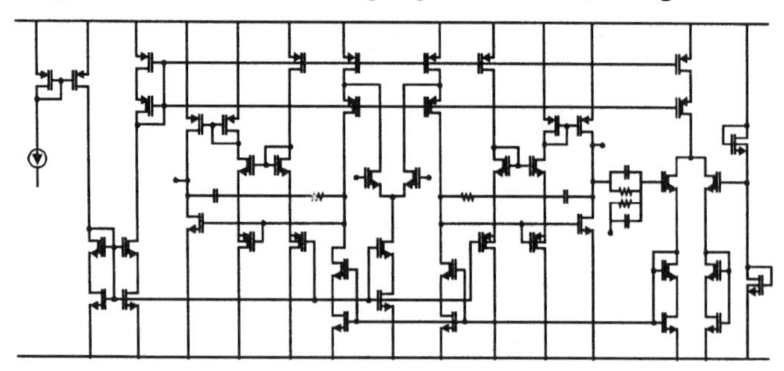

Fig. 5 The low noise operational amplifier used in the EQF.

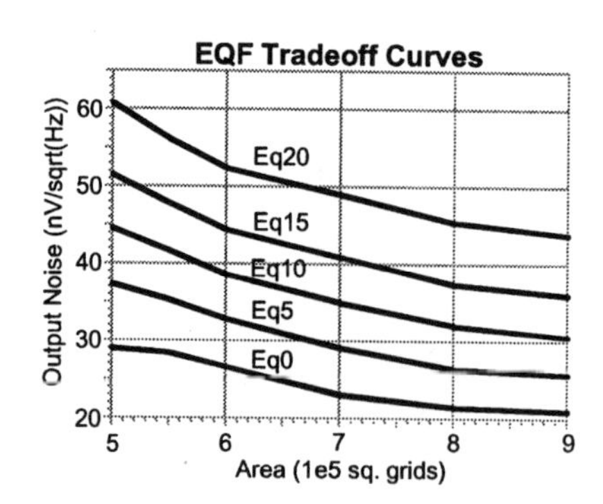

Fig. 6 Area vs. noise trade-off curves for EQF design exploration.

CMOS switch parasitics. This allowed us to estimate the absolute minimum noise for a given area.

After eliminating the dependent variables (*e.g.*, set by the transfer function) we were left with twelve independent variables that could be set by our optimization engine. We used the derived symbolic expressions as our "simulation" engine and instructed the optimization engine to minimize noise for a specified area. A double-sided penalty function was added to the cost function (2) to ensure that the specified area was obtained for each run. Using this approach we were able to generate the area versus noise trade-off curves shown in Fig. 6. These curves give us a good idea if the EQF system topology makes sense given the noise specifications for the design. Is the required area reasonable? Is it possible to meet the noise specification with the current topology? As it turns out, these results are reasonable but optimistic, as we shall see when we later compare this exploration to detailed synthesis results.

In our second approach we synthesized the EQF using a two step process. First, the top-level devices and components were sized, then the LNA was synthesized separately. Because we synthesized the top-level first, the simpler 2-pole model of opamps 1-5 in Fig. 2 was replaced with the more detailed circuit-level macromodel in Fig. 7. One drawback here is that if we wish only to synthesize the top-level components alone, the LNA macromodel parameters need to be set first. Some of the parameters, such as DC gain and the pole locations, can be calculated reasonably well during high-level design exploration. However, other parameters such as noise and input capacitance can only be approximated initially, then further refined once the LNA is actually synthesized. This is the reason this style of synthesis is ultimately iterative.

As with the exploration experiment, after pruning dependent variables, we were left with twelve independent designables for the top-level. Because we already had a finished hand design for EQF (*i.e.*, our "legacy IP"), we simply set the values for the LNA macromodel to the actual performance parameters measured from the hand design.

To test our approach we first synthesized the top-level several times to the same performance specifications as specified for the hand EQF design. Our synthesized results were *all* comparable to the hand design. More interestingly, the synthesized EQF's component values were very

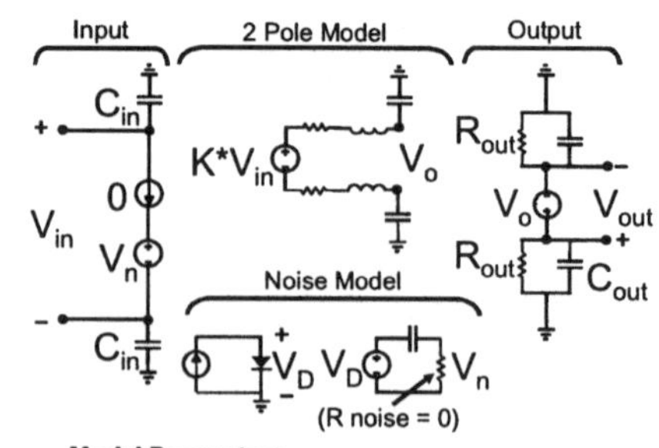

Fig. 7 LNA macromodel.

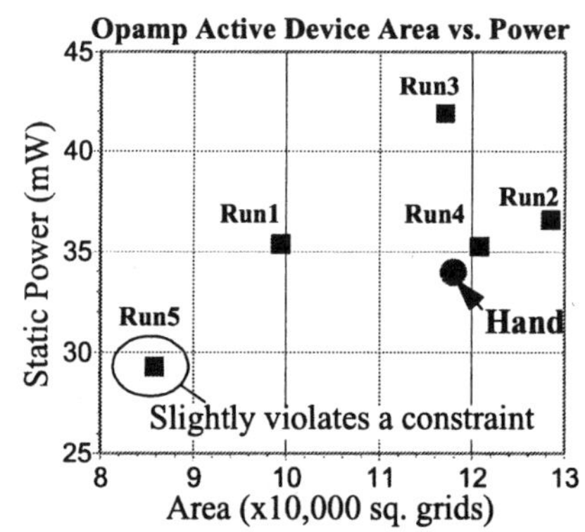

Fig. 8 Area vs. power scatter for 5 *consecutive* LNA syntheses. All met specs; Run5 had a few devices with an effective voltage smaller than desired.

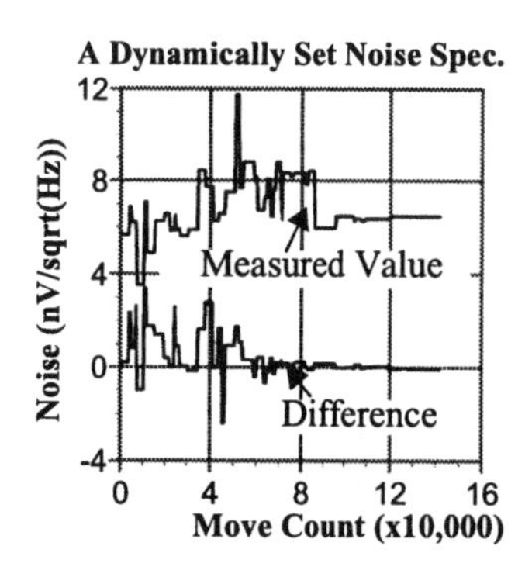

Fig. 9 Actual *measured value* from TISpice for input referred noise of the LNA over the lifetime of a synthesis run. The *difference* between the *measured value* and the evolving macromodel parameter value is also shown.

Table 1. Nominal performance specs for the LNA. R1-R5 correspond to nominal characteristics for the scatter plot points of Fig. 8.

Vdd = 3.3V

	Hand	R1	R2	R3	R4	R5
DC Gain (dB)	135	135	135	135	135	135
UGF (MHz)	110	121	126	132	118	118
PM (deg)	54	54	54	54	54	54
Noise ($nV/Hz^{-0.5}$)[1]	6.33	6.22	6.07	5.88	6.32	6.28
THD (%)[2]	0.16	0.13	0.12	0.14	0.15	0.15

1 Max input referred in range 25kHz-1104kHz

2 1MHz 2Vp-p input voltage, 5kΩ load

Table 2. - Derived performance characteristics for each of the three synthesis run.

Vdd = 3.3V

	Hand	Run1	Run2	Run3
DC Gain (dB)	135	131	132	130
UGF (MHz)	110	74	100	126
PM (deg)	54	64	53	56
Noise ($nV/Hz^{-0.5}$)[1]	6.33	8.35	7.16	6.44
THD (%)[2]	0.16	0.16	0.11	0.12

1 Max input referred in range 25kHz-1104kHz

2 1MHz 2Vp-p input voltage, 5kΩ load

close to being the *same* component values as the hand design. We regard these as good results because the hand design had been aggressively optimized by an experienced designer for several weeks.

Next we synthesized the LNA five *consecutive* times. The results are summarized as the area versus power scatter plot shown in Fig. 8 as well as the list of each opamp's nominal performance characteristics shown in Table 1. All five designs met the *nominal* performance constraints as defined by the hand design. Notice that four of the five synthesized results are within 20% of the hand design in terms of area and power. The circled result in Fig. 8 had a few devices with an effective voltage smaller than desired. This affected robustness across manufacturing variations; we verified each design across 3σ process, 0-100°C temperature and 10% supply voltage variations.

Finally, the entire EQF system was verified by replacing the symbolic macromodel with a synthesized opamp and simulating the complete EQF. This process was done for *each* of the four good opamp designs. *All* four met the EQF system level performance specifications.

A population size of 50 circuits, with 5 annealing streams, was used to synthesize the top-level. For each run approximately 25,000 design points were evaluated in roughly 2 hours. The LNA was synthesized using a population size of 100, with 10 annealers. Each run evaluated approximately 100,000 designs, resulting in a runtime of 10 hours.

C. Concurrent Synthesis for EQF

In our third experiment, we synthesized the top-level and LNA concurrently. The advantage of this approach is that the LNA's performance parameters are derived *dynamically* over the course of each synthesis run. An example of dynamically setting a performance parameter is shown in Fig. 9. The *difference* between the LNA's measured input referred noise and the macromodel's noise parameter is effectively zero when the synthesis run completes: the top-level and component synthesis tasks have *negotiated* the right specification here.

The number of independent variables for the entire system was 39: 12 top-level variables, 7 designable macromodel parameters and 20 LNA designables. Three *consecutive* synthesis runs were performed and, as before, several steps were taken to verify the results. First, each LNA was verified separately. Table 2 shows each LNA's nominal performance characteristics which were dynamically derived over the course of the synthesis runs. The LNAs were verified across manufacturing variations, and all had histograms with reasonable spreads when compared to the hand design. After verifying the LNAs, each was inserted into its respective top-level design, and the entire EQ was verified by simulation.

Fig. 10 shows the resulting passband responses with the equalizer gain set to 0dB. The two important observations to make are that the ripple in the passband and the cutoff frequency are nearly identical for the hand design and the three synthesized designs. When setting up this experiment, we decided to allow the gain to vary slightly to provide an extra degree of freedom to the optimization engine. The difference in gain between the hand design and the synthesized design can be compensated for using the programmable gain amplifier (see Fig. 1), which is the next stage in the CODEC receiver. Fig. 11 shows the EQF's response over the entire frequency range of interest. We observe from the figure that the cutoff frequency, stopband attenuation, and overall shape of the response are nearly identical for the hand design and the three synthesized EQF designs.

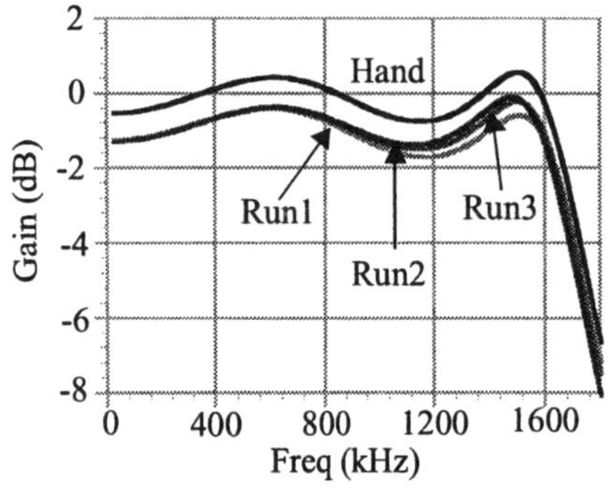

Fig. 10 Frequency response in the passband for the hand design and each of the three synthesis runs.

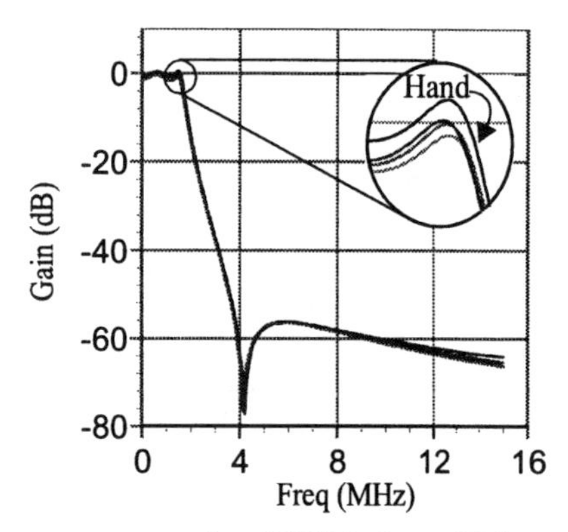

Fig. 11 Frequency response: hand EQF design and 3 three synthesis runs.

Table 3. - Noise and area results for three consecutive runs.

Nominal Noise Specs @25°C (nV/sqrt(Hz))
Max Output Noise 25kHz-1104kHz

	Hand	Run1	Run2	Run3
Eq0	36	35	33	33
Eq5	40	38	37	38
Eq10	44	41	40	43
Eq15	49	47	45	48
Eq20	55	54	52	54
Eq25	108	108	104	108
Area (1000 sq. grids)				
LNA	117	93.5	146	96.5
Top*	680	709	708	560
Total	1265	1177	1438	1043

(*) Area consumed by system level components.

Table 3 shows the final results for the EQF's area and output noise. Notice that the areas for the synthesized designs are close to the hand design: designs Run1 and Run3 have nearly identical noise, but with 7% and 17% *better* area, respectively; design Run2 has slightly better noise, but at a cost of 14% more area. It is also worth noting that the values for noise and area are worse than those predicted in our exploration experiment. For example, from Fig. 6, the predicted output noise value for a top-level area of 700,000 sq. grids was approximately $23\text{nV}/\sqrt{\text{Hz}}$. Table 3 shows that the final output noise actually obtained was $33\text{nV}/\sqrt{\text{Hz}}$ for Run2. This difference comes as no surprise: the symbolic macromodel ignored several effects, and was thus overly optimistic. Nevertheless, this does provide significant insight into the overall envelop of performance achievable by EQF in this technology. Any full synthesis flow will likely begin with high level exploration, followed by detailed synthesis to set final component values.

The population size used was 100, with 10 annealers. Fig. 12 shows how the cost of the population evolved over the course of each synthesis run. The log of the circuit population's mean and best cost are shown as a function of the move count. The move count is the total number of design points evaluated, about 150,000 total in this case, resulting in a total runtime of 12 hours.

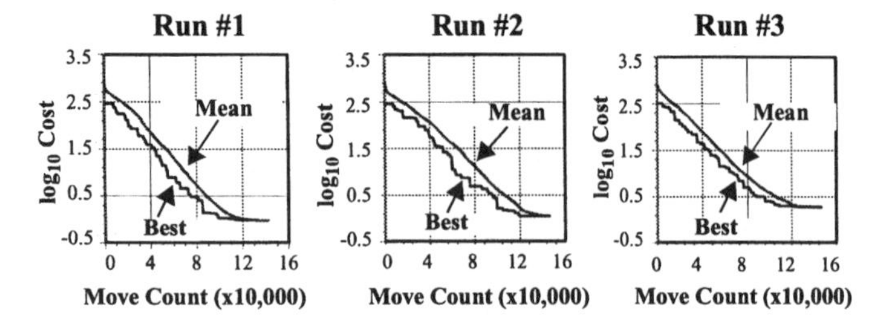

Fig. 12 Log cost versus the number of design points evaluated. The best circuit's cost in the population and the mean cost of the circuit population are shown over the lifetime of each synthesis run.

VI. Conclusions

To the best of our knowledge, this is the largest, most complex, most thorough controlled experiment ever undertaken to demonstrate that recent developments in analog synthesis have direct application to state-of-the-art industrial analog systems. We have successfully redesigned the EQF block in the ADSL frontend in several different ways, and examined the various trade-offs involved. Simulation-based synthesis, with a mix of macromodels, transistor-level detailed simulation, and vigorous global numerical search, can yield practical results on this important problem. We believe these ideas have significant potential in both fully custom analog design, and IP creation/reuse scenarios.

Acknowledgment: This work was funded by the Semiconductor Research Corp. and TI. We thank Felicia James and Gary Richey of TI for valuable discussions about this work.

References

[1] Steve Ohr, "Analog IP Slow to Start Trading", *EETimes*, Issue 1053, March 22 1999 (Also:.http://www.eet.com)
[2] W. Kruiskamp and D. Leenaerts, "DARWIN: CMOS Opamp Synthesis by Means of a Genetic Algorithm," *Proc. 32nd ACM/IEEE DAC*, 1995.
[3] P. C. Maulik, L. R. Carley, and R. A. Rutenbar, "Integer Programming Based Topology Selection of Cell Level Analog Circuits," *IEEE Trans. CAD*, vol. 14, no. 4, April 1995.
[4] M. Degrauwe *et al.*, "Towards an analog system design environment," *IEEE JSSC*, vol. sc-24, no. 3, June 1989.
[5] R. Harjani, R.A. Rutenbar and L.R. Carley, "OASYS: a framework for analog circuit synthesis," *IEEE Trans. CAD*, vol. 8, no. 12, Dec. 1989.
[6] B.J. Sheu, *et al.*, "A Knowledge-Based Approach to Analog IC Design," *IEEE Trans. Circuits and Systems*, CAS-35(2):256-258, 1988.
[7] J. P. Harvey, *et al.*, "STAIC: An Interactive Framework for Synthesizing CMOS and BiCMOS Analog Circuits," *IEEE Trans. CAD*, Nov. 1992.
[8] H.Y. Koh, C.H. Sequin, and P.R. Gray, "OPASYN: a compiler for MOS operational amplifiers," *IEEE Trans. CAD*, vol. 9, no. 2, Feb. 1990.
[9] G. Gielen, *et al.*, "Analog circuit design optimization based on symbolic simulation and simulated annealing," *IEEE JSSC*, vol. 25, June 1990.
[10] M. Hershenson, S. Boyd, T. Lee, "GPCAD: a Tool for CMOS Op-Amp Synthesis", *Proc. ACM/IEEE ICCAD*, pp. 296-303, 1998
[11] G. Gielen, P. Wambacq, W. Sansen, "Symbolic Analysis Methods and Applications for Analog Circuits: A Tutorial Overview," *Proc. IEEE*, vol. 82, no. 2, Feb., 1994.
[12] F. Medeiro, F.V. Fernandez, R. Dominguez-Castro and A. Rodriguez-Vasquez, "A Statistical Optimization Based Approach for Automated Sizing of Analog Cells," *Proc. ACM/IEEE ICCAD*, 1994.
[13] E. Ochotta, R.A. Rutenbar, L.R. Carley, "Synthesis of High-Performance Analog Circuits in ASTRX/OBLX," *IEEE Trans. CAD*, vol. 15, no. 3, March 1996.
[14] Y-C Ju, V.B. Rao and R. Saleh, "Consistency Checking and Optimization of Macromodels", *IEEE Transactions on CAD*, August 1991.
[15] B. Antao and A. Brodersen, "ARCHGEN: Automated Synthesis of Analog Systems", *IEEE Transaction on VLSI Systems*, June 1995.
[16] F. Medeiro, B. Pérez-Verdú, A. Rodríguez-Vázquez, J. Huertas, "A vertically-integrated tool for automated design of SD modulators," *IEEE Journal of Solid-State Circuits*, Vol. 30, No. 7, pp. 762-772, July 1995.
[17] A. Doboli, et al, "Behavioral synthesis of analog systems using two-layered design space exploration," *Proc. ACM/IEEE DAC*, June 1999.
[18] M. Krasnicki, R. Phelps, R.A. Rutenbar, L.R. Carley, "MAELSTROM: Efficient Simulation-Based Synthesis for Analog Cells," *Proc. ACM/IEEE Design Automation Conference*, June 1999.
[19] R. Phelps, M. Krasnicki, R.A. Rutenbar, L.R. Carley, J.R. Hellums, "ANACONDA: Robust Synthesis of Analog Circuits Via Stochastic Pattern Search," *Proc. IEEE Custom Integrated Circuits Conference*, May 1999.
[20] R. Hester, *et al.*, "CODEC for Echo-Canceling, Full-Rate ADSL Modems," *IEEE Int'l Solid-State Circuits Conference*, pages 242-243. 1999.
[21] P.J.M van Laarhoven and E.H.L. Aarts, *Simulated Annealing: Theory and Applications*, D. Reidel Pub. Co./Kluwer, Dordrecht, Holland, 1987.
[22] M. Krasnicki, *Generalized Analog Circuit Synthesis*, M.S. Thesis, Dept. of ECE, Carnegie Mellon University, Dec. 1997.
[23] R. Phelps, *Population-Based Synthesis for Analog Cells and Systems*, Ph.D. Thesis, Dept. of ECE, Carnegie Mellon University, June 2000 (expected).

WiCkeD: Analog Circuit Synthesis Incorporating Mismatch

K. Antreich[1], J. Eckmueller[2], H. Graeb[1], M. Pronath[1], F. Schenkel[1], R. Schwencker[1,3], S. Zizala[1,2]

[1] Institute of Electronic Design Automation, Technical University of Munich
[2] Infineon Technologies, Munich, Corporate Development Mixed-Signal
[3] Infineon Technologies, Munich, Design Automation LIB IO

Abstract

This paper presents a method to consider local process variations, which crucially influence the mismatch-sensitive analog components, within a new simulation-based analog synthesis tool called WiCkeD. WiCkeD includes tolerance analysis, performance optimization and design centering, and is a university tool used in industry for the design of analog CMOS circuits.

1 Introduction

Analog parts of integrated systems are important: in mixed-signal systems they contribute the same portion to the system as digital parts, in digital systems they provide vital functions like power-on reset or clock generation. As design automation of analog circuits still lags behind that of digital circuits, analog components often are a bottleneck in the design flow [1].

The analog synthesis task consists in designing a circuit topology and sizing circuit parameters such that the circuit performance is optimal with respect to given specifications (performance optimization) and to both manufacturing and operating tolerances (design centering).

Approaches for analog synthesis including topology design [2–4] often rely on declarative models and perform the circuit sizing process based on statistical methods. Approaches concentrating on performance optimization [5, 11, 13, 14] and design centering [1, 6–10, 12] rely on declarative models [12, 13] or, more often, numerical simulation. For performance optimization, statistical methods [1, 14], geometric programming [13] or nonlinear programming are applied. Statistical design centering methods are based on Monte Carlo analysis [7, 15]. Deterministic methods approximate the acceptance region [6] or use robustness objectives for individual performances, e.g. linearized performance penalties [9], worst-case distances [8] or performance scores [10].

This paper presents a new method to incorporate the important local process variations [16] into analog synthesis. While Monte Carlo analysis could be applied at once, it is expensive. Moreover, integrated circuits' statistical parameters usually are transistor model parameters (e.g. V_{th}) and thus Monte Carlo analysis does not provide yield gradients for the design parameters.

Deterministic approaches on the other hand have to cope with two challenges: the nominal performance is in a local extremum (Fig. 3), and the variances are a function of the design parameters due to the area law [16]. Current approaches do not look after these problems. Another problem is that automatic circuit sizing, which usually starts from incomplete specifications, frequently fails to converge or leads to pathological designs that violate basic rules.

The paper is organized as follows. An orientation about mismatch in Sec. 2 is followed by Sec. 3 and 4 presenting how to analyze and synthesize analog circuits in the presence of mismatch. Sec. 5 describes tool features required for practical applicability. Sec. 6 shows experimental results and Sec. 7 concludes this paper.

2 Mismatch

It is a key principle of analog circuit design to generate constant differences and ratios of currents or voltages with transistor pairs. These functional relationships are robust with respect to deviations of corresponding parameters of transistor pairs in the same direction, while being very sensitive to deviations of corresponding parameters in the opposite direction (mismatch). This can be illustrated with a current mirror (e.g. dashed box "P1" in Fig. 1). The drain-source currents $i_{ds1,2}$ through the two transistors should have a constant ratio, which can be approximated by $i_{ds2} = k_2\,(\sqrt{i_{ds1}/k_1} + V_{th1} - V_{th2})^2$.

One can see that local variations of the transistors' threshold voltages $V_{th1,2}$ are a major source of mismatch. For global process variations ($\Delta V_{th1} = \Delta V_{th2}$), $i_{ds2} = i_{ds1} \cdot k_2/k_1$ holds. On the other hand, for local process variations ($\Delta V_{th1} \neq \Delta V_{th2}$), i_{ds2} varies with the statistically independently varying $V_{th1,2}$.

This variation is illustrated in Fig. 2 for different technologies. For each technology, the current deviation of the current mirror is shown for the 3σ deviations of V_{th1} and V_{th2}. The disproportionate increase shows the growing significance of local process variations. Fig. 3 shows the mismatch effect on circuit level for the gain of the folded-cascode operational amplifier in Fig. 1. Like for the current mirror sub-block, the circuit performance is insensitive to equal changes $\Delta V_{th1} = \Delta V_{th2}$ and very sensitive to mismatch. For a 1σ mismatch deviation of V_{th}, the gain decreases by 50%.

These examples illustrate that it is very important to consider local process variations (e.g. of threshold voltages V_{th}) besides operating tolerances (e.g. of Temp or V_{DD}) and global

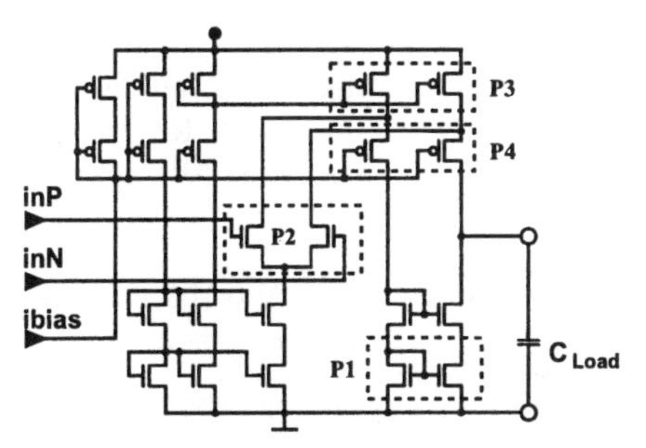

Figure 1: Folded-cascode operational amplifier

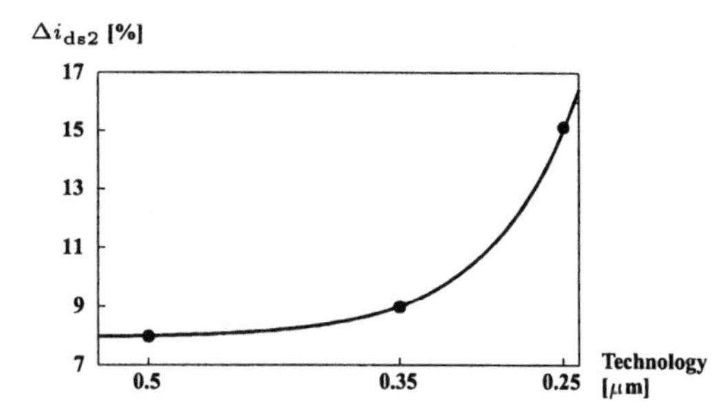

Figure 2: Effect of mismatch on the current of a current mirror

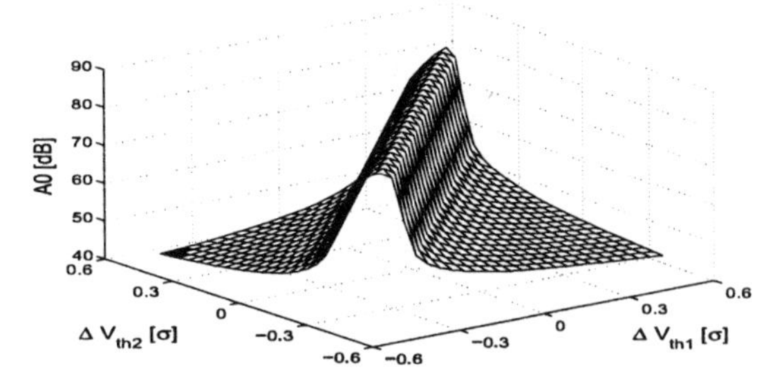

Figure 3: Effect of threshold voltage variations within the current mirror P1 on the gain of the operational amplifier (Fig. 1) before synthesis

process variations (e.g. of oxide thickness T_{ox} or threshold voltage V_{th}) in design centering.

Usually, local and global process variations are modeled by normal (Gaussian), log-normal or uniform distributions of statistical parameters $\mathbf{s}$. Without loss of generality, these distributions can be transformed into a Gaussian distribution with zero mean and unity covariance matrix $N(\mathbf{0}, \mathbf{I})$ [17]. Hence, $\mathbf{s} \sim N(\mathbf{0}, \mathbf{I})$ is assumed in the following.

3 Analysis

Robustness objectives have been proven to be accurate design objectives for analog synthesis [8–10]. This section presents how to calculate them in the presence of mismatch.

Deterministic approaches calculate a *geometric* approximation of the acceptance region with respect to specifications $\mathbf{f}(\mathbf{s}) \geq \mathbf{f}_b$ of the circuit performances $\mathbf{f}$ (e.g. gain, bandwidth). For each specification[1] f_b, a meaningful parameter set $\mathbf{s}_{wc}$ among all boundary parameter sets $\mathbf{s}_b \in \mathcal{S}_b = \{\mathbf{s}\,|\,f(\mathbf{s}) = f_b\}$ of the acceptance region is determined. $\mathbf{s}_{wc}$ is called a worst-case parameter set as it has highest probability density among all boundary parameter sets. It is determined by solving

$$\mathbf{s}_{wc} = \underset{\mathbf{s}}{\operatorname{argmin}} \left\{\beta^2(\mathbf{s})\,|\,f(\mathbf{s}) = f_b\right\}, \quad \beta^2(\mathbf{s}) = \mathbf{s}^T\mathbf{s}\,. \tag{1}$$

The robustness objective $\beta_{wc} = \pm\,|\beta(\mathbf{s}_{wc})|$ describes the performance distance to the spec in multiple of σ (with neg. sign if the spec is violated) and serves for yield estimation [8–10].

Circuit analysis including mismatch faces the following challenges, which cause problems for standard sequential quadratic programming (SQP [18, 20]) approaches to problem (1):

1. expensive performance evaluation by numerical simulation $\mathbf{s} \mapsto \mathbf{f}(\mathbf{s})$,
2. local quadratic performance behavior with semidefinite Hessian (Fig. 3), and
3. initial performance gradient is zero (Fig. 3).

[1] In this section, the index i indicating the spec is left out for simplicity.

For a robust and efficient solution of problem (1), a customized SQP approach with the following features has been developed:

- In each iteration j, the tangential step of the quadratic subproblem (with solution $\mathbf{s}_t^{(j+1)}$) is followed by a normal step *along* $\mathbf{s}_t^{(j+1)}$ starting from the nominal point in order to guarantee a solution $\mathbf{s}_b^{(j+1)}$ that satisfies the specification:
$$\mathbf{s}_b^{(j+1)} = \min_{\alpha} \left\{\alpha\mathbf{s}_t^{(j+1)} \,\middle|\, f(\alpha\mathbf{s}_t^{(j+1)}) = f_b\right\}. \tag{2}$$
Accurately solving problem (2) is crucial for fast convergence of the algorithm.
- The performance is approximated by a second-order model at the *boundary*:
$$f(\mathbf{s}) \approx f_b + \mathbf{g}^{(j)^T}(\mathbf{s} - \mathbf{s}_b^{(j)}) + \frac{1}{2}(\mathbf{s} - \mathbf{s}_b^{(j)})^T\mathbf{B}^{(j)}(\mathbf{s} - \mathbf{s}_b^{(j)}). \tag{3}$$
The approximate Hessian $\mathbf{B}^{(j)}$ is obtained by symmetric-rank-one updates SR1 [18]. The gradient $\mathbf{g}^{(j)} = \nabla_{\mathbf{s}} f(\mathbf{s}_b^{(j)})$ is calculated by finite differences.
- All subsequent tangential directions $\mathbf{q}^{(m)} = \mathbf{s}_t^{(m+1)} - \mathbf{s}_b^{(m)}$ are required to be orthogonal:
$$\mathbf{q}^{(m)^T}\mathbf{q}^{(n)} = 0, \;\; 0 \leq m, n \leq j-1, \; m \neq n. \tag{4}$$
This can be interpreted as a *conjugate-direction* property added to the quadratic subproblem.
- The Lagrange formulation of problem (1) including (3) and (4),
$$\begin{aligned} L(\mathbf{s}, \boldsymbol{\lambda}) = \mathbf{s}^T\mathbf{s} &+ \mu^{(j)}(\mathbf{s} - \mathbf{s}_b^{(j)})^T\mathbf{B}^{(j)}(\mathbf{s} - \mathbf{s}_b^{(j)}) \\ &+ 2\boldsymbol{\lambda}^T\mathbf{A}^{(j)}(\mathbf{s} - \mathbf{s}_b^{(j)}), \end{aligned} \tag{5}$$
with $\mathbf{A}^{(j)} = \left[\mathbf{q}^{(0)}, \ldots, \mathbf{q}^{(j-1)}, \mathbf{g}^{(j)}\right]^T$, leads to the linear equation system for the tangential step:
$$\begin{bmatrix} \mathbf{1} + \mu^{(j)}\mathbf{B}^{(j)} & \mathbf{A}^{(j)^T} \\ \mathbf{A}^{(j)} & \mathbf{0} \end{bmatrix} \cdot \begin{bmatrix} \mathbf{s}_t^{(j+1)} \\ \boldsymbol{\lambda}_t^{(j+1)} \end{bmatrix} = \begin{bmatrix} \mu^{(j)}\mathbf{B}^{(j)}\mathbf{s}_b^{(j)} \\ \mathbf{A}^{(j)}\mathbf{s}_b^{(j)} \end{bmatrix}. \tag{6}$$
$\mu^{(j)}$ is estimated e.g. by the jth component of $\boldsymbol{\lambda}_t^{(j)}$ [18].
- A regular SQP problem without conjugate-direction feature is appended after convergence for fine tuning.

Experimental results in Sec. 6 show superior convergence properties of this algorithm compared to standard SQP for solving problem (1). Please note that this accurate and efficient analysis is a prerequisite for a successful synthesis.

4 Synthesis

This paper concentrates on the sizing part of analog synthesis as it is most time-consuming in practice. Sizing means to maximize the robustness objectives over the design parameters $\mathbf{d}$, e.g. by minimizing the cost function

$$\varphi(\mathbf{d}) = \sum_i \exp\left(-a \cdot \beta_{wc,i}(\mathbf{d})\right), \qquad a > 0\,. \tag{7}$$

Function	Sizing Rules			
	Functional		Tolerance	
	Geometr.	Electr.	Geometr.	Electr.
Transistor as VCCS (T)	–	3	3	–
Current Mirror (C)	2	1	–	2
Level Shifter (L)	1	1	–	2
Differential Pair (D)	2	1	–	2
Cascode Curr. Mirr. (LC)	2	–	–	–

Table 1: Sizing rules for important transistor-pair elements

An efficient solution is based on a linearization of only the $\beta_{\mathrm{wc},i}$ and sizing rules $\mathbf{c}(\mathbf{d})$ [21] at $\mathbf{d}_0$ in each iteration within a trust-region algorithm [18, 19]:

$$\min_{\mathbf{d}}\{\widehat{\varphi}(\mathbf{d}) \mid \mathbf{c}(\mathbf{d}_0) + \nabla_{\mathbf{d}}^{\mathrm{T}}\mathbf{c} \cdot (\mathbf{d} - \mathbf{d}_0) \geq \mathbf{0}\}\,, \tag{8}$$

$$\widehat{\varphi}(\mathbf{d}) = \sum_i \exp\left(-a \cdot \left(\beta_{\mathrm{wc},i0} + \nabla_{\mathbf{d}}^{\mathrm{T}}\beta_{\mathrm{wc},i} \cdot (\mathbf{d} - \mathbf{d}_0)\right)\right)\,.$$

The sizing algorithm includes a cost-saving preprocessing performance-optimization stage [19].

4.1 Gradients

Local variations usually are uncorrelated and depend on design parameters (transistor geometries) [16]: $\sigma_{V_{\mathrm{th}}} = A/\sqrt{WL}$. The required optimization gradient can be calculated based on the performance gradient at the worst-case parameter set:

$$\nabla_{d_i}\beta_{\mathrm{wc}} = \frac{\beta_{\mathrm{wc}}}{\mathbf{g}^{\mathrm{T}}\mathbf{g}}\frac{g_i^2}{2d_i} - \frac{1}{\sqrt{\mathbf{g}^{\mathrm{T}}\mathbf{g}}}\nabla_{d_i} f|_{\mathbf{s}_{\mathrm{wc}}} \tag{9}$$

4.2 Sizing Rules

Sizing rules formalize intuitive design knowledge (e.g. to guarantee that a transistor functions as voltage controlled current source) on the level of transistor pairs for CMOS technology [21]. On circuit level, sizing rules are established for each circuit by structure recognition of transistor pairs.

Table 1 shows an overview of sizing rules for the most important transistor-pair elements. For each element, the number of sizing rules introduced on its hierarchical level is given, characterizing geometrical rules for transistor widths and lengths, electrical rules for currents and voltages, functional rules enforcing the required behavior, and tolerance rules for manufacturability. For the cascode current mirror, the total number of sizing rules (35) is determined by adding the number of rules of its building elements, i.e. current mirror, level shifter, single transistors.

Using these sizing rules enables an automatic synthesis that yields technically meaningful and robust results.

5 WiCkeD

The presented algorithms are implemented in a tool called WiCkeD (100k lines of code), which can be demonstrated and which is applied in industry. It is based on a mature and flexible simulation environment, which can be adapted to any underlying simulator. WiCkeD's CORBA-based simulation environment features simultaneous cross-platform simulations on a large number of host computers. Circuit performances are automatically extracted from simulator output files by Tcl scripts, so arbitrary circuit performances can be defined. Network traffic is reduced to a minimum by sourcing out the tasks of preparing the netlists and extracting the performances' values on the remote simulation hosts. The results of all performed simulations are kept in a database and are re-used by all algorithms to avoid redundant simulator runs. Sensitivities are estimated by means of finite differences by default in order to have a general-purpose tool, but sensitivities calculated by the simulator [5, 11] and performance extraction scripts can be used as well.

Performance	A_0 [dB]	f_t [MHz]	CMRR [dB]	SR_p [V/μs]	Power [mW]
Specification	> 35	> 40	> 75	>32	< 4.5
$f_i^{(0)}$	98	14	153	12	0.6
$\beta_i^{(0)}$ [σ]	2.2	< -6	3.3	< -6	> 6
$Y^{(0)}$	WiCkeD: 0%		Monte Carlo: 0%		
$\beta_i^{(2)}$ [σ]	1.8	0.6	2.9	2.9	2.7
$Y_i^{(2)}$ [%] (Monte Carlo)	96.0 (95.0)	72.6 (74.6)	99.8 (99.8)	99.8 (100)	99.6 (99.6)
$Y^{(2)}$	WiCkeD: 72.3%		Monte Carlo: 73.8%		
$f_i^{(9)}$	85	69	164	57	2.2
$\beta_i^{(9)}$ [σ]	4.1	3.8	5.5	4.1	3.8
$Y^{(9)}$	WiCkeD: 99.9%		Monte Carlo: 99.9%		

Table 2: Analysis results before ($^{(0)}$), during ($^{(2)}$) and after ($^{(9)}$) synthesis

WiCkeD features interactive and automatic circuit sizing with a comprehensive graphical user interface. Its batch mode enables the user to integrate WiCkeD into existing frameworks and to automate recurring design tasks to a large extent, as it provides a powerful scripting language based on Tcl.

6 Experimental Results

The presented tool is in use for the design of e.g. opamp cells, interface circuits, switched-capacitor filters. The opamp in Fig. 1 has been selected to present experimental results. For this circuit, 22 statistically independent threshold voltages with local variations, 44 design parameters (channel width, length of each transistor), 4 operational parameters and 5 performances with specifications (Table 2) are given. Introduction of sizing rules according to Sec. 4.2 leads to 165 constraint equations and a reduction to 9 independent design parameters.

Table 2 shows values of the robustness objectives (β) and estimated yield values (Y) (obtained according to Sec. 3) before, during and after synthesis according to Sec. 4. The performance values (f) before and after synthesis are also given.

From an initial design with violated specs and a yield of 0%, the method progresses to a centered design with 100% yield in 9 iteration steps. The design center is determined by the smallest, competing robustness objectives. In this case, nearly a 4σ design has been achieved.

In Table 2, the yield estimations obtained with 500-sample Monte Carlo simulations are added. The numbers show that the WiCkeD yield estimates are practically equal to those of a Monte Carlo analysis. Please note that after synthesis the gain A_0 has a higher robustness with respect to manufacturing and operating variations, although its nominal value is closer to the spec value. This result can only be achieved by design centering with regard to all types of tolerances.

Table 3 shows the computational costs obtained on a net-

#iterations	6 performance optimization +3 design centering
#simulations	**712** (simulator-internal sensitivity: ≈ 200)
wall clock time	**20min.** (1 computer: 65min.)

Table 3: Computational costs

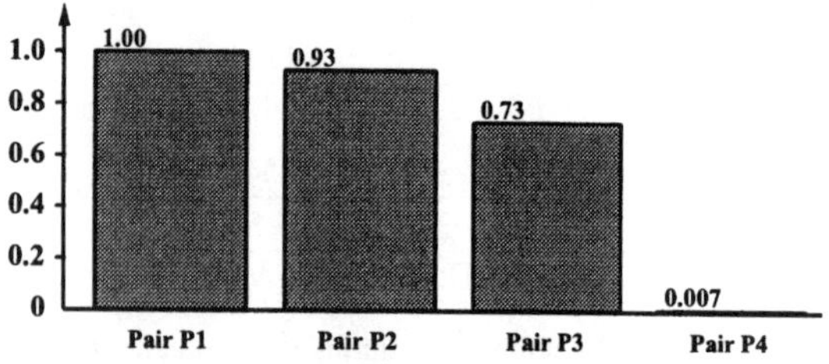

Figure 4: Mismatch-sensitive transistor pairs

work (100 Mbit/sec) of 5 computers (500MHz Pentium III) using the Infineon in-house simulator TITAN. One can see that the whole synthesis process takes 712 simulations (1 simulation includes DC, AC and transient simulation) and 20 minutes elapsed time (exclusive computer usage), which is comparable to the cost of a Monte Carlo analysis. The number in brackets describes the cost in units of simulations if a simulator with internal sensitivity analysis would be used (10% overhead for each parameter in sensitivity calculations [5]). The time in brackets describes the synthesis time if only one computer would be used. These numbers show that a fast, automatic synthesis has been achieved.

Fig. 4 shows a mismatch analysis of the circuit after synthesis based on the worst-case parameter sets and the sizing rules. There are three dominant transistor pairs concerning mismatch sensitivity. These pairs P1, P2 and P3 are marked in Fig. 1.

Table 4 compares the computational costs for a standard SQP algorithm [20] and for the presented algorithm in Sec. 3 (modSQP) at two steps ($^{(0)}$, $^{(2)}$) of the synthesis process. One can see that the total number of iterations for modSQP is 25% below SQP and, more important, SQP fails (ϟ) to find the solution in 4 of 10 cases, while modSQP provides reliable solutions.

7 Conclusion

A deterministic method for circuit analysis and synthesis in the presence of mismatch has been presented. Experimental results show that yield analysis for local parameter variations with the presented customized SQP is more efficient than a standard SQP or Monte Carlo approach and more reliable than a standard SQP approach. Unlike Monte Carlo-based yield analysis, the presented deterministic approach provides the required optimization gradient including the dependency of local parameter variations from design parameters. Parametric synthesis as presented can be performed automatically at costs comparable to a Monte Carlo analysis. The implemented tool is in industrial practice.

	A_0	f_t	CMRR	SR_p	Power
modSQP $^{(0)}$	5	7	9	5	5
SQP $^{(0)}$	11 ϟ	11	10 ϟ	5	2
modSQP $^{(2)}$	5	5	9	4	4
SQP $^{(2)}$	6	9 ϟ	20 ϟ	2	2

Table 4: #iterations for yield analysis: new modSQP vs. SQP

Acknowledgment

We would like to thank Dr. H. Eichfeld and Dr. U. Schlichtmann from Infineon Technologies AG for supporting this work and Th. Eichler from Infineon Technologies AG for discussion of results.

References

[1] E. Ochotta, R. Rutenbar, L. Carley, "Synthesis of high-performance analog circuits in ASTRX/OBLX," *IEEE Trans. CAD*-15:273–294, 1996.

[2] R. Harjani, R. Rutenbar, L. Carley, "OASYS: A framework for analog circuit synthesis," *IEEE Trans. CAD*-8:1247–1266, 1989.

[3] G. Gielen, H. Walscharts, W. Sansen, "Analog circuit design optimization based on symbolic simulation and simulated annealing," *IEEE J. SC*-25:707–713, 1990.

[4] A. Doboli, A. Nunez-Aldana, N. Dhanwada, S. Ganesan, R. Vemuri, "Behavioral synthesis of analog systems using two-layered design space exploration," in *ACM/IEEE DAC*, 1999.

[5] W. Nye, D. Riley, A. Sangiovanni-Vincentelli, A. Tits, "DELIGHT.SPICE: An optimization-based system for the design of integrated circuits," *IEEE Trans. CAD*-7:501–519, 1988.

[6] S. Director, W. Maly, A. Strojwas, *VLSI Design for Manufacturing: Yield Enhancement*, Kluwer Academic Publishers, USA, 1990.

[7] K. Antreich, R. Koblitz, "Design centering by yield prediction," *IEEE Trans. CAS*-29:88–95, 1982.

[8] K. Antreich, H. Graeb, C. Wieser, "Circuit analysis and optimization driven by worst-case distances," *IEEE Trans. CAD*-13:57–71, 1994.

[9] K. Krishna, S. Director, "The linearized performance penalty (LPP) method for optimization of parametric yield and its reliability," *IEEE Trans. CAD*-14:1557–1568, 1995.

[10] A. Dharchoudhury, S. Kang, "Worst-case analysis and optimization of VLSI circuit performances," *IEEE Trans. CAD*-14:481–492, 1995.

[11] A. Conn, P. Coulman, R. Haring, G. Morill, C. Visweswariah, Chai Wah Wu, "JiffyTune: Circuit optimization using time-domain sensitivities," *IEEE Trans. CAD*-17:1292–1309, 1998.

[12] G. Debyser, G. Gielen, "Efficient analog circuit synthesis with simultaneous yield and robustness optimization," in *IEEE/ACM ICCAD*, 1998.

[13] M. del Mar Hershenson, S. Boyd, T. H. Lee, "GPCAD: A tool for CMOS op-amp synthesis," in *IEEE/ACM ICCAD*, 1998.

[14] M. Krasnicki, R. Phelps, R. Rutenbar, L. Carley, "MAELSTROM: Efficient simulation-based synthesis for custom analog cells," in *ACM/IEEE DAC*, 1999.

[15] P. Feldmann, S. Director, "Integrated circuit quality optimization using surface integrals," *IEEE Trans. CAD*-12:1868–1879, 1993.

[16] M. Pelgrom, A. Duinmaijer, A. Welbers, "Matching properties of MOS transistors," *IEEE J. SC*-24:1433–1440, 1989.

[17] K. Eshbaugh, "Generation of correlated parameters for statistical circuit simulation," *IEEE Trans. CAD*-11:1198–1206, 1992.

[18] J. Nocedal, S. Wright, *Numerical Optimization*, Springer, 1999.

[19] R. Schwencker, F. Schenkel, H. Graeb, K. Antreich, "The generalized boundary curve – A common method for automatic nominal design and design centering of analog circuits," in *DATE*, 2000.

[20] K. Schittkowski "The nonlinear programming method of Wilson, Han, and Powell with an augmented Lagrangian type line search function,," *Numer. Math.*-38:83-114, 1981.

[21] S. Zizala, J. Eckmueller, H. Graeb, "Fast Calculation of Analog Circuits' Feasibility Regions by Low Level Functional Measures," in *IEEE ICECS*, 1998.

Optimal Design of a CMOS Op-Amp via Geometric Programming

Maria del Mar Hershenson, Stephen P. Boyd, *Fellow, IEEE*, and Thomas H. Lee

***Abstract*—We describe a new method for determining component values and transistor dimensions for CMOS operational amplifiers (op-amps). We observe that a wide variety of design objectives and constraints have a special form, i.e., they are *posynomial* functions of the design variables. As a result, the amplifier design problem can be expressed as a special form of optimization problem called *geometric programming*, for which very efficient *global optimization* methods have been developed. As a consequence we can efficiently determine *globally optimal* amplifier designs or globally optimal tradeoffs among competing performance measures such as power, open-loop gain, and bandwidth. Our method, therefore, yields completely automated sizing of (globally) optimal CMOS amplifiers, directly from specifications.**

In this paper, we apply this method to a specific widely used operational amplifier architecture, showing in detail how to formulate the design problem as a geometric program. We compute globally optimal tradeoff curves relating performance measures such as power dissipation, unity-gain bandwidth, and open-loop gain. We show how the method can be used to size *robust designs*, i.e., designs guaranteed to meet the specifications for a variety of process conditions and parameters.

***Index Terms*—Circuit optimization, CMOS analog integrated circuits, design automation, geometric programming, mixed analog–digital integrated circuits, operational amplifiers.**

I. Introduction

As THE demand for mixed-mode integrated circuits increases, the design of analog circuits such as operational amplifiers (op-amps) in CMOS technology becomes more critical. Many authors have noted the disproportionately large design time devoted to the analog circuitry in mixed-mode integrated circuits. In this paper, we introduce a new method for determining the component values and transistor dimensions for CMOS op-amps. The method handles a very wide variety of specifications and constraints, is *extremely fast*, and results in *globally optimal* designs.

The performance of an op-amp is characterized by a number of performance measures such as open-loop voltage gain, quiescent power, input-referred noise, output voltage swing, unity-gain bandwidth, input offset voltage, common-mode rejection ratio, slew rate, die area, and so on. These performance measures are determined by the design parameters, e.g., transistor dimensions, bias currents, and other component values. The CMOS amplifier design problem we consider in this paper is to determine values of the design parameters that optimize an objective measure while satisfying specifications or constraints on the other performance measures. This design problem can be approached in several ways, e.g., by hand or a variety of computer-aided-design methods, e.g., classical optimization methods, knowledge-based methods or simulated annealing. (These methods are described more fully below.)

In this paper, we introduce a new method that has a number of important advantages over current methods. We formulate the CMOS op-amp design problem as a very special type of optimization problem called a *geometric program*. The most important feature of geometric programs is that they can be reformulated as *convex optimization problems* and, therefore, *globally optimal* solutions can be computed with *great efficiency*, even for problems with hundreds of variables and thousands of constraints, using recently developed interior-point algorithms. Thus, even challenging amplifier design problems with many variables and constraints can be (globally) solved.

The fact that geometric programs (and, hence, CMOS op-amp design problems cast as geometric programs) can be globally solved has a number of important practical consequences. The first is that sets of infeasible specifications are unambiguously recognized: the algorithms either produce a feasible point or a proof that the set of specifications is infeasible. Indeed, the choice of initial design for the optimization procedure is completely irrelevant (and can even be infeasible); it has no effect on the final design obtained. Since the global optimum is found, the op-amps obtained are not just the best our method can design, but in fact the best *any* method can design (with the same specifications). In particular, our method computes the *absolute limit of performance* for a given amplifier and technology parameters.

The fact that geometric programs can be solved very efficiently has a number of practical consequences. For example, the method can be used to simultaneously optimize the design of a large number of op-amps in a single large mixed-mode integrated circuit. In this case, the designs of the individual op-amps are coupled by constraints on total power and area, and by various parameters that affect the amplifier coupling such as input capacitance, output resistance, etc. Another application is to use the efficiency to obtain *robust designs*, i.e., designs that are guaranteed to meet a set of specifications over a variety of processes or technology parameter values. This is done by simply replicating the specifications with a (possibly large) number of representative process parameters, which is practical only because geometric programs with thousands of constraints are readily solved.

All the advantages mentioned (convergence to a global solution, unambiguous detection of infeasibility, sensitivity anal-

Manuscript received November 10, 1997; revised March 9, 2000. This paper was recommended by Associate Editor K. Mayalam.

M. del Mar Hershenson is with Barcelona Design, Inc., Sunnyvale, CA 94086 USA.

S. P. Boyd and T. H. Lee are with the Electrical Engineering Department, Stanford University, Stanford, CA 94305 USA.

Publisher Item Identifier S 0278-0070(01)00348-7.

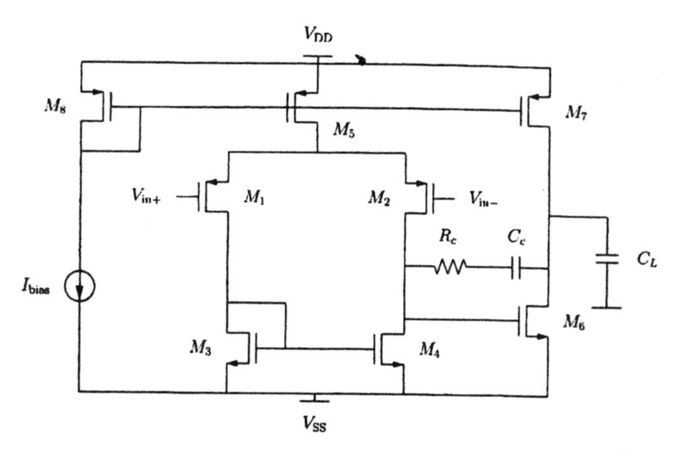

Fig. 1. Two-stage op-amp considered herein.

ysis, ...) are due to the formulation of the design problem as a *convex* optimization problem. Geometric programming (when reformulated as described in Section II-A) is just a special type of convex optimization problem. Although general convex problems can be solved efficiently, the special structure of geometric programming can be exploited to obtain an even more efficient solution algorithm.

The method we present can be applied to a wide variety of amplifier architectures, but in this paper, we apply the method to a specific two-stage CMOS op-amp. The authors show how the method extends to other architectures in [49] and [50]. A longer version of this paper, which includes more detail about the models, some of the derivations, and SPICE simulation parameters, is available at the authors' web site [51]. Related work has been reported in several conference publications, e.g., [48]–[50].

A. *The Two-Stage Amplifier*

The specific two-stage CMOS op-amp we consider is shown in Fig. 1. The circuit consists of an input differential stage with active load followed by a common-source stage also with active load. An output buffer is not used; this amplifier is assumed to be part of a very large scale integration (VLSI) system and is only required to drive a fixed on-chip capacitive load of a few picofarads. This op-amp architecture has many advantages: high open-loop voltage gain, rail-to-rail output swing, large common-mode input range, only one frequency compensation capacitor, and a small number of transistors. Its main drawback is the nondominant pole formed by the load capacitance and the output impedance of the second stage, which reduces the achievable bandwidth. Another potential disadvantage is the right half-plane zero that arises from the feedforward signal path through the compensating capacitor. Fortunately, the zero is easily removed by a suitable choice for the compensation resistor R_c (see [2]).

This op-amp is a widely used general purpose op-amp [88]; it finds applications, for example, in switched capacitor filters [23], analog-to-digital converters [60], [72], and sensing circuits [85].

There are 18 design parameters for the two-stage op-amp:

- The widths and lengths of all transistors, i.e., $W_1, \ldots, W_8$ and $L_1, \ldots, L_8$.
- The bias current I_{bias}.
- The value of the compensation capacitor C_c.

The compensation resistor R_c is chosen in a specific way that is dependent on the design parameters listed above (and described in Section V). There are also a number of parameters that we consider fixed, e.g., the supply voltages V_{DD} and V_{SS}, the capacitive load C_L, and the various process and technology parameters associated with the MOS models. To simplify some of the equations we assume (without any loss of generality) that $V_{\text{SS}} = 0$.

B. *Other Approaches*

There is a huge literature, which goes back more than 20 years, on computer-aided design (CAD) of analog circuits. A good survey of early research can be found in the survey [11]; more recent papers on analog-circuit CAD tools include [4], [12], [13]. The problem we consider in this paper, i.e., selection of component values and transistor dimensions, is only a part of a complete analog-circuit CAD tool. Other parts, which we do not consider here, include topology selection (see [66]) and actual circuit layout (see, e.g., ILAC [27], KOAN/ANAGRAM II [15]). The part of the CAD process that we consider lies between these two tasks; the remainder of the discussion is restricted to methods dealing with component and transistor sizing.

1) Classical Optimization Methods: General-purpose classical optimization methods, such as steepest descent, sequential quadratic programming, and Lagrange multiplier methods, have been widely used in analog-circuit CAD. These methods can be traced back to the survey paper [11]. The widely used general-purpose optimization codes NPSOL [39] and MINOS [71] are used in [25], [64], and [67]. LANCELOT [16], another general-purpose optimizer, is used in [22]. Other CAD approaches based on classical optimization methods, and extensions such as a minimax formulation, include the one described in [47], [61], and [63], OAC [78], OPASYN [56], CADICS [54], WATOPT [31], and STAIC [45]. The classical methods can be used with more complicated circuit models, including even full SPICE simulations in each iteration, as in DELIGHT.SPICE [75] (which uses the general-purpose optimizer DELIGHT [76]) and ECSTASY [86].

The main advantage of these methods is the wide variety of problems they can handle; the only requirement is that the performance measures, along with one or more derivatives, can be computed. The main disadvantage of the classical optimization methods is they only find *locally optimal* designs. This means that the design is at least as good as neighboring designs, i.e., small variations of any of the design parameters results in a worse (or infeasible) design. Unfortunately this does not mean the design is the best that can be achieved, i.e., globally optimal; it is possible (and often happens) that some other set of design parameters, far away from the one found, is better. The same problem arises in determining feasibility: a classical (local) optimization method can fail to find a feasible design, even though one exists. Roughly speaking, classical methods can get stuck at local minima. This shortcoming is so well known that it is often not even mentioned in papers; it is taken as understood.

The problem of nonglobal solutions from classical optimization methods can be treated in several ways. The usual approach is to start the minimization method from many different initial designs, and to take the best final design found. Of course, there are no guarantees that the globally optimal design has been found; this method merely increases the likelihood of finding the globally optimal design. This method also destroys one of the advantages of classical methods, i.e., speed, since the computation effort is multiplied by the number of different initials designs that are tried. This method also requires human intervention (to give "good" initial designs), which makes the method less automated.

The classical methods become slow if complex models are used, as in DELIGHT.SPICE, which requires more than a complete SPICE run at each iteration ("more than" since, at the least, gradients must also be computed).

2) Knowledge-Based Methods: Knowledge-based and expert-systems methods have also been widely used in analog circuit CAD. Examples include genetic algorithms or evolution systems like SEAS [74], DARWIN [58], [100]; systems based on fuzzy logic like FASY [46] and [92]; special heuristics-based systems like IDAC [29], [30], OASYS [44], BLADES [21], and KANSYS [43].

One advantage of these methods is that there are few limitations on the types of problems, specifications, and performance measures that can be considered. Indeed, there are even fewer limitations than for classical optimization methods since many of these methods do not require the computation of derivatives.

These methods have several disadvantages. They find a locally optimal design (or, even just a "good" or "reasonable" design) instead of a globally optimal design. The final design depends on the initial design chosen and the algorithm parameters. As with classical optimization methods, infeasibility is not unambiguously detected; the method simply fails to find a feasible design (even when one may exist). These methods require substantial human intervention either during the design process, or during the training process.

3) Global Optimization Methods: Optimization methods that are guaranteed to find the globally optimal design have also been used in analog-circuit design. The most widely known global optimization methods are branch and bound [103] and simulated annealing [94], [101].

A branch and bound method is used, e.g., in [66]. Branch and bound methods unambiguously determine the globally optimal design: at each iteration they maintain a suboptimal feasible design and also a lower bound on the achievable performance. This enables the algorithm to terminate nonheuristically, i.e., with complete confidence that the global design has been found within a given tolerance. The disadvantage of branch and bound methods is that they are extremely slow, with computation growing exponentially with problem size. Even problems with 10 variables can be extremely challenging.

Simulated annealing (SA) is another very popular method that can avoid becoming trapped in a locally optimal design. In *principle* it can compute the globally optimal solution, but in implementations there is no guarantee at all, since, for example, the cooling schedules called for in the theoretical treatments are not used in practice. Moreover, no real-time lower bound is available, so termination is heuristic. Like classical and knowledge-based methods, SA allows a very wide variety of performance measures and objectives to be handled. Indeed, SA is extremely effective for problems involving continuous variables and discrete variables, as in, e.g., simultaneous amplifier topology and sizing problems. SA has been used in several tools such as ASTR/OBLX [77], OPTIMAN [38], FRIDGE [68], SAMM [105], and [14].

The main advantages of SA are that it handles discrete variables well, and greatly reduces the chances of finding a nonglobally optimal design. (Practical implementations do not reduce the chance to zero, however.) The main disadvantage is that it can be very slow, and cannot (in practice) guarantee a globally optimal solution.

4) Convex Optimization and Geometric Programming Methods: In this section, we describe the general optimization method we employ in this paper: convex optimization. These are special optimization problems in which the objective and constraint functions are all convex.

While the theoretical properties of convex optimization problems have been appreciated for many years, the advantages in practice are only beginning to be appreciated now. The main reason is the development of extremely powerful interior-point methods for general convex optimization problems in the last five years (e.g., [73] and [102]). These methods can solve large problems, with thousands of variables and tens of thousands of constraints, very efficiently (in minutes on a small workstation). Problems involving tens of variables and hundreds of constraints (such as the ones we encounter in this paper) are considered small, and can be solved on a small current workstation in less than one second. The extreme efficiency of these methods is one of their great advantages.

The other main advantage is that the methods are truly global, i.e., the global solution is *always* found, regardless of the starting point (which, indeed, need not be feasible). Infeasibility is unambiguously detected, i.e., if the methods do not produce a feasible point they produce a certificate that proves the problem is infeasible. Also, the stopping criteria are completely nonheuristic: at each iteration a lower bound on the achievable performance is given.

One of the disadvantages is that the types of problems, performance specifications, and objectives that can be handled are far more restricted than any of the methods described above. This is the price that is paid for the advantages of extreme efficiency and global solutions. (For more on convex optimization, and the implications for engineering design, see [10].)

The contribution of this paper is to show how to formulate the analog amplifier design problem as a certain type of convex problem called geometric programming. The advantages, compared to the approaches described above, are extreme efficiency and global optimality. The disadvantage is less flexibility in the types of constraints we can handle, and the types of circuit models we can employ.

Aside from work we describe below, the only other application of geometric programming to circuit design is in transistor and wire sizing for Elmore delay minimization in digital circuits, as in TILOS [36] and other programs [81], [82], [87]. Their use of geometric programming can be distinguished from ours in several ways. First of all, the geometric programs that arise in Elmore delay minimization are very specialized (the only exponents that arise are 0 and ± 1). Second, the problems they encounter in practice are extremely large, involving up to hundreds of thousands of variables. Third, their representation of the problem as a geometric program is quite an approximation, since the actual circuits are nonlinear, and the threshold delay, not Elmore delay, is the true objective.

Convex optimization is mentioned in several papers on analog-circuit CAD. The advantages of convex optimization are mentioned in [65] and [66]. In [25] and [26], the authors use a supporting hyperplane method, which they point out provides the global optimum if the feasible set is convex. In [89], the authors optimize a few design variables in an op-amp using a Lagrange multiplier method, which yields the global optimum since the small subproblems considered are convex. In [95] and [96], convex optimization is used to optimize area, power, and dominant time constant in digital circuit wire and transistor sizing.

During the review process for this paper, the authors were informed of similar work that had been submitted to IEEE TRANSACTIONS ON COMPUTER-AIDED DESIGN OF INTEGRATED CIRCUITS AND SYSTEMS by Mandal and Visvanathan [24]. Mandal and Visvanathan show how geometric programming can be used to size another simple op-amp, and describe a simple method for iteratively refining monomial device models.

C. Outline of Paper

In Section II, we briefly describe geometric programming, the special type of optimization problem at the heart of the method, and show how it can be cast as a convex optimization problem. In Sections III–VI we describe a variety of constraints and performance measures, and show that they have the special form required for geometric programming. In Section VII we give numerical examples of the design method, showing globally optimal tradeoff curves among various performance measures such as bandwidth, power, and area. We also verify some of our designs using high fidelity SPICE models, and briefly discuss how our method can be extended to handle short-channel effects. In Section IX, we discuss robust design, i.e., how to use the methods to ensure proper circuit operation under various processing conditions. In Section X, we give our concluding remarks.

II. GEOMETRIC PROGRAMMING

Let $x_1, \ldots, x_n$ be n real, positive variables. We will denote the vector $(x_1, \ldots, x_n)$ of these variables as x. A function f is called a *posynomial* function of x if it has the form

$$f(x_1, \ldots, x_n) = \sum_{k=1}^{t} c_k x_1^{\alpha_{1k}} x_2^{\alpha_{2k}} \cdots x_n^{\alpha_{nk}}$$

where $c_j \geq 0$ and $\alpha_{ij} \in \boldsymbol{R}$. Note that the coefficients c_k must be nonnegative, but the exponents α_{ij} can be any real numbers, including negative or fractional. When there is exactly one nonzero term in the sum, i.e., $t = 1$ and $c_1 > 0$, we call f a *monomial* function. (This terminology is not consistent with the standard definition of a monomial in algebra, but it should not cause any confusion.) Thus, for example, $0.7 + 2x_1/x_3^2 + x_2^{0.3}$ is posynomial (but not monomial); $2.3(x_1/x_2)^{1.5}$ is a monomial (and, therefore, also a posynomial); while $2x_1/x_3^2 - x_2^{0.3}$ is neither. Note that posynomials are closed under addition, multiplication, and nonnegative scaling. Monomials are closed under multiplication and division.

A *geometric program* is an optimization problem of the form

$$\begin{array}{lll} \text{minimize} & f_0(x) & \\ \text{subject to} & f_i(x) \leq 1, & i = 1, \ldots, m \\ & g_i(x) = 1, & i = 1, \ldots, p \\ & x_i > 0, & i = 1, \ldots, n \end{array} \qquad (1)$$

where $f_0, \ldots, f_m$ are posynomial functions and $g_1, \ldots, g_p$ are monomial functions.

Several extensions are readily handled. If f is a posynomial and g is a monomial, then the constraint $f(x) \leq g(x)$ can be handled by expressing it as $f(x)/g(x) \leq 1$ (since f/g is posynomial). For example, we can handle constraints of the form $f(x) \leq a$, where f is posynomial and $a > 0$. In a similar way if g_1 and g_2 are both monomial functions, then we can handle the equality constraint $g_1(x) = g_2(x)$ by expressing it as $g_1(x)/g_2(x) = 1$ (since g_1/g_2 is monomial).

We will also encounter functions whose reciprocals are posynomials. We say h is *inverse posynomial* if $1/h$ is a posynomial. If h is an inverse posynomial and f is a posynomial, then geometric programming can handle the constraint $f(x) \leq h(x)$ by writing it as $f(x)(1/h(x)) \leq 1$. As another example, if h is an inverse posynomial, then we can maximize it, by minimizing (the posynomial) $1/h$.

Geometric programming has been known and used since the late 1960s, in various fields. There were two early books on geometric programming, by Duffin *et al.* [18] and Zener [106], which include the basic theory, some electrical engineering applications (e.g., optimal transformer design), but not much on numerical solution methods. Another book appeared in 1976 [9]. The 1980 survey paper by Ecker [19] has many references on applications and methods, including numerical solution methods used at that time. Geometric programming is briefly described in some surveys of optimization, e.g., [20, pp. 326–328] or [99, Ch. 4]. While geometric programming is certainly known, it is nowhere near as widely known as, say, linear programming. In addition, advances in general-purpose nonlinear constrained optimization algorithms and codes (such as the ones described above) have contributed to decreased use (and knowledge) of geometric programming in recent years.

A. Geometric Programming in Convex Form

A geometric program can be reformulated as a *convex optimization problem*, i.e., the problem of minimizing a convex function subject to convex inequality constraints and linear equality constraints. This is the key to our ability to globally and efficiently solve geometric programs. We define new variables $y_i = \log x_i$, and take the logarithm of a posynomial f to get

$$h(y) = \log(f(e^{y_1}, \ldots, e^{y_n})) = \log\left(\sum_{k=1}^{t} e^{a_k^T y + b_k}\right)$$

where $a_k^T = [\alpha_{1k} \cdots \alpha_{nk}]$ and $b_k = \log c_k$. It can be shown that h is a *convex* function of the new variable y: for all $y, z \in \boldsymbol{R}^n$ and $0 \leq \lambda \leq 1$ we have

$$h(\lambda y + (1 - \lambda)z) \leq \lambda h(y) + (1 - \lambda)h(z).$$

Note that if the posynomial f is a monomial, then the transformed function h is affine, i.e., a linear function plus a constant.

We can convert the standard geometric program (1) into a convex program by expressing it as

$$\begin{array}{lll} \text{minimize} & \log f_0(e^{y_1}, \ldots, e^{y_n}) & \\ \text{subject to} & \log f_i(e^{y_1}, \ldots, e^{y_n}) \leq 0, & i = 1, \ldots, m \\ & \log g_i(e^{y_1}, \ldots, e^{y_n}) = 0, & i = 1, \ldots, p. \end{array} \quad (2)$$

This is the so-called *convex form* of the geometric program (1). Convexity of the convex form geometric program (2) has several important implications: we can use efficient interior-point methods to solve them, and there is a complete and useful duality, or sensitivity theory for them; see, e.g., [10].

B. Solving Geometric Programs

Since Ecker's survey paper, there have been several important developments, related to solving geometric programming in the convex form. A huge improvement in computational efficiency was achieved in 1994, when Nesterov and Nemirovsky developed efficient interior-point algorithms to solve a variety of nonlinear optimization problems, including geometric programs [73]. Recently, Kortanek *et al.* have shown how the most sophisticated primal-dual interior-point methods used in linear programming can be extended to geometric programming, resulting in an algorithm approaching the efficiency of current interior-point linear programming solvers [57]. The algorithm they describe has the desirable feature of exploiting *sparsity* in the problem, i.e., efficiently handling problems in which each variable appears in only a few constraints. Other methods developed specifically for geometric programs include those described by Avriel *et al.* [7] and Rajpogal and Bricker [80], which require solving a sequence of linear programs (for which very efficient algorithms are known).

The algorithms described above are specially tailored for the geometric program (in convex form). It is also possible to solve the convex form problem using general purpose optimization codes that handle smooth objectives and constraint functions, e.g., LANCELOT [16], MINOS [71], LOQO [97], or LINGO-NL [83]. These codes will (in principle) find a globally optimal solution, since the convex form problem is convex. They will also determine the optimal dual variables (sensitivities) as a by-product of solving the problem. In an unpublished report [104], Xu compares the performance of the sophisticated primal-dual interior-point method developed by Kortanek *et al.*(XGP, [57]) with two general-purpose optimizers, MINOS and LINGO-NL, on a suite of standard geometric programming problems (in convex form). The general-purpose codes fail to solve some of the problems, and in all cases take substantially longer to obtain the solution.

For our purposes, the most important feature of geometric programs is that they can be *globally* solved with *great* efficiency. Problems with hundreds of variables and thousands of constraints are readily handled, on a small workstation, in minutes; the problems we encounter in this paper, which have on the order of ten variables and 100 constraints, are easily solved in under one second.

To carry out the designs in this paper, we implemented, in MATLAB, a simple and crude primal barrier method for solving the convex form problem. Roughly speaking, this method consists of applying a modified Newton's method to minimizing the smooth convex function

$$t \log f_0(e^{y_1}, \ldots, e^{y_n}) + \sum_{i=1}^{m} \log(-\log f_i(e^{y_1}, \ldots, e^{y_n}))$$

subject to the affine (linear equality) constraints $\log g_i(e^{y_1}, \ldots, e^{y_n}) = 0$, $i = 1, \ldots, p$, for an increasing sequence of values of t, starting from the optimal y found for the last value of t. It can be shown that when $t \geq m/\epsilon$, the optimal solution of this problem is no more than ϵ suboptimal for the original convex form geometric program (GP). The computational complexity of this simple method is $O(pn^3)$, where n is the number of variables, and p is the total number of terms in monomials and posynomials in the objective and constraints. For much more detail, see [10] and [35].

Despite the simplicity of the algorithm (i.e., primal only, with no sparsity exploited) and the overhead of an interpreted language, the geometric programs arising in this paper were all solved in approximately 1 or 2 s on an ULTRA SPARC1 running at 170 MHz. Since our simple interior-point method is already extremely fast on the relatively small problems we encounter in this paper, we feel that the choice of algorithm is not critical. When the method is applied to large-scale problems, such as the ones obtained for a robust design problem (see Section IX), the choice may well become critical.

C. Sensitivity Analysis

Suppose we modify the right-hand sides of the constraints in the geometric program (1) as follows:

$$\begin{array}{lll} \text{minimize} & f_0(x) & \\ \text{subject to} & f_i(x) \leq e^{u_i}, & i = 1, \ldots, m \\ & g_i(x) = e^{v_i}, & i = 1, \ldots, p \\ & x_i > 0, & i = 1, \ldots, n. \end{array} \quad (3)$$

If all of the u_i and v_i are zero, this modified geometric program coincides with the original one. If $u_i < 0$, then the constraint $f_i(x) \leq e_i^u$ represents a *tightened* version of the original ith constraint $f_i(x) \leq 1$; conversely if $u_i > 0$, it represents a *loosening* of the constraint. Note that u_i gives a logarithmic or fractional measure of the change in the specification: $u_i = 0.0953$ means that the ith constraint is loosened 10%, whereas $u_i = -0.0953$ means that the ith constraint is tightened 10%.

Let $f_0^*(u, v)$ denote the optimal objective value of the modified geometric program (3), as a function of the parameters $u = (u_1, \ldots, u_m)$ and $v = (v_1, \ldots, v_p)$, so the original objective value is $f_0^*(0, 0)$. In *sensitivity analysis*, we study the variation of f_0^* as a function of u and v, for small u and v. To express the change in optimal objective function in a fractional form, we use the *logarithmic sensitivities*

$$S_i = \frac{\partial \log f_0^*}{\partial u_i} \quad T_i = \frac{\partial \log f_0^*}{\partial v_i} \tag{4}$$

evaluated at $u = 0$, $v = 0$. These sensitivity numbers are dimensionless, since they express fractional changes per fractional change.

For simplicity we are assuming here that the original geometric program is feasible, and remains so for small changes in the right-hand sides of the constraints, and also that the optimal objective value is differentiable as a function of u_i and v_i. More complete descriptions of sensitivity analysis in other cases can be found in the references cited above, or in a general context in [10]. The surprising part is that the sensitivity numbers $S_1, \ldots, S_m$ and $T_1, \ldots, T_p$ come for free, when the problem is solved using an interior-point or Lagrangian-based method (from the solution of the dual problem; see [10]).

We start with some simple observations. If at the optimal solution x^* of the original problem, the ith inequality constraint is not active, i.e., $f_i(x^*)$ is strictly less than one, then $S_i = 0$ (since we can slightly tighten or loosen the ith constraint with no effect). We always have $S_i \leq 0$ since increasing u_i slightly loosens the constraints, and hence lowers the optimal objective value. The sign of T_i tells us whether increasing the right-hand side side of the equality constraint $g_i = 1$ increases or decreases the optimal objective value.

The sensitivity numbers are extremely useful in practice, and give tremendous insight to the designer. Suppose, for example, that the objective f_0 is power dissipation, $f_1(x) \leq 1$ represents the constraint that the bandwidth is at least 30 MHz, and $g_1(x) = 1$ represents the constraint that the open-loop gain is 10^5 V/V. Then $S_1 = -3$, say, tells us that a small fractional increase in required bandwidth will translate into a three times larger fractional increase in power dissipation. $T_1 = 0.1$ tells us that a small fractional increase in required open-loop gain will translate into a fractional increase in power dissipation only one-tenth as big. Although both constraints are active, the sensitivities tell us that the design is, roughly speaking, more tightly constrained by the bandwidth constraint than the open-loop gain constraint. The sensitivity information from the example above might lead the designer to reduce the required bandwidth (to reduce power), or perhaps increase the open-loop gain (since it would not cost much). We give an example of sensitivity analysis in Section VII-D.

III. Dimension Constraints

We start by considering some very basic constraints involving the device dimensions, e.g., symmetry, matching, minimum or maximum dimensions, and area limits.

A. *Symmetry and Matching*

For the intended operation of the input differential pair, transistors M_1 and M_2 must be identical and transistors M_3 and M_4 must also be identical. These conditions translate into the four equality constraints

$$W_1 = W_2 \quad L_1 = L_2 \quad W_3 = W_4 \quad L_3 = L_4. \tag{5}$$

The biasing transistors M_5, M_7, and M_8 must match, i.e., have the same length

$$L_5 = L_7 = L_8. \tag{6}$$

The six equality constraints in (5) and (6) have monomial expressions on the left- and right-hand sides and hence, are readily handled in geometric programming (by expressing them as monomial equality constraints such as $W_1/W_2 = 1$).

Note that (5) and (6) effectively reduce the number of variables from 18 to 12. We can, for example, eliminate the variables L_7 and L_8 by substituting L_5 wherever they appear. For clarity, we will continue to use the variables L_7 and L_8 in our discussion; for computational purposes, however, they can be replaced by L_5. (In any case, the number of variables and constraints is so small for a geometric program that there is almost no computational penalty in keeping the extra variables and equality constraints.)

B. *Limits on Device Sizes*

Lithography limitations and layout rules impose minimum (and possibly maximum) sizes on the transistors

$$\begin{aligned} L_{\min} &\leq L_i \leq L_{\max} \\ W_{\min} &\leq W_i \leq W_{\max}, \qquad i = 1, \ldots, 8. \end{aligned} \tag{7}$$

These 32 constraints can be expressed as posynomial constraints such as $L_{\min}/L_1 \leq 1$, etc. Since L_i and W_i are variables (hence, monomials), we can also fix certain devices sizes, i.e., impose equality constraints.

We should note that a constraint limiting device dimensions to a finite number of allowed values, or to an integer multiple of some fixed small value, *cannot* be (directly) handled by geometric programming. Such constraints can be approximately handled by simple rounding to an allowed value, or using more sophisticated mixed convex-integer programming methods.

C. *Area*

The op-amp die area A can be approximated as a constant plus the sum of transistor and capacitor area as

$$A = \alpha_0 + \alpha_1 C_c + \alpha_2 \sum_{i=1}^{8} W_i L_i. \tag{8}$$

Here $\alpha_0 \geq 0$ gives the fixed area, α_1 is the ratio of capacitor area to capacitance, and the constant $\alpha_2 > 1$ (if it is not one) can take into account wiring in the drain and source area. This expression for the area is a posynomial function of the design parameters, so we can impose an upper bound on the area, i.e., $A \leq A_{\max}$, or use the area as the objective to be minimized. This simple expresion does not take routing area into account; more accurate posynomial formulas for the amplifier die area could be developed, if needed.

D. *Systematic Input Offset Voltage*

To reduce input offset voltage, the drain voltages of M_3 and M_4 must be equal, ensuring that the current I_5 is split equally between transistors M_1 and M_2. This happens when the current densities of M_3, M_4, and M_6 are equal, i.e.,

$$\frac{W_3/L_3}{W_6/L_6} = \frac{W_4/L_4}{W_6/L_6} = \frac{1}{2}\frac{W_5/L_5}{W_7/L_7}. \tag{9}$$

These two conditions are equality constraints between monomials, and are therefore readily handled by geometric programming.

IV. Bias, Conditions, Signal Swing, and Power Constraints

In this section, we consider constraints involving bias conditions, including the effects of common-mode input voltage and output signal swing. We also consider the quiescent power of the op-amp (which is determined, of course, by the bias conditions). In deriving these constraints, we assume that the symmetry and matching conditions (5) and (6) hold. To derive the equations, we use a standard long-channel square-law model for the MOS transistors, which is described in detail in the Appendix. We refer to this model as the GP0 model; the same analysis also applies to the more accurate GP1 model, also described in the Appendix.

In order to simplify the equations, it is convenient to define the bias currents I_1, I_5, and I_7 through transistors M_1, M_5, and M_7, respectively. Transistors M_5 and M_7 form a current mirror with transistor M_8. Their currents are given by

$$I_5 = \frac{W_5 L_8}{L_5 W_8} I_{\text{bias}} \quad I_7 = \frac{W_7 L_8}{L_7 W_8} I_{\text{bias}}. \tag{10}$$

Thus I_5 and I_7 are monomials in the design variables. The current through transistor M_5 is split equally between transistor M_1 and M_2. Thus, we have

$$I_1 = \frac{I_5}{2} = \frac{W_5 L_8}{2 L_5 W_8} I_{\text{bias}} \tag{11}$$

which is another monomial.

Since these bias currents are monomials, we can include lower or upper bounds on them, or even equality constraints, if we wish. We will use I_1, I_5, and I_7 in order to express other constraints, remembering that these bias currents can simply be eliminated (i.e., expressed directly in terms of the design variables) using (10) and (11).

A. Bias Conditions

The setup for deriving the bias conditions is as follows. The input terminals are at the same dc potential, the common-mode input voltage V_{cm}. We assume that the common-mode input voltage is allowed to range between a minimum value $V_{\text{cm, min}}$ and a maximum value $V_{\text{cm, max}}$, which are given. Similarly, we assume that the output voltage is allowed to swing between a minimum value $V_{\text{out, min}}$ and a maximum value $V_{\text{out, max}}$ (which takes into account large signal swings in the output).

The bias conditions are that each transistor $M_1, \cdots, M_8$ should remain in saturation for all possible values of the input common-mode voltage and the output voltage. The derivation of the bias constraints given below can be found in the longer report [51]. The important point here is that the constraints are each posynomial inequalities on the design variables and, hence, can be handled by geometric programming.

- **Transistor** M_1. The lowest common-mode input voltage $V_{\text{cm, min}}$ imposes the toughest constraint on transistor M_1 remaining in saturation. The condition is

$$\sqrt{\frac{I_1 L_3}{\mu_n C_{\text{ox}}/2W_3}} \le V_{\text{cm, min}} - V_{\text{ss}} - V_{\text{TP}} - V_{\text{TN}}. \tag{12}$$

Note that if the right-hand side of (12) were negative, i.e., if $V_{\text{cm, min}} \le V_{\text{ss}} + V_{\text{TP}} + V_{\text{TN}}$, then the design is immediately known to be infeasible (since the left-hand side is, of course, positive).

- **Transistor** M_2. The systematic offset condition (9) makes the drain voltage of M_1 equal to the drain voltage of M_2. Therefore, the condition for M_2 being saturated is the same as the condition for M_1 being saturated, i.e., (12). Note that the minimum allowable value of $V_{\text{cm, min}}$ is determined by M_1 and M_2 entering the linear region.
- **Transistor** M_3. Since $V_{\text{gd},3} = 0$ transistor M_3 is always in saturation and no additional constraint is necessary.
- **Transistor** M_4. The systematic offset condition also implies that the drain voltage of M_4 is equal to the drain voltage of M_3. Thus, M_4 will be saturated as well.
- **Transistor** M_5. The highest common-mode input voltage $V_{\text{cm, max}}$, imposes the tightest constraint on transistor M_5 being in saturation. The condition is

$$\sqrt{\frac{I_1 L_1}{\mu_p C_{\text{ox}}/2W_1}} + \sqrt{\frac{I_5 L_5}{\mu_p C_{\text{ox}}/2W_5}} \le V_{\text{dd}} - V_{\text{cm, max}} + V_{\text{TP}}. \tag{13}$$

Thus, the maximum allowable value of $V_{\text{cm, min}}$ is determined by M_5 entering the linear region. As explained above, if the right-hand side of (13) is negative, i.e., $V_{\text{cm, max}} \ge V_{\text{dd}} + V_{\text{TP}}$, then the design is obviously infeasible.

- **Transistor** M_6. The most stringent condition occurs when the output voltage is at its minimum value $V_{\text{out, min}}$

$$\sqrt{\frac{I_7 L_6}{\mu_n C_{\text{ox}}/2W_6}} \le V_{\text{out, min}} - V_{\text{ss}}. \tag{14}$$

In this case the right-hand side of (14) will not be negative if we assume the minimum output voltage is above the negative supply voltage.

- **Transistor** M_7. For M_7, the most stringent condition occurs when the output voltage is at its maximum value $V_{\text{out, max}}$

$$\sqrt{\frac{I_7 L_7}{\mu_p C_{\text{ox}}/2W_7}} \le V_{\text{dd}} - V_{\text{out, max}}. \tag{15}$$

Here too, the right hand-side of (15) will be positive assuming the maximum output voltage is below the positive supply voltage.

- **Transistor** M_8. Since $V_{\text{gd},8} = 0$, transistor M_8 is always in saturation; no additional constraint is necessary.

In summary, the requirement that all transistors remain in saturation for all values of common-mode input voltage between $V_{\text{cm, min}}$ and $V_{\text{cm, max}}$, and all values of output voltage between $V_{\text{out, min}}$ and $V_{\text{out, max}}$, is given by the four inequalities (12)–(15). These are complicated, but *posynomial* constraints on the design parameters.

B. Gate Overdrive

It is sometimes desirable to operate the transistors with a minimum gate overdrive voltage. This ensures that they operate away from the subthreshold region, and also improves matching between transistors. For any given transistor this constraint can be expressed as

$$V_{\text{gs}} - V_{\text{T}} = \sqrt{\frac{I_D L}{\mu C_{\text{ox}}/2W}} \geq V_{\text{overdrive, min}}. \tag{16}$$

The expression on the left is a monomial, so we can also impose an upper bound on it, or an equality constraint, if we wish. (We will see in Section IX that robustness to process variations can be dealt with in a more direct way.)

C. Quiescent Power

The quiescent power of the op-amp is given by

$$P = (V_{\text{dd}} - V_{\text{ss}})(I_{\text{bias}} + I_5 + I_7) \tag{17}$$

which is a posynomial function of the design parameters. Hence, we can impose an upper bound on P or use it as the objective to be minimized.

V. Small–Signal Transfer Function Constraints

A. Small-Signal Transfer Function

We now assume that the symmetry, matching, and bias constraints are satisfied, and consider the (small-signal) transfer function H from a differential input source to the output. To derive the transfer function H, we use a standard small-signal model for the transistors, which is described in the Appendix, Section B. The standard value of the compensation resistor is used, i.e.,

$$R_c = 1/g_{\text{m6}} \tag{18}$$

(see [2]).

The transfer function can be well approximated by a four-pole form

$$H(s) = A_v \frac{1}{(1 + s/p_1)(1 + s/p_2)(1 + s/p_3)(1 + s/p_4)}. \tag{19}$$

Here, A_v is the open-loop voltage gain, $-p_1$ is the dominant pole, $-p_2$ is the output pole, $-p_3$ is the mirror pole, and $-p_4$ is the pole arising from the compensation circuit. In order to simplify the discussion in the sequel, we will refer to $p_1, \ldots, p_4$, which are positive, as the poles (whereas precisely speaking, the poles are $-p_1, \ldots, -p_4$).

We now give the expressions for the gain and poles. The two-stage op-amp has been previously analyzed by many authors [32], [53], [88]. The compensation scheme has also been analyzed previously, e.g., in [2].

- The open-loop voltage gain is

$$A_v = \left(\frac{g_{\text{m2}}}{g_{\text{o2}} + g_{\text{o4}}}\right)\left(\frac{g_{\text{m6}}}{g_{\text{o6}} + g_{\text{o7}}}\right) = \frac{2C_{\text{ox}}}{(\lambda_{\text{n}} + \lambda_{\text{p}})^2}\sqrt{\mu_{\text{n}}\mu_{\text{p}}\frac{W_2 W_6}{L_2 L_6 I_1 I_7}} \tag{20}$$

  which is monomial function of the design parameters.
- The dominant pole is accurately given by

$$p_1 = \frac{g_{m1}}{A_v C_c}. \tag{21}$$

  Since A_v and g_{m1} are monomials, and C_c is a design variable, p_1 is a monomial function of the design variables.
- The output pole p_2 is given by

$$p_2 = \frac{g_{\text{m6}} C_c}{C_1 C_c + C_1 C_{\text{TL}} + C_c C_{\text{TL}}} \tag{22}$$

  where C_1, the capacitance at the gate of M_6, can be expressed as

$$C_1 = C_{\text{gs6}} + C_{\text{db2}} + C_{\text{db4}} + C_{\text{gd2}} + C_{\text{gd4}} \tag{23}$$

  and C_L, the total capacitance at the output node, can be expressed as

$$C_{\text{TL}} = C_L + C_{\text{db6}} + C_{\text{db7}} + C_{\text{gd6}} + C_{\text{gd7}}. \tag{24}$$

  The meanings of these parameters, and their dependence on the design variables, is given in the Appendix, Section B. The important point here is that p_2 is an inverse posynomial function of the design parameters (i.e., $1/p_2$ is a posynomial).
- The mirror pole p_3 is given by

$$p_3 = \frac{g_{\text{m3}}}{C_2} \tag{25}$$

  where C_2, the capacitance at the gate of M_3, can be expressed as

$$C_2 = C_{\text{gs3}} + C_{\text{gs4}} + C_{\text{db1}} + C_{\text{db3}} + C_{\text{gd1}}. \tag{26}$$

  Thus, p_3 is also an inverse posynomial.
- The compensation pole is

$$p_4 = \frac{g_{\text{m6}}}{C_1} \tag{27}$$

  which is also inverse posynomial.

In summary: the open-loop gain A_v and the dominant pole p_1 are monomial, and the parasitic poles p_2, p_3, and p_4 are all inverse posynomials. Now we turn to various design constraints and specifications that involve the transfer function.

B. Open-Loop Gain Constraints

Since the open-loop gain A_v is a monomial, we can constrain it to equal some desired value A_{des}. We could also impose upper or lower bounds on the gain, as in

$$A_{\text{min}} \leq A_v \leq A_{\text{max}} \tag{28}$$

where $A_{\min}$ and $A_{\max}$ are given lower and upper limits on acceptable open-loop gain.

C. Minimum Gain at a Frequency

The magnitude squared of the transfer function at a frequency ω_0 is given by

$$|H(j\omega_0)|^2 = \frac{A_v^2}{\prod_{i=1}^{4}(1+\omega_0^2/p_i^2)}.$$

Since p_i are all inverse posynomial, the expressions ω_0^2/p_i^2 are posynomial. Hence, the whole denominator is posynomial. The numerator is monomial, thus we conclude that the squared magnitude of the transfer function, $|H(j\omega_0)|^2$, is inverse posynomial. (Indeed, it is inverse posynomial in the design variables and ω_0 as well.) We can, therefore, impose any constraint of the form

$$|H(j\omega_0)| \geq a$$

using geometric programming [by expressing it as $a^2/|H(j\omega_0)|^2 \leq 1$].

The transfer function magnitude $|H(j\omega)|$ decreases as ω increases (since it has only poles), so $|H(j\omega_0)| \geq a$ is equivalent to

$$|H(j\omega)| \geq a \qquad \text{for } \omega \leq \omega_0. \tag{29}$$

We will see below that this allows us to specify a minimum bandwidth or crossover frequency.

D. 3-dB Bandwidth

The 3-dB bandwidth $\omega_{3\,\text{dB}}$ is the frequency at which the gain drops 3 dB below the dc open-loop gain, i.e., $|H(j\omega_{3\,\text{dB}})| = A_v/\sqrt{2}$. To specify that the 3-dB bandwidth is at least some minimum value $\omega_{3\,\text{dB, min}}$, i.e., $\omega_{3\,\text{dB}} \geq \omega_{3\,\text{dB, min}}$, is equivalent to specifying that $|H(\omega_{3\,\text{dB, min}})| \geq A_v/\sqrt{2}$. This is turn can be expressed as

$$A_v/|H(\omega_{3\,\text{dB, min}})|^2 \leq 2 \tag{30}$$

which is a posynomial inequality.

In almost all designs p_1 will be the dominant pole, (see below) so the 3-dB bandwidth is very accurately given by

$$\omega_{3\,\text{dB}} = p_1 = \frac{g_{m1}}{A_v C_c} \tag{31}$$

which is a monomial. Using this (extremely accurate) approximation, we can constrain the 3-dB bandwidth to equal some required value. Using the constraint (30), which is exact but inverse posynomial, we can constrain the 3-dB bandwidth to exceed a given minimum value.

E. Dominant Pole Conditions

The amplifier is intended to operate with p_1 as the dominant pole, i.e., p_1 much smaller than p_2, p_3, and p_4. These conditions can be expressed as

$$\frac{p_1}{p_2} \leq 0.1 \quad \frac{p_1}{p_3} \leq 0.1 \quad \frac{p_1}{p_4} \leq 0.1 \tag{32}$$

where we (arbitrarily) use one decade, i.e., a factor of ten in frequency, as the condition for dominance. These dominant pole conditions are readily handled by geometric programming, since p_1 is monomial and p_2, p_3, and p_4 are all inverse posynomial. In fact these dominant pole conditions usually do not need to be included explicitly since the phase margin conditions described below are generally more strict, and describe the real design constraint. Nevertheless, it is common practice to impose a minimum ratio between the dominant and nondominant poles; see, e.g., [42].

F. Unity-Gain Bandwidth and Phase Margin

We define the unity-gain bandwidth ω_c as the frequency at which $|H(j\omega_c)| = 1$. The phase margin is defined in terms of the phase of the transfer function at the unity-gain bandwidth

$$\text{PM} = \pi - \angle H(j\omega_c) = \pi - \sum_{i=1}^{4} \arctan\left(\frac{\omega_c}{p_i}\right).$$

A phase margin constraint specifies a lower bound on the phase margin, typically between 30°–60°.

The unity-gain bandwidth and phase margin are related to the closed-loop bandwidth and stability of the amplifier with unity-gain feedback, i.e., when its output is connected to the inverting input. If the op-amp is to be used in some other specific closed-loop configuration, then a different frequency will be of more interest, but the analysis is the same. For example, if the op-amp is to be used in a feedback configuration with closed-loop gain +20 dB, then the critical frequency is the 20-dB crossover point, i.e., the frequency at which the open-loop gain drops to 20 dB, and the phase margin is defined at that frequency. All of the analysis below is readily adapted with minimal changes to such a situation. For simplicity, we continue the discussion for the unity-gain bandwidth.

We start by considering a constraint that the unity-gain bandwidth should exceed a given minimum frequency, i.e.,

$$\omega_c \geq \omega_{c,\,\text{min}}. \tag{33}$$

This constraint is just a minimum gain constraint at the frequency $\omega_{c,\,\text{min}}$ [as in (29)], and, thus, can be handled exactly by geometric programming as a posynomial inequality.

Here too we can develop an approximate expression for the unity-gain bandwidth which is monomial. If we assume the parasitic poles p_2, p_3, and p_4 are at least a bit (say, an octave) above the unity-gain bandwidth, then the unity-gain bandwidth can be approximated as the open-loop gain times the 3-dB bandwidth, i.e.,

$$\omega_{c,\,\text{approx}} = \frac{g_{m1}}{C_c} \tag{34}$$

which is a monomial. If we use this approximate expression for the unity-gain bandwidth, we can fix the unity-gain bandwidth at a desired value. The approximation (34) ignores the decrease in gain due to the parasitic poles and, consequently, overestimates the actual unity-gain bandwidth (i.e., the gain drops to 0 dB at a frequency slightly less than $\omega_{c,\,\text{approx}}$).

We now turn to the phase margin constraint, for which we can give a very accurate posynomial approximation. Assuming the open-loop gain exceeds ten or so, the phase contributed by the dominant pole at the unity-gain bandwidth, i.e., $\arctan(\omega_c/p_1)$, will be very nearly 90°. Therefore, the phase margin constraint can be expressed as

$$\sum_{i=2}^{4} \arctan\left(\frac{\omega_c}{p_i}\right) \leq \frac{\pi}{2} - \text{PM} \tag{35}$$

i.e., the nondominant poles cannot contribute more than 90° − PM total phase shift.

The phase margin constraint (35) cannot be exactly handled by geometric programming, so we use two reasonable approximations to form a posynomial approximation. The first is an approximate unity-gain bandwidth $\omega_{c,\,\text{approx}}$ [from (34)] instead of the exact unity-gain bandwidth ω_c as the frequency at which we will constrain the phase of H. As mentioned above, we have $\omega_{c,\,\text{approx}} \leq \omega_c$, thus, our specification is a bit stronger than the exact phase margin specification (since we are constraining the phase at a frequency slightly above the actual unity gain bandwidth). We will also approximate $\arctan(x)$ as a monomial. A simple approximation is given by $\arctan(x) \approx x$, which is quite accurate for $\arctan(x)$ less than 25°. Thus, assuming that each of the parasitic poles contributes no more than about 25° of phase shift, we can approximate the phase margin constraint accurately as

$$\sum_{i=2}^{4} \frac{\omega_{c,\text{approx}}}{p_i} \leq \frac{\pi}{2} - \text{PM}_{\text{min}} \tag{36}$$

which is a posynomial inequality in the design variables (since $\omega_{c,\,\text{approx}}$ is monomial). The approximation error involved here is almost always very small for the following reasons. The constraint (36) makes sure none of the nondominant poles is too near ω_c. This, in turn, validates our approximation $\omega_{c,\,\text{approx}} \approx \omega_c$. It also ensures that our approximation that the phase contributed by the nondominant poles is $\sum_{i=2}^{4} \omega_c/p_i$ is good.

Finally, we note that it is possible to obtain a more accurate monomial approximation of $\arctan(x)$ that has less error over a wider range, e.g., $\arctan(x) \leq 60°$. For example the approximation $\arctan(x) \approx 0.75x^{0.7}$ gives a fit around ±3° for angles between 0–60°, as shown in Fig. 2.

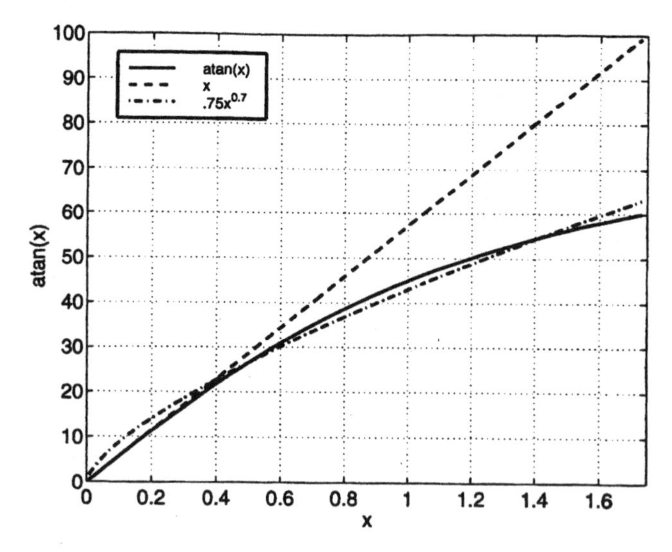

Fig. 2. Approximations of $\arctan(x)$.

VI. Other Constraints

In this section, we collect several other important constraints.

A. Slew Rate

The slew rate can be expressed [79] as

$$\text{SR} = \min\{2I_1/C_c,\ I_7/(C_c + C_{\text{TL}})\}.$$

In order to ensure a minimum slew-rate SR_{min} we can impose the two constraints

$$\frac{C_c}{2I_1} \leq \frac{1}{\text{SR}_{\text{min}}}, \qquad \frac{C_c + C_{\text{TL}}}{I_7} \leq \frac{1}{\text{SR}_{\text{min}}}. \tag{37}$$

These two constraints are posynomial.

B. Common-Mode Rejection Ratio

The common-mode rejection ratio (CMRR) can be approximated as

$$\begin{aligned}\text{CMRR} &= \frac{2g_{\text{m1}}g_{\text{m3}}}{(g_{\text{o3}} + g_{\text{o1}})g_{\text{o5}}} \\ &= \frac{2C_{\text{ox}}}{(\lambda_{\text{n}} + \lambda_{\text{p}})\lambda_{\text{p}}}\sqrt{\mu_{\text{n}}\mu_{\text{p}}\frac{W_1W_3}{L_1L_3I_5^2}}\end{aligned} \tag{38}$$

which is a monomial. In particular, we can specify a minimum acceptable value of CMRR.

C. Power-Supply Rejection Ratio

1) Negative Power-Supply Rejection Ratio: The negative power-supply rejection ratio (PSRR) is given by [52], [59]

$$\frac{g_{\text{m2}}g_{\text{m6}}}{(g_{\text{o2}} + g_{\text{o4}})g_{\text{o6}}}\,\frac{1}{(1 + s/p_1)(1 + s/p_2)}. \tag{39}$$

Thus, the low-frequency negative PSRR is given by the inverse posynomial expression

$$\frac{g_{\text{m2}}g_{\text{m6}}}{(g_{\text{o2}} + g_{\text{o4}})g_{\text{o6}}} \tag{40}$$

which, therefore, can be lower bounded.

The high-frequency PSRR characteristics are generally more critical than the low-frequency PSRR characteristics since noise in mixed-mode chips (clock noise, switching regulator noise, etc.) is typically high frequency. One can see that the expression for the magnitude squared of the negative PSRR at a frequency ω_0 has the form

$$|\text{PSRR}(j\omega_0)|^2 = \frac{A_p^2}{(1 + \omega_0^2/p_1^2)(1 + \omega_0^2/p_2^2)}$$

where A_p, p_1, and p_2 are given by inverse posynomial expressions. As in Section V-C, we can impose a lower bound on the negative PSRR at frequencies smaller than the unity-gain bandwidth by imposing posynomial constraints of the form

$$|\text{PSRR}(j\omega_0)| \geq a. \tag{41}$$

2) Positive Power-Supply Rejection Ratio: The low-frequency positive PSRR is given by

$$\text{PSRR} = \frac{2g_{m2}g_{m3}g_{m6}}{(g_{o2}+g_{o4})(2g_{m3}g_{o7}-g_{m6}g_{o5})} \tag{42}$$

which is neither posynomial nor inverse-posynomial, thus, it follows that constraints on the positive power supply rejection cannot be handled by geometric programming. However, this op-amp suffers from much worse negative PSRR characteristics than positive PSRR characteristics, both at low and high frequencies [40], [42]. Therefore, not constraining the positive PSRR is not critical.

We must at the least check the positive PSRR of any design carried out by the method described in this paper. (It is more than adequate in every design we have carried out.) However, if the positive PSRR specification becomes critical, it can be approximated (conservatively) by a posynomial inequality, e.g., using Duffin linearization [7], [17].

D. Noise Performance

The equivalent input-referred noise power spectral density $S_{in}(f)^2$ (in V²/Hz, at frequency f assumed smaller than the 3-dB bandwidth), can be expressed as

$$S_{in}^2 = S_1^2 + S_2^2 + \left(\frac{g_{m3}}{g_{m1}}\right)^2 (S_3^2 + S_4^2)$$

where S_k^2 is the input-referred noise power spectral density of transistor M_k. These spectral densities consist of the input-referred thermal noise and a $1/f$ noise

$$S_k(f)^2 = \left(\frac{2}{3}\right)\frac{4kT}{g_{m,k}} + \frac{K_f}{C_{ox}W_kL_kf}.$$

Thus, the input-referred noise spectral density can be expressed as

$$S_{in}(f)^2 = \alpha/f + \beta$$

where

$$\alpha = \frac{2K_p}{C_{ox}W_1L_1}\left(1 + \frac{K_n\mu_nL_1^2}{K_p\mu_pL_3^2}\right)$$

$$\beta = \frac{16KT}{3\sqrt{2\mu_pC_{ox}(W/L)_1I_1}}\left(1 + \sqrt{\frac{\mu_n(W/L)_3}{\mu_p(W/L)_1}}\right).$$

Note that α and β are (complicated) posynomial functions of the design parameters.

We can, therefore, impose spot noise constraints, i.e., require that

$$S_{in}(f)^2 \le S_{max}^2 \tag{43}$$

for a certain f, as a posynomial inequality. (We can impose multiple spot noise constraints, at different frequencies, as multiple posynomial inequalities.)

TABLE I
DESIGN CONSTRAINTS AND SPECIFICATIONS FOR THE TWO-STAGE OP-AMP

Specification/constraint	Type	Equation(s)
Symmetry and matching	Mono.	5, 6
Device sizes	Mono.	7
Area	Posy.	8
Systematic offset voltage	Mono.	9
Bias conditions		
Common-mode input range	Posy.	12, 13
Output voltage swing	Posy.	14, 15
Gate overdrive	Mono.	16
Quiescent power	Posy.	17
Open loop gain	Mono.	20
Dominant pole conditions	Posy.	32
3dB bandwidth	Mono.	31
Unity-gain bandwidth	Mono.	34
Phase margin	Posy.	36
Slew rate	Posy.	37
CMRR	Mono.	38
Neg. PSRR	Inv. Posy.	40, 41
Pos. PSRR	Neither	42
Input-referred spot noise	Posy.	43
Input-referred total noise	Posy.	44

The total rms noise level V_{noise} over a frequency band $[f_0, f_1]$ (where f_1 is below the equivalent noise bandwidth of the circuit) can be found by integrating the noise spectral density:

$$V_{noise}^2 = \int_{f_0}^{f_1} S_{in}(f)^2\, df = \alpha\log(f_1/f_0) + \beta(f_1 - f_0).$$

Therefore, imposing a maximum total rms noise voltage over the band $[f_0, f_1]$ is the posynomial constraint

$$\alpha\log(f_1/f_0) + \beta(f_N - f_0) \le V_{max}^2 \tag{44}$$

(since f_1 and f_0 are fixed, and α and β are posynomials in the design variables).

VII. OPTIMAL DESIGN PROBLEMS AND EXAMPLES

A. Summary of Constraints and Specifications

The many performance specifications and constraints described in the previous sections are summarized in Table I. Note that with only one exception (the positive supply rejection ratio), the specifications and constraints can be handled via geometric programming.

Since all the op-amp performance measures and constraints shown above can be expressed as posynomial functions and posynomial constraints, we can solve a wide variety of op-amp design problems via geometric programming. We can, for example, maximize the bandwidth subject to given (upper) limits on op-amp power, area, and input offset voltage, and given (lower) limits on transistor lengths and widths, and voltage gain, CMRR, slew rate, phase margin, and output voltage swing. The resulting optimization problem is a geometric programming problem. The problem may appear to be very complex, involving many complicated inequality and equality constraints, but in fact is readily solved.

TABLE II
SPECIFICATIONS AND CONSTRAINTS FOR DESIGN EXAMPLE

Constraint	Specification
Device length	$\geq 0.8\mu m$
Device width	$\geq 2\mu m$
Area	$\leq 10000\mu m^2$
Common-mode input voltage	fixed at $V_{DD}/2$
Output voltage range	$[0.1, 0.9]V_{DD}$
Quiescent power	$\leq 5mW$
Open-loop gain	$\geq 80dB$
Unity-gain bandwidth	Maximize
Phase margin	$\geq 60°$
Slew rate	$\geq 10V/\mu s$
Common-mode rejection ratio	$\geq 60dB$
Neg. power supply rejection ratio	$\geq 80dB$
Pos. power supply rejection ratio	$\geq 80dB$
Input-referred spot noise, 1kHz	$300nV/\sqrt{Hz}$

TABLE III
OPTIMAL DESIGN FOR DESIGN EXAMPLE

Variable	Value
$W_1 = W_2$	$232.8\mu m$
$W_3 = W_4$	$143.6\mu m$
W_5	$64.6\mu m$
W_6	$588.8\mu m$
W_7	$132.6\mu m$
W_8	$2.0\mu m$
$L_1 = L_2$	$0.8\mu m$
$L_3 = L_4$	$0.8\mu m$
L_5	$0.8\mu m$
L_6	$0.8\mu m$
L_7	$0.8\mu m$
L_8	$0.8\mu m$
C_c	3.5pF
I_{bias}	$10\mu A$

B. Example

In this section, we describe a simple design example. A 0.8 μm CMOS technology was used; see the longer report [51] for more details and the technology parameters. The positive supply voltage was set at 5 V and the negative supply voltage was set at 0 V. The load capacitance was 3 pF.

The objective is to maximize unity-gain bandwidth subject to the requirements shown in Table II. The resulting geometric program has 18 variables, seven (monomial) equality constraints, and 28 (posynomial) inequality constraints. The total number of monomial terms appearing in the objective and all constraints is 68. Our simple MATLAB program solves this problem in under one second real-time. The optimal design obtained is shown in Table III.

The performance achieved by this design, as predicted by the program, is summarized in Table IV. The design achieves an 86-MHz unity-gain bandwidth. Note that some constraints are tight (minimum device length, minimum device width, maximum output voltage, quiescent power, phase margin and input-referred spot noise) while some constraints are not tight (area, minimum output voltage open-loop gain, common-mode rejection ratio, and slew rate).

TABLE IV
PERFORMANCE OF OPTIMAL DESIGN FOR DESIGN EXAMPLE

Specification/Constraint	Performance
Minimum device length	$0.8\mu m$
Minimum device width	$2\mu m$
Area	$8200\mu m^2$
Output voltage range	$[0.03, 0.9]V_{DD}$
Minimum gate overdrive	130mV
Quiescent power	5mW
Open-loop gain	89.2dB
Unity-gain bandwidth	86MHz
Phase margin	60°
Slew rate	$88V/\mu s$
Common-mode rejection ratio	92.5dB
Negative power supply rejection ratio	98.4dB
Positive power supply rejection ratio	116dB
Maximum input-referred spot noise, 1kHz	$300nV/\sqrt{Hz}$

C. Tradeoff Analyses

By repeatedly solving optimal design problems as we sweep over values of some of the constraint limits, we can sweep out globally optimal tradeoff curves for the op-amp. For example, we can fix all other constraints, and repeatedly minimize power as we vary a minimum required unity-gain bandwidth. The resulting curve shows the globally optimal tradeoff between unity-gain bandwidth and power (for the values of the other limits).

In this section, we show several optimal tradeoff curves for the operational amplifier. We do this by fixing all the specifications at the default values shown in Table II, except two that we vary to see the effect on a circuit performance measure. When the optimization objective is not bandwidth we use a default value of minimum unity-gain bandwidth of 30 MHz.

We first obtain the globally optimal tradeoff curve of unity-gain bandwidth versus power for different supply voltages. The results can be seen in Fig. 3. Obviously the more power we allocate to the amplifier, the larger the bandwidth obtained; the plots, however, show exactly how much more bandwidth we can obtain with different power budgets. We can see, for example, that the benefits of allocating more power to the op-amp disappear above 5 mW for a supply voltage of 2.5 V, whereas for a 5 V supply the bandwidth continues to increase with increasing power. Note also that each of the supply voltages gives the largest unity-gain bandwidth over some range of powers.

In Fig. 4, we plot the globally optimal tradeoff curve of open-loop gain versus unity-gain frequency for different phase margins. Note that for a large unity-gain bandwidth requirement only small gains are achievable. Also, we can see that for a tighter phase margin constraint the gain bandwidth product is lower.

Fig. 5 shows the minimum input-referred spectral density at 1 kHz versus power, for different unity-gain frequency requirements. Note that when the power specification is tight, increasing the power greatly helps to decrease the input-referred noise spectral density.

In Fig. 6 we show the optimal tradeoff curve of unity-gain bandwidth versus area for different different power budgets.

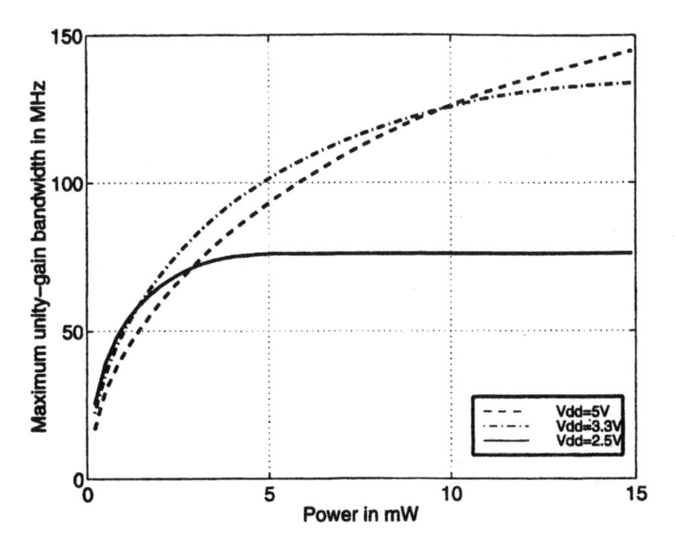

Fig. 3. Maximum unity-gain bandwidth versus power for different supply voltages.

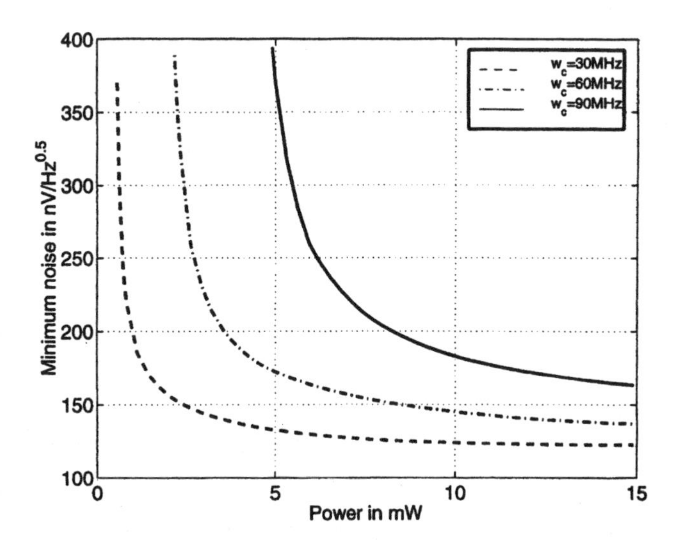

Fig. 5. Minimum noise density at 1 kHz versus power for different unity-gain bandwidths.

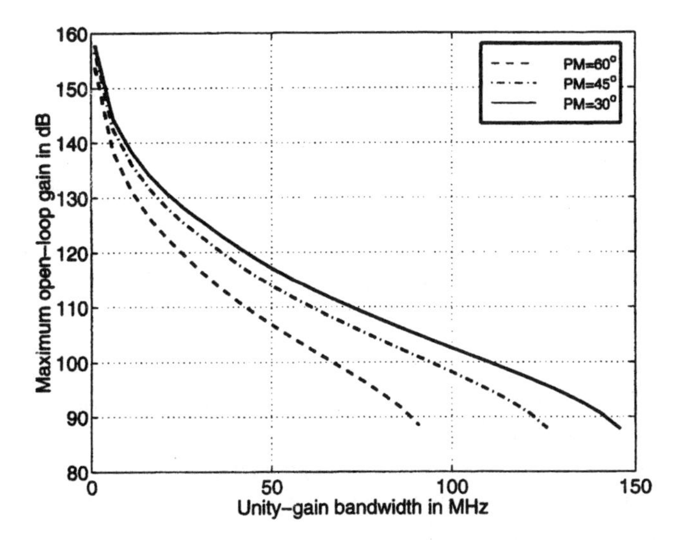

Fig. 4. Maximum open-loop gain versus unity-gain bandwidth for different phase margins.

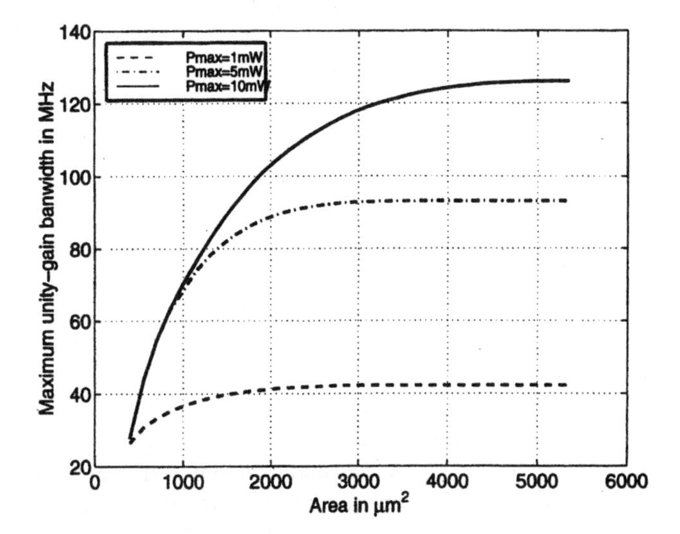

Fig. 6. Maximum unity-gain bandwidth versus area for different power budgets.

We can see that when the area constraint is tight, increasing the available area translates into a greater unity bandwidth. After some point, other constraints become more stringent and increasing the available area does not improve the maximum achievable unity-gain bandwidth.

Several other optimal tradeoff curves are given in the longer report [51].

D. Sensitivity Analysis Example

In this section, we analyze the information provided by the sensitivity analysis of the first design problem in Section VII-B (maximize the unity-gain bandwidth when the rest of specifications/constraints are set to the values shown in Table II). The results of this sensitivity analysis are shown in Table V. The column labeled "Sensitivity" (numerical) is obtained by tightening and loosening the constraint in question by 5% and resolving the problem. (The average from the two is taken.) The column labeled "Sensitivity" comes (essentially for free) from solving the original problem. Note that it gives an excellent prediction of the numerically obtained sensitivities.

There are six active constraints: minimum device length, minimum device width, maximum output voltage, quiescent power, phase margin, and input-referred spot noise at 1 kHz. All of these constraints limit the maximum unity-gain bandwidth. The sensitivities indicate which of these constraints are more critical (more limiting). For example, a 10% increase in the allowable input-referred noise at 1 kHz will produce a design with (approximately) 2.4% improvement in unity-gain bandwidth. However, a 10% decrease in the maximum phase margin at the unity-gain bandwidth will produce a design with (approximately) 17.6% improvement in unity-gain bandwidth. It is very interesting to analyze the sensitivity to the minimum device width constraint. A 10% decrease in the minimum device width produces a design with only a 0.05% improvement in unity-gain bandwidth. This can be interpreted as follows: even though the minimum device width constraint is binding, it can be considered not binding in a practical sense since tightening (or loosening) it will barely change the objective.

The program classifies the given constraints in order of importance from most limiting to least limiting. For this design

TABLE V
SENSITIVITY ANALYSIS FOR THE DESIGN EXAMPLE

Constraint	Spec.	Program	Sensitivity (numerical)	Sensitivity
Min. device length	$\geq 0.8\mu m$	$.8\mu m$	0.299	0.309
Min. device width	$\geq 2\mu m$	$2.0\mu m$	0.0049	0.0048
Area	$\leq 10000\mu m^2$	$8200\mu m^2$	0	0
Max. output voltage	4.5V	4.5V	−0.365	−0.349
Min. output voltage	0.5V	0.13V	0	0
Quiescent power	$\leq$ 5mW	4.99mW	−0.482	−0.483
Open-loop gain	$\geq$ 80dB	89.2dB	0	0
Phase margin	$\geq 60°$	60°	−1.758	−1.757
Slew rate	$\geq 10V/\mu s$	$88V/\mu s$	0	0
CMRR	$\geq$ 60dB	92.5dB	0	0
Neg. PSRR	$\geq$ 80dB	98.4dB	0	0
Pos. PSRR	$\geq$ 80dB	116dB	0	0
Spot noise, 1kHz	$\leq 300nV/\sqrt{Hz}$	$300nV/\sqrt{Hz}$	0.24	0.241

TABLE VI
DESIGN VERIFICATION WITH HSPICE LEVEL-1. THE PERFORMANCE MEASURES OBTAINED BY THE PROGRAM ARE COMPARED WITH THOSE FOUND BY HSPICE LEVEL 1 SIMULATION

Constraint	Spec.	Program	HSPICE 1
Max. output voltage	$\geq$ 4.5V	4.5V	4.5V
Min. output voltage	$\leq$ 0.5V	0.13V	0.13V
Quiescent power	$\leq$ 5mW	4.99mW	4.95mW
Open-loop gain	$\geq$ 80dB	89.2dB	89.4dB
Unity-gain bandwidth	maximize	86MHz	81MHz
Phase margin	$\geq 60°$	60°	64°
Slew rate	$\geq 10V/\mu s$	$88V/\mu s$	$92.5V/\mu s$
CMRR	$\geq$ 60dB	92.5dB	94dB
Neg. PSRR	$\geq$ 80dB	98.4dB	98.1dB
Pos. PSRR	$\geq$ 80dB	116dB	114dB
Spot noise, 1kHz	$\leq 300nV/\sqrt{Hz}$	$300nV/\sqrt{Hz}$	$280nV/\sqrt{Hz}$

the order is: phase margin, quiescent power, maximum output voltage, minimum device length, input-referred noise at 1 kHz, and minimum device width. The program also tells the designer which constraints are *not critical* (the ones whose sensitivities are zero or small). A small relaxation of these constraints will not improve the objective function, so any effort to loosen them will not be rewarded.

VIII. DESIGN VERIFICATION

Our optimization method is based on GP0 models, which are the simple square-law device models described in the Appendix, Section A. Our model does not include several potentially important factors such as body effect, channel length modulation in the bias equations, and the dependence of junction capacitances on junction voltages. Moreover we make several approximations in the circuit analysis used to formulate the constraints. For example, we approximate the transfer function with the four-pole form (19); the actual transfer function, even based on the simple model, is more complicated. As another example, we approximate $\arctan(a) \approx a$ in our simple version of the phase margin constraint.

While all of these approximations are reasonable (at least when channel lengths are not too short), it is important to *verify* the designs obtained using a higher fidelity (presumably nonposynomial) model.

A. HSPICE Level-1 Verification with GP0 Models

We first verify the designs generated by our geometric programming method (using GP0 or long-channel models) with HSPICE using a long-channel model (HSPICE level–1 model). We take the design found by the geometric programming method, and then use the HSPICE level-1 model to check the various performance measures. The level-1 HSPICE model is substantially more accurate (and complicated) than our simple posynomial models. It includes, for example, body effect, channel-length modulation, junction capacitance that depends on bias conditions, and a far more complex transfer function that includes many other parasitic capacitances. The unity gain bandwidth and phase margin are computed by solving the complete small-signal model of the op-amp. The results of such verification always show excellent agreement between our posynomial models and the more complex (and nonposynomial) HSPICE level-1 model. As an example, Table VI summarizes the results for the standard problem described above in Section VII-D. Note that the values of the performance specifications from the posynomial model (in the column labeled "Program") and the values according to HSPICE level 1 (in the right-hand side column) are in close agreement. Moreover, the deviations between the two are readily understood. The unity-gain frequency is slightly overestimated and the phase margin is slightly underestimated because we use the approximate expression (34), which ignores the effect of the parasitic poles on the crossover frequency. The noise is overestimated 7% because the open-loop gain has decreased 7% already at 1 kHz; this gain reduction translates into a reduction in the input-referred noise.

We have verified the geometric program results with the HSPICE level-1 model simulations for a wide variety of designs (with a wide variety of power, bandwidth, gain, etc.). The results are always in close agreement. Thus, our simple posynomial models are reasonably good approximations of the HSPICE level-1 models.

B. HSPICE BSIM Model Verification with GP1 Models

In this section, we show how the geometric programming method performs well even for short-channel devices, when more sophisticated GP1 transistor models are used, by verifying designs against sophisticated HSPICE level-39 (BSIM3v1) simulations. The GP1 model is described in the Appendix, Section B; the only difference is that we use an empirically found monomial expression for the output conductance of a MOS transistor instead of the standard long channel formula. Using these GP1 models, all of the constraints described above are still compatible with geometric programming.

Table VII shows the comparison between the results of geometric programming design, using GP1 models, and HSPICE level-39 simulation, for the standard problem described in Section VII-D. The predicted values are very close to the simulated values. The agreement holds for a wide variety of designs.

IX. DESIGN FOR PROCESS ROBUSTNESS

Thus far, we have assumed that parameters such as transistor threshold voltages, mobilities, oxide parameters, channel modulation parameters, supply voltages, and load capacitance are

IS USED TO SOLVE THE STANDARD PROBLEM, USING THE MORE SOPHISTICATED GP1 MODELS. THE RESULTS ARE COMPARED WITH HSPICE LEVEL-39 SIMULATION

Constraint	Spec.	Program	HSPICE 39
Max. output voltage	≥ 4.5V	4.55V	4.4V
Min. output voltage	≤ 0.5V	240mV	200mV
Quiescent power	≤ 5mW	4.99mW	5.2mW
Open-loop gain	≥ 80dB	80dB	83dB
Unity-gain bandwidth	maximize	75MHz	73MHz
Phase margin	$\geq 60°$	60°	62°
Slew rate	$\geq 10\text{V}/\mu\text{s}$	$97\text{V}/\mu\text{s}$	$95\text{V}/\mu\text{s}$
CMRR	≥ 60dB	86dB	88dB
Neg. PSRR	≥ 80dB	84dB	86dB
Pos. PSRR	≥ 80dB	93dB	92dB
Spot noise, 1kHz	$\leq 300\text{nV}/\sqrt{\text{Hz}}$	$300\text{nV}/\sqrt{\text{Hz}}$	$285\text{nV}/\sqrt{\text{Hz}}$

all known and fixed. In this section, we show to how to use the methods of this paper to develop designs that are *robust* with respect to variations in these parameters, i.e., designs that meet a set of specifications for a set of values of these parameters. Such designs can dramatically increase yield.

There are many approaches to the problem of robustness and yield optimization (see [5], [28]). The robust design problem can be formulated as a so-called semi-infinite programming problem, in which the constraints must hold for all values of some parameter that ranges over an interval, as in [75], which used DELIGHT.SPICE to do robust designs, or more recently, Mukherjee *et al.* [69], who use ASTRX/OBLX. These methods often involve very considerable run times, ranging from minutes to hours.

We also formulate the problem as a sampled version of a semi-infinite program. The method is practical only because geometric programming can readily handle problems with many hundreds, or even thousands, or constraints; the computational effort grows approximately linearly with the number of constraints.

The basic idea in our approach is to list a set of possible parameters, and to replicate the design constraints for all possible parameter values. Let $\alpha \in \boldsymbol{R}^k$ denote a vector of parameters that may vary. Then the objective and constraint functions can be expressed as functions of x (the design parameters) and α (which we will call the *process parameters*, even if some components, e.g., the load capacitance, are not really process parameters):

$$f_0(x, \alpha), \qquad f_i(x, \alpha), \qquad g_i(x, \alpha).$$

The functions f_i are all posynomial functions of x, for each α, and the functions g_i are all monomial functions of x, for each α. Let $\mathcal{A} = \{\alpha_1, \ldots, \alpha_N\}$ be a (finite) set of possible parameter values. Our goal is to determine a design (i.e., x) that works well for all possible parameter values (i.e., $\alpha_1, \ldots, \alpha_N$).

First we describe several ways the set $\mathcal{A}$ might be constructed. As a simple example, suppose there are six parameters, which vary independently over intervals $[\alpha_{\min,i}, \alpha_{\max,i}]$. We might sample each interval with three values (e.g., the midpoint and extreme values), and then form every possible combination of parameter values, which results in $N = 3^6$.

eter values, but only the ones likely to actually occur. For example, if it is unlikely that the oxide capacitance parameter is at its maximum value while the n-threshold voltage is maximum, then we delete these combinations from our set $\mathcal{A}$. In this way, we can model interdependencies among the parameter values.

We can also construct $\mathcal{A}$ in a straightforward way. Suppose we require a design that works, without modification, on several processes, or several variations of processes. $\mathcal{A}$ is then simply a list of the process parameters for each of the processes.

The robust design is achieved by solving the problem

$$\begin{array}{lll} \text{minimize} & \max_{\alpha \in \mathcal{A}} f_0(x, \alpha) & \\ \text{subject to} & f_i(x, \alpha) \leq 1, & i = 1, \ldots, m \quad \text{for all } \alpha \in \mathcal{A} \\ & g_i(x, \alpha) = 1, & i = 1, \ldots, p \quad \text{for all } \alpha \in \mathcal{A} \\ & x_i \geq 0, & i = 1, \ldots, n. \end{array} \tag{45}$$

This problem can be reformulated as a geometric program with N times the number of constraints, and an additional scalar variable γ [22]:

$$\begin{array}{llll} \text{minimize} & \gamma & & \\ \text{subject to} & f_0(x, \alpha_j) \leq \gamma, & j = 1, \ldots, N & \\ & f_i(x, \alpha_j) \leq 1, & i = 1, \ldots, m; & j = 1, \ldots, N \\ & g_i(x, \alpha_j) = 1, & i = 1, \ldots, p; & j = 1, \ldots, N \\ & x_i > 0, & i = 1, \ldots, n. & \end{array} \tag{46}$$

The solution of (45) [which is the same as the solution of (46)] satisfies the specifications for all possible values of the process parameters. The optimal objective value gives the (globally) optimal minimax design. (It is also possible to take an average value of the objective over process parameters, instead of a worst-case value.)

Equality constraints have to be handled carefully. Provided the transistor lengths and widths are not subject to variation, equality constraints among them (e.g., matching and symmetry) are likely not to depend on the process parameter α. Other equality constraints, however, can depend on α. When we enforce an equality constraint for each value of α, the result is (usually) an infeasible problem. For example suppose we specify that the open-loop gain equal exactly 80 dB. Process variation will change the open-loop gain, making it impossible to achieve a design that has an open-loop gain of *exactly* 80 dB for more than a few process parameter values. The solution to this problem is to convert such specifications into inequalities. We might, for example, change our specification to require that the open-loop gain exceed 80 dB, or require it to be between 80 and 85 dB. Either way the robust problem now has at least a chance of being feasible.

It is important to contrast a robust design for a set of process parameters $\mathcal{A} = \{\alpha_1, \ldots, \alpha_N\}$ with the optimal designs for each process parameter. The objective value for the robust design is worse (or no better) than the optimal design for each parameter value. This disadvantage is offset by the advantage that the design works for all the process parameter values. As a simple example, suppose we seek a design that can be run on two processes (α_1 and α_2). We can compare the robust design to the two optimal designs. If the objective achieved by the robust design is not much worse than the two optimal designs, then we have the advantage of a single design that works on two processes. On the other hand if the robust design is much worse (or even infeasible) we may elect to have two versions of the amplifier design, each one optimized for the particular process.

Thus far, we have considered the case in which the set $\mathcal{A}$ is finite. However, in most real cases it is infinite; e.g., individual parameters lie in ranges. We have already indicated above that such situations can be modeled or approximated by sampling the interval. While we believe this will always work in practice, it gives no guarantee, in general, that the design works for *all* values of the parameter in the given range; it only guarantees performance for the sampled values of the parameters.

There are many cases, however, when we *can* guarantee the performance for a parameter value in an interval. Suppose that the function $f_i(x, \alpha)$ is posynomial not just in x, but in x and α as well, and that α lies in the interval $[\alpha_{\min}, \alpha_{\max}]$. (We take α scalar here for simplicity.) It then suffices to impose the constraint at the endpoints of the interval, i.e.,

$$\begin{aligned} & f_i(x, \alpha_{\min}) \le 1, \qquad f_i(x, \alpha_{\max}) \le 1 \\ \Rightarrow \quad & f_i(x, \alpha) \le 1, \qquad \text{for all } \alpha \in [\alpha_{\min}, \alpha_{\max}]. \end{aligned}$$

This is easily proved using convexity of the log f_i in the transformed variables.

The reader can verify that the constraints described above are posynomial in the parameters C_{ox}, μ_{n}, μ_{p}, λ_n, λ_p, and the parasitic capacitances. Thus, for these parameters at least, we can handle ranges with no approximation or sampling, by specifying the constraints only at the endpoints.

The requirement of robustness is a real practical constraint, and is currently dealt with by many methods. For example, a minimum gate overdrive constraint is sometimes imposed because designs with small gate overdrive tend to be nonrobust. The point of this section is that robustness can be achieved in a more methodical way, which takes into account a more detailed description of the possible uncertainties or parameter variations. The result will be a better design than an *ad hoc* method for achieving robustness.

Finally, we demonstrate the method with a simple example. In Table VIII we show how a robust design compares to a nonrobust design. We take three process parameters: the bias current error factor, the positive power supply error factor, and oxide capacitance. The bias current error factor is the ratio of the actual bias current to our design value, so when it is one, the true bias current is what we specify it to be, and when it is 1.1, the true bias current is 10% larger than we specify it to be. Similarly, the positive power supply error factor is the ratio of the actual bias current to our design value. The bias current error factor varies between 0.9–1.1, the positive power supply error factor varies between 0.9–1.1, and the oxide capacitance varies ±10% around its nominal value. The three parameters are assumed independent, and we sample each with three values (midpoint and extreme values) so all together we have $N = 3^3$, i.e., 27 different process parameter vectors. In the third column, we show

TABLE VIII
ROBUST DESIGN

Constraint	Spec.	Rob. design	Std. design
Quiescent power	$\le$ 5mW	4.99mW	5.75mW
Open-loop gain	$\ge$ 80dB	89dB	87dB
Unity-gain bandwidth	maximize	72MHz	77MHz
Phase margin	$\ge 60°$	60°	55°
CMRR	$\ge$ 60dB	93dB	90dB
Neg. PSRR	$\ge$ 80dB	94dB	93dB
Pos. PSRR	$\ge$ 80dB	110dB	109dB
Spot noise, 1kHz	$\le 300\text{nV}/\sqrt{\text{Hz}}$	$300\text{nV}/\sqrt{\text{Hz}}$	$316\text{nV}/\sqrt{\text{Hz}}$

the performance of the robust design. For each specification, we determine the worst performance over all 27 process parameters. In the fourth column we show the performance of the nonrobust design. Again, only the worst-case performance over all 27 process parameters is indicated for each specification. The resulting geometric program involves 18 variables, seven monomial equality constraints (i.e., symmetry and matching) and 756 posynomial inequality constraints.

The new design obtains a unity-gain bandwidth of 72 MHz. The design in Section VII-B obtains a worst-case unity gain bandwidth of 77 MHz, but since it was specified only for nominal conditions, it fails to meet some constraints when tested over all conditions. For example, the power consumption increases by 15%, the open-loop gain decreases by 20%, the input-referred spot noise at 1 kHz increases by 5% and the phase margin decreases by 5°. The robust design, on the other hand, meets *all* specifications for all 27 sets of process parameters.

X. Discussion and Conclusion

We have shown how geometric programming can be used to design and optimize a common CMOS amplifier. The method yields globally optimal designs, is extremely efficient, and handles a very wide variety of practical constraints.

Since no human intervention is required (e.g., to provide an initial "good" design or to interactively guide the optimization process), the method yields completely automated sizing of (globally) optimal CMOS amplifiers, directly from specifications. This implies that the circuit designer can spend more time doing real design, i.e., carefully analyzing the optimal tradeoffs between competing objectives, and less time doing parameter tuning, or wondering whether a certain set of specifications can be achieved. The method could be used, for example, to do full custom design for each op-amp in a complex mixed-signal integrated circuit; each amplifier is optimized for its load capacitance, required bandwidth, closed-loop gain, etc.

In fact, the method can handle problems with constraints or coupling between the different op-amps in an integrated circuit. As simple examples, suppose we have 100 op-amps, each with a set of specifications. We can minimize the *total* area or power by solving a (large) geometric program. In this case, we are solving (exactly) the power/area allocation problem for the 100 op-amps on the integrated circuit. We can also handle direct coupling between the op-amps, i.e., when component values in one op-amp (e.g., input transistor widths) affect another (e.g., as load capacitance). The resulting geometric program will have perhaps hundreds of variables, thousands of constraints, and be

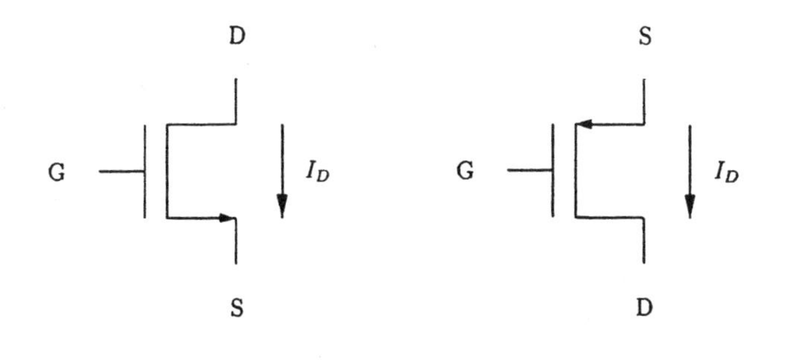

Fig. 7. Transistor symbols.

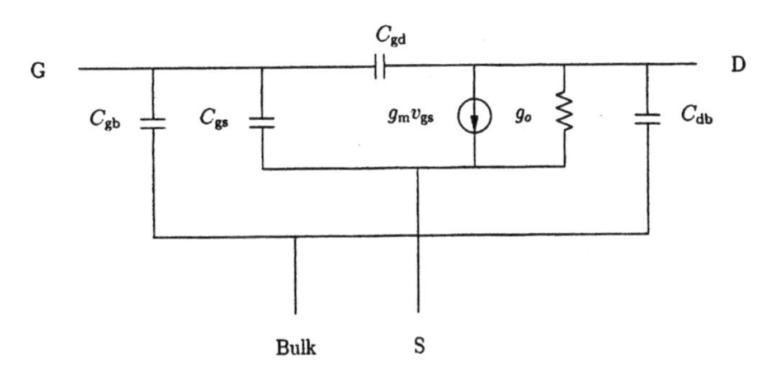

Fig. 8. Small signal model for a MOSFET.

For a PMOS transistor, the saturation condition is

$$V_{DS} \leq V_{GS} - V_{TP}. \quad (49)$$

The drain current is then given by

$$I_D = \frac{1}{2}\mu_p C_{ox}\frac{W}{L}(V_{GS} - V_{TP})^2(1 + \lambda_p V_{DS})$$

where

μ_p hole mobility;
V_{TP} PMOS threshold voltage;
λ_p channel-length modulation parameter.

Here too, we ignore the channel modulation effects and use the simplified expression

$$I_D = \frac{1}{2}\mu_p C_{ox}\frac{W}{L}(V_{GS} - V_{TP})^2. \quad (50)$$

2) Small-Signal Models: Fig. 8 shows the small-signal model around the operating point for a MOSFET transistor in saturation. The derivation of this model can also be found in [41]. The values of the various elements and parameters are described below.

The transconductance g_m is given by

$$g_m = \frac{\partial I_D}{\partial V_{GS}} = \sqrt{2\mu C_{ox} I_D \frac{W}{L}} \quad (51)$$

(where we ignore, with only small error, channel-length modulation effects). The output conductance g_o is given by

$$g_o = \frac{\partial I_D}{\partial V_{DS}} = \lambda I_D. \quad (52)$$

Note that we ignore channel-length modulation in our transconductance expression, but must include it in the output conductance expression (which would otherwise be zero).

The gate-to-source capacitance is given by the sum of the gate oxide capacitance and the overlap capacitance

$$C_{gs} = \tfrac{2}{3}WLC_{ox} + WL_D, C_{ox} \quad (53)$$

where L_D is the source/drain lateral diffusion length.

The source to bulk capacitance is a junction capacitance and can be expressed as

$$C_{sb} = \frac{C_{sb0}}{\left(1 + \frac{V_{SB}}{\psi_o}\right)^{1/2}} \quad (54)$$

where

$$C_{sb0} = C_j L_s W + C_{jsw}(2L_s + W) \quad (55)$$

ψ_o is the junction built-in potential, and L_s is the source diffusion length.

The drain-to-bulk capacitance is also a junction capacitance given by

$$C_{db} = \frac{C_{db0}}{\left(1 + \frac{V_{DB}}{\psi_o}\right)^{1/2}} \quad (56)$$

where $C_{db0} = C_{sb0}$ for equal source and drain diffusions.

The gate-to-drain capacitance is due to the overlap capacitance and is given by

$$C_{gd} = C_{ox} W L_D. \quad (57)$$

Equations (53), (55), and (57) are posynomial in the design variables and, therefore, are readily handled. The expressions for the junction capacitances (54) and (56) are not posynomial, except in the special case where V_{SB} and V_{DB} do not depend on the design variables. We can take two approaches to approximating these capacitances. One simple method is to take a worst-case analysis, and use the maximum values (which decreases bandwidth, slew rate, phase margin, etc.) This corresponds to the approximation $V_{SB} = 0$ or $V_{DB} = 0$. It is also possible to estimate the various junction voltages as constant, so (54) and (56) are constant.

In our op-amp circuit, the only junction capacitances that appear in the design equations (see Section V) are the drain-to-bulk capacitances of M_1, M_2, M_3, M_4, M_6, and M_7. We have estimated the drain-to-bulk voltages of transistors M_1, M_2, M_3, M_4, M_6, and M_7, and use these estimated voltages for calculating the junction capacitances.

The bulk terminal of the PMOS transistors is connected to the positive supply V_{DD} and that of the NMOS transistors is connected to the negative supply $V_{SS} = 0$. The drain voltages of M_1, M_2, M_3, and M_4 are the same as the gate voltage of M_6, namely, $V_{G,6}$. In most designs, $V_{G,6}$ is a few hundred millivolts above V_{TN} (recalling that we assume $V_{SS} = 0$). Thus, we can write $V_{G,6}$ as

$$V_{G,6} = V_{TN} + \Delta V_o \quad (58)$$

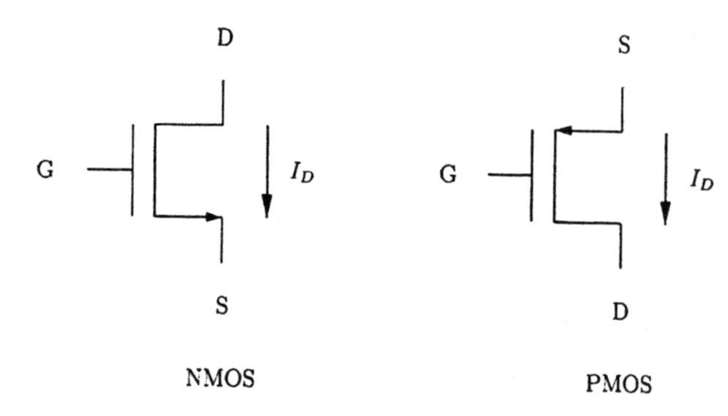

Fig. 7. Transistor symbols.

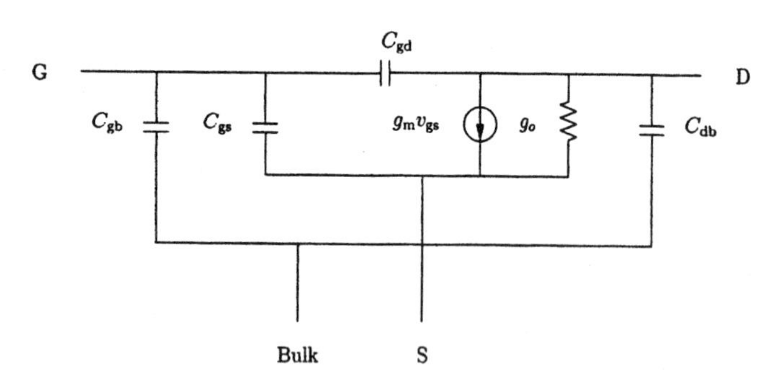

Fig. 8. Small signal model for a MOSFET.

For a PMOS transistor, the saturation condition is

$$V_{DS} \leq V_{GS} - V_{TP}. \tag{49}$$

The drain current is then given by

$$I_D = \frac{1}{2}\mu_p C_{ox}\frac{W}{L}(V_{GS} - V_{TP})^2(1 + \lambda_p V_{DS})$$

where

μ_p hole mobility;
V_{TP} PMOS threshold voltage;
λ_p channel-length modulation parameter.

Here too, we ignore the channel modulation effects and use the simplified expression

$$I_D = \frac{1}{2}\mu_p C_{ox}\frac{W}{L}(V_{GS} - V_{TP})^2. \tag{50}$$

2) Small-Signal Models: Fig. 8 shows the small-signal model around the operating point for a MOSFET transistor in saturation. The derivation of this model can also be found in [41]. The values of the various elements and parameters are described below.

The transconductance g_m is given by

$$g_m = \frac{\partial I_D}{\partial V_{GS}} = \sqrt{2\mu C_{ox} I_D \frac{W}{L}} \tag{51}$$

(where we ignore, with only small error, channel-length modulation effects). The output conductance g_o is given by

$$g_o = \frac{\partial I_D}{\partial V_{DS}} = \lambda I_D. \tag{52}$$

Note that we ignore channel-length modulation in our transconductance expression, but must include it in the output conductance expression (which would otherwise be zero).

The gate-to-source capacitance is given by the sum of the gate oxide capacitance and the overlap capacitance

$$C_{gs} = \tfrac{2}{3}WLC_{ox} + WL_D,\ C_{ox} \tag{53}$$

where L_D is the source/drain lateral diffusion length.

The source to bulk capacitance is a junction capacitance and can be expressed as

$$C_{sb} = \frac{C_{sb0}}{\left(1 + \frac{V_{SB}}{\psi_o}\right)^{1/2}} \tag{54}$$

where

$$C_{sb0} = C_j L_s W + C_{jsw}(2L_s + W) \tag{55}$$

ψ_o is the junction built-in potential, and L_s is the source diffusion length.

The drain-to-bulk capacitance is also a junction capacitance given by

$$C_{db} = \frac{C_{db0}}{\left(1 + \frac{V_{DB}}{\psi_o}\right)^{1/2}} \tag{56}$$

where $C_{db0} = C_{sb0}$ for equal source and drain diffusions.

The gate-to-drain capacitance is due to the overlap capacitance and is given by

$$C_{gd} = C_{ox} W L_D. \tag{57}$$

Equations (53), (55), and (57) are posynomial in the design variables and, therefore, are readily handled. The expressions for the junction capacitances (54) and (56) are not posynomial, except in the special case where V_{SB} and V_{DB} do not depend on the design variables. We can take two approaches to approximating these capacitances. One simple method is to take a worst-case analysis, and use the maximum values (which decreases bandwidth, slew rate, phase margin, etc.) This corresponds to the approximation $V_{SB} = 0$ or $V_{DB} = 0$. It is also possible to estimate the various junction voltages as constant, so (54) and (56) are constant.

In our op-amp circuit, the only junction capacitances that appear in the design equations (see Section V) are the drain-to-bulk capacitances of M_1, M_2, M_3, M_4, M_6, and M_7. We have estimated the drain-to-bulk voltages of transistors M_1, M_2, M_3, M_4, M_6, and M_7, and use these estimated voltages for calculating the junction capacitances.

The bulk terminal of the PMOS transistors is connected to the positive supply V_{DD} and that of the NMOS transistors is connected to the negative supply $V_{SS} = 0$. The drain voltages of M_1, M_2, M_3, and M_4 are the same as the gate voltage of M_6, namely, $V_{G,6}$. In most designs, $V_{G,6}$ is a few hundred millivolts above V_{TN} (recalling that we assume $V_{SS} = 0$). Thus, we can write $V_{G,6}$ as

$$V_{G,6} = V_{TN} + \Delta V_o \tag{58}$$

where we use a typical overdrive voltage of $\Delta V_o = 200$ mV. The drain-to-bulk capacitances of M_1, M_2, M_3, and M_4 are then given by the expressions

$$C_{\mathrm{db},1} = C_{\mathrm{db},2} = \frac{C_{\mathrm{dbo},1}}{\left(1+\dfrac{V_{\mathrm{DD}} - V_{\mathrm{TN}} - \Delta V_o}{\psi_o}\right)^{1/2}}$$

$$C_{\mathrm{db},3} = C_{\mathrm{db},4} = \frac{C_{\mathrm{dbo},3}}{\left(1+\dfrac{V_{\mathrm{TN}} + \Delta V_o}{\psi_o}\right)^{1/2}}.$$

The drain voltage of M_6 and M_7 is the output voltage of the amplifier. The quiescent output voltage is at mid-supply for an op-amp with small offset. Then, we can write $V_{\mathrm{D},6}$ as

$$V_{\mathrm{D},6} = \frac{V_{\mathrm{DD}}}{2}$$

and we obtain constant expressions for C_{db6} and C_{db7}

$$C_{\mathrm{db},6} = \frac{C_{\mathrm{dbo},6}}{\left(1+\dfrac{V_{\mathrm{DD}}}{2\psi_o}\right)^{1/2}} \qquad C_{\mathrm{db},7} = \frac{C_{\mathrm{dbo},7}}{\left(1+\dfrac{V_{\mathrm{DD}}}{2\psi_o}\right)^{1/2}}.$$

These approximations can be validated in several ways. First, we have observed that changing these typical voltages has very little effect on the final designs. And second, SPICE simulation (which includes the junction capacitances) reveals that we incur only small errors.

B. GP1 Models

The GP0 models described above are essentially the same as the standard long channel device models. It is also possible to derive device models that are more accurate than the long channel models, but at the same time are compatible with geometric programming based design.

Analysis of the errors incurred by the GP0 model shows that most of the modeling error comes from the expressions for transconductance and output conductance. By fitting monomial expressions to empirical data, or data obtained from a high-fidelity SPICE simulation, we obtain transistor models that are still compatible with geometric programming-based design. We refer to these models as GP1.

We found that the following simple models work very well. For NMOS devices, we use the monomial expression

$$g_{\mathrm{d,NMOS}} = 3.1 \cdot 10^{-2} W^{0.18} L^{-1.14} I_{\mathrm{D}}^{0.82}$$

where the output conductance is given in millisiemens, the bias current is in milliamps, and the width and length are in micrometers. This simple model provides a very good fit over a wide range of transistor width, length, and bias current. For PMOS devices, we find it useful to use two models, one model ($g_{\mathrm{d1,PMOS}}$) for devices operating at low drain-to-source voltage (M_5 and M_8) and another one ($g_{\mathrm{d2,PMOS}}$) for devices operating at high drain-to-source voltage (M_1, M_2, and M_7)

$$g_{\mathrm{d1,PMOS}} = 4.5 \cdot 10^{-1} W^{0} L^{-1.58} I_{\mathrm{D}}^{1.04}$$
$$g_{\mathrm{d2,PMOS}} = 8.9 \cdot 10^{-2} W^{0.13} L^{-1.97} I_{\mathrm{D}}^{0.87}$$

where again the output conductance is given in millisiemens, the bias current is in milliamps, and the width and length are in micrometers.

For all other circuit parameters, we used the GP0 model described above (although we could easily have improved the models using empirical fits to monomials and posynomials).

ACKNOWLEDGMENT

The authors would like to thank E. Waks, who wrote the geometric programming solver originally used for the numerical experiments shown in this paper. The authors are grateful to B. Fowler, A. E. Gamal, A. Hajimiri, and many anonymous reviewers, for very useful comments and suggestions.

REFERENCES

[1] M. A. Aguirre, J. Chávez, A. Torralba, and L. G. Franquelo, "Analog design optimization by means of a tabu search approach," in *Proc. IEEE Int. Symp. Circuits Syst.*, vol. 1, 1994, pp. 375–378.

[2] B. K. Ahuja, "An improved frequency compensation technique for CMOS operational amplifiers," *IEEE J. Solid-State Circuits*, vol. 18, pp. 629–633, Dec. 1983.

[3] P. E. Allen and D. R. Holberg, *CMOS Analog Circuit Design*, 1st ed, U.K.: Oxford, 1987.

[4] B. A. A. Antao, "Trends in CAD of analog ICs," *IEEE Circuits Devices Mag.*, vol. 12, no. 5, pp. 31–41, Sept. 1996.

[5] K. J. Antreich and H. E. Graeb, "Circuit optimization driven by worst-case distances," in *Proc. IEEE Int. Conf. Computer-Aided Design*, 1991, pp. 166–169.

[6] J. Assael, P. Senn, and M. S. Tawfik, "A switched-capapcitor filter silicon compiler," *IEEE J. Solid-State Circuits*, vol. 23, pp. 166–174, Feb. 1988.

[7] M. Avriel, R. Dembo, and U. Passy, "Solution of generalized geometric programs," *Int. J. Numer. Methods Eng.*, vol. 9, pp. 149–168, 1975.

[8] A. Barlow, K. Takasuka, Y. Nambu, T. Adachi, and J. Konno, "An integrated switched-capacitor filter design system," in *Proc. IEEE Custom Integrated Circuit Conf.*, 1989, pp. 4.5.1–4.5.5.

[9] C. S. Beightler and D. T. Phillips, *Applied Geometric Programming*. New York: Wiley, 1976.

[10] S. Boyd and L. Vandenberghe. (1997) Introduction to convex optimization with engineering applications. Stanford Univ., Stanford, CA. [Online]http://www.stanford.edu/class/ee364/

[11] R. K. Brayton, G. D. Hachtel, and A. Sangiovanni-Vincentelli, "A survey of optimization techniques for integrated-circuit design," *Proc. IEEE*, vol. 69, pp. 1334–1362, Oct. 1981.

[12] L. R. Carley, G. G. E. Gielen, R. A. Rutenbar, and W. M. C. Sansen, "Synthesis tools for mixed-signal ICs: Progress on front-end and back-end strategies," in *Proc. 33rd Annu. Design Automation Conf.*, 1996, pp. 298–303.

[13] L. R. Carley and R. A. Rutenbar, "How to automate analog IC designs," *IEEE Spectrum*, vol. 25, pp. 26–30, Aug. 1988.

[14] J. Chávez, M. A. Aguirre, and A. Torralba, "Analog design optimization: A case study," in *Proc. IEEE Int. Symp. Circuits Syst.*, vol. 3, 1993, pp. 2083–2085.

[15] J. M. Cohn, D. J. Garrod, R. A. Rutenbar, and L. R. Carley, "KOAN/ANAGRAM II: New tools for device-placement and routing," *IEEE J. Solid-State Circuits*, Mar. 1991.

[16] A. R. Conn, N. I. M. Gould, and PH. L. Toint, *LANCELOT: A Fortran Package for Large-Scale Nonlinear Optimization (Release A)*. New York: Springer-Verlag, 1992, vol. 17.

[17] R. J. Duffin, "Linearizing geometric programs," *SIAM Rev.*, vol. 12, pp. 211–227, 1970.

[18] R. J. Duffin, E. L. Peterson, and C. Zener, *Geometric Programming—Theory and Applications*: Wiley, 1967.

[19] J. Ecker, "Geometric programming: Methods, computations and applications," *SIAM Rev.*, vol. 22, no. 3, pp. 338–362, 1980.

[20] J. Ecker and M. Kupferschmid, *Introduction to Operations Research*. Melbourn, FL: Krieger, 1991.

[21] F. El-Turky and E. E. Perry, "BLADES: An artificial intelligence approach to analog circuit design," *IEEE Trans. Computer-Aided Design*, vol. 8, pp. 680–692, June 1989.

[22] A. R. Conn *et al.*, "Optimization of custom CMOS circuits by transistor sizing," in *Proc. IEEE Int. Conf. Computer-Aided Design*, 1996, pp. 174–180.

[23] D. G. Marsh *et al.*, "A single-chip CMOS PCM codec with filters," *IEEE J. Solid-State Circuits*, vol. SC-16, pp. 308–315, Aug. 1981.

[24] P. Mandal and V. Visvanathan, "An efficient method of synthesizing CMOS op-amps for globally optimal design," *IEEE Trans. Computer-Aided Design*, submitted for publication.

[25] H. Chang *et al.*, "A top-down constraint-driven design methodology for analog integrated circuits," in *Proc. IEEE Custom Integrated Circuit Conf.*, 1992, pp. 8.4.1–8.4.6.

[26] I. Vassiliou *et al.*, "A video-driver system designed using a top-down, constraint-driven, methodoly," in *Proc. 33rd Annu. Design Automation Conf.*, 1996.

[27] J. Rijmenants *et al.*, "ILAC: An automated layout tool for analog CMOS circuits," *IEEE J. Solid-State Circuits*, vol. 24, pp. 417–425, Apr. 1989.

[28] J. W. Bandler *et al.*, "Robustizing circuit optimization using Huber functions," in *IEEE MTT-S Int. Microwave Symp. Dig.*, 1993, pp. 1009–1012.

[29] M. G. R. Degrauwe *et al.*, "IDAC: An interactive design tool for analog CMOS circuits," *IEEE J. Solid-State Circuits*, vol. SC-22, pp. 1106–1115, Dec. 1987.

[30] M. G. R. Degrauwe *et al.*, "Toward an analog system design environment," *IEEE J. Solid-State Circuits*, vol. 24, pp. 1587–1597, Dec. 1989.

[31] R. Chadha *et al.*, "WATOPT: An optimizer for circuit applications," *IEEE Trans. Computer-Aided Design*, vol. CAD-6, pp. 472–479, May 1987.

[32] W. C. Black *et al.*, "A high performance low power CMOS channel filter," *IEEE J. Solid-State Circuits*, vol. SC-15, pp. 929–938, Dec. 1980.

[33] F. V. Fernández, A. Rodríguez-Vázquez, and J. L. Huertas, "Interactive ac modeling and characterization of analog circuits via symbolic analysis," *Analog Integrated Design and Signal Processing*, vol. 1, pp. 183–208, Nov. 1991.

[34] F. V. Fernández, A. Rodríguez-Vázquez, J. L. Huertas, and G. G. R. Girlen, *Symbolic Analysis Techniques*. Piscataway, NJ: IEEE Press, 1998.

[35] A. Fiacco and G. McCormick, *Nonlinear Programming: Sequential Unconstrained Minimization Techniques*. New York: Wiley, 1968.

[36] J. P. Fishburn and A. E. Dunlop, "TILOS: A posynomial programming approach to transistor sizing," in *Proc. ICCAD'85*, pp. 326–328.

[37] G. G. E. Gielen, H. C. C. Walscharts, and W. M. C. Sansen, "ISAAC: A symbolic simulator for analog integrated circuits," *IEEE J. Solid-State Circuits*, vol. 24, pp. 1587–1597, Dec. 1989.

[38] ——, "Analog circuit design optimization based on symbolic simulation and simulated annealing," *IEEE J. Solid-State Circuits*, vol. 25, pp. 707–713, June 1990.

[39] P. E. Gill, W. Murray, M. A. Saunders, and M. H. Wright, "User's guide for NPSOL (Version 4.0): A FORTRAN package for nonlinear programming," Operations Res. Dept., Stanford Univ., Standford, CA, Tech. Rept. SOL 86-2, Jan. 1986.

[40] P. R. Gray and R. G. Meyer, "MOS operational amplifier design—A tutorial overview," *IEEE J. Solid-State Circuits*, vol. SC-17, pp. 969–982, Dec. 1982.

[41] ——, *Analysis and Design of Analog Integrated Circuits*, 3rd ed. New York: Wiley, 1993.

[42] R. Gregorian and G. C. Temes, *Analog MOS Integrated Circuits for Signal Processing*, 1st ed. New York: Wiley, 1986.

[43] S. K. Gupta and M. M. Hasan, "KANSYS: A CAD tool for analog circuit synthesis," in *Proc. 9th Int. Conf. VLSI Design*, 1996, pp. 333–334.

[44] R. Harjani, R. A. Rutenbar, and L. R. Carley, "OASYS: A framework for analog circuit synthesis," *IEEE Trans. Computer-Aided Design*, vol. 8, pp. 1247–1265, Dec. 1989.

[45] J. P. Harvey, M. I. Elmasry, and B. Leung, "STAIC: An interactive framework for synthesizing CMOS and BiCMOS analog circuits," *IEEE Trans. Computer-Aided Design*, vol. 11, pp. 1402–1417, Nov. 1992.

[46] M. Hashizume, H. Y. Kawai, K. Nii, and T. Tamesada, "Design automation system for analog circuits based on fuzzy logic," in *Proc. IEEE Custom Integrated Circuit Conf.*, 1989, pp. 4.6.1–4.6.4.

[47] P. Heikkilä, M. Valtonen, and K. Mannersalo, "CMOS op-amp dimensioning using multiphase optimization," in *Proc. IEEE Int. Symp. Circuits Systems*, 1988, pp. 167–170.

[48] M. Hershenson, S. Boyd, and T. H. Lee, "CMOS operational amplifier design and optimization via geometric programming," in *Proc. 1st Int. Workshop Design Mixed-Mode Integrated Circuits Applicat.*, Cancun, Mexico, 1997, pp. 15–18.

[49] ——, "Automated design of folded-cascode op-amps with sensitivity analysis," in 5th IEEE Int. Conf. Electron., Circuits Syst., Lisbon, Portugal, Sept. 1998.

[50] ——, "GPCAD: A tool for CMOS op-amp synthesis," in *Proc. IEEE/ACM Int. Conf. Computer Aided Design*, San Jose, CA, 1998.

[51] ——, (2000) Optimal design of a CMOS op-amp via geometric programming. [Online]. Available: http://ww.stanford.edu/~boyd/opamp.html

[52] L. P. Huelsman and G. E. Gielen, *Symbolic Analysis of Analog Circuits: Techniques and Applications*. Norwell, MA: Kluwer, 1993.

[53] D. A. Johns and K. Martin, *Analog Integrated Circuit Design*, 1st ed. New York: Wiley, 1997.

[54] G. Jusuf, P. R. Gray, and A. Sangiovanni-Vincentelli, "CADICS: Cyclic analog-to-digital converter synthesis," in *Proc. IEEE Int. Conf. Computer-Aided Design*, 1990, pp. 286–289.

[55] S. Khorram, A. Rofougaran, and A. A. Abidi, "A CMOS limiting amplifier and signal-strength indicator," in *Symp. VLSI Circuits Dig. Tech. Papers*, 1995, pp. 95–96.

[56] H. Y. Koh, C. H. Séquin, and P. R. Gray, "OPASYN: A compiler for CMOS operational amplifiers," *IEEE Trans. Computer-Aided Design*, vol. 9, pp. 113–125, Feb. 1990.

[57] K. O. Kortanek, X. Xu, and Y. Ye, "An infeasible interior-point algorithm for solving primal and dual geometric progams," *Math. Programming*, vol. 76, pp. 155–181, 1996.

[58] W. Kruiskamp and D. Leenaerts, "DARWIN: CMOS op amp synthesis by means of a genetic algorithm," in *Proc. 32nd Annu. Design Automation Conf.*, 1995, pp. 433–438.

[59] K. R. Laker and W. M. C. Sansen, *Design of Analog Integrated Circuits and Systems*, 1st ed. New York: McGraw-Hill, 1994.

[60] G. F. Landsburg, "A charge-balancing monolithic A/D converter," *IEEE J. Solid-State Circuits*, vol. SC-12, pp. 662–673, Dec. 1977.

[61] F. Leyn, W. Daems, G. Gielen, and W. Sansen, "Analog circuit sizing with constraint programming modeling and minimax optimization," in *Proc. IEEE Int. Symp. Circuits Systems*, vol. 3, 1997, pp. 1500–1503.

[62] S. P. Boyd, M. Hershenson, S. S. Mohan, and T. H. Lee, "Optimization of inductor circuits via geometric programming," in *Design Automation Conf., Session 54.3*, June 1999, pp. 994–998.

[63] K. Madsen, O. Niedseln, H. Schjaer-Jakobsen, and H. Tharne, "Efficient minimax design of networks without using derivatives," *IEEE Trans. Microwave Theory Techn.*, vol. MTT-23, pp. 803–809, Oct. 1975.

[64] P. C. Maulik and L. R. Carley, "High-performance analog module generation using nonlinear optimization," in *Proc. 4th Annu. IEEE Int. ASIC Conf. Exhibit*, 1991, pp. T13-5.1–T13-5.2.

[65] P. C. Maulik, L. R. Carley, and D. J. Allstot, "Sizing of cell-level analog circuits using constrained optimization techniques," *IEEE J. Solid-State Circuits*, vol. 28, pp. 233–241, Mar. 1993.

[66] P. C. Maulik, L. R. Carley, and R. A. Rutenbar, "Integer programming based topology selection of cell-level analog circuits," *IEEE Trans. Computer-Aided Design*, vol. 14, pp. 401–412, Apr. 1995.

[67] P. C. Maulik, M. J. Flynn, D. J. Allstot, and L. R. Carley, "Rapid redesign of analog standard cells using constrained optimization techniques," in *Proc. IEEE Custom Integrated Circuit Conf.*, 1992, pp. 8.1.1–8.1.3.

[68] F. Medeiro, F. V. Fernández, R. Domínguez-Castro, and A. Rodríguez-Vázquez, "A statistical optimization-based approach for automated sizing of analog cells," in *Proc. 31st Annu. Design Automation Conf.*, 1994, pp. 594–597.

[69] T. Mukherjee, L. R. Carley, and R. A. Rutenbar, "Synthesis of manufacturable analog circuits," in *Proc. 31st Annu. Design Automation Conf.*, 1994, pp. 586–593.

[70] R. S. Muller and T. H. Kamins, *Device Electronics for Integrated Circuits*, 2nd ed. New York: Wiley, 1986.

[71] B. A. Murtagh and M. A. Saunders, "Minos 5.4 user's guide," Systems Optimization Lab., Stanford Univ., Stanford, CA, Tech. Rept. SOL 83-20R, Dec. 1983.

[72] F. H. Musa and R. C. Huntington, "A CMOS monolithic 3.5 digit A/D converter," in *Int. Solid-State Conf.*, 1976, pp. 144–145.

[73] Y. Nesterov and A. Nemirovsky, *Interior-Point Polynomial Methods in Convex Programming*. Philadelphia, PA: SIAM, 1994, vol. 13, Studies in Applied Mathematics.

[74] Z. Ning, T. Mouthaan, and H. Wallinga, "SEAS: A simulated evolution approach for analog circuit synthesis," in *Proc. IEEE Custom Integrated Circuit Conf.*, 1991, pp. 5.2.1–5.2.4.

[75] W. Nye, D. C. Riley, A. Sangiovanni-Vincentelli, and A. L. Tits, "DELIGHT.SPICE: An optimization-based system for the design of integrated circuits," *IEEE Trans. Computer-Aided Design*, vol. 7, pp. 501–518, Apr. 1988.

[76] W. T. Nye, E. Polak, and A. Sangiovanni-Vincentelli, "DELIGHT: An optimization-based computer-aided design system," in *Proc. IEEE Int. Symp. Circuits and Systems*, 1981, pp. 851–855.

[77] E. S. Ochotta, R. A. Rutenbar, and L. R. Carley, "Synthesis of high-performance analog circuits in ASTRX/OBLX," *IEEE Trans. Computer-Aided Design*, vol. 15, pp. 273–293, Mar. 1996.

[78] H. Onodera, H. Kanbara, and K. Tamaru, "Operational amplifier compilation with performance optimization," *IEEE J. Solid-State Circuits*, vol. 25, pp. 466–473, Apr. 1990.

[79] S. Rabii and B. A. Wooley, "A 1.8-V digital-audio sigma-delta modulator in 0.8-μm CMOS," *IEEE J. Solid-State Circuits*, vol. 32, pp. 783–795, June 1997.

[80] J. Rajpogal and D. L. Bricker, "Posynomial geometric programming as a special case of semi-infinite linear programming," *J. Optim. Theory. Appl.*, vol. 66, pp. 455–475, Sept. 1990.

[81] S. Sapatnekar, V. B. Rao, P. Vaidya, and S.-M. Kang, "An exact solution to the transistor sizing problem for CMOS circuits using convex optimization," *IEEE Trans. Computer-Aided Design*, vol. 12, pp. 1621–1634, 1993.

[82] S. S. Sapatnekar, "Wire sizing as a convex optimization problem: exploring the area-delay tradeoff," *IEEE Trans. Computer-Aided Design*, vol. 15, pp. 1001–1011, Aug. 1996.

[83] L. E. Schrage, *User's Manual for Linear, Integer and Quadratic Programming with LINDO*, 3rd ed. San Francisco, CA: Scientific, 1987.

[84] S. J. Seda, M. G. R. Degrauwe, and W. Fichtner, "A symbolic tool for analog circuit design automation," in *Dig. Tech. Papers IEEE Int. Conf. Computer-Aided Design*, 1988, pp. 488–491.

[85] J. Shieh, M. Patil, and B. J. Sheu, "Measurement and analysis of charge injection in MOS analog switches," *IEEE J. Solid-State Circuits*, vol. SC-22, pp. 277–281, Apr. 1987.

[86] J. Shyu and A. Sangiovanni-Vincentelli, "ECSTASY: A new environment for IC design optimization," in *Dig. Tech. Papers IEEE Int. Conf. Computer-Aided Design*, 1988, pp. 484–487.

[87] J.-M. Shyu, A. Sangiovanni-Vincentelli, J. P. Fishburn, and A. E. Dunlop, "Optimization-based transistor sizing," *IEEE J. Solid-State Circuits*, vol. 23, pp. 400–409, 1988.

[88] J. E. Solomon, "The monolithic op amp: A tutorial study," *IEEE J. Solid-State Circuits*, vol. SC-9, pp. 969–982, Dec. 1974.

[89] H. Su, C. Michael, and M. Ismail, "Statistical constrained optimization of analog MOS circuits using empirical performance tools," in *Proc. IEEE Int. Symp. Circuits Systems*, vol. 1, 1994, pp. 133–136.

[90] K. Swings, G. Gielen, and W. Sansen, "An intelligent analog IC design system based on manipulation of design equations," in *Proc. IEEE Custom Integrated Circuit Conf.*, 1990, pp. 8.6.1–8.6.4.

[91] S. M. Sze, *Physics of Semiconductor Devices*, 2nd ed. New York: Wiley, 1981.

[92] A. Torralba, J. Chávez, and L. G. Franquelo, "FASY: A fuzzy-logic based tool for analog synthesis," *IEEE Trans. Computer-Aided Design*, vol. 15, pp. 705–715, July 1996.

[93] L. Trontelj, J. Trontelj, T. Slivnik, R. Sosic, and D. Strle, "Analog silicon compiler for switched capacitor filters," in *Proc. IEEE Int. Conf. Computer-Aided Design*, 1987, pp. 506–509.

[94] P. J. M. van Laarhoven and E. H. L. Aarts, *Simulated Annealing: Theory and Applications*. Amsterdam, The Netherlands: Reidel, 1987.

[95] L. Vandenberghe, S. Boyd, and A. El Gamal, "Optimal wire and transistor sizing for circuits with nontree topology," in *Proc. 1997 IEEE/ACM Int. Conf. Computer Aided Design*, pp. 252–259.

[96] L. Vandenberghe, S. Boyd, and A. El Gamal, "Optimizing dominant time constant in RC circuits," *IEEE Trans. Computer-Aided Design*, vol. 2, pp. 110–125, Feb. 1998.

[97] R. J. Vanderbei, *Linear Programming: Foundations and Extensions*. Norwell, MA: Academic, 1997.

[98] F. Wang and R. Harjani, "Optimal design of opamps for oversampled converters," in *Proc. IEEE Custom Integrated Circuit Conf.*, 1996, pp. 15.5.1–15.5.4.

[99] D. Wilde and C. Beightler, *Foundations of Optimization*. Englewood Cliffs, NJ: Prentice-Hall, 1967.

[100] M. Wójcikowski, J. Glinianowicz, and M. Bialko, "System for optimization of electronic circuits using genetic algorithm," in *Proc. IEEE Int. Conf. Electronics, Circuits Syst.*, 1996, pp. 247–250.

[101] D. F. Wong, H. W. Leong, and C. L. Liu, *Simulated Annealing for VLSI design*. Norwell, MA: Kluwer, 1988.

[102] S. J. Wright, *Primal–Dual Interior-Point Methods*. Philadelphia, PA: SIAM, 1997.

[103] C. Xinghao and M. L. Bushnell, *Efficient Branch and Bound Search With Application to Computer-Aided Design*. Norwell, MA: Kluwer, 1996.

[104] X. Xu. (1995, May) XGP—An optimizer for geometric programming. Tech. Rep. [Online]. Available: ftp://col.biz.uiowa.edu/dist/xu/doc/home.html

[105] H. Z. Yang, C. Z. Fan, H. Wang, and R. S. Liu, "Simulated annealing algorithm with multi-molecule: An approach to analog synthesis," in *Proc. 1996 European Design & Test Conf.*, 1996, pp. 571–575.

[106] C. Zener, *Engineering Design by Geometric Programming*, New York: Wiley, 1971.

Maria del Mar Hershenson was born in Barcelona, Spain. She received the B.S.E.E. degree from the Universidad Pontificia de Comillas, Madrid, Spain, in 1995, and the M.S. and Ph.D. degrees in electrical engineering from Stanford University, Stanford, CA, in 1997 and 1999, respectively.

In 1994, she was an Intern at Linear Technology Corporation, Milpitas, CA, where she worked on low-power voltage regulators. In 1999, she co-founded Barcelona Design Inc., Sunnyvale, CA, where she currently designs analog circuits based on new optimization techniques. Her research interest are RF circuits and convex optimization techniques applied to the automated design of analog integrated circuits.

Dr. Hershenson is the recipient of a 1998 IBM Fellowship.

Stephen P. Boyd (S'82–M'85–SM'97–F'99) received the A.B. degree in mathematics from Harvard University, Cambridge, MA, in 1980, and the Ph.D. degree in electrical engineering and computer science from the University of California at Berkeley, in 1985.

In 1985, he joined the Electrical Engineering Department, Stanford University, Stanford, CA, where he is currently a Professor and Director of the Information Systems Laboratory. In 1999, he co-founded Barcelona Design Inc., Sunnyvale, CA. His interests include computer-aided control system design and convex programming applications in control, signal processing, and circuits.

Thomas H. Lee received the S.B., S.M., and Sc.D. degrees in electrical engineering from the Massachusetts Institute of Technology, Cambridge, in 1983, 1985, and 1990, respectively.

He joined Analog Devices in 1990 where he was primarily engaged in the design of high-speed clock recovery devices. In 1992, he joined Rambus Inc. Mountain View, CA, where he developed high-speed analog circuitry for 500 MB/s CMOS DRAMs. He has also contributed to the development of phased-locked loops (PLLs) in the StrongARM, Alpha, and K6/K7 microprocessors. Since 1994, he has been an Assistant Professor of electrical engineering at Stanford University, Stanford, CA, where his research focus has been on gigahertz-speed wireline and wireless integrated circuits built in conventional silicon technologies, particularly CMOS. He is also cofounder of Matrix Semiconductor. He holds 12 U.S. patents and has authored the textbook, *The Design of CMOS Radio—Frequency Integrated Circuits* (Cambridge, MA: Cambridge Univ. Press, 1998) and co-authored two additional books on RF circuit design.

Prof. Lee is a Distinguished Lecturer of the IEEE Solid-State Circuits Society and was recently named a Distinguished Microwave Lecturer. He twice received the "Best Paper" award at the International Solid-State Circuits Conference, was co-author of a "Best Student Paper" at ISSCC, and recently won a Packard Foundation Fellowship.

PART III

Symbolic Analysis

TECHNIQUES AND APPLICATIONS OF SYMBOLIC ANALYSIS FOR ANALOG INTEGRATED CIRCUITS: A TUTORIAL OVERVIEW

Georges G. E. Gielen

Katholieke Universiteit Leuven
Departement Elektrotechniek-ESAT, division ESAT-MICAS
Kasteelpark Arenberg 10, B-3001 Leuven
E-mail : georges.gielen@esat.kuleuven.ac.be

Abstract

This tutorial paper gives an overview of the state of the art in symbolic analysis techniques for analog integrated circuits. Symbolic analysis allows to generate closed-form analytic expres-sions for a circuit's small-signal characteristics with the circuit's elements and the frequency variable represented by symbols. Such analytic information complements the results from numerical simulations. The paper then describes the different application areas of symbolic analysis in the design of analog integrated circuits. Symbolic analysis is mainly used as a means to obtain insight into a circuit's behavior, to generate analytic models for automated circuit sizing, for behavioral model generation and in applications requiring the repetitive evaluation of circuit characteristics. Recent extensions of both the functionality (such as towards symbolic distortion or pole-zero analysis) and the efficiency of symbolic analysis for larger circuits (through new algorithmic developments) are discussed. Finally, an overview and comparison of existing tools is presented.

1. Introduction

Symbolic analysis of electronic circuits received strong attention during the late sixties and the seventies, where a lot of computer-oriented analysis techniques were proposed. Since the late eighties, symbolic analysis of electronic circuits has gained a renewed and growing interest in the electronic design community [1,2]. This is illustrated by the success of modern symbolic analyzers for analog integrated circuits such as ISAAC [3,4], ASAP [5,6], SYNAP [7,8], SAPEC [9], SSPICE [10], SCYMBAL [11], SCAPP [12], Analog Insydes [13], CASCA [14], SIFTER [15] and RAINIER [16].

In section 2, the technique of symbolic circuit analysis is first defined and illustrated for some practical examples. The basic methodology of how symbolic analysis works is explained. Section 3 then presents the different applications of symbolic analysis in the analog design world, and indicates the advantages and disadvantages of symbolic analysis compared to other techniques, especially numerical simulation. Section 4 then describes the present capabilities and limitations in symbolic analysis, and details recent algorithmic advances, especially for the analysis of large circuits. Finally, an overview of existing tools is presented in Section 5, and conclusions are provided in section 6.

2. What is symbolic analysis ?

In this section first a definition of symbolic analysis is provided and illustrated with some examples. Next, the basic methodology of how symbolic analysis works is explained..

2.1. Definition of symbolic analysis

Symbolic analysis of an analog circuit is a formal technique to calculate the behavior or a characteristic of a circuit with the independent variable (time or frequency), the dependent variables (voltages and currents) and (some or all of) the circuit elements represented by symbols. The technique is complementary to numerical analysis (where the variables and the circuit elements are represented by numbers) and qualitative analysis (where only qualitative values are used for voltages and currents, such as increase, decrease or no_change). A symbolic simulator is then a computer program that receives the circuit description as input and can automatically carry out the symbolic analysis and thus generate the symbolic expression for the desired circuit characteristic.

Almost all of the symbolic analysis research carried out in the past concerned the analysis of linear circuits in the frequency domain. For lumped, linear, time-invariant circuits, the symbolic network functions obtained are rational functions in the complex frequency variable x (s for continuous-time and z for discrete-time circuits) and the circuit elements p_i that are represented by a symbol (instead of a numerical value) :

$$H(x) = \frac{\sum_i x^i a_i(p_1,...,p_m)}{\sum_j x^j b_j(p_1,...,p_m)} \quad (1)$$

where $a_i(..)$ and $b_i(..)$ are symbolic polynomial functions in the circuit elements p_i. In general these symbolic equations can be expanded in sums-of-products form or can be in nested form.

In a fully symbolic analysis all circuit elements are represented by symbols. But a mixed symbolic-numerical analysis is also possible where only part of the elements are represented by symbols and the others by their numerical values. In the extreme case with no symbolic circuit elements, a rational function with numerical coefficients is obtained with the frequency variable (s or z) as only symbol.

Note that the symbols in these expressions represent the elements that are present in the linear circuit. In case of semiconductor circuits, the devices are typically linearized around their operating point and the symbols therefore represent the elements in the small-signal expansion (e.g. g_m, g_o, etc.) of these devices. As long as the topology of the small-signal device expansion remains the same, the resulting expressions (in terms of g_m, g_o, etc.) remain valid, independent of the particular device model used (e.g. SPICE level 2, BSIM, etc.) and to some extent also independent of the operating region of the device (although the results of course depend on the operating point when symbolic simplification techniques are used). Only when the equations are numerically evaluated or when the small-signal elements are further symbolically substituted by their describing equations, then a particular set of device model equations has to be used to relate the small-signal elements to the device sizes and biasing. For complicated models, the symbolic equations might soon become too cumbersome though to be interpretable by designers.

The technique of symbolic analysis is now illustrated for two practical examples, a continuous-time and a discrete-time filter.

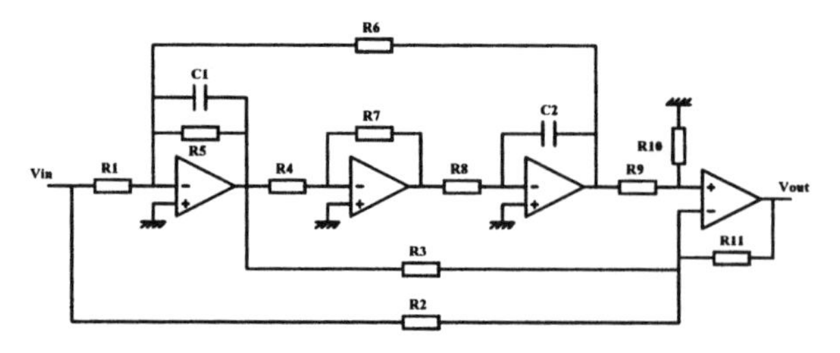

Fig. 1. Active RC filter.

Example 1

Consider the active RC filter of Fig. 1. Starting from the circuit description of this filter, a symbolic analysis program will return the following symbolic expression for the transfer function of this filter :

$$\frac{\begin{matrix} -G_4G_8(G_1G_2G_9+G_1G_3G_9+G_1G_9G_{11}+G_2G_6G_9 \\ +G_2G_6G_{10}) \\ +s\ G_7C_2(G_1G_3G_9+G_1G_3G_{10}-G_2G_5G_9-G_2G_5G_{10}) \\ -s^2\ G_2G_7C_1C_2(G_9+G_{10}) \end{matrix}}{G_{11}(G_9+G_{10})(G_4G_6G_8+s\ G_5G_7C_2+s^2\ G_7C_1C_2)} \quad (2)$$

where G_x is the conductance corresponding to resistor R_x in the schematic of Fig. 1. The operational amplifiers have been considered ideal in this case, but other, more realistic opamp models are possible as well.

Example 2

As a second example, consider the switched-capacitor biquad of Fleischer and Laker of Fig. 2. Starting from the circuit description of this filter and the timing information of the clock phases, a symbolic simulator automatically generates the following symbolic expression for the z-domain transfer function of this filter (with operational amplifiers considered as ideal) from the input (sampled at phase 1 and held constant during phase 2) to the output (at phase 1) :

$$\frac{\begin{matrix} DK+DJ-AL-AH \\ +z\,(-2DK+AL+AG-DJ-DI)+z^2\ D(K+I) \end{matrix}}{AE-DB+z\,(-AE+2DB-AC+DF)-z^2\ D(F+B)} \quad (3)$$

where the symbols A up to L represent the capacitors in the schematic of Fig. 2. When the input and output are considered during other phases, other transfer functions are obtained. Note that only a limited number of symbolic analysis tools can handle time-discrete circuits, for instance the tools described in [4,11,17,18].

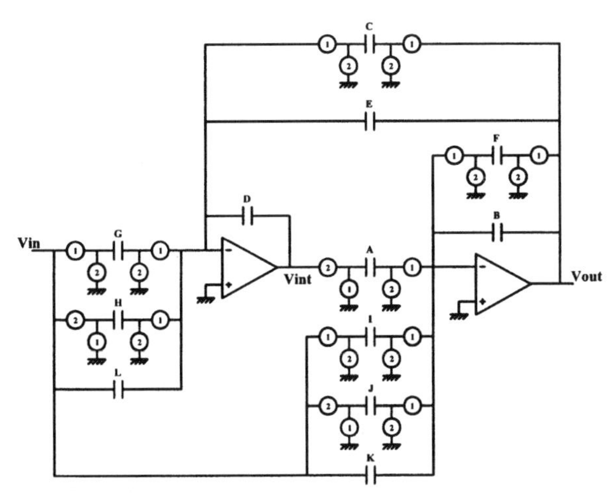

Fig. 2. Switched-capacitor biquad.

2.2. Basic methodology of symbolic analysis

The principle operation of symbolic analysis programs can now be explained. The basic flow is illustrated in Fig. 3 with the example of a bipolar single-transistor amplifier. The input is a netlist description of the circuit to be analyzed (for example in the well known SPICE syntax). Since symbolic analysis is primarily restricted to linear circuits, the first step for nonlinear analog circuits (such as the amplifier in Fig. 3) is to generate a linearized small-signal equivalent of the circuit. In the example of Fig. 3, only the dc model is shown in order not to overload the drawing. In general, circuits consisting of resistors, capacitors, inductors, independent and controlled sources can be handled by most symbolic analyzers.

After the user has indicated the network function (i.e. the input and output) that he wants, the analysis can start. For this, two basic classes of methods exist (as shown by the two paths in Fig. 3) [19,20] :

* in the class of ***algebraic (matrix- or determinant-based) methods***, the behavior of the linear(ized) circuit is described by a set of equations with symbolic coefficients. The required symbolic network function is obtained by algebraic operations on this set of equations (such as for instance a symbolic expansion of the determinant).
* in the class of ***graph-based (or topological) methods***, the behavior of the linear(ized) circuit is represented by one or two graphs with symbolic branch weights. The required symbolic network function is obtained by operations on these graphs (such as for instance the enumeration of loops or spanning trees).

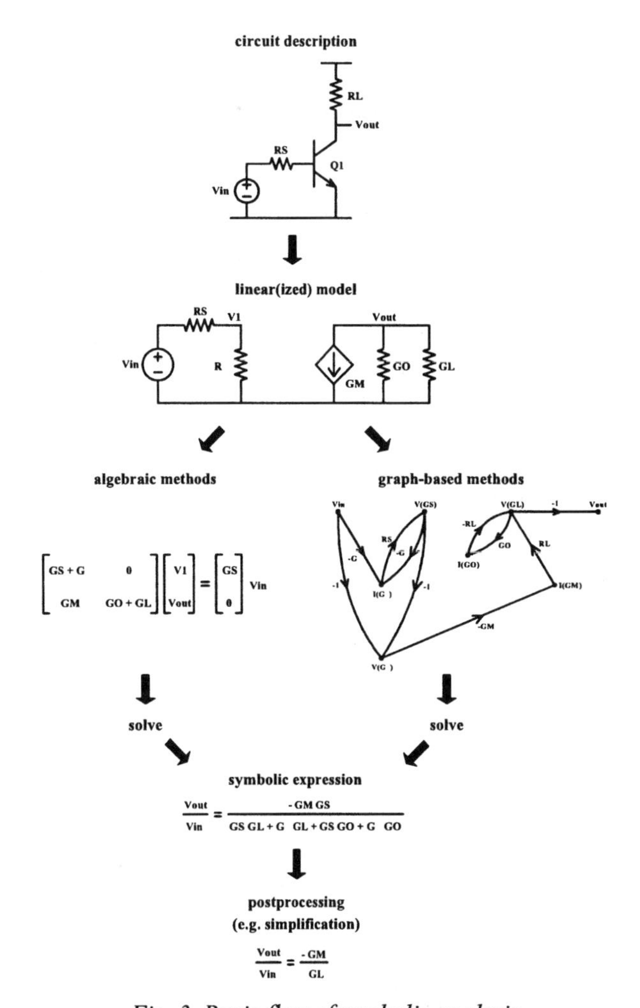

Fig. 3. Basic flow of symbolic analysis.

Whatever the symbolic analysis method followed, the expression generated can then further be postprocessed in symbolic format. For example and especially in the case of the small-signal analysis of semiconductor circuits, the network function can further be simplified or approximated. This means that information of the relative magnitudes of the circuit elements (of course depending on the operating point) is used to discard up to a maximum user-defined error smaller terms with respect to larger terms in order to obtain a less complicated symbolic expression with only the dominant contributions (see section 4 for details). Other symbolic postprocessing operations could be factorization of results, or symbolic substitutions of certain symbols by their expressions, etc. The result after postprocessing is then returned to the user or can be used in other design applications or for numerical postprocessing (e.g. evaluations).

3. Applications of symbolic analysis

This section now summarizes the major applications of symbolic analysis in the design of analog circuits [1,2,20].

3.1. Insight into circuit behavior

Symbolic analysis provides closed-form symbolic expressions for the characteristics of a circuit. It is an essential complement to numerical simulation, where a series of numbers is returned (in tabulated or plotted form). Although these numbers can accurately simulate the circuit behavior, they are specific for a particular set of parameter values. With numerical simulation, the functional behavior of the circuit can be verified in a very short time for the given parameter set. But no indication is given of which circuit elements determine the observed performance. No solutions are suggested when the circuit does not meet the specifications. A multitude of simulations are required to scan performance trade-offs and to check the influence of changes in the parameter values.

A symbolic simulator on the other hand returns first-time correct analytic expressions in a much shorter time and for more complex characteristics and circuits than would be possible by hand. The resulting symbolic expressions remain valid even when the numerical parameter values change (as long as all models remain valid). As such, symbolic analysis gives a different perspective on a circuit than that provided by numerical simulators, which is most appropriate for students and practicing designers to obtain real insight into the behavior of a circuit [20,21]. This is especially true if the generated expressions are simplified to retain the dominant contributions only, as already discussed before. With symbolic simulation, circuits can be explored without any hand calculation in a fast and interactive way. This is illustrated in the following example.

Example

Consider the CMOS three-stage amplifier of Fig. 4 with nested Miller compensation. (The stages are represented by equivalent circuits because we are interested in the stability properties of the overall amplifier in this example). In the presence of the two compensation capacitors, the 25%-maximum-error simplified transfer function with the dominant contributions is given by :

$$\frac{-g_{m1}(g_{m2}g_{m3} + sg_{m2}C_{C1} + s^2C_{C1}C_{C2})}{g_{o1}g_{o2}g_L + sg_{m2}g_{m3}C_{C2} + s^2(g_{m3} + g_L - g_{m2})C_{C1}C_{C2} + s^3C_{C1}C_{C2}C_L} \tag{4}$$

This expression clearly provides insight into the stability properties of this amplifier. The two compensation capacitors C_{C1} and C_{C2} can stabilize the circuit, provided that $g_{m3} + g_L$ is larger than g_{m2}. Also, C_{C1} and C_{C2} cause one left-half-plane zero and one right-half-plane zero, which influence the phase margin.

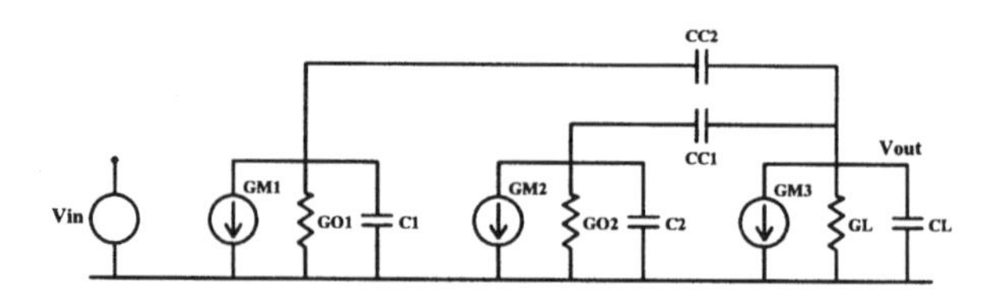

Fig. 4. Principle schematic of a 3-stage amplifier with nested Miller compensation.

In addition to improving the insight of students and novice designers, a symbolic simulator is also a valuable design aid for experienced designers, to check their own intuitive knowledge, and to obtain analytic expressions for second-order characteristics such as the power-supply rejection ratio or the harmonic distortion, which are almost impossible to calculate by hand, especially at higher frequencies and in the presence of device mismatches. On the downside, symbolic analysis is limited to much smaller circuit sizes and fewer analysis types (e.g. no transient simulation yet) than numerical simulation and inevitably takes much longer CPU times for the same circuit and type of analysis.

3.2. Analytic model generation for automated analog circuit sizing

A symbolic simulator can automatically generate all ac characteristics in the analytic model of a circuit. Such a model, that approximates the behavior of a circuit with analytic formulas, can then be used to efficiently size the circuit in an optimization program. An alternative is to use a numerical simulator to evaluate the circuit performance. Optimization time however is strongly reduced by replacing the full numerical simulation at each iteration with an evaluation of the analytic model equations. This approach is explicitly adopted in the OPTIMAN [22] and OPASYN [23] programs. The use of an equation manipulation program such as DONALD [24] even allows to automatically construct a design plan by ordering the equations of the analytic model such as to optimize the efficiency of each evaluation of this analytic model. To this end, DONALD constructs a computational path: it determines for the given set of independent variables the sequence in which the equations have to be solved, grouped into minimum subsets of simultaneous equations [24].

Other analog synthesis programs of course also rely on analytic information mixed with heuristics and design strategies. In IDAC [25] and OASYS [26] for example all this knowledge has to be converted manually into schematic-specific design plans. Tools like CATALYST [27] and SD-OPT [28] use analytic equations from the tool's library to perform the architectural-level design of specific classes of analog-to-digital converters in terms of their subblocks. The manual derivation of the analytic equations however is a time-consuming and error-prone process that has to be carried out for each circuit schematic. The use of a symbolic simulator largely reduces the effort required to develop the analytic model for a new circuit schematic. In this way, an open, non-fixed-topology analog CAD system can be created, in which the designer himself can easily include new circuit topologies. This approach is adopted in the AMGIE [20,29] and ADAM [30] analog design systems.

It can be objected that the exponential growth of the number of terms with the circuit size in expanded symbolic expressions makes the evaluation of such expressions for automatic circuit sizing less efficient than a simple numerical simulation. Therefore, for applications requiring the repetitive evaluation of symbolic expressions, one solution is to use expressions that have been simplified for typical ranges of the design parameters. If in addition to evaluation efficiency also full accuracy is required, then the non-simplified symbolic expressions have to be generated in the most compact form, demanding for nested formats. On the other hand, for interactive use and to obtain insight in the circuit behavior, the symbolic expressions are better expanded and simplified.

3.3. Interactive circuit exploration

Symbolic simulation can be used to interactively explore and improve new circuit topologies. The influence of topology changes (e.g. adding an extra component) can immediately be seen from the new resulting expressions. A powerful environment for the design and exploration of analog circuits then consists of a schematic editor, both a symbolic and a numerical simulator in combination with numerical and graphical postprocessing routines and possibly optimization tools. In this way, the interactive synthesis of new, high-performance circuits becomes feasible.

Moreover, in [31] a method is presented to automatically generate new circuit schematics with the combined aid of a symbolic simulator and the PROLOG language. A PROLOG program exhaustively generates all possible topologies with a predefined set of elements (for example one opamp and three capacitors). The switched-capacitor-circuit symbolic simulator SCYMBAL then derives a symbolic transfer function for each structure. A PROLOG routine finally evaluates this function to check whether it fulfils all requirements (for example a stray-insensitive switched-capacitor integrator function with low opamp-gain sensitivity).

3.4. Repetitive formula evaluation

The use of symbolic formulas is also an efficient alternative if a network function has to be evaluated repeatedly for multiple values of the circuit components and/or the frequency. For such applications, the technique of compiled-code simulation can be used. The symbolic expressions for the required network functions are generated once and then compiled. This same compiled code can then be evaluated many times for particular values of the circuit and input parameters. This technique can be more efficient than performing a full numerical simulation for each set of circuit and input parameters. The efficiency of this technique has been shown in [11] for the analysis of switched-capacitor circuits. Another obvious application is in statistical analysis (e.g. Monte Carlo simulations where the same characteristics have to be evaluated many times in order to assess the influence of statistical device mismatches and process tolerances on the circuit performance), large-signal sensitivity analysis, yield estimation and design centering. An application reported in [32] is the characterization of a semiconductor device as a function of technological and geometrical parameters.

If the values of all components of a circuit are known, then symbolic analysis can also be an efficient alternative to obtain frequency spectra of the characteristics of that circuit. In that case, the symbolic analysis has to be carried out with only the frequency variable represented by a symbol and the circuit parameters by their numerical value. This results in an exact, non-simplified rational function in s or z, which can be evaluated for many different frequency points. This technique can be more efficient than repeated numerical simulations in for example sensitivity analysis or distortion analysis. If in addition to the frequency variable also one or two circuit elements are retained as a symbol, the frequency spectra can be parameterized with respect to these symbolic elements. The same expressions in s or z can also be used to numerically derive poles and zeros, and generate pole-zero position diagrams as a function of one of the symbolic parameters. (These can then be compared to the results of any symbolic expressions derived for poles and zeros, if available). It has been shown in [33] that the CPU time needed to numerically evaluate such parametrized symbolic-numerical characteristics is much lower than that of a full numerical simulation, up to five orders of magnitude in one example, which is essential in highly iterative applications.

Example

Fig. 5 shows the position on the frequency axis of the gain-bandwidth (*), poles (straight lines) and zeros (dotted lines) for the three-stage amplifier of Fig. 4 as a function of the value of the compensation capacitor C_{C1} for a constant C_{C2} value. The expression for the transfer function was first derived semi-symbolically as a function of C_{C1} and s, and the roots were then solved numerically for different values of C_{C1}. The information of Fig. 5 can be used to select the optimum design values to stabilize this amplifier.

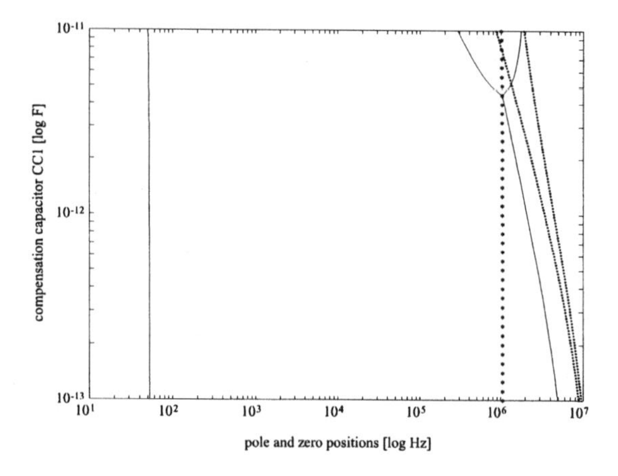

Fig. 5. Diagram of the frequency positions of the poles (straight lines) and zeros (dotted lines) for the three-stage amplifier as a function of C_{C1}.

3.5. Analog fault diagnosis

Interesting is also the application of the symbolic technique in fault diagnosis of linear and nonlinear analog circuits [34]. Using the symbolic network function, the circuit response is evaluated many times so as to determine those numerical values of the components that best fit the simulated response with the measured one. Based on the resulting component values, it can be decided which component is faulty. For nonlinear devices a piecewise-linear approximation is used. Reactive elements are replaced by their backward-difference models. The time-domain simulations are then carried out by means of a Katznelson-type algorithm.

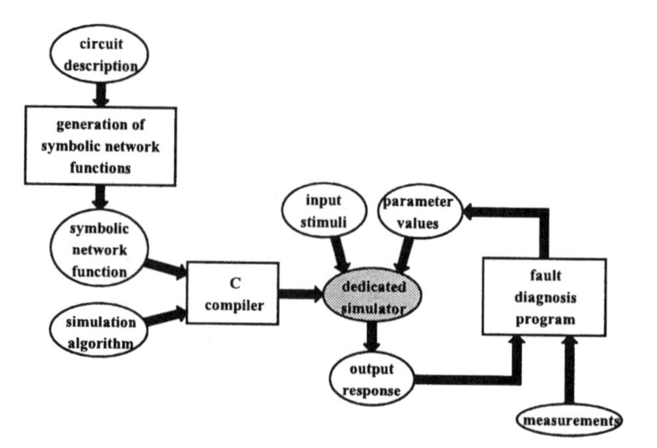

Fig. 6. Principle of analog fault diagnosis using dedicated compiled code simulators.

As shown in Fig. 6, an interesting aspect of the approach is that the symbolic expression is generated only once for the circuit and then together with the code of the simulation algorithm compiled into a dedicated simulator, specific for that circuit. During the fitting of the output response, this dedicated simulator is called many times for the given input signal with different numerical values for the components. Symbolic expressions have also been used in testability analysis to determine the observability of a circuit for a given number of observation nodes [35].

3.6. Behavioral model generation

Symbolic simulation can also easily be used to automatically generate behavioral models for certain analog blocks such as opamps and filters, which are most naturally described by transfer functions in s or z. Such behavioral models are required to simulate higher-level systems in acceptable CPU times, or in top-down design.

A big problem for general nonlinear circuits and general transient characteristics however is the generation of an accurate behavioral model. Therefore, in [36], an approach based on symbolic simplification techniques was explored. First the general nonlinear differntial-algebraic symbolic equations of the circuit are set up. These are then simplified using both global (e.g. complete elimination of a variable) and local (e.g. pruning of a term or expansion of a nonlinear function in a truncated series) approximations. The resulting simplified expressions form the behavioral model and are converted into the desired hardware description language. In this way simulation speed-ups between 4x and 10x have been reported [36]. The critical part of the method is the estimation of the approximation error. Since the circuit is nonlinear, this can only be done for a specific circuit response (e.g. DC transfer characteristic). For other responses, other models have to be generated or the error is not guaranteed.

3.7. Formal verification

Another recent application is analog formal verification [37]. The goal is to prove that a given circuit implementation implements a given specification. For linear(ized) circuits both the circuit implementation and the specification can be described with symbolic transfer functions. Algorithms have then been developed to numerically calculate an outer bound for the characteristic of the circuit implementation, taking into account the variations of process parameters and operating conditions, and an inner bound for the characteristic of the specification. If the former is enclosed by the latter, then it is formally proven that the circuit is a valid implementation of the specification. The algorithms are however very time-consuming and the extension to nonlinear circuits or characteristics is not at all obvious.

3.8. Summary of applications

We can now summarize the major applications of symbolic analysis in analog design as :

- knowledge acquisition and educational/ training purposes (insight)
- analytic model generation for automated circuit sizing in an open, flexible-topology analog design system
- design space exploration and topology generation
- repetitive formula evaluation, e.g. statistical analysis
- analog fault diagnosis and testability analysis
- analog behavioral model generation
- analog formal verification
- etc.

It can be concluded that symbolic analysis has a large potential for analog circuit analysis and design. It will now be described in the next section what the present capabilities and limitations of the symbolic analysis algorithms are. This will allow the reader to assess whether the above promising potential can be exploited by designers in real life or not.

4. Present capabilities and limitations of symbolic analysis

The different types of circuits that can be analyzed symbolically nowadays and the different types of symbolic analyses which are presently feasible on these circuits are shown in highlighted form in Fig. 7. For many years, symbolic simulation has only been feasible for the analysis of lumped, linear, time-continuous and time-discrete (switched-capacitor) analog circuits in the frequency domain (s or z). Nonlinear circuits, including both MOS and bipolar devices, are then linearized around the dc operating point, yielding the small-signal equivalent circuit. In this way, analytic expressions can be derived for the ac characteristics of a circuit, such as transfer functions, common-mode rejection ratio, power-supply rejection ratio, impedances, noise...

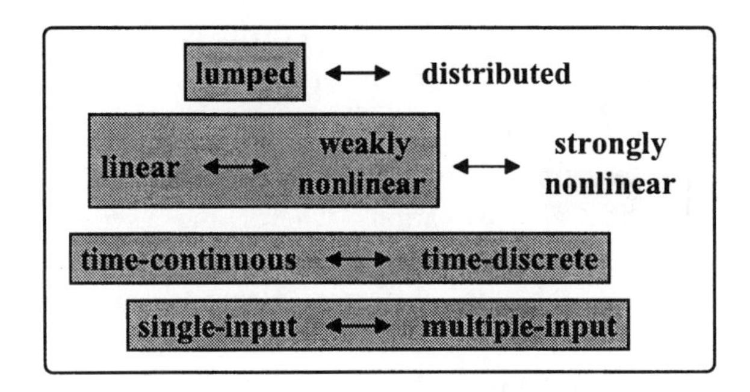

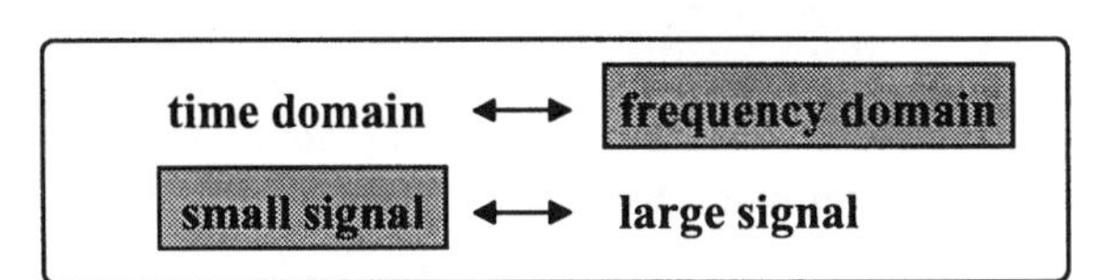

Fig. 7. Overview of present capabilities of symbolic analysis.

Originally, the size of the circuits that could be analyzed in a symbolic way was rather restricted (a few transistors) [38], partially because of the poor algorithms used, partially because of the complexity (large number of terms for even a small circuit) that make the resulting expressions intractable for designers. The introduction of symbolic simplification or approximation techniques in the late eighties [4,7] therefore was a first step to improve both the interpretability and extend the maximum analyzable circuit size. In recent years, other techniques have been developed to extend the capabilities of symbolic analysis both in terms of the functionality offered to the user and in terms of the computational efficiency for larger circuits. These improvements will now be described in detail.

4.1. Symbolic approximation

ISAAC [3,4] and SYNAP [7,8], later on followed by other tools like ASAP [5,6] and SSPICE [10], have introduced the idea of approximation (or simplification or truncation or pruning) of the symbolic expressions. Due to the exponential growth of the number of terms with the circuit size (e.g. more than 4.5 10^{17} terms in the system denominator of the well known μA741 opamp), the symbolic expressions for analog integrated circuits rapidly become too lengthy and complicated to use or interpret, rendering them virtually useless. Fortunately, in semiconductor circuits, some elements are much larger in magnitude than others. For example, the transconductance of a transistor is usually much larger than its output conductance. This means that the majority of the terms in the full symbolic expression are relatively unimportant, and that only a small number of dominant terms are important to really determine the circuit behavior and therefore also to gain insight into this behavior.

This observation has led to the technique of symbolic expression approximation where the symbolic expressions are simplified based on the relative magnitudes of the different elements and the frequency, defined in some nominal design point or over some range [3,7]. Conceptually, this means that the smaller-size terms are discarded with respect to the larger-size terms, so that the resulting simplified expression only contains a small and interpretable number of dominant contributions, that still describes the circuit behavior accurately enough.. Of course, this introduces some error ε equal to the contribution of the discarded terms, but the maximum error allowed ε_{max} and hence also the number of terms in the final expression is in most programs controllable by the user. The goal of simplification therefore is to find, based on the (order of) magnitude of the circuit elements (or their ranges) and the frequency variable (or its range), an approximating symbolic expression $h(\mathbf{p})$ for the original symbolic circuit characteristic $g(\underline{\mathbf{p}})$ such that [20] :

$$\left\|\frac{g(\mathbf{p})-h(\mathbf{p})}{g(\mathbf{p})}\right\| \leq \varepsilon_{max} \tag{5}$$

where the circuit parameters $\mathbf{p}$ (and the frequency) are evaluated over a certain range or in a nominal design point $\mathbf{p}_0$, and the error is measured according to some suitable norm. Symbolic expression approximation is thus a trade-off between expression accuracy (error) and expression simplicity (complexity). Note that in many tools the error ε_{max} is applied to each frequency coefficient separately, although in reality the changes in the magnitude and phase as well as in the poles and zeros of the overall network function should be controlled to within ε_{max} by the simplification algorithm [39]. Originally the error was only evaluated in one nominal design point (which could also be a typical set of default values), but then the error could be much larger outside this single design point. Therefore in [39] the error criterion was extended to consider ranges for all circuit elements, turning all calculations into interval arithmetic operations.

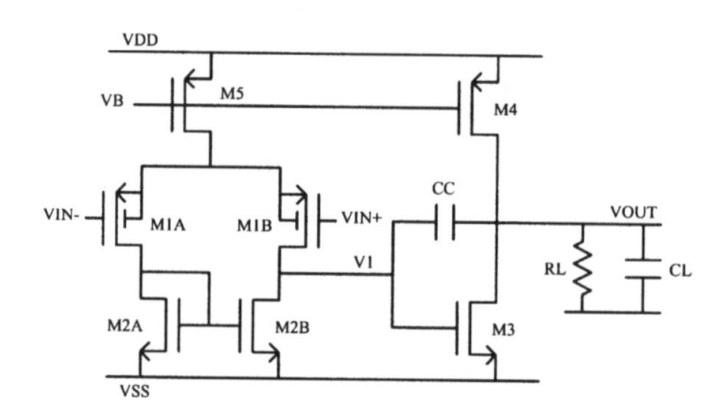

Fig. 8. CMOS two-stage Miller-compensated opamp.

Example

Consider the low-frequency value of the power-supply rejection ratio for the positive supply voltage of the CMOS two-stage Miller-compensated opamp of Fig. 8. For a 25% maximum approximation error, the denominator is reduced from 56 terms to 6 dominant terms. The simplified symbolic expression with the dominant contributions is given by :

$$\frac{2g_{m1}g_{m1}g_{m2}g_{m3}}{g_{m1}(g_{o1}+g_{o2})(2g_{m2}g_{o4}-g_{m3}g_{o5}) + (\Delta g_{m2BA}g_{m1}-\Delta g_{m1BA}g_{m2})g_{m3}g_{o5}} \tag{6}$$

where the Δ-symbols are the explicit symbolic representation of the statistical mismatch in transconductance between the matching transistors M1A-M1B and M2A-M2B respectively. Approximated results like (6) show the dominant contributions only, and are thus convenient for obtaining insight in the behavior of a circuit. In this example, the power-supply rejection ratio is analyzed, which is a complicated characteristic even for experienced designers. The partial cancellation between the nominal contributions of the first and the second stage, and the influence of mismatches can easily be understood from expression (6).

4.2. Improving computational efficiency

The main limitation inherent to symbolic analysis and hence to all symbolic analysis techniques, is the large computing time and/or memory storage, which increase very rapidly with the size of the network, especially for expanded expressions. This is mainly due to the exponential rise of the number of terms with the complexity of the circuit. As a result, this limits the maximum size of circuit that can be analyzed in a symbolic way [20]. In recent years, however, a real breakthrough was achieved with new algorithmic developments that have largely reduced the CPU time and memory consumption required for large circuits, and that have made possible the symbolic analysis of analog circuits of practical size (up to 40 transistors). These developments of course avoid the complete expansion of the exact expression.

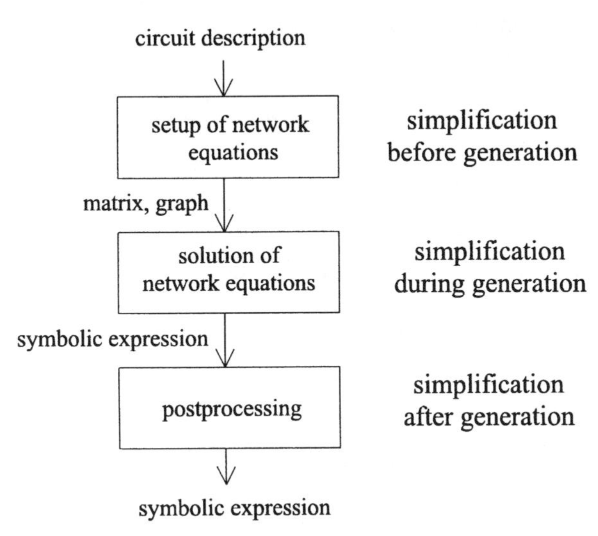

Fig. 9. Different simplification techniques in symbolic analysis.

Fig. 9 situates the different simplification techniques, depending on the place in the symbolic analysis process flow where the simplifications are introduced. The original simplification techniques first generated the complete exact expression in fully expanded format (in order to elaborate any term cancellations), after which the smallest terms were pruned repeatedly as long as the allowed error was not exceeded. This technique can be called **simplification after generation (SAG)**. Tools based on this technique like the original ISAAC [3,4], SYNAP [7] or ASAP [5,6] were restricted to circuits with maximum 10 to 15 transistors (depending on the actual topology configuration).

4.2.1. Simplification during generation

An improvement to save CPU time and memory was to generate the exact expression in nested format and to carry out a lazy expansion [8], where the initially nested expression is expanded term by term with the largest terms first until the error is below the required limit. With this technique, simplified results have been generated for circuits with up to 25 transistors. When using nested formulas, however, extensive care must be paid to term cancellations. Therefore, a more viable and inherently cancellation-free alternative are the recent **simplification during generation (SDG)** techniques [16,40,41] (see Fig. 9) that don't generate the exact expression but directly build the wanted simplified expression by generating the terms one by one in decreasing order of magnitude, until the approximation error is below the maximum user-supplied value. This problem can be formulated as finding the bases of the intersection of three matroids in decreasing order of magnitude, which however in general is an NP-complete problem. Applied to the two-graph method the three matroids involved are the graphic matroids corresponding to the voltage and current graph of the circuit and the partition matroid corresponding to the correct number of reactive elements needed to have a valid term of the frequency coefficient under generation. Fortunately polynomial-time algorithms have been developed to generate bases of the intersection of two matroids in descending order. For each base generated in this way it then has to be checked whether it is also a base in the third matroid. Such techniques have been implemented in SYMBA [41] and in RAINIER [16], and other developments are still going on. With these techniques large analog circuits can be analyzed with sizes corresponding to the current industrial practice (25 to 30 transistors). A benchmark example is the well known μA741 opamp.

Example

Another example, a fully differential BiCMOS opamp, is show in Fig. 10. The simplified expression of the gain of this opamp can (after factorization of the results) be obtained as :

$$A_{V0} = \frac{g_{m,M2}}{g_{m,M1}} \frac{g_{m,M4}}{\left(\frac{g_{o,M4} g_{o,M5}}{g_{m,M5} + g_{mb,M5}} + \frac{G_a + g_{o,M9} + g_{o,Q2}}{\beta_{Q2}} \right)} \quad (7)$$

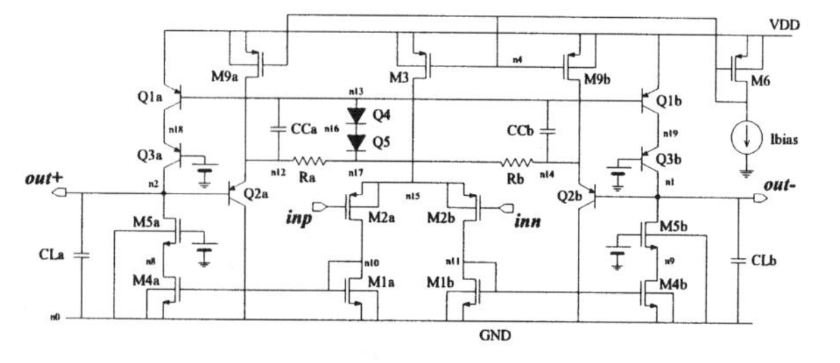

Fig. 10. BiCMOS fully differential opamp.

A more coarse approach in the same SDG line but that does not allow the user to accurately control the error was presented in [42]. The circuit elements are grouped in equivalence classes of the same order of magnitude, and the determinant is calculated by trying different combinations of equivalence classes in descending order of magnitude until the simplified symbolic expression is obtained as the first nonzero group of terms resulting from such combination.

AWEsymbolic [43], on the other hand, uses Asymptotic Waveform Evaluation (AWE) to produce a low-order symbolic approximation of the circuit response, both in time and frequency domain. Significant elements that have to be represented by a symbol are identified by means of numerical pole-zero sensitivity analysis in AWE. Numerical and symbolic computations are substantially decoupled by the use of moment-level partitioning. The method is most successful when only a small number of elements are represented by symbols.

4.2.2. Simplification before generation

In order to extend the capabilities of symbolic analysis to even larger circuits (40-50 transistors) that are still handled as one big flat circuit, the only possibility today is to introduce simplification on the circuit schematic, on the circuit matrix or the circuit graph(s) before the symbolic analysis starts. This technique is therefore called **simplification before generation (SBG)** (see Fig. 9) and corresponds to operations like a priori throwing away unimportant elements or shorting nodes, but also partial removals of elements become possible, for instance by removing only one or two of the entries out of the complete stamp of an element in an MNA-type matrix. Note that the principle of SBG is exactly the same as designers use during hand calculations, as the calculations become intractable otherwise anyway. The same also applies to a symbolic analysis tool. The golden rule to get interpretable results from a symbolic analysis run, is to simplify the circuit as much as possible in advance. A typical example is the replacement of biasing circuitry by a single voltage or current source with source impedance. Normally the devices in this biasing circuitry don't show up in the final simplified expressions, but they do complicate the symbolic calculations unnecessarily. The added value of SBG is that it quantifies the impact of every simplification. Yet controlling the overall error is still the most difficult part of these methods.

In Analog Insydes [13], the simplifications are performed on the system matrix and the Sherman-Morrisson theorem is used to calculate the influence of the simplifications. In SIFTER [15], a combination of SBG, SDG and SAG is used. First, circuit parameters are eliminated from the cofactor during determinant calculation provided that the removal causes an error smaller than the allowed margin. Next, dimension reduction is tried and heuristic row and column operations are applied to increase the sparseness of the matrix, after which the determinant is expanded and possibly factorized.

4.2.3. Hierarchical decomposition

If one wants to extend the capabilities of symbolic simulation to even larger circuits, it is clear that the generation of expanded expressions is no longer feasible or results in uninterpretable expressions, and that the only solution is to generate and keep the expressions in nested format. Due to the problem of term cancellations this approach is most useful only for circuits that consist of loosely coupled subcircuits, like for instance active filter structures.

An interesting method in this respect is the use of **hierarchical decomposition** [12,44,45,46]. The circuit is recursively decomposed into more or less loosely connected subcircuits. The lowest-level subcircuits are analyzed separately and the resulting symbolic expressions are combined bottom-up (in nested format, without expansion) according to the previously determined decomposition hierarchy, resulting in the global nested expression for the complete circuit. This is now illustrated in the following example.

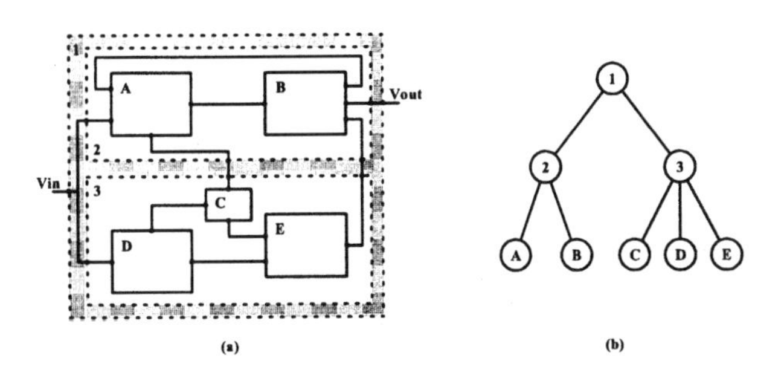

Fig. 11. (a) Hierarchical decomposition of a circuit, and (b) corresponding decomposition tree.

Example

Consider the theoretical example of Fig. 11a. The circuit (node 1 in the decomposition tree of Fig. 11b) is decomposed into two parts (corresponding to nodes 2 and 3), which are each decomposed again into the leaf subcircuits A and B (for node 2) and C, D and E (for node 3). The leaf subcircuits are then analysed by the symbolic simulator resulting in the following sets of symbolic transfer functions $\mathbf{H}_i$:

$$\mathbf{H}_A = f_A(s,\mathbf{p}_A) \tag{8}$$

$$\mathbf{H}_B = f_B(s,\mathbf{p}_B) \tag{9}$$

$$\mathbf{H}_C = f_C(s,\mathbf{p}_C) \tag{10}$$

$$\mathbf{H}_D = f_D(s,\mathbf{p}_D) \tag{11}$$

$$\mathbf{H}_E = f_E(s,\mathbf{p}_E) \tag{12}$$

where $\mathbf{p}_Y$ are the symbolic circuit parameters for leaf subcircuit Y. The transfer functions for the non-leaf subcircuits up the decomposition hierarchy can then be obtained without expansion in terms of the transfer functions of the composing subcircuits. For subcircuits 2 and 3, this is in terms of the above leaf transfer functions :

$$\mathbf{H}_2 = f_2(\mathbf{H}_A,\mathbf{H}_B) \tag{13}$$

$$\mathbf{H}_3 = f_3(\mathbf{H}_C,\mathbf{H}_D,\mathbf{H}_E) \tag{14}$$

The top-level transfer function of the complete circuit is then derived in terms of the transfer functions of the subcircuits 2 and 3 :

$$\mathbf{H}_1 = f_1(\mathbf{H}_2,\mathbf{H}_3) \tag{15}$$

and is therefore given as a sequence of small expressions having a hierarchical dependence on each other.

It is clear that the technique of hierarchical decomposition results in symbolic expressions in nested format, which are much more compact than expanded expressions. This results in a CPU time which increases about linearly with the circuit size [45], allowing the analysis of really large circuits, provided that the coupling between the different subcircuits is not too strong. Also the number of operations needed to numerically evaluate the symbolic expression is reduced because of the nested format. Hierarchical decomposition has been used in combination with a signal-flow-graph method in [45] and [46], and in combination with a matrix-based method in [12]. The calculation of senstivity functions in nested format has been presented in [47].

Following the use of binary decision diagrams in logic synthesis, determinant decision diagrams (DDD) have recently been proposed as a technique to canonically represent determinants in a compact nested format [48]. (Note that this has nothing to do with hierarchical decomposition as such, but merely with a compact representation format for the symbolic expressions.) The advantage is that all operations on these DDD's are linear with the size of the DDD, but the DDD itself is not always linear with the size of the circuit. This technique needs some further investigation before conclusions can be drawn about its usefulness.

None of the above mentioned programs, however, provides any approximation, which is essential to obtain insight in the behavior of semiconductor circuits. The human interpretation of nested expressions is also more difficult, especially if the nested expression is not fully factorized, which often is the case. In addition, the results of hierarchical decomposition are in general not cancellation-free, which further complicates the interpretation. A key issue here is to find efficient and reliable algorithms for carrying out the approximation of the nested expression but without complete expansion, also in the presence of term cancellations. Some attempts in this direction have already been presented. In [49], the symbolic expressions are generated in nested format whereby the terms are grouped and ordered in a way that eases the interpretation of the results. The decomposition of the circuit into preferably cascaded blocks becomes possible after a preceding rule-based simplification of the circuit structure (such as the lumping of like elements in series or parallel, or the application of the Norton-Thévenin transformation). Although the influence of neglected elements can be examined by means of the extra-element(s) theorem, this method does not offer a global view on the overal approximation error. Similarly, the lazy expansion technique [8] expands the nested expression term by term with the largest terms first until the error is below the required limit. Extensive care must be paid however to term cancellations. Another technique was presented in [50]. The numerically calculated contribution factors of each leaf-node expression to all frequency coefficients in the numerator and denominator of the total nested expression are used to determine with which individual error percentage each leaf node can be simplified on its own. The above approximation techniques are of course not needed for applications requiring only the numercial evaluation of the expressions, such as behavioral modeling or statistical analysis, where the compactness of the expression is needed to gain CPU time.

4.3. Symbolic pole/zero analysis

Algorithmic techniques to derive symbolic expressions for poles and zeros have recently been included in some symbolic analyzers, such as ASAP [5] or SANTAFE [51]. As closed-form solutions for poles and zeros can only be found for lower-order systems, the most straightforward approach is to make use of the pole-splitting hypothesis to come up with approximate symbolic expressions for the poles and zeros [5]. The use of such a capability together with advanced parametric graphical representations, however, is of significant help for interactive circuit improvement and design space exploration [6]. In SANTAFE symbolic Newton iterations are performed to calculate higher-order transfer functions in factorized pole/zero form [51].

4.4. Symbolic distortion analysis

Besides techniques for the symbolic analysis of linear(ized) circuits, also a technique has been developed and included in the ISAAC program [4] for the symbolic analysis of nonlinear circuits with multiple inputs but without hard nonlinearity (such as mixers and multipliers) [52], as well as for the symbolic analysis of harmonic distortion in weakly nonlinear circuits [53]. The method is based on the technique of Volterra series, where each higher-order response is calculated as a correction on the lower-order responses [53]. The function $i = f(v)$ that describes a nonlinear element is expanded into a power series around the operating point according to :

$$i = K_{1,f}v + K_{2,f}v^2 + K_{3,f}v^3 + \cdots \qquad (16)$$

where $K_{i,f}$ is the i th-order nonlinearity coefficient of f. This power series is truncated after the first three terms. The **probing method** [53] then allows to calculate the responses at the fundamental frequency, at the second harmonic and at the third harmonic by solving the same linear(ized) circuit but for different inputs. For the analysis of the second and third harmonic, the inputs are the nonlinearity-correction current sources associated with each nonlinear element. These current sources are placed in parallel with the linearized equivalent of the element and their value depends on the type of the nonlinearity, the order, the element's nonlinearity coefficients, the input frequencies and the lower-order solutions for the controlling voltage nodes. For example, the second-order correction current INL_2 for a capacitor C_x between nodes p and q is given by the following expression :

$$INL_{2,C_x} - 2j\omega\, K_{2,C_x}\left[H_{1,p}(j\omega) - H_{1,q}(j\omega)\right]^2 \qquad (17)$$

where K_{2,C_x} is the second-order nonlinearity coefficient for the capacitor and $H_{1,k}(j\omega)$ the 1st-order Volterra kernel for the voltage at node k (with respect to ground). This method has been extended towards the symbolic analysis of gain, noise and distortion of weakly nonlinear circuits with multiple inputs such as mixers and multipliers [52]. The expressions for the correction current sources are of course different and the solutions are a function of the frequencies ω_1 and ω_2 of the two input signals. The symbolic distortion analysis technique is now illustrated for a practical example.

Example

The second harmonic distortion at the output of the CMOS two-stage Miller-compensated opamp of Fig. 8 in open loop has been calculated by ISAAC [4] and is given by :

$$\frac{-A^2 K_{2,g_{m,M3}} g_{m,M1}^2 (g_{out}+sC_L)^2 (g_{o,M1}+g_{o,M2}+sC_C)}{(g_{out}(g_{o,M1}+g_{o,M2})+2sC_C g_{m,M3})(g_{out}(g_{o,M1}+g_{o,M2})+sC_C g_{m,M3})^2} \quad (18)$$

where A is the amplitude of the input signal. These analytic distortion expressions are almost impossible to obtain by hand, and provide invaluable information and insight to the designer. For a certain design point the results are also graphically plotted in Fig. 12 as a function of the frequency of the input signal. Fig. 12 shows the total harmonic distortion ratio (straight line), as well as the major contributions: $K_{2,gm3}$ (*), $K_{2,gm1}$ (+) and $K_{2,gm2}$ (o). All contributions are scaled to an input voltage amplitude of 1V (i.e. $A=1$).

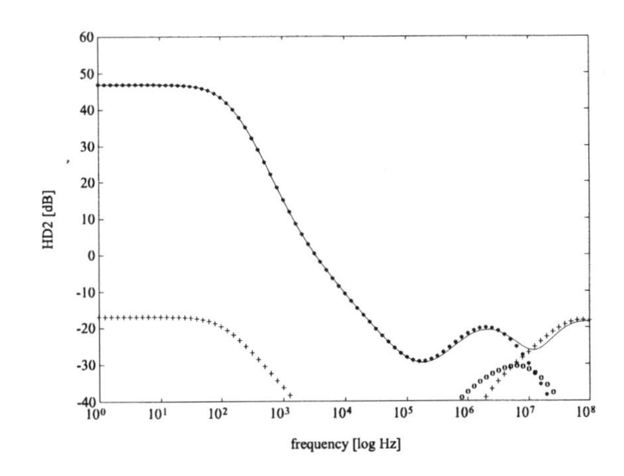

Fig. 12. Second harmonic distortion ratio at the output of the CMOS two-stage opamp in open loop versus the frequency of the input signal. The total harmonic distortion ratio is shown (straight line) as well as the major individual contributions.

4.5. Open research topics

At present and as far as functionality is concerned, the following topics are still open research areas in symbolic analysis of analog circuits (see Fig. 6). The analysis of switched-capacitor circuits has up till now only been carried out in the z-domain (such as for example in ISAAC [3,4], SCYMBAL [11], SSCNAP [17] or in [18]), excluding s-domain effects such as the finite opamp gain-bandwidth. Recently, a first step towards the combined symbolic analysis of s and z effects for a restricted class of circuits has been presented in [54].

Other unsolved topics at this moment are the symbolic analysis of large-signal behavior (e.g. slew rate), of time-domain behavior, and the symbolic analysis of strongly nonlinear circuits (if this will ever be feasible). A first approach towards symbolic time-domain simulation based on inverse Laplace transformation has been presented in [55]. But especially techniques for nonlinear analysis need to be developed in the near future.

Despite these yet unsolved topics, it can be concluded that the above reported algorithmic developments in symbolic analysis have recently created the breakthrough that was needed to make symbolic analysis feasible for circuits of practical size, at least for the linear(ized) behavior.

5. Comparison of symbolic simulators

The most prominent symbolic simulators for analog circuits existing nowadays are ISAAC [3,4], ASAP [5,6], SYNAP [7,8], SAPEC [9], SSPICE [10], SCYMBAL [11], SCAPP [12], Analog Insydes [13], CASCA [14] and GASCAP [21]. Although other tools have been published as well, the above programs are all available as stand-alone symbolic simulators. A comparison of the functionality offered by most of these programs and some implementation details are given in Table 1. ISAAC, ASAP, SYNAP and to some extent also SSPICE are targeted towards the symbolic analysis of analog integrated circuits, with a built-in small-signal linearization and symbolic approximation of the expressions. ISAAC and SCYMBAL offer the symbolic analysis of switched-capacitor circuits. SYNAP is the only program that offers an approximate dc analysis. SAPEC and GASCAP are more targeted towards the symbolic analysis of analog filters, whereas SCAPP is the only program that offers hierarchical analysis for large circuits. ISAAC is the only program that offers a symbolic distortion analysis for weakly nonlinear circuits, and ASAP and SSPICE offer an approximate symbolic pole/zero extraction.

Tools that include the most recent algorithmic developments such as simplification before and during generation offer similar functionality but can handle larger circuits. They include SIFTER [15], RAINIER [16] and SYMBA [56]. The SYMBA tool is currently reaching the commercial prototype stage and will offer a fully integrated environment for the symbolic analysis and modeling of analog circuits. It

	ISAAC	ASAP	SYNAP	SAPEC	SSPICE	SCYMBAL	SCAPP	GASCAP
analysis domains	s- & z-domain	s-domain	dc & s-domain	s-domain	s-domain	z-domain	s-domain	s-domain
primitive elements	complete	complete	complete	complete	limited	?	?	limited
small-signal linearization	yes	yes	yes	no	yes	no	no	no
mismatching	yes	yes	yes	no	no	no	no	no
approximation	yes	yes	yes	no	yes	no	no	no
weakly nonlinear analysis	yes	no	no	no	no	no	no	no
hierarchical analysis	no	no	no	no	limited	no	yes	no
symbolic pole/zero extraction	no	limited	no	no	limited	no	no	no
graphical interface	no	yes	no	yes	no	no	no	yes
formulation	CMNA	SFG	MNA	MNA	admittance matrix	SFG	RMNA & SFG	?
language	LISP	C	C++	LISP	C	FORTRAN	C	C

Table 1. Functional comparison between existing symbolic simulators for analog circuits.

will also be linked to a numerical simulator to extract the dc operating point information needed for the simplifications [56].

Besides functionality, a second criterion to compare symbolic analysis programs is their computational efficiency. A direct comparison, however, is extremely difficult here, because most CPU time values in the literature are published for different computing platforms (processor and memory configuration) and for different circuit examples. Therefore, a standard set of benchmark circuits should be agreed upon and used worldwide for all future comparisons. A second problem is that the different tools use different simplification techniques and expression formats (expanded versus nested) which make that reported CPU figures must be handled with care. Therefore no CPU times are reported in this tutorial, but the reader is advised to compare the efficiency of the different tools in the light of the accuracy and reliability of their symbolic results and in the light of his targeted application.

6. Conclusions

In this tutorial paper, an overview has been presented of the present state of the art in symbolic analysis of analog integrated circuits. The use for symbolic analysis as a complementary technique to numerical simulation has been shown. In the analog design world, the major applications of symbolic analysis are to obtain insight into the circuit's behavior, to generate analytic models for automated circuit sizing, in applications requiring the repetitive evaluation of characteristics (such as fault diagnosis) and in analog behavioral model generation and formal verification. For these applications, symbolic analysis is a basic technique that can be combined with powerful numerical algorithms to provide efficient solutions.

The basic methodology of symbolic circuit analysis has been described. The present capabilities and limitations of symbolic analysis, both in functionality and efficiency, have been discussed. Especially the algorithmic improvements realized over the past years (with the techniques of simplification before and during generation) have rendered symbolic analysis possible for circuits of practical size. Also the analysis of weakly nonlinear circuits has been developed.

Important future research topics in the field of symbolic analysis are the improvement of the postprocessing capabilities to enhance the interpretability of the generated expressions. Also, the potential of symbolic analysis in analog design automation, such as in analog design optimization, or in statistical and yield analysis, etc., has to be explored in much more detail. At the algorithmic level, analysis of nonlinear and time-domain characteristics are the key issues to be solved. If these issues can be solved, then commercial symbolic analysis tools might show up in the EDA marketplace in the near future.

Acknowledgements

The author acknowledges Prof. Willy Sansen and all Ph.D. researchers who have contributed to the progress in symbolic analysis research at the Katholieke Universiteit Leuven. Also support of several companies such as Philips Research Laboratories (NL) and Robert Bosch GmbH (D) as well as of ESPRIT AMADEUS is acknowledged.

References

[1] G. Gielen, P. Wambacq, W. Sansen, "Symbolic analysis methods and applications for analog circuits: a tutorial overview," Proceedings of the IEEE, Vol. 82, No. 2, pp. 287-304, February 1994.

[2] F. Fernández, A. Rodríguez-Vázquez, J. Huertas, G. Gielen, "Symbolic analysis techniques - Applications to analog design automation," IEEE Press, 1998.

[3] Willy Sansen, Georges Gielen, Herman Walscharts, "A symbolic simulator for analog circuits," proceedings International Solid-State Circuits Conference (ISSCC), pp. 204-205, 1989.

[4] G. Gielen, H. Walscharts, W. Sansen, "ISAAC: a symbolic simulator for analog integrated circuits," IEEE Journal of Solid-State Circuits, Vol. 24, No. 6, pp. 1587-1597, December 1989.

[5] F. Fernández, A. Rodríguez-Vázquez, J. Huertas, "A tool for symbolic analysis of analog integrated circuits including pole/zero extraction," proceedings European Conference on Circuit Theory and Design (ECCTD), pp. 752-761, 1991.

[6] F. Fernández, A. Rodríguez-Vázquez, J. Huertas, "Interactive AC modeling and characterization of analog circuits via symbolic analysis," Kluwer Journal on Analog Integrated Circuits and Signal Processing, Vol. 1, pp. 183-208, November 1991.

[7] S. Seda, M. Degrauwe, W. Fichtner, "A symbolic analysis tool for analog circuit design automation," proceedings International Conference on Computer-Aided Design (ICCAD), pp. 488-491, 1988.

[8] S. Seda, M. Degrauwe, W. Fichtner, "Lazy-expansion symbolic expression approximation in SYNAP," proceedings International Conference on Computer-Aided Design (ICCAD), pp. 310-317, 1992.

[9] S. Manetti, "New approaches to automatic symbolic analysis of electric circuits," IEE Proceedings part G, pp. 22-28, February 1991.

[10] G. Wierzba et al., "SSPICE - A symbolic SPICE program for linear active circuits," proceedings Midwest Symposium on Circuits and Systems, 1989.

[11] A. Konczykowska, M. Bon, "Automated design software for switched-capacitor IC's with symbolic simulator SCYMBAL," proceedings Design Automation Conference (DAC), pp. 363-368, 1988.

[12] M. Hassoun, P. Lin, "A new network approach to symbolic simulation of large-scale networks," proceedings International Symposium on Circuits and Systems (ISCAS), pp. 806-809, 1989.

[13] R. Sommer, E. Hennig, G. Drôge, E.-H. Horneber, "Equation-based symbolic approximation by matrix reduction with quantitative error prediction," Alta Frequenza - Rivista di Elettronica, Vol. 5, No. 6, pp. 29-37, November-December 1993.

[14] H. Floberg, S. Mattison, "Computer aided symbolic circuit analysis CASCA," Alta Frequenza - Rivista di Elettronica, Vol. 5, No. 6, pp. 24-28, November-December 1993.

[15] J. Hsu, C. Sechen, "DC small signal symbolic analysis of large analog integrated circuits," IEEE Transactions on Circuits and Systems, Part I, Vol. 41, No. 12, pp. 817-828, December 1994.

[16] Q. Yu, C. Sechen, "A unified approach to the approximate symbolic analysis of large analog integrated circuits," IEEE Transactions on Circuits and Systems, Part I, Vol. 43, No. 8, pp. 656-669, August 1996.

[17] B. Li, D. Gu, "SSCNAP: a program for symbolic analysis of switched capacitor circuits," IEEE Transactions on Computer-Aided Design, Vol. 11, No. 3, pp. 334-340, March 1992.

[18] M. Martins, A. Garção, J. Franca, "A computer-assisted tool for the analysis of multirate SC networks by symbolic signal flow graphs,"Alta Frequenza - Rivista di Elettronica, Vol. 5, No. 6, pp. 6-10, November-December 1993.

[19] P. Lin, "Symbolic network analysis," Elsevier, 1991.

[20] G. Gielen, W. Sansen, "Symbolic analysis for automated design of analog integrated circuits," Kluwer Academic Publishers, 1991.

[21] L. Huelsman, "Personal computer symbolic analysis programs for undergraduate engineering courses," proceedings International Symposium on Circuits and Systems (ISCAS), pp. 798-801, 1989.

[22] G. Gielen, H. Walscharts, W. Sansen, "Analog circuit design optimization based on symbolic simulation and simulated annealing," IEEE Journal of Solid-State Circuits, Vol. 25, No. 3, pp. 707-713, June 1990.

[23] H. Koh, C. Séquin, P. Gray, "OPASYN: a compiler for CMOS operational amplifiers," IEEE Transactions on Computer-Aided Design, Vol. 9, No. 2, pp. 113-125, February 1990.

[24] K. Swings, W. Sansen, "DONALD: a workbench for interactive design space exploration and sizing of analog circuits," proceedings European Design Automation Conference (EDAC), pp. 475-478, 1991.

[25] M. Degrauwe et al., "IDAC: an interactive design tool for analog CMOS circuits," IEEE Journal of Solid-State Circuits, Vol. 22, No. 6, pp. 1106-1116, December 1987.

[26] R. Harjani, R. Rutenbar, L. Carley, "OASYS: a framework for analog circuit synthesis," IEEE Transactions on Computer-Aided Design, Vol. 8, No. 12, pp. 1247-1266, December 1989.

[27] J. Vital, N. Horta, N. Silva, J. Franca, "CATALYST: a highly flexible CAD tool for architecture-level design and analysis of data converters," proceedings European Design Automation Conference (EDAC), pp. 472-477, 1993.

[28] F. Medeiro, B. Pérez-Verdú, A. Rodríguez-Vázquez, J. Huertas, "A vertically integrated tool for automated design of ΣΔ modulators," IEEE Journal of Solid-State Circuits, Vol. 30, No. 7, pp. 762-772, July 1995.

[29] G. Gielen et al., "An analog module generator for mixed analog/digital ASIC design," John Wiley International Journal of Circuit Theory and Applications, Vol. 23, pp. 269-283, July-August 1995.

[30] M. Degrauwe et al., "The ADAM analog design automation system," proceedings ISCAS, pp.†820-822, 1990.

[31] A. Konczykowska, M. Bon, "Analog design optimization using symbolic approach," proceedings International Symposium on Circuits and Systems (ISCAS), pp. 786-789, 1991.

[32] A. Konczykowska, P. Rozes, M. Bon, W. Zuberek, "Parameter extraction of semiconductor devices electrical models using symbolic approach," Alta Frequenza - Rivista di Elettronica, Vol. 5, No. 6, pp. 3-5, November-December 1993.

[33] A. Konczykowska, M. Bon, "Symbolic simulation for efficient repetitive analysis and artificial intelligence techniques in C.A.D.," proceedings International Symposium on Circuits and Systems (ISCAS), pp. 802-805, 1989.

[34] S. Manetti, M. Piccirilli, "Symbolic simulators for the fault diagnosis of nonlinear analog circuits," Kluwer Journal on Analog Integrated Circuits and Signal Processing, Vol. 3, No. 1, pp. 59-72, January 1993.

[35] R. Carmassi, M. Catelani, G. Iuculano, A. Liberatore, S. Manetti, M. Marini, "Analog network testability measurement: a symbolic formulation approach," IEEE Transactions on Instrumentation and Measurement, Vol. 40, pp. 930-935, December 1991.

[36] C. Borchers, L. Hedrich, E. Barke, "Equation-based behavioral model generation for nonlinear analog circuits," proceedings Design Automation Conference (DAC), pp. 236-239, 1996.

[37] L. Hedrich, E. Barke, "A formal approach to verification of linear analog circuit with parameter tolerances," proceedings Design and Test in European Conference (DATE), pp. 649-654, 1998.

[38] P. Lin, "A survey of applications of symbolic network functions," IEEE Transactions on Circuit Theory, Vol. 20, No. 6, pp. 732-737, November 1973.

[39] F. Fernández, A. Rodríguez-Vázquez, J. Martin, J. Huertas, "Formula approximation for flat and hierarchical symbolic analysis," Kluwer Journal on Analog Integrated Circuits and Signal Processing, Vol. 3, No. 1, pp. 43-58, January 1993.

[40] P. Wambacq, G. Gielen, W. Sansen, F. Fernández, "Approximation during expression generation in symbolic analysis of analog ICs," Alta Frequenza - Rivista di Elettronica, Vol. 5, No. 6, pp. 48-55, November-December 1993.

[41] P. Wambacq, F. Fernández, G. Gielen, W. Sansen, A. Rodríguez-Vázquez, "Efficient symbolic computation of approximated small-signal characteristics," IEEE Journal of Solid-State Circuits, Vol. 30, No. 3, pp. 327-330, March 1995.

[42] M. Amadori, R. Guerrieri, E. Malavasi, "Symbolic analysis of simplified transfer functions," Kluwer Journal on Analog Integrated Circuits and Signal Processing, Vol. 3, No. 1, pp. 9-29, January 1993.

[43] J. Lee, R. Rohrer, "AWEsymbolic: compiled analysis of linear(ized) circuits using Asymptotic Waveform Evaluation," proceedings Design Automation Conference (DAC), pp. 213-218, 1992.

[44] J. Smit, "A cancellation-free algorithm, with factoring capabilities, for the efficient solution of large sparse sets of equations," Proceedings SYMSAC, pp. 146-154, 1981.

[45] J. Starzyk, A. Konczykowska, "Flowgraph analysis of large electronic networks," IEEE Transactions on Circuits and Systems, Vol. 33, No. 3, pp. 302-315, March 1986.

[46] M. Hassoun, K. McCarville, "Symbolic analysis of large-scale networks using a hierarchical signal flowgraph approach," Kluwer Journal on Analog Integrated Circuits and Signal Processing, Vol. 3, No. 1, pp. 31-42, January 1993.

[47] P. Lin, "Sensitivity analysis of large linear networks using symbolic programs," proceedings International Symposium on Circuits and Systems (ISCAS), pp. 1145-1148, 1992.

[48] R. Shi, X. Tan, "Symbolic analysis of large analog circuits with determinant decision diagrams," proceedings International Conference on Computer-Aided Design (ICCAD), pp. 366-373, 1997.

[49] F. Dorel, M. Declercq, "A prototype tool for the design oriented symbolic analysis of analog circuits," proceedings Custom Integrated Circuits Conference (CICC), pp. 12.5.1-12.5.4, 1992.

[50] F. Fernández, A. Rodríguez-Vázquez, J. Martin, J. Huertas, "Approximating nested format symbolic expressions," Alta Frequenza - Rivista di Elettronica, Vol. 5, No. 6, pp. 29-37, November-December 1993.

[51] G. Nebel, U. Kleine, H.-J. Pfleiderer, "Symbolic pole/zero calculation using SANTAFE," IEEE Journal of Solid-State Circuits, Vol. 30, No. 7, pp. 752-761, July 1995.

[52] P. Wambacq, J. Vanthienen, G. Gielen, W. Sansen, "A design tool for weakly nonlinear analog integrated circuits with multiple inputs (mixers, multipliers)," proceedings Custom Integrated Circuits Conference (CICC), pp. 5.1.1-5.1.4, 1991.

[53] P. Wambacq, G. Gielen, W. Sansen, "Symbolic simulation of harmonic distortion in analog integrated circuits with weak nonlinearities," proceedings International Symposium on Circuits and Systems (ISCAS), pp. 536-539, 1990.

[54] Z. Arnautovic, P. Lin, "Symbolic analysis of mixed continuous and sampled-data systems," proceedings International Symposium on Circuits and Systems (ISCAS), pp. 798-801, 1991.

[55] M. Hassoun, B. Alspaugh, S. Burns, "A state-variable approach to symbolic circuit simulation in the time domain," proceedings International Symposium on Circuits and Systems (ISCAS), pp. 1589-1592, 1992.

[56] C. Baumgartner, "AMADEUS - Analog modeling and design using a symbolic environment," proceedings 4th international workshop on Symbolic Methods and Applications in Circuit Design (SMACD), Leuven, 1996.

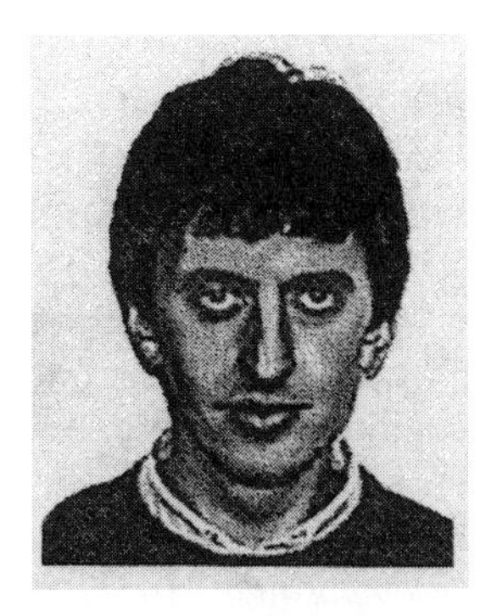

Georges G.E. GIELEN (Senior Member, IEEE) received the MSc and PhD degrees in Electrical Engineering from the Katholieke Universiteit Leuven, Belgium, in 1986 and 1990, respectively. In 1990, he was appointed as a postdoctoral research assistant and visiting lecturer at the department of Electrical Engineering and Computer Science of the University of California, Berkeley, U.S.A. From 1991 to 1993, he was a postdoctoral research assistant of the Belgian National Fund of Scientific Research at the ESAT-MICAS laboratory of the Katholieke Universiteit Leuven. In 1993, he was appointed as a tenure research associate of the Belgian National Fund of Scientific Research and at the same time as an assistant professor at the Katholieke Universiteit Leuven. In 1995 he promoted to associate professor and in 2000 to full-time professor at the same university.

His current research interests are in the design of analog and mixed-signal integrated circuits, and especially in analog and mixed-signal CAD tools and design automation (modeling, simulation and symbolic analysis, analog synthesis, analog layout generation, analog and mixed-signal testing). He has authored or coauthored two books and more than 100 papers in edited books, international journals and

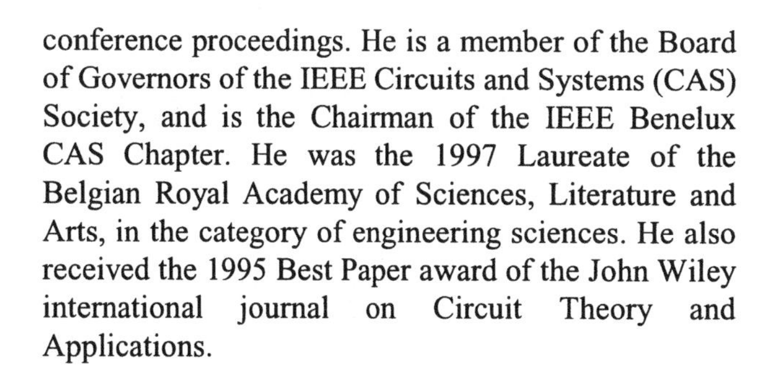

conference proceedings. He is a member of the Board of Governors of the IEEE Circuits and Systems (CAS) Society, and is the Chairman of the IEEE Benelux CAS Chapter. He was the 1997 Laureate of the Belgian Royal Academy of Sciences, Literature and Arts, in the category of engineering sciences. He also received the 1995 Best Paper award of the John Wiley international journal on Circuit Theory and Applications.

Flowgraph Analysis of Large Electronic Networks

JANUSZ A. STARZYK, SENIOR MEMBER, IEEE, AND A. KONCZYKOWSKA

Abstract —**The paper presents a new method for signal flowgraph analysis of large electronic networks. A hierarchical decomposition approach is realized using the so-called upward analysis of the decomposed network. This approach allows fully symbolic network formulas to be obtained in time linearly proportional to the size of the network. A multiconnection characterization, suitable for upward analysis, has been defined and used in topological formulas. Examples of large scale networks analysis are discussed. The approach can be used to obtain symbolic solutions of linear systems of equations.**

I. Introduction

THE NOTION of *topological analysis* of electrical networks is concerned with the determination of the network characteristics from the knowledge of elements and their connections (*network topology*) without applying numerical methods. As a result, for linear, lumped, stationary (LLS) networks, the transfer functions (defined as the ratio of the Laplace transform of the output to the input signals under zero initial state) are obtained.

Topological methods, independently of the graph representation used, allow a network transfer function to be obtained in a rational function form. The numerator and denominator of this function are expressed as a sum of products of edges weights [3]

$$K(s)=\frac{L(s)}{M(s)}=\frac{\sum\limits_{1}\prod\limits_{i} y_i}{\sum\limits_{m}\prod\limits_{j} y_j}. \tag{1}$$

These weights depend directly on the type and value of network elements.

Realization of topological formulas requires the knowledge of all graph connections [3]. To make the computations efficient, one should use the algorithms which generate connections rapidly and without duplication. Only in this case, it is possible not to check any new connection with all previously generated ones. There are many efficient algorithms to generate graph connections [20], [25], [27]. They form the basis of topological analysis programs intended for small linear networks [14], [16].

Direct application of topological formulas permits the analysis of networks with graphs having approximately 10 nodes [6]. This limitation is not the result of the low efficiency of generation algorithms but of great number of terms in topological formulas for determinant of a coefficient matrix [3]. Even if we could generate all terms in zero time, the time needed for weights evaluation would grow at least proportionally to the number of terms, and for relatively small networks (with about 20 nodes), will attain enormous values. In any case, it is obvious that application of topological formulas for networks having more than 10 nodes is much more time consuming than the methods of symbolic analysis based on the numerical techniques of determinant evaluation [2], [26].

For these reasons the methods of topological analysis were judged by McCalla and Pederson as completely inefficient [15]. Nevertheless, research in this area has been carried out [1], [17]–[19].

Attempts to introduce methods of graph reduction [4], [9], [11] or decomposition [5], [22] to the analysis did not provide universally efficient programs and were deemed unacceptable in a paper by Alderson and Lin [2].

An important development has been achieved with the introduction of *hierarchical decomposition*. In [24] the method of signal flowgraph analysis has been presented and in [23] the hierarchical analysis of directed graphs has been discussed. Based on both these methods and downward decomposition, a program for topological analysis of large networks has been successfully developed [12].

Further improvement was attained when the *upward hierarchical method* was introduced [13]. The details of the latter method will be presented in this paper. Our goal is to reduce the time consumption from the involution dependence for the previous (downward) decomposition to the linear dependency. We only consider Coates flowgraph representation of the network [3]. A similar approach is possible with other representations (e.g., unistor graph [8]).

II. Topological Formulas

The form of topological formulas is different for direct analysis and analysis with decomposition. It depends on the type of partition and on the kind of topological representation. In practice two-terminal immittances and two-port transfer functions are the most frequently calculated. As the basis for topological dependencies we consider evaluation of network immittances and transfer functions expressed by the determinant and cofactors of the nodal admittance matrix, as discussed in [3].

Manuscript received August 10, 1983; revised July 22, 1985.
J. A. Starzyk is with the Department of Electrical and Computer Engineering, Ohio University, Athens, OH 45701.
A. Konczykowska is with the Centre National d'Etudes des Telecommunication, 92220 Bagneux, France.
IEEE Log Number 8406682.

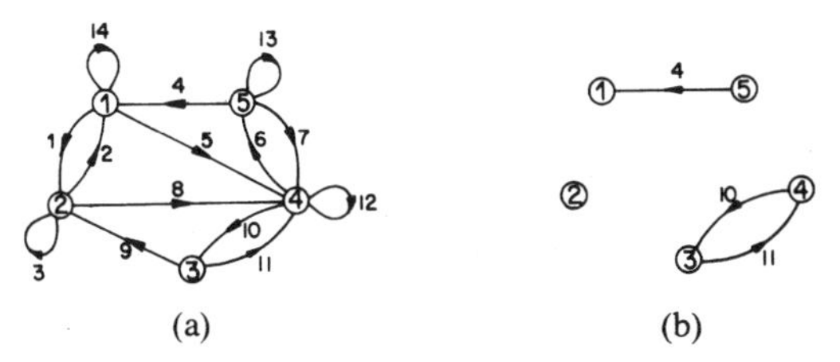

(a) (b)

Fig. 1. (a) Flowgraph. (b) 1-connection.

Let us denote W—a set of pairs of nodes in the Coates graph G_c

$$W=\{(v_1, r_1),\cdots,(v_k, r_k)\}, \qquad v_l \neq v_m,\ v_l \neq r_m,\ r_l \neq r_m \quad \text{for } l \neq m.$$

Definition 1

We call a *k-connection* (multiconnection) of graph G_c a subgraph p_W, composed of k node-disjoint directed paths and node-disjoint directed loops incident with all graph nodes. The initial node of ith path is v_i and the terminal node is r_i (pairs of nodes from the set W).

In Fig. 1 a flowgraph and its 2-connection p_W is presented. In this case $W=\{(5,1),(2,2)\}$. A 0-connection or simply a connection is denoted by p, because $W=0$. When $v_i=r_i$, a multiconnection has the isolated node v_i. A multiconnection is a natural generalization of terms "connection" and "1-connection" defined by Coates [7] and is useful for the topological analysis of a decomposed network. This notion corresponds to that of a *k-tree* (multitree) occurring in the analysis (with decomposition) when the representation with a pair of conjugated graphs or a directed graph is used. A tree can be obtained from the k-tree by adding $k-1$ edges. Similarly, a k-connection can be transformed into a connection by adding k edges. A set of all k-connections p_W will be denoted by P_W.

Definition 2

The weight function of $|P_W|$ of a multiconnection set P_W of a Coates graph with n nodes is defined as follows:

$$|P_W| = \sum_{p \in P_W} \operatorname{sign} \prod_{e \in p} y_e \tag{2}$$

where

$$\operatorname{sign} p = (-1)^{n+k+l_p} \operatorname{ord}(v_1,\cdots,v_k) \operatorname{ord}(r_1,\cdots,r_k)$$

$$\operatorname{ord}(x_1, x_2,\cdots,x_k) = \begin{cases} 1, & \text{when the number of permutations ordering the set is even} \\ -1, & \text{otherwise} \end{cases}$$

n number of graph nodes,
l_p number of loops in multiconnection p,
y_e weight of an element e.

Consider a flowgraph of a two-port network shown in Fig. 2. Let Y be an indefinite admittance matrix of the two-port. Denote Y_{uv} the first-order cofactor of Y as

$$Y_{uv} = (-1)^{u+v} \det \mathbf{Y}_{uv} \tag{3}$$

where $\mathbf{Y}_{uv}$ is the submatrix obtained from Y by deleting the uth row and vth column. The second order cofactor

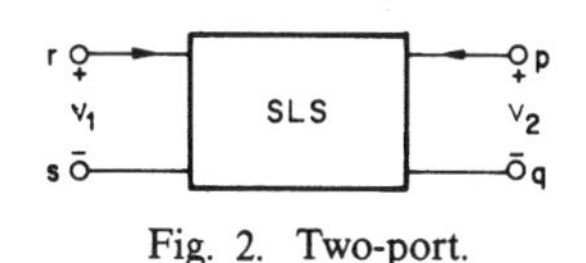

Fig. 2. Two-port.

$Y_{rp,ss}$ is defined as

$$Y_{rp,ss} = \operatorname{sgn}(r-s) \operatorname{sgn}(p-s) (-1)^{r+p} \det \mathbf{Y}_{rp,ss}, \quad r \neq s,\ p \neq s \tag{4}$$

where $\mathbf{Y}_{rp,ss}$ is the submatrix obtained from $\mathbf{Y}$ by deleting rows r and s and columns p and s. Similarly we define the third-order cofactor $Y_{pq,rr,ss}$ ($p \neq s$, $q \neq s$, $r \neq s$, $p \neq r$, $q \neq r$). Using these cofactors we can obtain formulas for transfer functions of the two-port (see [3]).

Theorem 1

Cofactors of the indefinite admittance matrix of a given multiterminal network can be expressed by the weight functions of multiconnection sets as follows:

$$Y_{uv} = |P_{\{(s,s)\}}| \tag{5}$$

$$Y_{rp,ss} = |P_{\{(r,p),(s,s)\}}| \tag{6}$$

$$Y_{pq,rr,ss} = |P_{\{(p,q),(r,r),(s,s)\}}|. \tag{7}$$

Proof:

If the Coates graph is based on $n \times n$ indefinite admittance matrix $\mathbf{Y}=[y_{ij}]$, then its edge directed from node x_j to node x_i has the weight equal to y_{ij}. We have

$$\mathbf{Y} = \lambda_- \mathbf{Y}_e \lambda_+^T \tag{8}$$

where the element ij of λ_- is equal to 1 if the jth edge is directed towards the ith node and zero otherwise, and the element ij of λ_+ is equal to 1 if the jth edge is directed away from the ith vertex and is zero otherwise, and $\mathbf{Y}_e$ is a diagonal matrix of element admittances.

The submatrix $\mathbf{Y}(A|B)$ obtained from $\mathbf{Y}$ by deleting rows represented by the set of nodes A and columns represented by the set B can be written in the form

$$\mathbf{Y}(A|B) = \lambda_{-A} \mathbf{Y}_e \lambda_{+B}^T \tag{9}$$

where $\lambda_{-A}(\lambda_{+B})$ is obtained from $\lambda_-(\lambda_+)$ by deleting rows $A(B)$, respectively. According to the Binet–Cauchy theorem [10] and relation (9), we have

$$\det \mathbf{Y}(A|B) = \sum \det \mathbf{C}^- \det \mathbf{C}^+ \tag{10}$$

where $\mathbf{C}^-$ is a major submatrix of $\lambda_{-A}\mathbf{Y}_e$ with order equal to (n-card A) and $\mathbf{C}^+$ is the corresponding major submatrix of λ_{+B}^T. A major determinant of $\lambda_{-A}\mathbf{Y}_e$ is different from zero if and only if there exists one nonzero element in every row of the chosen submatrix $\mathbf{C}^-$. This corresponds to the set of (n-card A) edges, such that every edge has a different terminal node from the set of nodes $(N-A)$, where N indicates the set of all nodes of the Coates graph. The corresponding submatrix $\mathbf{C}^+$ is different from zero if the same edges have different initial nodes from the set of nodes $(N-B)$. Now it is easy to check that these edges form a multiconnection p_W, such that if $(v_i, r_i) \in W$ then $v_i \in A$ and $r_i \in B$ (see Fig. 3). Formulas (5), (6), and (7) follow from this general observation. □

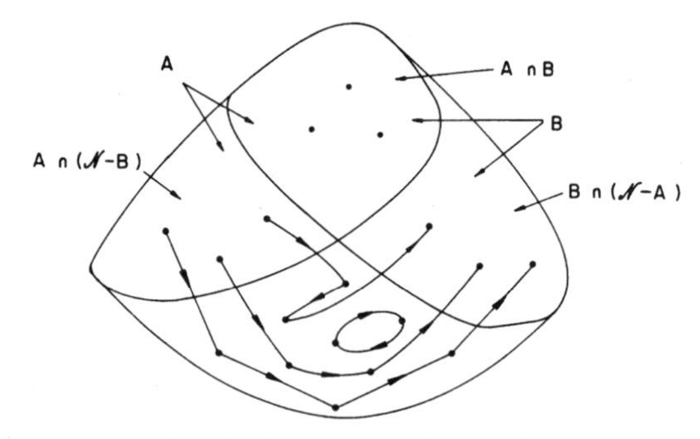

Fig. 3. Required multiconnections.

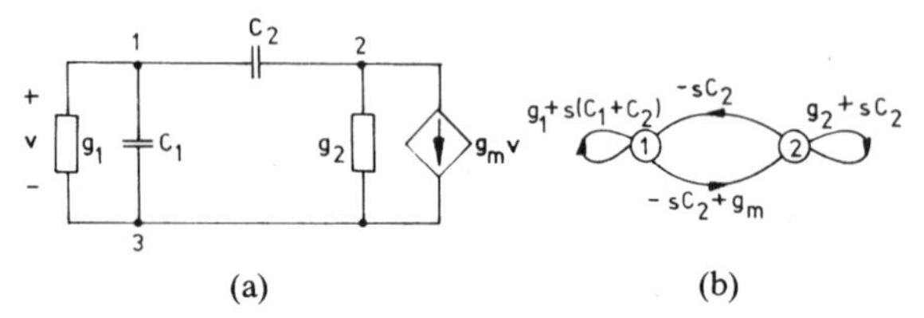

(a) (b)

Fig. 4. (a) Network. (b) Coates graph.

$\boldsymbol{Y}$ is a singular matrix and only its cofactors are used to evaluate network functions. From (5), (6) and (7) it can be seen that multiconnections used to evaluate cofactors of $\boldsymbol{Y}$ contain s as an isolated node. We can treat this node as a reference and delete all edges incident to it. From now on the Coates graph of a network will be assumed with the reference node deleted.

Example 1

An active linear network and its Coates graph with node 3 deleted are shown in Fig. 4. The indefinite admittance matrix is

$$\boldsymbol{Y}=\begin{bmatrix} g_1+s(C_1+C_2) & -sC_2 & -g_1-sC_1 \\ -sC_2+g_m & g_2+sC_2 & -g_2-g_m \\ -g1-sC_1-g_m & -g_2 & g_1+g_2+sC_1+g_m \end{bmatrix}. \tag{11}$$

It is easy to confirm that

$$\begin{aligned} \boldsymbol{Y}_{uv} &= |P_{\{(s,s)\}}| \\ &= (sC_2+g_2)\,(g_1+s(C_1+C_2)) \\ &\quad + sC_2(g_m-sC_2) \end{aligned}$$

for any u,v and

$$\boldsymbol{Y}_{12,33}=|P_{\{(1,2),(3,3)\}}|=-(g_m-sC_2).$$

Computer time needed for realization of direct topological formulas is proportional to the number of connections in a flowgraph. Let $\boldsymbol{D}(G)=[d_{ij}]$ be a matrix denoting the connection of a Coates flowgraph; d_{ij} is equal to the number of edges directed from the node i to the node j. $\boldsymbol{D}$ is a square matrix with the dimension equal to the number of nodes. The *number of connections* in a graph is equal [4] to

$$\text{card } P = \text{ per }[\boldsymbol{D}(g)] \tag{12}$$

where per $\boldsymbol{A}$ is a permanent of the matrix $\boldsymbol{A}$ [3].

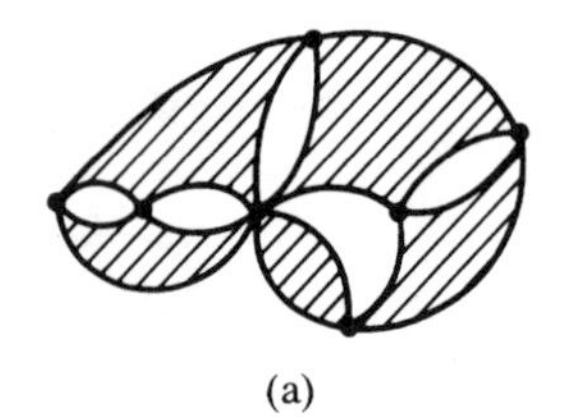
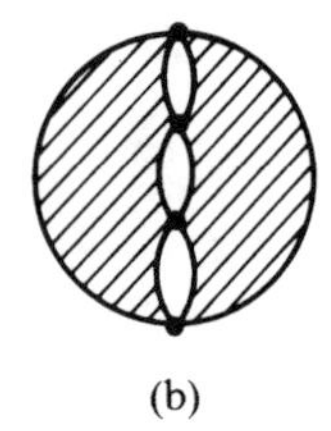

(a) (b)

Fig. 5. (a) Node decomposition. (b) Four terminal bisection.

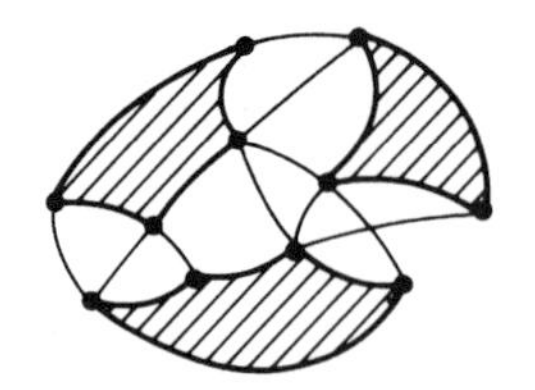

Fig. 6. Edge decomposition.

A very rough estimation for the number of connections for the graph with n nodes and k edges is given by [24]

$$\text{card } P \leqslant \left(\frac{k+1}{n}-1\right)^n. \tag{13}$$

Although (13) is only an upper estimate, it expresses correctly the rate of change in the number of terms. The exponential increase in the number of terms is observed in practice for direct topological analysis, which causes such analysis of large networks to be inexecutable.

III. The Graph Decomposition

The graph of an electrical network can be analyzed directly with the aid of (5)–(7), and the transfer function of the analyzed network can be obtained in all symbolic form. From the previous discussion, it is evident that the number of terms in the symbolic function is too large. As a result, the analysis of medium and large networks is a formidable task; network and graph decomposition becomes necessary.

The procedure of graph partition and determination of parts called *blocks* will be called decomposition.

A flowgraph can be decomposed in one of the three manners.

1) Node Decomposition: A graph is divided into edge disjoint subgraphs (blocks) (Fig. 5). Nodes common to two or more blocks are called *block nodes*. A particular case of node decomposition is bisection or decomposition into two subgraphs.

2) Edge decomposition: In a graph we isolate node disjoint blocks. Blocks are connected together by the means of edges which form cutsets of the graph (Fig. 6). These edges are called *cutting edges*. In the case of edge decomposition, nodes incident with cutting edges are called block nodes.

3) Hybrid Decomposition: This partition is a combination of two previous decompositions (Fig. 7). Nodes incident with cutting edges or common for more than one block are block nodes.

In both edge and hybrid decompositions, a bisection can be distinguished as a special case. We focus our attention

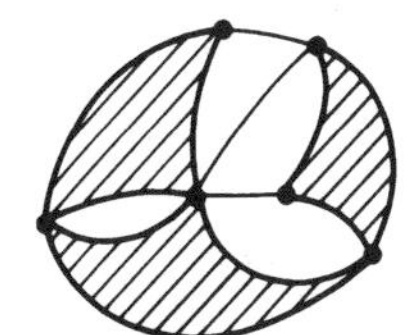

Fig. 7. Hybrid decomposition.

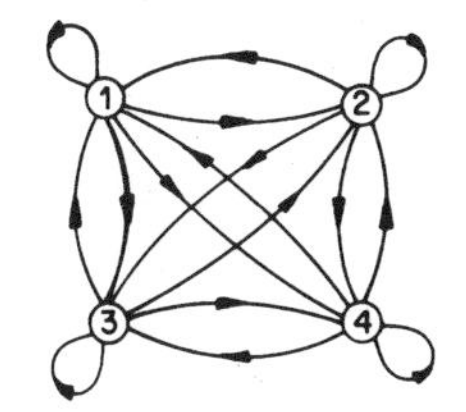

Fig. 8. Substitute graph spanned on four block nodes.

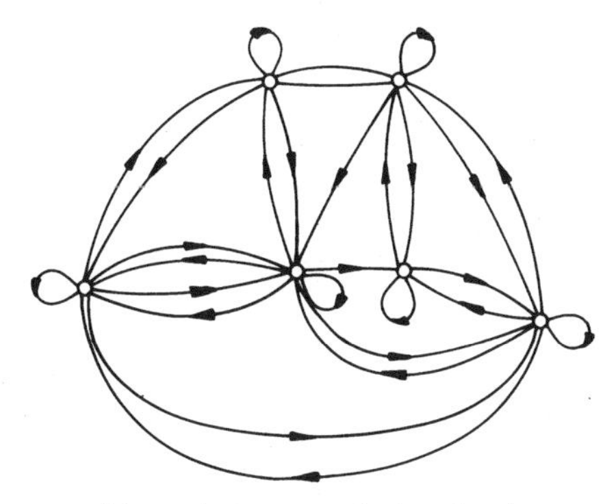

Fig. 9. Decomposition substitute graph for the decomposition from Fig. 7.

on bisection because of its special usefulness for the hierarchical decomposition. It is evident that any decomposition can be represented as a sequence of bisections, and for computer algorithms such an assumption produces simple data structures and simplifies organization of computations.

Definition 3

A complete symmetrical directed graph with self loops spanned on block nodes of subgraph G_i is called a *substitute graph* for that block and is denoted G_i^s (Fig. 8).

Definition 4

Graph G^d obtained when replacing blocks G_i by their substitute graphs is called a *decomposition substitute graph* (Fig. 9).

In the case of edge or hybrid decompositions, cutting edges belong to the decomposition substitute graph.

A decomposition substitute graph should not be too complex because the complexity of its analysis depends on the number of edges and nodes exactly as estimated for the case of proper graph (12), (13). Hence, it appears necessary to limit the number of blocks and block nodes. This limitation results in the *simple decomposition* method being ineffective for the case of very large networks. For large networks either the decomposition substitute graph G^d is too complex for analysis or blocks G_i are still too complex for direct topological analysis.

When simple decomposition is applied to subgraphs, we deal with *hierarchical decomposition*.

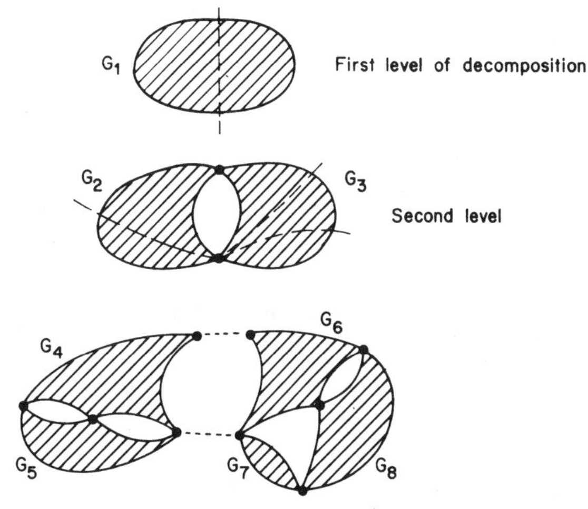

Fig. 10. Two level hierarchical decomposition.

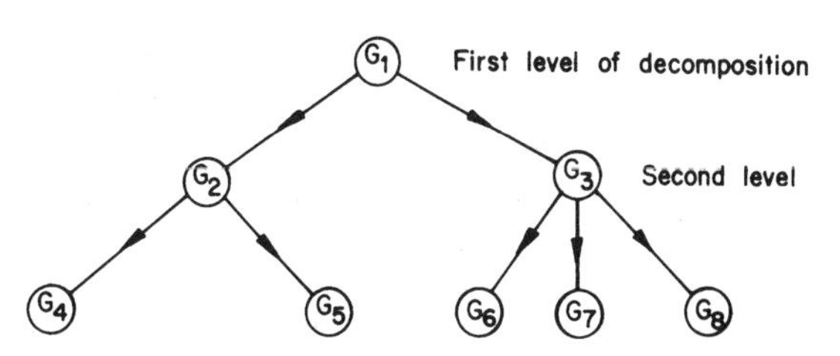

Fig. 11. Tree of decomposition shown in Fig. 10.

Decomposition of a network graph should be executed automatically. There are two reasons for this. First, the graph structure is not known when network data are provided and an *a priori* decision about block partition regarding only network structure could be nonoptimal. Second, elaboration of data would be cumbersome for the program user and would demand the knowledge of decomposition methods and calculations regardless of whether the partition is desirable or not.

The problem of graph decomposition is of the nonpolynomially bounded class. This means that time τ to find an optimal decomposition cannot be limited by a polynomial expressed in terms of nodes (n) or edges (k) number.

Taking the above into account we should not expect an efficient algorithm giving optimal solutions. Useful algorithms will provide a correct and nearly optimal solution in a short time. One such efficient algorithm has been presented in [21]. A modification of this algorithm gives the time of graph decomposition bounded linearly by the number of nodes.

IV. Hierarchical Analysis

Let us concentrate first on the case of node hierarchical decomposition. In Fig. 10 an example of hierarchical decomposition is presented. The hierarchical decomposition structure can be illustrated by a tree of decomposition. Nodes of the tree correspond to subgraphs obtained on different levels of decomposition. If a subgraph G_k was obtained during decomposition of subgraph G_i, then there is an edge from node G_i to node G_k. Fig. 11 shows the tree of decomposition from Fig. 10.

In the decomposition tree we have one *initial node* which is the root of the tree. *Terminal nodes* are leaves of the tree. All nodes that are not terminal nodes are *middle nodes*. For middle nodes we determine the *decomposition level* which is equal to the number of nodes in the path from the initial node to that node. *Range of hierarchical decomposition* is equal to the maximal decomposition level.

Every middle node has its *descendants* and every node except the initial one has its *ascendant*. If we limit ourselves to the bisection as the only graph partition, every middle node has exactly two descendants. As remarked previously, every decomposition can be considered as a sequence of bisections in hierarchical structure. Hence, without loss of generality, we shall examine this case only, obtaining a simpler expression of formulas and easier algorithm organization.

During the course of hierarchical decomposition analysis the following tasks are to be performed:

a) direct topological analysis of terminal blocks, and

b) analysis of middle blocks used to combine results from the higher level.

Analysis of Terminal Block

Let us consider a connection of a Coates graph. When we deal with a decomposed graph we can see that the part of the connection contained in a particular terminal block forms a multiconnection in this block.

The incidence of the block nodes determines the type of multiconnection. It means that topological analysis of terminal blocks will consist of enumeration of multiconnections, with paths linking different block nodes. Analysis of middle blocks will consist of combining together various types of multiconnections.

It is evident that combining multiconnections one by one will not reduce the computation time considerably. Multiconnections should be generated in groups and whole groups should be combined together. The larger the groups of multiconnections are the simpler the terminal block analysis is, and the more efficient middle block analysis is. One rule should be obeyed, namely, the resulting multiconnections should be generated without duplications.

The most detailed characterization is that presented in Definition 1, which is the generalization of Coates definition of 0- and 1-connections. For a block the different multiconnections may be grouped in sets P_W, of multiconnections characterized by the same set of nodes W.

However, it should be noted that a block with nb block nodes has

$$M(nb) = \sum_{i=0}^{nb} \binom{nb}{i}^2 i! \tag{14}$$

different types of multiconnection sets. This dependence could seriously limit the decomposition method. This led us to investigate other characterizations of multiconnection sets. After some trials [12], [13] the following type of characterization was chosen.

Definition 5

$P(B, E)$ is a set of multiconnections which have the following properties:

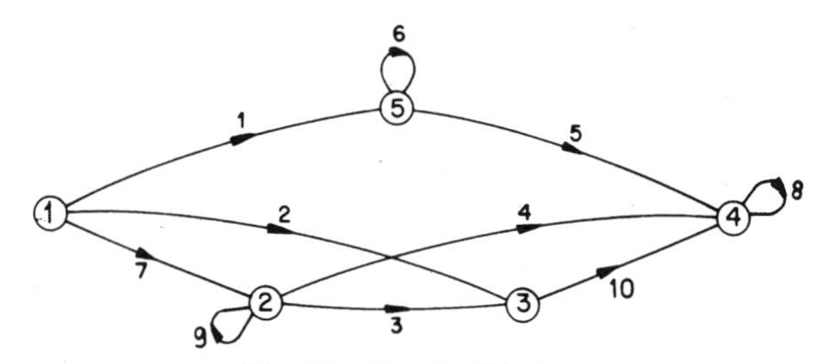

Fig. 12. Terminal block.

a) the incidence of block nodes is defined by sets B and E only—where B represents initial nodes and E represents terminal nodes of multiconnection edges;

b) all other nodes (internal nodes of a block) have full incidences, i.e., they are initial nodes as well as terminal nodes of multiconnection edges, where $B = \{b_1, b_2, \cdots, b_m\}$, $E = \{e_1, e_2, \cdots, e_m\}$ and $B \cup E \subset NB$—the set of block nodes.

Remark 1

Block nodes which are not included in $B \cup E$ are isolated nodes. Block nodes which are included in $B \cap E$ have full incidence.

Remark 2

In the sense of Definition 5 the set $P(B, E)$ contains k-connections with $k = \text{card}\,(NB - B \cap E)$.

Example 2

Let us consider the block shown in Fig. 12. If $NB = \{1,2,3,4\}$, $B = \{1,2\}$, $E = \{3,4\}$, then the set $P(B, E)$ is equal to $\{\{1,5,3\}, \{2,4,6\}\}$. According to the Definition 1 each set of these multiconnections has different characterizations by sets W

$$\{1,5,3\} \in P_{\{(1,4),(2,3)\}}$$
$$\{2,4,6\} \in P_{\{(1,3),(2,4)\}}.$$

For another pair $B = \{1,2\}$, $E = \{2,3\}$ with the common node 2, the set of multiconnections $P(B, E)$ is equal to $\{\{2,6,9\}, \{3,6,7\}\}$. In this case node 4 is isolated. The equivalent characterization by sets W, according to Definition 1, is as follows:

$$\{\{2,6,9\}, \{3,6,7\}\} = P\{(1,3),(4,4)\}.$$

The weight function of multiconnection set $P(B, E)$ is defined as in (2). Note that for a block with nb block nodes, the number of different types of multiconnections sets $P(B, E)$ is

$$MR(nb) = \sum_{i=0}^{nb} \binom{nb}{i}^2 \tag{15}$$

which means an important reduction in comparison with (14). It will be shown that multiconnections characterized by sets $P(B, E)$ can be generated without duplication.

Analysis of Middle Block

Analysis on an intermediate level consists of evaluation of multiconnections of a block that result from the association of two (or in general, more) blocks. Let us denote the sets of block nodes for both blocks and the resulting block by NB_1, NB_2 and NB, respectively. When connecting two blocks, some of their block nodes become internal nodes, which means that no other blocks are connected to these nodes on upper levels. These nodes will be called *reducible nodes*.

Let us denote

$COM = NB_1 \cap NB_2$, the set of common nodes

$RED = COM - NB$, the set of reducible nodes (16)

$P_1(B_1, E_1)$, $P_2(B_2, E_2)$, and $P(B, E)$—sets of multiconnections (as defined in Definition 5) for both blocks and resulting block, respectively.

An important result is presented in the following theorem.

Theorem 2

Any set of multiconnections $P(B, E)$ can be obtained according to the following rule:

$$P(B, E) = U\, P_1(B_1, E_1) \times P_2(B_2, E_2) \qquad (17)$$

where summation is performed over all sets of multiconnections $P_1(B_1, E_1)$ and $P_2(B_2, E_2)$ satisfying conditions

$$B_1 \cap B_2 = \emptyset, \quad E_1 \cap E_2 = \emptyset$$

$$RED = (B_1 \cup B_2) \cap (E_1 \cup E_2) \qquad (18)$$

and $\times$ is a Cartesian product [3] of sets $P_1(B_1, E_1)$ and $P_2(B_2, E_2)$. Sets B and E are in this case equal to

$$B = B_1 \cup B_2 - RED, \quad E = E_1 \cup E_2 - RED.$$

If all element weights are different, there are no duplicate terms in the formula (17). For every multiconnection $p \in P$, the sign of p can be calculated as follows:

$$\text{sign}\, p = \text{sign}\, p_1 \cdot \text{sign}\, p_2 \cdot (-1)^k \cdot \Delta \qquad (19)$$

where

$$p = p_1 \cup p_2, \quad p_1 \in P_1, \quad p_2 \in P_2$$

$$k = \min(\text{card}(E_1 \cap B_2 \cap COM), \text{card}(E_2 \cap B_1 \cap COM)) + \text{card}(COM)$$

$$\Delta = \text{ord}(b_{11}, b_{22}, \cdots, b_{1m_1})\, \text{ord}(e_{11}, e_{12}, \cdots, e_{1m_1})\, \text{ord}(b_{21}, b_{22}, \cdots, b_{2m_2})\, \text{ord}(e_{21}, e_{22}, \cdots, e_{2m_2})$$

$$B_1 = \{b_{11}, b_{12}, \cdots, b_{1m_1}\}, \quad E_1 = \{e_{11}, e_{12}, \cdots, e_{1m_1}\}$$

$$B_2 = \{b_{21}, b_{22}, \cdots, b_{2m_2}\}, \quad E_2 = \{e_{21}, e_{22}, \cdots, e_{2m_2}\}.$$

Proof of Theorem 2 is based on the observation that each element of $P_1(B_1, E_1) \times P_2(B_2, E_2)$ is a multiconnection of the type $P(B, E)$, and similarly, for each element $p \in P(B, E)$ there is a unique pair of elements $p_1 \in P_1(B_1, E_1)$ and $p_2 \in P_2(B_2, E_2)$ such that $p = p_1 p_2$. Therefore, multiconnection sets on both sides of (17) are equal. Since $P_1(B_1, E_1)$ and $P_2(B_2, E_2)$ are defined on edge disjoint subgraphs there will be no duplicate terms in (17). The sign of multiconnections is important in realization of formulas for transfer functions and (19) is to update the sign according to the topology of the association of two blocks.

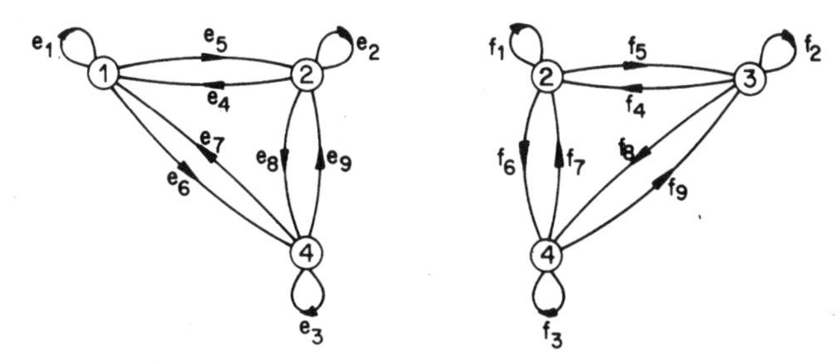

Fig. 13. Blocks to be connected.

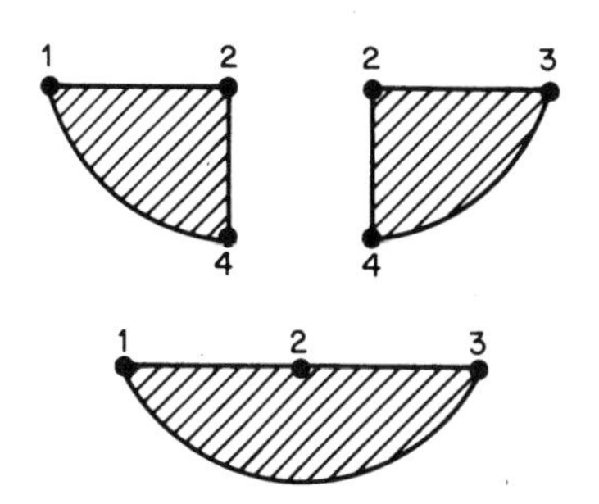

Fig. 14. Association of two blocks.

Remark

An important feature of (17) is the possibility of obtaining a set of multiconnections $P(B, E)$ by combining whole groups of multiconnections from the lower level. At the same time, from (19) we notice that the new sign is attributed simultaneously to the whole group of terms $P_1 \times P_2$, as k and Δ depend only on sets B_1, E_1, B_2, E_2. These features are of great importance in the computer realization because we do not have to deal with each multiconnection separately.

Example 3

Consider an association of two blocks presented in Figs. 13 and 14. We have $NB_1 = \{1,2,4\}$, $NB_2 = \{2,3,4\}$. $NB = \{1,2,3\}$, $COM = \{2,4\}$, $RED = \{4\}$. Let us calculate multiconnections of the type $P(\{1,2\},\{2,3\})$ of the resulting block. From the formula (17), with the condition (18), we obtain

$$\begin{aligned} P(\{1,2\},\{2,3\}) = {} & P_1(\{1,4\},\{2,4\}) \times P_2(\{2\},\{3\}) \\ & \cup P_1(\{1\},\{2\}) \times P_2(\{2,4\},\{3,4\}) \\ & \cup P_1(\{1,2\},\{2,4\}) \times P_2(\{4\},\{3\}) \\ & \cup P_1(\{1\},\{4\}) \times P_2(\{2,4\},\{2,3\}). \end{aligned}$$

A formula similar to that of Theorem 2 can be derived for the case of edge decomposition. Analysis of terminal blocks is realized in the same way as described previously. An edge bisection will be considered. We denote:

E_{cut}	a cutset of a graph G;
$G_1(E_1, V_1), G_2(E_2, V_2)$	two disconnected graphs obtained from G after removing edges E_{cut};
NB, NB_1, NB_2	sets of block vertices for G, G_1 and G_2, respectively,
$RED = (NB_1 \cup NB_2 - NB)$	the set of reducible nodes,

P_{cut} the set of multiconnections formed by edges E_{cut} only.

Theorem 3 [12]

Any set of multiconnections $P(B, E)$ can be obtained according to the following rule[1]

$$P(B, E) = \cup P_1(B_1, E_1) \times P_2(B_2, E_2) \times P_{cut}(B_c, E_c) \quad (20)$$

where summation is performed over all sets of multiconnections P_{cut}, with sets B_1, E_1, B_2, E_2 satisfying the following conditions:

$$B_c \cap B_1 = B_c \cap B_2 = E_c \cap E_1 = E_c \cap E_2 = \varnothing$$

$$RED \subset (B_c \cup B_1 \cup B_2) \cap (E_c \cup E_1 \cup E_2).$$

Sets B and E are then equal to

$$B = B_1 \cup B_2 \cup B_c - RED$$

$$E = E_1 \cup E_2 \cup E_c - RED.$$

If all element weights are different, there are no duplicate terms in the formula (20). For every multiconnection $p \in P$, the sign of p can be calculated as follows:

$$\text{sign } p = \text{sign } p_1 \cdot \text{sign } p_2 \cdot \text{sign } p_{cut}. \quad (21)$$

Downward and Upward Hierarchical Analysis

Now the method of analysis of terminal blocks and middle blocks is completed. The remaining step is the exploration of hierarchical structure to obtain a description of the initial network.

Two approaches are possible and are called the upward and downward methods of analysis. The upward method presents many advantages over the downward method, including savings of computer time and memory, so the latter will be only briefly outlined.

In the downward method, the analysis starts at the 1-level (initial block) and proceeds down to the next levels according to the connections in the hierarchical tree. The substitute graphs of blocks corresponding to the middle nodes are analyzed. On each intermediate level the type of necessary functions from the next level is determined. Arriving at the terminal node the analysis of the terminal block is executed to get the necessary function of this block. Then one proceeds upward. For each pass through the middle node, the multiplication of two functions from the lower level is executed. After arriving at the 1-level, we obtain a part of the function of the initial network. Many passes up and down the tree are necessary; much processing has to be performed. The formula (17) expresses a set of multiconnections of the middle block as a sum of products of multiconnection sets from lower level. Each term of this sum requires the above described up and down procedure.

The downward method presented in [12], permits the hierarchical analysis of large networks but has the following disadvantages:

a) multiple passes through the hierarchical structure causes multiple calculations of the same function;
b) complicated organization scheme;
c) problems with efficient storage of all-symbolic results.

For these reasons a new form of hierarchical tree exploration was elaborated. In the new method, only one pass along the hierarchical structure is necessary. The name *upward method* is due to the direction in which the decomposition tree is worked out—from the terminal nodes upward to the initial node.

Let us describe the upward hierarchical analysis in more detail. First, to facilitate the organization of the algorithm, a specific numeration of blocks is introduced. If N is the number of blocks (i.e., terminal and middle nodes), we shall number them from 1 to N in such way that each ascendant has lower number than its descendants. Such a numeration is easy to perform, e.g., we can number nodes starting from level 1 and move sequentially to the lowest level (as in Fig. 11). With this numeration the initial block has always number 1.

The upward method of analysis starts from the block having the number N and is performed sequentially to the number 1. When the terminal block is reached, the analysis presented in Section V is executed. When we arrive at the middle block, descendants of which have been previously analysed, the formula (17) is used. Two approaches to realize formula (17) are possible:

1) using the substitute graph, the combinations of sets B_1, E_1, B_2 and E_2 which satisfy conditions of the Theorem 2 are obtained directly, or
2) examining all possible combinations of multiconnections of descendant blocks, only those which satisfy conditions of the Theorem 2 are retained.

Since a simple test for combinations has been found (see Section V), the second approach was chosen for the algorithm and the program. The procedure ends after the initial block is analyzed. Then the functions of the original network are calculated.

V. Algorithm of Upward Hierarchical Analysis

As can be noted from the general presentation of the method, there are two distinct stages in the upward hierarchical analysis: analysis of terminal blocks and analysis of middle blocks. These two stages are resolved separately and each one can be ameliorated without affecting the other.

Analysis of the Terminal Block

An algorithm to generate multiconnections of the Coates graph will be presented. This part of the method corresponds to the methods of direct topological analysis of electrical circuits. Generation of 0-connections of a flowgraph can be converted to the problem of generation of disjoint cycles of a graph (see [6]).

[1] This form of the formula (20), a modification of the formula presented in [12], has been proposed by M. Bon.

Let us consider a Coates graph with n nodes. Let $\boldsymbol{M}$ be an incidence matrix defined as follows: $\boldsymbol{M}=[m_{ij}]_{n\times n}; m_{ij}$ = the set of edges starting from the ith node and ending at the jth node. The set of 0-connections of a flow-graph can be calculated from the formula

$$P=\bigcup_{(i_1,\cdots,i_n)\in I} m_{1i_1}\times m_{2i_2}\times\cdots\times m_{ni_n} \quad (22)$$

where I is the set of all permutations of numbers $(1,2,\cdots,n)$. There is no duplication in the formula (22). The sign of 0-connection $p\in m_{1i_1}\times m_{2i_2}\times\cdots m_{ni_n}$ is equal to $(-1)^{n+h}$, where h is a number of permutations necessary to order the set $i_1,\cdots,i_n$.

In the formulas for the hierarchical analysis, not only the set of all 0-connections is necessary but also sets of multiconnections characterized in Definition 5. This problem can be transformed to the generation of all 0-connections of the modified graph as follows.

Lemma 1:

The set of multiconnections $P(B,E)$ of a graph with an incidence matrix $\boldsymbol{M}$ is equal to the set of 0-connections of a graph described by a matrix $\boldsymbol{M}(B,E)$. The matrix $\boldsymbol{M}(B,E)$ is obtained from the matrix $\boldsymbol{M}$ by deleting:

all columns corresponding to nodes B;

all rows corresponding to nodes E.

Example 4

To generate the set of multiconnections $P(\{1,2\},\{3,4\})$ of the graph discussed in the Example 2, let us reduce the incidence matrix $\boldsymbol{M}$, where

$$\boldsymbol{M}=\begin{bmatrix}0&7&2&0&1\\0&9&3&4&0\\0&0&0&10&0\\0&0&0&8&0\\0&0&0&5&6\end{bmatrix}.$$

According to Lemma 1 the matrix $M(B,E)$ is obtained from $\boldsymbol{M}$ by deleting columns 1 and 2, and rows 3 and 4. Hence

$$\boldsymbol{M}(B,E)=\begin{bmatrix}2&0&1\\3&4&0\\0&5&6\end{bmatrix}.$$

Applying the formula (22) to $\boldsymbol{M}(B,E)$ we obtain the sets of multiconnections $P(B,E)$ so that

$$P(\{1,2\},\{3,4\})=\{\{2,4,6\},\{1,3,5\}\}$$

as expected.

The complete description of the block with nb block nodes is given by weight functions of all possible sets $P(B,E)$ with $B\cup E\subset NB$. Different sets B,E can be generated in the manner described below.

Let us numerate nodes NB from 1 to nb. For $i=0,\cdots,nb$, all i-element subsets of the set $\{1,\cdots,nb\}$ are generated. For a given i, let $\binom{nb}{i}$ such subsets form the set $K(i)$. Each pair of sets (m,k), where $m,k\in K(i)$ (note that m may be equal to k) has a corresponding set of potential multiconnections $P(B,E)$ with $B=m$ and $E=k$. Such sets of multiconnections are generated and stored. Each set may be identified by its type B,E. This type may be coded on a single computer word. The $2*nb$ bits would be occupied. Successive pairs of bits describe block nodes from 1 to nb. All elements b from B produce 1 on the position $2*b-1$ and elements e from E produce 1 on the position $2*e$. All other positions are equal to 0. The *identification code* C of a set of multiconnections $P(B,E)$ can be completely calculated from the formula

$$C=\sum_{b\in B}2**(2b-2)+\sum_{e\in E}2**(2e-1). \quad (23)$$

Example 5

For the set of block nodes $NB=\{1,2,3,4\}$, 8 bits are occupied to code different sets of multiconnections. If $B=\{1,2\}$ and $E=\{2,3\}$ the code for $P(B,E)$ is equal to

$$C=2^0+2^2+2^3+2^5=45.$$

This coding permits an easy identification of a multiconnection set by one interger number and a simple practical realization formula (17).

Analysis of the Middle Block

In the upward hierarchical method, the analysis of a middle block is performed at the moment when both its descendants have already been analyzed. The sets of multiconnections of these blocks are stored in the computer memory each having its identification code. The following rules of block nodes numeration are to be observed (renumerate if necessary):

first group is formed of reducible nodes *RED*;
second group is formed of other common nodes *COM-RED*;
third group is formed of other block nodes.

Both the first and second group should have the same numeration in blocks to be associated. We examine all possible combinations of functions describing two blocks. Let us denote the following bit fields in a computer word containing the code of a multiconnection:

RED_1, RED_2	corresponding to the nodes *RED* in both blocks (first group);
CR_1, CR_2	corresponding to the second group of nodes;
$REST_1, REST_2$	corresponding to the third group of nodes.

The following tests are performed

$$\text{AND}(RED_1,RED_2)=0$$
$$\text{AND}(CR_1,CR_2)=0$$
$$\text{OR}(RED_1,RED_2)=\text{ field having 1 on each bit.} \quad (24)$$

For the chosen code of multiconnection (23), conditions (24) are equivalent to (18). So if any of these conditions are not fulfilled, the combination is rejected. In the contrary case, we have the combination characterized by sets of nodes satisfying the formulas (18).

The code for resulting multiconnections can easily be composed from the parts of codes of component multiconnections. Since none of the first group of nodes remains a block node, there is no information concerning this group.

TABLE I
ORGANIZATION OF ALGORITHM OF HIERARCHICAL ANALYSIS

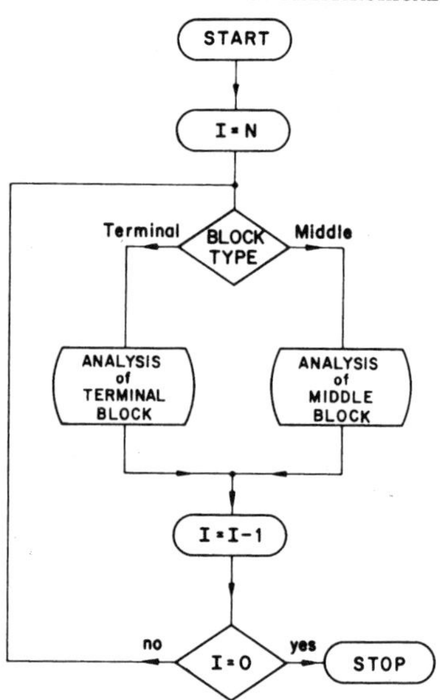

Nodes from the second group have code equal $\text{OR}(CR_1, CR_2) = CR_1 + CR_2$. As nodes from the third group are distinct in two blocks, their description remains $REST_1$, $REST_2$.

General Organization of the Algorithm

The general organization of the algorithm is presented in Table I. Once the proper numeration of block nodes is established, the analysis can be carried out as presented. With this numeration, analysis of any middle block is performed when both its descendants have been analyzed. The last analyzed block is the initial block.

The all-symbolic or semi-symbolic descriptions for large networks are intermediate results only. These results are used later in various types of network analysis.

The symbolic form of the transfer function for a large network contains a very large number of terms. To make possible the storage and to facilitate further work the decomposed form of results is preserved.

A terminal block is described by the weight functions of its multiconnection sets. Each term of a weight function has the form

$$t = r \cdot s^k \prod_i y_i \tag{25}$$

where $r =$ numerical factor; $s =$ Laplace variable; $y_i =$ symbolic admittances or symbolic element parameters. Any weight function is stored in the form of three vectors with successive elements equal: r, k and coded y_i. Each function can be recognized by its identification code (23).

From formula (17) we see that any function for a middle block is expressed as a sum of products of functions from the lower level. In the upward hierarchical method, the analysis of any middle block is performed after its descendants have been previously analyzed and resulting functions stored. The function of a middle block can be stored in an unexpanded form containing only addresses or functions to be multiplied. A term of such function is of the form

$$m = v \cdot F(i) \cdot F(k) \tag{26}$$

where $v =$ sign of term equal to ± 1, and $F(i), F(k) =$ functions describing descendants of the analyzed block.

The term m can be represented by three numbers: v and addresses of $F(i)$ and $F(k)$ stored previously.

The analysis is terminated by analyzing the initial block. Therefore, the whole hierarchical structure should be run through. From the functions of the initial block we only choose the necessary ones. The given addresses send us to next blocks. At the end we find functions of the terminal block. On these functions different kinds of operations can be performed, depending on what kind of analysis is required.

Example 6

Let us take a practical network to illustrate the algorithm. In Fig. 15 the scheme of an analyzed band-pass filter is shown. Operational amplifiers are considered ideal. The Coates flowgraph corresponding to this network is shown in Fig. 16. This graph has been decomposed into 5 terminal blocks (Fig. 17). The hierarchical structure of successive associations is presented in Fig. 18.

To illustrate the analysis of the terminal blocks let us consider the terminal block 8. Its flowgraph is indicated in Fig. 16 by the dashed line. In this block the only nonempty

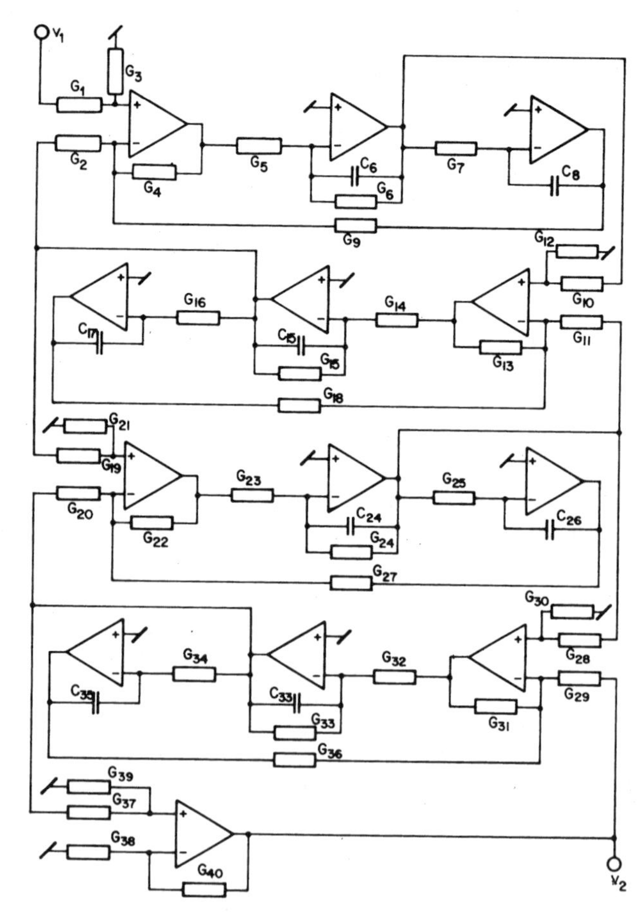

Fig. 15. Band-pass filter.

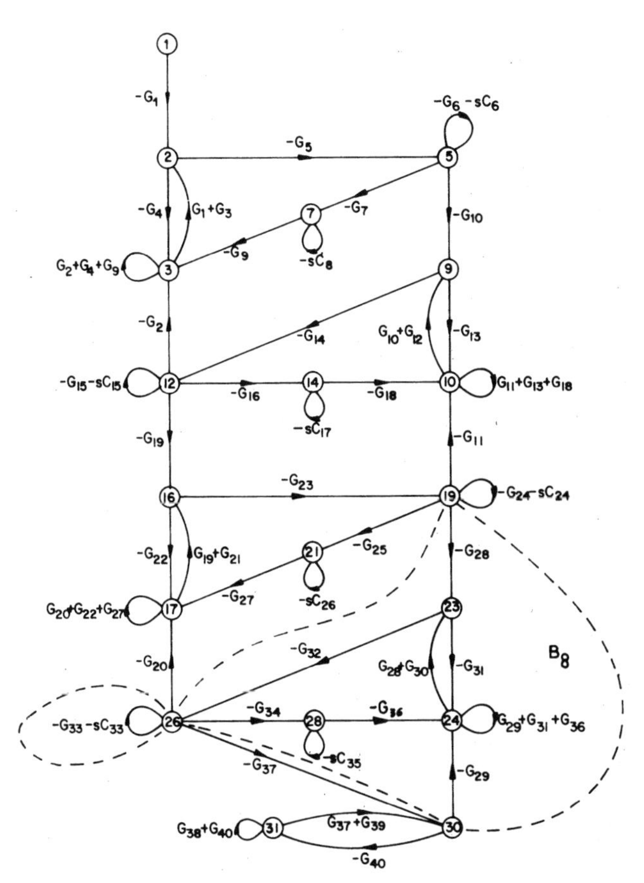

Fig. 16. Flowgraph for bandpass filter.

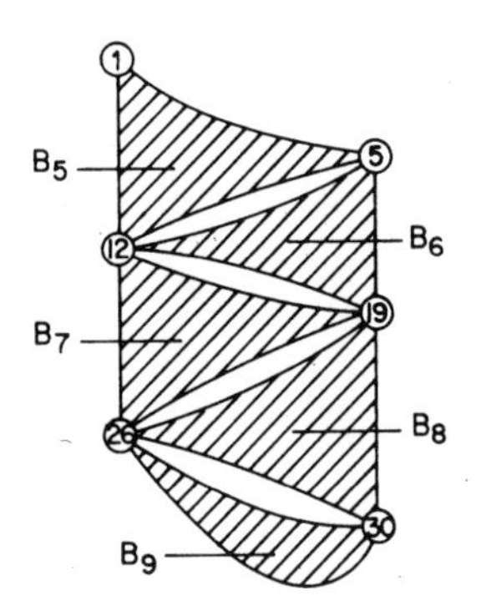

Fig. 17. Block graph.

types of sets of multiconnections $P_8(B, E)$ are

1) For $B = \{30\}$, $E = \{26\}$
$$|P_8(B, E)| = F(3) = G_{29}G_{32}sC_{35}(G_{28} + G_{30}).$$
2) For $B = \{26\}$, $E = \{26\}$
$$|P_8(B, E)| = F(4) = (G_{28} + G_{30}) \cdot [G_{32}G_{34}G_{36} + G_{31}sC_{35}(G_{33} + sC_{33})].$$
3) For $B = \{19\}$, $E = \{26\}$
$$|P_8(B, E)| = F(5) = G_{28}G_{32}sC_{35}(G_{29} + G_{31} + G_{36}).$$
4) For $B = \{0\}$, $E = \{0\}$
$$|P_8(B, E)| = sC_{35}G_{31}(G_{28} + G_{30}).$$

In Table II the first three types only are shown as they are required for voltage transfer function evaluation.

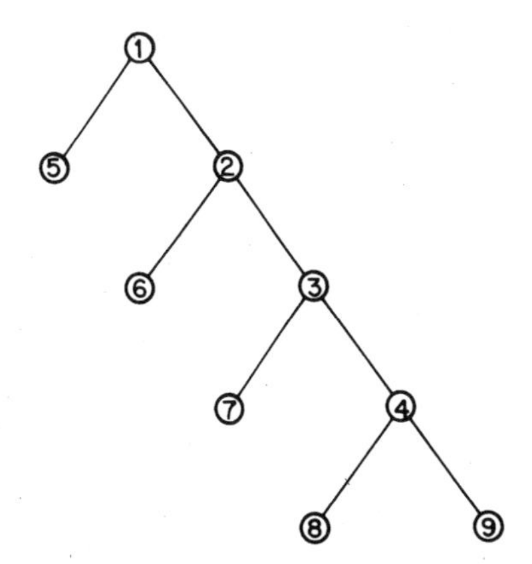

Fig. 18. Tree of the hierarchical structure.

TABLE II
RESULTS OF SYMBOLIC ANALYSIS

Block number	NB	B	E	Weight Function	
9	26,30	26	30	F(1)	$G_{37}(G_{38} + G_{40})$
		30	30	F(2)	$G_{40}(G_{37} + G_{39})$
8	19,26,30	30	26	F(3)	$G_{29}G_{32}sC_{35}(G_{28} + G_{30})$
		26	26	F(4)	$(G_{28} + G_{30})[G_{32}G_{34}G_{36} + G_{31}sC_{35}(G_{33} + sC_{33})]$
		19	26	F(5)	$G_{28}G_{32}sC_{35}(G_{29} + G_{31} + G_{36})$
7	12,19,26	26	19	F(6)	$G_{20}G_{23}sC_{26}(G_{19} + G_{21})$
		19	19	F(7)	$(G_{19} + G_{21})[G_{23}G_{25}G_{27} + G_{22}sC_{26}(G_{24} + sC_{24})]$
		12	19	F(8)	$G_{19}G_{23}sC_{26}(G_{20} + G_{22} + G_{27})$
6	5,12,19	19	12	F(9)	$G_{11}G_{14}sC_{17}(G_{10} + G_{12})$
		12	12	F(10)	$(G_{10} + G_{12})[G_{14}G_{16}G_{18} + G_{13}sC_{17}(G_{15} + sC_{15})]$
		5	12	F(11)	$G_{10}G_{14}sC_{17}(G_{11} + G_{13} + G_{18})$
5	1,5,12	12	5	F(12)	$G_2G_5sC_8(G_1 + G_3)$
		5	5	F(13)	$(G_1 + G_3)[G_5G_7G_9 + G_4sC_8(G_6 + sC_6)]$
		1	5	F(14)	$G_1G_5sC_8(G_2 + G_4 + G_9)$
4	19,26,30	26,30	30,30	F(15)	- F(3)F(1) + F(4)F(2)
		19,26	26,30	F(16)	- F(5)F(1)
		19,30	26,30	F(17)	F(5)F(2)
3	12,19,30	12,19	19,30	F(18)	F(8)F(16)
		19,30	19,30	F(19)	F(7)F(15) - F(6)F(17)
		12,30	19,30	F(20)	- F(8)F(15)
2	5,12,30	5,12	12,30	F(21)	F(11)F(18)
		12,30	12,30	F(22)	F(10)F(19) - F(9)F(20)
		5,30	12,30	F(23)	- F(11)F(19)
1	1,30	1	30	F(24)	F(14)F(21)
		30	30	F(25)	F(13)F(22) - F(12)F(23)

Association of blocks is performed according to the Theorem 2. For example the middle block 2 has $NB = \{5,12,30\}$ and is obtained as the association of block 6 with $NB_6 = \{5,12,19\}$ and block 3 with $NB_3 = \{12,19,30\}$ (see Fig. 18). For this association we have $COM = NB_6 \cap NB_3 = \{12,19\}$, $RED = COM - NB = \{19\}$. Considering only multiconnections necessary to obtain te required transfer function we evaluate:

$$\begin{aligned}
F(21) &= |P_2(\{5,12\},\{12,30\})| \\
&= |P_6(\{5\},\{12\})|\cdot|P_3(\{12,19\}\{19,30\})| \\
&= F(11)F(18), \\
F(22) &= |P_2(\{12,30\},\{12,30\})| \\
&= |P_6(\{12\},\{12\})|\cdot|P_3(\{19,30\},\{19,30\})| \\
&\quad - |P_6(\{19\},\{12\})|\cdot|P_3(\{12,30\},\{19,30\})| \\
&= F(10)F(19) - F(9)F(20), \\
F(23) &= |P_2(\{5,30\},\{12,30\})| \\
&= -|P_6(\{5\},\{12\})|\cdot|P_3(\{19,30\},\{19,30\})| \\
&= -F(11)F(19).
\end{aligned}$$

Symbolic results for the total network are shown in Table II. The results are presented in the unexpanded form as they are computed by the program. The voltage transfer function for the considered filter can be expressed as

$$K_v = \frac{v_2}{v_1} = \frac{|P_1(\{1\},\{30\})|}{|P_1(\{30\},\{30\})|} = \frac{F(24)}{F(25)}. \tag{27}$$

The obtained formula represents the symbolic transfer function, which can be used in compact form or expanded if necessary. This network has 44 elements and consequently 44 symbolic parameters in the symbolic results. Fully symbolic analysis of networks of this size can require considerable computer time when direct topological methods are applied. In the case of hierarchical analysis, it is even possible to obtain these results by hand calculations.

Notice that for this structure, the graphs of blocks 8, 7, 6, and 5 are isomorphic. If an isomorphism of graphs is detected, it is possible to execute block analysis only once since the symbolic descriptions of isomorphic blocks are identical. This permits reduction of computer time as well as the memory needed to store the results.

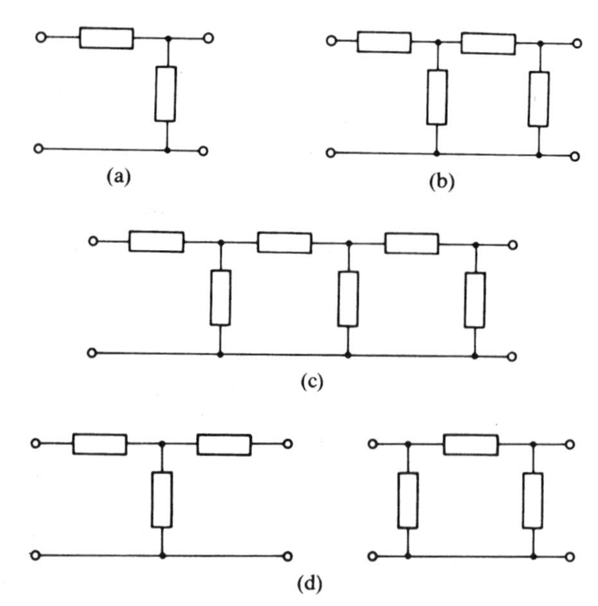

Fig. 19. Terminal blocks of ladder decomposition.

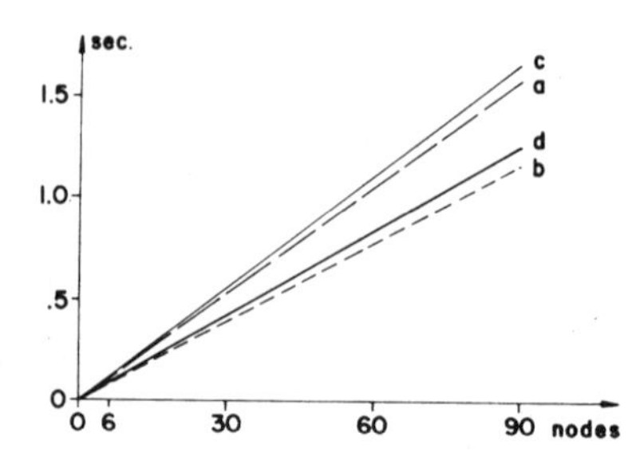

Fig. 20. Relationship between the analysis time and the number of nodes.

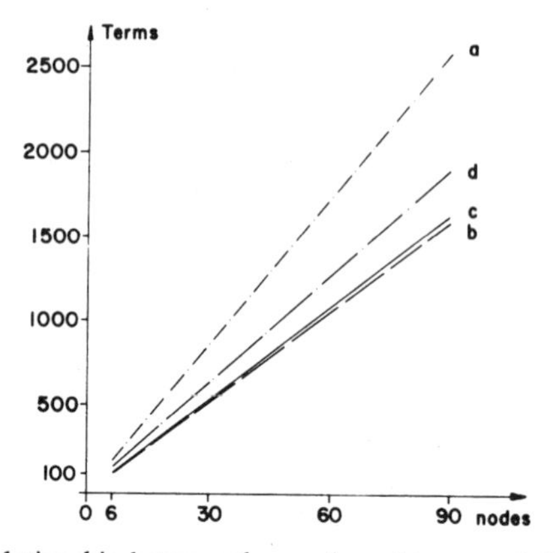

Fig. 21. Relationship between the number of terms and the number of nodes.

VI. Computer Realization and Results

Two computer programs were developed on the basis of presented algorithms. Programs FANES [12] realizes the downward analysis of hierarchical structure. The edge decomposition is used in this program. Some comparisons between SNAPEST, NAPPE, SNAP [14], [26] and FANES are presented in [12].

First results of the program FLOWUP realizing the upward hierarchical method were published in [13]. Program FLOWUP is written in Fortran and is implemented on CDC Cyber 73 and CIIHB DPS/8 computers. Memory demands for the program are not important and additionally two parts of the program, namely terminal and middle block analysis, can be separated and overlayed. The BASIC version for the minicomputer HP9835 (or HP9845) with standard memory has been realized.

Input data contains a node-to-node description of network elements. The program generates a signal flow-graph using element graphs obtained on the basis of modified admittance matrix (see Appendix). Then the hierarchical decomposition is automatically carried out and a structure of the decomposition tree is established by the program. Next the hierarchical analysis of the entire structure is

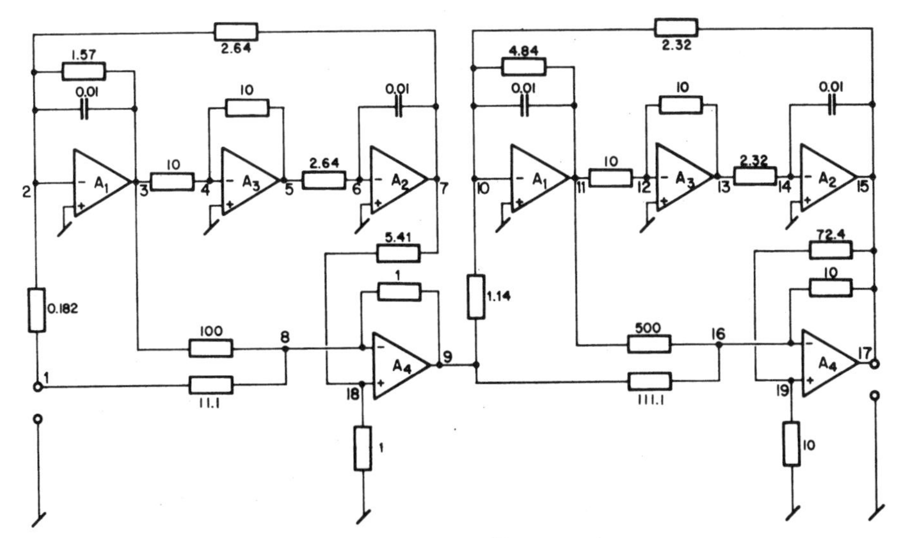

Fig. 22. Low-pass filter network.

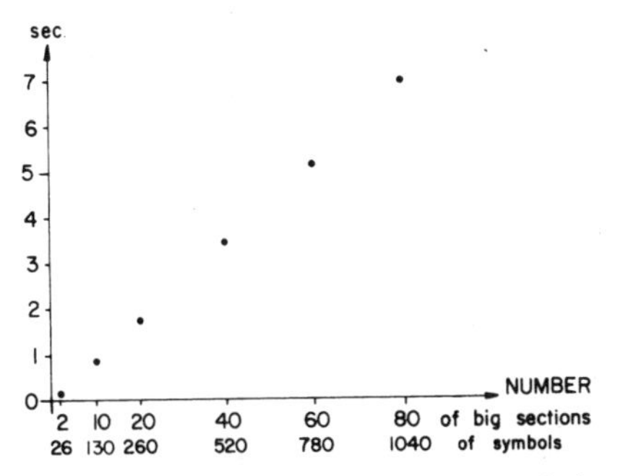

Fig. 23. Relationship between the analysis time and the size of the network.

performed starting from the graph of the highest number as shown in Table I. For the terminal block analysis, matrices of the range $n \times n$ are to be stored, where n-number of block nodes (in general not greater than 10). The demand for middle block analysis is due to the number of block nodes. In the case of analysis of large networks, the most important memory demand is due to the storage of symbolic results. Three vectors (25), each with length equal to the number of terms, are necessary. When the compact form is used the all-symbolic form for quite large networks can be calculated. In the minicomputer version the successive transfer of results to other memory supports may be performed during the program execution.

Let us present now some comparative results of analysis with the FLOWUP program. First let us examine the ladder structure decomposed into different terminal blocks, as shown in Fig. 19. Time of computer analysis and number of terms in the results are presented in Figs. 20 and 21, where the lines a, b, c, d represent the terminal blocks having the structure, as shown in Fig. 19(a), (b), (c), and (d), respectively. The isomorphism of the terminal blocks was not exploited. Both time and memory depend linearly on the number of nodes of the analyzed ladder. Linear dependence is typical for all cascade connections of blocks. Note that both time and memory depend on the kind of partition performed. These computations have been done on a CDC Cyber 73.

Analysis of the filter presented in Fig. 22 was executed on a DTS/8 GCOS. Analysis time for this filter was 0.165 s. In the case of cascade connection of many such sections, we have the linear growth of computer time as presented in Fig. 23. The isomorphism of sections has not been taken into account. When connection of blocks is more complicated than cascade, the analysis is expected to be more time consuming.

VII. Conclusion

We have discussed a new method that increases the computation power of topological analysis due to the reduction in the computer time needed for the analysis of large electronic networks. The approach will significantly affect the applications of topological methods to the analysis of large networks which was impossible even with the aid of the fastest computers.

Hence, network design problems requiring topological analysis can be solved with the help of the symbolic form of results. The method preserves the advantages of direct methods of topological analysis such as high accuracy of computations and possibility of generating fully symbolical results.

As the method is based on hierarchical decomposition, different blocks can be analyzed independently. Thus the use of parallel processing techniques is feasible and further reduction in computational time is possible.

The restriction of the presented method in its application to large networks lies in the number of block nodes in each block. This is usually overcome by using an effective decomposition algorithm which minimizes the number of partition nodes.

Appendix

In Table III flowgraph models for chosen network elements are shown. They are based on modified admittance matrices of elements and they can be directly connected resulting in a flowgraph of an analyzed circuit. Adding a new element does not alter the structure of an existing

TABLE III
ELEMENT'S MODELS

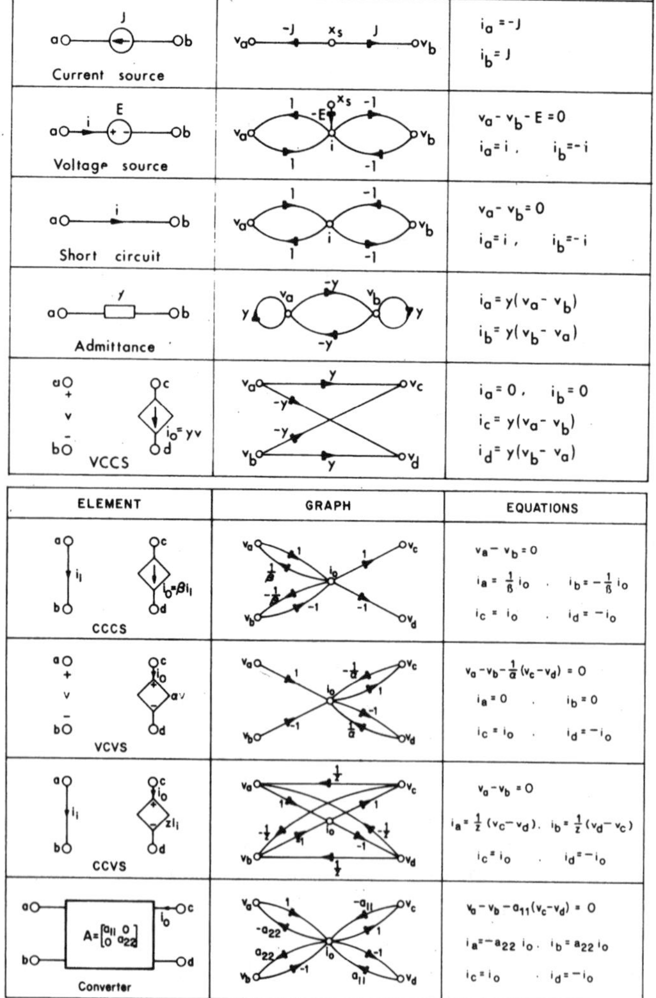

ELEMENT	EQUATIONS
Current source	$i_a = -J$ $i_b = J$
Voltage source	$v_a - v_b - E = 0$ $i_a = i$, $i_b = -i$
Short circuit	$v_a - v_b = 0$ $i_a = i$, $i_b = -i$
Admittance	$i_a = y(v_a - v_b)$ $i_b = y(v_b - v_a)$
VCCS	$i_a = 0$, $i_b = 0$ $i_c = y(v_a - v_b)$ $i_d = y(v_b - v_a)$
CCCS	$v_a - v_b = 0$ $i_a = \frac{1}{\beta} i_o$. $i_b = -\frac{1}{\beta} i_o$ $i_c = i_o$. $i_d = -i_o$
VCVS	$v_a - v_b - \frac{1}{\alpha}(v_c - v_d) = 0$ $i_a = 0$. $i_b = 0$ $i_c = i_o$. $i_d = -i_o$
CCVS	$v_a - v_b = 0$ $i_a = \frac{1}{z}(v_c - v_d)$. $i_b = \frac{1}{z}(v_d - v_c)$ $i_c = i_o$. $i_d = -i_o$
Converter	$v_a - v_b - a_{11}(v_c - v_d) = 0$ $i_a = -a_{22} i_o$. $i_b = a_{22} i_o$ $i_c = i_o$. $i_d = -i_o$

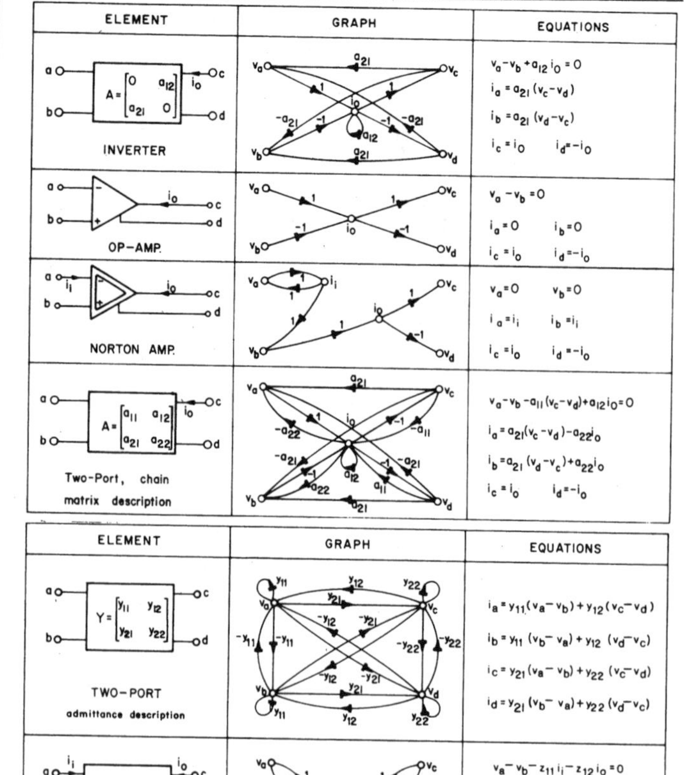

ELEMENT	EQUATIONS
INVERTER	$v_a - v_b + a_{12} i_o = 0$ $i_a = a_{21}(v_c - v_d)$ $i_b = a_{21}(v_d - v_c)$ $i_c = i_o$ $i_d = -i_o$
OP-AMP.	$v_a - v_b = 0$ $i_a = 0$ $i_b = 0$ $i_c = i_o$ $i_d = -i_o$
NORTON AMP.	$v_a = 0$ $v_b = 0$ $i_a = i_i$ $i_b = i_i$ $i_c = i_o$ $i_d = -i_o$
Two-Port, chain matrix description	$v_a - v_b - a_{11}(v_c - v_d) + a_{12} i_o = 0$ $i_a = a_{21}(v_c - v_d) - a_{22} i_o$ $i_b = a_{21}(v_d - v_c) + a_{22} i_o$ $i_c = i_o$ $i_d = -i_o$
TWO-PORT admittance description	$i_a = y_{11}(v_a - v_b) + y_{12}(v_c - v_d)$ $i_b = y_{11}(v_b - v_a) + y_{12}(v_d - v_c)$ $i_c = y_{21}(v_a - v_b) + y_{22}(v_c - v_d)$ $i_d = y_{21}(v_b - v_a) + y_{22}(v_d - v_c)$
Impedance description	$v_a - v_b - z_{11} i_i - z_{12} i_o = 0$ $v_c - v_d - z_{21} i_i - z_{22} i_o = 0$ $i_a = i_i$. $i_b = -i_i$ $i_c = i_o$. $i_d = -i_o$
Hybrid description	$v_a - v_b - h_{12}(v_c - v_d) - h_{11} i_i = 0$ $i_c = h_{22}(v_c - v_d) + h_{21} i_i$ $i_d = h_{22}(v_d - v_c) - h_{21} i_i$ $i_a = i_i$. $i_b = -i_i$

flowgraph—only new edges will be added according to the new element's model. In equations describing the models in Table III symbol i_x $(x = a, b, c, d)$ indicates current directed away from the node x.

REFERENCES

[1] C. Acar, "New expansion for signal-flow graph determinant," *Electron. Lett.*, Dec. 1971.

[2] G. E. Alderson and P. M. Lin, "Computer generation of symbolic network functions—a new theory and implementation," *IEEE Trans. Circuit Theory*, vol. CT-20, pp. 48–56, Jan. 1973.

[3] W. K. Chen, *Applied Graph Theory—Graphs and Electrical Networks*. Amsterdam, The Netherlands: North-Holland, 1976.

[4] W. K. Chen, "Flow graphs: Some properties and methods of simplifications," *IEEE Trans. Circuit Theory*, vol. CT-12, pp. 128–130, 1965.

[5] W. K. Chen, "Unified theory on the generation of trees of a graph," *Int. J. Electron.*; "Part I. The Wang algebra formulation," vol. 27, no. 2, 1969, "Part II. The Matrix Formulation," vol. 27, no. 4, 1969; Part III. Decomposition and Elementary Transformations," vol. 31, no. 4, 1971.

[6] L. O. Chua and P. M. Lin, *Computer Aided Analysis of Electronic Circuits—Algorithms and Computational Techniques*. Englewood Cliffs, NJ: Prentice-Hall, 1975.

[7] C. L. Coates, "Flow graph solutions of linear algebraic equations," *IRE Trans. Circuit Theory*, vol. CT-6, pp. 170–187, 1959.

[8] C. L. Coates, "General topological formulas for linear networks functions," *IRE Trans. Circuit Theory*, vol. 5, pp. 42–54, Mar. 1958.

[9] J. Cajka, "New formula for the signal-flow graph reduction," *Electron. Lett.*, pp. 437–438, July 1970.

[10] N. Deo, *Graph Theory with Applications to Engineering and Computer Sciences*. Englewood Cliffs, NJ: Prentice-Hall, 1974.

[11] W. R. Dunn and S. P. Chan, "Analysis of active networks by a subgraph-construction technique," presented at *Second Asilomar Conf. Circuits Sys.* 1968.

[12] A. Konczykowska and J. Starzyk, "Computer analysis of large signal flowgraphs by hierarchical decomposition method," in *Proc. European Conf. Circuit Theory Design*, (Warsaw, Poland), 1980, pp. 408–413.

[13] A. Konczykowska and J. Starzyk, "Computer justification of upward topological analysis of signal-flow graphs," in *Proc. European Conf. Circuit Theory Design*, (The Hague), 1981, pp. 464–467.

[14] P. M. Lin and G. E. Alderson, "SNAP-A computer program for generating symbolic network functions," School Elec. Eng., Purdue Univ. Lafayette, Ind. Tech. Rep. TR-EE, 70–16, Aug. 1970.

[15] W. J. McCalla and P. O. Pederson, "Elements of computer-aided analysis," *IEEE Trans. Circuit Theory*, vol. CT-18, pp. 14–26, Jan. 1971.

[16] O. P. McNamee and N. Potash, "A user's and programmer's manual for NASAP," Univ. California, Los Angeles, Rep. 63-68, Aug. 1968.

[17] R. R. Mielke, "A new signal flow graph formulation of symbolic network functions," *IEEE Trans. Circuits Sys.*, vol. CAS-25, pp. 334–340, June 1978.

[18] U. Ozguner, "Signal flow graph analysis using only loops," *Electron. Lett.* pp. 359–360, Aug. 1973.

[19] D. E. Riegle and P. M. Lin, "Matrix signal flow graphs and an optimum topological method for evaluating their gains," *IEEE Trans. Circuit Theory*, vol. CT-19, pp. 427–435, 1972.

[20] S. M. Roberts and B. Flores, "Systematic generation of Hamiltonian circuits," *Comm. ACM*, vol. 9, pp. 690–694, 1966.

[21] A. Sangiovanni-Vincentelli, L. K. Chen and L. O. Chua, "An efficient heuristic cluster algorithm for tearing large-scale networks,"

IEEE Trans. Circuits Syst., vol. CAS-24. pp. 709–717, Dec. 1977.
[22] S. D. Shieu and S. P. Chan, "Topological formulas of symbolic network functions for linear active networks," in *Proc. Int. Symp. Circuit Theory*, pp. 95–98, 1973.
[23] J. A. Starzyk and E. Sliwa, "Hierarchic decomposition method for the topological analysis of electronic networks," *Int. J. Circuit Theory Applic.*, vol. 8, pp. 407–417, 1980.
[24] J. A. Starzyk, "Signal flow-graph analysis by decomposition method," *IEEE Proc. Electronic Circuits Sys.*, no. 2, pp. 81–86, 1980.
[25] J. C. Tiernan, "An efficient search algorithm to find the elementary circuits of a graph," *Comm. ACM*, vol. 13, pp. 722–726, Dec. 1970.
[26] M. K. Tsai and B. A. Shenoi, "Generation of symbolic network functions using computer software techniques," *IEEE Trans. Circuits Sys.*, vol. CAS-26, pp. 344–346, June 1979.
[27] H. A. Weinblatt, "New search algorithm for finding the simple cycles of a finite directed graph," *J. ACM*, no. 1, pp. 43–56, 1972.

Janusz A. Starzyk (SM'83) was born in Rybnik, Poland, on June 26, 1947. He received the M.S. degree in applied mathematics and Ph.d. degree in electrical engineering from Technical University of Warsaw, Warsaw, Poland, in 1971 and 1976, respectively.

He became a Research and Teaching Assistant in 1972 and Assistant Professor in 1977 at the Institute of Electronics Fundamentals, Technical University of Warsaw, Poland. From 1981 to 1983 he was a Post-Docorate Fellow and research engineer at the McMaster University, Hamilton, Canada. He joined the Department of Electrical and Computer Engineering, Ohio University, Athens, OH, as Associate Professor in 1983. His current research is in the areas of large scale networks, graph theory and fault analysis.

Agnieszka Konczykowska received the M.S. degree in applied mathematics and the Ph.d. in electrical engineering from the Technical University of Warsaw, Poland, in 1971 and 1977, respectively.

From 1971 to 1981 she had research and teaching activities at the same University. She is now with the CNET-Bagneux, France. Her main research interests are circuit theory, computer aided design, particularly using symbolic design methods and decomposition methods, and switched-capacitor circuits. She has authored or coauthored about 25 publications.

ISAAC: A Symbolic Simulator for Analog Integrated Circuits

GEORGES G. E. GIELEN, STUDENT MEMBER, IEEE, HERMAN C. C. WALSCHARTS, AND WILLY M. C. SANSEN, SENIOR MEMBER, IEEE

Abstract —The symbolic simulator ISAAC (Interactive Symbolic Analysis of Analog Circuits) is presented. The program derives all ac characteristics for any analog integrated circuit (both time-continuous and switched-capacitor, CMOS, JFET, and bipolar) as symbolic expressions in the circuit parameters. This yields analytic formulas for transfer functions, CMRR, PSRR, impedances, noise, etc. Two novel features are included in the program. First, the expressions can be simplified with a heuristic criterion based on the magnitudes of the elements. This yields interpretable formulas showing only the dominant terms. Secondly, the explicit representation of mismatch terms allows the accurate calculation of second-order effects, such as the PSRR.

The symbolic simulator ISAAC provides analog designers with more insight into the circuit behavior than numerical simulators do and is a useful tool for instruction or designer assistance. Moreover, it generates complete analytic ac circuit models, which are used for automatic sizing in a non-fixed topology analog module generator. The program's capabilities are illustrated with several examples. The efficiency is established by a dedicated sparse-matrix algorithm.

I. Introduction

ADVANCES in VLSI technology nowadays allow the design of complex integrated circuits and systems. The use of computer-aided design tools has become indispensible to reduce the design time and cost. However, whereas for digital VLSI a lot of design tools, methodologies, and architectures have been developed, resulting now in complete digital silicon compilers, analog design tools and methodologies are just beginning to be developed. Most of the design is still done manually by analog experts, deriving the circuit expressions by hand and only using numerical circuit simulators as verification tools. This explains why the decreasing analog part takes relatively more and more of the overall design time and cost of the present mixed analog–digital chips. This is especially true in high-performance applications requiring high frequencies, low noise, and low distortion. Besides, due to the hand design and hand layout, errors regularly occur in the small analog section, resulting in several reruns.

Manuscript received May 3, 1989; revised August 25, 1989.
G. G. E. Gielen is a research assistant with the National Fund of Scientific Research (Belgium) at the Laboratory Elektronika, Systemen, Automatisatie en Technologie (ESAT), Departement Elektrotechniek, Katholieke Universiteit Leuven, B-3030 Heverlee, Belgium.

H. C. C. Walscharts and W. M. C. Sansen are with the Laboratory Elektronika, Systemen, Automatisatie en Technologie (ESAT), Departement Elektrotechniek, Katholieke Universiteit Leuven, B-3030 Heverlee, Belgium.
IEEE Log Number 8931433.

The design of high-performance analog circuits is known to be very knowledge-intensive. It relies strongly on the insight and expertise of the analog designer. These designers are rather rare and it takes a long time to instruct novice designers. This is mainly due to the complicated interaction between the device characteristics and the circuit performance. But also, insight into performance-degrading effects such as noise and distortion is only obtained after long and tedious calculations and simulations. So, there is a strong need for a CAD tool that provides the designer with insight into the circuit behavior.

To this purpose, the symbolic simulator ISAAC (Interactive Symbolic Analysis of Analog Circuits) has been developed [1]. The program analyzes lumped, linear, or linearized (small-signal) circuits in the frequency domain and returns analytic expressions with the complex frequency and the circuit elements represented by symbols. Time-continuous circuits are analyzed in the s-domain (Laplace domain), time-discrete circuits in the z-domain. Generally stated, the program derives network functions as symbolic rational forms in the complex frequency variable x (s or z), as given by

$$H(x)=\frac{N(x)}{D(x)}=\frac{\sum_i x^i \cdot a_i(p_1,\cdots,p_m)}{\sum_i x^i \cdot b_i(p_1,\cdots,p_m)} \tag{1}$$

where a_i and b_i are symbolic polynomial functions of the circuit elements p_j, which are represented by a symbol instead of a numerical value.

This is illustrated for the active RC filter of Fig. 1 [2]. The symbolic transfer function is given by

$$\frac{\begin{aligned}&-G4\,G8\,(G1\,G2\,G9+G1\,G3\,G9+G1\,G9\,G11\\&\quad+G2\,G6\,G9+G2\,G6\,G10)\\&+S\,G7\,C2\,(G1\,G3\,G9+G1\,G3\,G10\\&\quad-G2\,G5\,G9-G2\,G5\,G10)\\&+S^2\,(-1)\,G2\,G7\,C1\,C2\,(G9+G10)\end{aligned}}{\begin{aligned}&G11\,(G9+G10)\,(G4\,G6\,G8+S\,G5\,G7\,C2\\&\quad+S^2\,G7\,C1\,C2)\end{aligned}}. \tag{2}$$

The op amps are considered to be ideal. For semiconduc-

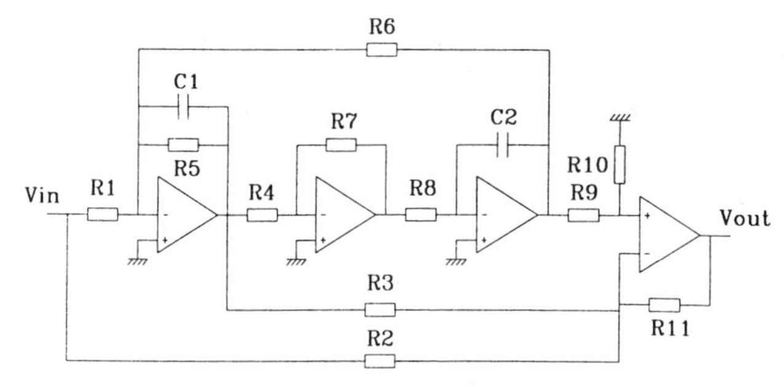

Fig. 1. Active *RC* filter.

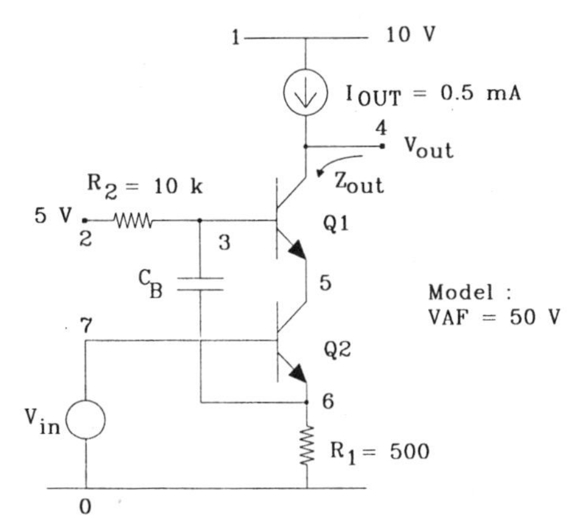

Fig. 2. Bipolar cascode stage with bootstrap capacitor.

tor circuits, however, the exact expressions are usually long and complicated. Therefore, a heuristic approximation algorithm has been built in. It simplifies the expressions based on the relative magnitudes of the elements and returns the dominant terms in the result.

Section II describes symbolic circuit analysis and its advantages over numerical methods. The main applications of symbolic simulation in analog circuit design are discussed. The ISAAC program can be used for instruction or designer assistance. It is also incorporated in the module generation part of an analog silicon compiler. Section III describes the internal structure and capabilities of the program. This is illustrated with several examples. In Section IV, algorithmic details are presented and the efficiency is compared to similar tools from literature. Section V provides final remarks and outlines future work.

II. Applications of Symbolic Analysis in Analog Design

A. Insight into Circuit Behavior: Symbolic Versus Numerical Simulation

Nowadays, many numerical circuit simulators (such as SPICE) exist. Although very efficient, they only return a collection of numbers, in tabular or plot form. The only conclusion that can be drawn from these numbers is whether the circuit meets the specified functional behavior, for the topology description and the technology and environmental parameters provided in the input file. No indication is given of which circuit elements essentially determine the observed performance. No potential problems are pointed out. No solutions are suggested when the circuit does not meet the specifications. To scan performance trade-offs and to check the influence of parameter variations, many simulations have to be carried out. Yet the results still cannot be extrapolated. This is especially the case for second-order phenomena such as the power-supply rejection ratio (PSRR), which strongly depends on component mismatching and layout overlap capacitances. It can be concluded that numerical circuit simulators are mainly useful for functional (nominal) verification, once the circuit has already been designed.

On the other hand, to really gain insight into the circuit's behavior, the circuit has to be analyzed with (all or part of) the circuit elements represented by symbols. This is the goal of a symbolic simulator. Such a symbolic simulator replaces the long and tedious hand calculations of the analog designer, especially for second-order effects such as distortion and PSRR. It returns first-time correct analytic expressions that are valid whatever the element values are. Assuming that all transistor small-signal models remain valid, the same expressions hold when the parameter values change. By examination of the resulting expressions, especially when they are simplified to retain only the dominant terms, insight can be gained into the circuit behavior. The analytic expressions clearly indicate the fundamental design variables. They show performance trade-offs and sensitivities to parameter variations and are suitable for pole–zero extraction, for sensitivity and tolerance analysis [3], and even for fault diagnosis [4].

Since a symbolic analysis program provides insight into the behavior of analog circuits, it is more useful for the instruction of students and novice designers than a numerical simulator such as SPICE. Circuits can be explored without any hand calculation in a fast and interactive way. In addition, a symbolic simulator is also a valuable aid for experienced designers. For example, the influence of the finite op amp gain and bandwidth on a filter characteristic can be examined. It can be concluded that a symbolic simulator is a valuable analog design aid and forms an essential complement to numerical simulators.

Example: The differences between symbolic and numerical simulation are now illustrated in the following example [1]. Consider the bipolar cascode stage, depicted in Fig. 2. A bootstrap capacitor C_B has been added, which is supposed to increase the output impedance at higher frequencies. A SPICE result of the output impedance versus frequency for a C_B value of 100 pF is shown in Fig. 3. The transistors are modeled by r_π, g_m, and g_o only. The plot confirms the expected impedance improvement. However, no indication is given regarding which elements determine the impedance levels at low and high frequencies or the two breakpoints. Even an experienced designer has difficulties formally predicting the influence of C_B on the output impedance, although it is a simple two-transistor circuit.

The input file to the ISAAC program is shown below. It is stated in the well-known SPICE format, except that for some elements, no numerical values have been provided.

```
BOOTCAS
*
VDD1 1 0  DC 10
VDD2 2 0  DC 5
Q1   4 3 5 0  NPN1
Q2   5 7 6 0  NPN1
R1   6 0  500
R2   2 3  10K
CB   3 6
VIN  7 0
IOUT 1 4
*
.MODEL NPN1 NPN VAF=50
```

The complete symbolic expression for the output impedance is given by

$$\frac{\begin{array}{c} G2\ GM1\ GM2+G1\ G2\ GM1+G2\ GM2\ G\pi1 \\ +G2\ GM1\ G\pi2+GM2\ GO1\ G\pi1+G1\ G2\ G\pi1 \\ +G2\ GM2\ GO1+G2\ GM1\ GO2+G1\ GO2\ G\pi1 \\ +G2\ G\pi1\ G\pi2+G1\ GO1\ G\pi1+G1\ G2\ GO2 \\ +G1\ G2\ GO1+GO2\ G\pi1\ G\pi2+GO1\ G\pi1\ G\pi2 \\ +G2\ GO2\ G\pi1+G2\ GO1\ G\pi2+G2\ GO2\ G\pi2 \\ +GO1\ GO2\ G\pi1+G2\ GO1\ GO2 \\ +S\ CB\ (GM1\ GM2 \\ +G1\ GM1+GM1\ G\pi2+G2\ GM1 \\ +G1\ G\pi1+GM2\ GO1+G\pi1\ G\pi2 \\ +G1\ GO2+G2\ G\pi1+G1\ GO1 \\ +GO1\ G\pi2+GO2\ G\pi2+GO1\ G\pi1 \\ +G2\ GO1+G2\ GO2+GO1\ GO2) \end{array}}{\begin{array}{c} GO1\ (G2\ GM2\ G\pi1+G1\ G2\ G\pi1+G2\ G\pi1\ G\pi2 \\ +G1\ GO2\ G\pi1+G1\ G2\ GO2+GO2\ G\pi1\ G\pi2 \\ +G2\ GO2\ G\pi2+G2\ GO2\ G\pi1 \\ +S\ CB\ (G1\ G\pi1 \\ +G\pi1\ G\pi2+G2\ G\pi1+G1\ GO2 \\ +GO2\ G\pi2+G2\ GO2)). \end{array}} \quad (3)$$

This expression is already long for such a simple circuit. However, for a bipolar transistor it usually holds that $g_m \gg g_\pi \gg g_o$. Exploiting this information, the expressions can be simplified, thereby introducing some error. The maximum error percentage is supplied by the user. The error definition and the simplification algorithm are described in Section IV-C. The simplified output impedance for a 10-percent error is given by

$$\frac{GM1\ (GM2+G1)\ (G2+S\ CB)}{GO1\ G\pi1\ (G2\ (GM2+G1)+S\ CB\ (G1+G\pi2))}. \quad (4)$$

The approximation can be carried out further on, yielding simpler and simpler formulas, at the expense of more and more error. If a 25-percent error is allowed, the impedance is given by

$$\frac{GM1\ GM2\ (G2+S\ CB)}{GO1\ G\pi1\ (G2\ GM2+S\ CB\ G1)}. \quad (5)$$

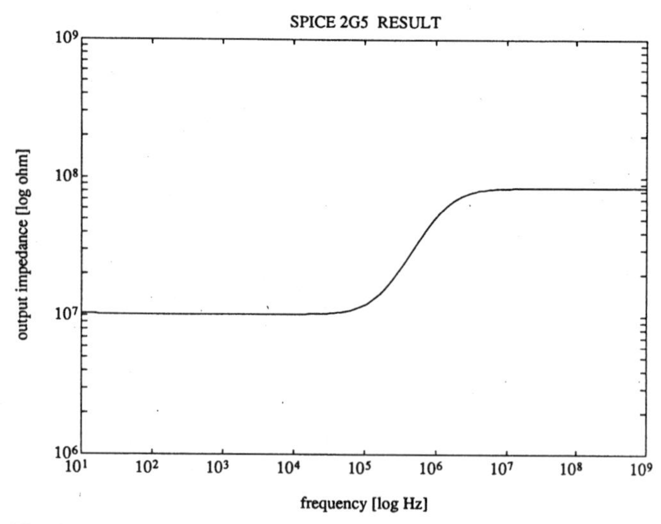

Fig. 3. SPICE plot of the output impedance of the bipolar cascode stage.

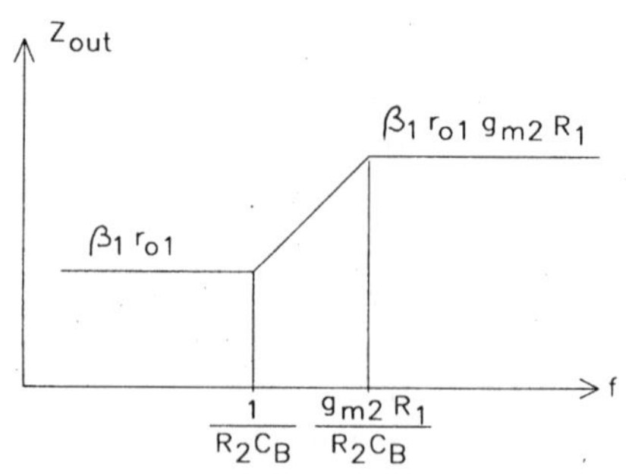

Fig. 4. ISAAC plot of the output impedance of the bipolar cascode stage.

From (5), the impedance levels and the pole and zero can easily be retrieved:

$$z_{\text{out}}(\text{low } f)=\frac{GM1}{G\pi1\ GO1} \quad (6)$$

$$z=-\frac{G2}{CB} \quad (7)$$

$$p=-\frac{GM2}{G1}\frac{G2}{CB}=\frac{GM2}{G1}z \quad (8)$$

$$z_{\text{out}}(\text{high } f)=\frac{GM1\ GM2}{G\pi1\ GO1\ G1}=\frac{GM2}{G1}z_{\text{out}}(\text{low } f). \quad (9)$$

At lower frequencies, this stage actually does not behave as a cascode stage, since the impedance at the emitter of $Q1$ is determined by $r_{\pi1}$ and not by $R_1 \cdot (g_{m2} r_{o2})$. However, at higher frequencies, the bootstrap capacitor C_B connects $r_{\pi1}$ and g_{o2} in parallel and the cascode formula holds again. The output impedance is now given by $R_1 \cdot (g_{m2} r_{\pi1}) \cdot (g_{m1} r_{o1})$. This behavior is summarized in Fig. 4. The symbolic simulator clearly gives more insight into the circuit than SPICE (Fig. 3) does.

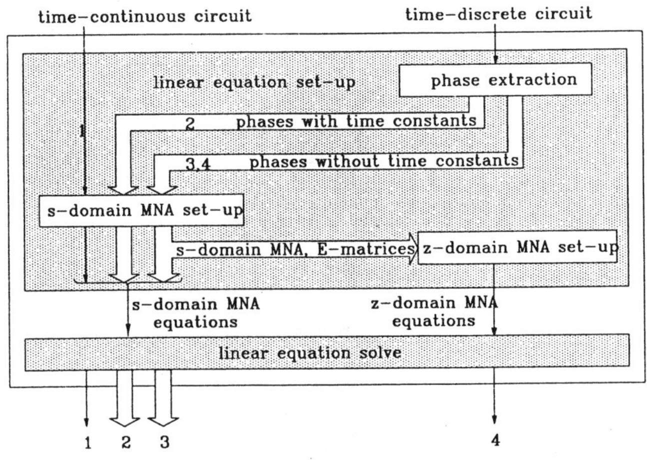

Fig. 5. ISAAC's four analysis modes.

B. Interactive Circuit Improvement

A symbolic simulator also enables interactive circuit improvement. The symbolic expressions immediately translate topology changes into performance changes. For example, the influence of adding an extra component can be examined in the expressions of the new circuit. In this way, the interactive synthesis of new high-performance circuits becomes feasible. Moreover, in [5], a method is described for the automatic generation of new circuit structures with the combined aid of a symbolic simulator and the Prolog language.

C. Repetitive Calculation of Formulas

Symbolic formulas are also useful if a circuit has to be evaluated repeatedly for several values of the components or the frequency [6]. It takes less time to evaluate the same analytic expression with one or a few parameters or the frequency varying over some range than to perform a full ac analysis each time. This can be useful in tolerance or large-scale sensitivity analysis.

D. Analytic Model Generation for Automatic Circuit Dimensioning

An important application of symbolic analysis is to integrate it as a CAD module into an analog silicon compiler. As opposed to digital circuits, the use of a fixed standard cell library is not optimal for analog circuits in terms of power, area, and overall performance. Depending on the specifications the transistor dimensions, power consumption and area occupation can widely vary for the same circuit. In addition, the range of circuit topologies used in analog design is quite broad. Too many cells are needed, which are not employed frequently enough.

Therefore, a more optimal solution to analog module generation is the use of a parameterized library. Circuits are characterized by a full analytic model, with free design variables left. These design variables are optimized according to the specifications of the actual application. In this way, the module generator can tailor all circuits to the actual application. However, as opposed to Degrauwe *et al.* [7] who create the formal circuit descriptions for IDAC after an in-depth analytic study, the circuit models should be generated automatically. This can be done by a symbolic simulator. It allows the fast inclusion of new circuits and does not restrict the module generator to a fixed set of topologies. Also, the use of simplified formulas can strongly speed up the optimization for large circuits. This design approach is adopted in the OPTIMAN program [8], which calls ISAAC to generate the ac characteristics of the circuit.

III. Structure and Capabilities of the ISAAC Program

A. General Program Description

The symbolic simulator ISAAC analyzes lumped, linear circuits in the frequency domain. Nonlinear circuits (CMOS, JFET, and bipolar) are linearized around the operating point, yielding the small-signal equivalent circuit. The program returns symbolic formulas for transfer functions, CMRR, PSRR, impedances, noise, etc. In this way, ISAAC can generate complete analytic models for building blocks such as amplifiers, filters, ladder networks, etc. The program has been written in COMMON LISP and is implemented on VAX, Texas Instruments EXPLORER, and Apollo DOMAIN computers.

ISAAC integrates the analysis of time-continuous and time-discrete circuits in one program. It currently provides the user with four analysis modes, depicted in Fig. 5 [9]. Time-continuous circuits, such as op amps and active *RC* filters, are analyzed in the *s*-domain (path 1 in Fig. 5). Time-discrete circuits, such as switched-capacitor filters, are analyzed with zero time constants in the *z*-domain (path 4 in Fig. 5). A typical example is the *z*-domain voltage gain from a full sample-and-hold input signal to the output sampled at phase 1 for the biquad of Laker and Fleischer (shown in Fig. 6):

$$\frac{\begin{array}{c} DK+DJ-AL-AH \\ +Z(-2DK+AL+AG-DJ-DI) \\ +Z^2D(K+I) \end{array}}{\begin{array}{c} AE-DB+Z(-AE+2DB-AC+DF) \\ +Z^2(-1)D(F+B) \end{array}} \tag{10}$$

In the presence of nonzero time constants, the analysis of a time-discrete circuit requires a combination of *s*- and *z*-domain analyses. This mode, however, is not yet provided in the program. On the other hand, each phase of a time-discrete circuit can be analyzed separately in the *s*-domain with nonzero time constants (path 2 in Fig. 5) or with zero time constants (path 3 in Fig. 5). As an example

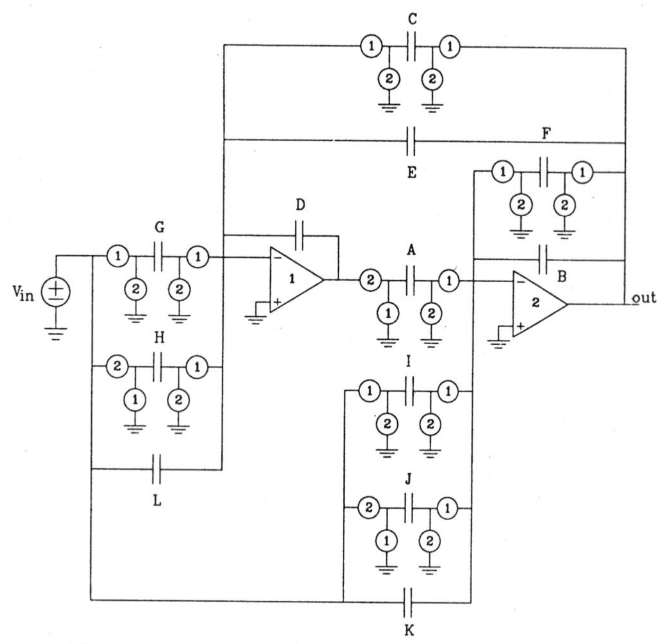

Fig. 6. Switched-capacitor biquad of Laker and Fleischer.

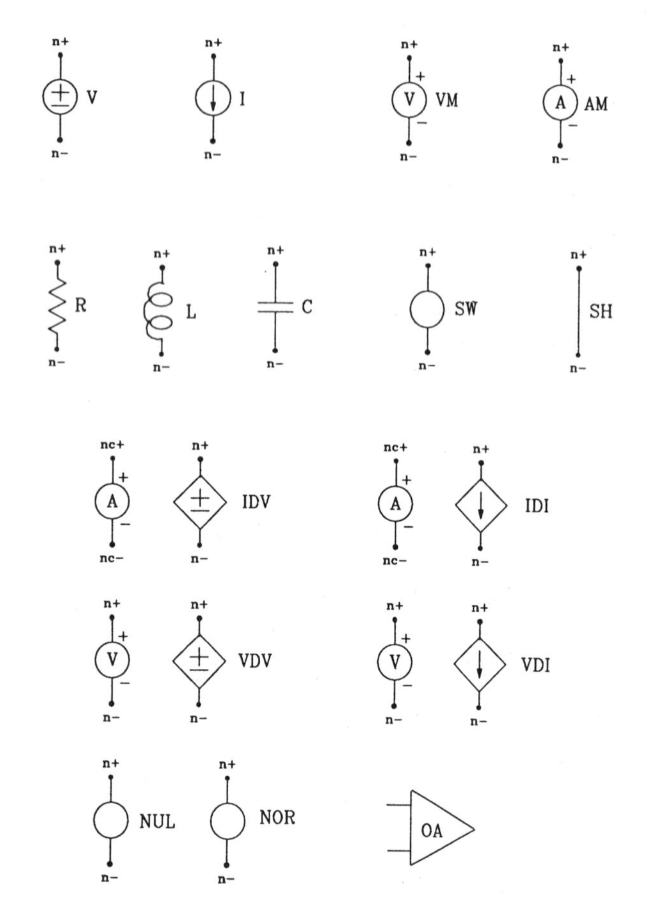

Fig. 7. Primitive elements available in ISAAC.

of an s-domain analysis without time constants, the load capacitance of the first op amp in the Laker and Fleischer biquad (Fig. 6) during phase 1 is given by

$$\frac{S\,D\,(C+E+G+H+L)}{C+D+E+G+H+L}. \tag{11}$$

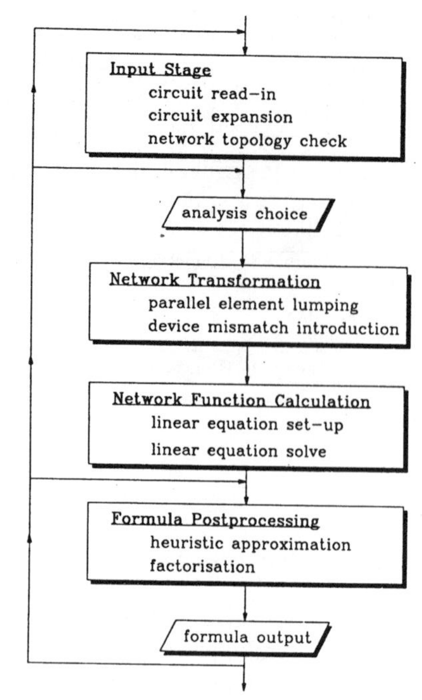

Fig. 8. Flowchart of the ISAAC program.

An example of an s-domain analysis with time constants is the noise transfer function during phase 2 from the noise source of one of the switches connected between ground and capacitor G to a voltage across capacitor G:

$$\frac{GON2}{GON2+S\,2\,G} \tag{12}$$

where $GON2$ is the on-conductance of the switch. This sampled noise voltage is then transferred to the biquad output during the next phase.

Fig. 7 shows the primitive elements available in ISAAC. These include all basic elements from circuit theory: independent voltage and current source, resistor, capacitor, inductor, current-controlled current and voltage source, voltage-controlled current and voltage source, nullator, and norator. Two measuring elements have been added: a current meter and a voltage meter. For time-discrete circuits a switch and a short have been introduced. For filters an ideal op amp has been added (in fact, a combination of nullator and norator). The user may define other nonprimitive elements by means of a subcircuit statement. These subcircuits are then expanded into their primitive components during the analysis.

As inputs, single and differential voltage and current sources can be selected. As outputs, single and differential node voltages and currents through elements can be selected. This enables, for example, the direct simulation of common-mode and differential-mode inputs and outputs.

B. Program Structure and Capabilities

Fig. 8 shows the flowchart of the ISAAC program. At the input stage, the circuit topology is read in by means of

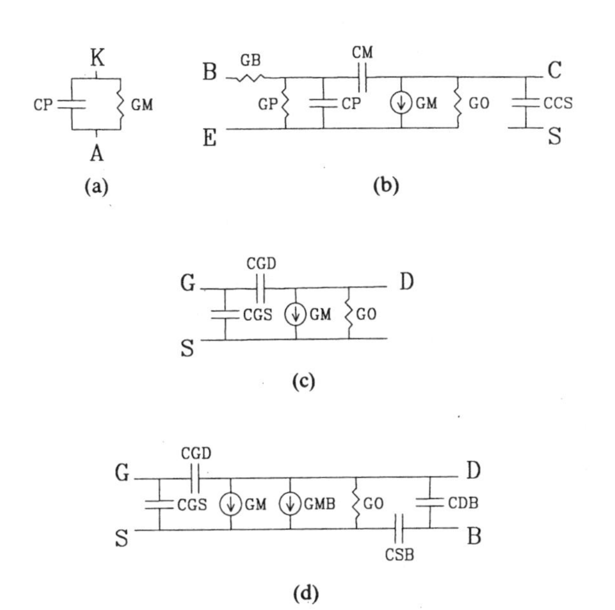

Fig. 9. Default small-signal models for (a) a diode, (b) a bipolar transistor, (c) a JFET transistor, and (d) a MOS transistor.

a conventional SPICE or SWAP [10] file for time-continuous and time-discrete circuits, respectively. In this way, compatibility with existing numerical simulators is established. Designers can use the same files and format as they did before. The only difference is that values do not have to be supplied.

Next, the circuit is expanded. Subcircuits are expanded into their primitive elements. Nonlinear devices such as diodes and transistors are replaced by their small-signal model. These models are controlled by the user through the model statements. As a general rule, the small-signal models are topologically the same as would be used in a SPICE run with the same input file. If no models are specified, default models are used. These default models are shown in Fig. 9(a)–(d) for a diode, a bipolar transistor, a JFET transistor, and a MOS transistor, respectively. These default models are quite general and independent of the technology process. The MOS model, for instance, is independent of the region of operation of the transistor. It is topologically valid both in the linear and saturation regions and in strong and weak inversion. Of course, the physical quantity represented by a particular symbol, its relation to geometrical aspects, and its magnitude do depend on the region of operation.

On the other hand, the more complex the small-signal expansion model of semiconductor devices becomes, the more terms the final expression contains. Most of these terms give only a minor contribution to the result, but they strongly increase the analysis time and storage requirements. This reveals the importance of the user-definable model facility. The user can select the appropriate model according to the application area he has in mind. High-frequency applications, for example, require a more sophisticated model than voice- or audio-band applications. So for the latter, the number of model parameters and the analysis time may be reduced without any loss of accuracy. Also, the small-signal model may differ from transistor to transistor. This allows unimportant parasitic capacitances to be dropped, which strongly speeds up the analysis.

A topology check is then performed on the expanded network. This prevents electrical and algorithmic problems afterwards. For example, loops of voltage sources are detected before any analysis is done.

Next, the analysis type is chosen and the excitation source, the input(s), and output(s) are interactively assigned by the user. The following possibilities are provided in the analysis menu:

ISAAC analysis menu	
Which analysis do you want to perform on circuit TEST?	
exit to main menu	: 0
signal transfer function	: 1
internal transfer function	: 2
transfer function ratio	: 3
rejection ratio	: 4
differential mode gain	: 5
common mode gain	: 6
CMRR	: 7
power supply gain	: 8
PSRR	: 9
noise transfer function	: 10
input-referred noise	: 11
Make your choice (0–11):	

Besides the general options (transfer function and transfer function ratio), some frequently used analyses for op amps are provided, such as differential-mode gain, input-referred noise, PSRR, etc.

Once the analysis has been chosen, two network transformations are performed. A first transformation is possible for matched components, such as transistors $M1A$ and $M1B$ in the CMOS Miller compensated two-stage op amp of Fig. 10. The small-signal components of these matching devices can be represented by different symbols:

$$\begin{array}{ll} g_{m1A} \rightarrow g_{m1A} & g_{m1B} \rightarrow g_{m1B} \\ C_{GS1A} \rightarrow C_{GS1A} & C_{GS1B} \rightarrow C_{GS1B} . \\ \cdots & \end{array} \quad (13)$$

However, since these devices are nominally identical, they can be represented by the same symbols:

$$\begin{array}{ll} g_{m1A} \rightarrow g_{m1} & g_{m1B} \rightarrow g_{m1} \\ C_{GS1A} \rightarrow C_{GS1} & C_{GS1B} \rightarrow C_{GS1} . \\ \cdots & \end{array} \quad (14)$$

This accelerates the network function calculation and improves the readability of the result. However, for second-order effects such as the PSRR, device mismatching becomes important. That is why mismatch terms are represented explicitly in ISAAC:

$$\begin{aligned} &g_{m1A} \to g_{m1} && g_{m1B} \to g_{m1} + \Delta g_{m1BA} \\ &C_{GS1A} \to C_{GS1} && C_{GS1B} \to C_{GS1} + \Delta C_{GS1BA} \cdot \\ &\cdots \end{aligned} \tag{15}$$

As a drawback, these mismatch terms slightly increase the analysis time.

As a second transformation, like parallel elements are lumped into one equivalent element and processed as one symbol. For example, the parallel capacitors C_L, C_{DB3}, C_{DB4}, and C_{GD4} in the Miller op amp of Fig. 10 can be replaced by one equivalent capacitor. This technique emulates the designer concept of node impedances. Technically, this transformation further reduces computer time and storage. Running ISAAC, the user interactively determines for each analysis whether mismatch terms are included and whether parallel elements are lumped or not. In this way, the analysis time can be reduced as much as possible, since, for instance, mismatch terms are of no importance for a differential-mode gain, but are for a PSRR. The influence of these transformations on the CPU time and storage is examined more thoroughly in Section IV-D.

A third option provided in ISAAC is to select between a fully symbolic analysis and a mixed numerical–symbolic analysis. In the latter case, all elements for which a numerical value is known are represented by this value. The other elements are represented by a symbol. If all elements have been assigned numerical values, the program returns expressions in symbolic form with respect to the complex frequency s or z only. If no elements have been assigned numerical values, the program returns fully symbolic expressions. Of course, all these options may be selected for each element separately, or for the whole circuit at once.

Example: These network transformations are now illustrated in the PSRR expression for the positive supply voltage for the op amp of Fig. 10. An approximation with 25 percent yields

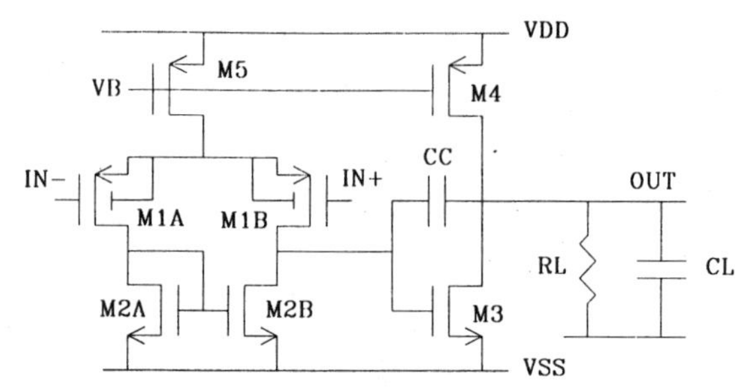

Fig. 10. CMOS Miller-compensated two-stage op amp.

with the following lumped elements being used:

$$\begin{aligned} CEQ1 &= CDB5 + CGD5 \\ CEQ2 &= CC + CGD3 \\ CEQ3 &= CDB2 + 2\ CGS2 \\ CEQ4 &= CDB4 + CGD4 \\ CEQ5 &= CDB2 + CGS3. \end{aligned} \tag{16}$$

At lower frequencies, a partial cancellation of two terms, $2\ g_{m1}\ g_{m2}\ g_{o4}\ (g_{o1} + g_{o2})$ and $-g_{m1}\ g_{m3}\ g_{o5}\ (g_{o1} + g_{o5})$, is noticed. These two terms correspond to the contribution of the first and the second stage. Also, the mismatch terms $\Delta g_{m2BA}\ g_{m1}\ g_{m3}\ g_{o5}$ and $-\Delta g_{m1BA}\ g_{m2}\ g_{m3}\ g_{o5}$ give a contribution which is of the same order of magnitude as the other terms and which may not be neglected. At higher frequencies, the PSRR behavior is determined by the lumped capacitors $CEQ1$, $CEQ2$, $CEQ3$ $CEQ4$, and $CEQ5$.

If all elements are represented by their value for a typical design (GBW = 1 MHz, C_L = 10 pF, PM = 60°), the

$$\frac{\begin{aligned} &2\ GM1\ GM1\ GM2\ GM3 \\ &+ S(-2)\ CEQ2\ GM1\ GM1\ GM2 \\ &+ S^2(-1)\ CEQ2\ GM1\ (CEQ3\ GM1 + 2\ CGS1\ GM2) \\ &+ S^3(-1)\ CEQ2\ GM1\ (CEQ3\ CGS1 + CDB1\ CEQ3 + 1/2\ CEQ1\ CEQ3 \\ &\qquad + CDB1\ CGS1) \\ &+ S^4 CEQ2\ CGD1\ (CEQ3\ CGS1 + CDB1\ CEQ3 + 1/2\ CEQ1\ CEQ3 \\ &\qquad + CDB1\ CGS1) \end{aligned}}{\begin{aligned} &2GM1\ GM2\ GO1\ GO4 + 2\ GM1\ GM2\ GO2\ GO4 + \Delta GM2BA\ GM1\ GM3\ GO5 \\ &\quad - \Delta GM1BA\ GM2\ GM3\ GO5 - GM1\ GM3\ GO1\ GO5 - GM1\ GM3\ GO2\ GO5 \\ &+ S\ GM1\ (2\ CEQ2\ GM2\ GO4 + 2\ CEQ5\ GM2\ GO4 - CEQ3\ GM3\ GO5 \\ &+ S^2(2\ CEQ2\ CEQ4\ GM1\ GM2 + 2\ CEQ4\ CEQ5\ GM1\ GM2 \\ &\quad - CEQ1\ CEQ3\ GM1\ GM3 + 2\ CEQ2\ CEQ3\ GM1\ GO4 \\ &\quad + CEQ2\ CEQ3\ GM1\ GO5 + 2\ CEQ2\ CGS1\ GM2\ GO4) \\ &+ S^3(2\ CEQ2\ CEQ3\ CEQ4\ GM1 + CEQ1\ CEQ2\ CEQ3\ GM1 \\ &\quad + 2\ CEQ2\ CEQ4\ CGS1\ GM2 + 2\ CEQ3\ CEQ4\ CEQ5\ GM1) \\ &+ S^4(2\ CEQ2\ CEQ3\ CEQ4\ CGS1 + 2\ CDB1\ CEQ2\ CEQ3\ CEQ4 \\ &\quad + 2\ CEQ3\ CEQ4\ CEQ5\ CGS1 + CEQ1\ CEQ2\ CEQ3\ CEQ4 \\ &\quad + CDB1\ CEQ1\ CEQ2\ CEQ3 + 2\ CDB1\ CEQ2\ CEQ4\ CGS1) \end{aligned}}$$

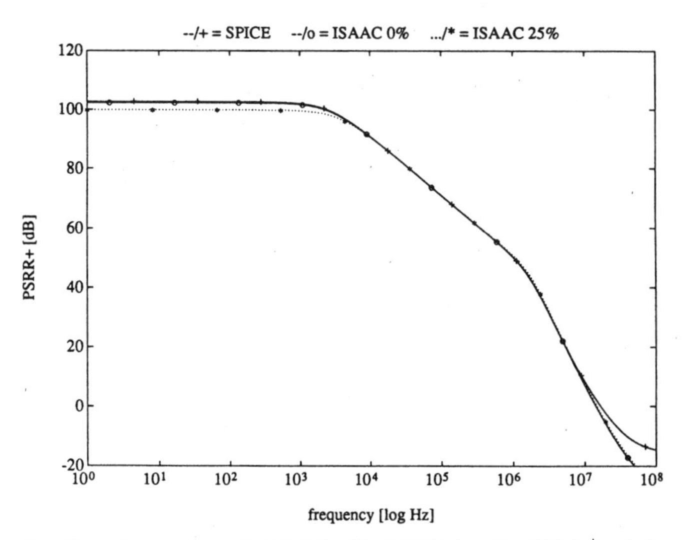

Fig. 11. Comparison of ISAAC with SPICE for the PSRR$^+$ of the Miller op amp.

exact expression in s for the positive PSRR of the Miller op amp becomes

$$\frac{\begin{array}{c}283587.56 + S\,10947.523 + S^2\,20.855215 \\ + S^3\,(-0.373244) + S^4\,0.000324661\end{array}}{\begin{array}{c}2.19817 + S\,127.54224 + S^2\,20.21014 \\ + S^3\,1.3461461 + S^4\,0.015403658\end{array}}. \tag{17}$$

After these transformations, the linear circuit equations are set up (see Fig. 8). The desired network functions are then calculated by a dedicated equation solution routine, which is described in Section IV-B. It returns exact expressions in the complex frequency s or z. Usually, these expressions are long and difficult to interpret, due to the large number of symbolic terms. However, taking into account the (order of) magnitude of all symbolic quantities, most of the terms provide only a minor contribution to the final result. Therefore, a heuristic approximation is provided, which simplifies the expressions up to a user-defined error percentage. The approximation algorithm is described in Section IV-C. The magnitude of a symbol itself can be determined in three ways. First, if the input file contains numerical values, these values are used. Secondly, if no value is provided in the input file, the user can supply an (order of) magnitude for the symbol during the interactive use of the program. Finally, ISAAC employs default values. These built-in defaults are average values, taking into account the relative magnitude of the different quantities. For example, taking the output conductance of a MOS transistor equal to 1.10^{-6}, the transconductance is set to 100.10^{-6}.

The simplified results are then factorized and returned to the user or stored in an output file. It is also possible to numerically evaluate the exact or simplified expression over a user-supplied frequency range. This yields plots which visualize the analytic behavior. These plots may be compared to SPICE output, to get an idea of the effective overall error introduced by the approximation. Fig. 11 compares the exact expression (17) and the 25-percent error approximated expression (16) with SPICE simulations of the PSRR$^+$ for the typical design mentioned above. In these analyses, mismatches of 1-percent were introduced. Notice how close the ISAAC curves follow the SPICE result up to frequencies far beyond the gain bandwidth (1 MHz). Notice also that although the maximum error percentage for (16) was 25 percent, the effective error for the magnitude of the PSRR$^+$ is much smaller over the whole frequency range. This is due to the error definition used, as will be explained in Section IV-C.

After each analysis (see Fig. 8), the user may try another error percentage on the same expressions. He can ask for another analysis on the same circuit, or he can read in a new circuit.

IV. Algorithmic Aspects

A. Equation Setup

In the literature, many symbolic analysis techniques have been presented for linear networks. These methods may be classified as follows:

1) matrix manipulation and determinant calculation [4], [11], [12];
2) signal-flow-graph method [2], [13];
3) tree-enumeration method [14], [15];
4) parameter-extraction method [16], [17]; and
5) numerical interpolation method [3], [18].

This has resulted in programs such as TAPLAN, SNAP, NAPPE [16], SNAPEST [12], and SAPEC [4]. Unfortunately, these systems generate exact expressions only and are not tuned to the needs of analog designers. Indeed, the two novel features, the heuristic approximation and the introduction of mismatch terms, make ISAAC especially useful for the instruction and assistance of analog designers and for analytic circuit modeling. Also, ISAAC is the first program which integrates s-domain and z-domain analyses within the same routines.

In the program, a determinant calculation method is implemented. The circuit equations are formulated according to a compacted modified nodal analysis (CMNA) method:

$$\begin{bmatrix} \underline{A} & \underline{B} \\ \underline{C} & \underline{D} \end{bmatrix} \begin{bmatrix} \underline{v} \\ \underline{i} \end{bmatrix} = \begin{bmatrix} \underline{E} \\ \underline{J} \end{bmatrix} \tag{18}$$

where $\underline{v}$ is the vector of node voltages, $\underline{i}$ is the vector of branch currents, and $\underline{E}$ and $\underline{J}$ contain the contributions of independent voltage and current sources. The CMNA method differs from the standard MNA method [19] in that branch currents are only incorporated for current meters and dependent current sources, and not for voltage sources. Also, several compactions are carried out for current meters, shorts, nullators, and independent voltage sources. The terminal voltages of these elements are equal or differ by a constant value. Hence, both node voltages are represented by one variable, thereby eliminating one row and column in the matrix [9]. Similar technique is used for the z-domain CMNA equation [9].

The CMNA method has the advantage that it reduces the matrix size and the number of term cancellations during the determinant calculation. As compared to the MNA method, a compaction from 11 to 3 equations is observed for the Vogel biquad [4]. A compaction from 49 to 2 equations is observed for both phases of the Laker and Fleischer biquad (Fig. 6).

B. Equation Solution Algorithm

The network functions are obtained by calculating two determinants with a dedicated algorithm. The efficiency is established by exploiting the sparse nature of the CMNA matrices and by sharing common subresults. Actually, the determinants are calculated by a sparse Laplace expansion algorithm with storage of minors [9]. The determinant $|D|$ is developed along the column j according to the following formula:

$$|D|_{nxn} = \sum_{\substack{i=1 \\ d_{ij} \neq 0}}^{n} (-1)^{i+j} d_{ij} |M_{ij}| \tag{19}$$

where M_{ij} is the matrix obtained from D by removing row i and column j. The summation in (19) is extended over the nonzero elements only. The minors $|M_{ij}|$ are recursively calculated by means of (19). A lot of time and memory is saved because term cancellations in every minor $|M_{ij}|$ are elaborated before the minor is used on a higher recursion level. In addition, every subminor is calculated only once and stored in a memo hashtable, from where it is retrieved when it is needed again. The CPU times, obtained with this algorithm, are comparable to topological methods, as will be illustrated in Section IV-D.

C. Approximation Algorithm

Approximating an expression means pruning insignificant terms based on the magnitude of these terms. Mathematically, it is trading off accuracy (error) against simplicity (complexity). The complexity of an expression is measured by the number of terms it contains in its canonical sum-of-products form. The error definition being used is the accumulated absolute error ϵ_A [9]:

$$\epsilon_A = \left. \frac{\sum |t_i(\underline{x})|}{|g(\underline{x})|} \right|_{\underline{x} = \underline{x}_o} \tag{20}$$

where $g(\underline{x})$ is the original expression, $t_i(\underline{x})$ are the pruned terms, and $\underline{x}_o$ is the point of evaluation. Notice that in the numerator absolute values are summed. In this way, the error ϵ_A used in ISAAC is a worst-case error, an overestimation of the effective error. However, this error definition is appropriate in the case of a nominal cancellation of large-magnitude terms, especially in the presence of large tolerances and covariances. An example is the dc value of the PSRR+ of the Miller op amp (16), where the contributions of the first and the second stage may nominally cancel. Hence, discarding these canceling terms does not change the effective (nominal) error, but strongly reduces the insight in the cancellation mechanism and yields an underestimation of the error if parameter values differ due to process tolerances. Actually, the error ϵ_A to be supplied by the user is more a parameter to control the information content of the resulting formula than a measure of the effective error. Further details on this approximation are presented in [20]. Also, from the error definition (20), it is clear that the explicit introduction of mismatch terms allows a much more accurate error estimation, since mismatch terms are smaller than the corresponding nominal parameters.

The approximation algorithm itself consists of two steps [9] and is consistent with error definition (20). First, all terms are discarded which are smaller than the fraction ϵ_{max} of the original expression's mean value. This means that all terms $t_i(\underline{x})$ are discarded for which

$$\frac{|t_i(\underline{x}_o)|}{|g(\underline{x}_o)|} \leqslant \frac{\epsilon_{max}}{n} \tag{21}$$

with n being the number of terms in the original expression $g(\underline{x})$ and ϵ_{max} the maximum error (as supplied by the user). The remaining terms are then sorted according to their magnitude and the smallest terms are removed as long as $\epsilon < \epsilon_{max}$. At the end, a small backtrace is done to reinclude terms of the same order as the last term which was not discarded. This approximation can be done for a specific value of the frequency or over the whole frequency range (coefficient per coefficient). The latter method works well for the magnitude response. However, in particular cases, rather large errors in zero or pole positions may occur, even with a small error tolerance on the individual power series coefficients.

D. ISAAC Performance

The performance of ISAAC is comparable to that of topological methods, although CPU time values may differ due to differences in computers. In this paper, CPU times are given for a Texas Instruments EXPLORER (1 MIPS, 8-Mbyte physical memory, 100-Mbyte virtual memory). The active *RC* filter of Fig. 1 requires 0.156 s of CPU time, which is of the same order as in [2]. The switched-capacitor biquad of Fig. 6 takes 0.718 s. However, typical CPU times are 8 and 26 s for the voltage gain of the Miller op amp (Fig. 10) and of a fifth-order switched-capacitor ladder filter, respectively.

The influence of lumping and matching on the CPU time and storage requirements is illustrated for the differential gain of the Miller op amp (Fig. 10). This gain has been analyzed with matching (14) and without matching

(13), and with lumping and without lumping. The CPU time and the total number of terms in the expression are summarized in the following table for the four analyses:

	without lumping	with lumping
without matching	77 s/7533 terms	19 s/982 terms
with matching	44 s/3469 terms	8 s/495 terms.

Clearly, both transformations, lumping and matching, strongly reduce the CPU time and the complexity of the result. Notice also that the fully symbolic expression contains 495 terms for a circuit with four internal nodes and seven transistors (33 symbolic elements).

For the operational transconductance amplifier (OTA), the CPU time for a fully symbolic analysis with all parasitic capacitors included is 76 s. This is far below the 32 min mentioned in [11]. However, for larger circuits the number of terms and the analysis time rapidly increase, especially for a fully symbolic analysis of linearized (small-signal) circuits. The expression for the OTA, for example, already contains 2866 terms. Clearly, incorporating all parasitic capacitances for large circuits becomes impractical.

V. Conclusions

A software package, ISAAC, is presented for the symbolic analysis of analog integrated circuits. ISAAC generates symbolic expressions for the ac characteristics of linear (or linearized) time-continuous and time-discrete circuits. This yields analytic formulas for transfer functions, PSRR, CMRR, impedances, noise, etc. In addition, the expressions can be simplified with a heuristic criterion based on the magnitude of the elements. This yields simple and interpretable formulas, which provide analog designers with insight into the circuit behavior. Also, the explicit representation of mismatch terms allows the accurate calculation of second-order effects, such as the PSRR. In this way, the program has proven itself to be a valuable design aid for experienced and novice designers, complementary to numerical simulators. Moreover, the program generates complete analytic circuit models, which are used for automatic sizing in a non-fixed-topology analog module generator.

In the future, ISAAC's routines will be improved to further reduce the analysis time for large circuits by hierarchical decomposition, exploitation of topological knowledge, and application of rules. The final circuit size aimed at with the symbolic simulator is of the order of practical building blocks, such as amplifiers, buffers, leapfrog filters, etc. Also, new functions being planned include an approximate symbolic pole–zero extraction, sensitivity, nonlinear distortion, and transient and large-signal analyses.

Acknowledgment

The authors are grateful to I. Rutten, W. Reijntjens, and T. Kostelijk for their cooperation and discussions, and Philips Industries, the Netherlands, for the logistic support. Special thanks are addressed to L. Callewaert, for the testing and evaluation of the program and the many useful suggestions.

References

[1] W. Sansen, G. Gielen, and H. Walscharts, "Symbolic simulator for analog circuits," in *ISSCC Dig. Tech. Papers*, 1989, pp. 204–205.

[2] J. A. Starzyk and A. Konczykowska, "Flowgraph analysis of large electronic networks," *IEEE Trans. Circuits Syst.*, vol. CAS-33, no. 3, pp. 302–315, Mar. 1986.

[3] K. Singhal and J. Vlach, "Symbolic analysis of analog and digital circuits," *IEEE Trans. Circuits Syst.*, vol. CAS-24, no. 11, pp. 598–609, Nov. 1977.

[4] A. Liberatore and S. Manetti, "SAPEC—A personal computer program for the symbolic analysis of electric circuits," in *Proc. ISCAS*, 1988, pp. 897–900.

[5] A. Konczykowska and M. Bon, "Automated design software for switched-capacitor IC's with symbolic simulator SCYMBAL," in *Proc. 25th ACM/IEEE Design Automation Conf.*, 1988, pp. 363–368.

[6] P. M. Lin, "A survey of applications of symbolic network functions," *IEEE Trans. Circuit Theory*, vol. CT-20, no. 6, pp. 732–737, Nov. 1973.

[7] M. G. R. Degrauwe *et al.*, "IDAC: An interactive design tool for analog CMOS circuits," *IEEE J. Solid-State Circuits*, vol. SC-22, no. 6, pp. 1107–1116, Dec. 1987.

[8] G. Gielen, H. Walscharts, and W. Sansen, "Analog circuit design optimization based on symbolic simulation and simulated annealing," in *Proc. ESSCIRC*, 1989, pp. 252–255.

[9] H. Walscharts, G. Gielen, and W. Sansen, "Symbolic simulation of analog circuits in *s*- and *z*-domain," in *Proc. ISCAS*, 1989, pp. 814–817.

[10] "SWAP 2.2 reference manual," Silvar Lisco, Heverlee, Belgium, Doc. M-037-2, 1983.

[11] S. J. Seda, M. G. R. Degrauwe, and W. Fichtner, "A symbolic analysis tool for analog circuit design automation," in *Proc. ICCAD*, 1988, pp. 488–491.

[12] M. K. Tsai and B. A. Shenoi, "Generation of symbolic network functions using computer software techniques," *IEEE Trans. Circuits Syst.*, vol. CAS-24, no. 6, pp. 344–346, June 1977.

[13] R. R. Mielke, "A new signal flowgraph formulation of symbolic network functions," *IEEE Trans. Circuits Syst.*, vol. CAS-25, no. 6, pp. 334–340, June 1978.

[14] J. T. Chow and A. N. Willson, Jr., "A microcomputer-oriented algorithm for symbolic network analysis," in *Proc. ISCAS*, 1985, pp. 575–577.

[15] S. D. Shieu and S. P. Chan, "Topological formulation of symbolic network functions and sensitivity analysis of active networks," *IEEE Trans. Circuits Syst.*, vol. CAS-21, no. 1, pp. 39–45, Jan. 1974.

[16] G. E. Alderson and P. M. Lin, "Computer generation of symbolic network functions—A new theory and implementation," *IEEE Trans. Circuit Theory*, vol. CT-20, no. 1, pp. 48–56, Jan. 1973.

[17] P. Sannuti and N. Nath Puri, "Symbolic network analysis—An algebraic formulation," *IEEE Trans. Circuits Syst.*, vol. CAS-27, no. 8, pp. 679–687, Aug. 1980.

[18] J. K. Fidler and J. I. Sewell, "Symbolic analysis for computer-aided circuit design—The interpolative approach," *IEEE Trans. Circuit Theory*, vol. CT-20, no. 6, pp. 738–741, Nov. 1973.

[19] L. O. Chua, C. A. Desoer, and E. S. Kuh, *Linear and Nonlinear Circuits*. New York: McGraw-Hill, 1987.

[20] H. Walscharts, G. Gielen, and W. Sansen, "Exact and approximated symbolic analysis of analog circuits in *s*- and *z*-domain," submitted to *IEEE Trans. Circuits Syst.*

Georges G. E. Gielen (S'87) was born in Heist-op-den-Berg, Belgium, on August 25, 1963. He received the engineering degree in electronics from the Katholieke Universiteit Leuven, Heverlee, Belgium, in 1986.

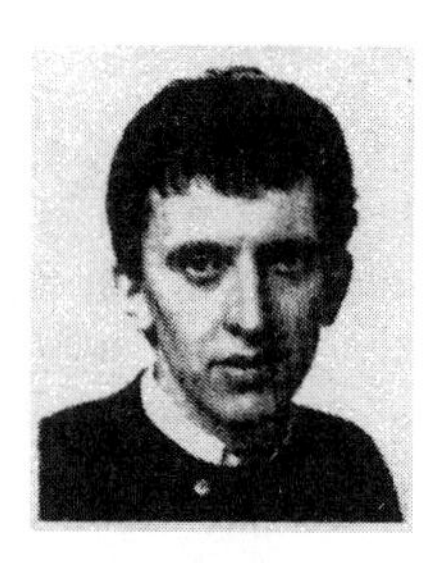

In the same year he was appointed by the National Fund of Scientific Research (Belgium) as a research assistant, at the Laboratory Elektronika, Systemen, Automatisatie en Technologie (ESAT), Katholieke Universiteit Leuven, where he is now working towards the Ph.D. degree in electronics. His research interests are in the design of analog integrated circuits and in analog design automation (synthesis, optimization and layout).

Herman C. C. Walscharts was born in Antwerpen, Belgium, on July 22, 1961. He received the engineering degree in electronics from the Katholieke Universiteit Leuven, Heverlee, Belgium, in 1984.

In the same year he joined the Laboratory Elektronika, Systemen, Automatisatie en Technologie (ESAT), Katholieke Universiteit Leuven, where he is currently working towards the Ph.D. degree in electronics. In the period from 1985 to 1988 he was supported by the Belgian I.W.O.N.L. Recently, he joined Bell Telephone M. C., Belgium. His interests are in design of analog integrated circuits and in analog design automation.

Willy M. C. Sansen (S'66–M'72–SM'86) was born in Poperinge, Belgium, on May 16, 1943. He received the engineering degree in electronics from the Katholieke Universiteit Leuven, Heverlee, Belgium, in 1967 and the Ph.D. degree in electronics from the University of California, Berkeley, in 1972. In 1968 he was employed as an Assistant at the Katholieke Universiteit Leuven. In 1971 he was employed as a Teaching Fellow at the University of California. In 1972 he was appointed by the N.F.W.O. (Belgian National Education) as a Research Associate, at the Laboratory Elektronika, Systemen, Automatisatie en Technologie (ESAT), Katholieke Universiteit Leuven, where he has been full Professor since 1981. Since 1984 he has been the head of the Department of Electrical Engineering. In 1978 he spent the winter quarter as a Visiting Assistant Professor at Stanford University, Stanford, CA; in 1981 at the Technical University of Lausanne; and in 1985 at the University of Pennsylvania, Philadelphia. His interests are in device modeling, the design of integrated circuits, and medical electronics and sensors.

Dr. Sansen is a member of the Koninklijke Vlaamse Ingenieurs Vereniging (K.V.I.V.), the Audio Engineering Society (A.E.S.), the Biotelemetry Society, and Sigma Xi. In September 1969 he received a CRB Fellowship from the Belgian American Educational Foundation, in 1970 a G.T.E. Fellowship, and in 1978 a NATO Fellowship.

Interactive AC Modeling and Characterization of Analog Circuits via Symbolic Analysis

F.V. FERNÁNDEZ, A. RODRÍGUEZ-VÁZQUEZ, AND J.L. HUERTAS
Department of Design of Analog Circuits, Centro Nacional de Microelectrónica, 41012-Sevilla, Spain

Received February 12, 1991; Revised April 10, 1991.

Abstract. An advanced symbolic analyzer, called ASAP, has been developed for the automatic ac modeling of analog integrated circuits. ASAP works on a data base of model primitives and provides error-free symbolic expressions for the different system functions of analog circuits composed by the primitives. Both complete and simplified expressions can be calculated. Two simplification criteria have been implemented. The basic one is based on pruning the least significant terms in the different system function coefficients. This may yield important errors in pole and zero locations. To avoid that, an improved criterion has been developed where pole and zero displacements are forced to remain bounded. Also implemented are routines for symbolic pole/zero extraction and parametric ac circuit characterization. ASAP uses the signal flow graph method for symbolic analysis and has been written in the C language for portability. Together with portability, efficiency and ability to manage complexity have been fundamental goals in the implementation of ASAP. These features are demonstrated in this paper via practical examples.

1. Introduction

Current achievable complexity levels for commercial VLSI technologies make it possible to implement specialized analog/digital systems into single application-specific chips (ASICs). Although the digital part of typical ASICs represents the largest percentage in terms of chip area and complexity (in the sense of number of transistors), it is widely accepted that most of the design effort for these circuits has to be devoted to the analog portion. One reason for that is that analog circuits, even elementary building blocks, display very complicated nonlinear relationships among performance specifications and circuit design parameters. Also, the number of specifications to deal with is much larger than in the digital case, the specifications themselves varying over wide ranges from one to another application.

Analog integrated circuits are typically designed by following an iterative approach based, on the one hand, upon an electrical numerical simulator for validation purposes and, on the other, on circuit knowledge for interpretation of the electrical simulator outputs and proper updating of the design parameters from one iteration to another. Whether the design procedure be manual or done by a knowledge-based CAD tool [1], [2], [3], [4], knowledge acquisition is recognized to be a fundamental cornerstone. However, the derivation of proper sets of design equations for analog building blocks is a difficult task, which requires intensive use of modeling and circuit analysis techniques. Even for experienced analog designers, increasing knowledge by modeling a new circuit topology may lead to a critical waste of time until an error-free set of equations is derived [2].

Recently it has been suggested to use symbolic analyzers to help in the process of generating design equations for analog circuits [5], [6]. A *symbolic analyzer* is a CAD tool for the automatic ac analysis of analog circuits in symbolic, as opposed to numerical, form. Automating the symbolic analysis of analog circuits is not a new idea. An excellent survey of the methods and applications of symbolic analyzers was already written by Lin in 1973 [7]. A more recent survey can be found in [8]. Starting with SNAP from Purdue University [9], several symbolic analyzers have been constructed during the last two decades [7], [8]. However, as stated in [6], old programs do not fit the needs of present analog designers. This fitting is carried out in two more recent tools, SYNAP [5] and ISAAC [6], via the incorporation of novel features such as matching, lumping, and simplification of expressions. In particular, simplifications are indispensable due to the large number of terms appearing in exact expressions corresponding to typical analog circuits, which makes these exact expressions useless for hand manipulations.

Both SYNAP and ISAAC are written in Common Lisp and use matrix manipulation methods [8] to solve the circuit equations. However, an important difference exists in that SYNAP uses a symbolic mathematics program, MACSYMA [10], to perform algebraic manipulations, while ISAAC is a self-contained tool, thereby yielding much higher efficiency. On the other hand, SYNAP and ISAAC share the following drawbacks: (1) They provide symbolic expressions for numerators and denominators of circuit system functions and do not give any additional information. Since for most practical circuits the associated system functions contain a huge number of terms, there is still a lot of work to be done by the user for interpretation of the results and, hence, for exploitation of the tool capabilities. (2) They use a simplification technique which may result in large pole/zero displacements and, hence, in important modeling errors.

In this paper a tool for automatic ac modeling of analog integrated circuits is presented which overcomes above drawbacks. This tool is based on a new advanced core symbolic analyzer, called ASAP (*A*nalog *S*ymbolic *A*nalysis *P*rogram). *Efficiency, portability, accuracy,* and *interpretability* have been the fundamental issues in the implementation of the proposed tool. For portability, ASAP has been written in the C language. Efficiency is achieved by proper exploitation of the adequacy of the C language to deal with the signal flow graph symbolic analysis method [8], [11]. Accuracy is guaranteed by providing mechanisms for careful control of pole/zero displacements when a simplification is carried out. For interpretability, a procedure for symbolic pole/zero calculation has been devised and incorporated in ASAP. Besides, different interfaces have been implemented allowing parametric analysis to be carried out and results to be displayed graphically which may help considerably for interpretation of the program results.

The paper is organized as follows. Section 2 describes the envisioned role of ASAP in the analog design procedure. An analysis example is used to illustrate program potential for providing useful information to the analog designer. ASAP inputs and outputs are summarized in Section 3, where the criteria used for simplification of expressions are discussed via their application to a CMOS Miller op amp. In Section 4 program performance is discussed via benchmark analog building blocks. Finally, Section 5 is devoted to discuss the more relevant ASAP algorithms and implementation details.

2. A Rationale for Symbolic Analyzers

Analog circuits are not typically designed by analytical methods, but by a semiempirical approach where design parameters are iteratively updated until specifications are achieved. Fundamental to this methodology is the availability of an accurate electrical numerical simulator to evaluate circuit performance after each new setting of the design parameters. However, electrical numerical simulators are simply validation tools, providing neither interpretation of the results nor guidelines for parameter updating. To illustrate that, consider the MOS inverter of figure 1a and assume we are interested in its voltage gain. Several numerical simulations are required to conclude that the voltage gain increases in module as $(W/L)_1$ increases and $(W/L)_2$ decreases. Many more simulations have to be run to assess the influence of the transistor channel lengths L_1 and L_2.

For efficiency of the heuristic analog design procedure, proper updating of the design parameters,

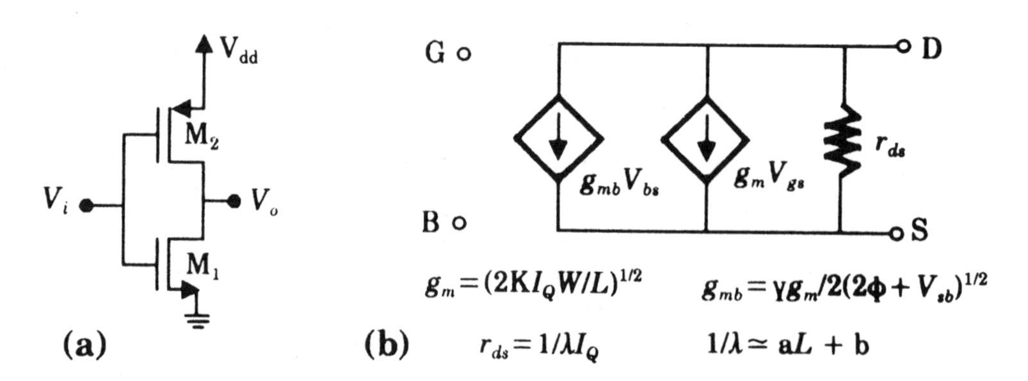

Fig. 1. (a) CMOS simple amplifier; (b) simplified MOS transistor model.

which requires a deep knowledge of the circuit operation, may be more critical than the availability of an accurate electrical numerical simulator. Coming back to the circuit of figure 1, a deeper insight about the properties of the voltage gain can be obtained via a small signal analysis using the transistor model of figure 1b. This gives

$$A_o = \frac{g_{m1} + g_{m2}}{g_{ds1} + g_{ds2}} \qquad (1)$$

from where, and taking into account the dependence of g_m and g_{ds} on W and L, it is possible to know which parameters influence the gain together with the degree of influence of each one of them. These pieces of information provide the designer with guidelines to interpret results from the numerical simulator and, hence, to yield efficiency in the semiempirical design approach.

In general, for a given analog circuit, analyzing it by keeping all or part of the circuit components as symbols provides insight into the circuit operation and, hence, is an invaluable help for the analog design procedure. Unfortunately, and although fig. 1 can be easily solved by hand to get (1), hand symbolic analysis of even elementary blocks containing a few transistors is a time-consuming, tedious, and error-prone task.

The role of symbolic analyzers (in particular, ASAP) in the analog design procedure can be conceptualized as illustrated in fig. 2. There, ASAP is to be interactively used by the designer to assess the influence of parameter variations or even the use of alternative circuit architectures and, hence, to provide support for optimum updating of the design parameters.[1] For this purpose the ASAP capabilities for pole/zero extraction and parametric analysis appear to be specially appealing, as can be seen in what follows with the help of an example.

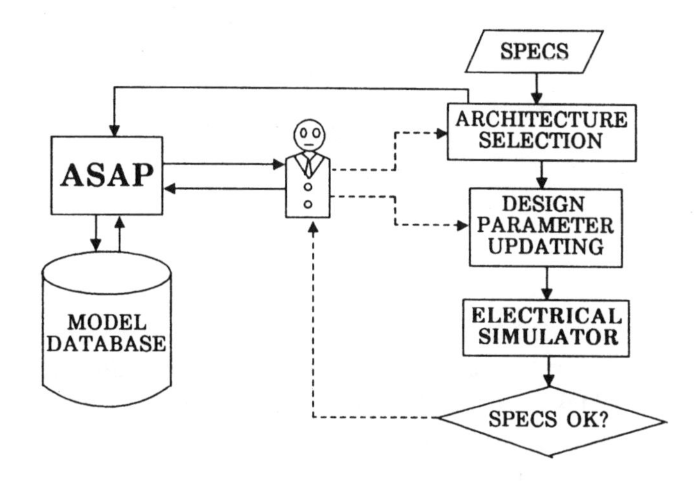

Fig. 2. Illustrating the role of ASAP in the analogue design procedure.

Consider the CMOS current-mirror structures shown in figure 3. Figure 4 shows, on the other hand, the four levels of default models implemented in ASAP for the MOS transistor. Assume we are interested in the current gain (I_o/I_i) of the simple mirror of figure 3a

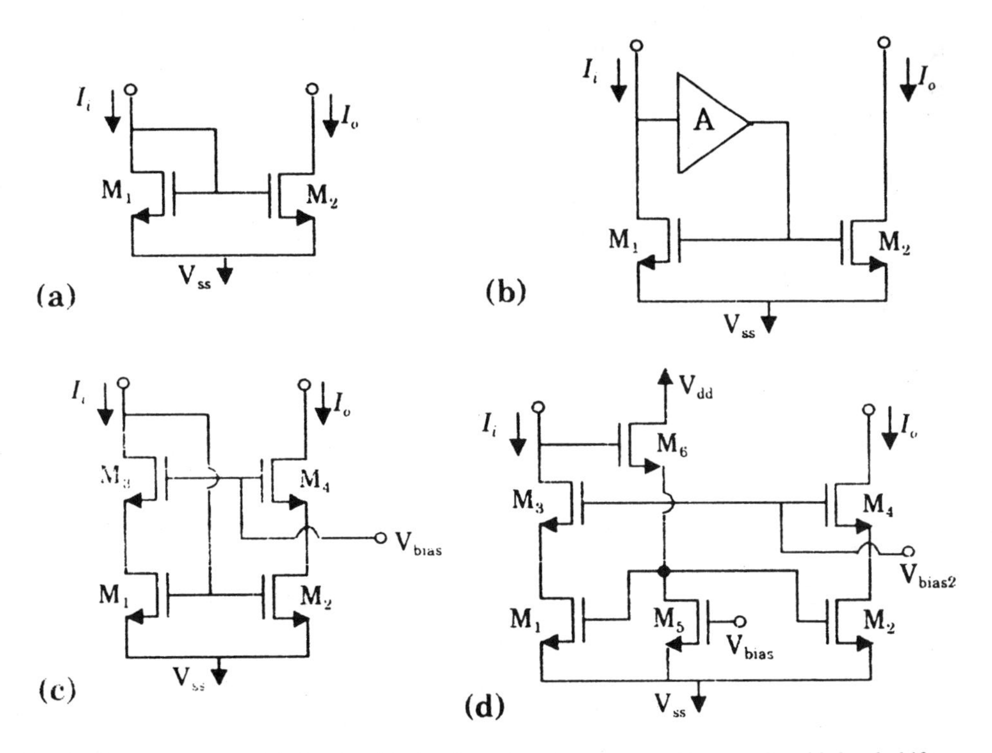

Fig. 3. Current mirrors: (a) simple; (b) simple active; (c) cascode; (d) cascode with level shifter.

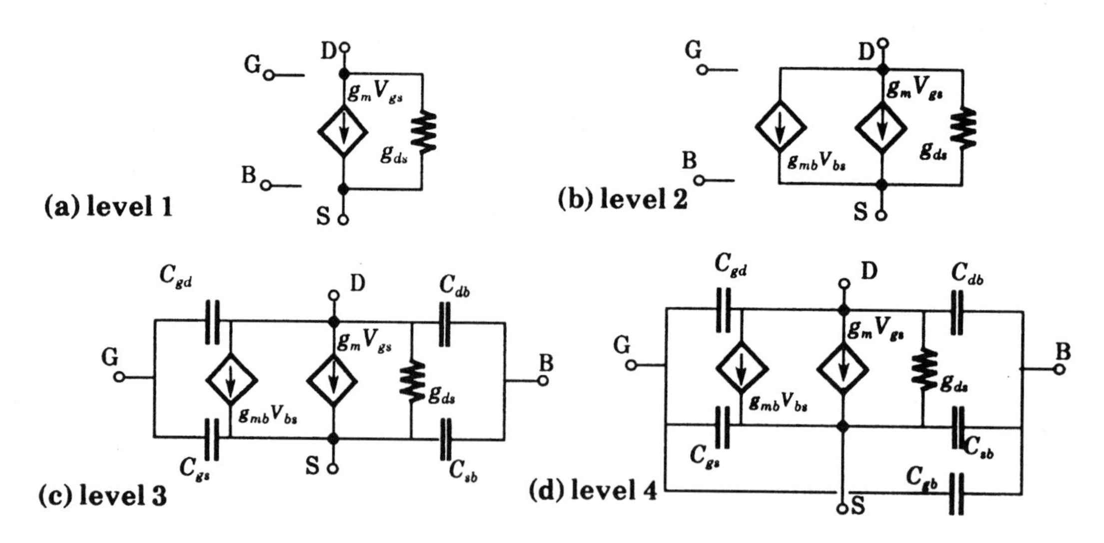

Fig. 4. Modeling levels for the MOS transistors.

when loaded by a resistor and a capacitor in parallel and using the level 3 transistor model of figure 4c. ASAP takes 0.6 s CPU time (measured on a SUN3/260 workstation with 4 MIPS and 8-Mbyte physical memory) in providing a symbolic expression for this gain,

$$\frac{I_o(s)}{I_i(s)} = \frac{N(s)}{D(s)} \tag{2}$$

where the numerator and denominator polynomials are given in table 1. There we can observe the ac response of the circuit is characterized by two poles and two zeros. We can ask ASAP to interactively extract these poles and zeros and to represent them versus the bias current, as it appears in figure 5a.[2] In figure 5a the upper diagram is for the real part of the roots, while the lower one is for the imaginary part. In both cases the Y axis corresponds to the bias current in logarithmic scale. Besides, a logarithmic scale has been used for the X axis. Positive (alternatively, negative) values on this axis correspond to logarithms of positive (alt., minus logarithm of modulus of negative) numbers whose moduli are greater than 1. If smaller-than-1 numbers in the modulus are present, a separate decimal representation of that area is provided by ASAP. It can be observed that pole p_1 and zero z_2 cancel each other in part of the current range. It remains a high frequency pole, p_2, which increases as current increases. We can hence conclude that this kind of mirror is very fast. Unfortunately, its output impedance is rather low and its input impedance is rather high [14], which may produce significant errors in case these circuits are cascaded, which occurs when using current-mode analog techniques [15]. Two solutions are a priori possible to overcome these errors: (a) keep the input nodes to a constant voltage, (b) increase the output impedance.

The circuit in figure 3b [16] addresses the first of these solutions. The amplifier decouples the drain and the gate of transistor M_1, introducing a negative feedback loop which reduces the input impedance and keeps the input node to a rather constant voltage. The numerical simulation of figure 3b with a bias current of some tens of microamps does not show any abnormal behavior. Thus we may conclude this circuit is acceptable. But, as figure 5b shows, the circuit is rather controversial. In this figure we can see a graphical representation (provided by ASAP) of the real and imaginary part of the poles and zeros of figure 5b as a function of the bias current. A simple macromodel was used for the amplifier. Amazingly, two complex conjugate poles appear in the right half of the s-plane for bias currents up to 7 μA. In fact, the numerical simulation using HSPICE [17] of this circuit for $I_{\text{bias}} = 5\ \mu$A showed

Table 1. Symbolic expression for the current gain of the simple current mirror of figure 3a.

Numerator =	Denominator =
$+ G_L g_{m2}$	$+(g_{ds1} + g_{m1})(G_L + g_{ds2})$
$+ s(C_L g_{m2} - C_{gd2} G_L)$	$+ s((G_L + g_{ds2})(C_{gs1} + C_{gs2} + C_{gd2} + C_{db1})$
$- s^2 C_L C_{gd2}$	$+ (g_{ds1} + g_{m1})(C_L + C_{gd2} + C_{db2}) + C_{gd2} g_{m2})$
	$+ s^2((C_L + C_{gd2} + C_{db2})(C_{gs1} + C_{gs2} + C_{db1}) + C_{gd2}(C_L + C_{db2}))$

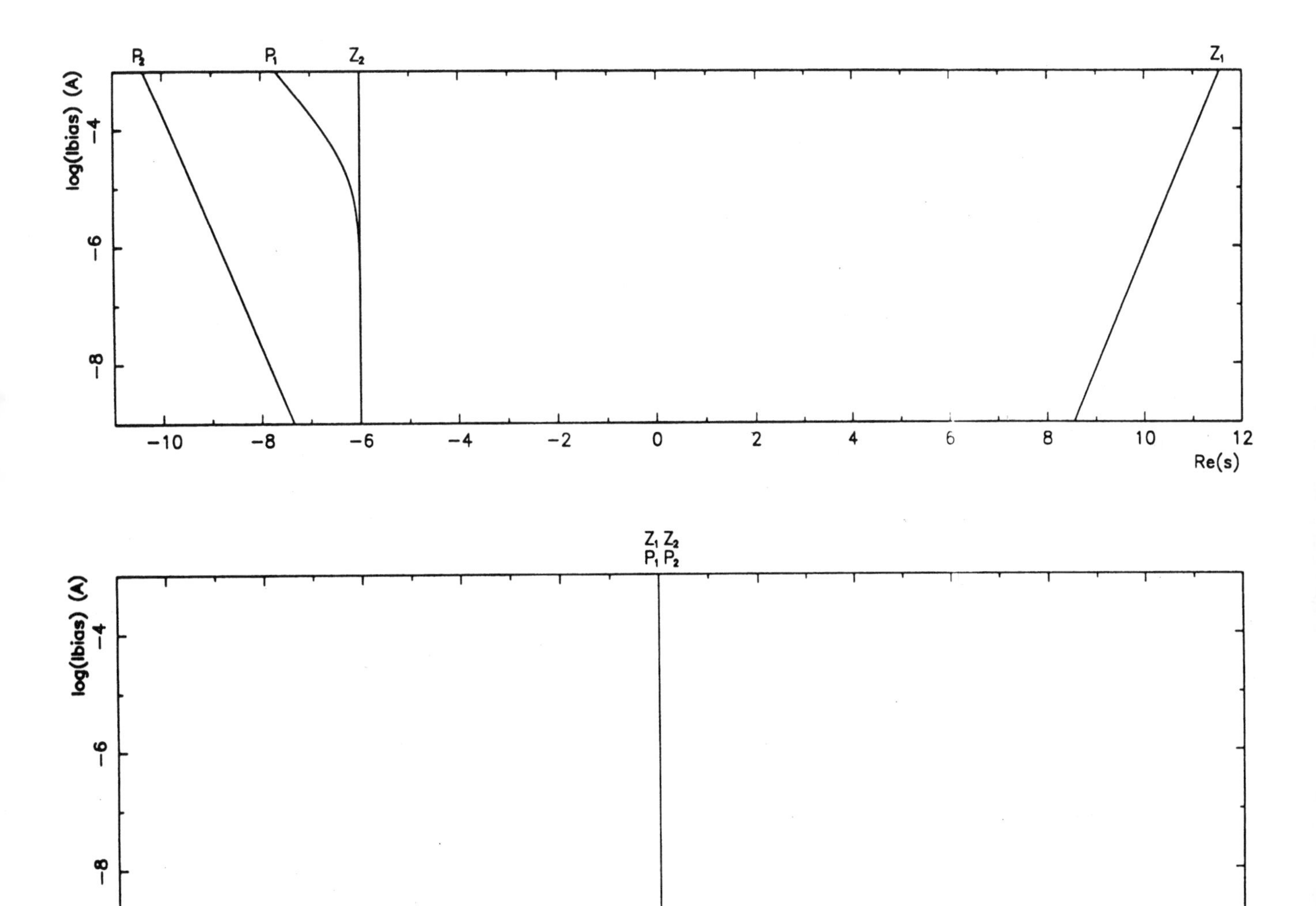

Fig. 5a. Root location of simple current mirror.

this instability. Without the information provided by ASAP, this problem could have kept hidden until the circuit test stage.

The solution of increasing the output impedance can be addressed by using cascode structures [14]. In particular, the circuit shown in figure 3c is frequently used in switched-current circuits [15]. Figure 5c represents the pole and zero locations of the current gain of this circuit as a function of the bias current. As can be seen, two real poles cancel with two zeros, with the result that the external behavior is determined by a positive zero and a pair of complex conjugate poles with negative real part. It can be observed that this circuit is slightly slower than the simple current mirror, although its output impedance can be shown, using ASAP for instance, to be considerably larger.

Such a circuit has an important problem in that it only allows for a small dynamic range of bias currents, the limits of the range being very sensitive to technological parameters. To increase this range, the circuit of figure 3d appears as a natural choice. The numerical simulation of this circuit for different circuit parameters and bias currents leads to instability without an easy way to find the boundary conditions. Again the analysis and the graphical representation using ASAP gives the key to the riddle. For a fixed set of circuit parameters figure 5d shows the real and imaginary parts of poles and zeros for different bias currents. A pair of complex conjugate poles appear, which move to the right half of the s-plane for a limited range of bias currents. Having this information helps the designer delimit the region of correct circuit operation.

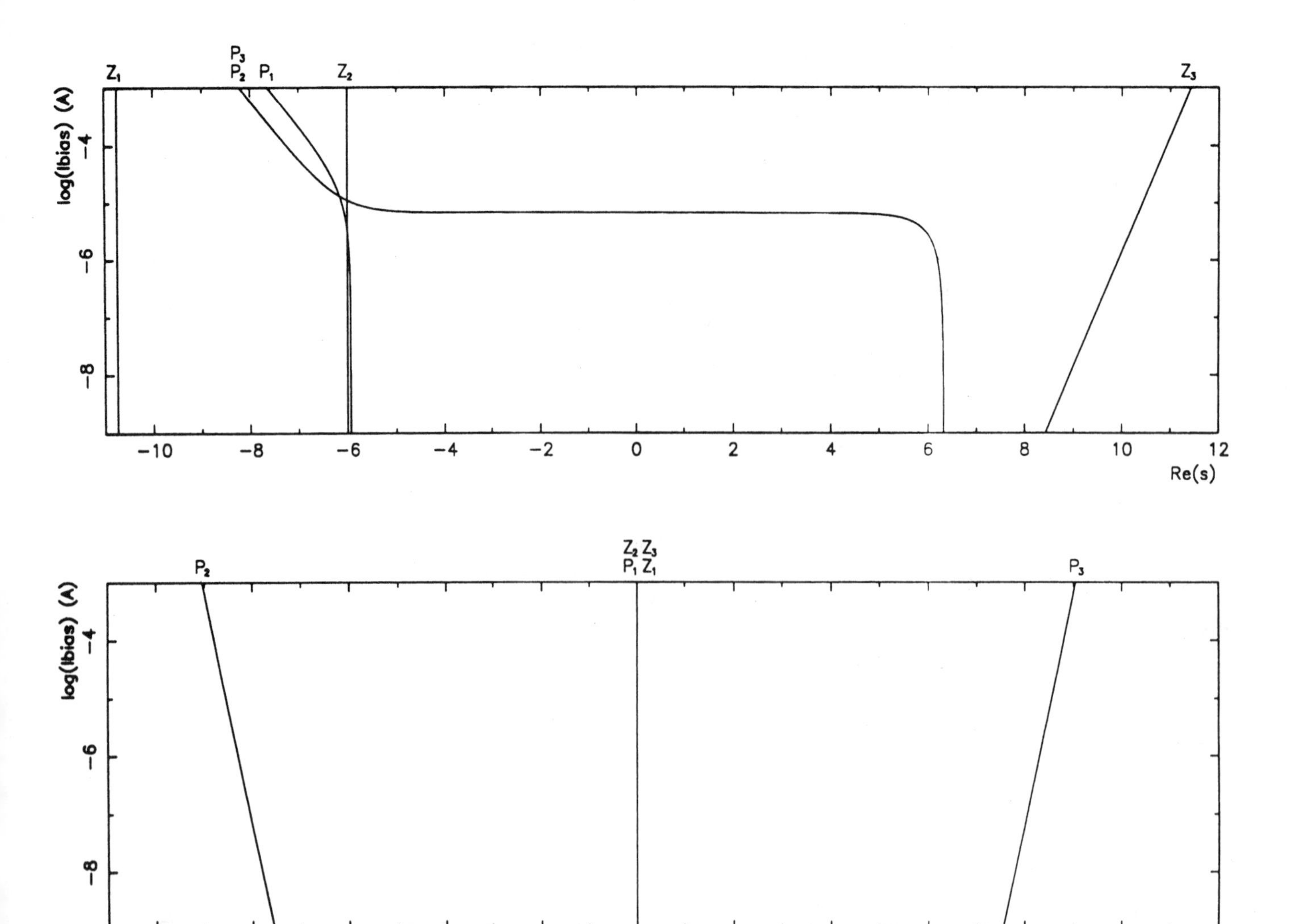

Fig. 5b. Root location of simple active current mirror.

Previous examples have illustrated how ASAP can be used as a valuable tool to study qualitative behavior of analog circuits. For all the cases CPU time required for the program is under 4 s, thereby allowing for interactive usage. It should be clear, on the other hand, that much more computer and human effort would have been needed to get the conclusions outlined above by just using hand analysis and a numerical simulation tool. Also many other examples can be found where the use of pole/zero plots versus some design variables can be a useful guide for design. Just to mention one case, this type of representation has been recently used to study feedforward compensation for CMOS op amps [18].

3. Program Description

3.1. ASAP Inputs and Outputs

Figure 6 is a conceptual block diagram showing the main operation flow of the herein reported symbolic analyzer. Inputs to the program include the following:

- *Description of the circuit topology* in SPICE format. Primitives for ASAP include resistors, capacitors, inductors, independent voltage and current sources, controlled sources (VCCS, VCVS, CCCS, and CCVS), semiconductor devices (diodes, BJTs, JFETs,

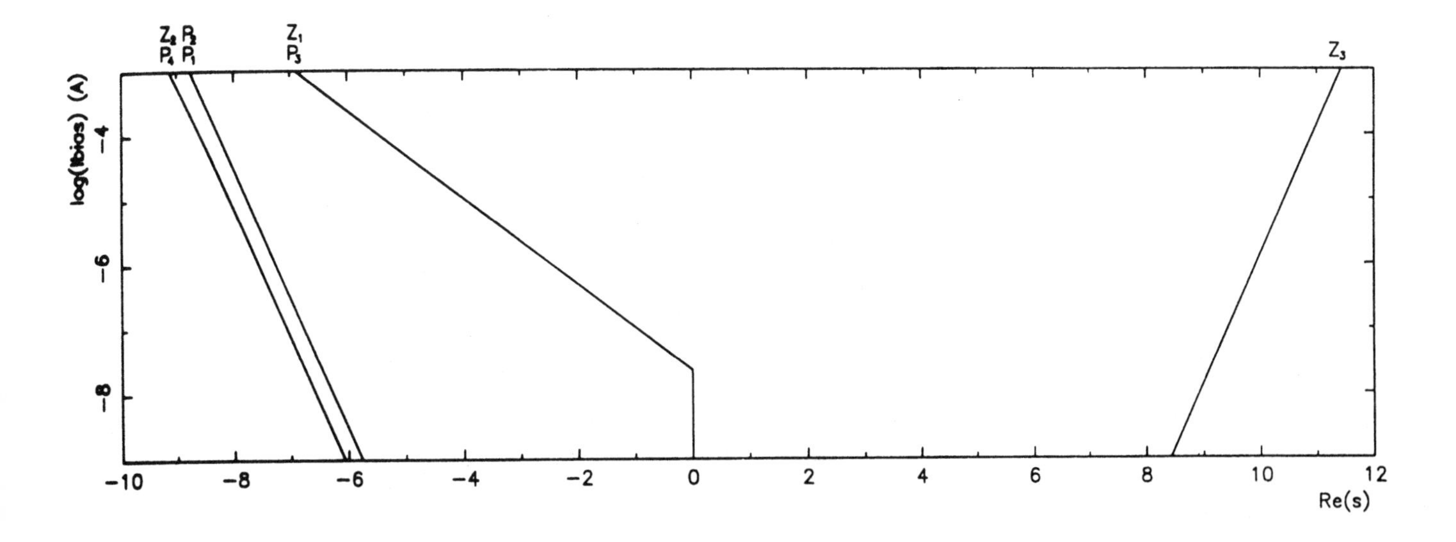

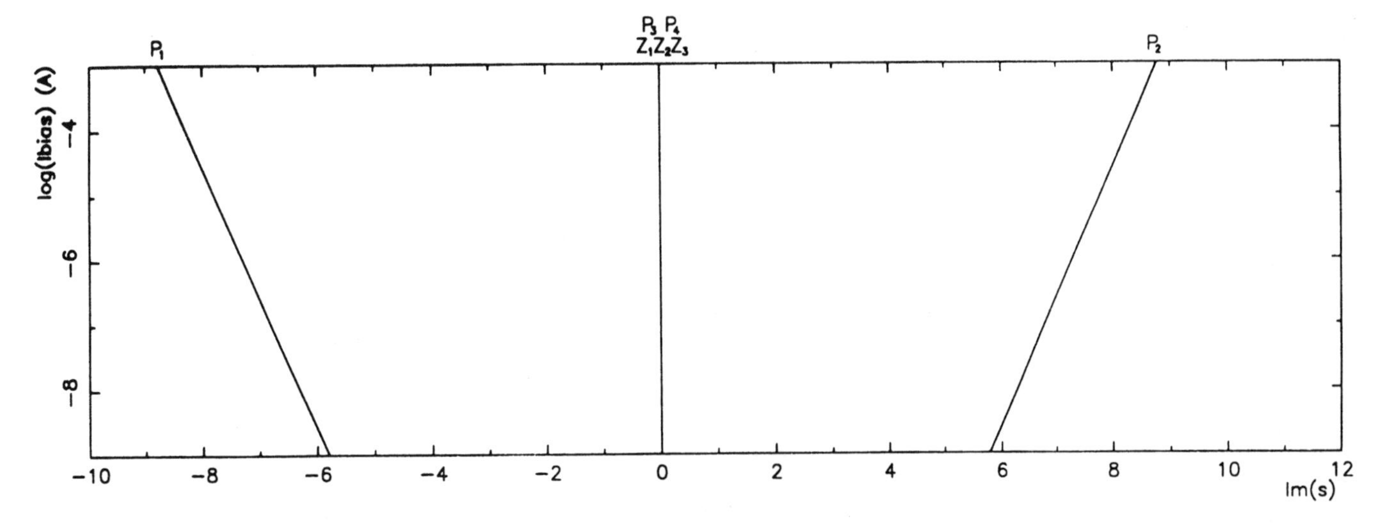

Fig. 5c. Root location of cascode current mirror.

MOSFETs, and GaAs MESFETs), op amps, and OTAs (see figure 7).

- *Level of modeling* for the semiconductor devices and for the amplifiers. Device model schematics are taken from a library where different level models are stored. For the sake of example, figure 4 shows the four levels considered for MOS transistors.
- *Relative parameter size* for formula simplifications. This is optional. In case this information is not provided, default values are used for this purpose.
- *Requested analyses and error margin* for formula simplification.
- *Matched devices information*. Mismatches can explicitly appear as symbols.

Outputs provided by the program are in the form of system functions [13],

$$H(s, \mathbf{x}) = \sum_{i=0}^{N} s^i f_i(\mathbf{x}) \Big/ \sum_{j=0}^{M} s^j g_j(\mathbf{x}) \qquad (3)$$

where $\mathbf{x}^{\mathrm{T}} = \{x_1, x_2, \ldots, x_Q\}$ is the vector of circuit symbolic parameters associated with the primitives (i.e., g_m, g_{ds}, g_{mb}, etc.) and $f_i(\mathbf{x})$ and $g_j(\mathbf{x})$ are polynomial functions in the parameter symbols.

Transfer characteristics (voltage and current gains, transimpedances, and transadmittances) and driving point immitances can be asked for as output system functions. Formulas for other ac specifications like

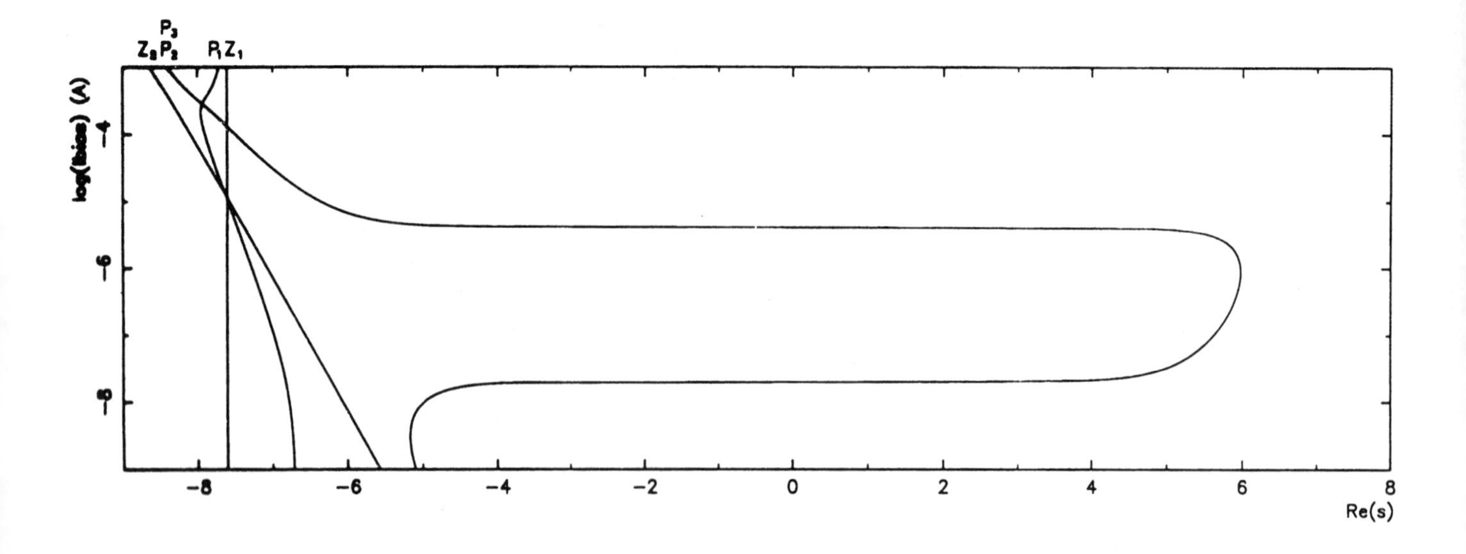

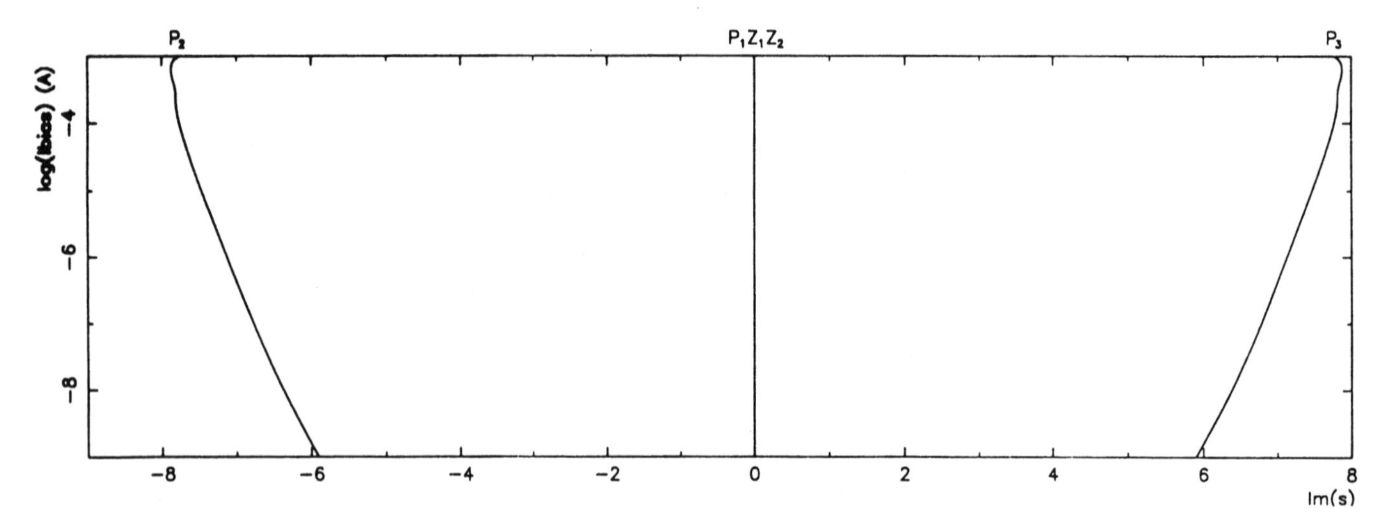

Fig. 5d. Root location of cascode current mirror with level shifter.

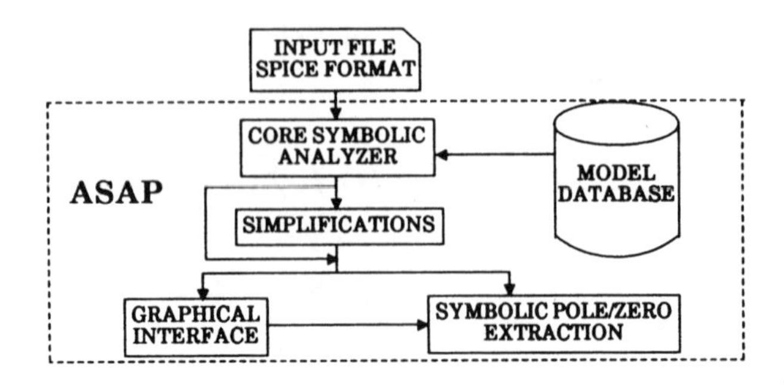

Fig. 6. Conceptual diagram of ASAP.

CMRR, PSRR, etc., can also be calculated. ASAP is able to provide complete or simplified expressions of the different network functions asked for in the input file. Besides, approximate symbolic expressions for poles and zeros can be calculated in an interactive way. To this purpose, and starting from numerical information about pole/zero positions the program uses heuristic rules to suggest possible approximations that are performed in case they are accepted by the user.

3.2. Formula Simplification in ASAP

As pointed out in Section 1, simplification of expressions are mandatory for hand analysis due to the large number of symbolic terms appearing even for low complexity analog blocks containing just a few transistors.

Recall the primary program outputs are in the form of rational system functions in the complex frequency s (see (3)), the coefficients of both the numerator and the denominator polynomials being in turn polynomial functions in the parameter symbols. In ASAP, simplifications are automatically performed by deleting the

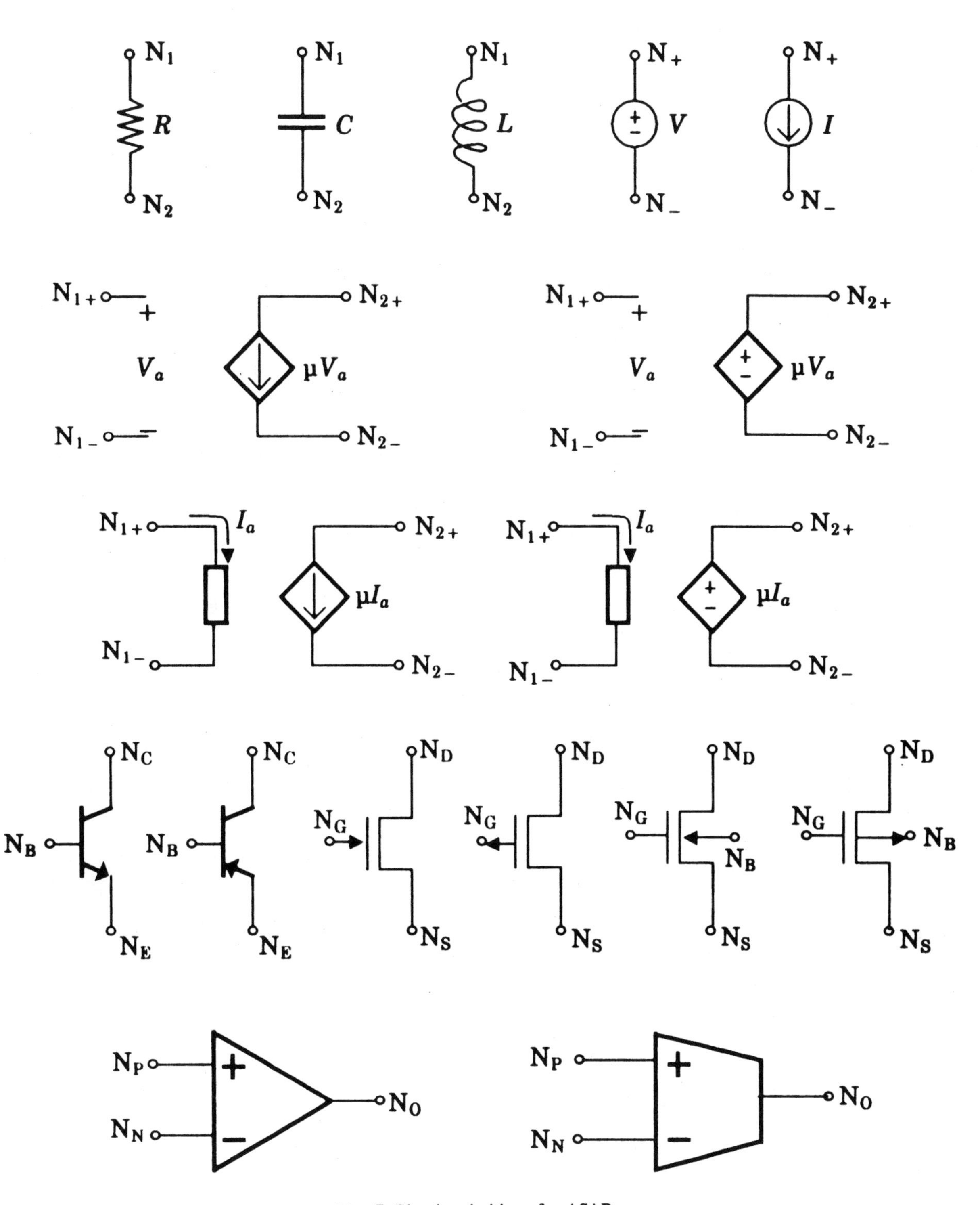

Fig. 7. Circuit primitives for ASAP.

least significant terms in the coefficient polynomials $f_i(\mathbf{x})$ and $g_j(\mathbf{x})$ ($1 \le i \le N$, $1 \le j \le M$) in (3). As stated, information about the relative size of the circuit parameters must be provided for simplifications to be carried out. Besides, the requested accuracy margin for dropping nonsignificant terms must be also given as an input. Assume an arbitrary symbolic polynomial,

$$h_k(\mathbf{x}) = \sum_{l=1}^{L} h_{kl}(\mathbf{x}) \tag{4}$$

where $h_k(\mathbf{x})$ represents either $f_i(\mathbf{x})$ or $g_j(\mathbf{x})$ in (3) and where the different terms $h_{kl}(\mathbf{x})$ are products of symbols. The basic technique used in ASAP for deleting the P least significant terms of (4) is

$$\sum_{l=1}^{P} |h_{kl}(\mathbf{x}_o)| \Big/ \sum_{l=1}^{L} |h_{kl}(\mathbf{x}_o)| < \epsilon \qquad (5)$$

where ϵ is the accuracy margin specified by the user.

Observe in (5) that a point $\mathbf{x}_0$ of the design parameter space has been considered for evaluation. Parameter values yielding this point can be provided by the user or left as defaults. In the more general case a finite set of evaluation points is required, since it is computationally impossible to go over the full range of variation of the different symbolic parameters. Notice also that summations in the numerator and denominator of (5) are made on the moduli of the different terms of $h_k(\mathbf{x})$. For the numerator, it is done to avoid problems in case there exist mutually canceling terms of significant magnitude, a common fact in calculating second-order characteristics. Concerning the denominator, summing the moduli of the terms instead of calculating the modulus of the sum of the signed terms avoids excessively conservative simplifications and disparities among the margin errors applied to the coefficients of the different powers of s in (3).

However, although the basic simplification criterion performs correctly for many cases, under some circumstances its use may lead to important errors in pole and zero locations. To illustrate that, let us consider the Miller op amp of figure 8a. Assume transistors in the differential pair are matched, and consider the level 3 MOS transistor model (repeated for convenience in figure 8b) to evaluate the op amp differential voltage gain. After simplification with a specified error margin $\epsilon = 20\%$,[3] table 2 results, giving the numerator and the denominator polynomials of the voltage gain.[4]

Figure 9 shows the pole locations as calculated by ASAP for the complete and the simplified expressions of the gain. As can be seen, 200% errors result in high frequency poles as a consequence of the simplification process.

To overcome this problem a new simplification criterion has been incorporated into ASAP. In this new procedure, elimination of the least significant terms of each symbolic polynomial is made via the iteration of (5), starting with a small value of ϵ and increasing it by $\Delta\epsilon$ from one iteration to another. At each iteration, pole and zero movements are monitored. Simplification

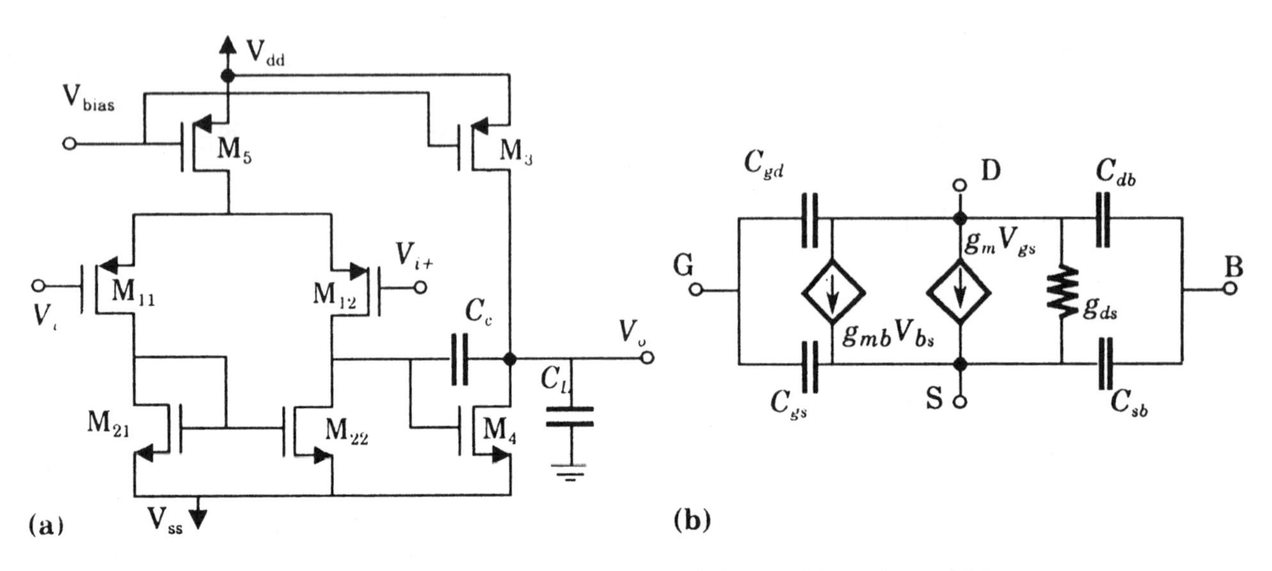

Fig. 8. (a) CMOS Miller op amp; (b) small signal MOS transistor model.

Table 2. Simplified expression for the differential voltage gain of figure 8a using the basic criterion.

Numerator =	Denominator =
$+ 2g_{m1}^{2}g_{m2}g_{m4}$	$+ 2g_{m1}g_{m2}(g_{ds1} + g_{ds2})(g_{ds3} + g_{ds4})$
$+ s(2C_{gs1}g_{m1}g_{m2}g_{m4} - 2g_{m1}^{2}g_{m2}C_c)$	$+ s2C_c g_{m1}g_{m2}g_{m4}$
$+ s^2(C_{gs1}g_{m1}g_{m4}(C_{db1} + 2C_{gs2})$	$+ s^2(2C_L g_{m1}g_{m2}(C_c + C_{gs4}) + 2C_cC_{gs1}g_{m2}g_{m4} + 4C_cC_{gs2}g_{m1}g_{m4})$
$- 2C_cC_{gs1}g_{m1}g_{m2} - 2C_cC_{gs2}g_{m1}^{2})$	$+ s^3(2C_{gs1}C_L g_{m2}(C_c + C_{gs4}) + 4C_{gs2}C_L g_{m1}(C_c + C_{gs4})$
$- s^3(C_cC_{gs1}g_{m1}(C_{db1} + C_{db2} + 2C_{gs2})$	$+ C_cC_{db1}C_L g_{m1} + 2C_cC_{db1}C_L g_{m2}$
$+ s^4C_cC_{dg1}C_{gs1}(C_{db1} + 2C_{gs2})$	$+ 2C_cC_{gs1}g_{m4}(C_{db1} + 2C_{gs2}) + 2C_cC_{gs1}C_{gs4}g_{m2})$
	$+ s^4(4C_cC_{gs2}C_L(C_{db1} + C_{gs1}) + 2C_cC_{gs1}C_L(C_{gd1} + C_{db1})$
	$+ 4C_{gs1}C_{gs2}C_{gs4}(C_c + C_L) + C_cC_{db1}C_LC_{db5})$

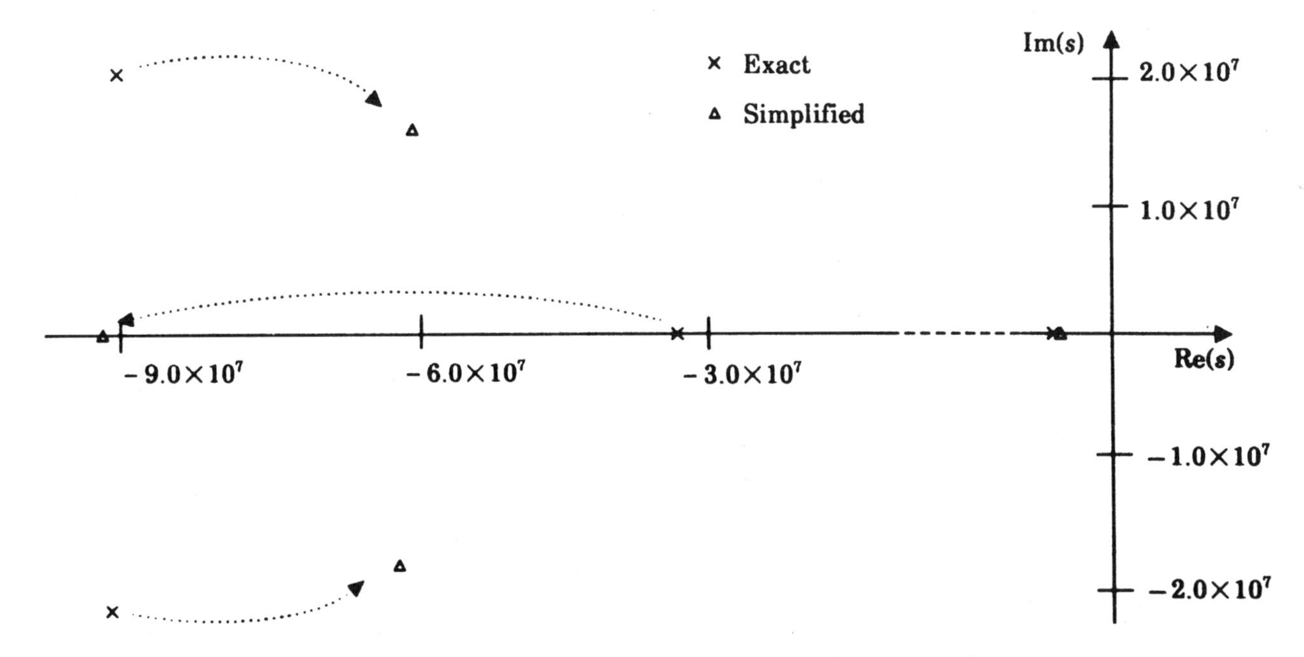

Fig. 9. Pole displacement for the Miller op amp of figure 8a using the basic simplification criterion.

stops when displacements are larger than a margin specified by the user. Consider the general case of a pair of complex conjugate roots. Assume for illustration purposes such critical frequencies are poles. Their contribution to the network function can be expressed as

$$H(s) = H'(s)\,\frac{1}{s^2 + 2\xi\omega_n\, s + \omega_n^2} \tag{6}$$

During simplification the roots are allowed to move provided ω_n and ξ keep inside a limited interval around the corresponding value for the exact expression. Control on ω_n and ξ instead of the real and imaginary part of the roots happens to be much more adequate (i.e., the imaginary part of high frequency poles with greater real part than the imaginary one is very sensitive to small variations in the polynomial coefficients, while its effect on circuit behavior is negligible).

When using this simplification criterion, it is also possible for the user to specify a frequency range where reducing the errors is specially important. In this case pole and zero displacements are more strictly limited in this band and constraints are gradually relaxed as root locations are far away from this region.

Table 3 has been calculated by ASAP for the Miller op amp of figure 8a when using the new simplification criterion. The corresponding poles are shown in figure 10. As can be seen, poles after simplification are kept much closer to the nominal positions than for the case of figure 9.

3.3. ASAP Graphical Interface

Symbolic expressions generated by ASAP can be formatted and dynamically loaded for graphical representation. Two basic graphic interfaces have been developed so that symbolic outputs of ASAP can be easily interpreted:

a. Representation of Bode diagrams
b. Representation of pole and zero loci

The first one allows us to draw magnitude and phase diagrams of any transfer function. Complete and simplified expressions can be represented as well as families of curves for different values of one generic parameter. By parameter we mean the value of any circuit component or transistor small signal parameter, any transistor dimension, or the current across any element.

The second allows us to find out pole and zero coordinates as a function of one parameter (in the same sense of parameter). As the symbolic expressions are dynamically loaded, compilers and machine characteristics impose limits on the expression size. Hence, simplified expressions must be used. Accuracy in graphic representations shows the importance of obtaining simplified expressions with small errors in zero and pole locations.

4. Illustrating ASAP Capabilities

Several examples are included to illustrate the topics of efficiency, ability to manage complexity, and exactness in the operation of ASAP. The

Table 3. Simplified expression for the differential voltage gain of figure 8a using the improved criterion.

Numerator =	Denominator =
$+ 2g_{m1}{}^2 g_{m2} g_{m4} + s(2C_{gs1} g_{m1} g_{m2} g_{m4}$ $\quad + 2g_{m1}{}^2 g_{m4} C_{gs2} - 2g_{m1}{}^2 g_{m2} C_c)$ $+ s^2(C_{gs1} g_{m1} g_{m4}(C_{db1} + 2C_{gs2})$ $\quad - 2C_c g_{m1} g_{m2}(C_{db1} + C_{gs1}) - 2C_c C_{gs2} g_{m1}{}^2)$ $- s^3(2C_c C_{gs2} g_{m1}(C_{db1} + C_{gs1})$ $\quad + C_c C_{gs1} g_{m1}(C_{db1} + C_{db2}))$ $+ s^4 C_c C_{gd1} C_{gs1}(C_{db1} + C_{db2} + 2C_{gs2})$	$+ 2g_{m1} g_{m2}(g_{ds1} + g_{ds2})(g_{ds3} + g_{ds4}) + s2C_c g_{m1} g_{m2} g_{m4}$ $+ s^2(2C_L g_{m1} g_{m2}(C_c + C_{gs4}) + 2C_c C_{gs1} g_{m2} g_{m4}$ $\quad + 2C_c C_{gs4} g_{m1} g_{m2} + C_c g_{m1} g_{m4}(C_{db1} + 4C_{gs2}))$ $+ s^3(2C_{gs1} C_L g_{m2}(C_c + C_{gs4}) + C_c C_L g_{m2}(2C_{db1} + C_{db5})$ $\quad + C_L g_{m1}(C_c + C_{gs4})(C_{db1} + 4C_{gs2})$ $\quad + 2C_c C_{gs1} g_{m2}(C_{db3} + C_{gs4}) + 2C_c C_{db2} C_L g_{m1}$ $\quad + 2C_c C_{gs1} g_{m4}(C_{db1} + C_{db2} + 2C_{gs2}) + 4C_c C_{gs2} C_{gs4} g_{m1})$ $+ s^4(4C_c C_{gs2} C_L (C_{db1} + C_{gs1}) + 2C_c C_{db3} C_{gs1}(C_{db1} + 2C_{gs2})$ $\quad + 2C_c C_{gs1} C_L (C_{gd1} + C_{db1} + C_{db2})$ $\quad + 2C_{gs1} C_{gs4}(C_{db1} + 2C_{gs2})(C_c + C_L) + 2C_{gs1} C_{gs4} C_{db2} C_L$ $\quad + C_c C_{db5} C_L (C_{db1} + 2C_{gs2})$

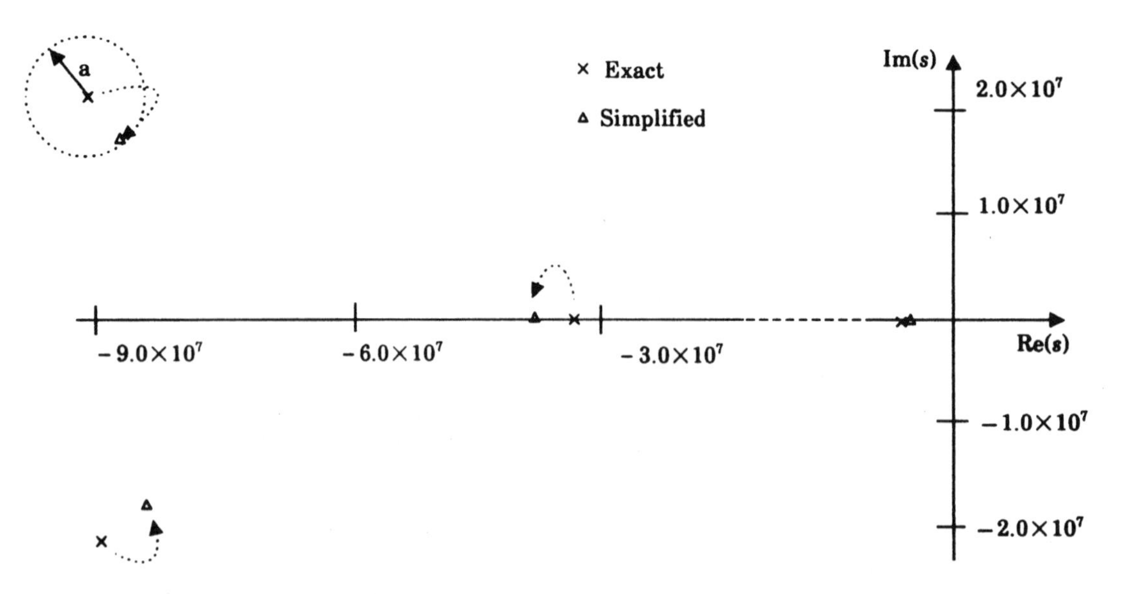

Fig. 10. Pole displacement for the Miller op amp of figure 8a using the improved simplification criterion.

five benchmark circuits shown in figure 11 will be used.

Efficiency can be assessed from table 4, which shows the measured CPU times (for a 4-MIPS machine) in computing exact expressions for different system functions of the circuits of figure 11 (A_v is voltage gain, Z_{out} is output impedance) and for the different MOS transistor model levels of figure 4. Observe that these CPU times are low enough to allow interactive use of ASAP for the analog circuit design process. Let us now consider in more detail the results provided by ASAP for the Miller op amp and the folded-cascode op amp.

4.1. Miller Op Amp

Consider the simplified expression for the voltage gain of the Miller op amp given in table 3. Figure 12 shows the result of graphically postprocessing the symbolic expressions of table 3. In figure 12a the four poles and four zeros of the system function are plotted (as done by ASAP) versus the bias current of the differential input stage (see Section 2 for details about the representation procedure). For the output stage a bias current 10 times greater than that of the differential pair was assumed. On the other hand, figures 12b and c represent Bode diagrams of the voltage gain of this circuit. The X axis corresponds to the frequency (in hertz) on the logarithmic scale. The Y axis in figure 12b represents the magnitude in decibels, and in figure 12c the phase in degrees. The families of curves are parametrized with respect to the bias current across M_5 ($10^{-7} \le I_{\text{Bias}} \le 10^{-4.5}$, on the logarithmic scale).

Several pieces of useful information for the analog design procedure can be extracted from a detailed examination of the graphic representations of figure 12. For instance, for the different bias currents a trade-off between gain and bandwidth is highlighted. Also, just to give another example, figure 12a shows how to choose the bias current for a given dominant and nondominant pole location. Moreover, since for a selected bias current poles and zeros can be plotted again as a

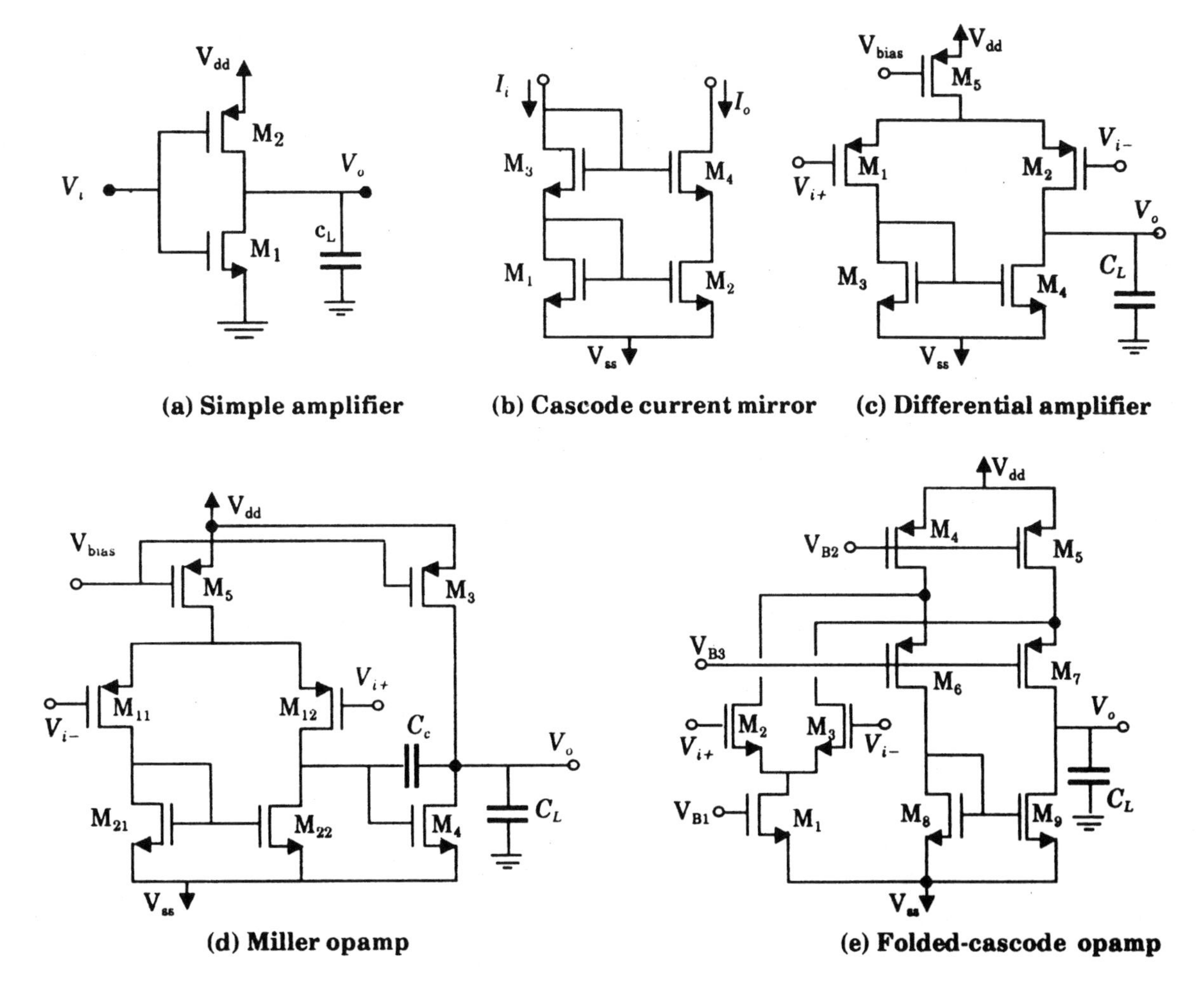

Fig. 11. Benchmark circuits for ASAP.

Table 4. CPU times (seconds) for different circuits and levels of modeling (measured in a SUN3/260 workstation).

	Level 1	Level 2	Level 3	Level 4
Simple amplifier (A_v)	0.2	0.2	0.3	0.3
Cascode current mirror (Z_{out})	0.4	0.4	2.4	3.7
Differential pair (A_v)	0.4	0.4	2.4	2.8
Miller op amp (A_v)	0.7	0.7	10.8	17.6
Folded cascode op amp (A_v)	2.3	8.6	232.6	299.8

function of a new parameter much deeper insight can be gained into the circuit operation.

4.2. *Folded-Cascode Op Amp*

The folded cascode op amp of figure 11e is used for further illustration of the program capabilities. The level 3 transistor model of figure 4c was again used to calculate the differential gain. The exact expression of this gain contains, for mismatched transistors, 97,953 terms and requires 232.6 s of CPU time. Table 5 shows a simplified expression of this gain for an error margin $\epsilon = 60\%$. A sized version of figure 11e was used, and matched transistors were considered to evaluate the simplified expression of table 5. At dc, a gain of 58.27 dB was obtained, the corresponding numerical value computed by HSPICE [17] being 57.32 dB. Observe the error is below 1 dB even for this so much simplified formula.

The exactness of the ASAP computation for higher frequencies can be assessed from figure 13, where a comparison to the results provided by the numerical simulator is made. The complete expression calculated by ASAP coincides exactly, on the other hand, with the results provided by the numerical simulator.

Table 6 shows a low frequency simplified formula for the positive power supply rejection ratio (PSRR+). In this analysis mismatches between transistors play a fundamental role [6]. Higher insight is thus gained by taking mismatches into account, namely g_{m3} =

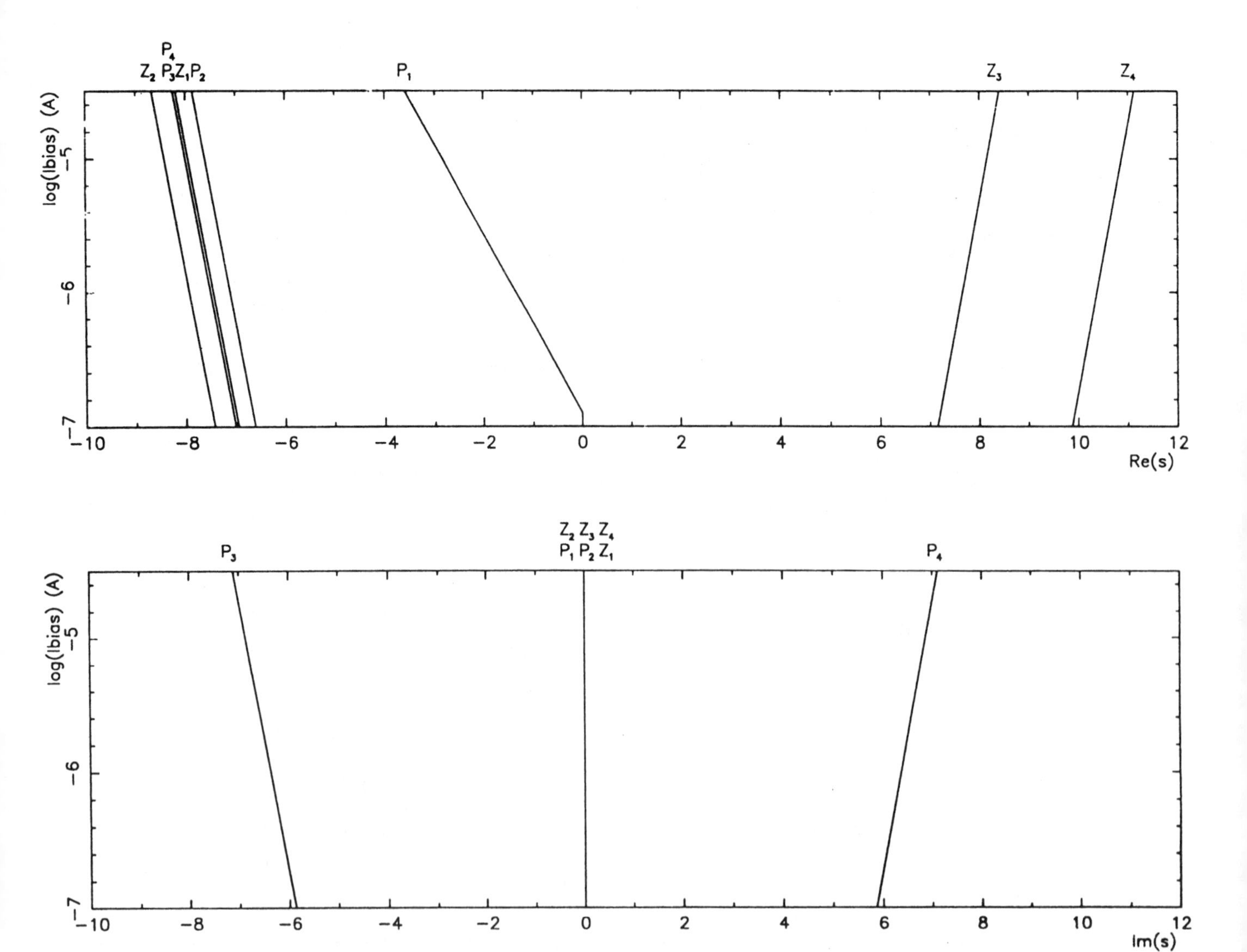

Fig. 12a. Graphical postprocessing of Miller opamp voltage gain: pole/zero location.

$g_{m2} + \Delta g_{m2}$, $g_{ds3} = g_{ds2} + \Delta g_{ds2}$, ... This possibility is included in the ASAP capabilities. The corresponding PSRR− is shown in table 7. Finally, table 8 shows a simplified expression for the CMRR. Mismatches are also explicitly shown in these latter expressions.

5. ASAP Algorithms and Implementation Details

As mentioned, ASAP uses signal flow graph methods to solve the circuit equations. The signal flow graph method is illustrated in figure 14 for the circuit of figure 1. Interrelationships among different variables in the small signal model of the circuit are first displayed in a flow graph diagram. Then, given an input node variable and an output one, the ratio between them can be calculated via Mason's rule, requiring the paths from the input node to the output one and the loops of any order existing in the graph [13].

Figure 15 is a block diagram showing the flow of operations in ASAP. Starting from the circuit description in SPICE format, the transistors and the amplifiers are expanded by ASAP using models taken from a library. A new description of the circuit is thus generated. This new description contains sources (controlled or not) and passive elements interconnected in a certain topology. Then this new circuit topology is reduced by substituting each group of shunted elements with admittance description by one corresponding element. Once the grouping is made, a *tree*[5] is selected

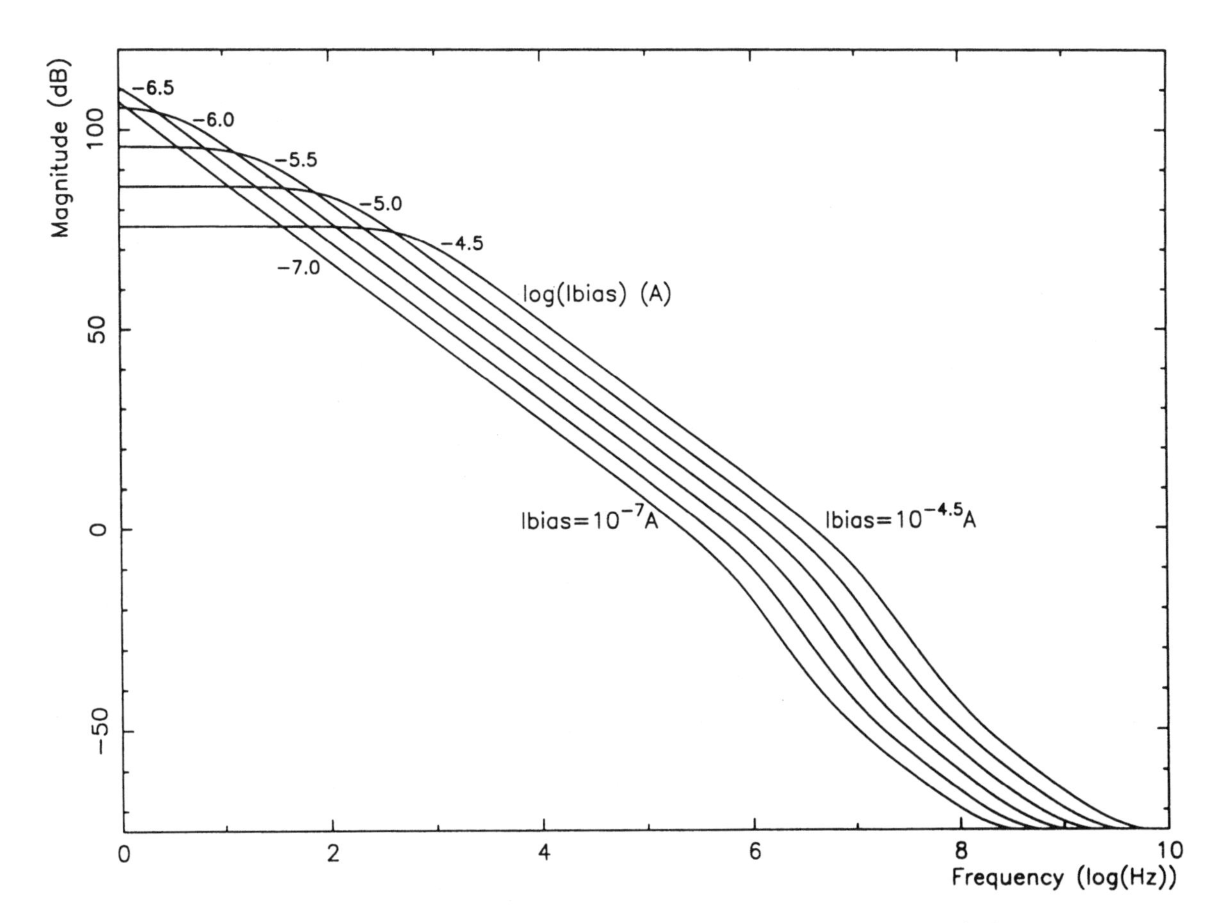

Fig. 12b. Graphical postprocessing of Miller opamp voltage gain: magnitude.

for the new circuit topology resulting after the grouping. A signal flow graph is then generated containing the information corresponding to all constitutive and topological equations of the circuit. Nodes in this graph correspond to the voltages in the tree branches and the currents in the cotree branches. Mason's rule [13] is then used to obtain relationships among the different nodes of the graph and the input nodes. The weights of the loops of different order of the graph are involved in these relationships. From them, the ac circuit characteristics are obtained by expanding the symbols associated with each weight.

The graph formulation in ASAP follows the technique proposed by Chua and Lin [11]. We assume the circuit contains neither loops of voltage sources nor cutsets of current sources. If it is the case, a tree is selected containing all voltage sources. The circuit topological equations can then be written as follows [19]:

$$\mathbf{V}_C = \mathbf{F}_T\mathbf{V}_T \qquad \mathbf{I}_T = \mathbf{F}_C\mathbf{I}_C \tag{7}$$

where $\mathbf{V}_T$ and $\mathbf{I}_T$ (alternatively $\mathbf{V}_C$ and $\mathbf{I}_C$) are column vectors for the voltages and currents in the tree branches (alternatively cotree branches).

In the flow graph formulation these topological equations have to be combined with the constitutive ones to yield all currents and voltages to be given as functions of the graph node variables, i.e., tree voltages and cotree currents. In case there are no controlled sources, each source (either voltage or current) directly defines a corresponding node in the signal flow graph. The other nodes can be defined by first writing the constitutive equations for the nonsource tree and cotree branches,

$$\mathbf{V}_T = \begin{bmatrix} \mathbf{E}_S \\ \mathbf{V}_Z \end{bmatrix} = \begin{bmatrix} \mathbf{U} & \mathbf{0} \\ \mathbf{0} & \mathbf{Z} \end{bmatrix} \begin{bmatrix} \mathbf{E}_S \\ \mathbf{I}_Z \end{bmatrix} \tag{8a}$$

$$\mathbf{I}_C = \begin{bmatrix} \mathbf{J}_S \\ \mathbf{I}_Y \end{bmatrix} = \begin{bmatrix} \mathbf{U} & \mathbf{0} \\ \mathbf{0} & \mathbf{Y} \end{bmatrix} \begin{bmatrix} \mathbf{J}_S \\ \mathbf{V}_Y \end{bmatrix} \tag{8b}$$

where $\mathbf{J}_S$ and $\mathbf{E}_S$ are the vector of current sources and the vector of voltage sources, respectively. Then these equations have to be combined with the topological ones in (7) to yield

$$\mathbf{V}_Z = \mathbf{Z}\,\mathbf{I}_Z = \mathbf{Z}\,\mathbf{F}_{CZ}\,\mathbf{I}_C \tag{9a}$$

$$\mathbf{I}_Y = \mathbf{Y}\,\mathbf{V}_Y = \mathbf{Y}\,\mathbf{F}_{TY}\,\mathbf{V}_T \tag{9b}$$

where $\mathbf{F}_{CZ}$ is the submatrix resulting of selecting those rows of $\mathbf{F}_C$ in (7) corresponding to $\mathbf{I}_Z$, and $\mathbf{F}_{TY}$ is formed from $\mathbf{F}_T$ by selecting the rows corresponding to $\mathbf{V}_Y$.

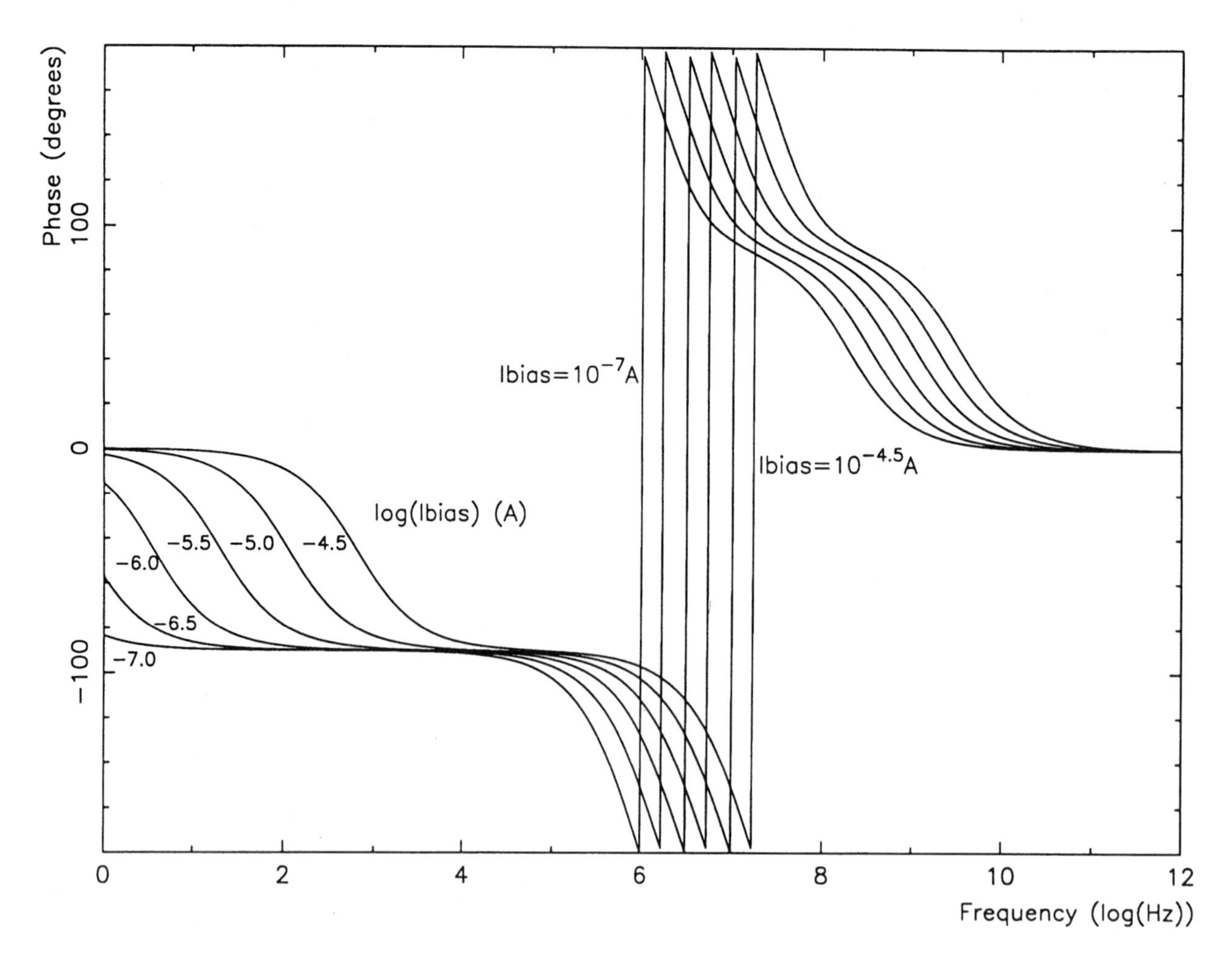

Fig. 12c. Graphical postprocessing of Miller opamp voltage gain: phase.

Table 5. Simpified expression for the differential voltage gain of figure 1le.

Numerator =	Denominator =
$+ 2g_{m8}g_{m2}^{2}g_{m6}^{2}$	$+ 2g_{m8}g_{ds8}g_{m2}g_{m6}^{2}$
$+ s(g_{m2}^{2}g_{m6}^{2}(2C_{gs8} + C_{db6})$	$+ s2g_{m8}C_{L}g_{m2}g_{m6}^{2}$
$\quad + 2C_{gs6}g_{m2}^{2}g_{m6}g_{m8})$	$+ s^{2}2C_{L}g_{m2}g_{m6}(g_{m8}C_{gs6} + g_{m6}(2C_{gs8} + C_{gd6} + C_{db6}))$
$+ s^{2}g_{m2}g_{m6}$	$+ s^{3}C_{L}(2C_{db6}g_{m6}^{2}(C_{gs2} + C_{sb2}) + 2g_{m8}C_{gs6}^{2}g_{m2}$
$\quad (C_{sb6}g_{m2}(2C_{gs8} + C_{db6})$	$\quad + 4g_{m2}g_{m6}(2C_{gs8} + C_{gd6} + C_{db6})(C_{gs6} + C_{sb6})$
$\quad + C_{gs6}g_{m2}(2C_{gs8} + C_{gd6})$	$\quad + 4C_{db6}g_{m2}g_{m6}(C_{db2} + C_{db4}))$
$\quad + C_{db6}g_{m2}(C_{db2} + C_{gs6}) + C_{gs2}C_{db6}g_{m6})$	$+ s^{4}2C_{L}(2C_{db6}C_{gs6}g_{m2}(C_{db2} + C_{db4})$
$+ s^{3}g_{m}2\,g_{m6}$	$\quad + 2C_{db6}g_{m6}(C_{gs6} + C_{sb6})(C_{gs2} + C_{sb2})$
$\quad (C_{gs2}C_{sb6}(2C_{gs8} + C_{db6})$	$\quad + 2C_{gs6}C_{sb6}g_{m2}(2C_{gs8} + C_{db6}) + C_{db1}C_{db6}g_{m6}(C_{gs6} + C_{sb6})$
$\quad + C_{gs2}C_{gs6}(2C_{gs8} + C_{gd6})$	$\quad + C_{gs6}^{2}g_{m2}(2C_{gs8} + C_{gd6} + C_{db6}) + C_{sb6}^{2}g_{m2}(2C_{db2} + C_{db6}))$
$\quad + C_{gs6}C_{db6}(C_{gs2} + C_{sb2}) + C_{gs2}C_{db2}C_{db6})$	$+ s^{5}C_{L}(C_{db6}(C_{gs6} + C_{sb6})^{2}(C_{db1} + 2C_{sb2})$
$- s^{4}C_{gd2}g_{m6}$	$\quad + 4C_{gs2}C_{gs6}C_{sb6}(2C_{gs8} + C_{db6}) + 2C_{gs2}C_{db6}C_{sb6}^{2}$
$\quad (C_{gs2}C_{gs6}(2C_{gs8} + C_{db6}) + C_{gs2}C_{db6}C_{sb6}$	$\quad + 4C_{gs2}C_{db6}C_{gs6}(C_{db2} + C_{db4})$
$\quad + C_{db6}(C_{gs6} + C_{sb6})(\tfrac{1}{2}C_{db1} + C_{sb2}))$	$\quad + 2C_{gs2}C_{gs6}^{2}(2C_{gs8} + C_{gd6} + C_{db6}))$

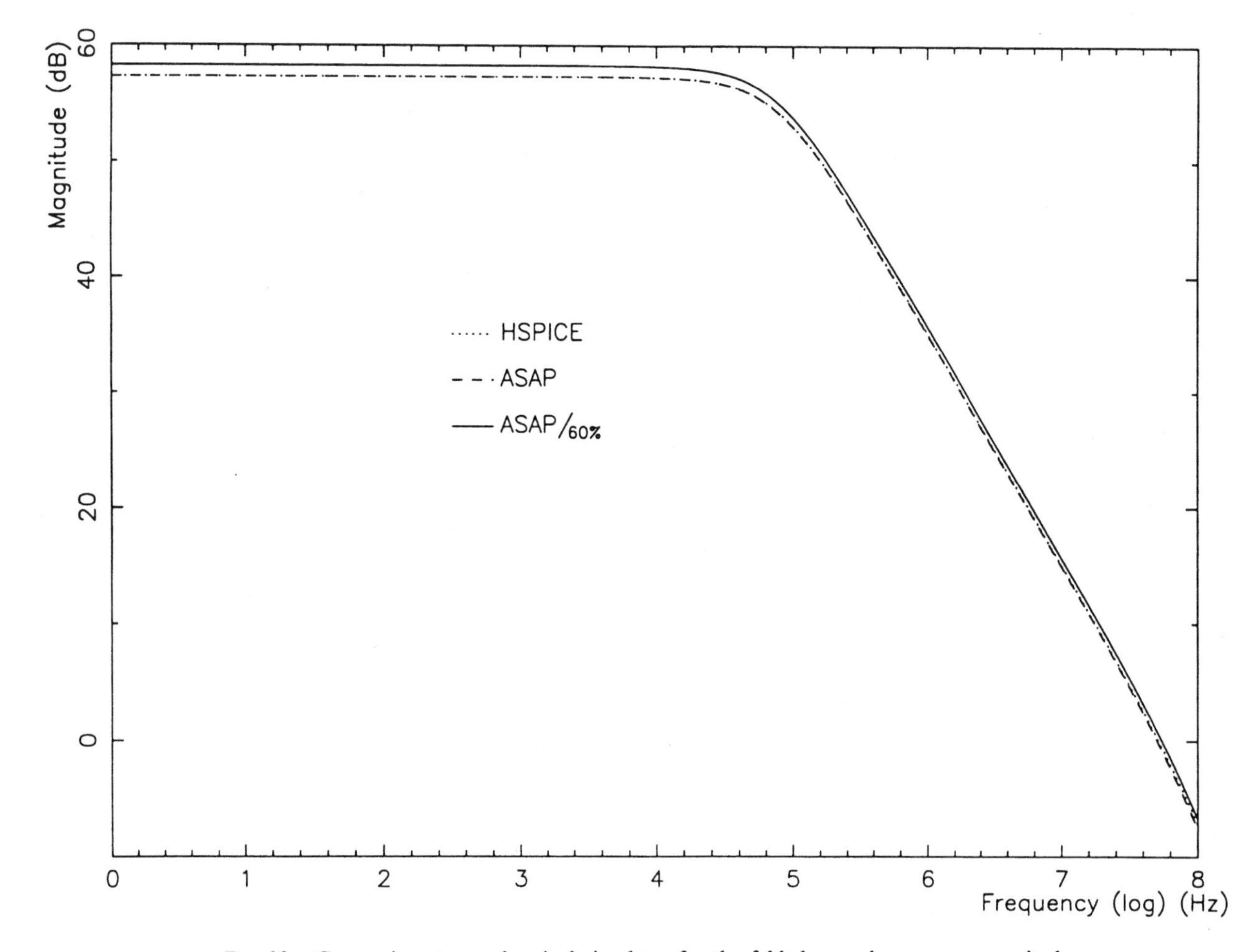

Fig. 13a. Comparison to an electrical simulator for the folded-cascode op amp: magnitude.

For a controlled source the corresponding node in the signal flow graph is not directly defined by the source value. In this case a procedure similar to that indicated above for $\mathbf{V}_Z$ and $\mathbf{I}_Y$ has to be followed to yield the source value to be given as a function of the flow graph node variables $\mathbf{V}_T$ and $\mathbf{I}_C$. For the sake of illustration, assume a VCCS whose controlling variables are included in the cotree. Consider the current source is the jth branch in the cotree and that the controlling voltages are associated with the kth and lth cotree branches. Then we get for the constitutive equation

$$|\mathbf{J}_S|_j = g_m\,|\mathbf{V}_C|_k - g_m\,|\mathbf{V}_C|_l \tag{10}$$

from which, after taking into account the topological equations, we obtain

$$|\mathbf{J}_S|_j = g_m(|\mathbf{F}_T|_k - |\mathbf{F}_T|_l)\,\mathbf{V}_T \tag{11}$$

where in both equations subscripts are used to specialize a row of the vectors and matrices involved.

Several innovative contributions have been made in ASAP to tailor the previously described network formulation method to the needs arising in the automatic modeling of analog integrated circuits. Increasing efficiency and ability to manage relatively large complexities have been the targets. To achieve these, special emphasis has been put on the following points:

- Codification of symbols
- Grouping of elements
- Tree selection
- Loop enumeration

5.1. Symbol Codification

The codification of symbols in ASAP is made by a combination of two systems:

a. *Prime number codification:* each symbol is represented by a prime number. Symbol operations reduce to integer number operations.
b. *Bit-to-bit codification:* each symbol is represented by a bit of a string. Symbol operations reduce to bit operations.

As symbol operations correspond to bit operations, the second system is faster. Also, it is usually able to

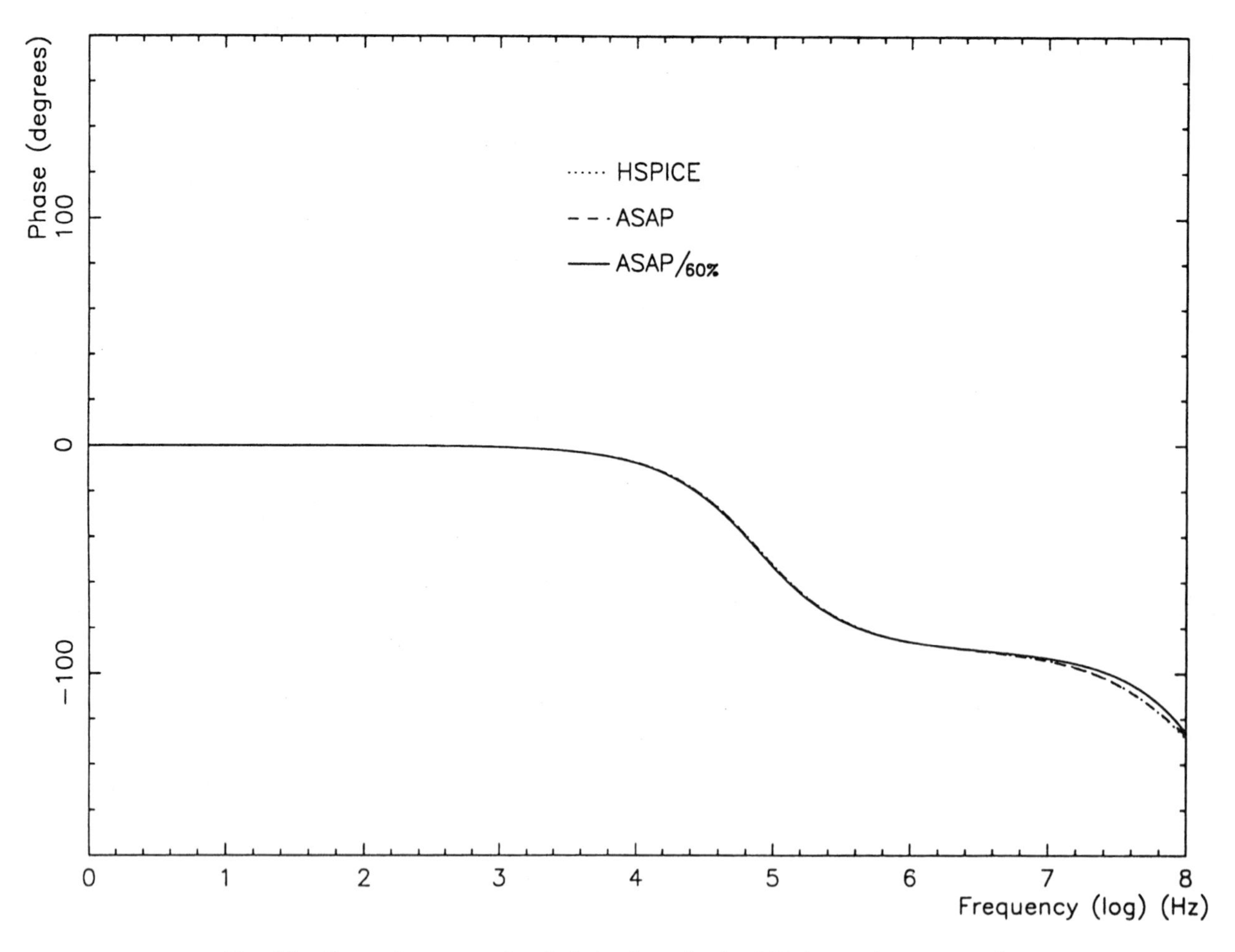

Fig. 13b. Comparison to an electrical simulator for the folded-cascode op amp: phase.

Table 6. Simplified expression of the PSRR+ for figure 11e.

Numerator =	Denominator =
$+\ 2g_{m2}^2 g_{m6} g_{m8}$	$-\ g_{m1} g_{m2} g_{m6} \Delta g_{m8} + g_{m1} g_{m6} g_{m8} \Delta g_{m2} - 2 g_{m2} g_{m6} g_{m8} \Delta g_{mb2}$
$+\ 4g_{m2}^2 g_{m8} g_{mb6}$	$+\ 2 g_{m6} g_{m8} g_{mb2} \Delta g_{m2} + 2 g_{m1} g_{m8} g_{mb6} \Delta g_{m2} - 2 g_{m1} g_{m2} g_{mb6} \Delta g_{m8}$
$+\ 2g_{m2} g_{m6} g_{m8} g_{mb2}$	$+\ g_{ds8} g_{m1} g_{m2} g_{m6}$

Table 7. Simplified expression of the PSRR− for figure 11e.

Numerator =	Denominator =
$+\ g_{m2} g_{m6} g_{m8} + 2 g_{m2} g_{m8} g_{mb6}$	$-\ g_{m4} g_{m6} \Delta g_{m8} + g_{m6} g_{m8} \Delta g_{m4} + 2 g_{m8} g_{mb6} \Delta g_{m4}$
$+\ g_{m6} g_{m8} g_{mb2}$	$-\ 2 g_{m4} g_{mb6} \Delta g_{m8} - g_{ds4} g_{m6} g_{m8} + g_{ds8} g_{m4} g_{m6}$

Table 8. Simplified expression of the CMRR for figure 11e.

Numerator =	Denominator =
$+\ g_{m2}^3 g_{m6}^2 + 3 g_{m2}^3 g_{m6} g_{mb6}$	$-\ g_{m2} g_{m6}^2 g_{mb2} \Delta g_{m2} + g_{m2}^2 g_{m6}^2 \Delta g_{mb2}$
$+\ 2 g_{m2}^2 g_{m6}^2 g_{mb2} + 4 g_{m2}^3 g_{mb6}^2$	$-\ 2 g_{m2} g_{m6} g_{mb2} g_{mb6} \Delta g_{m2} + 2 g_{m2}^2 g_{m6} g_{mb6} \Delta g_{mb2}$
$+\ 4 g_{m2}^2 g_{m6} g_{mb2} g_{mb6} + g_{m2} g_{m6}^2 g_{mb2}^2$	$-\ g_{m6}^2 g_{mb2}^2 \Delta g_{m2} + g_{m2} g_{m6}^2 g_{mb2} \Delta g_{mb2}$

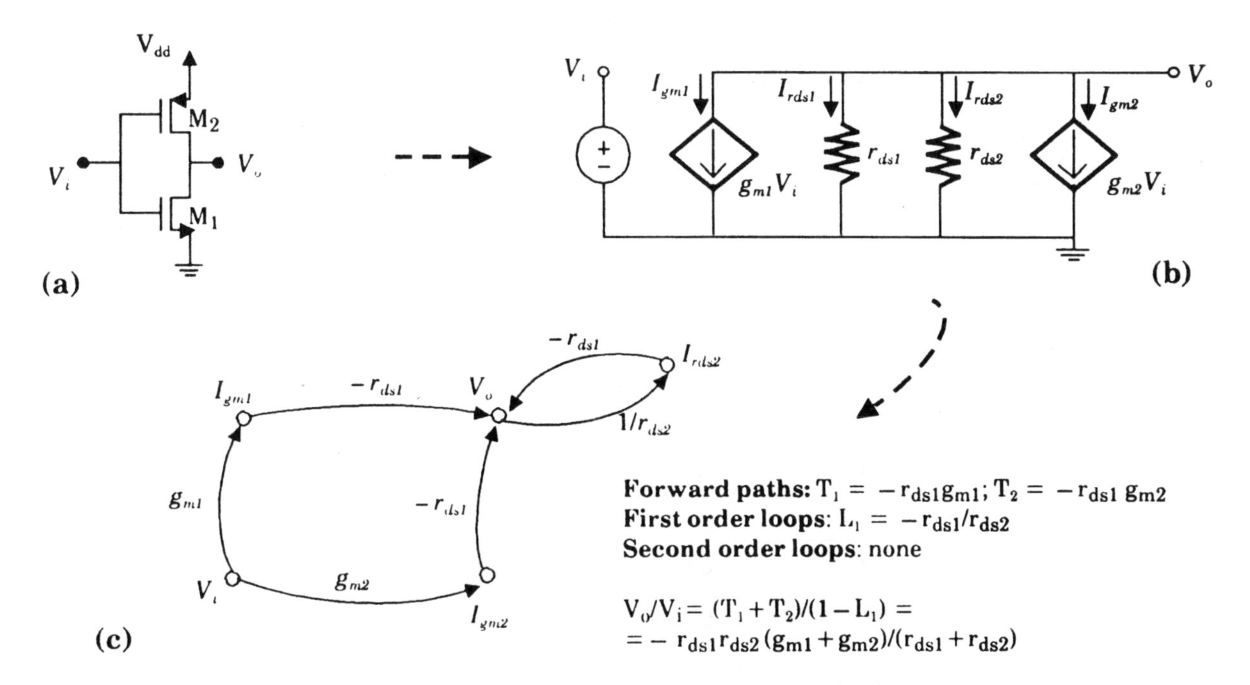

Fig. 14. (a) Simple amplifier; (b) circuit graph; (c) compact signal flow graph.

handle a larger number of symbols. On the contrary, the first system is advantageous in case there is repetition of symbols, corresponding to matched circuit elements.

In ASAP, both systems are not competitive but complementary. After the circuit description has been read, ASAP performs a symbol checking. The optimum system is then automatically chosen, depending upon the characteristics of the symbols to be manipulated.

5.2. Grouping of Elements

Computational cost of symbolic analysis of a given circuit is directly related to the number of elements of its small signal model. To minimize this cost, ASAP performs a transformation consisting in substituting each group of elements with admittance description by one generic element. This grouping is illustrated graphically in figure 16. Notice that the elements being grouped need not belong to the same type. The advantage of making these groupings is a very important reduction in CPU time. Besides, the program incorporates the feature of storing the information associated with each group of elements. Since this information is used for a final expansion, where all the original symbols are recovered, the process of grouping is, hence, completely transparent to the user, not obscuring results with symbols which represent different types of elements.

5.3. Tree Selection Heuristics

For a given graph, the selection of a tree can be made by using the algorithmic method proposed by Chua and Lin [11]. It is based on elementary row and column operations on the incidence matrix summarizing the circuit topology [19]. The purpose of the procedure is to reduce the incidence matrix to an echelon form. From this, the tree is directly provided. The problem of this method is that it does not guarantee an optimum tree will be found. On the contrary, one tree is selected in a rather arbitrary way, depending only on the ordering of the elements in the input file (notice that the algorithm selects the first set of branches which form a tree). Taking into account that the CPU time and memory resources required for symbolic analysis strongly depend on the selection of an appropriate tree, arbitrary selection of the tree is an important drawback.

To overcome this drawback of existing tree-finding algorithms, a novel heuristic technique has been developed by us and implemented in ASAP. For most cases, this technique yields a "good" tree, not far from the optimum one. Two facts are assumed for the development of this heuristic technique: (a) The voltage sources contain no loops, and the current sources contain no cutsets. (b) There are no elements in parallel with immittance description. (Observe that this is ensured by the grouping of elements performed by ASAP.)

The criterion used in the ASAP heuristic tree selection module is based on the fact that most of the CPU

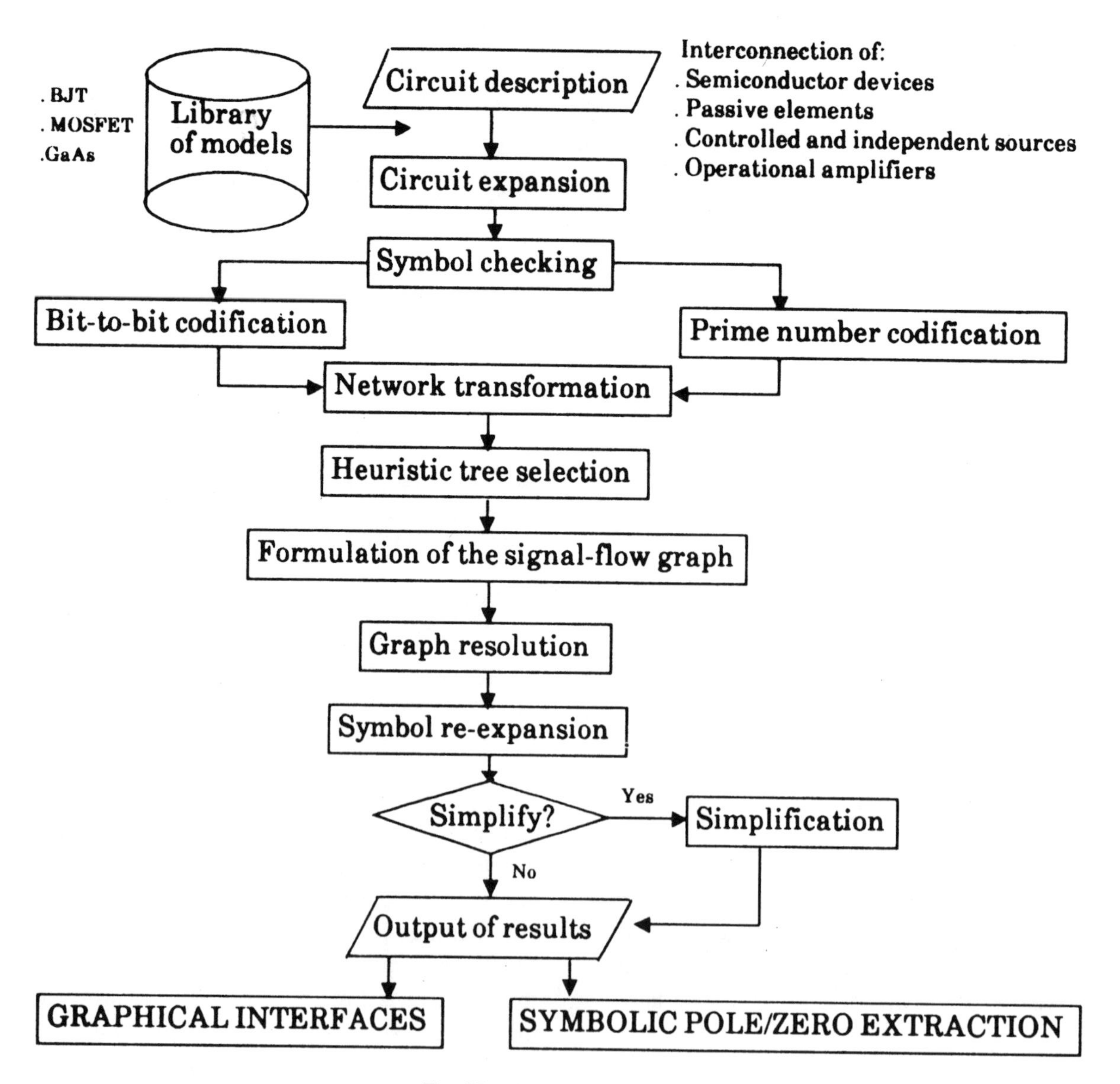

Fig. 15. ASAP operation flow.

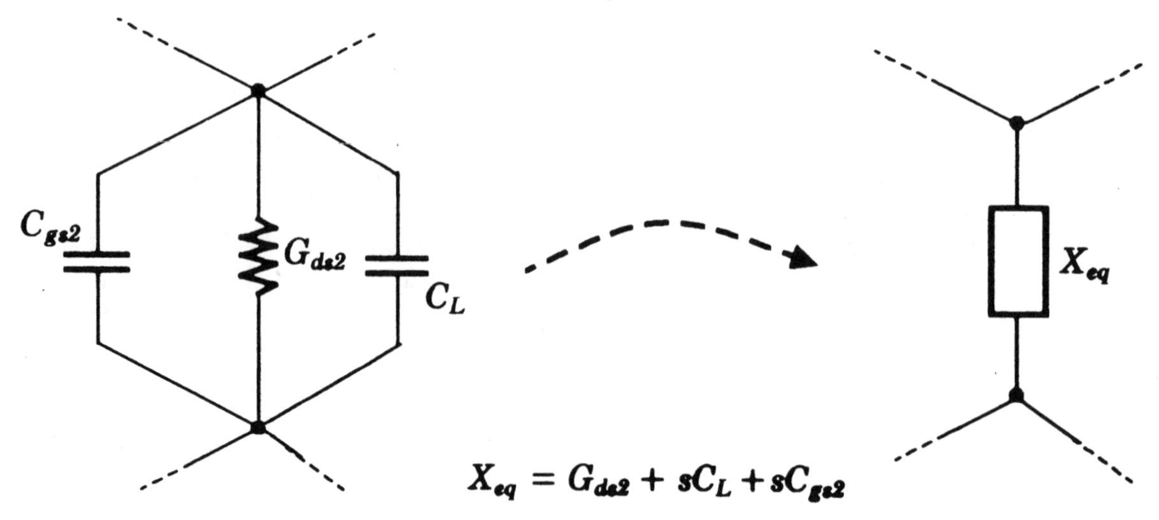

Fig. 16. Illustrating the grouping of elements.

time (about 90%) is spent in loop enumeration and term cancellations. Thus, to reduce the CPU time the number of loops must also be reduced. Since there is commonly a direct proportionality between the number of loops and the number of graph branches, it is clear that the selection of the tree must be directed toward minimizing the number of branches of the resulting signal flow graph. To achieve this, we exploit the fact that the number of branches in the signal flow graph is directly related to the number of nonzero entries in the part of the fundamental loop matrix corresponding to the tree branches, $\mathbf{F}_T$ (see (7)), and in the part of the fundamental cutset matrix corresponding to the cotree branches, $\mathbf{F}_C$. On the other hand, matrices $\mathbf{F}_T$ and $\mathbf{F}_C$ are related by [11], [19]

$$\mathbf{F}_T = -\mathbf{F}_C^T \tag{12}$$

from which we can see that the number of nonzero entries in both is equal. As a consequence, as long as our heuristics can minimize the number of nonzero entries in one of them it will automatically minimize it in the other.

With this basic criterion, two cases can be found in practice in trying to, by heuristic, select an appropriate tree:

a. A starlike tree[6] exists. This is the optimum case. It is so because for this kind of tree there are just two nonzero entries in each row of $\mathbf{F}_T$, as any pair of terminal nodes of a cotree branch is connected by just two tree branches. Hence, if the starlike tree were not the optimum one, there would be another one with at least one less branch in the corresponding signal flow graph. Thus, at least the terminal nodes of one cotree branch would be connected by just one tree branch. However, this is not possible, as we have assumed that there are no parallel branches.
b. A starlike tree cannot be found. In this case our heuristic module selects a tree, rooted at one common node, such that the sum of the number of branches from the remaining nodes to the common one is minimum. Notice that the starlike tree is a particular case in which there is only one branch from each node to the root node. In case several trees are found each one of them is evaluated, and the one with the smallest number of branches in the corresponding signal flow graph is selected.

Although the devised heuristic does not guarantee that the best tree will be found, practical results from a large number of different analog integrated circuit schematics confirm the correctness of the proposed method.

The histograms in figure 17 are intended to illustrate the performance of the tree selection module. Different trees have been considered in the symbolic analysis of a Miller op amp. Tree 2 is obtained by the heuristic module. The others are arbitrarily selected by using algorithmic tree selection methods. For the different trees, the histogram in figure 17a shows the CPU times spent in obtaining the voltage gain of the device. CPU times are measured in seconds and have been represented on a logarithmic scale for better understanding of differences among the trees. Numbers on top of each bar show the number of branches in the signal flow graph diagram for the corresponding tree. As can be seen from the figure, more than two orders of magnitude reduction in CPU time can be achieved by the use of the ASAP tree selection heuristic.

For further illustrating the performance of the heuristic tree selection module, figure 17b represents the total number of loops required in each case to calculate Mason's determinant. This number is directly related to the peak of memory resources needed along the execution of the program to perform the circuit analysis. As for the CPU time, memory resources can be considerably reduced by using the ASAP heuristic. Both features allow for moderate complexity blocks to be analyzed by ASAP using typical workstation configurations, as illustrated in Section 4.

5.4. Loop Enumeration

The signal flow graph method requires the calculation of all loops of any order, which is a very memory-intensive process. To reduce the need of memory resources, an algorithm has been developed by us which calculates the loops of $(n + 1)$st order by using only those of nth order. Thus, only the higher order loops must be stored at any instant of time. The algorithm is based in the comparison of pairs of loops against a second-order loop table.

Figure 18 illustrates graphically the algorithm operation. Given any pair of nth order loops composed of $n - 1$ identical first-order loops, a pair is built with the two remaining first-order loops. If a second-order loop can be found composed of the same pair of first-order loops, the corresponding $(n + 1)$st order loop is built. The different order loops are generated in a sorted way so that a limited number of comparisons is needed.

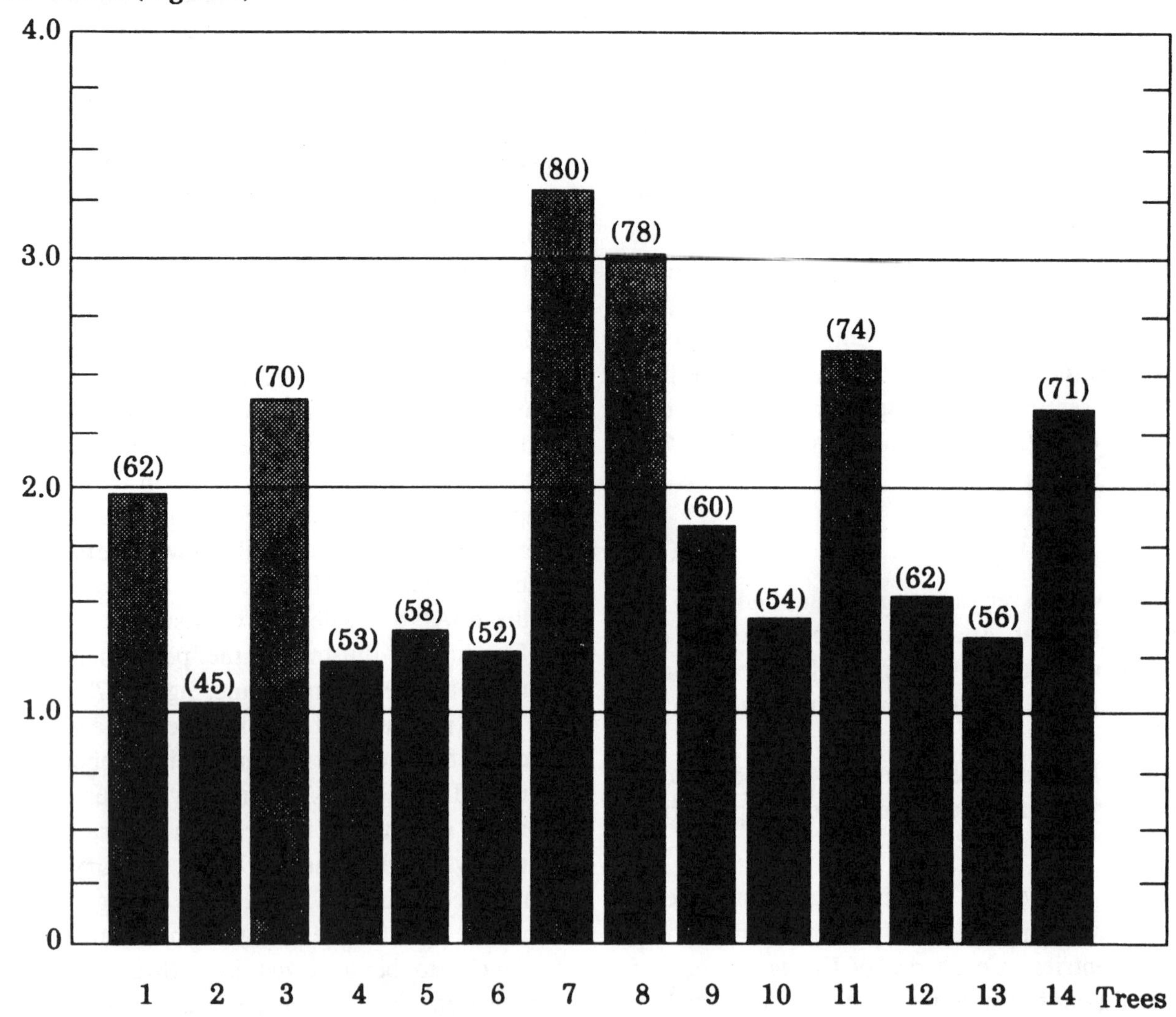

Fig. 17a. CPU time in seconds (on logarithmic scale) for the symbolic analysis of the Miller op amp of figure 8a using different trees.

The speed of the proposed algorithm is similar to others reported before [11]. However, since in the new algorithm the $(n - 1)$st order loops are no longer needed once the nth order loops have been calculated, the storage requirements are much lower. This is a very important feature for the symbolic analysis of analog integrated circuits. The ability of the C language to allocate and deallocate memory plays an important role here. The complexity levels ASAP is able to deal with (see Section 4) demonstrate the validity of the techniques used.

5.5. Symbolic Pole/Zero Calculation

Once the symbolic network functions have been generated, symbolic formulas for poles and zeros of either the complete or the simplified expressions can be calculated. Figure 19 shows the sequence of operations performed to calculate approximate symbolic expressions for poles and zeros. First, the program numerically calculates zero and pole locations using typical or user-specified numerical values. To this end, algorithmic methods are used in case the order of the polynomial is lower than 5. Laguerre's method is applied, on the other hand, for larger degree polynomials. Starting from this information, ASAP used heuristic rules to suggest possible approximations (i.e., either approximate expressions when two or more roots are split sufficiently or displacement to infinity for roots of very large magnitude). If they are rejected by the user, algorithmic methods are applied when possible. If they are accepted, the approximate symbolic poles and zeros are calculated, and a symbolic polynomial deflation is performed, reducing the order of the polynomial by one. Finally, if any root still remains, algorithmic methods are applied when possible. It is clear that the symbolic expressions for poles and zeros

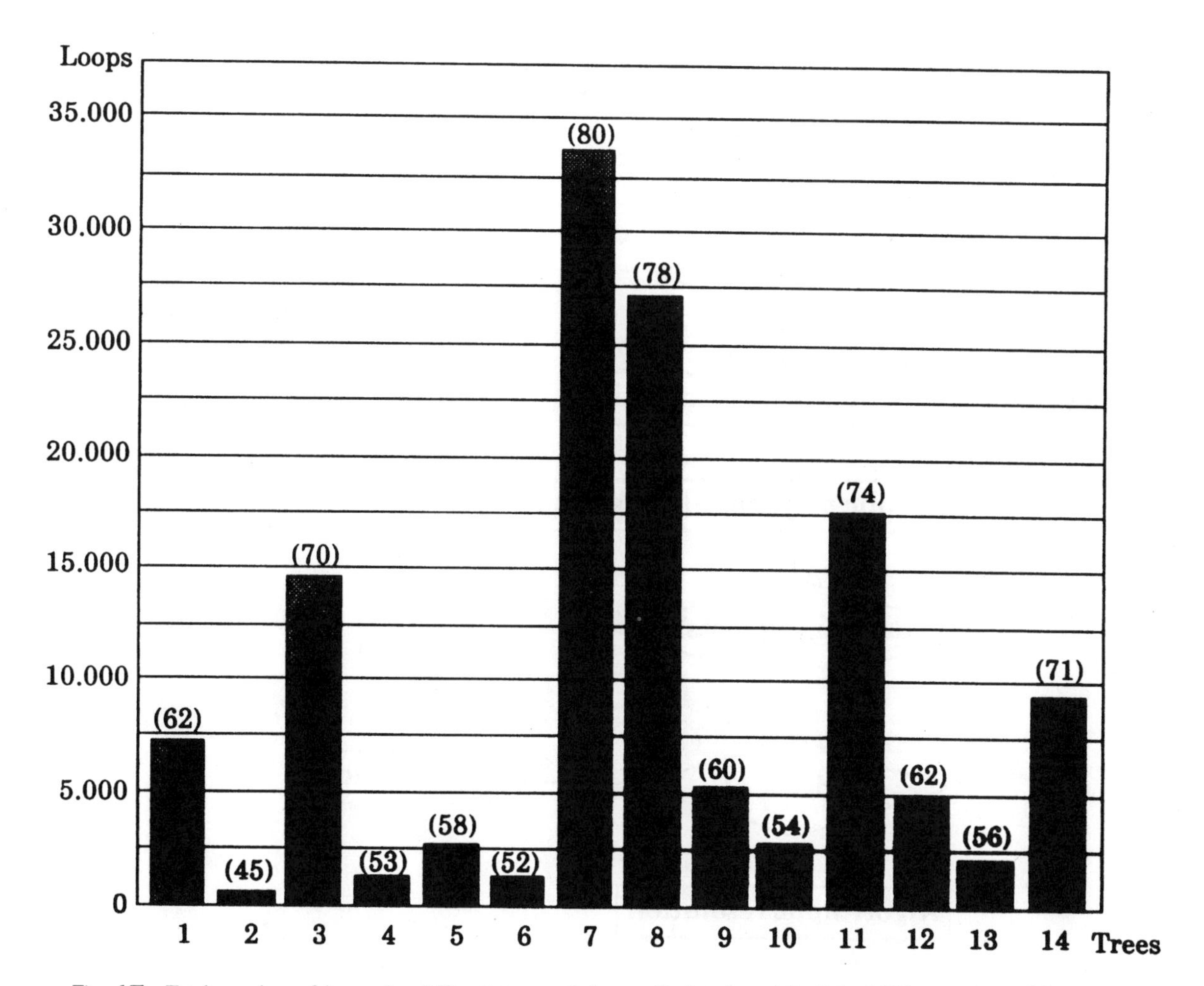

Fig. 17b. Total number of loops for different trees of the small signal model of the Miller op amp of figure 8a.

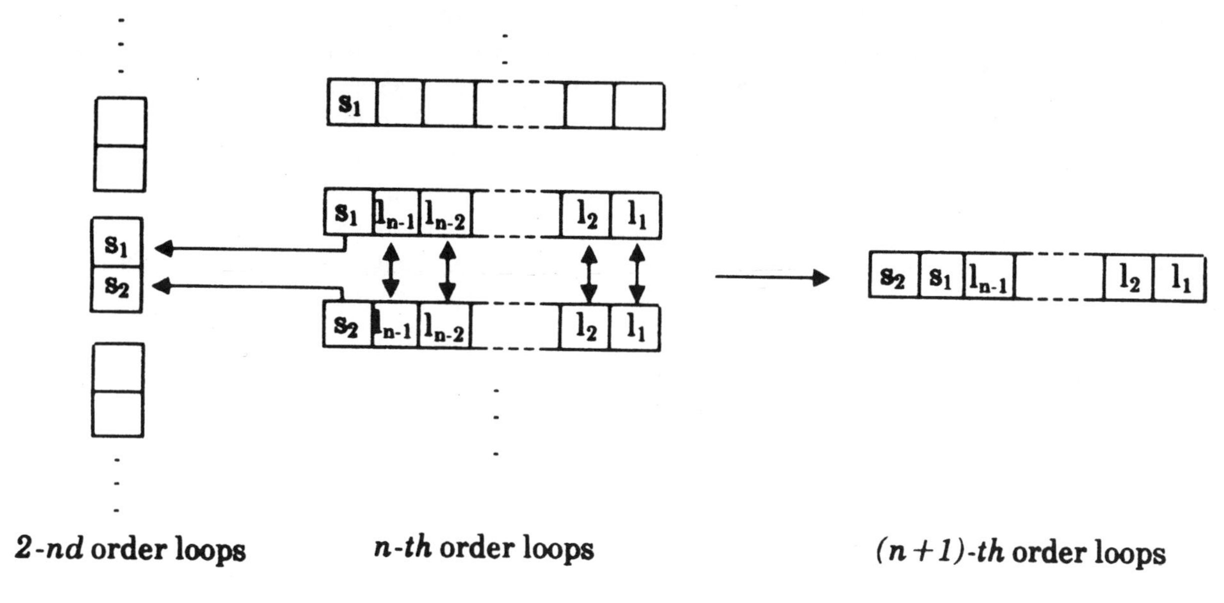

Fig. 18. Illustrating the loop enumeration algorithm.

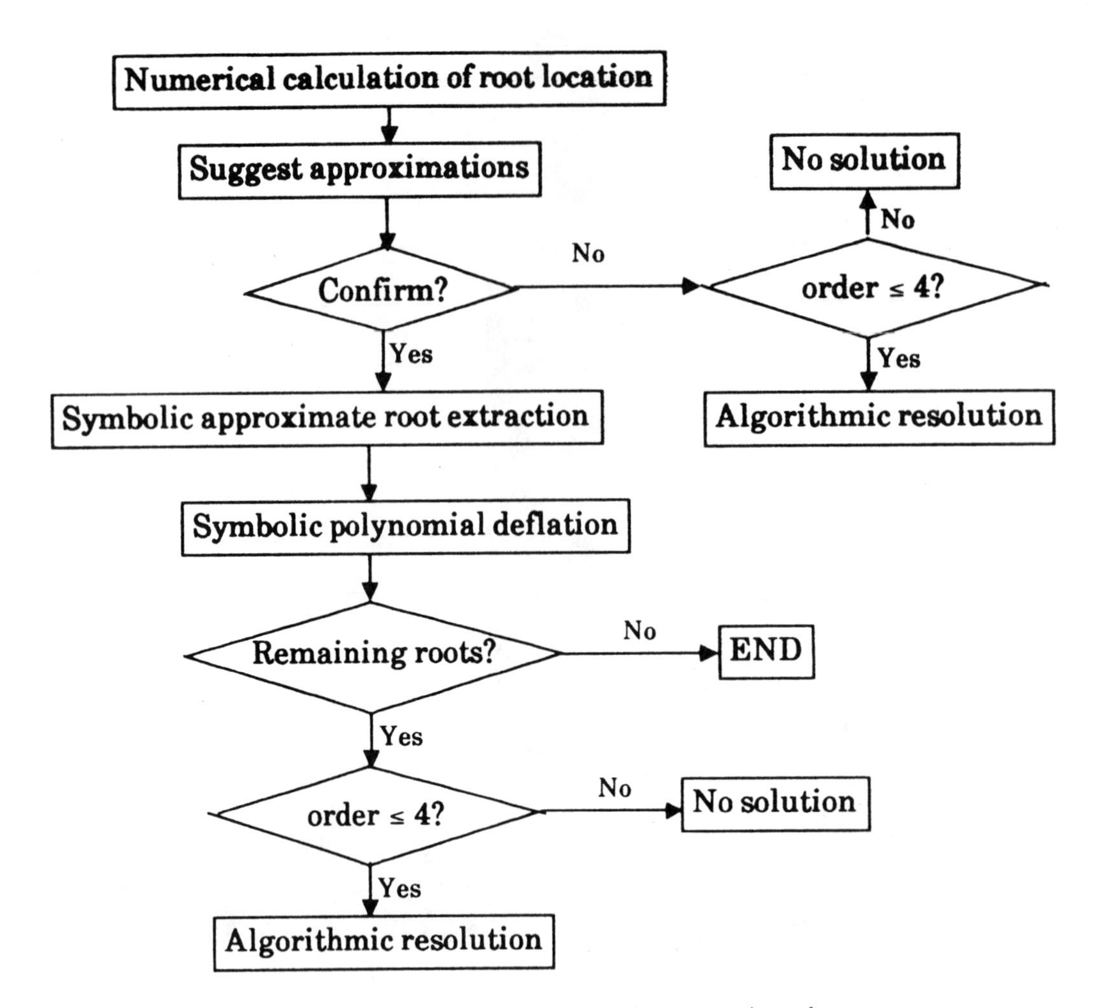

Fig. 19. Flow diagram of the symbolic pole/zero extraction subprogram.

of simplified formulas are only valid when the new simplification criterion has been used. It is useless to get simplified symbolic expressions of roots which are far from the real locations.

6. Conclusions

We have shown that by exploiting the adequacy of the C language to deal with flow graphs it is possible to implement a portable tool for highly efficient symbolic analysis of analog integrated circuits. This tool, which we call ASAP, provides exact and simplified symbolic expressions for the different system functions of circuits comprising the interconnection of linear circuit elements, semiconductor devices, and op amps. The performance of ASAP has been illustrated via selected examples where efficiency, complexity, and exactness of the computations are addressed. The observed performance allows interactive use of ASAP to help the designer in the automatic modeling of analog circuits. This application is exploited by embedding the new tool in a general CAD framework for analog circuits, including behavioral and electrical simulation.

In its current version, and running on a machine with 8 Mbytes central memory, ASAP is able to deal with the complexity levels arising in typical analog building blocks when described by device level models. More complex circuits, for instance, complete active *RC* filters, have to be described by macromodels. We are currently working on a new ASAP version which approaches the analysis of complex circuits in a hierarchical way. The ASAP modeling capabilities are exploited to automatically calculate macromodels for the components on the different levels of the hierarchy. These macromodels are then used for analysis at the next higher level. Other points to be addressed in the near future are the analysis of weakly nonlinear characteristics and the incorporation of routines for the analysis of discrete time circuits.

Acknowledgment

Research for this paper has been supported in part by Spanish CICYT under contract no. ME87-0004. The authors would also like to acknowledge the many contributions made by R. Domínguez-Castro and Juan D. Martín.

Notes

1. ASAP can be also used for automatic design procedures based on circuit analytical models and optimization routines [2], [12]. For these procedures the designer in figure 2 should be replaced by an optimization tool.
2. ASAP offers the user the possibility of drawing poles and zeros as a circuit variable (bias current, transistor geometries, etc.) changes. Either a conventional root locus diagram (where the root locations are drawn in the s-plane for different values of the parameter [13]) or diagrams like figure 5a (where the real and imaginary parts of the roots are separately drawn versus the parameter) can be provided. We have preferred to include one of the latter type because it more clearly displays the parameter values corresponding to different pole and zero locations.
3. Recall that this error refers to the elimination of the least significant terms of the different coefficients of the exact expression, according to (5).
4. ASAP took 10.8 s CPU time on a SUN3/260 (4 MIPS) workstation in calculating the exact expression for this gain, containing 6756 different terms. Simplifications took around 3 s extra CPU time.
5. For a given circuit topology containing a number of two-terminal circuit components connecting a number of nodes, a *tree* is a subcircuit containing all the original nodes and a number of components large enough to achieve all the nodes to be connected while avoiding the existence of any loop.
6. This is a tree where there is a common node which is connected to each of the remaining nodes by just one branch.

References

1. M.G.R. Degrauwe, R.A. Rutenbar, and L.R. Carley, "IDAC: An interactive design tool for analog CMOS circuits," *IEEE J. Solid-State Circ.* vol. 22, pp. 1106–1116, 1987.
2. H.Y. Koh, C.H. Séquin, and P.R. Gray, "OPASYN: A compiler for CMOS operational amplifiers," *IEEE J. Solid-State Circ.* vol. 9, pp. 113–125, 1990.
3. R. Harjani, O. Nys, E. Dijkstra et al., "OASYS: A framework for analog circuit synthesis," *IEEE Trans. Computer-Aided Design*, vol. 8, pp. 1247–1266, 1989.
4. F.M. El-Turky and E.E. Perry, "BLADES: An artificial intelligence approach to analog circuit design," *IEEE Trans. Computer-Aided Design*, vol. 8, no. 6, pp. 680–692, 1989.
5. S.J. Seda, M.G.R. Degrauwe, and W. Fichtner, "A symbolic analysis tool for analog circuit design automation," in *Proc. IEEE 1988 Int. Conf. Comp. Aided Design*, 1988, pp. 488–491.
6. G. Gielen, H. Walsharts, and W. Sansen, "ISAAC: A symbolic simulator for analog integrated circuits," *IEEE J. Solid-State Circ.* vol. 24, pp. 1587–1597, 1989.
7. P.M. Lin, "A survey of applications of symbolic network functions," *IEEE Trans. Circ. Theory*, vol 20, pp. 732–737, 1973.
8. T. Ozawa, ed., *Analog Methods for Computer-Aided Circuit Analysis and Diagnosis*, Marcel Dekker: New York, 1988.
9. P.M. Lin and G.E. Alderson, "SNAPDDA computer program for generating symbolic network functions," School Elec. Eng., Purdue University, Lafayette, Ind., Tech Rep. TR-EE70-16, Aug. 1970.
10. MACSYMA reference manual version 10, M.I.T. and Symbolics, 1984.
11. L.O. Chua and P.M. Lin, *Computer Aided Design of Electronic Circuits: Algorithms and Computational Techniques,* Prentice Hall: Englewood Cliffs, NJ, 1974.
12. G. Gielen, H. Walsharts, and W. Sansen, "Analog circuit design optimization based on symbolic simulation and simulated annealing," *IEEE J. Solid-State Circ.* vol. 25, no 3., pp. 707–713, 1990.
13. B.C. Kuo, *Automatic Control Systems*, Prentice Hall: Englewood Cliffs, NJ, 1987.
14. P.E. Allen, *CMOS Analog Circuit Design*, Holt, Rinehart and Winston: New York, 1987.
15. C. Toumazou, F.J. Lidgey, and D.G Haigh, eds., *Analog IC Design: The Current Mode Approach*, Peter Peregrinus: London, 1990.
16. D.G. Nairn and C.A.T. Salama, "High-resolution current mode A/D converters using active current mirror," *Elect. Lett.* vol. 24, pp. 1331–1332, 1988.
17. Meta-Software, *HSPICE User's Manual H8801*, 1988.
18. W. Sansen and Z.Y. Chang, "Feedforward compensation techniques for high-frequency CMOS amplifiers," *IEEE Solid-State Circ.* vol. 25, no 6, pp. 1590–1595, 1990.
19. N. Balabanian and T. Bickart, *Linear Network Theory*, Matrix: Beaverton, OR, 1981.

Francisco V. Fernández received the Licenciado en Física degree from the University of Seville, Spain, in 1988. He is currently working towards the Ph.D in the field of automatic modeling of analog integrated circuits. Since 1988 he has been working as a research assistant in the Departamento de Electrónica y Electromagnetismo at the University of Seville. His research interests are in design and modeling of analog integrated circuits.

Angel Rodríguez-Vázquez (M'80) received the Licenciado en Física degree in 1977, and the Doctor en Ciencias Físicas degree in 1983, both from the University of Seville, Spain. Since 1978 he has been with the Departamento de Electrónica y Electromagnetismo at the University of Seville where he is an associate professor. His research interests are analog/digital integrated circuit design, analog integrated neural and nonlinear networks, and modeling analog integrated circuits.

José L. Huertas received the Licenciado en Física degree in 1969 and the Doctor en Ciencias Físicas degree in 1973, both from the University of Seville, Spain. From 1970 to 1971 he was with the Philips International Institute, Eindhoven, The Netherlands, as a postgraduate student. Since 1971 he has been with the Departamento de Electrónica y Electromagnetismo at the University of Seville, where he is a professor. His research interests are multivalued logic, sequential machines, analog circuit design, and nonlinear network analysis and synthesis.

SSCNAP: A Program for Symbolic Analysis of Switched Capacitor Circuits

Bixia Li and Deren Gu, *Senior Member, IEEE*

***Abstract*—SSCNAP is a program developed on a microcomputer for symbolic analysis of ideal *K*-phase switched capacitor circuits. Symbolic transfer functions and sensitivity functions can be obtained by using SSCNAP. It can also calculate the poles and their sensitivities in an SC circuit. In the paper, an efficient computer method for symbolic analysis of linear SC circuits is described. No topological or duty-cycle constraints have been imposed on the circuits. An example is included to illustrate the capabilities of the SSCNAP program.**

I. Introduction

FOR the development of SC networks with new topological structures, it is very important to obtain closed-form solutions in the Z domain with the coefficients of the rational transfer function explicitly given as functions of the network elements. Several authors have implemented digital computer programs which can produce the transfer function when Z is the only symbolic variable [1]–[4]. Konczykowska and Bon [5] presented a topological method for obtaining totally symbolic transfer functions. As with all direct topological analyses, the method is limited to SC networks of very small size. Wehehahn [6] presented an equivalent circuit in which every capacitance in a two-phase SC circuit is represented by six analog elements. A linear network symbolic analysis program may then be used to obtain the desired result. Moschytz and Mulawaka [7] presented a new signal-flow graph method for closed-form analysis of two-phase SC networks. The method, however has the following restriction: every capacitor must possess both a source and a sink switch. Numerical methods [8] seem to have better chances for success.

This paper presents a symbolic SC circuit analysis program, called SSCNAP, which was developed on a microcomputer. The program is based on the method that uses numerical algorithms at all stages. Partially or totally symbolic transfer functions, large-change sensitivities, relative sensitivities, and poles and their sensitivities for arbitrary K-phase SC circuits can be obtained by using SSCNAP. No topological or duty-cycle constraints have been imposed on the circuits. Our discussion will be limited to ideal SC circuits that do not contain any resistances.

Manuscript received October 25, 1988; revised December 5, 1989, and December 13, 1990. This paper was recommended by the former Editor, M. R. Lightner.

The authors are with the Institute of Applied Physics, University of Electronic Science and Technology of China, Chengdu, 610054, People's Republic of China.

IEEE Log Number 9103042.

II. Symbolic Analysis When Z Is the Only Symbol and Calculation of Circuit Poles

The SSCNAP program has adopted the modified nodal two-graph formulation, which reduces the sets of equations in each phase of SC circuits to the smallest possible number [10]. Assuming that a unit excitation is applied, the modified nodal two-graph formulation in the Z domain for a K-phase SC circuit can be obtained directly as follows [3], [4]:

$$\begin{aligned} \boldsymbol{A}_1 \boldsymbol{V}_1 - Z^{-1}\boldsymbol{B}_1\boldsymbol{V}_K &= \boldsymbol{g}_1 \\ \boldsymbol{A}_i\boldsymbol{V}_i - \boldsymbol{B}_i\boldsymbol{V}_{i-1} &= \boldsymbol{g}_i, \qquad i = 2, 3, \cdots, K \end{aligned} \tag{1}$$

or

$$\boldsymbol{M}\boldsymbol{V} = \boldsymbol{H} \tag{2}$$

where

$$\boldsymbol{M} = \begin{bmatrix} \boldsymbol{A}_1 & & & -Z^{-1}\boldsymbol{B}_1 \\ -\boldsymbol{B}_2 & \boldsymbol{A}_2 & & \\ & \ddots & \ddots & \\ & & & -\boldsymbol{B}_K\boldsymbol{A}_K \end{bmatrix}$$

$$\begin{aligned} \boldsymbol{V}' &= [\boldsymbol{V}_1', \boldsymbol{V}_2', \cdots, \boldsymbol{V}_K'] \\ \boldsymbol{H}' &= [\boldsymbol{g}_1', \boldsymbol{g}_2', \cdots, \boldsymbol{g}_K']. \end{aligned} \tag{3}$$

The prime represents transpose, $\boldsymbol{V}_i$ is the solution vector in phase i, and $\boldsymbol{g}_i$ is the vector composed of zeros and ones, indicating the presence of the source in the appropriate row of the equations. $\boldsymbol{A}_i$ and $\boldsymbol{B}_i$ are constructed directly from contributions of the circuit. The output is a linear combination of the entries of the vector $\boldsymbol{V}$:

$$T = \boldsymbol{y}'\boldsymbol{V} = \boldsymbol{y}'\boldsymbol{M}^{-1}\boldsymbol{H} \tag{4}$$

with $\boldsymbol{y}$ is a constant column vector. The inverse of the system matrix $\boldsymbol{M}$ is determined by the Kader method [11] in SSCNAP:

$$\boldsymbol{M}^{-1} = [\boldsymbol{A} + Z^{-1}\boldsymbol{B}]^{-1} = \frac{\sum_{i=0}^{n} \boldsymbol{G}_i Z^{-i}}{\sum_{i=0}^{n} d_i Z^{-i}} = \frac{\sum_{i=0}^{n} \boldsymbol{G}_i Z^{-i}}{d(Z^{-1})} \tag{5}$$

where

$$d_0 = \det A = \prod_{i=1}^{K} \det A_i \qquad G_0 = \text{adj } A = d_0 A^{-1}$$

$$d_i = \frac{1}{i} \text{tr } (BG_{i-1}), \qquad i = 1, 2, \cdots, n$$

$$G_i = A^{-1}(d_i E - BG_{i-1}), \qquad i = 1, 2, \cdots, n$$

$$n = \text{rank } B_1$$

$$A = \begin{bmatrix} A_1 & & & \\ -B_2 & A_2 & & \\ & \ddots & \ddots & \\ & & -B_K & A_K \end{bmatrix}$$

$$B = [\quad -B_1] \tag{6}$$

and E is a unit matrix of appropriate size.

The circuit poles in the Z domain are determined by the roots of the equation

$$d(Z^{-1}) = 0.$$

III. Large-Change Sensitivity

Assume that m is the number of symbolic variable elements. Denote by $h_{10}, h_{20}, \cdots, h_{m0}$ the nominal values of the elements and permit changes by the amounts $\delta_1, \delta_2, \cdots, \delta_m$. the new values are then $h_i = h_{i0} + \delta_i$. The nominal value h_{i0} may be zero, indicating that the element is a parasitic. The program SSCNAP uses perturbation theory to obtain the symbolic functions in terms of $\delta_1, \delta_2, \cdots, \delta_m$. This theory was originally developed for analog networks [9]. At present, the modified nodal two-graph formulation for an SC circuit can be obtained as follows:

$$(A_1 + P_1X_1Q_1')V_1 - (B_1Z^{-1} + P_1X_1Q_K'Z^{-1})V_K = g_1$$

$$(A_i + P_iX_1Q_i')V_i - (B_i + P_iX_1Q_{i-1}')V_{i-1} = g_i$$

$$i = 2, 3, \cdots, K \tag{7}$$

or

$$(M + PXQ')V = H \tag{8}$$

$$P = \begin{bmatrix} P_1 & & & P_1 \\ & P_2 & & \\ & & \ddots & \\ & & P_K & O \end{bmatrix}$$

$$Q = \begin{bmatrix} Q_1 & -Q_1 & & & \\ & Q_2 & -Q_2 & & \\ & & \ddots & \ddots & \\ & & & Q_K & -Q_K \end{bmatrix}$$

$$P_i = [p_1^{(i)}, p_2^{(i)}, \cdots, p_m^{(i)}]$$

$$Q_i = [q_1^{(i)}, q_2^{(i)}, \cdots, q_m^{(i)}]. \tag{9}$$

Here $P_i(Q_i)$ is an $L_i \times m$ $(J_i \times m)$ matrix, which is obtained from the Q graph (V graph) of the ith phase. $L_i(J_i)$ is the number of nodes, except the reference node, of the Q graph (V graph) of the ith phase. As an example, consider the jth symbolic variable h_j with the corresponding Q graph (V graph) edge of the ith phase pointing from node $k_Q(k_v)$ to the node $k_Q'(k_v')$; then

$$p_j^{(i)} = e_{KQ}^{(i)} - e_{K_Q'}^{(i)} \qquad q_j^{(i)} = U_{K_v}^{(i)} - U_{K_v'}^{(i)}$$

with $e_l^{(i)}(U_l^{(i)})$ a column vector of zeros except for the lth entry, which is unity. The matrix X is defined as

$$X = \begin{bmatrix} X_1 & & & & \\ & X_2 & & & \\ & & \ddots & & \\ & & & X_K & \\ & & & & Z^{-1}X_K \end{bmatrix}$$

$$X_i = \begin{bmatrix} \delta_1 & & & \\ & \delta_2 & & \\ & & \ddots & \\ & & & \delta_m \end{bmatrix}, \qquad i = 1, 2, \cdots, K. \tag{10}$$

We introduce the following notation:

$$\hat{P} = [P, H] \qquad \hat{Q} = [Q, y]$$

$$\hat{F}(Z^{-1}) = \hat{Q}'G\hat{P} = \sum_{i=0}^{n} \hat{Q}'G_i\hat{P}Z^{-i} \tag{11}$$

$$a_0(Z^{-1}) = \sum_{i=0}^{n} y'G_iHZ^{-i} \qquad b_0(Z^{-1}) = d(Z^{-1})$$

$$a_{K_1K_2\cdots K_l}(Z_i^{-1}) = [d(Z_i^{-1})]^{-l} \det \hat{F}_i(k_1, k_2, \cdots, k_l, I + 1)$$

$$b_{K_1K_2\cdots K_l}(Z_i^{-1}) = [d(Z_i^{-1})]^{-l} \det \hat{F}_i(k_1, k_2, \cdots, k_l).$$

$$I = (K + 1) \times m \tag{12}$$

$\hat{F}_i(k_1, k_2, \cdots)$ stands for the submatrix of $\hat{F}(Z^{-1})$ formed by retaining only the rows and columns $k_1, k_2, \cdots$ at $Z^{-1} = Z_i^{-1}$. G_i $(i = 0, 1, 2, \cdots, n)$ and $d(Z^{-1})$ have been given in Section II. As the matrix

$$\frac{\partial^K(M + PXQ')}{\partial^K \delta i} \qquad (i = 1, 2, \cdots, m)$$

is of rank 1, the output function in terms of $\delta_1, \delta_2, \cdots, \delta_m$ can be obtained as

$$T_\delta(\delta_1, \delta_2, \cdots, \delta_m) = y'V = \frac{N_\delta(Z^{-1}, \delta_1, \delta_2, \cdots, \delta_m)}{D_\delta(Z^{-1}, \delta_1, \delta_2, \cdots, \delta_m)}$$

$$= \frac{A_0(Z^{-1}) + \sum_{i_1}^{m} A_{i_1}(Z^{-1})\,\delta_{i_1} + \sum_{i_1 \le i_2}^{m}\sum^{m} A_{i_1 i_2}(Z^{-1})\,\delta_{i_1}\,\delta_{i_2} + \sum_{i_1 \le i_2 \le i_3}^{m}\sum^{m}\sum^{m} A_{i_1 i_2 i_3}(Z^{-1})\,\delta_{i_1}\,\delta_{i_2}\,\delta_{i_3} + \cdots}{B_0(Z^{-1}) + \sum_{i_1}^{m} B_{i_1}(Z^{-1})\,\delta_{i_1} + \sum_{i_1 \le i_2}^{m}\sum^{m} B_{i_1 i_2}(Z^{-1})\,\delta_{i_1}\,\delta_{i_2} + \sum_{i_1 \le i_2 \le i_3}^{m}\sum^{m}\sum^{m} B_{i_1 i_2 i_3}(Z^{-1})\,\delta_{i_1}\,\delta_{i_2}\,\delta_{i_3} + \cdots}$$

$$= \frac{\sum_{j_1=0}^{K}\sum_{j_2=0}^{K} \cdots \sum_{j_m=0}^{K} \hat{A}_{j_1 j_2 \cdots j_m}(Z^{-1})\,\delta_1^{j_1}\,\delta_2^{j_2} \cdots \delta_m^{j_m}}{\sum_{j_1=0}^{K}\sum_{j_2=0}^{K} \cdots \sum_{j_m=0}^{K} \hat{B}_{j_1 j_2 \cdots j_m}(Z^{-1})\,\delta_1^{j_1}\,\delta_2^{j_2} \cdots \delta_m^{j_m}} \tag{13}$$

where

$$A_0(Z^{-1}) = a_0(Z^{-1}) \qquad B_0(Z^{-1}) = b_0(Z^{-1})$$

$$A_{i_1 i_2 \cdots i_l}(Z_i^{-1}) = \sum_{j_1}^{K+1}\sum_{j_2}^{K+1} \cdots \sum_{j_l}^{K+1} (a_{i_1+(j_1-1)m,\, i_2+(j_2-1)m, \cdots i_l+(j_l-1)m} Z_i^{-f(j_1-K-1)} Z_i^{-f(j_2-K-1)} \cdots Z_i^{-f(j_l-K-1)})$$

$$B_{i_1 i_2 \cdots i_l}(Z_i^{-1}) = \sum_{j_1}^{K+1}\sum_{j_2}^{K+1} \cdots \sum_{j_l}^{K+1} (b_{i_1+(j_1-1)m,\, i_2+(j_2-1)m, \cdots i_l+(j_l-1)m} Z_i^{-f(j_1-K-1)} Z_i^{-f(j_2-K-1)} \cdots Z_i^{-f(j_l-K-1)}) \tag{14}$$

with

$$f(x) \triangleq \begin{cases} 1, & x = 0 \\ 0, & x \neq 0. \end{cases}$$

We calculate $A_{i_1 i_2 \cdots i_l}(Z_i^{-1})$ and $B_{i_1 i_2 \cdots i_l}(Z_i^{-1})$ for 2^n distinct values of Z_i^{-1} that are uniformly distributed on the unit circle, 2^n being greater than and the nearest to rank $\boldsymbol{B}_1$. This will provide pairs of values $[Z_i^{-1}, A_{i_1 i_2 \cdots i_l}(Z_i^{-1})]$ and $[Z_i^{-1}, B_{i_1 i_2 \cdots i_l}(Z_i^{-1})]$. The polynomials $A_{i_1 i_2 \cdots i_l}(Z^{-1})$ and $B_{i_1 i_2 \cdots i_l}(Z^{-1})$ can be recovered by the FFT program.

The large change sensitivity function is obtained as

$$\Delta T_\delta(\delta_1, \delta_2, \cdots, \delta_m) \triangleq T_\delta(\delta_1, \delta_2, \cdots, \delta_m) - \frac{A_0(Z^{-1})}{B_0(Z^{-1})}$$

IV. Symbolic Network Functions

The program offers symbolic output functions in terms of element parameters:

$$T_h(h_1, h_2, \cdots, h_m) = y'V = \frac{N_h(Z^{-1}, h_1, h_2, \cdots, h_m)}{D_h(Z^{-1}, h_1, h_2, \cdots, h_m)}.$$

Since $h_i = h_{i0} + \delta_i$, $T_h(h_1, h_2, \cdots, h_m)$ is of the following form:

$$T_h(h_1, h_2, \cdots, h_m) = T_\delta(h_1 - h_{10}, h_2 - h_{20}, \cdots h_m - h_{m0})$$

$$= \frac{\sum_{j_1=0}^{K}\sum_{j_2=0}^{K}\sum_{j_m=0}^{K} R_{j_1 j_2 \cdots j_m}(Z^{-1}) h_1^{j_1} h_2^{j_2} \cdots h_m^{j_m}}{\sum_{j_1=0}^{K}\sum_{j_2=0}^{K}\sum_{j_m=0}^{K} S_{j_1 j_2 \cdots j_m}(Z^{-1}) h_1^{j_1} h_2^{j_2} \cdots h_m^{j_m}}. \tag{15}$$

The coefficients $R_{j_1 j_2 \cdots j_m}(Z^{-1})$ and $S_{j_1 j_2 \cdots j_m}(Z^{-1})$ are computed according to

$$\frac{1}{j_1! j_2! \cdots j_m!} \cdot \left.\frac{\partial^n N_h(Z^{-1}, h_1, h_2, \cdots, h_m)}{\partial^{j_1} h_1\, \partial^{j_2} h_2 \cdots \partial^{j_m} h_m}\right|_{h_i = h_{i0}} = \frac{1}{j_1! j_2! \cdots j_m!} \cdot \left.\frac{\partial^n N_\delta(Z^{-1}, \delta_1, \delta_2, \cdots, \delta_m)}{\partial^{j_1} \delta_1\, \partial^{j_2} \delta_2 \cdots \partial^{j_m} \delta_m}\right|_{\delta_i = 0} \equiv \hat{A}_{j_1 j_2 \cdots j_m}(Z^{-1})$$

$$\frac{1}{j_1! j_2! \cdots j_m!} \cdot \left.\frac{\partial^n D_h(Z^{-1}, h_1, h_2, \cdots, h_m)}{\partial^{j_1} h_1\, \partial^{j_2} h_2 \cdots \partial^{j_m} h_m}\right|_{h_i = h_{i0}} = \frac{1}{j_1! j_2! \cdots j_m!} \cdot \left.\frac{\partial^n D_\delta(Z^{-1}, \delta_1, \delta_2, \cdots, \delta_m)}{\partial^{j_1} \delta_1\, \partial^{j_2} \delta_2 \cdots \partial^{j_m} \delta_m}\right|_{\delta_i = 0} \equiv \hat{B}_{j_1 j_2 \cdots j_m}(Z^{-1}). \tag{16}$$

V. Sensitivity Analysis

The transfer function's sensitivity with respect to network elements is defined as

$$S_{h_i}^{T_h(h_i)} \triangleq \frac{h_i}{T_h(h_i)} \cdot \frac{\partial T_h(h_i)}{\partial h_i} = h_i \frac{\partial \ln T_h(h_i)}{\partial h_i}$$

where h_i may be a capacitor parameter or the gain of a controlled source. The function $T_h(h_i)$ can be obtained easily by using the method given in Section IV:

$$T_h(h_i) = \frac{\sum_{j=0}^{K} R_j(Z^{-1})h_i^j}{\sum_{j=0}^{K} S_j(Z^{-1})h_i^j}. \quad (17)$$

$S_{h_i}^{T_h(h_i)}$ is of the form

$$S_{h_i}^{T_h(h_i)} = \frac{\sum_{j=0}^{K} jR_j(Z^{-1})h_i^j}{\sum_{j=0}^{K} R_j(Z^{-1})h_i^j} - \frac{\sum_{j=1}^{K} jS_j(Z^{-1})h_i^j}{\sum_{j=1}^{K} S_j(Z^{-1})h_i^j}. \quad (18)$$

To calculate the pole sensitivity to network elements, we first find the denominator of the transfer function in terms of element parameter h_i:

$$D_h(h_i) = \sum_{n=0}^{K} S_n(Z^{-1})h_i^n$$

and the circuit pole Z_j when h_i is assigned its nominal value h_{i0}. The sensitivity of the pole is computed according to

$$\frac{\partial Z_j}{\partial h_i} - \frac{\sum_{n=1}^{K} nh_{i0}^{n-1}S_n(Z_j^{-1})}{\sum_{n=0}^{K} h_{i0}^n \cdot \left.\frac{\partial S_n(x)}{\partial x}\right|_{x=Z_j^{-1}}} \cdot \frac{1}{Z_j^{-2}}. \quad (19)$$

VI. Program Implementation and Performance

The SSCNAP program was written on the basis of the above results. Currently, it is in use on an MC68000 microcomputer.

The program can now process arbitrary switching phase SC circuits containing 50 capacitors, 100 ideal switches, 30 ideal operational amplifiers, 20 unity gain buffers, and 20 VCVS's with constant gains. The number of elements that the program can process could be increased easily. Even if it is not extended, the program can still analyze nearly all the SC circuits which have been described in the recent literature. The program is limited to at most ten symbolic elements.

Because the SSCNAP is a program developed on a microcomputer, attention should be given to three points. First, the computing time should be reasonable. Second, the memory of the microcomputer is limited, so the program must be installed very efficiently. Third, the program should be user friendly. User-friendly programs are needed on microcomputers because microcomputers are widely used and can be operated easily. To achieve reasonable operation time, the FFT method is used in polynomial interpolation, a fast algorithm established by the first author [14] is used to solve (16), and the rules for topological screening of symbol combinations in [12] are modified and integrated in the SSCNAP. To solve the problem of the shortage of memory, the modified nodal two-graph formulation is adopted; the sparse matrix technique is used in topological screening of symbol combinations and in solving (16). The program can now be run on a microcomputer with 260KB storage. To be user friendly, a free format input is provided. The user can input the circuit topology and element values in a manner which is similar to the language normally used by circuit designers. The input file of SSCNAP is similar to that of the well-known program SPICE. So, the user can define a circuit, choose the type of simulation, and obtain the required result without any difficulty. The results are printed in the form of tables.

The steps taking advantage of all possible savings are now summarized

1) Prepare matrices $\hat{F}(Z_i^{-1})$ and $d(Z_i^{-1})$. Detect valid symbol combinations.
   a) Formulate the matrices $\boldsymbol{A}_i$, $\boldsymbol{B}_i$, $\boldsymbol{P}_i$, and $\boldsymbol{Q}_i$, and vectors $\boldsymbol{g}_i$ and $\boldsymbol{y}$ using two-graph theory. $\boldsymbol{P}_i$, $\boldsymbol{Q}_i$, $\boldsymbol{g}_i$, and $\boldsymbol{y}$ are stored in tabular form.
   b) Detect valid symbol combinations [14]. Sparse matrix technique are used here.
   c) Calculate $\boldsymbol{G}_i$ and $d(Z^{-1})$ using (6).
   d) Select points Z_i^{-1} uniformly distributed on a unit circle. Calculate $\hat{F}(Z_i^{-1})$ and $d(Z_i^{-1})$ using (11).
2) Calculate the coefficients $\hat{A}_{j_1j_2\cdots j_m}(Z_i^{-1})$ and $\hat{B}_{j_1j_2\cdots j_m}(Z_i^{-1})$ corresponding to the valid symbol combinations using (12) and (14). The FFT is used in polynomial interpolation.
3) The transfer function in terms of Z^{-1} and network elements is obtained. Calculate the coefficients $R_{j_1j_2\cdots j_m}(Z^{-1})$ and $S_{j_1j_2\cdots j_m}(Z^{-1})$ corresponding to the valid symbol combinations using (16). Transfer function (15) is obtained.
4) The transfer function in terms of Z^{-1} and increments $\delta_1, \delta_2, \cdots, \delta_m$ (large change sensitivity) is obtained. Calculate all coefficients $\hat{A}_{j_1j_2\cdots j_m}(Z^{-1})$ and $\hat{B}_{j_1j_2\cdots j_m}(Z^{-1})$ using the relations

$$\frac{1}{j_1!j_2!\cdots j_m!} \cdot \left.\frac{\partial^n N_\delta(Z^{-1}, \delta_1, \delta_2, \cdots, \delta_m)}{\partial^{j_1}\delta_1\partial^{j_2}\delta_2\cdots\partial^{j_m}\delta_m}\right|_{\delta_i=h_{i0}}$$

$$\equiv R_{j_1j_2\cdots j_m}(Z^{-1})$$

$$\frac{1}{j_1!j_2!\cdots j_m!} \cdot \left.\frac{\partial^n D_\delta(Z^{-1}, \delta_1, \delta_2, \cdots, \delta_m)}{\partial^{j_1}\delta_1\partial^{j_2}\delta_2\cdots\partial^{j_m}\delta_m}\right|_{\delta_i=h_{i0}}$$

$$\equiv S_{j_1j_2\cdots j_m}(Z^{-1})$$

where $R_{j_1j_2\cdots j_m}(Z^{-1})\,(S_{j_1j_2\cdots j_m}(Z^{-1}))$ corresponding to invalid symbol combinations equals zero.

5) The transfer function's sensitivity with respect to network elements is obtained. Calculate the transfer function $T_h(h_i)$ and apply the results in (18).
6) Poles and their sensitivities to network elements are obtained.

   a) Calculate poles using $d(Z^{-1}) = 0$.
   b) Poles' sensitivities are obtained by using (19).

VII. Example

Many SC circuits have been analyzed by using the SSCNAP program. An example is given here.

Example: Fig. 1 is a fifth-order SC elliptic low-pass filter implemented by an inductor simulation ladder circuit [13]. Symbolic analysis was performed with the transfer function

$$T(Z^{-1}) = \frac{V_{\text{out}}^{(1)}(Z^{-1})}{V_{\text{in}}^{(2)}(Z^{-1})}.$$

The nominal values of C_i, $i = 1, 2, \cdots, 15$, are shown in Table I. The symbolic transfer function in terms of C_6, C_{10} and C_{13}:

$$T(C_6, C_{10}, C_{13}) = \frac{\sum_{i_1=0}^{2}\sum_{i_2=0}^{2}\sum_{i_3=0}^{2} \hat{A}_{i_1i_2i_3}(Z^{-1})C_6^{i_1}C_{10}^{i_2}C_{13}^{i_3}}{\sum_{i_1=0}^{2}\sum_{i_2=0}^{2}\sum_{i_3=0}^{2} \hat{B}_{i_1i_2i_3}(Z^{-1})\,C_6^{i_1}\,C_{10}^{i_2}\,C_{13}^{i_3}}$$

the large change sensitivity in terms of the increments δ_6, δ_{10}, δ_{13} $(C_i = C_{i0} + \delta_i)$:

$$\Delta T(\delta_6, \delta_{10}, \delta_{13}) = \frac{\sum_{i_1=0}^{2}\sum_{i_2=0}^{2}\sum_{i_3=0}^{2} \hat{A}_{i_1i_2i_3}(Z^{-1})\delta_6^{i_1}\delta_{10}^{i_2}\delta_{13}^{i_3}}{\sum_{i_1=0}^{2}\sum_{i_2=0}^{2}\sum_{i_3=0}^{2} \hat{B}_{i_1i_2i_3}(Z^{-1})\,\delta_6^{i_1}\,\delta_{10}^{i_2}\,\delta_{13}^{i_3}}$$

$$-\frac{A_{000}(Z^{-1})}{B_{000}(Z^{-1})}$$

and the sensitivity function in terms of C_2:

$$S_{C_2}^{T(C_2)} = \frac{\sum_{j=1}^{2} jr_j(Z^{-1})C_2^{j-1}}{\sum_{j=0}^{2} r_j(Z^{-1})C_2^j} - \frac{\sum_{j=1}^{2} jS_j(Z^{-1})C_2^{j-1}}{\sum_{j=0}^{2} S_j(Z^{-1})C_2^j}$$

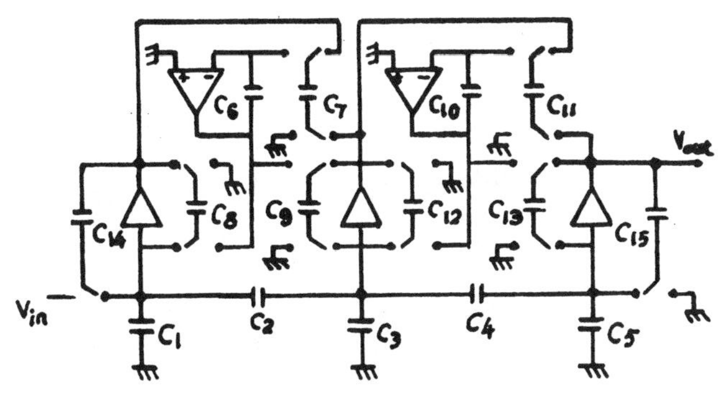

Fig. 1. Fifth-order SC elliptic low-pass filter.

TABLE I
Nominal Values of Elements

C1	5.892
C2	1.253
C3	9.160
C4	3.855
C5	4.222
C6	6.954
C7	1.0
C8	1.0
C9	1.0
C10	4.809
C11	1.0
C12	1.0
C13	1.0
C14	1.0
C15	1.0

are given in Tables II through IV. Poles in the Z domain and their sensitivities to C_6 and C_7 are shown in Table V. Those coefficients which are not listed in the tables are equal to zero. The computing time for the example is 5 min.

VIII. Conclusion

The SSCNAP program is a general K-phase symbolic SC circuit analysis program that is a useful tool for analyzing and designing SC circuits. For the case where a relatively small number of network elements are represented by symbols, network functions can be obtained for very large SC networks. In fact, with a comparable amount of computer storage, it is now possible to analyze networks of the order that can be handled by a numerical program such as WATSCAD [3]. SSCNAP further demonstrates a very significant improvement over topological programs based on tree enumeration or signal flow graphs. The most significant improvement is the ability to analyze much larger networks if many of the elements are characterized by nonsymbolic parameters. The SSCNAP has the capacity of large-change sensitivity analysis. The program can give symbolic transfer functions. It can also calculate poles and their sensitivities to network elements and is user friendly.

TABLE II

NUMERATOR AND DENOMINATOR COEFFICIENTS IN Z^{-1} OF SYMBOLIC TRANSFER FUNCTION IN TERMS OF C_6, C_{10}, AND C_{13}

	Z^0	Z^{-1}	Z^{-2}	Z^{-3}	Z^{-4}	Z^{-5}
$\hat{A}_{111}$	0	0	0	−7.04844	0	0
$\hat{A}_{120}$	0	0	−27.1600	54.3203	−27.1600	0
$\hat{A}_{211}$	0	0	−8.82796	17.6559	−8.82796	0
$\hat{A}_{220}$	0	−34.0318	136.127	−204.191	136.127	−34.0317
$\hat{B}_{110}$	0	0	−29.2525	22.6575	0	0
$\hat{B}_{111}$	0	0	−106.252	98.9975	0	0
$\hat{B}_{120}$	0	−971.180	2723.48	−2540.44	788.080	0
$\hat{B}_{210}$	0	−212.554	557.503	−484.508	139.458	0
$\hat{B}_{211}$	0	−513.127	1466.02	−1392.65	439.762	0
$\hat{B}_{220}$	−4963.85	23404.8	−44081.3	41453.5	−19463.1	3649.95

TABLE III

COEFFICIENTS IN Z^{-1} OF LARGE-CHANGE SENSITIVITY

	Z^0	Z^{-1}	Z^{-2}	Z^{-3}	Z^{-4}	Z^{-5}
A_{000}	0	−38.0597	145.817	−215.751	145.817	−38.0596
A_{001}	0	0	−2.05308	3.87025	−2.05308	0
A_{010}	0	−15.8285	61.0704	−90.5329	61.0704	−15.8285
A_{020}	0	−1.64571	6.39398	−9.49654	6.39398	−1.64571
A_{100}	0	−10.9461	42.5658	−63.2733	42.5658	−10.9461
A_{101}	0	0	0	1.14700	0	0
A_{110}	0	−4.55234	17.8253	−26.5530	17.8253	−4.55233
A_{120}	0	0	1.86610	−2.78556	1.86610	0
A_{200}	0	0	3.10569	−4.63730	3.10569	0
A_{210}	0	0	1.30044	−1.94625	1.30044	0
B_{000}	−5551.34	25849.9	−48394.4	45518.6	−21505.2	4081.95
B_{001}	0	−119.330	337.382	−320.556	102.269	0
B_{010}	−2308.73	10785.7	−20223.5	19020.5	−8971.74	1697.63
B_{011}	0	−24.8139	70.1563	−66.6576	21.2661	0
B_{020}	−240.042	1125.06	−2112.75	1986.95	−935.720	176.505
B_{100}	−1596.59	7457.00	−13980.8	13149.5	−6203.21	1173.99
B_{101}	0	−34.3198	97.5424	−92.6695	29.4129	0
B_{110}	−664.000	3111.36	−5842.43	5494.70	−2587.89	488.244
B_{111}	0	7.13657	20.2833	−19.2700	6.11621	0
B_{120}	−69.0372	324.543	−610.360	573.995	−269.905	50.7635
B_{200}	−114.796	537.781	−1009.72	949.647	−447.328	84.4106
B_{201}	0	−2.46763	7.05007	−6.69726	2.11482	0
B_{210}	−47.7423	224.382	−421.951	396.823	−186.617	35.1053
B_{211}	0	0	1.46602	−1.39265	0	0
B_{220}	−4.96385	23.4048	−44.0813	41.4535	−129.4631	3.64995

TABLE IV

COEFFICIENTS IN Z^{-1} OF SENSITIVITY FUNCTION

	Z^0	Z^{-1}	Z^{-2}	Z^{-3}	Z^{-4}	Z^{-5}
S_0	−3163.05	14606.9	−27140.3	25353.7	−11905.1	2247.53
S_1	−1639.77	7701.29	−14528.7	13759.0	−6539.83	1247.92
S_2	−212.519	1014.82	−1942.44	1862.96	−895.282	172.461
r_0	0	0	−3.29709	6.41639	−3.29709	0
r_1	0	−22.9280	89.6208	−133.432	89.6208	−22.9280
r_2	0	−5.94310	23.4518	−35.0174	23.4518	−5.94310

TABLE V

POLES IN Z DOMAIN AND THEIR SENSITIVITIES WITH RESPECT TO THE CAPACITORS C_6 AND C_7

Poles	Sensitivities to C_6	Sensitivities to C_7
0.9663977 $+j$ −0.1739691	0.1566609E−02 $+j$ 0.5787323E−02	−0.9699889E−02 $+j$ −0.3846329E−01
0.9663977 $+j$ 0.1739691	0.1566609E−02 $+j$ −0.5787323E−02	−0.9699889E−02 $+j$ 0.3846329E−01
0.8827388 $+j$ 0.0000000	−0.9982548E−03 $+j$ 0.0000000E+00	0.4398299E−02 $+j$ 0.0000000E+00
0.9204910 $+j$ −0.1289035	0.9433908E−03 $+j$ 0.6534982E−02	−0.6436553E−02 $+j$ −0.4877771E−01
0.9204910 $+j$ 0.1289035	0.9433908E−03 $+j$ −0.6534982E−02	−0.6436553E−02 $+j$ 0.48777771E−01

References

[1] C. F. Kurth and G. S. Moschytz, "Two-port analysis of switched-capacitor networks using four-port equivalent circuits in the Z-domain," *IEEE Trans. Circuits Syst.*, vol. CAS-26, pp. 166–180, Mar. 1979.

[2] Y. L. Kuo, M. L. Liou, and J. W. Kasinskas, "An equivalent circuit approach to the computer-aided analysis of switched-capacitor circuits," *IEEE Trans. Circuit Syst.*, vol. CAS-26, pp. 708–714, Sept. 1979.

[3] J. Vlach, K. Singhal, and M. Vlach: "Computer oriented formulation of equation and analysis of switched-capacitor networks," *IEEE Trans. Circuits Syst.*, vol. CAS-31, pp. 753–765, Sept. 1984.

[4] C. K. Pun and J. I. Sewell, "Symbolic analysis of ideal and non-ideal switched capacitor networks," in *Proc. Int. Symp. Circuits Syst.*, 1985, pp. 1165–1168.

[5] A. Konczykowska and M. Bon, "Topological analysis of switched-capacitor networks," *Electron. Lett.*, vol. 16, pp. 89–90, 1980.

[6] E. Wehehahn, "Evaluation of transfer functions of ideal SC networks in Z domain using standard linear symbolic or semisymbolic networks," *Electron. Lett.*, vol. 16, pp. 801–802, 1980.

[7] G. S. Moschytz and J. J. Mulawka, "New methods of direct closed form analysis of switched-capacitor networks," in *Proc. IEEE Int. Symp. Circuits Syst.*, 1986, pp. 369–372.

[8] Y. Cheng and P. M. Lin, "Symbolic analysis of general switched-capacitor networks—New methods and implementation," in *Proc. Int. Symp. Circuits Syst.*, 1987, pp. 55–59.

[9] K. Singhal and J. Vlach, "Symbolic circuit analysis," *Proc. Inst. Elec. Eng.*, vol. 128, pt G, pp. 81–86, Apr. 1981.

[10] J. Vlach and K. Singhal, *Computer Methods for Circuit Analysis and Design*. New York: Van Nostrand Reinhold, 1983.

[11] A. Kader and A. Tabot, "Inversion of polynominal network matrices with particular reference to sensitivity analysis," *Proc. Inst. Elec. Eng.*, vol. 128, pt. G, no. 4, pp. 170–172, Aug. 1981.

[12] K. Singhal and J. Vlach, "Two-graph tableau and nodal tableau formulation of networks with ideal elements," in *Proc. 1978 European Conf. Circuit theory and Design*, pp. 553–557.

[13] M. S. Lee and S. Chang, "Low-sensitivity switched-capacitor ladder filters," *IEEE Trans. Circuits Syst.*, vol. CAS-27, June 1980.

[14] B. Li, "Computer aided symbolic analysis of switched capacitor networks," Ph.D. thesis, University of Electronic Science and Technology of China, Mar. 1989.

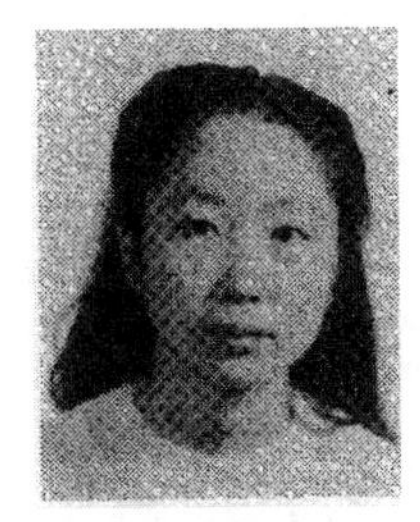

Bixia Li was born in Sichuan, People's Republic of China. She received the B.S., M.S., and Ph.D. degrees in electrical engineering from the University of Electronic Science and Technology of China in 1983, 1986, and 1989, respectively.

She is currently a Lecturer at the University of Electronic Science and Technology of China working on switched capacitor networks, computer-aided circuit design, and image processing.

Deren Gu (M'89–SM'90) received the B.Sc. degree in 1946 from Jiaotung University, Shanghai, China, and the M.S. degree from Cornell University, Ithaca, NY, in 1950, both in electrical engineering.

He was a former president of the University of Electronic Science and Technology of China. Prof. Gu's research currently focuses on switched capacitor networks, computer-aided circuit design, nonlinear filters, and image processing.

Efficient Symbolic Computation of Approximated Small-Signal Characteristics of Analog Integrated Circuits

Piet Wambacq, Francisco V. Fernández, Georges Gielen, Willy Sansen, and Angel Rodríguez-Vázquez

***Abstract*—A symbolic analysis tool is presented that generates simplified symbolic expressions for the small-signal characteristics of large analog integrated circuits. The expressions are approximated while they are computed, so that only those terms are generated which remain in the final expression. This principle causes drastic savings in CPU time and memory, compared with previous symbolic analysis tools. In this way, the maximum size of circuits that can be analyzed, is largely increased. By taking into account a range for the value of a circuit parameter rather than one single number, the generated expressions are also more generally valid. Mismatch handling is explicitly taken into account in the algorithm. The capabilities of the new tool are illustrated with several experimental results.**

I. Introduction

CURRENT tools for small-signal symbolic analysis of analog integrated circuits, like for instance ISAAC [1] and ASAP [2], are able to evaluate network functions in the s-domain with the complex frequency variable and the circuit parameters (capacitances, resistances, transconductances, etc.) kept as symbols. These functions are typically given as an expanded cancellation-free sum of products,

$$\frac{f_0(\boldsymbol{x}) + sf_1(\boldsymbol{x}) + s^2 f_2(\boldsymbol{x}) + \cdots + s^N f_N(\boldsymbol{x})}{g_0(\boldsymbol{x}) + sg_1(\boldsymbol{x}) + s^2 g_2(\boldsymbol{x}) + \cdots + s^M g_M(\boldsymbol{x})} \quad (1)$$

in which $\boldsymbol{x}^T = \{x_1, x_2, \cdots, x_Q\}$ is the vector of symbolic circuit parameters and the f_i and g_j are sums of products.

Since these expressions are calculated automatically, analog designers are released from the involved calculations needed to get insight into the ac behavior of circuits. Also, analog cells can be automatically sized for given ac specifications through the iterative optimization of the symbolic equations generated for their gain, poles, zeros, terminal impedances, PSRR, CMRR, etc. Other potential applications of symbolic analyzers, for synthesis, statistical optimization, testability, etc., exploit also the computational advantages to perform repetitive evaluations of precalculated models [3]. However, these applications can be realized at fully only if the automatic generation of symbolic expressions runs parallel to the automatic pruning of insignificant terms in these expressions—similar to what expert analog designers do when they analyze circuits by hand.

Manuscript received July 13, 1994; revised November 4, 1994

P. Wambacq, G. Gielen, and W. Sansen are with Katholieke Universiteit Lueven, Dep. Elektrotechniek, ESAT-MICAS, B-3001 Heverlee, Belgium.

F. V. Fernandez and A. Rodríguez-Vázquez are with Department of Analog Circuit Design, Centro Nacional de Microelectrónica, Edif. CICA, E-41012 Sevilla, Spain.

IEEE Log Number 9408739.

Although existing analyzers like ISAAC and ASAP incorporate such simplification feature, their algorithms have two important drawbacks: a) simplifications are performed only after the exact symbolic expression is generated in an expanded sum-of-product format; and b) the significance of each term in the sums-of-products is assessed on the basis of numerical evaluations using typical values of the circuit parameters, at a single point of the design parameter space. Since the size of the exact symbol expressions increases exponentially with the number of nodes and elements in the circuit, the first drawback puts an upper limit on the complexity of analyzable circuits; around ten transistors if each transistor is represented by a high-frequency model containing about nine circuit elements. On the other hand, approximating symbolic expressions by considering just a single point of the parameter space does not seem to be consistent with the very nature of the symbolic analysis procedure, where the exact numerical value of the parameters is, by definition, unknown *a priori*. Even in the case symbolic analysis is used to study critical parameter variations in an already sized schematic, simplifying by using just information about a nominal point may lead to important inaccuracies in mismatch-sensitive characteristics, as for instance PSRR or CMRR of operational amplifiers.

This paper presents a simplification algorithm to overcome both drawbacks above. First of all, the complexity limits of analyzable circuits are extended by generating only the dominant terms, without first computing the complete exact expression. In this approach, which is denoted as *simplification during generation*, the dominant terms are generated until the accuracy falls within a given user-supplied accuracy. As shown in Fig. 1, this is a much more efficient approach, both in terms of memory usage and CPU time, than the classical approach followed in [1], [2]. The idea of simplification during generation was first mentioned in [4], and later found also in [5] and [6]. However, simplifications in [4] and [6] are performed at a nominal point of the design space, and, consequently, any evaluation of the expressions in another operating point might cause large errors. To reduce these errors, this paper further elaborates the approach in [5] to combine the concept of simplification during generation with the use of ranges [7], instead of single values, for the circuit design parameters. This increases the compliance of generated expressions, while keeping the computation time and the memory resources needed for symbolic analysis of large analog circuits bounded. Also, the combination of both

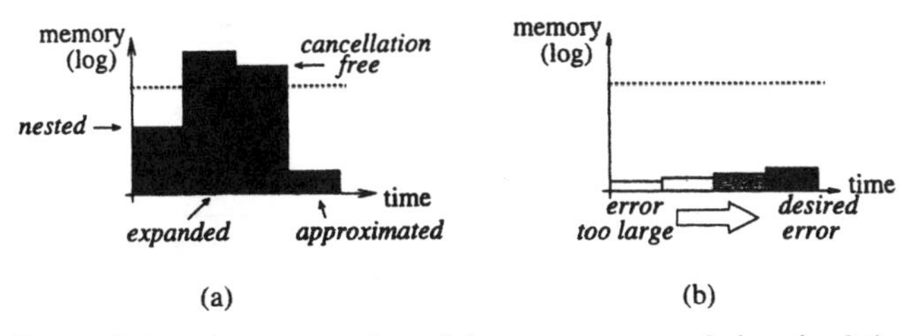

Fig. 1. Schematic representation of the memory usage during simulation time with the classical approach (a) and with the proposed approach of simplification during generation (b). Classically, a symbolic expression is first generated in a nested format. For a reliable approximation, the expression is then expanded and the cancelling terms are elaborated. This expansion can lead to a huge number of terms whose storage exceeds the memory limits (dashed line), while only a few terms are retained after approximation. This problem is circumvented if the expression is simplified when it is generated: the memory usage increases with the required accuracy or the number of terms of the symbolic expression.

techniques, simplification during generation and the use of ranges, demonstrate better results than previous approaches for the handling of matching between symbolic parameters and mismatches.

Section II presents the concept and outlines the algorithms used for simplification during generation. Section III explains how intervals are incorporated in the stopping criterion that controls the generation of terms. In Section IV it is explained how matching elements and corresponding mismatches are handled in the new approach. Finally, Section V presents examples that demonstrate the suitability of the techniques presented for analog cells containing more than 20 transistors, which approaches the size of practical circuits used in todays IC designs.

II. Simplification During Generation

The idea of simplification during generation needs a term by term generation mechanism, which finds the terms in decreasing order of magnitude, without skipping any term. This can be achieved with the undirected tree enumeration method [8]. This is a topological method that operates on two weighted graphs, the voltage graph and the current graph, which are easily derived from the given (small-signal) network. A term is valid only if its corresponding branches constitute a spanning tree in both graphs. The symbolic term is given by the product of the branch weights (admittances) in any of the graphs. The sign of a term is determined separately, using topological information of both graphs. By augmenting the network in a special way with fictitious elements, it is possible to generate the terms for both numerator and denominator at the same time [8].

The number of trees increases exponentially with the circuit size. Since we are interested only in the dominant terms and therefore not in all trees, the new algorithm enumerates spanning trees in the voltage graph in decreasing order. For every spanning tree, it is checked whether the corresponding branches in the current graph constitute a spanning tree as well. If so, a valid term is found and its sign is determined.

This technique is performed for every power of the frequency variable s in both the numerator and denominator of the network function. For a nonzero power of s, say the k-th power, spanning trees in decreasing order must be generated containing exactly k capacitance branches. This can be formulated as the following graph-theoretical problem: given a graph with n nodes and with red (corresponding to (trans)conductances) and blue (corresponding to capacitances) weighted branches, enumerate in decreasing order the spanning trees that contain exactly k blue branches and $n-k-1$ red branches, in which k can have a value between zero and $n-1$. For this problem, an algorithm [9] has been developed whose time complexity and memory requirements increase linearly with the number of generated spanning trees.

III. Generation of the Numerical Reference and Approximation Over Ranges

The tree enumeration procedure described above obviously needs a stopping criterion to know when enough terms have been generated. The generation of terms for a certain power k of s in the enumerator or denominator can stop when

$$\frac{|\sum \textit{num. evaluation of generated terms}|}{|\textit{num. value of coefficient of } s^k|} > (1-\epsilon_k). \quad (2)$$

In this equation, the numerical evaluation is performed in a nominal operating point of the circuit. The denominator in (2) represents the numerical value of the coefficient of s^k in either the numerator or denominator of the network function. The complete coefficients are never generated and, hence, their numerical value must be calculated in advance (without knowing the symbolic expressions). This is efficiently performed using the polynomial interpolation method [10].

For the extension of the stopping criterion of (2) to intervals, it is assumed that a symbolic parameter x can take a value inside a given interval determined by its lower bound x_L and its upper bound x_H.

The introduction of intervals for the symbolic circuit parameters gives rise to multidimensional intervals for the value of the coefficients f_i and g_j from (1). These are computed by an interval extension of the polynomial interpolation method. The resulting interval for a coefficient is usually a pessimistic overestimate. Therefore, intervals are narrowed using the algorithm described in [11].

Intervals for the small-signal circuit parameters are either determined by specifying a relative variation around a given nominal value, or they can be derived from intervals of the bias values and technological parameters. Intervals for symbolic terms or sums of products are then determined using the direct interval extension [11], or, more accurately, with the mean value form [11].

The stopping criterion given in (2) can now be reformulated as:

$$\frac{\mathcal{L}(|[G_L, G_H]|)}{\mathcal{U}(|[S_L, S_H]|)} > (1-\epsilon_k). \quad (3)$$

In this equation $[S_L, S_H]$ represents the interval of the coefficient of s^k obtained as described above. The interval $[G_L, G_H]$ denotes the interval of the sum of the significant terms that have already been generated. The symbols $\mathcal{L}$ and $\mathcal{U}$ denote the lower and upper bound of an interval, respectively.

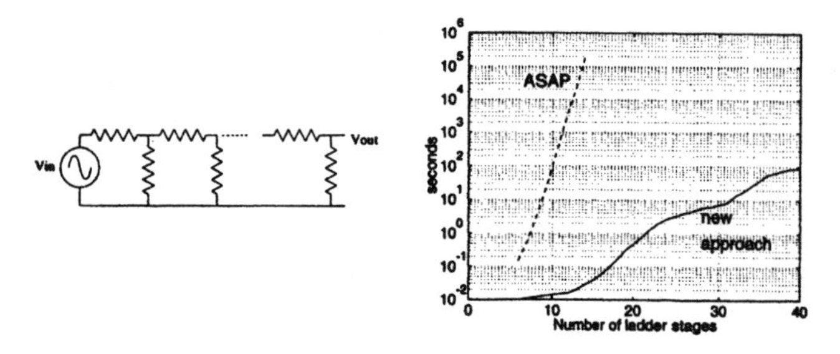

Fig. 2. CPU time on a SUN SPARC 10 for the symbolic computation with a 25% error of the voltage gain V_{out}/V_{in} of the resistive ladder network. The dotted line corresponds to times measured with ASAP (conventional symbolic analyzer). The solid line corresponds to the new approach.

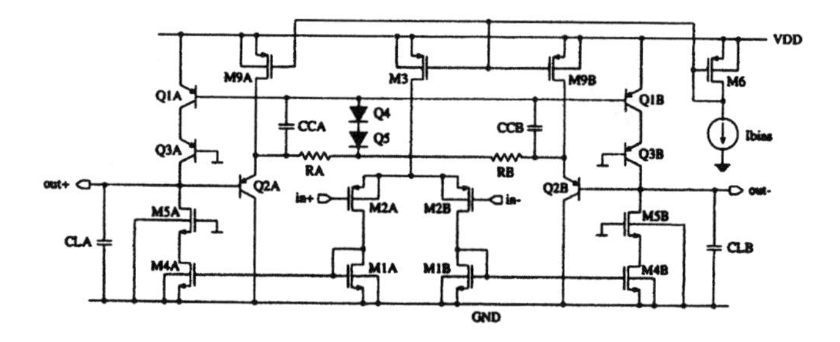

Fig. 3. A fully differential BiCMOS operational transconductance amplifier with common-mode feedback.

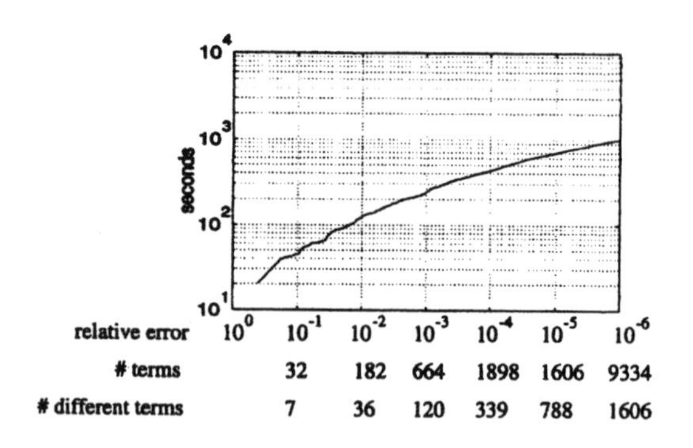

Fig. 4. CPU time (in seconds on a SUN SPARC 10 workstation) versus the relative error of the generated symbolic expression for the denominator of the low-frequency differential-mode gain of the BiCMOS amplifier (Fig. 3). The number of terms that corresponds to the accuracy is indicated as well.

The use of intervals provides a very good trade-off between accuracy and complexity. Obviously, more terms appear in the final result than when using fixed values, and if the intervals are taken too wide, then the interpretation of results can become complicated again.

IV. Matching Elements

Matching elements play an important role in analog and especially in differential integrated circuits. In symbolic calculations they are represented by the same nominal symbol. After doing so, product terms can occur with exactly the same symbols, so that they cancel or add, depending on their sign. The detection of matching terms requires a lot of overhead in CPU time and memory consumption in conventional symbolic analyzers [1], [2]. With the new technique, however, matching terms are easily detected: since they are equal in magnitude, they are generated one immediately after the other. Hence, the cancellations can be elaborated by looking only at the last few generated terms that have the same magnitude as the last generated term.

Mismatches are modeled explicitly by adding a small symbolic mismatch element in parallel with the nominal element. From that moment, both elements are handled independently, and with their own numerical magnitude. For example, the transconductances of two matching transistors M_{1A} and M_{1B} are written as $g_{m_{M1}}$ and $g_{m_{M1}} + \Delta g_{m_{M1BA}}$, respectively. In this way, product terms containing mismatch symbols are generated much later than the corresponding nominal terms and only when necessary.

In techniques previously used [1], [2] in symbolic analyzers, mismatch terms were always given a magnitude (the maximum deviation) and a sign. This is not realistic, since their sign is not known in advance. This problem is overcome here by representing a mismatch term by a symmetric interval around zero.

V. Examples

The new technique not only exceeds the limits of a conventional symbolic analyzer, it can also—for smaller circuits—generate an approximate expression in a CPU time that is up to several orders of magnitude smaller than with conventional analysis. This is shown with the symbolic analysis of the resistive ladder network shown in Fig. 2, which is often used as a benchmark circuit for symbolic analyzers. The CPU time is shown as a function of the number of stages for the symbolic computation of the voltage gain with a 25% error. The dramatic increase in CPU time with the number of stages for conventional symbolic analysis is due to the fact that the exact expression must be generated.

The efficiency of the simplification during generation technique in terms of CPU time is illustrated with the symbolic computation of the system determinant of the BiCMOS amplifier of Fig. 3. This circuit, containing twenty transistors, is far too complex to be analyzed with classical symbolic analyzers. Fig. 4 indicates how with the new technique the CPU time increases with the accuracy of the generated symbolic expression and hence with the number of terms, just as with the principle idea shown in Fig. 1. In contrast with the conventional symbolic analysis approaches, the less terms are generated (the larger the error), the less CPU time is required, which is a very "natural" way of generating terms that constitute a large expression.

For large circuits complicated expressions may be generated. This is illustrated in Fig. 5, which shows the symbolic expression of the low-frequency differential-mode gain of the amplifier of Fig. 3. This expression has been generated in 58 s on a SUN Sparc 10. The expression, however, can be further simplified without increasing the error by a symbolic postprocessing procedure, that factorizes the expressions and that takes advantage of the fact that the error on a ratio of two coefficients of a network function is often much smaller than the error on the coefficients individually. Doing so, the

$$
\begin{aligned}
\Big(& 4\,gm_{Q2}^2\,gm_{Q3}^2\,gm_{M6}\,gm_{M2}^2\,gm_{M5}^2\,gm_{M1}\,gm_{Q5}\,gm_{M4}\,gm_{Q4}\,Ga\,gm_{Q1} \\
+ & \;8\,gm_{Q2}^2\,gm_{Q3}^2\,gm_{Q5}\,gm_{M2}^2\,gmb_{M5}\,gm_{M5}\,gm_{M1}\,gm_{M6}\,gm_{M4}\,gm_{Q4}\,Ga\,gm_{Q1} \\
+ & \;4\,gm_{Q1}\,gm_{Q3}^2\,gm_{Q5}\,gm_{M2}^2\,gm_{Q2}^2\,gmb_{M5}^2\,gm_{M1}\,gm_{M6}\,gm_{M4}\,gm_{Q4}\,Ga \\
+ & \;4\,gm_{Q2}^2\,gm_{Q3}^2\,go_{M1}\,gm_{M2}^2\,gm_{Q5}\,gm_{M5}^2\,gm_{M6}\,gm_{M4}\,gm_{Q4}\,Ga\,gm_{Q1}\Big) \Big/ \\
\Big(& 8gm_{Q2}^2\,gm_{Q3}^2\,go_{M1}\,gm_{M2}\,gm_{M1}\,go_{M4}\,gm_{Q5}\,Ga\,gm_{M5}\,gm_{M6}\,go_{M5}\,gm_{Q4}\,gm_{Q1} \\
+ & \;4\,gm_{Q2}^2\,gm_{Q3}^2\,Ga\,gm_{M1}^2\,go_{M4}\,gm_{Q5}\,go_{M5}^2\,gm_{M6}\,gm_{M2}\,gm_{Q4}\,gm_{Q1} \\
+ & \;8\,g\pi_{Q2}\,gm_{Q2}\,gm_{Q3}^2\,Ga^2\,gm_{M2}\,gm_{M1}^2\,gmb_{M5}\,gm_{Q5}\,gm_{M5}\,gm_{M6}\,gm_{Q4}\,gm_{Q1} \\
+ & \;4\,g\pi_{Q2}\,gm_{Q2}\,gm_{Q3}^2\,Ga\,gm_{M1}^2\,gm_{Q5}\,gm_{M5}^2\,gm_{M6}\,geq2\,gm_{M2}\,gm_{Q4}\,gm_{Q1} \\
+ & \;4\,gm_{Q2}^2\,gm_{Q3}^2\,gm_{M1}^2\,gm_{M2}\,go_{M4}\,gm_{Q5}\,Ga\,gm_{M5}\,gm_{M6}\,go_{M5}\,gm_{Q4}\,gm_{Q1} \\
+ & \;4\,gm_{Q2}^2\,gm_{Q3}^2\,Ga\,gm_{M1}^2\,go_{M4}\,gmb_{M5}\,gm_{Q5}\,gm_{M6}\,go_{M5}\,gm_{M2}\,gm_{Q4}\,gm_{Q1} \\
+ & \;4\,g\pi_{Q2}\,gm_{Q2}\,gm_{Q3}^2\,Ga^2\,gm_{M2}\,gm_{M1}^2\,gm_{Q5}\,gm_{M5}^2\,gm_{M6}\,gm_{Q4}\,gm_{Q1} \\
+ & \;4\,gm_{Q2}^2\,gm_{Q3}^2\,gm_{M2}\,gm_{M1}^2\,go_{M4}^2\,gm_{Q5}\,Ga\,gm_{M6}\,go_{M5}\,gm_{Q4}\,gm_{Q1}\Big)
\end{aligned}
$$

Fig. 5. Approximated expression ($\epsilon = 20\%$) for the low-frequency differential-mode gain of the circuit of Fig. 3. The terms are sorted in decreasing order. Due to matching, several product terms occur more than once, which explains the occurance of integer coefficients 4 and 8. The element $geq2$ is a lumped element [1] consisting of the parallel conductances go_{M9} and go_{Q2}. Elements from the bias circuitry (like gm_{M6}) or from the common-mode feedback circuitry (like gm_{Q4}) that don't influence the differential-mode gain at all, disappear after factorization.

differential-mode gain reduces to

$$
\frac{gm_{M2}gm_{M4}}{gm_{M1}\left(\frac{go_{M4}go_{M5}}{gm_{M5}+gm_{M5}} + \frac{Ga+geq2}{\beta_{Q2}}\right)}. \tag{4}
$$

It is found that the output conductance of the bipolar cascode is too small, even with the inclusion of intervals, to contribute significantly to the conductance seen at the output node.

An even more complex circuit is the commercial μA741 opamp. This circuit contains 23 nodes, 22 transistors, and 13 resistors. The generation of a symbolic expression for the amplifier's transfer function with an error of 0.1% (110 terms) requires 38 seconds on a SUN Sparc 10 workstation.

VI. Conclusion

A new program has been presented that generates approximated symbolic expressions for small-signal characteristics of analog circuits. The approximation is performed during the generation of the expression. In this way, only the necessary terms of the simplified symbolic expressions are generated, which contrasts to approaches of conventional symbolic analyzers which require a lot of over head for the generation of the exact symbolic expression, which is then pruned. The new approximation technique also takes into account a range for the value of the symbolic circuit parameters rather than one single value. This extends the range of validity of the generated symbolic expressions. Moreover, the new technique allows an accurate control of the approximation error. The interpretability of the expressions can be enhanced by further postprocessing. Several examples have demonstrated that this approach enables the symbolic analysis of large analog integrated circuits of the size of practical analog cells, which were impossible to analyze properly before.

Acknowledgment

The authors wish to thank P. Eindhoven, The Netherlands, and the Human Capital and Mobility Program of the CEC for their support.

References

[1] G. Gielen and W. M. Sansen, *Symbolic Analysis for Automated Design of Analog Integrated Circuits.* Norwell, MA: Kluwer Academic, 1991.

[2] F. V. Fernández, A. Rodríguez-Vázquez, and J. L. Huertas, "Interactive ac modeling and characterization of analog circuits via symbolic analysis," *Kluwer Journal on Analog Integrated Circuits and Signal Processing*, vol. 1. Norwell, MA: Kluwer, 1991, pp. 183–208.

[3] G. Gielen, P. Wambacq, and W. M. Sansen, "Symbolic analysis methods and applications for analog circuits: A tutorial overview," *Proc. IEEE*, vol. 82, pp. 287–304, Feb. 1994.

[4] P. Wambacq, G. Gielen, and W. M. Sansen, "A cancellation-free algorithm for the symbolic analysis of large analog circuits," in *Proc. IEEE Int. Symp. Circuits Syst.*, 1992, pp. 1157–1160.

[5] P. Wambacq, F. V. Fernández, G. Gielen, and W. M. Sansen, "Efficient symbolic computation of approximated small-signal characteristics," in *Proc. CICC 1994*, 1994, pp. 21.5.1–21.5.4.

[6] Q. Yu and C. Sechen, "Generation of color-constrained spanning trees with application in symbolic circuit analysis," in *Proc. 4th Great Lakes Symp. VLSI*, Mar. 1994, pp. 252–255.

[7] F. V. Fernández, J. D. Martín, A. Rodríguez-Vázquez, and J. L. Huertas, "On simplification techniques for symbolic analysis of analog integrated circuits," in *Proc. IEEE Int. Symp. Circuits Syst.*, pp. 1149–1152, May 1992.

[8] P.-M. Lin, *Symbolic Network Analysis.* Amsterdam, The Netherlands: Elsevier, 1991.

[9] P. Wambacq, F. V. Fernández, G. Gielen, W. M. Sansen, and A. R odríguez-Vázquez, "An algorithm for efficient symbolic analysis of large analogue circuits," *IEE Electron. Lett.*, vol. 30, no. 14, pp. 1108–1109, July 1994.

[10] J. Vlach and K. Singhal, *Computer Methods for Circuit Analysis and design.* New York: Van Nostrand Reinhold, 1983.

[11] R. Moore, *Methods and Applications of Interval Analysis*, Studies in Applied Mathematics, Philadelphia, 1979.

Efficient Approximation of Symbolic Network Functions Using Matroid Intersection Algorithms

Qicheng Yu, *Member, IEEE*, and Carl Sechen, *Member, IEEE*

***Abstract*—An efficient and effective approximation strategy is crucial to the success of symbolic analysis of large analog circuits. In this paper we propose a new approximation strategy for the symbolic analysis of linear circuits in the complex frequency domain. The strategy directly generates common spanning trees of a two-graph in decreasing order of tree admittance product, using matroid intersection algorithms. The strategy reduces the total time for computing an approximate symbolic expression in expanded format to polynomial with respect to the circuit size under the assumption that the number of product terms retained in the final expression is polynomial. Experimental results are clearly superior to those reported in previous works.**

***Index Terms*—Approximation methods, matroid, matroid intersection, symbolic analysis, trees (graphs), two-graph.**

I. Introduction

THE conventional focus of symbolic analysis of analog circuits has been the generation of network functions in expanded nonhierarchical format for linear or linearized circuits in the complex frequency domain. An arbitrary network function of a continuous time circuit can be expressed in the cancellation-free expanded format as the ratio of two polynomials in the Laplace variable s

$$T(s) = \frac{N(s)}{D(s)} = \frac{b_0 + b_1 s^1 + \cdots + b_n s^n}{a_0 + a_1 s^1 + \cdots + a_d s^d} \tag{1}$$

where a_i and b_j are sums of product terms and each product term is the product of circuit element parameters [10]. It is well known that the number of product terms in an exact symbolic network function grows exponentially with the circuit size. For this reason symbolic network functions have to be approximated during or after its computation. Much energy has been directed in recent years to developing effective approximation strategies so that the computational effort and the size of resulting approximate expressions are kept under control. A survey of previous strategies and an overview of symbolic analysis methods and their applications can be found in [6]. Among the existing strategies, approximation-during-computation (ADC) strategies are superior because they avoid the burden of generating a huge amount of intermediate data related to the numerically negligible terms only to be discarded in the subsequent approximation stages.

A unified approach for approximate symbolic analysis has been recently proposed and implemented [19], [21]. Using novel ADC strategies, this approach has been able to produce approximate symbolic expressions in expanded format for the largest analog circuits reported to date. However, a polynomial bound on computation time could not be claimed for this approach or any other previously reported approaches generating expanded symbolic expressions.

In this paper, we propose a new approximation strategy that directly generates common spanning trees of a two-graph in decreasing order of tree admittance product using matroid intersection algorithms. The total time for computing an approximate symbolic expression in expanded format is reduced to polynomial with respect to the circuit size. Compared to the strategy of generating voltage graph spanning trees in decreasing order [15], [18], [19] used in the above-mentioned unified approach, the average time per product term grows much slower with the size of the two-graph for our new strategy. Experimental results show dramatic performance improvements.

II. Problem Formulation

The unified approach for approximate symbolic analysis in [19] is based on the classical two-graph tree enumeration method [10] and is outlined as follows: First, a two-graph consisting of a voltage graph and a current graph is constructed for both the numerator and the denominator of the network function whose approximate symbolic expression is to be computed. Each voltage graph edge and its counterpart in the current graph corresponds to a resistive or capacitive circuit element. The two-graphs are such that each product term in the cancellation-free symbolic expression corresponds to a common spanning tree of the voltage graph and the current graph and is called a tree admittance product. Second, a sensitivity-based simplification step is carried out in which the two-graphs are reduced by edge contraction and deletion based on the numerical contribution of corresponding circuit elements to the network function. Finally, only the product terms with the largest magnitude in the symbolic expression are computed from the two-graphs. The common spanning trees of the simplified two-graphs are generated in decreasing

Manuscript received October 20, 1995; revised December 30, 1996. This work was supported by the National Science Foundation under Grant MIP-940670 and by the National Science Foundation Center for Design of Analog and Digital Integrated Circuits (CDADIC). This paper was recommended by Associate Editor J. White.

Q. Yu is with Cirrus Logic, Inc., Crystal Semiconductor Product Division, Nashua, NH 03063 USA (e-mail: qiyu@crystal.cirrus.com).

C. Sechen is with the Department of Electrical Engineering, University of Washington, Seattle, WA 98195 USA.

Publisher Item Identifier S 0278-0070(97)09237-3.

order of their tree admittance product until certain error control criteria are satisfied. The generated common spanning trees are called significant common trees. For the symbolic expression to be valid over the whole frequency range of interest, spanning trees containing different numbers of capacitive edges are generated separately. This is because the admittance of a capacitor changes with frequency and so does the relative amplitude of tree admittance products with different numbers of capacitive edges. The problem involved in the final step above can be abstracted as the following, where resistive and capacitive edges are labeled green and red, respectively, and greater tree admittance product is translated to lower weight by definition.

Problem 1: Two undirected graphs G_V (voltage graph) and G_I (current graph) with n nodes each are given such that:

1) both graphs consist of the same set E of m weighted edges; each edge in E has one "copy" in G_V and another "copy" in G_I;
2) the topologies of G_V and G_I are different;
3) each edge in E has a color which is either red or green.

Find K_i best subsets of E in terms of lowest total weight, in increasing order of weight, such that each subset:

a) forms a spanning tree in G_V;
b) forms a spanning tree in G_I;
c) contains exactly i red edges.

This problem was "solved" by generating the spanning trees of the voltage graph with exactly i red edges in increasing order of weight and retaining only those that are also spanning trees of the current graph [15], [18], [19]. The simplified problem is the following.

Problem 2: Given an undirected graph G_V with n nodes and m edges where each edge has a color which is either red or green, find K_i^V lowest weight spanning trees of G_V containing exactly i red edges in increasing order of weight.

The advantage of generating voltage graph spanning trees is that very efficient algorithms are available for the purpose [18]. Every such spanning tree in the ordered sequence can be generated in time virtually linear with respect to the number of edges in the graph. However, the overall efficiency of the tree generation process is also determined by the ratio of common spanning trees obtained over the number of voltage-graph spanning trees generated. This ratio depends on the size and topological structure of the two-graph and therefore on the size and structure of the circuit and the network function to be computed. Not surprisingly, the ratio tends to decrease when the two-graph size increases. Therefore there are circuits and/or network functions for which generating voltage graph spanning trees is not efficient overall.

We therefore propose the approximation strategy of directly generating common spanning trees of a two-graph in decreasing order of tree admittance product, using the so-called *matroid intersection algorithms*. A *matroid* is defined by a finite set of elements, which is E in our case, and a set of subsets of the finite set, each of which is called *independent*, so that these subsets satisfy certain relations [9]. An independent set of maximum cardinality is called a *base*. In the view of combinatorial optimization theory, Problem 1 is defined with respect to three matroids on E: Condition 1.a is associated with a *graphic matroid* M_V so that each spanning tree of the graph G_V is a base of the matroid. Similarly, Condition 1.b is associated with another graphic matroid M_I. Condition 1.c is associated with a *partition matroid* M_p so that each subset of E satisfying Condition 1.c is a base of the matroid. Problem 1 is therefore to find K_i lowest weight common bases of the three matroids in increasing order of weight.

We first consider the simpler problem of finding a minimum weight common base of the three matroids.

Problem 3: Given undirected graphs G_V and G_I as defined in Problem 1, find a best subset of E in terms of lowest total weight such that it satisfies Conditions 1.a, 1.b, and 1.c.

In terms of combinatorial optimization, this is a case of the *weighted intersection problem* of three matroids. The weighted intersection problem of three general matroids is NP-hard [12] and there is no existing polynomial time algorithm for Problem 3 in particular. However, there are polynomial time algorithms for the weighted intersection problem of two general matroids and for the following problem of generating a minimum weight common spanning tree without color constraints in particular [1], [2], [5], [9].

Problem 4: Given undirected graphs G_V and G_I as defined in Problem 1, find a best subset of E in terms of lowest total weight such that it satisfies Conditions 1.a and 1.b.

Furthermore, polynomial time algorithms also exist for the following problem of generating uncolored common spanning trees in increasing order of weight [2].

Problem 5: Given undirected graphs G_V and G_I as defined in Problem 1, find K best subsets of E in terms of lowest total weight, in increasing order of weight, such that each subset satisfies Conditions 1.a and 1.b.

III. MATROID INTERSECTION ALGORITHMS[1]

In the following sections, we will show how matroid intersection algorithms, particularly an algorithm for Problem 5 by Camerini and Hamacher, which we refer to as the CH [2] algorithm, could be used in an efficient manner for computing approximate symbolic network functions valid over a given frequency range of interest. Before doing so, we briefly describe how matroid intersection algorithms operate in terms of matroids M_V and M_I associated with G_V and G_I.

Algorithms for finding a lowest weight common tree have been reported in [1], [5], [9], and other literature. One of the algorithms in [9] uses a data structure called a *border graph* to compute a lowest weight common tree in polynomial time. Another algorithm by Brezovec, Cornuéjols, and Glover [1], which we use as part of the CH algorithm and refer to as the BCG algorithm, has a similar data structure. The CH algorithm finds K lowest weight common trees without color constraints by employing a procedure of partitioning the solution space which is also used in [4], [8], and [18]. This procedure requires finding a second lowest weight common tree in each block of

[1] The relationship between matroids or matroid theory and electrical networks was studied in the past [7], [14]. The application of matroid intersection algorithms to approximate symbolic analysis further demonstrates the usefulness of matroids in practical engineering problems.

the partition of the solution space. A second lowest weight common tree in a block can be computed by further dividing this block and finding a lowest weight common tree in each subdivision. However, Camerini and Hamacher 1) proved a key property of the border graph which allowed the faster computation of a second lowest weight common tree directly from a lowest weight common tree and 2) introduced an auxiliary data structure called the *condensed border graph* that further reduced the run time.

IV. Approximation over a Frequency Range

For the approximate symbolic expression to be valid over the whole frequency range, we must be able to compare the relative significance of product terms with different powers of the Laplace variable s. This is achieved by generating uncolored common spanning trees at multiple frequency points using an algorithm for Problem 5, while at each frequency point product terms with different powers of s are compared by their magnitudes.

Let

$$P(s) = \sum_{i=0}^{n_p} p_i s^i = \sum_{i=0}^{n_p} \left(\sum_{l} p_{i,l} \right) s^i \tag{2}$$

be the expanded symbolic expression of the numerator or denominator corresponding to the two-graph after sensitivity-based simplification, where p_i is a sum of product terms and $p_{i,l}$ is an individual product term excluding the power term s^i.

According to our error control mechanism as proposed in [19], $p_{i,l}s^i$ is to be retained in the approximate expression if and only if

$$|p_{i,l}s^i|_{s=j2\pi f_k} \geq |\varepsilon P(s)|_{s=j2\pi f_k} \tag{3}$$

for at least one frequency f_k in a set of *sample frequencies*

$$F = \{f_k | k = 1, 2, \cdots, n_f\} \tag{4}$$

chosen in the range of interest and for a given error-control variable ε. The values of $P(s)$ at the frequency f_k are computed numerically. Equivalently, we may define a reference product term

$$p_i^R(\varepsilon) = \min_{1 \leq k \leq n_f} \left\{ \left| \frac{\varepsilon P(s)}{s^i} \right|_{s=j2\pi f_k} \right\} \tag{5}$$

for all product terms in p_i, so that $p_{i,l}s^i$ is retained if and only if $|p_{i,l}| \geq p_i^R(\varepsilon)$. Let f_{k_i} be a sample frequency that achieves the above minimum, then we only have to examine each product term in p_i at frequency f_{k_i} by comparing $|p_{i,l}s^i|$ with $|\varepsilon P(s)|$ to determine whether it should be retained. Here f_{k_i} does not depend on ε.

Therefore, we generate uncolored common spanning trees at every sample frequency in the set

$$F_g = \{f_{k_i} | i = 0, 1, \cdots, n_p\} \tag{6}$$

in decreasing order of tree admittance product using an algorithm for Problem 5, where "uncolored" means conductance and capacitance type edges are directly compared by their admittance at the frequency. Because each $p_{i,l}$ in $P(s)$ has its unique symbol combination, the same product term generated at more than one frequency can be easily detected. The set of product terms in the approximated symbolic expression for $P(s)$ is the union of the sets of product terms generated at every frequency in F_g. Experiments shows that while the number of sample frequencies in F could be large, the number of operationally distinct frequencies in F_g is typically very small, making the new approximation strategy efficient. If K is the number of product terms in the final approximated symbolic expression for $P(s)$, n_g is the cardinality of F_g, m and n are numbers of edges and vertices in the two-graph, and $C(K, m, n)$ is the time to run the algorithm for Problem 5 on the two-graph for $P(s)$ at one frequency, then the total time to obtain the approximate symbolic expression for $P(s)$ is $O(n_g C(K, m, n))$. Furthermore, n_g is bounded by $(n_p + 1)$, which is in turn bounded by n

$$n_g \leq \min\{n_p + 1, n_f\} \leq \min\{n, n_f\}. \tag{7}$$

Some network functions contain abrupt transitions in their magnitude or phase response. The choice of sample frequencies around an abrupt transition region may affect in a significant way the number of product terms retained in the final expression. For example, if the numerator of a transfer function has a zero on the imaginary axis and the zero frequency is chosen as a sample frequency, every product term in the numerator would have to be retained according to (3). Since the approximation strategy offers complete freedom in choosing the sample frequencies, the user is expected to choose according to the specific need of accuracy and simplicity of the network function rather than arbitrarily. In general, sample frequencies should be those frequencies where the accuracy of the approximate expression is critical. This is similar to the choice of passband and stopband edge frequencies in the design of a filter. In the above example, if it is not necessary for the approximate expression to be exactly accurate at the zero frequency, a pair of sample frequencies on the two sides of the zero could be chosen instead, such that the numerical value of the numerator is sufficiently small at these frequencies for the desired accuracy. An advantage of the new strategy is that no matter how many sample frequencies are chosen, the number of frequencies at which common spanning trees are generated never exceeds n. The user could select arbitrarily dense sample frequencies if necessary, knowing that common spanning tree generation will occur only on a limited subset of these frequencies.

The theoretical significance of the new strategy is that it reduces the total computation time for generating an approximate symbolic expression in expanded nonhierarchical format to *polynomial* with respect to circuit size, under the condition that the number of product terms retained in the final expression is polynomial with respect to circuit size. This has not been achieved by any previously reported approximation strategies.

TABLE I
STATISTICS OF APPROXIMATE SYMBOLIC EXPRESSIONS GENERATED USING STRATEGIES VT AND CT

Circuit name	**741**		**Cascode**		**Rail-to-rail**		**725**	
Network function	**voltage gain**		**voltage gain**		**voltage gain**		**input impedance**	
	N	**D**	**N**	**D**	**N**	**D**	**N**	**D**
# g type edges in the two-graph	7	14	11	14	19	24	9	9
# C type edges in the two-graph	7	9	6	9	9	9	5	5
# vertices in the two-graph	7	9	8	8	9	9	7	8
# sample frequencies: n_f	22		30		26		26	
# frequencies at which common trees are generated: n_g	2	3	2	4	2	3	4	5
# ν-graph spanning trees generated	64	182	179	454	117	3006	66	110
# common trees generated	20	37	57	188	11	25	20	19
run time using strategy VT (sec)	19.0		21.5		65.7		38.3	
run time using strategy CT (sec)	20.0		21.3		63.5		37.6	
# g type edges in the two-graph	32	41	16	19	25	33	22	22
# C type edges in the two-graph	24	30	10	13	10	11	19	19
# vertices in the two-graph	20	20	9	9	13	13	15	16
# sample frequencies: n_f	22		30		26		26	
# frequencies at which common trees are generated: n_g	4	4	3	4	3	4	8	9
# ν-graph spanning trees generated	4051	84819	8054	23663	11916	157448	79408	168807
# common trees generated	570	615	830	2840	53	64	1002	1225
run time using strategy VT (sec)	162.9		52.5		230.9		426.2	
run time using strategy CT (sec)	57.7		33.2		60.5		73.6	

V. IMPLEMENTATION

The above approximation strategy has been implemented in the symbolic analysis program RAINIER. Given the two-graph for either the numerator or the denominator of the network function, we apply the CH algorithm at every frequency in the corresponding frequency set F_g. According to the CH algorithm, the generation of lowest weight common trees (at one frequency) is carried out in two steps. The first step is to generate the minimum weight common tree. This is done using the BCG algorithm, whose run time is $O(m^2n)$ for our implementation. The second step is to generate the remaining lowest weight common trees in the sequence, which is carried out by the CH algorithm itself and takes $O(mn^2)$ time per common tree. Since there are n_g distinct frequencies in F_g and a total of K common trees are generated, the time complexity is $O(n_g mn(m + Kn))$. This procedure is performed on the two-graphs of both the numerator and the denominator.

In the following pseudocode, N_2 and D_2 are the numerator and denominator of the transfer function represented by the simplified two-graphs, respectively.[2] F_g^N and F_g^D are the frequency set F_g for N_2 and D_2, respectively. $N_{\rm appr}$ and $D_{\rm appr}$ are the approximate symbolic expressions for N_2 and D_2 consisting of all product terms already generated but excluding duplication, and the approximate network function is

$$T_{\rm appr} = \frac{N_{\rm appr}}{D_{\rm appr}}. \tag{8}$$

Strategy CT (Two-graphs for N_2 and D_2, set of sample frequencies $F, \varepsilon_{\max}, \varepsilon_{\min}, \xi$):

1) (numerically) compute the frequency sets F_g^N and F_g^D from F;
2) **for** each $f_{k_i} \in F_g^N$ **do** generate the largest product term in N_2 at frequency f_{k_i} with the BCG algorithm; **enddo**;
3) **for** each $f_{k_i} \in F_g^D$ **do** generate the largest product term in D_2 at frequency f_{k_i} with the BCG algorithm; **enddo**;
4) $\varepsilon \leftarrow \varepsilon_{\max}$;
5) **do**
6)   **for** each $f_{k_i} \in F_g^N$ **do**
7)   generate all product terms in N_2 that are larger in magnitude than $|\varepsilon N_2|$ at frequency f_{k_i} that have not been generated, with the CH algorithm;
8)   **enddo**;
9)   **for** each $f_{k_i} \in F_g^D$ **do**
10)   generate all product terms in D_2 that are larger in magnitude than $|\varepsilon D_2|$ at frequency f_{k_i} that have not been generated, with the CH algorithm;
11)   **enddo**;

[2] The sensitivity-based simplification algorithms for two-graphs were described in [19] and [21].

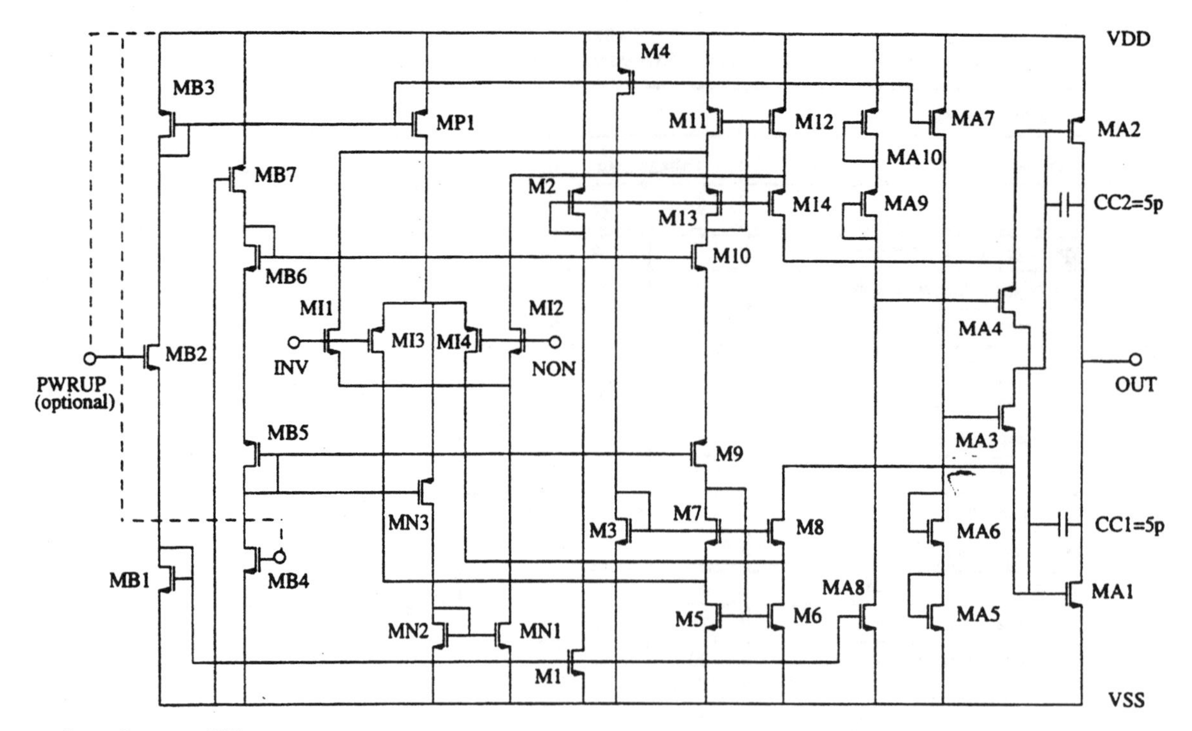

Fig. 1. A rail-to-rail opamp [17].

12) update N_{appr} and D_{appr} by adding newly generated product terms excluding duplications;
13) **if** T_{appr} is accurate enough **then return** (ok);
14) $\varepsilon \leftarrow \varepsilon \cdot \xi$;
15) **while** $\varepsilon \geq \varepsilon_{\min}$;
16) **return** (warning).

VI. Results

To evaluate the performance of the new approximation strategy (strategy CT), we ran the program with both the new strategy and the strategy of generating voltage graph spanning trees (strategy VT) for the circuit examples given in [19] and compared the central processing unit (CPU) time. The four examples are for the voltage gains of the μA 741 opamp, a complementary metal–oxide–semiconductor (CMOS) cascode opamp and a CMOS rail-to-rail opamp, and the input impedance of the μA 725 opamp. In each case sensitivity-based simplification was performed to reduce the sizes of the two-graphs before tree generation. Such simplification is controlled by user supplied error bounds. Larger error bounds mean more contracted and deleted edges and smaller two-graphs after simplification. In all examples given in this section, the sample frequencies in the set F were chosen to be logarithmically evenly spaced between the lower and upper edges of the frequency range of interest. Strategy CT was first applied to the same simplified two-graphs and generated the same approximate symbolic expressions as obtained in [19] with strategy VT. The results are shown in the upper part of Table I. All run times are total times including the time spent in parts of the program other than tree enumeration. Notice that n_g is very small in all examples and strategy CT takes comparable run time with strategy VT for these two-graphs. Then we tightened the error bounds controlling

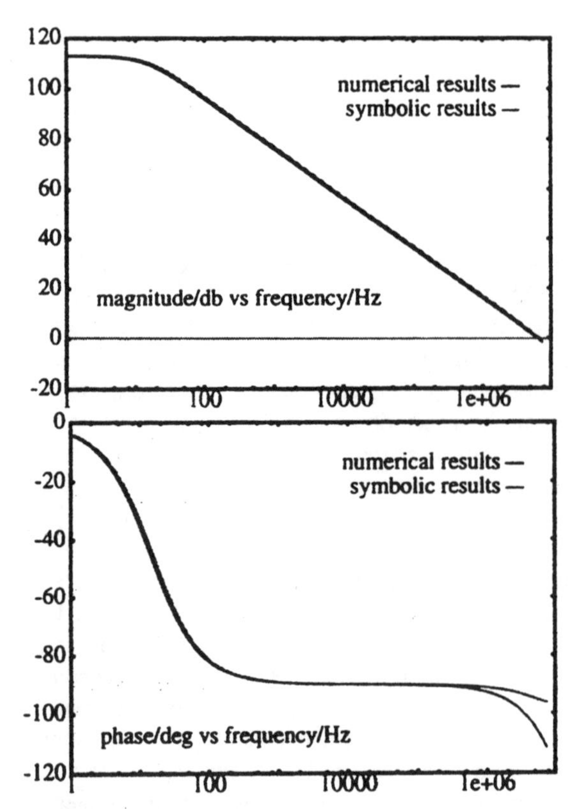

Fig. 2. Voltage gain of rail-to-rail opamp as given by the approximate symbolic expression.

the sensitivity-based simplification of the two-graphs and reran the program for each example. The consequence is "less simplified" two-graphs of larger size, which in turn means symbolic expressions with more product terms. Both strategy CT and strategy VT were applied to compute the same approximate symbolic expression in each example. The

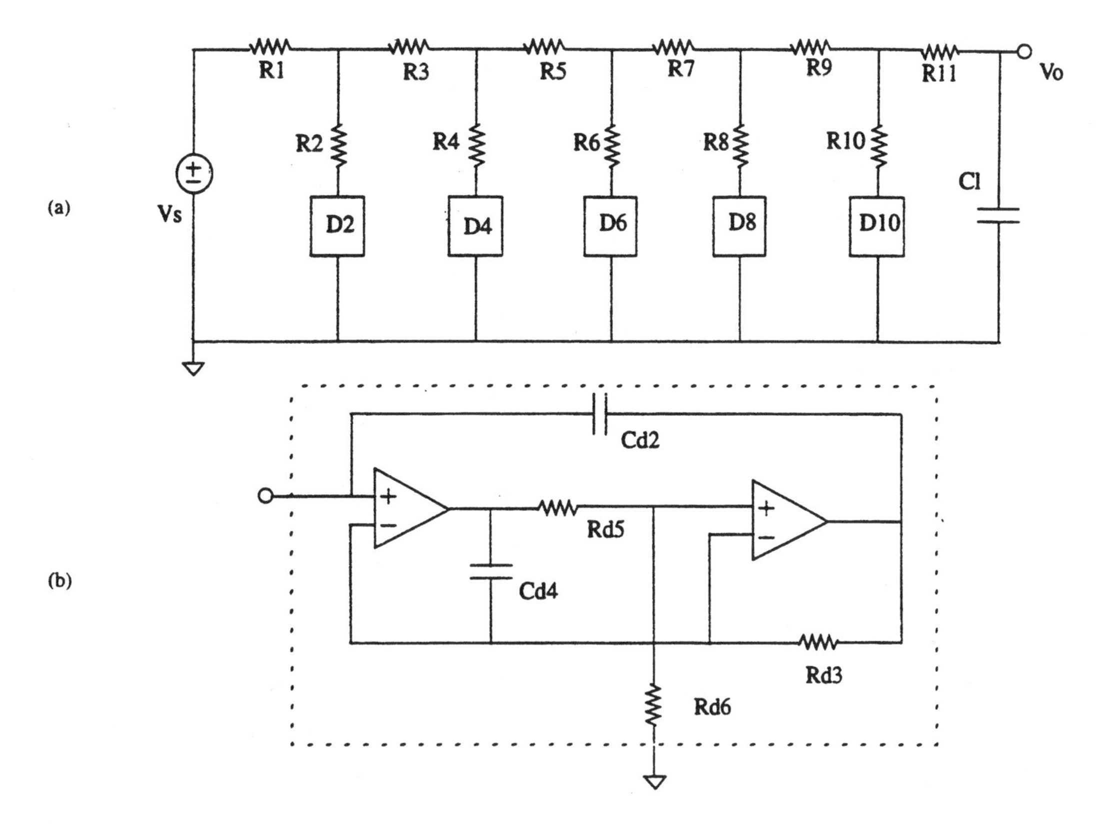

Fig. 3. An eleventh-order elliptic filter. (a) The filter. (b) Building block $D.x$.

TABLE II
STATISTICS OF SYMBOLIC EXPRESSIONS FOR VOLTAGE GAIN OF RAIL-TO-RAIL OPAMP AND ELEVENTH-ORDER ELLIPTIC FILTER

Circuit name	Rail-to-rail		Elliptic[a]	
Network function	voltage gain		voltage gain	
	N	D	N	D
# g type edges in the two-graph	79	81	26	26
# C type edges in the two-graph	58	58	10	11
# vertices in the two-graph	25	25	22	22
# sample frequencies: n_f	26		20	
# frequencies at which common trees are generated: n_g	5	6	2	1
# common trees generated	38	60	32	2048
run time using strategy CT (sec)	20.0		36.3	

[a]For a network function having zeros on the imaginary axis, such as the voltage gain of this elliptic filter, it is usually desirable to select more sample frequencies around the abrupt transition region of the frequency response in order to maintain local numerical accuracy. However, since the exact symbolic expression is being generated in this example, the choice of the sample frequencies affects only the CPU time but not the resulting expression. For simplicity, the sample frequencies are chosen to be logarithmically evenly spaced in the entire frequency range of interest.

results are shown in the lower part of Table I. For these larger two-graphs, strategy CT outperforms strategy VT with much shorter run times. In all cases, RAINIER was run on a SUN SparcStationII.

Next, strategy CT was directly applied to the unsimplified two-graphs for the voltage gain of the CMOS rail-to-rail opamp [17]. The numerator and denominator two-graphs contain 137 and 139 edges (circuit elements), respectively. The program computes a 98 product term approximate symbolic expression in 20.0 s, whose frequency response is compared to exact numerical simulation results in Fig. 2. The statistics[3] are shown in Table II. In contrast, strategy VT was not able to obtain the same expression after generating at least 1.2e6 voltage graph spanning trees in about 4500 s.

TABLE III
COMPONENT VALUES FOR ELEMENTS OF THE ELEVENTH-ORDER ELLIPTIC FILTER

Element name	Element value	Element name	Element value
R1 (ohm)	17043	R7 (ohm)	21363
R2 (ohm)	34213	R8 (ohm)	9561
R3 (ohm)	11868	R9 (ohm)	26748
R4 (ohm)	40211	R10 (ohm)	3347
R5 (ohm)	13991	R11 (ohm)	11575
R6 (ohm)	18728	Cl (pf)	2947

[3]Comparing statistics pertinent to the rail-to-rail opamp, one sees that the graph in Table II is larger than the graph in Table I, and also that n_g in Table II is larger than in Table I. However, strategy CT generates fewer common trees in Table II than in Table I and the total CPU time is less in Table II than in Table I. The reasons are as follows. First, sensitivity-based simplification is carried out for the case in Table I, but not for the case in Table II. For the examples in Table I, this step takes from 1/3 to 2/3 of the total CPU time. Second, the numerical accuracy specification on magnitude and phase of the approximate symbolic expression is different for the cases in Tables I and II. Third, even if two two-graphs have the same size, the time per common tree could be different due to different topological structures of the two-graphs and the number of common trees needed for the same approximation accuracy could also be different.

TABLE IV
COMPONENT VALUES FOR ELEMENTS IN EVERY Dx BLOCK OF THE ELEVENTH-ORDER ELLIPTIC FILTER

Block name	D2	D4	D6	D8	D10
Cd2, Cd4 (pf)	2200	2200	2700	3300	3300
Rd5, Rd6 (ohm)	20000	20000	20000	20000	20000
Rd3 (ohm)	18745	16646	19962	18348	19059

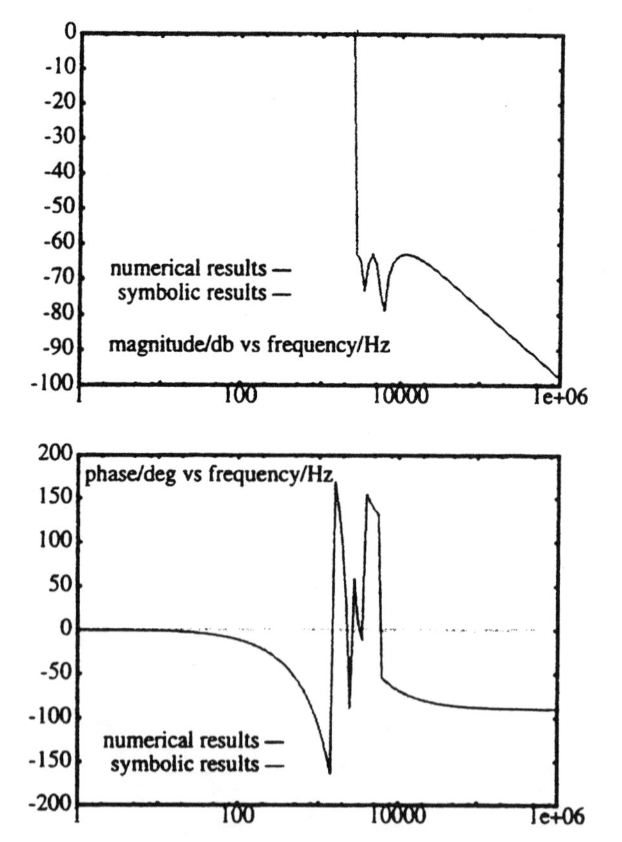

Fig. 4. Frequency response of eleventh-order elliptic filter as given by the exact symbolic expression.

Strategy CT was then applied to the unsimplified two-graphs for the voltage gain of an eleventh-order elliptic filter shown in Fig. 3, with component values in Tables III and IV. The numerator and denominator two-graphs contain 36 and 37 edges, respectively. The program computes the 2080 product term exact symbolic expression in 36.3 s, whose frequency response is compared to exact numerical simulation results in Fig. 4. The statistics are shown in Table II. In contrast, strategy VT was not able to obtain the same expression after generating at least 8.0e5 voltage graph spanning trees in about 3200 s.

Finally, we show a network function of a circuit for which the unified approach, even with direct generation of common trees, is not able to find a compact approximate symbolic expression over the desired frequency range. The circuit is the uncompensated μA 725 opamp as shown in [19]. The attempt was to generate an approximate symbolic expression for its voltage gain valid up to the unity gain frequency. The number of product terms needed for accuracy in both magnitude and phase turns out to be very large, rendering the approximate expression uninterpretable. Figs. 5–7 compare the frequency

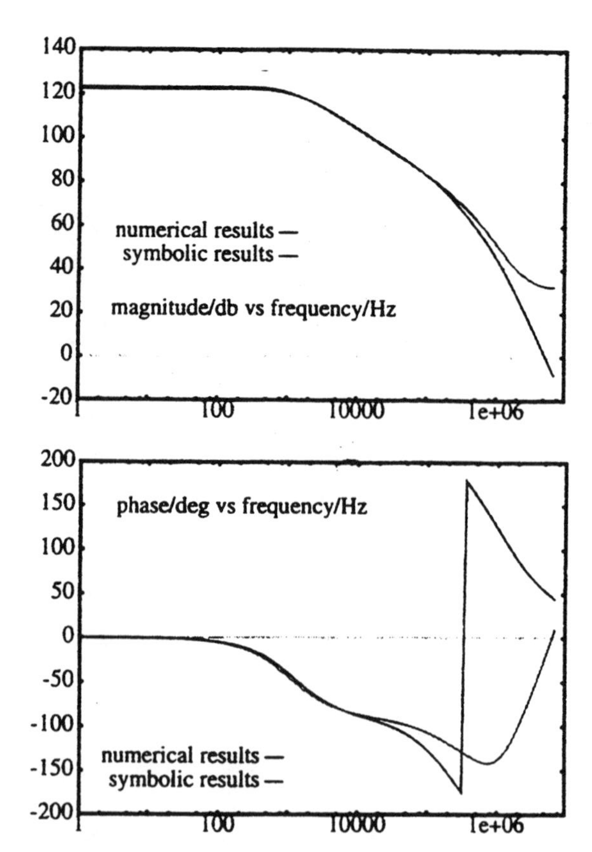

Fig. 5. Voltage gain of μA 725 as given by an approximate symbolic expression with 214 and 124 product terms in the numerator and denominator, respectively.

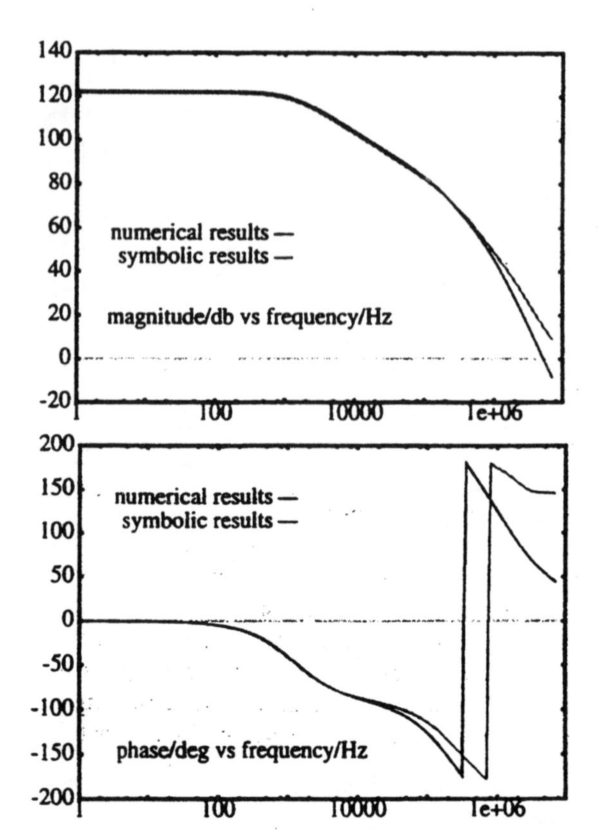

Fig. 6. Voltage gain of μA 725 as given by an approximate symbolic expression with 2824 and 3278 product terms in the numerator and denominator, respectively.

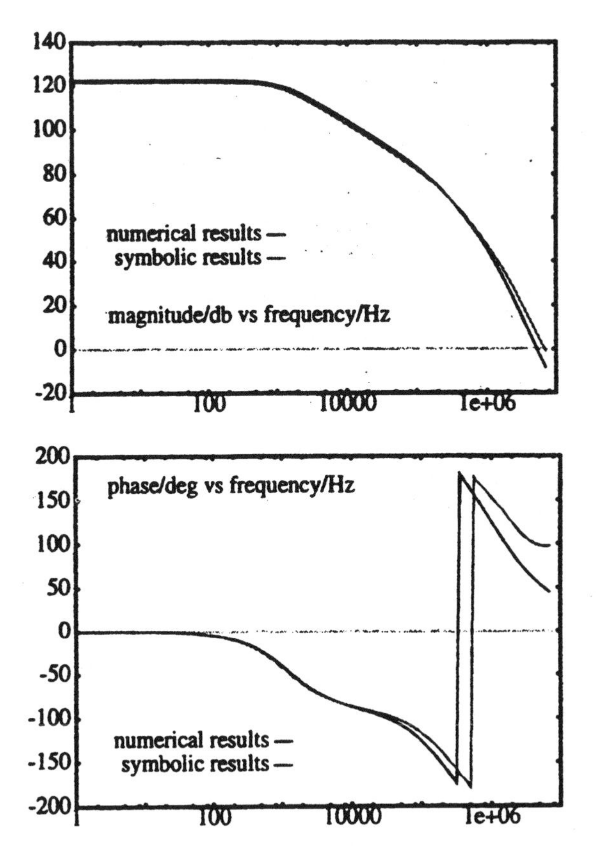

Fig. 7. Voltage gain of μA 725 as given by an approximate symbolic expression with 7244 and 22 454 product terms in the numerator and denominator, respectively.

responses given by approximate symbolic expressions containing 338, 3278, and 29 698 product terms before matching and algebraic postprocessing [21] to the exact numerical simulation results, respectively. The expressions are computed by RAINIER in 71.3, 142.7, and 423.8 s, respectively. With strategy VT, these expressions could not be obtained in a reasonable amount of computation time.

VII. Conclusion

We have presented a efficient approximation strategy for computing symbolic network functions in expanded format using matroid intersection algorithms. A polynomial time bound was achieved for the first time. Common trees only need to be generated at a small subset of all sample frequencies at which the error control criteria are applied. The performance improvement is drastic when the size of the two-graph is large.

Matroid intersection algorithms can also be used in sensitivity-based simplification of two-graphs to ensure that no numerically cancelling large product terms are removed from the symbolic expression when edges are contracted and deleted [19]. Every time a contraction or deletion is attempted, an algorithm for Problem 4 can be invoked to compute the largest product term removed. The contraction or deletion is rejected if this product term is of significant magnitude. Future work along this direction will minimize the distortion of information content during the approximation of a symbolic network function.

Finally, it is worth mentioning that as opposed to symbolic analysis, there has been research directly seeking a numerical reduced-order model of a large linear network. Such methods include asymptotic waveform evaluation [13], the PVL algorithm [3], and the Arnoldi method [11]. All these approaches add to a growing arsenal for analysis of larger analog circuits with better efficiency and deeper insight.

References

[1] C. Brezovec, G. Cornuéjols, and F. Glover, "Two algorithms for weighted matroid intersection," *Math. Programming*, vol. 36, pp. 39–53, 1986.

[2] P. M. Camerini and H. W. Hamacher, "Intersection of two matroids: (Condensed) border graph and ranking," *SIAM J. Disc. Math.*, vol. 2, pp. 16–27, Feb. 1989.

[3] P. Feldmann and R. W. Freund, "Efficient linear circuit analysis by Padé approximation via the Lanczos process," *IEEE Trans. Computer-Aided Design*, vol. 14, pp. 639–649, May 1995.

[4] H. N. Gabow, "Two algorithms for generating weighted spanning trees in order," *SIAM J. Comput.*, vol. 6, pp. 139–150, Mar. 1977.

[5] H. N. Gabow and Y. Xu, "Efficient algorithms for independent assignment on graphic and linear matroids," in *Proc. 30th Annu. Symp. Foundations Comput. Sci.*, 1989, pp. 106–111.

[6] G. Gielen, P. Wambacq, and W. Sansen, "Symbolic analysis methods and applications for analog circuits: A tutorial overview," *Proc. IEEE*, vol. 82, pp. 287–304, Feb. 1994.

[7] M. Iri, *Applications of Matroid Theory, Mathematical Programming: the State of Art*, A. Bachem, M. Grötschel, and B. Korte, Eds. Berlin, Germany: Springer-Verlag, 1983.

[8] E. L. Lawler, "A procedure for computing the K best solutions to discrete optimization problems and its application to the shortest path problem," *Manage. Sci.*, vol. 18, pp. 401–405, Mar. 1972.

[9] ——, *Combinatorial Optimization: Networks and Matroids*. New York: Holt, Rinehart and Winston, 1976.

[10] P. M. Lin, "Symbolic network analysis," in *Studies in Electrical and Electronic Engineering 41*. New York: Elsevier, 1991.

[11] L. Miguel-Silveira, M. Kamon, I. Elfadel, and J. White, "A coordinate-transformed Arnoldi algorithm for generating guaranteed stable reduced-order models of RLC circuits," in *IEEE/ACM ICCAD, Dig. Tech. Papers*, 1996, pp. 288–294.

[12] C. H. Papadimitriou and K. Steiglitz, *Combinatorial Optimization: Algorithms and Complexity*. Englewood Cliffs, NJ: Prentice-Hall, 1982.

[13] L. T. Pillage and R. A. Rohrer, "Asymptotic waveform evaluation for timing analysis," *IEEE Trans. Computer-Aided Design*, vol. 9, pp. 352–366, Apr. 1990.

[14] A. Recski, *Matroid Theory and its Applications in Electric Network Theory and Statics*. Berlin, Germany: Springer-Verlag, 1989.

[15] P. Wambacq, F. V. Fernandez, G. Gielen, and W. Sansen, "Efficient symbolic computation of approximated small-signal characteristics," in *Proc. Custom Integ. Circuits Conf.*, May 1994, pp. 461–464.

[16] P. Wambacq *et al.*, "A family of matroid intersection algorithms for the computation of approximated symbolic network functions," in *Proc. IEEE ISCAS*, 1996, pp. 806–609.

[17] W.-C. S. Wu *et al.*, "Digital-compatible high-performance operational amplifier with rail-to-rail input and output ranges," *IEEE J. Solid-State Circuits*, vol. 29, pp. 63–66, Jan. 1994.

[18] Q. Yu and C. Sechen, "Generation of color-constrained spanning trees with application in symbolic circuit analysis," in *Proc. 4th Great Lakes Symp. VLSI, IEEE Comput. Soc. Press*, pp. 252–255, Mar. 1994; also see *Int. J. Circuit Theory Appl.*, vol. 24, pp. 597–603, 1996.

[19] ——, "Approximate symbolic analysis of large analog integrated circuits," in *IEEE/ACM ICCAD, Dig. Tech. Papers*, 1994, pp. 664–671.

[20] ——, "Efficient approximation of symbolic network functions using matroid intersection algorithms," in *Proc. IEEE ISCAS*, 1995, pp. 2088–2091.

[21] ——, "A unified approach to the approximate symbolic analysis of large analog integrated circuits," *IEEE Trans. Circuits Syst. I*, vol. 43, pp. 656–669, Aug. 1996.

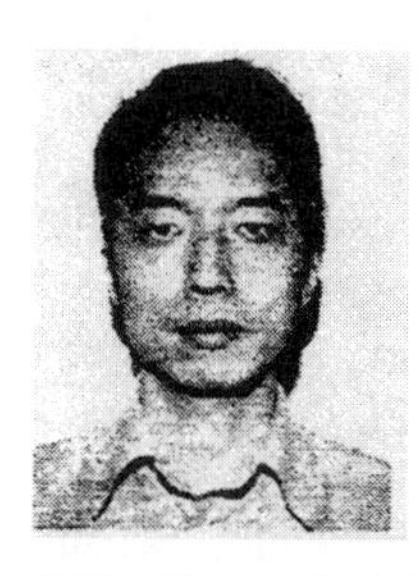

Qicheng Yu (M'93) received the B.S. degree from Fudan University, China, in 1985, the M.S. degree from the State University of New York at Stony Brook in 1991 and Yale University, New Haven, CT, in 1992, and the Ph.D. degree from the University of Washington in 1995.

From 1985 to 1988, he was in the graduate program of the Department of Electronic Engineering, Fudan University, while acting as a Research and Teaching Assistant. Since 1995, he has been with Cirrus Logic, Inc., Nashua, NH. His past and present fields of interest include computer-aided circuit analysis, digital system-level fault diagnosis, symbolic analysis of analog circuits, and analog and mixed signal circuit design.

Dr. Yu is a member of Eta Kappa Nu. He received the Sir Run Run Shaw Fellowship at SUNY at Stony Brook.

Carl Sechen (S'74–M'85) received the B.E.E. degree from the University of Minnesota, Minneapolis, St. Paul, the M.S. degree from the Massachusetts Institute of Technology, and the Ph.D. degree from the University of California, Berkeley.

From July 1986 through June 1992, he was an Assistant and then an Associate Professor at Yale University. Since July 1992, he has been an Associate Professor in the Department of Electrical Engineering at the University of Washington, Seattle. His primary research interests are the design and computer-aided design of high-speed digital integrated circuits. He was the initial developer of the TimberWolf placement and routing package. Together with W. Swartz, he cofounded TimberWolf Systems, Inc., a placement and routing software vendor.

High-Frequency Distortion Analysis of Analog Integrated Circuits

Piet Wambacq, Georges G. E. Gielen, Peter R. Kinget, *Member, IEEE,* and Willy Sansen, *Fellow, IEEE*

***Abstract*— An approach is presented for the analysis of the nonlinear behavior of analog integrated circuits. The approach is based on a variant of the Volterra series approach for frequency-domain analysis of weakly nonlinear circuits with one input port, such as amplifiers, and with more than one input port, such as analog mixers and multipliers. By coupling numerical results with symbolic results, both obtained with this method, insight into the nonlinear operation of analog integrated circuits can be gained. For accurate distortion computations, the accuracy of the transistor models is critical. A MOS transistor model is discussed that allows us to explain the measured fourth-order nonlinear behavior of a 1-GHz CMOS upconverter. Further, the method is illustrated with several examples, including the analysis of an operational amplifier up to its gain-bandwidth product. This example has also been verified experimentally.**

***Index Terms*—Analog integrated circuits, harmonic distortion, nonlinear distortion.**

I. Introduction

In the analog design community, many circuits are emerging in which nonlinear effects play an important role. For example, in switched-current filters [1] and in transconductance-C filters [2], [3], the amplifiers are open-loop circuits, which can cause considerable nonlinear distortion. Further, in analog radio frequency (RF) front-ends of integrated transceivers [4], [5], a knowledge of the nonlinear behavior of building blocks such as low-noise amplifiers and mixers is essential.

In this paper, it is described how nonlinear effects in weakly nonlinear, continuous-time analog integrated circuits can be analyzed with the inclusion of more than one nonlinearity and with high-frequency effects. Harmonic and intermodulation distortion are computed both in numerical and symbolic form. The numerical results provide extra insight since the method used can give the contribution to the output distortion of each nonlinearity separately. The symbolic results, on the other hand, are closed-form expressions as a function of one or two input frequencies and of the circuit elements. The symbolic expressions can be simplified with a criterion based on the relative magnitudes of the circuit elements, leading to interpretable expressions. In this way, circuit designers do not have to worry about how to calculate nonlinear effects. Instead they can obtain insight into those effects by concentrating upon the interpretation of the results.

For an analysis of the weakly nonlinear circuit behavior, the nonlinear circuit elements in the equivalent small-signal circuit of a transistor are described with power series which are broken down after the first few terms. The coefficients in these power series, referred to as *nonlinearity coefficients*, occur in the symbolic expressions of harmonics and intermodulation products. These coefficients can be computed by taking higher order derivatives from the transistor model equations as discussed in Section II. However, the value of nonlinearity coefficients is largely influenced by oversimplifications of the model equations or by errors on the model parameters. An accurate model for the drain current of a MOS transistor is presented in Section II. This model is used to explain the measured nonlinear behavior of a 1-GHz CMOS upconverter.

The method that is used here to compute harmonics and intermodulation products closely resembles the Volterra series approaches that have been used previously [6]–[9] to compute nonlinear responses numerically. Although the same results are obtained with the Volterra series approach, the method used here is simpler in the sense that it avoids redundant computations. The method is briefly explained in Section III.

The symbolic expressions for harmonics or intermodulation products are generated by the coupling of the computation method with a symbolic simulator that generates symbolic expressions for network functions of linearized analog circuits [10]. This is possible since the computation method described here reduces to a repeated solution of sets of linear equations. Modern symbolic analyzers [10]–[14] are able to generate approximate expressions, where the approximation is based upon the numerical values of the circuit parameters. Aspects of the coupling of the calculation method of harmonics, and intermodulation products with the symbolic network analysis approach for linearized circuits described in [12] are discussed in Section IV.

The analysis technique is illustrated in Section V with two examples: the second harmonic at the output of a CMOS Miller-compensated operational amplifier, and the third harmonic distortion of a single bipolar transistor amplifier.

II. Description of Nonlinearities

Most devices commonly used in analog integrated circuits can—for their small-signal behavior—be described with instances of nonlinear (trans)conductances and nonlinear ca-

Manuscript received July 31, 1997; revised June 15, 1998.

P. Wambacq was with the Department of Elektrotechniek, ESAT-MICAS, Katholieke Universiteit Leuven, Heverlee, Belgium. He is now with IMEC, Heverlee, Belgium.

G. G. E. Gielen is with the National Fund of Scientific Research, Belgium.

P. R. Kinget was with the Department of Elektrotechniek, ESAT-MICAS, Katholieke Universiteit Leuven, Heverlee, Belgium. He is now with Bell-Labs, Lucent Technologies, Murray Hill, NJ USA.

W. Sansen is with the Department of Elektrotechniek, ESAT-MICAS, Katholieke Universiteit Leuven, Heverlee, Belgium.

Publisher Item Identifier S 1057-7130(99)01767-X.

pacitances. A nonlinear conductance can be controlled by one or more voltages.

1) Nonlinear Conductance: For a nonlinear conductance or transconductance, the current through the element, i_{OUT}, is a nonlinear function f of up to three voltages u, v, and w. Using a power series expansion around the quiescent value, the total value of the current can be split into a quiescent part $I_{\mathrm{OUT}} = f(U_{CONTR}, V_{CONTR}, W_{CONTR})$ and an ac part. This ac part is given by

$$\begin{aligned} i_{\mathrm{out}} =& g_1 \cdot u + K_{2_{g_1}} \cdot u^2 + K_{3_{g_1}} \cdot u^3 + \cdots + g_2 \cdot v \\ &+ K_{2_{g_2}} \cdot v^2 + K_{3_{g_2}} \cdot v^3 + \cdots + g_3 \cdot w + K_{2_{g_3}} \cdot w^2 \\ &+ K_{3_{g_3}} \cdot w^3 + \cdots + K_{2_{g_1 \& g_2}} \cdot u \cdot v \\ &+ K_{3_{2g_1 \& g_2}} \cdot u^2 \cdot v + K_{3_{g_1 \& 2g_2}} \cdot u \cdot v^2 + \cdots \\ &+ K_{2_{g_1 \& g_3}} \cdot u \cdot w + K_{3_{2g_1 \& g_3}} \cdot u^2 \cdot w \\ &+ K_{3_{g_1 \& 2g_3}} \cdot u \cdot w^2 + \cdots + K_{2_{g_2 \& g_3}} \cdot v \cdot w \\ &+ K_{3_{2g_2 \& g_3}} \cdot v^2 \cdot w + K_{3_{g_2 \& 2g_3}} \cdot v \cdot w^2 + \cdots \\ &+ K_{3_{g_1 \& g_2 \& g_3}} \cdot u \cdot v \cdot w + \cdots. \end{aligned} \tag{1}$$

In this series, g_1, g_2, and g_3 denote the first-order derivative with respect to u, v, and w, respectively. These are nothing else but small-signal parameters. Further, this series implies the introduction of the following *nonlinearity coefficients*:

$$K_{n_{g_1}} = \frac{1}{n!} \frac{\partial^n f(u)}{\partial u^n} \tag{2}$$

$$K_{n_{jg_1 \& (n-j)g_2}} = \frac{1}{j!} \cdot \frac{1}{(n-j)!} \cdot \frac{\partial^n f(u, v)}{\partial u^j \partial v^{n-j}} \tag{3}$$

$$K_{n_{jg_1 \& kg_2 \& (n-j-k)g_3}} = \frac{1}{j!} \cdot \frac{1}{k!} \cdot \frac{1}{(n-j-k)!} \cdot \frac{\partial^n f(u, v, w)}{\partial u^j \partial v^k \partial w^{n-j-k}} \tag{4}$$

in which n is an integer larger than 1, and j and k are positive integers. If in $K_{n_{jg_1 \& (n-j)g_2}}$ the number j or $(n-j)$ equals one, then this number is omitted as a subscript, as in $K_{2_{g_1 \& g_2}}$:

$$K_{2_{g_1 \& g_2}} = \frac{\partial^2 f(u, v)}{\partial u \partial v}. \tag{5}$$

The same holds for the numbers j, k, and $(n-j-k)$ in (4) such as in $K_{3_{g_1 \& g_2 \& g_3}}$.

As a simple example, consider the simplified model of the collector current i_C of a bipolar transistor:

$$i_C = I_S \exp\left(\frac{v_{BE}}{V_t}\right) \tag{6}$$

in which I_S, v_{BE}, and V_t are the transistor saturation current, the base–emitter voltage, and the thermal voltage, respectively. The first derivative is the transistor transconductance $g_m = I_C/V_t$. The nonlinearity coefficients are found to be

$$K_{n_{g_m}} = \frac{I_C}{n! V_t^n} = \frac{g_m}{n! V_t^{n-1}} \qquad (n = 2, 3, \cdots). \tag{7}$$

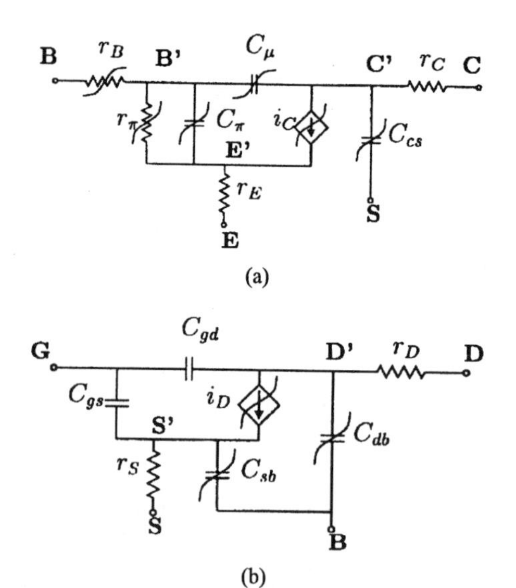

Fig. 1. Nonlinear model of a (a) bipolar and (b) MOS transistor.

2) Nonlinear Capacitance: A nonlinear capacitor is described with a nonlinear relationship between the charge upon the capacitor and the voltage over the capacitor in the same way as the relationship between the current and the controlling voltage of a one-dimensional conductance. The current through the capacitor is obtained by taking the time derivative of the charge.

A. Weakly Nonlinear Transistor Models

The nonlinearities described in the previous section can now be tailored together to construct nonlinear equivalent circuits for transistors. These are straightforward extensions of the linear equivalent small-signal circuits. The equivalent nonlinear circuit for a bipolar and a MOS transistor are shown in Fig. 1.

The nonlinearity coefficients that are needed to describe the different nonlinearities in a transistor are determined by taking second- and third-order derivatives, respectively, of the model equations. Existing circuit simulators provide values for the current and its first derivatives, which are the small-signal parameters, but the higher order derivatives are seldom computed. Therefore, C routines have been developed to compute those higher order derivatives. In these routines, the different derivatives are computed as a function of the bias voltages, the transistor dimensions, and the model parameters. The source code for these routines is derived in an automatic way, by computing symbolically with the symbolic algebra program MAPLE [15] the derivatives of the model equations. The generated expressions are then dumped to C format.

In this way, this approach is open to models that are different from the ones that are used in commercial circuit simulators. This has the advantage that physical effects that are poorly modeled in commercially used models can be modeled more accurately. For example, in MOS models such as the BSIM3 model [16], many effects are modeled with low-order polynomials rather than with more complicated transcendental functions. In this way, the accuracy for the computation of the dc values and the small-signal parameters is still acceptable, whereas the CPU time for the evaluation of the model equations is drastically reduced. However, the error on the higher order derivatives can be high.

With this in mind, the accurate models are only used for the computation of nonlinearity coefficients, whereas the dc bias point is computed with a numerical simulator using a model that can be efficiently evaluated, yielding only minor differences for the bias solution.

As an example, a drain current model for a MOS transistor in the triode region is considered. If velocity saturation and mobility reduction due to the vertical field are taken into account, then the drain current can be written as the product of three functions:

$$i_D = large(v_{GS}, v_{DS}, v_{SB}) \cdot mobred(v_{GS}, v_{DS}, v_{SB}) \cdot hot(v_{GS}, v_{DS}, v_{SB}). \tag{8}$$

The function $large(v_{GS}, v_{DS}, v_{SB})$ models the drain current in absence of mobility reduction and velocity saturation. An accurate expression for this function, derived in [17], is given by

$$large(v_{GS}, v_{DS}, v_{SB}) = \frac{\mu_0 C'_{ox} W}{L} \left\{ (v_{GS} - V_{FB} - \phi) v_{DS} - \frac{1}{2} v_{DS}^2 - \frac{2}{3}\gamma \left[(\phi + v_{SB} + v_{DS})^{3/2} - (\phi + v_{SB})^{3/2} \right] \right\} \tag{9}$$

in which μ_0 is the surface mobility in the absence of mobility reduction, C'_{ox} is the gate oxide capacitance per unit area, V_{FB} is the flat-band voltage, ϕ is the surface inversion potential, and γ is the body-effect coefficient. In order to have a function that can be evaluated more efficiently, the function *large* is very often approximated as follows [16], [18], [19]:

$$large = \frac{\mu_0 C'_{ox} W}{L} \left(v_{GS} - V_T - \frac{a}{2} v_{DS} \right) v_{DS} \tag{10}$$

in which various values are used for the parameter a. In several models, a is a function of v_{SB} only, and not of the other terminal voltages. As a result, the third-order nonlinearity coefficient $K_{3_{g_o}}$ is zero, since the dependence of the current on v_{DS} is quadratic. This is not correct as can be seen from the more accurate expression [see (9)] which contains a 3/2 power term that comprises v_{DS}. This illustrates that an oversimplification of a model can yield very large errors on nonlinearity coefficients.

The function $mobred(v_{GS}, v_{DS}, v_{SB})$ models the mobility reduction. It has the form

$$mobred(v_{GS}, v_{DS}, v_{SB}) = 1/(1 + \theta f_\mu). \tag{11}$$

The factor θ is the mobility reduction factor [17]. In many models, the parameter f_μ is often as simple as

$$f_\mu = v_{GS} - V_T. \tag{12}$$

In [20] a more accurate model is used:

$$f_\mu = (v_{GS} - V_{FB} - \phi) - \tfrac{1}{2} v_{DS} + \frac{2}{3} \frac{(\phi + v_{SB} + v_{DS})^{3/2} - (\phi + v_{SB})^{3/2}}{v_{DS}}. \tag{13}$$

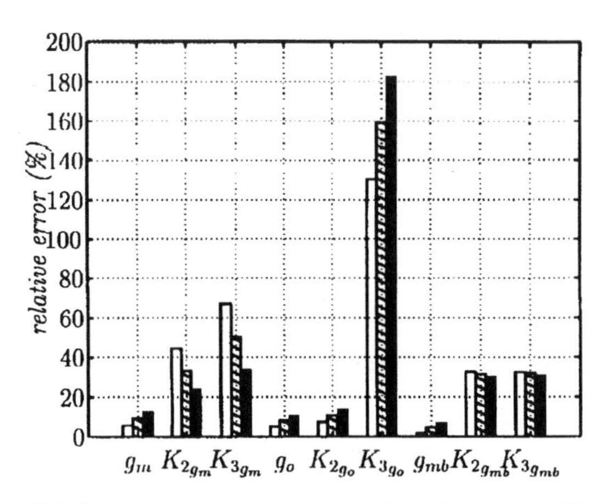

Fig. 2. Relative error in percentage between the nonlinearity coefficients of an n-MOS transistor in the triode region ($W = 320$ μm, $L = 14$ μm, $v_{DS} = 0.45$ V, $v_{GS} = 1.9$ V, $v_{SB} = 1.3$ V) computed with the simple mobility reduction model of (12) and the coefficients computed with the more exact model of (13). With the latter model, three values of θ have been considered: $\theta = 0.05$ V^{-1}, corresponding to the white bars, $\theta = 0.065$ V^{-1}, corresponding to the hatched bars, and $\theta = 0.079$ V^{-1}, corresponding to the black bars.

The difference between the nonlinearity coefficients of the drain current computed with the two models of f_μ given above increases with the order of the derivative. This is shown in Fig. 2.

Finally, the function $hot(v_{GS}, v_{DS}, v_{SB})$ models velocity saturation. In most MOS models that are commonly used, this is modeled with a function of the form

$$hot(v_{GS}, v_{DS}, v_{SB}) = 1 \Big/ \left(1 + \frac{v_{DS}}{LE_c} \right). \tag{14}$$

Here E_c is the critical electric field [17] that is approximately equal to v_{sat}/μ_{eff}, in which v_{sat} is the saturation velocity and $\mu_{\text{eff}} = \mu_0 \cdot mobred$ is the effective mobility. In [21] a more accurate model for the function *hot* is used:

$$hot(v_{GS}, v_{DS}, v_{SB}) = \frac{1}{\sqrt{1 + \left(\frac{v_{DB} - v_{SB}}{LE_c} \right)^2}}. \tag{15}$$

The problem of using more accurate expressions for the functions *large*, *mobred*, and *hot* than the expressions that are used in widely used MOS models, is that the drain current in saturation has to be computed iteratively, since the drain–source saturation voltage v_{DSAT} is given by an implicit equation. This would require too much CPU time for efficient circuit simulations. For the computation of nonlinearity coefficients, however, these models can be used without iteration, since it is possible to derive explicit expressions for the derivatives of the drain current as a function of the bias voltages and of v_{DSAT}. If, in the expressions for the derivatives, v_{DSAT} is evaluated with another, simpler MOS model that yields an explicit expression for v_{DSAT}, then the error is usually very small.

The expressions for the nonlinearity coefficients are usually very complicated. In order to get insight into the physical effects that mainly determine the value of a nonlinearity coefficient, an approximation procedure has been developed that can filter out the dominant terms in the expressions of the

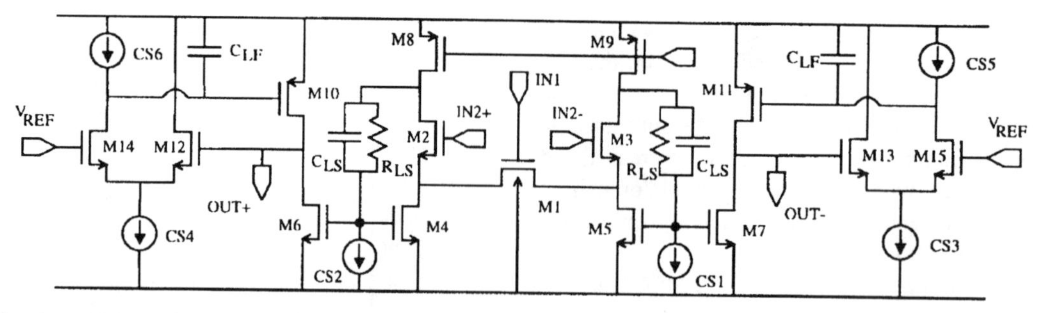

Fig. 3. Schematic of a 1-GHz CMOS upconversion mixer (22).

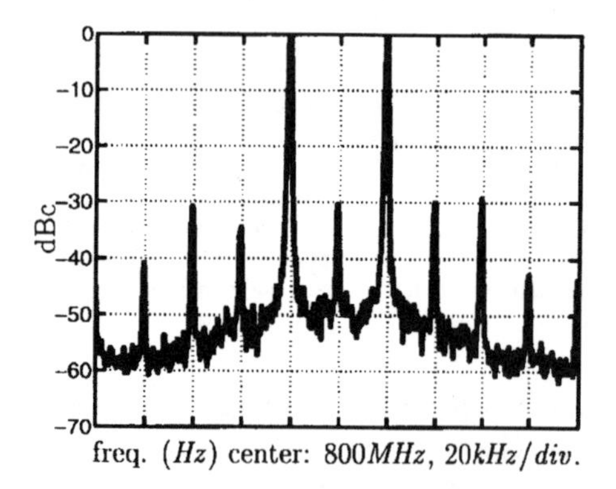

Fig. 4. Measured output spectrum of the upconverter of Fig. 3. The baseband signal is a 20-kHz differential sinusoidal voltage with an amplitude of 1 V. The LO signal is a 0 dBm sinewave at 800 MHz.

nonlinearity coefficients.[1] The procedure exploits the fact that the model equations from which the nonlinearity coefficients are computed by symbolic differentiation are products of functions that in turn are composed functions [see, for example, (8)]. Differentiating such products yields a sum, the dominant terms of which can be determined. This procedure is illustrated in the following example.

B. Example: A CMOS Upconverter

Fig. 3 depicts a 1-GHz CMOS upconverter that has been designed in a 0.7-μm CMOS technology [22]. The operation of the circuit is based on the mixing operation of transistor M_1 which is biased in the triode region with $v_{DS} = 0$ V. In this way, the drain and the source play an identical role. The local oscillator is applied at the gate of M_1, whereas the baseband signal is applied over the drain-source terminals via the source followers M_2 and M_3. The measured output spectrum for a 0-dBm 800-MHz local oscillator signal and a 1-V differential sinusoidal baseband signal at 20 kHz is shown in Fig. 4. The measured intermodulation products can be explained by looking at the mixing transistor M_1 only. The intermodulation products generated in that transistor propagate in a linear way through the rest of the circuit. Changes in the amplitude of the measured intermodulation products due to nonlinear behavior of the rest of the circuit are higher order effects, which can be neglected.

The wanted signal is at $f_{LO} \pm f_{BB}$, in which f_{LO} is the local oscillator frequency and f_{BB} is the frequency of the baseband signal. It is seen that the largest unwanted spectral component is at $f_{LO} \pm 3f_{BB}$. This component is due to fourth-order nonlinear behavior of transistor M_1. The amplitude of this component can be computed using the accurate expressions of the functions *large*, f_μ, and *hot* as given in (9), (13), and (15), respectively. The advantage of the used drain current model is that at $V_{DS} = 0$ V, it is perfectly symmetric with respect to source and drain, when the terminal voltages are referred to the bulk. This corresponds to the physical reality but it is very often not the case in widely used MOS models.

The high-frequency local oscillator signal is applied between the gate and the bulk of M_1. Its amplitude is A_{LO}:

$$v_{gb} = A_{LO}\,\sin(2\pi f_{LO}t). \tag{16}$$

The baseband signal with amplitude A_{BB} is applied between drain and source:

$$v_{db} = +\tfrac{1}{2}A_{BB}\,\sin(2\pi f_{BB}t) \tag{17}$$

$$v_{sb} = -\tfrac{1}{2}A_{BB}\,\sin(2\pi f_{BB}t). \tag{18}$$

1) Second-Order Response: The wanted component is caused by second-order nonlinear behavior of the drain current of M_1. This can be computed using the second-order nonlinearity coefficients that arise from the Taylor series expansion of the drain current around the bias point. At $V_{DS} = 0$ V, the derivatives of i_D with respect to v_{DB} are opposite to the derivatives with respect to v_{SB}. In this way, the wanted response is found as

$$|\text{drain current component at } (\omega_{LO} \pm \omega_{BB})| = \frac{1}{2} \cdot A_{LO} \cdot A_{BB} \cdot \frac{\partial^2 i_D}{\partial v_{GB} \partial v_{DB}}. \tag{19}$$

An approximate expression for the derivative in this equation can be obtained by using the routines explained above. In this way, it is found

$$|\text{drain current component at } (\omega_{LO} \pm \omega_{BB})| \approx \frac{1}{2} \cdot A_{LO} \cdot A_{BB} \cdot \mu_0 C'_{ox} \frac{W}{L} \cdot \frac{1}{1+\theta f_\mu}. \tag{20}$$

2) Fourth-Order Response: The fourth-order response at $f_{LO} \pm 3f_{BB}$ is caused by the fourth-order terms in a Taylor series expansion of the drain current that are proportional to v_{gb}. Using the symmetry of source and drain at $v_{DS} = 0$ V, and the fact that $v_{sb} = -v_{db}$, these fourth-order terms can be

[1] For the numerical computations, of course, the exact values are used.

combined in one term K_4:

$$K_4 = \frac{2}{3!} \frac{\partial^4 i_D}{\partial v_{GB} \partial v_{SB}^3} + \frac{\partial^4 i_D}{\partial v_{GB} \partial v_{SB} \partial v_{DB}^2}. \tag{21}$$

The fourth-order component in the drain current of M_1 now becomes

$$|\text{drain current component at } (\omega_{LO} \pm 3\omega_{BB})| = A_{LO} \cdot A_{BB}^3 \cdot \frac{K_4}{64}. \tag{22}$$

Using the routines described in Section II-A an approximation for the coefficient K_4 has been computed. With these routines, it is seen that the largest term is 62 times larger than the second largest term. Taking into account the largest term only for K_4, yields

$$K_4 \approx -\mu_0 C'_{ox} \frac{W}{L} \frac{2}{L^2 E_c^2 (1 + \theta f_\mu)} \tag{23}$$

with f_μ given by (13). The error on K_4, that is made by taking into account the largest term only is 2.8% for the given bias point in the given CMOS technology.

The factor $L^2 E_c^2$ arises from taking the derivative of the function *hot* that models velocity saturation. This means that the fourth-order response is almost completely determined by the variation of the drain current due to velocity saturation. Indeed, when the instantaneous drain current becomes high during operation, then the velocity of the carriers may saturate.

The above results can now be evaluated for the current design. With $\mu_0 = 0.047$ m^2/(V·s), $W = 16$ μm, $L = 0.7$ μm, $C'_{ox} = 0.002$ F/m^2, $v_{\text{sat}} = 10^5$ m/s, $\theta = 0.0503$ V^{-1}, and a mobility reduction factor f_μ of 0.146 in the given bias point, the effective mobility μ_{eff} is 0.041 m^2/(V·s) and $E_c = 2.439$ MV/m. The ratio of the fourth-order and the second-order component of the drain current with the given amplitudes is then found to be -33 dB, which is only 3 dB different from the measured ratio.

III. Calculation of Nonlinear Responses

In weakly nonlinear analog circuits, harmonics and intermodulation products can be computed using Volterra series [7]. In [23] and [24] a similar method is derived that circumvents the use of Volterra series. The analysis is limited to circuits with at most two input ports or with two input frequencies applied at one input port. In this way, harmonic and two-tone intermodulation distortion can be computed. The method leads to the same results as the Volterra series approach but it is simpler to use, especially for circuits with more than one input port, such as a mixer. For such circuits, the use of Volterra series requires tensors, which are not used here. Instead, only the components of the tensors are computed that are of interest for the required response. Further, the Volterra kernel transforms of order n, which are functions of n distinct frequency variables, are not computed completely. For example, when a harmonic needs to be computed, then only the value of the tensor element of interest needs to be computed for all frequency variables being equal.

TABLE I
Nonlinear Second-Order Current Sources for the Basic Nonlinearities to Compute Second-Order Intermodulation Products at $|\omega_1 \pm \omega_2|$ and Second Harmonics at $2\omega_1$. The Controlling Voltages Are v_i for the Nonlinear (Trans)Conductance and the Nonlinear Capacitor and v_i and v_j for the Two-Dimensional Conductance

type of nonlinearity	nonlinear current source for response at $\lvert\omega_1 \mp \omega_2\rvert$	nonlinear current source for response at $2\omega_1$
(trans)conductance	$K_{2_{g_1}} V_{i,1,0} V_{i,0,1}$	$\frac{K_{2_{g_1}}}{2} (V_{i,1,0})^2$
capacitor	$j(\omega_1 \pm \omega_2) K_{2_{C_1}} V_{i,1,0} V_{i,0,1}$	$j\omega_1 K_{2_{C_1}} (V_{i,1,0})^2$
twodimensional conductance (only crossterms)	$\frac{K_{2_{g_1 \& g_2}}}{2} V_{i,1,0} V_{j,0,1}$ $\frac{K_{2_{g_1 \& g_2}}}{2} V_{i,0,1} V_{j,1,0}$	$\frac{K_{2_{g_1 \& g_2}}}{2} V_{i,1,0} V_{j,1,0}$

It is assumed that a circuit is excited by at most two sinusoidal input signals at a frequency ω_1 and ω_2, respectively. Under steady-state conditions, every node voltage $v_i(t)$ consists of a sum of harmonic functions as shown in (24) at the bottom of the page. In this equation, $\text{Re}(\cdot)$ denotes the real part of its argument, $V_{i,x,y}$ is the phasor of the component of the voltage on node i at the frequency $x\omega_1 + y\omega_2$.

In a first step, the response of the linearized circuit to the external inputs is calculated. Together with the component of the output voltage at ω_1 and ω_2, the component at ω_1 and ω_2 of all voltages that control a nonlinearity are calculated as well.

In the next step, the second-order responses are calculated. Suppose, for example, that the intermodulation product at the difference frequency $|\omega_1 - \omega_2|$ has to be calculated. To this purpose, the physical inputs are first put to zero in the linearized circuit. Instead, the linearized circuit is excited with higher order correction current sources, denoted as *nonlinear current sources of order two*. Every nonlinearity in the original nonlinear circuit gives rise to such a current source in the linearized circuit. The sources are placed in parallel with each linearized element. The output of the circuit as a result of those current sources yields the required intermodulation product. Hereby, the transfer functions from the current sources to the output have to be evaluated at the frequency $|\omega_1 - \omega_2|$.

The value of the nonlinear current sources of order two (see Table I) depends on the type of the basic nonlinearity and on the first-order responses at the controlling voltages.

$$\begin{aligned}
v_i(t) = {} & \text{Re}(V_{i,1,0} e^{j\omega_1 t}) + \text{Re}(V_{i,0,1} e^{j\omega_2 t}) && \text{linear response} \\
& \left.\begin{aligned} & + \text{Re}(V_{i,2,0} e^{j2\omega_1 t}) + \text{Re}(V_{i,0,2} e^{j2\omega_2 t}) \\ & + \text{Re}(V_{i,1,1} e^{j(\omega_1+\omega_2)t}) + \text{Re}(V_{i,1,-1} e^{j(\omega_1-\omega_2)t}) \end{aligned}\right\} && \text{2nd-order response} \\
& + \text{Re}(V_{i,3,0} e^{j3\omega_1 t}) + \text{Re}(V_{i,0,3} e^{j3\omega_2 t}) && \text{3rd-order response} \\
& + \cdots
\end{aligned} \tag{24}$$

TABLE II

NONLINEAR THIRD-ORDER CURRENT SOURCES FOR THE BASIC NONLINEARITIES TO COMPUTE THE THIRD HARMONIC AT $3w_2$. THE CONTROLLING VOLTAGES ARE v_i FOR THE NONLINEAR (TRANS)CONDUCTANCE AND THE NONLINEAR CAPACITOR, v_i, AND v_j FOR THE TWO-DIMENSIONAL CONDUCTANCE, AND v_i, v_j, AND v_k FOR THE THREE-DIMENSIONAL CONDUCTANCE

type of nonlinearity	nonlinear current source for response at $3\omega_2$
(trans)conductance	$K_{2_{g_1}} V_{i,0,1} V_{i,0,2} + \frac{1}{4} K_{3_{g_1}} V_{i,0,1}^3$
capacitor	$3j\omega_2 \left[K_{2_{C_1}} V_{i,0,1} V_{i,0,2} + \frac{1}{4} K_{3_{C_1}} V_{i,0,1}^3 \right]$
twodimensional conductance (only crossterms)	$\frac{1}{2} K_{2_{g_1 \& g_2}} \left[V_{i,0,1} V_{j,0,2} + V_{i,0,2} V_{j,0,1} \right] + \frac{1}{4} K_{3_{2g_1 \& g_2}} V_{i,0,1}^2 V_{j,0,1} + \frac{1}{4} K_{3_{g_1 \& 2g_2}} V_{i,0,1} V_{j,0,1}^2$
threedimensional conductance (only crossterms)	$+\frac{1}{4} K_{3_{g_1 \& g_2 \& g_3}} V_{i,0,1} V_{j,0,1} V_{k,0,1}$

If one is interested in the second harmonic, a similar approach is followed. In this case, the transfer functions from the current sources to the output have to be evaluated at $2\omega_1$, and the values of the current sources, which are also given in Table I, are slightly different.

The method can be explained intuitively as follows. It is assumed that a second-order harmonic or intermodulation product is caused by nonlinearities of order two and not by higher-order nonlinearities. A second-order nonlinearity combines two first-order signals (signals at one of the two fundamental frequencies) at its controlling terminals and produces a second-order signal. This second-order signal propagates through the circuit to the output. When it is fed into other second-order nonlinearities, then higher order signals arise, which are neglected for the computation of second-order harmonics and intermodulation products. Hence, only the propagation of the second-order signal through the linearized network is important.

After the second-order signals of interest have been computed, the third-order responses can be calculated. Here again, they are computed as the response to nonlinear current sources. These are now determined by the second- and third-order nonlinearity coefficients and by the first- and second-order response at the controlling voltage(s). This can be seen in Table II, which lists the values of the nonlinear current sources for the computation of the third harmonic of ω_2. For other third-order responses, the values of the current sources are slightly different. If one is interested in responses of order higher than three, then a similar procedure can be followed.

A. Interpretation of the Results

With the calculation method explained above, the harmonics and intermodulation products of order $p > 1$ at the frequency $|\pm m\omega_1 \pm n\omega_2|$ with $m + n = p$ are computed as a sum of contributions:

$$\sum_{k=1}^{\#\text{ nonlinearities}} i_{NL_{p_k}} \cdot TF_{i_{NL_{p_k}} \rightarrow \text{output}}(\pm m\omega_1 \pm n\omega_2). \quad (25)$$

In this equation, $i_{NL_{p_k}}$ is the nonlinear current source of order p for nonlinearity k, and $TF_{i_{NL_{p_k}} \rightarrow \text{output}}(\cdot)$ denotes the transfer function from the applied nonlinear current source to the output. The sum is taken over all nonlinearities (conductances or capacitances) in the circuit.

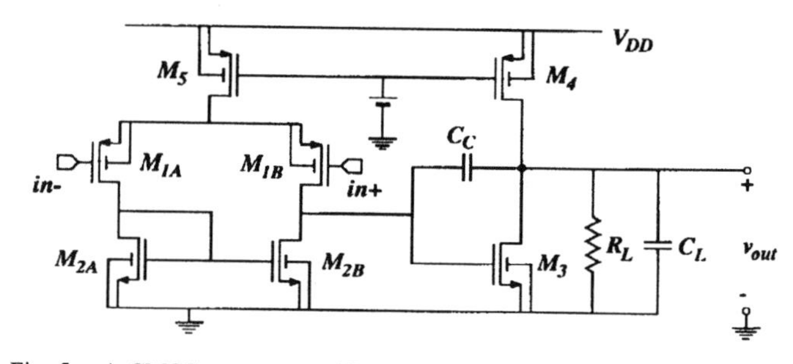

Fig. 5. A CMOS two-stage Miller-compensated opamp.

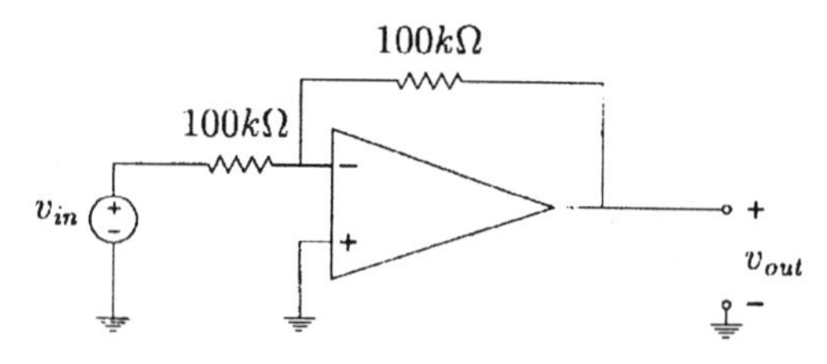

Fig. 6. An inverting amplifier.

The formulation of nonlinear responses in the form of (25) makes reasoning about distortion possible. Indeed, the transfer functions which determine the nonlinear current sources, on one hand (see Tables I and II), and the transfer functions from the current sources to the circuit's output, on the other hand, can be analyzed either numerically or symbolically and interpreted. Moreover, since the current sources are applied to a linear network, the effect of every current source, corresponding to one single nonlinearity, can be studied apart from the other ones.

IV. ALGORITHMIC ASPECTS

Responses of order higher than one are determined by analyzing the same linearized network that is used for the analysis of the linear behavior. This can be exploited in the symbolic computations. Indeed, the admittance matrix that arises from the network equations is always the same, apart from the value of the frequency variable. In a symbolic expression of a network function, the denominator corresponds to the determinant $\det(j\omega)$ of the admittance matrix. Since the expression of a nonlinear current source depends on lower-order responses, it is not difficult to see that the denominator of the value of a nonlinear current source is a combination of products of admittance matrix determinants, evaluated at a different frequency. For example, it can be shown that the denominator of any second-order response at $\omega_1 \pm \omega_2$ is $[\det(j\omega_1 \pm j\omega_2)\det(j\omega_1)\det(j\omega_2)]$, whereas the denominator of any second harmonic at ω_1 is $[(\det(j\omega_1))^2 \det(2j\omega_1)]$. This reasoning can be extended to order three.

From the previous paragraph, it is clear that a symbolic expression for the denominator of a harmonic or intermodulation product can be computed once an expression for $\det(j\omega)$ has been obtained, whereas the numerator is found by combining the numerators of several symbolic transfer functions, either from the input to a voltage that controls a nonlinearity or from a nonlinear current source to the output or to another controlling voltage. The final numerator is a nested expression, which can be expanded afterwards. However, huge expressions are generated, since a practical circuit contains a lot of basic nonlinearities and since each nonlinearity gives rise to a nonlinear current source, whose expression can already be quite complicated. In order to manage this complexity, two simplification procedures are used. In a first step, the contributions of the different nonlinearities are computed numerically as a function of one of the fundamental frequencies in a frequency range of interest, which is discretized into a set of frequency points. Next it is checked at every frequency point whether any contributions can be neglected. If at every frequency point a nonlinearity can be neglected, then the corresponding nonlinearity coefficient is set to zero. This elimination process is controlled by a user-defined error that should not be exceeded in a frequency range of interest. After this first step, usually very few nonlinearities remain with nonzero nonlinearity coefficients.

In a second step an approximate symbolic expression for the harmonic or intermodulation product of interest is computed with the nonlinearities that have not been eliminated in the previous step. The final symbolic expression will be a hierarchical one, in the form of (25). Even with just a few nonlinearities, the exact expression is very large for circuits of practical size. The reason is that the number of terms of a network function increases exponentially with the size of the circuit. For large circuits, it is even impossible to compute the exact expression due to a huge number of terms. This problem is solved by generating the dominant terms only for every subexpression in (25) without first computing the exact expression. This is performed with the so-called *simplification during generation* approach that is described in [12]. This approach uses numerical values of the symbolic circuit parameters for the approximation. The errors that must be used for the approximation of every subexpression are computed in advance from the user-specified error for the total expression. This is performed with the algorithm described in [25] that has been developed for the approximation of nested (hierarchical) expressions. The final simplified expression can be expanded on user demand.

It must be remarked that the intermediate result obtained in the first step, namely a knowledge of the significant nonlinearities, provides information which is already much more valuable than simulation results obtained from SPICE-like simulators with the .DISTO command [9], which do not select the significant nonlinearities at all. This knowledge, together with a plot of the different contributions and the total response as a function of frequency, can already yield enough insight such that the user does not need a symbolic analysis anymore.

Currently, symbolic network analysis programs are able to generate interpretable expressions for linearized circuits having about twenty transistors, each being represented by a nine elements equivalent circuit. The same limitation of course applies here. Since in many practical circuits only a few nonlinearities play a significant role, the extra memory usage and CPU time for symbolic analysis of nonlinear behavior is limited compared to symbolic analysis of thc linear behavior only.

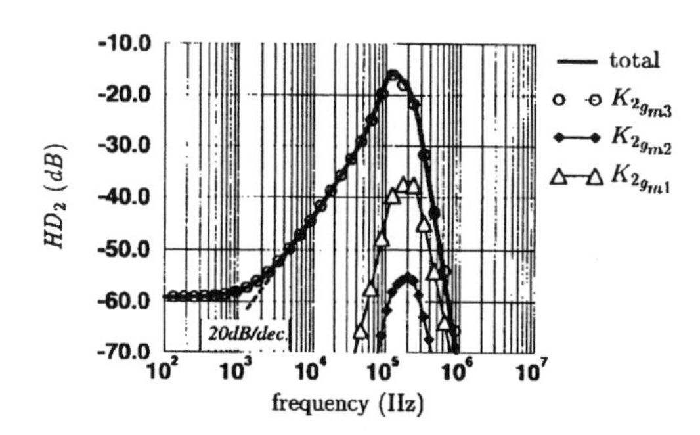

Fig. 7. The most important contributions to the second harmonic distortion of the output voltage of the CMOS two-stage Miller-compensated opamp of Fig. 5 in the feedback configuration of Fig. 6.

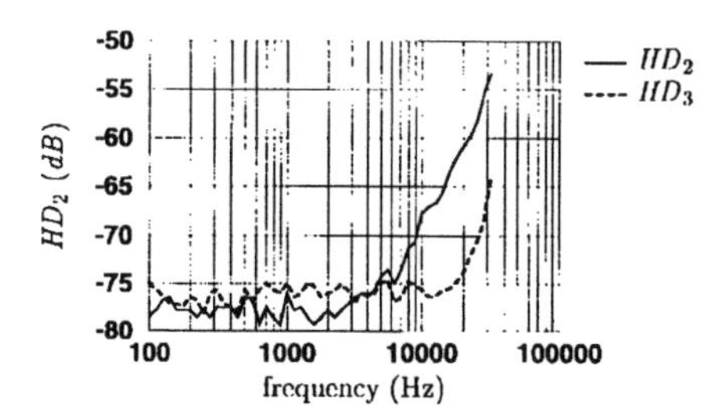

Fig. 8. Measured second and third harmonic distortion on the CMOS two-stage Miller-compensated opamp of Fig. 5 in the feedback configuration of Fig. 6.

V. Examples

A. Example 1: A Miller-Compensated Operational Amplifier

In this example, the harmonic distortion is analyzed at the output of the CMOS two-stage Miller-compensated operational amplifier of Fig. 5 up to its gain-bandwidth product (GBW). The amplifier is put in the inverting feedback configuration of Fig. 6. It is assumed that the transistors M_{1A} and M_{1B} match, just like M_{2A} and M_{2B}. Small-signal parameters and nonlinearity coefficients of such matching transistors are represented with one symbol. For example, g_{o1} represents the output conductance of both M_{1A} and M_{1B}.

The amplifier has a gain-bandwidth of 100 kHz, the load capacitance C_L is 10 pF, the load resistance R_L in addition to the load formed by the feedback resistors is 100 kΩ, and the compensation capacitance equals 1 pF.

The second harmonic distortion HD_2 has sixteen contributions. These are first calculated as a function of the fundamental frequency with the method explained in Section III. The most important contributions to HD_2 are shown in Fig. 7. The other contributions are below -70 dB.

It is seen that HD_2 starts to increase from 1 kHz ($\approx 0.01 GBW$) with 20 dB per decade, and from about 50 kHz ($\approx GBW/2$) with a steeper slope. Beyond the gain-bandwidth HD_2 decreases rapidly. This behavior is also seen in measurements, as shown in Fig. 8. For the third harmonic distortion, a similar increase in frequency is seen, but with a steeper slope. Differences in absolute value between the computed and measured distortion levels are mainly due to the poor modeling of the output conductance with the available SPICE level 2 models.

Clearly, only one nonlinearity dominates for frequencies below the gain-bandwidth product GBW (100 kHz), namely the second-order nonlinearity coefficient $K_{2_{gm}}$ of transistor M_3. This can be explained by the fact that the largest contributions to the nonlinear distortion at the output of an amplifier originate from the circuit elements close to or at the output, where signal swings are large.

An expression for HD_2 can be computed by considering the contribution of $K_{2_{gm3}}$ only. By omitting the other contributions, the error is never larger than 4%, which is the error at 100 kHz. At low frequencies, the error is much lower.

The contribution of $K_{2_{gm3}}$ is first computed in open loop. From (25) this is given by

$$\text{contribution of } K_{2_{gm3}} = i_{NL2_{gm3}} \cdot TF_{i_{NL2_{gm3}} \to \text{out}}(2\omega_1). \tag{26}$$

From Table I, the value of $i_{NL2_{gm3}}$ is found:

$$i_{NL2_{gm3}} = \frac{K_{2_{gm3}}}{2} V_{gs3,1,0}^2 \tag{27}$$

in which $V_{gs3,1,0}$ is the fundamental response of the gate-source voltage of M_3. At frequencies well below GBW, this is easily found to be

$$V_{gs3,1,0} \approx -\frac{g_{m1}(g_L + j\omega_1(C_L + C_C))}{g_{o1} + g_{o2} + j\omega_1 C_C(g_{m3}/g_L)} V_{\text{in}} \tag{28}$$

in which g_L is the sum of $1/R_L$ and the output conductances of M_3 and M_4.

The nonlinear current source of order two that corresponds to $K_{2_{gm3}}$ flows from the drain of M_3 to its source. The transfer function from this source to the output of the amplifier, which is the drain of M_3, is given by

$$TF_{i_{NL2_{gm3}} \to \text{out}}(2\omega_1) = -\frac{g_{o1} + g_{o2} + 2j\omega_1 C_C}{g_L(g_{o1} + g_{o2} + 2j\omega_1(g_{m3}/g_L))}. \tag{29}$$

In order to know the second harmonic in closed loop, the second harmonic in open loop needs to be divided by $[(1 + T(j\omega_1))^2(1 + T(2j\omega_1))]$, T being the loop gain [27]. The second harmonic distortion is obtained by dividing the second harmonic by the fundamental response. Doing so, the poles in (28) and (29) are cancelled, and HD_2 is computed with the above method as shown in (30) at the bottom of the page. It is seen that HD_2 increases with 20 dB per decade from the frequency $(g_{o1} + g_{o2})/(4\pi C_C))$. For the given design, this frequency is computed to be 2.3 kHz. At the frequency $g_L/(2\pi(C_L + C_C)$, which equals 31 kHz here, the increase is with 60 dB per decade.

The error between HD_2 given in (30) and the exact numerical value (without neglecting any contribution of a nonlinearity coefficient) is smaller than 0.1% at low frequencies and reaches a maximum of 37% at GBW.

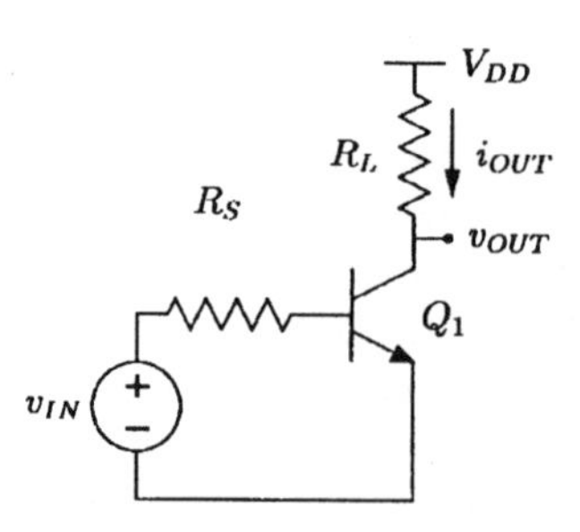

Fig. 9. A single bipolar transistor amplifier with a voltage source excitation.

B. Example 2: A Single Bipolar Transistor Amplifier

Fig. 9 shows a common-emitter amplifier loaded with a resistor R_L. The different contributions to the third harmonic at the collector of Q_1 will be analyzed at high frequencies.

The ac equivalent circuit of the common emitter amplifier is shown in Fig. 10, while the linearized equivalent is shown in Fig. 11. It is seen in Fig. 10 that the collector current is described as a function of two voltages in order to model both the exponential dependence on the base–emitter voltage and the Early effect. The different nonlinearities are modeled with the Gummel–Poon model, with the extension that the nonlinearity of the Early effect is modeled as well [28]. However, this nonlinearity plays a negligible role in this example. Further, the base resistance has been considered as a linear element for simplicity. This is a reasonable assumption for transistors with a large emitter area and at moderate bias currents. At high base currents effects such as current crowding, base pushout and/or base conductivity modulation make the base resistance dependent on the base current [26], such that its nonlinear behavior needs to be taken into account.

The most important transistor parameters are: $I_C = 0.834$ mA, $\beta_{AC} = 90$, $r_B = 213\ \Omega$, $\tau_F = 10$ ps, $V_{AF} = 30$ V.

Fig. 12 shows the third harmonic distortion HD_3 as a function of frequency together with the most important contributions. It is seen that at low frequencies two contributions play a role, namely the contributions of $K_{3_{gm}}$ and $K_{2_{gm}}$. They partially compensate, such that the total value of HD_3 is smaller than the value of the largest contribution.

The total value of HD_3 begins to increase from about 10 MHz. This can be explained as follows. At low frequencies, the imaginary part of the contributions of $K_{3_{gm}}$ and $K_{2_{gm}}$ is zero. At about 10 MHz, however, a zero occurs in the contribution of $K_{2_{gm}}$ and the phase of this contribution goes to 90°. On the other hand, the frequency behavior of the contribution of $K_{3_{gm}}$ is still flat, which means that the phase is still about zero. As a result, the two contributions will not cancel anymore; and when their magnitude is equal, at about 80 MHz, then the sum of the two contributions is not zero.

The third harmonic distortion reaches a maximum value around 200 MHz. This maximum is about 10 dB higher than the low-frequency value of HD_3. Beyond 200 MHz, HD_3 falls off.

$$HD_2 = \frac{1}{2} V_{\text{in}} \frac{(R_1 + R_2)R_2}{R_1^2} \frac{|(g_L + j\omega_1(C_L + C_C)^2(g_{o1} + g_{o2} + 2j\omega_1 C_C)|}{g_{m1} g_{m3}^3} K_{2_{gm3}} \tag{30}$$

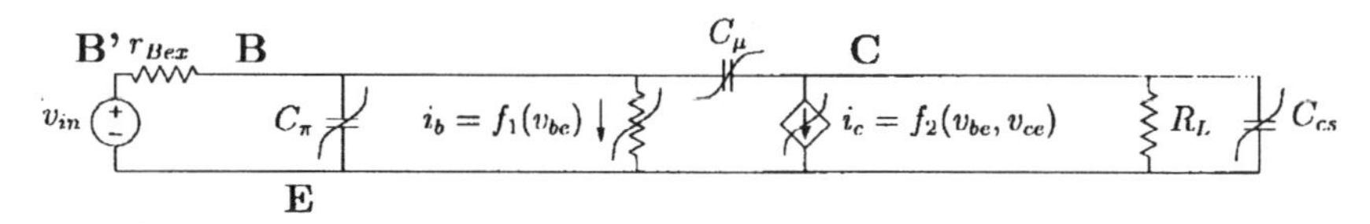

Fig. 10. AC-equivalent circuit of the one-transistor amplifier of Fig. 9 driven by a voltage source. The nonlinear capacitances C_π, C_μ, C_{cs}, and the nonlinear base and collector current are included.

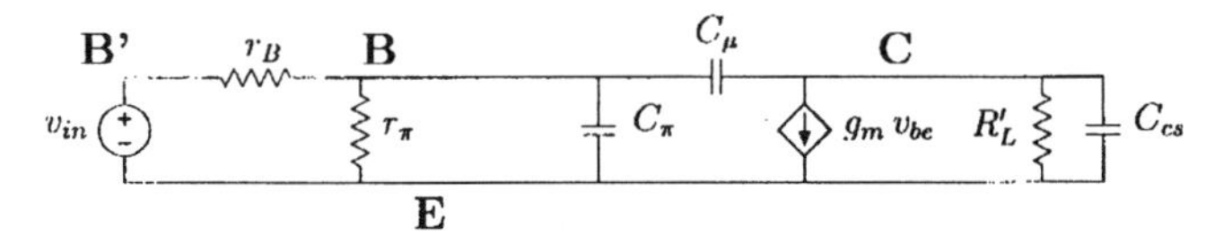

Fig. 11. Linearized equivalent circuit of Fig. 10. R'_L is the parallel connection of R_L and the early resistance of Q_1.

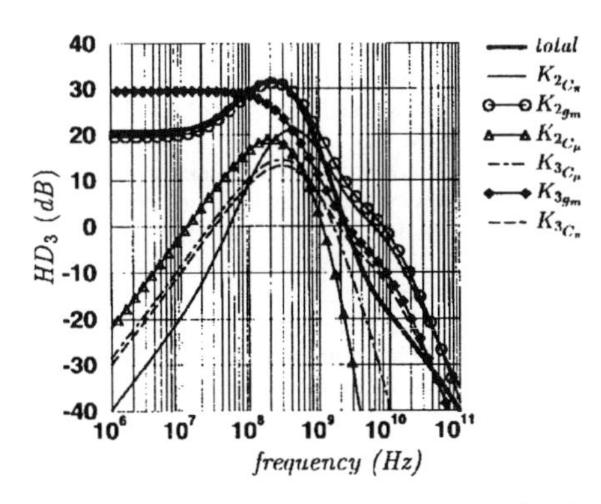

Fig. 12. Third harmonic distortion at the output of the single-transistor amplifier of Fig. 10, as a function of frequency together with its contributions. HD_3 has been normalized to an input amplitude of 1 V. Only the most important contributions have been shown.

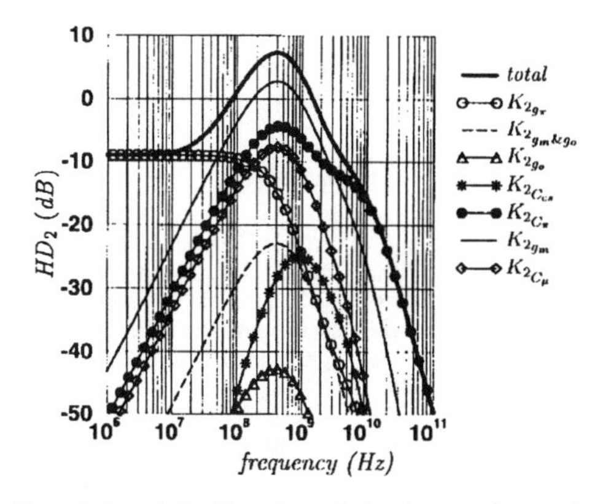

Fig. 13. Second harmonic distortion of the base–emitter voltage in the single-transistor amplifier of Fig. 10, as a function of frequency together with its most importatnt contributions. HD_2 has been normalized to an input amplitude of 1 V.

Around the maximum value in the vicinity of 200 MHz, the most important contribution is due to $K_{2_{gm}}$. From Table II it is seen that this contribution is proportional to the second harmonic $V_{be,2,0}$ at the base–emitter voltage. This second harmonic, together with its most important contributions, is shown in Fig. 13. It is seen that around 200 MHz, the most important contribution to $V_{be,2,0}$ comes from $K_{2_{gm}}$. The nonlinear current source that corresponds to $K_{2_{gm}}$ flows from the collector to the emitter. At low frequencies, the transfer function from this source to the base–emitter voltage is zero, but at high frequencies there is a path from the collector to the base along C_μ.

An approximate expression for HD_3 that only takes into account $K_{2_{gm}}$ has been computed with the above method, resulting in (31) at the bottom of the page, in which $\det(s)$ is the determinant of the admittance matrix, given by

$$\det(s) = (g_B + g_\pi)G'_L + s(C_\mu(g_B + g_\pi) + C_\mu g_m + C_{cs}(g_B + g_\pi) + C_\pi G'_L + C_\mu G'_L) + s^2(C_{cs}C_\pi + C_{cs}C_\mu + C_\mu C_\pi). \quad (32)$$

In the given operating point, and around 200 MHz, this expression can be simplified to

$$\det(s) \approx g_B G'_L + s(C_\mu(g_m + g_B) + C_{cs}g_B). \quad (33)$$

VI. Conclusions

In this paper, the analysis of weakly nonlinear behavior of analog integrated circuits has been addressed. The method makes use of a variant of the Volterra series approach to calculate harmonics and intermodulation products in the frequency domain. However, it is simpler although equally accurate for circuits with more than one input port, such as mixers and multipliers. It has been shown how numerical and symbolic results obtained with this method can be combined in order to obtain more insight in the nonlinear operation of analog integrated circuits. The numerical results yield the different contributions to the harmonic or intermodulation products of interest, and the most important ones can be retrieved. The symbolic expressions are generated by a symbolic network analysis method that can generate approximate expressions of the most important contributions with a user-definable error.

The results depend on nonlinearity coefficients that are proportional to higher order derivatives of model parameters. Routines have been presented for the computation of these coefficients, not only based on widely used transistor models but also on more accurate models that can be defined by the

$$HD_3 \approx \left| \frac{g_B^2(G'_L + j\omega_1(C_{cs} + C_\mu))^3 K_{2_{gm}}^2 C_\mu(g_B + g_\pi + 3j\omega_1(C_\pi + C_\mu))}{(\det(j\omega_1))^2 \det(2j\omega_1)\det(3j\omega_1) g_B(-g_m + j\omega_1 C_\mu)} \right| V_{\text{in}}^2 \quad (31)$$

user. The capabilities of these routines have been illustrated with the analysis of the fourth-order nonlinear behavior in a 1-GHz CMOS upconverter.

The complete approach has been illustrated with several examples: the harmonic distortion in an operational amplifier in closed loop has been analyzed as a function of frequency and compared to experimental results; next, the main contributions to the third harmonic distortion in a single bipolar transistor amplifier at high frequencies have been analyzed.

REFERENCES

[1] T. Fiez, G. Liang, and D. Allstot, "Switched-current circuit design issues," *IEEE J. Solid-State Circuits,* vol. 26, pp. 192–202, Mar. 1991.

[2] V. Gopinathan, Y. Tsividis, K.-S. Tan, and R. Hester, "Design considerations for high-frequency continuous-time filters and implementation of an antialiasing filter for digital video," *IEEE J. Solid-State Circuits,* vol. 25, pp. 1368–1378, Dec. 1990.

[3] G. De Veirman, S. Ueda, J. Cheng, S. Tam, K. Fukahori, M. Kurisu, and E. Shinozaki, "A 3.0 V 40 Mb/s hard disk drive read channel IC," *IEEE J. Solid-State Circuits,* vol. 30, pp. 788–799, July 1995.

[4] A. Abidi, "Direct conversion radio transceivers for digital communications," *IEEE J. Solid-State Circuits,* vol. 30, pp. 1399–1410, Dec. 1995.

[5] R. Meyer and W. Mack, "A 2.5 GHz BiCMOS transceiver for wireless LAN," in *Proc. ISSCC 1997,* 1997, pp. 310–311.

[6] Y. Kuo, "Distortion analysis of bipolar transistor circuits," *IEEE Trans. Circuit Theory,* vol. CT-20, pp. 709–716, Nov. 1973.

[7] J. Bussgang, L. Ehrman, and J. Graham, "Analysis of nonlinear systems with multiple inputs," *Proc. IEEE,* vol. 62, pp. 1088–1118, Aug. 1974.

[8] L. O. Chua and C.-Y. Ng, "Frequency-domain analysis of nonlinear systems: Formulation of transfer functions," *IEE J. Electron. Circuits Syst.,* vol. 3, no. 6, pp. 257–269, Nov. 1979.

[9] J. Roychowdhury, "SPICE3 distortion analysis," Memo. no. UCB/ERL M89/48, Univ. Calif., Berkeley, Apr. 1989.

[10] G. Gielen, H. Walscharts, and W. Sansen, "ISAAC: A symbolic simulator for analog integrated circuits," *IEEE J. Solid-State Circuits,* vol. 24, pp. 1587–1597, Dec. 1989.

[11] F. V. Fernandez, A. Rodnguez-Vazquez, and J. L. Huertas, "Interactive ac modeling and characterization of analog circuits via symbolic analysis," *Kluwer J. Analog Integrated Circuits Signal Processing,* vol. 1, pp. 183–208, Nov. 1991.

[12] P. Wambacq, F. Fernandez, G. Gielen, W. Sansen, and A. Rodnguez-Vázquez, "Efficient symbolic computation of approximated small-signal characteristics," *IEEE J. Solid-State Circuits,* vol. 30, pp. 327–330, Mar. 1995.

[13] J. J. Hsu and C. Sechen, "DC small signal symbolic analysis of large analog integrated circuits," *IEEE Trans. Circuits Syst.—I,* vol. 41, pp. 817–828, Dec. 1994.

[14] Q. Yu and C. Sechen, "Approximate symbolic analysis of large analog integrated circuits," in *Proc. IEEE/ACM ICCAD,* 1994, pp. 664–671.

[15] *MapleV Language Reference Manual.* New York: Springer-Verlag, 1991.

[16] *BSIM3v3 User's Manual,* Dept. Elec. Eng. Computer Sci., Univ. Berkeley, 1995.

[17] Y. Tsividis, *Operation and Modeling of the MOS Transistor.* New York: McGraw-Hill International Editions, 1988.

[18] S. Liu and L. Nagel, "Small-signal MOSFET models for analog circuit design," *IEEE J. Solid-State Circuits,* vol. SC-17, pp. 983–998, Dec. 1982.

[19] B. Sheu, D. Sharfetter, P.-K. Ko, and M.-C. Jeng, "BSIM: Berkeley short-channel IGFET model for MOS transistors," *IEEE J. Solid-State Circuits,* vol. SC-22, pp. 558–566, Aug. 1987.

[20] M. White, F. Van De Wiele, and J.-P. Lambot, "High-accuracy MOS models for computer-aided design," *IEEE Trans. Electron Devices,* vol. ED-27, pp. 899–906, May 1980.

[21] T. Grotjohn and B. Hoefflinger, "A parametric short-channel MOS transistor model for subthreshold and strong inversion current," *IEEE Trans. Electron Devices,* vol. ED-31, pp. 234–246, Feb. 1984.

[22] P. Kinget and M. Steyaert, "A 1-GHz CMOS up-conversion mixer," *IEEE J. Solid-State Circuits,* vol. 32, pp. 370–376, Mar. 1997.

[23] P. Wambacq, G. Gielen, and W. Sansen, "Symbolic simulation of harmonic distortion in analog integrated circuits with weak nonlinearities," in *Proc. ISCAS,* May 1990, pp. 536–539.

[24] P. Wambacq and W. Sansen, *Distortion Analysis of Analog Integrated Circuits.* Norwell, MA: Kluwer, 1998.

[25] F. V. Fernández, A. Rodríguez-Vázquez, and J. L. Huertas, "Formula approximation for flat and hierarchical symbolic analysis," *Kluwer J. Analog Integrated Circuits Signal Processing,* vol. 3, no. 1, pp. 43–58, Jan. 1993.

[26] J.-S. Yuan, J. Liou, and W. Eisenstadt, "A physics-based current-dependent base resistance model for advanced bipolar transistors," *IEEE Trans. Electron Devices,* vol. 35, pp. 1055–1062, July 1988.

[27] S. Narayanan, "Application of Volterra series to intermodulation distortion analysis of transistor feedback amplifiers," *IEEE Trans. Circuit Theory,* vol. CT-17, pp. 518–527, Nov. 1970.

[28] C. Andrew and L. Nagel, "Early effect modeling in SPICE," *IEEE J. Solid-State Circuits,* vol. SC-31, pp. 136–138, Jan. 1996.

Piet Wambacq was born in Asse, Belgium, in 1963. He received the M.Sc. degree in electrical and mechanical engineering in 1986 from the Katholieke Universiteit Leuven, Belgium. From 1986 to 1996, he worked as a Research Assistant at the ESAT-MICAS Laboratory of the Katholieke Universieit Leuven, where he received the Ph.D. degree in 1996 on symbolic analysis of large and weakly nonlinear analog integrated circuits.

Since 1996, he has been working at IMEC on design methodologies for mixed-signal integrated circuits. His research interests are design and CAD of analog and mixed-signal integrated circuits. He has authored or coauthored more than 30 papers in edited books, international journals, and conference proceedings. He is the author of the book *Distortion Analysis of Analog Integrated Circuits.*

Georges G. E. Gielen received the M.Sc. and Ph.D. degrees in electrical engineering from the Katholieke Universiteit Leuven, Belgium, in 1986 and 1990, respectively.

From 1986 to 1990, he was appointed as a Research Assistant by the Belgian National Fund of Scientific Research for carrying out his Ph.D. research in the ESAT-MICAS Laboratory of the Katholieke Universieit Leuven. In 1990, he was appointed as a Postdoctoral Research Assistant and Visiting Lecturer at the Department of Electrical Engineering and Computer Science of the Unviersity of California, Berkeley. From 1991 to 1993, he was a postdoctoral Research Assistant of the Belgian National Fund of Scientific Research at the ESAT-MICAS Laboratory of the Katholieke Universiteit Leuven. In 1993, he was appointed as a tenure Research Associate of the Belgian National Fund of Scientific Research, and at the same time as an Assistant Professor at the Katholieke Universieit Leuven. In 1995, he was promoted to Associate Professor at the same university. His research interests are in the design of analog and mixed-signal integrated circuits, especially analog and mixed-signal CAD tools and design automation (modeling, simulation, symbolic analysis, analog synthesis, analog layout generation, and analog and mixed-signal testing). He is coordinator or partner of several (industrial) research projects in this area. He has authored or coauthored one book and more than 100 papers in edited books, international journals, and conference proceedings.

Dr. Gielen is a regular member of the program committees of international conferences (ICCAD, ED&TC, etc.), he has served as Associate Editor of the IEEE TRANSACTIONS ON CIRCUITS AND SYSTEMS PART I, and is a member of the editorial board of the *Kluwer International Journal on Analog Integrated Circuits and Signal Processing*. He is a member of ACM.

Peter R. Kinget (S'88–M'90) received the engineering degree in electrical and mechanical engineering and the Ph.D. degree in electrical engineering from the Katholieke Universiteit Leuven, Belgium, in 1990 and 1996, respectively.

From 1991 to 1995, he received a fellowship from the Belgium National Fund for Scientific Research (NFWO) that allowed him to work as a Research Assistant at the ESAT-MICAS Laboratory of the Katholieke Universiteit Leuven. In October 1996, he joined Bell Laboratories, Lucent Technologies, in Murray Hill, NJ, as a Member of the Technical Staff. His research interests are in analog and telecommunications integrated circuits.

Willy Sansen (S'66–M'72–SM'86–F'95) was born in Poperinge, Belgium, on May 16, 1943. He received the masters degree in electrical engineering from the Katholieke Universieit Leuven in 1967 and the Ph.D. degree in electronics from the University of California, Berkeley, in 1972.

In 1968, he was employed as a Research Assistant at K.U. Leuven. In 1971, he was employed as a Teaching Fellow at U.C. Berkeley. In 1972, he was appointed by the National Fund of Scientific Research (Belgium) as a Research Associate at the ESAT Laboratory of the K.U. Leuven, where he has been a Full Professor since 1981. During the period 1984–1990, he was head of the Electrical Engineering Department. In 1978, he spent the winter quarter at Stanford University as a Visiting Assistant Professor. In 1981, he was a Visiting Professor at the Federal Technical University Lausanne, and in 1985 at the University of Pennsylvania, Philadelphia. He has been involved in design automation and in numerous analog integrated circuit designs for telecom, consumer electronics, medical applications, and sensors. He has been supervisor of 25 Ph.D. theses in the same field. He has authored and coauthored five books and 300 papers in international journals and conference proceedings.

Dr. Sansen is a member of several editorial committees of journals such as the IEEE JOURNAL OF SOLID-STATE CIRCUITS, *Sensors and Actuators*, *High Speed Electronics*, etc. He serves regularly on the program committees of conferences such as ISSCC, ESSCIRC, ASICTT, EUROSENSORS, TRANSDUCERS, EDAC, etc.

Canonical Symbolic Analysis of Large Analog Circuits with Determinant Decision Diagrams

C.-J. Richard Shi, *Senior Member, IEEE* and Xiang-Dong Tan, *Member, IEEE*

***Abstract*—Symbolic analysis has many applications in the design of analog circuits. Existing approaches rely on two forms of symbolic-expression representation: expanded sum-of-product form and arbitrarily nested form. Expanded form suffers the problem that the number of product terms grows exponentially with the size of a circuit. Nested form is neither canonical nor amenable to symbolic manipulation. In this paper, we present a new approach to exact and canonical symbolic analysis by exploiting the *sparsity* and *sharing* of product terms. It consists of representing the symbolic determinant of a circuit matrix by a graph—called a determinant decision diagram (DDD)—and performing symbolic analysis by graph manipulations. We show that DDD construction, as well as many symbolic analysis algorithms, takes time almost linear in the number of DDD vertices. We describe an efficient DDD-vertex-ordering heuristic and prove that it is optimum for ladder-structured circuits. For practical analog circuits, the numbers of DDD vertices are several orders of magnitude less than the numbers of product terms. The algorithms have been implemented and compared respectively to symbolic analyzers *ISAAC* and *Maple-V* in generating the expanded sum-of-product expressions, and *SCAPP* in generating the nested sequences of expressions.**

***Index Terms*—Analog symbolic analysis, circuit simulation, determinant decision diagrams (DDD's), symbolic matrix determinant, zero-suppressed binary decision diagram (ZBDD).**

I. Introduction

SYMBOLIC analysis calculates the behavior or characteristic of a circuit in terms of symbolic parameters. In contrast to numerical simulators such as *SPICE* [32] that only provide numerical results, symbolic simulators can explicitly express which circuit parameters determine the circuit behavior. They can offer more advantages than numerical simulators in many applications such as optimum topology selection, design space exploration, behavioral model generation, and fault detection; these have been summarized and illustrated by a recent survey paper of Gielen *et al.* [22].

Despite its advantages, symbolic analysis has not been widely used by analog designers and was once judged as completely inefficient [28]. The root of the difficulty is apparent: the number of product terms in a symbolic expression may increase exponentially with the size of a circuit. For example, for a BiCMOS amplifier that has about 15 nodes and 25 devices (transistors, diodes, resistors, and capacitors), the determinant of the circuit matrix contains more than 10^{11} product terms [47]. Any manipulation and evaluation of sum-of-product-based symbolic expressions will require CPU time, at best, linear in the number of terms and, therefore, have both time and space complexities exponential in the size of a circuit.

To cope with large analog circuits, modern symbolic analyzers[1] rely on two techniques—hierarchical decomposition and symbolic simplification. Hierarchical decomposition generates symbolic expressions in the nested instead of expanded form [23], [24], [40]. Symbolic simplification discards those insignificant terms based on the relative magnitudes of symbolic parameters and the frequency defined at some nominal design points or over some ranges. It can be performed before/during the generation of symbolic terms [25], [37], [45], [51] or after the generation [17], [21], [47]. Exploitation of these techniques has enabled the use of symbolic simulators in several university research projects [9], [19], [21], [50]; however, both techniques have some major deficiencies. Symbolic manipulation (other than numerical evaluation) of a nested expression usually requires complicated and time-consuming procedures; e.g., sensitivity calculation in [27] and lazy approximation in [37]. On the other hand, simplified expressions only have a sufficient accuracy at some points or frequency ranges. Even worse, simplification often loses certain information, such as sensitivity with respect to parasitics, which is crucial for circuit optimization and testability analysis.

In this paper, we present a new approach to exact symbolic analysis, which is capable of analyzing analog integrated circuits substantially larger than those previously handled. Our approach is based on two observations concerning symbolic analysis of large analog circuits: 1) the circuit matrix is sparse and 2) a symbolic expression often shares many subexpressions. Under the assumption that all the matrix elements are distinct, each product term can be viewed as a subset of all the symbolic parameters. Therefore, we adapt a special data structure called ZBDD's[2] introduced originally for representing sparse subset systems [29]. This leads to a new graph representation of symbolic determinants, called DDD's. This representation has several advantages over both the expanded and arbitrarily nested

Manuscript received June 4, 1997; revised April 10, 1999. This work was supported in part by the U.S. Defense Advanced Research Projects Agency (DARPA) under Grant F33615-96-1-5601, in part by United States Air Force, Wright Laboratory, Manufacturing Technology Directorate, and in part by Conexant Systems. A preliminary version of this paper was presented at the IEEE/ACM Int. Conf. Computer-Aided Design, San Jose, CA, November 9–13, 1997 [38]. This paper was recommended by Associate Editor M. Fujita.

C.-J. R. Shi is with the Department of Electrical Engineering, University of Washington, Seattle, WA 98195 USA (e-mail: cjshi@ee.washington.edu).

X.-D. Tan is with the Department of Electrical Engineering, University of Washington, Seattle, WA 98195 USA.

Publisher Item Identifier S 0278-0070(00)01381-6.

[1]Some are surveyed in a paper by F. V. Fernández and A. Rodríguez-Vázquez [19].

[2]We remark that ZBDD's is a variant of binary decision diagrams (BDD's) introduced by Akers [1] and popularized by Bryant [4]. Our work is inspired by the success of BDD's as an enabling technology for industrial use of symbolic analysis and formal verification in digital logic design [5].

forms of a symbolic expression. First, similar to the nested form, our representation is compact for a large class of analog circuits. A ladder-structured network can be represented by a diagram with the number of vertices (called its *size*) equal to the number of symbolic parameters. As indicated by our experiments, the size of a DDD is usually dramatically smaller than the number of product terms. For example, 5.71×10^{20} terms can be represented by a diagram with 398 vertices. Second, similar to the expanded form, our representation is canonical; i.e., every determinant has a **unique** DDD representation. The representation canonicity facilitates efficient symbolic analysis and may provide a potential tool to formally verify analog circuits. Finally, derivation, manipulation and evaluation of the DDD representations of symbolic determinants have time complexity proportional to the DDD sizes.

This paper is organized as follows: Section II introduces the background and basic notation for the rest of the paper. Section III presents the notion of determinant decision diagrams as an application of ZBDD's to represent symbolic determinants. Section IV describes an effective heuristic for ordering DDD vertices so that the resulting DDD has a smallest or near-smallest size. DDD-based algorithms for symbolic analysis and applications are described in Section V. Experimental results are presented in Section VI. The proposed approach is compared to some related work in Section VII. Section VIII concludes the paper.

II. Notations and Problem Statement

In this section, we introduce some basic notation and concepts that will be used in the rest of the paper. Since these come from several different research areas, an attempt has been made to choose a self-consistent set of notations.

A. Subset Systems and ZBDD's

Let V be a *set* of elements. The number of elements in V is called the *cardinality* of V, denoted by $|V|$. The set of all *subsets* of V is called the *power set* of V, denoted by 2^V. A subset X of the power set, written as $X \subseteq 2^V$, is called a *subset system* of V.

A subset system X of V can be decomposed with respect to an element v in V into two unique subset systems, X_v and $X_{\overline{v}}$, where X_v is the set of subsets of V belonging to X that contain v, from which v has been removed, and $X_{\overline{v}}$ is the set of subsets of V belonging to X that do not contain v. For instance, let $X = \{\{v_1, v_2\}, \{v_1, v_3, v_5\}, \{v_3, v_4, v_5\}\}$. Then we have $X_{v_1} = \{\{v_2\}, \{v_3, v_5\}\}$, and $X_{\overline{v}_1} = \{\{v_3, v_4, v_5\}\}$. This decomposition can be represented graphically by a decision *vertex*. It is labeled by v_1, and represents the subset system X. As illustrated in Fig. 1(a), the vertex has two outgoing edges: one points to X_{v_1} (called *1-edge*), and the other to $X_{\overline{v}_1}$ (called *0-edge*). We say that the edges are *originated* from the vertex. If X is recursively decomposed with respect to all the elements of V, one obtains a binary decision tree whose leaves are $\{\{\}\}$ and $\{\}$, respectively. For convenience, we denote leaf $\{\{\}\}$ by the 1-terminal, and $\{\}$ by the 0-terminal.

When a subset system X is decomposed with respect to an element that does not appear in X, then its 1-edge points to

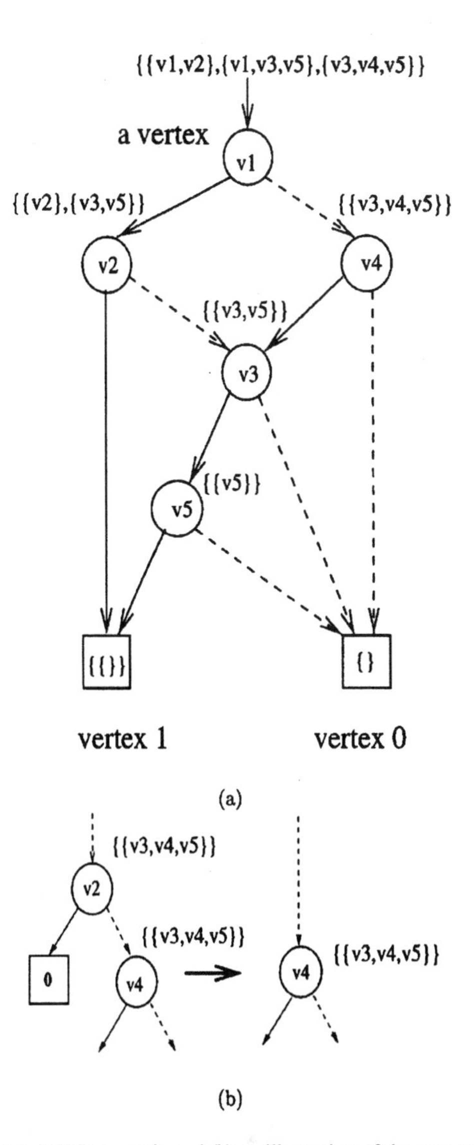

Fig. 1. (a) A ZBDD example and (b) an illustration of the zero-suppression rule.

the 0-terminal. This is illustrated in Fig. 1(b) for decomposing $X = \{\{v_3, v_4, v_5\}\}$ with respect to v_2. To make the diagram compact, Minato suggested the following *zero-suppression* rule for representing sets of *sparse* subsets: eliminate all the vertices whose 1-edges point to the 0-terminal vertex and use the subgraphs of the 0-edges, as shown in Fig. 1(b) [29]. A ZBDD is such a zero-suppressed graph with the following two rules due to Bryant [4]: *ordered*—all elements of V, if one appears, will appear in a fixed order in all the paths of the graph; and *shared*—all equivalent subgraphs are shared. ZBDD's are a canonical representation of subset systems; i.e., every subset system has a unique ZBDD representation under a given vertex ordering. For example, Fig. 1(a) is a unique ZBDD representation for the subset system $\{\{v_1, v_2\}, \{v_1, v_3, v_5\}, \{v_3, v_4, v_5\}\}$ with respect to ordering v_1, v_2, v_4, v_3 and v_5. For convenience, every nonterminal vertex is indexed by an integer number greater than those of its descendant vertices [29]. The process of assigning indexes to the nonterminal vertices is called *vertex ordering*.

A path from a nonterminal vertex to the 1-terminal is called *1-path*. It defines a subset of V. The subset consists of all the elements of V from which the 1-edges in the 1-path originate. A

1-path from the root is a *rooted* 1-path. The number of vertices in a ZBDD is called its *size*.

B. Matrix, Determinant, and Cofactors

Let $e = \{1, \cdots, n\}$ be a set of integers. Let A denote a set of m elements, called *symbolic parameters* or simply *symbols*, $\{a_1, \cdots, a_m\}$, where $1 \le m \le n^2$ and each symbol is labeled by a unique pair (r, c), where $r \in e$ and $c \in e$. Often, we write A as an $n \times n$ (square) *matrix*, denoted by $\boldsymbol{A}$, and use $a_{r,c}$ to denote the element of matrix $\boldsymbol{A}$ at row r and column c. We sometimes use $r(a)$ and $c(a)$ to denote, respectively, the row and column indexes of element a

$$\boldsymbol{A} = \begin{pmatrix} a_{1,1} & a_{1,2} & \cdots & a_{1,n} \\ a_{2,1} & a_{2,2} & \cdots & a_{2,n} \\ \cdots & \cdots & \cdots & \cdots \\ a_{n,1} & a_{n,2} & \cdots & a_{n,n} \end{pmatrix}.$$

If $m = n^2$ the matrix is said to be *full*. If $m \ll n^2$ the matrix is said to be *sparse*. The *determinant* of $\boldsymbol{A}$, denoted by $\det(\boldsymbol{A})$, is defined by

$$\begin{vmatrix} a_{1,1} & a_{1,2} & \cdots & a_{1,n} \\ a_{2,1} & a_{2,2} & \cdots & a_{2,n} \\ \cdots & \cdots & \cdots & \cdots \\ a_{n,1} & a_{n,2} & \cdots & a_{n,n} \end{vmatrix} = \sum_{j_1 \ne j_2 \ne \cdots \ne j_n} (-1)^p \cdot a_{1,j_1} \cdot a_{2,j_2} \cdots a_{n,j_n}. \quad (1)$$

Here

$(j_1, j_2, \cdots, j_n)$ permutation of e;

p number of permutations needed to make the sequence $(j_1, j_2, \cdots, j_n)$ monotonically increasing.

The right-hand side of (1) is a symbolic expression of $\det(\boldsymbol{A})$ in the *expanded* form, more precisely, the *sum-of-product* form, where each *term* is an algebraic product of n symbolic parameters. We note that each symbol can be assigned a real or complex value for analog circuit simulation.

Let p, $p \subseteq e$, and q, $q \subseteq e$ such that $|p| = |q|$. The square matrix obtained from the matrix $\boldsymbol{A}$ by deleting those rows not in p and columns not in q forms a *submatrix* of $\boldsymbol{A}$, and is represented by $\boldsymbol{A}(p, q)$. It has dimension $|p|$ by $|q|$.

Let $a_{r,c}$ be the element of $\boldsymbol{A}$ at row r and column c. Let $\boldsymbol{A}_{a_{r,c}}$ be the $(n-1) \times (n-1)$-matrix obtained from the matrix $\boldsymbol{A}$ by deleting row r and column c, and let $\boldsymbol{A}_{\overline{a}_{r,c}}$ be the $n \times n$-matrix obtained from $\boldsymbol{A}$ by setting $a_{r,c} = 0$. Then, the determinant of matrix $\boldsymbol{A}$ can be *expanded* in a way similar to Shannon expansion for Boolean functions [8]

$$\det(\boldsymbol{A}) = a_{r,c}(-1)^{r+c}\det(\boldsymbol{A}_{a_{r,c}}) + \det(\boldsymbol{A}_{\overline{a}_{r,c}}) \quad (2)$$

where

$(-1)^{r+c}\det(\boldsymbol{A}_{a_{r,c}})$ referred to as the *cofactor* of $\det(\boldsymbol{A})$ with respect to $a_{r,c}$;

$\det(\boldsymbol{A}_{\overline{a}_{r,c}})$ *remainder* of $\det(\boldsymbol{A})$ with respect to $a_{r,c}$;

determinant $\det(\boldsymbol{A}_{a_{r,c}})$ *minor* of $\det(\boldsymbol{A})$ with respect to $a_{r,c}$.

We note that the following two special cases of the expansion above are well known as *Laplace expansions* along row r and column c, respectively

$$\det(\boldsymbol{A}) = \sum_{r=1}^{n} a_{r,c}(-1)^{r+c}\det(\boldsymbol{A}_{a_{r,c}}) \quad (3)$$

$$\det(\boldsymbol{A}) = \sum_{c=1}^{n} a_{r,c}(-1)^{r+c}\det(\boldsymbol{A}_{a_{r,c}}). \quad (4)$$

C. Symbolic Analysis Problem of Analog Circuits

Consider a linear(ized) time-invariant analog circuit. Its system of equations can be formulated by, for example, the modified nodal analysis (MNA) approach in the following general form [44]:

$$\boldsymbol{Tx} = \boldsymbol{w}. \quad (5)$$

The *circuit unknown vector* $\boldsymbol{x} \in \mathcal{R}^n$ may be composed of node voltages and branch currents, and the *circuit matrix* $\boldsymbol{T}()$: $\mathcal{R}^n \rightarrow R^{n \times n}$ is a large **sparse** symbolic matrix, typically with just a few nonzero entries per row/column.

Symbolic analysis of analog circuits can be stated as the problem of solving the systems of symbolic equation (5), i.e., deriving the closed-form expression of a circuit unknown in terms of symbolic parameters in $\boldsymbol{T}$ and symbolic excitations expressed by $\boldsymbol{w}$. According to Cramer's rule, the kth component x_k of the unknown vector $\boldsymbol{x}$ is obtained as follows:

$$x_k = \frac{\sum_{i=1}^{n} w_i(-1)^{i+k}\det(\boldsymbol{T}_{t_{i,k}})}{\det(\boldsymbol{T})}. \quad (6)$$

Most symbolic simulators are targeted at finding various network functions, each function being defined as the ratio of an output unknown from $\boldsymbol{x}$ to an input from $\boldsymbol{w}$. These are special cases of (6) or the ratios of the two expressions in the form of (6).

Note that $(-1)^{i+k}\det(\boldsymbol{T}_{t_{i,k}})$ in (6) is the cofactor of $\det(\boldsymbol{T})$ with respect to element $t_{i,k}$ of matrix $\boldsymbol{T}$ at row i and column k. Therefore, the central issue in determinant-based symbolic analysis is how to find symbolic expressions of $\det(\boldsymbol{T})$ and the cofactors of $\det(\boldsymbol{T})$. In the rest of the paper, we focus on how to represent a symbolic determinant (Sections III and IV), and how to compute, manipulate, and evaluate a symbolic determinant (Section V). For simplicity, we assume in the paper that all the entries in $\boldsymbol{T}$ are distinct. This assumption has been used in previous symbolic analysis approaches; methods have been proposed to formulate the symbolic equations to meet (or closely meet) this assumption [21], [44].

III. ZBDD Representation of Symbolic Matrix Determinant

In this section, we apply the notation of ZBDD's to represent a symbolic matrix determinant. This leads to a new interpreted graph called DDD's. DDD's are a canonical representation for matrix determinants, similar to BDD's for representing *binary functions* and ZBDD's for representing *subset systems*.

A key observation is that the circuit matrix is sparse, and many times, a symbolic expression may share many subexpressions. For example, consider the following determinant:

$$\det(\boldsymbol{M}) = \begin{vmatrix} a & b & 0 & 0 \\ c & d & e & 0 \\ 0 & f & g & h \\ 0 & 0 & i & j \end{vmatrix} = adgj - adhi - aefj - bcgj + cbih. \tag{7}$$

We note that subterms ad, gj, and hi appear in several product terms, and each product term involves a subset (four) out of ten symbolic parameters. Therefore, we view each symbolic product term as a subset, and use a ZBDD to represent the subset system composed of all the subsets each corresponding to a product term. Fig. 2 illustrates the corresponding ZBDD representing all the subsets involved in $\det(\boldsymbol{M})$ under ordering $a > c > b > d > f > e > g > i > h > j$. It can be seen that subterms ad, gj, and ih have been shared in the ZBDD representation.

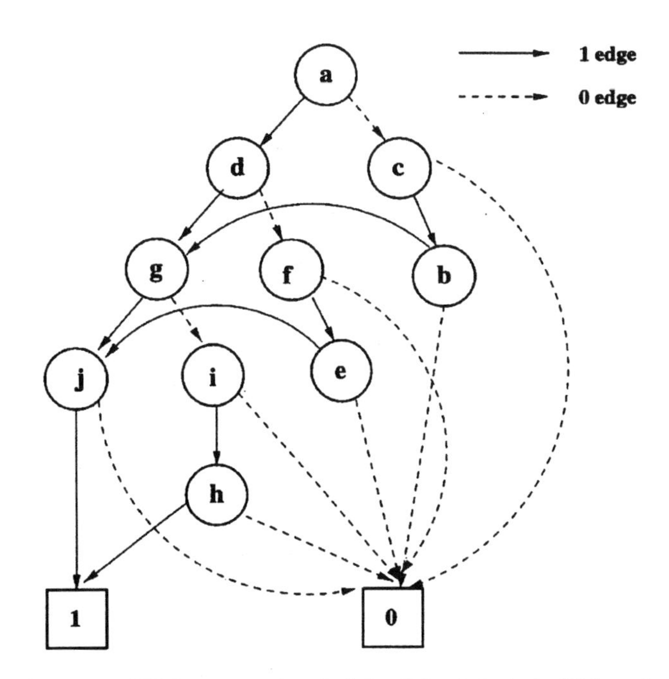

Fig. 2. A ZBDD representing $\{adgj, adhi, afej, cbgj, cbih\}$ under ordering $a > c > b > d > f > e > g > i > h > j$.

Following directly from the properties of ZBDD's, we have the following observations. First, given a fixed order of symbolic parameters, all the subsets in a symbolic determinant can be represented uniquely by a ZBDD. Second, every rooted 1-path in the ZBDD corresponds to a product term, and the number of 1-edges in any rooted 1-path is n. The total number of rooted 1-paths is equal to the number of product terms in a symbolic determinant.

We can view the resulting ZBDD as a graphical representation of recursive application of determinant expansion formula (2) with the expansion order a, c, b, d, f, e, g, i, h, j. Each vertex is labeled with a matrix entry, and represents all the subsets contained in the corresponding submatrix determinant. The 1-edge points to the vertex representing all the subsets contained in the cofactor of the current expansion, and 0-edge points to the vertex representing all the subsets contained in the remainder.

To embed the signs of the product terms of a symbolic determinant into its corresponding ZBDD, we consider one step of matrix expansion with respect to $a_{r,c}$ as defined by (2). The sign is $(-1)^{r+c}$. Note that r and c are, respectively, the row and column indexes of the element $a_{r,c}$ in the submatrix before this step of expansion, say $\boldsymbol{A}'$. Let the *absolute* row and column indexes of the element $a_{r,c}$ in the original matrix $\boldsymbol{A}$ before any expansion be $r(a_{r,c})$ and $c(a_{r,c})$, respectively. Then, we observe that $(-1)^{r+c} = (-1)^{r+c-2}$ and $r + c - 2$ is equal to the number of rows in $\boldsymbol{A}'$ with absolute indexes less than $r(a_{r,c})$ plus the number of columns in $\boldsymbol{A}'$ with absolute indexes less than $c(a_{r,c})$. We also note that all the rows and columns in $\boldsymbol{A}'$ except that of $a_{r,c}$ are represented in the subgraph rooted at the vertex pointed to by the 1-edge of vertex $a_{r,c}$. Therefore, the sign of a nonterminal vertex v, denoted by $s(v)$, can be defined recursively as follows.

1) Let $P(v)$ be the set of ZBDD vertices that originate the 1-edges in any 1-path rooted at v. Then

$$s(v) = \prod_{x \in P(v)} \operatorname{sign}(r(x) - r(v)) \operatorname{sign}(c(x) - c(v)) \tag{8}$$

where $r(x)$ and $c(x)$ refer to the absolute row and column indexes of vertex x in the original matrix, and u is an integer so that

$$\operatorname{sign}(u) = \begin{cases} 1, & \text{if } u > 0, \\ -1, & \text{if } u < 0. \end{cases}$$

2) If v has an edge pointing to the 1-terminal vertex, then $s(v) = +1$.

This is called the *sign rule*. For example, in Fig. 3, shown aside by each vertex are the row and column indexes of that vertex in the original matrix, as well as the sign of that vertex obtained by using the sign rule above. We note that all the paths rooted at the same vertex yield the same vertex sign.

It can be verified that the product of all the signs in a rooted 1–path is exactly the sign of the corresponding product term. For example, consider the 1-path *acbgih* in Fig. 3. The vertices that originate all the 1-edges are c, b, i, h, their corresponding signs are –, +, –, and +, respectively. Their product is +. This is the sign of the symbolic product term *cbih*.

With ZBDD's and the sign rule as two foundations, we are now ready to introduce formally our representation of a symbolic determinant. Let $\boldsymbol{A}$ be an $n \times n$ sparse matrix with a set of distinct m symbolic parameters $\{a_1, \cdots, a_m\}$, where $1 \le m \le n^2$. Each symbolic parameter a_i is associated with a unique pair $r(a_i)$ and $c(a_i)$, which denote, respectively, the row index and column index of a_i. A DDD is a signed, rooted, directed, acyclic graph with two terminal vertices, namely the 0-terminal vertex and the 1-terminal vertex. Each nonterminal vertex is labeled with a symbolic parameter a_i and the sign $s(a_i)$ determined by the sign rule defined by (8). It has two outgoing edges, called 1-edge and 0-edge. A vertex labeled by a_i represents a matrix determinant D defined recursively as follows.

1) If the vertex is the 1-terminal vertex, then $D = 1$.
2) If the vertex is the 0-terminal vertex, then $D = 0$.

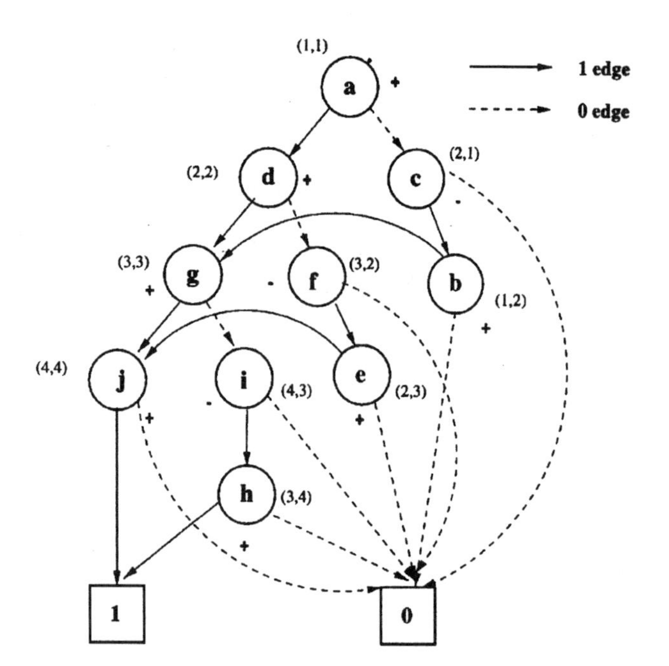

Fig. 3. A signed ZBDD for representing symbolic terms.

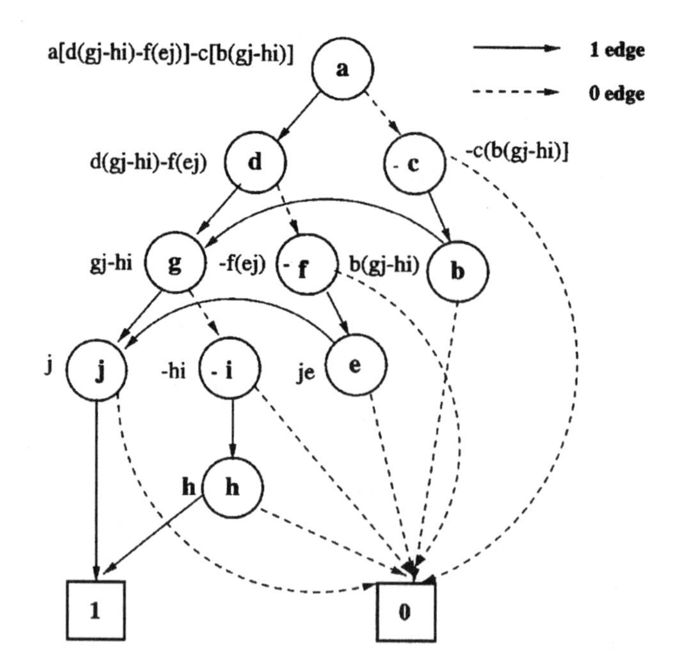

Fig. 4. A determinant decision diagram for matrix $\boldsymbol{M}$.

3) If the vertex is a nonterminal vertex, then $D = a_i s(a_i) D_{a_i} + D_{\overline{a}_i}$, where D_{a_i} ($D_{\overline{a}_i}$) is the matrix determinant represented by the vertex that was pointed to by the 1-edge (0-edge) of the vertex labeled by a_i.

Note that $s(a_i)D_{a_i}$ is the *cofactor* of D with respect to a_i, D_{a_i} is the *minor* of D with respect to a_i, $D_{\overline{a}_i}$ is the *remainder* of D with respect to a_i, and operations are algebraic multiplications and additions. For example, Fig. 4 shows the DDD representation of $\det(\boldsymbol{M})$ under ordering $a > c > b > d > f > e > g > i > h > j$.

To enforce the uniqueness and compactness of the DDD representation, the three rules of ZBDD's, namely, zero-suppression, ordered, and shared, described in Section II-A are adopted. This leads to DDD's having the following properties.

- Every 1-path from the root corresponds to a product term in the fully expanded symbolic expression. It contains exactly n 1-edges. The number of 1-paths from the root is equal to the number of product terms.
- For any determinant D, there is a unique DDD representation under a given vertex ordering.

We use |DDD| to denote the *size of* a DDD, i.e., the number of vertices in the DDD.

IV. An Effective Vertex-Ordering Heuristic

A key problem in many decision diagram applications is how to select a vertex ordering, since the size of the resulting decision diagram strongly depends on the chosen ordering. For example, if we choose vertex order $a > c > d > f > g > h > b > e > i > j$ for $\det(\boldsymbol{M})$ in Section III, then the resulting DDD is shown in Fig. 5. It has 13 vertices, in comparison to ten in Fig. 4, although they represent the same determinant. In this section, we describe an efficient heuristic for selecting a good vertex ordering, and show that it is optimal for a class of circuit matrices.

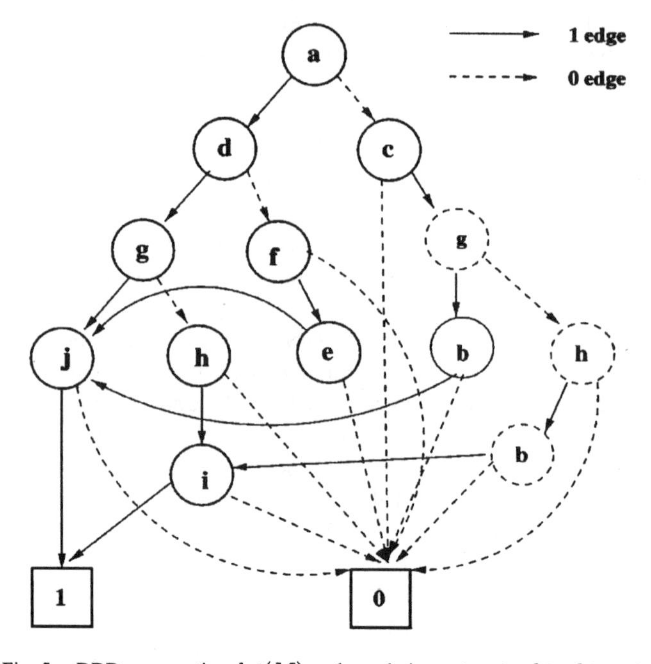

Fig. 5. DDD representing $\det(\boldsymbol{M})$ under ordering $a > c > d > f > g > h > b > e > i > j$.

We propose to select the vertex ordering for a DDD by examining the structure of the original matrix. Suppose that $\boldsymbol{A}$ is an $n \times n$ matrix with m nonzero elements (entries, or symbols). The vertex-ordering problem is how to *label* all the nonzero elements in $\boldsymbol{A}$ using integers one to m so that the resulting DDD constructed with the chosen order has a small size. As stated in Section II-A, those elements labeled by smaller integers will appear close to the leaves, or the bottom of the DDD, and the root is labeled by index m.

We propose a vertex-ordering heuristic, based on the refinement of a well-known strategy for Laplace expansion of a sparse matrix [21]. The basic idea is to label with larger indexes those columns or rows containing fewer nonzero

```
MATRIX_GREEDY_LABELING(A(p, q))
1   select column j (row i) that has least nonzero elements
2   s ← all the rows i (columns j) so that a_{i,j} ≠ 0
3   for all row i (column j) in s from the one with least nonzero elements
4       if there exist unlabeled elements in A(p − {i}, q − {j})
5           MATRIX_GREEDY_LABELING(A(p − {i}, q − {j})
6   for all row i (column j) in s from the one with most nonzero elements
7       if a_{i,j} has not been labeled
8           label a_{i,j} by k and then increment k
```

Fig. 6. A DDD vertex-ordering heuristic.

elements. Elements in those dense rows and columns will be labeled using small indexes. Intuitively, this strategy increases the possibility of DDD subgraph sharing. Fig. 6 describes the proposed heuristic MATRIX_GREEDY_LABELING ($\boldsymbol{A}(p, q)$) for labeling all the elements in matrix $A(p, q)$. In the algorithm, $A(p - \{i\}, q - \{j\})$ denotes the matrix obtained from $A(p, q)$ by removing row i and column j. We keep a global counter k. Initially k is set to 1, and $p = q = e$.

As an example, consider how to apply MATRIX_GREEDY_LABELING to label the matrix $\boldsymbol{M}$ defined in Section III. The complete process is illustrated in Fig. 7. First, columns 1 and 4 of $\boldsymbol{M}$, as well as rows 1 and 4, have the least number of nonzero elements (2). We arbitrarily select column 1. Then the set of rows that have a nonzero element at column 1 are one and two, i.e., $s = \{1, 2\}$. Since row 1 has one nonzero element, and row 2 has two nonzero elements, lines 3–5 invoke first MATRIX_GREEDY_LABELING on matrix $\boldsymbol{M}$ after removing row 1 and column 1, then on matrix $\boldsymbol{M}$ after removing row 2 and column 1. The process is applied recursively on the resulting submatrices, and is illustrated in Fig. 7 from the top to the bottom. Then, elements are labeled in the Fig. 7 from the bottom to the top in the reverse order of expansion. These labels are marked in Fig. 7 at the top-right corner of each element. If we summarize all the labels assigned to the matrix elements using the original matrix structure, we have

$$\begin{pmatrix} a & b & 0 & 0 \\ c & d & e & 0 \\ 0 & f & g & h \\ 0 & 0 & i & j \end{pmatrix} \rightarrow \begin{pmatrix} 10 & 8 & 0 & 0 \\ 9 & 7 & 5 & 0 \\ 0 & 6 & 4 & 2 \\ 0 & 0 & 3 & 1 \end{pmatrix}.$$

The heuristic leads to compact DDD's for a large class of circuit matrices, as observed in our experiments described in Section VI. In the rest of this section, we show that the heuristic is optimal for a class of circuit matrices, called *tridiagonal matrices*. Tridiagonal matrices are matrices that have only nonzero elements at positions (i, i), $(i-1, i)$, and $(i+1, i)$. They have the structure as shown in (8a), found at the bottom of the page.

$$\begin{pmatrix} a & b & 0 & 0 \\ c & d & e & 0 \\ 0 & f & g & h \\ 0 & 0 & i & j \end{pmatrix}$$

$$a^{(10)}\begin{pmatrix} d & e & 0 \\ f & g & h \\ 0 & i & j \end{pmatrix} \qquad c^{(9)}\begin{pmatrix} b & 0 & 0 \\ f & g & h \\ 0 & i & j \end{pmatrix}$$

$$d^{(7)}\begin{pmatrix} g & h \\ i & j \end{pmatrix} \quad f^{(6)}\begin{pmatrix} e & 0 \\ i & j \end{pmatrix} \quad b^{(8)}\begin{pmatrix} g & h \\ i & j \end{pmatrix}$$

$$g^{(4)}(j^{(1)}) \quad i^{(3)}(h^{(2)}) \qquad e^{(5)}(j^{(1)})$$

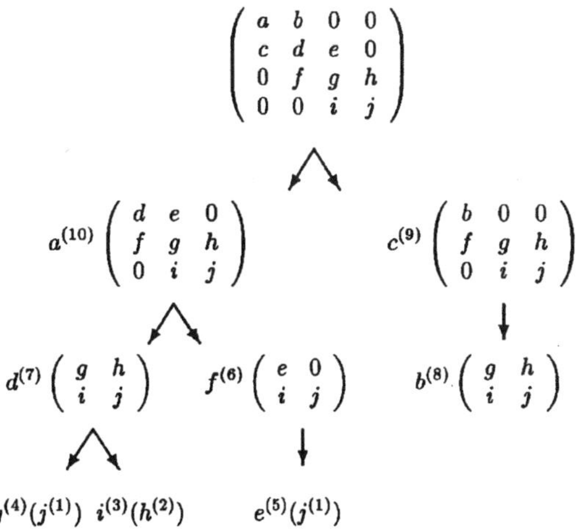

Fig. 7. An illustration of DDD vertex-ordering heuristic.

The number of nonzero elements in $\boldsymbol{A}_{n+1,n+1}$ is $3n+1$. Matrix $\boldsymbol{M}$ used throughout this paper is a 4×4 tridiagonal matrix.

To show that the algorithm MATRIX_GREEDY_LABELING yields an optimum ordering, we note that the lower bound on the number of DDD vertices is equal to the number of matrix entries, i.e., each vertex appears only once in the final DDD. We will show that, for tridiagonal matrices, the ordering given by MATRIX_GREEDY_LABELING results in a DDD with the number of vertices equal to the number of nonzero matrix elements. This is proved by induction. It is easy to see that the result is true for 1×1 and 2×2 matrices. Now we assume that it is true for $\boldsymbol{A}_{n,n}$, i.e., the number of DDD vertices is $3n - 2$. We prove that it is also true for $\boldsymbol{A}_{n+1,n+1}$: the number of DDD vertices is $3(n + 1) - 2$. Let vertex $D_{n-1,n-1}$ represent the DDD of $\det(\boldsymbol{A}_{n-1,n-1})$ and $D_{n,n}$ represent the DDD of $\det(\boldsymbol{A}_{n,n})$. Matrix $\boldsymbol{A}_{n+1,n+1}$ has an extra row and column with three nonzero elements $a_{n+1,n+1}$, $a_{n,n+1}$, and $a_{n+1,n}$. Algorithm MATRIX_GREEDY_LABELING assigns integer labels $3n + 1$, $3n$ and $3n - 1$ to elements $a_{n+1,n+1}$, $a_{n,n+1}$, and $a_{n+1,n}$, respectively. This gives ordering $a_{n+1,n+1} > a_{n,n+1} > a_{n+1,n} > a_{n,n}$. We, thus, first create a DDD vertex labeled by $a_{n+1,n+1}$. Its 1-edge points to vertex $D_{n,n}$, and its 0-edge points to the vertex that corresponds to $a_{n,n+1}$. The 0-edge of vertex $a_{n,n+1}$ points to the 0-terminal. Its 1-edge points to the vertex that represents the determinant of matrix $\boldsymbol{A}_{n+1,n+1}$ after removing the first column and the second row. Since the first row in the resulting matrix contains only one nonzero element $a_{n+1,n}$, we can create a DDD vertex for $a_{n+1,n}$ with its 1-edge pointing to

$$\boldsymbol{A}_{n+1,n+1} = \begin{pmatrix} a_{n+1,n+1} & a_{n+1,n} & 0 & \cdots & \cdots & \cdots & 0 \\ a_{n,n+1} & a_{n,n} & a_{n,n-1} & 0 & \cdots & \cdots & 0 \\ 0 & a_{n-1,n} & a_{n-1,n-1} & a_{n-1,n-2} & 0 & \cdots & 0 \\ 0 & 0 & \cdots & \cdots & \cdots & \cdots & \cdots \\ 0 & \cdots & 0 & a_{3,4} & a_{3,3} & a_{3,2} & 0 \\ 0 & \cdots & \cdots & 0 & a_{2,3} & a_{2,2} & a_{2,1} \\ 0 & \cdots & \cdots & \cdots & 0 & a_{1,2} & a_{1,1} \end{pmatrix} \tag{8a}$$

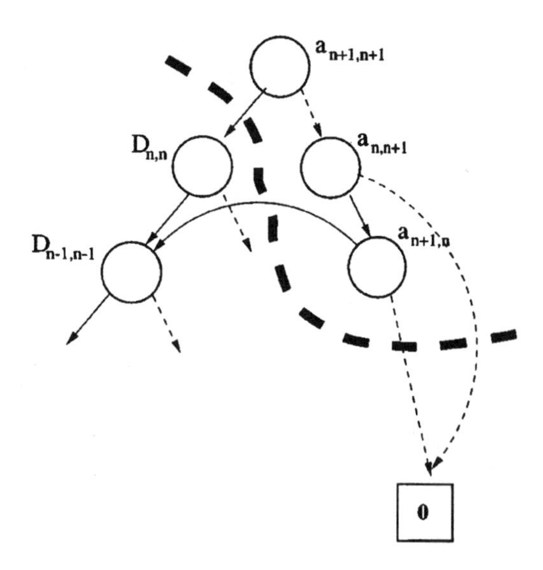

Fig. 8. An illustration of DDD construction for tridiagonal matrices.

$D_{n-1,\,n-1}$, and its 0-edge pointing to the 0-terminal. This is illustrated in Fig. 8. Only three vertices are added for representing $\boldsymbol{A}_{n+1,\,n+1}$. The total number of DDD vertices for $\boldsymbol{A}_{n+1,\,n+1}$ is, thus, $3n-2+3=3(n+1)-2$. Our conjecture is, therefore, proved.

We have proved that for tridiagonal matrix $\boldsymbol{A}_{n,\,n}$, the number of DDD vertices is $3n-2$. As a comparison, the number of expanded product terms in $\det(\boldsymbol{A}_{n,\,n})$ is $F(n+1)$, where $F(i)$ is the ith Fibonacci number defined by

$$F(n)=F(n-1)+F(n-2) \quad n>2,$$
$$F(2)=F(1)=1.$$

Each product term involves n symbols. To store all the product terms without considering sharing, the memory requirement is proportional to $nF(n+1)$. Further, any symbolic manipulation using the expended form would have time complexity at best proportional to $nF(n+1)$. In Fig. 9, the number of DDD vertices is plotted against the number of product terms.

The practical relevance of tridiagonal matrices is that they correspond to the circuit matrices for ladder networks—an important class of circuit structures in analog design. A three-section ladder circuit is shown in Fig. 10. The system of equations can be formulated as shown in (9) at the bottom of the page. If we view each entry as a distinct symbolic parameter, the resulting circuit matrix is a tridiagonal matrix. We note that many practical circuits have the structure of ladders or close to ladders.

We emphasize that just like the BDD representation for Boolean functions, in the worst case, the number of DDD vertices can grow exponentially with the size of a circuit. Nevertheless, as we have observed (in Section VI) that with the proposed vertex-ordering heuristic, the numbers of vertices in the resulting DDD's are reasonable for practical analog circuits.

V. Manipulation and Construction of Determinant Decision Diagrams

In this section, we show that, using determinant decision diagrams, algorithms needed for symbolic analysis and its applications can be performed with the time complexity proportional to the size of the diagrams being manipulated, **not the number of rooted 1-paths in the diagrams, i.e., product terms in the symbolic expressions**. Hence, as long as the determinants of interest can be represented by reasonably small graphs, our algorithms are quite efficient.

A basic set of operations on matrix determinants is summarized in Table I. Most operations are simple extensions of subset operations introduced by Minato on ZBDD's [29]. These few basic operations can be used directly and/or combined to perform a wide variety of operations needed for symbolic analysis. In this section, we first describe these operations, and then use an example to illustrate the main ideas of these operations and how they can be applied to compute network function sensitivities—a key operation needed in optimization and testability analysis. We also show that the generation of significant product terms can be casted as the k-shortest path problem in a DDD and solved elegantly in time $O(k \cdot |\mathrm{DDD}|)$.

A. Implementation of Basic Operations

We summarize the implementation of these operations in Fig. 11. For the clarity of the description, the steps for computing the signs associated with DDD vertices, using the sign rule defined in Section III, are not shown.

As the basis of implementation, we employ two techniques originally developed by Brace, Rudell and Bryant for implementing decision diagrams efficiently [7]. First, a basic procedure GETVERTEX(top, D_1, D_0) is to generate (or copy) a vertex for a symbol top and two subgraphs D_1 and D_0. In the procedure, a hash table is used to keep each vertex unique; vertex elimination and sharing are managed mainly by GETVERTEX. With GETVERTEX, all the major operations for DDD's in Table I are described in Fig. 11.

$$\begin{pmatrix} \frac{1}{R_1} & -\frac{1}{R_1} & 0 & 0 \\ -\frac{1}{R_1} & \frac{1}{R_1}+\frac{1}{R_2}+\frac{1}{R_3} & -\frac{1}{R_3} & 0 \\ 0 & -\frac{1}{R_3} & \frac{1}{R_3}+\frac{1}{R_4}+\frac{1}{R_5} & -\frac{1}{R_5} \\ 0 & 0 & -\frac{1}{R_5} & \frac{1}{R_5}+\frac{1}{R_6} \end{pmatrix} \begin{pmatrix} v_1 \\ v_2 \\ v_3 \\ v_4 \end{pmatrix} = \begin{pmatrix} I_{in} \\ 0 \\ 0 \\ 0 \end{pmatrix} \tag{9}$$

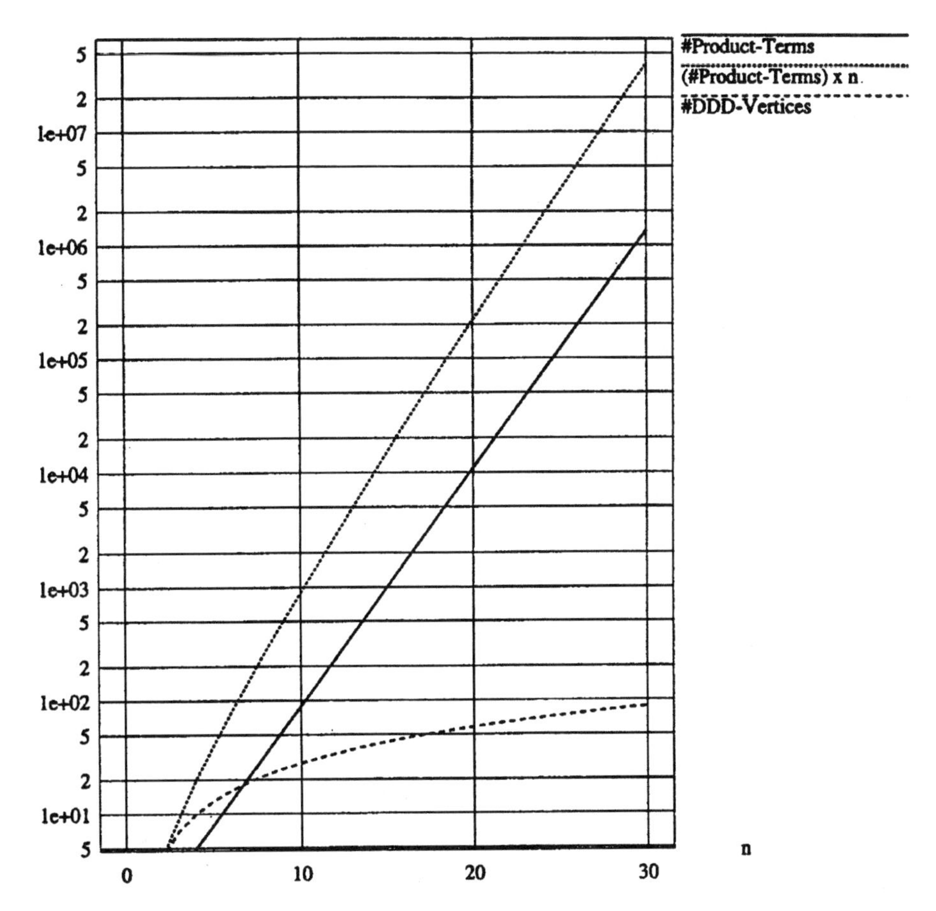

Fig. 9. A comparison of DDD sizes versus numbers of product terms for tridiagonal matrices.

TABLE I
SUMMARY OF BASIC OPERATIONS

Determinant operation	Result	Subset operation
VERTEXONE()	return 1	Base()
VERTEXZERO()	return 0	Empty()
COFACTOR(D, s)	return the cofactor of D wrt s	Subset1(D, s)
REMAINDER(D, s)	return the remainder of D wrt s	Subset0(D, s)
MULTIPLY(D, s)	return $s \times D$	Change(D, s)
TERMSUBTRACT(D, P)	return $D - P$ where P is a product term in D	Diff(D, P)
DDD_OF_MATRIX$(\mathbf{A})$	construct the DDD for matrix $\mathbf{A}$	
EVALUATE(D)	return the numerical value of D	

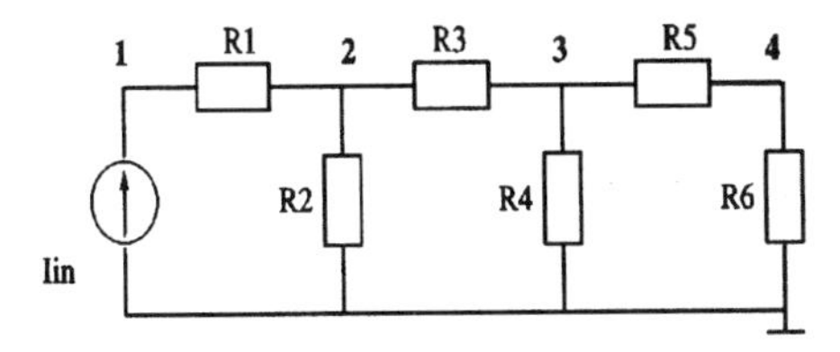

Fig. 10. A three-section ladder circuit.

Second, similar to conventional BDD's, we use a cache to remember the results of recent operations, and refer to the cache for every recursive call. In this way, we can avoid duplicate executions for equivalent subgraphs. This enables us to execute these operations in a time that is (almost) linearly proportional to the size of a graph.

Evaluation: Given a determinant decision diagram rooted at a vertex D and a set of numerical values for all the symbolic parameters, EVALUATE(D) computes the numerical value of the corresponding matrix determinant. EVALUATE(D) naturally exploits subexpression sharing in a symbolic expression, and has time complexity (almost) *linear* in the size of the diagram.

Construction: Let $\boldsymbol{A}(e, e)$ be an n-by-n symbolic matrix, where e is a set of integers from one to n. DDD_OF_MATRIX $(\boldsymbol{A}(p, q))$ constructs a determinant decision diagram of a submatrix of $\boldsymbol{A}$ with row set $p \subseteq e$ and column set $q \subseteq e$ such that $|p| = |q|$ for a given ordering of symbolic parameters. It can be viewed as a generalized Laplace expansion procedure of matrix determinants. In line 3 of the procedure DDD_OF_MATRIX, a nonzero element s is selected, and the determinant is expanded. Due to the canonicity of DDD's, s can be any nonzero matrix element, and the resulting DDD is always the same. However, the best expansion order is to use the element with the largest integer label in line 3 of the procedure DDD_OF_MATRIX.

```
COFACTOR(D, s)
1   if (D.top < s) return VERTEXZERO()
2   if (D.top = s) return D_1
3   if (D.top > s) return GETVERTEX(D.top, COFACTOR(D_0, s), COFACTOR(D_1, s))

REMAINDER(D, s)
1   if (D.top < s) return D
2   if (D.top = s) return D_0
3   if (D.top > s) return GETVERTEX(D.top, REMAINDER(D_0, s), REMAINDER(D_1, s))

MULTIPLY(D, s)
1   if (D.top < s) return GETVERTEX(s, 0, D)
2   if (D.top = s) return GETVERTEX(s, D_1, D_0)
3   if (D.top > s) return GETVERTEX(D.top, MULTIPLY(D_0, s), MULTIPLY(D_1, s))

TERMSUBTRACT(D, P)
1   if (D = 0) return VERTEXZERO()
2   if (P = 0) return D
3   if (D = P) return VERTEXZERO()
4   if (D.top > P.top) return GETVERTEX(D.top,TERMSUBTRACT(D_0, P), D_1)
5   if (D.top < P.top) return TERMSUBTRACT(D, P_0)
6   if (D.top = P.top)
        return GETVERTEX(D.top, TERMSUBTRACT(D_0, P_0), TERMSUBTRACT(D_1, P_1))

DDD_OF_MATRIX(A(p, q))
1   if (A = 0) return VertexZero()
2   if (A = 1) return VertexOne()
3   let s be a nonzero element at row i and column j of A
4       return GETVERTEX(s, DDD_OF_MATRIX(A(p - {i}, q - {j})), DDD_OF_MATRIX(A|s=0))

EVALUATE(D)
1   if (D = 0) return 0
2   if (D = 1) return 1
3   return EVALUATE(D_0) + s(D) * D.top * EVALUATE(D_1)
```

Fig. 11. Implementation of basic operations for symbolic analysis and applications.

Cofactor and Derivative: COFACTOR(D, s) is to compute the cofactor of a symbolic determinant represented by a DDD vertex D with respect to symbolic parameter s. COFACTOR is perhaps the most important operation in symbolic analysis of analog circuits. For example, the network functions can be obtained by first computing some cofactors, and then combining these cofactors according to some rules (Cramer's rule).

B. Illustration of Basic Operations and its Use in Circuit Sensitivity

In this subsection, we use an example to show how the network function sensitivity can be computed using DDD-based COFACTOR. We also use COFACTOR to exemplify the main ideas of a typical DDD-based operation.

Consider the ladder circuit shown in Fig. 10. Its system of equations has been formulated in (9). The input impedance Z_{in} is defined as

$$Z_{\text{in}} = \frac{v_1}{I_{\text{in}}}.$$

If each matrix entry is viewed as a distinct symbol, the determinant of the circuit matrix can be rewritten as (7). We redraw its DDD in Fig. 12(a), where the 0-terminal and all the 0-edges pointing to the 0-terminal are suppressed. In Fig. 12(a), for each vertex labeled by a lower-case letter, we use the corresponding upper-case letter to denote the determinant represented by that vertex. The root of the DDD represents the determinant, denoted by A, of the circuit matrix. Note that $a = 1/R_1$ and $d = (1/R_1) + (1/R_2) + (1/R_3)$. From Cramer's rule

$$v_1 = \frac{I_{\text{in}} \times \text{COFACTOR}(A,\ a)}{A}.$$

Thus

$$Z_{\text{in}} = \frac{\text{COFACTOR}(A,\ a)}{A}.$$

We consider the normalized sensitivity of the input impedance Z_{in} with respect to resistor R_2

$$\begin{aligned} S_{R_2}^{Z_{\text{in}}} &= \left(\frac{R_2}{Z_{in}}\right)\left(\frac{\partial Z_{in}}{\partial R_2}\right) \\ &= \left(\frac{R_2 A}{\text{COFACTOR}(A,\ a)}\right)\frac{\partial}{\partial d}\left(\frac{\text{COFACTOR}(A,\ a)}{A}\right) \\ &\quad \cdot\left(\frac{\partial d}{\partial R_2}\right) \\ &= \left(\frac{R_2 A}{\text{COFACTOR}(A,\ a)}\right) \\ &\quad \cdot\left(-\frac{\text{COFACTOR}(A,\ a)}{A^2}\frac{\partial A}{\partial d}\right. \\ &\quad \left.+\frac{1}{A}\frac{\partial \text{COFACTOR}(A,\ a)}{\partial d}\right)\left(-\frac{1}{R_2^2}\right). \end{aligned}$$

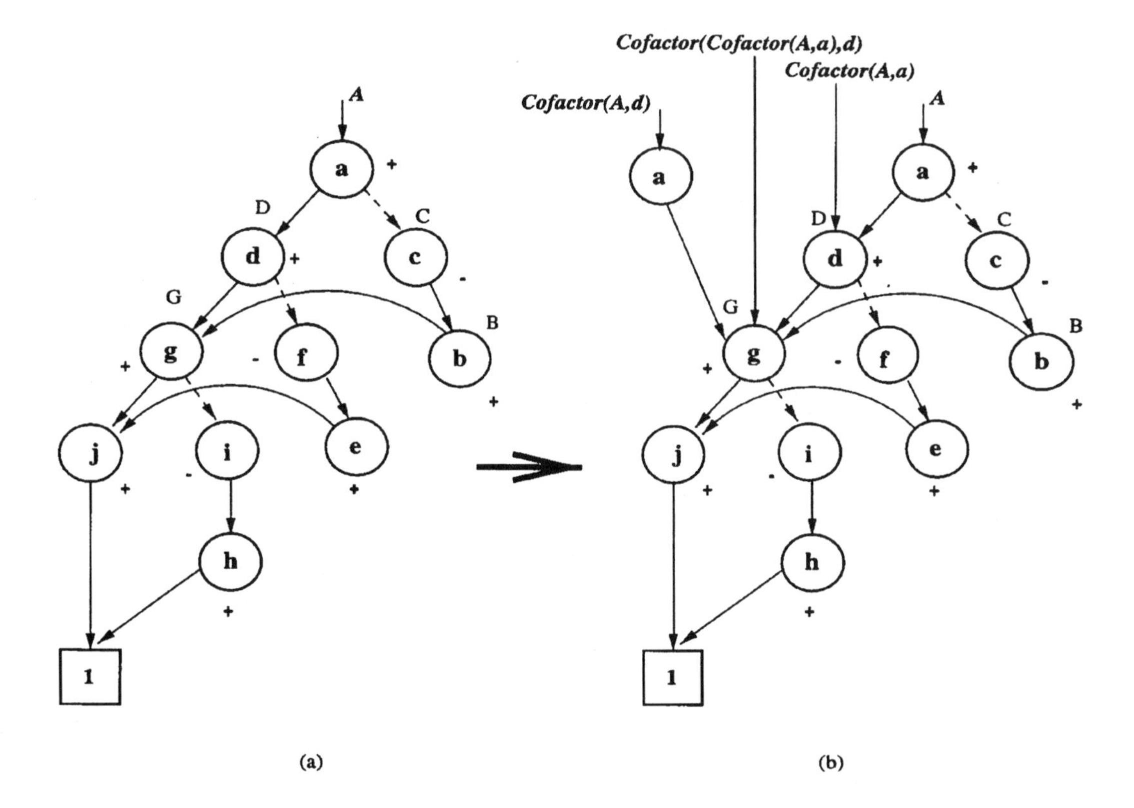

Fig. 12. DDD-based derivation of cofactors.

Note that

$$\frac{\partial A}{\partial d} = \text{Cofactor}(A,\, d)$$

and

$$\frac{\partial \text{Cofactor}(A,\, a)}{\partial d} = \text{Cofactor}(\text{Cofactor}(A,\, a),\, d)$$

we have

$$S_{R_2}^{Z_{\text{in}}} = \left(\frac{1}{R_2}\right)\left(\frac{\text{Cofactor}(A,\, d)}{A} - \frac{\text{Cofactor}(\text{Cofactor}(A,\, a),\, d)}{\text{Cofactor}(A,\, a)}\right).$$

The three cofactors COFACTOR(A, a), COFACTOR(A, d), and COFACTOR(COFACTOR(A, a), d) in the expression above can be computed elegantly using algorithm COFACTOR in Fig. 11 on the DDD shown in Fig. 12(a). Recall that the vertex ordering in Fig. 12(a) is $a > c > b > d > f > e > g > i > h > j$.

First, consider how to compute COFACTOR(A, a). Since $A.\text{top} = a = s$, $A_1 = D$ is returned. COFACTOR(A, a) points to D. Similarly, COFACTOR(COFACTOR(A, a),d) points to G. They are shown in Fig. 12(b).

Next consider COFACTOR(A, d). Since $A.\text{top} = a > d$, line 3 of the algorithm is executed with $A_0 = C$ and $A_1 = D$; i.e., GETVERTEX(a, COFACTOR(C, d), COFACTOR(D, d)). Then the procedure is invoked recursively, respectively, for COFACTOR(C, d) and COFACTOR(D, d)). This process is shown in Fig. 13, where top-down solid arrows illustrate the recursive invocation of the procedure COFACTOR, bottom-up dashed arrows show how the final result is synthesized, and each step is labeled by a number that indicates its order of execution. Note that GETVERTEX(s, 0, 0) returns the 0-terminal based on the zero-suppression rule. Eventually, COFACTOR(A, d) returns GETVERTEX(a, 0, G). Since no vertex exists with label a, the 0-edge pointing to zero and the 1-edge pointing to d, GETVERTEX(a, 0, G) will create a new vertex as shown in Fig. 12(b). During the recursive process, COFACTOR(0,d) is first calculated at Steps 3 and 4. Later on at Step 6, its return value has been cached and is used directly. This avoids the duplicate execution of COFACTOR on the same subgraph.

All three cofactors and the original determinant are compactly represented in a **single** four-root DDD as shown in Fig. 12(b). From this DDD, the sum-of-product expressions of cofactors and determinants can be generated efficiently by enumerating all its corresponding rooted 1-paths. For this example, we have $\text{Cofactor}(A,\, d) = agj - aih$, $\text{Cofactor}(\text{Cofactor}(A,\, a),\, d) = G = gj - ih$, and $\text{Cofactor}(A,\, a) = D = dgj - dih - fej$. The DDD representation enables efficient computation of exact sum-of-product symbolic expressions and their sensitivities. We can also generate the sequence-of-expression representations by introducing one intermediate symbolic symbol for each vertex. In comparison, sensitivity computation using directly the sequence-of-expressions approach requires grammar-driven compilation [27].

C. Generation of Significant Terms

Many small-signal characteristics are dominated by a small number of product terms. This has been observed and exploited previously in the context of transfer function approximation. Many times, analog designers are interested in the symbolic expressions of the first few dominating terms. In our framework, the extraction of significant product terms can be transformed to

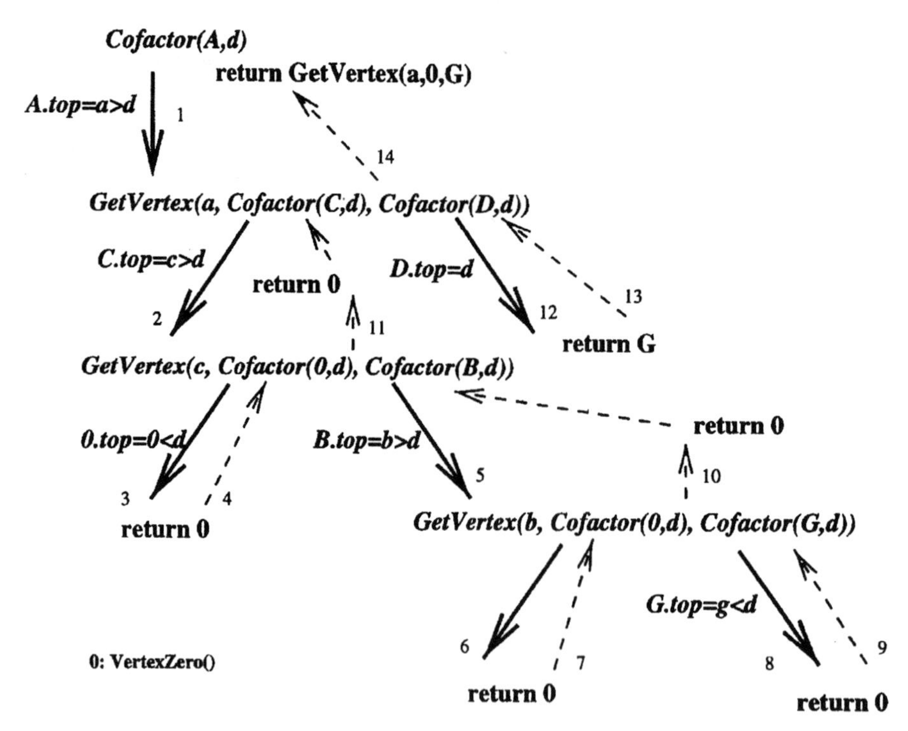

Fig. 13. Illustration of DDD-based cofactoring.

the problem of finding k shortest paths in a DDD, and solved elegantly by an $O(k|\mathrm{DDD}|)$ algorithm. We note that the problem itself can be solved efficiently by matroid-intersection-based methods [45], [46], [51], [52].

We adapted the weighting scheme of Yu and Sechen [51]. A 1-edge originated from vertex a_i is assigned weight $-\log|a_i|$, where $|a_i|$ is the numerical value of symbolic parameter a_i. All the 0-edges are assigned weight 0. Then the *cost* of a path in a DDD is defined to be the total weights of all the edges along the path. With this, the most significant product term in a symbolic determinant D corresponds to the minimum cost (shortest) path between the DDD root and the 1-terminal. The shortest path in a DDD can be obtained by a depth-first search, which has the time complexity $O(|\mathrm{DDD}|)$ [15].

A nice property of DDD's is that after we find the shortest path from a DDD, we can subtract it from the DDD using DDD operation TERMSUBTRACT. We can find the next shortest path in the resulting DDD. In this manner, we can find the k shortest paths in time $O(k \cdot |\mathrm{DDD}|)$.

VI. Experimental Results

We have implemented in C++ the proposed symbolic analysis algorithms, and tested our program on a set of circuits varying from RLC filters to bipolar and MOS integrated circuits. The set of test circuits includes

- *millerOpamp*, a two-stage miller compensated MOS opamp from [21];
- μA741, a bipolar opamp containing 26 transistors and 11 resistors, with the schematic in Fig. 15;
- *cascodeOpamp*, a CMOS cascode opamp containing 22 transistors with the schematic in Fig. 16,
- *ladder7, ladder21, ladder100*, 7-, 21-, and 100-section cascade resistive ladder networks;
- *rctree1, rctree2*, two RC tree networks;
- some RLC filters, named *butter, rlctest, vcstst, ccstest*, and *bigtst*.

For nonlinear integrated circuits, DC analysis is first performed using *SPICE*, and the resulting small-signal models from the output of *SPICE* are used in symbolic analysis. For the completeness, the small-signal models used for bipolar and MOS transistors are described in Fig. 14(a) and (b), respectively. The MNA approach as used in *SPICE* is employed to formulate the circuit equations. Exact symbolic expressions for (voltage) transfer functions ($V_{\mathrm{out}}/V_{\mathrm{in}}$) are computed using Cramer's rule and shared DDD representations of symbolic determinants and cofactors. The transfer function is in the form of the ratio of two DDD's, which are represented compactly using a **single shared** DDD with two roots; this is referred to as the DDD representation of the transfer function.

Table II[3] describes the statistics of all the test circuits, the resulting DDD sizes, and *SCAPP* results. Columns 2–6 describe, respectively, the number of nodes in each circuit, the number of nonzero elements in each circuit matrix, the number of product terms in the transfer function and the numbers of vertices in the DDD representation of the transfer function without and with the use of the vertex-ordering heuristic described in Fig. 6. For each circuit, the number of DDD vertices without vertex-ordering is the average of that of ten randomly generated orderings. *SCAPP* is used to generate the sequence of expressions (in the C program) for each transfer function. The number of multiplications (divisions are counted as multiplications), the number of additions, and the number of intermediate expressions used for computing each transfer function are reported in Columns 7–9. If the sequence of expressions is generated from the DDD

[3]In [38], only the statistics of the determinants of the circuit matrices are reported. Here, the statistics are collected for the transfer functions in order to compare with *SCAPP* which calculates the transfer functions.

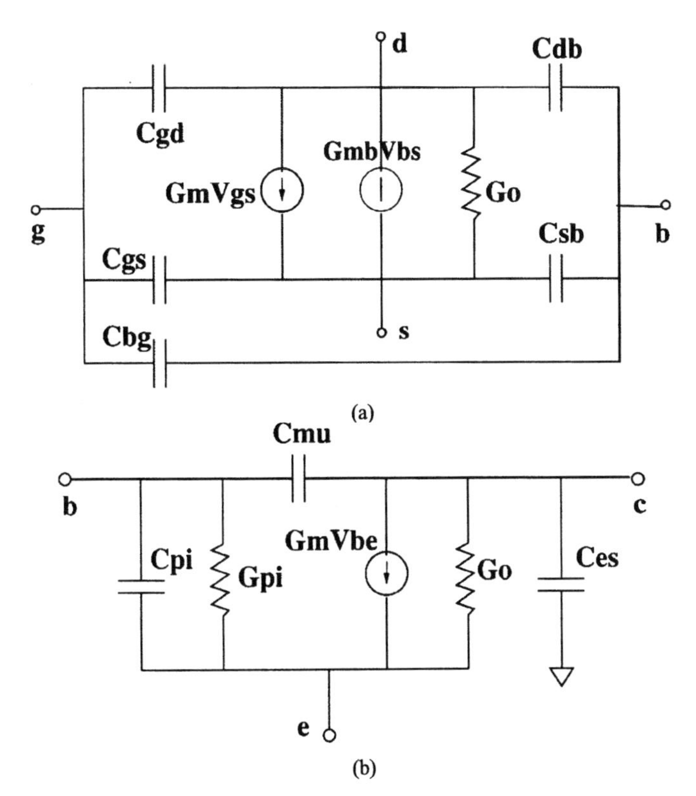

Fig. 14. (a) MOSFET small-signal model and (b) bipolar transistor model.

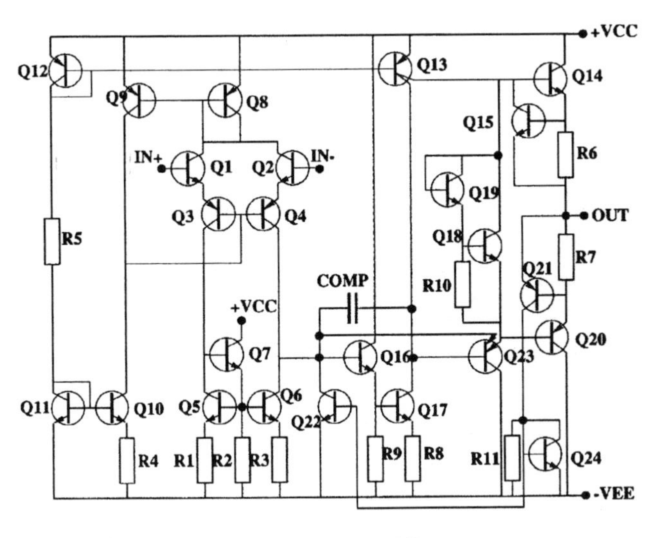

Fig. 15. The circuit schematic of bipolar μA741.

representation of the transfer function, each vertex will introduce one multiplication and one addition (if its 0-edge does not point to the 0-terminal) and, hence, the total number of multiplications and the total number of additions are bounded by the number of DDD vertices.

From Table II, we can make several observations.

1) Ordering leads to DDD's significantly smaller than that of nonordering. For ladder networks *ladder7*, *ladder21*, and *ladder100*, the numbers of DDD vertices are exactly the numbers of nonzero matrix elements.
2) For a small circuit, the number of DDD vertices may be greater than the number of product terms. This is not surprising, since the number of DDD vertices without exploiting sharing would be the number of product terms times the number of symbolic parameters in each product term.
3) For large circuits, the numbers of DDD vertices can be several orders of magnitude smaller than the numbers of product terms. Further, the difference becomes more dramatic with the increase of the circuit size.
4) The sequences of expressions generated from DDD's use about two thirds of multiplications required by *SCAPP* for ladder-structured circuits (*ladder7, ladder21,* and *ladder100*), and use less than half the multiplications of *SCAPP* for tree-structured circuits (*rctree1* and *rctree2*). It may use much more multiplications than *SCAPP* for those circuits that do not have ladder- or tree-like structures (*cascodeOpamp* and *μA*741).[4]

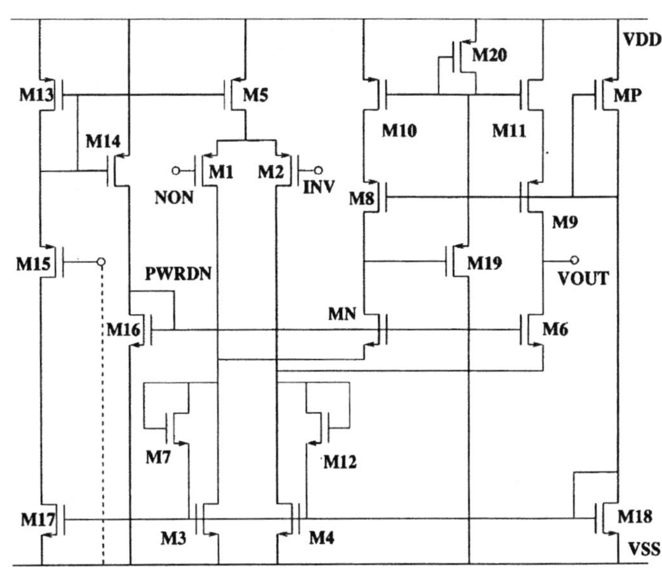

Fig. 16. The circuit schematic of MOS cascode Opamp.

We then compare our program with *ISAAC* [21] and *Maple-V* [10] for generating the complete sum-of-product expressions of the transfer functions. *ISAAC* is a well-known special-purpose symbolic analyzer designed for analog integrated circuits. *Maple-V* is a general-purpose mathematic package capable of solving linear equations symbolically. To use *Maple-V*, we use our program to set up the circuit equations and then feed the circuit matrices to *Maple-V*. The results are described in Table III. All data are obtained using a SUNsparc 20 with 32M memory. We can observe that our program runs significantly faster than both *ISAAC* and *Maple-V*, and uses much less memory. For slightly large circuits, both *ISAAC* and *Maple-V* ran out of memory. The symbolic expressions generated by *ISAAC*, *Maple-V* and the DDD-based approach are the same for each circuit.

We have implemented a frequency-domain circuit simulator by evaluating the computed DDD-represented transfer functions and compared the results with *SPICE*. For all the test circuits, our simulator produced the same numerical outputs as those of *SPICE*. Note that *SPICE* employs pivoting to improve the nu-

[4]Very recently, we showed that by exploiting the design hierarchy and automated partitioning, the DDD-based approach can generate even more compact representations than that of *SCAPP* [41], [42].

TABLE II
COMPARISON OF DDD SIZES WITH/WITHOUT VERTEX-ORDERING AND *SCAPP*

ckt name	#nodes	#symbols (matrix)	#terms	|DDD| w/o ordering	|DDD| w/ ordering	*SCAPP* #mul	#add	#expr
millerOpamp	7	21	29	95.2	51	36	60	44
butter	8	19	14	57.1	22	30	21	39
ladder7	9	22	22	85.6	26	36	25	46
rlctest	10	37	200	634.2	152	208	161	219
ccstest	10	35	176	510.8	97	208	162	219
vcstst	11	46	120	435.8	70	238	169	252
ladder21	23	64	17712	53434.4	84	116	79	140
cascodeOpamp	15	77	119395	150868.1	1994	444	596	460
$\mu A741$	24	89	119011	-*	6654	198	365	233
bigtst	33	112	1.6×10^7	-	951	996	715	1030
ladder100	102	301	5.7×10^{20}	-	398	594	398	697
rctree1	41	119	7.1×10^7	-	208	508	311	550
rctree2	54	158	3.0×10^{10}	-	299	660	407	715

-*: out of memory.

TABLE III
COMPARISON OF THE PROPOSED ALGORITHM AGAINST *ISAAC* AND *Maple-V*

circuit	*ISAAC* CPU time (seconds)	*ISAAC* Memory (bytes)	*Maple-V* CPU time (seconds)	*Maple-V* Memory (bytes)	Proposed Algorithm CPU time (seconds)	Proposed Algorithm Memory (bytes)
butter	0.61	493k	0.03	10.9k	0.033	8.1K
ladder7	1.05	980k	0.63	250k	0.033	8.1k
millerOpamp	0.49	640k	0.6	250k	0.066	24.5k
rlctest	-*		17.3	15.2M	0.10	49.1k
vcstst	-	-	104.3	23.1M	0.15	112.1k
ladder21	-	-	-	-	0.066	32.7k
cascodeOpamp	-	-	-	-	7.62	4.9M
$\mu A741$	-	-	-	-	9.13	5.2M
bigtst	-	-	-	-	2.81	2.6M

-*: out of memory.

merical accuracy of the solution of a system of linear equations, whereas no special consideration has been given to the numerical aspect in our DDD-based approach. We further observe that, in comparison with *SPICE* which uses numerical **LU** decomposition, and numerical evaluation with the sequences of expressions generated by *SCAPP*,[5] our DDD-based simulator has the following interesting features.

- Evaluation of DDD determinants and cofactors uses only multiplications, additions, subtractions, and **no** divisions. The notorious "divided-by-zero" problem does not occur in DDD evaluation.
- All multiplications in DDD evaluation are performed between a derived value (the value of a cofactor) and one from the original problem (matrix entry). Further, the *depth* of derived operation, i.e., the number of (nested) multiplications required to obtain the final value from a value in the original problem, is at most n. In contrast, in **LU** decomposition, most times, operations are performed among two derived values, and the depth of derived operation is at least $2n$.
- DDD's are constructed in such a way to achieve maximal sharing of subexpressions. As a consequence, those values close in their magnitudes are likely to be manipulated together.

Table IV shows the comparison of our DDD-based simulator with *SPICE* in terms of CPU time for repetitive numerical evaluation. For each circuit, 1000 frequency points were simulated.

TABLE IV
COMPARISON OF FREQUENCY ANALYSIS BY THE PROPOSED ALGORITHM AND BY *SPICE*

circuit	*SPICE* (seconds)	DDD-based numerical evaluation (seconds)
butter	0.96	0.26
millerOpamp	0.36	0.28
rcltest	0.71	0.65
ladder21	0.96	0.75
bigtst	3.68	2.78
cascodeOpamp	3.16	11.8

At each frequency point, the DDD-represented transfer function was evaluated by first computing the numerical value of each matrix entry from the values of circuit parameters and frequency, and then substituting the computed value of each entry to compute the DDD values. The proposed algorithm is actually faster than *SPICE* for small circuits, but is slower for large circuits. We note that the complexity of such *repetitive numerical evaluation* is linearly proportional to the number of DDD vertices, but the number of DDD vertices may grow exponentially with the size of a circuit. Sparse-matrix-based numerical **LU** decomposition has been observed to run in $O(n^{1.1-1.5})$ for typical circuits [33]. Therefore, the straightforward use of exact symbolic expressions—DDD-based or even *SCAPP*'s sequences of expressions—for numerical evaluation may not offer

[5]We note that hierarchical symbolic analysis employed in *SCAPP* is essentially partial symbolic LU decomposition by Gaussian elimination [44].

any speed advantage over fine-tuned numerical simulators such as *SPICE*. However, DDD-based symbolic analysis may still be attractive for repetitive numerical evaluation, since it allows the other "latency" properties of symbolic expressions to be exploited efficiently. For example, for the frequency-domain simulation of time-invariant circuits, all the circuit parameters but the complex variable s remain unchanged for all the frequency points, and, therefore, the use of s-expanded symbolic expressions can provide a significant speedup over *SPICE*. We have shown that s-expanded symbolic expressions can be derived very efficiently using DDD's and then only one DDD evaluation is needed for all the frequency points; this speeds up *SPICE* significantly [39].

VII. Related Work

The proposed approach is an application of decision diagram concepts to symbolic network analysis. In this section, we summarize some closely related work in these two areas. We refer the reader to [21] and [26] for comprehensive surveys of symbolic analysis techniques and applications, and [30] and [36] for decision diagrams.

A. Comparison with Existing Determinant-Based Symbolic Techniques

Previously, determinant expansion has been exploited for symbolic analysis of analog circuits. The work includes the parameter extraction method [2], the algebraic formulation method [35], and recursive Laplace expansion with minor storage and row/column ordering as implemented in *ISAAC* [21]. Parameter extraction was developed for handling large sparse matrices with a few symbolic entries and many numerical entries. The key idea is to apply a refined form of determinant expansion in (2) on all the symbolic entries first, then to use any standard numerical method to evaluate the values of minors that contain only numerical entries. This idea can be naturally incorporated into DDD's where symbolic entries are labeled first and numerical entries are labeled after symbolic entries (with small indexes). With this, numerical entries will appear at the bottom of a DDD, which can be evaluated and condensed. The incorporation of parameter extraction into DDD's will lead to a new method capable of handling large sparse matrices with both numerical and (potentially many) symbolic entries. In contrast, it is generally difficult to combine parameter extraction with other symbolic methods.

The algebraic formulation method of Sannuti and Puri exploits the structure of determinants and circuits to establish the condition for valid nonzero product terms as expressed in (1). Then the valid product terms are enumerated.

The DDD-based approach inherits many ideas found in recursive Laplace expansion with minor storage and row/column ordering as implemented in *ISAAC* [21]. For the derivation of sum-of-product expressions, both approaches are based on determinant expansion, follow the same order of expansion, and both use the cache to store the minors. But the DDD-based approach has the following several subtle differences from *ISAAC*. First, we impose the fixed ordering rule so that the structure of expansion is canonical. Further, the structure of expansion is formalized as a binary decision diagram, where each time only one matrix entry is considered. Then, the generation of sum-of-product expressions for a symbolic determinant is broken into three separate steps: 1) entry labeling (vertex ordering), 2) construction of the diagram with the chosen order, and 3) generation of the product terms from the diagram. These considerations enable us to exploit the understandings and implementation of BDD's developed mainly in the past decade in the area of formal verification and logic synthesis. We further show that symbolic manipulation on sum-of-product expressions can be performed much more efficiently on the proposed diagrams.

B. Comparison with Hierarchical Symbolic Analysis

DDD's can be viewed as a special form of sequences of expressions as used in hierarchical symbolic analysis [24], [40]. They differ in how they are created and the canonicity of the representation, which lead to several fundamental differences in their performance.

First, the DDD representation is unique. For a given circuit, regardless of which algorithms to use, the generated code based on the DDD representation must always be the same (should be able to be compared simply using UNIX shell command *diff*). The canonicity property may be useful for formal analog verification.

Second, symbolic manipulation with DDD's is simpler than with arbitrarily nested sequences of expressions. From the DDD's, the expanded sum-of-product expressions can be generated by a simple DDD traversal; the significant terms can be generated efficiently by finding k-shortest DDD paths; the s-expanded expressions can be derived in linear time in the size of a DDD [39]. Very recently, we have shown that the problem of deriving simplified, reliable, and interpretable symbolic network functions can be performed effectively and efficiently with DDD's [43]. In contrast, manipulation and approximation of arbitrarily nested symbolic expressions are known to be more difficult and involved [18], [27], [37].

Third, in comparison with numerical evaluation using hierarchical symbolic analyzer *SCAPP*, which is the compiled partial **LU** decomposition by Gaussian elimination [44], numerical evaluation with DDD's is division-free, manipulates fewer derived values, and generally adds/subtracts the values with less differing magnitudes. Furthermore, since pivoting is not employed in *SCAPP*, repetitive numerical evaluation with the generated code may be subject to the numerical accuracy problem. We note that compiled-code simulation has been studied in the area of numerical circuit simulation [48], and partial **LU** decomposition by Gaussian elimination itself has been exploited in the context of circuit tearing, for example as in [49]. If targeted at circuit simulation, **LU** decomposition by Gaussian elimination has been observed to be less preferred over the Crout method or other methods of **LU** decomposition [44].

Finally, it is worth noting that efficient linear(ized) circuit analysis can be accomplished via numerical reduced-order modeling techniques such as asymptotic waveform evaluation [34], the PVL algorithm [16], and the Arnoldi method [31]. It is very intriguing to combine DDD's and these techniques for possible **symbolic** reduced-order modeling.

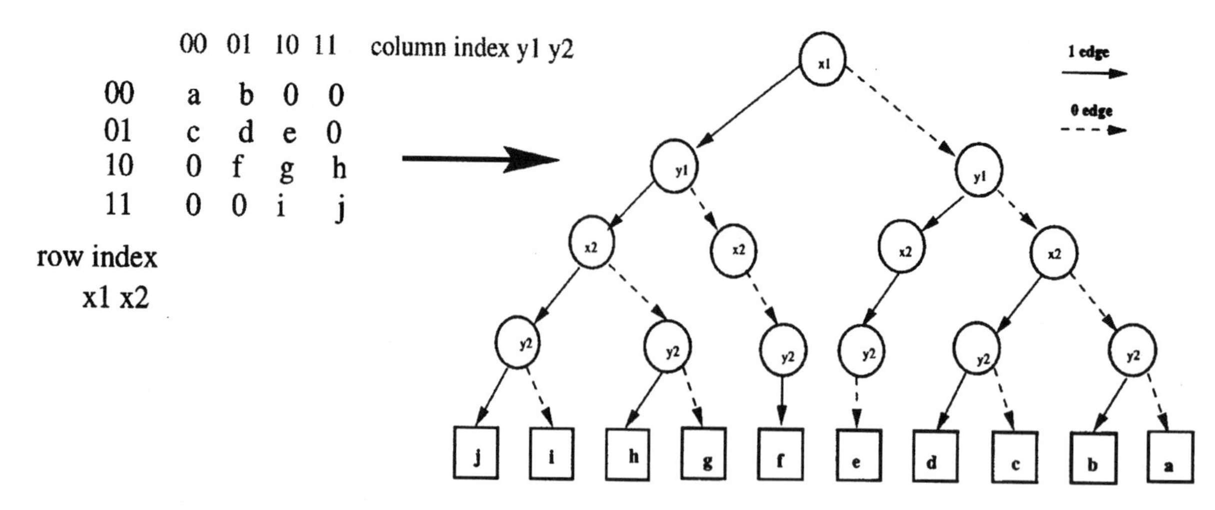

Fig. 17. An MTBDD representation of matrix $\boldsymbol{M}$.

C. Relevance to Other Decision Diagram Concepts

The notion of DDD's comes from the application of ZBDD's, a variant of BDD's, to represent symbolic matrix determinants. Each vertex in the DDD represents a symbolic determinant and is defined by the determinant expansion rule (2), where operations are the addition and multiplication in normal algebra. We note that several other extensions of BDD's for multivalued functions and arithmetic applications have been made; for example, multiterminal binary decision diagrams [11]–[13], [20] (also called algebraic decision diagrams [3]), hybrid decision diagrams [14], binary moment diagrams (BMD's) [6], and several others as described in the book edited by Sasao and Fujita [36]. Among them, Multiterminal BDD's (MTBDD's) have been explored for the implementation of matrix algebra.

MTBDD's is an extension of BDD's with multiple terminals, each of which as a real value [11]–[14], [20]. MTBDD's can be used to represent matrices by observing that the row and column indexes of a matrix can be encoded as Binary vectors, say $\{x_1 x_2 \cdots x_k\}$ and $\{y_1 y_2 \cdots y_l\}$ ($k = l$ for square matrices), and then the matrix can be conceptually viewed as a function $f\colon B^{k+l} \to R$ where R is the set of nonzero matrix entries. In the resulting MTBDD for representing this multivalued Boolean function f, each terminal represents a nonzero matrix entry, and each nonterminal vertex is labeled by either a row or a column encoding bit (x_i or y_i). Each path from the root to a terminal defines the row and column position at which the matrix entry—represented by the terminal—locates. For example, Fig. 17 shows an MTBDD representation of the matrix $\boldsymbol{M}$. As shown by Fujita, McGeer and Yang, using MTBDD's, many matrix algebraic operations such as Strassen matrix multiplication, spectral transforms, and **LU** decomposition can be performed elegantly [20]. As a representation of **matrices**, MTBDD's differ from DDD's, which is a representation of **matrix determinants**. Although it may be argued that a matrix determinant may be calculated explicitly or implicitly based on the determinant definition (1) in Section II-B from the MTBDD representation of the matrix and basic MTBDD operations described in [20]. However, the computational procedure and results would be significantly complicated, and MTBDD's have not been used for determinant computation. To the best of our knowledge, the introduction of DDD's represents the first effort in exploring BDD's and their variations for representing and manipulating matrix determinants and cofactors. On the other hand, DDD's represent matrix determinants, not matrices, and are not adequate for implementing general matrix algebra. Nevertheless, matrices and determinants are intriguingly related. It is interesting to explore more connections between MTBDD's and DDD's.

BMD's[6] are more closely connected to DDD's. BMD's provide a canonical representation for multilinear functions. They are a variation of BDD's where the expansion rule is the following decomposition rule of function f

$$f = f_{\overline{x}} + x(f_x - f_{\overline{x}})$$

where f_x (respectively, $f_{\overline{x}}$) denotes the positive (respectively, negative) cofactor of f with respect to x, i.e., the function resulting when constant one (respectively, zero) is substituted for x. Note that a determinant is a multilinear function in its entries. Recall the DDD expansion rule (2)

$$\det(\boldsymbol{A}) = a_{r,c}(-1)^{r+c}\det(\boldsymbol{A}_{a_{r,c}}) + \det(\boldsymbol{A}_{\overline{a}_{r,c}}).$$

Here, det $(\boldsymbol{A}_{\overline{a}_{r,c}})$ is the remainder of $\det(\boldsymbol{A})$, which is equal to $\det(\boldsymbol{A})|_{a_{r,c}=0}$, i.e., the value of $\det(\boldsymbol{A})$ by substituting $a_{r,c}$ by zero. It can be verified that

$$(-1)^{r+c}\det(\boldsymbol{A}_{a_{r,c}}) = \det(\boldsymbol{A})|_{a_{r,c}=1} - \det(\boldsymbol{A})|_{a_{r,c}=0}.$$

Therefore, the DDD representation can be viewed as a special case of the BMD representation.

However, the matrix determinants as a special form of multilinear functions have several special properties. For example, the sign can be determined nicely using the sign rule (Section III) and be attached as part of a vertex, whereas in BMD's, signs are encoded as edge weights. The BMDs' objective is for formal verification of arithmetic digital circuits, where matrix

[6]The connection of BMD's to ZBDD's and MTBDD's has been noted by several researchers [6], [14], [30].

operations such as multiplication are emphasized, whereas in our application, solving a system of linear equations is the objective and the representation and manipulation of symbolic determinants and cofactors of **sparse** matrices are of primary interest. Nevertheless, many ideas developed in the area of decision diagrams such as using multiterminals and attributed edges to represent polynomials and numeric coefficients can be adapted to enhance the power of DDD's.

VIII. Conclusion

In this paper, a new graph representation, called DDD's, for symbolic matrix determinants is introduced and symbolic analysis algorithms for analog circuits are presented. Unlike previous approaches based on either the expanded form or the nested form representations of symbolic expressions, DDD-based symbolic analysis exploits the sparsity and sharing in a canonical manner. We described an efficient vertex-ordering heuristic and proved that it is optimum for ladder-structured circuits; in this case, the number of DDD vertices is equal to the number of nonzero matrix entries. We emphasize that the DDD size depends on the size of a circuit, its structure and sparsity, as well as the chosen vertex ordering. In the worst case, the DDD size can grow exponentially with the size of a circuit. Fortunately, for practical circuits, we have observed that with the proposed vertex-ordering heuristic, the numbers of DDD vertices are quite small—usually several orders of magnitude smaller than the numbers of product terms in the expanded form. Generating the complete sum-of-product symbolic expressions from the DDD representation offers orders-of-magnitude improvement in both CPU time and memory usages over symbolic analyzers *ISAAC* and *Maple-V* for large analog circuits. It enables the exact and canonical symbolic analysis of such large analog circuits as μA741 for the first time.

We also compared the DDD-based approach with the state-of-art hierarchical symbolic analyzer *SCAPP* in generating the sequences of expressions for network transfer functions. For ladder-structured circuits, the DDD-based approach uses only two thirds of multiplications as required by *SCAPP*. For large circuits that do not have ladder-like structures such as μA741 opamp, the sequences of expressions generated by the DDD-based approach are generally manageable by modern computers, but can be substantially longer than those from *SCAPP*. However, the DDD representation is unique. That is, for a given circuit, regardless of which algorithms to use, the generated code based on the DDD representation must always be the same (should be able to be compared simply using UNIX shell command *diff*). The canonicity property may be useful for formal analog verification.

In contrast to the nested form as used in *SCAPP*, the DDD representation is more amenable to efficient symbolic manipulation. As shown in this paper, symbolic analysis algorithms such as driving cofactor computation, network function construction, sensitivity calculation, and generating significant terms, can be performed in time almost linear in the number of DDD vertices. Very recently, we have shown other important manipulations such as deriving s-expanded symbolic expressions and symbolic approximation can be accomplished in a similar manner [39], [43].

In comparison with numerical **LU** decomposition as in *SPICE* and symbolic **LU** decomposition as in *SCAPP*, DDD evaluation is division-free, manipulates fewer derived values, and generally adds/subtracts the values with less differing magnitudes. Inspired by these features, as well as the canonicity and compactness of the DDD representation, research is being extended to exploit the full potential of canonical symbolic analysis for the design and test automation of analog circuits, as well as in general symbolic algebra.

Acknowledgment

The authors would like to thank Prof. F. Brewer and his research group at the University of California, Santa Barbara, for providing them with their *HomeBrew* BDD package, which accelerated significantly their initial DDD implementation, Prof. G. Gielen of Katholieke Universiteit Leuven, Belgium, and Prof. M. Hassoun of Iowa State University, Ames, for providing them their symbolic circuit analysis packages *ISAAC* and *SCAPP*. They are also grateful to Prof. T. Sasao of Kyushu Institute of Technology, Japan, and the anonymous reviewers for valuable comments that improve the presentation of this paper.

References

[1] S. B. Akers, "Binary decision diagrams," *IEEE Trans. Comput.*, vol. C-27, pp. 509–516, June 1976.

[2] G. E. Alderson and P. M. Lin, "Computer generation of symbolic network functions—a new theory and implementation," *IEEE Trans. Circuit Theory*, vol. CT-20, pp. 48–56, Jan. 1973.

[3] R. I. Bahar, E. A. Frohm, C. M. Gaona, G. A. Hachtel, E. Macii, A. Pardo, and F. Somenzi, "Algebraic decision diagrams and their applications," in *Proc. IEEE/ACM Int. Conf. Computer-Aided Design*, Nov. 1993, pp. 188–191.

[4] R. E. Bryant, "Graph-based algorithms for Boolean function manipulation," *IEEE Trans. Comput.*, vol. C-37, pp. 677–691, Aug. 1986.

[5] R. E. Bryant, "Binary decision diagrams and beyond: Enabling technologies for formal verification," in *Proc. Int. Conf. Computer-Aided Design*, Nov. 1995, pp. 236–243.

[6] R. E. Bryant and Y. A. Chen, "Verification of arithmetic functions with binary moment diagrams," in *Proc. 32nd IEEE/ACM Design Automation Conf.*, San Francisco, CA, June 1995, pp. 535–541.

[7] K. S. Brace, R. L. Rudell, and R. E. Bryant, "Efficient implementation of a BDD package," in *Proc. 27th IEEE/ACM Design Automation Conf.*, June 1990, pp. 40–45.

[8] J. G. Broida and S. G. Williamson, *A Comprehensive Introduction to Linear Algebra*. Reading, MA: Addison-Wesley, 1989.

[9] R. Carmassi, M. Catelani, G. Iuculano, A. Liberatore, S. Manetti, and S. Marini, "Analog network testability measurement: A symbolic formulation approach," *IEEE Trans. Instrum. Meas.*, vol. 40, pp. 930–935, Dec. 1991.

[10] B. W. Char, *et al.*, *Maple V: Language Reference Manual*. Berlin, Germany: Springer-Verlag, 1991.

[11] E. M. Clarke, X. Zhao, M. Fujita, Y. Matsunga, P. C. McGeer, and J. Yang, "Fast Walsh transform computation using binary decision diagrams," presented at the IFIP WG 10.5 Workshop on Applications of Reed–Muller Expansion in Circuit Design, 1993.

[12] E. M. Clarke, M. Fujita, P. C. McGeer, J. Yang , and X. Zhao, "Multiterminal binary decision diagrams: An efficient data structure for matrix representation," presented at the Int. Workshop on Logic Synthesis (IWLS), Tahoe City, CA, May 23–26, 1993.

[13] E. M. Clarke, M. Fujita, P. C. McGeer, X. Zhao , and J. Yang, "Fast spectrum computation for logic functions using binary decision diagrams," in *Proc. IEEE Int. Symp. Circuits and Systems*, May 1994, pp. 275–278.

[14] E. M. Clarke, M. Fujita, and X. Zhao, "Multi-terminal binary decision diagrams and hybrid decision diagrams," in *Representations of Discrete Functions*, T. Sasao and M. Fujita, Eds. Norwell, MA: Kluwer Academic, 1996, pp. 93–108.

[15] T. Cormen, C. E. Leiserson, and R. L. Rivest, *Introduction to Algorithms*. Cambridge, MA: MIT Press, 1990.

[16] P. Feldmann and R. Freund, "Efficient linear circuit analysis by Pade approximation via the Lanczos process," *IEEE Trans. Computer-Aided Design*, vol. 14, pp. 639–649, May 1995.

[17] F. V. Fernández, J. D. Martín, A. Rodríguez-Vázquez, and J. L. Huertas, "On simplification techniques for symbolic analysis of analog integrated circuits," in *Proc. IEEE Int. Symp. Circuits and Systems*, 1992, pp. 1149–1152.

[18] F. V. Fernández, A. Rodríguez-Vázquez, J. D. Martín, and J. L. Huertas, "Accurate simplification of large symbolic formulae," in *Proc. IEEE Int. Conf. Computer Aided Design (ICCAD)*, 1992, pp. 318–321.

[19] F. V. Fernández and A. Rodríguez-Vázquez, "Symbolic analysis tools—the state of the art," in *Proc. IEEE Int. Symp. Circuits and System*, 1996, pp. 798–801.

[20] M. Fujita, P. C. McGeer, and J. C.-Y. Yang, "Multi-terminal binary decision diagrams: An efficient data structure for matrix representation," *Formal Meth. Syst. Design*, vol. 10, pp. 149–169, 1997.

[21] G. Gielen and W. Sansen, *Symbolic Analysis for Automated Design of Analog Integrated Circuits*. Norwell, MA: Kluwer Academic, 1991.

[22] G. Gielen, P. Wambacq, and W. Sansen, "Symbolic analysis methods and applications for analog circuits: A tutorial overview," *Proc. IEEE*, vol. 82, no. 2, pp. 287–304, Feb. 1994.

[23] M. M. Hassoun and P. M. Lin, "A new network approach to symbolic simulation of large-scale network," in *Proc. IEEE Int. Symp. Circuits and Systems*, 1989, pp. 806–809.

[24] ——, "A hierarchical network approach to symbolic analysis of large scale networks," *IEEE Trans. Circuits Syst.*, vol. 42, pp. 201–211, Apr. 1995.

[25] J.-J. Hsu and C. Sechen, "DC small-signal symbolic analysis of large analog integrated circuits," *IEEE Trans. Circuits Syst.*, vol. 41, pp. 817–828, Dec. 1994.

[26] P. M. Lin, *Symbolic Network Analysis*. Amsterdam, the Netherlands: Elsevier Science, 1991.

[27] ——, "Sensitivity analysis of large linear networks using symbolic program," in *Proc. IEEE Int. Symp. Circuits and Systems*, 1992, pp. 1145–1148.

[28] W. J. McCalla and D. O. Pederson, "Elements of computer-aided analysis," *IEEE Trans. Circuit Theory*, vol. CT-18, pp. 14–26, Jan. 1971.

[29] S. Minato, "Zero-suppressed BDD's for set manipulation in combinatorial problems," in *Proc. 30th IEEE/ACM Design Automation Conf.*, Dallas, TX, June 1993, pp. 272–277.

[30] ——, *Binary Decision Diagrams and Applications for VLSI CAD*. Norwell, MA: Kluwer Academic, 1996.

[31] L. Miguel-Silveira, M. Kamon, I. Elfadel, and J. White, "A coordinate-transformed Arnoldi algorithm for generating guaranteed stable reduced-order models of RLC circuits," in *Proc. IEEE/ACM Int. Conf. Computer-Aided Design*, 1996, pp. 288–294.

[32] L. W. Nagel, "SPICE2: A computer program to simulate semiconductor circuits," Ph.D. dissertation, Univ. California, Berkeley, CA, May 1975.

[33] A. R. Newton and A. L. Sangiovanni-Vincentelli, "Relaxation-based electrical simulation," *IEEE Trans. Computer-Aided Design*, vol. CAD-3, pp. 308–331, Oct. 1984.

[34] L. T. Pillage and R. A. Rohrer, "Asymptotic waveform evaluation for timing analysis," *IEEE Trans. Computer-Aided Design*, vol. 9, pp. 352–366, Apr. 1990.

[35] P. Sannuti and N. N. Puri, "Symbolic network analysis—an algebraic formulation," *IEEE Trans. Circuits Syst.*, vol. CAS-27, pp. 679–687, Aug. 1980.

[36] T. Sasao and M. Fujita, *Representations of Discrete Functions*. Norwell, MA: Kluwer Academic, 1996.

[37] S. J. Seda, M. G. R. Degrauwe, and W. Fichtner, "Lazy-expansion symbolic expression approximation in SYNAP," in *Proc. IEEE Int. Conf. Computer Aided Design (ICCAD)*, 1992, pp. 310–317.

[38] C.-J. Shi and X. Tan, "Symbolic analysis of large analog circuits with determinant decision diagrams," in *Proc. IEEE/ACM Int. Conf. Computer-Aided Design*, San Jose, CA, Nov. 1997, pp. 366–373.

[39] ——, "Efficient derivation of exact s-expanded symbolic expressions for behavioral modeling of analog circuits," in *Proc. IEEE Custom Integrated Circuits Conf. (CICC'98)*, San Clara, CA, May 1998, pp. 463–466.

[40] J. A. Starzky and A. Konczykowska, "Flowgraph analysis of large electronic networks," *IEEE Trans. Circuits Syst.*, vol. CAS-33, pp. 302–315, Mar. 1986.

[41] X. Tan and C.-J. Shi, "Hierarchical symbolic analysis of large analog circuits with determinant decision diagrams," in *Proc. IEEE Int. Symp. Circuits and Systems*, vol. VI, May 1998, pp. 318–321.

[42] ——, "Balanced multilevel multiway partitioning of large analog circuits for hierarchical symbolic analysis," in *Proc. Asia and South Pacific Design Automation Conf. (ASP-DAC)*, Jan. 18–21, 1999, pp. 1–4.

[43] ——, "Interpretable symbolic small-signal characterization of large analog circuits using determinant decision diagrams," in *Proc. Design, Automation, and Test in Europe Conf. and Exhibition (DATE)*, Mar. 10–13, 1999, pp. 448–453.

[44] J. Vlach and K. Singhal, *Computer Methods for Circuit Analysis and Design*, New York: Van Nostrand Reinhold, 1994.

[45] P. Wambacq, G. Gielen, and W. Sansen, "A cancellation-free algorithm for the symbolic simulation of large analog circuits," in *Proc. IEEE Int. Symp. Circuits and Systems*, 1992, pp. 1157–1160.

[46] P. Wambacq, *et al.*, "A family of matroid intersection algorithms for the computation of approximate symbolic network functions," in *Proc. IEEE Int. Symp. Circuits and Systems*, 1996, pp. 806–809.

[47] P. Wambacq, G. Gielen, and W. Sansen, "A new reliable approximation method for expanded symbolic network functions," in *Proc. IEEE Int. Symp. Circuits and Systems*, 1996, pp. 584–587.

[48] W. T. Weeks, A. J. Jimenez, G. W. Mahoney, D. Mehta, H. Wassemzadeh, and T. R. Scott, "Algorithms for ASTAP—A network analysis program," *IEEE Trans. Circuit Theory*, vol. CT-20, pp. 628–634, Nov. 1973.

[49] P. Yang, "An investigation of ordering, tearing, latency algorithms for the time-domain simulation of large circuits," Ph.D. dissertation, Univ. Illinois at Urbana-Champaign, Urbana, IL, 1980.

[50] Z. You, E. Sánchez-Sinencio, and J. P. de Gyvez, "Analog system-level fault diagnosis based on a symbolic method in the frequency domain," *IEEE Trans. Instrum. Meas.*, vol. 44, no. 1, pp. 28–35, Jan. 1995.

[51] Q. Yu and C. Sechen, "A unified approach to the approximate symbolic analysis of large analog integrated circuits," *IEEE Trans. Circuits Syst.*, vol. 43, pp. 656–669, Aug. 1996.

[52] ——, "Efficient approximation of symbolic functions using matroid intersection algorithms," *IEEE Trans. Computer-Aided Design*, vol. 16, pp. 1073–1081, Oct. 1997.

C.-J. Richard Shi (M'91–SM'99) received the B.S. and M.S. degrees in electrical engineering from Fudan University, Shanghai, China, in 1985 and 1988, respectively, the M.A.Sc. degree in electrical engineering and the Ph.D. degree in computer science from the University of Waterloo, Waterloo, Ont., Canada, in 1991 and 1994, respectively.

He is currently an Assistant Professor in the Department of Electrical Engineering, University of Washington, Seattle. His research interests include methodologies and tools for systems-on-a-chip design, with the particular emphasis on analog, mixed-signal, and deep-submicron design and test automation. He has published more than 70 technical papers, and has been a principal investigator of more than $2M in research funding from DARPA, NSF, USAF, CDADIC, and industry since 1995. He is a consultant to several semiconductor and EDA companies.

Dr. Shi co-founded IEEE/ACM/VIUF International Workshop on Behavioral Modeling and Simulation, and served as its Technical Program Chair from 1997 to 1999. Having been involved in IEEE DASC 1076.1 VHDL-AMS Working Group since 1994, he is one of the contributors to, and promoters of, IEEE std 1076.1-1999 standard language (VHDL-AMS) for the description and simulation of mixed-signal/mixed-technology systems. He has delivered tutorials on VHDL-AMS and behavioral modeling at several conferences including DAC, EuroDAC, and ASP-DAC. He has been a recipient or co-recipient of several awards including the T. D. Lee Physics Award for excellence in graduate study from Fudan, University of Waterloo Outstanding Achievement in Graduate Studies Award, the Natural Sciences and Engineering Research Council of Canada Doctoral Prize, a National Science Foundation CAREER Award, four Best Paper Awards (including the 1999 IEEE/ACM Design Automation Conference Best Paper Award and the 1999 IEEE VLSI Test Symposium Best Paper Award), and three other Best Paper Award Nominations (ASP-DAC'98, EuroDAC'96, and ASP-DAC'95). He is a member of IEEE Design Automation Standards Committee. He is an Associate Editor of IEEE TRANSACTIONS ON CIRCUITS AND SYSTEMS–II.

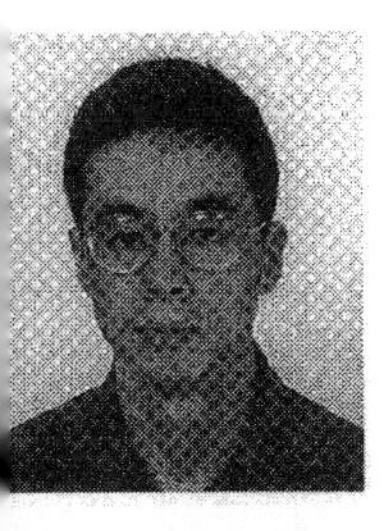

Xiang-Dong Tan (S'96–M'99) received the B.S. and M.S. degrees in electrical engineering from Fudan University, Shanghai, China, in 1992 and 1995, respectively, and the Ph.D. degree in electrical and computer engineering from the University of Iowa, Iowa City, in 1999.

He is currently a Member of Technical Staff at Monterey Design Systems, Monterey, CA. He worked with Rockwell Semiconductor Systems in the summer of 1997, and Avant! Corporation in the summer of 1998. He was a Research Assistant in the Department of Electrical Engineering, University of Washington, Seattle, from September 1998 to April 1999. His current research interests include very large scale integration (VLSI) physical design automation, symbolic analysis of large analog circuits, layout optimization for performance, timing, power, and clock tree synthesis.

Dr. Tan received a Best Paper Award from the 1999 IEEE/ACM Design Automation Conference in 1999 and the First-Place Student Poster Award from the 1999 Spring Meeting of the Center for Design of Analog Digital Integrated Circuits (CDADIC). He received a Best Graduate Award in 1992 and a number of Excellent College Student Scholarships from 1988–1992, all from Fudan University.

PART IV

Analog Layout

Layout Tools for Analog ICs and Mixed-Signal SoCs: A Survey

Rob A. Rutenbar
Dept. of ECE, Carnegie Mellon University
Pittsburgh, Pennsylvania, 15213
rutenbar@ece.cmu.edu

John M. Cohn
IBM
Essex Junction, Vermont, 05477
johncohn@us.ibm.com

Abstract—*Layout for analog circuits has historically been a time-consuming, manual, trial-and-error task. The problem is not so much the size (in terms of the number of active devices) of these designs, but rather the plethora of possible circuit and device interactions: from the chip substrate, from the devices and interconnects themselves, from the chip package. In this short survey we enumerate briefly the basic problems faced by those who need to do layout for analog and mixed-signal designs, and survey the evolution of the design tools and geometric/electrical optimization algorithms that have been directed at these problems.*

1 Introduction

Layout for digital integrated circuits is usually regarded as a difficult task because of the *scale* of the problem: millions of gates, kilometers of routed wires, complex delay and timing interactions. Analog designs and the analog portions of mixed-signal systems-on-chip (SoCs) are usually vastly smaller—up to 100 devices in a cell, usually less than 20,000 devices in a complete sub-system—and yet they are nothing if not *more* difficult to lay out. Why is this? The answer is that the complexity for analog circuits is not so much from the sheer number of devices, as from the complex *interactions* among the devices, among the various continuous-valued performance specifications, and with the fabrication process and the operating environment.

This would be less of a problem if analog circuits and subsystems were rare or exotic commodities, or if they were sufficiently generic that a few stable designs could be easily retargeted to each new application. Unfortunately, neither is true. The markets for application-specific ICs (ASICs), application-specific standard parts (ASSPs) and high-volume commodity ICs are characterized by an increasing level of integration. In recent years, complete systems that before occupied separate chips are being integrated on a single chip. Examples of such "systems on a chip" include telecommunications ICs such as modems, wireless designs such as components in radio frequency receivers and transmitters, and networking interfaces such as local area network ICs. Although most functions in such integrated systems are implemented with digital (or especially digital signal processing) circuitry, the analog circuits needed at the interface between the electronic system and the "real" world are now being integrated on the same die for reasons of cost and performance.

The booming market share of mixed-signal ASICs in complex systems for telecommunications, consumer, computing, and automotive applications is one direct result of this. But along with this increase in achievable complexity has come a significant increase in design complexity. And at the same time, many present ASIC application markets are characterized by shortening product life cycles and time-to-market constraints. This has put severe pressure on the designers of these analog circuits, and especially on those who lay out these designs. If cell-based library methodologies were workable for analog (as they are for semi-custom digital designs) layout issues would be greatly mitigated. Unfortunately, such methodologies fare poorly here. Most analog circuits are one-of-a-kind, or at best few-of-a-kind on any given IC. Today, they are usually designed by hand, and laid out by hand. The problem recurs at the system level: the discipline of row-based layout as the basis for large function blocks that is so successful in digital designs is not (yet) as rigorously applied in analog designs.

Despite the problems here, there is a thriving research community working to make custom analog layout tools a practical reality. This brief survey attempts to describe the history and evolution of these tools; it extends two earlier reviews [1,2]. Our survey is organized as follows. We begin with a brief taxonomy of problems and strategies for analog layout in Sec 2. Then, in Sec. 3 we review attacks on the cell-level layout problem. In Sec. 4 we review mixed-signal system-level layout problems. In Sec. 5 we review recent work on field programmable analog arrays. Sec. 6 offers some concluding remarks. We end with an extensive, annotated bibliography.

2 Analog Layout Problems and Approaches

Before trying to categorize the various geometric strategies that have been proposed for analog layout, it is essential first to understand what are the electrical problems that affect analog design. We enumerate here briefly the salient effects that layout can have on circuit performance. References Cohn [35], Verghese [76], and Stanisic [101] together comprise a fairly complete treatment of these issues. There are really three core problems, which we describe first below. We then briefly survey solution strategies here.

2.1 Loading problems

The non-ideal nature of inter-device wiring introduces capacitive and resistive effects which can degrade circuit performance. At sufficiently high frequencies, inductive effects arise as well. There are also parasitic RLC elements associated with the geometry of the devices themselves, e.g., the various capacitances associated with MOS diffusions. All these effects are remarkably sensitive to detailed (polygon-level) layout For example, in MOS circuits, layout designers have a useful degree of freedom in that they can fold large devices (i.e., devices with large width-to-length ratios), thus altering both their overall geometric shape and detailed parasitics. Folding transforms a large device (large channel width) into a parallel connection of smaller devices. The smaller devices are *merged*: parallel side-by-side alignment allows a single source or drain diffusion to be shared between adjacent gate regions, thus minimizing overall capacitance. Every diffused structure also has an associated parasitic resistance which varies with its shape. These resistances can be reduced by minimizing the aspect ratio of all diffusions (reducing the width of the device), merging diffusions when possible, and strapping diffusion with low resistance layers such as metal where possible. Of course, this "strapping" may interfere with signal routing. Cohn [35] offers a careful treatment of the layout issues here for MOS devices.

One of the distinguishing characteristics of analog layout, in comparison to digital layout, is the amount of effort needed to create correct layouts for atomic devices. MOS devices with large width-to-length rations, bipolar devices with large emitter areas, etc., are common in analog designs and require careful attention to detail. Passive components that implement resistors or capacitors or inductors are also more frequent in analog design, and require careful layout. (See, for example, Bruce [15] as an example of procedural generation of complex MOS devices.) Extremely low-level geometric details of the layout of individual devices and passives can have a significant circuit-level impact.

2.2 Coupling problems

Layout can also introduce unexpected signal coupling into a circuit which may inject unwanted electrical noise or even destroy its stability through unintended feedback. At lower frequencies, coupling may be introduced by a combination of capacitive, resistive, or thermal effects. At higher frequencies inductive coupling becomes an issue. Especially in the modern deep submicron digital processes in which analog systems are being integrated, coupling is an increasing problem. Metal conductors couple capacitively when two metal surfaces are sufficiently close, e.g., if wires run in parallel on a single layer or cross on adjacent layers. If a parallel run between incompatible signals is unavoidable, a neutral wire such as a ground or reference line can be placed between them as a coupling shield.

Current flowing through a conductor also gives rise to a fluctuation in the voltage drop across the conductor's finite resistance. This fluctuation is then coupled into all devices attached to the conductor. This effect is particularly problematic in power supply routing for analog cells on digital ICs. Sensitive analog performance often depends on an assumption of moderate stability in the power rails--this may not be true if the power distribution network is improperly laid out. See Stanisic [99, 100, 101] for a detailed treatment of the issues in mixed-signal power distribution.

Signals can also be coupled through the silicon substrate or bulk, either through capacitive, resistive, or thermal effects. Because all devices share the same substrate, noise injected into the substrate is capacitively or resistively coupled into every node of the circuit. This is particularly problematic when analog circuits must share a substrate with inherently noisy high speed digital logic. On mixed-signal ICs conventional solutions focus on *isolation*: either by locating the sensitive analog far away from the noise-injecting source, or surrounding the noise-sensitive circuits with a low impedance diffusion guard-ring to reduce the substrate noise level in a particular area. Unfortunately, the structure of the substrate and details of the layout of the power supply network that biases the substrate greatly affect even the qualitative behavior of this coupling. For example, epitaxial substrates have such low resistivity that injected noise can "reappear" far from its origin on the chip surface; simple isolation schemes do not always work well here. Bulk substrates are somewhat more resistive and noise may remain more local. Evolving silicon-on-insulator processes (SOI) may dramatically reduce this problem, but these are not yet in widespread use for commodity mixed-signal designs. References [68, 85, 100] are a good starting point for analysis of the substrate coupling problem in mixed-signal design.

In addition, since silicon is an excellent conductor of heat, local temperature variations due to current changes in a device can also cause signal coupling in nearby thermally sensitive devices. This phenomenon is most prevalent in Bipolar or Bi-CMOS processes. Placing thermally sensitive devices far away from high-power thermally dissipating devices can reduce this effect. Placing matching devices symmetrically about thermally "noisy" sources can also be effective in reducing the effects of thermal coupling.

2.3 Matching problems

Unavoidable variations which are present in all fabrication processes lead to small mismatches in electrical characteristics of identical devices. If these mismatches are large enough, they can effect circuit performance by introducing electrical problems such as offsets. Four major layout factors can affect the matching of identical devices: *area*, *shape*, *orientation*, and *separation*.

Device area is a factor because semiconductor processing introduces unavoidable distortions in the geometry which make up devices. Creating devices using identical geometry (identical shape) improves matching by insuring that both devices are subject to the same (or at least *similar*) geometric distortions. Similarly, since the proportional effect of these variations tends to decrease as the size of the device increases, matching devices are usually made as large as the circuit performance and area constraints will allow. Since many processing effects, *e.g.*, ion-implantation, introduce anisotropic geometric differences, devices which must match should also be placed in the same orientation. Finally, spatial variations in process parameters tend to degrade the matching characteristics of devices as their separation increases. This is largely due to process induced gradients in parameters such as mobility or oxide thickness. Sensitivity to these effects can be reduced by placing devices which must match well in close proximity. Devices which must be *extremely* well matched may be spatially interdigitated in an attempt to cancel out the effects of global process gradients.

Device matching, particularly of bipolar devices, may also be degraded by thermal gradients. Two identical devices at different points on a thermal gradient will have slight differences in VBE for a given collector current. To combat this, it is common practice to arrange thermally sensitive matching devices symmetrically around thermally generating noise sources. Parasitic capacitive and resistive components of interconnect can also introduce problems of matching in differential circuits, i.e., those circuits which are comprised of two matching halves. A mismatch in the parasitic capacitance and resistance between the two matching halves of the circuit can give rise to offsets and other electrical problems. The most powerful technique used to improve interconnect parasitic matching is layout symmetry, in which the placement and wiring of matching circuits are forced to be identical, or in the case of differential circuits, mirror symmetric.

2.4 Layout solution strategies

We mentioned a variety of solution techniques in the above enumeration of layout problems: careful attention to atomic device layout, MOS merging, substrate noise isolation, symmetric layout, etc. However, these are really low-level *tactics* for dealing with specific problems for specific circuits in specific operating or fabrication environments. The more general question we wish to address next is how the analog portion of a large mixed-signal IC is laid out—what are the overall geometric *strategies* here?

We note first that, like digital designs, analog designs are attacked hierarchically. However, analog systems are usually significantly smaller than their digital counterparts: 10,000 to 20,000 analog devices versus 100,000 to 1,000,000 digital gates. Thus, analog layout hierarchies are usually not as deep as their digital counterparts. The need for low-level attention to coupling and interaction issues is another force that tends to flatten these hierarchies. The typical analog layout hierarchy comprises two fundamentally different types of layout tasks:

- **Cell-level layout:** The focus here is really device-level layout, placement and routing of individual active and passive devices. At this level, many of the low level matching, symmetry, merging, reshaping, and proximity-management tactics for polygon-level optimization are applied. The goal is to create cells that are suitably insulated not only from fluctuations in fabrication and operating environment, but from probable coupling with neighboring pieces of layout.
- **System-level layout:** The focus here is cell composition, arranging and interconnecting the individual cells to complete an analog subsystem. At this level, isolation is one major concern: from nearby noisy digital circuits coupled in via substrate, power grid, or package, The other concern is signal integrity. Some number of digital signals from neighboring digital blocks need to penetrate into any analog regions, and these signals may be fundamentally incompatible with some sensitive analog signals.
- **Programmable layout:** The focus here is applying field programmable gate array (FPGA, see for example, Rose [114]) ideas to analog designs. The idea is to bypass completely the need to do custom cells. Rather, a set of programmable active and passive elements is connected with programmable wiring to achieve low-performance analog functions. This is a relatively recent implementation strategy, but we expect to see it grow.

We visit the ideas behind each of these layout strategies in the following three sections.

3 Analog Cell Layout Strategies

For our purposes a "cell" is a small analog circuit, usually comprising not more than about 100 active and passive devices which is designed and laid out as a single atomic unit. Common examples include operational amplifiers, comparators, voltage references, analog switches, oscillators, and mixers.

The earliest approaches to custom analog cell layout relied on procedural module generation. These approaches are a workable strategy when the analog cells to be laid out are relatively static, *i.e.*, necessary changes in device sizing or biasing result in little need for global alterations in device layout, orientation, reshaping, *etc*. Procedural generation schemes usually start with a basic geometric template (sometimes called a *topology* for the circuit), which specifies all necessary device-to-device and device-to-wiring spatial relationships. Generation completes the template by correctly sizing the devices and wires, respacing them as necessary. References [6-11] are examples dedicated mainly to opamps. The mechanics for capturing the basic design specifications can often be as familiar as a common spreadsheet interface, e.g., [12]. Owen [12] is a more recent example focused on opamp generation, and describes both a language for specifying these layouts and several optimized layout results. The system at Philips [13] is another good example of practical application of these ideas on complex circuits. Bruce [15] shows an example of module generation useful for atomic MOS devices.

Often, however, changes in circuit design require full custom layout, which can be handled with a *macrocell-style* strategy. The terminology is borrowed from digital floorplanning algorithms, which manipulate flexible layout blocks, arrange them topologically, and then route them. For analog cells, we regard the flexible blocks as devices to be reshaped and reoriented as necessary. Module generation techniques are used to generate the layouts of the individual devices. A placer then arranges these devices, and a router interconnects them—all while attending to the numerous parasitics and couplings to which analog circuits are sensitive.

The earliest techniques used a mixed of knowledge-based and constructive techniques for placement, with routers usually adapted from common semicustom digital applications [18,20,21,24]. For example, a common constructive placement heuristic is to use the spatial relationships in a drawn circuit schematic as an initial basis from mask-level device placement [23]. Unfortunately, these techniques tended to be rather narrow in terms of which circuits could be laid out effectively.

The ILAC tool from CSEM was an important early attempt in this style [17,22]. It borrowed heavily from the best ideas from digital layout: efficient slicing tree floorplanning with flexible blocks, global routing via maze routing, detailed routing via channel routing, area optimization by compaction. The problem with the approach was that it was difficult to extend these primarily-digital algorithms to handle all the low-level geometric optimizations that characterize expert manual design. Instead, ILAC relied on a large, very sophisticated library of device generators.

ANAGRAM and its successor KOAN / ANAGRAM II from CMU kept the macrocell style, but reinvented the necessary algorithms from the bottom up, incorporating many manual design optimizations [19,25,27,35]. For example, the device placer KOAN relied on a very small library of device generators, and migrated important layout optimizations into the placer itself. KOAN could dynamically fold, merge and abut MOS devices, and thus discover desirable optimizations to minimize parasitic capacitance during placement. KOAN was based on an efficient simulated annealing algorithm [3]. (Recent placers have also extended ideas from sequence-pair module-packing representations [5] to handle analog layout tasks [38].) KOAN's companion, ANAGRAM II, was a maze-style detailed area router capable of supporting several forms of symmetric differential routing, mechanisms for tagging compatible and incompatible classes of wires (*e.g.*, noisy and sensitive wires), parasitic crosstalk avoidance, and over-the-device routing. Other device placers and routers operating in the *macrocell-style* have appeared (e.g., [26, 31, 32, 33, 34]), confirming its utility.

In the next generation of cell-level tools, the focus shifted to quantitative optimization of performance goals. For example, KOAN maximized MOS drain-source merging during layout, and ANAGRAM II minimized crosstalk, but without any specific, quantitative performance targets. The routers ROAD [29] and ANAGRAM III [30] use improved cost-based schemes that route instead to minimize the deviation from acceptable parasitic bounds derived from designers or sensitivity analysis. The router in [39] can manage not just parasitic sensitivities, but also basic yield and testability concerns. Similarly, the placers in [28, 36, 37] augment a KOAN-style model with sensitivity analysis so that performance degradations due to layout parasitics can be accurately controlled. Other tools in this style include [40].

In the newest generation of CMOS analog cell research, the device placement task has been separated into two distinct phases: device *stacking*, followed by *stack placement*. By rendering the circuit as an appropriate graph of connected drains and sources, it is possible to identify natural clusters of MOS devices that ought to be merged—called *stacks*—to minimize parasitic capacitance. Malavasi [43, 44, 47] gave an exact algorithm to extract all the optimal stacks, and the placer in [45, 46] extends a KOAN-style algorithm to dynamically choose the right stacking and the right placement of each stack. Basaran [48] offers another variant of this idea: instead of extracting all the stacks (which can be time-consuming since the underlying algorithm is exponential), this technique extracts one optimal set of stacks very fast (in linear time). The technique is useful in either the inner loop of a layout algorithm (to evaluate quickly a merging opportunity) or in an interactive layout editor (to stack a set of devices optimally, quickly).

The notion of using sensitivity analysis to quantify the impact on final circuit performance of low-level layout decisions (*e.g.*, device merging, symmetric placement / routing, parasitic coupling due to specific proximities, *etc.*) has emerged as a strategy to link the various approaches being taken for cell level layout and system assembly. Several systems from U.C. Berkeley are notable here. The influential early formulation of the sensitivity analysis problem was Choudhury [59] which not only quantified layout impacts on circuit performance, but also showed how to use nonlinear programming techniques to map these sensitivities into constraints on various portions of the layout task. Charbon [61] extended these ideas to handle constraints involving nondetermininistic parasitics of the type that arise from statistical fluctuations in the fabrication process. In related work, Charbon [60] also showed how to extract critical constraints on symmetry and matching directly from a device schematic.

One final problem in the macrocell style is the separation of the placement and routing steps. In manual cell layout, there is no effective difference between a rectangle representing a wire and one representing part of a device: they can each be manipulated simultaneously. In a place-then-route strategy, one problem is estimating how much space to leave around each device for the wires. One solution strategy is *analog compaction*, *e.g.*, [49,51,52], in which we leave extra space during device placement and then compact. A more radical alternative is *simultaneous device place-and-route*. An experimental version of KOAN [61] supported this by iteratively perturbing both the wires and the devices, with the goal of optimally planning polygon-level device and wire layout interactions.

As wireless and mobile design has proliferated, more radio frequency designs have appeared. These offer yet another set of challenges at the cell level. Radio frequency (RF) circuits and higher frequency microwave circuit have unique properties which make their automated layout impossible with the techniques developed for lower frequency analog cells. Because every geometric property of the layout of an individual wire—its length, bends, proximity to other wires or devices—may play a key role in the electrical performance of the overall circuit, most RF layouts are optimized for performance first and density second.

Most layout work targeting RF circuits comprises interactive tools that aid the designer to speed manual design iterations [54, 55]. Other work in the area includes semi-automated approaches that rely on knowledge of the relative position of all cells [53]. However, these template-based approaches with predefined cells do limit the design alternatives possible. Recently, Charbon introduced a performance-driven router for RF circuits [56]. In their approach, sensitivity analysis is employed to compute upper bounds for critical parasitics in the circuit, which the router then attempts to respect. Aktuna [57,58] recently introduced the idea of device-level floorplanning for these circuits; using a genetic algorithm, the tool simultaneously evolves placement and detailed routing under constraints on length, bends, phase, proximity and planarity.

There are several open problems in cell-level layout. For example, the optimal way to couple the various phases of cell layout—stacking, placement, routing, compaction—to each other and back to circuit design (or redesign) remains a challenging problem. However, there is a maturing base of workable transistor-level layout techniques now to build on. We note that commercial tools offering device-level analog layout synthesis have just recently appeared, and are in now in production use. Figure 1, an industrial analog cell produced by a commercial analog layout synthesis tool, is one such automatic layout example [41,42].

4 Mixed-Signal System Layout

A mixed-signal *system* is a set of custom analog and digital functional blocks. At the system-level the problem is really an *assembly* problem [2]. Assembly means block floorplanning, placement, global and detailed routing (including the power grid). As well as parasitic sensitivities, the two new problem at the chip level are *coupling* between noisy signals and sensitive analog signals, and *isolation* fro digital switching noise that couples through the substrate, power grid, or package.

Just as at the cell level, procedural generation remains a viable alternative for well-understood designs with substantial regularity. Many signal processing applications have the necessary highly stylized, regular layout structure. Procedural generation has been successful for many switched capacitor filters and data converters [13,62-66], and especially regular, array-style blocks [16].

More generally though, work has focussed on custom placement and routing at the block level, with layout optimization aimed at not only area, but also signal coupling and isolation. Trnka [89] offered one very early attempt aimed at bipolar array-style (i.e., fixed device image) layout, adapting semicustom digital layout tools to this analog layout image. For row-based analog standard cell layout, an early elegant solution to the coupling problem was the *segregated channels* idea of [86,87] to alternate noisy digital and sensitive analog wiring channels in a row-based cell layout. The strategy constrains digital and analog signals never to be in the same channel, and remains a practical solution when the size of the layout is not too large. However, in modern multi-level interconnect technologies, this rigorous segregation can be overly expensive in area.

For large designs, analog channel routers were developed. In Gyurscik [90], it was observed that a well-known digital channel routing algorithm [88] could be easily extended to handle critical analog problems that involve varying wire widths and wire separations needed to isolate interacting signals. Work at Berkeley substantially extended this strategy to handle complex analog symmetries, and the insertion of shields between incompatible signals

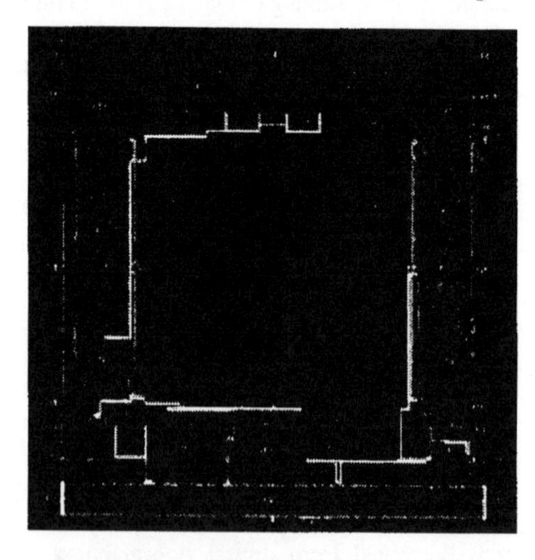

Figure 1. Bootstrapped fully differential amplifier in 0.25um CMOS process. (Courtesy Rocketchips, and NeoLinear, Inc.)

[91,92,94]. This work also introduced the idea of *constraint mapping*, which begins with parasitic sensitivities available from analysis of the system (or the cell) to be laid out, and transforms these into hard bounds on the allowable parasitics of each wire in each channel. The mapping process is itself a nonlinear programming problem, in this case a quadratic programming formulation. These tools are particularly effective for stylized row-based layouts such as switched capacitor filters, where complex routing symmetries are necessary to balance subtle parasitics, and adjoint simulation methods can yield the necessary sensitivities.

The WREN [93] and WRIGHT [95,96] systems from CMU generalized these ideas to the case of arbitrary layouts of mixed functional blocks. WREN comprises both a mixed-signal global router and channel router. WREN introduced the notion of *SNR-style* (signal-to-noise ratio) constraints for incompatible signals, and both the global and detailed routers strive to comply with designer-specified noise rejection limits on critical signals. WREN incorporates a constraint mapper (influenced by [59]) that transforms input noise rejection constraints from the across-the-whole-chip form used by the global router into the per-channel per-segment form necessary for the channel router (as in [94]). WRIGHT uses a KOAN-style annealer to floorplan the blocks, but with a fast substrate noise coupling evaluator so that a simplified view of substrate noise influences the floorplan. WRIGHT used a coarse resistive mesh model with numerical pruning to capture substrate coupling; the approach in Charbon [83] uses semi-analytical substrate modeling techniques which allow fast update when blocks move, and can also support efficient noise sensitivity analysis.

The substrate coupling problem is an increasingly difficult one as more and faster digital logic is placed side-by-side with sensitive analog parts. One avenue of relevant work here seeks to model the substrate accurately, efficiently extract tractable electrical models of its conduction of overall chip noise, and understand qualitatively how various isolation mechanisms (e.g., separation, guard rings) will work. This has been an active area of late. References [67,69-71,73-80] address basic computational electromagnetics attacks on modeling and analysis of substrate coupling. The approaches vary in their discretization of the substrate, their numerical technique to solve for the point-to-point resistance between two devices in the substrate, and their model-order reduction techniques to reduce potentially large, extracted circuit-level substrate models to smaller, more efficient circuit models. Su [68] offers experimental data from test circuits on the mechanisms of substrate noise conduction for CMOS mixed-signal designs in epitaxial substrates. Charbon and Miliozzi [82,83] address substrate coupling in the context of linking substrate modeling with the generation of constraints on allowable noise in the synthesis and layout process. Mitra [72] and Miliozzi [81] address the problem of estimating substrate current injection; Mitra [72] uses a circuit-level switching model and circuit simulation, and transforms simulation results into an equivalent single-tone sinusoid with the same total energy as the original random switching waveform. Miliozzi [81] uses a digital simulator to capture simple digital switching waveforms, which are then combined with precharacterized circuit-level injection models to estimate block-level injection. Tsukada [84] suggests an active guard ring structure to mitigate substrate noise, based on some of these modeling ideas. Verghese [85] offers a recent survey of substrate modeling, extraction, reduction, and injection work, along with a review of how substrate issues are dealt with in current mixed-signal design methodologies

Another important task in mixed-signal system layout is power grid design. Digital power grid layout schemes usually focus on connectivity, pad-to-pin ohmic drop, and electromigration effects. But these are only a small subset of the problems in high-performance mixed-signal chips which feature fast-switching digital systems next to sensitive analog parts. The need to mitigate unwanted substrate interactions, the need to handle arbitrary (non-tree) grid topologies, and the need to design for transient effects such as current spikes are serious problems in mixed-signal power grids. The RAIL system [97-101] addresses these concerns by casting mixed-signal power grid synthesis as a routing problem that uses fast AWE-based [4] linear system evaluation to electrically model the entire power grid, package and substrate during layout. By allowing changes in both grid topology (where segments are located, where power pins are located, where substrate contacts are located) and grid segment sizing, the tool can find power grid layouts to optimize ac, dc, and transient noise constraints. Techniques such as [72, 81] are useful to estimate the digital switching currents needed here for power grid optimization. Chen [102] discusses a similar power distribution formulation applied to digital circuits.

Most of these system layout tools are fairly recent, but because they often rely on mature core algorithms from similar digital layout problems, many have been prototyped both successfully and quickly. Several full, top-to-bottom prototypes have recently emerged, *e.g.*, [83, 103-106].

There are many open problems here. Signal coupling and substrate/package isolation are still addressed via rather *ad hoc* means overall. There is still much work to be done to enhance existing constraint mapping strategies and constraint-based layout tools to handle the full range of industrial concerns, and to be practical for practicing designers.

5 Field-programmable analog arrays

We mention finally one very recent analog layout style which is radically different from those mentioned above. In the digital realm, field programmable gate arrays (FPGAs) have revolutionized digital prototyping and rapid time-to-market designs [114]. Integrating programmable logic elements and programmable interconnect, these allow rapid customization with no fabrication steps. An obvious question is: can a similar technology be adapted for analog and mixed-signal designs. The apparent answer is a qualified "yes".

Early work such as [107-109] mimicked directly the FPGA style of small primitive functions connectable by programmable interconnect. However, the loading which is already problematic in digital designs proved even more deleterious here. Later designs such as [110-113] moved up to higher-level buildinng blocks (e.g., opamps, switches, capacitor arrays) and also focussed new energy on sensitive analog design of the programmable interconnect. Field programmable analog arrays (FPAAs) are just now beginning to be commercially available. They are currently so small, however, that layout is not really a problem. It will be interesting to see if these designs become larger (e.g., larger digital blocks with smaller analog blocks), and if so, if automatic layout becomes a requirement.

6 Conclusions

There has been substantial progress on tools for custom analog and mixed-signal circuit layout. Cast mostly in the form of numerical and combinatorial optimization tasks, linked by various forms of sensitivity analysis and constraint mapping, leveraged by ever faster workstations, these tools are beginning to have practical--even commercial--application. There remain many open problems here, and some newly created problems as analog circuits are increasingly embedded in unfriendly digital deep submicron designs. Given the demand for mixed-signal ICs, we expect no diminishment in the interest in various layout tools to speed the design of these important circuits.

Acknowledgments

We are grateful to our colleagues at Carnegie Mellon for their assistance with the preparation of this manuscript. Rick Carley, Mehmet Aktuna and Brian Bernberg in particular helped with early drafts. Sachin Sapatnekar of U. Minnesota also offered helpful comments on the final version of the paper.

References

For clarity here, we have grouped the references by topic, in roughly the order in which each topic area is visited in our review.

General references

[1] R.A. Rutenbar, "Analog design automation: Where are we? Where are we going?" *Proc. IEEE Custom IC Conference*, May 1993.

[2] L.R. Carley, G.G.E. Gielen, R.A. Rutenbar, W.M.C. Sansen, "Synthesis tools for mixed-signal ICs: Progress on frontend and backend strategies," *Proc. ACM/IEEE Design Automation Conference*, June 1996.

[3] S. Kirkpatrick, C.D. Gelatt, M.P. Vecchi, "Optimization by simulated annealing," *Science*, vol. 220, no. 4598, 13 May 1983.

[4] L. T. Pillage, R. A. Rohrer, "Asymptotic waveform evaluation for timing analysis", *IEEE Trans. CAD*, Vol. 9, No. 4, Apr. 1990.

[5] H. Murata, K. Fujiyoshi, S. Nakatake, Y. Kajitani, "VLSI module placement based on rectangle-packing by the sequence-pair method," *IEEE Transactions on CAD*, vol. 15, no. 2, Dec. 1996.

Cell-level module generators

[6] J. Kuhn, "Analog Module Generators for Silicon Compilation," *VLSI Systems Design*, May 1987.

[7] E. Berkan, M. d'Abreu, W. Laughton, "Analog compilation based on successive decompositions," *Proc. ACM/IEEE Design Automation Conference*, 1988.

[8] H. Koh, C. Séquin, P. Gray, "OPASYN: a compiler for CMOS operational amplifiers," *IEEE Trans. CAD*, Vol. 9, No. 2, pp. 113-125, February 1990.

[9] H. Onodera, *et al.*, "Operational amplifier compilation with performance optimization," *IEEE JSSC*, Vol. SC-25, No. 2, pp. 466-473, Apr. 1990.

[10] J.D. Conway and G.G. Schrooten, "An Automatic Layout Generator for Analog Circuits," *Proc. EDAC*, pp. 513-519, March 1992.

[11] J. P. Harvey, *et al.*, "STAIC: An Interactive Framework for Synthesizing CMOS and BiCMOS Analog Circuits," *IEEE Trans. CAD*, pp. 1402-1417, Nov. 1992.

[12] R. Henderson, *et al.*, "A spreadsheet interface for analog design knowledge capture and re-use," *Proc. IEEE CICC*, 13.3, May 1993.

[13] G. Beenker, J. Conway, G. Schrooten, A. Slenter, "Analog CAD for consumer ICs," chapter 15 in "Analog circuit design" *Analog Circuit Design* (J. Huijsing, R. van der Plassche and W. Sansen. eds.), Kluwer Academic Publishers, pp. 347- 367, 1993.

[14] B.R. Owen, R. Duncan, S. Jantzi, C. Ouslis, S. Rezania, K. Martin, "BALLISTIC: An analog layout language," *Proc. IEEE Custom IC Conference*, 1995.

[15] J.D. Bruce, H.W. Li, M.J. Dallabetta, and R.J. Baker, "Analog layout using ALAS!", *IEEE JSSC*, vol. 31, no. 2, Feb. 1996.

[16] G. van der Plas, J. Vandenbussche, G. Gielen, W. Sansen, "Mondriaan: a tool for automated layout of array-type analog blocks," *Proc.IEEE Custom IC Conference*, May 1998.

Device-Level Placement and Routing

[17] J. Rijmenants, T.R. Schwarz, J.B. Litsios, R. Zinszner, "ILAC: An automated layout tool for CMOS circuits," *Proc. IEEE Custom IC Conference*, 1988.

[18] M. Kayal, S. Piguet, M. Declerq, B. Hochet, "SALIM: A layout generation tool for analog ICs," *Proc. IEEE Custom IC Conference*, 1988.

[19] D. J. Garrod, R. A. Rutenbar, L. R. Carley, "Automatic layout of custom analog cells in ANANGRAM", *Proc. ICCAD*, pp. 544-547, Nov. 1988.

[20] M. Ayal, S. Piguet, M. Declerq, B. Hochet, "An interactive layout generation tool for CMOS ICs," *Proc. IEEE ISCAS*, 1988.

[21] M. Mogaki, *et al.*, "LADIES: An automatic layout system for analog LSI's," *Proc. ACM/IEEE ICCAD*, pp. 450-453, Nov. 1989.

[22] J. Rijmenants, J.B. Litsios, T.R. Schwarz, M.G.R. Degrauwe, "ILAC: An automated layout tool for analog CMOS circuits," *IEEE JSSC*, Vol. 24, No. 4, pp. 417-425, April 1989.

[23] S.W. Mehrenfar, "STAT" A schematic to artwork translator for custom analog cells," *Proc. IEEE Custom IC Conference*, 1990.

[24] S. Piguet, F. Rahali, M. Kayal, E. Zysman and M. Declerq, "A new routing method for full-custom analog ICs," *Proc. IEEE Custom IC Conference*, 1990.

[25] J. M. Cohn, D. J. Garrod, R. A. Rutenbar, L. R. Carley, "New algorithms for placement and routing of custom analog cells in ACACIA," *Proc. IEEE Custom IC Conference*, 1990.

[26] E. Malavasi, *et al.*, "A routing methodology for analog integrated circuits," *Proc. ACM/IEEE ICCAD*, pp. 202-205, Nov 1990.

[27] J. M. Cohn, D. J. Garrod, R. A. Rutenbar, L. R. Carley, "KOAN/ANAGRAM II: New tools for device-level analog placement and routing," *IEEE JSSC*, Vol. 26, No. 3, March, 1991.

[28] E. Charbon, E. Malavasi, U. Choudhury, A. Casotto, A. Sangiovanni-Vincentelli, "A constraint-driven placement methodology for analog integrated circuits", *Proc. IEEE CICC*, pp. 28.2/1-4, May 1992.

[29] E. Malavasi, A. Sangiovanni-Vincentelli, "Area routing for analog layout," *IEEE Trans. CAD*, Vol. 12, No. 8, pp. 1186-1197, Aug, 1993.

[30] B. Basaran, R. A. Rutenbar, L. R. Carley, "Latchup-aware placement and parasitic-bounded routing of custom analog cells", *Proc. ACM/IEEE ICCAD*, Nov. 1993.

[31] M. Pillan, D. Sciuto, "Constraint generation and placement for automatic layout design of analog integrated circuits," *Proc. IEEE ISCAS*, 1994.

[32] G. J. Gad El Karim, R. S. Gyurcsik, G. L. Bilbro, "Sensitivity driven placement of analog modules," *Proc. IEEE ISCAS*, 1994.

[33] J.A. Prieto, J.M. Quintana, A.Rueda, J.L. Huertas, "An algorithm for the place-and-route problem in the layout of analog circuits," *Proc. IEEE ISCAS*, 1994.

[34] E. Malavasi, J.L. Ganley and E. Charbon, "Quick placement with geometric constraints," *Proc. IEEE Custom IC Conf.*, 1997.

[35] J.M Cohn, D. J. Garrod, R. A. Rutenbar, L. R. Carley, *Analog Device-Level Layout Automation*, Kluwer Acad. Publ., 1994.

[36] K. Lampaert, G. Gielen and W. Sansen, "Direct performance-driven placement of mismatch-sensitive analog circuits," *Proc. ACM/IEEE Design Automation Conference*, 1995.

[37] K. Lampaert, G. Gielen, W.M. Sansen, "A performance-driven placement tool for analog integrated circuits," *IEEE JSSC*, Vol. 30, No. 7, pp. 773-780, July 1995.

[38] F. Balasa, K. Lampaert, "Module placement for analog layout using the sequence pair representation," *Proc. ACM/IEE Design Automation Conference*, June 1999.

[39] K. Lampaert, G. Gielen, W. Sansen, "Analog routing for manu-

facturability," *Proc. IEEE CICC*, May 1996.
[40] C. Brandolese, M. Pillan, F. Salice, D. Sciuto, "Analog circuits placement: A constraint driven methodology," *Proc. IEEE ISCAS*, 1996.
[41] Stephan Ohr, "Electronica: Cell-builder tool anticipates analog synthesis," *EE Times*, November 9, 1998.
[42] NeoCell Layout Synthesis User Manual, 1999. NeoLinear, Inc., Pittsburgh, PA. http://www.neolinear.com.

Optimal MOS Device Stacking

[43] E. Malavasi, D. Pandini, V. Liberali, "Optimum stacked layout for analog CMOS ICs," *Proc. IEEE Custom IC Conference*, 1993.
[44] V. Liberali, E. Malavasi, D. Pandini, "Automatic generation of transistor stacks for CMOS analog layout," *Proc. IEEE ISCAS*, 1993.
[45] E. Charbon, E. Malavasi, D. Pandini, A. Sangiovanni-Vincentelli, "Simultaneous placement and module optimization of analog ICs," *Proc. ACM/IEEE Design Automation Conference*, 1994.
[46] E. Charbon, E. Malavasi, D. Pandini, A. Sangiovanni-Vincentelli, "Imposing tight specifications on analog IC's through simultaneous placement and module optimization," *Proc. ACM/IEEE Design Automation Conference*, 1994.
[47] E. Malavasi, D. Pandini, "Optimum CMOS stack generation with analog constraints", *IEEE Trans. CAD*, Vol. 14, No. 1, pp. 107-12, Jan. 1995.
[48] B. Basaran, R. A. Rutenbar, "An O(n) algorithm for transistor stacking with performance constraints", *Proc. ACM/IEEE DAC*, June 1996.

Device-Level Compaction and Layout Optimization

[49] R. Okuda, T. Sato, H. Onodera, K. Tamuru, "An efficient algorithm for layout compaction problem with symmetry constraints," *Proc. IEEE ICCAD*, pp. 148-151, Nov. 1989.
[50] J. Cohn, D. Garrod, R. Rutenbar, L. R. Carley, "Techniques for simultaneous placement and routing of custom analog cells in KOAN/ANAGRAM II," *Proc. ACM/IEEE ICCAD*, pp. 394-397, Nov. 1991.
[51] E. Felt, E. Malavasi, E. Charbon, R. Totaro, A. Sangiovanni-Vincentelli, "Performance-driven compaction for analog integrated circuits," *Proc. IEEE Custom IC Conference*, 1993.
[52] E. Malavasi, E. Felt, E. Charbon, A. Sangiovanni-Vincentelli, "Symbolic compaction with analog constraints," *Int. J. Circuit Theory and Applic.*, Vol. 23, No. 4, pp. 433-452, Jul/Aug 1995.

Radio Frequency Cell Layout

[53] J.F. Zurcher, "MICROS- A CAD/CAM Program for Fast Realization of Microstrip Masks," *IEEE Journal MTT-S*, pp. 481-484, 1985.
[54] R. H. Jansen, "LINMIC: A CAD Package for the Layout-Oriented Design of Single- and Multi-Layer MICs/MMICs up to mm-Wave Frequencies," *Microwave Journal*, pp. 151-161, Feb 1986.
[55] R. H. Jansen, R. G. Arnonld and I. G. Eddison, "A Comprehensive CAD Approach to Design of MMICs up to MM-Wave Frequencies," *IEEE Journal MTT-T*, vol. 36, no. 2, pp. 208-219, Feb. 1988.
[56] E. Charbon, G. Holmlund, B. Donecker and A. Sangiovanni-Vincentelli, "A Performance-Driven Router for RF and Microwave Analog Circuit Design," *Proc. IEEE Custom Integrated Circuits Conference* pp. 383-386, 1995.
[57] M. Aktuna, R. A. Rutenbar, L. R. Carley, "Device Level Early Floorplanning for RF Circuits," Proc. 1998 ACM International Symposium on Physical Design, April 1998.
[58] M. Aktuna, R. A. Rutenbar, L. R. Carley, "Device Level Early Floorplanning for RF Circuits," *IEEE Trans. CAD*, vol. 18, no. 4, April 1999.

Constraint Generation and Mapping to Physical Design

[59] U. Choudhury, A. Sangiovanni-Vincentelli, "Automatic generation of parasitic constraints for performance-constrained physical design of analog circuits", *IEEE Trans. CAD*, Vol. 12, No. 2, pp. 208-224, February 1993.
[60] E. Charbon, E. Malavasi, A. Sangiovanni-Vincentelli, "Generalized constraint generation for analog circuit design", *Proc. IEEE/ACM ICCAD*, pp. 408-414, Nov. 1993.
[61] E. Charbon, P. Miliozzi, E. Malavasi, A. Sangiovanni-Vincentelli, "Generalized constraint generation in the presence of non-deterministic parasitics," *Proc. ACM/IEEE Int'l Conf. on CAD*, 1996.

System-Level Mixed-Signal Module Generators

[62] W. J. Helms and K. C. Russel, "Switched Capacitor Filter Compiler," in *Proc. IEEE CICC.*, 1986.
[63] H. Yaghutiel, A. Sangiovanni-Vincentelli, P. R. Gray, "A methodology for automated layout of switched capacitor filters," in *Proc. ACM/IEEE ICCAD*, Nov. 1986.
[64] G. Jusef, P.R. Gray and A. Sangiovanni-Vincentelli, "CADICS–Cyclic Analog-to-Digital Converter Synthesis," *Proc. IEEE ICCAD*, pp. 286-289, Nov. 1990.
[65] H. Chang, A. Sangiovanni-Vincentelli, F. Balarin, E. Charbon, U. Choudhury, G. Jusuf, E. Liu, E. Malavasi, R. Neff and P. Gray, "A top-down, constraint-driven methodology for analog integrated circuits," Proc. IEEE Custom IC Conference, 1992.
[66] R. Neff, P. Gray and A. Sangiovanni-Vincentelli, "A module generator for high speed CMOS current output digital/analog converters," *Proc. IEEE Custom IC Conference*, 1995.

Substrate Modeling, Extraction, and Coupling Analysis

[67] T.A. Johnson, R.W. Knepper, V. Marcello, and W. Wang, "Chip substrate resistance modeling technique for integrated circuit design," *IEEE Trans. CAD*, vol. CAD-3, no. 2, April 1984.
[68] D.K. Su, M. Loinaz, S. Masui and B. Wooley, "Experimental results and modeling techniques for substrate noise in mixed-signal integrated circuits," *IEEE JSSC*, vol. 28, no. 4, April 1993.
[69] N. Verghese, D. Allstot and S. Masui, "Rapid simulation of substrate coupling effects in mixed-mode ICs," *Proc. IEEE Custom IC Conference*, 1993.
[70] F. Clement, E. Zysman, M. Kayal and M. Declerq, "LAYIN: Toward a global solution for parasitic coupling modeling and visualization," *Proc. IEEE Custom IC Conference*, 1994.
[71] R. Gharpurey and R.G. Meyer, "Modeling and analysis of substrate coupling in ICs," *Proc IEEE Custom IC Conference*, 1995.
[72] S. Mitra, R.A. Rutenbar, L.R. Carley and D.J. Allstot, "A methodology for rapid estimation of substrate-coupled switching noise," *Proc. IEEE Custom IC Conference*, 1995.
[73] N.K. Verghese, D.J. Allstot, M.A. Wolfe, "Fast parasitic extraction for substrate coupling in mixed-signal ICs," *Proc. IEEE Custom IC Conference*, 1995.
[74] I.L. Wemple and A.T. Yang, "Mixed signal switching noise analysis using Voronoi-tessellated substrate macromodels," *Proc. ACM/IEEE Design Automation Conference*, 1995.
[75] N. Verghese and D. Allstot, "SUBTRACT: A program for efficient evaluation of substrate parasitics in integrated circuits," *Proc. ACM/IEEE ICCAD*, 1995.
[76] N. Verghese, T. Schmerbeck, D. Allstot, *Simulation Techniques and Solutions for Mixed-Signal Coupling in Integrated Circuits*, Kluwer Academic Publishers, Norwell MA, 1995.

[77] T. Smedes, N.P. van der Meijs and A.J. van Genderen, "Extraction of circuit models for substrate cross-talk," *Proc. ACM/IEEE ICCAD*, 1995.

[78] K.J. Kerns, I.L. Wemple, and A.T. Yang, "Stable and efficient reduction of substrate model networks using congruence transforms," *Proc. ACM/IEEE ICCAD*, 1995.

[79] N.K. Verghese, D.J. Allstot, M.A. Wolfe, "Verification techniques for substrate coupling and their application to mixed signal IC design," *IEEE JSSC*, vol. 31, no. 3, March 1996.

[80] R. Gharpurey and R.G. Meyer, "Modeling and analysis of substrate coupling in integrated circuits," *IEEE JSSC*, vol. 31, no. 3, March 1996.

[81] P. Milliozzi, L. Carloni, E. Charbon and A. L. Sangiovanni-Vincentelli, "SUBWAVE: A methodology for modeling digital substrate noise injection in mixed-signal ICs," *Proc. IEEE Custom IC Conference*, 1996.

[82] P. Miliozzi, I. Vassiliou, E. Charbon, E. Malavasi, A. Sangiovanni-Vincentelli, "Use of sensitivities and generalized substrate models in mixed-signal IC design," *Proc. ACM/IEEE Design Automation Conference*, 1996.

[83] E. Charbon, R. Gharpurey, R.G. Meyer and A. Sangiovanni-Vincentelli, "Semi-analytical techniques for substrate characterization in the design of mixed-signal ICs," *Proc. ACM/IEEE ICCAD*, 1996.

[84] T. Tsukada, K. M. Makie-Fukuda, "Approaches to reducing Digital Noise Coupling in CMOS Mixed Signal LSIs," *IEICE Transactions on Fundamentals of Electronics, Communications, and Computer Sciences*, vol. E80-A, No. 2, Feb. 1997.

[85] N. K. Verghese and D.J. Allstot, "Verification of RF and mixed-signal integrated circuits for substrate coupling effects," *Proc. IEEE Custom IC Conference*, 1997.

System-Level Mixed-Signal Placement and Routing

[86] C. D. Kimble, *et al.*, "Analog autorouted VLSI," *Proc. ACM/IEEE CICC.*, June 1985.

[87] A.E. Dunlop et al, "Features in the LTX2 for analog layout," *Proc. IEEE ISCAS*, 1985.

[88] H. H. Chen and E. Kuh, "Glitter: A Gridless Variable Width Channel Router," *IEEE Trans. CAD*, vol. CAD-5, no. 4, pp. 459-465, Oct. 1986.

[89] J. Trnka, R. Hedman, G. Koehler and K. Lading, "A device level auto place and wire methodology for analog and digital masterslices," *IEEE ISSCC Digest of Technical Papers*, 1988.

[90] R. S. Gyurcsik, J. C. Jeen, "A generalized approach to routing mixed analog and digital signal nets in a channel," *IEEE JSSC*, Vol. 24, No. 2, pp. 436-442, Apr. 1989.

[91] U. Chowdhury and A. Sangiovanni-Vincentelli, "Use of Performance Sensitivities in Routing Analog Circuits," *Proc. IEEE ISCAS*, pp. 348-351, May 1990.

[92] U. Chowdhury and A. Sangiovanni-Vincentelli, "Constraint Generation for Routing Analog Circuits," *Proc. ACM/IEEE DAC*, pp. 561-566, June 1990.

[93] S. Mitra, S. Nag, R. A. Rutenbar, and L. R. Carley, "System-level routing of mixed-signal ASICs in WREN," *Proc. ACM/IEEE ICCAD*, Nov. 1992.

[94] U. Choudhury, A. Sangiovanni-Vincentelli, "Constraint-based channel routing for analog and mixed analog/digital circuits," *IEEE Trans. CAD*, Vol. 12, No. 4, pp. 497-510, Apr. 1993.

[95] S. Mitra, R. A. Rutenbar, L. R. Carley, D.J. Allstot, "Substrate-aware mixed-signal macrocell placement in WRIGHT," *Proc. IEEE Custom IC Conference*, 1994.

[96] S. Mitra, R. A. Rutenbar, L. R. Carley, D.J. Allstot, "Substrate-aware mixed-signal macrocell placement in WRIGHT," *IEEE JSSC*, Vol. 30, No. 3, pp. 269-278, Mar. 1995.

Mixed-Signal Power Distribution Layout

[97] B.R. Stanisic, R.A. Rutenbar and L.R. Carley, "Power distribution synthesis for analog and mixed signal ASICs in RAIL," *Proc. IEEE Custom IC Conference*, 1993.

[98] B.R. Stanisic, R.A. Rutenbar and L.R. Carley, "Mixed-Signal Noise Decoupling via simultaneous power distribution design and cell customization in RAIL," *Proc. IEEE Custom IC Conference*, 1994.

[99] B. R. Stanisic, N. K. Verghese, R. A. Rutenbar, L. R. Carley, D. J. Allstot, "Addressing substrate coupling in mixed-mode IC's: simulation and power distribution synthesis", *IEEE JSSC*, Vol. 29, No. 3, Mar. 1994.

[100]B.R. Stanisic, R.A. Rutenbar and L.R. Carley, "Addressing noise decoupling in Mixed-Signal ICs: Power distribution design and cell customization," *IEEE JSSC*, vol. 30, no. 3, March 1995.

[101]B.R. Stanisic, R.A. Rutenbar and L.R. Carley, *Synthesis of Power Distribution to Manage Signal Integrity in Mixed-Signal ICs*, Kluwer Academic Publishers, Norwell MA, 1996.

[102]H.H. Chen and D.D. Ling, "Power supply noise analysis methodology for deep submicron VLSI chip design," *Proc. ACM/IEEE Design Automation Conference*, June 1997.

Examples of Complete Analog Layout Flows

[103]R. Rutenbar *et al.*, "Synthesis and layout for mixed-signal ICs in the ACACIA system," in *Analog Circuit Design*, (J.H. Huijsing, R. J. van de Plassche and W.M.C Sansen, eds.), Kluwer Acad. Publishers, pp. 127-146, 1996.

[104]I. Vassiliou, H. Chang, A. Demir, E. Charbon, P. Miliozzi, A. Sangiovanni-Vincentelli, "A video driver system designed using a top-down constraint-driven methodology," *Proc. ACM/IEEE ICCAD*, 1996.

[105]E. Malavasi, E. Felt, E. Charbon and A. Sangiovanni-Vincentelli, "Automation of IC Layout with Analog Constraints," *IEEE Trans. CAD*, vol. 15, no. 8, August 1996.

[106]H. Chang, E. Charbon, U. Choudhury, A. Demir, E. Felt, E. Liu, E. Malavasi, A. Sangiovanni-Vincentelli, I. Vassiliou, *A Top-Down, Constraint-Driven Design Methodology for Analog Integrated Circuits*, Kluwer Academic Publishers, Norwell, MA, 1997.

Field Programmable Analog Arrays

[107]M. Sivilotti, "A dynamically configurable architecture for prototyping analog circuits," *Proc. Decennial Caltech Conference*, Cambridge MA, 1988.

[108]E.K.F. Lee and P.G. Gulak, "A field programmable analog array based on MOSFET transconductors," *Electronics Letters*, Vol. 28, no. 1, Jan. 2, 1992.

[109]E.K.F. Lee and P.G. Gulak, "A transconductor-based field programmable analog array," *IEEE ISSCC Digest of Technical Papers*, Feb., 1995

[110]H.W. Klein, "Circuit development using EPAC technology: an analog FPGA," *Proc. of the SPIE, International Society for Optical Engineering*, vol. 2607, 1995.

[111]A. Bratt and I. Macbeth, "Design and implementation of a field programmable analogue array," *Proc. ACM International Symposium on FPGAs*, 1996.

[112]P. Chow and P.G. Gulak, "A field programmable mixed analog digital array," *Proc. ACM International Symposium on FPGAs*, 1995.

[113]C.A. Looby and C. Lyden, "A CMOS continuous time field programmable analog array," *Proc. ACM International Symposium on FPGAs*, 1997.

[114]S.D. Brown, R.J. Francis, J. Rose and Z.G. Vranesic, *Field Programmable Gate Arrays*, Kluwer Academic Pub., Norwell MA, 1992.

ILAC: An Automated Layout Tool for Analog CMOS Circuits

JEF RIJMENANTS, MEMBER, IEEE, JAMES B. LITSIOS, MEMBER, IEEE, THOMAS R. SCHWARZ, MEMBER, IEEE, AND MARC G. R. DEGRAUWE, MEMBER, IEEE

Abstract —**ILAC (Interactive Layout of Analog CMOS Circuits) is a process-independent tool that automatically generates geometrical layout for analog CMOS cells from a circuit description. ILAC handles typical analog layout constraints such as device matching, symmetry, and distance and coupling constraints. ILAC supports user-specified constraints on cell height and input/output pin locations. Together with the design tool IDAC (Interactive Design for Analog CMOS Circuits), ILAC makes a fully functional analog CMOS cell compiler that automatically produces geometrical layout from functional specifications for a library of circuits including amplifiers, voltage and current references, comparators, oscillators, and A/D converters.**

I. Introduction

The TREND to replace boards with discrete electronic circuitry by a single application-specific IC (ASIC) leads to the integration of analog functions together with digital circuitry on a single chip. Although the analog circuits take only a minor part of most ASIC's, their design is very often a bottleneck. Until recently CAD tools for synthesis and layout of analog circuits were rare or nonexistent, but because of the ASIC revolution they have received a great amount of interest lately [1]–[9].

An analog CMOS cell compiler has been developed at the Centre Suisse d'Electronique et de Microtechnique (CSEM) which automatically produces geometrical layout from functional specifications for a library of circuits including amplifiers, voltage and current references, comparators, and oscillators. As shown in Fig. 1, the analog cell compiler is built around two subsystems: the design system IDAC (Interactive Design for Analog CMOS Circuits) [1] and the layout system ILAC (Interactive Layout of Analog CMOS Circuits) [2].

The design system IDAC sizes analog CMOS circuits from a library of proven schematics given a set of building-block specs and technological parameters. The IDAC cell library contains over 40 schematics such as amplifiers, voltage and current references, quartz oscillators, and comparators. The output of IDAC consists of a complete data sheet for the cell, SPICE files, and the layout description file from which ILAC generates the geometrical layout.

Manuscript received August 19, 1988; revised December 8, 1988.
The authors are with the Centre Suisse d'Electronique et de Microtechnique S.A. (CSEM), Maladière 71, CH-2007 Neuchâtel, Switzerland.
IEEE Log Number 8826159.

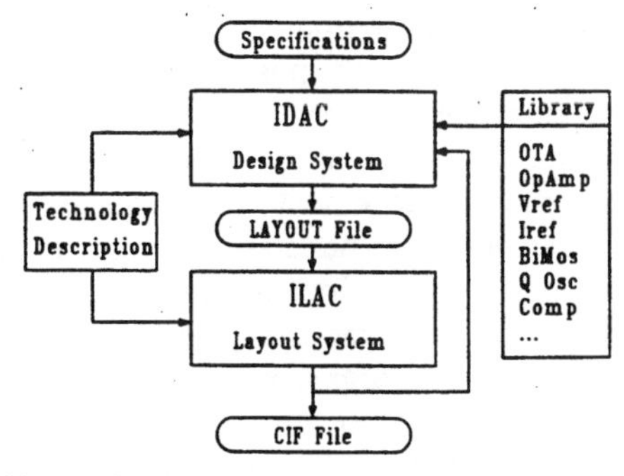

Fig. 1. Block diagram of the IDAC/ILAC system.

The layout system ILAC automatically generates geometrical layout for analog CMOS cells from net-list information under user-specified constraints on cell height and input/output pin positions. ILAC handles typical analog layout constraints such as device matching, symmetry, and distance and coupling constraints. ILAC uses the block (or macrocell) place-and-route approach to generate the layout. The blocks can be simple primitive cells such as a single MOS transistor or fairly complex subcircuits such as a lateral bipolar centroïdal structure or a programmable current mirror. Specialized layout generators are used to generate layout for the blocks. ILAC is not limited to the circuits in the IDAC library and, unlike existing analog silicon compilers that use some predefined or template placement for a specific type of circuit [8], [9], ILAC determines an optimized layout for any circuit and any set of input parameters.

Most of the placement and routing algorithms used by ILAC were inspired by algorithms used for digital design that were adapted to satisfy the typical constraints imposed on analog circuit layout, such as device matching, symmetry, and signal uncoupling. Among the techniques adopted from the digital design world are: the block place-and-route approach, the slicing-tree placement, the floorplan optimization technique based on simulated annealing, and the routing in channels [10]–[15].

The block place-and-route approach is used to be independent of circuit topology and to handle the enormous disparity in device sizes typically found in analog circuits. Shape functions [12] are used extensively to exploit the usually large number of possible ratios of the blocks.

Device matching is handled in ILAC at the device level by using specialized block layout generators (for example, for current mirrors and matched centroïdal bipolar transistors) and at the circuit level by placing blocks in matched groups where all elements have the same orientation and geometrical form (for example, for matched resistors). Symmetry is handled at the device level by using specialized layout generators (for example, for differential pairs) and at the circuit level by placing blocks in symmetry groups (for example, for differential gain stages). Distance and coupling constraints between blocks are handled by weighted penalties in the cost function of the placement optimization process. These constraints allow undesired electrical and/or thermal couplings to be avoided between, for example, input and output stages.

Routing of analog circuits poses the additional problems of avoiding couplings between sensitive and noisy nets and of minimizing parasitic capacitance and/or series resistance to get the desired circuit performance. Nets in analog circuits can therefore be classified into four categories: 1) sensitive nets such as high-impedance nodes, 2) noisy nets such as output nodes and high swing nodes (clock signals), 3) noncritical signal nets, and 4) power supply nets that require special treatment for planarity and wire sizing. The global routing algorithm used in ILAC is based on best-first maze search, but instead of just searching for a minimal interconnect length as most of its digital counterparts, it handles net couplings, undesired crossings and planarity (power nets), and channel congestion. The channel router used in ILAC is a novel type of router that handles analog routing constraints. Digital channel routers usually try only to minimize the channel size and are designed to run fast by simplifying the routing task using grids, fixed layers for horizontal and vertical segments, etc. The analog channel router used in ILAC tries to find routing paths that minimize the parasitics on sensitive nodes and the couplings between sensitive and noisy nodes.

The next section gives a brief overview of the ILAC system. Section III describes circuit partitioning and block layout generation. Section IV describes block placement and routing. Section V describes layout compaction. Further, in Section VI practical results obtained with the system are presented. Finally, future developments and concluding remarks are given in the last section.

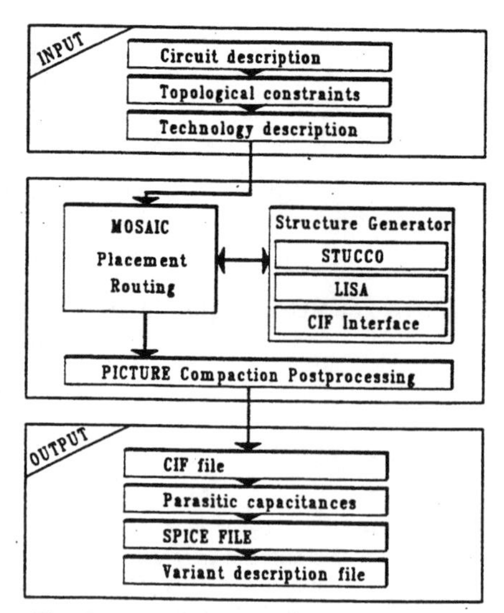

Fig. 2. Block diagram of the ILAC system.

II. Overview of the ILAC Layout System

A block diagram of the ILAC system is shown in Fig. 2. The input for ILAC consists of a technology description file, a circuit description file, and an optional file with topological constraints. Starting from this information the program generates a geometrical layout in CIF [23] or Calma GDS II [24] format and a parasitic capacitance file from which IDAC can generate the complete data sheet of the cell.

To be process tolerant, ILAC internally uses generic layers that are mapped to the physical mask layers using Boolean mask operations in a postprocessing step. The technology description file contains all process-dependent information such as process type, design rules, electrical characteristics, and mask layer definitions. The syntax allows description of all the rules found in CMOS processes and can be extended for BiCMOS processes. Fig. 3 shows an example of the technology description file: it specifies basic characteristics such as types of tubs (or wells), the number of interconnection layers, definitions of additional layers to help specify rules, aliases for condensed specification, actual design rules that can have minimum as well as maximum values, electrical characteristics such as maximum current density for wires, and definitions of additional and CIF mask layers as functions of others.

The layout description input file for ILAC contains device information, interconnections, and matching and distance constraints. The input description language is general enough so that ILAC is not limited to the circuits in the IDAC library. Fig. 4 shows the schematic of a voltage reference circuit with lateral bipolar transistors [20], [21] and Fig. 5 shows the layout description file for this circuit generated by IDAC. The circuit contains MOS transistors, lateral bipolar transistors, resistors, and a capacitor. These elements are specified here with their interconnections, their size, and variant information. The file also contains specifications for structures in the circuit, the nodes that are external connections, and some matching constraints between elements.

The user can specify constraints on cell height and input/output pin positions. Pin positions can be free or on a specific side of the cell. Pins can be ordered relatively or placed at absolute positions with respect to the cell border. This makes it possible to generate cells for a standard cell environment or a general cell environment. By default the program generates a standard-cell-like layout with the power supply pins on the left and right side, and the other signal pins on the top or bottom side. Topological con-

```
VERSION( 1, 1 );
TECHNOLOGY techno_x;

UNITS
LAMBDA = 1u;
RESOLUTION = 2/1u;

CHARACTERISTICS
n_bulk; p_tub;
metal( 2 ); poly( 1, MONODOPED );

DEFINE
ccap; nchan_gate; pchan_gate;
diff_tub; inco;
...

ALIAS
a_cont = <cpol, cdiff, cbulk, cbutt>;
a_ptub = <ptub, lnpn>;
a_ndiff = <ndiff, ntr, nguard, ndnt >;
...

RULES
# p-well
WIDTH    [a_ptub] >= 3;
EQUIPOT [a_ptub, a_ptub] >= 6;
SPACING [a_ptub, a_ptub] >= 9;
OVERLAP [a_ptub, a_diff] >= 6;
SPACING [a_ptub, a_tdiff] >= 6;
INTERSECT [a_diff, a_ptub] >= 3;
EXTENSION [a_diff, a_ptub] >= 3;
# contact cuts
WIDTH    [a_cont] >= 3;
         [a_cont] <= 9;
SPACING [a_cont, a_cont] >= 3;
...
```

```
ELECTRICAL
CURRENT [metal] := 0.5m/1u;
        [via] := 0.07m/1u;
        [smetal] := 1m/1u;

EVALUATE
nchan_gate := ntr AND poly;
pchan_gate := ptr AND poly;
ptub := ptub OR lnpn OR (ntr OR nguard)
  OVERSIZED OVERLAP[ptub, ntr] OR
  pdpt OVERSIZED OVERLAP[ptub, pdpt];
diff_tub := a_diff AND ptub;
ntr := ntr ANDNOT nchan_gate;
ptr := ptr ANDNOT pchan_gate;
...

CIF
cpw := ptub;
cnd := a_ndiff OR nchan_gate;
cpd := a_pdiff OR pchan_gate;
cp := poly OR cpol OVERSIZED 0.2;
cpi :=
  ((pdiff OR ptr OR pchan_gate) ANDNOT
    ndnt OVERSIZED OVERLAP[pimpl,pdiff])
  OVERSIZED OVERLAP[pimpl,pdiff] OR
  (pdpt ANDNOT
   a_ndiff OVERSIZED OVERLAP[pimpl,pdpt])
  OVERSIZED OVERLAP[pimpl,pdpt];
cm := metal OR inco OVERSIZED 0.2 OR bond;
cod := cdiff OR cbulk OR cbutt;
ce := cpol OVERSIZED OVERLAP[ce, cpol];
...

COLOURS
...
END.
```

Fig. 3. Sample technology description file.

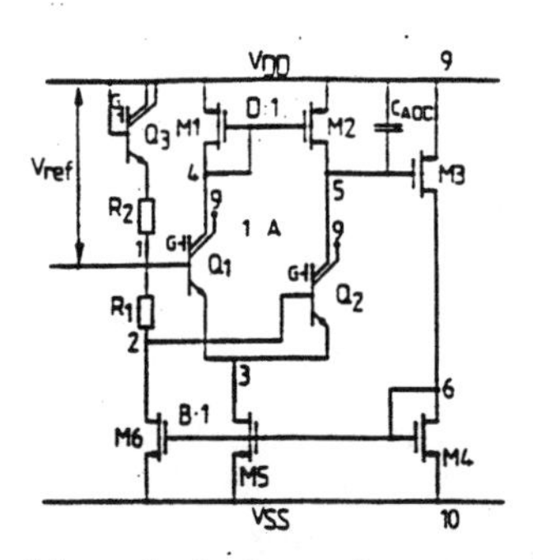

Fig. 4. Schematic of voltage reference source *V*REF2.

```
VERSION( 1, 3 );
CELL( 'VREF2' );
TECHNOLOGY( 'TECHNO3U' );
CONSUMPTION( 4.767E-04 );

PMOS( M01, PIN( 4, 4,  9,  9 ),   129u, 5u, FREE, 1, 1 );
PMOS( M02, PIN( 5, 4,  9,  9 ),   129u, 5u, FREE, 1, 1 );
PMOS( M03, PIN( 6, 5,  9,  9 ), 737.5u, 5u, FREE, 1, 1 );
NMOS( M04, PIN( 6, 6, 10, 10 ),   155u, 3u, FREE, 1, 1 );
NMOS( M05, PIN( 3, 6, 10, 10 ),   155u, 3u, FREE, 1, 1 );
NMOS( M06, PIN( 2, 6, 10, 10 ),   155u, 3u, FREE, 1, 1 );

LATNPN( QL01, PIN( 4, 9, 1,  3, 99 ), 1 );
LATNPN( QL02, PIN( 5, 9, 2,  3, 99 ), 8 );
LATNPN( QL03, PIN( 9, 9, 9, 13, 99 ), 1 );

RES( R01, PIN(  1, 2 ), SQUARES( 388 ), POLY, FREE );
RES( R02, PIN( 13, 1 ), SQUARES( 1669 ), POLY, FREE );
CAP( CADD, PIN( 5, 9 ), AREA( 10418uu ), METAL, POLY, FREE );

MIRROR( MIR01, M04, M05, M06 );
CENTER( CEN01, QL01, QL02 );

SUPPLY_VDD( VDD, 9 );
SUPPLY_VSS( VSS, 10 );
OUTPUT( VREF, 1 );
INPUT( VG, 99 );

MATCH( MAT01, M01, M02 );
MATCH( MAT02, R01, R02 );

END.
```

Fig. 5. Layout description file for cell *V*REF2 (generated by IDAC).

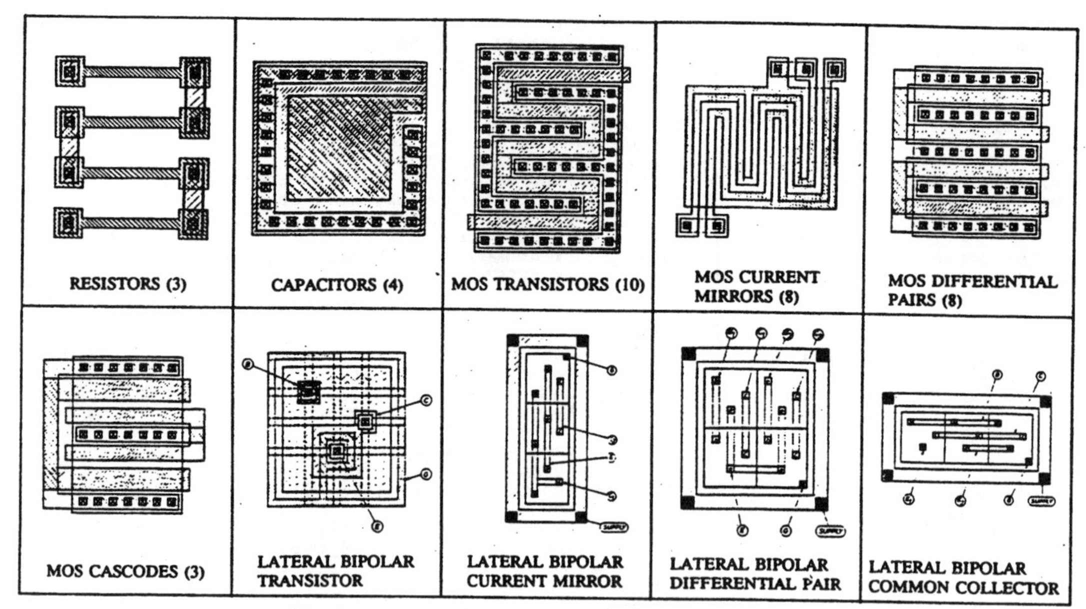

Fig. 6. Overview of the STUCCO block generator library.

straints can be added and modified interactively from a menu.

Starting from the above-mentioned information, the ILAC system generates the geometrical layout in CIF or GDS II format, a file with routing parasitic capacitances in SPICE format, and a variant description file that allows the cell generated to be reused in a higher hierarchical level by means of the CIF file interface.

The layout is generated in three steps: 1) partitioning the circuit into a number of blocks and generating a layout for the blocks; 2) placement and routing of these blocks; and 3) layout compaction. The next sections describe these three steps in more detail.

III. Circuit Partitioning and Block Layout Generation

The first step in layout generation is the partitioning of the circuit into blocks and selecting how the blocks will be laid out according to the input specifications. The blocks can be basic devices such as transistors, resistors, capacitors, or structures such as current mirrors or differential pairs. The blocks can be placed together in clusters, matched groups, or symmetry groups.

Each block can have a number of possible physical realizations, called variants. Typical MOS transistor variants are the following: rectangular, snake, interdigit, circular, concentric, S-shape, and waffle. Each variant can have a number of different ratios or shapes. For example, an interdigit MOS transistor has a number of different ratios or shapes depending on its W, L, and the number of fingers of its interdigital gate. All the shapes of a variant are encoded in a shape function, which is a curve that delimits the region containing all possible rectangles that can accommodate a given block [12].

The block layout generator serves two functions. For each possible variant of a block it provides the list of possible shapes with their corresponding ports. Secondly it generates the geometrical layout for a particular shape. ILAC currently supports the following three ways for block generation: 1) the STUCCO procedural layout generator module, 2) the LISA layout language system, and 3) the CIF file interface for externally generated blocks. A brief description of these three possibilities is given below.

A. The STUCCO Procedural Block Layout Generator

The STUCCO module is a set of parameterized procedural layout generators for the most commonly used blocks. Fig. 6 shows an overview of the blocks supported by STUCCO. The STUCCO library currently contains more than 40 process-independent and parameterized blocks. There are three types of resistors (snake polysilicon, segmented polysilicon, segmented tub), four capacitors (rectangular metal-poly, low-resistance metal-poly, interdigit metal-poly, and rectangular poly-poly2), ten MOS transistors (rectangular, interdigit, U-shape, snake, S-shape, concentric, waffle, $\cdots$), eight MOS current mirrors, eight MOS differential pairs, three MOS cascodes, and one tub diode. There are a unitary lateral bipolar transistor [20], a lateral bipolar current mirror, a differential pair, and a common collector and a centroïdal structure. STUCCO has specialized procedures to get the best possible device matching for structures.

Fig. 7 shows a centroïdal structure of two lateral bipolar transistors with a ratio of 1 : 8. The block layout generator always minimizes the interconnect resistance for emitters and bases by using aluminum as interconnection level. Furthermore a guard ring is designed around the structure in order to collect the vertical current. The block generator STUCCO automatically assures a good matching as well as a proper functioning of the structure by "copying" proven "hand" layouts in a parameterizable way.

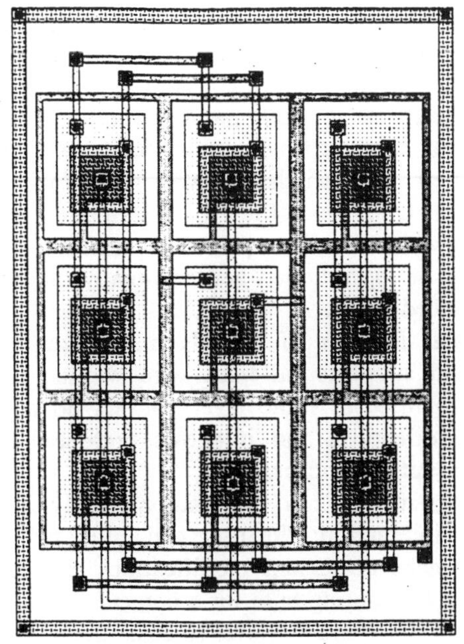

Fig. 7. Lateral bipolar centroïdal structure generated by STUCCO.

B. The LISA Procedural Layout System

An alternative way to generate layout for blocks is the LISA system [16]. LISA permits the interactive layout of technology-independent structures. LISA differs from other procedural systems (i [17], L [25]) in its nonimperative language syntax and its focus on the generation of analog structures. The LISA system is hierarchical, incremental, and constantly shows the graphical layout of the structure being edited.

The LISA language is declarative and nonimperative, that is, it is line oriented and sentence-position independent. This makes it not only possible to keep the power of a interactive layout design with its incremental and graphics features, but also to embed behavioral and structural information algorithmically. Elementary language statements allow the creation of basic geometrical primitives and their placement with relational technology-independent constraints. Using sets of elements, conditional statements, and hierarchical features, it is possible to implement complex algorithms and generate parameterizable structures.

Although LISA has been successfully used to create digital structures, it is especially suited for analog structures because of its simple geometrical elements (rectangles, doughnuts, wires, ports, cells). Unlike transistor-level primitives, these low-level geometric primitives allow the direct implementation of complex analog structures.

The LISA language can be seen as a means to define a formal technology-independent specification of a structure. The LISA system contains a compactor that will lay out the structures for a given technology. The compactor contains efficient algorithms for overconstraint detection and area optimization [17], [19]. This results in a fast change–compile–check cycle for the creation of technology-independent analog structures.

C. The CIF File Interface

Blocks generated outside the ILAC system can be imported by providing a variant description file that describes the possible variants, the shape functions, and the port lists. The geometrical layout for each shape is stored in a separate CIF file that is referenced from the variant description file. This feature also allows a basic form of hierarchy since one of the outputs of the placement and routing program is a variant description file of the cell generated.

IV. Block Placement and Routing

The second step in layout generation is block placement and routing, which is done by MOSAIC. Placement and routing of the blocks is done in a three step process: 1) a floorplan preoptimization phase to find an initial floorplan; 2) the actual placement optimization phase which includes global routing (or route planning); and 3) the detailed routing phase.

The placement algorithm is based on slicing structures [12], [13] and handles individual blocks as well as groups of blocks. Groups can be matched groups, symmetry groups, or cluster groups. Functional blocks can have a set of possible physical realizations, called variants, each with their own shape function.

Placement optimization is based on the technique of simulated annealing [10], [11]. During the optimization process, a large number of possible slicing structures is generated by making small modifications, called moves, at random to the current placement. For every placement generated a given cost function is evaluated and the newly generated placement is accepted as the current one if its cost is lower than the current cost. If its cost is higher, the new placement can be accepted with a probability that decreases during the process.

The first step in MOSAIC is to find an initial floorplan using a very rough routing space estimate. This is done in a simulated annealing optimization loop. The starting solution is constructed by putting the blocks in alternately horizontal and vertical slices. The set of moves to generate new placements during the annealing process includes swapping two components, reversing the cut type for a subslice, and changing the orientation or the variant for a component. Moves treat groups as an entity or are local to the group. Blocks in a symmetry group are kept symmetrical two-by-two around some axis. Blocks in a matched group always have the same variant and orientation. Blocks in a cluster group are just kept together in one subtree. Channel sizes are estimated from the number of nets that have ports in the channel. This usually overestimates the channel size, but the results are better than with the fixed channel size approach. The cost function has three factors: cell area, center-point distance between connected blocks, and penalties for constraint violations. This initial floorplan is just a starting solution for the next step, which is more CPU-intensive.

The next step is to optimize the floorplan and global routing simultaneously. This is also done in a simulated annealing loop using the floorplan from the previous step as a starting solution. The set of moves is the same as for the first step. The cost function also has three factors: cell area, total wire length, and penalties for constraint violations. For every floorplan generated the following operations have to be carried out: a) a first slicing tree evaluation to find positions for the blocks and ports; b) planar global power and ground routing; c) global routing for the other nets; d) tub routing; e) channel size estimation based on net and tub data; and f) a second slicing tree evaluation with the estimated channel sizes.

Slicing tree evaluation consists of: a) determining the shape function for each slice by adding up the shape functions of its child slices in a bottom-up traversal of the tree (the size of the routing channels that correspond to the cut lines of the slicing structure are added in as offsets); b) selecting the bounds of the cell from the global shape function to satisfy the user-specified cell height constraint or to obtain a minimal area; and c) calculating slice bounds and center positions from the parent's bounds and center position in a top-down traversal of the tree [13].

The global routing algorithm determines for each net the path to follow through the routing channels. The algorithm is best-first search based [14], [15] and handles net sensitivities, planarity (for the power nets), and channel congestion. Ports are not restricted to grid points but are Manhattan intervals that can be anywhere inside the blocks as long as they can be contacted from the outside by wires perpendicular to the port interval without causing design-rule violations. Tub (or well) routing consists in scanning the channels to find out whether or not tubs can be merged across the channel and to add tub contacts to make sure that all tubs are properly anchored to some potential. After global and tub routing, the size of the routing channels is estimated based on the number of nets and the type of tubs in the channels.

The last step is to flesh out the interconnections. The channels are routed one by one from the bottom upwards in the slicing structure. If the estimated size and actual channel size are different, the blocks and channels are repositioned accordingly. If the channel router fails to complete a net (typically because the spacing between ports is less than the contact-to-contact spacing), it attempts to route the nets in another order; if these attempts fail too, the placement is rejected.

Channel routing is done by a novel scan-line-based incremental channel router specially developed to handle analog routing constraints such as net coupling and minimization of parasitics. The router routes net-by-net in a given priority order; power nets are normally routed first to ensure planarity, followed by the sensitive nets to ensure the shortest path on the preferential layer to minimize parasitics. The noisy nets are routed last, after the noncritical nets that together with the supply nets act as shields between the sensitive and the noisy nets. The router scans the channel along its spine and evaluates all possible paths between a source and multiple target points. Spacings between already routed wire segments that run parallel to the spine of the channel are considered to be stretchable. The cost function to rank paths includes distance and penalties for crossing or running adjacent to noisy or sensitive nets, for switching layers, and for increasing channel size. Once an optimal path is found, the path is embedded, which usually requires that segments of already routed nets be repositioned and eventually that the channel is stretched.

Additional features of the channel router are: it is gridless and supports different wire widths; it can be extended to three or more layers; layers do not have a predefined direction; and it allows prerouted nets and interactive routing modification.

The program can run fully automatically but allows some user interaction: the user can enter his preferred floorplan, make placement modifications (moves) manually, change the routing path for a net, etc. All interactions are done on the symbolic level (blocks, ports, wires), not on the geometrical (polygon) level. Additional features of the user interface allow the user, for example, to highlight objects such as a sensitive net to verify its path and to acquire valuable information such as the total parasitic capacitance for the net.

V. Geometrical Postprocessing and Compaction

The final step in layout generation is compaction and geometrical postprocessing, which is done by PICTURE. The geometrical layout generated by MOSAIC contains only a limited set of generic layers. PICTURE is used to convert this set of process-independent layers into the final set of process-specific layers. Integrated in this conversion process, to insure the validity of the layout, is a design rule check (DRC). This operation is usually followed by a compaction. In view of the fact that it is impossible to be absolutely sure that the layout generated by the block layout generators is correct, a DRC is always performed in order to enhance the reliability of the tool. The compactor is useful to minimize the layout size. The size of the layout after routing is not optimal because the routing does not interact with the inside of the modules. The compactor eliminates these extra spaces, but does not modify the internals of the structures. Constraints on cell height and input/output pin positions (pitch matching) are also taken into account.

All of the internal PICTURE operations are done with the scan-line algorithm. The program supports hierarchical Manhattan geometries and uses an "on the fly" flattening for the scan line. All basic sizing and logical layer operations are supported, as are equipotential node extraction, wire length minimization [18], overconstraint detection [19], and multigrid compaction.

The geometrical layout is output in a CIF format (compatible with the format used by VLSI Technology Inc.) or in a Calma GDS II stream format.

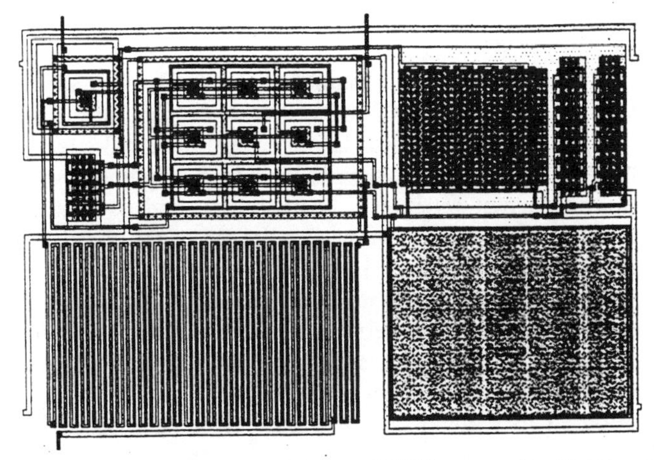

Fig. 8. Layout of voltage reference VREF2 generated by ILAC.

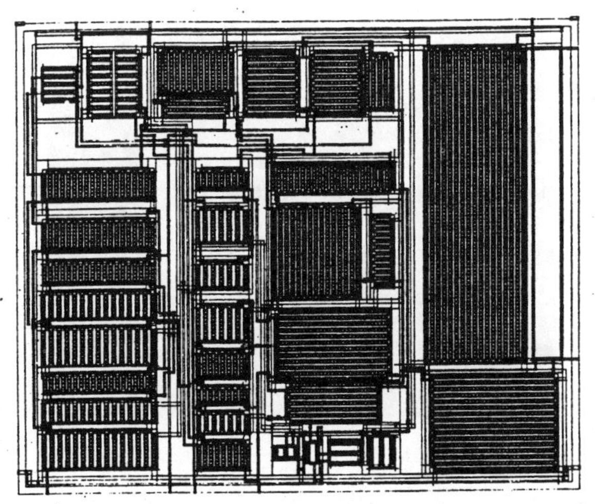

Fig. 10. Layout of an operational amplifier generated by ILAC.

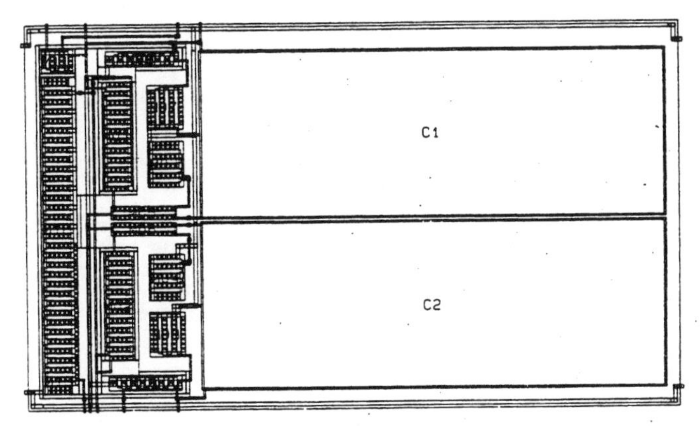

Fig. 9. Symmetrical layout of comparator COMP1 generated by ILAC.

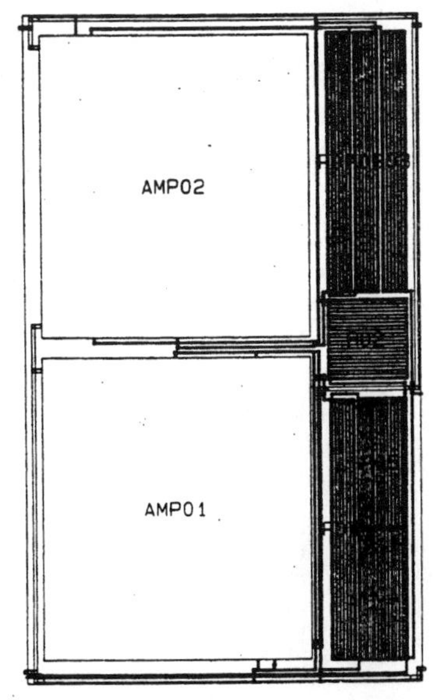

Fig. 11. Hierarchical layout example generated by ILAC.

VI. Practical Results

This section shows some layouts generated by the ILAC system. The run times mentioned are for a DEC VAX 8600, which is a 4 MIPS machine. The bulk of the CPU time is spent in the second phase of MOSAIC, i.e., the simultaneous optimization of floorplan and global routing.

Fig. 8 shows the layout generated by ILAC for the voltage reference circuit VREF2 of Figs. 4 and 5. The program was asked to minimize the total circuit size. Run time for this circuit was 3 minutes of CPU time.

Fig. 9 shows the layout generated by ILAC for a comparator circuit (COMP1). The circuit is laid-out symmetrically according to the input specifications. Run time for this circuit was 3.5 minutes of CPU time.

Fig. 10 shows the layout generated by ILAC for an operational amplifier taken from literature [22]. The program took 15 minutes of CPU time to generate the layout for this circuit which contains 36 transistors. The layout shown is not compacted. Clearly visible in this layout are the different cluster groups, with the output stage at the right and the input differential pairs in the top left corner.

Fig. 11 shows an hierarchical example generated by ILAC. The circuit, a gyrator, consists of two identical op amps and a number of matched resistor groups. The op amp was laid out by ILAC from an input description generated with IDAC. The program took 1.5 minutes of CPU time to generate the op amp, and 2 minutes to generate the gyrator itself.

A chip photograph of an operational transconductance amplifier (OTA2) dimensioned by IDAC and laid out by an early version of ILAC is shown in Fig. 12. From the comparison between predicted and measured characteristics shown in Table I one can see that the performances are adequately predicted.

Compared to handcrafted layouts made at CSEM for cells of the IDAC library, the layouts generated by ILAC are on the average 25 percent larger. The density of the resulting layout is mainly determined by the availability of block generators to realize a circuit. For a worst-case circuit, composed of separate transistors of minimum size, the layout generated by ILAC is not dense, because of the slicing structure placement and the routing in channels. Even compaction does not eliminate most of the white space left by MOSAIC for such a circuit, as MOSAIC treats the blocks as black boxes and uses block spacings

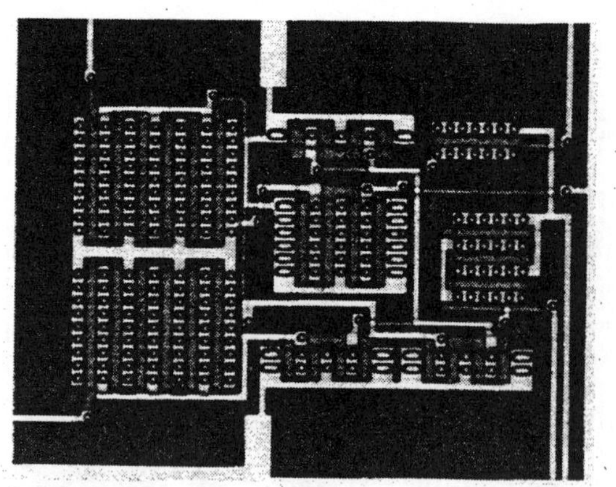

Fig. 12. Chip photograph of operational transconductance amplifier OTA2.

TABLE I
PREDICTED AND MEASURED CHARACTERISTICS OF OTA2

Parameters	Measurement results				IDAC			Units
	MIN	MAX	MEAN	SIGMA	MIN	MAX	SLOW-SLOW	
DC gain	87	90.7	88.8	1.38	88	95.6	88.6	dB
GBW	1440	1600	1520	45	1170	1543	1221	kHz
Phase margin	70.7	72.4	71.7	0.5	75.7	82.2	81	deg
Slew-rate +	0.666	0.704	0.682	0.014	0.89	0.99	0.94	V/usec
Slew-rate −	0.447	0.484	0.471	0.011	−	−	−	V/usec
Out swing +	1.162	1.362	1.254	0.056	1.22	1.33	−	V
Out swing −	−1.81	−1.63	−1.76	0.054	−1.32	−1.29	−	V
Offset	−	−	−6.2	1.68	2.1	3.3	−	mV
White noise	−	−	−	−	14.6	20.6	17.6	nV/√(Hz)
1/f noise(1kHz)	219	262	239	13.8	274	359	333	nV/√(Hz)
Itot	31.6	33.6	32.48	0.62	29.1	32.1	30.6	uA
Power	158	168.1	162.4	3.07	145.5	160.5	−	uW
Gm	150.6	161.8	156.3	3.44	−	−	120.9	uA/V
CMRR (10Hz)	63.9	77	68	3.9	−	−	−	dB
CMRR (1kHz)	63.3	75.2	67.2	3.6	−	−	−	dB
PSRR + (10Hz)	51	52.9	52.1	0.75	−	−	−	dB
PSRR + (1kHz)	50.1	51.9	51.2	0.73	−	−	−	dB
PSRR − (10Hz)	47.7	49.6	48.8	0.76	−	−	−	dB
PSRR − (1kHz)	46.8	48.8	47.9	0.8	−	−	−	dB

that are "on the safe side." For manual layout it is, however, also a nontrivial task to get a very dense layout for such a circuit.

VII. IMPLEMENTATION AND FUTURE DEVELOPMENTS

ILAC is written in Pascal (90 000 lines of code for the complete system, of which 20 000 are for placement and routing and 18 000 are for block generation) and runs on DEC VAX/VMS machines. Interactive graphics are supported on Tektronix 41xx graphics terminals and DEC microVAX-GPX workstations.

Work is currently being done to port the system to UNIX-based workstations. Future developments of the ILAC system include: a) automatic wire sizing in routing and block generation from current information passed by IDAC; b) an improved hierarchical version with extensions for mixed digital/analog layout—this version will be able to handle more complex circuits such as A/D converters; and c) layout generators for vertical bipolar devices and support for BiCMOS technology.

VIII. CONCLUSIONS

Together with its companion tool IDAC, ILAC makes a fully functional analog CMOS cell compiler that automatically produces geometrical layout from functional specifications for a library of circuits including amplifiers, voltage and current references, comparators, and oscillators. The ILAC system is, however, not limited to the circuits in the IDAC library.

ILAC handles typical analog layout constraints such as device matching, symmetry, and distance and coupling constraints, and generates geometrical layout automatically in a modest amount of CPU time. The area penalty compared to manual layout is acceptable because of the gain in time over manual layout and the fact that the resulting layout is error-free and device matching rules are respected.

ACKNOWLEDGMENT

The authors wish to thank their colleagues of the CSEM Circuits Department for the many interesting discussions and valuable suggestions, and for volunteering to test the software. They also wish to express their gratitude to M. Schwoerer for measuring the test circuits, as well as to R. Zinszner who helped to start up this work.

REFERENCES

[1] M. Degrauwe *et al.*, "IDAC: An interactive design tool for analog CMOS circuits," *IEEE J. Solid-State Circuits*, vol. SC-22, no. 6, pp. 1106–1116, Dec. 1987.

[2] J. Rijmenants, T. Schwarz, J. Litsios, and R. Zinszner, "ILAC: An automated layout tool for analog CMOS circuits," in *Proc. Custom Integrated Circuits Conf.*, May 1988, Paper 7.6.

[3] M. Kayal, S. Piguet, M. Declercq, and B. Hochet, "SALIM: A layout generation tool for analog ICs," in *Proc. Custom Integrated Circuits Conf.*, May 1988, Paper 7.5.

[4] R. Harjani, R. Rutenbar, and L. Carley, "A prototype framework for knowledge based analog circuit synthesis," in *Proc. 24th Des. Automation Conf.*, June 1987, pp. 42–49.

[5] H. Koh, C. Sequin, and P. Gray, "Automatic synthesis of operational amplifiers based on analytic circuit models," in *Proc. Int. Conf. Computer Aided Des.*, Nov. 1987, pp. 502–505.

[6] R. J. Bowman and D. J. Lane, "An knowledge based system for analog integrated circuit design," in *Proc. Int. Conf. Computer Aided Des.*, Nov. 1985, pp. 210–212.

[7] F. M. El-Turky and R. A. Nordin, "Blades: An expert system for analog circuit design," in *Proc. IEEE ISCAS*, June 1986, pp. 552–555.

[8] P. E. Allen and P. R. Barton, "A silicon compiler for successive approximation A/D and D/A converters," in *Proc. Custom Integrated Circuits Conf.*, June 1986, pp. 552–555.

[9] J. Kuhn, "Analog module generators for silicon compilation," *VLSI Syst. Des.*, pp. 74–80, May 4, 1987.

[10] S. Kirkpatrick, C. D. Gelatt, and M. P. Vecchi, "Optimization by simulated annealing," *Science*, vol. 220, pp. 671–680, 1983.

[11] P. J. M. van Laarhoven and E. H. L. Aarts, *Simulated Annealing: Theory and Applications*. Dordrecht, The Netherlands: Reidel (Kluwer), 1987.

[12] R. H. J. M. Otten, "Efficient floorplan optimization," in *Proc. ICCD*, 1983, pp. 499–502.

[13] D. F. Wong and C. L. Liu, "A new algorithm for floorplan design," in *Proc. 23rd Des. Automation Conf.*, June 1986, pp. 101–107.

[14] C. Y. Lee, "An algorithm for path connections and its applications," *IRE Trans. Electron. Comput.*, vol. EC-10, pp. 346–365, Sept. 1961.

[15] G. W. Clow, "A global routing algorithm for general cells," in *Proc. 21st Des. Automation Conf.*, June 1984, pp. 45–51.

[16] M. van Swaay, "LISA: A declarative language for interactive layout generation," CSEM, Neuchâtel, Switzerland, Internal Rep. 193, Apr. 1988.

[17] D. D. Hill, J. P. Fishburn, and M. D. P. Leland, "Effective use of virtual grid compaction in macro-module generators," in *Proc. 22nd Des. Automation Conf.*, June 1985, pp. 777–780.

[18] W. L. Shiele, "Improved compaction by minimized length of wires," in *Proc. 20th Des. Automation Conf.*, June 1983, pp. 121–127.

[19] J. L. Burns and A. R. Newton, "SPARCS: A new constraint-based IC symbolic layout spacer," in *Proc. Custom Integrated Circuits Conf.*, May 1986, pp. 534–539.

[20] E. Vittoz, "MOS transistors operated in the lateral bipolar mode and their application in CMOS technology," *IEEE J. Solid-State Circuits*, vol. SC-18, no. 6, pp. 273–279, June 1983.

[21] M. Degrauwe, O. Leuthold, E. Vittoz, H. Oguey, and A. Descombes, "CMOS voltage references using lateral bipolar transistors," *IEEE J. Solid-State Circuits*, vol. SC-20, no. 6, pp. 1151–1157, Dec. 1985.

[22] J. A. Fisher and R. Koch, "A highly linear CMOS buffer amplifier," in *Proc. ESSCIRC* (Delft, The Netherlands), Sept. 1986, pp. 80–82.

[23] C. Mead and L. Conway, *Introduction to VLSI Systems*. Reading, MA: Addison-Wesley, 1980, pp. 115–127.

[24] *GDS II System Manager's Manual*, Release 5.2, Appendix B, Calma Co., Milpitas, CA, 1985.

[25] *GDT IC Design Guide*, Silicon Compiler Systems Corp., Warren, NJ, 1987.

Jef Rijmenants (S'80–M'81) was born in Nijlen, Belgium, on May 30, 1956. He received the degree of Burgerlijk Ingenieur Electronica (electrical engineer) from the Katholieke Universiteit Leuven (KUL), Leuven, Belgium, in 1979.

After graduating he spent two years at ESAT Laboratories, Leuven, Belgium, as a Research Assistant working in the field of micropower sampled data circuits. Later he joined Silvar-Lisco N.V., Leuven, Belgium, where he worked as a Software Engineer on a symbolic layout editor integrated with a compactor, a graphics postprocessor, and database interfaces. He joined the Centre Suisse d'Electronique et de Microtechnique (CSEM), Neuchâtel, Switzerland, in 1985, where he is currently working on automated layout techniques for analog integrated circuits.

James B. Litsios (S'85–M'87) was born in Boston, MA, on January 6, 1963. He received the engineering degree in electronics from the Ecole d'Ingénieur de Genève in 1982, and in computer science from the Ecole Polytechnique Fédéral de Lausanne (EPFL) in 1987.

After graduating he joined the Centre Suisse d'Electronique et de Microtechnique (CSEM), Neuchâtel, Switzerland, where he is currently working in the Circuits Department. His current research involves geometrical algorithms for postprocessing analog circuits and automated synthesis techniques for analog circuits.

Thomas R. Schwarz (S'85–M'86) was born in Basle, Switzerland, on October 12, 1962. He received the engineering degree in electronics from the Swiss Federal Institute of Technology (EPFL), Lausanne, Switzerland, in 1986.

After graduating he was a Research Assistant at the EPFL. Later he joined the Centre Suisse d'Electronique et de Microtechnique (CSEM), Neuchâtel, Switzerland, where he is currently involved in the development of a technology interface for process independence, a procedural analog structures generator, and a floorplan editor for a gate-matrix layout system.

Marc G. R. Degrauwe (S'78–M'84) was born in Brussels, Belgium, on August 16, 1957. He received the engineering degree in electronics and the Ph.D. degree in applied sciences from the Katholieke Universiteit Leuven (KUL), Leuven, Belgium, in 1980 and 1983, respectively.

During the summer of 1980, he was on leave at the Centre Electronique Horloger S.A. (CEH), Neuchâtel, Switzerland. From autumn 1980 to 1983 he was associated with KUL where he worked on the design of micropower amplifiers and sampled data filters. In July 1983 he returned to CEH. In 1984 when CEH was reorganized into the Swiss Centre of Electronics and Microtechnics (CSEM), he became Head of the Circuits Department. His actual field of interest is design automation of analog circuits. He is also lecturing on analog circuit design and supervising student work at the University of Neuchâtel.

Dr. Degrauwe received the 1987 ESSCIRC Conference Best Paper Award for a paper on crystal oscillators he co-authored.

KOAN/ANAGRAM II: New Tools for Device-Level Analog Placement and Routing

John M. Cohn, *Member, IEEE*, David J. Garrod, *Student Member, IEEE*, Rob A. Rutenbar, *Senior Member, IEEE*, and L. Richard Carley, *Senior Member, IEEE*

***Abstract*—This paper describes KOAN and ANAGRAM II, new tools for device-level analog placement and routing. A block place-and-route style from macrocell digital IC's has recently emerged as a viable methodology for the automatic layout of custom analog cells. In this *macrocell* style, parameterized module generators produce geometry for individual devices, a placer arranges these devices, and a router embeds the wiring. However, analog layout tools that merely apply known digital macrocell techniques fall far short of achieving the density and performance of handcrafted analog cells. KOAN and ANAGRAM II differ from existing approaches by employing general algorithmic techniques to find critical device-level layout optimizations rather than relying on a large library of fixed-topology module generators. New placement algorithms implemented in KOAN handle complex layout symmetries, dynamic merging and abutment of individual devices, and flexible generation of wells and bulk contacts. New routing algorithms implemented in ANAGRAM II handle arbitrary gridless design rules in addition to over-the-device, crosstalk avoiding, mirror-symmetric, and self-symmetric wiring. Examples of CMOS and BiCMOS analog cell layouts produced by these tools are presented.**

I. Introduction

TWO recent trends have exposed custom analog layout as a bottleneck in the path from specifications to silicon for analog cells. The first trend is the growing importance of mixed-signal ASIC's. Although an increasing fraction of ASIC designs requires integration of analog and digital components, the analog standard cell libraries used to implement them often cannot supply all the analog cells necessary for a given design. The second trend is the emergence of cell-level analog circuit synthesis tools [1]–[4] which can quickly transform circuit specifications into sized schematics for some important classes of cells. In both of these design scenarios, custom analog cell layout is required, which can be a time-consuming, critical bottleneck.

The problem of custom analog layout has generated considerable interest in the last few years. The critical problems involve handling analog-specific constraints that render the layout much more sensitive to low-level geometric choices than digital cells of similar size. Work to date on analog layout has included knowledge-based approaches [4]–[6], algorithmic techniques for placement [7]–[11], routing [7]–[13], compaction [7], [14], procedural device and module generation [3], [9], [15]–[17], and performance constraint generation [18], [19]. Our interest is algorithmic placement and routing for custom analog cells. A block place-and-route approach, which we refer to as the macrocell layout style [3], [7], [8], has emerged as a popular candidate for recent analog cell layout tools: from a netlist, critical primitives (such as matched devices or complex folded structures) are produced by parameterized module generators; these primitive blocks are placed, and then routed. This style is derived from techniques for layout of digital IC's. Unsurprisingly, many existing analog cell layout tools borrow heavily from the digital macrocell style, adapting digital layout ideas to analog problems.

Our central argument is that certain attributes of this digital layout style limit its ability to achieve high-quality analog cell layouts. Some common assumptions that help manage the complexity of large digital layouts, e.g., a slicing-style placement [20], the restriction that signal routing be confined to channels between placed objects, and the exclusive emphasis on minimizing wire length and area, are not essential for attacking analog device-level layout. In our experience, these assumptions actually interfere with the type of low-level optimizations common in manual analog cell layout. For example, much of the creativity displayed by analog layout experts involves shaping, folding, placing, and merging individual devices to achieve dense layouts. In such high-quality layouts, many connections are achieved by abutment rather than explicit wires, and some fraction of the remaining wires is routed directly over devices. These optimizations reduce not only layout area, but more importantly, the device parasitics themselves. In current analog macrocell systems, such optimizations appear inside procedurally generated subcircuits, but not between the modules involved in placement and wiring. Indeed, these systems usually require a large library of device generator programs, each implementing some common arrangement of basic devices, to achieve even moderately dense layouts. None of the analog layout systems of which we are aware can support the more free-form style of device layout characteristic of expert designs.

This paper presents an alternative macrocell layout style that permits more of the low-level layout optimizations described above. Specifically, we have designed new device placement and routing algorithms to support the following

Manuscript received August 1, 1990; revised November 9, 1990. This work was supported in part by the National Science Foundation under Grants MIP-8657369 and MIP-8451496, by the Semiconductor Research Corporation under Contract 90-DC-068, and by Harris Corporation. J. M. Cohn was supported by the IBM Resident Study Program.

The authors are with the Department of Electrical and Computer Engineering, Carnegie Mellon University, Pittsburgh, PA 15213.

IEEE Log Number 9041735.

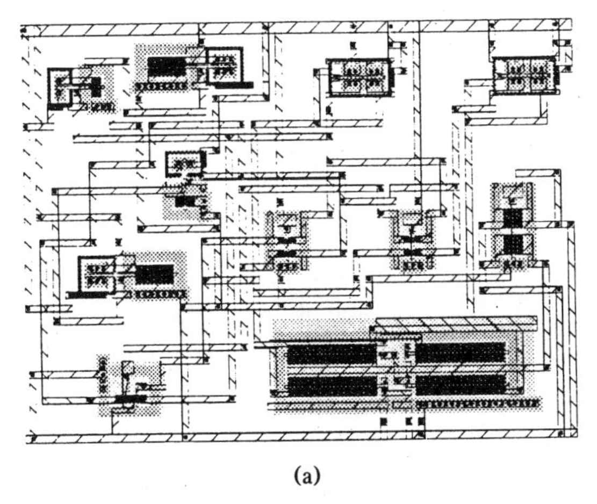
(a)

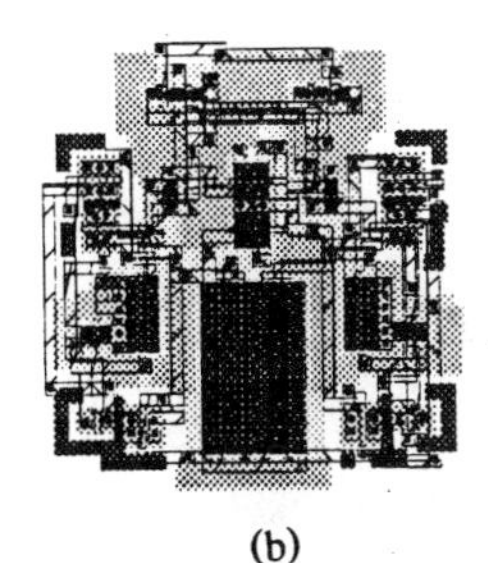
(b)

Fig. 1. Comparison of (a) ANAGRAM I comparator layout versus (b) KOAN/ANAGRAM II layout.

for analog cells:

- *Layout symmetries*: we handle both symmetric placement and symmetric detailed routing for differential circuits, including those with nonsymmetric components. This supports, for example, the layout requirements for a symmetric differential signal path and its associated nonsymmetric biasing circuitry.
- *Device merging / abutment*: we permit individual devices to be merged and abutted during placement. This not only increases cell performance, because of reduced parasitics, but also increases layout density, because the placer can now arrange devices into complex merged structures that are unlikely to be present in a library of module generators.
- *Well merging and bulk contacts*: we generate merged wells, and well and substrate contacts. This is required since the placed no longer assumes that these structures are completely fixed at the time of device generation.
- *Over-the-device wiring*: we do not restrict signal wires to be routed in channels between placed devices; instead, we allow wires to traverse devices at designer discretion. The router can handle arbitrary design rules on wires, use portions of placed devices as wiring, and does not require electrical terminals to be restricted to the perimeter of devices.
- *Crosstalk avoidance*: we model elementary capacitive coupling between signal nets, including simple shielding effects, and use these in the router to coerce embedded wires to avoid potentially damaging crossings or adjacencies.
- *Integrated rip-up / reroute*: we allow the router to rip up existing paths to improve wiring in densely placed, highly merged/abutted cells. This significantly increases the router's reliability by nearly eliminating its sensitivity to the order in which nets are routed.

We arrived at this set of layout requirements based on fabrication experiences with high-performance CMOS cells designed using our first-generation analog layout system, ANAGRAM I [8], [21]. These features are implemented in a new placement tool called KOAN, and a new router called ANAGRAM II. The layout results in Fig. 1 illustrate most clearly the differences between these first- and second-generation tools. Shown are two cell layouts for the same CMOS comparator design, one done by ANAGRAM I, which we regard as fairly typical of first-generation macrocell-style tools, and the other done by KOAN and ANAGRAM II. The layout generated by our new place-and-route tools is approximately one third the size of the earlier layout.

The remainder of the paper describes the architecture of these new layout tools. Section II describes the new device generation and placement algorithms used in KOAN. Section III next describes the new routing and path optimization algorithms used in ANAGRAM II. Some program implementation details for these tools are given in Section IV. Example results, including automatically generated CMOS and BiCMOS cells, appear in Section V. Finally, Section VI offers some concluding remarks.

II. Analog Device Placement in KOAN

A. Basic Architecture

KOAN is a device-level analog placement tool that supports symmetric placement, device merging, and abutment routing. KOAN consists of a set of procedural device generators, a device placer, and a well/substrate generator. In this, it architecturally resembles other analog macrocell layout systems [3], [4], [7]–[9]. It differs from these tools, however, in its ability to selectively overlap modules to reduce parasitic capacitance and cell area by appropriate sharing of geometry. Our strategy contrasts with other schemes in which such geometry sharing can only occur within the boundaries of procedurally-generated modules. This allows KOAN to exploit a potentially richer variety of geometry sharing alternatives than the few static topologies embodied in a typical procedural module generator library.

This strategy also strongly affects the design of the placement algorithms in KOAN: details about design rules, electrical connectivity, parasitic minimization, wells, etc., must be dealt with during placement, since they are not fixed during procedural device generation. Our placement strategy, like some device placers [7], [8], is based on simulated annealing. However, the annealing formulation is considerably more complex, given the new density and electrical performance optimizations we require it to support. KOAN employs a post-placement well generation strategy based on elementary computational geometry methods. These are described below.

B. Procedural Device Generation

KOAN has a library of procedural device generators, distinguished mainly by its small size. The library currently consists of two generators, one for folded FET's and the other for nonprecision capacitors. In an attempt to match the performance of hand crafted layouts, the common subcircuit layout structures that make up the bulk of typical generator libraries, e.g., cascode structures, matched differential pairs, and so forth, are created, *during* placement, by combining primitive devices. Since device generators are tedious to construct and maintain, we hope this approach will reduce the size of generator libraries, and encourage designers to implement new generators only for a smaller set of special-purpose structures unlikely to be found algorithmically, e.g., interdigitated cascode structures [7].

For each given set of electrical requirements, our generators produce several layout variants. For example, MOS devices are generated with a varying number of folds, varying contact locations, etc. The placer then chooses the variant that best fits the geometric and electrical requirements of the evolving layout.

Allowing the placer to merge and abut individual devices has several consequences on the design of device generators. First, the resultant device layouts are no longer *opaque* to the placer; they cannot be represented as simply a rectangle with terminals on its perimeter. Our generators supply to the placer detailed geometry for all electrical terminals that may subsequently participate in merge/abutment decisions. These include, for example, MOS drain, source, gate, and well contact geometry. Another consequence is that device generators cannot generate fully specified wells. If these are fixed at generation time, they adversely limit the flexibility of the placer to merge and abut devices. Hence, we generate those features that do not compromise the placer, for example, abutting substrate contacts on well-tied or rail-connected devices. The wells themselves are handled separately, in a post-processing phase after device placement or routing as described in subsection D.

C. Placement by Annealing

KOAN borrows from ANAGRAM I the idea of using a flat, nonslicing annealing model [22], [23] for device placement. Alternative approaches [3], [7] that adopt a slicing constraint have employed existing algorithms to handle block placement [24] and block shape (generated device variant) selection [25]. However, the slicing assumption is undesirable in our application for two reasons. First, it limits the set of reachable layout topologies, an effect most easily seen when the layout consists of many small objects, or objects with a wide variation in size [8]. Second, and more importantly, it prohibits us from reaching layouts where devices overlap in desirable ways. In the flat annealing style, intermediate states of the evolving layout can have arbitrary overlaps among movable devices. We use this as the mechanism to discover arbitrarily complex merged structures. One consequence of the flat style is that our placer produces an absolute, and not a relative (topological) placement. Hence, the placer is responsible for ensuring that there is sufficient wiring space for the router.

Simulated annealing [26] is a general optimization strategy based on iterative improvement with controlled hill climbing. This hill climbing allows annealing strategies to avoid many local minima in a complex cost surface, and reach better global solutions. To characterize KOAN's device placement strategy, we need to describe the four components of any annealing-based optimizer: 1) the *representation* for intermediate states of the layout visited during iterative improvement; 2) the set of allowable *moves* that transform one intermediate state of the layout to the next; 3) the *cost function* used to evaluate the quality of each intermediate layout; and 4) the *cooling schedule* used to control hill climbing.

Because it does not assume a slicing structure, KOAN uses a simple representation for evolving layouts. KOAN manipulates a set of rectilinear objects moving among arbitrary locations in a two-dimensional plane. Placeable objects can overlap in arbitrary ways as annealing proceeds. A bin hashing scheme is used to improve the performance of the overlap calculation as in [23].

One trade-off in any annealing-based layout algorithm is whether the responsibility for "good" layouts is embedded primarily in the choice of moves, called the *moveset*, or in the choice of cost function. KOAN relies mostly on its cost function to find good analog layouts. We do not employ moves that specifically seek to merge devices, to abut them, to share well contacts, etc. In our preliminary experiments, move sets that emphasized such specialized moves were consistently inferior to schemes that favored more random moves and a sophisticated cost function. The one important exception is that device symmetry and matching are supported directly in the move set. KOAN supports three classes of moves: *relocation* moves, *reshaping* moves, and *group* moves.

Relocation moves can translate, rotate, mirror, or swap devices. Three classes of analog constraints are maintained during all relocation moves.

1) Symmetry constraints reduce the effect of parasitic mismatch in differential circuits. Devices with symmetry constraints are always relocated to new symmetric positions. Single devices with symmetry constraints must slide along a symmetry line that bisects the evolving placement. Pairs of devices with symmetry constraints are always relocated in mirrored positions about this symmetry line. Such constraints have been handled elsewhere by adopting a slicing style [7]. However, because of its flat annealing formulation, arbitrary symmetries on arbitrary devices are especially easy to handle in KOAN.

2) Matching constraints force a common gate orientation (and overall device shape; see the discussion on reshaping moves below) on different devices. These help to reduce the effect of processing-induced mismatches.

3) Topological constraints allow the circuit designer to fix some aspects of the placement, while the placer handles the remainder. For example, individual device locations can be fixed in either one or both dimensions, or constrained to one of the edges of the layout or to its symmetry line. Likewise, cell terminals can be fixed in location, allowed to slide along one of the layout edges, or constrained to be symmetric to allow the resulting layouts to be abutment routed when placed side to side.

Reshaping moves replace one procedurally generated variant of a device with a different variant. This is how malleable devices are refolded, terminal contacts realigned, and so forth. To our knowledge, these sorts of alterations have only been used to date for cell area minimization [3], [7], [25]. In contrast, we allow the iterative improvement process to select any variant, not just the area-minimizing variant, because such perturbations may improve many different aspects of the layout. For example, a reshaped device might better merge with another device, thus reducing a critical nodal capacitance. As with the relocation moves, symmetry and matching are also enforced. If one device in a matched or symmetric group is reshaped, its companion devices are similarly reshaped.

Group moves extend the idea of moves for individual devices to moves for complex merged structures incorporating an arbitrary number of devices. Because we seek to encourage formation of dense merged/abutted device groups, it is essential that merged objects, once formed, be free to participate as a single unit in the same placement transformations as individual devices. It is also important to be able to dissolve such structures, by relocating one or more constituent devices sufficiently far to prevent interaction. The rationale here is the need to explore many different possible group mergings to prevent suboptimal groups from freezing too early in the annealing process. Group moves and single-device moves are handled uniformly. The process can be summarized as follows:

1) Randomly select a placed device D_i. Check to see if D_i is part of a larger group structure. Any set of devices $G = \{D_1, D_2, \cdots, D_i, \cdots, D_k\}$ that can be reached by traversing merged or abutted geometry constitutes such a group.
2) Randomly select one of the possible relocation moves {*translate*, *rotate*, *mirror*, *swap*}, or the reshaping move.
3) If reshaping was selected, replace device D_i with one of its generated variants. If D_i was part of a group, make a random binary decision either to a) align the new device variant so that it preserves the same overlap with its previously merged neighbors, or b) make no local adjustment.
4) If instead a relocation move was selected, randomly choose a subset of devices in the group, $M \subseteq G$. All devices in M participate in this move. Note that M might contain only device D_i, it might contain several devices, or it might be all of group G.
5) Apply the selected relocation to all devices in M. Note, if a *swap* was selected, identify another device E_i, and repeat step 4 to find a target set of devices with which to swap M. Maintain symmetry/matching if any members of the group have these constraints.

Step 3 is the one exception to the general rule that moves do not target specific optimizations. Since reshaping a device in a merged structure is highly disruptive, we have found that occasional attempts to properly align the new device in its local environment are more likely to cause beneficial shape changes to be accepted. This allows variants with similar contact placement to be interchanged without disturbing existing merges.

Much of the sophistication required to reach dense highly merged placements is embodied in KOAN's cost function, which is given by the following weighted sum:

$$cost = w_0 Overlap + w_1 Area + w_2 AspectRatio + w_3 NetLength + w_4 Proximity + w_5 Merge$$

where the w_i are experimentally chosen weights. The annealing process searches among different layout configurations to minimize this cost function. Some of these terms are familiar from digital macrocell placers. However, other terms are new, and all have been reformulated to handle analog-specific concerns.

The *Overlap* term penalizes illegal overlaps (measured as overlap areas) among devices. Recall that the placer has access to detailed geometry in each movable device. Part of each generated device is a protection frame that determines how closely distinct devices can be placed. The protection frame for each generated device accounts for the design-rule distances that must be maintained around the perimeter of each device. Space for wires to be embedded is also maintained, using a simple variant of the adaptive halos mechanism from [23]. Wiring space is allocated around each device depending on its size and number of wiring terminals. As devices are moved during annealing, an illegal overlap occurs if protection frames overlap, i.e., two pieces of electrically distinct geometry overlap, or are closer than design rules or wire space estimates allow. This term of the cost function must be driven to zero to ultimately produce a feasible layout. The *Area* term penalizes the total bounding box area of the placement. The *AspectRatio* term penalizes deviation from the desired aspect ratio R, and has the form $(R_{current} - R_{desired})^2$.

The *NetLength* term is the familiar sum of estimated lengths for each net. However, it is critical to use the right length estimator. Note that the detailed geometry representing terminals of nets to be wired may be large relative to individual devices, or the overall cell itself. Hence, estimators such as the half-perimeter of the least bounding rectangle of these terminals, or the length of a minimum rectilinear spanning tree connecting the centers of these terminals, can be extremely inaccurate. Instead, we construct the minimum rectilinear spanning tree that touches any piece of geometry in each electrically distinct terminal, and use its length as our estimator. This estimator also has the essential property that when the placement process merges the geometry of previously distinct terminals, the predicted net length to connect those terminals is zero.

The *Proximity* term allows designers to improve matching by encouraging arbitrary (possibly unconnected) devices to cluster. Devices in such proximity groups are modeled as having a dummy net connecting their respective centers. The *Proximity* term is the weighted sum of these dummy net lengths.

The *Merge* term is perhaps the most interesting part of the cost function. It can be regarded as the complement of the *Overlap* term: whereas *Overlap* penalizes overlaps that cause electrical violations, *Merge* rewards overlaps that are electrically beneficial. KOAN supports a flexible model of what geometry can be merged: roughly speaking, if two pieces of geometry on different devices are electrically common, and on *compatible* layers, they can be overlapped as far as the other prevailing design rules allow. Note that the notion of compatible layers is technology specific, and must be specified in the technology file. Fig. 2 shows examples of possible geometry sharing for a typical BiCMOS process.

The *Merge* term is formulated as

$$Merge = C_{area}(TotalMergeableArea - TotalMergedArea) + C_{perim}(TotalMergeablePerimeter - TotalMergedPerimeter).$$

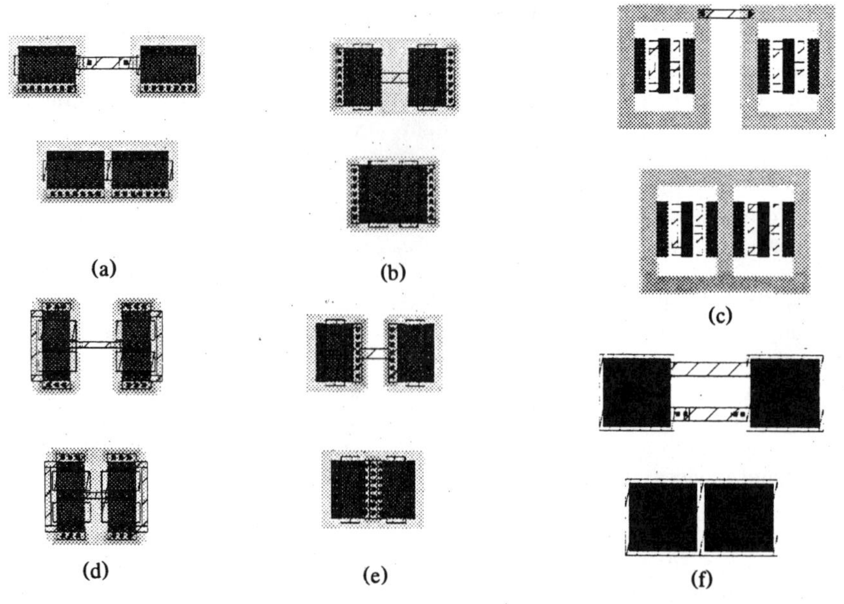

Fig. 2. Geometry sharing supported for typical BiCMOS process. (a) MOS gate abutment. (b) MOS diffusion sharing. (c) BJT guard-ring merging. (d) MOS metal abutment routing. (e) MOS substrate contact sharing. (f) Capacitor contact sharing.

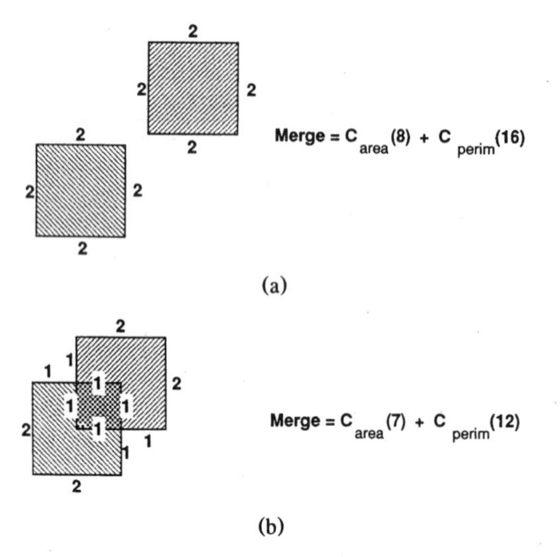

Fig. 3. Merge metric calculation. (a) Unmerged geometry, with more area and larger perimeter. (b) Merged geometry, with smaller area and perimeter.

Overlaps and abutments among compatible and electrically connected pieces of geometry do not count toward the *Overlap* penalty, since they are not electrically illegal. The total area of these mergable electrical terminals is the constant *TotalMergeableArea*. Similarly, the total perimeter of these terminals is the constant *TotalMergeablePerimeter*. *TotalMergedArea* and *TotalMergedPerimeter*, in contrast, change as devices are relocated to share differing amounts of geometry. Note that in the presence of no merges or abutments, *TotalMergedArea* and *TotalMergedPerimeter* are each zero, hence the *Merge* term is simply a constant overhead on the cost function. However, when merges occur, the terms measuring merged area/perimeter increase, decreasing the *Merge* penalty, and the overall cost of the layout. This is illustrated in Fig. 3. The overall effect of minimizing the *Merge* term is to maximize the sharing of geometry by maximizing allowable overlaps and shared perimeter. Since parasitic capacitance is directly proportional to such overlaps and perimeters, this term has the direct effect of minimizing such parasitics, when the per-unit-area (C_{area}) and per-unit-perimeter (C_{Perim}) capacitance scaling factors are set appropriately.

The cooling schedule is the policy that controls hill climbing during annealing. The placement must be "heated" to a high temperature to allow many random placements to evolve, then carefully "cooled" to allow the desired structure of the placement to freeze out. In the hot regime of annealing, moves that substantially increase the cost function are tolerated (again, to jump out of the local minima), but as the placement is cooled, fewer disruptive moves are permitted. There has been considerable progress of late in automating the decisions involved during cooling. We employ the automatic schedule from [27]. It is worthwhile to note that the individual terms in the overall cost function each dominate a different phase of the cooling process. In other words, the cost function itself is fixed, but the terms comprising it tend to freeze at different temperatures. Early in the annealing, the *Overlap* term forces illegal random overlaps to disappear. Later, the *NetLength* and *Proximity* terms encourage a good relative arrangement of devices. As the placement cools further, the *Merge* term starts to coerce desirable sharing of geometry, and finally the *Area* term causes the overall layout to shrink as much as possible.

Fig. 4 shows how all the new features of KOAN come into play. Three KOAN layouts are shown for a differential op amp with 11 devices and 12 nets. Fig. 4(a) shows a placement generated without symmetry, and with merging disabled. Although fairly dense, the nonsymmetric placement of the input-stage devices would likely increase the offset voltage. Fig. 4(b) shows a placement with symmetry, but again no merging. Notice, however, that not all components had symmetry constraints. This is not much smaller than the previous layout, but essential symmetries are now enforced for those devices that require it. This result is typical of other

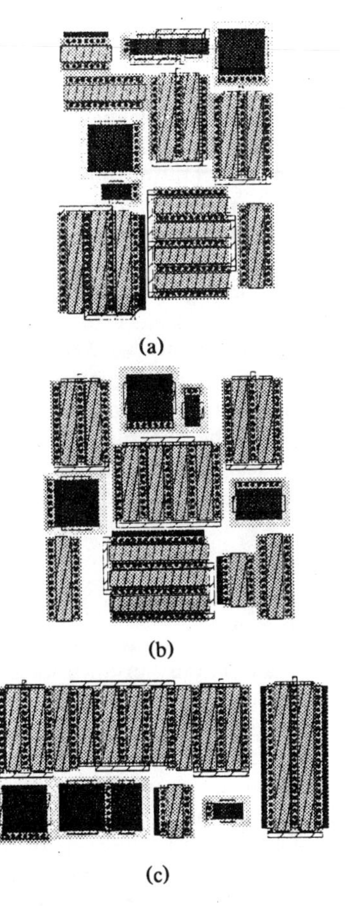

Fig. 4. Impacts of symmetry, device merging in a simple op amp. (a) Symmetry and device merge/abutment disabled. (b) Symmetry enabled, device merge/abutment disabled. (c) Both symmetry and device merge/abutment enabled.

macrocell systems [8]. Fig. 4(c) shows a placement generated when both symmetry and merging are enabled. This layout is appreciably smaller (about 10%) and also much more typical of a handcrafted placement. In particular, the dense area at the top left is a symmetric, folded, five-way merged group of devices, which is a very compact solution. We believe it is unlikely that such a complex structure would be available as a single parameterized cell in a typical macrocell library. Nevertheless, KOAN was able to discover and optimize this structure automatically.

D. Well Generation

Correct handling of wells and associated geometry presents problems during both device generation and device placement. In most analog placement systems, wells and associated geometry are created when devices are generated. For example, in ANAGRAM I, each device was created with an appropriate bulk contact, well, and guard ring. In addition to being overly conservative, this approach wasted space, created extra bias routing, and precluded dynamic device merging. Hence, we have adopted an alternative approach in which most well geometry is generated after devices have been placed or routed. An exception to this is well- and bulk-connected devices. By examining the input netlist during device generation, we detect when the source of an FET is connected to the same potential as its bulk. For these devices, we generate appropriate bulk contacts and abut them to the device's source. In this way, no additional routing is required for bulk biasing. We then allow bulk contacts to participate in contact merging during placement. This allows bulk contacts on different devices to merge into one contact shared by the merged devices. While merging bulk contacts does not present a significant parasitic advantage, it does improve layout density and simplify device wiring. Note also that while we do not force a segregation of n- and p-channel devices during placement (as is done in some other layout systems [3]), this type of segregation tends to occur automatically because of the device connectivity.

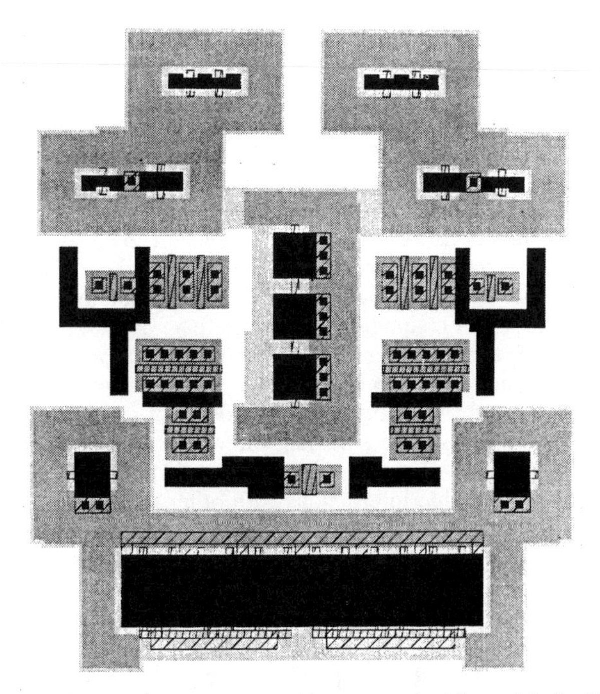

Fig. 5. Symmetric comparator with generated wells and bulk diffusions.

Well generation proceeds as a series of simple computational geometry steps such as shrinks, expands, unions, and intersections on the appropriate layers. For example, wells are produced approximately as follows. First, we expand the geometry of the well-bound devices (e.g., the n devices in a p-well process), find the union of these expanded shapes, and intersect this unioned region with an expansion of the geometry of the devices outside this type of well. The complement of this intersection defines all well regions; wells merge wherever possible in this scheme, which is essentially that used in [28]. Any well regions that remain floating after this shapes processing are contacted during routing by ANAGRAM II. To reduce the series resistance from the bulk contacts to other parts of the well and substrate, similar sequences of steps fill all unused space with the appropriate low-resistance diffusion straps. It is worth noting that although wells themselves can be generated before or after routing, we have found it beneficial to generate these low-resistance diffusion straps after routing, since they can interfere with wire paths that must be routed in polysilicon. Fig. 5 shows a placement for a high-speed CMOS comparator, along with its automatically generated wells and substrate contacts.

III. Analog Wire Routing in ANAGRAM II

A. Basic Architecture

ANAGRAM II is a detailed general-area router for analog cells. It borrows two critical ideas from the router in its predecessor, ANAGRAM I: a general-area routing strategy instead of a channel-routing model, and capacitive crosstalk penalty functions to encourage intelligent path decisions about net crossings and adjacencies. Although the sparser placements produced by first-generation layout tools could accommodate a channel-routing style (e.g., consider Fig. 1), the dense, highly merged/abutted placements produced by KOAN completely preclude such a stylized approach. Our area-routing strategy incorporates models of capacitive coupling, including simple shielding effects, in its basic evaluation mechanism for paths, allowing path selections to be coerced by possible interactions with other wired nets.

ANAGRAM II also has supports several new features missing in ANAGRAM I, and the other analog area routers of which we are aware. The first is the use of a tile-plane representation [29] for the to-be-routed layout, instead of a simple coarse grid. This representation supports essentially arbitrary wiring rules, in particular, over-the-device wiring. Second, new algorithms have been devised for line-expansion wire routing in this framework. The major contribution here is an algorithm for embedding geometrically matched, crosstalk-avoiding symmetric paths for differential signals. Third, a new integrated ripup/rerouting strategy has been designed. During the search for individual segments of wiring paths, the router can choose at any time to remove an existing wire. For the dense, highly merged/abutted placements produced by KOAN (or manually by layout experts), this integrated rip-up and rerouting turns out to be essential: without the ability to remove embedded nets on demand, it is often the case that there is no way to embed the next net. We describe below the tile-based representation used in ANAGRAM II, the new line-expansion algorithms for wire embedding in this framework, and the integrated rip-up/rerouting strategy.

B. Tile Representation

ANAGRAM II represents all placed devices, free wiring space, and embedded wires in a single-tile-plane data structure [29]. The tile representation frees us from some of ANAGRAM I's unnatural grid-based limitations on the location of devices and their terminals, and on the width and pitch of individual wires. Because there is no difference between unused space, wires, and device geometry, the router can "see" the internal details of devices. This has several advantages: over-the-device wiring incurs no overhead; pieces of devices can now themselves be used for wiring paths; electrical terminals can appear as arbitrary collections of geometry; and the same crosstalk penalties that accrue to wire segments can be applied to pieces of placed devices. A layout representation that supports careful over-the-device wiring is essential for routing dense placements. For example, terminals in some KOAN-generated placements can appear inside complex merged structures, and simply cannot be reached without extending a wire over some device geometry.

We use a single tile plane in which each individual tile represents a unique combination of mask layers. This facilitates many operations required by the ANAGRAM II's routing algorithm, most notably the enumeration of all geometric objects within a small region to check for design-rule and crosstalk violations. During initialization, ANAGRAM II constructs its tile-plane representation from an input device placement. It then determines the precise location of all routing terminals. Tile planes, which naturally support connectivity propagation, prove useful here by allowing the designer to tag only a small piece of a terminal and rely on ANAGRAM II to maximally expand the terminal to all connected geometry. During actual wire routing, the tile plane is employed primarily as a database, i.e., ANAGRAM II continually queries the tile plane for information about geometric objects within small regions. During routing, new wires are embedded and existing wires deleted from the tile plane as needed.

C. Line-Expansion Routing

ANAGRAM II is a line-expansion router. For each net to be embedded, a set of partial paths is maintained, sorted by cost. The least-cost path is selected, and expanded by probing incrementally outward from the head of the path. Each such probe results in a new partial path with a new cost, which is inserted back into the data structure used to sort the paths, in our case a *heap* [30]. The cycle of selecting, expanding, evaluating, and saving partial paths continues until all electrical terminals of the net have been connected, and no partial path can be expanded to make the same set of connections with less cost. The two critical components of the ANAGRAM II router are its expansion strategy and its path cost function.

Our expansion scheme relies heavily on the underlying tile-plane representation. ANAGRAM II always deals with exact wiring geometry, hence, each new path segment must conform to the prevailing design rules concerning width, pitch, interaction, via spacing, electrical interaction, etc. Again, the advantage of the tile plane is that it permits us to support essentially arbitrary wiring rules easily. However, the tile plane is used mainly for design-rule and electrical connectivity checking. Partial paths themselves are not embedded in the tile plane as they evolve (in contrast to [31]); they are simply stored in the heap. Only when a final path is found is the tile plane itself modified to embed the found path.

Another important algorithmic decision is the length of each probe attempted during path expansion. Long probes improve efficiency by allowing large distances to be traversed in a single step. However, long probes require much more complex computations to determine the optimal probe length, especially when an arbitrary number of crosstalk-interacting wiring layers is allowed. (An efficient scheme for whole-chip tile-based routing in this style is discussed in [32].) Short probes are less time efficient, but easily handle the small jogs, bends, etc., usually needed to reach dense terminals and avoid crosstalk problems. Our experiments with both styles favored short, unit-distance probes because of their simpler computational requirements, and the fact that successful probes in dense device placements tend to be short anyway. The following algorithm summarizes the expansion process for finding a path from a set of *source* terminals to one *target* terminal:

```
find_path(source, target)
{
  initialize heap to EMPTY;
  add source geometry to heap;/*begin with sources as
      partial paths*/
  /*main routing loop*/
  While (least-cost path in heap does not contact target) {
     expand(path);
  }
}
expand(partial_path)
}
  /*expand in all connected layers*/
  foreach (L in {connected, legal routing layers}){
    foreach (D in {each of 3 nonbackward probe
    directions}){
      p_new =partial_path + new geometry to expand it
                  one unit in direction D on layer L;
      if (p_new is design rule correct){
         compute cost of p_new;
         add p_new to heap;
      }
    }
  }
  /*expand possible contacts*/
  foreach (C in {connected, legal contact layers}){
    p_new =partial_path + new geometry to add
              a contact to layer C;
    if (p_new is design rule correct){
       compute cost of p_new;
       add p_new to heap;
    }
  }
}
```

Multipoint nets are handled in the usual fashion by decomposing them into a minimum spanning tree of two-point nets. The full geometry of the current, partially routed multipoint net serves as the *source* for the expansion cycle to find the next two-point connection.

The cost of a partial path P has two components:

$$PathCost(P) = \sum_{segments\ s \in P} SegmentCosts(s) + Distance(s\ to\ target).$$

The Σ *SegmentCost*(s) term sums the costs of P's constituent rectangular segments. The *Distance* term estimates distance from the head of the path P to the final target geometry, and is used to bias the expansion preferentially toward the target in the manner of conventional best-first search approaches [33]. After each expansion, the cost of the evolving partial path is updated by adding the cost of the newly expanded segment, and recomputing the *Distance* estimate.

Each *SegmentCost* has the form

$$Segment\_Cost = Wire + Direction + Crosstalk + Rip\text{-}up.$$

The *Wire* term is proportional to the area of the probed segment and the designer-specific layer cost. The effect is to favor short (low area) segments on preferred wiring layers,

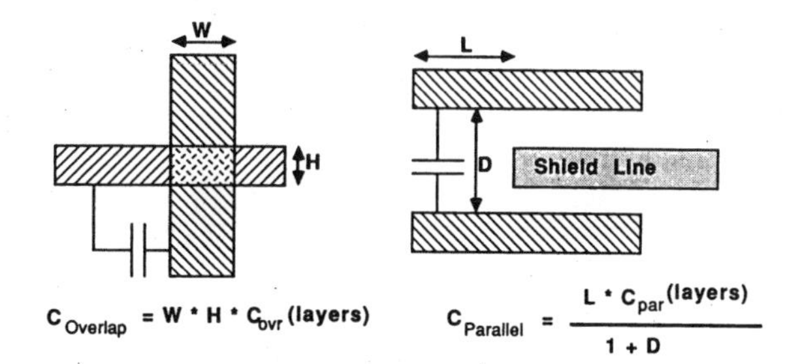

Fig. 6. Crosstalk models for routing cost function.

e.g., metal instead of polysilicon. The *Direction* term can be used to enforce preferred directions on individual wiring layers. Some routers use this to reduce congestion; our experience suggests this can be done more effectively with integrated rip-up and reroute. The *Crosstalk* term is unique to both ANAGRAM II and its predecessor, ANAGRAM I. By penalizing partial paths that contain undesirable crosstalk, we can force the router to search for less costly paths without undesirable net interactions. Both KOAN and ANAGRAM II allow the designer to specify arbitrary *compatibility classes* for electrical nodes. These classes specify which subsets of nets, when wired, may cross or be closely adjacent. For example, designers can define different classes of sensitive nodes (such as charge storage nodes) and noisy nodes (such as clocks) that should not interact. In addition, designers can specify which nodes can serve as potential shields for these unwanted interactions, for example, supply lines and some dc bias lines. This generalizes the simple noisy/sensitive/neutral classification scheme introduced by ANAGRAM I to handle arbitrary classes of net interactions. The crosstalk cost estimates the capacitance from a given probe segment to other *unshielded* interacting nodes. Simple capacitance models are used to penalize overlapping nets and non-shielded parallel runs, as shown in Fig. 6. The result of using this crosstalk cost during routing is that nets will attempt to take detours that avoid expensive crosstalk violations. The final path selected is the one which properly balances crosstalk cost and wiring cost. The *Rip-up* term is used during rip-up and reroute path optimization, and is described in the following section.

The router also has the unique capability to find symmetric paths for differential signals with symmetrically placed terminals, even in the presence of arbitrary asymmetric blockages. Although we are aware of approaches that can support perfectly symmetric device placement, e.g., [7], we are unaware of any routers that can complete these placements with perfectly symmetric differential wiring. We model a differential net as two wires, each of which is exactly mirrored about an assumed symmetry line bisecting the layout. Embedding such nets is accomplished by routing one net of the symmetric pair using the same line-expansion algorithm described above, but constraining the search process to consider only partial paths that would also be legal if reflected across the symmetry line. A differential net is expanded twice during the probe process: once for the existing net's proposed probe, and once for the mirror reflection of the net and the probe on the opposite side of the symmetryline. Of course, the router must be sensitive not only to design-rule violations on the symmetric net's reflection, but to crosstalk violations as well. Thus, when routing a symmet-

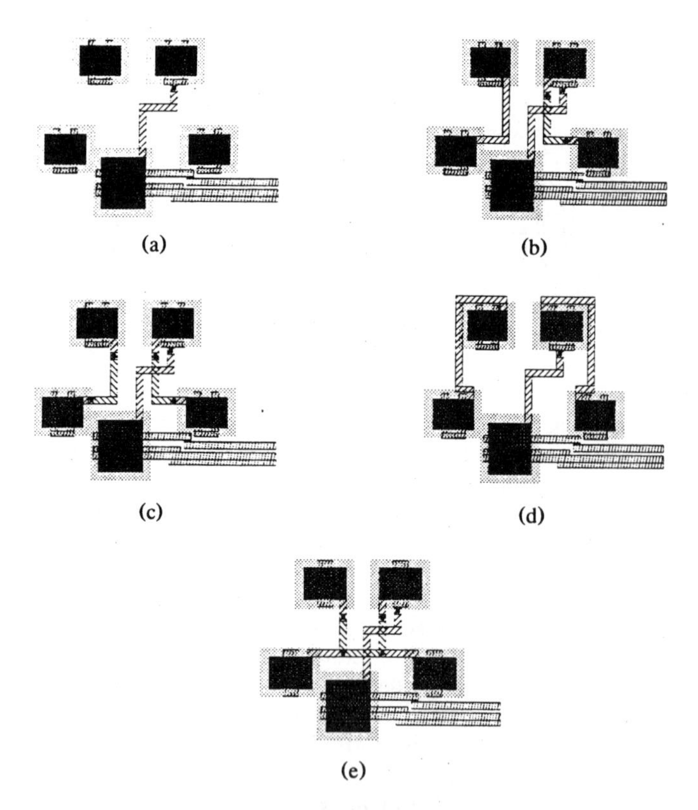

Fig. 7. Symmetric crosstalk-avoiding routes. (a) Four symmetric devices, with one asymmetric device and wire. (b) Routed with no symmetry, no crosstalk avoidance. (c) Routed with symmetry, but no crosstalk avoidance. (d) Routed with symmetry and crosstalk avoidance. (e) Routed as a single self-symmetric net, no crosstalk avoidance.

ric net, the crosstalk cost of a segment becomes the sum of the crosstalk cost of that segment and the crosstalk cost of its reflection across the symmetry line. Particularly good results can be achieved when all critically interacting nets are symmetric because any crosstalk violations that cannot be eliminated through detours in the routing will be identically matched on both sides of the signal path. Other approaches, such as [13], attempt to balance crossing of differential lines by adding equal-area crossings where necessary. However, we believe it is likely that superior matching occurs when the crossings/adjacencies are geometrically identical.

To further exploit the advantages of symmetric crossings. we have also added a new class of net called *self-symmetric*. A self-symmetric net is a single net whose pattern of pins is identical on both sides of the symmetry line. For example, this is frequently true of the clock nets in sampled-data circuits. With the knowledge that a net is self-symmetric, the router can then apply the symmetric routing algorithm to find a symmetric route for the individual halves and can connect the two halves in a symmetric manner by treating the symmetry line itself as a pin. Thus, the benefits of symmetric routing can be maintained for a single net whose function is identical on both sides of the symmetry line.

Fig. 7 shows how all the above features of ANAGRAM II come into play. A simple routing problem with a differential signal path (two pairs of symmetric terminals) is shown, along with an asymmetric noisy blocking wire. If no differential symmetry or crosstalk avoidance is enabled, the unbalanced paths in Fig. 7(b) result; note that one wire changes layers, while its companion does not. If differential symmetry is enabled, the matched paths in Fig. 7(c) are produced; both wires now make identical layer changes, one to avoid an obstacle, the other to match it to its companion. If both symmetry and crosstalk avoidance are enabled, the matched paths in Fig. 7(d) result; now both wires make a long detour to avoid the crosstalk violation that results if the rightmost wire crosses the noisy obstacle wire. Fig. 7(e) shows all four pins routed as a single self-symmetric net. Note how the two halves of the net are routed symmetrically and are connected at the center line.

D. Integrated Rerouting

Routers that route one net at time are often highly sensitive to the order in which nets are routed. For example, the currently evolving route cannot predict whether it is using space that will critically impact an unrouted net, nor can it determine that a small change in a previously embedded net might greatly help the current routing task. This is especially critical for analog routers that strive to avoid unwanted parasitic interactions among wired nets. Different net ordering schemes have been tried, e.g., routing short sensitive signals before clock nets, etc., but in general these techniques are highly unreliable, both in achieving 100% net completions, or perfect crosstalk avoidance. ANAGRAM II instead avoids this problem by relying on an aggressive iterative improvement strategy for routing. The router in ANAGRAM I routed all nets once, then randomly removed and rerouted nets until no further improvement was seen. In contrast, ANAGRAM II integrates net rip-up directly into the path-search mechanism used during line-expansion routing. The ability to remove any existing net as a new net is being routed proves to be essential for embedding wires in dense, KOAN-generated placements. In our experience, such placements are only wirable after a considerable amount of net rip-up and rerouting.

This is implemented as follows. Suppose an existing partial path P is being expanded. Suppose also that a probe with new wire segments s creates a design-rule violation with a previously embedded net N. Without any rip-up capability, we would simply reject segment s, since the new partial path $P+s$ is infeasible. With integrated rip-up, we allow the path $P+s$, but add to it a cost, *Rip-up*(N), associated with removing the obstacle net N. This is the *Rip-up* term in the *SegmentCost* for segment s. All partial paths that evolve from path $P+s$ are penalized by this amount, since they all include this segment s. Note that a partial path accrues a cost *Rip-up*(N_i) once for each net N_i it needs to remove. In particular, if some descendant of path $P+s$ also needs to use space occupied by net N, no additional rip-up penalty is assessed because it is assumed that the net has already been removed. The search otherwise proceeds exactly as previously described. During the search of individual wiring paths, a wide variety of different net rip-ups may be evaluated. However, only those associated with the final chosen path are actually removed. When a final path is determined, the nets that must be deleted to embed this path are removed from the tile plane, and scheduled for later rerouting. After their removal, the newly found path is embedded in the tile plane.

At any time, nets can be in one of two states: unrouted or routed. Unrouted nets are ordered in a queue. Nets are routed (or rerouted) in the order they are appear in the

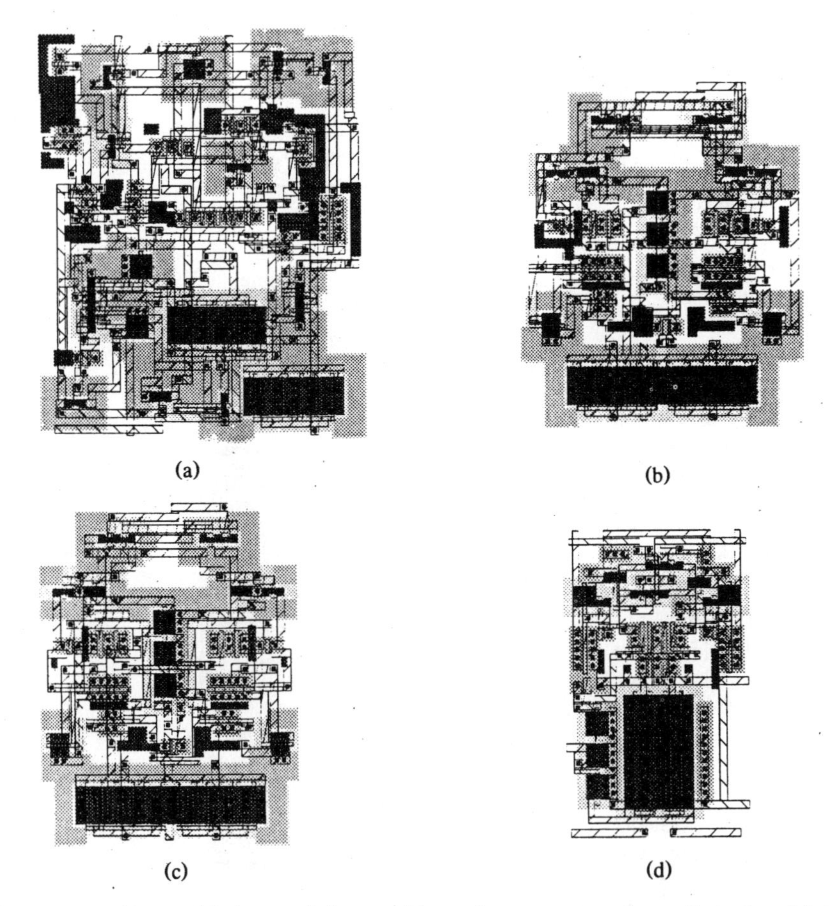

Fig. 8. CMOS comparator layouts. (a) Automatic layout with no placement or routing optimization. (b) Automatic layout with placement optimization only. (c) Automatic layout with both placement and routing optimization. (d) Manual layout.

queue. Initially, the queue is filled with all nets to be routed, usually ordered with critical nets at the front of the queue. As nets are routed, the queue empties. As newly routed nets remove embedded nets, the ripped-up nets are reinserted in the queue. The *reroute policy* determines whether such nets are replaced at the head or tail of the queue. If an *early* policy is used, ripped nets are placed at the head of the queue, and thus immediatly rerouted. This has the effect of allowing a small set of interacting nets to negotiate their final placement, rerouting themselves in tight loop. If a *late* policy is used, ripped nets are placed at the end of the queue, and thus all other unrouted nets are routed first. This has the effect of allowing a removed net to reroute itself against the background of space and crosstalk constraints imposed by currently routed nets. Our experience favors the late scheme when many crosstalk interactions and terminal blockages are present in the placement, although with particularly critical nets it can be desirable to apply the early policy to these critical nets to ensure that they remain routed.

Because the routing/rerouting process might never terminate, the cost *Rip-up*(N) associated with removing net N cannot remain constant, but must increase as the routing cycle proceeds. This is handled by associating with each net N an aging factor $A(N)$. Each time net N is ripped up, its rip-up cost is multiplied by its aging factor $A(N)$. As net N ages—by being ripped up—its *Rip-up* cost increases, and it becomes more costly to remove. This suffices to guarantee eventual termination of the routing process, even when no final solution is found. In the beginning of the routing cycle, net rip-up costs are low, since nets embedded early in the routing process are more likely to cause congestion problems for future nets. In difficult routing problems a small set of critically interacting nets usually emerges and the route optimization process iterates among these nets for several cycles. It is during these critical negotiation cycles that the importance of individual net aging factors becomes apparent: the net with the larger aging factor emerges from this cycle with the favorable route, while the other net(s) must compromise.

Route blockages usually occur very near a net's source or target. ANAGRAM II's integrated rip-up/reroute scheme can quickly find a path out of a blocked source pin: after a small amount of search, the router is forced to accept a path which removes the net, despite its *Rip-up* cost. However, a net which blocks the target pin causes difficulties. Here, a *horizon effect* [33] problem forces the router to explore a nearly endless variety of slightly less desirable paths before choosing one that removes the relatively expensive blocking net(s); the router can easily waste a great deal of time or run out of memory. Fortunately, the designation of source and target is arbitrary. Thus, ANAGRAM II solves its horizon effect problem by reversing these assignments. Because this horizon effect often dominates search time, ANAGRAM II will actually make three attempts at routing a net. The first will terminate after a very small amount of search under the assumption that a horizon effect is impeding the search. The second attempt will reverse the terminals and search until

TABLE I
CMOS COMPARATOR PERFORMANCE COMPARISON
Speed is measured as decision time with overdrive corrected for offset.

Comparator Layout	KOAN Place Optimized	ANAGRAM II Route Optimized	Area	Speed (at 3 mV Overdrive)	Systematic Offset
Automatic (Fig. 8(a))	No	No	34 272 μm^2	25 ns	+3.5 mV
Automatic (Fig. 8(b))	Yes	No	24 768 μm^2	21 ns	+2.7 mV
Automatic (Fig. 8(c))	Yes	Yes	22 100 μm^2	20 ns	−220 μV
Manual (Fig. 8(d))	—	—	15 092μm^2	23 ns	−680μV

the maximum search depth is reached. If this fails, a final full depth search using the original terminal order is performed.

IV. IMPLEMENTATION DETAILS

KOAN and ANAGRAM II together comprise approximately 20 000 lines of C code. The only input to the tools is 1) a common SPICE deck netlist [34] with annotations to control place/route options, and 2) a process-specific technology file. The SPICE netlist annotations are in the form of one-line comments which specify device and net symmetries, matchings, and sensitivities. The technology file is a text file containing line-by-line keyword/value specifications of layer-wise spacings, connectivities, extensions, merge compatibilities, etc., and is used commonly by both KOAN and ANAGRAM II. All communication of layout information between the tools is in MAGIC format [28], which allows designers to examine or modify intermediate layout results. Results presented in the following section were run on a DECstation 3100 under ULTRIX. Typical placement times for KOAN average 1 to 45 minutes of elapsed time, depending on the number of devices and amount of device merging optimization. Typical routing times for ANAGRAM II average 1 to 45 minutes of elapsed time, depending on the amount of crosstalk interaction to be managed, and the density of placed device terminals.

V. LAYOUT RESULTS

To demonstrate the effectiveness of KOAN and ANAGRAM II in custom analog cell layout, we present four sets of placed and routed layout examples produced by these tools. The first example is a high-performance CMOS comparator design. This circuit is particularly difficult to lay out, and was chosen to illustrate how the tools can optimize for electrical performance. The second and third examples are typical CMOS op-amp designs that illustrate how circuits with many large malleable devices can be aggressively optimized for density by our tools. The final example is a BiCMOS op amp, which illustrates how the tools can be easily retargeted to different technologies.

Example layouts for the comparator appear in Fig. 8. Table I summarizes these layouts and the results after parasitic extraction and simulation with HSPICE(tm) [35]. The circuit is a high-speed regenerative comparator designed in MOSIS 2-μm p-well CMOS. The circuit has 26 devices and 21 nets, and is difficult to lay out because it has many small devices (not much smaller than wires themselves), a relatively large number of interconnections, and many potential

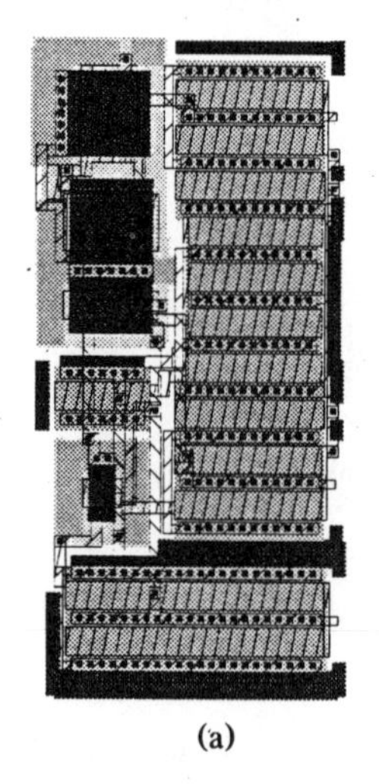

(a)

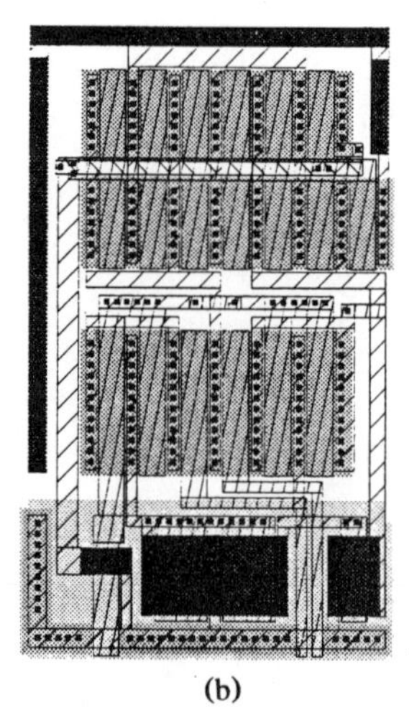

(b)

Fig. 9. Small CMOS op-amp layout. (a) Automatic layout with placement and routing optimization. (b) Manual layout.

crosstalk interactions between clocks and sensitive nodes. Fig. 8(a) shows a very poor automatic layout that makes no use of the optimization features in either KOAN or ANAGRAM II. No symmetry, merging, or abutment was encouraged during placement, and no symmetry or crosstalk avoidance was attempted during routing. As expected, the result is a slower comparator with a large systematic offset. Fig. 8(b) shows a better automatic layout. The device placement is highly optimized, because the placer was allowed to enforce symmetry and matching, and to merge/abut devices as necessary. However, the routing is as before—no symmetry, no crosstalk avoidance. The result is better: speed is improved considerably because of device merges, but there is still a large systematic offset due to asymmetric routing. Fig. 8(c) shows a fully optimized result. This is the same placement as Fig. 8(b), but with fully symmetric, crosstalk-avoiding routing. This result is the best in terms of speed, and has negligible systematic offset due to careful routing. Fig. 8(d) shows a comparable manual layout. The manual layout was aggressively optimized for density, and is 32% smaller than our best automatic layout. Nevertheless, the best automatic layout has somewhat higher performance due to well-chosen device merges and careful routing. This example illustrates the need to consider detailed electrical optimizations during both device placement and routing.

Layouts for a small CMOS op-amp example appear in Fig. 9. The circuit is a differential op amp designed in MOSIS 2-μm p-well CMOS. The circuit has 11 devices and 12 nets.

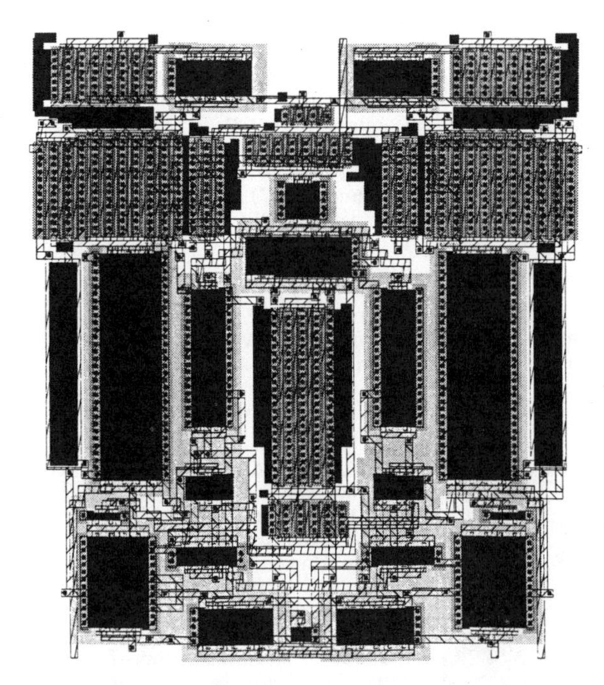

Fig. 10. Larger CMOS op-amp automatic layout.

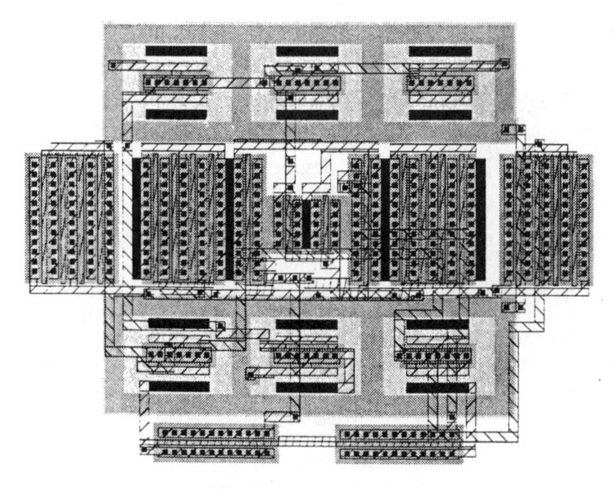

Fig. 11. Automatic BiCMOS op-amp layout with placement and routing optimization.

Fig. 9(a) shows an automatic layout exploiting all the capabilities of the placer and router. Fig. 9(b) shows for comparison a manual layout done by an industrial layout technician. Both layouts have essentially identical performance after extraction and simulation. However, the automatic layout made a different set of choices for device merging, abutment, and over-the-cell wiring, resulting in a cell that is actually 20% smaller than the manual design. The areas of the automatically generated cell and the manually generated cell were 25 872 and 32 430 μm^2 respectively.

An automatic layout for a much more complex op amp appears in Fig. 10. The circuit is also a differential op amp designed in MOSIS 2-μm p-well CMOS. However, this circuit has 31 devices and 24 nets. The automatic layout occupies 105 876 μm^2, and is only 13% larger than a high-quality manual layout.

As a final example, Fig. 11 shows an automatic layout of a folded-cascode op amp, now in a 2-μm n-well MOSIS BiCMOS process. The circuit has 16 devices and 15 nets. The current mirrors in this op amp are bipolar. A set of characterized bipolar device layouts was imported for use in this example. The layout tools can import arbitrary manual device layouts, and combine them with procedurally generated devices during placement and routing. We also altered the merging rules in KOAN to recognize and allow merging of the guard rings around each bipolar device. (Merging of collectors could also be supported, but does not provide any advantage in this layout.) Note the high degree of merging, both guard rings and MOS devices, in the final result. The layout occupies 41 888 μm^2.

VI. Conclusions

We have described new algorithms for analog device placement and routing, and their implementation in the tools KOAN and ANAGRAM II. Together, these tools support a more detailed model of analog device layout than other analog macrocell systems. Several layout capabilities are unique to these tools, in particular, their reliance on a very small library of procedural device generators, dynamic merging and abutment of devices during placement, over-the-device routing, mirror-symmetric and self-symmetric routing, and crosstalk avoiding area routing. Preliminary results are very encouraging. Our recent layouts are considerably smaller than our earlier attempts, and are beginning to approximate the density and aesthetics of expert manual designs. More importantly, our layout algorithms can reliably produce high-performance layouts. Moreover, we believe the flexible placement and routing models that underly these tools will allow us to target new technologies as they emerge. The extension from CMOS to BiCMOS cell layout, for example, required only the modification of a few technology rules in the placer and router. Our current efforts are focused on improving the communication between the placer and the router, since it is clear that division of the analog layout task into sequential placement and routing steps is artificial and often problematic in very dense layouts.

References

[1] M. G. R. DeGrauwe *et al.* "IDAC: An interactive design tool for analog CMOS circuits," *IEEE J. Solid-State Circuits*, vol. SC-22, no. 6, pp. 1106–1116, Dec. 1987.

[2] R. Harjani, R. A. Rutenbar, and L. R Carley, "OASYS: A framework for analog circuit synthesis," *IEEE Trans. Computer-Aided Design*, vol. 8, no. 12, pp. 1247–1266, Dec. 1989.

[3] H. Koh, C. Sequin, and P. Gray, "OPASYN: A compiler for CMOS operational amplifiers," *IEEE Trans. Computer-Aided Design*, vol. 9, no. 2, pp. 113–125, Feb. 1990.

[4] A. H. Fung, D. J. Chen, Y. N. Lai, and B. J. Sheu, "Knowledge-based analog circuit synthesis with flexible architecture," in *Proc. IEEE Int. Conf. Computer Design*, Oct. 1988.

[5] M. Mogaki, N. Kato, Y. Chikami, N. Yamada, and Y. Kobayashi, "LADIES: An automatic layout system for analog LSI," in *IEEE Int. Conf. CAD*, Nov. 1989.

[6] M. Ayal, S. Piguet, M. Declercq, and B. Hochet, "An interactive layout generation tool for CMOS analog IC's," in *Proc. IEEE Int. Symp. Circuits Syst.*, June 1988.

[7] J. Rijmenants *et al.*, "ILAC: An automated layout tool for analog CMOS circuits," *IEEE J. Solid-State Circuits*, pp. 417–425, no. 2, Apr. 1989.

[8] D. Garrod, R. A. Rutenbar, and L. R. Carley, "Automatic layout of custom analog cells in ANAGRAM," in *Proc. IEEE Int. Conf. CAD*, Nov. 1988.

[9] E. Berkcan, M. d'Abreu, and W. Laughton, "Analog compilation based on successive decompositions," in *Proc. Design Automation Conf.*, June 1988.

[10] J. Trnka, R. Hedman, G. Koehler, and K. Lading, "A device level auto place and wire methodology for analog and digital masterslices," in *ISSCC Dig. Tech. Papers*, Feb. 1988.

[11] S. W. Mehranfar, "STAT: A schematic to artwork translator for custom analog cells," in *Proc. IEEE Custom Integrated Circuits Conf.*, May 1990.

[12] R. S. Gyurcsik and J. C. Jeen, "A generalized approach to routing mixed analog and digital signals net in a channel," *IEEE J. Solid-State Circuits*, vol. 24, no. 2, pp. 436–442, Apr. 1989.

[13] S. Piguet, F. Rahali, M. Kayal, E. Zysman, and M. Declercq, "A new routing method for full custom analog IC's," in *Proc. IEEE Custom Integrated Circuits Conf.*, May 1990.

[14] R. Okuda, T. Sato, H. Onodera, and K. Tamaru, "An efficient algorithm for layout compaction problem with symmetry constraints," in *Proc. IEEE Int. Conf. CAD*, Nov. 1989.

[15] H. Onodera, H. Kanbara, and K. Tamaru, "Operational amplifier compilation with performance optimization," in *Proc. IEEE Custom Integrated Circuits Conf.*, May 1989.

[16] J. Kuhn, "Analog module generators for silicon compilation," *VLSI Design*, May 1987.

[17] R. J. Bowman, "Analog macrocell layout generation," in *Proc. 2nd Annual IEEE ASIC Seminar and Exhibit*, Sept. 1989.

[18] U. Choudhury and A. Sangiovanni-Vincentelli, "Constraint generation for routing analog circuits," in *Proc. Design Automation Conf.*, June 1990.

[19] J. Y. Lee, "Efficient pole zero sensitivity calculation using asymptotic waveform evaluation (AWE)," M.S. thesis, Dept. Electrical Comput. Eng., Carnegie-Mellon Univ., Pittsburgh, PA, May, 1990.

[20] R. H. J. M. Otten, "Automatic floor-plan design," in *Proc. 19th ACM/IEEE Design Automation Conf.*, June 1982, pp. 261–267.

[21] L. R. Carley, "ACACIA, The CMU analog design system," CMUCAD Tech. Rep. CMUCAD-89-64, Carnegie Mellon Univ., Pittsburgh, PA, Nov., 1989.

[22] D. W. Jepsen and C. D. Gelatt Jr., "Macro placement by Monte Carlo annealing," in *Proc. IEEE Int. Conf. Computer Design*, Nov. 1984, pp. 495–498.

[23] C. Sechen, "Chip-planning, placement and global routing of macro/custom cell integrated circuits using simulated annealing," in *Proc. 25th ACM/IEEE Design Automation Conf.*, June 1988, pp. 73–80.

[24] D. F. Wong and C. L. Liu, "A new algorithm for floorplan design," in *Proc. 23rd ACM/IEEE Design Automation Conf.*, June 1986, pp. 101–107.

[25] L. Stockmeyer, "Optimal orientations of cells in slicing floorplan designs," *Inform. Contr.*, vol. 59, pp. 91–101, 1983.

[26] S. Kirkpatrick, C. D. Gelatt, and M. P. Vecchi, "Optimization by simulated annealing," *Science*, vol. 220, no. 4598, pp. 671–680, May 1983.

[27] M. D. Huang, F. Romeo, and A. Sangiovanni-Vincentelli, "An efficient general cooling schedule for simulated annealing," in *Proc. 1986 IEEE Int. Conf. CAD*, Nov. 1986.

[28] J. K. Ousterhout *et al.*, "Magic: A VLSI layout system," in *Proc. 21st ACM/IEEE Design Automation Conf.*, June 1984.

[29] J. K. Ousterhout, "Corner stitching: A data-structuring technique for VLSI layout tools," *IEEE Trans. Computer-Aided Design*, vol. CAD-3, 1984.

[30] E. M. Reingold. J. Nevergelt, and N. Deo, *Combinatorial Algorithms: Theory and Practice*. Englewood Cliffs, NJ: Prentice Hall, 1977.

[31] A. Margarino *et al.*, "A tile-expansion router," *IEEE Trans. Computer-Aided Design*, vol. CAD-6, no 4, Apr. 1987.

[32] M. H. Arnold and W. S. Scott, "An interactive maze router with hints," in *Proc. 25th ACM/IEEE Design Automation Conf.*, June 1988.

[33] A. Barr and E. Feigenbaum, Ed., *The Handbook of Artificial Intelligence, Vol. I*. Los Altos, CA: W. Kaufman, 1981.

[34] L. W. Nagel, "SPICE2: A computer program to simulate semiconductor circuits," ERL Memo ERL-M520, Univ. of Calif., Berkeley, May 1975.

[35] *HSPICE User's Manual*, Meta-Software Inc., Campbell, CA, 1987.

John M. Cohn was born in New York City in 1959. He received the B.S.E.E. degree from the Massachusetts Institute of Technology, Cambridge, in 1981.

From 1981 until the present he has worked at IBM's General Technologies Division, Essex Junction, VT, in the area of analog CAE tool development. In 1988 he was admitted to the IBM Resident Study Program, which has allowed him to attend Carnegie Mellon University, Pittsburgh, PA. He is currently a Ph.D. candidate in the Department of Electrical Engineering. His current research is in the area of analog layout automation. He is the author of the KOAN analog placer.

David J. Garrod was born in New York City in 1963. He received the B.S.E.E. degree from the University of California at Davis in 1985 and the M.S.E.E. degree from Carnegie Mellon University, Pittsburgh, PA, in 1987, where he is currently completing work for the Ph.D. degree in electrical and computer engineering. His research has concentrated on the area of layout algorithms for analog circuits. Specifically, he is the author of ANAGRAM I and II.

Rob A. Rutenbar (S'77–M'84–SM'90) received the B.S. degree in electrical and computer engineering from Wayne State University, Detroit, MI, in 1978, and the M.S. and Ph.D. degrees in computer engineering (CICE) from the University of Michigan, Ann Arbor, in 1979 and 1984, respectively.

In 1984 he joined the faculty of Carnegie Mellon University, Pittsburgh, PA, where he is currently an Associate Professor of Electrical and Computer Engineering, and of Computer Science. His research interests include VLSI layout algorithms, parallel CAD algorithms, and applications of automatic synthesis techniques to VLSI design, in particular, synthesis of analog integrated circuits, and synthesis of CAD software.

Dr. Rutenbar received a Presidential Young Investigator Award from the National Science Foundation in 1987. At the 1987 IEEE-ACM Design Automation Conference, he received a Best Paper Award for work on analog circuit synthesis. In 1989, he was Guest Editor of a special issue of *IEEE Design & Test*. In 1990 he received the Benjamin Teare Award for Excellence in Teaching from the College of Engineering at CMU. He is a member of ACM, Eta Kappa Nu, Sigma Xi, and AAAS.

L. Richard Carley (S'77–M'84–SM'90) received the S.B. degree from the Massachusetts Institute of Technology (MIT), Cambridge, in 1976 and was awarded the Guillemin Prize for the best EE undergraduate thesis. He remained at MIT where he received the M.S. degree in 1978 and the Ph.D. degree in 1984.

He has worked for MIT's Lincoln Laboratories and has acted as a consultant in the area of analog circuit design and design automation for Analog Devices and Hughes Aircraft among others. In 1984 he joined Carnegie Mellon University, Pittsburgh, PA, where he is currently an Associate Professor of Electrical and Computer Engineering. His research is in the area of analysis, design, automatic synthesis, and simulation of mixed analog/digital systems.

Dr. Carley received a National Science Foundation Presidential Young Investigator Award in 1985, and a Best Paper Award at the 1987 Design Automation Conference.

Automatic Generation of Parasitic Constraints for Performance-Constrained Physical Design of Analog Circuits

Umakanta Choudhury and Alberto Sangiovanni-Vincentelli, *Fellow, IEEE*

***Abstract*—A new design methodology for the physical design of analog circuits is proposed. The methodology is based on the automatic generation of constraints on parasitics introduced during the layout phase from constraints on the functional performance of the circuit. In this novel performance-constrained approach, the parasitic constraints drive the layout tools to reduce the need for further layout iterations. Parasitic constraint generation involves 1) generation of a set of bounding constraints on the critical parasitics of a circuit to provide maximum flexibility to the layout tools while meeting the performance constraints; and 2) deriving a set of matching constraints on the parasitics from matched-node-pair and matched-branch-pair information in differential circuits.**

Constraint generator PARCAR is described and results presented for test circuits.

I. Introduction

DIGITAL processing has become more and more pervasive in the design of electronic circuits and systems because of its favorable properties in terms of accuracy and noise immunity. However, the analog nature of *real-world signals* makes it impossible to eliminate analog circuits completely. Common analog operations include A–D and D–A conversion, filtering, and amplification. Moreover, for high-frequency applications, analog processing may often be preferred over digital processing. Recently, attention has been devoted to analog signal processing with particular emphasis on massively parallel architectures such as neural networks. In any case, for a wide variety of applications there is a clear cost advantage in integrating both digital and analog functions on the same chip. But while CAD tools for digital design are well developed, CAD tools for analog design are still in their infancy. Even if the analog portion occupies a small fraction of the total area on the chip, it can become the design bottleneck when designed without the aid of CAD tools.

Manuscript received June 7, 1991; revised February 17, 1992. This work was supported in part by DARPA under Grant N00039-87-C-0182 and by AT&T, MICRO, Harris, Hewlett-Packard, and Rockwell. This paper was recommended by Associate Editor R. K. Brayton.

A. Sangiovanni-Vincentelli is with the Department of Electrical Engineering and Computer Sciences, University of California at Berkeley, Berkeley, CA 94720.

U. Choudhury was with the Department of Electrical Engineering and Computer Sciences, University of California at Berkeley. He is now with AT&T Bell Laboratories, Murray Hill, NJ 07974.

IEEE Log Number 9201140.

Analog circuit designers spend a significant fraction of their design time in the physical design phase. Since the performance of analog circuits can be very sensitive to parasitic effects, existing CAD tools for digital circuits that do not control parasitics cannot be used for their layout design. Examples of performance functions of analog circuits are bandwidth, phase margin, slew rate, gain, and noise figure. Traditionally, the layout of analog circuits is carried out in an iterative and trial-and-error fashion. At each iteration, all the layout parasitics are extracted from the layout and the circuit is simulated to check if performance specifications are met. If not, the layout is modified in an attempt to meet the specifications. Several problems arise when using this trial-and-error approach.

- The extracted list of parasitics is very large.
- If performance specifications are not met, no information is systematically extracted to guide the next iteration (for example, the problem may not necessarily be due to a single parasitic, but can be due to the combined effect of several parasitics).
- The number of iterations needed cannot be estimated and may be rather large for commonly designed analog circuits.

A new *performance-constrained* approach towards physical design (placement and routing) of analog circuits which reduces the need for layout iterations arising from parasitics is proposed in this paper.

The parasitics which are controlled during placement and routing are the interconnect parasitics. After the electrical design phase, it is possible to estimate the device parasitics from the device sizes. However, the interconnect parasitics depend entirely on the details of placement and routing.

Interconnect parasitics can be line resistances, capacitances, inductances, or line-to-line capacitances and inductances (all of these parasitics can be considered as linear circuit elements). For monolithic circuits, these lumped approximations are valid for frequencies up to a few gigahertz (above which transmission-line effect becomes important).

The interconnect parasitics can play an important role in altering circuit behavior if proper care is not exercised in layout design. For example, in switched-capacitor circuits unwanted capacitive coupling between interconnects can destroy ratio accuracy of precision capacitors. In layouts containing both digital and analog blocks, noise signal fed from digital to analog lines can impair precision of analog circuits in a serious way, besides limiting their dynamic range. In an amplifier circuit, even a small capacitive coupling can degrade significantly the frequency response due to the Miller effect. Stray coupling which gives rise to positive feedback may lead to oscillations. Interconnect resistances in poly lines can cause voltage drop. For high frequencies of operation, even inductive effects can be important. Often analog circuits employ fully differential architecture to achieve higher performance. This results in an additional need for the interconnect parasitics associated with appropriate nodes or branches to match nominally, for impedance matching and noise cancellation purposes.

The effects of interconnect capacitances are illustrated in Fig. 1 for a fully differential circuit which uses an opamp and precision-ratioed capacitors C_1 and C_2 to realize a precision gain. The ratio accuracy of the capacitors can be of the order of 0.1%, depending on the technology. The precision of the gain can be critical when the opamp is used in an A–D converter, since it affects the resolution of the conversion process. If the precision capacitances are of the order of 0.1 – 1 pF, even a few femtofarad of coupling capacitance between input and output terminals of the opamp can degrade the gain accuracy substantially. Coupling can arise due to crossing lines or adjacent lines running parallel to each other (particularly significant for technologies employing aggressive design rules). Also, in the example shown, if the interconnects associated with the output nets are long and the output nodes happen to be the high-impedance nodes of the opamp, the associated self capacitances will degrade the bandwidth of the circuit. This is pictorially illustrated in the layout portion of Fig. 1, where not only do the interconnects have to be symmetrical with respect to net pairs (1, 2) and (3, 4), but it also may be necessary to bound the coupling and self-capacitances of the interconnects shown.

Although device capacitances *to ground* are usually larger then the interconnect capacitances, the *line-to-line* coupling capacitances due to routing can be extremely critical. It should be noted that internode *device* capacitances can be present only between specific pairs of nodes, and the circuit designer is aware of them. On the other hand, *interconnect* capacitance can be present between any pair of nodes, and such a coupling, which is not anticipated during electrical design when present in the layout, may sometimes be dangerous.

There have been previous attempts to automate the physical design of analog circuits. A standard-cell placement strategy to eliminate coupling between *zero-swing-sensitive* nets and *large-swing-insensitive* nets has been described in [1]. Placement and routing which include parasitics in their cost function have been reported in [2], [3]. Several other layout systems have been reported in

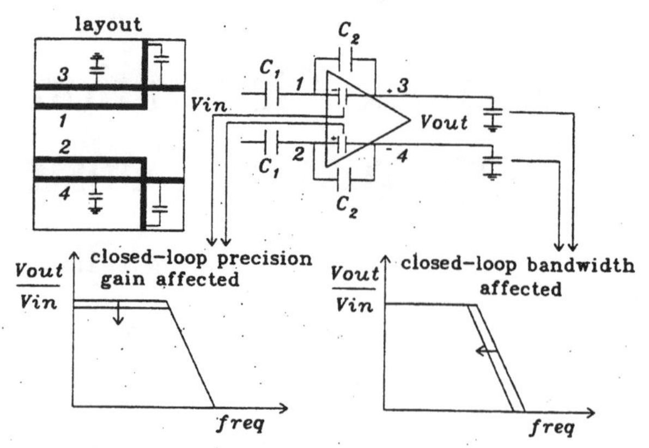

Fig. 1. Effects of parasitics on circuit performance.

[4]–[11]. Moreover, work on specific problems such as routing and compaction has also been reported. Channel-routing for analog circuits considering constraints on routing capacitances has been reported in [12]–[14]. Compaction considering symmetry constraints has been reported in [15] and [16]. One common limitation of all the previous approaches towards physical design is that no attempt is made to quantify performance degradation during layout design. The approaches above are based on layout algorithms that can take into consideration parasitics, but no attention is given to how to obtain constraints on parasitics from the specifications. Our new approach is to automatically extract bounds (constraints) on parasitics by analyzing the circuit and its performance, thus tying the functional specifications to the actual layout parasitics.

We assume that electrical design of the circuit has been done in such a way that the performance specifications have been met. For meeting the performance specifications of the circuit after physical design, finite degradation can always be allowed in the performance functions during the layout design, as long as the specifications are below certain thresholds. Performance constraints during layout design are the maximum changes allowed in performance functions because of the layout parasitics. Hence, in our approach, layout design is treated as a *constraint-driven optimization* problem, where the objective function is the chip area, and the constraints are the performance constraints of the circuit.

All possible *combinations* of parasitics which meet performance constraints define a *feasible region* in the space described by the parasitics. Since the high-level performance constraints of a circuit are too abstract for the layout tools to handle directly, a set of constraints on the parasitics are generated from the performance constraints, so that when the parasitic constraints are met, the values of the parasitics always remain in the feasible region of the space described by the parasitics. These parasitic constraints can then be used to drive the layout tools. This results in a much more reliable procedure that should yield a circuit meeting the performance constraints in the very first pass of layout if all the parasitic constraints are met. Even if some of the performance constraints are not met,

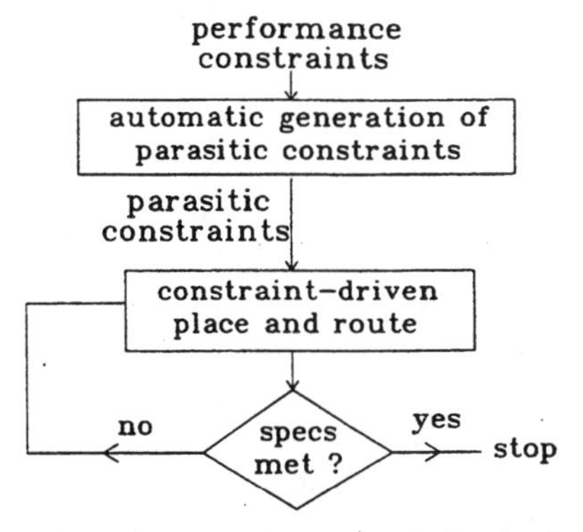

Fig. 2. Performance-constrained approach for physical design.

the responsible parasitics can be detected easily as they are the ones which violated the constraints. Hence tighter control can be exercised on those parasitics in the next iteration, resulting in a well-guided approach for physical design.

The parasitic constraints which are imposed on the layout are of two types: *bounding constraints* and *matching constraints*. Since in general, there will be many possible combinations of bounding constraints which meet performance constraints, a unique algorithm is used to generate a set of constraints which maximize the flexibility provided to the layout tools. An overview of the approach is shown in Fig. 2.

This paper is organized as follows: Section II lists the notations used in this paper. Section III provides a description of performance constraints and their specifications. In Section IV, the parasitic-constraint-generation problem is formulated and algorithms are presented to generate parasitic constraints from performance constraints. In Section V, a prototype parasitic constraint generator PARCAR is presented. Section VI contains results for some test circuits.

II. Notations

The notations used in this paper are listed below.

- α — Fraction (small compared to 1) used to set the limit for performance degradation caused by the noncritical parasitics.
- a_j, b_j, c_j — Coefficients of the polynomial modeling the layout flexibility f_j.
- b — $N_{cp} \times 1$ column vector whose jth element is $-b_j$.
- $C(n_i, n_j)$ — Interconnect capacitance between n_i and n_j.
- $C(n_i)$ — Interconnect capacitance from n_i to ground.
- f — Total layout flexibility associated with all the parasitics.
- f_j — Layout flexibility associated with parasitic p_j.
- M — Set of matched pairs of nodes in the circuit.
- N_{cp} — Number of critical parasitics.
- N_{mc} — Number of matching constraints.
- N_i^+ — Number of parasitics noncritical for positive constraint on W_i.
- N_i^- — number of parasitics noncritical for negative constraint on W_i.
- N_p — Number of parasitics.
- N_q — Number of process parameters.
- N_w — Number of performance functions.
- p — $N_{cp} \times 1$ column vector having p_{j_bound} as jth element.
- p_0 — Fictitious parasitic which replaces a matched pair of parasitics.
- p_{j_bound} — Bound on parasitic p_j to satisfy performance constraints.
- p_{j_max} — Conservative estimate of maximum value of p_j.
- p_{j_min} — Conservative estimate of minimum value of p_j.
- p_{max} — $N_{cp} \times 1$ column vector having p_{j_max} as jth element.
- p_{min} — $N_{cp} \times 1$ column vector having p_{j_min} as jth element.
- p_{thresh} — Threshold value for the parasitics (below which any parasitic can be ignored).
- q — Process parameter.
- Δq_{j_max} — Maximum variation of q_j around its nominal.
- Q — $N_{cp} \times N_{cp}$ matrix used in the cost function of quadratic programming.
- R — Set of parasitics noncritical for all the performance functions.
- R_i — Set of parasitics noncritical for performance function W_i.
- r — Additional degradation in precision due to parasitics allowed as a fraction of nominal precision.
- r_{max} — Maximum fractional process variation of a parasitic.
- r_{mis} — Maximum fractional mismatch between matched parasitics.
- S_0 — Sensitivity of a performance W with respect to fictitious parasitic p_0.
- S_{ij} — Sensitivity of performance W_i with respect to parasitic p_j.
- S_j — Sensitivity of performance W with respect to parasitic p_j.
- S_{ij}^- — $-S_{ij}$ if $S_{ij} \leq 0$, otherwise zero.
- S_{ij}^+ — S_{ij} if $S_{ij} \geq 0$, otherwise zero.
- S_i^- — $1 \times N_{cp}$ row vector whose jth element is S_{ij}^-.
- S_i^+ — $1 \times N_{cp}$ row vector whose jth element is S_{ij}^+.
- S_{lo} — Lower bound on sensitivity S_{ij} due to process variations.
- S_{up} — Upper bound on sensitivity S_{ij} due to process variations.
- W — Performance function.
- W^+ — Set of performance functions having constraint in the positive direction from nominal.
- W^- — Set of performance functions having con-

	straint in the negative direction from nominal.
W_{ideal}	Ideal value of W for a precision specification.
W_{max}	Maximum value of W with process variations and without parasitics.
W_{min}	Minimum value of W with process variations without parasitics.
W_{nom}	Nominal value of W (value without process variations and parasitics).
W_{prec}	Precision associated with W.
W^0_{prec}	Precision associated with W without parasitics.
ΔW_i	Change in performance function W_i due to parasitics.
ΔW^-_{max}	Maximum allowed change in W in the negative direction due to parasitics.
ΔW^+_{max}	Maximum allowed change in W in the positive direction due to parasitics.
ΔW_{prec}	Degradation in precision of W caused by parasitics.
ΔW^-_{proc}	Maximum change in W in the negative direction due to process variations.
ΔW^+_{proc}	Maximum change in W in the positive direction due to process variations.
X	Set of nodes in the circuit.
X_s	Set of first nodes in the matched pairs of nodes.
$\overline{X}_s$	Set of second nodes in the matched pairs of nodes.
X_q	Set of unmatched nodes.

III. Performance Constraints

Before formally describing parasitic constraint generation in Section IV-A, a description of performance constraints and their specification is provided.

Performance functions of analog circuits can be dc, ac, or transient. For an opamp, offset voltage, bandwidth, and slew rate are examples of dc, ac, and transient performance functions, respectively. When both digital and analog circuits are on the same chip, the noise signal fed from a digital line to any analog circuit can also be described as a performance function of the analog circuit.

As mentioned earlier, the performance constraints during layout design are defined to be the maximum changes (because of the interconnect parasitics) allowed in the performance functions from their nominal values to satisfy a set of performance specifications. The nominal value W_{nom} of a performance function W is defined as the performance realized without any interconnect parasitics and process variations. W_{nom} is known before placement and routing commences. Let ΔW^+_{max} and ΔW^-_{max}, respectively, denote the maximum positive and negative changes allowed in W from W_{nom} to satisfy the specifications on W. Hence, ΔW^+_{max} and ΔW^-_{max} represent the performance constraints associated with W.

Let W^+ and W^-, respectively, denote the sets of performance functions, which have constraints in the positive and negative directions from their nominal values. Each performance function belonging to W^+ will have a positive finite value for $\Delta W^+_{i\_max}$ and each performance function belonging to W^- will have a positive finite value for $\Delta W^-_{i\_max}$. If a performance function has a two-sided constraint from its nominal, it will belong to both W^+ and W^-. If a performance function has one sided constraint, it will belong to only one of the two sets.

The performance specifications of analog circuits can be categorized into two classes, a) nonprecision specifications, and b) precision specifications. The manner in which the performance constraints should be specified for a performance function depends on whether the associated specification is a nonprecision or a precision specification. Specification of performance constraints for each of these two cases is now described.

A. Nonprecision Specifications

A nonprecision specification allows the performance functions to vary over a wide range, the most commonly encountered ones being those having one-sided performance constraints (the function is required only to be larger than some minimum W_{min}, or smaller than some maximum value W_{max}). Examples of such specifications are those associated with bandwidth and phase margin of an opamp to be used in a custom design. Both bandwidth and phase margin will belong to W^- as there will be a constraint on the maximum decrease in each of these performance functions. Often for such specifications, it is possible to overdesign the circuit, allowing substantial margins for performance degradation due to interconnect parasitics (which are not known during electrical design), and process variations. However, it is to be realized that overdesign may require sacrifice of objective functions like chip area. Hence, it should only be used as means to compensate for factors such as process variations, over which the designer does not have any control, but whose effects can be statistically predicted in advance. Thus, for optimal design it is a good idea to make the performance degradation caused by interconnect parasitics small compared to that caused by process variations.

B. Precision Specifications

A precision specification is one which requires a performance function W to stay in a very narrow range around an ideal value W_{ideal} (e.g., the precision closed-loop gain of an opamp). Hence such performance functions belong to both W^+ and W^-. Because of the tight two-sided constraint on the performance, there is no room for overdesign.

The nominal value W_{nom} (W without interconnect parasitics and process variations) will be different from W_{ideal} because of the various nonidealities associated with the circuit. For example, in the circuit shown in Fig. 1, the nominal value of closed-loop gain of the circuit which has a precision specification associated with it will deviate from its ideal value set by nominal ratio of capacitors C_1 and C_2, due to the finite open-loop gain of the opamp.

In addition to the nonidealities, further change in performance functions will occur due to process variations and interconnect parasitics. The precision W_{prec} associated with the performance function is defined as its maximum deviation from the ideal value, i.e.,

$$W_{prec} = \max(|W - W_{ideal}|). \quad (1)$$

where W_{prec} is always positive, and the lower its value, the higher is the accuracy realized. Let W^0_{prec} denote the precision of the performance function in the absence of interconnect parasitics. The critical interconnect parasitics which impair nominal precision W^0_{prec} can be controlled very well, if proper care is taken during placement and routing. Hence, it is important to keep the degradation in precision caused by these layout phases (denoted by ΔW_{prec}) small compared to W^0_{prec}.

In order to compute the nominal precision W^0_{prec}, one should be able to estimate worst-case change in performance function due to process variations. The maximum changes in performance W (due to process variations) about its nominal value W_{nom} in the positive and negative directions are denoted by ΔW^+_{proc} and ΔW^-_{proc}, respectively. Let W depend on N_q process parameters, say $q_1, \cdots, q_{N_q}$, and each process parameter q_j has a maximum variation of $\pm\Delta q_{j\_max}$ around its nominal value. If the performance function varies monotonically with respect to each q_j in the range described by process variations, a fast way of estimating ΔW^+_{proc} and ΔW^-_{proc} is to first compute sensitivities of performance with respect to the process parameters at their nominal values. Then, based on the signs of the individual sensitivities, the value of each process parameter can be set at one of its extreme values and the circuit simulated for a worst-case analysis. For precision specifications, since the performance is made to depend on parameters which are tightly controlled (such as ratios of capacitors or resistors), even linear analysis using sensitivities should yield a good estimate of the worst-case variation in performance. In that case, a simple expression for ΔW^+_{proc} and ΔW^-_{proc} is given by

$$\Delta W^+_{proc} = \Delta W^-_{proc} = \sum_{j=1}^{N_q} |S_j| \Delta q_{j\_max} \quad (2)$$

where S_j is the sensitivity (gradient) of W with respect to q_j. However, if linear approximations are not valid, ΔW^+_{proc} and ΔW^-_{proc} may not be equal.

The nominal precision W^0_{proc} of the performance can be computed in terms of the defined quantities as depicted by an example in Fig. 3. When all the routing parasitics are zero, the performance function W lies in the range (W_{min}, W_{max}) as a result of process variations, where

$$W_{max} = W_{nom} + \Delta W^+_{proc} \quad (3)$$

$$W_{min} = W_{nom} - \Delta W^-_{proc}. \quad (4)$$

The maximum deviations of W from W_{ideal} in positive and negative directions are $W_{max} - W_{ideal}$, and $W_{ideal} - W_{min}$, respectively. Hence, from definition (1),

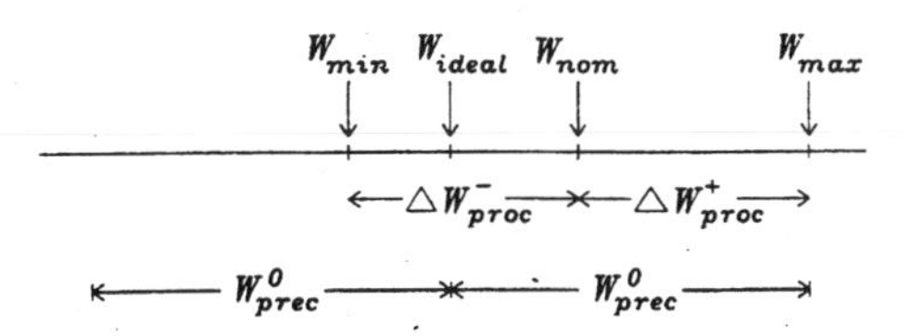

Fig. 3. Parameters determining nominal precision.

$$W^0_{prec} = \max(W_{max} - W_{ideal}, \quad W_{ideal} - W_{min}) \quad (5)$$

$$= \max(W_{nom} - W_{ideal} + \Delta W^+_{proc}, W_{ideal} - W_{nom} + \Delta W^-_{proc}). \quad (6)$$

If degradation in precision due to the parasitics should be limited to $\Delta W_{prec} = rW^0_{prec}$, $r << 1$, then the performance function with the parasitics should stay in the range ($W_{ideal} - (1 + r)W^0_{prec}$, $W_{ideal} + (1 + r)W^0_{prec}$). It is assumed that the relative process variations ΔW^+_{proc} and ΔW^-_{proc} do not change significantly because of shift in the operating point of the performance due to the parasitics. This should be a reasonable assumption, since in this analysis one is interested in small shifts caused while the parasitics meet the performance constraints. Also, the process variations in the parasitics themselves are taken care of by using worst-case sensitivities for modeling performance constraints (Section IV-B). Thus, the maximum positive and negative deviations (ΔW^+_{max} and ΔW^-_{max}) in performance W from its nominal value because of parasitics should satisfy the following relations:

$$W_{nom} + \Delta W^+_{max} + \Delta W^+_{proc} = W_{ideal} + (1 + r)W^0_{prec} \quad (7)$$

$$W_{nom} - \Delta W^-_{max} - \Delta W^-_{proc} = W_{ideal} - (1 + r)W^0_{prec} \quad (8)$$

from which the general expressions for ΔW^+_{max} and ΔW^-_{max} are obtained as follows:

$$\Delta W^+_{max} = (1 + r)W^0_{prec} - (W_{nom} - W_{ideal}) - \Delta W^+_{proc} \quad (9)$$

$$\Delta W^-_{max} = (1 + r)W^0_{prec} - (W_{ideal} - W_{nom}) - \Delta W^-_{proc}. \quad (10)$$

As seen from (2), $\Delta W^+_{proc} = \Delta W^-_{proc}$, when linear analysis is used for predicting these values. For this useful practical case, simple expressions will be derived for performance constraints ΔW^+_{max} and ΔW^-_{max}.

Case 1: $W_{nom} \geq W_{ideal}$:
From (6)

$$W^0_{prec} = W_{nom} - W_{ideal} + \Delta W^+_{proc}. \quad (11)$$

Substituting this expression in (9) and (10), and identifying rW^0_{prec} as ΔW_{prec}, one gets

$$\Delta W^+_{max} = \Delta W_{prec} \quad (12)$$

$$\Delta W^-_{max} = \Delta W_{prec} + 2(W_{nom} - W_{ideal}). \quad (13)$$

Case 2: $W_{nom} < W_{ideal}$:
Again, from (6),

$$W^0_{prec} = W_{ideal} - W_{nom} + \Delta W^-_{proc} \quad (14)$$

and substituting this in (9) and (10), one gets

$$\Delta W^+_{\max} = \Delta W_{\text{prec}} + 2(W_{\text{ideal}} - W_{\text{nom}}) \quad (15)$$

$$\Delta W^-_{\max} = \Delta W_{\text{prec}}. \quad (16)$$

To summarize, if linear analysis is valid for studying process variations of a precision function, then (12), (13), (15), and (16) can be used for specifying performance constraints. The maximum allowed change in a precision performance due to parasitics is a) ΔW_{prec} in the *offset direction* of W_{nom} with respect to W_{ideal}, and b) an additional amount of twice this offset in the opposite direction, where ΔW_{prec} (degradation in precision which can be tolerated) should be chosen small, compared to the nominal precision of the circuit given by (11) and (14). When linear analysis is not valid, (6), (9), and (10) have to be used.

IV. Parasitic Constraint Generation

A. Problem Definition

Parasitic constraint generation is the process of generating parasitic constraints for the layout tools from the high-level performance constraints of the circuit. It is formally defined below.

For a set of performance functions $\{W_i\}, i = 1, \cdots, N_w$ and a set of parasitics $\{p_j\}, j = 1, \cdots, N_p$, *parasitic constraint generation* is defined as generation of the following two types of constraints on a subset of $\{p_j\}$.

$$p_j = p_k \text{ (matching constraint)} \quad (17)$$

$$p_j \leq p_{j\_\text{bound}} \text{ (bounding constraint)} \quad (18)$$

to ensure that

$$\Delta W_i \leq \Delta W^+_{i\_\max}, \qquad \forall W_i \in W^+ \quad (19)$$

$$\Delta W_i \geq -\Delta W^-_{i\_\max}, \qquad \forall W_i \in W^-. \quad (20)$$

Matching constraints are necessary to match parasitics associated with certain pairs of nodes and branches (necessary in differential circuits and circuits employing matched devices). Bounding constraints limit the maximum values of the parasitics to meet the performance constraints. Constraints are imposed only on a subset of the parasitics as others may be noncritical.

B. Modeling Performance Constraints

In general, the parasitic constraint-generation problem seems to be extremely difficult, as the performance functions are nonlinear functions of the parasitics and no analytical expressions for those functions may be available. However, since one wants to limit variations in performance functions (because of the additional interconnect parasitics) to small values around their nominal values, linear approximations using sensitivities can be used to model dependence of performance on the parasitics, when the parasitics are within the bounds specific by the bounding constraints. It is to be emphasized that the performance degradation will not be automatically small, if parasitic constraints are not imposed.

The sensitivity S_{ij} of a performance function W_i with respect to a parasitic p_j at the nominal value of W_i is defined as

$$S_{ij} = [\partial W_i / \partial p_j]_{p_j = 0}. \quad (21)$$

A brief description of methods for sensitivity computation is now provided. One way to compute sensitivity is by using the perturbation method. In this method, a parameter is perturbed and the performance reevaluated using circuit simulation. However, this method is very time consuming since a resimulation of the circuit is required for each parameter of interest. This method can also be very error-prone, particularly for transient simulations, since the circuit simulator output has some inherent error, and while taking difference between two close numbers, the percentage error gets larger. Sensitivities can be computed much more efficiently and accurately, compared to the perturbation method, by using direct or adjoint techniques of sensitivity computation. For a detailed description of these methods, refer to [19] and [22]. Sensitivity analysis is supported in standard circuit simulators such as SPICE3 [17], SPICE2 [19] and TISPICE [20]. In SPICE3 for the direct method of sensitivity analysis the additional CPU time required for each parameter is about 8% for transient sensitivity, 4% for ac sensitivity and 1% for dc sensitivity. The sensitivities available from the circuit simulator are those associated with circuit variables (node voltages and branch currents). Intermediate calculations have to be performed to compute the sensitivities of the performance functions as described in [27].

The matched node pair information is then: In the case of matched parasitics, the value of only one parasitic can be independently controlled. Hence, the matched pair is treated like a fictitious single parasitic for the generation of bounding constraints. The bounding constraint generated on that fictitious parasitic imposes the same constraint on both the parasitics in the matched pair. The expression the for sensitivity associated with that single fictitious parasitic is derived later in this section.

In general, for each performance function some of the sensitivities will be positive and some negative. However, when the layout is not available, one cannot take advantage of the possible cancellation effects while generating bounds on the various parasitics, except for the special case when a matching constraint is imposed on a pair of parasitics (handled by defining the fictitious parasitic). Hence, for modeling the constraint on a performance function in positive (negative) direction, only the parasitics which have positive (negative) sensitivities with respect to that performance function are considered.

Approximations to performance constraints can thus be modeled by the following inequalities:

$$\sum_{j=1}^{N_p} S^+_{ij} p_j \leq \Delta W^+_{i\_\max} \ \forall \ W_i \in W^+ \quad (22)$$

$$\sum_{j=1}^{N_p} S_{ij}^- p_j \le \Delta W_{i\_max}^- \ \forall\ W_i \in W^- \tag{23}$$

where

$$S_{ij}^+ = S_{ij}, \quad \text{if } S_{ij} \ge 0;\ S_{ij}^+ = 0, \quad \text{if } S_{ij} < 0 \tag{24}$$

$$S_{ij}^- = -S_{ij}, \quad \text{if } S_{ij} \le 0;\ S_{ij}^- = 0, \quad \text{if } S_{ij} > 0. \tag{25}$$

Note that S_{ij}^+ and S_{ij}^- are always nonnegative.

Sensitivity computation for matched parasitics is now addressed. Let two parasitics p_j and p_k be matched, such that $p_j = p_k$, and process variations are ignored. For the purpose of generation of bounding constraints, the two parasitics are replaced by a single fictitious parasitic p_0 whose value is defined to be the same as the value of p_j and p_k without any process variations.

Then the sensitivity of a performance function W_i with respect to p_0 is

$$S_0 = [\partial W_i/\partial p_j][\partial p_j/\partial p_0] + [\partial W_i/\partial p_k][\partial p_k/\partial p_0]. \tag{26}$$

$\partial p_j/\partial p_0$ and $\partial p_k/\partial p_0$ are both unity since $p_j = p_0$ and $p_k = p_0$. Hence,

$$S_0 = S_{ij} + S_{ik}. \tag{27}$$

However, due to finite *mismatch* because of process variations, use of this expression for sensitivity may give meaningless results, particularly when S_{ij} and S_{ik} tend to cancel each other.

Expressions for the upper and lower bounds of sensitivity have been derived in Appendix A1, taking worst-case process variations into account, and are listed below.

Case I: S_{ij} and S_{ik} are of the same sign:

$$(1 \pm r_{max}) S_0. \tag{28}$$

Case II: S_{ij} and S_{ik} are of opposite signs:

$$S_0 \pm [r_{max}(|S_{ik}| - |S_{ij}|) + r_{mis}|S_{ij}|], \quad \text{if } |S_{ij}| \le |S_{ik}| \tag{29}$$

$$S_0 \pm [r_{max}(|S_{ij}| - |S_{ik}|) + r_{mis}|S_{ik}|], \quad \text{if } |S_{ij}| > |S_{ik}| \tag{30}$$

where r_{max} is the maximum fractional process variation of a parasitic, and r_{mis} is the maximum fractional mismatch between the matched parasitics.

For the important special case when $S_{ij} = -S_{ik}$ (refer to Appendix A1 for an example where this situation may arise), the bounds on the sensitivity become $+r_{mis}|S_{ij}|$ and $-r_{mis}|S_{ij}|$. This is compared to "0" sensitivity predicted if mismatch is not taken into account. The magnitude of these bounds can be significant when $|S_{ij}|$ is large, since r_{mis} can be as large as 0.2 (as the values of interconnect parasitics are poorly controlled).

Even for parasitics which are not matched, process variations should be considered to define the worst-case sensitivities. The worst-case sensitivity with respect to a single unmatched parasitic p_j is $S_{ij} = (1 + r_{max}) S_{ij\_0}$, where S_{ij_0} is the sensitivity without any process variations (easy to show).

For worst-case modeling of performance constraints in the positive and negative directions, the upper bound and lower bound on the sensitivities should be used, respectively. Let the upper bound on the sensitivity S_{ij} is denoted by S_{up}, and the lower bound by S_{lo}. Hence, following substitutions have to be made for S_{ij} in (22) and (23) for worst-case modeling:

$$S_{ij} = S_{up} \text{ in (22)};\ S_{ij} = S_{lo} \text{ in (23)}. \tag{31}$$

C. Matching Constraints on Parasitics

In this section, derivation of capacitive matching constraints from matched-node-pair information in differential circuits will be shown (inductive and resistive matching constraints can be derived in a similar manner from matched-branch-pair information).

Let X denote the set of nodes present in a circuit. In each differential circuit, there exists a set of matched pairs of nodes:

$$M = \{(n_i, \overline{n_i}), i = 1, \cdots, N_m\} \qquad n_i, \overline{n_i} \in X. \tag{32}$$

The sets of nodes $\{n_i\}$, $\{\overline{n_i}\}$ are denoted by X_s and $\overline{X}_s$, respectively. The set of remaining nodes which are not matched is denoted by X_q. The interconnect capacitance present from node n_i to ground is denoted by $C(n_i)$ and that between nodes n_i and n_j by $C(n_i, n_j)$.

Usually, the pair of nodes in each matched pair of a differential circuit have equal impedance levels, and equal and opposite transfer functions to the output. For impedance matching, the capacitances to ground associated with each matched pair of nets should be equal. Also, when a net which is not matched comes close to a matched pair of nets in the layout, it is required to match the coupling capacitances between that net and the pair of matched nets. This causes equal loading on the pair of nets, and in addition any noise in the third net is equally coupled to the pair, resulting in noise cancellation at the output. When *two pairs* of matched nets come close to each other, it is necessary to match the direct-coupling capacitances (C_5 in Fig. 4) and cross-coupling capacitances (C_6 in Fig. 4). Besides causing symmetrical loading, this ensures that equal level of noise on the two nodes of one matched pair causes the same on the other pair if any coupling is present. The capacitive constraints can thus be derived from matched-node-pair information of the circuit using the following relations:

Algorithm 1 (Derivation of Matching Constraints):

$$C(n_i) = C(\overline{n_i}), \qquad \forall n_i \in X_s. \tag{33}$$

$$C(n_i, n_k) = C(\overline{n_i}, n_k), \qquad \forall\ n_i \in X_s, \forall n_k \in X_q. \tag{34}$$

$$C(n_i, \overline{n_j}) = C(\overline{n_i}, n_j), \qquad \forall n_i, n_j \in X_s. \tag{35}$$

$$C(n_i, n_j) = C(\overline{n_i}, \overline{n_j}), \qquad \forall\ n_i, n_j \in X_s. \tag{36}$$

Matching constraints have been pictorially illustrated in Fig. 4, where capacitances to be matched have the same label. The matching constraints which are derived from Algorithm 1 not only impose the corresponding symmetry

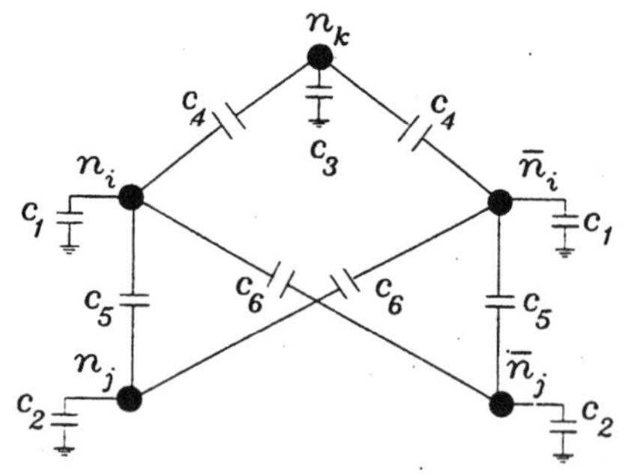

Fig. 4. Matching constraints.

requirements on the layout tool, but also affect the worst case modeling of performance constraints as described in Section IV-B.

D. Bounding Constraints on Parasitics

Although the actual bounding constraints are on performance functions represented by (22) and (23), it may be difficult to develop layout tools to handle them directly. For example, at any stage of routing one does not have a global picture of all the parasitics affecting the various performance functions. On the other hand, it is much more convenient to develop constraint-driven routing algorithms to meet bounding constraints on the individual parasitics which are derived from (22) and (23). Such constraint-driven algorithms have already been developed by us in the channel router ART [25]. The process for obtaining the bounding constraints on individual parasitics from the performance constraints will now be described.

1) Limits for parasitics: During layout design, often it is possible to estimate reasonable limits (not to be confused with bounding constraints) for the various parasitics. For example, for a given placement the minimum possible values of capacitances for various nets are determined by the half perimeters of bounding boxes associated with the pins. When the routing algorithm to be used is known, then it may be possible to estimate lower and upper limits for coupling capacitances also. For the constraint-driven channel router ART, algorithms for making conservative estimates of maximum and minimum values of coupling capacitances have been described in [25]. When such limits are estimated for a given placement before routing, they can be used for generating bounding constraints on the parasitics for routing as described later. When constraints have to be generated for placement, only very rough estimates for limits on the parasitics can be estimated, based on the estimated dimensions of the chip.

2) Selecting the Critical Parasitics: If N_{mc} matching constraints are imposed, then there are $(N_p - N_{mc})$ *independent* parasitics available, on which bounding constraints can be imposed. However, in practice a significant fraction of these parasitics will have very low sensitivities. If constraints are generated on all the independent parasitics, a large number of noncritical constraints will unnecessarily overload the layout tools. For any given circuit, the sensitivities of each performance function with respect to the parasitics follow a particular distribution from the high to the low end. The nature of this distribution depends on the circuit and the performance function considered. Determination of a cutoff point for selecting the critical parasitics will now be discussed.

Initially, sensitivity computation has to be performed with respect to all the parasitics which may possibly exist in the layout. For example, if capacitances are considered, then capacitances to ground from all nodes and coupling capacitances between all possible pairs of nodes have to be considered. If the *adjoint* technique of sensitivity analysis [22] is used, the analysis time will be almost independent of the number of parasitics considered. The adjoint method for transient sensitivity computation may not be supported in standard circuit simulators such as SPICE. However, the direct method of transient sensitivity analysis takes only about 8% additional CPU time for each parameter (in SPICE3), and in our experience does not take a prohibitive amount of time for all the parasitics in typical analog circuits.

For each performance function $W_i \in W^+$, let the parasitics with nonnegative sensitivities be sorted in increasing value, and for each $W_i \in W^-$, let the parasitics with negative sensitivities be sorted in increasing magnitude. Suppose p_{j_max} is the conservative estimate of the maximum value that p_j can attain in the layout (as discussed in Section IV-D-1). Note that the order of sorting may be different for different performance functions. If only performance function W_i is considered, then the first N_i^+ and N_i^- parasitics in the sorted lists which, respectively, make positive and negative contribution to W_i can be considered noncritical, if N_i^+ and N_i^- are the largest integers such that

$$\sum_{j=1}^{N_i^+} S_{ij}^+ p_{j\_max} \le \alpha \Delta W_{i\_max}^+ \qquad \text{if } W_i \in W^+ \tag{37}$$

$$\sum_{j=1}^{N_i^-} S_{ij}^- p_{j\_max} \le \alpha \Delta W_{i\_max}^- \qquad \text{if } W_i \in W^- \tag{38}$$

where the threshold α should be set to a value small compared to 1, say 0.01. Effective values of $\Delta W_{i\_max}^+$ and $\Delta W_{i\_max}^-$ available for critical parasitics can then be reduced by the same fraction. Hence, this procedure separates the parasitics which can collectively cause only a small performance degradation, compared to the maximum allowed. Let R_i denote the intersection of sets of these N_i^+ and N_i^- parasitics. The set of parasitics which can be considered noncritical when all the performance are considered is given by

$$\bigcap_{i=1}^{N_w} R_i. \tag{39}$$

The noncritical parasitics so determined are eliminated from further analysis. Let the remaining *critical* parasitics be indexed from 1 to N_{cp}. If parasitics which have different dimensions (capacitances, resistances, and inductances) are considered, then different sorted lists have to be maintained for each kind (since the sensitivities are also of different dimensions), and elimination carried out separately. But after this elimination, the critical parasitics can all be considered together for the generation of bounding constraints.

Even for the critical parasitics chosen for further consideration, if the value of a parasitic is so small that it falls below a predefined threshold, it can be considered to be zero. Hence, during final extraction (used for layout verification), one can avoid overloading the netlist by ignoring such small parasitics. The threshold value p_{thresh} of the parasitic of a given kind, say capacitance, can be estimated by solving for the maximum value all the critical parasitics can attain simultaneously while resulting in negligible degradation (compared to the maximum allowed) in each performance function. Note that p_{thresh} is different from (usually much less than) the bounding constraint p_{j_bound} associated with a parasitic p_j.

3) Modeling layout flexibility: In general, there are several parasitics affecting several performance functions. Hence, there are several possible combinations of bounding constraints on parasitics which will satisfy the performance constraints. However, they may not all be equally easy for layout tools to meet. This is particularly the situation when the parasitics have substantially different sensitivities. In that case, given one set of constraints it may be possible to *tighten* the constraints on *more sensitive* parasitics by *small* amounts and *relax* the constraints on *less sensitive* ones by *large* amounts to obtain another set of constraints easier to meet. As an example, let there be two parasitics p_1 and p_2 and one performance function W of interest. Let W be ten times more sensitive to p_2 than to p_1. Suppose $p_1 \leq 1$ fF and $p_2 \leq 1$ fF form one possible set of bounding constraints which satisfy the specified constraint on the performance. Then if the bound on p_1 is reduced by a small amount (say 0.1 fF), the bound on p_2 can be relaxed by a much larger amount (about 1 fF), since p_1 is ten times more sensitive than p_2. The second set of constraints may be easier to meet as it increases the total allowable capacitance during layout design.

The general problem is formalized by introducing the notion of *layout flexibility* associated with the bounding constraints, and the maximizing this flexibility subject to the performance constraints while generating the bounding constraints on the parasitics.

As described in Section IV-D-1, often it is possible to make estimates of maximum and minimum values ($p_{j\_\min}$ and $p_{j\_\max}$) a parasitic p_j can attain in the layout. Such information can be used to model the layout flexibility associated with the parasitic constraints. A bounding constraint p_{j_bound} should not be less than $p_{j\_\min}$, otherwise that constraint cannot be met. Also, the constraint should not be pushed above $p_{j\_\max}$, the reason for this being that bounding constraints are generated by maximizing overall layout flexibility. If the constraint is allowed to increase above $p_{j\_\max}$, no additional flexibility is provided to the tool, but it might require unnecessary tightening of constraint on some other parasitic (possible when the two parasitics have different sensitivities, as illustrated earlier). Hence, each bounding constraint should satisfy

$$p_{j\_\min} \leq p_{j\_\text{bound}} \leq p_{j\_\max}. \tag{40}$$

The total layout flexibility is modeled as the sum of flexibilities associated with individual bounding constraints. f_j is defined as the layout flexibility associated with bounding constraint on p_j, and is assumed to be a number ranging from 0 (minimum flexibility), to 1 (maximum flexibility). f_j is not defined when $p_{j\_\text{bound}} < p_{j\_\min}$, since it is not possible to meet the constraint. f_j should be 0 (minimum flexibility) when $p_{j\_\text{bound}} = p_{j\_\min}$; and it should be 1 (maximum flexibility) when $p_{j\_\text{bound}} = p_{j\_\max}$. Also, when p_{j_bound} increases beyond $p_{j\_\max}$, since no additional flexibility is given to the layout tool, the slope of the f_j versus p_{j_bound} curve should be zero at $p_j = p_{j\_\max}$, if one wants to fit a smooth curve. The question of matching of derivatives at $p_{j\_\text{bound}} = p_{j\_\min}$ does not arise, as flexibility is not defined for $p_{j\_\text{bound}} < p_{j\_\min}$. Since one cannot use a linear relationship to model this, a second-order polynomial, as depicted in Fig. 5, is used.

It can be shown that the second-order polynomial in p_{j_bound} satisfying the above three conditions is

$$f_j = a_j + b_j p_{j\_\text{bound}} + c_j p_{j\_\text{bound}}^2 \tag{41}$$

where

$$c_j = -1/(p_{j\_\max} - p_{j\_\min})^2 \tag{42}$$

$$b_j = -2c_j p_{j\_\max} \tag{43}$$

$$a_j = -b_j p_{j\_\min} - c_j p_{j\_\min}^2. \tag{44}$$

The total layout flexibility is

$$f = \sum_1^{N_{cp}} f_j \tag{45}$$

$$= \sum_1^{N_{cp}} a_j + \sum_1^{N_{cp}} b_j p_{j\_\text{bound}} + \sum_1^{N_{cp}} c_j p_{j\_\text{bound}}^2. \tag{46}$$

4) The main algorithm: The problem of maximizing the total flexibility can be converted into the standard form of minimization by changing the sign of the objective function (denoted by *cost*). Hence,

$$\text{cost} = -f \tag{47}$$

$$= -\sum_1^{N_{cp}} a_j - \sum_1^{N_{cp}} b_j p_{j\_\text{bound}} - \sum_1^{N_{cp}} c_j p_{j\_\text{bound}}^2. \tag{48}$$

For minimization, the first term in (48), which is a constant, can be ignored. Hence, the cost function can be more compactly written as $p^T Q p + b^T p$, where

- p — $N_{cp} \times 1$ column vector having jth element equal to p_{j_bound},
- Q — $N_{cp} \times N_{cp}$ positive-definite diagonal matrix whose jth diagonal element is $-c_j (c_j < 0)$,
- b — $N_{cp} \times 1$ column vector whose jth element is $-b_j$.

The minimization has to be subject to the linear performance constraints represented by (22) and (23) involv-

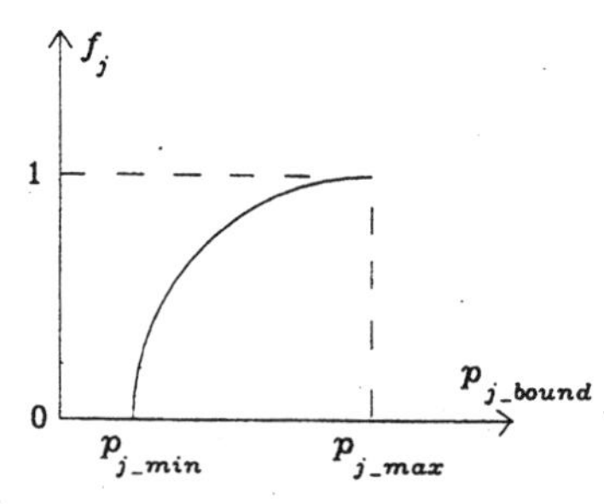

Fig. 5. Flexibility model.

ing the *critical* parasitics (refer to Section IV-D-2), and the imposed limit $(p_{\min}, p_{\max})$, where $p_{\min}$ and $p_{\max}$ are the vectors of the minimum and maximum estimated values of the individual parasitics, respectively. The main algorithm for generation of bounding constraints now follows.

Algorithm 2 (Generation of Bounding Constraints):

$$\text{minimize } p^T Qp + b^T p \tag{49}$$

such that

$$S_i^+ p \leq \Delta W_{i\_\max}^+ \ \forall \ W_i \in W^+ \tag{50}$$

$$S_i^- p \leq \Delta W_{i\_\max}^- \ \forall \ W_i \in W^- \tag{51}$$

$$p_{\min} \leq p \leq p_{\max}. \tag{52}$$

where S_i^+ and S_i^- are $1 \times N_{cp}$ row vectors whose jth elements are S_{ij}^+ and S_{ij}^-, respectively.

This is a quadratic programming problem (referred to as the *constraint-generation problem*) in N_{cp} variables, with a quadratic cost function and linear constraints. Due to positive definiteness of Q, it can be solved by standard QP packages. By definition, a *feasible solution* of this problem is a solution which satisfies (50), (51), and (52). An *optimum solution* p_{opt} is a feasible solution which minimizes the cost function given by (49). A vector p_c is called a *bounding constraint*, if for all $p \leq p_c$ (where this inequality implies term by term inequality for the two vectors), ((50) and (51)) are satisfied.

Lemma 1: A vector p_c is a bounding constraint if and only if it satisfies (50) and (51).

Proof: If p_c is a bounding constraint, by definition, any $p \leq p_c$ satisfies the performance constraints described by (50) and (51), and so does p_c. To show the reverse, it is observed that the coefficients and RHS of (50) and (51) are all nonnegative (refer to Sections IV-A and -B). So if a solution is found which meets performance constraints (50) and (51), then the parasitics, whose values are less than those of the solution vector, also meet the performance constraints. It follows from the above lemma that a feasible solution of the constraint-generation problem is a bounding constraint, but the reverse is not necessarily true.

Theorem 1: The necessary and sufficient condition for finding a feasible solution of the constraint-generation problem is that $p_{\min}$ be a feasible solution.

It is obviously a sufficient condition, since it is possible to find at least one feasible solution, i.e., $p_{\min}$. Using the argument of proof in Lemma 1, it can be easily shown that if p_{sol} is any feasible solution, $p_{\min}$ is also a feasible solution and, hence, a necessary condition for finding a feasible solution.

Lemma 2: There is a unique optimum solution p_{opt} for the constraint-generation problem.

This follows from the properties of quadratic programming, and from the fact that Q matrix is strictly positive definite.

Theorem 2: The optimum solution p_{opt} of the constraint-generation problem yields a maximal bounding constraint in the closed interval $[p_{\min}, p_{\max}]$, i.e., for no other bounding constraint p_c in this interval, $p_c \geq p_{\text{opt}}$.

Proof: The result will be shown for a more general case, when f_j is a strictly monotonically increasing function of p_{j_bound} in the closed interval $[p_{j\_\min}, p_{j\_\max}]$ (the flexibility model which is used for f_j satisfies this condition). Suppose there exists a bounding constraint p_c ($p_{\min} \leq p_c \leq p_{\max}$), which is different from p_{opt}, and $p_c \geq p_{\text{opt}}$. p_c satisfies (52) according to the assumption made.

Using (47) and (45),

$$\text{cost}\,(p_c) - \text{cost}\,(p_{\text{opt}}) = \sum_{1}^{N_{cp}} [f_j(p_{j\_\text{opt}}) - f_j(p_{j\_c})] \tag{53}$$

where p_{j_opt} and p_{j_c} are the jth elements of p_{opt} and p_c, respectively.

Since $p_{\text{opt}} \leq p_c$, and p_c is different from p_{opt}, $p_{j\_\text{opt}} < p_{j\_c}$ for at least one j. Due to the strictly monotonically increasing nature of f_j, it follows that

$$\text{cost}\,(p_c) < \text{cost}\,(p_{\text{opt}}). \tag{54}$$

This is a contradiction, since p_c is a feasible solution and cannot have a cost function less than that associated with the optimum solution p_{opt}.

The results of Lemma 1 and Theorem 1 are valid, irrespective of how the layout flexibility is modeled. As shown, Theorem 2 applies to any flexibility model for f_j which is strictly monotonically increasing with the bounding constraint in the desired range. The way this flexibility is modeled is going to determine how easily the generated bounding constraints can be met by the layout tool.

The constraint generation will be successful if $p_{\min}$ is a feasible solution. If the minimum possible values of some of the parasitics (like capacitance to ground) due to placement are such that a feasible solution does not exist for routing, then the placement has to be changed. Shielding can also be used to eliminate unavoidable coupling capacitances, and reduce the effective values of minimum capacitances to obtain a feasible solution, as discussed in Appendix A2.

The minimum estimate $p_{j\_\min}$ of a parasitic can always be set to threshold p_{thresh} (defined in Section IV-D-2) if it falls below p_{thresh}. Hence, the bounding constraint on any parasitic is always forced to lie above a certain nonzero threshold. This makes sure that the layout tool has nonzero flexibility associated with each critical parasitic without affecting the chances of performance constraints being met (because of the way p_{thresh} is defined).

V. PARCAR

The constraint-generation algorithms which were presented are implemented in PARCAR, a parasitic constraint generator. It has approximately 7000 lines of C code. An overall flow diagram of PARCAR is shown in Fig. 6. It generates constraints on interconnect capacitances (both line-to-ground and line-to-line). Currently it is being expanded to generate constraints on resistances and inductances also. It can be used either for generating constraints on routing (for a given placement) or for generating constraints on placement itself.

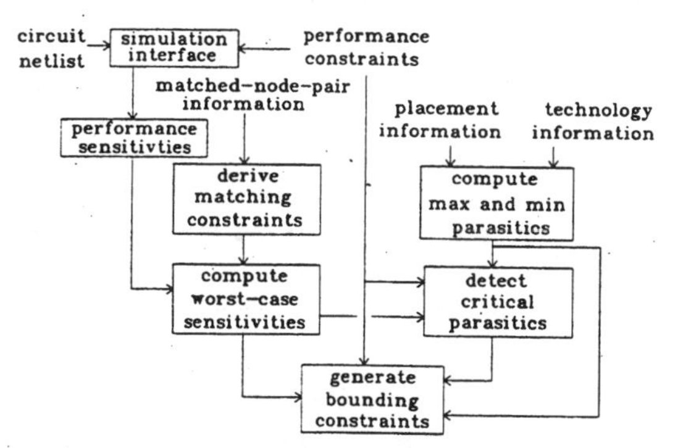

Fig. 6. Flow diagram of PARCAR.

A. Program Input

The input data PARCAR needs are: a) circuit netlist, b) names of matched-node pairs and c) performance constraints. Besides these inputs, some additional inputs may be necessary depending on the mode the program is used in. They are a) placement information, b) technology information, and c) maximum and minimum limits for parasitics.

The circuit netlist can be a SPICE netlist or a netlist for the switched-capacitor simulator SWAP [21]. The matched pairs of nodes in the circuit have to be specified. The performance constraints are specified first by defining the performance functions of interest for the circuit. This can be done by choosing one or more among a set of standard specifications such as gain bandwidth, phase margin, etc. If a performance function not in the standard list is to be defined, then a special interface can be used [28] in which the constraint functions are described procedurally. The constraints on the performance functions of interest are then specified as maximum allowed changes from their nominal values in either positive or negative or both directions (due to interconnect parasitics).

When the program is used to generate constraints on routing for a given placement, the placement and technology information are fed to the program. They are used to estimate maximum and minimum values for the individual parasitics which are used in the constraint generation. The technology information consists of the coefficients used for computation of line-to-ground and coupling capacitances between interconnects. These coefficients can be generated automatically by using the model generator CAPMOD [29]. The technology information also consists of worst-case process variation of parasitics and worst-case mismatch between two matched parasitics (expressed in percent). When constraints have to be generated for placement, then conservative estimates of minimum and maximum values for all the parasitics can be specified, simply based on the estimates of chip dimensions, as mentioned in Section IV-D-1.

B. Program Operation

The program has two modes of operation, a normal mode and a special mode. In the normal mode, the PARCAR can be interfaced to any layout tool which can meet bounding and matching constraints on parasitics. A constraint-driven area router RoAD has been interfaced to PARCAR in this mode [26]. The placement information in this mode has to be provided in OCT format. A special mode has been developed in which the program has been interfaced to a constraint-driven channel router ART described in [25]. In this mode, the user can directly input the dimensions of the unrouted channels, the pin positions, and the global routing information. The constraint generator generates constraints for each channel separately, as the channel router routes one channel at a time. A parasitic may have a contribution coming from each channel. For the purpose of constraint generation, these contributions for a parasitic are treated as different parasitics having the same performance sensitivities.

The sequence of steps followed by PARCAR is now described. The program first inserts zero-valued capacitances from all the nodes to ground and between all possible pairs of nodes in the circuit. However, to generate constraints on resistances and inductances, very small values of the resistances and inductances have to be used in the branches to avoid nonconvergence in simulation and simultaneously to make sure that circuit operation is not altered significantly by the addition of these elements. Interface has been developed for the simulators SPICE3 [17], [18], and SWAP [21], for accessing the sensitivities of the performance functions with respect to all the parasitics. The matched node pair information is then used to impose matching constraints on the parasitics and reduce the number of independent parasitics to be considered for generating bounding constraints. Then worst-case sensitivities with respect to the independent parasitics are computed as described in Section IV-B, using the technology information provided. When running in the special mode with the constraint-driven channel router ART described in [25], the program automatically computes the maximum and minimum values of the parasitics in each channel. When running in the normal mode with the constraint-driven area routing RoAD [26], a special interface program is used to perform this task. The maximum values for parasitics are used along with the worst-case sensitivities to select the critical parasitics (Section IV-D-2). The bounding constraints on these critical parasitics are then generated using the algorithm presented in Section IV-D-4.

The output of the program consists of the critical pairs of nodes between which bounding constraints are generated, the associated bounds on the capacitances, and the associated performance sensitivities. Intermediate files are created to store the sensitivities of each performance function with respect all the parasitics (critical and noncritical), in decreasing order. Hence, the user can get information about the noncritical parasitics as well.

VI. Results

Results for automatic constraint generation and constraint-driven routing are presented in this section. For results of constraint-driven placement, refer to [30].

A. SC Filter Example

Results are now presented for a fifth-order switched-capacitor filter circuit, as shown in Fig. 7. The numbers in the figure refer to the node numbers. The filter circuit was simulated using the switch-cap simulator SWAP [21]. The simulated response has a nominal ripple of about ± 0.1 dB in the passband of 0–1.6 kHz. The nominal magnitude response assumes maximum positive and negative peak in the passband at about 0.1 and 1.6 kHz, respectively. Hence, we have considered responses at these two frequencies to model the performance constraints, although more samples can be taken for better accuracy. The ideal value of magnitude response in the passband has been taken as the value about which peak deviation in nominal response curve is symmetric. It is the relative flatness of response curve in the passband which is the performance of interest in this example. After following the procedure suggested in Section III-B $\Delta W^{+}_{\max}$ and $\Delta W^{-}_{\max}$ in the magnitude response were set to 0.01 dB and -0.17 dB at 0.1 kHz, and vice versa at 1.6 kHz. The worst-case process variation was calculated by considering a maximum variation of 0.1% in the ratios between the integrating capacitor and the other capacitors connected to the input of each opamp (directly or through switches). The nominal frequency response of a switched capacitor filter is designed to depend on these ratios only, to the first order.

The placement and global routing for the layout of the filter was generated using the program ADORE [23]. The program follows a standard-cell type of placement. It places all the modules in three rows: switches in the top row, opamps in the bottom row, and capacitors in the middle row. There is a channel available for routing above each row. The switches are oriented in such a fashion that the top channel contains all the clock lines and no analog net. Hence, constraints have been generated for the other two channels only.

The circuit has 25 nodes. Sensitivity analysis was performed using SWAP with respect to all the parasitic capacitances (self and coupling). The sensitivities of magnitude of frequency response of the filter at 0.1 and 1.6 kHz have been shown in Tables I and II, respectively, for five of the most sensitive parasitics. *net1* and *net2* refer to the two nets between which the parasitic capacitance is considered for sensitivity analysis. Based on the sensitivities and the performance constraints, 30 capacitances were detected to be critical, and bounding constraints were generated on them. The channels were routed by the constraint-driven channel router ART. The bounding constraints and the extracted values for five of the 30 critical capacitances have been shown in Table III. The first row in Table III states that there is a bounding constraint of 0.0015 fF between nets 18 and 21, but 0 coupling was extracted after constraint-driven routing. 21 is a sensitive net as it switches between ground and the virtual ground node 8. 18 is a large-swing net as it switches between ground and node 9, which is output of an opamp. In each of the rows of Table III, one of the nets can be identified as a sensitive net and the other a large-swing net. Although two out of the 30 parasitic constraints could not be met by the router, since the other parasitic constraints were met with some margins, the circuit with the extracted parasitics met the performance constraints (note that the additional degradation caused by some parasitics violating the constraints may be balanced by less degradation due to others which met the constraints with some margins). When the circuit was routed without the con-

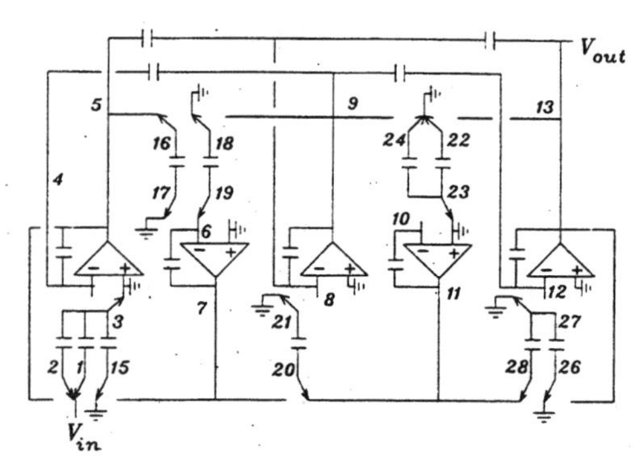

Fig. 7. Fifth-order low-pass switched-capacitor filter.

TABLE I
SENSITIVITIES OF MAGNITUDE RESPONSE AT 0.1 kHz W.R.T. FIVE OF THE MOST SENSITIVE CAPACITANCES

net1	*net2*	*sensitivity at* 0.1 kHz
18	21	−0.0062 dB/fF
22	23	−0.0044 dB/fF
18	23	0.0044 dB/fF
16	19	−0.0043 dB/fF
17	18	0.0043 dB/fF

TABLE II
SENSITIVITIES OF MAGNITUDE RESPONSE AT 1.6 KHz W.R.T. FIVE OF THE MOST SENSITIVE CAPACITANCES

net1	*net2*	*sensitivity at* 1.6 kHz
23	28	0.0184 dB/fF
18	21	−0.0156 dB/fF
18	23	0.0104 dB/fF
27	28	0.0062 dB/fF
17	18	0.0058 dB/fF

TABLE III
FIVE OF THE BOUNDING CONSTRAINTS ON CAPACITANCES AND THE ACTUAL ESTIMATED CAPACITANCES AFTER ROUTING

net1	*net2*	C_{j_bound}	C_{j_actual}
18	21	0.0015 fF	0 fF
12	28	3.3 fF	1.3 fF
4	15	0.01 fF	0 fF
27	28	0.5 fF	0.01 fF
23	28	0.0015 fF	0.007 fF

straints, there was 170% violation in performance constraint. One of the channels of the filter routed with and without the constraints by the channel router ART has been shown in Fig. 8. The noticeable differences resulting from the imposition of the constraints are the changes in positions of the segments of some of the nets, and the use of a shield net to control the couplings. More detailed results for constraint-driven channel routing are provided in [25].

The CPU time for generating the constraints was about 5 s on VAX 8650. Error due to linear approximations (used is modeling performance constraint) was found to be less than 0.5%.

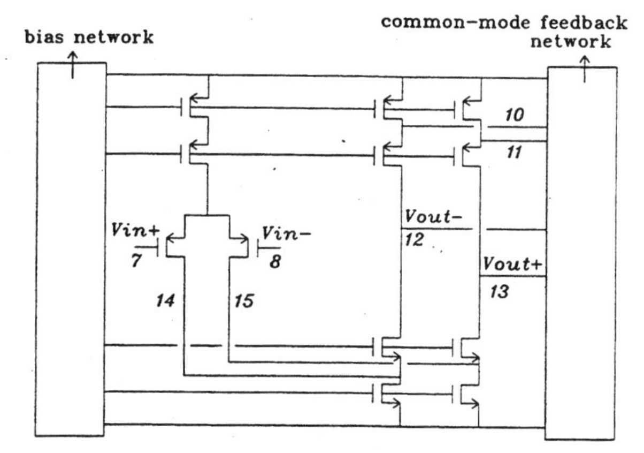

Fig. 8. One of the channels of the filter routed with and without the constraints.

Routed without the bounding constraints:

Routed with the bounding constraints:

shield net

crossover shielding

Fig. 9. Fully differential CMOS opamp.

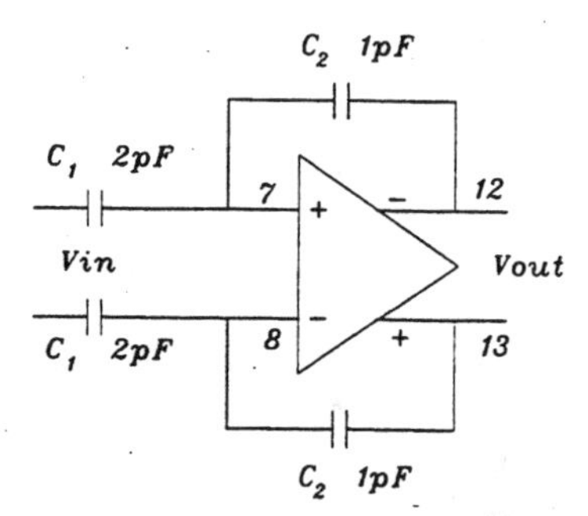

Fig. 10. Closed-loop configuration.

B. CMOS Opamp Example

This section presents the results of automatic constraint generation for a CMOS fully-differential folded-cascode opamp. The schematic of the gain stage of the opamp is shown in Fig. 9. SPICE3 was used for simulation and sensitivity computation. The performance functions of this opamp considered are a) unity-gain bandwidth (UGB), b) phase margin (PM), and c) closed-loop gain (CLG). Performance functions UGB and PM are considered for the open-loop configuration and CLG for the configuration shown in Fig. 10 (this configuration is often used in A–D conversion). The circuit has nominal UGB of 13.4 MHz, and PM of 79.5°. These two performance functions have one-sided constraints (in the negative direction) and are nonprecision specifications (as explained in Section III A). We decided to keep ΔUGB and ΔPM smaller than 0.1 MHz and 0.5°, respectively, in the negative direction (less than about 1% of their nominal values). The closed-loop gain of the circuit is, however, a precision specification having two-sided performance constraints. The ideal value of gain W_{ideal} is 2, and the nominal value W_{nom} is 1.994 (without any interconnect parasitics), mainly because of the finite open-loop gain of the opamp. The process variations ΔW^+_{proc} and ΔW^-_{proc} are 0.002 (computed as suggested in Section III-B by considering a worst-case mismatch of 0.1% between capacitors C_1 and C_2). Hence, the nominal precision W^0_{proc} without any interconnect parasitics is $(2 - 1.994) + 0.002 = 0.008$. The maximum allowed degradation ΔW_{prec}, in precision due to the parasitics has been set to 0.0005. Hence, from (15) and (16), $\Delta W^+_{max} = 0.0165$ and $\Delta W^-_{max} = 0.0005$ for the closed-loop gain.

The placement and global routing for the layout of the opamp have been done manually. The opamp has 37 nodes. Labeling is done only for those nodes which appear in Table V. The circuit has seven matched pairs of nodes, including the pairs (7, 8), (10, 11), (12, 13), and (14, 15), which are shown in Fig. 9. Matching constraints were directly derived from matched-node-pair information as illustrated in Section IV-C. Sensitivity analysis was performed with respect to all the self and coupling capacitances using SPICE3. The UGB was found most sensitive to capacitances involving nets 12 and 13, since they contribute the dominant poles. Nodes 14 and 15 on the other hand contribute the second dominant poles, making the phase margin most sensitive to the capacitances involving these nodes. The CLG, as expected, is highly sensitive to coupling between input (7, 8) and output (12, 13) nodes of the opamp. The sensitivities of the performance functions with respect to some of the critical parasitics have been presented in Table IV–VI. *net*1 and *net2* refer to the two nets between which the parasitic capaci-

TABLE IV
SENSITIVITIES OF UNITY-GAIN-BANDWIDTH W.R.T. FIVE OF THE MOST SENSITIVE CAPACITANCES

net1	net2	UGB sensitivity
12	13	5.5 kHz/fF
12	15	2.85 kHz/fF
10	13	2.75 kHz/fF
7	13	1.7 kHz/fF
12	0	1.4 kHz/fF

TABLE V
SENSITIVITIES OF PHASE MARGIN W.R.T. FIVE OF THE MOST SENSITIVE CAPACITANCES

net1	net2	PM sensitivity
14	15	0.015°/fF
7	14	0.01°/fF
10	15	0.0075°/fF
7	10	0.0065°/fF
14	0	0.0038°/fF

TABLE VI
SENSITIVITIES OF CLOSED-LOOP-GAIN W.R.T. TWO OF THE MOST SENSITIVE CAPACITANCES

net1	net2	CLG sensitivity
7	12	0.002/fF
7	13	−0.002/fF

TABLE VII
FIVE OF THE BOUNDING CONSTRAINTS ON CAPACITANCES AND THE ACTUAL ESTIMATED CAPACITANCES AFTER ROUTING

net1	net2	C_{j_bound}	C_{j_actual}
7	12	0.001 fF	0.0 fF
14	15	0.924 fF	0.036 fF
4	13	0.015 fF	0.053 fF
12	14	9.6 fF	0.035 fF
3	15	5.7 fF	0.4 fF

tance is considered for sensitivity analysis. Sensitivities with respect to matched parasitics have been computed taking the matching constraints into account. Bounding constraints were generated on 47 critical capacitances detected after sensitivity analysis. The opamp has been routed by the constraint-driven area router RoAD. Five of the bounding constraints as well as the actual capacitances after routing are shown in Table VII. Again, although five of the 47 constraints could not be met, the performance constraints could be met. More detailed results regarding the constraint-driven area routing are provided in [26].

The CPU time taken by the constraint generator to generate the constraints was about 25 s on a VAX 8650. Worst-case error due to linear approximations were found to be less than 2.0%, which means that the total performance degradation caused (when the parasitics are below their bounding constraints) may exceed specified values by 2%. But, since only conservative values need to be specified for ΔW^+_{max} and ΔW^-_{max}, these errors are not critical, unless they are too large.

VII. CONCLUSIONS AND FUTURE WORK

A performance-constrained approach for physical design of analog circuits was proposed to reduce the number of time-consuming layout iterations necessary for analog circuits. Algorithms for generating parasitic constraints from the specified performance constraints were presented. A parasitic constraint generator PARCAR, which implements these algorithms was reported. The output of the constraint generator can be used to drive constraint-driven layout tools. The feasibility of this approach was illustrated for automated routing. The constraint generator in its own right can be a valuable design aid, as it provides useful insight and guidelines to layout designers.

APPENDIX A1
DERIVATION OF EXPRESSIONS FOR WORST CASE SENSITIVITY WITH PROCESS VARIATIONS

The expressions for upper and lower bounds for sensitivity of performance function W_i with respect to p_0 (which represents matched pair (p_j, p_k) and has the value of p_j and p_k without any process variations) are now derived, taking process variations into account. Let r_{max} be the maximum fractional process variation associated with a parasitic. Then, the actual values of the matched parasitics in terms of p_0 are

$$p_j = (1 + r_j)p_0, \; p_k = (1 + r_k)p_0 \tag{55}$$

where

$$|r_j|, \; |r_k| \leq r_{max}. \tag{56}$$

If the values of the two parasitics p_j and p_k were statistically independent of each other, then the maximum possible mismatch between the parasitics would have been $2r_{max}$. But the actual mismatch will be less than $2r_{max}$, since the two parasitics will tend to track each other at least to some extent on the same chip.

$$|r_j - r_k| \leq r_{mis} \tag{57}$$

where

$$r_{mis} \leq 2r_{max}. \tag{58}$$

Both r_{mis} and r_{max} are positive numbers.

Differentiating (55) as in (26), the sensitivity with process variation becomes

$$S = (1 + r_j)S_{ij} + (1 + r_k)S_{ik} \tag{59}$$

$$= S_0 + r_jS_{ij} + r_kS_{ik} \tag{60}$$

where S_0 is given by (27). Our goal is to compute the upper and lower bounds on the last two terms of (60) (contributions of process variation). This is a linear programming problem in r_j and r_k with a cost function (denoted by *cost*) equal to

$$r_jS_{ij} + r_kS_{ik} \tag{61}$$

with (56) and (57) as the linear constraints.

The problem has been illustrated with the aid of Fig. 11. The linear inequalities represented by (56) define a

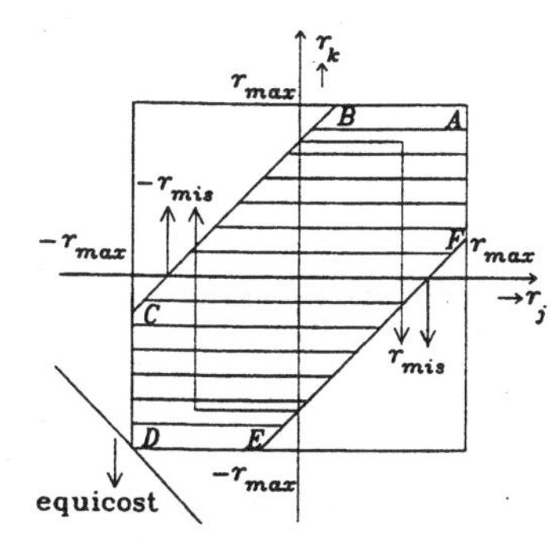

Fig. 11. Sensitivity with process variations.

square box of dimension $r_{\max}$ in the two-dimensional space described by r_j and r_k. The additional mismatch inequalities represented by (57) cut off the shaded region of the square as shown. The two inclined lines (BC and EF) have slopes of 1. The cost function defined by (61) is constant on a straight line (referred to as *equicost* in Fig. 11) with a slope of $-S_{ij}/S_{ik}$. Note that (58) always restricts the points B, C, E, and F on the respective edges of the bounding square, since the point at which BC and EF meet the axes do not go beyond the range $(2r_{\max}, -2r_{\max})$.

Case I: S_{ij} and S_{ik} are of the same sign.

Case (Ia): $S_{ij}, S_{ik} \geq 0$:

$$-r_{\max} \leq r_j, r_k \leq r_{\max} \tag{62}$$

$$\Rightarrow -r_{\max} S_{ij} \leq r_j S_{ij} \leq r_{\max} S_{ij} \tag{63}$$

$$-r_{\max} S_{ik} \leq r_k S_{ik} \leq r_{\max} S_{ik} \tag{64}$$

$$\Rightarrow -r_{\max}(S_{ij} + S_{ik}) \leq r_j S_{ij} + r_k S_{ik} \leq r_{\max}(S_{ij} + S_{ik}). \tag{65}$$

Thus $r_j S_{ij} + r_k S_{ik}$ attains its maximum and minimum at $(r_{\max}, r_{\max})$ and $(-r_{\max}, -r_{\max})$, respectively, (which correspond to points A and D in Fig. 11, respectively).

Case (Ib): $S_{ij}, S_{ik} < 0$:

Since S_{ij} and S_{ik} are negative, the inequalities in (63), (64), and (65) now go in opposite directions. The maximum and minimum in cost function are achieved at $(-r_{\max}, -r_{\max})$ and $(r_{\max}, r_{\max})$, respectively.

Hence, we see that, when S_{ij} and S_{ik} are of the same sign, the two limits for the range of sensitivity S are given by $(1 + r_{\max})S_0$ and $(1 - r_{\max})S_0$. The maximum mismatch of parasitics does not come into the picture in determining the range of the sensitivity.

Case II: S_{ij} and S_{ik} are of the opposite sign.

Case (IIa): $|S_{ij}| = |S_{ik}|$:

Since $S_{ik} = -S_{ij}$, cost $= S_{ij}(r_j - r_k)$. But $|r_j - r_k| \leq r_{\text{mis}}$ implies that upper and lower limits of *cost* are $|S_{ij}|\, r_{mis}$ and $-|S_{ij}|r_{mis}$ attained at $|r_j - r_k| = \pm r_{\text{mis}}$, which correspond to points on EF and BC in Fig. 11, respectively. Although the expressions for this case can be derived by setting $|S_{ij}| = |S_{ik}|$ in the expressions for case IIb and IIc, this case has been treated separately since both the maximum and minimum occur over a continuous set of points in the (r_j, r_k) space, rather than single points as in the other two cases.

Case (IIb): $|S_{ij}| < |S_{ik}|$:

Claim: *cost* attains its extreme values at $(r_{\max} - r_{\text{mis}}, r_{\max})$ and $(r_{\text{mis}} - r_{\max}, -r_{\max})$ (points B and E, respectively, in Fig. 11).

Proof: At point B,

$$\text{cost } (B) = (r_{\max} - r_{\text{mis}})S_{ij} + r_{\max} S_{ik} \tag{66}$$

$$= \pm[(r_{\max} - r_{\text{mis}})|S_{ij}| - r_{\max}|S_{ik}|] \tag{67}$$

"+" sign when $S_{ij} \geq 0$ and $S_{ik} < 0$; "−" sign when $S_{ij} < 0$ and $S_{ik} \geq 0$.

$$\text{cost } (B) = \pm[r_{\max}(|S_{ij}| - |S_{ik}|) - r_{\text{mis}}|S_{ij}|] \tag{68}$$

$$= \mp[r_{\max}(|S_{ik}| - |S_{ij}|) + r_{\text{mis}}|S_{ij}|] \tag{69}$$

$$= \mp C_{\max} \tag{70}$$

where $C_{\max} = r_{\max}(|S_{ik}| - |S_{ij}|) + r_{\text{mis}}|S_{ij}| > 0$, since $|S_{ik}| > |S_{ij}|$.

Similarly, at point E

$$\text{cost } (E) = (r_{\text{mis}} - r_{\max})S_{ij} - r_{\max} S_{ik} \tag{71}$$

$$= \pm[(r_{\text{mis}} - r_{\max})|S_{ij}| + r_{\max}|S_{ik}|] \tag{72}$$

$$= \pm[r_{\max}(|S_{ik}| - |S_{ij}|) + r_{\text{mis}}|S_{ij}|] \tag{73}$$

$$= \pm C_{\max}. \tag{74}$$

Now we shall show that

$$-C_{\max} \leq r_j S_{ij} + r_k S_{ik} \leq C_{\max}. \tag{75}$$

Let $\text{dif}^+ = C_{\max} - r_j S_{ij} - r_k S_{ik}$. Substituting expression for $C_{\max}$,

$$\text{dif}^+ = r_{\max}(|S_{ik}| - |S_{ij}|) + r_{mis}|S_{ij}| - r_j S_{ij} - r_k S_{ik} \tag{76}$$

$$= r_{\max}(|S_{ik}| - |S_{ij}|) + r_{\text{mis}}|S_{ij}| - (r_j - r_k + r_k)S_{ij} - r_k S_{ik} \tag{77}$$

$$= r_{\max}(|S_{ik}| - |S_{ij}|) + r_{\text{mis}}|S_{ij}| - (r_j - r_k)S_{ij} - r_k(S_{ij} + S_{ik}). \tag{78}$$

If $S_{ij} \geq 0$ and $S_{ik} < 0$, then $S_{ij} = |S_{ij}|$, $S_{ik} = -|S_{ik}|$. Thus,

$$\text{dif}^+ = r_{\max}(|S_{ik}| - |S_{ij}|) + r_{\text{mis}}|S_{ij}| - (r_j - r_k)|S_{ij}| + r_k(|S_{ik}| - |S_{ij}|) \tag{79}$$

$$= (r_{\max} + r_k)(|S_{ik}| - |S_{ij}|)| + [r_{\text{mis}} - (r_j - r_k)]\,|S_{ij}|. \tag{80}$$

The first term in the right hand side of the above equation is nonnegative since $|S_{ik}| > |S_{ij}|$ and $|r_k| \leq r_{\max}$. The second term is nonnegative since $|r_j - r_k| \leq r_{\text{mis}}$. Hence, $\text{dif}^+ \geq 0$.

If $S_{ij} < 0$ and $S_{ik} \geq 0$, then $S_{ij} = -|S_{ij}|$, $S_{ik} = |S_{ik}|$.

$$\text{dif}^+ = (r_{max} - r_k)(|S_{ik}| - |S_{ij}|)| + [r_{\text{mis}} + (r_j - r_k)]\,|S_{ij}|. \tag{81}$$

Again, $\text{dif}^+ \geq 0$ due to the same reasons.

$$\text{dif}^+ \geq 0 \Rightarrow r_j S_{ij} + r_k S_{ik} \leq C_{\max}. \quad (82)$$

Let $\text{dif}^- = r_j S_{ij} + r_k S_{ik} - (-C_{\max})$. Substituting expression for $C_{\max}$,

$$\text{dif}^- = r_j S_{ij} + r_k S_{ik} + r_{\max}(|S_{ik}| - |S_{ij}|) + r_{\text{mis}}|S_{ij}| \quad (83)$$

$$= (r_j - r_k + r_k) S_{ij} + r_k S_{ik} + r_{\max}(|S_{ik}| - |S_{ij}|) + r_{\text{mis}}|S_{ij}| \quad (84)$$

$$= (r_j - r_k) S_{ij} + r_k (S_{ij} + S_{ik}) + r_{\max}(|S_{ik}| - |S_{ij}|) + r_{\text{mis}}|S_{ij}|. \quad (85)$$

If $S_{ij} \geq 0$ and $S_{ik} < 0$ ($S_{ij} = |S_{ij}|$, $S_{ik} = -|S_{ik}|$),

$$\text{dif}^- = (r_{\max} - r_k)(|S_{ik}| - |S_{ij}|)| + [r_{\text{mis}} + (r_j - r_k)]|S_{ij}|. \quad (86)$$

Arguing as before for dif^+, $\text{dif}^- \geq 0$.

If $S_{ij} < 0$ and $S_{ik} \geq 0$ ($S_{ij} = -|S_{ij}|$, $S_{ik} = |S_{ik}|$),

$$\text{dif}^- = (r_{\max} + r_k)(|S_{ik}| - |S_{ij}|) + [r_{\text{mis}} - (r_j - r_k)]|S_{ij}|. \quad (87)$$

Again, $\text{dif}^- \geq 0$.

$$\text{dif}^- \geq 0 \Rightarrow r_j S_{ij} + r_k S_{ik} \geq -C_{\max}. \quad (88)$$

From (82) and (88), we conclude that the upper and lower bounds of cost are $\pm C_{\max}$, i.e., $\pm[r_{\max}(|S_{ik}| - |S_{ij}|) + r_{\text{mis}}|S_{ij}|]$, respectively.

Case (IIc): $|S_{ij}| > |S_{ik}|$:

Claim: *cost* attains its extreme values at $(r_{\max}, r_{\max} - r_{\text{mis}})$ and $(-r_{\max}, r_{\text{mis}} - r_{\max})$ (points F and C, respectively, in Fig. 11).

Proof: Proof is the same as for Case IIB, with the roles of subscripts j and k interchanged in S_{ij}, S_{ik}, r_j, and r_k. The upper and lower bounds of *cost* are $\pm[r_{\max}(|S_{ij}| - |S_{ik}|) + r_{\text{mis}}|S_{ik}|]$, respectively.

One has to add S_0 to the bounds for cost to get the bounds for the sensitivity (see (60)).

We shall now illustrate the need for taking process variations into account while computing sensitivities with a practical example. Consider a differential amplifier circuit having two input nodes which are extremely sensitive to external noise. If a large-swing net comes close to these two nodes, there will be a matching constraint imposed on the coupling capacitances as in (34). Without any process variations, the sensitivities of the amplifier output to the two coupling capacitances will be equal in magnitude and opposite in sign, i.e., the overall sensitivity is zero if computed using (27). This will result in *no bounding constraint* on the two capacitances. But in practice, the two coupling capacitances may have a mismatch as large as 20%, since the routing capacitances are poorly controlled. As each of the capacitances gets bigger (when the large-swing net comes closer), the absolute value of the maximum mismatch also grows proportionately, increasing the worst-case noise voltage at the amplifier output (since the two sensitivities act in opposite directions). This gives rise to the need for bounding constraints on these coupling capacitances.

When we want to compute sensitivity with respect to a single parasitic which is not to be matched with any other parasitic, then proceeding in similar fashion it can be shown that the two bounds for the actual sensitivity S are $(1 + r_{\max}) S_0$ and $(1 - r_{\max}) S_0$, where S_0 is the sensitivity computed without taking into account any process variations. Hence $(1 + r_{\max}) S_0$ can be used as the worst-case sensitivity.

Appendix A2
Shielding Decision

It is easy to see that there can be nonzero value of minimum crossover capacitance between two nets decided by pin positions of the nets (for example, when pins of a net are present both on top and bottom edges of a channel, and another net has to go right through the channel from the left to the right edge). For a given placement and global routing, the minimum crossover capacitance for channel routing algorithm used in ART (which is interfaced to the parasitic constraint generator PARCAR) can be estimated as described in [25]. It was earlier mentioned in Section IV-D-4 that the minimum values of parasitics should form a feasible solution for a successful generation of bounding constraints. If that is not the case, we have to *shield* some of the crossovers to reduce the minimum possible capacitances. Shielding can be done when a third layer of interconnection is available (like POLY in addition to MET1 and MET2), which need not be used frequently for routing (since it is a high-resistivity material), but can be used for running short stretches of interconnects if necessary. In that case, one can shift the interconnect in one of the layers to the third layer for a short length, and use a grounded plate in the intermediate layer to shield the crossover (for more detailed description of crossover shielding; refer to [23]). Shielding should be done only when it is not possible to meet performance constraints without it, since shielding requires additional ground wires to be brought to the shielding site and, hence, increases the area occupied by the chip. Also, it increases the capacitive loading of neighboring lines.

Shielding decisions are made before constraint generation. To cause minimum possible shielding, for each performance function for which (50) or (51) is not satisfied by the minimum capacitances, the capacitances are sorted in decreasing magnitudes of performance sensitivities and eliminated successively (marked as shielded) until (50) and (51) are satisfied. Once a feasible solution is assured, the bounding constraints can be generated. The shielding decisions can be passed to the constraint-driven router, along with the parasitic constraints. However, the router is free to go for additional shielding if necessary. During routing it is also possible to run shielding wires between two parallel adjacent lines.

Acknowledgment

The author would like to thank Dr. Bruce Donecker and Dr. Gary Homeland of Hewlett-Packard, Santa Rosa, CA, for providing guidance in the initial phase of the project. They thank Nicholas Weiner for many useful discussions regarding issues involving interface of the constraint generator with the router. They acknowledge the help of Prof. Jan Rabaey, Dr. Hormoz Yaghutiel, and Dr. Gani Jusuf in connection with SWAP, ADORE, and the CMOS opamp example.

References

[1] C. D. Kimble *et al.*, "Autorouted analog VLSI," in *Proc. IEEE Custom Integrated Circuits Conf.*, 1985, pp. 72–78.

[2] J. Rijmenants, J. B. Litsios, T. R. Schwarz, and M. G. R. Degrauwe, "ILAC: An automated layout tool for analog CMOS circuits," *IEEE J. Solid State Circ.*, vol. 24, pp. 436–442, Apr. 1989.

[3] J. M. Cohn, D. J. Garrod, R. A. Rutenbar, and L. R. Carley, "KOAN/ANAGRAM II: New tools for device-level analog placement and routing," *IEEE J. Solid State Circ.*, vol. 26, pp. 330–342, Mar. 1991.

[4] J. Tronteli, L. Tronteli, T. Pletersek, G. Shenton, M. Robinson, K. Floyd, and C. Jungo, "Expert. system. for automated mixed analog/digital layout compilation," in *Proc. IEEE Custom Integrated Circuits Conf.*, 1987, pp. 165-167.

[5] M. Kayal, S. Piguet, M. Declercq, and B. Hochet, "SALIM: A layout generation tool for analog IC's," in *Proc. IEEE Custom Integrated Circuits Conf.*, 1988, pp. 7.5.1–7.5.4.

[6] D. J. Chen, J. C. Lee, and B. J. Sheu, "SLAM: A smart analog module generator for mixed analog-digital VLSI design," in *Proc. IEEE Int. Conf. Computer Design*, 1989, pp. 24–27.

[7] M. Mogaki, N. Kato, Y. Chikami, N. Yamada, and Y. Kobayashi, "An automatic layout system for analog LSI's," in *Proc. IEEE ICCAD*, 1989, pp. 450–453.

[8] S. K. Hong and P. E. Allen, "Performance-driven analog layout compiler," in *Proc. IEEE Int. Symp. Circ. Syst.*, 1990, pp. 835–838.

[9] M. Itoh and H. Mori, "A layout generating and editing system for analog LSI's," in *Proc. IEEE Int. Symp. Circ. Syst.*, 1990, pp. 843–846.

[10] Y. Shiraishi, M. Kimura, K. Kobayashi, T. Hino, M. Seriuchi, and M. Kusaoke, "A high-packing density module generator for bipolar analog LSI's," in *Proc. IEEE ICCAD*, 1990, pp. 194–197.

[11] S. W. Mehranfar, "A technology-independent approach to custom analog cell generation," *IEEE J. Solid-State Circuits*, vol. 26, pp. 386–393, Mar. 1991.

[12] R. S. Gyurcsik and J.-C. Jeen, "A generalized approach to routing mixed analog and digital signal nets in a channel," *IEEE J. Solid-State Circuits*, vol. 24, pp. 436–442, Apr. 1989.

[13] S. Piguet, F. Rahali, M. Kayal, E. Zysman, and M. Declercq, "A new routing method for full custom analog IC's," in *Proc. Custom Integrated Circ. Conf.*, 1990, pp. 27.7.1–27.7.4.

[14] I. Harada, H. Kitazawa, and T. Kaneko, "A routing system for mixed A/D standard cell LSI's," in *Proc. IEEE ICCAD*, 1990, pp. 378–381.

[15] J. L. Burns and A. R. Newton, "SPARCS: A new constraint-based IC symbolic layout spacer," in *Proc. IEEE Custom Integrated Circ. Conf.*, May 1986, pp. 534–539.

[16] R. Okuda, T. Sato, H. Onodera, and K. Tamaru, "An efficient algorithm for layout compaction problem with symmetry constraints," in *Proc. ICCAD*, Nov. 1989, pp. 148–151.

[17] U. Choudhury, "Sensitivity computation in SPICE3," Master's thesis, Univ. California, Berkeley, Dec. 1988.

[18] T. H. Nguyen and H. Guo, "Performance sensitivity computation for interconnect parasitics," Berkeley, CA, Univ. CA, Berkeley, Project Rep. EECS 2441, Dec. 1990.

[19] W. T. Nye, D. Riley, A. Sangiovanni-Vincentelli, and A. L. Tits, "DELIGHT.SPICE: An optimization-based system for the design of integrated circuits and systems," *IEEE Trans. Computer-Aided Design*, vol. 7, pp. 501–519, Apr. 1988.

[20] D. E. Hocevar and P. Yang, "Practical issues for implementing transient sensitivity computation," in *Proc. IEEE ICCAD*, 1984, pp. 1-3.

[21] "The switched capacitor network simulator SWAP reference manual" Release 2.0, Silvar-Lisco, Nov. 1983.

[22] S. W. Director and R. A. Rohrer, "The generalized adjoint network sensitivities," *IEEE Trans. Circuit Theory*, vol. CT-16, pp. 318–323, Aug. 1969.

[23] H. Yaghutiel, A. Sangiovanni-Vincentelli, and P. R. Gray, "A methodology for automated layout of switched-capacitor filters," in *Proc. IEEE ICCAD*, 1986, pp. 444–447.

[24] H. Y. Koh, C. H. Sequin, and P. R. Gray, "Automatic layout generation for CMOS operational amplifiers," in *Proc. IEEE ICCAD*, 1988, pp. 548–551.

[25] U. Choudhury and A. Sangiovanni-Vincentelli, "Constraint-based channel routing for analog and mixed-analog digital circuits," in *Proc. ICCAD*, Nov. 1990, pp. 198–201.

[26] E. Malavasi, U. Choudhury, and A. Sangiovanni-Vincentelli, "A routing methodology for analog integrated circuits," in *Proc. IEEE ICCAD*, Nov. 1990, pp. 202–205.

[27] W. Nye, "Techniques for using SPICE sensitivity computations in DELIGHT.SPICE optimization," in *Proc. IEEE ICCAD*, 1986, pp. 92–95.

[28] E. Malavasi, "A user interface for performance constraint specification," Project Rep., Univ. California, Berkeley, Dec. 1991.

[29] U. Choudhury and A. Sangiovanni-Vincentelli, "An analytical-model generator for interconnect capacitances," in *Proc. IEEE Custom Integrated Circ. Conf.*, May 1991, pp. 8.6.1–8.6.4.

[30] E. Charbon, E. Malavasi, U. Choudhury, A. Casotto, and A. Sangiovanni-Vincentelli, "A constraint-driven placement methodology for analog integrated circuits," to be published in *Proc. IEEE Custom Integrated Circ. Conf.*, 1992.

Umakanta Choudhury received the B.Tech. degree in electrical engineering from the Indian Institute of Technology, Kharagpur, in 1986, and the M.S. and Ph.D. degrees in electrical engineering from the University of California, Berkeley, in 1988 and 1992, respectively.

Currently he is with AT&T Bell Laboratories as a Member of the Technical Staff. In the summer of 1988, he was with Hewlett-Packard, Santa Rosa, CA, working on interconnect modeling for high-frequency circuits. His research interests include CAD for mixed analog/digital and high frequency circuits.

Dr. Choudhury received the Tong Leong Lim Predoctoral Prize from the University of California, Berkeley, in 1988.

Alberto Sangiovanni-Vincentelli (M'74–SM'81–F'83) received the Dr. Eng. degree from the Politecnico di Milano, Italy, in 1971.

From 1971 to 1977, he was with the Politecnico di Milano, Italy. In 1976, he joined the Department of Electrical Engineering and Computer Sciences, University of California, Berkeley, where he is presently a Professor. His research interests are in various aspects of computer-aided design of integrated circuits, with particular emphasis on VLSI, simulation, and optimization. He was an Associate Editor of the IEEE Transactions on Circuits and Systems, and is currently an Associate Editor of the IEEE Transactions on Computer-Aided Design and a member of the Large-Scale Systems Committee of the IEEE Circuits and Systems Society and the Computer-Aided Network Design (CANDE) Committee. He was Executive Vice-President of the IEEE Circuits and Systems Society in 1983.

Dr. Sangiovanni-Vincentelli received the Distinguished Teaching Award from the University of California in 1981. At both the 1982 and 1983 IEEE-ACM Design Automation Conference, he was given a Best Paper and a Best Presentation Award. In 1983, he received the Guillemin-Cauer Award for the best paper published in the IEEE Transactions on Circuits and Systems and Computer-Aided Design in 1981-1982. He is a member of ACM and Eta Kappa Nu.

PERFORMANCE-DRIVEN COMPACTION FOR ANALOG INTEGRATED CIRCUITS

Eric Felt, Enrico Malavasi†, Edoardo Charbon, Roberto Totaro†,
and Alberto Sangiovanni-Vincentelli

Department of Electrical Engineering
and Computer Sciences
University of California
Berkeley, CA 94720

† Dip. di Elettronica e Informatica
Università di Padova
via Gradenigo, 6/A
35131, Padova, Italy

ABSTRACT

This paper describes a new approach to layout compaction of analog integrated circuits which respects *all* of the performance and technology constraints necessary to guarantee proper analog circuit functionality. Our approach consists of two stages: a *fast* constraint graph critical path algorithm followed by a *general* linear programming algorithm. Circuit performance is guaranteed by mapping high-level performance constraints to low-level bounds on parasitics and then to minimum spacing constraints between adjacent nets.

1. INTRODUCTION

The automatic generation of circuit layout consists of three major phases: placement, routing, and compaction. The compaction phase can be viewed as an optimization problem in which one wishes to minimize layout area subject to a set of constraints necessary to insure proper circuit functionality. These constraints are of three general types: technology constraints, topology constraints, and performance constraints.

In digital systems most approaches to layout compaction consider only technology constraints, i.e. object-to-object minimum and maximum spacing requirements. This approach is sufficient because in digital layouts the dependence of electric performances on the details of physical implementation is limited to logic functions and delay requirements. As long as connectivity is not disrupted and interconnection length is not significantly increased, timing specifications are preserved.

In analog systems, however, all three types of functionality constraints are critical issues during all phases of layout synthesis. Topological constraints and performance constraints are essential because analog circuit performances are extremely sensitive to parasitics. Variations in device symmetries (topological constraints) or interconnect length (performance constraints), for example, affect cross-coupling capacitances and stray resistances, which can dramatically degrade circuit performance.

The symbolic layout compaction problem is NP-hard [1]. However, heuristics have been developed which provide good quality layouts. In particular, the use of mono-dimensional compaction phases alternately performed in orthogonal directions is one of the most popular approaches. The mono-dimensional topological compaction problem is typically formulated as a longest path problem on a directed graph, which is solvable in polynomial time. However, not all analog circuit constraints can be expressed in a form which can be solved by a constraint-graph algorithm.

Several approaches to specific portions of the general analog compaction problem have been previously described. In [2] symmetries are enforced by using an iterative graph perturbation technique. Unfortunately false overconstraints frequently arise with this approach because of the order in which the symmetric objects are processed, especially when multiple symmetry axes are present. In [3] a linear programming (LP) algorithm was proposed. Objects not directly interacting with symmetric items in the layout are collapsed, thus generating "super-constraints" which are solved using linear programming. This technique is particularly appropriate when hierarchical compaction is performed [4]. Even after reducing the set of constraints considered in the LP problem, however, this algorithm is severely limited in the maximum circuit complexity which can be considered and in the classes of constraints which can be accommodated. In [5] symmetries are enforced with a combination graph and LP algorithm which uses constraint graph longest path techniques to arrive at a good starting point for an LP solver.

These prior approaches address the problem of enforcing topological constraints (symmetry requirements, in particular), but no provision is made for performance constraints. In this paper we propose an efficient approach to symbolic compaction which accounts for *all* analog constraints. The size and complexity of the circuits with which it can cope are large enough to make the algorithm suitable for industrial-strength applications. The work presented in [5] served as a basis for the results presented here, with the primary contributions of this paper being: 1) a new aggressive technique for controlling parasitics and 2) an original technique for *global* wire length minimization.

Our approach is part of a general top-down, constraint-driven analog circuit design methodology [6]. Top-level constraints are propagated down the design hierarchy from step to step until the final layout is reached. Since early verification techniques are employed to insure that the performance specifications are satisfied at each stage of the layout, the layout which is fed into the compactor is guaranteed to be feasible in the sense that it meets the performance specifications. This methodology has been successfully used for placement [7] and routing [8, 9]. In this paper we extend the same methodology to compaction, thus ensuring that the results obtained from constraint-driven placement and routing are not spoiled during compaction. Not using constraint-driven compaction would cause the whole methodology to fail by permitting uncontrolled constraint violations.

We believe that compaction is an essential step in any analog layout synthesis framework. The synthesis step immediately prior to compaction is usually routing. While the layout from the router is guaranteed to meet all high-level performance specifications, compaction is still essential to insure that the final layout is: 1) as small as possible and 2) design rule correct. Area minimization is an important target and it is difficult to achieve with only placement and routing, especially when area routing is used. Moreover, constraint-driven compaction makes it possible to relax the requirements on routing and accept slight design rule spacing violations by relying on the possibility of fixing the violations during compaction. Thus compaction is not merely an optional post-processing step; in addition to minimizing the area of the chip, compaction also improves the entire analog synthesis process by permitting more aggressive placement and routing.

The algorithm we propose to accommodate the analog performance constraints during compaction is as follows:

1. Generate a set of minimum spacing constraints between wire segments which satisfy all high-level performance constraints. Add these additional constraints to the constraint graph.
2. Solve the modified graph.
3. Introduce shield and symmetry constraints and use the solution achieved in step 2 as the starting point for an LP solver. The LP solver resolves all constraints and produces a final layout of minimum width (height) and minimum interconnect length.

The algorithm is illustrated in Figure 1. If overconstraints are detected in steps 2 or 3 then feedback is provided to step 1 so that an alternative set of spacing constraints can be generated.

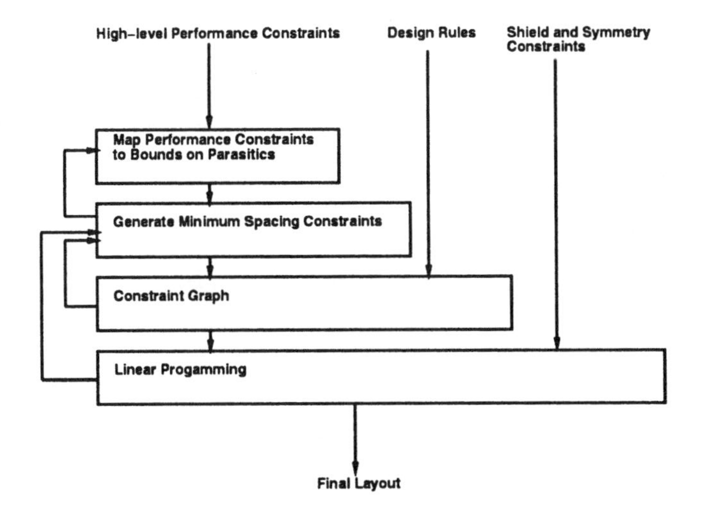

Figure 1: *Compaction algorithm which accommodates all analog constraints.*

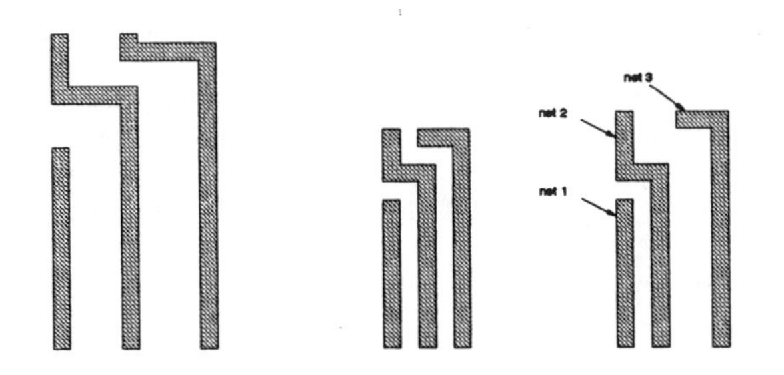

Figure 2: *Coupling constraints. From left to right: original layout, layout compacted without coupling constraints, layout compacted with coupling constraints.*

The use of the graph solution obtained in step 2 as the starting point for the LP solver in step 3 is the key to obtaining a significant speed-up in the solution of the LP, compared to previous approaches using pure LP [3]. In most observed practical cases the configuration yielded by step 2 is close enough to the final solution to require the LP solver to carry out only a small number of steps, yielding 20% to 50% reductions in cpu time over pure LP algorithms. The range of cases that can be managed with acceptable computational complexity is therefore significantly expanded. The details of using the constraint graph and LP solver together to enforce symmetries are reported in [5]. With an LP solver it is also possible to perform global wire length minimization, which is described in section 4.5.

2. ANALOG CONSTRAINTS

Compaction is one of the last steps in the layout synthesis of analog integrated circuits, so the methodology used to synthesize the circuit layout determines the set of constraints which the analog layout compactor must resolve. Under our top-down, constraint-driven methodology, high-level performance specifications are mapped onto low-level constraints which are made available to the compactor. The design methodology insures that the compactor will be

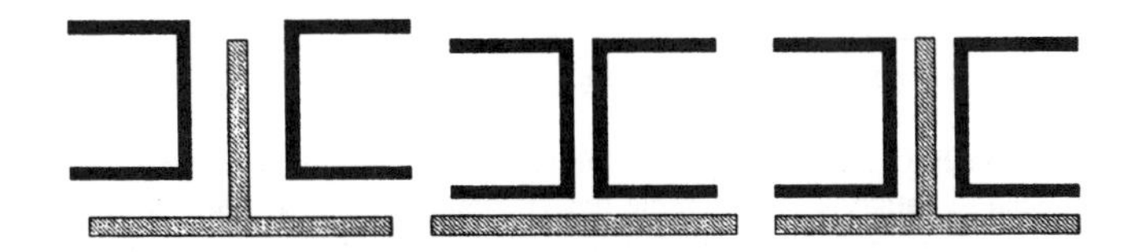

Figure 3: *Shield requirements. From left to right: original layout, layout compacted without shield preservation constraints, layout compacted with shield-preservation constraints.*

provided with *all* of the necessary layout constraints and that no information except interconnects must be inferred directly from the layout. The methodology also insures that the layout and routing tools will provide a reasonable starting point for the compactor, and that if all of the constraints can be satisfied then the circuit will operate within all of its high-level specifications.

Low-level constraints fall into the following categories: 1) maximum cross-coupling, 2) shield preservation, and 3) symmetry enforcement. Each of these constraints will be considered in turn.

- **Maximum Cross-Coupling**

The minimum design rule spacing requirements used when compacting digital layouts are often not sufficient for analog circuits because of capacitive cross-coupling between adjacent wires. When two critical wires are routed closely together, the capacitive coupling between them can seriously degrade the circuit performance. To prevent this detrimental coupling, additional constraints must be added during compaction to insure that the signal coupling does not cause the circuit to violate high-level performance specifications. An example is shown in Figure 2. Without maximum cross-coupling constraints the compacted circuit may not meet the high-level performance requirements because of excessive coupling between nets 2 and 3.

- **Shield Preservation**

When routing an analog circuit it is sometimes impossible to reduce the coupling between two critical lines below the required maximum because of the circuit topology. In these cases the router may introduce a shield between the critically coupled nets. Special constraints must be passed from the router to the compactor to insure that shields are properly preserved. An example is shown in Figure 3.

- **Symmetry Requirements**

Analog circuit designers frequently introduce topological symmetries in differential circuits to optimize offset, differential gain, and noise. Special care must be taken during the compaction step to insure that those symmetries are preserved.

3. ALGORITHM

3.1. Technology Constraints

If only technology constraints are considered, mono-dimensional compaction can be solved efficiently with the constraint graph longest path algorithm [10] [1, Ch. 10]. The pattern of component connectivity and minimum separations required by the technology is described as a weighted, directed graph. Each component is represented in the graph by a node. Maximum and minimum spacing constraints of the form:

$$x_2 - x_1 \geq K \tag{1}$$

are represented as edges of weight K. Note that K is negative for maximum spacing constraints.

A simple plane sweep algorithm is used to generate the minimum spacing constraints. The longest path in this graph provides the minimum width (height).

3.2. Coupling Constraints

High-level performance specifications can be mapped efficiently onto a set of low-level bounds on parasitics [11]. If every parasitic is kept below its bound, all performance constraints are guaranteed to be met. Bounds are set so as to be feasible, on the basis of reasonable estimates of parasitic values drawn from analysis of the layout.

The algorithm to determine the minimum spacings between parallel wire segments is an iterative procedure:

1. Compute the longest path on the current graph.
2. Extract actual parasitics and compare to bounds. If all bounds are met, exit. Otherwise
3. Compute the actual performance degradations, based on extracted parasitics and sensitivities. If performance constraints are met, exit. Otherwise

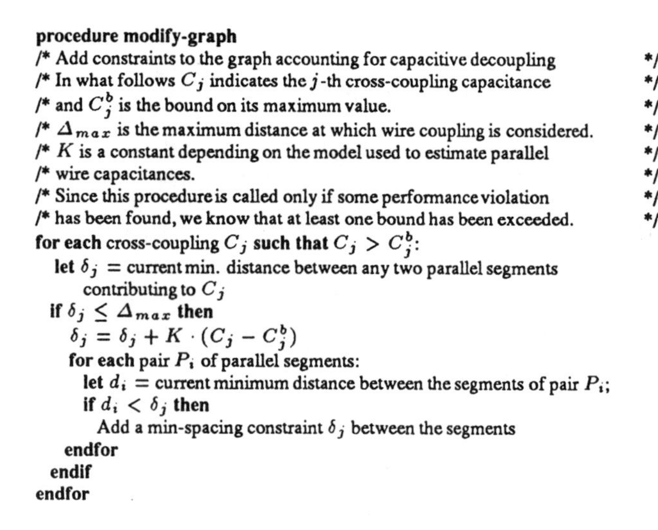

```
procedure modify-graph
/* Add constraints to the graph accounting for capacitive decoupling      */
/* In what follows C_j indicates the j-th cross-coupling capacitance      */
/* and C_j^b is the bound on its maximum value.                           */
/* Δ_max is the maximum distance at which wire coupling is considered.    */
/* K is a constant depending on the model used to estimate parallel       */
/* wire capacitances.                                                     */
/* Since this procedure is called only if some performance violation      */
/* has been found, we know that at least one bound has been exceeded.     */
for each cross-coupling C_j such that C_j > C_j^b:
  let δ_j = current min. distance between any two parallel segments
      contributing to C_j
  if δ_j ≤ Δ_max then
     δ_j = δ_j + K · (C_j − C_j^b)
     for each pair P_i of parallel segments:
       let d_i = current minimum distance between the segments of pair P_i;
       if d_i < δ_j then
         Add a min-spacing constraint δ_j between the segments
     endfor
  endif
endfor
```

Figure 4: *Constraint generation for capacitive decoupling.*

4. Modify the graph by increasing the spacings between critically coupled wire segments. Then go to step 1.

In step 3 performance degradations are computed based on their sensitivities to each parasitic. This step substantially improves the quality of the algorithm's output by permitting the layout to violate some parasitic bounds as long as high-level performance constraints are still satisfied. This situation can occur when one or more parasitics lay below their upper bounds, since the resultant performance improvement can offset the performance degradation due to the parasitics that violate their upper bounds. The algorithm thus guarantees that if a feasible spacing exists then it will be found, even if a non-feasible set of bounds is used by the previous steps in the synthesis process. In addition, area minimization is not artificially constrained by the exact set of parasitic bounds used in the previous steps; if an alternative set of bounds which yields more area improvement exists, it will be found and used.

Pseudo-code for the graph modification procedure implemented at step 4 is shown in Figure 4. For each cross-coupling C_j exceeding its bound C_j^b, δ_j represents the distance between the closest parallel wire segments, namely the segments whose coupling provides the largest contribution to the overall capacitance. The distance between such segments is increased by an amount proportional to the parasitic bound violation. Notice that the distance is increased not only for the closest segments, but also for all the segment pairs whose distances are below δ_j, after this parameter has been increased.

The spacing step implemented by procedure **modify-graph** can introduce overconstraints, making the graph unsolvable. When an overconstraint is detected, a *pruning* procedure is invoked which relaxes the newly-added spacing constraints contained in positive weight loops. In such situations the task of decoupling the two nets is left to the remaining segment pairs. If a feasible solution involving the remaining segment pairs does not exist, then an error is reported because in that event at least one high-level performance constraint cannot be met.

3.3. Shield Constraints

Figure 5 illustrates a lateral and a vertical shield created during routing to reduce coupling between critical nets. The spacing requirement for the lateral shield between two nets labeled 1 and 2 is:

$$x_{shield} \geq x_{net1} \ AND \ x_{shield} \geq x_{net2}. \tag{2}$$

where

x_{shield} = location of top of shield
x_{net1} = location of top of net 1
x_{net2} = location of top of net 2

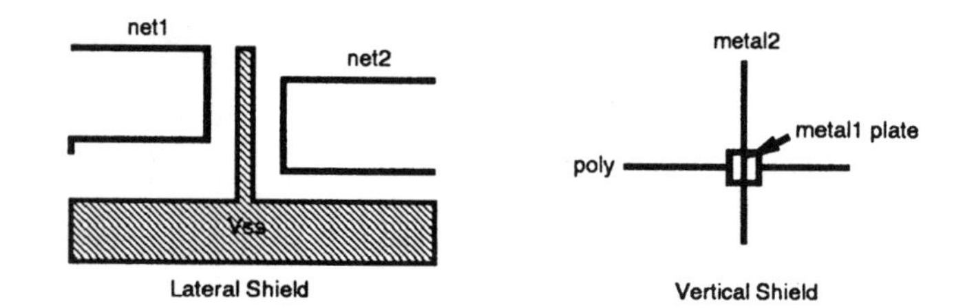

Figure 5: *Horizontal and vertical shields.*

Vertical shield constraints are expressed as:

$$x_{m1} = x_{m2} \tag{3}$$

$$y_{m1} = y_{poly} \tag{4}$$

where

x_{m1}, y_{m1} = center of shielding sheet (usually on metal1).
x_{m2} = location of upper conductor (usually metal2) in horizontal direction.
y_{poly} = location of lower conductor (usually poly) in vertical direction.

Note that shields are inserted between two nets by the router only when absolutely necessary, i.e. when increased spacing between the nets cannot sufficiently reduce the coupling. If the compactor were to remove those shields then the coupling control algorithm outlined in section 4.2 would report an overconstraint and fail.

3.4. Symmetry Constraints

Consider two devices a and b between which a symmetry constraint is required with respect to a vertical axis s (the case with respect to a horizontal axis is perfectly dual). The constraints which must be enforced are:

$$x_a - x_s = x_s - x_b \tag{5}$$

$$y_a = y_b \tag{6}$$

where

x_a = horizontal position of a
x_b = horizontal position of b
y_a = vertical position of a
y_b = vertical position of b

and x_s is the axis position.

While the vertical constraint (6) can be easily resolved in the constraint graph, the horizontal constraint (5) has no graph representation and hence cannot be enforced. It is therefore introduced as a linear programming constraint:

$$x_a + x_b - 2x_s = 0. \tag{7}$$

Symmetry constraints between complete wire segments can be reduced to object symmetry constraints on the wire endings. The horizontal constraints for the two wires labeled 1 and 2 in Figure 6 are formulated as:

$$x_{2,left} - x_s = x_s - x_{1,right} \tag{8}$$

and

$$x_{2,right} - x_s = x_s - x_{1,left} \tag{9}$$

where

$x_{1,right}$ = rightmost ending of wire segment 1
$x_{1,left}$ = leftmost ending of wire segment 1
$x_{2,right}$ = rightmost ending of wire segment 2
$x_{2,left}$ = leftmost ending of wire segment 2.

In this fashion the problem of enforcing one wire symmetry constraint is reduced to satisfying two simple object symmetries. The wire symmetry constraints can thus be directly incorporated into the LP solver and handled in the same way as the device symmetry constraints.

The formulation of the problem and the nature of the starting point allow us to achieve the optimum integer solution using linear programming (LP)

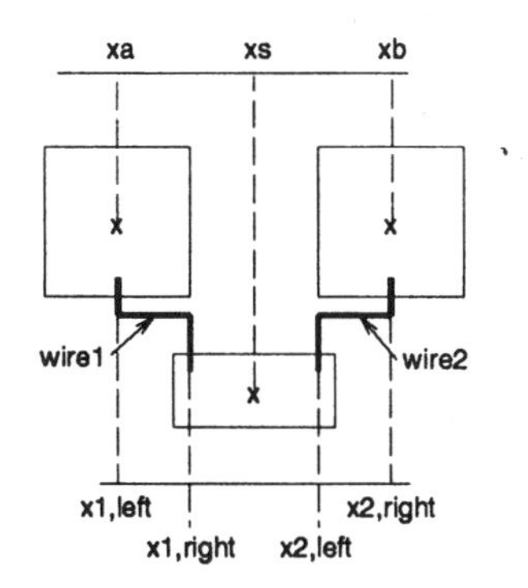

Figure 6: *Wire symmetry constraints.*

rather than integer programming (IP). This simplification is possible because if the nodes of each symmetric pair are also constrained by design-rule minimum spacing requirements, all design-rule minimum spacing requirements are integers, and the leftmost node is located at an integer coordinate. Under these conditions the optimum solution to the LP problem contains only integer coordinates [5].

The LP solver used in this implementation is a simplex algorithm which finds the optimum solution by sequentially visiting the vertices of the polytope bounding the feasible region. Because of the nature of this search algorithm, providing the LP solver with a good initial solution significantly reduces the number of iterations with respect to the case of a random initial basis.

4. WIRE LENGTH MINIMIZATION

Once a minimum area solution which satisfies all of the performance requirements is obtained, a secondary optimization is performed to minimize total interconnect length. This wire length minimization is performed by the LP solver so that the complete set of compaction constraints (including shield and symmetry constraints) can be considered during the optimization. The objective function is formulated as follows:

$$minimize\ L = \sum_{i \in \{all\ wire\ segments\}} (x_{i,right} - x_{i,left}) \quad (10)$$

where

$x_{i,right}$ = rightmost ending of wire segment i
$x_{i,left}$ = leftmost ending of wire segment i
$x_{i,right} \geq x_{i,left}$

This formulation of the objective function permits the LP solver to find the global minimum horizontal (vertical) wire length. While not required by the analog design methodology, optimizing on this secondary objective improves circuit performance by further reducing wire resistances and parasitic capacitances.

One alternative approach [12] performs heuristic local wire length minimization, but no previous approach addresses the issue of global wire length minimization. This important secondary optimization is only possible in this implementation because of the use of the LP solver as a post-processor of the constraint-graph solution.

5. RESULTS

Three example circuits were compacted with the new tool, SPARCS-A. Two are folded cascode CMOS operational transconductance amplifiers (OTA) and the third is a CMOS comparator. For "ota1," high-level performance constraints were specified for unity gain bandwidth and low frequency gain. For "ota2," constraints were specified for unity gain bandwidth, low frequency gain, and offset. For "comp," a high-performance clocked comparator, the high-level constraints were decision time and offset. Circuit performance was measured by extracting all capacitances and resistances from the layout and running SPICE3 with level 2 transistor models. The "comp" circuit, which

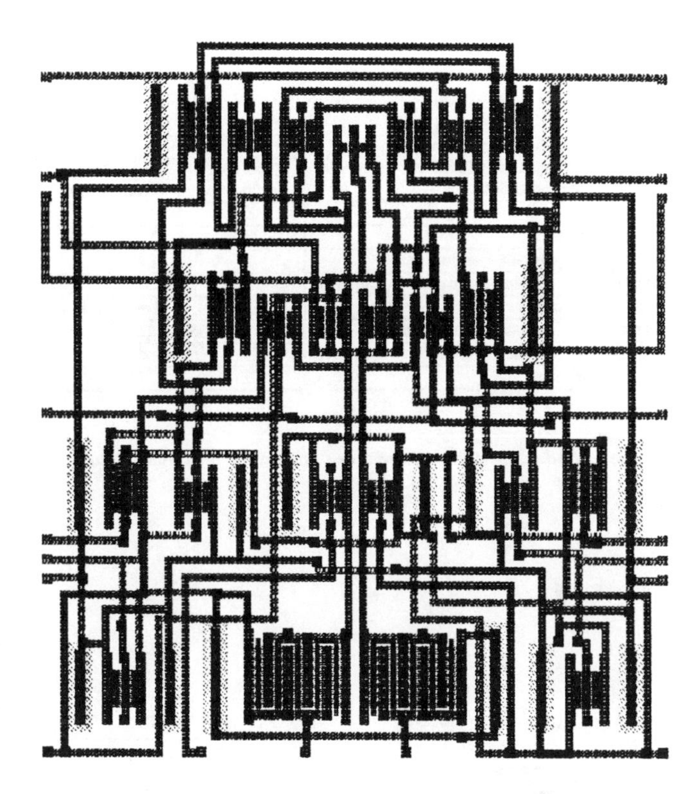

Figure 7: *Layout of the circuit "ota" after compaction.*

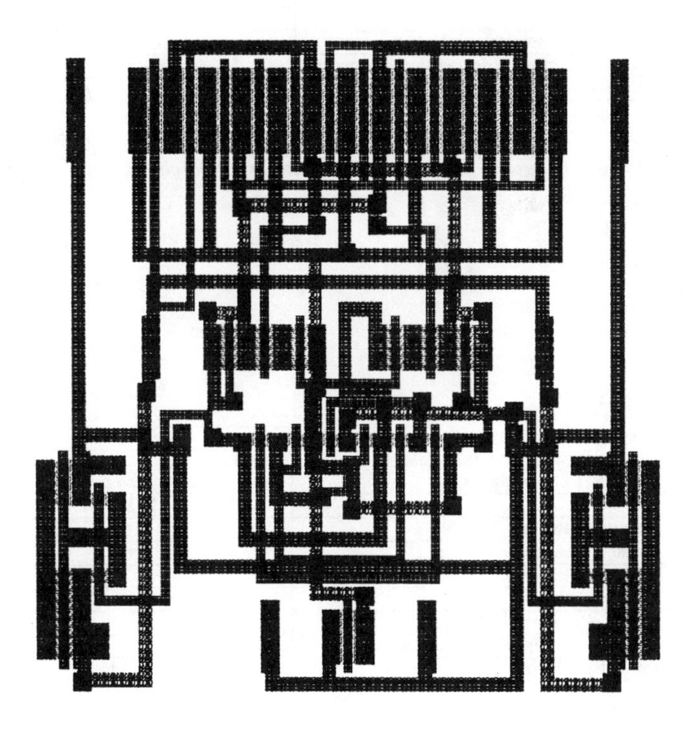

Figure 8: *Layout of the circuit "comp" after compaction.*

ota1, 39 objects, 61 critical parasitics
cpu time for compaction: **106.2 sec**
2 performance constraints

	ota1	
	before compaction	after compaction
area	102480	69918 (68.2%)
wiring length	7333	5964 (81.3%)
unity gain bandwidth–nominal value	$6.015\ MHz$	
maximum permissible degradation (constraint)	$-180\ kHz$	
actual degradation	$-180\ kHz$	$-165\ kHz$
low frequency gain–nominal value	$63.42\ dB$	
maximum permissible degradation (constraint)	$+/-1\ dB$	
actual degradation	$+0.27\ dB$	$+0.32\ dB$

Table 1: *Reduction and performance statistics for the circuit "ota1."*

ota2, 23 objects, 69 critical parasitics
cpu time for compaction: **44.3 sec**
3 performance constraints

	ota2	
	before compaction	after compaction
area	40086	27250 (68.0%)
wiring length	3870	2870 (74.2%)
unity gain bandwidth–nominal value	$6.015\ MHz$	
maximum permissible degradation (constraint)	$-180\ kHz$	
actual degradation	$-188\ kHz$	$-143\ kHz$
low frequency gain–nominal value	$63.42\ dB$	
maximum permissible degradation (constraint)	$+/-1\ dB$	
actual degradation	$+0.04\ dB$	$+0.04\ dB$
offset voltage–nominal value	$0\ \mu V$	
maximum permissible degradation (constraint)	$+100\ \mu V$	
actual degradation	$+81\ \mu V$	$+74\ \mu V$

Table 2: *Reduction and performance statistics for the circuit "ota2."*

comp, 26 transistors, 8 critical parasitics
cpu time for compaction: **36.2 sec**
2 performance constraints

	comp	
	before compaction	after compaction
area	38982	20338 (52.5%)
wiring length	2758	2144 (77.7%)
decision time–nominal value	$6.80\ ns$	
maximum permissible degradation (constraint)	$+3.00\ ns$	
actual degradation	$+2.25\ ns$	$+2.17\ ns$
offset voltage–nominal value	$756\ \mu V$	
maximum permissible degradation (constraint)	$+10\ \mu V$	
actual degradation	$+8\ \mu V$	$-13\ \mu V$

Table 3: *Reduction and performance statistics for the circuit "comp."*

first appeared in [13], was simulated with a 1 pF load capacitance. Note that the final area which SPARCS-A reports for "comp" is almost identical to the area reported in [13].

Tables 1 and 2 report the area reduction and relevant performance statistics for "ota1" and "comp." Figure 7 depicts the final layout of "ota1" and Figure 8 depicts the final layout of "comp." From the tables it is apparent that performance-driven compaction respects the circuits' high-level performance constraints and that significant reduction of chip area can be obtained in a very short CPU time.

5. CONCLUSIONS

A performance-driven compaction algorithm which accommodates *all* layout constraints associated with analog design has been presented. Because of the manner in which the set of constraints is derived, the layout produced by the compactor is guaranteed to meet all of the circuit's high-level analog performance requirements. The algorithm has been implemented and displays remarkable completeness and efficiency. When compared to previous approaches, the algorithm is the first to consider all of the analog constraints necessary to meet high-level circuit performance specifications. Furthermore, we have applied the algorithm to circuits far larger than those shown by previous approaches. The methodology performs well on practical circuits and is suitable for industrial applications.

REFERENCES

[1] T. Lengauer, *Combinatorial Algorithms for Integrated Circuit Layout*, Applicable Theory in Computer Science. John Wiley & Sons, New York, 1990.

[2] J. R. Burns and A. R. Newton, "SPARCS: A New Constraint-Based IC Symbolic Layout Spacer", in *Proc. IEEE Custom Integrated Circuits Conference*, pp. 534–539, 1986.

[3] R. Okuda, T. Sato, H. Onodera and K. Tamaru, "An Efficient Algorithm for Layout Compaction Problem with Symmetry Constraints", in *Proc. IEEE ICCAD*, pp. 148–151, November 1989.

[4] C. S. Bamji and R. Varadarajan, "Hierarchical Pitchmatching Compaction Using Minimum Design", in *Proc. Design Automation Conference*, pp. 311–317, June 1992.

[5] E. Felt, E. Charbon, E. Malavasi and A. Sangiovanni-Vincentelli, "An Efficient Methodology for Symbolic Compaction of Analog IC's with Multiple Symmetry Constraints", in *Proc. European Design Automation Conference*, pp. 148–153, September 1992.

[6] H. Chang, A. Sangiovanni-Vincentelli, F. Balarin, E. Charbon, U. Choudhury, G. Jusuf, E. Liu, E. Malavasi, R. Neff and P. Gray, "A Top-down, Constraint-Driven Design Methodology for Analog Integrated Circuits", in *Proc. IEEE Custom Integrated Circuits Conference*, pp. 841–846, May 1992.

[7] E. Charbon, E. Malavasi, U. Choudhury, A. Casotto and A. Sangiovanni-Vincentelli, "A Constraint-Driven Placement Methodology for Analog Integrated Circuits", in *Proc. IEEE Custom Integrated Circuits Conference*, pp. 2821–2824, May 1992.

[8] E. Malavasi, U. Choudhury and A. Sangiovanni-Vincentelli, "A Routing Methodology for Analog Integrated Circuits", in *Proc. IEEE ICCAD*, pp. 202–205, November 1990.

[9] U. Choudhury and A. Sangiovanni-Vincentelli, "Constraint-Based Channel Routing for Analog and Mixed-Analog Digital Circuits", in *Proc. IEEE ICCAD*, pp. 198–201, November 1990.

[10] A. Mlynsky and C.-H. Sung, "Layout Compaction", in *Layout Design and Verification*, ch. 6, pp. 199–235. T.Ohtsuki Ed., North Holland, 1986.

[11] U. Choudhury and A. Sangiovanni-Vincentelli, "Constraint Generation for Routing Analog Circuits", in *Proc. Design Automation Conference*, pp. 561–566, June 1990.

[12] J. L. Burns, "Techniques for IC Symbolic Layout and Compaction", Memorandum UCB/ERL M90/103, UCB, November 1990.

[13] J. M. Cohn, D. J. Garrod, R. A. Rutenbar and L. R. Carley, "KOAN/ANAGRAM II: New Tools for Device-Level Analog Placement and Routing", *IEEE Journal of Solid State Circuits*, vol. 26, n. 3, pp. 330–342, March 1991.

Automation of IC Layout with Analog Constraints

Enrico Malavasi, Edoardo Charbon, Eric Felt, and Alberto Sangiovanni-Vincentelli, *Fellow, IEEE*

***Abstract*—A methodology for the automatic synthesis of full-custom IC layout with analog constraints is presented. The methodology guarantees that all performance constraints are met when feasible, or otherwise, infeasibility is detected as soon as possible, thus providing a robust and efficient design environment. In the proposed approach, performance specifications are translated into lower-level bounds on parasitics or geometric parameters, using sensitivity analysis. Bounds can be used by a set of specialized layout tools performing stack generation, placement, routing, and compaction. For each tool, a detailed description is provided of its functionality, of the way constraints are mapped and enforced, and of its impact on the design flow. Examples drawn from industrial applications are reported to illustrate the effectiveness of the approach.**

I. Introduction

THE LAYOUT of analog circuits is intrinsically more difficult than the digital one. High performance can be achieved by taking advantage of the physical characteristics of integrated devices and of the correlation between electrical parameters and their variations due to statistical fluctuations of the manufacturing process. Device matchings, parasitics, thermal, and substrate effects must all be taken into account. The nominal values of performance functions are subject to degradation due to a large number of parasitics which are generally difficult to estimate accurately before the actual layout is completed. With severe performance degradation, some specifications may not be satisfied, thus jeopardizing the functionality of larger designs of which the circuit is a relevant component.

At the system level, analog silicon compilers have reached satisfactory results with systems characterized by regular hierarchical structures. Examples are programs for the automatic synthesis of opamps and comparators [1]–[3], switched-capacitor filters [4]–[6] and data converters [7]–[9]. Although these generators cover a substantial fraction of the analog circuits needed in most industrial applications, a more general approach, able to cope with arbitrary architecture and full custom layout, is often needed. A variety of approaches inherited from the digital CAD world, with placement based on slicing structures and channel routing, have been proposed: in macro-cell [10] and standard-cell [11]–[13] approaches, capacitive coupling between interconnections is minimized during global routing by allocating sensitive nets to separate channels. In ILAC [14], the layout generator for the analog synthesis system IDAC [15], the layout is based on the generation of specialized predefined modules. Knowledge-based approaches, such as the ones presented in SALIM [16], LADIES [17], and BLADES [18], rely heavily on the user's expertise. In [19], a gridless channel-router is described, where great importance is given to parasitic control during routing; capacitance bounds between nets are preserved by setting the minimum separation between horizontal segments in a channel, and by ordering them to avoid crossovers (if possible). In [20], cross-coupling minimization is the routing target, while stray resistances are controlled by means of variable wire-segment widths. In STAT [21] and in KOAN/ANAGRAM [22], placement and routing rely on weighted parasitic minimization and matching constraint enforcement. Area routing and unconstrained placement with abutment capability provide the layout with high flexibility and good area performance. However, no clear strategy is indicated for the definition of parasitic weights, of routing schedule and matching constraints. This information must be supplied by the user on the ground of his/her experience and knowledge of the circuit behavior. Finally, most of the systems cited above produce noncompacted layouts. Unconstrained compaction can degrade parasitics by modifying the spacings between interconnections and matched devices. So far, few approaches [23] have been reported which face, at least in part, the multiconstrained analog compaction problem.

A set of tools able to guarantee that constraints are met in a reasonable number of applications would have a considerable impact, since analog designers need to trust a tool to meet their specs before using it. Analog CAD tools, like their digital counterparts, must guarantee to meet all specs, or otherwise to detect as soon as possible infeasibility and its causes. Only recently, constraint-driven layout generation tools [24]–[27] have been proposed, generally based on sensitivity analysis of circuit performance [28], [29]. In this paper, a methodology and the supporting tools [30] for performance-driven layout synthesis are presented. In the methodology, high-level constraints are automatically translated into a set of low-level bounds on the parameters (i.e., parasitics and geometry) that can be effectively controlled during layout synthesis. Design choices are taken trying to detect infeasible configurations as early as possible. After each stage of the design, further elaboration is allowed only if the partial design can meet all performance specifications. While the tools supporting the methodology have been presented [27], [28], [31]–[34], so

Manuscript received February 15, 1995; revised December 18, 1995. This work was supported in part by SRC Grant 91-DC-008, by ARPA Grant J-FBI-90-073, by FUJITSU, by the MICRO Program of the State of California, by the Italian National Council of Research, and by Asea Brown Boveri, Baden, Switzerland. This paper was recommended by Associate Editor R. Otten.

E. Malavasi is with Cadence Design Systems, San Jose, CA. USA, on leave from the Dipartimento di Elettronica e Informatica, University of Padova, Italy.

E. Charbon, E. Felt, and A. Sangiovanni-Vincentelli are with the Department of Electrical Engineering and Computer Science, University of California, Berkeley, CA 94720 USA.

Publisher Item Identifier S 0278-0070(96)05721-1.

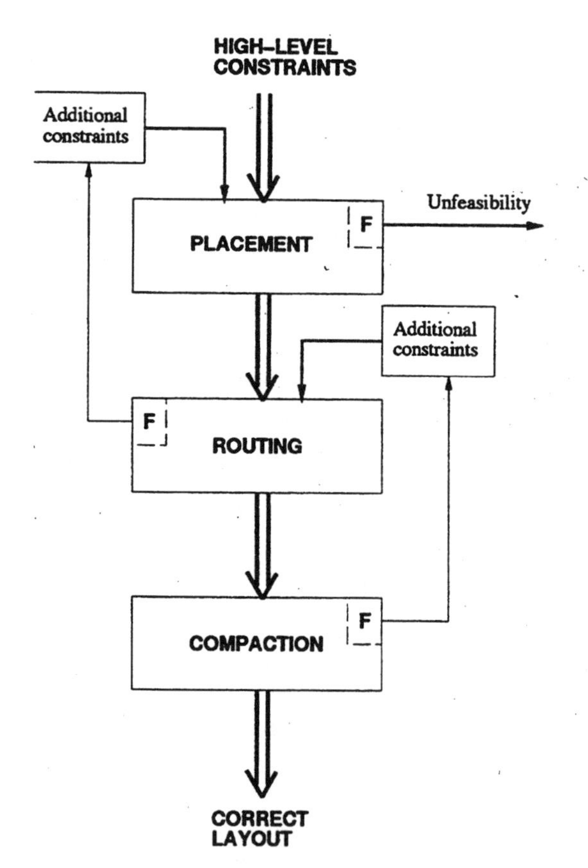

Fig. 1. The traditional design partition into tasks has been modified by adding information paths between layout phases.

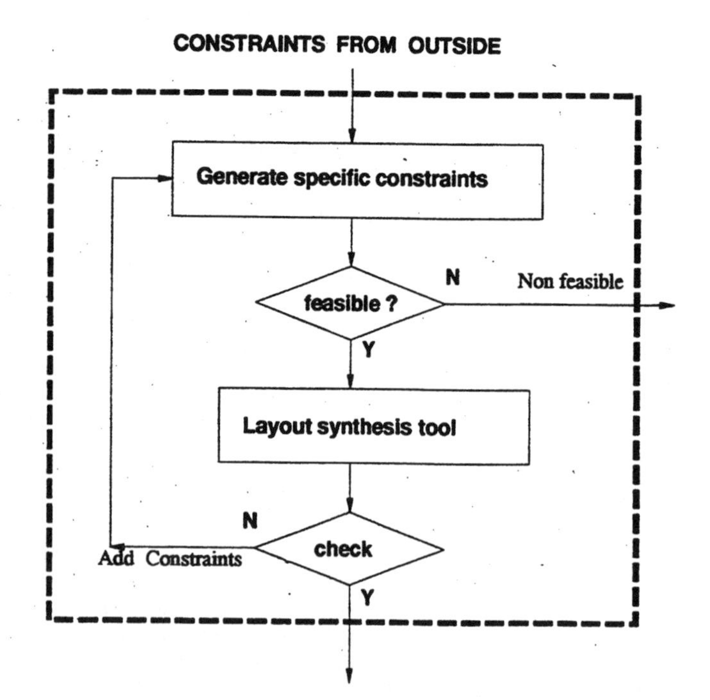

Fig. 2. The organization of each layout phase. Internal feedback paths provide information to the constraint generator. External feedback paths provide information to the previous layout phases about the reason of failure to meet performance specifications.

far, no paper has been published to present an overall view of the layout methodology. Moreover, to the best of our knowledge, this paper is the first comprehensive presentation of a fully integrated performance-driven analog layout system and of the tools supporting it, targeted toward a general set of applications.

The top-down layout design flow is illustrated in Fig. 1. At each step, the existence of a feasible configuration is checked and high-level constraints are translated into a set of bounds on low-level parameters. Sensitivity analysis and parasitic estimates are used to determine feasible bounds. Among all the possible sets of bounds, the one maximizing the *flexibility* of the tool to be used is chosen. Flexibility is a function which measures how easily the tool is able to meet the given set of constraints. In feedback paths, infeasible solutions are analyzed to increase the accuracy of the parameter estimates used for bound generation.

Each layout phase is organized as illustrated in Fig. 2. The design task is constrained by a set of input specifications, which are either high-level performance specifications or additional design constraints introduced by other layout phases. Constraints are translated into a set of bounds on parasitics by a *constraint generator*, based on estimates of the feasible values of each parasitic. These bounds drive each tool independently. The resulting layout is then analyzed to check whether performance specifications have actually been met. If some constraint has been violated, the values of the extracted parasitics can provide more accurate estimates to the constraint generator. The constraint generator also executes the feasibility check. In fact, low-level bounds must be feasible, i.e., they must lay between the minimum and maximum possible values estimated for the parameters. Such early detection of infeasibility provides an efficient control of design iterations, thus minimizing overall computation time. Feedback control paths provide previous design phases with information on those critical parasitics for which it was not possible to determine feasible bounds with the current configuration.

In this paper, several tools supporting the methodology are described. The basic algorithms employed by the tools are described, while detailed discussion on each tool can be found in the referenced literature. Emphasis is given to techniques and algorithms for the management of analog constraints, and to their coordination in the design flow. The organization of this paper is as follows. Section II presents an overview of sensitivity analysis and on the techniques for constraint translation to low-level bounds. The constraint generator PARCAR is described in Section III. In the following sections, the constraint-driven tools are described. The placement tool PUPPY-A is described in Section IV, the routers ART and ROAD are presented in Section V, and the compactor SPARCS-A is in Section VI. Experimental results on industrial-strength benchmarks are reported in Section VII, followed by conclusions in Section VIII.

II. Sensitivity Analysis and Constraint Generation

Constraint generation is the translation of high-level performance specifications into bounds on low-level layout parameters, such as parasitics, wire and device spacing, and symmetries.

High-level performance constraints are expressed as maximum allowed degradations from nominal values, due to process variance and to the parasitics introduced in the definition of layout details. Both absolute parasitic values and mismatch play a role in the deviation of performance functions from their nominal behavior. If some regularity assumptions are satisfied, the relative importance of each parameter can be expressed by the *sensitivity* of performance functions with respect to the parameters.

We denote by N_p the number of layout parameters, by $\mathbf{p} = [p_1 \dots p_{N_p}]^T$ the array of all such parameters, and by $\mathbf{p}^{(0)} = [p_1^{(0)} \dots p_{N_p}^{(0)}]^T$ the array of their *nominal* values. Each performance K_i is a nonlinear continuously differentiable function of all parasitics $K_i = K_i(\mathbf{p})$, and the array of the N_k performance functions will be indicated as $\mathbf{K} = \mathbf{K}(\mathbf{p}) = [K_1(\mathbf{p}) \dots K_{N_k}(\mathbf{p})]^T$. If all parasitics are subject to variations with respect to their nominal values, let $\mathbf{\Delta K}(\mathbf{p}) = \mathbf{K}(\mathbf{p}) - \mathbf{K}(\mathbf{p}^{(0)})$ be the corresponding degradation of $\mathbf{K}$ due to such variations.

A generalized expression for the computation of sensitivities from a set of arbitrary performance functions has been derived in [35], [36]. With this formulation, all performance functions can be represented in a compact and rigorous way, as long as they are continuous and sufficiently regular in an interval around their nominal value, The sensitivity of K_i with respect to p_j is defined as[1]

$$S_{i,j} = \left. \frac{\partial K_i(\mathbf{p})}{\partial p_j} \right|_{\mathbf{p}^{(0)}}.$$

The array of all sensitivities is

$$\mathbf{S} = \begin{bmatrix} S_{1,1} & \dots & S_{1,N_p} \\ \dots & \dots & \dots \\ S_{N_k,1} & \dots & S_{N_k,N_p} \end{bmatrix}.$$

Sensitivities are computed for each performance function, with respect to all the parameters that may be introduced or modified by the layout phase, i.e., parasitics and geometric parameters. The *adjoint technique* of sensitivity analysis [37] has been used in the AC, DC, and time [38] domain. Performance degradations are approximated by linearized expressions using sensitivities [39], which is acceptable as long as we assume that degradations are small compared to the nominal values. The array of all degradations of performance functions due to parasitic variations is

$$\mathbf{\Delta K}(\mathbf{p}) = \mathbf{S}[\mathbf{p} - \mathbf{p}^{(0)}]. \tag{1}$$

Before the definition of layout details, one cannot take advantage of the possible cancellation effects due to positive and negative sensitivities for different parasitics. Hence, each performance constraint is modeled only with respect to the parasitics whose sensitivity is either positive or negative, depending on the sign of the constraint itself. In the general problem formulation, performance constraints are modeled by the following inequalities[2]

$$\mathbf{\Delta K}(\mathbf{p}) - \overline{\mathbf{\Delta K^+}} \le \mathbf{0} \tag{2}$$
$$\mathbf{\Delta K}(\mathbf{p}) + \overline{\mathbf{\Delta K^-}} \ge \mathbf{0} \tag{3}$$

where $\overline{\mathbf{\Delta K^+}}$ and $\overline{\mathbf{\Delta K^-}}$ are the vectors of constraints, in absolute value, on the degradation of performance functions $\mathbf{K}(\mathbf{p})$ in the positive and negative direction, respectively. They can be different and one of them can eventually be infinite. By substituting the linearized expression (1) in inequalities (2) and (3), the general problem can be rewritten as

$$\mathbf{S}^+[\mathbf{p} - \mathbf{p}^{(0)}] - \overline{\mathbf{\Delta K^+}} \le \mathbf{0} \tag{4}$$
$$\mathbf{S}^-[\mathbf{p} - \mathbf{p}^{(0)}] - \overline{\mathbf{\Delta K^-}} \le \mathbf{0} \tag{5}$$

where $\mathbf{S}^+$ is the matrix of the worst-case positive sensitivities and $\mathbf{S}^-$ is the matrix of the absolute values of the worst-case negative sensitivities

$$\mathbf{S}^+\{i,j\} = \max(0, S_{i,j})$$
$$\mathbf{S}^-\{i,j\} = \max(0, -S_{i,j}).$$

In the remainder of this paper, the '+' and '−' signs have been omitted in the notations of sensitivities and constraints. Expressions (4) and (5) are given for positive and negative directions, and the general problem formulation becomes

$$\mathbf{S}[\mathbf{p} - \mathbf{p}^{(0)}] - \overline{\mathbf{\Delta K}} \le \mathbf{0}. \tag{6}$$

We want to determine an array of *bounds* $\mathbf{p}^{(b)} = [p_1^{(b)} \dots p_{N_p}^{(b)}]^T$ for all parasitics, such that inequality (6) holds as long as each parasitic remains below its bound, i.e.,

$$\mathbf{S}\left[\mathbf{p}^{(b)} - \mathbf{p}^{(0)}\right] - \overline{\mathbf{\Delta K}} = \mathbf{0}. \tag{7}$$

All bounds must be *feasible* and *meaningful*, i.e., they must be within the range of values that the parasitics can assume in practice. Let $p_j^{(\min)}$ and $p_j^{(\max)}$ be, respectively, the minimum and maximum possible values which can be assumed by parasitic p_j, and let $\mathbf{p}^{(\min)} = [p_1^{(\min)} \dots p_{N_p}^{(\min)}]^T$ and $\mathbf{p}^{(\max)} = [p_1^{(\max)} \dots p_{N_p}^{(\max)}]^T$. The array of bounds $\mathbf{p}^{(b)}$ must satisfy the following inequalities

$$\begin{cases} \mathbf{p}^{(b)} - \mathbf{p}^{(\min)} \ge \mathbf{0} \\ \mathbf{p}^{(b)} - \mathbf{p}^{(\max)} \le \mathbf{0} \end{cases}. \tag{8}$$

The solution of (7), subject to the feasibility constraints (8), is called the *constraint-generation problem.*

[1] Here and in what follows, the nonnormalized notation, first used in [28], is used for sensitivities, without loss of generality.

[2] The notation $\mathbf{A} \le \mathbf{0}$ means that every element of array $\mathbf{A}$ is a real number not greater than 0. Similarly, $\mathbf{A} \ge \mathbf{0}$ indicates that every element of $\mathbf{A}$ is a nonnegative real number.

TABLE I
NOTATION FOR PARASITICS AND PERFORMANCE FUNCTIONS

symbol	meaning
$Rs_{\_i}$	degeneration resistance at the source of transistor M_i
$Rs_{\_i,j}$	mismatch between $Rs_{\_i}$ and $Rs_{\_j}$
C_i	substrate capacitance of net i
$C_{i,j}$	cross-coupling capacitance between nets i and j
V_{t_i}	voltage threshold of transistor M_i.
$V_{t\_i,j}$	mismatch between V_{t_i} and V_{t_j}
V_{dd}	Supply voltage
ω_0	Unity-gain bandwidth
A_v	Low-frequency gain
V_{off}	Systematic offset
ϕ_M	Phase margin
τ_D	Switching delay

The notation adopted in this paper, for frequently used parasitics and performance functions, is reported in Table I.

III. THE CONSTRAINT GENERATOR

In general, an infinite number of solutions exist for the constraint-generation problem. PARCAR [28] is a constraint-generator, namely a tool able to find a solution to the constraint-generation problem under particular assumptions. Among all solutions, PARCAR chooses the one maximizing the layout tool *flexibility*, which is a measure of how easily the tool is able to meet the constraints. To explain this concept, suppose that the bound for a given parasitic p_j is close to its lower limit $p_j^{(\min)}$, and far from its upper limit $p_j^{(\max)}$. Then the tool is required to maintain p_j within a bound which imposes a tight limit to its variation. If, on the contrary, the bound is close to $p_j^{(\max)}$, the effort required is lower, and the constraint is easier to meet. Therefore, flexibility is defined as

$$F = 1 - \frac{\|\mathbf{p}^{(\max)} - \mathbf{p}^{(b)}\|_2}{\|\mathbf{p}^{(\max)} - \mathbf{p}^{(\min)}\|_2}.$$

A discussion of this definition and of the quadratic norm choice can be found in [28]. In PARCAR, a geometric norm is used, and the constraint-generation problem is solved by minimizing a quadratic function (the geometric norm) subject to linear constraints (7) and (8), using a standard quadratic programming (QP) package.

The quality of the result depends on the estimates of parasitic limits $\mathbf{p}^{(\min)}$ and $\mathbf{p}^{(\max)}$, which become more and more accurate as layout details are defined during the design. The values of $\mathbf{p}^{(\min)}$ and $\mathbf{p}^{(\max)}$ are generally not known *a priori*. However, it is possible to compute suitable estimates that depend on the layout algorithm used. For example, the minimum value of the cross-coupling capacitance between unrouted nets can be set either to zero, or to the crossover capacitance due to unavoidable crossings. The latter estimate, however, is possible only if the router is able to detect unavoidable net crossings. This is the case for a channel router, where wire paths have been predefined in the global routing phase. With maze routing, on the contrary, the minimum value is always set to zero.

A substantial speed-up of the QP solver is achieved by removing from the problem those parasitics whose cumulative contribution to performance degradation is negligible. A threshold value $\alpha < 1$ is defined (in PARCAR we set $\alpha = 0.01$). For each performance function K_i, all parasitics are sorted by increasing value. The first n_i parasitics in the sorted list such that

$$\sum_{j=1}^{n_i} S_j^i p_j^{\max} \leq \alpha \overline{\Delta K_i} \tag{9}$$

are considered noncritical with respect to the threshold α. This procedure detects the parasitics whose cumulative contribution to performance degradation is small. To compensate for this simplification in the constraint-generation problem, (7) is modified by replacing $\Delta\mathbf{K}$ with $(1-\alpha)\Delta K$. Notice that the sorting order may be different for different performance functions. Let P_i denote the set of n_i critical parasitics sorted according to performance K_i. When all performance functions are considered simultaneously, the set of noncritical parasitics is

$$P = \bigcap_{i=1}^{N_k} P_i.$$

The set P of noncritical parasitics determined in this way is eliminated from further analysis. Different sorted lists are maintained for each kind of parasitics (capacitances, resistances, and inductances) and elimination is carried out separately. This simplification can be very effective, since in most cases it allows to eliminate a relevant number of negligible parasitics.

A. Matching Constraints

Matching constraints are drawn from high-level constraints through sensitivity analysis. Matching of devices or interconnections can be defined as a correlation enforced between their electrical parameters, by means of a proper layout setup minimizing the effect of technological gradients and random mask errors.

Consider two parasitics p_1 and p_2. Within the limits of linear approximation (1), their contribution to the degradation of performance K_i is

$$\Delta K_i|_{1,2} = S_{i,1}p_1 + S_{i,2}p_2 = 2S_{i,p}p + \frac{S_{i,\Delta}}{2}\Delta p \tag{10}$$

where

$$p = \frac{p_1 + p_2}{2} \quad S_{i,p} = S_{i,1} + S_{i,2}$$
$$\Delta p = p_1 - p_2 \quad S_{i,\Delta} = \frac{S_{i,1} - S_{i,2}}{2}.$$

It is evident that if

$$\left|\frac{S_{i,\Delta}}{S_{i,p}}\right| \gg 1 \tag{11}$$

the contribution of p_1 and p_2 to the degradation of K_i can be significantly reduced by increasing the correlation between the two parasitics, i.e., by enforcing matching between them. Inequality (11) determines quantitatively the benefit deriving from matching enforcement. For each pair of parasitics, their mismatch and average sensitivities are computed. If relation (11) holds, the mismatch Δp and the average value p replace p_1 and p_2 in the list of parasitics. In our approach, the magnitude requested to ratio $|\frac{S_{i,\Delta}}{S_{i,p}}|$ is user-defined. In our tests, we have obtained good results by requiring the ratio to be at least 10, and this value has been used in all the examples of this paper. If we assume that all parameters in $\mathbf{p}$ (mismatches as well as parasitics) are independent random variables with zero mean, the variance of the degradation of performance function K_i with respect to the variances of all mismatches is

$$\sigma^2(\Delta K_i) = \sum_{j \neq l} |S_{i,\Delta_{j,l}}|^2 \sigma^2(\delta p_{j,l}). \tag{12}$$

In [40], and more recently in [41], relations have been determined between variances and the relative orientation and distance between device pairs. This information can be used to translate the maximum allowed performance degradation into constraints on the physical separation and relative orientation between devices. This procedure has been described in detail in [42].

B. Symmetry Constraints

A quantitative approach to the determination of all parasitic and device symmetry constraints has been developed and is used to generate automatically symmetry constraints. Symmetry is recognized as a particular case of matching between devices or interconnections belonging to distinct differential signal paths, which become effective when the circuit is operated in differential mode. A graph-based search algorithm, described in detail in [42], has been designed for the automated detection of all critical symmetry constraints. First, a graph is built, with a node for each circuit net, and an edge for each device, to represent the circuit connectivity. Then all virtual grounds are detected by comparing the common- and differential-mode gains of all nets. The search algorithm recognizes all the subgraphs whose structure has the following characteristics:

1) symmetric topology;
2) matching constraints between symmetric graph elements;
3) the two halves of the structure are connected with one another by one or more real or virtual ground nets.

Each of these subgraphs is a differential structure, and the symmetry constraints are all the matching constraints recognized at Step 2.

C. Example

As a practical example, consider the clocked comparator COMPL, whose schematic is shown in Fig. 3. This comparator has been used as a benchmark in several recent works on analog CAD [32], [43], [44], due to its relevant performance sensitivity to layout details. Consider the following stray resistances (see Table I for notation) and the corresponding sensitivities of systematic offset V_{off} with respect to each of them

$$\mathbf{p} = \begin{bmatrix} R_{S\_1} \\ R_{S\_2} \\ R_{S\_3} \\ R_{S\_4} \\ R_{S\_6} \\ R_{S\_7} \\ R_{S\_20} \\ R_{S\_21} \\ R_{S\_22} \\ R_{S\_23} \end{bmatrix} \quad \mathbf{S} = \begin{bmatrix} 56.53 \\ -56.53 \\ 0.202 \\ -0.202 \\ 11.83 \\ -11.83 \\ 16.76 \\ -16.76 \\ -16.72 \\ 16.72 \end{bmatrix}^T.$$

Offset sensitivities to resistances are expressed in $\mu V/\Omega$. They were computed by SPICE-3 [38] with a precision within the third digit. Therefore, for each of the pairs $R_{S\_1,2}$, $R_{S\_3,4}$, $R_{S\_6,7}$, $R_{S\_20,21}$, $R_{S\_22,23}$, $R_{S\_21,23}$, $R_{S\_20,22}, R_{S\_21,22}$, the ratio (11) is $|\frac{S_{i,\Delta}}{S_{i,p}}| \geq 10^3$, i.e., the resistive mismatch is at least 10^3 times more important for offset than the absolute values of these resistances. By simplification (10), offset sensitivities with respect to mismatches become

$$\mathbf{p} = \begin{bmatrix} R_{S\_1,2} \\ R_{S\_20,23} \\ R_{S\_21,23} \\ R_{S\_20,22} \\ R_{S\_21,22} \\ R_{S\_6,7} \\ R_{S\_3,4} \end{bmatrix} \quad \mathbf{S} = \begin{bmatrix} 56.53 \\ 16.76 \\ 16.74 \\ 16.74 \\ 16.72 \\ 11.83 \\ 0.201 \end{bmatrix}^T.$$

The cumulative effect of all average values on performance degradation is negligible according to (9) and, therefore, they are all eliminated from $\mathbf{p}$.

The symmetry-constraint graph-search algorithm detected the following symmetric net pairs

$$(52, 53), (15, 16), (10, 11), (13, 14), (55, 56)$$

and the following device pairs

$$(M_1, M_2), (M_{20}, M_{22}), (M_{21}, M_{23}),$$
$$(M_{25}, M_{26}), (M_6, M_7), (M_{10}, M_{11}), (M_8, M_9).$$

Performance constraints are enforced on the max switching delay τ_D and on systematic offset V_{off}

$$\tau_D \leq 7 \text{ ns}$$
$$|V_{\text{off}}| \leq 1 \text{ mV}. \tag{13}$$

In the first steps of layout, we assume that the nominal value of all parasitics is 0, i.e., $\mathbf{p}^{(0)} = [0 \ldots 0]^T$. Simulation yields a nominal value of the switching delay $\tau_D^{(0)} = 4$ ns and null offset. Therefore

$$\mathbf{K} = \begin{bmatrix} \tau_D \\ V_{\text{off}} \\ -V_{\text{off}} \end{bmatrix} \quad \mathbf{K}(\mathbf{p}^{(0)}) = \begin{bmatrix} 4.0 \text{ ns} \\ 0.0 \\ 0.0 \end{bmatrix} \quad \overline{\Delta \mathbf{K}} = \begin{bmatrix} 3.0 \text{ ns} \\ 1 \text{ mV} \\ 1 \text{ mV} \end{bmatrix}.$$

As expected, sensitivity analysis shows that delay is sensitive to stray capacitances, while resistances and mismatch affect

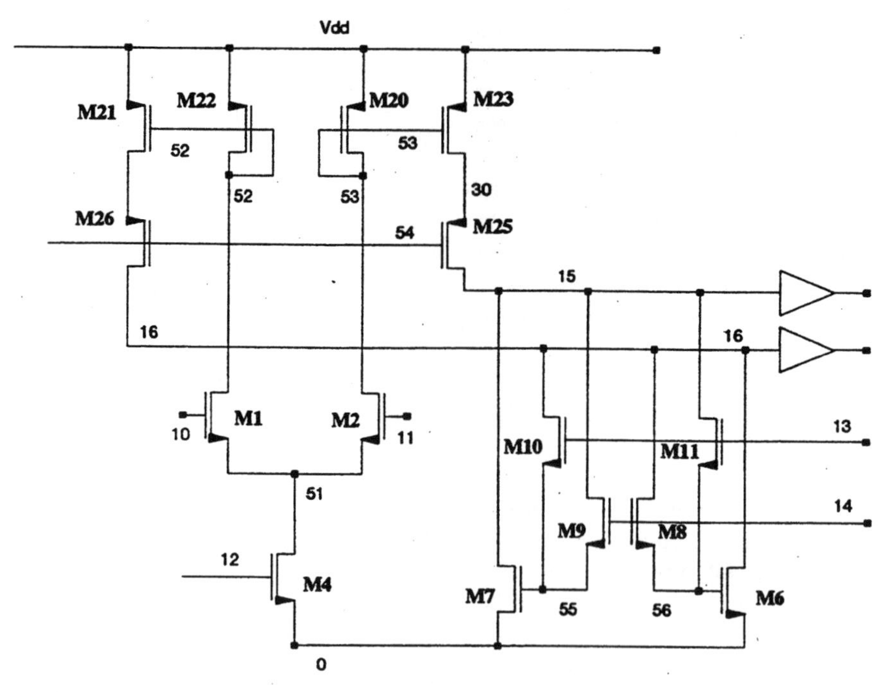

Fig. 3. Clocked comparator COMPL.

only offset

$$\mathbf{p} = \begin{bmatrix} C_{15} \\ C_{16} \\ C_{55} \\ C_{56} \\ R_{S\_1,2} \\ R_{S\_20,23} \\ R_{S\_21,23} \\ R_{S\_20,22} \\ R_{S\_21,22} \\ R_{S\_6,7} \\ R_{S\_3,4} \end{bmatrix}$$

$$\mathbf{S} = \begin{bmatrix} 36 \text{ ps/fF} & 0.0 & 0.0 \\ 36 \text{ ps/fF} & 0.0 & 0.0 \\ 47 \text{ ps/fF} & 0.0 & 0.0 \\ 47 \text{ ps/fF} & 0.0 & 0.0 \\ 0.0 & 0.056 \text{ mV}/\Omega & 0.056 \text{ mV}/\Omega \\ 0.0 & 0.016 \text{ mV}/\Omega & 0.016 \text{ mV}/\Omega \\ 0.0 & 0.016 \text{ mV}/\Omega & 0.016 \text{ mV}/\Omega \\ 0.0 & 0.016 \text{ mV}/\Omega & 0.016 \text{ mV}/\Omega \\ 0.0 & 0.016 \text{ mV}/\Omega & 0.016 \text{ mV}/\Omega \\ 0.0 & 0.011 \text{ mV}/\Omega & 0.011 \text{ mV}/\Omega \\ 0.0 & 0.201\ \mu\text{V}/\Omega & 0.201\ \mu\text{V}/\Omega \end{bmatrix}^T$$

Because of symmetries, and since the nominal value of mismatch is zero, offset sensitivities in the positive and negative direction are equal. We use the following conservative minimum and maximum parasitic estimates

$$\begin{aligned} \min C &= 1 \text{ fF} \\ \max C &= 100 \text{ fF} \\ \min R &= 0 \\ \max R &= 50\ \Omega. \end{aligned}$$

With these estimates, PARCAR computed the following set of parasitic bounds

$$\mathbf{p}^{(\mathbf{b})} = \begin{bmatrix} 71.96 \text{ fF} \\ 71.96 \text{ fF} \\ 78.52 \text{ fF} \\ 78.52 \text{ fF} \\ 1.0\ \Omega \\ 7.4\ \Omega \\ 7.4\ \Omega \\ 7.4\ \Omega \\ 7.5\ \Omega \\ 19.9\ \Omega \\ 49.5\ \Omega \end{bmatrix}.$$

Here, the relation between sensitivity and tightness of bounds is evident. Only a few parameters critically affect the performance of this circuit and, therefore, need to be bounded tightly. In practice, only the mismatch between the source resistances in the differential pair and the mismatch between the two current mirrors (M_{20}, M_{23}) and (M_{21}, M_{22}) are responsible for offset.

IV. Placement with Analog Constraints

PUPPY-A [27] is a macro-cell-style placement tool based on Simulated Annealing (SA) [45]. In PUPPY-A, the cost function is a weighted sum of nonhomogeneous parameters controlling parasitics, symmetries, and device matching. Let s be a placement configuration, i.e., the set of the positions and rotation angles of all layout modules. The cost function is given by the following expression

$$\begin{aligned} f(\mathbf{s}) = {} & \alpha_{\text{wl}} f_{\text{wl}}(\mathbf{s}) + \alpha_a f_a(\mathbf{s}) + \alpha_{\text{ov}} f_{\text{ov}}(\mathbf{s}) \\ & + \alpha_{\text{sy}} f_{\text{sy}}(\mathbf{s}) + \alpha_{\text{ma}} f_{\text{ma}}(\mathbf{s}) + \alpha_{\text{we}} f_{\text{we}}(\mathbf{s}) + \alpha_{\text{co}} f_{\text{co}}(\mathbf{s}) \end{aligned}$$

where

- $f_{\text{wl}}(\mathbf{s})$ is the sum of wire length estimates over all the modules. Two estimation methods are available, one based on semiperimeter and the other on pseudo-Steiner tree technique;
- $f_a(\mathbf{s})$ is the total area of the circuit. Space for routing is estimated with the *halo* mechanism described in [46];
- $f_{\text{ov}}(\mathbf{s})$ is the total overlapping area between cells;
- $f_{\text{we}}(\mathbf{s})$ is a measure of the discontinuity of well regions. This parameter is used only in device-level placement. It is given by the sum of the distances between devices that should lay within the same well or substrate region;
- $f_{\text{sy}}(\mathbf{s})$ is a measure of the *distance* between placement $\mathbf{s}$ and a symmetric configuration, given by the following expression

$$f_{\text{sy}}(\mathbf{s}) = \sum (d(\mathbf{s})_i + \rho_i) \tag{14}$$

where the sum is extended to all symmetric devices. Item $d(\mathbf{s})_i$ is the translation needed to bring the ith cell to a symmetric position. The value of ρ_i is zero if mirroring and/or rotation are not needed to enforce symmetry, otherwise it is set to ten;
- $f_{\text{ma}}(\mathbf{s})$ is a measure of the mismatch between circuit devices. Its definition is similar to the one of $f_{\text{sy}}(\mathbf{s})$

$$f_{\text{ma}}(s) = \sum \left(d_i^{(m)} + \rho_i\right)$$

where the sum is extended to all matched devices. Item $d_i^{(m)}$ is the translation needed to bring the ith device inside an area of adequate matching characteristics with the other matched devices. This area can be user-specified or automatically computed as explained in Section III. Parameter ρ_i and has the same meaning as in (14);
- $f_{\text{co}}(\mathbf{s})$ is a penalty function accounting for performance constraint violations. Its computation is key to our performance-driven approach. Estimates $\mathbf{p}^{(\min)}$ and $\mathbf{p}^{(\max)}$ of minimum and maximum interconnect capacitances and resistances are obtained on the ground of net length estimates and of the available routing layers. Using the linearized expression (1), performance degradation can be computed at each annealing iteration, and one of the following cases can apply:
  1) if the maximum degradation is within the specifications, that is
  $$\mathbf{S}[\mathbf{p}^{(\max)} - \mathbf{p}^{(0)}] - \overline{\mathbf{\Delta K}} \leq \mathbf{0} \tag{15}$$
  no cost function penalty is imposed. In fact, in this case, constraints (7) and (8) are met whatever values the parasitics assume;
  2) otherwise, $f_{\text{co}}(\mathbf{s})$, is a function of the constraint violation $\mathbf{\Delta K}(\mathbf{p})$.

  In case 2, the penalty term is computed as follows. Let $\Delta K_i^{(\min)}$ and $\Delta K_i^{(\max)}$ be, respectively, the minimum and maximum values that the degradation of performance K_i can assume with different values of parasitics

  $$f_{\text{co}} = \sum_{i=1}^{N_k} C_i$$

  where C_i is given by the equation at the bottom of this page. ρ_c is the ratio between the maximum and minimum value of the minimum-width unit-length substrate capacitance of interconnections on the available routing layers. If $S_r \gg 1$, then the values of C_i for feasible and infeasible placements differ by at least one order of magnitude. In our implementation, $S_r = 10$.

  In case deterministic values for $\mathbf{p}^{(\max)}$ are not available, the degradation variance $\sigma^2(\Delta K_i)$ is computed with the model (12), and then compared with specification $\overline{\sigma^2(\Delta K_i)}$, using (15). The added measure of violation is then treated like any other performance violation and used to drive directly the annealing algorithm;
- $\alpha_{\text{wl}}, \alpha_a, \alpha_{\text{ov}}, \alpha_{\text{we}}, \alpha_{\text{sy}}, \alpha_{\text{ma}}$, and α_{co} are nonnegative weights. Their initial default values are adjusted dynamically during the algorithm using heuristics so that, at the beginning of the annealing, area and wire length dominate in the expression of the cost function, then their importance decreases progressively, until at low temperatures overlaps, symmetries, and constraint violations become dominating.

A. Abutment and Control of Junction Capacitances

Device abutment during placement is useful to reduce interconnect and junction capacitances, and to obtain substantial gain in area. It can also be used to merge the diffusion regions of MOS transistors or of other components, such as capacitors, BJT's, etc. In PUPPY-A, abutment is obtained in two different ways. The first is by dynamic device abutment (similar to the approach in KOAN [22]), performed by PUPPY-A during the annealing algorithm. In PUPPY-A, dynamic abutment is driven by parasitic constraints, as well as by area and wiring considerations. Instead of randomly choosing the devices to merge, the algorithm operates first on the nets whose parasitics are critical for performance constraints. The second is through the *stack generator* LDO, which efficiently builds stacks containing transistors all with the same width.

LDO [34] implements stacks of folded or interleaved MOS transistors sharing their drain and source diffusions. Dense layouts can be achieved with this approach, the junction capacitances associated with shared diffusions being minimized. Moreover, matching between transistors decomposed into elements stacked together is usually good, in particular if

$$C_i = \begin{cases} 0, & \text{if } \Delta K_i^{(\max)} \leq \overline{\Delta K_i} \\ \Delta K_i^{(\max)} - \overline{\Delta K_i}, & \text{if } \Delta K_i^{(\min)} < \overline{\Delta K_i} \leq \Delta K_i^{(\max)} \\ (S_r + 1)(\Delta K_i^{(\max)} - (S_r \rho_c + 1)\overline{\Delta K_i}), & \text{if } \overline{\Delta K_i} \leq \Delta K_i^{(\min)} \end{cases}$$

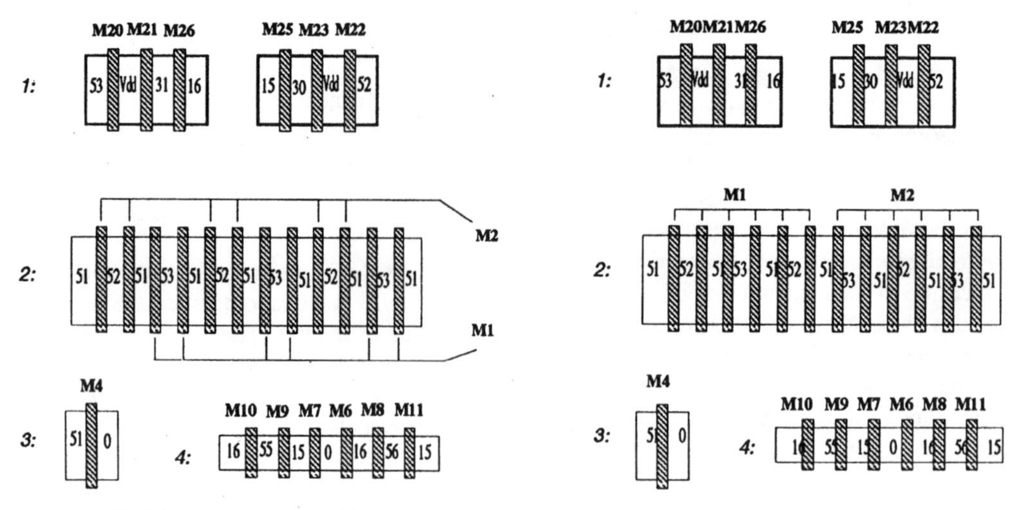

Fig. 4. Clocked comparator COMPL. Two alternative full-stacked implementations.

the elements are interleaved. Because of the regularity of these structures, routing is usually dense with this layout style.

The target of the stack generator can be summarized as follows:

1) obtain maximally compact stacks, so that the area occupied by the devices is minimum;
2) keep all critical capacitances at their minimum value by exploiting the abutment of source/drain diffusion areas;
3) provide control over device matching, so that critically matched devices can be decomposed into interleaved elements, and common-centroid structures are obtained when symmetry constraints are enforced;
4) provide control over net length, by conveniently distributing the elements within the stacks.

LDO is based on an algorithm exploiting the equivalence between stack generation and path partitioning in the circuit graph. The algorithm is guaranteed to find all optimum stacked configurations, according to an optimality criterion defined by a cost function, which takes into account parasitic criticality, matching constraints, and device area. The stack generation algorithm is based on a two-phase approach, working on the *circuit graph*, i.e., a graph whose nodes are circuit nets, and whose edges are MOS transistors. In the first phase, a dynamic programming procedure generates all possible paths in the circuit graphs, namely in the connected subgraphs whose nodes have no more than two adjacent edges. The second phase explores the compatibility between all paths. By solving a clique problem, an optimum set of paths is selected, which minimizes the cost function and contains all the transistors of the circuit. More details on the algorithm and its implementation can be found in [34].

In LDO, not only are symmetries fully taken into account, but they have proved effective to reduce the computational complexity by limiting the size of the search space, while preserving the admissibility of the algorithm (i.e., the optimum solution is always found). In practice, the higher the number of symmetry constraints, the faster the algorithm runs.

By abutting elemental transistors into one stack, their source/drain regions are merged, thus effectively reducing their junction capacitances. The cost function driving LDO tries to minimize the most critical capacitances, according to the tightness of their bounds. Junction capacitances are the only parasitics that can be directly controlled by LDO, because they are directly influenced by the shape and inner organization of the stacks. Routing parasitics, such as interconnect stray resistances and capacitances, can be controlled effectively only after the placement phase. This limitation can be overcome by simultaneously generating and placing the stacks. This has been achieved by means of an annealing move-set extension, to include a move called "alternative solution swap," which randomly selects a module in the circuit and swaps it with one of its alternative implementations found by LDO. The criterion whether to accept the move is based on the usual annealing scheme [33].

B. Example

Consider the clocked comparator COMPL. Many possible stack implementations exist for this circuit. Two of such possible solutions are shown in Fig. 4. All transistors have been grouped in four subcircuits, according to their channel widths, their matching requirements and bulk nets. Only transistors belonging to the same subcircuit can belong to the same stack. The two solutions only differ by the implementation of the stack containing the input differential pair. In the first realization, they are interleaved in a common-centroid pattern, which minimizes device mismatch, but usually requires a considerable area overhead, due to the complex routing required. The second solution is symmetric, but without the common-centroid structure. The choice between such alternative realizations is left to the user or it can be made automatically during the placement phase on the ground of area and routing considerations. In both solutions, critical nets 55, 56, 15, 16, whose capacitance toward the substrate strongly influences the comparator speed, have been kept in internal positions when possible. Their capacitances are reported in Table II. In both cases stack abutment yielded a reduction of net capacitance. Such a reduction can be exploited to improve the flexibility of the routing stage. For example, consider nets 55 and 56. Abutment allowed each of them to be reduced by more than 6.6 fF, which in our process is the capacitance

TABLE II
CAPACITANCES IN THE STACKS GENEREATED FOR THE CLOCKED COMPARATOR COMPL

nets	without abutment	with abutment	% reduction
15,16	41.2 fF	34.4 fF	(16.5%)
55,56	13.4 fF	6.6 fF	(51%)

of a 136 μm-long minimum-width metal-1 wire. Therefore, the router is allowed to draw longer wires for the sensitive nets, thus increasing the success rate and the robustness of the entire layout synthesis.

These capacitance values constitute new nominal values and better lower limits, and can be used to compute a new set of bounds. By using these values

$$\begin{aligned} &C_{15}^{(\min)} = C_{15}^{(\mathrm{nom})} = C_{16}^{(\min)} = C_{16}^{(\mathrm{nom})} = 34.4 \text{ fF} \\ &C_{55}^{(\min)} = C_{55}^{(\mathrm{nom})} = C_{56}^{(\min)} = C_{56}^{(\mathrm{nom})} = 6.6 \text{ fF} \\ &\max C = 100 \text{ fF} \\ &\min R = 0 \\ &\max R = 50\ \Omega \end{aligned} \tag{16}$$

we obtain the following arrays

$$\mathbf{K}(\mathbf{p}^{(0)}) = \begin{bmatrix} 5.5 \text{ ns} \\ 0.0 \\ 0.0 \end{bmatrix} \qquad \overline{\Delta \mathbf{K}} = \begin{bmatrix} 1.5 \text{ ns} \\ 1 \text{ mV} \\ 1 \text{ mV} \end{bmatrix}.$$

Here, the delay degradation, due to the insertion of junction capacitances, is apparent. The next set of bounds found by PARCAR is the following:

$$\mathbf{p}^{(b)} = \begin{bmatrix} 67.1 \text{ fF} \\ 67.1 \text{ fF} \\ 48.9 \text{ fF} \\ 48.9 \text{ fF} \\ 1.0\ \Omega \\ 7.4\ \Omega \\ 7.4\ \Omega \\ 7.4\ \Omega \\ 7.5\ \Omega \\ 19.9\ \Omega \\ 49.5\ \Omega \end{bmatrix}. \tag{17}$$

Notice that all bounds on critical capacitances have been lowered, because the degradation allowed to delay is smaller than in the previous step. In fact half of the degradation allowed at the beginning of the layout design has been introduced by junction capacitances alone, and the remaining half will be available to the remaining tools (i.e., placement and routing tools). The placement of Fig. 5 was obtained with the set of bounds (17). After placement, estimates of the minimum values of all critical parasitics can be drawn, taking into account the junction capacitances of all terminals and the estimated minimum length of interconnections between terminals

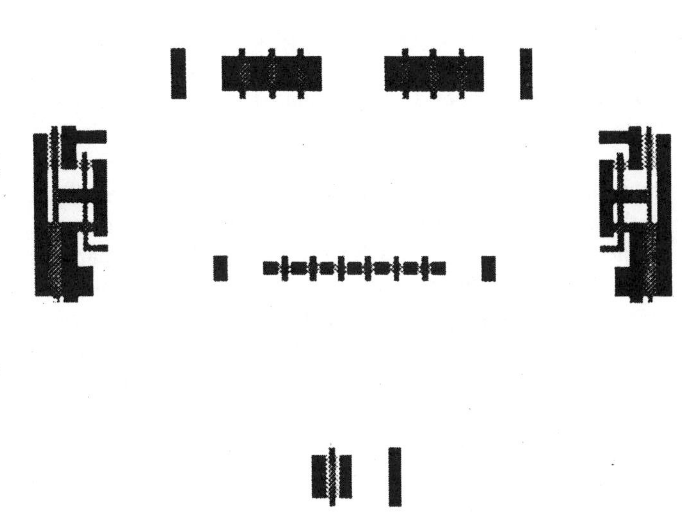

Fig. 5. Placement of comparator COMPL obtained with PUPPY-A.

$$\mathbf{p}^{(0)} = \mathbf{p}^{(\min)} = \begin{bmatrix} 10.1 \text{ fF} \\ 10.1 \text{ fF} \\ 51.0 \text{ fF} \\ 51.0 \text{ fF} \\ 0.0 \\ 0.0 \\ 0.0 \\ 0.0 \\ 0.0 \\ 0.0 \\ 0.0 \end{bmatrix}. \tag{18}$$

V. ROUTING WITH ANALOG CONSTRAINTS

A. Channel Routing

In the channel router ART [24], the two-layer gridless channel-routing problem [47] is represented by a *vertical-constraint graph* (VCG) whose nodes correspond to the horizontal segments of a net or subnet. An undirected edge links two nodes if the associated segments have a common horizontal span. A directed edge links two nodes if one segment has to be placed above the other because of pin constraints. The weight of an edge is the minimum distance between the center lines of two adjacent segments. Hence, the channel-routing problem is formulated as the problem of directing all the undirected edges to minimize the longest directed path in the VCG. The length of such path corresponds to the channel width. Over-constraints can be solved by assigning to each net more than one node in the VCG. However, with this approach, we introduce additional capacitive couplings due to the wire jogs. In the current implementation of ART, one VCG node is supported for each net.

In ART, all parasitic bounds are mapped into constraints for the VCG. Within a channel, nets provide two different contributions to cross-coupling capacitance: crossover capacitances between overlapped orthogonal wire segments, and capacitances between segments running parallel to each other. Both depend on the distribution of terminals along the channel edges. Unavoidable crossovers can be determined directly on the ground of the terminal positions. If such a crossover is detected, it introduces a lower bound for the cross-coupling capacitance between the corresponding nets. The coupling between horizontal adjacent edges is controlled by their minimum separation and, therefore, it is proportional to the weight of the corresponding edge in the VCG. This contribution can be theoretically reduced at will by inserting sufficient space, or by exploiting the shielding effect due to other wires. Shielding nets can be inserted on purpose, if the presence of a further wire segment in the channel is more convenient in terms of area than extra spacing. In ART, this is automatically carried out by adding a new node and edges to the VCG.

Perfect mirror symmetry can be achieved when symmetric nets are restricted to different sides of the symmetry axis (i.e., they don't cross). If the horizontal spans of a pair of symmetric nets intersect the symmetry axis, perfect mirror symmetry cannot be achieved. However, good parasitic matching can be obtained between the nets with the technique illustrated in Fig. 6. A "connector" allows two symmetric segments to cross over the axis. Resistances and capacitances of the two nets are matched, because for each one, the connector introduces the same interconnect length, the same number of corners on each layer, and the same number of vias. Only coupling capacitances with other nets running close to the connector will suffer slight asymmetries.

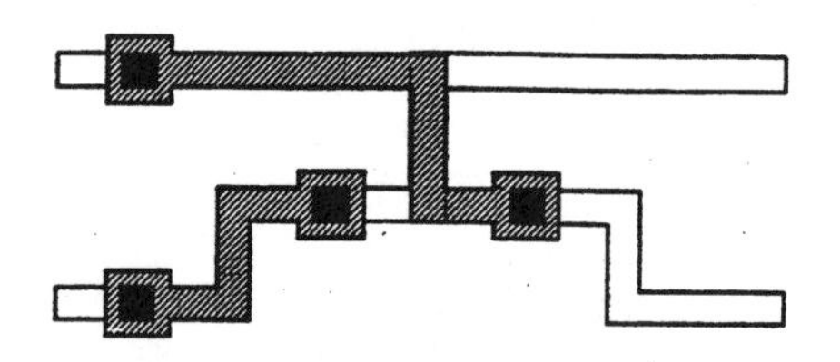

Fig. 6. Connector for symmetric nets spanning across the symmetry axis.

B. Area Routing

ROAD [31] is a maze router based on the A^* algorithm [48], using a relative grid with dynamic allocation. For each net, the path found by the maze router is the one of minimum length. If a cost function is defined on the edges of the grid, the path found is the one minimizing the integral of the cost function. In ROAD, the cost function is a weighted sum of several nonhomogeneous items. Let $\mathcal{N}$ be the set of all nets. On a given grid edge x with length $L(x)$, on layer l, the cost function for a net $N \in \mathcal{N}$ has the following form

$$F(x) = L(x) \cdot \left(1 + \frac{\mathrm{Cr}(x)}{Cr_0} + w_R \frac{R_u(l)}{R_0} + w_{C_u} \frac{C_u(l)}{C_0} + \frac{1}{C_P} \sum_{n \in \mathcal{N}-\{N\}} w_{C_n} C_n(x)\right) \quad (19)$$

where

- $\mathrm{Cr}(x)$ is a measure of local area crowding. It is computed in a simplified form, by giving over-congested areas steep cost function "hills," which prevent future wires from crossing these areas. Area crowding $\mathrm{Cr}(x)$ is given by

$$\mathrm{Cr}(x) = \begin{cases} 0, & \text{if } R \leq 1 \\ \mathrm{Cr}_{\max} & \text{if } R > 1 \end{cases} \quad (20)$$

where $\mathrm{Cr}_{\max}$ is a large constant (the height of the "hills"), and R is the ratio between the needed room for the new wire which has to be built, and the room available on the sides of edge x

$$R = \frac{\text{needed room}}{\text{available room}};$$

- $R_u(l)$ is the resistance of a minimum-width unit-length wire segment on layer l;
- $C_u(l)$ is the capacitance to bulk of a minimum-width unit-length wire segment on layer l. The model is described in Appendix;
- $C_n(x)$ is the capacitance between a unit-length wire segment located across edge x and the wire implementing net n. The model is described in Appendix;
- w_{R_u}, w_{C_u}, and w_{C_n} are weights regulating the relative importance of each item;
- $\mathrm{Cr}_0, R_0, C_0, C_P$ are reference parameters providing dimensional homogeneousness to the addenda and a meaning to their comparison.

Weights provide an efficient way to limit the magnitude of critical parasitics. Performance sensitivities to parasitics are used to generate the weights for the cost function driving the area router. The contribution of a parasitic to performance degradation is proportional to the sensitivity and inversely proportional to the maximum variation range allowed to that performance. The weight w_j associated to parasitic p_j is defined as follows:

$$w_j = \sum_{i=1}^{N_k} \left(\frac{S_{i,j}^-}{\Delta K_i^-} + \frac{S_{i,j}^+}{\Delta K_i^+}\right) P_0$$

where P_0 is a normalization factor, such that, if sensitivities are not all zero, at least one weight is set to one, and the others are all between zero and one. The dimensional unit of $P_0(\Omega, V, F, \ldots)$ depends on the parasitic type. With this definition, each item in (19) can be interpreted as the contribution to performance degradation due to one of the parasitics introduced by the wire segment routed along edge x of the grid.

The routing schedule is determined with a set of heuristic rules set up and tuned with experimental tests. The higher the number of constraints on a net, the higher is its priority. We define a number of *properties* that a net can have, for instance symmetry, belonging to the class of supply nets, of clocks, etc.

The priority of net n is given by the following expression

$$\mathrm{Pr}(n) = \sum_{j} a_j \,\mathrm{Prop}_j(n) + \sum_{i=1}^{N_p} a_{p_i} w_{p_i}$$

where $\mathrm{Prop}_j(n)$ is one if net n has the jth property, and zero otherwise. Parameters w_{p_i} are the same parasitic weights used to define the cost function (19), while a_j and a_{p_i} are *priority weights* expressing the importance of each property. Priority weights are assigned in such a way that maximum priority is given to symmetric nets, followed by supply nets, and then by nets with tight electrical requirements. The most difficult nets are routed first, and the unrouted ones are less and less critical as the circuit crowding increases. If two nets have the same priority, the shorter one is routed first.

After performing the weight-driven routing, parasitics are extracted and performance degradation is estimated and compared with its specifications. If constraints (6) are not met, the weights of the most sensitive parasitics are raised and routing is repeated. When the weights of all sensitive parasitics hit their maximum value (that is one), iterations stop. This means that even considering maximum criticality for the sensitive parasitics, routing is not possible on the given placement, without constraint violations. In this case, the circuit placement needs to be generated again, using a wider range of variation for the detected sensitive parasitics.

Symmetries: ROAD is able to find symmetric paths for differential signals with symmetric placements, even in the presence of a nonsymmetric distribution of terminals. The algorithm, described in detail in [31], is illustrated in Fig. 7. Let us assume without loss of generality that the symmetry axis is vertical and it splits the circuit into a left half and a right half. If the placement is not perfectly symmetric, we consider the outline determined by the union of real obstacles and *virtual obstacles* obtained by mirroring each obstacle with respect to the symmetry axis. First, every net is built considering only the terminals located on the left side of the symmetry axis or on the border of the wiring space, that is not contained by any virtual image of an obstacle. The wire segments defined in this way are called *left-side segments*. Then, each left-side segment is mirrored with respect to the symmetry axis. Next, the routing is extended to cover the portions of area occupied by virtual obstacles, but not by real obstacles. The segments whose existence is not required for the full net connectivity are pruned. Only those branches of nonsymmetric nets should be pruned, that don't cross or run close to symmetric nets.

Electrostatic Shields: Decoupling based only on wire spacing can increase excessively area, and this can be avoided by inserting wire stubs, connected to a virtual ground, shielding critically coupled wires. Shields are built after all wires have been routed. Given two wires to be decoupled, first a grid node between each pair of parallel segments of the two wires is found, or generated by dynamic grid allocation. On each of these nodes, area congestion is computed with expression (20), and a terminal is defined wherever congestion is sufficiently low, i.e., where $\mathrm{Cr}(x) < \mathrm{Cr}_{\max}$. Next, a new wire is routed through all these terminals, and connected (with null weight on resistive constraints) to the proper ground node. If local congestion doesn't allow to create a suitable shield, then we pass to the reroute phase as described above.

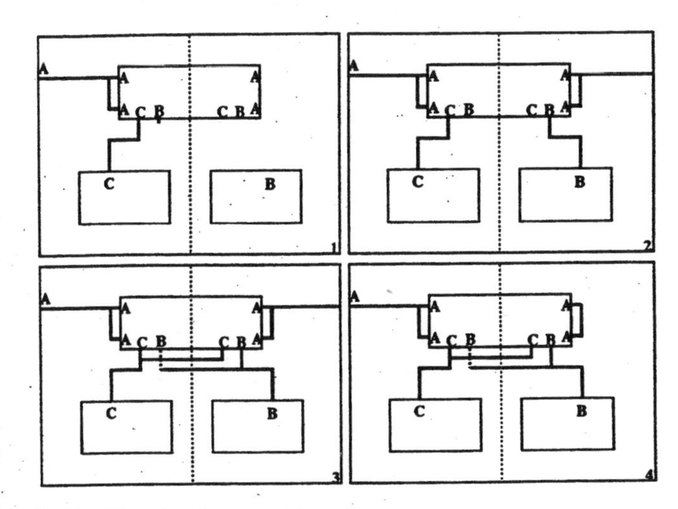

Fig. 7. Algorithm for symmetric routing.

C. Example

Consider the clocked comparator COMPL. After the placement step, the nominal values of all parasitics have been updated as shown in (18). With these new nominal values, the least capacitive interconnect among the terminals of the critical nets would give (by simulation) a total delay of 5.9 ns

$$\mathbf{K}(\mathbf{p}^{(0)}) = \begin{bmatrix} 5.9\text{ ns} \\ 0.0 \\ 0.0 \end{bmatrix} \quad \overline{\Delta\mathbf{K}} = \begin{bmatrix} 1.1\text{ ns} \\ 1\text{ mV} \\ 1\text{ mV} \end{bmatrix}.$$

With the high-level constraints (13), now only 1.5 ns of delay degradation are allowed to the router, of which 0.4 ns have been recognized as being unavoidable with this placement. Therefore, tight bounds will have to be enforced by the router. In fact, PARCAR now requires the following bounds

$$\underline{\mathbf{p}}^{(b)} = \begin{bmatrix} 76.5\text{ fF} \\ 76.5\text{ fF} \\ 39.3\text{ fF} \\ 39.3\text{ fF} \\ 1.0\ \Omega \\ 7.4\ \Omega \\ 7.4\ \Omega \\ 7.4\ \Omega \\ 7.5\ \Omega \\ 19.9\ \Omega \\ 49.5\ \Omega \end{bmatrix}.$$

Comparing the capacitive bounds with the junction capacitances (16) computed after module generation, it is evident that 42.5 fF are available for routing nets 15 and 16, and 32.7 fF are available for nets 55 and 56. The layout routed by ROAD is shown in Fig. 8.

Extraction results for capacitances are the following:

$$C_{15} = 70.3\text{ fF}$$
$$C_{16} = 70.4\text{ fF}$$
$$C_{55} = 20.6\text{ fF}$$
$$C_{56} = 18.0\text{ fF}.$$

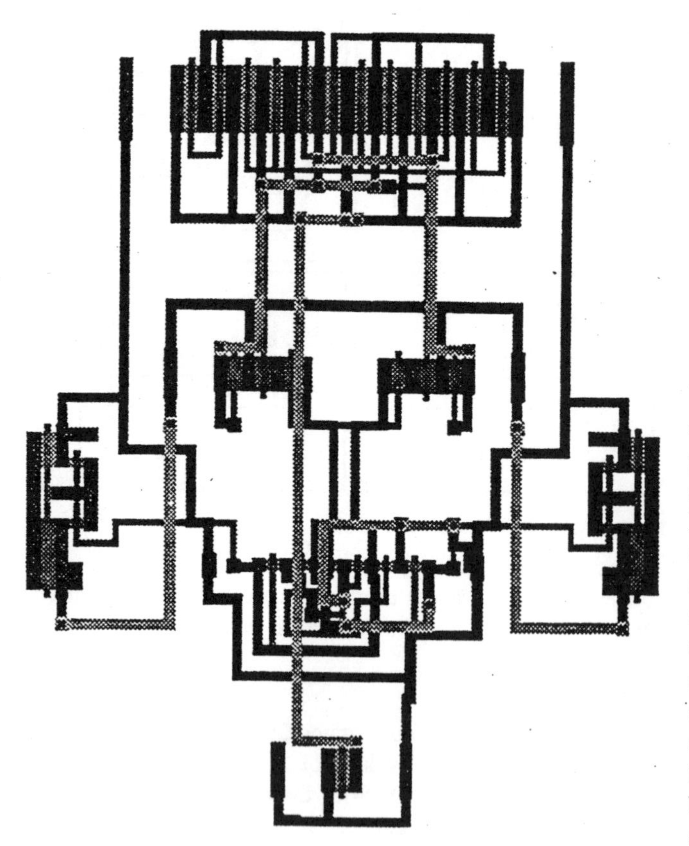

Fig. 8. Comparator COMPL after automatic placement and routing with ROAD.

Simulation results after extraction of the routed layout give a delay of 6.5 ns and an offset of 756 μV.

VI. Compaction

SPARCS-A [49] is a monodimensional constraint-graph (CG) longest-path compactor, implementing algorithms to enforce symmetry and parasitic constraints. The role of compaction in the constraint-driven approach is important for the following two reasons:

1) constraints enforced by the previous layout steps, such as parasitic bounds, symmetries, and shields, should not be disrupted for the sake of area minimization. The compactor must be able to respect, and if necessary, to enforce such constraints;
2) the compactor can recover design-rule errors and constraint violations. Hence, the requirements on placement and routing in terms of constraint enforcement can be relaxed. Since compaction has generally higher computational efficiency than routing, the overall CPU cost of layout design can be substantially reduced. Thus, in addition to reducing chip area, compaction also improves the efficiency and robustness of the entire analog synthesis process by permitting the use of more aggressive techniques during placement and routing.

The algorithm implemented in SPARCS-A takes advantage of the high speed of the CG-based technique to provide a good starting point to a Linear Programming (LP) solver. Monodimensional compaction is iterated alternatively in the two orthogonal directions, until no area improvement is achieved. The algorithm used in each iteration is the following:

1) with the CG technique, solve the spacing problem without symmetry constraints;
2) use the simplex LP algorithm to solve the symmetry constraints, using the CG solution as initial starting point;
3) round off the coordinates of the elements laying out of the critical path;
4) verify that all constraints are satisfied.

Using the CG solution as starting point is key to a significant speed-up in the solution of the linear problem. Compared to previous approaches [23] solving compaction using an LP solver, this algorithm represents a substantial improvement. In fact, in our case, the LP solver starts from a feasible configuration that is already close to the final solution. The range of cases that can be managed with acceptable computational complexity is therefore significantly broadened [50]. Control over cross-coupling capacitances is enforced by modifying the constraint-graph before computing the longest path. Proper distances between parallel interconnection edges are kept to maintain cross-coupling capacitances below their bounds. This is achieved by employing a heuristic which adds extra spacing between wire segments, based on the need for decoupling and on their length, which has a direct impact on the overall area. The procedure implementing this heuristic is the following:

procedure modify-graph
/* Purpose: Add constraints to improve capacitive decoupling
/* Since this procedure is called only if some performance violation
/* has been found, we know that at least one bound has been exceeded.
for each cross-coupling C_j **such that** $C_j > C_j^b$:
  let δ_j = current min. distance between any two parallel segments
      contributing to C_j
  $\delta_j = \delta_j + mindist(C_j, C_j^b)$
  for each pair P_i of parallel segments:
    let d_i = current minimum distance between the segments of pair P_i;
    if $d_i < \delta_j$ **then**
      Add constraint to graph requiring $d_i \geq \delta_j$ between the segments
  end
end

C_j indicates the jth cross-coupling capacitance, and C_j^b is the bound on its maximum value. Function *mindist* (C_j, C_j^b), which depends on the model used for capacitances (see Appendix), returns the minimum distance increment to add between parallel segments of the jth pair of wires, to reduce their cross-coupling capacitance from C_j to C_j^b. For each cross-coupling C_j exceeding its bound, C_j^b, δ_j is the minimum distance to be kept between parallel segments of the jth pair of wires. The distance increment is a function of the parasitic bound violation: the bigger the violation, the wider the extra spacing

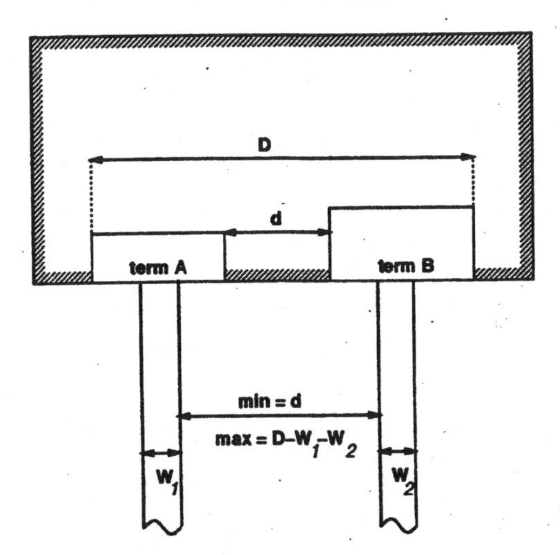

Fig. 9. Minimum and maximum wire spacing constraints deriving from connections to fixed-distance terminals.

added. Notice that in procedure *modify-graph*, spacing is added not only between the nearest segments, but also between all the segment pairs whose distance is less than δ_j.

The spacing step implemented by procedure *modify-graph* can introduce over-constraints making the graph unsolvable. An example where this situation might occur is illustrated in Fig. 9. Two wire segments are connected to terminals A and B, whose relative position is fixed with respect to the instance of a subcell. An over-constraint, due to a positive loop in the CG, is generated if the spacing required between the segments is $\delta_j > D - W_1 - W_2$. When a positive loop is detected, a *pruning* procedure is invoked, which removes the newly-added spacing constraints contained in the positive-weight loops. In such situation, the task of decoupling the two nets is left to the remaining segment pairs. If a feasible solution involving the remaining segment pairs does not exist, an error is reported because the constraint cannot be met.

One of the main advantages of the longest-path compaction algorithm is that right after each compaction step, it provides the exact value of the minimum layout pitch. This can be exploited to add geometric constraints together with electrical performance specifications. Let us consider, without loss of generality, a horizontal compaction step. The pitch W of the longest path can be checked against the maximum size $W_{\max}$ allowed to the circuit width. If it is smaller, the difference between them is the maximum amount by which the vertical parallel wire segments belonging to the longest path itself can be brought apart from each other. Otherwise, all such pairs of wire segments must be kept at their minimum distance. This corresponds to an additional constraint

$$\sum_i \delta_i \leq \min(0, W_{\max} - W_{\text{lp}}) \tag{21}$$

where the sum is extended to all the vertical segment pairs laying on the longest path. Aspect ratio constraints can be reduced to absolute-size constraints by considering, at each step, the pitch of the layout in the orthogonal direction as fixed.

A. Example

Consider once more the clocked comparator COMPL. As shown at the end of Section V, after routing the problem instances are the following:

$$\mathbf{p}^{(0)} = \begin{bmatrix} 70.3\ \text{fF} \\ 70.4\ \text{fF} \\ 20.6\ \text{fF} \\ 18.0\ \text{fF} \\ 0 \\ 0 \\ 0 \\ 0 \\ 0 \\ 0 \\ 0 \end{bmatrix} \quad \mathbf{K}(\mathbf{p}^{(0)}) = \begin{bmatrix} 6.2\ \text{ns} \\ 756\ \mu\text{V} \\ -756\ \mu\text{V} \end{bmatrix}$$

$$\overline{\Delta\mathbf{K}} = \begin{bmatrix} 0.8\ \text{ns} \\ 244\ \mu\text{V} \\ -244\ \mu\text{V} \end{bmatrix}.$$

The solution to the constraint-generation problem is

$$\mathbf{p}^{(\mathbf{b})} = \begin{bmatrix} 87.7\ \text{fF} \\ 87.8\ \text{fF} \\ 32.8\ \text{fF} \\ 28.4\ \text{fF} \\ 1.0\ \Omega \\ 1.0\ \Omega \\ 1.0\ \Omega \\ 1.0\ \Omega \\ 1.0\ \Omega \\ 5.2\ \Omega \\ 49.3\ \Omega \end{bmatrix}.$$

Notice that now resistive mismatch has become critical because of the shrunk margin allowed to offset degradation. Of the capacitive parasitics, C_{55} and C_{56} have been recognized as more critical than the others, and their bounds have been further tightened with respect to their previous values used to drive the router. Other bounds on C_{15} and C_{16} have been relaxed as a consequence. The layout produced with this set of bounds is shown in Fig. 10. Capacitive extraction from this layout yields the following values

$$C_{15} = 73.9$$
$$C_{16} = 75.2$$
$$C_{55} = 18.8$$
$$C_{56} = 17.2.$$

Simulation showed that in the compacted layout performance specifications were met with an offset of 743 μV and a delay of 6.7 ns.

VII. Results

All the tools described in this paper have been implemented within the OCTTOOLS framework of the University of California at Berkeley. This has allowed us to test the described algorithms, and to validate the methodological approach on a large set of test circuits. All sensitivity computation and simulations have been done using SPICE-3 [38].

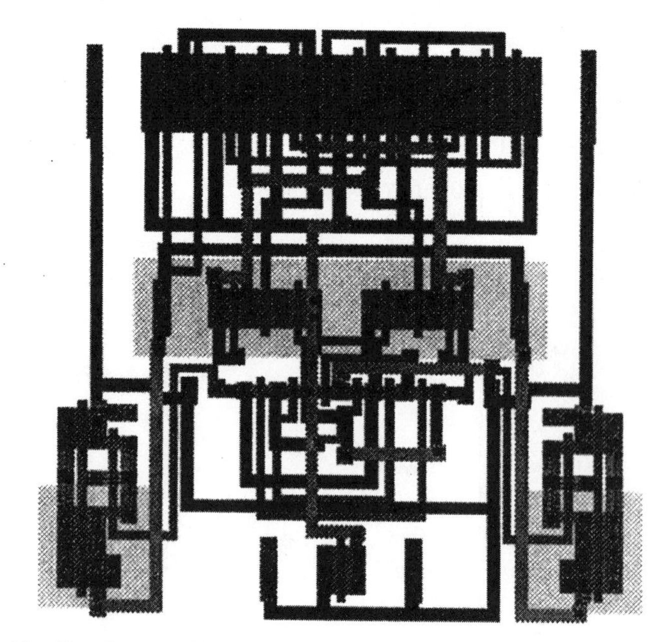

Fig. 10. Compacted layout of COMPL with all analog constraints enforced.

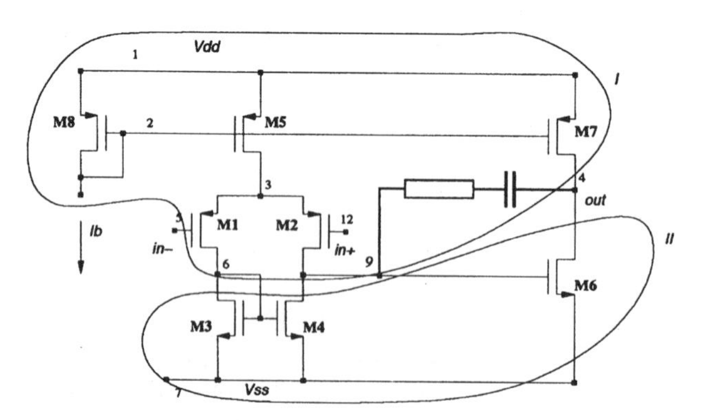

Fig. 11. Two-stage CMOS opamp. The bubbles represent the partition operated by LDO on the ground of transistor polarity.

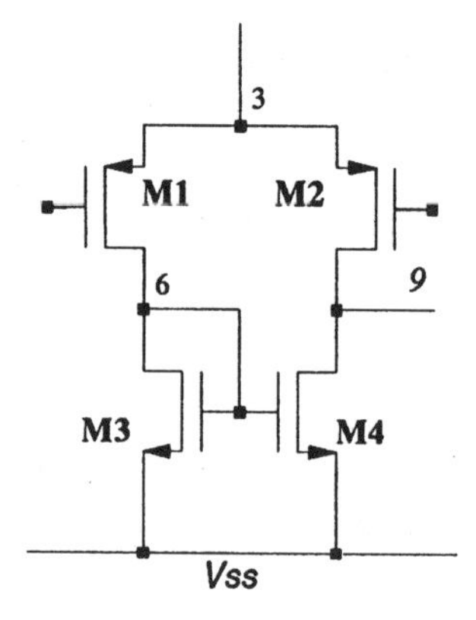

Fig. 12. Differential pair and its active load.

This section reports a few circuit examples to illustrate the methodology and the results which can be achieved.

A. Example: Two-Stage CMOS Opamp

Consider the two-stage CMOS opamp shown in Fig. 11, with specification constraints on offset V_{off} and on unity-gain

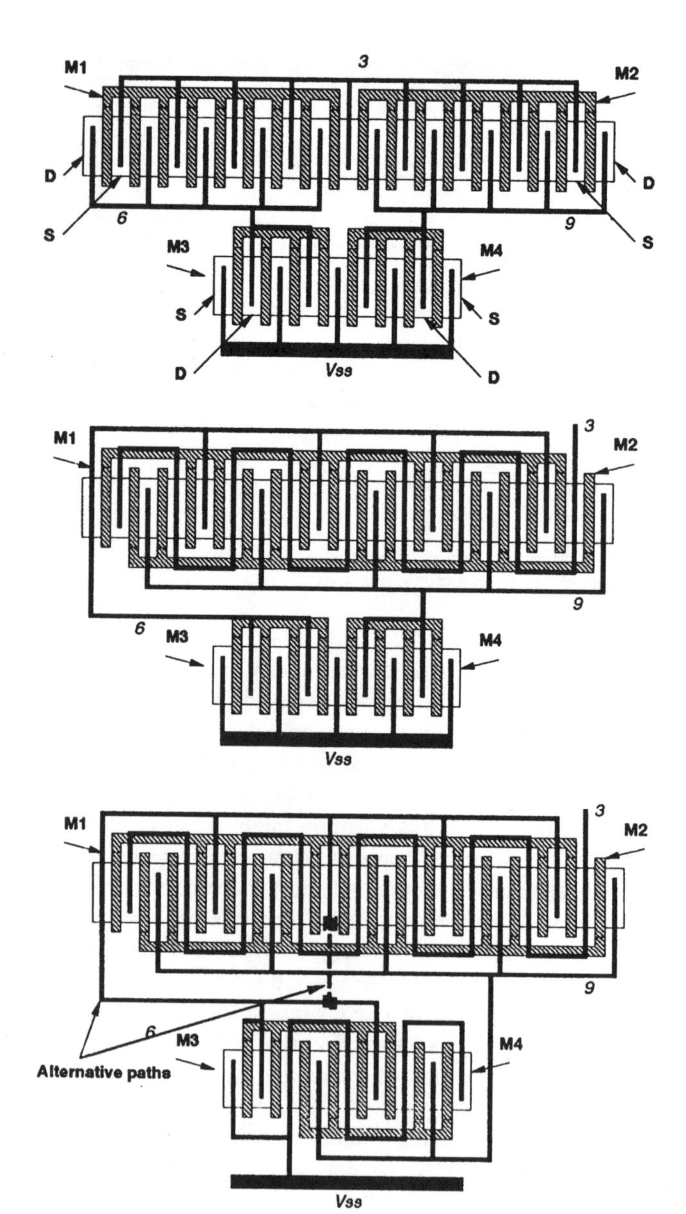

Fig. 13. Alternative implementations of the differential pair and its active load.

bandwidth ω_0

$$|V_{\text{off}}| < 2.6 \text{ mV}$$
$$\omega_0 > 6.5 \text{ MHz}.$$

Simulation results confirm that with no parasitics the nominal values of systematic offset and bandwidth would be $V_{\text{off}}^{(0)} = 2.4$ mV, and $\omega_0^{(0)} = 6.6$ MHz, respectively. Hence

$$\mathbf{K}(\mathbf{p}^{(0)}) = \begin{bmatrix} 2.4 \text{ mV} \\ -2.4 \text{ mV} \\ -6.6 \text{ MHz} \end{bmatrix} \quad \overline{\Delta \mathbf{K}} = \begin{bmatrix} 0.2 \text{ mV} \\ 5.0 \text{ mV} \\ 0.1 \text{ MHz} \end{bmatrix}.$$

For sake of clarity only a portion of the circuit in Fig. 11 is quantitatively analyzed in terms of the effects on performance by interconnect parasitics. In Fig. 12, the input differential pair and its active load are shown. Three different solutions

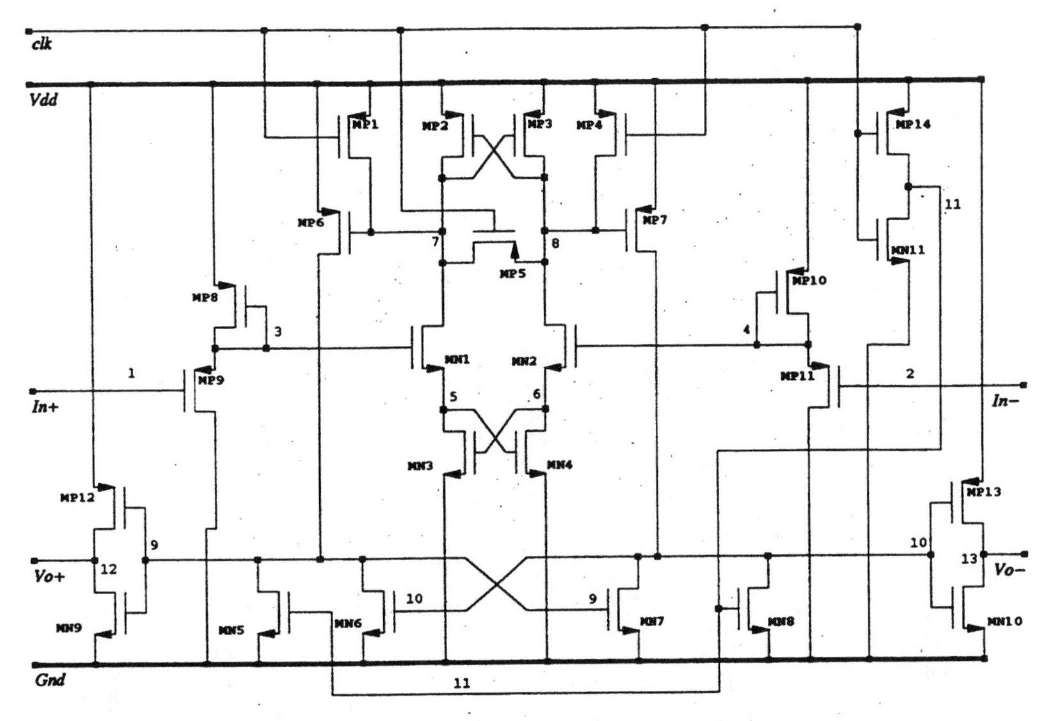

Fig. 14. Schematic of the clocked comparator FASTCOMP.

found by LDO are shown in Fig. 13. All meet matching and symmetry requirements have the same values for all critical junction capacitances, and require the same area for active devices, i.e., for LDO the costs of the three implementations are the same. According to the notation defined in Table I, the critical parasitic array and the matrix of sensitivities are

$$\mathbf{p} = \begin{bmatrix} R_{S\_1} \\ R_{S\_2} \\ R_{S\_2,1} \\ R_{S\_3,4} \\ V_{t\_2,1} \\ V_{t\_3,4} \\ C_{6,9} \\ C_6 \\ C_9 \end{bmatrix}$$

$$\mathbf{S} = \begin{bmatrix} 0.0 & 0.0 & 0.46\ \text{kHz}/\Omega \\ 0.0 & 0.0 & 0.51\ \text{kHz}/\Omega \\ 0.020\ \text{mV}/\Omega & 0.020\ \text{mV}/\Omega & 0.0 \\ 0.013\ \text{mV}/\Omega & 0.013\ \text{mV}/\Omega & 0.44\ \text{kHz}/\Omega \\ 5.00 \times 10^{-4} & 5.00 \times 10^{-4} & 0.0 \\ 1.25 \times 10^{-4} & 1.25 \times 10^{-4} & 0.0 \\ 0.0 & 0.0 & 1.21\ \text{kHz/fF} \\ 0.0 & 0.0 & 1.03\ \text{kHz/fF} \\ 0.0 & 0.0 & 0.84\ \text{kHz/fF} \end{bmatrix}^T$$

Notice that offset sensitivities are not null only w.r.t. mismatch parameters. Consider now the three alternative implementations shown in Fig. 13. Realization a) requires the smallest routing area for nets 6 and 9. This implies low interconnect resistances and capacitances and, therefore, low ω_0 degradation. Estimations of parasitics would yield a $\Delta\omega_0$ of -74 kHz and a ΔV_{off} of 0.3 mV. Clearly, only one

TABLE III
FASTCOMP: BOUNDS ON THE RESISTIVE MISMATCH

capacitive mismatch		
Mismatch	bound	$C^{(max)}$
(C_7, C_8)	27.15fF	100fF
$(C_{7,V_{dd}}, C_{8,V_{dd}})$	27.15fF	100fF
$(C_{1,5}, C_{2,6})$	68.32fF	100fF
resistive mismatch		
Mismatch	bound	$R^{(max)}$
(R_{S_N3}, R_{S_N4})	0.100Ω	1Ω
(R_{S_P2}, R_{S_P3})	7.708Ω	50Ω
(R_{S_P8}, R_{S_P10})	42.90Ω	50Ω

specification can be met with such a configuration, unless the specification on offset is relaxed. This is mainly due to the role played by the threshold voltage mismatch. In realization c) both the differential pair and the active load are tightly interleaved and, therefore, the threshold voltage mismatch is minimized, though at the expenses of capacitive cross-coupling and of substrate capacitance of nets 6 and 9. In fact, ω_0 exceeds the specifications (-173.98 kHz), while V_{off} becomes acceptable (150.33 μV). If both tight performance constraints are specified simultaneously, a tradeoff configuration must be chosen. For instance, in realization b), the differential pair is interleaved in a common-centroid pattern, while the active load is implemented in a simpler way. With this configuration, both constraints are satisfied. In fact, ΔV_{off} and $\Delta\omega_0$ are 148.0 μV and 97.2 kHz, respectively. Notice that this particular layout cannot be found with any tool relying only on automatic abutment during placement, because of the interleaved pattern in the differential pair. Nor could it be generated with standard module generators [51], unless a detailed knowledge of the circuit structure was known *a priori*.

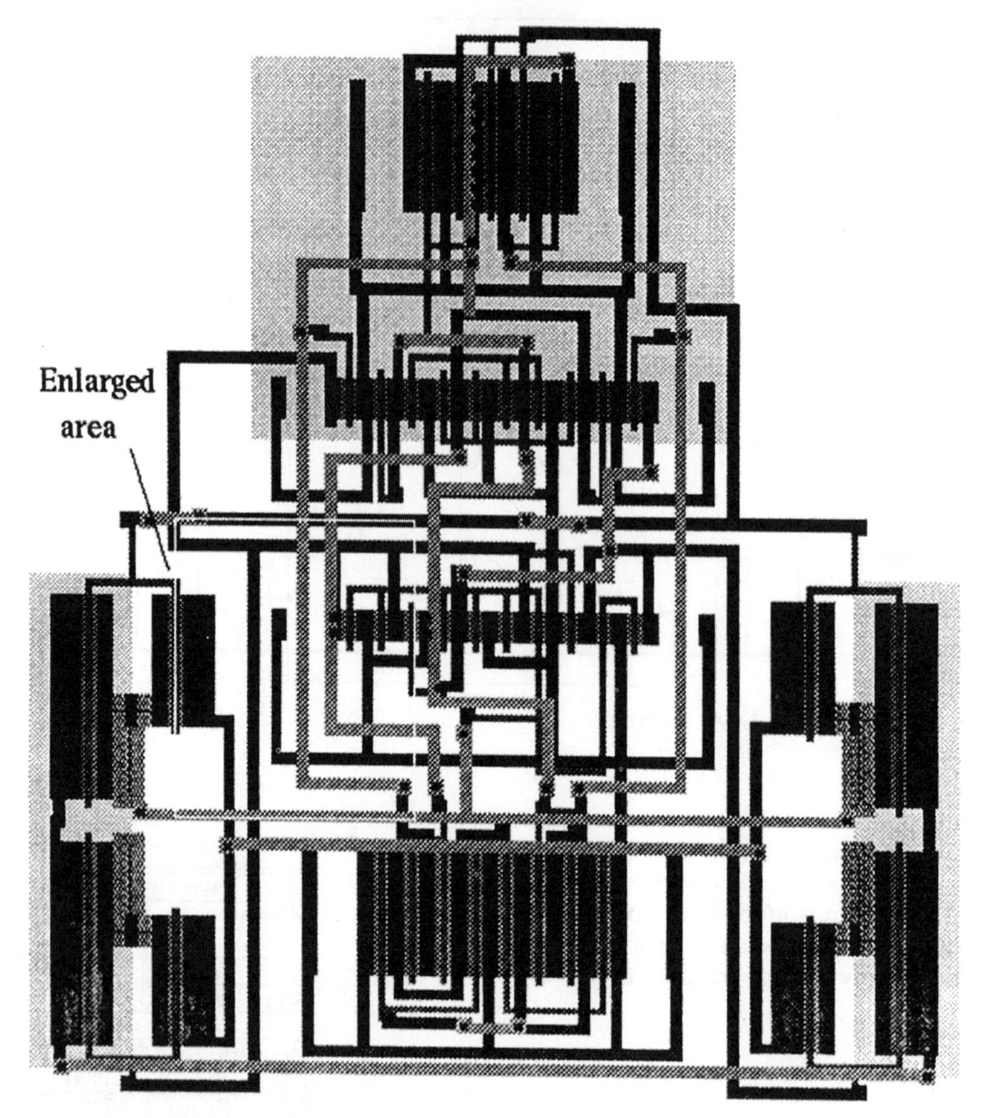

Fig. 15. Complete layout of FASTCOMP, with enforcement of all analog constraints.

B. Example: FASTCOMP

Fig. 14 shows the schematic of a clocked comparator named FASTCOMP. For this circuit we consider specifications on voltage offset and switching speed. The nominal values are

$$V_{\text{off}} = 0.0 \text{ mV}$$
$$\tau_D(H \to L) = 2.42 \text{ ns}$$
$$\tau_D(L \to H) = 2.49 \text{ ns}.$$

The constraint specifications are

$$|V_{\text{off}}| \leq 2.0 \text{ mV}$$
$$|\Delta\tau_D(H \to L)| \leq 0.25 \text{ ns}$$
$$|\Delta\tau_D(L \to H)| \leq 0.25 \text{ ns}.$$

Therefore

$$\mathbf{K} = \begin{bmatrix} V_{\text{off}} \\ -V_{\text{off}} \\ \tau_D(H \to L) \\ -\tau_D(H \to L) \\ \tau_D(L \to H) \\ -\tau_D(L \to H) \end{bmatrix} \quad \mathbf{K}(\mathbf{p}^{(0)}) = \begin{bmatrix} 0.0 \\ 0.0 \\ 2.42 \text{ ns} \\ -2.42 \text{ ns} \\ 2.49 \text{ ns} \\ -2.49 \text{ ns} \end{bmatrix}$$

$$\overline{\boldsymbol{\Delta}\mathbf{K}} = \begin{bmatrix} 2 \text{ mV} \\ 2 \text{ mV} \\ 0.25 \text{ ns} \\ 0.25 \text{ ns} \\ 0.25 \text{ ns} \\ 0.25 \text{ ns} \end{bmatrix}$$

Table III shows some of the most critical parasitic constraints found by PARCAR. As expected, the main contribution to voltage offset is due to parasitic resistances responsible for source degeneration of the input pair. The input source followers (MP10-11 and MP8-9) are less critical than the high-gain pairs (MN3-4, MN1-2 and MP2-3).

The complete layout of FASTCOMP is shown in Fig. 15. Fig. 16 shows two details of the layout area highlighted in Fig. 15, respectively with and without parasitic and topological constraint enforcement. In the right-hand side example, large capacitive couplings between critical nets are clearly visible. In particular, a considerable mismatch is present between nets 7 and 8. The capacitance of nets 3, 4, 9, 10 is large, thus slowing down the signal path. These capacitances are much smaller in the example shown in the left-hand side. Notice

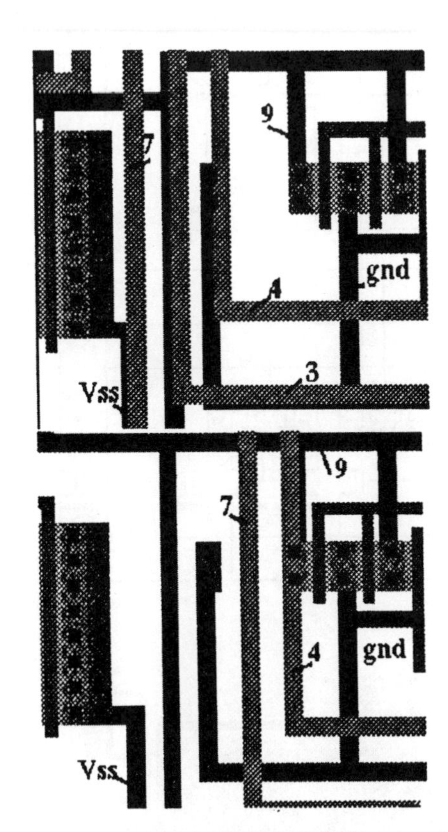

Fig. 16. Details of the routing of FASTCOMP. (Left) no parasitic constraints enforced. (Right) all parasitic constraints successfully enforced.

TABLE IV
FASTCOMP: RESULTS

	Unconstrained	Constrained
V_{off}	8.0mV	2.0 mV
$\tau_D(H \rightarrow L)$	3.11ns	2.73ns
$\tau_D(L \rightarrow H)$	2.92ns	2.65ns

TABLE V
FASTCOMP: CPU TIME FOR EACH LAYOUT PHASE

Layout phase	Unconstrained	Constrained
Constraint generation (PARCAR)	-	10.9 sec
Placement (PUPPY-LDO)	246.4 sec	1466.4 sec
Routing (ROAD)	2086.3 sec	2086.3 sec
Compaction (SPARCS-A)	2.5 sec	49.6 sec
TOTAL	2353.2 sec	3602.3 sec

that relatively large cross-couplings between nets 3, 4, and 9 were accepted due to their low criticality. A performance comparison of both the constrained and the unconstrained layouts is summarized in Table IV. Table V lists the CPU times required on a DEC-station 5000/240 for each layout phase.

C. Example: MPH

Fig. 17 shows the schematic of a micropower amplifier. This is an example that shows how the layout methodology described in this paper fits also tight constraint specifications on relatively large circuits. The nominal performance values for this circuit are the following:

$$V_{\text{dd}} = 1.5 \text{ V}$$
$$\omega_0 = 6.0 \text{ MHz}$$
$$A_v = 120 \text{ dB}$$
$$\phi_M = 60^\circ.$$

The following constraints have been specified

$$|\Delta V_{\text{dd}}| \leq 150 \text{ mV}$$
$$\Delta\omega_0 \geq -100 \text{ kHz}$$
$$\Delta A_v \geq -0.1 \text{ dB}$$
$$|\Delta\phi_M| \leq 10^\circ.$$

Therefore

$$\mathbf{K} = \begin{bmatrix} V_{\text{dd}} \\ -V_{\text{dd}} \\ -\omega_0 \\ -A_v \\ \phi_M \\ -\phi_M \end{bmatrix} \quad \mathbf{K}(\mathbf{p}^{(0)}) = \begin{bmatrix} 1.5 \text{ V} \\ -1.5 \text{ V} \\ -6.0 \text{ MHz} \\ -120 \text{ dB} \\ 60^\circ \\ -60^\circ \end{bmatrix}$$

$$\overline{\Delta\mathbf{K}} = \begin{bmatrix} 150 \text{ mV} \\ 150 \text{ mV} \\ 100 \text{ kHz} \\ 0.1 \text{ dB} \\ 10^\circ \\ 10^\circ \end{bmatrix}.$$

The complete layout of MPH is shown in Fig. 18.

Results for this layout are reported in Table VI, compared with the data from a handmade implementation of the same circuit, made by an experienced designer. The comparison between the layouts shows the usefulness of the constraint-driven approach. Table VII shows the CPU times required by each phase of the design, referred to a DEC-station 5000/240. Due to the very large set of critical parasitics and the tightness of constraints, PARCAR required a considerably longer CPU time than with FASTCOMP and the previous examples.

TABLE VI
MPH: RESULTS

	Constrained	Manual design
V_{dd}	full range	full range
ω_0	7.2MHz	6.0
A_v	136dB	120dB
ϕ_M	56°	63°

Table VIII summarizes the results obtained with the tools described in this paper on a set of benchmarks of industrial strength. In each of these examples, all the performance specifications have been met. CPU times refer to a DEC-station 5000/240.

VIII. CONCLUSION

In this paper, we have presented a constraint-driven methodology for the design of analog layout, supported by a set of

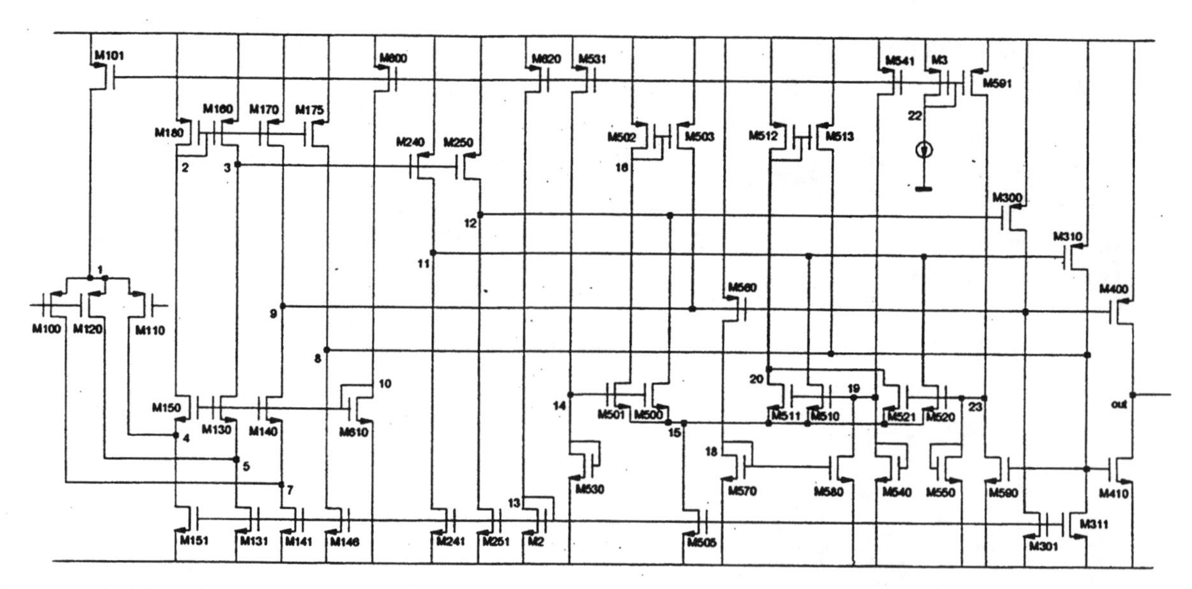

Fig. 17. Schematic of MPH.

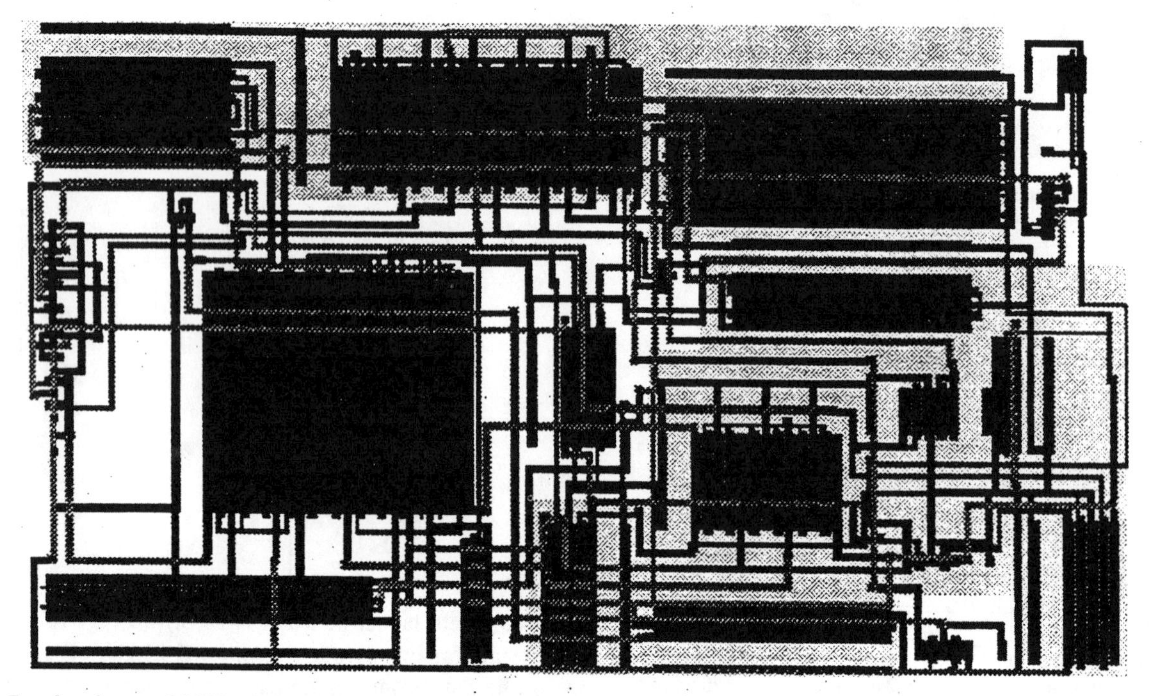

Fig. 18. Complete layout of MPH, obtained enforcing all analog constraints.

TABLE VII
MPH: CPU TIME FOR EACH LAYOUT PHASE

Synthesis phase	Constrained	Manual design
Constraint generation (PARCAR)	12,129 sec	-
Placement (PUPPY - LDO)	6096.7 sec	-
Routing (ROAD)	6625.0 sec	-
Compaction (SPARCS-A)	10.9 sec	-
Total	6.9 hrs	2 weeks

specialized tools. The key points of the methodology can be summarized as follows:

- We apply a rigorous methodology to translate high-level performance specifications into the set of constraints that the tools are able to control. The constraint generation technique guarantees that if we can satisfy the low-level constraints, all high-level specifications will be met.
- At each step of the layout design the tools are able to enforce constraints on all low-level parameters of the circuit.
- Infeasibility is detected as soon as possible in the design flow. A quantitative analysis allows us to determine the causes of infeasibility and to address a redesign strategy.

The tools presented cover all the major steps of layout synthesis, namely placement, routing, and compaction. The presence of a constraint-aware compactor allows the routing phase a more aggressive approach, thus improving the success rate and the robustness of the entire synthesis. All tools have been

TABLE VIII
PERFORMANCE ON A SET OF INDUSTRIAL BENCHMARKS

Circuit name	ω_0	A_v	ϕ_M	V_{off}	Decis. time	Power	CPU (sec.)
NEWOTA	*	*					2510
AB	*	*	*	*			1552
COMPL				*	*		378
OTA731	*	*	*				2629
FASTCOMP				*	*		3604
FCPHIL	*	*	*	*			3631
OPAMP1	*	*	*	*			315
MPH	*	*	*			*	24900

integrated in an environment where they share the database, the constraint representation, the parasitic models, and the performance analysis methods.

The impact of each layout step on the flexibility of the entire design flow has been analyzed in detail. The examples shown are benchmarks of industrial strength, and validate the effectiveness of our methodology.

APPENDIX

A. Capacitive Models for Interconnections

This appendix reports the capacitive models adopted to compute the stray capacitances of interconnections. In all our tools, the capacitive models described by [52] have been used. The dependence of the capacitance between a net and the substrate, or between two nets, is expressed as a polynomial in terms of wire widths and spacings. The coefficients are technology dependent, and can be computed, for each process, by accurate three-dimensional (3-D) simulation as in [52], by interpolation on experimental measurement, or by solving the Laplace equation when geometries are sufficiently regular.

The capacitance between a unity-length wire segment and the substrate is given by

$$C_u = k_0 + k_1 w$$

where w is the wire segment width.

The capacitance between unity-length parallel wire segments is given by

$$C_{\text{parallel}} = k_0' + k_1' w_1 + k_2' w_2 + \frac{k_3'}{d} + \frac{k_4'}{d^2} \tag{22}$$

where d is the distance between the wire segments, and w_1, w_2 are their respective widths. Model (22) holds for wires on the same layer or on different layer. In the former case, the model is symmetric, i.e., $k_1' = k_2'$, in the latter case k_1' and k_2' may differ.

The capacitance between orthogonal wire segments crossing each other is given by

$$C_{\text{cross}} = k_0'' + k_1'' w_1 + k_2'' w_2 + k_3'' w_1 w_2 \tag{23}$$

where w_1 and w_2 are the widths of the two segments. The fringe effect is accounted for by the two linear terms, while the quadratic term corresponds to the parallel-plate contribution, proportional to the crossing area $(w_1 w_2)$.

ACKNOWLEDGMENT

The authors wish to thank the students, researchers, and professors of the Electronics Research Laboratory of the Dept. of EECS, University of California, for many discussions on the topics of this paper. This paper could not have been possible if the authors had not been surrounded by such a stimulating human environment. Some of the examples shown in this paper have been provided by M. Gandini of CSELT Laboratories, Torino, Italy, J. Cohn of IBM Corp., Essex Jct. VT, J. H. Huijsing and R. G. H. Eschauzier of Delft University of Technology, The Netherlands, whom the authors gratefully acknowledge.

REFERENCES

[1] H. Y. Koh, C. H. Séquin, and P. R. Gray, "OPASYN: A compiler for CMOS operational amplifiers," *IEEE Trans. Computer-Aided Design*, vol. 9, pp. 113–126, Feb. 1990.

[2] H. Onodera, H. Kanbara, and K. Tamaru, "Operational amplifier compilation with performance optimization," in *Proc. IEEE CICC*, May 1989, pp. 17.4.1–17.4.6.

[3] R. Harjani, R. A. Rutenbar, and L. R. Carley, "OASYS. A framework for analog circuit synthesis," Carnegie Mellon Univ., Pittsburgh, PA, Res. Rep. CMUCAD-89-65, Nov. 1989.

[4] H. Yaghutiel, A. Sangiovanni-Vincentelli, and P. R. Gray, "A methodology for automated layout of switched-capacitor filters," in *Proc. IEEE ICCAD*, 1986, pp. 444–447.

[5] Y. Therasse, L. Reynders, R. Lannoo, and B. Dupont, "A switched-capacitor filter compiler," *VLSI Syst. Des.*, vol. 8, no. 10, pp. 85–88, Sept. 1987.

[6] J. Assael, P. Senn, and M. S. Tawfik, "A switched-capacitor filter silicon compiler," *IEEE J. Solid-State Circuits*, vol. 23, pp. 166–174, Feb. 1988.

[7] P. E. Allen and P. R. Barton, "A silicon compiler for successive approximation A/D and D/A converters," in *Proc. IEEE CICC*, 1986, pp. 552–555.

[8] G. Jusuf, P. R. Gray, and A. Sangiovanni-Vincentelli, "CADICS—Cyclic analog-to-digital converter synthesis," in *Proc. IEEE ICCAD*, Nov. 1990, pp. 286–289.

[9] H. Chang, E. Liu, R. Neff, E. Felt, E. Malavasi, E. Charbon, A. Sangiovanni-Vincentelli, and P. R. Gray, "Top-down, constraint-driven methodology based generation of n-bit interpolative current source D/A converters," in *Proc. IEEE CICC*, May 1994, pp. 369–372.

[10] G. Winner *et al.*, "Analogue macrocell assembler," *VLSI Syst. Des.*, vol. 8, no. 5, pp. 68–71, May 1987.

[11] C. D. Kimble, A. E. Dunlop, G. F. Gross, V. L. Hein, M. Y. Luong, K. J. Stern, and E. J. Swanson, "Autorouted analog VLSI," in *Proc. IEEE CICC*, 1985, pp. 72–78.

[12] T. Pletersek *et al.*, "High-performance designs with CMOS analogue standard cells," *IEEE J. Solid-State Circuits*, vol. 21, pp. 215–222, Apr. 1986.

[13] L. D. Smith *et al.*, "A CMOS-based analog standard cell product family," *IEEE J. Solid-State Circuits*, vol. 24, pp. 370–379, Apr. 1989.

[14] J. Rijmenants, J. B. Litsios, T. R. Schwarz, and M. G. R. Degrauwe, "ILAC: An automated layout tool for analog CMOS circuits," *IEEE J. Solid-State Circuits*, vol. 24, pp. 417–425, Apr. 1989.

[15] M. G. R. Degrauwe *et al.*, "Toward an analog system design environment," *IEEE J. Solid-State Circuits*, vol. 24, pp. 659–671, June 1989.

[16] M. Kayal, S. Piguet, M. Declercq, and B. Hochet, "SALIM: A layout generator tool for analog IC's," in *Proc. IEEE CICC*, May 1988, pp. 751–754.

[17] M. Mogaki, N. Kato, Y. Chikami, N. Yamada, and Y. Kobayashi, "LADIES: An automatic layout system for analog LSI's," in *Proc. IEEE ICCAD*, Nov. 1989, pp. 450–453.

[18] F. M. Turky and E. E. Perry, "BLADES: An A. I. approach to analog circuit design," *IEEE Trans. Computer-Aided Design*, vol. 8, pp. 680–692, June 1989.

[19] R. S. Gyurcsik and J.-C. Jeen, "A generalized approach to routing mixed analog and digital signal nets in a channel," *IEEE J. Solid-State Circuits*, vol. 24, pp. 436–442, Apr. 1989.

[20] I. Harada, H. Kitazawa, and T. Kaneko, "A routing system for mixed A/D standard cell LSI's," in *Proc. IEEE ICCAD*, Nov. 1990, pp. 378–381.

[21] S. W. Mehranfar, "A technology-independent approach to custom analog cell generation," *IEEE J. Solid-State Circuits*, vol. 26, pp. 386–393, Mar. 1991.

[22] J. M. Cohn, D. J. Garrod, R. A. Rutenbar, and L. R. Carley, "KOAN/ANAGRAM II: New tools for device-level analog placement and routing," *IEEE J. Solid-State Circuits*, vol. 26, pp. 330–342, Mar. 1991.

[23] R. Okuda, T. Sato, H. Onodera, and K. Tamaru, "An efficient algorithm for layout compaction problem with symmetry constraints," in *Proc. IEEE ICCAD*, Nov. 1989, pp. 148–151.

[24] U. Choudhury and A. Sangiovanni-Vincentelli, "Constraint-based channel routing for analog and mixed-analog digital circuits," in *Proc. IEEE ICCAD*, Nov. 1990, pp. 198–201.

[25] E. Malavasi, U. Choudhury, and A. Sangiovanni-Vincentelli, "A routing methodology for analog integrated circuits," in *Proc. IEEE ICCAD*, Nov. 1990, pp. 202–205.

[26] S. K. Hong and P. E. Allen, "Performance driven analog layout compiler," in *Proc. IEEE Int. Symp. Circuits Syst.*, 1990, pp. 835–838.

[27] E. Charbon, E. Malavasi, U. Choudhury, A. Casotto, and A. Sangiovanni-Vincentelli, "A constraint-driven placement methodology for analog integrated circuits," in *Proc. IEEE CICC*, May 1992, pp. 2821–2824.

[28] U. Choudhury and A. Sangiovanni-Vincentelli, "Constraint generation for routing analog circuits," in *Proc. IEEE/ACM DAC*, June 1990, pp. 561–566.

[29] G. Gad-El-Karim and R. S. Gyurcsik, "Use of performance sensitivities in analog cell layout," in *Proc. IEEE Int. Symp. Circuits Syst.*, June 1991, pp. 2008–2011.

[30] H. Chang, A. Sangiovanni-Vincentelli, F. Balarin, E. Charbon, U. Choudhury, G. Jusuf, E. Liu, E. Malavasi, R. Neff, and P. Gray, "A top-down, constraint-driven design methodology for analog integrated circuits," in *Proc. IEEE CICC*, May 1992, pp. 841–846.

[31] E. Malavasi and A. Sangiovanni-Vincentelli, "Area routing for analog layout," *IEEE Trans. Computer-Aided Design*, vol. 12, pp. 1186–1197, Aug. 1993.

[32] E. Felt, E. Malavasi, E. Charbon, R. Totaro, and A. Sangiovanni-Vincentelli, "Performance-driven compaction for analog integrated circuits," in *Proc. IEEE CICC*, May 1993, pp. 1731–1735.

[33] E. Charbon, E. Malavasi, D. Pandini, and A. Sangiovanni-Vincentelli, "Simultaneous placement and module optimization of analog IC's," in *Proc. IEEE/ACM DAC*, June 1994, pp. 31–35.

[34] E. Malavasi and D. Pandini, "Optimum CMOS stack generation with analog constraints," *IEEE Trans. Computer-Aided Design*, vol. 14, pp. 107–122, Jan. 1995.

[35] W. Nye, D. C. Riley, A. Sangiovanni-Vincentelli, and A. L. Tits, "DELIGHT-SPICE: An optimization-based system for the design of integrated circuits," *IEEE Trans. Computer-Aided Design*, vol. 7, pp. 501–519, Apr. 1988.

[36] J.-M. Shyu, "Performance optimization of integrated circuits," Univ. California, Berkeley, CA, Memo. UCB/ERL M88/74, Nov. 1988.

[37] S. W. Director and R. A. Rohrer, "The generalized adjoint network and network sensitivities," *IEEE Trans. Circuit Theory*, vol. 16, pp. 318–323, Aug. 1969.

[38] U. Choudhury, "Sensitivity computation in SPICE3," M.S. thesis, Univ. California, Berkeley, CA, 1988.

[39] U. Choudhury and A. Sangiovanni-Vincentelli, "Use of performance sensitivities in routing of analog circuits," in *Proc. IEEE Int. Symp. Circuits Syst.*, May 1990, pp. 348–351.

[40] M. J. M. Pelgrom, A. C. J. Duinmaijer, and A. P. G. Welbers, "Matching properties of MOS transistors," *IEEE J. Solid-State Circuits*, vol. 24, pp. 1433–1440, Oct. 1989.

[41] C. Guardiani, A. Tomasini, J. Benkoski, M. Quarantelli, and P. Gubian, "Applying a submicron mismatch model to practical IC design," in *Proc. IEEE CICC*, May 1994, pp. 297–300.

[42] E. Charbon, E. Malavasi, and A. Sangiovanni-Vincentelli, "Generalized constraint generation for analog circuit design," in *Proc. IEEE ICCAD*, Nov. 1993, pp. 408–414.

[43] B. Basaran, R. A. Rutenbar, and L. R. Carley, "Latchup-aware placement and parasitic-bounded routing of custom analog cells," in *Proc. IEEE ICCAD*, Nov. 1993, pp. 415–421.

[44] Z. Daoud and C. J. Spanos, "DORIC: Design of optimal and robust integrated circuits," in *Proc. IEEE CICC*, May 1994, pp. 361–364.

[45] E. H. L. Aarts and P. J. M. van Laarhoven, *Simulated Annealing: Theory and Applications*. Dordrecht, The Netherlands: D. Reidel Publishing, 1987.

[46] C. Sechen and A. Sangiovanni-Vincentelli, "Chip-planning, placement and global routing of macro/custom cell IC's using simulated annealing," in *Proc. IEEE/ACM DAC*, June 1988, pp. 73–80.

[47] M. Burstein, "Channel routing," in *Layout Design and Verification*, T. Ohtsuki, Ed. Amsterdam, The Netherlands: North Holland, 1986, ch. 4, pp. 133–167.

[48] G. W. Clow, "A global routing algorithm for general cells," in *Proc. IEEE/ACM DAC*, 1984, pp. 45–51.

[49] E. Felt, E. Charbon, E. Malavasi, and A. Sangiovanni-Vincentelli, "An efficient methodology for symbolic compaction of analog IC's with multiple symmetry constraints," in *Proc. Euro. Design Automation Conf.*, Sept. 1992, pp. 148–153.

[50] E. Malavasi, E. Felt, E. Charbon, and A. Sangiovanni-Vincentelli, "Symbolic compaction with analog constraints," *Int. J. Circuit Theory Appl.*, vol. 23, no. 4, pp. 433–452, July–Aug. 1995.

[51] J. Kuhn, "Analog module generators for silicon compilation," *VLSI Syst. Des.*, vol. 8, no. 5, pp. 74–80, May 1987.

[52] U. Choudhury and A. Sangiovanni-Vincentelli, "An analytical-model generator for interconnect capacitances," in *Proc. IEEE CICC*, May 1991, pp. 861–864.

Enrico Malavasi received the Laurea degree (summa cum laude) in electrical engineering from the University of Bologna, Italy, in 1984, and the M.S. degree in electrical engineering from the University of California, Berkeley, in 1993.

Between 1986 and 1989, he was with the Department of Electrical Engineering and Computer Science (DEIS), University of Bologna, working on computer-aided-design for analog circuits. In 1989, he joined the *Dipartimento di Elettronica ed Informatica*, University of Padova, Italy, as an Assistant Professor. Since 1990, he has collaborated with the CAD group of the Department of EECS, University of California, where he has carried out research on performance-driven CAD tools and methodologies for analog design. In 1995, he joined Cadence Design Systems Inc., San Jose, CA, as architect for physical design automation. His research interests include several areas of analog design automation: layout, design methodologies, optimization, and circuit analysis.

Edoardo Charbon received the Diploma in electrical engineering from the Swiss Federal Institute of Technology (ETH), Zurich, Switzerland, in 1988, the M.S. degree in electrical and computer engineering from the University of California, San Diego, in 1991, and the Ph.D. degree in electrical engineering and computer sciences from the University of California, Berkeley, in 1995.

Between 1988 and 1989, he was with the Department of Electrical Engineering, ETH, where he designed CMOS A/D converters for integrated sensor applications. In 1989, he visited the Dept. of Electrical Engineering, University of Waterloo, Canada, where he was involved in the design and fabrication of ultra low-noise, nano-Tesla magnetic sensors. His research interests include CAD for analog and mixed-signal IC's, RF design, microwave and superconducting parasitic analysis, and micromachined sensor design.

Eric Felt received the B.S.E. degree in electrical engineering, summa cum laude, with distinction, from Duke University, Durham, NC, in 1991, the M.S. degree in electrical engineering and computer sciences from the University of California, Berkeley, in 1993, and is currently working toward his Ph.D. degree at the same department and institution.

His current research topic is CAD for the testing and characterization of analog IC's.

Alberto Sangiovanni-Vincentelli (M'74–SM'81–F'83) for a photograph and biography, see p. 505 of the May 1996 issue of this TRANSACTIONS.

A Performance-Driven Placement Tool for Analog Integrated Circuits

Koen Lampaert, Georges Gielen, *Member, IEEE,* and Willy M. Sansen, *Senior Member, IEEE*

***Abstract*—This paper presents a new approach toward performance-driven placement of analog integrated circuits. The freedom in placing the devices is used to control the layout-induced performance degradation within the margins imposed by the designer's specifications. This guarantees that the resulting layout will meet all specifications by construction. During each iteration of the simulated annealing algorithm, the layout-induced performance degradation is calculated from the geometrical properties of the intermediate solution. The placement tool inherently handles symmetry constraints, circuit loading effects and device mismatches. The feasibility of the approach is demonstrated with practical circuit examples.**

I. Introduction

Generating the layout of high-performance analog circuits is a difficult and time-consuming task which has a considerable impact on circuit performance. Asymmetries and device mismatches can easily upset the critical precision of component values and together with the parasitics associated with the interconnections they can introduce intolerable performance degradation. Since these parasitics are unavoidable, the main concern in computer-aided analog layout synthesis is to control the effects of the parasitics on circuit performance and to keep the layout-induced performance degradation within user-defined margins, as specified in the circuit's performance specifications. In this way, it can be guaranteed that the resulting layout will meet the specifications by construction.

Many existing analog layout programs, like KOAN/ANAGRAM [1] or ALSYN [2], however, try to optimize the layout without quantifying the performance degradation. They therefore cannot guarantee that the resulting layout will also meet the specifications. In recent years therefore, a performance-driven methodology has been introduced [3] which does try to achieve this. The approach in [3] and [4] first maps the performance specifications onto a set of constraints for critical parasitics which are then used to drive the layout tools (see Fig. 1(a)). Using sensitivity information and quadratic optimization, the set of constraints that maximizes the flexibility of the layout tools is computed. If the layout tools fail to meet one of these parasitic constraints, one or more iterations with another set of constraints is needed. This approach suffers from a number of drawbacks. First, the flexibility of layout tools is something which is very hard, if not impossible to quantify with any reasonable accuracy, for use as cost function in the optimization. Second, by imposing *one* set of constraints only, several valid alternative solutions are rejected, unnecessarily increasing the number of time-consuming iterations.

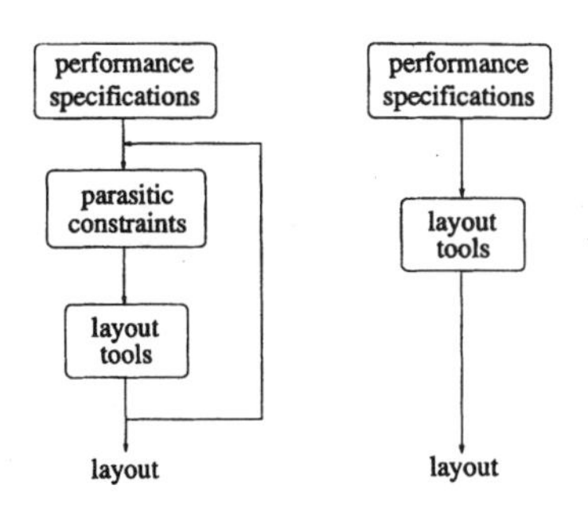

Fig. 1. Performance-driven layout methodologies: (a) with an intermediate constraint generation step, and (b) performance specifications directly taken into account by the layout tools.

To overcome these drawbacks, we therefore propose an alternative solution which eliminates the intermediate constraint generation step, while still guaranteeing a fully functional layout that meets all performance specifications. In our approach, the layout tools are driven *directly* by the performance constraints (see Fig. 1(b)). This has several advantages. First, by directly taking into account the high-level performance specifications, a complete and sensible trade-off between the different alternative solutions can be made. Second, since the performance degradation is calculated at run-time, it is not only possible to keep all performance characteristics within their limits, but also to optimize the layout with respect to a subset of the specifications. Finally, while the methodology described in [3] and [4] can lead to a number of iterations with different constraint values, our algorithm will either yield a correct layout or will flag the specifications as being impossible to meet, without iterations. This approach therefore eliminates the feedback route between constraint derivation, placement and layout extraction.

In this paper we present an analog placement tool based on this approach. The tool uses a simulated-annealing-based optimization algorithm, with a cost function designed to drive a random start solution to a placement where all performance constraints are satisfied. Our placement tool differs from other analog placement approaches [1], [2], [5] in that it takes into account symmetry constraints, performance degradation due to interconnect parasitics as well as due to device mismatches,

Manuscript received December 20, 1994; revised April 11, 1994. This work was supported in part by the ESPRIT Project ADMIRE and a Contract with ESA-ESTEC.

The authors are with the Katholieke Universiteit Leuven, Department of Elektrotechniek, ESAT-MICAS, Kardinaal Mercierlaan 94, B-3001 Heverlee, Belgium.

IEEE Log Number 9412408.

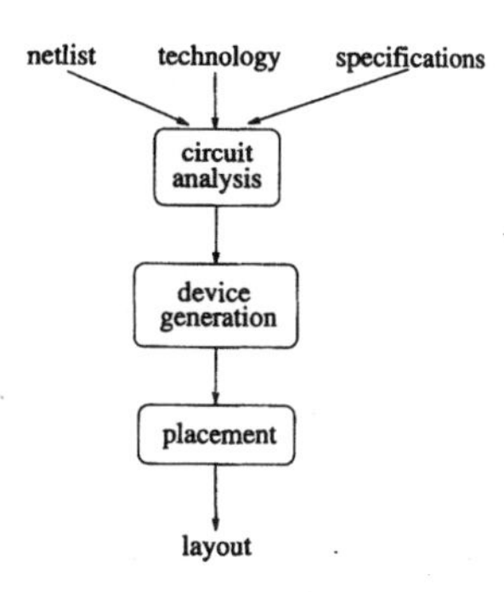

Fig. 2. Program flow of the analog placement tool.

and combines this with the aggressive geometrical optimization techniques (devices merges, abutment, etc.) introduced in [1]. In addition, a number of features have been added to make the tool suitable for use with submicron technology processes in a mixed-signal context.

This paper is organized as follows: Section II describes the general flow of the analog placement tool. Circuit analyses that have to be performed first are discussed in Section III. Section IV describes the device generators that are used in the placement program. The placement algorithm itself is then discussed in Section V and details of how the performance degradation due to parasitics and mismatches are estimated, are presented in Section VI. Examples of practical placements produced with the placement tool are then presented in Section VII. Finally, conclusions of our research are formulated in Section VIII.

II. Overview

The basic components of the placement program are illustrated in Fig. 2. The input to the tool is the circuit netlist after sizing as well as the list of performance specifications that the circuit has to meet (e.g., phase margin $\geq 60°$). The difference between these specified values and the actual performance values obtained by the circuit after sizing are the margins that can be taken up by layout-induced performance degradation. The latter is evaluated using the numerically calculated sensitivities of the considered performances toward the considered layout-dependent performance degrading effects

$$\Delta P_j = \sum_i S^j_{x_i} x_i \tag{1}$$

where ΔP_j is the degradation of the jth performance, $S^j_{x_i} = (\delta P_j)/(\delta x_i)$ the sensitivity of performance P_j to layout effect x_i, evaluated only once in the parasitic-free design point. This linear approximation is valid for small values of the parasitic effect. The placement tool is driven such that the total performance degradations ΔP_j for all performances are below their maximum margins as given by the input specifications

$$\Delta P_j \leq \Delta P_{j,\max} \quad j = 1 \cdots k \tag{2}$$

where k is the number of specifications.

This performance-driven principle can be applied to any layout-dependent performance degrading effect x_i for which the relation between the actual layout and the impact on the performance can be modeled and simulated. In this paper, this will be illustrated for parasitic wire capacitances and resistances, as well as for device mismatches. Note also that although this is only a placement program, the effect of the wiring introduced in the later routing phase is already estimated during the placement, and routing will therefore introduce only an incremental additional degradation of the performance to the extent of the difference between the actual and the estimated wire parasitics.

The first step in the execution of the placement program consists of a number of numerical simulations which result in the set of required performance sensitivities and operating-point information (branch currents, node voltages) of the circuit. This information, together with the circuit netlist, is then used as input for a set of device generators that construct a list of geometrical variants for every device (see Section IV). Only the information needed for the optimization of the placement is generated.

Next, a simulated-annealing algorithm is used to create a placement which respects all circuit specifications. The specific analog features are implemented in the following way:

1) *Symmetry:* Symmetry is considered to be an absolute constraint. If the user specifies a number of devices as being symmetric and/or selfsymmetric, the devices have to be symmetric in the resulting placement. Consequently, symmetry constraints are handled in the move set. Groups of symmetric devices are moved simultaneously such that their symmetry is preserved at all times during the optimization and also in the final result.
2) *Matching:* The impact of mismatches is included in the cost function. The user can specify a pair of devices as being matched without specifying the degree of matching. Matching devices are always generated identically and with identical orientations but it is up to the placement tool to determine the positions and therefore also the distance between the matched devices such that the circuit performance constraints are met. Since it is not always possible in an analog circuit layout to, at the same time, meet all symmetry requirements, put all matching devices directly next to each other and obtain a fairly compact layout, the matching degree of a pair of devices has to be selected in view of its influence on the performance of the circuit. Mismatch is therefore included in the set of parasitic effects for which the degradation of the performance characteristics is calculated and included in the placement cost function.
3) *Circuit Loading:* The performance degradation due to parasitic wire and node capacitance and resistance is also a factor which is included in the placement cost function.
4) *Device Merging:* When it is possible to merge two device terminals, the total junction capacitance of the net to which the terminals belong decreases with an amount proportional to the area and the perimeter of the overlap region. This decreases the total performance degradation and lowers the cost function. In this way, device geometry sharing is promoted specifically for sensitive nets.

After the optimization, the device generators are called again to create the actual layout for the selected geometrical device variants and the final layout is constructed. The output of the program is then the final layout, together with information about the performance degradation in this final layout and an identification of the most important contributions to this degradation. In case the degradation exceeds the required performance specifications, this information can be used by the designer to see the failing performance(s) and to identify the critical effects. This allows him to improve his design when desired.

The relevant details of the different steps will now be discussed in the following sections.

III. Circuit Analysis

An operating-point analysis is first performed on the sized circuit to extract all branch currents and node voltages. The current through a branch is used to determine the necessary width of the routing wires. For small currents, minimal width wires can be used. For larger currents, the wire width has to be adjusted in order to avoid electromigration. This information is used to make accurate estimations for the values of parasitic node capacitances and resistances and to estimate the area which has to be reserved for routing.

Another set of simulations is needed to extract the sensitivities of the circuit specifications with respect to device mismatches and parasitic node capacitances and resistances. The concept of sensitivities and how they are used to calculate performance degradation will be explained in detail in Section VI.

IV. Device Generation

With respect to device generation, we follow the approach outlined in [1]: we only provide device generators for basic circuit devices (transistors, resistors and capacitors) and we rely as much as possible on dynamic geometry sharing techniques to obtain merged structures for multi-device subcircuits (differential pairs, current mirrors, etc.). This drastically reduces the number of generators that needs to be developed and maintained. These device generators can also be used interactively during manual layout.

Before the execution of the optimization algorithm, the device generators are called in interface mode to generate a list of possible geometrical variants for each device, based on the actual device parameters, the technology process and a number of user-specified options. For each variant, the geometrical and electrical information relevant for the execution of the placement algorithm is generated: the bounding box, protection frames for each terminal and the values for parasitic terminal capacitances. The terminal protection frames are used to detect device merges as described in [1]. The parasitic terminal capacitances are needed to calculate performance degradation due to interconnect parasitics (see Section VI). They can be calculated accurately since the dc voltages of the connecting nodes are known.

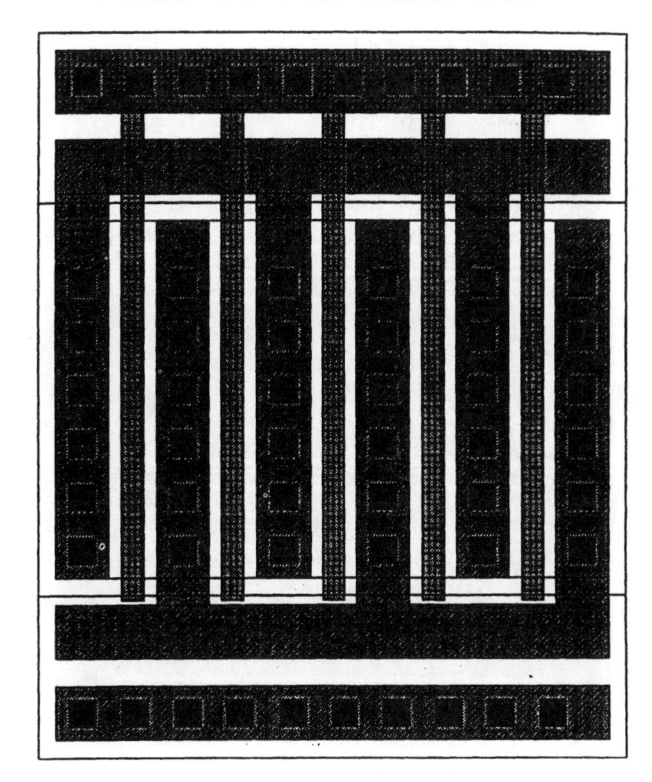

Fig. 3. A regular finger-style layout for a MOS transistor.

The protection box of a device is calculated by taking the physical bounding box of the device and expanding it on each side by a certain amount to reserve space for routing. The necessary routing space is estimated for each side based on the width of the wires which must be connected to the terminal on that side. By taking into account variable wire widths, we are able to make fairly accurate routing space estimates.

We have added special features to the device generators in order to make them suitable for use in a mixed-signal context. Switching transients in digital MOS circuits can perturb analog circuits integrated on the same die by means of coupling through the substrate. An effective way to protect sensitive devices against coupling noise is to make sure that no part of a MOS transistor is more than a specified minimum distance away from a bulk contact. This minimum distance can be a design rule imposed by the foundry or can be specified by the designer. Our MOS transistor layout generator can handle this kind of constraint. When necessary, the MOS transistor will be split in a number of parallel transistors and bulk straps will be added between the different parts. Fig. 3 shows a regular MOS finger style transistor, while Fig. 4 shows a large transistor which has been split into parallel parts. Also, the wire width of the source and drain wires of the last transistor has been automatically adapted to the large current.

After the execution of the placement algorithm, the device generator is called in layout mode for the selected variant. This time, the generator returns the actual layout of the selected variant of the device together with a model that includes all second-order parasitic elements. This model can then be backannotated to the netlist to carry out more accurate simulations.

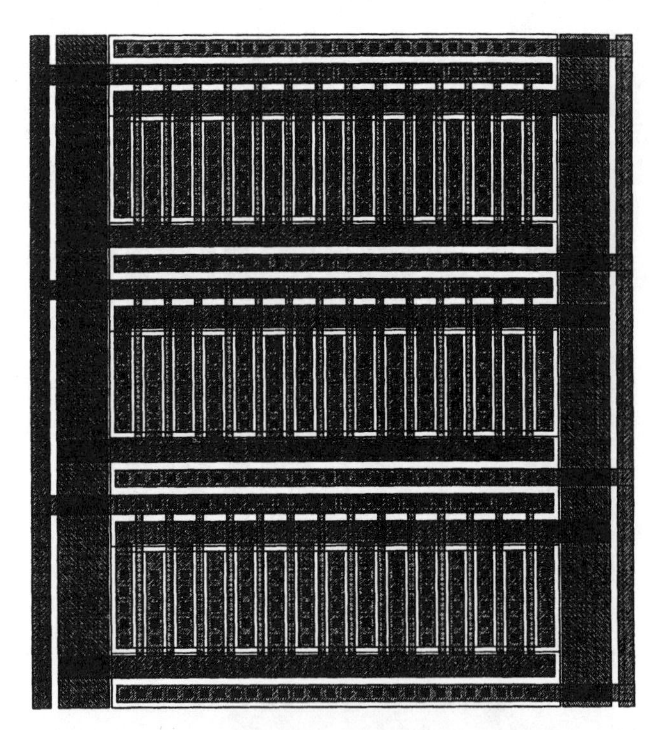

Fig. 4. A special layout structure for a large MOS transistor. The transistor is split into parallel parts and width of the source and drain wires has been adapted to large currents.

V. The Analog Placement Algorithm

A. The Simulated Annealing Algorithm

The optimization algorithm used for the performance-driven placement program is simulated annealing [6], a general and robust optimization technique that uses stochastically controlled hill-climbing to avoid the many local minima in complex cost surfaces and thus to find better global solutions. The algorithm works by generating a large number of randomly selected perturbations on an initial random placement. For each proposed move, the effect that the move has on the cost of the circuit is evaluated. If the proposed move decreases the cost function, it is accepted. If, however, the proposed move increases the cost function C, it is accepted with a probability that is affected by the temperature T, a control parameter that is decreased gradually in the course of the annealing process, in the following manner:

$$P[uphill] \sim e^{(-\Delta C)/T}. \tag{3}$$

The important aspects of the annealing algorithm are discussed in the following subsections.

B. Placement Representation

We use a flat, nonslicing placement representation with the absolute coordinates, the orientation and the geometrical variant of each device (or device cluster) as annealing variables. For analog performance-driven placement this model is to be preferred over the alternative slicing-style representation for several reasons. The main reasons are that an accurate estimation of parasitic circuit loading and mismatch effects requires the knowledge of absolute device coordinates, which are not known in a slicing structure, and secondly that a slicing-style placement model limits the set of reachable device configurations, prohibits geometry-sharing optimizations and makes it difficult to implement symmetry constraints.

C. Move Set

The following moves are executed during placement optimization:

1) *Device Translation:* A device is chosen at random and its center position is translated to a randomly selected coordinate.
2) *Device Reorientation:* A device is chosen at random and its orientation is changed to one of the eight orthogonal rotation and/or mirror values possible by combining 90° rotations with mirroring in the X and Y direction.
3) *Device Swap:* Two devices are chosen at random and their center coordinates are interchanged.
4) *Device Reshape:* A device is chosen at random and casted into a new geometrical variant, chosen from the list created by the device generator.

In addition to this classic single-device move set, a number of moves which operate on clusters of matched, merged or symmetric devices have also been implemented. These moves simultaneously change the position, orientation or the variant of a set of clustered devices.

D. Cost Function

The search for an optimum placement is driven by the cost function. This cost function C is calculated for each intermediate placement result and is a weighted sum of 4 terms

$$C = \alpha C_{area} + \beta C_{aspect\ Ratio} + \gamma C_{overlap} + \delta C_{perf\ Degr} \tag{4}$$

where

1) C_{area}: the area of the bounding box of the intermediate placement. This term is used to minimize the area.
2) $C_{aspectRatio}$: this term is proportional to the deviation of the aspect ratio of the intermediate placement to the aspect ratio specified by the user.
3) $C_{overlap}$: this term is proportional to the total amount of illegal overlap present in the intermediate placement and is driven to zero in the final placement.
4) $C_{perfDegr}$: this term is used to keep the performance degradation induced by interconnect parasitics and device mismatches within user-specified limits. This term is extremely important for our placement approach and is therefore described in detail in the next section.

The weighting coefficients α, β, γ, and δ are adapted dynamically during the placement optimization.

VI. Estimating Performance Degradation

We now describe how the performance degradation of an analog circuit due to interconnect parasitics and device mismatches, is estimated in the program. Note, however that the same technique can be applied to other layout-dependent

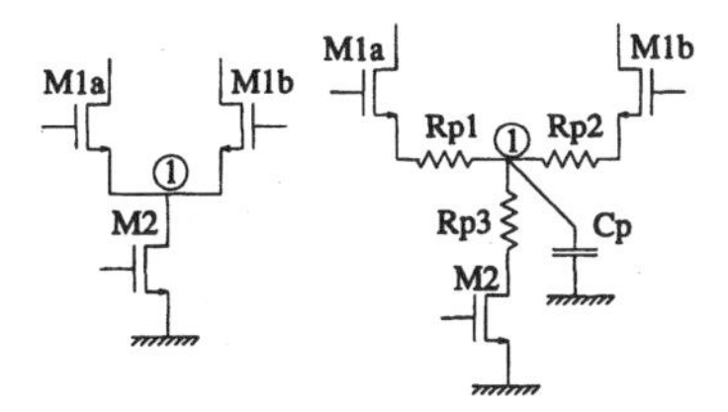

Fig. 5. Modeling of interconnect parasitics on node 1. (a) Differential pair without parasitics. (b) Differential pair with parasitic capacitor and resistors.

performance-degrading effects, provided that the effect can be quantified.

In each case we follow the same three-step methodology. First we extract the relevant geometrical information out of the intermediate placement solution. Based on this information we calculate the value of the parasitic circuit elements which model the parasitic effect during simulation. Finally, we calculate the influence of the parasitic circuit elements on the performance characteristics of the circuit using a linear approximation (valid for small values of the parasitic effect) based on performance sensitivities, which have been calculated in advance before the start of the placement optimization. This methodology will now be applied to interconnect parasitics and device mismatches.

A. Interconnect Parasitics

An n-terminal net can be modeled by a starred configuration of n parasitic resistors and 1 parasitic capacitor (see Fig. 5). The values of the capacitor and the resistors can only be approximated since the actual geometry of the wire is unknown before routing. A minimum spanning tree connecting all terminals is calculated and the sum of the lengths of its paths is used as an approximation for the total net length L.

The parasitic capacitance C_p is the sum of two components

$$C_p = C_j + C_w. \tag{5}$$

C_j is caused by the sum of the junction capacitances of all connecting terminals and C_w is caused by the capacitance to the substrate of the actual wire.

The parasitic resistances $R_{p,i}, i = 1, \cdots, n$ of the wire segments are calculated as follows:

$$R_{pi} = \rho_{square} \frac{\frac{L}{n}}{W_{wire,\,i}} \tag{6}$$

where ρ_{square} is the sheet resistance of the primary routing layer, L is the estimated total net length, and n the number of terminals connected to the net. $W_{wire,\,i}$ is the width of the ith wire segment which can be calculated from the current flowing through the ith terminal.

Once the value of C_p and $R_{p,i}$, $i = 1, \cdots, n$ is known for every net, the performance degradation ΔP_j for the jth performance characteristic due to interconnect parasitics can be determined using the precalculated sensitivity information

$$\Delta P_j = \sum_{k=1}^{m} \left(S^j_{C_{p,\,k}} C_{p,\,k} + \sum_{i=1}^{n_k} S^j_{R_{p,\,ki}} R_{p,\,ki} \right) \tag{7}$$

where m is the number of nodes minus the ground node and n_k is the number of terminals of net k. $S^j_{C_{p,\,k}} = [(\delta P_j)/(\delta C_{p,\,k})]$ and $S^j_{R_{p,\,ki}} = [(\delta P_j)/(\delta R_{p,\,ki})]$ are the sensitivities of performance characteristic P_j to small changes in the parasitic capacitance $C_{p,\,k}$ and the parasitic resistances $R_{p,\,ki}$. These sensitivities are determined in advance by simulation.

B. Mismatches

We follow a comparable approach with respect to mismatches. $\sigma^2(V_{TO})$ and $\sigma^2(\beta)$ for the V_{TO} and β mismatch of MOS transistors can be estimated using the following equations [7]:

$$\sigma^2(V_{TO}) = \frac{A^2_{V_{TO}}}{WL} + S^2_{V_{TO}} D^2 \tag{8}$$

$$\sigma^2(\beta) = \frac{A^2_{\beta}}{WL} + S^2_{\beta} D^2 \tag{9}$$

where $A_{V_{TO}}$, $S_{V_{TO}}$, A_β, and S_β are constants depending on the process. The area of the devices WL and the distance D are known for each intermediate placement. Based on this information, the standard deviations of V_{TO} and β can be calculated with (8) and (9).

Using sensitivity information, the effect on the performance degradation can be estimated as follows:

$$\Delta P_j = \sum_{k=1}^{m} \{ S^j_{\Delta V_{TO,\,k}} [3\sigma(V_{TO})_k] + S^j_{\Delta\beta_k} [3\sigma(\beta)_k] \} \tag{10}$$

where $S^j_{\Delta V_{TO,\,k}} = [(\delta P_j)/\ (\delta \Delta V_{TO,\,k})]$ and $S^j_{\Delta\beta_k} = [(\delta P_j)/\ (\delta\Delta\beta_k)]$ are the sensitivities of performance characteristic P_j to small changes in ΔV_{TO} and $\Delta\beta$ of matching transistor pair k. The sum is taken over all pairs of matching devices.

Equation (10) can be rewritten as

$$\Delta P_j = \Delta P_{j,\,area} + \Delta P_{j,\,distance}. \tag{11}$$

$\Delta P_{j,\,area}$ represents the degradation of the jth performance characteristic due to area effects. This term can be computed after sizing and remains constant during placement.

$\Delta P_{j,\,distance}$ represents the degradation of the jth performance characteristic due to distance effects and therefore depends on the actual layout. This term can be computed as follows:

$$\Delta P_{j,\,distance} = \sum_{k=1}^{m} [S^j_{D_k} (D_k)] \tag{12}$$

where D_k represents the distance between the transistors of matching pair k and $S^j_{D_k}$ is the sensitivity of performance characteristic j to small variations in distance D_k. This term must be recomputed for every new placement.

VII. Experimental Results

The algorithm described in this paper has been implemented using the C++ language in the UNIX environment and has been integrated in a complete synthesis environment for analog

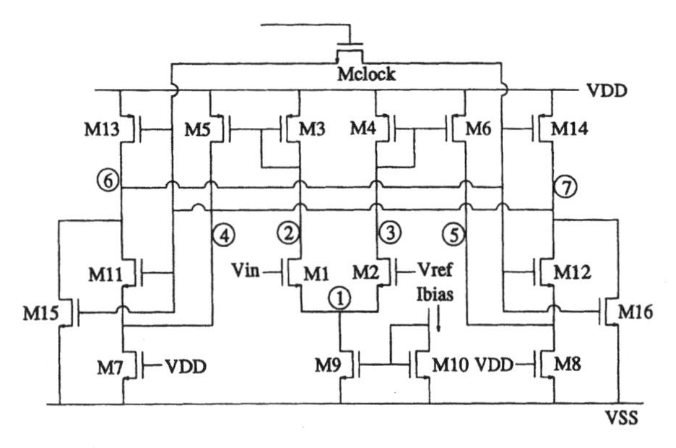

Fig. 6. High speed CMOS comparator.

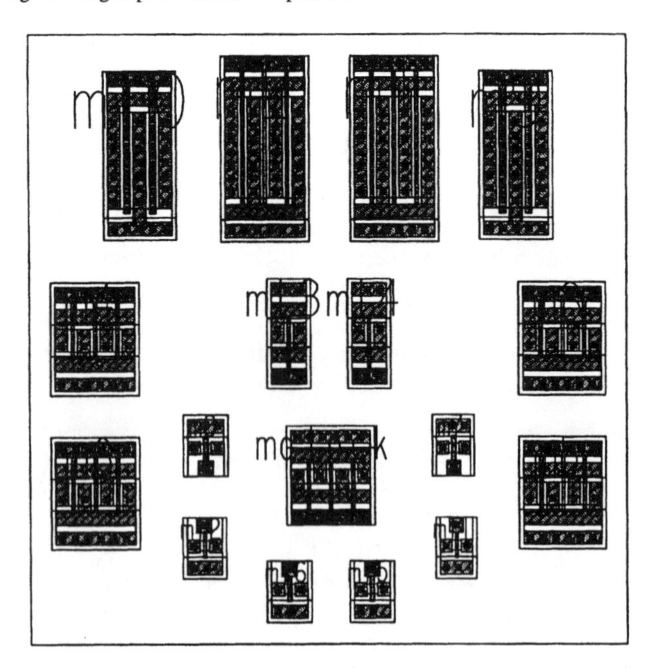

Fig. 7. CMOS comparator: placement 1: generated automatically by the performance-driven placement program.

TABLE I
COMPARISON OF THE PERFORMANCE OF TWO AUTOMATICALLY GENERATED COMPARATOR PLACEMENTS

performance				
Performance	Spec	Plac 1	Plac 2	Unit
offsetvoltage	< 5	3.7	6.9	*mV*
delay	< 5	2.8	5.4	*nsec*

TABLE II
PLACEMENT 1 (PERFORMANCE DRIVEN): OFFSET VOLTAGE DEGRADATION DUE TO DISTANCE EFFECTS

comparator offset voltage degradation			
Transistor Pair	Distance	Sensitivity	Degradation
	μm	$\frac{\mu V}{\mu m}$	mV
M1 – M2	60	12	.720
M3 – M5	70	2.9	.203
M4 – M6	70	2.9	.203
M7 – M8	118	.2	.024
M11 – M12	119	1.93	.230
M13 – M14	38	3.8	.145
M15 – M16	40	3	.120
total			1.645

TABLE III
PERFORMANCE CHARACTERISTICS OF THE FULLY DIFFERENTIAL CMOS OPAMP

opamp performance				
Performance	Specification	Nominal Value	After Placement	Unit
A_v	> 100	107	104	*dB*
GBW	> 200	205	202	*MHz*
PM	> 70	77	74	deg
CMRR@10Hz	> 70	∞	78	*dB*
PSRR@10Hz	> 80	∞	86	*dB*

circuits [11]. The program was tested on a number of practical circuits. Three of them are presented in this section.

The first example is a high-speed CMOS comparator [8]. The circuit is used in a CMOS A/D converter and its performance is a limiting factor for the performance of the overall A/D converter. The specifications imposed upon the circuit are a propagation delay of less than 5 ns and an offset voltage of less than 5 mV. The circuit schematic is shown in Fig. 6. To demonstrate the effectiveness of our direct performance-driven approach we have generated two placements for this circuit. Placement 1 (see Fig. 7) was generated with the presented performance-driven placement tool, while placement 2 was generated in the traditional way, with the same placement tool but with the performance-driven mechanism disabled. It can be seen from Table I that the simulated performance of placement 1 is significantly better than that of placement 2. Placement 1 has both performance characteristics within the user-specified ranges, while for placement 2 both specifications are violated. The optimized distances between the matching transistor pairs together with the resulting offset voltage degradation due to distance effects are shown in Table II for placement 1. The nominal values are the values obtained after sizing of the circuit without parasitic layout effects (no parasitic node capacitances and no mismatch). It can be seen that the performance-driven algorithm selectively minimizes the distances for the most sensitive transistor pairs, which results in a lower offset voltage. CPU times were 106 and 93 s for placement 1 and 2, respectively, on a SUN SPARC 10 workstation, which means that the performance-driven mechanism significantly improves the circuit performance at only a small increase in CPU time.

As a second example, a fully differential CMOS operational amplifier [9] (see Fig. 8) was used to test the efficiency of the algorithm for larger circuits. The placement of the opamp is shown in Fig. 9. Note the clear symmetry axis in this fully differential circuit. The circuit's specifications together with the obtained performance after placement are given in Table III. The degradation of all performances clearly remains within the specified margins. This placement required a CPU time of 163 s (less than 3 min) on a SUN SPARC 10 workstation.

The third example is a high-speed CMOS operational amplifier [10]. The schematic of the circuit is shown is Fig. 10. The placement that was generated for this circuit is shown in Fig. 11. The symmetry axis of the circuit is clearly visible in the placement. This circuit contains some very large transistors which have been split into parallel parts for reasons

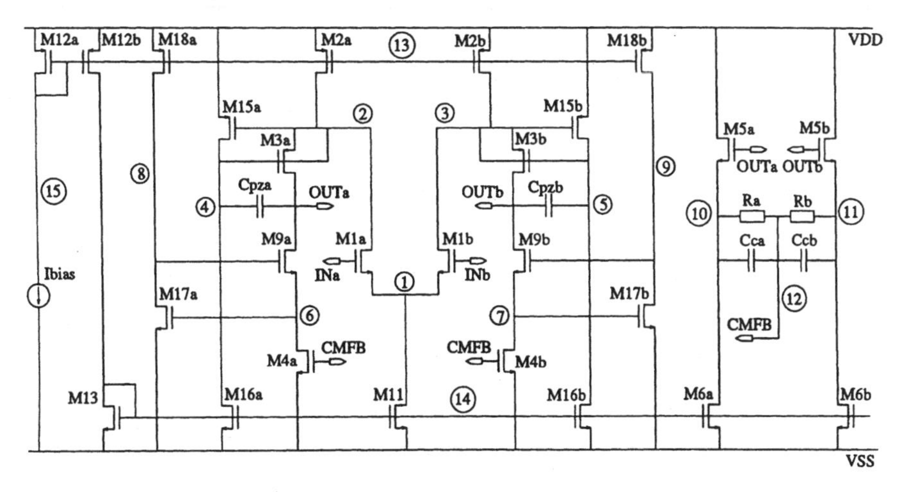

Fig. 8. CMOS fully differential opamp.

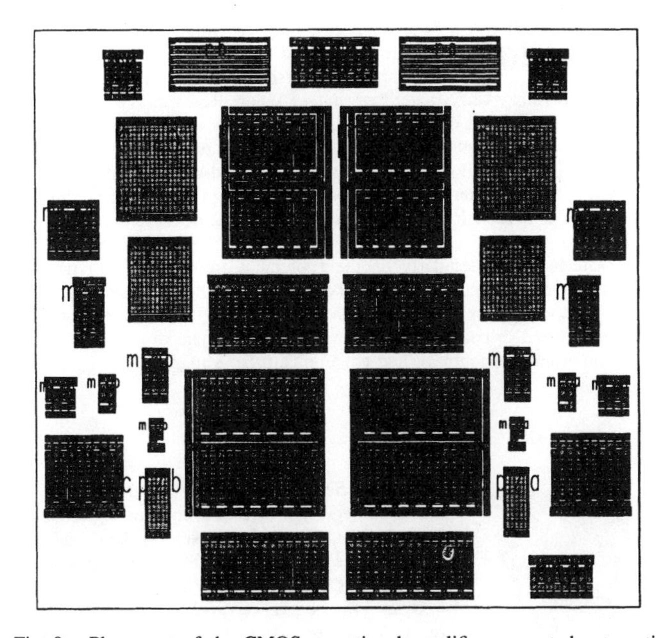

Fig. 9. Placement of the CMOS operational amplifier generated automatically by the performance-driven placement program.

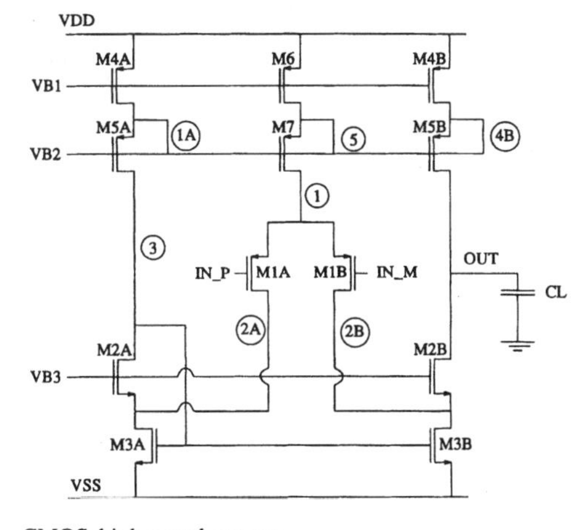

Fig. 10. CMOS high-speed opamp.

explained in Section IV. As shown in Table IV all performance characteristics for this circuit are within the specifications. All specifications were met in one pass of the program (CPU time 83 s).

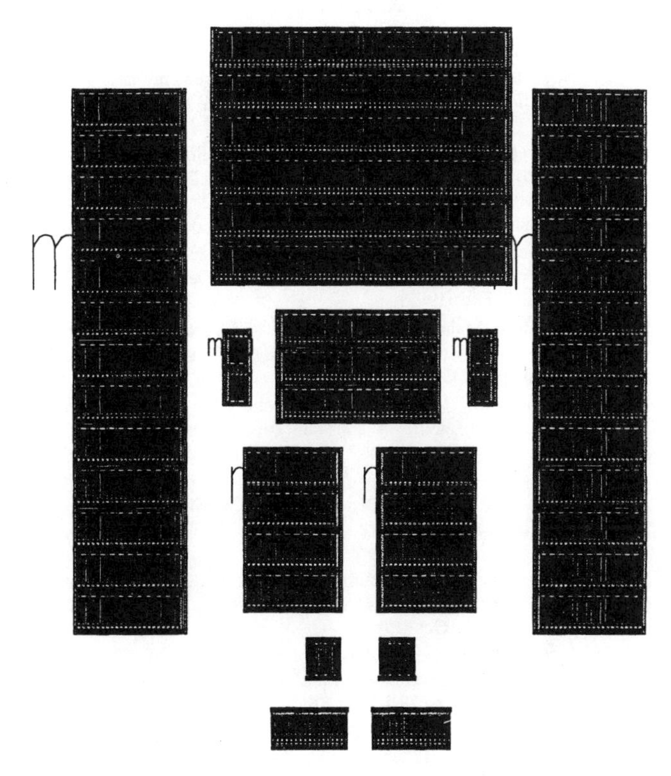

Fig. 11. Placement of the high-speed CMOS operational amplifier generated automatically by the performance-driven placement program.

TABLE IV
PERFORMANCE CHARACTERISTICS OF THE HIGH-SPEED CMOS OPAMP

opamp performance				
Performance	Specification	Nominal Value	After Placement	Unit
GBW	> 225	228	226	MHz
A_v	> 60	67	67	dB
PM	> 60	61	60.2	deg
$slew-rate$	> 150	163	163	$V/\mu sec$
$Voffset$	< 5	4.5	4.8	mV

VIII. CONCLUSIONS AND FUTURE WORK

A new direct performance-driven placement tool for analog integrated circuits has been presented. A simulated annealing algorithm is used to optimize a placement while keeping the layout-induced performance degradation within the margins imposed by the designer's specifications. The cost assigned to an intermediate placement is based on an accurate estimation of the performance degradation caused by interconnect parasitics and by device mismatches. This technique can easily be extended to other effects as well. Experimental results have demonstrated the feasibility and efficiency of our approach.

Future work includes the extension of the methodology to analog routing and the integration of performance-driven placement and routing for analog circuits, to avoid possible layout iterations due to the additional performance degradation caused by the actual wiring.

REFERENCES

[1] J. M. Cohn, R. A. Rutenbar, and L. R. Carley, "KOAN/ANAGRAM II: New tools for device-level analog placement and routing," *IEEE J. Solid-State Circuits,* vol 26, no. 3, pp. 330–342, Mar. 1991.

[2] V. K. Meyer zu Bexten, C. Moraga, and R. Klinke, "ALSYN: Flexible rule-based layout synthesis for analog IC's," *IEEE J. Solid-State Circuits,* vol. 28, no. 3, pp. 261–268, Mar. 1993.

[3] U. Choudhury and A. Sangiovanni-Vincentelli, "Automatic generation of parasitic constraints for performance-constrained physical design of analog circuits," *IEEE Trans. Computer-Aided Design,* vol. 12, no. 2, pp. 208–224, Feb. 1993.

[4] E. Charbon, E. Malavasi, and A. Sangiovanni-Vincentelli, "Generalized constraint generation for analog circuit design," in *Proc. IEEE ICCAD,* Nov. 1993, pp. 408–414.

[5] E. Charbon, E. Malavasi, U. Choudhury, A. Casotto, and A. Sangiovanni-Vincentelli, "A constraint-driven placement methodology for analog integrated circuits," in *Proc. IEEE CICC,* May 1992, pp. 28.2.1–28.2.4.

[6] S. Kirkpatrick, C. D. Gelatt, and M. P. Vecchi, "Optimization by simulated annealing," *Science,* vol. 220, no. 4598, pp. 671–680, May 1983.

[7] M. J. M. Pelgrom, A. C. J. Duinmaijer, and A. P. G. Welbers, "Matching properties of MOS transistors," *IEEE J. Solid-State Circuits,* vol. 24, no. 5, pp. 1433–1440, Oct. 1989.

[8] M. Steyaert, R. Roovers, and J. Craninckx, "A 110 MHz 8 bit CMOS interpolating A/D converter," in *Proc. IEEE Custom Integrated Circuits Conf.,* May 1993, pp. 28.1.1–28.1.4.

[9] E. Peeters and K. Ghafoor, "Design of a fully differential high-speed CMOS amplifier," Master's thesis, Katholieke Universiteit Leuven, Dept. Elektrotechniek, Heverlee, Belgium, June 1993.

[10] J. Fisher and R. Koch, "A highly linear CMOS buffer amplifier," *IEEE J. Solid-State Circuits,* vol. SC-22, no. 3, pp. 330–334, June 1987.

[11] G. Gielen *et al.,* "An analog module generator for mixed analog/digital ASIC design," to be published in *Int. J. Circuit Theory Applicat.,* Special Issue on Analog Tools for Circuit Design, 1995.

Koen Lampaert was born in Roeselare, Belgium, in 1968. He received the M.Sc. degree in electrical and mechanical engineering in 1992 from the Katholieke Universiteit Leuven, Belgium.

He is currently with the ESAT-MICAS laboratories of the Katholieke Universiteit Leuven as a Research Assistant. He is working towards the Ph.D. degree on automated analog layout. His research interests are in analog design automation.

Georges Gielen (M'92) received the M.Sc. and Ph.D. degrees in electrical engineering from the Katholieke Universiteit Leuven, Belgium, in 1986 and 1990, respectively.

In 1990, he was with the Department of EECS of the University of California at Berkeley as a Postdoctoral Research Assistant and Visiting Lecturer. From 1991 to 1993, he was a Postdoctoral Research Assistant of the Belgian National Fund of Scientific Research at the ESAT laboratory of the Katholieke Universiteit Leuven. In 1993, he was with the Katholieke Universiteit Leuven as a Tenure Research Associate of the Belgian National Fund of Scientific Research and as an Assistant Professor. His current research interests are in the design of analog and mixed-signal integrated circuits, and especially in analog and mixed-signal CAD (numerical and symbolic simulation, synthesis, layout, and design for manufacturability) and test. He is a Technical Coordinator of several industrial research projects in this area.

Dr. Gielen serves regularly on the Program Committee of international conferences and he is currently an Associate Editor of the IEEE TRANSACTIONS ON CIRCUITS AND SYSTEMS—PART I: FUNDEMENTAL THEORY AND APPLICATIONS. He has authored or coauthored one book and more than 50 papers in edited books, international journals, and conference proceedings.

Willy M. Sansen (S'66–M'72–SM'86) received the M.S. degree in electrical engineering from the Katholieke Universiteit Leuven in 1967 and the Ph.D. degree in electronics from the University of California at Berkeley in 1972.

Since 1981 he has been with the ESAT laboratory of KU Leuven as a Full Professor. Between 1984–1990, he was the Head of the Electrical Engineering Department. He was a Visiting Professor with Stanford University, Stanford, CA, in 1978, at the Federal Technical University Lausanne in 1983, at the University of Pennsylvania, Philadelphia in 1985, and at the Technical University Ulm in 1994. He has been involved in design automation and in numerous analogue integrated circuit designs for telecom, consumer electronics, medical applications, and sensors. He has been supervisor of 30 Ph.D. theses in that field.

Dr. Sansen has authored and coauthored more than 300 papers in international journals and conference proceedings and six books. He has also coauthored (with K. Laker) a textbook, *Design of Analog Integrated Circuits and Systems.*

Optimum CMOS Stack Generation with Analog Constraints

Enrico Malavasi and Davide Pandini

Abstract—**An algorithm for the automatic generation of full-stacked layouts in CMOS analog circuits is described in this paper. The set of stacks obtained is optimum with respect to a cost function which accounts for critical parasitics and device area minimization. Device interleaving and common-centroid patterns are automatically introduced when possible, and all symmetry and matching constraints are enforced. The algorithm is based on operations performed on a graph representation of circuit connectivity, exploiting the equivalence between stack generation and path partitioning in the circuit graph. Path partitioning is carried out in two phases: In the first phase, all paths are generated by a dynamic programming procedure. In the second phase, the optimum partition is selected by solving a clique problem. Original heuristics have been introduced, which preserve the optimality of the solution, while effectively improving the computational efficiency of the algorithm. The algorithm has been implemented in the "C" programming language. Many test cases have been run, and the quality of results is comparable to that of hand-made circuits. Results also demonstrate the effectiveness of the heuristics employed, even for relatively complex circuits.**

Index Terms—**CMOS, module generation, analog layout.**

I. Introduction

ANALOG AND MIXED analog/digital circuits find applications in several fields of current IC industry. Due to the tight performances needed, to the fast technology innovation rate and to strict time-to-market requirements, the design of analog parts in mixed systems has become an expensive bottleneck. So far, computer-aided design has helped reduce the development time of digital circuits. Analog circuits, on the contrary, still suffer from a lack of adequate tools.

The layout of analog circuits is intrinsically more difficult than the digital one. Matching and symmetries add constraints to the layout, and parasitics are sources of performance degradation. Different approaches have been investigated for the automation of IC physical implementation with analog constraints. Some methodologies are derived from the digital style, based on macro-cell [1] or standard-cell [2] paradigms, or on knowledge-based approaches [3], [4], [5], which are often critically dependent on the user's expertise. In ILAC [6], the layout is based on the generation of modules, tailored on specific library subcircuits generated by IDAC [7], a knowledge-based optimization system. Approaches more targeted toward analog design issues are the ones adopted in KOAN/ANAGRAM [8], which employs sophisticated layout algorithms for placement and routing, and in STAT [9], where a more relevant role is assigned to interactive options. In these tools, all issues related to device and interconnect matching, symmetries, and parasitic control are dealt with effectively. In the tools by UC-Berkeley [10], [11] emphasis is given to a constraint-driven approach consistent with a top-down design methodology [12]. In all existing approaches, placement and routing operate on modules which are either provided by the user, or automatically generated with fixed patterns. Transistor modules are often implemented according to the "full-stacked" design style [13]. In this style, the Source and Drain (S/D) regions of MOS transistors are systematically abutted to reduce area occupation and stray capacitances. The resulting "stacks" often offer convenient size and aspect ratios for macrocell-like placement. The full-stacked style yields compact layouts and good matching, and it is widely used in MOS analog applications. Elemental stacks are usually built by dedicated module-generators [14], on a device-by-device basis. Such modules are later arranged by the placement tools. This approach is unsatisfactory, since it yields layouts whose quality is not comparable to that of hand-made circuits. The major shortcomings are:

1) By operating with pre-constructed modules, common-centroid patterns and interleaved configurations cannot be built. These configurations are useful to achieve compact layouts characterized by very low dependence on technological gradients.
2) Junction capacitances are not taken into account. By properly abutting S/D regions, critical capacitances can be effectively reduced, with beneficial effects on performance. This is systematically done by skilled designers in hand-made layouts.

To our knowledge, no approaches have been proposed so far to automate stacked module generation for full-custom layout, overcoming the two limits outlined above. An interesting methodology has been adopted in KOAN [8] and PUPPY-A [10], where an automatic abutment procedure is used to merge the S/D regions of devices connected to the same net, so that area and junction capacitances are conveniently reduced. In both tools abutment is aimed at local area minimization, and in PUPPY-A parasitics are also considered, on the ground of their impact on performance degradation. The problem of optimal transistor chaining has been studied by Uehara and vanCleemput [15] for digital cells, and later extended by Wimer *et al.* [16] for series/parallel CMOS networks. More recently, an approach for non-series/parallel circuits has been

Manuscript received September 14, 1993; revised August 16, 1994. This work was supported in part by Philips Research, Eindhoven, NL. This paper was recommended by Associate Editor Satoshi Goto.

The authors are with the Dipartimento di Elettronica ed Informatica, University of Padova, Padova, Italy.

IEEE Log Number 9506359.

described by Zhang and Asada [17]. In these approaches a mono-dimensional stacked layout paradigm is considered. All circuit devices are aligned along two rows, one of N-channel transistors and one of P-channel transistors. In [18] it was pointed out how this layout pattern, although useful for digital circuits, is not flexible enough for analog circuits, where parasitic control is a major issue, and transistors are not paired as in standard CMOS logic cells. Matching and symmetry constraints are often present, and therefore a bidimensional layout pattern must be considered.

In this paper an algorithm for the automatic generation of full-stacked layouts in CMOS analog circuits is described. The set of stacks is optimum with respect to a cost function accounting for critical parasitics and area minimization. Device interleaving and common-centroid patterns are automatically found when possible, and all symmetry and matching constraints are always met. The algorithm stems from the approach described by Wimer *et al.* [16] for digital cells. Graph operations are performed on the *circuit graph*, which represents the connectivity of the circuit. The algorithm exploits the equivalence between stack generation and path partitioning in the circuit graph. Path partitioning is performed in two phases. In the first phase all paths are generated, while in the second phase the optimum set of paths is selected. A cost function defines the optimality criterion, according to parasitic criticality and matching constraints. In our approach, the original algorithm proposed in [16] has been extended to accommodate a bidimensional layout pattern. Transistor pairing constraints have been removed, and all matching and symmetry constraints are enforced. Tight matching requiring current direction control is enforced by means of *dummy devices* inserted into the stacks. The computational complexity of the algorithm has been drastically reduced in two ways:

- *Symmetries* are used to reduce the cost of the path set selection phase. Configurations violating symmetry constraints are discarded, thus reducing effectively the size of the path partitioning problem.
- A *branch-and-bound* heuristic has been introduced into the path set selection phase, based on the cost function. The configurations leading to costly solutions are discarded, while the cheapest solutions are preserved.

Symmetries and branch-and-bound have proved effective in reducing the complexity of the algorithm, as confirmed by experimental results. At the same time, they preserve the admissibility of the approach, i.e. the optimum solution is always found.

The aim of the algorithm is to create stacks containing transistor modules of the same size. Merging S/D regions of modules with different widths is not considered, but it is left to the abutment capability of the placement phase. Such feature is available in particular in PUPPY-A, the placement tool used in our framework, and employed to produce all the example layouts shown in this paper.

A detailed outline of the entire algorithm is presented in Section II. The cost function is described in Section III. Path generation and optimum path set selection are described in Sections IV and V respectively. Section VI focuses on matching enforcement. Experimental results are reported in Section VII, showing the suitability and computational efficiency of our algorithm for industrial-strength circuits. Conclusions and remarks on the future developments of this research are presented in Section VIII.

II. The Algorithm for Stack Generation

We start by giving a synopsis of the basic definitions which will be used throughout this paper. These definitions are included only for establishing the notation. Details and examples can be found in [19, ch. 5].

Consider an undirected graph $G(E, V)$, where E denotes the edge set, and V is the set of vertices.

Definition 1 (Paths): A *chain* or *path* p from vertex v_0 to vertex v_k in G is a sequence of vertices $v_0, v_1, v_2, \ldots, v_k \in V$ and edges $e_1, e_2, \ldots, e_k \in E$ such that e_i links v_{i-1} and v_i and $e_i \neq e_j, \forall i, j, i \neq j$. Vertices v_0 and v_k are called *endpoints* of p. The *length* of a path is the number of edges it contains. An *n-path* is a path whose length is n.

Definition 2 (Connected graphs): A connected graph is a graph $G(V, E)$ where for all $v, w \in V$ there is a path in G whose endpoints are v, w.

Definition 3 (Complete components): A complete component of G is a subgraph of G where all pairs of vertices are adjacent, i.e. they are linked by an edge. The size of a complete component is the number of its vertices. A complete component C is maximal when its size cannot be increased, i.e. when for each $v \in V$, either v is already in C, or v is not adjacent to some vertex of C. A maximal complete component is called clique.

Given a CMOS circuit $\mathcal{C}$, we create a *circuit graph* $G(E, V)$ where each vertex is a net in $\mathcal{C}$. For each MOS transistor in $\mathcal{C}$, an edge exists in E, linking the vertices associated to the nets connected to the source and drain terminals of the transistor. A stack of n transistors in $\mathcal{C}$ can be created if the corresponding edges in G form an n-path. The source and drain (S/D) regions corresponding to the path endpoints are said to be in *external* positions in the stack. The others are said to be in *internal* positions. Each full-stacked implementation of the layout corresponds to a *path partition* of the circuit graph, namely a set $\mathcal{P}$ of paths such that:

$$p \cap q = \phi \qquad \forall p, q \in \mathcal{P} \tag{1}$$

$$\bigcup_{p \in \mathcal{P}} p = E \tag{2}$$

where ϕ is the empty set and E is the set of all the edges in G. Operators $\bigcap$ and $\bigcup$ are applied to the edge sets of the paths. Therefore the intersection between paths means the set of those edges that belong to both paths at the same time, and the union represents the set of all the edges of the paths. Condition (1) is the *non-overlapping* condition that no two stacks in the layout contain the same transistor. Condition (2) is the *covering* condition, that each transistor must appear in a stack.

Notice that in each circuit at least one trivial partition $\mathcal{P}_0$ exists, where each path has exactly one edge. Such partition of 1-paths corresponds to separate elemental modules and it

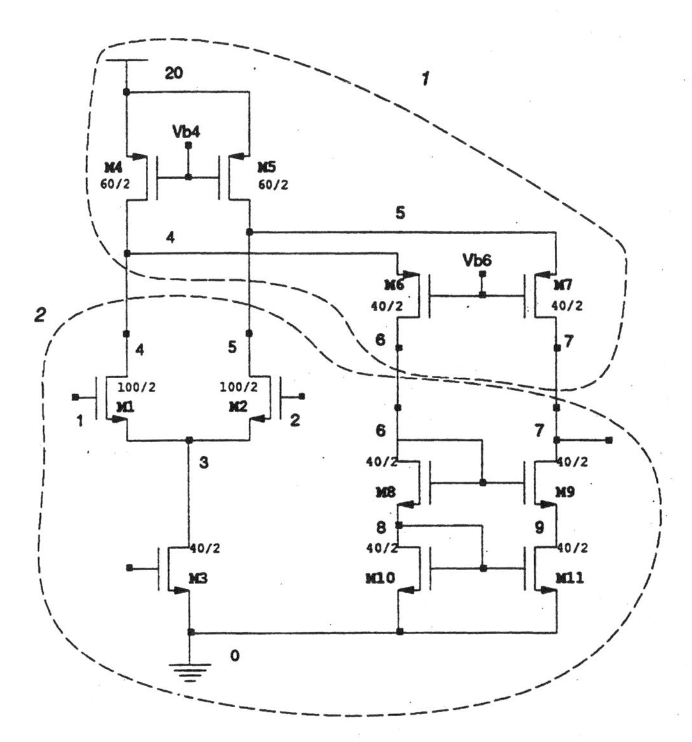

Fig. 1. OTA folded cascode.

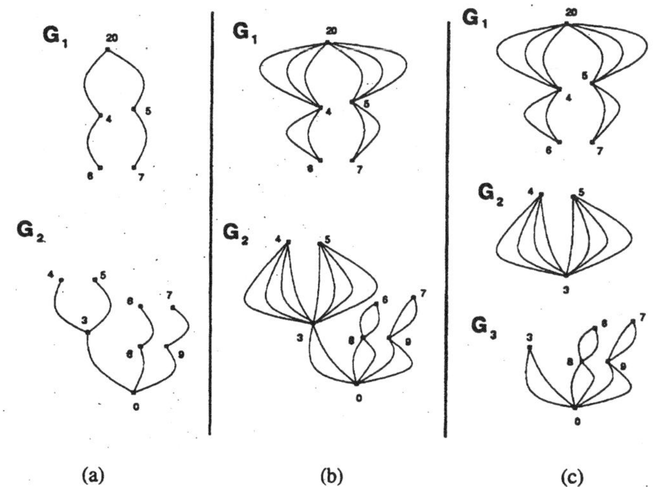

(a) (b) (c)

Fig. 2. OTA folded cascode. (a) The circuit subgraphs after step 1. (b) The circuit subgraphs after step 2. (c) The circuit subgraphs after step 3.

is often the starting configuration for placement tools with automatic abutment capability [8], [10].

Our stack generation algorithm is as follows.

The Stack-Generation Algorithm:

1) The circuit graph is split into two or more subgraphs $G_i, i = 1, \ldots, k$ in such a way that each subgraph contains only transistors with the same bulk bias net.
2) Large transistors are split into smaller modules connected in parallel.
3) Each subgraph is split into smaller connected subgraphs, containing only edges corresponding to modules with the same channel width.
4) Optimum path partition is carried out on each subgraph independently.

Step 1 yields separated subgraphs which can be processed independently from each other. In fact, transistors with different bulk nets cannot be contained in the same stack. The problem size is therefore significantly reduced, without modifying the set of solutions. As an example, consider the simple OTA folded cascode shown in Fig. 1. The dashed lines show the two parts in which the circuit is split, according to the bias net splitting operated in step 1 of the algorithm. After step 1, two subgraphs G_1, G_2 are created, as shown in Fig. 2(a).

In step 2, device splitting is carried out in two phases. In the first phase, the user's explicit requirements are enforced. In the second phase, matched transistors are split into modules with the same width, so that matching is improved by enforcing the same fringe effects on all modules. The new width is the greatest common divider (GCD) of the widths of all matched transistors. A lower bound applies, based on layout rules and user-specified minimum device width. Splitting is not applied if the GCD is smaller than this bound. After a transistor is split, the new modules become different matched devices, each introducing a different edge into the circuit graph. The new modules take the place of the split transistor. Step 2 is illustrated in Fig. 2(b). Here the user required M_1 and M_2 to be split into modules with channel width equal to 25 μm. The other devices were split into modules with channel width equal to 20 μm. Notice that the matched transistors whose width GCD is either lower than the bound or equal to their width are not split, and therefore they would not benefit from folding. Specifications about how these devices, as well as all the unmatched ones, must be split, are usually related to aspect-ratio considerations, and derive from floor-planning design, hence they could hardly be inferred from the input spice netlist. In our approach, we require these specifications to be explicitly provided by the user.

After step 3, each subgraph contains only transistors with the same channel width. Stacks containing modules with different widths can still be built with the automatic abutment procedure in the placement phase [8], [10]. The subgraphs obtained are independent from each other, with the exception of the ones containing devices related by symmetry and/or matching constraints. Such subgraphs cannot be processed independently, therefore they are merged together to form larger non-connected subgraphs. Let $G_1(E_1, V_1)$, $G_2(E_2, V_2)$ be two such interdependent subgraphs. They are merged into a non-connected subgraph $G'(E', V')$, where $E' = E_1 \bigcup E_2$, and $V' = V_1 \bigcup V_2$. As a consequence, at the end of step 3 each subgraph is not necessarily connected, but it is totally independent of the others. Hence, all the resulting subgraphs can be processed in parallel. In the example reported above, step 3 yields the three subgraphs G_1, G_2, G_3 shown in Fig. 2(c). In step 4, path partitioning is based on a two-phase algorithm, illustrated in Fig. 3.

First all the existing paths in the circuit graph are generated by a dynamic programming procedure (Fig. 3(a)). Then the problem of finding a path partition is transformed into a *clique* problem [20, p. 194]. Each path becomes a vertex for a *chain-graph* G_c, whose edges link two vertices if and only if the corresponding paths are *mutually compatible*, that is if they

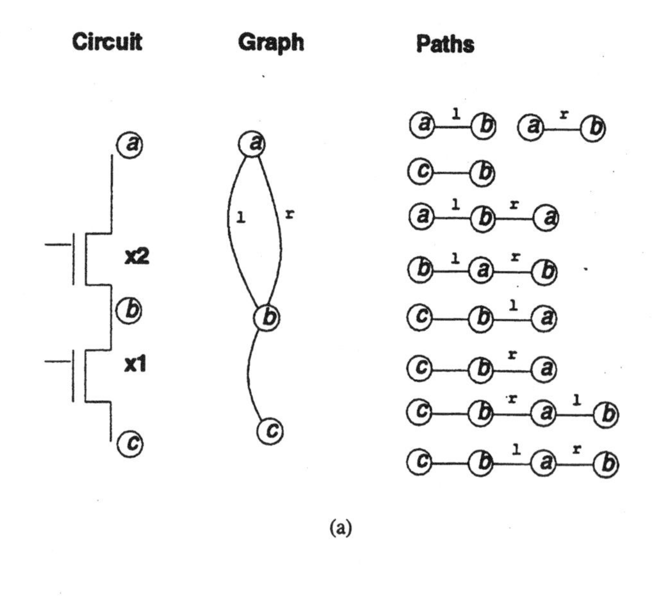

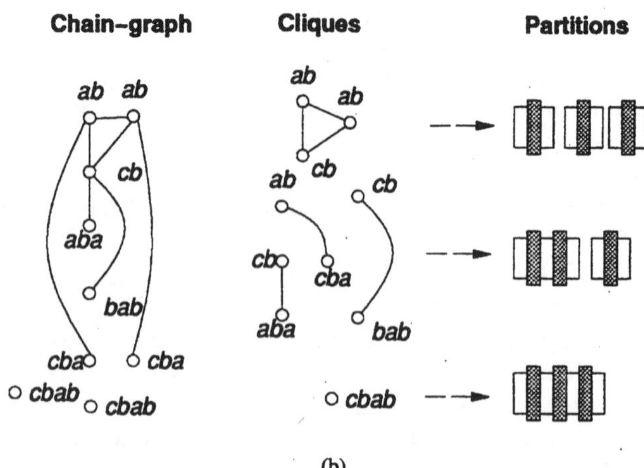

Fig. 3. (a) All the existing paths in the circuit graph are generated. (b) A clique problem is solved in the *chain-graph*.

can coexist in the same partition. Since the non-overlapping condition (1) is necessary for mutual compatibility, every partition is a complete component in G_c. In Section V we prove that such complete components are maximal. Therefore each partition is a clique in G_c. The cliques found are checked against the coverage condition (2) to determine whether they constitute a partition, otherwise they are discarded. A cost function evaluates the cost of each clique. The cheapest clique found is the optimum solution to the partitioning problem.

The approach followed in step 4 stems from the algorithm described in [16] for CMOS logic cells. The original algorithm cannot be applied to analog circuits. Because of the shortcomings discussed in Section I, this algorithm lacks flexibility to deal with analog constraints. Moreover, the original algorithm is computationally inefficient when applied to graphs where the number of edges is large compared to the number of vertices. In this case, the number of paths generated by the dynamic programming procedure grows almost factorially with the number of edges. The original algorithm has been modified to account for analog specifications and symmetries and to deal more efficiently with circuit graphs with a large

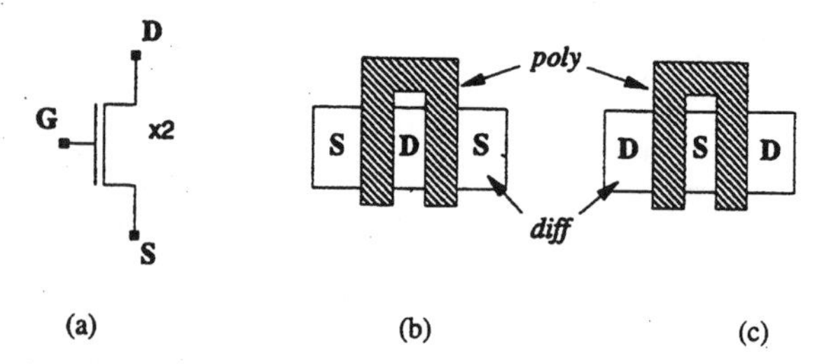

Fig. 4. (a) A transistor split in two modules. (b) Layout minimizing the capacitance of net D. (c) Layout minimizing the capacitance of net S.

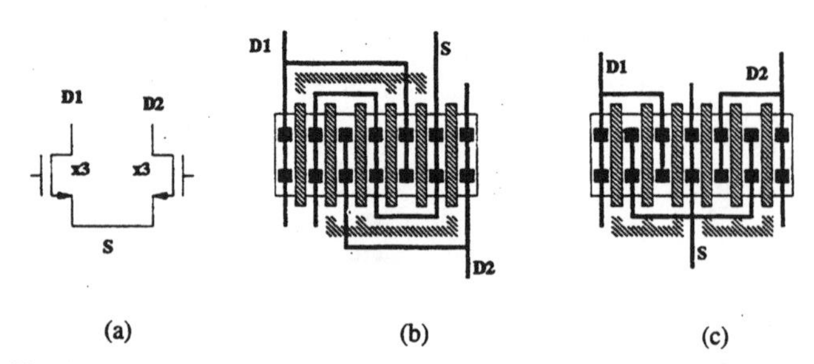

Fig. 5. (a) A differential pair with transistors split into 3 modules. (b) Implementation with common-centroid pattern. (c) Implementation minimizing routing area.

number of edges. A cost function has been introduced to choose among different solutions the ones minimizing critical parasitics.

III. The Cost Function

The quality of a stack arrangement is evaluated by a cost function that takes into account critical parasitics and area. For each module width W we denote by $C_{\text{ext}}(W)$ and $C_{\text{int}}(W)$ the values of the junction capacitances of S/D regions respectively in external and internal positions in the stack. In general, due to the side-wall junction capacitances associated to S/D regions, $C_{\text{ext}}(W) > C_{\text{int}}(W)$ for all values of W [21]. These capacitances are voltage-dependent, therefore upper bounds have been computed for them with a worst-case analysis.

Let n_j be a net, connected to M_j transistor modules. Let $C(n_j)$ be the total capacitance of net n_j towards the substrate. The value of $C(n_j)$ is maximum when all the S/D regions connected to n_j are in external positions. It is given by

$$C^{\max}(n_j) = \sum_{k=1}^{M_j} C_{\text{ext}}(W_k).$$

$C(n_j)$ is minimum when all the S/D regions connected to n_j are in internal positions. If all modules have the same width W:

$$C^{\min}(n_j) = \begin{cases} \dfrac{M_j}{2} C_{\text{int}}(W) & \text{if } M_j \text{ is even} \\ \dfrac{M_j - 1}{2} C_{\text{int}}(W) + C_{\text{ext}}(W) & \text{if } M_j \text{ is odd.} \end{cases} \tag{3}$$

Notice that nets connected to an odd number of transistor modules must have at least one S/D region in external position.

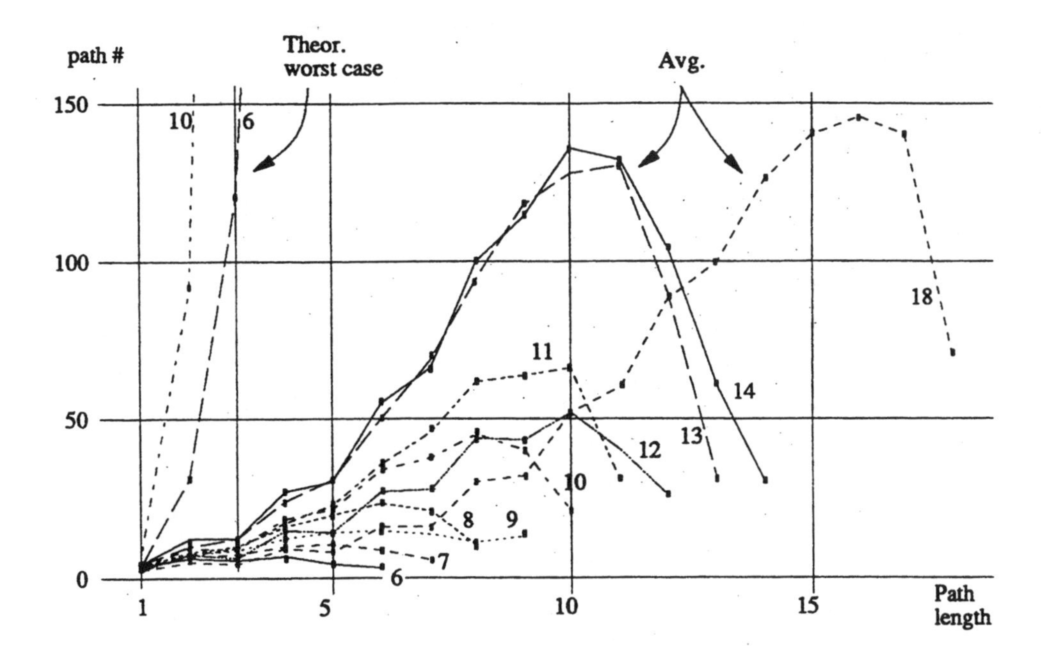

Fig. 6. Path lengths for different graph sizes. Numbers indicate the graph size. Lengths are average values drawn from more than 100 tests.

If the modules have different widths, $C^{\min}(n_j)$ is the sum of the capacitances computed with expression (3) for each width value. These limits are used by the constraint generator PARCAR [22] to generate bounds $C^{(b)}$ for all capacitances, in such a way that

$$C^{\min}(n_j) \leq C^{(b)}(n_j) \leq C^{\max}(n_j).$$

Bounds are defined in such a way that the circuit specifications are met, provided all parasitics are kept below their bounds. These bounds are used to define a set of *criticality weights* for stray capacitances:

$$w(n_j) = \left[\frac{C^{\max}(n_j) - C^{(b)}(n_j)}{C^{\max}(n_j) - C^{\min}(n_j)}\right]^2.$$

All criticality weights are between 0 and 1. Weights are close to 1 for capacitances with tight bounds ($C^{(b)} \approx C^{\min}$), they are close to 0 for capacitances with loose bounds ($C^{(b)} \approx C^{\max}$).

Let s_i be a stack containing M_i modules of width W_i. Its $(M_i + 1)$ S/D regions are connected to $(M_i + 1)$ nets n_j, $j = 0, \ldots, M_i$. Nets n_0 and n_{M_i} are in external positions. If $M_i > 1$, there are $(M_i - 1)$ other nets in internal positions. The cost $F(s_i)$ of stack s_i is

$$F(s_i) = \sum_{j=0}^{M_i} k_j(W_i) \cdot w(n_j) \tag{4}$$

where k_j are the *position weights*, defined as follows:

$$k_j(W_i) = \begin{cases} C_{\text{ext}}(W_i)/C_{\text{int}}(W_i) & \text{if } j = 0 \text{ or } j = M_i \\ 1 & \text{otherwise.} \end{cases}$$

The position weights account for the different junction capacitances due to net positions in the stack. The cost of a partition $\mathcal{P}$ is defined as the sum of the costs of all its stacks:

$$F(\mathcal{P}) = \sum_{s_i \in \mathcal{P}} F(s_i)$$

This formulation of the cost function implicitly accounts for area minimization. In fact, suppose the criticality weights were all approximately the same. Then the partitions made of long stacks would be cheaper than partitions made of shorter ones, because in the former case both the total number of S/D regions and the number of S/D regions in external positions would be smaller. In general, the cost function is cheaper for partitions made of long stacks, with most critical nets located in internal positions.

As an example, consider the 2-module transistor shown in Fig. 4(a) and its two implementations 4(b) and 4(c). Assume $C_{\text{ext}}/C_{\text{int}} = 1.6$, and let the criticality weights for nets S and D be respectively $w(S) = 0.8$ and $w(D) = 0.4$. The costs of the two implementations are:

$$F(s_i) = \begin{cases} 1.6 \cdot 0.8 + 1 \cdot 0.4 + 1.6 \cdot 0.8 = 2.96 & \text{(layout 4.b)} \\ 1.6 \cdot 0.4 + 1 \cdot 0.8 + 1.6 \cdot 0.4 = 2.08 & \text{(layout 4.c)} \end{cases}$$

Solution 4(c) is 30% cheaper than solution 4(b). Notice that the trivial partition made of two stacks of length 1 (separated elemental modules) would cost $2 \times 1.6 \times (0.8 + 0.4) = 3.84$, i.e. more than either 2-path solution.

The choice between partitions with the same cost is taken based on area or matching criteria. If area is a critical issue, routability is locally optimized by selecting the solution with minimum component spread. Otherwise, matching is optimized by solutions with maximum device interleaving. This strategy automatically generates common-centroid patterns when symmetry constraints are present. As an example, consider the differential pair shown in Fig. 5(a). Both solutions 5(b) and 5(c) are symmetrical. However, solution 5(b) optimizes matching with a common-centroid pattern, while solution 5(c) minimizes the amount of external routing between S/D regions and gates.

TABLE I
PATH LENGTHS FOR DIFFERENT GRAPH SIZES. FOR EACH GRAPH, THE NUMBER OF EDGES IS THE MAXIMUM THEORETICAL PATH LENGTH

graph size	6		7		8		9		10	
no. of tests	39		17		20		16		10	
path length	avg.	max	avg.	max	avg.	max	avg.	max	avg.	max
1	3.28	6	3.24	5	3.75	4	3.94	5	3.40	6
2	6.00	9	6.36	10	8.10	11	7.31	13	9.10	15
3	5.38	14	7.41	20	9.85	17	8.31	15	9.20	20
4	6.53	20	9.76	13	16.20	33	12.43	24	18.70	47
5	4.23	18	10.23	18	19.65	52	13.94	26	21.60	78
6	3.02	10	8.24	14	23.35	71	14.81	32	33.70	144
7	–	–	5.18	14	20.65	75	13.56	34	37.60	210
8	–	–	–	–	9.95	30	11.19	38	44.90	272
9	–	–	–	–	–	–	12.75	18	40.40	300
10	–	–	–	–	–	–	–	–	21.20	150

The subgraphs, generated by splitting the circuit graph in steps 1 to 3 of the stack-generation algorithm, should not be considered totally independent as assumed above. In fact, global routability considerations should be taken into account when choosing among solutions with the same cost. However, global routability can be evaluated only after placement. Moreover, area is taken into account only indirectly, by maximizing abutment between device S/D regions, but the area required by interconnects cannot be estimated at this stage. Therefore the simple cost function (4) adopted is not fully adequate to evaluate the quality of the final layout. A more sophisticated cost function could be used, and the concept of optimality defined in this section (minimizing the cost function) would still hold, as long as we can determine a suitable optimistic estimate to preserve admissibility with branch-and-bound. Taking care of global routability, however, would make it hard to determine such an optimistic estimate, accurate enough to keep low the computational cost of the algorithm. Hence, we perform only a local routability optimization, limited to the interconnections linking different S/D regions and/or gates of the same stack. Notice that the complete set of all optimal solutions (optimality being defined with respect to the given cost function) is *always* found by our algorithm. Therefore, strategies can be applied *after* path partitioning, during the placement phase, to choose among all solutions on the ground of area, global routability and aspect-ratio constraints [23].

IV. THE PATH-FINDING ALGORITHM

All the existing paths in the graph are found by a dynamic programming procedure. First all 1-paths are generated. At the i-th step, for $i > 1$, all i-paths are generated by augmentation of the paths of length $(i-1)$. Augmentation is carried out by adding to each $(i-1)$-path one of the edges connected to its endpoints. In the worst case, the number of paths generated at each step grows factorially with their length. In fact, the maximum number of paths of length k is the number of k-permutations on the edge set:

$$\frac{|E|!}{(|E|-k)!}$$

where $|E|$ is the number of edges in the graph. However, the average case is much less costly than factorial. Experimental data drawn from more than 100 circuit graphs built on opamps, OTA's and comparators, show the behavior summarized in Table I. The number of paths generated at each step does not grow factorially with the path length. It grows smoothly for short paths, then it decreases for the longest ones. This behavior is graphically illustrated in Fig. 6 for the average path length. A similar behavior has been observed for the maximum path length. The sharp discrepancy between the worst theoretical case and the average case can be explained with the following observations:

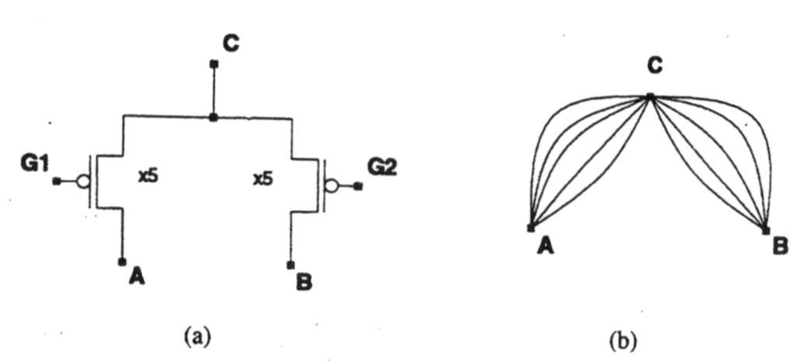

Fig. 7. (a) A differential pair with transistors split into 5 modules each. (b) Its circuit graph.

1) The worst case is achieved only with complete graphs. Circuit graphs, at least the ones obtained from useful circuits, are generally sparse. With the possible exception of supply and ground lines, each vertex is usually connected only with a few other vertices. As an example, consider the simple case of the differential pair shown in Fig. 7. Although it cannot be considered a sparse graph, yet it is sparse enough to contain considerably fewer paths than the worst case. Each transistor is split into five elemental modules, and the circuit graph has 3

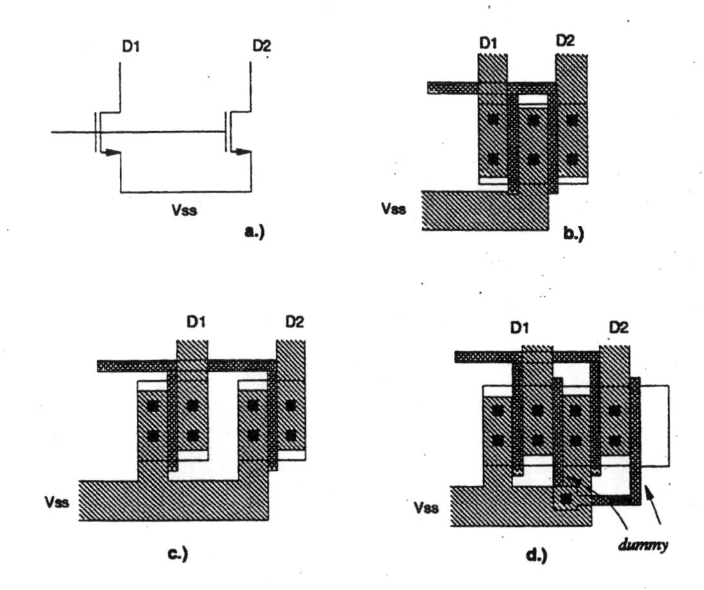

Fig. 8. (a) Matched transistors generating nominally equal currents. (b) Layout implementation minimizing area. (c) Layout accounting for channel orientation matching. (d) Alternative layout optimized for maximum matching, using dummy transistors.

```
Recursive_Dummy_Generator( N_d )
/* N_d = number of dummy transistors inserted so far (initially 0) */
IF F = 0 THEN return F;
LOOP on all available stack positions;
    insert a dummy transistor in the next available stack position;
    SET f = Recursive_Dummy_Generator( N_d + 1 );
    IF ( f < f_min ) THEN
        SET f_min = f
        save all dummy positions;
    end IF;
end LOOP;
return ( f_min );
```

Fig. 9. Recursive procedure for dummy generation.

vertices and 10 edges. Considering the paths of length 2, the max theoretical number of paths is 90. However, the actual number of 2-paths is 65. Similarly, the worst-case number of 3-paths is 720, while only 140 exist in the graph. The total number of paths is about 65 000, much fewer than $\sum_{k=1}^{10} \frac{10!}{(10-k)!} = 6,235,300$.

2) The edges generated from the same transistor by module splitting are mutually interchangeable. Two paths that can be obtained from each other by exchanging one or more mutually interchangeable edges in the same position are not distinct. Paths containing the same set of edges in different orders are distinct, because their costs can be different, due to the parasitics associated with S/D regions. In the example of Fig. 7, the number of distinct 2-paths is only 8. The number of distinct 3-paths is only 4. Of the about 65 000 paths in the graph, only 84 are distinct, all the others are discarded.

V. The Clique-Finding Algorithm

All circuit partitions are found by solving a clique problem on the chain-graph G_c. We have exploited the existence of interchangeable edges to reduce the computational complexity. As a consequence, the path-finding algorithm does not generate all possible paths, but only a set of distinct paths. However, the equivalence between circuit partitions and cliques in G_c still holds, provided the mutual compatibility of paths is defined correctly.

Definition 4 (Mutual Compatibility): Two paths are mutually compatible, if they have no common edges.

In the chain-graph, two vertices are linked by an edge if and only if their corresponding paths in the circuit graph are mutually compatible. Notice that paths containing equivalent edges are mutually compatible, provided the edges are not the same. Now we can prove the following

Theorem 1: If the chain-graph G_c is built using Definition 4, then each partition that can be built with the paths associated to the vertices of G_c, and satisfying conditions (1) and (2), is a clique in G_c.

Proof: A partition is a subset $\mathcal{P}$ of paths satisfying the non-overlapping (1) and the covering (2) conditions. By the nonoverlapping condition, the subgraph in G_c corresponding to such a subset must be a complete component. If it were not maximal, $\mathcal{P}$ would not satisfy the covering condition. In fact, suppose that $\mathcal{P}$ is not maximal, i.e., there exists a path p such that $p \notin \mathcal{P}$, and $\mathcal{P} \bigcup \{p\}$ is still a complete component. Since p must be compatible with all the paths in $\mathcal{P}$, each edge of G contained in p is not contained in any of the paths in $\mathcal{P}$. Therefore $\mathcal{P}$ does not satisfy the covering condition. □

Not every clique in G_c is necessarily a partition. In fact, while all cliques satisfy the nonoverlapping condition, some might contain less than $|E|$ edges, and thus violate the covering condition. Furthermore, symmetry constraints must be checked in all solutions. The cliques violating the covering condition or symmetry are discarded.

Notice that by discarding equivalent paths some of the partitions are lost. For instance, consider the circuit shown in Fig. 3. Only one of the two 1-paths (a)–(b) is kept, the other is discarded. As a consequence, the trivial partition {(a)–(b), (a)–(b), (c)–(b)} is lost. However, when equivalent paths are simplified, an alternative solution is always present, given by longer paths including all the discarded path segments. These longer paths constitute in general a better solution than the lost one. For instance, paths (a)–(b)–(a) and (b)–(a)–(b) can be used to make partitions with two paths, both less costly than the lost trivial partition. The existence of such alternative solutions is guaranteed by the fact that two equivalent paths have the same endpoints. Therefore a longer path always exists, that links them by one of the endpoints.

To solve the clique problem, we apply the Bron-Kerbosch algorithm [24]. This is an iterative algorithm, where at each step an attempt is made to augment a complete component, i.e. to find a vertex which does not belong to that component, but is adjacent to all its vertices. If such attempt fails, the complete component is maximal, and therefore by definition 3 it is a clique. Branch-and-bound has been applied to the complete-

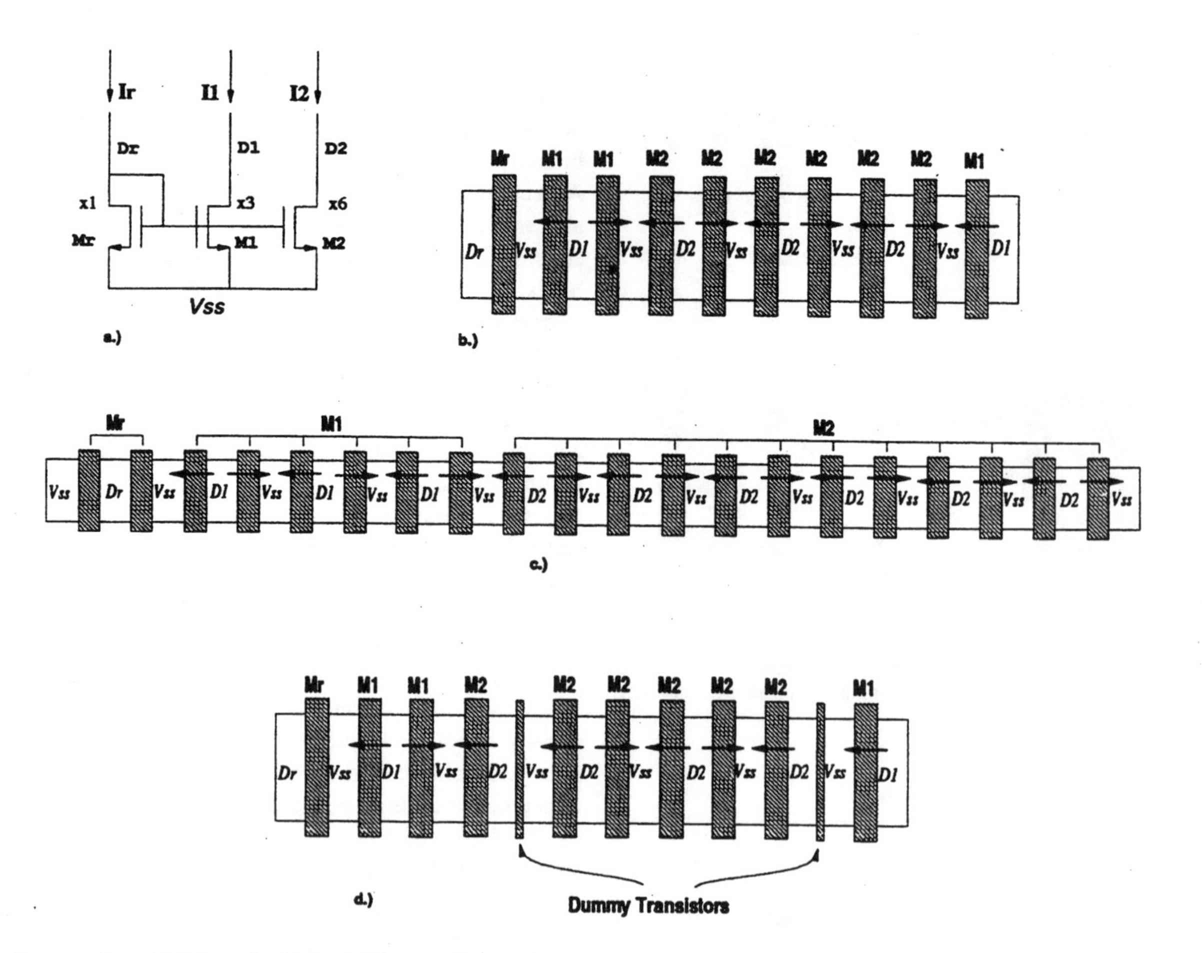

Fig. 10. Current mirror. (a) Schematic. (b) Stacked layout without tight matching constraints. (c) Stacked layout where each transistor is split into an even number of modules. (d) Stacked layout with dummy transistors to enforce same current direction in all modules.

component augmenting step. Let F be the cost of the cheapest clique yielded by augmentation of a complete component of G_c. If an estimate $\tilde{F}$ of F can be provided, it is possible to decide whether augmentation is worthwhile. If $\tilde{F}$ is larger than the cost of the cheapest clique found so far, augmentation is not carried out and the component is discarded. If $\tilde{F}$ is an optimistic estimate, i.e., it is $\tilde{F} \leq F$, then *admissibility* is preserved, that is the cheapest clique in G_c is always found.

Let $\mathcal{P} = \{s_i\}$ be the set of paths corresponding to the vertices of a complete component of G_c. The estimate has the following expression:

$$\tilde{F} = \sum_{s_i \in \mathcal{P}} F(s_i) + \tilde{h} \tag{5}$$

where $F(s_i)$ is the cost of path s_i given by expression (4) and $\tilde{h}$ is an estimate of the cheapest combination of paths needed to complete a clique. Let M be the total number of edges contained in the paths of the complete component. By the covering condition (2), exactly $(|E|-M)$ edges are needed to augment the complete component to a clique satisfying the covering condition. For sake of simplicity suppose that all the missing edges correspond to modules with the same width W. The cheapest arrangement for them is one path with $(|E|-M-1)$ nets in internal positions and 2 nets in external positions. The expression of $\tilde{h}$ is the following:

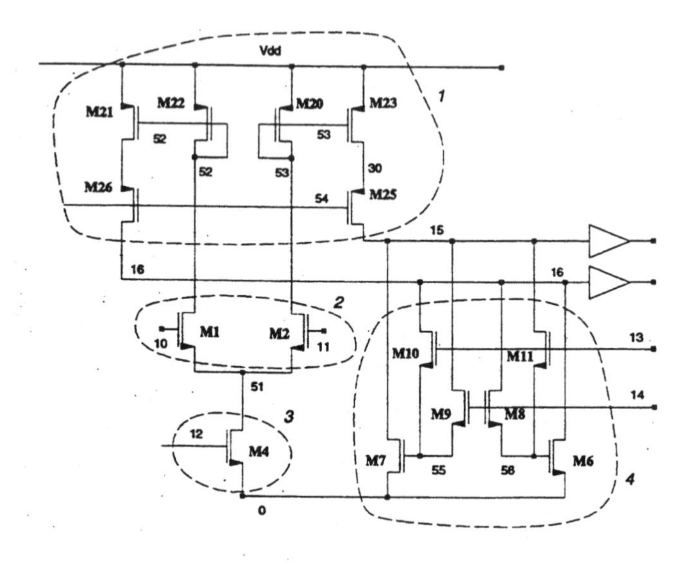

Fig. 11. Clocked comparator CMP2.

$$\tilde{h} = \sum_{j=1}^{(|E|-M+1)} \tilde{k}_j \cdot w(n_j).$$

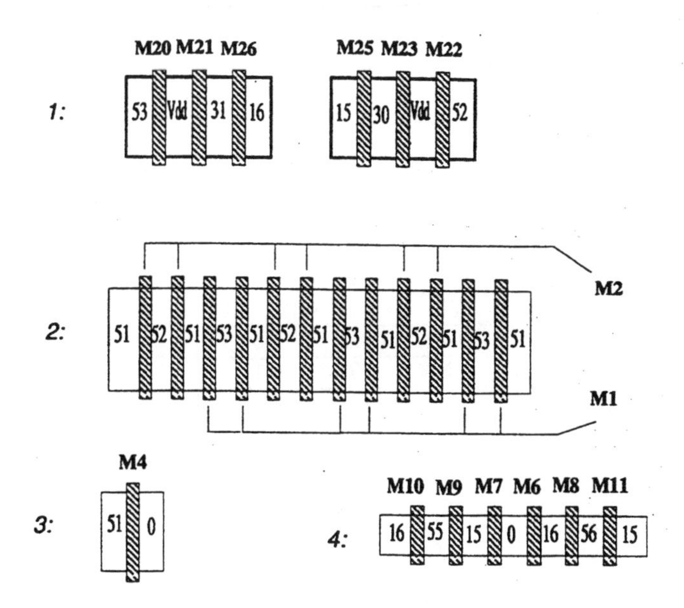

Fig. 12. Clocked comparator CMP2—optimum stack partition.

TABLE II
JUNCTION CAPACITANCE WITHOUT AND WITH ABUTMENT FOR SOME CRITICAL NETS, AND CORRESPONDING EQUIVALENT INTERCONNECT LENGTH

circuit	net(s)	caps (fF) without abutment	caps (fF) with abutment	equiv. length (μm)
CMP2	15,16	41.2	34.4	136
	55,56	13.4	6.6	136
OPAMP2	1	629	310.4	6372
	2,3	240	137.6	2048
	5	80	28.8	1024
CMP3	3,4	106	53	1060
	5,6	124	45.6	1568
	7,8	287	170.4	2332
	9,10	82.5	56.9	512
AB1	6,7,8,9	59.2	24	704
	10,11	179.6	103.4	1524
OPAMP5	4,9	264.5	185.7	1576
	6,13	215.9	171.5	888

where $\tilde{k}_j$ is an optimistic estimate of k_j. Let M_j be the number of modules connected to net n_j. The value of $\tilde{k}_j$ is $M_j/2$ if M_j is even, it is $(M_j - 1)/2 + C_{\text{ext}}(W)/C_{\text{int}}(W)$ if M_j is odd. At this stage, no information is available about net positions in the stack. Therefore, expression (5) corresponds to the path cost definition (4), with all position weights set to 1, except the ones corresponding to nets connected to an odd number of modules (in fact such nets must be connected to at least one S/D region in external position). By construction this estimate is optimistic and the branch-and-bound heuristic satisfies the admissibility condition. The generalization to the case where there are modules with different channel widths is straightforward. The cheapest implementation is a layout with one stack for each value of width and $\tilde{h}$ is the sum of the costs of such stacks, each computed using expression (5).

A further reduction in the CPU time required by the clique solver is achieved with a heuristic exploiting the structure of the cost function. Even when the capacitances involved are not critical, longer paths are always cheaper than the shorter ones, because of the area saving provided by abutments. The clique solver is preceded by a check on the existence of maximum-length paths, i.e. paths containing all the edges of the subgraph. If such a path exists and if it is consistent with matching and symmetry requirements, we can discard all shorter paths. In fact each partition containing two or more paths is going to be more costly than a solution containing one maximum path. In this case the clique problem is reduced to choosing the cheapest maximum paths. The entire path partitioning is reduced to the path-finding algorithm, and the clique solver is not even run.

Long paths, corresponding to long stacks, are not always desirable, due to aspect-ratio limitations stemming from floorplanning considerations. However, as pointed out in Section II, circuit geometry information is not available in the SPICE netlist (the only input available to our tool), hence this information must be explicitly provided by the user. Constraints on circuit size and aspect ratio determine the maximum length a stack can achieve. If such constraints are specified, the dynamic programming procedure generating all paths will stop as soon as the max bound is reached. Obviously, in this case the heuristic described above, based on the check on maximum-length paths, does not apply.

VI. MATCHING CONSTRAINTS

Device mismatch can be minimized by means of a careful layout of the transistors. For instance, common-centroid configurations are effective in reducing the mismatch due to parameters varying linearly across the chip surface. As explained in Section III, it is possible to obtain stacks where transistor module interleaving is maximum, and common centroid is automatically found when possible. At the same time, by setting a bound on maximum stack length, the distance between matched devices can be kept conveniently short. According to [25], the mismatch between MOS transistors, under certain assumptions, is inversely proportional to device channel area ($W \cdot L$), and directly proportional to the spacing between them. The mismatch between transistors is also dependent on their relative channel orientation, as shown in Fig. 8. Due to the non-orthogonal ion implantation angle on silicon surface and to silicon crystal anisotropy, channel orientation affects the current factor of mismatch [25]. The effective importance of mismatch depends on the performance degradation it can induce. Given a set of performance constraints, the bounds on maximum distance and relative orientation between matched devices can be automatically derived from a sensitivity analysis of the circuit [26].

Consider two MOS transistors Mi and Mj, respectively split into n_i and n_j modules, all in the same stack. Let $n_i^r(n_j^r)$ be the number of modules of $Mi(Mj)$ oriented in the right direction and $n_i^l(n_j^l)$ the number of modules oriented in the left direction. Obviously $n_i = n_i^r + n_i^l$ and $n_j = n_j^r + n_j^l$. We will denote by Δn_i and Δn_j the differences $n_i^r - n_i^l$ and $n_j^r - n_j^l$

TABLE III

RESULTS WITH SOME TEST CIRCUIT EXAMPLES OF OPAMPS, OTAS, AND COMPARATORS. THE NUMBER OF OPTIMUM SOLUTIONS STEMS FOR THE COMBINATION OF THE ALTERNATIVES FOR EACH SUB-GRAPH

Circuit name	No. of modules	No. of sub-graphs	No. of opt. solutions		Max G_c size vertices	Max G_c size edges	Time (sec.)
AB1	29	6	2	(2×1^5)	61	168	0.1
OPAMP1	56	8	384	$(16 \times 6 \times 2^2 \times 1^4)$	990	85	5.9
OPAMP2	32	5	112	$(7 \times 4 \times 2^2 \times 1)$	66	1228	3.0
OPAMP3	30	3	32	$(4^2 \times 2)$	200	1400	0.8
OPAMP4	30	2	8	(8×4)	1756	43789	3412 (11.0)
OPAMP5	40	9	288	$(36 \times 8 \times 1^7)$	218	1080	1.5
CMP1	25	5	4	(4×1^4)	21	90	0.1
CMP2	25	4	8	(4×1^2)	21	90	0.1
CMP3	29	4	320	$(20 \times 4^2 \times 1)$	194	830	7.5

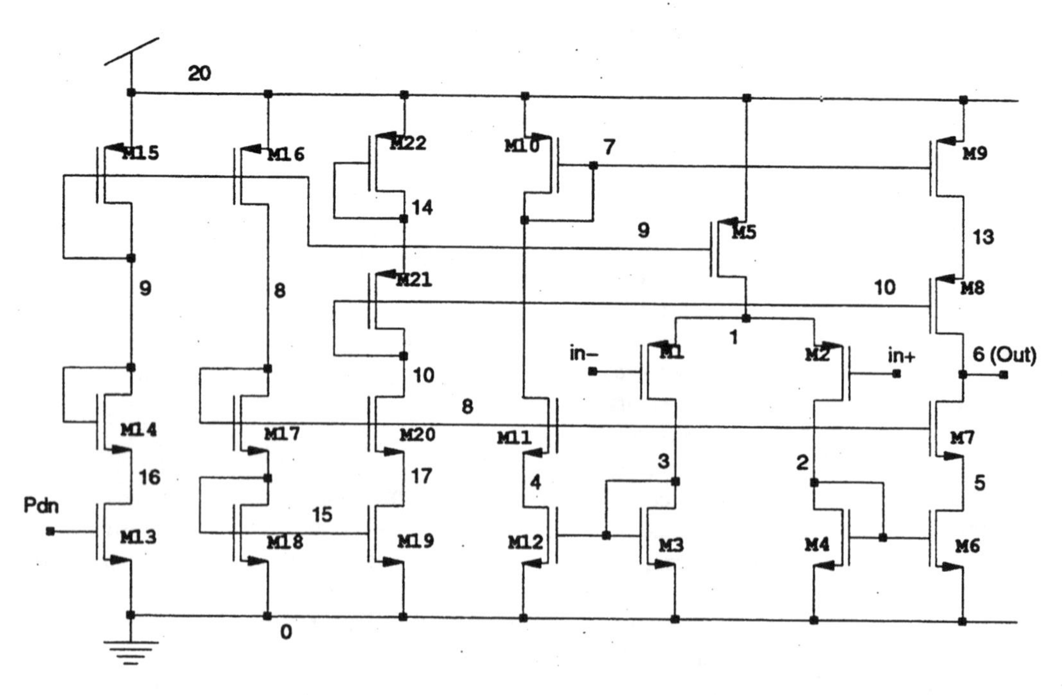

Fig. 13. Schematic of OPAMP2.

respectively. We will assume that Mi and Mj operate in their saturation region. With the same nominal voltages, all modules carry the same nominal current I. However, currents I^r, I^l flowing through channels with opposite orientations differ by a max error

$$\epsilon_I = |I^r - I^l|.$$

The value of ϵ_I for a given process depends on the module size and can be determined experimentally [27]. The *current offset* of transistor Mi is defined as follows:

$$\Delta I_i = (\Delta n_i)\epsilon_I$$

and similarly the current offset of Mj is

$$\Delta I_j = (\Delta n_j)\epsilon_I.$$

The *current mismatch* $\mathcal{F}_{ij}$ between transistors Mi and Mj is defined as:

$$\mathcal{F}_{ij} \triangleq \left|\frac{\Delta I_i}{n_i I} - \frac{\Delta I_j}{n_j I}\right| = \frac{\epsilon_I}{I}\left|\frac{\Delta n_i}{n_i} - \frac{\Delta n_j}{n_j}\right|.$$

For each stack a cost function is defined as follows:

$$\mathcal{F} = \max_{i,j} \mathcal{F}_{ij} \quad \forall i, j$$

and all matching constraints are satisfied when condition

$$\mathcal{F} = 0 \tag{6}$$

holds.

The mismatch due to channel orientation can always be avoided by splitting each transistor into an even number of elemental modules. Such modules can be oriented half in one direction and half in the opposite direction. However, this solution often leads to small elemental modules. As a consequence, mismatch problems due to small channel area arise. An alternative solution is to assign all modules the same orientation. However, often this solution over-constrains the layout, it reduces the possibility of interleaving matched transistors and it requires excessive area for module placement and routing.

Tight matching can be enforced with an alternative strategy, based on a recursive procedure reducing the cost of function $\mathcal{F}$ by inserting "dummy" transistors in the stack, kept permanently OFF by proper gate polarization. The advantage deriving from this strategy is the reduction of sidewall capacitances in S/D regions, and a reduction of area, due to the fact that the minimum channel length is usually shorter than the minimum distance between S/D regions.

A. Dummy Modules Strategy

By inserting a dummy transistor, a stack is split in two disconnected *sub-stacks*, which are free to rotate independently by 180 degrees. The recursive dummy generation function is shown in Fig. 9. The solution found by the recursive procedure satisfies condition (6) with the minimum number N_d of dummy elements. At the first step of the recursion, the value of N_d is set at 0. When the optimum solution is found, the stack is split into $N_d + 1$ smaller stacks, which are rotated and abutted to N_d dummy modules. In a stack with N_s modules, the maximum number of dummy transistors which can be inserted is $(N_s - 1)$. This is the trivial solution, where all transistor modules are oriented in the same direction. Hence, the recursive procedure always terminates in no more than N_s steps.

Additional dummy modules can be added to the ends of the stack to reduce the capacitance of the external S/D regions. Moreover, such dummy devices provide the external modules of the stack with the same fringe parameters as all internal modules. Consequently, they further improve matching among them.

As an example, consider the current mirror shown in Fig. 10(a). The mirror generates two currents $I_1 = 3I_r$ and $I_2 = 6I_r$. In order to achieve a good matching between transistors M_1 and M_2, they are split into 3 and 6 modules, respectively. All modules have the same channel width as the reference transistor M_r, and they are placed in one stack with M_r as shown in Fig. 10(b). With this configuration, $n_1 = 3$, $\Delta n_1 = -1$, $n_2 = 6$, $\Delta n_2 = 0$. The current mismatch between M_1 and M_2 is

$$\mathcal{F}_{1,2} = \frac{\epsilon_I}{I}\left|\frac{-1}{3} - \frac{0}{6}\right| = \frac{1}{3}\frac{\epsilon_I}{I} \neq 0$$

By splitting all transistors into an even number of modules, the stack shown in Fig. 10(c) is obtained. Half the modules are oriented in one direction and half in the opposite direction, and $\mathcal{F}_{1,2} = 0$. However, all transistors have smaller channel area and there is considerable distance between modules at opposite ends of the stack. An alternative solution, shown in Fig. 10(d), is obtained by inserting two dummy elements. Now $\Delta n_1 = -1$ and $\Delta n_2 = -2$, therefore $\mathcal{F}_{1,2} = 0$. Notice that with this solution the channel area of each module is twice that of the case of Fig. 10(c). Therefore the mismatch contribution due to fringe effects on channel gates is less relevant. The stack is only slightly longer than the original one and therefore its aspect ratio is also much more convenient.

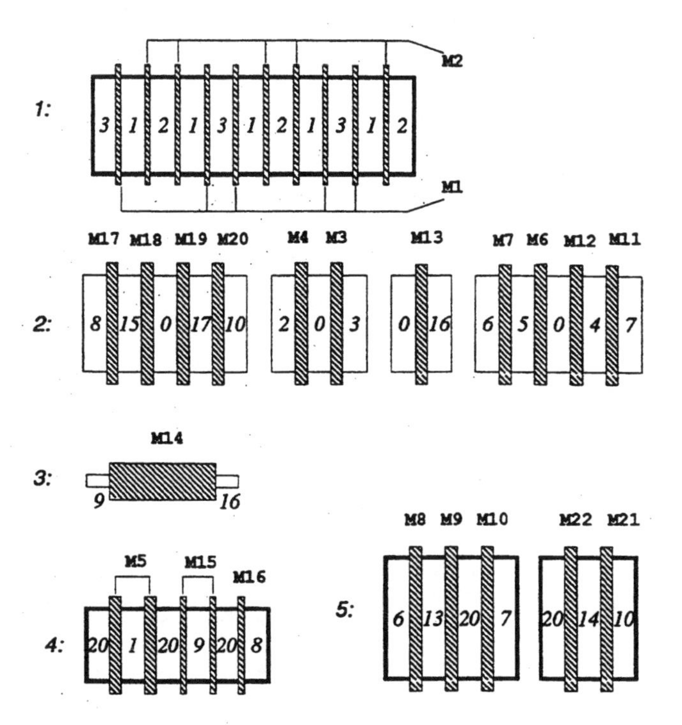

Fig. 14 Operational amplifier OPAMP2—optimum stack partition.

B. Same-Stack Requirement

Given a group of tightly matched devices, an optional requirement is that the entire group be contained in the same stack. As a consequence, each path containing some but not all the devices of the group is discarded. The clique-finding algorithm operates on much fewer paths and therefore it becomes considerably faster. However, this requirement prevents some solutions from being found, and the optimality of the result is not guaranteed. This option has been introduced because in some cases the user expects a particular configuration. An example is a differential pair with large devices, to be interleaved in one stack with a common-centroid pattern. In such cases, a hint on the layout configuration can provide the algorithm with a significant speed-up (see the example OPAMP4 in Table III), while still finding the optimum solution.

VII. Results

The algorithm for optimum stack partitioning has been implemented in a program of about 7000 lines in the "C" language. The program, running in the UNIX environment, has been tested on numerous circuits of industrial strength. All the examples have been compared to hand-made layouts made by experienced designers. Invariably, the hand-made solution was one of the optimum partitions found by the algorithm. When two or more equivalent solutions were found, they were all of comparable objective quality in terms of area, parasitics, and matching with the original one. Some of those examples are reported in this section.

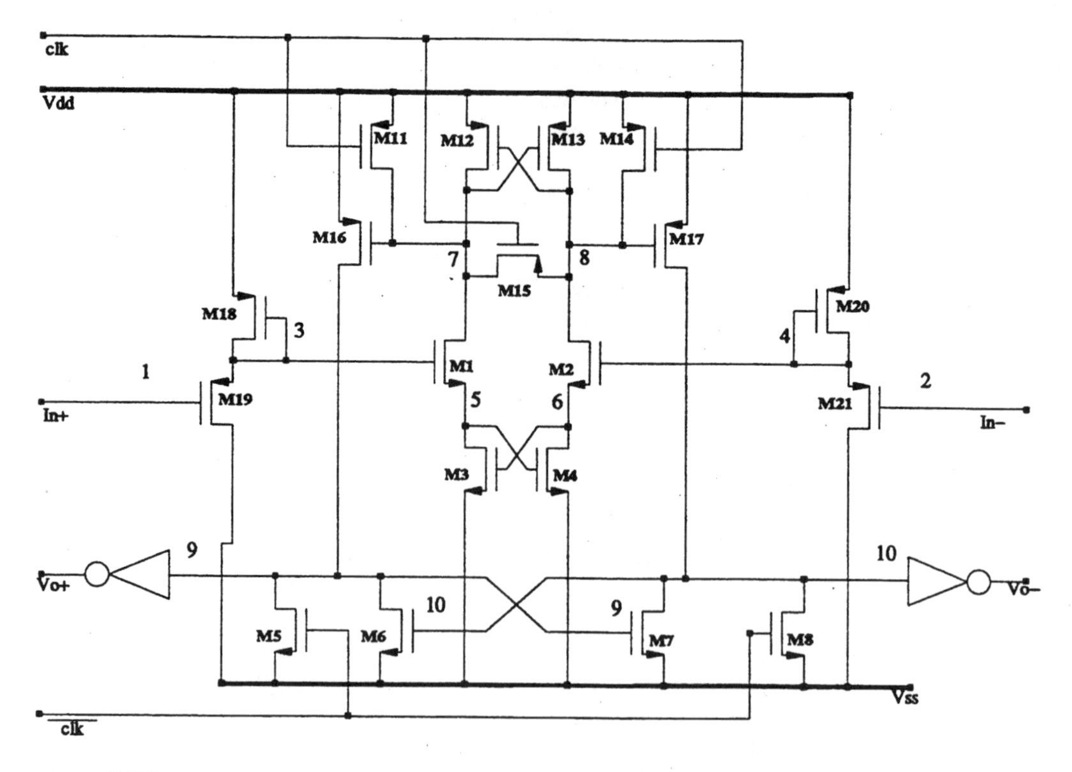

Fig. 15. Fast comparator CMP3.

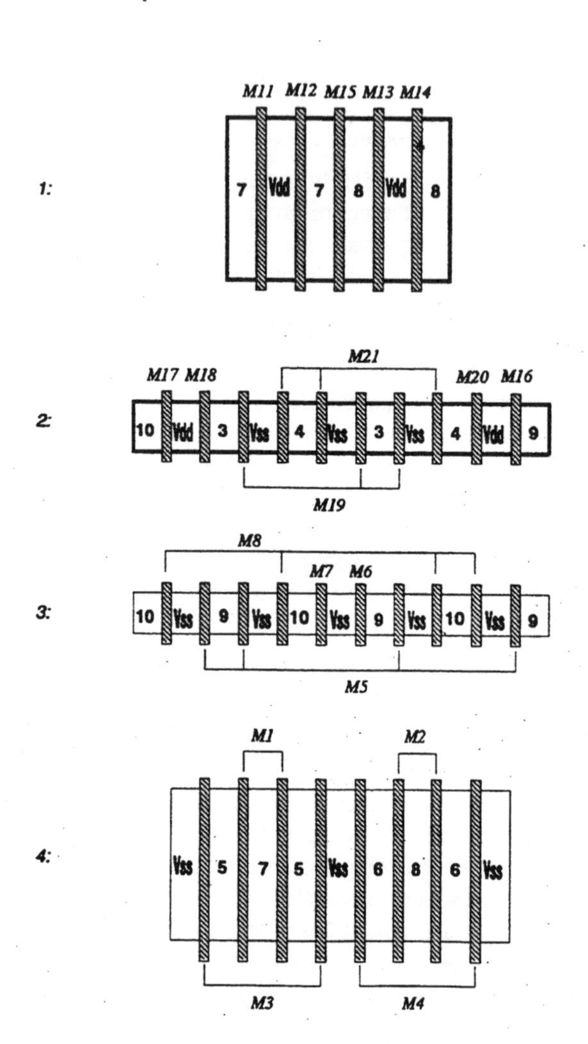

Fig. 16. Fast comparator CMP3—optimum stack partition.

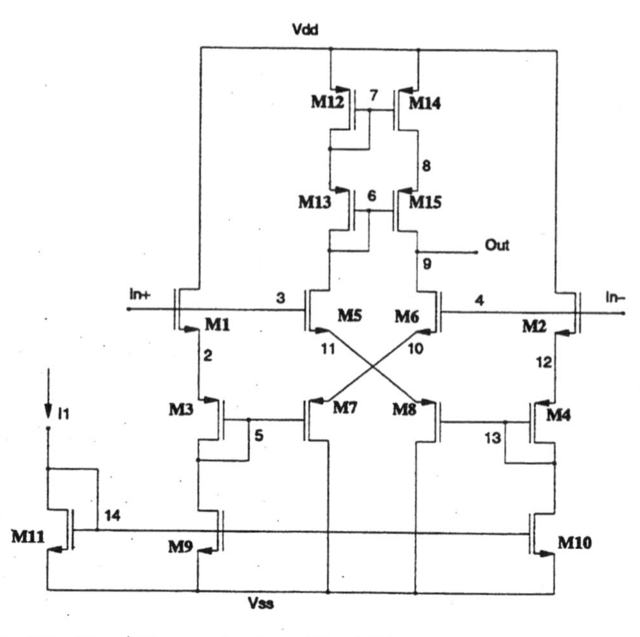

Fig. 17. Class-AB operational amplifier AB1.

Fig. 11 shows the schematic of a clocked comparator CMP2 [8]. In steps 1 to 3 of the stack generation algorithm, the circuit was split into four subgraphs as indicated by dashed lines in Fig. 11, on the ground of bulk bias nets, of matching requirements, and of user-defined transistor folding constraints. A solution for this circuit, consistent with all specifications on matching and symmetry, is reported in Fig. 12. Notice that for subgraphs 2, 3 and 4, partitions made of one maximum stack were found, and therefore no clique problem had to be solved. For subgraph 1, on the contrary, the optimum solution has two stacks with three modules each.

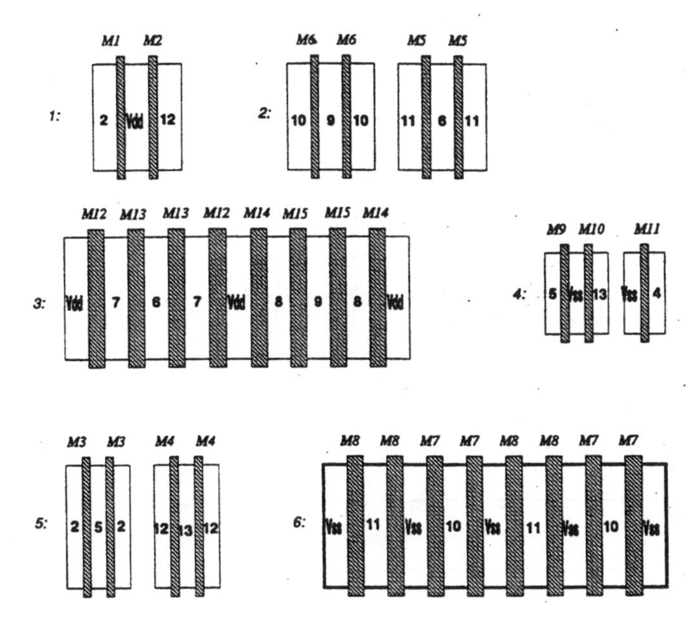

Fig. 18. Class-AB operational amplifier AB1—optimum stack partition.

Fig. 13 shows the schematic of an OTA named OPAMP2. In steps 1 to 3 of the stack generation algorithm, the circuit graph was split into 5 subgraphs, corresponding to the following circuit partition:

$$
\begin{aligned}
1:&\quad \{M_1, M_2\}\\
2:&\quad \{M_3, M_4, M_6, M_7, M_{11}, M_{12}, M_{13}, M_{17},\\
&\quad M_{18}, M_{19}, M_{20}\}\\
3:&\quad \{M_{14}\}\\
4:&\quad \{M_5, M_{15}, M_{16}\}\\
5:&\quad \{M_8, M_9, M_{10}, M_{21}, M_{22}\}.
\end{aligned}
$$

Constraints on GBW and phase margin were used to set weight specifications on critical capacitances. With the corresponding cost function, 112 optimum solutions were found in 3 CPU s. One of such solutions, with maximum transistor interleaving and common-centroid configuration for the input devices M_1 and M_2, is shown in Fig. 14. As in the case of circuit CMP2, maximum stacks were found for some of the subgraphs (1, 3, and 4), so that the clique solver was needed only for the remaining two subgraphs.

The comparator CMP3, shown in Fig. 15, is very sensitive to parasitics. Its speed depends critically on the junction capacitances on nets $\{3,4\}$ and $\{7,8\}$. Offset specifications require the perfect preservation of symmetries. Devices M_5, M_8, M_9, M_{21} were decomposed to provide a more compact layout. The partition generated is the following:

$$
\begin{aligned}
1:&\quad \{M_{11}, M_{12}, M_{13}, M_{14}, M_{15}\}\\
2:&\quad \{M_{16}, M_{17}, M_{18}, M_{19}, M_{20}, M_{21}\}\\
3:&\quad \{M_5, M_6, M_7, M_8\}\\
4:&\quad \{M_1, M_2, M_3, M_4\}
\end{aligned}
$$

Fig. 16 shows one of the 320 equivalent optimum solutions found. The two inverters were implemented with a specific generator and are not included in this figure. Common-centroid configurations have been used whenever possible to increase matching between symmetric devices. The layout turns out very compact and critical capacitances are small. The capacitances on nets $\{7,8\}$ could be further reduced by adding dummy modules to the sides of the upper stack. In this case it was not deemed necessary with the given constraints.

Fig. 17 shows the class-AB operational amplifier AB1. Transistors M_3 and M_4 are not connected, but they are linked by a symmetry constraint. In order to enforce such constraint, they must belong to the same subgraph, which is therefore not connected. The partition generated has 9 stacks and is shown in Fig. 18.

In the folded-cascode OPAMP5[1] [28], shown in Fig. 19, tight matching with channel orientation control is required among the transistors of the following two groups

$$
\begin{aligned}
&\{M_{16}, M_{20}, M_{17}, M_{23}\}\\
&\{M_3, M_4\}.
\end{aligned}
$$

An optimum partition is shown in Fig. 20. Some channel directions had to be changed with the help of dummy transistors. The arrows indicate the current direction in the matched devices.

The partitions obtained have been used as starting configurations by analog layout tools for automatic placement [10], routing [11] and compaction [29], with excellent results. The layouts for CMP1 and CMP3 are shown in Figs. 21 and 22, respectively, for a MOSIS-2 μm process. The area, speed and offset of these layouts are competitive with the best hand-made implementations known to us for these circuits.

A quantitative analysis of the effect of abutment on parasitic capacitances is shown in Table II. For each of the examples shown above, junction capacitances are reported for the most sensitive nets. Design rules and technology parameters refer to a 1.0 μm CMOS process by ORBIT. Junction capacitance is shown, for each sensitive net, without abutment in the second column, and with abutment in the third column. The reduction of parasitic capacitance due to abutment provides the router with improved flexibility, so that a bigger quota of capacitance is available for interconnections. Such quota is shown in the fourth column of Table II, as the length of minimum-width interconnection on metal-1 (the *equiv. length*), which would contribute the same amount of capacitance saved with abutment.

Table III summarizes the results obtained with these and other test examples. The number of optimum solutions is given by the product of the number of alternative solutions for each subgraph, reported between parentheses in the table. For instance, in the case of OPAMP2, 5 subgraphs were solved. The number of solutions is reported as 112 $(7 \times 4 \times 2^2 \times 1)$, which means that 1 subgraph yielded 7 solutions, two yielded 2 solutions each, and the remaining two gave 4 solutions and 1 solution respectively. Their combinations provide 112 different layouts. The maximum chain-graph size is also reported in the table. Such size is significant to express the cost of path generation. It is not always significant for the cost of the clique solver though. In fact the solver is not required for some subgraphs. In CMP3, for example (see Figs. 15 and 16), no

[1] This circuit is shown here by courtesy of Philips Research, Eindhoven, NL.

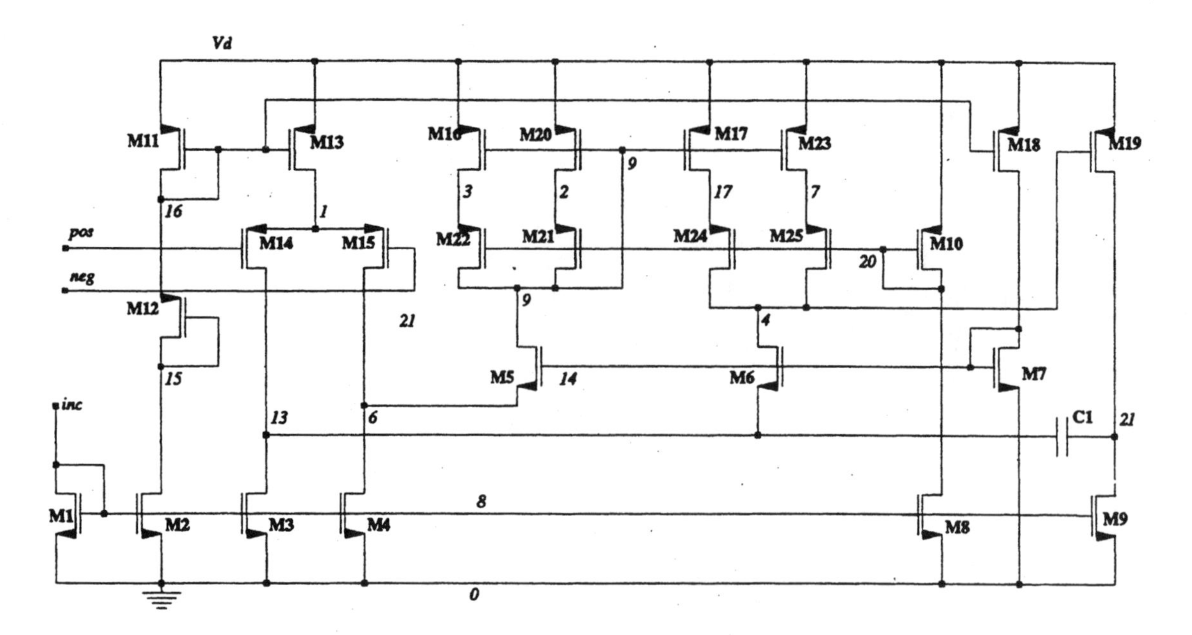

Fig. 19. Folded-cascode opamp OPAMP5 (courtesy of Philips Research, Eindhoven, NL).

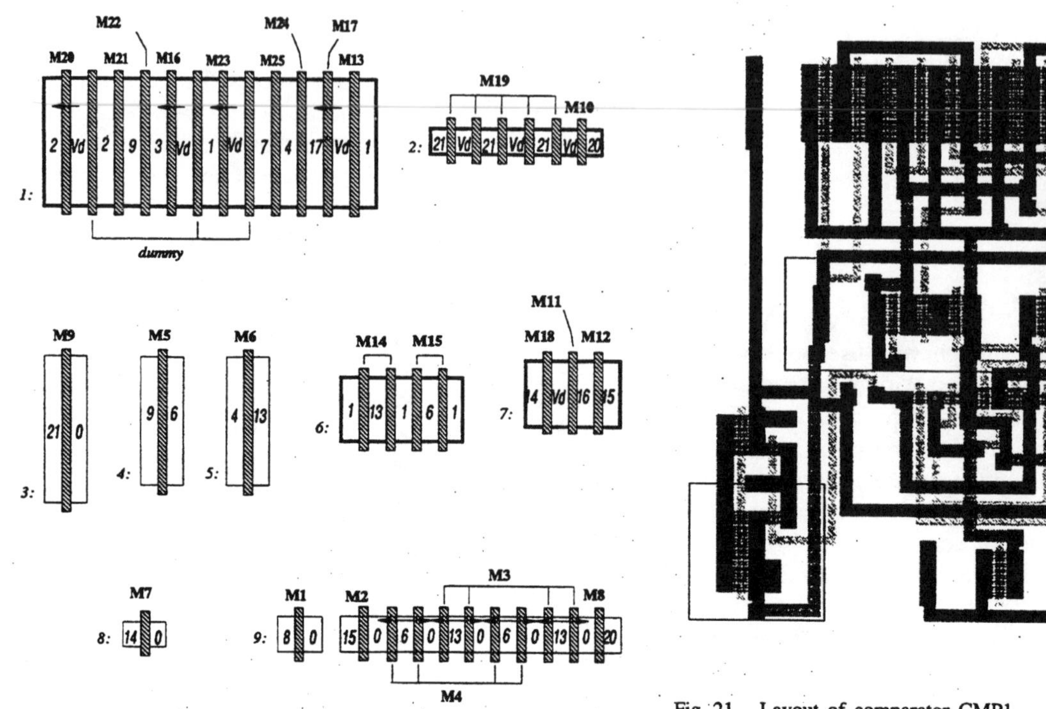

Fig. 20. Folded-cascode opamp OPAMP5—optimum stack partition.

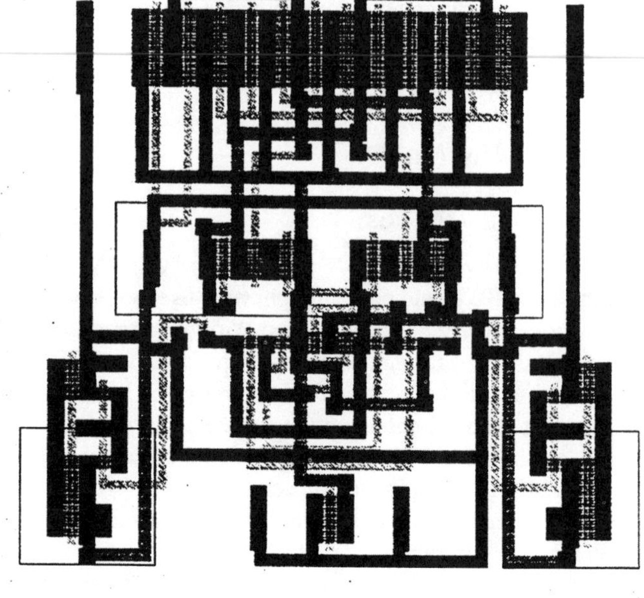

Fig. 21. Layout of comparator CMP1.

clique problems needed to be solved, since for all subgraphs solutions with maximum-length stacks were found.

CPU times are referred to a DECstation 5000/125, and include all parts of the algorithm, from circuit graph generation to clique solving. Notice that, in the case of OPAMP4, a subgraph with high degree of parallelism in its edges was found. As a consequence, a clique problem had to be solved on a large chain-graph with more than 40 000 edges, requiring almost an hour of CPU. The time between parentheses (11.0 s.) refers to the case with a "hint" requiring the input differential pair to be in the same stack. This option was so effective in reducing the size of the chain-graph, that the CPU time was decreased by 2 orders of magnitude, while still yielding the same optimum solution.

VIII. Conclusion

In this paper a general and rigorous methodology for maximally stacked layouts in CMOS analog integrated circuits has been proposed. Our algorithm always finds an optimum solution according to a cost function which takes into account parasitic effects, matching, and area. By means of dummy

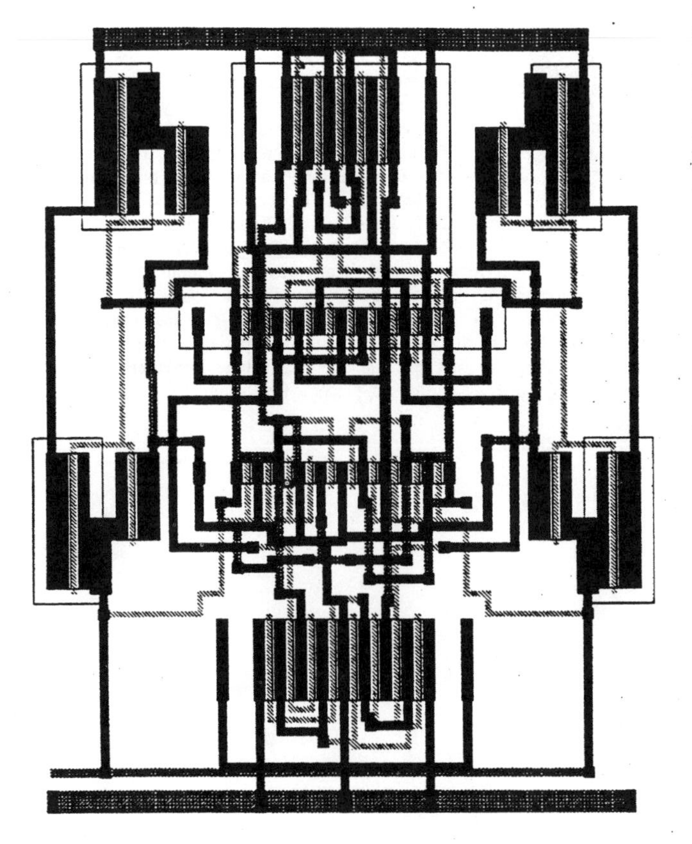

Fig. 22. Layout of comparator CMP3.

transistors, elemental modules laying in the same stack can be oriented so as to eliminate the mismatch due to different channel orientations.

Heuristics have been introduced, which have proved effective in reducing the computational complexity of the algorithm. Analog constraints and symmetries provide a key to a branch-and-bound procedure restricting the size of the problem while preserving the admissibility of the algorithm.

When combined with sensitivity analysis and automatic constraint generation, this methodology provides a suitable performance-driven approach to analog layout module generation in a full-stacked design style. The algorithm has been implemented in the "C" programming language and used as module generator in a framework for automated layout of analog circuits. The solutions found are equal or competitive with the ones obtained by hand by experienced designers. The results are very encouraging.

In the future, dedicated module generators will be integrated in the tool, in order to solve efficiently important and frequent configurations, such as differential pairs and current mirrors. Further work will be dedicated to a parallel implementation and to strategies for global routability optimization, integrating the module generation phase with placement.

Acknowledgment

The authors would like to thank A. Slenter and M. Pelgrom (Philips Research), V. Liberali (Univ. of Pavia), E. Charbon (UC-Berkeley), and R. Guerrieri (Univ. of Bologna) for many discussions and for their expert advice on layout and CAD topics. Some of the circuits shown in this paper have been provided by Dr. V. Liberali (Univ. of Pavia), Dr. J. Cohn (IBM), Prof. R. G. Meyer (UC Berkeley), Dr. M. Pelgrom (Philips Res.) and Dr. M. Gandini (CSELT), whom we gratefully acknowledge.

References

[1] G. Winner *et al.*, "Analog macrocell assembler," *VLSI System Design*, pp. 68–71, May 1987.

[2] T. Pletersek *et al.*, "High-performance designs with CMOS analog standard cells," *IEEE J. Solid-State Circ.*, vol. 21, no. 2, pp. 215–222, Apr. 1986.

[3] M. Kayal, S. Piguet, M. Declercq, and B. Hochet, "SALIM: A layout generator tool for analog ICs," in *Proc. IEEE Custom Integr. Circ. Conf.*, May 1988, pp. 751–754.

[4] M. Mogaki, N. Kato, Y. Chikami, N. Yamada, and Y. Kobayashi, "LADIES: An automatic layout system for analog LSI's," in *Proc. IEEE ICCAD*, Nov. 1989, pp. 450–453.

[5] F. M. Turky and E. E. Perry, "BLADES: An A.I. approach to analog circuit design," *IEEE Trans. Computer-Aided Design*, vol. 8, no. 6, pp. 680–692, June 1989.

[6] J. Rijmenants, J. B. Litsios, T. R. Schwarz, and M. G. R. Degrauwe, "ILAC: An automated layout tool for analog CMOS circuits," *IEEE J. Solid-State Circ.*, vol. 24, no. 2, pp. 417–425, Apr. 1989.

[7] M. Degrauwe *et al.*, "IDAC: An interactive design tool for analog CMOS circuits," *IEEE J. Solid- State Circ.*, vol. SC-22, no. 6, pp. 1106–1116, Dec. 1987.

[8] J. M. Cohn, D. J. Garrod, R. A. Rutenbar, and L. R. Carley, "KOAN/ANAGRAM II: New tools for device-level analog placement and routing," *IEEE J. Solid-State Circ.*, vol. 26, no. 3, pp. 330–342, Mar. 1991.

[9] S. W. Mehranfar, "A technology-independent approach to custom analog cell generation," *IEEE J. Solid- State Circ.*, vol. 26, no. 3, pp. 386–393, Mar. 1991.

[10] E. Charbon, E. Malavasi, U. Choudhury, A. Casotto, and A. Sangiovanni-Vincentelli, "A constraint-driven placement methodology for analog integrated circuits," in *Proc. IEEE Custom Integr. Circ. Conf.*, May 1992, pp. 2821–2824.

[11] E. Malavasi and A. Sangiovanni-Vincentelli, "Area routing for analog layout," to be published in *IEEE Trans. Computer-Aided Design*, vol. 12, no. 8, pp. 1186–1197, Aug. 1993.

[12] E. Malavasi *et al.*, "A top-down, constraint-driven design methodology for analog integrated circuits," in *Analog Circuit Design*, (J. H. Huijsing, R. J. van der Plassche and W. Sansen Ed.). Norwell, MA: Kluwer, 1993, pp. 285–324.

[13] U. Gatti, F. Maloberti, and V. Liberali, "Full stacked layout of analogue cells," in *Proc. IEEE Int. Symp. Circ. Syst.*, 1989, pp. 1123–1126.

[14] J. Kuhn, "Analog module generators for silicon compilation," *VLSI Systems Design*, pp. 74–80, May 1987.

[15] T. Uehara and W. M. vanCleemput, "Optimal layout of CMOS functional arrays," *IEEE Trans. Comput.*, vol. C-30, no. 5, pp. 305–312, May 1981.

[16] S. Wimer, R. Y. Pinter, and J. A. Feldman, "Optimal chaining of CMOS transistors in a functional cell," *IEEE Trans. Computer-Aided Design*, vol. CAD-6, no. 5, pp. 795–801, Sep. 1987.

[17] H. Zhang and K. Asada, "An improved algorithm of transistors pairing for compact layout of non-series-parallel cmos networks," in *Proc. IEEE Custom Integr. Circ. Conf.*, May 1993, pp. 1721–1724.

[18] E. Malavasi, D. Pandini, and V. Liberali, "Optimum stacked layout for analog CMOS ICs," in *Proc. IEEE Custom Integr. Circ. Conf.*, pp. 1711–1714, May 1993.

[19] T. H. Cormen, C. E. Leiserson, and R. L. Rivest, *Introduction to Algorithms*. Cambridge, MA: MIT Press, 1991.

[20] M. R. Garey and D. S. Johnson, *Computers and Intractability. A Guide to the Theory of NP-Completeness*. New York: W. H. Freeman, 1979.

[21] N. Weste and K. Eshraghian, *Principles of CMOS VLSI Design*. Reading, MA: Addison-Wesley, 1985.

[22] U. Choudhury and A. Sangiovanni-Vincentelli, "Constraint generation for routing analog circuits," in *Proc. Design Auto. Conf.*, pp. 561–566, June 1990.

[23] E. Charbon, E. Malavasi, D. Pandini, and A. Sangiovanni-Vincentelli, "Simultaneous placement and module optimization for analog ICs," in *Proc. Design Auto. Conf.*, June 1994, pp. 31–35.

[24] C. Bron and J. Kerbosch, "Algorithm 457—Finding all cliques of an undirected graph," *Comm. ACM*, vol. 16, no. 9, pp. 575–577, Sep. 1973.

[25] M. J. M. Pelgrom, A. C. J. Duinmaijer, and A. P. G. Welbers, "Matching properties of MOS transistors," *IEEE J. Solid- State Circ.*, vol. 24, pp. 1433–1440, Oct. 1989.

[26] E. Charbon, E. Malavasi, and A. Sangiovanni-Vincentelli, "Generalized constraint generation for analog circuit design," in *Proc. IEEE ICCAD*, Nov. 1993, pp. 408–414.

[27] R. W. Gregor, "On the relationship between topography and transistor matching in an analog CMOS technology," *IEEE Trans. Electron Devices*, vol. 39, no. 2, pp. 275–282, Feb. 1992.

[28] M. J. M. Pelgrom, "A 10-b 50-MHz CMOS D/A converter with 75-Ω buffer," *IEEE J. Solid-State Circ.*, vol. 25, pp. 1347–1352, Dec. 1990.

[29] E. Felt, E. Malavasi, E. Charbon, R. Totaro, and A. Sangiovanni-Vincentelli, "Performance-driven compaction for analog integrated circuits," in *Proc. IEEE Custom Integr. Circ. Conf.*, May 1993, pp. 1731–1735.

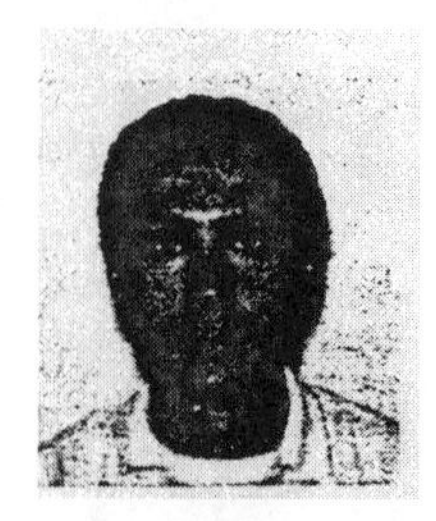

Davide Pandini received the Laurea degree in electrical engineering from the University of Bologna, Italy. He is currently a Ph.D. candidate in the area of computer-aided design for integrated circuits at University of California, Berkeley.

From 1991 to 1992, he was with the Technical Direction of Sirti S.p.A., Milano, Italy, where he was involved in research on network management systems. In 1992, he was a visiting researcher at Philips Research Laboratories, Eindhoven, The Netherlands, in the *CAD for VLSI Circuits* group. In 1992, he joined the Department of Electronics and Computer Science at the University of Padova, Italy. From 1993 to 1994, he joined the Department of EECS, University of California, Berkeley, as a graduate student. He is also a research intern at Digital Equipment Corporation, Western Research Laboratory, Palo Alto. His research interests include computer-aided design for analog IC's and algorithms for placement problems.

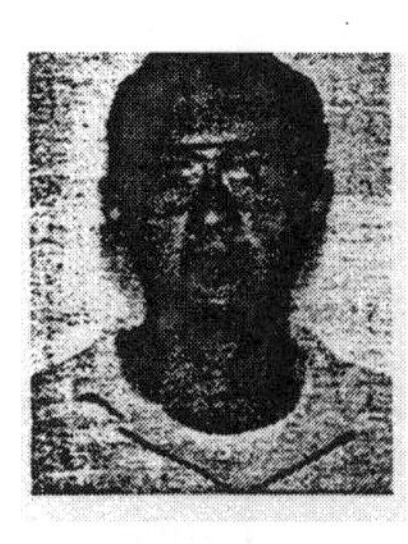

Enrico Malavasi received the B.S. degree in electrical engineering from the University of Bologna, Italy, in 1984. In 1993, he received the M.S. degree in electrical engineering from the University of California at Berkeley.

Between 1986 and 1989, he was with the Department of Electrical Engineering and Computer Science (DEIS) of the University of Bologna on research topics related to computer-aided design for analog circuits. In 1989, he joined the *Dipartimento di Elettronica ed Informatica* of the University of Padova, Italy, as Assistant Professor. Since 1990, he has collaborated with the computer-aided design group of the Department of EECS of the University of California at Berkeley, where he is working on the development of performance-driven computer-aided design tools and methodologies for analog design. His research interests include several areas of analog design automation, including layout, design methodologies, module generators, optimization and circuit analysis.

Substrate-Aware Mixed-Signal Macrocell Placement in WRIGHT

Sujoy Mitra, *Student Member, IEEE,* Rob A. Rutenbar, *Senior Member, IEEE,* L. Richard Carley, *Senior Member, IEEE,* and David J. Allstot, *Fellow, IEEE*

***Abstract*— We describe a set of placement algorithms for handling substrate-coupled switching noise. A typical mixed-signal IC has both sensitive analog and noisy digital circuits, and the common substrate parasitically couples digital switching transients into the sensitive analog regions of the chip. To preserve the integrity of sensitive analog signals, it is thus necessary to electrically isolate the analog and digital. We argue that optimal area utilization requires such isolation be designed into the system during first-cut chip-level placement. We present algorithms that incorporate commonly used isolation techniques within an automatic placement framework. Our substrate-noise evaluation mechanism uses a simplified substrate model and simple electrical representations for the noisy digital macrocells. The digital/analog interactions determined through these models are incorporated into a simulated annealing macrocell placement framework. Automatic placement results indicate these substrate-aware algorithms allow efficient mixed-signal placement optimization.**

I. Introduction

THE ability to implement mixed-signal systems by integrating both digital and analog portions on the same silicon substrate [1], [2] has paved the way for a new generation of reliable, cost-effective single-chip solutions. However, this integration has also introduced a host of new problems that were previously solved by simply isolating the analog and digital circuity in separate packages. Most of these problems are a result of parasitic interactions—in particular, substrate interactions—between the digital and analog subsystems [3]. CAD tools for mixed-signal system design must model and manage such parasitic interactions to be of practical use to the design community. Recent efforts here have included new noise models [4], the first practical, large-scale substrate simulation techniques [5]–[7], and physical design techniques [5]–[11]. Our focus is on physical design, in particular chip-level macrocell placement. We argue that gross coupling problems are best addressed as early as possible in the physical design of a mixed-signal IC, i.e., during actual placement of noisy and sensitive blocks. Given a good placement, downstream layout/verification tools such as noise-sensitive routers [10], [11], power grid synthesis [7]–[9], and full-chip substrate simulators [5], [7] can be used to maximum advantage.

Manuscript received August 1, 1994; revised October 20, 1994. This work was supported by the Semiconductor Research Corporation under Contract DC-068-94.

The authors are with the Department of Electrical and Computer Engineering, Carnegie Mellon University, Pittsburgh, PA 15213 USA.

IEEE Log Number 9408751.

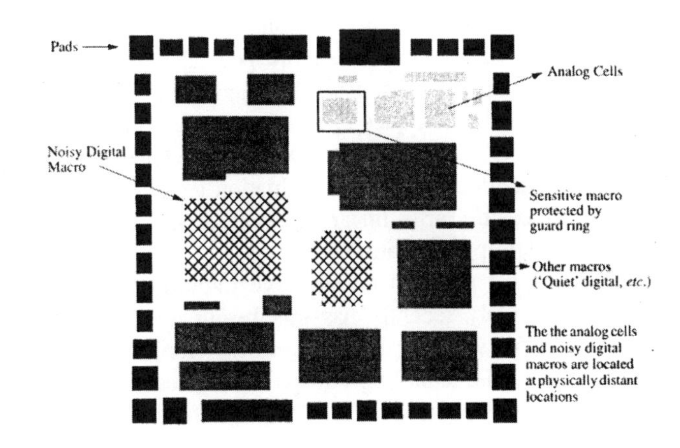

Fig. 1. Typical isolation mechanisms in mixed-signal systems.

This paper describes algorithms for automatic placement that are specifically targeted toward mitigating the effects of noise coupling through the substrate. The key innovation is the inclusion of efficient models for the chip substrate, noise sources and receivers on the macrocells, and mitigation measures such as guard rings *inside* the placement algorithm. We refer to this ability to make placement choices in light of their consequences for substrate noise coupling as *substrate-aware* placement.

II. Background

Prior work in this area has focused on chip-level layout methodologies, modeling and simulation for substrate coupling, and physical design algorithms.

Most methodologies for handling substrate coupling advocate the physical separation [12] of analog and digital circuitry in clearly delineated parts of the substrate. This is adequate in most cases, but some especially sensitive macros require additional isolation. This is achieved by inserting *guard rings* around these blocks (Fig. 1). These guard rings are usually low-impedance ties to a quiet potential and hence they have the effect of creating (ideally) a zero potential ring around the sensitive macro, thereby electrically isolating it from the rest of the chip. Unfortunately, most existing design methodologies remain *ad hoc*, and make no attempt either to quantify the level of interaction or formalize a clear methodology for separation or isolation. Often enough, it is possible for the experienced designer to manually manage the digital–analog interaction. But the growing complexity of these chips is making this task progressively more difficult and time consuming.

Modeling and simulation techniques for predicting substrate noise have been addressed in [4]–[6] and [13]. In [4], Su *et al.* address the problem of modeling the substrate. The active devices (transistors) are resistively coupled to a substrate represented by a single node. This model is adequate for epitaxial processes but rather inaccurate for the more common bulk processes. (Epitaxial processes compare poorly against bulk processes from the economic standpoint.) In the bulk case, the substrate is represented as a mesh or grid of appropriate linear elements, as in [5]. Noise couples into and out of the substrate through well diffusions and contacts modeled through appropriate linear approximations. The advantage of the single-node model is its computational simplicity for simulation; the advantage of the more general finite element-style substrate mesh is its generality and greater accuracy. The first practical techniques to simulate noise with complex substrate meshes appeared in [5], which introduced the idea of using Asymptotic Waveform Evaluation techniques (AWE [14]) to reduce the large linear substrate model to a small, tractable modal approximation.

CAD tools for physical design have traditionally ignored the substrate. Only recently has this begun to change. For example, the power distribution synthesis tool, RAIL [7], [8], considers noise coupling into power busses through the substrate. However, to our knowledge, the effects of substrate coupling have never been considered quantitatively in the context of automatic placement.

III. Strategy

We outline here the key ideas behind our basic placement strategy.

- *Macrocell Formulation:* We target placement problems composed of rectilinear analog and digital macrocells. We assume that all individual analog cells are placeable, and that many digital cells will already be grouped into larger functional blocks (e.g., via row-based placement inside the blocks). We are mostly concerned with placement problems as they appear during chip-level floorplanning when major functional blocks have already been laid out.
- *Simulated Annealing Optimization:* We adopt the basic iterative-improvement placement formulation of [15], and extensively refined in [16] and [17]. Cells are iteratively and arbitrarily relocated across the chip, without concern about whether they overlap or not. A cost function is evaluated after each placement perturbation; the goal of placement is to minimize this function. The cost function has two types of terms: *objective* terms which should be minimized (e.g., area and wirelength) and *penalty* terms which must be driven to zero to obtain a legal placement. Hard constraints are mapped into these penalty terms. For example, illegal overlaps among modules are evaluated and transformed into one such penalty term. In our substrate-aware placer, the key innovation is the formulation of penalty terms for hard electrical constraints, e.g., substrate-coupled switching noise above acceptable limits at any designer-specified location. This is the same overall strategy used successfully in [7] and [8].
- *Simplified Substrate Modeling:* We update a model of the substrate noise profile after *each* annealing placement perturbation. Since an annealer must visit many placement configurations (10^5 to 10^6) we need extremely fast, though necessarily approximate noise evaluation. The appropriate analogy here is to wirelength estimators used in the inner loop of a placer: we can afford simple estimators such as bounding-box length or simple spanning tree estimators [18], but we cannot afford to invoke a real router. Similarly, we can afford a simplified noise estimator, but not a full-chip simulator such as [5], [7], [19]. We rely on a coarse, lumped substrate grid, preprocessing of the grid to a smaller n-port model, and adaptive thresholding of the n-port matrix to render it sparse (and thus faster to solve) during the early phases of the placement.
- *Simplified Noise Injection and Noise Receiver Modeling:* Just as with the substrate model, we need only an extremely coarse model for substrate noise injectors and sensitive analog circuits. Because substrate coupling occurs through both capacitors and resistors, it is frequency dependent. However, in order to simplify the analysis which must occur in the inner loop of an annealer, we assume that there is a single user selected frequency (e.g., the clock frequency) that will provide a good first-order estimation for signal coupling. Further, all of the resistive and capacitive couplings to the substrate within a module are collected together into one or a few geometrically localized patches (the number would be determined by the cell size compared with the substrate grid size). Then, a magnitude must be assigned to the sinusoidal noise current injected across the admittance at each of these patches at the fundamental frequency selected. For example, this magnitude can be derived from circuit simulations of functional blocks with representative or worst-case input waveforms, or from the power estimates of the individual macrocells [20]. Finally, the designer must provide constraints on the magnitude of substrate voltage variations and on the gradients in substrate voltage that will ensure correct operation of sensitive analog cells.

Our basic goal is first-cut cell placement sensitive to significant substrate effects. We focus on an early phase of the layout, prior to signal and power grid routing, and thus prior to the availability of full-chip geometry for full-chip substrate simulation. The accuracy available in these tools is unaffordable within an iterative placement framework that must visit thousands of candidate placement configurations. Our central argument is that this accuracy is also unnecessary: simpler models can produce reasonable results more reliably than trial-and-error manual techniques. These automatic layouts can adequately serve as the starting point for further layout optimizations with more sophisticated noise coupling models.

IV. Annealing-Based Macrocell Placement

In this section we will briefly describe our use of simulated-annealing for substrate-aware placement. We chose this optimization method because it has been successful for many different classes of placement problems [15]–[17] and because its characteristics allow seamless incorporation of the features necessary for substrate-awareness. Annealing is a global optimization method whose principal advantage is a controlled hill-climbing mechanism by which many local optima can be avoided. Annealing is itself a stochastic iterative improvement strategy: the perturbations (called *moves*) that evolve the solution, and their acceptance/rejection are random, although controlled by a parameter usually called *temperature*. Analogous to physical annealing, the problem starts at a high temperature in which large uphill moves are highly probable; in this high-temperature regime we see aggressive global search of the problem space. As annealing progresses, the temperature cools and uphill moves decrease both in size and in likelihood. Eventually, only minor local improvements are possible, and the problem is said to be *frozen*.

Any annealing algorithm can be specified by describing its four key components: the representation for each *state* (in our case, placement configuration) visited during iterative improvement; the *move-set* which specifies the list of allowable perturbations to these states; the *cost-function* that measures the quality of the evolving solution; and the *cooling schedule* that controls the rate of hill-climbing as the solution evolves from randomness to a final "frozen" solution.

Our algorithm uses the so-called flat (nonhierarchical) Gelatt–Jepsen annealing formulation of [15], and extensively refined in [17] and [16]. Macrocells are arbitrarily relocated about the surface of the chip. Illegal overlaps are permitted, but penalized. Hence, the state representation is just the coordinate locations and instance selections (each cell can have a set of alternate instances) for each movable macrocell in the current placement. The size and shape of each macrocell is input to the placer in terms of its *footprint*. The location and shape of each terminal in the cell is indicated on this footprint. A list of alternate *instances*—if any—for each macro is also provided. The algorithm starts with a randomly generated placement and iteratively improves the solution by applying moves from the move-set. Moves include *translating* a cell to new location, *reorienting* a cell by rotating or reflecting it, and *reinstancing* a cell by substituting a different instance for the cell. Efficient, general-purpose adaptive cooling schedules are available to match the rate of cooling (and the choice of which move to make) to the problem [21]. See [22] for an extensive discussion of the implementation details for this basic placement strategy. This placement flow is summarized in Fig. 2.

The key innovations for substrate-aware placement are in the move-set and in the cost-function and its evaluation. Most macrocell placement algorithms allow the selection of alternate instances to handle cells available in a range of different shapes and sizes. Our algorithm also incorporates this feature, but uses it not only for geometric packing but also for electrical optimization. We model the insertion of a guard ring around a cell as an "alternative instance" of that cell, i.e., we simply add to the palette of alternatives for this macrocell a version of this cell with a guard ring. This allows the placer to select a move that inserts or deletes a guard ring. The advantage is that

```
Algorithm Placer(Netlist, Constraints, Macro definitions) {
    Create random placement instance, P;
    Initialize annealing temperature T (see White's Algorithm [32]);
    do {
        Perturb P to P + ΔP using move from move-set;
        Evaluate cost-function for placement instance P + ΔP;
        Use Metropolis Criterion to determine acceptance or rejection of instance P + ΔP;
        If ( Accept ) P := P + ΔP;
        Use cooling schedule (e.g. [21]) to modulate T;
    } while ( not frozen );
    Output placement instance, P, as solution;
} /* Placer */
```

Fig. 2. A simulated annealing-based macrocell-placement algorithm.

the placer now sees directly how the guard ring changes the shape and size of the cell, and also its new demands for routing area in its vicinity. We refer to this as *active mitigation*, in the sense that it is the placer's task to determine the appropriate set of mitigation measures required to satisfy the specified constraints.

Absent any concern for substrate coupling, the classical cost function includes estimated area (the smallest bounding rectangle of the cells) and wirelength (sum of half-perimeter of bounding box for each net's pins) terms to minimize. The penalty term measures illegal cell overlaps (area of overlapping footprints) and is driven to zero. To incorporate substrate-awareness, we allow the user to specify the maximum allowable switching noise at any location on any sensitive analog macro. Violations give rise to new penalty terms that must also be driven to zero in the cost function. Thus we need to model the substrate and the noise sources that inject noise into the substrate. We update an approximate substrate noise profile *after each placement move*, and then use this profile to determine the noise at any designer-specified locations on each sensitive macro. Each possible constraint violation is then checked and penalized as appropriate. Note that in contrast to simulation applications, we must update this substrate noise profile thousands of times as the movable cells converge to their final locations. Hence, our real problem is fast noise evaluation with "just enough" accuracy.

V. Modeling the Substrate

In this section, we develop a method for estimating substrate-coupled switching noise. Various tradeoffs among accuracy, generality, and efficiency are possible. For example, one can determine the substrate noise-profile by evaluating a two-dimensional Fourier series expression derived by solving the Laplace equation for noise in an epitaxial substrate using the approach described in [23]; such a model is very accurate, but computationally expensive, and inapplicable to the commonly used bulk case. For generality, most substrate simulators model the substrate in terms of an electrical equivalent: a grid of linear electrical elements derived using box-integration [24] techniques. For example, [6] and [8] use a three-dimensional resistive grid for each substrate layer; [25] uses a grid of resistances and capacitances for higher accuracy; and [4] represents the substrate as a single node (i.e., a degenerate grid) for the special case of an epitaxial process.

In the following, we develop a highly simplified gridded substrate model with accuracy sufficient for our placement application.

Despite its generality, the gridded substrate model has one major drawback: its accuracy is adversely affected by the coarseness of the grid. To improve accuracy, it is necessary to refine the grid in all three dimensions. Many substrate simulators try to overcome the efficiency issue by modulating the coarseness of the grid across the surface. Thus, a fine grid is used around the "interesting" areas of the substrate; for example, near transistors specified to be sensitive by the user. Unfortunately, this approach cannot be used in our substrate model because the location of the interesting objects—the noise receivers—*varies* after each annealing move, and it is computationally untenable to redefine the grid after *every* placement perturbation. Thus, we must use a static substrate grid with an adequate degree of refinement in each dimension.

Let us assume that the substrate is discretized to have $N = N_x \times N_y \times N_z$ grid elements. It has been observed empirically that the complexity of solving an electrical system with N nodes is about $N^{1.3}$ when sparse matrix techniques are used. Such substrate models has been used previously [5], [7], [8], but we believe that a more abstract model is desirable in this case. For example, the substrate can be fairly complex: it may have backside contacts and the lead-frame will certainly affect the noise profile. We describe such an abstract model subsequently. Such models have been used previously with some success [26].

To begin, we make the observation that only one part of the substrate is relevant during placement: the *active* area or the *face* of the substrate where placement is allowed. The placeable objects—noise sources and receivers—interact with the substrate through the grid points located on this surface. Therefore, we can characterize the entire substrate in terms of its *n-port admittance* (*y-parameter*) equivalent for the n grid points accessible on the surface of the substrate. Of course, any substrate, even the epitaxial type, can generally be represented as a grid and characterized as an n-port. The electrical equivalents of the macrocells can then be *plugged into* the nodes at the surface of the substrate. Each move of a single macrocell now entails disconnecting it from its nearest substrate nodes, relocating it, then reconnecting to the nearest nodes. Once this is done, the entire model (cells and substrate) can be solved for the substrate noise profile using common matrix solution techniques. This profile can then be used to determine the noise levels at the noise receivers.

An n-port equivalent is a multiport transfer function (where each element is a function of frequency) allowing evaluations of responses to transient stimuli. For example, [26] renders the substrate as a multiport s-domain transfer function using AWE [14] before computing thermal transients. But this degree of accuracy is not affordable in the context of automatic placement. We need merely characterize the n-port in terms of its admittance parameters at one fundamental frequency. In addition, the complexity of the substrate model has decreased in that the admittance model has many fewer nodes. The downside, however, is that the n-port admittance matrix is not sparse.

These issues are best examined in the context of a simple example. Consider a single-layer substrate and a user specification of $N = N_x \times N_y \times N_z$ grid points in the three dimensions. For a resistive, homogeneous substrate, the gridded equivalent circuit will have N nodes. A substrate simulator would convert this gridded equivalent and the circuitry on the chip to an admittance matrix of size N or more (reflecting the nodes in the circuitry on chip). This admittance matrix will be banded and sparse allowing the solution complexity to be marginally superlinear. The matrix is sparse because the elements are located only between nodes that are physically adjacent.

Assuming that the macros are placed on the $z = 0$ plane, the n-port equivalent circuit will have $N' = N_x \times N_y$ nodes and hence is a much smaller matrix. The diagonal terms in this equivalent matrix represent the equivalent admittance from the appropriate node to the reference node in the circuit. The off-diagonal terms represent the equivalent admittance between the appropriate nodes in the original network. This observation implies that this n-port equivalent will not be sparse because there is an electrical path between every pair of ports in the original circuit. Thus, it is likely that what we gained by reducing the size of the substrate model from N to N' was lost to reduced sparsity. This is not unexpected because the circuits are exact electrical equivalents of each other.

However, we can actually overcome this problem by exploiting the structure of the annealing process. In any annealing optimizer, convergence toward a final solution (a placement) is gradual and incremental. Early in the process (at hotter temperatures), large, uphill moves are allowed, which vigorously explore the solution space. In these temperature regimes, great accuracy in the solution of the substrate noise profile is unnecessary. What we seek instead, for efficiency, is a variable accuracy model of the substrate which can increase in detail, accuracy—and computational expense—as we converge to a final solution.

Fortunately, there is a computationally simple method to achieve this variable accuracy for an n-port substrate model: we *threshold* the nondiagonal terms in the matrix. The idea is that by removing small elements from the matrix, we render it sparse and thus fast to solve. At the beginning of annealing, only the diagonal terms are present in the matrix; as annealing proceeds toward a solution, more and more off-diagonal terms appear. Near a final solution, the entire substrate model is employed. This can be achieved by establishing a threshold and monotonically reducing it to zero across the optimization process. If $\boldsymbol{Y_s} = [y_{ij}]$ is the n-port matrix, a suitable formulation is to set $y_{ij} = 0$ if it satisfies

$$ty^{max} + (1-t)y^{min} \geq |y_{ij}|, \qquad t = [1, 0]. \qquad (1)$$

Here, y^{max} and y^{min} are the maximum and minimum nondiagonal terms in the matrix. Parameter t varies from 1 to 0 across the annealing process; at the start, $t = 1.0$ and all nondiagonal terms are excluded.

We illustrate the value of thresholding through a series of experiments summarized in Fig. 3. A placement instance is shown in Fig. 3(a), and its noise contours (without thresholding) are shown in Fig. 3(b). Note that the noise level outside of the noise-source regions is about 200 mV. The effects of

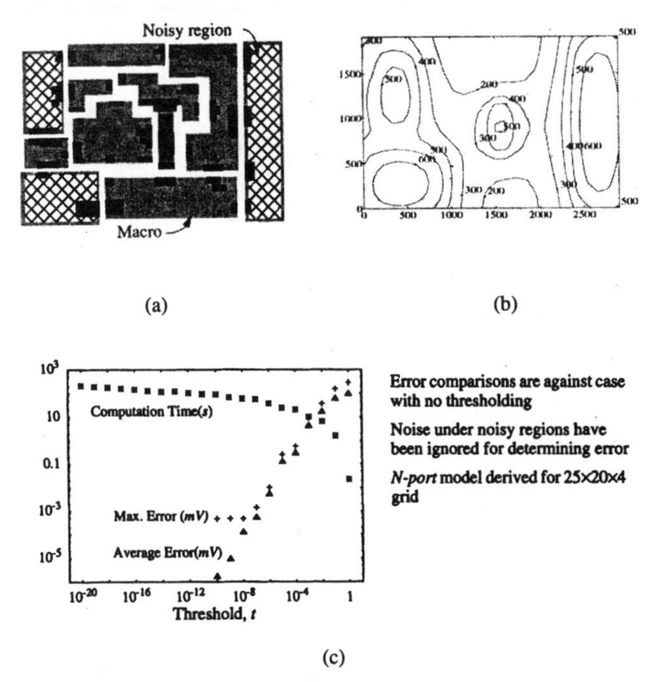

Fig. 3. Effect of thresholding on error and evaluation time. (a) Placement instance. (b) Noise contours (mV), no thresholding. (c) Effect of thresholding.

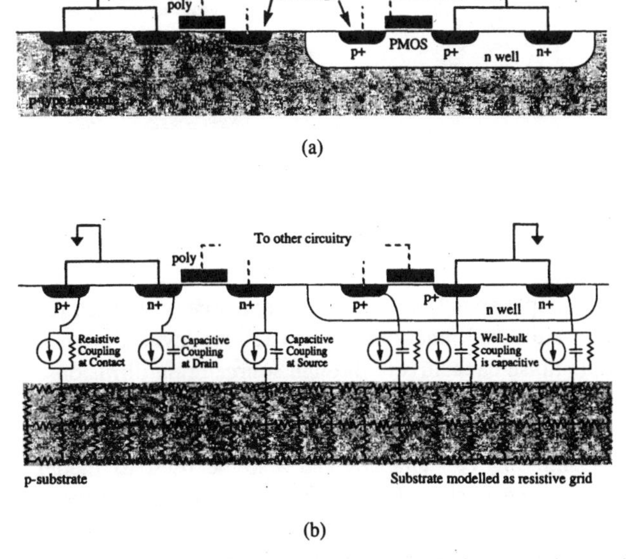

Fig. 4. Noise injection in CMOS logic circuits. (a) Noise generation mechanisms in CMOS. (b) Norton equivalents for substrate coupling.

thresholding are depicted in Fig. 3(c). Naturally, we expect the error to be the highest for $t = 1.0$. For this case, the average and the maximum errors are about 100 and 300 mV, respectively. These large errors decrease to about 4 and 10 mV, respectively, when the threshold is down to $t = 10^{-3}$. Within the same threshold range, the time per evaluation increases from 20 ms to 20 s. Thus, great improvements in efficiency can be obtained by thresholding the model when the error in the noise levels can be tolerated. A sparse matrix solver [27] running on a DECStation 5000™/200 was used in this experiment.

Note, however, that the times involved are still rather large given that repetitive evaluations would be required. One of the reasons is that the grid is quite fine—25 × 20—which implies that the *n-port* model can have as many as one-quarter of a million elements. The sparse matrix solver requires as much as 250 s per evaluation when the threshold is very low. Thus, the solver should employ a dense matrix technique when the matrix is no longer sparse.

VI. Modeling Substrate Interaction

As mentioned in Section II, the interaction of the macros and the substrate can be abstracted into a simple model comprised of the noise sources, the substrate, and the noise receivers. In this section, we will define simple models for noise sources and receivers, and indicate how they can be combined with the substrate model for evaluating switching noise.

A. Noise Sources

As mentioned in Section III, a simplified model is used for noise generation. Because substrate coupling occurs through both capacitors and resistors, it is frequency dependent. However, in order to simplify the analysis which must occur in the inner loop of an annealer, we assume that there is a single user-selected frequency (e.g., the clock frequency) that will provide a good first-order estimation for signal coupling. Although the assumption that all the noise in the system is at one (the fundamental) frequency only is simplistic, it has the advantage that it greatly simplifies the inputs that must be provided by the designer concerning the nature of the noise sources. To characterize a more complex time domain noise source might be extremely difficult. More accurate models would necessitate transient simulation and temporal description of the noise waveforms, both of which would be infeasible and expensive within an iterative placement framework, and, we argue, unnecessary at the very early stages of the layout process which we target.

In addition, all of the resistive and capacitive couplings to the substrate within a module are collected together into one or a few geometrically localized patches (the number would be determined by the cell size compared with the substrate grid size). These Norton equivalent *regions* are characterized by a complex current source and an admittance. Many such regions may be present on a digital macrocell. Then, a magnitude must be assigned to the sinusoidal noise current injected across the admittance at each of these patches at the fundamental frequency selected. For example, this magnitude can be derived from circuit simulations of functional blocks with representative or worst-case input waveforms. Finally, the designer must provide constraints on the magnitude of substrate voltage variations and on the gradients in substrate voltage that will ensure correct operation of sensitive analog cells.

How does the designer determine the admittance and current source values for a noise generating cell? Generally, there are three mechanisms that inject noise into the substrate. These are shown in Fig. 4(a). Power buses couple noise arising out of switching transients into the substrate through ohmic contacts;

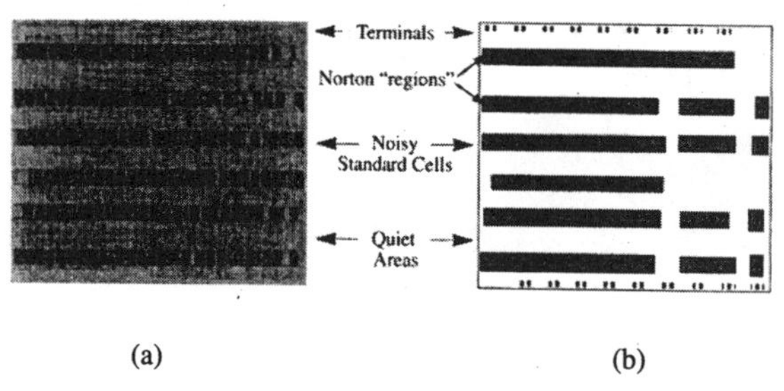

Fig. 5. Modeling a digital macro. (a) Digital macrocell. (b) Footprint.

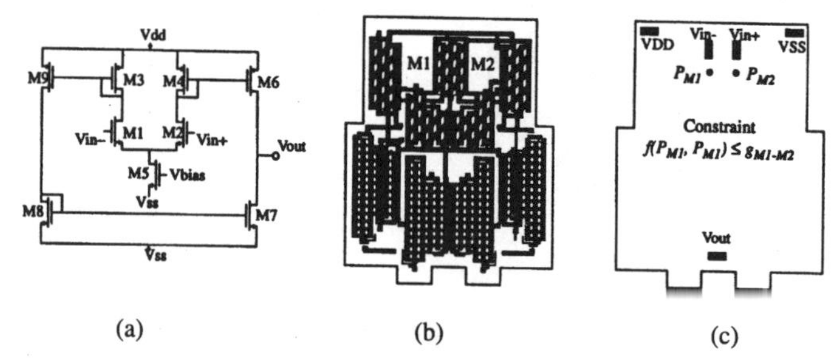

Fig. 6. The model for a two-stage op-amp. (a) Schematic. (b) Layout. (c) Footprint.

wells couple switching transients capacitively through reverse biased bulk/well junctions; and switching devices (transistors) couple noise capacitively through source/drain diffusions. The noise sources at the contacts can be modeled with an equivalent current source across a contact resistance. Current sources with shunting capacitances can also be used to model the effect of the charge injected into the substrate at the wells and the source/drain diffusions of the switching devices. A simplified model of such substrate interactions is shown in Fig. 4(b): a more accurate model of such substrate interactions (for example, one including the series resistance of the reverse biased source/drain diffusion region "diodes") would involve more elements to model some of the second-order effects. In a circuit simulator, the admittance between the circuit and the substrate can be determined at the selected frequency. Next, a transient circuit simulation of one or more functional blocks with representative or worst-case input waveforms can be carried out. Then, by measuring the currents flowing into the substrate from all elements in each patch (e.g., body contact of transistors) during the transient simulation, a magnitude and approximate sinusoidal frequency can be chosen for the current waveform.

For our placer, the geometry of a digital macrocell is specified in terms of its *footprint.* An example of a digital macro and its footprint appears in Fig. 5. Besides the outline and the terminal description, areas generating noise are indicated on the footprint and tagged with their Norton equivalent. For example, the noisy areas would correspond to the location of the noisy standard cells in a row-based functional block.

B. Noise Receivers

Substrate simulators usually couple sensitive (high impedance) signals or transistors to the substrate through parasitic capacitors and resistors. However, we require a much simpler coupling model. Thus, we will assume that the designer is able to provide *constraints* on the noise levels and the gradients in noise level at sensitive locations on analog macros. This scenario is best explained through an example.

Consider the two-stage operational amplifier (op-amp) depicted in Fig. 6. The schematic is given in Fig. 6(a) (some biasing circuitry is omitted for clarity). Fig. 6(b) is the layout for this macro and Fig. 6(c) shows its footprint. The footprint encodes the outline of the cell and the location, shape, and layers of its terminals. In addition, the cell may now be tagged with a substrate-noise constraint. In practice, substrate noise-induced mismatches between any pair of matched devices would disturb the output, but here we focus on the mismatches of the input devices only for reasons of clarity and because their effect is the greatest. For example, the impact of the mismatches between $M3$ and $M4$ in Fig. 6(a) is reduced by the gain across the input devices $M1$ and $M2$ when referred to the input.

The input devices are susceptible to substrate noise via two different mechanisms. First, due to manufacturing mismatches between their body effect coefficients ($\Delta\gamma$), variations in substrate voltage will be manifested as variations in input offset voltage. Second, even if the body effect coefficients match perfectly, a gradient in substrate potential will still result in a noise-dependent input offset voltage. Thus, if $M1$ is located at $P1$ and $M2$ is located at $P2$, the user may specify a constraint (2) aimed at keeping the potential gradient within a limit, $g_{M1\text{-}M2}$:

$$f(P_{M1}, P_{M2}) = \frac{|\text{Noise}\,(P_{M1}) - \text{Noise}\,(P_{M2})|}{|P_{M1} - P_{M2}|} \leq g_{\text{M1-M2}}. \tag{2}$$

Such gradients can easily be determined from potential profiles in the substrate. The placer will attempt to locate this macro such that this constraint is satisfied. The op-amp designer is usually aware of such substrate sensitivities, and is therefore capable of tagging such constraints onto the footprint of the cell shown in Fig. 6(c). Extremely sensitive analog macros are often surrounded by guard rings. Such guard rings can also be modeled by our Norton equivalent (i, y). In this case, i is zero and the admittance y is entirely resistive.

A simple example of this substrate-interaction model is shown in Fig. 7. A placement instance is shown in Fig. 7(a). The corresponding noise profile on a 10 × 10 substrate grid is shown in Fig. 7(b). This profile is determined from the electrical equivalent of substrate interaction depicted in Fig. 7(c). A bulk substrate without a backside contact is assumed. (This is typical for inexpensive packages like the plastic quad flat pack.) Noise couples in through direct coupled current sources at the grid points located under the noise source indicated in Fig. 7(a). (For clarity of illustration, we make the simplifying assumption that the substrate contacts are uniformly distributed across the substrate and hence characterized into the substrate model.) A guard-ring and its electrical equivalent are also indicated. Note that the profile in Fig. 7(b) flattens out in the region surrounded by the guard ring. Naturally, the guard ring serves as a low impedance path (typically, the series combination of the diffusion and the

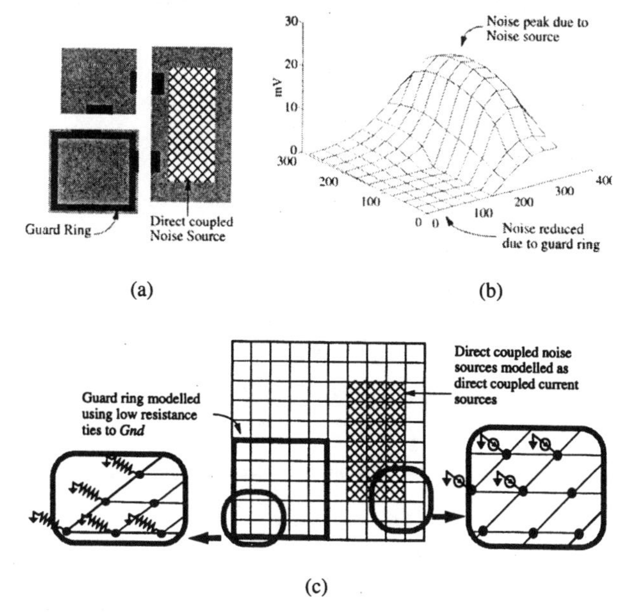

Fig. 7. Substrate interactions for a given placement instance. (a) Placement instance. (b) Noise profile. (c) Electrical equivalent of placement instance.

wiring) to a quiet potential, thereby flattening out the noise profile by "intercepting" the substrate noise before it reaches the protected area. Usually, this ring would be connected to the low-noise analog ground supplying the analog cells on the chip.

C. Evaluating Substrate Coupled Switching Noise

A placement instance combined with the models described above can be easily solved to determine the substrate noise profile. Note that noise sources and receivers have been modeled such that they do not add any extra nodes to the electrical equivalent. Let the n-port admittance (y-parameter) model of the substrate be $\boldsymbol{Y_S}$. This is a matrix of size $N' \times N'$, where N' is the number of grid points on the active area of the substrate. The electrical interaction of the placement instance and the substrate is represented by a diagonal matrix, $\boldsymbol{dY}$ of size $N' \times N'$. A diagonal element in this matrix is nonzero if there is a macro with a nonzero Norton admittance located at the substrate node corresponding to that element. Let $\boldsymbol{J}$ be a vector of size N' such that an element therein is nonzero if there is a macro with a nonzero Norton current source located at the substrate node corresponding to that element. Also, let $\boldsymbol{V}$ be the vector (of size N') representing the potential profile across the substrate measured against an arbitrarily selected reference potential for the entire system. Then from basic circuit theory [28]

$$[\boldsymbol{Y_S} + \boldsymbol{dY}]\boldsymbol{V} = \boldsymbol{J}. \tag{3}$$

In general, this equation would be complex and dense. However, it may be rendered sparse by applying the thresholding methods described in Section V to the substrate model $\boldsymbol{Y_S}$. Also, the designer may choose to use a simplified substrate interaction model without any capacitances. In this case, (3) is realvalued and a solver tuned to solving real matrices can be used. The matrix $[\boldsymbol{Y_S} + \boldsymbol{dY}]$ is also diagonally dominant,

TABLE I
SUMMARY OF EXPERIMENTS WITH EXAMPLE $C1$

Expt.	Constraint (V)	Normalized		Time (min)
		Area	WireLength	
1	-	1	1	4
2	$V_n(A) \leq 0.6$	1	2.33	139
3	$V_n(A) \leq 0.1$	1.1	1.166	200

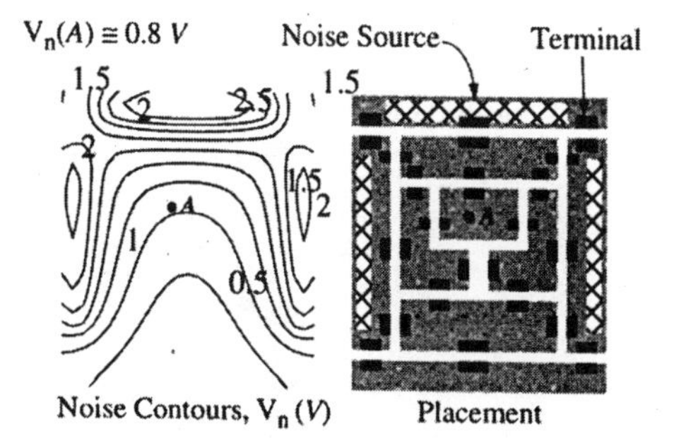

Fig. 8. Placement without considering substrate noise.

implying that efficient iterative solution schemes like the *Incomplete Choleski Conjugate Gradient* [29] can be used to achieve further gains in speed.

VII. Experimental Results

We will now describe and analyze a small set of experiments that demonstrate the viability of the placement techniques presented here. The algorithms described in the preceding sections have been implemented in a prototype substrate-aware mixed-signal placement tool called WRIGHT. This program is about 14 000 lines of C code. A program to generate n-port substrate models using a circuit simulator [30] and simple structural and electrical descriptions of the substrate was also implemented. WRIGHT incorporates a sparse [27] and a dense [31] matrix package. All examples decribed here were executed on a DECStation 5000™/200 workstation.

To begin, we ran a set of controlled experiments on a simple synthetic example, $C1$. This has 9 blocks, 20 nets, 40 terminals, 3 noisy macros, and 1 sensitive macro. This example is essentially a jigsaw puzzle: the macros here are rectilinear slices out of a rectangle. A placement that finds the global optimum with respect to area and wire length reassembles the pieces back into the original solid rectangle. These experiments are summarized in Table I, and the results are depicted in Figs. 8–10. The noise contours and the final placement are given for each. In these figures, each gray box is a macro, the darker boxes are terminals, and the hatched boxes are noise sources. A bulk substrate abstracted from a $10 \times 10 \times 2$ grid is used.

In experiment 1, we place to minimize area and wire length *without* any regard to substrate noise. The solution (Fig. 8) has both minimum wire length and area. As expected, WRIGHT

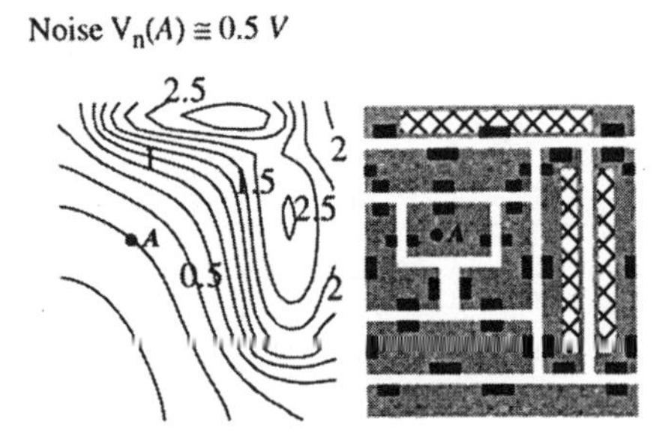

Fig. 9. Constrain $V_n(A) \leq 0.6$ V.

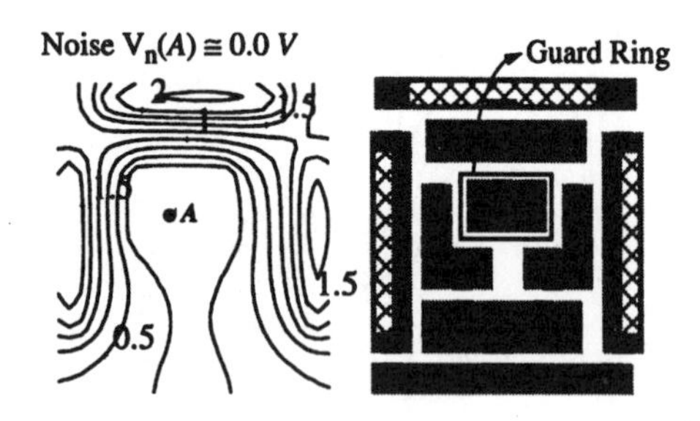

Fig. 10. Constrain $V_n(A) \leq 0.1$ V.

just reassembles the puzzle when ignoring the substrate noise. Note that the noise level at location A is predicted as 0.8 V.

In experiment 2, we impose a constraint on the acceptable substrate noise level at A, a location on the sensitive macro. The new placement is shown in Fig. 9. Here WRIGHT attempts to satisfy this constraint by moving the sensitive macro away from the noisy ones. WRIGHT satisfies the constraints by appropriately choosing the relative locations of sensitive and noisy macros.

Finally, we tighten the noise constraint further in experiment 3. The placer now decides that the area penalty involved in further separating the macro containing A is higher compared to putting a guard ring around that macro (Fig. 10). This is an example of the placer making a noise/area tradeoff in favor of actively mitigating a noise constraint violation.

From the results given in Table I, we can draw several conclusions. Not unexpectedly, the CPU time increases when noise constraints are activated. Note, however, only one matrix solve *per* placement iteration is required to evaluate the noise at all receivers: the number of noise constraints or noise receivers does not affect the efficiency of this evaluation. However, like all other annealing placement algorithms, the time required to complete placement is a function of the *size* of the problem as quantified by the number of macros.

In Fig. 11, we place a larger example, $C2$, which is a simplified version of the mixed-signal routing benchmark from [10] and [8]. This example has 25 macros, 381 terminals, 163 nets, 3 noisy digital macros and 12 sensitive analog macros with receivers located at their centers. The analog macros have constraints similar to example $C1$. The constraints and the results of running WRIGHT are tabulated in Table II. We note that WRIGHT satisfies a greater number of constraints when the consideration of substrate noise is enabled. Moreover, the degree of constraint violation is also minimized. WRIGHT uses a simple weighted sum of constraint violations (noise voltages), $\sum_i w_i v_i$, as the penalty term to be driven toward zero here. By design here, the noise constraints for M5 and M9 cannot be met, so WRIGHT attempts to minimize the constraint violation for each macro. However, the constraint on M9 has greater weight attached to it, so the placer attempts to reduce the violation for M9 more aggressively. So, as Fig. 11 shows, M9 is placed at a quieter location a greater distance away from the noisy digital macros compared to M5, thereby reducing the substrate noise constraint penalty term. Overall, we see that wire length is traded off to achieve these gains. (The area does not change because the bounding box of the chip is fixed.) This more complex example requires about 7 hours to place when using a substrate model derived from a rather fine 15 × 20 × 3 grid. Improvements in time efficiency can be expected with a coarser grid. This example was run with thresholding enabled.

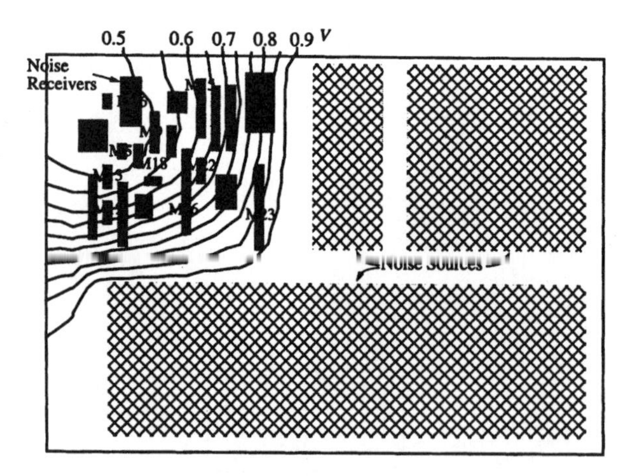

Fig. 11. Noise contours for example $C2$ placed with 12 noise constraints.

TABLE II
PLACING A LARGE EXAMPLE $C2$

Receiver, i	Constraints		Substrate Noise			
	$V_i^{noise} \leq V_i^{max}$		Ignored		Considered	
	V^{max}_i (V)	Weight, w_i	V_i^{noise} (V)	Violation, v_i	V_i^{noise} (V)	Violation, v_i
M1	0.9	1	0.58		0.76	
M5	0.2	100	0.88	0.68	0.46	0.26
M12	0.8	1	0.57		0.69	
M13	0.8	1	0.61		0.73	
M15	0.8	1	0.80		0.69	
M9	0.2	10	0.69	0.49	0.51	0.31
M18	0.7	1	0.66		0.56	
M16	0.9	1	0.63		0.48	
M22	0.9	1	0.75		0.89	
M26	0.6	1	0.89	0.28	0.71	0.11
M23	0.7	1	0.92	0.22	0.84	0.14
M24	0.7	1	0.83	0.13	0.48	

Placement metrics	Ignored	Considered
$\sum w_i v_i$	73.53	29.35
Wire-Length (μm)	252449	282461
Time (*min*)	2.87	433
Area (μm²)	3.6x10⁷	3.6x10⁷

VIII. Conclusions

In this paper we presented a strategy for substrate-aware mixed-signal macrocell placement. This approach incorporates simplified switching noise estimation into a simulated anealing placement algorithm. Preliminary results suggest our algorithms generate placements that are intuitively correct. This satisfies our goal of generating reasonable first-cut placements that consider the most significant effects. One of the basic principles underlying the selection and design of our substrate and substrate interaction models has been the tradeoff between efficiency and accuracy. This is necessary because the models are meant to be used during a part of the design cycle when little information about the completed chip is available and efficient evaluation is critical. We expect that our placements will be incrementally tuned based on results from downstream layout steps (e.g., power grid and signal routing) and full-chip substrate simulation. We also expect that further work on estimation of power dissipation and switching activity [20] for digital circuitry will be directly applicable to our simple noise injection models.

Our current work focuses on improving the speed and the accuracy of our models and on extending this method for handling thermal layout problems [26].

Acknowledgment

The authors would like to thank W. Getchell of Bell-Northern Research, B. Stanisic of IBM, G. Gad-Elkarim of Cadence Design Systems, and N. K. Verghese of Carnegie Mellon University for their comments and the highly stimulating discussions during the initial phases of this project.

References

[1] D. A. Hodges, P. R. Grey, and R. W. Brodersen, "Potential for MOS technologies for analog integrated circuits," *IEEE J. Solid-State Circuits,* vol. SC-13, pp. 285–294, June 1978.

[2] J. R. Lineback, "ASIC houses rush to add analog functions to libraries," *Electron.* vol. 60, no. 22, p. 31, Oct. 29, 1987.

[3] J. A. Olmstead and S. Vulih, "Noise problems in mixed analog/digital integrated circuits," in *Proc. Custom Integrated Circuits Conf.,* 1987, pp. 659–662.

[4] D. K. Su, M. J. Loinaz, S. Masui, and B. A. Wooley, "Experimental results and modeling techniques for substrate noise in mixed-signal integrated circuits," *IEEE J. Solid-State Circuits,* vol. 28, pp. 420–430, Apr. 1993.

[5] N. K. Verghese, D. J. Allstot, and S. Masui, "Rapid simulation of substrate coupling effects in mixed-mode IC's," in *Proc. IEEE Custom Integrated Circuits Conf.,* May 1993, pp. 18.3.1–18.3.4.

[6] N. K. Verghese, T. J. Schmerbeck, and D. J. Allstot, *Simulation Techniques and Solutions for Mixed-Signal Coupling in Integrated Circuits.* Norwell, MA: Kluwer Academic, 1994.

[7] B. R. Stanisic, N. K. Verghese, R. A. Rutenbar, L. R. Carley, and D. J. Allstot, "Addressing substrate coupling in mixed-mode IC's: Simulation and power distribution synthesis," *IEEE J. Solid-State Circuits,* vol. 29, pp. 226–238, Dec. 1994.

[8] B. R. Stanisic, R. A. Rutenbar, and L. R. Carley, "Power distribution synthesis for analog and mixed-signal ASIC's in RAIL," in *Proc. IEEE Custom Integrated Circuits Conf.,* May 1993, pp. 17.4.1–17.4.5.

[9] B. R. Stanisic, R. A. Rutenbar, and L. R. Carley, "Mixed-signal noise-decoupling via simultaneous power distribution design and cell customization in RAIL," in *Proc. IEEE Custom Integrated Circuits Conf.,* May 1994, pp. 24.3.1–24.3.4.

[10] S. Mitra, S. K. Nag, R. A. Rutenbar, and L. R. Carley, "System-level routing of mixed-signal ASIC's in WREN," in *Proc. IEEE Int. Conf. Comput.-Aid. Design,* 1992, pp. 394–399.

[11] U. Chowdhury and A. Sangiovanni-Vincentelli, "Constraint-based channel routing for analog and analog/digital circuits," in *Proc. IEEE Int. Conf. Computer-Aided Design,* Nov. 1990, pp. 198–201.

[12] A. E. Dunlop, G. F. Gross, C. D. Kimble, M. Y. Luong, K. J. Stern, and E. J. Swanson, "Features in LTX2 for analog layout," in *Proc. ISCAS-85,* 1985, pp. 21–23.

[13] T. J. Schmerbeck, R. A. Richetta, and L. D. Smith, "A 27MHz mixed analog/digital magnetic recording channel DSP using partial response signalling with maximum likelihood detection," in *Proc. IEEE Int. Solid-State Circuits Conf.,* 1991, pp. 136–137.

[14] L. T. Pillage and R. A. Rohrer, "Asymptotic waveform evaluation for timing analysis," *IEEE Trans. Computer-Aided Design,* vol. 9, pp. 352–366, Apr. 1990.

[15] D. W. Jepsen and C. D. Gelatt, Jr., "Macro placement by Monte Carlo annealing," in *Proc. 1984 IEEE Int. Conf. Comput. Design,* Nov. 1984, pp. 495–498.

[16] J. M. Cohn, D. J. Garrod, R. A. Rutenbar, and L. R. Carley, "KOAN/ANAGRAM II: New tools for device-level analog placement and routing," *IEEE J. Solid-State Circuits,* vol. 26, pp. 330–342, Mar. 1991.

[17] C. Sechen and A. Sangiovanni-Vincentelli, "The TimberWolf placement and routing package," *IEEE J. Solid-State Circuits,* vol. SC-20, pp. 510–522, Apr. 1985.

[18] B. T. Preas and M. J. Lorenzetti, Eds., *Physical Design Automation of VLSI Systems.* Menlo Park, CA: Benjamin/Cummins, 1988.

[19] S. Masui, "Simulation of substrate-coupling in mixed-signal MOS Circuits," in *Proc. VLSI Circuits Symp.,* 1992, pp. 42–43.

[20] A. Yang and K. Mayaram, "Estimating power dissipation in VLSI circuits," *IEEE Circuits Devices,* pp. 11–19, July 1994.

[21] M. D. Huang, F. Romeo, and A. Sangiovanni-Vincentelli, "An efficient general cooling schedule for simulated annealing," in *Proc. IEEE Int. Conf. Computer-Aided Design,* 1986, pp. 381–384.

[22] J. M. Cohn, D. J. Garrod, R. A. Rutenbar, and L. R. Carley, *Analog Device-Level Layout Automation.* Norwell, MA: Kluwer Academic, 1994.

[23] A. G. Kokkas, "Thermal analysis of multiple-layer structures," *IEEE Trans. Electron. Devices,* vol. ED-21, pp. 674–681, Nov. 1974.

[24] K. Fukahori and P. R. Grey, "Computer simulation of integrated circuits in the presence of electrothermal interaction," *IEEE J. Solid-State Circuits,* vol. SC-11, pp. 834–846, Dec. 1976.

[25] N. K. Verghese, S.-S. Lee, and D. J. Allstot, "A unified approach to simulating electrical and thermal substrate coupling interactions in IC's," in *Proc. Int. Conf. Comput.-Aid. Design,* Nov. 1993.

[26] S.-S. Lee and D. J. Allstot, "Electrothermal simulations of integrated circuits," *IEEE J. Solid-State Circuits,* vol. 28, pp. 1283–1293, Dec. 1993.

[27] K. S. Kundert and A. Sangiovanni-Vincentelli, *Sparse User's Guide,* Version 1.3a ed., Dep. Elec. Eng. Comput. Sci., Univ. California, Berkeley, 1988.

[28] J. Vlach and K. Singal, *Computer Methods for Circuit Analysis and Design.* New York: Van Nostrand Reinhold, 1983.

[29] J. A. Meijerink and H. A. van der Vorst, "An iterative solution method for linear systems of which the coefficient matrix is a symmetric M-matrix," *Math. Computat.,* vol. 31, pp. 148–162, Jan. 1977.

[30] Meta-Software, *HSPICE User's Manual,* Meta-Software Inc., Campbell, CA, 1987.

[31] J. J. Dongarra *et al., LINPACK User's Guide,* Soc. Ind. Appl. Math., Philadelphia, 1979.

[32] S. R. White, "Concepts of scale in simulated annealing," in *Proc. IEEE Int. Conf. Comput. Design,* 1984, pp. 646–651.

Sujoy Mitra (S'94) received the B.Tech. degree in electronics and electrical communication engineering from the Indian Institute of Technology, Kharagpur, India, in 1987, and the M.S. degree in electrical and computer engineering from Carnegie Mellon University in 1992, where he is currently working toward the Ph.D. degree.

He has held software and design engineering positions at Texas Instruments, Bangalore, India, during 1987–1990. His research interest include physical design algorithms and design automation of mixed-signal systems.

Rob A. Rutenbar (S'77–M'78–SM'90) received the Ph.D. degree in computer engineering (CICE) from the University of Michigan, Ann Arbor, in 1984.

He then joined the Faculty of Carnegie Mellon University, Pittsburgh, PA, where he is currently Professor of Electrical and Computer Engineering, and of Computer Science, and Director of the Semiconductor Research Corporation-CMU Center of Excellence in CAD and IC's. His research interests focus on circuit and layout synthesis for mixed-signal ASIC's, for high-performance digital IC's, and for FPGA's.

Dr. Rutenbar received a Presidential Young Investigator Award from the National Science Foundation in 1987. At the 1987 IEEE-ACM Design Automation Conference, he received a Best Paper Award for work on analog circuit synthesis. He is currently on the Executive Committee of the IEEE International Conference on CAD, and Program Committees for the ACM/IEEE Design Automation Conference and European Design & Test Conference. He is on the Editorial Board of IEEE SPECTRUM, and chairs the Analog Technical Advisory Board for Cadence Design Systems. He is a member of ACM, Eta Kappa Nu, Sigma Xi, and AAAS.

L. Richard Carley (S'77–M-81–SM'90) received the S.B. degree from the Massachusetts Institute of Technology, Cambridge, MA, in 1976 and was awarded the Guillemin Prize for the best EE Undergraduate Thesis; he received the M.S. degree in 1978 and the Ph.D. degree in 1984, also from MIT.

He is currenly a Professor of Electrical and Computer Engineering at Carnegie Mellon University. He has worked for MIT's Lincoln Laboratories and has acted as a Consultant in the area of analog circuit design and design automation for Analog Devices and Hughes Aircraft, among others. In 1984 he joined Carnegie Mellon, and in 1992 he was promoted to Full Professor. His current research interests include the development of CAD tools to support analog circuit design, the design of high-performance analog signal processing IC's, and the design of low-power high-speed magnetic recording channels.

Dr. Carley received a National Science Foundation Presidential Young Investigator Award in 1985, a Best Technical Paper Award at the 1987 Design Automation Conference, and a Distinguished Paper Mention at the 1991 International Conference on Computer-Aided Design.

David J. Allstot (S'72–M'72–SM'83–F'92) received the B.S. degree in engineering science from the University of Portland, Portland, OR, the M.S. degree in electrical and computer engineering from Oregon State University, Corvallis, and the Ph.D. degree in electrical engineering and computer science from the University of California at Berkeley in 1979. His Ph.D. work dealt with the analysis, design, and implementation of switched-capacitor filters.

He has held industrial positions with Tektronix, Texas Instruments, and MOSTEK, and academic positions with UC Berkeley (Acting Assistant Professor and Visiting MacKay Lecturer), Southern Methodist University, Dallas, TX (Adjunct Assistant Professor), and Oregon State University (Associate Professor). He is currently Professor of Electrical and Computer Engineering at Carnegie Mellon University, Pittsburgh, PA, and Associate Director of the SRC-CMU Research Center for Computer-Aided Design. He has advised about 40 M.S. and Ph.D. students and has published more than 100 papers with students and colleagues.

Dr. Allstot was a co-recipient of the 1980 IEEE W.R.G. Baker Prize Paper Award, and has received excellence in teaching awards from SMU, OSU, and CMU. He has served as Associate Editor of the IEEE TRANSACTIONS ON CIRCUITS AND SYSTEMS, Guest-Editor for the IEEE JOURNAL SOLID-STATE CIRCUITS, Technical Program Committee member of the IEEE Custom IC Conference, the IEEE International Symposium on Circuits and Systems, the 1994 Symposium on Low Power Electronics, and member of the IEEE Circuits and Systems Society Board of Governors. He is currently Editor of the IEEE TRANSACTIONS ON CIRCUITS AND SYSTEMS II: ANALOG AND DIGITAL SIGNAL PROCESSING and a member of the Technical Program Committee of the IEEE International Solid-State Circuits Conference. He is member of Eta Kappa Nu and Sigma Xi.

System-level Routing of Mixed-Signal ASICs in WREN

Sujoy Mitra, Sudip K. Nag, Rob A. Rutenbar and L. Richard Carley

Department of Electrical and Computer Engineering
Carnegie Mellon University
Pittsburgh, Pennsylvania 15213

Abstract†

This paper presents new techniques for global and detailed routing of the macrocell-style analog core of a mixed-signal ASIC. We combine a comparatively simple geometric model of the problem with an aggressive simulated annealing formulation that selects paths while accommodating numerous signal-integrity constraints. Experimental results demonstrate that it is critical to attack such constraints both globally (system-level) and locally (channel-level) to meet designer-specified performance targets.

1 Introduction

An increasing fraction of new application-specific integrated circuit (ASIC) designs require some core of analog functionality. Although the die area of these mixed-signal ASICs devoted to analog circuitry is usually modest, the lack of mature synthesis and layout tools renders their design a difficult, time-consuming process. This paper focuses on system-level analog routing, specifically, global and detailed routing of the analog core of a mixed-signal IC, assuming the core to be composed of arbitrary rectangular analog macrocells. For mixed-signal applications, robust methods for these basic problems are lacking. Indeed, it is not uncommon in industrial practice to route such designs with digital tools and then resolve the analog problems later, by hand.

The key problem in analog routing is signal integrity, *i.e.*, ensuring that critical wires do not encounter performance-degrading parasitics. Several schemes for analog cell routing have appeared [4][12]. At the system level, most attempts focus on row-based layouts rather than the more general 2-d macrocell style. (One exception is [10] which supports 2-d analog macrocells, but optimizes primarily for area instead of signal integrity.) In row-based schemes, global routing is either done implicitly, by segregrating channels to carry exclusively analog or digital signals [7], or explicitly, by priority schemes that give sensitive signals right-of-way in mixed-signal channels [11]. Detailed routing is done typically with constraint-graph-based channel routing techniques [8][1]. The graph is augmented with analog-specific spacing constraints to avoid deleterious signal crossings or proximities [6][3].

A related problem in system-level routing is the form of the electrical constraints that can be accommodated. For some limited but important application domains, *e.g.*, linear or mostly-linear circuits such as filters, elegant techniques exist for transforming performance goals into accurate parasitic bounds on individual nets [2]. In practice, these rely on sensitivities obtained via fast adjoint methods. For some special circuits, *e.g.*, those with perfect row-wise bilateral symmetry, some constraints can be met constructively by symmetric detailed routing of signals [3].

Unfortunately, many large industrial designs violate the assumptions that underlie these prior routing methods. Sophisticated analog subsystems are often composed of custom cells not limited to rows, are often too large for channel-wise segregation of interacting signals, and are often missing any compliant chip-level symmetrical placement of pins. The functions implemented by these subsystems–and hence the performance goals to be met during routing–are often nonlinear. Also, because such systems are simulated at the behavioral level, it is hence difficult to extract low-level sensitivities. Accordingly, many critical performance goals of these subsystems are difficult to abstract into simple constraints that *guarantee* system correctness.

In industrial practice, such systems are successfully laid out by relying on their designers to specify critical signal interactions. Such specifications may be *structural*, *e.g.*, they define classes of signals that must not interact, or *quantitative*, *e.g.*, they prescribe a minimum acceptable signal-to-noise ratio to bound the coupling among signals with known relative magnitude and frequency.

This paper presents new techniques for global and detailed routing of analog subsystems on mixed-signal ASICs. The key ideas are: (1) we restrict ourselves to comparatively simple geometric models for the routing; (2) but we use aggressive optimization techniques to balance among competing electrical tradeoffs among paths. Specifically, we use simulated annealing to select paths and optimize signal integrity for both global and channel routing. We support structural and quantitative constraints of the type mentioned above. We are unaware of any previous system-level routing techniques that handle this full spectrum of practical constraints for 2-d macrocell-style analog subsystems.

The remainder of the paper is organized as follows. Sec. 2 summarizes the overall strategy and assumptions that underlie the approach. Secs. 3 and 4 then describe our global and detailed routing algorithms, respectively. Sec. 5 describes the implementation of these ideas in WREN, and gives experimental results on a realistic example. Finally, Sec. 6 offers concluding remarks.

2 Solution Strategy and Constraints

We outline here briefly the major design decisions that underlie our mixed-signal routing strategy.

†. This work was supported in part by the Semiconductor Research Corporation, and by IBM.

Focus on signal integrity: our principal goal is to ensure that signals do not interact, via coupling parasitics, beyond arbitrary designer-specified electrical limits. Note this is quite distinct from the typical goals of digital system routing: minimize area, wirelength, delay. Given that the analog core of a mixed-signal ASIC occupies, for example, 25% of overall silicon area, a 20% area penalty paid to ensure that coupling constraints are *not* violated will impact the overall chip by a negligible 5%. We explicitly expect to pay such a penalty to achieve *working* silicon.

Scale: note that the size of a typical analog core is modest. Tens of blocks and hundreds of nets is common. Mixed-signal wiring channels are also typically sparser than comparable digital-only channels, to no surprise: signals in analog form require fewer wires than, say, a digitized signal routed as a wide parallel bus. We conclude that we may employ more aggressive optimization techniques because the amount of geometry we have to manipulate is not burdensome.

Simple geometric model: we assume a fixed, slicing-style placement of rectangular macrocells, and assume a simple 2-layer gridless channel routing model to facilitate the use of combinatorial optimization techniques for both global and detailed routing. We do not support the complex net symmetries of [3] since these are rarely useful outside of perfectly-symmetric, module-generated placements. We do, however, support special routing of matched differential net pairs.

Optimization: we use annealing for both global and detailed routing. Such techniques have been proposed in the past, but were not widely regarded as successful [17][18]. One reason is that when the goal is just to pack geometry as tightly as possible, opportunistic heuristics do work well, and quickly. However, when the goal is to *balance* a large set of *competing* electrical constraints determined by fine-grain geometric choices, optimization strategies like annealing become much more attractive. Further, this early work was hampered by choices for geometric layout models (*e.g.*, 2-point nets, gridded channels) and annealing control mechanisms that are primitive by the standards of recent work [9][12].

Partitioning of concerns: it is important to understand *where* and *how* signal integrity constraints are handled in the global router and channel router. The global router seeks to avoid fundamental crossings of incompatible signals, a primary source of parasitic coupling. Because it works with entire nets and handles junction pin ordering, it can detect such interactions. But because the global router does not work with detailed net geometry, it is blind to parasitics caused by the mere proximity of nets, *e.g.*, crosstalk between parallel lines. Conversely, given a complete global routing, the channel router has relatively few degrees of freedom to avoid fundamental net crossings, but it can easily respace proximate signals to avoid unwanted coupling. Thus, the channel router strives not to add crossings beyond those known to appear in the globally routed layout, and not to violate proximity constraints.

Performance constraints: we assume designers specify arbitrary *functions* of measurable properties of the evolving routes (e.g., length, crossing/parallel-run capacitance, etc.) and *bounds* on these functions. Because the global router manipulates complete (though somewhat simplified) nets, it selects paths to minimize the deviation from these bounds. However, because the channel router sees only channel-wise portions of nets, we must transform the net-level performance constraints into per-channel constraints. We rely on a linear programming formulation here (in contrast with [2],which uses an alternate quadratic-programming solution). It is worth emphasizing that we strive to satisfy *designer*-specified constraint functions. Guaranteed parasitic targets for system-level performance goals are rarely available or affordable.

3 The Global Router

The global router assigns nets to channels by selecting from a finite palette of alternate feasible topologies for each net. Our annealing-based scheme is most similar to [13], though the general style has also been attempted via integer-programming [14], and other heuristic means [16]. The major departure in our work is the emphasis on signal integrity rather than area and wirelength. By perturbing all nets simultaneously (as opposed to sequential schemes [10]) we can evaluate all global signal interactions and select alternative paths to avoid these interactions.

The major technical difference between our formulation and previous formulations is the use of two-point segments as the atomic, movable objects during annealing. Similar to [12], we decompose each multi-point net into a minimum spanning tree, each of the connections of which we refer to as a *segment*. We then use Lawler's algorithm [15] to create set of alternate topologies, called *paths*, for each two-point *segment* in the spanning tree. The annealing process assembles multi-point nets by choosing a path for each segment of each net. By correctly accounting for where paths of the same net share channels, we allow the paths to create a Steiner tree. Specifically, we avoid double counting portions of different paths that will become a single piece of geometry in a single net after detailed routing.

The motivation for this scheme is to maximize the set of topological alternatives for each net, while managing the complexity of the annealing move set. When two incompatible nets cross during global routing, we want to be able to incrementally reshape individual two-point segments of each net, *i.e.*, to detour dramatically small sections of each net to avoid the problem. Digital routers in this style usually select from M good (short) paths for each net. In contrast, if we select M paths for *each* of the n-1 two-point segments of an n-point net, we can potentially create M^{n-1} different net topologies, and allow each net segment to detour widely to avoid other incompatible nets.

For efficiency, a table of path-to-path interactions is compiled in a pre-processing step before annealing. A subtle point is that this determination must be made in a way that is ultimately compatible with the final junction pin ordering performed after annealing. If pin ordering fails to respect (as far as possible) path interaction decisions made during annealing, pin ordering can seriously degrade the quality of the global routing. For each pair of electrically common paths, we determine where, if at all, these segments overlap. For electrically distinct paths, we use the pla-

Figure 1. *Path exploration during global routing*

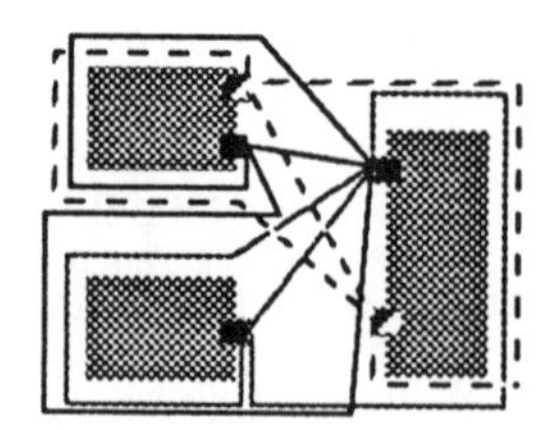

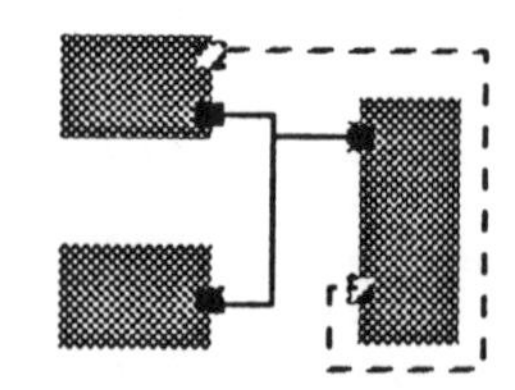

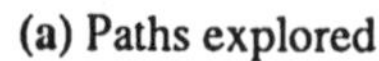

(a) Paths explored

(b) Paths accepted; 3-pt net is composed of two 2-pt paths

nar tree-embedding scheme of [5] to predict if, after final junction pin ordering, these signals *must* cross. We also employ some simple heuristics to deal with the problem of multi-point nets, *i.e.*, to avoid double counting crossings if one path of net *i* intersects two paths of net *j* in a location *common* to both net *j* paths.

The annealer perturbs nets by selecting random paths for each segment. Though we strive to account carefully for all net crossings, crossings *per se* are *not* part of the cost function. Rather, the annealer minimizes the *deviation* from designer-supplied performance functions that depend on geometric characteristics such as crossings. The cost function also includes a secondary netlength term to encourage short paths where possible. Weights in the cost function are initially set by performing a quick quench to determine the relative magnitudes of its component terms. Weights are thereafter modulated dynamically to "keep pressure" on the hard constraints (see, *e.g.*, [9]). The cooling schedule is as in [12].

4 The Detailed Router

To date, the total constraints graph (TCG) model of [8][1] has been the common choice for analog channel routers [6][3]. Analog proximity constraints are easily added to this model[6]. We also employ a variant of this model, but depart from tradition in that we substitute simulated annealing for the greedy heuristics commonly used to resolve the necessary topological constraints. Our early experiments suggested that annealing can better trade off among competing electrical constraints than opportunistic schemes that cannot undo poor routing decisions. This obviously impacts efficiency: channels require a few minutes, rather than seconds, to route; but we regard this as an acceptable trade-off.

In the TCG model, a vertex represents a horizontal segment of a net in a channel, a directed edge represents a *vertical* constraint and an undirected edge represents a *horizontal* constraint. The weights of these edges reflect the design rules. The assignment of directions to undirected edges constitutes a complete routing solution. Note that a routing solution can be considered to be a *topological sorting* of these vertices. We use this fact as shown in Fig. 2 in our annealing formulation. We maintain at all times an ordered array representing the nets as visited in *some* solution (*i.e.*, *some* topological sorting) of the TCG. Annealing randomly perturbs the order of the nets in this array. Moves that violate vertical constraints in the TCG are rejected. Feasible moves perform a depth-first longest-path search of the graph, which uniquely locates every piece of channel geometry. Thereafter, the exact values of all parasitics are determined through simple computational geometry methods.

Figure 2. *Channel routing model (nets B,D incompatible)*

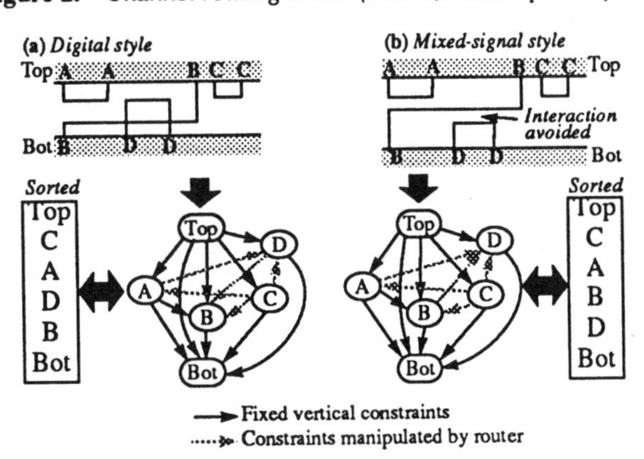

Of course, an acyclic TCG is required. To break the cycles, we rank all nets as to their suitability for doglegs, create a small palette of dogleg topologies for *each* net on this list, then dogleg nets in rank order until the TCG is cycle-free. We then exhaustively evaluate all possible combinations of these dogleg net topologies and choose the best, by means of a cost function that measures potential incompatible net crossings, and potential incompatible proximities (*i.e.*, common horizontal span of incompatible signals). This scheme works well in practise, since (1) cycles can usually be broken with a handful of good dogleg choices, and (2) careful dogleg choices are necessary to comply with the junction pin ordering specified during global routing.

The annealer's cost function is a weighted sum of four terms: *ExcessParasitic*, which varies in proportion to the violations of the parasitic budget allocated to each net in this channel; *ChannelHeight*, *ViaArea*, and *PinArea*, which we minimize as secondary objectives. Weighting and cooling are similar to the global router.

Another critical part of the channel routing process is the mapping scheme by which our constraint functions (used directly by the global router) are transformed into per-channel per-net parasitic budgets. Assume we are given a set of F constraint functions of the form $\varphi_i(P) \in [L_i, U_i]$, where $P = (p_1, p_2, \ldots, p_E)$ is the set of E potential system-wide parasitics. Recognizing that detailed routing is the incremental addition of parasitics, and that each system-wide p_e must be modeled as a channel-wise sum of local parasitics p_{ec} visible over C different channels, we linearize each $\varphi_i(P)$ constraint to obtain:

$$L_i \le \varphi_i(P) \le U_i \Rightarrow L_i \le \sum_{e=1}^{E} S_e^i \left(\sum_{c=1}^{C} p_{ec} \right) \le U_i \,,\; S_e^i = \frac{\partial \varphi_i}{\partial p_e} \qquad (1)$$

We have, in addition, one more set of constraints. Global routing places hard constraints on how much of each system parasitic p_e *must* appear in each channel c, namely:

$$p_{ec}(\min) \le p_{ec} \le p_{ec}(\max) \qquad (2)$$

The goal is to budget a maximum p_{ec}^b for each parasitic p_{ec} in each of the C channels that does not violate constraints (1) or (2) above. Detailed routing is, of course, most flexible when the parasitic budgets are as *large* as the constraining equations allow. We choose to maximize, via standard linear programming, a linear combination of these parasitics weighted by the maximum of the absolute values of the sensitivities that affect each parasitic:

$$\operatorname*{Max}_{p_{ec}^b} \left\{ \sum_{e=1}^{E} \max_i \left| S_e^i \right| \cdot \sum_{c=1}^{C} p_{ec}^b \right\} \qquad (3)$$

The excess over these computed bounds p_{ec}^b comprises the *ExcessParasitic* penalty term in the router's annealing cost function

5 Experimental Results

We have implemented these routing ideas in a package called WREN. The global and detailed channel routers that comprise this package are called WRENGR and WRENDR, respectively. We begin with two small examples that explain in more detail the interaction between global and detailed routing, and the precise man-

Figure 3. *SNR-style model of parasitic coupling constraints*

V_j — *Noisy*; C_{ij} coupling parasitic; V_i — *Sensitive*; G_{is}, C_{is}; Substrate

$$y_{ij} = j\omega C_{ij}$$

$$y_{is} = G_{is} + j\omega C_{is}$$

$$p_{ij} = \frac{y_{ij}}{y_{ij} + y_{is}}$$

ner in which electrical constraints are handled. We then describe a more realistic system-level mixed-signal ASIC example.

Fig. 3 illustrates the *style* of the performance constraints chosen for all the examples in this section. We seek to bound the coupling of a noisy signal j onto a sensitive signal i, assuming we know their relative voltage swings (V_i, V_j) and frequency (ω). Observe that this models a typical mixed-signal interaction such as a noisy digital clock interfering with a high-impedance analog signal as though all the energy in the clock is at its fundamental harmonic. This formulates parasitic bounds for signal i roughly in the style of a *signal-to-noise* ratio (SNR) constraint. Given a user-supplied bound SNR_i, we strive to enforce:

$$SNR_i \le \frac{|V_i|}{\sum_{j \ne i} |V_j|\,|p_{ij}|}, \qquad \sum_{j \ne i} |V_j|\,|y_{ij}| \le \frac{|V_i|\,|y_{is}|}{SNR_i} \tag{4}$$

WRENGR uses the constraint on the left side of (4); WRENDR uses the right side of (4) during parasitic budgeting. Observe that each constraint is a function of many parasitics.

Let us examine how the global and detailed routers use this electrical model of performance constraints. Consider first the simple global routing problem of Fig. 4. The problem has five signal nets, labeled r, s, t, u, v. Assume signal s is sensitive to interactions with its neighboring nets. These interactions are shown as parasitic capacitors between nets. Accordingly, the designer provides a target SNR to ensure the integrity of signal s. WRENGER generates a set of topological alternatives for signal s, in this case, three alternatives labeled s_1, s_2 and s_3 in the figure. The job of the global router is to select the optimal alternative, *i.e.*, the one that minimizes any violation of the SNR constraint, as well as net length. These three choices gives rise to three different *circuits* modeling the effect of crossing-induced parasitics. For example, if path s_1 is chosen, a somewhat stylized version of

Figure 4. *Simple global routing example*

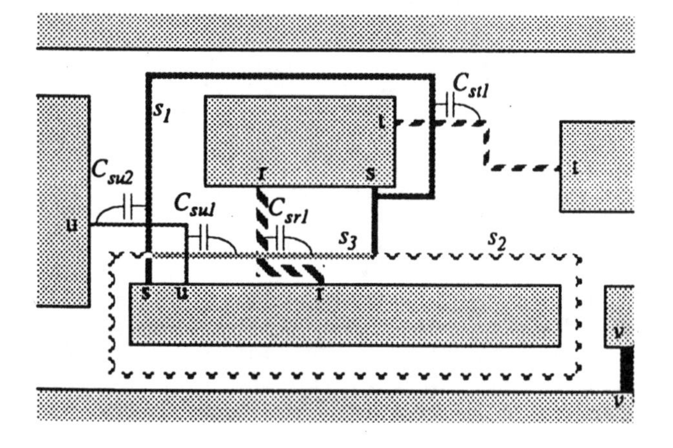

Figure 5. *Parasitic coupling for path alternative s_1*

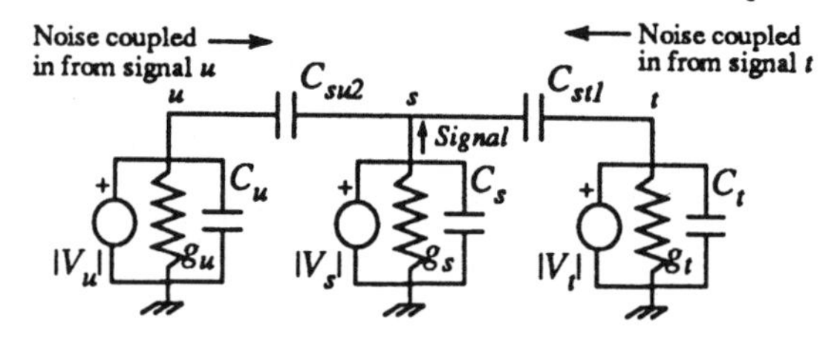

the equivalent electrical circuit is shown in Fig. 5:

The expressions for the signal-to-noise ratios for each of the three alternative paths can be easily derived:

$$snr_s(\mathrm{dB}) = 20\log\frac{|V_s|}{|N_i|},\ i = 1, 2, 3$$

$$N_1 = V_u \frac{j\omega C_{su2}}{g_u + j\omega(C_u + C_{su2})} + V_t \frac{j\omega C_{st1}}{g_t + j\omega(C_t + C_{st1})}$$

$$N_2 = 0 \tag{5}$$

$$N_3 = V_u \frac{j\omega C_{su1}}{g_u + j\omega(C_u + C_{su1})} + V_r \frac{j\omega C_{sr1}}{g_r + j\omega(C_r + C_{sr1})}$$

As these equations indicate, WREN needs to know certain electrical information for each signal k. These include the expected voltage swing for each signal V_k, the capacitances C_k and conductances g_k from the terminals of each signal to the substrate, the operating frequency f, and the width w_k of the net implementing each signal and capacitance per unit area of the various layer-to-layer crossings. Assuming the electrical data specified in Table 1, Table 2 gives precise values for the various possible crossing capacitances the global router will encounter for this example, and the SNR values for each of the three possible topological alternates for signal s.

Assuming a SNR target of *at least* 75 dB for signal s, observe that choice s_3 is unacceptable. Both s_1 and s_2 are acceptable, but

Table 1. *Electrical specifications (assuming f = 30 MHz, crossing capacitance =0.06 fF/μ^2)*

Signal k	w_k (μ)	$\|V_k\|$ (V)	C_k (pF)	g_k (mS)
r	5	3	2	0.1
s	2	5	2	0.1
t	2	2	2	0.1
u	2	5	2	0.1
v	10	5	2	0.1

Table 2 *Resulting SNR for path choices for signal s*

Crossing Cap	fF
C_{stl}	0.24
C_{su2}	0.24
C_{sul}	0.24
C_{srl}	0.60

Path for signal s	SNR (*dB*)
s_1	75.81
s_2	∞
s_3	70.75

Figure 6. *Simple detailed routing example*

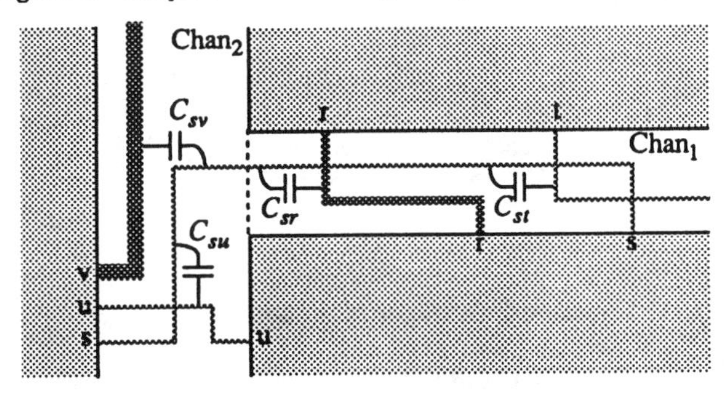

WRENGR chooses s_1 because its length is shorter than s_2. If, however, the SNR target was 100 dB, WRENGR would choose path s_2, despite its length.

Of course, only *crossing* capacitance is accommodated in the global router on the assumption that all *proximate* capacitance can be removed during detailed routing by appropriately spacing out parallel runs of sensitive wires. To illustrate this, consider the simple two-channel detailed routing problem with five signal nets *r, s, t, u, v* shown in Fig. 6. We will again use the electrical specifications of Table 1. As before, assume signal *s* is sensitive to interactions with its neighbors. The expression for the signal-to-noise ratio on signal *s* is:

$$snr_s = 20\log\frac{|V_s|}{\left|\sum_{i=r,t,u,v} N_i\right|}$$
$$N_i = V_i\frac{j\omega C_{is}}{g_s + j\omega(C_s + C_{is})} \quad (6)$$

The designer specifies a constraint $snr_s = SNR_s$. The parasitic budgeting step in WRENDR attempts to determine budgets for routing-induced parasitics so that these constraints are satisfied. The budgeting process relies on linear programming, so it is necessary to linearize the expression for snr_s. In linearizing, we make two observations: all proximate capacitance can be reduced to negligible values by appropriately spacing out parallel runs of sensitive wires, and the crossing capacitances (≈ 0.6 fF in this example) are negligible *compared* to the capacitance at the terminals (≈ 2 pF here) of the cells. This implies that the parasitic capacitance terms in the *denominator* of N_i in (6) can be neglected, thereby linearizing the expression for snr_s.

Table 3 shows how WRENDR will manage these specific parasitics. Note that parasitics C_{su} and C_{sr} are unavoidable by the detailed router, but that C_{st} and C_{sv} can be mitigated by appropriate spacing (between *v* and *s* in $Chan_2$) and topological ordering (place *t* above *s* in $Chan_1$).

Depending on the value of the SNR_s constraint specified for signal *s*, different budgets will be created for each parasitic in each channel. The four columns of Table 3 illustrate the budgets that must be enforced by the detailed router for each parasitic. For example, if a relatively small value of SNR_s is specified, less than 58 dB, WRENDR will allow signals *s* and *t* to cross in $Chan_1$, and will allow signals *s* and *v* to be adjacent in $Chan_2$ to reduce channel height. However, as the SNR_s constraint is tightened, note that the budgets for the avoidable parasitics are reduced. For example, if more than 73 dB of rejection is required, then WRENDR will not cross *s* and *t* in $Chan_1$, and will add extra space between *s* and *v* in $Chan_2$. Note this will obviously increase the height of each channel, but that this is unavoidable if the signal integrity constraints are to be met.

To illustrate the use of WRENGR and WRENDR on a more realistic example, we have routed the analog core of a *pseudo*-industrial mixed-signal ASIC: the netlist was abstracted (and sanitized) from a production industrial ASIC, but retains similar density of wiring and critical signal interactions. This analog core has 25 macrocells and 270 nets. Of these nets, 110 are digital control signals, 28 are sensitive analog signals, and the remainder are relatively insensitive. Using the same SNR-style performance constraints introduced at the beginning of this section creates 28 SNR-style constraints for this ASIC. For this chip, the noisy signals we are concerned about operate at $\omega = 2\pi \cdot 30MHz$.

Fig. 7 shows the successfully routed analog core of this ASIC. Table 4 shows the results of a controlled experiment in which both the global and detailed routers were run with signal-integrity constraints *disabled* (labeled "*Digital*" in table) and *enabled* ("*Mix-Sig*"). Each row of the table counts the number of signals violating SNR constraints by the listed amount. The important conclusion is that it is essential to consider parasitic interactions during *both* global and detailed routing; neither tool can succeed in avoiding numerous SNR violations alone. As expected, the penalty for allowing signals to detour or to add excess wire-to-wire spacing is added area, about 5% here. Not shown in the table is that the wire-length penalty is about 15%.

It is worth mentioning that in the case of failure to meet constraints, we rerun the global router, but allow it to create *more* distinct paths per segment, to explore net topologies that detour more widely. The results of Table 4 were done with 6 paths/segment. Interestingly, if we reduce this to 5 or fewer paths/segment, we

Table 1 *Parasitic management during detailed routing*

Parasitic Cap	Computed Parasitic Budget (fF) depending on specified SNR_s			
	58 dB	59 dB	71 dB	73 dB
C_{st}	0.48	0	0.48	0
C_{sr}	0.60	0.60	0.60	0.60
C_{su}	0.24	0.24	0.24	0.24
C_{sv}	2.40	2.40	0	0

Figure 7. *Analog core routed by* WREN

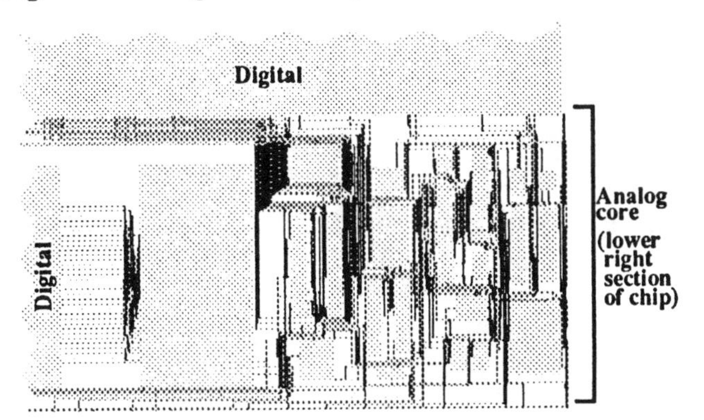

Table 3. WREN *ASIC benchmark results*

Global Route Style	Chan Route Style	SNR Violations (dB) / Count						Norm Area
		< 0	0-3	3-10	10-30	30-60	>60	
Digital	*Digital*	9	3	7	6	1	2	1.0
Digital	*MixSig*	11	3	9	2	3	0	1.016
MixSig	*Digital*	10	1	8	9	0	0	1.034
MixSig	*MixSig*	25	3	0	0	0	0	1.049

generate several unacceptable SNR violations. Finally, it is also worth noting that the CPU times for these examples are reasonable, about 3 CPU hours end-to-end on a DECstation^{tm} 3100.

6 Conclusions

We have presented new algorithms for global and detailed routing of the macrocell-style analog core of a mixed-signal ASIC. A comparatively simple geometric model, coupled with novel annealing techniques to minimize violations of signal-integrity constraints, produces excellent results. The area and wirelength penalties (~15%), and overall CPU time (~hours) necessary to meet these signal integrity constraints, are an acceptable price for working silicon.

Acknowledgments: We are grateful to Bob Stanisic of IBM for help with the benchmark, and formulation of the SNR-style constraint functions.

References

[1] H. H. Chen & E. Kuh, "Glitter: A Gridless Variable Width Channel Router," *IEEE Trans. CAD*, vol. CAD-5, no. 4, pp. 459-465, Oct. 1986.

[2] U. Chowdhury & A. Sangiovanni-Vincentelli, "Constraint Generation for Routing Analog Circuits," *Proc. DAC*. pp. 561-566, June 1990.

[3] U. Chowdhury and A. Sangiovanni-Vincentelli, "Constraint Based Channel Routing for Analog and Analog/Digital Circuits" *Proc. IEEE ICCAD*, pp. 198-201, Nov 1990.

[4] J. M. Cohn, *et al.*, "KOAN/ANAGRAM II: New Tools for Device-Level Analog Placement and Routing," *IEEE JSSC*, vol. 26, no. 3, March, 1991.

[5] P. Groeneveld, "Wire Ordering for Detailed Routing," *IEEE Design & Test of Computers*, pp. 6-17, Dec. 1989.

[6] R. S. Gyurcsik and J. C. Jeen, "A Generalized Approach to Routing Mixed Analog and Digital Signal Nets in a Channel," *IEEE JSSC*, vol. CAD-24, no. 2, pp. 436-442, April 1989.

[7] C. D. Kimble, *et al.*, "Analog Autorouted VLSI," *Proc. IEEE CICC.*, June, 1985.

[8] H. Kimura, *et al.*, "An Automatic Routing Scheme for General Cell LSI," *IEEE Trans. CAD*, vol. 2, pp. 285-292, Oct. 1983.

[9] E. S. Ochotta, *et al*, "Equation-Free Synthesis of High Performance Linear Analog Circuits," *Proc. 1992 Brown/MIT Conference on Advanced VLSI*, The MIT Press.

[10] J. Rijmenants, *et al*, "ILAC: An Automated Layout tool for Analog CMOS Circuits," *IEEE JSSC*, pp. 417-425, vol. 24, no. 2, April 1989.

[11] I. Harada, H. Kitazawa, and T. Kaneko, "A Routing System for Mixed A/D Standard Cell LSI's," *Proc. IEEE ICCAD*, pp.378-381, 1990.

[12] J. M Cohn, D. J. Garrod, R. A. Rutenbar and L. R. Carley, "Techniques for Simultaneous Placement and Routing of Custom Analog Cells in KOAN/ANAGRAM II, "*Proc. IEEE ICCAD*, pp.394-397, 1991.

[13] C. Sechen, "Placement and Global Routing of Macro/Custom Cell Integrated Circuits using Simulated Annealing", *Proc. 25th ACM/IEEE DAC*, pp. 73-80, June 1988.

[14] T. Lengauer, *Combinatorial Algorithms for Integrated Circuit Layout*, Wiley-Taubner, 1990.

[15] E. Lawler, *Combinatorial Optimization: Networks and Matroids*, Holt, Rinehart, and Winston, 1976.

[16] D. Chen and C. Sechen, "Mickey: A Macro Cell Global Router," *Proc. SRC TECHCON '90*, pp. 241-244, 1990.

[17] H. W. Leong, D. F. Wong and C. L. Liu, "A Simulated Annealing Channel Router," *Proc. IEEE ICCAD*, 1985, pp. 226-228.

[18] M. P. Vecchi and S. Kirkpatrick, "Global Wiring by Simulated Annealing," *IEEE Trans. Computer-Aided Design*, vol CAD-2, No. 4, pp. 215-222, Oct 1983.

Addressing Substrate Coupling in Mixed-Mode IC's: Simulation and Power Distribution Synthesis

Balsha R. Stanisic, *Member, IEEE,* Nishath K. Verghese, *Student Member, IEEE,* Rob A. Rutenbar, *Senior Member, IEEE,* L. Richard Carley, *Senior Member, IEEE,* and David J. Allstot, *Fellow, IEEE*

Abstract—**This paper describes new techniques for the simulation and power distribution synthesis of mixed analog/digital integrated circuits considering the parasitic coupling of noise through the common substrate. By spatially discretizing a simplified form of Maxwell's equations, a three-dimensional linear mesh model of the substrate is developed. For simulation, a macromodel of the *fine* substrate mesh is formulated and a modified version of SPICE3 is used to simulate the electrical circuit coupled with the macromodel. For synthesis, a *coarse* substrate mesh, and interconnect models are used to couple linear macromodels of circuit functional blocks. Asymptotic Waveform Evaluation (AWE) is used to evaluate the electrical behavior of the network at every iteration in the synthesis process. Macromodel simulations are significantly faster than device level simulations and compare accurately to measured results. Synthesis results demonstrate the critical need to constrain substrate noise and simultaneously optimize power bus geometry and pad assignment to meet performance targets.**

I. Introduction

The push for reduced cost, more compact circuit boards, and added customer features has provided incentives for the inclusion of analog functions on primarily digital MOS integrated circuits. This has prompted the development of a new generation of electrically cognizant analysis and synthesis tools. To date however, these tools have largely ignored the critical effect of the common chip substrate.

The common substrate couples noise between the on-chip digital and analog circuits that corrupts low-level analog signals, impairing the performance of mixed-signal IC's. The potential crosstalk problem of noise finding its way into sensitive analog circuitry has traditionally been handled effectively by judicious use of multiple power bus layouts and desensitized analog circuitry [1]. However, in the future, digital speeds will increase, additional analog circuitry will be included, chips will become more densely packed, interconnect layers will be added and analog resolution will be increased. In consequence, the noise crosstalk problem will worsen and designer skills will be severely taxed.

As an illustration of the significance of the problem, consider the performance of single-bit sigma-delta D/A converters which have traditionally been limited to a resolution of 16 b. Recently [2], 18-b resolution was achieved with the same topology by separating the analog and digital functional blocks onto two different IC's, with the aim of alleviating the substrate coupling problem. As technologies continue to scale and compactness of design becomes more important, analog designers will not have the leeway to fabricate their functional blocks on independent IC's. Performance degradation due to substrate noise will become difficult to control and even more difficult to predict. The need for high performance simulation and synthesis tools to identify and help avoid the problem is increasingly evident in the industry today [3].

Accurate simulation of substrate coupling has only recently begun to receive attention [4]–[6]. A device simulation program has been used to study the mechanism of substrate coupling noise [4], [5]; however, its applicability for simulating integrated circuits is limited by the long simulation times required. A single node substrate model has also previously been used to simulate substrate coupling [3], [4]. This approach, however, is applicable only to technologies employing a lightly doped epitaxial layer on a heavily doped substrate. To date, we are unaware of any process-independent tool efficient enough to simulate substrate coupling effects in large integrated circuits, or any tool able to synthesize power distribution cognizant of these same substrate effects.

In this paper we present such a process-independent simulation strategy for substrate coupling effects that is not only valid for all silicon IC technologies but also fast and accurate. We also present a power distribution synthesis strategy that incorporates substrate coupling effects and simultaneously optimizes power bus geometry and power I/O pad assignment. Section II introduces the substrate coupling problem and develops the basic substrate model. Section III describes the simulation techniques used: macromodeling and circuit simulation with macromodels. Section IV compares simulation results to those from a device simulation program and also compares simulation results with reported measurements on a test chip [4]. Section V introduces the power distribution synthesis strategy, Section VI describes the simultaneous power I/O pad assignment and power bus synthesis formulation, and Section VII describes the optimization approach. Sections VIII and IX describe the electrical modeling and evaluation strategies in the synthesis process, respectively. Section X shows synthesis results for several mixed-signal examples, and compares synthesis and simulation results to those reported in [4].

Manuscript received July 28, 1993; revised October 3, 1993. This work was supported by International Business Machines and the Semiconductor Research Corporation under Contract 93-DC-068.

The authors are with the Department of Electrical and Computer Engineering, Carnegie Mellon University, Pittsburgh, PA 15213.

IEEE Log Number 9214786.

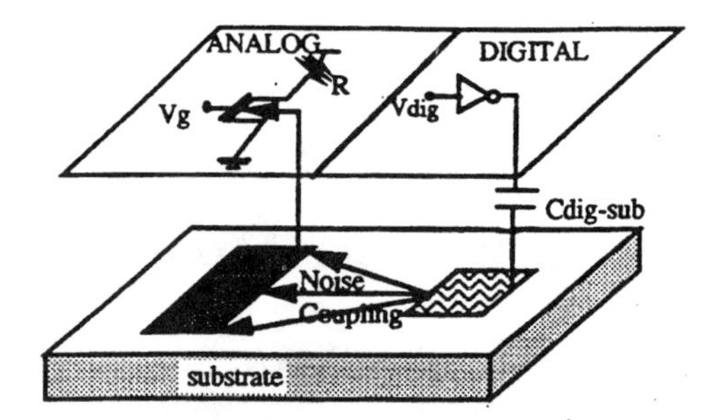

Fig. 1. The substrate coupling problem.

II. Modeling the Substrate

The general substrate coupling problem is illustrated in Fig. 1 in which digital switching nodes are capacitively coupled to the substrate through junction capacitances and interconnect/bonding pad capacitances, causing fluctuations in the underlying voltage. As a result of these fluctuations, a substrate current pulse flows between the switching node and the surrounding substrate contacts. The switching-induced current flow causes the substrate potential underlying critical transistors in the path of this flow to change. As a result of the body effect and the junction capacitances of a sensitive transistor, changes in the backgate voltage induce noise spikes in its drain current and consequently its drain voltage. To simulate this phenomenon, a necessary first step is the development of a suitable model for the substrate.

Outside the diffusion/active areas and contact areas, the substrate can be treated as consisting of layers of uniformly doped semiconductor material of varying doping densities. In these layers, a simplified form of Maxwell's equations can be formulated, ignoring the influence of magnetic fields and using the identity $\nabla \cdot (\nabla \times a) = 0$:

$$\epsilon \cdot \frac{\partial}{\partial t}(\nabla \cdot E) + \frac{1}{\rho}\nabla \cdot E = 0 \tag{1}$$

where E is the electric field intensity vector, and ρ and ϵ are the sheet resistivity and dielectric constant of the semiconductor, respectively.

There are two general approaches to solve (1): *analytical* and *numerical*. The *analytical* approach involves a search for an exact solution for structures with mathematically tractable (usually rectangular) geometries. Although it is possible to obtain a closed-form analytical solution of (1) for a simple structure, analytical solutions of complicated geometries do not generally exist. Hence *numerical* techniques must be used to solve the geometries encountered in typical integrated circuits. To employ numerical techniques, the substrate is spatially discretized using a simple box integration technique [7].

From Gauss' law,

$$\nabla \cdot E = k \tag{2}$$

where E is the electric field intensity and $k = \rho'/\epsilon$ where ρ' is the charge density of the material. Integrating $\nabla \cdot E$ over a volume Ω_i surrounding node i as shown in Fig. 2,

$$\int_{\Omega_i} \nabla \cdot E d\Omega = \int_{\Omega_i} k d\Omega. \tag{3}$$

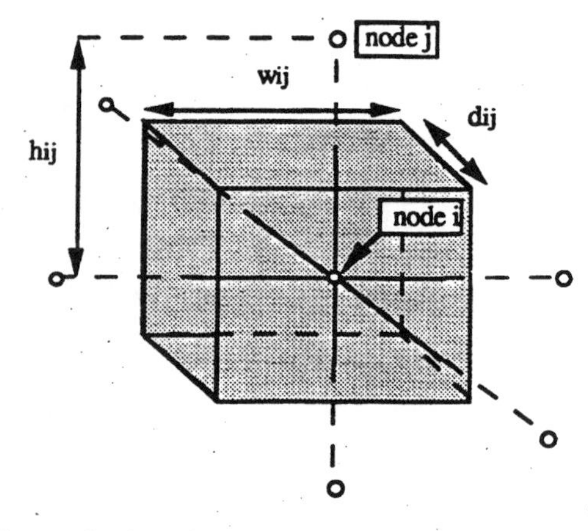

Fig. 2. A control volume in the box integration technique.

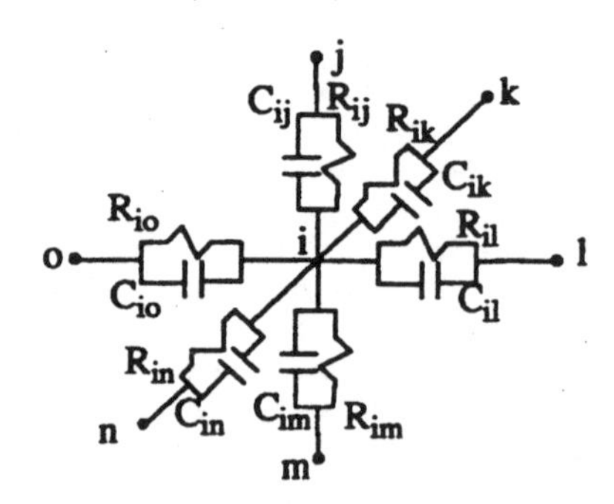

Fig. 3. Resistances and capacitances around a mesh node in the electrical substrate mesh.

From the divergence theorem,

$$\int_{S_i} E dS = \int_{\Omega_i} k d\Omega \tag{4}$$

where S_i is the surface area of the cube shown in Fig. 2. The integral on the left side of (4) can be approximated as

$$\sum_j E_{ij} \cdot S_{ij} = \sum_j E_{ij} \cdot w_{ij} d_{ij} = k \cdot \Omega_i \tag{5}$$

and hence,

$$\nabla \cdot E = k = \frac{1}{\Omega_i} \sum_i E_{ij} \cdot w_{ij} d_{ij}. \tag{6}$$

Using (6) in (1) and noting that $E_{ij} = V_i - V_j/h_{ij}$ in Fig. 2, (1) reduces to

$$\sum_j \left[\frac{(V_i - V_j)}{R_{ij}} + C_{ij}\left(\frac{\partial V_i}{\partial t} - \frac{\partial V_j}{\partial t} \right) \right] = 0 \tag{7}$$

where $R_{ij} = \rho h_{ij}/w_{ij} d_{ij}$ and $C_{ij} = \epsilon w_{ij} d_{ij}/h_{ij}$ as modeled with lumped circuit elements in Fig. 3.

Since the relaxation time of the substrate (outside of active areas and well diffusions) given by $\tau = \rho\epsilon$ is of the order of $10^{-11} s$ (with $\rho = 15\Omega - cm$ and $\epsilon_0 = 11.9$), it is reasonable to neglect intrinsic substrate capacitances for operating speeds of up to a few GHz and switching times of the order of 0.1 ns. Moreover, if the capacitances to the substrate introduced

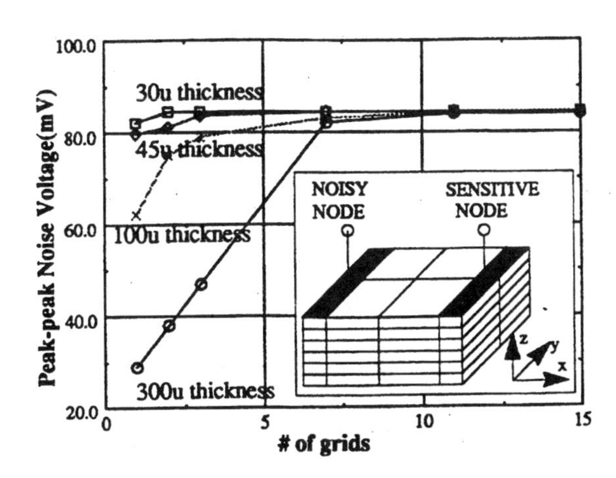

Fig. 4. The peak–peak noise voltage at the sensitive node as a function of the number of grids in the z direction (with the grid density in the x and y directions fixed) for different substrate thicknesses.

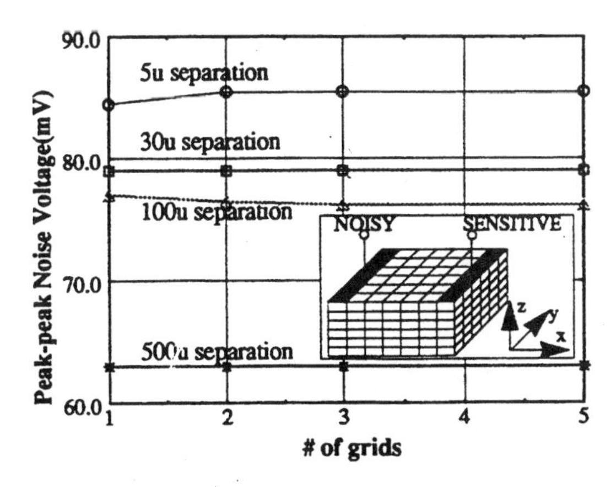

Fig. 5. The peak–peak noise voltage at the sensitive node as a function of the number of grids in the x and y directions (with the grid density in the z direction fixed) for different amounts of lateral separation.

by the depletion regions of well diffusions and interconnects overlying field oxide can be accurately modeled as lumped circuit elements outside the mesh, the substrate can be modeled as a purely resistive mesh. (Note that junction capacitances of active devices are already modeled outside the mesh as lumped capacitances.)

Although the electric field varies nonlinearly as a function of distance, the box integration method approximates this variation as a piecewise constant function. In regions where the gradient of the electric field is high, it is necessary to use fine grids to accurately approximate the nonlinearity of the electric field. Elsewhere, coarse grids can be used to reduce the overall number of grids. However, since the field intensity cannot be determined before discretization, the density of grids needed is not known *a priori*.

We determine the density of grids for a substrate with either contacts/diffusions on the surface or a backside contact using the setup of Fig. 4. It consists of a noisy node and a sensitive node separated by a fixed lateral distance. With a fixed number of uniformly distributed grids in the x and y directions between these nodes, the peak–peak noise voltage at the sensitive node is determined as a function of the grid density in the z direction for different substrate thicknesses. For the geometry shown in Fig. 4, noise-coupling is a strong function of grid density in the z direction as expected, since the gradient of the electric field is high in that direction. The number of grids needed (in the z direction) is also a strong function of the substrate thickness.

Fig. 5 shows the effects of grid density in the x and y directions with a fixed grid density in the z direction. Since the fixed boundaries are in the xy plane, the noise coupling is not as sensitive to the number of grids in the x and y directions; consequently, coarse grids are used in these directions. Thus, we empirically determine a grid density based on total substrate thickness from the results of Figs. 4 and 5. While a large body of research on the subject of automatic mesh generation exists [28], we have avoided it in our work since we have found that in the case of a linear substrate model with contacts/diffusions at the surface or a backside contact, it is relatively easy to determine the density of grids required *a priori* using the technique mentioned above.

III. Simulation Techniques

Once the substrate is discretized into a 3-D mesh it is necessary to solve it in conjunction with the electrical circuit. However, solving a 3-D mesh using traditional variable time-step trapezoidal integration techniques is prohibitive in terms of cpu time and memory since the substrate mesh has in general many more nodes than the rest of the electrical circuit. By macromodeling the mesh and performing transient analysis with the macromodel, computations for the mesh need to be performed only on those nodes that physically connect the electrical circuit to the substrate. Since the objective is to use SPICE3 [8] as a basic framework that is modified to perform substrate coupling simulations, an admittance (Y - parameter) macromodel of the substrate mesh is formulated. In computing the admittance macromodel, the matrix solution technique utilized plays a crucial role in determining the cpu time requirement. To optimize the simulation time required, specialized matrix solution techniques are used as described below.

A. Matrix Solution

The matrix solution of a large 3-D mesh network using the traditional direct solution method is known to require a large amount of cpu time and memory for LU factorization even when reordering and sparse matrix techniques are used. Since the matrix resulting from the substrate mesh is both strictly diagonally dominant and symmetric (with a nodal analysis formulation), the incomplete Choleski conjugate gradient (ICCG) iterative method [13] with an ILU (1) (incomplete LU factorization with diagonal correction only) preconditioner has been adopted as the matrix solver. Unlike other iterative methods, the ICCG method always converges to the exact solution in at most n iterations where n is the order of the matrix. Experience with ICCG shows typical iteration counts to be far fewer because of preconditioning.

B. AWE Admittance Macromodel

If the substrate is modeled as an RC mesh, Asymptotic Waveform Evaluation (AWE) can be used to determine its admittance macromodel. AWE is a technique to approximate the behavior of a linear(ized) circuit using a few dominant poles and residues in either the time or the frequency domains [9]. The AWE technique involves the computation of the circuit moments in an efficient and recursive manner. A reduced-order pole-residue model of the circuit transfer function is determined from the circuit moments using a form of Pade approximation [10]. Similarly, AWE approximations to the admittance parameters of an RC mesh can be determined in a simple manner. The admittance parameter macromodel can be simulated together with the nonlinear portions of the circuit in the time domain using the inverse Laplace transform symbolically, on a term-by-term basis [11].

C. DC Admittance Macromodel

If the substrate is modeled as a purely resistive mesh, the macromodel consists of only the steady-state/DC values of the admittance parameters, the higher-order mesh moments being zero. In contrast to the AWE macromodel where $2q+1$ matrix inversions (using the ICCG method) have to be performed for every port in the mesh to determine the qth order approximation to the admittance parameters, the DC macromodel requires only one matrix inversion per port for the computation of its admittance parameters. Moreover, transient simulation of the DC macromodel along with the nonlinear portions of the circuit is trivial and requires only the introduction of each admittance parameter into its corresponding location in the global admittance matrix generated by SPICE3 at every time-point in the simulation run.

IV. Simulation Results

To validate the macromodeling strategy, simulation results have been compared both to results from the device simulation program MEDICI [12], and also to results from measurements reported on an experimental chip [4].

A. Comparisons to Device Simulation

The device simulation program models two-dimensional distributions of potential and carrier concentrations in a device to predict its electrical characteristics for any bias condition. It solves Poisson's equation and both the electron and hole continuity equations using numerical simulation techniques to analyze devices such as diodes, BJT's, MOSFET's, etc. for dc, steady-state, or transient operating conditions. Since mixed-mode IC's are generally fabricated in processes with either a heavily doped bulk with an epitaxial layer or a lightly doped substrate (without an epitaxial layer), circuits representative of both processes have been verified using the simulation tool. The doping profile of a 2 μm BiCMOS technology [5] is used for both simulations with MEDICI and the macromodel.

Fig. 6 shows the experimental setup used to simulate substrate coupling in a heavily doped substrate with a lightly doped epitaxial layer [5]. It consists of a diffused region equivalent to the drain of a switching transistor and a single NMOS transistor current source considered to be part of a sensitive analog circuit separated by a distance of 30 μm. Several established shielding techniques to reduce the coupling from the switching node to the sensitive transistor,

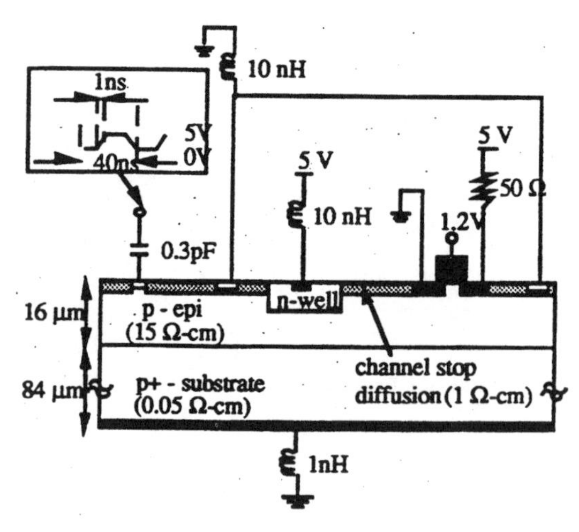

Fig. 6. Circuit schematic/layout profile for simulations with a heavily doped bulk and a lightly doped epitaxial layer.

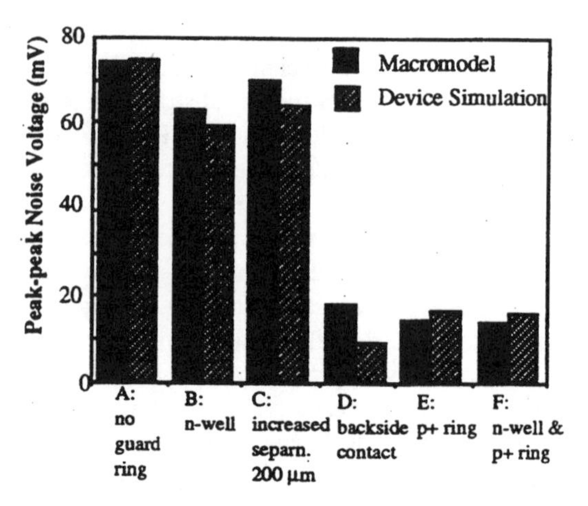

Fig. 7. Effect of various shielding techniques on peak–peak noise voltage at the sensitive node in Fig. 6.

including increased separation, an n-well diffusion, a p+ ring and a p+ contact strapping the backside of the substrate to ground potential [4] have been tested with our macromodeling technique. Figs. 7 and 8 compare the results obtained with the macromodeling technique to those obtained with the device simulation program and plots the drain noise voltage (peak-to-peak) and settling time behavior of the sensitive transistor as functions of the shielding technique used. The separation between the switching and sensitive nodes is 30 μm unless otherwise specified. The substrate contact on the far right of Fig. 6 is present in all the cases of Fig. 7 and Fig. 8. In case A there is no guard ring or backside contact between the switching and sensitive nodes. Case C shows the effect of simply increasing the separation between the two nodes to 200 μm. In case D a backside contact is used to additionally bias the substrate. In case B an 8 μm wide n-well is placed midway between the switching and sensitive nodes while in case E an additional p+ contact is placed between the nodes.

Fig. 9 is the setup used for simulations with a uniformly lightly doped substrate (without an epitaxial layer) [5] and is otherwise identical to Fig. 6. Fig. 10 shows the peak–peak voltage and Fig. 11 the settling time behavior of the result-

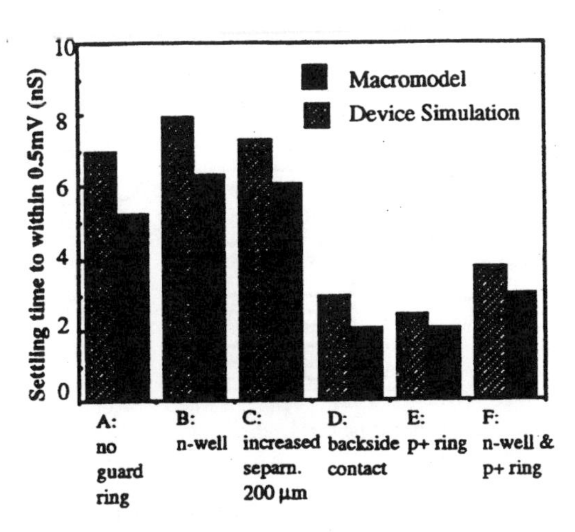

Fig. 8. Effect of various shielding techniques on settling time of the noise voltage at the sensitive node in Fig. 6.

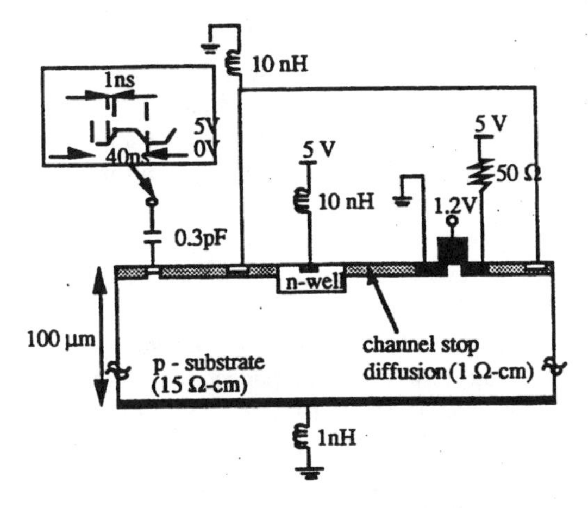

Fig. 9. Circuit schematic/layout profile for simulations with a lightly doped substrate.

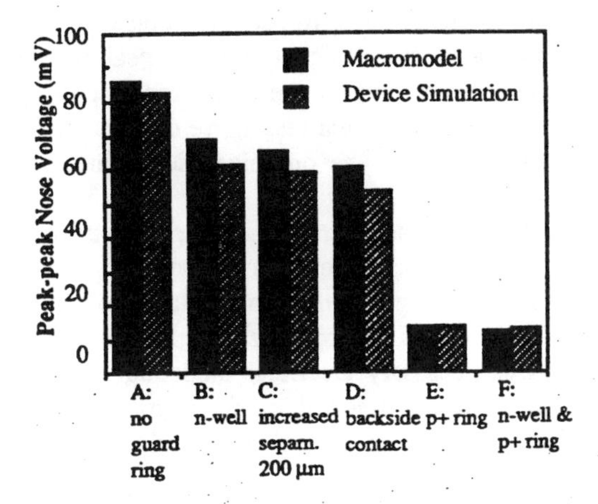

Fig. 10. Effect of various shielding techniques on peak–peak noise voltage at the sensitive node in Fig. 9.

ing noise voltage waveforms using both the macromodeling technique and the device simulation program.

Table I compares the cpu time requirements for the device simulation program and the AWE (second order) and DC macromodeling techniques. With the DC macromodeling technique, the n-well diffusions of Fig. 6 and Fig. 9 were

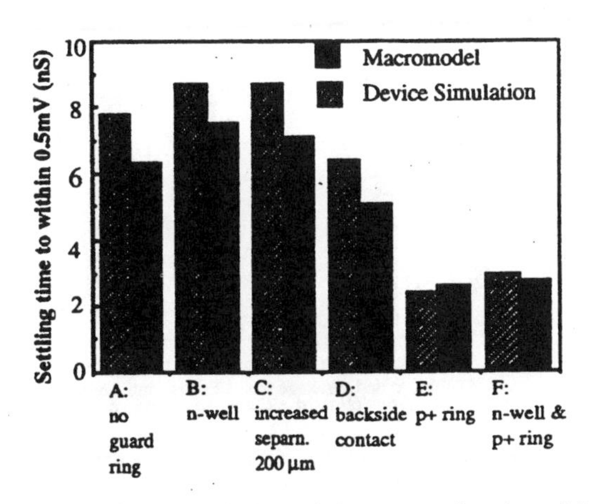

Fig. 11. Effect of various shielding techniques on settling time of the noise voltage at the sensitive node in Fig. 9.

TABLE I
RUN TIME (ON DECSTATION 5000) COMPARISON BETWEEN THE DEVICE SIMULATION PROGRAM AND THE MACROMODELING TECHNIQUES

Number of Mesh Nodes	Device Simulation cpu time (s)	AWE Macromodel cpu time(s)	dc Macromodel cpu time(s)
2940	4375	60.6	5.5
3716	7192.8	92.3	7.2
6605	16 882.2	202.2	20.3
8712	23 732.2	294.3	28.3

extracted as ports and connected to lumped capacitances equivalent to the depletion capacitance of the n-well. Both the AWE and DC macromodeling techniques produced identical results for the examples shown. As exemplified in Table I, the main advantage in using the macromodeling techniques is the significantly shorter cpu times required in their analysis as compared to the device simulation program.

B. Comparisons to Measured Results

The DC macromodeling technique has been used to simulate substrate coupling on the experimental chip reported in [4]. The 2 mm × 2 mm test chip realized in a 2 μm BiCMOS n-well process consists of transistors fabricated in a 15 μm lightly-doped epitaxial layer over a heavily-doped bulk. An on-chip ring oscillator drives a block of 12 CMOS inverters with each inverter output capacitively coupled to the substrate. The switching noise introduced into the substrate is measured by ten single transistor NMOS current sources distributed across the chip. The substrate is biased using a combination of several p+ contacts on the die surface. Seven of the current sources are shielded from the substrate noise using guard rings placed at varying distances (6 μm or 22 μm) from the sources and biased either with a dedicated package pin or connected to two large substrate contacts (one located at the chip center and one diffusion ring surrounding the chip).

The measured results [4] and simulated results are in good agreement as shown in Fig. 12 although only approximate values have been used for bonding pad and chip-to-package capacitance and bond-wire/package pin inductances. Simulations with the macromodel confirm the observation in [4] that is a substrate with a heavily doped bulk, lateral current flow in

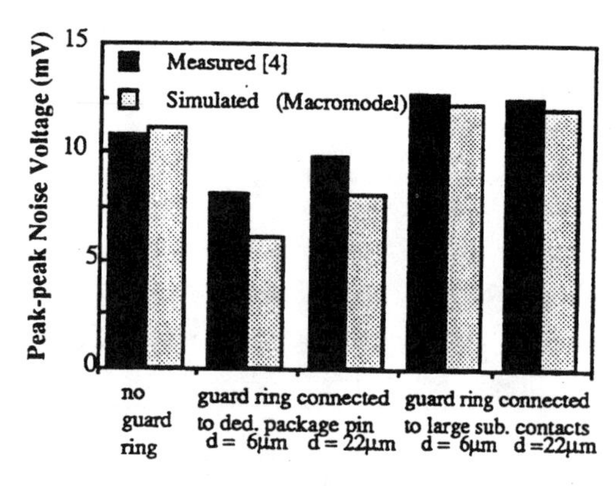

Fig. 12. Effect of various guarding configurations on the drain noise voltage of an NMOS transistor on the test chip [4].

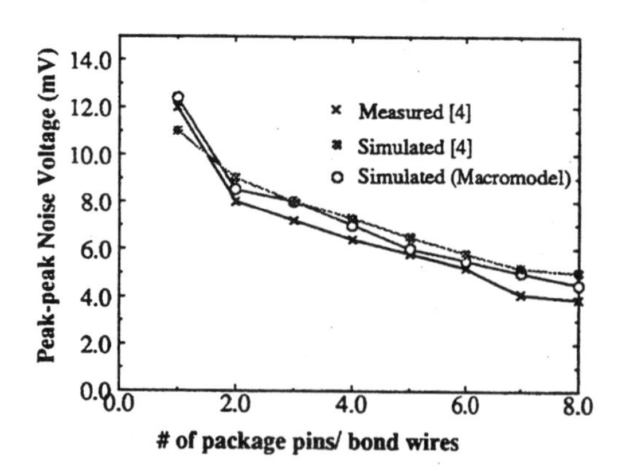

Fig. 13. Effect of multiple substrate bias package pins/bond wires on noise voltage.

the epitaxial layer is negligible as long as the devices and/or contacts are separated by more than 4 times the thickness of the epitaxial layer. A single node substrate model [3], [4] can be justified in these circumstances. When devices/contacts are closer than the critical distance, additional resistances must be introduced into the single node model to represent the resulting lateral current flow. The values of these resistances are easily obtained from the macromodel.

Fig. 13 shows the effect of reducing the power supply inductance on the noise voltage of a sensitive transistor. By increasing the number of package pins used to bias the substrate power supply, noise voltage is reduced dramatically. Fig. 13 compares the measured results [4], simulated results with a single node model [4], and simulated results with the macromodel. The settling time behavior of the noise voltage waveforms are shown in Fig. 22(b).

V. Power Distribution Synthesis Strategy

As can be seen from the previous sections, efficient simulation tools allow designers to devise and analyze power distribution techniques to reduce the noise problem for key analog circuits. The next step is to create tools that automate the design of the power distribution network, and that accommodate such design techniques. However, existing power grid layout techniques focus mainly on the geometric problem—assuring connectivity to all power pins [14], [15]—while allowing only rather simple electrical constraints, e.g., pad-to-pin ohmic drop [17]–[19]. Yet to handle realistic mixed-signal design problems, we must be able simultaneously to optimize the topology of the power grid, the sizing of individual segments, and the choice of I/O pad number and location, under tight dc, ac, and transient electrical constraints arising from the interaction of the power grid with the rest of the IC—notably via substrate coupling effects.

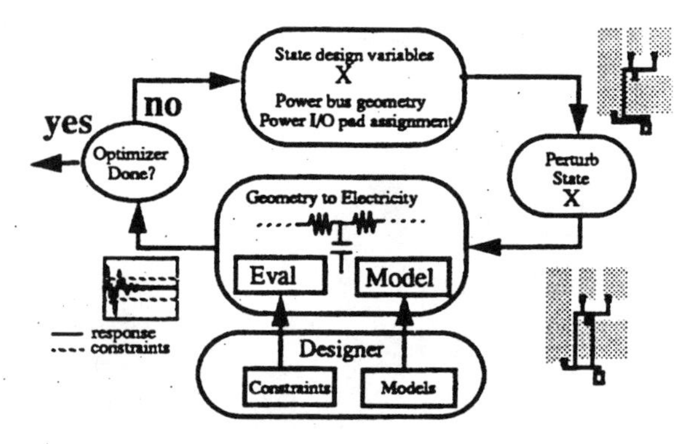

Fig. 14. Optimization-based power synthesis.

We suggest a power distribution synthesis strategy based on combinatorial optimization techniques that allows us to attack all these concerns simultaneously. We employ an iterative improvement approach illustrated in Fig. 14.

To design a power grid, we begin with an initial (rather arbitrary) "state" for the power bus geometry and power I/O pad configuration. We then perturb this geometry to create a new candidate power grid or pad configuration and update the electrical models for the busses and I/O pads. We next combine those models with designer-supplied circuit macromodels for blocks being supplied by the power grid, and for the substrate. With this complete electrical model—power grid, blocks, pads, substrate—we evaluate the resulting electrical performance, and compare against designer constraints. For example, we might evaluate the coupled noise waveform at a sensitive node against a designer-supplied peak-to-peak noise amplitude constraint. Finally, the optimizer accepts or rejects the perturbation based on the result. We continue the iterative improvement loop until the optimizer determines no further improvement is possible.

Our optimization-based strategy comprises four major components which we describe in the following four sections. Section VI describes the geometric representation of the problem. Section VII describes the optimization formulation. Section VIII discusses necessary electrical modeling techniques. Finally, Section IX discusses efficient electrical evaluation of the performance of the evolving power grid.

VI. Geometric Representation

The highest level physical design decision that affects the geometric design of the power grid is the overall design style selected for the IC. As with previous methods, we focus on the custom 2-D macrocell design style [14], [17], [18], which can appear in either a flat or a slicing representation. We select the slicing representation to avoid the channel identification preprocessing step associated with flat representations.

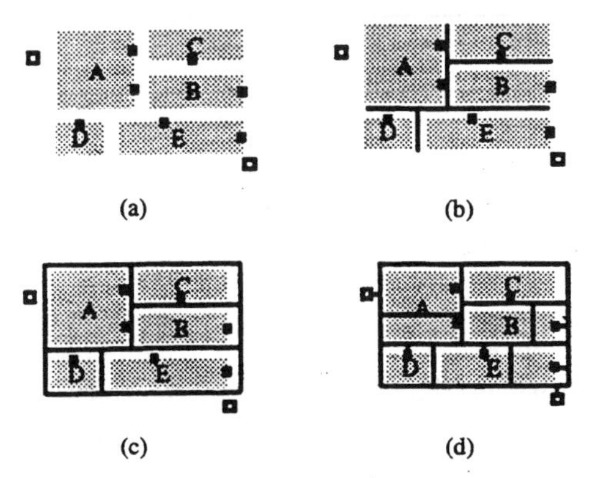

Fig. 15. Creating the general grid. (a) Slicing tree placement. (b) Power bus segments for slicing channels. (c) Power bus segments for I/O channels. (d) Power bus segments for macros (includes over-the-cell).

A. Power Busses

To represent power bus geometry, we offer a new formulation which we term a *general grid*. Unlike previous synthesis methods which made at most one single bus available for each power net in *many* channels [14]–[17], we allow a single bus, at most, for each power net in *every* channel and *every* designer-supplied over-the-cell power feedthrough. This formulation was inspired by work done for signal nets in gate-arrays [20] and was independently rediscovered for standard cell power grids in [19]. The idea is essentially subtractive: we allocate power grid segments everywhere, and formulate as an optimization problem which to resize, and which to remove. The idea is illustrated in Fig. 15.

Starting with a slicing tree macrocell placement in the layout [Fig. 15(a)] we generate power bus nets in all available channels [Fig. 15(b)]. Then, to connect to power I/O pads, we build power bus segments in the four I/O channels that surround the macrocells [Fig. 15(c)]. Finally, we introduce power bus segments perpendicular to these channel segments to handle over-the-cell feedthroughs and to connect to macros and pads [Fig. 15(d)]. For maximum flexibility, we provide power bus geometry for *all* power nets in *each* feedthrough, but later ensure that at most one net uses the resource.

This collection of segments constitutes the *general grid*, the master topology of which all final topologies are a subset. Each individual grid segment's width is an independent variable. These segments may be sized to one of several discrete widths ranging from the minimum width allowable in the technology to a designer-specified maximum. In addition to sizing the master topology, we add *zero* width to the sizing options to remove segments for topological selection. Now, both sizing and topological optimization can proceed in the guise of sizing a master topology. We avoid singularities in the associated electrical model by modeling the zero width segment (ideally an electrical "open") as a minimum conductance. Similarly, a maximum conductance is used to model a "short," as we shall see in the next subsection. This formulation has several advantages: it can handle over-the-cell power routing at designer discretion; it can discover power layouts in both tree and grid topologies; and it can partition quiet and noisy pins onto separate nets/pads.

B. Reconfigurable I/O Blocks

In prior work, the power busses have been the major focus of attention and optimization methods have synthesized *either* power bus topologies [16], [19] *or* power bus sizings [17], [18]. Our general grid formulation allows us to handle these concerns simultaneously. However, Fig. 13 clearly demonstrated the importance of proper power I/O pad assignment as another component of this problem. We can easily extend the idea of discretely-selected power bus segment widths to handle power I/O pad assignment. Each I/O pad is connected to each power bus via a resistor. We refer to this as a *reconfigurable* I/O block. By discretely selecting between on/off switch values for resistors inside each I/O pad, we can effectively reassign the I/O to the appropriate power net, or even no net.

VII. Optimization

Most previous methods assume a fixed power I/O pad assignment, then determine power bus topologies, then determine the sizing. In contrast, we reformulate the problem to allow pad assignment, topology selection, and segment sizing to be optimized simultaneously. We argue that all these decisions must be handled simultaneously to make optimal tradeoffs under tight constraints. To attack this combined problem we employ simulated annealing [24], a general optimization strategy based on iterative improvement with controlled hill climbing.

Hill climbing allows annealing strategies to avoid local minima in a complex cost surface, and reach better global solutions. To characterize our power distribution synthesis strategy, we need to describe the four components of any annealing-based optimizer: 1) the *representations* for intermediate states of the power distribution visited during iterative improvement; 2) the *moves* that transform one intermediate state of the layout to the next; 3) the *cost function* used to evaluate the quality of each intermediate power distribution; and 4) the *cooling schedule* used to control hill climbing.

A. Representation

We represent power bus segments with the *general grid*, updating their widths as the power distribution evolves. Similarly, we represent reconfigurable I/O blocks with their user-defined switch configurations and associated geometries.

B. Move Set

Controlled random perturbations, or *moves*, in conjunction with the cost function, are responsible for generating a "good" final power distribution. Our move set supports two classes of perturbations: changes to paths of connected segments and changes to interdependent sets of reconfigurable I/O blocks.

For on-chip power nets, moves alter segment paths between macrocells and pads (see Fig. 16). We first compute dc currents in each power bus segment. These currents assign a direction to each segment, used for path finding between macros and pads. A move begins by selecting a random segment in a random power net. It then traverses a random number of segments, going either toward a pad or macro. The resulting

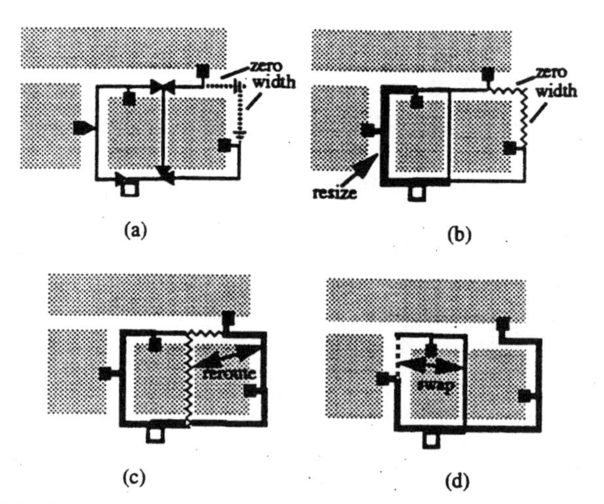

Fig. 16. Power bus moveset. (a) Segments with current directions. (b) Resize a path. (c) Reroute a path. (d) Swap two paths.

set of connected segments forms our starting point. The move then randomly selects from among three possible changes to make. A *resizing* move alters the width of segments on this path [Fig. 16(b)]. A *rerouting move* selects a second random path through zero-width segments, removes the first path, and distributes its area over the zero-width segments. [Fig. 16(c)]. A *swapping* move selects a second random path of similar area, and redistributes the segment widths of the first path over the second, and vice versa [Fig. 16(d)]. Note that the rerouting and swapping moves can alter the connectivity of the grid. Finally, we adaptively weight our move selection to maximize the probability of selecting successful moves [26].

For reconfigurable I/O blocks, the key is to maintain specified dependencies among blocks during perturbations. A move randomly selects an independent I/O block, connects it to either a power bus or the substrate, then randomly selects compatible connections for all dependent blocks. For example, a move which connects the package pin to VDD will also force its associated chip pad to connect to the chip VDD bus.

C. Cost Function

Our success in generating a "good" final power distribution is measured by a weighted-sum cost function, updated incrementally after each move:

$$Cost = w_1 Area + w_2 dc + w_3 ac + w_4 Tran \quad (8)$$

The weights are determined empirically. Our goal is to minimize power bus area consumed (the $Area$ term) while satisfying interacting electrical constraints (the *dc*, *ac*, and $Tran$ terms). The annealer searches among the power bus and pad configurations and selects one minimizing this function.

The $Area$ term sums the area consumed by each power bus within its assigned routing area. Each channel is partitioned into signal and power net routing areas. Power bus area consists not only of the physical area of each segment, but the spacing halo around each segment dictated by design rules. Minimizing an accurately estimated $Area$ term ensures we reserve maximal space for subsequent signal routing.

The remaining terms are all penalty terms, i.e., they increase with violations from a hard electrical constraint, but contribute zero to the cost when the constraint is met. Each constraint is normalized and includes an absolute tolerance, $ATOL$, obtained as input, to handle zero specifications. All constraint costs are given in the following form:

$$Constraint\,cost_i = \max\left(\frac{performance_i - spec_i}{spec_i + ATOL}, 0\right) \quad (9)$$

The *dc* term penalizes violations of dc constraints. For electromigration, we constrain power bus maximum current density and power pad maximum current. For maintaining local power supply levels at macros, we constrain voltage drops at the macro between power nets. To limit voltage offsets and indirectly control power dissipation and circuit bias, we constrain ground shift between macros and pads. To directly constrain chip power dissipation and macrocell bias currents, we constrain macrocell currents.

The *ac* term penalizes violations of ac constraints. Particularly, we focus on chip resonance frequencies and design power distribution to ensure they are well above clock frequencies present on chip. We also examine the driving point impedance at some key digital switching blocks to gauge the amplitude of noise generated.

The $Tran$ term penalizes violations in the transient constraints. Designers typically characterize noise effects in terms of peak-to-peak noise seen at a sensitive analog node [3], [4]. Thus, we also constrain peak-to-peak noise amplitude at these nodes. Since the noise problem originates in the digital portion of the chip, we also can constrain that peak-to-peak noise amplitude.

We should also note here the interaction between the cost function, the move set, and the general grid representation. Because segment widths and pad assignments are discrete, the cost surface is highly discontinuous. Moreover, we permit infeasible intermediate configurations of the power grid, i.e., segments can be sized to zero width, disconnecting parts of the grid. Such a formulation poses severe problems for traditional gradient optimizers, but poses no problems for an annealer. Indeed, the advantage of this strategy is that it simplifies the design of the move set, permits simultaneous optimization of pad assignment, grid topology and segment sizes, and allows large-scale perturbations of the entire power distribution configuration which lead to more aggressive search among tradeoffs.

D. Cooling Schedule

The *cooling schedule* controls the hill climbing during annealing. The power distribution is "heated" to a high enough temperature to allow any perturbation in power distribution to be accepted. As the power distribution is carefully "cooled," pad configurations solidify, power bus topologies crystallize, and power bus segment sizes fine-tune to generate the final power distribution. We employ an efficient, automatic cooling schedule [25].

VIII. Electrical Modeling

Good design practice provides near ideal power supplies on the card so we need only consider the electrical aspects of

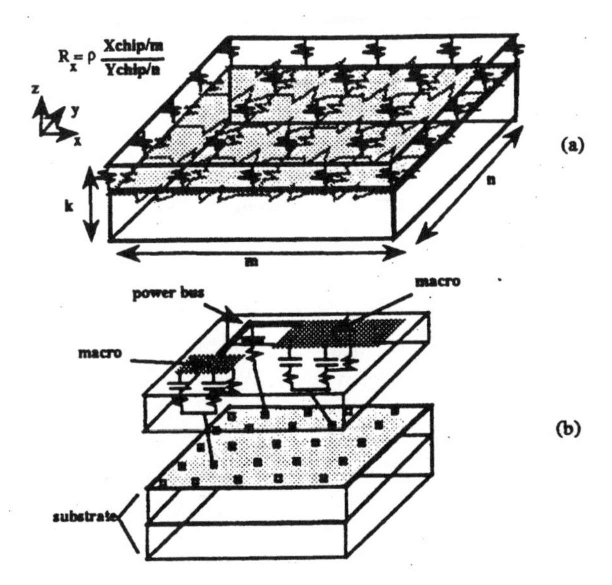

Fig. 17. Substrate coupled macros and busses on-chip. (a) Coarse substrate mesh couples. (b) Macros connected to it via arbitrary circuit models.

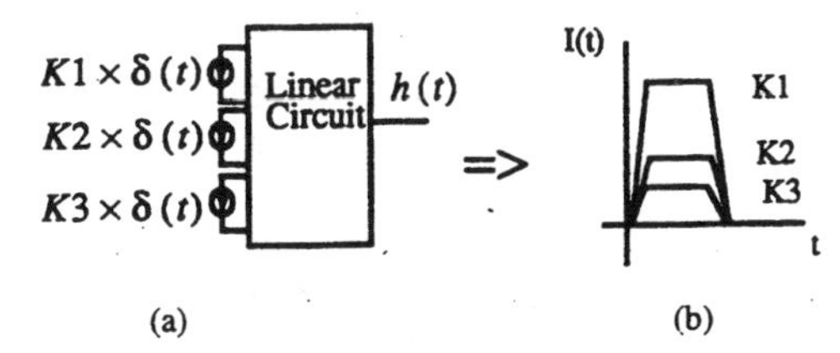

Fig. 18. Simultaneous switching of multiple sources. (a) Weighted impulses generate (b) weighted pulses.

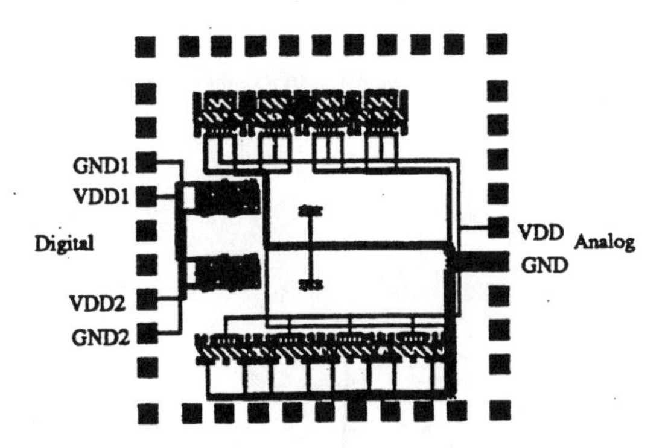

Fig. 19. Mixed1 sample/hold, placement and power distribution.

the chip and package. We model the principal components of chip and package power distribution. This includes not only models for the chip macrocells and power busses, but models for the chip substrate and chip-to-package interconnect. With each annealing perturbation we map the attempted geometric changes to their electrical equivalents. In contrast to the simple resistor and dc current source modeling of previous methods [17]–[19], we support arbitrary independent and controlled sources, frequency and time-varying sources, and complete RLC models for all components of the power distribution network. We provide parameterizable default models, but to handle the expected variety of realistic models, we also support input of designer-specified models.

There are four modeling problems: 1) chip macrocells, 2) power bus segments, 3) chip-package interconnect, and 4) chip substrate. For macrocells we accept user input of a linear model arbitrarily interconnected between macro terminal pins and any additional substrate pins defined in the macro. For each power bus segment, we generate a parasitic RC π model. For chip-package interconnect, we estimate wire bond inductance. For the substrate, we either accept a user-supplied model, or automatically generate from technology data a *coarse* resistive mesh model with the designer-specified number of grids. For each substrate pin in macrocells and net segments, we assume connection to the nearest substrate mesh node. Fig. 17 illustrates the resulting model.

IX. Electrical Evaluation

For optimization problems of this scale, an annealer will typically visit 10^4–10^5 design configurations. Since each move requires us to reevaluate the cost function, we require efficient techniques for the electrical evaluation of our power distribution models. Because we allow arbitrary power grid topologies, and complex macro, substrate, and chip-package models, the simple analytical formulations of early approaches [17], [18] cannot capture the behavior of the resulting networks. To deal with realistic design constraints, we must handle not only dc analysis (see, e.g., [19]), but also ac and transient analyses.

A. Evaluating dc Performance

To evaluate circuit behavior we use modified-nodal analysis. This method is general, and allows us to analyze the dc behavior of the arbitrary linear network resulting from each annealing move. It is also efficient, requiring 1 LU factorization and forward-and-back substitution. Further, matrix reformulation is never required and matrix reordering is rarely needed since each new topology visited by the annealer is a subset of the master topology created by the general grid and reconfigurable block formulations. The computed dc currents and voltages are used for cost function evaluation; the currents are saved for each power bus segment to provide the directions needed to trace paths in the next move.

B. Evaluating ac and Transient Performance

We again employ AWE [9] for efficient ac and transient evaluation. For this synthesis application, AWE is typically 100–1000× faster than SPICE. An ac response to an input is available directly from AWE's computed pole-zero approximation. A transient response for a typical input waveform comprised of steps and ramps is also easily generated. An interesting aspect of our use of AWE is in how we model simultaneous switching. Most noise analysis tools based on CMOS switching sources use trapezoidal, triangular, or saw-tooth waveforms to model current switching [21]–[23], all of which can be described with a superposition of steps and ramps. The transient response from multiple input sources of varying strengths switching simultaneously can be efficiently estimated with one AWE evaluation. Using the linearity of the system and the constant amplitude scaling for ramps and steps, the multiple source response is obtained by weighting

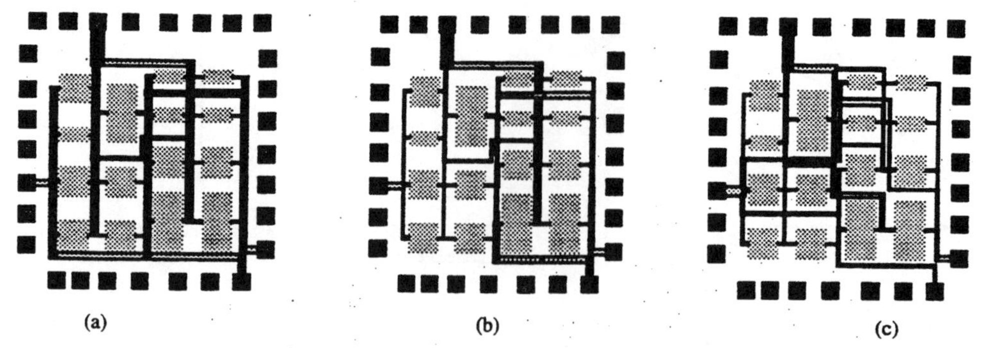

Fig. 20. Analog1 benchmark layouts for VDD and GND nets. (a) Manual layout. (b) Topology fixed and sized. (c) Topology selected and sized.

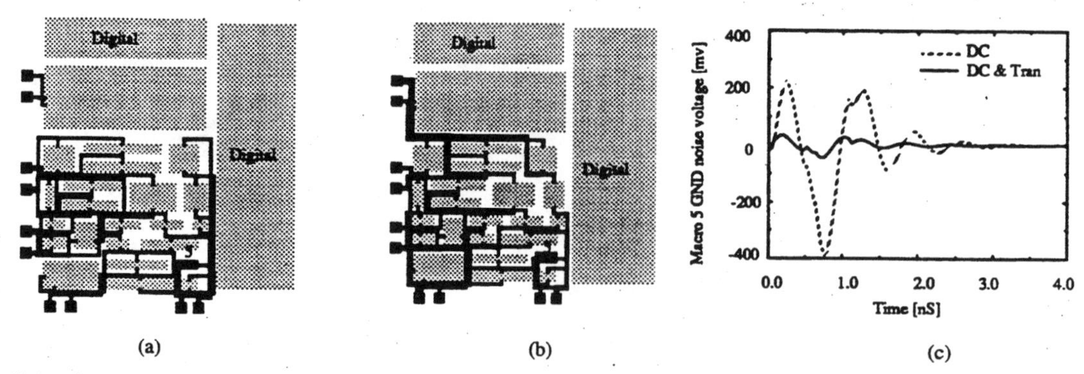

Fig. 21. Mixed3 benchmark layouts and simulated noise comparisons. (a) Layout for dc constraints. (b) Layout for dc and transient constraints. (c) SPICE GND noise comparison between layouts for macro 5 (highlighted).

TABLE II
SYNTHESIZED POWDER BUS AREA AND RUN TIMES FOR BENCH MARK PROBLEM SIZES

Example	Power Bus Segments/Pin Blocks Configured	Circuit Nodes/Substrate Grid Size	Power Bus Area [μm^2]	CPU Time IBM RS6000/550 [hours]
Mixed 1	252/–	588/2 × 2 × 1	2.39E5	1.7
Analog 1(a)	142/–	—	4.74E5	manual
Analog 1(b)	142/–	330/6 × 6 × 1	3.82E5	2.2
Analog 1(c)	142/–	330/6 × 6 × 1	3.25E5	3.4
Mixed 3(a)	294/–	618/10 × 10 × 1	7.83E5	5.3
Mixed 3(b)	294/–	618/10 × 10 × 1	9.06E5	14.6
Noise test	132/9	560/2 × 2 × 1	6.6E4	1.3

the impulses in the original moment generation phase of the AWE algorithm as shown in Fig. 18. In other words, with one superposition of weighted impulses, AWE approximates the effect of multiple, independently switching sources.

X. SYNTHESIS RESULTS

We have implemented our ideas in a tool called RAIL. RAIL has generated power distribution for several analog and mixed-signal examples [27]. We have included three mixed-signal and one analog example in this section. The power bus area, circuit size, and CPU time for these power distribution synthesis examples are shown in Table II.

The first example in Fig. 19 is a sample and hold circuit with on-chip clock generation. Three sets of power nets (one analog, two digital) were synthesized by RAIL. The most interesting feature is the automatic tapering in the analog ground bus and the minimizing of the analog VDD bus since much of the current from the VDD rail flows through signal pads connected to NMOS drains.

The next example in Fig. 20 is an industrial analog bipolar IC. This example demonstrates the importance of simultaneous topology selection and sizing. Fig. 20(a) shows the conservative manual design for the power busses. Fig. 20(b) demonstrates the benefit from automatically sizing this given (manual) topology. Finally, Fig. 20(c) shows the further benefit of simultaneous power bus topology selection and sizing.

The third example is a larger mixed-signal chip which illustrates the importance of ac and transient noise constraints during power distribution synthesis. Fig. 21(a) shows the synthesized layout when ac and transient constraints are ignored. Short routes which meet dc electromigration and voltage constraints are created. However the noise is intolerable at 600 mV as shown in Fig. 21(c). After applying transient noise constraints on sensitive analog power net connections, the layout in Fig. 21(b) is generated and the noise is reduced to 75 mV.

Our last example demonstrates the accuracy obtainable through linear macromodeling, and the importance of simultaneous optimization of power busses and power I/O pad assignment. We again use the test chip of [4], for which detailed noise measurements have been published. From the

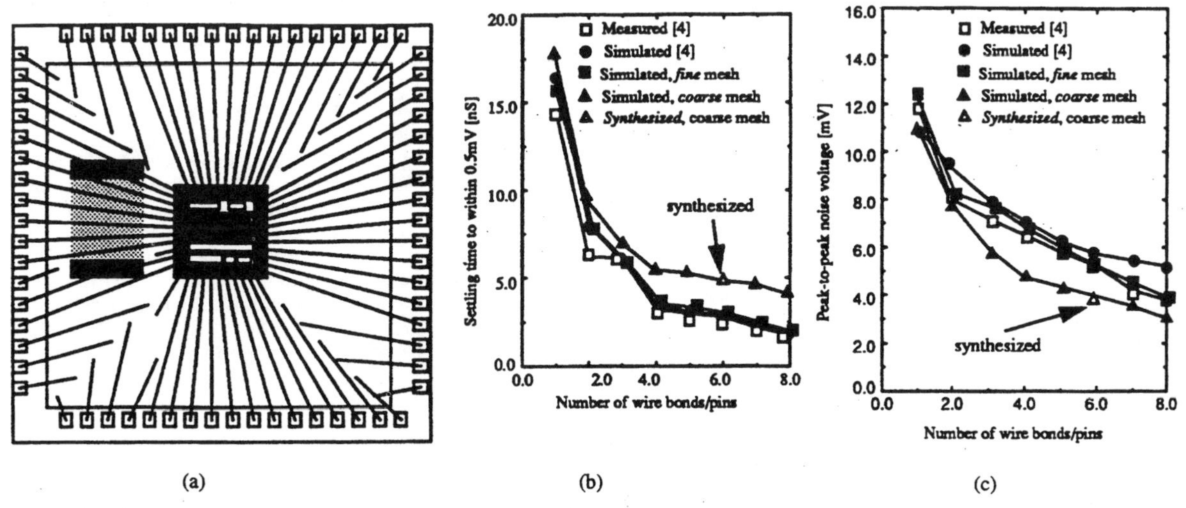

Fig. 22. Results for 68 pin PLCC package with ceramic decoupling capacitor and noise test chip [4]. (a) PLCC cavity with chip, decoupling capacitor, and wire bonds to lead frame. (b) Settling time at drain for number of wire bonds. (c) Noise voltage (peak-to-peak) at drain for number of wire bonds.

description of the chip-package environment, Fig. 22(a) shows our rendering of the chip-package physical design. The 68 pin PLCC cavity contains the 2 × 2 mm chip and a ceramic decoupling capacitor. First, we created linear macromodels of the analog current sources (noise receivers) and the oscillator-based digital switching logic (noise generators). Next, we synthesized power distribution with the constraint that the noise coupled at the drain of one current source must not exceed 4 mV. Further, only one substrate I/O pad was allowed. As expected, meeting both the dc constraints and the transient 4 mV noise constraint was unachievable. However, when RAIL was allowed to perform power pad I/O assignment simultaneously, it synthesized the power busses to meet dc constraints, and added five substrate contact pins to meet the transient noise constraint, thus completing the power distribution.

Fig. 22(b) and (c) show measured and simulated settling times and noise voltages for this experiment, respectively. We believe this result justifies the use of coarse mesh substrate approximations in heavily doped substrates.

XI. Conclusions

We have described simulation and power distribution synthesis techniques to address the parasitic coupling of noise through the substrate in mixed-signal integrated circuits. The simulation tool uses a modified version of SPICE3 which spatially discretizes the substrate from the layout information, develops its macromodel and performs circuit simulation with the macromodel. Simulation results indicate a significant reduction in cpu time while preserving the accuracy of device simulations. Results also compare favorably with measurements made on a test chip. The synthesis tool, RAIL, creates analog power distribution while minimizing noise coupling to sensitive analog circuits. RAIL simultaneously optimizes power bus topology, segment sizing, and power I/O pad assignment. Its circuit evaluation efficiency is derived from a synergy between our geometric formulations and AWE. Power distribution synthesis results demonstrate the critical need to model the substrate and constrain the noisy ac and transient behavior it introduces. The simulator can be valuable as a verification tool for analog and mixed-mode IC designs and the complementary synthesis tool can be valuable as a power distribution floorplanner for these same designs.

Acknowledgment

The authors thank T. J. Schmerbeck of International Business Machines, Dr. S. Kumashiro of NEC, Dr. V. Raghavan, J. E. Bracken of Performance Signal Integrity, and S. Masui of Nippon Steel Corporation for several useful discussions, and T. Mukherjee of Carnegie Mellon University for first using RAIL. M. J. Loinaz, T. Blalack, D. K. Su, and Prof. B. A. Wooley of Stanford University provided useful information on their test chip.

References

[1] J. A. Olmstead and S. Vulih, "Noise problems in mixed analog/digital integrated circuits," in *Proc. Custom Integrated Circuits Conf.*, 1987, pp. 659–662.

[2] B. M. J. Kup, E. C. Dijkmans, P. J. A. Naus, and J. Sneep, "A bit stream digital-to-analog converter with 18-b resolution," *IEEE J. Solid State Circuits*, vol. 26, no. 12, pp. 1757–1763, Dec. 1991.

[3] T. Schmerbeck, R. Richetta, and L. Smith, "A 27 MHz mixed analog/digital magnetic recording channel DSP using partial response signalling with maximum likelihood detection," in *Tech. Dig. IEEE Int. Solid State Circuits Conf.*, Feb. 1991, pp. 136–137.

[4] D. K. Su, M. J. Loinaz, S. Masui, and B. A. Wooley, "Experimental results and modeling techniques for substrate noise in mixed-signal integrated circuits," *IEEE J. Solid State Circuits*, vol. 28, no. 4, pp. 420–430, Apr. 1993.

[5] S. Masui, "Simulation of substrate coupling in mixed-signal MOS circuits," in *Tech. Dig. Symp. VLSI Circuits*, 1992, pp. 42–43.

[6] N. K. Verghese, D. J. Allstot, and S. Masui, "Rapid simulation of substrate coupling effects in mixed-mode ICs," in *Proc. IEEE Custom Integrated Circuits Conf.*, May 1993, pp. 18.3.1–18.3.4.

[7] S. Kumashiro, "Transient simulation of passive and active VLSI devices using asymptotic waveform evaluation," Ph.D. dissertation, Dep. Elec. Comput. Eng., Carnegie Mellon Univ., Aug. 1992.

[8] T. L. Quarles, "The SPICE3 implementation guide," Univ. California, Berkeley, Tech. Rep. ERL-M89/44, Apr. 1989.

[9] L. T. Pillage and R. A. Rohrer, "Asymptotic waveform evaluation for timing analysis," *IEEE Trans. Comput.-Aided Design Integrated Circuits Syst.*, vol. 9, no. 4, pp. 352–366, Apr. 1990.

[10] X. Huang, V. Raghavan, and R. A. Rohrer, "AWEsim: A program for the efficient analysis of linear(ized) circuits," in *Proc. IEEE Int. Conf. Computer Aided Design*, 1990, pp. 534–537.
[11] V. Raghavan, J. E. Bracken, and R. A. Rohrer, "AWEspice: A general tool for the accurate and efficient simulation of interconnect problems," in *Proc. 29th Design Automation Conf.*, 1992, pp. 87–92.
[12] *TMA MEDICI: Two Dimensional Device Simulation Program*, Technology Modeling Associates, Inc., Vols. 1, 2, Mar. 1992.
[13] J. A. Meijerink and H. A. Van der Vorst, "An iterative solution method for linear systems of which the coefficient matrix is a symmetric M-matrix," *Math. Computat.*, vol. 31, pp. 148–162, Jan. 1977.
[14] H.-J. Rothermel and D. A. Mlynski, "Computation of power supply nets in VLSI layout," in *Proc. 18th Design Automation Conf.*, June 1981, pp. 37–42.
[15] Z. A. Syed, A. El Gamal, and M. A. Breuer, "On routing for custom integrated circuits," in *Proc. 19th Design Automation Conf.*, June 1982, pp. 887–893.
[16] A. S. Moulton, "Laying the power and ground wires on a VLSI chip," in *Proc. 20th Design Automation Conf.*, June 1983, pp. 745–755.
[17] S. Chowdhury, "An automated design of minimum-area IC power/ground nets," in *Proc. 24th Design Automation Conf.*, June 1987, pp. 223–229.
[18] R. Kolla, "A dynamic programming approach to the power supply net sizing problem," in *Proc. European Design Automation Conf.*, Mar. 1990, pp. 600–604.
[19] T. Mitsuhashi and E. Kuh, "Power and ground network topology optimization for cell based VLSIs," in *Proc. 29th Design Automation Conf.*, June 1992, pp. 524–529.
[20] A. M. Patel, N. L. Soong, and R. K. Korn, "Hierarchical VLSI routing—an approximate routing procedure," *IEEE Trans. Comput.-Aided Design*, vol. CAD-4, no. 2, Apr. 1985.
[21] R. Burch, F. Najm, P. Yang, and D. Hocevar, "Pattern-independent current estimation for reliability analysis of CMOS circuits," in *Proc. 25th Design Automation Conf.*, June 1988, pp. 294–299.
[22] S. Chowdhury and J. Barkatullah, "Current estimation in MOS IC logic circuits," in *Proc. IEEE Int. Conf. Computer-Aided Design*, Nov. 1988, pp. 212–215.
[23] J. Hall, D. Hocevar, *et al.*, "SPIDER-A CAD system for modeling VLSI metallization patterns," *IEEE Trans. Comput.-Aided Design*, vol. CAD-6, no. 6, pp. 1023–1031, Nov. 1987.
[24] S. Kirkpatrick, D. Gelatt, and M. Vecchi, "Optimization by simulated annealing," *Science*, vol. 220, 1983.
[25] M. Huang, F. Romeo, and A. Sangiovani-Vincentelli, "An efficient general cooling schedule for simulated annealing," in *Proc. Int. Conf. Computer-Aided Design*, Nov. 1986, pp. 381–384.
[26] S. Hustin and A. Sangiovani-Vincentelli, "TIM, a new standard cell placement program based on simulated annealing algorithm," presented at the IEEE Physical Design Workshop Placement and Floorplanning, Hilton Head, SC, Apr. 1987.
[27] B. R. Stanisic, R. A. Rutenbar, and L. R. Carley, "Power distribution synthesis for analog and mixed-signal ASICs in RAIL," in *Proc. IEEE Custom Integrated Circuits Conf.*, May 1993, pp. 17.4.1–17.4.5.
[28] Z. J. Cendes and D. N. Shenton, "Adaptive mesh refinement in the finite element computation of magnetic fields," *IEEE Trans. Magnetics*, vol. 21, pp. 1221–1228, 1985.

Balsha R. Stanisic (M'84) was born in Milwaukee, WI, in 1962. He received the B.S. and M.S. degrees in electrical and computer engineering from the University of Wisconsin at Madison in 1984 and 1987, respectively, and the Ph.D. degree from Carnegie Mellon University in 1993.

From 1984 to present he has worked at IBM's AS/400 division, Rochester, MN, in the area of analog circuit CAE and design. His research interests include VLSI layout algorithms and synthesis techniques. His current research is in the area of analog power distribution synthesis. He is the author of RAIL.

Nishath K. Verghese (S'91) received the B.Eng. (Hons.) degree from the Birla Institute of Technology and Science, Pilani, India, in 1990 and the M.S. degree from Carnegie Mellon University in 1993.

From 1990 to 1991 he was with the VLSI Design Laboratory, McGill University, where he worked on new circuit implementations of sigma-delta A/D converters and on estimation techniques for power bus current in CMOS logic circuits. He is currently working toward the Ph.D. degree at Carnegie Mellon University where his research focuses on the simulation of substrate coupling in mixed-mode IC's. His research interests include design and verification of analog and mixed-mode circuits.

Rob A. Rutenbar (S'78–M'84–SM'90) received the Ph.D. degree in computer engineering (CICE) from the University of Michigan, Ann Arbor, in 1984.

He subsequently joined the faculty of Carnegie Mellon University, Pittsburgh, PA. He is currently Professor of Electrical and Computer Engineering, and of Computer Science, and Director of the SRC-CMU CAD Center. His research interests focus on circuit and layout synthesis for mixed-signal ASIC's.

In 1987, Dr. Rutenbar received a Presidential Young Investigator Award from the National Science Foundation. At the 1987 IEEE-ACM Design Automation Conference, he received a Best Paper Award for work on analog circuit synthesis. He is currently on the Executive Committee of the IEEE International Conference on CAD, Program Committees for the ACM/IEEE Design Automation Conference and European Design and Test Conference, and a member of the Editorial Board of IEEE SPECTRUM. He is on the Strategic Advisory Council for Cadence Design Systems. He is a member of ACM, Eta Kappa Nu, Sigma Xi, and AAAS.

L. Richard Carley (S'77–M'84–SM'90) received the S.B., M.S., and Ph.D. degrees from the Massachusetts Institute of Technology, Cambridge, in 1976, 1978, and 1984, respectively.

He is a Professor of Electrical and Computer Engineering at Carnegie Mellon University. He has worked for MIT's Lincoln Laboratories and has acted as a consultant in the area of analog circuit design and design automation for Analog Devices and Hughes Aircraft among others. In 1984 he joined Carnegie Mellon, and in 1992 he was promoted to Full Professor. His current research interests include the development of CAD tools to support analog circuit design, the design of high performance analog signal processing IC's, and the design of low-power high-speed magnetic recording channels.

Dr. Carley was awarded the Guillemin Prize for the best EE Undergraduate thesis in 1976, a National Science Foundation Presidential Young Investigator Award in 1985, a Best Technical Paper Award at the 1987 Design Automation Conference, and a Distinguished Paper Mention at the 1991 International Conference on Computer-Aided Design.

David J. Allstot (S'72–M'78–SM'83–F'92) received the B.S.E.S. degree from the University of Portland, Portland, OR, the M.S.E.E. degree from Oregon State University, Corvallis, OR, and the Ph.D. degree in electrical engineering and computer science from the University of California, Berkeley, in 1979.

He has held industrial positions with Tektronix, Texas Instruments, and MOSTEK, and academic positions with the University of California, Berkeley, Southern Methodist University, Dallas, TX, and Oregon State University. He is currently Professor of Electrical and Computer Engineering at Carnegie-Mellon University Pittsburgh, PA, and Associate Director of the SRC-CMU Research Center for Computer-Aided Design.

Dr. Allstot's professional service positions have included: Associate Editor of the IEEE TRANSACTIONS ON CIRCUITS AND SYSTEMS, Guest Editor for the IEEE JOURNAL OF SOLID-STATE CIRCUITS, Editorial Board Member of the *Analog Integrated Circuits and Signal Processing Journal*, Technical Program Committee member of the IEEE Custom IC Conference and the IEEE International Symposium on Circuits and Systems, and elected member of the IEEE Circuits and Systems Society Board of Governors. He is currently Editor of the IEEE TRANSACTIONS ON CIRCUITS AND SYSTEMS II: ANALOG AND DIGITAL PROCESSING. He was a corecipient of the 1980 IEEE W. R. G. Baker Award, and has received excellence in teaching awards from SMU, OSU and CMU. He is a member of Eta Kappa Nu and Sigma Xi.

Mondriaan : a Tool for Automated Layout Synthesis of Array-type Analog Blocks

G. Van der Plas, J. Vandenbussche, G. Gielen*, W. Sansen

Department of Electrical Engineering, Katholieke Universiteit Leuven, ESAT-MICAS
Kardinaal Mercierlaan 94, 3001 Heverlee, Belgium
E-mail : georges.gielen@esat.kuleuven.ac.be

Abstract

A tool set, Mondriaan, is proposed which targets the physical design automation of all array-type analog blocks. The approach takes in consideration typical analog constraints and is very flexible. A three step procedure (floorplan generation, symbolic place & route and technology mapping) solves the layout synthesis problem in a fast and technology independent way. Industrial strength examples show the applicability and usefulness of the proposed solution.

I. Introduction

The ever shortening life-cycles of electronic products causes a continuous improvement of the product design cycle. In digital design the productivity in terms of number of gates has increased steadily over the last decade. This productivity gain has been realized both in logic synthesis and physical design. The world is however still analog and the interface with the digital cores is therefore formed by analog blocks (as for instance drivers, D/A and A/D converters). Unfortunately the design productivity for analog blocks has not kept track with the digital productivity improvement. The sizing synthesis and physical design of analog blocks both require high expertise. In [1] an overview is given of the recent advances in the automation of the analog design process.

The automation of the physical design task for analog circuits has received a lot of attention in the past. Device generators now are standard components in commercial frameworks as Cadence or Mentor Graphics. More complex device generators, which are capable of generating pairs of devices in an interdigitated form, have been reported in [2]. In [2,3,4] layout environments have been presented which place and route device-level analog circuits. With these tools higher-level blocks can be created as for instance has been done in [5] for a $\Sigma\Delta$ D/A converter and a video driver. One important class of layout generation tasks has however received less attention. In analog blocks, very often a highly regular architecture of basic cells is used. Examples of this are flash type A/D converters (ADC), Cellular Neural Networks (CNN) or current-steering D/A converters (DAC). In [5,6] the automatic generation of current-steering DACs has been realized using a stretch and tile approach. In this paper a new approach will be presented which targets the general class of highly regular array types of analog architectures. This approach has been implemented in a tool called Mondriaan**.

The paper is organized as follows. In section II the currently available methods for layout generation of regular structures will be reviewed. In section III the new approach will be explained in detail. Experimental results will be presented in section IV. Conclusions are given in section V.

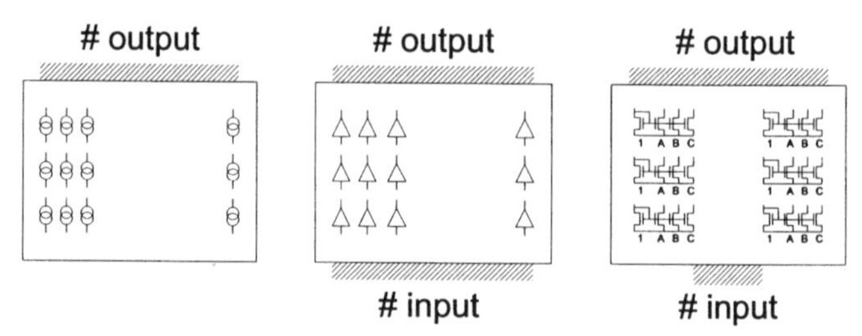

Fig. 1: Three analog array types: signal generation (current source array), signal processing (amplification) and signal multiplication and processing.

II. Overview of physical design methods

Typically, the regular layout structures used in analog blocks contain an array of unit cells (potentially with slightly different versions), which process in a parallel way one or more input signals and steer one or more output signals. Three different possibilities (each represented by a typical example) are shown in Fig. 1. Two methods for generating such array types of layout have been used up till now :

1. The layouts are drawn manually in a hierarchical way, reusing as much as possible of the regularity of the array. If a number of columns is identical, only one column is drawn and the resulting layout is instantiated the requested number of times. The remaining columns are drawn separately and also instantiated. Connections to neighbors and distribution of power and biasing are in this case often realized through abutment of cells. In general a completely regular layout can be generated quite rapidly with this method. However it is very unlikely that an architecture is completely regular. Almost always a number of cells are special (a reference, a dummy, etc.)

* research associate of the National Fund of Scientific Research (Belgium)
** a Dutch painter of the 20th century famous for his "array-type" paintings

and these cases destroy the gracefulness of the hierarchical method, since one ends up with more levels and exceptions than can be overseen. This increases the chance of an error and eventual small changes often lead to an important hierarchical reorganization which takes more time than would be expected. In addition, the whole work has to be redone if the unit cells or the array structure are changed or another technology process is used.

2. In [5,6] the generation of current-steering D/A converters has been automated, including the layout generation. Essentially the layout is generated by stretching and tiling a number of basic cells [6] which are basic circuit cells on the one hand and pure connectivity cells on the other hand. The stretching of the cells allows a tuning of the basic cells without having to regenerate them. This is important since the cells need to fit in order to realize the complete array connectivity by abutment of the basic cells. This approach however requires extensive preparation of the basic cells in order that they can be stretched and that the connections remain correct by abutment. Furthermore the complete architecture needs to be mapped on a cell array which is completely connected through abutment. Although this approach automates the layout generation, the flexibility and ease of use are severely limited, especially when the array structure itself can change.

Therefore a new, more flexible and general approach is proposed in the next section.

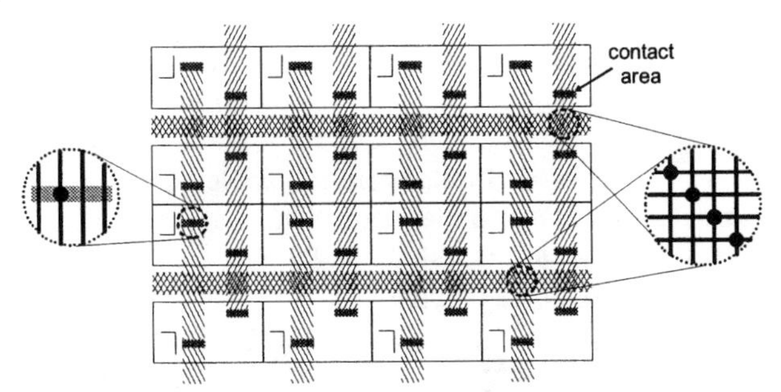

Fig. 2: Symbolic view of cells and routing channels : vertical routing across cells, horizontal routing in between cells. Vertical wires connect to the contact areas in the cell, horizontal wires connect to the vertical wires

III. Novel Approach to Array-type Layout Generation

In many applications a cell array is used to implement the circuit layout. The connectivity of this cell array is not easily realized through abutment, except for ground, biasing or power supply connections. The other connections in and out of the array or internal to the array are more easily realized through routing channels across or between the cells. The basic cells contain contact areas underneath the routing channels. In these areas a connection to the underlying cells can be made. This is symbolically shown in Fig. 2. Busses passing through these channels connect to the cells in the matrix. The vertical direction is used to connect wires to the cells, the horizontal direction is used to connect vertical wires. Cells can be flipped upsidedown (shown in Fig. 2) or sideways every next row or column to share lines by abutment. The thus offered placement and routing functionality is much more powerful than the stretch and tile approach and covers the requirements of a large variety of analog circuits as will be shown by the examples.

Note that the placement and routing of Field Programmable Gate Arrays somewhat resembles this approach [7]. An essential difference is that in FPGA routing the majority of the connections is internal to the logic array, while in analog applications the majority of the connections is to pins at the edge of the matrix. Furthermore the placement and routing of FPGAs is faced with a fixed number of wires and blocks and the critical delay (caused by routing) is to be minimized. This is not the case in analog applications: the number of wires is variable, but should be minimized, and the performance depends on equal capacitance, resistance or surrounding (matching) rather than the critical delay.

The layout generation process will now be discussed in more detail. The overall flow is shown in Fig. 3 and consists of three basic steps.

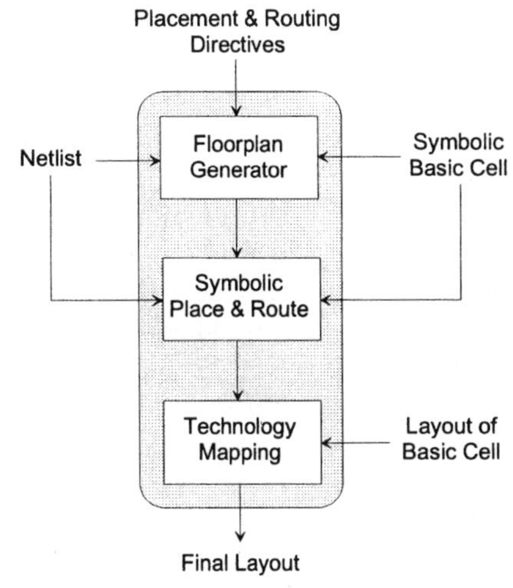

Fig. 3: Layout generation flow

A. The Floorplan Generator

The inputs to the floorplan generator are the netlist of the architecture, the symbolic basic cell (containing the contact areas) and placement directives. Two cases can be distinguished. In some analog blocks interdigitated or common centroid placements are required for matching purposes. These requirements are entered as directives. In this case the layout process is placement driven, and the initial placement generated during floorplanning does not change in further steps. In other architectures, the final cell placement can be deferred to the symbolic place & route and only a fixed IO pin placement is available. In this case the floorplan data is limited to the size of the array and the IO pin assignment. The layout generation is now routing driven. The number of wires in the busses passing the contactable areas is determined by the tool or can be given by the user and is passed to the symbolic place & route step.

B. Symbolic Place & Route

The symbolic place & route algorithm can be summarized as follows:

```
1. determine # of vertical wires for every column
2. forall IO pins
       select free wire, connect
       propagate connectivity (net of IO pin -> wire)
       if (connected cell isn't placed)
           place cell in free slot
3. forall cells (column-wise)
       forall contact areas
           if (wire passing with correct net)
               connect with wire
           else
               select free wire, connect with contact area
               propagate connectivity (net of cell -> wire)
4. determine # of horizontal wires
5. forall vertical wires
       find all wires to be connected to the same net
       select free horizontal wire and connect
```

From the floorplan, netlist and symbolic basic cell, the symbolic placement and routing is generated. To accomplish this a search algorithm is used to propagate the placement and connectivity information across the array. First the number of vertical wires for every column is determined (if not specified by the user). Next, for all fixed IO pins free vertical wires are selected and the net of the wire is updated. If the cell connected to this net isn't placed, it is placed in a free array slot. When all fixed IO pins have been connected and all cells have been placed, the cells can be scanned column-wise, to propagate their connectivity. As a last step, the number of horizontal wires is determined (if it is not given by the user). This is done by counting the number of vertical wires which need to be connected. Then the wires are scanned and free horizontal wires are selected to connect the vertical wires. Of course if no vertical wires need to be connected no horizontal wires are created.

C. Technology Mapping

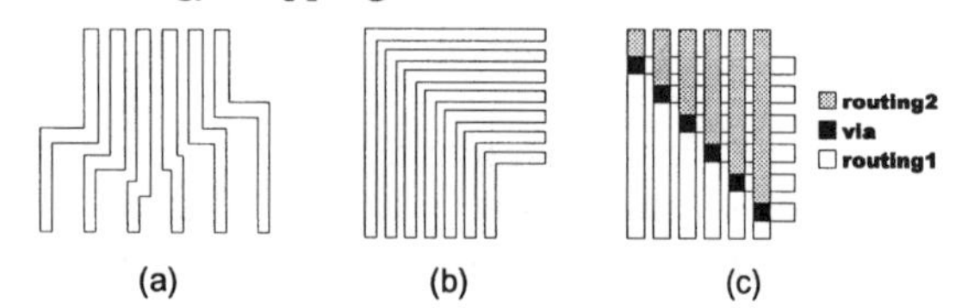

Fig. 4: Bus device generators: Y-type for pitch change (a), L-type for corner (b) and T-type for bus splitting (c).

The technology mapping stretches the cells to accommodate for the passing wires. The cells are tiled and all horizontal routing by abutment is realized. The user can specify to insert routing cells between the columns to connect to the basic cells horizontally and route vertically out of or into the array. This is useful when clocks or other signals need to be applied vertically, since the width of the array is too large. The symbolic layers are mapped on physical routing layers, then the physical vertical and horizontal routing busses are inserted and contacts are added to generate the connectivity. The result is a mask layout which is DRC error free and the final placement of the IO pins (the coordinates).

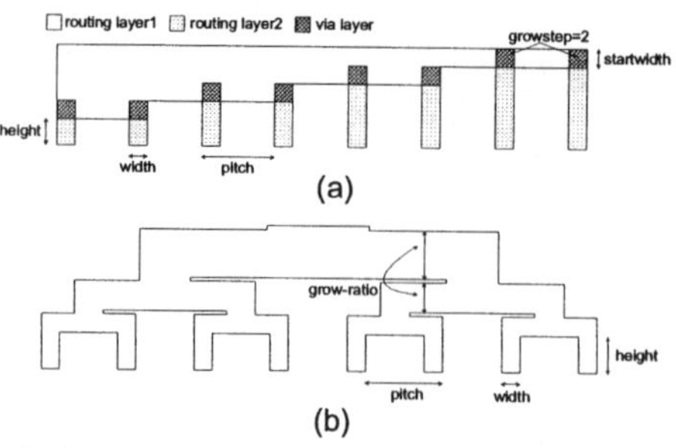

Fig. 5: Tree device generators: a unary tree, tapered to have a linear voltage drop along the length (a) and a binary tree (b).

In this way the cell arrays used in ADCs, DACs and other regular structures can automatically be generated in a technology independent way. The connections between the cell arrays are easily realized by the use of bus generators (see Fig. 4). The distribution of clock signals and biasing is realized by the use of tree generators (see Fig. 5). A set of tree and bus generators is part of the Mondriaan layout tool set.

D. Illustrative example

Let us illustrate this methodology for a 4-bit unary current source array [8]. The different steps are as follows. First the floorplan is generated from the user specifications: 4 unary bits, spiral switching sequence, mirrored on both axes, surrounding dummies. This implies that every current source is implemented as 4 MOS transistors in parallel, see Fig. 6(a). This results in a symbolic layout specification of which the floorplan is shown in Fig. 6(b). The IO pins are to be placed on top of the array, on a regular grid.

j = jA jB jC jD

dummies

-	4A	5A	6A	6B	5B	4B	-
14A	3A	0A	7A	7B	0B	3B	14B
13A	2A	1A	8A	8B	1B	2B	13B
12A	11A	10A	9A	9B	10B	11B	12B
12D	11D	10D	9D	9C	10C	11C	12C
13D	2D	1D	8D	8C	1C	2C	13C
14D	3D	0D	7D	7C	0C	3C	14C
-	4D	5D	6D	6C	5C	4C	-

dummies

(a) (b)

Fig. 6: (a) One unit current source is implemented as 4 parallel MOS transistors. (b) Floorplan of the 4-bit unary current source array.

The search algorithm determines the minimum number of wires required to connect all cells in both vertical and horizontal direction. It scans all IO pins and connects them to wires. Next all cells are scanned column-wise to connect them to wires. In a last pass the horizontal wires are used to connect the vertical wires. The output is a task file containing the symbolic final placement and routing for the technology mapping. From the DRC rules the minimal width of the wires is determined. The cells are stretched to accommodate the routing and tiled. Physical wires are created and contacts inserted to realize the connectivity. Depending on the number of routing levels, the internal routing of the cell and the required connectivity by abutment, the routing is performed on top of the cells or in between the cells.

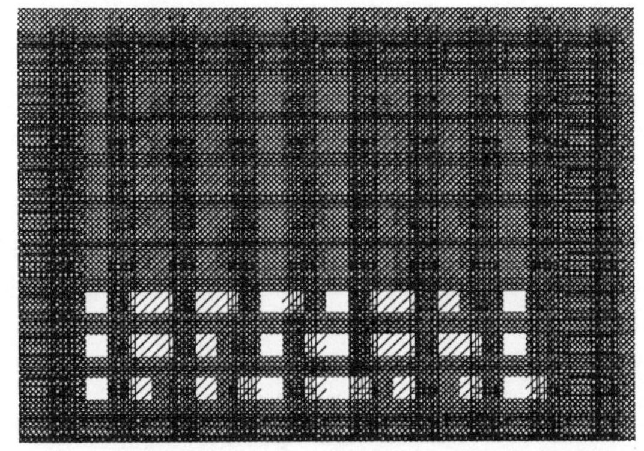

Fig. 7: Automatically generated 4-bit unary current source array. For reasons of clarity all current source cells have not been shown except for the dummy cells surrounding the array. The IO pins are shown on top of the figure.

The final layout in a two-metal process is shown in Fig. 7: the cells have been left out to show the routing. The IO pin assignment is shown on top of the figure. In the Y direction the second metal routing level was used, since metal level one was used internally in the X direction to distribute the power and biasing of the cell by abutment. The parallel devices of each current source have been connected in the X direction by wires in between the cells since metal level one had already been used internally. Of course, for this illustrative 4-bits example no tool is needed. But this is no longer true for higher resolutions (which are encountered in practice).

IV. Experimental Results

The proposed methodology has been implemented in a tool called Mondriaan and has been applied to the physical design of several high-resolution current steering DACs [8]. In Fig. 8 the layout of a 12-bit analog core (the current source array and the switch array) is shown. The current source array (lower part of Fig. 8) contains all the current sources of the segmented DAC (both the unary and the binary current sources). The binary sources have been realized by putting the current source transistors in series. Although the array consists of identical transistors, the routing of the binary sources is different. The resulting pin placement drives the placement of the switches (upper part of Fig. 8). The binary switches have different widths and are placed vertically above the binary current sources (this is a result of the routing driven placement).

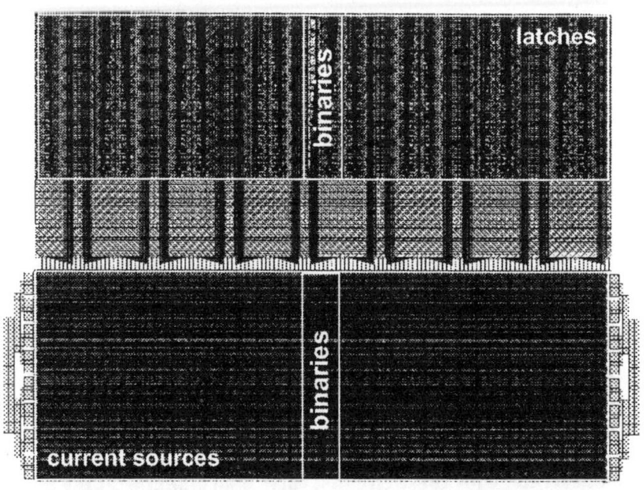

Fig. 8: Synthesized analog core of a current-steering D/A converter.

V. Conclusions

A tool set, Mondriaan, has been implemented which targets the layout synthesis of all array-type analog blocks. The approach takes in consideration typical analog constraints and is very flexible. A three step procedure (floorplan generation, symbolic place & route and technology mapping) solves the layout synthesis problem in a fast and technology independent way. Industrial strength examples show the applicability and usefulness of the implemented approach.

Acknowledgments

This work has been supported in part by ESA/Estec.

References

[1] L. Carley, G. Gielen, R. Rutenbar, W. Sansen, "Synthesis tools for mixed-signal ICs: progress on front-end and back-end strategies," *Proc. DAC*, pp.298-303 (1996)

[2] J. D. Bruce, H. W. Li, M. J. Dallabetta, and R. J. Baker, "Analog Layout Using ALAS!," *IEEE JSSC*, vol. 31, no. 2, February 1996, pp. 271-274.

[3] K. Lampaert, G. Gielen, and W. Sansen, "A performance-driven placement tool for analog integrated circuits," *IEEE JSSC*, vol. 30, no. 7, July 1995, pp. 773-780.

[4] J. Cohn, D. Garrod, R. Rutenbar, and L. Carley, "KOAN/ANAGRAM II: New tools for device-level analog placement and routing," *IEEE JSSC*, vol. 26, no. 3, March 1993, pp. 330-342.

[5] H. Chang, "A Top-Down, Constraint-Driven Design Methodology for Analog Integrated Circuits," PhD. Dissertation Electronics Research Laboratory, College of Engineering, UCB, CA 94720, 1995.

[6] R. Neff, "Automatic Synthesis of CMOS Digital/Analog Converters," PhD. Dissertation Electronics Research Laboratory, College of Engineering, UCB, CA 94720, 1995 (available on the WWW).

[7] S. D. Brown, R. J. Francis, J. Rose, Z. G. Vranesi, *Field-programmable gate arrays*, Kluwer Academic Publisher, 1997, Edition 5, ISBN 0-7923-9248-5.

[8] R. van de Plassche, *Integrated Analog-to-Digital and Digital-to-Analog Converters*, Kluwer Academic Publishers, 1994.

Device-Level Early Floorplanning Algorithms for RF Circuits

Mehmet Aktuna, Rob A. Rutenbar, *Fellow, IEEE*, and L. Richard Carley, *Fellow, IEEE*

***Abstract*—High-frequency circuits are notoriously difficult to lay out because of the tight coupling between device-level placement and wiring. Given that successful electrical performance requires careful control of the lowest-level geometric features-wire bends, precise length, planarity, etc., we suggest a new layout strategy for these circuits: early floorplanning at the device level. This paper develops a floorplanner for radio-frequency circuits based on a genetic algorithm (GA) that supports fully simultaneous placement and routing. The GA evolves slicing-style floorplans comprising devices and planned areas for wire meanders. Each floorplan candidate is fully routed with a gridless, detailed maze-router which can dynamically resize the floorplan as necessary. Experimental results demonstrate the ability of this approach to successfully optimize for wire planarity, realize multiple constraints on net lengths or phases, and achieve reasonable area in modest CPU times.**

***Index Terms*—Algorithms, integrated circuit layout, routing.**

I. Introduction

The growing market for wireless technologies has increased the need for design tools for high-frequency circuits. Most work to date in this area has focused on the difficult problems of verification and simulation for such designs, e.g., [13], [19], [20], and [32]. As the number of designs proliferates, however, other phases of the design process are becoming bottlenecks. Layout is a notorious problem for these designs because of the tight coupling between device placement and wiring, and the potentially significant impact of even small geometric perturbations on the overall performance of the circuit.

Radio-frequency (RF) circuits have unique properties which make their automated layout impossible with standard techniques developed for lower frequency analog and digital circuits. Because every geometric property of the layout of an individual wire—its length, bends, proximity to other wires or devices—may play a key role in the electrical performance of the overall circuit, most RF layouts are optimized for performance first and density second. Worse, in some cases the crossing of two wires creates an unacceptable level of signal degradation and parasitic coupling, requiring a completely planar layout for some high-performance circuits.

Given this level of electrical and geometric coupling, we suggest in this paper a new layout strategy: device-level early floorplanning. The central idea, borrowed from chip-level floorplanning, is to resolve as early as possible all problematic device/wiring interactions by correctly planning the placement and the wiring of the full circuit. The scale of these problems admits an aggressive optimization-based attack. In our approach, a genetic algorithm (GA) evolves a population of device-level candidate floorplans; the location of not only the active devices but also the necessary extra space for planned wire meanders (extra detours taken by individual wires to control total length or phase) are managed by this floorplanning process. Each candidate floorplan is evaluated by completely routing it with a fast, gridless, detailed maze router which can dynamically resize the floorplan as necessary. We extend here our earlier treatments of [1] and [2]. The idea is similar to Cohn *et al.* [8]: for maximum control over performance, we need simultaneous placement and routing so that we may evaluate subtle performance issues correctly.

Much of the related computer-aided design (CAD) work for layout here has focused on lower-speed CMOS analog designs, e.g., [7], [9], [21], [23], [24], and [30], and is not directly applicable at higher frequencies. There is some recent RF circuit synthesis, e.g., [10], which focuses on efficient representations of these circuits for use in numerical optimization. Most CAD work targeting RF circuits comprises interactive tools that aid the designer to speed manual design iterations [15], [16]. Other work in the area includes semi-automated approaches that rely on knowledge of the relative position of all cells [14], [35]. However, these template-based approaches with predefined cells strongly limit the design alternatives possible. Recently, Charbon *et al.* introduced a performance-driven router for RF circuits [5]. In their approach, sensitivity analysis is employed to compute upper bounds for critical parasitics in the circuit, which the router then attempts to respect. Nagao *et al.* [25] also recently targeted a subset of the RF layout problem, using a variant of the sequence pair method and introducing a planarity-preserving routing algorithm. None of these techniques plan simultaneously for both device placement and wiring, and none of them can target difficult constraints such as planar wiring with precise length control.

Our goal in this work is to create a basic substrate of geometric algorithms that can manage the complex geometric interactions that determine performance for an RF layout. We assume here that the critical electrical concerns can be

Manuscript received June 10, 1998; revised October 29, 1998. This work was supported by HP-EEsof and the Semiconductor Research Corporation. This paper was recommended by Associate Editor M. Wong.

M. Aktuna is with the Department of Electrical and Computer Engineering, Carnegie Mellon University, Pittsburgh, PA 15213 USA (e-mail: aktuna@ece.cmu.edu).

R. A. Rutenbar is with the Department of Electrical and Computer Engineering, Carnegie Mellon University, Pittsburgh, PA 15213 USA (e-mail: rutenbar@ece.cmu.edu).

R. Carley is with the Department of Electrical and Computer Engineering, Carnegie Mellon University, Pittsburgh, PA 15213 USA (e-mail: carley@ece.cmu.edu).

Publisher Item Identifier S 0278-0070(99)02316-7.

reduced—at least approximately—to a set of purely geometrical constraints that guide the device-level floorplanning task. Automatic, sensitivity-based constraint-mapping techniques have been demonstrated in [4], [6], and [24]. In practice, we expect designers to use a mix of expertise and extraction/simulation to guide this floorplanning process.

The remainder of the paper is organized as follows. Section II revisits the general floorplanning strategy outlined here, and describes more carefully our assumptions. Section III describes our device-level floorplanner. Section IV describes the device-level router used to evaluate each floorplan. Section V offers experimental results to demonstrate the merits of the approach. Section VI offers concluding remarks.

II. Assumptions and Strategy

RF cells are significantly different from either digital or lower-speed analog cells of similar size. Fig. 1 shows a simple side-by-side comparison of the three cell types. We can observe the following.

- Digital cells have mostly small devices, and for MOS technologies are often laid out in a stylized, row-dominated fashion. MOS devices are aggressively merged for performance and density, and over-the-device wiring is common. This cell, from [3] is a CMOS dynamic logic gate.
- Analog cells feature a wider typical range of device sizes and device shapes, and must manage precision issues not found in digital cells, such as required device/wiring alignment, matching, proximity, or symmetry. These cells are typically optimized first for performance, and then for density. This cell, from [26] is a CMOS operational amplifier.
- RF cells, unlike either low-speed analog or digital cells, are mostly wire dominated. There is no MOS-like device merging. There are no wires routed over devices. There are new precision issues, involving precise control of wire topology and length; indeed these layouts are often required to be planar. These cells are aggressively optimized for performance with density being more of a secondary concern. This cell, from [22] is an RF limiting amplifier. More precisely, we can enumerate the specific layout issues that make RF cell synthesis difficult.
  1) *Performance:* Every geometric property of a wire is a performance concern. Signal degradation can be caused by bends and airbridges (the three-dimensional structures used to allow wires to cross with an insulated air-space in between) on a particular net, and may impact the functionality of the overall design.
  2) *Routing:* Wiring often dominates the layout area. Wire width plus wire spacing is often quite large, causing substantial area consumption for routing. More importantly, wire detours which result from explicit length constraints on nets often take up a large fraction of the layout area. This can be seen in the manual RF layout in Fig. 2.

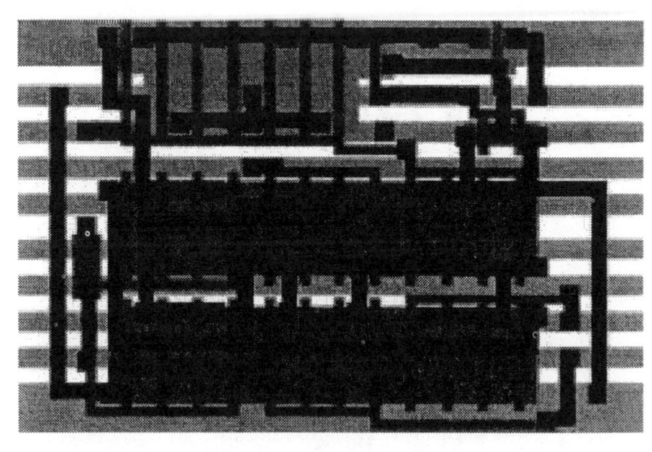

(a)

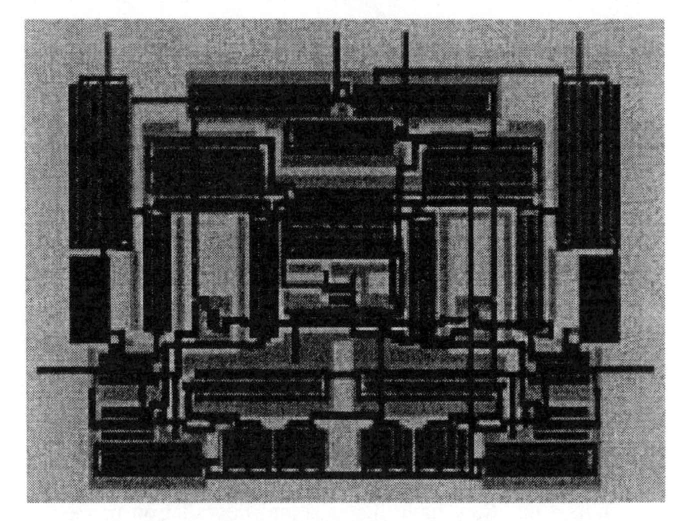

(b)

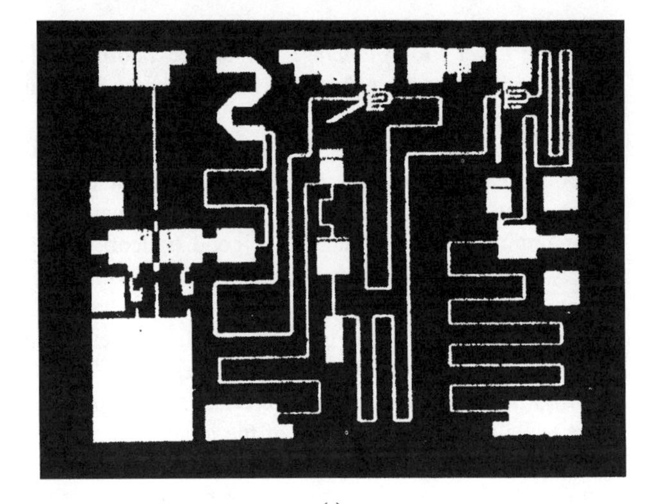

(c)

Fig. 1. Typical digital, analog, and RF cell layouts. (a) Dynamic logic gate [3]. (b) CMOS op amp [26]. (c) Layout of RF limiting amplifier [22].

  3) *Optimization:* Area minimization is not the primary concern. Optimizing the planarity of the routing—necessary when crossing introduces unacceptable coupling and desirable to reduce expensive airbridges—and meeting length constraints on

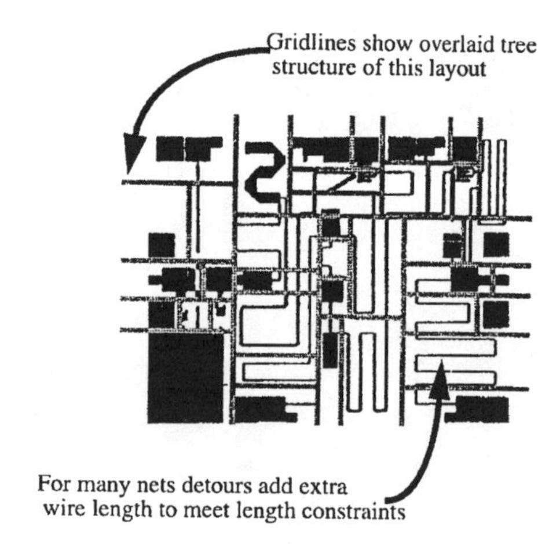
Gridlines show overlaid tree
structure of this layout
For many nets detours add extra
wire length to meet length constraints

Fig. 2. Layout of RF limiting amplifier [22] with slicing tree floorplan overlaid.

performance-critical nets take precedence. These wirespecific constraints directly determine the functionality of the circuit. Given these RF-specific layout issues, we propose a new approach to layout for RF circuits: early floorplanning. Before going into specific algorithmic and implementation details in the following sections, we summarize here our basic strategy and the critical engineering decisions on which it depends.

- *Target Technology:* We assume a one-layer signal routing technology with airbridges for net crossings. Even when multiple metal layers are available, despite their area cost, airbridges can be more desirable for performance-critical nets since they have better signal degradation properties compared to vias that are needed to switch layers. We assume multipoint nets are allowed.
- *Device Level Floorplanning:* We evolve floorplans that specify the locations of both active and passive individual devices of an RF circuit.
- *Representation:* We use slicing trees to represent the floorplans. This restricts the floorplans to some extent, but since slicing trees can efficiently be manipulated for optimization, we find this a good tradeoff. In fact, very complex floorplans can be realized with slicing trees. Fig. 2 shows a slicing tree overlaid on top of the manual RF layout from Fig. 1, and illustrates the practical example on the usability of slicing trees.
- *Wire Meanders as Placeable Objects:* Length constraints on wires are specified as part of the input. In a particular floorplan, wire detours or meanders may be needed to meet this exact length. In our approach, wire meanders are located in the same floorplan room as one of the devices to which the net connects. The size of the room in the floorplan is adjusted by the router to accommodate extra meandering if needed.
- *Geometric Problem Abstraction:* Rather than evaluating the exact performance characteristics of our layouts by using expensive circuit simulation or electromagnetic field analysis, we focus on optimizing the geometric properties of the layout. This is primarily due to the high computational cost of field analysis which we cannot afford for the thousands of layouts we evaluate. Circuit simulation needs the results of field analysis to determine element values and is also computationally intensive. Our strategy strives to provide the designer with fully routed layout alternatives that meet essential geometric concerns such as net length constraints and planarity. The designer, using sensitivity-based analysis tools, can then further adjust the layout to resolve subtle performance issues.
- *Stochastic Optimization of Layouts:* We evolve floorplans using a GA formulation. The GA creates new solution candidates from promising floorplans and invokes the router for their evaluation. In the following two sections, we discuss our device-level floorplanning and routing algorithms in detail.

III. Device-Level Floorplanning

In this section we describe the details of our device-level floorplanning strategy for RF cells. Despite the fact that there is no explicit placement stage after floorplanning, the floorplanner does not produce fixed device locations for the router to work on. The result of the floorplanning is a starting point for the router to work on, laying out the relative placements of devices and the wire detours that are adjacent to them. The router will further expand this "seed" placement and dynamically create a "sized" floorplan that can accommodate the optimized routing. Because of this, the device-level floorplanning strategy introduced here should be regarded as a preliminary stage before the floorplan is finalized. Due to the tight links between the floorplanner and the GA optimizer around it, genetic optimization issues are also discussed here.

A. Floorplanning by Genetic Optimization

We recast the floorplan optimization problem as a stochastic optimization using a GA. The critical components of any genetic algorithm are as follows.

- *A Representation for Individual Solutions:* In our case this is a slicing tree representation of floorplans.
- *A Population of Solutions:* We evolve a population of device-level RF cell floorplans, with typical population size of a few hundred.
- *A Selection Scheme:* We use tournament selection as described in [17] with a tournament size of two.
- *Evolution Operators:* We use a new subtree-driven crossover scheme and mutation operations adapted from simulated annealing of slicing trees.
- *An Evaluation Method:* We use our router for evaluation of the floorplans. Our specific genetic algorithm implementation uses a continuous population model that replaces a user-controllable fraction of the population in every generation. The default is replacing the 30% of the population with the worst scores. Keeping the best individuals of the population after a generation is called elitism [12]. It should be noted that the continuous population model implements elitism implicitly, since the individuals with better scores will always be preserved.

The score of the best individual in the population is tracked during the course of evolution. This is used to determine the stopping criteria, which is to stop if the score of the best answer found has not changed in a user-defined number of generations more than a tolerance percentage. The default is to stop if the best individual has not changed more than 1% in the last 100 generations.

B. Floorplan Representation

In our strategy, we represent the floorplan using canonical polish expressions of slicing trees [28], [34], and the genetic algorithm evolves the polish expressions directly. Slicing trees capture the relative placement of objects in a compact way. More importantly, the optimizer can produce a new floorplan from a given one with little computational effort, allowing efficient search of the design space. The choice of slicing trees for representation will not allow the realization of some nonslicing floorplans. With the target application of at most 50 devices—where for us a "device" is any active or passive component that must be placed and wired in an RF cell—this does not impact area significantly. We believe efficient traversal of the slicing tree design space can more than make up for this restriction.

Each of the objects in the polish expression is either a device or a device with planned space for wire meandering next to it. The floorplanner chooses an aspect ratio for every module using the Stockmeyer algorithm [34]. When there is no user defined constraint on the overall aspect ratio of the layout, the choice of aspect ratio for each module is optimal for area with respect to the given slicing tree. This optimization allows the router to start with the best packing possible for each slicing tree.

When there is a constraint on the aspect ratio for the overall layout, the floorplanner will look for aspect ratios that are within given upper and lower bounds. If there are multiple aspect ratios within the allowed range the one with minimum area is selected. If none of the aspect ratios are within the allowed range, the one closest is selected. Note that after routing, the final aspect ratio of the layout may be different than what the Stockmeyer algorithm determined after placement. After routing, a penalty is imposed if the final aspect ratio is not within specified bounds. This guides the evolution of floorplans toward solutions with more desirable aspect ratios.

C. Evolution Operators

In their floorplanner that uses simulated annealing to evolve canonical polish expressions of slicing trees, Wong *et al.* [34] used three moves to perturb the current floorplan. These were as follows.

M1) Swap two adjacent objects.

M2) Flip every cut in a chain in the polish expression, where a chain is a maximal series of operators not delimited by objects.

M3) Swap an adjacent operator operand pair.

Any new slicing floorplan of n rectangles can be reached from another with some sequence comprised solely of these three moves. Our genetic algorithm uses the same three moves as mutation operators to introduce diversity into the population. However, for efficiency we also need to mate pairs of floorplans, with the goal of propagating the best components of each. For this purpose, we introduce a new crossover operator based on subtrees. Subtrees are a good choice of building blocks for slicing trees since they encapsulate the adjacency relations among subsets of nearby devices. The crossover operator preserves the subtrees in parents as much as possible with the hope of preserving the adjacency relations that allowed the parents to have a good score. We call this strategy subtree-driven evolution. Two parents chosen for mating by the genetic algorithm then go through mutation with a fixed mutation probability.

We can describe this mating process qualitatively as follows. There are two basic cases. In the simplest, the crossover operator picks a random location in the first parent tree. If the crossover location holds a device, the child is obtained by swapping two devices in the second parent tree to enforce the same location for the crossover module in the second parent. This resembles an M1-type mutation operator, but it is influenced by the other parent.

The more interesting, second type of crossover occurs when the crossover location in the first parent, P1, contains an operator. The subtree is implanted into a suitable place in the other parent tree, P2. The primary goal while doing this is to introduce a subtree from P1 into P2 with minimal disruption. This is done in two main ways.

- The crossover algorithm rigorously searches for a subtree of comparable size in P2 for implantation, in order not to destroy a significant portion of the organization of P2.
- Further conservation is sought even within this tree of comparable size, by looking for subtrees that can be preserved.

Fig. 3 shows a simple crossover example where nodes labeled "|" mark operators whose children are horizontally adjacent, and nodes labeled "-" delimit vertically adjacent children. Note how the subtree with f-k-g-h is implanted in the final offspring, and how the a-c subtree is preserved. Fig. 4 has simple psuedocode for the operation. Together, these evolution operators can efficiently find dense, low-area slicing floorplans that can meet the geometric constraints imposed as input. But to evaluate these layouts accurately in their context as RF circuit designs, we need to route them.

IV. Detailed Routing of Floorplans

We use a novel router as the evaluation tool for evolving floorplans generated by the genetic optimizer. However, in our overall layout strategy this router has responsibilities beyond those of a traditional router. Our router completes the placement process by determining exact device locations and placing airbridges, which may occupy substantial area in RF circuits.

We use a detailed, one-wire-at-a-time area router. The router does not have separate global and detailed routing stages, instead it takes wiring details into account while selecting paths. There are two major reasons for this. The first concern

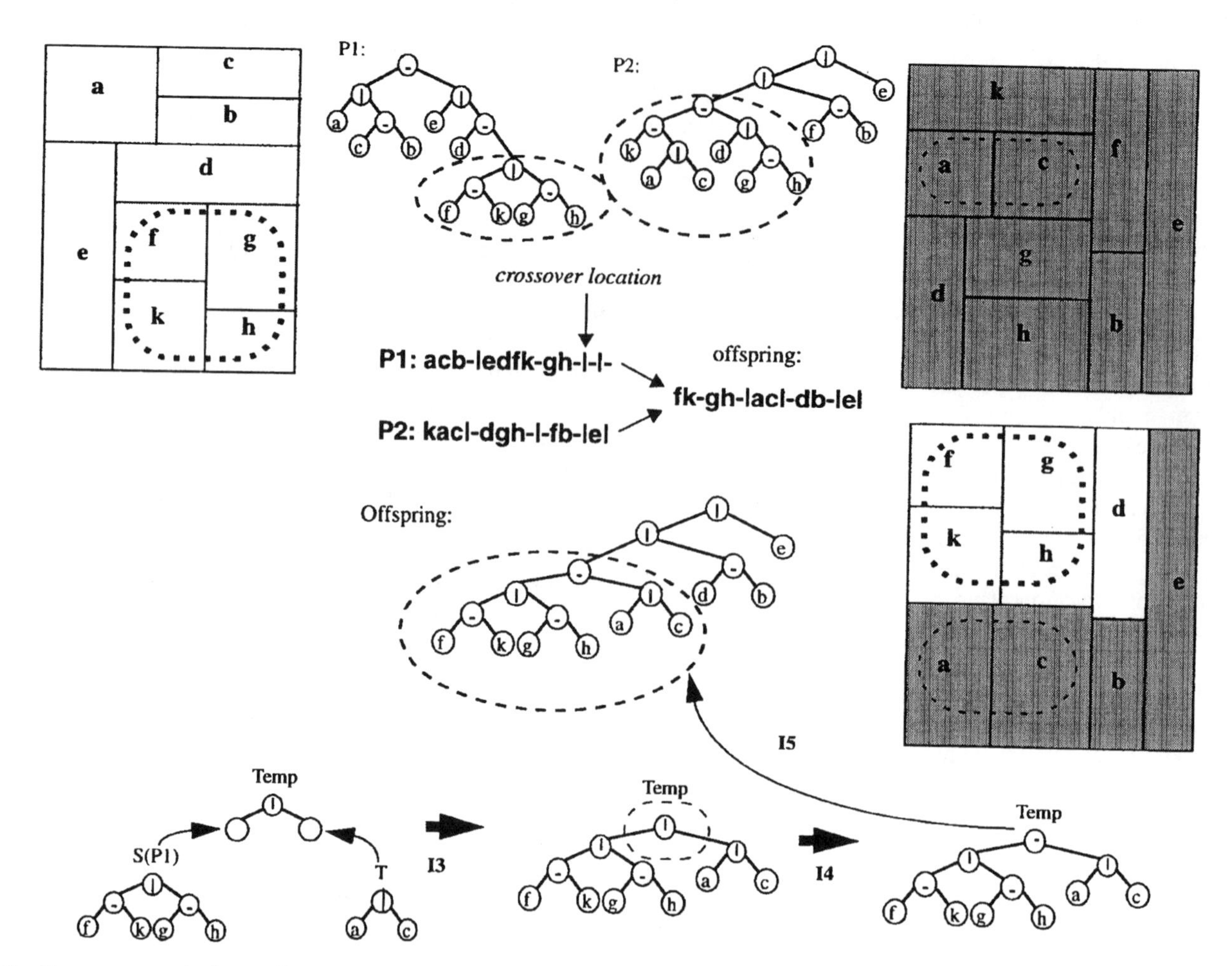

Fig. 3. Crossover example for combining two floorplans with implanting. Parent and resulting offspring floorplans appear at the top; steps of the implant process appear at the bottom.

Crossover algorithm:

C1.Pick a random subtree S(P1) in parent 1

C2.Find the subtree S(P2) rooted at S(P1)'s root in parent 2

C3.If S(P2) has fewer modules than S(P1):

  C3.1Search around S(P2)'s root for subtree with at least as many modules as S(P1), make it S(P2).

C4.Call subroutine *Implant(S(P1),S(P2))*.

Subroutine *Implant (S(P1),S(P2))*

I1.Let n1= number of modules in S(P1)

  n2= number of modules in S(P2) .

I2.Pick T={(n2-n1) modules from S(P2)} from modules that are not in S(P1)

I3.Create random tree Temp composed of S(P1) and modules in T.

I4.Correct operator chains in Temp in parent 2

I5.Swap Temp and S(P2).

I6.Correct duplicate module conflicts by swapping modules from outside S(P2) in parent 2

Fig. 4. Algorithm for crossover with subtree implantation.

is planarity. Without taking routing details into account, the number of net crossings will not be optimized. This is very important since the number of airbridges has a major impact on the quality of a global path. Maximizing planarity is important not only because it decreases area due to fewer airbridges, but also because it reduces signal degradation at airbridges. Similarly, bends require attention to detail for minimization. The second concern is the need to update channel dimensions (i.e., the wiring areas between our placed devices, where all routing occurs) as routing progresses, which can only be done with detailed routing information. In our dynamic sizing formulation, channel dimensions change dramatically as wires and airbridges are embedded, and global routing decisions have to be made taking this into account.

A. Routing Strategy

We use a graph-based router with the cost function described in pseudocode form as

$$\begin{aligned}&\Sigma\ (\text{length of floorplan edges in path})\cdot(\text{net width})\\&\quad+\Sigma\ (\text{length of routing in floorplan nodes for path})\\&\quad\cdot(\text{net width})\\&\quad+\Sigma\ \text{airbridge penalty for crossings in expanded path}\\&\quad+\Sigma\ \text{bend penalty for bends in expanded path}\\&\quad+\Sigma\ (\text{number of other nets in edge})\\&\quad\cdot(\text{length of floorplan edge})\cdot(\text{net width}).\end{aligned}$$

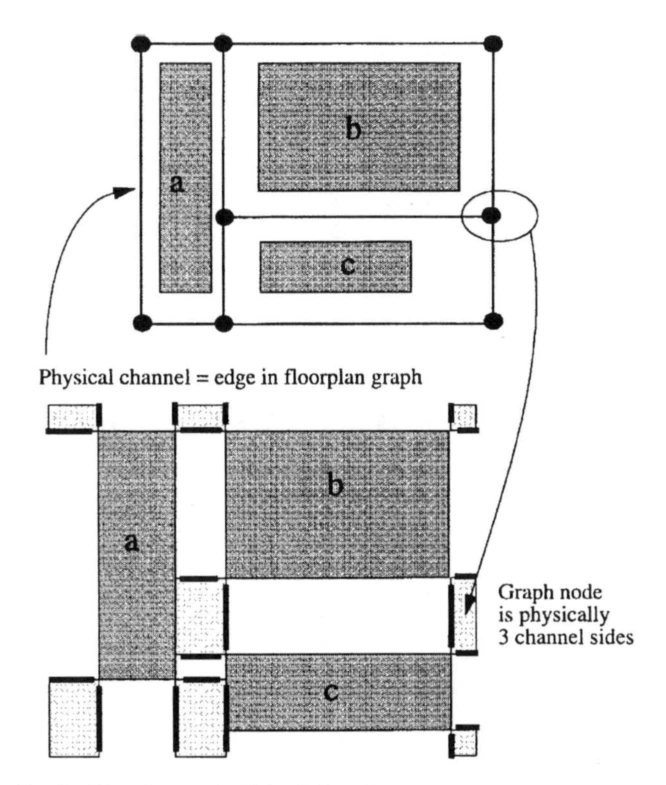

Fig. 5. Floorplan graph of simple layout.

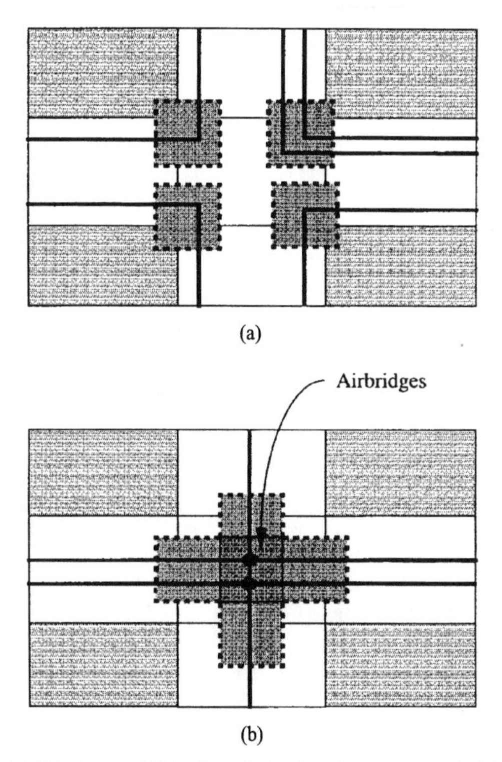

Fig. 6. (a) The four net-lists for nets turning the corners and (b) the two net-lists for nets crossing a node vertically.

This cost-based method is a practical method for capturing the effects of routing details during path selection. It also gives the user the flexibility to choose the criticality of airbridges and bends on a net-by-net basis, since the corresponding penalties are proportional to user-set coefficients. The last term in the cost function penalizes parallel runs of nets in crowded regions, proportional to the length of the parallel run. This is introduced to control congestion in channels and nodes with the aim of reducing long parallel runs of wires. This—at least qualitatively—reduces some crosstalk problems.

One key mechanism of the router is dynamic floorplan resizing. The router starts off on a floorplan with fixed-aspect-ratio devices placed with no extra routing space between them. To ensure sufficient space, the layout is dynamically enlarged while each net is embedded. By avoiding fixed, predetermined channel widths, the layout quality is substantially improved in RF circuits since the variation of channel widths is higher than for lower-frequency analog or digital circuits. Having a dynamic resizing mechanism also means that a channel is never blocked due to congestion. The dynamic resizing ability is also key to meeting length constraints since we can resize as needed to create extra wire detour space.

B. Routing Data Structures and Basic Routing Engine

The routing engine is a graph based maze router that minimizes channel congestion, number of air bridges and bends. A floorplan graph is a commonly used tool for describing the topological relation of routing regions and placed blocks [11]. Our gridless router works on an extended version of the floorplan graph. Fig. 5 shows a simple layout and its associated floorplan graph. Each edge of the graph corresponds to a routing channel with device pins on at most two opposing sides. Each node in the graph corresponds to the intersection of channels. Our maze expands nets along one edge and through one node as its atomic expansion step. In our extended floorplan graph, nodes and edges are rectangular regions and are not necessarily perfectly aligned. The router keeps track of the sizes and the alignment offsets for each edge or node.

Further extending the floorplan graph, the router relies on topological enumerations of nets in each floorplan graph edge and node. Each enumeration tracks how the net moves through the routing region associated with the edge or node: from which side it entered, its pin order on this entering side, whether it goes straight through or chooses to turn in the region, its exiting side, its pin order on this side. Given that we currently route in just one layer with a focus on maintaining planarity, this topological representation of each routing region is critical. Each node in the graph can have several of these topological entry/order/exit lists. At every graph node, nets are grouped into lists using as the unique key the entry–exit channels for the net. Therefore there is a net list for every combination of two edges incident on a node. Since a node may have two, three, or four edges incident on it, it may have up to $C(4, 2) = 6$ net-lists. Two typical cases are illustrated in Fig. 6. Fig. 6(a) shows five nets changing direction as they traverse a node; the shaded boxes highlight the four topological entry/order/exit lists that track nets that change direction through this node. Similarly, in Fig. 6(b) we highlight the two other lists that track nets traversing the node vertically or horizontally. If a node has nets in both of these latter two lists, they are nonplanar in the node and will require airbridges to cross each other in the node as shown in Fig. 6(b).

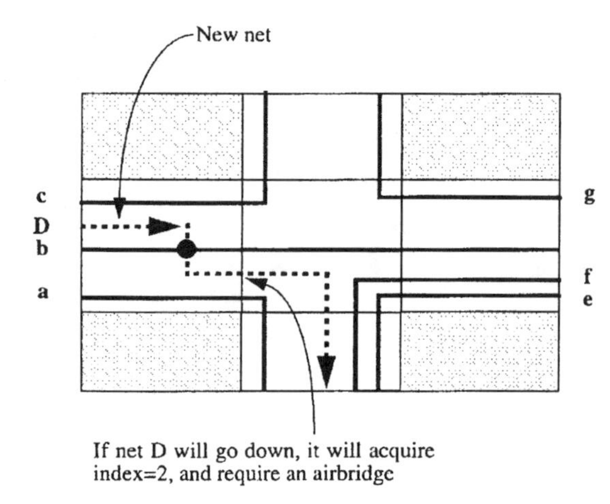
New net
c
D
b
a
g
f
e
If net D will go down, it will acquire index=2, and require an airbridge

Fig. 7. Wavefront expansion for a new net among previously routed nets.

Nets in these topological lists are ordered via their pin-order on both their entering and exiting sides. For example, counting bottom-up, net f in Fig. 7 will be the second net in the bottom-right-corner topological list for the node. Given any edge or node in the graph, we can combine the appropriate topological lists from the adjacent node/edges in the graph and create an ordered topological list of all nets on any side of any routable region in the layout. We will call this list the pin ordering for brevity. We also assign an index to each net in pin orderings. For the channel to the left of the node in Fig. 7 the bottom-up pin ordering is $(a\,b\,c)$ and the indexes for a, b, and c are one, two, and three, respectively. Pin orderings allow the router to deal straightforwardly with planarity issues in a channel. While expanding potential paths, the router considers the net as a pin to be inserted into the existing pin orders on the entering and exiting sides of the routing region being traversed. The new pin's index entering a channel is known since the order in the previous routing region has already been determined and nets do not change order in a node. Referring again to Fig. 7, the index for the new net D is three on the left side of the channel. Given the entering index, the router computes the index leaving the channel to minimize airbridges required. For net D in our example, the index will be two on the right side of the channel. Once this is determined, the index entering the next channel on the bottom side of the node is also determined to be two since the pin order $(a\,D\,f\,e)$ is known. This index depends on the direction the net will proceed in the succeeding node, after traversing this channel; this is illustrated in Fig. 8. Therefore, every channel is expanded in all possible directions in which the net can next proceed. More details of this process will be discussed in the next section on wavefront expansion.

C. Wavefront Expansion

Routing of signals is done in a single layer of metal with air bridges inserted to resolve nonplanarities. The path search algorithm is an extended form of general maze routing [29]. However, the cost of a net location on the next node-side is rather more complex than costs in general cell expansion. The cost of adding a net location to the evolving wavefront is the sum of the cost to cross a node and an edge. Both costs

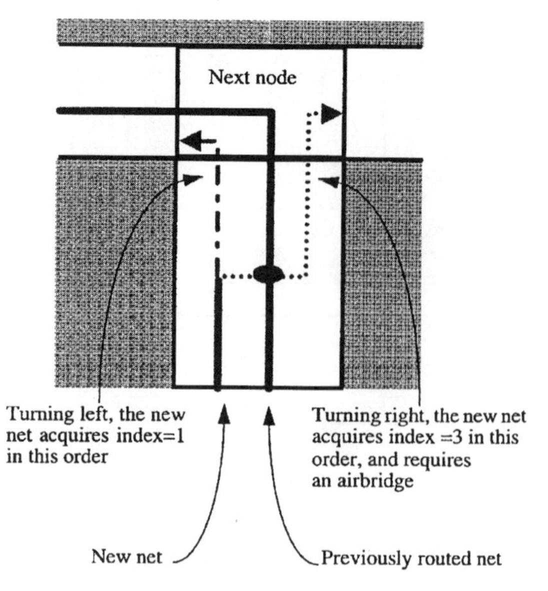
Next node
Turning left, the new net acquires index=1 in this order
Turning right, the new net acquires index =3 in this order, and requires an airbridge
New net
Previously routed net

Fig. 8. Wavefront expansion for a net in two directions.

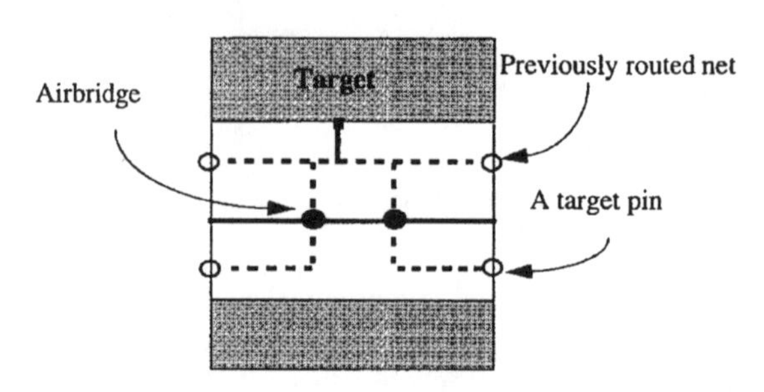
Target
Airbridge
Previously routed net
A target pin

Fig. 9. Four target pins for a new net, two of which require airbridges.

include bend and airbridge costs. After a path is selected by the maze router, it is embedded into the current layout by inserting the required air bridges and resizing the channels as needed. We discuss details of the expansion process in this section.

In our formulation, device terminals generally have fixed positions as part of the circuit description. That is, for every terminal, the designer specifies a device, one of the device's four edges, and the distance from the left or bottom corner. Another option we provide the designer is specifying multiple edges without a coordinate. This is intended for devices like capacitors and inductors where the connections can be made anywhere on the perimeter of the device. For such nets the router will choose the center of the cheapest edge among the ones available.

Maze routing requires a source and a set of targets before expansion can start. Given the floorplan graph, we propagate device terminals as pins onto the escaping sides of the routing region we are expanding through. While doing this, all distinct possible pin locations have to be represented so as not to restrict the router. For example, as illustrated in Fig. 9, for the target one of the two device terminals being connected by a net is chosen. Then all possible net indexes on the two ends of the channel are initially recorded as target pins. It should be noted that the target pins have different costs to the terminal since they may have different distance or different number of airbridges required. The goal of path search is to reach one of these target pins with minimum cost. Sources are generated with a similar procedure.

The cost of expansion depends on the following:

1) the length and congestion of the channel involved;
2) the number of airbridges and bends required to cross it;
3) the relevant physical dimensions: the width of the channel, the dimensions of the node at its end, and the width of the wire being routed.

Since the number of airbridges required to cross an edge depends on the direction in which the net will proceed, each new routing region entry/exit being added to the wavefront may have a different set of airbridges required, and hence a different cost. Therefore, the router expands each of these topological path alternatives separately. To be precise, what is pushed to the wavefront by expansion during path search for net n is a record that has: 1) one side of one node in the floorplan graph and 2) the index for net n in the pin order for that side the router proposes as an entry point for net n. If different paths lead to different net indexes (i.e., pin orders) on the same side of one region for the net being routed, they will coexist on the wavefront.

The following three causes of wire bends are tracked and penalized during expansion:

1) bends caused by connecting a vertical edge to a horizontal edge;
2) bends caused by airbridges;
3) bends caused by connecting two edges that are both horizontal or both vertical, but not aligned.

Bend costs are user-specified on a per-net basis, giving the user finer control of bends on critical nets.

While expanding the wavefront, it is not sufficient to account for just the edge lengths, i.e., the simple distance of traversal across each physical channel, to compute the exact routing length of a net. Edge widths as well as detours in nodes have to be taken into account. These are accounted for by a simple algorithm that keeps track of turns at node corners during expansion.

D. Resolving Planarity During Expansion

The maze router computes the number of air bridges required to insert an expanding net into each channel using locally stored pin order information; this is a critical component of the routing cost. In our model, evolving nets may need to take nonplanar paths through channels to make the turns necessary to reach their targets. However, we do not allow nets to make nonplanar turns through floorplan graph nodes. During expansion through a channel, we use a simple heuristic to determine if a planar embedding is feasible. This is also made tractable by the fact that we restrict ourselves to a single wiring layer. The number of net crossings will also change with the order nets are routed. We do not explore different orders in this version and use the net ordering the designer provided.

If two nets are nonplanar in a channel, an air bridge has to be inserted at the point they become adjacent along the channel. The number of airbridges a new net requires to cross a channel depends on two pieces of information which are known for the expanding net. We call the following the planarity data set:

1) the entrance index of the net among previously routed nets in the edge;
2) the direction in which the net will proceed after it exits this edge.

The simple idea is that a net n will fail to find a planar embedding if its entry pin order to the channel requires it to insert between other previously embedded wires, and net n needs to make a turn that crosses some of these wires. For example, net D in Fig. 7 has to cross net b in order to make a turn toward the bottom of the node because it entered the channel above net b. Comparison of the planarity data set for the new net n and each other net in the channel suffices to determine if an airbridge is required. By doing this for all nets in the channel, a set of nets which must each be crossed, and the index in which the new net leaves the edge, are computed. This determines the index of entry into the successor routing region, along with the exact cost of airbridges in the edge just traversed. We use a simple linear scan down the spine of the channel, and compare the evolving net to each previously embedded net. This operation is linear in the number of nets embedded in the channel. The approach is similar to the pin-order net-combing technique introduced by Groeneveld in [18] for selecting an ordering for all pins at junction edges, given the global routing. However unlike their problem, we already have the ordering for previously routed nets and need to make a decision only for the new net being expanded. Once a path is decided for a new net, it will be embedded to the graph edges and nodes on the path. The topological lists in each edge or node are updated at embedding time.

E. Interaction Between the Router and the Floorplanner

It is important to note in our strategy how precise net lengths are achieved. We do not require the maze router to embed each controlled net at a precise length; rather, we rely on evolution to create floorplans which can be routed with nets of the correct length. This is less random than it might initially appear: the floorplanner plans space for meanders on individual nets, and the router then negotiates with the floorplan to ensure that the combined length of the embedded wires and the flexible wiring in the meandered spaces meets the length constraints. This is illustrated in Fig. 10. This is critical to the success of the overall approach since, on a net-by-net basis, the router is constantly resizing the floorplan. It is not possible to embed a net once, early, at a specified length and then maintain this as an invariant as subsequent routes embed.

After embedding all nets, length constraints are checked. If a net is shorter than it has to be, it is detoured in the space adjacent to its source. When that device does not have enough unused space in its floorplan room, appropriate floorplan resizers are called to create enough space in the floorplan for meandering this wire.

When the net is longer than its constraint, this shows that the current slicing tree is not suitable to meet the constraint, and we impose a penalty to the score of the slicing tree, reducing its chance of survival into the next generation of layout solutions evolved by the genetic algorithm. RF signal phase constraints

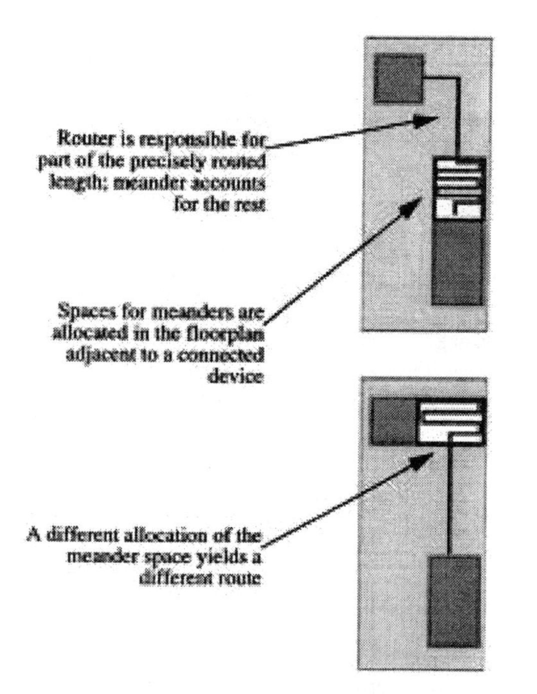

Router is responsible for part of the precisely routed length; meander accounts for the rest
Spaces for meanders are allocated in the floorplan adjacent to a connected device
A different allocation of the meander space yields a different route

Fig. 10. Interaction between floorplanner and router for precise-length control.

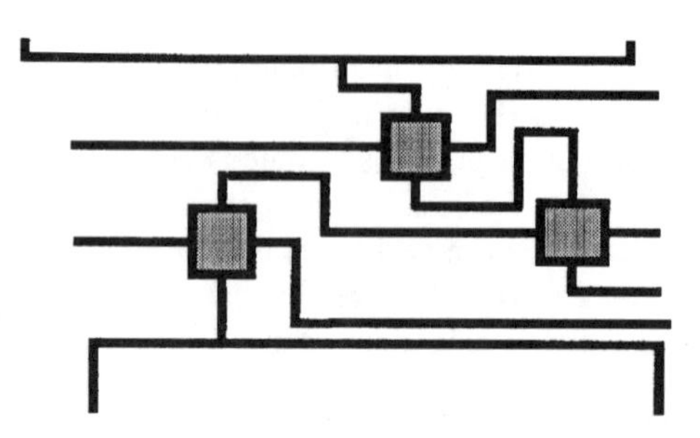

Fig. 11. A set of airbridges placed in a channel.

can be mapped to length constraints and satisfied with the same mechanism. The only difference is that a penalty is not needed since a net can always be made longer until it reaches the required phase.

The router makes fine adjustments on the floorplan by resizing it. Embedding of nets is done after the maze router finds a path. Channel heights and lengths are adjusted starting from the source while embedding the net with its air bridges. While embedding, the ordered list of airbridges in each channel is updated with the new airbridges. The airbridge list defines the placement order of airbridges as well; in our model, all airbridges will be placed along the length of the channel, as illustrated in Fig. 11. The height and length of the channel are also adjusted to accommodate the new net with resizing operations. Qualitatively, there are five major causes for the resizing operations the router invokes.

1) *Insufficient Channel Height to Place Wires and Airbridges:* The router makes sure that all wires can fit side by side for the current channel height. The required channel height is also a function of the airbridges in the channel. A net is not necessarily present in the whole channel—it may exit to a terminal. Furthermore, airbridges may have different sizes, therefore the maximum height will depend on the placement of airbridges and the current set of nets around them.
2) *Insufficient Channel Length to Place Airbridges and the Bends They Require:* While placing the airbridges the router keeps track of the space they take side by side. As illustrated in Fig. 11, this depends not only on the width of the airbridges but the width of the nets that have to bend around to get in and out of the airbridges.
3) *Improper Airbridge-Device Pin Alignment in the Channel:* The router makes sure that there is enough room in the channel for required airbridges before a net exits. For example in Fig. 11, there has to be enough space to the right of the net exiting to a terminal above to fit the two airbridges to the right of the terminal.
4) *Insufficient Wire Meandering Space Next to a Device:* When a net needs a large meander, new space may have to be created next to a device to fit the meander properly.
5) *Insufficient Space for Airbridges at a Channel Intersection (a Floorplan Graph Node):* When there are airbridges in a node as in Fig. 6, the router makes sure the node is large enough to contain them.

These resize operations are handled by tracking at all times the available space or slacks around devices and wires. Managing slacks around modules for floorplanning is common; unusual here is that we also track slacks for wire segments in each routing region. Recall that our goal is fully simultaneous placement and routing, to explore a layout space with tight coupling between device placement and device routing. Each proposed floorplan is fully routed, and each new wire can embed so as to move all other devices and previously embedded nets. We track slacks to allow fast, incremental resizing across all layout geometry as each wire embeds.

To accomplish this, we keep a detailed record of all geometry and slacks around them. For every device the slacks on its four sides are tracked. For example, in Fig. 12 device A has slacks $s1$, $s2$, $s3$, and $s4$ in its slicing floorplan room. Besides devices, channels and nodes are first-class objects whose sizes are tracked. Channels have both horizontal and vertical slacks. Channel 1 in Fig. 12 has horizontal slacks $s5$ and $s7$ and the vertical slack $s6$. Slacks originate from unused space created by slicing during the Stockmeyer detailed floorplan sizing algorithm [34], or previous resizing operations. Fig. 13(a) shows slacks above device A and to the left of device B created by the slicing tree algorithm. The first of these two slacks increases when a net is routed where the second one decreases as shown in Fig. 13(b). Fig. 13(c) shows a new slack created to the right of device B by resizing operations when another net gets routed. When a space request originates, devices, channels and nodes will try to satisfy it from their own slacks as much as possible and then propagate the remainder across the layout. Eventually all objects will have enough space. To allow the communication for this propagation, every device has pointers to nodes on its sides which in turn have pointers to all channels incident on them. The channels in turn have pointers to devices and nodes around them. For example, in Fig. 12 device A keeps track of the first and last nodes on its right, nodes 1 and 3. Since nodes and channels are stitched

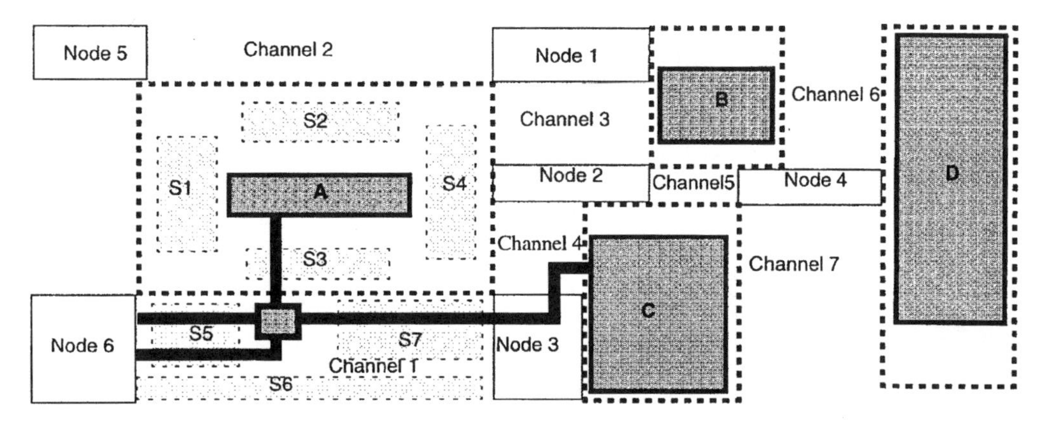

Fig. 12. Slacks around devices and channels in a floorplan.

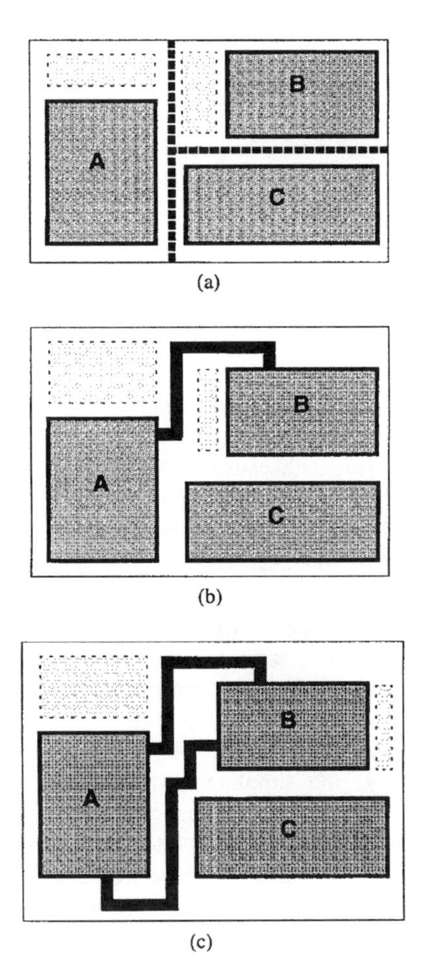

Fig. 13. Evolution of slacks in a floorplan as routing progresses.

with pointers, when device A has to move right it can probe nodes 1, 2, and 3 to move objects to its right further by passing requests to all these nodes.

Resizing of channels is done by moving the devices defining the channel. Once a vertical channel requests more width, the device to the right is passed a request for extra space, where a horizontal channel would probe the device on top of it. Resizing moves are always to the "right" or "up," in the plane of the device-level floorplan. For example, if channel 1 in Fig. 12 needs more vertical space it will pass a request to device A for more space. Device A will first try to satisfy the request from $s3$ then $s2$. If devices can not satisfy the request from their own slacks, extra space requests are passed to all channels on the opposite side through nodes on the opposite side. In this case, device A would probe nodes 1 and 5 to move channel 2 as needed. After a device moves, the required slack is added to its neighboring channels on the side from which the request originated. For example, if channel 6 makes device D move right, after the move device D adds slack to channel 7. Channel 6, being the originator for the move, does not get extra slack here.

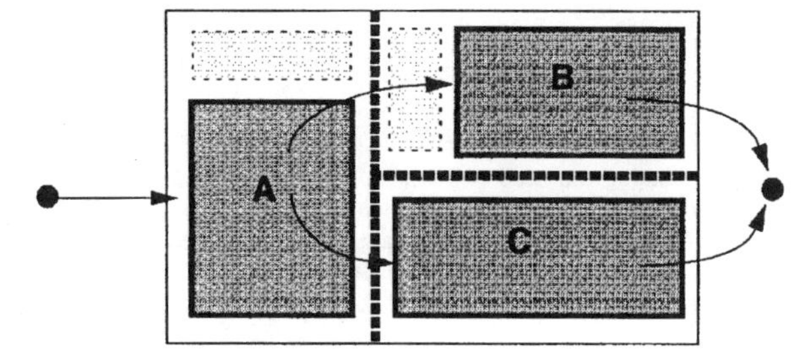

Fig. 14. Horizontal constraint graph overlaid on a floorplan with shown slicing tree.

This resizing method is equivalent to finding and maintaining critical paths on the floorplan graph and computing slacks in the noncritical paths. Fig. 14 shows the horizontal constraint graph for a simplified layout where the path that goes through devices A and C is critical, therefore the path that goes through devices A and B has a horizontal slack at B. However our approach does not need to identify explicit critical paths across the floorplan for every resizing; in practice, the incremental updates do not propagate across the whole floorplan for most resizing operations.

After devices move, they also check that the neighboring channels and nodes can still be routed properly. The new location of the device may leave insufficient room for airbridges and wires around them or may make aligning them with device terminals impossible. To fix this, extra resizing requests may be generated.

The router is also responsible for the placement of pads and other fixed objects. We allow the designer to specify how far a device should be from one or multiple edges with given tolerances. For a pad this is usually a specification to place the

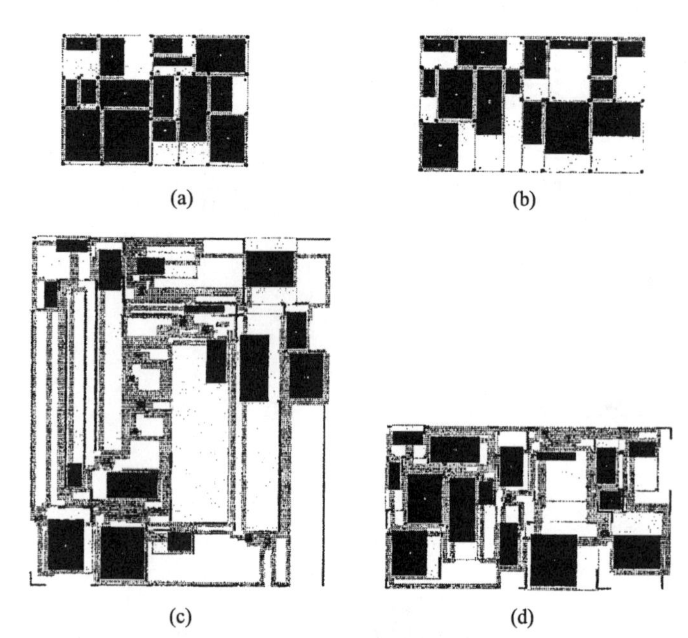

Fig. 15. Validating the need to evaluate device-level floorplans with complete, detailed routing. (a) Floorplan from optimization with sizeless wires; area: 1612. (b) Floorplan from optimization with real wires. area: 2119. (c) Routing the floorplan from (a); area: 7236; airbridges: 12. (d) Routing the floorplan from (b); area: 3311; airbridges: four.

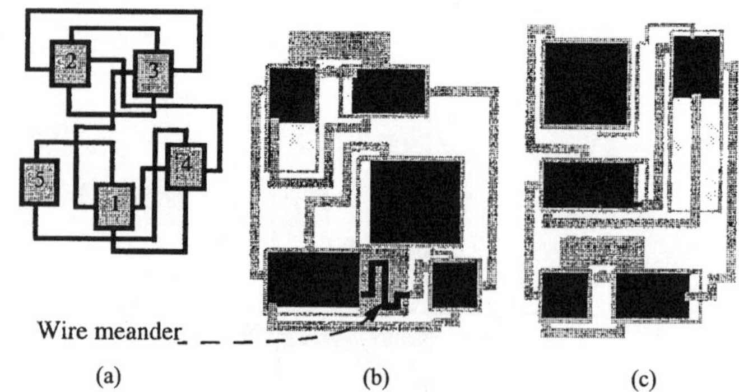

Fig. 16. Impact of bend costs on optimization.

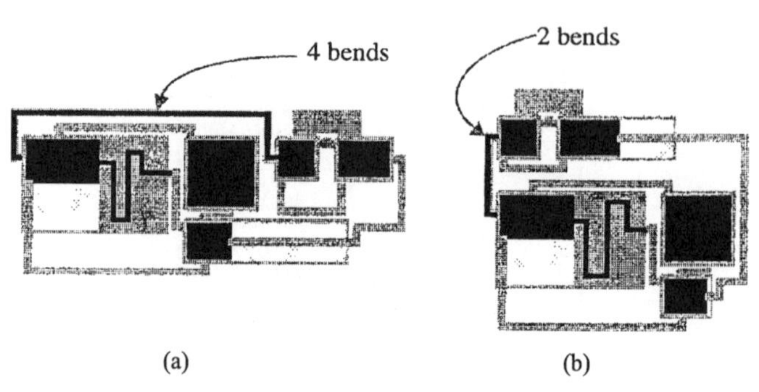

Fig. 17. Manual and automatic layouts for limiting amplifier of Fig. 2.

pad at zero distance from a specific edge of the layout with zero tolerance. We also allow a specification to place pads on any one of the four edges without the designer choosing a particular one. The router moves pads to the specified locations if there is no obstruction and imposes a cost penalty if the specification is not met. This simple approach is reasonably functional for pad support at the cost of increased runtime since we need larger populations to accommodate the pad constraints.

V. Results

The algorithms we presented in the preceding sections have been implemented in roughly 18 000 lines of $C{+}{+}$ code. We employed a modified version of a genetic algorithm library, GAlib, available from MIT [33].

To begin, let us validate one of the core assumptions of the overall strategy: the need to have detailed routing geometry to evaluate the quality of each proposed device-level floorplan. We use a synthetic netlist with 15 devices, 20 nets, and no length constraints. The netlist is evolved twice. First, we use a "sizeless" wire routing without airbridges; the idea is that each route is of zero-width and so does not require any negotiation with the floorplan for resizing. In effect, this minimizes a simplified wire length for each net. This floorplan is optimized for idealized estimates of area and wirelength. We then evolve another floorplan with the tools full real-geometry routing capabilities.

It is possible to find very dense—indeed superior—floorplans if we ignore the details of routing. Unfortunately, when we then actually route these floorplans, the results can be dramatically inferior, as illustrated in Fig. 15. Without real wires, the floorplan at the top left offers a better packing of the devices. But when routed, it is clear this is a poor solution candidate: it has three times the number of airbridges and more than twice the area compared to the layout resulting from full optimization. This effect is especially pronounced in our RF circuit layouts because of the need to route in a single wiring layer under length constraints, and the significant area penalty (much larger than a conventional via) of each air bridge to resolve nonplanar connections. We believe that this simple result demonstrates to the need to capture fine details of the routing simultaneous to the device placement. Next, we shall highlight two specific capabilities of our floorplanning strategy: the ability to control length precisely, and the ability to optimize wire bends.

First, we show the impact of optimizing for precise wire lengths. We use the simple netlist shown in Fig. 16. The floorplan at left is optimized first without taking the length constraint into account, and then (at the right) with the length constraint. When we ignore precise net length requirements during device-level floorplan evolution, we cannot guarantee that subsequent routing can meet the constraints. In this case we can—by adding a meander as shown at the left in the figure—but at increased area. The layout at the right is simply better planned to meet the constraint, and thus saves area.

Next, we show the impact of controlling bends. We use the same synthetic netlist as from Fig. 16. Fig. 17 shows the results. The layout at left optimizes with uniform bend penalties on all nets. Increasing the bend cost of the highlighted net results in a different floorplan which allows the net to embed with two rather than with four bends, which is minimum for this design. To give a better view of the capabilities of the approach, we turn finally to a larger and more realistic layout with several interacting constraints. We compare an automatic floorplan with a manual layout we

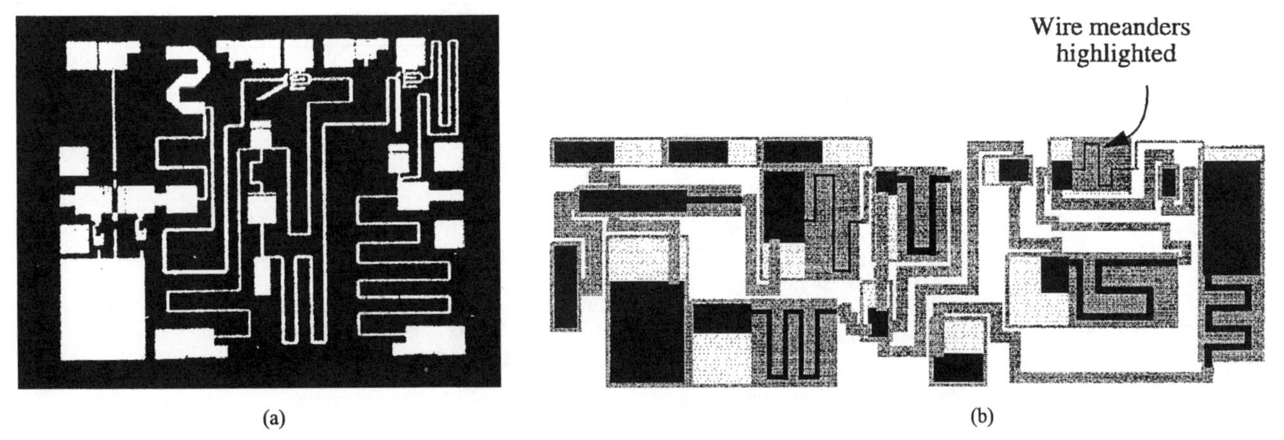

Fig. 18. Manual and automatic layouts for limiting amplifier of Fig. 2. (a) Manual layout from [22]; area: 3.8 × 3.0 mm; norm. area: 1.0; 15 devices, 15 nets, eight length constraints; zero airbridges. (b) Automatic layout; area: 6.75 × 2.33 mm; norm. area: 1.38; meets all eight length constraints; zero airbridges; runtime: 25.7 min.

TABLE I
EXECUTION TIMES TO FLOORPLAN OUR EXAMPLE CIRCUITS

Example Circuit (Figure ref.)	Number Devices /Nets	GA Population Size	Number of GA Generations	CPU Time (sec)	CPU Time per generation (sec)
Fig. 16b)	5/9	100	65	28.8	0.44
Fig. 15c) (with sizeless nets)	15/20	200	102	336	3.29
Fig. 15d)	15/20	200	172	637	3.67
Fig. 18b)	15/15	1000	286	1542[a]	5.39

a. This result was generated on a 233MHz IBM 604 CPU; others in the table used a 100MHz IBM 604 CPU.

extracted from [22]. The manual layout comprises 15 devices and 15 nets, eight of which have precise net length constraints. Since the manual layout is planar, it has no airbridges. We ran our tool on a simplified version with approximately sized rectangles for each device. We imposed the same net length constraints. As in the manual layout, we restricted three pads to the upper edge, two pads to the lower edge, the input to the left edge and the output to the right edge. Running the tool produced the layout at the right of Fig. 18. Our tool was able to evolve a planar layout. More importantly, all length constraints were met. Total runtime for this layout was roughly 26 min on an IBM 233-MHz PowerPC604 workstation.

The automatic layout is roughly 38% larger than the manual layout. This is primarily because wire meanders of the same net are spread across multiple floorplan "rooms" in the manual layout, many of which are dedicated solely to meandering. Currently, our meandering space model does not support this, resulting in inferior density. This suggests the need for a more sophisticated model of how meanders can distribute themselves across a layout. Nevertheless, this is the first time to our knowledge that any RF circuit with these sorts of tightly interacting placement and routing constraints has been automatically floorplanned.

Table I gives the runtimes for the tool with a termination criteria of 1% change tolerance at 50 generations.

In general, the runtime goes up rapidly as the number of devices and nets is increased. This is due to the larger population size required for larger problems, the larger number of generations necessary for convergence, and the longer evaluation times (routing time) for each circuit. However, the tool is capable of optimizing typical designs in 1–15 min. The second row of Table I gives the statistics for one layout optimization with sizeless wires and airbridges to show an example of the incremental cost of floorplan resizing operations. A rough analysis using the decrease in the time spent per generation on examples run with sizeless optimization shows that sizing operations take about 10%–35% of the total runtime, depending on the particular circuit.

One final issue to examine is the robustness of the basic genetic algorithm. We are solving a difficult, constrained problem using a stochastic optimization attack that is essentially an unconstrained minimization (i.e., our GA seeks to find a routed floorplan with a minimal cost score). A reasonable question is that of the robustness, by which mean both the likelihood that the GA can find a feasible solution, and that upon repeated runs of the algorithm feasible solutions dominate. As a test here, we ran 22 separate trials of the layout experiment from Fig. 18; the GA starts from a different random seed in each. Results appear in Fig. 19. The plot on the left shows the distributions of the best layout cost and mean layout cost for each of the 22 final populations at the end of genetic optimization. Despite the wide variation in the mean cost (which is influenced by the existence of some relatively poor layouts in the final populations) there is surprising uniformity of results for the best layout in each population. The scatter plot at the right then sorts these 22 best final layouts with respect to the number of

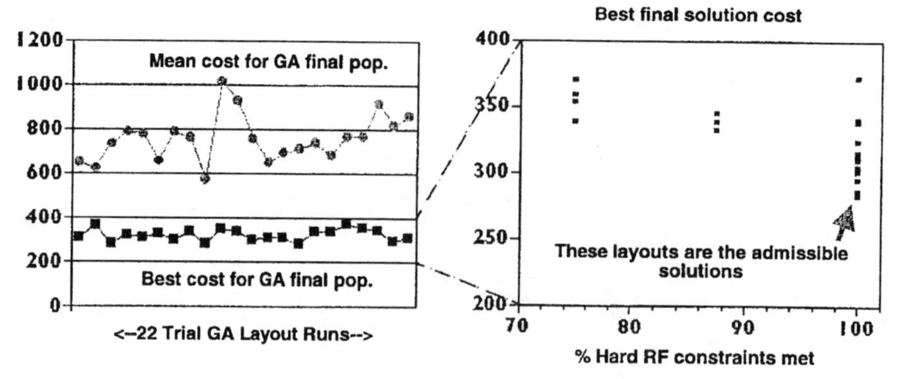

Fig. 19. Analysis of GA layout algorithm repeatability and robustness.

hard RF constraints met. Unsurprisingly, not all final solutions meet all constraints. However, roughly two-thirds of these solutions do meet all constraints. The practical conclusion here is that a few separate runs of our GA RF layout tool will suffice to find a good quality, feasible solution.

Overall, we regard this as a very satisfactory set of results for a first attempt at this difficult, tightly constrained, geometrically complex layout task.

VI. Conclusions

In this paper we suggested that the tight interaction between performance and layout for RF circuits could be addressed by device-level early floorplanning. We developed new algorithms for device-level floorplanning which integrate simultaneous detailed routing. The key idea is to use a complete—though rough—circuit layout to evaluate the low-level geometric interactions that must be carefully controlled in high-frequency designs. One of the more novel features of the approach is the integration of the placement and routing algorithms: the floorplanner plans space for large wire meanders, and the router negotiates fine-grain space for individual nets one segment at a time. This ensures that all layouts can be routed, and that both placement and wiring can be adjusted to optimize for constraints.

A preliminary implementation of these ideas works well on small designs. Our prototype can handle multiple constraints on precise net length, wire bends (and congestion; see [1]), and optimize for overall area, wirelength and—especially critical for RF cells—planarity. For these circuits, the floorplanning process is computationally reasonable.

Preliminary comparison to manual layout suggests the need for a more sophisticated model for embedding wire meanders to achieve density comparable to manual designs. The other obvious extension is to incorporate more direct evaluation of electrical interactions (e.g., local parasitics) on top of the geometric abstractions we introduced in this paper. This will allow us to take into account subtle electromagnetic interactions and make more accurate quantitative tradeoffs to optimize performance of the designed circuits. This should lead the way to a more complete layout optimization strategy for RF circuits, allowing us to move from device-level floorplans to more complete device-level layouts.

Acknowledgment

The authors are grateful to M. Mlinar of HP-EEsof for several insightful discussions of the potential role of genetic optimization in device-level RF layout.

References

[1] M. Aktuna, "A framework for simultaneous placement and routing of radio-frequency circuits," Masters thesis, Elect., Comput. Eng. Dept., Carnegie Mellon Univ., Pittsburgh, PA, , Dec. 1996.

[2] M. Aktuna, R. A. Rutenbar, and L. R. Carley, "Device-level early floorplanning algorithms for RF circuits," in *Proc. ACM Int. Symp. Physical Design,* 1998.

[3] B. Basaran, "Optimal diffusion sharing in digital and analog CMOS layout," Ph.D. thesis, Dept. of Elect., Comput. Eng., Carnegie Mellon Univ., Pittsburgh, PA, Mar. 1997.

[4] E. Charbon, E. Malavasi, and A. Sangiovanni-Vincentelli, "Generalized constraint generation for analog circuit design," in *Proc. IEEE/ACM ICCAD,* Nov. 1993, pp. 408–414.

[5] E. Charbon, G. Holmlund, B. Donecker, and A. Sangiovanni-Vincetelli, "A performance-driven router for RF and microwave analog circuit design," in *Proc. IEEE Custom Integrated Circuits Conf.* 1995, pp. 383–386.

[6] U. Choudhury and A. Sangiovanni-Vincentelli, "Constraint-based channel routing for analog and mixed analog/digital circuits," *IEEE Trans. Computer-Aided Design,* vol. 12, no. 4, pp. 497–510, Apr. 1993.

[7] J. M. Cohn, D. J. Garrod, R. A. Rutenbar, and L. R. Carley, "KOAN/ANAGRAM II: New tools for device-level analog placement and routing," *IEEE JSSC,* vol. 26, Mar. 1991.

[8] ——, "Techniques for simultaneous placement and routing of custom analog cells in KOAN/ANAGRAM II," in *Proc. ACM/IEEE ICCAD,* Nov. 1991, pp. 394–397.

[9] ——, *Analog Device-Level Layout Automation.* Norwell, MA; Kluwer Academic, 1994.

[10] J. Crols, S. Donnay, M. Steyaert, and G. Gielen, "A high-level design and optimization tool for analog RF receiver front-ends," presented at *ACM/IEEE ICCAD,* Nov. 1995.

[11] W. Dai and E. S. Kuh, "Simultaneous floor planning and global routing for hierarchical building-block layout," *IEEE Trans. Computer-Aided Design,* vol. CAD-6, no. 5, Sept. 1987.

[12] K. A. De Jong, "An analysis of the behavior of a class of genetic adaptive systems," Doctoral dissertation, Univ. Michigan, Ann Arbor, 1975.

[13] P. Feldmann, "Computation of circuit waveform envelopes using an efficient, matrix decomposed harmonic balance algorithm," presented at *ACM/IEEE ICCAD,* Nov. 1991.

[14] *Hewlett-Packard Microwave and RF Design System Manuals,* Santa Rosa Systems Division, Santa Rosa, CA, Dec. 1992.

[15] R. H. Jansen, "LINMIC: A CAD package for the layout-oriented design of single- and multi-layer MIC's/MMIC's up to mm-wave frequencies," *Microwave J.,* pp. 151–161, Feb. 1986.

[16] R. H. Jansen, R. G. Arnold, and I. G. Eddison, "A comprehensive CAD approach to design of MMIC's up to MM-wave frequencies," *IEEE J. Microwave Theory Tech.,* vol. 36, pp. 208–219, Feb. 1988.

[17] D. E. Goldberg, K. Deb, and B. Korb, "Do not worry, be messy," in *Proc. 4th Int. Conf. Genetic Algorithms,* 1991, pp. 24–30.

[18] P. Groeneveld, "Wire ordering for detailed routing," *IEEE DTC,* pp. 6–17, Dec. 1989.

[19] K. S. Kundert, J. K. White, and A. Sangiovanni-Vincentelli, *Steady-State Methods for Simulating Analog and Microwave Circuits.* Norwell, MA: Kluwer Academic, 1990.

[20] K. S. Kundert and D. Sharrit, "Simulation methods for RF circuits," in *Proc. ACM/IEEE ICCAD,* Nov. 1997.

[21] K. Lampaert, G. Gielen, and W. M. Sansen, "A performance-driven placement tool for analog integrated circuits," *IEEE JSSC,* vol. 30, pp. 773–780, July 1995.

[22] G. K. Lewis, I. J. Bahl, E. L. Griffin, and E. R. Schineller, "GaAs MMIC's for digital radio-frequency memory (DRFM) subsystems," *IEEE Trans. Microwave Theory Tech.,* vol. MTT-35, no. 12, p. 1478, Dec. 1987.

[23] E. Malavasi and A. Sangiovanni-Vincentelli, "Area routing for analog layout," *IEEE Trans. Computer-Aided Design,* vol. 12, no. 8, pp. 1186–1197, Aug. 1993.

[24] E. Malavasi, E. Felt, E. Charbon, and A. Sangiovanni-Vincentelli, "Automation of IC layout with analog constraints," *IEEE Trans. Computer-Aided Design,* 1996.

[25] A. Nagao, I. Shirakawa, and T. Kambe, "A layout approach to monolithic microwave IC," presented at *ACM Int. Symp. Physical Design,* 1998.

[26] K. Nakamura and L. R. Carley, "An enhanced fully differential folded-cascode operational amplifier," *IEEE JSSC,* vol. 27, Apr. 1992.

[27] R. H. J. M. Otten, "Automatic floor-plan design," in *Proc. 19th ACM/IEEE Design Automation Conf.,* 1982, pp. 261–267.

[28] ——, "Efficient floorplan optimization," in *Proc. IEEE Int. Conf. Computer Design,* 1983, pp. 499–502.

[29] F. Rubin, "The Lee path connection algorithm," *IEEE Trans. Computer-Aided Design,* vol. 3, no. 4, pp. 308–318, Oct. 1974.

[30] R. A. Rutenbar, L. R. Carley, P. C. Maulik, E. S. Ochotta, T. Mukherjee, B. Basaran, S. Mitra, S. K. Nag, and B. R. Stanisic, "Synthesis and layout for analog and mixed signal IC's in the ACACIA system," in *Advances in Analog Circuit Design.* Norwell, MA: Kluwer Academic, 1996.

[31] N. Sherwani, *Algorithms for Physical Design Automation.* Norwell, MA: Kluwer Academic, 1993, pp. 215–223.

[32] R. Telichevesky, K. S. Kundert, and J. K. White, "Efficient steady-state-analysis based on matrix-free Krylov-subspace methods," presented at *ACM/IEEE DAC,* June 1995.

[33] M. Wall, Massachusetts Inst. Technol, available: http://lancet.mit.edu/ga.

[34] D. F. Wong and C. L. Liu, "A new algorithm for floorplan design," in *Proc. 23rd ACM/IEEE Design Automation Conf.,* 1986, pp. 101–107.

[35] J. F. Zurcher, "MICROS—A CAD/CAM program for fast realization of microstrip masks," *IEEE Trans. Microwave Theory Tech.,* vol. MTT-S, pp. 481–484, 1985.

Mehmet Aktuna received the S.B. degree in electrical computer and systems engineering from Harvard University, Cambridge, MA, in 1994 and the M.S. degree in electrical and computer engineering from Carnegie Mellon University, Pittsburgh, PA, in 1996, where he is currently working towards the Ph.D. degree.

During the summer of 1996, he worked on microprocessor routing at Intel Corporation, Hillsboro, OR. He worked with the Physical Design Group of HP-EEsof, Santa Rosa, CA, during the summer of 1997. His current research interests include routing and placement for analog and digital circuits, and high-frequency interconnect modeling.

Rob A. Rutenbar (S'77–M'84–SM'90–F'98) received the Ph.D. degree from the University of Michigan in 1984 and subsequently joined the faculty of Carnegie Mellon University, Pittsburgh, PA.

He is currently Professor of Electrical and Computer Engineering, and (by courtesy) of Computer Science. His research interests focus on circuit and layout synthesis algorithms for mixed-signal ASIC's, for high-speed digital systems and FPGA's.

In 1987, Dr. Rutenbar received a Presidential Young Investigator Award from the National Science Foundation (NSF). He has been on the program committees for the IEEE International Conference on CAD, the ACM/IEEE Design Automation Conference, the European Design and Test Conference, and the ACM International Symposium on FPGA's. From 1992–1995, he was on the Editorial Board of IEEE SPECTRUM. He was General Chair of the 1996 ICCAD and is currently on the Program Committee of the ACM International Symposium on Physical Design. He chaired the Analog Technical Advisory Board for Cadence Design Systems from 1992–1996. He was Director of Carnegie Mellon's Center for Electronic Design Automation from 1993–1998. He is a member of the ACM.

L. Richard Carley (S'77–M'81–SM'90–F'97) received the S.B. degree in 1976, the M.S. degree in 1978, and the Ph.D. degree in 1984 from the Massachusetts Institute of Technology (MIT), Cambridge.

He has worked for MIT's Lincoln Laboratories and has acted as a Consultant in the areas of analog and mixed analog/digital circuit design, analog circuit design automation, and signal processing for data storage. In 1984, he joined Carnegie Mellon University, and in 1992 he was promoted to the rank of Full Professor of Electrical and Computer Engineering. His research interests include the development of computer-aided design (CAD) tools to support analog circuit design, the design of high-performance analog/digital signal processing IC's, and the design of integrated microelectromechanical systems (MEMS). Since joining Carnegie Mellon he has been granted eight patents and authored or coauthored more than 120 technical papers.

Dr. Carley is a member of program committees of the Custom IC Conference (CICC), the Magnetic Recording Conference (TMRC), and the International Symposium on Low-Power Electronics Design (ISLPED). He has served as an Associate Editor for the IEEE TRANSACTIONS ON CIRCUITS AND SYSTEMS and on the Editorial Board of the *Analog Signal Processing Journal.* He received a National Science Foundation (NSF) Presidential Young Investigator Award in 1985, a Best Technical Paper Award at the 1987 Design Automation Conference, a Distinguished Paper Mention at the 1991 International Conference on Computer-Aided Design, and a Best Panel Award at the 1993 International Solid-State Circuits Conference. He was awarded the Guillemin Prize for best undergraduate thesis in the electrical engineering department at MIT.

PART V

Analog Modeling and Analysis

Macromodeling of Integrated Circuit Operational Amplifiers

GRAEME R. BOYLE, BARRY M. COHN, DONALD O. PEDERSON, FELLOW, IEEE, AND JAMES E. SOLOMON, MEMBER, IEEE

Abstract—**A macromodel has been developed for integrated circuit (IC) op amps which provides an excellent pin-for-pin representation. The model elements are those which are common to most circuit simulators. The macromodel is a factor of more than six times less complex than the original circuit, and provides simulated circuit responses that have run times which are an order of magnitude faster and less costly in comparison to modeling the op amp at the electronic device level.**

Expressions for the values of the elements of the macromodel are developed starting from values of typical response characteristics of the op amp. Examples are given for three representative op amps. In addition, the performance of the macromodel in linear and nonlinear systems is presented. For comparison, the simulated circuit performance when modeling at the device level is also demonstrated.

I. Introduction

INTEGRATED circuit (IC) simulators have proven to be a useful tool to the IC design engineer. Nonetheless, their widespread acceptance in the design of large-scale integrated circuits and IC subsystems has been impeded by excessive simulation costs and increasing convergence problems. Present simulators model semiconductor devices at the p-n junction and 2-terminal element level. Because of the large number of these devices in large-scale IC systems, the analysis can surpass the computer's memory capability, simulator circuit-size capability, or the inherent numerical accuracy of the computer. Even if an adequate simulator and computer are available, the required simulation time makes the analysis financially impractical. This paper describes one solution to this problem: macromodels which have been developed for IC's such as operational amplifiers and comparators.

The idea and use of macromodels in electronic circuit design is very common at the system level. For example, in developing an analog signal processor, one might utilize a number of ideal voltage amplifiers, integrators, and other subsystem blocks. In effect, a variety of zero-order circuit models are used. To determine the actual system performance, a prototype circuit is constructed and tested at the device level. The size and complexity of today's inexpensive IC's are large; therefore, the cost of using present simulators for design and evaluation can be very large. The cost for large IC's can only be justified if very large manufacture is anticipated. The costs and other problems can be relieved by the development of macromodels for IC's which provide an adequate pin-for-pin representation of the IC. For digital IC's, logic simulation and macromodels have been developed for digital logic blocks [1], [8]. For analog IC's, this paper describes a very effective macromodel that has been developed for IC op amps [2], [3], [9].

The aim of macromodeling is to obtain a circuit model of an IC or a portion of an IC which has a significantly reduced complexity to provide for smaller, less costly simulation time, or to permit the simulation of larger IC's or IC systems for the same time and cost. In the macromodel for IC op amps shown in Fig. 1, a reduction of approximately 6 in branch and node count has been achieved while providing a very close approximation to the actual performing op amp, i.e., accurate modeling of the input and output characteristics, differential- and common-mode gain versus frequency characteristics, quiescent dc characteristics, offset characteristics, and large-signal characteristics, such as slew rate, output voltage swing, and short-circuit current limiting. Further, since much of a simulation run is involved with iterative analysis to an equilibrium circuit solution, the reduction of 60 to 80 p-n junctions in an actual op amp to the 8 junctions in the macromodel of Fig. 1 indicates better how much faster and cheaper can be the simulation using the macromodels instead of device-level models. The results with amplifiers, timers, and filters that are cited in this paper show that a reduction in time of 6 to 10 is typical.

In many design or evaluation situations, it is not necessary to model an op amp in all of its performance characteristics. For example, maximum short-circuit current limiting may not be of interest. If the elements in the macromodel which provide this feature are eliminated, further simplification of the macromodel is obtained. As an example, the simulation time of the filter in Section IV is reduced by a factor of 1.4 if the current and voltage limiters are omitted.

Manuscript received August 2, 1974; revised August 16, 1974. This research was sponsored in part by the Joint Services Electronics Program under Contract F44620-71-C-0087 and by the National Science Foundation under Grant GK-17931. This paper was presented at the International Solid-State Circuits Conference, Philadelphia, Pa., February 1974.

G. R. Boyle and D. O. Pederson are with the Department of Electrical Engineering and Computer Sciences and the Electronics Research Laboratory, University of California, Berkeley, Calif.

B. M. Cohn is with Intel Corporation, Santa Clara, Calif.

J. E. Solomon is with National Semiconductor Corporation, Santa Clara, Calif.

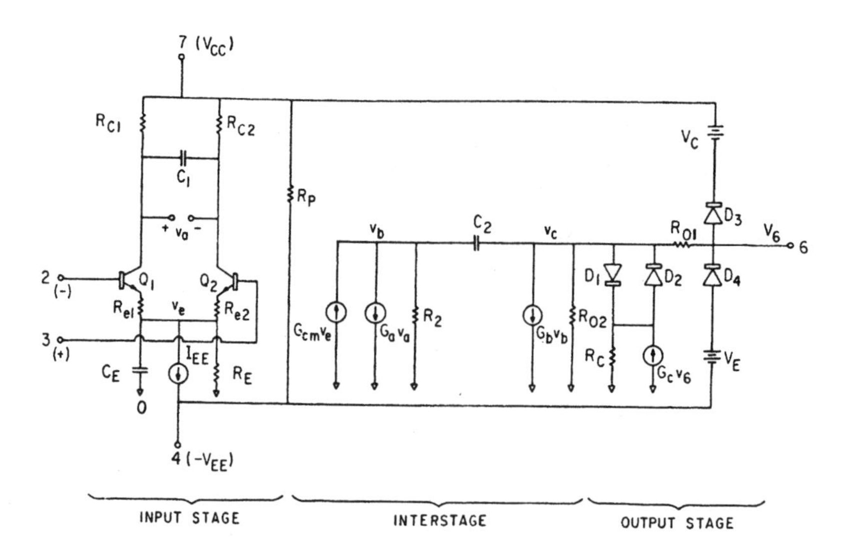

Fig. 1. Circuit diagram of the op amp macromodel.

II. Macromodel Development

The circuit model for an IC op amp which is developed in this paper is shown in Fig. 1. The configuration, with a suitable choice of parameters and elements, accurately models a broad class of IC op amps. For a given op amp, the model provides an essentially pin-for-pin correspondence with the op amp, and accurately represents the circuit behavior for nonlinear dc, ac, and large-signal transient responses.

The circuit of Fig. 1 is subdivided into three stages. The input stage consists of ideal transistors Q_1 and Q_2 and the associated sources and passive elements. This stage produces the necessary linear and nonlinear differential-mode (DM) and common-mode (CM) input characteristics. For convenience, the stage is designed for unity voltage gain. The stage can be designed to provide desired voltage and current offsets. As brought out in the next section, the capacitor C_E is used to introduce a second-order effect for the slew rate [4], and the capacitor C_1 introduces a second-order effect to the phase response.

The DM and CM voltage gains of the op amp are provided by the linear interstage and output stage elements consisting of G_{cm}, G_a, R_2, G_b, and R_{02}. The function of each element is presented in the next section. The dominant time constant of the op amp is produced with the internal feedback capacitor C_2. A feedback connection in the macromodel is used for C_2 in order to provide the necessary ac output resistance change with frequency. In addition, the two nodes of C_2 can be made available to the outside world in order that the circuit designer can introduce the same compensation modification as might be added to the actual op amp. Notice the complete isolation that exists between the input and the interior stages. This leads to a simplification of the frequency and the slew rate performances.

The output stage provides the proper dc and ac output resistance of the op amp. The elements D_1, D_2, R_C, and G_C produce the desired maximum short-circuit current. The elements D_3, V_C and D_4, V_E are voltage-clamp circuits to produce the desired maximum voltage excursion.

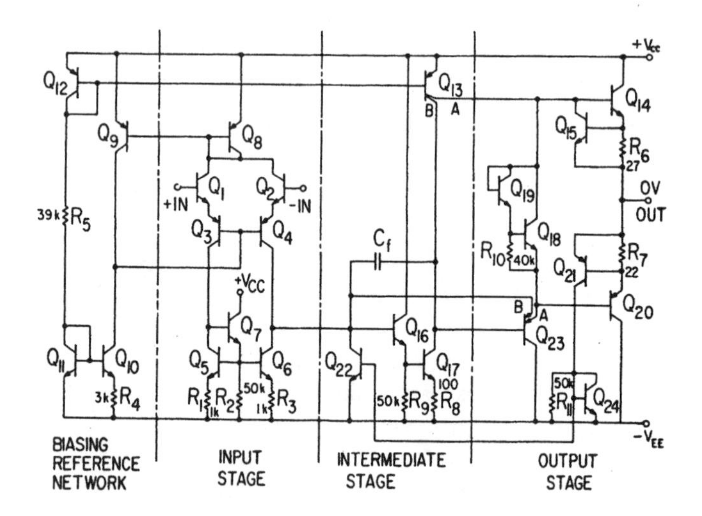

Fig. 2. Circuit diagram of the ICL8741 op amp.

The circuit model of Fig. 1 has been developed using two basic macromodeling techniques: simplification and build-up. In the simplification technique, representative portions of op amp circuitry are successively simplified by using simple ideal elements to replace numerous real elements. Thus, the final model using this approach bears a strong resemblance to the real circuit. In Fig. 1, the input stage design is an example of the simplification technique. In the build-up technique, a circuit configuration composed of ideal elements is proposed to meet certain external circuit specifications without necessarily resembling a portion of an actual op amp circuit configuration. The build-up technique is employed in the development of the output stage.

To illustrate these aspects further, consider the schematic diagram shown in Fig. 2 of the 741-type op amp

TABLE I
DESIGN EQUATIONS FOR THE OP AMP MACROMODEL

$$V_T = \frac{kT}{q} = 25.85 \text{ mV for } 300 \text{ K}$$

$$I_{S1} = I_{SD3} = I_{SD4} = 8 \cdot 10^{-16} \text{ A}$$

$$R_2 = 100 \text{ k}\Omega$$

$$I_{C1} = I_{C2} = \frac{C_2}{2} S_R^+$$

$$C_E = \frac{2I_{C1}}{S_R^-} - C_2$$

$$I_{B1} = I_B + \frac{I_{Bos}}{2}$$

$$I_{B2} = I_B - \frac{I_{Bos}}{2}$$

$$\beta_1 = I_{C1}/I_{B1}$$

$$\beta_2 = I_{C2}/I_{B2}$$

$$I_{EE} = \left(\frac{\beta_1 + 1}{\beta_1} + \frac{\beta_2 + 1}{\beta_2}\right) I_{C1}$$

$$R_E = 200/I_{EE}$$

$$I_{S2} = I_{S1}\left(1 + \frac{V_{os}}{V_T}\right)$$

$$\frac{1}{g_{m1}} = V_T/I_{C1}$$

$$R_{c1} = 1/2\pi f_{0 \text{ dB}} C_2$$

$$R_{e1} = \left(\frac{\beta_1 + \beta_2}{\beta_1 + \beta_2 + 2}\right)\left(R_{C1} - \frac{1}{g_{m1}}\right)$$

$$C_1 = \frac{C_2}{2} \tan \Delta\phi$$

$$R_p = (V_{CC} + V_{EE})^2/(P_d - V_{CC}(2I_{C1}) - V_{EE}I_{EE})$$

$$G_a = 1/R_{c1}$$

$$G_{cm} = \frac{1}{R_{c1}\,(\text{CMRR})}$$

$$R_{01} = R_{0-ac}$$

$$R_{02} = R_{out} - R_{01}$$

$$G_b = \frac{a_{VD}R_{c1}}{R_2 R_{02}}$$

$$I_X = (2I_{C1})G_b R_2 - I_{SC}$$

$$I_{SD1} = I_{SD2} = I_X \exp - \frac{R_{01}I_{SC}}{V_T}$$

$$R_C = \frac{V_T}{100 I_X} \ln \frac{I_X}{I_{SD1}}$$

$$G_C = 1/R_C$$

$$V_C = V_{CC} - V_{out}^+ + V_T \ln \frac{I_{SC}^+}{I_{SD3}}$$

$$V_E = V_{CC} + V_{out}^- + V_T \ln \frac{I_{SC}^-}{I_{SD4}}$$

(Intersil ICL8741). This type is the most common, general purpose IC op amp. In developing a macromodel using the simplification technique, the circuitry employed for biasing can be replaced with ideal passive elements (pure current and voltage sources).

Similarly, the active load and balance-to-unbalance converter in the input stage can be replaced with ideal elements. Finally, it is not necessary to use composite transistors in the input stage. Thus, as shown in Fig. 1, a simple differential stage can be proposed to model accurately the nonlinear input characteristic of the op amp.

The op amp macromodel is developed keeping in mind existing IC simulators. Therefore, the model contains only elements which are common to most IC simulators (i.e., resistors, capacitors, inductors, dependent current sources, independent sources, diodes, and bipolar transistors). In addition, effort is made to minimize the number of p-n junctions. These nonlinear elements make necessary iterative analysis to obtain the equilibrium state of the circuit. A reduction of the number of nonlinear elements leads to smaller simulation time.

For the input stage, our investigations showed that at least four ideal junctions were necessary to provide the needed balanced, nonlinear behavior in the macromodel. It was determined that the simplest arrangement is that of Fig. 1 where the four ideal junctions were obtained with two ideal transistors, each modeled with the lowest order Ebers–Moll (E–M) transistor model which includes two ideal p-n junctions and two dependent current sources.

For the output stage, a simplified model of an actual op amp does not provide the best solution. A stripped-down class-AB stage with ideal transistors leads to a branch count of over 13 in comparison with 11 branches in the output stage of Fig. 1. In addition, the class-AB stage must be augmented with voltage limiters in the drive circuitry to limit the voltage excursion at the transistor bases to the supply potentials. It was found that the idealized built-up procedure provides an output stage which is considerably simpler.

III. Parameters and Element Values of the Macromodel

In this section, expressions are developed to relate the performance of the op amp and the macromodel to the parameters and elements of the macromodel. A summary of all design equations is presented in Table I. The determination of the element values of the macromodel proceeds from the input, transfer, and output characteristics of the op amp.

The Input Stage: I_{C1} and C_E

The value of the necessary collector current of the first stage is established by the slew rate of the op amp. If the op amp is connected as a voltage follower, the positive going slew rate S_R^+ is

$$S_R^+ = \frac{2I_{C1}}{C_2} \tag{1}$$

where an n-p-n stage has been assumed [4]. From a rearrangement of this expression

$$I_{C1} = \tfrac{1}{2}C_2S_R^+. \tag{2}$$

For a quiescent situation, equal collector currents are used in the input stage $I_{C2} = I_{C1}$.

The negative going slew rate S_R^- is smaller because of the charge-storage effects in the input stage which is modeled by C_E [4].

$$S_R^- = \frac{2I_{C1}}{C_2 + C_E} \tag{3}$$

or

$$C_E = \frac{2I_{C1}}{S_R^-} - C_2. \tag{4}$$

If $S_R^+ < S_R^-$, the macromodel should be modified to use p-n-p transistors in the input stage. In the equations above S_R^+ and S_R^- should then be interchanged.

In addition to the transient slew rate effects, the element C_E also introduces a desirable modification to the ac response of the CM gain of the macromodel.

The Transistor Parameters

The values of β_1 and β_2 for the two ideal transistors are obtained from the specifications for the average input bias current I_B and the desired level of input current offset I_{Bos}.

$$I_{B1} = I_B + \frac{I_{Bos}}{2}, \qquad I_{B2} = I_B - \frac{I_{Bos}}{2}. \tag{5}$$

$$\beta_1 = \frac{I_{C1}}{I_{B1}}, \qquad \beta_2 = \frac{I_{C2}}{I_{B2}}. \tag{6}$$

The voltage offset V_{os} for the macromodel is produced by specifying different saturation currents I_S for the two transistors. Assume a given value for I_{S1} of Q_1

$$I_{C1} = I_{S1} \exp \frac{V_{BE1}}{V_T} \tag{7}$$

where $V_T = kT/q = 0.02585$ V at $T = 300$ K. A similar expression holds for $I_{C2} = I_{C1}$

$$I_{C2} = I_{S2} \exp \frac{V_{BE2}}{V_T}. \tag{8}$$

The offset voltage is

$$\begin{aligned} V_{os} &= V_{BE1} - V_{BE2} \\ &= V_T \ln \frac{I_{S1}}{I_{S2}}. \end{aligned} \tag{9}$$

This leads to

$$I_{S2} = I_{S1} \exp \frac{V_{os}}{V_T} \cong I_{S1}\left[1 + \frac{V_{os}}{V_T}\right]. \tag{10}$$

The Input Stage: R_{c1} and R_{c1}

Values for the resistors $R_{c1} = R_{c2}$ are derived from the required value of the 0 dB frequency $f_{0\,\mathrm{dB}}$ of the fully compensated op amp. The 0 dB frequency is approximately the product of the DM voltage going a_{VD} and the -3 dB corner frequency $f_{3\,\mathrm{dB}}$ of the gain function

$$f_{0\,\mathrm{dB}} \simeq a_{VD} f_{3\,\mathrm{dB}}. \tag{11}$$

The corner frequency can be estimated using a Miller-effect approximation in the interior stage.

$$\begin{aligned} f_{3\,\mathrm{dB}} &\simeq \frac{1}{2\pi R_2 C_2(1 + G_b R_{02})} \\ &\simeq \frac{1}{2\pi R_2 C_2 G_b R_{02}}. \end{aligned} \tag{12}$$

The DM voltage gain at very low frequencies is

$$a_{VD} = (G_a R_2)(G_b R_{02}). \tag{13}$$

G_a is chosen to be equal to $1/R_{c1}$ in order to obtain a convenient slew rate expression as in (1). The last three expressions lead to

$$f_{0\,\mathrm{dB}} = \frac{1}{2\pi R_{c1} C_2} \tag{14}$$

or

$$R_{c1} = \frac{1}{2\pi f_{0\,\mathrm{dB}} C_2}. \tag{15}$$

Alternately, a relationship between $f_{0\,\mathrm{dB}}$ and S_R^+ can be written using (1).

$$f_{0\,\mathrm{dB}} = \frac{S_R^+}{2\pi R_{c1}(2I_{C1})}. \tag{16}$$

The value of R_{c1} is usually small, of the order of $2/g_m$. R_{c1} and R_{c2} should be small in order that saturation of the input stage (and concomittent latchup of the op amp model) is avoided with maximum input. The resistances R_{e1} and R_{e2} in the input stage are introduced to provide a degree of freedom with respect to slew rate and 0 dB frequency, and to simulate better certain op amps which use emitter resistors for slew rate enhancement, e.g., the LM118. R_{e1} is found from the DM voltage gain of the first stage, which for convenience is taken to be unity.

$$\frac{v_a}{v_{\text{in}}} = \frac{\beta_1 R_{c1} + \beta_2 R_{c2}}{\dfrac{\beta_1}{g_{m1}} + (\beta_1 + 1)R_{e1} + \dfrac{\beta_2}{g_{m2}} + (\beta_2 + 1)R_{e2}} = 1. \tag{17}$$

If $I_{C1} = I_{C2}$, then $g_{m1} = g_{m2}$. If also $R_{c1} = R_{c2}$ and $R_{e1} = R_{e2}$,

$$R_{e1} = \frac{\beta_1 + \beta_2}{\beta_1 + \beta_2 + 2}\left[R_{c1} - \frac{1}{g_{m1}}\right]. \tag{18}$$

The Input Stage: I_{EE} and R_E

The value of the dc current source in the input stage for equal collector currents is

$$I_{EE} = \left(\frac{\beta_1 + 1}{\beta_1} + \frac{\beta_2 + 1}{\beta_2}\right) I_{C1}. \tag{19}$$

The resistor R_E is added to provide a finite CM input resistance. Because the current source I_{EE} is often realized with an n-p-n transistor, the resistance R_E is taken as its output resistance

$$R_E \simeq \frac{V_A}{I_C} = \frac{V_A}{I_{EE}} \tag{20}$$

where V_A is the early voltage of the device. V_A for a small n-p-n transistor is typically 200 V.

The Input Stage: C_1

To introduce excess phase effects in the DM amplifier response, another capacitor C_1 is added in the input stage. The second pole of the DM gain function is located at

$$p_2 = -1/2R_{e1}C_1. \tag{21}$$

Notice that there is no interaction amongst the three capacitors because of the use of unilateral devices and stages.

The excess phase at $f = f_{0\,\text{dB}}$ due to the nondominant pole p_2 is

$$\Delta\phi = \tan^{-1}\frac{2\pi f_{0\,\text{dB}}}{|p_2|} = \tan^{-1}(2\pi f_{0\,\text{dB}})(2R_{e1}C_1) = \tan^{-1}\frac{2C_1}{C_2}. \tag{22}$$

The phase margin of the DM open-loop response is then

$$\phi_m = 90° - \Delta\phi. \tag{23}$$

The necessary value of C_1 to produce the excess phase is

$$C_1 = \frac{C_2}{2}\tan\Delta\phi. \tag{24}$$

DC Power Drain

To model the actual dc power dissipation of an op amp, a resistor R_p is introduced into the macromodel. For the circuit of Fig. 1, in a quiescent state, the power dissipation is

$$P_d = V_{CC}2I_{C1} + V_{EE}I_{EE} + \frac{(V_{CC} + V_{EE})^2}{R_p}. \tag{25}$$

The necessary value of R_p to produce this dissipation is

$$R_p = \frac{(V_{CC} + V_{EE})}{P_d - V_{CC}2I_{C1} - V_{EE}I_{EE}}. \tag{26}$$

In a typical op amp, most of the current drain from the voltage supplies is due to diode-resistor current defining paths. Therefore, as the supply voltages are changed, the current drain varies almost linearly and R_p will continue to model accurately the power dissipation.

The Interstage: G_a, R_2, and G_{cm}

As indicated earlier, the coefficient G_a of the voltage dependent current source $G_a v_a$ is chosen equal to $1/R_{c1}$ for convenience. Similarly, the value of R_2 or G_b can be arbitrarily chosen. Only the product is determined by the DM gain. For "active region" considerations, the choice of R_2 is not important. However, it must be kept in mind that the voltage response at node b is linear with R_2. If too large a value of v_b is developed during a transient excursion through the active region of the op amp, a considerable discharge or recovery time can be encountered after the active region excursion. To prevent these discharge delays in relation to actual op amp behavior, a small value of R_2 should be used. Empirically, a value of 100 kΩ is found to be appropriate.

If a second voltage-controlled current source is introduced across R_2, the CM voltage gain response can be introduced. The CM voltage gain in the input stage from v_{in} to v_e is approximately unity since R_E is large. The CM voltage gain from the input to v_b is then approximately

$$\frac{v_{b\text{CM}}}{v_{\text{inCM}}} \cong G_{cm}R_2. \tag{27}$$

The differential voltage gain from the input to v_b is

$$\frac{v_{b\text{DM}}}{v_{\text{inDM}}} = G_a R_2 = \frac{1}{R_{c1}} R_2. \tag{28}$$

The CM rejection ratio (CMRR) is the ratio of the two gains [5]

$$\text{CMRR} = \frac{a_{VD}}{a_{VC}} = \frac{1}{R_{c1}G_{cm}}. \tag{29}$$

Therefore,

$$G_{cm} = \frac{1}{(\text{CMRR})R_{c1}}. \tag{30}$$

The dominant behavior of the CM frequency response will be approximately the same as the DM frequency response except that the presence of the capacitor C_E in the input stage introduces a transmission zero in the CM gain function at $-1/R_E C_E$.

The Output Stage: R_{01}, R_{02}, and G_b

The output stage provides the desired dc and ac output resistances and the output current and voltage limitations. From Fig. 1, it is seen that the output resistance at very low frequencies for the quiescent state is

$$R_{\text{out}} = R_{01} + R_{02}. \tag{31}$$

At high frequencies, R_{02} is shorted out by the (current) Miller-effect output capacitance across it due to C_2. The effective shunt capacitance is $C_{sh} \simeq C_2(1 + R_2 G_b)$. For the situation where a large load resistance is presented to the macromodel, the corner frequency of the output impedance is

$$f_c = \frac{1}{2\pi R_{02}C_2(1 + R_2G_b)}. \tag{32}$$

For frequencies well above this value, the output resistance is R_{01}. Therefore,

$$R_{01} = R_{0-ac}. \tag{33}$$

With this value established, R_{02} from (31) and G_b from (13) are

$$R_{02} = R_{out} - R_{01} \tag{34}$$

and

$$G_b = \frac{a_{VD}R_{c1}}{R_2R_{02}}. \tag{35}$$

The Output Stage: Current Limiting

In the output stage of Fig. 1, the desired output-current limiting is provided by the elements G_CV_6, R_C, D_1, D_2, and R_{01}. The R_C, G_CV_6 combination is an equivalent to a voltage-controlled voltage source (which is not available in simulators such as program SPICE). Thus, $V_{out} = V_6$ also appears across R_C. If both of the voltage-clamp diodes D_3 and D_4 are off, the maximum current to the output is the ratio of the potential across D_1, D_2, and R_{01}

$$I_{SC} \simeq \frac{V_D}{R_{01}} \tag{36}$$

where

$$V_D = V_T \ln \frac{I_X}{I_{SD1}} \tag{37}$$

I_X Maximum current through D_1 or D_2.

I_{SD1} Saturation current of diodes D_1, D_2.

Since R_{01} is known, I_{SD1} can be established once I_X is determined. In Fig. 3, a reduced portion of the output stage is shown which applies for a positive output excursion, and where a very small load resistance is assumed. A Thévenin equivalent for G_bv_b and R_{02} is used. The Thévenin open-circuit available voltage is $a_{VD}v_{in}$. An ideal voltage-controlled voltage source v_0 is also used in place of G_CV_6 and R_C. Assume first that the voltage $a_{VD}v_{in}$ is not large. The output current flowing through the resistor R_{01} then produces only a small voltage drop. The polarity of the voltage v_c is such as to forward bias diode D_1. If the voltage drop across R_{01} is small, the current through D_1 is very small and can be neglected.

As $a_{VD}v_{in}$ increases, so does I_L and the voltage drop across R_{01}. As the latter approaches the "ON" voltage of D_1, i.e., the voltage for appreciable current through the diode, the increasing current from the source I_{D1} flows through the diode. I_L is then approximately limited because of the exponential increase of I_{D1} with respect

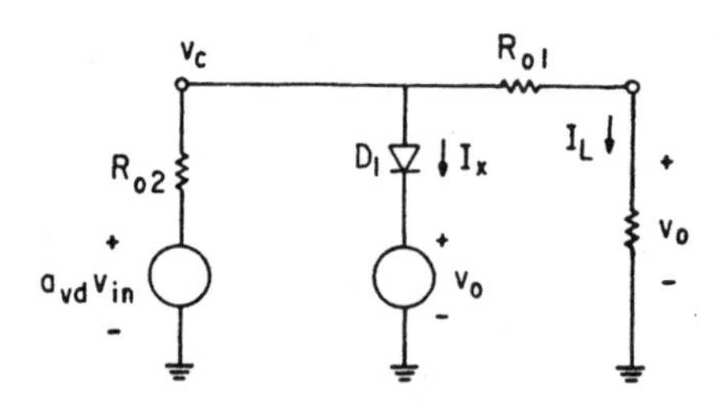

Fig. 3. Simplified circuit diagram of the output stage.

to I_LR_{01}. The approximate limiting condition is found from

$$I_X = I_{SD1} \exp \frac{I_{SC}R_{01}}{V_T}. \tag{38}$$

The limiting value of I_X is determined by an overdrive condition at the input. The short-circuit available current from G_bv_b is I_{max}

$$I_{max} = I_X + I_{SC} = 2I_{C1}R_2G_b. \tag{39}$$

A typical value of I_{max} is 100 A.

From the equation above,

$$I_{SD1} = I_{SD2} = I_X \exp\left(-\frac{R_{01}I_{SC}}{V_T}\right). \tag{40}$$

For large required values of R_{01}, the value of I_{SD1} can be extremely small which may lead to numerical difficulty. In many applications when the output resistance is not critical, a smaller value of R_{01} can be used neglecting the exact realization of R_{0-ac}, e.g., if $I_{sD1} = I_{s1}$,

$$R_{01} \simeq \frac{V_T}{I_{SC}} \ln \frac{I_X}{I_{S1}}. \tag{41}$$

R_{02} is then increased to $R_{out} - R_{01}$, and G_b is decreased to maintain the same value of the $G_b\,R_{02}$ product.

In order to approximate well a voltage-controlled voltage source, R_C must be very small. If the voltage drop across R_C is to be only 1 percent of V_{D1} or V_{D2},

$$R_C = \frac{V_T}{100I_X} \ln \frac{I_X}{I_{SD1}}. \tag{42}$$

The necessary value for the voltage-controlled current source G_CV_6 is

$$G_C = \frac{1}{R_C}. \tag{43}$$

The Output Stage: Voltage Limiting

The output voltage excursion is limited by the voltage source-diode clamp combinations V_C, D_3 and V_E, D_4, shown in Fig. 1. With a large, positive output voltage such as to forward bias D_3

$$\begin{aligned} V_{out}^+ &= V_{CC} - V_C + V_{D3} \\ &= V_{CC} - V_C + V_T \ln \frac{I_{SC}^+}{I_{SD3}}. \end{aligned} \tag{44}$$

As indicated above, the diode current is limited to the

TABLE II

CIRCUIT DATA AND GUMMEL–POON TRANSISTOR PARAMETERS FOR THE ICL8741 OP AMP

Circuit Data			Gummel-Poon Transistor Parameters						
-Element	Nodes	Value	.MODEL	BNP1	NGP	BFM=209.	BRM=2.5	RB= 670.
R1	02 17	1.0K	+		RC=300.	CCS=1.417P	TF=1.15N	TR= 405.N
R2	02 16	50.K	+		CJE=0.65P	CJC=0.36P	IS= 1.26E-15	VA=178.6
R3	02 18	1.0K	+		C2= 1653.	IK=1.611M	NE=2.0	PE=0.60
R4	02 08	3.0K	+		ME=3.	PC=0.45	MC=3.	
R5	04 05	39.K	.MODEL	BNP2	NGP	BFM=400.	BRM=6.1	RB= 185.
R6	12 26	27.	+		RC= 15.	CCS=3.455P	TF=0.76N	TR= 243.N
R7	12 25	22.	+		CJE=2.80P	CJC=1.55P	IS= 0.395E-15	VA=267.0
R8	02 23	100.	+		C2= 1543.	IK=10.00M	NE=2.0	PE=0.60
R9	02 21	50.K	+		ME=3.	PC=0.45	MC=3.	
R10	24 27	40.K	.MODEL	BPN1	PGP	BFM= 75.	BRM=3.8	RB= 500.
R11	02 22	50.K	+		RC=150.	CCS=2.259P	TF=27.4N	TR=2540.N
C	15 19	30.P	+		CJE=0.10P	CJC=1.05P	IS= 3.15E-15	VA=55.11
Q1	10 07 13	BNP1	+		C2= 1764.	IK=270.0U	NE=2.0	PE=0.45
Q2	10 06 11	BNP1	+		ME=3.	PC=0.45	MC=3.	
Q3	14 09 13	BPN1	.MODEL	BPN2	PGP	BFM=117.	BRM=4.8	RB= 80.
Q4	15 09 11	BPN1	+		RC=156.		TF=26.5N	TR=2430.N
Q5	14 16 17	BNP1	+		CJE=4.05P	CJC=2.80P	IS= 17.6E-15	VA=57.94
Q6	15 16 18	BNP1						
Q7	01 14 16	BNP1						
Q8	10 10 01	BPN1	+		C2= 478.4	IK=590.7U	NE=2.0	PE=0.60
Q9	09 10 01	BPN1	+		ME=4.	PC=0.60	MC=4.	
Q10	09 05 08	BNP1	.MODEL	BPN3	PGP	BFM=13.8	BRM=1.4	RB=100.
Q11	05 05 02	BNP1	+		RC= 80.	CCS=2.126P	TF=27.4N	TR= 55.N
Q12	04 04 01	BPN1	+		CJE=0.10P	CJC=0.30P	IS= 2.25E-15	VA=83.55
Q13A	20 04 01	BPN3	+		C2=84.37K	IK=5.000M	NE=2.0	PE=0.45
Q13B	19 04 01	BPN4	+		ME=3.	PC=0.45	MC=3.	
Q14	01 20 26	BNP2	.MODEL	BPN4	PGP	BFM=14.8	BRM=1.5	RB=160.
Q15	20 26 12	BNP1	+		RC=120.	CCS=2.126P	TF=27.4N	TR= 220.N
Q16	01 15 21	BNP1	+		CJE=0.10P	CJC=0.90P	IS= 2.25E-15	VA=83.55
Q17	19 21 23	BNP1	+		C2=84.37K	IK=171.8U	NE=2.0	PE=0.45
Q18	20 27 24	BNP1	+		ME=3.	PC=0.45	MC=3.	
Q19	20 20 27	BNP1	.MODEL	BPN5	PGP	BFM= 80.	BRM=1.5	RB=1100.
Q20	02 24 25	BPN2	+		RC=170.		TF=26.5N	TR=9550.N
Q21	22 25 12	BPN1	+		CJE=1.10P	CJC=2.40P	IS= 0.79E-15	VA=79.45
Q22	15 22 02	BNP1	+		C2= 1219.	IK=80.55U	NE=2.0	PE=0.60
Q23A	02 19 24	BPN5	+		ME=4.	PC=0.60	MC=4.	
Q23B	02 19 15	BPN6	.MODEL	BPN6	PGP	BFM= 19.	BRM=1.0	RB= 650.
Q24	22 22 02	BNP1	+		RC=100.		TF=26.5N	TR=2120.N
			+		CJE=1.90P	CJC=2.40P	IS=0.0063E-15	VA=167.1
			+		C2=57.49K	IK=80.55U	NE=2.0	PE=0.60
			+		ME=4.	PC=0.60	MC=4.	

short current I_{SC}^{+}. The necessary bias voltage is

$$V_C = V_{CC} - V_{\text{out}}^{+} + V_T \ln \frac{I_{SC}^{+}}{I_{SD3}}. \qquad (45)$$

Similarly,

$$V_E = V_{EE} + V_{\text{out}}^{-} + V_T \ln \frac{I_{SC}^{-}}{I_{SD4}}. \qquad (46)$$

The Complete Model

A summary of the design equations for the parameters of the macromodel is given in Table I. An example of the use of these equations is given in the Appendix. The starting point is op amp performance data.

The particular IC used to illustrate the design procedure was the object of an earlier study [6]. The configuration of this IC was established to be that of Fig. 2 and the transistors were characterized by the Gummel–Poon (G–P) parameters of Table II.[1]

Program SPICE has been used to establish the performance and characteristics of the op amp [7]. The results are summarized in Table III, column 1. These values are used in the Appendix to develop the element and parameter values of the macromodel. As brought out in the Appendix, several parameters of the macromodel could be chosen arbitrarily:

[1] A slight modification of the G–P parameters of the output transistors has been made to produce a typical level of maximum short-circuit available current.

TABLE III

OP AMP PERFORMANCE CHARACTERISTICS

	8741 Device-Level Model	8741 Macromodel	LM741 Data Sheet	LM118 Data Sheet
C_2 (pF)	30	30	30	5
S_R^+ (V/μs)	0.9	0.899	0.67	100
S_R^- (V/μs)	0.72	0.718	0.62	71
I_B (nA)	256	255	80	120
I_{Bos} (nA)	0.7	<1	20	6
V_{os} (mV)	0.299	0.298	1	2
a_{VD}	$4.17\cdot10^5$	$4.16\cdot10^5$	$2\cdot10^5$	$2\cdot10^5$
a_{VD} (1 kHz)	$1.219\cdot10^3$	$1.217\cdot10^3$	10^3	$16\cdot10^3$
$\Delta\phi$ (°)	16.8	16.3	20	40
CMRR (dB)	106	106	90	100
R_{out} (Ω)	566	566	75	75
$R_{o\text{-ac}}$ (Ω)	76.8	76.8	—	—
I_{SC}^+ (mA)	25.9	26.2	25	25
I_{SC}^- (mA)	25.9	26.2	25	25
V^+ (V)	14.2	14.2	14.0	13
V^- (V)	−12.7	−12.7	−13.5	−13
P_d (mW)	59.4	59.4	—	—

$$T = 300\ \text{K}(V_T = 25.85\ \text{mV}),$$

$$I_{S1} = I_{SD3} = 8\cdot10^{-16}\ \text{A},$$

$$R_2 = 0.1\ \text{M}\Omega,$$

where I_{S1}, I_{SD3} are the saturation currents of the first transistor and the voltage-limiter diodes, respectively. In addition, the major compensation capacitor is fixed

TABLE IV
MACROMODEL PARAMETERS

	8741	LM 741	LM 118
T (K)	300	300	300
I_{S1} (A)	$8\cdot10^{-16}$	$8\cdot10^{-16}$	$8\cdot10^{-16}$
I_{SD3} (A)	$8\cdot10^{-16}$	$8\cdot10^{-16}$	$8\cdot10^{-16}$
R_2 (kΩ)	100	100	100
C_2 (pF)	30	30	5
C_E (pF)	7.5	2.41	2.042
β_1	52.6726	111.67	$2.033\cdot10^3$
β_2	52.7962	143.57	$2.137\cdot10^3$
I_{EE} (μA)	27.512	20.26	500
R_E (mΩ)	7.2696	9.872	0.40
I_{S2} (A)	$8.0925\cdot10^{-16}$	$8.309\cdot10^{-16}$	$8.619\cdot10^{-16}$
R_{c1} (Ω)	4352	5305	1989
R_{e1} (Ω)	2391.9	2712	1884
C_1 (pF)	4.5288	5.460	2.098
R_p (kΩ)	15.363	—	—
G_a (μmho)	229.774	188.6	502.765
G_{CM} (nmho)	1.1516	6.28	5.028
R_{01} (Ω)	76.8	32.13	32.13
R_{02} (Ω)	489.2	42.87	42.87
G_b (mho)	37.0978	247.49	92.792
I_X (A)	100.138	—	—
I_{SD1} (A)	$3.8218\cdot10^{-32}$	$8\cdot10^{-16}$	$8\cdot10^{-16}$
R_C (Ω)	$0.1986\cdot10^{-3}$	$0.02129\cdot10^{-3}$	$0.00279\cdot10^{-3}$
G_C (mho)	5034.3	46 964	358 000
V_C (V)	1.6042	1.803	2.803
V_E (V)	3.1042	2.303	2.803

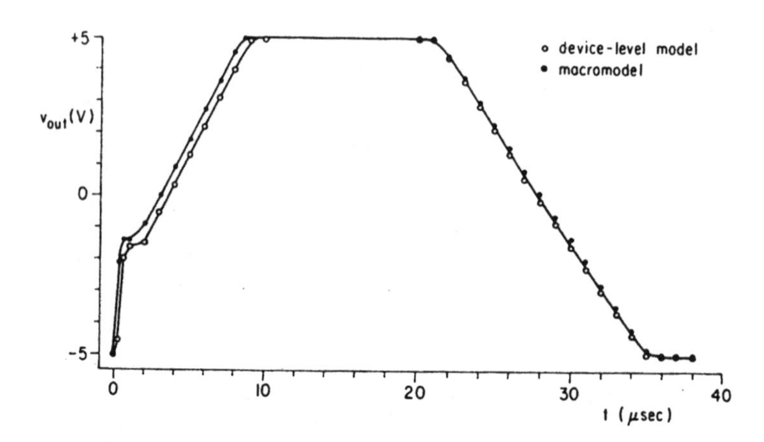

Fig. 4. Simulated voltage follower slew rate performance using both device-level models and macromodels.

by the type of op amp under study or is chosen appropriately. For the case at hand, $C_2 = 30$ pF.

The remaining values of the parameters of the macromodel are presented in Table IV, column 1.

IV. COMPARISON WITH DEVICE-LEVEL MODELS

Basic Macromodel Performance

The values for the macromodel of Table IV, column 1 were used to define an external model in program SPICE. The same set of computer runs was made as lead to the op amp performance results of Table III, column 1. The results for the macromodel are presented in column 2 of this table. It is seen that the comparison is excellent for both small-signal and large-signal experiments.

To provide a further comparison, the large-signal, slew rate performance for a voltage follower is shown Fig. 4 for both the device-level model and the macromodel. It is seen that the responses are very similar.

The presence of C_E produces a step in the initial response of the voltage follower. From simple theory [4], the jump should be approximately

$$\Delta V_{out} = \frac{C_E}{C_2}\,\Delta V_{in}\,.$$

For this example,

$$\Delta V_{in} = 10\text{ V} \quad\text{and}\quad \Delta V_{out} = \frac{7.5}{30}(10) = 2.5\text{ V}.$$

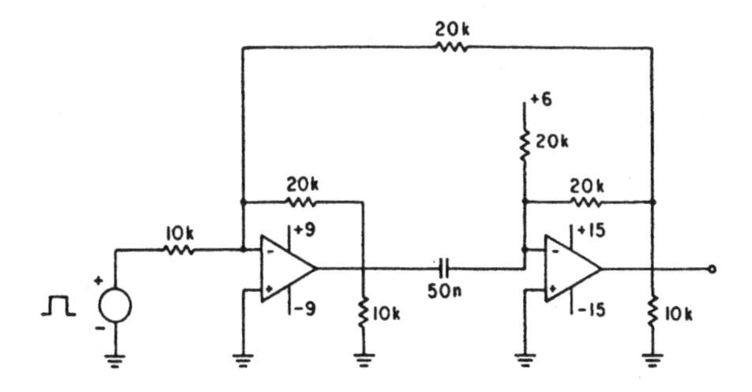

Fig. 5. A monostable time delay circuit.

From Fig. 4, the observed jump for the macromodel is 3.6 and 3.4 V for the device-level model.

A measure of the complexity of the two op amp models can be obtained by comparing the node and branch counts of each circuit. For the device-level model, where each G–P transistor model has 2 internal nodes and 7 branches, the totals are 81 nodes including the datum node, and 193 branches. For the macromodel, where each E–M transistor model has no internal nodes and 4 branches, the totals are 16 nodes and 28 branches. The ratios for the two models are 5.1 for the nodes and 6.9 for the branches. The number of p-n junctions in the device-level model is 52 and 8 in the macromodel, a ratio of 6.5.

The total computer central processing unit (CPU) time on a CDC 6400 to simulate the voltage follower slew rate performance is 39.2 s for the device-level model and 4.0 s for the macromodel, a ratio of 9.8. An alternate comparison is obtained if only the simulation times for the initial state and the transient analyses are used. The improvement ratio is then 12.0.

For the dc and ac simulations, the device-level model to macromodel CPU times, for the analyses only, have the ratios of 3.9 and 6.0, respectively. Note that the improvement ratio is less for the dc analyses.

A Regenerative Timer

The monostable time delay circuit of Fig. 5 provides a good test of the ability of the macromodel to perform as desired when demanding nonlinear performance is required. The voltage and timing levels of the circuit of Fig. 5 have been chosen to provide a maximum stress on the op amps with respect to voltage limits, critical voltage switching levels and speed of response, slew-rate limitations, etc. The output waveforms of the circuit as predicted by program SPICE using both the device-level model and the macromodel are shown in Fig. 6. It is seen that the responses compare closely. The leading and trailing edges of the output pulse differ in timing by less than 1 time step of the computer output, i.e., better than 3 percent of the overall pulsewidth. The total CPU times for the simulations using the two models are in the ratio of 8.9. If the common output time is deleted, the improvement ratio is 9.6.

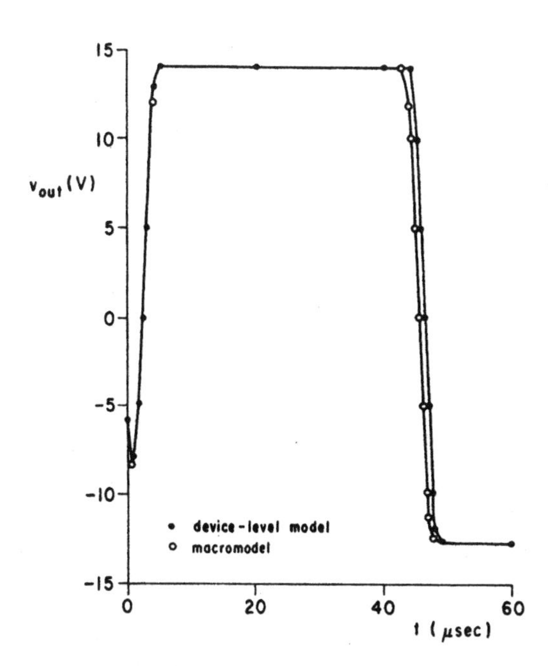

Fig. 6. Simulated output pulse response of time delay circuit using both device-level models and macromodels.

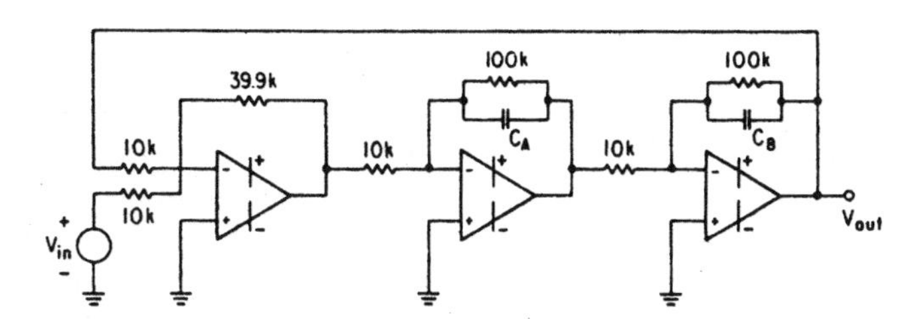

Fig. 7. "Ring of Three" bandpass filter.

TABLE V
RESPONSE DATA FOR THE "RING OF THREE" BANDPASS FILTER

Design Center Frequency f_{od} (kHz)	Actual Center Frequency f_{oa} (kHz)	Gain Magnitude at f_{oa}	Gain Magnitude at $0.9 f_{od}$	Gain Magnitude at $1.1 f_{od}$
1	0.998	11.01	4.806	4.311
	(0.998)	(11.00)	(4.806)	(4.311)
2	1.996	12.24	4.898	4.368
	(1.996)	(12.23)	(4.896)	(4.367)
10	9.934	112.1	5.547	4.398
	(9.934)	(107.1)	(5.542)	(4.401)

Numbers in parentheses refer to results with device-level model.

An Active RC Filter

To further check the second-order ac response of the macromodel, the simple "Ring of Three" op amp filter of Fig. 7 was designed for a center frequency of 1 kHz and a Q of 10. The frequency response from program SPICE for the two models is summarized in Table V. Again, it is seen that the comparison is very close.

At higher frequencies, the phase response of the (1 MHz) 741-type op amps comes into effect. A response comparison for the two models as shown in rows 2 and 3 indicates that the macromodel is providing the proper phase response.

The total CPU simulation times to determine the dc state and the frequency response using the two models have the ratio of 5.8. If the common output time is omitted, the ratio becomes 6.8.

In this application, the nonlinear performance of the op amp is not of major interest. In order to check on the improvement of computer run time for a reduced macromodel, the voltage and current limiting circuitry of Fig. 1 was omitted. The simulated response of the filter did not change, of course; however, the total CPU run time was reduced by a factor of 1.4.

V. Macromodel Parameters for Other IC Op Amps

The detailed, precise performance characteristics for an individual op amp as obtained from the use of the device-level models are usually not available. The precision used in the numerical example of this paper is employed in order to obtain an accurate estimate of the performance of a macromodel in relation to a known reference. Experimental results with actual op amps could include significant measurement inaccuracies.

Typically, one has a data sheet or averaged experimental data for a type of op amp which is to be included in a system. As an example of the choice of macromodel parameters in this situation, two further examples are given. In Table III, columns 3 and 4, measured typical op amp data are given for both the LM741 and the LM118. The macromodel parameters corresponding to these data are given in columns 2 and 3 of Table IV.

It is possible to introduce programming into a simulator to determine automatically the macromodel parameters. This has been done at one location for program SPICE. In this situation, all that is necessary to define a specific op amp model is to list its characteristics on an op amp "model card" in much the same way that one currently defines a transistor model by specifying its characteristics on a transistor model card. When an op amp characteristic is not inputed, a default value is used.

Appendix

The 8741 Macromodel

In this Appendix, a numerical example is used to illustrate the development of the parameters of the op amp macromodel. For the example, the response characteristics of the 8741 op amp are used as determined by several simulator runs using device-level modeling. The circuit of Fig. 2 together with the transistor parameters of Table II has been analyzed to obtain the characteristics which are summarized in column 1, Table III.

The development procedure follows the sequence of expressions of Table I. The final results are presented in column 1, Table IV.

From the slew rate performance of the 8741 as given in Table III

$$S_R^+ = 0.90 \text{ V}/\mu\text{s} \quad \text{and} \quad S_R^- = 0.72 \text{ V}/\mu\text{s}.$$

For the given compensation capacitor of $C_2 = 30$ pF, these values lead to

$$I_{C1} = \tfrac{1}{2} S_R^+ C_2 = 13.50 \ \mu\text{A}$$

$$C_E = \frac{2I_{C1}}{S_R^-} - C_2 = 7.50 \text{ pF}.$$

The average base current is 256 nA and the desired base current offset is 0.7 nA.

$$I_{B1} = 256.3 \text{ nA}, \qquad I_{B2} = 255.7 \text{ nA}$$

$$\beta_1 = 52.6727, \qquad \beta_2 = 52.7962.$$

A high level of precision is used in this example in order to obtain an accurate comparison of macromodel performance in relation to that of the op amp modeled at the device level.

The necessary emitter current source for the input stage is $I_{EE} = 27.512$ μA. The value of the CM emitter resistor is

$$R_E = 200/I_{EE} = 7.2696 \text{ M}\Omega$$

where a value of $V_A = 200$ V has been used.

For Q_1, the assumed value of saturation current is $8\cdot 10^{-16}$ A which is a typical value for a small n-p-n IC transistor. To produce the desired input offset voltage of 0.299 mV

$$I_{S2} = 8\cdot 10^{-16}\left(1 + \frac{0.299}{25.85}\right) = 8.0925\cdot 10^{-16} \text{ A}.$$

For a fully compensated op amp with a rolloff of −6 dB/octave, the 0 dB frequency can be calculated from the product of the gain and the value of the frequency at which it is measured providing that the frequency is well above the corner frequency of the gain characteristic. From the data of column 1, Table III,

$$f_{0\text{ dB}} = (1.219\cdot 10^3)(10^3) \text{ Hz} = 1.219\cdot 10^6 \text{ MHz}.$$

The value of the collector resistors of the first stage is

$$R_{c1} = \frac{1}{2\pi f_{0\text{ dB}} C_2} = 4352 \ \Omega.$$

The value of the reciprocal of g_m for the first stage is

$$\frac{1}{g_{m1}} = 1915 \ \Omega.$$

The required value of the emitter resistor is

$$R_{e1} = 0.9814(4352 - 1915) = 2392 \ \Omega.$$

The final element for the input stage is C_1, which produces the nondominant pole of the gain function. For $\Delta\phi = 16.80°$

$$C_1 = \frac{C_2}{2} \tan 16.80° = 4.529 \text{ pF}.$$

The value of the resistor R_p to simulate power dissipation for ±15 V supplies and a dissipation of 59.4 mW is

$$R_p = 15.363 \text{ k}\Omega.$$

For the interstage, R_2 is taken to be 100 kΩ and

$$G_a = \frac{1}{R_{c1}} = 229.774 \ \mu\text{mho}.$$

For a CMRR of 106 dB ($199.5 \cdot 10^3$)

$$G_{CM} = \frac{G_a}{\text{CMRR}} = 1.1516 \text{ nmho}.$$

In the output stage, the desired dc and ac output resistances are 566 Ω and 76.8 Ω, respectively. Therefore,

$$R_{01} = 76.8 \ \Omega, \ R_{02} = 489.2 \ \Omega.$$

The value of G_b to provide the correct DM voltage gain of $417 \cdot 10^3$ is

$$G_b = \frac{a_{VD} R_{c1}}{R_2 R_{02}} = 37.097.$$

The maximum current through the diode D_1 or D_2 is

$$I_X = 2 I_{C1} G_b R_2 - I_{SC} = 100.14 \text{ A}$$

where the desired value of I_{SC} is 25.9 mA. With these values, the saturation currents of D_1 and D_2 are

$$I_{SD1} = I_X \exp - \frac{R_{01} I_{SC}}{V_T} = 3.822 \cdot 10^{-32} \text{ A}.$$

The values for the approximate voltage-controlled voltage source are

$$R_C = \frac{V_T}{100 I_X} \ln \frac{I_X}{I_{SD1}} = 0.199 \text{ m}\Omega$$

and

$$G_C = \frac{1}{R_C} = 5034 \text{ mho}.$$

For the voltage-clamp circuits, the saturation currents of diodes D_3 and D_4 are chosen to be $8 \cdot 10^{-16}$ A. The voltage sources should be

$$V_C = 15 - 14.2 + 0.02585 \ln \frac{0.0259}{8 \cdot 10^{-16}} = 1.604 \text{ V}$$

and

$$V_E = 15 - 12.7 + 0.804 = 3.104 \text{ V}.$$

Acknowledgment

The authors are pleased to acknowledge the aid and discussion on this topic with C. Battjes, R. Bohlman, S. Taylor, R. Dutton, and I. Getreu.

The staffs at the Computer Centers at the University of California, Berkeley, and at Tektronix, Inc., Beaverton, Oreg., have been extremely generous and helpful in the support given for the numerous computer runs necessary for this project.

References

[1] J. R. Greenbaum, "Digital-IC models for computer-aided design," *Electronics*, vol. 46, 25, pp. 121–125, Dec. 6, 1973.

[2] B. M. Cohn, D. O. Pederson, and J. E. Solomon, "Macromodeling of operational amplifiers," in *ISSCC Dig. Tech. Papers*, Feb. 1974, pp. 42–43.

[3] D. O. Pederson and J. E. Solomon, "The need and use of macromodels in IC subsystem design," in *Proc. 1974 IEEE Symp. Circuits and Systems*, p. 488.

[4] J. E. Solomon, "The monolithic op amp: a tutorial study," *IEEE J. Solid-state Circuits*, this issue, pp. 314–332.

[5] P. R. Gray and R. G. Meyer, "Recent advances in monolithic operational amplifier design," *IEEE Trans. Circuits and Syst.*, vol. CAS-21, pp. 317–327, May 1974.

[6] B. A. Wooley, S.-Y. J. Wong, and D. O. Pederson, "A computer-aided evaluation of the 741 amplifier," *IEEE J. Solid-State Circuits*, vol. SC-6, pp. 357–366, Dec. 1971.

[7] L. W. Nagel and D. O. Pederson, "Simulation program with integrated circuit emphasis (SPICE)," Electron. Res. Lab., Univ. of California, Berkeley, Memo ERL-M382, and in *Proc. 16th Midwest Symp. Circuit Theory*, 1973.

[8] D. N. Pocock and M. G. Krebs, "Terminal modeling and photocompensation of complex microcircuits," *IEEE Trans. Nucl. Sci.*, vol. NS-19, pp. 86–93, Dec. 1972.

[9] D. H. Treleaven and F. N. Trofimenkoff, "Modeling operational amplifiers for computer-aided circuit analysis," *IEEE Trans. Circuit Theory* (Corresp.), vol. CT-18, pp. 205–207, Jan. 1971.

Graeme R. Boyle was born in Echuca, Victoria, Australia, on October 26, 1949. He received the B.E. and M.Eng.Sc. degrees in electrical engineering from the University of Melbourne, Melbourne, Australia, in 1972 and 1974, respectively.

He is currently at the University of California, Berkeley, working toward the Ph.D. degree in the field of modeling and computer simulation of integrated circuits.

Barry M. Cohn (S'68) was born in Seattle, Wash., on April 8, 1949. He received the B.S.E.E. degree, graduating Magna Cum Laude and with College Honors, from the University of Washington, Seattle, in 1971. He received the M.S. degree in electrical engineering from the California Institute of Technology, Pasadena, in 1972.

He has done power engineering for Seattle City Light during the summers of 1968–1970. From 1972–1974, while at the University of California, Berkeley, he did research and published on the topic of macromodeling integrated circuits for computer-aided design. During the summer of 1973, he was employed as an Engineer with National Semiconductor Corporation, where he did research on the topic of macromodeling operational amplifiers. From 1973–1974 he has been a Research Assistant at the Electronic Research Lab and a Teaching Assistant at the University of California, Berkeley, from which he is currently on leave and presently employed by Intel Corporation, Santa Clara, Calif., where he directs CAD operations.

Mr. Cohn is a member of Tau Beta Pi and Phi Eta Sigma. He is the recipient of a National Science Foundation Traineeship, a General Telephone Electronics Graduate Fellowship, and numerous undergraduate scholarships.

Donald O. Pederson (S'49–A'51–M'56–F'64) was born in Hallock, Minn., on September 30, 1925. He received the B.S. degree from North Dakota Agricultural College (now North Dakota State University), Fargo, in 1948 and the M.S. and Ph.D. degrees from Stanford University, Stanford, Calif., in 1949 and 1951, respectively.

From 1951 to 1953 he was a Research Associate in the Electronics Research Laboratory, Stanford University. From 1953 to 1955 he worked at Bell Telephone Laboratories, Inc., Murray Hill, N.J., and was also a Lecturer at Newark College of Engineering, Newark, N.J. In 1955 he joined the Electrical Engineering Department, University of California, Berkeley, where is is now a Professor and engaged in research in integrated circuits and computer-aided circuit analysis and design. From 1960 to 1964 he was Director of the Electronics Research Laboratory.

Dr. Pederson is a member of Sigma Xi and Eta Kappa Nu. He and three coauthors were awarded a Best Paper Award for a paper presented at the 1963 International Solid-State Circuits Conference. He was a Guggenheim Fellow in 1964 and was the recipient of the IEEE Education Medal in 1969. In 1974 he was elected to the National Academy of Engineering.

James E. Solomon (S'57–M'61), for a photograph and biography, please see p. 332 of this issue.

A Macromodeling Algorithm for Analog Circuits

Giorgio Casinovi, *Member, IEEE,* and Alberto Sangiovanni-Vincentelli, *Fellow, IEEE*

***Abstract*—The advantages of using macromodels of functional blocks in the simulation of large scale circuits have long been pointed out. However, all the methods for generating macromodels that have so far appeared in the literature make very stringent assumptions about either the type of circuits to be modeled or the shape of the waveforms involved. The models generated in this way are, therefore, of limited use in general circuit simulation. In this paper we propose a general purpose macromodeling algorithm applicable to any type of circuit without any restrictions on the waveforms. Starting from a user-supplied template, the algorithm modifies the values and the branch characteristics of macromodel elements with the goal of optimizing the macromodel's time-domain accuracy. Experimental results obtained so far indicate that the algorithm requires reasonable amounts of CPU time and that the models are accurate enough to be used for standard circuit simulation.**

I. Introduction

The word macromodel usually refers to a compact representation of a circuit that captures those features that are useful for a particular purpose while discarding redundant information. The sheer size of most VLSI circuits forces the use of a modular design style, in which multiple instances of a limited number of cells are used repeatedly in different parts of the circuit. In this context macromodels are extremely useful both in the synthesis and in the verification phases of the design of complex circuits, because they provide a simple, effective representation of the functional units contained in the circuit. These reasons, coupled with the need to reduce circuit simulation times for large circuits, prompted the development of macromodels for frequently used functional blocks: in [1], techniques known as "circuit simplification" and "circuit build-up" were used to generate a macromodel for operational amplifiers. In the same paper it was shown that the macromodel could be simulated up to ten times faster than the original circuit, while still giving accurate results within a few percent. The same ideas were later used to develop macromodels for comparators [2] and MOS operational amplifiers [3]. Although successful in generating fast and accurate macromodels, the techniques proposed in those papers are not easily automated, because they rely on human understanding of the functionality of the circuits being modeled.

A great deal of attention was also given to the development of macromodels to be used for logic, switch-level, or timing simulation of digital circuits [4]–[6]: these types of macromodels provide very limited information about the behavior of the circuits they represent. For instance, a macromodel of a logic gate used in timing analysis will typically contain only information about delay, output rise, and fall times, etc. For this reason, those macromodels cannot be used when a more accurate description is needed (e.g., circuit simulation). To overcome these limitations, some effort was put into the development of better macromodels that could take into account the effects of waveforms on the behavior of the circuit being modeled [7], [8]; however, very restrictive assumptions were made about the shapes of the waveforms, essentially limiting the scope of those methods to the macromodeling of digital circuits. An alternative, more general approach was proposed in [9]: although the example presented in that paper is a macromodel of a NAND gate, the method suggested by the authors is applicable to any circuit and does not rely on any specific assumptions about the waveforms involved. Unfortunately the macromodels are derived under static conditions, so that their accuracy during transients cannot be predicted. More recent work [10] has focused specifically on modeling analog blocks: the technique suggested by the authors consists in generating automatically C code describing KCL and KVL equations for the circuit to be modeled. This is a simple and flexible approach, which, however, does not solve the problem of reducing model complexity and cannot be used to improve the accuracy of an existing model.

A new theoretical approach to macromodeling was proposed by the first author in [11]. The purpose of this paper is to show how those ideas can be used in practice to modify and optimize analog macromodels. Here the word macromodel refers to an electrical network containing fewer devices and/or fewer nodes than the circuit it represents: as an example, a voltage-controlled current source, possibly including an output resistance, could be taken as a very simple macromodel for an operational amplifier. The proposed algorithm compares a circuit and its macromodel under dynamic conditions; however, no *a priori* assumptions are made about the waveforms involved. As a consequence, the macromodels thus generated can be used for general purpose circuit simulation, where accurate modeling under widely varying conditions is required.

II. Comparison of Circuit Dynamics

Most VLSI circuits are designed following a modular style, in which multiple instances of functional blocks are interconnected to achieve certain design specifications. The behavior of the whole circuit is determined by how the individual blocks interact with each other, while what happens inside each of them is unimportant. In brief, the relevant feature of each functional block in this context is its input–output behavior. The accuracy of a macromodel must, therefore, be defined in terms of how closely its input–output behavior matches the one of the original circuit. Thus a precise measure of the accuracy of a macromodel requires the capability of quantifying the intuitive notion

Manuscript received August 29, 1989; revised January 17, 1990. This work was supported in part by JSEP-AFOSR under Grant F49620-87-C-0041 and by MICRO under Grant 532491-19900. This paper was recommended by Associate Editor A. Ruehli.

G. Casinovi was with the Department of Electrical Engineering and Computer Science, University of California, Berkeley, CA. He is now with the School of Electrical Engineering, Georgia Institute of Technology, Atlanta, GA 30332-0250.

A. Sangiovanni-Vincentelli is with the Department of Electrical Engineering, University of California, Berkeley, CA 94720.

IEEE Log Number 9040259.

that two circuits behave in a similar way: this problem is addressed in the remainder of this section, and a solution is proposed.

Throughout the rest of this paper it will be assumed that all the circuits under consideration satisfy the following assumptions.

- All the devices contained in the circuit can be modeled by independent current sources, resistors, capacitors, and voltage-controlled current sources.
- Resistors, capacitors, and controlled sources are not required to be linear. However, it is assumed that they are voltage controlled, i.e., that they can be described by branch equations of the form $i_d = f_d(v_d)$ or $q_d = f_d(v_d)$, where i_d is the current flowing through the device, q_d is the charge on the capacitor, f_d is a function of class C^1 which depends on the device and v_d is the controlling branch voltage. Furthermore, it is assumed that there exists a constant $c_{\min} > 0$ such that $(dq_d/dv_d) \ge c_{\min}$ for *all* voltages v_d and for *all* capacitors in the circuit.
- There are capacitors connecting the ground node to all other nodes of the circuit.

The conditions listed above are satisfied by most VLSI circuits. Moreover, the theoretical results presented in this and in the following sections remain valid even if some of those restrictions are lifted, at the expense of more complex mathematical proofs: for instance, the requirement that each node be connected to ground by a capacitor can be dropped altogether.

The first step to be taken in approaching the macromodeling problem is to obtain an appropriate description of the dynamics of the circuits involved. For this purpose, consider an elementary cell (e.g., an operational amplifier), and let n be the total number of its nodes, m of which are to be considered *inputs* and p *outputs*. Input and output nodes can overlap partially or totally, or be completely distinct. For notational convenience, input nodes will be assigned numbers 1–m. Let $q = (q_1, \cdots, q_n)^T$ be the vector of the node charges and let $i = (i_1, \cdots, i_m)^T$ be the vector of the currents flowing into the input nodes from the outside (see Fig. 1). Then the node equations can be written as

$$\dot{q} = -f(v) + Bi \tag{1}$$

where $v = (v_1, \cdots, v_n)$ is the vector of the node voltages, f is a C^1 function and B is a rectangular matrix given by

$$B = \begin{pmatrix} I_{m\times m} \\ O_{(n-m)\times m} \end{pmatrix}.$$

Under the assumptions made at the beginning of this section, the node charges are a function of the node voltages: $q = q(v)$.[1] Moreover, $q(v)$ is *uniformly monotone* according to the following definition.

Definition 1: A function $f: R^n \to R^n$ is uniformly monotone if there exists a $\gamma > 0$ such that

$$[f(x) - f(y)]^T(x - y) \ge \gamma(x - y)^T(x - y), \quad \forall x, y \in R^n.$$

Uniformly monotone functions can be inverted as a consequence of the following theorem [12, p. 143].

[1]This is a compact way of denoting a functional relationship between q and v, i.e., the fact that there exists a function f such that $q = f(v)$.

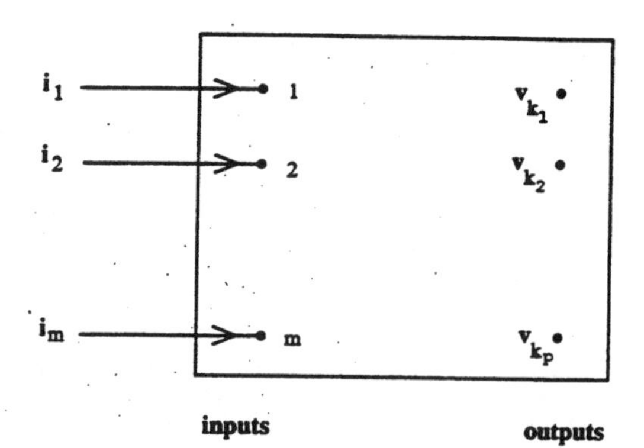

Fig. 1. Abstract representation of the block to be modeled.

Theorem 1: If $f: R^n \to R^n$ is of class C^1 and uniformly monotone on R^n, then f has a C^1 inverse $f^{-1}: R^n \to R^n$.

The following criterion [12, p. 142] can be used to decide whether a function is uniformly monotone.

Theorem 2: f is uniformly monotone on R^n if and only if there is a $\gamma > 0$ such that:

$$\Delta x^T \frac{\partial f}{\partial x} \Delta x \ge \gamma \|\Delta x\|^2, \quad \forall x \in R^n, \forall \Delta x \in R^n.$$

The function $q(v)$ satisfies this criterion, because its derivative $(\partial q/\partial v)$ is the capacitance matrix of a linear network obtained from the original circuit by replacing every capacitor whose branch equation is $q_d = q_d(v_d)$ with a linear capacitor of value (dq_d/dv_d). The quantity $(1/2)\Delta v^T(\partial q/\partial v)\Delta v$ represents the energy stored in this network when voltages $\Delta v_1, \cdots, \Delta v_n$ are applied to its nodes. This is certainly larger than the energy stored just on the grounded capacitances, which is equal to $(1/2)\sum_{i=1}^n c_{ii}\Delta v_i^2$. However, $c_{ii} \ge c_{\min}$, so that the following chain of inequalities is satisfied:

$$\frac{1}{2}\Delta v^T \frac{\partial q}{\partial v}\Delta v \ge \frac{1}{2}\sum_{i=1}^n c_{ii}\Delta v_i^2 \ge \frac{1}{2} c_{\min}\|\Delta v\|^2.$$

According to Theorem 1, $q(v)$ can be inverted, i.e., the node voltages can be expressed as a function of the node charges: $v = v(q)$. In particular, the vector of the output node voltages $v^0 = (v_{k_1}, \cdots, v_{k_p})$ is a function of q, so that the dynamical behavior of a circuit satisfying the assumptions enumerated at the beginning of this section is completely characterized by the following equations:

$$\dot{q} = -f(v) + Bi$$
$$v^0 = g(q). \tag{2}$$

It is easily recognized that these equations describe a dynamical system whose input, state, and output are i, q, and v^0, respectively.

Assume now that a macromodel M for the circuit described by (2) is available: according to the definition given in the introduction, this is a circuit containing fewer nodes and devices than the original circuit. The dynamics of both circuits are described by the following set of equations:

$$\dot{q}_1 = -f_1(v_1) + B_1 i$$
$$v_1^0 = g_1(q_1) \tag{3}$$
$$\dot{q}_2 = -f_2(v_2) + B_2 i$$
$$v_2^0 = g_2(q_2). \tag{4}$$

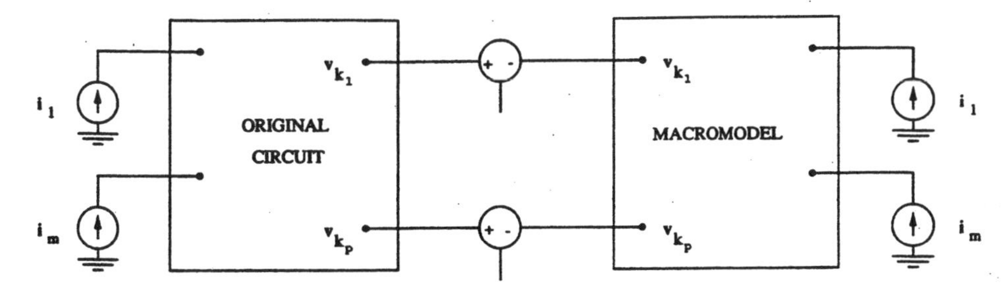

Fig. 2. Comparing the dynamics of two circuits.

Note that the original circuit and its macromodel must have matching input and output nodes, but the number of internal nodes in general will not be the same. As a consequence, the dimensions of v_1^0 and v_2^0 are equal, but the dimension of q_2 is in general smaller than the dimension of q_1, i.e., fewer differential equations must be solved to simulate (4) than (3). Thus it is an advantage to replace the original circuit with its macromodel, provided that the difference in the external dynamical behavior of the two circuits is within limits prescribed by the user. This difference can be quantified in the following way: fix a time interval $[0, T]$, initial conditions $q_{1,0}$ and $q_{2,0}$ and a subset $\mathcal{U}$ of admissible input waveforms: this means that the dynamics of the circuit and of its macromodel are compared only for those inputs that belong to $\mathcal{U}$. At first sight this seems to render such a comparison almost meaningless; in fact, it will be shown later that $\mathcal{U}$ can be taken large enough to include all waveforms of practical interest. For the time being the only condition imposed upon the admissible inputs is that they must be bounded on the interval $[0, T]$; such limitation makes perfect sense because of physical reasons, and in mathematical terms it translates into the requirement that $\mathcal{U}$ be a subset of $L^\infty([0, T], R^m)$. Standard theorems about the solutions of differential equations ensure that for every $i(t) \in \mathcal{U}$ there exist unique functions $q_1(t)$, $q_2(t)$ satisfying (3), (4) and the initial conditions $q_1(0) = q_{1,0}$, $q_2(0) = q_{2,0}$. Let $v_1^0(t)$, $v_2^0(t)$ be the corresponding outputs, and define:

$$c(M, i_w) = \frac{1}{2}\int_0^T \left\| v_2^0(t) - v_1^0(t) \right\|^2 dt. \tag{5}$$

Here the subscript w in i_w denotes the fact that the value of the integral depends on the whole current waveform on the interval $[0, T]$ (as opposed to the instantaneous value i appearing, for instance, in (3), (4)). The notation $c(M, i_w)$ stresses the point that c depends both on the equations describing macromodel M as well as on the input waveform i_w through (3), (4). From a practical point of view, the above equations can be interpreted in the following way (see Fig. 2): two identical sets of current generators $i_1(t), \cdots, i_m(t)$ are connected to the inputs of the original circuit and of the macromodel. The resulting voltages at the output nodes are compared and the L^2 norm of their difference is the quantity $c(M, i_w)$ defined in (5). The largest value of c over all admissible inputs i_w can then be taken as a measure of the difference in the input-output behavior of the two circuits; in formulas:

$$D(M) = \sup_{i_w \in \mathcal{U}} c(M, i_w). \tag{6}$$

The considerations that follow assume the following choice of the set of admissible inputs:

$$\mathcal{U} = \left\{ i(t) \in L^\infty([0, T], R^m) : \left| i_j(t) \right| \le M_j, \; t \in [0, T], \quad j = 1, \cdots, m \right\}.$$

This choice means that the shapes of $i_1(t), \cdots, i_m(t)$ can be arbitrary, as long as they never exceed certain specified bounds. Obviously $\mathcal{U}$ is large enough to contain all waveforms of practical interest. For this particular choice of $\mathcal{U}$ it can be proven that there exists $i_{\max} \in \mathcal{U}$ such that $D(M) = c(M, i_{\max})$. It is immediately recognized that the computation of $i_{\max}$ requires solving the following optimal control problem:

$$\begin{aligned} &\max_{i_w \in \mathcal{U}} \frac{1}{2}\int_0^T \left\| v_2^0(t) - v_1^0(t) \right\|^2 dt \\ &\begin{pmatrix} v_1^0 \\ v_2^0 \end{pmatrix} = \begin{pmatrix} g_1(q_1) \\ g_2(q_2) \end{pmatrix} \\ &\begin{pmatrix} \dot{q}_1 \\ \dot{q}_2 \end{pmatrix} = \begin{pmatrix} -f_1(v_1) + B_1 i \\ -f_2(v_1) + B_2 i \end{pmatrix}. \end{aligned} \tag{7}$$

The Hamiltonian for this problem is

$$\begin{aligned} H(v, \lambda, i) = \tfrac{1}{2}\left\| v_2^0 - v_1^0 \right\|^2 &+ \lambda_1^T[-f_1(v_1) + B_1 i] \\ &+ \lambda_2^T[-f_2(v_2) + B_2 i]. \end{aligned}$$

Because H is a linear function of i, Pontryagin's maximum principle [13, p. 108] can be used to give the following characterization of $i_{\max}$: *let* $i_{\max} = (i_1(t), \cdots, i_m(t))^T$. *For every* $j = 1, \cdots, m$ *and for almost all* $t \in [0, T]$ *either* $i_j(t) = M_j$ *or* $i_j(t) = -M_j$, provided *that* $i_{\max}$ *is not a singular control [13, p. 246].*

This result is somewhat surprising, because it goes against the natural expectation that the worst-case input could be almost any function, depending on the specific structure of the two circuits. Instead, *because the systems under consideration are linear in the input*, only a much smaller class of functions need be examined, namely those that are piecewise constant in the interval $[0, T]$; only the number and the location of the switching points remain unknown. From a practical point of view this is a big advantage, because it allows the set of admissible inputs to be parametrized by a finite dimensional vector in the following way [15]: arbitrarily fix the maximum number of switching points that each component can have; this is a parameter (call it N) that can be set by the user. Let $(t_{j,1}, \cdots, t_{j,N})$ and $M_j > 0$ be the switching points and the bound of $i_j(t)$, respectively (see Fig. 3); for notational convenience define $t_{j,0} = 0$ and $t_{j,N+1} = T$ (obviously we must have $t_{j,0} \le t_{j,1} \le \cdots \le t_{j,N+1}$). For every $t \in [0, T]$ the value of $i_j(t)$ is computed as follows: let k be that index such that $t_{j,k} \le t < t_{j,k+1}$; then $i_j(t) = (-1)^k M_j$. This completely determines the input, which can be, therefore, parametrized by a vector $t = (t_{1,1}, \cdots, t_{m,N})$ containing mN components. It is natural to ask what an appropriate value for N is. Unfortunately this question cannot be answered

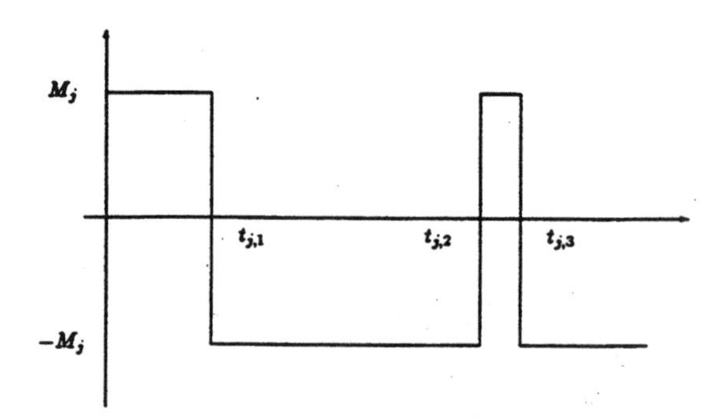

Fig. 3. Input parametrization according to switching points.

in general terms, even in the case of linear systems [14, p. 15]. Therefore, the value of N must be chosen by the user on a case-by-case basis.

III. Macromodeling as a Min–Max Problem

The measure of the difference of the dynamical behavior of two circuits introduced in the previous section makes it possible to quantify the accuracy of a macromodel. In particular, the best macromodel in a (finite or infinite) collection $\mathfrak{M}$ is the one that minimizes such a difference. Mathematically speaking, the best macromodel in $\mathfrak{M}$ is a solution of the following equation:

$$\min_{M \in \mathfrak{M}} \left(\max_{i_w \in \mathfrak{U}} c(M, i_w) \right). \tag{8}$$

Thus the task of finding the optimal macromodel is equivalent to solving a min-max problem.

In many instances the elements of $\mathfrak{M}$ are identified by a multidimensional parameter $x = (x_1, \cdots, x_r)$ ranging over a set $\mathcal{P}$ of admissible parameter values. The x_i's can be the values of elements contained in the macromodel, such as resistances or capacitances, or coefficients appearing in the transfer characteristics of nonlinear devices. The parameter x affects the behavior of the macromodel through (4), so that any $M \in \mathfrak{M}$ is described by a set of equations of the following type:

$$\begin{aligned} \dot{q}_2 &= -f(v_2, x) + B_2 i \\ v_2^0 &= g_2(q, x). \end{aligned} \tag{9}$$

Moreover, because each macromodel M is uniquely identified by x, (8) can be rewritten as:

$$\min_{x \in \mathcal{P}} \left(\max_{i_w \in \mathfrak{U}} c(x, i_w) \right). \tag{10}$$

As for actually solving the above equation, it must be pointed out that computing the global optimum of a general real-valued function is almost always impossible. Because solving (10) effectively requires finding a global minimum of the function $D(x) = \min_{i_w \in \mathfrak{U}} c(x, i_w)$, in most cases it will not be possible to find the optimal model. The best that can be hoped for is to compute a local minimum of $D(x)$: the corresponding macromodel is the best among all those that can be obtained from it by small parameter variations. Most of the algorithms that can be used for this purpose require the computation of the gradient of the objective function. It will be shown next that the gradient of c can in fact be computed numerically using formulas very similar to those appearing in optimal control theory. For this purpose it is convenient to let $q = (q_1, q_2)$, $v = (v_1, v_2)$ and $\lambda = (\lambda_1, \lambda_2)$; then (3), (9) can be rewritten compactly as

$$\dot{q} = f(v, i, x) \tag{11}$$

where

$$f = \begin{pmatrix} -f_1(v_1) + B_1 i \\ -f(v_2, x) + B_2 i \end{pmatrix}. \tag{12}$$

Computation of the differential δc caused by a small change δi in the input function has been extensively dealt with in optimal control theory: it is well known [13] that δc is given by

$$\delta c = \int_{\gamma_0} \frac{\partial H}{\partial i} \delta i \, dt = \int_{\gamma_0} \lambda^T \frac{\partial f}{\partial i} [v(t), i(t), x] \delta i(t) \, dt \tag{13}$$

where H is the Hamiltonian associated to (11):

$$\begin{aligned} H(v, \lambda, i, x) &= \lambda^T f(v, i, x) \\ &= \tfrac{1}{2} \| v_2^0 - v_1^0 \|^2 + \lambda_1^T [-f_1(v_1) + B_1 i] \\ &\quad + \lambda_2^T [-f_2(v_2, x) + B_2 i] \end{aligned}$$

and λ^T is the vector of the Lagrange multipliers, which satisfy the equations:

$$\begin{aligned} \dot{\lambda}^T &= -\frac{\partial H}{\partial q} \\ \lambda(T) &= 0. \end{aligned} \tag{14}$$

In the particular case being considered here (14) splits into two equations:

$$\dot{\lambda}_i^T = -\frac{\partial H}{\partial q_i} = -\frac{\partial H}{\partial v_i} \frac{\partial v_i}{\partial q_i}, \qquad i = 1, 2. \tag{15}$$

However, $(\partial v_i / \partial q_i)$ is the inverse of the capacitance matrix $(\partial q_i / \partial v_i)$, and (15) can be rewritten in the following form:

$$\frac{\partial q_1^T}{\partial v_1} \dot{\lambda}_1 - \frac{\partial f_1^T}{\partial v_1} \lambda_1 = v_2^0 - v_1^0 \tag{16}$$

$$\frac{\partial q_2^T}{\partial v_2} \dot{\lambda}_2 - \frac{\partial f_2^T}{\partial v_2} \lambda_2 = -v_2^0 + v_1^0. \tag{17}$$

The computation of the sensitivities of c with respect to changes in x can be tackled in a similar way. In fact it can be proven that sensitivity computations of any kind can be dealt with in a unified setting, whose final result is the following theorem.

Theorem 3: Let x be an arbitrary parameter appearing in (11). Then

$$\frac{\partial c}{\partial x} = \int_{\gamma_0} \lambda^T \frac{\partial f}{\partial x} dt = \int_{\gamma_0} \frac{\partial H}{\partial x} dt \tag{18}$$

where $\gamma_0 = (q(t), \lambda(t))$ is the trajectory described by the solutions of (11) and (14).

The similarity between (13) and (18) is manifest; it is also important to stress the fact that x is *arbitrary:* it can indifferently be a parameter appearing in the input waveform or in the circuit equations.

A rigorous proof of the validity of (18) is beyond the scope of this paper and can be found in [16], [17]; a different proof is given in [11], [18]. However, a less correct but more intuitive explanation might be of interest. For this purpose, let $Q(t, q_0)$ be the general integral of (11), i.e., the value of the solution of (11) with initial condition q_0 at time t. The Lagrange multipliers give the derivative of c with respect to variations in the trajectory [13, p. 49]:

$$\lambda(t) = \frac{\partial c\big(Q[T - t, q(t)]\big)}{\partial q(t)}.$$

Now let $q(t)$ be the trajectory of (11) for a given choice of the parameter x; a small change in x causes a small perturbation δf to f and we obtain a nearby trajectory, say $q(t) + \delta q(t)$. The corresponding change in the value of c can be obtained by adding up all the contributions of each individual perturbation $\delta q(t)$, as t varies from 0 to T:

$$\delta c = \int_0^T \frac{\partial c\big(Q[T - t, q(t)]\big)}{\partial q(t)} \delta q(t) = \int_0^T \lambda^T(t)\, \delta q(t). \tag{19}$$

However, $\delta q(t) = \delta \dot{q}\, dt \approx \delta f\, dt$, and we get

$$\delta c = \int_0^T \lambda^T(t)\, \delta f\, dt = \int_{\gamma_0} \lambda^T \frac{\partial f}{\partial x} \delta x\, dt \tag{20}$$

which is precisely (18).

As mentioned before, (18) holds for any dynamical system described by a set of equations similar to (11) and for any type of parameter. Some special cases relevant to macromodeling are examined below in more detail: the first one occurs when x is a parameter appearing in the branch equations of some memoryless element. For example, suppose that the element which depends on x is a resistor or a current source connected between nodes i and j, and denote its branch current by I_{ij}. In this case x does not affect $\dot{v}_2^0(q)$ and $v_2(q)$, because the functional relationship between voltages and charges is determined exclusively by the capacitive elements in the circuit. Therefore, H depends on x only through I_{ij}, which is one of the terms contained in f_2. Hence:

$$\frac{\partial c}{\partial x} = -\int_0^T \lambda_2^T \frac{\partial f_2}{\partial x}(v_2, x)\, dt = -\int_0^T (\lambda_i - \lambda_j) \frac{\partial I_{ij}}{\partial x} dt. \tag{21}$$

In the case of a linear resistor, $I_{ij} = G\Delta v_{ij}$ where $\Delta v_{ij} = v_i - v_j$. If the resistor conductance is taken as parameter we have $(\partial I_{ij}/\partial G) = \Delta v_{ij}$, so that the derivative of the cost function with respect to variations in G is

$$\frac{\partial c}{\partial G} = -\int_{\gamma_0} (\lambda_i - \lambda_j) \Delta v_{ij}\, dt \tag{22}$$

which is the same formula given in [19].

A more interesting case occurs when the branch equation describing the element is nonlinear and the type of the nonlinearity (polynomial, exponential, etc.) is not known *a priori*. Then the most appropriate way of describing the element characteristic is to use a table lookup model: the values of the function are given explicitly on a finite set of nodes, and an interpolation scheme is used to compute the function elsewhere. To make a concrete example, take an interpolation scheme based on B-splines [20, p. 108], and let $v^* = (v_1^*, \cdots, v_l^*)$ be the spline knot sequence; then the branch equation has the following expression:

$$I_{ij} = \sum_k x_k B_k(\Delta v_{ij}, v^*) \tag{23}$$

where $B_k(\Delta v_{ij}, v^*)$ is a B-spline. Here the parameters x_k are approximately the values of the branch current at the interpolation knots: $x_k \cong I_{ij}(v_k^*)$ [20, p. 123]. Proceeding as in the case of a linear resistance, it is easy to prove that $(\partial c/\partial x_k)$ is given by

$$\frac{\partial c}{\partial x_k} = -\int_{\gamma_0} (\lambda_i - \lambda_j) B_k(\Delta v_{ij}, v^*)\, dt. \tag{24}$$

It is in this case that the generality of (18) proves to be most valuable, because it relieves the user of the burden of having to guess *a priori* whether a certain nonlinearity is best approximated by a polynomial, or by a sum of exponentials, or by some other type of function. Instead, the user simply enters the values of the transfer characteristic at the interpolation nodes. Those values will then be adjusted automatically by the algorithm until the best fit between the macromodel and the original circuit is found.

The case of an element with memory is slightly more complex, because x affects the equations relating the charges to the currents, i.e., we have $q_2 = q_2(v_2, x)$ and $v_2 = v_2(q_2, x)$. Because these two functions are one of the inverse of the other, the function $q_2[v_2(q_2, x), x]$ does not depend on x, so taking its total derivative with respect to x yields the following equation:

$$0 = \frac{\partial q_2}{\partial v_2}\frac{\partial v_2}{\partial x} + \frac{\partial q_2}{\partial x}. \tag{25}$$

Moreover, from (15) we obtain:

$$\frac{\partial H}{\partial v_2} = -\dot{\lambda}_2^T \left(\frac{\partial v_2}{\partial q_2}\right)^{-1} = -\dot{\lambda}_2^T \frac{\partial q_2}{\partial v_2}.$$

$(\partial H/\partial x)$ can now be computed as follows:

$$\frac{\partial H}{\partial x} = \frac{\partial H}{\partial v_2}\frac{\partial v_2}{\partial x} = -\dot{\lambda}_2^T \frac{\partial q_2}{\partial v_2}\frac{\partial v_2}{\partial x} = \dot{\lambda}_2^T \frac{\partial q_2}{\partial x} \tag{26}$$

because $(\partial q_2/\partial v_2)(\partial v_2/\partial x) = -(\partial q_2/\partial x)$ from (25). To make a concrete example, let x be the capacitance C of a linear capacitor connected between nodes i and j. Then the charge on the capacitor is $q_{ij} = C(v_i - v_j)$, and the charge equation at node i contains the term $+q_{ij}$, while the charge equation at node j contains the term $-q_{ij}$. Then:

$$\dot{\lambda}_2^T \frac{\partial q_2}{\partial C} = (\dot{\lambda}_i - \dot{\lambda}_j)(v_i - v_j).$$

Substituting this expression in (21) gives the following equation for $(\partial c/\partial C)$:

$$\frac{\partial c}{\partial C} = \int_{\gamma_0} (\dot{\lambda}_i - \dot{\lambda}_j)(v_i - v_j)\, dt \tag{27}$$

which coincides with the one found in [19].

As a final example of the generality of (18), we use it to compute the derivatives of c with respect to the input switching times $t_{k,j}$ introduced at the end of the previous section. Because the model parameters x are kept constant throughout the computations that follow, explicit dependence on x will be dropped, in order to simplify notation; moreover, only one parameter $t_{k,j}$ at a time need be considered. With these simplifications (18) becomes:

$$\frac{\partial c}{\partial t_{k,j}} = \lim_{\Delta t_{k,j} \to 0} \int_0^T \frac{H[v(\tau), \lambda(\tau), i(\tau; t_{k,j} + \Delta t_{k,j})] - H[v(\tau), \lambda(\tau), i(\tau; t_{k,j})]}{\Delta t_{k,j}} d\tau \tag{28}$$

(here the notation $i(\tau; t_{k,j})$ is used to emphasize the difference between the variable τ and the parameter $t_{k,j}$). $t_{k,j}$ affects only one component of the input vector i, namely i_k; let i_k be connected to node n_1 in the original circuit and to node n_2 in the macromodel, and let λ_{n_1} and λ_{n_2} be the corresponding Lagrange multipliers. Then:

$$\begin{aligned}
&H[v(\tau), \lambda(\tau), i(\tau; t_{k,j} + \Delta t_{k,j})] - H[v(\tau), \lambda(\tau), i(\tau; t_{k,j})] \\
&= \lambda_1^T(-f_1[v_1(\tau)] + B_1 i(\tau; t_{k,j} + \Delta t_{k,j})) \\
&\quad - \lambda_1^T(-f_1[v_1(\tau)] + B_1 i(\tau; t_{k,j})) \\
&\quad + \lambda_2^T(-f_2[v_2(\tau)] + B_2 i(\tau; t_{k,j} + \Delta t_{k,j})) \\
&\quad - \lambda_2^T(-f_2[v_2(\tau)] + B_2 i(\tau; t_{k,j})) \\
&= \lambda_1^T B_1[i(\tau; t_{k,j} + \Delta t_{k,j}) - i(\tau; t_{k,j})] \\
&\quad + \lambda_2^T B_2[i(\tau; t_{k,j} + \Delta t_{k,j}) - i(\tau; t_{k,j})] \\
&= (\lambda_{n_1} + \lambda_{n_2})[i_k(\tau; t_{k,j} + \Delta t_{k,j}) - i_k(\tau; t_{k,j})].
\end{aligned}$$

However, $i_k(\tau; t_{k,j} + \Delta t_{k,j}) - i_k(\tau; t_{k,j})$ is equal to zero, unless $\tau \in [t_{k,j}, t_{k,j} + \Delta t_{k,j}]$, in which case it is equal to $(-1)^{j+1}M_k$. Therefore, (28) becomes:

$$\begin{aligned}
\frac{\partial c}{\partial t_{k,j}} &= (-1)^{j+1} M_k \lim_{\Delta t_{k,j} \to 0} \frac{1}{\Delta t_{k,j}} \int_{t_{k,j}}^{t_{k,j}+\Delta t_{k,j}} [\lambda_{n_1}(\tau) + \lambda_{n_2}(\tau)]\, d\tau \\
&= (-1)^{j+1} M_k[\lambda_{n_1}(t_{k,j}) + \lambda_{n_2}(t_{k,j})].
\end{aligned}$$

This gives the same expression for $(\partial c/\partial t_{k,j})$ as the one obtained in [15] following a different line of proof.

IV. Saddle Point Location

This section explores the issues involved in solving (10) numerically. As pointed out in the previous section, finding the global minimum of a general real-valued function is a very challenging task, so that trying to solve (10) exactly is unrealistic. However, local minima are normally considered acceptable solutions of optimization problems; because global minima are also local minima, the new set of acceptable solutions includes the old one as a special case. In the same spirit, it can be noted that any solution of (10) must be a saddle point of the function c. This observation suggests the following idea: instead of trying to compute an exact solution, look for a saddle point of c, i.e., for a local solution of (10). This section presents an algorithm that can be used for this purpose.

It has been shown that the set $\mathcal{U}$ of admissible inputs can be parametrized by a finite dimensional vector $t = (t_{1,1}, \cdots, t_{m,N})$. Thus c depends in fact on two vector variables: x, the vector of macromodel parameters and t. Throughout this section it will be assumed that $c(x, t)$ is twice continuous differentiable; we introduce the joint variable $z = (x, t)^T$, and we denote the Hessian matrix of c by

$$F(z) = \frac{\partial c^2}{\partial z^2} = \begin{pmatrix} \dfrac{\partial c^2}{\partial x^2} & \dfrac{\partial c^2}{\partial x \partial t} \\ \dfrac{\partial c^2}{\partial t \partial x} & \dfrac{\partial c^2}{\partial t^2} \end{pmatrix} = \begin{pmatrix} P(z) & R(z) \\ R(z)^T & -Q(z) \end{pmatrix}. \tag{29}$$

It will be assumed that there exists a constant $m > 0$ such that $\langle P(z)w_1, w_1 \rangle \geq m\|w_1\|^2$, $\langle Q(z)w_2, w_2 \rangle \geq m\|w_2\|^2$ for all z, w_1, w_2, and $g(z)$ will denote the gradient of c: $g(z) = (\partial c/\partial z)^T$.

We also introduce the matrix:

$$U = \begin{pmatrix} I & 0 \\ 0 & -I \end{pmatrix}$$

and we note that the following inequality is satisfied:

$$\begin{aligned}
\langle w, UFw \rangle &= \tfrac{1}{2}\langle w, (UF + FU)w \rangle \\
&= (w_1^T \;\; w_2^T)\begin{pmatrix} P & 0 \\ 0 & Q \end{pmatrix}\begin{pmatrix} w_1 \\ w_2 \end{pmatrix} \\
&\geq m(\|w_1\|^2 + \|w_2\|^2) = m\|w\|^2.
\end{aligned}$$

In other words, the symmetric matrix $(1/2)(UF + FU)$ is positive definite uniformly in (x, t).

The algorithm used for locating a saddle point of c is a modification of one first proposed by Broyden as an iterative method to solve nonlinear systems of equations [21]. It generates a sequence of points $\{z_k\}$ and a sequence of matrices $\{H_k\}$ which under certain conditions converge to $\hat{z}$ and to $F^{-1}(\hat{z})$, respectively, where $\hat{z}$ is a stationary point of c. The justification behind the equation used to generate the matrices H_k is the following approximate equality, which can be obtained from the first-order Taylor expansion of $g(z)$ at the point z_{k+1}:

$$g(z_{k+1}) - g(z_k) \cong F(z_{k+1})(z_{k+1} - z_k).$$

This approximate equation is the reason why many algorithms used to generate successive approximations B_k to the Hessian of a function require that the following equation, known as the *secant relation*, be satisfied:

$$\Delta g_k = B_{k+1}\Delta z_k \tag{30}$$

where $\Delta g_k = g(z_{k+1}) - g(z_k)$ and $\Delta z_k = z_{k+1} - z_k$. Broyden's algorithm, in its original formulation, falls in that category; the matrices B_k are generated through the following recurrence relation, commonly known as *Broyden's update:*

$$B_{k+1} = B_k + \frac{(\Delta g_k - B_k \Delta z_k)\Delta z_k^T}{\langle \Delta z_k, \Delta z_k \rangle}. \tag{31}$$

It is important to point out that any sequence of matrices generated from (31) satisfies the secant relation (30): no additional assumptions on Δg_k and Δz_k need be made.

Our implementation of the saddle point location algorithm is based on a modified version of Broyden's algorithm, in which F^{-1} is approximated instead of F. This is obtained by letting $H_k = B_k^{-1}$ and by making use of the following result, known as the Sherman–Morrison formula [22].

Proposition 1: If the matrix A is nonsingular and $\langle v, A^{-1}u \rangle \neq -1$, the matrix $A + uv^T$ is also nonsingular, and its inverse is given by

$$(A + uv^T)^{-1} = A^{-1} - \frac{A^{-1}uv^T A^{-1}}{1 + \langle v, A^{-1}u \rangle}.$$

Applying this formula to (31) with $v = (\Delta z_k / \langle \Delta z_k, \Delta z_k \rangle)$ and $u = \Delta g_k - B_k \Delta z_k$ leads to the following updating relation for the matrices H_k [23, p. 345]:

$$H_{k+1} = H_k - \frac{(H_k \Delta g_k - \Delta z_k)\Delta z_k^T H_k}{\langle \Delta z_k, H_k \Delta g_k \rangle}. \tag{32}$$

The complete definition of the algorithm used to locate saddle points of c is the following.

Saddle Point Algorithm

Data: $H_0 = U$, z_0, ϵ.

Step 0: Set $k = 0$.

Step 1: Set $g_k = g(z_k)$, $d_k = -H_k g_k$.

Step 2: Compute $z_{k+1} = z_k + t_k d_k$ such that the equation $\langle U g_{k+1}, d_k \rangle = 0$ is satisfied.

Step 3: If $\| g_{k+1} \| < \epsilon$ stop. Else set:

$$\Delta z_k = z_{k+1} - z_k$$

$$\Delta g_k = g_{k+1} - g_k$$

$$H_{k+1} = H_k - \frac{(H_k \Delta g_k - \Delta z_k) \Delta z_k^T H_k}{\langle \Delta z_k, H_k \Delta g_k \rangle}$$

$$k = k + 1$$

and go to Step 1.

Gradient computations in Steps 1 and 2 are performed by solving (11) and (14) (forward and backward integration), and then using (18) to compute the derivatives of c with respect to changes in the input waveform and in the macromodel parameters as explained in the previous section.

The behavior of this algorithm can be summarized in the following theorem.

Theorem 4: Let $\{z_k\}$ be a sequence generated by the algorithm, and suppose that $\lim_{k\to\infty} z_k = \hat{z}$ and $\lim_{k\to\infty} H_k = F^{-1}(\hat{z})$. Then $\hat{z}$ is a stationary point for c.

Proof: Let $\hat{F} = F(\hat{z})$, $\hat{g} = g(\hat{z})$. It follows that:

$$0 = \lim_{k\to\infty} \langle U g_{k+1}, d_k \rangle = \lim_{k\to\infty} \langle U g_{k+1}, H_k g_k \rangle = \langle U\hat{g}, \hat{F}^{-1}\hat{g} \rangle$$
$$= \tfrac{1}{2} \langle \hat{g}, (U\hat{F}^{-1} + \hat{F}^{-1}U)\hat{g} \rangle.$$

$U\hat{F} + \hat{F}U > 0$ implies $U\hat{F}^{-1} + \hat{F}^{-1}U > 0$, as can be readily seen from the identity:

$$\hat{F}^{-1}(U\hat{F} + \hat{F}U)\hat{F}^{-1} = (U\hat{F}^{-1} + \hat{F}^{-1}U)$$

and from the fact that $A > 0$ implies $C^T A C > 0$ for any nonsingular matrix C. Therefore, we must have $\hat{g} = 0$. □

It should be pointed out again that in general $\hat{z}$ is not a global solution of (10). As a consequence, the macromodel arrived at by this algorithm is guaranteed to be optimal only locally, i.e., with respect to small variations in the input. Other algorithms (e.g., Monte Carlo methods) could conceivably be used to overcome this problem, but at the price of increasing the computational cost by several orders of magnitude.

V. Numerical Results

The macromodeling algorithm outlined in the previous sections has been implemented in a computer program that can handle both linear and nonlinear circuits. The starting point is a SPICE-like description of the circuit to be modeled, and a similar description of a user-supplied template for the macromodel. Here the word "template" means a circuit containing a number of elements (identified by the user) whose characteristics the program is allowed to modify. Because the program cannot change the topology of the macromodel, and because it is not guaranteed to find global optima for the values of the macromodel elements, the user's choice of a template can and in general does affect the accuracy of the final macromodel. Thus it is desirable to choose a template that captures the basic functionality of the circuit to be modeled. For this purpose some of the techniques described in the papers referenced in the introduction can be useful. For instance, the templates for two of the macromodels introduced later in this section are based on a linear single-pole approximation of an operational amplifier. The advantage of using the algorithm proposed in this paper is that element values and, more importantly, element nonlinearities need not be given accurately in the template: they will be adjusted automatically by the program in order to optimize the macromodel's performance. This fact is also exemplified in the templates for the operational amplifier macromodels that will be introduced later: the initial characteristics of some elements were linear, and a certain amount of nonlinearity was introduced by the program in order to match the behavior of the original circuit.

Modifications to the user-supplied template are carried out according to the algorithm described in the previous section. This requires the computation of time-domain sensitivities, which is performed using the adjoint network algorithm described briefly in Section III and more extensively in [17]. The program is written entirely in C, and the user interface is through NUTMEG, a front-end written for SPICE3 [24]; in this way, circuits can be described in the same format used by SPICE3. The input file contains a description of the circuit to be modeled, as well as a description of a template for the macromodel. In addition, the user specifies what macromodel parameters the program is allowed to modify: as mentioned in Section III, these can include the values of certain elements contained in the macromodel, such as resistances and capacitances, as well as interpolation points to be used in table lookup models for nonlinear devices. The final values of the macromodel parameters are saved in a file of the same format as the input deck, so that the macromodel can be used by any simulator understanding SPICE3 syntax.

To illustrate the performance of the algorithm, we report the results obtained in three test cases: the first example is a linear network which is a part of more complex circuit used to simulate the behavior of biological neurons on a digital computer [25]. One of the components of the neuron, the dendrite, is usually represented by a distributed RC line. Because distributed elements are difficult to simulate using standard techniques, the RC line itself is replaced by an RC ladder containing N identical cells (see Fig. 4). The element values are determined by the equations $G_1 = NG_a$, $G_2 = G_l/N$, $C_1 = C/N$, where G_a is the axial conductance of the distributed line, G_l is the longitudinal conductance, and C is the total capacitance. Obviously the larger N is, the closer the ladder approximates a distributed line. However, it is conceivable that improved accuracy could also be obtained by varying the element values in each cell, while keeping the number of cells fixed. This hypothesis was verified on the RC ladder shown in Fig. 4 with $N = 40$. The values of G_a, G_l, and C are 4.71e-2 mho, 1.13e-2 mho, and 0.452 F, respectively, giving values for G_1, G_2, and C of 1.88 mho, 2.83e-4 mho, and 1.13e-2 F. This ladder was in turn modeled with a smaller one containing only 3 cells (Fig. 5). The initial element values were computed using the above formulas with $N = 3$, giving an initial model with 3 identical cells, but all the element values were allowed to vary, for a total of 9 model parameters to be optimized. Both end nodes were inputs as well as outputs, with node 1 in Fig. 4 corresponding to node $m1$ in Fig. 5 and node $N + 1$ corresponding to node

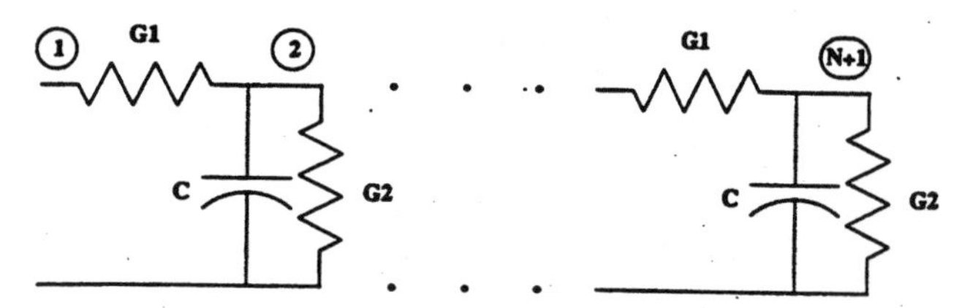

Fig. 4. *RC* ladder model for a neuron.

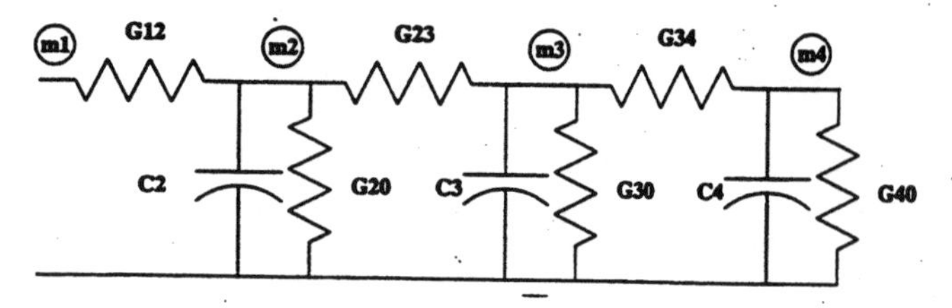

Fig. 5. Reduced *RC* ladder model for a neuron.

TABLE I
ELEMENT VALUES FOR THE OPTIMIZED NEURON MODEL

G_{12}	1.81e-1	mho
G_{20}	3.82e-3	mho
C_2	1.62e-2	F
G_{23}	6.15e-1	mho
G_{30}	3.78e-3	mho
C_3	1.59e-1	F
G_{34}	7.19e-2	mho
G_{40}	3.69e-3	mho
C_4	1.33e-1	F

*m*4. Optimizing this model took approximately 670 s of CPU time on a Microvax II running Ultrix™ V2.0. The element values of the optimized model are reported in Table I. The performance of the optimized model as a part of a complete neural circuit is displayed in Fig. 6, which shows the voltage waveforms at one end of the neuron modeled in three different ways: with an *RC* ladder of 40 identical cells (continuous line), with the optimized 3-cell model computed above (dotted line), and with a *RC* ladder of 3 identical cells (dashed line). The improvement is noticeable, both in the amplitude of the spike and in the shape of the descending transient: apart from a small amount of time lead, the waveform corresponding to the optimized model is almost an identical copy of the waveform generated by the 40 cell ladder. The difference in time is explained by the fact the spikes are generated by two highly nonlinear elements placed at both ends of the neuron: when the voltage at one end of the neuron reaches a certain threshold the spike is triggered. Because this threshold is reached by the 3-cell model before the 40-cell ladder, the corresponding spikes occur at different times.

The second example is a one stage CMOS operational amplifier containing nine transistors shown in Fig. 7; it is used as a comparator in a A/D converter [26]. The macromodel chosen to represent it is shown in Fig. 8: the voltage-controlled current source is nonlinear, and its characteristic is described by a table lookup model containing a total of 9 points. The program was allowed to vary the transfer characteristic of the controlled source, and also the values of the output resistance and capacitance. Initial values for all the elements were taken from a

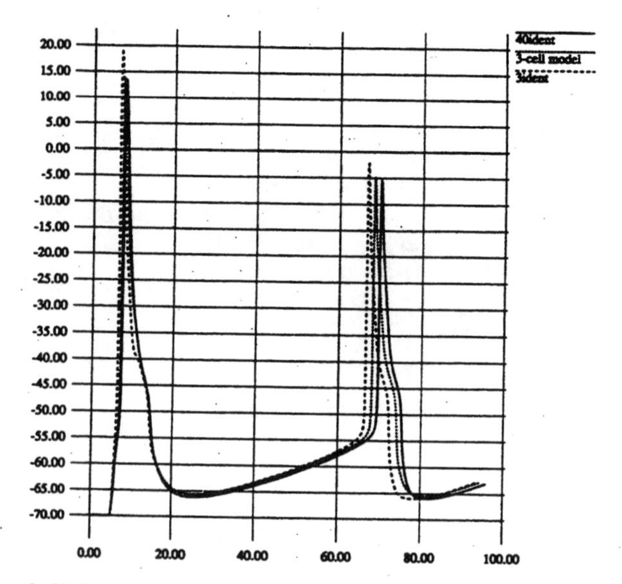

Fig. 6. Voltage waveforms corresponding to three different neuron models.

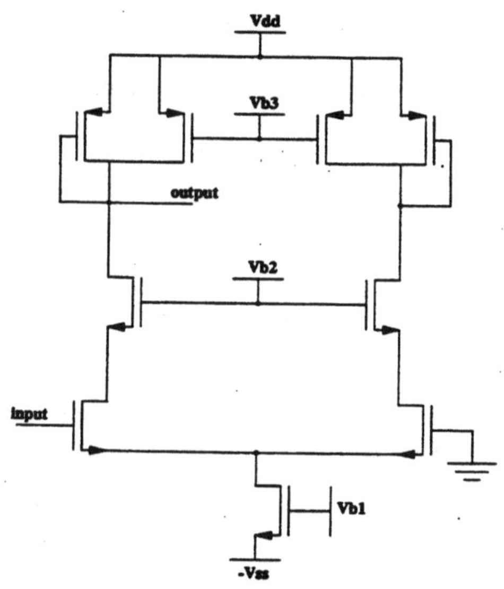

Fig. 7. Schematic of a CMOS operational amplifier.

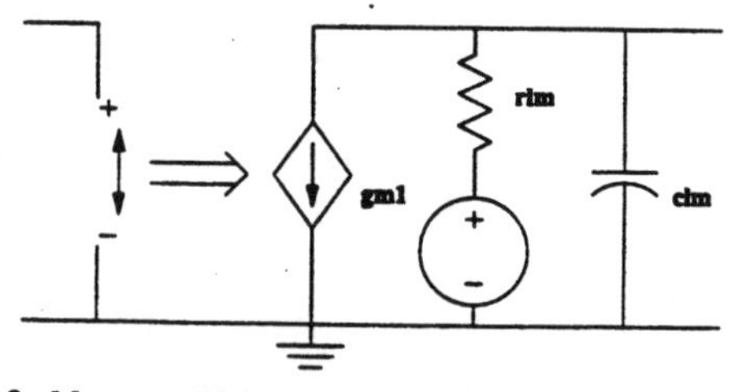

Fig. 8. Macromodel for the operational amplifier of Fig. 7.

small-signal analysis of the circuit performed by SPICE3. In particular, the initial transfer characteristic of the controlled source was linear, with a slope equal to the small-signal transconductance. The program used approximately 60 s of CPU time on a VAX 8800 running Ultrix V3.0 to modify the macromodel. The value of the output resistance was modified to 55 kΩ from 65 kΩ, and the output capacitance decreased to 35 pF from 45 pF. However, it is more interesting to compare the transfer characteristic of the current source before and after optimization

™Ultrix is a trademark of Digital Equipment Corporation.

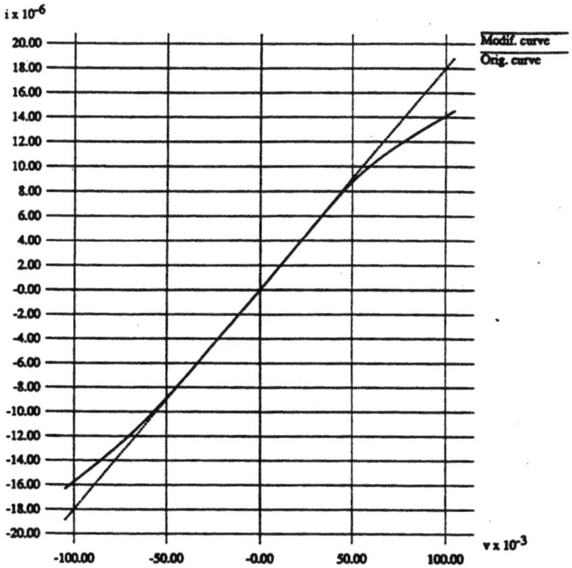

Fig. 9. Current source transfer characteristic.

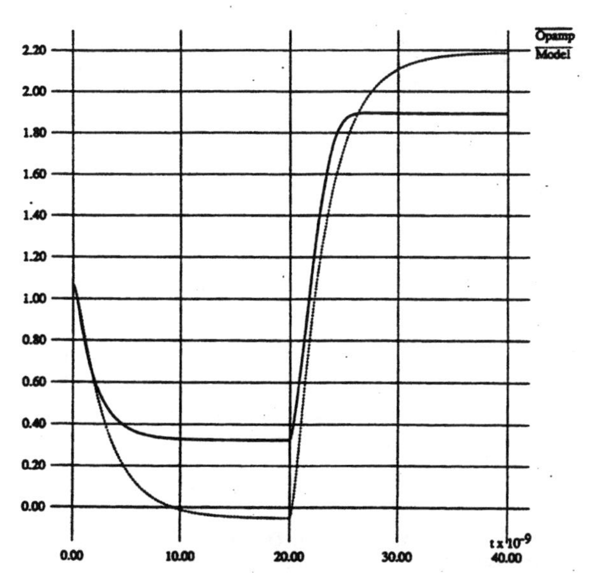

Fig. 10. Macromodel and operational amplifier outputs before optimization.

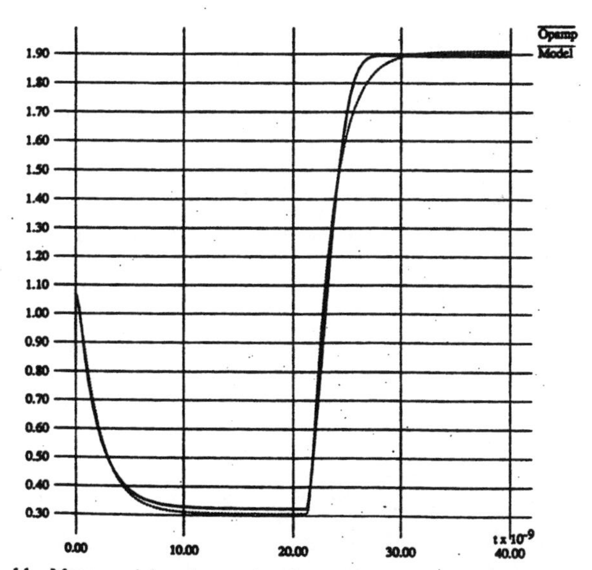

Fig. 11. Macromodel and operational amplifier outputs after optimization.

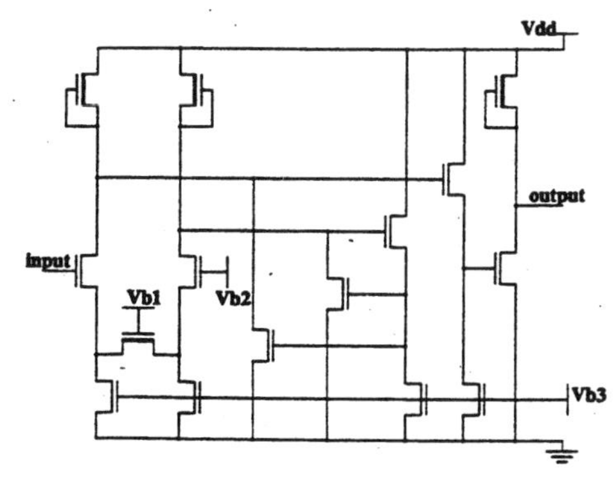

Fig. 12. Schematic of an NMOS operational amplifier.

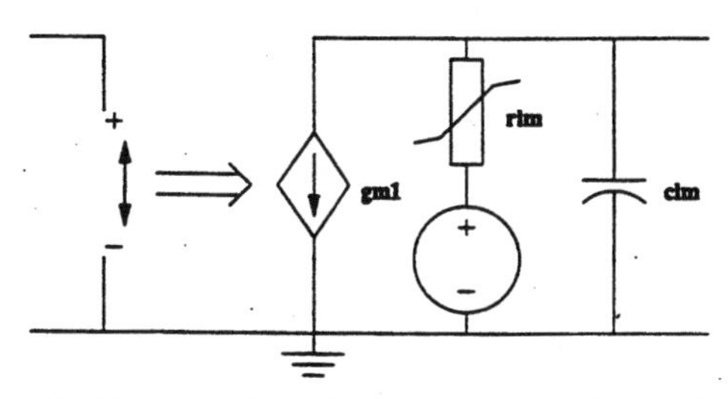

Fig. 13. Macromodel for the operational amplifier of Fig. 12.

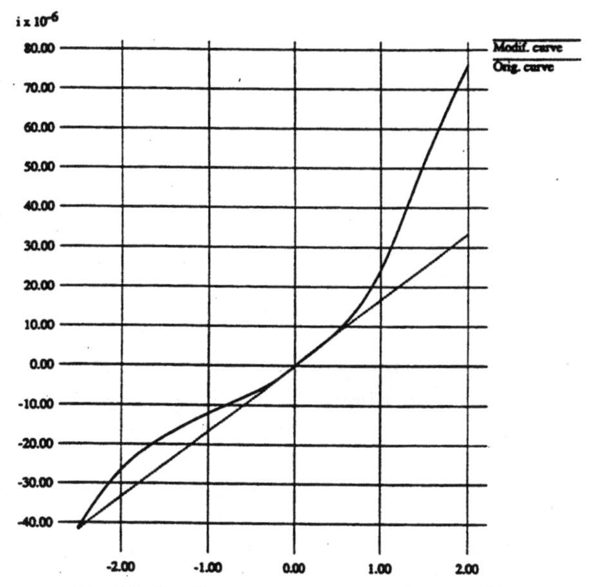

Fig. 14. Nonlinear output resistance characteristic.

(see Fig. 9): as expected, a certain amount of nonlinearity was introduced at both ends of the curve. Note that trying to compute an analytical expression for the final characteristic starting from the original circuit is practically impossible. The improvement in the accuracy of the macromodel can be evaluated by comparing Figs. 10 and 11. The first shows the outputs of the operational amplifier and of the macromodel before optimization: the input is a square waveform chosen arbitrarily as an initial guess to the worst-case input. The second refers to the optimized macromodel: its output is compared to the operational amplifier output when both are driven by the same input, which in this case is chosen by the program as the one that gives the largest possible difference.

The third example is a three-stage NMOS operational amplifier containing fifteen transistors, and it is shown in Fig. 12.

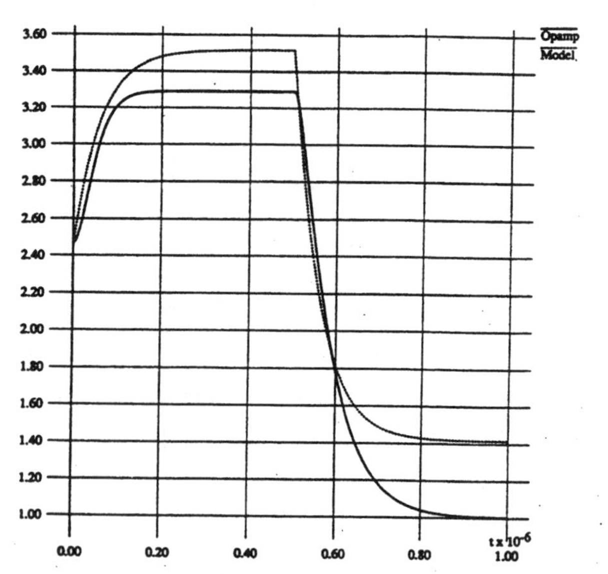

Fig. 15. Macromodel and operational amplifier outputs before optimization.

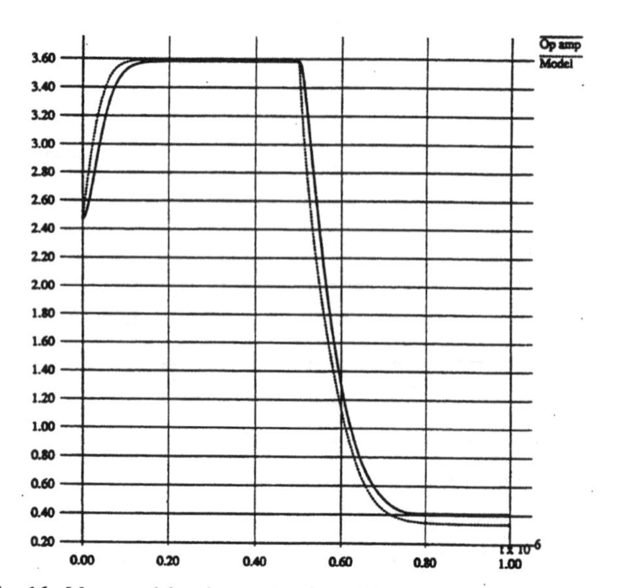

Fig. 16. Macromodel and operational amplifier outputs after optimization.

The macromodel is similar to the one used in the previous case: the only difference is that the output resistance is nonlinear (see Fig. 13). The macromodel parameters were again initialized to their small-signal values, and the program was allowed to change the value of the output capacitance and the characteristics of the current source and of the output resistance. Modeling took approximately 364 s of CPU time on a VAX 8800 running Ultrix 3.0: the initial and final characteristics of the nonlinear output resistance can be compared in Fig. 14. Comparisons of the macromodel output and of the operational amplifier output before and after optimization are shown in Figs. 15 and 16, respectively; as in the previous case, the improvement in the accuracy of the macromodel is clearly noticeable.

VI. Conclusions

The goal of this paper was to present a general purpose algorithm for the generation of macromodels suitable for circuit simulation. The algorithm is based exclusively on a comparison of the input–output behavior of the macromodel with that of the circuit to be modeled. Because no reliance on any particular properties of the circuit is made, the algorithm can be used to model a very wide class of circuits; in particular, even if we focused on the more challenging analog case, our macromodeling technique can also be applied to digital circuits. From a mathematical point of view, the approach followed here can be regarded as an extension of previous work on parameter optimization and sensitivity computation in the time domain [27], [16]. The main differences are that in this case the input is not determined *a priori* and that the formulas for the sensitivities are derived in a different way, based on an extension of the Hamiltonian formulation of a classical optimal control problem. This approach provides a unified setting for the computation of all sensitivities; in particular, it was shown that sensitivities with respect to changes in element values, in table lookup model parameters, and in input waveforms can all be obtained as special cases of a single equation.

The computer implementation of the algorithm uses the same interface as SPICE3, providing the user with a simple and familiar way to describe both the circuit to be modeled and the macromodel. The optimized macromodel is described in the same format, so that it can be used by any simulator understanding SPICE3 input files. Three examples were presented to demonstrate the algorithm's performance, two of which were chosen specifically to test the algorithm's ability to approximate nonlinearities in the circuits to be modeled. In all cases the accuracy of the macromodels was improved substantially using reasonable amounts of CPU time.

Acknowledgment

The first author would like to thank Hermann K. Gummel, Ajoy K. Bose, Hao N. Nham, and Chin F. Chen of AT&T Bell Laboratories, where he began to work on this research topic.

References

[1] G. R. Boyle, B. M. Cohn, D. O. Pederson, and J. E. Solomon, "Macromodeling of integrated circuit operational amplifiers," *IEEE J. Solid-State Circuits*, vol. SC-9, pp. 353–363, 1974.

[2] I. E. Getreu, A. D. Hadiwidjaja, and J. M. Brinch, "An integrated-circuit comparator macromodel," *IEEE J. Solid-State Circuits*, vol. SC-11, pp. 826–833, 1976.

[3] C. Turchetti and G. Masetti, "A macromodel for integrated all-MOS operational amplifiers," *IEEE J. Solid-State Circuits*, vol. SC-18, pp. 389–394, 1983.

[4] J. Ousterhout, "Switch-level delay models for digital MOS VLSI," in *Proc. 21st Design Automation Conf.*, June 1984, pp. 542–548.

[5] T. Lin and C. A. Mead, "Signal delay in general RC networks with application to timing simulation of digital integrated circuits," in *Proc. Conf. on Advanced Research in VLSI*, MIT, Jan. 1984, pp. 93–99.

[6] T. Tokuda *et al.*, "Delay-time modeling for ED MOS logic LSI," *IEEE Trans. Computer-Aided Design*, vol. CAD-2, pp. 129–134, 1983.

[7] M. D. Matson, "Macromodeling of digital MOS VLSI circuits," MIT Tech. Rep. VLSI Memo No. 84-212, Nov. 1984.

[8] S. R. Nassif and S. W. Director, "WASIM: A waveform based simulator for VLSICs," in *Proc. 1985 Int. Conf. on Computer-Aided Design*, Nov. 1985, pp. 29–31.

[9] W. M. Coughran, Jr., E. Grosse, and D. J. Rose, "CAzM: A circuit analyzer with macromodeling," *IEEE Trans. Electron. Devices*, vol. ED-30, pp. 1207–1213, 1983.

[10] C. Visweswariah, R. Chadha, and C. Chen, "Model development and verification for high level analog blocks," in *Proc. 25th Design Automation Conf.*, June 1988, pp. 376–382.

[11] G. Casinovi, "Macromodeling for the simulation of large scale analog integrated circuits," Ph.D. dissertation, Univ. of California, Berkeley, June 1988.

[12] J. M. Ortega and W. C. Rheinboldt, *Iterative Solution of Nonlinear Equations in Several Variables*. New York: Academic, 1970.

[13] A. E. Bryson, Jr. and Y. Ho, *Applied Optimal Control*. Waltham, MA: Ginn, 1969.

[14] G. M. Smirnov, *Oscillation Theory of Optimal Processes*. New York: Wiley, 1984.

[15] R. Gonzalez and E. Rofman, "On bang-bang control policies," in *Optimization Techniques, Part 2*, J. Cea, ed. New York: Springer-Verlag, 1976, pp. 587–602.

[16] G. D. Hachtel and R. A. Rohrer, "Techniques for the optimal design and synthesis of switching circuits," *Proc. IEEE*, vol. 55, pp. 1864–1877, 1967.

[17] R. K. Brayton and R. Spence, *Sensitivity and Optimization*. Amsterdam, The Netherlands: Elsevier, 1980.

[18] G. Casinovi, "Input-output approximations of dynamical systems," presented at 1989 IFAC Nonlinear Control Systems Design Symposium, Capri, Italy, June 14–16, 1989.

[19] R. A. Rohrer, "Fully automated network design by digital computer: Preliminary considerations," *Proc. IEEE*, vol. 55, pp. 1929–1939, 1967.

[20] C. de Boor, *A Practical Guide to Splines*. New York: Springer-Verlag, 1978.

[21] C. G. Broyden, "A class of methods for solving nonlinear simultaneous equations," *Math. Comput.*, vol. 19, no. 92, pp. 577–593, 1965.

[22] A. S. Householder, *The Theory of Matrices in Numerical Analysis*. New York: Blaisdell, 1964.

[23] M. Avriel, *Nonlinear Programming: Analysis and Methods*. Englewood Cliffs, NJ: Prentice-Hall, 1976.

[24] T. L. Quarles, "SPICE3 version 3c1 user's guide," Tech. Rep. UCB/ERL M89/46, Univ. of California, Berkeley, 1989.

[25] L. J. Borg-Graham, "Modelling the somatic electrical response of hippocampal pyramidal neurons," Master's thesis, Massachusetts Institute of Technology, May 1987.

[26] J. Doernberg, "High speed analog-to-digital conversion using 2-step flash architectures," Tech. Rep. UCB/ERL M89/46, Univ. of California, Berkeley, 1989.

[27] S. W. Director and R. A. Rohrer, "The generalized adjoint network and network sensitivities," *IEEE Trans. Circuit Theory*, vol. CT-16, pp. 318–323, 1969.

Giorgio Casinovi (M'89) received the B.S. degrees in electrical engineering and in mathematics from the University of Rome, Italy, in 1980 and 1982, respectively. In 1984 he received the M.S.E.E. degree and in 1988 the Ph.D. degree in electrical engineering, both from the University of California, Berkeley.

He joined the faculty of the School of Electrical Engineering of the Georgia Institute of Technology as an Assistant Professor in 1989. His research interests include computer-aided design of analog integrated circuits, numerical analysis, optimal control theory, and optimization.

Alberto Sangiovanni-Vincentelli (M'74–SM'81–F'83) for a photograph and biography please see page 3 of the January 1991 issue of this TRANSACTIONS.

Consistency Checking and Optimization of Macromodels

Yun-Cheng Ju, Vasant B. Rao, *Member, IEEE,* and Resve A. Saleh, *Member, IEEE*

Abstract—**Circuit macromodeling has been in use for many years to improve the efficiency of simulation at the expense of some accuracy. While this approach has been found to be quite effective, the success or failure of the method depends critically on how closely the macromodels approximate the actual behavior of the functional blocks. This paper describes a systematic methodology for automatic consistency checking and optimization of parameterized macromodels. Initially, an overview of the entire macromodel verification system, iMAVERICK, is provided. An efficient stochastic algorithm to optimize the parameters of a given macromodel is then described and a number of characteristics of the algorithm are presented. Next, a waveform comparison technique for evaluating the performance of the proposed configuration in the optimization process is described. Finally, several examples of macromodel optimization using the iMAVERICK system are presented.**

I. Introduction

CIRCUIT simulation has been proven to be an extremely useful CAD tool for circuit designers. However, time-domain simulation of certain types of circuits such as switched-capacitor filters (SCF) and phase-locked loops (PLL) can be very time consuming. The excessive simulation costs make such a simulation impractical. Clearly, more efficient methods are required to verify the performance of these types of circuits, especially if they are to be simulated often. One common approach is to break the circuit up into functional blocks, such as the voltage-controlled oscillator, low-pass filter, phase comparator, amplifier, etc., (in the case of a PLL) and to replace each functional block by a corresponding higher level model called a *macromodel* [1], [2], [3]. The simulation is then performed using the macromodels to greatly reduce the overall simulation time at the expense of some accuracy. While this approach has been found to be quite effective, the success of the method depends critically on how closely the macromodels approximate the actual behavior of the functional blocks they represent. Therefore, each macromodel must somehow be verified against the corresponding transistor level circuit to establish both consistency and accuracy before it is used. This process, until very recently [4], was usually performed manually. The goal of our research is to develop a systematic methodology for automatic consistency checking and optimization of time-domain macromodels for efficient and accurate simulation.

Manuscript received April 18, 1990. This work was supported by the Digital Equipment Corporation and the Semiconductor Research Corporation. This paper was recommended by Associate Editor A.E. Ruehli.

Y.-C. Ju is with the Coordinated Science Laboratory, Department of Computer Science, University of Illinois, Urbana, IL 61801.

V.B. Rao and R.A. Saleh are with the Coordinated Science Laboratory, Department of Electrical and Computer Engineering, University of Illinois, Urbana, IL 61801.

IEEE Log Number 9143312.

This paper begins with a brief overview of our macromodels optimization and verification system called iMAVERICK. The optimization loop in iMAVERICK involves a performance comparison of a macromodel against its transistor level circuit and the adjustment of the macromodel parameters until the desired level of accuracy is obtained. This procedure ensures that the macromodels are both reliable and accurate and gives the designer a higher level of confidence in the simulation results.

The remainder of this paper focuses on two key aspects of the system: the optimization algorithm and consistency checking. A new stochastic optimization scheme has been developed that is well suited for our application. This new scheme is shown to be effective and efficient for a number of test problems. A consistency checking approach based on waveform comparison techniques is then presented. It is used to derive the ''cost'' of an associated configuration defined by the parameters selected in a given step of the optimization process. A new waveform operator is developed for this special application and its properties are presented.

II. Overview of the iMAVERICK System

2.1. The Macromodeling Approach

Several techniques have been used over the years to build general and/or special-purpose macromodels. Of recent interest is the automatic generation of macromodels [5], but this work is still in progress. Since circuit designers are usually capable of producing reasonable macromodels based on their understanding of the circuit operation and function, a more practical solution is to build a complete library of macromodels. Ideally, a macromodel library should offer a variety of different configurations for each circuit block. They should range from very simple models to increasingly complex ones in which critical second-order effects are modeled. Each macromodel should be flexible enough so that its parameters can be adjusted to modify its performance and to improve its accuracy for a given application. Of course, the accuracy of the macromodel relative to the transistor level circuit depends on how much time is spent on optimizing these parameters, a rather tedious and time-consuming task if done manually. The main issue addressed in this paper is not the automatic construction of a macromodel but rather the choice of parameters for a given macromodel to accurately model the behavior of the corresponding transistor level circuit.

2.2. *Overview of the Macromodel Verification and Optimization System*

Visweswariah *et al.* [4] described an open loop system called ACME to verify the correctness of a high-level analog model. A graphical comparison was used to compare the waveforms generated from both the transistor level circuit and the corresponding macromodel. However, no optimization capabilities were provided in this open-loop system. Casinovi [3] developed a general-purpose algorithm to automatically optimize a given macromodel. Briefly, the given optimization problem is transformed into an approximation problem in a space whose elements are mappings between function spaces. While this approach is efficient and rigorous, only a local optimum is guaranteed. However, the basic idea is attractive. In fact, our work can be viewed as combination of the two approaches in [3] and [4].

We now describe a closed-loop verification and optimization system called iMAVERICK, which not only ensures the validity of each macromodel provided, but also has the capability of optimizing the macromodels to improve accuracy. The system, as illustrated in Fig. 1, consists of two paths. The circuit designer initially provides a transistor level description of the circuit block under consideration, a time-domain macromodel with a number of adjustable parameters, and a set of target specifications. The input excitations used to verify the transistor circuit and a reasonable range for each parameter are also provided by the designer. It is assumed that these input excitations, which are used for circuit simulation, capture all the important performance characteristics needed in the macromodel so that it can be used in place of the real circuit in its particular application. Casinovi [3] has developed an algorithm to identify and generate the worst-case input excitations which can also be used for this purpose.

The two sets of waveforms, generated by a circuit simulator and a macromodel simulator, are used as input to a waveform consistency checker. Any number of waveforms and target specifications can be provided to the system. The output waveforms can be time-domain and/or frequency-domain responses of the circuit. The consistency checker performs a set of comparisons between two sets of waveforms and returns a value indicating the relative proximity of the waveforms. If the macromodel does not compare favorably with the transistor level circuit based on these waveforms, the parameters are adjusted and the simulation and consistency checking cycle is repeated. This optimization process continues until the desired level of accuracy is achieved or the maximum allowable CPU-time is exceeded.

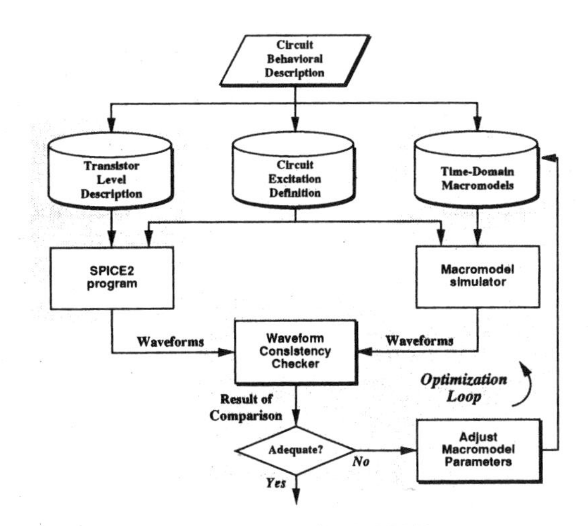

Fig. 1. Overview of the iMAVERICK system.

In order to make the procedure efficient and accurate, the following issues must be addressed.

1) For efficiency: The optimization loop attempts to find the best configuration by generating and evaluating a number of new candidate configurations. However, in order to evaluate these candidates, at least one simulation and a set of waveform comparisons must be performed on each new macromodel configuration. Since simulation and waveform comparison are both expensive, the primary goal is to *reduce the number of evaluations* while preserving the capability of finding the best configuration.

2) For accuracy: The performance and accuracy of each new macromodel configuration generated is evaluated by checking the proximity of the waveforms. Poor error measures will generate inaccurate and inconsistent results and may mislead the optimization process. A suitable waveform error measuring method for this application must be derived. Previous approaches in this area have only used direct comparison methods which are not effective in our application.

We show in the remaining sections of this paper that iMAVERICK achieves these two goals.

III. A Parameter Optimization Algorithm

3.1. *Motivation for a New Algorithm*

Given a macromodel and a reasonable min/max range for every parameter by the circuit designer, the optimization problem in our case is to determine the set of macromodel parameters that provide the most accurate representation of a corresponding transistor circuit. Several characteristics of the optimization problem are as follows.

1) It is constrained nonlinear programming problem with (possibly) multiple optimal solutions.

2) There is no closed form analytic expression available for either the cost function or its gradients (although it could be computed numerically at some additional expense).

3) Evaluation of the cost function is a fairly expensive process and should be avoided as much as possible.

4) The objective function is smooth in the sense that a very small change in a single parameter does not result in a large change in the value of the objective function.

Conventional deterministic algorithms for nonlinear problems such as the steepest descent method or Newton's method [6] are not appropriate for our application due to observation 2) above. An exhaustive search is too computationally prohibitive since the search space is a continuous n-dimensional space, where n is the number of parameters to be adjusted in the macromodel. An algorithm known as simulated annealing [7] has proven to be a powerful and general-purpose algorithm for solving a variety of combinatorial problems such as the *traveling salesman problem*, the *VLSI placement problem* [8], and certain nonlinear problems [9]. However, it is very slow and inefficient for our application where the evaluation of the performance of a macromodel configuration is fairly expensive since each evaluation of the cost function requires one or more complete simulations of the macromodel and waveform post-processing operations. Unless the number of cost function evaluations can be significantly reduced, the optimization procedure will be quite impractical for use by circuit designers. Therefore, because of the characteristics of our problem and the lack of an appropriate optimization method, we developed and implemented a new approach in iMAVERICK.

3.2. Conjectural Cost Function

A new heuristic algorithm has been developed to solve the macromodel optimization problem described above. The algorithm is iterative and stochastic in nature and attempts to evaluate the objective function at as few points as possible (to save computation time) while maintaining the ability to explore all possible points in order to find the globally optimal point. It also has the feature that only the value of the objective function is computed at each sample point; the calculation of its derivatives is not required. It uses past sample points to *predict* the behavior of the objective function and to decide whether a given region of the search space is worth revisiting. A heuristic measure of the likelihood that a cost evaluation will provide useful information or a better solution is introduced in the algorithm.

Fig. 2 illustrates the basic idea behind our new algorithm. The algorithm dynamically constructs a piecewise constant *conjectural cost function* and then uses this function to partition the search space into several regions. Note that each region has only one actual sample point (solid dot). In this example, we have three regions based on three sampled points. Region 3 is the "best," followed by Region 1, and finally Region 2 is the "worst." For each region, a probability, P_{Evaluate}, can be computed based on many factors including the cost at the associated sample point. This probability is used to decide whether or not a cost evaluation at a new sample point in that region should

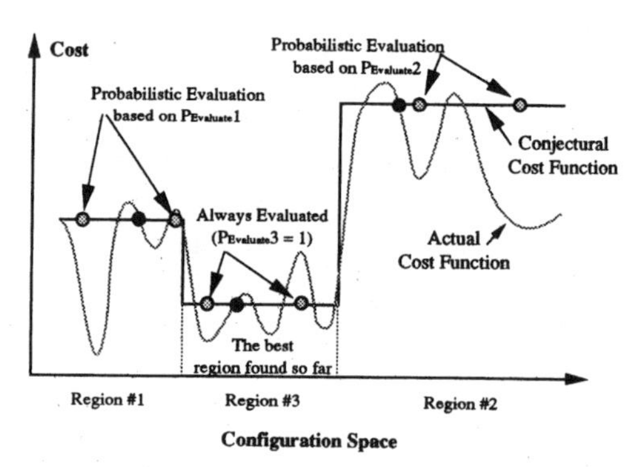

Fig. 2. Conjectural cost function.

be performed (shaded dots). For example, in Fig. 2 we may find that $P_{\text{Evaluate}}3 > P_{\text{Evaluate}}1 > P_{\text{Evaluate}}2$. We require that $P_{\text{Evaluate}}3 = 1.0$ since Region 3 is the "best" found so far. Thus any new configuration generated in Region 3 will always be evaluated while those in Region 1 and Region 2 may be evaluated depending on the value of P_{Evaluate} for those regions. New configurations in regions with higher probabilities have a better chance of being evaluated. This method is different from the basic approach in simulated annealing where a new configuration is always *evaluated* and then a probabilistic decision is made as to whether or not the new configuration should be *accepted* or *rejected*. Our cost evaluations are too expensive to do this! However, our new optimization algorithm could be viewed as a variation of simulated annealing with a directed search in which the method is applied to a *conjectural cost function*, i.e., an approximate cost function, built by our program incrementally using past sample points. The conjectural cost is used to *accept* or *deny* actual cost function evaluations (simulations and waveform comparisons) for new configurations.

3.3. A Heuristic Stochastic Algorithm

We now describe the details of our new stochastic algorithm and the probability function associated with it. Consider the problem of seeking the global minimum of an objective function $f\colon \Omega \subset R^n \to R$, where n is the dimension of the search space and Ω is a user-defined compact region. In our case, Ω is assumed to be an n-dimensional *hypercube*. For example, when $n = 2$, Ω is simply a rectangle. The algorithm proceeds as follows: first, a random sample point x_1 is selected in Ω and the objective function is evaluated to be $c_1 = f(x_1)$. Next, as iteration 1 begins, a point x_2 is selected at random and the region Ω is divided into two subregions Ω_1 and Ω_2 such that each Ω_i is represented by its sample point x_i for $i = 1, 2$. The cost at $c_2 = f(x_2)$ is evaluated at x_2. Let n_k denote the number of existing subregions at the beginning of iteration k. Hence, so far $n_1 = 1$ and $n_2 = 2$. Eventually, region Ω becomes a collection of disjoint subregions $\Omega_1, \Omega_2, \cdots, \Omega_{nk}$, where each region Ω_i is associated with a sample point x_i of cost $c_i = f(x_i)$. Let $c_{\min}$ and $c_{\max}$ denote the minimum and maximum cost, respectively,

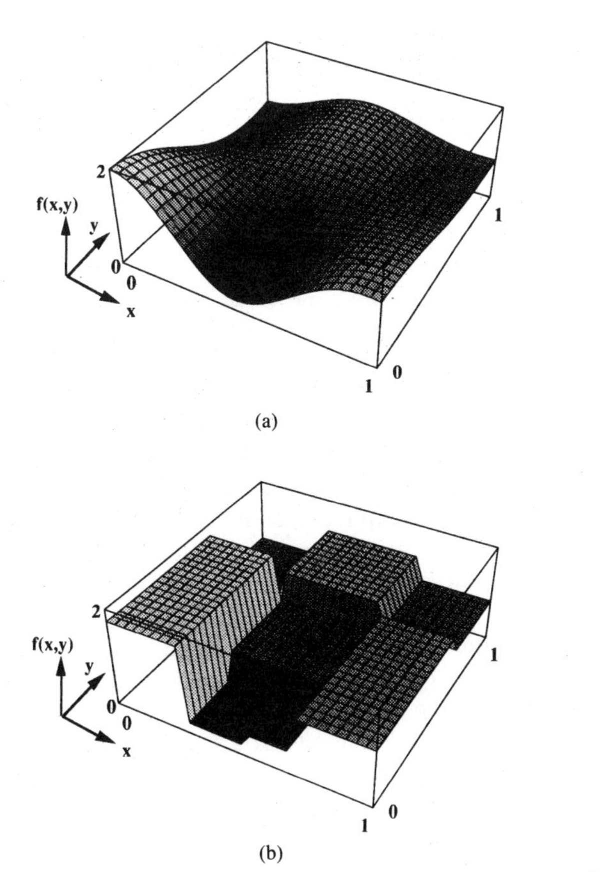

Fig. 3. Conjectural cost function for $f(x, y)$. (a) $f(x, y) = 1 + \exp(-x^2 - y^2) \cos(2\pi x) \cos(\pi y)$. (b) Conjectural cost function generated after 8 function evaluations.

among the $c_1, c_2, \cdots, c_{n_k}$ values evaluated thus far. Fig. 3(a) shows a test cost function

$$f(x, y) = 1 + \exp(-x^2 - y^2) \cos(2\pi x) \cos(\pi y) \quad (1)$$

and the conjectural cost function obtained by our algorithm after the first 8 function evaluations is shown in Fig. 3(b).

The kth iteration proceeds as follows. A new point x is selected randomly from Ω. Suppose the new point falls in region Ω_i. The algorithm now computes a probability P_{Evaluate} associated with the region Ω_i based on the values of c_i, $c_{\min}$, $c_{\max}$, n_k, and the number of times the algorithm has attempted to explore this region but has failed. Details on exactly how this probability P_{Evaluate} is computed are given in the next section. This probability P_{Evaluate} is then compared against a random number $r \in [0,1]$ to *decide whether* or *not to perform a cost evaluation* (i.e., simulation) at the new point $x \in \Omega_i$. If $r < P_{\text{Evaluate}}$, a cost evaluation is performed at x and Ω_i is subdivided into two new disjoint subregions such that x and x_i fall into separate regions. Since there will now be a total of $n_k + 1$ subregions we set $n_{k+1} = n_k + 1$. The subregion containing x_i remains labeled as Ω_i, while the other region is labeled as $\Omega_{n_{k+1}}$ with a sample point $x_{n_{k+1}} = x$ of cost $c_{n_{k+1}}$. The algorithm then updates $c_{\min}$, $c_{\max}$ and proceeds to the next iteration. If, however, $r > P_{\text{Evaluate}}$, a cost evaluation is NOT performed at the new point leaving the subregions unmodified. In this case $n_{k+1} = n_k$, and the algorithm proceeds to the next iteration by choosing another random sample point. Clearly, the number of cost evaluations performed by this algorithm in k iterations is $n_k \leq k$. In practice we have observed that $n_k << k$ and this provides the key advantage of our approach. For example, in one test case only $n_k = 3000$ cost evaluations were performed in $k = 200\,000$ trials, which means that 98.5% time saving was achieved, thus satisfying observation (3) stated in Section III-3.1.

The overall heuristic search algorithm is outlined in the following, where c_x* is the cost at random starting point

while (($c_x* > \epsilon$) and (max CPU-time not exceeded)) {
  Randomly generate a new point $x \in \Omega$;
  Find subregion Ω_i containing x;
  Compute probability P_{Evaluate} for Ω_i;
  if (random [0,1] $< P_{\text{Evaluate}}$) {
    Evaluate the cost function at x; /* *perform simulation* */
    if ($c_x < c_{x*}$) $x* \leftarrow x$; /* *keep best solution so far* */
    Subdivide region Ω_i;
  else { /* *skip this point* */ }
}
return ($x*$).

3.4. *The Probability Function,* P_{Evaluate}

The behavior of the above algorithm is essentially controlled by the choice of the probability, P_{Evaluate}. Forcing $P_{\text{Evaluate}} = 1$ simply evaluates the configuration at every trial point or, equivalently, leads to a blind search. This gives $n_k = k$ and no savings in the number of cost evaluations. Recall that the overall goal is to avoid excessive evaluations of the cost function without affecting the quality of the final solution. It is, therefore, desirable to choose P_{Evaluate} intelligently to avoid the evaluation of sample points in "bad" regions. In other words, the regions around the global minimum should ideally be sampled more frequently than other regions. With this goal in mind, the following assertions can be made.

1) Since the only information we have about the behavior of the objective function is the set of sample points found so far, a region is assumed to be "good" if the sample point in that region happens to have a "small" cost.

2) Larger probabilities should be assigned to "good" regions to improve the likelihood of revisiting them.

3) Large regions should be assigned larger probabilities even though they may be considered "bad" because our knowledge about these regions may not be accurate. The chances are high that there may be some small, "good" regions hidden in those large, "bad" regions.

4) The probability associated with any region that has not been explored for a long time should be high (increased). The heuristic algorithm may underestimate the potential of having "good" solutions in current "bad"

regions. However, by increasing the probability value associated with those regions, the algorithm is guaranteed to eventually explore any region later no matter how "bad" that region is currently.

Since there is a CPU-time limit on the optimization process, a limit is placed on the maximum number of samples that iMAVERICK can perform. Therefore, one other important property to include in our probability function is the ability to accept many random samples initially, and then increase its selectivity as it proceeds. This is because little is known about the configuration space initially, but the knowledge improves as more and more samples are taken. Therefore, it can afford to be greedy about the locations of the samples near the end of the optimization process. The probability function used in simulated annealing has this property and forms the basis for probability used in iMAVERICK.

The precise computation of the probability P_{Evaluate} is now presented. Suppose the algorithm is at iteration k and has chosen a random point x in region Ω_i with an existing sample point x_i of cost c_i. Let η_i be the number of times the algorithm has tried to explore region Ω_i but has failed, and let *max_rejection_allowed* be the maximum number of times a region can be visited without being evaluated at a trial sample point. The probability is then computed as

$$P_{\text{Evaluate}} = \begin{cases} \exp \dfrac{-\alpha(c_i - c_{\min})}{c_{\max} - c_{\min}}, & \text{if } \eta_i < \textit{max\_rejection\_allowed} \\ 1, & \text{if } \eta_i \geq \textit{max\_rejection\_allowed}. \end{cases} \quad (2)$$

For the reader familiar with simulated annealing for combinatorial optimization, the parameter α plays the role of $1/T$ where T is the temperature parameter. In a similar way, our algorithm starts by setting α to be a small value at the beginning of the optimization process and increases this parameter gradually as the process evolves like the cooling schedule of simulated annealing. A small value of α forces many cost evaluations to be performed whereas a large value of α chooses only the best regions for cost evaluations. This is fairly intuitive because at the beginning of the search, information about the behavior of our objective function is very limited, and therefore, the algorithm should explore new regions to improve its knowledge. However, the accuracy of the prediction increases as the number of sample points increases.

The predefined limit *max_rejection_allowed* is used as a safety factor to ensure that those regions not visited recently will be explored later if a sample point is randomly chosen in those regions, even if those regions are currently considered to be "bad." Our algorithm keeps track of the number of times the algorithm selects a point in a region Ω_i but fails to evaluate the cost function. When the number associated with Ω_i exceeds this limit, the probability P_{Evaluate} will be set to 1, which will force the algorithm to further explore that region. This satisfies requirement 4) listed earlier. Even though the size of each subregion is not explicitly shown in (2), our algorithm automatically favors larger subregions because larger subregions are more likely to be explored since each sample point is chosen in region Ω randomly.

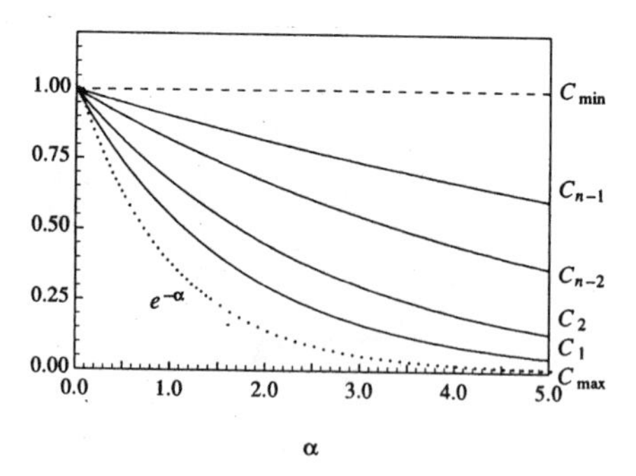

Fig. 4. Probability curves for a range of costs as a function of α.

The other term in the exponent, namely $(c_i - c_{\min})/(c_{\max} - c_{\min})$, is a normalized weighting factor for α that lies in the range [0,1]. It is used to favor "good" regions over "bad" regions. If a sample point in the best region is chosen, the exponent evaluates to 0 since $c_i = c_{\min}$, and hence, $P_{\text{Evaluate}} = 1$. Therefore, any point picked in this region Ω_i will always be evaluated. However, if $c_i = c_{\max}$, the exponent evaluates to $-\alpha$. Clearly, intermediate regions will generate probabilities that are bounded above by 1 and below by $e^{-\alpha}$. Fig. 4 shows these probability curves for a range of costs as a function of α. As α increases from 0 to ∞, the optimization procedure becomes more and more selective and prefers "good" regions over "bad" regions.

3.4.1. The Role of α: As an example of the influence of α, consider the following test function:

$$f(x, y) = 1 + \exp(-x^2 - y^2) \cos(10\pi x) \cos(5\pi y) \quad (3)$$

over the region $\Omega = [0,1] \times [0,1]$. This function is plotted in Fig. 5 and has a unique global minimum near the point (0.1, 0.0). It has, however, several local minima near points $(0.2k + 0.1, 0.4j + 0.2)$ where $k = 0, 1, 2, 3, 4$, and $j = 0, 1, 2$. The global minimum is very difficult for any deterministic algorithm to find.

The results of four different optimization runs after 5000 cost evaluations are shown in Fig. 6 for different choices of the parameter α in each case. A small α assigns similar probabilities to both good and bad regions. In the first two cases of $\alpha = 0$ and $\alpha = 2.0$, the search space is sampled almost uniformly. Therefore, the search scheme may be much slower, but it is more likely to identify the region containing the globally optimal solution after a sufficiently large number of samples. If α is chosen to be large (i.e., the probabilities assigned to "good" regions are much higher than those assigned to "bad" regions), the search scheme is faster in finding "good" solutions but it has higher risk of missing the globally optimal solution in the first few visits because the globally optimal solution

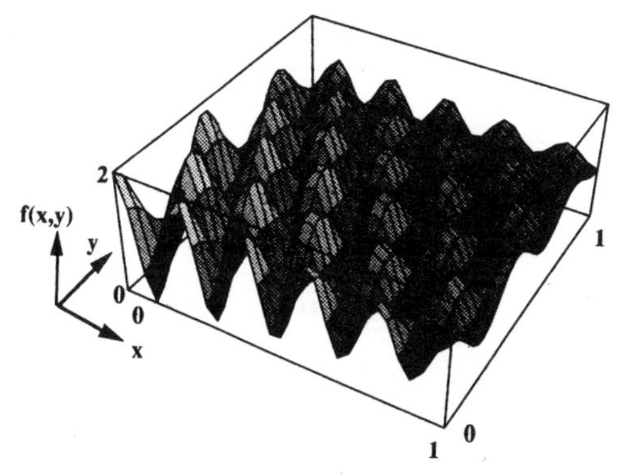

Fig. 5. $f(x, y) = 1 + \exp(-x^2 - y^2) \cos(10\pi x) \cos(5\pi y)$.

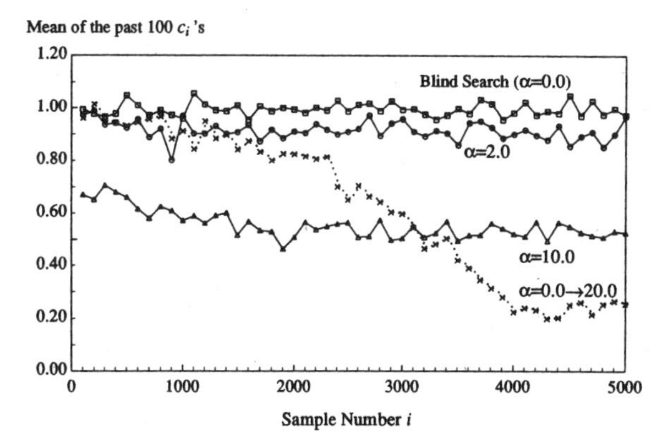

Fig. 7. Quality of sample points for each scheme versus sample number i for test function (3).

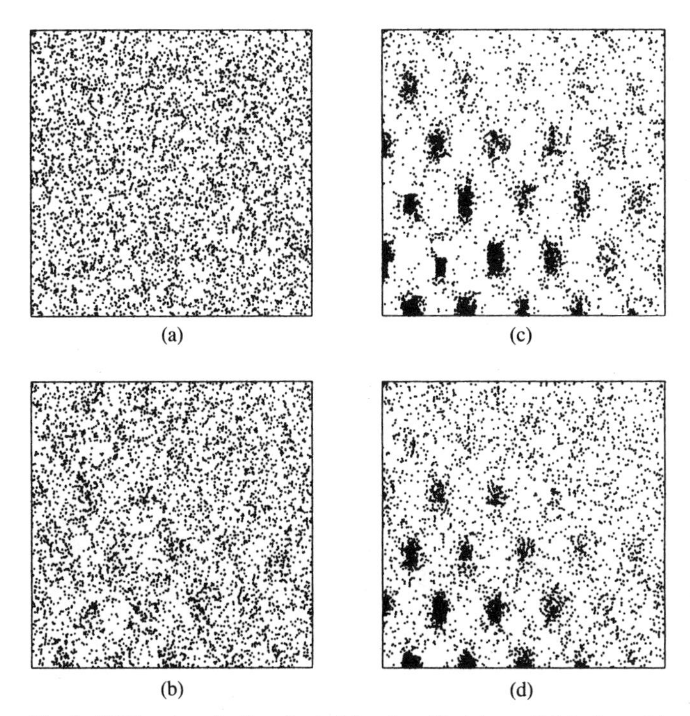

Fig. 6. 5000 cost evaluations for test function (3). (a) $\alpha = 0$ (blind search). (b) $\alpha = 2.0$. (c) $\alpha = 10.0$. (d) $\alpha = 0.0 \rightarrow 20.0$.

may hide in large, "bad" regions. For $\alpha = 10.0$, the sampling is significantly increased around the global minimum and the local minima at which the value of the objective function is close to that of the global minimum.

Based on the above cases, the most effective procedure is to adjust α as the optimization progresses from a small value to a large value. This is analogous to a cooling schedule in simulated annealing. A variety of schemes can be used to vary α. Perhaps the simplest scheme is to hold $\alpha = 0$ as part of a blind search initially, and then vary it linearly after a sufficient number of samples have been obtained:

$$\alpha_k = \begin{cases} 0, & \text{if } n_k \leq \dfrac{n_{\max}}{10} \\ \alpha_{\max} \dfrac{(n_k - 0.1 * n_{\max})}{(0.9 * n_{\max})}, & \text{if } n_k > \dfrac{n_{\max}}{10} \end{cases} \qquad (4)$$

where n_k is the number of subregions at iteration k and $n_{\max}$ is the maximum number of cost function evaluation allowed. This scheme used 10% of the samples to build the knowledge base and the remainder of the samples to find the best solution. More sophisticated sampling schemes such as Hadamard sampling [10] and Latin Hypercube sampling [11] can be used instead of the blind search, but this is not implemented in iMAVERICK at present. The value $\alpha_{\max}$ is the final value of α used after $n_{\max}$ samples. Fig. 6 shows an example that varies α from 0 to a maximum of 20.[1]

One way of illustrating the effectiveness of a particular scheme involving α is to examine the quality of the sample points picked by plotting the cost, c_i, associated with sample points, x_i. The higher the quality of the sample points, the more effective the scheme in locating "good" regions. The mean values for 100 consecutive costs are plotted in Fig. 7 as a function of the sample number i. It can be seen that a blind search with $\alpha = 0$ has the highest (i.e., worst) mean value of 1.0, which is the same as the mean of the test function in Ω. By contrast, for $\alpha = 10.0$, a much better set of sample points are found. More importantly, the key feature of the adaptive scheme that uses (4) is observed in the figure; namely, the quality of its sample points becomes better and better as the number of sample points becomes larger.

IV. Consistency Checking and Waveform Comparison

4.1. *Preliminary Observations*

Consistency checking of a macromodel against a transistor level circuit involves a comparison of a set of scalar quantities and a set of output waveforms. This process is used to compute the cost function for the optimization loop described in previous section. The complete cost function has the following form

$$\text{cost} = \sum_i \beta_i |X_i - Y_i| + \sum_j \gamma_j \|x_j(\cdot) - y_j(\cdot)\| \qquad (5)$$

[1] Although it would be worthwhile to investigate logarithmic schedules, we have found that a linear scheme has been adequate for our purposes thus far. The best scheme to use may, in fact, be problem dependent.

where the β's and γ's are the weights associated with each target specification. The specifications for the transistor circuit are represented by X and $x(\cdot)$, and those for the macromodel by Y and $y(\cdot)$. In general, the total cost function of a macromodel is derived from the waveform differences $\|x_i(\cdot) - y_i(\cdot)\|$ as well as the difference in the scalar performance measures $|X_i - Y_i|$ (such as gain, CMRR, input resistance, and output resistance, for an operational amplifier). However, we only focus on the waveform comparison portion in this paper because the others are computed simply as a weighted sum of scalar differences. During each cost function evaluation, two sets of waveforms are compared using a *distance metric* to measure the *error* between them. This error is taken to be a measure of the performance in the macromodel relative to the transistor level circuit. The waveforms may be either time- or frequency-domain responses, and there may be as many of them as needed by the user.

Consider two transient output waveforms, $x(t)$ and $y(t)$, defined over an interval $[0,T]$ to be compared in order to verify a time-domain macromodel, where $x(t)$ is the reference waveform at some output node of a circuit and $y(t)$ is the waveform produced by the corresponding output node of the macromodel. First, the direct difference waveform $z(t) = x(t) - y(t)$ is generated. Then a conventional waveform *norm* such as

$$\|z\|_1 = \int_{t=0}^{T} |z(t)| \, dt \tag{6}$$

or

$$\|z\|_\infty = \max_{t \in [0,T]} |z(t)|. \tag{7}$$

may be applied to the difference signal $z(t)$. Each of the above norms induces a *metric* that measures the error between the two waveforms $x(t)$ and $y(t)$ as follows:

$$d_p(x, y) = \|x - y\|_p. \tag{8}$$

In this paper, we will only consider the conventional metrics d_1 and d_∞ for use as distance measure. Physically, d_1 is the area between two functions and d_∞ is the largest difference between the two functions over all t.

The use of the above conventional metrics is not appropriate for our application because they may occasionally mislead the optimization process. Fig. 8(a) shows a reference waveform, $x(t)$, and two voltage waveforms, $y_1(t)$ and $y_2(t)$. If the d_∞ metric is used to select the best waveform, it would select $y_2(t)$ over $y_1(t)$ as shown by the difference waveforms in Fig. 8(b). The better choice would, of course, be $y_1(t)$ as it only has a small phase error. The d_1 metric would correctly select $y_1(t)$ in this example, but it may be too pessimistic in other cases. For example, in Fig. 9(a), the reference waveform, $x(t)$, contains some glitches due to point-to-point ringing, possibly from numerical integration using the trapezoidal method. As illustrated in Fig. 9(b), the d_1 metric indicates a large error between $x(t)$ and $y(t)$ when, in fact, they may be equivalent waveforms obtained by the use of different integration methods. In order to properly guide the optimization loop, we require a metric that can tolerate small phase errors and ignore glitches in the waveforms.

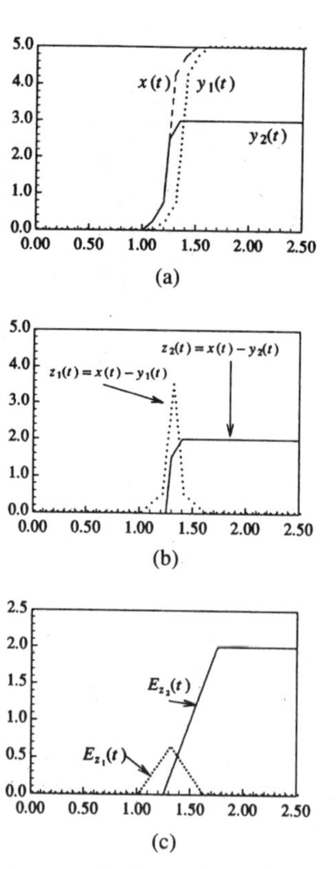

Fig. 8. An example illustrating deficiency of d_∞.

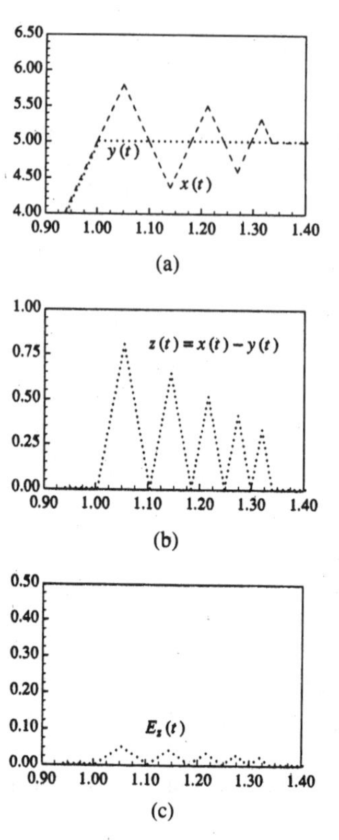

Fig. 9. An example illustrating deficiency of d_1.

4.2. *A New Waveform Operator*

We now describe a class of distance metrics for time-domain waveform comparison that are better suited for our application if circuit designers are not interested in the high frequency behavior of their macromodels. Consider a piecewise differentiable (smooth) waveform $z(t)$ over a finite time-interval $[0,T]$. Define a *Euclidean operator* (similar to the one described in [12]) that produces a new waveform $E_z(t)$ as follows:

$$E_z(t_0) = \inf_{t\in[0,T]} \sqrt{z^2(t) + s^2(t - t_0)^2} \tag{9}$$

where s is the scaling factor which can be used to scale time into an "equivalent" voltage. Graphically, $E_z(t_0)$ is the *shortest Euclidean distance* between the point $(t_0,0)$ and any point $(t,z(t))$ on the waveform which is the radius of the circle in Fig. 10 instead of the length of the shaded arrow. Clearly, for all $t \in [0,T]$, we have $0 \leq E_z(t) \leq |z(t)|$. Also, $E_z(t_k) = 0$ if and only if $z(t_k) = 0$. The new waveform operator defines a family of distance metrics given by

$$d_p^E(x, y) = \|E_{x-y}(t)\|_p \tag{10}$$

but we typically use $d_1^E(x,y)$ in iMAVERICK.

Figs. 8(c) and 9(c) illustrate $E_z(t)$ for the two cases examined in the previous section. Note that the peaks appearing in $z(t)$ are attenuated in the waveform $E_z(t)$. Also, it can be shown that the waveform $E_z(t)$ is always continuous even if $z(t)$ is not. Thus $E_z(t)$ appears to be a "filtered" version of $z(t)$. The operation can be characterized as a *nonlinear low-pass filter* that suppresses any undesired high frequency behavior in the original signal $z(t)$. The nonlinearity is due to the fact that, in general, $E_{\alpha z} \neq |\alpha| E_z$, i.e., the Euclidean operation does not satisfy the principal of homogeneity.

The role of s in (9) is to set the range of useful frequencies for the filter and provides the scale factor between time and voltage. Equation (9) requires that s be in units of volts/seconds. Typically, s is chosen by determining the time resolution that is equivalent to a given voltage resolution. For example, if 0.5 V of resolution is equivalent to 1 ns of resolution, then $s = 0.5\ \text{V}/1\ \text{ns} = 5 \times 10^8 \text{V/s}$. As s approaches infinity, the behavior of $E_z(t)$ approaches $z(t)$. However, for finite values of s, the waveform operator acts as a nonlinear low-pass filter with a −3-dB frequency that can be easily derived (for sinusoidal waveforms with amplitude A) to be:

$$f_{-3\text{dB}} \approx 2s/A.$$

Therefore, for a given application with a signal frequency of f, s should be set such that

$$Af/2 < s < 5Af$$

so that the necessary frequencies are captured but the high frequencies (10 times larger than the signal frequencies) are rejected.

In summary, our new waveform operator can tolerate small phase errors between x and y and has the capability of filtering out glitches and high frequency "noise" present in one of the waveforms. However, if the designer is interested in preserving these features, the d_1 metric can be used. In fact, we use the d_1 metric for all frequency-domain waveform comparisons and d_1^E for time-domain comparisons.

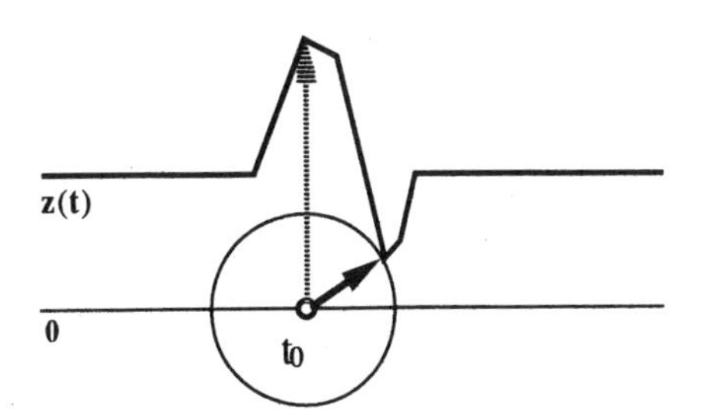

Fig. 10. Example of the calculation of Euclidean distance.

V. Experimental Results

5.1. *Example I: Operational Amplifier 741*

The operational amplifier is one of the most important building blocks utilized in analog integrated circuits. Therefore, accurate macromodels for operational amplifiers are very important to circuit designers and much work has been done in this area. However, little emphasis has been placed on optimizing existing operational amplifier macromodels. To demonstrate the usefulness of iMAVERICK, a general-purpose operational amplifier, the 741 [13], and the macromodel of Chua and Lin [14], as shown in Fig. 11, will be used. The SPICE3 program [15] was modified slightly to handle the nonlinear voltage-controlled current source used in the Chua–Lin macromodel. Several experiments were performed to optimize the macromodel based on different criteria. The experiments and the results are presented below.

The first experiment was to optimize the macromodel to match the frequency response of the 741 operational amplifier. In Fig. 11, stage N_3 consists of the $R_1 C_1$ parallel combination and is used to set the dominant-pole corner frequency. Higher frequency poles are specified in the fourth stage N_4 which is a grounded, unity-gain, linear, two-port *RC* network. The initial macromodel configuration given by Chua and Lin has the following parameters: $g_m = 1$, $I_m = 68$ mA, $C_1 = 0.136\ \mu$F, and $R_1 = 835$ kΩ. Fig. 12 shows a plot of the gain and phase characteristics of the actual 741 operational amplifier obtained using SPICE3 simulations (curve labeled "741") and the initial macromodel configuration (curve labeled "*C–L*"). In this experiment, C_1 and the slope g_m in stage N_2 were automatically determined by iMAVERICK to match the frequency response of the operational amplifier 741. 1646 sample points out of 8547 possible trials were evaluated (simulated) by iMAVERICK to generate the best macromodel configuration having the closest frequency response to the real operational amplifier 741. The optimized configuration has parameters $g_m = 0.33$ and $C_1 = 0.068\ \mu$F. Its frequency response is also plotted in Fig.

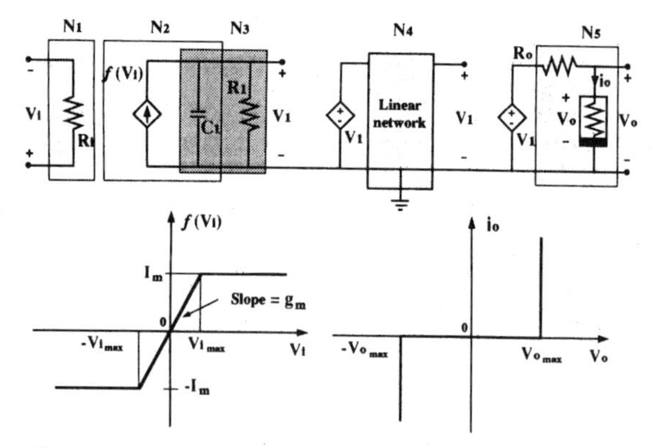

Fig. 11. Chua-Lin macromodel for operational amplifier 741.

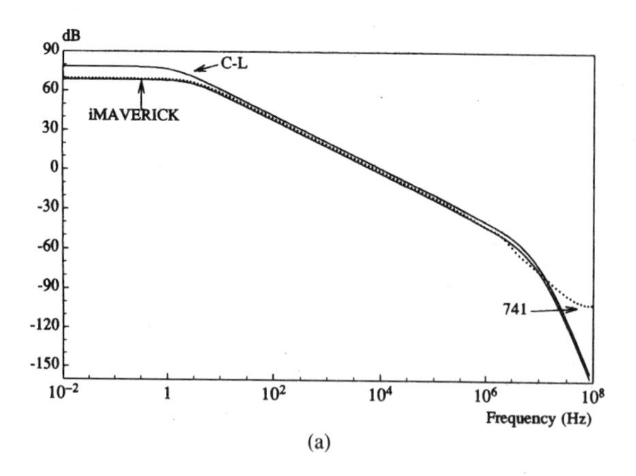

(a)

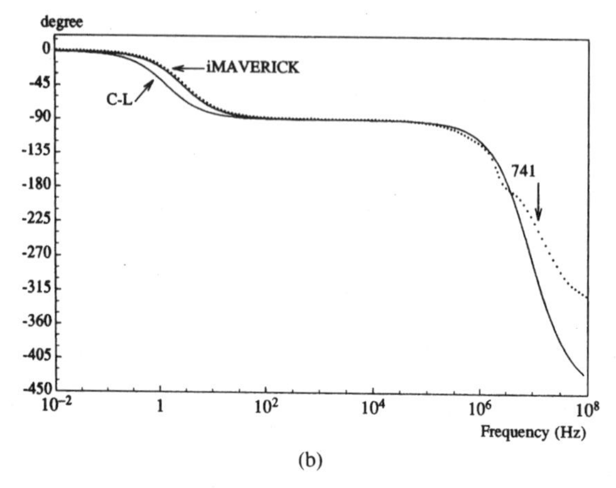

(b)

Fig. 12. Comparison of frequency responses. (a) Magnitude. (b) Phase.

12 (curve labeled "iMAVERICK"). Clearly, the optimized configuration has a frequency response that is closer to the real circuit in the range 10 mHz $\leq f < 1$ MHz. Note that this macromodel is not capable of modeling the true high frequency behavior due to the fact that it only provides poles at high frequency whereas the actual 741 has a number of poles and zeros. However, this experiment does illustrate the ability of iMAVERICK in preforming macromodel optimization in the frequency domain.

A common test of the large-signal performance of an amplifier is to apply a step input voltage from 0 to +5 V

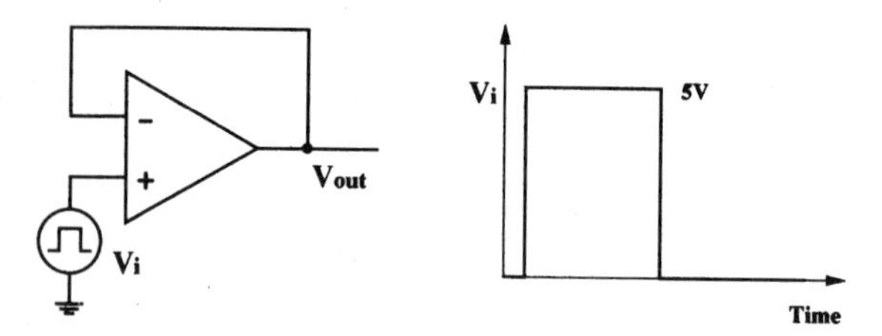

Fig. 13. Circuit for testing slew rate performance.

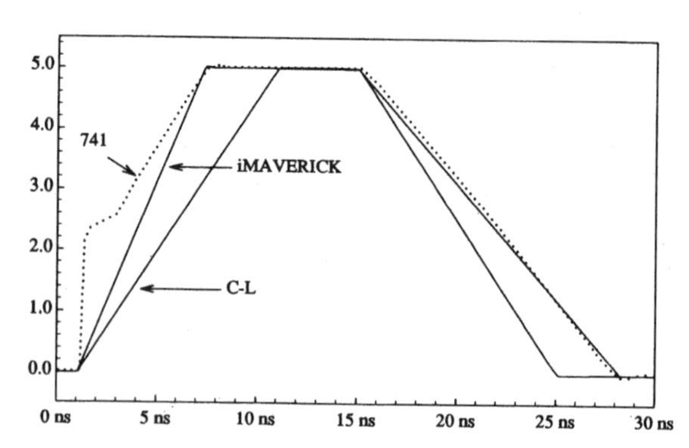

Fig. 14. Simulated voltage follower slew rate performances.

and then back again from +5 V to 0 as illustrated in Fig. 13. This was the basis for the second experiment. The figure shows an operational amplifier in a unity-gain feedback configuration, which can be used to observe the slew rate performance. Fig. 14 shows the simulated voltage follower slew rate performances using both the 741 and the macromodels. Unfortunately, the step (jump) in the initial response of the voltage follower is not captured by the Chua-Lin macromodel. This is another deficiency of this macromodel.

Attempts to accurately match both the frequency response and the time-domain response failed due to the fact that the Chua-Lin macromodel allows only symmetric slewing action. Therefore, the nonlinear voltage-controlled current source in the Chua-Lin macromodel (see Fig. 11) was modified slightly by allowing different slopes in positive and negative regions in order to obtain different rising in the falling slew rates. The positive slope g_m^+, negative slope g_m^-, and maximum output current of the controlled source I_m were the parameters to be determined by iMAVERICK. The optimized configuration (labeled "iMAVERICK") with parameters $g_m^+ = 1.4$, $g_m^- = 0.68$, and $I_m = 76.82$ mA has the same rise and fall times as the operational amplifier 741 as it was optimized to produce similar waveforms.

5.2. *Example II: Wien-Bridge Oscillator*

The second example is a Wien-bridge oscillator shown in Fig. 15 that is to be simulated using the modified *C-L* macromodel. The design parameters $k = 1 + R_3/R_4$ and $1/R_1C_1$ control the oscillating frequency and are set to be $k = 3.5$ and $1/R_1C_1 = 11915s^{-1}$ in this experiment. Fig. 16 shows the waveforms of the output node V_{out}. The dot-

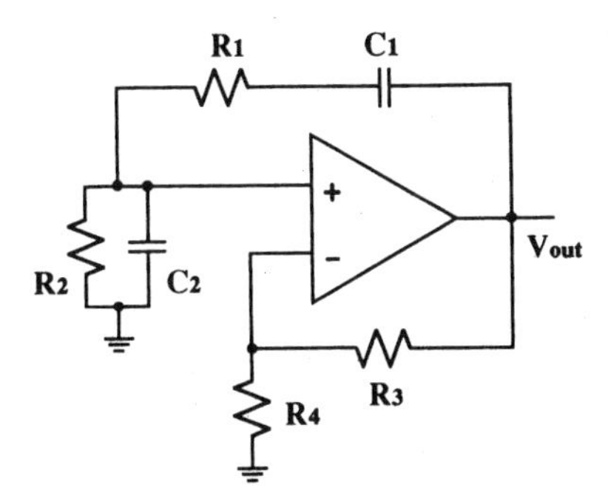

Fig. 15. Wien-bridge oscillator.

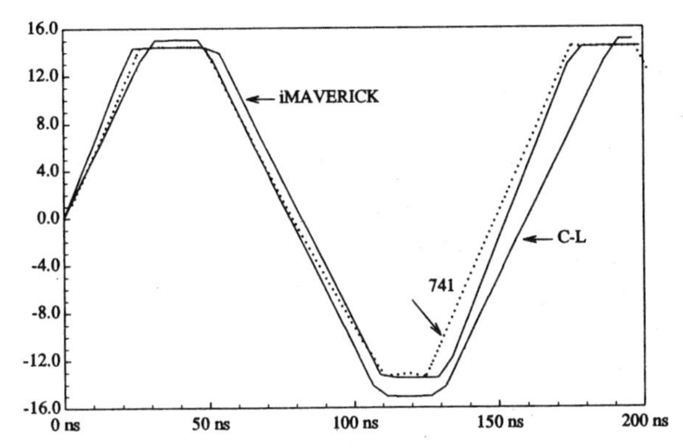

Fig. 16. Simulated macromodel performances for the Wien-bridge oscillator.

ted curve labeled "741" is the waveform of the real operational amplifier. The *C–L* macromodel using the original parameters is also shown in this figure with a label "*C–L*." It can be seen from this figure that the configuration with the optimized parameters performs better than the configuration that uses the original *C–L* parameters.

VI. Conclusions

A closed-loop verification and optimization system called iMAVERICK has been developed to eliminate the time spent in manual optimization of macromodel parameters. The key contributions of this paper are: a new stochastic optimization algorithm and a waveform comparison technique. This stochastic algorithm is suitable for applications where the "cost" evaluations are expensive. A heuristic measure of the likelihood that a cost evaluation will provide useful information or a better solution is used in the algorithm based on the information accumulated from the previous cost evaluations. This heuristic measure is used to reject poor sample points without sacrificing the quality of solutions. The new waveform comparison techniques compare two waveforms more effectively than a direct difference approach. They are sensitive to phase errors between waveforms and have the capability of filtering out glitches and high frequency noise present in the waveforms.

One point worth emphasizing is that the overall accuracy of macromodel-based simulation depends critically on the nature of the macromodels used and the type of input provided by the designer to the optimization loop. If the macromodel is not able to capture a particular desired feature, such as the high frequency response of the 741 operational amplifier in the Chua–Lin macromodel, obviously the use of optimization will not improve it. Furthermore, if a complete set of objectives are not provided to the optimization loop, the parameters generated may not be adequate for all applications. However, iMAVERICK provides a framework for experimentation to determine the features and limitations of a given macromodel, and to generate the best parameters for each intended application. iMAVERICK is written in C and can be obtained by writing to the authors at the University of Illinois.

Acknowledgment

The authors would like to thank Dr. Y. T. Yen at Digital Equipment Corporation for suggesting this area of research.

References

[1] G. R. Boyle, B. M. Cohn, D. O. Pederson, and J. E. Solomon, "Macromodeling of integrated circuit operational amplifiers," *IEEE J. Solid-State Circuits*, vol. SC-9, pp. 353–363, Dec. 1974.

[2] B. Perez-Verdu, J. Huertas, and A. Rodriguez-Vazquez, "A new nonlinear time-domain op-amp macromodel using threshold functions and digitally controlled network elements," *IEEE J. Solid-State Circuits*, vol. 23, pp. 959–970, Aug. 1988.

[3] G. Casinovi, "Macromodeling for the simulation of large scale analog integrated circuit," Ph.D. dissertation, Univ. California, Berkeley, Aug. 1988.

[4] C. Visweswariah, R. Chadha, and C-F Chen, "Model development and verification for high level analog blocks," in *Proc. 25th ACM/IEEE Design Automation Conf.*, June 1988, pp. 376–382.

[5] G. Casinovi, "Automatic generation of models for circuit simulation," in *Proc. Midwest Symp. Circuits and Systems*, Aug. 1989, pp. 993–996.

[6] D. A. Wismer and R. Chattergy, *Introduction to Nonlinear Optimization: A Problem Solving Approach*. New York: North-Holland, 1978.

[7] S. Kirkpatrick, C. D. Gelatt, and M. P. Vecchi, "Optimization by simulated annealing," *Sci.*, vol. 220, no. 4598, pp. 671–680, May 1983.

[8] C. Sechen, *VLSI Placement and Global Routing Using Simulated Annealing*. Boston, MA: Kluwer Academic, 1988.

[9] R. Otten and L. van Ginneken, *The Annealing Algorithm*. Boston, MA: Kluwer Academic, 1989.

[10] Hedayat and Wallis, "Hadamard matrices and their applications," *Ann. Statistics*, 1978.

[11] Iman and Shortencarier, "A FORTRAN 77 Program and user's guide for the generation of latin hypercube and random samples for use with computer models," Sandia National Labs Rep., Mar. 1984.

[12] J. Kleckner, "Advanced mixed-mode simulation techniques," Ph.D. dissertation, Univ. California, Berkeley, Apr. 1984.

[13] P. R. Gray and R. G. Meyer, *Analysis and Design of Analog Integrated Circuits*, 2nd ed. New York: Wiley, 1983.

[14] L. O. Chua and P. Lin, *Computer-Aided Analysis of Electronic Circuits*. Englewood Cliffs, NJ: Prentice-Hall, 1975.

[15] T. Quarles, "SPICE3 version 3cl users guide," ERL Memo. No. M89/46, Univ. of California, Berkeley, CA, 1989.

Yun-Cheng Ju received the B.S. degree in electrical engineering from the National Taiwan University, Taiwan, Republic of China, in 1984, and the M.S. degree in computer science from the University of Illinois at Urbana-Champaign, Urbana, IL, in 1990. He is currently working towards a Ph.D. degree in computer science at the University of Illinois at Urbana-Champaign, Urbana, IL.

Since 1988 he has been a Research Assistant at the Coordinated Science Laboratory, University of Illinois, Urbana, IL. His research interests include computer-aided design and mixed-mode simulation of VLSI circuits, parallel processing, and graph algorithms.

Vasant B. Rao (S'81–M'85), for a photograph and biography, please see page 535 of the April 1991 of this TRANSACTIONS.

Resve A. Saleh (S'85–M'86) received the B.Eng. degree in (electrical) engineering from Carleton University, Ottawa, Canada, in 1979, and the M.S. and Ph.D. degrees from the University of California at Berkeley in 1983 and 1986, respectively.

He has worked in industry for Mitel Corporation, Kanata, Ontario, Canada, Tektronix, Beaverton, OR, Toshiba Corporation, Kawasaki, Japan, and Shiva Multisystems, Menlo Park, CA. He joined the University of Illinois in 1986 where he is currently an Assistant Professor directing research in mixed-mode simulation and parallel processing. His research interests also include analog CAD and synthesis.

Dr. Saleh has served on the technical committees of the Custom Integrated Circuits Conference and the Design Automation Conference since 1987, and was a member of the organizing committee of the MidWest Symposium on Circuits and Systems in 1989. He received a Presidential Young Investigator Award in 1990.

Computer-Aided Design Considerations for Mixed-Signal Coupling in RF Integrated Circuits

Nishath K. Verghese, *Member, IEEE* and David J. Allstot, *Fellow, IEEE*

***Abstract*— This paper reviews computer-aided design techniques to address mixed-signal coupling in integrated circuits, particularly wireless RF circuits. Mixed-signal coupling through the chip interconnects, substrate, and package is detrimental to wireless circuit performance as it can swamp out the small received signal prior to amplification or during the mixing process. Specialized simulation techniques for the analysis of periodic circuits in conjunction with semi-analytical methods for chip substrate modeling help analyze the impact of mixed-signal coupling mechanisms on such integrated circuits. Application of these computer-aided design techniques to real-life problems is illustrated with the help of a design example. Design techniques to mitigate mixed-signal coupling can be determined with the help of these modeling and analysis methods.**

***Index Terms*—CAD, mixed-signal noise, modeling, noise, simulation, substrate coupling.**

I. Introduction

THE rapid growth of personal wireless communication promises to continue well into the next century. The recent FCC allocation of new personal communication system (PCS) channels in the 1.8–2 GHz band will further this growth for the next generation of wireless systems [1]. The recent increase in wireless applications brings a new set of aggressive design goals: low-power for portability, lower-cost, and higher integration of RF components. A single-chip transceiver is already a reality for certain relatively undemanding applications, such as RF ID, but IC manufacturers are not close to producing a similar device capable of operating in the more demanding application world of cellular telephony. In the meantime, incremental changes are helping provide consumers with handsets that offer major performance and size gains.

The biggest obstacles to integrating cellular telephone components are in the receiver channel. As more and more channels are packed into the same range of frequencies, techniques must be found to eliminate interfering signals. This is particularly difficult in cellular telephony due to the high dynamic range required [2]. For acceptable performance, a cellular phone has to receive a signal clearly from a base station miles away, while interfering signals blast away on adjacent channels. Interfering signals on wireless IC's can also arise due to the integration of analog and digital components on the same chip [3]. The goal of putting mixed-signal "systems on a chip" has been a difficult challenge that is only recently beginning to be met. Some of the more prevalent problems in integrating mixed-signal (analog and digital) circuits are the noise coupling mechanisms on-chip. Switching currents induced by logic circuits work in tandem with the chip/package parasitics to cause ringing in the power-supply rails and in the output driver circuitry. This in turn couples through the common substrate to corrupt sensitive analog signals on the same chip. Compounding the problems of wireless circuit designers dealing with this mixed-signal noise coupling problem is that received signals on-chip are often in the range of only a few tens of microvolts. At microvolt signal levels, RF receiver circuits are sensitive to fundamental device noise, let alone mixed-signal substrate-coupled noise.

Several design considerations must be given to the realization of integrated mixed-signal RF circuits. The potential of computer simulation in aiding these design decisions is becoming increasingly evident. These simulation techniques are a result of recent progress in the development of steady-state methods to simulate periodically time-varying circuits [4], [5] and of semi-analytical extraction methods for the modeling of the chip substrate [6], [7]. The application of these techniques to the computer-aided design of mixed-signal/wireless IC's will be described in this paper. Design considerations to mitigate noise coupling mechanisms and to successfully integrate mixed-signal/wireless IC's can be arrived at with the help of these modeling and simulation techniques.

Section II discusses some of the problems in integrating mixed-signal wireless circuits. Section III reviews the mixed-signal coupling problem and the use of substrate, package, interconnect, and suitable device models for its analysis. Section IV illustrates the use of simplified circuit macromodels for the efficient simulation of noise coupling in mixed-signal/wireless IC's. It includes a review of various macromodeling techniques for switching noise introduced by on-chip logic circuits. A macromodeling technique for substrate coupling in RF circuits using specialized simulation techniques for periodically time varying circuits is also described. Section V illustrates the application of these modeling and simulation techniques to a real-world design problem.

II. Mixed-Signal Wireless IC Problems

During signal propagation and reception in a wireless channel, there exist many sources of noise such as interferers, channel noise, and device nonlinearity (due to mixers and other front-end RF circuitry) that can affect the integrity of

Manuscript received August 5, 1997; revised October 15, 1997.

N. K. Verghese was with Cadence Design Systems, San Jose, CA 95134 USA. He is now with Apres Technologies Inc., Cupertino, CA 95014 USA.

D. J. Allstot is with the Department of Electrical Engineering, Arizona State University, Tempe, AZ 85287 USA.

Publisher Item Identifier S 0018-9200(98)01712-0.

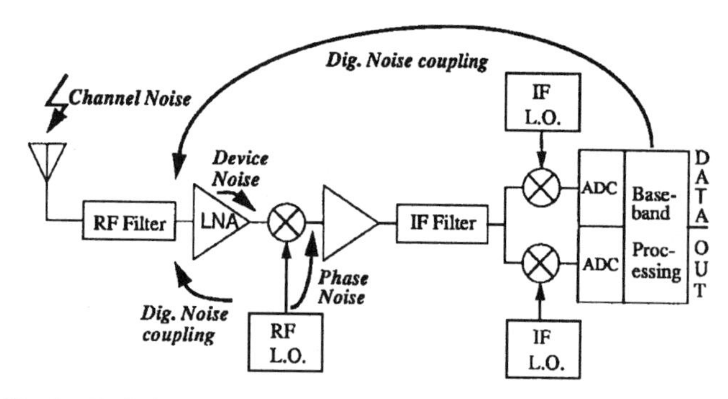

Fig. 1. Typical problems in integrated mixed-signal wireless IC's.

the demodulated signal. Fig. 1 shows some of the possible sources of noise during signal transmission and reception. These interferers and noise mechanisms can be categorized as channel noise sources, front-end receiver nonidealities, and mixed-signal noise coupling.

Channel noise sources include background white Gaussian noise, atmospheric noise, radio propagation losses, Doppler effect, fast fading, multipath signal losses, and adjacent channel interferers [8], [9]. The wireless channel is constantly changing over time due to the uncertainty in the channel terrain, signal path propagation, and the mobility of the transceiver unit. Hence, the RF signal received at the antenna is a function of several factors including distance from the base station, signal path geography, obstacles between the transmitter and the receiver, and multipath propagation of reflective signals. For a typical receiver, the incoming RF signal can range from 20 to -120 dBm; i.e., with 50-Ω termination, the incoming signal strength can be as low as 0.22 μV_{RMS} modulated from 900 MHz to 2 GHz [8].

Furthermore, the receiver adds noise to the incoming signal due to RF circuit nonlinearity, electronic device noise, and mixed-signal noise coupling. Receiver circuit nonlinearity results in spurious responses or apparent on-channel responses to undesired interfering signals. Intermodulation products during the mixing operation can create spurious responses in the signal frequency range. Other common spurious responses (spurs) include image, half-IF, and Able–Baker spurs [3]. Also, phase noise in the frequency synthesizer can cause undesired interferers to be mixed down, thereby corrupting the IF signal. Phase noise in the frequency synthesizer is caused both by device noise and by mixed-signal substrate-coupled noise mechanisms.

With greater emphasis on integration in modern wireless circuits, mixed-signal coupling is becoming a significant bottleneck in wireless IC development. On-chip integration of sensitive analog RF circuits with the noisy local oscillator (LO), frequency synthesizer, and other high-speed digital circuits can be detrimental to the weak incoming RF signal. The synthesizer consists of a high-speed divider/counter circuit which can produce significant switching noise. Any digital switching noise is of significant importance since cumulatively it can reach several hundred millivolts and propagate to the RF section via the substrate and I/O pins. The common substrate can also act as a dc and small-signal feedback path causing variations in dc operating conditions and small-signal parameters such as gain and bandwidth [6]. In direct conversion receivers, substrate coupling can cause the LO signal to feedthrough to the low noise amplifier (LNA) input ("LO leakage"), where it gets amplified and subsequently mixed with itself ("self mixing") resulting in dc offsets. A similar effect can occur if a large interferer leaks from the LNA or mixer input to the LO port and is multiplied by itself [10]. In the rest of this paper, the focus will be on the mixed-signal coupling problem in wireless integrated circuits.

III. Modeling Mixed-Signal Coupling

The received RF signal can range from 20 to -120 dBm; i.e., the incoming signal can be as low as 0.22 μV_{RMS}. Considering that the receive signal may be at 1.8–2 GHz, any internal on-chip noise can destroy the desired signal. The main source of on-chip noise in mixed-signal IC's is the digital switching noise. In purely digital applications, the CMOS static logic family offers several attractive features including zero static power dissipation, high packing densities, wide noise margins, high operating frequencies, etc. For high-frequency wireless applications however, its major drawback is the generation of a large amount of digital switching noise [11]–[14]. When many static gates change states, a large cumulative current spike flows through parasitic resistances and inductances creating power supply noise voltage spikes known as "V_{DD} *bounce*" or "G_{nd} *bounce*." Some fraction of this noise inevitably propagates to the sensitive analog circuitry through the substrate, power supply lines, bonding wires, package pins, etc., as shown in Fig. 2, where it often limits the achievable accuracy. The switching noise current per gate can range from 0.1 mA to several mA depending on the frequency of operation and device sizes. Besides being a coupling path for switching noise, the common substrate is also a signal path. Impact ionization in MOS device channels can cause currents to be injected into the substrate even under dc operating conditions, causing substrate biases to vary, which in turn cause variations in MOS threshold voltages, depletion capacitances, and other circuit bias and performance quantities. Signal coupling through the substrate can also lead to gain and bandwidth variations in LNA's and self-mixing problems in direct conversion receivers. Moreover, signal loss in the substrate is a significant concern both in the design of receiver front-end circuitry and in the realization of on-chip high-Q inductors [15].

Several mechanisms must be understood and suitably modeled in an electrical circuit in order to analyze it for mixed-signal coupling problems. These include the effects of impact ionization, device/interconnect capacitance, package/bondwire inductance, and substrate resistance (and capacitance).

A. Impact Ionization Model

With increasing speeds of operation and decreasing technology feature sizes, impact ionization is becoming a primary cause of substrate current injection in integrated circuits. When the electric field in the depleted drain end of a MOS transistor becomes large enough to cause impact ionization, electron-hole pairs are created causing a current flow to the substrate.

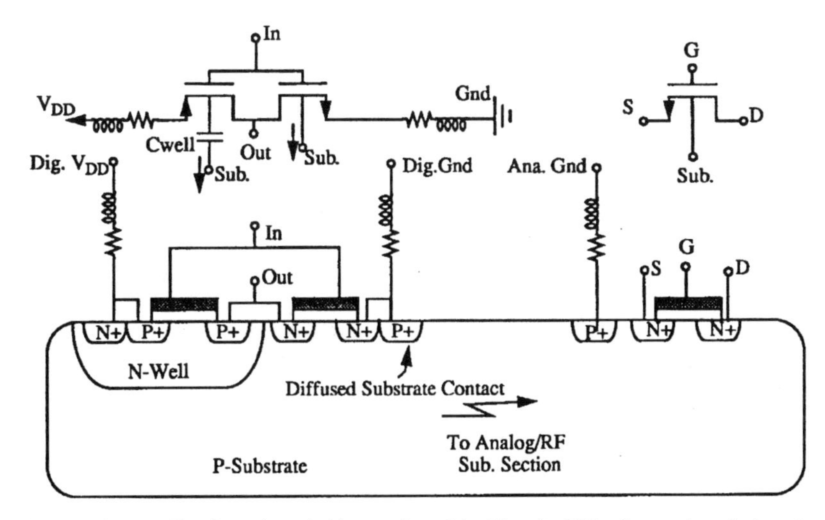

Fig. 2. The mixed-signal problem of noise coupling from the switching portion of the IC to the RF/analog portion via the substrate, package, and supply lines.

The hot-electron induced substrate current can be expressed in semi-analytical form as [16]

$$I_{sub} = C_1(V_{\text{ds}} - V_{\text{dsat}})I_d \exp\left(-\frac{C_2 t_{ox}^{1/3} \cdot x_j^{1/2}}{V_{\text{ds}} - V_{\text{dsat}}}\right) \tag{1}$$

where C_1 and C_2 are process-related, empirically determined parameters, t_{ox} is the oxide thickness, and x_j is the junction depth. Using results from device simulations or measurements, it is possible to determine the empirical coefficients C_1 and C_2 and to incorporate impact ionization induced substrate currents into existing device models for circuit simulation.

B. Device/Interconnect Capacitance Model

Every transistor on an IC die is coupled capacitively to the substrate through its p-n junction depletion capacitances. Moreover, every interconnect routed on an IC has some capacitance to substrate. Capacitively coupled substrate current is of significant consequence in mixed-signal circuits due to the presence both of a large number of switching digital nodes that inject current into the substrate and of high impedance analog nodes that are affected by this injected current. Since the amount of injected current is directly proportional to the slew rate of the switching voltage, at higher rates of circuit operation the substrate coupling problem is greatly aggravated. Moreover, with decreasing technology feature sizes, the interconnect capacitances to substrate are becoming increasingly important contributors of injected current. In order to account for capacitively coupled substrate currents, it is necessary to perform a parasitic capacitance extraction on the design to determine all significant capacitances to substrate in the circuit.

C. Package Model

The effect of nonideal (inductive) power supplies has a tremendous impact on the amount of substrate coupled switching noise in an IC design. Since the bondwires and package pins associated with the substrate supplies have finite and often large inductances, any substrate current picked up by

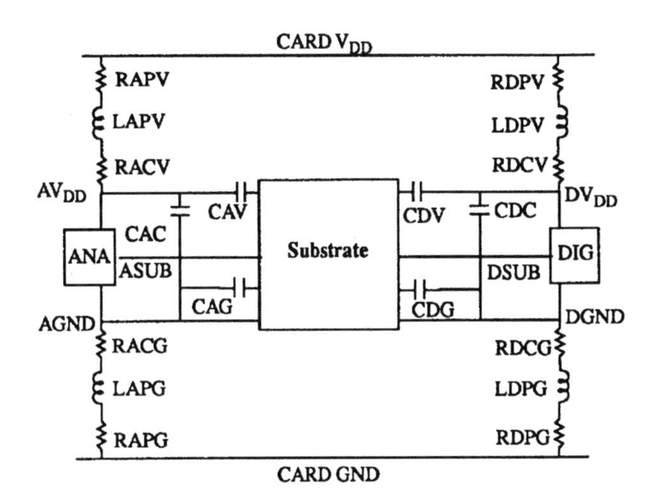

Fig. 3. An illustrative parasitic model of a chip in its package [13].

these supplies can cause large glitches in the value of the substrate supply voltage bias. This phenomenon is referred to as inductive or Ldi/dt noise. The presence of parasitic inductances in the substrate supplies[1] can severely aggravate the noise coupling problem and much of current mixed-signal IC design methodology focuses on techniques to minimize their effect. For simulation purposes, it is necessary to use suitable package inductance models in the supply leads to accurately analyze substrate-coupled switching noise [13], [17].

A simple chip-package model [13] is illustrated in Fig. 3 where RAPV, LAPV, and RACV represent the package resistance, package inductance, and on-chip resistance, respectively, in the analog V_{DD} line, while RAPG, LAPG, and RACG represent similar parasitics in the analog ground line. CAC represents the chip V_{DD} to G_{ND} capacitance, while CAV and CAG represent the capacitances from the analog V_{DD} and G_{ND} lines to substrate. Similar parasitics are also illustrated for the digital G_{ND} and V_{DD} lines.

[1] There are typically several separate digital and analog supplies connected to substrate.

D. Substrate Model

Outside of the active areas formed by devices and substrate contacts, the substrate can be treated as consisting of layers of uniformly doped semiconductor material. Neglecting the effects of magnetic fields on-chip, a simplified form of Maxwell's equations can be applied to the substrate yielding

$$\frac{1}{\rho}\nabla^2 V(r,t) + \varepsilon \frac{\partial}{\partial t}\left[\nabla^2 V(r,t)\right] = -\dot{q}(r,t) \quad (2)$$

where ρ is the resistivity and ε the permittivity of the uniformly doped semiconductor. $V(r, t)$ is the transient voltage vector and $\dot{q}(r, t)$ is the rate of charge generation per unit volume at location $r = (x, y, z)$ on the substrate. Assuming a three-dimensional (3-D) semi-infinite substrate that goes to infinity in all but one of the six spatial directions, the solution to (2) in the Laplace domain for the voltage at any point on the substrate due to a unit current injected into the substrate a distance r away, is given by

$$\nu_2(s) = \frac{\rho}{2\pi r} \cdot \frac{i_1(s)}{s(\rho\varepsilon)+1}. \quad (3)$$

Consequently, the substrate impedance $z_{21}(s) = v_2(s)/i_1(s)$ has a single pole response with a -3 dB frequency given by the reciprocal of the relaxation time constant, $\tau = \rho\varepsilon$, of the substrate. The net substrate admittance between the two points $y(s)$ can be determined as

$$y(s) = \frac{i_1}{2[v_1(s) - v_2(s)]} = \frac{1}{2[z_{11}(s) - z_{12}(s)]} \quad (4)$$

where the self impedance term $z_{11}(s) = v_1(s)/i_1(s)$ also has a single pole response with the same -3 dB frequency. Note that $y(s)$ in (4) can equivalently be expressed as $y(s) = [y_{11}(s) - y_{12}(s)]/2$ where z_{11}, z_{12}, y_{11}, and y_{12} are related as in (5). Consequently, the admittance $y(s)$ has a zero in its response at the same 3 dB frequency. Moreover, this frequency scales with doping and is nearly 150 GHz for a 1-Ω-cm substrate. Consequently, for most frequencies of interest, the substrate can be assumed to behave as a purely resistive medium.

The analysis conducted above assumed two points on a simplistic homogeneous substrate. When the substrate profile consists of several layers of different doping density and we consider physical dimensions of the "points" on the substrate, the analysis becomes more complicated. Techniques like the numerical finite difference method [13] or the semi-analytical boundary element method [6], [7] can be applied to the problem to determine reduced-order R(C) models between the devices in the circuit to capture the behavior of the substrate. To determine such a reduced order substrate model, begin by defining every device and every contact location in the physical layout of the design as an equipotential *port*. The substrate, then, acts to resistively couple these different ports as shown in Fig. 4. Note that the substrate is represented by vertically stacked layers of uniform doping density, with or without a metallized backside contact. Of the different techniques that can be employed to solve the governing equations of this problem, and to represent it as an equivalent matrix of admittances interconnecting the ports, the boundary element method is more efficient.

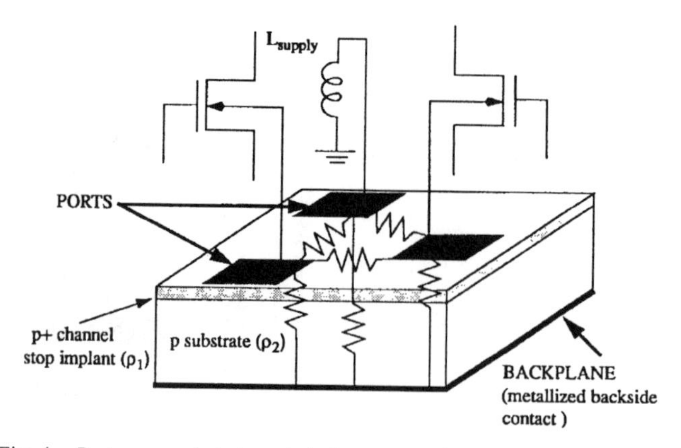

Fig. 4. Ports on a substrate resistively connected to one another.

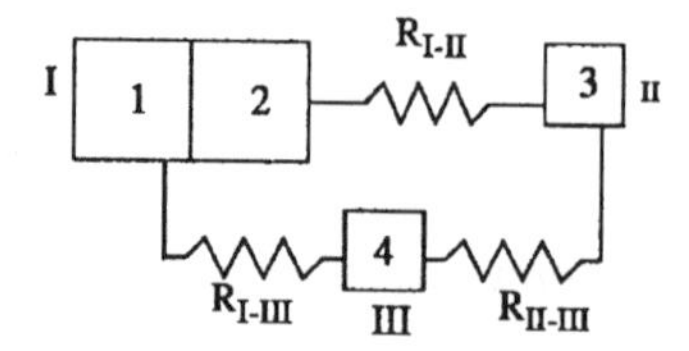

Fig. 5. Determining resistive coupling between ports using the boundary element method.

The boundary element method of extracting the substrate resistances consists of discretizing each port on the substrate into a collection of *panels*, formulating an impedance matrix using Green's function for the medium, and inverting the resulting matrix to determine the corresponding coefficients of the admittance matrix. The Green's function for a medium is defined as the potential at any point in a medium with suitable boundary conditions due to a unit current injected at any point also within the medium. Thus, for the ports labeled I, II, and III in Fig. 5, the substrate resistances are computed as

$$\begin{bmatrix} y_{11} & y_{12} & y_{13} & y_{14} \\ y_{12} & y_{22} & y_{23} & y_{24} \\ y_{13} & y_{23} & y_{33} & y_{34} \\ y_{14} & y_{24} & y_{34} & y_{44} \end{bmatrix} = \begin{bmatrix} z_{11} & z_{12} & z_{13} & z_{14} \\ z_{12} & z_{22} & z_{23} & z_{24} \\ z_{13} & z_{23} & z_{33} & z_{34} \\ z_{14} & z_{24} & z_{34} & z_{44} \end{bmatrix}^{-1} \quad (5)$$

$$R_{\text{I-II}} = \frac{1}{y_{13}+y_{23}}; \quad R_{\text{II-III}} = \frac{1}{y_{34}}; \quad R_{\text{I-III}} = \frac{1}{y_{14}+y_{24}}. \quad (6)$$

The panel-to-panel impedance z_{ij} is analytically determined using the Green's function for the substrate in a preprocessing phase [6], [7]. The dense matrix inversion in (5) using standard Gaussian elimination, being roughly $O(n^3)$ in the number of panels, is overly expensive and consequently, iterative matrix solution methods like GMRES [18] must be employed. The most expensive part of the iterative method is the matrix vector product ($Zi = v$) requiring $O(n^2)$ operations (for n panels) at every iteration of the method until convergence is achieved. The overall complexity of iterative methods for extraction therefore approaches $O(n^3)$ in complexity for a large number of ports since the iterative solution must be repeated for each port. Fortunately, several different techniques can be employed

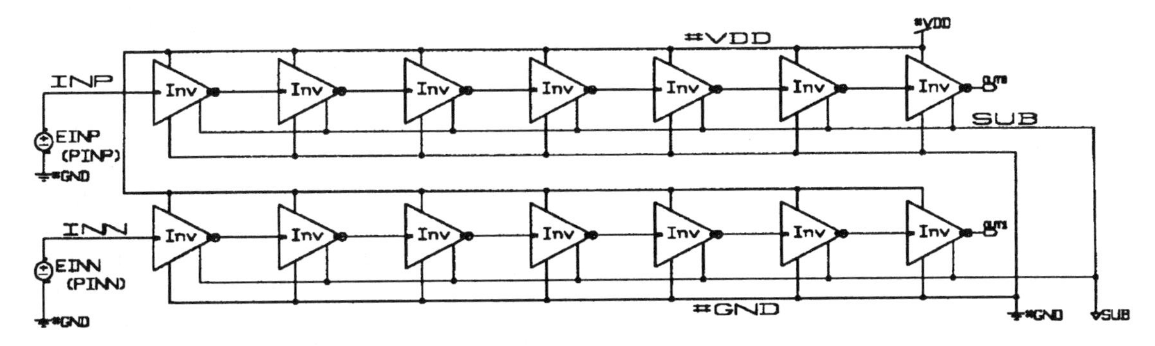

Fig. 6. Macromodeling logic switching activity using inverter chains [13].

to accelerate the matrix-vector product inherent to iterative matrix solution methods for the substrate model extraction problem. Choice of the appropriate acceleration technique for substrate extraction depends on the nature of the substrate profile (the presence/absence of a backplane or a heavily doped bulk). These techniques allow the matrix-vector product to be calculated to within a user-defined error tolerance in linear $[O(n)]$ or near-linear time [7], [19].

If the RC effects of the substrate need to be considered, the Green's function can be formulated using a complex valued dielectric constant $\sigma + j\omega$ where ω is the frequency at which the computation is performed [6]. At every frequency of interest, the resulting complex impedance matrix can be effectively inverted using an iterative method with accelerated matrix vector product. The resulting y-parameters can either be used directly in frequency domain analysis or converted to equivalent lumped-circuit models for time-domain simulation [20].

IV. Circuit Macromodeling

The overall verification strategy for mixed-signal coupling effects in RF IC's is to combine substrate models, chip, and package parasitic models and suitable macromodels of the circuit for fast yet accurate simulation. The chip, substrate, and package can be modeled efficiently as outlined in Section III. Also, as seen in Section III, impact ionization effects must be contained in device models for accurate simulation of substrate currents. However, simulating an entire mixed-signal or RF circuit at a transistor level is beyond the scope of traditional circuit simulators even without the complex device models necessary. Therefore, wherever possible, circuit behavior has to be abstracted to a higher level while retaining the circuit's relationship to the substrate.

A. Logic Circuit Macromodels

It is evident that one of the bottlenecks of simulating mixed-signal coupling in large IC's is the simulation of large digital blocks of the circuit at a transistor level. It is important to characterize the switching noise injected into the substrate by these logic circuit blocks as accurately as possible using a simple macromodel. Several macromodeling techniques for switching current injected by logic circuits have been proposed in literature [13], [21]–[23].

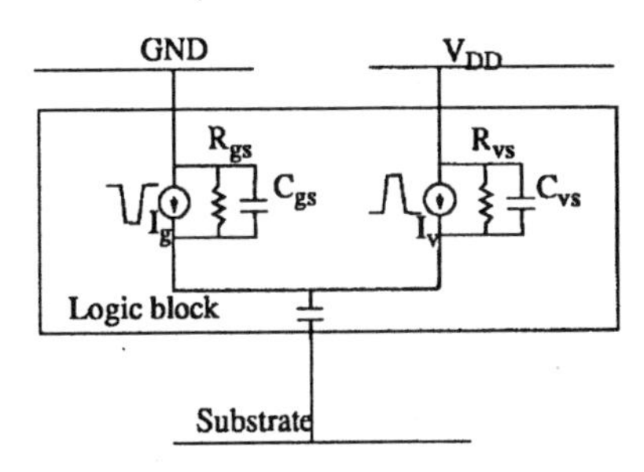

Fig. 7. Simple linear macromodel of logic block for switching noise [21].

One possible macromodel proposed by Smith and Schmerbeck [13] involves the use of chains of inverters to approximate logic switching activity as shown in Fig. 6. There are several inverters in each string to take into account the fact that the output switching occurs over a number of gate delays after the clock edge. The bottom inverter string is driven by a signal 180° out of phase relative to the top inverter string to emulate simultaneous out-of-phase switching. Each inverter model includes a multiplication factor to increase the amount of output switching power without modeling large numbers of inverters, and each inverter in the string can have a different multiplication factor [13]. The multiplication factor for each inverter can be determined by simulating the logic circuit involved using an event-driven simulator to determine switching instants. Since switching activity is not guaranteed to be constant, the weighting factors may have to change from clock edge to clock edge.

A simple linear macromodel as shown in Fig. 7 has been used to quickly evaluate switching noise effects for power distribution synthesis [21]. The current sources I_v and I_g model the current switching on the digital V_{DD} and ground rails. In addition, the finite resistance of the sources is modeled along with an equivalent capacitance across the sources. The capacitance to substrate represents the net junction and interconnect capacitances in the logic block.

A modeling methodology for digital switching noise introduced in [22] involves the precharacterization of the substrate current injected by each logic component as a function of its input switching pattern. Each signature noise signal is stored in a library consisting of all such signatures for every logic component of the circuit. For a given input pattern to the logic circuit, an event driven simulator is used to record every transition (event) in the circuit. Utilizing this event information

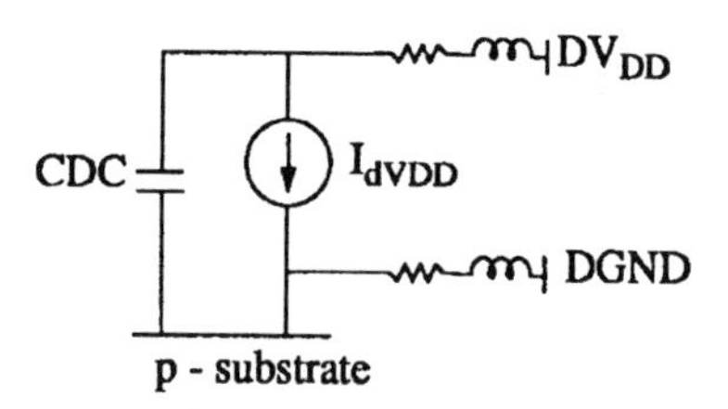

Fig. 8. Macromodel of switching noise in the digital substrate ground line.

and the precomputed signature noise signals for each gate and assuming spatial independence of all the noise sources, the net noise in the substrate is computed as a convolution of the impulse train of events and the corresponding noise signatures.

Although the assumption that the noise sources are spatially independent may fail in the absence of a heavily doped bulk substrate, the aforementioned methodology is quite useful in determining simple linear macromodels of logic blocks as in Fig. 7. The current models I_v and I_g for a given logic block can effectively be determined by convolving its noise signature with the event train associated with that block.

A simpler macromodeling strategy can be arrived at by assuming that coupling from the digital blocks to the substrate occurs solely through the substrate supply line [23]. Then, a macromodel as shown in Fig. 8 is constructed for all logic circuits connected to that supply. Since typically multiple digital supplies are used to bias the substrate, one such macromodel is required for each substrate supply used. The current $I_{dV_{DD}}$ flowing from V_{DD} to ground in the macromodel of Fig. 8 can be computed using a suitable power analysis tool. CDC is the net chip V_{DD} to ground capacitance, while the resistors and inductors represent package parasitics. It should be noted that in using such a macromodeling approach, appropriate modifications must be made in designs where a Kelvin ground is used to bias the substrate (i.e., a quiet ground different from the circuit ground), since in such cases the short-circuit displacement component of $I_{dV_{DD}}$ does not flow directly to the substrate as suggested in Fig. 8.

B. RF/Analog Circuit Macromodels

A characteristic of RF circuits is that they often process signals that contain closely spaced large frequencies. Transient simulation of such circuits using a traditional circuit simulator can be extremely expensive because the simulator must take very small time steps to follow the large frequency carrier (LO) signal and simulate over a long period of time in order to allow the circuit to reach steady state. More efficient simulation techniques have emerged recently with the advent of efficient matrix-free iterative methods [4], [5], [24]–[26] for the periodic analyses of RF circuits. Characterizing a given RF circuit with periodic analyses is a two-step process. First, a periodic steady-state (PSS) analysis with only the large periodic stimulus applied is performed to set the periodic operating point of the given circuit. In a mixer for example, only the LO is applied, whereas in a switched capacitor filter only the clock is applied. Finding the periodic steady-state solution of a circuit means finding the initial condition for the circuit's associated system of differential equations such that the solution at the end of the period matches the initial condition. Once the periodic operating point is computed, a periodic small-signal analysis is performed with the second tone applied to the circuit. This is done by linearizing the circuit about its periodic operating point and computing the steady state response of the periodically varying linear circuit assuming that it is driven with a small sinusoid of arbitrary frequency. The time required to compute the response of the linearized circuit is about the same as that required to compute the periodic steady-state response to the large stimulus alone, independent of the frequency of the sinusoid. Note that when applying a small sinusoid to a linear periodically time-varying circuit, the circuit responds with sinusoids at many frequencies. Hence, the periodic small-signal analysis essentially computes a series of transfer functions, one for each frequency where the input and output frequencies are offset by the harmonics of the LO [25], [26].

Two periodic small-signal analyses are possible. A periodic ac (PAC) analysis computes the small-signal response from a single stimulus at a single frequency to every node in the circuit at every output frequency (i.e., input frequency and all frequencies offset from it by a harmonic of the LO). A periodic transfer function (PXF) analysis, on the other hand, computes transfer functions to a single output at a single frequency from every source in the circuit at every input frequency (i.e., output frequency and all frequencies offset from it by a harmonic of the LO). The latter analysis is particularly useful in determining suitable macromodels of an RF circuit for the simulation of substrate coupling effects. Using a PXF analysis, it is possible to compute transfer functions from the bulk node of every device in the RF circuit to a specific output at a given frequency of interest. Note that for a given output frequency of interest, this results in a set of transfer functions for each bulk node, where each transfer function models the frequency conversion effect inherent in the RF circuit. Hence, each transfer function models the gain of the circuit in converting a sideband frequency component of the bulk node voltage to the specified frequency component of the output voltage.

Once the transfer functions have been computed for the RF circuit, it no longer needs to be represented at the transistor level. A transient simulation can be performed on the logic circuit using the switching noise macromodels and the parasitics coupling the RF circuit and the logic circuits through the substrate and package. The transient noise responses at the bulk nodes of the RF devices can be postprocessed to determine their equivalent noise spectra, and the frequency components of interest can be multiplied with the transfer functions determined earlier to compute the net spurious response at the prespecified output.

A similar macromodeling approach can be employed for analog linear time invariant systems where standard small-signal transfer function analysis is used to determine transfer functions from the substrate to the circuit outputs over all frequencies of interest. Once the transfer functions are computed, a behavioral macromodel of the circuit can be formulated, taking into account both the circuit's functional and substrate coupling behavior.

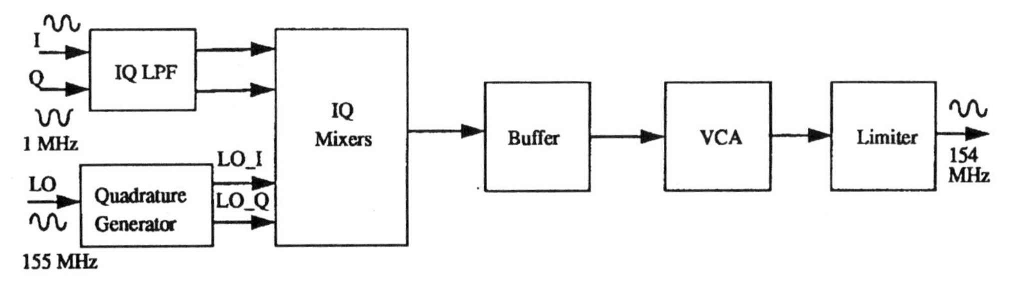

Fig. 9. Block-level schematic of the up-conversion mixer.

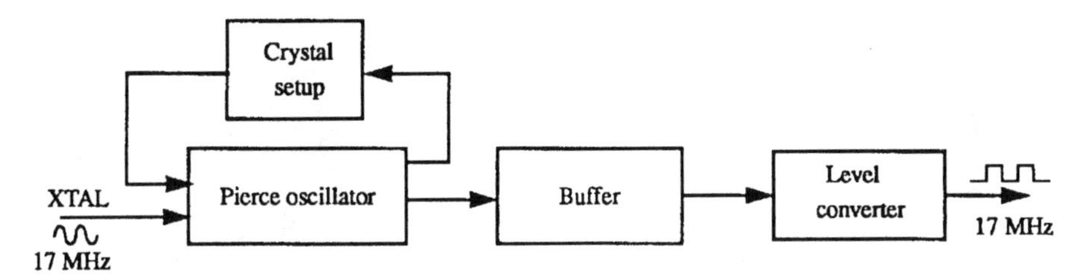

Fig. 10. Block-level schematic of the reference frequency generator.

V. A Design Example

The methodology described in this paper has been applied to the verification of the transmit section of a portable radio front-end IC. Measured results on the fabricated IC indicated an RF spur (undesired or spurious signal) at the output of an up-conversion mixer (modulator) in the transmit section of the circuit. A 310-MHz carrier signal is divided by two and drives the quadrature modulator. The modulating signals are at 1 MHz, so the expected modulator output is at 154 or 156 MHz depending on whether I leads Q or vice-versa. Also present on-chip is a crystal buffer amplifier that is used to generate a clock signal at the reference frequency of 17 MHz. The measured RF spur at the output of the modulator is at 153 MHz, which corresponds to the ninth harmonic of the 17-MHz clock frequency. Since this spur is present at the modulator output even in the absence of the 1-MHz modulating signal, it was evident that this signal had coupled through the substrate and package to the output of the modulator.

Block diagrams of the modulator and reference frequency generator are shown in Figs. 9 and 10, respectively. The modulator has an image-reject architecture where the in-phase and quadrature-phase modulating signals are mixed with the in-phase and quadrature-phase LO signals, respectively. The desired frequency component at the outputs of the mixers is the same while the undesired component differs in phase. A simple addition of the two mixer outputs cancels the undesired component. Each of the I and Q mixers is a conventional bipolar doubly balanced commutating mixer as shown in Fig. 11 where the LO signal is sufficiently large to make the input transistors behave as switches. Therefore, each differential pair alternately steers the modulating signal (plus any bias I_Q) into the output. The modulator and frequency generator blocks are part of a 16 mm $\times$ 16 mm single-chip radio front-end fabricated in a BiCMOS technology on a substrate of 5-Ω-cm resistivity and 25 mils thickness. The

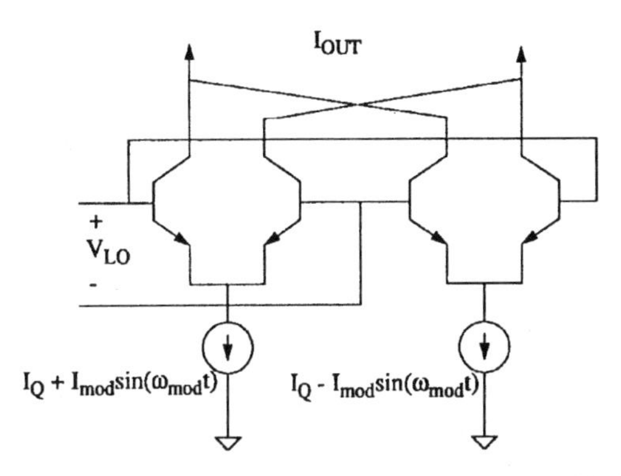

Fig. 11. Active doubly balanced commutating mixer employed in the modulator.

backside of the substrate is connected to a flag/heat sink through a nonconductive epoxy. The substrate is grounded via substrate contacts and guard rings on the substrate surface, surrounding the logic and analog circuitry. To isolate the sensitive circuits, different substrate supplies were used to bias the modulator circuit and the reference frequency generator. For further isolation, these two blocks are separated by a large distance on the lightly doped substrate as shown in Fig. 12 which shows the layout of the frequency generator on the left and the modulator on the right separated by almost 5 mm on the 5-Ω-cm substrate. An extraction of the chip substrate showed that the most significant coupling between the two circuits is approximately 93 Ω between the two substrate ground lines. Together with the package/bond-wire inductance of 8 nH, at 153 MHz the substrate provides an attenuation of approximately 20 dB from the noisy digital ground line to the modulator ground. Note that neither substrate supply is a Kelvin ground (i.e., each substrate ground is also a circuit ground)

The periodic operating point of the modulator under the application of only the carrier signal was first determined by performing a periodic steady-state analysis on the modulator

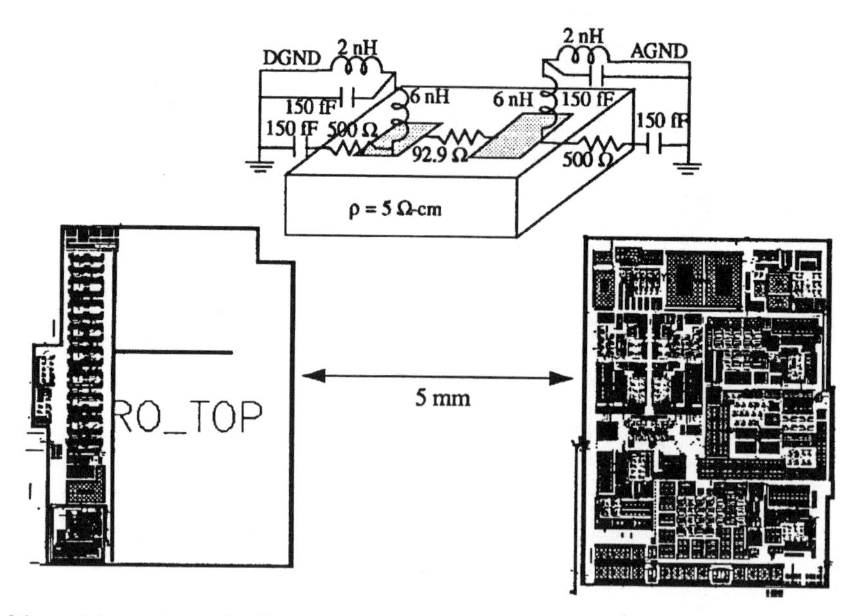

Fig. 12. Physical design of the modulator and crystal buffer amplifier on the chip substrate. Inset shows a simplified model of chip substrate and package.

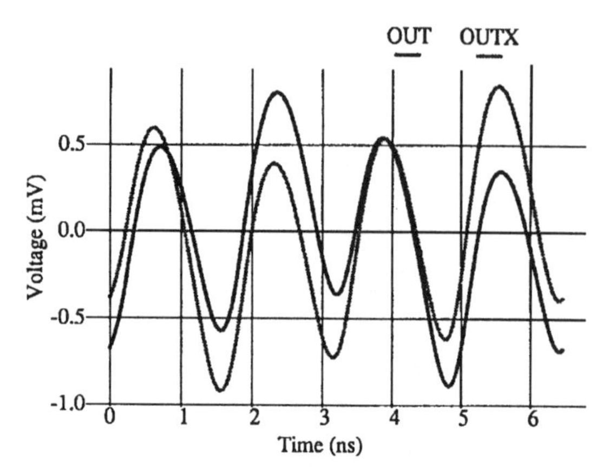

Fig. 13. Periodic steady state analysis output showing carrier feedthrough of modulator under application of the carrier signal only.

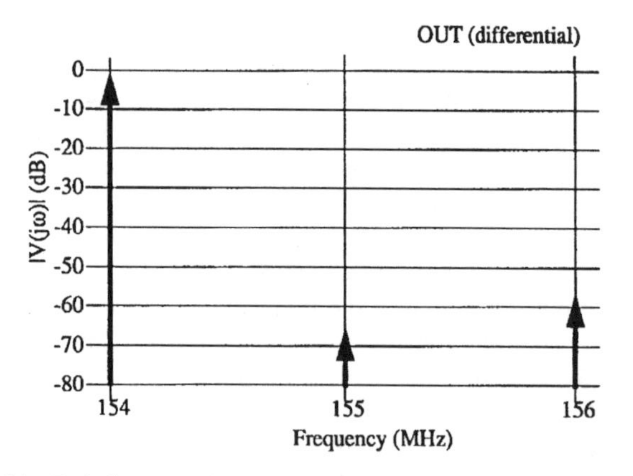

Fig. 14. Periodic ac analysis of up-conversion mixer under application of carrier and 1-MHz modulating signal showing carrier feedthrough and undesired upper sideband at output relative to lower sideband.

at 155 MHz. Each of the differential outputs of the modulator under the large-signal excitation of the carrier is plotted in Fig. 13. To verify operation of the mixer, a periodic ac analysis was performed with a small 1 MHz signal applied to the circuit at the modulating inputs of the circuit linearized about its previously computed periodic operating point. The PAC analysis results in output responses at frequencies $f_{\text{out}} = f_{\text{mod}} + kf_{\text{LO}}$ where f_{LO} is the LO frequency (155 MHz), $k = \cdots -2, -1, 0, 1, 2, \cdots$ is the harmonic number of the LO signal, and f_{mod} is the modulating signal frequency (1 MHz). Results of the PAC analysis showing the carrier feedthrough (155 MHz) and undesired upper sideband (156 MHz) relative to the desired lower sideband (154 MHz) are plotted in Fig. 14. The mixed-signal noise coupling analysis strategy to determine the output spur at 153 MHz for this circuit is illustrated in Fig. 15. First, a macromodel of the RF circuit is determined to characterize the coupling from the modulator substrate to the modulator output. To do so, a periodic small-signal transfer function analysis is performed on the modulator circuit linearized about its previously computed periodic operating point. For a given output frequency component of interest f_{out} (in this case $f_{\text{out}} = 153$ MHz), the PXF analysis results in transfer functions to the modulator output at f_{out} from the substrate at frequencies $f_{sub} = f_{\text{out}} + kf_{\text{LO}}$ where k and f_{LO} are the harmonic number and frequency of the LO signal as before. The computed transfer functions inherently model the conversion of spectral components in the substrate at 153 MHz, 2 MHz, 157 MHz, etc., to the output at 153 MHz due to mixing with the fundamental, first, and second harmonics of the 155-MHz carrier frequency, respectively. Next, to determine the actual noise at the modulator substrate, a transistor-level circuit simulation was performed on just the reference frequency generator and the substrate and package parasitics. (Since the reference frequency generator was small enough to be simulated at the transistor-level, no logic circuit macromodeling was employed.) Fig. 16 shows the resulting transient noise waveform at the substrate ground line of the modulator. A Fourier transform of the transient noise waveform at the modulator substrate was then performed and the resulting spectral components at 153 MHz, 2 MHz, 157 MHz, etc., were multiplied with the previously computed transfer functions to the modulator output and summed to

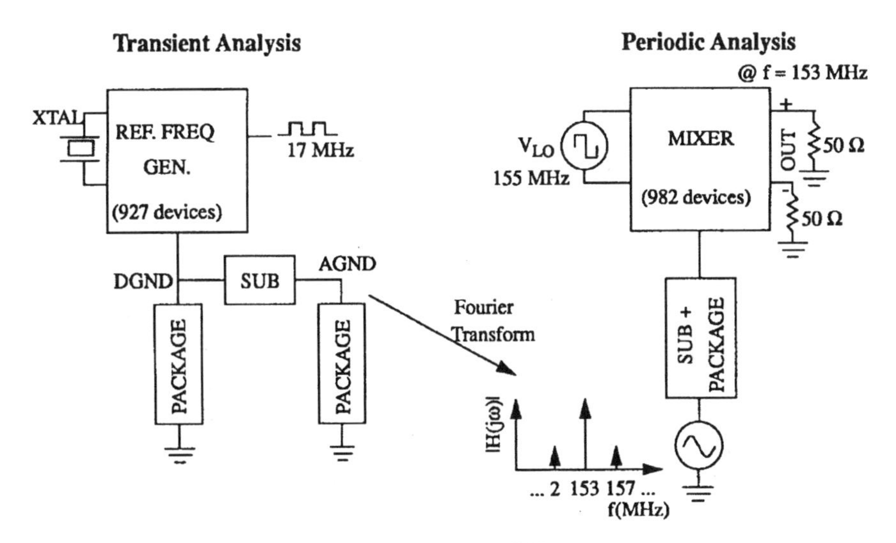

Fig. 15. The mixed-signal noise coupling analysis strategy for the radio front-end IC.

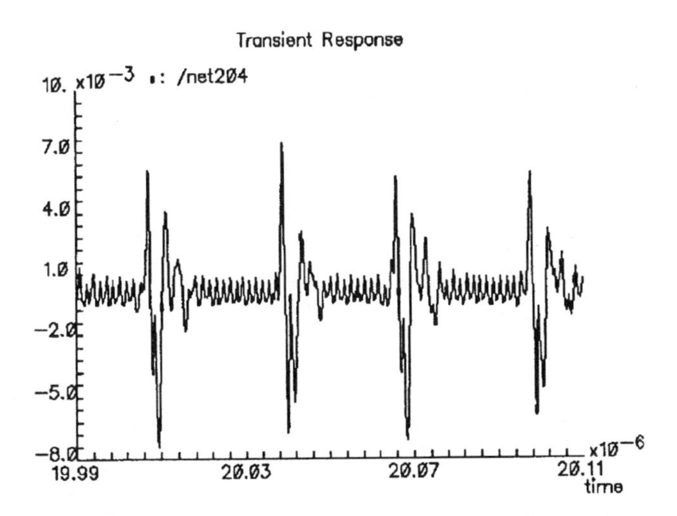

Fig. 16. Simulated transient noise waveform on the modulator substrate ground line.

determine the net output spur at 153 MHz. Under application of only the LO (carrier) signal, the simulated differential spurious output of the modulator at 153 MHz is −72 dBm into a 50-Ω load which matches fairly well with the measured result of −75 dBm. Note that the spur is both measured and simulated in the absence of the 1-MHz modulating signal to isolate the mixed-signal coupling component at 153 MHz from the third-order intermodulation product of the 155-MHz LO with the 1-MHz signal.

With roughly 1900 devices, 717 nodes, and 3234 equations to solve, a transient analysis of the modulator and reference frequency generator together would have required nearly two days of computation to simulate the 20 periods that are required for the modulator to attain steady state. Using periodic analysis instead, a macromodel of the modulator (containing 982 devices, 438 nodes, and 1445 equations) was obtained in under an hour of CPU time. Another hour was required to simulate the transient noise coupling from the reference frequency generator.

VI. Conclusions

Computer-aided design techniques to address mixed-signal coupling in RF integrated circuits were reviewed in this paper. Efficient simulation of mixed-signal coupling is possible using macromodels of the logic circuits, RF circuits, substrate, and package. Several methods for the computation of substrate models and for the macromodeling of logic switching noise were reviewed. Additionally, a technique to determine macromodels of RF circuits for mixed-signal crosstalk using periodic analyses was described. A design example illustrated an application of such simulations to mixed-signal RF IC design.

Acknowledgment

Thanks are due to A. Zocher of Motorola Cellular Research Labs for providing the design example and to K. Kundert, K. Nabors, D. Dumlugol, J. White, T. Schmerbeck, B. Stanisic, and S. Mitra for several helpful discussions.

References

[1] D. G. Steer, "Coexistence and access etiquette in the United States unlicensed PCS bands," *IEEE Personal Commun.*, vol. 1, pp. 36–43, 1994.
[2] C. Huang, "RF research chases elusive one-chip radio," *Op-Ed*, EE Times, p. 45, June 23, 1997.
[3] S. Kiaei, D. J. Allstot, K. Hansen, and N. K. Verghese, "Noise considerations for mixed-signal RF IC transceivers," *ACM J. Wireless Networks*, Mar./Apr. 1998.
[4] R. Telichevesky, K. Kundert, I. Elfadel, and J. White, "Fast simulation algorithms for RF circuits," in *Proc. IEEE Custom IC Conf.*, May 1996, pp. 437–444.
[5] R. Melville, P. Feldmann, and J. Roychowdhury, "Efficient multi-tone distortion analysis of analog integrated circuits," in *Proc. IEEE Custom IC Conf.*, May 1995, pp. 241–244.
[6] R. Gharpurey and R. G. Meyer, "Modeling and analysis of substrate coupling in integrated circuits," *IEEE J. Solid-State Circuits*, vol. 31, pp. 344–353, Mar. 1996.
[7] N. K. Verghese, D. J. Allstot, and M. A. Wolfe, "Verification techniques for substrate coupling and their application to mixed-signal IC design," *IEEE J. Solid-State Circuits*, vol. 31, pp. 354–365, Mar. 1996.
[8] G. C. Hess, *Land Mobile Radio System Engineering.* Boston, MA: Artech House, 1993.

[9] T. S. Rappaport, *Wireless Communications, Principles and Practice.* Upper Saddle River, NJ: Prentice Hall PTR, 1996.
[10] B. Razavi, "Design considerations for direct-conversion receivers," *IEEE Trans. Circuits Syst.–II,* vol. 44, pp. 428–435, June 1997.
[11] G. H. Warren and C. Jungo, "Noise, crosstalk and distortion in mixed analog/digital integrated circuits," *Proc. IEEE Custom IC Conf.,* pp. 12.1.1–12.1.4, May 1988.
[12] B. J. Hosticka and W. Brockherde, "The art of analog circuit design in a digital VLSI world," in *Proc. IEEE Int. Symp. Circuits Systems,* Apr. 1990, pp. 1347–1350.
[13] N. K. Verghese, T. J. Schmerbeck, and D. J. Allstot, *Simulation Techniques and Solutions for Mixed-Signal Coupling in Integrated Circuits.* Boston, MA: Kluwer, 1995.
[14] D. J. Allstot, S. Kiaei, and R. H. Zele, "Analog logic techniques steer around the noise," *IEEE Circuits Devices Mag.,* vol. 9, pp. 18–21, Sept. 1993.
[15] J. R. Long and M. A. Copeland, "The modeling, characterization, and design of monolithic inductors for silicon RF IC's," *IEEE J. Solid-State Circuits,* vol. 32, pp. 357–369, Mar. 1997.
[16] C. Hu, *VLSI Electronics: Microstructure Science.* New York, NY: Academic, 1981, vol. 18, ch. 3.
[17] D. K. Su, M. J. Loinaz, S. Masui, and B. A. Wooley, "Modeling techniques and experimental results for substrate noise in mixed-signal integrated circuits," *IEEE J. Solid-State Circuits,* vol. 28, pp. 420–430, Apr. 1993.
[18] Y. Saad and M. H. Schultz, "GMRES: A generalized minimum residual algorithm for solving nonsymmetric linear systems," *SIAM J. Scientific Statistical Comput.,* vol. 7, pp. 856–859, July 1986.
[19] N. K. Verghese and D. J. Allstot, "Verification of RF and mixed-signal integrated circuits for substrate coupling effects," in *Proc. IEEE Custom IC Conf.,* May 1997, pp. 363–370.
[20] A. M. Niknejad and R. G. Meyer, "Analysis and optimization of monolithic inductors and transformers for RF IC's," in *Proc. IEEE Custom IC Conf.,* May 1997, pp. 375–378.
[21] B. R. Stanisic, R. A. Rutenbar, and L. R. Carley, "Addressing noise decoupling in mixed-signal IC's: Power distribution design and cell customization," *IEEE J. Solid-State Circuits,* vol. 30, pp. 321–326, Mar. 1995.
[22] P. Miliozzi, L. Carloni, E. Charbon, and A. L. Sangiovanni-Vincentelli, "SUBWAVE: A methodology for modeling digital substrate noise injection in mixed-signal IC's," in *Proc. IEEE Custom IC Conf.,* May 1996, pp. 385–388.
[23] S. Mitra, R. A. Rutenbar, L. R. Carley, and D. J. Allstot, "A methodology for rapid estimation of substrate-coupled switching noise," *Proc. IEEE Custom IC Conf.,* May 1995, pp. 129–132.
[24] R: Telichevesky, K. Kundert, and J. White, "Efficient steady-state analysis based on matrix-free Krylov-subspace methods," in *Proc. Design Automation Conf.,* June 1995, pp. 480–484.
[25] R. Telichevesky, K. Kundert, and J. White, "Efficient AC and noise analysis of two-tone RF circuits," in *Proc. Design Automation Conf.,* June 1996, pp. 21-3.1–21-3.6.
[26] ——, "Receiver characterization using periodic small-signal analysis," in *Proc. IEEE Custom IC Conf.,* May 1996, pp. 449–452.

Nishath K. Verghese (S'91–M'95) received the B.E. (Hons.) degree from Birla Institute of Technology and Science, Pilani, India, in 1990 and the M.S. and Ph.D. degrees from Carnegie Mellon University, Pittsburgh, PA, in 1993 and 1995, respectively.

His research at Carnegie Mellon focused on extraction and simulation techniques for substrate-coupled noise in mixed-signal IC's. In 1994, he worked at the Mixed-Signal Design Department, Texas Instruments, Dallas, TX, applying these techniques to the design of a video A/D converter. From 1995 to 1997, he was with Cadence Design Systems, San Jose, CA, where he worked on algorithms and methodologies for mixed-signal/RF IC design and verification. He is currently with Apres Technologies, Cupertino, CA, developing software tools for the verification and optimization of IC layouts for noise crosstalk, manufacturability, and reliability concerns. He has several publications in the areas of crosstalk analysis, layout verification, and synthesis. He is co-author of the book *Simulation Techniques and Solutions for Mixed-Signal Coupling in Integrated Circuits* (Kluwer, 1995). His research interests include design and verification of analog and mixed-signal circuits.

David J. Allstot (S'72–M'78–SM'83–F'92) received the B.S., M.S., and Ph.D. degrees from the University of Portland, OR, Oregon State University, Corvallis, and the University of California, Berkeley, respectively.

He held industrial positions with Tektronix, Texas Instruments, and MOSTEK, and academic positions with Oregon State and Carnegie Mellon Universities. He has also served as an Adjunct Professor at Southern Methodist University and a Visiting Professor at the University of California, Berkeley. He is currently Professor of Electrical Engineering at Arizona State University, Tempe.

Dr. Allstot served as Editor of the IEEE TRANSACTIONS ON CIRCUITS AND SYSTEMS and Guest Editor of the IEEE JOURNAL OF SOLID-STATE CIRCUITS. He is a member of the ISSCC Technical Program and Executive Committees. He was co-recipient of the 1980 IEEE W. R. G. Baker Award, the 1995 IEEE Circuits and Systems Society Darlington Award, the 1998 ISSCC Beatrice Winner Award for Editorial Excellence, and the recipient of several teaching awards. He is a member of Eta Kappa Nu and Sigma Xi.

Integration and Electrical Isolation in CMOS Mixed-Signal Wireless Chips

ROBERT C. FRYE

Invited Paper

The technological trends underlying the application of CMOS technology in wireless applications are leading to the integration of the RF analog functions and the digital baseband processing into a single chip. Two key technical requirements for this integration are the capability to fabricate high Q passive components and the need to maintain electrical isolation between analog and digital components in the resulting mixed-signal chip. Some basic arguments that illustrate the technological conflict between these two important demands are presented, focusing on their implications for the structure of the IC substrate. This structure and the characteristics of the device package play important roles in determining the levels of coupled ground noise that will be present in the mixed-signal IC. A simple, high-level model for coupled ground noise is presented and used to illustrate the impact of design alternatives for the package and for the IC substrate.

Keywords*—*CMOS, coupled noise, mixed signal, wireless.

I. Technological Trends

Portable telecommunications (cordless phones, cellular phones, pagers) and wireless LANs have emerged as commercially important applications of RF wireless technology. The more technically demanding of these applications (e.g., cellular phones) are likely for some time to continue to require mixed technology implementations in order to satisfy their stringent performance requirements. Some less demanding wireless applications (e.g., cordless phones), however, operating at similar frequencies, may operate satisfactorily with predominantly CMOS integrated circuits [1].

Fig. 1 shows the projections for $f_{\max}$ versus technology node (gate length) from the International Technology Roadmap for Semiconductors [2]. A typical RF gain stage requires 20 dB of power gain. Based on a single-pole roll-off in transistor characteristics, devices are useful for frequencies below about 1/10 of $f_{\max}$. CMOS performance

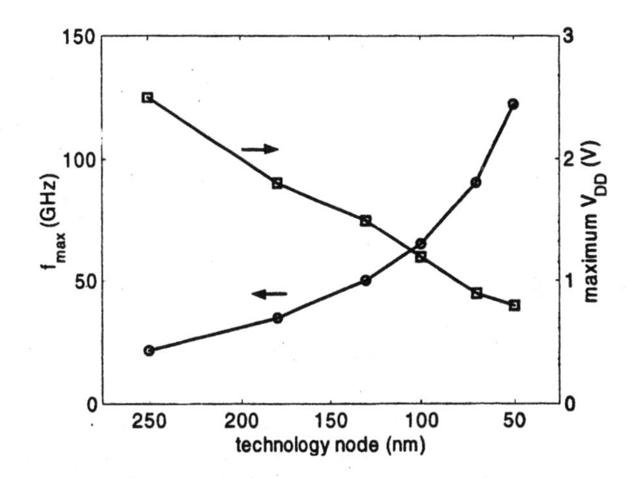

Fig. 1. International technology roadmap for semiconductors projections for maximum operating frequency $f_{\max}$ and maximum operating supply voltage V_{DD} of analog MOS transistors [2].

has recently begun to meet the needs of the 2.4-GHz ISM band, and within a few years it can be expected to be useful in the 5.7-GHz band, as well. This promises to result in low-cost electronic hardware that can access more than 200 MHz of allocated spectrum for new and existing wireless applications.

The projected improvements in CMOS performance are mainly driven by digital applications. Improved transistor speed is a convenient by-product of device scaling, but the important driver behind this trend is economic. To exploit these advantages in RF analog circuits and to realize the goals of highly integrated radios, technologists have focused not only on transistor performance, but also on the fabrication of passive components. For analog circuits, and especially for RF wireless ICs, these passive components dominate the chip area. In contrast to digital circuits that scale in size with new generations of technology, the size of reactive passive components is mainly determined by their component value, which scales inversely with frequency [3]. Consequently, the size of RF analog ICs is nearly constant, despite new generations of transistor technology.

Manuscript received June 11, 2000; revised October 2, 2000.
The author is with Bell Laboratories, Murray Hill, NJ 07974 USA.
Publisher Item Identifier S 0018-9219(01)03197-8.

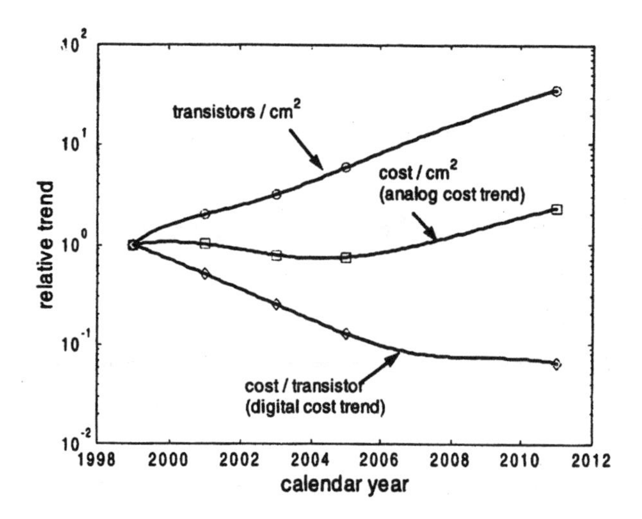

Fig. 2. Projected trends in CMOS technology. Cost/transistor and transistors/cm^2 are projections from ITRS [2], cost/cm^2 is derived.

Fig. 2 shows the projected trends with time in CMOS technology for the number of transistors per unit area and the cost per transistor [2]. In particular, the cost per transistor, which for digital circuits is equivalent to cost per unit functionality, is a strongly decreasing function of time. From these two trends, we can find the projected cost per unit area. RF wireless analog circuits take advantage of the transistor speed improvements that accompany new CMOS generations, but their physical size is roughly constant, so they follow a very different, and less favorable, cost trend.

Clearly, there is a strong economic incentive to move functionality into the digital regime. This is being reflected in the design of a new generation of direct conversion [4] and low IF (intermediate frequency) [5] radio architectures that eliminate analog IF stages. These radios have a minimal amount of analog circuitry in the radio front-end and shift the system complexity into the digital baseband circuits. They take advantage of high-performance analog-to-digital converters [6] and faster digital circuits that are capable of processing at higher baseband frequencies and rely on the continued trends toward lower size and cost for these circuits. This approach has been successfully used to make integrated transceiver front-end ICs for low-end wireless applications [7]–[9] and more recently for mixed-signal integration of the analog front-end with digital baseband circuitry [10].

Two conflicting requirements impede the extension of this approach to higher performance radios. Projections for a direct-conversion "highly integrated" radio in cellular applications suggest that they will require about 75 external components, in addition to the IC itself [11]. The inability to integrate certain key components of the radio stems from fundamental limits in the quality factor, Q, of the passive devices in the IC. Implementing more of the radio's functionality in digital circuitry is motivated, to a large extent, by the desire to replace these components by more cost-effective digital equivalents. However, the integration of both digital and analog components in a common chip inevitably leads to problems of noise coupling from the large signal digital circuits into the sensitive analog circuits. As we will show, physical design choices that result in higher Q passive components also have consequences for electrical isolation.

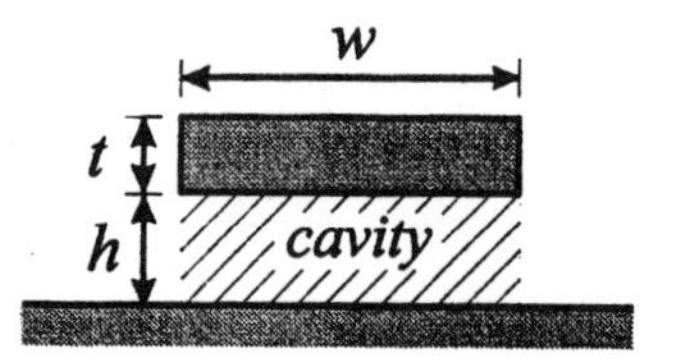

Fig. 3. Cross section of a basic microstrip transmission line.

II. Q VERSUS ISOLATION

A central problem for highly integrated radios is the fundamental conflict between the requirements for electrical isolation and quality factor, Q. This conflict can be illustrated by considering a simple example of a microstrip resonator, whose basic geometric parameters are shown in Fig. 3.

If the dielectric medium surrounding the line has a dielectric constant of ε, then a lower bound for its capacitance per unit length C_0 is given by

$$C_0 > \varepsilon w/h. \tag{1}$$

The expression in (1) is a lower bound because it does not include fringing or sidewall capacitance of the line. If the magnetic permeability of the dielectric medium is μ, then the speed of light in the medium relates the capacitance and inductance per unit length

$$c = 1/\sqrt{\mu\varepsilon} = 1\Big/\sqrt{L_0 C_0}\,. \tag{2}$$

From (1) and (2), we find an upper bound for the inductance per unit length

$$L_0 < \mu h/w. \tag{3}$$

Resistance per unit length at low frequencies is determined by the cross-sectional area and resistivity of the wire

$$R_0 = \rho/(wt) \tag{4}$$

where ρ is the resistivity of the metal. Using (3) and (4), the quality factor of a resonator made from a length of microstrip transmission line is

$$Q \equiv \omega L_0/R_0 < \omega\mu h t/\rho \tag{5}$$

where ω is the angular frequency $2\pi f$.

This expression gives an upper bound on Q as a function of geometric and material parameters. It is clearer if we introduce the skin depth

$$\delta = \sqrt{2\rho/(\omega\mu_C)} \tag{6}$$

where μ_C is the permeability of the conductor. Expressing (5) in terms of the skin depth δ, we find

$$Q < \frac{\mu}{\mu_C}\left(\frac{2ht}{\delta^2}\right) \quad \text{(low frequency)}. \tag{7}$$

At high frequency, electrical conduction is confined to a skin depth and (4) is no longer valid. Instead, the resistance per unit length at very high frequencies is given by

$$R_0 = \rho/(2w\delta) \tag{8}$$

where the metal thickness t has been replaced by 2δ to account for conduction on the upper and lower surface of the line. In this case, (7) becomes

$$Q < \frac{\mu}{\mu_C}\left(\frac{4h}{\delta}\right) \quad \text{(high frequency).} \tag{9}$$

This particular relationship between microstrip dimensions and Q is a special case of a more fundamental, well-known relationship for microwave cavities [12]

$$Q < \frac{\mu}{\mu_C}\left(\frac{V}{A\delta}\right) \times \text{(geometrical factor).} \tag{10}$$

In (10), μ is the permeability of the cavity of volume V and μ_C the permeability of the conducting walls of area A. The geometrical factor depends on cavity shape and is of order unity. For the structure sketched in Fig. 3, if we consider the resonant cavity to be the region between the line and the ground plane, as indicated, and apply the formalism in [12] we arrive at an equivalent result.

A basic implication of (9) is that, barring the use of highly permeable magnetic materials in the cavity volume, high Q is unobtainable in thin physical structures, such as CMOS circuits. In conventional digital CMOS technology, the height of the conductors from the silicon (i.e., the thickness of the interlevel dielectrics) is usually a few micrometers. To minimize the possibility of latch-up, the silicon substrate often consists of a highly conductive bulk region, essentially a ground plane, with a high resistivity epitaxial layer that is on the order of 10 μm thick. Overall, in this structure, the effective separation of the metal conductors from the ground plane is comparable to the skin depth in the metal, and the resulting Q of spiral inductors fabricated in this technology is quite low, on the order of 3 to 5 [13]. An improvement often used for RF CMOS circuits is to use a high-resistivity bulk substrate, on the order of 10 Ω·cm. This makes the separation between the conductor and ground plane (i.e., the underside of the chip) much greater, typically about 200 to 300 μm. This can raise the Q of the devices to values as high as 15, limited mainly by resistive losses in the metal and in the substrate.

One of the key system elements in a radio that depends critically on the Q of its components is the local oscillator. The Q of its resonant tank limits the phase noise of an ideal oscillator. At an offset $\Delta\omega$ from its center frequency ω_0, the phase signal-to-noise ratio is given by [14]

$$N \text{ (dBc/Hz)} = 10 \cdot \log\left[\frac{2kT}{P_{osc}} \cdot \left(\frac{\omega_0}{2Q\Delta\omega}\right)^2\right] \tag{11}$$

where P_{osc} is the power dissipated in the oscillator tank circuit. From this, we see that low phase noise in the local oscillator can always be obtained if we are willing to allocate an arbitrarily large amount of power. However, the desirable combination of low power and low phase noise can only by obtained in conventional, resonator-based oscillator circuits by using high Q passive components. Many of the key applications for low-cost CMOS-based radios are in battery operated portable electronic products for which power consumption is a major concern. Consequently, because of the strong economic incentives to eliminate external components and need for low-power consumption, the current trend in RF CMOS technology is to use high-resistivity bulk substrates and low-resistance metal interconnect layers to help maximize the Q of on-chip inductors.

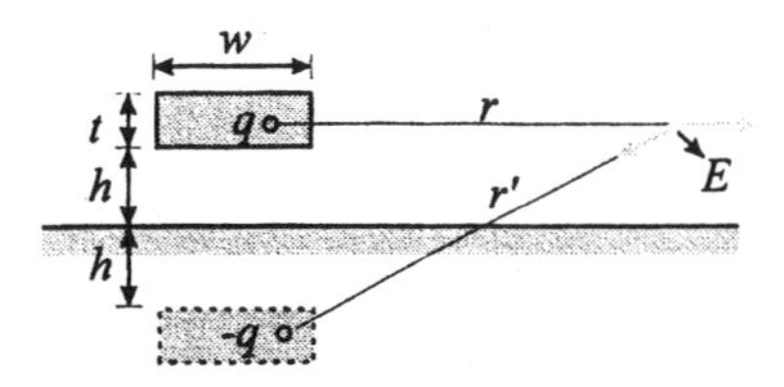

Fig. 4. Induced electric field E at a lateral distance r from a microstrip structure.

In addition to Q, isolation is a key requirement for densely packed highly integrated circuits. In contrast to Q, however, electromagnetic isolation between adjacent components improves with thinner structures [15]. Consider the same microstrip geometry as above in terms of isolation. This is illustrated for the case of induced electric field in Fig. 4. A point charge q in the line induces charge of the opposite sign on the surface of the ground plane. This induced charge can be modeled by an image charge of value $-q$ that is the mirror of q.

At a distance from the charge, the net field is the vector sum of the induced fields from both the charge and its image

$$\vec{E} = \varepsilon q\left(\frac{\langle\vec{r}\rangle}{|r|} - \frac{\langle\vec{r}'\rangle}{|r'|}\right). \tag{12}$$

In the far field (i.e., $|r| \gg |h|$) and neglecting the thickness of the metal

$$\langle\vec{r}\rangle \approx \langle\vec{r}'\rangle$$

and

$$|r'| = \sqrt{|r|^2 + |h|^2} \approx |r|\left(1 + \frac{|h|^2}{2|r|^2}\right) \tag{13}$$

so, in the far field

$$\begin{aligned}\vec{E} &\approx \frac{\varepsilon q}{|r|}\langle\vec{r}\rangle\left(1 - \left(1 + \frac{|h|^2}{2|r|^2}\right)^{-1}\right)\\ &\approx \frac{\varepsilon q}{|r|}\langle\vec{r}\rangle\left(1 - \left(1 - \frac{|h|^2}{2|r|^2}\right)\right).\end{aligned} \tag{14}$$

Or

$$\vec{E} \approx \vec{E}_0 \frac{|h|^2}{2|r|^2} \quad \text{(far field)} \tag{15}$$

where $\vec{E}_0$ is the field in the absence of the shielding ground plane. From this, we can see that the electric field falls off much more quickly (as r^{-3}) in the presence of a ground plane. This effect, however, only applies in the far field, i.e., at lateral distances that are large compared to the vertical separation from the ground plane.

Magnetic coupling has the same spatial dependence in the presence of a ground plane. The analysis is similar, with the fundamental source of the magnetic field being the magnetic dipole moment, rather than charge.

For thin structures like silicon ICs built on epitaxial substrates, the fields in the plane of the devices fall off in a very short distance, making it possible to densely integrate components with good isolation. (Note that this only considers direct electromagnetic coupling. Isolation is also affected by common impedance paths, considered in the section that follows.) However, the same thin structure that confines the fields for high electromagnetic isolation necessarily, as expressed in (10), leads to low Q passive components. Although it is generally recognized that both high Q and electromagnetic isolation are desirable properties for RF circuit technologies, it is not generally recognized that one is obtained at the expense of the other.

III. Coupled Noise in Mixed-Signal ICs

Analysis of electrical isolation in mixed-signal circuits that integrate both the analog radio front-end and the digital baseband processing circuits poses a different set of problems from simple electromagnetic coupling discussed above. This is a result of the extremely dissimilar signal levels in the RF front-end, particularly at the receiver input, and in the digital logic circuits. Digital logic signals that are typically on the order of volts may couple unwanted noise into a receiver that is designed to detect signals on the order of microvolts. To some extent, the frequency selectivity of the receiver mitigates this, but the difference in signal levels makes the mixed-signal coupling problem especially difficult.

Early analyses of coupled noise in mixed-signal ICs attempted to understand and model its physical mechanisms, focusing primarily on the role of the semiconductor substrate. The field-coupling mechanism discussed above treats the resistive part of the silicon substrate as an insulator and the conductive bulk or back gate of the chip as a perfect ground. In reality, neither of these is true. The silicon forms an imperfectly grounded common impedance path that is shared by all circuits on the chip. Very small potential fluctuations in the substrate that would be negligible in the large-signal digital circuits can be significant compared with small-signal levels in the analog circuits.

Su *et al.* [16] examined the effects of circuit placement in both high-resistivity bulk and epitaxial substrates. They showed that circuit placement has some effect on noise, particularly in higher resistivity substrates. Significantly, they found that for circuits separated by a distance greater than about four times the thickness of the high-resistivity layer (i.e., in the far field discussed above) the noise was dominated by voltage fluctuations on the imperfectly grounded back-gate node. This observation is consistent with the simple electromagnetic model discussed above, but also indicates that there are other, more subtle coupling mechanisms that come into play at low coupled noise levels. This work motivated a number of subsequent studies [17]–[19] focused on detailed modeling of current flows in the semiconductor substrate and simulation of coupled noise effects in analog circuits.

Gharpurey and Meyer [20] used a Green's function analysis of the distributed impedance characteristics of an IC substrate to examine electrical isolation as a function of circuit separation. Interestingly, this study considered also variations in the ground connection to the back-gate by varying its inductance in the model. Their simulations showed that for epitaxial substrates, isolation improves with increased separation between circuits, as expected. Importantly, however, they also showed that the degree of isolation that can be obtained is ultimately limited by the package's ground inductance. In high-resistivity bulk substrates, the dependence on the inductance of the substrate contact was much weaker.

These conclusions are further supported in measurements reported by Blalack *et al.* [21], [22]. The main focus of this work was on coupling in analog-to-digital converters. Two important features of this problem are the time-sampled nature of the analog circuitry and the inherent close proximity of the digital and analog transistors. This and subsequent studies [23], [24], seek to understand and characterize the mechanisms of digital switching noise in the time domain and, by using appropriately synchronized sample times, avoid much of the coupled noise.

Although mixed-signal ICs for wireless systems usually contain A/D converters, these components are used at a point in the gain stage of the receiver where the signals are relatively large and noise is less of a problem. For the most part, the nature of noise coupling in wireless applications is different than in A/D converters. The digital and sensitive analog blocks of the circuitry are usually in well-defined separate blocks and can be physically segregated to avoid most of the direct short-range coupling. Furthermore, the system is generally required to receive signals at carrier frequencies that bear no fixed relation to the digital baseband frequency, making synchronization between the two infeasible. Also, the analog front-end of mixed-signal wireless systems is particularly sensitive to some frequencies and insensitive to others. Consequently, investigation and characterization of coupled noise in wireless systems are more naturally concerned with their spectral properties [25]–[27]. Also, since the digital circuits are located far from the analog circuits, it is possible to use macromodels [16], [28] to account for their collective behavior.

IV. A Macromodel for Estimating Coupled Noise

Fig. 5 shows a simple circuit model of the coupling between the digital and analog ground nodes through a conductive back-gate. This model is based on the assumption that the lateral separation between the analog and digital circuits is sufficiently large compared with the thickness of the substrate h so that there is minimal direct coupling between the grounds. In this case, the coupling is predominantly vertical, through the conductive backside of the substrate. Generally,

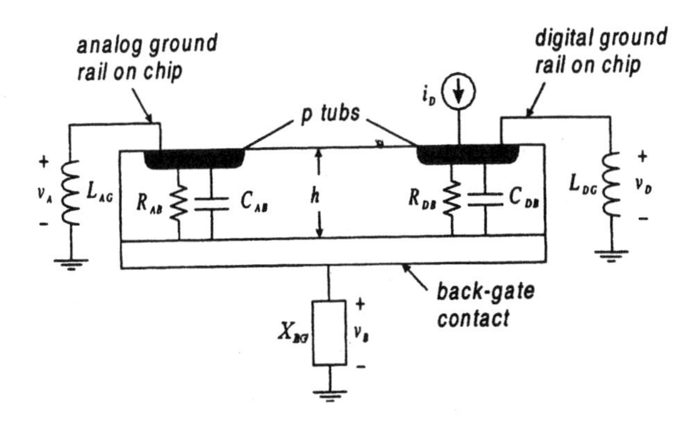

Fig. 5. Model for noise voltage coupling between the digital and analog rails on a chip through the common impedance formed by the substrate.

the n-channel transistors in CMOS ICs are formed in implanted p-type "tubs" having a lower resistivity than the surrounding p-type substrate or epitaxial material. If the overall area of the digital circuitry is A_D, and a fraction α_D is occupied by the p tubs, then the resistance and capacitance coupling the digital ground to the back-gate are given by

$$R_{DB} = \rho h/(\alpha_D A_D) \tag{16}$$

and

$$C_{DB} = \varepsilon \alpha_D A_D / h. \tag{17}$$

Here, ρ and ε are the substrate resistivity and permittivity. Note that the RC time constant is independent of geometry, depending only on the substrate doping level. For typical resistivity, on the order of 10 Ω·cm, the coupling is predominantly resistive below 10 GHz. Similar expressions describe the resistance and capacitance coupling the analog ground rail to the back-gate.

The remaining parameters in this model depend on the characteristics of the package. We assume that the ground reference for the circuit is on the system circuit board where it is typically implemented with a very low impedance conductive plane. The ground return inductance in the digital circuit L_{DG} and in the analog circuit L_{GA} represent the composite inductance of the package ground pins. Commonly, circuits are designed using several pins to lower the overall inductance in this return. This model does not account for local variations in the ground potential, which can generally arise in real circuits having nonuniform current flows. The final element in the model is the reactance that couples the back-gate to ground, X_{BG}. This parameter is strongly dependent on the package structure, as we will discuss below, and may be especially important for mixed-signal noise performance.

The source of the coupled noise in the model is the digital switching current i_D which couples through the impedance of the chip and package to produce voltage disturbances on the ground rails and on the back-gate. Digital switching current, depicted in Fig. 6, exhibits features that make it especially troublesome in mixed-signal radio transceivers. It mainly consists of short duration transients that have wide-band and high-frequency power spectral content. The way this power is distributed over the spectrum is especially problematic. Simulation results for the switching current

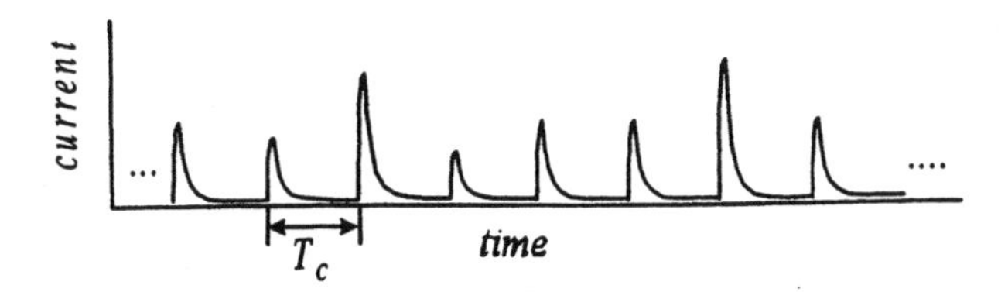

Fig. 6. Switching current transients in a digital circuit.

waveforms in digital circuits have generally shown that their spectrum consists of two parts: a continuous spectrum and a discrete spectrum with its components at multiples of the digital clock frequency [25], [26]. The discrete spectrum causes particular problems because one of its elements may unavoidably coincide with one of the communication frequencies in the radio, or it may occur at a nearby frequency and effectively act like a strong blocker.

Our purpose in the analysis that follows is to estimate the severity of coupled noise in a typical wireless mixed-signal IC and to evolve some guidelines for managing it. Consequently, we seek a high-level description of the switching current waveform that captures the important features of its behavior with a minimum of detailed characterization of the digital circuitry or system. A common high-level method for analyzing the spectral properties of digital waveforms is to treat the waveform as the outcome of a stochastic process. This is often used, for example, to find the spectrum of a digital signal waveform for which the modulation scheme is known, but the information content is not (see, e.g., [29]). The Appendix contains an analysis that uses a similar approach. The analysis decomposes the switching current, as depicted in Fig. 6, into a sequence of pulses each of duration T_C

$$i_D(t) = \sum_{n=-\infty}^{\infty} i_n(t - nT_C) \tag{18}$$

where the current pulse in each period $i_n(t)$ is a stochastic variable. The key result of the analysis is to show that the discrete spectrum of the switching current waveform is given by

$$P_d(f) = F\left[\sum_{k=-\infty}^{\infty} (R_\mu(\tau - kT_C))\right] \tag{19}$$

where

$$\begin{aligned} R_\mu(\tau) &= \frac{1}{T_C}\int_{-\infty}^{\infty} \overline{i_n(t)}\;\overline{i_n(t-\tau)}\,dt \\ &= \frac{1}{T_C}\int_{-\infty}^{\infty} \mu(t)\mu(t-\tau)\,dt. \end{aligned} \tag{20}$$

This result shows that the discrete spectrum is determined entirely by the average behavior of the digital switching current pulses. The higher order statistics of the waveform (contained in its covariance) contribute to the continuous spectrum of the noise. The importance of this result is that it suggests that we can use a high-level description of the average switching current $\mu(t)$ to estimate the levels of noise in the discrete part of the spectrum. This description, illustrated in

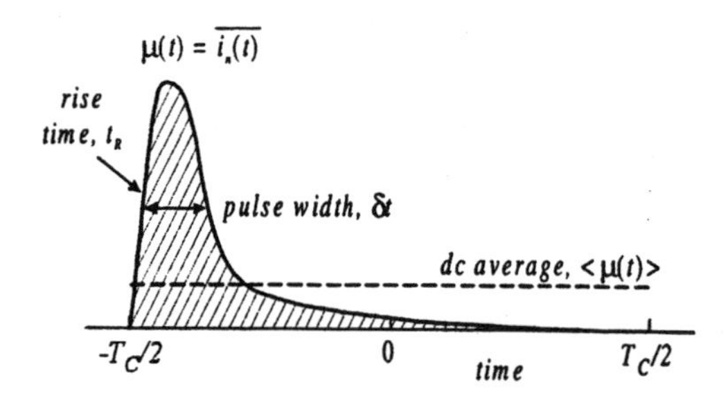

Fig. 7. High-level description of the ensemble average current pulse $\mu(t)$.

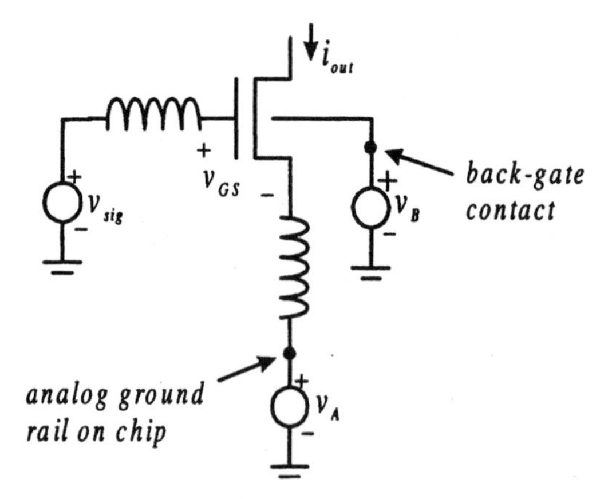

Fig. 8. Typical inductively degenerated common source LNA, showing added noise voltage sources from coupled ground noise.

Fig. 7, is typically characterized by a sharp initial rise in current coinciding with the start of the digital clock cycle, followed by a more gradual decay as signals propagate through the logic levels. The key parameters that describe the waveform are readily estimated from the known or estimated properties of the digital circuit. Descriptions and estimates of this sort are especially useful as a basis for comparison of technological or structural alternatives in the design, as illustrated by the following examples.

V. Example Noise Estimates

To appreciate the significance of coupled noise levels in mixed-signal wireless applications, it is helpful to focus on a specific circuit. Additive noise has the greatest impact on a wireless signal at the input to the receiver, where the signal is weakest. In subsequent stages of the circuit, the signal has been amplified and the same levels of additive noise have much less effect on the overall signal-to-noise ratio (SNR). As an example circuit, consider the inductively degenerated common source low noise amplifier (LNA) circuit shown in Fig. 8. LNA circuits like this one are commonly used in CMOS RF IC designs [13]. The figure also shows the added coupled noise sources at the back-gate, v_B, and on the analog ground rail, v_A. The input signal, which generally arrives from an antenna and input filter has the system circuit board ground plane as its reference. The noise sources, as shown in Fig. 5, are also referred to this same potential.

The output current i_{out} in this circuit is driven by the NMOS transistor's gate-to-source voltage, v_{GS}, characterized by its transconductance, g_m. It is clear from this circuit configuration that the noise voltage on the analog ground rail effectively adds in series with the signal voltage to modulate v_{GS}, so it is a direct source of additive noise. The noise on the back-gate, however, modulates the output current by changing the channel potential of the transistor. Typically, the transconductance for back-gate modulation g_{mb} is lower than g_m. Moreover, this circuit is resonant at its designed operating frequency, and the modulating signal v_{GS} is larger than the input signal by a factor equal to the quality factor of the resonant circuit. These two factors reduce the effects of the back-gate noise on the output current compared with the input signal. We can define an effective back-gate noise voltage referred to the input

$$v'_B = \frac{g_{mb}}{g_m Q_{\text{LNA}}} v_B. \tag{21}$$

With this definition, the overall noise referred to the input of this circuit is the sum of the noise on the analog ground rail and the effective back-gate noise. However, because these two noise voltages arise from a common source, they are correlated and do not add in the same RMS fashion as thermal noise.

Table 1 lists the parameters for the example noise estimates. Entries in the table are grouped into four sections: the first contains parameters that describe the overall structure and physical parameters of the chip. The second contains the high-level description of the digital switching current waveform. The third section describes parameters of the package and the last section describes the LNA circuit parameters.

The physical parameters of the chip are chosen only to be representative of a mixed-signal radio, which may vary considerably from design to design. The main important features are that the digital part is usually physically larger, and a significant fraction of its area is occupied by p-tubs containing transistors. Passive components make up most of the area of the analog circuitry, so the overall p-tub area is much smaller. The following estimates will compare, among other things, the coupled noise in both bulk and epitaxial substrates, for which we will assume the thicknesses h_{bulk} and h_{epi}, respectively.

Comparisons of the spectra of various mathematical representations of $\mu(t)$ show that, if they share the same rise time, pulsewidth and dc average, their spectral properties are remarkably similar. A convenient method for modeling the average digital current waveform is to use the convolution of two exponential waveforms

$$\mu(t) = i_0(u_0(t)e^{-t/t_R}) \otimes (u_0(t)e^{-t/t_F}),$$
$$0 < t < T_C,\ t_R < t_F \tag{22}$$

where $u_0(t)$ is the unit step function and t_R and t_F are the resulting waveform's rise and fall times. The normalization constant i_0 is chosen to give the appropriate dc current. The width of the pulse is

$$\delta t \approx t_R + t_F. \tag{23}$$

The parameters that describe the package pin inductance are mainly dictated by overall package size. Physically larger

Table 1
Model Parameters used in the Example Noise Comparisons.

parameter	value	description
A_D	0.25cm^2	area of digital circuitry
α_D	25%	fraction of digital circuit occupied by p-tub
A_A	0.1cm^2	area of analog circuitry
α_A	5%	fraction of analog circuit occupied by p-tub
h_{bulk}	300μm	thickness of lightly doped bulk substrate
h_{epi}	10μm	thickness of lightly doped epi layer
ρ	10Ωcm	substrate resistivity
t_R	1ns	noise current waveform rise time
δt	5ns	noise current pulse width
T_C	100ns	digital clock period
$< i_D >$	20mA	DC average digital current
L_P	3nH	package pin inductance (including wire bond)
L_{GD}	0.6nH	digital ground composite inductance (5 pins)
L_{GA}	0.6nH	analog ground composite inductance (5 pins)
C_{BG}	3.5pF	substrate ground capacitance (conventional package)
L_{BG}	0.05nH	substrate ground inductance (enhanced package)
g_m/g_{mb}	5	front to back-gate transconductance ratio
Q_{LNA}	10	quality factor of the resonant LNA circuit
R_{in}	50Ω	LNA input resistance

packages generally have longer electrical leads from the chip bond pad to the package's external connection to the board and consequently higher values of inductance. The value of 3 nH per pin used in the estimates is typical of the small plastic molded packages (e.g., TQFP or TSSOP) commonly used in portable wireless products. These packages are low in cost, and their size is typically about 1 cm^2 (see, e.g., [30]). The estimates of the composite inductance in the digital and analog ground connections have assumed a total of five pins allocated to each. The common structure of plastic molded packages is to mount the chip on a conductive pad and electrically connect it to the package leads by wire bonds. This assembly is then encapsulated in molded plastic. The result is to entirely surround the back-gate node of the chip in plastic. Consequently, the reactive connection from the back-gate to the circuit board ground (X_{BG} in Fig. 5) is capacitive. The value of 3.5 pF for C_{BG} listed in the table is based on a pad area of 0.5 cm^2, a separation from the board of 0.5 mm and an average dielectric constant of 4.

An interesting recent development in the technology of molded packages is the use of packages with exposed metal substrate pads. The original motivation for this was mainly to enhance heat transfer from the package to the board, and a variety of package configurations have been developed for this purpose [29], [31]. Originally, these packages were mainly for large, high-pin-count ICs, but more recently a variety of package configurations for small outline packages have been developed [29], [32]. Packages with exposed substrate pads are now available with the same form factor as the small TQFP and TSSOP packages commonly used in low cost portable wireless applications. It is especially interesting to compare the electrical performance of these "enhanced" packages to conventional ones with respect to coupled ground noise. The main electrical difference is in the reactance coupling the substrate pad (back-gate contact) to ground. For a direct solder contact between the substrate pad and ground, the reactance X_{BG} is inductive rather than capacitive. For the very low structural inductance of such a connection, the series impedance is dominated by the inductance of the power/ground plane supply network. The inductance value L_{BG} of 50 pH in the table is estimated from the finite inductance of a power-ground plane radial transmission line [33], assuming an exposed pad size of 5 mm by 5 mm.

Fig. 9 shows the calculated noise referred to the input for a bulk substrate, showing the contributions from the back-gate node v_B and from the analog ground rail v_A. The results compare the conventional package with the enhanced package. Note in these results that the calculated result is not a power density. Values of the components of the discrete spectrum give the total power at each of the harmonics of the clock frequency, 10 MHz in this example. The results as depicted in Fig. 9 assume that one of these components falls within the receiver's band of sensitivity. To place these results in perspective, the range of common receiver sensitivities for terrestrial telecommunications applications is from about -70 to -90 dBm. Noise power comparable to these signal levels would have a significant impact on receiver performance.

An interesting feature of this comparison is that for frequencies of interest in most wireless communication

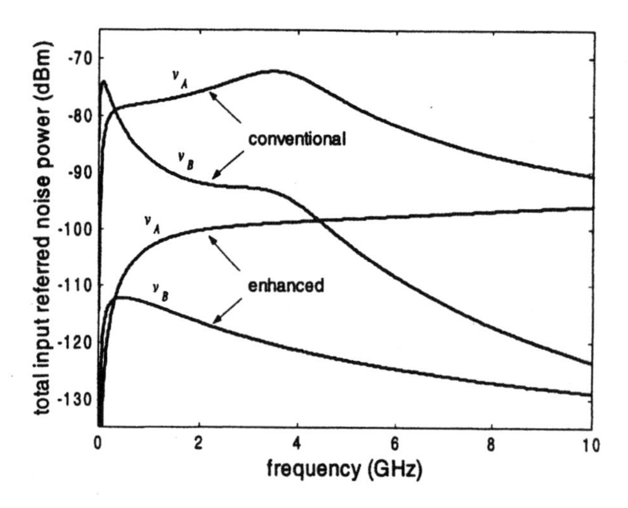

Fig. 9. Calculated values of input referred coupled noise power on the back-gate node (v_B) and on the analog ground rail (v_A) for conventional and enhanced packages with bulk substrates.

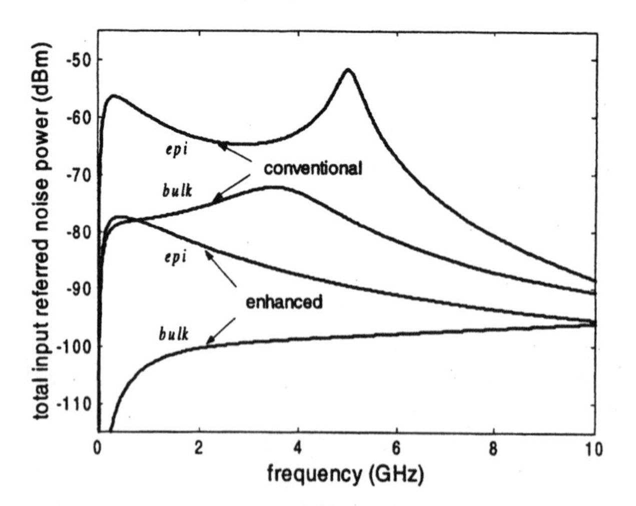

Fig. 10. Calculated values of input referred coupled noise power on analog ground rail (v_A) for conventional and enhanced packages, comparing noise levels for epitaxial and bulk substrates.

applications, ranging from about 1 to 3 GHz, noise coupled on the analog ground rail is more important than noise arising from back-gate modulation of the signal current. For the impedance network depicted in Fig. 5, it is generally the case that the noise voltage level on the back-gate is larger, but the LNA circuit in this example is less sensitive to back-gate coupled noise than to noise on the analog voltage rail. The other main feature of this comparison that is of more practical interest is that a slight modification of the package structure in the case of the enhanced package results in a noise reduction of more than 20 dB in the main frequency range of interest.

Fig. 10 shows a similar comparison between the noise performance of circuits built in epitaxial and in bulk substrates. This example focuses on the noise on the analog ground rail v_A, which was shown in the above comparison to be the dominant noise source. For this particular noise coupling mechanism, we see a large difference between the two substrate types. In contrast to the direct electromagnetic coupling discussed above, for which epitaxial substrates provide better isolation, ground noise coupling through the package and

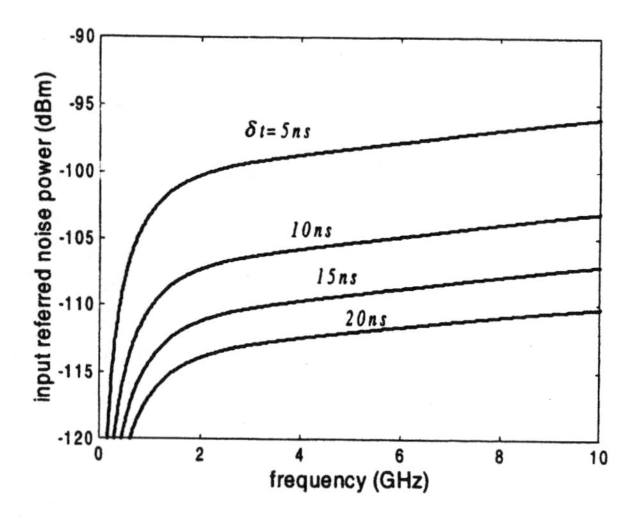

Fig. 11. Calculated values of input referred coupled noise power on analog ground rail (v_A) for enhanced packages, comparing noise levels various values of the switching current pulsewidth.

substrate is less severe in bulk substrates. This is easy to understand by reference to the model circuit in Fig. 5, where it is clear that the higher resistance values in the bulk substrates help to block digital switching current from flowing in the analog ground return path.

The final example, shown in Fig. 11, considers the effect of increasing the average current pulsewidth. This can be realized in practice by adding on-chip decoupling capacitance between the digital power and ground rails to smooth out the switching pulse transients. This lowers the high-frequency spectral content of the current waveform, but leaves its dc average unchanged. The calculations shown in the figure are for the noise on the analog ground rail, using the parameters of the enhanced package. We see in this example that increasing the width of the switching noise current pulse lowers the coupled noise, as we would expect.

VI. Discussion and Conclusion

It is important to keep in mind that the estimates presented above consider only one of a number of coupled noise mechanisms. However, because of the particular spectral nature of this noise source and its magnitude, it is one of the most important mechanisms for mixed-signal wireless integrated circuits. An interesting aspect of this problem, well illustrated in Fig. 10, is that simple and seemingly minor changes in the structure of the package and IC substrate can cause the noise levels arising from common impedance coupling in the ground to change by several orders of magnitude. Additionally, compared with the sensitivity levels of wireless receivers in typical applications, these noise levels that result from various technological choices range from overwhelmingly large to manageably small.

A fortunate property of coupled ground noise is that the adoption of differential design techniques can mitigate some of its effects on circuit performance. Since it is likely that mixed-signal wireless ICs will require differential front-ends, the main benefit to be derived from reducing the noise levels is to relax the demands on the common mode rejection ratio. This is generally limited by the degree of component matching that is available within the ICs manufacturing tolerances. Ultimately, improved ground noise isolation results in relaxed matching requirements and higher device yield.

The development of highly integrated CMOS radios is mainly driven by economic considerations. An important contribution to the cost effectiveness of this technology is the replacement of large numbers of discrete components that characterizes current designs with a single low cost integrated circuit. As we have seen, integration of high Q resonant passive components, especially spiral inductors, dictates the use of high resistivity bulk silicon. Past studies of the effects of circuit layout in high resistivity substrates, as well as simple electromagnetic isolation considerations as presented above, show that this requires more careful attention to placement than in epitaxial substrates. Assuming that the digital and analog circuits are physically segregated, however, an encouraging result of these noise calculations is that high resistivity substrates help to minimize common impedance path coupling through the ground network.

The other significant observation is the beneficial effects that can be obtained by using packages with exposed substrate pads for mixed-signal wireless ICs. Providing a low inductance path from the back-gate of the IC to the system ground reduced the noise by more than 20 dB in the examples that were presented and strongly argues for the adoption of these kinds of packages in wireless designs.

Finally, it is interesting to note that the noise-coupling mechanism described above leads to induced noise voltages in the analog circuits that are proportional to the digital switching current. Consequently, the magnitude of the equivalent input referred noise power increases quadratically with the digital current or power. So, all other factors being equal a twofold decrease in the digital power results in a fourfold reduction in equivalent noise power. This further emphasizes the desirability of low-power digital design methods in mixed-signal wireless applications.

Appendix
Power Spectral Density of Digital Switching Current Waveforms

A digital switching current waveform, $i(t)$ like the one sketched in Fig. 6 may be generally described by

$$i_D(t) = \sum_{n=-\infty}^{\infty} i_n(t - nT_C) \tag{A1}$$

where $i_n(t)$ is the current waveform in the nth clock cycle. With this definition, $i_n(t)$ is zero outside the interval

$$-\frac{T_C}{2} < t < \frac{T_C}{2}. \tag{A2}$$

Lacking detailed knowledge of the digital system that generates the current, we will consider $i_n(t)$ to be a random process and each clock cycle to be a trial whose outcome is $i_n(t)$. In general, because the mean value of $i_n(t)$ is clearly time-dependent, neither $i_n(t)$ nor $i(t)$ is a stationary process.

We can represent a truncated version of the waveform over a time interval T by

$$i_T(t) = \sum_{n=-N}^{N} i_n(t - nT_C) \tag{A3}$$

where the interval T and the summation index N are related by

$$T = (2N+1)T_C. \tag{A4}$$

The autocorrelation function for this truncated current waveform is given by

$$R_i(t, t+\tau) = \overline{i_T(t) i_T(t+\tau)} \tag{A5}$$

or, using (A3)

$$\begin{aligned} R_i(t, t+\tau) &= \overline{\sum_{n=-N}^{N} i_n(t-nT_C) \sum_{m=-N}^{N} i_m(t+\tau-mT_C)} \\ &= \sum_{n=-N}^{N} \sum_{m=-N}^{N} \overline{i_n(t-nT_C) i_m(t+\tau-mT_C)} \end{aligned} \tag{A6}$$

where the bars over the right-hand sides of (A5) and (A6) denote the ensemble average over many samples of the waveform $i_T(t)$.

For a nonstationary process, the power spectral density can be evaluated using the time average of the autocorrelation function in the limit of an infinitely long sample [28]

$$\begin{aligned} R_i(\tau) &= \langle R_i(t, t+\tau) \rangle \\ &= \lim_{T\to\infty} \left[\frac{1}{T} \int_{-T/2}^{T/2} R_i(t, t+\tau)\, dt \right]. \end{aligned} \tag{A7}$$

Here, the time interval T is used in the same sense as in (A4) and (A5). Note that, in the limit of large T, the truncated current waveform approaches the true waveform. The power spectral density is given by the Fourier transform of the time-averaged autocorrelation

$$P(f) = F[R_i(\tau)]. \tag{A8}$$

So, in terms of the function $i_n(t)$, the power spectral density of the switching current $i(t)$ is given by

$$\begin{aligned} P(f) = F\Bigg[\lim_{T\to\infty} \frac{1}{T} \int_{-T/2}^{T/2} \\ \cdot \left(\sum_{n=-N}^{N} \sum_{m=-N}^{N} \overline{i_n(t-nT_C) i_m(t+\tau-mT_C)} \right) dt \Bigg]. \end{aligned} \tag{A9}$$

We introduce the notation

$$R_{nm}(\tau) \equiv \frac{1}{T_C} \int_{-\infty}^{\infty} \overline{i_n(t) i_m(t+\tau)}\, dt \tag{A10}$$

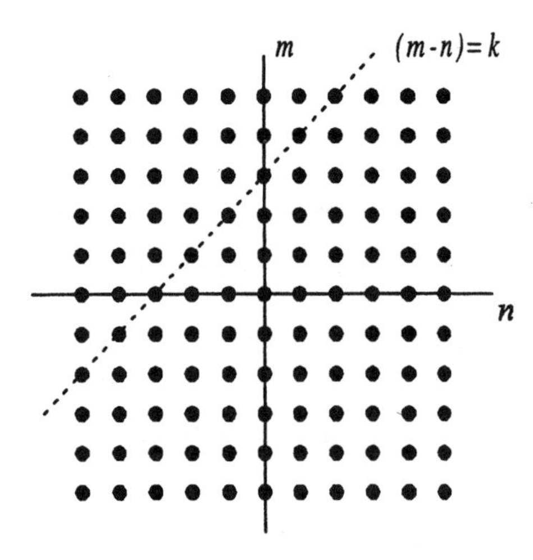

Fig. 12. Region of summation in (A12), shown here for $N = 5$.

and identify $R_{nm}(\tau)$ as the cross-correlation function of the waveforms in the mth and nth clock cycles. Using this notation, (A9) can be rewritten as

$$P(f) = F\left[\lim_{T\to\infty}\left(\frac{T_C}{T}\right)\sum_{n=-N}^{N} \cdot\left(\sum_{m=-N}^{N} R_{nm}(\tau+(n-m)T_C)\right)\right]. \quad \text{(A11)}$$

Making use of (A4)

$$P(f) = F\left[\lim_{N\to\infty}\left(\frac{T_C}{(2N+1)T_C}\sum_{n=-N}^{N} \cdot\sum_{m=-N}^{N} R_{nm}(\tau+(n-m)T_C)\right)\right]. \quad \text{(A12)}$$

The double summation in (A12) can be simplified by noting that its argument depends only on the difference in the indices, $m - n$. We will denote this difference by k. Fig. 12 illustrates the region of summation, shown in this example for $N = 5$. The dots represent the values of the argument in (A12) to be summed. Since this argument only depends on the difference k, all of the points lying along the dotted line for a particular value of k have the same value. The number of points as a function of k is given by

$$p(k) = (2N+1) - |k|. \quad \text{(A13)}$$

So, the summation in (A12) becomes

$$\frac{T_C}{(2N+1)T_C}\sum_{n=-N}^{N}\sum_{m=-N}^{N} R_{nm}(\tau+(n-m)T_C) = \sum_{k=-2N}^{2N}\frac{(2N+1)-|k|}{(2N+1)} R_{nm}(\tau+kT_C). \quad \text{(A14)}$$

Taking the limit for large N, we obtain

$$P(f) = F\left[\sum_{k=-\infty}^{\infty} R_{nm}(\tau+kT_C)\right]. \quad \text{(A15)}$$

The general behavior of the argument on the right-hand side of (A15) is easier to interpret if we define the function $\delta_n(t)$ to describe the departure of the current from its mean, i.e.,

$$i_n(t) = \mu(t) + \delta_n(t) \quad \text{(A16)}$$

where

$$\mu(t) = \overline{i_n(t)}. \quad \text{(A17)}$$

This notation emphasizes that $\mu(t)$ is deterministic and independent of the clock cycle. Note, too, that for a given time t, the ensemble average of $\delta_n(t)$ is zero. An additional consequence of this representation, made apparent in (A17), is that the dc current $\langle\overline{i_n(t)}\rangle$ is entirely specified by $\mu(t)$. With this change in notation, the cross-correlation function becomes

$$\begin{aligned} R_{nm}(\tau) &= \frac{1}{T_C}\int_{-\infty}^{\infty}\overline{(\mu(t)+\delta_n(t))(\mu(t-\tau)+\delta_m(t-\tau))}\,dt \\ &= \frac{1}{T_C}\int_{-\infty}^{\infty}\mu(t)\mu(t-\tau)\,dt + \frac{1}{T_C}\int_{-\infty}^{\infty}\mu(t)\overline{\delta_m(t-\tau)}\,dt \\ &+ \frac{1}{T_C}\int_{-\infty}^{\infty}\mu(t-\tau)\overline{\delta_n(t)}\,dt \\ &+ \frac{1}{T_C}\int_{-\infty}^{\infty}\overline{\delta_n(t)\delta_m(t-\tau)}\,dt. \end{aligned} \quad \text{(A18)}$$

Since $\delta_n(t)$ is a zero-mean variable, the middle two terms on the right-hand side of (A18) are zero, leaving

$$\begin{aligned} R_{nm}(\tau) &= \frac{1}{T_C}\int_{-\infty}^{\infty}\mu(t)\mu(t-\tau)\,dt \\ &+ \frac{1}{T_C}\int_{-\infty}^{\infty}\overline{\delta_n(t)\delta_m(t-\tau)}\,dt \\ &= R_\mu(\tau) + C_{nm}(\tau). \end{aligned} \quad \text{(A19)}$$

In (A19), we identify $R_\mu(\tau)$ as the autocorrelation function of the mean current waveform and $C_{nm}(\tau)$ as the cross-covariance function. Equation (A15) then becomes

$$P(f) = F\left[\sum_{k=-\infty}^{\infty}(R_\mu(\tau-kT_C)+C_{nm}(\tau-kT_C))\right]. \quad \text{(A20)}$$

If we separate this expression into the two main constituents of the spectrum, we find

$$P_c(f) = F\left[\sum_{k=-\infty}^{\infty}(C_{nm}(\tau-kT_C))\right] \quad \text{(A21)}$$

and

$$P_d(f) = F\left[\sum_{k=-\infty}^{\infty}(R_\mu(\tau-kT_C))\right]. \quad \text{(A22)}$$

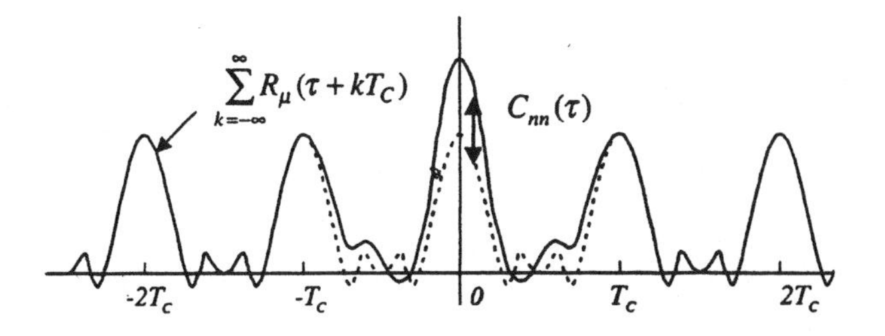

Fig. 13. Example behavior of the autocorrelation function of the switching current.

In (A21), the covariance function can be expected to be a decreasing function of k, the time separation between digital clock cycles. Consequently, we identify $P_c(f)$ as corresponding to the continuous part of the current switching waveform's spectrum. In contrast, in (A22), the autocorrelation of the average switching current is independent of k, so the argument of the Fourier transform is a periodic function. Thus, we can identify $P_d(f)$ as corresponding to the current switching waveform's discrete spectrum.

Fig. 13 further illustrates the general nature of the argument of (A20) for a particular case in which the current waveforms in each clock cycle are uncorrelated. In this case the covariance is zero for nonzero k. Because R_μ is independent of k, the summation describes a periodic component that is determined by the ensemble average behavior of current pulses. Around $\tau = 0$, however, the waveform departs from its periodic behavior because $C_{nm}(\tau)$ is nonzero for $n = m$, and over the interval $-T_C < \tau < T_C$. The power spectral density of this function, then, consists of a continuous spectrum arising from the aperiodic part of this function near $\tau = 0$, and a discrete spectrum arising from periodic part [25].

In the more general case where correlations are present in the digital data, and hence in the current waveform, $C_{nm}(\tau)$ will be nonzero over a wider interval. We would generally expect, however, that for sufficiently large values of τ, i.e., for clock periods that are far apart in time, the waveforms will be uncorrelated. In any case, however, the discrete part of the power spectrum is entirely specified by the $R_\mu(t)$, which only depends on the ensemble mean of the switching current waveform.

Acknowledgment

The author is grateful to J. Glas, A. Demir, and P. Feldmann for helpful discussions and comments, and to the students and faculty of the Berkeley Wireless Research Center for valuable insights into the problems of CMOS radio design.

References

[1] L. E. Larson, "Integrated circuit technology options for RFIC's—Present status and future directions," *IEEE J. Solid State Circuits*, vol. 33, p. 387, 1998.
[2] Semiconductor Industry Association, *The International Technology Roadmap for Semiconductors*, 1999 ed. Austin, TX: International SEMATECH.
[3] R. Frye, "MCM-D implementation of passive RF components: Chip/package tradeoffs," in *Proc. IEEE Symp. IC/Package Design Integration*, Santa Cruz, CA, Feb. 2–3, 1998, pp. 100–104.
[4] A. Abidi, "Direct-conversion radio transceivers for digital communications," *IEEE J. Solid State Circuits*, vol. 30, p. 1399, 1995.
[5] J. Crols and M. Steyaert, "A single-chip 900MHz CMOS receiver front-end with a high performance low-IF topology," *IEEE J. Solid State Circuits*, vol. 30, pp. 1483–1492, 1995.
[6] R. Walden, "Performance trends for analog-to-digital converters," *IEEE Commun. Mag.*, pp. 96–101, Feb. 1999.
[7] M. Steyaert, M. Borremans, J. Janssens, B. De Muer, N. Itoh, J. Craninckx, J. Crois, E. Morifuji, H. Sasaki, and W. Sansen, "A single-chip CMOS transceiver for DCS-1800 wireless communications," in *ISSCC Dig. Tech. Papers*, 1998, pp. 48–49.
[8] A. Rofougarian, G. Chang, J. Rael, J. Chang, M. Rofogarian, P. Chang, M. Djafari, M.-K. Ku, E. Roth, A. Abidi, and H. Samueli, "A single-chip 900 MHz spread-spectrum wireless transceiver in 1μm CMOS, I. Architecture and transmitter design,," *IEEE J. Solid State Circuits*, vol. 33, pp. 515–534, 1998.
[9] A. Rofougarian, G. Chang, J. Rael, J. Chang, M. Rofogarian, P. Chang, M. Djafari, J. Min, E. Roth, A. Abidi, and H. Samueli, "A single-chip 900 MHz spread-spectrum wireless transceiver in 1μm CMOS, II. Receiver design," *IEEE J. Solid State Circuits*, vol. 33, pp. 535–547, 1998.
[10] T. Cho, E. Dukatz, M. Mack, D. MacNally, M. Marringa, S. Mehta, C. Nilson, L. Plouvier, and S. Rabii, "A single-chip CMOS direct-conversion transceiver for 900MHz spread-spectrum digital cordless phones," in *ISSCC Dig. Tech. Papers*, 1999, pp. 228–229.
[11] P. Collander and P. Alinikula, "Cellular phones pull packaging technology," in *Wireless Syst. Des.*, May 1999, vol. 15.
[12] J. Jackson, *Classical Electrodynamics*. New York: Wiley, 1962, p. 258.
[13] L. E. Larson, "Integrated circuit technology options for RFIC's—Present status and future directions," *IEEE J. Solid State Circuits*, vol. 33, pp. 387–399, 1998.
[14] T. Lee, *The Design of CMOS Radio-Frequency Integrated Circuits*. Cambridge: Cambridge University Press, 1998.
[15] R. Frye, C. Hui, and M. Johnson, "Crosstalk noise in a mixed-signal silicon-on-silicon multichip module," in *Proc. Int. Symp. Microelectronics*, Boston, MA, 25–30, 1994.
[16] D. K. Su, M. J. Loinaz, S. Masui, and B. A. Wooley, "Experimental results and modeling techniques for substrate noise in mixed-signal integrated circuits," *IEEE J. Solid State Circuits*, vol. 28, pp. 420–430, 1993.
[17] B. R. Stanisic, N. K. Verghese, R. A. Rutenbar, R. Carley, and D. J. Allstot, "Addressing substrate coupling in mixed mode ICs: Simulation and power distribution synthesis," *IEEE J. Solid State Circuits*, vol. 29, pp. 226–238, 1994.
[18] S. Mitra, R. A. Rutenbar, L. R. Carley, and D. J. Allstot, "A methodology for rapid estimation of substrate-coupled switching noise," in *Proc. IEEE Custom Integrated Circuits Conf.*, 1995, pp. 129–132.
[19] N. K. Verghese, D. J. Allstot, and M. A. Wolfe, "Verification techniques for substrate coupling and their application to mixed-signal IC Design," *IEEE J. Solid State Circuits*, vol. 31, pp. 354–365, 1996.
[20] R. Gharpurey and R. Meyer, "Modeling and analysis of substrate coupling in integrated circuits," *IEEE J. Solid State Circuits*, vol. 31, pp. 344–353, 1996.
[21] T. Blalack and B. A. Wooley, "The effects of switching noise on an oversampling A/D converter," in *Proc. IEEE Int. Solid State Circuits Conf.*, 1995, pp. 199–200.
[22] T. Blalack, J. Lau, F. J. R. Clement, and B. A. Wooley, "Experimental results and modeling of noise coupling in a lightly doped substrate," in *Proc. IEEE Int. Electron Devices Meeting*, 1996, pp. 623–626.
[23] K. Makie-Fukuda, T. Kikuchi, T. Matsuura, and M. Hotta, "Measurement of digital noise in mixed-signal integrated circuits," *IEEE J. Solid State Circuits*, vol. 30, pp. 87–92, 1995.
[24] K. Makie-Fukuda, T. Anbo, T. Tsukuda, T. Matsuura, and M. Hotta, "Voltage-comparator-based measurements of equivalently sampled substrate noise waveforms in mixed-signal integrated circuits," *IEEE J. Solid State Circuits*, vol. 31, pp. 726–731, 1996.
[25] M. van Heijningen, J. Compiet, P. Wambacq, S. Donnay, and I. Bolsens, "A design experiment for measurement of the spectral content of substrate noise in mixed-signal integrated circuits," in *Proc. Southwest Symp. Mixed-Signal Design*, Tucson, AZ, Apr. 11–13, 1999, pp. 27–32.
[26] D. J. Ciplickas and R. A. Rohrer, "Expected current distributions for CMOS circuits," in *Proc. Int. Conf. Computer Aided Design*, San Jose, CA, Nov. 10–14, 1996, pp. 589–592.

[27] A. Demir and P. Feldmann, "Modeling and simulation of the interference due to switching in mixed-signal ICs," in *IEEE/ACM Int. Conf. Computer-Aided Design Dig. Tech. Papers*, San Jose, CA, Nov. 1999, pp. 70–74.
[28] N. K. Verghese and D. J. Allstot, "Computer-aided design considerations for mixed-signal coupling in integrated circuits," *IEEE J. Solid State Circuits*, vol. 33, pp. 314–323, 1998.
[29] L. Couch, *Digital and Analog Communication Systems*, 5th ed., 1997, ch. 6.
[30] JEDEC Solid State Products Engineering Council, *Publication 95*, Jan. 1997.
[31] R. Waterman-Masey, "Thermally enhanced quad flat packages," Motorola Application Note AN1094.
[32] B. Guenin, A. Chowdhury, R. Groover, and E. Derian, "Analysis of thermally enhanced SOIC packages," *IEEE Trans. Comp., Packag., Manufact.Technol. A*, vol. 19, pp. 458–468, 1996.
[33] Y. Chen, Z. Wu, A. Agrawal, Y. Liu, and J. Fang, "Modeling of delta-I noise in digital electronics packaging," in *Proc. IEEE Conf. Multichip Modules*, Santa Cruz, CA, Mar. 15, 1994, pp. 126–131.

Robert C. Frye received the B.S. degree in electrical engineering from the Massachusetts Institute of Technology (MIT), Cambridge, in 1973.

From 1973 to 1975, he was with the Central Research Laboratories of Texas Instruments, where he worked on charge-coupled devices for analog signal processing. In 1975 he returned to MIT, and received the Ph.D. degree in electrical engineering in 1980. Since then, he has been with Bell Laboratories, Murray Hill, NJ. His research activities there have included thin-film semiconductor devices and neural network implementation and applications. More recently, his work has focused on advanced electronic interconnection technology, multichip modules and integrated passive components for RF applications. He is currently a Distinguished Member of Technical Staff in the Design Principles Research Department, and a part-time Industrial Member Researcher at the Berkeley Wireless Research Center.

PART VI

Specialized Analog Simulation

Simulating and Testing Oversampled Analog-to-Digital Converters

BERNHARD E. BOSER, STUDENT MEMBER, IEEE, KLAUS-PETER KARMANN, HORST MARTIN, AND BRUCE A. WOOLEY, FELLOW, IEEE

Abstract—**Quantities such as peak error and integral or differential nonlinearity are commonly used to characterize the performance of analog-to-digital converters. However, these measures are not readily applicable to converter architectures that employ feedback and oversampling. An alternative set of parameters for characterizing the linear, nonlinear, and statistical properties of A/D converters is suggested, and a new algorithm, referred to as the sinusoidal minimum error method, is proposed to estimate the values of these parameters. The suggested approach is equally suited to examining the performance of A/D converters by means of either computer simulations or experimental measurements on actual circuits.**

I. Introduction

SIGNAL PROCESSING systems can be divided into data acquisition and data processing components. While modern VLSI technology greatly simplifies implementation of the processing function by digital means, the same is not true for data acquisition, where analog signals must typically be conditioned and then converted to digital codes. A large number of transistors of small size and high speed are needed for digital processing, whereas conventional means of implementing the analog-to-digital (A/D) conversion function call for a variety of high-precision components. As a result, A/D converters are often implemented using special integrated circuit processes and are fabricated as separate chips. This approach is inefficient in that it both fails to take full advantage of VLSI technology and complicates system implementation by requiring multiple processes and chip sets. There is thus a pressing need for robust A/D conversion techniques that are insensitive to component variations and are compatible with VLSI technology. Oversampling is one approach to meeting this need.

In oversampled A/D converters coarse quantization at a high sampling rate is combined with negative feedback and digital filtering to achieve increased resolution at a lower sampling rate [1]–[3]. These converters thus exploit the speed and density advantages of VLSI while at the same time reducing the requirements on component accuracy. Digital speech processing systems and voice-band telecommunications codecs with A/D converters based on oversampling have already been realized [4]–[10], and the extension of the technique to performance levels required for digital audio has been demonstrated [11].

The performance of signal processing systems is typically specified by *mean squared error* (MSE) or *signal-to-noise ratio* (SNR), rather than by quantities such as peak error or integral and differential linearity that are commonly used to characterize classical Nyquist rate A/D converters. In this paper appropriate definitions of measures for both dynamic range and harmonic distortion based on the MSE are derived, and an algorithm, called the sinusoidal minimum error method, for estimating these parameters is suggested. In this algorithm the error at the output of a converter is determined by minimizing the power of the difference between the output signal and a template comprising a sinusoid at the frequency of the input signal, a dc offset, and explicitly evaluated harmonics. The applicability of the method to both the simulation and testing of oversampled converters is considered, and the algorithm is compared with alternative methods of evaluating A/D converter performance.

The sinusoidal minimum error method offers two advantages over spectral estimation techniques traditionally used for the simulation and testing of A/D converters [12], [13]. First, it is computationally more efficient because amplitude and phase of the converter output are computed only at the signal frequency and, optionally, any harmonics of interest. Second, no windowing of the data is necessary even in cases where the sampling frequency is not an integral multiple of the signal frequency. Thus, the difficult problem posed by spectral estimation of trading off accuracy against spectral resolution is avoided [14].

II. Performance of Oversampled A/D Converters

In oversampled A/D converters, such as that shown in Fig. 1, the analog input $x(t)$ is sampled at a rate well above the Nyquist frequency and the amplitude is coarsely quantized [15]. The error introduced by the quantizer is spread out over the entire frequency band from zero to the sampling frequency f_s. Quantization noise above the signal band is then removed with a digital decimation filter wherein the signal is resampled at a rate $1/T$.

Manuscript received January 10, 1987; revised July 28, 1987, and January 4, 1988. The review of this paper was arranged by A. J. Strojwas, Editor.

B. E. Boser and B. A. Wooley are with the Center for Integrated Systems, Stanford University, Stanford, CA 94305.

K.-P. Karmann is with Siemens AG, Communication and Information Systems, Otto-Hahn-Ring 6, 8000 Munich 83, West Germany.

H. Martin is with Siemens AG, EWSD Basic Development System Peripherie, ETP 33, Boschetsriederstrasse 133, 8000 Munich 70, West Germany.

IEEE Log Number 8820200.

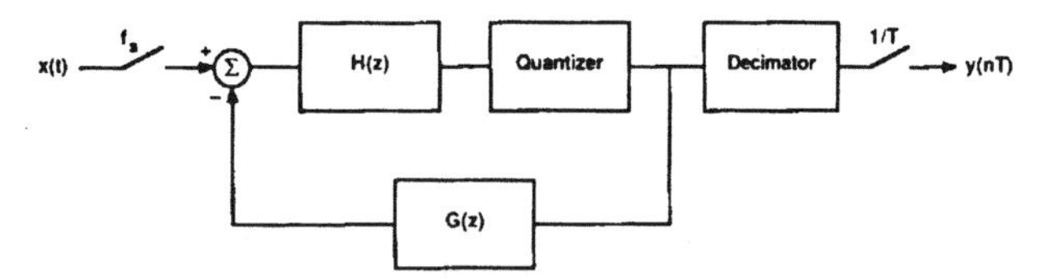

Fig. 1. Oversampled A/D converter.

Due to the feedback in the modulator, every digital output sample is a function of a sequence of analog input samples. Measures, such as peak error, that do not account for the history of the input signal are therefore meaningless for oversampled converters. Consequently, performance evaluation methods based on these parameters [16]–[18] generally cannot be used to characterize such converters. However, the mean squared error within the signal bandwidth of interest is an appropriate alternative basis for performance specification.

For every input signal $x(t)$, an A/D converter produces a sequence of digital output codes $y(nT)$. Errors in this conversion can be attributed to a variety of sources, including quantization in time (sampling) and amplitude, which are intrinsic to any A/D converter, linear errors such as gain and delay, nonlinear errors such as harmonic distortion, and additive errors such as thermal noise.

Errors due to the ideal sampling process itself are avoided simply by ensuring that the input signal is sampled at a rate that is at least twice the signal bandwidth. Quantization in amplitude maps the continuous amplitude range of the analog input signal onto a finite number of digital codes corresponding to a discrete set of amplitude levels. The linear errors describe the deviation of the gain and phase shift of the converter from ideal—usually unity gain and linear phase—and are often a function of the frequency of the input signal. Phase linearity and variation of the gain with frequency are important parameters in many applications, while the absolute value of the gain is often of less concern. Harmonic distortion is caused by nonlinearities in the A/D converter, and thermal noise is present in any analog circuit.

It is crucial that performance evaluation methods discriminate among the different types of error rather than simply lump them into a single parameter. For example, harmonic distortion has very different consequences for the operation of a system than a constant delay of the signal. Another important criterion for an evaluation procedure is its applicability to both computer simulations and measurements of actual devices, so as to simplify the comparison of expected with measured performance.

III. Sinusoidal Minimum Error Method

In order to measure the performance of an A/D converter, test signals must be easy to define and generate. Sinusoids fulfill these requirements and are widely used for system characterization [19]. For a sinusoidal input

$$x(t) = A \cos 2\pi f_x t \tag{1}$$

the output of the A/D converter is a sinusoidal signal at the input frequency, f_x, together with the error introduced by the quantization. In distortion measurement instrumentation [20], the total noise and distortion are commonly estimated by measuring the power of the difference between the output signal and a sinusoid at the frequency f_x that has been fitted to the output so as to minimize the difference power. The *sinusoidal minimum error* algorithm is a generalization of this same principle. Specifically, the difference between the system output, $y(nT)$, and a template consisting of a sinusoid at the input frequency, a dc offset term, and harmonics that are to be explicitly evaluated is minimized. The template, $\hat{y}(nT)$, and a residual error, $e(nT)$, are defined such that

$$y(nT) = \hat{y}(nT) + e(nT) \tag{2}$$

where

$$\hat{y}(nT) = \underbrace{a_0}_{\text{offset}} + \underbrace{a_1 \cos(2\pi f_x nT + \phi_1)}_{\text{signal}} + \underbrace{\sum_{k=2}^{K} a_k \cos(2\pi k f_x nT + \phi_k)}_{\text{harmonics}} \tag{3}$$

and $e(nT)$ is an additive error comprising errors due to both amplitude quantization and thermal noise, as well as any harmonics not included in $\hat{y}(nT)$.

The offset, signal, and harmonics constitute spectral components of the system output and are combined in $\hat{y}(nT)$. In general, the number of harmonics present in the output is infinite. However, in practice only the power in the lower order harmonics is significant. Thus, an explicit parameter K has been introduced in (3) to reflect that only a limited number of harmonics relevant to a particular application need be considered.

The amplitudes, a_k, and phases, ϕ_k, of the output signal and its harmonics can be determined by fitting $\hat{y}(nT)$ to the system output $y(nT)$ so as to minimize the mean square (power) of the error $e(nT)$, σ_{ee}^2, where

$$\sigma_{ee}^2 = E\{e^2(nT)\} \triangleq \frac{1}{N} \sum_n e^2(nT). \tag{4}$$

In this notation $E\{u(nT)\}$ is the expectation of the sequence $\{u(nT)\}$ and N is the number of samples over which the expectation is evaluated.

The sequence $\hat{y}(nT)$ is a linear function of all a_k, but is nonlinear in ϕ_k. However, if (3) is rewritten as

$$\hat{y}(nT) = \sum_{k=-K}^{+K} \hat{Y}(k) W^{kn} \tag{5}$$

where

$$W = e^{j(2\pi f_x T)} \tag{6}$$

and

$$\hat{Y}(k) = \frac{a_k}{2} e^{j\phi_k} \tag{7}$$

then the minimization of σ_{ee}^2 becomes linear in the complex variables $\hat{Y}(k)$. This can be demonstrated as follows.

A necessary condition for minimizing the power in $e(nT)$ is that

$$\frac{\partial}{\partial \hat{Y}(l)} \sigma_{ee}^2 = 0, \qquad -K \le l \le +K. \tag{8}$$

From substituting (2) and (4) for σ_{ee}^2 it follows that

$$\frac{\partial}{\partial \hat{Y}(l)} E\left\{\left[y(nT) - \hat{y}(nT)\right]^2\right\} = 0. \tag{9}$$

Because the order of differentiation and taking the expectation is interchangeable, it is apparent from (5) and (9) that

$$E\left\{-2\left(y(nT) - \sum_{k=-K}^{+K} \hat{Y}(k) W^{kn}\right) W^{ln}\right\} = 0 \tag{10}$$

and thus

$$E\left\{y(nT) W^{ln}\right\} = E\left\{W^{ln} \sum_{k=-K}^{+K} \hat{Y}(k) W^{kn}\right\}, \qquad -K \le l \le +K. \tag{11}$$

Equation (11) is a system of $2K + 1$ linear equations in the $2K + 1$ complex unknowns $\hat{Y}(k)$. The unknowns a_k and ϕ_k are in turn related to the $\hat{Y}(k)$ by

$$a_k = \begin{cases} \hat{Y}(0) & \text{if } k = 0 \\ 2\left|\hat{Y}(k)\right|, & 1 \le k \le K \end{cases} \tag{12}$$

and

$$\phi_k = \arg \hat{Y}(k), \qquad 1 \le k \le K. \tag{13}$$

IV. Performance Specification

The linear properties of an A/D converter can be characterized by its gain, $G = a_1/A$, and phase, ϕ_1, both possibly functions of the signal frequency f_x. The signal power at the output of the A/D converter is

$$\sigma_{\text{out}}^2 = \frac{a_1^2}{2} \tag{14}$$

and the power in the harmonics is

$$\sigma_{hh}^2 = \frac{1}{2} \sum_{k=2}^{K} a_k^2. \tag{15}$$

Since $e(nT)$ is uncorrelated with $\hat{y}(nT)$, as proven in the Appendix, its power is given by

$$\sigma_{ee}^2 = \sigma_{yy}^2 - \sigma_{\text{out}}^2 - \sigma_{hh}^2 - a_0^2 \tag{16}$$

where

$$\sigma_{yy}^2 = E\left\{y^2(nT)\right\}. \tag{17}$$

Additive errors are commonly expressed in terms of signal-to-noise ratio. Of the many possible definitions of this quantity, the following are particularly useful. The *signal-to-noise ratio*,

$$SNR = \frac{\sigma_{\text{out}}^2}{\sigma_{ee}^2} \tag{18}$$

accounts for the quantization error and thermal noise, and the *total signal-to-noise ratio* includes the harmonic distortion resulting from nonlinearities in the converter:

$$TSNR = \frac{\sigma_{\text{out}}^2}{\sigma_{ee}^2 + \sigma_{hh}^2}. \tag{19}$$

These formulas exclude the dc offset, a_0, which can be included as part of either the signal or error terms.

V. Practical Considerations

Only a finite number, N, of output samples $y(nT)$ can be obtained from simulations or measurements on an A/D converter. Consequently, the expectation in (11) must be approximated by a sum over N samples:

$$\sum_{n=0}^{N-1} y(nT) W^{ln} = \sum_{n=0}^{N-1} W^{ln} \sum_{k=-K}^{+K} \hat{Y}(k) W^{kn}, \qquad -K \le l \le +K. \tag{20}$$

The output $y(nT)$ and the signal frequency, f_x, relative to the sampling rate, $1/T$, can be determined either by a simulation or an actual measurement on the system. The amplitude A of the input signal $x(t)$ of the converter need not be known other than to evaluate the gain G.

As an example, Fig. 2 presents simulation results obtained for an oversampled A/D converter of the form shown in Fig. 1, where $H(z)$ performs a double integration [1]. Both the *SNR* and the *TSNR* of the A/D converter are shown as a function of the normalized input signal power. For linear $H(z)$, the power in the harmonics is zero and the *SNR* and *TSNR* do not differ. Nonlinearities in $H(z)$ introduce harmonic distortion at high signal powers, and the *TSNR* thus decreases relative to the *SNR*.

The sinusoidal minimum error method presented here is equally suited to both simulations and measurements on actual circuits. A signal source with high stability and low harmonic distortion is the only precise analog component needed for experimental measurements. Consequently, very high signal-to-noise ratios can be determined accurately.

For simulation purposes, computational errors can be made arbitrarily small in digital algorithms and are therefore of no concern for the method presented here. However, both the finite record length, N, and the precision with which the signal frequency f_x is known introduce systematic errors. Generally, these errors depend on the spectrum of the converter output $y(nT)$. The magnitude of the errors can easily be estimated by means of an example where, for the sake of simplicity, the output $y(nT)$ is assumed to be the sum of two sinusoids at frequencies f_1 and f_2:

$$y(nT) = A_1 \cos 2\pi f_1 nT + A_2 \cos 2\pi f_2 nT. \tag{21}$$

Such an output would be generated by a converter with a quadratic nonlinearity in response to a sinusoidal input with frequency f_1. When the sinusoidal minimum error

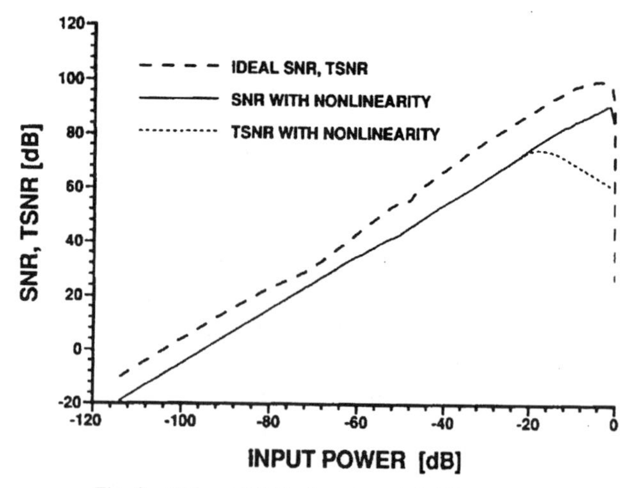

Fig. 2. *SNR* and *TSNR* of oversampled A/D converter.

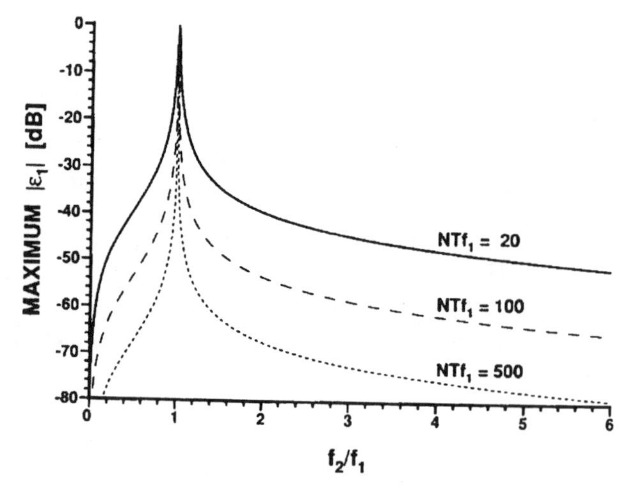

Fig. 3. Minimum sidelobe suppression for $NTf_1 = 20$ 100 and 500.

method is applied to determine the amplitude A_1 of the signal at f_1,

$$\hat{y}(nT) = a_1 \cos 2\pi f_1 nT. \tag{22}$$

The estimated amplitude a_1 can be expressed in terms of A_1 and A_2 as

$$a_1 = A_1 + \epsilon_1 A_2. \tag{23}$$

Ideally, the estimation of the amplitude, a_1, of the output signal at frequency f_1 is independent of the sine wave at frequency f_2; i.e., $a_1 = A_1$. For an infinite number of samples, $\epsilon_1 = 0$ for $f_1 \neq f_2$; that is, a signal at frequency f_2 is completely suppressed. For finite N, ϵ_1 is bounded by

$$|\epsilon_1| \leq \left| \frac{1}{\pi NTf_1} \frac{f_2/f_1}{(f_2/f_1)^2 - 1} \right|, \quad NTf_1 \gg 1. \tag{24}$$

Fig. 3 shows the upper bound on $|\epsilon_1|$ as a function of f_2/f_1 for various NTf_1. For example, with $N = 200$ and $f_1 = 1/10T$ ($NTf_1 = 20$), ϵ_1 is less than -38 dB when $f_2 = 2f_1$.

Fig. 4 shows the estimation error that results when the signal frequency is not accurately known. For this calculation, the system output $y(nT)$ is assumed to be sinusoidal with frequency $f_2 = f_1 + \Delta f$; that is,

$$y(nT) = A_1 \cos 2\pi(f_1 + \Delta f)nT \tag{25}$$

where Δf is the inaccuracy of the signal frequency. If A_1 is to be estimated with the sinusoidal minimum error method, but due to measurement errors the frequency at which the estimation is performed is f_1 rather than $f_1 + \Delta f$, then

$$\hat{y}(nT) = a_1 \cos 2\pi f_1 nT. \tag{26}$$

In this case, the estimated amplitude a_1 is given by

$$a_1 = (1 - \epsilon_2)A_1. \tag{27}$$

The correct amplitude, $a_1 = A_1$, is obtained when $\epsilon_2 = 0$. The error in computing a_1 increases with N but is negligible as long as $\Delta fNT \ll 1$. From Fig. 4 it can be seen

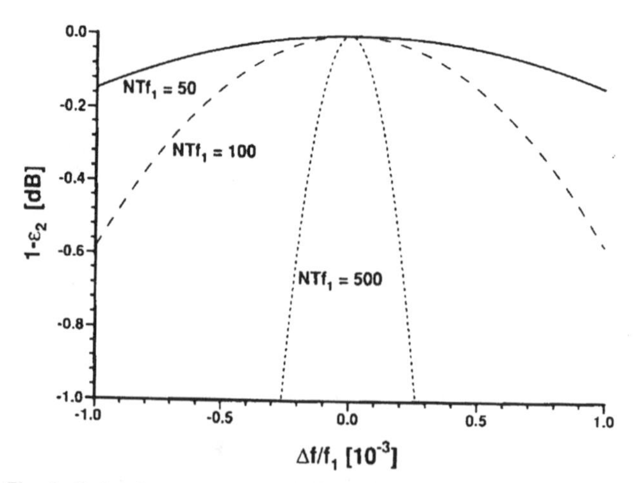

Fig. 4. Estimation error as a function of frequency error ($NTf_1 =$ 50 100 500).

that with $f_1T = 1/10$, $N = 500$, and $\Delta f/f_1 = 0.05$ percent, $1 - \epsilon_2 = 0.996$; that is, the estimation error is less than 0.04 dB.

VI. Comparison with Other Techniques

Several other techniques have been suggested to measure distortion in A/D converters. In this section the minimum sinusoidal error method is compared with two of these approaches.

Spectral techniques based on the Fourier transform [12], [21] are commonly used to determine the signal-to-noise ratio of A/D converters. One approach is to estimate the entire power density spectrum of the converter output and then separate the signal and the errors in the frequency domain [13]. The sinusoidal minimum error method is computationally more efficient than spectral methods based on estimating the entire spectrum since only the amplitude and phase of the signal and, optionally, the harmonics of interest need be estimated, rather than the entire spectrum. This savings in computational effort has proven crucial in simulation tools intended for the design and optimization of A/D converter structures.

While being computationally more efficient than estimating the entire spectrum, the sinusoidal minimum error method is otherwise very similar to spectral estimation techniques. In fact, the left-hand side of (20) is exactly the discrete Fourier transform (DFT) of $y(nT)$, whereas the right-hand side is the spectrum $\hat{Y}(k)$ multiplied by the matrix

$$\left[\sum_{n=0}^{N-1} W^{n(l+k)}\right]_{kl}, \quad -K \le k, l \le +K. \tag{28}$$

If the duration, NT, of the sequence $y(nT)$ is a multiple of the signal period $1/f_x$, (28) is a unity matrix, and the sinusoidal minimum error method and the DFT give the same results. If this condition is violated—for example when the sampling frequency $1/T$ is not an integral multiple of the signal frequency—a window must be used for the DFT. Windows call for a tradeoff between sidelobe suppression and frequency resolution for a given number of samples, N. The sinusoidal minimum error method does not call for such tradeoffs. Consequently, in many cases fewer samples of $y(nT)$ are needed with this method than with the DFT.

An important reason for the popularity of the DFT is the existence of the FFT [12]. Because of the similarity of the sinusoidal minimum error method and the DFT, the same means of speeding up computations is applicable to both techniques. Equation (20) can be rewritten for real arithmetic and the matrix (28) can be precomputed and decomposed for a series of computations.

A computationally efficient technique for estimating only uncorrelated errors (which exclude, for example, harmonic distortion) has been suggested by Candy [22]. The same signal $x(t)$ is applied to two identical A/D converters as illustrated in Fig. 5. The signal and correlated errors at the outputs of the two systems will then be equal. If we further make sure that the uncorrelated errors are orthogonal by initializing the two oversampled converters to different states, the power in the difference of the two system outputs equals twice the power of the uncorrelated errors. While this procedure does not reveal any information about linear errors or harmonic distortion in an A/D converter, it is computationally more efficient than either the sinusoidal minimum error method or the discrete Fourier transform approach. Moreover, it is not restricted to periodic test signals, but allows for random input signals as well. The critical prerequisite for using this algorithm is the availability of two *identical* A/D converters. For example, only slight variations of the gains of the two systems could cause a difference of the signal at the outputs which far exceeds the uncorrelated errors. Consequently, the technique is practical only for computer simulations.

As pointed out previously, many of the techniques used to characterize classical A/D converters are not applicable to oversampled converters. Conversely, however, the sinusoidal minimum error method presented here can be used to evaluate both classical and oversampled A/D converters, thus making possible direct comparison between different conversion architectures.

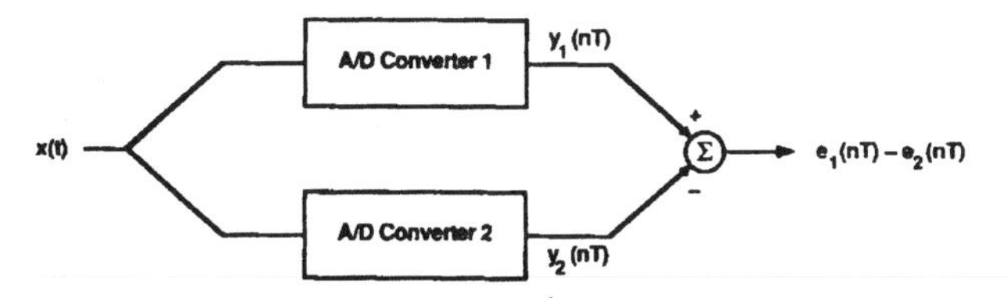

Fig. 5. Estimation of uncorrelated noise.

VII. Conclusions

A method for estimating the performance of A/D converters under real-time conditions has been proposed. The technique is applicable to both classical and oversampled converters and does not entail analog measurements that could limit its resolution. The tradeoffs between measurement time and accuracy of the algorithm have been explored, and estimates of the precision have been presented.

Appendix

First it will be shown that $e(nT)$ and $\hat{y}(nT)$ are uncorrelated. This result will then be used to derive (16).

Lemma:

$$E\{e(nT)W^{ln}\} = 0. \tag{29}$$

Proof: It follows from (5) and (10) that

$$E\{[y(nT) - \hat{y}(nT)]W^{ln}\} = 0, \quad -K \le l \le +K \tag{30}$$

and then from (2) that

$$E\{e(nT)W^{ln}\} = 0, \quad -K \le l \le +K \quad \text{QED.} \tag{31}$$

Proposition: The cross-correlation of $e(nT)$ and $\hat{y}(nT)$ is zero:

$$E\{e(nT)\,\hat{y}(nT)\} = 0. \tag{32}$$

Proof: Substitution of (5) for $\hat{y}(nT)$ leads to

$$E\{e(nT)\,\hat{y}(nT)\} = E\left\{e(nT) \sum_{k=-K}^{+K} \hat{Y}(k)W^{kn}\right\}. \tag{33}$$

Upon exchanging the summation and expectation operators,

$$E\{e(nT)\,\hat{y}(nT)\} = \sum_{k=-K}^{+K} \hat{Y}(k)\,E\{e(nT)W^{kn}\}. \tag{34}$$

The value of the expectation in this expression is zero according to the above Lemma; hence

$$E\{e(nT)\,\hat{y}(nT)\} = 0 \quad \text{QED.} \tag{35}$$

From (35) it follows that the power in the output sequence $y(nT)$ is

$$\begin{aligned}\sigma_{yy}^2 &= E\{y^2(nT)\} \\ &= E\{[\hat{y}(nT) + e(nT)]^2\} \\ &= E\{\hat{y}^2(nT)\} + E\{\hat{y}(nT)\,e(nT)\} \\ &\quad + E\{e(nT)\,\hat{y}(nT)\} + E\{e^2(nT)\} \\ &= E\{\hat{y}^2(nT)\} + E\{e^2(nT)\} \\ &= \sigma_{\hat{y}\hat{y}}^2 + \sigma_{ee}^2. \end{aligned} \tag{36}$$

The power of $e(nT)$ is thus

$$\sigma_{ee}^2 = \sigma_{yy}^2 - \sigma_{\hat{y}\hat{y}}^2. \tag{37}$$

Since sine waves at different frequencies are orthogonal, the power of $\hat{y}(nT)$ is

$$\sigma_{\hat{y}\hat{y}}^2 = a_0^2 + \sigma_{\text{out}}^2 + \frac{1}{2}\sum_{k=2}^{K} a_k^2. \tag{38}$$

This last expression is exact only if either the sampling frequency $1/T$ is an integer multiple of the signal frequency, f_x, or the number of samples N is infinite. The case when N is finite is discussed in Section V.

The fact that $e(nT)$ is uncorrelated with $\hat{y}(nT)$ does not imply that $e(nT)$ and the system input $x(nT)$ are uncorrelated. Specifically, correlation between the input and the noise has been observed in some oversampled A/D converter architectures.

Acknowledgment

The authors would like to thank Dr. J. Tiemann and Prof. R. Gray for their comments and suggestions.

References

[1] J. C. Candy, "A use of double integration in sigma delta modulation," *IEEE Trans. Commun.*, vol. COM-33, pp. 249–258, Mar. 1985.

[2] J. C. Candy, B. A. Wooley, and O. J. Benjamin, "A voiceband codec with digital filtering," *IEEE Trans. Commun.*, vol. COM-29, pp. 815–830, June 1981.

[3] J. C. Candy, "Decimation for sigma delta modulation," *IEEE Trans. Commun.*, vol. COM-34, pp. 72–76, Jan. 1986.

[4] G. L. Baldwin and S. K. Tewksbury, "Linear delta modulator integrated circuit with 17-MBit/s sampling rate," *IEEE Trans. Commun.*, vol. COM-22, pp. 977–985, July 1974.

[5] B. A. Wooley and J. L. Henry, "An integrated per-channel PCM encoder based on interpolation," *IEEE J. Solid-State Circuits*, vol. SC-14, pp. 14–20, Feb. 1979.

[6] J. D. Everhard, "A single-channel PCM codec," *IEEE J. Solid-State Circuits*, vol. SC-14, pp. 25–37, Feb. 1979.

[7] T. Misawa, J. E. Iwersen, L. J. Loporcaro, and J. G. Ruch, "Single-chip per channel codec with filters utilizing sigma-delta modulation," *IEEE J. Solid-State Circuits*, vol. SC-16, pp. 333-341, Aug. 1981.

[8] H. Kuwahara *et al.*, "An interpolative PCM codec with multiplexed digital filters," *IEEE J. Solid-State Circuits*, vol. SC-15, pp. 1014–1021, Dec. 1980.

[9] M. W. Hauser, P. J. Hurst, and R. W. Brodersen, "MOS ADC-filter combination that does not require precision analog components," in *IEEE Int. Solid-State Circuits Conf. Dig. Tech. Papers*, Feb. 1985, pp. 80–81.

[10] K. Yamakido, S. Nishita, M. Kokubo, H. Shirasu, K. Ohwada, and T. Nishihara, "A voiceband 15b interpolative converter chip set," in *IEEE Int. Solid-State Circuits Conf. Dig. Tech. Papers*, Feb. 1986, pp. 180–181.

[11] U. Roettcher, H. Fiedler, and G. Zimmer, "A compatible CMOS-JFET pulse density modulator for interpolative high-resolution A/D conversion," *IEEE J. Solid-State Circuits*, vol. SC-21, pp. 446-452, June 1986.

[12] A. V. Oppenheim and R. W. Schafer, *Digital Signal Processing*. Englewood Cliffs, NJ: Prentice Hall, 1975.

[13] F. Irons, "Dynamic characterization and compensation of analog to digital converters," in *Proc. IEEE Int. Symp. Circuits Syst.*, May 1986, pp. 1273–1277.

[14] F. J. Harris, "On the use of windows for harmonic analysis with the discrete Fourier transform," *Proc. IEEE*, vol. 66, pp. 51–83, Jan. 1978.

[15] D. J. Goodman, "Delta modulation granular quantization noise," *Bell Syst. Tech. J.*, vol. 48, pp. 1197–1218, May–June 1969.

[16] Hewlett Packard, "Dynamic performance testing of A-to-D converters," *Hewlett Packard Product Note 5180A-2*.

[17] J. Doernberg, H.-S. Lee, and D. A. Hodges, "Full-speed testing of A/D converters," *IEEE J. Solid-State Circuits*, vol. SC-19, pp. 820–827, Dec. 1984.

[18] M. Bossche, J. Schoukens, and J. Renneboog, "Dynamic testing and diagnostics of A/D converters," *IEEE Trans. Circuits Syst.*, vol. CAS-33, pp. 775–785, Aug. 1986.

[19] CCITT, "Digital networks—Transmission systems and multiplexing equipment," Tech. Report III.3, The International Telegraph and Telephone Consulative Committee, Nov. 1980.

[20] Hewlett-Packard, *Distortion Measurement Set 339A*, HP, 1984.

[21] L. R. Rabiner and B. Gold, *Theory and Application of Digital Signal Processing*. Englewood Cliffs, NJ: Prentice Hall, 1975.

[22] J. C. Candy, "Simulation of sigma-delta modulators based on uncorrelated noise," unpublished.

*

Bernhard E. Boser (S'80) received the diploma in electrical engineering in 1984 from the Swiss Federal Institute of Technology in Zurich, Switzerland, and the M.S. degree from Stanford University in 1985. He is currently a Ph.D. candidate in electrical engineering at Stanford University.

His research interests include the simulation and design of analog integrated circuits for signal processing applications. At the Swiss Federal Institute of Technology he designed and fabricated a CMOS analog multiplier and also contributed to the development and installation of CAD tools for VLSI circuit design. His doctoral research is directed toward the study of the modeling, simulation, and design of oversampled analog-to-digital converters for audio applications.

*

Klaus-Peter Karmann was born in 1954 in Bielefeld, West Germany. He received the diploma degree in physics from the University of Bielefeld in 1979 and the Dr.rer.nat. degree from the Darmstadt Institute of Technology in 1984.

From 1984 to 1986 he was associated with the Siemens Corporate Research Laboratory in Munich, where he worked on modeling, simulation, and computer-aided design of electronic devices and circuits. From April to September 1986 he visited the Center for Integrated Systems at Stanford University, where he was engaged in a project on oversampled A/D conversion. Since then he has been with Siemens Computer Division in Munich and his work concentrates on circuit design and design automation.

*

Horst Martin was born in Hagen, West Germany, in 1958. He received the diploma in physics from the University of Münster, West Germany, in 1984.

He worked on a VLSI design project with Siemens Erlangen, West Germany, from 1984 to 1986. With a scholarship from his employer he worked at Stanford University in the summer of 1986. He is now with Siemens Munich, where he is working on hardware for telecommunications (ISDN).

Bruce A. Wooley (S'64–M'70–SM'76–F'82) was born in Milwaukee, WI, on October 14, 1943. He received the B.S., M.S., and Ph.D. degrees in electrical engineering from the University of California, Berkeley, in 1966, 1968, and 1970, respectively.

From 1970 to 1984 he was a member of the research staff at Bell Laboratories in Holmdel, NJ. In 1980 he was a Visiting Lecturer at the University of California, Berkeley. In 1984 he assumed his present position as Professor of Electrical Engineering at Stanford University. His research is in the field of integrated circuit design and technology, where his interests have included monolithic broad-band amplifier design, circuit architectures for high-speed arithmetic, analog-to-digital conversion, and digital filtering for telecommunications, tactile sensing for robotics, and circuit techniques for video A/D conversion and broad-band fiber-optic communications.

Dr. Wooley is a member of the IEEE Solid-State Circuits Council, and he is the Editor of the IEEE JOURNAL OF SOLID-STATE CIRCUITS. He was the Chairman of the 1981 International Solid-State Circuits Conference. He is also a past Chairman of the IEEE Solid-State Circuits and Technology Committee and has served as a member of the IEEE Circuits and Systems Society Ad Com. He is a member of Sigma Xi, Tau Beta Pi, and Eta Kappa Nu. In 1966 he was awarded the University Medal by the University of California, Berkeley, and he was the IEEE Fortescue Fellow for 1966–67.

Simulation of Mixed Switched-Capacitor/Digital Networks with Signal-Driven Switches

KEN SUYAMA, MEMBER, IEEE, SAN-CHIN FANG, MEMBER, IEEE, AND YANNIS P. TSIVIDIS, FELLOW, IEEE

***Abstract*—This paper considers the simulation of mixed switched-capacitor/digital (SC/D) networks containing capacitors, independent and linear-dependent voltage sources, switches, comparators, and logic gates using a new simulator, SWITCAP2. The switches, in particular, may be controlled by not only periodic waveforms but also by nonperiodic waveforms from the outputs of comparators and logic gates. The signal-dependent modification of network topology through the comparators, logic gates, and signal-driven switches makes the modeling of various nonlinear switched-capacitor circuits possible. Simulation results for a PCM voice encoder, a sigma-delta modulator, a neural network, and a phase-locked loop (PLL) are presented to demonstrate the flexibility of the approach.**

I. Introduction

An IMPORTANT class of mixed analog–digital VLSI circuits can be termed mixed switched-capacitor/digital (SC/D) networks. These contain capacitors, sources, switches driven by clocks or by signals, and digital logic. Many modern mixed analog–digital chips, in production or in development, consist wholly or partly of such networks [1]–[11], [32]. Examples include charge-redistribution A/D and D/A converters and PCM codecs, certain oversampled sigma-delta coders, switched-capacitor VCO's, rectifiers and phase-locked loops, and switched-capacitor neural networks. The complexity of some of these circuits is such that simulation is a must. Yet traditional simulation tools for time-invariant networks such as SPICE [12] and ASTAP [13] are not adequate for accurately computing the sharp waveform transitions generated by switching efficiently. Simulation algorithms for specifically analyzing SCN's [14]–[17], [20], [21] are not applicable to mixed SC/D networks either, since these algorithms deal only with linear switched-capacitor networks (SCN's). Although several methods have been reported for simulating general mixed analog/digital networks [24]–[28], they, too, fail to take full advantage of the algebraic nature of network equations in the case of mixed SC/D networks. An event-driven simulator such as SPLICE [29] can also be modified to handle mixed-mode sample-data systems [30]. However, this simulator requires the circuit to be represented by a behavioral model using difference equations. For general mixed SC/D networks, the modeling procedure becomes complex. Furthermore, it is difficult to interpret the effect of the original circuit elements once the circuit is modeled in difference equations.

This paper describes a unified linear SCN and mixed SC/D network simulator, SWITCAP2, and its applications to several widely used and novel nonlinear SCN's. This simulation program has evolved from the general linear SCN analysis program SWITCAP [18], [19], which has been enhanced to accommodate the recent trend of mixed-signal systems. The most notable enhancement is the transient analysis of mixed SC/D networks [23], [36]. It incorporates comparators, switches controlled by periodic or nonperiodic logic signals, and logic gates with multiple inputs in addition to all of SCN's regular network elements which are supported in the original program. The comparators and switches allow for signal-dependent modification of circuit topology. Therefore, many interesting nonlinear circuits can be modeled. In the paper, the overall architecture of mixed SC/D networks as defined in SWITCAP2 is first described, followed by a discussion of simulation facilities and user interface for transient analysis of mixed SC/D networks. Finally, simulation examples of a pulse-code modulation (PCM) voice encoder, a sigma–delta modulator, a switched-capacitor neural network (SCNN), and a switched-capacitor phase-locked loop (SC PLL) are presented.

II. Mixed Switched-Capacitor/Digital Network

A mixed SC/D network is shown in Fig. 1. The network is divided into four groups: SCN, THRESHOLD, TIMING, and LOGIC, which are interconnected as shown in the figure. The feedback loop from the SCN to itself through the THRESHOLD and LOGIC allows signal-dependent modification of the SCN topology. The program provides various sources such as function generators and dc sources as the input to the network and an oscilloscope to measure the output. This division of the network can be effectively used to model nonlinear SCN's, and at the same time facilitates the efficient implementation of computer algorithms for time-domain analysis of

Manuscript received May 1, 1990; revised July 19, 1990.
K. Suyama and Y. P. Tsividis are with the Department of Electrical Engineering and Center for Telecommunications Research, Columbia University, New York, NY 10027.
S.-C. Fang is with AT&T Bell Laboratories, Murray Hill, NJ 07974.
IEEE Log Number 9038971.

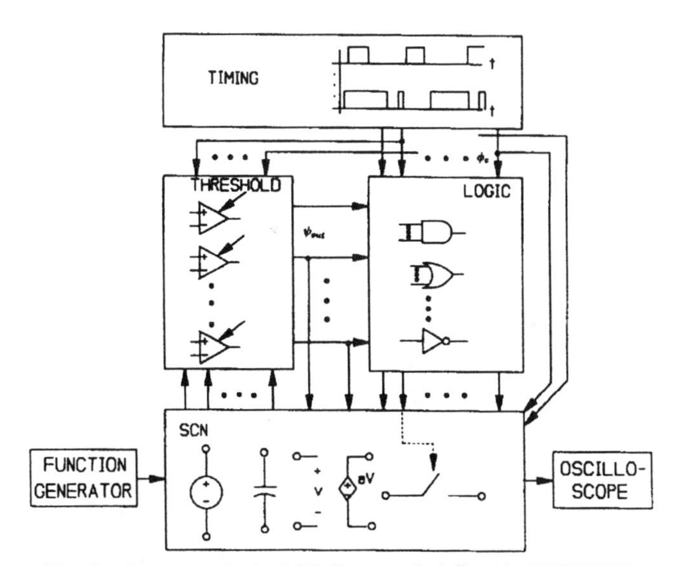

Fig. 1. Structure of mixed SC/D network defined in SWITCAP2.

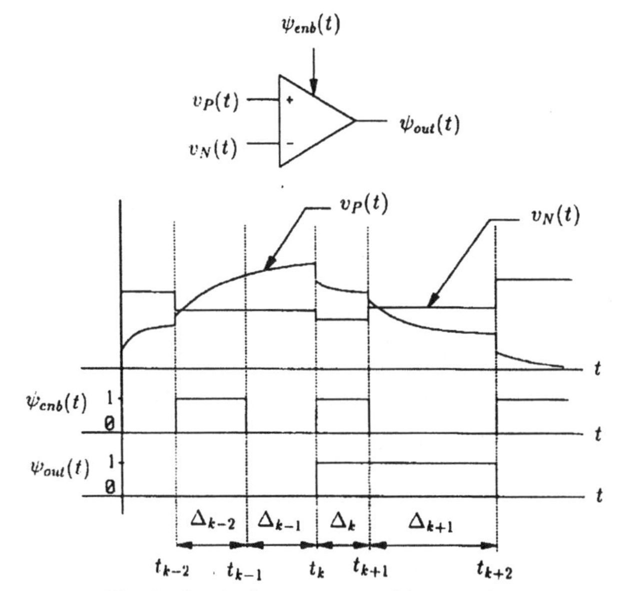
Fig. 2. Latched comparator and its operation.

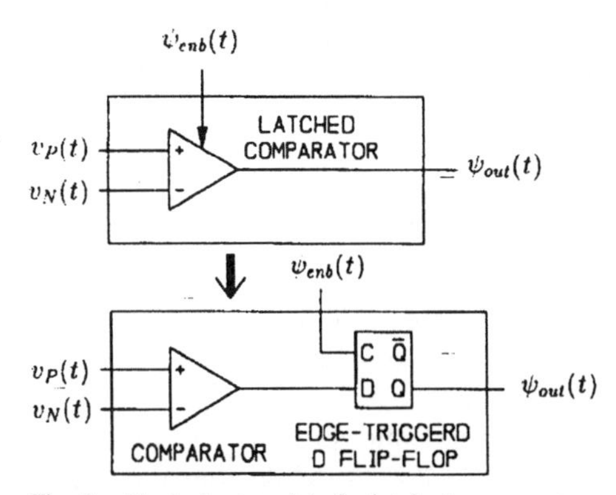

Fig. 3. Equivalent model of a latched comparator.

the network [23], [36]. Each network group is described as follows.

A. SCN Block

The SCN may be composed of linear capacitors, independent voltage sources, linear voltage-controlled voltage sources, and ideal switches which are controlled by periodic or nonperiodic switch control signals. The SCN receives voltage inputs through independent voltage sources which model function generators and dc sources. The switch control signals may be periodic or nonperiodic binary signals and they are supplied by the TIMING, THRESHOLD, and LOGIC blocks. The output variables of the SCN are node voltages; a subset of these voltages is monitored by the THRESHOLD block as shown in Fig. 1. The topology of the network within the SCN can be set arbitrarily except for obvious pathological cases such as loops of voltage sources or disjoint networks. The size of the SCN is only limited by the available memory size of the user's computer. The minimum memory requirements for several example circuits are given in Section IV.

B. THRESHOLD Block

The THRESHOLD block contains latched comparators that act as the interface from the SCN block to the LOGIC block. A latched comparator (Fig. 2) is a comparator with noninverting and inverting input terminals (signals at the terminals are denoted as $v_P(t)$ and $v_N(t)$, respectively, as shown in Fig. 2), an enable terminal for a binary input $\psi_{enb}(t)$, and a binary output $\psi_{out}(t)$. The latched comparator samples its input signals $v_P(t)$ and $v_N(t)$ at the moment immediately before a switching instant t_k and, if warranted, changes its output value accordingly only when the enable signal $\psi_{enb}(t)$ changes from 0 to 1 at t_k. Then, it latches the output value until the next rising edge is observed at the enable terminal. For example, in Fig. 2, the output $\psi_{out}(t)$ of the latched comparator can change only at switching instants $t = t_{k-2}$, t_k, and t_{k+2}. Although the inputs $v_P(t)$ and $v_N(t)$ in Fig. 2 intersect at time instants within the interval Δ_{k-2} and Δ_{k+1}, the output $\psi_{out}(t)$ does *not* change at these instants. The enable signal must be periodic and is generated by the TIMING block. It should be emphasized that the latched comparator can be modeled by a composite circuit, which consists of an ideal comparator and an edge-triggered *D*-type flip-flop as shown in Fig. 3.

A continuous comparator by itself, wherein its input terminals are monitored continuously and the output is allowed to change at the instant that the inputs cross over, is not directly implemented in the program to avoid the time-consuming search for the instant (such a search involves multiple network evaluations at many fine time steps). Another potential problem associated with the continuous comparator is the *race-around* condition where a change in the output state of a comparator propagates through a set of switches in the SCN and, in turn, forces a change of the same comparator again immediately after the instant. This problem exists in actual circuits and is

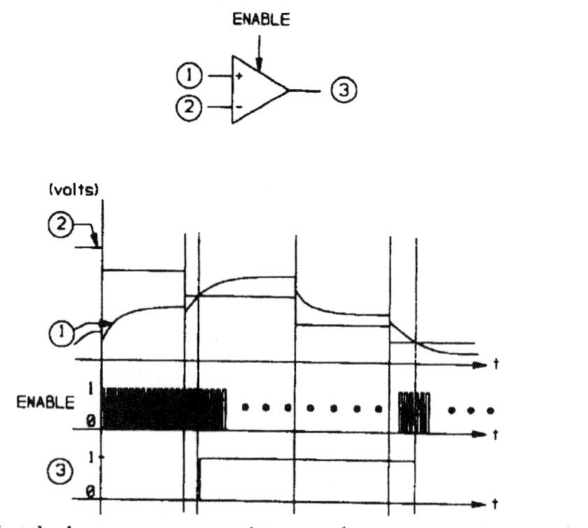

Fig. 4. Latched comparator used as an (approximately) continuous comparator.

often avoided by inserting a latch at the output of the comparator. However, if a continuous comparator is needed for some applications, it is possible to drive the enable terminal with a high-frequency clock as shown in Fig. 4, so that the latched comparator behaves practically like a continuous one. Obviously, this incurs a sacrifice in simulation speed.

C. TIMING Block

This block generates periodic clock waveforms which may be used as enable signals for latched comparators in the THRESHOLD block, inputs to the LOGIC block, or switch control signals directly. Each clock is allowed to have its own period and multiple ON–OFF phases within each period under the condition that the period and ON and OFF switching instants must be normalized to one common master clock period and expressed as a rational number times the master clock period. Although the TIMING block can be implemented by an oscillator and digital circuit (flip-flops and a few logic gates) in actual implementations, here it is architecturally separated from the rest of the digital circuits in the LOGIC block since the computer implementation of the periodic waveforms can be efficiently carried out by constructing an event table of all predetermined switching instants.

D. LOGIC Block

The LOGIC block realizes Boolean logic functions without internal clocks. It consists of basic logic gates such as AND, OR, NOT, NAND, NOR, XOR, and XNOR. Multiple inputs are allowed where applicable. Arbitrary combinational and sequential logic circuits (with external clock signals supplied by the TIMING block for the latter) can be designed using the available logic gates. All gates are assumed to have zero gate delay and, therefore, the outputs from the LOGIC block are determined immediately at the instants that the inputs change their states. The inputs to the LOGIC block may consist of the outputs of latched comparators in the THRESHOLD block and of periodic clock waveforms generated by the TIMING block as shown in Fig. 1. It should be emphasized that switching instants are determined by the TIMING block and are thus periodic since the LOGIC block as well as the THRESHOLD block are also regulated by periodic signals from the TIMING block. However, not all output state sequences from the LOGIC and THRESHOLD blocks are periodic because the signal path through the THRESHOLD block is dependent on analog signals. This signal path characterizes the unique feature of the mixed SC/D network. If this loop is absent, then the whole network becomes a periodically switched linear SCN.

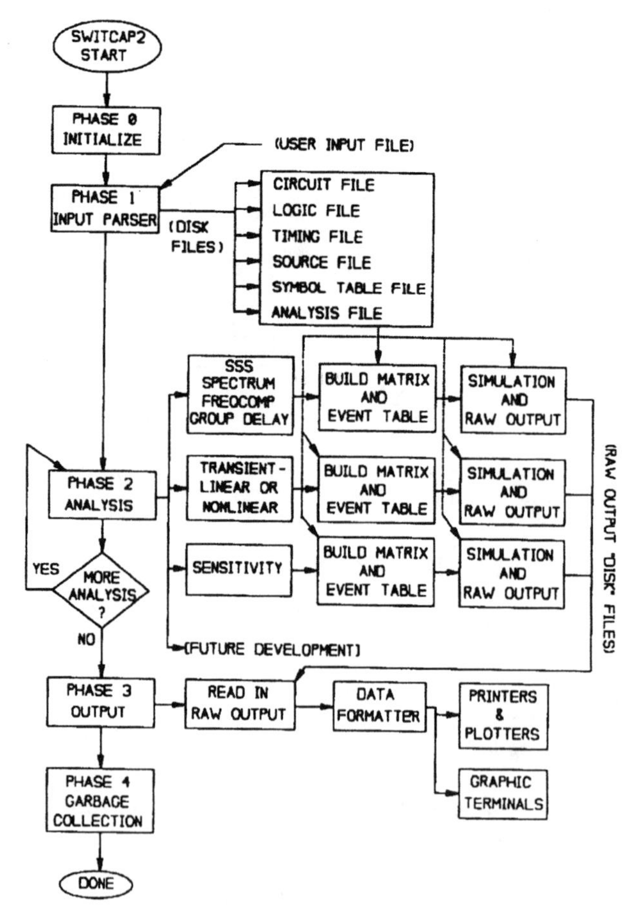

Fig. 5. SWITCAP2 structure.

III. The Program

The program can perform frequency-domain analyses for linear SCN's and transient analysis for linear SCN's and mixed SC/D networks as shown in Fig. 5. After the input parser translates the user input file, several intermediate data files, which have a data structure common to all analysis modes, are created. Necessary matrices and event tables are built from the data files for each analysis since each analysis may require different sets of matrices and event tables to ensure optimized performance. Although the mixed SC/D network capability of the program is emphasized in this paper, the program also provides linear SCN analyses such as group delay and frequency-domain sensitivities in addition to all features found in the original version of SWITCAP. Detailed discussions of the features associated with the original SWITCAP are given in [18] and [19], and the formulation used in the group delay and sensitivity analyses is described in [22] and [23].

A. Transient Analysis of Mixed SC/D Networks

For the transient analysis of mixed SC/D networks, each block (i.e., SCN, THRESHOLD, TIMING, and LOGIC) is formulated separately to take full advantage of its own properties. The SCN block is formulated using a version of the modified nodal analysis (MNA) method [14] where a switching matrix is the only time-varying matrix to be computed at each switching instant while the rest of the network matrices are precomputed and stored at the beginning of the simulation [23], [36]. This is particularly important for mixed SC/D networks since a new set of network matrices must be computed at each switching instant due to the nonperiodic feedback signals through the THRESHOLD block.

After the final MNA matrix within a switching interval is computed from the switching matrix and the time-invariant precomputed network matrices, one LU decomposition and forward and backward substitutions per switching interval are needed for the transient analysis to move forward in time. In order to reduce the computation time required to solve the network matrices, a matrix cache [23], [36] is implemented to store the frequently used set of LU decomposed MNA matrices, since several network matrices are repeatedly used during the simulation of many nonlinear SCN's. Therefore, the analysis becomes more efficient as the size of available computer memory increases.

A simple zero-delay event-driven logic simulator called EZSIM [23], [36] is implemented for the LOGIC block. It handles basic logic gates (Section II-D) that are needed to realize arbitrary combinational or sequential logic circuits without internal clocks. The periodic clock waveforms needed for sequential circuits are generated by a simple event table in the TIMING block.

Latched comparators are used for the interface from the SCN to the LOGIC block. Unlike regular comparators, latched comparators operate only at the known switching instants. Therefore, the formulation of a comparator simply becomes a comparison of two node voltages at switching instants (as opposed to modeling the comparator into the MNA matrix). This interface scheme in combination with the zero gate delay nature of EZSIM ensures that the network equations in the SCN stay algebraic, and thus the transient analysis can *jump* from one switching instant to the next without having to undergo a time-consuming numerical integration. Detailed discussions of the formulation and algorithms used for mixed SC/D networks are presented in [23] and [36], and a summary of the overall algorithm used for the transient analysis is given in the Appendix.

B. User Interface

Nested subcircuits, in which a subcircuit calls another subcircuit, may be defined in the program for logic gates and latched comparators as well as for regular SCN elements with the exception of independent voltage sources. Symbolic values and clocks are allowed in subcircuits. The actual values are assigned when the subcircuits are called. It is also possible to declare some logic nodes and clock names as global variables.

For the transient analysis of a mixed SC/D network, a set of the initial condition of analog and digital nodes at $t = 0^-$ may be specified. Unspecified nodes are set to 0 V for analog nodes and logic ZERO for digital nodes. SWITCAP2 makes an attempt to propagate the given initial condition to the rest of the network in order to avoid inconsistent states. If the inconsistency persists, a warning is given. Various signal sources such as dc, sinusoidal, pulse, pulse train, exponential, and piecewise linear waveforms are available through independent voltage sources.

IV. Applications

In this section, we illustrate how the mixed SC/D networks defined above are capable of modeling a wide variety of nonlinear SCN's through well-known circuits, as well as some "exotic" ones.

A. PCM Voice Encoder

The PCM CODEC (enCOder-DECoder) is probably one of the most widely used mixed SC/D networks to date. A companding CODEC using the segmented μ-255 encoding law is simulated by SWITCAP2 in this section. Fig. 6 shows the analog section of the complete encoder [31]. It consists of two binary-weighted capacitor arrays for segment and step determination, a comparator for an interface from the analog to the logic section, and switches controlled by the logic. The digital logic used in the simulation consists of 32 edge-triggered D-type flip-flops and approximately 180 gates (not shown). The flip-flops are designed using seven NAND gates. Sixteen periodic clock waveforms, which in an actual implementation would be realized using flip-flops and gates driven from a single clock, are specified in the TIMING block in Fig. 1.

The node voltage at the input of the comparator, V_x, versus the conversion time steps for a given input voltage of 4 V is shown in Fig. 7. The reference voltage V_{ref} used in Fig. 6 is 5 V. The result matches exactly with the theoretical operation described in [31]. The simulation time is 0.045 s per requested time point on a Sun 4/280

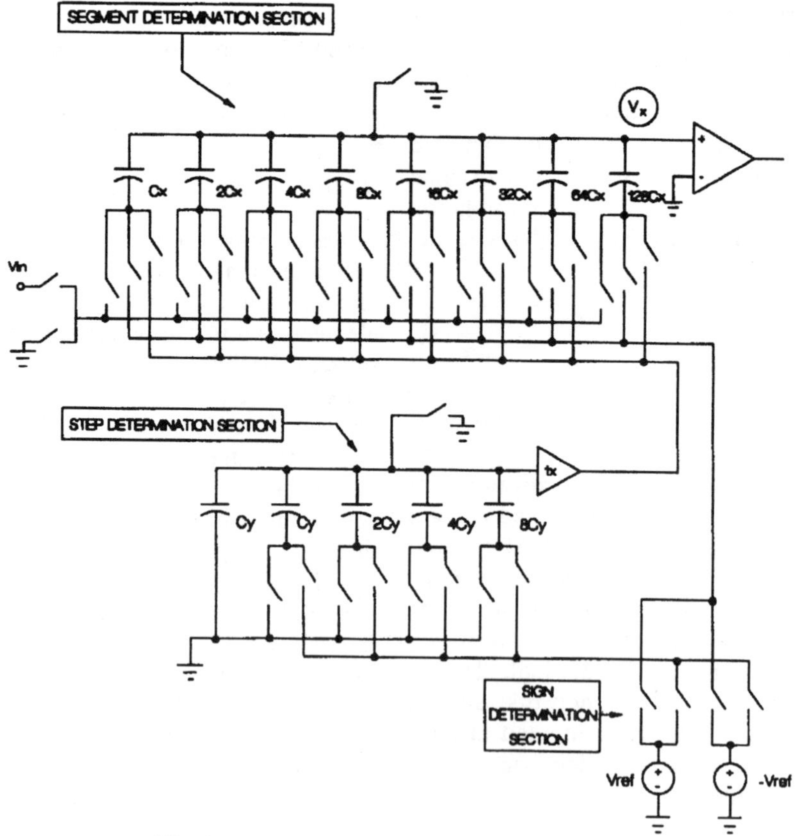

Fig. 6. Analog section of a complete PCM encoder.

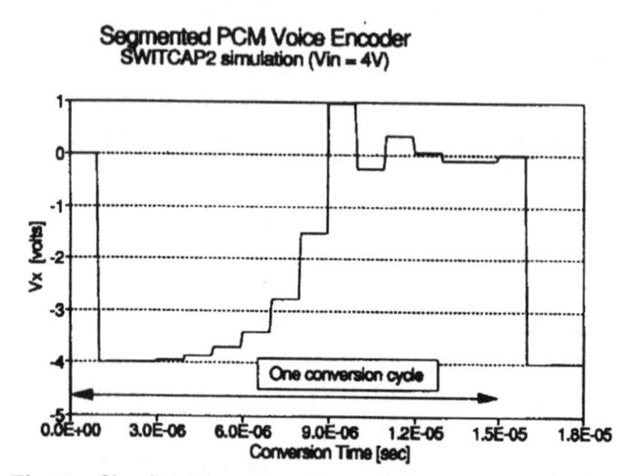

Fig. 7. Simulated input waveform of the comparator in Fig. 6.

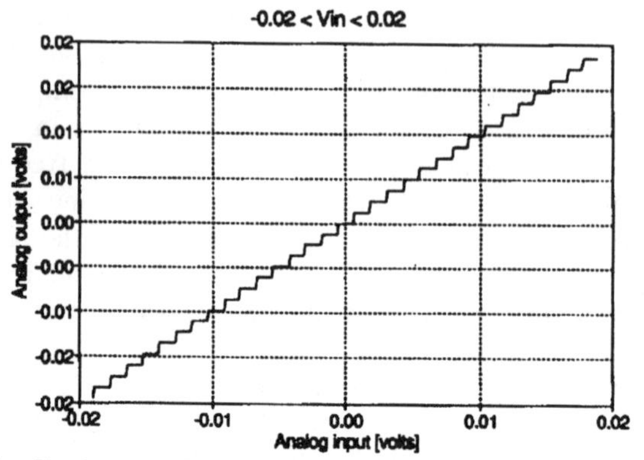

Fig. 8. Simulated dc transfer characteristics (the first segment) of the encoder–decoder combination.

and the minimum required memory allocation[1] to simulate the circuit is 43 336 words.

When an encoder is followed by a decoder, we can obtain the dc transfer characteristics of the encoder–decoder combination as it was done in [31]. The decoder is implemented with a structure very similar to that in Fig. 6. The plot of the decoder output versus the encoder input for the first segment is shown in Fig. 8. It should be noted that the well-known half-step adjustment around the origin is incorporated in the encoder (not shown).

[1]The minimum required memory allocation reflects the smallest size of memory needed for the matrices and tables for the SCN, THRESHOLD, and TIMING blocks to perform the analysis without the matrix cache.

B. *Sigma–Delta Modulator*

An SCN implementation of a second-order sigma–delta modulator used in high-precision A/D conversion [32] is shown in Fig. 9. The modulator consists of two fully balanced SC integrators, a comparator used as a 1-b quantizer, and an inverter. A fully balanced op amp is modeled as shown in the inset of Fig. 9(c). A careful distinction between a "balanced" output op amp and a "differential" output op amp must be observed to model

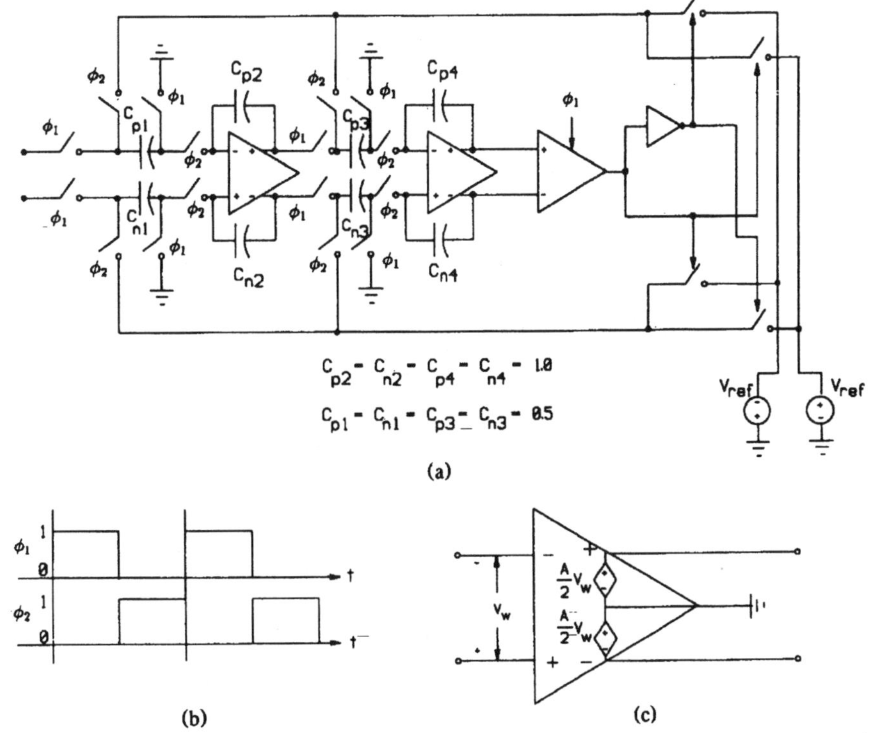

Fig. 9. (a) SCN implementation of a second-order sigma–delta modulator [32], (b) timing waveforms, and (c) a balanced op-amp model.

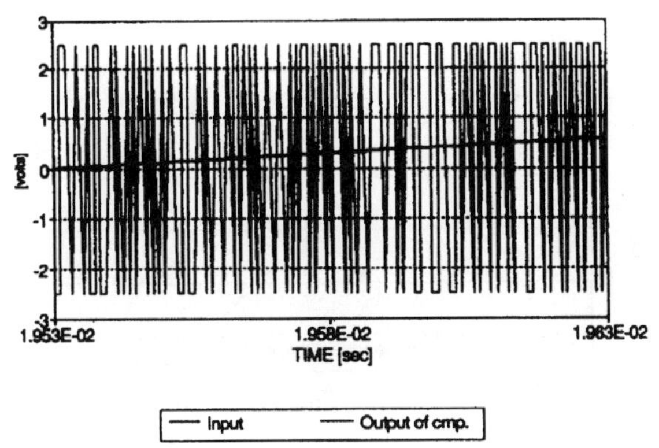

Fig. 10. Simulated transient characteristics of the sigma–delta modulator.

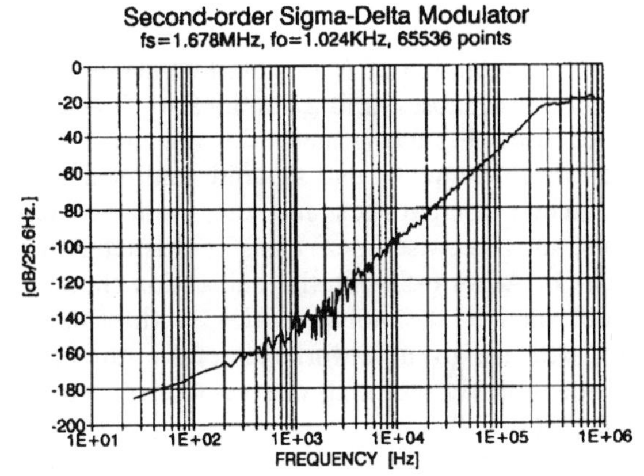

Fig. 11. Spectrum of the comparator output.

a fully balanced op amp [33]. The clock waveform ϕ_1 in the timing diagram is used for the enable signal of the latched comparator. Therefore, the comparison of the integrator outputs is done at the rising edge of ϕ_1. The transient analysis of the circuit has been performed with a switching frequency of 1.678 MHz and an input frequency of 1.02 kHz. The transient characteristics of the modulator are shown in Fig. 10. Fig. 11 shows the spectrum of the comparator output computed by taking an FFT over 65 536 points. The FFT routine from OSIM [34] is used here. Care must be taken while computing the comparator output spectrum since the spectrum can differ significantly depending on the FFT algorithm used and its parameters (e.g., the number of transient points) [37]. The simulation time is 0.088 s per sampling period on a Sun 4/110. The minimum memory requirement for this circuit is 19 548 words.

It should be emphasized that some of the nonidealities mentioned in [32] (e.g., integrator gain variations, the dc gain of the integrator, comparator hysteresis, etc.) can also be investigated using SWITCAP2 despite the fact that it only allows ideal elements. For example, the integrator gain variations can be simulated by perturbing the ratio of the integrating and switching capacitors.

C. Switched-Capacitor Neuron

The SCN implementation of electronic neural networks has been considered by several research groups [6], [7], [9]–[11]. Essential elements of a typical neural network

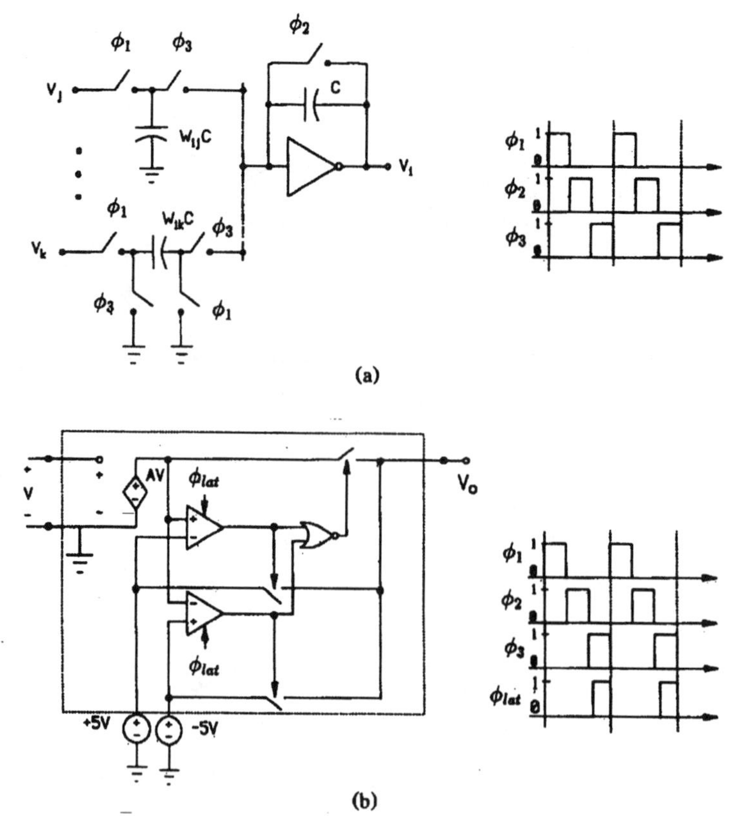

Fig. 12. Neuron-synapse circuit. (a) SC implementation. (b) A model of the inverter for simulation.

model are a summation of N weighted inputs, where N is the number of synapses, and a nonlinearity. Typical nonlinearities used in neural networks are hard limiters, threshold logic elements, and sigmoid nonlinearities [35]. Using latched comparators and the available network elements, any piecewise linear transfer characteristics can be modeled. As an example, this subsection shows a way to model one of the nonlinearities, a threshold logic element, with the given network elements in the program.

The equation representing one version of neural networks is [35]

$$x_i(nT) = f\left[\sum_{j=1}^{N} w_{ij} x_j(nT - T)\right] \tag{1}$$

where $x_i(nT)$ is the output of the ith neuron at discrete time instant nT, w_{ij} are real numbers representing weights from the jth neuron to the ith neuron, and $f[\cdot]$ denotes a nonlinearity.

An SC implementation of (1), which is shown in Fig. 12, has been proposed in [6]. The nonlinearity associated with the inverter in Fig. 12(a) can be modeled using a mixed SC/D network shown in Fig. 12(b). It consists of a voltage-controlled voltage source, two latched comparators, three switches, two independent sources, and a logic gate. The duration of ϕ_{lat} is insignificant since the rising edge of the clock is of interest. Note that there is no gap between clock waveforms in Fig. 12 to avoid disjoint networks which cause numerical problems, although the gap is necessary in an actual implementation. Fig. 13

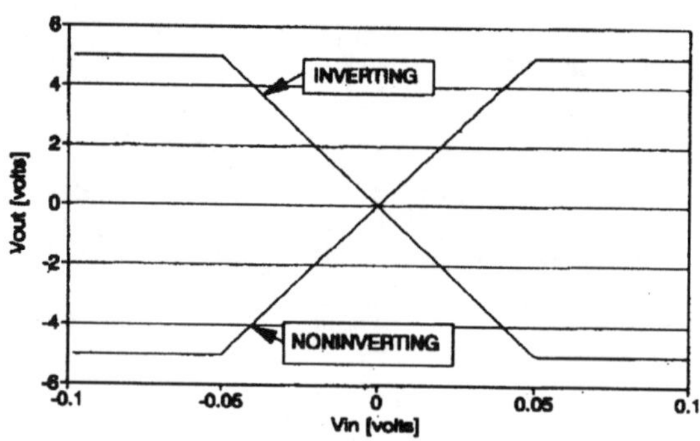

Fig. 13. Simulated dc transfer characteristics of SC neuron. Gain = 100.

shows the transfer characteristics of the neuron model. The slope of the linear region observed in Fig. 13 can be changed simply by adjusting the ratio of the feedforward and feedback capacitors (Fig. 12(a)).

D. Switched-Capacitor Neural Network

By interconnecting SC neurons as shown in Fig. 14, a 20-neuron Hopfield net with fixed weights was simulated in SWITCAP2. The weights were computed in such a way that the network can classify three patterns of the outputs. The way to compute each weight is well documented in [35].

The input to each neuron is entered by storing an initial value to the feedback capacitor of the neuron (Fig. 12(a)). The transient waveforms of the output of neurons 5–8 with a set of initial values are shown in Fig. 15. The

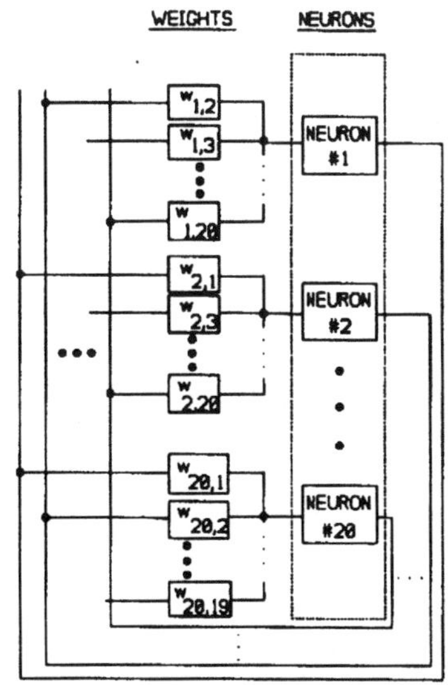

Fig. 14. 20-neuron Hopfield net.

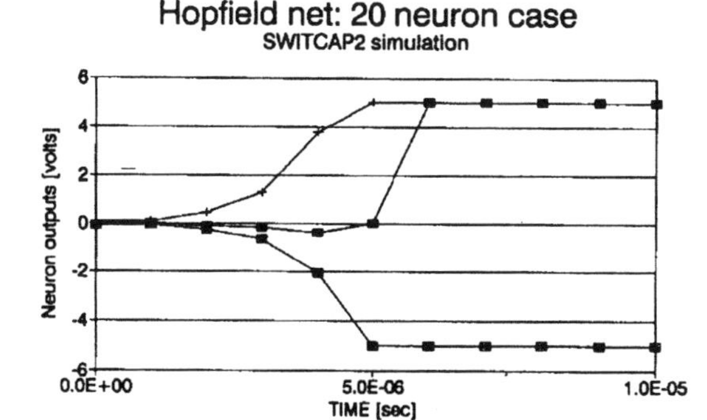

Fig. 15. Transient waveforms of neurons 5, 6, 7, and 8.

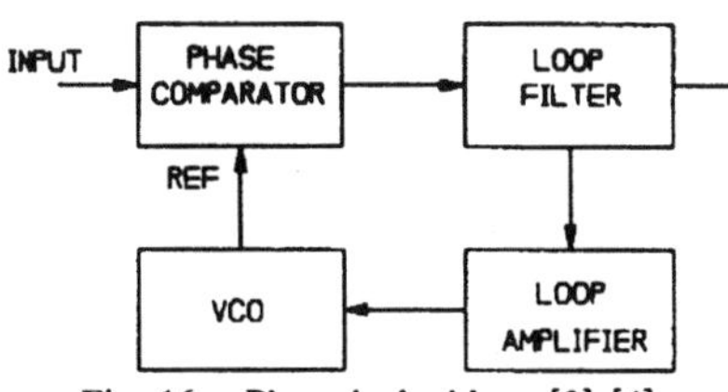

Fig. 16. Phase-locked loop [3], [4].

network contains 1172 analog circuit elements (switches, capacitors, etc.), 442 analog network nodes, 120 digital gates, and 166 digital nodes. The computation time required by SWITCAP2 is approximately 10 min per sampling period on a Sun 4/280 and the minimum memory requirement is 425 128 words. The results match exactly with the theory published in [35].

It should be emphasized that the purpose of this example is not to promote the Hopfield net for the implementation of SCNN's, but rather to demonstrate the robustness of SWITCAP2 for a large nonlinear network (even with feedback), by deliberately using a computationally intensive neural network model.

E. Switched-Capacitor Phase-Locked Loop

An SC PLL, reported elsewhere [2]–[4], consists of a phase comparator, a low-pass filter, a loop amplifier, and a voltage controlled oscillator (VCO), connected as shown in Fig. 16. This subsection shows how SWITCAP2 can model the nonlinearities necessary for the proper operation of the SC PLL. An SC implementation of a VCO is proposed in [2] and [3]. It consists of an SC Schmitt-trigger circuit, an SC integrator, and a phase reverser [2], [3]. A modified version of the VCO, which is suitable for simulation, is shown in Fig. 17. For the SC Schmitt trigger shown in Fig. 17(b), the output of the latched comparator must be converted to an analog value using two dc sources and an inverter as shown in the figure, since a latched comparator gives a Boolean output. The SC phase detector [3] and SC loop filter can easily be implemented in the program. The PLL parameters used in this example are obtained from [2]–[4].

Fig. 18 shows a plot from the transient analysis of the SC PLL. The input starts as a cosine waveform of 1 kHz, and at $t = 12$ ms changes to a 1.5-kHz cosine waveform. The computation time is approximately 0.2 s per switching period on Sun 4/110 and the minimum memory requirement is 20 344 words.

V. Conclusions

We have presented a new methodology for simulating mixed switched-capacitor/digital networks which include capacitors, independent and linear-dependent voltage sources, switches controlled by either periodic or nonperiodic Boolean signals, latched comparators, and logic gates. The signal-dependent modification of network topology provided by the comparators, logic gates, and switches is shown to be useful for modeling essential nonlinearities for widely used and novel nonlinear SCN's. The method is incorporated into a general SCN analysis program called SWITCAP2. The simulation of several nonlinear SCN/digital circuits such as a PCM codec, a sigma–delta modulator, a switched-capacitor neural network, and a switched-capacitor phase-locked loop has been illustrated.

Appendix

The overall algorithm for the transient analysis of mixed SC/D networks is summarized here. More detailed discussions of algorithms are presented elsewhere [23], [36]. We begin with some definitions.

Definition

$\boldsymbol{\Psi}_{\text{out},k}$ is an $n_T \times 1$ latched comparator output vector, the ith row of which represents the logical state of output of the ith latched comparator during the switching interval Δ_k.

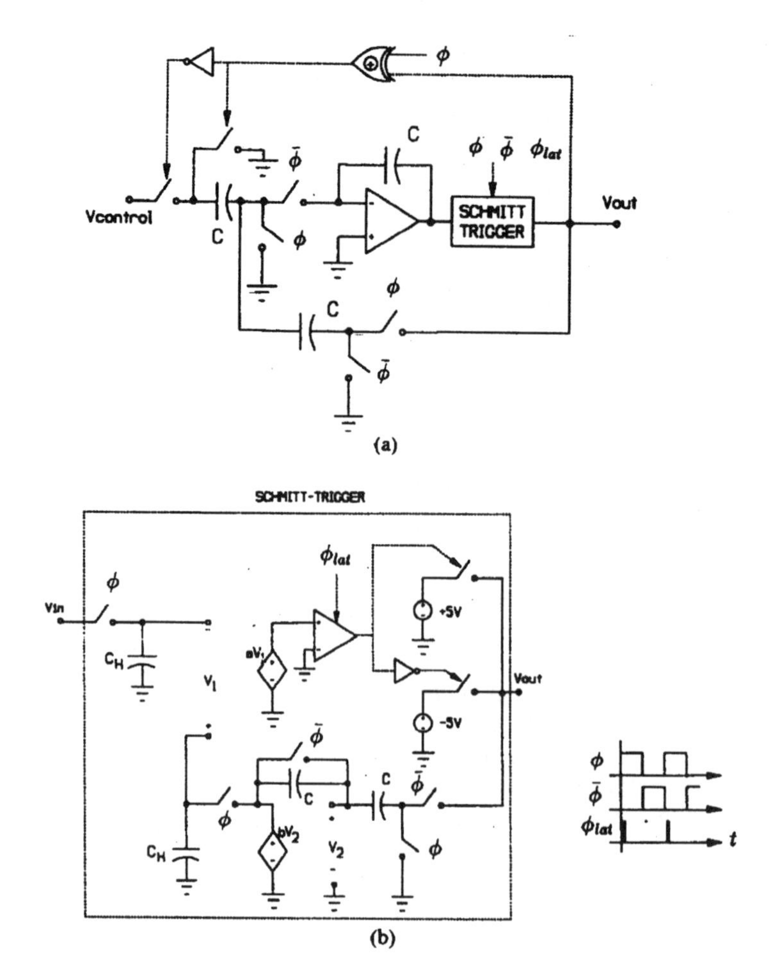

Fig. 17. (a) SC implementation of VCO for simulation. (b) SC Schmitt trigger for simulation.

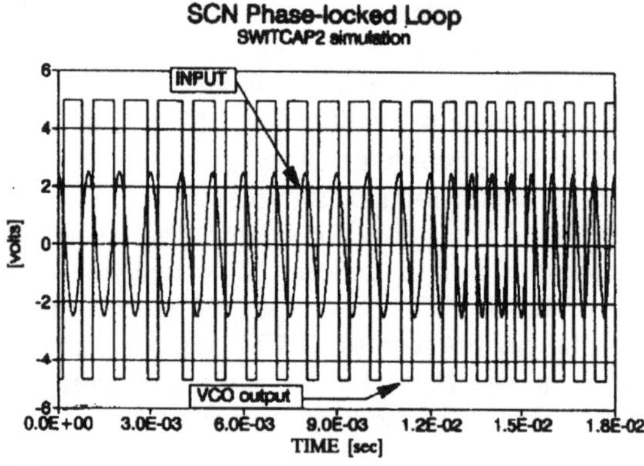

Fig. 18. Simulated transient characteristics of the SC VCO.

$\Psi_{lat,k}$ is an $n_T \times 1$ latch state vector, the ith row of which represents the logical state of the latch control signal seen by the ith latched comparator during the switching interval Δ_k.

$v(t)$ is an $n \times 1$ node voltage vector, the ith element of which is the voltage of node i with respect to the ground node at t.

n is the number of nodes excluding the ground node in the SCN, n_T is the number of latched comparators, and $\Delta_k = [t_k, t_{k+1})$.

Algorithm: Transient

[Build time invariant matrices and store them]
  /*These matrices include node-to-branch incident matrices of capacitors, independent sources, linear VCVS's, switches, and latched comparators, the branch-admittance matrix for the capacitors, and the control voltage coefficient matrix for the VCVS's*/
[Build an event table for periodic clocks generated by the TIMING block]
[Build a source list and a vector for independent voltage sources]
[Initialize EZSIM: define inputs and outputs, parse logic operations]
[Initialize a matrix cache: allocate memory for cache table and nodes]
[$k = -1$]/*initialize an index k*/
[Initialize $v(0^-)$, $\Psi_{lat,-1}$, and $\Psi_{out,-1}$]
while ($k < k_{end}$) **do**
  [$k = k + 1$]
  [Do Algorithm THRESHOLD to determine $\Psi_{out,k}$ from $v(t_k^-)$, $\Psi_{lat,k-1}$, and $\Psi_{lat,k}$]
    /*Algorithm THRESHOLD compares two analog inputs of each latched comparator and determines the output state of the comparators.*/

[Do EZSIM: $\mathbf{\Phi}_{EZO,k} \leftarrow \phi_c, \mathbf{\Psi}_{\text{out},k}$]

/*ϕ_c is a clock vector and $\mathbf{\Phi}_{EZO,k}$ is the output vector from EZSIM.*/

[Generate a switching matrix from $\Phi_{EZO,k}$]

[Check whether the LU decomposed matrices for the switching matrix exist in the matrix cache. If they exist, use them. If not, build the MNA matrix and LU decompose it]

[Evaluate the desired output for $t_k \leqslant t < t_{k+1}$ using the decomposed matrices]

[Evaluate $v(t_{k+1}^-)$ using the decomposed matrices]

end

end

ACKNOWLEDGMENT

The authors would like to thank I. Yusim for his help with the programming of the input parser for SWITCAP2, S. Satyanarayana for his help with neural networks, and Dr. T. Cataltepe and H. Wang for their help in the sigma–delta example.

REFERENCES

[1] K. Martin "Non-filtering applications of switched-capacitor circuits: A tutorial overview emphasizing technological constraints," in *Proc. IEEE Int. Symp. Circuits Syst.* (Montreal, Canada), May 1984, pp. 162–165.

[2] B. J. Hosticka, W. Brockherde, U. Kleine, and R. Schweer, "Design of nonlinear analog switched-capacitor circuits using building blocks," *IEEE Trans. Circuits Syst.*, vol. CAS-31, no. 4, pp. 345–368, Apr. 1984.

[3] B. J. Hosticka, W. Brockherde, U. Kleine, and G. Zimmer, "Switched-capacitor FSK modulator and demodulator in CMOS technology," *IEEE J. Solid-State Circuits*, vol. SC-19, no. 3, pp. 389–396, June 1984.

[4] Y. Tsividis and P. Antognetti, Eds., *Design of MOS VLSI Circuits for Telecommunications.* Englewood Cliffs, NJ: Prentice-Hall, 1985.

[5] R. Gregorian and G. C. Temes, *Analog MOS Integrated Circuits for Signal Processing.* New York: Wiley, 1986.

[6] Y. Tsividis and D. Anastassiou, "Switched capacitor neural networks," *Electron. Lett.*, vol. 23, pp. 958–959, Aug. 27, 1987.

[7] A. Rodriguez-Vazquez, A. Rueda, J. L. Huertas, and R. Dominguez-Castro, "Switched-capacitor neural networks for linear programming," *Electron. Lett.*, vol. 24, pp. 496–498, Apr. 14, 1988.

[8] S. R. Norsworthy and I. G. Post, "A 13-bit, 160 kHz sigma–delta A/D converter for ISDN," in *Proc. IEEE 1988 CICC* (Rochester, NY), May 1988, pp. 21.3.1–21.3.4.

[9] Y. Horio, S. Nakamura, H. Miyasaka, and H. Takase, "Speech recognition network with SC neuron-like components," in *Proc. IEEE Int. Symp. Circuits Syst.* (Espoo, Finland), June 1988, pp. 495–498.

[10] T. Bernard, P. Garda, A. Reichart, B. Zavidovique, and F. Devos, "Design of a half-toning integrated circuit based on analog quadratic minimization by nonlinear multistage switched capacitor network," in *Proc. IEEE Int. Symp. Circuits Syst.* (Espoo, Finland), June 1988, pp. 1217–1220.

[11] J. E. Hansen, J. K. Skelton, and D. J. Allstot, "A time-multiplexed switched-capacitor circuit for neural network applications," in *Proc. IEEE Int. Symp. Circuits Syst.* (Portland, OR), May 1989, pp. 2177–2180.

[12] L. W. Nagel, "SPICE2: A computer program to simulate semiconductor circuits," Electron. Res. Lab., Univ. of Calif., Berkeley, Rep., ERL-M520, May 1975.

[13] W. T. Weeks *et al.*, "Algorithms for ASTAP—A network-analysis program," *IEEE Trans. Circuit Theory*, vol. CT-20, pp. 628–634, Nov. 1973.

[14] Y. P. Tsividis, "Analysis of switched capacitive networks," *IEEE Trans. Circuits Syst.*, vol. CAS-26, no. 11, pp. 935–947, Nov. 1979.

[15] H. De Man, J. Rabaey, G. Arnout, and J. Vandewalle, "Practical implementation of a general computer aided design technique for switched capacitor circuits," *IEEE J. Solid-State Circuits*, vol. SC-15, no. 2, pp. 190–200, Apr. 1980.

[16] J. Vandewalle, H. J. De Man, and J. Rabaey, "Time, frequency, and z-domain modified nodal analysis of switched-capacitor networks," *IEEE Trans. Circuits Syst.*, vol. CAS-28, pp. 187–195, Mar. 1981.

[17] R. D. Davis, "Computer analysis of switched capacitor filters including sensitivity and distortion effects," Ph.D. dissertation, Dept. Elect. Eng., University of Illinois, Urbana, Oct. 1981.

[18] S. C. Fang, Y. P. Tsividis, and O. Wing, "SWITCAP: A switched-capacitor network analysis program. Part I: Basic features," *IEEE Circuits Syst. Mag.*, vol. 5, no. 3, pp. 4–10, Sept. 1983.

[19] S. C. Fang, Y. P. Tsividis, and O. Wing, "SWITCAP: A switched-capacitor network analysis program. Part II: advanced features," *IEEE Circuits Syst. Mag.*, vol. 5, no. 4, pp. 41–46, Dec. 1983.

[20] S. C. Fang, Y. P. Tsividis, and O. Wing, "Time- and frequency-domain analysis of linear switched-capacitor networks using state charge variables," *IEEE Trans. Computer-Aided Design*, vol. CAD-4, pp. 651–661, Oct. 1985.

[21] J. Vlach, K. Singhal, and M. Vlach, "Computer oriented formulation of equations and analysis of switched-capacitor networks," *IEEE Trans. Circuits Syst.*, vol. CAS-31, no. 9, pp. 753–765, Sept. 1984.

[22] K. Suyama and S. C. Fang, "Efficient state charge variable analysis of group delay and sensitivity for switched capacitor networks," in *Proc. 29th Midwest Symp. CAS* (Lincoln, NE), Aug. 1986, pp. 789–792.

[23] K. Suyama, "Analysis, simulation, and application of linear and nonlinear switched-capacitor and mixed switched-capacitor/digital networks," Ph.D. dissertation, Columbia Univ., New York, NY, 1989.

[24] G. Arnout and H. J. De Man, "The use of threshold functions and boolean-controlled network elements for macromodeling of LSI circuits," *IEEE J. Solid-State Circuits*, vol. SC-13, no. 3, pp. 326–332, June 1978.

[25] Z. Masňaay, Z. Oliva, and H. Mann, "DAVID5—A program for programmable analog-digital system simulation," in *Proc. ECCTD* (Prague, Czechoslovakia), Sept. 1985, pp. 121–124.

[26] J. Buddefeld, P. Richert, W. Brockherde, B. J. Hosticka, and G. Zimmer, "CAD tools for the design and analysis of switched-capacitor networks," in *Proc. ESSCIRC* (Delft, The Netherlands), Sept. 1986, pp. 140–142.

[27] E. S. Lee and T. F. Fang, "A mixed-mode analog-digital simulation methodology for full custom designs," in *Proc. IEEE 1988 CICC* (Rochester, NY), May 1988, pp. 3.5.1–3.5.4.

[28] E. L. Acuna, J. P. Dervenis, A. J. Pagones, F. L. Yang, and R. A. Saleh, "Simulation techniques for mixed analog/digital circuits." *IEEE J. Solid-State Circuits*, vol. 25, pp. 353–363, Apr. 1990.

[29] R. Saleh, "Iterated timing analysis and SPLICE 1," M.S. thesis, Dept. Elec. Eng. Computer Sci., Univ. of Calif., Berkeley, June 1983.

[30] M. A. Copeland, G. P. Bell, and T. A. Kwasniewski, "A mixed-mode sampled-data simulation program," *IEEE J. Solid-State Circuits*, vol. SC-22, no. 6, pp. 1098–1105, Dec. 1987.

[31] Y. P. Tsividis, P. R. Gray, D. A. Hodges, and J. Chacko, Jr., "A segmented μ-255 law PCM voice encoder utilizing NMOS technology," *IEEE J. Solid-State Circuits*, vol. SC-11, pp. 740–747, Dec. 1976.

[32] B. E. Boser and B. A. Wooley, "The design of sigma–delta modulation analog-to-digital converters," *IEEE J. Solid-State Circuits*, vol. 23, pp. 1298–1308, Dec. 1988.

[33] Y. Tsividis, M. Banu, and J. Khoury, "Continuous-time MOSFET-C filters in VLSI," *IEEE Trans. Circuits Syst.*, vol. CAS-33, no. 2, pp. 125–140, Feb. 1986.

[34] K. C. Chao, "Limitations of interpolative modulators for oversampling A/D converters," M.S. thesis, Mass. Inst. Technol., Cambridge, 1988.

[35] R. P. Lippman, "An introduction to computing with neural nets," *IEEE ASSP Mag.*, pp. 4–20, Apr. 1987.

[36] K. Suyama, S. C. Fang, and Y. P. Tsividis, "Simulation methods for mixed switched-capacitor/digital networks," to be published.

[37] T. Cataltepe, private communication, 1990.

Ken Suyama (S'81–M'88) was born in Tokyo, Japan, in 1955. He received the B.S. degree in electrical engineering from the University of California, Davis, in 1980, and the M.S. and Ph.D. degrees in electrical engineering from Columbia University, New York, NY, in 1982 and 1989, respectively.

Since 1989 he has been an Associate Research Scientist at the Center for Telecommunications Research, Columbia University, New York, NY. His research interests are computer-aided analysis and design, analog and mixed analog/digital integrated circuits, and circuit implementations of neural networks and fuzzy logic.

San-Chin Fang (S'80–M'82) received the B.S. degree from Princeton University, Princeton, NJ, in 1974, the M.S. degree from Stanford University, Stanford, CA, in 1977, and the Ph.D. degree from Columbia University, New York, NY, in 1983.

In 1982 he joined AT&T Bell Laboratories in Murray Hill, NJ, as a Member of the Technical Staff to work on analog CAD tools and high-speed analog circuit design. Currently he is a Supervisor in the VLSI Technology Laboratory. From June 1985 to May 1986 he was on leave at the Center for Telecommunications Research, Columbia University, as an Associate Research Scientist. He also taught several courses in analog IC design at Columbia University from 1982 to 1984.

Yannis P. Tsividis (S'71–M'76–SM'81–F'86) was born in Greece and received the Ph.D. degree from the University of California, Berkeley, in 1976.

He is a Professor at Columbia University, New York, NY, and is the author of *Operation and Modeling of the MOS Transistor* (McGraw-Hill, 1987) and co-editor and co-author of *Design of MOS VLSI Circuits for Telecommunications* (Prentice-Hall, 1985).

Dr. Tsividis has been a member of the Administrative Committee of the IEEE Circuits and Systems Society and of the United Nations Advisory Committee on Science and Technology for Development. He is the recipient of the 1984 IEEE W. R. G. Baker Best Paper Award and the 1986 European Solid-State Circuits Conference Best Paper Award, and co-recipient of the 1987 IEEE Circuits and Systems Society Darlington Best Paper Award.

Behavioral Simulation for Analog System Design Verification

Brian A. A. Antao, *Member, IEEE,* and Arthur J. Brodersen, *Senior Member, IEEE*

***Abstract*— Synthesis of analog circuits is an emergent field, with efforts focused at the cell level. With the growing trend of mixed ASIC designs that contain significant portions of analog sections, compatible design methodologies in the analog domain are necessary to complement those in the digital domain. The synthesis process requires an associated verification process to ensure that the designs meet performance specifications at the onset. In this paper we present a behavioral simulation methodology for analog system design veri,1fication and design space exploration. The verification task integrates with analog system-level synthesis for an integrated synthesis-verification process that avoids expensive post synthesis simulation by invoking external simulators. Thus rapid redesign at the architectural level can be undertaken for design parameter variation and during optimization. The verification suite is composed of a repertoire of analysis modes that include time and frequency domain analysis, sensitivity analysis and distortion analysis. Besides verification of design specifications, these analysis modes are also used to generate metrics for comparison of various architectural choices that could realize a given set of specifications. The implementation is in the form of a behavioral simulator, ARCHSIM.**

I. Introduction and Motivation

THE DESIGN of integrated circuits is an enormous and iterative process, with numerous CAD tools being essential to deal with the ever growing complex nature of designs. Simulators are an essential component of the VLSI designers tool kit, replacing the discrete circuit breadboard.[1] *Circuit simulators* such as SPICE [1] were widely used in the early days [2] where the number of transistors were few; however, with the increasingly dense designs newer simulators and simulation methodologies are required. The SPICE like simulators are used for detailed simulation of smaller circuits, and have been the only option available till now to the analog designer. Since the major portion of VLSI designs are digital in nature, much of the effort went into developing methodologies that expedited the digital design and simulation process. For digital circuits, one of the key tasks in simulation is timing analysis, for which special-purpose simulators such as MOTIS [3] and the more recent AWE [4] and SPECS [5] that consider the characteristics of digital MOS circuits were developed. *Logic-level simulators* were introduced to simulate circuits directly using boolean values instead of continuous values for signals. One class of logic simulators are *switch-level simulators* which model the transistor with a simple switch model incorporating delays, thus speeding up the simulation of digital circuits at the transistor-level. The switch-level also facilitates symbolic analysis of digital networks [6]. A second category of logic simulators are *Gate-level simulators*, which are a higher-level simulators that model an entire gate as a functional element, instead of a transistor. The individual gates can be verified by using detailed SPICE simulations, and their delays extracted and incorporated in the gate models. Since it was necessary to simulate critical-paths of a digital system to a greater detail, *Mixed-mode simulators* were introduced to handle this task in an unified manner [7]. DIANA [8] and SPLICE [9] were notable developments which mixed circuit or *electrical-level* simulation and logic-level simulation. High level simulators have also been used for system level simulation of digital systems at the register-transfer level [10]. *Behavioral simulators* in the digital domain can directly simulate the functional behavior utilizing behavioral models and hardware descriptions in widely used languages such as VHDL [11] and Verilog[2].

The analog domain however is not as well structured as the digital domain. Digital behavior is easily represented using boolean expressions or higher-level programming language constructs. The notion of the *Analog Behavioral Level* has only recently been introduced by developments such as the SABER[3] [12] and the ATTSIM[4] [13], [14] simulators. Efforts are also underway towards defining a companion AHDL to accompany VHDL of the digital domain [15]. In order to support system-level design of analog systems, it is essential to have a supporting verification methodology. Some of the recent general purpose analog behavioral simulators can be utilized for this purpose, which however are limited for the most part to time domain analysis. However these behavioral simulators focus on standalone simulation, driven by the need for expediting simulation. These developments do not directly address system verification from the synthesis perspective.

Among other developments in the efficient simulation of analog systems are the ATTSIM simulator that can simulate a system comprised of analog as well as digital blocks; a mixed mode simulator, M^3, that enhances the MOTIS digital

Manuscript received April 25, 1994; revised August 2, 1994. This work was supported in part by the Semiconductor Research Corporation under Grants 93-DP-109 and 94-DP-109.

B. Antao was with the Department of Electrical Engineering, Vanderbilt University, Nashville, TN 37235 USA. He is now with Coordinated Science Laboratory, University of Illinois at Urbana-Champaign, Urbana, IL 61801 USA.

A. Brodersen is with the Department of Electrical Engineering, Vanderbilt University, Nashville, TN 37235 USA.

IEEE Log Number 9413460.

[1] Breadboarding integrated circuits—i.e., fabricating intermediate prototypes is an extremely expensive affair, seldom undertaken.

[2] Verilog is a trademark of Cadence Design Systems Inc.

[3] SABER is a trademark of Analogy Inc.

[4] ATTSIM is a trademark of AT&T Bell Laboratories.

timing simulator to simulate analog-digital systems [16]; and iMACSIM [17, 18] another multilevel analog simulation tool. These developments expedite simulation through the use of various modeling techniques. Other techniques used to speed up simulation are *event-driven simulation* [5], [19] that exploit *latency, piecewise approximations* [5] and *waveform relaxation* [20]. *Macromodeling*, which replaces detailed circuit blocks such as opamps with simpler equivalent circuits, is often used to expedite the simulation of analog circuits at the circuit-level. However these behavioral simulators do not address system synthesis and verification, external postprocessors such as FFT routines often need to be used to generate different responses.

Most of the analog design automation research efforts have been focussed at the circuit-level or cell level [10–50 devices]. Cell-level synthesis involves circuit-topology design and device sizing; and layout synthesis [21], [22]. Since the focus of these synthesis efforts are at the circuit-level, verification of the synthesized circuits is carried out using conventional SPICE-like circuit level simulators. While these efforts are slowly beginning to mature, the design and realization of mixed analog-digital systems also requires design automation efforts at the *analog systems level* that are compatible with the system-level design efforts in the digital domain. The high-level synthesis task begins with analog behavioral specifications. The analog behavioral level is characterized in terms of mathematical expressions in the time and frequency domains. These behavioral specifications are synthesized into functional architectures independent of any specific circuit implementation style. The functional architectures are then synthesized into an implementation-specific intermediate architecture. The implementation styles can be transconductance-capacitor, switched-capacitor, switched-current. etc. A behavioral architecture is composed of high-level functional elements that are independent of a circuit technology. An *intermediate architecture* is a synthesized behavioral architecture in a specific circuit implementation technology.

To support the above described synthesis paradigm, we have formulated a customized behavioral modeling and simulation framework. The motivation being:

1) To analyze various high-level architectures to determine an appropriate architectural solution which is based on meeting performance specifications. Sensitivity is a key metric used in differentiating architectures.
2) To simulate at the systems level prior to the design of the constituent sub-circuits.
3) To provide an integrated modeling, analysis and synthesis environment. Additionally, simulation description for general purpose time-domain mixed-mode simulators can be generated that can be included in the simulation of a complete mixed-signal ASIC.
4) To support the top-down design methodology, through the use of various levels of simulation models.
5) To provide a rapid simulation technique that can be invoked iteratively for design parameter variations and optimization.

The behavioral simulator, ARCHSIM (ARCHitecture SIMulator) integrates with the architectural synthesis module, ARCHGEN [24], [25]. The synthesis module and the behavioral simulator share a common internal design representation which can be directly verified instead of generating an external description and invoking an external simulator. This tight coupling also increases the efficiency of the verification process, facilitating the easy use of iterative optimization methods in the synthesis process. External designs at the behavioral and intermediate architecture levels can be specified for verification in a C++ input description.

The paper is organized across the various sections as follows: In Section II we present an overview of the analog system synthesis process, in Section III we present the verification methodology along with the model definitions. In Section IV we describe the formulations for the different analysis modes for analog system verification. Details of the simulator architecture and system description are presented in Section V. In Section VI we present simulation examples and experimental results, followed by conclusions in Section VII.

II. Overview of Analog System Synthesis

The analog behavioral level is characterized in terms of mathematical expressions in the time and frequency domains. In the time domain, continuous systems are mathematically modeled by differential equations and discrete systems by difference equations. In the frequency domain, s and z transfer functions are used to characterize systems. These behavioral representations serve as a set of high-level specifications for the analog architectural synthesis process. From the transfer function specifications, a state variable model is first derived in a canonical form, which then serves as a basis for generating a functional architecture. The functional architectures are independent of any circuit design style, and are composed of functional blocks such as adders, integrators etc. A given set of behavioral specifications can be realized into different possible functional architectures. The choice of a specific architectural configuration depends on the various design goals that need to be satisfied. Thus system level verification and analysis plays a crucial role in the synthesis process—1) to verify that the synthesized architectures meet the behavioral specifications, and 2) for comparison of various performance characteristics of these architectures to explore the design space.

General behavioral level representation for continuous systems is expressed as a s-domain transfer function

$$H(s) = K\frac{b_m s^m + b_{m-1}s^{m-1} + \cdots + b_1 s + b_0}{a_n s^n + a_{n-1}s^{n-1} + \cdots + a_1 s + a_0} \quad m \leq n. \tag{1}$$

The corresponding general state-space model is,

$$\dot{\mathbf{x}}(t) = \mathbf{A}(t)\mathbf{x}(t) + \mathbf{B}(t)\mathbf{u}(t) \tag{2}$$

$$\mathbf{y}(t) = \mathbf{C}(t)\mathbf{x}(t) + \mathbf{D}(t)\mathbf{u}(t). \tag{3}$$

Here, $\mathbf{x}(t)$ is the internal state vector. Equation (2) is the internal description of the system and (3) specifies the system output in terms of the internal states and input. Which for example can be realized as a controllable architecture whose

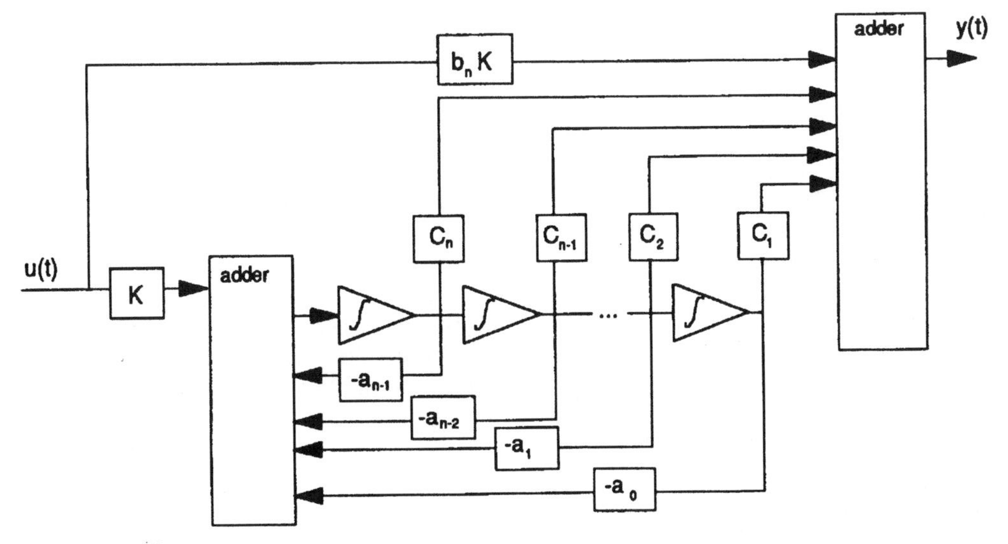

Fig. 1. A controllable form architecture.

state-space model has the following form

$$\begin{bmatrix} \dot{x}_1(t) \\ \dot{x}_2(t) \\ \vdots \\ \vdots \\ \dot{x}_n(t) \end{bmatrix} = \begin{bmatrix} 0 & 1 & 0 & \cdots & 0 \\ 0 & 0 & 1 & \cdots & 0 \\ \vdots & \vdots & \vdots & \cdots & \vdots \\ 0 & 0 & 0 & \cdots & 1 \\ -a_0 k^n & -a_1 k^{n-1} & -a_2 k^{n-2} & \cdots & -a_{n-1} k \end{bmatrix} \times \begin{bmatrix} x_1(t)/k \\ x_2(t)/k \\ \vdots \\ \vdots \\ x_n(t)/k \end{bmatrix} + \begin{bmatrix} 0 \\ 0 \\ \vdots \\ \vdots \\ k^n K \end{bmatrix} u(t) \tag{4}$$

$$y(t) = \left[(b_0 - b_n a_0) \quad \frac{(b_1 - b_n a_1)}{k} \quad \cdots \quad \frac{(b_{n-1} - b_n a_{n-1})}{k^{n-1}} \right] \times \frac{\bar{X}(t)}{k} + [b_n K]u(t). \tag{5}$$

Fig. 1 shows a controllable architecture corresponding to the derived state-space model.

Once a functional architecture has been synthesized, it can be implemented into one of the many circuit implementation styles, such as switched-capacitor, transconductance-capacitor, switched current etc. The choice of the implementation style would depend on another set of design constraints such as the fabrication technology of the overall system in which the analog section needs to be embedded, and the performance limitations of the various styles. Implementation style specific intermediate architectures are synthesized from the behavioral architecture. Fig. 2 shows two intermediate architecture realizations for an integrator functional block. The intermediate architecture serves the purpose of design space exploration—evaluating various implementation styles; and serves as a bridge between system-level design and detailed circuit-level synthesis. Again at the intermediate architecture stage, verification and design space exploration are necessary to ensure that specifications are met as the design evolves through the hierarchy.

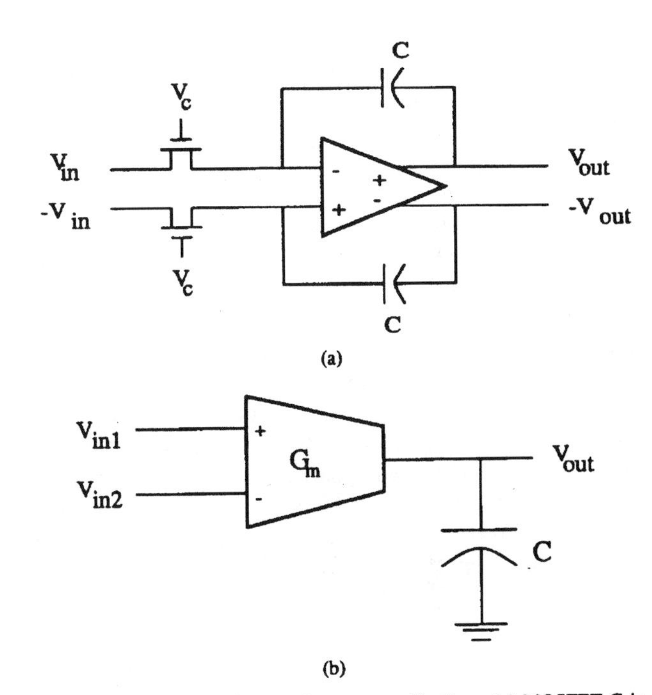

Fig. 2. Intermediate architecture integrator realizations. (a) MOSFET-C integrator. (b) $G_m - C$ integrator.

III. Verification Methodology

The synthesized architectures have to be analyzed to verify that they meet the behavioral specifications, which requires analysis in the time and frequency domains. Besides these analysis, sensitivity analysis, distortion analysis, and dynamic range measurements are useful in comparing the performance of the various architectures. The time and frequency domain responses of the various architectures are likely to be

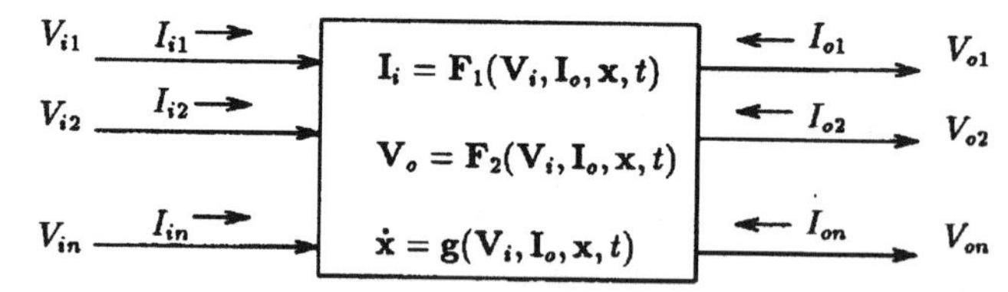

Fig. 3. A general model definition.

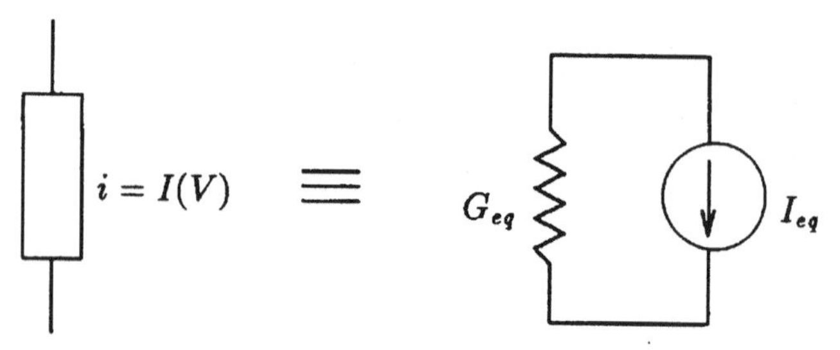

Fig. 4. Linearized companion model.

identical since they are synthesized from the same set of behavioral specifications. Whereas sensitivity and distortion analysis provide a set of metrics that aid in the design space exploration task by providing metrics to distinguish between the various possible architectures. At the intermediate architecture level, the various simulation metrics are useful in choosing between implementation styles, and to check if the fabrication technology limits are not pushed beyond feasible bounds.

System verification is implemented in a circuit simulation paradigm. Central to this approach is the formulation of the system equations in an analysis regime and the solution of these equations. The system of equations are generated in the Modified Nodal Analysis (MNA) formulation [26]. The models for the various components are decoupled from the analysis algorithms. Each model is implemented as an MNA stencil. At each time or frequency step, the model evaluation phase results in updating the system MNA matrix.

A. Model Definition

In general, the model for any functional block is formulated using a hybrid port definition. Fig. 3 shows a general model. This model includes the port node voltages, controlling currents as well as port currents that simulate loading effects. Each model also has internal states $\mathbf{x}$, used to model storage elements such as capacitor voltages or inductor currents in terms of equivalent circuits.

A linear n-port can be modeled using the following hybrid-port equations

$$I_i = h_{11} V_i + h_{12} I_o$$
$$V_o = h_{21} V_i + h_{22} I_o.$$

Introducing internal states $\mathbf{x}$ in the model, the relations for the general nonlinear case are as follows

$$\mathbf{I}_\mathrm{i}(\mathbf{t}) = \mathbf{F}_1(\mathbf{V}_\mathrm{i}(\mathbf{t}), \mathbf{I}_\mathrm{o}(\mathbf{t}), \mathbf{x}(\mathbf{t}), \mathbf{t}) \quad (6)$$
$$\mathbf{V}_\mathrm{o}(\mathbf{t}) = \mathbf{F}_2(\mathbf{V}_\mathrm{i}(\mathbf{t}), \mathbf{I}_\mathrm{o}(\mathbf{t}), \mathbf{x}(\mathbf{t}), \mathbf{t}) \quad (7)$$
$$\dot{\mathbf{x}}(\mathbf{t}) = \mathbf{g}(\mathbf{V}_\mathrm{i}(\mathbf{t}), \mathbf{I}_\mathrm{o}(\mathbf{t}), \mathbf{x}(\mathbf{t}), \mathbf{t}). \quad (8)$$

The above set of model equations result in a system of algebraic-differential equations in the time domain of the form

$$\mathbf{f}(\mathbf{v}(\mathbf{t}), \mathbf{i}(\mathbf{t}), \mathbf{x}(\mathbf{t}), \dot{\mathbf{x}}(t)) = 0 \quad (9)$$

which are solved using numerical integration and iterative nonlinear equation methods during transient analysis.

In general any component characterized by a closed form expression $i = I(V)$ is linearized and replaced by an equivalent *companion* element during simulation. The linearized equivalent at each iteration is derived through the application of the Newton–Raphson method, or a Taylor's series expansion. In the companion model,

$$G_{\mathrm{eq}}^{j+1} = \left[\frac{\partial i}{\partial V}\right]_j$$
$$I_{\mathrm{eq}}^{j+1} = i^j - \left[\frac{\partial i}{\partial V}\right]_j V^j$$

For models without a closed form expression numerical models are used to arrive at a companion model.

In the general case, a behavioral model can be characterized using nonlinear integro-differential equations expressed as

$$\mathbf{F}(\mathbf{x}(\mathbf{t}), \dot{\mathbf{x}}(\mathbf{t}), \mathbf{t}) = 0. \quad (10)$$

Here $\mathbf{x}(\mathbf{t})$ represents a set of variables associated with the models, these include port voltages and currents and internal states. Given this set of equations, we next derive the corresponding MNA stencil. The entries corresponding to the location of the variables $\mathbf{x}(\mathbf{t})$ in the system MNA matrix are filled in accordingly. At time point $t = t_{n+1}$

$$\mathbf{F}(\mathbf{x}_{n+1}, \dot{\mathbf{x}}_{n+1}, t_{n+1}) = 0. \quad (11)$$

The system of (10) include the state equations of the form

$$\dot{\mathbf{x}} = \mathbf{h}(\mathbf{x}, t).$$

Using a general integration formula, the backward differentiation formula (BDF), to find an approximate value to a desired precision, for $\dot{\mathbf{x}}_{n+1}$ at time $t = t_{n+1}$. The BDF is expressed as

$$\dot{\mathbf{x}}_{n+1} = -\frac{1}{h}\sum_{i=0}^{k} \alpha_\mathrm{i} \mathbf{x}_{\mathbf{n+1-i}} \triangleq \mathbf{g}(\mathbf{x}_{n+1}). \quad (12)$$

$\alpha_0, \alpha_1, \ldots \alpha_k$ are the BDF constants, and $h \triangleq t_{n+1} - t_n$ is the present time step. In this system of equations, $\mathbf{x}_{n+1}$ is still unknown. Application of the BDF to (12) results in the following system of nonlinear algebraic equations

$$\mathbf{F}(\mathbf{x}_{n+1}, \mathbf{g}(\mathbf{x}_{n+1}), t_{n+1}) = 0. \quad (13)$$

The solution $\mathbf{x}_{n+1}$ can now be found by solving (13) using nonlinear iterative solution methods such as the Newton-Raphson algorithm through the use of the recursive relation:

$$\mathbf{x}_{n+1}^{(j+1)} = \mathbf{x}_{n+1}^{(j)} - \left[\mathbf{J}\left(\mathbf{x}_{n+1}^{(j)}\right)\right]^{-1} \mathbf{F}\left(\mathbf{x}_{n+1}^{(j)}, \mathbf{g}\left(\mathbf{x}_{n+1}^{(j)}\right), t_{n+1}\right) \quad (14)$$

$\mathbf{x}_{n+1}^{(j+1)}$ is the value of $\mathbf{x}_{n+1}$ at the $(j+1)$th iteration. $\mathbf{J}(\mathbf{x}_{n+1}^{(j)})$ is the *Jacobian matrix* of $\mathbf{F}(\mathbf{x}_{n+1}, \mathbf{g}(\mathbf{x}_{n+1}), t_{n+1})$ evaluated at $\mathbf{x}_{n+1} = \mathbf{x}_{n+1}^{(j)}$ and is defined as

$$\mathbf{J}(\mathbf{x}_{n+1}^{(j)}) \triangleq \left.\frac{\delta \mathbf{F}(\mathbf{x}, \dot{\mathbf{x}}, t)}{\delta \mathbf{x}}\right|_{\substack{x=x_{n+1}^{(j)} \\ \dot{x}=g(x_{n+1}^{(j)})}} - \left.\frac{\alpha_0}{h}\frac{\delta \mathbf{F}(\mathbf{x}, \dot{\mathbf{x}}, t)}{\delta \dot{\mathbf{x}}}\right|_{\substack{x=x_{n+1}^{(j)} \\ \dot{x}=g(x_{n+1}^{(j)})}}. \tag{15}$$

Equation (14) can be expressed as

$$[\mathbf{J}(\mathbf{x}_{n+1}^{(j)})]\mathbf{x}_{n+1}^{(j+1)} = [\mathbf{J}(\mathbf{x}_{n+1}^{(j)})]\mathbf{x}_{n+1}^{(j)} - \mathbf{F}(\mathbf{x}_{n+1}^{(j)}, \mathbf{g}(\mathbf{x}_{n+1}^{(j)}), t_{n+1}). \tag{16}$$

From the system of (16), the entries of the Jacobian matrix at each row associated with each output port voltage variable and unknown current or state variable are correspondingly filled into the MNA matrix, and the quantities on the RHS are added to the RHS vector. We further illustrate this method through the use of two specific examples next.

The analog multiplier is an example of a functional block modeled by a nonlinear algebraic equation

$$V_{\text{out}} = KV_1 \cdot V_2.$$

Applying Taylor's series expansion and neglecting second order and higher terms, at the $(j+1)$th iteration,

$$\begin{aligned} V_{\text{out}}^{(j+1)} &= V_{\text{out}}^{(j)} + KV_2^{(j)}(V_1^{(j+1)} - V_1^{(j)}) \\ &\quad + KV_1^{(j)}(V_2^{(j+1)} - V_2^{(j)}) \\ &= KV_2^{(j)}V_1^{(j+1)} + KV_1^{(j)}V_2^{(j+1)} \\ &\quad + V_{\text{out}}^{(j)} - 2KV_1^{(j)}V_2^{(j)}. \end{aligned} \tag{17}$$

The linearized MNA model for a *multiplier* is

$$\begin{matrix} \\ e_{vout} \end{matrix} \begin{bmatrix} V_1^{(j+1)} & V_2^{(j+1)} & V_{\text{out}}^{(j+1)} \\ -KV_2^{(j)} & -KV_1^{(j)} & 1 \end{bmatrix} \begin{bmatrix} \text{RHS} \\ V_{\text{out}}^{(j)} - 2KV_1^{(j)}V_2^{(j)} \end{bmatrix}. \tag{18}$$

Next consider the formulation of the time-domain MNA stencil for an analog voltage controlled oscillator (VCO). The behavior of a VCO is characterized by an algebraic-differential equation. The angular frequency is given by the expression

$$\omega_o = \omega_c + K_o V_{\text{in}}(t). \tag{19}$$

The phase angle is the integral of the angular frequency,

$$\theta(t) = \int [\omega_c + K_o V_{\text{in}}(t)]dt. \tag{20}$$

The output of the VCO is

$$V_{\text{out}}(t) = V_m \sin(\theta(t)). \tag{21}$$

The instantaneous phase angle is modeled by introducing the state x

$$\dot{x} = \omega_c + K_o V_{\text{in}}(t).$$

```
Algorithm 1: Time-domain sensitivity analysis

time_domain_sensitivity_analysis()
{
  t = t0;
  compute_initial_conditions();
  pn = number_of_variable_parameters();
  stoptime = end_of_simulation;
  while (t < stoptime ) {
    do_tran_iteration(t);            /* do a transient analysis iteration at current time-point */
    foreach (parameter p in pn ) {
      construct_RHS_vector();
      compute_sensitivity_network_response();        /* reuses MNA matrix from do_tran_iteration */
      }
    store_single_parameter_sensitivities();
    compute_multiparameter_sensitivity();
    timestep = compute_time_step();
    update_time();
  }
}
```

Fig. 5. Time domain sensitivity analysis algorithm.

Application of one of the integration methods, backward euler (BE), or Trapezoidal method to the above state equations, and a Taylor series expansion to the VCO output equation results in a linearized equation at each iteration of the transient simulation given by

$$\begin{aligned} V_{\text{out}(n+1)}^{(j+1)} &= V_{\text{out}(n+1)}^{(j)} + hK_o V_m \cos \\ &\quad \times (x_n + h[\omega_c + K_o V_{\text{in}(n+1)}^{(j)}]) \\ &\quad \times (V_{\text{in}(n+1)}^{(j+1)} - V_{\text{in}(n+1)}^{(j)}). \end{aligned} \tag{22}$$

Equation (22) is a linearized equation at each iteration and represents an MNA stencil in the form of an equivalent conductance in parallel with a current source. The equivalent conductance is given by

$$G_{\text{eq}} = hK_o V_m \cos(x_n + h[\omega_c + K_o V_{\text{in}(n+1)}^{(j)}])$$

and the equivalent current source

$$\begin{aligned} I_{\text{eq}} &= V_{\text{out}(n+1)}^{(j)} - hK_o V_m \cos \\ &\quad \times (x_n + h[\omega_c + K_o V_{\text{in}(n+1)}^{(j)}]) V_{\text{in}(n+1)}^{(j)}. \end{aligned}$$

IV. Analysis Modes

Architecture verification requires the simulation of the synthesized architectures in the time and frequency domains to ensure that the behavioral specifications are met. An additional design task is design space exploration for which suitable metrics such as sensitivity and distortion analysis are required to compare and choose from a set of possible synthesis solutions. To cater to these diverse analysis and verification requirements, the various analysis modes are formulated and implemented within a common framework.

A. Time-Domain Analysis

Conventional time-domain analysis is implemented in the form of iterative time step analysis. Newton iterations are carried out for evaluating nonlinear models. At each time point, the algorithm exercises each component instance; i.e., the model evaluation task that results in the generation of a new set of values for the time-domain MNA stencils. MNA stencils for the various functional elements that compose the behavioral and intermediate architectures are derived and implemented as part of the common synthesis-verification objects. The analog

Algorithm 2: Frequency-domain sensitivity analysis (adjoint method)

```
frequency_domain_sensitivity_analysis()
{
  startfreq = starting frequency;
  stopfreq = end frequency;
  h_ω = frequency step ;
  while (freq < stopfreq ) {
    freq_domain_model_evaluation(freq);
    assemble_system_equations();
    LU_decompose();
    compute_original_response();
    compute_adjoint_response();
    compute_single_parameter_sensitivity();
    compute_cumulative_sensitivity();
    update_frequency();
  }
}
```

Fig. 6. Frequency domain sensitivity analysis algorithm.

Algorithm 3: Time-domain steady-state analysis

```
time_domain_steady-state_analysis()
{
  /* x_0 is a set of initial conditions to be computed */
  x_0 = initial_guess();
  T = signal_frequency_period;
  repeat {
    do_tran(t = 0 to t = T);
    sensitivity_analysis();      /* compute jacobian F'(x_0^(j)) = ∂F(x_0)/∂x_0 |_{x_0=x_0^(j)} */
    x_0^(j+1) = x_0^(j) - [1 - F'(x_0^(j))]^{-1} [x_0^(j) - F(x_0^(j))]      /* update initial conditions */
  until (||x_0(T, x_0^(j)) - x_0^(j)|| < ε and ||x_0^(j+1) - x_0^(j)|| < δ)
  }
  set_initial_conditions( x_0^(n));
  do_tran(steady-state_simulation_time);
}
```

Fig. 7. Time domain steady-state analysis algorithm.

Algorithm 4: Frequency-domain steady-state analysis

```
frequency_domain_steady-state_analysis()
{
  startfreq = starting frequency;
  stopfreq = end frequency;
  h_ω = frequency step ;
  while (freq < stopfreq ) {
    T = period of current signal frequency;
    time_domain_steady-state_analysis(T);
    assemble_coefficients();            /* compute the matrix Γ^{-1} */
    compute_Fourier_coefficients();
    compute_nodal_spectra();
    update_frequency();
  }
}
```

Fig. 8. Frequency domain steady-state analysis algorithm.

multiplier and VCO MNA stencils presented in the previous section are two examples.

B. Frequency-Domain Analysis

A set of linearized MNA stencils in the frequency domain are defined for each component. The core of the frequency domain algorithm evaluates each of the frequency domain MNA stencils and the system of equations is solved for the frequency domain unknowns. This mode of frequency domain analysis is used for linear and linearized systems.

C. Sensitivity Analysis

The synthesized architectures and circuits are likely to undergo deviations in their component parameters by the time the circuit is fabricated. These deviations can result in degradation in performance. Thus one of the goals of the design process is to choose architectures that meet the specifications with minimum deviation and whose performance is least sensitive to variations. Sensitivity analysis provides a set of metrics that indicate how sensitive a design is to certain parameter variations. Sensitivity analysis also provides gradients of performance functions with respect to design parameters which are required during optimization and in other analysis such as time-domain steady-state analysis. Both time domain and frequency domain sensitivity analysis methods have been formulated and implemented for behavioral level simulation. Sensitivities are computed with respect to the model variable parameters. Frequency domain sensitivity analysis is used as a metric for design space exploration of possible architectural choices during synthesis. Time domain sensitivity analysis finds application in optimization and steady-state analysis.

1) Time Domain Sensitivity Analysis: The direct sensitivity analysis method [28] is adapted for time-domain behavioral sensitivity analysis. This method requires the construction of an associated sensitivity network by direct differentiation of the components. Each branch in the sensitivity network is of the same type as that in the original network. In the time domain, the direct method is preferred, as the simulation is less involved as compared to the adjoint method which requires the adjoint network to be simulated backward in time, and convolving variables across the two networks [30].

For an adder,

$$v_o = a_1 v_1 + a_2 v_2 + \cdots + a_k v_k.$$

Let p be the parameter of interest:

$$\begin{aligned}\frac{dv_o}{dp} &= \frac{da_1}{dp}v_1 + a_1\frac{dv_1}{dp} + \frac{da_2}{dp}v_2 + a_2\frac{dv_2}{dp} \\ &\quad + \cdots + \frac{da_k}{dp}v_k + a_k\frac{dv_k}{dp} \\ &= a_1\hat{v}_1 + a_2\hat{v}_2 + \cdots + a_k\hat{v}_k + \sum_{i=1}^{k}\hat{V}_i.\end{aligned}$$

Here,

$$\hat{v} = \frac{dv}{dp}, \qquad V_i = \frac{da_i}{dp}v_i.$$

Thus the component corresponding to an adder in the sensitivity network is an adder with an additional driving source. The additional driving source is stamped in the RHS vector. In simulating the sensitivity network, the MNA matrix for the original network is reused with a new RHS vector that includes the driving source in the sensitivity branch of interest. For each parameter of interest, sensitivity computation only requires a new RHS vector, with the LU factors of the original MNA

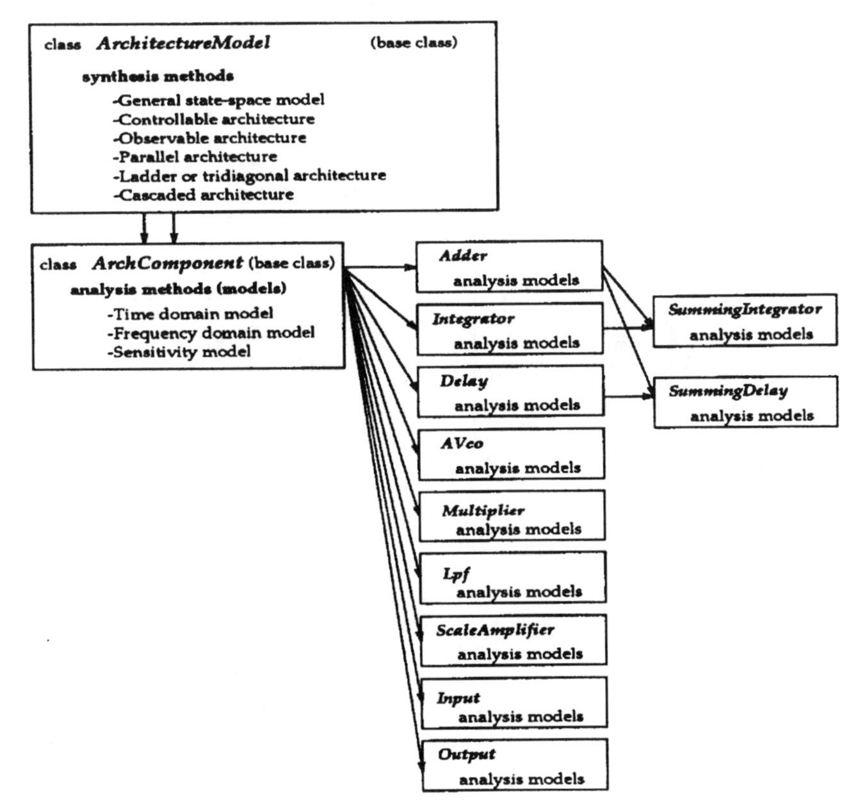

Fig. 9. Architecture representation objects.

matrix being reused. The algorithm is depicted in pseudocode form in Fig. 5.

2) Frequency Domain Sensitivity Analysis: For frequency domain sensitivity computation at the behavioral level, we have devised an algorithm based on the adjoint method [29]. A formulation of the Tellegen's theorem at the behavioral level [30] is used for this purpose, which is stated as follows: If V_1 is the set of variables associated with the original network A, and V_2, the set associated with $\hat{A}$, x_i, γ_i being the internal variables and x_p, γ_p the external variables. Tellegen's theorem is expressed as

$$\sum_i \Delta x_i^{V_1} \gamma_i^{V_2} = \sum_p \Delta x_p^{V_1} \gamma_p^{V_2}. \tag{23}$$

By applying the above expression to the model equations at the behavioral level, the adjoint models are generated. The procedure for sensitivity computation in the frequency domain involves constructing the adjoint network in the frequency domain and analyzing the original and adjoint networks at each frequency step to determine the sensitivities of a performance function with respect to all the component parameters. The pseudocode for the algorithm is described next in Fig. 6.

3) Multiparameter Sensitivity: The sensitivities obtained are usually with respect to a single model parameter, referred to as *single parameter sensitivities*. Often two architectures synthesized from the same set of behavioral specifications differ significantly in the topology and components. Thus single parameter sensitivities may not provide a realistic measure for comparison. Multiparameter sensitivity provides a cumulative sensitivity measure obtained by combining the single parameter sensitivities of the performance functions. Multiparameter sensitivity is computed as the complex congugate scalar product of the single parameter sensitivities [30].

D. Steady-State Analysis

1) Time-Domain Steady-State Analysis: Time-domain steady-state analysis can be viewed as finding a solution to a two point boundary value problem [31]. Since the time domain solution starting from zero initial conditions contains the transient component, a suitable set of initial conditions is sought that have the effect of cancelling out the transient component. Consider the system of equations that characterize a system in the time domain,

$$\dot{\mathbf{x}} = \mathbf{f}(\mathbf{x}, \mathbf{t}). \tag{24}$$

Let $\mathbf{x}(\mathbf{t})$ be a solution with initial state $\mathbf{x}(\mathbf{t_0}) = \mathbf{x_0}$ such that

$$\mathbf{x}(\mathbf{t}) = \int_0^t \mathbf{f}(\mathbf{x}(\mathbf{t}), \mathbf{t})\mathbf{dt} + \mathbf{x_0} \triangleq \mathbf{x}(\mathbf{t}, \mathbf{x_0}). \tag{25}$$

Here $\mathbf{x}(\mathbf{t}, \mathbf{x_0})$ signifies that the solution $\mathbf{x}(\mathbf{t})$ at any time t depends on the initial conditions $\mathbf{x_0}$. For periodic inputs, the periodic solution is such that $\mathbf{f}(\mathbf{x}, \mathbf{t}) = \mathbf{f}(\mathbf{x}, \mathbf{t} + \mathbf{T})$. We need to find $\mathbf{x_0}$, the initial conditions that exist when the transient component dies out and the steady-state periodic response begins. Thus $\mathbf{x_0}$ represents the state at the beginning of the first steady-state period such that

$$\mathbf{x}(\mathbf{T}, \mathbf{x_0}) = \mathbf{x_0}. \tag{26}$$

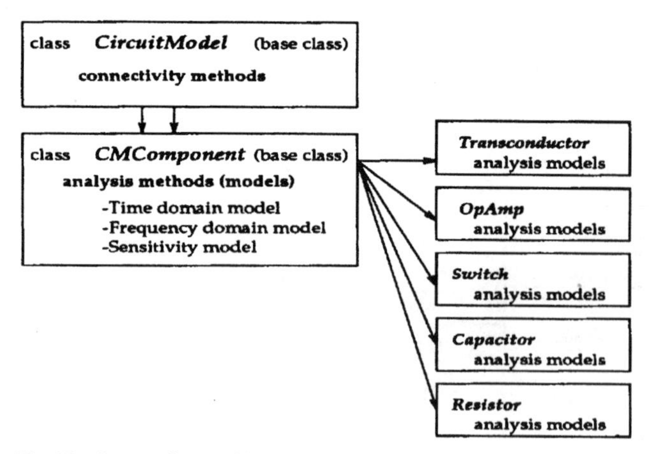

Fig. 10. Intermediate architecture representation objects.

Define a function, $\mathbf{F}(\mathbf{x_0})$ by fixing the time $t = T$, such that $\mathbf{F}(\mathbf{x_0})$ is a function of only $\mathbf{x_0}$

$$\mathbf{F}(\mathbf{x_0}) \triangleq \mathbf{x}(\mathbf{T},\mathbf{x_0}) = \int_0^{\mathbf{T}} \mathbf{f}(\mathbf{x}(\mathbf{t}),\mathbf{t})\mathbf{dt} + \mathbf{x_0}. \tag{27}$$

Then the initial condition that represents the state with no transient component must satisfy the equation

$$\mathbf{x_0} = \mathbf{F}(\mathbf{x_0}). \tag{28}$$

The solution to the above equation provides the desired steady-state initial conditions. This equation is solved by applying an iterative solution method, specifically the Newton–Raphson method, for which we define the equation

$$\hat{\mathbf{F}}(\mathbf{x_0}) = \mathbf{x_0} - \mathbf{F}(\mathbf{x_0}). \tag{29}$$

At each Newton–Raphson iteration, we have

$$\mathbf{x_0^{(j+1)}} = \mathbf{x_0^{(j)}} - \left[1 - \mathbf{F}'(\mathbf{x_0^{(j)}})\right]^{-1}\left[\mathbf{x_0^{(j)}} - \mathbf{F}(\mathbf{x_0^{(j)}})\right] \tag{30}$$

$$\mathbf{F}'(\mathbf{x_0^{(j)}}) = \left.\frac{\partial \mathbf{F}(\mathbf{x_0})}{\partial \mathbf{x_0}}\right|_{x_0 = x_0^{(j)}}. \tag{31}$$

$\mathbf{F}'(\mathbf{x_0})$ is the Jacobian matrix with respect to the initial conditions of each branch or component in the network, defined as

$$\mathbf{F}'(\mathbf{x_0}) = \begin{bmatrix} \frac{\partial x_1(T,\mathbf{x_0})}{\partial \mathbf{x_{0_1}}} & \frac{\partial x_1(T,\mathbf{x_0})}{\partial \mathbf{x_{0_2}}} & \cdots & \frac{\partial x_1(T,\mathbf{x_0})}{\partial \mathbf{x_{0_n}}} \\ \frac{\partial x_2(T,\mathbf{x_0})}{\partial \mathbf{x_{0_1}}} & \frac{\partial x_2(T,\mathbf{x_0})}{\partial \mathbf{x_{0_2}}} & \cdots & \frac{\partial x_2(T,\mathbf{x_0})}{\partial \mathbf{x_{0_n}}} \\ \vdots & \vdots & & \vdots \\ \frac{\partial x_n(T,\mathbf{x_0})}{\partial \mathbf{x_{0_1}}} & \frac{\partial x_n(T,\mathbf{x_0})}{\partial \mathbf{x_{0_2}}} & \cdots & \frac{\partial x_n(T,\mathbf{x_0})}{\partial \mathbf{x_{0_n}}} \end{bmatrix}. \tag{32}$$

The Jacobian matrix is computed via a transient sensitivity analysis over the time interval $t = 0$ to $t = T$. The direct sensitivity method, described in the previous section, is utilized with the sensitivity networks constructed by direct differentiation of the BCEs with respect to the initial conditions. Once the initial conditions are obtained, the transient simulation method described earlier is applied with this set of initial conditions to obtain the responses over the desired simulation time. The pseudocode for the algorithm is shown in Fig. 7.

```
#include <archModel.h>
class LowPassFilter4 :public ArchitectureModel {
  public:
    Node N1, N2, N3, N4, N5, N6;
    Adder add1;
    Integrator integ1, integ2, integ3, integ4;
    ExternInput in1;
    ExternOutput out1;
    LowPassFilter4()
      : ArchitectureModel( "LowPassFilter4" )
      , N1( "1" ), N2( "2" ), N3( "3" ), N4( "4" ), N5( "5" )
      , N6( "6" )
      , in1( "in1", N1)
      , out1( "out1", N6)
      , add1( "add1", 5, N1, N3, N4, N5, N6, N2,
                       1, -2.61313, -3.41421, -2.61313, -1)
      , integ1( "integ1", N2, N3, 3.97887e-05)
      , integ2( "integ2", N3, N4, 3.97887e-05)
      , integ3( "integ3", N4, N5, 3.97887e-05)
      , integ4( "integ4", N5, N6, 3.97887e-05)
    { compile(); }
};
```

Fig. 11. Architecture Specification Language (ASL) description of a 4thorder low pass filter behavioral architecture.

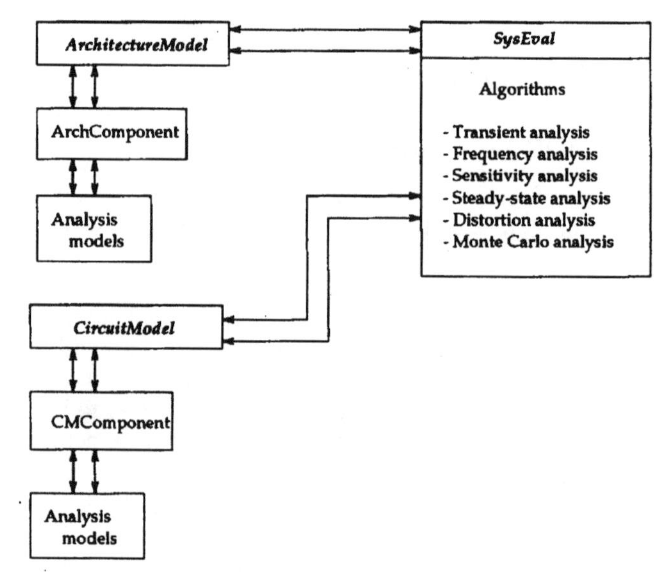

Fig. 12. Architecture of ARCHSIM.

2) Frequency Domain Steady-State Analysis: Frequency domain steady-state analysis in required for generating the frequency response of nonlinear systems, and for noise and distortion analysis of systems in the presence of higher order nonlinearities. In the frequency domain, steady-state analysis is formulated as follows: A periodic function $v(t)$ with fundamental frequency ω and period T can be expressed as a Fourier series by the expression truncated to K harmonics. At each fundamental frequency we need $(2K + 1)$ time domain samples that span a complete period. If $t_0, t_1, \ldots t_n$ are the $n = (2K + 1)$ time steps at the fundamental frequency with time steps $h = \frac{T_0}{n+1}$, then

$$v(t_l) = V_0 + \sum_{k=1}^{K} \left(V_k^C \cos k\omega t_l + V_k^S \sin k\omega t_l\right). \tag{33}$$

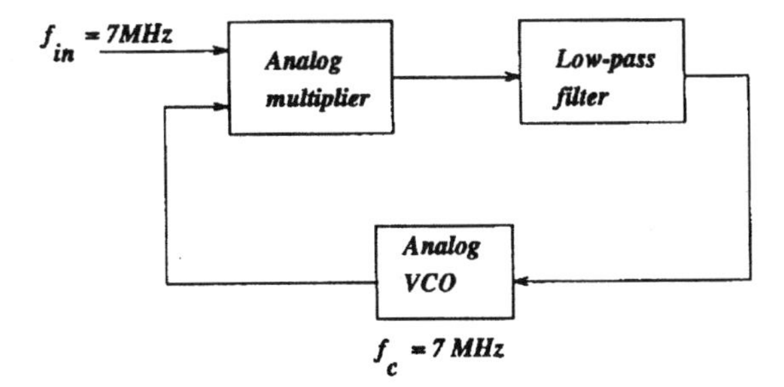

Fig. 13. Analog phase-locked loop.

```
class pllarch :public ArchitectureModel {
  public:
      Node N1, N2, N3, N4;
      AVco vco1;
      Multiplier mult;
      Lpf lpf;
      ExternInput in1;
      ExternOutput out1;
      pllarch()
        : ArchitectureModel( "pllarch" )
          , N1( "1" ), N2( "2" ), N3( "3" )
          , N4( "4" )
          , in1( "in1", N1)
          , out1( "out1", N3)
          , vco1( "vco1", N4, N2, 30000, 7e6)
          , mult( "mult1", N1, N2, N3, 3.72)
          , lpf( "lpf1", N3, N4, 7e5)
      { compile(); }
};
```

Fig. 14. ASL input description of PLL.

We now solve the $(2K + 1)$ equations at each period to compute the Fourier series coefficients V_k^C and V_k^S. The system of equations to be solved are formulated as

$$\Gamma^{-1} \begin{bmatrix} V_0 \\ V_1^C \\ V_1^S \\ \vdots \\ V_K^C \\ V_K^S \end{bmatrix} = \begin{bmatrix} v(t_0) \\ v(t_1) \\ \vdots \\ v(t_n) \end{bmatrix} \tag{34}$$

Γ^{-1} is given by

$$\begin{bmatrix} 1 & \cos\omega t_0 & \sin\omega t_0 & \cdots & \cos K\omega t_0 & \sin K\omega t_0 \\ 1 & \cos\omega t_1 & \sin\omega t_1 & \cdots & \cos K\omega t_1 & \sin K\omega t_1 \\ 1 & \cos\omega t_2 & \sin\omega t_2 & \cdots & \cos K\omega t_2 & \sin K\omega t_2 \\ \vdots & \vdots & \vdots & \cdots & \vdots & \vdots \\ 1 & \cos\omega t_n & \sin\omega t_n & \cdots & \cos K\omega t_n & \sin K\omega t_n \end{bmatrix}. \tag{35}$$

$V(k\omega) = \frac{1}{2}(V_k^C - jV_k^S)$ is the complex frequency domain component of $v(t)$ at the kth harmonic frequency. In algorithmic form, frequency-domain steady-state analysis or nonlinear frequency domain analysis is implemented as shown by the pseudocode in Fig. 8.

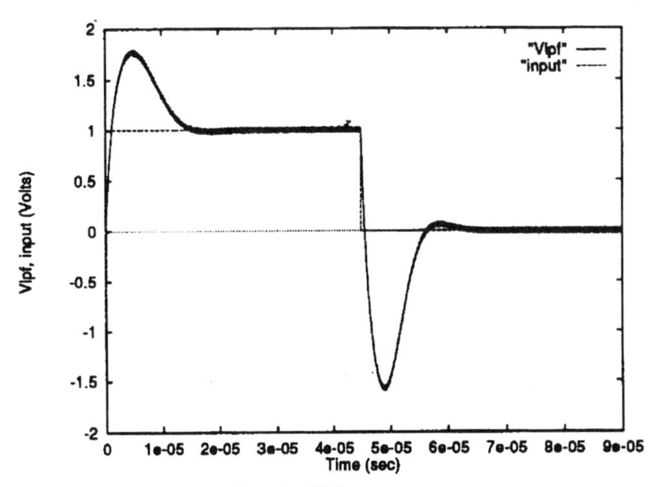

Fig. 15. Step response of analog PLL.

V. Simulator Architecture

The simulator, **ARCHSIM** is implemented in the C++ programming language, using a compiled approach. External system (nonsynthesized) descriptions are specified in a customized C++ description format which are directly compiled and linked to the simulator code. Taking advantage of inheritance and polymorphism provided by the C++ object-oriented environment [27], a common set of objects that represent the various functional blocks required for synthesis and verification are defined. The MNA stencils are directly included in the object definitions. This feature facilitates the decoupling of the analysis algorithms from the models.

A. System Representation

An architecture at the behavioral level is represented by the base class ***ArchitectureModel*** composed of functional blocks. Each functional block is implemented by a generic base class ***ArchComponent***. Specific functional blocks such as an *Adder* or an ***Integrator*** are derived from this base class. Fig. 9 shows the behavioral architecture representation objects. The base class ***CircuitModel*** represents a circuit-implementation style specific intermediate architecture. The base class ***CMComponent*** defines a generic component in the intermediate architecture, various intermediate architecture components such as OTA's, OPAMP's etc., are represented as derived classes of this base class. Fig. 9 shows the data objects for representing the intermediate architecture. The various analysis models are encapsulated within each component object definition.

B. Architecture Representation

Hardware description languages using programming language constructs have long been in use in the digital domain. Programming languages have been directly applicable to the task of behavioral description of digital systems as the behavior of digital systems is naturally expressed in an algorithmic form. We extend the object-oriented capabilities of the C++ programming language for analog hardware description. An architecture is described using a C++ class description that

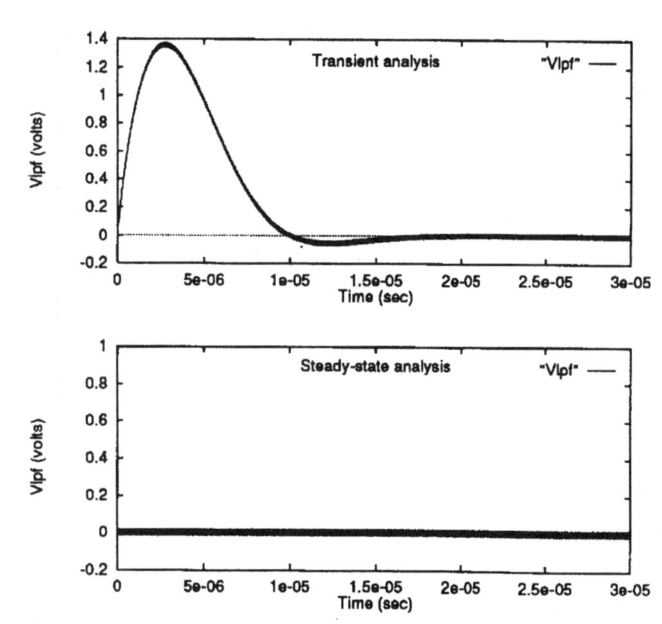

Fig. 16. Steady state simulation of analog phase-locked loop.

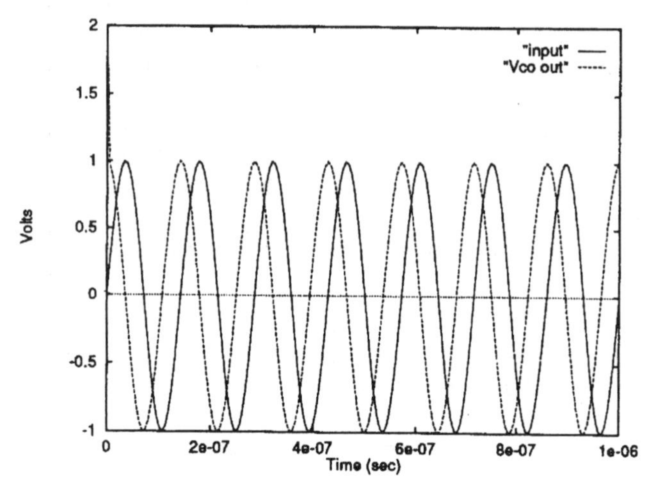

Fig. 17. Steady-state locked signals of analog PLL.

specifies the connectivity of the various components and their parameters. The format of the description is shown in Fig. 11, which is the hardware description of a behavioral architecture of a fourth order continuous time filter. The architecture description is in the Architecture Specification Language (ASL)[5] which follows the conventions in defining a C++ class [27]. This representation was chosen, since the overall framework is implemented in C++, and hence no additional input parsers need to be defined. By utilizing a C++ based description, a compiled approach is used that relies on the programming language compiler's syntax checking abilities.

In the ASL description format, first the nodes and the component instances to be used are declared as variables. The connectivity and the instance specific parameters such as the adder coefficients are defined in the *constructor*. A *constructor*

[5] We refer to the C++ based hardware description as an ASL description for notational convenience.

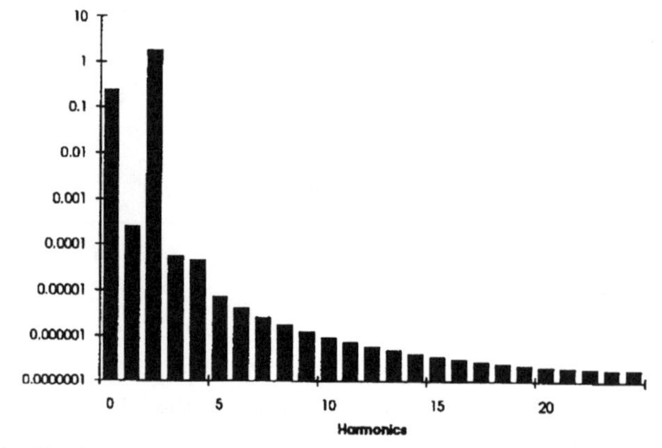

Fig. 18. Frequency spectrum of analog multiplier output of analog PLL.

is a C++ method invoked each time a new instance of a class is created. A common node between two components signifies a connection between the corresponding component terminals. Thus each time a new instance of an architecture is created, the constructor is executed instantiating the various components. A call to the method *compile*() is necessary within the body of the constructor to assemble the various component instances into the target architecture.

Besides the direct link to the synthesis module ARCHGEN, the simulator ARCHSIM can also simulate external designs in a stand-alone mode of operation. An input architecture description is compiled and linked with the ARCHGEN or ARCHSIM object code, which is then executed directly. An intermediate architecture description in the format shown in Fig. 11 in a specific technology such $g_m - C$ is generated as an outcome of a synthesis run, the input is a s-domain transfer function. The intermediate architecture description is annotated to indicate the corresponding higher-level functional blocks realized by circuit-level components.

C. Simulator Core

The core of each analysis algorithm involves the solution of a set of linearized equations. These linearized system of equations are usually very sparse requiring the use of sparse matrix methods for greater efficiency. The **Sparse** linear equation solver [33] is integrated within the *ARCHSIM* software framework, and the MNA matrices are constructed using the sparse matrix data structures. Fig. 12 shows the functional organization or the architecture of *ARCHSIM*. The various algorithms, and the associated data structures for MNA matrix definitions, etc. are implemented as an encapsulated object class ***SysEval***, which in essence can be viewed as the simulation engine. The behavioral and intermediate architectures are represented by the base classes ***ArchitectureModel*** and *CircuitModel*, in which the interface to the simulation algorithms is defined.

VI. Examples and Experimental Results

Phase-locked loops (PLL's) are a widely used class of circuits in many applications, both in the design of digital, analog and mixed analog-digital systems. PLL's are also a

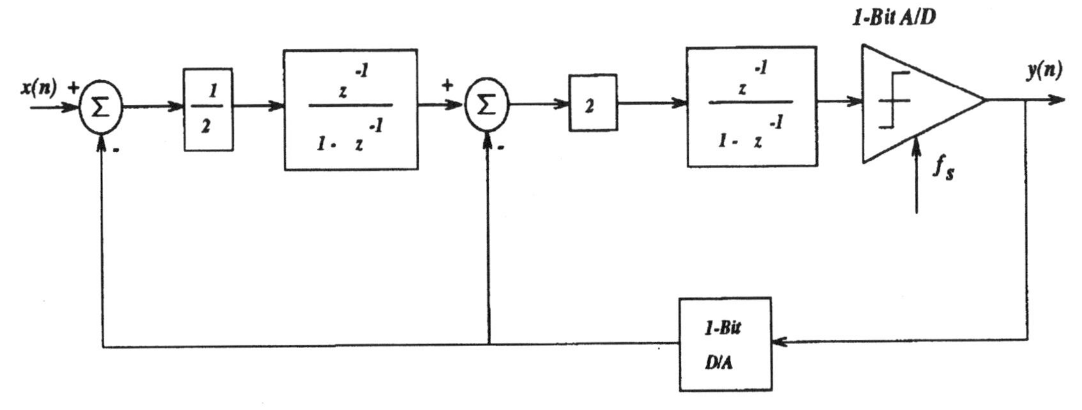

Fig. 19. A single stage second-order ΣΔ modulator.

difficult class of systems to simulate [14]. Fig. 13 shows an analog PLL configuration. The models for the multiplier and analog VCO were derived earlier in this paper, on similar lines, a direct model for the low-pass filter is derived, or a complete filter architecture can be used if a higher order filter is desired. This PLL design has the following component parameters:

Analog multiplier gain $K_m = 3.72$

Low-pass filter pole at $\omega_o = 7 \times 10^5$ rad/s.

VCO center frequency $f_c = 7$ MHz.

VCO gain $K_o = 30$ kHz/V = 188.5 rad/s/V.

Fig. 14 shows the ASL input description of this PLL configuration.

During transient analysis, the various models that compose the PLL are evaluated, resulting in the corresponding model MNA stencils being stamped in the global system matrix. Each model instance is provided with a set of pointers to its contributing entries in the global system matrix. These entries are updated locally by each model at each simulation iteration. As a result of the initial processing, the system of equations, in symbolic form are shown in (36).

$$\begin{bmatrix} 1 & 0 & 0 & 0 \\ 0 & 1 & 0 & -hK_oV_m\cos\left(x_{n+1}^{(j)}\right) \\ -K_mV_{2(n+1)}^{(j)} & -K_mV_{1(n+1)}^{(j)} & 1 & 0 \\ 0 & 0 & -1 & \left(1+\frac{1}{\omega_o h}\right) \end{bmatrix} \cdot \times \begin{bmatrix} V_{1(n+1)}^{(j+1)} \\ V_{2(n+1)}^{(j+1)} \\ V_{3(n+1)}^{(j+1)} \\ V_{4(n+1)}^{(j+1)} \end{bmatrix} = \begin{bmatrix} U_{n+1} \\ V_{2(n+1)}^{(j)} - hK_oV_m\cos\left(x_{n+1}^{(j)}\right)V_{4(n+1)}^{(j)} \\ V_{3(n+1)}^{(j)} - 2K_mV_{1(n+1)}^{(j)}V_{2(n+1)}^{(j)} \\ \frac{1}{\omega_o h}V_{3(n+1)}^{(j)} \end{bmatrix}. \tag{36}$$

Fig. 15 shows the transient step response of this PLL, where $Vlpf$ is the input to the VCO. A complete time domain simulation of a PLL is necessary to estimate the lock-in time. The lock-in time can run from a few microseconds to seconds [14]. Once the lock-in characteristics have been established, simulation of the PLL within a larger system can be carried out using time domain steady-state analysis. By arriving at the steady state directly other characteristics such as stability to frequency variations, jitter and noise effects can be measured effectively. Fig. 16 shows the transient and steady-state simulation of the analog PLL example that was simulated earlier using transient analysis. Fig. 17 shows the locked signals of the PLL in steady-state.

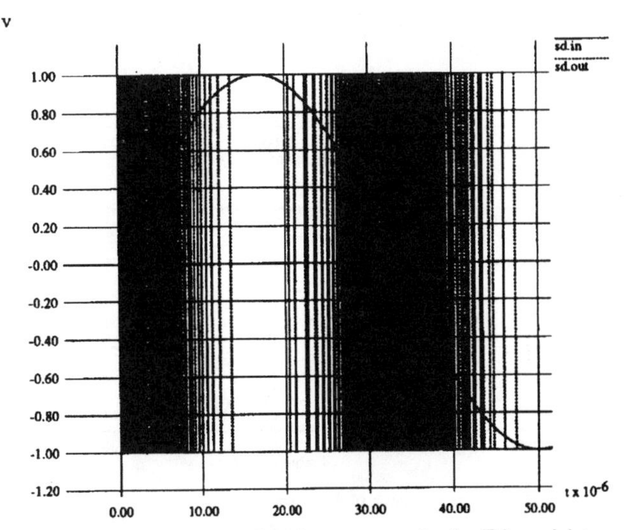

Fig. 20. Simulation response of single stage second order ΣΔ modulator.

Fig. 18 shows the frequency spectrum of the output of the multiplier obtained through a nonlinear frequency domain analysis of this PLL example. The frequency spectrum is measured with the reference signal to the PLL being at the VCO center frequency, and the simulation was carried out using the nonlinear frequency analysis algorithm.

Another widely used class of data converters in mixed-signal designs are ΣΔ modulators, whose architectures consist of a single quantizer and a noise shaping filter. Most modulators, single-stage or cascade use a small set of functional blocks, discrete integrators, single-bit or multi-bit quantizers, simple D/A converter, and adders. ΣΔ modulators can be simulated in the ARCHSIM using discrete models. Fig. 19 shows a second-order ΣΔ modulator architecture used in the design published in [34]. This configuration has a sampling

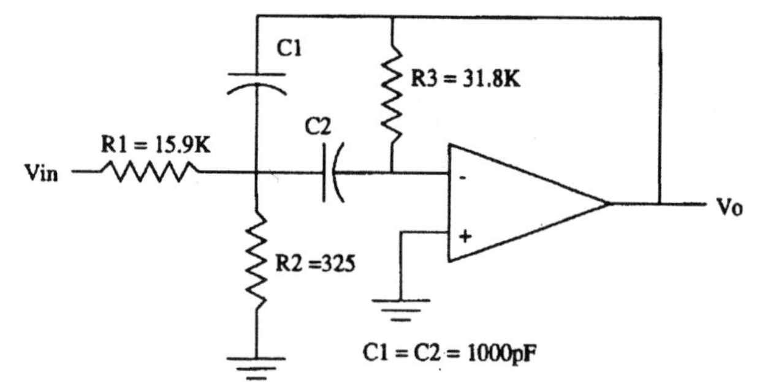

Fig. 21. Bandpass filter, modeled at the intermediate architecture level.

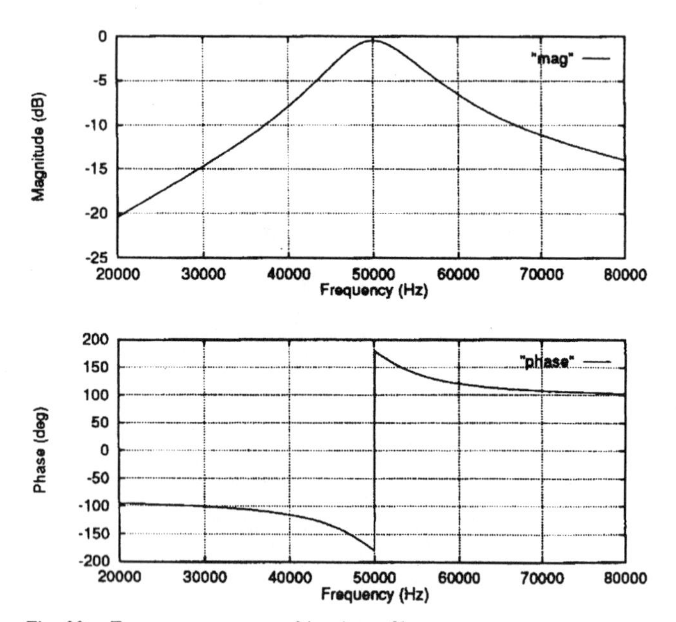

Fig. 22. Frequency response of bandpass filter.

rate of 10.24 MHz, and was simulated using an input signal of 15 kHz. The output of the modulator is ± 1 V. Fig. 20 shows the simulation results of the modulator.

Fig. 21 shows a bandpass filter, modeled using components (which include the frequency domain nonideal opamp model) at the intermediate architecture stage. Fig. 22 shows the frequency response of the bandpass filter.

One of the primary applications of sensitivity analysis is to measure the sensitivity or tolerance of a synthesized design to component parameter variations. Four architectures synthesized by ARCHGEN were simulated in the frequency domain and the cumulative sensitivity measured. Fig. 23 shows the cumulative or multiparameter sensitivity of the four architectures synthesized from the following s-domain transfer function (behavioral) specification of a fourth order Butterworth low pass filter. In this example sensitivity analysis is carried out at the behavioral architecture level (see (37) at the bottom of this page).

Fig. 24 shows the multiparameter time domain sensitivity of the output of the controllable architecture synthesized from the above filter specifications.

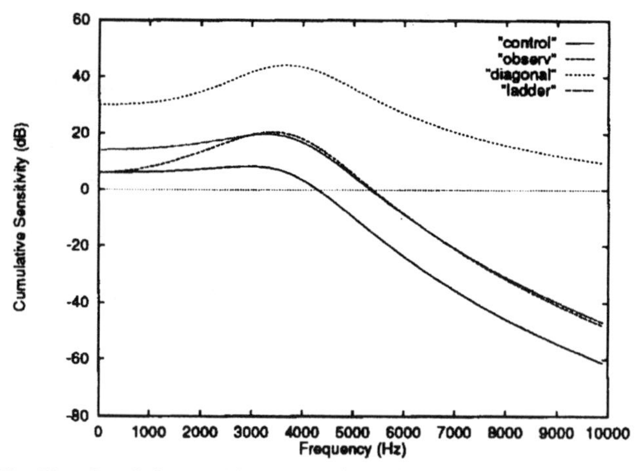

Fig. 23. Cumulative sensitivity comparison of low-pass filter architectures.

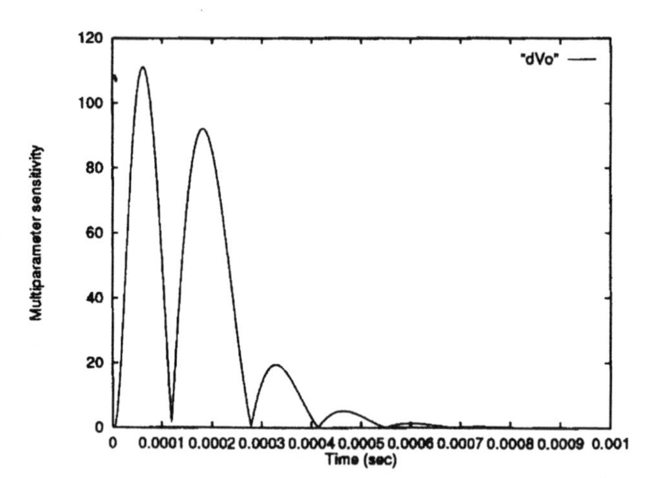

Fig. 24. Multiparameter time domain sensitivity of low-pass filter architecture.

Fig. 25 shows the Monte Carlo simulation results of the frequency response of the low pass filter, the probability distributions assigned to each component parameter at the behavioral level span a 10% tolerance range.

VII. Summary and Conclusion

Analog systems-level design is slowly evolving to fill in the need for higher-level analog design solutions to aid in the design process of mixed-signal ASICS. In this paper we have presented a methodology for analysis and verification of analog systems. The verification provides for the two critical design tasks, verification of design specifications and design space exploration. A behavioral simulation paradigm is developed that integrates with the synthesis process for quick verification of the synthesized designs, thus allowing rapid redesign and optimization. The simulator is general in nature and is not tied to a specific type of a circuit or system architecture. The suite of analysis regimes, formulated and

$$H(s) = \frac{3.989876e17}{s^4 + 6.567502e4s^3 + 2.156604e9s^2 + 4.148393e13s + 3.989876e17}. \tag{37}$$

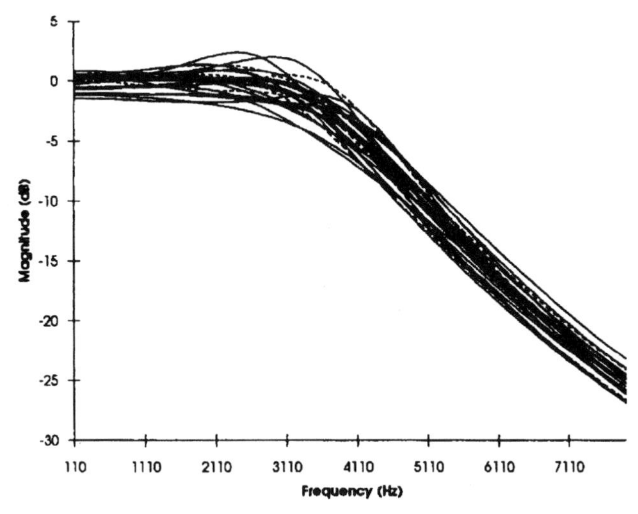

Fig. 25. Monte Carlo analysis of low pass filter.

implemented at the behavioral level, provide a comprehensive analysis of analog systems. Integration of the analysis models with synthesis opens a new avenue for design space exploration and optimization at the systems level. In addition a customized system description language in the C++ language format makes it possible to simulate systems designed outside the synthesis path as well. The implementation is open-ended, thus providing a framework that can be further extended to include, for example, higher-order nonlinearities and models. Another extension to this research is to include specialized modes for stability analysis of higher-order $\Sigma\Delta$ systems.

ACKNOWLEDGMENT

The authors would like to thank P. Feldmann, F. El-Turky, B. Leonowich, and R. Saleh for discussions and suggestions during the course of this research, and the referees whose comments have helped in making a better presentation of this effort.

REFERENCES

[1] L. W. Nagel, "SPICE2: A computer program to simulate semiconductor circuits," Univ. Calif., Berkeley, Electron. Res. Lab., Memo UCB/ERL M520, May 1975.

[2] R. A. Rohrer, "Circuit simulation—The early years," *IEEE Circ., Devices Mag.*, pp. 32–37, May 1992.

[3] B. R. Chawla, H. K. Gummel, and P. Kozak, "MOTIS—A MOS timing simulator," *IEEE Trans. Circuit Syst.*, vol. CAS-22, pp. 901–910, Dec. 1975.

[4] L. W. Pillage and R. A. Rohrer, "Asymptotic waveform evaluation for timing analysis," *IEEE Trans. Computer-Aided Design*, vol. 9, pp. 352–366, Apr. 1990.

[5] C. Vishweswariah and R. A. Rohrer, "Piecewise approximate circuit simulation," *IEEE Trans. Computer-Aided Design*, vol. 10, pp. 861–870, July 1991.

[6] R. E. Bryant, "Algorithmic aspects of symbolic switch network analysis," *IEEE Trans. Computer-Aided Design*, pp. 618–633, July 1987.

[7] R. A. Saleh and A. R. Newton, *Mixed-Mode Simulation*. Norwell, MA: Kluwer, 1990.

[8] H. J. De Man *et al.*, "Practical implementation of a general computer-aided design technique for switched capacitor circuits," *IEEE J. Solid-State Circuits*, vol. SC-15, pp. 190–200, Apr. 1980.

[9] R. A. Saleh, J. E. Kleckner, and A. Richard Newton, "Iterated timing analysis in SPLICE1," Dig. Tech. Papers, in *IEEE ICCAD*, 1983, pp. 139–140.

[10] M. R. Barbacci, "Instruction set processor specifications (ISPS): The notation and its applications," *IEEE Trans. Comput.*, vol. C-30, pp. 24–40.

[11] R. Lipsett, E. Marschner, and M. Shahdad, "VHDL—The Language," *IEEE Design Test Comput.*, pp. 28–41, Apr. 1986.

[12] I. E. Getreu, "Behavioral modeling of analog blocks using the SABER simulator," in *Proc. Midwest Symp. Circ., Syst.*, Aug. 1989, pp. 977–980.

[13] ADAMS Team, "The ABCDL: A robust environment for analog circuit behavioral modeling," AT&T Bell Labs. Internal Tech. Memo., Mar. 1991.

[14] B. A. A. Antao, F. M. El-Turky, and R. H. Leonowich, "Behavioral modeling phase-locked loops for mixed-mode simulation," in *Analog Integrated Circuits and Signal Processing.*, to

[15] R. A. Saleh, D. L. Rhodes, E. Christen, and B. A. A. Antao, "Analog harware description languages," *IEEE Custom Integr. Circ. Conf. (CICC)*, May 1994.

[16] R. Chadha, C. Visweswariah, and C. Chen, "M^3—A multilevel mixed-mode mixed A/D simulator," *IEEE Trans. Computer-Aided Design*, vol. 11, pp. 575–584, May 1992.

[17] J. Singh and R. Saleh, "iMACSIM: A program for multi-level analog circuit simulation," *IEEE ICCAD*, Nov. 1991, pp. 16–19.

[18] R. A. Saleh, B. A. A. Antao, and J. Singh, "Multi-level and mixed-domain simulation of analog circuits and systems," *IEEE Trans. Computer-Aided Design*, to be published.

[19] K. A. Sakallah and S. W. Director, "SAMSON2: An event driven VLSI circuit simulator," *IEEE Trans. Computer-Aided Design*, vol. CAD-4, pp. 668–684, Oct. 1985.

[20] A. R. Newton and A. L. Sangiovanni-Vincentelli, "Relaxation-Based electrical simulation," *IEEE Trans. Computer-Aided Design*, vol. CAD-3, pp. 308–331, Oct. 1984.

[21] B. A. A. Antao and A. J. Brodersen, "Techniques for synthesis of analog integrated circuits," *IEEE Design Test Comput.*, pp. 8–18, Mar. 1992.

[22] R. A. Rutenbar, "Analog design automation: Where are we? where are we going?," *IEEE Custom Integr. Circuits Conf. (CICC)*, May 1993, pp. 13.1.1–13.1.8.

[23] G. Jusuf, P. R. Gray, and A. L. Sangiovanni-Vincentelli, "CADICS—Cyclic analog-to-digital converter synthesis," *IEEE ICCAD*, 1990, pp. 286–289.

[24] B. A. A. Antao and A. J. Brodersen, "ARCHGEN: Automated synthesis of analog systems," *IEEE Trans. VLSI Syst.*, vol. 3, pp. 231–244, June 1995.

[25] ——, "A framework for synthesis and verification of analog systems," to appear in *Analog Integ. Circuits, Signal Processing.*

[26] C. W. Ho *et al.*, "The modified nodal approach to network analysis," *IEEE Trans. Circuits Syst.*, vol. CAS-22, pp. 504–508, June 1973.

[27] M. A. Ellis and B. Stroustrup, *The Annotated C++ Reference Manual*. Reading, MA: Addison-Wesley, 1990.

[28] D. E. Hocevar, P. Yang, T. N. Trick, and B. D. Epler, "Transient sensitivity computation for MOSFET circuits," *IEEE Trans. Computer-Aided Design*, vol. CAD-4, pp. 609–620, Oct. 1985

[29] S. W. Director and R. A. Rohrer, "The generalized adjoint network and network sensitivities," *IEEE Trans. Circ. Theory*, vol. CT-16, pp. 318–323, Aug. 1969.

[30] B. A. A. Antao and A. J. Brodersen, "A formulation for analog behavioral-level sensitivity and steady-state analysis," to be published.

[31] T. J. Aprille and T. N. Trick, "Steady-state analysis of nonlinear circuits with periodic inputs," *Proc. IEEE*, vol. PROC-60, pp. 108–114, Jan. 1972.

[32] *MHDL Language Reference Manual.* Intermetrics Inc., and Army Res. Lab., Mar. 1994.

[33] K. S. Kundert and A. L. Sangiovanni-Vincentelli, "Sparse user's guide: A sparse linear equation solver," Dep. of EECS, Univ. Calif., Berkeley, Apr. 1988.

[34] S. R. Norsworthy, I. G. Post, and H. S. Fetterman, "14-bit sigma-delta A/D converter: Modeling, design and performance evaluation," *IEEE J. Solid-State Circuits*, vol. 24, pp. 256–266, Apr. 1989.

Brian A. A. Antao (S'85–M'94), for a photograph and biography, see the June 1995 issue of this TRANSACTIONS, p. 231.

Arthur J. Brodersen (SM'76), for a photograph and biography, see the June 1995 issue of this TRANSACTIONS, p. 231.

Multilevel and Mixed-Domain Simulation of Analog Circuits and Systems

Resve A. Saleh, *Senior Member, IEEE,* Brian A. A. Antao, *Member IEEE,* and Jaidip Singh

Abstract— **Integrated circuit design has evolved to the stage where large, complex, analog and digital functionalities are implemented on a single chip or as an integrated chip set. Besides the mixed signal nature of the designs, the analog sections also include continuous-time and discrete-time components. Thus, for analysis of these integrated modules, all encompassing simulation capabilities are required that address, not only transient analysis of mixed analog–digital circuits, but frequency domain analysis as well. In this paper, we present mechanisms for multilevel and mixed-domain simulation of analog systems tied in with mixed analog–digital simulation. We then describe the implementation in the form of an open-ended and expandable simulation framework, iMACSIM. A simulation backplane is used to provide a general event-processing and scheduling framework that ties together the various algorithms necessary for simulation of various classes of circuits in different analysis regimes.**

I. Introduction and Background

RECENT generations of integrated circuits and application specific integrated circuits (ASIC's) include both analog and digital sections with the proportion of the analog content growing significantly as designers strive to achieve higher levels of integration to realize complete systems on a single chip. Furthermore, complete analog systems are being designed using a combination of discrete-time components and continuous-time components. A complete system includes digital circuitry as well as some combination of continuous-time and discrete-time analog sections. Verification and analysis of these designs are crucial to ensure functionality and manufacturability. With the advent of newer integrated modules, the need arises for simulation and analysis tools that allow mixed analog–digital simulation as well as multidomain analog simulation. That is, the analysis of complex analog–digital systems requires that the simulation platform provide an all encompassing integrated analysis environment.[1] A simulation environment that allows these newer generation of systems composed of digital and analog sections, both continuous and discrete to be evaluated, is implemented as the tool iMACSIM and described in this paper.

In digital design, the analysis of large circuits has been made tractable through the use of higher-level models. For example, switch-level models can be used for transistors and simulated in switch-level simulators [1]. Higher level logic models can be used to directly represent the functionality of the various logic components instead of a transistor-level description. The system would then be simulated using logic simulators [2]. Entire digital systems are also simulated at the register-transfer level and behavioral simulation [3] with the use of hardware description languages such as VHDL and Verilog [4]. Thus, a multilevel simulation paradigm has evolved from the digital environment that allows a digital system to be modeled and simulated across different levels, trading off speed versus accuracy. Typically, noncritical sections of the design are modeled at a higher level, whereas detailed transistor-level representations are used for the critical sections.

Similarly, a multilevel paradigm is emerging for analog circuits and systems, and one of the goals of the research described in this paper is to further develop this aspect of analog simulation. To support the analysis of complex analog–digital systems requires analysis tools that go beyond the capabilities of SPICE [5], [6] to support multilevel and mixed-domain simulation. Multilevel models should be supported across the analog–digital interface and mixed-domain representation across the continuous, discrete, and frequency domains. In this paper, we describe multilevel and mixed-domain representations of analog circuits and the simulation algorithms needed for analog–digital systems as implemented in the simulation program *iMACSIM*. We discuss various issues that need to be addressed when simulating mixed-signal circuits represented across multilevels and in different domains. The iMACSIM framework includes the well-established digital multilevel simulation algorithms developed in the iSPLICE program [7], [8] along with the integration of the various analog time and frequency analysis algorithms for various types of analog circuits, thus providing a truly integrated, multilevel and mixed-domain simulation capability.

While newer generations of analog behavioral simulators are coming into prominence, they are limited in some respects in their analysis capabilities. True multilevel and mixed-domain simulation with analysis capabilities in both time and frequency domains is still lacking. The developments reported in this paper address multilevel and mixed-domain simulation and bridge the gaps between SPICE and SWITCAP [9] and integrate multilevel digital simulation as well. Many specialized

Manuscript received August 19, 1994; revised October 27, 1995. This work was supported in part by Semiconductor Research Corporation under Grant 93-DP-109 and Grant 94-DP-109, Analog Devices Inc., and by the National Science Foundation under Grant MIP-0957887 (PYI). This paper was recommended by Associate Editor J. White.

R. A. Saleh and B. A. A. Antao are with the Coordinated Science Laboratory, University of Illinois at Urbana-Champaign, Urbana, IL 61801 USA.

J. Singh was with the Coordinated Science Laboratory, University of Illinois at Urbana-Champaign, Urbana, IL 61801 USA. He is now with the Wharton School, University of Pennsylvania, Philadelphia, PA 19103 USA.

Publisher Item Identifier S 0278-0070(96)01346-2.

[1] An integrated approach ties simulation algorithms across the analog–digital boundaries as opposed to the glued approach where two distinct simulators are used for the analog and digital sections.

techniques and tools have been developed to meet the needs specific to discrete-time analog systems such as SWITCAP2 [9] and AWEswit [10] for linearized analysis of switched capacitor circuits and MIDAS [11] for $\Sigma\Delta$ modulator-based systems. These programs resort to macromodeling techniques and charge-conserving algorithms to simulate switched-capacitor circuits and $\Sigma\Delta$ circuits. However, these programs do not encompass all the necessary levels of continuous-time simulation in a tightly integrated manner. For example, MIDAS was originally targeted towards $\Sigma\Delta$ modulators and handles simulation at the behavioral level. SWITCAP2 uses a fixed discrete-time charge formulation and is specifically targeted at the simulation of switched-capacitor circuits with some logic analysis capability. AWEswit is based on asymptotic waveform evaluation (AWE) [12] and addresses specific limitations of SWITCAP in handling resistive nonlinearities. The efforts in developing the M^3 simulator [13] concentrated on incorporating analog behavioral blocks modeled using Laplace transforms and z-transforms for time domain simulation of analog–digital systems. This effort however did not directly address frequency domain issues for multilevel and mixed-domain descriptions and the ability to simulate discrete-time circuits. A more recent analog behavioral simulator, ARCHSIM [14] provides a full suite of analysis for analog behavioral simulation and is tied to analog system synthesis for integrated analog system synthesis and verification. The GTSIM effort [15] explores a mixed Broyden–Newton algorithm for analog behavioral simulation. Commercial analog behavioral simulators include SABER [16], ATTSIM [17], and ELDO [18]. Again, the above programs do not incorporate the discrete-time techniques of SWITCAP2 in their simulation capabilities. However, most of these simulators do provide a template in a customized modeling language for modeling analog systems at the behavioral level using transfer functions and a general set of equations. Some of the commercial template description languages are MAST[2] in SABER, ABCDL[3] in ATTSIM, and FAS.[4] Additionally, efforts overseen by IEEE standardization committee on analog hardware description languages (AHDL) are underway at developing AHDL standards [19], [20].

We begin this paper by addressing the critical issues in multilevel and mixed-domain simulation. Section II is devoted to a description of the various simulation levels in the analog and digital domain, and in Section III, we describe transformations and interface models necessary for integrated simulation across the levels. In Section IV, we provide details on the transient simulation issues and the event-driven simulation methodology that makes it possible to simulate analog–digital circuits across multilevels and in mixed domains. In Section IV.3, a novel mixed relaxation/MNA formulation for transient analysis across multiple levels and mixed-domains is presented. In Section V, we describe methods to obtain the frequency response analysis of various classes of circuits. In Section VI, we describe the simulator architecture, followed by simulation examples in Section VII, and conclusions in Section VIII.

[2] MAST is a trademark of Analogy Inc.

[3] Analog Behavior Circuit Description Language, ABCDL is a trademark of AT&T Bell Laboratories

[4] FAS is a trademark of ANACAD Inc.

TABLE I
DIGITAL-ANALOG SIMULATION LEVELS

Digital	Analog
Behavioral	Behavioral
RTL/Gate	Ideal functional
Timing	Non-ideal functional
Electrical	Electrical

II. SIMULATION LEVELS

A mixed analog–digital system can be partitioned across the analog and digital boundaries and represented at different levels of hierarchy. The various levels are shown in Table I, with equivalent levels for the analog and digital domains.

Thus, given that a system can be represented across these various levels, **multilevel** simulation refers to the simulation of a system whose parts are represented at different levels. In multilevel simulation, the entire system need not be represented at one consistent level as would be required by a circuit simulator such as SPICE.

A. *Multilevel Digital Simulation*

In iMACSIM, the digital simulation is carried out using algorithms implemented in the iSPLICE3 simulator [7], [8]. iSPLICE3 has three levels of simulation, LOGIC, ELOGIC, and *Electrical* levels. The LOGIC level is for gate-level logic simulation, the ELOGIC level for switch-level timing simulation, and the *Electrical* level for circuit-level simulation. At the LOGIC level, the system is modeled using boolean logic gate models such as NAND's, NOR's, AND's, OR's, XNOR's, XOR's, inverters, and MOS pass transistors. The LOGIC level provides first-order timing simulation. In the simplest case, a two-state logic model is used, and for greater accuracy, a multistate logic model can be used. Accuracy and computational cost of the LOGIC level simulation is controlled by the number of states and the delay models. In converting a block from a transistor-level representation to the discrete LOGIC level, loss of some circuit characteristics may impact the overall simulation results, more so in the case of analog–digital systems. In such situations, switch-level simulation or the ELOGIC level is recommended. The ELOGIC level allows more detailed timing analysis to be obtained using electrical device models in the switch-level analysis. As part of the ELOGIC modeling process, a number of discrete voltage levels are selected. These levels need not be equally spaced, but these levels and their values have an impact on performance and accuracy. The electrical level is the circuit-level simulation using detailed electrical components and models such as MOS transistors, capacitors, resistors, etc. Since digital blocks have loosely coupled feedback paths and high latency, a more efficient form of circuit analysis called *iterated timing analysis* (ITA) [7] is used. The iMACSIM program also includes behavioral level logic elements but does not include RTL simulation.

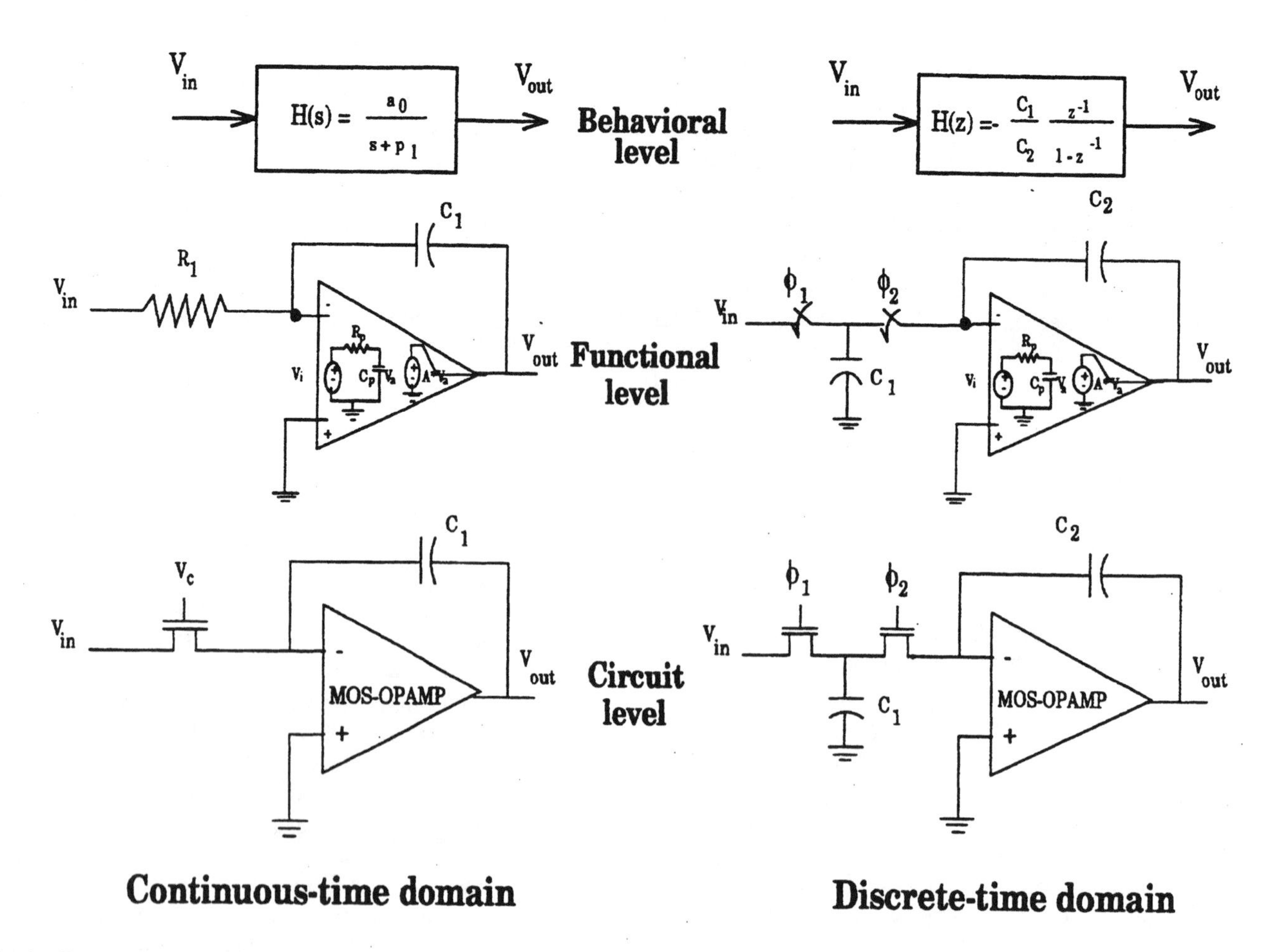

Fig. 1. Equivalent analog representations across the various levels.

B. Multilevel Analog Simulation

Similar to the simulation levels for the digital domain, the various analog simulation levels are shown on the right side of Table I. Here, the analog levels are shown in correspondence to those established in the digital domain so that the relationship between the digital and analog domains is clear. We further develop these levels in greater detail in this paper. The behavioral level can be used where the function of a block is known, but its detailed structure is unavailable or is in the process of being designed. The functional level in the analog domain is composed of macromodels, and depending on the amount of nonidealities included in the models, a first-order ideal functional level or a more detailed or nonideal functional level model may be defined. At the electrical level, identical components are used across both sections. The analog sections may be further characterized by additional modeling constraints [21]. Improvements on electrical level simulation using the relaxation formulation for analog–digital electrical simulation are presented in Section IV.

C. Mixed-Domain Analog Modeling

As analog circuits can contain both continuous-time or discrete-time blocks, a system can have a mixed-domain representation requiring simulation capabilities in both domains. Thus, **mixed-domain** simulation refers to the simulation of an analog system composed of sections defined in the different domains, simulated together. The details of the analog simulation levels, both for continuous-time and discrete-time systems, are shown in Table II. Fig. 1 shows continuous-time and discrete-time representations for integrator functions across the different levels in the hierarchy. At the behavioral level, the representation is in the form of transfer functions. At the functional level, the discrete integrator is simulated using models for switches and a functional macromodel for the operational amplifier; circuit-level simulation involves using a detailed MOS transistor realization of switches and opamps. We now describe these levels in greater detail below.

At the highest level in the simulation hierarchy is behavioral simulation, as described earlier. At this level, the individual blocks are modeled in terms of s-domain transfer functions or differential equations or other high-level functions and their interactions in terms of signal flow diagrams that include summers and multipliers. Hardware description languages can be used to define the behavioral blocks that are parsed and linked to the simulator. Often, analog processors have a continuous time front-end for coarse analog filtering followed by sampling and A/D conversion. The sampling and data conversion stages are discrete time systems, often realized in a switched capacitor technology. These stages can be modeled in the discrete time functional level, z-domain behavioral models, and the electrical level. Thus,

TABLE II
ANALOG SIMULATION LEVELS

	Continuous Time	Discrete Time
Behavioral Level	S-domain functions, Differential equations, Logic gates, A/D, D/A, opamps, comparators	Z-domain functions, Discrete-time equations, Discrete-time filters, Sampled-data blocks
Functional Level	Linear and nonlinear controlled sources, macromodels	Switches, Voltage-controlled Voltage sources
Circuit Level	Transistors, Diodes, Capacitors, Resistors, Inductors	

the simulation of such a system requires the mixed-domain simulation capability.

At the functional level, macromodels are used to represent the circuit blocks. Complex models may be specified at this level, including nonlinear properties, limiting effects, inductive effects, capacitive effects, and input/output loading characteristics. Functional relationships among the components in the macromodel are defined using controlled elements such as voltage-controlled voltage sources (VCVS) and current-controlled voltage sources (CCVS). At the other extreme, the macromodels can be relatively simple descriptions of the intended operation of a block. If the functionality of the entire block can be specified in a single description, it is usually considered as a behavioral block. Typical circuit blocks at this level include opamps, comparators, and voltage-controlled oscillators. In the case of an opamp, a simple model would use a VCVS. A detailed macromodel would include at least finite gain, finite bandwidth, input loading effects, output loading effects, and output limiting behavior. An advanced model would also incorporate input offset current, input offset voltage, nonlinear slewing properties, common-mode rejection ratio (CMRR), power supply rejection ratio (PSRR), and output drive capability. However, there is a distinction between the functional-level macromodels and the Boyle *et al.* [22] type macromodels composed of physical elements such as transistors and diodes and are circuit-level simplifications.

At the circuit level, the simulator has the capability of carrying out detailed electrical simulation. Physical elements such as transistors, capacitors, inductors, resistors, and diodes are used and have a one-to-one correspondence with the actual circuit. To fit the event-driven framework and exploit latency, *relaxation methods* [23], [24] are used. In addition, through the use of partitioning, independent subcircuits can be processed using direct methods.

III. MULTILEVEL AND MIXED-DOMAIN TRANSFORMATIONS

As described in the previous section, analog blocks can be modeled using different representations. Table III shows the various representations along with the terminology to be used in this paper. Essentially, an analog block can have

TABLE III
DEFINITIONS FOR ANALOG BLOCK REPRESENTATION

Domain	View	Representation
discrete	time	difference/algebraic equations.
continuous	time	differential/algebraic equations.
discrete	frequency	z-transform.
continuous	frequency	Laplace transform.
discrete	circuit	structural, continuous time circuit elements.
continuous	circuit	structural, discrete time circuit elements.

continuous-time or discrete-time behavior, which we define as the **domain** of the analog functionality shown in column 1 of Table III. Thus, a system composed of both continuous-time and discrete-time components is a mixed-domain system. Furthermore, from the simulation point-of-view, the block needs to have its behavior defined for transient response or frequency response analysis. Thus, we define a **view** to be the characterization of an analog block for a specific type of analysis. A view can also be interpreted as the domain of analysis. The *time* view is required for transient analysis, *frequency* view for frequency response analysis, and for detailed simulation, a *circuit* view. For each view, the analog block is defined using specific **representations**, such as differential equations, transfer functions, or the circuit-level structural description. Thus, each block can be represented in the continuous or discrete domains with behavioral views that provide necessary information for generating the transient or frequency responses. Additionally, the blocks can have functional-level or circuit-level views in the form of equivalent circuit macromodels or actual circuit realizations.

From this discussion, the discrete integrator of Fig. 1 in the discrete-time domain block can be modeled at the behavioral level by the following set of equations:

1) Discrete-time view

$$V_{out}(t) = V_{out}(t_{n-1}) - \frac{C_1}{C_2} V_{in}(t_{n-1}).$$

2) Discrete-frequency view

$$H(z) = \frac{V_{out}(z)}{V_{in}(z)} = -\frac{C_1}{C_2}\frac{z^{-1}}{1 - z^{-1}}.$$

The discrete circuit view at the functional and circuit levels is shown in Fig. 1. Both of the forms shown above can be generated from the circuit view, as described next.

A. Transformations Across Domains

A problem arises when models that are defined in a number of different domains are connected together. Fig. 2 illustrates an interconnection of circuit elements and models defined across different domains with different views. This is a representation of a worst-case interdomain interconnection, and the ability to process such an interconnection characterizes a mixed-domain simulator. The simulation of such a system in one of the analysis modes, either a transient analysis or a frequency response, requires that a suitable set of transformations be applied such that a consistent representation of the entire

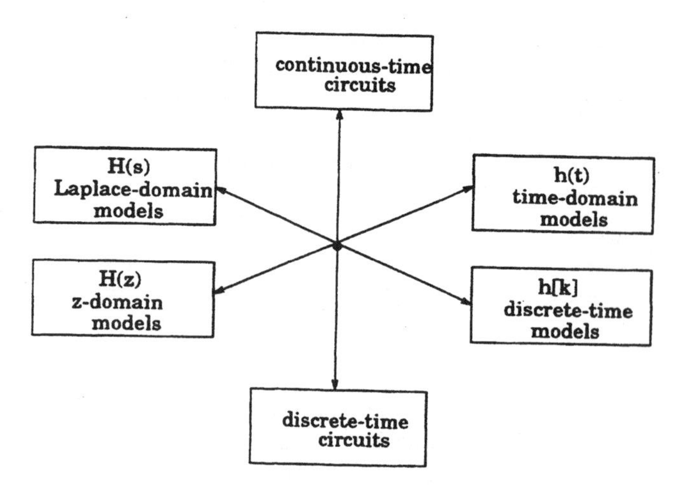

Fig. 2. Worst-case interconnection of mixed-domain analog models.

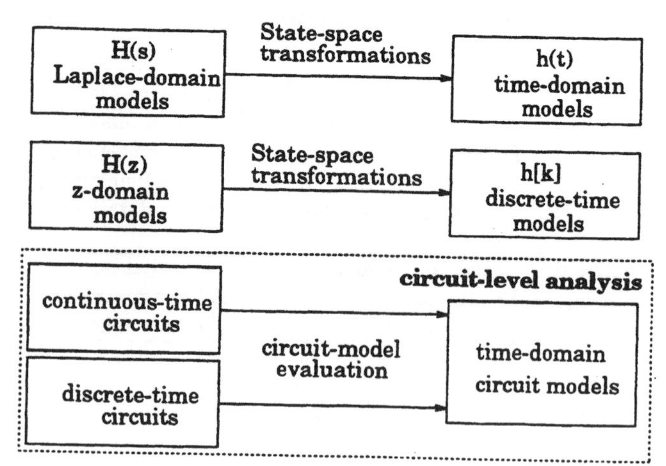

Fig. 3. Transformations for transient simulation.

system is obtained in an analysis regime. One of the goals of the research presented in this paper is to resolve this problem of multilevel and mixed-domain interconnections. In case of circuit-level or electrical simulation, the system is composed of devices such as transistors, capacitors, and resistors. Typically, device models are defined for transient analysis as well as frequency dependent models in the form of linearized small-signal models. Depending on the simulation mode specified, the simulation algorithms choose the type of models to be used in the analysis. Given a circuit-level representation, a higher-level model in the form of a transfer function can be extracted automatically through the application of symbolic analysis [29]. Using symbolic analysis, analytic expressions or models for various circuit blocks are generated, which can then be used for simulation at higher levels.

The transformations arising at the higher or behavioral levels are mathematical in nature. The set of transformations for linearized frequency domain transfer functions described in [25], [26] generate time-domain state-space models. For a general s-domain transfer function

$$H(s) = \frac{a_0 + a_1 s + a_2 s^2 + \cdots + a_p s^p}{b_0 + b_1 s + b_2 s^2 + \cdots + b_{q-1} s^{q-1} + s^q}, \qquad p < q \tag{1}$$

the equivalent state-space model is

$$\dot{\boldsymbol{x}} = \boldsymbol{A}\boldsymbol{x} + \boldsymbol{b}u; \quad y = \boldsymbol{c}\boldsymbol{x} \tag{2}$$

where $\boldsymbol{x}$ is the vector of state variables, u is the input, y is the output, and the matrices $\boldsymbol{A}, \boldsymbol{b}$, and $\boldsymbol{c}$ are defined in [27]. A model defined in the continuous time-domain can be converted into a frequency domain model through the use of the Laplace transform. Models with nonlinearities can be linearized and transformed to the frequency domain, or explicit simulation methods such as nonlinear frequency domain analysis [28] can be used. For the state equations shown in (2), the corresponding frequency domain model is

$$H(s) = \boldsymbol{c}(s\boldsymbol{1} - \boldsymbol{A})^{-1}\boldsymbol{b}. \tag{3}$$

For a general z-domain transfer function

$$H(z) = \frac{Y(z)}{U(z)} = \frac{a_0 + a_1 z^{-1} + \cdots + a_p z^{-p}}{1 + b_1 z^{-1} + \cdots + b_q z^{-q}}. \tag{4}$$

The corresponding state-space model is

$$\boldsymbol{x}(k) = \boldsymbol{A}\boldsymbol{x}(k-1) + \boldsymbol{b}u(k); \quad y(k) = \boldsymbol{c}\boldsymbol{x}(k). \tag{5}$$

Discrete-time models are converted to the frequency domain through the application of the z-transform or the Fourier transform. For the state equations shown in (5), the frequency domain model is

$$H(z) = \boldsymbol{c}(\boldsymbol{1} - \boldsymbol{A}z^{-1})^{-1}\boldsymbol{b}. \tag{6}$$

Yet another set of transforms are applicable when combining discrete circuits with behavioral models. Though a multilevel simulation can be carried out mixing circuit and behavioral representations, the simulation can be further expedited by extracting behavioral models from the circuit-level representation. One approach is to apply symbolic analysis with pruning to the circuit blocks [29]. Symbolic analysis has been used to generate analytic models of circuits. These symbolic models can be parameterized and used in place of the actual circuit block and processed by the behavioral-level simulation algorithms. These provide transformations from the circuit-level to behavioral models. Using behavioral models in place of circuits expedites the simulation as evaluation of a behavioral model is computationally less expensive than the simulation of the entire circuit. Fig. 3 summarizes the transformations for transient simulation. A similar set of transformations are needed for the frequency response as shown in Fig. 4.

The transformations outlined in Figs. 3 and 4 that generate a representation of a mixed-domain system in an analysis regime can be applied explicitly or implicitly. The explicit transformations are in the form of model generation programs such as described in [13] and [26], which are applied before invoking the simulator. These generate models in a code native to the simulator.

Alternately, the transformations can be integrated into the simulator. In this case, the transformations are applied implicitly as part of the model-processing algorithms. The behavioral simulation algorithms that process or evaluate models represented at the behavioral level as implemented in iMACSIM apply these transformations implicitly during simulation. Depending on the simulation regime specified for the entire

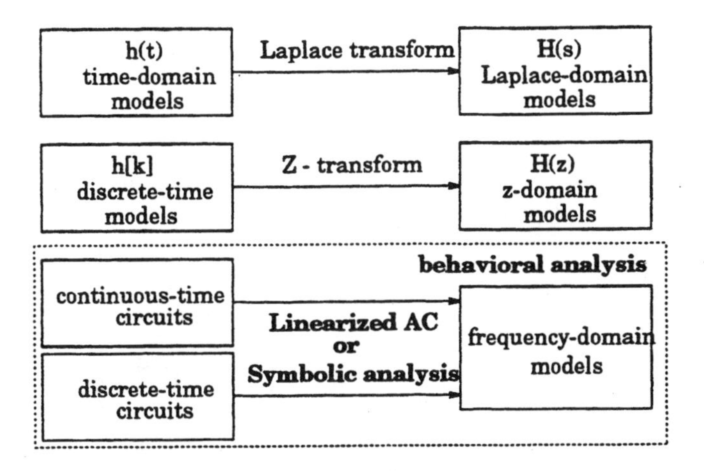

Fig. 4. Transformations for frequency response analysis.

system, the transformations are applied during model evaluation. The implicit transformation technique offers greater flexibility and a true mixed-domain capability to a simulator allowing it to process interconnections of the type shown in Fig. 2. The transformation process is transparent outside the simulator. Explicit transformations are necessary when the underlying simulator has limited simulation capabilities. For example, in the M^3 simulator, a pure time-domain simulation requires all simulation input to be transformed into time-domain models.

A simple example is used to illustrate the concepts described thus far. Fig. 5 shows the interconnection of continuous-time and discrete-time behavioral models. The physical representation is then simulated in an analysis regime. For transient analysis, the transformations shown in Fig. 3 are applied to each model to construct a time-domain representation of the entire system. This time-domain representation is then simulated using transient analysis algorithms. Fig. 6(a) shows the transient response, and Fig. 6(b) shows the frequency response. Here, **sz_s** is the response at the interface of the two models, and **sz_out** is the output. For the transient simulation, the input signal has an added high frequency noise component. The signal seen at the output of the continuous-time analog block is a filtered signal that is integrated through the discrete integrator. **sz_out** is a continuous-time filtered and discrete-time integrated signal. For frequency response analysis, no transformation is required, as the models are defined in the frequency domain. In Fig. 6(b), the cumulative effect of the poles in the two blocks can be observed at the output **sz_out**. It is important to note that any of the representations shown in Fig. 1 could have been used in the above example. Regardless of which view of the model is utilized, the simulation results for all first-order models should be identical. Slight variations in responses can be expected with addition of higher-order nonlinearities.

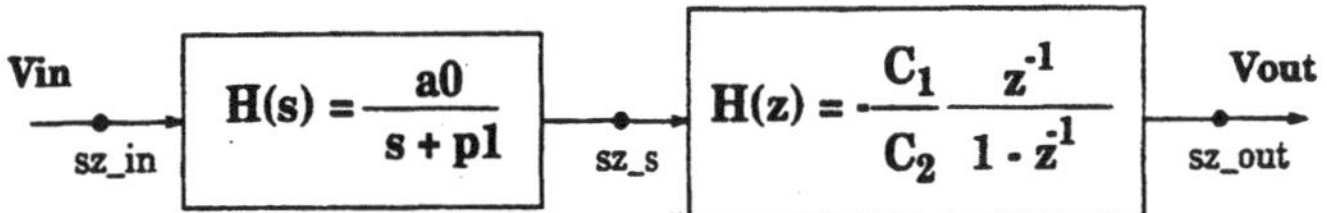

Fig. 5. Mixed-domain example.

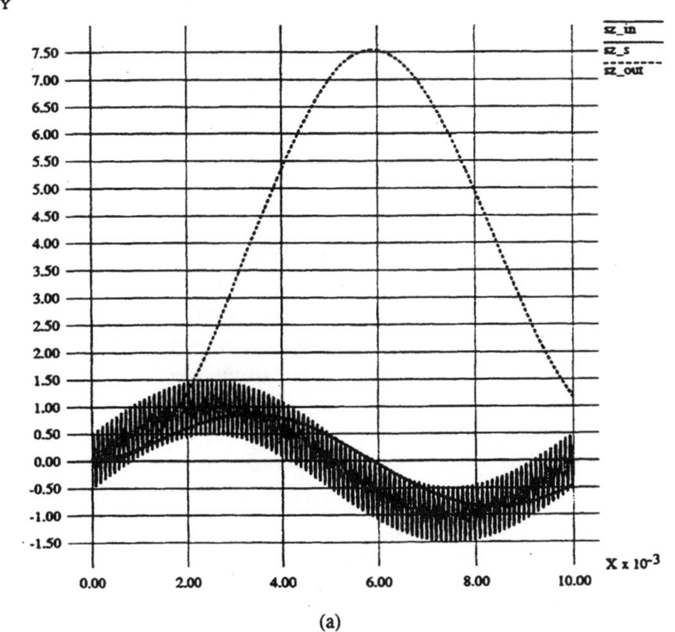

(a)

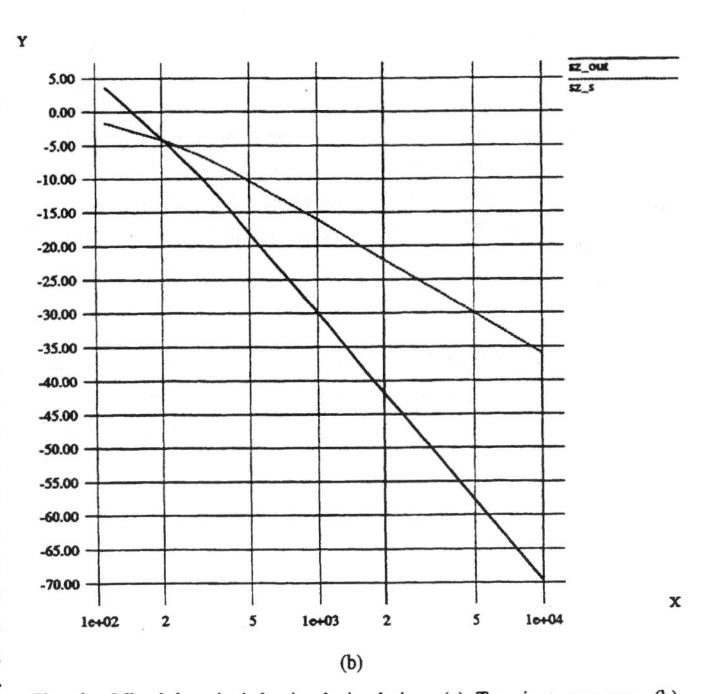

(b)

Fig. 6. Mixed-domain behavioral simulation. (a) Transient response. (b) Frequency response.

IV. Transient Simulation Issues

In order to coordinate the specialized analysis algorithms required to process the various blocks that comprise analog–digital systems, an integrated simulation framework is necessary to carry out the analysis. First, the simulator must partition the circuit into electrical, functional, and behavioral blocks and, during simulation, different types of blocks are processed by different algorithms. For example, an electrical-level block is processed by an electrical algorithm and a behavioral block by a corresponding behavioral algorithm. To improve the efficiency of the overall simulator, an event-driven paradigm is used that ties together simulation across the different levels and domains. An event is defined as a change in state of a node in the circuit that would cause the fan-outs of that node to be affected. The fan-out nodes could belong to blocks of a different type other than the type of the scheduling node. The event-driven paradigm permits latency to be exploited. At a given timepoint, only those blocks driven by nodes whose state has changed will be scheduled. Thus, inactive portions of the circuit will not be processed, resulting in substantial reductions in computation cost.

A. Analog Event Scheduling

The event-driven paradigm is well understood for digital circuits, where a logic transition constitutes an event. However, in analog simulation, waveforms span across the signal ranges, thus the scheduling process is somewhat complicated compared to that for standard logic simulators.

A.1 Scheduling Analog Fan-Out Components

Consider Fig. 7 where the output node Y of an analog component drives fan-out components that are electrical, logical, switch-level, and behavioral blocks. Node Y is a continuous function of time, and this presents two scheduling problems. The first problem is that each of these fan-out components may have different scheduling requirements at their input pins, A, I, D, and P, respectively, due to the fact that they are associated with different algorithms. That is, the electrical block should be scheduled continuously as long as node Y changes. On the other hand, the logic block should be scheduled only when its input is beyond the threshold voltage needed to trigger a change at the inverter output. Devices with multiple inputs may have different scheduling requirements at their inputs, as is the case for the pass transistor (with inputs G and D) shown in the figure. Finally, the scheduling of a block may be based on a condition on a pair of inputs as is the case for the comparator. If each gate is processed whenever Y changes, the fanout blocks may be excessively processed whenever Y undergoes a transition. To overcome this efficiency problem, iMACSIM attaches a scheduling function to each input of a block, which is executed whenever the voltage at the pin changes. This scheduling function determines if the block should be scheduled in the queue.

Currently, the following scheduling functions are implemented in iMACSIM. These functions are associated with the input pins of each subcircuit and are used to determine the type of scheduling to be performed. These functions along with examples are:

Continuous scheduling for circuit-level components, which causes the device to be scheduled for evaluation until the magnitude of the change is small. (node A)

Delta scheduling occurs when a voltage changes by a predefined amount. (for ELOGIC)

Threshold scheduling occurs whenever a voltage crosses a predefined voltage level in either direction. (node G)

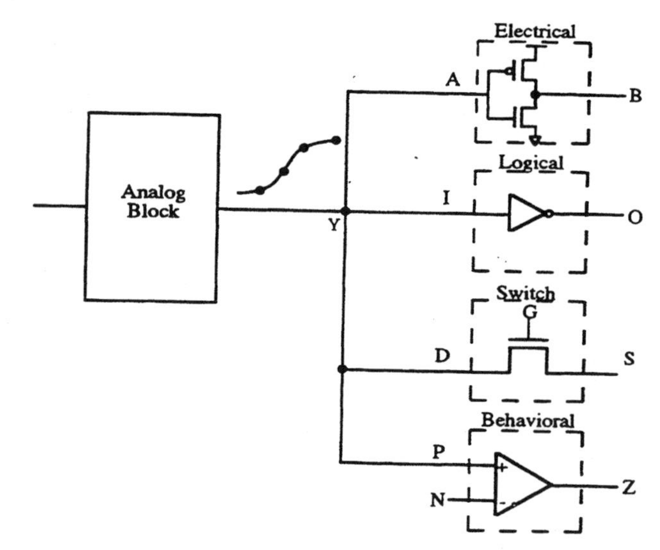

Fig. 7. Scheduling at a mixed-mode interface node.

Signal direction-based scheduling is the above scheduling functions with a directional assignment, indicating the rising or falling nature of the signal. (node I)

A.2 Accurate Threshold Event Scheduling

The second important issue is the computation of the time when a given threshold voltage level has been reached at a given node. Two possible scenarios must be considered for threshold scheduling. In the simpler of the two cases, node Y of Fig. 7 is completely known a priori if it depends only on the function of the analog block and a delay value. In this case, the transition can be predicted and the time at which the threshold will be crossed can be exactly computed. The corresponding fanout blocks can be scheduled in the queue at a precise time.

In the second case, the value of node Y does not follow a simple, predictable pattern if it is being computed incrementally as time flows forward. In this case, some form of root finding is needed to find the threshold. Consider the situation shown in Fig. 8 where the threshold time $t_{threshold}$ for a given control voltage lies between the present time-point, $t_{present}$ and the newly selected time-point t_{new}. The maximum timing error created on this step, if simulation proceeds at t_{new}, is on the order of $t_{new} - t_{present}$. If the circuit has a large number of threshold elements such as switches or logic gates, the timing errors accumulate, causing inaccurate final results. Therefore, it is crucial to detect $t_{threshold}$ and to force the simulator to evaluate the threshold element at $t_{threshold}$ before advancing the time-wheel any further.

One method [33] computes a new time-step, independent of threshold constraints, and predicts if a threshold crossing will occur during the time-step. That is, at $t_{present}$, compute $\Delta t = t_{new} - t_{present}$ using time-step control. Now, given the control voltage $v_{t_{present}}$ predict if $v_{threshold}$ will be crossed during Δt. If a threshold crossing will occur, reduce Δt so that t_{new} = predicted $t_{threshold}$. The prediction mechanism is a by-product of the numerical integration method being used for the electrical-level subcircuits. For example, if the

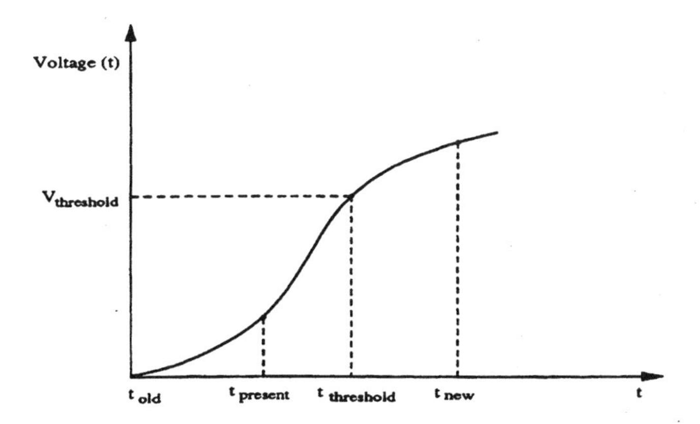

Fig. 8. Threshold event scheduling.

Backward Euler method is being used, then the formula for predicting if $V_{threshold}$ will be crossed during Δt is given as follows. Let t_{old} be the time-point immediately preceding $t_{present}$ and compute Δt from

$$V_{threshold} = V_{present} + \frac{(V_{present} - V_{old})}{(t_{present} - t_{old})}(\Delta t). \quad (7)$$

This approach has the drawback that many small time-steps may have to be taken before the exact threshold is reached. When many time-points are clustered closely together, it becomes necessary to tighten the local truncation error controls since the error in a very tiny step may be small in an absolute sense but very large in a relative sense. One heuristic is to discard the results at $t_{present}$ if the time-point is very close to the threshold and to compute the local truncation error based on an intermediate point generated between t_{old} and $t_{present}$. Another heuristic is to add a small $\Delta\tau$ to the predicted Δt to slightly overshoot the predicted threshold time and thereby avoid the clustering problem. The amount of inaccuracy introduced due to this overshoot must be controlled to prevent timing errors from building up. However, efficiency is the overriding consideration in this approach.

Another possibility is to accept the given time-step without imposing any threshold constraints and to evaluate V_{new}. If a threshold crossing has occurred between $V_{present}$ and V_{new}, the problem of finding the threshold can be formulated as a root-finding problem, and $t_{threshold}$ can be determined. This has an advantage over the predictive method in that there is no "creeping" up to the threshold [34], [35]. The only aspect affected by the numerical integration method is the root-finding routine. Typical root-finding routines used are Hermite interpolation [34] and Newton-Raphson [35], and polynomial interpolation in multistep integration methods. One disadvantage with this approach is that a backup mechanism must be incorporated into the scheduler to discard all events that have been processed after the threshold was passed, and this may be expensive.

B. Scheduling Mixed-Domain Circuits

In the transient simulation of mixed-domain circuits, two other issues need to be considered. The first issue is the design of a mechanism to maintain a consistent solution for

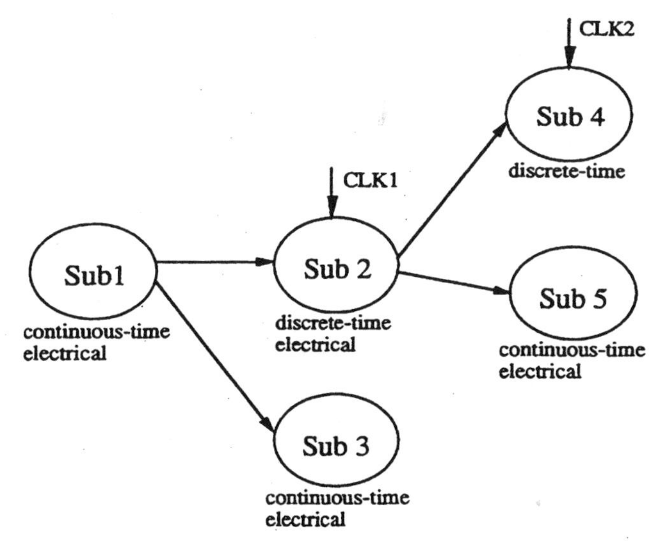

Fig. 9. Combined simulation of discrete- and continuous-time circuits.

the circuit at any given time-point in the simulation interval across the boundaries of the continuous-time subcircuit and the discrete-time subcircuit. A given circuit is partitioned by the simulator into discrete-time and continuous-time subcircuits based on the models specified by the user. For a continuous-time subcircuit, an event is defined to occur when the state of one or more of its nodes changes. This change of state is usually defined as a voltage change. When an event occurs, a subcircuit will try to schedule all subcircuits that fan out from it to be processed by the simulator. As shown in Fig. 9, a continuous-time subcircuit can fan out to either discrete-time or continuous-time subcircuits. With reference to the figure, a continuous-time subcircuit can schedule another continuous-time subcircuit, but a discrete-time subcircuit can be scheduled only by a specially designated clock, in this case, CLK1 for subcircuit Sub 2. On the other hand, whenever the output of a discrete-time subcircuit changes, it can schedule any continuous-time subcircuits on its fan-out list. This event-driven mechanism ensures that a discrete-time subcircuit will not be scheduled at inappropriate times and also permits latency in the circuit to be exploited.

The second issue concerns convergence to a solution in the presence of discontinuities. An element such as an ideal switch can have a very sharp transition from an on-state to an off-state, or vice-versa, and the voltages at its terminals can change almost instantaneously. In some cases, these voltages could function as inputs to continuous-time electrical or functional-level subcircuits. The numerical integration methods used at the electrical and functional levels are extremely prone to error in the presence of sharp transitions or discontinuities. Furthermore, the local truncation error (LTE) checking schemes used in conjunction with the integration methods may have to be modified to take into account the fact that voltages at the affected circuit nodes prior to transition cannot be used as predictors of the voltages at the circuit nodes after the transition. This is because the relative voltage change may be large and too abrupt to assure continuity, which is implicit in the LTE scheme.

The proposed methodology is to detect the exact time at which the transition occurs and to schedule a pseudo-dc solution of the circuit at that time. In the case of an ideal switch, one terminal of the switch may be labeled as a controlling node and the other terminal as a controlled node. At the threshold instant when a switch has just closed, the controlling nodes are used to set the values of the controlled nodes. Convergence to a dc-like solution at the transition time is not guaranteed under this scheme and further investigation of the problem is needed.

An important issue that must also be considered is that a clocking signal for a discrete-time subcircuit may not be known in advance. The user should have the option of either completely specifying the clocking signal in advance before the simulation, or deriving it from nodes in the circuit, whose behavior is not known a priori. In general, this problem may exist even in purely continuous-time circuits containing elements whose behavior changes at some threshold (usually a threshold voltage). If constraints are not imposed, the time-step selection algorithm may never schedule a threshold element to be processed at the exact time that its threshold is reached, and this would give rise to a timing error in the simulation, as described previously.

Algorithm: (Partitioning MNA circuits)

```
partition()
{
  nodelist ← {all nodes};
  foreach(MNA element i) {
    place all its nodes in subcircuit i;
    nodelist ← {nodelist} - {MNA nodes};
  }
  G_partitioner(nodelist);             /* conductance based */
  C_partitioner(nodelist);             /* capacitance based */
  Build_Subcircuit_Data(sublist);            /* tentative subcircuits */
  Levelize(sublist);          /* rank */
  foreach Level i, (i=n, ..., 1) {           /* process in reverse-order */
    foreach (subcircuit j in Level i) {
      Levelize(sub_nodelist(j));           /* order nodes in subcircuit */
    }
  }
  /*Check feedforward and feedback nodes for capacitances*/
  foreach Level i, (i=n, ..., 1) {           /*rank*/
    foreach (subcircuit j in Level i) {          /*subcircuit*/
      foreach (node k in subcircuit j), (k=p_j, ..., 1) {            /*node*/
        if (k = unvisited) {
          foreach ((x ∈ fanout_list(k)) and (x = unmarked)) {
            Mark(x);
            if (check_cap_feedback(x,k) = NO or check_cap_feedforward(x,k) = NO) {
              /*no feedback/feedforward capacitance*/
              Coalesce(j, find_subcircuit(x));           /*combine subcircuits*/
              Update(j, find_subcircuit(x));
              Continue;
            }
          }
        }
        k = visited;
      }
    }
  }
  Build_subcircuits();
  foreach Level i, (i=n, ..., 1) {
    Partial_order();           /* establish final ordering */
  }
}
```

Fig. 10. Algorithm for partitioning MNA circuit elements.

C. Mixed Relaxation/MNA Formulation

Conventional electrical simulation uses direct integration methods implemented in the SPICE type of circuit simulators. However, while direct methods are successful in simulating tightly coupled circuits, these methods are not very amenable to event-driven processing. Thus, relaxation-based nonlinear methods such as the iterated timing analysis (ITA) have been developed [36]. Relaxation methods offer significant advantages over direct methods in their ability to exploit latency inherent in many classes of circuits [24]. However, for guaranteed convergence, these methods require that circuits meet a set of convergence criteria. The sufficiency conditions for convergence that have been shown for the various relaxation-based solution methods [37] require that a grounded capacitance exist for all boundary nodes in the various subcircuits. This requirement is often not satisfied by many analog circuits when they are modeled at the functional level with elements such as floating voltages, inductances, and controlled sources. These elements necessitated the use of Modified Nodal Analysis (MNA) [38] and are referred to as the MNA elements. In order to handle MNA elements within the relaxation framework, a partitioning algorithm is used that embeds an MNA element in a subcircuit such that *the boundary nodes of the subcircuit have grounded capacitances.* This partitioning process ensures that the convergence criteria are satisfied. The subcircuits with MNA elements are then solved using direct methods within the general context of the relaxation formulation. Similarly, behavioral blocks are grouped in separate subcircuits and processed by the behavioral model evaluation algorithms. Thus, a system composed of circuit-level components, functional-level elements, and behavioral blocks are processed in an integrated manner using the overall relaxation-based simulation framework. Fig. 10 shows the partitioning algorithm implemented in iMACSIM for handling MNA elements in a relaxation framework.

D. Discrete-Time Circuits

Switched-capacitor circuits are discrete-time circuits that require special simulation techniques, as direct methods prove to be computationally expensive on account of the high frequency switching clocks used as compared to the lower frequency signals being processed. At the functional level, a combination of switch models, controlled sources, and capacitors are sufficient to model these circuits. Charge conservation algorithms that solve a set of charge-conservation equations once per clock phase prove to be orders of magnitude faster than direct methods. A MNA formulation is used to formulate the charge-based equations. In addition to the node voltages, variables for the charge flowing across each node is modeled as an additional variable. An underlying assumption is that complete charge transfer is achieved instantaneously after each clock transition. Thus, charge-based MNA matrices are constructed for each clock-phase and switch setting for later reuse.

V. Frequency-Response Issues

Frequency-response analysis is currently limited to linear or linearized circuits in iMACSIM. Behavioral representations, in the form of transfer functions or the equivalent frequency-domain representation obtained from the time domain models, are simulated at the behavioral level. Circuit-level or functional-level continuous-time circuits are simulated using linear or linearized ac analysis. Here, the phasor analysis technique as implemented in SPICE-like simulators is used. Nonlinear devices such as transistors are linearized about the dc operating point. The linearized frequency response analysis of switched or discrete time circuits requires a specialized formulation described next, followed by the approaches for simulating mixed-domain circuits that require additional consideration.

A. Frequency Response of Switched-Capacitor Circuits

The switched capacitor analysis program SWITCAP2 uses an algorithm proposed by Sun to obtain closed-form expressions for the frequency response of idealized switched capacitor circuits [40]. Starting with a set of time-domain state equations for each circuit phase, the frequency response is computed through the solution of a closed-form expression. The closed-form solution requires the computation of a frequency-response matrix of the form $(Ie^{j\omega T} - M_K)^{-1}Q$ for every frequency ω at which the response is to be calculated. Here, $M_K \in \Re^{n\times n}$ is derived from the companion matrix of the state-space formulation at each phase, and $Q \in \Re^{n\times m}$ represents an output for an input $ue^{j\omega t}$ with zero initial conditions, and m is the number of independent sources in the circuit. The integer n is related to the number of nodes in the circuit. The matrix inversion for each frequency ω is $O(n^3)$. Using a modified version of Laub's algorithm [41], [42], significant speedup in the frequency-response analysis is obtained. In this method, the matrix M_K is first transformed to an upper Hessenberg form, thus enabling subsequent LU factorizations and inversions to be carried out efficiently. However, this method assumes that the matrix Q is independent of ω. To circumvent this assumption, the implementation of this algorithm in iMACSIM, uses a mixed numeric-symbolic representation for Q, thus avoiding the recomputation of Q at each frequency. With these modifications, the iMACSIM implementation demonstrates a speedup of approximately 9.7× over SWITCAP2 for a fifth-order lowpass filter with identical results [43].

B. Frequency Response of Mixed-Domain Circuits

When discrete-time and continuous-time subcircuits are interconnected, a problem arises with computing the frequency response of the composite system. This interconnection could be processed by partitioning the system, computing the frequency response of the continuous-time block, and using the response to drive the analysis of the discrete-time block. However, this approach is limited to interconnections with little or no feedback elements across the partitioned blocks. Thus, for generality and for processing tightly coupled mixed-domain systems, a more integrated solution is required. This problem is addressed by generating a common set of frequency-dependent equations in the MNA formulation [14]. The continuous-time components are stenciled in the MNA matrix by evaluating

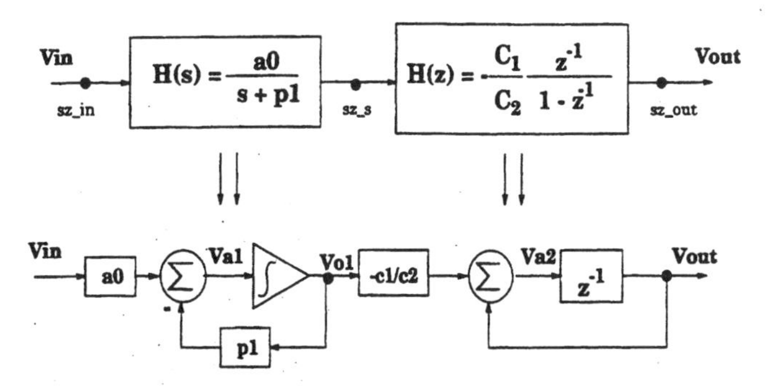

Fig. 11. Frequency response of mixed-domain example.

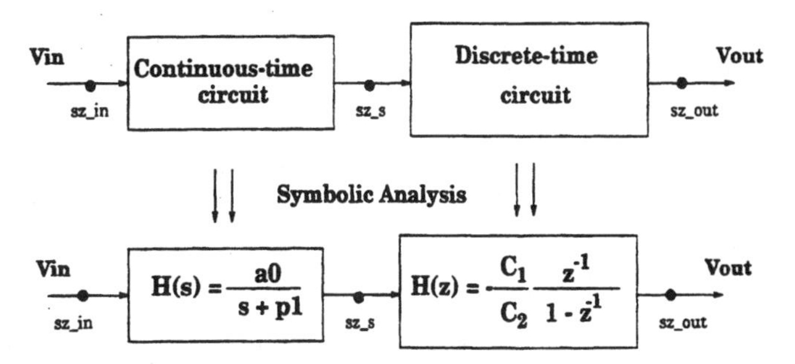

Fig. 12. Frequency response of mixed-domain circuits.

the models using the continuous-time frequency variable $j\omega$, and the discrete components are stenciled using the discrete-time frequency variable $j\omega T$. Revisiting the example shown in Fig. 5, the behavioral blocks represented by the Laplace and z-domain transfer functions are transformed into their equivalent physical representations. From the physical representation, a composite MNA matrix is constructed for the entire system. Continuous-time components are represented by the equivalent Laplace domain representation and discrete-time components by the corresponding z-domain representation. Fig. 11 shows the equivalent physical representation, and the composite MNA matrix is given by (8)

$$\begin{bmatrix} 1 & p_1 & 0 & 0 \\ -1/s & 1 & 0 & 0 \\ 0 & C_1/C_2 & 1 & -1 \\ 0 & 0 & -z^{-1} & 1 \end{bmatrix} \begin{bmatrix} V_{a1} \\ V_{o1} \\ V_{a2} \\ V_{out} \end{bmatrix} = \begin{bmatrix} a_0 \\ V_{in} \\ 0 \\ 0 \\ 0 \end{bmatrix}. \tag{8}$$

In the case when two circuit-level descriptions, from those shown in Fig. 1 are connected, symbolic analysis is used to construct the behavioral representation in the form of transfer functions $H(s)$ and $H(z)$, which are then simulated. Fig. 12 shows the application of symbolic analysis to construct the frequency-response analysis representation.

VI. Simulator Architecture

The overall architecture of iMACSIM is shown in Fig. 13. It consists of an interface called the *algorithmic backplane*, which contains sparse matrix routines, a scheduler package, waveform processing routines, and input/output routines. The backplane coordinates the simulation across the different levels

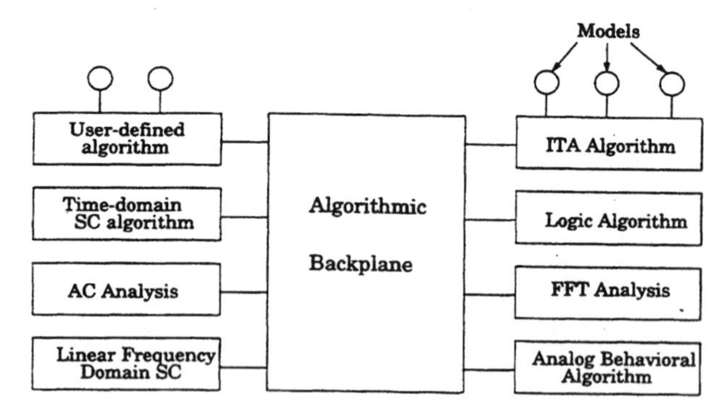

Fig. 13. Simulator architecture.

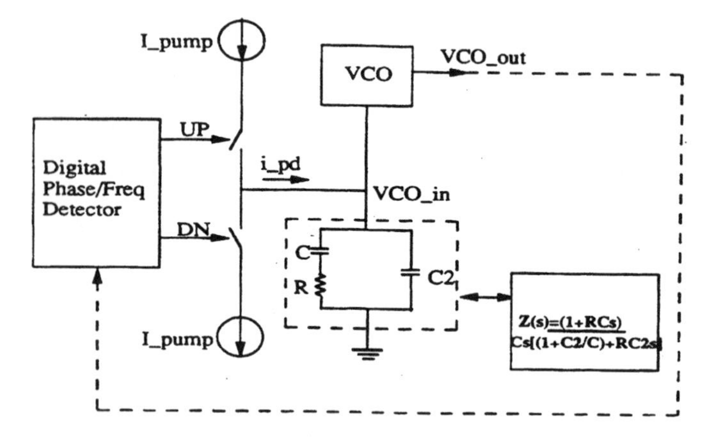

Fig. 14. PLL clock generator circuit.

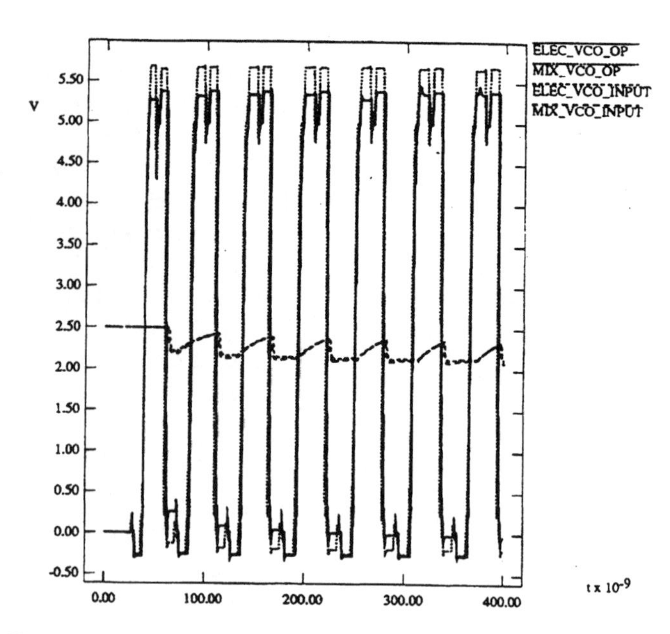

Fig. 15. Comparison of oscillator outputs as a function of $V_{control}$.

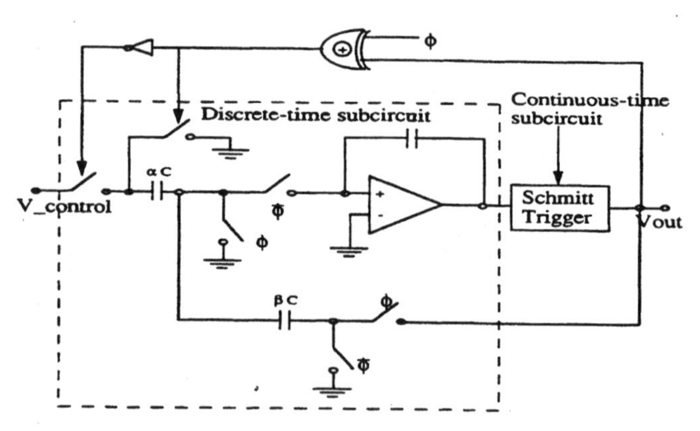

Fig. 16. Switched-capacitor voltage-controlled oscillator.

and domains. Several different algorithms are shown that plug directly into the backplane. Each algorithm has a corresponding set of dedicated models. That is, a given model is processed by its associated algorithm. This modular design provides the simulator with an open, expandable, and reconfigurable architecture. Thus, the simulator can be tailored to very specific needs, such as switched capacitor circuit analysis or provide the entire range of analysis across all the analog and digital levels. In addition, other analysis algorithms, which are in the process of development, can be added to this backplane. The open architecture also provides an accessible platform for experimenting with different algorithms and solution methods, for example, different variants of integration methods and ITA methods can be incorporated. Knowing the limitations of these algorithms in dealing with certain kinds of circuits, the most appropriate and efficient method that suits a given circuit or system characteristics can be chosen.

The algorithmic backplane is a procedural interface to the simulation algorithms and consists of a set of macros and subroutine calls that provide transparent access to the central data structures and operations. Since the various algorithms are decoupled from each other, development on one aspect of the simulation can proceed independent of the other modes. The backplane contains various packages that are shared by the various simulation algorithms.

Fig. 13 shows some of the simulation algorithms and techniques that are presented in this paper thus far which have been attached to the backplane. The iterated timing analysis (ITA) algorithm is a relaxation-based algorithm for electrical and functional simulation, as described in Section IV.3. The associated models include resistors, capacitors, inductors, transistors, and controlled sources. The analog behavioral algorithm controls both discrete-time and continuous-time models. The logic algorithm processes behavioral descriptions of logic gates. Currently, the iMACSIM library includes over 40 models for logic gates. The SC algorithm processes switched-capacitor subcircuits in both the time and the frequency domains. Its associated models include a capacitor model, ideal switches, and ideal VCVS's. The ac analysis algorithm carries out SPICE-like small signal analysis for linearized frequency response. An FFT algorithm allows the user to obtain the frequency-domain characteristics of nonlinear circuits from a steady-state transient solution. The FFT algorithm has no associated models as it is used as a postprocessing routine. Since iMACSIM uses an event-driven paradigm, the inner loop of the program is an event processor that simply sequences through an ordered list of simulation events such as *dc_solution()*, *pick_time_step()*, *process_analog_event()*, *process_digital_event()*, and *roll_back_time()*.

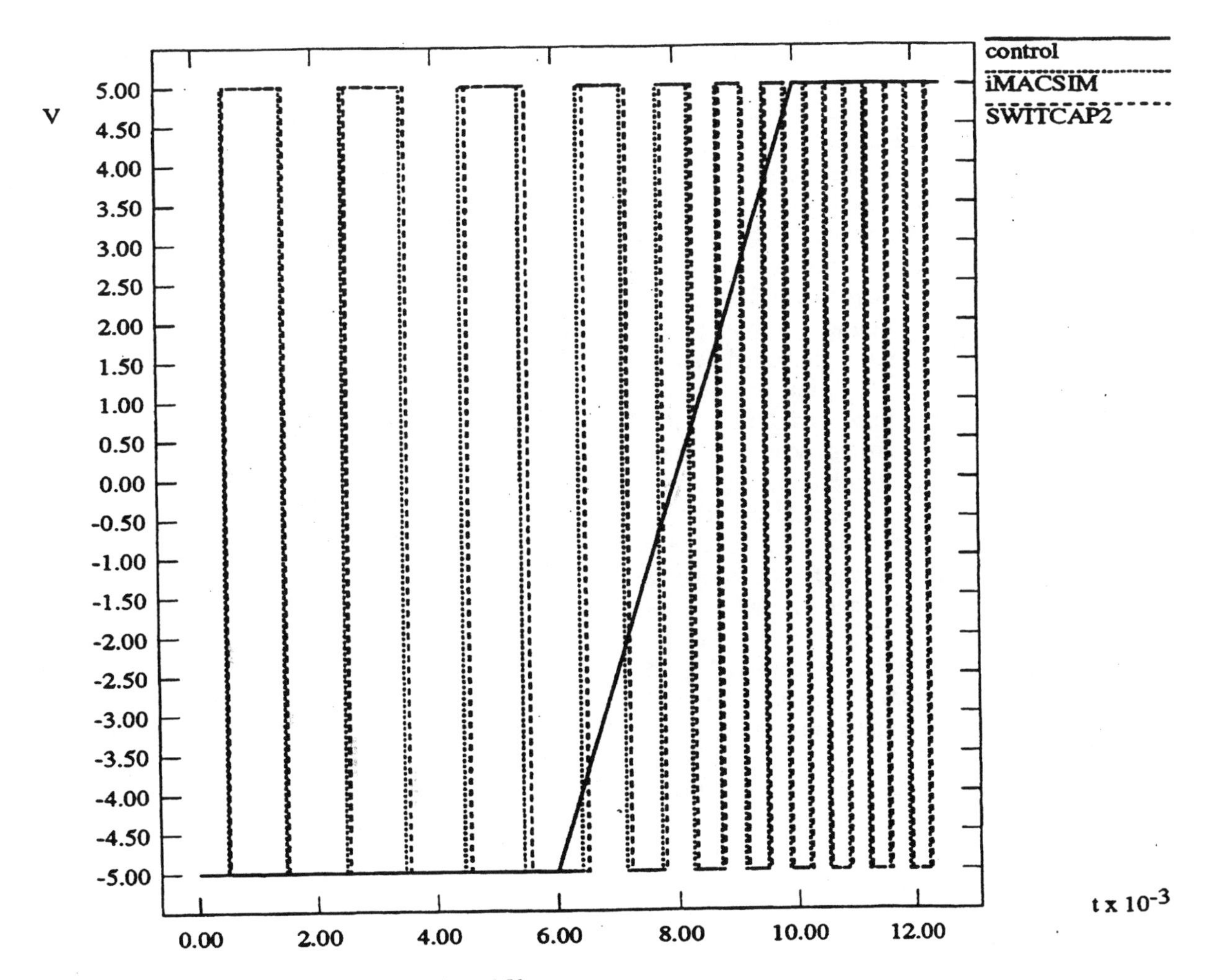

Fig. 17. Comparison of oscillator outputs as a function of $V_{control}$.

A. Adding Algorithms

When a new analysis capability is to be added to the program, the user must write the procedure using a combination of C and predefined macros. To give an idea of the program flow, the sequence of steps that occur after the scheduler picks up a *process_analog_event* is described. This simulation event is associated with a subcircuit. The first step is to obtain the input voltages for that subcircuit. The subcircuit is then processed by the *process_analog* algorithm and, if a change occurs at the designated circuit variables, which could be voltages, currents, or charges, the subcircuit will be scheduled for processing at a later time. This later time is calculated by a time-step routine. Also, fan-out subcircuits will be scheduled as appropriate at the current time. For example, some subcircuits on the fan-out list may require immediate scheduling whenever their input voltages change. Others may require scheduling only when the input voltage crosses a designated threshold, and discrete-time subcircuits would be scheduled independently at fixed times, irrespective of their input voltage. For illustration, the pseudocode for an analog behavioral event is provided

```
process_analog_event(subckt, time, step)
{
        get_inputs(subckt, time);
        analog_solve(subckt);
        if(output_changed(subckt)) {
            schedule_self(subckt);
            schedule_fanouts(subckt);
        }
}
```

In the pseudocode above, *subckt* is the portion of the circuit being processed, *time* is the current time, and *step* is the time-step used for the numerical integration. Since the rest of the simulator must be made aware of this routine, the name of the routine and other information about the type of analysis must be stored in globally accessible tables in the simulator for use during simulation. The information would include the models associated with the analysis, a definition of event scheduling for this analysis, and a definition of how subcircuits are to be constructed for models associated with this analysis. This information will be accessed by processing functions as needed during simulation.

VII. Examples

In this section, we present examples of various circuits that have multilevel, mixed-mode, and multidomain representations. These are representative examples that validate the various concepts and implementation issues that are presented in this paper. SPICE-like simulators have limitations and are proving to be inadequate for many analog simulation tasks. The special characteristics of many analog circuits, such as stiffness and high switching speeds of the switched-capacitor circuits pose severe problems. Phase-locked loops (PLL's) are an example of stiff systems that are very time consuming to evaluate using SPICE-like simulators as the PLL lock-in time is very large compared to the period of the operating signal. Similarly, switched-capacitor circuits are driven by high-frequency clocks and process lower frequency signals, resulting in the conventional simulation to be carried at very small integration time-steps within the clock period over the simulation time.

The first example is a multilevel mixed-mode circuit, a PLL clock generator circuit with a mix of analog and digital components [45]. A complete MOS transistor-level implementation of this circuit was simulated at the electrical level. Next, the digital phase detector was represented at the LOGIC level using gate models, the loop filter at the analog behavioral level using a Laplace domain transfer function, and a circuit-level implementation of the VCO. Fig. 14 shows the PLL configuration, and the simulation results are shown in Fig. 15, which compares the simulations in iMACSIM at the electrical level and the multilevel. The same results are obtained with a speedup of a factor of 4.

The next example is a switched capacitor voltage-controlled oscillator circuit shown in Fig. 16. Two seperate simulations were conducted in iMACSIM; in the first case, the Schmitt trigger was modeled at the behavioral level and in the second case at the circuit-level using the transistor implementation. In both cases, a mixed continuous-time/discrete-time simulation was performed. The results are shown in Fig. 17 with the control voltage $V_{control}$ swept from -5 to $+5$ V. Additionally, this circuit was modeled using wholly discrete-time components in SWITCAP2, and the simulation results from the two simulators are compared.

For frequency response analysis of switched capacitor circuits, an efficient method is implemented in iMACSIM described earlier in Section VI. A fifth-order switched capacitor low pass filter is shown in Fig. 18 [46] along with the frequency response.

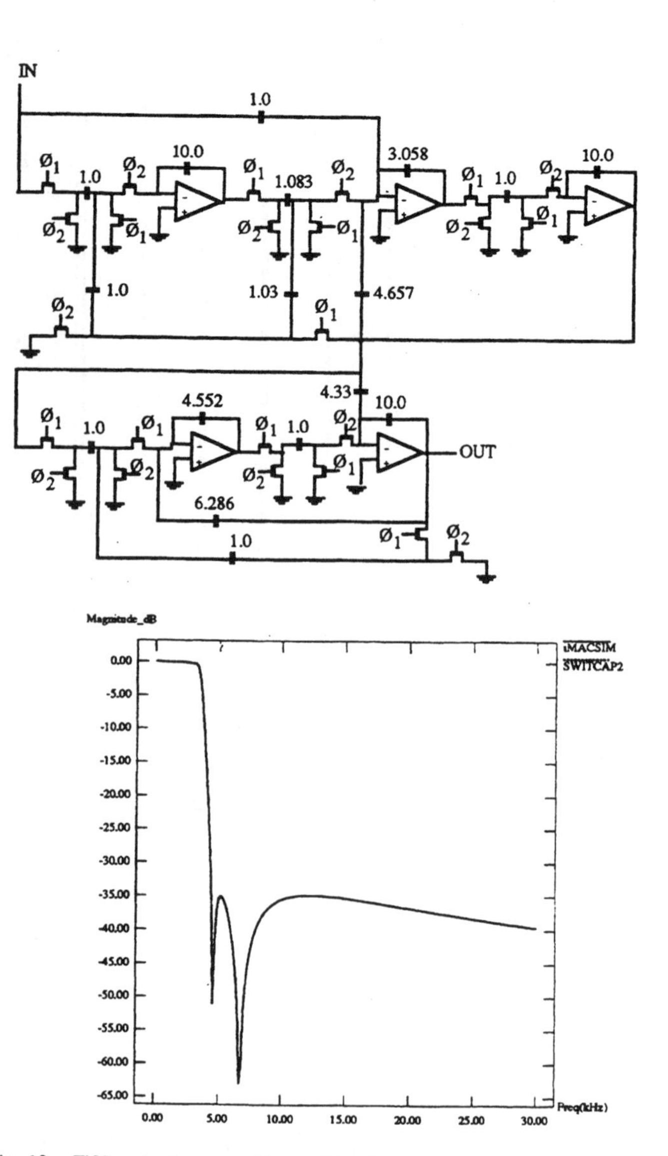

Fig. 18. Fifth-order low-pass filter and its frequency response.

VIII. Conclusions and Future Research

Simulation of analog–digital systems across different levels, domains, and analysis regimes requires a special set of mechanisms and techniques to make such a simulation possible in an integrated manner. Previous approaches have resorted to the use of special purpose simulators for analysis of each of the circuit types separately. Though some succesful efforts have been reported in simulation of mixed analog–digital simulation, their limitations include simulation across domains and frequency-response analysis. Integrated simulation across multilevels and mixed-domains requires addressing issues of interfaces, interdomain transformations, and algorithms for various analysis regimes; these issues have been presented in this paper. The implementation of iMACSIM demonstrates the feasibility of undertaking such an all-encompassing simulation with analysis in the time and frequency regimes. The implementation is an open-ended framework that allows additional algorithms to be added to include analysis of systems that require very specialized methods.

Future research efforts are directed towards developing advanced simulation algorithms for steady-state analysis and sensitivity analysis. A new method for efficient nonlinear frequency analysis of switched-capacitor circuits and for parameter variation studies that can be used to measure manufacturability yields is in the stage of being developed and added to this framework [47]. Additional efforts are also directed towards developing algorithms for steady-state analysis of mixed analog–digital circuits, using efficient steady-state simulation algorithms, instead of the brute-force transient analysis until steady-state method.

References

[1] R. E. Bryant, "Algorithmic aspects of symbolic switch network analysis," *IEEE Trans. Computer-Aided Design*, vol. 6, pp. 618–633, July 1987.

[2] J. M. Beardslee, "Implementation of a logic simulator and mixed-level simulation for SAMSON2," Carnegie Mellon Univ., M. S. Rep. CMUCAD-86-12, May 1986.

[3] *Digital System Design Automation: Languages, Simulation and Data Base*, M. A. Breuer, Ed. New York: Computer Science Press, 1975.
[4] R. Lipsett, C. Schaefer, and C. Ussery, *VHDL: Hardware Description and Design*. Boston, MA: Kluwer, 1990.
[5] L. W. Nagel, "SPICE2: A computer program to simulate semiconductor circuits," Univ. California, Berkeley, Electronics Research Laboratory, Memo UCB/ERL M520, May 1975.
[6] R. A. Rohrer, "Circuit simulation—the early years," *IEEE Circuits and Devices Mag.*, pp. 32–37, May 1992.
[7] R. A. Saleh, J. E. Kleckner, and A. Richard Newton, "Iterated timing analysis in SPLICE1," in *Dig. Tech. Papers, IEEE ICCAD*, pp. 139–140, 1983.
[8] E. L. Acuna, J. P. Dervenis, A. J. Pagones, F. L. Yang, and R. A. Saleh, "Simulation techniques for mixed analog/digital circuits," *IEEE J. Solid-State Circuits*, vol. 25, pp. 353–362, Apr. 1990.
[9] K. Suyama, S. C. Fang, and Y. P. Tsividis, "Simulation of mixed switched-capacitor/digital networks with signal-driven switches," *IEEE J. Solid-State Circuits*, vol. 25, pp. 1403–1412, Dec. 1990.
[10] R. J. Trihy and R. A. Rohrer, "A switched capacitor circuit simulator: AWEswit," *IEEE J. Solid-State Circuits*, vol. 29, pp. 217–225, Mar. 1994.
[11] L. A. Williams, B. E. Boser, E. W. Y. Liu, and B. A. Wooley, "MIDAS User Manual," Center for Integrated Systems, Stanford Univ., 1989.
[12] L. W. Pillage and R. A. Rohrer, "Asymptotic waveform evaluation for timing analysis," *IEEE Trans. Computer-Aided Design*, vol. 9, pp. 352–366, Apr. 1990.
[13] R. Chadha, C. Visweswariah, and C. Chen, "M^3—A multilevel mixed-mode mixed A/D simulator," *IEEE Trans. Computer-Aided Design*, vol. 11, pp. 575–584, May 1992.
[14] B. A. A. Antao and A. J. Brodersen, "Behavioral simulation for analog system design verification," *IEEE Trans. VLSI Systems*, vol. 3, pp. 417–429, Sept. 1995.
[15] G. Casinovi and J. Yang, "Simulation of analog behavioral models," in *Proc. Custom Integrated Circuits Conf., CICC*, pp. 12.4.1–12.4.4, 1992.
[16] I. E. Getreu, "Behavioral modeling of analog blocks using the SABER simulator," in *Proc. Midwest Symp. Circuits Syst.*, Aug. 1989 pp. 977–980,.
[17] B. A. A. Antao, F. M. El-Turky, and R. H. Leonowich, "Behavioral modeling phase-locked loops for mixed-mode simulation," to appear in *Analog Integrated Circuits and Signal Processing*, Special issue on Modeling and Simulation of Mixed Analog–Digital Systems.
[18] H. El Tahawy, A. Chianale, and B. Hennion, "Functional verification of analog blocks in Fideldo: A unified mixed-mode simulation environment," in *Int. Symp., Circuits Syst.*, 1989, pp. 2012–2015.
[19] R. A. Saleh, D. L. Rhodes, E. Christen, and B. A. A. Antao, "Analog hardware description languages," presented at the IEEE Custom Integrated Circuits Conf. (CICC), May 1994.
[20] B. Antao, "Analog hardware description language requirements: A snapshot," presented at the Int. Conf. Hardware Description Languages (ICEHDL), Jan. 1995.
[21] Y. P. Tsividis and K. Suyama, "MOSFET modeling for analog circuit CAD: Problems and prospects," *IEEE J. Solid-State Circuits*, vol. 29, pp. 210–216, Mar. 1994.
[22] G. R. Boyle, B. M. Cohn, D. O. Pederson, and J. E. Solomon, "Macromodeling of integrated circuit operational amplifiers," *IEEE J. Solid-State Circuits*, vol. SC-9, pp. 353–364, Dec. 1974.
[23] A. R. Newton and A. L. Sangiovanni-Vincentelli, "Relaxation-based electrical simulation," *IEEE Trans. Computer-Aided Design*, vol. CAD-3, pp. 308–331, Oct. 1984.
[24] R. A. Saleh and A. R. Newton, "The exploitation of latency and multirate behavior using nonlinear relaxation for circuit simulation," *IEEE Trans. Computer-Aided Design*, vol. 8, Dec. 1989.
[25] C. Visweswariah, R. Chadha, and C-F. Chen, "Model development and verification for high level analog blocks," in *Proc. 25th ACM/IEEE DAC*, pp. 376–382, June 1988.
[26] B. A. A. Antao and F. M. El-Turky, "Automatic analog model generation for behavioral simulation," in *Proc. IEEE Custom Integrated Circuits Conf. (CICC)*, May 1992, pp. 12.2.1–12.2.4.
[27] C. T. Chen, *Linear System Theory and Design*. New York: Holt, Rinehart, and Winston, 1984.
[28] B. A. A. Antao, "Synthesis and verification of analog integrated circuits," Ph.D. dissertation, Dept. Elect. Eng., Vanderbilt Univ., Nashville, TN, Dec. 1993.
[29] G. Gielen and W. Sansen, *Symbolic Analysis for Automated Design of Analog Integrated Circuits*. Norwell, MA: Kluwer, 1991.
[30] V. M. Ma, J. Singh, and R. A. Saleh, "Modeling, simulation and optimization of analog macromodels," in *Proc. CICC*, 1992, pp. 12.1.1–12.1.4.
[31] A. R. Newton, "The simulation of large-scale integrated circuits," Ph.D. dissertation, Univ. California, Berkeley, ERL Memo ERL-M78/52, July 1978.
[32] H. J. De Man *et al.*, "Practical implementation of a general computer-aided design technique for switched capacitor circuits," *IEEE J. Solid-State Circuits*, vol. SC-15, pp. 190–200, Apr. 1980.
[33] M. B. Carver, "Efficient integration over discontinuities in ordinary differential equation simulations," in *Mathematics and Computers in Simulation XX*, pp. 190–196.
[34] D. G. Bedrosian, "Analysis of networks with internally controlled switches," Ph.D. dissertation, Univ. Waterloo, Ontario, Canada, 1991.
[35] D. Ellison, "Efficient automatic integration of ordinary differential equations with discontinuities," in *Mathematics and Computers in Simulation XXIII*. pp. 12–20.
[36] J. K. White and A. Sangiovanni-Vincentelli, *Relaxation Techniques for the Simulation of VLSI Circuits*. Boston, MA: Kluwer, 1987
[37] M. P. Desai and I. N. Hajj, "On the convergence of block relaxation methods for circuit simulation," *IEEE Trans. Circuits Syst.*, vol. 36, pp. 948–958, July 1989.
[38] C. W. Ho *et al.*, "The modified nodal approach to network analysis," *IEEE Trans. Circuits Syst.*, vol. CAS-22, no. 6, pp. 504–508, June 1973.
[39] R. A. Saleh and A. R. Newton, *Mixed-Mode Simulation*. Boston, MA: Kluwer, 1990.
[40] Y. Sun, "Direct analysis of time-varying continuous and discrete difference equations with application to nonuniformly switched-capacitor circuits," *IEEE Trans. Circuits Syst.*, vol. CAS-28, pp. 93–100, Feb. 1981.
[41] A. J. Laub, "Efficient multivariable frequency response computations," *IEEE Trans. Automat. Contr.*, vol. AC-26, pp. 407–408, 1981.
[42] ——, "Algorithm 640: Efficient calculation of frequency response matrices from state space models," *ACM Trans. Math. Software*, vol. 12, no. 1, pp. 26–33.
[43] J. Singh and R. A. Saleh, "Frequency-domain analysis of switched-capacitor circuits containing nonidealities," CSL, Internal Rep.
[44] K. S. Kundert and A. L. Sangiovanni-Vincentelli, "Sparse User's Guide: A sparse linear equation solver," Dept of EECS, Univ. California, Berkeley, Apr. 1988.
[45] J. Sherred, "A phase-locked clock generator for VLSI application," M.S. thesis, MIT, Cambridge, MA., June 1988.
[46] E. Sanchez-Sinencio, P. E. Allen, A. W. T. Ismail, and E. R. Klinkovsky, "Switched capacitor filters with partial positive feedback," *AEU-Electron. and Commun.*, vol. 38, no. 5, pp. 331–339, Sept.–Oct. 1984.
[47] J. G. Mueller, B. A. A. Antao, and R. A. Saleh, "Rapid estimation of the nonlinear frequency response of switched-capacitor circuits," presented at the IEEE Custom Integrated Circuits Conf., May 1995.

Resve A. Saleh (S'78–M'79–SM'95) received the B. Eng. degree (electrical) from Carleton University, Ottawa, Canada, in 1979, and the M.S. and Ph.D. degrees from the University of California at Berkeley in 1983 and 1986, respectively.

He has worked in industry for a number of companies in Canada, U.S., and Japan. He joined the Department of Electrical and Computer Engineering at the University of Illinois in 1986 where he directed research in mixed-mode simulation, analog simulation, and parallel processing. His research interests also include analog CAD and synthesis. He coauthored a book called *Mixed-Mode Simulation* (Norwell, MA: Kluwer) in 1990 and a second edition entitled *Mixed-Mode Simulation and Analog Multilevel Simulation* in 1994. He has served on the technical committee of the Custom Integrated Circuits Conference as the Technical Program Chair (1993), Conference Chair (1994), and General Chair (1995), the Design Automation Conference since 1987, and was a member of the organizing committee of the Midwest Symposium on Circuits and Systems in 1989. He is a member of the IEEE Circuits and Systems Society and a member of the Solid-State Circuits Council. He served as an Associate Editor on the IEEE TRANSACTIONS ON COMPUTER-AIDED DESIGN OF CIRCUITS AND SYSTEMS from 1993 to 1995 and Chairman of the SCC-30 committee involved in standardizing an analog HDL from 1992 to 1995.

Dr. Saleh received a National Science Foundation Presidential Young Investigator Award in 1990.

Brian A. A. Antao (S'86–A-86–M'93) received the B.E. degree (honors) in electrical engineering from the University of Bombay (V.J.T.I.) in 1986 and the M.S. and Ph.D. degrees in electrical engineering from Vanderbilt University, in 1988 and 1993, respectively.

Currently, he is a member of the research faculty in the Coordinated Science Laboratory at the University of Illinois at Urbana-Champaign and will be joining the Advanced Design Technology Group at Motorola Inc. in Austin, TX. In addition, he has held summer research positions at AT&T Bell Laboratories in 1991 and 1992 working on behavioral modeling and mixed-mode simulation. His research is in the design and synthesis of high performance integrated systems and mixed analog–digital systems through an interdisciplinary effort synergizing various aspects of architectural design, circuit design, and computer-aided design techniques. Specific areas of focus at present include high-level analog synth+esis and optimization, modeling, and mixed-mode simulation. He is the guest editor of the forthcoming special issue of the *Analog Integrated Circuits and Signal Processing Journal* on modeling and simulation of mixed analog–digital systems.He is also a member of the Technical Program Committee of the IEEE Custom Integrated Circuits Conference.

Dr. Antao is a member of the IEEE Computer and the Circuits and IEEESystems Societies, ACM, and Tau Beta Phi.

Jaidip Singh received the B.S. degree in electronics engineering from the Indian Institute of Technology at Madras in 1986, the M.S. degree in electrical and computer engineering from Rensselaer Polytechnic Institute in 1988, and the Ph.D. degree in electrical engineering from the University of Illinois at Urbana-Champaign in 1994. He is currently in the MBA program at the Wharton School of the University of Pennsylvania..

He has worked at Meta-Software, Analog Devices, and IBM in the areas of circuit simulation and computer performance analysis.

Simulation Methods for RF Integrated Circuits

Ken Kundert

Cadence Design Systems, San Jose, California, USA

Abstract — **The principles employed in the development of modern RF simulators are introduced and the various techniques currently in use, or expected to be in use in the next few years, are surveyed. Frequency- and time-domain techniques are presented and contrasted, as are steady-state and envelope techniques and large- and small-signal techniques.**

I. RF Circuits

The increasing demand for low-cost mobile communication systems has greatly expanded the need for simulation algorithms that are both efficient and accurate when applied to RF communication circuits.

RF circuits have several unique characteristics that are barriers to the application of traditional circuit simulation techniques. Over the last decade, researchers have developed many special purpose algorithms that overcome these barriers to provide practical simulation for RF circuits, often by exploiting the very characteristic that represented the barrier to traditional methods.

Despite dramatic progress, the average design cycle of an RFIC is still twice the length of that for other ICs found in a communication system, such as a cellular phone. This represents a significant practical barrier to integration of the RF and baseband sections of a transceiver onto a single chip. Clearly, more progress is necessary.

This paper is a overview of RF simulation methods that seeks to provide an understanding of how the various methods address the RF simulation problem, and how they relate to each other.

It begins by describing the unique characteristics of RF circuits. The basic solution methods of transient analysis, harmonic balance, and shooting methods are presented and contrasted. Small-signal analysis versions of both harmonic balance and shooting methods are covered. Composite methods are next. These methods apply the base methods in layers to provide dramatic new capabilities. The composite methods are divided into two groups. The first is based on a multidimensional representation of time. The second is based on sampling the RF signal and finding its envelope. The paper concludes with comparisons of the methods.

Published in the *Proceedings of ICCAD'97*, November 9-13, 1997 in San Jose, California. Manuscript received August 11, 1997.

This work was supported by the Defense Advanced Research Projects Agency under the MAFET program.

K. Kundert can be reached via e-mail at kundert@cadence.com.

A. Narrowband Signals

RF circuits process narrowband signals in the form of modulated carriers. Modulated carriers are characterized as having a periodic high-frequency carrier signal and a low-frequency modulation signal that acts on either the amplitude, phase, or frequency of the carrier. For example, a typical cellular telephone transmission has a 10-30 kHz modulation bandwidth riding on a 1-2 GHz carrier. In general, the modulation is arbitrary, though it is common to use simple periodic or quasiperiodic modulations constructed from a small number of sinusoids for test signals.

The ratio between lowest frequency present in the modulation and the frequency of the carrier is a measure of the relative frequency resolution required of the simulation. General purpose circuit simulators, such as SPICE, use transient analysis to predict the nonlinear behavior of a circuit. Transient analysis is inefficient when it is necessary to resolve low modulation frequencies in the presence of a high carrier frequency because the high-frequency carrier forces a small time step while a low-frequency modulation forces a long simulation interval.

Passing a narrowband signal though a nonlinear circuit results in a broadband signal whose spectrum is relatively sparse, as shown in Figure 1. In general, this spectrum

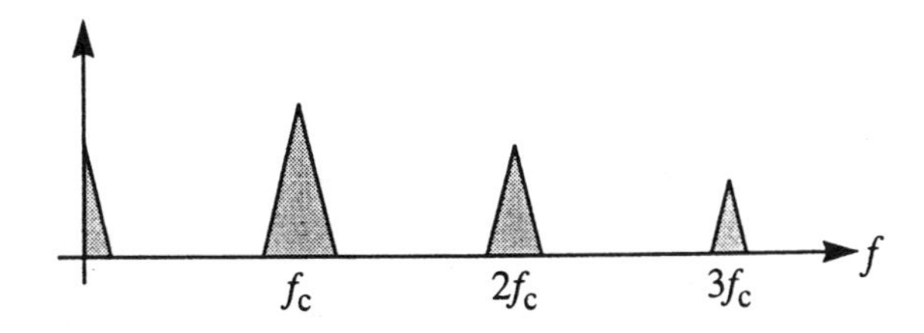

Fig. 1. Spectrum of a narrowband signal centered at a carrier frequency f_c after passing though a nonlinear circuit.

consists of clusters of frequencies near the harmonics of the carrier. These clusters take the form of a discrete set of frequencies if the modulation is periodic or quasiperiodic, and a continuous distribution of frequencies otherwise.

RF simulators exploit the "sparse" nature of this spectrum in various ways and with varying degrees of success. Steady-state methods are used when the spectrum is discrete, and transient methods are used when the spectrum is continuous.

B. Time-Varying Linear Nature of the RF Signal Path

Another important but less appreciated aspect of RF circuits is that they are generally designed to be as linear as possible from input to output to prevent distortion of the *modulation* or *information signal.* Some circuits, such as mixers, are designed to translate signals from one frequency to another. To do so, they are driven by an additional signal, the LO, a large periodic signal the frequency of which equals the amount of frequency translation desired. For best performance, mixers are designed to respond in a strongly nonlinear fashion to the LO. Thus, mixers behave both near-linearly (to the input) and strongly nonlinearly (to the LO).

Since *timing* or *synchronization signals*, such as the LO or the clock, are not part of the path of the information signal, they may be considered to be part of the circuit rather than an input to the circuit. This simple change of perspective allows the mixer to be treated as having a single input and a near-linear, though periodically time-varying, transfer function. As an example, consider a mixer made from an ideal multiplier and followed by a low-pass filter. A multiplier is nonlinear and has two inputs. Applying an LO signal of $\cos(\omega_{LO}t)$ consumes one input and results in a transfer function of

$$v_{out}(t) = \mathrm{LPF}\{\cos(\omega_{LO}t)v_{in}(t)\}, \tag{1}$$

which is clearly time-varying and is easily shown to be linear with respect to v_{in}. If the input signal is

$$v_{in}(t) = m(t)\cos(\omega_c t), \tag{2}$$

then

$$v_{out}(t) = \mathrm{LPF}\{m(t)\cos(\omega_c t)\cos(\omega_{LO}t)\} \tag{3}$$

and

$$v_{out}(t) = m(t)\cos((\omega_c - \omega_{LO})t). \tag{4}$$

This demonstrates that a linear periodically-varying transfer function implements frequency translation.

Often we can assume that the information signal is small enough to allow the use of a linear approximation of the circuit from its input to its output. Thus, a small-signal analysis can be performed, as long as it accounts for the periodically varying nature of the signal path, which is done by linearizing about the periodic operating point. This is the idea behind the small-signal analyses of Section III. Traditional simulators such as SPICE provide several small-signal analyses, such as the AC and noise analyses, that are considered essential when analyzing amplifiers and filters. However, they start by linearizing a nonlinear time-invariant circuit about a constant operating point, and so generate a linear time-invariant representation, which cannot exhibit frequency translation. By linearizing a nonlinear circuit about a periodically varying operating point, we extend small-signal analysis to circuits that must have a periodic timing signal present to operate properly, such as mixers, switched filters, samplers, and oscillators (for oscillators the timing signal is the desired output of the oscillator, while the information signal is generally an undesired signal, such as the noise). In doing so, a periodically varying linear representation results, which does exhibit frequency translation.

All of the traditional small-signal analyses can be extended in this manner, enabling a wide variety of applications (some of which are described in [36,42]). In particular, a noise analysis that accounts for noise folding and cyclostationary noise sources can be implemented [23,29], which fills a critically important need for RF circuits. When applied to oscillators, it also accounts for phase noise.

The linearity of the RF signal path can also be exploited to improve the convergence of the methods, as will be presented later.

C. Linear Passive Components

At the high frequencies present in RF circuits, the passive components, such as transmission lines, spiral inductors, packages (including bond wires) and substrates, often play a significant role in the behavior of the circuit. The nature of such components often make them difficult to include in the simulation.

Generally the passive components are linear and are modeled with phasors in the frequency-domain. This greatly simplifies the modeling of distributed components such as transmission lines, using either analytical expressions or tables of S-parameters. Large distributed structures, such as packages, spirals, and substrates, often interface with the rest of the circuit through a small number of ports. Thus, they can be easily replaced by a N-port macromodel that consists of the N^2 transfer functions. These transfer functions are found by reducing the large systems of equations that describe these structures, leaving only the equations that relate the signals at their ports. The relatively expensive reduction step is done once for each frequency as a preprocessing step. The resulting model is one that is efficient to evaluate if N is small. This is usually true for transmission lines and spirals, and less often true for packages and substrates.

Time-domain simulators are formulated to solve sets of first-order ordinary-differential equations (ODE). However, distributed components, such as transmission lines, are described with partial-differential equations (PDE) and so are problematic for time-domain simulators. Generally, the PDEs are converted to a set of ODE using some form of discretization [20]. Such approaches suffer from bandwidth limits. A more general approach is to compute the impulse response for a distributed component from a frequency domain description and use convolution to determine the response of the component in the circuit [11,31]. Evaluating lossy or dispersive transmission line models or tables of S-parameters with this approach is generally expensive and error-prone [33]. Packages, substrates and spirals can be modeled with large lumped networks, but such systems are too large to be efficiently incorporated into a time domain simulation, and so some form of reduction is necessary [6,25].

II. Numerical Solution of DAEs

If we ignore distributed components for the moment, the basic idea behind circuit simulation is to solve a nonlinear set of differential-algebraic equations (DAEs). We start by proposing the form of the solution as a linear combination of basis functions. A primary requirement of the basis functions is that we must know their derivatives. This allows us to analytically evaluate the derivatives in the DAE, thereby converting the DAE to a system of nonlinear algebraic equations that can be solved with Newton's method. Examples of possible basis functions are polynomials and sinusoids. The efficiency and accuracy of the simulation is dependent on the choice of basis functions. A careful choice of basis functions can result in tremendous performance advantages.

The choice of basis functions defines the solution space for the method. Note that this space may not contain good approximations to all of the solutions to the original DAE, in which case the method based on that choice of basis functions will not be able to find those solutions. This is used to advantage to avoid particular solutions that are either uninteresting or undesirable. For example, if the basis functions are chosen to be sinusoids, then only steady-state behavior is representable and so the initial transient behavior is avoided. Furthermore, the circuit can be simulated even if it is unstable.

The other way to avoid undesired solutions is to apply constraints on the solution in the form of initial or boundary conditions. In general, DAEs have an infinite continuum of solutions, and so constraints must be applied before they can be solved for a particular solution unless the solution space of the method contains only isolated solutions of the DAE.

A. *Transient Analysis*

Transient analysis breaks the time continuum into a series of adjacent short intervals and uses low-order polynomials as the basis functions over each interval (the time step) with the constraint that the solution must be continuous across interval boundaries (the time points). For example, consider the nonlinear DAE

$$f(v(t), t) = i(v(t)) + \frac{dq(v(t))}{dt} + u(t) = 0. \qquad (5)$$

While this equation is capable of modeling any lumped time-invariant nonlinear system, it is convenient to think it as being generated from nodal analysis, and so representing a statement of Kirchhoff's Current Law for a circuit containing nonlinear conductors, nonlinear capacitors, and current sources. In this case, $v(t) \in \mathcal{R}^N$ is the node voltage, $i(v(t)) \in \mathcal{R}^N$ represents the current out of the node from the conductors, $q(v(t))$ represents the charge out of the node from the capacitors, and $u(t)$ represents the current out of the node from the sources.

Consider trapezoidal rule. This is a second-order method that assumes that the solution and its first derivative are continuous at the interval boundaries. As such, they would be known at the start of the interval from the solution of the previous interval. With a small amount of algebra, one can show that if t_{s-1} is the initial point in the interval, t_s is the final point, and $h_s = t_s - t_{s-1}$, then for a second-order polynomial,

$$\frac{dq(v(t_s))}{dt} = \frac{2}{h_s}(q(v(t_s)) - q(v(t_{s-1}))) - \frac{dq(v(t_{s-1}))}{dt}. \qquad (6)$$

Substituting (6) into (5) converts it into a sequence of nonlinear algebraic systems of equations that can be solved with Newton's method to build up a pointwise approximation to the solution from some initial state.

Under certain mild conditions, it is possible to show that Newton's method will converge at every time point of a transient analysis. Newton's method is an iterative procedure that converges to the solution of a nonlinear system of equations, *if* the initial starting point is close to the final solution. As long as the solution trajectory for the DAE is continuous, which it must be except at a set of distinct points (usually small in number and often zero), and if the starting point used for the Newton iteration is an extrapolation from the values at previous time points, then one can always take a time step small enough for Newton's method to converge.

B. Harmonic Balance

Harmonic balance [15,21,26] uses harmonically related sinusoids for basis functions. As noted earlier, such basis functions cannot represent the initial transient, so harmonic balance directly computes the steady-state behavior of the circuit.

For simplicity, we will use phasors (complex exponentials) as basis functions rather than sines and cosines. We will also treat the time-domain signals as being complex valued and simply ignore the imaginary parts. While this is not as efficient as if we assumed real-valued signals, it results in a simpler presentation. For more efficient implementations, see [15,18,27].

With phasors as basis functions, harmonic balance has a natural ability to incorporate the frequency-domain descriptions of distributed components. Thus, we formulate a new test problem that is similar to (5), except that the signals are complex valued and another term is added to model distributed components. In addition, we assume that the circuit is driven (i.e. that $u(t)$ is not constant).

$$i(v(t)) + \frac{dq(v(t))}{dt} + \int_{-\infty}^{t} y(t-\tau)v(\tau)d\tau + u(t) = 0 \quad (7)$$

or

$$f(v, t) = 0, \quad (8)$$

where $v(t)$, $u(t)$, $y(t)$, $i(v(t))$, $q(v(t))$, and $f(v, t) \in C^N$, $v(t)$ and $u(t)$ are assumed to be T-periodic, $y(t)$ is the impulse response of the linear components, while i and q now represent only the nonlinear components.

Harmonic balance assumes that both v and f of (8) are formulated as Fourier series,

$$x(t) = \sum_{k=-\infty}^{\infty} X(k)e^{j\omega_k t}, \quad (9)$$

where $\omega_k = k\lambda$ and $\lambda = 2\pi/T$ is the fundamental frequency. Now (8) is rewritten as

$$\sum_{k=-\infty}^{\infty} F(V, k)e^{j\omega_k t} = 0, \quad (10)$$

where

$$F(V, k) = j\omega_k Q(V, k) + I(V, k) + Y(k)V(k) + U(k) \quad (11)$$

Since $e^{j\omega_{k_1}}$ and $e^{j\omega_{k_2}}$ are linearly independent if $k_1 \neq k_2$, $F(V, k) = 0$ for each k individually. In vector form,

$$F(V) = I(V) + \Omega Q(V) + YV + U = 0, \quad (12)$$

where $F(V), I(V), Q(V), V, U \in C^{(2K+1)N}$ are vectors of vectors. Each is composed of N vectors that represent the spectrum at each node. Ω, Y are block matrices. Ω is block diagonal and

$$\Omega_{nn} = \mathrm{diag}\{-j\omega_K, \ldots, 0, \ldots, j\omega_K\} \quad (13)$$

is the frequency-domain differentiation operator and $j = \sqrt{-1}$. Y_{mn} is the Laplace transform of y_{mn} evaluated at $j\omega_k$ for each k. It is diagonal because the components modeled by y are time-invariant.

While it is possible to model some types of nonlinearities directly in the frequency domain [34], this is not practical or desirable in all cases. Instead, a procedure is employed where V is first converted into the time domain at evenly spaced sample points using the Inverse Discrete Fourier Transform (IDFT),

$$x(s) = \frac{1}{K}\sum_{k=-K}^{K} X(k)e^{j2\pi ks/S}. \quad (14)$$

where $s = 0, 1, \ldots, S-1$ and $S \geq S_{\min} = 2K+1$. While S may be set equal to $S_{\min}$, typically it is chosen in the range $2S_{\min} < S < 10S_{\min}$ to reduce aliasing. For each time point, both $i(v(s))$ and $q(v(s))$ are evaluated, and the result converted back into the frequency domain using the forward Discrete Fourier Transform (DFT),

$$X(k) = \sum_{s=0}^{S-1} x(s)e^{-j2\pi ks/S}. \quad (15)$$

Both the DFT and the IDFT can be written a matrix operations [15]:

$$X = \mathcal{F}x, \quad (16)$$

$$x = \mathcal{F}^{-1}X, \quad (17)$$

where $\mathcal{F}$ represents the DFT and $\mathcal{F}^{-1}$ of represents the IDFT. (16) is a restatement of (15) and (17) is a restatement of (14). If $S > S_{\min}$, these matrices are not square. Now

$$v = [v_n] = [\mathcal{F}^{-1}V_n], \quad (18)$$

$$I_m(V) = \mathcal{F}i_m(v), \quad (19)$$

$$Q_m(V) = \mathcal{F}q_m(v). \quad (20)$$

Applying Newton-Raphson to solve (12) results in the iteration

$$J(V^{(r)})(V^{(r+1)} - V^{(r)}) = -F(V^{(r)}), \quad (21)$$

where r is the iteration number and

$$J(V) = \frac{\partial F(V)}{\partial V} = \frac{\partial I(V)}{\partial V} + \Omega\frac{\partial Q(V)}{\partial V} + Y \quad (22)$$

is the *harmonic Jacobian*.

$$J(V) = [J_{mn}(V)] = \left[\frac{\partial F_m(V)}{\partial V_n}\right], \quad (23)$$

where

$$\frac{\partial F_m(V)}{\partial V_n} = \frac{\partial I_m(V)}{\partial V_n} + \Omega_{mm}\frac{\partial Q_m(V)}{\partial V_n} + Y_{mn} \qquad (24)$$

is a *conversion matrix*. The derivation of $\partial I_m(V)/\partial V_n$ follows from the chain rule.

$$I_m(V) = \mathcal{F} i_m(v) \qquad (25)$$

$$\frac{\partial I_m(V)}{\partial V_n} = \mathcal{F}\frac{\partial i_m(v)}{\partial v_n}\frac{\partial v_n}{\partial V_n} \qquad (26)$$

Since $i(v)$ is algebraic, $\partial i_m/\partial v_n$ is diagonal.

$$\frac{\partial i_m(v)}{\partial v_n} = \operatorname{diag}\{g_{mn}(v(t_s))\}, \qquad (27)$$

where $g_{mn}(v(t)) = \partial i_m(v(t))/\partial v_n(t)$ is the conductance waveform for the nonlinear resistors.

From $v_n = \mathcal{F}^{-1}V_n$,

$$\frac{\partial I_m(V)}{\partial V_n} = \mathcal{F}\frac{\partial i_m(v)}{\partial v_n}\mathcal{F}^{-1}. \qquad (28)$$

$\partial Q_m/\partial V_n$ is derived and constructed from c_{mn} in a similar manner, where $c_{mn}(v(t)) = \partial q_m(v(t))/\partial v_n(t)$ is the capacitance waveform for the nonlinear capacitors.

Efficiency: $J(V)$ is big and relatively dense, and so expensive to store and factor. To reduce the cost of solving (21), matrix-implicit methods are used. In these methods, only the component pieces of the Jacobian (g, c, and Y) are stored and iterative linear equation solvers such as the Krylov-subspace methods [30] are employed. These methods solve this linear system of equations by evaluating a sequence of matrix-vector products (MVP) that involve the Jacobian. The MVPs can be evaluated with nearly linear time and storage by exploiting the structure of (28) and by using fast algorithms such as the FFT to implement $\mathcal{F}$ and $\mathcal{F}^{-1}$ [19]. Unfortunately, these iterative linear solvers are not guaranteed to converge and require the use of a preconditioner to improve their convergence. The original linear system of equations

$$Ax = b \qquad (29)$$

is preconditioned by multiplying both sides by $\hat{A}^{-1}$. One then applies the iterative solver to the preconditioned system

$$\hat{A}^{-1}Ax = \hat{A}^{-1}b. \qquad (30)$$

Generally, the preconditioner $\hat{A}$ is chosen to be a close approximation to A that is also easy to invert. For mildly nonlinear problems, constructing $\hat{A} = J(V_{DC})$ by linearizing about the DC operating point and performing a simple AC analysis at each ω_k is an effective and efficient choice [9,19,37]. In this case, $\hat{A}$ is the Jacobian generated by the Gauss-Jacobi-Newton harmonic relaxation algorithm [15]. This preconditioner is not sufficient for strongly nonlinear problems. To handle these problems, it is necessary to adaptively prune the full harmonic Jacobian as in the harmonic-relaxation Newton algorithm [15,18].

When using the matrix-implicit methods, harmonic balance requires roughly $O(NK)$ operations, where N is the number of circuit equations and K is the number of frequencies required to accurately represent both v and f. This does not include the operations required to precondition the system of equations, which on strongly nonlinear problem may be far from negligible.

Extensions: An extremely important application of harmonic balance is determining the steady-state behavior of oscillators. However, as presented, harmonic balance is not suitable for autonomous circuits such as oscillators. The method was derived assuming the circuit was driven, which made it possible to know the operating frequency in advance. Instead, it is necessary to modify harmonic balance to directly compute the operating frequency [15].

Applications: Harmonic balance is generally used to predict the distortion of RF circuits. It is also used to compute the operating point about which small-signal analyses are performed (presented later). When applied to oscillators, it is used to predict the operating frequency and power, and can also be used to determine how changes in the load affect these characteristics (load pull).

Its space of application is similar to that of shooting methods (presented next). It is preferred over shooting methods when the circuit includes distributed components.

C. Shooting Methods

Transient analysis solves initial-value problems. A shooting method is an iterative procedure layered on top of transient analysis that is designed to solve boundary-value problems. Boundary-value problems play an important role in RF simulation. For example, assume that (5) is driven with a non-constant T-periodic stimulus. The T-periodic steady state solution is the one that also satisfies the two-point boundary constraint,

$$v(T) - v(0) = 0. \qquad (31)$$

If the initial state $v(t_0)$ is known, then transient analysis can solve (5) and compute the state as some later time t_1. In general, one writes

$$v(t_1) = v(t_0) + \phi(v(t_0), t_0, t_1) \qquad (32)$$

where ϕ is the state transition function for the differential equation. Shooting methods reformulate (5) and (31) as

$$\phi_T(v(0), 0) = 0, \qquad (33)$$

where $\phi_T(v(t_0), t_0) = \phi(v(t_0), t_0, T + t_0)$, which is a nonlinear algebraic problem and so standard Newton methods can be used to solve for $v(0)$. We refer to the combination of the Newton and shooting methods as the shooting-Newton algorithm.

When applying Newton's method directly to (33), it is necessary to compute both the response of the circuit over one period and the sensitivity of the final state $v(T)$ with respect to changes in the initial state $v(0)$. The sensitivity is used to determine how to correct the initial state to reduce the difference between the initial and final state.

Applying Newton's method to (33) results in the iteration

$$v_0^{(r)} = v_0^{(r-1)} - [J_\phi(v_0^{(r-1)}) - 1]^{-1}[\phi_T(v_0^{(r-1)}, 0) - v_0^{(r-1)}] \tag{34}$$

where r is the iteration number, $v_0 = v(0)$, 1 is the identity matrix, and

$$J_\phi(v_0) = \frac{d\phi_T(v_0, 0)}{dv_0} = \frac{dv(T)}{dv_0}. \tag{35}$$

There are two important pieces to the computation of the Newton iteration given in (34): evaluating the state-transition function $\phi_T(v_0, 0)$, and forming and factoring the matrix $J_\phi(v_0)$, which is a dense matrix in general.

The state-transition function is computed by integrating (5) numerically over the shooting interval. The derivative of the state-transition function, referred to as the sensitivity matrix, is computed simultaneously because there are several quantities that are common to both computations [1,15,35,37].

Efficiency: Forming J_ϕ requires $O(N^2S)$ operations where S is the number of time points used to evaluate ϕ_T. Factoring $J_\phi - 1$ requires $O(N^3)$ operations. As a result, forming and factoring $J_\phi - 1$ becomes intractable when N exceeds several hundred. As with harmonic balance, matrix-implicit Krylov-subspace methods are used to avoid forming and factoring $J_\phi - 1$ [35]. Again, the component pieces of $J_\phi - 1$ are saved and the matrix-vector products are performed on the fly. The component pieces are

$$J_f(v(t_s)) = G(v(t_s)) + \frac{C(v(t_s))}{h_s} \tag{36}$$

and $C(v(t_s))$ at each time point s, where

$$G(v(t)) = \frac{di(v(t))}{dv(t)}, \tag{37}$$

$$C(v(t)) = \frac{dq(v(t))}{dv(t)}, \tag{38}$$

and $h_s = t_s - t_{s-1}$. A natural preconditioner is applied simply by saving and applying J_f in LU factored form [37]. This preconditioner has proven itself to be extremely robust.

Convergence: Newton's method is applied both in the outer loop to solve (33) and in the inner loop to solve (5) at each time point, making this a multi-level Newton method. As described before, the ability to adjust the time step during transient analysis results in Newton's method being extremely reliable in the inner loop. The outer loop is also quite reliable because ϕ_T is generally near linear as a direct result of RF circuits having near linear signal paths. Thus, shooting-Newton represents a well-designed multi-level Newton method where the inner loop is robust and shields the outer loop from the nonlinearity inherent in the problem.

Extensions: As with harmonic balance, it is extremely important to be able to determine the steady-state behavior of oscillators. To do so it is necessary to modify shooting methods to directly compute the period of the oscillator [15].

Applications: Shooting methods are applied in the same situations as harmonic balance. It is generally preferred if the circuit is driven with strongly discontinuous signals (pulses as opposed to sinusoids). As such, shooting methods are well suited for simulating switching mixers, switched filters, samplers, frequency dividers, and relaxation oscillators as long as the circuits do not contain distributed components.

III. Small-Signal Analysis

Consider a circuit whose input is the sum of two periodic signals, $u(t) = u_L(t) + u_s(t)$, where $u_L(t)$ is an arbitrary periodic waveform with period T_L and $u_s(t)$ is a sinusoidal waveform of radial frequency ω_s whose amplitude is small.

Let $v_L(t)$ be the steady-state solution waveform when $u_s(t)$ is zero. Then allow $u_s(t)$ to be small, but nonzero. We can consider the new solution $v(t)$ to be a perturbation $v_s(t)$ on $v_L(t)$, as in $v(t) = v_L(t) + v_s(t)$. The small-signal solution $v_s(t)$ is computed by linearizing the circuit about $v_L(t)$ and applying one of the methods for finding the steady-state solution already described. From the theory of periodically time-varying systems [5,40], it is known that for

$$u_s(t) = U_s e^{j\omega_s t} \tag{39}$$

the steady-state response is given by

$$v_s(t) = \sum_{k=-\infty}^{\infty} V_s(k) e^{j(\omega_s + k\lambda)t} \tag{40}$$

where $\lambda = 2\pi/T_L$ is the large signal fundamental frequency. $V_s(k)$ represents the sideband for the kth harmonic of V_L. In this situation, there is only one sideband per harmonic because U_s is a single frequency complex exponential and the circuit is linear. This representation has terms at negative frequencies. If these terms are mapped to positive frequencies, then the sidebands with $k < 0$ become lower sidebands of the harmonics of v_L and those with $k > 0$ become upper sidebands.

$V_s(k)/U_s$ is the transfer function for the input at ω_s to the output at $\omega_s + k\lambda$. Notice that with periodically-varying linear systems there are an infinite number of transfer functions between any particular input and output. Each represents a different frequency translation.

Extensions: In the next few sections, the basic techniques are introduced that are used to compute the small-signal steady state response of a circuit linearized about a time varying operating point. This is sufficient for performing a time-varying AC analysis and can be extended to other types of small-signal analyses, such as computing the S-parameters of the circuit. These small-signal analyses are also extendable to cyclostationary noise analysis [3,7], which is an extremely important capability for RF designers [36].

These methods can also be extended so as to allow a small-signal analysis about a quasiperiodic operating point or small-signal analysis of autonomous circuits.

Applications: Small-signal analyses are tremendously useful for computing transfer functions (such as conversion gain and supply rejection) and predicting noise performance. In this way, they are similar to the AC and noise analyses in SPICE, but they can be applied to all kinds of circuits that the traditional small-signal analyses cannot, such as mixers, switched-filters, and samplers. Because they compute transfer functions in the presence of large signals, they can be used to determine the degradation of gain and noise when there is a large interfering signal on an adjacent channel. They can also be used to estimate intermodulation distortion [36].

When performing noise analysis, they accurately capture noise folding present in all circuits that have time-varying operating points, but which is especially important in mixers, samplers, and oscillators. In addition, they can be used to predict the phase noise of oscillators.

A. Transient Small-Signal Analysis

Though beyond the scope of this paper, there has been recent progress on performing small-signal noise analysis about a transient operating point [3]. However, in most cases, it is the small-signal behavior when the circuit is in steady state that is of interest. Using these methods generally requires integrating through any initial transient.

B. Harmonic Balance Small-Signal Analysis

Consider (12) where $U = U_L + U_s$ and $V = V_L + V_s$. Initially set $U_s = 0$ and solve (12) for V_L such that

$$F(V_L) = I(V_L) + \Omega Q(V_L) + YV_L + U_L = 0. \quad (41)$$

Since U_s is expected to be small, we can expand about V_L using a Taylor series truncated to first order.

$$\frac{\partial I(V_L)}{\partial V}V_s + \Omega_s\frac{\partial Q(V)}{\partial V}V_s + YV_s + U_s \approx 0 \quad (42)$$

or

$$J(V_L)V_s \approx -U_s \quad (43)$$

where $J(V_L)$ is computed as in (22) and Ω_s is as in (13) except with $\omega_k = \omega_s + k\lambda$.

Solving (43) for V_s gives the sidebands generated by U_s.

These ideas are extended to cyclostationary noise analysis in [10,12,29].

C. Shooting Method Small-Signal Analysis

Consider the circuit described by (5). Assume $u_s(t) = 0$ and let $v_L(t)$ be the resulting solution that also satisfies the two-point constraint (31). Linearizing (5) about $v_L(t)$ yields a time-varying linear system

$$\frac{dC(v_L(t))}{dt}v_s(t) + G(v_L(t))v_s(t) + u_s(t) = 0 \quad (44)$$

that can be solved for $v_s(t)$ if $u_s(t)$ is small.

From (40) it follows that

$$v_s(t + T_L) = v_s(t)e^{j\omega_s T_L}. \quad (45)$$

Equation (45) in the periodically time-varying linear steady-state problem is analogous to (31) in the standard steady-state problem. It is solved using a modified shooting method [22,38]. Note that (45) implies that the entire small-signal steady-state response of the periodic time-varying system is determined by the behavior of $v_s(t)$ on any interval of length T_L.

Since the solution is computed by performing a time-domain simulation, there is an upper bound on the analysis frequency that is imposed by the largest time step used. The period of analysis frequency must be much larger than the largest time step. Typically, the time steps are the same ones used when computing the large signal operating point.

Standard shooting methods can be used if we form $x_s(t) = v_s(t)e^{-j\omega_s t}$. Then, (45) becomes

$$x_s(t + T_L) = x_s(t), \quad (46)$$

which is the same form as (31) if $t = 0$. Now conventional shooting methods are applied to find x_s. Once known, Fourier analysis [16] is applied to find X_k, from which V_k is trivially computed. This is a different, but equivalent, way of deriving the method given in [22,38].

This procedure is applied to cyclostationary noise analysis in [23,38].

IV. Multi-Variate Methods

An interesting approach to solving narrowband problems uses a *multi-variate* representation of the signals and reformulates (5) as a partial-differential equation. Consider a simple two-tone quasiperiodic signal

$$x(t) = \cos(\omega_1 t)\cos(\omega_2 t), \tag{47}$$

where ω_1 is much smaller than ω_2. A large number of time points would be needed to accurately approximate such a signal. For example, if 15 points per period were needed to accurately represent the highest frequency, and if $\omega_2 = 1000\,\omega_1$, then 15,000 points are needed to accurately represent $x(t)$. Now consider a bi-variate representation obtained by replacing t with t_1 in the slowly varying parts of the expression, and t with t_2 in the rapidly varying parts. Then

$$\hat{x}(t_1, t_2) = \cos(\omega_1 t_1)\cos(\omega_2 t_2). \tag{48}$$

$x(t)$ is easily recovered from $\hat{x}(t_1, t_2)$ simply by setting $t = t_1 = t_2$. In this case, a grid of $15 \times 15 = 225$ points are needed to accurately represent $\hat{x}(t_1, t_2)$.

Using these ideas, we can replace (5) with

$$\frac{\partial q(\hat{v}(t_1, t_2))}{\partial t_1} + \frac{\partial q(\hat{v}(t_1, t_2))}{\partial t_2} + i(\hat{v}(t_1, t_2)) + \hat{u}(t_1, t_2) = 0 \tag{49}$$

or

$$f(\hat{v}(t_1, t_2), t_1, t_2) = 0. \tag{50}$$

A wide variety of methods are formulated from this equation by proposing different forms of the solution and different boundary conditions along each time axis. One can apply transient, harmonic balance, and shooting methods in layers and customize methods to particular classes of problems [2,28].

A. Quasiperiodic Harmonic Balance

If a circuit is driven with two signals at unrelated fundamental frequencies, it generally responds in steady state by generating quasiperiodic (also known as polyperiodic) signals. Quasiperiodic signals have the form of a Fourier series that is generalized in that the frequencies of the sinusoids are not just integer multiples of a single fundamental frequency. Instead, they are the linear combination of integer multiples of several fundamentals. For example, if there are two fundamentals, then

$$x(t) = \sum_{k_1=-\infty}^{\infty} \sum_{k_2=-\infty}^{\infty} X(k_1, k_2) e^{j(k_1\lambda_1 + k_2\lambda_2)t}, \tag{51}$$

where λ_1 and λ_2 are the fundamental frequencies.

The response contains components at multiples of each fundamental frequency as well as at the sums and differences of these frequencies. Rearranging (51) allows us to see this as being equivalent to constructing the waveform as a conventional Fourier series where the sinusoidal frequencies are at integer multiples of λ_1, except that the Fourier coefficients themselves are time-varying. In particular, the coefficient $\tilde{X}(t, k_1)$ is periodic with period $T_2 = 2\pi/\lambda_2$ and can itself be represented as a Fourier series in t_2.

$$x(t) = \sum_{k_1=-\infty}^{\infty} \underbrace{\sum_{k_2=-\infty}^{\infty} X(k_1, k_2) e^{jk_2\lambda_2 t}}_{\tilde{X}(t, k_1)} e^{jk_1\lambda_1 t} \tag{52}$$

Convert to a bi-variate representation by associating t_1 with λ_1 and t_2 with λ_2.

$$\hat{x}(t_1, t_2) = \sum_{k_1=-\infty}^{\infty} \sum_{k_2=-\infty}^{\infty} X(k_1, k_2) e^{jk_2\lambda_2 t_2} e^{jk_1\lambda_1 t_1} \tag{53}$$

This is a two-dimensional Fourier series, and so $\hat{x}$ and X are related by a two-dimensional Fourier transform.

Assuming v and f of (50) take the form of (53) (a linear combination of periodically AM modulated sinusoids) results in

$$\sum_{k_1=-\infty}^{\infty} \sum_{k_2=-\infty}^{\infty} F(V, k_1, k_2) e^{jk_1\lambda_1 t_1 + jk_2\lambda_2 t_2} = 0, \tag{54}$$

where

$$F(V, k_1, k_2) = j(k_1\lambda_1 + k_2\lambda_2)Q(V, k_1, k_2) + I(V, k_1, k_2) + U(k_1, k_2) \tag{55}$$

As with periodic harmonic balance, the terms in (54) are linearly independent, so $F(V, k_1, k_2) = 0$ for each k_1, k_2. In vector form,

$$F(V) = I(V) + \Omega Q(V) + U = 0, \tag{56}$$

where Ω is generalized such that $\omega_k = k_1\lambda_1 + k_2\lambda_2$. This becomes finite-dimensional by bounding $|k_j| \le K_j$. This is similar to (12), except the term for the distributed linear components (YV) is missing. This term can be included in (56) without difficulty because the spectrum of V is discrete. When evaluating I, and Q the multidimensional discrete Fourier transform is used.

Using a multidimensional Fourier transform is just one way of formulating harmonic balance for quasiperiodic problems. It is used here because of its simple derivation and because it represents an early application of multivariate methods [27,43]. An alternate approach that is generally preferred in practice is the false frequency method, which is based on a one-dimensional Fourier transform [8,15].

Parametric Harmonic Balance: Harmonic balance as described above exploits the sparse spectrum of modulated carrier signals and the linear nature of the passive components. However, it only incidentally exploits the nature of RF circuits to respond in a near-linear manner to the input signal (more effective preconditioner, fewer Newton iterations). Parametric harmonic balance [39] is a variation that retains the advantages of standard harmonic balance, but also exploits the near-linear nature of quasiperiodic RF circuits in a more deliberate manner. It splits harmonic balance into two phases, one to compute the response to the large periodic drive signal alone (ex., the LO in a mixer), and a second phase computes the response from the input as a perturbation of the solution computed in the first phase. The advantage of this method is that it solves for many fewer frequencies in the first phase when convergence is a struggle. In addition, information generated in the first phase can be used to accelerate the second phase. In this regard, parametric harmonic balance is similar to several of the methods presented later. Unfortunately, further description of this method is beyond the scope of this article.

Extensions: One obvious extension is to allow more than two fundamentals, though the method becomes expensive fast as the number of fundamentals increases, so in practice only two or three fundamentals are used. Another important extension would be to support autonomous or semi-autonomous circuits [43]. An example of an autonomous quasiperiodic circuit is one that consists of two coupled oscillators. An example of an semi-autonomous circuit is a free-running oscillator connected to a mixer with a driven input.

Applications: Quasiperiodic harmonic balance is used predict intermodulation distortion of narrowband circuits and both harmonic and intermodulation distortion of mixers. Both it and the mixed frequency-time method are used in similar situations. Quasiperiodic harmonic balance is preferred when the circuit includes distributed components.

B. Circuit Envelope Method

In the last section, the concept of harmonic balance with time-varying Fourier coefficients was introduced. In that case, the Fourier coefficients were assumed to be periodic, with the result that signals themselves were quasiperiodic. With circuit envelope [32], the Fourier coefficients are again time varying, but are not necessarily periodic. Instead, the Fourier coefficients $\tilde{X}(t, k)$ are taken to be transient waveforms. Thus, signals take the form

$$x(t) = \sum_{k=-\infty}^{\infty} \tilde{X}(t_1, k)e^{j\omega_k t_2}, \tag{57}$$

where $\omega_k = k\lambda$ and λ is the fundamental frequency of the base Fourier series. $\tilde{X}(t_1, k)$ represents the complex modulation of the kth harmonic.

Now, (50) can be rewritten in the form of (57),

$$\sum_{k=-\infty}^{\infty} \tilde{F}(\tilde{V}(t_1), t_1, k)e^{j\omega_k t_2} = 0, \tag{58}$$

where

$$\tilde{F}(\tilde{V}(t_1), t_1, k) = \frac{d\tilde{Q}(\tilde{V}(t_1), k)}{dt} + j\omega_k\tilde{Q}(\tilde{V}(t_1), k) + \tilde{I}(\tilde{V}(t_1), k) + \tilde{U}(t_1, k) \tag{59}$$

Because $e^{j\omega_{k_1}t_2}$ is linearly independent of $e^{j\omega_{k_2}t_2}$ if $k_1 \neq k_2$, and because $\tilde{F}(\tilde{V}(t_1), t_1, k)$ is independent of t_2, each term in (58) is zero independently of the others. In other words, $\tilde{F}(\tilde{V}(t), t, k) = 0$ for each k, or in vector form

$$\tilde{F}(\tilde{V}(t), t) = \frac{d\tilde{Q}(\tilde{V}(t))}{dt} + \Omega\tilde{Q}(\tilde{V}(t)) + \tilde{I}(\tilde{V}(t)) + \tilde{U}(t) = 0 \tag{60}$$

As with transient analysis, discretization methods such as trapezoidal rule or the backward difference formulae replace dQ/dt with a finite-difference approximation, converting (60) to a system of nonlinear algebraic equations that is solved with Newton's method. For example, applying backward Euler converts (60) to

$$\frac{\tilde{Q}(\tilde{V}(t_s)) - \tilde{Q}(\tilde{V}(t_{s-1}))}{t_s - t_{s-1}} + \Omega\tilde{Q}(\tilde{V}(t_s)) + \tilde{I}(\tilde{V}(t_s)) + \tilde{U}(t_s) = 0 \tag{61}$$

One of the important strengths of harmonic balance is its ability to easily incorporate frequency-domain models for the linear components such as lossy or dispersive transmission lines. Unfortunately, this is not true with the circuit envelope method. The transient nature of the modulations $\tilde{V}(t, k)$ introduces the same difficulties that are present with distributed components in transient analysis, which are addressed using similar techniques. In particular, one can use convolution [11,31], or the model for the distributed component can be separated into delay and dispersion, with the dispersion being replaced by a lumped approximation [33].

As the distributed components are linear, the sidebands for each harmonic k can be treated individually. Thus, a separate model is generated for each harmonic k, which greatly reduces the bandwidth requirements on the models. The model for each harmonic must only be valid over the bandwidth of the sidebands associated with that harmonic. In RF circuits, the bandwidths of the sidebands are usually small relative to the carrier frequency, and so generating models of distributed components for use in the circuit envelope method is much easier than for conventional transient analysis. In fact, it is not uncommon for the bandwidth of an RF circuit to be so small that the transfer function of a distributed component does not change appreciably over the bandwidth of the sidebands. In this case, the transfer function is taken to be constant. In other words, $Y_{mn}(\omega, k)$ is simply replaced with $Y_{mn}(\omega_k, k)$ where $\omega = \omega_k + \delta\omega$.

Extensions: One can build circuit envelope on top of a quasiperiodic harmonic balance rather than on top of periodic harmonic balance as has been done here. One can also build it on top of autonomous harmonic balance.

One can also wrap shooting around circuit envelope to find a periodic or quasiperiodic modulation waveform, which results the multi-variate mixed frequency-time method [28].

Applications: Circuit envelope has two primary applications. The first is predicting the response of a circuit when it is driven with a complex digital modulation. An important problem is to determine the interchannel interference that results from intermodulation distortion. Simple intermodulation tests involving a small number of sinusoids as can be performed with quasiperiodic harmonic balance is not a good indicator of how the nonlinearity of the circuit couples digitally modulated signals between adjacent channels. Instead, one must apply the digital modulation, simulate with circuit envelope, and then determine how the modulation spectrum spreads into adjacent channels.

Another important application of circuit envelope is to predict the long term transient behavior of certain RF circuits. Examples include the turn-on behavior of oscillators, power supply droop or thermal transients in power amplifiers, and the capture and lock behavior of phase-locked loops. Another important example is determining the turn-on and turn-off behavior of TDMA transmitters. In TDMA (time-division multiple access), transmitters broadcast during a narrow slice of time. During that interval the transmitter must power up, stabilize, send the message, and then power down. If it powers up and down too slowly, the transmitter does not work properly. If it powers up and down too quickly, the resulting spectrum will be too wide to fit in the allotted channel. Simulating with traditional transient analysis would be prohibitively expensive because the time slice lasts on the order of 100 ms and the carrier frequency is typically at 1 GHz or greater.

The capabilities of circuit envelope are similar to envelope following (presented next). Circuit envelope is preferred when the circuit contains distributed components.

V. Sampling Methods

RF circuits are generally influenced by one periodic timing signal, often referred to as the LO or the clock, and one or more information signals. For oscillators, the timing signal is the oscillation signal itself and the information signal is generally noise. With sampling methods, we designate the one timing signal as the *clock*. If there is more than one timing signal, then usually the largest and fastest is chosen to be the clock. The discrete sample-envelope for a signal x is defined as signal x_{env} that results when x is sampled with a period equal to that of the clock, as shown in Figure 2. The continuous sample-envelope is the trajectory that is traced out if the phase of both the clock and the sampling is allowed to drift relative to the other signals present. The sampling is assumed to always occur at the same phase of the clock.

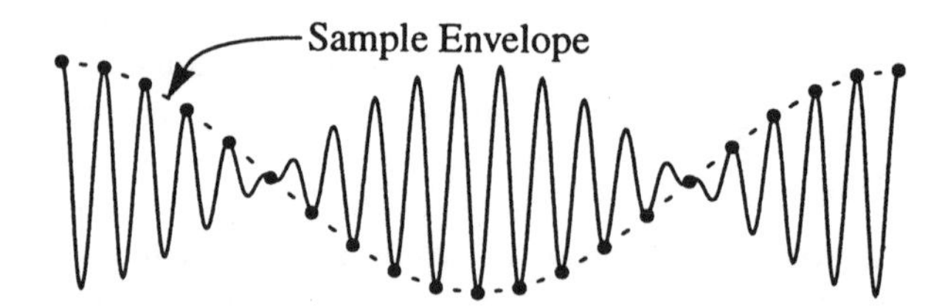

Fig. 2. Sample envelope is the waveform traced out when signal is sampled with a period equal to that of the clock.

The sample-envelope methods apply shooting methods over one or more clock cycles using boundary conditions that are formulated based on assumptions about the sample envelope. One can apply a wide variety of constraints on the sampled envelope, which results in a plethora of methods. For example, assuming the sampled envelope is constant results in simple shooting with a periodic boundary constraint. Assuming the circuit is linear and the sample envelope is sinusoidal results in periodic small-signal analysis. However, more interesting methods result when we make different assumptions about the sample envelope.

A. Envelope Following

Envelope Following approximates the sample envelope as a piecewise polynomial [13,24] in a manner that is analogous to conventional transient analysis. This approach is efficient if the sequence formed by sampling the state at the beginning of each clock cycle, $v(0)$, $v(T)$, $v(2T)$, ..., $v(mT)$, changes slowly as a function of m. A "differential-like" equation is formed from (32)

$$\Delta v(mT) = \phi_T(v(mT), mT), \quad (62)$$

where $\Delta v(mT) = v((m+1)T) - v(mT)$ is a measure of the time-derivative of the sample envelope at mT. We can apply traditional integration methods to compute an approximation to the solution using a procedure that involves solving (62) at isolated time points. If the sample envelope is accurately approximated by a low-order polynomial, then this procedure should allow us to skip many cycles, and so find the solution over a vast number of cycles in an efficient manner. For RF circuits, (62) is stiff and so requires implicit integration methods such as trapezoidal rule, which can be written as

$$\Delta v(mT) \approx \frac{2}{l}(v(mT) - v((m-l)T)) - \Delta v((m-l)T), \quad (63)$$

where l is the time step, which is measured in terms of cycles. This equation represents a 2-point boundary constraint on (62), and so together they can be solved with shooting-Newton methods to find $v(mT)$. As with transient analysis, once $v(mT)$ is computed, it is necessary to check that the trajectory is following the low-order polynomial as assumed. If not, the point should be discarded and the step l should be reduced. If desired, other integration methods can be used, such as the backward-difference formulae.

Extensions: Envelope following can be extended to support autonomous circuits such as oscillators.

One can wrap shooting around envelope following to find a periodic or quasiperiodic envelope, which results the hierarchical shooting method [28].

Applications: The applications of envelope following are the same as circuit envelope. Envelope following is preferred when the carrier signal is strongly discontinuous (consists of pulses rather than sinusoids). As such, envelope following is suitable for simulating switched filters and switching power supplies in addition to the traditional RF circuits.

B. Mixed Frequency-Time Method

The Mixed Frequency-Time (or MFT) Method [14,15] makes the assumption that the sample envelope can be accurately approximated by a Fourier series with K terms (excluding DC), where K is presumed to be small. If true, then once the value of $S = 2K+1$ distinct points along the sample envelope are known, then all points can be found. Specifically, once the S points are known, then the $2K+1$ Fourier coefficients can be determined using the DFT, and then the resulting Fourier series can be evaluated to determine the value of any point. In particular, let

$$\mathcal{V}_0 = [v(t_1), v(t_2), \ldots, v(t_S)] \quad (64)$$

be the value of S points, then we can find the value of the S points that follow them by one cycle,

$$\mathcal{V}_T = [v(t_1 + T), v((t_2) + T), \ldots, v(t_S + T)] \quad . (65)$$

Since the DFT is a linear operator $\mathcal{F}$, there exists a linear operator $D_T = \mathcal{F}^{-1} e^{j\omega_k T} \mathcal{F}$ that maps $\mathcal{V}_0$ to $\mathcal{V}_T$.

$$\mathcal{V}_T = D_T \mathcal{V}_0 \quad (66)$$

This is SN equations with $2SN$ unknowns. They were formulated purely from the constraints on the sample envelope. It represents a boundary condition on solution to (5) in a way analogous to (31) and (45). Designate Φ_T as the collection of S state transition functions from t_s to t_s+T. Then

$$\mathcal{V}_T = \Phi_T(\mathcal{V}_0) \quad (67)$$

Applying (66) to (67) gives

$$D_T \mathcal{V}_0 - \Phi_T(\mathcal{V}_0) = 0, \quad (68)$$

which can be solved using Newton's method for $\mathcal{V}_0$. As pointed out earlier, from $\mathcal{V}_0$ one can calculate any point on the sample envelope. Any point $v(t)$ on the original waveform is found by integrating (5) from the appropriate point on the sample envelope.

Extensions: As with quasiperiodic harmonic balance, the mixed frequency-time method can be extended to an arbitrary number of fundamentals, though in practice is limited to two or three. It can also be extended to handle autonomous and semi-autonomous circuits.

Applications: The applications of the mixed frequency-time method are the same as quasiperiodic harmonic balance. The mixed frequency-time method is preferred when the carrier signal is strongly discontinuous (consists of pulses rather than sinusoids).

VI. Comparisons

The methods presented can be grouped into two broad families, those methods based on harmonic balance, and those based on shooting methods. Table I shows how the various methods are related. to baseband methods, and to each other.

Most of the differences between the methods emanate from the attributes of the base methods (harmonic balance and shooting methods). So only the base methods will be compared.

Progress on developing RF simulation methods has been rapid over the last decade and is continuing. As a result,

TABLE I
RELATING THE RF SIMULATION METHODS

Baseband (SPICE)	HB Family	SM Family
DC	HB	SM
AC, Noise	Sm Sig HB	Sm Sig SM
Transient	Ckt Env	Env Follow
HB, SM	QPHB	MFT

any comparisons based on currently implemented versions of the methods will quickly become dated. So instead, I will try to extrapolate from current trends and compare what is likely to be the eventual attributes of the methods.

A. Strengths of Harmonic Balance

The main strength of harmonic balance is its natural support for frequency-domain models — both linear and nonlinear. Distributed components such as lossy and dispersive transmission lines and interpolated tables of S-parameters from either measurements or electromagnetic simulators are examples of linear models that are handled easily and efficiently with harmonic balance. Nonlinear frequency domain models are based on Volterra series and are derived either from simulation or from measurement [41]. In addition, it is becoming more common for abstract behavioral models to be written directly in the frequency domain. There is, however, one caveat. It is relatively easy to generate nonphysical models in the frequency domain.

Harmonic balance is extremely efficient and accurate if both v and f in (8) are nearly sinusoidal. However this is not a feature that finds much application in practice. It is generally only significant when trying to determine the distortion of very low distortion amplifiers and filters. It does not help analyzing mixers, amplifiers, and samplers because they contain signals that are far from sinusoidal.

B. Strengths of Shooting Methods

The strengths of shooting methods stem from the properties of its underlying transient analysis. In particular, it chooses nonuniform timesteps in order to control error, and it has excellent convergence properties.

The ability of transient analysis, and so shooting methods, to place time points in a nonuniform manner allows it to accurately and efficiently follow abruptly discontinuous waveforms. Small time steps can be used to accurately resolve rapid transitions without taking small steps everywhere. This is very important for circuits such as mixers, relaxation oscillators, switched-capacitor and switched-current filters, samplers, sample-and-holds, and chopper stabilized amplifiers. In addition, the timestep is automatically chosen to control error. Though not an inherent issue, it is a failing of existing harmonic balance simulators that they do not automatically control error.

The strong convergence properties of shooting methods result from its implementation as a multilevel Newton method, and not from the fact it is a time-domain method. Indeed, it is possible to formulate harmonic balance as a time-domain method [15,37], yet its convergence properties do not fundamentally change.

Harmonic balance can be made as robust as shooting methods by incorporating them in a carefully designed continuation or homotopy method [15]. However, because continuation methods end up calling harmonic balance tens, or perhaps hundreds, of times, they can be slow. The ability of shooting methods to converge on a large class of strongly nonlinear circuits without the need for continuation methods represents a significant advantage in efficiency over harmonic balance.

With shooting methods, it is natural to perform transient analysis for a while before starting the shooting iteration in order to generate a good starting point. This is usually sufficient to get convergence even on troublesome circuits except when the time constants in the circuit are much larger than the period of the signal. If this is not sufficient, one can also use continuation methods with shooting methods.

Finally, the preconditioner available in shooting methods seems to be more robust and less burdensome than the preconditioners available with harmonic balance. The preconditioner is used to implement the matrix-implicit linear solvers that allow both shooting methods and harmonic balance to handle large problems. This is currently an area of innovation, and so this situation may change.

A significant disadvantage of shooting methods is that they do not support distributed components. While it is conceivable that shooting methods can be extended to handle distributed components, doing so will likely compromise their strong convergence properties and their preconditioner.

The fundamental strengths of shooting methods and harmonic balance are compared in Table II.

TABLE II
STRENGTHS OF SHOOTING METHODS AND HARMONIC BALANCE

Shooting Methods	Harmonic Balance
• Convergence • Nonuniform timesteps • Robust preconditioner	• Frequency domain models • No frequency limit on "small" signals

VII. Conclusion

There has been a tremendous amount of innovation and progress in RF simulation methods in the past decade, with the result being the wide variety of methods available today. Each method carries with it limiting assumptions that it exploits to perform efficiently when those assumptions are satisfied. However, the assumptions also prevent each method from being used in a general setting. There is no universal method, and it is unlikely there will ever be one. This is mainly a problem in that it prevents the whole RF section of a transceiver from being simulated together. While it is starting to be possible to simulate the whole signal path of a receiver or a transmitter, incorporating the frequency synthesizer and any digital signal processing is still beyond reach.

While the existing methods and their obvious extensions do not solve all RF simulation problems for individual blocks, their coverage is pretty good. As such, work in developing new methods is expected to gradually decline. A new area of effort is expected to be in using the existing methods to generate high level models of individual blocks that will allow more general methods to simulate the whole system. Examples of this new trend include using Volterra series to model the nonlinear behavior of RF blocks [41] and using the small-signal time-varying noise analyses on the individual blocks in a phase-locked loop to generate behavioral models that include jitter [3,17], which allows accurate and efficient noise prediction for frequency synthesizers.

Acknowledgments

I would like to thank Dave Sharrit of HP and Jacob White of MIT for their direct and significant contributions to this paper.

References

[1] T. Aprille and T. Trick. Steady-state analysis of nonlinear circuits with periodic inputs. *Proceedings of the IEEE*, vol. 60, no. 1, pp. 108-114, January 1972.

[2] H. Brachtendorf, G. Welsch, R. Laur and A. Bunse-Gerstner. Numerical steady state analysis of electronic circuits driven by multi-tone signals. *Electrical Engineering*, Springer-Verlag, vol. 79, pp. 103-112, 1996.

[3] A. Demir. *Analysis and Simulation of Noise in Nonlinear Electronic Circuits and Systems.* To be published by Kluwer Academic Publishers in 1997.

[4] C. Dragone. Analysis of thermal and shot noise in pumped resistive diodes. *The Bell System Technical Journal*, vol. 47, no. 9, pp. 1883-1902, November 1968.

[5] S. Egami. Nonlinear, linear analysis and computer-aided design of resistive mixers. *IEEE Transactions on Microwave Theory and Techniques*, vol. MTT-22, no. 3, pp. 270-275, March 1974.

[6] P. Feldmann and R. Freund. Efficient linear circuit analysis by Padé Approximation via the Lanczos Process. *IEEE Transactions on Computer-Aided Design of Integrated Circuits and Systems*, vol. 14, no. 5, pp. 639-649, May 1995.

[7] W. Gardner. *Introduction to Random Processes: With Applications to Signals and Systems.* McGraw-Hill, 1989.

[8] D. Hente and R. Jansen. Frequency-domain continuation method for the analysis and stability investigation of nonlinear microwave circuits. *IEE Proceedings*, part H, vol. 133, no. 5, pp. 351–362, October 1986.

[9] P. Heikkilä. *Object-Oriented Approach to Numerical Circuit Analysis.* Ph. D. dissertation. Helsinki University of Technology, January 1992.

[10] D. Held and A. Kerr. Conversion loss and noise of microwave and millimeter-wave mixers: part 1 — theory. *IEEE Transactions on Microwave Theory and Techniques*, vol. MTT-26, no. 2, pp. 49-55, February 1978.

[11] S. Kapur, D. Long and J. Roychowdhury. Efficient time-domain simulation of frequency-dependent elements. *IEEE/ACM International Conference on Computer-Aided Design: Digest of Technical Papers*, November 1996.

[12] A. Kerr. Noise and loss in balanced and subharmonically pumped mixers: part 1 — theory. *IEEE Transactions on Microwave Theory and Techniques*, vol. MTT-27, no. 12, pp. 938-950, December 1979.

[13] K. Kundert, J. White and A. Sangiovanni-Vincentelli. An envelope-following method for the efficient transient simulation of switching power and filter circuits. *IEEE International Conference on Computer-Aided Design: Digest of Technical Papers*, November 1988.

[14] K. Kundert, J. White and A. Sangiovanni-Vincentelli. A mixed frequency-time approach for distortion analysis of switching filter circuits. *IEEE Journal of Solid-State Circuits*, April 1989, vol. 24, no. 2, pp. 443-451.

[15] K. Kundert, J. White and A. Sangiovanni-Vincentelli. *Steady-State Methods for Simulating Analog and Microwave Circuits.* Kluwer Academic Publishers, 1990.

[16] K. Kundert. Accurate Fourier analysis for circuit simulators. *Proceedings of the IEEE Custom Integrated Circuits Conference*, May 1994.

[17] K. Kundert. Modeling and simulation of jitter in phase-locked loops. To appear in *Analog Circuit Design* by Sansen, Huijsing, and van de Plassche (editors), Kluwer Academic Publishers, 1997.

[18] D. Long, R. Melville, K. Ashby and B. Horton. Full-chip harmonic balance. *Proceedings of the IEEE Custom Integrated Circuits Conference*, May 1997.

[19] R. Melville, P. Feldmann and J. Roychowdhury. Efficient multi-tone distortion analysis of analog integrated circuits. *Proceedings of the IEEE Custom Integrated Circuits Conference*, May 1995.

[20] K. Nabors, T. Fang, H. Chang, K. Kundert, and J. White. Lumped Interconnect Models Via Gaussian Quadrature.

Proceedings of the 34th Design Automation Conference, June 1997.

[21] M. Nakhla and J. Vlach. A piecewise harmonic balance technique for determination of periodic response of nonlinear systems. *IEEE Transactions of Circuits and Systems*. vol. CAS-23, no. 2, pp. 85-91, February 1976.

[22] M. Okumura, T. Sugawara and H. Tanimoto. An efficient small signal frequency analysis method for nonlinear circuits with two frequency excitations. *IEEE Transactions of Computer-Aided Design of Integrated Circuits and Systems*, vol. 9, no. 3, pp. 225-235, March 1990.

[23] M. Okumura, H. Tanimoto, T, Itakura and T. Sugawara. Numerical noise analysis for nonlinear circuits with a periodic large signal excitation including cyclostationary noise sources. *IEEE Transactions on Circuits and Systems — I. Fundamental Theory and Applications*, vol. 40, no. 9, pp. 581-590, September 1993.

[24] L. Petzold. An efficient numerical method for highly oscillatory ordinary differential equations. *SIAM Journal of Numerical Analysis*, vol. 18, no. 3, June 1981.

[25] J. Phillips, E. Chiprout and D. Ling. Efficient full-wave electromagnetic analysis via model-order reduction of fast integral transforms. *Proceedings of the 33nd Design Automation Conference*, June 1996.

[26] V. Rizzoli and A. Neri. State of the art and present trends in nonlinear microwave CAD techniques. *IEEE Transactions on Microwave Theory and Techniques*, vol. 36, no. 2, pp. 343-365, February 1988.

[27] V. Rizzoli, C. Cechetti, A. Lipparini and F. Mastri. General-purpose harmonic balance analysis of nonlinear microwave circuits under multitone excitation. *IEEE Transactions on Microwave Theory and Techniques*, vol. 36, no. 12, pp. 1650–1660, December 1988.

[28] J. Roychowdhury. Efficient methods for simulating highly nonlinear multi-rate circuits. *Proceedings of the 34th Design Automation Conference*, June 1997.

[29] J. Roychowdhury, D. Long and P. Feldmann. Cyclostationary noise analysis of large RF circuits with multi-tone excitations. *Proceedings of the IEEE Custom Integrated Circuits Conference*, May 1997.

[30] Y. Saad. Iterative *Methods for Sparse Linear Systems*. PWS Publishing, 1996.

[31] J. Schutt-Aine and R. Mittra. Scattering parameter transient analysis of transmission lines loaded with nonlinear terminations. *IEEE Transactions on Microwave Theory and Techniques*, vol. MTT-36, no. 3, pp. 529-536, March 1988.

[32] D. Sharrit. New method of analysis of communication systems. *MTTS'96 WMFA: Nonlinear CAD Workshop, June 1996.*

[33] M. Silveira, I. El-Fadel, J. White, M. Chilukuri, and K. Kundert. Efficient frequency-domain modeling and circuit simulation of transmission lines. *IEEE Transactions on Components, Packaging, and Manufacturing Technology — Part B: Advanced Packaging*, November 1994.

[34] G. Rhyne, M. Steer and B. Bates. Frequency-domain nonlinear circuit analysis using generalized power series. *IEEE Transactions on Microwave Theory and Techniques*, vol. 36, no. 2, pp. 717-720, February 1988.

[35] R. Telichevesky, K. Kundert and J. White. Efficient steady-state analysis based on matrix-free Krylov-subspace methods. *Proceedings of the 32nd Design Automation Conference*, June 1995.

[36] R. Telichevesky, K. Kundert and J. White. Receiver characterization using periodic small-signal analysis. *Proceedings of the IEEE Custom Integrated Circuits Conference*, May 1996.

[37] R. Telichevesky, K. Kundert, I. El-Fadel and J. White. Fast simulation algorithms for RF circuits. *Proceedings of the 1996 IEEE Custom Integrated Circuits Conference*, May 1996.

[38] R. Telichevesky, K. Kundert and J. White. Efficient AC and noise analysis of two-tone RF circuits. *Proceedings of the 33rd Design Automation Conference*, June 1996.

[39] Y. Thodesen and K. Kundert. Parametric harmonic balance. *IEEE MTT-S International Microwave Symposium Digest*, June 1996.

[40] H. Torrey and C. Whitmer. *Crystal Rectifiers*. McGraw-Hill, 1948.

[41] F. Verbeyst and M. Vanden Bossche. VIOMAP, the S-parameter equivalent for weakly nonlinear RF and microwave devices. *IEEE Transactions on Microwave Theory and Techniques*, vol. 42, no. 12, pp. 2531-2535, December 1994.

[42] N. Verghese and D. Allstot. Verification of RF and mixed-signal integrated circuits for substrate coupling effects. *Proceedings of the IEEE Custom Integrated Circuits Conference*, May 1997.

[43] A. Usihda, L. Chua and T. Sugawara. A substitution algorithm for solving nonlinear circuits with multi-frequency components. *International Journal on Circuit Theory and Application*, vol. 15, pp. 327–355, 1987.

VHDL-AMS—A Hardware Description Language for Analog and Mixed-Signal Applications

Ernst Christen, *Member, IEEE,* and Kenneth Bakalar

(Invited Paper)

***Abstract*— This paper provides an overview of the VHDL-AMS hardware description language for analog and mixed-signal applications, by describing the major elements of the language and illustrating them by examples.**

***Index Terms*— Analog simulation, hardware description languages, mixed-signal simulation, VHDL-AMS.**

I. Introduction

HARDWARE description languages (HDL's) are programming languages designed for describing the behavior of physical devices and processes, a task commonly called modeling. Models written in an HDL are used as input to a suitable simulator to analyze the behavior of the devices. HDL's have been used since the 1960's to model and simulate applications as diverse as digital and analog electronic systems, fluid concentrations in chemical processes, and parachute jumps.

Modern HDL's support the description of both behavior and structure. The structural mechanisms of an HDL allow a user to compose the model of a complete system from reusable model components stored in a library. Stored components are assembled into a design hierarchy that often closely resembles the decomposition of the system into subsystems and sub-subsystems. The behavioral mechanisms of an HDL allow a user to express the operation of a subsystem at various levels of abstraction: very detailed, highly abstract, or anything in between. The designer can proceed using a top-down methodology, first performing conceptual studies using less detailed models, and then continually refining the design until each subsystem has been completed in sufficient detail for implementation. Top-down design and indefinite extensibility using stored components are the major advantages of an HDL-based design methodology over traditional approaches.

HDL's can be divided into digital, analog, and mixed-signal HDL's, depending on the available language constructs. Digital HDL's, such as VHDL [1] or Verilog [2], are based on event-driven techniques and a discrete model of time. They support the modeling of digital hardware at abstraction levels, from system level down to gate level. Analog HDL's support the description of systems of differential and algebraic equations (DAE's) whose solution varies continuously with time. Languages following the continuous system simulation language (CSSL) specification [3] focus on the description of behavior and provide only rudimentary facilities for hierarchical composition of a design. They do not automatically support conservative connections that satisfy Kirchhoff's laws, which makes them less suited for modeling electronic systems. Modern analog HDL's support both structural composition and conservation semantics, in addition to behavioral descriptions. Examples of such languages are FAS (Anacad, 1988), SpectreHDL (Cadence, 1994), and Verilog-A (Open Verilog International, 1997). Mixed-signal languages support both event-driven techniques and differential/algebraic equations (DAE's). Languages in this category are MAST (Analogy, 1986), HDL-A (Mentor Graphics, 1992), VHDL-AMS (IEEE, 1999), and Verilog-AMS (Open Verilog International, under development).

In this contribution, we give an overview of VHDL-AMS, a new hardware description language for analog, digital, and mixed-signal applications. VHDL-AMS is an informal name for the combination of two IEEE standards: VHDL 1076-1993 [1] and VHDL 1076.1-1999 [4]. In Section II, we describe the foundations of VHDL-AMS: the design objectives that guided the development of VHDL-AMS, the base language VHDL 1076-1993, and the mathematical foundation in the theory of differential/algebraic equations. Section III introduces the elements of the language, focusing on the static semantics of the extensions introduced by IEEE Std. 1076.1-1999. Each language element is illustrated by examples. Finally, Section IV describes the dynamic semantics of VHDL-AMS, including initialization and the simulation cycle.

II. Foundations of VHDL-AMS

A. Design Objectives

The requirements for VHDL-AMS were described in a design objective document [5], compiled from the raw requirements submitted by a group of people interested in the development of the language. That document, together with the design objective rationale [6], defines the scope of VHDL-AMS.

Manuscript received June 26, 1999; revised July 29, 1999. This work was supported by a grant from the Joint European Systems on Silicon Initiative, the U.S. Air Force—Rome Laboratory, a grant from Wright-Patterson Laboratory, and by the employers of participating engineers.

E. Christen is with Analogy, Inc., Beaverton, OR 97008 USA.

K. Bakalar is with Mentor Graphics Corporation, Rockville, MD 20850 USA.

Publisher Item Identifier S 1057-7130(99)08720-0.

In summary, VHDL-AMS is required to be a superset of VHDL 1076-1993, supporting the hierarchical description and simulation of continuous and mixed-continuous/discrete system with conservative and nonconservative semantics. The language must support modeling at various abstraction levels in electrical and nonelectrical energy domains. The systems to be modeled are lumped systems that can be described by ordinary differential equations and algebraic equations; the solution of the equations describing the behavior of the system may include discontinuities. Interactions between the discrete part of a model and its continuous part, are to be supported in a flexible and efficient manner, and support for small-signal frequency-domain analysis and noise analysis must be provided.

B. VHDL 1076-1993

VHDL-AMS is not a new language. Rather, it is a superset of VHDL 1076-1993. It builds on the solid foundations of VHDL 1076-1993: its syntactic and semantic framework, including structural and functional decomposition, separate compilation, a powerful sequential notation, and the strong type system of a modern programming language. VHDL-AMS adds new objects and new kinds of statements, while preserving the scope and visibility rules, the requirement of declaration before use, the way names are formed, and other essential elements of digital VHDL.

C. Differential and Algebraic Equations

The continuous aspects of the behavior of the lumped systems targeted by VHDL-AMS can be described by a system of ordinary DAE's of the form

$$\underline{F}(\underline{x},\, d\underline{x}/dt,\, t) = 0 \tag{1}$$

where $\underline{F}$ is a vector of expressions, $\underline{x}$ is a vector of unknowns, $d\underline{x}/dt$ is a vector of derivatives of the unknowns with respect to time, and t is time. There is no alternative systematization with equivalent power and scope. Most such systems of equations have no analytic solution, so in practice the solution must be approximated using numerical techniques.

DAE's have been studied extensively by numerical mathematicians over the past 25 years [7]. DAE's can be classified based on their structure. Roughly speaking, the structural index describes how close a system of DAE's is to a system of ordinary differential equations; a system of ordinary differential equations has index 0. Several numerical solution methods exist for DAE's with low index. The algorithms are not suited for high index DAE systems, but methods have been developed that allow the structural index of such systems to be reduced by augmenting the system of equations. In spite of a few open issues, the theory of DAE's is sufficiently mature to justify existing techniques and to provide a sound foundation for the development of new algorithms.

The VHDL-AMS definition provides a notation for DAE's and describes precisely what system of equations is implied at each simulation time by the text of a VHDL-AMS model. It does not specify a technique for their solution, leaving the selection of a suitable numerical method to the implementor of a compliant simulator. The solution algorithm itself is referred to as the "analog solver." Only the results that the analog solver must achieve, and not its algorithm, are characterized in the language definition.

III. Language Elements of VHDL-AMS

This section introduces the language elements of VHDL-AMS for the description of continuous time and mixed continuous/discrete time systems.

A. Overview of VHDL-AMS Models

A VHDL-AMS model consists of an *entity* and one or more *architectures*. The entity specifies the interface of the model to the outside world. It includes the description of the *ports* of the model (the points that can be connected to other models) and the definition of its generic parameters. The architecture contains the implementation of the model. It may be coded using a structural style of description, a behavioral style, or a style combining structural, and behavioral elements. A structural description is a netlist; it is a hierarchical decomposition of the model into appropriately connected instances of other models. A behavioral description consists of concurrent statements to describe event-driven behavior and simultaneous statements to describe continuous behavior. Concurrent statements include the concurrent signal assignment for data flow modeling and the process statement for more general event-driven modeling. Simultaneous statements are discussed in Section III-C.

When a VHDL-AMS model is instantiated in a structural description, the designer can specify which of several architectures to use for each instance. Alternatively, the decision can be postponed until immediately prior to the simulation. This allows for an easy and flexible reconfiguration of the model. For example, in top-down design, one architecture can describe a subsystem behaviorally with little detail, while another can add parasitics and a third can decompose the subsystem into lower level components.

B. Quantities

The unknowns in the collection of DAE's implied by the text of a model are analytic functions of time; that is, they are piecewise continuous with a finite number of discontinuities. The analog solver solves for the values of all unknowns over time by first converting, at specific values of time, the differential part of the DAE's to algebraic equations using appropriate discretization methods, and then solving the algebraic equations simultaneously.

VHDL-AMS introduces a new class of objects, the *quantity*, to represent the unknowns in the DAE's. Quantities can be scalar or composite (arrays and records), but must have scalar subelements of a floating-point type. A quantity object can appear anywhere a value of the type is allowed, in particular in an expression. In the remainder of this section, we describe the characteristics of scalar quantities. The characteristics of a composite quantity are simply the aggregation of the characteristics of its scalar subelements. The behavior of each scalar subelement is independent of the others.

```
entity summer is
   port (quantity in1, in2: in REAL;
         quantity sum: out REAL);
end entity summer;
```

Fig. 1. Entity declaration of a signal-flow model.

Quantities can be declared anywhere a signal can be declared except in a VHDL package. The following statement declares three quantities q1, q2, and q3 of type REAL:

quantity q1, q2, q3: REAL

where bold text indicates reserved words and upper-case text indicates predefined concepts.

A quantity can also be declared as an interface element in a port list of a model. Interface quantities support signal flow modeling. Each has a mode, similar in concept to the mode of an interface signal, indicating the direction of signal flow. For instance, Fig. 1 shows the entity declaration of a signal flow model with two interface quantities of mode **in** and one interface quantity of mode **out**. When this model is instantiated each interface quantity is associated with a quantity declared in the instantiating model. For example

a1: **entity** summer **port map**
(in1 ⇒ q1, in2 ⇒ q2, sum ⇒ q3);

The effect of a quantity association is to constrain the two quantities to be equal.

In addition to quantities declared explicitly, some quantities are implicitly declared by using their name in the text of a model. For example, Q'Dot is a quantity that holds the derivative of quantity Q with respect to time. Implicit quantities also exist for the integral of a quantity over time, ideal delay, Laplace, z-domain transfer functions, and ideal sample-and-hold. Other implicit quantities related to conservative systems and mixed-signal modeling are described in Sections III-D and III-G.

C. Simultaneous Statements

Simultaneous statements are a new class of statements in VHDL-AMS for notating differential and algebraic equations. Simultaneous statements contain ordinary VHDL expressions that can be evaluated in the ordinary way. Simultaneous statements can appear anywhere a concurrent signal assignment is allowed. The basic form is the *simple simultaneous statement*, which has the following syntax:

$$\text{[label:] expression} = \text{expression} \tag{2}$$

where the square brackets indicate that this part of the statement is optional. For instance, the constitutive equation of a signal flow model related to Fig. 1 could be written as

$$\texttt{sum} = \texttt{in1} + \texttt{in2}.$$

The expressions may have composite values, in which case there must be a matching subelement on the left for each subelement on the right. The expressions may refer to signals, quantities, constants, literals, and functions. Each scalar subelement of a simple simultaneous statement is mapped to one expression in the vector $\underline{F}$ in (1) by subtracting the right expression in (2) from the left expression. When the analog solver has properly established the value of each quantity, the matching subelements of the expressions will be (approximately) equal.

Several additional forms of the simultaneous statement have been defined. The *simultaneous case statement* and *simultaneous if statement* are analogous to their sequential counterparts and allow the description of piecewise defined behavior. Each contains an arbitrary list of simultaneous statements in its statement parts, including nested simultaneous case and if statements. The analog solver considers only the simultaneous statements selected by the case expressions and chosen by the conditional expressions. The *simultaneous procedural statement* is merely a way to rewrite the function body f in the simultaneous statement $f(q, x) = q$ "in line," where q is a collection of quantities and x is an arbitrary collection of other objects.

Fig. 2 shows the VHDL-AMS model of a signal flow amplifier that brings together many of the new language concepts described so far. The entity declares a model parameter gain with type REAL and a default value of REAL'High, the largest value representable for type REAL. The model uses this value to indicate infinite gain. There is one interface quantity with mode **in**, input, and one interface quantity with mode **out**, output. The architecture includes three simultaneous statements: two simple simultaneous statements and a simultaneous if statement that selects one of the two simple simultaneous statements, depending on the value of the parameter gain. The right-hand side of the equation "input = 0.0" is nothing more than $\lim_{\text{gain}\to\infty}$ output/gain.

D. Conservative Systems

Systems with conservation semantics—for example, electrical systems obeying Kirchhoff's laws—merit separate treatment because they are so commonly encountered. Special purpose syntax and semantics can provide a simplified notation, that reduces the risk of errors and thereby improves productivity. In VHDL-AMS, equations describing the conservative aspects of such a system need not be explicitly notated by the modeler. Only the so-called constitutive equations remain the modeler's responsibility.

The description of conservative systems uses a graph-based conceptual model. Consider, for example, the bipolar inverter circuit and its equivalent graph, shown in Fig. 3. The vertices of the graph represent equipotential nodes in the circuit, and the edges represent branches of the circuit through which current flows. Similar graphs can be created for systems in other energy domains such as the thermal or fluidic domains. This conceptual model does not dictate any particular implementation for the analog solver. In particular, it does not force the selection of a formulation technique with the word "nodal" in its name (for example, the modified nodal formulation).

Branch quantities represent the unknowns in the equations describing conservative systems. There are two kinds

```
entity sfg_amplifier is
   generic (gain: REAL := REAL'High);     -- model parameter
   port (quantity input: in REAL;         -- interface quantities
         quantity output: out REAL);
end entity sfg_amplifier;

architecture one of sfg_amplifier is
begin
   if gain = REAL'High use                -- simult. if statement
      input == 0.0;                       -- simple simult. statement
   else
      output == gain * input;
   end use;
end architecture one;
```

Fig. 2. Entity and architecture of a signal-flow amplifier.

of branch quantities: *across quantities* and *through quantities*. Across quantities represent effort like effects such as voltage, temperature, or pressure. They correspond to the potential difference between two vertices in the graph. Through quantities represent flow like effects such as current, heat flow rate, or fluid flow rate. They correspond to the edges in the graph. The constitutive equations of conservative systems are expressed by relating the across and through quantities of one or several branches using simultaneous statements. For example, a resistor has a single branch, and its constitutive equation (Ohm's law) relates the voltage across (the across quantity) and the current through the resistor (the through quantity): $i = v/r$.

A branch quantity is declared with reference to two *terminals*. The terminal is the second new object of the extended language. A terminal is declared to be of some *nature*. Natures can be scalar or composite (arrays and records). Each scalar nature represents a distinct energy domain—electrical, thermal, fluidic, etc. Its definition includes the types of across and through quantities incident to a terminal of the nature, and the common *reference terminal* (e.g., electrical ground, or mechanical anchor) shared by all terminals with elements of that scalar nature. These concepts are illustrated in Fig. 4, where the declarations of two subtypes `voltage` and `current`, and a scalar nature `electrical` are shown. For reuse, these declarations are contained in a package `electrical_system`. In the following, we will assume that this package has been compiled into a library `disciplines`. The definition of natures of this kind for various energy domains and the definition of other infrastructure for continuous modeling is the topic of an ongoing standardization effort at the time of this writing.

Using the declarations in Fig. 4, the following statements declare two terminals `t1` and `t2` of nature `electrical`, an across quantity `v`, and two through quantities `i1` and `i2` between the terminals. The across quantity represents the potential difference between the terminals and the through quantities represent two parallel current-carrying branches

```
terminal t1, t2: electrical;
quantity v across i1, i2 through t1 to t2;
```

The type of a branch quantity is not explicitly declared. Rather, it is derived from the nature of its terminals. It may be a composite type. In the example the across quantity `v` is of

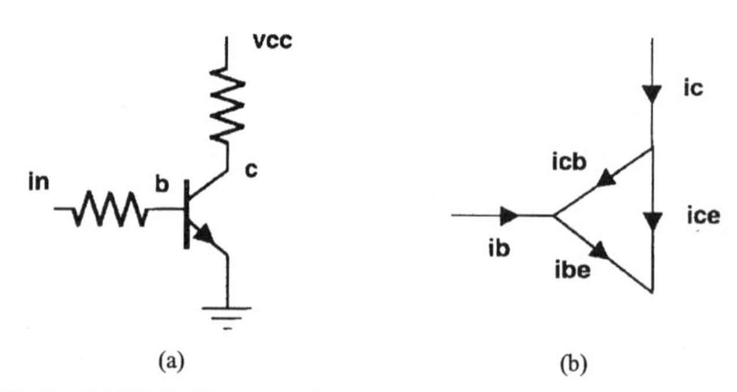

Fig. 3. (a) Bipolar inverter and (b) its equivalent graph.

```
package electrical_system is
   subtype voltage is REAL;
   subtype current is REAL;
   nature electrical is
      voltage across
      current through
      ground  reference;
end package electrical_system;
```

Fig. 4. Declaration of nature electrical in package electrical_system.

type `voltage`, and the type of the two through quantities `i1` and `i2` is `current`. As in the case of ordinary quantities, the characteristics of the composite are just the aggregate of the characteristics of its scalar subelements. The terminals must have elements of the same scalar nature and must agree in other specified ways. The terminals of a branch quantity are called the plus terminal and minus terminal, and the direction of the branch is plus to minus—in an electrical system, the direction of positive current flow.

A terminal may be declared anywhere a signal declaration is allowed. In particular, a terminal can be an interface element in a port list. For instance, the following statement declares the interface terminals of a diode:

```
port (terminal anode, cathode: electrical).
```

When a model is instantiated, the association of interface terminals is used to construct nodes in hierarchical descriptions in a fashion paralleling the use of interface signals to construct nets in digital hierarchies.

The declaration of a terminal T creates two implicit quantities.

```
library ieee, disciplines;
use ieee.math_real.all;
use disciplines.electrical_system.all;
entity diode is
   generic (iss: REAL := 1.0e-14;       -- saturation current
            af:  REAL := 1.0;           -- flicker noise coeff.
            kf:  REAL := 0.0);          -- flicker noise exponent
   port (terminal anode, cathode: electrical);
end entity diode;

architecture ideal of diode is
   quantity v across i through anode to cathode;
   constant vt: REAL := 0.0258;         -- thermal voltage at 300K
begin
   i == iss * (exp(v/vt) - 1.0);
end architecture ideal;
```

Fig. 5. VHDL-AMS model of an ideal diode.

1) The *reference quantity* T'Reference is an across quantity with T as its plus terminal and the reference terminal of the nature of T as its minus terminal (e.g., voltage to ground).
2) The *contribution quantity* T'Contribution is a through quantity whose value is equal to the sum, with appropriate sign, of all through quantities incident to T (for example, the current flowing into an electrical terminal T).

Both T'Reference and T'Contribution are composite if T is composite. In that case, the rules apply to each scalar subelement of T.

The conservation equations of the system (in an electrical system, those equations due to Kirchhoff's laws) are extracted from the graph created by the declared branch quantities and terminals and the association of terminals into nodes. A *node* is a set of scalar terminals created by a tree of terminal associations. The value of each scalar across quantity is constrained to be equal to the difference of the reference quantities of its terminals. All the reference quantities of the terminals of a node are constrained to be equal, and the contribution quantity of the terminal at the root of the tree is constrained to zero.

With these definitions the model of an ideal diode can be written as shown in Fig. 5. The library clause and the use clause make all declarations in the packages `math_real` and `electrical_system` visible in the model. This is necessary, because the model uses nature `electrical` from package `electrical_system` and function `exp` from package `math_real`. The entity declares the saturation current `iss` and the flicker noise parameters `af` and `kf` as generics (i.e., parameters) and `anode` and `cathode` as two interface terminals of nature `electrical`. We will use `af` and `kf` in an alternative architecture of the diode model in Section III-H. The architecture declares `v` as across quantity and `i` as through quantity between `anode` and `cathode`, and a constant `vt` representing the thermal voltage. It includes a simple simultaneous statement describing the behavior of the ideal diode.

A single model may have terminals of several natures. For example, the model of an electrical motor might have, in addition to its electrical terminals, a terminal of a rotational nature representing the shaft of the motor. Similarly, the model of a diode with self-heating might have terminals of a thermal nature. There are no limitations on the number of different natures that can be used in a model.

E. Tolerances

When the analog solver solves the DAE's of a model it determines the value of each quantity such that the values of all expressions in (1) are close to zero. To allow a user to control how close to zero the solution must be, the language defines the concept of a *tolerance group*: each quantity and each expression in (1) belongs to a tolerance group. The tolerance group of a quantity is defined by the subtype of the quantity. For example, the subtype declarations in Fig. 4 would more properly be written as

```
subtype voltage is REAL tolerance "default_voltage";
subtype current is REAL tolerance "default_current";
```

where the string following the reserved word **tolerance** defines the tolerance group of the subtype. The default tolerance group of a simple simultaneous statement whose left- or right-hand side is the name of a quantity is the tolerance group of that quantity. However, the tolerance group of a simple simultaneous statement may also be specified explicitly. For example

```
i = iss * (exp(v/vt) - 1.0) tolerance "low_voltage";
```

It is the responsibility of an implementation to associate suitable simulator control parameters with each tolerance group and to give the user control over those parameters. For example, the parameters might be the numerical values for absolute tolerance and relative tolerance for a SPICE-like analog solver. This approach keeps the language definition completely independent of any solution algorithm, recognizing that different algorithm use different means to control the accuracy of a solution.

F. A/D Interaction

When during simulation the value of a quantity Q crosses a threshold E an event occurs on the Boolean signal Q'Above(E). The value of Q'Above(E) is TRUE if $Q > E$ and FALSE if $Q < E$. The value changes at the instant

```
library IEEE, disciplines;
use IEEE.std_logic_1164.all;
use disciplines.electrical_system.all;
entity comparator is
   generic (vlo, vhi: REAL;
            timeout: DELAY_LENGTH);
   port (terminal ain, ref: electrical;
         signal dout: out std_logic);
end entity comparator;

architecture hysteresis of comparator is
   type states is (unknown, zero, one, unstable);
   quantity vin across ain to ref;
   function level(vin, vlo, vhi: REAL) return states is
   begin
      if    vin < vlo then return zero;
      elsif vin > vhi then return one;
      else                 return unknown;
      end if;
   end function level;
begin
   process
      variable state: states := level(vin, vlo, vhi);
   begin
      case state is
         when zero =>
            dout <= '0';
            wait on vin'above(vlo);
            state := unstable;
         when one =>
            dout <= '1';
            wait on vin'above(vhi);
            state := unstable;
         when unknown =>
            dout <= 'X';
            wait on vin'above(vlo), vin'above(vhi);
            state := level(vin, vlo, vhi);
         when unstable =>
            wait on vin'above(vlo), vin'above(vhi) for timeout;
            state := level(vin, vlo, vhi);
      end case;
   end process;
end architecture hysteresis;
```

Fig. 6. VHDL-AMS model of a comparator with hysteresis.

of the threshold crossing. Threshold crossing can be used for A/D conversions. For example, the statement

$$\texttt{s} \Leftarrow \text{`1'}\ \textbf{when}\ \text{q' above } (0.0)\ \text{else `0'};$$

where `q` is a quantity and `s` is a signal, implements the behavior of an ideal comparator.

A more complete version of a comparator that includes hysteresis and outputs an unknown state if the input remains in the transition region for more than a specified time can be implemented using a finite-state machine. Its implementation is shown in Fig. 6. The branch quantity declaration declares `vin` to be an across quantity between terminals `ain` and `ref`. Note that no through quantity is declared; none is needed in this model. Function `level` returns one of three values, depending on the value of the input voltage `vin` with respect to the two thresholds `vlo` and `vhi`. The case statement defines the action to be taken for each state: changing the value of the output signal when entering the state, defining the condition when the finite state machine leaves the state, and determining the next state. For example, the wait statement in the unstable state suspends the process until there is an event on either `vin'above (vlo)` or `vin'above(vhi)`, or until the specified timeout expires, whichever comes first. It is instructive to reconstruct the state transition diagram of this model from the code shown in Fig. 6.

G. Digital-to-Analog (D/A) Interaction

The name of a signal may be used directly in a simple simultaneous statement to influence the continuous portion of a model, provided that the type rules for VHDL expressions are obeyed. For example, the statement

$$\texttt{q} = \texttt{s};$$

where `q` is a quantity and `s` is a signal of a floating-point type makes `q` follow `s` at all times. However, since signal values change instantaneously, this statement introduces a discontinuity in the value of `q` whenever the value of `s` changes. Since algorithms for the numerical solution of DAE's rely on the solution being continuous, a model introducing a discontinuity must notify the analog solver exactly when the discontinuity occurs. The break statement provides this notice.

```
library IEEE, disciplines;
use IEEE.std_logic_1164.all;
use disciplines.electrical_system.all;
entity dac is
   generic (tt: REAL := 1.0e-9);        -- transition time
   port (signal din: in std_logic;
         terminal aout, ref: electrical);
end entity dac;
architecture simple of dac is
   type real_table is array(std_logic) of REAL;
   constant v_table: real_table :=
      (2.5, 2.5, 0.2, 4.8, 2.5, 2.5, 0.2, 4.8, 2.5);
   constant r_table: real_table :=
      (0.1, 0.1, 0.1, 0.1, 1.0e9, 1.0e4, 1.0e4, 1.0e4, 1.0e9);
   quantity vout across iout through aout to ref;
   signal veff: REAL := v_table(din);  -- effective voltage
   signal reff: REAL := r_table(din);  -- effective resistance
begin
   veff <= v_table(din);
   reff <= r_table(din);
   vout == veff'ramp(tt) - iout * reff'ramp(tt);
end architecture simple;
```

Fig. 7. VHDL-AMS model of a one-bit D/A converter.

For example, the statement

break on s;

notifies the analog solver of a discontinuity in the model when an event occurs on s. This statement must complement the above simple simultaneous statement.

An alternative mechanism to influence the continuous solution with a signal is to use one of the predefined quantities S'Ramp(trise, tfall) or S'Slew(rising_slope, falling_slope), where S is a signal of a floating-point type

$$q = s'\mathtt{ramp}\,(1.0e-9,\ 1.2e-9);$$

S'Ramp ramps linearly over the specified rise and fall time from the previous value of S to its new value, starting at the time of the event. S'Slew does the same, but with specified slopes. This approach is usually closer to the physical behavior of a device, and no break statement is required.

As an example, the model shown in Fig. 7 implements a one-bit D/A converter with a nine-state signal input and a voltage source output with a series resistor. For simplicity, the voltage and resistance values corresponding to each state have been coded directly in the model. The concurrent signal assignment statements assign values to the local signals `veff` (the effective voltage) and `reff` (the effective resistance) by indexing the corresponding tables with the value of the input signal `din`. Their values change whenever the value of `din` changes. The output voltage of the DAC is the effective voltage minus the voltage drop across the series resistor. Both voltage and resistance ramp linearly from their previous value to their new value with rise and fall time `tt`.

Discontinuities can also be introduced by simultaneous if statements. A break statement is always required in the model if, for any reason, there is a discontinuity in a quantity or in its first derivative with respect to time, except if the discontinuity is caused by a predefined quantity, such as S'Ramp or S'Slew.

```
library IEEE, disciplines;
use IEEE.math_real.all;
use disciplines,electrical_system.all;
entity isrc is
   generic (ampl, freq: REAL;
            ac_mag, ac_phase: REAL);
   port (terminal p, m: electrical);
end entity isrc;
architecture sine of isrc is
   quantity i through p to m;
   quantity ac: current spectrum ac_mag, ac_phase;
begin
   i == ampl * sin(math_2_pi * freq * NOW) + ac;
end architecture sine;
```

Fig. 8. Spectral source quantity in a current source model.

H. Small-Signal Frequency-Domain and Noise Modeling

To support small-signal frequency-domain (AC) simulation and noise simulation, VHDL-AMS provides a third kind of quantity: the *source quantity*. Source quantities allow a model writer to define small-signal stimulus in the frequency domain. They come in two varieties: spectral source quantities and noise source quantities. Source quantities have special semantics that give them a zero value except during a small-signal frequency-domain simulation or a noise simulation, respectively. Their value may depend on frequency and on the biasing of a model.

The declaration and use of a spectral source quantity is illustrated with the VHDL-AMS model of a sinusoid current source in Fig. 8. The source quantity declaration declares `ac` as a spectral source quantity of subtype `current` whose magnitude is `ac_mag` and whose phase (in radians) is `ac_phase`. The name of this quantity is then used in the simple simultaneous statement that constrains the current `i` to be a sine wave with specified amplitude and frequency. NOW is a predefined function that returns the current simulation time. The definition of the small-signal model has the effect of eliminating the time-dependent portion of the statement in any frequency-domain simulation, and the source quantity `ac` then

```
architecture noisy of diode is
   quantity v across i through anode to cathode;
   quantity flns: current noise kf * i**af / FREQUENCY;
   constant vt: REAL := 0.0258;      -- thermal voltage at 300K
begin
   i == iss * (exp(v/vt) - 1.0) + flns;
end architecture noisy;
```

Fig. 9. Flicker noise modeling in a diode.

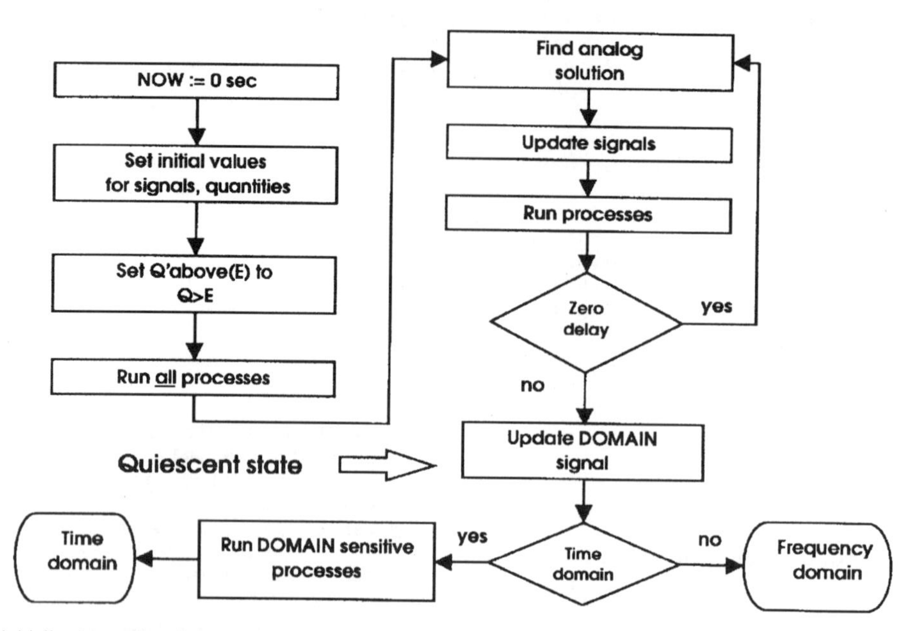

Fig. 10. VHDL-AMS initialization and quiescent state.

has the specified magnitude and phase. Therefore, during a small-signal frequency-domain simulation the current i has the specified spectrum.

Noise source quantities are used in a similar fashion, as illustrated in Fig. 9 by a diode architecture with bias-dependent flicker noise. The expression following the reserved word **noise** specifies the noise power; its value depends on the current flowing through the diode and on frequency, indicated by a call to the predefined function FREQUENCY. The definitions guarantee that at the time the noise power expression is evaluated the value of i equals the value of the diode current at the quiescent point of the model (see Section IV-A). As with spectral source quantities, the noise source quantity is simply added to the simple simultaneous statement defining the constitutive equation of the diode.

VHDL-AMS does not support general frequency-domain modeling, because DAE's are not sufficient to describe such models. Consequently, function FREQUENCY is restricted to expressions that are part of the declaration of a source quantity.

IV. Dynamic Semantics of VHDL-AMS

This section describes the semantics of VHDL-AMS related to the simulation of analog and mixed-signal systems.

A. Elaboration, Initialization, and the Quiescent Point

When a VHDL-AMS simulator reads a design, it first compiles the text of a description, including entities, architectures,

```
library disciplines;
use disciplines.electrical_system.all;
entity sampler is
   generic (period: TIME);
   port (terminal ain, ref: electrical;
         signal dout: out REAL := 0.0);
end entity sampler;

architecture ideal of sampler is
   quantity vin across ain to ref;
begin
   process  begin
      wait on DOMAIN;
      loop
         dout <= vin;
         wait for period;
      end loop;
   end process;
end architecture ideal;
```

Fig. 11. Use of the DOMAIN signal in a sampling device.

configurations, packages and package bodies, into a library. Then, it creates the so-called simulatable model by a process called *elaboration*, which includes, among other things, the creation of the data structures used by the simulator and the evaluation of certain expressions, in particular expressions that determine the size of objects such as the length of a signal vector.

The elaborated model is initialized before simulation begins. Each signal and each quantity is given a well-defined initial value, and each process executes until it suspends itself. Initialization is illustrated by the sequence of steps on the left of Fig. 10.

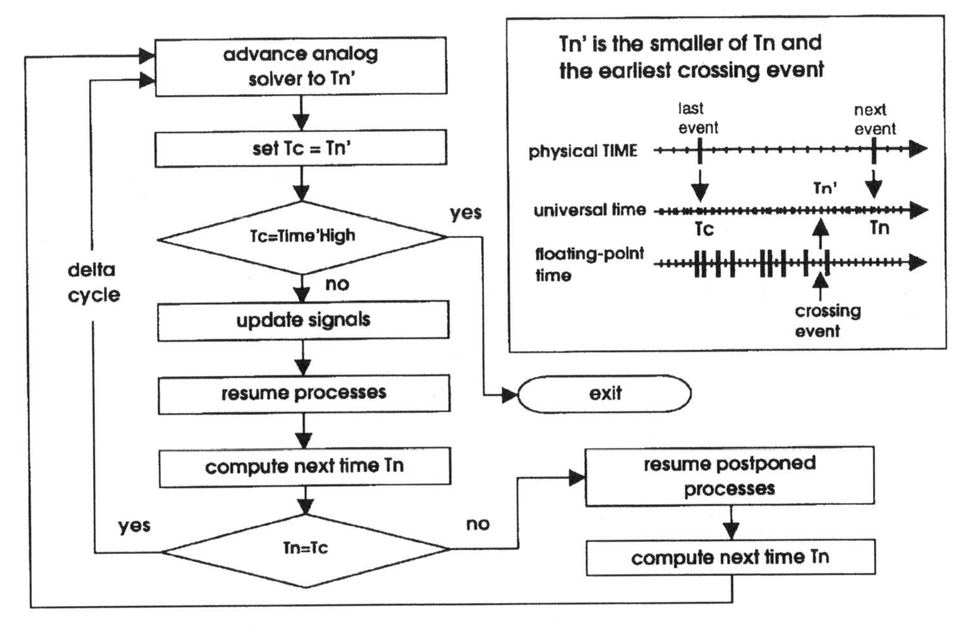

Fig. 12. The VHDL-AMS simulation cycle.

After initialization the simulator enters an abbreviated simulation cycle, shown at the top right of Fig. 10. While executing the abbreviated simulation cycle, the simulator alternatingly finds an analog solution, determines threshold crossings, and processes zero delay events. The equations in (1) are temporarily modified to constrain all explicit time derivatives to zero for the duration of the abbreviated cycle. When all zero delay events have been processed, a predefined signal DOMAIN whose initial value was QUIESCENT_DOMAIN gets assigned a new value: TIME_DOMAIN if a time domain simulation follows, FREQUENCY_DOMAIN if a small-signal frequency domain or a noise simulation follows. The state of the simulatable model at that point is called the *quiescent point* of the model. In purely continuous models it corresponds to the dc operating point.

The DOMAIN signal can be used to write models that behave differently in different domains. It is most commonly used to initialize mixed-signal models. As an example, consider the model of a sampling device that does not start sampling until after the quiescent state has been reached. The corresponding model is shown in Fig. 11.

B. VHDL-AMS Simulation Cycle

During a time domain simulation, the simulator repeatedly executes the simulation cycle shown in Fig. 12. In this figure, T_c is the current time, T_n is the next time, and T'_n is the adjusted next time. T'_n is equal to T_n unless there are crossing events in the interval $[T_c, T_n]$. In such a case, T'_n is the time of the earliest crossing event in the interval. The analog solver suspends at T'_n because at this time, there is either a "regular" event or a crossing event to be processed. The simulation cycle supports discontinuities, and additional rules define the initial conditions after a discontinuity, such that by default the system conserves charge and flux or their equivalents in other energy domains.

The inset in Fig. 12 shows the relationship of the time used by the event-driven kernel and the time used by the analog solver. The former is an integer multiple of a minimum time resolution (by default 1 fs) and is represented by the predefined type TIME, the latter approximates a continuous time and is represented by a floating-point value. The simulation cycle is defined in terms of a union of the two definitions, called universal_time. Note that some values of type universal_time cannot be represented exactly in type TIME, while others cannot be represented exactly by a floating-point value. The definitions ensure that any delays in signal assignments and timeouts in wait statements are interpreted correctly, while providing for an efficient simulation.

V. CONCLUSION

We have given an overview of VHDL-AMS, a hardware description language for analog, digital, and mixed-signal applications. We have focused on the main aspects of the language and illustrated the introduction of the language concepts by examples. VHDL-AMS includes much more, and the interested reader is encouraged to explore the language by studying the language reference manual [4] and other available literature.

REFERENCES

[1] *VHDL Language Reference Manual,* IEEE Standard 1076-1993.
[2] *Standard Description Language Based on the Verilog(TM) Hardware Description Language,* IEEE Standard 1364-1995.
[3] J. C. Strauss *et al.,* "The SCI continuous system simulation language (CSSL)," *Simulation,* pp. 281–303, Dec. 1967.
[4] *Definition of Analog and Mixed Signal Extensions to IEEE Standard VHDL,* IEEE Standard 1076.1-1999.
[5] 1076.1 Design Objective Document V 2.3 (1995). [Online]. Available HTTP: http://vhdl.org/vi/analog
[6] 1076.1 Design Objective Rationale V 1.2 (1995). [Online]. Available HTTP: http://vhdl.org/vi/analog
[7] K. E. Brenan, S. L. Campbell, and L. R. Petzold, *Numerical Solution of Initial-Value Problems in Differential-Algebraic Equations.* Philadelphia, PA: SIAM, 1996.

Ernst Christen (S'75–M'82) received the diploma in electrical engineering and the Dr.Sc.Techn. degree, both from the Swiss Federal Institute of Technology, Zurich, Switzerland, in 1975 and 1985, respectively.

From 1985 to 1987, he was with the Department of Electrical Engineering, University of Waterloo, Ontario, Canada, as a postdoctoral Fellow, working on worst-case optimization of active *RC* and switched-capacitor filters. Since 1987, he has been with Analogy, Inc., Beaverton, OR, where he is currently a Principal Engineer in simulation technology. His interests include hardware description languages and simulation and design of circuits and systems.

Dr. Christen has made major contributions to the definition of IEEE Std. 1076.1-1999.

Kenneth Bakalar received the B.S. degree in computer science and the M.D. degree in clinical engineering from The Johns Hopkins University and Medical School, Baltimore, MD, in 1973 and 1977, respectively.

After being engaged for nine years in laboratory instrumentation, data analysis, and system modeling in the Department of Biomedical Engineering, Johns Hopkins University, he joined VHDL pioneer CAD Language Systems, Inc., where he began teaching and writing about VHDL in 1987. Since 1997, he has been Program Manager for analog and mixed-signal hardware description languages at Mentor Graphics Corporation, Rockville, MD.

Dr. Bakalar is one of the principal Language Architects of IEEE Std. 1076.1-1999.

PART VII

Analog Centering and Yield Optimization

DELIGHT.SPICE: An Optimization-Based System for the Design of Integrated Circuits

WILLIAM NYE, MEMBER, IEEE, DAVID C. RILEY,
ALBERTO SANGIOVANNI-VINCENTELLI, FELLOW, IEEE,
AND ANDRÉ L. TITS, MEMBER, IEEE

***Abstract*—DELIGHT.SPICE is the union of the DELIGHT interactive optimization-based computer-aided-design system and the Spice circuit analysis program. With the DELIGHT.SPICE tool, circuit designers can take advantage of recent powerful optimization algorithms—and a methodology that emphasizes designer *intuition* and man-machine *interaction* in an approach in which designer and computer are complementary—to automatically adjust parameters of electronic circuits in order to improve their performance. They may optimize arbitrary performance criteria as well as study complex tradeoffs between multiple competing objectives, while simultaneously satisfying multiple constraint specifications. The optimization runs much more efficiently than previously by dint of the fact that the Spice program used has been enhanced to perform dc, ac, and transient sensitivity computation. Industrial analog and digital circuits have been redesigned using this tool, yielding substantial improvement in circuit performance.**

I. Introduction

FOR OUR PURPOSES, circuit design can be considered a two-step iterative process. The designer first selects an initial circuit configuration and then determines values of circuit parameters (e.g. resistor and capacitor values, and device geometries such as bipolar transistor areas and MOS transistor lengths and widths) that satisfy a set of specifications and optimize a set of possibly competing design objectives. This process is repeated until a satisfactory design has been achieved. The most creative part of the design process is, in general, the selection of the circuit topology. For large nonlinear circuits, the selection of values of the design parameters is often time consuming, and usually stops before even a local optimum of the design objectives is reached. This is due to the complex dependencies of specifications and objectives on ac, dc, and transient responses, and in turn of these responses on many design parameters. Thus it is usually difficult for designers to predict the effect of parameter changes on circuit performance without numerous circuit simulations.

Manuscript received November 13, 1986; revised October 2, 1987. This work was supported by DARPA under Grant N00039-C-0107, by a grant from the Semiconductor Products Division of the Harris Corporation, by a grant from MICRO, and by the National Science Foundation under Grant ECS-82-04452. The review of this paper was arranged by A. J. Strojwas, Editor.

W. Nye is with Epsilon Active Inc., Berkeley, CA.

D. C. Riley and A. Sangiovanni-Vincentelli are with the Department of Electrical Engineering and Computer Sciences, University of California at Berkeley, Berkeley, CA 94720.

A. L. Tits is with the Electrical Engineering Department and Systems Research Center, University of Maryland, College Park, MD.

IEEE Log Number 8718822.

Optimization algorithms coupled with circuit simulation programs are obvious candidates to aid in the selection of design parameters. In fact, the idea of using optimization to design circuits—which we now survey—dates back to the 1950's and 1960's, beginning, perhaps, with the least squares curve fitting or matching problems discussed by Aaron [5] and Calahan [12]. Examples of engineering problems which were formulated as matching problems include model parameter determination for modeling system performance for computer simulation, black box techniques [38] (which have now evolved into present-day macromodeling), and the design of electronic filters and microwave integrated circuits. Early optimization techniques were most widely used in the electronic filter problems, as is suggested by the large number of papers on the subject. In these problems, the filter response functions to be matched against desired curves were usually the magnitude and phase of transfer functions (including input and output impedances) discretized over the independent parameter frequency.

A recent microwave circuit design system is COMPACT [3], a question/answer style interactive optimization program. It allows designers to minimize a weighted scalar error function consisting of the sum of the squared deviations between several frequency-domain properties of circuits. Control of this fixed problem formulation is accomplished by adjusting the weights or simply setting them to zero to remove their respective terms from the error function. However the inability of COMPACT to compute transient circuit responses or treat MOS circuits severely limits its range of application.

Extension of the use of optimization in electronic circuit design to consider both dc biasing effects and frequency-domain matching was tackled by Dowell [17] and later by McCalla [36] at Berkeley. The optimization performance function was again the sum of the squared errors between actual and desired responses summed over frequency, and only passive element values were allowed as design parameters.

The next step in the evolution of optimization in circuit design was the formulation of certain design problems as general nonlinear programming problems. This approach required the formulation of a single performance function (cost) to minimize subject to a set of inequality constraints, as in the following standard form:

$$\underset{x}{\text{minimize}} \left\{ f(x) \mid g(x) \leq 0 \right\}.$$

Little use was made, however, of the few constrained optimization algorithms that existed; in many cases [30], [54] the constrained problem was transformed into an unconstrained minimization problem using, for example, the penalty function approach of Fiacco and McCormick [19]

In parallel with the increased use of nonlinear programming in design came great improvements in the efficiency of calculating the gradients (network sensitivities) necessary for the most useful optimization algorithms [15], [16], [23], [53] (and more recently [28], [31], [41]).

The maturation of simulation and optimization techniques has led to the introduction of several computer-aided design systems over the past several years. A2OPT [24] is an optimization-based CAD system for electronic circuits which added constrained optimization features to the ASTAP [1] network analysis program. It was probably the first system to allow designers to conveniently specify a wide range of objectives and constraints as arbitrary expressions of circuit outputs. The program combined the weighted sum of several objective or constraint functions, the latter using penalty functions, into a single scalar function which was minimized using the variable metric, rank-one update method due to Cullum [14]. The choice of weights was somewhat ad hoc—they were adjusted experimentally by the designer after observing initial optimization performance [24].

Several years ago the INTEROPTDYN design package [10] was put together to enhance the use of optimization algorithms for engineering design in general and to study the methodology needed for man–machine interaction and graphical display. It combined a particularly powerful *method of feasible directions* optimization code called OPTDYN [9], INTRAC-C, an extension of the INTRAC [57] language-interpreter construction module developed at the Lund Institute of Technology, and various application-dependent simulation codes. In one version, [49] the objective and constraints of the design problem formulation were fixed, with the designer only providing numeric parameter values for his performance goals. Versions of INTEROPTDYN were used successfully for several different design applications but the difficulty in coding arbitrary objectives or constraints limited their use.

The APLSTAP system [25], [26], developed at IBM, is an interactive design system in which algorithms written in APL exercise control over the ASTAP [1] circuit analysis program. A novel linear programming (LP) step reveals tradeoffs between multiple objective and constraint functions to the designer, who then attempts to select a maximally effective LP step based on his experience about the various functions. Emphasis is placed on getting the most from the first few optimization steps rather than on a completely convergent sequence of steps.

In spite of the many different approaches mentioned here, there has been little use of parametric optimization in the circuit design community. In some cases, such as in electronic filter design, early efforts were successful. But for more complex present-day design situations, both early and recent optimization-based design systems have many shortcomings which limit their value. These include the use of primitive and nontunable optimization algorithms, the use of rigid classical optimization problem formulations [55], a lack of adequate interaction in the software for circuit optimization (a review of the literature reveals that optimization techniques were intended to completely *automate* the design process), the inadequacy of computing facilities [11], and deficiencies of circuit simulators. The most notable simulator deficiency is that most do not compute sensitivities. As a result, gradient calculations are performed using finite differences, leading to a large increase in the time needed to perform an optimization and to the possibility of numerical errors. Efforts such as the recent work of Trimberger and Matson [33] provide significant speedups and can be very useful but deal with very particular types of circuits and rigid performance specifications.

DELIGHT.SPICE, the union of the DELIGHT interactive optimization-based computer-aided design system [40], [42] and the Spice circuit analysis program [39], attempts to alleviate the above limitations by (1) placing optimization in the framework of a new application-oriented design problem formulation, (2) placing heavy emphasis on designer interaction, (3) utilizing an enhanced Spice that includes sensitivity computation, and (4) allowing arbitrary specifications on just about any type of circuit that can be handled by Spice.

As pointed out previously, circuit performance and specifications are, in general, functions of the design parameters through circuit responses. For example, the design of a wide-band amplifier could have as a *performance objective* to maximize the bandwidth of the amplifier and as a *constraint specification* that the dc power be less than some value. The design parameters could be a capacitor value and a BJT area. Neither the performance objective nor the constraint specification is an explicit function of the design parameters; this dependence is implicit through analyses of the circuit equations. In particular, the bandwidth can be evaluated by finding the -3 dB point of the frequency response from an ac analysis, while the dc power can be computed as the product of the dc current through the power supply times the supply voltage. This dc current would be computed by a dc analysis. Thus the performance and specification evaluations required to perform optimization of a circuit often involve expensive circuit simulations. The DELIGHT.SPICE system computes these circuit responses using the simulation program Spice.

The organization of this paper is as follows. In Section II we provide an overview of the optimization-based design methodology used in DELIGHT.SPICE. Our multiobjective problem formulation, an algorithm for solving problems formulated as such, and the DELIGHT.SPICE user interface are covered. The sensitivity computation for dc, ac, and transient analysis modes, added to Spice since our initial report on DELIGHT.SPICE [43], is discussed in Sections III-A and III-B while its computational requirements are discussed in Section III-C. In Section IV we present two detailed examples—the design of a wide-band operational amplifier and the design of a fast comparator for use in an analog-to-digital converter—both carried out using the optimization and design methodology provided by DELIGHT.SPICE. Finally, concluding remarks are given in Section V.

II. An Application-Oriented Design Methodology

DELIGHT.SPICE makes use of a design methodology recently proposed in the broader context of engineering design [44]. In Section II-A we motivate and exhibit an intuitive, application-oriented formulation of circuit design problems as optimization problems. In Section II-B, an algorithm is proposed capable of solving problems thus formulated. Section II-C briefly describes various aspects of the corresponding user interface.

A. Problem Formulation

The use of DELIGHT.SPICE for optimization requires the formulation of the design problem as a certain standard mathematical programming problem. Fortunately, formulating circuit design problems in this way is almost always possible. The discussion here begins with several simple but restrictive formulations and progresses to our general problem formulation.

The simplest nonlinear mathematical programming problem is the *unconstrained* nonlinear programming problem:

$$\underset{x}{\text{minimize:}}\ M(x)$$

in which the minimum (or maximum) value of some scalar function M of the design parameters viewed as elements of a vector x is sought. In engineering design, however, it is very likely that there are inequality constraints (e.g., power dissipation less than 1.5 W) which must be met. This can be accommodated in the standard *constrained* nonlinear programming problem:

$$\underset{x}{\text{minimize:}}\ M(x)$$

subject to:

$$I_1(x) \leq SpecValue_1$$

$$I_2(x) \leq SpecValue_2$$

$$\cdots$$

$$I_n(x) \leq SpecValue_n$$

where $I_1, I_2, \cdots, I_n$ are, in general, nonlinear functions of the design parameter vector x, and $SpecValue_1, \cdots, SpecValue_n$ are scalars representing the limits on the specifications that the circuit must satisfy.

Many commonly occurring constraints require that some specification be met *over a range of an independent parameter*, such as time, temperature, frequency, or even the voltage of an independent voltage source. These constraints are called *functional* inequality constraints and must be handled in a special way by optimization algorithms [35], [46], [48]. An example of a functional constraint is "Maintain the common mode rejection ratio within prespecified limits for every frequency in the interval 100 kHz to 10 MHz." DELIGHT.SPICE allows designers to specify these functional constraints in addition to the ordinary constraints and objective function introduced above. By adding functional inequality constraints to the problem formulation one arrives at the following *semi-infinite*[1] nonlinear programming problem:

$$\underset{x}{\text{minimize:}}\ M(x)$$

subject to:

$$I_1(x) \leq SpecValue_1$$

$$\cdots$$

$$I_n(x) \leq SpecValue_n$$

and

$$FI_1(x, \omega_1) \leq SpecCurve_1(\omega_1)\ \forall\ \omega_1 \in [\omega_{1_0}, \omega_{1_c}]$$

$$\cdots$$

$$FI_p(x, \omega_p) \leq SpecCurve_p(\omega_p)\ \forall\ \omega_p \in [\omega_{p0}, \omega_{pc}]$$

where $FI_1, \cdots, FI_p$ are functional inequality constraints and the *SpecCurve*'s are the specifications for the functional constraints that are in general given in the form of a function (possibly a constant) of the independent parameters $\omega_1, \cdots, \omega_p$.

When we first applied DELIGHT.SPICE in an industrial environment, we observed that capturing the intents of circuit designers and formulating them as outlined above were not such easy tasks. For example, it was often not clear how the objective function should be chosen or what precise values to give to each $SpecValue_i$. Also, designers' ideas of a good design seemed to evolve as possible solutions were exhibited by the system.

The way DELIGHT.SPICE now deals with these aspects is the result of this close interaction with circuit designers. We observed that designers often want to choose values of the design parameters so that a *set of objectives* rather than a single objective are "optimized" subject to a set of ordinary and functional constraints. Consequently, in DELIGHT.SPICE the previous single objective formulation is extended to accept specifications of the type

$$\underset{x}{\text{minimize}}\ \{M_1(x), \cdots, M_r(x)\}$$

[1]This name stems from the fact that the second argument to FI in the formulation may be viewed as leading to an infinite number of constraints.

where the minimization is interpreted in the minimax sense; i.e., the maximum of the r objective function values is minimized [32].

In addition, we observed that not all constraints are perceived by designers in the same way, leading us to classify constraints as either *hard* or *soft*. By *hard constraints* we mean constraints the designer considers most essential to have satisfied. He does not want them to take part in any subsequent design tradeoff. Obviously, any constraint required for physical realizability, such as resistance being positive, must be treated as a hard constraint. A stability constraint might also be treated as a hard constraint. *Soft constraints*, on the other hand, are those that the designer is interested in trading off against one another and against the performance objectives during intermediate iterations of an optimization run. In the terminology of [44], however, soft constraints are to be distinguished from design objectives in that when their specified thresholds have been achieved, no effort is made to make them progress further—the thresholds indicate the "point of diminishing return" for soft constraints. DELIGHT.SPICE allows the user to specify whether a constraint is to be treated as hard or soft.

Since objectives and soft constraints may be traded off by the designer, it is important to specify their relative importance to the optimization algorithm of DELIGHT.SPICE. For example, a constraint on power dissipation in a circuit such as power $\leq$ 400 mW might be very important to prevent chip overheating whereas the constraint $Z_{in} \geq 10$ MΩ on input impedance may be less important since often a considerably lower input impedance is acceptable. DELIGHT.SPICE provides the user with what we consider a natural way of describing the relative importance of soft constraints and objectives. The user specifies two values for each: a *good* value and a *bad* value. The meaning of these values is limited to the following understanding—the so-called *uniform satisfaction/dissatisfaction rule*: having each of the various objectives and soft constraints individually achieve their corresponding good values should provide the same level of "satisfaction" to the designer, while individually achieving the bad values should provide the same level of "dissatisfaction." Good and bad values are used in defining *normalized* objectives and soft constraints according to the transformation

$$f_i' \triangleq \frac{f_i - good_i}{bad_i - good_i}$$

where f_i is the "raw" value. Note that f_i' takes on values 0 and 1 when f_i equals its good and bad values, respectively.[2] Analogously, for functional specifications, DELIGHT.SPICE provides for good and bad curves, which are functions of the independent parameters. The corresponding normalized specification is

$$f_i'(\omega) \triangleq \frac{f_i(\omega) - good_i(\omega)}{bad_i(\omega) - good_i(\omega)}$$

for all values of ω in the specified range $[\omega_0, \omega_c]$ of the independent parameter. Here $f_i(\omega)$ is the raw value of an objective or constraint for the value ω of the independent parameter.

Good and bad values provide a very simple way to do tradeoff analyses: if a designer is unhappy with the performance level achieved by a particular objective or constraint, he simply changes what he considers to be satisfactory or unsatisfactory by adjusting the good and bad values, and resumes execution of the optimization. This is demonstrated later in the design examples of Section IV.

B. Optimization Algorithm

DELIGHT has a large library of optimization algorithms (see [40]). This paper concentrates on an algorithm introduced in [44] developed specifically for engineering design problems, and handling the problem formulation outlined in Section II-A.[3] This algorithm is an extension of the phase-1/phase-2 method of feasible directions with functional constraints (see, e.g., [48]). Methods of feasible directions were chosen for the following reasons:

1) They always produce a feasible design, if one exists, in a finite number of iterations (while the faster methods do not), and once a feasible design has been obtained, all subsequent iterates remain feasible.
2) They improve the design at each iteration, so that an acceptable solution may be achieved in the early steps. This is important when dealing with a time-consuming simulator such as Spice since rapid design improvement is sought in a few optimization iterations rather than after the near convergence of an infinite sequence.
3) They are easily extended to the case of semi-infinite constraints, i.e., constraints to be maintained over a range of an independent parameter [21], [46], [47].
4) They involve only first-order derivatives of objective and constraint functions (in fact, they require the computation of the gradients of only a subset of all the constraints: those which are almost active).
5) They are robust; i.e., they behave satisfactorily for any initial guess, and versions of this method are available which approximate derivatives and function evaluations adaptively [29] without losing their convergence properties.

Also, work has been done recently to improve the computational behavior of methods of feasible directions in an engineering design environment [55].

[2]Hence it is always desirable to *decrease* the normalized objectives, even if the original specification is a maximization (good value larger than bad value).

[3]The user does have the possibility of selecting a different algorithm but in that case the problem formulation and interactive features of the system would be different from what is explained in this paper.

The optimization algorithm used in DELIGHT.SPICE (called the phase I–II–III method of feasible directions [44]) may be viewed as consisting of three distinct phases. In phase I, each iteration tries to decrease the hard constraint violations, hopefully achieving a design where all hard constraints are satisfied. Phase II then takes over, in which each iteration progressively improves the worst normalized value among objectives and soft constraints while maintaining the hard constraints satisfied. Phase II terminates when (and if) all the good values have been achieved. Phase III then begins, in which the worst normalized objective value is progressively improved, while both hard and soft constraints are maintained satisfied.

In each of the three phases, one is confronted with a constrained (or unconstrained) minimax optimization problem of the form[4]

$$\min_{x} \left\{ \max_{j \in \boldsymbol{n}_f} f^j(x) \,\middle|\, g^j(x) \le 0 \;\forall\, j \in \boldsymbol{n}_g \right\} \tag{1}$$

where we have used the notation

$$\boldsymbol{n}_f \triangleq \{1, 2, \cdots, n_f\} \tag{2}$$

$$\boldsymbol{n}_g \triangleq \{1, 2, \cdots, n_g\} \tag{3}$$

and where n_g may be zero (in phase I it always is since only hard constraints are considered and these appear as f^j's in (1)) and at the beginning of each phase, a design vector x is known which satisfies the corresponding constraints. If and when a given phase is completed as described above, the overall design algorithm automatically switches to the problem associated with the next phase. Hence the central problem is solving (1) in such a way that none of the iterates ever violates the constraints.

Before considering a few properties of problem (1), we need to introduce some notation. For simplicity, we assume that no functional specifications are present. The *feasible set* F for problem (1) is defined as

$$F \triangleq \left\{ x \in \mathbb{R}^n \,\middle|\, g^j(x) \le 0 \;\forall\, j \in \boldsymbol{n}_g \right\} \tag{4}$$

and any x in F is called a *feasible* point for (1). For any x in F we then define

$$\psi_f(x) \triangleq \max_{j \in \boldsymbol{n}_f} f^j(x). \tag{5}$$

For any x in F and $\epsilon \ge 0$ we define the ϵ-*active* index sets as

$$J_{f,\epsilon}(x) \triangleq \left\{ j \in \boldsymbol{n}_f \,\middle|\, f^j(x) \ge \psi_f(x) - \epsilon \right\}, \tag{6}$$

the indices of f^j functions that are within ϵ of the maximum, and

$$J_{g,\epsilon}(x) \triangleq \left\{ j \in \boldsymbol{n}_g \,\middle|\, g^j(x) \ge -\epsilon \right\}, \tag{7}$$

the indices of g^j constraints that are within ϵ of being violated. Later it is the ϵ-*active* functions that steer an optimization iteration.

Our goal now is to obtain local solutions to (1) according to the following natural definition.

Definition 1: A point $\hat{x}$ in F is a *local solution* to (1) if there exists a positive scalar ρ such that

$$\psi_f(\hat{x}) \le \psi_f(x) \;\forall\, x \in B(\hat{x}, \rho) \cap F \tag{8}$$

where $B(\hat{x}, \rho)$ is the open ball of radius ρ centered at $\hat{x}$. ■

The following definition of a stationary point is analogous to that of a Fritz John point for a classical constrained optimization problem. A similar concept has been used in multiobjective optimization [37]. It is a particular case of the notion of stationary point in nondifferentiable optimization [13], [50].

Definition 2: A point $\hat{x}$ in F is called a *stationary point* for (1) if there exist two vectors $\mu_f \in \mathbb{R}^{n_f}$ and $\mu_g \in \mathbb{R}^{n_g}$, not both zero, such that the following conditions hold:

$$\sum_{j=1}^{m} \mu_f^j \nabla f^j(\hat{x}) + \sum_{j=1}^{l} \mu_g^j \nabla g^j(\hat{x}) = 0 \tag{9}$$

$$\mu_f \ge 0 \qquad \mu_g \ge 0 \tag{10}$$

$$\mu_f^j(\psi_f(\hat{x}) - f^j(\hat{x})) = 0 \;\forall\, j \in \boldsymbol{n}_f \tag{11}$$

$$\mu_g^j g^j(\hat{x}) = 0 \;\forall\, j \in \boldsymbol{n}_g. \tag{12}$$

■

The following statement of necessary conditions is easily proven.

Proposition 1: If $\hat{x}$ is a local solution to (1), then it is a stationary point for (1).

■

Proposition 1 is important because our optimization algorithm (and most others) can only hope to meet the *necessary* (as opposed to *sufficient*) conditions for a *local* (as opposed to *global*) optimum solution.

The optimization algorithm used in DELIGHT.SPICE [44] makes use of a search direction function $h_\epsilon(x)$ and of an optimality function $\theta_\epsilon(x)$ defined as

$$h_\epsilon(x) \triangleq \frac{h}{\|h\|} \tag{13}$$

$$\theta_\epsilon(x) \triangleq -\|h\| \tag{14}$$

where h is defined as

$$h = -\text{Nr co}\left\{ \left\{ \nabla f^j(x), j \in J_{f,\epsilon}(x) \right\} \cup \left\{ \nabla g^j(x), j \in J_{g,\epsilon}(x) \right\} \right\} \tag{15}$$

and "Nr co" designates the nearest point to the origin in the convex hull of a set. Here we see that it is only the ϵ-*active* functions that take part in determining the search direction.

[4]Functional specifications are handled by considering the worst value over the given range, i.e., $f^j(x)$ stands for $\max_{\omega} \phi^j(x, \omega)$, where $\phi^j(x, \omega)$ is the corresponding normalized function.

Algorithm 1
(with comments after the "#" character)

Parameters. $\delta > 0$, $\epsilon_0 > 0$, $\alpha, \beta \in (0, 1)$.
Data. $x_0 \in F$.
$i = 0$
repeat {
  $\epsilon = \epsilon_0$
  if $(\theta_0(x_i) = 0)$ **stop** # Stop if x_i is a stationary point.
  while $(\theta_\epsilon(x_i) \geq -\delta\epsilon)$ **do**
    $\epsilon = \epsilon/2$ # Find a sufficiently large search direction h.
  obtain $h_\epsilon(x)$ using (13)–(15)
  $k = 0$
  repeat { # Find a sufficient decrease in the maximum
    if $(\psi_g(x_i + \beta^k h_\epsilon(x_i)) \leq 0$ # f^j satisfying all the constraints.
      and $\psi_f(x_i + \beta^k h_\epsilon(x_i)) - \psi_f(x_i) \leq \alpha\beta^k\theta_\epsilon(x_i))$ **break**[5]
    $k = k + 1$
    }
  forever
  $x_{i+1} = x_i + \beta^k h_\epsilon(x_i)$
  $i = i + 1$
  }
forever
end

■

In the above algorithm, β^k means β to the power of k. Thus the stepsize starts at unity and is halved after each pass through the inner *repeat* loop. We usually use $\beta = 0.5$, with $\delta = 0.3$, $\epsilon_0 = 1$, and $\alpha = 0.3$ (but the user can modify these default values).

The algorithm can be proven to converge [44]:

Theorem 1: All the x iterates constructed by Algorithm 1 are feasible for (1). Moreover, if the sequence $\{x_i\}$ constructed by Algorithm 1 is finite, then the last point is stationary for (1). Otherwise, any accumulation point of the sequence $\{x_i\}$ is stationary for (1).
■

Theorem 1 says that Algorithm 1 in general produces a feasible point satisfying the necessary conditions for a local solution of (1). Now if, in a given phase of the solution of the design problem, Algorithm 1 constructs a sequence with an accumulation point $\hat{x}$ such that the corresponding cost $\psi_f(\hat{x})$ is negative, the sequence will be truncated after finitely many iterations, when the first point of negative cost will be reached, and a switch to the next phase will occur.

Various extensions and refinements of Algorithm 1 are in order if one wants to use it effectively, or even at all, in engineering design problems. It is crucial, for instance, to be able to handle functional specifications such as the common mode rejection constraint. To this effect, there is no conceptual difficulty in grafting onto Algorithm 1 the scheme developed by Mayne, Polak, Gonzaga, and Trahan, [21], [47] as well as the improved discretization scheme described in [46]. Also, the various enhancements introduced in [55] to improve the efficiency of methods of feasible directions in engineering applications can easily be fitted on Algorithm 1 as well. Among these enhancements is a special scheme for efficiently handling hard *box* constraints, i.e., constraints that consist of simple bounds on some of the design parameters.

[5] *break* means "exit the *repeat-forever* loop."

C. DELIGHT.SPICE User Interface

As pointed out in the Introduction, interaction is essential for effectively using optimization in design. We have developed a front end for easily describing the problem to be solved and capturing the knowledge and experience of the designer.

Describing a circuit design problem consists of defining the circuit structure, specifying the design parameters and their respective nominal variations and initial guesses, and expressing the various specifications. Defining the circuit structure is accomplished by creating a file that is virtually identical to a normal Spice input file, and further details are omitted from this paper. In the remainder of this subsection we give a feel for our description language for design parameters and performance specifications.

1) Design Parameters: Design parameters and possibly corresponding box constraint specifications are declared using, for example, the statement *sim_design-parameter M37_W variation=20u min=(10u,20u)*, which declares the width of MOS transistor M37 to be a design parameter. These statements not only identify the design parameters but also allow their names to be used in other objective and constraint expressions, and put the names in 1-1 correspondence with the optimization unknown vector x (see Section II-A). The nominal variation argument is introduced below. The optimal *min* (and *max*) pair in parentheses specify the good and bad value for a soft box constraint on the design parameter. (A hard box constraint can also be created.)

Parameter scaling is very important in achieving good optimization performance in DELIGHT.SPICE. To achieve it, the designer must specify a positive *nominal variation* for each design parameter, estimated according to the following *uniform parameter influence rule*: a change in each parameter by its nominal variation should influence the objectives and constraints to roughly the same degree. For example, the frequency response of an amplifier might be influenced to nearly the same degree by a 1 pF change in a 5 pF capacitor as by a 0.5 k change in a 10 k resistor.

2) Track Parameters: Another type of design parameter comprises *track parameters*. An important property of integrated circuits is the fact that the values of many circuit parameters can be made to match to a fairly high degree. The accuracy with which two identical transistors or resistors can be matched has a first-order effect on the attainable performance in monolithic operational amplifiers as well as other types of analog integrated circuits such as voltage regulators, analog multipliers, analog–digital converters, and voltage comparators [22]. Thus a designer need only consider one out of such a matched set of parameters as a design parameter; the others in the set (called "track parameters") simply track the first. For example, to make BJT areas in a differential pair matched, one is declared a design parameter while the other is declared a track parameter, the latter being set to an arbitrary expression of the former.

3) Objectives and Constraints: Specifications are first classified into objectives and constraints, resulting in two very similar types of declarations. For each objective the designer defines the actual expression to be minimized or maximized, the good and bad values, and any other statements which are necessary in the computation of the objective expression. These optional statements, which might be used for, say, setting up test conditions or loading or extracting results from Spice, appear after the keyword *using* as in the example

```
objective 3 'DC Power' minimize VDD* idc(VDD) good=300mw bad=400mw using {
   set RSWITCH1 = 10meg
   set RSWITCH2 = 1
   }
```

This is an objective to minimize the dc power in a circuit, measured *after* setting up the proper test conditions with two "switch" resistors. Objectives can be defined as an arbitrary expression of the design parameters, usually an implicit dependence through Spice current or voltage outputs from the various analysis modes (explained further below). Declarations for *functional* objectives and constraints differ from the above in that the interval range over which the constraint is to be satisfied must be given. For example, to constrain the output voltage of a voltage reference to lie within a 0.05 V band around 1.2 V over temperature we might have

```
constraint 1 'Upper Band' vdc(101) <= good=1.25 bad=1.3 soft
for_every TEMPDC from -55 to 125 initially by 10

constraint 2 'Lower Band' vdc(101) >= good=1.15 bad=1.1 soft
for_every TEMPDC from -55 to 125 initially by 10
```

where *TEMPDC* is the Spice simulation temperatures in degrees centigrade. The purpose of the keyword *initially* is to stress the fact that the *for_every* line specifies only an initial set of sample points; to guarantee convergence, our optimization algorithm adaptively modifies the distribution of sample points and automatically increases their number as an optimal point is approached using the techniques developed in [46].

4) Simulation Outputs: To allow the expressions in *objective* and *constraint* declarations to depend on output results of the various Spice analysis types (dc, ac, and transient), the desired output circuit voltage or currents must be declared *simulation outputs*. For the dc voltage at node 101, *simulation_output vdc(101)* would be used while magnitude and transient voltages would require *vm(101)* and *vtr(101)*.

A small digression on the use of these Spice outputs is in order. *vdc* can be used in any DELIGHT.SPICE expression and returns the dc node voltage from the last Spice dc analysis. If any Spice circuit parameters or test circuit "configuration" parameters have changed since the last analysis, *vdc* automatically triggers a new dc analysis; this rule applies to the ac and transient Spice simulation modes as well. The result is that the designer can specify objectives and constraints as *arbitrary* functions of simulation outputs with no need to indicate when the various Spice analyses should be performed—they are performed automatically as needed.

5) Pcomb Graphical Display: In the progress of a multiple-objective optimization computation, it is very desirable to have a display of objective and constraint values at each iteration which facilitates subjective evaluation of the design associated with that iteration. In DELIGHT.SPICE this purpose is served by the *Pcomb* performance comb, a graphical display which shows the designer how close each of his multiple objectives and soft constraints are to their corresponding good and bad values. Since most designer interaction with DELIGHT.SPICE occurs in phase II of the algorithm, hard constraints are not displayed. However, if any are violated in phase I, a message is printed.

The *Pcomb* presents the raw objective and constraint values (see Section II-A) numerically but the normalized values graphically. Referring to Fig. 1, the display consists of a vertical good line to the left and a vertical bad line to the right, with *G* appearing above the good line and *B* appearing above the bad, as shown. The normalized value of each objective or soft constraint is displayed by two horizontal bars, or *teeth*, on the comb, one for the previous comb drawn and one for the current comb. The previous comb teeth are in a lighter color or shade (and each slightly above the current one). The goal of the optimization algorithm is to move the tips of all the comb teeth to the left (in the direction of the good line). By minimizing the maximum of the (normalized) objectives or constraints, the algorithm in effect tries to move the rightmost tip to the left, even if other tips move slightly to the right; i.e., their performance becomes worse. If an objective or constraint value is such that the tip of its comb tooth should be drawn off the *Pcomb* display, an arrow is drawn to show that the tooth is out of the comb range.

The performance comb may be output automatically during each optimization iteration or manually after, say, adjusting the good or bad values for a particular objective or soft constraint. Since the comb display shows the previous comb teeth as well as the present one, the designer can easily see the results of such an adjustment of good or bad values as well as the improvement made by an optimization iteration. Extensive illustration of these uses of the *Pcomb* are shown later in Section IV.

6) Performing Tradeoffs: Tradeoffs between competing objectives or constraints are explored by adjusting good and bad values between optimization iterations. Basically, after several optimization iterations have been carried out with a set of good and bad values, the designer displays a performance comb and decides whether he is happy with the present values of his objectives and constraints. (In a tightly constrained design problem most of the algorithm execution will probably be spent in phase II. Recall that in phase II objectives and soft constraints are competing equally to be improved by the algorithm and the meaning of the good and bad values applies consistently to both.) If he is not happy with the present performance, he adjusts good or bad values to reflect his feelings and resumes the optimization.

For example, suppose dc power is a performance objective in an optimization design problem and good and bad values of 30 mW and 50 mW, respectively, have been specified. Suppose that at the current iteration of the optimization, the power is 40 mW. For these values, the tip of the associated comb tooth would appear midway between the good and bad vertical lines. Suppose that the designer is unhappy with the way several objectives (and/or constraints) have traded off and in particular wants to reduce the power further, at the expense of other objectives. This means that he now considers the value 40 mW to be worse than he previously did relative to other objectives. This means that the dc power objective bad value should be closer to 40 mW than it is presently. Thus, setting the bad value to, say, 45 mW or even 40 mW is the proper action. He then redisplays the comb and proceeds with additional iterations.

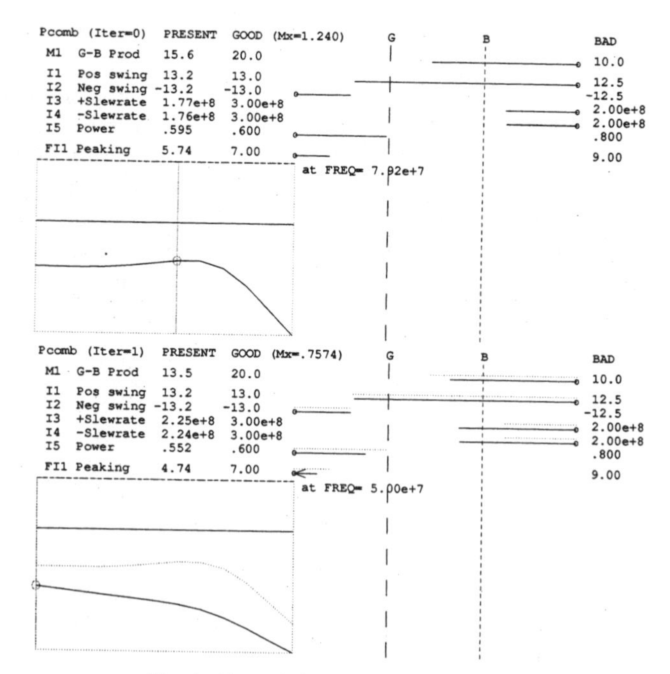

Fig. 1. Example Pcomb performance comb.

III. Computation of Sensitivities

In Section II-B we discussed an algorithm which has proven effective for the problem formulation which is the basis of DELIGHT.SPICE optimization. As was clear in that section, the algorithm requires the numerical computation of the gradients with respect to all design parameters of all the *ε-active* objectives and constraints occurring in the problem formulation for the circuit of interest. As was already mentioned in Section II-C, the objectives and constraints in general depend on dc, ac, and transient voltage and current *simulation_outputs* of Spice, denoted here by ψ. Calculation of the needed objective and constraint gradients requires the calculation of the gradients of these simulation outputs with respect to all design parameters. These gradients comprise the rows of the matrix $\partial\psi/\partial x$ needed to calculate $\nabla_x f(x, \psi)$ according to the (simplified) chain rule $\nabla_x f(x, \psi) = \partial f/\partial x + (\partial f/\partial\psi)(\partial\psi/\partial x)$. The elements of $\partial\psi/\partial x$ are conventionally referred to, with minor abuse of terminology, as sensitivities. Note that in the transient case, ψ is a function of time and hence so is $\partial\psi/\partial x$. Once the user has expressed the design objectives and constraints, DELIGHT.SPICE evaluates the corresponding chain rules *completely automatically*. Thus, as seen in Fig. 4 in Section IV below, the user does not have to provide any partial derivative information whatsoever. For more on how the chain rule is actually assembled inside DELIGHT.SPICE, see [45].

The use of finite differences to compute sensitivities is easily implemented, requiring only repeated calls to Spice with parameter values perturbed from the nominal. However unacceptable errors may occur and the computational cost of the approach, particularly in computing derivatives of transient responses, is prohibitive in most cases. The purpose of this section is to present the theoretical basis for the methods that have been implemented in DELIGHT.SPICE to calculate the gradients of Spice outputs with improved accuracy and acceptable computational efficiency.

The sensitivities of the dc outputs are computed using a special case of the computation of the transient sensitivities. The latter computation bears only a token conceptual similarity to the computation of the ac sensitivities. In view of this, the transient and dc cases are discussed together, then the ac case.

A. Transient and DC Sensitivities

The first objective here is to discuss the method for computing the transient sensitivities. The specialization which yields the dc sensitivity computation is then described.

To discuss these computations using a fully detailed form of the circuit equations underlying Spice would obscure the important concepts of the computations, while adding considerably to the length of the discussion. Instead, the presentation here is based on a very general formulation of the circuit equations, of which the Spice equation formulation is a special case. Specifically, this general formulation is written for an arbitrary lumped nonlinear network in terms of a vector $\boldsymbol{y}$ of network voltages, currents, capacitor charges, and fluxes spanning the solution space of the network. For prescribed source excitation (which are in general functions of time), $\boldsymbol{y}$ is a function of the design parameter vector $\boldsymbol{x}$ and of time t, i.e.,

$$\boldsymbol{y} = \boldsymbol{y}(\boldsymbol{x}, t). \tag{16}$$

The equations of the general formulation are written in terms of a differentiable vector-valued function $\boldsymbol{h}$ as

$$\boldsymbol{h}\big(\boldsymbol{y}(\boldsymbol{x}, t), \dot{\boldsymbol{y}}(\boldsymbol{x}, t), \boldsymbol{x}, t\big) = 0 \tag{17}$$

where a dot appearing over a variable denotes partial differentiation with respect to t.

Let the elements of the $\boldsymbol{y}$ vector be indexed by i from 1 to n_y and the elements in the $\boldsymbol{x}$ vector be indexed by j from 1 to n_x. The immediate goal here is to discuss the computation of the $n_y \times n_x$ matrix of time functions called the *sensitivity of* $\boldsymbol{y}$ *with respect to* $\boldsymbol{x}$, denoted $\boldsymbol{s}_{yx}$ and defined by

$$\boldsymbol{s}_{yx} = \frac{\partial \boldsymbol{y}(\boldsymbol{x}, t)}{\partial \boldsymbol{x}} \tag{18}$$

the (i, j) element of which is $\partial y_i(\boldsymbol{x}, t)/\partial x_j$.

The method for computing the sensitivity matrix, based on ideas in [31], [41], and [56], is as follows. Differentiating (17) with respect to x_j gives

$$\frac{\partial \boldsymbol{h}(*)}{\partial \boldsymbol{y}} \frac{\partial \boldsymbol{y}(\boldsymbol{x}, t)}{\partial x_j} + \frac{\partial \boldsymbol{h}(*)}{\partial \dot{\boldsymbol{y}}} \frac{\partial}{\partial x_j} \frac{\partial \boldsymbol{y}(\boldsymbol{x}, t)}{\partial t} + \frac{\partial \boldsymbol{h}(*)}{\partial x_j} = 0 \tag{19}$$

where $(*)$ is a symbol representing arguments of $\boldsymbol{h}$, specifically,

$$(*) = \big(\boldsymbol{y}(\boldsymbol{x}, t), \dot{\boldsymbol{y}}(\boldsymbol{x}, t), \boldsymbol{x}, t\big). \tag{20}$$

Interchanging the order of the differentiation in the second term gives

$$\frac{\partial \boldsymbol{h}(*)}{\partial \boldsymbol{y}} \frac{\partial \boldsymbol{y}(\boldsymbol{x}, t)}{\partial x_j} + \frac{\partial \boldsymbol{h}(*)}{\partial \dot{\boldsymbol{y}}} \frac{\partial}{\partial t} \frac{\partial \boldsymbol{y}(\boldsymbol{x}, t)}{\partial x_j} + \frac{\partial \boldsymbol{h}(*)}{\partial x_j} = 0. \tag{21}$$

If for $j = 1, 2, \cdots, n_x$ the *sensitivity of* $\boldsymbol{y}$ *with respect to* x_j, denoted $\boldsymbol{s}_{yx_j}$, is defined as

$$\boldsymbol{s}_{yx_j}(\boldsymbol{x}, t) = \frac{\partial \boldsymbol{y}(\boldsymbol{x}, t)}{\partial x_j} \tag{22}$$

then (21) becomes a differential equation in $\boldsymbol{s}_{yx_j}$, that is,

$$\frac{\partial \boldsymbol{h}(*)}{\partial \boldsymbol{y}} \boldsymbol{s}_{yx_j}(\boldsymbol{x}, t) + \frac{\partial \boldsymbol{h}(*)}{\partial \dot{\boldsymbol{y}}} \dot{\boldsymbol{s}}_{yx_j}(\boldsymbol{x}, t) + \frac{\partial \boldsymbol{h}(*)}{\partial x_j} = 0. \tag{23}$$

$\boldsymbol{s}_{yx_j}$ is the jth column of the desired sensitivity matrix $\boldsymbol{s}_{yx}$. At the highest level the determination of this matrix is carried out column by column, by solving (23) with an appropriate intial condition for $\boldsymbol{s}_{yx_j}$, $j = 1, 2, \cdots, n_x$.

Equation (23) is a linear differential equation with time-varying coefficients dependent on the solution ($\boldsymbol{y}$ and $\dot{\boldsymbol{y}}$) of the circuit equations through the arguments symbolized by $(*)$ (see (20)). Spice solves the circuit response differential equations numerically by discretizing the time axis and using the backward Euler and trapezoidal numerical integration formulas. The same approach can be applied in solving (23).

In one implementation of this approach to computing the sensitivity matrix, the computation of the circuit responses for the entire simulation time interval of interest could precede the sensitivity computations. However, this requires that in general most of the elements of $\boldsymbol{y}$ and $\dot{\boldsymbol{y}}$ over the simulation interval be stored. This can be avoided if all the sensitivity column vectors at each time point are computed as soon as $\boldsymbol{y}$ and $\dot{\boldsymbol{y}}$ for that time point are known. If a variable-timestep integration is implemented, this approach offers the additional advantage that adjustment of the timestep can be controlled in part by errors occurring in the sensitivity solution.

Another important feature of the sensitivity computation is that at each time point the coefficient matrix needed to solve for all of the sensitivity column vectors is the Jacobian at convergence of the Newton–Raphson solution of the circuit equations at that timepoint. The availability of this coefficient matrix in LU-factored form represents a significant computational cost savings.

The special case of the method discussed above which describes the Spice sensivity computation is obtained as a particular choice of the y vector, and a particular structure of (17). Regarding the choice of y vector, first note that the circuit equations which are the basis for Spice circuit analyses are formulated using modified nodal analysis [2]. Among the unknown circuit variables in this formulation are currents through inductors, independent and controlled voltage sources, and current-controlled current sources. In the context of the Spice implementation, the y vector can be considered to be of the form

$$y = [v \quad q]' \tag{24}$$

where v consists of the node voltages, and currents through the device types just described, treated as unknowns in Spice, and q consists of the capacitor charge and inductor flux variables treated as unknowns in Spice. Note that the computation of s_{yx}, and hence of its "upper" submatrix, $[\partial y(x, t)]/\partial x$, suffices to yield the partial derivatives of all current and voltage outputs of Spice and thus the elements of $\partial\psi/\partial x$.

Regarding the structure of (17) in terms of v and q and differentiable functions w and ϕ, these circuit equations can be written for circuits simulated by Spice in the form

$$w(v(x, t), \dot{q}(x, t), x, t) = 0 \tag{25}$$

$$q(x, t) - \phi(v(x, t), x) = 0. \tag{26}$$

The first equation is not explicitly dependent on q, nor is the second on $\dot{q}$ or t. Neither equation is explicitly dependent on $\dot{v}$. More importantly, the second set provides an explicit expression for q. This allows a transformation of the solution equations in which the unknown vector of the simultaneous linear equations which must be solved to compute the sensitivities is v rather than the larger y vector. Carrying out this transformation is relatively straightforward, and the details are omitted here.

The dc outputs of Spice are the results of the analysis of the circuit with constant excitation and under the condition that the capacitors are replaced by open circuits and the inductors by short circuits. In this case, the unknown vector y is a function only of the parameter vector x, and it should be clear from the discussion of the transient case that the result desired from the dc sensitivity computation is the matrix $[\partial v(x)]/\partial x$. That this can be computed as a special case of the transient sensitivity computation can be argued from (25). With excitations fixed and with no time-varying circuit elements, there is no dependence of the circuit equations on time. Without loss of generality, t can be set to 0, and hence (25) can be modified to

$$w(v(x), 0, x, 0) = 0. \tag{27}$$

Therefore the dc circuit and sensitivity solutions can be obtained from the transient solution equations by setting the second and fourth arguments of w to zero (and likewise for the term $\partial w/\partial \dot{q}$ which appears in the solution equations).

In the discussion of the transient sensitivity computation, the determination of initial conditions for the sensitivity differential equation of (23) was not addressed. Having discussed the dc sensitivity computation facilitates returning to this issue. In the transformed formulation underlying the Spice implementation, what is needed specifically is the value of the sensitivity of the q vector at $t = 0$, which, in agreement with notational practice here, is written as the matrix

$$s_{qx}(x, 0) = \frac{\partial q(x, 0)}{\partial x}. \tag{28}$$

The jth column of this matrix serves as the initial condition for the d.e. pertaining to the computation of s_{yx_j}. Although there are of course an infinity of values to which $s_{qx}(x, 0)$ can be set, there are two discrete approaches which stand out for their usefulness. First, if the application in which the circuit being optimized is to be used is such that initial conditions are *imposed* on the circuit independent of the values of its design parameters, $s_{qx}(x, 0)$ should be set to the 0 matrix. If, as is more commonly the case, no such initial conditions are imposed on the circuit, the circuit solution is initialized in Spice by the quiescent conditions of the circuit. These are in general a function of design parameter values. In this case it is not difficult to see that $s_{qx}(x, 0)$ should be set to the value of the dc q vector sensitivity matrix which corresponds to the source excitation values at $t = 0$. Hence, in this second method of setting initial conditions, the transient sensitivity solution is preceded by a dc solution. The treatment of sensitivity initial conditions presents examples of the extensive parallelism between Spice circuit analysis and the sensitivity computation described in this paper.

B. AC Sensitivity

The ac outputs of which sensitivities are needed in DELIGHT.SPICE can be represented by the complex amplitude vector, say V, analogous to the real vector v of unknown currents and voltages introduced above. Given the circuit to be optimized, the response vector V depends on the frequency and complex amplitudes of the independent sources in the circuit, and on design parameter values. However as far as the circuit response and sensitivity calculations are concerned, the frequency and complex amplitudes of the independent sources for each of the circuit conditions needed to compute the objectives and constraints are constants prescribed by the problem formulation. (If functional constraints are present, the optimization algorithm also plays a role in determining the frequencies of interest.) Therefore in the context of this discussion, V can be considered to depend only on the design parameter vector x.

In Spice, V is computed as the solution of the linear equation which results from the modified nodal analysis formulation of the ac circuit equations. Letting W represent the complex coefficient matrix of this equation, and

J the right-hand side, the equation is $WV = J$. J depends only on prescribed source amplitudes, but W depends on circuit quantities such as incremental capacitances, conductances, and transconductances of semiconductor devices. These circuit quantities depend on design parameter values, in general both explicitly and through their dependence on dc quiescent conditions (which depend in turn on design parameter values). For the purposes of this discussion, let $\boldsymbol{v}$ denote the vector of dc quiescent currents and voltages of the circuit in question, and let the elements of $\boldsymbol{v}$ be indexed by i from 1 to n_v. Then the functional dependencies in the circuit equations can be represented according to

$$W(x, v_1(x), v_2(x), \cdots, v_{n_v}(x))\, V(x) = J. \quad (29)$$

As in the transient sensitivity case, the derivation of the equations for ac sensitivity is basically carried out by differentiation of the circuit equations with respect to the design parameters taken one at a time. Noting that J is independent of the design parameter vector x, the derivation is a straightforward application of the chain rule and the result for the jth design parameter x_j can be written in the form

$$W(*)\frac{\partial V(x)}{\partial x_j} = -\left[\sum_{i=1}^{n_v} \frac{\partial W(*)}{\partial x_j} + \frac{\partial W(*)}{\partial v_i}\frac{\partial v_i(x)}{\partial x_j}\right] V(x)$$

where $(*)$ represents the argument list appearing in (29), $(x,\ v_1(x),\ v_2(x),\ \cdots,\ v_{n_v}(x))$. Recognizing that $[\partial v_i(x)]/\partial x_j$ is just the dc sensitivity $s_{v_i x_j}(x)$, and letting s_{Vx_j} denote $\partial V/\partial x_j$, this can be rewritten as

$$W(*)\, s_{Vx_j} = -\left[\sum_{i=1}^{n_v} \frac{\partial W(*)}{\partial x_j} + \frac{\partial W(*)}{\partial v_i} s_{v_i x_j}(x)\right] V(x). \quad (30)$$

Given x, $V(x)$ from the ac circuit solution, and $s_{v_i x_j}(x)$ from the dc sensitivity solution, the only unknown quantity in (30) is the desired sensitivity vector s_{Vx_j}. This is computed from this equation. Note that as in the transient and dc cases, the coefficient matrix of the sensitivity equation is the same as the coefficient matrix of the circuit equations, affording significant computational savings.

There is in the ac sensitivity case an additional type of computational savings. The technique is loosely analogous to one used by Spice in the ac circuit solution. Note that for each particular value of x, the dependence of the coefficient matrix W on the frequency of the independent sources has the particularly simple form given by

$$W(\omega) = P + \hat{j}\omega Q \quad (31)$$

where $\hat{j}^2 = -1$, ω is radian frequency, and P and Q are real matrices independent of ω. (Inductance elements do not give rise to an imaginary term inversely proportional to ω, on the right of (31), because in the Spice modified nodal analysis formulation, equations for the voltages across all inductors take the place of inductance entries in the nodal admittance matrix.) Computation of the elements of P and Q is expensive. Consequently when ac responses are desired at more than one frequency, as is usually the case, Spice in effect stores P and Q, so that only the computation of (31) is required to obtain W at a new frequency.

A computationally similar circumstance occurs in the ac sensitivity calculation. The calculation of the right-hand side of (30), which usually dominates the cost of the ac sensitivity calculation, entails the calculation of a total derivative of an admittance matrix, which is affine in $(\hat{j}\omega)$, as is W in (31). Accordingly, at each new x value, the coefficient matrices of this affine relationship are stored to speed computations when sensitivity analysis is needed at more than one frequency.

C. Computational Requirements of Spice Sensitivity

The computational cost of the sensitivity method of the previous section is directly proportional to the number of design parameters, since the solution of (23) (calculation of a right-hand-side vector and a forward/backward substitution with an existing L/U factored circuit equation Jacobian) is repeated for each parameter. To quantify this dependency, we performed several tests on our two circuit examples of the next section. For the high-speed op-amp, we measured the CPU time required for independently performing dc, ac, and transient analyses with sensitivity computation for 0, 1, 2, 5, 7, and 10 resistor design parameters and again for these numbers of BJT area design parameters. For the comparator, the same tests were performed but only for transient sensitivity. The results are shown in Table I.

Below each column is the linear least squares formula for computing the CPU time versus N, the number of design parameters in the third column. For example, for the op-amp, the table shows that each additional BJT area design parameter costs 2 additional seconds for dc analysis, 0.58 for ac, and 19 for transient. Overall it appears that the additional cost for each added design parameter is *very roughly* 10 percent of the analysis cost without any sensitivity computation, a conclusion similar to the 11 percent value reached by Hocevar *et al.* [28] for the transient sensitivity of MOSFET circuits. Although this is small in comparison with an entire analysis required by a finite difference gradient approach, it is still reason to choose design parameters with care—just ten design parameters approximately double the analysis time. It is for this reason that in our design examples we choose only circuit parameters that might have considerable effect on the design objectives and constraints.

To illustrate the benefit of using Spice sensitivity over finite differences, we measured the CPU times for a comparator circuit (the same circuit used later in Section IV-B). After two complete optimization iterations, we obtained the following results:

	Total CPU time (seconds)	Number of Transient Analyses
Sensitivity	1362	7
Finite Difference	5566	31

TABLE I
CPU TIME FOR SPICE SENSITIVITY

Circuit	Design Parameter	Number of Design Parameters	Cpu Times in Seconds		
			DC	AC	Transient
Opamp	Resistor	0	27	2.1	158
		1	28	3.0	223
		2	33	3.8	225
		5	39	5.6	278
		7	40	7.0	317
		10	45	8.6	369
			28+1.8N	3.3+.53N	182+19N
	Bjt Area	0	26	2.2	158
		1	26	3.1	216
		2	33	4.0	225
		5	39	6.0	279
		7	40	7.7	311
		10	45	8.8	364
			27+2.0N	3.2+.58N	180+19N
Comparator	Resistor	0	-	-	54
		1	-	-	64
		2	-	-	70
		5	-	-	87
		7	-	-	99
		10	-	-	113
					57+5.8N
	Bjt Area	0	-	-	55
		1	-	-	66
		2	-	-	72
		5	-	-	91
		7	-	-	105
		10	-	-	122
					58+6.6N

The finite difference case did 24 analyses for two iterations of 12 design parameters, plus the same seven analyses in the stepsize computation as done by the sensitivity case, for a total of 31 analyses. In the sensitivity case, all transient analyses occurred during the stepsize computation (sensitivity is computed during *every* analysis) so there were only the seven analyses total. The extreme computational cost of the direction-finding step of our algorithm has been reduced to practically zero!

IV. DESIGN EXAMPLES

All of the examples introduced in this paper are derived from actual product development or redesign activities at Harris Semiconductor; the results of our efforts are presently being incorporated into several of the products discussed. The values of the design parameters with which the optimization sessions started were the end products of manual design procedures by experienced designers. Thus, the improvements in performance that we report are indeed significant.

A. Design of a High-Speed Operational Amplifier

The circuit shown in Fig. 2 is an operational amplifier currently in production at Harris Semiconductor. This amplifier was to offer maximum bandwidth and stability at a reasonable closed-loop gain (5), along with a minimum settling time. The following were the initial design goals:

gain–bandwidth product	$\geq$ 300 MHz
output voltage swing	$\geq$ 13 V
$\pm$ slew rate	$\geq$ 300 V/μs
dc power dissipation	$\leq$ 600 mW
offset voltage	$\leq$ 5 mV

"stable" at a closed-loop gain of 5

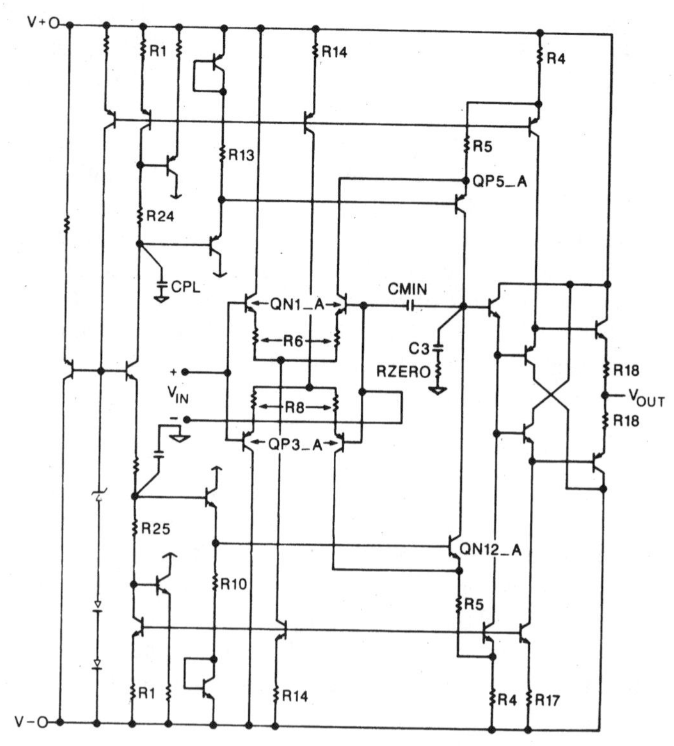

Fig. 2. High-speed operational amplifier of the first design example. The values of all of the labeled elements are optimization design parameters.

The dc limits, while important, were not of primary concern and could be relaxed to achieve better ac performance since the relative marketability of these specifications was very subjective. For example, if significant gain–bandwidth product could be maintained and stability increased at the cost of a small increase in power, the product might be more desirable. The design methodology of DELIGHT.SPICE is well suited to this sort of subjective tradeoff.

1) Op-Amp Problem Formulation: The first step in using DELIGHT.SPICE is the choice of design parameters and their initial values. Ideally, to explore the full potential of a design, one should include all the parameters of a circuit. However, since as shown in Section III-C the CPU time spent by the system increases with the number of design parameters, this strategy would be impractical. Thus, only circuit parameters that have considerable effect on the design objectives and specifications should be considered. For this example, the parameters chosen are the ones labeled in Fig. 2.

After the selection of a set of design parameters, it was necessary to translate the goals listed above into the mathematical programming formulation of DELIGHT.SPICE. In general, this consists of four steps. The first step is to express them as explicit functions of the design parameters or of quantities that can be computed by Spice. For example, consider the unity gain–bandwidth product GBW. We determine it by measuring the gain at a relatively low frequency (on the dominant pole portion of the Bode plot) and extrapolating. Thus $GBW = FREQ \cdot |Vout/Vin|$, where $FREQ$ is the frequency at which the open-loop gain is measured.

Second, the designer must decide whether each performance goal is to be treated as an objective or a constraint. Both of these require the specification of desired values. However, objectives continue to improve throughout the optimization while constraints "stop being pushed" after they achieve their desired good (target) values. For our op-amp example we decided to consider the gain–bandwidth product as the only objective; all other goals were treated as constraints.

The third step is to decide whether each constraint is to be treated as *hard* or *soft*. For the present example we wanted all of the constraints to take part in tradeoffs and hence we indicated (with the keyword *soft* as shown in Fig. 3) that all were soft.

The last step in setting up the problem formulation is to indicate the relative importance of the objectives and constraints by providing a *good* and a *bad* value for each. The precise values specified need only be "ball-park" values since DELIGHT.SPICE contains commands for modifying them during the optimization session. For example, the goal of having the gain–bandwidth product greater than 300 MHz reflects our objective to improve it as much as possible but at least have it as large as that of the initial design. We chose initial good and bad values of 500 MHz and 150 MHz. These translated into gains of 500 MHz/50 MHz = 10 and 150 MHz/50 MHz = 3 at frequency 50 MHz, leading to good and bad values for this objective of approximately 20 dB and 10 dB, respectively. Good and bad values were determined for the various soft constraints in a similar manner, keeping in mind the uniform satisfaction/dissatisfaction rule of Section II-A.

2) Op-Amp Problem Description: A complete example of the problem-oriented input language in DELIGHT.SPICE is given in our problem description for the op-amp in Fig. 3. It is important to mention that *vm* and *vp* return the ac magnitude and phase of the specified node at the value of global variable *FREQ*. Similarly, *vtr* returns the transient output voltage of the specified node at the value of global variable *TIME*.

File bopampM

```
prob_function multicost

  objective 1 'G-B Prod' maximize retval good=20 bad=10 using {
    FREQ = 50meghz
    retval = db(vm(24)/vm(80))
    }

end_multicost
```

File bopampI

```
prob_function ineq

  constraint 1 'Pos swing' vdc(13) >= good=13 bad=12.5 soft

  constraint 2 'Neg swing' vdc(16) <= good=-13 bad=-12.5 soft

  constraint 3 '+Slewrate' retval >= good=300e6 bad=200e6 soft using
    retval = vdc(78,15) / (C3+CMIN+6.4p)

  constraint 4 '-Slewrate' retval >= good=300e6 bad=200e6 soft using
    retval = -vdc(79,15) / (C3+CMIN+6.4p)

  constraint 5 'Power' power <= good=.6 bad=.8 soft using
    power = ( vdc(222,2)*15 + vdc(111,1)*(-15) )

end_ineq
```

File bopampFI

```
prob_function fineq

  constraint 1 'Peaking' vm(24) <= good=7 bad=9 soft
  for_every FREQ from 50me to 120me initially dec 30

end_fineq
```

Fig. 3. Problem description for the high-speed operational amplifier.

The ordinary (*ineq*) constraints show that the output voltage swing was formulated as a separate inequality constraint for positive and negative swing. Similarly, the slew rate specification also resulted in two constraints. The slew rate was determined in an approximate way to avoid expensive transient simulations. In particular, it was computed by dividing the maximum charging current by the total compensation plus parasitic capacitance. (Although we avoid an expensive transient analysis here, we *do* use transient analysis with sensitivity computation, an important result of this research, in Section IV-B for the second design example.) The offset voltage specification does not appear as a constraint. V_{os} is made up of a preferential term and a statistical term. Since V_{os} was not of vital importance to this product, we avoided Monte Carlo analysis by considering only the preferential term. To ensure that V_{os} was centered at zero, we simply forced certain nominally matched op-amp parameters to track one another using track parameters, introduced in Section II-C. To quantify the stability goal at a gain of 5, we constrained the closed-loop frequency response to be less than 7.0 for every frequency in an interval. The interval range end points were frequency values that we decided would surely contain the peak. Thus, stability was handled as a *functional* inequality constraint over frequency (see Section II-A). Another approach that would be much more expensive computationally is to measure stability using the settling time or overshoot of a transient simulation output waveform. In general, care must be used in choosing how to compute the quantities involved in the specifications. Note that a similar problem arises when a designer simply wants to *examine* the performance of his/her design using circuit simulation, e.g., with batch Spice.

The complete problem description is shown in Fig. 3.

3) Op-Amp Optimization Sessions: After completing the description of the design problem, entering the DELIGHT.SPICE environment, and issuing a command to process the problem description files, we were ready to begin optimization. The command *run 3* requested that three iterations be performed, each automatically fol-

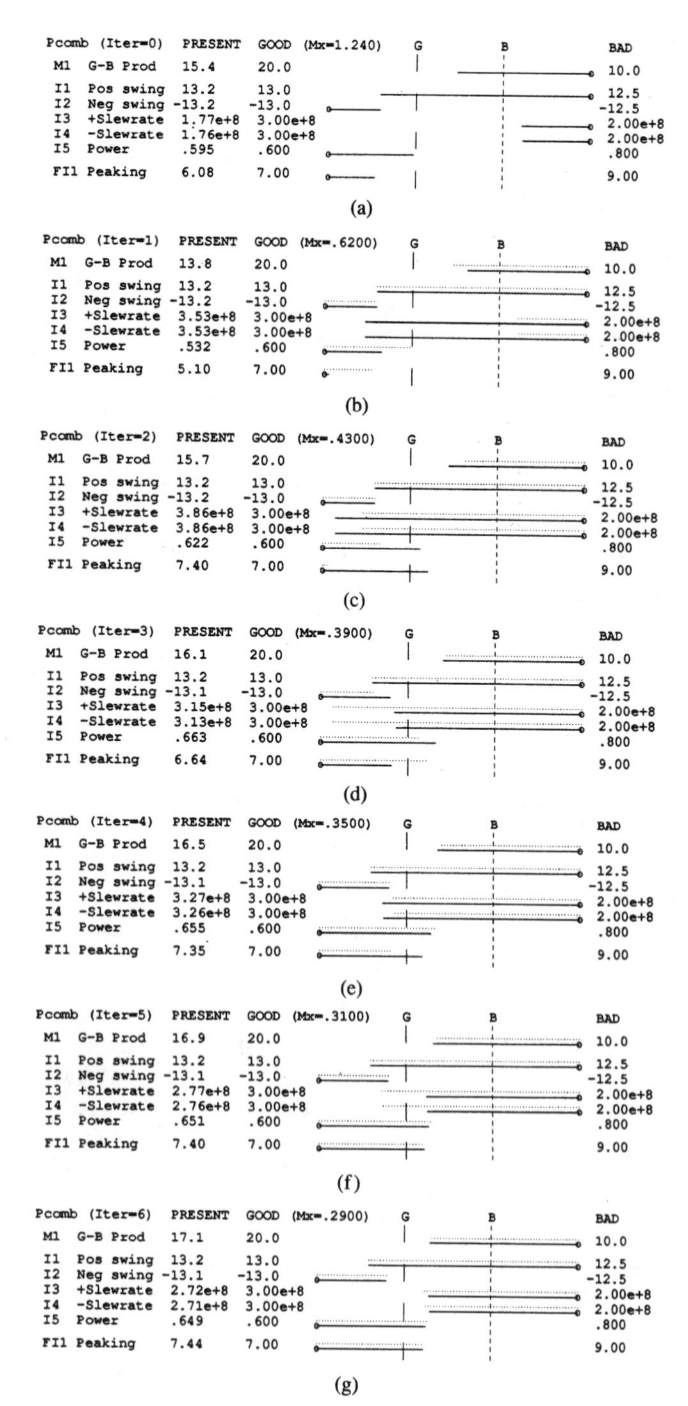

Fig. 4. Performance combs for the first six optimization iterations.

lowed by output of the *Pcomb* performance comb. In Fig. 4(a), the initial (Iter = 0) Pcomb is shown. It immediately illuminates the fact that most of the objectives and constraints are fairly satisfactory—better than or equal to their good values—but that the two slew rate soft constraints are quite unsatisfactory—worse than their respective bad values.

After one rather spectacular iteration, the Pcomb in Fig. 4(b) shows that the slew rate constraints and others have made a large improvement, but at a slight expense of the gain-bandwidth product, labeled as *G–B Prod*, or objective M1. To give an example of how the algorithm displays the current design parameters after each iteration,

	Parameter	Value	%wrt 0	Prev
1	C3	6.213e − 13	211%	211%
2	CPL	3.000e − 13	0%	0%
3	CMIN	4.105e − 12	3%	3%
4	R1	1.395e + 3	0%	0%
5	R14	2.364e + 2	24%	24%
6	R4	1.227e + 2	−6%	−6%
7	R17	3.300e + 2	0%	0%
8	R10	2.400e + 3	0%	0%
9	R13	2.400e + 3	0%	0%
10	R5	5.877e + 1	−41%	−41%
11	R24	7.115e + 2	2%	2%
12	R25	7.104e + 2	1%	1%
13	R6	4.000e + 1	0%	0%
14	R8	4.000e + 1	0%	0%
15	R18	2.000e + 1	0%	0%
16	QN1_A	.9999	0%	0%
17	QP3_A	1.000	0%	0%
18	QN12_A	1.016	2%	2%
19	QP5_A	1.015	2%	2%
20	RZERO	1.000	0%	0%

Fig. 5. Parameter values after the first optimization iteration.

Fig. 5 shows this output after the first iteration. Shown in four columns are each parameter name, its current value, and the percentage change with respect to the original (Iter = 0) guess and with respect to the previous iteration. The top line also shows the iteration number, that we are in phase II of the phase I–II–III algorithm, and that the **Max**imum scaled value of all **M**ultiple objectives plus **S**oft **C**onstraints (indicated *MxM+SC*) is 0.6231. This value can also be seen in the Pcomb in Fig. 4(b), where the rightmost present comb tooth (M1) is just about six tenths of the way between the zero-valued vertical good line and the unity-valued bad line.

Our original *run 3* command completed after iterations 2 and 3, the Pcomb's of which are shown in Fig. 4(c) and 4(d), respectively. Since the previous and present comb teeth in these figures show that the optimization had hardly settled down, i.e., approached a stationary point in the sense of Definition 2, we decided to give another *run 3* command. This produced the Pcomb's for iterations 4, 5, and 6, shown in Fig. 4(e), (f), and (g). Finally, in Fig. 4(g) we saw that almost all of the objectives and soft constraints were competing and had almost reached what for this problem might be called a *maximally regular point* (see [25] for a discussion of regular points). In fact, the small values in the "percentage change with respect to previous" column of the iteration 6 output shown in Fig. 6 also testified to the fact that the optimization was nearing a stationary point. The total computational cost of these six iterations was 6513 seconds on a Masscomp MC500 workstation (with 68010 and floating point hardware). Even though the op-amp problem was formulated *without* any transient analyses, this rather high cost was due to the 64.4 seconds per dc analysis and the 15.4 seconds per ac analysis at a single frequency point required by this moderately sized, 28-transistor circuit. (Also, each transistor was modeled with the full Spice Gummel–Poon transistor model.)

	Parameter	Value	%wrt 0	Prev
1	C3	0.000	−100%	****%
2	CPL	1.306e − 12	335%	10%
3	CMIN	4.561e − 12	14%	3%
4	R1	1.420e + 3	1%	0%
5	R14	2.436e + 2	28%	0%
6	R4	1.003e + 2	−23%	1%
7	R17	3.291e + 2	0%	0%
8	R10	2.398e + 3	0%	0%
9	R13	2.399e + 3	0%	0%
10	R5	8.038e + 1	−20%	0%
11	R24	7.032e + 2	0%	0%
12	R25	7.014e + 2	0%	0%
13	R6	3.673e + 1	−8%	−1%
14	R8	3.316e + 1	−17%	−3%
15	R18	1.705e + 1	−15%	−3%
16	QN1_A	1.216	22%	3%
17	QP3_A	1.214	21%	3%
18	QN12_A	.9040	−10%	−2%
19	QP5_A	.6844	−32%	−4%
20	RZERO	1.708	71%	0%

Fig. 6. Parameter values after the sixth optimization iteration.

Although we were quite content with the results of the first six iterations above, we decided to change our major emphasis for the product to one of low power consumption. Thus, the good and bad values for the dc power constraint were lowered from 600 mW and 800 mW to 300 mW and 500 mW, respectively. Before continuing the optimization, we first output the Pcomb shown in Fig. 7(a) so we could see the effect of these changes. Note that the *Power* constraint (I5) was now not very satisfactory, as shown by its comb tooth protruding far to the right of the vertical bad line. After issuing a *run 3* and then later another *run 4* command, we were entertained by the optimization iteration Pcomb's shown in Fig. 7(b) through (h). The Pcomb in Fig. 7(b) shows very clearly the tradeoff between dc power and gain-bandwidth product—power had dropped significantly at the slight cost of a lower gain-bandwidth product. The Pcomb in Fig. 7(h) again suggests that we had reached a stationary point. As before, this fact was also backed up by the fourth column of the parameter value output in Fig. 8. With these seven additional iterations, a low-power version of the op-amp was obtained with the following characteristics:

$$\begin{aligned} \text{gain-bandwidth product} &= 300 \text{ MHz} \\ \text{peaking} &= 20\% \\ \text{dc power} &= 398 \text{ mW} \\ \text{slew rate} &= 253 \text{ V}/\mu\text{s}. \end{aligned}$$

As demonstrated above, the ability to give different emphasis to objectives and constraints can result in a variety of attractive solutions, all of which can be realized with only minor changes in a few IC masks. Thus the designer can offer the marketing department several product options, each of which represents the circuit's best performance for the particular emphasis given. (A methodology for accurately identifying distinct product options desired

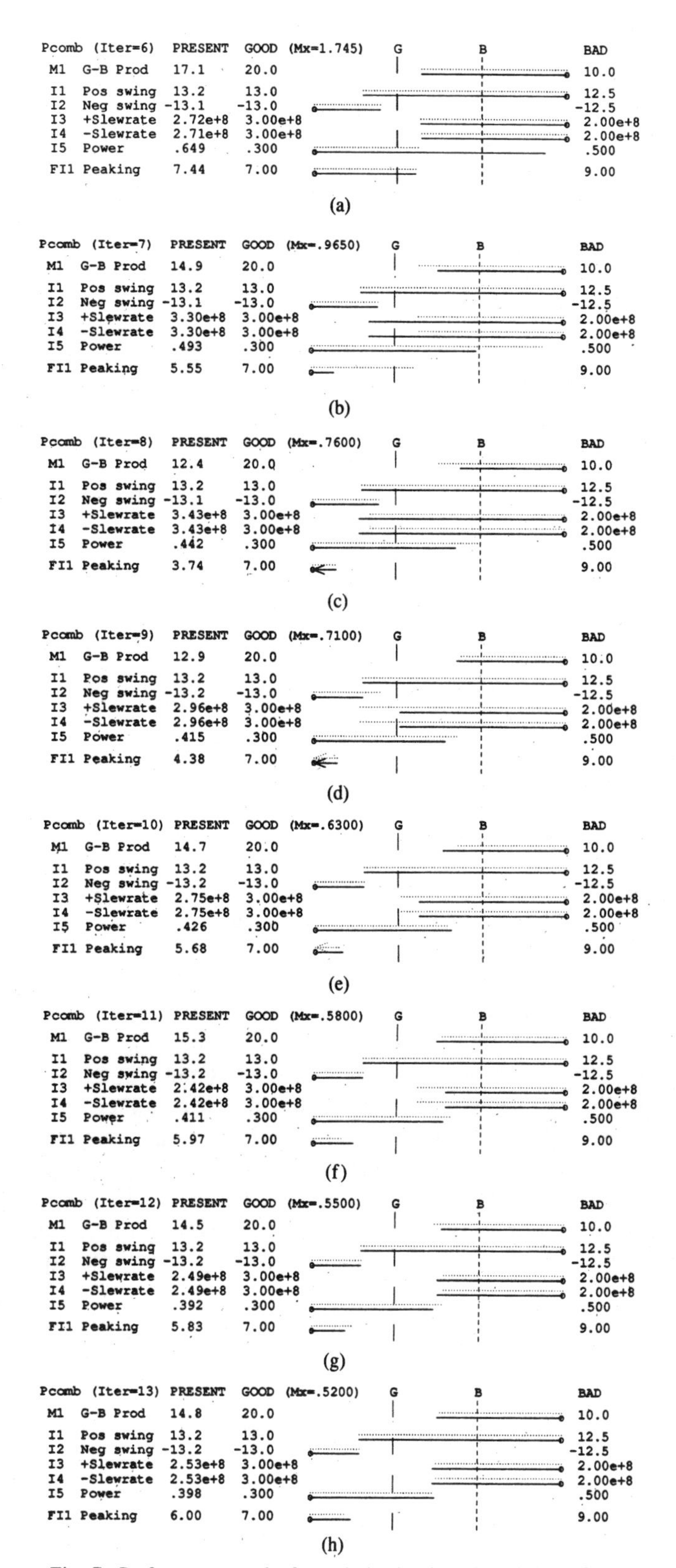

Fig. 7. Performance combs for optimization iterations 7 through 13.

by potential customers has in fact been developed as part of an evolving statistical design methodology [52].)

B. Design of an A/D Comparator

A critical part of an analog-to-digital converter system is the comparator section, which must resolve the polarity

	Parameter	Value	%wrt 0	Prev
1	C3	4.371e − 13	119%	0%
2	CPL	1.352e − 12	351%	0%
3	CMIN	4.567e − 12	14%	0%
4	R1	1.358e + 3	−3%	0%
5	R14	3.332e + 2	75%	1%
6	R4	1.881e + 2	45%	−3%
7	R17	3.422e + 2	4%	0%
8	R10	2.399e + 3	0%	0%
9	R13	2.401e + 3	0%	0%
10	R5	7.108e + 1	−29%	2%
11	R24	7.224e + 2	3%	0%
12	R25	6.948e + 2	−1%	0%
13	R6	3.640e + 1	−9%	0%
14	R8	3.240e + 1	−19%	0%
15	R18	1.964e + 1	−2%	0%
16	QN1_A	1.221	22%	0%
17	QP3_A	1.230	23%	0%
18	QN12_A	.8480	−15%	0%
19	QP5_A	.6850	−31%	−3%
20	RZERO	1.708	71%	0%

Fig. 8. Parameter values after the 13th optimization iteration.

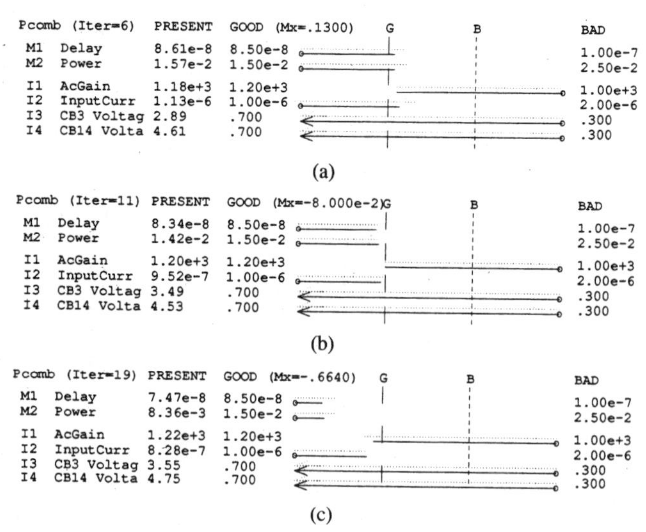

Fig. 9. Performance combs for several of the first 25 iterations.

of the difference between the binary divided input signal and a reference voltage. The conversion time of the particular A/D converter considered here is significantly increased by the delay time through the comparator portion of the circuit. Because for the most part the A/D converter was considered improved if this delay was made smaller, the time delay was selected as an objective rather than a constraint. We also made dc power an objective to be minimized, with a tentative good value of 15 mW for the cell. There was a requirement that certain key transistors not saturate under input overdrive conditions. This stemmed from the lightly doped collectors obtained from the linear fabrication process used and from the fact that the storage time under saturated switching conditions could increase by as much as an order of magnitude. We implemented this constraint by constraining the dc collector–base voltage on two key transistors with the input overdriven, thus avoiding a second (expensive) transient analysis. These objectives and this constraint along with other "soft" constraints are summarized in the following table:

delay	target 85 ns or better
power	target 15 mW or better
ac small signal gain	$\geq$ 1200
input bias current $\lvert Ibias \rvert$	$\leq$ 1 μA
collector–base voltage 3	$\geq$ 0.7 V
collector–base voltage 14	$\geq$ 0.7 V

Notice that these specifications are based on the results of *transient analyses* as well as both dc and ac analyses from Spice.

After setting up the problem description files, we were ready to begin the optimization. At this point it should suffice to simply point out the Pcomb's from optimization iterations 6, 11, and 19 in Fig. 9(a) through 9(c).

We then decided to try to make the comparator even faster and so lowered the good and bad values of the delay objective from 8.5 ns and 100 ns to 60 ns and 80 ns, respectively. As in the previous design example, before continuing the optimization, we first output the performance comb in Fig. 10(a) so we could calibrate ourselves to the effect of these changes.

Proceeding with optimization, we obtained the performance comb of Fig. 10(b). The next five iterations produced a slow decrease in the most active objective, the time delay. We decided to stop when the Pcomb in Fig. 10(c) showed that negligible improvement was obtained for iteration 31. Since the algorithm was in phase II and thus all objectives and constraints on the comb were competing equally for improvement, we concluded that the delay objective was sitting in an unconstrained local minimum.

The performance combs in Fig. 9(b) and (c) deserve some additional comment. In 9(b), all of the objectives and constraints first became "satisfied," i.e., equal to or better than their respective good values. Recall from Section II-B that this is precisely what causes the algorithm to switch from phase II to phase III, in which the worst objective value is now progressively improved, while the soft (and hard) constraints are maintained satisfied. This is shown in Fig. 9(c), where objectives M1 and M2 improve while the *AcGain* soft constraint I1 becomes slightly worse—but not worse than its good value, as guaranteed by the algorithm when in phase III.

Several other industrial circuits have been optimized using DELIGHT.SPICE. The performance of each of them was significantly improved, as summarized in Table II.

V. Concluding Remarks

We have illustrated that DELIGHT.SPICE can be a powerful tool for the design of analog circuits and digital cells. Industrial IC's have been considerably improved with the use of the system. The system is presently used

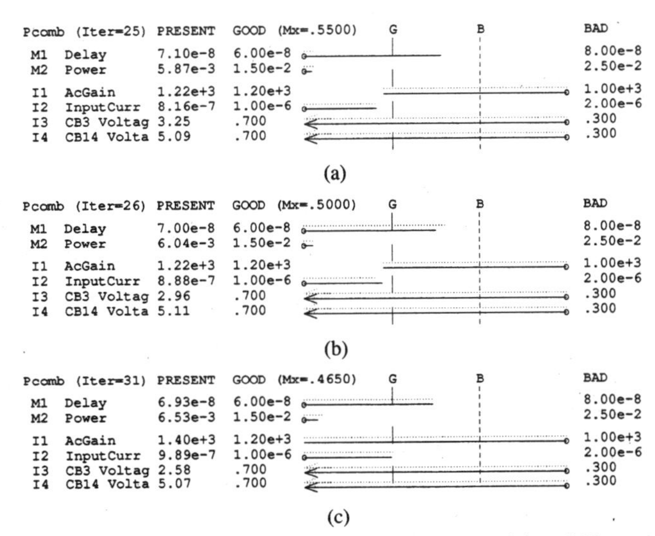

(c)

Fig. 10. Performance combs for iterations 25, 26, and 31.

TABLE II
DELIGHT.SPICE OPTIMIZATION RESULTS ON OTHER INDUSTRIAL DESIGNS

Circuit	Performance Objectives	Before	After	Number Iterations
DAC Control Amplifier	Settling Time Power	545ns 100mw	395ns 95mw	12
10'th Order Modem Filter	Group Delay Pass Band Stop Band	$.15^0/Hz$ $\pm 5db$ -57db	$.025^0/Hz$ $.65db$ -48db	55
Bus Precharge Circuit	Bus_1 Delay Bus_2 Delay Area	38ns 50ns 150	36ns 38ns 158	6

by experts in optimization theory for developing new algorithms as well as by circuit designers with little background in optimization methods. However several drawbacks still limit its application. First, while the formulation of the optimization is very general and applies to a wide range of engineering design areas, it would be helpful for the circuit designer to have at his disposal shorthand ways of entering often-used circuit specifications such as dc power minimization or slew rate maximization. Forms or a menu-driven problem definition would achieve this and have the additional advantage of greatly helping the beginning user. Second, it is clear that for specific problems the general nature of the optimization algorithms in the DELIGHT system can yield inefficient solutions; special-purpose algorithms can be added to deal with important cases such as timing optimization in digital circuit design (see, e.g., the works of Matson [33], [34] and of Fishburn *et al.* [20]). Finally, the complexity of the current system, especially the interface to Spice2, makes it difficult to maintain and thus hinders its general use. The new circuit simulator Spice3 [51], written in C and (hopefully) including additional routines such as those needed for sensitivity calculation, will permit a dramatic reduction of this complexity. A new system currently under development is aimed at eliminating or significantly alleviating the drawbacks just pointed out. This system, similar in philosophy to DELIGHT.SPICE, will be based on the C programming language and Spice3 and on the general design environment that has been built over the past few years [27]. The system is expected to be more flexible and efficient.

As hinted above in regard to Spice3, the basic DELIGHT "kernel" can be interfaced to other versions of Spice besides the Spice2G.6 version that we use. One such example is DELIGHT.SLICE, interfaced to the in-house enhanced Spice version called SLICE [4] at Harris Semiconductor in Melbourne, FL. But, since to date SLICE does not compute sensitivities, the optimization run times are much larger (see Table I at the end of Section III-C). The design methodology of [44] has also been applied with success in other engineering contexts: the design of multi-input, multi-output control systems in DELIGHT.MaryLin [18] and in the design of earthquake-resistant building structures in DELIGHT.STRUCT [6]–[8]. These other systems present the designer with a similar application-oriented problem formulation and run-time user feedback but simply use underlying simulators other than Spice.

ACKNOWLEDGMENT

The authors are pleased to thank the Harris Corporation, in particular, J. Cornell, J. Spoto, and T. Coston, for providing an ideal industrial environment in which to perform this research. They are grateful to many experienced Harris designers, particularly P. Hernandez, J. Lazar, T. Guy, and B. Webb, for their close collaboration. The authors also thank Prof. P. Gray and Prof. B. Meyer for discussing the use of DELIGHT.SPICE in integrated circuit design. A profound influence on the DELIGHT system in general is due to Dr. M. K. H. Fan, Dr. T. L. Wuu, and Prof. L. E. Polak.

REFERENCES

[1] "ASTAP advanced statistical analysis program," IBM Program Product Document SH20-1118-0, IBM Data Processing Division, White Plains, NY, 1973.
[2] C.-W. Ho, "The modified nodal approach to network analysis," *IEEE Trans. Circuits Syst.*, vol. CAS-22, pp. 504–509, June 1975.
[3] *COMPACT Version 4.8, United Computing Systems*, Compact Engineering, Inc., Los Altos, CA, Oct. 1978.
[4] *SLICE User's Guide*, Harris Semiconductor, Melbourne, FL, 1986.
[5] M. R. Aaron, "The use of least squares in system design," *IRE Trans. Circuit Theory*, vol. CT-3, pp. 224–231, Dec. 1956.
[6] M. Austin, "A methodology for the computer-aided design of earthquake-resistant steel structures," Ph.D. thesis, Department of Civil Engineering, University of California, Berkeley, CA, Nov. 1985.
[7] R. Balling, "DELIGHT.STRUCT: A computer-aided design environment for structural engineering," Ph.D. thesis, Department of Civil Engineering, University of California, Berkeley, CA, Dec. 1981.
[8] R. J. Balling, K. S. Pister, and E. Polak, "DELIGHT.STRUCT: A computer-aided design environment for structural engineering," Memo No. UCB/EERC-81/19, Earthquake Engineering Research Center, University of California, Berkeley, CA, Dec. 1981.
[9] M. A. Bhatti, K.S. Pister, and E. Polak, "OPTDYN—A general purpose optimization program for problems with or without dynamic constraints," Report No. UCB/EERC-79/16, Earthquake Engineering Research Center, University of California, Berkeley, CA, July 1979.
[10] M. A. Bhatti, T. Essebo, W. T. Nye, K. S. Pister, E. Polak, A. Sangiovanni-Vincentelli, and A. L. Tits, "A software system for optimization-based interactive computer-aided design," Memo No.

UCB/ERL M80/14, Electronics Research Laboratory, University of California, Berkeley, CA, Apr. 1980.
[11] R. K. Brayton, G. D. Hachtel, and A. L. Sangiovanni-Vincentelli, "A survey of optimization techniques for integrated circuit design," *Proc. IEEE*, vol. 69, pp. 1334–1362, 1981.
[12] D. A. Calahan, "Computer design of linear frequency selective networks," *Proc. IEEE*, vol. 53, pp. 1701–1706, 1965.
[13] F. H. Clarke, "A new approach to Lagrange multipliers," *Math. Oper. Res.*, vol. 1, pp. 165–174, 1976.
[14] J. Cullum, "An algorithm for minimizing a differentiable function that uses only function values," in *Techniques of Optimization*, A. V. Balakrishnan, Ed. New York: Academic Press, pp. 117–127.
[15] S. W. Director and R. A. Rohrer, "Automated network design—The frequency-domain case," *IEEE Trans. Circuit Theory*, vol. CT-16, pp. 330–337, 1969.
[16] S. W. Director and R. A. Rohrer, "The generalized adjoint network and network sensitivities," *IEEE Trans. Circuit Theory*, vol. CT-16, pp. 318–323, 1969.
[17] R. I. Dowell, "Automated biasing of integrated circuits," Ph.D. dissertation, Department of Electrical Engineering and Computer Science, University of California, Berkeley, CA, Apr. 1972.
[18] M. K. H. Fan, W. T. Nye, and A. L. Tits, "DELIGHT.MaryLin user's guide," Electrical Engineering Department, University of Maryland, College Park, MD, in preparation.
[19] A. V. Fiacco and G. P. McCormick, "The sequential unconstrained minimization technique for nonlinear programming. Algorithm II, Optimum gradients by Fibonacci search," Technical Paper RAC-TP-123, Research Analysis Corporation, McLean, VA, June 1964.
[20] J. P. Fishburn and A. E. Dunlop, "TILOS: A posynomial programming approach to transistor sizing," in *Proc. 1985 IEEE Int. Conf. Computer-Aided Design*, Nov. 1985, pp. 326–328.
[21] C. Gonzaga, E. Polak, and R. Trahan, "An improved algorithm for optimization problems with functional inequality constraints," *IEEE Trans. Automat. Contr.*, vol. AC-25, no. 1, pp. 49–54, 1980.
[22] P. R. Gray and R. G. Meyer, *Analysis and Design of Analog Integrated Circuits*. New York: Wiley, 1977.
[23] G. D. Hachtel and R. A. Rohrer, "Techniques for the optimal design and synthesis of switching circuits," *Proc. IEEE*, vol. 55, pp. 1864–1877, 1967.
[24] G. D. Hachtel, M. R. Lightner, and H. J. Kelly, "Application of the optimization program AOP to the design of memory circuits," *IEEE Trans. Circuits Syst.*, vol. CAS-22, pp. 496–503, June 1975.
[25] G. D. Hachtel, T. R. Scott, and R. P. Zug, "An interactive linear programming approach to model parameter fitting and worst case circuit design," *IEEE Trans. Circuits Syst.*, vol. CAS-27, pp. 871–881, Oct. 1980.
[26] G. D. Hachtel and P. Zug, "APLSTAP—Circuit design and optimization system—User's guide," IBM Yorktown Research Facility, Yorktown, NY, 1981.
[27] D. Harrison, P. Moore, A. R. Newton, A. Sangiovanni-Vincentelli, and C. H. Sequin, "Data management and graphics editing in the Berkeley design environment," in *Proc. 1986 IEEE Int. Conf. Computer-Aided Design*, 1986.
[28] D. E. Hocevar, P. Yang, T. N. Trick, and B. D. Epler, "Transient sensitivity computation for MOSFET circuits," *IEEE Trans. Computer-Aided Design*, vol. CAD-4, no. 4, pp. 609–620, Oct. 1985.
[29] R. Klessig and E. Polak, "An adaptive precision gradient method for optimal control," *SIAM J. Contr.*, vol. 11, no. 1, pp. 80–93, 1973.
[30] L. S. Lasdon and A. D. Waren, "Optimal design of filters with bounded, lossy elements," *IEEE Trans. Circuit Theory*, vol. CT-13, pp. 175–187, 1966.
[31] E. Lelarasmee and A. L. Sangiovanni-Vincentelli, *Time Domain Sensitivity Computation by the Perturbation Method*. Berkeley, CA: University of California, 1981.
[32] M. R. Lightner and S. W. Director, "Multiple criterion optimization for the design of electronic circuits," *IEEE Trans. Circuits Syst.*, vol. CAS-28, pp. 169–179, Mar. 1981.
[33] M. D. Matson, "Macromodeling and optimization of digital MOS VLSI circuits," Ph.D. thesis, Department of Electrical Engineering and Computer Science, Massachusetts Institute of Technology, Cambridge, MA, 1985.
[34] M. D. Matson and L. A. Glasser, "Macromodeling and optimization of digital MOS VLSI circuits," *IEEE Trans. Computer-Aided Design*, vol. CAD-5, no. 4, pp. 659–678, Oct. 1986.
[35] D. Q. Mayne, E. Polak, and A. Sangiovanni-Vincentelli, "Computer-aided design via optimization: A review," *Automatica*, vol. 18, no. 2, pp. 147–154, 1982.
[36] W. J. McCalla, "Computer-aided design of integrated bandpass amplifiers," Ph.D. dissertation, Department of Electrical Engineering and Computer Science, University of California, Berkeley, CA, June 1972.
[37] M. Minami, "Weak Pareto optimality of multiobjective problems in a locally convex linear topological space," *J. Optimization Theory Appl.*, vol. 34, no. 4, pp. 469–484, 1981.
[38] M. A. Murray-Lasso, "Analysis of linear integrated circuits by digital computer using black-box techniques," in *Computer-Aided Integrated Circuit Design*, G. J. Herskowitz, Ed. New York: McGraw-Hill, 1968, pp. 113–159.
[39] L. W. Nagel, "SPICE2: A computer program to simulate semiconductor circuits," Memo No. ERL-M520, Electronics Research Laboratory, University of California, Berkeley, CA, May 1975.
[40] W. T. Nye, E. Polak, A. Sangiovanni-Vincentelli, and A. L. Tits, "DELIGHT: An optimization-based computer-aided design system," in *Proc. 1981 IEEE Int. Symp. Circuits Syst.*, Apr. 1981, pp. 851–855.
[41] W. T. Nye and D. C. Riley, "Transient sensitivity in SPICE," EECS 290H course project report, Department of Electrical Engineering and Computer Science, University of California, Berkeley, CA, June 1982.
[42] W. T. Nye, "DELIGHT: An interactive system for optimization-based engineering design, Ph.D. thesis, Department of EECS, University of California, Berkeley, CA, 1983.
[43] W. T. Nye, A. Sangiovanni-Vincentelli, J. P. Spoto, and A. L. Tits, "DELIGHT.SPICE: An optimization-based system for the design of integrated circuits," in *Proc. 1983 Custom Integrated Circuits Conf.*, May 1983, pp. 233–238.
[44] W. T. Nye and A. L. Tits, "An application-oriented, optimization-based methodology for interactive design of engineering systems," *Int. J. Contr.*, vol. 43, no. 6, pp. 1693–1721, 1986.
[45] W. T. Nye, "Techniques for using Spice sensitivity computations in DELIGHT.SPICE optimization," in *Proc. 1986 IEEE Int. Conf. Computer-Aided Design*, 1986.
[46] H. Parsa and A. L. Tits, "Nonuniform, dynamically adapted discretization for functional constraints in engineering design problems," in *Proc. 22nd IEEE Conf. Decision and Control*, Dec. 1983, pp. 410–411.
[47] E. Polak and D. Q. Mayne, "An algorithm for optimization problems with functional inequality constraints," *IEEE Trans. Automat. Contr.*, vol. 21, pp. 184–193, 1976.
[48] E. Polak, "Algorithms for a class of computer-aided design problems: A review," *Automatica*, vol. 15, pp. 795–813, Sept. 1979.
[49] E. Polak, K. J. Astrom, and D. G. Mayne, "INTEROPTDYN-SISO: A tutorial," Memo No. UCB/ERL M81/99, Electronics Research Laboratory, University of California, Berkeley, CA, Dec. 1981.
[50] E. Polak, D. Q. Mayne, and Y. Wardi, "On the extension of constrained optimization algorithms from differentiable to nondifferentiable problems," *SIAM J. Cont. Optimization*, vol. 21, no. 2, pp. 179–203, Mar. 1983.
[51] T. Quarles, "SPICE3 preliminary report," in progress at the Electronics Research Laboratory, University of California, Berkeley, CA, 1982.
[52] D. C. Riley and A. L. Sangiovanni-Vincentelli, "Models for a new profit-based methodology for statistical design of integrated circuits," *IEEE Trans. Computer-Aided Design*, vol. CAD-5, no. 1, pp. 131–169, Jan. 1986.
[53] R. A. Rohrer, "Fully automated network design by digital computer: Preliminary considerations," *Proc. IEEE*, vol. 55, pp. 1929–1939, 1967.
[54] P. O. Scheibe and E. A. Huber, "The application of Carroll's optimization technique to network synthesis," in *Proc. Third Annual Allerton Conf. Circuits Syst.*, 1965, pp. 182–191.
[55] A. L. Tits, W. T. Nye, and A. Sangiovanni-Vincentelli, "Enhanced methods of feasible directions for engineering design problems," to appear in *J. Optimization Theory Appl.*
[56] T. N. Trick, "Computation of capacitor voltage and inductor current sensitivities with respect to initial conditions for the steady-state analysis of nonlinear periodic circuits," *IEEE Trans. Circuits Syst.*, vol. CAS-22, pp. 391–396, May 1975.
[57] J. Wieslander and H. Elmquist, "INTRAC: A communication module for interactive programs: Language manual," CODEN LUTFD2/(TFRT-3149)/1-060/(1978), Dept. of Automatic Control, Lund Institute of Technology, Lund, Sweden, Aug. 1978.

William Nye (S'80–M'83) was born on September 18, 1951, in Miami, FL. He received the B.S. and M.S. degrees in electrical engineering from the University of Florida, Gainesville, in 1973 and 1974, respectively.

From 1976 to 1978 he was an analog integrated circuits designer at Harris Semiconductor in Melbourne, FL. From 1978 to 1983 he was a research assistant at the University of California, Berkeley, where he completed the Ph.D. degree in electrical engineering, and continued with two additional years of postdoctoral work. Currently he is president of Epsilon Active Inc. in Berkeley, CA. His areas of interest include software engineering, interactive software for optimization-based computer-aided design with particular applications to integrated circuit design, the application of DELIGHT optimization to many different disciplines, design centering, and, recently, surface fitting and approximation theory.

Dr. Nye is a member of the Association for Computing Machinery and the Society for Industrial and Applied Mathematics.

*

David C. Riley was born in Batavia, NY, on August 11, 1946. He received the B.E.E. (with distinction) and M.S. degrees from Cornell University, Ithaca, NY, in June and September 1969, respectively.

He then joined NASA at Goddard Space Flight Center, Greenbelt, MD, where he worked on problems in analog circuit design, simulation, and systems analysis and design. He is now completing Ph.D. studies in the Department of Electrical Engineering and Computer Science at the University of California, Berkeley. His research interests include analog circuit design, statistical circuit design, optimization, and the man–machine interface.

Mr. Riley is a member of Tau Beta Pi and Eta Kappa Nu.

Alberto Sangiovanni-Vincentelli (M'74–SM'81–F'83) received the Dr.Eng. degree (summa cum laude) from the Politecnico di Milano, Italy, in 1971.

From 1971 to 1977, he was with the Istituto di Elettrotecnica ed Elettronica, Politecnico di Milano, Italy, where he held the positions of Research Associate, Assistant, and Associate Professor. In 1976, he joined the Department of Electrical Engineering and Computer Sciences of the University of California at Berkeley, where he is presently Professor. He is a consultant in the area of computer-aided design to several industries. His research interests are in various aspects of computer-aided design of integrated circuits, with particular emphasis on VLSI simulation and optimization. He was Associate Editor of the IEEE Trans-actions on Circuits and Systems, and is Associate Editor of the IEEE Transactions on Computer-Aided Design of Integrated Circuits and Systems and a member of the Large-Scale Systems Committee of the IEEE Circuits and Systems Society and of the Computer-Aided Network Design (CANDE) Committee. He was the Guest Editor of a special issue of the IEEE Transactions on Circuits and Systems on CAD for VLSI. He was Executive Vice-President of the IEEE Circuits and Systems Society in 1983.

In 1981, Dr. Sangiovanni-Vincentelli received the Distinguished Teaching Award of the University of California. At the 1982 IEEE-ACM Design Automation Conference, he was given a Best Paper and a Best Presentation Award. In 1983, he received the Guillemin-Cauer Award for the best paper published in the IEEE Transactions on CAS and CAD in 1981-1982. At the 1983 Design Automation Conference, he received a Best Paper Award. He is a member of ACM and Eta Kappa Nu.

*

André L. Tits (S'76–M'80) was born in Verviers, Belgium, on April 13, 1951. He received the 'Ingénieur Civil' degree from the University of Liège, Belgium, and the M.S. and Ph.D. degrees from the University of California, Berkeley, all in electrical engineering, in 1974, 1979, and 1980, respectively.

He held a visiting position at the University of California, Berkeley. Since 1981 he has been with the Electrical Engineering Department at the University of Maryland, College Park, where he is currently an Associate Professor. His research interests include numerical methods in optimization and their application to the design of engineering systems.

Dr. Tits was the recipient of a 1985 NSF Presidential Young Investigator Award.

Design Centering by Yield Prediction

KURT J. ANTREICH, SENIOR MEMBER, IEEE, AND RUDOLF K. KOBLITZ

Abstract—Design centering is an appropriate design tool for all types of electrical circuits to determine the nominal component values by considering the component tolerances. A new approach to design centering will be presented, which starts at the initial nominal values of the circuit parameters and improves these nominal values by maximizing the circuit yield step by step with the aid of a yield prediction formula. Using a variance prediction formula additionally, the yield maximization process can be established with a few iteration steps only, whereby a compromise between the yield improvement and the decrease in statistical certainty must be made in each step. A high-quality interactive optimization method is described which allows a quantitative problem diagnosis. The yield prediction formula is an analytical approximation based on the importance sampling relation. This relation can also be used to reduce the sample size of the necessary Monte Carlo analyses. Finally the efficiency of the presented algorithm will be demonstrated on a switched-capacitor filter.

Introduction

FOR A DESIGN of electrical circuits, particulary of semiconductor integrated circuits, the statistical variations of the circuit parameter values must be considered.

These statistical fluctuations due to manufacturing tolerances will involve a variation of the circuit's response. In this case the manufacturing yield—which is the proportion of manufactured circuits fulfilling the desired performance specifications—is a very useful design criterion.

Design centering is the process of defining a set of nominal parameter values to maximize the yield for known but unavoidable statistical fluctuations of the circuit parameters.

During the last years, several proposals for this task have been published which can be classified into geometric and statistical techniques.

The geometric approaches give a meaningful insight into the design centering problems [1], [2], but the proposed algorithms assume some properties (e.g., convexity) of the acceptance region and are difficult to deal with a large number of circuit parameters [3], [4]. In [5] efforts to overcome these difficulties are made.

On the other hand, the significant amount of computer time is a frequently mentioned drawback of the statistical approaches. Nevertheless, only the Monte Carlo method—used in the statistical techniques [6]–[19]—enables the designer to estimate the yield of a given design in a reliable way.

Manuscript received October 9, 1980; revised June 25, 1981.
The authors are with the Department of Electrical Engineering, Technical University of Munich, D-8000 Munich 2, Germany.

One of the first published approaches concerning statistical network optimization is the method by Kjellström *et al.* [6], [7] which has been successfully demonstrated on large examples.

There are also other approaches applicable for practical examples [10], [11]. Other methods are especially suited for linear circuits [12], [13]. The calculation of yield, its gradient and Hessian matrix based on a single Monte Carlo analysis, and the application to some simple optimization methods are derived in [14] first.

In this context, a design procedure based on "parametric sampling" [15] should be mentioned. This procedure is derived from the importance sampling concept [16]. With this approach, the manufacturing yield and the yield gradient can be computed for different tolerance situations and different nominal values. Thus well-known optimization methods are applicable.

In this paper, a design centering approach is proposed which is also based on the concept of importance sampling and which was first published in [17] and [18]. Here, we derive an analytical yield prediction formula which contains the yield-gradient and the Hessian matrix.

Due to the quadratic form, this yield prediction is valid in an extensive region around the nominal parameter values. In order to control the statistical uncertainty of the yield prediction, we deduce a variance prediction in addition and formulate design centering as a constrained optimization problem. An eigenvalue decomposition is proposed, to solve the optimization problem (in the case of a nonpositive definite system matrix) in an interactive manner.

Problem Formulation

Consider a set of circuit performance functions $\boldsymbol{\varphi}(\boldsymbol{p})$ which depends on a set of parameter values $\boldsymbol{p}$. The acceptance of a circuit by a specification test can be expressed as follows:

$$\boldsymbol{\varphi}(\boldsymbol{p}) \in \Phi. \tag{1}$$

Φ is the region of acceptable performance specifications in the performance space. With

$$\boldsymbol{\varphi}(\boldsymbol{p}) \in \Phi \rightarrow \boldsymbol{p} \in R \tag{2}$$

the acceptance region R in the parameter space is defined.

This region R can be described by the acceptance function $\delta(\boldsymbol{p})$ such that

$$\delta(\boldsymbol{p})=\begin{cases}1, & \text{if } \boldsymbol{\varphi}(\boldsymbol{p})\in\Phi\text{: circuit accepted} \rightarrow \boldsymbol{p}\in R\\ 0, & \text{if } \boldsymbol{\varphi}(\boldsymbol{p})\in\bar{\Phi}\text{: circuit rejected} \rightarrow \boldsymbol{p}\in \bar{R}.\end{cases} \tag{3}$$

The parameter values $\boldsymbol{p}$ are statistically varying with a probability density function (pdf) $g(\boldsymbol{p})$. Thus the manufacturing yield Y of a circuit can be formulated by

$$Y=\int_R\cdots\int g(\boldsymbol{p})\,d\boldsymbol{p}. \tag{4}$$

With the aid of the acceptance function $\delta(\boldsymbol{p})$, Y can be expressed as an expectation with respect to $g(\boldsymbol{p})$:

$$Y=\int_{-\infty}^{+\infty}\cdots\int \delta(\boldsymbol{p})\cdot g(\boldsymbol{p})\cdot d\boldsymbol{p}=\underset{g}{E}\{\delta(\boldsymbol{p})\}. \tag{5}$$

An unbiased estimator of Y is

$$\hat{Y}=\frac{1}{N}\cdot\sum_{\nu=1}^{N}\delta(\boldsymbol{p}^{(\nu)})=\frac{n}{N} \tag{6}$$

where n is the number of accepted circuits which pass the specification test and N is the total number of considered circuits.

Hence, the manufacturing yield can be estimated by repeating network analyses and performance specification tests (Monte Carlo analysis).

The proposed approach to design centering starts at the initial nominal values $\boldsymbol{\xi}_0$ of the circuit parameters. This initial situation is determined by, the initial pdf $g_0(\boldsymbol{p},\boldsymbol{\xi}_0,\boldsymbol{C}_0)$, containing the mean value vector $\boldsymbol{\xi}_0$ and for instance the variance-covariance matrix $\boldsymbol{C}_0$, and by the initial circuit yield Y_0

$$\left.\begin{aligned} Y_0&=\underset{g_0}{E}\{\delta(\boldsymbol{p})\}\\ \boldsymbol{\xi}_0&=\underset{g_0}{E}\{\boldsymbol{p}\}\\ \boldsymbol{C}_0&=\underset{g_0}{E}\{(\boldsymbol{p}-\boldsymbol{\xi}_0)(\boldsymbol{p}-\boldsymbol{\xi}_0)^T\}\end{aligned}\right\}\ \text{expectations with respect to } g_0(\boldsymbol{p}). \tag{7}$$

For further use, an additional pdf $h(\boldsymbol{p})$ is introduced, which is obtained from $g_0(\boldsymbol{p})$ by truncation with $\delta(\boldsymbol{p})$

$$h(\boldsymbol{p},\boldsymbol{\xi}_H,\boldsymbol{C}_H)=Y_0^{-1}\cdot\delta(\boldsymbol{p})\cdot g_0(\boldsymbol{p},\boldsymbol{\xi}_0,\boldsymbol{C}_0). \tag{8}$$

The corresponding mean value vector and variance-covariance matrix can also be expressed as expectations with respect to $g_0(\boldsymbol{p})$:

$$\boldsymbol{\xi}_H=\underset{g_0}{E}\{Y_0^{-1}\cdot\boldsymbol{p}\cdot\delta(\boldsymbol{p})\} \tag{9}$$

$$\boldsymbol{C}_H=\underset{g_0}{E}\{Y_0^{-1}\cdot(\boldsymbol{p}-\boldsymbol{\xi}_H)(\boldsymbol{p}-\boldsymbol{\xi}_H)^T\cdot\delta(\boldsymbol{p})\}. \tag{10}$$

Note that the corresponding values $\hat{Y}_0$, $\hat{\boldsymbol{\xi}}_H$, $\hat{\boldsymbol{C}}_H$, estimated on the basis of random samples according to $g_0(\boldsymbol{p})$, give information about the acceptance region R in the parameter space.

Yield Prediction

We develop a prediction of the circuit yield Y assuming a change of the probability density function from g_0 to g. For this purpose we use the relation known as the method of importance sampling [16] so that the yield Y can be expressed as an expectation with respect to $g_0(\boldsymbol{p})$

$$\begin{aligned}Y&=\int_{-\infty}^{+\infty}\cdots\int\delta(\boldsymbol{p})\cdot g_0(\boldsymbol{p})\frac{g(\boldsymbol{p})}{g_0(\boldsymbol{p})}d\boldsymbol{p}\\ &=\underset{g_0}{E}\left\{\delta(\boldsymbol{p})\frac{g(\boldsymbol{p})}{g_0(\boldsymbol{p})}\right\}.\end{aligned} \tag{11}$$

Note that the importance sampling relation—familiar in connection with variance reduction [15]—is used to formulate the expectation with respect to $g_0(\boldsymbol{p})$.

We now assume multinormal distributions for $g_0(\boldsymbol{p},\boldsymbol{\xi}_0,\boldsymbol{C}_0)$ and $g(\boldsymbol{p},\boldsymbol{\xi},\boldsymbol{C})$. With regard to practical requirements as well as to an especially clear illustration of the new approach this assumption is very useful. Under this condition we obtain from (11)

$$Y=\underset{g_0}{E}\left\{\delta(\boldsymbol{p})\cdot\frac{\sqrt{\det\boldsymbol{C}_0}\cdot\exp\left[-\frac{1}{2}(\boldsymbol{p}-\boldsymbol{\xi})^T\cdot\boldsymbol{C}^{-1}\cdot(\boldsymbol{p}-\boldsymbol{\xi})\right]}{\sqrt{\det\boldsymbol{C}}\cdot\exp\left[-\frac{1}{2}(\boldsymbol{p}-\boldsymbol{\xi}_0)^T\cdot\boldsymbol{C}_0^{-1}\cdot(\boldsymbol{p}-\boldsymbol{\xi}_0)\right]}\right\}. \tag{12}$$

Design centering implies only a change $\Delta\boldsymbol{\xi}$ from $\boldsymbol{\xi}_0$ to $\boldsymbol{\xi}$ whereas the initial variance-covariance matrix $\boldsymbol{C}_0$ remains unchanged:

$$\Delta\boldsymbol{\xi}=\boldsymbol{\xi}-\boldsymbol{\xi}_0\qquad \boldsymbol{C}=\boldsymbol{C}_0. \tag{13}$$

Hence, (12) can be written in the form

$$Y=\underset{g_0}{E}\left\{\delta(\boldsymbol{p})\cdot\exp\left[(\boldsymbol{p}-\boldsymbol{\xi}_0)^T\cdot\boldsymbol{C}_0^{-1}\cdot\Delta\boldsymbol{\xi}-\tfrac{1}{2}\Delta\boldsymbol{\xi}^T\cdot\boldsymbol{C}_0^{-1}\cdot\Delta\boldsymbol{\xi}\right]\right\}. \tag{14}$$

Using a Taylor series expansion of the exponential form in (14) about $\boldsymbol{\xi}_0$ and applying the expectation to the single terms we get

$$\begin{aligned}Y\approx Y_0\Big[1&+(\boldsymbol{\xi}_H-\boldsymbol{\xi}_0)^T\cdot\boldsymbol{C}_0^{-1}\Delta\boldsymbol{\xi}-\tfrac{1}{2}\Delta\boldsymbol{\xi}^T\cdot\boldsymbol{C}_0^{-1}\\ &\cdot\left(\boldsymbol{C}_0-\boldsymbol{C}_H-(\boldsymbol{\xi}_H-\boldsymbol{\xi}_0)(\boldsymbol{\xi}_H-\boldsymbol{\xi}_0)^T\right)\cdot\boldsymbol{C}_0^{-1}\Delta\boldsymbol{\xi}\Big].\end{aligned} \tag{15}$$

Thereby the cubic and higher order terms of $\Delta\boldsymbol{\xi}$ are neglected. In [18] this derivation is explained in detail. This yield prediction formula is an analytical approximation function with respect to the changes $\Delta\boldsymbol{\xi}$ of the nominal parameter values.

In order to get a better global yield-approximation, we propose the following, somewhat heuristical conversion into an exponential form

$$\begin{aligned}Y_p(\Delta\boldsymbol{\xi})=Y_0\cdot\exp\Big[&(\boldsymbol{\xi}_H-\boldsymbol{\xi}_0)^T\cdot\boldsymbol{C}_0^{-1}\cdot\Delta\boldsymbol{\xi}\\ &-\tfrac{1}{2}\Delta\boldsymbol{\xi}^T\cdot\boldsymbol{C}_0^{-1}(\boldsymbol{C}_0-\boldsymbol{C}_H)\cdot\boldsymbol{C}_0^{-1}\Delta\boldsymbol{\xi}\Big].\end{aligned} \tag{16}$$

Note that the Taylor series expansion of (16) about $\Delta\boldsymbol{\xi}=\boldsymbol{0}$ considering the first- and second-order terms of $\Delta\boldsymbol{\xi}$ is identical with (15).

It is worth mentioning that the step from (15) to (16) is not essential for the yield prediction. But due to this transformation, the system matrix of (16) is "more positive definite" than those of (15). Thus difficulties of the design centering procedure are reduced as pointed out below.

VARIANCE PREDICTION

An estimation $\hat{Y}_0$ of the initial circuit yield Y_0

$$\hat{Y}_0 = \frac{1}{N_0} \sum_{\nu=1}^{N_0} \delta(\boldsymbol{p}^{(\nu)}) \tag{17}$$

is characterized by a binominal random process. For a sufficient large number N_0 of random samples $\boldsymbol{p}^{(1)} \cdots \boldsymbol{p}^{(\nu)} \cdots \boldsymbol{p}^{(N_0)}$ the binominal distribution of $\hat{Y}_0$ can be approximately described by the normal distribution ($N_0 \gtrsim 100$). Hence, the variance $V(\hat{Y}_0)$ is a well-known measure of the expected error in the estimate $\hat{Y}_0$,

$$V(\hat{Y}_0) = \frac{1}{N_0} \cdot \underset{g_0}{E}\left\{\left[\delta(\boldsymbol{p}) - \underset{g_0}{E}\{\delta(\boldsymbol{p})\}\right]^2\right\}. \tag{18}$$

Since the expectation $E_{g_0}\{\delta(\boldsymbol{p})\}$ is Y_0, (18) leads to

$$V(\hat{Y}_0) = \frac{1}{N_0} \cdot Y_0 \cdot (1 - Y_0). \tag{19}$$

In order to determine the uncertainty of the yield prediction we must consider the expectation Y in (11). Hence, an estimator of Y with respect to $g_0(\boldsymbol{p})$ is

$$\hat{Y} = \frac{1}{N_0} \cdot \sum_{\nu=1}^{N_0} \delta(\boldsymbol{p}^{(\nu)}) \cdot \frac{g(\boldsymbol{p}^{(\nu)})}{g_0(\boldsymbol{p}^{(\nu)})}. \tag{20}$$

From the considerations above it is evident that the variance $V(\hat{Y})$ of Y is given by

$$V(\hat{Y}) = \frac{1}{N_0} \cdot \underset{g_0}{E}\left\{\left[\delta(\boldsymbol{p}) \cdot \frac{g(\boldsymbol{p})}{g_0(\boldsymbol{p})} - Y\right]^2\right\}. \tag{21}$$

Under the same conditions as the yield prediction (16) is derived from (11) we can obtain a variance prediction from (21):

$$V_P(\hat{Y}) = V(\hat{Y}_0)\left(\frac{Y_P(\boldsymbol{\Delta\xi})}{Y_0}\right)^2 \cdot \left(\frac{\exp\left[\boldsymbol{\Delta\xi}^T \cdot C_0^{-1} \cdot C_H \cdot C_0^{-1} \cdot \boldsymbol{\Delta\xi}\right] - Y_0}{1 - Y_0}\right) \tag{22}$$

where $Y_P(\boldsymbol{\Delta\xi})$ is given by (16).

In this variance prediction formula, derived in [19], the first- and second-order terms of $\boldsymbol{\Delta\xi}$ are considered.

Since in (22) C_H is positive definite, and Y_P is greater than Y_0 for a design centering step, we see that $V_P(\hat{Y})$ is greater than $V(\hat{Y}_0)$ as expected.

An approximation of $\hat{Y}$ in (20) is obtained from (16)

$$\hat{Y}_{P1}(\boldsymbol{\Delta\xi}) = \hat{Y}_0 \cdot \exp\Big[(\hat{\xi}_H - \hat{\xi}_0)^T \cdot C_0^{-1} \cdot \boldsymbol{\Delta\xi} - \tfrac{1}{2}\boldsymbol{\Delta\xi}^T \cdot C_0^{-1} \cdot (C_0 - \hat{C}_H) \cdot C_0^{-1} \cdot \boldsymbol{\Delta\xi}\Big] \tag{23}$$

where $\hat{Y}_0$, $\hat{\xi}_H$, and $\hat{C}_H$ can be estimated via Monte Carlo analysis. Hence, we conclude that the expression in (22) is also a quadratic approximation of the variance $V(\hat{Y}_{P1})$.

Note that the square root of the variance $V_P(\hat{Y}) \approx V_P(\hat{Y}_{P1})$ is a measure of the statistical uncertainty in the yield prediction $\hat{Y}_{P1}$.

In order to improve the statistical certainty of the yield prediction $\hat{Y}_{P1}$ in (23) we use the following yield prediction $\hat{Y}_{P2}$:

$$\hat{Y}_{P2}(\boldsymbol{\Delta\xi}) = \hat{Y}_0 \cdot \exp\Big[(\hat{\xi}_H - \hat{\xi}_0)^T \cdot C_0^{-1} \cdot \boldsymbol{\Delta\xi} - \tfrac{1}{2}\boldsymbol{\Delta\xi}^T \cdot C_0^{-1} \cdot (\hat{C}_0 - \hat{C}_H) \cdot C_0^{-1} \cdot \boldsymbol{\Delta\xi}\Big]. \tag{24}$$

Though, practical experiences and tedious derivations of regression estimates [20] show that normally

$$V(\hat{Y}_{P2}) < V(\hat{Y}_{P1}) \tag{25}$$

the improvement of the statistical certainty is not very significant.

Therefore, we use $V_P(\hat{Y})$ in (22) as a worst-case prediction for $V(\hat{Y}_{P2})$.

DESIGN CENTERING BY CONSTRAINED OPTIMIZATION

The proposed statistical approach to design centering can be characterized by maximizing the circuit yield Y directly with the aid of the yield prediction formula, whose parameters are estimated via Monte Carlo analysis.

$$Y(\boldsymbol{\Delta\xi}) \approx \hat{Y}_{P2}(\boldsymbol{\Delta\xi}) = \hat{Y}_0 \cdot \exp\left[\Gamma_\xi(\boldsymbol{\Delta\xi})\right] \overset{!}{=} \max. \tag{26}$$

Considering $Y_{P2}(\boldsymbol{\Delta\xi})$ as shown in (24), the generalized problem is to maximize the quadratic function

$$\Gamma_\xi(\boldsymbol{\Delta\xi}) = (\hat{\xi}_H - \hat{\xi}_0)^T \cdot C_0^{-1} \cdot \boldsymbol{\Delta\xi} - \tfrac{1}{2}\boldsymbol{\Delta\xi}^T \cdot C_0^{-1}(\hat{C}_0 - \hat{C}_H) \cdot C_0^{-1} \cdot \boldsymbol{\Delta\xi} \overset{!}{=} \max. \tag{27}$$

This optimization problem seems to be very simple. Unfortunately the following difficulties must be considered:

the restricted validity of the yield prediction formula;

the ill conditioning due to nearly linear dependent circuit parameters;

the uncertainties due to the statistical estimation of the elements $\hat{\xi}_0$, $\hat{\xi}_H$, $\hat{C}_0$, and $\hat{C}_H$;

the occasional appearance of saddle-point behavior in the yield prediction formula ($\hat{C}_0 - \hat{C}_H$) is not positive definite.

In order to consider the restricted validity of the objective function and the ill-conditioned maximization problem, the design centering task is to be formulated by constrained optimization which limits the magnitudes of the parameter changes $\boldsymbol{\Delta\xi}$.

As a constraint we, therefore, propose the normalized magnitude $\|\boldsymbol{\Delta\eta}\|$ of $\boldsymbol{\Delta\xi}$.

$$\|\boldsymbol{\Delta\eta}\| = \sqrt{\boldsymbol{\Delta\eta}^T \cdot \boldsymbol{\Delta\eta}} = \sqrt{\boldsymbol{\Delta\xi}^T \cdot C_0^{-1} \cdot \boldsymbol{\Delta\xi}} = \text{const.} \tag{28}$$

This normalization due to the matrix C_0^{-1}, is very useful with regard to the statistical uncertainty of the quadratic form (27), as pointed out below.

The relation between $\Delta\xi$ and $\Delta\eta$

$$\Delta\eta = R_0 \cdot \Sigma_0^{-1} \cdot \Delta\xi \tag{29}$$

is obtained by decomposition of the matrix C_0^{-1}

$$C_0^{-1} = (\Sigma_0 \cdot K_0 \cdot \Sigma_0)^{-1} = \Sigma_0^{-1} \cdot R_0^T \cdot R_0 \cdot \Sigma_0^{-1} \tag{30}$$

with

$$R_0^T \cdot R_0 = K_0^{-1}. \tag{31}$$

Σ_0 is the diagonal matrix containing the standard deviations and K_0 is the correlation matrix involving the correlation coefficients. Since K_0^{-1} is real, symmetric and positive definite, R_0 can easily be calculated, e.g., via Cholesky decomposition and becomes a real upper triangular matrix.

With the aid of these considerations, we get the following constrained maximization problem for each design centering step:

$$\Gamma_\eta(\Delta\eta) = (\hat{\xi}_H - \hat{\xi}_0)^T \cdot \Sigma_0^{-1} \cdot R_0^T \cdot \Delta\eta - \tfrac{1}{2} \Delta\eta^T \cdot R_0 \cdot \Sigma_0^{-1} (\hat{C}_0 - \hat{C}_H) \cdot \Sigma_0^{-1} \cdot R_0^T \cdot \Delta\eta \overset{!}{=} \max$$

$$\text{subject to} \quad \|\Delta\eta\| = \sqrt{\Delta\xi^T \cdot C_0^{-1} \cdot \Delta\xi} = \text{const}. \tag{32}$$

Considering the statistical certainty of the yield prediction, it is evident that with increased changes $\Delta\xi$, the yield prediction viz. (27) gives information about a region with a decreased number of random samples. The variance prediction (22) confirms this fact.

Since the random samples were chosen according to $g_0(p)$, the proposed constraint (28) corresponds to elliptical contours of constant densities of the random samples. Using this constraint the statistical certainty of the yield-prediction formula can be controlled by (22) in an appropriate way.

Now we perform an eigenvalue decomposition of the system matrix

$$R_0 \cdot \Sigma_0^{-1} (\hat{C}_0 - \hat{C}_H) \cdot \Sigma_0^{-1} \cdot R_0^T = U \cdot D \cdot U^T \tag{33}$$

where U is an orthogonal matrix, containing all eigenvectors ($U \cdot U^T = I$, I is the unit matrix) and D is a diagonal matrix involving all eigenvalues

$$D = \text{diag}(d_1, \cdots, d_\rho, \cdots, d_r). \tag{34}$$

Using the linear mappings

$$x = U^T \cdot \Delta\eta; \; b = U^T \cdot R_0 \cdot \Sigma_0^{-1} (\hat{\xi}_H - \hat{\xi}_0) \tag{35}$$

we get the constrained optimization problem

$$\Gamma(x) = b^T \cdot x - \tfrac{1}{2} x^T \cdot D \cdot x \overset{!}{=} \max$$

$$\text{subject to } x^T \cdot x = \text{const}. \tag{36}$$

With the aid of the positive parameter λ ($0 < \lambda < \infty$), we transform (36) into the unconstrained optimization problem

$$\Gamma_m(x) = b^T \cdot x - \tfrac{1}{2} x^T \cdot D \cdot x - \tfrac{1}{2} \lambda \cdot x^T \cdot x \overset{!}{=} \max. \tag{37}$$

Thus the stationary point of Γ_m is given by

$$x_s = \left[\frac{b_1}{\lambda + d_1}, \cdots, \frac{b_\rho}{\lambda + d_\rho}, \cdots, \frac{b_r}{\lambda + d_r} \right]^T \tag{38}$$

with

$$[b_1, \cdots, b_\rho, \cdots, b_r]^T = b. \tag{39}$$

Due to this eigenvalue decomposition, a saddle-point behavior of the single components of x_s can be recognized by their negative eigenvalues. In order to obtain a proper maximization step, a selection of a well-conditioned subsystem is proposed in [18].

In this paper an alternative treatment of the negative eigenvalues should be mentioned.

Corresponding to a proposal in [21], we change the signs of the negative eigenvalues in (38) and get a somewhat suboptimal solution:

$$x_0 = \left[\frac{b_1}{\lambda + |d_1|}, \cdots, \frac{b_\rho}{\lambda + |d_\rho|}, \cdots, \frac{b_r}{\lambda + |d_r|} \right]. \tag{40}$$

Let us suppose, that λ is determined in an appropriate way. Then the result of maximizing Γ_m yields the desired mean value vector ξ_1 as an improvement of ξ_0 considering (29) and (35)

$$\Delta\xi_0 = \Sigma_0 \cdot R_0^{-1} \cdot U \cdot x_0 \rightarrow \xi_1 = \xi_0 + \Delta\xi_0. \tag{41}$$

The corresponding value of the maximized circuit yield can be derived from (40), (36), and (26):

$$\hat{Y}_{P\max}(\lambda) = \hat{Y}_0 \exp\left[\sum_{\rho=1}^{r} b_\rho^2 \frac{\lambda + |d_\rho| - \frac{1}{2} d_\rho}{(\lambda + |d_\rho|)^2} \right]. \tag{42}$$

Thus the normalized magnitude of $\Delta\xi_0$ is given by

$$\sqrt{\Delta\xi_0^T \cdot C_0^{-1} \cdot \Delta\xi_0} = \|x_0\| = \sqrt{\sum_{\rho=1}^{r} \frac{b_\rho^2}{(\lambda + |d_\rho|)^2}}. \tag{43}$$

The parametric representation with respect to λ in (42) and (43) leads to the maximized yield prediction as a function of $\|x_0\|$:

$$\hat{Y}_{P\max} = f_1(\|x_0\|). \tag{44}$$

From (40), (41), and (43) we obtain the variance prediction according (22) as a function of $\|x_0\|$:

$$V(\hat{Y}_{P\max}) = f_2(\|x_0\|). \tag{45}$$

Note that the simple calculation of x_0 due to the eigenvalue decomposition (see (40)) allows the determination of $\hat{Y}_{P\max}$ and $V(\hat{Y}_{P\max})$ for a large number of λ so that a graphic representation of (44) and (45) can easily be obtained, as shown in Figs. 2 and 3.

For an appropriate interpretation, we prefer the square root of the variance as a measure of the expected error in the estimate $\hat{Y}_{P\max}$. The representation of both figures on a graphic terminal serves as a suitable criterion to control an interactive maximization process. With the aid of these graphic representations, the circuit designer chooses an appropriate normalized magnitude $\|x_0\|$, to get an acceptable compromise between the yield improvement and the decrease in statistical certainty. For this decision, elementary statistical relations can be used, for example, the single-sided confidence level.

If the variance prediction curve indicates an uncertainty which is too large and no compromise can be accepted, additional random samples must be carried out, to get a smaller value of the variance.

Hence, these curves are an appropriate assessment of the necessary sample size so that a minimum number of random samples can be effected in each design centering step. After calculating the improved nominal values according to (41), this design centering step is finished. In order to get a further improvement of the nominal values, the next design centering step with the changed nominal values including the Monte Carlo analysis is to be carried out. The end of the design centering process is reached when the graphic figures indicate that no further yield improvement can be achieved.

Reduction of the Sample Size

In any design centering step the use of the yield prediction and variance prediction formulas (24) and (22) requires the calculation of the estimates $\hat{Y}_k$, $\hat{\xi}_k$, $(\hat{\xi}_H)_k$, $\hat{C}_k$, and $(\hat{C}_H)_k$, where k is the kth step.

For a given accuracy of the estimate $\hat{Y}_k$, the necessary number N_k of random samples is obtained from (19) if the random samples of the kth Monte Carlo analysis are only used.

In this section it is shown that using the relation of importance sampling we can exploit random samples of past steps for the calculation of the estimates in the kth step. Doing this, the number N_k of random samples in the kth Monte Carlo procedure can be reduced while retaining accuracy of estimation [23].

For the kth design centering step with the probability density function $g_k(p)$, the yield estimate using the ith Monte Carlo analysis can be expressed with the aid of the importance sampling relation:

$$\hat{Y}_k^{(i)} = \frac{1}{N_i} \sum_{\nu=1}^{N_i} \delta_i(p^{(\nu)}) \frac{g_k(p^{(\nu)})}{g_i(p^{(\nu)})}. \tag{46}$$

The variance of the estimate $\hat{Y}_k^{(i)}$ is

$$V(\hat{Y}_k^{(i)}) = \frac{1}{N_i - 1} \left[\frac{1}{N_i} \sum_{\nu=1}^{N_i} \delta_i(p^{(\nu)}) \left(\frac{g_k(p^{(\nu)})}{g_i(p^{(\nu)})} \right)^2 - \hat{Y}_k^{(i)^2} \right]. \tag{47}$$

Exploiting all Monte Carlo procedures $(i = 0, 1, \cdots, k)$, the estimate $\hat{Y}_k$ can be formulated as

$$\hat{Y}_k = \sum_{i=0}^{k} \alpha_i \cdot \hat{Y}_k^{(i)}. \tag{48}$$

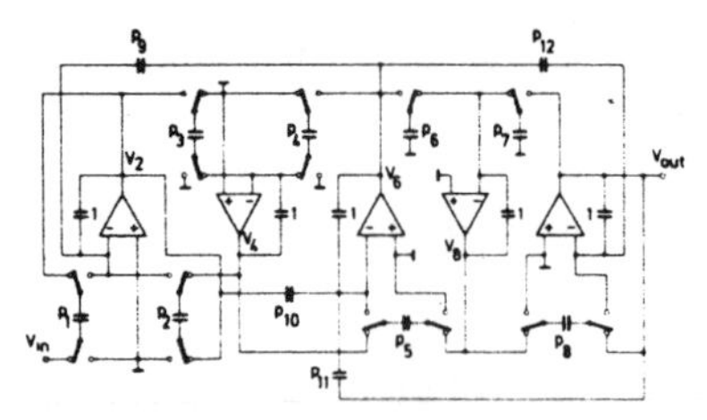

Fig. 1. Switched-capacitor filter with 12 design parameters.

The weights α_i must be chosen so that $V(Y_k)$ is minimized and the constraint

$$\sum_{i=0}^{k} \alpha_i = 1$$

is satisfied. Thus we obtain

$$\alpha_i = \frac{V(\hat{Y}_k)}{V(\hat{Y}_k^{(i)})} \tag{49}$$

where

$$\frac{1}{V(\hat{Y}_k)} = \sum_{i=0}^{k} \frac{1}{V(\hat{Y}_k^{(i)})}. \tag{50}$$

Using (19), the necessary number of random samples in the kth Monte Carlo analysis is

$$N_k = \frac{Y_k(1 - Y_k)}{V(\hat{Y}_k^{(k)})} \tag{51}$$

and with (50), we get

$$N_k = Y_k(1 - Y_k) \left[\frac{1}{V(\hat{Y}_k)} - \sum_{i=0}^{k-1} \frac{1}{V(\hat{Y}_k^{(i)})} \right]. \tag{52}$$

From (52) it is evident that the exploitation of the past Monte Carlo procedures $(i = 0, 1, \cdots, k-1)$ always reduces the sample size N_k in the kth design centering step.

The estimates for the past Monte Carlo procedures $(i = 0, 1, \cdots, k-1)$ are obtained without additional network analyses, because the results of the performance specification tests $[\delta_i(p^{(\nu)})]$ can be stored.

In accordance with (46) and (47) the estimates $\hat{\xi}_k^{(i)}$, $(\hat{\xi}_H)_k^{(i)}$, $\hat{C}_k^{(i)}$, $(\hat{C}_H)_k^{(i)}$ and the corresponding variances must be calculated in the same manner as $\hat{Y}_k^{(i)}$ and $V(\hat{Y}_k^{(i)})$. Note, that in general the calculation of the estimates $\hat{\xi}_k$, $(\hat{\xi}_H)_k$, $\hat{C}_k$, and $(\hat{C}_H)_k$ leads to different values of the corresponding weights α_i.

For determining the sample size N_k, we use only (52) with a desired variance $V(\hat{Y}_k)$ of the circuit yield estimate $\hat{Y}_k$.

Experimental Results

The yield prediction and the design centering algorithms have been implemented using ANS Fortran (approximately 5000 source statements). This program is called TACSY (Tolerance Analysis and Design Centering System). The circuit analysis and the performance specification test must be carried out via subroutine call. TACSY is highly portable and has been tested on several computers.

This program was applied to the switched capacitor filter, shown in Fig. 1. The z-transform of the node voltages can be described with (53):

$$\begin{bmatrix} 1+\frac{p_8}{z-1} & 0 & 0 & p_{12} & \frac{-p_8}{z-1} \\ 0 & 1+\frac{(p_1-p_2)}{z-1} & \frac{p_2}{z-1} & p_9 & 0 \\ 0 & \frac{-p_3 z}{z-1} & 1 & \frac{-p_4 z}{z-1} & 0 \\ p_{11} & p_{10} & \frac{p_5}{z-1} & 1 & \frac{-p_5}{z-1} \\ \frac{p_7 z}{z-1} & 0 & 0 & \frac{p_6 z}{z-1} & 1 \end{bmatrix} \begin{bmatrix} V_{out} \\ V_2 \\ V_4 \\ V_6 \\ V_8 \end{bmatrix} = \begin{bmatrix} 0 \\ \frac{p_1}{z-1} V_{in} \\ 0 \\ 0 \\ 0 \end{bmatrix}. \tag{53}$$

With $z=\exp(j2\pi f/f_c)$ we obtain the frequency-domain behavior where f is the frequency parameter and $f_c = 128$ Hz stands for the clock frequency. The insertion loss relative to the loss at $f=800$ Hz is given by

$$a(f) = 20 \cdot \log_{10} \frac{V_{out}(800)}{V_{out}(f)}. \tag{54}$$

Equation (54) has to fulfill the PCM performance-qualification listed in Table I. For checking up these specifications, the circuit was tested at the frequencies $f=100$ kHz, $f=10.0$ kHz, and from 1.2 to 4.7 kHz in steps of 100 Hz. In the critical frequency range range from 3.0 to 3.4 kHz, the step size was reduced to 40 Hz.

For solving (53) and checking the performance-specifications, a Fortran subroutine was written and loaded to the TACSY program.

The 12 parameters $p_1 \cdots p_\rho \cdots p_{12}$, as shown in Fig. 1 and in (53), are assumed to have normal probability density functions with mean values ξ_ρ listed in Table II and with standard deviations of $\sigma_\rho = 0.02 \cdot \xi_\rho$. Correlation coefficients between the parameters are supposed to be zero.

With the mean values of the initial design a Monte Carlo analysis using 500 random samples was carried out. From this statistical simulation we get the estimates $\hat{Y}_0$, $\hat{\xi}_H - \hat{\xi}_0$, and $\hat{C}_0 - \hat{C}_H$ considering (7), (9), and (10).

TABLE I
PERFORMANCE SPECIFICATIONS FOR THE SWITCHED-CAPACITOR FILTER

Frequency range f(KHz)	Loss specification a(f) [dB]	
	Minimum	Maximum
f < 3.0	- 0.25	+ 0.25
3.0 < f < 3.4	- 0.25	+ 0.9
3.4 < f < 3.6	- 0.25	-
3.6 < f < 4.6	+ 14·sin[π·(f-4)/1.2] + 14	-
4.6 < f < 10.0	+ 28	-

TABLE II
MEAN VALUES OF THE CIRCUIT PARAMETERS

circuit parameters	mean values initial design	mean values centered design	mean value changes from the initial to the centered design
p_1	0.177	0.191	7.9 %
p_2	0.0887	0.0911	2.7 %
p_3	0.307	0.320	4.2 %
p_4	0.307	0.305	- 0.65%
p_5	0.0527	0.0537	1.9 %
p_6	0.168	0.163	- 3.0 %
p_7	0.168	0.174	3.6 %
p_8	0.0994	0.103	3.6 %
p_9	0.684	0.623	- 8.9 %
p_{10}	0.406	0.394	- 3.0 %
p_{11}	0.117	0.120	2.6 %
p_{12}	0.220	0.217	- 1.4 %

Solving the optimization task, we get the graphic representations of the yield prediction and the variance prediction formulas with regard to (44) and (45) for this first step, as shown in Figs. 2 and 3.

With the yield and variance prediction curves the step size of each optimization step can be determined. Thereby a compromise between an increase in yield and a decrease in statistical certainty must be considered. In the two figures it is shown that for small values of $\|x_0\|$ the predictions are almost identical to the "correct" values estimated via Monte Carlo analyses.

For the first maximization step a value of $\|x_0\| = 1.0$ and an increase of yield from 0.24 to 0.36 seems to be an acceptable compromise. With (40) and (41) the new mean values of the circuit parameters were calculated. With these mean values a new Monte Carlo procedure and the next

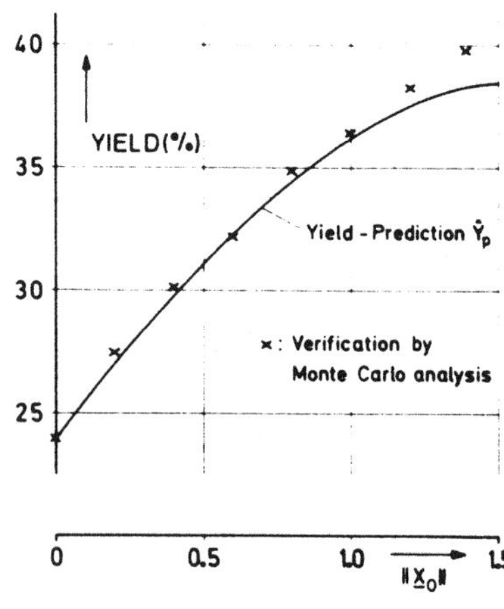

Fig. 2. Yield prediction.

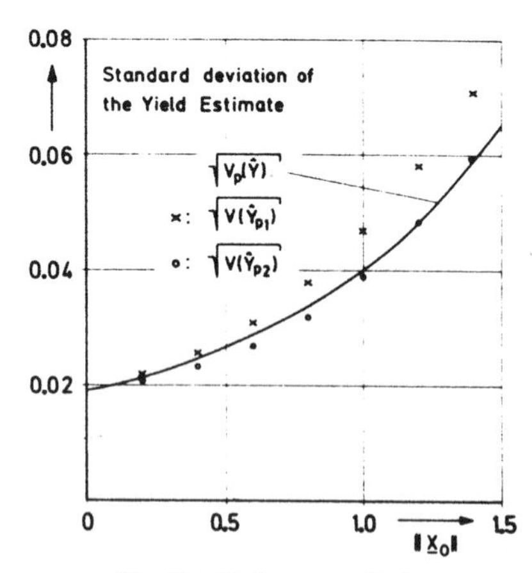

Fig. 3. Variance prediction.

maximization step were performed. After four steps a circuit yield of 0.54 as attainable. Now, near the design center the yield prediction curve becomes very flat and the corresponding variance prediction indicates relative large values in comparison to the increase in yield. To get a certain improvement of the circuit yield near the design center, the variance of yield estimate must normally be reduced. After several steps with increased sample size the circuit yield of 0.60 was attained. In Table II the change of the mean values from initial to the centered design are listed.

Note that the design center has always to be reached with a circuit yield $\hat{Y} < 1$. For $\hat{Y} = 1$ there is no guarantee that the design center of the acceptance region R with respect to chosen parameter tolerances will be attained. If the situation $\hat{Y} \approx 1$ occurs within the design centering process a change to $\hat{Y} < 1$ must be carried out, either by increasing the parameter tolerance or by restricting the circuit performance specifications.

For this example, in Fig. 4 the circuit's response due to the initial and the centered design are shown. Comparing the insertion loss of the initial design with the centered one and regarding the performance specifications, it is evident that the design centering procedure equalizes the attenuation reserves in the pass and stopband.

For an obvious illustration of the circuit improvement by design centering we propose a presentation according to

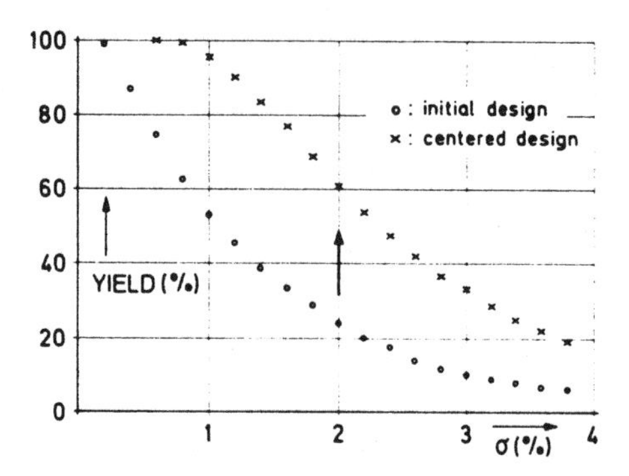

Fig. 4. Performance specifications and insertion loss.

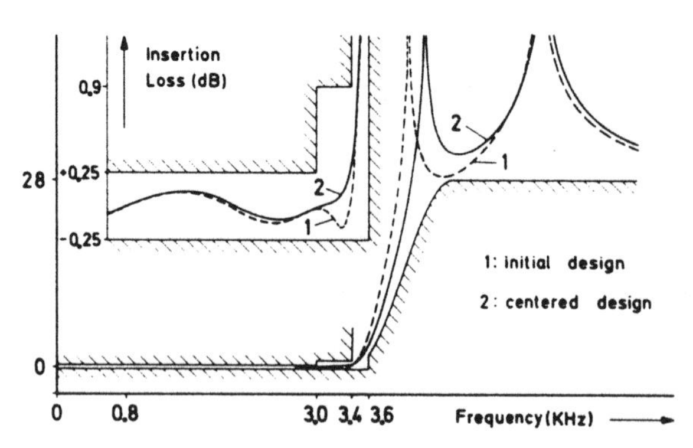

Fig. 5. Circuit yield for different values of σ.

Fig. 5. There, the yield for the initial and the centered circuit design is shown for different values of the standard deviations of the circuit parameters. Note, that the design centering procedure leads to a better design for a wide range of tolerance values, though the maximizing of yield is carried out for the relative standard deviation $\sigma_\rho = 2$ percent of all parameters ($\sigma_\rho = \sigma \cdot \xi_\rho$) only.

It is worth mentioning that these curves can be received with the aid of the previous Monte Carlo simulations, using the importance sampling relation without repeated network analyses.

Conclusions and Extensions

Finally some of the main aspects concerning the proposed approach will be listed.

1) Maximum yield with respect to given specifications is a very practical objective. A circuit, optimized via design centering for a specific set of tolerances will be close the optimum for a wide range of tolerance values (Fig. 5).

2) Note that the yield prediction formula is an analytical approximation based on the importance sampling relation. This relation can also be used to reduce the sample size in each design centering step. The complete Monte Carlo procedure must only be performed for the initial step.

3) With regard to the described difficulties typical for the yield maximization steps, a high-quality interactive optimization method which allows a quantitative problem diagnosis is necessary.

4) The proposed method of yield prediction and interactive optimization can also be used for tolerance assignment.

References

[1] E. Polak and A. Sangiovanni-Vincentelli, "Theoretical and computational aspects of the optimal design centering, tolerancing, and tuning problem," *IEEE Trans. Circuits Syst.*, vol. CAS-26, pp. 795–813, Sept. 1979.
[2] J. W. Bandler, P. C. Liu, and H. Tromp, "A nonlinear programming approach to optimal design centering, tolerancing, and tuning," *IEEE Trans. Circuits. Syst.*, vol. CAS-23, pp. 155–165, Mar. 1976.
[3] S. W. Director and G. D. Hatchtel, "The simplicial approximation approach to design centering," *IEEE Trans. Circuits Syst.*, vol. CAS-24, pp. 363–372, July 1977.
[4] J. W. Bandler and H. L. Abdel-Malek, "Optimal centering, tolerancing, and yield determination via updated approximations and cuts," *IEEE Trans. Circuits Syst.*, vol. CAS-25, pp. 853–871, Oct. 1978.
[5] L. M. Vidigal and S. W. Director, "Design-centering: The quasi-convex, quasi-concave performance function case," in *Proc. IEEE Int. Symp. on Circuits and Systems*, (Houston, TX), pp. 43–46, Apr. 1980.
[6] G. Kjellstroem, "Optimization of electrical networks with respect to tolerance costs," Ericsson Technics no. 3, pp. 157–175, 1970.
[7] G. Kjellstroem and L. Taxen, "On the efficient use of stochastic optimization in network design," in *Proc. IEEE Int. Symp. on Circuits and Systems*, (Munich, Germany) pp. 714–717, Apr. 1976.
[8] J. F. Pinel and K. Singhal, "Efficient Monte Carlo computation of circuit yield using importance sampling," in *Proc. IEEE Int. Symp. on Circuits and Systems*, (Phoenix, AZ), pp. 575–578, Apr. 1977.
[9] A. R. Thorbjornsen and E. R. Armbruster, "Computer-aided tolerance/correlation design of integrated circuits," *IEEE Trans. Circuits Syst.*, vol. CAS-26, pp. 763–767, Sept. 1979.
[10] N. J. Elias, "New statistical methods for assigning device tolerances," in *Proc. IEEE Int. Symp. on Circuits and Systems*, (Newton, MA), pp. 329–332, Apr. 1975.
[11] R. S. Soin and R. Spence, "Statistical design centering for electrical circuits," *Electron. Lett.*, vol. 14, no. 24, pp. 772–774, Nov. 1978.
[12] K. S. Tahim and R. Spence, "A radial exploration approach to manufacturing yield estimation and design centering," *IEEE Trans. Circuits Syst.*, vol. CAS-26, pp. 768–774, Sept. 1979.
[13] K. S. Tahim and R. Spence, "A radial exploration algorithm for the statistical analysis of linear circuits," *IEEE Trans. Circuits Syst.*, vol. CAS-27, pp. 421–425, May 1980.
[14] B. V. Batalov, Yu. N. Belyakov, and F. A. Kurmaev, "Some methods for statistical optimization of integrated microcircuits with statistical relations among the parameters of the components," *Microelectronica* (USA, translated from Russian), pp. 228–238, 1978.
[15] K. Singhal and J. F. Pinel, "Statistical design centering and tolerancing using parametric sampling," in *Proc. IEEE Int. Symp. on Circuits and Systems*, (Houston, TX), pp. 882–885, Apr. 1980.
[16] J. M. Hammersley and D. C. Handscomb, *Monte Carlo Methods.* Norwich, England: Fletcher, 1975.
[17] R. Koblitz, "Design centering and tolerance assignment of electrical circuits with gaussian-distributed parameter values," *Archiv Elek. Übertragung.* vol. 34, pp. 30–37, Jan. 1980.
[18] K. J. Antreich and R. K. Koblitz, "A new approach to design centering based on a multiparameter yield-prediction formula," in *Proc. IEEE Int. Symp. on Circuits and Systems*, (Houston, TX), pp. 886–889, Apr. 1980.
[19] K. J. Antreich and R. K. Koblitz, "An interactive procedure to design centering," in *Proc. IEEE Int. Symp. on Circuits and Systems*, (Chicago, IL), pp. 139–142, Apr. 1981.
[20] W. G. Cochran, *Sampling Techniques.* New York: Wiley, 1977.
[21] J. Greenstadt, "On the relative efficiencies of gradient methods," *Math. Comput.*, vol. 21, pp. 360–367, July 1967.
[22] B. J. Karafin, "The general component tolerance assignment problem in electrical networks," Ph.D. dissertation, Univ. of Pennsylvania, Philadelphia, 1974.
[23] K. Singhal and J. F. Pinel, "Statistical design centering and tolerancing using parametric sampling," *IEEE Trans. Circuits Syst.*, vol. CAS-28, pp. 692–702, July 1981.

Kurt J. Antreich (M'75–SM'78) was born in Chomutov, Czechoslovakia, on December 7, 1934. He received the Dipl. Ing. degree from the Technical University of Munich, Munich, Germany, in 1959, and the Dr. Ing. degree from the Technical University Fridericiana Karlsruhe, Karlsruhe, Germany, in 1966.

He joined the company AEG-Telefunken in 1959, where he was involved in computer-aided circuit design for telecommunication systems. From 1968 to 1975 he was head of the Advanced Development Department of AEG-Telefunken, Backnang, Germany. Since 1975, he has been a Professor of Electrical Engineering at the Technical University Munich, Munich, West Germany. His research interests are in computer-aided design of electronic circuits and systems, with particular emphasis on circuit optimization, layout synthesis, and fault diagnosis.

Dr. Antreich is a member of the NTG (Nachrichtentechnische Gesellschaft, Germany, the German Association of Communication Engineers). He received the NTG prize paper award in 1976. He was chairman of the NTG Circuit and System Group from 1972 to 1974. Since 1979, he has been member of the executive board of the NTG.

Rudolf K. Koblitz was born in Worms, Germany on September 12, 1950. He received the Dipl. Ing. degree from the Technical University Darmstadt, Darmstadt, Germany, in 1976.

Since May 1976 he has been with the Department of Electrical Engineering, Technical University of Munich, Munich, Germany, working in the field of CAD, statistical design, and design centering of integrated circuits.

Mr. Koblitz is a member of the German Nachrichtentechnische Gesellschaft (NTG). He received the 1981 prize paper award from the German Nachrichtentechnische Gesellschaft.

Statistical Integrated Circuit Design

Stephen W. Director, *Fellow, IEEE,* Peter Feldmann, *Member, IEEE,* and Kannan Krishna, *Student Member, IEEE*

***Abstract*— IC manufacturers have become increasingly interested in maximizing yield as a way to maximize profits. As a result, there has recently been renewed interest in computer aids for maximizing yield. While statistical design methods for maximizing circuit yield have been available for more than two decades, it is only recently that such methods have become practical. In this paper, we review some of these methods and illustrate the usefulness of one of these, the boundary integral method, with several examples.**

I. Introduction

TRADITIONAL or deterministic design focuses on finding a set of circuit parameter values such that all circuit performance specifications are met. It is usually assumed that this set of values can be accurately realized when the circuit is built. Unfortunately, this assumption is not realistic for mass-produced integrated circuits where fluctuations in the manufacturing process cause deviations in the actual values of the parameters from their nominal values. The statistical variability of the circuit parameters, in turn, causes the circuit performances to show a spread of values. Thus, if an amplifier was designed for a nominal dc gain of 50 dB, we could expect to see a distribution of actual gains for a population of manufactured chips to be similar to the one shown in Fig. 1.

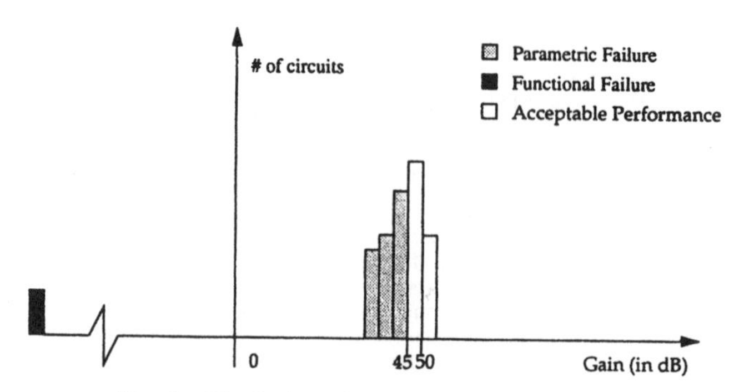

Fig. 1. Distribution of manufactured performances.

The ratio of the number of chips that meet specifications to the number of chips that are manufactured is referred to as the yield. Yield of less than 100% is due to two types of disturbances: local and global. Local disturbances, caused by spot defects in the manufacturing process, are the primary cause of catastrophic failures or functional failures which often cause a change in the basic functionality of the circuit. Global disturbances, primarily caused by phenomena that affect most devices on a chip (e.g., mask misalignment, variations in diffusion temperatures, or implant doses), are the primary cause of parametric failures. Parametric failures result when the chip is "functional" but specifications on such quantities as speed and power are not met.

Because of the close correlation between high yield and high profits, IC manufacturers strive to maximize yield. As the complexity of VLSI designs continues to increase, the need for computer aids for maximizing yield becomes more important. While statistical design methods for maximizing the circuit yield have been an active area of research for more than two decades, it is only relatively recently that such methods have been taken seriously by the IC design community. Among the reasons for this situation are that a number of practical yield maximization methods specific to integrated circuits have recently been developed. Second, it is only relatively recently that computers powerful enough to use these methods in a reasonable amount of time have appeared. Finally, the economics of chip manufacture are such that maximization of yield, rather than merely optimization of performance, has become of prime importance.

In this paper, we review several statistical design methods that have been developed to minimize the effects of IC manufacturing process disturbances on circuit performance. We also illustrate the effectiveness of the most recent of these, the boundary integral method, with several circuit design examples.

II. Parametric Circuit Yield Maximization

In this section, we will formally define "parametric yield" and formulate the "parametric yield maximization problem" for integrated circuits.

A circuit to be designed is characterized in terms of a set of n_φ performances, which are the components of the n_φ-dimensional vector $\varphi = (\varphi_1, \cdots, \varphi_{n_\varphi})$. For example, the performances of interest, for the circuit shown in Fig. 2, are the dc gain, the unity gain frequency, and the phase margin.

These performances are determined by a set of parameters that is denoted by the n_p-dimensional vector $\boldsymbol{p} = (p_1, \cdots, p_{n_p})$. Note that $\varphi = \varphi(\boldsymbol{p})$. For VLSI circuits, there are two categories of parameters comprising $\boldsymbol{p}$ that affect circuit performances: process-related parameters like the oxide thicknesses and length reduction of MOS devices, and circuit-related parameters such as device sizes and their placement. During the normal design process, nominal values are assumed for the parameters $\boldsymbol{p}_0$. During manufacture, however, inherent process disturbances cause the actual parameter values to deviate randomly from their nominal values causing actual

Manuscript received August 26, 1992; revised October 27, 1992. This work was supported in part by the Semiconductor Research Corporation.

S. W. Director and K. Krishna are with the Department of Electrical and Computer Engineering, Carnegie Mellon University, Pittsburgh, PA 15213.

P. Feldmann is with AT&T Bell Telephone Laboratories, Murray Hill, NJ 07974.

IEEE Log Number 9206123.

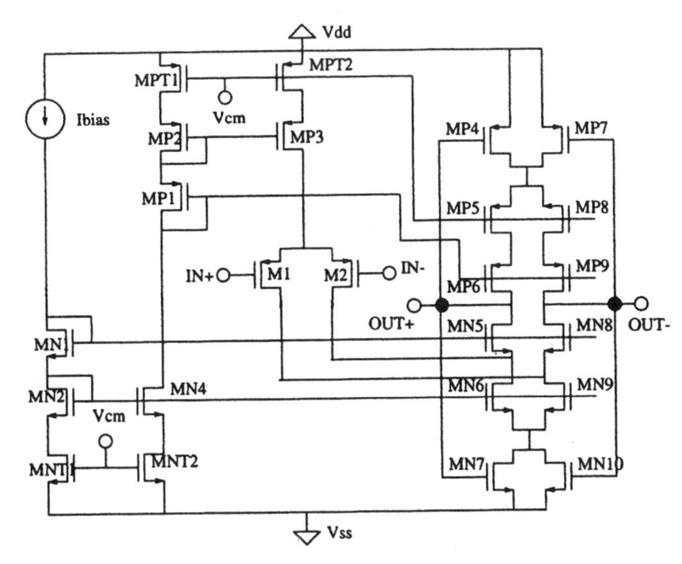

Fig. 2. Fully differential folded-cascode amplifier.

performances to vary as well. In particular, the actual layout geometries are affected by effects such as mask misalignment, lateral diffusion, and over (under) etching, while process-related parameters are primarily dependent on effects such as variations in diffusion times, temperatures, energies, doses of implants, etc.

If the random n_ξ-dimensional vector ξ denotes process disturbances, the vector $\boldsymbol{p}$ represents actual values for the parameters with nominal values $\boldsymbol{p}_0$, subjected to the disturbances ξ. Mathematically, the actual parameter values can be expressed as functions of their nominal values and of the disturbances: $\boldsymbol{p} = \boldsymbol{p}(\boldsymbol{p}_0, \xi)$. Note that the number of parameters is, in general, not related to the number of disturbances, i.e., a single process disturbance can simultaneously affect multiple integrated circuit parameters and a circuit parameter may be affected by multiple disturbances. The nominal values of the process-related parameters are not under the direct control of the circuit designer, who can only specify parameters such as device sizes and layout. Therefore, only a subset of the nominal parameter values $\boldsymbol{p}_0$ are truly designable in the circuit design phase. For the purpose of yield maximization, it is convenient to express circuit performances as functions of the deterministic nominal parameter values $\boldsymbol{p}_0$ and the random vector ξ: $\varphi = \varphi(\boldsymbol{p}_0, \xi)$. For the circuit in Fig. 2, the designable parameters are the device geometries and the bias current. Although there are 24 transistors and consequently 48 widths and lengths that are designable, only 12 of the widths and lengths are independent designables, due to matching constraints, e.g., $M1$ and $M2$ should have identical dimensions. The designable part of the $\boldsymbol{p}_0$ vector of this circuit therefore has 13 elements: 12 widths and lengths and a bias current.

The random variables that characterize the disturbances, $\xi = (\xi_1, \cdots, \xi_{n_\xi})$, may represent physically based effects at the process level. These variables, such as variations of temperatures, impurity diffusivities, etc., mimic the random variations that occur in a manufacturing line. A process and device simulator [1] can be used to predict their effect on circuit parameters. We can make reasonable approximations of the *jpdf* of ξ. One frequently made assumption is that the components of the random vector ξ, which model variations of controlling parameters at successive fabrication steps, are, at least to first order, mutually independent. A different approach to the modeling of disturbances is to use statistical techniques on data collected from the fabrication line. These techniques, such as principal factor analysis [20], model the highly dependent set of random variables as functions of a smaller set of independent random variables. While these independent factors may not necessarily be related to any physical quantities, they can be used to model the circuit parameter variations.

III. VLSI Circuit Optimization

A manufactured circuit will be considered acceptable if all of its actual performances fall within acceptable limits, i.e., if

$$\varphi_k^L \le \varphi_k \le \varphi_k^U, \qquad k = 1, \cdots, n_\varphi. \tag{1}$$

In the case of our amplifier circuit in Fig. 2, acceptable performances might be expressed as

- dc gain $\ge$ 55 dB,
- unity gain frequency $\ge$ 45 MHz,
- phase margin $\ge 30°$.

The system of inequalities, (1), defines a region of acceptability, denoted by A_φ, in the n_φ-dimensional performance space, which is a cartesian space spanned by the n_φ performances $\varphi_i, i = 1, \cdots, n_\varphi$.

We can now formally state the yield of a design as the probability that a manufactured circuit has acceptable performance, i.e., that the vector φ belongs to the acceptability region A_φ:

$$Y = \text{Prob}\{\varphi \in A_\varphi\} = \int_{A_\varphi} f_\varphi(\boldsymbol{p}_0, \varphi)\, d\varphi.$$

where $f_\varphi(\boldsymbol{p}_0, \varphi)$ is the joint probability density function (*jpdf*) of the circuit performances that constitute the vector φ. The *jpdf* is clearly dependent on the nominal parameter values $\boldsymbol{p}_0$. Yield maximization is performed by determining a set of nominal values of the designable parameters, $\boldsymbol{p}_0$, that maximizes the probability mass of the random performances that lies within A_φ. Consequently, the yield maximization problem, in the performance space, is formulated as

$$\max_{\boldsymbol{p}_0} \left\{ Y = \int_{A_\varphi} f_\varphi(\boldsymbol{p}_0, \varphi)\, d\varphi \right\}. \tag{2}$$

It is important to observe that the *jpdf* $f_\varphi(\boldsymbol{p}_0, \varphi)$ is not explicitly available to us and, in general, no easy means is available for obtaining it. Therefore, employing (2) as a basis for maximizing yield is difficult and we are motivated to consider alternative formulations to the yield maximization problem.

We can also formulate the yield optimization problem in the space of independent statistical disturbances, viz., the disturbance space. The disturbance space is a Cartesian space in which the axes correspond to independent statistical disturbances. The region of acceptability in the disturbance space, denoted by $\boldsymbol{A}_\xi(\boldsymbol{p}_0)$, consists of all possible combinations of

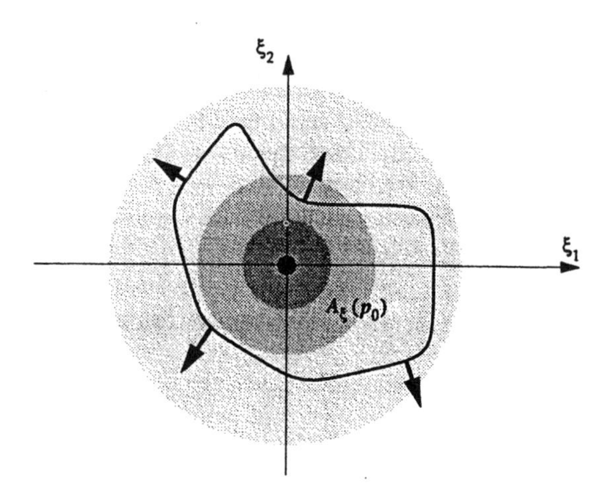

Fig. 3. Yield optimization by moving the acceptability region boundary.

disturbances that can occur in the manufacturing of a circuit, which, for specific nominal parameter values, do not result in unacceptable performance. It is given by

$$A_\xi(\boldsymbol{p}_0) = \{\xi | \varphi^L \leq \varphi(\boldsymbol{p}_0, \xi) \leq \varphi^U\}. \tag{3}$$

In the disturbance space, the acceptability region is the inverse image of the parametrized mapping $\xi \rightarrow \varphi$ which depends on the nominal parameter values $\boldsymbol{p}_0$.

Parametric yield can be expressed in the disturbance space as the probability of occurrence of a combination of acceptable disturbance values and can be formalized as

$$Y = \text{Prob}\,\{\xi \in A_\xi(\boldsymbol{p}_0)\} = \int_{A_\xi(\boldsymbol{p}_0)} f_\xi(\xi)\, d\xi. \tag{4}$$

Thus, the resulting optimization problem is

$$\max_{\boldsymbol{p}_0} \left\{ Y = \int_{A_\xi(\boldsymbol{p}_0)} f_\xi(\xi)\, d\xi \right\}. \tag{5}$$

Yield optimization is, therefore, performed by modifying the acceptability region in a way so as to increase the coverage of a fixed probability distribution. This is illustrated in Fig. 3 for a two-dimensional disturbance space.

For the case of the differential amplifier in Fig. 2, we assumed that the processing variations were adequately modeled by the variations in length and width of the n- and p-type devices, flat-band voltage variation of the n- and p-type devices, and the variation in oxide thickness. Thus, we have a seven-dimensional disturbance space.

We have found that the disturbance space formulation is the most adequate formulation for the case of integrated circuits where different components and devices are subject to variations caused by the same random disturbances. Some other statistical design methods that assume a different formulation cannot easily accommodate this type of situation.

IV. Statistical Design Issues

A. Alternate Optimization Formulations

So far, we have considered only the adjustment of nominal parameter values as a means for optimizing yield. Another possible way to improve yield is to adjust component tolerances. For the case of VLSI circuits, reducing tolerances means tighter control of the process, thereby minimizing the spread in the device parameter values. This can be achieved by applying statistical design techniques at the process level and through the use of more precise, and, in most cases, more expensive manufacturing equipment. Since tightening tolerances is likely to increase the cost of production, it is desirable to identify the components, device parameters, and processing steps, for which better tolerances result in savings that offset the cost. Alternatively, one can identify parameters for which relaxing the tolerance has only minor effects on the variability of performances. This may allow the use of a less expensive processing step, or a simpler circuit topology, thus reducing manufacturing cost.

Mathematically, tolerances can be modeled as parameters, $\boldsymbol{p}_f$, that control some measure of spread, such as variances, of the statistical distributions, $f_\xi(\xi, \boldsymbol{p}_f)$, of disturbances. Let $C(\boldsymbol{p}_f)$ be a function that models the manufacturing cost associated with a given set of tolerances. The tolerancing problem can then be formulated as

$$\max_{\boldsymbol{p}_f} Y = \int_{A_\xi} f_\xi(\xi, \boldsymbol{p}_f)\, d\xi$$
$$\text{such that } C(\boldsymbol{p}_f) < C_{\max}. \tag{6}$$

The tolerancing problem can also be formulated as a cost minimization problem with a minimum yield constraint:

$$\min_{\boldsymbol{p}_f} C(\boldsymbol{p}_f)$$
$$\text{such that } \left(Y = \int_{A_\xi} f_\xi(\xi, \boldsymbol{p}_f)\, d\xi \right) > Y_{\min}.$$

Such a formulation may be more appropriate in some cases.

Sometimes a low yield results from very stringent specifications. In many cases, especially when designing a subcomponent of a larger design, specifications may be negotiable. It may be the case that one or several specifications can be relaxed and compensated for by other subcomponents at a much lower cost. The problem here is to identify specification constraints responsible for significant yield loss and to predict to what extent yield can be improved by relaxing them. Of course, such an exercise is not possible when specifications result from standards or published specification sheets.

Specifications can be modeled, mathematically, as deterministic parameters that control the acceptability region. If the function $C(\varphi^B)$ quantifies the cost, on the entire system, of modifying the specifications φ^B, which represent lower or upper bounds on circuit or device performances, the specification assignment problem becomes

$$\max_{\varphi^B} Y = \int_{A_\xi(\varphi^B, \boldsymbol{p}_0)} f_\xi(\xi)\, d\xi$$
$$\text{such that } C(\varphi^B) < C_{\max}. \tag{7}$$

As in the case of the tolerancing problem, this can also be formulated as a cost minimization problem with a minimum yield constraint.

Finally, while parametric yield is the most frequently used measure of statistical behavior of mass produced circuits, other measures such as quality loss or signal-to-noise ratios [10], [26] are sometimes more accurate measures of performance variability. Several statistical design methodologies [10], [17] use these as the objective for optimization.

B. Analytical Models

The statistical design problem can be formulated as an optimization problem in which the objective function and/or constraint functions contain statistical averages, such as yield or another measure of performance variability. Optimization of a function is an iterative process that requires several evaluations of the objective and constraint functions and their gradients. Since statistical objective functions are, in general, very expensive to evaluate, it is important to choose an optimization strategy that requires a minimum number of objective function evaluations. State-of-the-art gradient-based optimization algorithms, such as [4], can significantly reduce the number of required iterations. Therefore, it is very important to be able to compute both the statistical objective function and its gradient efficiently.

In general, to determine if a circuit satisfies a given specification, a circuit simulation must be performed. Circuit simulation, especially of large circuits, is computationally expensive and has, consequently, been the bane of practical use of yield maximization strategies. Fortunately, techniques have been developed to overcome this bottleneck through the use of analytical approximations of performances. The efficiency of these techniques is determined by the cost of building sufficiently accurate approximations. These techniques require that some number of circuit simulations be performed on a sample of circuits according to some experimental plan [24], [25]. From these simulations, the analytical approximations are determined using either response surface or interpolation techniques.

Response surface models are constructed by computing the coefficients of linear or quadratic polynomials through linear regression from the simulation results at the experimental points. This requires a number of simulations that is larger than the number of coefficients to be computed, the extra degrees of freedom being used to estimate the modeling error. For quadratic polynomials, the number of coefficients increases as the square of the number of model parameters and, therefore, may quickly become prohibitively large. In order to reduce the number of coefficients that need to be computed, techniques such as parameter screening or stepwise regression [23] can be used. Some researchers have gone so far as to ignore all quadratic cross-products. This is, however, not a good compromise since the effect of this would be to neglect all interactions among parameters, information that is crucial for variability reduction [19].

A different way of generating simple analytic expressions for performances as a function of their parameters is through interpolation. In contrast to response surfaces, interpolation models pass exactly through all the experimental points, but provide no means for estimating the resulting modeling error. Interpolation methods used in statistical design methodologies range from simple linear interpolation [6], [7] to more complicated techniques such as maximally flat interpolation [15]. A method that combines the maximally flat interpolation with regression-based approximation was presented in [16]. A particularly interesting interpolation approach has been proposed in [31].

V. Yield Maximization Methodologies for IC's

As mentioned above, a large number of circuit yield maximization methodologies have been developed for discrete circuits. Integrated circuit yield optimization is fundamentally different from discrete circuit yield optimization in that for IC's a, typically small, number of disturbances affect a large number of devices causing variations in the devices to be highly dependent on each other. Here we discuss some important methodologies that are suitable for integrated circuit yield optimization.

A. The Method of Random Perturbations

One category of methods for IC yield optimization is based on stochastic approximation. Stochastic approximation is a general method for dealing with optimization of regression functions, $E[h(\boldsymbol{p})]$ ($E[\cdot]$ denotes expectation), where only noise corrupted observations of $h(\boldsymbol{p})$ can be obtained. Several stochastic approximation methods have been developed for both unconstrained and constrained statistical optimization problems.

Yield can be expressed as the expectation of the acceptability region indicator function

$$h_\xi(\xi, \boldsymbol{p}_0) = \begin{cases} 1, & \text{if } \xi \in A_\xi(\boldsymbol{p}_0) \\ 0, & \text{otherwise} \end{cases}$$

with respect to the *jpdf* of the statistical variables.

$$Y = \int_{A_\xi(\boldsymbol{p}_0)} f_\xi(\xi)\, d\xi = \int_\infty h_\xi(\xi, \boldsymbol{p}_0) f_\xi(\xi)\, d\xi \qquad (8)$$

Stochastic approximation methods can, therefore, be used for maximization of the yield function.

Styblinski and Ruszczynski [28] use a variant of the simplest stochastic approximation algorithm which is akin to the steepest ascent method. The main feature of this method is the fact that the Monte Carlo estimation of yield gradient is blended within the optimization algorithm. The expensive Monte Carlo analysis is not performed at every iteration to determine the yield gradient. The estimate of the gradient used in the stochastic approximation approach can be based on a very small number of samples, even one, and the direction of movement of the designable variables, at any iteration, is a convex combination of that of the previous iteration and the gradient determined during the current iteration. The rate of convergence is, however, dependent on the accuracy of the yield estimate.

Yield in expression (8) can be estimated through Monte Carlo using a sample of points in the disturbance space, produced by a random number generator that mimics the *jpdf* f_ξ. However, since designable parameters appear in h_ξ, which is not differentiable, there is no straightforward expression for the Monte Carlo estimate of the yield gradient. The method of random perturbation is used to cope with this situation.

This method involves perturbing the original problem by adding a random component η to the deterministic parameters $\boldsymbol{p}_0$. The variance of this random component is controlled by a parameter β. As β is decreased to 0, the random component η becomes a deterministic zero vector and the parameters become deterministic again. One of the possible choices for the distribution, $f_\eta(\eta, \beta)$, to be assigned to these perturbation parameters is a multivariate normal distribution with its covariance matrix scaled by β. With this random perturbation all parameters become statistical and can be decomposed into the sum of a deterministic vector (which is partially designable) and a random vector.

$$\tilde{\boldsymbol{p}} = \begin{bmatrix} \boldsymbol{p}_0 \\ 0 \end{bmatrix} + \begin{bmatrix} \eta \\ \xi \end{bmatrix} = \tilde{\boldsymbol{p}}_0 + \tilde{\xi}. \tag{9}$$

Using the indicator function $\tilde{h}$ (similar to the one introduced before), the "perturbed yield" can be expressed as

$$\tilde{Y}(\boldsymbol{p}_0, \beta) = \int_\infty \tilde{h}(\tilde{\boldsymbol{p}}_0 + \tilde{\xi}) \tilde{f}_{\tilde{\xi}}(\tilde{\xi}, \beta)\, d\tilde{\xi} \tag{10}$$

where $\tilde{f}_{\tilde{\xi}}(\tilde{\xi}, \beta) = f_\eta(\eta, \beta) f_\xi(\xi)$. By performing a simple change of variables $\zeta = \tilde{\boldsymbol{p}}_0 + \tilde{\xi}$, the designable parameters appear only in the *jpdf*, which is differentiable.

$$\tilde{Y}(\boldsymbol{p}_0, \beta) = \int_\infty \tilde{h}(\zeta) f_\zeta(\zeta - \tilde{p}_0, \beta)\, d\zeta. \tag{11}$$

In this formulation the gradient of the perturbed yield can be estimated using a Monte Carlo procedure. For the stochastic approximation algorithm, however, only one or a few samples are used at every iteration. During the course of optimization β is also swept to 0, therefore, at convergence the "perturbed yield" becomes the original yield.

This method was used in a system for production yield optimization with respect to fundamental fabrication parameters and mask layout dimensions. A statistical process simulator (FABRICS [1]) is run in conjunction with a circuit simulator (SPICE [2]) to determine the effect of fabrication process variations on circuit performance.

The main drawback of this method is the fact that the optimization algorithm is stochastic and may require a very large number of iterations to achieve convergence. Since simulations must be performed at each iteration, stochastic approximation may turn out to be prohibitively expensive. Analytic models can be employed to reduce the cost of simulation. The approximation must be done, however, in terms of both deterministic and statistical variables over the entire space of interest, and therefore the cost of modeling can increase rapidly as the circuit size increases.

B. The Texas Instruments Method

Researchers at Texas Instruments have proposed a yield optimization method tailored for digital MOS circuits. They have observed that both current and capacitance in MOS FET's are primarily sensitive to four variables [5]–[7]: width reduction, length reduction, oxide capacitance, and flat-band voltage (for both n-channel and p-channel devices for the case of CMOS). Effects of variables such as doping profile have been found to be significantly smaller. Since these parameters are determined by different steps in the manufacturing process, they can, to first order, be assumed to be independent of each other. All circuit variations are, therefore, assumed to be caused by the variations of these statistical variables. The aforementioned variables define the disturbance space.

To reduce the cost of circuit simulation, circuit performances are approximated as linear functions of the four statistical variables. A linear interpolation model for each performance is constructed from five (six for the case of CMOS circuits) circuit simulations. Performance constraints resulting from specifications applied to these interpolation models define a polytope in the disturbance space that approximates the acceptability region. Yield is defined as the integral of the disturbance *jpdf* over this approximation to the acceptability region. It can easily be estimated by Monte Carlo.

Having a means for evaluating yield, the next step is to have some means for optimizing it with respect to the designables. Two yield optimization strategies are proposed. The gradient-based approach involves computation of the yield gradient. The other is a geometric approach.

The gradient-based approach involves the computation of the yield and the yield gradient by multidimensional integration. Yield can be evaluated by quadrature techniques, by integrating the *jpdf* of the statistical variables over the approximating polytope. The gradient can be evaluated in a similar way, only that the integration is performed over the facets of the approximated acceptability region. A gradient-based algorithm using a quasi-Newton procedure is then used to optimize the circuit yield.

The geometric approach solves the problem in which the minimum distance of the mean point of the disturbance *jpdf* (the projection of the nominal point from the parameter space onto the disturbance space) from the acceptability region boundaries is maximized, i.e., the boundaries are pushed as far as possible from the center of the *jpdf* (Fig. 4). It uses an iterative approach to solve this problem in which, at each iteration, the distance functions, to the approximated acceptability region boundaries, are linearized. A linear programming problem is then formulated and solved with this linearization.

The main features of this method are approximation of the acceptability region boundaries by hyperplanes and yield maximization, done both directly with the yield and yield gradient and, indirectly, using a geometric approach.

The method described above makes use of a number of good ideas, such as the choice to compute yield in the disturbance space, the use of a small number of statistical variables to account for most of the performance variability, and the computation of the yield gradient as surface integrals. However, the method uses some algorithms and assumptions that restrict its application as a general yield optimization strategy. The method builds linear models for the performances in terms of the statistical variables which may not be accurate enough in other processes. Other aspects of the method, for example, using quadrature integration to determine the

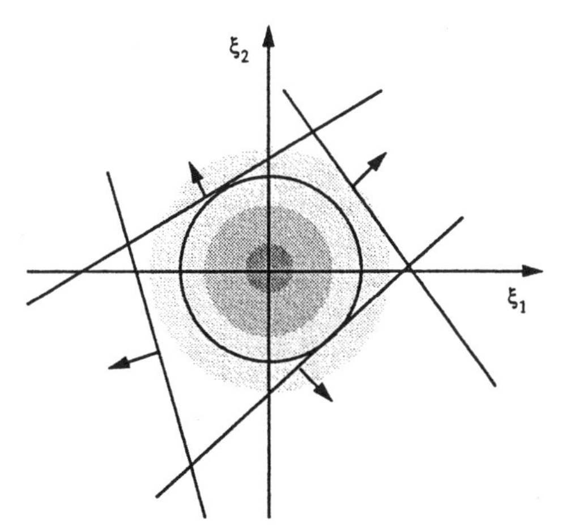

Fig. 4. Moving the acceptability region boundaries to improve parametric yield.

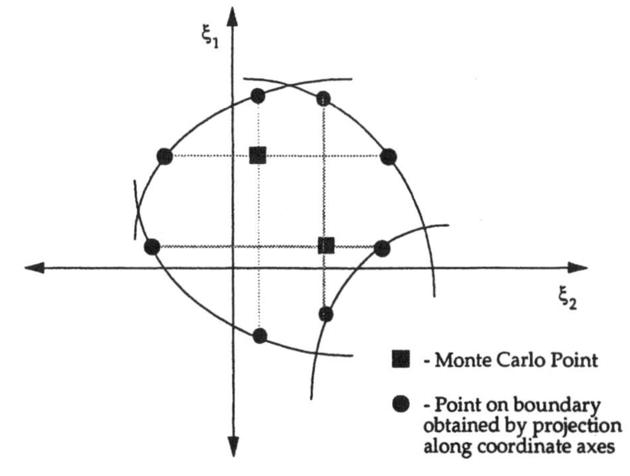

Fig. 5. Generation on points on acceptability region boundary in the boundary integral method.

yield gradient, can be applied only to acceptability regions determined by such linear models and even then only if the number of statistical variables is restricted to about four or five. Therefore, this methodology is limited in its application to the particular class of digital MOSFET circuits described by the authors.

C. The Boundary Integral Method

The boundary integral method [8]–[11], which may be viewed as a generalization of the Texas Instruments approach, uses a formulation of yield involving the acceptability region in the space of independent statistical disturbances. However, the boundary integral method is unique in the sense that through the application of Stokes' theorem it reformulates yield itself as a surface integral on the boundary of the acceptability region in this disturbance space. The advantages of this reformulation is that yield and derivatives of yield, with respect to deterministic nominal parameters, parameters of the disturbance distribution, and performance constraints, can all be expressed as surface integrals on the acceptability region boundary, and can be computed through Monte Carlo based on the same set of sample points. This sample is obtained by generating points within the acceptability region and then projecting them onto the acceptability region boundaries, as illustrated in Fig. 5.

Suppose we generate such a sample of points on the acceptability region boundary. $S = \{\zeta^{(k)}\}_{k=1,\cdots,N_\zeta}$. According to the boundary integral method the Monte Carlo estimate of yield is

$$\hat{Y} = \frac{1}{nN} \cdot \sum_{k=1}^{N_\zeta} s_k F_{\xi,q}(\xi_q)_{\xi=\zeta^{(k)}} \tag{12}$$

where n is the dimensionality of the disturbance space, N is the number of original Monte Carlo points, q is the projection direction used to generate point $\zeta^{(k)}$, $F_{\xi,q}(\xi_q)$ is the marginal probability function corresponding the qth component of $\zeta^{(k)}$, and s_k is a sign factor.

The expression for the derivative of the yield with respect to a designable circuit parameter, say x_i, is

$$\frac{\partial \hat{Y}}{\partial x_i} = \frac{1}{nN} \cdot \sum_{k=1}^{N_\zeta} f_{\xi,q}(\xi_q) \cdot \left. \frac{\partial \varphi_a / \partial x_i}{|\partial \varphi_a / \partial \xi_q|} \right|_{\xi=\zeta^{(k)}} \tag{13}$$

where φ_a is the performance that determines the boundary at point $\zeta^{(k)}$.

Similarly, the expression for the derivative of the yield with respect to a specification constraint, $\varphi_a^B, (\varphi_a - \varphi_a^B \geq 0)$ is

$$\frac{\partial \hat{Y}}{\partial \varphi_a^B} = \frac{1}{nN} \cdot \sum_{k=1}^{N_{\zeta,a}^B} f_{\xi,q}(\xi_q) \cdot \left. \frac{1}{|\partial \varphi_a / \partial \xi_q|} \right|_{\xi=\zeta_a^{(k)}} \tag{14}$$

where the terms correspond the subset of the boundary points $S_a^B = \{\zeta^{(k)}\}_{k=1,\cdots,N_{\zeta,a}^B}$ for which the constraint is active. The derivative values are the sensitivities of the yield to various constraint specifications.

In addition, expressions can be obtained for derivatives of yield with respect to parameters of the disturbance distribution. The sensitivity of the yield with respect to a parameter p that may affect the distribution of more than one disturbance is given by

$$\frac{\partial Y}{\partial p} = \frac{1}{nN} \sum_{k=1}^{N_\zeta} s_k \left[\frac{\partial}{\partial p} F_{\xi,q}(p, \xi_q)_{\xi=\zeta^{(k)}} + \sum_{i=1, i\neq q}^{n} \left[F_{\xi,q}(p, \xi_q) \frac{\partial}{\partial p} \log f_{\xi,i}(p, \xi_i) \right]_{\xi=\zeta^{(k)}} \right]. \tag{15}$$

This expression can be used to determine the effect of the variances or spreads of different process disturbances on the performance of the circuit. This expression is also used to obtain the yield gradient component contributed by disturbances that have distributions dependent on the designable parameters. Although a yield optimization procedure will only make use of the information regarding the gradient of the yield with respect to the designable parameters, the aforementioned sensitivities can provide valuable information about the circuit being designed.

The boundary integral method provides a way to estimate yield, yield gradients, and sensitivities that is completely decoupled from the larger statistical optimization problem. Yield, its gradient with respect to the designable parameter vector, and its derivatives with respect to performance specifications, disturbance distribution parameters, etc. can be computed without any reference to the kind of optimization that needs to be performed. One is not tied down to any one particular optimization formulation and the choice of an optimization algorithm is completely orthogonal.

The same orthogonality applies to the choice of the analytic approximation method. In [8]–[11], in order to save on the cost of performance evaluation by circuit simulation at every Monte Carlo point, the performances of the circuit are approximated by macromodels (response surfaces) that are quadratic in the disturbance variables. However, one can even choose to estimate performance employing a smaller number of boundary points obtained directly through line searches and avoid approximations altogether.

The advantages of using the surface integral formulation of yield are multiple. First, the computation involved in determining separate sets of Monte Carlo points for the yield and the yield gradient is avoided. Most importantly, though, the Monte Carlo estimate of the surface integral yield expression is consistent with the Monte Carlo estimate of the yield gradient integral, i.e., the gradient of the surface integral Monte Carlo estimate of the yield is the Monte Carlo estimate of the yield gradient. This consistency is absent if the yield is estimated as a volume integral based on a different sample. As a consequence of the consistency a gradient-based optimization using very poor estimates, even based on a few points, will still converge.

VI. Circuit Optimization Examples with Boundary Integrals

We now illustrate the effectiveness of the boundary integral method by several examples. As a first example, we used the boundary integral method to maximize the yield of the differential amplifier example shown in Fig. 2. The specifications, disturbances, and designables were as described earlier. The circuit yield improved from 39% to 100%. Figs. 6, 7, and 8 show distributions of the three performances of interest before and after optimization.

These figures show the trade-off that has been achieved between the gain and unity gain frequency (ugf) specs, and the phase margin spec. The final spreads of all three performance measures have been reduced. However, the nominal gain and the nominal ugf have been reduced and a corresponding increase in the nominal phase margin has been obtained.

The boundary integral method affords us an efficient way to compute circuit yield and its gradient. Traditionally, yield maximization examples tend to focus on improving circuit yield using this gradient information. However, as we have discussed in Section IV, statistical design need not restrict itself to maximizing yield. In this example we illustrate how we can increase a performance specification while maintaining a minimum yield specification. The problem formulation is of the form

$$\max_{\text{spec}\,\boldsymbol{p}_0} \text{spec}$$
$$\text{such that } Y(\boldsymbol{p}_0, \text{spec}) \geq Y_{\min}.$$

We chose to maximize the ugf specification (spec) while

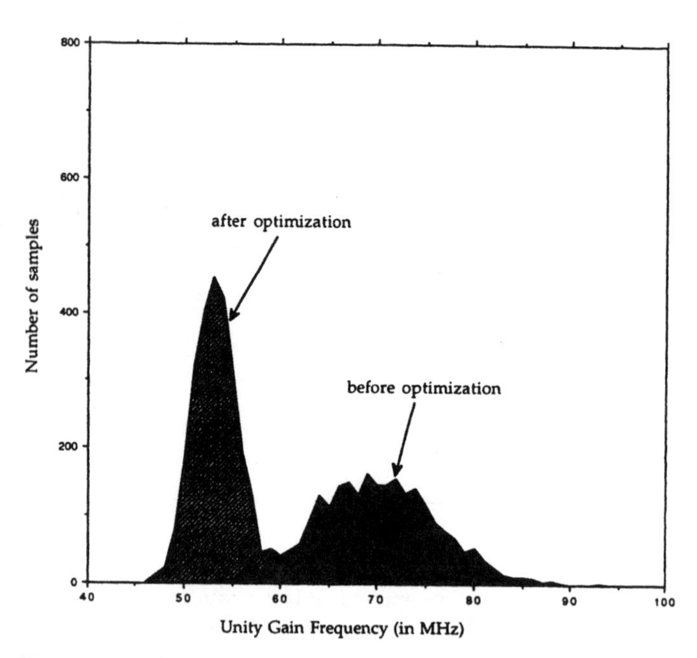

Fig. 6. Distributions of ugf (in megahertz) in a population of 2500 samples before and after optimization.

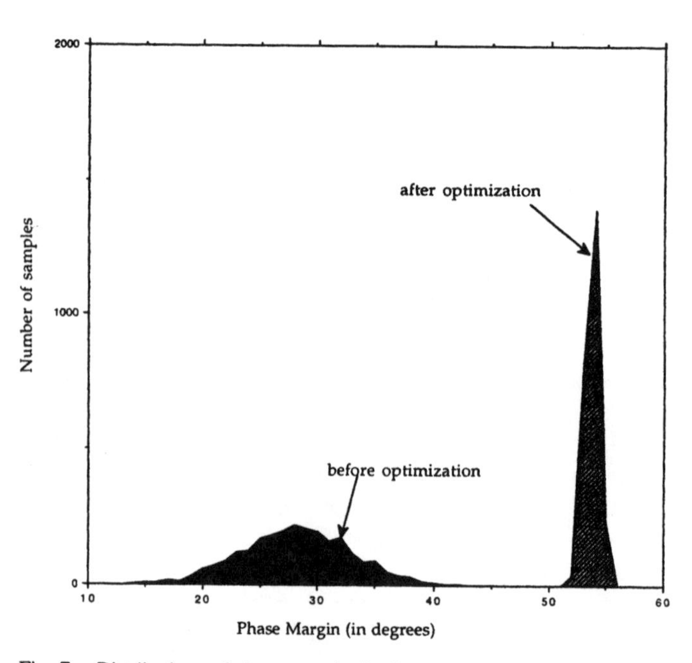

Fig. 7. Distributions of phase margin (in degrees) in a population of 2500 samples before and after optimization.

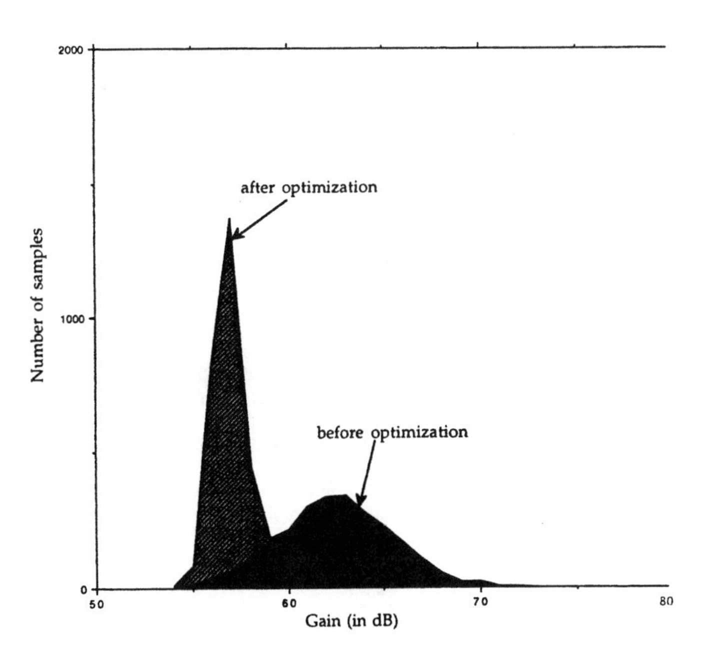

Fig. 8. Distributions of gain (in decibels) in a population of 2500 samples before and after optimization.

TABLE I
MARGINAL YIELDS AT INITIAL POINT

Performance	Marginal Yield (in%)
Bandwidth	88
DC gain	100
Ripple	54
Roll-off	100

maintaining a minimum yield ($Y_{\min}$) of 80%. The initial point had a yield of 100%. The final yield was decreased to 98% and the ugf spec was raised from 45 to 55 MHz.

Next consider the five-pole current mode filter described in [32]. The circuit has a total of 175 MOS transistors. The desired specifications for this circuit are

- dc gain ≥ -1.5 dB,
- bandwidth $\geq$ 35 MHz,
- passband ripple $\leq$ 0.5 dB,
- rolloff $\geq$ 100 dB/decade.

Disturbances include variations in the width, length, oxide thickness, and flat-band voltage of the n- and p-type devices. We found that linear macromodels in the disturbance variables were sufficiently accurate for all four performances. The designables in this circuit were five independent transistor widths. The initial yield of the circuit was 42%.

The marginal yield of a certain circuit with respect to a particular performance is defined as the fraction of circuits that satisfies that particular performance specification. For this example, the marginal yields at the initial point are given in Table I.

It is not entirely atypical for such a situation to occur with a design. In many cases the design is such that some performance specifications are comfortably satisfied and other performances are quite close to the specification. In other cases some performances that may nominally be well within the specification bounds are extremely sensitive to process variations and cause the circuit yield to be low. This, coupled with the fact that performances are usually competing against each other (e.g., an increase in gain usually causes an decrease in bandwidth), may cause the circuit to have low parametric yield even though it may seem to have good nominal performances and good marginal yields. Since the performances are not independent of each other, the overall parametric yield is not equal to the product of the marginal yields.

For this particular circuit, at the initial point the sensitivity of the yield to the dc gain and rolloff performance specification was 0.0. This does not come as a surprise to us after considering that these specifications are comfortably satisfied at the initial point. The normalized sensitivities of the yield to the disturbance variances at the initial point are given in Table II.

TABLE II
SENSITIVITIES OF YIELDS WITH RESPECT TO DISTURBANCE VARIANCES

Disturbance	Sensitivity of yield wrt variance
n flat-band voltage variation	0.007466858
n-device t_{ox} variation	-0.993829417
n-device length variation	0.060171695
n-device width variation	-0.009230657
flat-band voltage variation	0.030542683
p-device t_{ox} variation	-0.058411000
p-device length variation	-0.059349247
p-device width variation	0.025970466

We note that variance of the oxide thickness variation of the n-type devices has the most detrimental effect on the yield. Since the variance of a disturbance is a measure of its spread, we can determine which processing step needs to be controlled most stringently to ensure good circuit yield. While such control may require costly equipment or may not be possible for a particular circuit, we might find such information useful when designing newer processing equipment.

Upon optimization, which required only 288 circuit simulations, the yield of this circuit was increased to 98%. The designable parameters changed as given in Table III.

TABLE III
PERCENTAGE CHANGE IN DESIGNABLES

Width #	% Change
1	21.54
2	-23.55
3	50.76
4	2.21
5	95.53

These numbers indicate that the designable parameters may

TABLE IV
FINAL MARGINAL YIELDS

Performance	Marginal Yield (in %)
DC gain	100
Bandwidth	99
Ripple	99
Roll-off	100

undergo a significant percent change during optimization. A very small change is noticed in Width 4 whereas Width 5 has almost doubled during the course of the optimization. The marginal yields at the final point were as given in Table IV.

VII. CONCLUSIONS

We have shown that statistical design problems can be expressed as optimization problems in which either the objective function or the constraint functions depend on expectations of random variables. Traditionally, one of the deterrents to the use of statistical design methods has been the high cost of circuit/process/device simulation. With the advent of fast computers and the development of methods that work to minimize the cost of simulation, yield maximization methods are gaining acceptance among the circuit design community. In fact, the scaling down of device sizes without a corresponding scaling down of processing variations has made such tools invaluable for circuits designed for manufacture.

Except for an attempt to analyze several methods developed to handle the case of discrete circuit yield optimization [30], no comparative performance studies exist on various IC circuit yield maximization methodologies. In the absence of objective data, we offer our, somewhat biased, conclusion.

The boundary integral method offers the most flexible formulation of the IC statistical design problem. By decoupling the computation of the yield and its gradients with respect to a variety of parameters from the actual optimization formulation, it allows the flexibility of using the objective and constraint functions that are most suited for the problem at hand.

ACKNOWLEDGMENT

The authors would like to thank M. Anderson for helping with the examples.

REFERENCES

[1] S. R. Nassif, A. J. Strojwas, and S. W. Director, "FABRICS-II: A statistical based IC fabrication process simulator," *IEEE Trans. Computer-Aided Design*, vol. CAD-3, Jan. 1984.

[2] L. Nagel, "SPICE2: A computer program to simulate semiconductor circuits," Univ. of California, Berkeley, ERL Memo. ERL-M520, May 1975.

[3] J. M. Hammersley and D. C. Handscomb, *Monte Carlo Methods*. London: Metheun, 1964.

[4] P. E. Gill, W. Murray, M. A. Saunders, and M. H. Wright, "User's guide for NPSOL (version 4.0)," Systems Optimization Lab., Stanford Univ., Stanford, CA, Tech. Rep. SOL 86-2, Jan. 1986.

[5] P. Cox, P. Yang, S. S. Mahant-Shetti, and P. Chatterjee, "Statistical modeling for efficient parametric yield estimation of MOS VLSI circuits," *IEEE Trans. Electron Devices*, vol. ED-32, pp. 471–478, Feb. 1985.

[6] P. Yang, D. E. Hocevar, P. F. Cox, C. Machala, and P. K. Chatterjee, "An integrated and efficient approach for MOS VLSI statistical circuit design," *IEEE Trans. Computer-Aided Design*, vol. CAD-5, pp. 5–14, Jan. 1986.

[7] D. E. Hocevar, P. F. Cox, and P. Yang, "Parametric yield optimization for MOS circuit blocks," *IEEE Trans. Computer-Aided Design*, vol. 7, pp. 645–658, June 1988.

[8] P. Feldmann and S. W. Director, "Accurate and efficient evaluation of circuit yield and yield gradients," in *Proc. IEEE Int. Conf. Computer-Aided Design*, Nov. 1990, pp. 120–123.

[9] P. Feldmann, "Statistical integrated circuit design," Ph.D. dissertation, Carnegie Mellon Univ., Pittsburgh, PA, Jan. 1991.

[10] P. Feldmann and S. W. Director, "Improved methods for IC yield and quality optimization using surface integrals," in *Proc. IEEE Int. Conf. Computer-Aided Design*, Nov. 1991.

[11] P. Feldmann and S. W. Director, "Integrated circuit quality optimization using surface integrals," submitted to *IEEE Trans. Computer-Aided Design*.

[12] P. Feldmann and S. W. Director, "A Macromodeling approach for increasing efficiency of IC yield optimization," presented at the Int. Symp. Circuits and Syst., 1991.

[13] R. Y. Rubinstein, *Simulation and the Monte Carlo Method*. New York: Wiley, 1981.

[14] M. A. Styblinski and L. J. Opalski, "Algorithms and software tools of IC yield optimization based on fundamental fabrication parameters," *IEEE Trans. Computer-Aided Design*, vol. CAD-5, pp. 79–89, Jan. 1986.

[15] R. Biernacki and M. Styblinski, "Efficient performance function interpolation scheme and its application to statistical circuit design," *Int. J. Circuit Theory Application*, vol. 19, pp. 403–422, 1991.

[16] M. A. Styblinski and S. A. Aftab, "Efficient circuit performance modeling using a combination of interpolation and self organizing approximation techniques," in *1990 IEEE Int. Symp. Circuits Syst. Proc.*, 1990.

[17] L. J. Opalski and M. A. Styblinski, "Generalization of yield optimization problem: The maximum income approach," *IEEE Trans. Computer-Aided Design*, vol. CAD-5, pp. 346–360, Apr. 1986.

[18] M. A. Styblinski and J. C. Zhang, "Orthogonal array approach to gradient based yield optimization," in *1990 IEEE Int. Symp. Circuits Syst. Proc.*, 1990.

[19] M. A. Styblinski and J. C. Zhang, "Circuit performance variability reduction: Principles, problems and practical solutions," presented at the IEEE Int. Conf. Computer-Aided Design, Nov. 1991.

[20] N. R. Draper and H. Smith, *Applied Regression Analysis* (Wiley Series in Probability and Mathematical Statistics). New York: Wiley, 1980.

[21] S. W. Director and R. A. Rohrer, "The generalized adjoint network and network sensitivities," *IEEE Trans. Circuit Theory*, vol. CT-16, pp. 318–323, Aug. 1969.

[22] P. Feldmann, T. V. Nguyen, S. W. Director, and R. A. Rohrer, "Sensitivity computation in piecewise approximate circuit simulation," *IEEE Trans. Computer-Aided Design*, vol. 10, Feb. 1991.

[23] *SAS User's Guide: Statistics*, Version 5 ed., SAS Institute, Cary, NC, 1985.

[24] J. Sacks, S. B. Schiller, and W. J. Welch, "Designs for computer experiments," *Technometrics*, vol. 31, no. 1, pp. 41–47, Feb. 1989.

[25] M. D. McKay, R. J. Beckman, and W. J. Conover, "A comparison of three methods for selecting values of input variables in the analysis of output from a computer code," *Technometrics*, vol. 21, no. 2, pp. 239–245, May 1979.

[26] M. S. Phadke, *Quality Engineering Using Robust Design*. Englewood Cliffs, NJ: Prentice-Hall, 1989.

[27] M. J. M. Pelgrom, AAD C. J. Duinmaijer, and A. P. G. Welbers, "Matching properties of MOS transistors," *IEEE J. Solid-State Circuits*, vol. 24, no. 5, pp. 1433–1440, Oct. 1989.

[28] M. A. Styblinski and A. Ruszczynski, "Stochastic approximation approach to statistical circuit design," *Electron. Lett.*, vol. 19, no. 8, pp. 300–302, 1983.

[29] A. Strojwas, Ed., *Selected Papers on Statistical Design of Integrated Circuits*. New York: IEEE Press, 1987.

[30] E. Wehrhahn and R. Spence, "The performance of some design centering methods," in *Proc. IEEE Int. Symp. Circuits Syst.*, May 1984, pp. 1424–1438.

[31] W. J. Welch, T.-K. Yu, S.-M. Kang, and J. Sacks, "Computer experiments for quality control by parameter design," *J. Quality Technology*, vol. 22, pp. 15–22, 1990.

[32] S.-S. Lee *et al.*, "A 40 MHz CMOS continuous-time current-mode filter," in *Proc. IEEE 1992 CICC*, pp. 24.5.1–24.5.4.

Stephen W. Director (S'65–M'69–SM'75–F'78) received the B.S. degree from the State University of New York at Stony Brook in 1965 and the M.S. and Ph.D. degrees in electrical engineering from the University of California, Berkeley, in 1967 and 1968, respectively.

From 1968 until 1977 he was with the Department of Electrical Engineering at the University of Florida, Gainesville. From Sept. 1974 to Aug. 1975 he was a Visiting Scientist in the Mathematical Sciences Department at IBM's T. J. Watson Research Center, Yorktown Heights, NY. He joined Carnegie Mellon University, Pittsburgh, PA, in 1977 and served as Head of the Department of Electrical and Computer Engineering from 1982 to 1991. In 1982 he founded the SRC-CMU Research Center for Computer-Aided Design and served as its Director from 1982 to 1989. He is now Dean of the College of Engineering and U. A. and Helen Whitaker University Professor of Electrical and Computer Engineering.

Dr. Director has served as President of the IEEE Circuits and Systems Society, as Chairman of the CAS Technical Committee on Computer-Aided Network Design (CANDE), and as Associate Editor of the IEEE TRANSACTIONS ON CIRCUITS AND SYSTEMS. He has authored or coauthored four texts. In 1970 and 1985 he received Best Paper Awards from the IEEE Circuits and Systems Society; in 1976 he received the Frederick Emmons Terman Award from the American Society of Engineering Education; and in 1978 he received the W. R. G. Baker Prize Paper Award from the IEEE. In 1984 he received an IEEE Centennial Medal and was named a Distinguished Alumnus for the State University of New York, Stony Brook. He was elected to membership in the National Academy of Engineering in 1989. In 1992 he received the Society Award from the Circuits and Systems Society and a Best Paper Award from the ACM/IEEE Design Automation Conference.

Peter Feldmann (S'88–M'91) was born in Timisoara, Romania, in 1958. He received the B.Sc. degree in computer engineering in 1983 and the M.Sc. degree in electrical engineering in 1987, both from the Technion in Haifa, Israel, and the Ph.D. degree in 1991 from Carnegie Mellon in Pittsburgh, PA.

From 1985 through 1987 he designed digital signal processors at Zoran Microelectronics in Haifa. Currently, he is on the technical staff at Bell Labs in Murray Hill, NJ. His research interests include CAD for VLSI circuits, more specifically circuit simulation, optimization, and design for manufacturability.

Kannan Krishna (S'89) received the B.Tech. degree in electrical engineering from the Indian Institute of Technology, New Delhi, India, in 1990 and the M.S. degree from the Department of Electrical and Computer Engineering, Carnegie Mellon University, Pittsburgh, PA, in 1991, where he is currently working toward the Ph.D. degree. His research interests are in the areas of statistical design, device modeling, and circuit simulation.

Currently he is holding a temporary position as Member of Technical Staff at AT&T Bell Laboratories, Murray Hill, NJ.

Yield Optimization of Analog IC's Using Two-Step Analytic Modeling Methods

Carlo Guardiani, *Member, IEEE*, Primo Scandolara, Jacques Benkoski, *Member, IEEE*, and Germano Nicollini

***Abstract*—Innovative methods for statistical design optimization have been applied to the development of analog IC blocks. The most important feature of these methods is the derivation of an analytic function representing the yield surface in the design parameter space. Using this analytic model it is possible to optimize the yield accurately and efficiently. All the required operations are implemented in an integrated and fully automated CAD system. A comparison between simulated and measured data on several wafer lots demonstrates the validity of the approach.**

I. Introduction

THE capability of taking into account the natural random variability of the fabrication process at design stage is of strategic importance to decrease VLSI IC's manufacturing cost. Unfortunately the statistical nature of the problem generates a number of tasks of formidable complexity, starting from the introduction of reliable and significant measurements and data collection in the fabrication line to the practical feasibility of performing statistical circuit simulation. The final goal of automatically optimizing the parametric yield with respect to the design parameters is even more difficult because it requires the iterative evaluation of both yield and yield gradient for several values of the input parameters [1], [2]. To circumvent these difficulties a number of techniques have been introduced which either try to model the generally unknown region of the input parameters space that results in acceptable circuit performances (simplicial approximation [3]), or to reduce the number of samples required for the Monte-Carlo analysis by decreasing the variance of the yield estimate (importance sampling [4], [5]). More recently, a number of papers have been focusing on the use of analytic models as inexpensive computational surrogates of circuit simulation in the Monte-Carlo sampling process [6]–[9]. These analytic models are obtained following the prescriptions of a technique known as the response surface method (RSM) [10]. In [11] we described an innovative technique for IC yield maximization based on the derivation of a polynomial representation of the yield as a function of the design parameters. This goal was achieved with two different steps of RSM modeling. In the first step the design objectives were characterized as simple analytic functions of both design and statistical parameters. In the second step, these analytic functions were used to build a model of the yield in the design parameter space. This paper describes how this technique can be successfully applied to the yield optimization of analog IC's. We will also consider an alternative technique that leads to the same analytic formulation of the yield by building multiple analytic functions of the design objectives with respect to the statistical process variables in the design parameter space. We will denote these two techniques as the "global" and the "local" two-step analytic modeling approach, respectively. We will show that it is possible to automatically and accurately optimize the yield together with other design objectives using both techniques. A practical example will evidence the characteristic aspects of both methods. Finally, we will compare the computed results with the measurements on several lots of a manufactured circuit.

Manuscript received November 30, 1992; revised March 2, 1993. This work was supported in part by the Joint European Submicron Silicon (JESSI) AC12: Analog Expert Design System.

C. Guardiani, P. Scandolara, and G. Nicollini are with SGS-Thomson Microelectronics, 20041 Agrate Brianza, Italy.

J. Benkoski is with SGS-Thomson Microelectronics, 38019 Grenoble Cedex, France.

IEEE Log Number 9209009.

II. The Two-Step Yield Analytic Modeling Approach

The yield optimization methods described in this paper are based on the assumption that it is possible to model the fundamental performance of every IC through simple analytic functions of a set of design parameters $\{x\}$ and a set of statistical random variables $\{\delta\}$ representing the effect of the IC process randomness within a predefined and limited space. These simple analytic functions can accurately reproduce the response of the circuit in a limited range of variations of the independent variables and therefore can be used as computationally inexpensive substitutes of electrical simulation. Let Φ represent the set of the relevant circuit performances with lower and upper restrictions $\Phi_L < \Phi < \Phi_U$ on their acceptable values. We assume that it is possible to find a low-order polynomial approximation η such that $\eta = \Phi + \epsilon$, where ϵ, the lack of fit of η, is small compared with $|\Phi|$ in the range of $\{x, \delta\}$ explored. The subspace defined by

$$\Xi = \{x, \delta | \Phi_L < \Phi < \Phi_U\} \tag{1}$$

is called the acceptability region. The yield is defined as

$$\text{Yield} = \int_{\Xi} \varphi(\delta)\, d\delta \tag{2}$$

where φ is the joint probability distribution function of the random variables δ.

A common way to compute the multidimensional integral in (2) is to first extend the integration to the entire space by means of a support function $\zeta(x, \delta)$, defined as

$$\zeta(x, \delta) = \begin{cases} 0, & \text{if } (x, \delta) \notin \Xi \\ 1, & \text{if } (x, \delta) \in \Xi. \end{cases} \tag{3}$$

Using (3) in (2) it is possible to write

$$\text{Yield} = \int \zeta(x,\delta)\varphi(\delta)\,d\delta. \tag{4}$$

The Monte-Carlo method approximates the value of the integral in (4) with the following sum:

$$\text{Yield} \approx \lim_{N\to\infty} \frac{\sum_{i=1}^{N} \zeta(x,\delta_i)}{N} \tag{5}$$

where the sample points δ_i are randomly extracted with distribution φ. The explicit knowledge of the acceptability region is not necessary to compute $\zeta(x,\delta_i)$ if the inequalities in (1) can be evaluated efficiently and accurately. The simplified models $\{\eta\}$ are generated for this purpose in the first analytic modeling step using the RSM prescriptions as described, for instance, in [7], [8], and [11]. In the second analytic modeling step the polynomial approximations $\{\eta\}$ are used in (5) to compute $\zeta(x,\delta_i)$ and ultimately to evaluate the yield on a selected set of points $\{x^1,\cdots,x^m\}$ in the design parameters space, as prescribed by a suitable experimental design. This can be done in two ways: it is possible to either build the response surfaces (first step) in the complete parameters space or to produce several "local" models (one for each point in the $\{x\}$ space discretization defined by the second step) that are functions of the vector $\{\delta\}$ only. If we define with η_g^i and η_l^i, respectively, the value of the global and local polynomial approximation vectors in x^j, then in the first case one has

$$\eta_g^j = \eta(x^j,\delta), \qquad j = 1,\cdots,m \tag{6}$$

while in the second case it is

$$\eta_l^j = \eta^j(\delta), \qquad j = 1,\cdots,m. \tag{7}$$

In both cases one obtains a set of yield values $\{Y^1, Y^2,\cdots, Y^m\}$, where

$$\text{(global)} \qquad Y^j = \lim_{N\to\infty} \frac{\sum_{i=1}^{N} \zeta_g(x^j,\delta_i)}{N}$$

$$\text{(local)} \qquad Y^j = \lim_{N\to\infty} \frac{\sum_{i=1}^{N} \zeta_l^j(\delta_i)}{N} \tag{8}$$

and $\zeta_g(x^j,\delta_i)$ and $\zeta_l^j(\delta_i)$ represent, respectively, the support functions computed using the global and the local models.

The set of yield values $\{Y^1, Y^2,\cdots, Y^m\}$ thus obtained is used to derive an analytic model of the yield as a function of the vector $\{x\}$ by regression analysis. Finally, the maximum of the yield in the design parameter space is trivially found by numeric optimization of the yield model.

In general, the random variables in the set $\{\delta\}$ are correlated, so that their joint probability distribution function is difficult to obtain. In practice, however, it is possible to find a subset of independent factors that account for most of the natural process variance and whose distribution can be estimated from measurements. Alternatively, it is also possible to derive the statistical distribution of the $\{\delta\}$ from statistical process and device simulation as shown in [12] and [13].

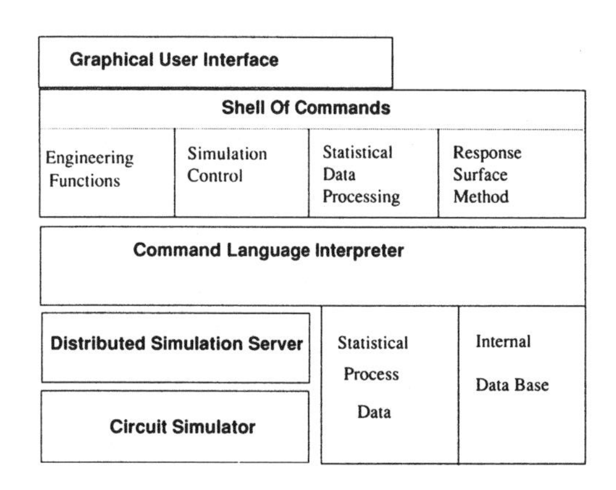

Fig. 1. The architecture of PLUTO.

III. Algorithm Implementation

The above-described yield optimization methods have been implemented in a CAD system named PLUTO [6]. The general architecture of the system is shown in Fig. 1. The core is the Command Language Interpreter (CLI), which allows any circuit performance to be described in terms of node voltages and currents by means of a structured programming language. The CLI also implements and automates all the steps of the RSM algorithm: starting from the designable and statistical parameters data base, which can be built by interacting with a schematic editor, PLUTO is able to generate several kinds of experimental designs (Plackett Burmann, fractional factorial, central composite design [14], [15] and Latin hyper-cube [16]), run simulations, and build the response surfaces. It is possible to take advantage of the intrinsic parallelism of the process that builds the response surfaces by distributing the simulations over a computers network. A module of PLUTO is able to arrange the simulations through the available computing resources, balancing the load and the memory usage. The language also includes commands that generate random points in the statistical parameters space. These points can be distributed according to the probability density functions that are most frequently observed in practice (Gaussian, log-normal, uniform, etc.) It is possible to define either exact or statistical matching between different parameters to account for tracking or correlation phenomena. Finally, the CLI implements a set of commands to evaluate the polynomial models in randomly generated points, to compute the yield as defined in (5), and to optimize a multiple-objective yield-performance cost function either using the weighted sum or the minmax approach [17]. Since the cost function is a linear combination of second-order polynomials, its stationary points can be directly found by solving the linear system that one obtains by equating the cost function gradient to zero. Alternatively, it is also possible to use one of the three different numerical minimization algorithms implemented in PLUTO, i.e., Levemberg–Marquardt, sequential quadratic programming [18], or modified random search [19].

IV. An Application Example: Yield Optimization of a CMOS Amplifier

Both the "local" and "global" methods were exploited in the design of the CMOS differential difference amplifier (DDA)

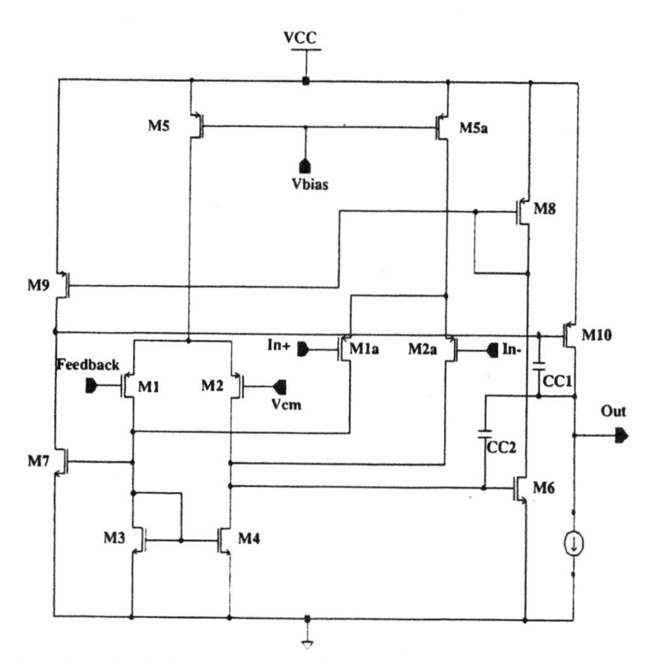

Fig. 2. Differential difference amplifier used in the application example.

TABLE I
INDEPENDENT STATISTICAL PARAMETERS

Name	Description
LP1[DSURF]	Surface Doping (P-Ch)
EN1[DSURF]	Surface Doping (N-Ch)
EN1[EOX]	Gate Oxide Thickness
EN1[DL]	Channel Lenght Mask Error
EN1[DW]	Channel Width Mask Error
EN1[NAB]	Implant Doping (N-Ch)
LP1[NAB]	Implant Doping (P-Ch)
EN1[NB]	Substrate Doping (N-Ch)
LP1[NB]	Substrate Doping (P-Ch)
LP1[Uo]	Carriers Mobility (P-Ch)
EN1[Uo]	Carriers Mobility (N-Ch)
POLY[Dimp]	Poly Doping
CAP[EOX]	Capacitor Oxide Thickness
Voff1[DC]	Voltage Offsets
Voff3[DC]	
Voff5[DC]	
Voff6[DC]	

TABLE II
CRITICAL DESIGN SPECIFICATIONS

Specification Name	Value
Bandwidth (20dB gain)	> 800 KHz
Phase margin	> 55 degrees
Total Harmonic Distortion	< -57 dB

[20] shown in Fig. 2. It consists of two input differential stages ($M1, M2, M5$ and $M1a, M2a, M5a$) that share a common current mirror ($M3, M4$), followed by a gain stage ($M6$–$M9$) and a class A output stage. A nested Miller compensation ($CC1$ and $CC2$) is used because of the presence of three high-impedance nodes in the signal path. If an input signal is connected between In+ and In–, V_{cm} is tied to a fixed reference value (about half the power supply), and Feedback is used to implement a resistive feedback from the output node, this amplifier can be used as an efficient substitute for an instrumentation amplifier. In fact, the inherent circuit simplicity of the former leads to a much lower power consumption and silicon area if compared with the latter. The drawback of a DDA is mainly related to the large harmonic distortion that arises in the case of mismatches in the input stage and/or in the presence of a relatively large common-mode input [20] which can bring some of the input stage devices out of their saturation region. Therefore, process-related parameters affect the amplifier linearity in such a way that an accurate tuning of the design variables taking into account the process spread is necessary to reach an acceptable parametric yield.

In our example, the 13 independent process parameters shown in Table I were chosen as statistical variables in a 1.5-μm CMOS n-well technology with double metal layers and implanted capacitors. In addition, four random offset voltages have been introduced to take into account the mismatch effects between $M1$ and $M2$, $M1a$ and $M2a$, $M5$ and $M5a$, $M3$ and $M4$, respectively (see Fig. 2).

Process parameters and mismatches are assumed to be independent random variables with truncated Gaussian distribution that reproduce the data collected in the fabrication pilot line. The total harmonic distortion (THD) of the DDA, connected in a 20-dB gain configuration, together with its open-loop 20-dB bandwidth and phase margin were assumed as design constraints for the yield evaluation, as shown in Table II. The widths of the six MOS transistors $M1, M1a, M2, M2a, M5$, and $M5a$ were chosen as design parameters; however, only two of these six parameters are actually independent because of device matching.

The first step of both global and local methods was completed by evaluating the design performances in N_{obs} different points of the input space chosen according to a Latin hyper-cube experimental design plan. This was done using an electrical circuit simulator [21] and taking advantage of the distributed computing capabilities of PLUTO by running all the simulations in parallel on a network of six CPU's. The number of simulations to be performed has been deduced from the following considerations. First, a second-order polynomial in the input variables was chosen as a suitable model for all the design performances. This choice represented a trade-off between accuracy and efficiency. In fact, a linear model is not able to account for curvature effects and interactions between the input variables [22] while third- and higher order polynomial models require a greater number of experimental points (i.e., simulations). On the other hand, a second-order polynomial in N variables has $N_c = [(N(N+3))/2] + 1$ coefficients, which represent the unknown parameters of the model. Therefore, at least N_c independent equations are needed to determine univocally these unknowns, which can be obtained, for instance, by applying the least squares conditions with respect to a set of at least $N_{obs} = N_c$ observations.

TABLE III
YIELD EVALUATION RESULTS

M1[w], M5[W]	Yield loss vs Design Specs (local) Bw - Pm - Thd	Yield (local)	Yield (global)
580 μ, 150 μ	100% - 99% - 97.56%	96.62	94.76
360 μ, 195 μ	100% - 100% - 87.08%	87.08	89.30
865 μ, 60 μ	97.72% - 99.58% - 0.0%	0.0	1.42
615 μ, 155 μ	100% - 96.22% - 96.04%	92.32	92.16
250 μ, 70 μ	0.0% - 99.74% - 100%	0.0	0.0
760 μ, 105 μ	100% - 96.94% - 85.32%	82.46	69.72
690 μ, 90 μ	99.94% - 99.98% - 64.08%	64.06	56.46
970 μ, 135 μ	100% - 35.36% - 84.38%	27.16	30.02
430 μ, 175 μ	100% - 100% - 92.14%	92.14	94.30

Second, a certain number of extra observations are needed to increase the degree of freedom $(N_{obs} - N_c)$ and consequently to improve the overall predictive power of the model.

Finally, some other extra observations are required to measure the accuracy of the models by means of one of the R^2, R^2_{adj} [15], or R^2_{press} indicators [8].

We decided to take $N_{obs} = 2 \times N_c$, with $N_c = [(19(19+3))/2] = 209$, for the global approach and $N_c = [(17(17+3))/2] = 170$ for each of the nine different models of the local approach. About two-thirds of these observations were used to compute the coefficients of the models and the remaining were used to verify the fitting accuracy. The total number of observations can be minimized, if necessary, by using iterative incremental techniques, as in [8]. It is evident that the local approach involves a much larger computational effort than the global method; nonetheless, local models give more accurate results because smaller ranges of variations are explored. The polynomial models (η) were obtained using least-squares regression analysis and fast Given transformations [23] for all the objectives with the exception of THD, which was best fitted using least absolute values because of the presence of outliers. The figures in Table III show the results of yield evaluations obtained with the two methods; the yield was evaluated for the nine different combinations of $M1$ and $M5$ channel widths generated in the second step using another Latin hyper-cube experimental design. These nine yield values are sufficient to compute an accurate second-order polynomial model of the yield as a function of the two designable parameters. The yield surface in this three-dimensional space can be easily visualized as shown by the contour plots of Figs. 3 (global) and 4 (local). In this case, the accuracy of the global technique is sufficient, as evidenced by the results of the local technique. The distribution of the yield loss among the different design specifications in each region is also shown. Finally, the yield was optimized together with die area and power consumption using a sequential quadratic programming algorithm [24]. The optimization results summarized in Table IV show a yield improvement from 47% to 99%. The described circuit was integrated on silicon as shown by the microphotograph in Fig. 5. The histograms in Fig. 6, comparing the expected and measured distributions of the total harmonic distortion in the optimum, demonstrate the satisfying accuracy of the algorithm predictions.

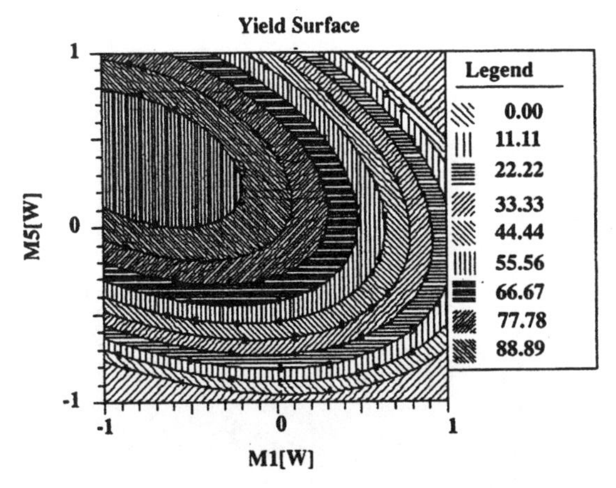

Fig. 3. Contour plot of the yield surface obtained with the global technique. The widths of the two MOS devices are scaled between −1 and 1.

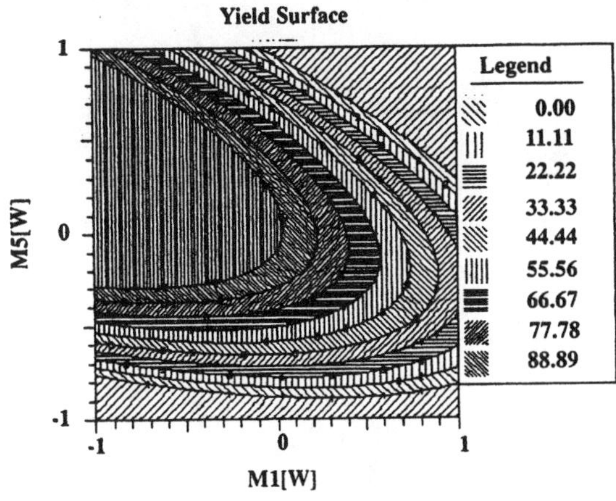

Fig. 4. Contour plot of the yield surface obtained with the local technique. The widths of the two MOS devices are scaled between −1 and 1.

TABLE IV
YIELD OPTIMIZATION RESULTS

Design Objectives/Parameters	Initial value	Final Value (local)	Final Value (global)
Bandwidth	1.2Mhz	1.1 Mhz	1.0 MHz
Phase margin	54 Deg	55 Deg	55 Deg
Total Harmonic Distortion	-59 dB	- 65 dB	-67 dB
Yield	47.18 %	99.97	99.50
M1[W]	1000 μ	365μ	376 μ
M5[w]	125 μ	167 μ	175 μ

V. CONCLUSION

The application of a new class of algorithms to the yield optimization of IC's has been presented in this paper. These algorithms are based on the derivation of an analytic formula expressing the yield as a function of the design parameters of the circuit. This allows a great simplification in the combined yield-performance optimization process. Another considerable advantage is the possibility of obtaining a direct graphical representation of the parametric yield in the space of the design variables, which gives a deeper insight and a better control on the statistical behavior of the circuit. In the application

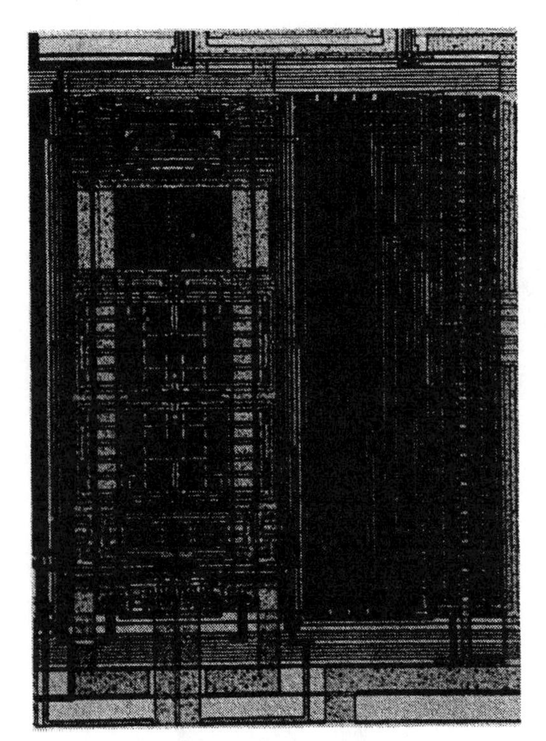

Fig. 5. Microphotograph of the integrated DDA used in the application example.

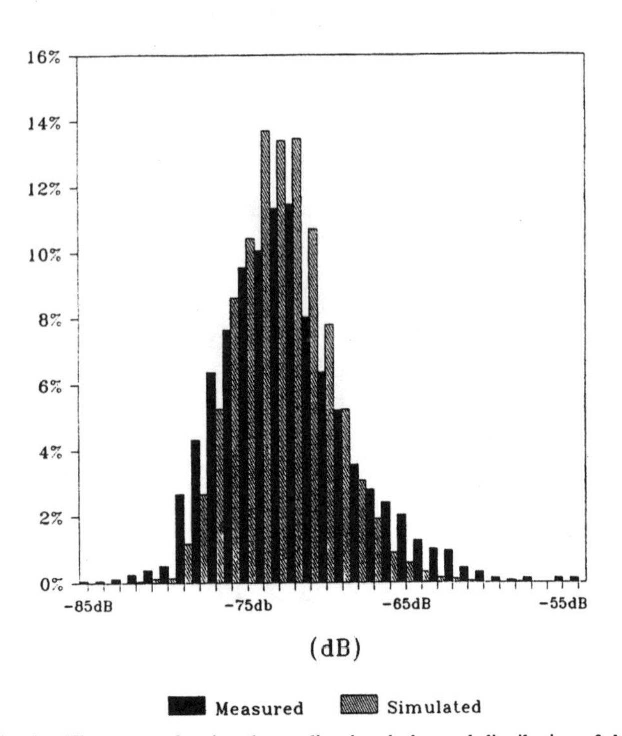

Fig. 6. Histograms showing the predicted and observed distribution of the THD of the DDA. The design parameters values have been fixed at their optimized values.

example shown in Section IV, the yield has been improved by over 50%. The described algorithms have been fully integrated in a CAD system automating most of the necessary operations, allowing accurate parametric yield optimization to become a standard piece of the design flow. The comparison between empirical and theoretical results reinforces the confidence in the accuracy of the predictions.

ACKNOWLEDGMENT

The authors are grateful to S. Mariani for the careful layout, to Dr. B. Franzini for his help in the realization of the application example using BASIS, and to Dr. G. Espinosa for the many helpful discussions on the subject. A special thanks goes to Dr. E. Profumo, who originally started and inspired the PLUTO project.

REFERENCES

[1] S. W. Director, P. Feldman, and K. Krishna, "Optimization of parametric yield: A tutorial," Carnegie-Mellon Univ., Pittsburgh, PA, Res. Rep. CMUCAD-92-10, Feb. 1992.

[2] R. Spence and R. S. Soin, *Tolerance Design of Electronic Circuits.* Reading, MA: Addison-Wesley, 1988 (ISBN 0-201-18242-4.

[3] S. W. Director, G. D. Hachtel, and L. M. Vidigal, "Computationally efficient yield estimation procedures based on simplicial approximation," *IEEE Trans. Circuits Syst.*, vol. CAS-25, pp. 121–130, Mar. 1978.

[4] D. E. Hocevar, M. R. Lightner, and T. N. Trick, "A study of variance reduction techniques for estimating circuit yield," *IEEE Trans. Computer-Aided Design,* vol. CAD-3, no. 3, pp. 279–287, 1984.

[5] K. Singhal and J. F. Pinel, "Statistical design centering and tolerancing using parametric sampling," *IEEE Trans. Circuits Syst.*, vol. CAS-28, no. 7, pp. 692–701, July 1981.

[6] C. Guardiani and M. Amadori, "PLUTO: An RSM analog circuit optimizer," in *Proc. EDAC* (Amsterdam), Feb. 1991, p. 596.

[7] A. R. Alvarez *et al.*, "Application of statistical design and response surface methods to computer-aided design," *IEEE Trans. Computer-Aided Design,* vol. 7, pp. 272–288, Feb. 1988.

[8] L. Milor and A. L. Sangiovanni-Vincentelli, "Computing parametric yield accurately and efficiently," in *Proc. ICCAD,* Nov. 1990, pp. 116–119.

[9] K. J. Antreich and R. K. Koblitz, "Design centering by yield prediction," *IEEE Trans. Circuits Syst.*, vol. CAS-29, no. 2, pp. 88–96, Feb. 1982.

[10] R. H. Myers, A. I. Khuri, and W. H. Carter, Jr., "Response surface methodology: 1966–1988," *Technometrics,* vol. 31, no. 2, pp. 137–157, May 1989.

[11] C. Guardiani, B. Franzini, and G. Nicollini, "Design for manufacturability: A two step analytic modeling approach," in *Proc. ISCAS* (San Diego, CA), May 1992, pp. 1997–2000.

[12] M. Cecchetti, M. Lissoni, C. Lombardi, and A. Marmiroli, "Process analysis using RSM and simulations," in *Proc. ESSDERC* (Leuven), Sept. 1992, pp. 511–514.

[13] M. Cecchetti, M. Lissoni, and A. Marmiroli, "ST-SPICE worst cases model cards for T6 process from statistical simulation," SGS-Thomson Microelectronics, Tech. Rep. 37.1049.92, Nov. 1992.

[14] G. E. P. Box, W. G. Hunter, and J. S. Hunter, *Statistics for Experimenters.* New York: Wiley, 1978.

[15] G. E. P. Box and N. R. Draper, *Applied Regression Analysis,* 2nd ed. New York: Wiley, 1987.

[16] M. D. McKay *et al.*, "A comparison of three methods for selecting values of input variables in the analysis of output from a computer code," *Technometrics,* vol. 21, no. 2, pp. 233–245, May 1979.

[17] R. K. Brayton, G. D. Hachtel, and A. L. Sangiovanni-Vincentelli, "A survey of optimization techniques for integrated-circuit design," *Proc. IEEE,* vol. 69, no. 10, pp. 1334–1360, Oct. 1981.

[18] D. G. Luenberger, *Linear and Nonlinear Programming,* 2nd ed. Reading, MA: Addison-Wesley, 1984.

[19] J. F. Tang, Q. Zheng, "Automatic design of optical thin-film systems—Merit function and numerical optimization method," *J. Opt. Soc. Amer.*, vol. 72, no. 11, pp. 1522–1528, Nov. 1982.

[20] E. Säckinger and W. Guggenbühl, "A versatile building block: The CMOS differential difference amplifier," *IEEE J. Solid-State Circuits,* vol. SC-22, no. 2, pp. 287–294, Apr. 1987.

[21] L. W. Nagel, "SPICE 2: A computer program to simulate semiconductor circuits," Electron. Res. Lab., Univ. of California, Berkeley, Rep. ERL-M520, 1975.

[22] W. J. Diamond, *Practical Experiment Designs for Engineers and Scientists.* New York: Van Nostrand Reinhold, 1981, pp. 41–46.

[23] W. M. Gentleman, "Basic procedures for large, sparse or weighted linear least squares problems," *Appl. Statistics,* vol. 23, pp. 448–454, 1974.

[24] J. Stoer, *Principles of Sequential Quadratic Programming Problems* (NATO ASI series), vol. 15. Berlin: Springer-Verlag, 1985.

Carlo Guardiani was born in Piacenza, Italy, in 1961. He received the engineering degree in theoretical physics (cum laude) from the University of Parma in 1986.

Since November 1986 he has been with SGS-Thomson Microelectronics Central R&D in Agrate, Italy, where he has been involved in statistical process control and in the development of tools for analog simulation. Since 1991 he has been responsible for the development of parametric and statistical design optimization tools.

Primo Scandolara was born in Milano, Italy, in 1962. He received the engineering degree in computer science from the University of Milano in 1986.

Since January of 1986 he has been with SGS-Thomson Microelectronics, Agrate, Italy. From 1986 to 1990 he worked on the development of tools for analog simulation. Since 1990 he has been involved in the development of parametric and statistical design optimization tools.

Jacques Benkoski (S'86–M'89) was born in Belgium in 1963. He received the B.Sc. degree in electrical and computer engineering (cum laude) from the Technion Israel Institute of Technology in 1985, having been distinguished in both the Dean's and the University President's Lists. He received the M.Sc. and Ph.D. degrees in electrical and computer engineering from Carnegie Mellon University, Pittsburgh, PA, in 1987 and 1989, respectively, for his work on statistical timing simulation.

He stayed at the IBM Scientific Center in Haifa, Israel, in the spring of 1985 and in the Interuniversity Micro-Electronics Center (IMEC) in Leuven, Belgium in the spring of 1987, where he worked on false path elimination algorithms in timing analysis. In 1989 he joined SGS-Thomson Microelectronics Central R&D in Grenoble, France, where he successively held the positions of Senior Member of Technical Staff in charge of timing analysis and mixed-mode simulation developments, Group Leader for analog, mixed, and cell-level design system, and is today heading the Design System Group. In 1990 he was elected to the Scientific and Technical Council of Thomson S.A. In 1992 he was appointed Project Leader for the Joint European Submicron Silicon (JESSI) project on HDL Component Modeling and Libraries. He has published over 30 articles in international journals and conferences.

Dr. Benkoski is on the technical committees of EDAC and Euro-DAC. He is a member of the ACM.

Germano Nicollini was born in Piacenza, Italy, in 1956. He received the degree in electronic engineering from the Universitá di Pavia in 1981.

Since 1982 he has been with SGS-Thomson Microelectronics, Milano, Italy, where he is involved in the design of analog and mixed integrated circuits for telecommunications. He was the Project Leader of the M5913/14/16/17 ST COMBO's and then he became responsible of the design of acoustic front ends for ISDN terminals and digital telephone sets.

At the present time, he is in charge of a group for the development of acoustic front ends for cordless and cellular telephones. During part of the academic years 1990–1991 and 1991–1992 he was a Visiting Professor at the Universita' di Genova. He also holds several pending patents and 14 patents granted abroad.

Circuit Analysis and Optimization Driven by Worst-Case Distances

Kurt J. Antreich, *Senior Member, IEEE,* Helmut E. Graeb, and Claudia U. Wieser

***Abstract*— In this paper, a new methodology for integrated circuit design considering the inevitable manufacturing and operating tolerances is presented. It is based on a new concept for specification analysis that provides exact worst-case transistor model parameters and exact worst-case operating conditions. Corresponding worst-case distances provide a key measure for the performance, the yield, and the robustness of a circuit. A new deterministic method for parametric circuit design that is based on worst-case distances is presented: It comprises nominal design, worst-case analysis, yield optimization, and design centering. In contrast to current approaches, it uses standard circuit simulators and at the same time considers deterministic design parameters of integrated circuits at reasonable computational costs. The most serious disadvantage of geometric approaches to design centering is eliminated, as the method's complexity increases only linearly with the number of design variables.**

I. INTRODUCTION

PREDICTING and improving the design quality in terms of performance, robustness, and yield, prior to manufacturing, is a central concern of computer-aided circuit design. Decreasing device sizes and the resulting increase in performance sensitivity require circuit designs that ensure acceptable robustness and yield under all manufacturing variations. Circuit design comprises nominal and tolerance design. While *nominal design* aims at satisfying the different *performance specifications*, the goal of *tolerance design* is the circuit's robustness against the inevitable statistical variations in the manufacturing process. This implies a high *yield*, i.e., a large percentage of manufactured circuits satisfying the specifications. This paper deals with *parametric circuit design*. For a given circuit topology, parametric circuit design is concerned with adjusting the design parameters. Parametric tolerance design investigates parametric manufacturing disturbances, such as mask misalignment, which cause a loss in *parametric yield*. Structural disturbances in the manufacturing process, such as spot defects, which lead to total performance failures and *functional yield* losses [25], are not considered. Circuit designers generally apply *circuit simulation* tools to verify the circuit behavior with selected parameter sets. In practice, the tools for circuit optimization [8], [10], [12], [14], [22], [28], [32] not only have to deal with ill-conditioned problems but suffer from expensive simulation costs. Performance macromodeling techniques have been developed to reduce costs [11], [23], [24], [33], [36], [9]. In order to determine the tolerance behavior at minimal simulation costs, methods for *worst-case analysis* operate with selected worst-case parameter sets [27], [20], [26], [30]. This paper continues from the worst-case analysis method as formulated in [26] and the design centering method as formulated in [24]. Section II presents a new specification analysis procedure that provides exact worst-case parameter sets for all specifications individually, using standard circuit simulators. Thereby, worst-case operating conditions are provided systematically, instead of scanning the whole operating range [9]. Macromodeling techniques may replace the circuit simulator in order to profit from their specific advantages. Corresponding worst-case distances enable a comparison of performance properties of any units and orders of magnitude. Moreover, they provide a key measure for the worst-case behavior (Section 2.1), the yield (Section 2.2), and the robustness (Section 2.3) of a circuit. Section III develops a new deterministic method for circuit optimization based on worst-case distances. It is shown that this method includes nominal design (Section 3.1), yield optimization (Section 3.2), and design centering (Section 3.3). Statistical approaches to tolerance design [5], [31], [34], [32], [35] usually involve high simulation costs, which depend on the sample size of the Monte-Carlo analysis, but which are independent of the dimension of the design space. Geometric methods [1], [14] usually work at lower simulation costs, but have a complexity that increases critically with the number of parameters. This serious disadvantage is eliminated by the new method. Another major advantage is that it treats all types of circuit parameters in a unified manner and enables the use of circuit simulation in the yield optimization procedure at reasonable simulation costs. In contrast, current methods for yield optimization handle deterministic design parameters of integrated circuits at reasonable simulation costs by using macromodeling techniques [15], [21], [34]. Section IV introduces planning aids for circuit optimization that provide the circuit designer with a systematic control of the design process. Section V demonstrates the efficiency of the new approach with the experimental results of three digital and analog, integrated and discrete circuits with statistical and deterministic design parameters.

II. BASIC RELATIONSHIPS

Parametric circuit design investigates *performance properties* $\boldsymbol{f} \in \mathcal{R}^{n_f}$, such as power dissipation or gate delays. The performance properties depend on the circuit parameters $\boldsymbol{p} \in \mathcal{R}^{n_p}$, such as transistor geometries or oxide thickness. In

Manuscript received May 4, 1992; revised June 16, 1993. This paper recommended by Associate Editor D. Scharfetter.

The authors are with the Institute of Electronic Design Automation, Technical University of Munich, Munich Germany.

IEEE Log Number 9212352.

general, the performance for a given parameter set has to be evaluated by numerical *circuit simulation*

$$f\colon p \mapsto f(p). \tag{1}$$

An adequate formulation of practical design problems requires the consideration of different types of circuit parameters. To this purpose, the following classification of parameters is proposed:

$$p = \begin{cases} d \in \mathcal{R}^{n_d} & \text{deterministic parameters} \\ s \in \mathcal{R}^{n_s} & \text{statistical parameters} \\ \theta \in \mathcal{R}^{n_\theta} & \text{operating parameters.} \end{cases} \tag{2}$$

Integrated circuit design features all three types of parameters. The transistor geometries are deterministic design parameters. The transistor model parameters are statistical parameters that reflect the inevitable manufacturing fluctuations. The performance has to be guaranteed for a specified region T_θ of operating parameters like temperature. Fig. 1 shows the structure of a digital library cell, a CMOS inverter. The performance properties considered are the gate delays of falling and rising output slopes t_f, t_r. The transistor widths W_p, W_n are deterministic design parameters. Their statistical variation is summarized in the width reduction ΔW, which is one of the statistical transistor model parameters. The tolerance region T_θ of operating parameters θ is determined by lower and upper bound θ_L, θ_U

$$T_\theta(\theta_L, \theta_U) = \{\theta \in \mathcal{R}^{n_\theta} | \theta_L \le \theta \le \theta_U\}. \tag{3}$$

With regard to practical requirements, the manufacturing tolerances are modeled by a normal probability density function *pdf* of the transistor model parameters s, which is defined by the mean value s_0 and the covariance matrix C, which includes parameter variances and correlations [2].

$$pdf(s, s_0, C) = \frac{\exp\left(-0.5(s - s_0)^T \cdot C^{-1} \cdot (s - s_0)\right)}{\sqrt{2\pi}^{n_s} \sqrt{\det C}}. \tag{4}$$

Other distributions like the unimodal are commonly approximated by normal *pdf*'s. In particular, skew distributions are approximated by lognormal *pdf*'s which are transformed into normal *pdf*'s. The equidensity contours of a normal *pdf* are n_s-ellipsoids with center s_0 and shape and orientation determined by C. An n_s -ellipsoid defines a *tolerance body* T_s [14]

$$T_s(\beta, s_0, C) = \{s \in \mathcal{R}^{n_s} | (s - s_0)^T \cdot C^{-1} \cdot (s - s) \le \beta^2\}. \tag{5}$$

The tolerance bodies given by (3) and (5) are examples of various tolerance bodies. In what follows, the presentation is based on these types of tolerance bodies and distributions.

The performance properties are supposed to satisfy certain *specifications*, which are given as bounds $B_i, i = 1, \cdots, n_B$. The index i denotes the ith specification B_i, which is either a lower bound L_i or an upper bound U_i. n_B denotes the number of specifications. f_i denotes the performance function that corresponds to the specification B_i. Then, a lower bound L_i and an upper bound U_j may correspond to the same

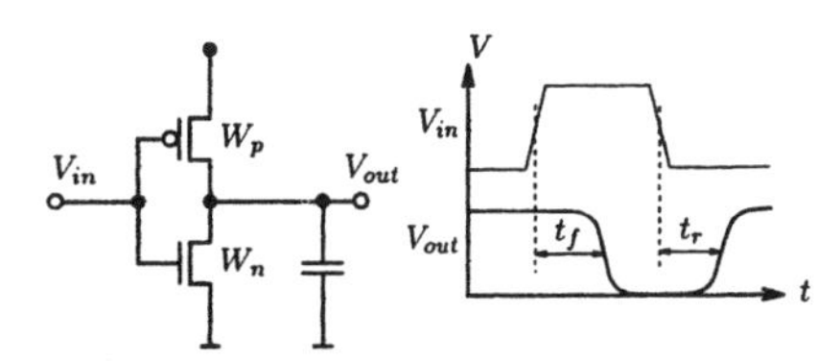

Fig. 1. CMOS inverter with design parameters W_p, W_n and performance properties t_f, t_r.

performance function $f_i = f_j$. Each specification B_i defines a region $A_{f,i}$ containing all performance values satisfying the ith specification

$$A_{f,i} = \left\{ f \in \mathcal{R}^{n_f} \middle| f_i \begin{cases} \ge L_i \text{ if lower bound} \\ \le U_i \text{ if upper bound} \end{cases} \right\}. \tag{6}$$

Performance sets satisfying *all* specifications form the *performance acceptance region* A_f

$$A_f = \bigcap_{i=1,\cdots,n_B} A_{f,i}. \tag{7}$$

Circuit simulation (1) provides nominal performance values $f(p)$ and *performance distances* α_i between specification values B_i and values of corresponding performance properties $f_i(p)$

$$\alpha_i(p) = \begin{cases} f_i(p) - L_i & \text{if lower bound} \\ U_i - f_i(p) & \text{if upper bound} \end{cases}. \tag{8}$$

Note that α_i has a negative value if specification B_i is violated. Figs. 2 and 3 illustrate the tolerance design problem for two statistical parameters and two performance properties with respective lower and upper specification bounds. In practice, performance tolerances cannot be calculated explicitly from parameter tolerances, and parameter acceptance regions cannot be calculated explicitly from performance specifications. The *parameter acceptance region* $A_s(d, \theta)$ is formulated implicitly using (6) and (7):

$$A_{s,i}(d, \theta) = \{s \in \mathcal{R}^{n_s} | f(d, s, \theta) \in A_{f,i}\} \tag{9}$$

$$A_s(d, \theta) = \bigcap_{i=1,\cdots,n_B} A_{s,i}(d, \theta). \tag{10}$$

Whether or not a parameter set lies inside the acceptance region can only be checked using circuit simulation (1) [5]. Using (6) and the acceptance function $\delta(d, s, \theta)$, we obtain

circuit accepted:

$$f(d, s, \theta) \in A_f \Leftrightarrow s \in A_s(d, \theta) \Leftrightarrow \delta(d, s, \theta) = 1$$

circuit rejected:

$$f(d, s, \theta) \notin A_f \Leftrightarrow s \notin A_s(d, \theta) \Leftrightarrow \delta(d, s, \theta) = 0 \tag{11}$$

The circuit *yield* Y is defined as the percentage of manufactured circuits satisfying all specifications

$$Y = \int_{A_s(d,\theta)} pdf(s) \cdot ds = \int_{-\infty}^{+\infty} \delta(d, s, \theta) \cdot pdf(s) \cdot ds = E\{\delta(d, s, \theta)\} \quad pdf(s). \tag{12}$$

$E\{\cdot\}$ denotes an expectation value and: $ds = ds_1 \cdot ds_2 \cdot \cdots \cdot ds_{n_s}$. With the *Monte-Carlo* technique, the parametric yield is

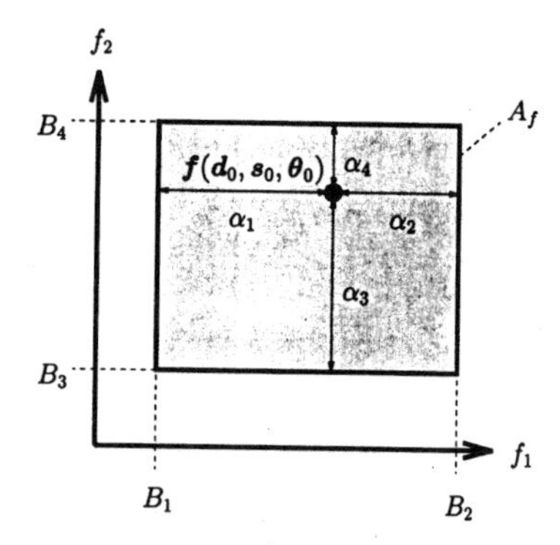

Fig. 2. Acceptance region A_f for 2 performance properties f_1, f_2 and 4 specifications $B_i, i = 1, \cdots, 4$; nominal performance $f(d_0, s_0, \theta_0)$; performance distances $\alpha_i, i = 1,,4$.

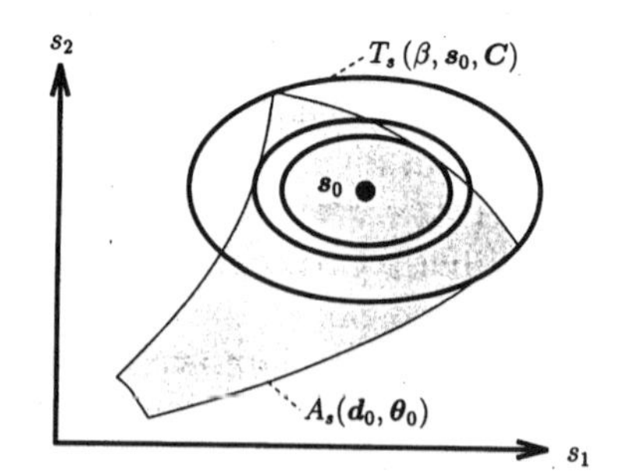

Fig. 3. Tolerance bodies $T_s(\beta, s_0, C)$ of the probability density function *pdf* for 2 statistical parameters s_1, s_2. Acceptance region A_s depends on deterministic and operating parameters d, θ.

estimated by simulating and testing the circuit for N parameter sets $s^\nu, \nu = 1, \cdots, N$, which are randomly sampled according to the *pdf* (n_{accept} is the number of accepted parameter samples):

$$\hat{Y} = \frac{1}{N} \sum_{\nu=1}^{N} \delta(d, s^{(\nu)}, \theta) = \frac{n_{\text{accept}}}{N}. \tag{13}$$

III. Specification Analysis

Circuit simulation provides nominal performance values and performance distances. Circuit design considering manufacturing tolerances additionally requires an accurate characterization of the acceptance region A_s, i.e., a definite mapping of performance specifications and performance distances into the parameter space. However, the inverse of the mapping (1) is ambiguous as shown in Fig. 4, where the curved lines mark *all* parameter sets satisfying $f_i(d_0, s, \theta) = B_i$. We now define a definite mapping of a specification B_i into the parameter space. We propose a so-called *specification analysis*, that maps each individual performance specification B_i on one worst-case statistical parameter set $s_{w,i}$ with minimal distance $\beta_{w,i}$ from the nominal set s_0, for any operating parameter set θ inside a predefined tolerance body T_θ. This worst-case distance $\beta_{w,i} = ||s_{w,i} - s_0||$ is measured in a well-defined norm that corresponds to the parameter distribution. Fig. 4 illustrates the procedure for two parameters. The operating range of parameters θ leads to a band of curves $f_i(d_0, s, \theta) = B_i$ that determine the boundary of an acceptance region $A_{s,i}(d_0, \theta)$ according to (9). The worst-case operating parameter set $\theta_{w,i}$

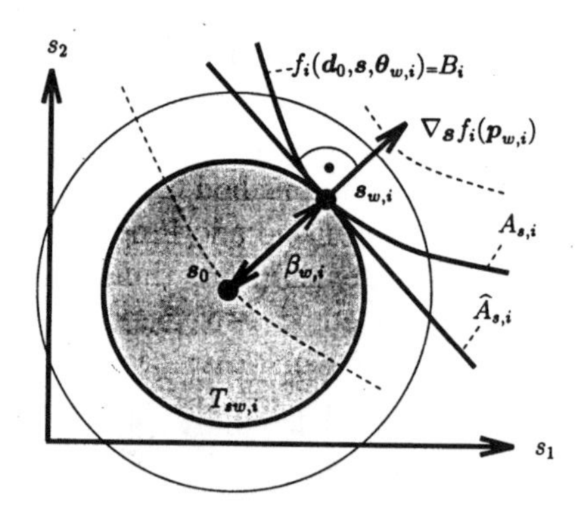

Fig. 4. Maximal tolerance body $T_{sw,i}$, worst-case parameter set $s_{w,i}$, worst-case distance $\beta_{w,i} = (s_{w,i} - s_0)^T \cdot C^{-1} \cdot (s_{w,i} - s_0)$, related performance gradient $\nabla_s f_i(p_{w,i})$, approximate acceptance region $\hat{A}_{s,i}$ for one specification B_i. ($C = I$: Equidensity contours of *pdf* are circles.)

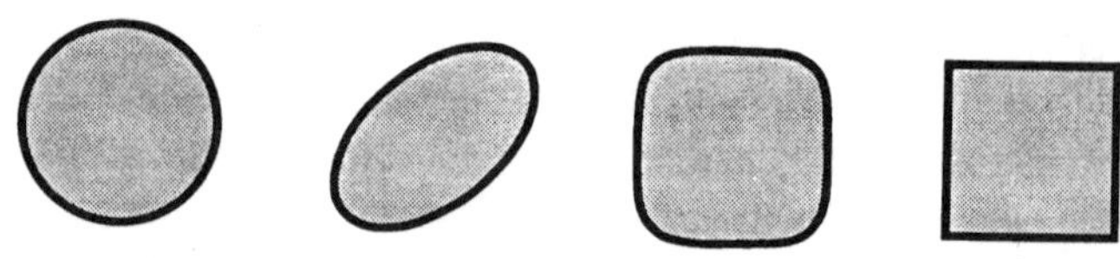

Fig. 5. Tolerance bodies for l_2-norm, weighted l_2-norm, l_4-norm, l_∞-norm.

determines the "orbit" $f_i(d_0, s, \theta_{w,i}) = B_i$ of the worst-case statistical parameter set $s_{w,i}$. The specification analysis performs a *specific backward evaluation* of one *specification* B_i and the corresponding performance distance α_i. It maps the specification onto one *worst-case set of statistical and operating parameters*, $B_i \mapsto s_{w,i}, \theta_{w,i}$, and, correspondingly, the performance distance onto one *worst-case distance*, $\alpha_i \mapsto \beta_{w,i}$. The worst-case distance $\beta_{w,i}$ defines a maximal tolerance body $T_{sw,i}$ inside the acceptance region $A_{s,i}(d_0, \theta_{w,i})$ for worst-case operating conditions $\theta_{w,i}$ as illustrated in Fig. 4.

Given the nominal values of deterministic and statistical parameters d_0, s_0, the covariance matrix C, the operating range T_θ, and the performance specifications $B_i, i = 1, \cdots, n_B$, we obtain worst-case statistical parameter sets $s_{w,i}$, worst-case operating conditions $\theta_{w,i}$ and worst-case distances $\beta_{w,i}$ by solving (14) [3], [4]

$$\min_{s,\theta} \beta(s, s_0, C) \quad \text{s.t.} \quad f_i(d_0, s, \theta) = B_1 \quad \text{and} \quad \theta \in T_\theta \tag{14}$$

individually for all specifications $B_i, i = 1 \cdots, N_B$:

$$d_0, s_0, C, T_\theta, B_i, \mapsto s_{w,i}, \theta_{w,i}, \beta_{w,i}, i = 1, \cdots, n_B.$$

The specification analysis (14) can be performed for any norm to measure the distance β. Fig. 5 sketches level contours of usual norms. The traditional worst-case analysis is adopted by using tolerance intervals according to (3), which corresponds to the l_∞-norm. An l_∞-norm is also adequate if the statistical parameters are modeled by a unimodal distribution.

In case of tolerance bodies (5) according to a normal *pdf* (4), as in Fig. 5 second from left, the ellipsoidal norm is used:

$$\begin{aligned} \beta^2 &= (s - s_0)^T \cdot C^{-1} \cdot (s - s_0), \\ \beta_{w,i}^2 &= (s_{w,i} - s_0)^T \cdot C^{-1} \cdot (s_{w,i} - s_0). \end{aligned} \tag{15}$$

Using (15), problem (14) describes an optimization problem with a quadratic objective function, a nonlinear equality constraint, and linear inequality constraints. To solve (14), an efficient deterministic solution method has been implemented, that is based on Sequential Quadratic Programming. Appendix A describes the implementation, simulation costs, and practical problems. The method also yields the performance gradient $\nabla \boldsymbol{p} f_i(\boldsymbol{p}_{w,i})$ at the worst-case parameter set $\boldsymbol{p}_{w,i}^T = [\boldsymbol{d}_0^T \boldsymbol{s}_{w,i}^T \boldsymbol{\theta}_{w,i}^T]$. It is worth mentioning that problem formulation (14) generally results in a unique worst-case distance and, correspondingly, in a unique maximal tolerance body, but might result in several equivalent worst-case parameter sets. However, practical investigations indicate that for technical problems (14) results in a unique worst-case parameter set.

For a normal *pdf* according to (4), parameter sets of equal distance β according to (15) represent parameter sets of equal probability density. In this case, problem formulation (14) is equivalent to (16) [3], [4]:

$$\max_{\boldsymbol{s},\boldsymbol{\theta}} pdf(\boldsymbol{s}, \boldsymbol{s}_0, \boldsymbol{C}) \quad \text{s.t.} \quad f_i(\boldsymbol{d}_0, \boldsymbol{s}, \boldsymbol{\theta}) = B_i \quad \text{and} \quad \boldsymbol{\theta} \in T_\theta. \tag{16}$$

Equation (16) defines the worst-case parameter set $\boldsymbol{s}_{w,i}$ as the parameter set with highest probability density among the parameter sets just about satisfying the specification B_i and therefore yields another adequate interpretation of the worst-case parameter set for a specification.

Sometimes, specifications B_i are not given. Then, a worst-case analysis has to determine the performance bounds that can be guaranteed for a given minimal circuit yield (i.e., a given tolerance body T_s). A new formulation of the worst-case analysis procedure is obtained by exchanging optimization objective and constraint in (14) [19]

$$\min_{\boldsymbol{s},\boldsymbol{\theta}} / \max f_i(\boldsymbol{d}_0, \boldsymbol{s}, \boldsymbol{\theta}) \quad \text{s.t.} \quad \boldsymbol{s} \in T_s(\beta, \boldsymbol{s}_0, \boldsymbol{C}) \quad \text{and} \quad \boldsymbol{\theta} \in T_\theta. \tag{17}$$

Given the nominal values of deterministic and statistical parameters $\boldsymbol{d}_0, \boldsymbol{s}_0$, the covariance matrix $\boldsymbol{C}$, the operating range T_θ, and the range of statistical parameters T_s determined by β according to (15), the solution of (17) yields all specified minimal/maximal worst-case performance values $f_{w,i}, i = 1, \cdots, n_B$, corresponding worst-case statistical parameter sets $\boldsymbol{s}_{w,i}$ and worst-case operating conditions $\boldsymbol{\theta}_{w,i}$:

$$\boldsymbol{d}_0, \boldsymbol{s}_0, \boldsymbol{C}, T_\theta, \beta \mapsto f_{w,i}, \boldsymbol{s}_{w,i}, \boldsymbol{\theta}_{w,i}, \qquad i = 1, \cdots, n_B.$$

Current methods for worst-case analysis select worst-case parameter sets based on performance gradients for the nominal design or macromodeling techniques, regardless of operating parameters [20], [26], [27]. E.g., in [26], the concept of (16) has been introduced to compute worst-case parameter sets for linear performance functions. In [20], realistic worst-case parameter sets are computed by first determining the tolerance body (5) that refers to a given yield value, and then selecting a worst-case parameter set within that tolerance body, based on user-defined model functions and the concept of (14).

Equations (14) and (17) provide a universal methodology for specification analysis, that is applicable to any performance functions (including standard circuit simulation) and any tolerance bodies, and that appropriately considers all types of circuit parameters. In the following, it will be shown that worst-case parameter sets and worst-case distances are well-suited to characterize worst-case behavior, circuit yield and robustness.

2.1 Worst-Case Behavior

The worst-case behavior of a circuit with respect to a specification B_i is characterized by the worst-case statistical parameter set $\boldsymbol{s}_{w,i}$ and the worst-case operating conditions $\boldsymbol{\theta}_{w,i}$. The worst-case statistical parameters $\boldsymbol{s}_{w,i}$ represent the smallest parameter distortion (14) or, alternatively, the parameter set with highest probability (16) that will cause violation of specification B_i for the specified range T_θ of operating parameters. In contrast to traditional "slow" and "fast" parameter sets that give a vague description of the worst-case behavior, the worst-case parameter sets for all specifications according to (14) determine the worst-case behavior with an accuracy according to the underlying method for performance evaluation, i.e., exactly when using numerical simulation or approximately when using performance macromodeling techniques.

2.2 Parametric Yield Using Worst-Case Distances

The worst-case parameter sets and corresponding performance gradients for one specification provide a tangential hyperplane of the parameter acceptance region $A_{s,i}$ as illustrated in Fig. 4. This hyperplane represents the boundary of an approximate acceptance region $\hat{A}_{s,i}$.

$$\hat{A}_{s,i}(\boldsymbol{d}_0, \boldsymbol{\theta}_{w,i}) = \left\{ \boldsymbol{s} \middle| \nabla_s^T f_i(\boldsymbol{p}_{w,i}) \cdot (\boldsymbol{s} - \boldsymbol{s}_{w,i}) \cdot \begin{cases} \geq 0 & \text{if lower bound} \\ \leq 0 & \text{if upper bound} \end{cases} \right\}. \tag{18}$$

In contrast to the geometric approach in [14], that involves special algorithms for nonconvex acceptance regions, the specification analysis (14) is applicable to both convex and nonconvex regions. As each specification defines one boundary plane of A_s, the complexity for the specification analysis and the approximation of the acceptance region increases only linearly with the number of specifications. In contrast, other geometric approaches have a complexity that increases critically with the number of parameters. Using $\hat{A}_{s,i}$, the *parametric yield Y_i for a single specification B_i*,

$$Y_i = \int_{A_{s,i}} pdf(\boldsymbol{s}) \cdot d\boldsymbol{s} \tag{19}$$

is approximated by $Y_i \approx \hat{Y}_i$,

$$\hat{Y}_i = \int_{\hat{A}_{s,i}} pdf(\boldsymbol{s}) \cdot d\boldsymbol{s} = \int_{-\infty}^{\text{sign}(\alpha_i)\cdot\beta_{w,i}} \frac{1}{\sqrt{2\pi}} \cdot \exp\left(-\frac{t^2}{2}\right) \cdot dt. \tag{20}$$

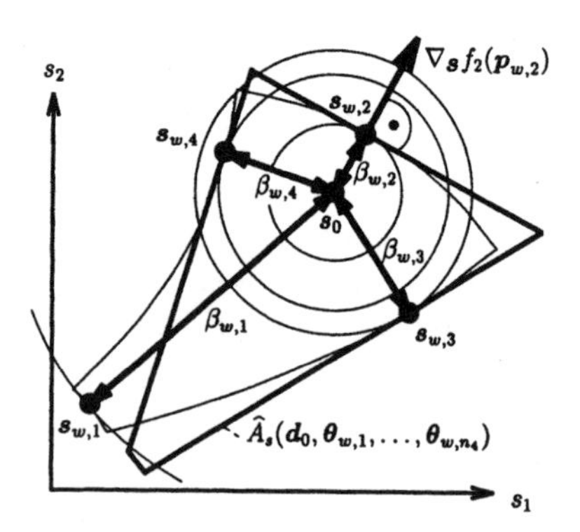

Fig. 6. Specification analysis of 4 specifications according to Fig. 2. ($C = I$: Equidensity contours of *pdf* are circles.)

Equation (20) shows that the yield $\hat{Y}_i$ determined by $\hat{A}_{s,i}$ can be looked up in a statistical table for a one-dimensional marginal distribution (Appendix C outlines the proof).

A yield of 50% is obtained for $\beta_{w,i} = 0$, i.e., for the case that the nominal performance value is equal to the specified bound. Yield values bigger than 50% indicate that the specification is met for nominal parameter values, i.e., $\text{sign}(\alpha_i) > 0$. If the specification is violated for nominal parameter values, i.e., $\text{sign}(\alpha_i) < 0$, the corresponding yield is less than 50%.

Moreover, it can be shown that $\hat{A}_{s,i}$ represents the most accurate among all possible approximations of convex regions $A_{s,i}$ or $\mathcal{R}^{n_s} A_{s,i}$ with tangential hyperplane boundaries [17].

According to (16), the boundary of $\hat{A}_{s,i}$ is exact for the worst-case parameter set with highest probability density and has a decreasing accuracy for decreasing values of the *pdf*. Even if $A_{s,i}$ or $\mathcal{R}^{n_s} \backslash A_{s,i}$ is not convex, there always exists a convex environment of the worst-case parameter set. Therefore, (20) provides very good yield estimates for any shapes of $A_{s,i}$. This is confirmed by the examples.

Considering all specifications simultaneously, an approximate parameter acceptance region $\hat{A}_s$ is then provided as shown in Fig. 6

$$\hat{A}_s(\boldsymbol{d}_0, \boldsymbol{\theta}_{w,1}, \cdots, \boldsymbol{\theta}_{w,n_B}) = \bigcap_{i=1,\cdots,n_B} \hat{A}_{s,i}(\boldsymbol{d}_0, \boldsymbol{\theta}_{w,i}). \quad (21)$$

The *circuit yield* is computed at no additional circuit simulation costs based on the approximate acceptance region (21). Like for Monte-Carlo analysis, a sample of stochastic parameter sets is chosen randomly. Circuit simulation however is substituted by the evaluation of (18) for every specification, the cost of which is negligible when compared to circuit simulation.

2.3 Circuit Robustness and Comparison of Performance Properties

According to (14), the worst-case distance $\beta_{w,i}$ is a unit-free quantity. The worst-case distances therefore allow for a comparison of all performance properties of different units and orders of magnitude with respect to *yield.* While performance distances measured in MHz and V/μs, e.g., cannot be compared, the corresponding worst-case distances, e.g., 2.0 $\rightarrow$ 99.7% and 3.5 $\rightarrow$ 99.97%, enable a straightforward comparison. Fig. 7 illustrates that the yield has a limited range

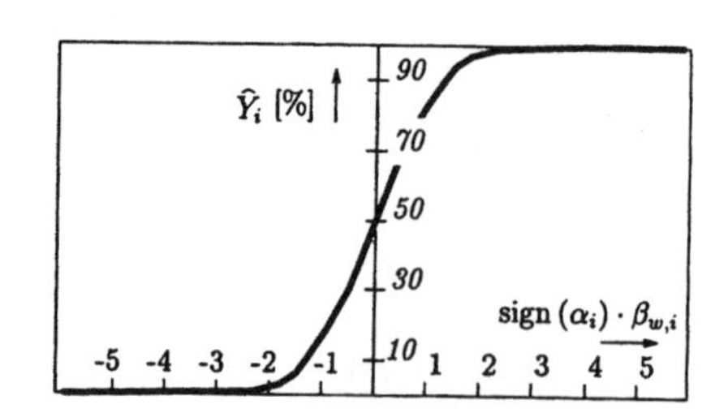

Fig. 7. Yield values for different worst-case distances.

of values between 0% and 100%. Therefore, yield is a very insensitive measure for bad or for good designs. Worst-case distances have an unlimited range of values. According to (20), they can be interpreted as a multiple of an imaginary process variance. In addition to the yield value, worst-case distances therefore provide a global measure for a circuit's *robustness* with respect to manufacturing tolerances.

To summarize: The specification analysis of the individual performance specifications provides exact worst-case sets of statistical and operating parameters. Corresponding worst-case distances provide very good yield estimates, they indicate the circuit's robustness and make different performance properties comparable with respect to yield and robustness.

IV. Circuit Optimization Using Worst-Case Distances

As worst-case distances are a unified measure for the performance, the yield, and the robustness of a circuit, it is obvious that performance, yield, and robustness are optimized by maximizing the worst-case distances over the nominal values of design parameters. The parametric circuit optimization procedure is now formulated based on the worst-case distances $\beta_{w,i} = \beta_{w,i}(\boldsymbol{d}_0, \boldsymbol{s}_0, \boldsymbol{C}, T_\theta, B_i)$ that are provided by a specification analysis (14):

$$\max_{\boldsymbol{d}_0, \boldsymbol{s}_0} \text{sign}\,(\alpha_i) \cdot \beta_{w,i}$$
$$\text{for all}\, i = 1, \cdots, n_B \,\text{simultaneously}. \quad (22)$$

Given the covariance matrix $\boldsymbol{C}$, the operating range T_θ, and the performance specifications $B_i, i = 1, \cdots, n_B$, the solution of problem (22) yields optimal (maximal) worst-case distances $\beta^*_{w,i}, i = 1, \cdots, n_B$, and an optimal nominal parameter set $\boldsymbol{d}^*_0, \boldsymbol{s}^*_0$, :

$$\boldsymbol{C}, T_\theta, B_i, \mapsto \beta^*_{w,i}, \boldsymbol{d}^*_0, \boldsymbol{s}^*_0, \qquad i = 1, \cdots n_B.$$

As (22) considers all worst-case distances simultaneously, a Multiple Criterion Optimization (MCO) problem has to be solved [22]. The MCO task has to take account of the trade-off between the individual optimization objectives $\beta_{w,i}, i = 1, \cdots, n_B$. A solution of (22) is obtained by solving the following scalar optimization problem, which corresponds to that in [24]

$$\min_{\boldsymbol{d}_0, \boldsymbol{s}_0} \sum_{i=1}^{n_B} k_i \cdot \exp(-a \cdot \text{sign}\,(\alpha_i) \cdot \beta_{w,i}). \quad (23)$$

Equations (22) and (23) represent general formulations of nominal and tolerance design. With $a = 2, k_i = 1, i =$

$1, \cdots, n_B$ in (23), the least squares formulation (24) is obtained

$$\min_{\boldsymbol{d}_0, \boldsymbol{s}_0} \gamma^T \cdot \gamma,$$
$$\gamma = [\cdots \exp(-\mathrm{sign}(\alpha_i) \cdot \beta_{w,i} \cdots]^T,$$
$$i = 1, \cdots, n_B. \quad (24)$$

Problem (24) is solved with an interactive trust region method [6], which was originally applied to nominal circuit design. This method is especially suited for ill-conditioned optimization problems. It is well-tailored to practical requirements: The individual view of performance properties during the optimization process is maintained, and a thorough insight into the optimization process enables a systematic control.

From (24) we see: If the performance specification is satisfied, i.e., $\mathrm{sign}(\alpha_i) = +1$, the corresponding component of γ has a value between 0 and 1. If the specification is violated, i.e., sign $(\alpha_i) = -1$, it has a value greater than 1. In that way, all violated specifications obtain an adequate penalty weight. The optimization procedure may start from an arbitrary initial design, as satisfied and violated specifications are simultaneously optimized. Starting from an unacceptable initial design, (24) performs a nominal design taking into account performance sensitivities. Continuing from an acceptable nominal design, it maximizes the yield. After yield values near 100% (saturation region in Fig. 7) have been obtained, (24) maximizes the circuit's robustness thus performing design centering. *Circuit optimization based on* (24) *therefore covers nominal design, yield optimization and design centering.* In the following, it is explained in detail how the new approach for circuit optimization incorporates these three tasks.

3.1 Nominal Design Using Worst-Case Distances

The goal of nominal design is to maximize or minimize performance values. Analogously to (22), we formulate the nominal design problem:

$$\alpha_i = \begin{cases} f_i(\boldsymbol{p}_0) - f_{0,i}, & \text{to maximize a performance value} \\ f_{0,i} - f_i(\boldsymbol{p}_0), & \text{to minimize a performance value} \end{cases}$$

$$\max_{\boldsymbol{d}_0, \boldsymbol{s}_0} \alpha_i(\boldsymbol{d}_0, \boldsymbol{s}_0, \boldsymbol{\theta}_0, f_{0,i})$$
$$\text{for all } i = 1, \cdots, n_B \text{ simultaneously.} \quad (25)$$

The nominal design procedure maximizes the performance distances $\alpha_i, i = 1, \cdots, n_B$, from given bias performance values $f_{0,i}, i = 1, \cdots, n_B$, over the nominal values of design parameters. Given the bias performance values $f_{0,i}, i = 1, \cdots, n_B$, and nominal operating conditions $\boldsymbol{\theta}_0$, the solution of problem (25) yields the optimal (maximal) performance distances $\alpha_i^*, i = 1, \cdots, n_B$, and an optimal nominal parameter set $\boldsymbol{d}_0^*, \boldsymbol{s}_0^*$:

$$\boldsymbol{\theta}_0, f_{0,i}, \mapsto \alpha_i^*, \boldsymbol{d}_0^* \boldsymbol{s}_0^*, \qquad i = 1, \cdots, n_B.$$

Obviously, (25) represents an MCO problem analogous to (22). Usually, MCO-methods for nominal design apply weighted minmax, least-squares, or generalized norms to calculate efficient tradeoff solutions [8], [10], [12], [22], [28]. If the worst-case distances are computed approximately based on the performance sensitivities at the nominal parameter set, it follows from (45), (55), and (54) in Appendix B, that the worst-case distances are equal to the performance distances weighted with the performance sensitivity: $\mathrm{sign}(\alpha_i) \cdot \beta_{w,i} = \alpha_i / \sqrt{\nabla_{\boldsymbol{s}}^T f_i(\boldsymbol{p}_0) \cdot \boldsymbol{C} \cdot \nabla_{\boldsymbol{s}} f_i(\boldsymbol{p}_0)}$. With one sensitivity calculation per design step, (24) now performs a nominal design according to (25) taking into account performance sensitivities.

3.2 Yield Optimization Using Worst-Case Distances

The general formulation of yield optimization is

$$\max_{\boldsymbol{d}_0, \boldsymbol{s}_0} \min_{\boldsymbol{\theta} \in T_\theta} Y(\boldsymbol{d}_0, \boldsymbol{s}_0, \boldsymbol{\theta}, \boldsymbol{C}, B_1, \cdots, B_{n_B}). \quad (26)$$

Given the covariance matrix $\boldsymbol{C}$, the operating range T_θ, and the performance specifications $B_i, i = 1, \cdots, n_B$, (26) determines the optimal (maximal) yield Y^*, optimal nominal parameter values $\boldsymbol{d}_0^*, \boldsymbol{s}_0^*$, and the operating conditions $\boldsymbol{\theta}^*$ that determine Y^*:

$$\boldsymbol{C}, T_\theta, B_i \qquad i = 1, \cdots, n_B \mapsto Y^*, \boldsymbol{d}_0^* \boldsymbol{s}_0^*, \boldsymbol{\theta}^*.$$

Equation (26) is mainly solved without consideration of operating parameters by Monte-Carlo based methods [31], [5], [32], [35], which usually have a high computational complexity due to the large number of circuit simulations. Methods based on (26) have great difficulties in handling deterministic design parameters of integrated circuits. In [34], random perturbations are introduced to consider deterministic parameters in a statistical optimization method. [21] and [15] define yield as a surface integral in order to consider deterministic parameters. [21] uses performance sensitivities to approximate the yield gradient, and [15] uses performance macromodels to keep the simulation costs low. In [15], [17], the yield gradient for deterministic parameters is determined by an expectation value. It is estimated with a Monte-Carlo analysis of N samples that requires at least $2 \cdot N$ one-dimensional line searches to compute boundary points of the parameters acceptance region and $2 \cdot N$ performance sensitivity calculations with regard to deterministic parameters. Therefore, methods for yield optimization based on the yield gradient handle deterministic design parameters by using macromodeling techniques for performance evaluation.

The presented method (24) maximizes the yield by maximizing all worst-case distances simultaneously. It is based on the gradients of the worst-case distances, which are computed according to (56) in Appendix B. In each optimization step, it performs a specification analysis and a performance sensitivity calculation for deterministic parameters at all worst-case parameter sets, which has a complexity of $n_B \cdot n_{grad}(n_d)$ [see (33)]. The simulation costs are low when compared to the cost of a Monte-Carlo analysis and extremely low when compared to the cost for the aforementioned yield gradient estimation for deterministic parameters. *In contrast to current approaches, the presented method is based on numerical circuit simulations and*

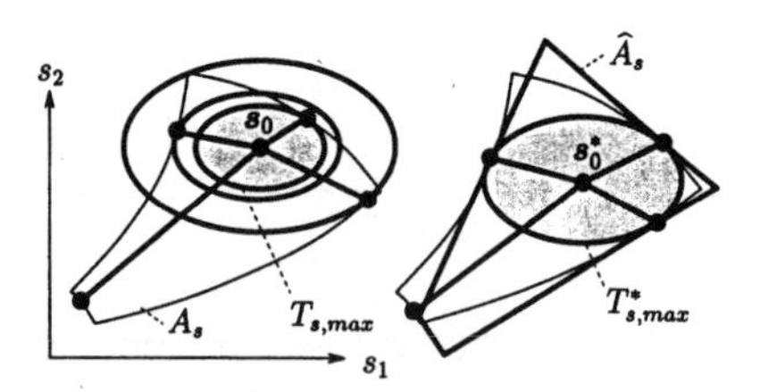

Fig. 8. Design centering (22) of the example of Fig. 6 using an ellipsoidal norm maximizes the maximal tolerance body inside the acceptance region $\hat{A}_s$ obtained by the specification analysis.

at the same time considers deterministic design parameters at reasonable computational costs.

As the gradients of the worst-case distances are derived with respect to any circuit or process parameter (Appendix B), (24) is also suitable for process optimization and tolerance assignment.

3.3 Design Centering Using Worst-Case Distances

Geometric approaches for design centering usually maximize a tolerance body inside the acceptance region [14], [7], [1], as illustrated in Fig. 8. This is equivalent to maximizing the minimal worst-case distance, i.e., to using a l_∞-norm in (22)

$$\max_{\boldsymbol{d}_0, \boldsymbol{s}_0} \min_{i=1,\cdots,n_B} \operatorname{sign}(\alpha_i) \cdot \beta_{w,i}. \tag{27}$$

Equation (27) is equivalent to (23) for $a \to \infty$. Obviously, the tradeoff between the optimization objectives of the MCO task (22) corresponds to a closed parameter acceptance region as in Fig. 8. While in [24], a corresponding formulation of problem (23) has been interpreted as an approximation to (27), we interpret (26) and (27) as specific tradeoff solutions to the original tolerance design problem (22). Current design centering methods do not consider operating parameters. Their computational complexity critically depends on the dimension of the design space due to the approximation of the parameter acceptance region. As the specification analysis has a complexity that increases only linearly with the number of specifications and design parameters, *the presented method eliminates the most serious disadvantage of geometric design centering methods compared to Monte-Carlo based approaches.*

To summarize: A deterministic circuit optimization methodology has been presented that comprises nominal design, yield optimization, and design centering. It is based on numerical circuit simulations and works at low computational costs that increase linearly with the number of design variables. In particular, deterministic design parameters are considered at reasonable computational costs.

V. Planning Aids for Circuit Design and Optimization

Circuit simulations being the crucial cost factor of the design process, an efficient optimization procedure has to be developed. In [24], a method for partitioning the design problem into disjoint subproblems of lower complexity is presented, that leads to a reduction in the simulation costs. In the following, a method for a detailed diagnosis of the design problem is presented, that is based on the results of the specification analysis, i.e., worst-case distances and gradients according to (56) and (57) (Appendix B). It analyzes the impact of circuit parameters and specifications on circuit quality and enables the designer to carefully plan the optimization steps. Selecting an independent set of relevant specifications that limit circuit quality most severely, selecting an independent set of relevant parameters, as well as predicting the amount of change in the different worst-case distances will produce a maximal effect with a minimal parameter change in each optimization step and lead to a reduction in the simulation costs.

4.1 Selection of Relevant Specifications

Relevant specifications are selected according to their impact on circuit yield and robustness. Comparing two specifications B_k and B_i, with corresponding worst-case-distances $\operatorname{sign}(\alpha_i) \cdot \beta_{w,i} \gg \operatorname{sign}(\alpha_k) \cdot \beta_{w,k}$, it is obvious, that B_k is relevant because circuit yield and robustness is mainly limited by B_k. B_i *depends* on B_k if:

- $\operatorname{sign}(\alpha_i) \cdot \beta_{w,i} \gg \operatorname{sign}(\alpha_k) \cdot \beta_{w,k}$, i.e., the impact of a specification B_i on circuit yield is negligible when compared to B_k, or
- $\angle(\nabla_{\boldsymbol{p}}(\operatorname{sign}(\alpha_k) \cdot \beta_{w,k}), \nabla_{\boldsymbol{p}}(\operatorname{sign}(\alpha_i) \cdot \beta_{w,i}))$ small, i.e., specifications B_i and B_k are linearly dependent with respect to the circuit parameters $\boldsymbol{p}$.

These two criteria are combined in (28): Specification B_i depends on B_k, if

$$W\left(\operatorname{sign}(\alpha_i) \cdot \beta_{w,i} - \operatorname{sign}(\alpha_k) \cdot \beta_{w,i}\right) \\ \angle(\nabla_{\boldsymbol{p}}(\operatorname{sign}(\alpha_k) \cdot \beta_{w,k}), \nabla_{\boldsymbol{p}}(\operatorname{sign}(\alpha_i) \cdot \beta_{w,i})) < \epsilon. \tag{28}$$

W is a monotonic weight function with $W(0) = 1$, e.g., $W(z) = \exp(-z)$ and $\epsilon = 20°$ in the examples. A set of relevant specifications is determined with:

1. The set F contains all performance specifications.
2. The specification $B_k \in F$ with minimal worst-case distance is relevant. Delete B_k from F.
3. Determine the specifications in F, that *depend* on B_k according to (28) and delete them from F.
4. Repeat 1 and 2 until F is empty.

4.2 Selection of Relevant Parameters

The selection of parameters is performed for the designable transistor geometries, in order to plan a design step. It is also performed for the statistical transistor model parameters, in order to determine a reduced set of dominant statistical parameters [21]. And it is performed for the variances of statistical parameters using (57), in order to tune or plan the manufacturing process and statistical measurements during manufacturing. A set of n_r relevant parameters $\boldsymbol{p}_r$ is selected with respect to a suitable cost function C. Relevant parameters enable a large decrease in cost at small parameter changes. Appropriate choices for C are $C = \Sigma_{i=1}^{n_r} \exp\left[-2 \cdot \operatorname{sign}(\alpha_i) \cdot \beta_{w,i}\right]$ as in (24) or $C = \max_i(-\operatorname{sign}(\alpha_i) \cdot \beta_{w,i})$. The impact of circuit parameters on circuit quality is mirrored in the gradient $\nabla_{\boldsymbol{p}} C$. Ordering the gradient elements according to their absolute values corresponds to ordering the parameters

TABLE I
CHARACTERISTIC PERFORMANCE PROPERTIES WITH SPECIFIED LOWER BOUNDS L_i, NOMINAL VALUES $f_i(\mathbf{p}_0)$ AND PERFORMANCE DISTANCES α_i FOR THE INITIAL DESIGN OF THE AMPLIFIER

	Gain	*GBW*	φ_M	SR_+	SR_-	ΔV_{out}
i	1	2	3	4	5	6
L_i	$69dB$	$2.5MHz$	60°	$4.0V/\mu s$	$4.0V/\mu s$	$3.8V$
$f_i(\mathbf{p}_0)$	$75dB$	$3.5MHz$	88°	$5.1V/\mu s$	$5.2V/\mu s$	$4.3V$
α_i	$6dB$	$1.0MHz$	18°	$1.1V/\mu s$	$1.2V/\mu s$	$0.5V$
$\beta_{w,i}$; $\hat{Y}_i$	?	?	?	?	?	?

according to their relevance. A minimal set $\mathbf{p}_r$ or relevant parameters is determined by selecting the parameters that correspond to the largest components of $\nabla_{\mathbf{p}}\mathcal{C}$, such that $\angle(\nabla_{\mathbf{p}}\mathcal{C}, \nabla_{\mathbf{p}_r}\mathcal{C}) < \epsilon$, with $\epsilon = 20^\circ$ in the examples.

4.3 Prediction of the Change in Circuit Quality for a Design Step

The change in circuit quality with respect to a certain specification is mirrored in the according worst-case distance, the change in the overall circuit quality is mirrored in the cost function. The change in worst-case distances and cost function due to a design step $\Delta\mathbf{p}$ is approximated by (29)

$$\Delta\beta_{w,i} = \nabla_{\mathbf{p}}^T(\text{sign}(\alpha_i)\cdot\beta_{w,i})\cdot\Delta\mathbf{p}, \quad \Delta\mathcal{C} = \nabla_{\mathbf{p}}^T\mathcal{C}\cdot\Delta\mathbf{p}. \quad (29)$$

4.4 Extended Circuit Analysis

An extended circuit analysis comprises

- *circuit simulation* to determine nominal performance values,
- *specification analysis* to provide worst-case parameter sets and worst-case distances,
- an *assessment of the actual circuit quality* determined by worst-case behavior, robustness and yield,
- an *analysis of the impact of the different performance specifications and parameters* on circuit quality.

The above listed steps have been described in the previous sections.

The extended circuit analysis enables a deep insight into the design problem. It provides the means for a straightforward rating and a comparison of the influence of the different performance specifications on circuit quality. It may be used to judge the overall quality of a circuit and to compare different circuit topologies with respect to yield and performance bounds. For these reasons the extended circuit analysis is well-tailored to the design process. It enables a careful planning of an optimization step, which is essential in order to make that step most efficient.

VI. EXAMPLES

5.1 CMOS Buffer Amplifier

The integrated CMOS buffer amplifier in Fig. 9, described in [16], features 66 circuit parameters—12 statistical transistor model parameters, 2 operating parameters, and 52 deterministic design parameters. The circuit's performance is characterized by the open loop gain at dc (*Gain*), the unity gain bandwidth (GBW), the phase margin (φ_M), the slew rates (SR_+, SR_-) for rising and falling output slope, respectively, and the output voltage swing (ΔV_{out}). For these 6 performance properties lower bounds L_i are specified as in Table I.

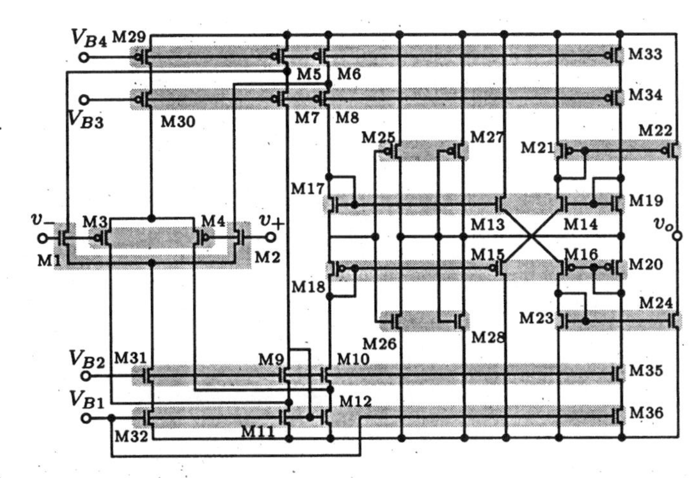

Fig. 9. CMOS buffer amplifier: $R_{load} = 50\,\Omega$. The lengths of the transistors surrounded by one shaded box are matched, e.g., $L1 = L2$. *Circuit parameters p: 52 deterministic parameters d*: transistor widths and lengths of the amplifier and the bias stage generating the bias voltages V_{B1} to V_{B4}. *12 statistical parameters s* with variations and correlations as in [26]: perturbations in transistor geometries ΔW and ΔL, oxide thickness t_{ox}; mobility $\mu_{0,p/n}$, threshold voltage $V_{T0,p/n}$, bulk threshold parameter $\gamma_{p/n}$, junction capacitance $c_{j,p/n}$ for p- and n-type transistors; bias current I. *2 operating parameters* $\boldsymbol{\theta}$: temperature $T(-40°C \le T \le 100°C)$, supply voltage V_{DD} $(4.75\,V \le V_{DD} \le 5.25\,V)$.

A *circuit simulation with nominal parameter values* $\mathbf{p}_0$ yields the nominal performance values $f_i(\mathbf{p}_0)$ and the performance distances α_i in Table I. All the specifications are met for the nominal design ($\alpha_i > 0$). As the performance distances are given in different units and orders of magnitude, like e.g., 1 MHz for the gain bandwidth or 0.5 V for the output voltage swing, they cannot be used to compare the different specifications with respect to their impact on circuit quality. Therefore, a *specification analysis* has to be carried out to determine the worst-case distances and corresponding yield values for the individual specifications as indicated by the last row in Table I.

The worst-case parameter sets are calculated with an accuracy of 1% in (30) and 20° in (31). This leads to an average of $n_{iter} = 3.5$ iterations per specification. The cost of the specification analysis for all $n_B = 6$ specifications therefore amounts to 21 sensitivity analyses for the 14 statistical and operating parameters [see (33)]

$$n_{SQP} = 21 \cdot n_{grad}(14).$$

These costs are low when compared to a Monte-Carlo analysis. In contrast to Monte-Carlo analysis, the specification analysis now provides worst-case statistical parameter sets $\mathbf{s}_{w,i}, i = 1, \cdots, 6$, the worst-case operating conditions $\boldsymbol{\theta}_{w,i}, i = 1, \cdots, 6$ (Table II) and the worst-case distances $\beta_{w,i}$ (Fig. 10) for all specifications individually.

TABLE II
WORST-CASE VALUES FOR THE RELEVANT STATISTICAL PARAMETERS IN DEVIATION FROM THE NOMINAL VALUE; WORST-CASE VALUES FOR THE OPERATING CONDITIONS T AND V_{DD}

	ΔV_{out}	$Gain$	GBW	φ_M	SR_+	SR_-
ΔL	$0.04\mu m$	$0.11\mu m$	$0.16\mu m$	$3.0\mu m$	$0.19\mu m$	$0.13\mu m$
$\mu_{0,p}$	−1.1%	−2.9%	−3.3%	17.0%	−1.9%	−2.5%
$\mu_{0,n}$	.	.	.	−30.0%	.	−1.0%
t_{ox}	.	1.9%	1.3%	−37.0%	0.7%	.
I	.	1.5%	−5.3%	50.0%	−10.0%	−13.0%
$V_{T0,p}$	.	−1.6%	.	45.0%	.	.
T	100°C	100°C	100°C	100°C	100°C	100°C
V_{DD}	4.75V	4.75V	4.75V	4.75V	4.75V	4.75V

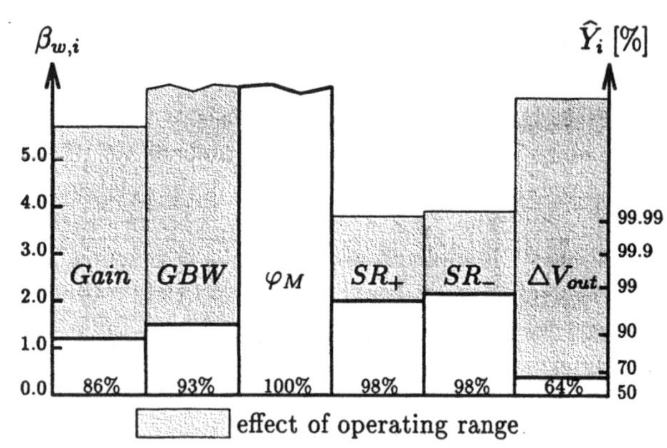

Fig. 10. Worst-case distances $\beta_{w,i}$ and yield estimates $\hat{Y}_i$ for the initial design of the amplifier.

The data provided by specification analysis are used to *judge the actual circuit quality.* The bars in Fig. 10 correspond to the worst-case distances that have been obtained for the 6 different performance specifications. The worst-case distances immediately provide yield estimates $\hat{Y}_i$ for the individual specifications according to (20). In Fig. 10 this is made evident by the left-hand axis indicating the worst-case distance and the right-hand axis indicating the corresponding yield value. A bar of height 0 corresponds to a worst-case distance of 0 and to a yield of 50%. Here, the smallest worst-case distance $\beta_{w,6} = 0.33$ for the output voltage swing ΔV_{out} implies a yield of 64% for ΔV_{out}. Now the designer can immediately compare the different specifications with respect to their impact on circuit quality, which he was not able to do just given the worst-case performance deviations. The *relevancy of a specification* is first of all determined by the according yield value. In Fig. 12, the specifications are ordered according to increasing worst-case distances, i.e., according to decreasing impact on circuit quality. Obviously, circuit yield is mostly restricted by the output voltage swing ΔV_{out}, whereas the phase margin φ_M has negligible impact. The shaded bars demonstrate the effect of the operating range of parameters T and V_{DD}. For the output voltage swing ΔV_{out}, e.g., the worst-case distance varies from $\beta_{w,6} = 0.33\,(\hat{Y}_6 = 64\%)$ for

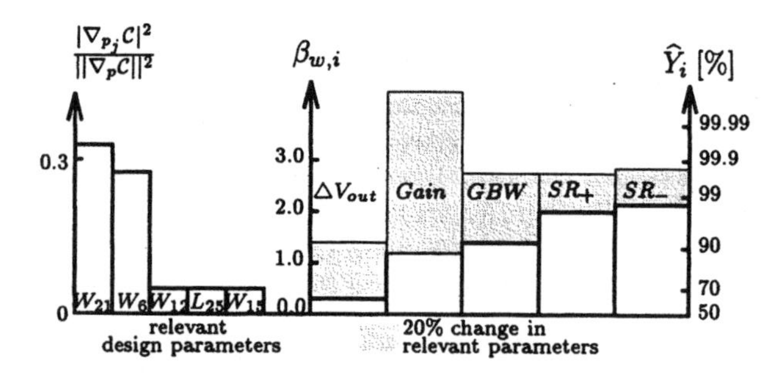

Fig. 11. Planning a design step.

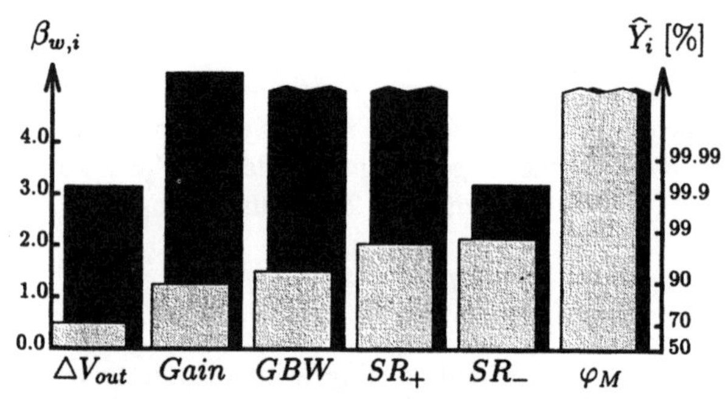

Fig. 12. Worst-case-distances $\beta_{w,i}$ and yield estimates $\hat{Y}_i$ for the initial (light shaded bars) and the optimal design (dark shaded bars) of the amplifier.

$T = 100°\text{C}, V_{DD} = 4.75$ V to $\beta_{w,6} = 6.2\,(\hat{Y}_6 = 100\%)$ for $T = -40°\text{C}, V_{DD} = 5.25$ V.

Based on the approximate acceptance region $\hat{A}_s$ according to (21) an estimate of the total yield is obtained as $\hat{Y} = 64\%$. The yield estimates obtained based on the results of the specification analysis differ less than 1% from the yield values obtained with a Monte-Carlo analysis based on a sample size of 1000 circuit simulations.

The *impact of the different specifications and circuit parameters on circuit quality* is investigated according to Section IV. The reduced worst-case parameter sets in Table II represent the relevant statistical parameters obtained for this circuit. Note that, unlike in [21], dominant model parameters are not presupposed. The relevant transistor model parameters are determined for each investigated circuit individually. In contrast to previous works, worst-case operating conditions are not presupposed. Worst-case sets of both operating and statistical parameters are systematically determined for each investigated performance property individually. This is especially useful as worst-case operating conditions cannot always be determined *a priori* and can differ for different performance properties [19].

For circuit optimization, ΔV_{out} and *Gain* are determined as the *most relevant specifications*. SR_+, e.g., is not a relevant specification because it is automatically optimized when ΔV_{out} is optimized.

Then, relevant design parameters are chosen. The left diagram in Fig. 11 orders the designable transistor geometry parameters according to the corresponding element of the cost function gradient. Considering that W_{22} may not be increased, and L_{19} and L_{20} may not be decreased because of their respective upper and lower bounds, the widths W_{21} and W_6 are the most relevant out of 52 design parameters. They allow for a large increase in the worst-case yield values at small

TABLE III
PARAMETER CHANGES AND NOMINAL VALUES FOR THE OPTIMAL DESIGN OF THE AMPLIFIER

	Gain	GBW	φ_M	SR_+	SR_-	ΔV_{out}
$f_i(p_0^*)$	90 dB	6.0 MHz	86°	9.5 V/μs	5.6 V/μs	4.5 V
	W_{21}: -50%, W_6: -40%, L_{25}: -20%					

parameter changes. For a 20% change in these relevant design parameters, the right diagram in Fig. 11 shows the predicted changes in the worst-case yield values. The smallest worst-case distance for ΔV_{out} is increased implying an increase in the corresponding yield from 64% to 92%.

This is verified by the first *optimization* step. An optimal design of the circuit is obtained after 6 optimization steps. In the first step, relevant design parameters are selected according to Fig. 11. Apart from improving the numerical condition of the optimization problem, this selection results in a considerable reduction in simulation costs. In the following optimization steps, only the relevant deterministic parameters are considered for the gradient evaluation. (A selection of relevant design parameters is performed once more in an intermediate optimization step.) The total optimization costs are given by the costs of 7 specification analyses, $2 \cdot n_B$ gradient evaluations for all 52 deterministic design parameters and $5 \cdot n_B$ gradient evaluations for the 3 relevant design parameters [see (33)], i.e., in total 189 sensitivity analyses.

$$n_{opt} = 147 \cdot n_{grad}(14) + 12 \cdot n_{grad}(52) + 30 \cdot n_{grad}(3).$$

The example demonstrates that the optimization of circuits with a large number of design parameters is performed efficiently by selecting the most relevant parameters. The worst-case values of the operating parameters $T$and V_{DD} do not change in the course of optimization. This confirms the practical proceeding of circuit designers, that use fixed, empirically derived worst-case values for the operating parameters.

The optimization progress is illustrated by the increase of all the worst-case distances (and corresponding yield estimates) in Fig. 12. The smallest worst-case distances for the slew rate SR_- and the output range ΔV_{out} are equal to 3.1, indicating a satisfactory robustness. The total yield of the optimal design computed with a specification analysis is 100%. A Monte-Carlo analysis with 1000 circuit simulations yields the same value. A further optimization is not possible because an increase in the worst-case distance for SR_- is coupled to a decrease in the worst-case distance for ΔV_{out}. The parameter change for the optimal design is given in Table III. Transistor widths and lengths are decreased, i.e., area is not increased. A circuit simulation yields the nominal values for the optimal design in Table III.

After tuning the circuit as in Fig. 12, the designer investigates a possible further improvement in circuit quality by reducing the variations of the underlying manufacturing process. Identifying critical process variances is an important aid for the tuning and planning of the manufacturing process and measurements during manufacturing. According to Section IV, the variances in transistor lengths ΔL and in the mobility of the p-type transistors are determined as the critical variances of the manufacturing process. For a 10% reduction in these variances, the increase in the worst-case distance for ΔV_{out} is predicted as 0.02. This shows that on the one hand circuit quality cannot be increased much without drastically reducing process variations, and that on the other hand circuit quality would not decrease much for a 10% increase in process variations.

5.2 CMOS Inverter

The integrated CMOS inverter of Fig. 1 is a typical element of a standard cell library. As high-level and layout synthesis tools are based on cell libraries, the "cell characterization" is a fundamental part of the chip design. This example illustrates that circuit optimization according to (24) incorporates nominal design, yield optimization, and design centering. The inverter is optimized with respect to the following specifications for the gate delays t_r and t_f: $t_r^L \leq t_r \leq t_r^U t_f^L \leq t_f \leq t_f^U$, where $t_r^L = t_f^L = 2$ ns and $t_r^U = t_f^U = 4$ ns. The lower bounds for the gate delays imply an upper bound for the gate area. The design parameters are the transistor widths W_p and W_n, the statistical transistor model parameters are the same as for the previous example. The initial design $(W_p = W_n = 4\,\mu\text{m}), (t_r = 7.9\,\text{ns}), (t_f = 3.5\,\text{ns})$, is chosen so that the specification t_r^U is violated and the yield is practically zero. The corresponding worst-case distances and yield estimates are represented by the dark bars in Fig. 13. During optimization, a preliminary design is obtained $(W_p = 11.6\,\mu\text{m}, W_n = 4.9\,\mu\text{m}), (t_r = t_f = 3.0\,\text{ns})$, for which the performance values lie midway between the lower and upper specification bounds, and for which the yield is 100%. Thus, nominal design and yield optimization have been performed. The corresponding worst-case distances have evened out considerably, but as the worst-case distances to the upper bounds are still smaller than those to the lower bounds, we see that the optimum with respect to robustness has not yet been reached. For the optimized design $(W_p = 12.4\,\mu\text{m}, W_n = 5.4\,\mu\text{m}), (t_r = 2.9\,\text{ns}), (t_f = 2.8\,\text{ns})$, the worst-case distances $\beta_{w,i}$ (light bars in Fig. 13) have been equalized as far as possible. Thus, design centering has been performed. It is interesting that the performance values for the optimal design are not centered with respect to the specification bounds, and that the rising delay has smaller worst-case distances than the falling delay and thus a higher sensitivity to manufacturing fluctuations.

The optimization was carried out in four steps. In each optimization step, a specification analysis was performed that took on average $n_{iter} = 2$ iteration steps per specification. The total optimization costs are therefore given by $n_{opt} = 40 \cdot n_{grad}(14) + 20 \cdot n_{grad}(2)$, i.e., in total 60 sensitivity analyses.

5.3 Switched Capacitor Filter

The switched-capacitor-filter of Fig. 14 features 12 statistical designable parameters and 73 specifications for the gain-frequency range. The initial yield is 24%. The performance of this circuit is given by an analytical, very ill-conditioned equation system described in [5]. Therefore, it serves as a

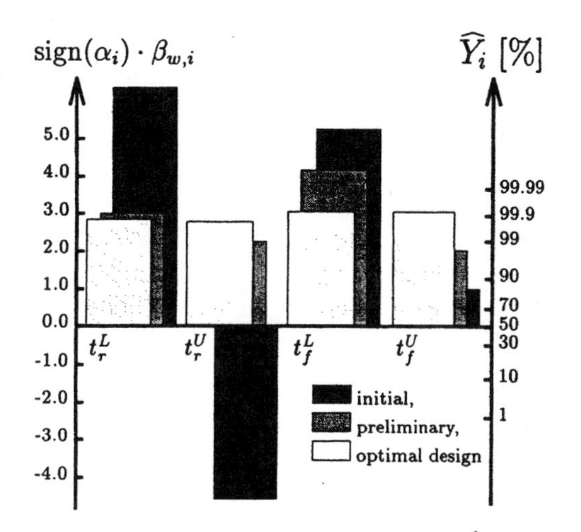

Fig. 13. Worst-case-distances $\beta_{w,i}$ and yield estimates $\hat{Y}_i$ for the initial, preliminary, and optimal design of the inverter.

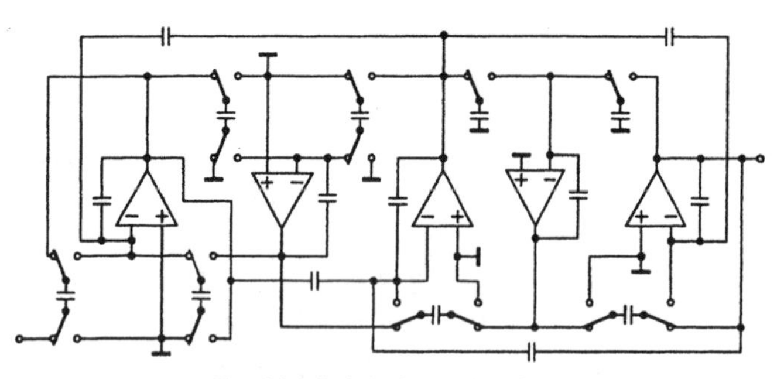

Fig. 14. Switched capacitor filter.

benchmark circuit to compare methods for yield optimization. In [5], it is optimized using a sophisticated Monte-Carlo based method, which results in an optimized yield of 60% after 5000 circuit simulations. Applying our new method we compute an optimized design with the optimal parameter values given by the values below. For this example an average of six relevant specifications is selected in every optimization step. For these six specifications, a specification analysis is performed. For the remaining 66 specifications, the worst-case distances are approximated using the sensitivities at the nominal parameter set. In this way, the simulation costs are significantly reduced. The relevant specifications are dynamically exchanged during the optimization process, so that only in the first optimization step a complete specification analysis of all 72 specifications is performed at an expense of 1872 simulations. After eight optimization steps, a yield of 64% at a total expense of 3120 simulations is attained. Thereby, the sensitivity calculation is performed using the more expensive finite difference method. Thus, the presented method yields a better result than statistical methods at lower simulation costs. All given (accurate) yield estimates are obtained with a Monte-Carlo analysis with 20 000 samples. The yield estimates provided by the new method differ less than 1% from these exact yield

VII. Conclusion

A generalized method for worst-case analysis of performance specifications and parameter tolerances has been presented, that adequately considers the integrated circuit's geometry, model, and operating parameters. The specification analysis provides exact worst-case parameter sets for the individual performance specifications. Worst-case parameter sets and related performance gradients provide an approximate acceptance region with a complexity that increases only linearly with the number of specifications. The examples demonstrate that this approximation of the acceptance region results in a very good estimate of the circuit yield at no additional simulation costs. Moreover, worst-case distances measure the yield and the robustness of the individual performance specifications, independently of differing units and orders of magnitude, and therefore enable a comparison of the specifications. Based on worst-case distances and related gradients, a new deterministic methodology for circuit design and optimization with respect to performance, robustness, and yield has been developed. It incorporates nominal design, worst-case analysis, yield optimization, and design centering. In contrast to current approaches to yield optimization, it is based on numerical circuit simulations and at the same time considers deterministic design parameters at reasonable computational costs. It eliminates the most serious disadvantage of geometric approaches to design centering, as its complexity increases only linearly with the number of design variables.

Acknowledgment

The authors wish to thank Dr. R. Koch, Dr. G. Mueller-Liebler, and Mr. H. Schinagel of SIEMENS Corporation for stimulating discussions and for providing examples of integrated circuits.

Appendix

A. Specification Analysis—Implementation and Practical Problems

(In this section, the index i is dropped for simplicity.) For a performance specification B the specification analysis determines a worst-case parameter set by solving problem (14). At the solution $(\boldsymbol{s}_w, \boldsymbol{\theta}_w)$ of problem (14), the performance value is equal to the specified bound and the gradient $\nabla_{\boldsymbol{s}} f(\boldsymbol{s}_w, \boldsymbol{\theta}_w)$ of the performance specification is parallel to the gradient $\nabla_{\boldsymbol{s}} \beta^2(\boldsymbol{s}_w, \boldsymbol{\theta}_w)$, i.e.:

$$f(\boldsymbol{d}_0, \boldsymbol{s}, \boldsymbol{\theta}) - B = 0 \tag{30}$$

$$\angle(\nabla_{\boldsymbol{s}} f(\boldsymbol{s}_w, \boldsymbol{\theta}_w), \nabla_{\boldsymbol{s}} \beta^2(\boldsymbol{s}_w, \boldsymbol{\theta}_w)) = 0. \tag{31}$$

Practical experience shows that a solution method for problem (14) has to carefully consider simulator inaccuracies and numerical ill-conditioning. To solve problem (14), we use the

---

$$[\cdots \boldsymbol{p}_i \cdots] = [0.1972 \quad 0.874 \quad 0.3141 \quad 0.3052 \quad 0.0508 \quad 0.1771 \quad 0.1976 \quad 0.0990 \quad 0.5518 \quad 0.3593 \quad 0.1269 \quad 0.2124].$$

Sequential Quadratic Programming (SQP) Method [29] with the Watchdog Technique [13] to calculate an appropriate step length for each iteration step. The SQP Method converges to a local minimum of (14). All the examples investigated so far have shown that the local approach is no impediment to the solution of practical problems. However, they showed that the crucial factor for the number of iteration steps needed to solve (14) is the second condition (31).

The SQP Method has been implemented in a program called EXCALIBUR. Any circuit simulator and macromodels may be used in EXCALIBUR to perform the function evaluations $f(\boldsymbol{d}_0, \boldsymbol{s}, \boldsymbol{\theta})$. For all the examples included here, the worst-case parameter sets were determined to an accuracy better than 1% and 20° in (30) and (31).

Appropriate scaling of the variables $\boldsymbol{s}$ and $\boldsymbol{\theta}$, the objective function $\beta(\boldsymbol{s}, \boldsymbol{\theta})$ and the constraints is essential for an efficient performance of the SQP Method. In EXCALIBUR, a statistical variable s is scaled according to its variance σ, an operating variable θ according to its tolerance interval $\theta_U - \theta_L$, and a performance f according to the amount of change that may occur during the iteration process:

$$s := s/\sigma, \quad \theta := \theta/(\theta_U - \theta_L), \quad f := f/|f(\boldsymbol{s}_0) - B|. \tag{32}$$

The computational cost for the solution of (14) is mainly determined by the number of required sensitivity calculations. In every iteration step the gradient $\nabla_{\boldsymbol{s},\boldsymbol{\theta}} f(\boldsymbol{d}_0, \boldsymbol{s}, \boldsymbol{\theta})$ has to be determined. No additional function evaluations are required for the approximations to second derivatives calculated with the Broyden–Fletcher–Goldfarb–Shanno (BFGS) update formula. The cost of an iteration step is therefore given by the cost of a sensitivity calculation. The number n_{grad} of circuit simulations required for the evaluation of the sensitivity $\nabla_{\boldsymbol{s},\boldsymbol{\theta}} f(\boldsymbol{d}_0, \boldsymbol{s}, \boldsymbol{\theta})$ depends on the number of statistical and operating parameters: $n_{grad} = n_{grad}(n_s + n_\theta)$. It is approximately determined using (33) below. If an average of n_{iter} iteration steps is needed per specification, the number n_{SQP} of circuit simulations required to solve (14) for all n_B specifications is given by

$$n_{SQP} = n_B \cdot n_{iter} \cdot n_{grad}(n_s + n_\theta). \tag{34}$$

The practical investigations showed that typically $n_{iter} = 3$ iterations are needed to solve (14). Our method for circuit analysis and optimization is cheap when compared to one Monte-Carlo analysis if the sensitivities are provided by the simulator as in [28]. If sensitivities are calculated by finite differences our method requires simulation costs for the complete analysis and optimization process which are comparable to those of one Monte-Carlo analysis.

B. Worst-Case Distance Gradient

Theorem B: The gradient of the worst-case distance according to (14) and (15) with respect to circuit parameters of any type (2) is given by

$$\nabla_{\boldsymbol{p}} \beta_{w,i}(\boldsymbol{p}_0) = \frac{-\text{sign}\,(B_i - f_i(\boldsymbol{p}_0))}{\sqrt{\nabla_{\boldsymbol{s}}^T f_i(\boldsymbol{p}_{w,i}) \cdot \boldsymbol{C} \cdot \nabla_{\boldsymbol{s}} f_i(\boldsymbol{p}_{w,i})}} \cdot \nabla_{\boldsymbol{p}} f_i(\boldsymbol{p}_{w,i}), \begin{cases} \beta_{w,i} \neq 0 \\ \nabla_{\boldsymbol{s}} f_i(\boldsymbol{p}_{w,i}) \neq \boldsymbol{0}. \end{cases}$$

Proof: For the proof, $\beta_{w,i} \neq 0$ and $\nabla_{\boldsymbol{s}} f_i(\boldsymbol{p}_{w,i}) \neq \boldsymbol{0}$ is assumed. For simplicity, the index i denoting the ith specification is dropped. Then, the Lagrange formulation of (14) is

$$\begin{aligned} &\mathcal{L}(\boldsymbol{s}, \boldsymbol{\theta}, \boldsymbol{\lambda}, \boldsymbol{\mu}, \boldsymbol{\nu}, \boldsymbol{\zeta}, \boldsymbol{\eta}) \\ &\quad = (\boldsymbol{s} - \boldsymbol{s}_0)T \cdot \boldsymbol{C}^{-1} \cdot (\boldsymbol{s} - \boldsymbol{s}_0) \\ &\quad\quad - 2\lambda \cdot (f(\boldsymbol{d}_0, \boldsymbol{s}, \boldsymbol{\theta}) - B) \\ &\quad\quad + 2\boldsymbol{\mu}^T \cdot (\boldsymbol{M}_\theta \cdot (\boldsymbol{\theta} - \boldsymbol{\theta}_0) - 1 + \zeta^2) \\ &\quad\quad + 2\boldsymbol{\nu}^T \cdot (\boldsymbol{M}_\theta \cdot (\boldsymbol{\theta}_0 - \boldsymbol{\theta}) - 1 + \boldsymbol{\eta}^2). \end{aligned} \tag{35}$$

Thereby, the inequality constraints of operating parameters have been included with the l_∞-norm using

$$\|\boldsymbol{M}_\theta \cdot (\boldsymbol{\theta} - \boldsymbol{\theta}_0)\|_\infty \leq 1, \quad \boldsymbol{\theta}_0 = \frac{1}{2}(\boldsymbol{\theta}_U + \boldsymbol{\theta}_L),$$
$$\boldsymbol{M}_\theta = \boldsymbol{diag}^{-1}\left(\frac{\boldsymbol{\theta}_U - \boldsymbol{\theta}_L}{2}\right). \tag{36}$$

$\boldsymbol{diag}(\boldsymbol{a})$ denotes a diagonal matrix containing the elements of $\boldsymbol{a}$. The optimality conditions for a solution $\boldsymbol{p}_w^T = [\boldsymbol{d}_w^T \boldsymbol{s}_0^T \boldsymbol{\theta}_w^T], \lambda_w, \boldsymbol{\mu}_w, \boldsymbol{\nu}_w, \zeta_w, \boldsymbol{\eta}_w$ of (35) are

$$\nabla_{\boldsymbol{s}} \mathcal{L} = \boldsymbol{0}: \quad \boldsymbol{s}_w - \boldsymbol{s}_0 = \lambda_w \cdot \boldsymbol{C} \cdot \nabla_{\boldsymbol{s}} f(\boldsymbol{p}_w) \tag{37}$$
$$\nabla_{\boldsymbol{\theta}} \mathcal{L} = \boldsymbol{0}: \quad \lambda_w \cdot \nabla_{\boldsymbol{\theta}} f(\boldsymbol{p}_w) = \boldsymbol{M}_\theta \cdot (\boldsymbol{\mu}_w - \boldsymbol{\nu}_w) \tag{38}$$
$$\nabla_\lambda \mathcal{L} = \boldsymbol{0}: \quad f(\boldsymbol{p}_w) = B \tag{39}$$
$$\nabla_{\boldsymbol{\mu}} \mathcal{L} = \boldsymbol{0}: \quad \boldsymbol{M}_\theta \cdot (\boldsymbol{\theta}_w - \boldsymbol{\theta}_0) = 1 - \zeta_w^2 \tag{40}$$
$$\nabla_{\boldsymbol{\nu}} \mathcal{L} = \boldsymbol{0}: \quad \boldsymbol{M}_\theta \cdot (\boldsymbol{\theta}_0 - \boldsymbol{\theta}_w) = 1 - \boldsymbol{\eta}_w^2 \tag{41}$$
$$\nabla_{\zeta^2} \mathcal{L} = \boldsymbol{0}: \quad \mu_{w,j} \cdot \zeta_{w,j} = 0, \quad i = 1, \cdots, n_\theta \tag{42}$$
$$\nabla_{\boldsymbol{\eta}^2} \mathcal{L} = \boldsymbol{0}: \quad \nu_{w,j} \cdot \eta_{w,j} = 0, \quad i = 1, \cdots, n_\theta. \tag{43}$$

From (37) and (15), we get:

$$\beta_w = \lambda_w \cdot \text{sign}\,(\lambda_w) \cdot \sqrt{\nabla_{\boldsymbol{s}}^T f(\boldsymbol{p}_w) \cdot \boldsymbol{C} \cdot \nabla_{\boldsymbol{s}} f(\boldsymbol{p}_w)}. \tag{44}$$

Equation (44) describes the worst-case distance as a function of the performance gradient at the worst-case parameter set and of λ_w. In the following, an explicit formula of λ_w as a function of $\boldsymbol{p}_0$ is derived. Insertion in (44) then concludes the proof. Towards this, a linearization at a worst-case parameter set $\tilde{\boldsymbol{p}}_w^T = [\tilde{\boldsymbol{d}}_0^T \tilde{\boldsymbol{s}}_w^T \tilde{\boldsymbol{\theta}}_w^T]$ that has been computed for a nominal point $\tilde{\boldsymbol{p}}_0^T = [\tilde{\boldsymbol{d}}_0^T \tilde{\boldsymbol{s}}_0^T \tilde{\boldsymbol{\theta}}_0^T]$, is inserted into (35) and into the optimality conditions:

$$\tilde{f}(\boldsymbol{p}) = B + \nabla_{\boldsymbol{p}}^T f \cdot (\boldsymbol{p} - \tilde{\boldsymbol{p}}_w), \quad \nabla_{\boldsymbol{p}} f \equiv \nabla_{\boldsymbol{p}} f(\tilde{\boldsymbol{p}}_w). \tag{45}$$

$$n_{grad}(k) \approx \begin{cases} 1 + 0.1 \cdot k, & \text{if the circuit simulator provides a sensitivity calculation [28].} \\ 1 + k, & \text{if a finite difference gradient approach is used.} \end{cases} \tag{33}$$

Equations (45) and (35) show that this linearization satisfies the optimality conditions. Equations (39), (45), (37), and $\boldsymbol{d}_w = \boldsymbol{d}_0$ result in

$$\lambda_w \cdot \nabla_{\boldsymbol{s}}^T f \cdot \boldsymbol{C} \cdot \nabla_{\boldsymbol{s}} f + \nabla_{\boldsymbol{\theta}}^T f \cdot (\boldsymbol{\theta}_w - \boldsymbol{\theta}_0) = \nabla_{\boldsymbol{p}}^T f \cdot (\tilde{\boldsymbol{p}}_w - \boldsymbol{p}_0). \quad (46)$$

Equations (40), (41), (42), and (43) yield three cases for an operating parameter θ_j:

Case 1: active upper bound $(\mu_{w,j} > 0, \nu_{w,j} = 0, \nu_{w,j} = 0, \eta_{w,j} > 0)$:

$$M_{\theta,j} \cdot (\theta_{w,j} - \theta_{0,j}) = 1 \quad (47)$$

$$\lambda_w \cdot \nabla_{\theta_j} f = M_{\theta,j} \cdot \mu_{w,j} \Leftrightarrow \text{sign}(\lambda_w) = \text{sign}(\nabla_{\theta_j} f). \quad (48)$$

Case 2: active lower bound $(\mu_{w,j} = 0, \zeta_{w,j} > 0, \nu_{w,j} > 0, \eta_{w,j} = 0)$:

$$M_{\theta,j} \cdot (\theta_{0,j} - \theta_{w,j}) = 1 \quad (49)$$

$$\lambda_w \cdot \nabla_{\theta_j} f = -M_{\theta,j} \cdot \nu_{w,j} \Leftrightarrow \text{sign}(\lambda_w) = -\text{sign}(\nabla_{\theta_j} f). \quad (50)$$

Case 3: inactive bounds $(\mu_{w,j} = 0, \zeta_{w,j} > 0, \nu_{w,j} = 0, \eta_{w,j} > 0)$:

$$\lambda_w \cdot \nabla_{\theta_j} f = 0 \Leftrightarrow \nabla_{\theta_j} f = 0. \quad (51)$$

From (47), (49), and (51) follows:

$$\nabla_{\boldsymbol{\theta}}^T f \cdot (\boldsymbol{\theta}_w - \boldsymbol{\theta}_0) = \sum_{\substack{\text{all active} \\ \text{upper bounds } j}} \frac{\nabla_{\theta_j} f}{M_{\theta,j}} - \sum_{\substack{\text{all active} \\ \text{lower bounds } j}} \frac{\nabla_{\theta_j} f}{M_{\theta,j}}. \quad (52)$$

Due to $a = |a| \cdot \text{sign}(a), \boldsymbol{M}_\theta$ being a diagonal matrix, and with (52), (48), (50), (51), we get

$$\nabla_{\boldsymbol{\theta}}^T f \cdot (\boldsymbol{\theta}_w - \boldsymbol{\theta}_0) = \text{sign}(\lambda_w) \cdot ||\boldsymbol{M}_\theta^{-1} \cdot \nabla_{\boldsymbol{\theta}} f||. \quad (53)$$

$|| \cdot ||_1$ denotes the l_1-norm. With (53), (46), and (44), the following explicit formula of the worst-case distance as a function of $\boldsymbol{p}_0$ is derived (using again the index i for the ith specification).

$$\beta_{w,i} = \text{sign}(\lambda_{w,i}) \cdot \frac{\nabla_{\boldsymbol{p}}^T f_i(\tilde{\boldsymbol{p}}_{w,i} - \boldsymbol{p}_0) - \text{sign}(\lambda_{w,i}) \cdot ||M_\theta^{-1} \cdot \nabla_{\boldsymbol{\theta}} f_i||_1}{\sqrt{\nabla_{\boldsymbol{s}}^T f_i(\tilde{\boldsymbol{p}}_{w,i}) \cdot \boldsymbol{C} \cdot \nabla_{\boldsymbol{s}} f_i(\tilde{\boldsymbol{p}}_{w,i})}}. \quad (54)$$

Using (54) and

$$\begin{aligned} \text{sign}(\lambda_{w,i}) &= \text{sign}(\nabla_{\boldsymbol{s}}^T f_i(\boldsymbol{p}_{w,i}) \cdot (\boldsymbol{s}_{w,i} - \boldsymbol{s}_0)) \\ &= \text{sign}(B_i - f_i(\boldsymbol{p}_0)) \end{aligned} \quad (55)$$

finally concludes the proof. ∎

In a similar manner, the worst-case distance gradient is derived for *any* norm that has been used to measure the worst-case distance. In this general case, the denominator of the gradient includes the corresponding dual norm of the weighted performance gradient.

In order to define a continuous worst-case distance gradient, that includes the case $\beta_{w,i} = 0$, definition (56) is introduced below.

Practical problems may arise for the case $\nabla_{\boldsymbol{p}} f_i(\boldsymbol{p}_{w,i}) = 0$, which has not been observed for the investigated circuits so far.

In a similar manner, the gradient of a worst-case distance is derived with respect to parameter variances:

$$\nabla_{\boldsymbol{\sigma}} \beta_{w,i}(\boldsymbol{\sigma}_0) = \frac{-\beta_{w,i}}{\nabla_{\boldsymbol{s}}^T f_i \cdot \boldsymbol{C} \cdot \nabla_{\boldsymbol{s}} f_i} \cdot \boldsymbol{vec}\{\boldsymbol{diag}^{-1}(\boldsymbol{\sigma}_0) \cdot \nabla_{\boldsymbol{s}} f_i \cdot \nabla_{\boldsymbol{s}}^T f_i\} \quad (57)$$

where $\boldsymbol{vec}(\boldsymbol{A}_{(n,n)}) = [A_{11} \quad A_{22} \cdots A_{nn}]^T$ and $\nabla_{\boldsymbol{s}} f_i \equiv \nabla_{\boldsymbol{s}} f_i(\boldsymbol{s}_{w,i})$.

C. Yield of a Specification

Theorem C: The yield $\hat{Y}_i$ of a parameter acceptance region (18), that is bounded by a hyperplane P_i,

$$P_i = \{\boldsymbol{s} | \nabla_{\boldsymbol{s}}^T f_i(\boldsymbol{p}_{w,i}) \cdot (\boldsymbol{s} - \boldsymbol{s}_{w,i}) = 0\} \quad (58)$$

is given by

$$\begin{aligned} \hat{Y}_i &= \int_{\hat{A}_{s,i}} pdf(\boldsymbol{s}) \cdot d\boldsymbol{s} = \int_{-\infty}^{\text{sign}(\alpha_i) \cdot \beta_{w,i}} \frac{1}{\sqrt{2\pi}} \cdot \exp\left(-\frac{t^2}{2}\right) dt \\ &= G(-\infty, \text{sign}(\alpha_i) \cdot \beta_{w,i}). \end{aligned}$$

$\beta_{w,i}$ is the worst-case distance according to (14) between nominal parameter set $\boldsymbol{s}_0$ and the worst-case parameter set $\boldsymbol{s}_{w,i}$, where P_i touches a tolerance body of the underlying normal distribution (Fig. 15(a)).

$$\nabla_{\boldsymbol{p}}(\text{sign}(\alpha_i) \cdot \beta_{w,i}(\boldsymbol{p}_0)) = \begin{cases} \text{sign}(\alpha_i) \cdot \nabla_{\boldsymbol{p}} \beta_{w,i}, & \beta_{w,i} \neq 0, \quad \nabla_{\boldsymbol{s}} f_i \neq \boldsymbol{0} \\ \dfrac{1}{\sqrt{\nabla_{\boldsymbol{s}}^T f_i \cdot \boldsymbol{C} \cdot \nabla_{\boldsymbol{s}} f_i}} \cdot \nabla_{\boldsymbol{p}} f_i, & \beta_{w,i} = 0, B = L, \quad \nabla_{\boldsymbol{s}} f_i \neq \boldsymbol{0} \\ \dfrac{-1}{\sqrt{\nabla_{\boldsymbol{s}}^T f_i \cdot \boldsymbol{C} \cdot \nabla_{\boldsymbol{s}} f_i}} \cdot \nabla_{\boldsymbol{p}} f_i, & \beta_{w,i} = 0, B = U, \quad \nabla_{\boldsymbol{s}} f_i \neq \boldsymbol{0} \end{cases} \quad (56)$$

using the abbreviations: $\nabla_{\boldsymbol{p}} f_i \equiv \nabla_{\boldsymbol{p}} f_i(\boldsymbol{p}_{w,i}), \nabla_{\boldsymbol{s}} f_i \equiv \nabla_{\boldsymbol{s}} f_i(\boldsymbol{p}_{w,i})$.

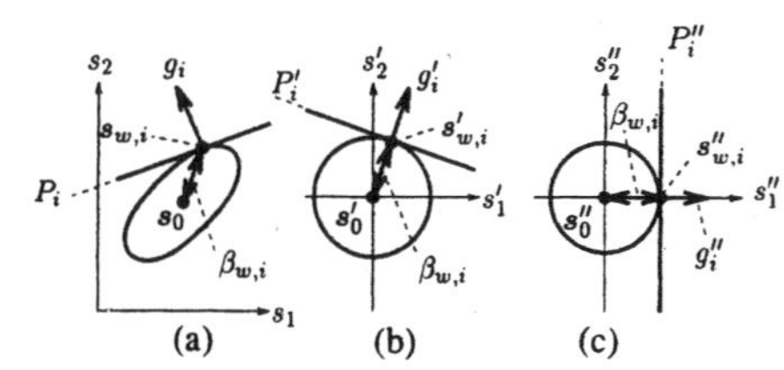

Fig. 15. Bounding hyperplane P_i according to (58), (59) for 2 statistical parameters s_1, s_2, worst-case parameter set $s_{w,i}$ with worst-case distance $\beta_{w,i}$: (a) for a (s_0, C)-normal distribution; (b) transformed into a $(0, I)$-normal distribution; (c) rotated.

Proof: The following abbreviation shall be used:

$$g_i \equiv \nabla_s f_i(p_{w,i}). \tag{59}$$

Using

$$s' = M \cdot (s - s_0), \quad C = M^{-1} \cdot M^{-1^T} \tag{60}$$

the (s_0, C)-normally distributed parameters s are transformed into $(0, I)$-normally distributed parameters s'. Then, from (37) follows:

$$s'_{w,i} = \lambda_{w,i} \cdot M^{-1^T} \cdot g_i. \tag{61}$$

Inserting (44), (61), and (55) in (58) yields the transformed bounding hyperplane P'_i and acceptance region $\hat{A}'_s$ [Fig. 15(b)]:

$$P'_i = \{s' | g_i'^T \cdot s' = \beta_{w,i} \cdot \text{sign}\,(B_i - f_i(p_0))\} \tag{62}$$

$$\hat{A}'_{s,i} = \left\{ s' \middle| g_i'^T \cdot s' \cdot \begin{cases} \geq \beta_{w,i} \cdot \text{sign}\,(B_i - f_i(p_0)), & B = L \\ \leq \beta_{w,i} \cdot \text{sign}\,(B_i - f_i(p_0)), & B = U \end{cases} \right\} \tag{63}$$

$$g'_i = \frac{M^{-1^T} \cdot g_i}{||M^{-1^T} \cdot g_i||}, \quad ||g'_i|| = 1. \tag{64}$$

Parameters s' are now orthogonally rotated, such that the first coordinate s_1 lies on the axis determined by s'_w [Fig. 15(c)]

$$s = H \cdot s', \quad \det H = 1, \quad H^{-1} = H^T. \tag{65}$$

The transformed worst-case parameter set $s_{w,i}$ and the bounding hyperplane P_i are now given by

$$s_{w,i} = [X \quad 0 \quad 0 \cdots 0]^T = \lambda_{w,i} \cdot H \cdot M^{-1^T} \cdot g_i \tag{66}$$

$$P_i = \{s | (H \cdot g_i'^T)^T \cdot s = \beta_{w,i} \cdot \text{sign}\,(B_i - f_i(p_0))\}. \tag{67}$$

With (64) and the orthogonal matrix H we get: $||H \cdot g'_i|| = ||g'_i|| = 1$. Using (66), we obtain: $H \cdot g'_i = [1 \quad 0 \quad 0 \cdots 0]^T$. The parameter acceptance region $\hat{A}_s$ is then given by

$$\hat{A}_{s,i} = \left\{ s \middle| s_1 \begin{cases} \geq \beta_{w,i} \cdot \text{sign}\,(B_i - f_i(p_0)), & B = L \\ \leq \beta_{w,i} \cdot \text{sign}\,(B_i - f_i(p_0)), & B = U \end{cases} \right\}. \tag{68}$$

From (68), the yield is derived in the following way:

$$\begin{aligned} \hat{Y}_i &= \int_{\hat{A}_{s,i}} pdf(s) \cdot ds = \int_{\hat{A}_{s,i}}^{\hat{A}} pdf(s) \cdot ds \\ &= \int_{s_1^U}^{s_1^O} \frac{1}{\sqrt{2\pi}} \cdot \exp\left(-\frac{s_1^2}{2}\right) \cdot ds_1 \\ &\quad \cdot \underbrace{\prod_{i=2,\cdots,n_s} \int_{-\infty}^{\infty} \frac{1}{\sqrt{2\pi}} \cdot \exp\left(-\frac{s_i^2}{2}\right) \cdot ds_i}_{1} \\ &= G(s_1^U, s_1^O). \end{aligned} \tag{69}$$

Thereby:

$$[s_1^U, s_1^O] = \begin{cases} [\beta_{w,i} \cdot \text{sign}\,(B_i - f_i(p_0)), \infty], & B = L \\ [-\infty, \beta_{w,i} \cdot \text{sign}\,(B_i - f_i(p_0))], & B = U. \end{cases} \tag{70}$$

$G(s_1^U, s_1^O) = G(-s_1^O, -s_1^U)$ and (69), (70), (8) conclude the proof. ■

REFERENCES

[1] H. Abdel-Malek and A. Hassan, "The ellipsoidal technique for design centering and region approximation," *IEEE Trans. Computer-Aided Design*, vol. 10, pp. 1006–1013, 1991.

[2] T. Anderson, *An Introduction to Multivariate Statistical Analysis*. New York: Wiley, 1958.

[3] K. Antreich and H. Graeb, "Circuit optimization driven by worst-case distances," in *Proc. IEEE ICCAD*, 1991, pp. 166–169.

[4] ——, "A unified approach towards nominal and tolerance design," in *Proc. IMACS World Congress on Appl. Math.*, 1991, pp. 176–181.

[5] K. Antreich and R. Koblitz, "Design centering by yield prediction," *IEEE Trans. Computer-Aided Design*, vol. 29, pp. 88–95, 1982.

[6] K. Antreich, P. Leibner, and F. Poernbacher, "Nominal design of integrated circuits on circuit level by an interactive improvement method," *IEEE Trans. Circuits Syst.*, vol. 35, pp. 1501–1511, 1988.

[7] J. Bandler and H. Abdel-Malek, "Optimal centering, tolerancing, and yield determination via updated approximations and cuts," *IEEE Trans. Circuits Syst.*, vol. 25, pp. 853–871, 1978.

[8] J. Bandler and S. Chen, "Circuit optimization: The state of the art," *IEEE Trans. Microwave Theory Tech.*, vol. 36, pp. 424–442, 1988.

[9] M. Bernardo, R. Buck, L. Liu, W. Nazaret, J. Sacks, and W. Welch, "Integrated circuit design optimization using a sequential strategy," *IEEE Trans. Computer-Aided Design*, vol. 11, pp. 361–372, 1992.

[10] R. Brayton, G. Hachtel, and A. Sangiovanni-Vincentelli, "A survey of optimization techniques for integrated-circuit design," *Proc. IEEE*, vol. 69, pp. 1334–1363, 1981.

[11] G. Casinovi and A. Sangiovanni-Vincentelli, "A macromodeling algorithm for analog circuits," *IEEE Trans. Computer-Aided Design*, vol. 10, pp. 150–160, 1991.

[12] R. Chadha, K. Singhal, J. Vlach, and E. Christen, "WATOPT—An optimizer for circuit applications," *IEEE Trans. Computer-Aided Design*, vol. 6, pp. 472–479, 1987.

[13] R. Chamberlain, M. Powell, C. Lemarechal, and H. Pedersen, "The watchdog technique for forcing convergence in algorithms for constrained optimization," *Mathematical Programming Study*, vol. 16, pp. 1–17, 1982.

[14] S. Director, W. Maly, and A. Strojwas, *VLSI Design for Manufacturing: Yield Enhancement*. Boston, MA: Kluwer Academic, 1990.

[15] P. Feldmann and S. Director, "Improved methods for IC yield and quality and optimization using surface integrals," in *Proc. IEEE ICCAD*, 1991, pp. 158–161.

[16] J. Fisher and R. Koch, "A highly linear CMOS buffer amplifier," *IEEE J. Solid-State Circuits*, vol. SC-22, pp. 330–334, 1987.

[17] H. Graeb, "Circuit optimization with worst-case distances as optimization targets," Ph.D. dissertation (in German), Technical Univ. of Munich, 1993.

[18] H. Graeb, C. Wieser, and K. Antreich, "Design verification considering manufacturing tolerances by using worst-case distances," in *Proc. EURO-DAC*, 1992, pp. 86–91.

[19] ——, "Improved methods for worst-case analysis and optimization incorporating operating tolerances," *Trans. ACM/IEEE DAC*, 1993.
[20] N. Herr and J. Barnes, "Statistical circuit simulation modeling of CMOS VLSI," *IEEE Trans. Computer-Aided Design*, vol. 5, pp. 15–22, 1986.
[21] D. Hocevar, P. Cox, and P. Yang, "Parametric yield optimization for MOS circuit blocks," *IEEE Trans. Computer-Aided Design*, vol. 7, pp. 645–658, 1988.
[22] M. Lightner, T. Trick, and R. Zug, "Circuit optimization and design," *Circuit Analysis, Simulation and Design, Part 2 (A. Ruehli). Advances in CAD for VLSI 3*. Amsterdam: North-Holland, 1987, pp. 333–391.
[23] K. Low and S. Director, "An efficient methodology for building macromodels of IC fabrication processes," *IEEE Trans. Computer-Aided Design*, vol. 8, pp. 1299–1313, 1989.
[24] ——, "A new methodology for the design centering of IC fabrication processes," *IEEE Trans. Computer-Aided Design*, vol. 10, pp. 895–903, 1991.
[25] W. Maly, "Computer-aided design for VLSI circuit manufacturability," *Proc. IEEE*, vol. 78, pp. 356–392, 1990.
[26] G. Mueller-Liebler, "Limit-parameters: The general solution of the worst-case problem for the linearized case," *Proc. IEEE ISCAS*, 1990.
[27] S. Nassif, A. Strojwas, and S. Director, "A methodology for worst-case analysis of integrated circuits," *IEEE Trans. Computer-Aided Design*, vol. 5, pp. 104–113, 1986.
[28] W. Nye, D. Riley, A. Sangiovanni-Vincentelli, and A. Tits, "DELIGHT.SPICE: An optimization-based system for the design of integrated circuits," *IEEE Trans. Computer-Aided Design*, vol. 7, pp. 501–519, 1988.
[29] K. Schittkowski, "The nonlinear programming method of Wilson, Han, and Powell with an augmented Lagrangian type line search function," *Numer. Math.*, vol. 38, pp. 83–114, 1981.
[30] H. Schjaer-Jacobsen and K. Madsen, "Algorithms for worst case tolerance optimization," *IEEE Trans. Circuits Syst.*, vol. 26, pp. 775–783, 1979.
[31] K. Singhal and J. Pinel, "Statistical design centering tolerancing using parametric sampling," *IEEE Trans. Circuits Syst.*, vol. 28, pp. 692–702, 1981.
[32] R. Spence and R. Soin, *Tolerance Design of Electronic Circuits*. England: Addison-Wesley, 1988.
[33] M. Styblinski and S. Aftab, "Efficient circuit performance modeling using a combination of interpolation and self organizing approximation techniques," in *Proc. IEEE ISCAS*, 1990, pp. 2268–2271.
[34] M. Styblinski and L. Opalski, "Algorithms and software tools for IC yield optimization based on fundamental fabrication parameters," *IEEE Trans. Computer-Aided Design*, vol. 5, pp. 79–89, 1986.
[35] K. Tahim and R. Spence, "A radial exploration approach to manufacturing yield estimation and design centering. *IEEE Trans. Circuit Syst.*, vol. 26, pp. 768–774, 1979.
[36] T. Yu, S. Kang, I. Hajj, and T. Trick, "Statistical performance modeling and parametric yield estimation of MOS VLSI," *IEEE Trans. Computer-Aided Design*, vol. 6, pp. 1013–1022, 1987.

Kurt J. Antreich (M'69–SM'78) received the Dipl.-Ing. degree from the Technical University of Munich, Munich, Germany, in 1959, and the Dr.-Ing. degree from the Technical University Fridericiana Karlsruhe, Karlsruhe, Germany, in 1966.

He joined AEG-Telefunken in 1959, where he was involved in computer-aided circuit design for telecommunication systems. From 1968 to 1975 he was head of the Advanced Development Department of AEG-Telefunken, Backnang, Germany. In 1983 the name AEG-Telefunken changed to ANT Bosch Telecom. Since 1975, he has been Professor and head of the Institute of Electronic Design Automation, Technical University of Munich, Munich, Germany. His research interests are in computer-aided design of electronic circuits and systems, with particular emphasis on circuit optimization, layout synthesis, testing, and synthesis of digital systems.

Dr. Antreich was Chairman of the NTG Circuit and Systems Group from 1972 to 1974, and a member of the executive board of the NTG from 1979 to 1982, and Chairman of the NTG CAD Group from 1985 to 1989. He received the NTG Prize Paper Award in 1976. In 1986 the name of NTG changed to ITG (Informationstechnische Gesellschaft, Germany). He is a member of the ITG and the GI (Gesellschaft für Informatik, Germany).

Helmut E. Graeb received the Dipl.-Ing. and the Dr.-Ing. degrees in electrical engineering from the Technical University of Munich, Munich, Germany, in 1986 and 1993, respectively. From 1986 to 1987, he was at Siemens Corp., Munich, where he was involved in the design and development of 4M dynamic RAM's.

Since 1987, he has been a research assistant at the Institute of Electronic Design Automation, Technical University of Munich. Since 1992 he has also been supervising the analog activities within the Institute. His current research interests are in computer-aided design of analog circuits, with particular emphasis on nominal and tolerance design.

Claudia U. Wieser received the Dipl.-Ing. degree in electrical engineering from the Technical University of Munich, Munich, Germany, in 1989. Currently, she is working towards the Dr.-Ing. degree at the Technical University of Munich.

Since 1989, she has been a research assistant at the Institute of Electronic Design Automation, Technical University of Munich. Her current research interests are in computer-aided design of analog circuits, with particular emphasis on nominal and tolerance design.

Efficient Analog Circuit Synthesis with simultaneous Yield and Robustness Optimization

Geert Debyser, Georges Gielen*

Katholieke Universiteit Leuven, Belgium

[geert.debyser, georges.gielen]@esat.kuleuven.ac.be

Abstract

This paper presents an efficient statistical design methodology that allows simultaneous sizing for performance and optimization for yield and robustness of analog circuits.

The starting point of this methodology is a declarative analytical description of the circuit. An equation manipulation program based on constraint satisfaction converts this declarative model into an efficient design plan for optimization based sizing. The efficiency is due to the use of an operating point driven DC formulation, so that the design plan avoids the calculation of simultaneous sets of nonlinear equations. From the same declarative analytical description also a direct symbolic yield estimation plan is generated. The parametric yield is estimated by propagating the spread of the technological variables through the analytical model towards the performance variables of the circuit. The design plan and the yield estimation plan are then combined together in the inner loop of a global optimization routine. The strength of this methodology lies in the low CPU times needed to perform yield estimation compared to the hours of simulation batches with Monte Carlo simulations, while the accuracy is comparable.

I. Introduction

Designing an analog circuit is one thing, producing it is another. Real-life technology parameter variations make the circuits fail for some or all of the specifications if no precautions are taken. The ratio of the number of successful circuits over the number of produced circuits is the total yield. The total yield consists of yield due to production faults and yield due to soft faults. In this paper we concentrate on the yield based on the soft faults, generated by the technology parameter variations: the parametric yield.

The hard way to make the design more robust is to run multiple batch jobs of Monte Carlo simulations in the inner loop of an optimization routine. Because a Monte Carlo simulation consists of typically hundreds of SPICE simulations, the computational effort is so large that only a post-design yield optimization is considered. This is the known design centering method. Starting from the nominal design a local optimization tries to push the performances away from the specification boundaries in order to make the circuit more robust against technology parameter variations [1].

The same is true for analog circuit sizing or synthesis programs [2]. Most of them concentrate on the nominal design only, without considering the process parameter or operating condition variations. Performing hours of simulated annealing for just a nominal design is only half of the game. Only in [3] a first approach towards analog synthesis for manufacturability was presented that combines nominal circuit optimization with variation analysis in an outer optimization loop. The results of the variation analysis are used to change the cost function of the inner circuit optimization by penalizing design solutions that do not meet the specifications over the entire operating range. The approach however is extremely time consuming.

Therefore, in this paper, an alternative approach is presented towards simultaneous circuit sizing for performance and yield/robustness optimization. The approach results in a drastic reduction of the required CPU time, without sacrifying too much accuracy. It is based on the use of symbolic techniques to capture the behavior of a circuit in a declarative model [4]. Constraint statisfaction techniques implemented in the tool DONALD [5] are then used to derive an efficient sizing plan as well as an efficient yield estimation plan. Both plans are then simultaneously evaluated in the inner loop of a global optimization routine. The outcome of the optimization is a circuit design point that fulfills all specifications and that at the same time has pushed away the performances from the specification boundaries under the influence of the yield and/or Cpk measure.

The paper is organized as follows. Section 2 explains the construction of the sizing plan by the DONALD tool starting from the declarative behavioral model of the circuit. Section 3 describes how we used symbolic techniques to construct the yield/robustness estimation plan. Section 4 explains the yield optimization strategy. Section 5 shows experimental results. Section 6 draws some conclusions.

II. Sizing plan

A declarative analytical model of an analog circuit is mainly obtained by 2 sources: by applying symbolic methods on the circuit's graph topology and by hand. A simple illustrative example of the outcome for an inverter is given in Fig. 1, where 7 equations describe the DC operating point.

The 7 equations are presented in a bipartite graph form (see

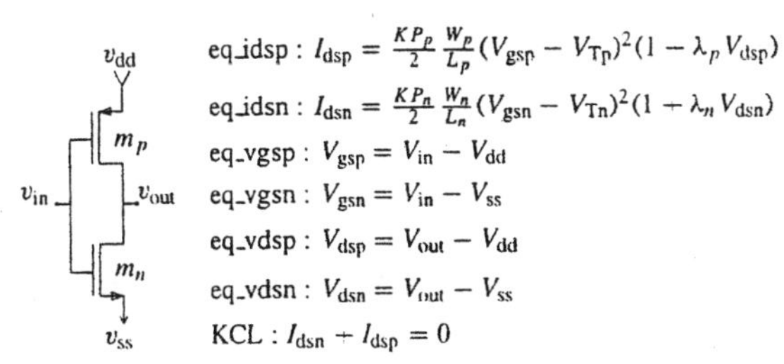

$$\text{eq\_idsp}: I_{dsp} = \frac{KP_p}{2}\frac{W_p}{L_p}(V_{gsp} - V_{Tp})^2(1 - \lambda_p V_{dsp})$$

$$\text{eq\_idsn}: I_{dsn} = \frac{KP_n}{2}\frac{W_n}{L_n}(V_{gsn} - V_{Tn})^2(1 + \lambda_n V_{dsn})$$

$$\text{eq\_vgsp}: V_{gsp} = V_{in} - V_{dd}$$

$$\text{eq\_vgsn}: V_{gsn} = V_{in} - V_{ss}$$

$$\text{eq\_vdsp}: V_{dsp} = V_{out} - V_{dd}$$

$$\text{eq\_vdsn}: V_{dsn} = V_{out} - V_{ss}$$

$$\text{KCL}: I_{dsn} + I_{dsp} = 0$$

Fig. 1. Inverter and its simplified behavioral DC model.

* research associate of the Belgian Fund of Scientific Research

ICCAD98, San Jose, CA, USA

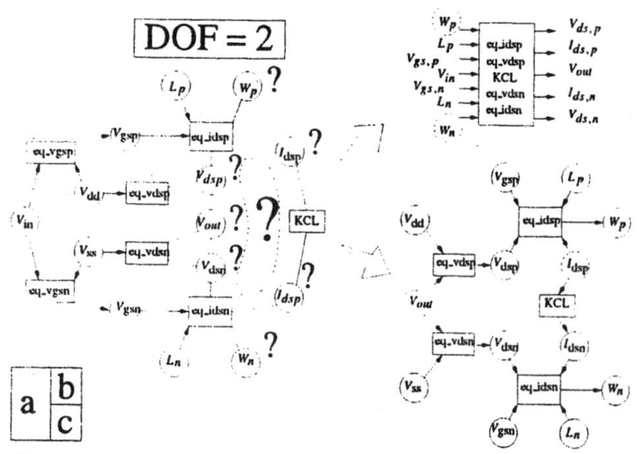

Fig. 2. Depending on the input variables, DONALD has to solve a 5x5 cluster or 2 simple equations.

Fig. 2a), where the square boxes represent the equations and the circles represent the variables. We assume that the user can give a value for V_{dd}, V_{ss}, V_{in} (typically half-way V_{dd} and V_{ss}). L_n and L_p can be chosen to be minimal size. The equation manipulation tool DONALD propagates these inputs throught the graph, which is then partially directed. The traditional SPICE-like inputs (choosing W_p and W_n, see Fig. 2b), lets DONALD simultaneously solve a set of 5 (possibly nonlinear) equations with a nonlinear root solver. In this case convergence cannot be guaranteed and the solution also depends on the starting point. The operating point driven DC formulation [6], however, constrains node voltages and branch currents (choosing V_{nout} and I_{dsp}, as in Fig. 2c). The DONALD program only has to solve 2 one-dimensional nonlinear equations to obtain W_n and W_p, which can be executed much faster. The computational effort is drastically reduced since, for a circuit with N nodes and M MOS-transistors, solving M one-dimensional nonlinear equations is of the order $O(M\ \mathrm{Root}(1))$ (where Root(x) is the effort to solve a cluster of size x). A simulator needs an effort of the order $O(\mathrm{Root}(N))$, with a risk of divergence. This approach has already been used for optimization based nominal sizing [6], [7]. In this paper we extend this to yield optimization.

III. Yield/robustness estimation plan

Our approach replaces the Monte Carlo simulations with a direct yield estimation technique. We start with a statistical transistor model, which gives us a reduced set of quasi-independent technology parameters $\boldsymbol{\theta}$. Then we calculate the nominal design point and then we calculate the variances of all performance parameters **y** with respect to the reduced set of technology parameters $\boldsymbol{\theta}$. Using these variances, we construct an efficient representation of "yield" based on C_p/C_{pk} indices. Both the sizing plan and the yield estimation plan are then placed in the inner loop of the optimization routine as explained in section 4. The flow diagram in Fig. 3 with the setup for the yield estimation plan is explained in the following subsections (x stands for designable parameters and e stands for simulator variables).

A. *Statistical transistor model*

The default transistor models have to be replaced by statistical models in order to take the correlations of the technology parameters into account. This is necessary to estimate the yield in a statistically correct way. The statistical model describes all technological variables as a function of only 7 quasi-uncorrelated input variables, which are TOX, NSUB_n, NSUB_p, CJ_n, CJ_p, LD_n and RSH_n. A method for deriving such a statistical transistor model has been discussed in [8]. The Monte Carlo routine perturbes only these 7 parameters and extracts performances from the SPICE output to construct the yield estimation figures based on a pass/fail mechanism. This way of calculating the yield is very costly: a large number (e.g. 300) of SPICE simulations have to be executed. Our symbolic yield estimator also starts from the same 7 parameters $\boldsymbol{\theta}$.

B. *Direct yield estimation*

As the pass/fail measure used in Monte Carlo simulation is very rough and gives little or no information about which performance does not meet its specification, another quality measure is preferred. The Taguchi quality measure gives a much better idea of the quality of a circuit, because it takes absolute variability and bias to the target spec into account:

$$M_{\mathrm{TAG}} = var\{\mathbf{y}(x)\} + (\bar{\mathbf{y}} - \mathbf{Spec}^T)^2 \qquad (1)$$

To ease the calculation of M_{TAG}, two capability indices were introduced in [9]: the capability *potential* index C_p:

$$C_p = \frac{\mathrm{Spec}^U - \mathrm{Spec}^L}{6\sigma_y} \qquad (2)$$

and the capability *performance* index C_{pk}:

$$C_{pk} = \min\{\frac{\mathrm{Spec}^U - \bar{y}}{3\sigma_y}, \frac{\bar{y} - \mathrm{Spec}^L}{3\sigma_y}\} \qquad (3)$$

The first index represents the variability, the second index the bias. As can be noticed from formulae (2,3), these indices strongly depend on the variances of the performances. The variances of the 7 technological parameters are propagated through the computational path by means of sensitivities. In case θ_1, θ_2, ... $\theta_{n=7}$ are not correlated, the variability of the performances can be written as follows:

$$\sigma_{y_i} = \sqrt{Var(y_i)} \approx \sqrt{\sum_{j=1}^{n=7} (S_{\theta_j}^{y_i})^2 \sigma_{\theta_j}^2} \qquad (4)$$

(assuming that $\sigma_{\theta_j}/\bar{\theta}_j$ is sufficiently small).

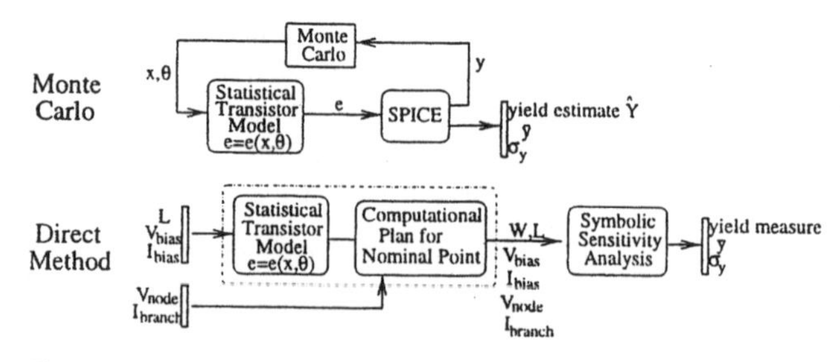

Fig. 3. Yield estimation methods. Monte Carlo and SPICE versus direct method.

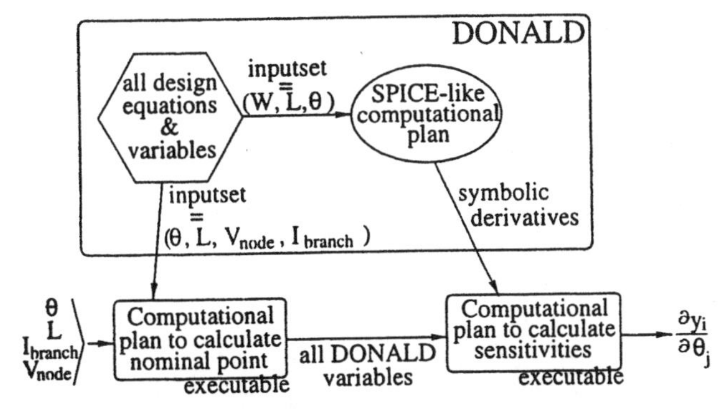

Fig. 4. Creation and use of direct yield estimation model.

An extension has been made to DONALD (see Fig. 4), so that it also builds a computational plan to calculate all sensitivities from the output variables w.r.t. the input variables. The sensitivities have to be known for *every* operating point x. Our Jacobians are in symbolic form, so the Jacobian updating for every x is relatively cheap.

This computational path is derived from the same set of equations used to determine the nominal point, but W and L must be chosen as input variables. This time however, we do *not* solve the set of nonlinear equations, but we reuse the values previously obtained in the nominal point calculation (see Fig. 4). Symbolic derivatives from all equations are automatically derived. Out of this a computational path results, which is a chain of one-dimensional and more-dimensional subsystems of equations. The local sensitivities are calculated using Jacobians in symbolic form. The global sensitivity matrix S_{θ}^{y} with elements $S_{\theta_j}^{y_i} = \partial y_i / \partial \theta_j$, is calculated by applying the chain rule according to the computational plan:

$$S_{\theta}^{y} = S_{z_n}^{y} S_{z_{n-1}}^{z_n} \cdots S_{\theta}^{z_1} \tag{5}$$

where z_k are the internal variables along the calculation path.

The calculation of the local sensitivities is complicated but straightforward. First, at each square subsystem with equations $f_1, \ldots, f_n$, a distinction is made between those variables of the subsystem that have been solved by the subsystem, and those variables that are solved by previous subsystems. The former are called the *output variables* $\mathbf{z^{out}} = (\mathbf{z_1^{out}}, \ldots, \mathbf{z_q^{out}})^T$ of the subsystem, the others are called the *input variables* $\mathbf{z^{in}} = (z_1^{in}, \ldots, z_p^{in})^T$. For each square subsystem the change $\delta \mathbf{z^{out}}$ is then calculated of the output variables with respect to a unit change $\delta \mathbf{z^{in}}$ of the input variables. The local sensitivity $S_{z_j^{in}}^{z_i^{out}} = \delta z_i^{out}$ is solved from the following system of linear equations:

$$\nabla \mathbf{F}^{out} \delta \mathbf{z^{out}} = -\nabla \mathbf{F^{in}} \delta \mathbf{z^{in}} \tag{6}$$

where $\delta \mathbf{z^{in}} = \mathbf{1}$ and

$$\nabla \mathbf{F}^{out} = \begin{bmatrix} \frac{\partial f_1}{\partial z_1^{out}} & \cdots & \frac{\partial f_1}{\partial z_q^{out}} \\ \vdots & \ddots & \vdots \\ \frac{\partial f_n}{\partial z_1^{out}} & \cdots & \frac{\partial f_n}{\partial z_q^{out}} \end{bmatrix} \tag{7}$$

and

$$\nabla \mathbf{F}^{in} = \begin{bmatrix} \frac{\partial f_1}{\partial z_1^{in}} & \cdots & \frac{\partial f_1}{\partial z_p^{in}} \\ \vdots & \ddots & \vdots \\ \frac{\partial f_n}{\partial z_1^{in}} & \cdots & \frac{\partial f_n}{\partial z_p^{in}} \end{bmatrix} \tag{8}$$

where all partial derivatives are in symbolic form. By multiplying local sensitivity values along the computational path between two variables r and s, a *global sensitivity* value $S_s^r = \partial r / \partial s$ can be calculated.

IV. Yield Optimization

We can now perform a nominal point analysis and to calculate all sensitivities and hence the C_p/C_{pk} values. The step towards simultaneous nominal and statistical optimization is small by combining the approach of Fig. 4 in a optimization loop. The optimization is done stepwise, as can be seen in Fig. 5, where the solution is gradually narrowed by enforcing more and more constraints on the design [4].

The initial solvability space is narrowed by adding the following sequence of constraints to the design plan. The manufacturability space Ω_M is the set of circuits that can be produced within a given technology. The operationality space Ω_O contains the circuits whose transistors are in the correct operating region. The functionality space Ω_F contains the designs that fulfill the design requirements (such as first order behavior for an OTA). The applicability space Ω_A contains the circuits that fulfill all specifications. The robustness space Ω_R contains the designs that take all performance variations into account by retreating from the design space boundaries. Once all specifications fulfilled, there is still room left for trade-off between different performance parameters.

With each of these spaces inequality functions can be associated, which in their turn correspond to penalty functions h_j:

$$h_j = \begin{cases} \leq 0 & \text{if inequality is fulfilled} \\ > 0 & \text{else} \end{cases}$$

The subdivision of the solvability region Ω_S is reached by adding a weighted sum of penalties to the total cost function of the optimization routine. To assure a *sequential design space pruning process*, the weights W are chosen as follows: the steering function for robustness gets the reference weight W_R. The penalties of the applicability space Ω_A get weights of $10W_R$. The weights belonging to the other spaces are $10^2 W_R$ for Ω_F,

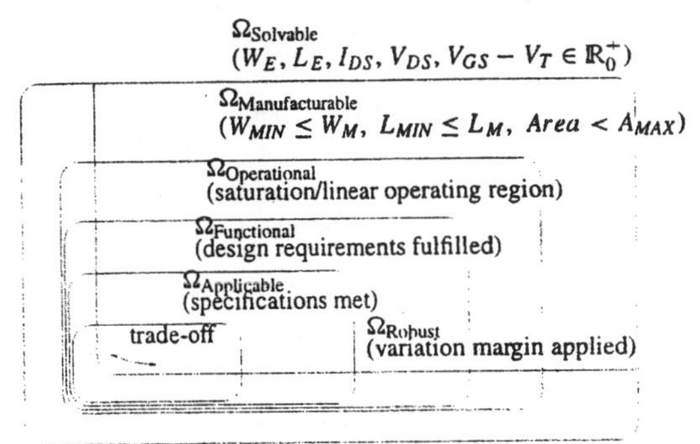

Fig. 5. Spaces encountered in optimization based sizing of analog circuits.

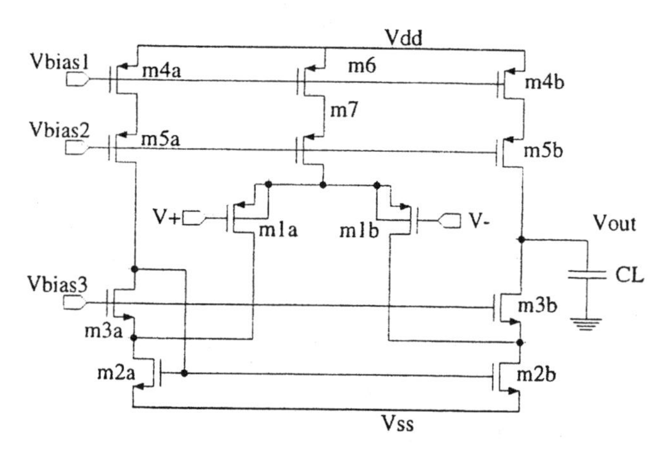

Fig. 6. Schematic of current buffer OTA.

$10^3 W_R$ for Ω_O and, $10^4 W_R$ for Ω_M. This way the optimizer tackles the sizing problem in a stepwise way. First it assures the manufacturability, then the operationality, the functionality and the applicability. In the final stage the yield estimation model is activated and the cost belonging to the not-yet optimal yield and variability is added to the global cost function of the optimization routine. We chose to leave out the yield estimation till the optimizer was in the applicability space Ω_A, because a yield estimation in an earlier stage of the optimization is computational expensive and doesn't change the problem that much.
The quality measure for robustness used in the cost function is:

$$\Phi_i^{\pm}(x) = C_{p,i} \pm \lambda \frac{\frac{\text{Spec}_i^U + \text{Spec}_i^L}{2} - \overline{y}_i}{3\sigma_{y_i}} \qquad (9)$$

$$\Phi_i(x) = \min\{\Phi_i^+(x), \Phi_i^-(x)\} \qquad (10)$$

The weight factor λ acts as a penalty term on the bias, with respect to target Spec_i^T. If $\lambda = 0$ then $\Phi_i = C_{p,i}$, if $\lambda = 1$ then $\Phi_i = C_{pk,i}$. A value of $\lambda = 0.8$ has been chosen experimentally. The optimization problem to be solved is then

$$\Phi(x) = \sum_{\substack{x \in R_x \\ j=\Omega_M,\Omega_O,\Omega_F,\Omega_A}} W_j h_j(x) + \max_{x \in R_x} \min_i \{\Phi_i(x)\} + \alpha f(\text{trade-off}) \qquad (11)$$

where R_x is a hyperbox of constraints.

V. Experimental Results

The circuit is a CMOS current buffer OTA (see Fig. 6) in a 0.7u CMOS process. An arbitrary starting point (formulated as a DC operating point) has been chosen for the optimizer. Then a simultaneous sizing and optimization is performed (in 2h15' time) with both the sizing plan and the yield/robustness estimation plan (see Fig. 4) in the inner loop of of the annealing routine. In the resulting optimized point a Monte Carlo is run for verification.
The yield/robustness estimation plan is extremely fast (10s) compared to Monte Carlo (300 samples: 2h20' on a Sparc I). Such an optimization with Monte Carlo in the inner loop would take approximately 20 days! The first two columns of Table 1 contain the mean and the variance of the performance variables,

TABLE I
FINAL RESULTS OF YIELD OPTIMIZATION.

Specs	After optimization			
	Yield Model		Monte Carlo	
	$\overline{y}$	σ_y	$\overline{y}$	σ_y
GBW>100MHz	165	23.1	172	21.4
A_{v0}>60dB	78.0	1.68	78.1	1.74
PM>60°	60.2	0.2	63.6	0.4
OR>3.0V	3.1	0.05	3.27	0.058
V_{off}<5mV	4.8	0.14	3.5	0.7
I_{tot}<3.0mA	2.6	0.4	2.6	0.56

which come out of the yield/robustness estimation model. The last two columns in Table 1 give the estimate mean and variance of the performances from the Monte Carlo run.

VI. Conclusions

A new statistical design method using symbolic techniques has been presented. A direct yield estimation model has automatically been derived from the whole set of symbolic design equations by constructing a computational plan to calculate symbolically the sensitivities of the performances w.r.t. the technology parameters. By propagating the variances of 7 quasi-independent technology parameters the quality measures C_p and C_{pk} for the performances were obtained. This yield model was directly used in the inner loop of a yield optimization routine to perform simultaneous nominal and yield/robustness optimization. The experimental results show that the model is accurate enough to steer the optimization in the direction of a "better" design in a much faster way than using a simulator. Research is still to be continued to improve the yield model.

Acknowledgments

The authors acknowledge the financial support for parts of this work from the Flemish IWT under the MEDEA SADE project.

References

[1] K. Antreich, H. Graeb, C. Wieser, "Circuit analysis and optimization driven by worst-case distances," IEEE Transactions on Computer-Aided Design, Vol. 13, No. 1, pp. 57-71, January 1994.
[2] C. Borchers, L. Hedrich, E. Barke, "Equation-based behavioral model generation for nonlinear circuits," proc. IEEE/ACM DAC, pp. 236-239, 1996.
[3] T. Mukherjee, L.R. Carley, "Rapid yield estimation as a computer aid for analog circuit design," IEEE Journal of Solid-State Circuits, Vol. 26, No. 3, pp. 291-199, March 1991.
[4] F. Leyn, W. Daems, G. Gielen, W. Sansen, "A behavioral signal path modeling methodology for qualitative insight in and efficient sizing of CMOS opamps," Proc. IEEE ICCAD, pp. 374-381, Nov. 1997.
[5] K. Swings, W. Sansen, G. Gielen, "DONALD : An intelligent analog IC design system based on manipulation of design equations," Proc. IEEE CICC, pp. 8.6.1-4, May 1990.
[6] F. Leyn, G. Gielen, W. Sansen, "An efficient DC root solving algorithm with guaranteed convergence for analog integrated CMOS circuits," submitted to ICCAD 1998.
[7] G. Gielen, G. Debyser, et al., "Use of symbolic analysis in analog circuit synthesis," Proc. IEEE ISCAS, pp. 2205-2208, May 1995.
[8] J. Chen, M. Styblinski, "A systematic approach of statistical modeling and its applications to CMOS circuits," Proc. IEEE ISCAS, pp. 1805-1808, May 1993.
[9] M. Styblinski, S. Aftab, "IC variability minimization using a new Cp and Cpk based variability/performance measure," Proc. IEEE ISCAS, May-June 1994.

Efficient Handling of Operating Range and Manufacturing Line Variations in Analog Cell Synthesis

Tamal Mukherjee, *Member, IEEE*, L. Richard Carley, *Fellow, IEEE*, and Rob A. Rutenbar, *Fellow, IEEE*

***Abstract*—We describe a synthesis system that takes operating range constraints and inter and intracircuit parametric manufacturing variations into account while designing a sized and biased analog circuit. Previous approaches to computer-aided design for analog circuit synthesis have concentrated on nominal analog circuit design, and subsequent optimization of these circuits for statistical fluctuations and operating point ranges. Our approach simultaneously synthesizes and optimizes for operating and manufacturing variations by mapping the circuit design problem into an Infinite Programming problem and solving it using an annealing within annealing formulation. We present circuits designed by this integrated synthesis system, and show that they indeed meet their operating range and parametric manufacturing constraints. And finally, we show that our consideration of variations during the initial optimization-based circuit synthesis leads to better starting points for post-synthesis yield optimization than a classical nominal synthesis approach.**

I. Introduction

Although one-fifth the size of the digital integrated circuit (IC) market, mixed-signal IC's represent one of the fastest-growing segments of the semiconductor industry. This continuing integration of analog functionality into traditionally digital application-specific ICs (ASIC's) coupled with the time-to-market pressures in this segment of the semiconductor industry has demanded increasing amounts of support from design automation tools and methodologies. While digital computer-aided design tools have supported the rapid transition of design ideas into silicon for a generation of designers, the analog portion of the mixed-signal IC, although small in size, is still typically designed by hand [1], [2].

A wide range of methodologies has emerged to design these circuits [3]. These methodologies start with *circuit synthesis* followed by *circuit layout*. This paper focuses on the *circuit synthesis* portion, in which performance specifications are translated into a schematic with sized transistors, a process involving topology selection as well as device sizing and biasing. Existing approaches tend to synthesize circuits considering only a nominal operating point and a nominal process point. At best, existing approaches allow the expert synthesis tool creator to preselect specific operating and process points for performance evaluation.

Manuscript received July 22, 1998; revised August 23, 1999. This work was supported in part by the Semiconductor Research Corporation (SRC) under Contract DC 068. This paper was recommended by Associate Editor M. Sarrafzadeh.

The authors are with the Electrical and Computer Engineering Department, Carnegie Mellon University, Pittsburgh, PA 15213 USA.

Publisher Item Identifier S 0278-0070(00)06421-6.

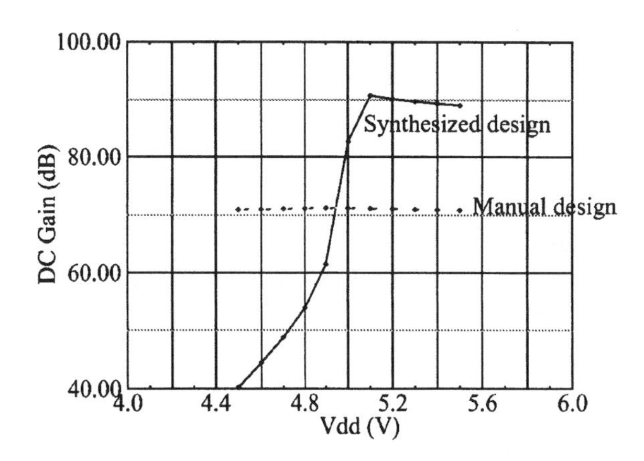

Fig. 1. Variation of dc gain of manual and synthesized design with V_{dd}.

Because ICs are sensitive to parametric fluctuations in the manufacturing process, design with a nominal set of manufacturing process parameters is insufficient. In addition, all circuits are sensitive to fluctuations in their operating conditions (e.g., power supply voltages and temperature). The importance of these variations leads to the two phase approach common in manual design: first, generate a nominal design using topologies known to be relatively tolerant of operating range and parametric manufacturing variations, then improve the design's manufacturablility using worst case methods.

An automated circuit design flow that combines analog circuit synthesis for nominal design and yield optimization for manufacturability can mimic this manual approach. The goal of analog circuit synthesis tools is to decrease design time. The goal of yield optimization tools is to improve the manufacturability of an already well-designed circuit. While both sets of tools are aimed at helping the designer, they both solve half the problem. Synthesis tools often create designs that are at the edge of the performance space, whereas a good human designer using a familiar topology can conservatively over-design to ensure adequate yield. We have empirically observed that yield optimization tools can improve the yields of good manual designs, but automatically synthesized circuits are often a bad starting point for gradient-based post-synthesis yield optimization.

Two examples make these issues concrete. Fig. 1 shows the impact of variation in a single environmental operating point variable on a small CMOS amplifier design. The variable is the supply voltage, V_{dd}, which has a nominal value of 5.0 V. Fig. 1 plots the dc gain performance over varying supply voltage for

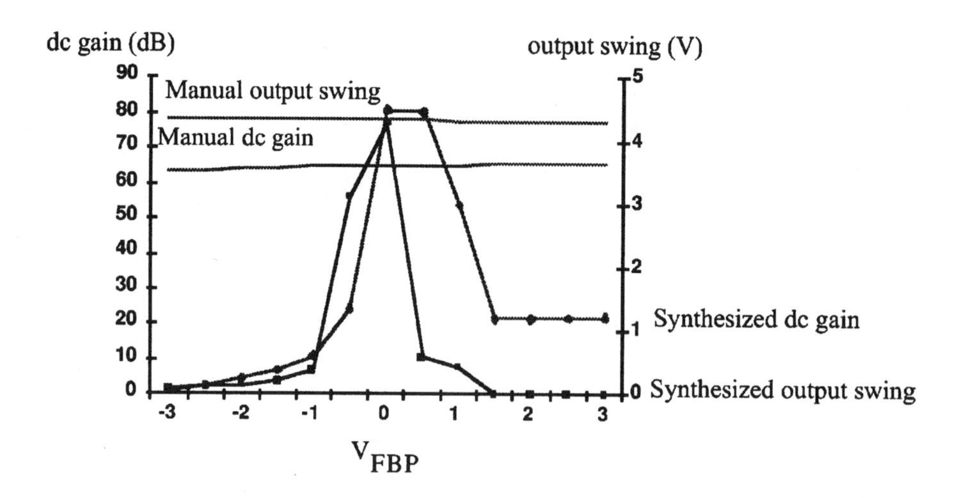

Fig. 2. Effect of manufacturing variation on circuit performance.

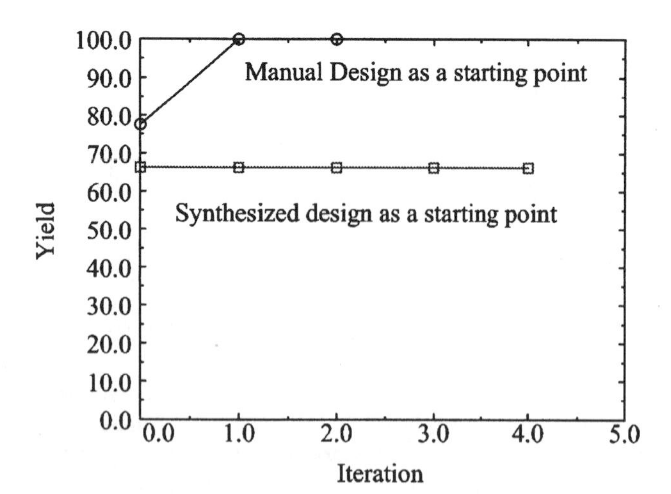

Fig. 3. Post-design yield optimization on manual and nominal designs.

two designs: one done manually, one done using synthesis [4]. Both designs meet the nominal specification of 70 dB of gain at 5.0 V, but the synthetic design shows large sensitivity to even small changes in V_{dd}. (One reason in this particular design is too many devices biased too close to the edge of saturation.) Fig. 2 shows the impact of manufacturing variation on a second small CMOS circuit. We consider here statistical variation in just one parameter: V_{FBP}, the PMOS flat-band voltage, ranging over $\pm 3\sigma$ variation. Fig. 2 plots two performance parameters, dc gain (on left y-axis) and output swing (on right y-axis), and also shows curves for both a manual, and a synthesized [4] design. Again, both designs meet nominal specifications in the presence of no V_{FBP} variation, but the synthesized design shows large sensitivities to any variation. Fig. 3 illustrates an attempt to use post-design statistical yield maximization on the two designs from Fig. 2. The figure plots circuit yield, determined via Monte Carlo analysis, after each iteration of a gradient-based yield optimization algorithm [5]. It can be seen that the manual design ramps quickly from 75% to near 100% estimated yield. However, the synthetic design makes no progress at all; its yield gradients are all essentially zero, locating this design in an inescapable local minimum.

Our goal in this paper is to address this specific problem: how to avoid nominal analog synthesis solutions that are so sensitive to operating/manufacturing variation that they defeat subsequent optimization attempts. To do this, we add simplified models of operating point and manufacturing variation to the nominal synthesis process. It is worth noting explicitly that we do not propose to obsolete post-design statistical optimization techniques. Our aim is not to create, in a single numerical search, a "perfect" circuit which meets all specifications and is immune to all variation. Rather, the goal is the practical one of *managing* the sensitivity to variation of these synthesized circuits, so that they are compatible with existing post-design yield optimization.

We achieve this goal by developing a unified framework in which interchip and intrachip parametric fluctuations in the fabrication process, environmental operating point fluctuations, and the basic numerical search process of sizing and biasing an analog cell are all dealt with concurrently. We show how the problem can be formulated by adapting ideas from Infinite Programming [6], and solved practically for analog cells using simulated annealing [7] for the required nonlinear optimizations. An early version of these ideas appeared in [8]. As far as we are aware, this is the first full circuit synthesis formulation which treats both operating and manufacturing variations, and the first application of infinite programming to this synthesis task.

The remainder of the paper is organized as follows. Section II reviews prior analog circuit synthesis approaches, with emphasis on the nominal synthesis approach that is our own numerical starting point. Section III shows how operating and manufacturing variations can be simplified so as to allow a unified treatment of these as constraints using a basic infinite programming formulation. Section IV introduces a highly simplified, idealized infinite programming algorithm, which allows us to highlight critical points of concern in problems of this type. Section V then develops a practical algorithm for application to analog synthesis, and also highlights the heuristics we employ when the assumptions that underly the idealized infinite programming algorithm are no longer valid. Section VI offers a variety of experimental synthesis results from a working implementation of these ideas. Finally, Section VII offers concluding remarks.

II. Review of Analog Circuit Synthesis

The advent of computer-based circuit simulation led to the first studies of computer-based circuit design, e.g., [9]. Since then, a variety of circuit synthesis approaches have been developed that range from solving both the topology selection and device sizing/biasing problems simultaneously [10] to solving them in tandem; from using circuit simulators for evaluating circuit performance [11], to behavioral equations predicting circuit performance [12], [13]; from searching the design space with optimization [14], [15], to using a set of inverted behavioral equations with a restricted search space [1]. See [3] for a recent survey of analog circuit synthesis.

The system presented in this paper is based on ASTRX/OBLX [4], which first translates the design problem into a cost function whose minimum is carefully crafted to be the best solution to the design problem, and then minimizes this cost function to synthesize the required circuit. It has been successful at designing the widest range of analog circuits to date [4], [16]–[18]. The key ideas in the ASTRX/OBLX synthesis strategy are summarized below:

- **Automatic Compilation:** ASTRX compiles the input circuit design into a performance prediction module that maps the circuit component parameters (such as device lengths and widths) to the performance metrics (such as the circuit's dc gain) specified by the designer. If the designer specifies an equation for the circuit performance metric as a function of the component parameters, the mapping is trivial. For the remaining performance metrics, ASTRX links to the device modeling and circuit evaluation approaches below to create a cost function, or single metric that indicates how well the current component parameters can meet the circuit design goals and specifications.
- **Synthesis via Optimization:** The circuit synthesis problem is mapped onto a constrained optimization formulation that is solved in an unconstrained fashion, as will be detailed below.
- **Device Modeling via Encapsulation:** A compiled database of industrial models for active devices is used to provide the accuracy of a general-purpose circuit simulator, while making the synthesis tool independent of low-level modeling concerns [4], [19]. The results presented in this paper use the BSIM1 model from this library.
- **Circuit Evaluation by Model Order Reduction:** Asymptotic waveform evaluation (AWE) [20] is augmented with some simple, automatically generated analytical analyses to convert AWE transfer functions into circuit performances. AWE is a robust, efficient approach to analysis of arbitrary linear RLC circuits that for many applications is several orders of magnitude faster than SPICE. The circuit being synthesized is linearized, and AWE is used for all linear circuit performance evaluation.
- **Efficiency via Relaxed-dc numerical formulation:** Efficiency is gained by avoiding a CPU-intensive dc operating point solution after each perturbation of the circuit design variables [4], [19]. Since encapsulated models must be treated numerically, as in circuit simulation, an iterative algorithm such as Newton Raphson is required to solve for the nodal voltages required to ensure the circuit obeys Kirchhoff's Current Law. For synthesis, Kirchhoff's Current Law is implicitly solved as an optimization goal by adding the list of nodal variables to the list of circuit design variables. Just as optimization goals are formulated, such as meeting gain or bandwidth constraints, now dc-correctness is formulated as yet another goal that needs to be met.
- **Solution by Simulated Annealing:** The optimization engine which drives the search for a circuit solution is simulated annealing [7]; it provides robustness and the potential for global optimization in the face of many local minima. Because annealing incorporates controlled *hill-climbing* it can escape local minima and is essentially starting-point independent. Annealing also has other appealing properties including: the inherent robustness of the algorithm in the face of discontinuous cost functions, and the ability to optimize without derivatives, both of which are taken advantage of later in this paper.

Two of these ideas are particularly pertinent to our implementation (which will be described in Section V), and are reviewed in more detail. They include the optimization formulation, and the use of simulated annealing for solving the optimization problem. Our work extends this *synthesis-via-optimization* formulation. In ASTRX/OBLX, the synthesis problem is mapped onto a constrained optimization formulation that is solved in an unconstrained fashion. As in [10], [11], and [14], the circuit design problem is mapped to the nonlinear constrained optimization problem (NLP) of (1), where is $\underline{u}$ the vector of independent variables—geometries of semiconductor devices or values of passive circuit components—that are changed to determine circuit performance; $\underline{f}(\underline{u})$ is a set of objective functions that codify performance specifications the designer wishes to optimize, e.g., area or power; and $\underline{g}(\underline{u}) \leq 0$ is a set of constraint functions that codify specifications that must be beyond a specific goal, e.g., 60 dB-Gain$(\underline{u}) \leq 0$. Scalar weights, w_i, balance competing objectives. The decision variables can be described as a set $\underline{u} \in U_P$, where U_P is the set of allowable values for $\underline{u}$

$$\left.\begin{array}{lc} \min_{\underline{u}} & z = \sum_{i=1}^{k} w_i \cdot f_i(\underline{u}) \\ \text{s.t.} & \underline{g}(\underline{u}) \leq 0 \\ & \underline{u} \in U_P \end{array}\right\} \text{(NLP)}. \tag{1}$$

To allow the use of simulated annealing, in ASTRX/OBLX this constrained optimization problem is converted to an unconstrained optimization problem with the use of additional scalar weights (assuming l constraint functions)

$$C(\underline{u}) = \sum_{i=1}^{k} w_i f_i(\underline{u}) + \sum_{i=1}^{l} w_i g_i(\underline{u}). \tag{2}$$

As a result, the goal becomes minimization of a scalar cost function, $C(\underline{u})$, defined by (2). The key to this formulation is that the minimum of $C(\underline{u})$ corresponds to the circuit design that best matches the given specifications. Thus, the synthesis task is divided into two subtasks: evaluating $C(\underline{u})$ and searching for its minimum.

In the next section, we will revisit the general NLP optimization formulation of (1) with the aim of extending it to handle operating range and manufacturing line variations. Then, in Section V we will revisit the unconstrained formulation of (2) used in ASTRX/OBLX, when we discuss our annealing-based implementation.

III. Infinite Programming Approach to Analog Circuit Synthesis

In this section we will expand the nonlinear constrained optimization formulation in ASTRX/OBLX to a nonlinear infinite programming formulation that considers operating range and parametric manufacturing variations. Our goal for this formulation is a complete model of the design problem, thereby solving the operating point and manufacturing variations in a unified manner. A complete model is required in order to use an optimization-based synthesis algorithm; partially modeled problems ignore practical constraints, hence, they let the optimization have freedom to search in areas that lead to impractical designs. The resulting unified formulation simultaneously synthesizes a circuit while handling the two independent variations.

IC designers often guess worst case range limits from experience for use during the initial design phase. Once they have generated a sized schematic for the nominal operating point (i.e., bias, temperature) and process parameters, they test their design across a wide range of process and operating points (*worst case set*). Typically, the initial design meets specifications at several worst case points, but needs to be modified to ensure that the specifications are met at the rest of the worst case points. We will now first consider the types of worst case points found in circuits before proposing a representation that mathematically models them.

Let us begin with a typical example of an environmental operating point specification: the circuit power supply voltage, V_{dd}. Most designs have a $\pm 10\%$ range specification on the power supply voltage, leading to an operating range $V_{dd} \in [4.5, 5.5]$ V in a 5-V process. Graphs similar to the one shown in Fig. 1 for the performances in many analog circuits (from simulation, as well as from data books) show the following.

- Low V_{dd} designs are the ones most likely to fail to meet the performance specifications, so there is a need to consider the additional worst case point of $V_{dd} = 4.5$ V in the mathematical program used to design the circuit.
- Not all of the performance parameters are monotonic functions of the operating point. Therefore, a mechanism to find the worst case point in the operating range for each performance function is needed. Even when the performance parameters are monotonic functions of the operating point, a mechanism to determine the worst case corner is needed since this corner may not be *a priori* known.

To investigate the effect of introducing operating ranges to the NLP model of (1), let us consider the example of dc gain, *a circuit performance metric*, and the power supply voltage range, an *operating range*: the dc gain needs to be larger than 60 dB for every power supply voltage value in the range $4.5 \leq V_{dd} \leq 5.5$. This can be written as in (1), where $\underline{u}$ is the vector of designable parameters, and ξ is a new variable (in this example only a scalar), to represent the operating range

$$60 \text{ dB} - \text{Gain}(\underline{u}, \xi) \leq 0 \quad \forall \xi \in [4.5, 5.5]\text{V} \tag{3}$$

where ξ is considered to be a *range variable*.

Since every single voltage in the given range needs to be investigated, this single operating range constraint adds an *infinite* number of constraints to the mathematical program. Heittich, Fiacco, and Kortanek present several papers in [6], [21] which discuss nonlinear optimization problems where some constraints need to hold for a given range of an operating range variable. These problems, and the one just presented, are called *semi-infinite programs* due to their finite number of objectives and infinite number of constraints. When there is an infinite number of objective functions (due to the presence of a range variable in the objective function), the mathematical program is called an *infinite program*. The complete mathematical program can now be re-written as the *nonlinear infinite program* (NLIP) shown in (4) where Ξ is the vector set of operating point ranges and statistical manufacturing fluctuations

$$\left.\begin{array}{ll} \min_{\underline{u}} & z = \sum_i^k w_i \cdot f(\underline{u}, \underline{\xi}) \\ \text{s.t.} & \left(\begin{array}{c} \underline{g}(\underline{u}, \underline{\xi}) \leq 0 \\ \underline{u} \in U_P \\ \forall \underline{\xi} \in \Xi \end{array}\right. \end{array}\right\} \text{(NLIP).} \tag{4}$$

Equation (4) reduces into the NLP formulation of (1) if the range variables are considered to be fixed at their nominal values (this is how ASTRX/OBLX solves the nominal synthesis problem).

So, we can see that environmental operating point variables that can vary over a continuous range can be incorporated directly into this nonlinear infinite programming formulation. However, for statistical fluctuations, such as those characterizing the manufacturing process, this formulation cannot be used directly. We need some suitable transformation of these statistical variations to treat them as in (4).

Consider how designers empirically deal with statistical process variations in manual circuit design. Once the circuit meets the performance specifications at a few token worst case points, the designer begins improving the circuit's yield by statistical design techniques [22]–[25]. Some of these techniques also include operating ranges [26], [27]. Such yield maximization algorithms can be used to determine the best design for a given set of specifications and a joint probability distribution of the fluctuations in the fabrication line. Our unified formulation attacks the problem of manufacturing line fluctuations, but in a different way than the yield maximization approach. Specifically, we convert the statistical variation problem into a *range problem*. This critical simplification allows us to treat both operating range variation and statistical manufacturing variation in a unified fashion.

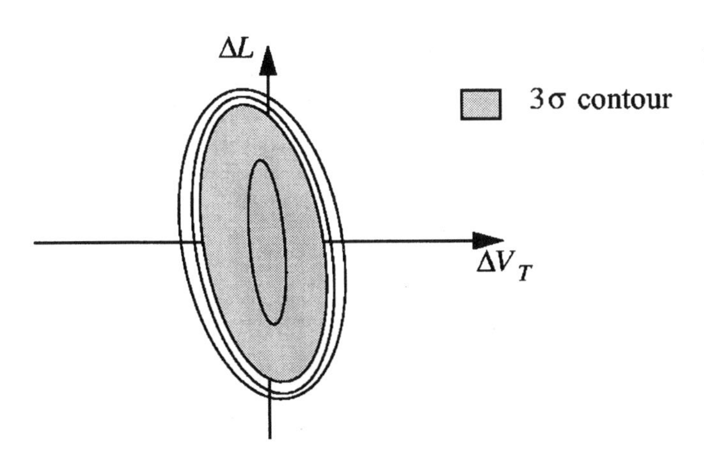

Fig. 4. Contours of jpdf of fabrication line model.

It has been shown that relatively few device model variables capture the bulk of the variations in a MOS manufacturing line [28], [29], with the most significant ones being threshold voltage, oxide thickness and the length and width lithographic variations. In practice, the oxide thickness variation is very tightly controlled in modern processes. In addition, for analog cells, the devices widths are much larger than the lengths. Therefore, the circuit performances will be more sensitive to length variations than to width variations. Therefore, two sources of variation, intrachip threshold voltage variation, ΔV_T and intrachip length reduction variation, ΔL dominate most analog designs. ΔV_T is the difference in the threshold voltage between two identical devices on the same chip. ΔL is the difference in the lengths of the identical devices on the same chip. This limited number of range variables is crucial to the run times of our approach, as we shall see later. To explain our modeling approach, let us consider an example where the remaining parameters do not vary.

What we want is a model of the fabrication line that allows us to determine sensible ranges for the constraints on ΔV_T and ΔL. While a simple strategy is to use the wafer acceptance specifications provided by the process engineers, we will consider a joint-probability distribution function (jpdf) that a fabricated chip has an intrachip threshold voltage difference of ΔV_T and an intrachip length difference of ΔL: $p(\Delta V_T, \Delta L)$. This will help in comparing and contrasting with the approaches previously used for yield optimization. This jpdf can be determined using measurements from test-circuits fabricated in the manufacturing line or via statistical process and device simulation [30]. We can draw the contours of $p(\Delta V_T, \Delta L)$ as shown in Fig. 4 with the 3σ contour highlighted. In our model of the fabrication line variations, we will treat this contour in exactly the same way as a range variable. We, therefore, specify that the circuit being designed should work for every possible value of ΔV_T and ΔL within the 3σ contour. More exactly, we are constraining the yield to be at least 3σ.

Stepping back from this specific example to consider the general approach, we see that this is *not* a statistical IC design formulation of the synthesis problem. Instead it is a unified formulation that converts the problem of manufacturing line variations into the same numerical form as the operating range variations. We do this by converting these statistical fluctuations into a range variable which spans a *sufficiently probable* range of these statistical distributions. Our formulation in (4) reduces to the fixed tolerance problem (FTP) of [31] if we only consider only the design objectives (and ignore the design constraints), and if we replace each objective with the maximum in the range space. An alternative approach, followed by [32] considers the integral of the objective over the jpdf (a metric for yield) by discretizing it into a summation, and approximating the continuous jpdf with discrete probabilities. Such an approach more correctly weights the objective with the correct probabilities of the range variable capturing the manufacturing variation, but prevents the development of the unified approach. A unified approach is pursued in this work to let the optimization simultaneously handle all the constraints, thereby having a full model of the design problem. However, the price here is the simplified range model we must use for statistical variation.

We have shown how we can represent both operating range variation and (with some simplification) statistical manufacturing variation as constraints in a nonlinear infinite programming formulation for circuit synthesis. The next problem is how we might numerically solve such a formulation. This is discussed next.

IV. A Conceptual Infinite Programming Algorithm

In this section, we will review the solution of a simple nonlinear infinite program. We begin with a simplified form of the nonlinear infinite program to illustrate all the critical points of concern. Only a single design variable u, and a single range variable ξ are considered. Also, only a single objective, and a single constraint are considered, and functions f and g are assumed to be convex and continuous and the set Ξ is assumed to be convex. So, the problem is

$$\begin{array}{lll} \min_{u \in U_P} & f(u, \xi) & \forall \xi \in \Xi \\ \text{subject to} & g(u, \xi) \leq 0 & \forall \xi \in \Xi. \end{array} \tag{5}$$

In this problem, a point u is feasible if $g(u, \xi)$ (as a function of ξ) is less or equal to zero on the *whole* of Ξ. If Ξ is a finite set, then $g(u, \xi) \leq 0$, $\xi \in \Xi$, represents a finite number of inequalities, and $f(u, \xi)$ represents a finite number of objectives, the problem reduces to a common multiobjective nonlinear programming problem. On the other hand, if Ξ is an infinite set (as in the case of a continuous variable ξ), this problem is considered to be an *infinite program*.

Three existing approaches to solving this problem are now outlined. The first approach uses nondifferentiable optimization, while the second and third both use some discretizing approximation of the initial infinite set Ξ, based on gridding and cutting planes.

In the first approach, the *for all* term in the objective and constraint are re-written into a nondifferentiable optimization problem. The infinite number of objectives can be replaced with a single objective function as determined by the problem $\max_{\xi \in \Xi} f(u, \xi)$. The objective now minimizes, in u-space, the worst case value of the objective function (which is in ξ-space). This approach is akin to the one followed for solving multiobjective optimization problems. Similarly, the infinite number of constraints can be replaced by a single constraint function as determined by the problem $\max_{\xi \in \Xi} g(u, \xi)$. Note that the worst case value for $g(u, \xi)$ is the maximum value in ξ-space due to the direction of the inequality (a greater-than inequality would lead to a minimization in ξ-space). In essence, the infinite constraints have been replaced by a single worst case constraint. Thus, the problem in (5) is equivalent to

$$\begin{array}{ll} \min_{u \in U_P} & \max_{\xi \in \Xi} f(u, \xi) \\ \text{subject to} & \max_{\xi \in \Xi} g(u, \xi) \le 0 \end{array}. \quad (6)$$

The resulting objective and constraint are both no longer differentiable and can be solved by applying methods of nondifferentiable optimization [33].

A second approach to solving the original problem, (5), is to overcome the infinite nature of the set Ξ by discretizing it to create a finite, approximate set via gridding. The simplest approach is to create a fine grid and, therefore, a large but finite number of objectives and constraints, and solve for the optimum using standard nonlinear programming approaches. A more elegant method is to create a coarse grid and solve a nonlinear program with a few constraints and objectives. Then, the information gained from this solution can be used to further refine the grid in the region of ξ-space where this is necessary [34].

Our solution approach is a derivative of the cutting plane-based discretization originally proposed by [35]. The starting point is, for simplicity, a semi-infinite program, as shown in (7). In addition to the assumptions leading to (5), only a single (finite) objective is considered. We call this *Problem P*, and will refer back to this as we develop the conceptual algorithm

$$P: \quad \begin{array}{ll} \min_{\underline{u}} & f(\underline{u}) \\ \text{s.t.} & g(\underline{u}, \underline{\xi}) \le \underline{\xi} \in \Xi \\ & \underline{u} \in U_P \end{array} . \quad (7)$$

A more detailed discussion of the cutting plane-based discretization is presented to introduce notation needed for the annealing-within-annealing formulation that will be described in Section V. The infinite set Ξ is replaced with a worst case constraint. Since we are trying to ensure that $g(\underline{u}, \underline{\xi}) \le 0 \; \forall \xi \in \Xi$, the *worst case* or *critical point* is the maximum value of g with respect to $\underline{\xi}$, as shown in Fig. 5. If the constraint value at the critical point is less than zero, then all points in $\underline{\xi}$-space will be less than zero, and the constraint due to the critical point is satisfied for the entire infinite set. Using this observation, we now break the infinite programming problem into two subproblems.

Let $I(\underline{u})$ denote the problem

$$I(\underline{u}): \quad \max_{\underline{\xi} \in \Xi} g(\underline{u}, \underline{\xi}). \quad (8)$$

A simple conceptual algorithm begins with a $\underline{\xi}_0 \in \Xi$ and solves

$$P_0: \quad \begin{array}{ll} \min_{\underline{u}} & f(\underline{u}) \\ \text{s.t.} & g(\underline{u}, \underline{\xi}_i) \le 0 \quad i = 0 \\ & \underline{u} \in U_P \end{array} \quad (9)$$

to obtain $\underline{u}_0$. Then, $\underline{\xi}_1$ is computed by solving $I(\underline{u}_0)$. If $g(\underline{u}_0, \underline{\xi}_i) \le 0$ for $i = 0, 1$, we are done. Note that since $\underline{u}_0$ was computed with the constraint $g(\underline{u}_0, \underline{\xi}_0) \le 0$, this constraint is trivially satisfied. Also, the critical point for $\underline{u} = \underline{u}_0$ is at $\underline{\xi}_1$,

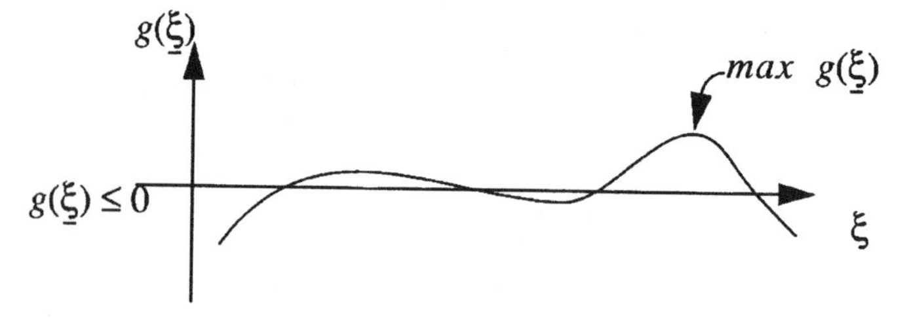

Fig. 5. The worst case point.

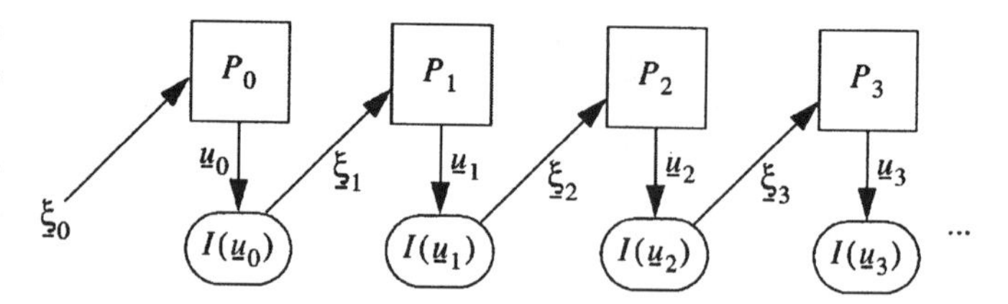

Fig. 6. Flow of conceptual cutting plane algorithm.

and if we meet the constraint at the critical point $\underline{\xi}_1$, we meet it within the entire range Ξ. So, if the inequality constraint $g(\underline{u}_0, \underline{\xi}_1) \le 0$ is met, we have met the constraint, and are at the minimum value of the objective, implying we have solved the optimization problem, P. If $g(\underline{u}_0, \underline{\xi}_1) > 0$ then we have not met the constraint, and we have to go on to solve another subproblem

$$P_1: \quad \begin{array}{ll} \min_{\underline{u}} & f(\underline{u}) \\ \text{s.t.} & g(\underline{u}, \underline{\xi}_i) \le 0 \quad i = 0, 1 \\ & \underline{u} \in U_P \end{array} \quad (10)$$

for $\underline{u}_1$, etc. The reason we consider both of the critical points $\underline{\xi}_0$ and $\underline{\xi}_1$ in P_1 is to ensure eventual convergence. The subscript in the problem label indicates the iteration number. We can generalize this sequence of problems, labeled as P_ρ, in which there are only $\rho + 1$ constraints; they can be written as

$$P_\rho: \quad \begin{array}{ll} \min_{\underline{u}} & f(\underline{u}) \\ \text{s.t.} & g(\underline{u}, \underline{\xi}_i) \le 0 \quad i = 0, 1, \cdots, \rho. \\ & \underline{u} \in U_P \end{array} \quad (11)$$

This sequence is interspersed with the problems $I(\underline{u}_i)$ with $i = 0, 1, \cdots, \rho$. The flow of this conceptual algorithm is, therefore, a sequence of two alternating optimization problems as shown in Fig. 6. Note that P_ρ only has a finite number of constraints. Therefore, we have been able to convert the semi-infinite programming problem into a sequence of finite nonlinear optimization problems.

Each optimization, $I(\underline{u}_i)$, adds a critical point, $\underline{\xi}_i$, resulting in a cutting plane $g(\underline{u}, \underline{\xi}_i) \le 0$, which reduces the feasible design space to that of Problem P, at which point the algorithm converges to determine $\underline{u}^*$

$$f(\underline{u}_0) \le f(\underline{u}_1) \le \cdots \le f(\underline{u}_\rho) \le \cdots \le f(\underline{u}^*) \quad (12)$$

where $\underline{u}^*$ is any solution to Problem P and $f(\underline{u}_i$ is the objective value of the problem P_i.

This overall approach, alternating an *outer* optimization which drives the basic design variables, with an *inner* optimization, which adds critical points to $\widehat{\Xi}$ approximating the infinite

```
ρ = 0, ξ_0 = nominal value, Ξ = {ξ_0}
repeat
        Solve outer optimization P_ρ to obtain u_ρ
        for i = 1...l
                Solve inner optimization I_i(u_ρ) to obtain critical point ξ_i
                if ξ_i ∉ Ξ then
                        add ξ_i to Ξ̂
                        ρ = ρ + 1
until g_j(u_ρ, ξ_i) ≤ 0, i = 0, ..., ρ, j = 1, ..., l
```

Fig. 7. Complete conceptual algorithm for solving NLIP.

set Ξ. The above conceptual solution can be generalized to handle more range variables and constraints, as well as to an objective that is a function of range variables. Pseudocode for the general algorithm appears in Fig. 7 where the set $\widehat{\Xi}$ contains all the critical points determined by prior inner optimizations, $I_i(\underline{u}_\rho)$ is identically (8) with the subscript i since there are now l constraints, g_i, $i = 1 \cdots l$ and problem P_ρ is a generalization of (11) for multiple objective and constraint functions, which, for completeness, is

$$\begin{aligned} \min_{\underline{u} \in U_P} \quad & z = \sum_i^k w_i \cdot \sum_j^\rho f(\underline{u}, \underline{\xi}_j) \\ \text{s.t.} \quad & g(\underline{u}, \underline{\xi}_j) \leq 0 \quad j = 0, \cdots, \rho. \end{aligned} \tag{13}$$

One of the problems of using this approach, like any other discretization approach, is that as ρ increases, the number of constraints in Problem P_ρ increases, therefore, we need to consider removing or pruning constraints for efficiency. This is often referred to as *constraint-dropping*. Eaves and Zangwill, in [35] show that under assumptions of convexity of the objective and constraint functions f and g, the constraint $g(\underline{u}, \underline{\xi}_i) \leq 0$ can be dropped from Problem P_ρ if:

- for the current solution $\underline{u}_{\rho-1}$, $f(\underline{u}_{\rho-1})$, is larger than $f(\underline{u}_i)$;
- $g(\underline{u}_{\rho-1}, \xi_i) \leq 0$;
- the next solution $\underline{u}_\rho$ satisfies $f(\underline{u}_\rho) \geq f(\underline{u}_{\rho-1})$.

Intuitively, $f(\underline{u}_i) < f(\underline{u}_{\rho-1})$ implies that an operating point other than $\underline{\xi}_i$ is contributing more significantly to the objective than $\underline{\xi}_i$ which had led to the optimum $\underline{u}_i$. Also, $g(\underline{u}_{\rho-1}, \xi_i) \leq 0$ means that the problem is feasible with the $\underline{\xi}_i$ critical point. Finally $f(\underline{u}_\rho) \geq f(\underline{u}_{\rho-1})$ ensures that the feasible set F_ρ is a subset of $F_{\rho-1}$, satisfying (12). This constraint-dropping scheme suggests that the growth of complexity of Problem P_ρ can be slowed if the objective is nondecreasing.

This study of the conceptual algorithm has shown that a solution to the nonlinear infinite program is possible using a sequence of alternating subproblems. These subproblems are finite and can be solved by traditional optimization techniques. We have also seen that in the case of nondecreasing $\{f(\underline{u}_\rho)\}$, the algorithm will converge, and constraint dropping may be used to improve the computational complexity of the algorithm.

Unfortunately, in our circuit synthesis application, the assumptions of convexity of either the objective function or the feasibility region do not hold, thus, the convergence results such as those in [6], [21], and [35] also cannot be guaranteed here. Nevertheless, we can still employ the basic algorithm of Fig. 7 as an empirical framework, if we can find practical methods to solve the individual inner and outer nonlinear optimization problems created as this method progresses. We discuss how to do this next.

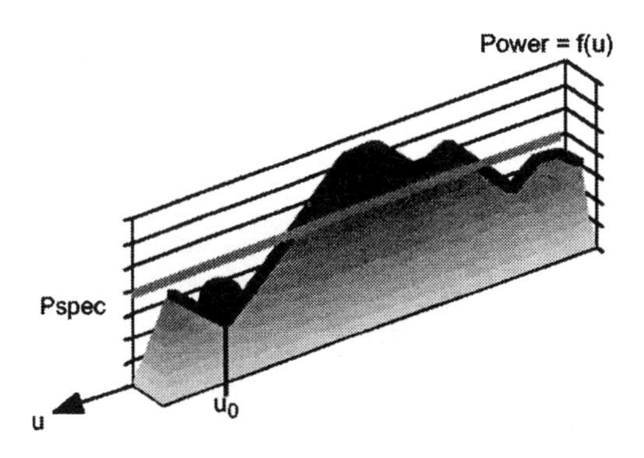

Fig. 8. The nominal cost function.

V. Infinite Programming in ASTRX/OBLX

In this section we extend the nominal synthesis formulation in ASTRX/OBLX to a nonlinear infinite programming formulation. We first outline the approach using an illustrative example. In the domain of circuit synthesis, complicated combinations of nonlinear device characteristics make it extremely difficult, if not impossible, to make any statements about convexity. Hence, like in ASTRX/OBLX, we use a simulated annealing optimization method, and exploit its hill-climbing abilities in the solution of the nonlinear infinite programming problem. Our basic approach is to use a sequence of optimization subproblems as described in Fig. 7.

For purposes of illustration, let us consider the example of (14), with one (infinite) objective function which is power$(u, V_{dd})\ \forall V_{dd} \in [4.5\,\text{V}, 5.5\,\text{V}]$, one (infinite) constraint which is power$(u, V_{dd} - P\text{spec} \leq 0\ \forall V_{dd} \in [4.5\,\text{V}, 5.5\,\text{V}]$, a single design variable, and a single range variable. For simplicity we assume a single design variable (u). So, our problem is formally:

$$\begin{aligned} \min_u \quad & \text{power}(u, V_{dd}) & \forall V_{dd} \in [4.5, 5.5] \\ \text{s.t.} \quad & \text{power}(u, V_{dd}) - P\text{spec} \leq 0 & \forall V_{dd} \in [4.5, 5.5] \\ & u \in [U_{\min}, U_{\max}]. \end{aligned} \tag{14}$$

In our conceptual algorithm, we started off with $\underline{\xi}_0 \in \Xi$, which in the domain of circuit synthesis maps to the nominal operating point and device parameters for the manufacturing process (in this example $V_{dd} = 5.0$ V). The first step (solving P_0) maps to the nominal synthesis problem, since it ignores all the other values of V_{dd} except for $V_{dd} = 5.0$ V. Graphically, we can depict this problem as in Fig. 8, with the optimal solution u_0.

Including the range variable leads to the cost function in Fig. 9, where an additional V_{dd} dimension for the *for all* constraint is shown. The next step in the sequence of alternating optimization subproblems requires us to periodically freeze the circuit design variables u, thus stopping ASTRX/OBLX, and go solve a new optimization problem to determine the worst case value of the range variable. This is the first inner

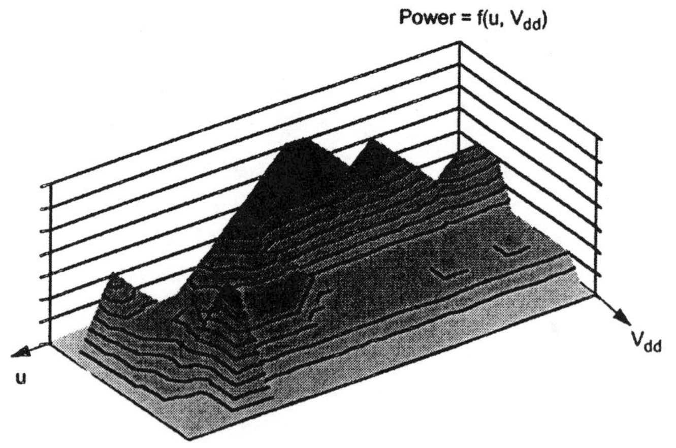

Fig. 9. The meaning of $\forall V_{dd} \in [4.5, 5.5]$.

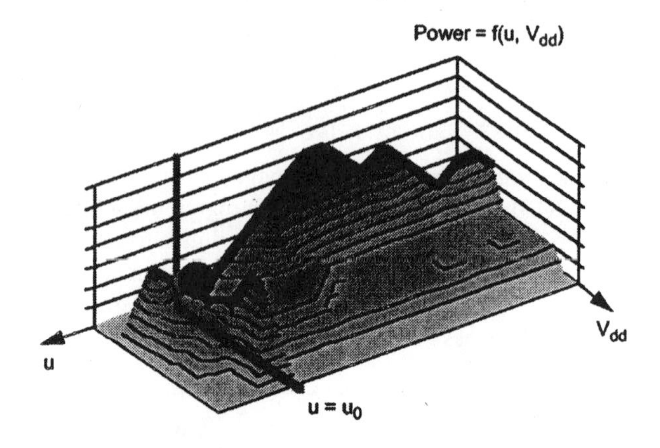

Fig. 10. Inner optimization cost function.

optimization problem, $I(u_0)$ whose cost function is described in Fig. 10. The original outer optimization problem, P_0, had found the minimum u_0. We freeze the design variables at $u = u_0$ and take the slice of power($u = u_0$, V_{dd}) as the cost function for the inner optimization. Since we want to ensure that power $\leq$ Pspec, this corresponds to a maximization problem. If the maximum value of power in the power($u = u_0$, V_{dd}) cost function meets the specification, then all points across the V_{dd} axis will meet the specification. Thus, the responsibility of the first inner optimization is to find the critical point, V_{dd}^* as in Fig. 11.

In the conceptual algorithm, the link back to the second outer optimization involves adding another constraint to the finite problem P_0 to create another finite optimization problem P_1. Since ASTRX/OBLX solves the constrained optimization problem by converting it to an unconstrained optimization problem, the addition of V_{dd}^* is handled by altering the ASTRX/OBLX cost function. The effect of adding this information must modify the cost function of the outer optimization in such a way to prevent it from searching around the areas that are optimal in the *nominal* sense, but not in the *for all* sense implied by the range variables. Fig. 12 shows this effect of adding the result of the inner optimization to the outer optimization cost function for our simple example about circuit power. The effect of this critical point is to make the original outer optimization result (u_0) suboptimal. Thus, we need to re-run the outer optimization on this new cost function. If the

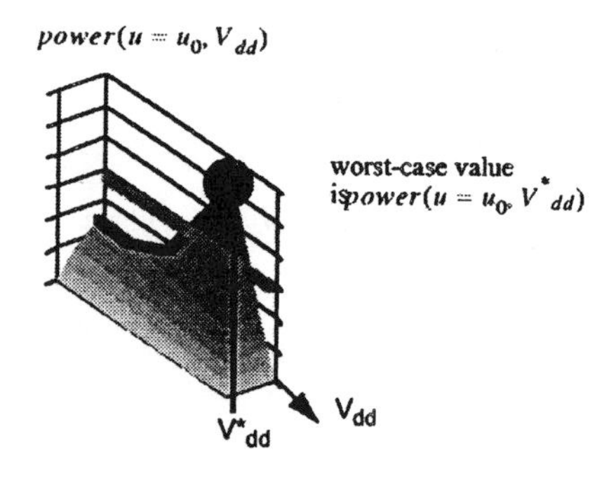

Fig. 11. Solution of the inner optimization.

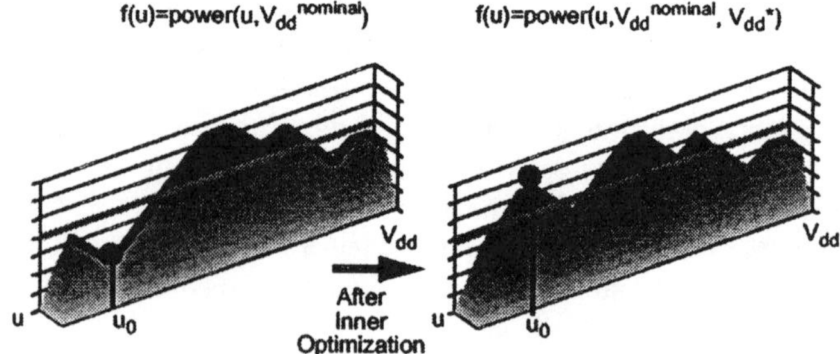

Fig. 12. Effect of inner optimization on outer optimization cost function.

inner optimization does not add any new critical points, then the outer optimization cost function remains the same, hence, no further outer optimizations are necessary.

We generate the second outer optimization cost function shown on the right graph of Fig. 12 using the critical points collected so far: V_{dd}^{nominal} and V_{dd}^*. This cost function can be written as

$$\begin{aligned} C(u) = \max\left\{\text{power}\left(u,\, V_{dd}^{\text{nominal}}\right),\, \text{power}\left(u,\, V_{dd}^*\right)\right\} \\ + \max\left\{0,\, \text{power}(u,\, V_{dd}^*)\right\} \\ + \max\left\{0,\, \text{power}\left(u,\, V_{dd}^{\text{nominal}}\right)\right\}. \end{aligned} \tag{15}$$

The first term in the outer optimization cost function relates to the discretization of the infinite objective. In the nominal case, we were trying to minimize power(u), and now we are minimizing the worst case value of the power at the discretized points. The second and third terms relate to the discretization of the infinite constraint. In the nominal cost function we have only one term for the constraint; now we have a term for every discretized value of V_{dd}. Solving this cost function, in terms of ASTRX/OBLX, is like designing a single circuit that is simultaneously working at two different operating points, V_{dd}^{nominal} and V_{dd}^*. Equation (15) reduces to the nominal synthesis formulation in ASTRX/OBLX described in (2) if we assume that the only possible value V_{dd}^* can take is V_{dd}^{nominal}.

Now let us consider implementing the inner/outer optimization approach in the ASTRX/OBLX framework. First, let us consider the direct approach. OBLX currently solves the problem P_0, and can easily be extended to solve the outer optimization problems P_i. Each OBLX run takes from a few minutes to a few hours, making it prohibitive to consider this alternative. A possible second approach would be to iterate on the range space ξ during every iteration of the design space u. In other words, after each perturbation in the current

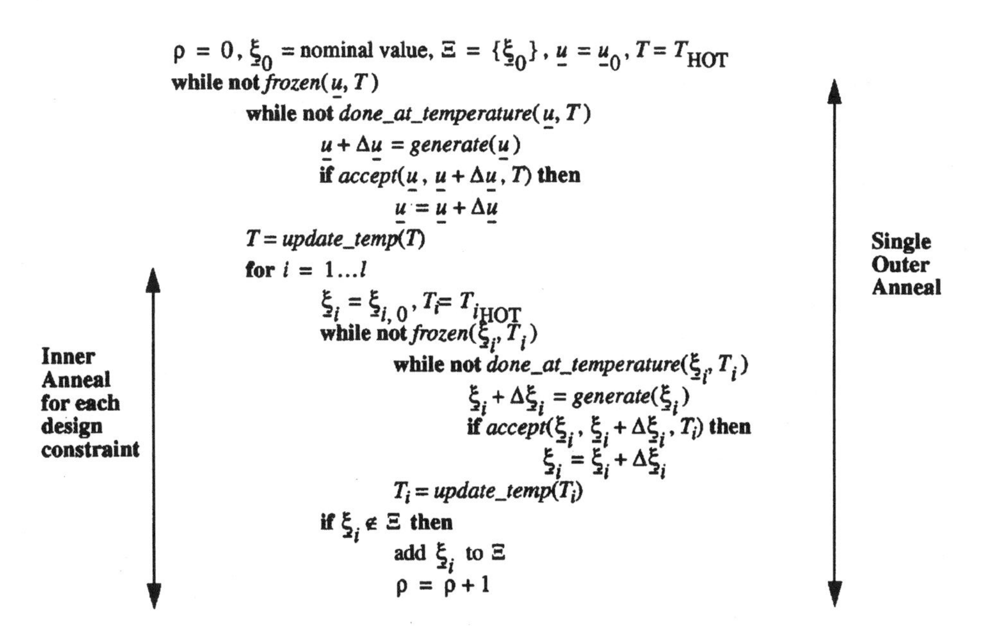

Fig. 13. Annealing-within-annealing algorithm.

ASTRX/OBLX annealing-based approach, the inner optimizations for each of the l constraints, $I_i(\underline{u}_\rho)$, $i = 1 \cdots l$, will be launched to update the list of critical points. Such a fine grained interaction between the inner and outer optimizations is also not reasonable. Note that each inner optimization involves several iterations, or perturbations of the $\underline{\xi}_i$ range variables; thus, we would have one nominal perturbation followed by several perturbations of the range variables, which is clearly unbalanced. The circuit design variable space is the larger and more complicated space, hence, more effort should be expended in determining the optimum in that space rather than determining more and more critical points.

Instead, we solve for the critical points in the middle of the annealing run, specifically, at each point when the annealing temperature related to the outer problem is reduced. This leads to a single annealing run to solve all the outer optimization problems P_i (albeit slightly longer than simply solving the nominal synthesis problem, P_0). Inside this single annealing run, at every change in the annealing temperature, the number of critical points increases depending on the inner optimizations. This heuristic approach is the middle ground between solving the inner optimization problems ($I_i(\underline{u}_\rho)$, $i = 1 \cdots l$) at each perturbation of the outer annealing problem (P_ρ), and solving the inner optimization between each annealing run as suggested by the direct implementation of the conceptual scheme presented in the previous section. Furthermore, empirical testing shows that this scheme tends to converge in a reasonable time period.

Given this overview, we can specify more precisely the overall algorithm we employ. Fig. 13 gives pseudocode for this annealing-in-annealing approach, in which the inner-optimizations are performed at each temperature decrement in the outer-optimization. The functions *frozen, done_at_temperature, generate, accept* and *update_temp* are required for the annealing algorithm and are detailed in [4] and [7]. We find empirically that these temperature decrement steps form a natural set of discrete stopping points in the outer optimizations which are numerous enough to update the critical points frequently, but not so numerous as to overwhelm the actual circuit synthesis process with critical-point-finding optimizations.

Although Fig. 13 gives the overall algorithm in its general form, there are several practical observations which we can apply to reduce the overall amount of computation required. For example, Fig. 1 shows that it is not always necessary to do an inner optimization, since the function $g_i(\underline{u}, \underline{\xi})$ is often in practice a one-dimensionally monotonic function of $\underline{\xi}$. Thus, the first part of the solution of $I_i(\underline{u}_\rho)$ should involve a test for monotonicity. We use a large-scale sensitivity computation to determine monotonicity, and pick the appropriate corner of the operating point and/or manufacturing range from there. This test can be applied to operating point variables which have box constraints on them ($\Xi_{jL} \leq \xi_j \leq \Xi_{jU}$, where Ξ_{jL} is the lower bound and Ξ_{jU} is the upper bound for the dimension in which $g_i(\underline{u}, \underline{\xi})$ is one dimensionally monotonic). Applying such bounds to the statistical variables will lead to conservative designs. It will be left up to the user to tradeoff between applying these bounds and getting conservative designs quickly, or actually doing an inner optimization over the space of statistical design variables to get a more aggressive design.

Constraint pruning or constraint dropping can also be employed to reduce the total run times of the combined outer/inner optimization. Recall that each inner optimization has the effect of indicating to the outer optimization that it should simultaneously design the circuit for another value of the range variables. For the outer optimization, this means that each configuration perturbation (change in the design variables, u) needs one more complete circuit simulation for each additional critical point. This is the primary source of growth in terms of run-time for this approach. To control this growth in run-time, a critical point pruning method can be applied. Unfortunately, in our nonconvex domain, the Eaves and Zangwill [35] approach cannot work.

Instead, we consider dropping critical points using a *least recently* inactive heuristic. (Note in our conceptual discussion there was a single inequality, so additional critical points added just one constraint, which is why it was called *constraint drop-*

ping by Eaves and Zangwill; in our problems the NLIP formulation has more than one constraint, so we prefer to use the term *critical point dropping* instead). During the early stages of the annealing, the Metropolis criterion encourages search across the entire design space, leading to the addition of several critical points that are not useful during the later stages of more focused searching. Our approach is to collect statistics for each critical point (each time the outer annealing decrements temperature) to indicate how often it is active. A critical point is considered to be active if a constraint evaluated at the critical point for the current design is violated. In the example of power minimization, we can look at (15). Each time a critical point is active, its $\max(\,)$ function will return a positive number to increase the cost at that design point. If the critical point is inactive, its $\max(\,)$ function returns a zero, leaving the cost unaffected by that term. A critical point becomes inactive when one of two situations occurs. Either another critical point has been added that is more strict than the inactive critical point, or the search in u-space has evolved to a region in this space where this critical point is no longer active.

In either case, we can drop this inactive critical point without affecting the cost function. There is a danger to doing this: the critical point may become active at a later point. We, therefore, need to regularly check for this, which we do during each per temperature iteration of the outer annealing. Therefore, our approach is not to drop a critical point *completely*, but rather to *suppress* the evaluation of the circuit at that critical point (hence, the critical point effectively does not contribute to the cost function). This saves us the time-consuming component of the each worst case point, the performance evaluation needed in each outer perturbation. In addition, late in the annealing, we unsuppress all evaluations ensuring that the final answer will meet all the critical points generated and will, therefore, be a synthesized manufacturable analog cell.

When the number of design constraints is large, critical points tend to always remain active, since at least one of the design constraints is active at each critical point. To overcome this, the total cost function contribution of each critical point is computed at each outer annealing temperature decrement. Critical points that contribute more than 15% (determined empirically) of the total cost function are always evaluated, while those contributing less are marked as inactive.

While various convergence proofs for the mathematical formulation of the infinite programming problem exist [6], [21], most require strict assumptions (e.g., linear constraint, convex objective). The analog circuit synthesis problem, on the other hand, tends to be highly nonlinear because of the mixture of discrete and continuous variables, our use of a penalty function for handling the constraints in the annealing formulation, and the inherent nonlinearities in the transistor models. The example set of cross sections of an OBLX cost surface showing a myriad of sharp peaks and valleys from [36] confirms that we cannot rely on any convexity assumptions.

Practical implementations of nondeterministic algorithms tend to have control parameters that are tuned to ensure quality solutions in the shortest amount of time. Our implementation of annealing uses the parameters determined in [4], with the run times for both the inner and outer annealing growing as

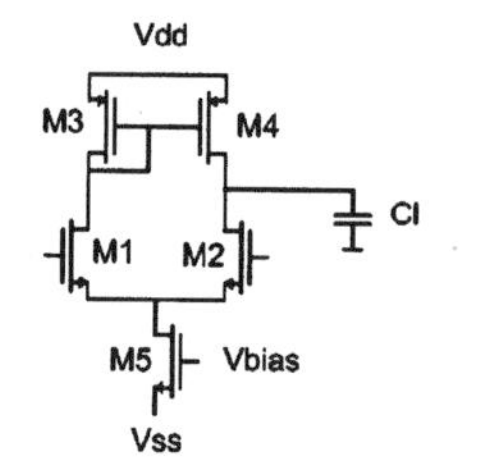

Simple OTA Circuit

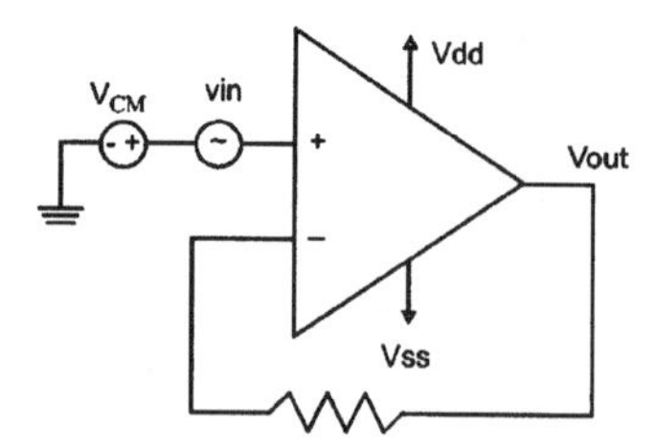

Test jig for Simple OTA

Fig. 14. An operational transconductance amplifier.

$O(n^{1.4})$ where n is the number of synthesis variables for the outer annealer, and the number of range variables for the inner annealer. Of course, each of the $O(1)$ evaluations of our cost function is proportional to the number of critical points that are active during the optimization, which can be an exponential function of the number of range variables. As the number of range variables tends to be small, this exponential nature has not been observed to be a critical issue in our experiments. Furthermore, the constraint dropping scheme can control this growth, as is shown in the results presented in Section VI.

VI. Results

We applied the above annealing approach to solving the nonlinear infinite programming formulation of the analog circuit synthesis problem to a small operational transconductance amplifier (OTA) cell, and to a large folded cascode amplifier cell. We compare these results with the original ASTRX/OBLX [4] designs to show that it is indeed important to take parametric manufacturing variations and operating point variations into account during analog circuit synthesis. In both circuits, we first added the V_{dd} operating range as the only worst case variable to consider. In addition, we also considered global variations in transistor geometries, global parametric variations in threshold voltages, and intrachip parametric variations in the threshold voltages of matched devices on the OTA to show that the NLIP formulation can incorporate both inter and intrachip statistical fluctuations in the form of yield as a constraint.

Fig. 14 shows the Simple OTA circuit and the test-jig used to simulate the circuit in HSPICE [37] for the results presented below. There are six design variables: the width and length of the differential-pair transistors (M1 and M2), the width and length of the current mirror transistors (M3 and M4), the width and length of the tail current source transistor (M5), and the V_{bias} voltage. For the NLIP formulation, we added the V_{dd} operating point variable as a worst case variable. We compare the designs generated by looking at the performance graphs across

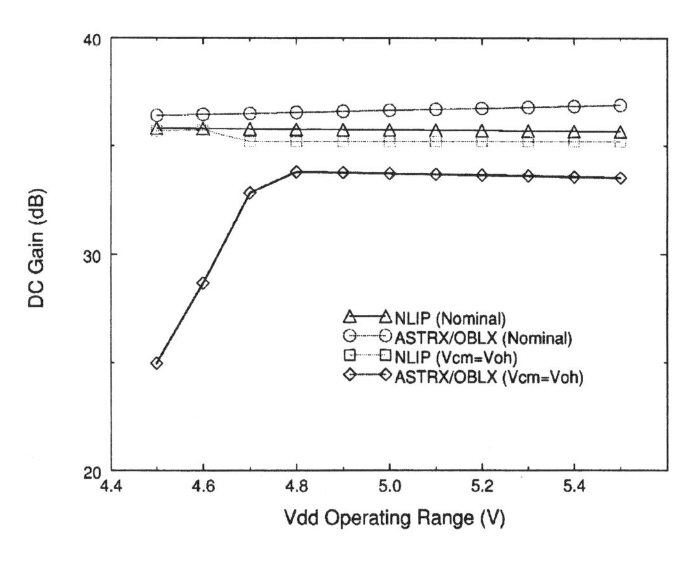

Fig. 15. OTA dc gain performance.

the $4.5 \leq V_{dd} \leq 5.5$ operating range. Fig. 15 shows the dc gain performance of the ASTRX/OBLX and NLIP formulation design, simulated at $V_{CM} = 2.5$ V (labeled Nominal) and $V_{CM} = V_{OH}$, the specified output high voltage (3.75 V). Note that at the nominal operating point ($V_{dd} = 5.0$ V, $V_{CM} = 2.5$ V) the dc gain of the ASTRX/OBLX design is no more sensitive to the operating point than is the nominal NLIP design. This illustrates that even adding small-change sensitivity constraints at the nominal operating point would not improve the design. The actual worst case gain of this circuit will occur when the common mode of the input voltage (called V_{CM} here) is at its highest specified value, in this case the highest output voltage (V_{OH}) since the test-jig is configured for unity-gain feedback, and V_{dd} is at its lowest value. It is clear from the graph that it is necessary to use the NLIP formulation to ensure that the dc gain is relatively insensitive to the operating range. Since the critical point is an operating range corner, the designer can actually ask ASTRX/OBLX to design for that corner by prespecifying $V_{dd} = 4.5$ V, and $V_{CM} = V_{OH}$ instead of their nominal values. However, it is not always possible *a priori* to identify the worst case corner in a larger example (with more worst case variables), and in some cases, the critical point can occur within the operating range.

The same experiment was repeated with the folded cascode amplifier shown in Fig. 16. In this design, there are 12 designable widths and lengths, a designable compensation capacitance and two dc bias voltages. Again the single operating point variable V_{dd} was used. We simulated the ASTRX/OBLX and the NLIP designs' output swing across the V_{dd} operating range (shown in Fig. 17). The output swing of the amplifier is a strong function of the region of operation of the output transistors (M3, M4, M5, and M6). The output swing is obtained by using a large-signal ac input, and determining the output voltage at which the output transistors move out of saturation (which will cause clipping in the output waveform). Compared to the OTA, this is a much more difficult design, hence, the output swing specification of 2.0 V is just met by both the ASTRX/OBLX design (at the nominal power supply voltage of 5.0 V) and the NLIP design (across the entire operating range).

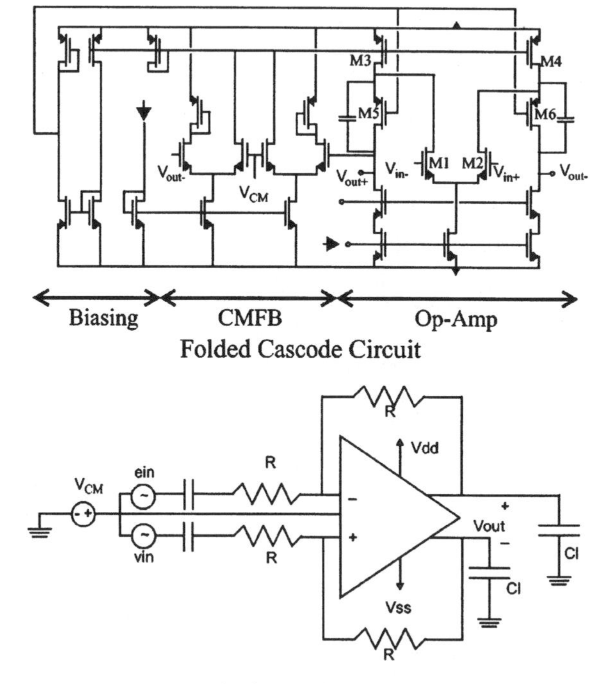

Fig. 16. The folded cascode amplifier.

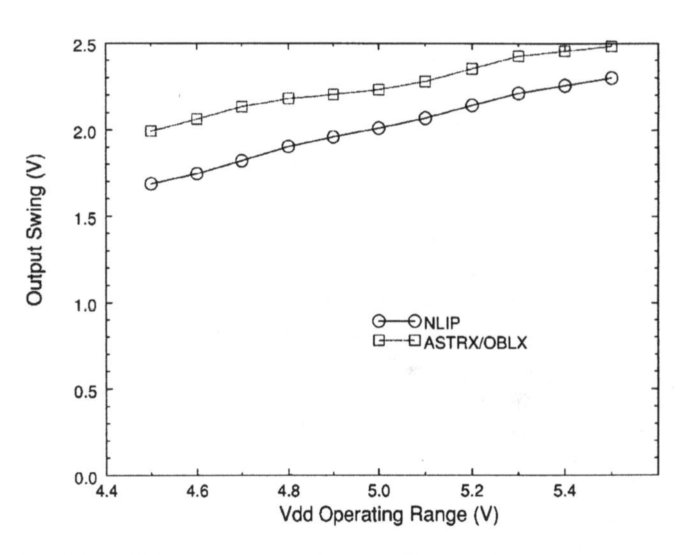

Fig. 17. Folded cascode output swing across V_{dd}.

The ASTRX/OBLX design fails to meet the output swing specification for the lower half of the operating range ($V_{dd} < 5.0$ V). This is a common problem of nominal synthesis tools. For an optimal design, it biased the circuit so that the output transistors were at the edge of saturation, and a slight decrease in the V_{dd} voltage resulted in their moving out of saturation, hence, the output swing falls below the 2.0-V specification. Again the NLIP formulation overcomes this by more completely modeling the design problem.

Stable voltage and current references are required in practically every analog circuit. A common approach to providing such a reference is the bandgap reference circuit. Bandgap reference circuits are some of the most difficult circuits to design. Precision bandgap circuits are still solely designed by

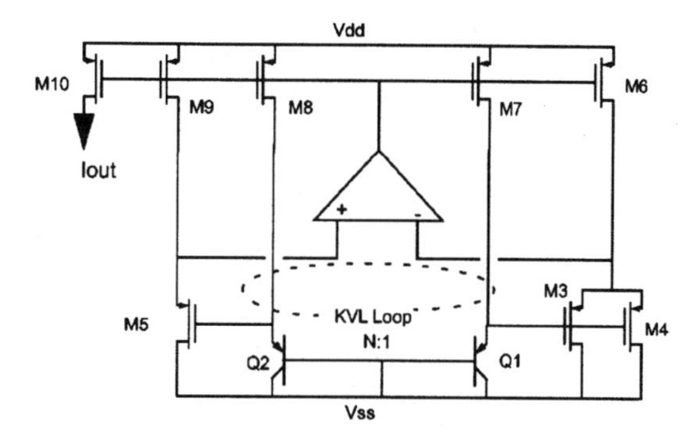

Fig. 18. Simple bandgap reference circuit.

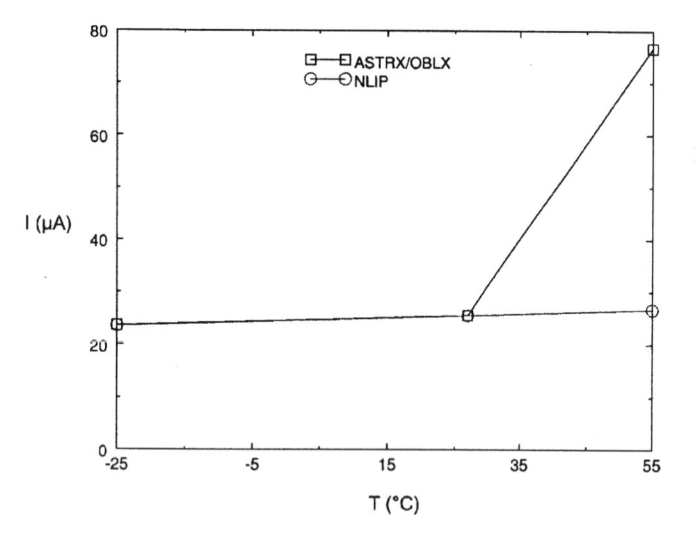

Fig. 19. Temperature sensitivity of bandgap reference.

highly experienced expert designers in industry. In comparison, the circuit we have chosen for our next experiment is a simple first-order bandgap reference circuit.

Fig. 18 shows the schematic of the bandgap reference circuit we synthesized for this experiment. It is based on the CMOS voltage reference described in [38], but has been modified to generate a bias current instead. Analyzing this circuit using the Kirchhoff's Voltage Law equation set by the op-amp feedback loop indicated by the dotted loop in Fig. 18, and first-order equations for the base-emitter and gate-source voltages as a function of bias current we can solve for the bias current

$$I = 100 \cdot (V_t)^2 \cdot k' \frac{W}{L}. \tag{16}$$

when the emitter ratio of $Q1$ to $Q2$ is $1:20$, the device width and lengths so that M3, M4, and M5 are identical, that is, M3 and M4 share the same current that M5 sees. In this equation, $V_t = kT/q$, and $k' = \mu_p C_{ox}$ where $\mu_p \sim T^{-1.5}$, thus the output current varies as $I_{\text{out}} \sim T^{-0.5}$, which is the best we'll be able to achieve if we synthesize this circuit.

In our bandgap synthesis experiment, we started off with the schematic shown in Fig. 18. While the schematic shows hierarchy, with an op-amp to control the feedback, we flattened the entire schematic in our input to the synthesis system, replacing the op-amp with the simple OTA shown in Fig. 14. The matching considerations described above were used to determine which subset of the device geometries (MOS width and length and

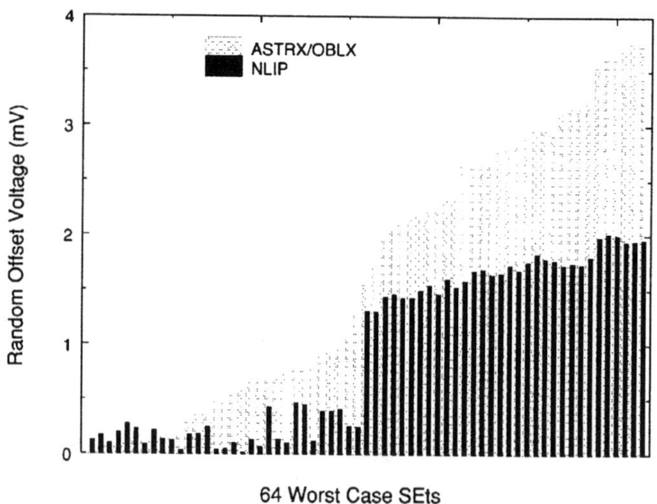

Fig. 20. OTA random offset voltage.

TABLE I
DEVICE SIZES (IN MICRONMETERS) OF THE OTA AND PARAMETRIC MANUFACTURING VARIATIONS

Design	ASTRX/ OBLX	NLIP
W_{M1}	26	71
L_{M1}	2.7	3.3
W_{M3}	4.8	15
L_{M3}	4.6	8.6
W_{M5}	6.6	58
L_{M5}	2.1	15
CPU Time on RS6000/58H	10 min	80 min

bipolar emitter areas) were the design variables. The output current was specified to be 25 μA, and a test-jig was placed around the op-amp devices to code gain, unity gain frequency and phase margin specifications to ensure those devices behaved like an op-amp. It took about 10 min to synthesize a circuit to these specifications on a 133-MHz PowerPC 604-based machine.

Fig. 19 shows the graph of the output current with respect to temperature. As can be seen from the ASTRX/OBLX design, a nominal synthesis of this circuit is extremely unstable with respect to temperature. In comparison, the NLIP formulation has a gentle slope with respect to temperature. As we have seen from the analysis above, this slope is to be expected (curve-fitting the data shows that the exponent is -0.42). We would need to expand the circuit topology to include curvature compensation resistors to ensure that dI_{out}/dT has zero slope at room temperature.

In our next experiment we reconsider the OTA circuit of Fig. 14. This time we introduce global variations in transistor geometries (e.g., ΔL of the p and n devices); global parametric variations in threshold voltages (V_{FB}, the flat-band voltage of the p and n devices); and intrachip parametric variations in the threshold voltages of matched devices (ΔV_{FB} of the

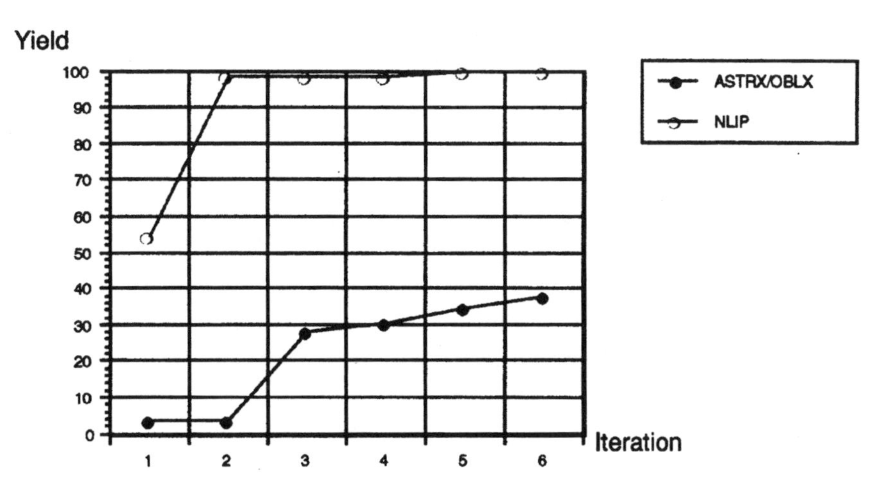

Fig. 21. Post-design yield optimization.

p and n devices, using a simple mismatch model proposed by [39]). Note, the intrachip parametric variations are particularly challenging because their amplitude depends on the design variables—these variations are roughly proportional to $1/(\sqrt{WL})$.

In this run there were six circuit design variables, and six worst case variables. We expect to see that the device sizes will be larger to minimize the effect of the mismatch in geometry and threshold voltage. It should be obvious that larger geometries reduce the sensitivity to the ΔL variation. In addition, larger devices are less sensitive to the ΔV_{FB} mismatch than are smaller devices [39], again pushing for larger designs.

Fig. 20 shows the random component of the input referred offset voltage (the systematic component of both designs is less than 0.5 mV), at the 64 worst case corners of the six variables used in the design. The ASTRX/OBLX design can have a random offset voltage of up to 4 mV. The nonlinear infinite programming formulation of the analog circuit synthesis problem was set up with a random offset voltage specification of 2 mV. In the graph we have sorted the corners by the size of the random offset voltage in the ASTRX/OBLX design. From the graph it is clear that about half of the 64 corners need to be considered by the formulation. It is also clear that the optimization has reduced the random offset voltage only as much as was needed to meet the specification. For the half of the worst case corners that are not active constraints, the NLIP optimization returns a design whose random offsets are actually greater than that of the ASTRX/OBLX design for that corner. While these corners are currently inactive, at a different sizing and biasing for the circuit, those corners can easily become active. This prevents the designer using ASTRX/OBLX from *a priori* defining the list of critical points. In an incompletely specified problem, the optimization will find a solution that might violate a constraint that had not been specified.

Our approach can dynamically determine the actual critical points, using the inner annealing and the large-scale sensitivity of the cost function terms to the worst case variables, hence, it limits the designer's responsibility to just providing the worst case ranges. As expected, the device sizes from the NLIP formulation are larger to reduce sensitivity to mismatch. This is shown in Table I, with execution times of both runs. Note that the execution time of this run for this prototype implementation is only eight times more than the nominal design done by ASTRX/OBLX. The total number of critical points determined during this run was 12, thus, the outer optimization was evaluating 12 circuits simultaneously for each perturbation it made during the last stages of the synthesis. In addition, since the search space is more complicated due to the additional constraints, the annealer for the outer optimization has to do more total perturbations to ensure that it properly searches the design space. The least recently active critical point pruning approach accounts for the less than 12× run-time degradation when the 12 critical points are added. Without the use of the runtime selection of critical points, all 64 "known" corners of the range-space can be searched exhaustively, which is unnecessarily conservative. Alternatively, a designer specified list of limited critical points would have been much less than 12—some of the points were added by the annealing due to the sequence of designs the synthesis tool considered.

Fig. 21 shows the effect of the inclusion of the simplified manufacturing variations into the synthesis process. We use the same metric used in Section I for this comparison, namely, post-design yield optimization. More precisely, we start off with two nominal designs: one synthesized by ASTRX/OBLX and one synthesized by the NLIP formulation. We use yield optimization to improve the yield of both these circuits. Instead of the traditional yield optimization algorithms, we pursue an indirect method detailed in [5]. The actual yield optimization approach used was a penalty function method that penalizes designs that do not meet the desired circuit performances, while at the same time trying to improve the circuit's immunity to manufacturing line variations. At every iteration in this gradient-based approach we evaluate the yield using a Monte Carlo analysis, and this yield is plotted on the graph in Fig. 21. The nonlinear infinite programming formulation provides a good starting point for the yield optimization, much like the manual design did in our earlier example. As in the previous case, the nominally synthesized design is a difficult starting point for yield optimization. The run with the nominal synthesis starting point terminates due to an inability to find a direction of improvement. Therefore, the sequence of NLIP-based synthesis and yield optimization truly leads to an automated environment for circuit synthesis with optimal yield, while the combination of nominal synthesis and yield optimization is inadequate in addressing the needs of automated analog cell-level design.

VII. Conclusion

In this paper, we have integrated analog circuit synthesis with worst case analysis of both parametric manufacturing and operating point variations. This unified approach maps the circuit design task into an infinite programming problem, and solves it using an annealing within annealing formulation. This integration has been used to design several manufacturable circuits with a reasonable CPU expenditure. By showing that an automated system can generate circuits that can meet some of the critical concerns of designers (operating range variations and parametric manufacturing variations), we believe that we have taken a significant step toward the routine use of analog synthesis tools in an industrial environment.

Acknowledgment

The authors would like to acknowledge Dr. E. Ochotta for extending ASTRX/OBLX to allow for an efficient implementation of NLIP within ASTRX/OBLX. The authors would also like to acknowledge Prof. S. W. Director for his direction in the early stages of this work.

References

[1] R. Harjani, R. A. Rutenbar, and L. R. Carley, "OASYS: A framework for analog circuit synthesis," *IEEE Trans. Computer-Aided Design*, vol. 8, pp. 1247–1266, Dec. 1989.

[2] P. R. Gray, "Possible analog IC scenarios for the 90's," in *VLSI Symp.*, 1995.

[3] L. R. Carley, G. G. E. Gielen, R. A. Rutenbar, and W. M. C. Sansen, "Synthesis tools for mixed-signal ICs: Progress on frontend and backend strategies," in *Proc. IEEE/ACM Design Automation Conf.*, June 1996, pp. 298–303.

[4] E. S. Ochotta, R. A. Rutenbar, and L. R. Carley, "Synthesis of high performance analog circuits in ASTRX/OBLX," *IEEE Trans. Computer-Aided Design*, vol. 15, pp. 273–294, Mar. 1996.

[5] K. Krishna and S. W. Director, "The linearized performance penalty (LPP) method for optimization of parametric yield and its reliability," *IEEE Trans. Computer-Aided Design*, vol. 14, pp. 1557–1568, Dec. 1995.

[6] A. V. Fiacco and K. O. Kortanek, Eds., *Semi-Infinite Programming and Applications*. ser. Lecture Notes in Economics and Mathematical Systems 215. Berlin, Germany: Springer-Verlag, 1983.

[7] S. Kirkpatrick, C. D. Gelatt, and M. P. Vecchi, "Optimization by simulated annealing," *Science*, vol. 220, no. 4598, pp. 671–680, May 1983.

[8] T. Mukherjee, L. R. Carley, and R. A. Rutenbar, "Synthesis of manufacturable analog circuits," in *Proc. IEEE/ACM Int. Conf. Computer-Aided Design*, Nov. 1994, pp. 586–593.

[9] R. A. Rohrer, "Fully automated network design by digital computer, preliminary considerations," *Proc. IEEE*, vol. 55, pp. 1929–1939, Dec. 1967.

[10] P. C. Maulik, L. R. Carley, and R. A. Rutenbar, "Integer programming-based topology selection of cell-level analog circuits," *IEEE Trans. Computer-Aided Design*, vol. 14, pp. 401–412, Apr. 1995.

[11] W. Nye, D. C. Riley, A. Sangiovanni-Vincentelli, and A. L. Tits, "DELIGHT.SPICE: An optimization-based system for design of integrated circuits," *IEEE Trans. Computer-Aided Design*, vol. 7, pp. 501–518, Apr. 1988.

[12] M. G. R. DeGrauwe, B. L. A. G. Goffart, C. Meixenberger, M. L. A. Pierre, J. B. Litsios, J. Rijmenants, O. J. A. P. Nys, E. Dijkstra, B. Joss, M. K. C. M. Meyvaert, T. R. Schwarz, and M. D. Pardoen, "Toward an analog system design environment," *IEEE J. Solid-State Circuits*, vol. 24, pp. 659–672, June 1989.

[13] G. E. Gielen, H. C. Walscharts, and W. C. Sansen, "Analog circuit design optimization based on symbolic simulation and simulated annealing," *IEEE J. Solid-State Circuits*, vol. 25, pp. 707–713, June 1990.

[14] H. Y. Koh, C. H. Sequin, and P. R. Gray, "OPASYN: A compiler for MOS operational amplifiers," *IEEE Trans. Computer-Aided Design*, vol. 9, Feb. 1990.

[15] J. P. Harvey, M. I. Elmasry, and B. Leung, "STAIC: An interactive framework for synthesizing CMOS and BiCMOS analog circuits," *IEEE Trans. Computer-Aided Design*, vol. 11, pp. 1402–1418, Nov. 1992.

[16] E. S. Ochotta, R. A. Rutenbar, and L. R. Carley, "Equation-free synthesis of high-performance linear analog circuits," in *Proc. Joint Brown/MIT Conf. Advanced Research in VLSI and Parallel Systems*, Providence, RI, Mar. 1992, pp. 129–143.

[17] ——, "ASTRX/OBLX: Tools for rapid synthesis of high-performance analog circuits," in *Proc. IEEE/ACM Design Automation Conf.*, June 1994, pp. 24–30.

[18] ——, "Analog circuit synthesis for large, realistic cells: Designing a pipelined A/D convert with ASTRX/OBLX," in *Proc. IEEE Custom Integrated Circuit Conf.*, May 1994, pp. 15.4/1–4..

[19] P. C. Maulik and L. R. Carley, "Automating analog circuit design using constrained optimization techniques," in *Proc. IEEE/ACM Int. Conf. Computer-Aided Design*, Nov. 1991, pp. 390–393.

[20] V. Raghavan, R. A. Rohrer, M. M. Alaybeyi, J. E. Bracken, J. Y. Lee, and L. T. Pillage, "AWE inspired," in *Proc. IEEE Custom Integrated Circuit Conf.*, May 1993, pp. 18.1/1–8.

[21] R. Heittich, Ed., *Semi-Infinite Programming*. ser. Lecture Notes in Control and Information Sciences 15. Berlin, Germany: Springer-Verlag, 1979.

[22] S. W. Director, P. Feldmann, and K. Krishna, "Optimization of parametric yield: A tutorial," *Proc. IEEE Custom Integrated Circuit Conf.*, pp. 3.1/1–8, May 1992.

[23] P. Feldmann and S. W. Director, "Integrated circuit quality optimization using surface integrals," *IEEE Trans. Computer-Aided Design*, vol. 12, pp. 1868–1879, Dec. 1993.

[24] D. E. Hocevar, P. F. Cox, and P. Yang, "Parametric yield optimization for MOS circuit blocks," *IEEE Trans. Computer-Aided Design*, vol. 7, pp. 645–658, June 1988.

[25] M. A. Styblinski and L. J. Opalski, "Algorithms and software tools of IC yield optimization based on fundamental fabrication parameters," *IEEE Trans. Computer-Aided Design*, vol. 5, Jan. 1996.

[26] K. J. Antriech, H. E. Graeb, and C. U. Wieser, "Circuit analysis and optimization driven by worst case distances," *IEEE Trans. Computer-Aided Design*, vol. 13, pp. 57–71, Jan. 1994.

[27] A. Dharchoudhury and S. M. Kang, "Performance-constrained worst case variability minimization of VLSI circuits," in *Proc. IEEE/ACM Design Automation Conf.*, June 1993, pp. 154–158.w.

[28] P. Cox, P. Yang, S. S. Mahant-Shetti, and P. Chatterjee, "Statistical modeling for efficient parametric yield estimation of MOS VLSI circuits," *IEEE Trans. Electron Devices*, vol. ED-32, pp. 471–478, Feb. 1985.

[29] K. Krishna and S. W. Director, "A novel methodology for statistical parameter extraction," in *Proc. IEEE/ACM Int. Conf. Computer-Aided Design*, 1995, pp. 696–699.

[30] S. R. Nassif, A. J. Strojwas, and S. W. Director, "FABRICS II: A statistically based IC fabrication process simulator," *IEEE Trans. Computer-Aided Design*, vol. CAD-3, pp. 20–46, Jan. 1984.

[31] K. Madsen and H. Schjaer-Jacobsen, "Algorithms for worst case tolerance optimization," *IEEE Trans. Circuits Syst.*, vol. CAS-26, Sept. 1979.

[32] I. E. Grossmann and D. A. Straub, "Recent developments in the evaluation and optimization of flexible chemical processes," in *Proc. COPE-91*, 1991, pp. 49–59.

[33] E. Polak, "On mathematical foundations of nondifferentiable optimization in engineering design," *SIAM Rev.*, vol. 29, pp. 21–89, 1987.

[34] H. Parsa and A. L. Tits, "Nonuniform, dynamically adapted discretization for functional constraints in engineering design problems," in *Proc. 22nd IEEE Conf. Decision and Control*, Dec. 1983, pp. 410–411.

[35] B. C. Eaves and W. I. Zangwill, "Generalized cutting plane algorithms," *SIAM J. Control*, vol. 9, pp. 529–542, 1971.

[36] M. Krasnicki, R. Phelps, R. A. Rutenbar, and L. R. Carley, "MAELSTROM: Efficient simulation-based synthesis for custom analog cells," in *Proc. IEEE/ACM Design Automation Conf.*, June 1999, pp. 945–950.

[37] *Avant! Star-HSPICE User's Manual*. Sunnyvale, CA: Avant! Corp., 1996.

[38] K.-M. Tham and K. Nagaraj, "A low supply voltage high PSRR voltage reference in CMOS process," *IEEE J. Solid-State Circuits*, vol. 30, pp. 586–590, May 1995.

[39] M. J. M. Pelgrom, A. C. J. Duinmaijer, and A. P. G. Welbers, "Matching properties of MOS transistors," *IEEE J Solid-State Circuits*, vol. 24, pp. 1433–1439, Oct. 1989.

Tamal Mukherjee (S'89–M'95) received the B.S., M.S., and Ph.D. degrees in electrical and computer engineering from Carnegie Mellon University, Pittsburgh, PA, in 1987, 1990, and 1995 respectively.

Currently he is a Research Engineer and Assistant Director of the Center for Electronic Design Automation in Electrical and Computer Engineering Department at Carnegie Mellon University. His research interests include automating the design of analog circuits and microelectromechanical systems. His current work focuses on the developing computer-aided design methodologies and techniques for integrated microelectromechanical systems, and is involved in modeling, simulation, extraction, and synthesis of microelectromechanical systems.

L. Richard Carley (S'77–M'81–SM'90–F'97) received the S.B. degree in 1976, the M.S. degree in 1978, and the Ph.D. degree in 1984, all from the Massachusetts Institute of Technology (MIT), Cambridge.

He joined Carnegie Mellon University, Pittsburgh, PA, in 1984. In 1992, he was promoted to Full Professor of Electrical and Computer Engineering. He was the Associate Director for Electronic Subsystems for the Data Storage Systems Center [a National Science Foundation (NSF) Engineering Research Center at CMU] from 1990–1999. He has worked for MIT's Lincoln Laboratories and has acted as a Consultant in the areas of analog and mixed analog/digital circuit design, analog circuit design automation, and signal processing for data storage for numerous companies; e.g., Hughes, Analog Devices, Texas Instruments, Northrop Grumman, Cadence, Sony, Fairchild, Teradyne, Ampex, Quantum, Seagate, and Maxtor. He was the principal Circuit Design Methodology Architect of the CMU ACA-CIA analog CAD tool suite, one of the first top-to-bottom tool flows aimed specifically at design of analog and mixed-signal IC's. He was a co-founder of NeoLinear, a Pittsburgh, PA-based analog design automation tool provider; and, he is currently their Chief Analog Designer. He holds ten patents. He has authored or co-authored over 120 technical papers, and authored or co-authored over 20 books and/or book chapters.

Dr. Carley was awarded the Guillemin Prize for best Undergraduate Thesis from the Electrical Engineering Department, MIT. He has won several awards, the most noteworthy of which is the Presidential Young Investigator Award from the NSF in 1985 He won a Best Technical Paper Award at the 1987 Design Automation Conference (DAC). This DAC paper on automating analog circuit design was also selected for inclusion in 25 years of Electronic Design Automation: A Compendium of Papers from the Design Automation Conference, a special volume, published in June of 1988, including the 77 papers (out of over 1600) deemed most influential over the first 25 years of the Design Automation Conference.

Rob A. Rutenbar (S'77–M'84–SM'90–F'98) received the Ph.D. degree from the University of Michigan, Ann Arbor, in 1984.

He joined the faculty of Carnegie Mellon University (CMU), Pittsburgh, PA, where he is currently Professor of Electrical and Computer Engineering, and (by courtesy) of Computer Science. From 1993 to 1998, he was Director of the CMU Center for Electronic Design Automation. He is a cofounder of NeoLinear, Inc., and served as its Chief Technologist on a 1998 leave from CMU. His research interests focus on circuit and layout synthesis algorithms for mixed-signal ASIC's, for high-speed digital systems, and for FPGA's.

In 1987, Dr. Rutenbar received a Presidential Young Investigator Award from the National Science Foundation. He has won Best/Distinguished paper awards from the Design Automation Conference (1987) and the International Conference on CAD (1991). He has been on the program committees for the IEEE International Conference on CAD, the ACM/IEEE Design Automation Conference, the ACM International Symposium on FPGA's, and the ACM International Symposium on Physical Design. He also served on the Editorial Board of IEEE SPECTRUM. He was General Chair of the 1996 ICCAD. He chaired the Analog Technical Advisory Board for Cadence Design Systems from 1992 through 1996. He is a member of the ACM and Eta Kappa Nu.

The Generalized Boundary Curve - A Common Method for Automatic Nominal Design and Design Centering of Analog Circuits

R. Schwencker[1,2], F. Schenkel[1], H. Graeb[1], K. Antreich[1]

[1]Institute of Electronic Design Automation
Technical University of Munich
81609 Munich, Germany

[2]Infineon Technologies
P.O. Box 80 09 49
81617 Munich, Germany

Abstract

In this paper, a new method for analog circuit sizing with respect to manufacturing and operating tolerances is presented. Two types of robustness objectives are presented, i.e. parameter distances for the nominal design and worst-case distances for the design centering. Moreover, the generalized boundary curve is presented as a method to determine a parameter correction within an iterative trust region algorithm. Results show that a significant reduction in computational costs is achieved using the presented robustness objectives and generalized boundary curve.

1 Introduction

In the era of System-on-Chip (SoC), mixed-signal ICs gain an ever growing market share. For the digital part of those mixed-signal ICs, an established design flow exists, while for the analog designer the circuit simulator is still the most important design tool. In order to keep up with the ever growing demands on the designers' productivity and time-to-market, algorithms and tools for the automatic nominal design and design centering of analog cells are needed to design robust analog cells in acceptable design times.

For the nominal design of analog circuits, big improvements were made in the area of symbolic [5, 12] and simulation-based [4, 10, 14, 16] algorithms. On the other hand, for the design centering statistical [13, 17] and deterministic methods based e.g. on simplicial approximation [6] or ellipsoidal techniques [1] were presented.

More recently, deterministic unified approaches for nominal design and design centering were published, that are based on multiple robustness objectives (MROs) for individual performances e.g. linearized performance penalties (LPP) [11] or worst-case distances (WCD) [2]. But these algorithms suffer from high computational cost that are determined by the number of simulations.

On the one hand the simulation effort is caused by the calculation of the MROs. These MROs give an estimation of the performance-oriented yield of the circuit, but at the beginning of an optimization, the performances are usually far away from their specification and the primary goal is to fulfill the specification. To reduce the simulation cost at that stage, robustness objectives are needed that describe the robustness of the individual performances and can be calculated with as little effort as possible.

On the other hand, the published algorithms transform the MROs into a scalar cost function with a sum of exponential functions. This results in a strongly nonlinear function, while the original MROs are often only weakly nonlinear in the region of interest. This in turn leads to increased simulation costs, if a standard optimization algorithm is used.

In this paper, the *parameter distances* are defined as new robustness objectives for the nominal design, that cause no additional simulation effort compared to a simple sensitivity analysis while they can be used with the same cost function as the WCDs and LPPs. Thus they are a significant progress compared to the state-of-art, since they allow to apply the same algorithms as for the design centering at a significantly lower simulation effort. Additionally, they may also be used if no process statistic is available.

Furthermore, the *generalized boundary curve* (GBC) is presented as a method to determine a step length within an iterative trust-region algorithm. Compared to a trust-region algorithm, that uses the linearized cost function to determine a step length (e.g. [16]), the GBC is based on the full nonlinear cost function calculated with the linearized objectives. Thus the "linearization error" for the strongly nonlinear cost function is kept small. As shown in the results (Section 4), using the full nonlinear cost function based on the linearized objectives significantly reduces the total number of iterations in the optimization. This is a significant improvement compared to gradient-based sizing algorithms like e.g. [16].

The remainder of this paper is structured as follows. In

the following section 2, the parameter distances and WCDs are introduced as robustness objectives for the nominal design and the design centering. Based on these objectives the optimization problem is defined. Section 3 gives a brief overview over the GBC. Section 4 presents results, and section 5 concludes the paper.

2 Problem definition

For a fixed topology, an analog circuit can be described by its parameters and performances. The circuit parameters can be divided into three different classes:

- The *design parameters* $\mathbf{d}$ (e.g. transistor widths and lengths) are tuned by the circuit designer in order to improve the circuit performances and yield.
- For the *operating parameters* $\boldsymbol{\theta}$ (e.g. supply-voltage and temperature), a tolerance region T_θ is defined by the upper bounds $\boldsymbol{\theta}_u$ and lower bounds $\boldsymbol{\theta}_l$:

  $$T_\theta = \{\boldsymbol{\theta} \mid \boldsymbol{\theta}_l \leq \boldsymbol{\theta} \leq \boldsymbol{\theta}_u\} \quad (1)$$

  The circuit must fulfill the given specifications for all operating parameters within this tolerance-region.
- The fluctuations of the manufacturing process are described by the *statistical parameters* $\mathbf{s}$ (e.g. oxide thickness). These parameters cannot be tuned by the circuit designer. They are characterized by their mean values $\mathbf{s}_0$, covariance matrix $\mathbf{C}$, and probability density function $\mathrm{pdf}(\mathbf{s})$. In the rest of the paper the statistical parameters are assumed to be normally distributed. As shown in [8], this is no serious limitation.

For given parameters, the *circuit performances* $\mathbf{f}$ can be calculated by simulation. For these performances, *specifications* $\mathbf{f}_b$ are given and can generally be written as upper bounds:

$$\mathbf{f}(\mathbf{d}, \boldsymbol{\theta}, \mathbf{s}) \leq \mathbf{f}_b \quad (2)$$

2.1 Parameter distances

Goal of the nominal design is to satisfy all specifications with as much safety margin as possible for process variations. This poses the problem to compare different performances with different units (e.g. slew-rate and gain) with each other.

Each performance f_i is linearized with respect to the design and operating parameters:

$$\bar{f}_i(\mathbf{d}, \boldsymbol{\theta}) = \mathbf{g}_{\mathbf{d},i}^{\mathrm{T}} \cdot (\mathbf{d} - \mathbf{d}_0) + \mathbf{g}_{\boldsymbol{\theta},i}^{\mathrm{T}} \cdot (\boldsymbol{\theta} - \boldsymbol{\theta}_0), \quad (3)$$
$$\text{with} \quad \mathbf{g}_{\mathbf{d},i} = \nabla_{\mathbf{d}} f_i, \qquad \mathbf{g}_{\boldsymbol{\theta},i} = \nabla_{\boldsymbol{\theta}} f_i$$

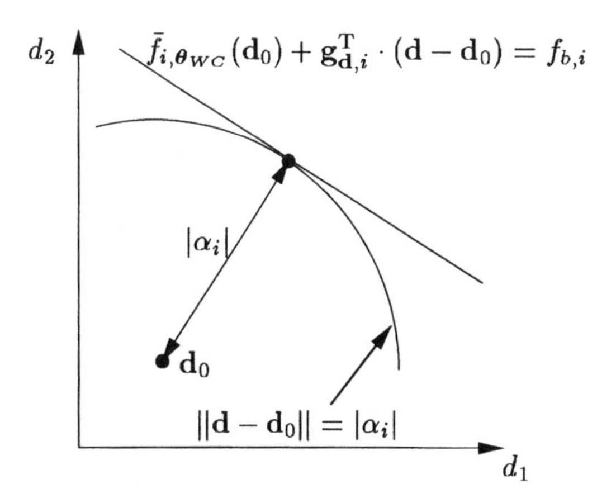

Figure 1. Parameter distance

Based on this linearization the influence of the operating parameters can be approximated as:

$$\boldsymbol{\theta}_{WC_i} = \underset{\boldsymbol{\theta}}{\operatorname{argmax}} \{\bar{f}_i(\mathbf{d}_0, \boldsymbol{\theta}) \mid \boldsymbol{\theta} \in T_\theta\} \quad (4)$$

$$\bar{f}_{i,\boldsymbol{\theta}_{WC}}(\mathbf{d}_0) = \bar{f}_i(\mathbf{d}_0, \boldsymbol{\theta}_{WC_i}) \quad (5)$$

The value $\bar{f}_{i,\boldsymbol{\theta}_{WC}}(\mathbf{d}_0)$ is the performance value for the approximated worst-case corner $\boldsymbol{\theta}_{WC_i}$ of the operating parameters.

The sensitivity of a performance can be determined as the norm of its gradient with respect to the design parameters. Taking this into account, the minimum deviation $||\Delta\mathbf{d}||$ of the design parameters that is needed to shift the performance $\bar{f}_i$ in the linear model from the value $\bar{f}_{i,\boldsymbol{\theta}_{WC}}(\mathbf{d}_0)$ to the specification $f_{b,i}$ is an appropriate measure to characterize the distance of the performance from its specification:

$$||\Delta\mathbf{d}|| = \min_{\mathbf{d}} \{||\mathbf{d} - \mathbf{d}_0|| \mid \bar{f}_i(\mathbf{d}, \boldsymbol{\theta}_{WC_i}) = f_{b,i}\} \quad (6)$$

The design parameters are assumed to be scaled, such that the different parameters are comparable. Eq. (6) can be solved exactly using a Lagrange formulation:

$$||\Delta\mathbf{d}|| = \frac{|f_{b,i} - \bar{f}_{i,\boldsymbol{\theta}_{WC}}(\mathbf{d}_0)|}{||\mathbf{g}_{\mathbf{d},i}||} \quad (7)$$

The value $||\Delta\mathbf{d}||$ in eq. (7) is defined as the *unsigned parameter distance*. A signed parameter distance is introduced, such that fulfilled specifications get a positive sign and violated specifications a negative one. Hence, the *signed parameter distance* $\alpha_i(\mathbf{d}_0)$ is defined as:

$$\alpha_i(\mathbf{d}_0) = \frac{f_{b,i} - \bar{f}_{i,\boldsymbol{\theta}_{WC}}(\mathbf{d}_0)}{||\mathbf{g}_{\mathbf{d},i}||} \quad (8)$$

The calculation of the parameter distance is illustrated in Figure 1. The parameter distance α_i considers both, the distance of the performance from its specification, and its sensitivity with respect to the design and operational parameters. Considering eq. (8) the parameter distance can be interpreted as the distance of the specification from its performance measured in units of the reciprocal value of its sensitivity.

Thus the parameter distance is well-suited to measure the "error" of one performance and compare different performances with each other. Compared to other norms with the same purpose (e.g. generalized l_p-norm [3]) it has the advantage of being differentiable, if the performance is differentiable. This is a prerequisite for many optimization algorithms. The goal of the nominal design is to tune the design parameters, such that the smallest parameter distances are improved and all parameter distances are as great as possible. This yields a good starting point for design centering.

2.2 Worst-case distances

Goal of the design centering is to adjust the design parameters, such that the yield of the circuit is maximized. Generally the overall yield cannot be calculated analytically. As shown in [2] the worst-case distances (WCD) β_i can be used to estimat and improve the yield. The worst-case distances have the following properties:

- The worst-case distances take operating parameters and variations of the statistical parameters, including correlations into account.

- A worst-case distance is negative, if the specification is violated for the worst-case operating parameters. It is positive, if the specification is fulfilled.

- The yield $Y_i(\mathbf{d}_0)$ with respect to the specification $f_{b,i}$ can be estimated based on the WCD β_i:

$$Y_i = \frac{1}{\sqrt{2\pi}} \int_{-\infty}^{\beta_i} \exp(-\xi^2/2) d\xi \qquad (9)$$

- Even for a high yield, the worst-case distances still have reasonable variations and thus can be used to further improve the yield and robustness of the circuit.

It can be seen, that the worst-case distances have similar properties as the parameter distances. So, except for their significantly higher simulation effort, they could also be used for a nominal design. In the approach presented in this paper, the parameter distances are used in the nominal design to calculate a good starting point for the design centering and thus reduce the overall number of simulations.

The design centering is based on the worst-case distances as objectives. The aim is to size the circuit, such that all worst-case distances are as great as possible.

2.3 Cost function

In both cases the goals for the design centering and nominal design are to maximize all objectives but especially the smallest ones. This means, that from a mathematical point of view, the two optimization tasks are the same. For the rest of the paper, the objectives are denoted with $\gamma_i(\mathbf{d}_0)$. For the design centering, γ_i is equal to the WCD β_i, for the nominal design, γ_i denotes the parameter distance α_i. It is important to note that the proposed method is not limited to WCDs and parameter distances. In fact it can be applied to any differentiable robustness objectives and cost function, as long as the cost function is convex for the linearized robustness objectives.

As proposed in [2, 11], a suitable cost function φ for the proposed optimization task is:

$$\varphi(\mathbf{d}) = \sum_{i=1}^{n_\gamma} \exp(-a \cdot \gamma_i(\mathbf{d})), \quad a > 0 \qquad (10)$$

The constant factor a in this equation is a scaling factor for the objectives. The cost function (10) is constructed, such that small or negative objectives have a high contribution to the overall cost and positive ones only have a minor contribution.

An automatic circuit sizing is only feasible if the principal functionality of the circuit is guaranteed during the whole sizing process by considering functional constraints (e.g. saturation condition for the MOS-transistors of a current mirror) [7, 16, 18]. These constraints can be divided into equality constraints for design parameters and inequality constraints for parameters and simulated properties.

Every equality constraint reduces the number of free design parameters by one and therefore reduces the design space and speeds up the sizing. The inequality constraints $\mathbf{u}(\mathbf{d})$ are formulated such that fulfilled constraints are positive:

$$\mathbf{u}(\mathbf{d}) \geq \mathbf{0} \qquad (11)$$

These constraints are either constraints on parameters or on simulated properties, that can be extracted from a DC-simulation, that must be done anyway. Thus they cause no additional simulation effort.

Combining the constraints (11) and the cost function (10), the sizing problem can be formulated as:

$$\min_{\mathbf{d}} \{\varphi(\mathbf{d}) \,|\, \mathbf{u}(\mathbf{d}) \geq \mathbf{0}\} \qquad (12)$$

3 The generalized boundary curve

The minimization problem (12) could be solved using a standard optimization algorithm. But as stated in the introduction, due to the strongly nonlinear cost function this may result in a huge number of iterations.

On the other hand, trust-region algorithms have proven to be very effective within circuit design (e.g. [16]). As key task in each iteration of these algorithms a parameter correction has to be determined, that has a good ratio between error reduction and norm of the parameter correction, such that the linearization is still valid for the chosen step-size.

The CBC [16] turned out to be very well suited for this job in presence of linearized objectives. Therefore in this paper the GBC is presented as a generalization of the CBC, that inherits its most important properties but is also suited for the proposed strongly nonlinear cost function. The sizing algorithm itself is the same as the one presented in [16], except that the GBC is used to determine a parameter correction.

The main idea of the GBC is, not to use the linearized cost function but only the linearized objectives to calculate a boundary curve similar to the characteristic boundary curve. The objectives γ and the constraints $\mathbf{u}$ are linearized at the linearization point $\mathbf{d}_0$:

$$\underbrace{\gamma(\mathbf{d})}_{\bar{\gamma}(\mathbf{x})}=\underbrace{\gamma(\mathbf{d}_0)}_{\gamma_0}+\underbrace{\nabla_{\mathbf{d}}\gamma|_{\mathbf{d}=\mathbf{d}_0}}_{\mathbf{S}}\cdot\underbrace{(\mathbf{d}-\mathbf{d}_0)}_{\mathbf{x}}+\ldots \tag{13}$$

$$\underbrace{\mathbf{u}(\mathbf{d})}_{\bar{\mathbf{u}}(\mathbf{x})}=\underbrace{\mathbf{u}(\mathbf{d}_0)}_{\mathbf{u}_0}+\underbrace{\nabla_{\mathbf{d}}\mathbf{u}|_{\mathbf{d}=\mathbf{d}_0}}_{\mathbf{U}}\cdot\underbrace{(\mathbf{d}-\mathbf{d}_0)}_{\mathbf{x}}+\ldots \tag{14}$$

Based on the linearized objectives, the approximated objective function $\bar{\varphi}(\mathbf{x})$ is set up:

$$\bar{\varphi}(\mathbf{x}) = \sum_{i=1}^{n_\gamma} \exp(-a \cdot \bar{\gamma}_i(\mathbf{x})) \tag{15}$$

Similarly to the CBC the cost function (15) is modified in order to find a good compromise between the error reduction and the norm of the parameter correction:

$$\bar{\Phi}(\mathbf{x}, \lambda) = \bar{\varphi}^2(\mathbf{x}) + \lambda \cdot ||\mathbf{x}||^2, \quad \lambda \geq 0 \tag{16}$$

The factor λ is a weight for the parameter correction. Based on (16) the modified optimization problem is set up:

$$\mathbf{x}_c(\lambda) = \underset{\mathbf{x}}{\operatorname{argmin}} \{\bar{\Phi}(\mathbf{x})|\bar{\mathbf{u}}(\mathbf{x}) \geq \mathbf{0}\} \tag{17}$$

$$\bar{\varphi}_c(\lambda) = \bar{\varphi}(\mathbf{x}_c(\lambda)) \tag{18}$$

To solve this problem, no simulations are necessary, because it relies on the linearized objectives. As suggested in [16] the resulting parameter correction $||\mathbf{x}_c||$ and the cost function $\bar{\varphi}_c$ are transformed, such that all solutions are in the interval $[0, 1]$:

$$a_c(\lambda) = \frac{||\mathbf{x}_c(\lambda)|| - ||\mathbf{x}_c(\lambda \to \infty)||}{||\mathbf{x}_c(0)|| - ||\mathbf{x}_c(\lambda \to \infty)||} \tag{19}$$

$$r_c(\lambda) = \frac{\bar{\varphi}_c(\lambda) - \bar{\varphi}_c(\lambda \to \infty)}{\bar{\varphi}_c(0) - \bar{\varphi}_c(\lambda \to \infty)} \tag{20}$$

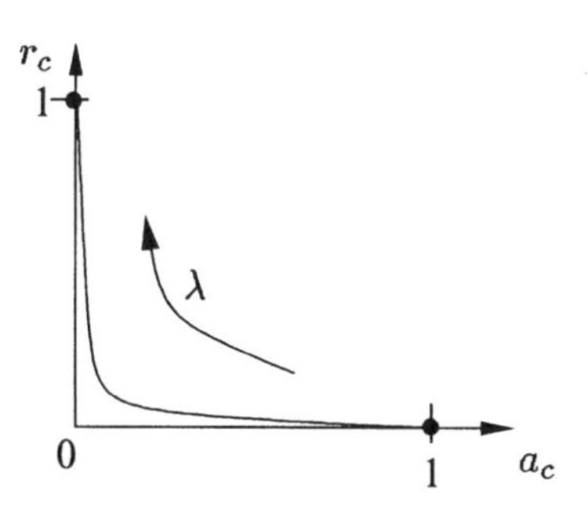

Figure 2. Typical example for a GBC.

A typical example of a GBC is shown in Fig. 2, where the GBC is the generalization of the CBC. The CBC can only be applied to linear objectives transformed into a scalar quadratic cost function. Whereas the GBC is suitable for any cost function, that is convex for the linearized objectives. For the GBC following theorems are true (proofs see [15]):

Theorem 1 *If $\bar{\varphi}(\mathbf{x})$ is convex and the active constraints are linearly independent, $\mathbf{x}_c(\lambda)$ is the parameter correction with minimum $\bar{\varphi}(\mathbf{x})$ for all $\mathbf{x} \in \mathcal{D}$ and $||\mathbf{x}|| \leq c = ||\mathbf{x}_c(\lambda)||$.*

□

Theorem 2 *If the cost function $\bar{\varphi}(\mathbf{x})$ is convex, then the resulting parametric curve $[\,a_c(\lambda), r_c(\lambda)\,]^{\mathrm{T}}$ is convex.*

□

Theorem 3 *The slope m_c of the GBC is given by:*

$$m_c = -\lambda \frac{||\mathbf{x}_c(\lambda)||}{\bar{\varphi}_c(\lambda)} \cdot \frac{||\mathbf{x}_c(0)|| - ||\mathbf{x}_c(\lambda \to \infty)||}{\bar{\varphi}_c(0) - \bar{\varphi}_c(\lambda \to \infty)} \tag{21}$$

□

These theorems show, that the GBC inherits all important properties from the CBC. Theorem 1 and 2 mean, that the GBC is suitable to determine a parameter correction in an automatic sizing algorithm. Theorem 3 guarantees that the GBC can be approximated with the same algorithms as the CBC and an identical algorithm can be used to search the kink of the curve. This kink represents a parameter correction, that on the one hand has a reasonable step size and on the other hand a significant error reduction.

4 Results

The introduced sizing method was used to design a folded cascode operational amplifier (Fig. 3) and an output buffer (Fig. 4).

The folded-cascode operational amplifier consists of 22 transistors, which result in 44 possible design parameters. Based on the structural constraints, the number of these can

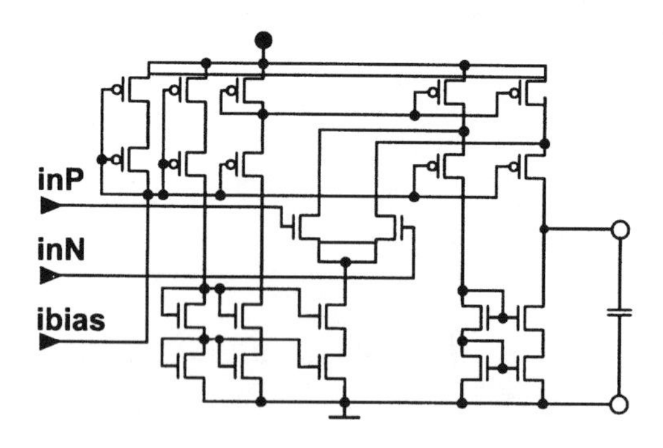

Figure 3. Folded-cascode operational amplifier

be reduced from 44 to 9 design parameters. The operating conditions are described by 4 operating parameters and the process statistic is modeled with technology data from Infineon Technologies. The detailed sizing results are shown in Table 1.

Perf.	Spec	Initial Val/WCD	Nominal Val/WCD	Centered Val/WCD
A_0[dB]	>65	95/0.2σ	76/2.7σ	76/4.2σ
f_t[MHz]	>30	21/-9.7σ	68/8.0σ	58/4.5σ
PHM[°]	>60	39/-6.6σ	67/1.6σ	71/3.9σ
$Slew_p$[V/μs]	>32	20.8/—	67/6.4σ	58/3.9σ
Power[mW]	<3.5	1.1/6.0σ	2.6/0.9σ	2.3/4.2σ
Yield (WCD)	—	0.0%	76.2%	99.9%
Yield (MC)	—	0.0%	77.3%	100%

Table 1. Sizing results and yield estimations based on worst-case distances (WCD) and a 1000 sample Monte Carlo analysis (MC) for the OP

All sensitivities were calculated with forward finite differences with Infineon's in-house simulator TITAN [9] on a Pentium II/450MHz. A further reduction in simulation time can be expected if simulator-built-in sensitivities are used.

Especially remarkable are the results for the gain A_0 of the operational amplifier. The design centering improves the WCD from 2.5 to 4.2 in 3 iterations, but the performance value is 76dB in both cases. This shows, that a parameter set is calculated where the circuit is significantly less sensitive to variations of the production process and the operating conditions. Such a solution can only be calculated by a design centering, that takes both statistical and operational parameters into account.

For the output buffer a technology transfer was done. This means, that the values for the design parameters are taken from an old technology and are scaled down. Afterwards a resizing is necessary in order to fulfill the specifications. Table 2 summarizes the results for the output buffer.

In both cases a good starting point for the design centering was calculated in the nominal design. This is enabled by the parameter distances as robustness objectives. Thus the number of iterations in the subsequent design centering

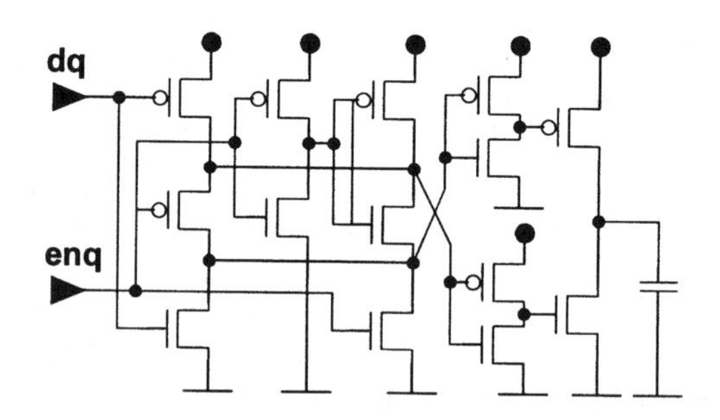

Figure 4. Output buffer

Perf.	Spec	Initial Val/WCD	Nominal Val/WCD	Centered Val/WCD
$Delay_{fall}$[ns]	<4	3.2/2.5σ	2.4/5.4σ	2.6/4.8σ
$Delay_{rise}$[ns]	<4	3.9/-0.2σ	2.8/2.9σ	2.7/3.4σ
$Slope_{fall}$[ns]	<4.5	5.0/-1.7σ	3.0/5.4σ	3.1/3.7σ
$Slope_{rise}$[ns]	<4.5	6.7/-3.4σ	3.3/2.5σ	3.3/3.4σ
$Noise_{gnd}$[mV/nH]	<25	12.3/6.8σ	14.9/4.1σ	14.4/3.9σ
$Noise_{vdd}$[mV/nH]	<25	12.3/6.8σ	14.4/4.3σ	15.3/3.4σ
Yield (WCD)	—	0.0%	99.2%	99.9%
Yield (MC)	—	0.0%	98.9%	100%

Table 2. Sizing results and yield estimations based on worst-case distances (WCD) and a 1000 sample Monte Carlo analysis (MC) for the output buffer

is kept small, which is essential for an industrial applicability of a design centering algorithm. For both circuits it was possible to calculate a 3σ design or better.

As stated in section 3, the circuit sizing could also be done with a standard optimization algorithm. But due to the strongly nonlinear cost function, this may result in a significantly larger number of iterations.

The results archived with the GBC-based algorithm are compared to the standard gradient-based algorithm discussed in [16]. Both algorithms resulted in nearly the same solution (performance differences less than 2%, WCD differences less than 0.1). But as shown in Table 3 the simulation effort for the gradient-based algorithm was significantly higher.

	OpAmp		Buffer	
	Grad.	GBC	Grad	GBC
Iterations nom. design	14	6	6	3
CPU nom. design	769s	305s	337s	147s
Iterations design cent.	5	3	6	2
CPU design centering	2812s	1567s	3310s	1160s
Average lin. error	95%	28%	58%	15%
Average $\|\|x\|\|$	0.13	0.35	0.05	0.11
Reduction in sim. time	48%		64%	

Table 3. Comparison of GBC and gradient-based minimization algorithm

For the gradient-based algorithm, due to the high linearization error, only small steps were possible. Thus, more iterations were needed to achieve the same results.

For the GBC based algorithm the linearization error was rather small. This shows, that the robustness objectives are only weakly nonlinear and thus the estimation of the error on linearized objectives gives a good idea of the real error.

5 Conclusion

In this contribution an automatic sizing method for analog cells based on robustness objectives was presented. The proposed method is suitable for both nominal design and design centering. Robustness objectives for the design centering are the worst-case distances (WCD); for the nominal design, the parameter distances are introduced as robustness objectives.

The chosen sizing algorithm requires the transformation of the robustness objectives into a cost function via a sum of exponential functions. This results in a strongly nonlinear cost function, even for linear objectives. To overcome this problem, the generalized boundary curve (GBC) is introduced. This GBC is based only on the linearized objectives themselves and not the strongly nonlinear cost function. Thus the linearization error is kept small. In every iteration step of the sizing algorithm, the GBC is used to calculate a parameter correction with a good ratio between error reduction and norm of the parameter correction.

Results show that a reduction in computational costs of about 50% with respect to [16] is achieved by the presented robustness objectives and generalized boundary curve.

Acknowledgements

The authors would like to thank Dr. U. Schlichtmann, Dr. H. Eichfeld and Dr. C. Sporrer from Infineon Technologies AG for supporting this work and Dr. J. Eckmüller from Infineon Technologies AG for the discussion of results.

References

[1] H. Abdel-Malek and A. Hassan. The ellipsoidal technique for design centering and region approximation. *IEEE Trans. on CAD*, 10:1006–1013, 1991.

[2] K. Antreich, H. Graeb, and C. Wieser. Circuit analysis and optimization driven by worst-case distances. *IEEE Trans. on CAD*, 13(1):57–71, 1994.

[3] J. Bandler and S. Chen. Circuit optimization: The state of the art. *IEEE Trans. on Microwaves Theory Techniques (MTT)*, 36:424–442, 1988.

[4] A. R. Conn, P. K. Coulman, R. A. Haring, G. L. Morill, C. Visweswariah, and C. W. Wu. JiffyTune: Circuit optimization using time-domain sensitivities. *IEEE Trans. on CAD*, 17(12):1292–1309, Dec. 1998.

[5] M. del Mar Hershenson, S. P. Boyd, and T. H. Lee. GPCAD: A tool for CMOS op-amp synthesis. In *IEEE/ACM Int. Conf. on CAD (ICCAD)*, 1998.

[6] S. Director, W. Maly, and A. Strojwas. *VLSI Design for Manufacturing: Yield Enhancement*. Kluwer Academic Publishers, USA, 1990.

[7] J. Eckmueller, M. Groepl, and H. Graeb. Hierarchical characterization of analog integrated CMOS ciruits. In *Design, Automation and Test in Europe (DATE)*, pages 636–643, Paris, France, Feb. 1998.

[8] K. Eshbaugh. Generation of correlated parameters for statistical circuit simulation. *IEEE Trans. on CAD*, 11:1198–1206, 1992.

[9] U. Feldmann, U. Wever, Q. Zheng, R. Schultz, and H. Wriedt. Algorithms for modern circuit simulation. *Archiv für Elektronik und Übertragungstechnik (AEÜ)*, 46:274–285, 1992.

[10] M. Krasnicki, R. Phelps, R. A. Rutenbar, and L. R. Carley. MAELSTROM: Efficient simulation-based synthesis for custom analog cells. 1999.

[11] K. Krishna and S. Director. The linearized performance penalty (LPP) method for optimization of parametric yield and its reliability. *IEEE Trans. on CAD*, 14(12):1557–1568, Dec. 1995.

[12] F. Leyn, W. Daems, G. Gielen, and W. Sansen. A behavioral signal path modeling methodology for qualitative insight in and efficient sizing of cmos opamps. In *IEEE/ACM Int. Conf. on CAD (ICCAD)*, 1997.

[13] M. D. Meehan and J. Purviance. *Yield and Reliablility in Microwave Circuit and System Design*. Artech House Boston/London, 1993.

[14] E. S. Ochotta, R. A. Rutenbar, and L. R. Carley. Synthesis of high-performance analog circuits in ASTRX/OBLX. *IEEE Trans. on CAD*, 15(3):273–294, March 1996.

[15] R. Schwencker. Automatic design centering of analog integrated circuits based on the generalized boundary curve of multiple robustness objectives. Technical Report TUM-LEA-99-1, Technical University Munich, 1999.

[16] R. Schwencker, J. Eckmueller, H. Graeb, and K. Antreich. Automating the sizing of analog cmos-circuits by consideration of structural constraints. In *Design, Automation and Test in Europe (DATE)*, Munich, Mar. 1999.

[17] J. C. Zhang and M. A. Styblinski. *Yield and Variability Optimization of Integrated Circuits*. Kluwer Academic Publishers, 1995.

[18] S. Zizala, J. Eckmueller, and H. Graeb. Fast calculation of analog circuits' feasibility regions by low level functional measures. In *IEEE Int. Conf. on Electronics, Circuits and Systems*, pages 85–88, Sept. 1998.

PART VIII

Analog Test

Metrics, Techniques and Recent Developments in Mixed-Signal Testing

Gordon W. Roberts
Microelectronics and Computer Systems Laboratory
McGill University
Montreal, CANADA H3A 2A7
Tel: 514-398-6029 Fax: 514-398-4470
http://www.macs.ee.mcgill.ca/~roberts/

Abstract

This paper presents a tutorial on mixed-signal testing. Our focus is on testing the analog portion of the mixed-signal device, as the digital portion is handled in the usual way. We begin by first outlining the role of test in a manufacturing environment, and its impact on product cost and quality. We will look at the impact of manufacturing defects on the behavior of digital and analog circuits. Subsequently, we will argue that analog circuits require very different test methods than those presently used to test digital circuits. We will then describe four common analog test methods and their measurement setups. We will also describe how analog testing can be accomplished using digital sampling techniques. Finally, we shall close this tutorial with a brief description of several developments presently underway on the design of testable mixed-signal circuits.

1. Introduction

With the growing importance of analog circuits in commercial mixed-signal ICs and systems, combined with the demand for shorter design and manufacturing cycles, the need for economical, fast and accurate test methods is readily apparent. At present, mixed-signal ICs are tested using *ad hoc* or unstructured test methods on a wide assortment of expensive analog and digital test equipment. Moreover, test of the analog portion of the mixed-signal circuit is usually considered as an afterthought of design, eliminating any influence that the test engineer may have over the test process, e.g., access to a particular node. In contrast, digital circuits are designed using methodologies that consider up front during the design phase how the circuit will be tested and interfaced to the test equipment. The advantage of such an approach is that the best compromise between functionality, performance, and test can be established, as test requirements become another design constraint. Furthermore, by standardizing the components of the test apparatus, possible test solutions can be incorporated into existing CAD tools to aid the designer in finding the optimum design.

While mixed-signal designs can benefit from similar design and test methodologies, none seems to be prevalent at this time. This stems largely from the nature of an analog circuit, its purpose, and how manufacturing errors influence their desired behavior. It is therefore the intent of this tutorial paper to describe the underlying reason for the difference between analog and digital test, followed by the metrics used to judge an analog or mixed-signal device, and their typical test set-ups. Subsequently, this paper will go on to describe some of the more recent developments presently underway in various research laboratories, including fault analysis, IEEE the proposed IEEE 1149.4 mixed-signal test bus, and methods for built-in self test.

2. The role of test

Consumers today demand high performance and quality in any of the electronic components that they may buy. Low prices and years of problem-free operation with minimal maintenance are now the norm. In order for manufacturers to deliver such products, an extensive testing program must be in place [1], [2]. This is to ensure that only good products are delivered to the consumer and that bad parts are either sent for repair or discarded. Distributing the testing throughout each stage of manufacture (i.e., at wafer, die, board and system assembly) will minimize the cost incurred by testing [3]. A commonly mentioned rule of thumb of test is the *rule of ten* which suggests that the cost of detecting a bad component in a manufactured part increases tenfold at each level of assembly. Thus, discovering its presence early is most desirable.

The question that emerges at this point is what kind of test should be performed, after all, testing does consume resources and takes time to perform. The answer lies with the observation that design errors are unrelated to those caused by manufacturing. Thus, the tests required

This work was supported by NSERC and by MICRONET, a Canadian federal network of centers of excellence dealing with microelectronic devices, circuits and systems for ultra large scale integration.

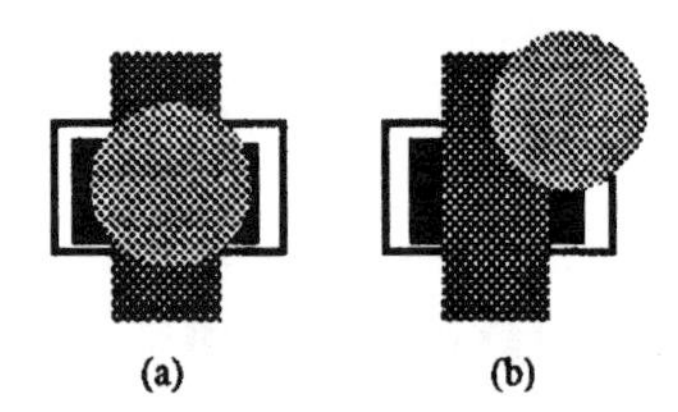

Figure 1: A spot defect creating (a) catastrophic and (b) parametric failure modes in an MOS transistor.

to determine whether a design is acceptable can be very different from those required to determine whether the design has been manufactured correctly. The same is true for analog circuits. However, as we shall describe next, the effects of manufacturing errors on analog circuits are quite different than those which occur with digital circuits. Thus, requiring very different test techniques.

3. Manufacturing defects

In this work we refer to manufacturing errors, or what is commonly referred to as a defects, as any error that leads to a device failure that is caused during the manufacturing phase. There are generally two types of manufacturing defects: those caused by spot or bridging defects, or what we shall refer to as environmental defects, and those caused by process variations. One example of a spot defect is a piece of dust or debris landing on the surface of a wafer of an integrated circuit during its fabrication. The result can then have two different effects classified as catastrophic or parametric. A catastrophic failure is one in which the component is destroyed or uncontrollable. For example, the gate of a transistor is completely removed from the channel region of the transistor by a spot defect as shown in Figure 1(a). A parametric failure, on the other hand, is one in which the component appears to function but may not be within the desired tolerance limits. A transistor that may turn on and off but carries less current than normal is one example.

Component Type	Absolute Tolerance	Matching Tolerance
NPN transistor:		
β	±20%	±5%
V_{BE}	±20 mV	±1 mV
NMOS transistor:		
V_T	±100 mV	±10 mV
k_p	±20%	±1%
Capacitor:		
MOS	±20%	±0.1% - ±1%
poly-poly	±20%	±0.1% - ±1%
Resistor:		
p-type diffused	±20%	±1%
epitaxial	±20%	±5%

Table 1: Absolute value and mismatch tolerances.

This situation may arise by a spot defect removing only a portion of the gate as illustrated in Figure 1(b).

Process variation is generally the result of equipment fluctuations in alignment and performance. In IC manufacturing it leads to an uneven layer deposition across the surface of the wafer. On account of their independent effects on electronic circuits in general, process variation are categorized into two types: global and local. Global variation refers to the systematic variation of a parameter that occurs between the extremities of the device, e.g. transistor threshold voltage may vary systematically from one side of the die to the other. Local variation refers to the small ($< 1\ \mu m^2$) random differences that occur between physically adjacent components. Using the IC example given previously, local variation gives rise to the mismatch error between two physically adjacent transistors. It is interesting to note that no two transistors, resistors, capacitors, etc., will have the same behavior; mismatch will always be present. Device mismatches are one of the fundamental performance limitations in analog circuits, as well as a major reason behind the difficulty encountered with mixed-signal testing. To shed some light on the magnitude of expected process tolerances, we provide in Table 1 a list of absolute value and matching tolerances for several standard IC components. As is evident, the effects of global variations on device behavior is much larger than that experienced by any local variation.

3.1 Impact of defects on digital circuits

To observe the effects that manufacturing defects have on digital circuits, let us consider the CMOS inverter circuit shown in Figure 2(a). Assume that the NMOS and PMOS devices have current gain factors, k_n and k_p, respectively, resulting in the DC transfer characteristic shown in Figure 2(b). Furthermore, let us assume that its pulse response is that shown in Figure 2(c). Hand analysis [4] reveals that the DC transfer characteristic depends on the ratio of k_n to k_p and not on any single current gain factor. So, provided the two transistors are physically close to one another, their current gains will be similar, differing by some small mismatch error. The result is a slight change in the DC transfer characteristic, but nothing very significant. As long as the input signals are within their intended range, the logic function of the inverter circuit will remain the same. The logic function of the inverter is therefore unchanged by the presence of the defect. In contrast, the rise and fall times of the inverter's pulse response is directly dependent on the absolute value of the current gain factors. Thus, to maintain the same logic function, it is important that the current gain factors are large enough to drive the logic level from one state to the next in the time required, i.e. clock period. Digital IC manufacturing ensures that this is achieved by testing the current gains of a small sampling of transistors at various locations on the wafer. If any device fails to meet the required current gain levels, the entire wafer is thrown out.

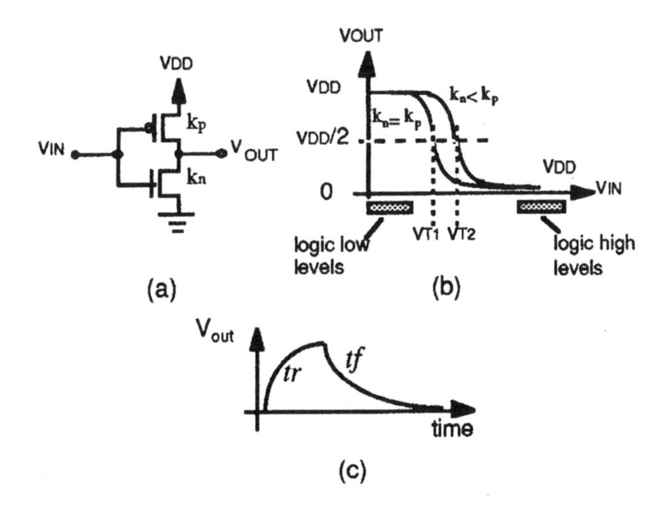

Figure 2: CMOS inverter circuit.

The effect of an environmental defect on a digital circuit is random. A dust particle knocking out the gate of a transistor will cause the logic function of the inverter to change. However, if it only removes a portion of the gate, the inverter may continue to perform the desired logic function as if no defect was present. This is a unique self-correcting property and one that analog circuits do not possess. The bulk of the testing performed on digital circuits is to identify those circuits whose logic function is altered by the presence of an environmental defect.

3.2 Impact of defects on analog circuits

Unlike digital circuits, the function of an analog circuit is sensitive to device mismatches. Analog circuits are also affected by other manufacturing defects, but in ways that are very similar to that described for digital circuits. To illustrate the dependence on device mismatches, consider the op amp inverter circuit shown in Figure 3(a). Assuming an ideal op amp, straightforward circuit analysis reveals that its gain is equal to $-R_2/R_1$. So, with equal resistors, the gain of the amplifier is expected to be -1 V/V. Unfortunately, due to local process variations, R_1 and R_2 will differ. Thus, the gain of the amplifier will not be equal to -1 V/V, but instead, it may equal, say -1.1 V/V. Moreover, repeating the same design at different locations on the same die will be subject to similar effects, however, not identical ones. That is, the gain of each amplifier may become, say -0.91, -0.94, or -1.05 V/V. In essence, the effects of local process variations on each resistor pair results in a level of uncertainty in the actual gain that can be achieved, see Figure 3(b). The same can be said for the function of any analog circuit, thus the expected level of uncertainty is an important design parameter, and not one that is left for chance. So, from a test point-of-view, it is meaningless to talk about measuring a function without assigning a range of acceptability. Table 2 summarizes the effect that various manufacturing defects have on the behavior of both digital and analog circuits.

The astute reader may be wondering why one does not directly measure the mismatch error in the circuit, and avoid the complexity associated with measuring the circuit's function. After all, this approach worked quite well for digital circuits. The answer to this question has two parts. Firstly, there is no obvious way in which to measure all the mismatch errors present in a circuit and, secondly, mismatch errors do not necessarily add to create a larger error; it is possible for errors to cancel. As an example, a circuit that realizes a gain of 1 V/V ± 5% can be constructed from a cascade of two op amp inverter circuits having a nominal gain of -1 V/V. If one of the inverter circuits has an actual gain of -0.7 V/V and the other -1.4 V/V then their combined gains would be 0.98 V/V, or an error of less than -2%. Clearly, on an individual basis, each stage when compared against the 5% tolerance band would be considered unacceptable. Test decisions based directly on measurements of individual component variations lead to an unacceptable number of good parts being rejected (i.e., false alarms) resulting in reduced yields.

Conversely, the acceptability of the system should not be determined from a test that is based on the performance of several analog circuits whose normal behaviors are independent of one another. This stems from the fact that a substantial number of defective parts can appear acceptable during the test, as errors in different circuits can mask one another. For example, a test decision based on, say, the power supply current (I_{DDQ}) of a mixed-signal device consisting of an acceptable A/D and D/A converter and a defective filter circuit can collectively appear as a good part. To minimize the risk, elementary statistical analysis suggests the only way to improve the situation is to reduce the tolerance band around individual components. Of course, this results in an increase in the

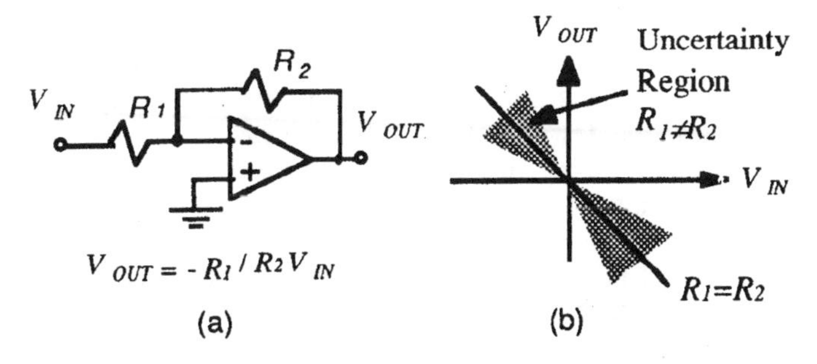

Figure 3: Op amp inverter circuit.

Manufacturing Errors	Impact on Circuit	
	Digital	Analog
Environmental	random	random
Process Variations:		
global	none provide within bound	none provide within bound
local	none	random

Table 2: Effects of manufacturing defects.

cost of the part and is generally frowned upon. Analog and mixed-signal circuits are usually pushing the envelop of technology and it is generally felt that any improvement in the tolerance bands can be used more effectively by improving the performance of the device rather than easing the test requirements. Of course, better understanding of the economic break-even point could better serve the electronics' manufacturer.

The conclusion from the above discussion is that there does not appear to be any easy way out of the mixed-signal test quandary. It appears at this time that the most economical method of testing analog and mixed-signal devices is to test their function directly. If particular catastrophic failures are expected, similar to those described for digital circuits, then a prescreening test may be called for. Here the power supply current monitoring method alluded to above may play a useful role [5]. However, in the end, some form of functional testing is necessary before the part can be accepted. Today, most analog or mixed-signal circuits exist at the interface between the analog world and the digital compute engine, and are intended to have high-performance capabilities. Thus, testing their functionality requires very expensive test equipment, performing lengthy and elaborate test routines. It is therefore not surprising that only a small number of mixed-signal devices are available from commercial electronic vendors when so much effort must be devoted to their quality assurance. Section 7 will consider some possible ways of improving the present situation.

4. Analog circuit functions

Of the many roles that analog circuits play in electronic systems, probably the most important is that in signal processing. In much the same way that a numerical analyst writes a computer program, analog circuit designers configure electronic sub-circuits so that a particular function is realized. They are usually guided by a mathematical description of the function they want, expressed in terms of algebraic, integral, and differential operations. For example, an eighth-order switched-capacitor elliptical filter would be described by an eighth-order difference equation. Similarly, at the core of an algorithmic A/D converter would be a divide-by-two circuit that would be used to aid in the conversion of the floating point (analog) number representation to its binary number equivalent. Owing to the nonlinear behavior of the underlying components, e.g. transistors, only an approximation of the function is actually realized. Through the application of negative feedback, the approximation is improved and made less sensitive to process variations. It is therefore the goal of the circuit designer to obtain the desired function within the acceptable error subject to the underlying process variations. What constitutes acceptable error depends on the application and is impossible to generalize. Thus, every analog circuit is characterized by its own special set of measurements that attempt to quantify this approximation in the environment that it is intended to be used.

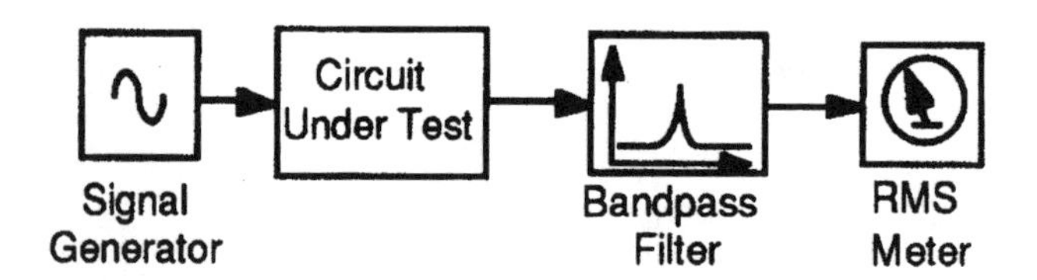

Figure 4: Typical analog test setup.

5. Test set-up and measurements

The most basic analog measurement setup consists of a signal generator exciting the circuit-under-test and an instrument to extract the appropriate parameter from the circuit's output response. Depending on the purpose of the test, the signal generator may be generating a DC, sinusoid, square-wave, or some arbitrary waveform shape. Which signal is used depends on the type of measurement that is to be taken. There are four main measurement categories:

1) *DC measurements*: they measure the static behavior of the circuit such as leakage currents, output resistance, transfer characteristics and offsets.
2) *AC measurements*: they measure both the small- and large-signal frequency response behavior of the circuit. Distortion measurements are also included in this test.
3) *Transient or time-domain measurements*: they measure the behavior of the circuit subject to signal shapes that the circuit will experience in its intended application.
4) *Noise measurements*: they measure the variation in the signal that appears at the circuit's output when the input is set to zero.

A common setup that is able to perform several of the above measurements (1, 2 and 4) is shown in Figure 4. It consists of a sinusoidal signal generator with variable amplitude and frequency control. The output of the circuit-under-test is then filtered by a narrowband bandpass filter. The center frequency of the filter is tunable, and may or may not track the frequency of the input signal. Finally, the power associated with the filtered output signal, once settled, is then measured using a true-RMS power meter. Transient-type measurements are not more complicated, but usually require very specialized equipment to generate and capture the appropriate test signal, e.g., bit-error rate.

Tests involving sinusoidal excitation is probably the most common among linear circuits, such as amplifiers, data converters and filter circuits. Amidst all waveforms, the sinusoid is unique in that its shape is not altered by its transmission through a linear circuit, only its magnitude and phase are changed. In contrast, a non-linear circuit will alter the shape of a sinusoidal input. The more non-linear the circuit is, the greater the change in the shape of the sinusoid. One means of quantifying the extent of the non-linearity present in a circuit is by observing the power distributed in the frequency components contained in the output signal using a Fourier Analysis. Using the setup shown in Figure 4, this would be obtained by exciting the circuit-under-test using a sinusoid and by measuring the power appearing at the output of the bandpass filter as it is tuned to discrete frequencies across the frequency band of interest. Figure 5 illustrates a typical power spectral density plot obtained using these methods. The fundamental component of the output signal is clearly visible, followed by several harmonics. Also shown in the plot is the noise floor of the circuit-under-test. This floor represents the smallest signal that can be distinguished from the noise generated by the circuit. By comparing the power contained in the harmonics to that in the fundamental signal, a measure of Total Harmonic Distortion (THD) is obtained. By comparing the fundamental power to the noise power over a specified bandwidth, one obtains the signal-to-noise ratio (SNR). By altering the frequency and amplitude of the input sinusoidal signal, or by adding an additional tone with the input signal, other transmission parameters can be derived from the power spectral density plot [6].

6. DSP-Based Testing

Since the early eighties, digital signal processing (DSP) has altered the traditional test setup shown in Figure 4 to that shown in Figure 6. Through the application of analog-to-digital (A/D) and analog-to-digital (D/A) data converters, and a very fast compute engine for performing vector manipulations, the function of each analog instrument can be emulated by a software program [7], [8].

The basic idea behind the DSP-based test station is that a signal, possibly sinusoidal, is numerically computed by the digital compute engine and then applied to the D/A block for conversion into analog form. The resulting analog signal is then applied to the circuit-under-test from which its response is digitized by the A/D converter and passed on to the digital compute engine for further processing. Depending on the measurement that is required, the appropriate software would be loaded in place.

The first obvious advantage that comes from this

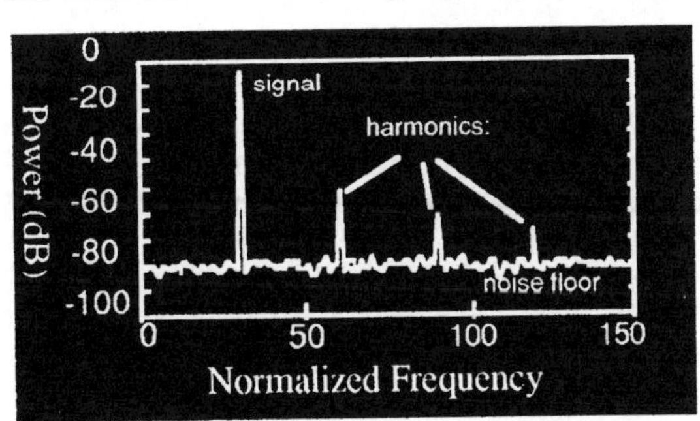

Figure 5: Power spectral density plot.

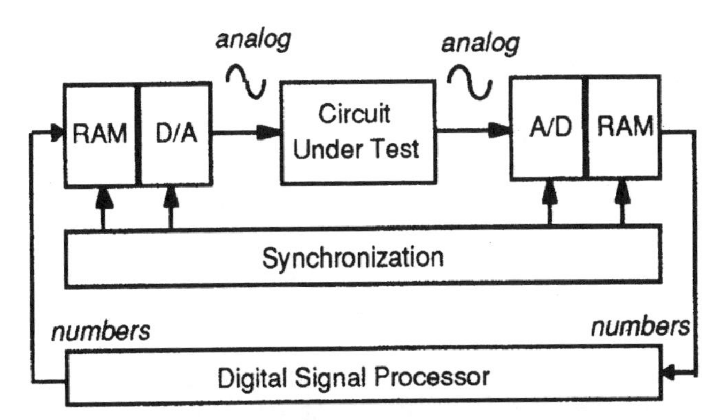

Figure 6: DSP-based measurement system.

approach is the flexibility that programmability provides. The same hardware can be used to perform a multitude of test functions. Secondly, the correction factors associated with a system calibration can be easily incorporated into the routine of any emulated instrument. Thus, correcting for the effects of drift and aging that comes from running test equipment continuously for 24 hours a day, 7 days a week. A less obvious advantage, but equally important, is the ability to pipeline the different phases of the test procedure. On account of the discrete events that are taking place, operations that do not need to run in real-time can be delayed and run in parallel with the operations of the next device. Finally, coherent testing provides a means in which to gather information using the least number of samples. Owing to its importance to DSP-based testing, a brief description is given below.

6.1 Coherent testing

One may be wondering at this point whether digitizing the excitation and response of the circuit-under-test somehow degrades the accuracy of the measurement or loses information. The answer is simply no, as Shannon so succinctly pointed out [9]. His observation suggests that the DSP-based test station has the same amount of information to work with as does the test station shown in Figure 4. More recently, Mahoney [7] pointed out that a DSP-based test station using an N-point Fast Fourier Transform (FFT) with rectangular windowing (i.e., no additional post-processing) can perform a much faster and more accurate frequency-selective power measurement than the setup of Figure 4. This is a result of coherent sampling and by eliminating the analog implementation of the squaring operation in the true-RMS power meter. He coined this test arrangement as *coherent testing*. As test time is of primary importance in a production environment, this observation was significant and lead to the creation of a new family of mixed-signal testers. Basically, coherent testing establishes which input test frequencies can be used to perform an accurate sinusoidal test using an N-point FFT. According to [8], the test frequency F_T is selected according to the following

$$F_T = \frac{M}{N} F_S , \tag{1}$$

where F_S is the sampling rate and M represents an arbitrary integer, usually less than $N/2$ and has no factors common with N, i.e., M and N are relatively-prime. The latter ensures that each point is unique in the N-point set, thereby maximizing the information content. An alternative description of the relationship given in Eqn. (1) is that the test frequency should only be selected as a harmonic of F_S/N, the so-called primitive frequency.

6.2 Multi-tone testing

By combining several sinusoids harmonically related to the primitive frequency of the test sequence, a very effective signal for probing the frequency characteristics of a circuit in a single measurement is obtained. Such a signal is known as a multi-tone signal [8]. Multi-tone signals are also very important signals for probing the nonlinear behavior of narrowband circuits, as they provide intermodulation distortion components that lie in-band. Multi-tone signals can also be used to generate a pseudo-random noise signal, by including a fairly broad selection of tones [10]. With independent control over both the amplitude and phase of each tone, energy in the signal can be distributed over time in an optimum manner thereby avoiding a high crest factor.

6.3 RF measurements

Through the application of heterodyning or undersampling, DSP-based test stations can extend their frequency measurement capabilities to the IF or RF frequency bands [11]. The basic idea is to excite a circuit-under-test using an IF or RF signal generated by a frequency synthesizer and translate its response back down into the baseband range of the digitizer for DSP processing. By doing so, most of the advantages of DSP for test are maintained.

6.4 Noise effects

Noise is present in all signals captured by the digitizer. Noise, in general, has many different sources: circuit-generated, static and man-made. We are usually only concerned about circuit noise, as it sets the lower limit of the maximum dynamic range available in a circuit. The influence from the other two sources of noise is minimized through good board layout and power supply decoupling. For the most part, circuit noise creates a randomness in the captured signal, resulting in a different measurement each time the test is run. Thus, noise adds an additional uncertainty with the circuit's function and must be accounted for when selecting the test limits. Statistical theory reveals that the amount of variation in a measurement will decrease inversely with the square-root of the number of samples in the measurement set. However, increasing the number of samples, increases the test time. Therefore a good compromise between acceptable variation and test time must be established. An in-depth discussion of the effects of noise on DSP-based measurements can be found in [12].

7. New test developments

It should be clear from the above discussion that mixed-signal devices will not benefit from the same test advancements that digital circuits presently experience. However, that is not to say that all is lost on the mixed-signal test front. Much can be done and should be done if the cost of mixed-signal testing is to be reduced. As in digital test, most of the mixed-signal research community advocates some form of *design-for-testability* as it is believed that the best compromise between functionality, performance, and test can only be achieved early in the design cycle. It is therefore the purpose of this section to briefly describe the directions and recent results of several advancements in this area.

7.1 Analog fault analysis

An area of research that is gaining some ground in mixed-signal testing is the concept of fault modeling [13] [14]. Fault modeling is a concept that has its roots in digital testing [15]. Fault modeling for analog circuits serves to identify the test conditions that will expose the presence of a fault in a circuit with the least amount of test effort. For catastrophic defects, fault modeling is a very effective way of identifying the optimum test setup. Unfortunately, for parametric failures, it quickly becomes unwieldy due to the amount of simulation time required. For a moderately sized analog circuit, such as a second-order delta-sigma modulator used in data conversion applications, it is not uncommon to run a SPICE simulation continuously for a week on a SUN SPARC 10 workstation before a single measure of the THD is available. With that said, effort is underway attempting to rank-order various tests for popular mixed-signal circuits. By assigning probabilities to potential failure modes, the order of the tests are arranged so that, on average, the least amount of time is spent searching for defective parts [16].

7.2 Circuit schemes for test

With such high levels of integration possible today using submicron VLSI technologies, it is both feasible and beneficial to consider placing all or part of the test circuitry directly on the same die as the desired circuit. Referring back to Figure 4 this would include the test stimulus, parameter extraction or measurement circuitry, and equally important, the interconnect and control circuits. Some of the benefits are: (1) it facilitates design-for-test, (2) provides a hierarchical test solution, as the test circuits can be used at all levels of the system, from the IC-level to the board and system-levels, thereby maximizing the return on the test hardware investment, and (3) standardization which simplifies automation and the integration of test into present day CAD facilities.

The following is a brief description of some of the circuit techniques that have been proposed for making mixed-signal circuits more easily testable. We limit our discussion to those schemes that have been reported to be prototyped or bench tested.

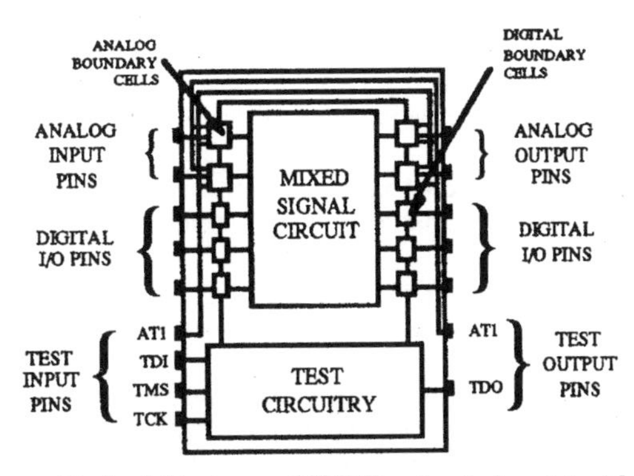

Figure 7: Architecture of IEEE mixed-signal test bus.

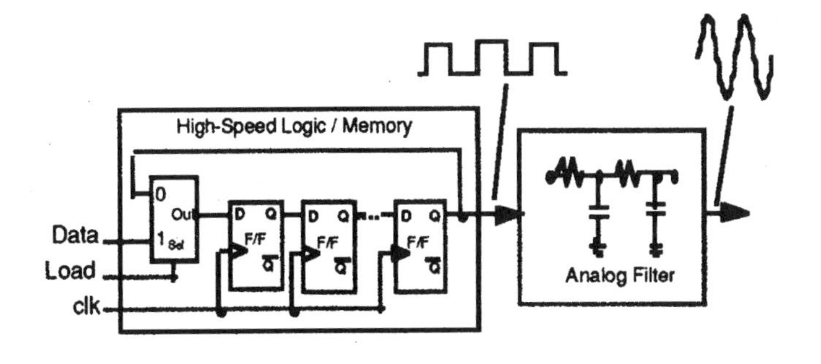

Figure 8: Analog test signal generation using a one-bit digital bit-stream.

Analog test bus

One of the most significant advancements in mixed-signal test is the proposed IEEE P1149.4 mixed-signal test bus standard. Over the past four years a group of international companies and R&D institutions have been working together to define the standard and discuss its compatibility with the IEEE 1149.1 digital test bus standard. The basic idea of the mixed-signal test bus is the inclusion of a set of analog boundary cells and two analog buses connected to two dedicated pins (AT1 and AT2) that allow the analog portions of the mixed-signal device to be tested in much the same manner as with the digital boundary-scan technique [17]. Figure 7 illustrates the basic architecture of the proposed IEEE 1149.4 mixed-signal test bus. Interested readers can learn more about this proposal and its current status by referring to the most recent proceedings of the International Test Conference.

Scan-based signal generation

Very recently a method [18] has been developed which makes use of the memory elements (i.e., flip-flops) in the digital boundary cells of the IEEE 1149.1 test standard to provide storage for a short periodic one-bit sequence that when filtered provides a high-quality analog test stimulus (see Figure 8). For an N-bit sequence being clocked at a rate of F_S, tones harmonically related to the primitive frequency F_S/N are available for excitation in much the same way as that described in Section 6. Similarly, if on-chip RAM is available then it can be used to store and play-back the appropriate bit-pattern. Except for the parameter extraction circuitry, a simple RC filter circuit and some interconnect, the digital boundary scan configuration provides the rest.

Built-in self test

One of the earliest proposal for a fully integrated built-in self-test was that made by a group of AT&T engineers for verifying the monotonicity of an Nyquist-rate A/D converter circuit [19]. An illustration of the proposal is shown in Figure 9. A linear ramp voltage is generated on chip and applied to the input of the A/D converter during test. The output codes are then checked for monotonicity by comparing the present output code with the past code. If true, the output counter is incremented. If false, the test is terminated and a fail flag is set. A counter keeps track of the number of successful comparisons. The final count is then checked against the expected value and a go/no-go type decision is then made. The final count can also be retrieved for possible diagnostics. A straightforward extension is to add additional registers whereby a histogram of the output codes can be obtained over an extended period of test time. The data in this histogram can then be used to determine the linearity of the data converter according to the integral nonlinearity error (INL) and differential linearity error (DLE) type tests.

Another self-test scheme is the so-called MADBIST scheme. This technique is applicable to devices containing A/D and D/A data converters, and some computing resources [20], although the technique is equally valid for devices containing only A/Ds. The basic idea is that an all-digital on-chip $\Delta\Sigma$ modulation oscillator circuit [21] generates a single-bit digital sequence as the test stimulus for the A/D converter. Within this binary sequence is a well-behaved sinusoid (or multi-tone) and a residual signal whose spectral properties are orthogonal to one another. This signal is then applied to the input of the A/D circuit whose pre-aliasing filter (AAF) suppresses the residual signal and allows the sinusoid to pass unattenuated. The A/D is then excited in the usual way and its output response is then processed using an FFT, if available on-chip, or a narrowband digital filter technique together with a peak detector. A go/no-go type decision is then made. Subsequently, another phase of the test is run whereby the D/A circuit is tested by exciting it with a similar type of digital test signal and whose output is then digitized by the A/D circuit. Once the D/A is considered functional,

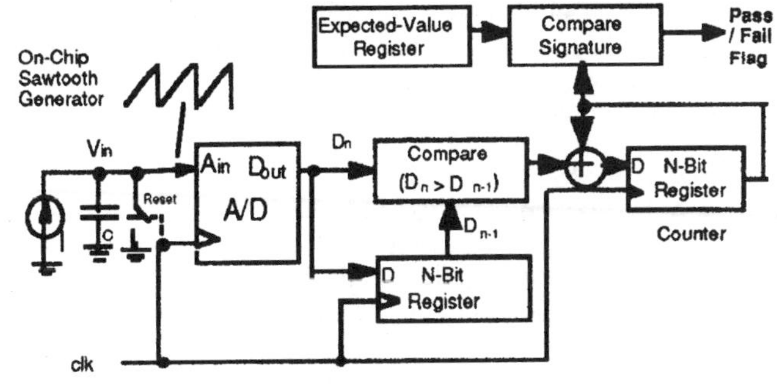

Figure 9: BIST for an A/D converter.

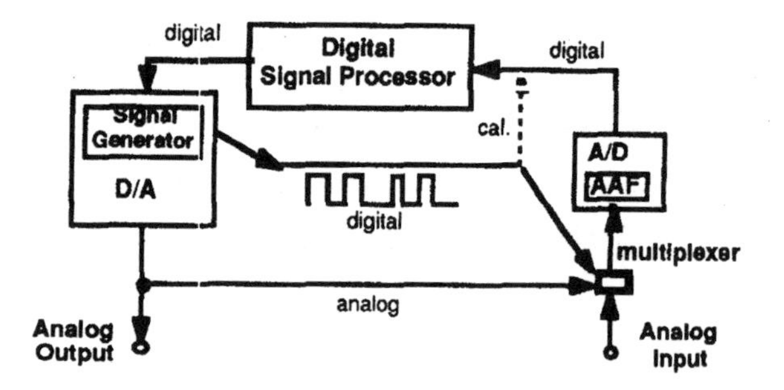

Figure 10: The MADBIST scheme.

both data converters can then be used to measure other analog circuits, in much the same way as the DSP-based test setup shown in Figure 6. The latter approach has been proposed as a means of testing bandpass-type mixed-signal devices such as those used in wireless communication systems [22].

Concurrent error detection

Another important research activity is in the area of concurrent error detection for analog and mixed-signal circuits. These techniques are used for detecting the presence of a defect that has manifested itself after the product was tested and sent to the customer for normal operation. A survey paper describing these techniques is provided in [23].

8. Conclusions

The function of an analog circuit is sensitive to local process variations whereas digital circuits are not. At the time of this writing, the most economical way of testing analog and mixed-signal circuits is to measure their function directly. A level of uncertainty caused by local process variations must be accounted for, as well as the uncertainty created by circuit generated noise. It is also important to realize that the function of an analog circuit is only approximately realized, resulting in another level of uncertainty on account of the difficulties in which to track them. This tutorial paper described several common test techniques, with an emphasis on spectral-based measurements using digital sampling techniques. New developments that help to make mixed-signal circuits more testable were also described.

References

[1] B. Davis, *The Economics of Automatic Testing*, McGraw-Hill, London, UK, 1982.

[2] S. D. Millman, "Improving quality: yield versus test coverage," Journal of Electronic Testing: Theory and Applications, Vol. 5, pp. 253-261, 1994.

[3] B. Davis, "Economic modeling of board test strategies," Journal of Electronic Testing: Theory and Applications, Vol. 5, pp. 157-170, 1994.

[4] D. A. Hodges and H. G. Jackson, *Analysis and Design of Digital Integrated Circuits*, 2nd Ed, McGraw-Hill Publishing Company, New York, 1988.

[5] S. Bracho, M. Martínez and J. Argüelles, "Current test methods in mixed-signal circuits," Proc. Midwest Symposium on Circuits and Systems, Rio de Janeiro, Brazil, pp. 1162-1167, Aug. 1995.

[6] K. I. Feher and Engineers of Hewlett-Packard, Telecommunications Measurements, Analysis and Instrumentation, Prentice-Hall, Englewood Cliffs, NJ, 1987.

[7] M. V. Mahoney, "New techniques for high speed analog testing," Proc. International Test Conference, pp. 589-597, Oct. 1983.

[8] M. V. Mahoney, *DSP-Based Testing of Analog and Mixed-Signal Circuits*, IEEE Computer Society Press, 1987.

[9] R. M. Gray and J. W. Goodman, *Fourier Transforms; An Introduction for Engineers*, Kluwer Academic Publishers, Norwell, Massachusetts, 1995.

[10] H. Kitayoshi, S. Sumida, K. Shirakawa and S. Takeshita, "DSP synthesized signal source for analog testing stimulus and new test method," Proc. International Test Conference, pp. 825-834, Oct. 1985.

[11] LTX Product Literature, *Synchro Test System*, 1995.

[12] M. F. Toner, *MADBIST: A Scheme for Built-In Self-Test of Mixed Analog-Digital Integrated Circuits*, Ph.D. Dissertation, McGill University, Montreal, Canada, Aug. 1996.

[13] M. Sachdev and B. Atzema, Industrial relevance of analog IFA: A fact or a fiction," Proc. International Test Conference, Oct. 1995.

[14] A. Meixner and W. Maly, "Fault modeling for the testing of mixed-integrated circuits," Proc. International Test Conference, pp. 564-572, Oct. 1991.

[15] W. Maly, "Realistic fault modeling for VLSI testing," Proc. Design Automation Conference, pp. 173-180, 1987.

[16] L. Milor and A. Sangiovanni-Vincentelli, "Optimal test set design for analog circuits," Proc. International Conference on Computer-Aided Design, pp. 294-297, 1990.

[17] A. Osseiran, "Getting to a test standard for mixed-signal boards," Proc. Midwest Symposium on Circuits and Systems, Rio de Janeiro, Brazil, pp. 1157-1161, Aug. 1995.

[18] E. M. Hawrysh and G. W. Roberts, " An integration of memory-based analog signal generation into current DFT architectures," Proc. International Test Conference, Washington, Oct. 1996.

[19] M. R. DeWitt, G. F. Gross and R. Ramachandran, "Built-in self-test for analog to digital converters," AT&T Bell Laboratories, Murray-Hill, NJ, US Patent, No. 5,132,685, filed Aug. 9, 1991; granted Jul. 21, 1992.

[20] M. F. Toner and G. W. Roberts, " A BIST scheme for an SNR test of a sigma-delta ADC," Proc. International Test Conference, Baltimore, Maryland, pp. 805-814, Oct. 1993.

[21] G. W. Roberts and A. K. Lu, *Analog Signal Generation For Built-In Self-Test Of Mixed-Signal Integrated Circuits*, Kluwer Academic Publishers, Norwell, MA, USA, 1995.

[22] B. R. Veillette and G. W. Roberts, "A built-in self-test strategy for wireless communication systems," Proc. International Test Conference, Washington, pp. 930-939, Oct. 1995.

[23] M. Lubaszewski, S. Mir, A. Rueda and J. L. Huertas, "Concurrent error detection in analog and mixed-signal integrated circuits," Proc. Midwest Symposium on Circuits and Systems, Rio de Janeiro, Brazil, pp. 1151-1156, Aug. 1995.

A Tutorial Introduction to Research on Analog and Mixed-Signal Circuit Testing

Linda S. Milor, *Member, IEEE*

Abstract—**Traditionally, work on analog testing has focused on diagnosing faults in board designs. Recently, with increasing levels of integration, not just diagnosing faults, but distinguishing between faulty and good circuits has become a problem. Analog blocks embedded in digital systems may not easily be separately testable. Consequently, many papers have been recently written proposing techniques to reduce the burden of testing analog and mixed-signal circuits. This survey attempts to outline some of this recent work, ranging from tools for simulation-based test set development and optimization to built-in self-test (BIST) circuitry.**

Index Terms—**Analog circuits, analog system fault diagnosis, analog system testing, built-in testing, integrated circuit testing, mixed analog–digital integrated circuits, testing.**

I. Introduction

HISTORICALLY, electronic circuits were almost exclusively analog and were designed with discrete components. The components were mounted on printed circuit boards and tested with a "bed of nails" tester, allowing access to all input and output voltages of components. Since the components of an electronic system could be individually tested, speed in identifying the cause of failures was more of a problem. Testing research focused on the development of methods to rapidly diagnose component failures and assembly errors during field servicing of weapons, navigation, and communication systems.

The advent of integrated circuit (IC) technology and the scaling of transistor sizes have allowed the development of much larger electronic systems. Digital design techniques have become predominant because of their reliability and lower power consumption. However, although large electronic systems can be constructed almost entirely with digital techniques, many systems still have analog components. This is because signals emanating from storage media, transmission media, and physical sensors are often fundamentally analog. Moreover, digital systems may have to output analog signals to actuators, displays, and transmission media. Clearly, the need for analog interface functions like filters, analog-to-digital converters (ADC's), phase-locked loops, etc., is inherent in such systems. The design of these interface functions as integrated circuits has reduced their size and cost, but in turn, for testing purposes, access to nodes is limited to primary inputs and outputs, making it more difficult to locate component failures when circuit specifications are not satisfied. Nevertheless, algorithms aimed at diagnosing component failures on boards can be applied to identifying faulty components in analog IC's. And in fact, given limited accessibility to internal nodes in analog integrated circuits, a large number of algorithms and theoretical findings for fault diagnosis and test signal selection have been developed throughout the 1970's and 1980's [1], [2].

Manuscript received November 7, 1996; revised December 23, 1997. This paper was recommended by Associate Editor G. W. Roberts.

The author is with the Submicron Development Center, Advance Micro Devices, Sunnyvale, CA 94086 USA.

Publisher Item Identifier S 1057-7130(98)07529-6.

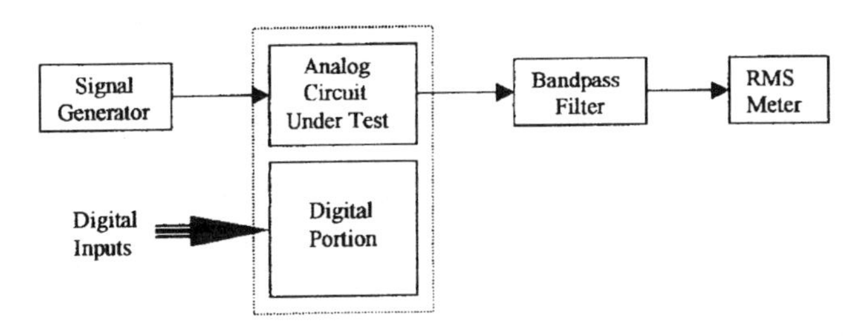

Fig. 1. Test setup for a mixed-signal device [5].

Recently, due to the exploding telecommunications market, as well as markets for consumer and automotive electronics, more and more mixed-signal devices are being designed, integrating digital and analog components on a single chip in order to improve performance and reduce board size and cost. In the production of mixed-signal circuits, test can be a limiting factor, contributing significantly to manufacturing cost [3]. A typical strategy for testing a mixed-signal chip involves, when possible, first testing the digital and analog components, followed by some system tests to check the at-speed interaction among components. In this case, the digital parts would be tested with standard methods, aided by software for automatic test pattern generation, scan chains, and built-in self-test (BIST), which has become mature and cost effective. Testing the analog parts and the combined system is less well understood, where test sets are typically based on a designer's experience and specifications on a circuit's functionality.

A typical test setup for testing the analog components is shown in Fig. 1. Such a setup involves applying digital inputs to the digital block, inputting a signal, which excites the analog portion of the mixed-signal circuit with a dc, sinusoid, square-wave, or some random signal, having a known probability distribution function, and measuring the response with an rms power meter, operating over a narrow and tunable frequency band. Sinusoidal inputs are commonly used to test linear analog circuits, such as amplifiers, data converters, and filters, to verify the magnitude and phase of an output signal as a function of the input frequency. Additionally, sinusoidal inputs are also used to quantify the extent of nonlinearity in an output signal by comparing the power contained in the harmonics or noise to that of the fundamental signal, referred to as total harmonic distortion [4]. As an alternative to deterministic input signals, random inputs, with a known spectral distribution, may be used to test an analog circuit's transfer characteristic, via analyzing the spectrum of the response.

Many factors limit the straightforward application of such an approach to testing mixed-signal circuits. First, market pressures require very efficient development of bug-free test sets, which not only check the functionality of the analog components, but also the at-speed operation of the entire system. Especially for at-speed testing, the interactions between the digital and analog portions of the chip can be complex and unique to the application, and as a result need to be fully understood. This is certainly nontrivial and requires extensive labor-intensive engineering work. It turns out that for many mixed-signal circuits, a significant contribution to time-to-market comes from the time required for test program development and debugging. Test program development often begins after a design is complete due to limited CAD tool support for test set development, and it usually takes many iterations between design and test to realize a testable design which satisfies specifications. This contrasts strongly with test program development for digital circuits. For digital circuits, CAD tools are extensively used to generate test patterns, which are then tested on prototypes of the circuit, including register transfer level software descriptions, gate-level software descriptions, and field-programmable gate array (FPGA) prototypes. This both validates the circuit design and the test program prior to the availability of silicon. Hence, this tutorial will begin with a discussion of recent tools that have been developed to automate the test development cycle so that test sets can be created concurrently with the design phase.

Then, even if mixed-signal test programs can be developed efficiently, complete testing of some analog circuit specifications can be very costly. Consider, for example, measuring the integral nonlinearity (INL) of an ADC. For a 13-bit ADC, this would require locating 8192 (2^{13}) input voltages which cause transitions in the output between codes, at multiple temperatures. Such a large number of long tests can limit throughput during production testing unless numerous test stations are in simultaneous operation. And given the high cost of high-speed mixed-signal test equipment, coupled with the time that each device spends on a tester, testing can add several dollars to the cost of a device. For example, locating all codes can add more than a dollar to the cost of a 13-bit ADC, which typically sells for $15 [6]. Therefore, in Section III, methods for reducing the cost to production testing will be presented. During production testing, the goal is to distinguish good circuits from faulty ones with minimum cost, where cost is influenced by test time, throughput, and the cost of test equipment. Unlike with board designs, fault location is not a target because it is not possible to repair or replace faulty components. On the other hand, during design characterization, if a circuit has been identified as faulty, it is desirable to find the cause of failure. Hence, in Section IV, approaches to fault location and identification are discussed. This problem involves selecting input signals and measurements, in addition to the decision algorithm.

Finally, the inputs to the analog components of a mixed-signal circuit may not be accessible to the tester. Moreover, it is not feasible for a designer to bring all of the analog inputs and outputs out to the package pins, and probe loading effects can degrade measurements made on naked die. Consequently, extra components are often required to access internal nodes through primary inputs and outputs. But in this case, the parasitics introduced when accessibility is augmented can degrade some circuit performances. In Section V, design techniques for improving the testability of embedded analog components will be summarized. Specifically, circuits for increasing controllability and observability of internal signals at analog component nodes will be presented. Additionally, work on analog and mixed-signal BIST will be discussed. Analog and mixed-signal BIST goes beyond simply improving controllability and observability of internal nodes by attempting to reduce the need for high-performance test equipment through implementing test signal generators and analyzing test results on chip. Analog and mixed-signal BIST allows a designer increased flexibility to make the tradeoff between the increased silicon area needed for BIST circuitry and external tester requirements. Lastly, Section VI concludes this paper with a summary.

II. Test Program Development

Creating the test programs to run mixed-signal testers is a major bottleneck in the product delivery cycle for many mixed-signal circuits. Unlike digital test program development, which is automated with the support of CAD tools for test program generation and verified with the help of software and hardware descriptions of the circuit prior to the availability of silicon, analog and mixed-signal test program development is labor-intensive, time-consuming, and must be done using fabricated devices and on the tester. Hence, because the test engineer works with prototypes of a circuit and the tester hardware and software to develop and debug test programs, test program development begins after a design is complete and prototypes have been manufactured. In fact, the delays due to waiting for prototypes and the lack of automation for mixed-signal test program development leads to significant increases in the product development cycles, which can potentially be reduced if some debugging of the tester-circuit interface (load board) and the test program can be done prior to the availability of first silicon.

Recently, work on simulation tools to address this need has begun, targeting emulating the testing of mixed-signal devices. The idealistic goal of this work is depicted in Fig. 2. Clearly, such an ideal situation is not possible, due to inaccuracies in simulation models. Nevertheless, the diagram points out that any reduction in test debug time after first silicon directly translates into a reduction in time-to-market. Consequently, it is worthwhile to target such a goal. To achieve this goal, the test engineer must create a software description of the test program, including models of tester hardware, the test board, code to program the tester to perform various tests in the

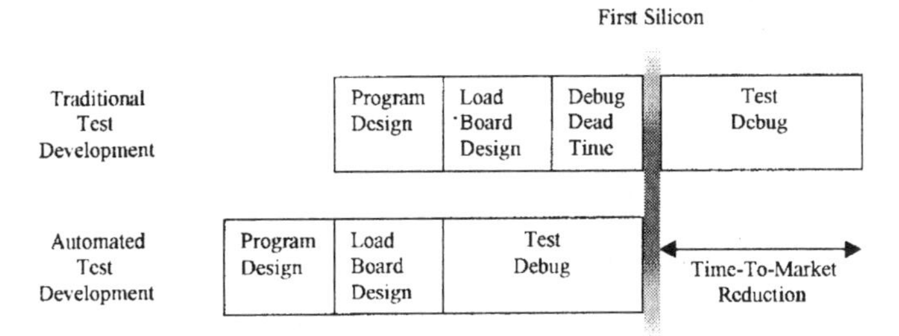

Fig. 2. Ideal potential time-to-market reduction due to the use of electronic design automation (EDA) tools for test program development.

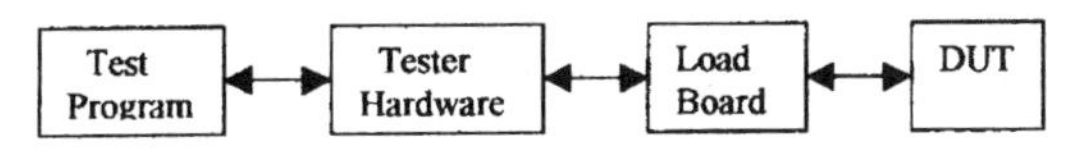

Fig. 3. The test environment.

language of the target tester, and, of course, the circuit. Under these conditions the test engineer may then simulate the testing process and evaluate the impact of parasitics on test results, check the synchronization between analog and digital tester resources, check impedance matching and transient effects, etc., in order to evaluate the effectiveness of a test program prior to the availability of prototypes. And, if simulation models are reasonably accurate, significant portions of the test program can be debugged before first silicon.

Writing a test program involves defining input stimuli, tester resources, output responses, the postprocessing calculations performed, and the load board. Tools have been proposed to automate several components in this process. These include the design of the load board, verification of each test, generation of tester source code, and the complete emulation of the entire test program.

The load board connects the device under test (DUT) to the tester resources (Fig. 3). Designing a load board is a time-consuming task in test program development, since a test program combines a number of tests, each of which may require different connections and components on the load board. Kao and Xia have proposed a tool to automatically synthesize the load board [8]. This tool considers connectivity data from test schematics for the individual tests, determines shared tester resources, inserts switches for different setups, and generates a net list for the final load board circuit. In addition, a printed circuit board place and route tool is used to automatically place and route load board components, after which parasitic information is extracted.

Simulating a test may help identify problems with, say, settling times, impedance matching, load board parasitics, and instrument precision. This requires the test engineer to evaluate the performance of a device as measured on a piece of commercial automatic test equipment with realistic approximations to actual measurement and instrumentation techniques used in the tester hardware. Some such models for tester resource simulation have been implemented and are described in [9]–[11]. Given tester resource models, in [9]–[11] the simulation of a test is performed in an environment which combines a traditional approach to simulation, i.e., SPICE, for smaller analog components, with behavioral models for measurement instruments and larger analog blocks. If the circuit is a mixed-signal device, mixed-signal simulation techniques are used, combining circuit-level, analog behavioral, switch-level, gate-level logic, and behavioral logic verification capabilities. Simulation of a test not only makes it possible to investigate DUT-test interactions, but also makes it possible to identify critical components through sensitivity analysis, whose tolerances, if too large, may degrade test results.

In [12], a source code generator is proposed which automatically creates a test program in the language of the target tester hardware. This source code generator provides tester setup conditions and connectivity information, sets control bits, and controls the movement and analysis of data.

Finally, a tool for checking test program synchronization is proposed in [7] and [11]. The software tools described in [7] and [11] are automatic test equipment (ATE) program emulators. Test program emulators are proposed to verify instrument settings, instrument precision, measurement synchronization, etc., using a workstation. In particular, during emulation, the test program interacts with simulation models of the DUT, load board, and tester resources through a communication channel, where, as the test program executes, it sets up the virtual instruments (simulation models of the tester resources) and sources the test patterns. These patterns are then simulated, and the simulation responses are collected and postprocessed. The emulator then performs the decision-making and branching, as is done in the real test program. Based on the results, more test patterns are fed back to the simulator, until testing is complete. The whole process works as if the test program is verified on-line on real tester hardware. Consequently, both fault-free and faulty simulation prototypes of the circuit can be tested to verify that the test program both passes good devices and rejects faulty ones. Given sufficient accuracy in the simulation models, if test program and design bugs are caught by the emulator, significant debug time would clearly be saved, which would otherwise have been spent using the tester hardware, after the availability of silicon.

III. Reducing the Cost of Production Test

Some analog and mixed-signal circuits require large numbers of specifications to be verified, where checking all specifications can result in prohibitive testing times on expensive automated test equipment [6], [13]. In this section the problem of minimizing the cost of production testing is considered, assuming that the circuit being tested has a predefined set of specifications that need to be measured, with prespecified binning limits, indicating limits for failed die and limits for various performance grades of good die. The aim of techniques presented in this section is to minimize testing time by optimally ordering the tests (Section III-A) and by dropping some specification tests without degrading fault coverage (Section III-B). The methods presented are best suited to large production runs where the cost of evaluating fault coverage using a sample of circuits that are exhaustively and nonoptimally tested can be amortized over large numbers of circuits that are subsequently optimally tested. Therefore, these methods can be applied without the use of fault models, since fault coverage of subsets of a test set can be computed based on historical pass/fail data. Nevertheless, it may be desirable to optimally order and/or select a best subset for production testing for a new product prior to the availability of a large database of historical pass/fail data. In this case, fault models are needed to generate such a database using simulation data. Hence, in Section III-C, fault modeling is discussed, together with its application to evaluating the effectiveness of a test set.

A. Optimal Ordering of Tests

Typical industrial practice in production testing involves performing groups of tests, where if any set is failed, the die is assigned a failure bin number and testing is terminated. The number of die in various bins provides information about the common failure modes of the chip. The order of the groups of tests with a common bin number often depends on the complexity of the test. In other words, usually gross failures, like shorts between power and ground, are checked first, followed by, say, tests for defects in digital components. The last set of tests usually relates to checking the performance of the entire system. This approach to test ordering intuitively maximizes the information attained about failure modes.

Because testing is terminated as soon as a test is failed, average production testing time varies depending on the order of the tests. Specifically, if tests for gross failures are performed first and a circuit frequently fails such tests, then the average production testing time will be shorter compared to another circuit which mostly fails the system tests which are performed last.

If groups of tests are ordered to maximize information about failure modes, they may not be optimized for minimizing production testing time. Production testing time not only depends on the time to complete each test, but also on the probability that a particular test will be performed on a circuit, given its position in the test set. Consequently, if a large number of die fail a certain system test and if this system test also fails die with shorts between power and ground, production test time may be less if such a test is performed first, at the expense of separating die with gross failures from die that just fail a particular system test. Hence, there is clearly a tradeoff between optimizing failure bin information and test time, where the former is usually more important during the early stages of production, while the latter is more important for mature products. On the other hand, if testing for each failure bin is terminated as soon as a test is failed, the order of tests can certainly be optimized within the group corresponding to each failure bin to reduce test time, while retaining as much failure information.

Mathematically, the order of tests may be optimized as follows. Suppose a test set has n tests which are ordered from the first position (O_1) to the last (O_n), requiring test times of T_{O_i}, $i = 1, \cdots, n$. The probability P_{O_i} that the ith test is performed is

$$P_{O_i} = \prod_{j=1}^{i-1} Y_{O_j}$$

where Y_{O_j} is the yield of the test in the jth position. Average test time is then

$$\text{Average Test Time} = \sum_{i=1}^{n} T_{O_i} P_{O_i}.$$

Hence, minimizing production testing time involves gathering pass/fail data for each of the circuit specifications using a sample of fabricated chips in order to calculate Y_{O_j}, the yield of the test in the O_j position, given previous tests in positions O_1 to O_{i-1}. Then Dijkstra's Algorithm can be used to optimize the order [14]. Specifically, the test selection problem is formulated as a shortest path problem in a directed graph, where the computational complexity is dominated by the number of possible subsets of the test set, 2^n. In order to cut the computational cost by avoiding the evaluation of all possible 2^n subsets of the test set, two heuristic approaches to test ordering have also been proposed [15].

B. Selecting a Subset of Specification Tests

The easiest way to reduce the number of tests in a test set is to drop the tests that are never failed. It is possible that a test will not be failed if it is designed to detect a processing problem that has not yet occurred. Moreover, it is likely that some tests will never be failed when there are many more circuit specifications that have to be measured than independent sources of variability in the manufacturing process. It turns out that the order in which tests are performed will influence which tests are never failed. For example, a redundant test, placed early in a test set, may detect some faulty circuits, which could be detected by a combination of tests performed later. For example, a power supply short problem most certainly will be detected by a system performance test, and therefore a test for such a short would be redundant. Hence, as there is a tradeoff between minimizing production test time, achieved by optimally ordering a test set, and maximizing failure information, there is also a tradeoff between achieving minimal production test time through eliminating tests and maximizing failure information. Clearly, failure information is more important early in the product cycle, while reducing test time is more important for mature products. And, if a group of tests is assigned a single failure bin, redundant tests from that group can be eliminated at little cost.

In [15] and [16], an algorithm has been developed that orders tests so that the number of tests that have no dropout is maximized. In other words, this algorithm identifies and maximizes the number of redundant tests. It uses historical pass/fail data to identify those tests that detect some faulty circuits which are detected by no other test. If the sample size is large enough, all necessary tests will detect some such faulty circuits that are detected by no other test. In this way redundant tests are identified. On the other hand, if the sample size is too small, some necessary tests may be wrongly identified as redundant, resulting in reduced fault coverage, defined as the probability of correctly rejecting a faulty circuit. This may occur if some process corners have not yet been exercised. And in fact, for circuits with high yield, large sample sizes of nonoptimally tested circuits are needed in order to achieve

TABLE I
SAMPLE SIZE AS A FUNCTION OF YIELD AND FAULT COVERAGE AT 95% CONFIDENCE

Yield	Desired Fault Coverage				
	80%	90%	95%	99%	99.9%
99%	1700	3000	6000	30,000	300,000
90%	170	300	600	3000	30,000
75%	62	120	240	1200	12,000
50%	29	59	120	600	6000
25%	19	39	79	400	4000

high fault coverages, since if large sample sizes are available, presumably all realistic process corners have been exercised (Table I).

A number of papers have attempted to go beyond just dropping tests that are never failed. These papers have proposed using a limited set of measurements to predict the results of other measurements. Supposing that the number of measurements is much larger than the number of primary sources of variation in the manufacturing process, the number of primary sources of variation in the manufacturing process limits the dimension of the set of basis vectors that span the measurement space. In order to characterize the measurement space, these methods rely on linear models of deviations in circuit performances as a function of variations in parameters characterizing the manufacturing process. These linear models can be constructed by simulation, where the basis vectors of the measurement space are sensitivities of measurements to changes in process parameters [17]. Alternatively, basis vectors for the measurement space can be found empirically using a sample of devices coming off the production line, assuming they manifest all of the sources of variability of the manufacturing process [18]. Using an empirical set of basis functions eliminates the need to simulate often very complex circuits, which can either lead to results of limited accuracy or can be extremely computationally intensive. But empirical basis functions include noise, which can be minimized by averaging repeated measurements of the same device. And basis functions can be missed if all process corners have not yet been exercised.

Mathematically, for the case when the span of the measurement space is characterized by simulation [17], let y denote a vector of changes in n measurement responses from nominal, and let x denote a vector of m changes in parameter values from nominal. Then, given a sensitivity matrix A

$$y = Ax.$$

Alternatively, for an empirical model, the columns of A would denote changes from nominal of the n measurements for m different devices [18]. The problem addressed is to find a set of measurements which can be used to predict the responses of other measurements not made. Suppose that the measurements y are divided into two sets, where y_1 is a vector of the n_1 measurements that are made and y_2 is a vector of n_2 measurements that need to be predicted. The sensitivity matrix can correspondingly be divided into two matrices, A_1 and A_2, where

$$\begin{bmatrix} y_1 \\ y_2 \end{bmatrix} = \begin{bmatrix} A_1 \\ A_2 \end{bmatrix} x.$$

Then, if $n_1 \geq m$ and A_1 has rank m, the measurements y_1 can be used to predict parameters deviations

$$x = (A_1^T A_1)^{-1} A_1^T y_1$$

where A_1^T is the transpose of A_1. And the remaining measurements are predicted as follows:

$$y_2 = A_2 (A_1^T A_1)^{-1} A_1^T y_1.$$

For a circuit, the results of measurements y_1 and the predicted results for measurements y_2 can be used to sort devices into performance bins.

One runs into problems with this approach when A is of rank less than m or nearly so. In this case, for any choice of measurements y_1, $A_1^T A_1$ cannot be inverted, and all of the parameters x cannot be predicted. This would happen if sensitivities for different sources of variation in the manufacturing process are not unique. In such a case, it would not be possible to determine from measurements the values of those parameters x with linearly dependent sensitivities. In fact, such parameters, which cannot be predicted unless the values of other parameters are known, are said to belong to the same ambiguity group. Since the purpose of this work is to find a set of basis vectors that span the measurement space, the problem of ambiguities in x is solved by dropping a column of A for one member of each ambiguity or approximate ambiguity group. One way of identifying dependent columns is QR factorization of A with column pivoting, where a set of unambiguous columns is sequentially selected until no more can be found [19]. In another approach, columns are sequentially selected by choosing the largest element, corresponding to a high sensitivity of a measurement y to a parameter x, so that it is possible to accurately solve for the parameters x that are selected. This process also continues until no more columns with large sensitivities can be found [20]. An alternative approach is to test all small subsets of columns for dependency, sequentially eliminating ambiguous columns [19]. The problem with this last approach is that the number of subsets to be tested grows exponentially with m. A fourth approach to identifying ambiguity groups from which ambiguous columns are identified begins by computing the null space of A using singular value decomposition or Gaussian elimination [21]. The rows of the matrix that span the null space correspond to different ambiguity groups only if they are orthogonal. Hence the ambiguity groups are found by checking the rows of the null space for pairwise orthogonality, which can be performed in polynomial time.

Once columns corresponding to ambiguity groups have been pruned from A, an optimal set of measurements y_1 can be chosen, where the rank of A determines the minimum number of measurements required. The method suggested in [17] involves finding a set of measurements which minimizes the average prediction variance of the model coefficients x. When the number of measurements is large, the maximum variance is minimized with the D-Optimality criterion, i.e., maximizing $\det(A_1^T A_1)$. This approach is computationally costly for large problems, and hence an approximate solution is found in [17] using QR factorization of A^T with pivoting. The resulting pivots correspond to the selected measurements y_1. In an alternative approach, the measurements are selected by minimizing the average standard error of the predicted output [22]. This algorithm is based on I-Optimal experimental design.

For both approaches to test selection, the minimum number of test vectors may not be sufficient to conclusively verify the performance specifications because of the size of the confidence bounds on the predicted test results, due to measurement noise and model inaccuracies. In fact, adding more measurements reduces the size of the confidence bounds on the predicted test results y_2, making it possible to conclusively verify if predicted test results are inside or outside of binning limits. In [22], the I-Optimal criteria is used to select additional tests, and in [23] the measurement with the highest prediction variance is iteratively selected.

These approaches to test selection based on linear modeling have been proven to be effective for ADC's [6]. Specifically, 50 commercial 13-bit ADC's were exhaustively tested to reveal that 18 measurements of code transitions out of 8192 sufficed to model all randomness in the manufacturing process for a batch of devices and to predict the remaining code transitions. The remaining 77 devices were tested with this reduced test set.

C. Evaluating the Effectiveness of a Test Set

How can we determine if a test set is effective for a given circuit? Fault coverage is a good measure of the effectiveness of a test set, i.e., the probability that a faulty circuit fails tests. One way to evaluate the fault coverage of a test set is to exhaustively test a sample of manufactured circuits. This approach would require a very large sample in order to evaluate high fault coverages (Table I). If such historical data is not available, an alternative is to use simulation data to determine the faults that are detectable by a given test set, optimize the order of a test set, and eliminate redundant tests in a test set. For this to be possible, one needs an accurate and efficient simulation methodology and a fault model. Moreover, an efficient statistical simulation strategy is needed because of the very high computational cost associated with simulating large systems and the very large number of random variables needed to model manufacturing processes.

Faults in analog circuits can be classified into two categories: catastrophic and parametric [24]. Catastrophic faults include open nodes, shorts between nodes, and other topological changes in a circuit. For integrated circuits, they usually result from local defect mechanisms like particles on the wafer surface generated by equipment during processing, particles that block exposure of a local area during masking, oxide defects which short out transistors, severe misalignment of layers, etc. For printed circuits boards, catastrophic faults can come from excess solder, resulting in bridging between pins, lack of solder, a broken pin of a component, missing components, improperly oriented components, or use of a wrong component (i.e., a resistor rather than a capacitor) [11]. For integrated circuits, one way to produce a catastrophic fault list is to mimic the defect size and frequency distribution of the manufacturing process using a Monte Carlo defect simulator that places missing or extra material in a given layer of a layout and extracts the impact on circuit topology [25], [26]. Alternatively, prior to the availability of a layout, a fault list may be generated from a schematic, i.e., broken wires, gate-drain shorts, etc. [27], [28]. For printed circuit board technology, a fault list may include open and shorted pins, missing components, improperly oriented components, or use of the wrong component [11].

Parametric faults refer to changes in a circuit that do not affect its connectivity. Such changes may be global or local. Global variation of parameters is due to imperfect parametric control in IC manufacturing. Such variations affect all transistors and capacitors on a die, and if large, result in circuits that fail performance specifications or are binned as lower performance parts. Such variations have a large component due to lot-to-lot fluctuations, because of processing in different equipment at different times, a smaller component due to wafer-to-wafer fluctuations, because of variations in the performance of individual pieces of equipment as a function of time, a smaller component due to within-wafer fluctuations, often because of temperature gradients during etching, and a smaller component due to within-reticle fluctuations, often due to lens distortion which affects lithography [29]. For example, transistor channel length will vary both systematically and randomly across a die, reticle, wafer, and between wafers and lots. In order to characterize global parametric variations, a set of independent factors is usually identified that explain lot-to-lot, wafer-to-wafer, and die-to-die variations in a process, often by principal components methods [30], [31].

Local variations of parameters are due to local defect mechanisms, like particles, which, say, enlarge a single transistor's channel length or capacitor locally, or local variations in parameters across a die because of imperfect process control, for example, due to local temperature gradients during etching or local variations in the lens of a stepper. Local variations in parameters due to defects are often modeled as large changes in a single device parameter, like a change in a single resistor value. Local variations due to imperfect process control have little effect on digital components, but result in random differences between physically adjacent devices in analog components, which gives rise to mismatch, to which many analog designs are very sensitive. Modeling mismatch entails supplementing global statistical process models with additional variables indicating the extent of mismatch, possibly as a function of device area and spacing [30], [32], [33]. Incidentally, the data needed to characterize mismatch is harder to come by due to the lack of test structures on scribe line monitors that can be used to measure mismatch.

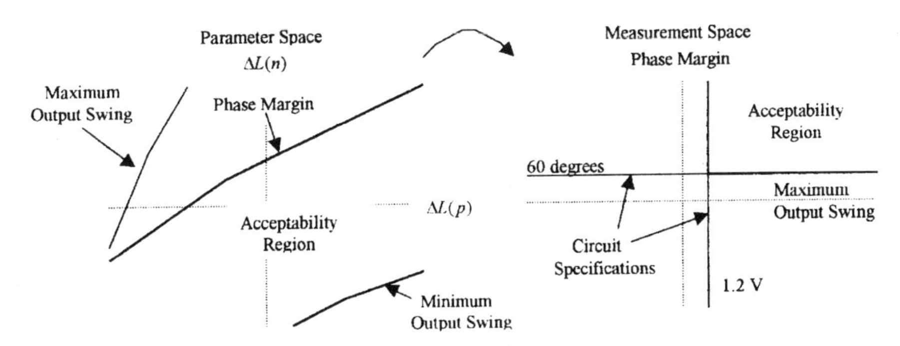

Fig. 4. The map between the parameter space and the measurement space. Parameters and measurements not in the acceptability region correspond to faculty circuits [15].

Simulating both catastrophic and parametric faults can be cumbersome because of the computational cost of simulating large analog systems. A hierarchical approach to fault simulation reduces the computational cost. Hierarchical simulation involves partitioning a system into blocks. The blocks are then electrically simulated and their responses are stitched together with a behavioral simulator in order to evaluate system performances. This is the approach taken in [11] and [33]–[37] for catastrophic fault simulation. And if a catastrophic fault model is used, fault coverage (fc) is typically defined as the ratio of faults detected (fd) over faults simulated (fs) [11], [28], [38]. If faults are weighted by their likelihood p_i, then

$$fc = \sum_{i \in fd} p_i \bigg/ \sum_{i \in fs} p_i.$$

Several authors use the same methodology for the simulation of parametric faults [34], [35], [39]. This corresponds to a parametric fault model involving only local variations in geometries due to defects. Such a fault model does not include global parametric variations resulting from imperfect process control. Given such a local parametric fault model, in order to simulate a fault, a circuit parameter is set to an out-of-tolerance value and the resulting circuit is simulated. But how much out of tolerance should the parametric faults be? Models of defect size frequency indicate that small defects are much more likely than large defects [40]. And very small defects result in only minor changes in circuit performances. Hence, such small defects may not cause a circuit to fail specifications. Clearly, the definition of a parametric fault needs to be related to the circuit specifications, and specifically defining a parametric fault involves determining parameter limits such that a circuit fails specifications, which may or may not coincide with parameter tolerances.

Similarly, for global parametric faults, which result from imperfect control in manufacturing, parameters closer to nominal values are much more likely than parameters which are far from nominal, while parameter values that are far from nominal are much more likely to cause a circuit to fail specifications. And, as with local parametric faults, circuits with parameter values that are close to tolerance limits may not fail specifications, and consequently may not be faulty. Hence, also for global parametric faults, determining if a parameter deviation results in a parametric fault involves determining the map between the random variables describing the manufacturing process and circuit performances [15], [16]. Typically, this is done using statistical modeling methods, where, based on a limited set of simulations, an equation is constructed for each circuit performance as a function of parameters modeling global variations in the manufacturing process and as a function of circuit parameters modeling local variations due to defects [33], [41]–[46]. Given models for each circuit performance and their corresponding specification, the set of parameters where all specifications are satisfied may be determined, called the acceptability region. Conversely, the set of parameters where at least one specification is failed is likewise determined (Fig. 4). Moreover, if the parameters characterizing the variations of the IC manufacturing process are described by a probability density function, parametric faults may similarly be characterized by a probability density function.

What is fault coverage for parametric faults? For local parametric faults, created by defects, a fault list can be generated from a layout using tools similar to those designed for creating catastrophic fault lists [25]. Such tools mimic the defect size and frequency distribution for each layer of a manufacturing process by placing extra or missing material in a given layer of a layout and extracting the resulting circuit changes. Those circuit changes that result in faults can be identified using the map relating parameters to measurements. Fault coverage is consequently the fraction of those circuit changes that result in faults that can be detected by a given test set.

Computing fault coverage for global parametric faults is more complex. The digital/catastrophic fault coverage definition (fd/fs) does not apply for such faults since such faults are characterized by a continuous distribution rather than a discrete one. Nevertheless, the ratio of the likelihood of faults detected over the likelihood of faults simulated is an equivalent definition of parametric fault coverage [15], [16], [47]

$$fc = \frac{\displaystyle\int_{x \in fd \cap fs} f(x)\,dx}{\displaystyle\int_{x \in fs} f(x)\,dx}$$

where $f(x)$ is the probability density function of parameters x modeling the manufacturing process, fs is the complement of the acceptability region, i.e., the set of all parametric faults, and fd is the set of parameters which correspond to circuits that fail a given test set.

A straightforward way to evaluate the integrals in the above equation is to use Monte Carlo analysis, where a sample of parameters x from the probability density function describing the manufacturing process is simulated. At each x, it is first determined if x is a fault, in which case $x \in fs$, and if so, it is determined if x is detected, and if so $x \in fd \cap fs$. The evaluation of whether or not x is a fault and if it is detected may be determined directly, using circuit simulation, or based on regression models of circuit performances, defining parametric faults. However, applying the Monte Carlo algorithm directly by simulating a circuit with a sample of parameters representing the manufacturing process may not lead to accurate results. Specifically, if a small sample size is used, results will be inaccurate because of the sample size. In fact, unless a test is highly inaccurate, it may be hard to find a sample of faulty parameters x which is not detected by the test set, i.e., $x \in fs$ and $x \notin fd$. Alternatively, if the sample size is large, the computational cost of simulating just the blocks, i.e., op amps, if not the whole system, hundreds of times, can be very high, unless very inaccurate simulation models are used. Importance sampling can reduce this cost of simulation [48]. Nevertheless, when applying the Monte Carlo algorithm, the use of regression models of block performances as a function of process parameters and a hierarchical simulation strategy reduces the computational cost most effectively with accurate results [33], [41], [42], [46]. In other words, a limited set of simulations is performed to construct regression models, and then the regression models are used to evaluate if tests are passed or failed for the much larger random sample of parameters representing manufacturing process variations. In this case, the most significant sources of inaccuracy will come from a combination of the accuracy of circuit simulation, which is used to construct the regression models, and the ability of the regression models to mimic the simulator.

If all circuit specifications are tested, the parametric fault coverage may still not be 100%, due to measurement noise [15]. Moreover, if tests other than the specification tests are used, as proposed in [28], [39], [47], and [49], there may be a systematic loss of fault coverage (Fig. 5). Similarly good circuits may fail tests due to measurement noise and if tests other than the specification tests are used. Yield coverage (yc) has been proposed as a parameter to quantify the problem of discarding good circuits [47]

$$yc = \frac{\displaystyle\int_{x \notin fs \cup fd} f(x)\,dx}{\displaystyle\int_{x \notin fs} f(x)\,dx}.$$

Yield coverage may be computed in the same way as fault coverage.

Given the three definitions of fault coverage above, i.e., catastrophic fault coverage, local parametric fault coverage, and global parametric fault coverage, which one should be used when evaluating a test set? Sachdev, in [50], considered

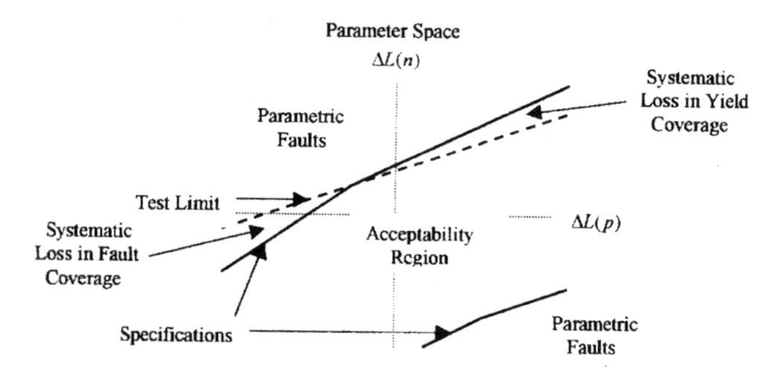

Fig. 5. Systematic loss in fault coverage and yield coverage.

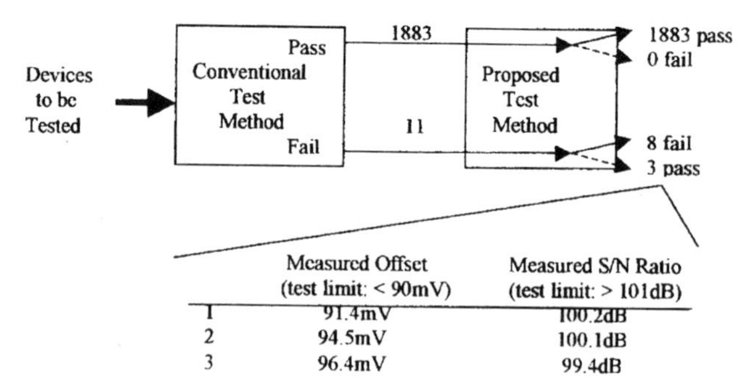

Fig. 6. Test results for the Class AB amplifier and the measured performances for the three devices that failed specification tests but passed the proposed test set for catastrophic faults [50].

generating test sets for just catastrophic faults. The test set he proposed for a Class AB amplifier was derived based on realistic catastrophic faults and demonstrated high catastrophic fault coverage of modeled faults by simple stimuli, i.e., simple dc, ac, and transient stimuli. This test program was then appended to the existing conventional (specification-based) test program, in order to judge its effectiveness in a production test environment. The results are shown in Fig. 6. As can be seen from the figure, the yield of the device was very high (99.5%), and the fault coverage of the proposed test set was only 73%. The performances of the three devices which passed tests for catastrophic faults but failed specification tests are also shown in Fig. 6. Because the proposed test set was designed to detect catastrophic faults and because distributions of both local and global parametric faults have higher frequencies of parameter values that correspond to circuit performances close to specification limits, it appears from these results that these three devices failed due to parametric faults.

Because the sample of failed devices was so small, Sachdev [50] followed up this experiment with a larger one using the same Class AB amplifier. The results of this second experiment are shown in Fig. 7. It can be seen that the fault coverage of the test set designed solely for catastrophic faults was 87%. The 433 circuits that failed the proposed test set but had passed the conventional test set were then retested by the conventional method. Of these 433 circuits, 51 passed the conventional test set, indicating that measurement results are very close to specification limits, causing the circuit to pass or fail based on noise levels. The remaining 382 circuits mostly failed specifications on the input offset voltage, total harmonic distortion, and the signal-to-noise ratio. All of

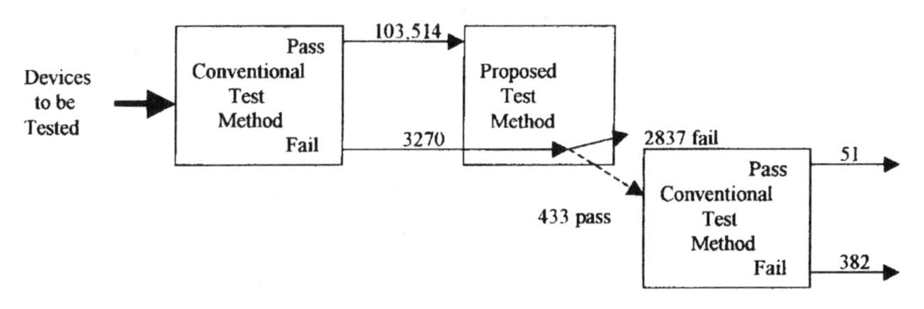

Fig. 7. Test results for the Class AB amplifier.

TABLE II
CATASTROPHIC FAULTS FOR A HIGH-PASS FILTER BLOCK [33]

Frequency	Block Behavior	Impact of System Performance
93	Pairs of shorted nodes cause a fixed output voltage	Fails all specifications
161	Pairs of shorted nodes cause $gain \leq 10$	Fails all specifications
3	Pairs of shorted nodes cause $gain \leq 4300$ and $Vos \geq 200mV$	Fails offset specification
8	Pairs of shorted nodes cause $gain = 4500$ and $Vos \geq 20mV$	Passes specifications at nominal parameter values, but has degraded yield due to process variations
6	Pairs of shorted nodes cause $gain = 700$ and $Vos = 10mV$	Passes specifications at nominal parameter values, but has degraded yield due to process variations
9	Shorted capacitors	Fails all specifications
6	Clock short to a node	Fails all specifications
3	Power supply short to a node	Fails all specifications

these specifications are very sensitive to transistor matching in the differential amplifiers. Since poor transistor matching is a parametric fault, it is likely that these circuits failed for parametric faults. Clearly, it can be concluded that a test based solely on process defects (catastrophic faults) is not sufficient for ensuring that specifications are satisfied, and consequently catastrophic fault coverage is insufficient in quantifying the quality of a test set for analog circuits. This is likely because, unlike digital circuits, which tend to have less tight performance requirements but more functional complexity, optimal analog circuit performance is often only achievable under optimal fabrication and operating conditions. Hence, the parametric fall-out for analog circuits is likely to be more significant compared to digital designs.

Suppose, on the other hand, a test set is designed for high parametric fault coverage. Would such a test set be able to detect catastrophic faults? In [15], [16], and [33], algorithms for selecting optimal sets of specification tests based on parametric fault coverage have been presented. Specifically in [33], a subset of 1024 frequency measurements and measurements of the system offset, dynamic range, and total harmonic distortion for a bandpass filter was selected. This bandpass filter is a switched-capacitor design, composed of five blocks, one high-pass filter, three biquads, and one sum-gain amplifier. For each of the blocks, local defects were generated in the layout using VLASIC [25] in order to obtain a fault list. Each fault was simulated in order to compute the resulting distortion of parameters characterizing each of the blocks, i.e., gain, offset, etc. The results for the high-pass block are shown in Table II, where modifications in block performances could be classified in eight groups, the most common two being low gain and fixed output voltage. System simulation using the behavioral model was then performed in order to determine if specification tests are passed or failed. It can be seen from Table II that almost all of the catastrophic faults failed all of the specification tests. The 5% of the catastrophic faults that did not fail all specifications resulted in circuits that under nominal processing conditions would pass or almost pass all specification tests. Hence, for these faults, the impact of these catastrophic faults is to lower yield. And, it turned out that the specification tests needed to achieve high parametric fault coverage were sufficient to detect these catastrophic faults when they were combined with variations in parameters and resulted in circuits that failed specifications. Consequently, in this example, a test set designed for high parametric fault coverage also achieved 100% catastrophic fault coverage.

It seems that it can be concluded that test sets should be evaluated both in terms of parametric and catastrophic fault coverage. Moreover, based on the limited set of experiments that have been done to date, there seems to be some evidence that test sets designed for high parametric fault coverage are more likely to detect catastrophic faults compared to the ability of test sets designed for high catastrophic fault coverage to detect parametric faults.

IV. FAULT DIAGNOSIS

If an integrated circuit has been found to be faulty during design characterization, before it is in high volume production, it may be useful to diagnose the cause of the failure. If faults

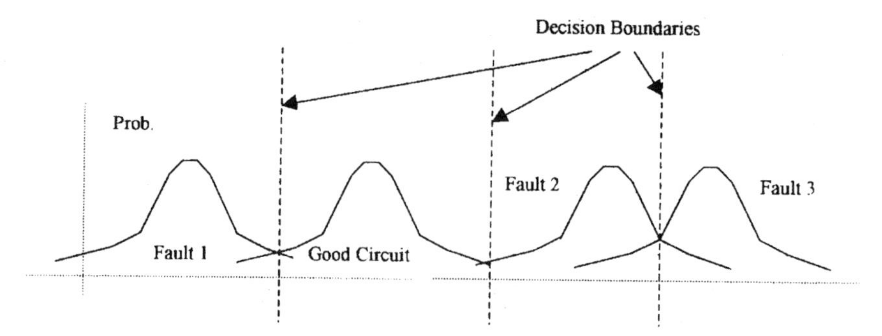

Fig. 8. Decision boundaries between the good circuit and three faulty circuits for measurement ϕ assuming equal prior probabilities.

are identified and located, a circuit can be redesigned to be less sensitive to common failure mechanisms. Alternatively, if an analog or mixed-signal system with components that have been separately tested fails system specifications, it is also useful to find the cause. In this case, problems in system performance can occur due to assembly errors and the degradation of components with time. It is therefore desirable to have a methodology to rapidly identify component failures. Two distinct strategies have been proposed for analog fault diagnosis: simulation-before-test and simulation-after-test. Simulation-before-test approaches begin with a fault list. The faults are then simulated to determine the corresponding responses to predetermined stimuli. Faults are consequently diagnosed by comparing simulated and observed responses. Simulation-after-test approaches, on the other hand, begin with the failed responses, which are then used to estimate faulty parameter or component values. For a comprehensive survey of these approaches see Bandler and Salama [2]. The sections below are intended to give a brief overview.

Fault diagnosis techniques need fault models. Simulation-before-test techniques are better suited for detecting catastrophic faults and local parametric faults, while they may perform less well in detecting global parametric faults, since for such faults the separation between good performances and faulty performances is less wide. On the other hand, simulation-after-test techniques are better suited for detecting problems with global parametric variations and mismatch, and are not well suited for detecting catastrophic faults.

A. Simulation-Before-Test

Simulation-before-test algorithms are based on a fault dictionary. In particular, the most likely faults (usually catastrophic) are anticipated based on a fault model, and a set of input stimuli and measurements are selected to detect faults. The measurements may be dc responses [51], ac responses at circuit outputs [52]–[55], ac responses at the power supply node [56], transient responses at circuit outputs [57], [58], or transient responses at the power supply node [59]. Then for the set of potential faults, the circuit's response for each stimulus is evaluated for each fault with all but the faulty parameter set to nominal values, i.e., the parameters for which the circuit was designed. The responses for each fault are typically evaluated using circuit simulation, except for the special case when the circuit is linear, where efficient techniques exist for computing the faulty responses [55], [60], [61]. In particular, in [55] a symbolic simulator is proposed for evaluating the impact of faults on the frequency response of a circuit, greatly reducing the computational cost needed to construct a fault dictionary. It is also possible to construct a fault dictionary using measured data, based on previously observed and diagnosed faults.

After evaluating the responses corresponding to the faults, the faults and their corresponding responses are stored in a dictionary. To diagnose a fault, the measurements of the circuit being tested are compared with measurements in the dictionary. The fault is identified by determining the closest simulated fault using inspection [53], [58], the Euclidean norm [51], [54], [55], pattern matching [52], fuzzy distance [62], or a neural network [59].

If parameters are assumed to lie exactly at their nominal values when the fault dictionary is constructed, errors in fault identification can occur. Parameters are usually not at nominal values because of uncontrollable fluctuations in circuit fabrication. Consequently, it is unlikely that measurements will have values exactly equal to those stored in the fault dictionary. Furthermore, different faults may result in exactly the same measurement because of variations in components. Approaches which account for distributions of parameters [63], [64] typically begin with Monte Carlo simulations of all anticipated faulty circuits and the good circuit. The use of Monte Carlo analysis, although greatly increasing the computational cost, provides estimates of the mean vectors μ_k and covariance matrices $\sum_k$ of the distributions for the good and every type of faulty circuit. Maximum likelihood is taken as the measure of distance to determine the most likely fault class. Specifically, let p_k be the prior probability of the kth fault type, i.e., p_k is the likelihood of the kth fault type based on past data. Then for an observation ϕ, the corresponding fault is identified as belonging to fault class k if the quadratic discrimination score is minimum

$$d_k(\phi) = \ln|\Sigma_k| + (\phi - \mu_k)^T \Sigma_k^{-1}(\phi - \mu_k) + \ln(p_k).$$

Fig. 8 shows the decision boundaries which separate fault classes, assuming equal prior probabilities. If it is known that the good circuit is much more likely than, say, fault 1, the decision boundary would shift to the left, making it more likely to identify the good circuit compared to fault 1. Moreover, if during testing the prior probabilities p_k are updated, the decision boundaries between faults would change accordingly.

Some faults may result in measurements that are very close to each other (Fig. 8). In this case, the probability that a fault will be wrongly diagnosed is high. Some algorithms attempt to aggregate faults into ambiguity groups, assuming parameters are within tolerance. Faults are said to belong to the same ambiguity group if they cannot be distinguished by a given set of measurements. To rigorously aggregate faults into ambiguity groups, it is necessary to determine the set of measurements that can result from each fault, given that parameters are within tolerance. Specifically, for the example in Fig. 8 it would be necessary to determine the sets of measurements ϕ which may result from, say, faults 2 and 3. And, it can be seen from the figure that these sets clearly overlap, and hence the faults should belong to a common ambiguity group. Rigorous algorithms for the computation of the sets of measurements resulting from each fault are presented in [62] for linear circuits and in [28] for nonlinear circuits.

The success of an algorithm in correctly diagnosing faults depends on its choice of measurements. The best choice of measurements should result in a maximum separation between the good circuit and faulty circuits, and among faulty circuits, as indexed, for example, by

$$\frac{\mu_0 - \mu_k}{\sigma_0} \quad \text{and} \quad \frac{\mu_j - \mu_k}{\sigma_0}$$

where μ_0 and σ_0 are the mean and standard deviation for a measurement for the good circuit, and μ_k and μ_j are the means for faulty circuits. Algorithms for selecting an optimal set of measurements have been presented in [28] and [65]. For the special case where testing is restricted to the frequency domain, theoretical guidelines for frequency selection exist [52], [54]. In particular, a set of frequencies should be chosen so that at least one is between each break point in the frequency response, plus one before the first breakpoint and one after the last breakpoint.

B. Simulation-After-Test

Simulation-after-test algorithms have been designed to solve for values of component parameters, given a set of measured responses and knowledge of the circuit topology. In other words, given that a simulator can be used to find measured responses for a set of component parameter values, the simulation-after-test problem is to determine the inverse map. A fault is identified if one or more parameter values are found to be outside of tolerance. Hence a fault can be diagnosed if it is possible to uniquely solve for all parameters from the given measurements. Clearly there need to be enough independent measurements to identify all parameters. And, under these conditions, faults can be diagnosed by solving a nonlinear set of equations. However, the solution is only locally unique for both linear [66]–[68] and nonlinear [69]–[72] circuits, except under the special case where testing of a linear network is performed at a single frequency [73], [74]. Nevertheless, to identify all circuit parameters, a formidable set of nonlinear equations has to be solved, and to reduce computation, algorithms have been developed based on the assumption that only a few parameters are faulty, for both linear [75], [76] and nonlinear [70], [77], [78] circuits. Solving for only a few faulty parameters also reduces test point requirements. But since the faulty parameters are not known in advance, a search is required. Since these algorithms solve for only a few parameters, when searching for the set of faulty parameters, all other parameters are assumed to be at nominal values, and this raises the question of robustness. Because it can only be guaranteed that good parameters are in tolerance, it is possible that wrong faults will be identified, as can occur with simulation-before-test algorithms if faults belong to a common ambiguity group. Moreover, because of the amount of computation required on-line, these algorithms are practically restricted to small circuits (50–100 components). Nevertheless, parameter identification algorithms are equally applicable to modules, reducing the size of the problem for large circuits, or to detecting process control problems which result from large variations in the global and mismatch parameters that model variations in the manufacturing process. On the other hand, common fault mechanisms, like catastrophic faults, cannot easily be diagnosed without substantially increasing the complexity of the numerical analysis. They can cause ill-conditioning and difficulty in convergence of the numerical solution.

Under the conditions that measurements vary approximately linearly as a function of changes in parameters, and faults are not catastrophic, but rather moderate changes in system component parameter values or parameters modeling fluctuations in the manufacturing process, matrix techniques can be used to solve for parameter deviations from nominal. Solving for parameter deviations may help identify problems with component aging. Alternatively, if a circuit has previously been screened for catastrophic faults, using an algorithm such as the ones suggested in [28] and [50], solving for global parameter values which model the manufacturing process may help identify those parameters that are not well controlled by the manufacturing process. In other words, such an analysis may help us identify critical mismatch parameters, for example, as in [21].

Specifically, if y is an n-dimensional vector of measurements, x is an m-dimensional vector of parameters, and A is a sensitivity matrix

$$y = Ax.$$

If A has rank m, it is straightforward to solve for parameter deviations from nominal using least squares

$$x = (A^T A)^{-1} A^T y.$$

A will have rank m if all of the sensitivity vectors are unique, i.e., all of the columns of A are independent. If there are dependencies or near dependencies in the sensitivity vectors, the above equation is less solvable, and the component values of the circuit or the process parameters are said to be less diagnosable. In fact, small amounts of measurement noise can induce wide variations in the computed parameter values x. The parameters that correspond to nearly dependent columns are said to belong to ambiguity groups. This means that these parameter deviations cannot be separately identified without additional measurements. Algorithms for finding such ambiguity groups have been developed in [19]–[21] and have been discussed in Section III-B. In addition, classifications of types of ambiguity, i.e., fault masking, fault dominance, and fault equivalence, are presented in [79]. However, even if all parameters can be identified, their values cannot be precisely determined, due to measurement inaccuracies and unmodeled variables. Specifically, given variations of component or process parameter values x within tolerance and measurement uncertainty, the minimum parameter changes that can be diagnosed can be determined and the corresponding best measurements can be identified [80]. Moreover, given measurement uncertainty, in [20] the accuracies of parameter estimates are discussed, for both the cases where parameters can be separately computed and where there are ambiguities among parameters.

V. Design for Testability Techniques

The increased complexity of analog circuits and the reduced access to internal nodes has made it not only more difficult to diagnose and locate faulty components, but also the functions of embedded components may be difficult to measure. Design-for-testability techniques at the very least aim to improve the controllability and observability of internal nodes, so that embedded functions can be tested. Such techniques include analog test busses and scan methods. Testability can also be improved with BIST circuitry, where signal generators and analysis circuitry are implemented on chip. A brief overview of these techniques will be presented in the following sections. A more detailed discussion can be found in [5] and [81].

A. Improving the Controllability and Observability of Internal Nodes

Improving the testability of increasingly complex digital designs has resulted in the widespread use of scan chains. Scan chains involve shift registers, where data is serially shifted in through a test data input pin and through the scan chain to reach internal nodes. The test is performed and the results are captured in registers. The results are then serially shifted to the output through the registers to reach the test data output pin. Scan chains not only make it possible to improve fault coverage, but also make it possible to link failures to specific circuit blocks. In addition, at the board level, new high-density packaging technologies have made it more difficult to identify failing components in a system. With such packaging technologies, once a chip is attached to a package, its pins are no longer accessible, and the ability to remove a part reliably and without damage for testing purposes may be limited. Boundary scan has been proposed to address this controllability and observability problem for system debug, where all chip pins for all components in a system are connected in a scan chain (Fig. 9). A key advantage of boundary scan has been its ability to detect opens and shorts in the board's wiring, since 80–90% of board failures result from wiring problems. Hence, boundary scan cells combine the shift registers that are used to input and output voltages to and from the core circuitry with components that a) disconnect the IO pin from the core, b) set the IO pin to a logic level, and c) detect the logic level on another IO pin.

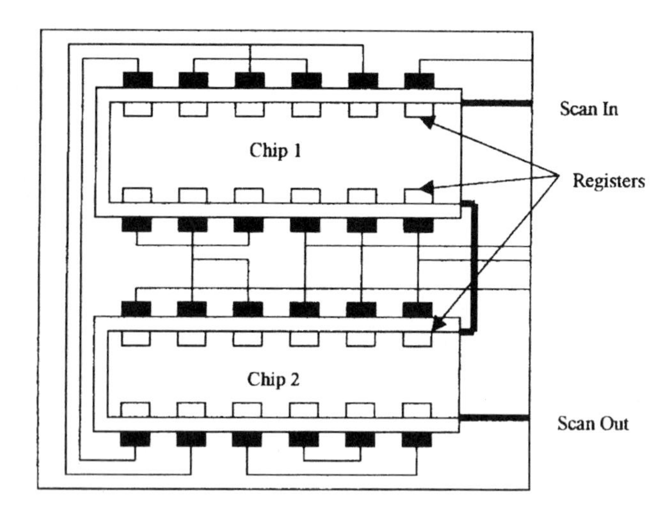

Fig. 9. Boundary scan chain on a board.

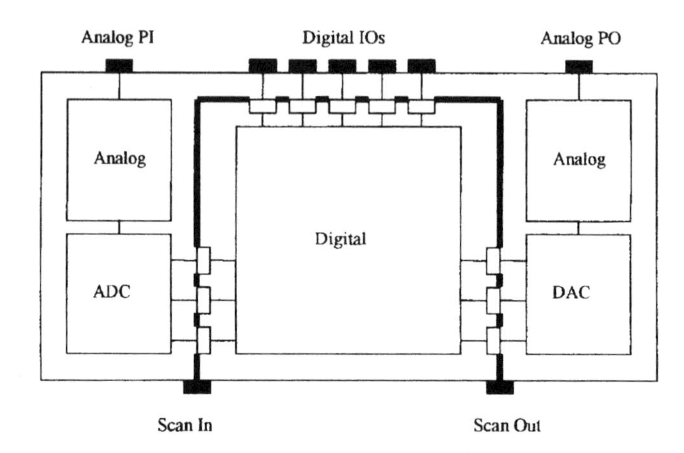

Fig. 10. Modified boundary scan for a mixed-signal component.

In the mixed-signal domain, Fasang [82] has proposed the use of boundary scan for mixed-signal designs. He proposes taking advantage of the fact that many mixed-signal designs have ADC's and DAC's on chip. These can be used to digitize all analog outputs before they are stored in scan cell registers and shifted to external pins and to convert digitized analog input signals which are shifted into the scan path into analog inputs for the analog components. The proposed modified boundary scan path for a mixed-signal component is shown in Fig. 10.

This configuration provides some controllability and observability of both digital and analog components. However, this configuration does not make it possible to isolate failures within the analog blocks. In order to enhance our ability to diagnose failures within analog blocks, it may be desirable to separately test the components of the analog blocks, i.e., ADC's, DAC's, operational amplifiers, oscillators, phase-locked loops, filters, etc. Given a partition of a complex chip, component test requires the isolation of the blocks, the control of component inputs, and the observation of component outputs. In [83], a set of nodes to be accessed and component tests to be performed, together with their test conditions, are outlined for some common analog building blocks. Isolation of the blocks from digital circuitry can be achieved through

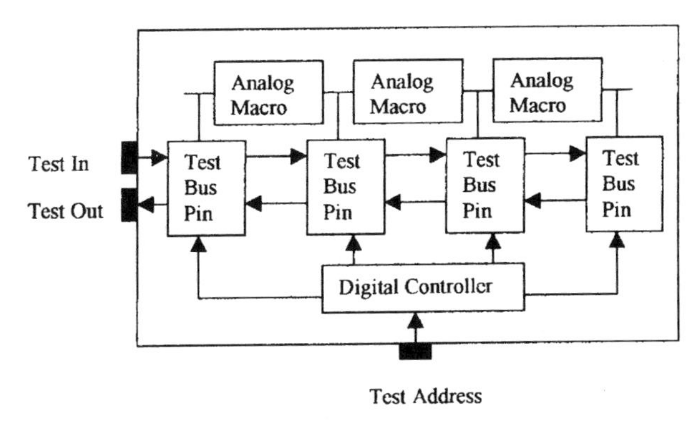

Fig. 11. Analog test bus architecture.

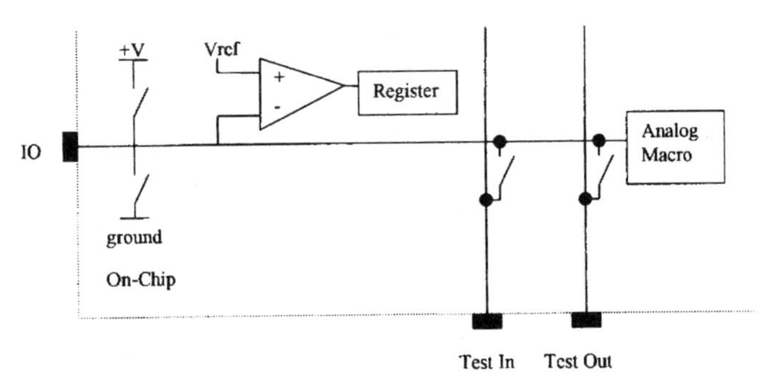

Fig. 12. Simplified schematic of a P1149.4 test bus pin [85].

interface storage elements, i.e., parallel or serial shift registers organized in a scan chain [84]. Isolation between analog blocks may be achieved through buffers, controlled by digital circuitry [84]. In addition, it may be necessary to disable feedback loops in order to measure open-loop parameters, since feedback reduces input controllability.

Given a set of nodes that need to be accessible in order to test the blocks, it is generally neither feasible nor practical to bring all of these nodes of an integrated circuit out to the package pins. One way to enhance observability is to add a small metal contact connected to the node. The signal can then be measured with either buffered high-impedance micro probes or an electron beam tester. The chip area required for such a contact is significantly less than what is needed for a bond, but if many such contracts are used, the area overhead may become significant. Alternatively, analog inputs and outputs may be combined by multiplexing and routed to primary inputs and outputs. Such a configuration is called an analog test bus [85] (Fig. 11). In this case any block input and/or output would be externally addressable so that real-time data may be input and observed. Even nodes which are shielded for electromagnetic interference can be made observable. The analog test bus cell shown in Fig. 12 has been designed as an analog boundary scan cell, although it can also be used to access internal analog nodes. It therefore has components designed to check board connectivity, i.e., to a) disconnect the IO pin from the analog macro, b) set the IO pin to a logic level, c) detect the logic level on another pin, and d) connect the IO pin to the two-wire analog test bus.

Clearly the design of the test bus must be done with care. Specifically, the test bus pin should not distort the

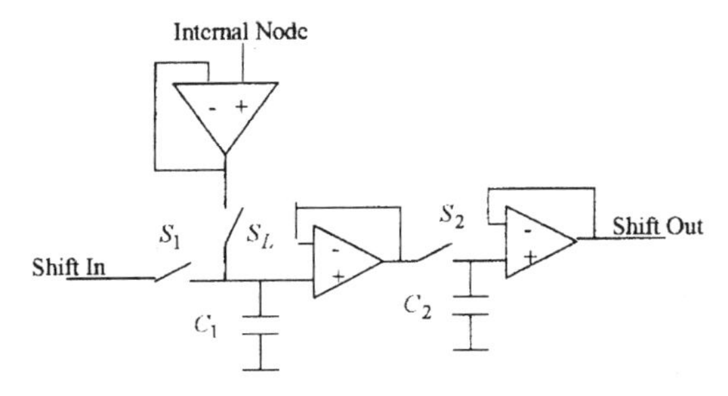

Fig. 13. Voltage-based scan cell.

real-time signals entering and leaving the blocks. Except for very high frequency nodes, the capacitance load added by, as a minimum, the two test bus transmission gates, the multiplexer/comparator, and the tri-state digital inverter is unlikely to cause significant distortions of internal signals [85]. On the other hand, the bus capacitance, combined with the high impedance of the transmission gates, may distort the signal that is observed at or driven by the output pin, by increasing delays and reducing bandwidth. Nevertheless, transmission gates or buffers can be designed to allow for higher output signal bandwidth at the expense of additional capacitive loading and/or area [85].

For high-speed testing of many analog signals simultaneously, the two signals at a time controllability and observability capability of the analog test bus may not be sufficient. Instead, Wey [86], [87] and Soma [88] have proposed an entirely analog scan path implementation. In [86], voltages are stored in the scan cells, which are composed of sample-and-hold circuits, each built with a switch for sampling, a capacitor for storage, and a voltage follower for impedance buffering between capacitors (Fig. 13). In order to minimize the influence of the analog shift register test circuit on the node being observed, a high input impedance and low output impedance buffer is placed between the node and the sampling switch. When a test is performed, data at various test points is simultaneously loaded to the holding capacitors C_1 by closing switch S_L. To scan out, the switch S_L is opened and a two-phase clock is used to scan out voltages, like for digital scan chains. Specifically, first, S_2 is closed so that the voltage on C_1 is copied to C_2. Then, S_1 is closed to copy the voltage to the next cell. This process continues until all voltages reach the output.

In [87] and [88], currents, instead of voltages, are passed through the scan chain. The shift register is composed of current mirrors and switches. A possible implementation is shown in Fig. 14. In this approach, if a node voltage is being observed, it is first converted to a current and isolated from the test circuitry with a V/I converter. During loading, the switch S_L is closed, and the current flows through transistor M_{1A} and sets the voltage on capacitor C_1 to the level needed for M_{1A} to support I_{in}. The transistor M_{1B} is now capable of sinking a current I_{in} when connected to a load. Then to scan out the stored current, S_L is opened and the switches S_2 and S_1 are alternately closed. When S_2 is closed, transistor M_{1B} is connected to a load, and consequently sinks I_{in}. As a result, I_{in} flows through M_{2A} and charges up C_2, so that I_{in} will flow through M_{2B} when it is connected to a load. Then, in

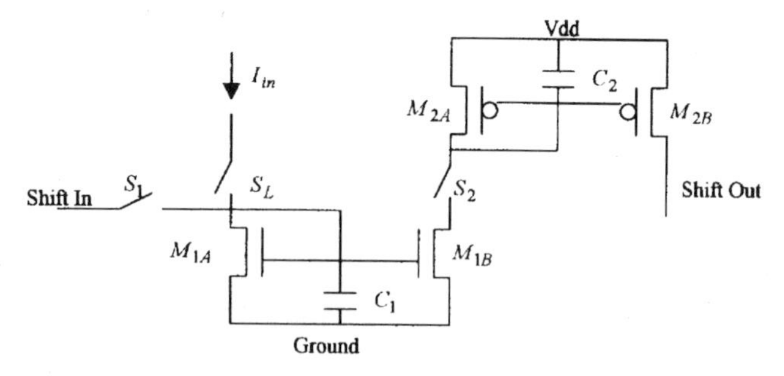

Fig. 14. Current-based scan cell.

the next clock phase, S_2 is opened and S_1 is closed, so that I_{in} is then copied from M_{2B} to M_{1A} of the following cell. As with the voltage scan chain, this process continues until cell currents reach the output. It should be noted that in both the current-based and the voltage-based approaches, scan chain length is limited by accuracy requirements. For example, clock feedthrough and mismatch in the current mirror transistors can limit the accuracy of the scanned currents and voltages.

B. Analog Built-In Self Test

When test busses and scan circuitry are used to enhance testability, signals have to be transmitted through long wires and/or have to pass through transmission gates before they can be measured. Hence, analog signals can be corrupted, and distortion may occur before measurements are made due to parasitic loading and coupling. BIST circuitry helps to overcome this problem by going beyond simply controlling and observing component inputs and outputs. Instead, signal generators and analysis circuitry are implemented on chip, and since signals do not have to be routed off chip, it is likely that there will be less distortion of these signals when they are measured. Consequently, dynamic tests can be performed at full speed even during wafer probe and in the field, without external test equipment. And the only signal that needs to be routed off chip is a pass/fail bit indicating the test results. On the other hand, one major problem faced by BIST circuit designs is the area overhead. Moreover, the design of high-quality signal generators and analysis circuitry can be complex and time-consuming.

The hardware overhead for BIST is minimized if the test circuitry is used by much more than one of the analog components. In addition, some BIST designs only attempt to perform on-chip analysis of test results and rely on external signal sources, thereby minimizing the hardware overhead. These designs are mainly targeted for on-line test of high safety systems, i.e., testing during operation. They signal errors due to component degradation, electromagnetic interference, or heat. Such circuits either rely on area redundancy (multiple copies of the same hardware) or time redundancy (the same hardware is used to carry out repeated operations).

In [89], a BIST circuit has been proposed for switched-capacitor filters, relying on partial replication of the circuit being tested (area redundancy). Using this technique, on-line testing is performed by first decomposing the filter into its component biquads. Multiplexing makes the input and output terminal of each biquad accessible. The circuit relies on a programmable biquad to implement a copy of each of the component biquads (Fig. 15). The programmable biquad can implement any of the basic filter types, i.e., low-pass, bandpass, and high-pass, and many different frequency specifications by changing the capacitance values for each node. During testing, the same signal is input to the programmable biquad as the component biquad being tested, and the resulting continuous signals are compared in real-time using a voter circuit. The voter circuit indicates an error if the outputs of the two filters differ by more than a specified tolerance margin at any time. An absolute acceptance window is implemented in [89], but in [90] it is noted that an absolute tolerance window can be too restrictive for signals with a large swing. Therefore, as an alternative, a circuit signaling an error based on a relative tolerance window has been proposed. In addition, instead of comparing signals in the time domain, the signal of the circuit being tested can be compared against a reference signal in the frequency domain as well, using a gain detector, phase detector, and a window comparator [91]. Note that for all techniques, the programmable biquad is used by all filters on the chip, and as a result, the area overhead is essentially limited to the area required for the programmable biquad.

Another approach to concurrent testing relying upon area redundancy involves using a continuous checksum [92]. In this case a circuit must be approximately linear. As a result, the time domain response can be described by state equations. A check variable is defined as equal to a linear combination of the state variables, and in this approach additional circuitry is proposed which outputs the check variable. Consequently, the additional circuitry produces a continuous nonzero signal when the signals corresponding to the state variables deviate from nominal.

Finally, in contrast to the previously mentioned approaches, time redundancy is exploited for concurrent testing of ADC's in [93]. Time redundancy employs a single piece of hardware to carry out repeated operations. In order to avoid producing the same erroneous result twice, in the repeated cycle, the input operand is coded and the result decoded, after which a comparison is made with the result obtained in the previous cycle.

All of the above techniques should be effective in detecting local catastrophic and some local parametric faults, since, given an appropriate input signal, there should be a clear discrepancy between the test circuit's output and that of the component being tested. On the other hand, because of the area overhead, the yield of the die will decrease [40]. This decrease, nevertheless, is likely to be small, if the area overhead of the test circuit is small. Such circuitry is likely to be less effective in detecting global parametric faults and component degradation since component variations in the test circuitry are likely to track those in the circuit being tested. Hence, only major changes in component values are likely to be detected.

C. BIST for Mixed-Signal Integrated Circuits

When designing BIST circuitry for mixed-signal integrated circuits, it is often possible to exploit existing on-chip hard-

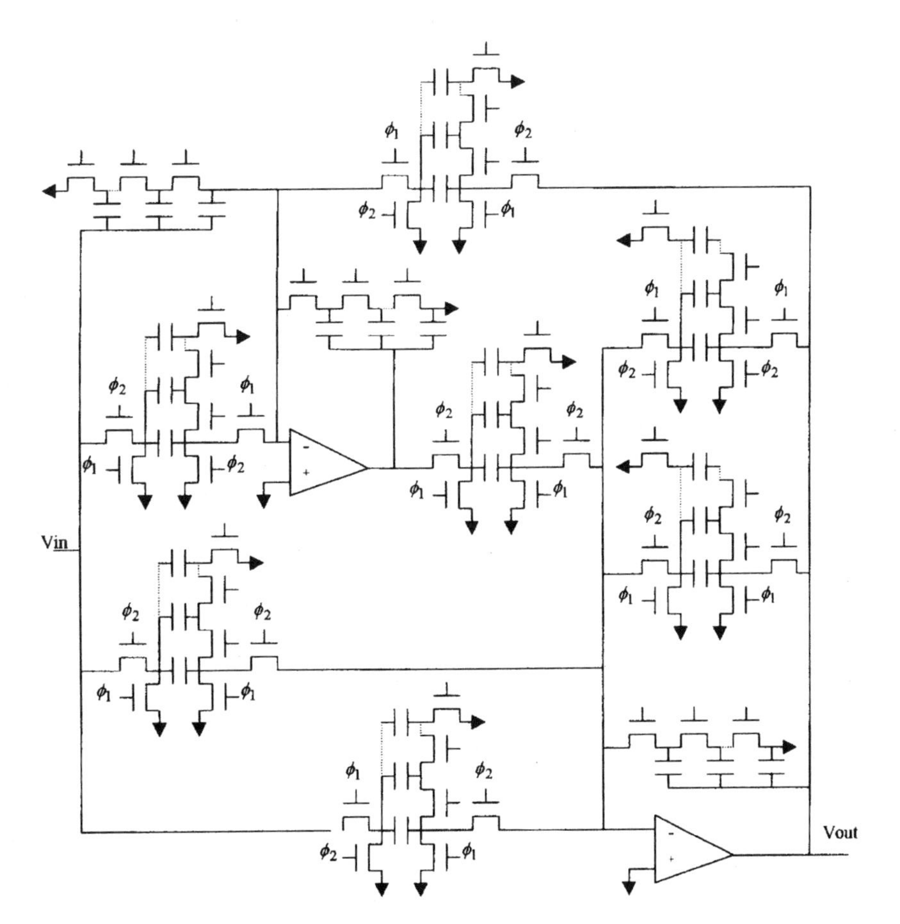

Fig. 15. A programmable biquad [89].

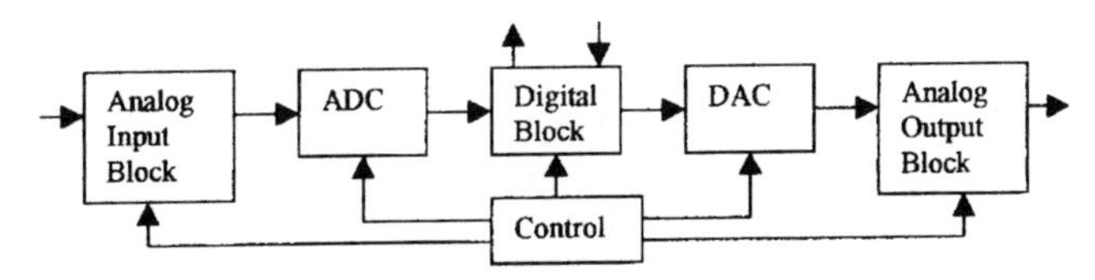

Fig. 16. An example of a mixed-signal circuit architecture.

ware and consequently reduce the area overhead needed for the on-chip generation of test signals and analysis of test results. In particular, such circuits have DAC's and ADC's which may be used for testing through reconfiguring the connections between blocks. By taking advantage of ADC's and DAC's, which are already part of a design, mixed-signal circuits can be tested with digital testers, components may be tested in parallel, and testing is more easily performed in the field.

A common architecture for a mixed-signal circuit is shown in Fig. 16. This architecture assumes that a mixed-signal circuit is composed of analog input components, connected to a large digital section by an ADC, which in turn is connected to analog output components by a DAC. Given the on-chip DAC, a digital test stimulus may be implemented on-chip, in order to test the analog output block. Specifically, Ohletz [49] has proposed a pseudorandom piecewise-constant input signal with different amplitudes, generated with a linear feedback shift register (LFSR), the DAC, and an output amplifier (Fig. 17). Alternatively, input stimuli could come from a ROM or DSP circuitry, rather than an LFSR. All of these approaches keep

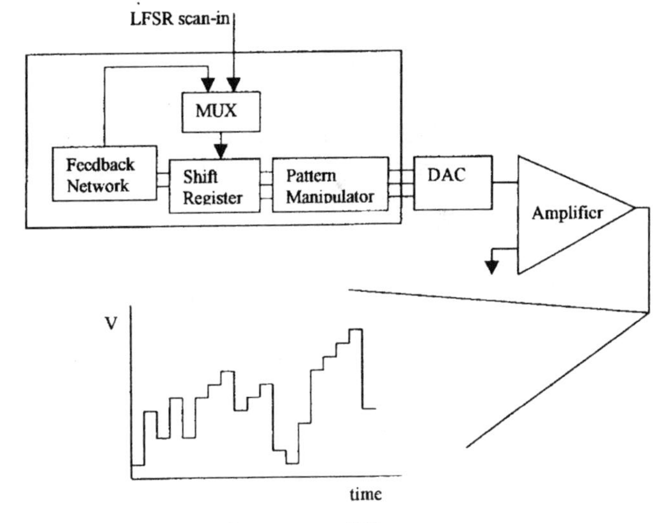

Fig. 17. An input stimulus generator [49].

the hardware overhead to a minimum by reconfiguring and reusing existing circuitry on chip during the test mode.

During test mode, outputs of the analog output block may be measured or routed to the analog input pins (Fig. 18). The outputs of the analog input block are embedded on chip and therefore not accessible. However, during the test mode, these outputs may be measured after conversion of the signal to digital by the on-chip ADC. One way to capture the output is by built-in logic block observers (BILBO) [49],

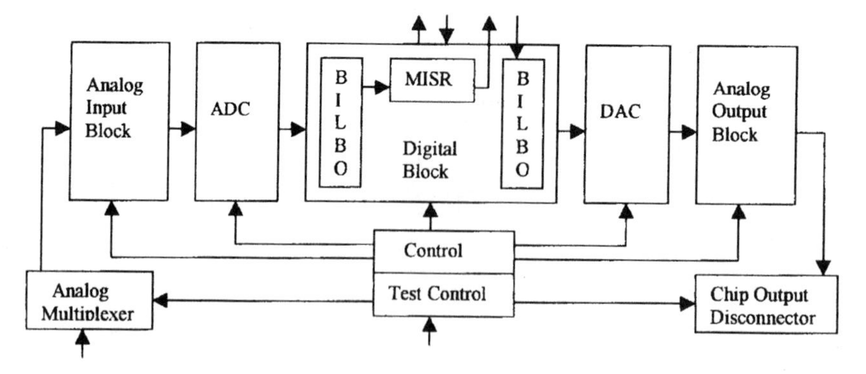

Fig. 18. Implementation of the BIST scheme in [49].

[94], which are also used for digital test. The signals stored in the BILBO registers may then be fed to a multiple input signature register (MISR), which performs the task of on-chip data compaction using signature analysis [49]. Hence, the analog test results are consequently evaluated in the digital domain, using the same techniques as used for on-chip evaluation of the digital response. Nevertheless, signature analysis is not the only way in which the digitized response from an analog block can be analyzed. The response could be compared against a known good response, stored in a ROM, before compaction, or postprocessing may be done based on the functional characteristics of the analog blocks. Specifically, in [39], given a pseudorandom piecewise-constant input signal, circuitry for computing the auto-correlation and cross-correlation of the impulse response is proposed.

In both [39] and [49], the effectiveness of pseudorandom inputs in detecting catastrophic faults in analog components has been demonstrated. However, the effectiveness of such approaches in detecting both local and global parametric faults still needs to be determined. Because the circuit performances that are tested are different than the circuit specifications, there may be significant systematic losses in fault coverage and/or yield coverage for parametric faults.

In [95], a BIST circuit is proposed for looking at abnormal changes in the power supply current. The proposed circuit involves an upper limit detector, for detecting an abnormally high power supply current, a lower limit detector, for detecting an abnormally low power supply current, and some logic to signal if there is a fault. The idea behind this approach is that faults will either increase or decrease the power supply current compared to the fault-free circuit. When using this power supply current monitor to test ADC's, the input voltage is varied, so that all states of the ADC are exercised. A reasonable fault coverage of catastrophic faults has been demonstrated by simulation. Nevertheless, as with using random inputs to test analog blocks, the effectiveness of this approach in detecting local and global parametric faults is still unknown.

A more traditional signal generator is proposed for BIST of ADC's in [4] and [96]. Specifically, tests are designed to measure the signal-to-noise ratio, gain tracking, and the frequency response of a sigma–delta ADC (Fig. 19). The stimulus is a precise multitone oscillator designed for an uncalibrated environment [97], [98]. The design of the oscillator is fully digital, except for an imprecise low-pass

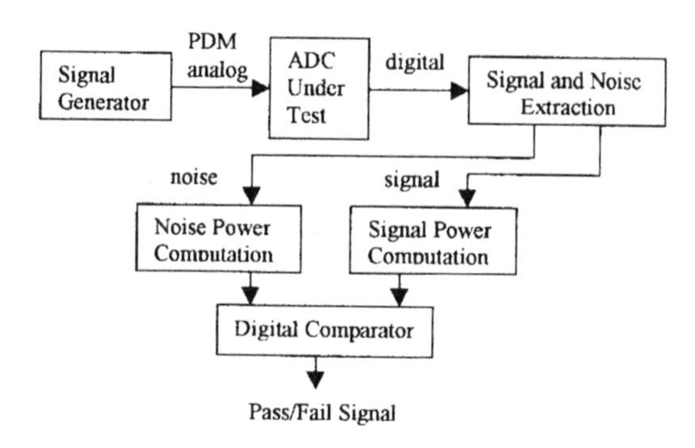

Fig. 19. BIST circuit for testing ADC's [4].

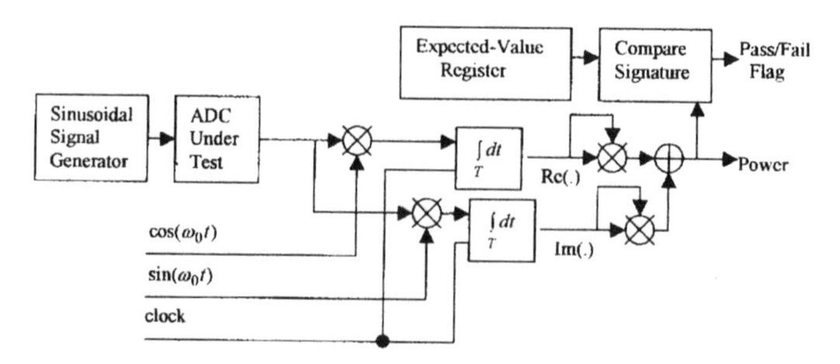

Fig. 20. Signal-to-noise ratio test using a correlator circuit [5].

filter, and it is digitally programmable for multiple amplitudes, frequencies, and phases. Three digital methods are proposed for analyzing test results: the fast Fourier transform, the IEEE standard 1057, and a narrow-band digital filter. The fast Fourier transform approach finds the signal and noise powers using a correlator circuit, like the one shown in Fig. 20. Hence, the on-chip computing resources that are required to implement this approach include registers to store the samples, plus circuitry to either compute or look-up values for the sine and cosine functions. The IEEE Standard 1057 finds the signal and noise powers by fitting a sinusoid using regression and requires similar on-chip resources. And lastly, the narrow-band digital filter approach relies on on-chip bandpass and notch filters, as shown in Fig. 21. In [4] and [96], it is argued that the area overhead for the narrow-band digital filter is the least of the above three methods, but the test results are biased, since some of the noise power may be mistakenly included in the signal power. However, the bias may be minimized through proper design of the filter.

It is not only possible to build BIST circuitry for ADC's, but also DAC's can be tested with on-chip circuitry. In [99],

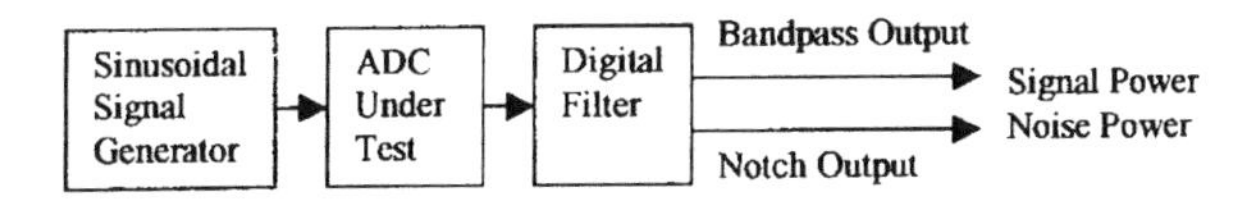

Fig. 21. Signal-to-noise ratio test using a digital filter [4].

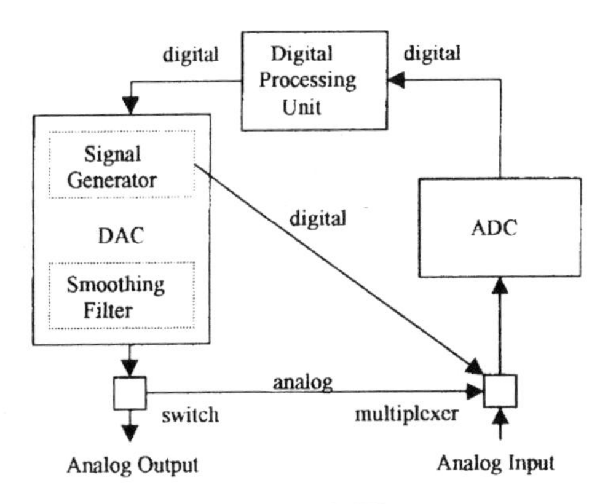

Fig. 22. Mixed-signal BIST scheme [4], [5].

a BIST circuit has been designed to test offset, gain, integral linearity, and differential linearity of DAC's. For this circuit, the input is a digital sequence generated on chip, and the output is analyzed with sample-and-hold circuitry, various reference voltages, and a comparator. A modified version of this circuit has also been proposed for testing successive approximation ADC's, since they involve DAC's.

Finally, since many circuits contain both ADC's and DAC's, a sequential strategy for verifying both is proposed in [4] (Fig. 22). First, all digital circuitry is tested, including the digital components of the DAC, ADC, and the on-chip signal generator. Then, the ADC is tested using the on-chip digital $\Delta\Sigma$ oscillator. Next, the smoothing filter of the DAC is tested using signals from the digital signal processing unit. Its output is digitized by the ADC, and its response is analyzed by the digital circuitry. This verifies the functionality of the smoothing filter, and the combination of the digital signal generator and smoothing filter can now be used as a calibrated analog signal source. This also completes the testing of the DAC and ADC, since all other digital circuitry has been tested. Once the DAC, ADC, and signal generator are considered functional, the converters and the analog signal generator can be used to test other on-chip analog functions.

The fault coverage of this approach to on-chip BIST still needs to be verified. However, because the test methods are very similar to traditional functional testing, it is likely that fault coverage will be high, even for global parametric faults. This is because, although global parametric variations in the test circuit will track those in the circuit being tested, the functions of the test circuit and the circuit being tested are different, and hence process sensitivities are likely to differ. As a result, one can expect less systematic loss in fault coverage and yield coverage for parametric faults.

VI. Summary

Recently, work on analog testing has evolved from its early focus on diagnosing failures and degradation in analog board designs toward a focus on production test, including CAD tools for test set design and optimization, and circuit design techniques targeted to explore the tradeoff between on-chip versus external testers. Clearly, diagnosing failures is still important. Many circuits are still designed with discrete components, and failures in the field should be quickly corrected. Moreover, integrated circuits must be characterized before production runs, and this involves determining any systematic sources of yield loss, including components whose performances may fluctuate too much due to variations in manufacturing. However, the new areas of research in analog and mixed-signal testing are motivated by new concerns. Specifically, as analog and mixed-signal circuits become more complex and have shorter product cycles, they frequently cannot be tested using methods developed in the past, due to the longer testing times needed for high precision analog components and the lack of accessibility of analog components embedded in large mixed-signal chips. Research addressing these problems is still preliminary and is likely to evolve rapidly in the coming years.

Acknowledgment

The author would like to thank M. Ohletz, W. Kao, and the four reviewers for their helpful comments on this manuscript.

References

[1] P. Duhamel and J. C. Rault, "Automatic test generation techniques for analog circuits and systems: A review," *IEEE Trans. Circuits Syst.*, vol. CS-26, pp. 411–439, July 1979.
[2] J. W. Bandler and A. E. Salama, "Fault diagnosis of analog circuits," *Proc. IEEE*, vol. 73, Aug. 1985.
[3] A. Rappaport *et al.*, "Panel discussion: Impediments to mixed-signal IC development," in *Proc. ISSCC*, 1991, pp. 200–201.
[4] M. F. Toner and G. W. Roberts, "A BIST SNR, gain tracking and frequency response test of a sigma–delta ADC," *IEEE Trans. Circuits Syst. II*, vol. 42, pp. 1–15, Jan. 1995.
[5] G. W. Roberts, "Improving the testability of mixed-signal integrated circuits," in *Proc. CICC*, 1997, pp. 214–221.
[6] T. M. Souders and G. N. Stenbakken, "Cutting the high cost of testing," *IEEE Spectrum*, pp. 48–51, Mar. 1991.
[7] J. Q. Xia, T. Austin, and N. Khouzam, "Dynamic test emulation for EDA-based mixed-signal test development automation," in *Proc. Int. Test Conf.*, 1995, pp. 761–770.
[8] W. H. Kao and J. Q. Xia, "Automatic synthesis of DUT board circuits for testing of mixed signal IC's," in *VLSI Test Symp.*, 1993, pp. 230–236.
[9] B. Webster, "An integrated analog test simulation environment," in *Proc. Int. Test Conf.*, 1989, pp. 567–571.
[10] S. C. Bateman and W. H. Kao, "Simulation of an integrated design and test environment for mixed-signal integrated circuits," in *Proc. Int. Test Conf.*, 1992, pp. 405–414.
[11] P. Caunegre and C. Abraham, "Achieving simulation-based test program verification and fault simulation capabilities for mixed-signal systems," in *European Design and Test Conf.*, 1995, pp. 469–477.
[12] W. Kao, J. Xia, and T. Boydston, "Automatic test program generation for mixed signal IC's via design to test link," in *Proc. Int. Test Conf.*, 1992, pp. 860–865.
[13] L. Bonet *et al.*, "Test features of the MC145472 ISDN U-transceiver," in *Proc. Int. Test Conf.*, 1990, pp. 68–79.
[14] S. D. Huss and R. S. Gyurcsik, "Optimal ordering of analog integrate circuit tests to minimize test time," in *Proc. DAC*, 1991, pp. 494–499.
[15] L. Milor and A. L. Sangiovanni-Vincentelli, "Minimizing production test time to detect faults in analog circuits," *IEEE Trans. Computer-Aided Design*, vol. 13, pp. 796–813, June 1994.
[16] ——, "Optimal test set design for analog circuits," in *Proc. ICCAD*, 1990, pp. 294–297.
[17] G. N. Stenbakken and T. M. Souders, "Test-point selection and testability measures via QR factorization of linear models," *IEEE Trans. Instrum. Measur.*, vol. 36, pp. 406–410, June 1987.

[18] ———, "Linear error modeling of analog and mixed-signal devices," in *Proc. Int. Test Conf.,* 1991, pp. 573–581.

[19] G. N. Stenbakken, T. M. Souders, and G. W. Stewart, "Ambiguity groups and testability," *IEEE Trans. Instrum. Measur.,* vol. 38, pp. 941–947, Oct. 1989.

[20] G. J. Hemink, B. W. Meijer, and H. G. Kerkhoff, "Testability analysis of analog systems," *IEEE Trans. Computer-Aided Design,* vol. 9, pp. 573–583, June 1990.

[21] E. Liu *et al.*, "Analog testability analysis and fault diagnosis using behavioral modeling," in *Proc. CICC,* 1994, pp. 413–416.

[22] E. Felt and A. L. Sangiovanni-Vincentelli, "Testing of analog systems using behavioral models and optimal experimental design techniques," in *Proc. ICCAD,* 1994, pp. 672–678.

[23] T. M. Souders and G. N. Stenbakken, "A comprehensive approach for modeling and testing analog and mixed-signal devices," in *Proc. Int. Test Conf.,* 1990, pp. 169–176.

[24] W. Maly, A. J. Strojwas, and S. W. Director, "VLSI yield prediction and estimation: A unified framework," *IEEE Trans. Computer-Aided Design,* vol. 5, pp. 114–130, Jan. 1986.

[25] D. M. H. Walker, *Yield Simulation for Integrated Circuits.* Norwell, MA: Kluwer, 1987.

[26] A. Jee and F. J. Ferguson, "Carafe: An inductive fault analysis tool for CMOS VLSI circuits," in *Proc. IEEE VLSI Test Symp.,* 1993, pp. 92–98.

[27] M. Ohletz, "L^2RFM—Local layout realistic faults mapping scheme for analogue integrated circuits," in *Proc. CICC,* 1996, pp. 475–478.

[28] L. Milor and V. Visvanathan, "Detection of catastrophic faults in analog integrated circuits," *IEEE Trans. Computer-Aided Design,* vol. 8, pp. 114–130, Feb. 1989.

[29] C. J. B. Spanos and S. W. Director, "Parameter extraction for statistical IC process characterization," *IEEE Trans. Computer-Aided Design,* vol. 5, pp. 66–79, Jan. 1986.

[30] C. Michael and M. Ismail, "Statistical modeling of device mismatch for analog MOS integrated circuits," *IEEE J. Solid-State Circuits,* vol. 27, pp. 154–165, Jan. 1992.

[31] K. Eshbaugh, "Generation of correlated parameters for statistical circuit simulation," *IEEE Trans. Computer-Aided Design,* vol. 11, pp. 1198–1206, Oct. 1992.

[32] N. Salamina and M. R. Rencher, "Statistical bipolar circuit design using MSTAT," in *Proc. ICCAD,* 1989, pp. 198–201.

[33] C. Y. Chao, H. J. Lin, and L. Milor, "Optimal testing of VLSI analog circuits," *IEEE Trans. Computer-Aided Design,* vol. 16, pp. 58–77, Jan. 1997.

[34] N. Nagi, A. Chatterjee, and J. A. Abraham, "Fault simulation of linear analog circuits," *J. Electron. Testing: Theory and Applicat.,* vol. 4, pp. 345–360, 1993.

[35] N. Nagi and J. Abraham, "Hierarchical fault modeling for linear analog circuits," *Analog Integrated Circuits and Signal Processing,* vol. 10, pp. 89–99, June/July 1996.

[36] R. J. A. Harvey *et al.*, "Analogue fault simulation based on layout dependent fault models," in *Proc. Int. Test Conf.,* 1994, pp. 641–649.

[37] A. Meixner and W. Maly, "Fault modeling for the testing of mixed integrated circuits," in *Proc. Int. Test Conf.,* 1991, pp. 564–572.

[38] C. Sebeke, J. P. Teixeira, and M. J. Ohletz, "Automatic fault extraction and simulation of layout realistic faults for integrated analogue circuits," in *European Design and Test Conf.,* 1995, pp. 464–468.

[39] C. Y. Pan and K. T. Cheng, "Pseudo-random testing and signature analysis for mixed-signal circuits," in *Proc. ICCAD,* 1995, pp. 102–107.

[40] C. H. Stapper and R. J. Rosner, "Integrated circuit yield management and yield analysis: Development and implementation," *IEEE Trans. Semiconduct. Manufact.,* vol. 8, pp. 95–102, May 1995.

[41] C. M. Kurker *et al.*, "Hierarchical yield estimation of large analog integrated circuits," *IEEE J. Solid-State Circuits,* vol. 28, pp. 203–209, Mar. 1993.

[42] C. Y. Chao and L. Milor, "Performance modeling of circuits using additive regression splines," *IEEE Trans. Semiconduct. Manufact.,* vol. 8, pp. 239–251, Aug. 1995.

[43] L. Milor and A. Sangiovanni-Vincentelli, "Computing parametric yield accurately and efficiently," in *Proc. ICCAD,* 1990, pp. 116–119.

[44] T. K. Yu *et al.*, "Statistical performance modeling and parametric yield estimation of MOS VLSI," *IEEE Trans. Computer-Aided Design,* vol. CAD-6, pp. 1013–1022, Nov. 1987.

[45] K. K. Low and S. W. Director, "An efficient methodology for building macromodels of IC fabrication processes," *IEEE Trans. Computer-Aided Design,* vol. 8, pp. 1299–1313, Dec. 1989.

[46] M. Li and L. Milor, "Computing parametric yield adaptively using local linear models," in *Proc. DAC,* 1996, pp. 831–836.

[47] W. M. Lindermeir, H. E. Graeb, and K. J. Antreich, "Design based analog testing by characteristic observation inference," in *Proc. ICCAD,* 1995, pp. 620–626.

[48] K. Singhal and J. F. Pinel, "Statistical design centering and tolerancing using parametric sampling," *IEEE Trans. Circuits Syst.,* vol. CS-28, pp. 692–701, July 1981.

[49] M. J. Ohletz, "Hybrid built-in self test (HBIST) for mixed analogue/digital integrated circuits," in *Proc. European Test Conf.,* 1991, pp. 307–316.

[50] M. Sachdev and B. Atzema, "Industrial relevance of analog IFA: A fact or a fiction," in *Proc. ITC,* 1995, pp. 61–70.

[51] W. Hochwald and J. D. Bastian, "A DC approach for analog fault dictionary determination," *IEEE Trans. Circuits Syst.,* vol. CS-26, pp. 523–529, July 1979.

[52] S. Seshu and R. Waxman, "Fault isolation in conventional linear systems—A feasibility study," *IEEE Trans. Reliab.,* vol. 15, pp. 11–16, 1966.

[53] G. O. Martens and J. D. Dyck, "Fault identification in electronic circuits with the aid of bilinear transformation," *IEEE Trans. Reliab.,* vol. 21, pp. 99–104, 1972.

[54] K. C. Varghese, J. H. Williams, and D. R. Towill, "Computer-aided feature selection for enhanced analogue system fault location," *Pattern Recogn.,* vol. 10, pp. 265–280, 1978.

[55] Z. You, E. Sanchez-Sinencio, and J. Pineda de Gyvez, "Analog system-level fault diagnosis based on a symbolic method in the frequency domain," *IEEE Trans. Instrum. Measur.,* vol. 44, pp. 28–35, Feb. 1995.

[56] J. Beasley *et al.*, "I_{dd} pulse response testing: A unified approach to testing digital and analogue IC's," *Electron. Lett.,* vol. 29, pp. 2101–2103, Nov. 25, 1993.

[57] C. J. Macleod, "System identification using time-weighted pseudorandom sequences," *Int. J. Contr.,* vol. 14, pp. 97–109, 1971.

[58] H. H. Schreiber, "Fault dictionary based upon stimulus design," *IEEE Trans. Circuits Syst.,* vol. CS-26, pp. 529–537, July 1979.

[59] S. S. Somayajula, E. Sanchez-Sinencio, and J. Pineda de Gyvez, "Analog fault diagnosis based on ramping power supply current signature clusters," *IEEE Trans. Circuits Syst. II,* vol. 43, pp. 703–712, Oct. 1996.

[60] G. C. Temes, "Efficient methods for fault simulation," in *Proc. 20th Midwest Symp. Circuits and Systems,* Lubbock, TX, Aug. 1977, pp. 191–194.

[61] A. T. Johnson, Jr., "Efficient fault analysis in linear analog circuits," *IEEE Trans. Circuits Syst.,* vol. CS-26, pp. 475–484, July 1979.

[62] J. H. Lee and S. D. Bedrosian, "Fault isolation algorithms for analog electronic systems using the fuzzy concept," *IEEE Trans. Circuit Syst.,* vol. CS-26, pp. 518–522, July 1979.

[63] S. Freeman, "Optimum fault isolation by statistical inference," *IEEE Trans. Circuits Syst.,* vol. CS-26, pp. 505–512, July 1979.

[64] B. R. Epstein, M. Czigler, and S. R. Miller, "Fault detection and classification in linear integrated circuits: An application of discrimination analysis and hypothesis testing," *IEEE Trans. Computer-Aided Design,* vol. 12, pp. 101–113, Jan. 1993.

[65] G. Devarayanadurg and M. Soma, "Analytical fault modeling and static test generation for analog IC's," in *Proc. ICCAD,* 1994, pp. 44–47.

[66] N. Sen and R. Saeks, "Fault diagnosis for linear systems via multifrequency measurements," *IEEE Trans. Circuits Syst.,* vol. CS-26, pp. 457–465, July 1979.

[67] L. Rapisarda and R. A. Decarlo, "Analog multifrequency fault diagnosis," *IEEE Trans. Circuits Syst.,* vol. CS-30, pp. 223–234, Apr. 1983.

[68] G. Iuculano *et al.*, "Multifrequency measurement of testability with application to large linear analog systems," *IEEE Trans. Circuit Syst.,* vol. CS-33, pp. 644–648, June 1986.

[69] J. A. Starzyk and H. Dai, "Sensitivity based testing of nonlinear circuits," in *Proc. ISCAS,* 1990, pp. 1159–1162.

[70] V. Visvanathan and A. Sangiovanni-Vincentelli, "Diagnosability of nonlinear circuits and systems—Part 1: The DC case," *IEEE Trans. Circuits Syst.,* vol. CS-28, pp. 1093–1102, Nov. 1981.

[71] R. Saeks, A. Sangiovanni-Vincentelli, and V. Visvanathan, "Diagnosability of nonlinear circuits and systems—Part II: Dynamical systems," *IEEE Trans. Circuits Syst.,* vol. CS-28, pp. 1103–1108, Nov. 1981.

[72] V. Visvanathan and A. Sangiovanni-Vincentelli, "A computational approach for the diagnosability of dynamical circuits," *IEEE Trans. Computer-Aided Design,* vol. CAD-3, pp. 165–171, July 1984.

[73] R. Saeks, S. P. Singh, and R. W. Liu, "Fault isolation via component simulation," *IEEE Trans. Circuit Theory,* vol. CT-19, pp. 634–640, 1972.

[74] T. N. Trick, W. Mayeda, and A. A. Sakla, "Calculation of parameter values from node voltage measurements," *IEEE Trans. Circuits Syst.,* vol. CS-26, pp. 466–474, July 1979.

[75] Z. F. Huang, C. S. Lin, and R. W. Liu, "Node-fault diagnosis and a design of testability," *IEEE Trans. Circuits Syst.,* vol. CS-30, pp. 257–265, 1983.

[76] R. M. Biernacki and J. W. Bandler, "Multiple-fault location of analog circuits," *IEEE Trans. Circuits Syst.,* vol. CS-28, pp. 361–367, May 1981.

[77] C. C. Wu *et al.*, "Analog fault diagnosis with failure bounds," *IEEE Trans. Circuits Syst.*, vol. CS-29, pp. 277–284, May 1982.
[78] A. E. Salama, J. A. Starzyk, and J. W. Bandler, "A unified decomposition approach for fault location in large analog circuits," *IEEE Trans. Circuit Syst.*, vol. CS-31, pp. 609–622, July 1984.
[79] M. Slamani and B. Kaminska, "Fault observability analysis of analog circuits in frequency domain," *IEEE Trans. Circuits Syst. II*, vol. 43, pp. 134–139, Feb. 1996.
[80] N. B. Hamida and B. Kaminska, "Multiple fault analog circuit testing by sensitivity analysis," *J. Electron. Testing: Theory and Applicat.*, vol. 4, pp. 331–343, 1993.
[81] H. Kerkhoff, "Design for testability," in *Analog VLSI Signal and Information Processing*, M. Ismail and T. Fiez, Eds. New York: McGraw-Hill, 1994, pp. 547–584.
[82] P. P. Fasang, "Boundary scan and its application to analog-digital ASIC testing in a board/system environment," in *Proc. CICC*, 1989, pp. 22.4.1–22.4.4.
[83] ——, "Analog/digital ASIC design for testability," *IEEE Trans. Ind. Electron.*, vol. 36, pp. 219–226, May 1989.
[84] K. D. Wagner and T. W. Williams, "Design for testability of analog/digital networks," *IEEE Trans. Ind. Electron.*, vol. 36, pp. 227–230, May 1989.
[85] S. Sunter, "The P1149.4 mixed signal test bus: Costs and benefits," in *Proc. Int. Test Conf.*, 1995, pp. 444–450.
[86] C. L. Wey, "Built-in self-test (BIST) structure for analog circuit fault diagnosis," *IEEE Trans. Instrum. Measur.*, vol. 39, pp. 517–521, June 1990.
[87] C. L. Wey and S. Krishnan, "Built-in self test (BIST) structure for analog circuit fault diagnosis with current test data," *IEEE Trans. Instrum. Measur.*, vol. 41, 1992.
[88] M. Soma, "Structure and concepts for current-based analog scan," in *Proc. CICC*, 1995, pp. 517–520.
[89] J. L Huertas, A. Rueda, and D. Vazquez, "Testable switched-capacitor filters," *IEEE J. Solid-State Circuits*, vol. 28, pp. 719–724, July 1993.
[90] V. Kolarik *et al.*, "Analog checkers with absolute and relative tolerances," *IEEE Trans. Computer-Aided Design*, vol. 14, pp. 607–612, May 1995.
[91] M. Slamani, B. Kaminska, and G. Quesnel, "An integrated approach for analog circuit testing with a minimum number of detected parameters," in *Proc. Int. Test Conf.*, 1994, pp. 631–640.
[92] A. Chatterjee, "Concurrent error detection and fault-tolerance in linear analog circuits using continuous checksums," *IEEE Trans. VLSI Syst.*, vol. 1, pp. 138–150, June 1993.
[93] S. Krishnan, S. Sahli, and C. L. Wey, "Test generation and concurrent error detection in current-mode A/D converters," in *Proc. Int. Test Conf.*, 1992, pp. 312–320.
[94] B. Konemann, J. Mucha, and G. Zwiehoff, "Built-in logic block observation techniques," in *Proc. Int. Test Conf.*, 1979, pp. 37–41.
[95] Y. Miura, "Real-time current testing for A/D converters," *IEEE Design, Test Comput.*, vol. 13, pp. 34–41, Summer 1996.
[96] M. F. Toner and G. W. Roberts, "On the practical implementation of mixed analog-digital BIST," in *Proc. CICC*, 1995, pp. 525–528.
[97] A. K. Lu and G. W. Roberts, "An analog multi-tone signal generator for built-in-self-test applications," in *Proc. Int. Test Conf.*, 1994, pp. 650–659.
[98] E. M. Hawrysh and G. W. Roberts, "An integrated memory-based analog signal generation into current DFT architectures," in *Proc. Int. Test Conf.*, 1996, pp. 528–537.
[99] K. Arabi, B. Kaminska, and J. Rzeszut, "A new built-in self-test approach for digital-to-analog and analog-to digital converters," in *Proc. ICCAD*, 1994, pp. 491–494.

Linda S. Milor (S'86–M'90) received the Ph.D. degree in electrical engineering from the University of California, Berkeley, in 1992.

She is currently the Device Engineering Manager in the Yield Department in the Submicron Development Center at Advance Micro Devices, Sunnyvale, CA, where she has been working since 1995. Before joining Advanced Micro Devices, she was an Assistant Professor with the Electrical Engineering Department at the University of Maryland, College Park. She has published in the areas of yield modeling, yield enhancement, circuit performance prediction, testing of analog integrated circuits, and statistical modeling.

About the Editors

Rob A. Rutenbar received the Ph.D. in Computer Engineering in 1984 from the University of Michigan, and subsequently joined the faculty of Carnegie Mellon University, where he is currently the Stephen J. Jatras Professor of Electrical and Computer Engineering. In 1985 he began work on the first comprehensive approaches for circuit and physical synthesis of full-custom analog designs. He has worked extensively on both circuit-level and system-level CAD problems for analog and mixed-signal ICs. His research interests continue to focus on synthesis algorithms for custom circuits and for high-performance layout.

He has been actively involved in the semiconductor design/CAD community. He was on the editorial board of IEEE Spectrum from 1992 to 1994. He chaired the Semiconductor Research Corporation's University Advisory Council from 1997 to 1998. In 1995 he coauthored the IC layout portions of the U.S. Semiconductor Industries Association National Technology Roadmap for Semiconductors. He was General Chair of the 1996 International Conference on CAD. He chaired the Analog Technical Advisory Board for Cadence Design Systems from 1992 to 1996. In 1997 he co-founded Neolinear, Inc., to commercialize full-custom analog synthesis tools. He served as CTO of Neolinear during a leave of absence in 1998, and currently holds the position of Chief Scientist. He has received a variety of awards for his research and teaching. He won Best Paper awards at the 1987 Design Automation Conference (for analog synthesis), 1991 International Conference on CAD (for analog layout), 1993 SRC TECHCON Conference (for analog synthesis), and 2000 SRC TECHCON Conference (for custom digital layout). He received a Presidential Young Investigator Award from the National Science Foundation in 1987. He was elected a Fellow of the IEEE in 1998. He was co-winner of the 2001 SRC Aristotle Award for excellence in teaching.

Georges G. E. Gielen received the M.Sc. and Ph.D. degrees in Electrical Engineering from the Katholieke Universiteit Leuven, Belgium, in 1986 and 1990, respectively. After being visiting lecturer at the department of Electrical Engineering and Computer Science of the University of California, Berkeley, U.S.A, he became a faculty member at the ESAT-MICAS laboratory of the EE department of the Katholieke Universiteit Leuven where he is now a full-time professor. His research interests are in the design of analog and mixed-signal integrated circuits, and especially in analog and mixed-signal CAD tools and design automation (modeling, simulation and symbolic analysis, analog synthesis, analog layout generation, analog and mixed-signal testing). He has authored or coauthored two books and more than 150 papers in edited books, international journals and conference proceedings. He regularly is a member of the program committees of international conferences and a member of the editorial board of international journals. He is a senior member of IEEE, and currently a member of the Board of Governors of the IEEE Circuits and Systems (CAS) Society, and the Chairman of the IEEE Benelux CAS Chapter. He was the 1997 Laureate of the Belgian Royal Academy of Sciences, Literature and Arts, in the category of engineering sciences. He also received the 1995 Best Paper award of the John Wiley international journal on Circuit Theory and Applications. He was elected a Fellow of the IEEE in 2001.

Brian A. Antao received a Ph.D. in Electrical Engineering from Vanderbilt University in 1993. His doctoral research was in Synthesis and Verification of Analog Integrated Circuits. While a graduate student he did two summer internships at the AT&T Bell Laboratories working on mixed-signal behavioral modeling and simulation. He spent two years on the research faculty at the Department of Electrical and Computer Engineering at the University of Illinois at Urbana-Champaign, doing research on Computer-aided Design for analog and mixed-signal circuits. During this time he also worked on the standardization efforts for Analog Hardware Description Languages. Since then he has worked in a variety of industrial positions in CAD for mixed-signal ICs. Currently he is with Silicon Metrics Corp. in Austin, Texas developing advanced HDL modeling solutions do go along with the Very Deep Sub-Micron (VDSM) characterization tools that Silicon Metrics develops. Brian has also been active in the EDA/CAD and IEEE professional circles, publishing numerous journal and conference papers, reviewing technical paper submissions, as well as assisting in the organization of conferences.